MGM

CO.

NEW YORK
BEVERLY HILLS
NASHVILLE
LONDON
ROME

WILLIAM MORRIS AGENCY, INC

ARTISTS MANAGER — ESTABLISHED 1898

The Agency Of The Entertainment World

NEW YORK—1350 Avenue Of The Americas, New York, N.Y. 10019 • (212) 586-5100, Telex 620165

BEVERLY HILLS—151 El Camino, Beverly Hills, Calif. 90212 • (213) 274-7451, (213) 272-4111, Telex 673102

NASHVILLE—2325 Crestmoor Road, Nashville, Tenn. 37125 • (615) 385-0310, Telex 554472

LONDON—147/149 Wardour Street, London W1, England • Phone 01-734-9361
Cable Address—"Willmorris London," Telex 851-27928

ROME—Via Giosue' Carducci, 10-00187, Roma, Italy • Phones: 474-1123; 474-1891
Cable Address—"Willmorris Rome," Telex 84368192

1982

INTERNATIONAL TELEVISION ALMANAC

Editor: RICHARD GERTNER

Associate Editor: WILLIAM PAY

Quigley Publishing Company, Inc.

New York • London

(212) 247-3100 (0268 42824)

Foreword

In this 26th edition of the INTERNATIONAL TELEVISION ALMANAC there is introduced a new section devoted to cable television—the industry within an industry whose growth has been nothing short of phenomenal. The statistics are impressive: revenues climbed 20 per cent in 1979 to over $1.8 billion, for instance, and in the same period some 1.6 new subscribers were added. This and other interesting data will be found in the new Almanac section.

—RICHARD GERTNER

INTERNATIONAL
TELEVISION ALMANAC
For 1982
ISSN: 0539-0761
ISBN: 0-900610-27-1

PRINTED IN THE UNITED STATES OF AMERICA

Table of Contents

STATISTICS

Network and Station Operations

There were 748 commercial stations operating in the United States as of January, 1981 located in over 390 markets and serving some 77,800,000 families

TOTAL REVENUE OF TELEVISION NETWORKS AND STATIONS

Year	Revenue	Year	Revenue
1959	1,335,400,000	1970	2,809,000,000
1960	1,456,200,000	1971	2,750,000,000
1961	1,514,300,000	1972	3,180,000,000
1962	1,704,700,000	1973	3,460,000,000
1963	1,836,400,000	1974	3,800,000,000
1964	2,067,800,000	1975	4,094,100,000
1965	2,265,900,000	1976	5,200,000,000
1966	2,557,600,000	1977	5,889,000,000
1967	2,634,400,000	1978	6,913,000,000
1968	2,916,100,000	1979	7,875,000,000
1969	3,235,500,000	1980	8,808,000,000

BROADCAST REVENUES, EXPENSES, AND INCOME OF TELEVISION NETWORK AND STATIONS

(Millions dollars)

	1980	1979	1978	1977	1976
Revenues	8,808.0	7,875.0	6,913.0	5,889.0	5,198.5
Expenses	7,154.0	6,185.0	5,265.9	4,488.0	3,948.3
Income (before Federal tax)	1,654.0	1,690.0	1,647.1	1,401.0	1,250.2

IN 1980 THE American television industry showed an 11.8 per cent gain in net revenue but profits declined 2.2 per cent. Data is contained in the annual report on the industry made by the Federal Communications Commission.

The FCC analysis covers the three major commercial networks—ABC, CBS and NBC; the 15 TV stations owned directly by those networks and all the other TV stations operating around the country.

Advertisers spent a record $10.28 billion on TV in 1980, up 11.9 per cent over the $9.2 billion spent in 1979. After deducting $1.67 billion in advertising commissions, the industry was left with net revenue of $8.8 billion in 1980, an 11.8 per cent increase over 1979 that pushed the total above the $8 billion mark for the first time ever.

While revenues were up, however, broadcast expenses jumped almost 16 per cent in 1980 to $7.15 billion, compared with $6.18 billion in 1979. This left total pretax income of $1.65 billion, down 2.2 per cent from the 1.69 billion in 1979.

According to the FCC breakdown of the industry, expenses climbed 25.9 per cent for UHF stations as a group, compared with 14.8 per cent for the three networks and 14.7 per cent for all VHF stations. Thus, while network profits dropped 12 per cent to $325.6 million and the profits of VHF stations together rose a modest 3 per cent to 1.29 billion, the pretax profits of all UHF stations together dropped a surprising 42.6 per cent to $39.7 million in 1980.

Of the 748 TV stations in operation in 1980, 516 were VHF and 232 UHF.

VIEWING HABITS

TELEVISION USAGE

Number of TV homes	77,800,000
Numbers of color TV homes	66,300,000
Homes with two or more sets	39,800,000

	1978	1979	1980
Avg. Hours per Home	6 hrs. 10 min.	6 hrs. 28 min.	6 hrs. 36 min.

COMMERCIAL TELEVISION STATIONS	1980	1975	1973	1971
Total	734	706	696	682
VHF	516	514	510	503
UHF	218	192	186	179

COLOR TELEVISION	1970	1975	1981
Homes	23,400,000	48,500,000	66,250,000
Per Cent of TV Home	39.2	70.8	85.2

Source: A. C. Nielsen

TELEVISION USAGE

A BREAKDOWN OF TELEVISION VIEWERS

	Avg. Viewing Time Per TV Time Per Day (Hours and Minutes)
Total U.S.	7:02
By Household Income	
Under $10,000	6:15
$10,000–$14,999	7:13
$15,000 and Over	7:31
$20,000 and Over	7:20
$30,000 and Over	6:54
By Age of Lady of House	
18–34	7:19
35–54	8:01
55 and Over	6:36
By Household Size	
1 person	4:37
2 persons	6:20
3 persons	8:38
4 or more	9:08
By Education of Head of House	
Under 4 Yrs. H.S.	7:28
4 Yrs. H.S.	7:22
1 + Yrs. College	6:21
4 + Yrs. College	6:08
By Territory	
Northeast	7:13
East Central	7:01
West Central	6:50
South	7:14
Pacific	6:39
By County Size	
A	6:56
B	6:58
C & D	7:12

Source: A. C. Nielsen

TV EXPENDITURES

TOTAL ADVERTISING VOLUME

	Millions of Dollars			
	1979	1980	% Change	% of Total
Total Advertising	$49,520	$54,750	+10.6	100.0
TELEVISION	10,154	11,330	+11.6	20.7
Newspapers	14,493	15,615	+ 7.7	28.5
Magazines	2,932	3,225	+10.0	5.9
Radio	3,277	3,690	+12.6	6.7
Outdoor	540	610	+13.0	1.1
Business Papers	1,575	1,695	+ 7.6	3.1
All Others	16,549	18,585	+12.3	34.0

Source: McCann-Erickson 1/81

NATIONAL ADVERTISING INVESTMENTS IN MAJOR MEDIA

	Millions of Dollars			
	1979	1980	% Change	% of Total
TELEVISION	$ 7,472	$ 8,365	+12.0	49.3
Network	4,599	5,105	+11.0	30.1
Spot	2,873	3,260	+13.5	19.2
Newspapers	2,085	2,335	+12.0	13.8
Magazines	2,932	3,225	+10.0	19.0
Radio (Network & Spot)	820	935	+14.0	5.5
Outdoor	355	400	+12.7	2.4
Business Papers	1,575	1,695	+ 7.6	10.0
Total Major Media	$15,239	$16,955	+11.3	100.0

Source: McCann-Erickson 1/81

LOCAL ADVERTISING INVESTMENTS IN MAJOR MEDIA

	Millions of Dollars			
	1979	1980	% Change	% of Total
TELEVISION	$ 2,682	$ 2,965	+10.5	15.4
Newspapers	12,408	13,280	+ 7.0	69.1
Retail	7,801	8,600	+10.2	44.8
Classified	4,607	4,680	+ 1.6	24.3
Radio	2,457	2,755	+12.0	14.4
Outdoor	185	210	+12.5	1.1
Total Major Media	$17,732	$19,210	+ 8.3	100.0

Source: McCann-Erickson 1/81

LEADING LOCAL TV CATEGORIES

	add (000)		
	1979	1980	% Change
1. Restaurants & Drive-Ins	$263,290.4	$287,873.1	+ 9
2. Food Stores & Supermarkets	133,802.4	148,708.6	+11
3. Banks, Savings & Loans	139,760.5	140,948.2	+ 1
4. Department Stores	118,002.8	131,241.1	+11
5. Furniture Stores	103,519.8	111,420.0	+ 8
6. Movies	94,203.3	102,401.4	+ 9
7. Auto Dealers*	122,229.3	99,106.1	−19
8. Discount Department Stores	68,458.4	73,505.4	+ 7
9. Radio Stations & Cable TV	53,364.3	69,264.2	+30
10. Amusement & Entertainment	57,532.7	68,024.5	+18

Source: BAR * Dealer Associations Not Included

LENGTH OF ANNOUNCEMENTS

Announcement Length	Non-Network 1965	1970	1975	1980	Network 1965	1970	1975	1980
10 seconds	15%	11%	8%	8%	—%	—%	—%	1%
20 seconds	13	3	—	—	—	—	—	—
30 seconds	1	54	83	87	—	32	83	87
60's piggybacked	7	9	1	—	25	45	12	10
60 sec. or more	64	23	8	5	75	23	5	2
	100%	100%	100%	100%	100%	100%	100%	100%

Source: BAR, November of each year

TELEVISION SET SALES TOP 17.5 MILLION IN 1980

	Total	Color	Monochrome
1975	10,637,000	6,219,000	4,418,000
1976	14,131,000	8,194,000	5,937,000
1977	15,431,000	9,341,000	6,090,000
1978	17,406,000	10,674,000	6,732,000
1979	16,619,000	10,043,000	6,576,000
1980	17,508,000	10,779,000	6,729,000

Source: EIA, Domestic & Imports

CABLE TV OPERATING SYSTEMS AND SUBSCRIBERS

	1960	1965	1970	1975	1980	1981 (pre.)
Operating Systems	640	1,325	2,490	3,366	4,048	4,300
Subscribers (millions)	0.7	1.3	4.5	9.8	15.5	17.0
% of TV Homes	1.4%	2.4%	7.6%	14.3%	20.5%	22.0%

Source: Television Digest, Jan. 1 of each year

ALPHABETICAL INDEX OF SUBJECTS

A

B

C

P

Q

R

S

T

U

V

W

Z

Emmy Award Winners

ACADEMY OF TELEVISION ARTS AND SCIENCES

1980–81 TELEVISION ACADEMY AWARDS Categories and Areas of Achievement For the Emmy Award FOR THE PERIOD FROM July 1, 1980 THROUGH JUNE 30, 1981

ENTERTAINMENT CATEGORIES

Outstanding Comedy Series—"Taxi," James L. Brooks, Stan Daniels, Ed. Weinberger, executive producers, Glen Charles, Les Charles, producers.

Outstanding Drama Series—"Hill Street Blues," Steven Bochco, Michael Kozoll, executive producers, Gregory Hoblit, producer.

Outstanding Variety, Music or Comedy Program—"Lily: Sold Out," Lily Tomlin, Jane Wagner, executive producers, Rocco Urbisci, producer, CBS.

Outstanding Limited Series—Shogun, James Clavell, executive producer, Eric Bercovici, producer.

Outstanding Drama Special—"Playing for Time," Linda Yellen, executive producer, Linda Yellen, producer, John E. Quill, co-producer, CBS.

Outstanding Lead Actor in a Drama Series—Daniel J. Travanti, "Hill Street Blues," NBC.

Outstanding Lead Actor in a Limited Series or a Special—Anthony Hopkins, "The Bunker," CBS.

Outstanding Lead Actress in a Drama Series—Barbara Babcock, "Hill Street Blues," NBC.

Outstanding Lead Actress in a Limited Series or a Special—Vanessa Redgrave, "Playing for Time," CBS.

Outstanding Achievement in Hairstyling—Shirley Padgett, "Madame X," NBC.

Outstanding Achievement in Lighting Direction—Ralph Holmes, "Nureyev and the Joffrey Ballet/in Tribute to Nijinsky Dance in America," PBS.

Outstanding Lead Actor in a Comedy Series—Judd Hirsh, "Taxi," ABC.

Outstanding Lead Actress in a Comedy Series—Isabel Sanford, "The Jeffersons," CBS.

Outstanding Supporting Actor in a Comedy or Variety or Music Series—Danny DeVito, "Taxi," ABC.

Outstanding Supporting Actor in a Drama Series—Michael Conrad, "Hill Street Blues," NBC.

Outstanding Supporting Actress in a Drama Series—Nancy Marchand, "Lou Grant," CBS.

Outstanding Supporting Actress in a Comedy or Variety or Music Series—Eileen Brennan, "Private Benjamin," CBS.

Outstanding Supporting Actor in a Limited Series or a Special—David Warner, "Masada," ABC.

Outstanding Supporting Actress in a Limited Series or a Special—Jane Alexander, "Playing for Time," CBS.

Outstanding Writing in a Limited Series or a Special—Arthur Miller, "Playing for Time," CBS.

Outstanding Writing in a Comedy Series—Michael Leeson, "Taxi,"—"Tony's Sister and Jim," ABC.

Outstanding Directing in a Comedy Series—James Burrows, "Taxi"—"Elaine's Strange Triangle," ABC.

Outstanding Directing in a Limited Series or a Special—James Goldstone, "Kent State," NBC.

Outstanding Directing in a Drama Series—Robert Butler, "Hill Street Blues"—"Hill Street Station," NBC.

Outstanding Writing in a Drama Series—Michael Kozoll, Steven Bochco, "Hill Street Blues"—"Hill Street Station," NBC.

Outstanding Writing in a Variety, Music or Comedy Program—Jerry Juhl, David Odell, Chris Langham, Jim Henson, Don Hinkley, "The Muppet Show With: Carol Burnett," SYN.

Outstanding Directing in a Variety, Music or Comedy Program—Don Mischer, "The Kennedy Center Honors: A National Celebration of the Performing Arts," CBS.

Outstanding Animated Program—"Life Is a Circus, Charlie Brown," Lee Mendelson executive producer; Bill Melendez, producer, CBS.

CREATIVE ARTS IN ENTERTAINMENT CATEGORIES

Outstanding Art Direction for a Series—Howard E. Johnson, production designer; John M. Dwyer, Robert George Freer, set decorators, "The Gangster Chronicles Chapter Eleven," NBC.

Outstanding Art Direction for a Limited Series or a Special—Ray Storey, art director; Dennis Peeples, David Love, set decorators, John Steinbeck's "East of Eden," Episode 3, ABC.

Outstanding Art Direction for a Variety or Music Program—Roy Christopher, art director, "53rd Annual Academy Awards," ABC.

Outstanding Achievement in Choreography—Walter Painter, "Lynda Carter's Celebration," CBS.

Outstanding Individual Achievement-Special Class—Sarah Vaughan, performer, "Rhapsody and Song—A Tribute to George Gershwin," PBS.

Outstanding Costume Design for a Series—Shin Nishida, "Shogun," Episode 5, NBC.

Outstanding Costume Design for a Special—Willa Kim, "The Tempest Live With the San Francisco Ballet Dance in America," PBS.

Outstanding Achievement in Makeup—Albert Paul Jeyte; James Kail, "Peter and Paul," CBS.

Outstanding Children's Program—"Donahue and Kids, Project Peacock," Walter Bartlett, executive producer; Don Mischer, producer; Jan Cornell, co-producer, NBC.

Outstanding Film Editing for a Series—M. Pam Blumenthal; Jack Michon, "Taxi—Elaine's Strange Triangle," ABC.

Outstanding Film Editing for a Limited Series or a Special—John A. Martinelli, A.C.E., "Murder in Texas," NBC.

Outstanding Video Tape Editing for a Series—Andy Ackerman, "WKRP in Cincinnati—Bah, Humbug," CBS.

Outstanding Video Tape Editing for a Limited Series or a Special—Marco Zappia, "Perry Como's Christmas in the Holy Land," ABC.

Outstanding Technical Direction and Electronic Camerawork—Heino Ripp, technical director; Peter Basil, Al Camoin, Tom Dezendorf, Vince Di Pietro, Gene Martin, camerapersons, "Live From Studio 8H: An Evening of Jerome Robbins' Ballets with Members of the New York City Ballet, NBC.

Outstanding Individual Achievement—Creative Technical Crafts—Jon Allison, Adolf Schaller, Don Davis, Rick Sternbach, John Lomberg, astronomical artists, "Cosmos, the Shores of the Cosmic Ocean," PBS.

Carey Melcher, Technical designer; Bob Buckner, Steve Burum, Jim Dow, John Gale, Larry Heider, Mike Johnson, Robert C. King, Cleve Landsberg, Joseph Matza, George C. Reilly, Joe Wolcott, Magicam crew, "Cosmos, the Shores of the Cosmic Ocean," PBS.

Outstanding Achievement in Tape Sound Mixing—Jerry Clemans, Doug Nelson, Donald Worsham, "John Denver With His Special Guest George Burns, Two of a Kind," ABC.

Outstanding Achievement in Film Sound Mixing—William R. Teague, production mixer; Robert L. Harman, William L. McCaughey, Howard Wollman, re-recording mixers, "Evita Peron, Part I," NBC.

Outstanding Achievement in Film Sound Editing—Samuel Horta, supervising editor; Robert Cornett, Denise Horta, Eileen Horta, editors, "Hill Street Blues, Hill Street Station," NBC.

Outstanding Classical Program in the Performing Arts—"Live From Studio 8H: An Evening of Jerome Robbins' Ballets With

Members of the New York City Ballet," Alvin Cooperman, Judith De Paul, producers, NBC.

Outstanding Achievement in Music and Lyrics—Ken Welch, composer; Mitzie Welch, lyricist, "Lynda in Wonderland," Song: "This Is My Night," CBS.

Outstanding Achievement in Music Direction—Ian Fraser, music director; Billy Byers, Chris Boardman, Bob Florence, arrangers, "Lynda in Wonderland," CBS.

Outstanding Achievement in Music Composition for a Limited Series—Jerry Goldsmith, "Masada, Episode 2," ABC.

Outstanding Achievement in Music Composition for a Series (Dramatic Underscore)—Bruce Broughton, "Buck Rogers," "The Satyr," NBC.

Outstanding Graphic Design and Title Sequences—Phil Norman, "Shogun, Episode 1," NBC.

Outstanding Cinematography for a Series—William H. Cronjager, "Hill Street Blues, Hill Street Station," NBC.

Outstanding Cinematography for a Limited Series or a Special—Arthur F. Ibbetson, B.S.C., "Little Lord Fauntleroy," CBS.

CREATIVE ARTS IN ENTERTAINMENT AREAS (Possibility of one award, more than one award or no award)

Outstanding Individual Achievement—Informational Programming—Kent Gibson, Gerald Zelinger, tape sound mixers, "Cosmos, Blues for a Red Planet," PBS.

Dick Rector, production mixer; Gary Bourgeois, Dave Dockendorf, John Mack, re-recording mixers, "Gorilla, National Geographic Special," PBS.

Outstanding Informational Series—Steve Allen's "Meeting of Minds," Loring d'Usseau, producer, PBS.

Outstanding Informational Special—"The Body Human: The Bionic Breakthrough," Thomas W. Moore, executive producer; Alfred R. Kelman, Robert E. Fuisz, M.D., producers; Charles A. Bangert, Nancy Smith, co-producers, CBS.

Syndicated Shows

Listed below are titles available for syndicated runs. Asterisk denotes barter program; double asterisk denotes barter or cash.

Title	Number	Type	Syndicator
After School Specials	14 1 hrs. (11 off-network)	Children's	Time-Life TV
American Express Presents*	24 ½ hrs.	Instructional	Vitt Media
Animaland	104 4 mins.	Children's	Laff-A-Bits
Animated Holiday Specials	5 ½ hrs.	Children's	Four Star
Backlot	24 ½ hrs.	Talk/Interview	Silverbach/Lazarus
Backstage At The Grand Old Oprey*	26 ½ hrs.	Music/Variety	Show Biz
Battle For The Ballot*	34 1 mins.	Instructional	Carter-Grant Prods.
B.B. Beegle Show	24 ½ hrs.	Comedy	Y & R Program Services
Beachcombers	150 ½ hrs.	Adventure	CBC Program Sales
Beatlemania	Weekly ½ hrs.	Documentary	Lexington Broadcast Services
Bert Convy Show	60 or 90 mins. strip	Music/Variety	Colbert TV Sales
Bet Your Bottom Dollar	½ hr strip	Game	MPC
Bill Dance Outdoors*	39 ½ hrs.	Sports	Advanswers
Body Break	130 1-mins.	Instructional	Rhodes Prods.
Bounce Back	65 ½-hr strip	Game	Claster TV Prods.
Bullseye	½-hr strip	Game	Colbert TV Sales
Burbon Street Parade	26 1 hrs.	Music/Variety	William F. Frank TV Programs
Butcher, Baker	260 1-mins.	Women's	Rhodes Prods.
Casino Game, The	240 ½-hr strip	Game	MPC
Camouflage	26 ½ hrs.	Game	Firestone Program Syndication
Canned Film Festival	26 90-mins.	Comedy	Lexington Broadcast Services
Card Sharks	½-hr strip	Game	Firestone Program Syndication
Champions: The Competitive Edge	39 ½ hrs.	Sports	Hollywood Distributors
Children's Specials	12 ½ hrs.	Children's	Video Voice, Inc.
Chuck Connors' Great Western Theatre	26 90/60-mins. (off-network)	Adventure	Leo A. Gutman
Circle Square	26 ½ hrs.	Children's	Janus TV Corp.
Connection	5 ½ hrs.	Documentary	D.L. Taffner
Coors Western Outdoorsman*	26 ½ hrs.	Sports	Advanswers
Countryside	104 2-mins.	Talk/Interview	Compro
Dear Tony	150 1-mins.	Instructional	Rhodes Prods.
Don Lane Show	26 1 hrs.	Music/Variety	Worldvision
Doug Llewelyn Entertainment Report	260 1½-mins.	Talk/Interview	Woodlake Prods.
Eyewitness To The Past	39 ½ hrs.	Instructional	Hollywood Distributors
Face The Music	½-hr strip (5 per week)	Game	Sandy Frank
'50s Connection	24 ½ hrs.	Music/Variety	Visual Prods.
Fight Back! With David Horowitz	26 ½ hrs.	Instructional	Group W
Five Star Whozit?	½-hr strip	Game	Rhodes Prods.
Flappers	26 ½ hrs.	Comedy	CBC Program Sales
Food, Folklore and Fun	26 ½ hrs.	Women's	Gerber/Carter
Fourth Dimension, The	7 1-hr specials	Science Fiction	Con Hartsock & Co.
Fantasies Fulfilled*	½-hr strip	Game	Air Time
Fun & Fitness	118 ½ hrs.	Instructional	Hollywood Distributors
Future Flipper	26 ½ hrs.	Adventure	NTA
Garden Gazette	260 1½ mins.	Instructional	Mighty Minute Programs
Glitter	78 ½ hrs.	Talk/Interview	Visual Prods.
Good Taste	½-hr strip	Magazine	Mighty Minute Programs
Hank Grant—Hollywood Reporter	260 90-secs.	Talk/Interview	Mighty Minute Programs
Health Field, The*	26 ½ hrs.	Instructional	Lexington Broadcast Services
Hour Magazine	26 1 hrs.	Talk/Interview	Group W
Incredible World	5 ½ hrs.	Documentary	Lexington Broadcast Services
Inside Track	29 ½ hrs.	Children's (teens)	Janus TV Corp.
Instant Comedy	½-hr strip	Game	MPC
Invitation To Dance**	26 1 hrs.	Music/Variety	Tele-Tran Prods.

Irish Shower Half-Hour	24 ½ hrs.	Music/Variety	Visual Prods. TV
It's A Fact	130 1-mins.	Instructional	Rhodes Prods.
Jerry Van Dyke Show	90-min. strip	Music/Variety	MPC
John Davidson Show	90-min. strip	Music/Variety	Group W
Johnny's Oasis Bar & Grill	½-hr strip	Music/Variety	TVS TV Network
Just Between Us	130 ½ hrs.	Instructional	Rhodes Prods.
Kid Break	130 1-mins.	Children's	Rhodes Prods.
Kidoozle	26 ½ hrs.	Children's	TV Syndication Group
Kid Stuff**	26 ½ hrs.	Children's	Syndicast
Let's Make A Deal*	½-hr strip	Game	Rhodes Prods.
Little Rascals	195 20-secs.	Instructional	King World Prods.
Littlest Hobo, The	26 ½ hrs.	Children's	Silverbach/Lazarus
Lively Country	78 ½ hrs.	Music/Variety	Visual Prods.
Lively Specials	26 ½ hrs.	Music/Variety	Visual Prods.
Look What They've Done To My Song*	26 ½ hrs.	Music/Variety	Lexington Broadcast Services
Magic Star Traveller**	86 ½ hrs.	Children's	Joseph Green Pictures
Malu, A Woman	13 1 hrs.	Drama	TV Globo of Brazil
Masters Of The Martial Arts	26 ½ hrs.	Instructional	TV Syndication Group
Meet The Lipits	26 ½ hrs.	Music/Variety	Gerber/Carter
Melody Ranch	38 ½ hrs.	Music/Variety	Jeralt TV Distribution
Merry Morning	104 2-mins.	Comedy	Compro
Mismatch	25 ½ hrs.	Game	Time-Life TV
Monte Carlo Show, The*	24 1 hrs.	Music/Variety	20th-Fox TV
Morning After*	90-min strip	Magazine	ICM TV Marketing
Motor Race Of The Week	11 ½ hrs.	Sports	Lexington Broadcast Services
Mouthtrap	½-hr strip	Game	MPC
Ms. Fixer Upper	130 3½ mins.	Instructional	Laff-A-Bits
Music Country, U.S.A.	20 1 hrs.	Music/Variety	NTA
Music World	78 1 hrs. or 156 ½ hrs.	Music/Variety	Hollywood Distributors
Nashville Swings	26 ½ hrs.	Music/Variety	William F. Cooke TV Programs
Nellie, Daniel, Emma & Ben	12 ½ hrs.	Comedy	Media Lab TV
New Ed Allen Show*	260 ½-hr strip	Instructional	American TV Syndication
New Original Amateur Hour	24 1 hrs.	Music/Variety	Brad Marks
The New Quiz Kids	24 ½ hrs.	Game	Columbia Pictures TV
Nutz & Boltz Theatre	26 ½ hrs.	Children's	Hollywood Distributors
Olympic Gold*	7 ½ hrs.	Sports	Syndicast
Open House Theatre	26 ½ hrs.	Music/Variety	Hollywood Distributors
Operation Prime Time	24 1 hrs.	Drama	Various
Our Incredible World	½ hrs.	Documentary	Lexington Broadcast Services
Pandemonium	½-hr strip	Game	Silverbach/Lazarus
Phrase It	½-hr strip	Game	Paramount TV
Picture Of Health, The	65 ½ hrs.	Instructional	Bardon & Associates
Pink Panther	26 ½ hrs.	Children's	UA-TV
Play The Percentages	½ hr strip	Game	Colbert TV Sales
P.M. Magazine Special Edition	6 1 hrs.	Magazine	Group W
Portrait—The New Breed	24 ½ hrs.	Talk/Interview	Four Star Entertainment
Prisoner: Cellblock H	260 ½ hrs.	Drama	Firestone Syndication
Press Your Luck	½-hr strip	Game	CPT
Punchlines	½-hr strip	Game	MPC
Rex Reed's Movie Guide	26 ½ hrs.	Talk/Interview	Group IV
Roaring Doughnuts	26 ½ hrs.	Children's	United Children's TV
Rock A Mania (Let's Rock)	24 ½ hrs.	Music/Variety	ITC
Rock World	26 1 hrs.	Music/Variety	NTN Entertainment
Sensational Seventies	52 ½ hrs. or 12 1 hrs.	Documentary	American TV
Snickle Tickle Library	½-hr strip	Children's	Osmond TV
Sounds Good	78 ½ hrs.	Music/Variety	Visual Prods.
Sports Extra	26 ½ hrs.	Sports	Lexington Broadcast Services
Sports Stars	26 ½ hrs.	Sports	Jeralt TV Distribution
Sports Whirl	26 ½ hrs.	Sports	Coe Films
Spy	6 1 hrs.	Drama	Viacom
Star Tracks	26 ½ hrs.	Music/Variety	Fremantle Corp.
Steve Edwards Show	60- or 90-min strip	Talk/Interview	CPT
Stranger Than Fiction	26 ½ hrs.	Science Fiction	NTA
Studio Two Revue	25 ½ hrs.	Music/Variety	Gerber/Carter
Success	104 ½ hrs.	Talk/Interview	Janus TV Corp.
Successworld	25 ½ hrs.	Magazine	Southbury Prods.
Sullivans, The	600 ½-hr strip	Drama	TAT
Superspecials	8 1 hrs.	Music/Variety	CBC Program Sales

Superstar Profile	13 ½ hrs.	Documentary	Jerry Dexter Syndication
Suspense Theatre Of The Air	52 1 hrs or ½ hrs.	Drama	Cine Centrum
Teamwork	36 ½ hrs.	Talk/Interview	TV Syndication Group
Tennis Break	130 1-mins.	Sports	Rhodes Prods.
That Good Ole Gospel Music	78 1 hrs.	Music/Variety	Hollywood Distributors
That Joan Rivers Show	260 ½ hrs.	Comedy	TV National
T.H.E. Hospital	½-hr strip	Comedy	20th-Fox TV
This Was America	12 ½ hrs.	Documentary	BBI Communications
Three On The Go*	25 ½ hrs.	Adventure	JWT
Toni Tennille Show	60- or 90-min. strip	Music/Variety	MCA-TV
Top Of The Hill	4 1 hrs.	Drama	Paramount TV
To Tell The Truth	½-hr strip	Game	Viacom
Wayne & Shuster	70 ½ hrs.	Comedy	D.L. Taffner
Wayne Thomas Show	26 1 hrs.	Music/Variety	Visual Prods.
Whatever Happened To?	26 ½ hrs.	Magazine	Deepwood Prods.
White Escape, The	13 ½ hrs.	Sports	Dick Barrymore
Wide World Of People	200 ½-hr strip	Magazine	Victory TV
Wide World of Women	26 ½ hrs.	Women's	Coe Films
Wilderness Alive	4 1 hrs.	Documentary	Time-Life TV
World Beyond, The	120 1 hrs.	Documentary	Hollywood Distributors
World Tennis	17 1 hrs.	Sports	Hollywood Distributors
Yellow Woodpecker Ranch	80 ½ hrs.	Children's	TV Globo of Brazil
You Bet Your Life	26 ½ hrs.	Game	MCA-TV
You Can Do It*	20 ½ hrs.	Instructional	Blair Advertising
Your New Day	130 ½ hrs.	Women's	SFM
Your Tax Man	65 45-sec.	Instructional	Virpo
ZAP	½-hr strip	Game	MPC

Recent additions:

Barnaby Jones	178 1 hrs.	Drama	Worldvision
Barney Miller available-fall 1980)	128 ½ hrs.	Comedy	CPT
Best Of Sullivan Returns To TV edited to ½ hrs.)	52 1 hrs.	Music/Variety	Syndicast
Charlie's Angels (available-fall 1981)	100 1 hrs.	Drama	CPT
Dallas	1 hrs.	Drama	Lorimar
Dan August	5 2 hrs. or 10 1 hrs.	Drama	Goldwyn TV
Dean Martin Show	26 1 hrs.	Music/Variety	NTA
Eight Is Enough	100 1 hrs.	Comedy	Lorimar
Flip Wilson Show, The	26 1 hrs.	Music/Variety	NTA
I, Claudius	13 1 hrs.	Drama	MPC
Jeffersons, The	159 ½ hrs.	Comedy	TAT
Laugh-In	130 ½ hrs.	Comedy	Silverbach/Lazarus
Little House On The Prairie	170 1 hrs.	Drama	Worldvision
One Day At A Time	161 ½ hrs.	Comedy	TAT
Return Of The Saint	22 1 hrs.	Drama	ITC
Sonny & Cher Comedy Hour*	26 1 hrs.	Music/Variety	Air Time
Three's Company	130 ½ hrs.	Comedy	D.L. Taffner
Washington: Behind Closed Doors	12 1 hrs.	Drama	Paramount TV
What's Happening	89 ½ hrs.	Comedy	CPT
Wonder Woman	61 1 hrs.	Adventure	WB-TV

INDEX OF ADVERTISERS

Who's Who

In MOTION PICTURES and TELEVISION

A

AARON, ROY H.: Executive. Attorney in L.A. law firm of Pacht, Ross, Warne, Bernhard & Sears. Joined Plitt Companies in 1978 as snr. v.p. & gen. counsel. In 1980 named pres. & chief operating officer of Plitt Theatres, Inc. and related Plitt companies.

ABARBANEL, SAM X.: Producer, writer, publicist. b. Jersey City, N.J. Mar. 27, 1914. e. Cornell U & U. of Illinois, B.S. 1935. Newspaperman Chicago before joining N.Y. exploitation dept. Republic, then to studio as ass't. publicity director. World War II in Europe with 103rd Div. After war independent publicist and producer. Co-prod. Argyle Secrets, 1948. Co-wrote orig. s.p., co-produced U.A.'s Prehistoric Women, 1950. Exec. Prod. U.A.'s Golden Mistress. In 1963 to Spain as associate prod. MGM's Gunfighters of Casa Grande, and Son of Gunfighter. Formed own co. in Spain, 1966. Produced Narco Men. Co-authored, prod. Last Day of War. Orig. & co-s.p. Avco's Summertime Killer, 1972.

ABEL, WALTER: Actor, b. St. Paul, Minn.; e. Amer. Acad. Dram. Art, N.Y. Role in touring co of N.Y. prod. "Come Out of the Kitchen," then stock, vauderville; stage mgr. touring cos. Minnie Maddern Fiske & others; on Broadway in "A Square Peg," "Long Watch"; m.p. sound debut in "Three Musketeers" 1935.

PICTURES INCLUDE: Witness Chair, Racket Busters; Arise, My Love; Holiday Inn, Star Spangled Rhythm, 13 Rue Madeleine, Kid from Brooklyn, Hal Roach Comedy Carnival, So This Is Love, Island in the Sky, Night People, Indian Fighter, The Steel Jungle, Bernadine, Raintree County, Handle with Care, Mirage.

ABELES, ARTHUR: Chairman, Filmartketeers, Ltd. b. 1914. Educated Duke University, Columbia University, U.S. 1939–1941: general manager, West Indies, Warner. 1941: assistant general manager, Brazil, Universal. 1942–5: general manager, Uruguay, Warner. 1945–7: supervisor, Argentina, Chile, Uruguay, Paraguay, Warner. 1948–68: Managing director, Warner Bros. Pictures, Warner Theatre, Warner Bros. Production; Director Henderson Film Laboratories; Vice-president Warner Bros. International Cor.; in charge of Europe, the Near East and North Africa. 1968–70: Managing director, Universal Pictures Ltd. Vice-president Universal International in charge of U.K., Continent and Middle East. 1970–7: Co-chairman, Cinema International Corporation. 1978: Chairman, Filmarketeers, Ltd.

ABEND, SHELDON: Executive. b. New York, N.Y. June 13, 1929. Maritime Labor-Rel. Negotiator, 1954–56; chmn, Maritime Union, 1954–56; head, exec. dir. Research Co. rep. estates decreased authors, est. 1957. indep. literary negotiator, C&C Films, A.A.P., RKO General Inc., David O. Selznick, 1959–65; pres. American Play Co. Inc., Century Play Co. Inc., 1961–72; est. Million Dollar Movie Play Library, 1962; pres. American Literary Consultants est. 1965; exec. v.p. Chantec Enterprises Inc., 1965; Literary prod. consultant Zane Grey Corp. Inc. 1969–72. Marketing Literary Consultant for Damon Runyon Estate. Copyright analyst and literary rights negotiator, United Artists Corp. Founder and chrm. of Guild for Author's Heirs, 1970–72, Literary negotiator and prod. consultant for Robert Fryer, 1972. Founder of the Copyright Royalty Co. for Authors Heirs, 1974; copyright consultant for Films Inc. 1975; literary agent for Bway. play, Chicago, 1975. Owner of 53 classic RKO motion pictures for the non-theatrical markets. Distributed by Films, Inc. Revived publishing of Damon Runyon stories in quality paperback in conjunction with Runyon-Winchell cancer fund. Published six mystery stories written by Cornell Woolrich—all of which were produced by Alfred Hitchcock for TV & motion pictures, 1976. 1978—assoc. prod. of film, Why Would I Lie?; Originator of Million Dollar Movie Book Theatre and Million Dollar Movie Sound Track Co., 1980; assoc. prod. of Bdwy. revival, Shanghai Gesture, 1981.

ABRAHAMS, GARY: Executive. e. Univ. of Arizona (BA); in 1968 joined Paramount TV as asst. to v.p.; 1974–75, asst. to prod., The Movies; 1971–78 assoc. dir., Los Angeles Intl. Film Exposition (Filmex); 1978, joined Marble Arch Productions, as exec. asst. to pres.; named dir. of corporate relations & merchandising, 1979; project coordinator, Los Angeles Film Center 1981.

ABRAHAMS, MORT: Producer. b. New York, N.Y., March 26, 1916; e. NYU, A.B., 1936; Columbia U., M.A., 1940. Prod. shorts, documentaries, Paul Falkenberg Prod., N.Y., 1940–42; Columbia Pictures, 1942–48 TV prod., Rockhill Radio, N.Y., 1950; film consultant, Bank of America, N.Y. 1951; TV prod. George F. Foley, 1952–53; TV prod., Music Corp. of America, 1953; v.p. in charge of East Coast TV prod., 1956; exec. prod. Producers Showcase, NBC-TV, June 1956–57; Suspicion, 1958; dir. programming, prod., NTA, 1958–1960; exec. prod. Cooga Mooga Prod. Inc., 1960–1961. Producer: Target, The Corruptors 1961, Route 66, 1962–63; writer T.V. Shows, 1963–64; prod., Kraft Suspense Theatre, 1965; prod. Man from U.N.C.L.E., 1965–66; Executive Vice-President, AJAC Prod. 1966. Assoc. prod. "Doctor Doolittle," "Planet of the Apes," "Goodbye, Mr. Chips." Prod., "The Chairman." Assoc. producer & writer, "Beneath the Planet of the Apes"; 1969, Vice President in charge of production, Rastar Prods.: 1971–74 exec. prod. American Film Theatre & v.p. Ely Landay Org. in chg. West Coast prod. exec. prod. on Luther, Homecoming, Man in the Glass Booth; 1975-exec. chg. prod., Transworld Entertainment; 1977–78, exec. prod., The Greek Tycoon, Hopscotch, exec. prod., 1980; exec. in chg. prod., The Chosen, Beatlemania, 1981.

ACKERMAN, BETTYE: Actress. b. Cottageville, S.C., Feb. 28, 1928. e. Columbia Univ., 1948–52. Taught dancing 1950–54.

PLAYS INCLUDE: No Count Boy (1954), Tartuffe (1956), Sophocles' Antigone, and "Oedipus at Colonus" (1956). The Merchant of Venice (1971).

PICTURES INCLUDE: Face of Fire (1959), Companions in Nightmare, Rascal.

TV: Alcoa Premiere, Hitchcock, Perry Mason, Breaking Point, Dr. Maggie Graham on Ben Casey series for five yrs., Hope-Chrysler Theatre, Bonanza, FBI Story, Mannix, Ironsides, "Medical Center," regular on "Bracken's World," Colombo, Sixth Sense, Heat of Anger, Return to Peyton Place, (6 months 1972) Rookies, Barnaby Jones, Police Story, Gunsmoke, Harry O, Streets of San Francisco, S.W.A.T.

LP RECORDS: Salome & School for Scandal.

ACKERMAN, HARRY S.: Executive. b. Albany, N.Y., Nov. 17, 1912, e. Dartmouth Coll. Began as writer, actor, producer. Joined Young & Rubicam adv. agcy. 1936; became v.p. of program operations, 1946. Exec. prod. CBS, 1948. Dir. of network programs—Hollywood June, 1948; v.p. charge network radio programs, 1950; v.p. chge. network TV programs CBS-TV, Hollywood, June, 1951; exec. dir. spec. progs. CBS network, 1956; formed independent company 1957; v.p. and exec. prod. Screen Gems Inc., 1958–1973; pres. Harry Ackerman Prods. na'l. pres., Academy TV Arts and Sciences, two terms; now v.p. in chg. TV, Capitol Pictures.

ACKLAND, JOSS: Actor. b. London, England, Feb. 29, 1928. Member of old Vic; toured Russia and U.S. with them. Over 400 TV appearances.

PICTURES INCLUDE: Seven Days to Noon, Royal Flash, Crescendo, The House That Dripped Blood, Villain, England Made Me, The Black Windmill, Great Expectations, The Greek Tycoon, Someone is Killing the Great Chefs of Europe, Saint Jack, The Apple, Rough Cut.

ADAM, KEN: Art dir. b. Berlin, Germany, 1921. e. St. Pauls School, London, and London Univ., student of architecture. 6 years war service as RAF pilot. Ent. m.p. ind. as draughtsman 1947. Art dir. Around the World in 80 Days, Trials of Oscar Wilde, Dr. No, Sodom and Gomorrah, Dr. Strangelove, Goldfinger, Thunderball, Ipcress File, You Only Live Twice, Funeral in Berlin, Chitty Chitty Bang Bang, Goodbye Mr. Chips, The Owl and the Pussycat, Diamonds Are Forever, Sleuth, The Last of Sheila, Barry Lyndon, Salon Kitty, 7% Solution, The Spy Who Loved Me, Moonraker.

ADAMS, BROOKE: Actress. b. New York City. e. High School of Performing Arts; Institute of American Ballet; Lee Strasberg. Made professional debut at age of six in Finian's Rainbow. Worked steadily in summer stock and TV until age 18. After hiatus resumed acting career.
PICTURES: Car Wash, Days of Heaven, Invasion of the Body Snatchers, Cuba, a Man, a Woman and a Bank.
TELEVISION: James Dean: Portrait of a Friend, The Last of the Belles, The Lords of Flatbush, The Daughters of Joshua Cabe, Murder on Flight 502, The Bob Newhart Show, Police Woman.

ADAMS, EDIE: Actress, singer. r.n. Edith Adams Kovacs. b. Kingston, Pa., April 16, 1931. e. Juilliard School of Music, Columbia School of Drama.
STAGE: Wonderful Town, 1952–54; Lil Abner, 1956–57.
TV: The Chevy Show, 1958: Take a Good Look, 1960–61; Here's Edie, 1962–63.
PICTURES INCLUDE: The Apartment, Mad, Mad, Mad, Mad World, Call Me Bwana, Under the Yum, Yum Tree, Love With A Proper Stranger, Lover Come Back, The Best Man, Maid in Paris, The Honey Pot, Up in Smoke.

ADAMS, GERALD DRAYSON: Writer. b. Winnipeg, Manitoba; e. Oxford University. Export exec. 1925–30; literary agt. 1931–45. Member: Screen Writers' Guild.
PICTURES INCLUDE: (jt. orig. screen story) Magnificent Rogue, (jt. s.p.) Plunderers, (s.p.) Gallant Legion, (jt. s.p.) Big Steal; collab. story, Dead Reckoning; Orig. s.p. Battle of Apache Pass, Son of Ali Baba, Flaming Feather, Flame of Araby; collab. s.p. Lady from Texas, Steel Town, Untamed, Frontier; story, collab. s.p., Duel at Silver Creek; st., s.p., Princess of the Nile, Three Young Texans, Gambler from Natchez; st., Wings of the Hawk, Between Midnight and Dawn; st., adapt., Taza Son of Cochise; collab. s.p. story, Gambler from Natchez, Chief Crazy Horse, s.p. Golden Horde; adapt. s.p. Prince Who Was a Thief; orig. s.p., Sea Hornet; s.p., Three Bad Sisters; orig., s.p., Duel on the Mississippi; orig. Black Sleep, s.p., War Drums; story, collab. s.p., Gun Brothers; story, Affair in Reno; story, s.p., Frontier Rangers; story, s.p., Gold, Glory & Custer; orig. story, collab. s.p., Kissin Cousins. Orig. s.p. Harem Scarem.
TV: Maverick, G.E. Theatre, Northwest Passage, Broken Arrow, Cheyenne, 77 Sunset Strip.

ADAMS, JULIE: Actress. r.n. Betty May Adams. b. Waterloo, La., Oct. 17, 1928; e. Jr. Coll., Little Rock, Ark., m. Ray Danton, Actor; Coll. dramatic; m.p. debut in Red Hot and Blue, Bright Victory; as Betty Adams in: The Dalton Gang, Marshal of Heldorado, West of the Brazos; since in Hollywood Story, Finders Keepers, Bend of the River, Treasure of Lost Canyon, Horizons West, Lawless Breed, Mississippi Gambler, Man from the Alamo, Wings of the Hawk, Star of Tomorrow, 1953.
PICTURES INCLUDE: Tickle Me, The Last Movie, MCQ, The Wild McCullochs, Killer Force, The Fifth Floor.
TV: Go Ask Alice.

ADDAMS, DAWN: Actress. b. Felixstowe, Suffolk, England, Sept. 21, 1930. e. Royal Acad. of Drama, London. Acted in rep. in Europe & on English stage; to Hollywood 1950.
PICTURES INCLUDE: Night into Morning, Singin' in the Rain, Hour of 13, Plymouth Adventure, Young Bess, Unknown Man, Moon is Blue, The Robe, Riders to the Stars, Return to Treasure Island, Khyber Patrol, Mizar, The Bed, Treasure of Rommel, London Calling North Pole, King in New York, The Silent Enemy, Secret Professional, Long Distance, The Two Faces of Dr. Jekyll, Follow That Man, The Liars, Lessons In Love, Come Fly With Me, The Black Tulip, Ballad in Blue, Vampire Lovers, Zeta One, The Vault of Horror.
PLAYS: London: Peter Pan, The Coming Out Party, The Wild Duck.
TV: The Saint, Danger Man, Romans and Friends, Emergency Ward 10, Dept. S. International Lawyer, Star Maidens.

ADDISON, JOHN: Composer. b. West Chobham (Surrey) Eng., March 16, 1920; e. Wellington and Royal Coll. of Music. Entered m.p. ind. in 1948, Professor, Royal College of Music.
PICTURES INCLUDE: (Music scores) The Honey Pot, Torn Curtain, A Fine Madness, I Was Happy Here, The Loved One, Guns At Batasi, The Uncle, Girl With Green Eyes, Tom Jones, The Loneliness of the Long Distance Runner, A Taste of Honey, The Charge of the Light Brigade, Smashing Time, Start The Revolution Without Me, Country Dance (in U.S., Brotherly Love), Forbush and the Penguins, Sleuth, Luther, Dead Cert, Ride A Wild Pony, Seven-Per-Cent Solution, Swashbuckler, Joseph Andrews, A Bridge Too Far, The Pilot, High Point.
BALLETS INCLUDE: Carte Blanche (at Sadler Wells and Edinburgh Festival).
PLAYS INCLUDE: The Entertainer, The Chairs, Luther, Midsummer Nights Dream, The Broken Heart, The Workhouse Donkey, Hamlet, Semi-Detached, The Seagull, Saint Joan of the Stockyards, Listen to the Mockingbird, Cranks (revue), Antony & Cleopatra, I Claudius, Twelfth Night, Bloomsbury.
TV: Sambo and the Snow Mountains, Detective, Hamlet, The Search for Ulysses, Way of the World, Back of Beyond, Black Beauty, The Bastard, Deadly Price of Paradise, Centennial, Pearl, Love's Savage Fury, Like Normal People, The French Atlantic Affair.
STAGE MUSICALS: The Amazons, Popkiss.

ADELMAN, JOSEPH A.: Executive. b. Winnipeg, Manitoba, Canada, Dec. 27, 1933. e. New York University, B.A., 1954; Harvard Law School, J.D., 1957, graduated cum laude, Attorney, United Artists Corp., New York, 1958; named west coast counsel, Hollywood, 1964; named executive assistant to the vice-president in charge of production 1968; named vice-president, west coast business and legal affairs, 1972; named executive vice-president, Association of Motion Pictures and Television Producers, 1977. Admitted to New York, California and U.S. Supreme Court bars; member, Phi Beta Kappa; American Bar Association; Los Angeles Copyright Society; Academy of Motion Picture Arts and Sciences; legislative liaison committee, Motion Picture Association of America; board of directors, AMPTP, 1969– ; board of trustees, Theatre Authority, 1970– .

ADLER, BOB: Actor, r.n. Robert Fick Adler. b. Hoboken, N.J., Mar. 24, 1906. Ranch Foreman, polo player, horse trainer, riding instructor; horseman in D. W. Griffiths "America" 1924, and others; in many Westerns since; member 20th-Fox Stock Co. from 1944.
PICTURES INCLUDE: My Darling Clementine, Smoky, Captain from Castile, Green Grass of Wyoming, Yellow Sky, Broken Arrow, Two Flags West, Lure of the Wilderness, Inferno, Hell and High Water, Prince Valiant, Broken Lance, Violent Saturday, Untamed, The Tall Men, Bottom of the Bottle, Proud Ones, The Silver Ship, Valerie, Fury at Showdown, Yellow Sky, Rope Law, Untamed, Rios Conchos, Whiskey Renegades, Stagecoach, Bandalero, The Undefeated.
TV: Flicka, Broken Arrow, Have Gun Will Travel, Cheyenne, Laramie, Rawhide, Bonanza, Daniel Boone, Gunsmoke, Batman, Judd, Big Valley, Lancer, Outcasts, Land of the Giants, Medical Center.

ADLER, LUTHER: Actor. b. New York City, 1903. On stage: Saigon, Loves of Carmen, Wake of the Red Witch.
PICTURES INCLUDE: House of Strangers, D.O.A., South Sea Sinner, Under My Skin, Kiss Tomorrow Goodbye, M. Magic Face, Desert Fox, Hoodlum Empire, Tall Texan, Miami Story, Girl in the Red Velvet Swing, Crashout, Hot Blood, Crazy Joe, Murph the Surf.

ADRIAN, IRIS: Actress, r.n. Iris Adrian Hostetter. b. Los Angeles, May 29, 1913; p. prof. Dancer, on stage N.Y. and abroad in "Follies of 1931"; m.p. debut in MGM two-reel color films.
PICTURES INCLUDE: Smart Woman, Out of the Storm, Paleface, Sky Dragon, There's a Girl in My Heart, Tough Assignment, Once a Thief, Always Leave Them Laughing, Stop That Cab, My Favorite Spy, Highway Magnet, Fast & The Furious, Crime Wave, Devil's Harbor, Carnival Rock, The Buccaneer, Blue Hawaii, The Errand Boy, Fate is the Hunter, That Darn Cat, The Odd Couple, The Love Bug.

AFTON, RICHARD LORD: Actor, director, producer, writer, sen. prod/dir. BBC Television since 1947. Prod. over 1200 shows on TV and many stage, film productions. Resigned BBC Oct. 1965 to form Freelance Co. Former man. dir. Ardmore Studios, Ireland. TV Columnist, London Evening News.

AGAR, JOHN: Actor. b. Chicago, Ill., Jan 31, 1921. In service World War II. Has appeared in Fort Apache, Adventure in Baltimore.
PICTURES INCLUDE: I Married a Communist, Sands of Iwo Jima, She Wore a Yellow Ribbon, Breakthrough, Magic Carpet, Along the Great Divide, Woman of the North Country, Man of Conflict, Bait, Rocket Man, Shield for Murder, Golden Mistress, Revenge of the Creature, Hold Back Tomorrow, Tarantula, Star in the Dust, Joe Butterfly, St. Valentine's Day Massacre, Waco, Johnny Reno, Women of the Prehistoric Planet.

AHERNE, BRIAN: Actor. b. Worcestershire, Eng., May 2, 1902; e. Malvern Coll., U. London 1920. In Brit. m.p. since 1924; on stage in London, on Broadway in Barretts of Wimpole Street, Lucrece, Romeo and Juliet, St. Joan, Othello, French Touch, The Constant Wife, She Stoops to Conquer, Quad-

rille, Escapade, On Tour, My Fair Lady, Dear Liar. LL.D (Hon., Baylor Univ., Texas)
PICTURES INCLUDE: Song of Songs, Constant Nymph, What Every Woman Knows, Beloved Enemy, I Live My Life, What a Woman, Hired Wife, Skylark, Man Who Lost Himself, The Great Garrick, Captain Fury, Smiling Through Juarez, My Son My Son, My Sister Eileen, Forever and a Day, Smart Woman, The Locket, I Confess, Titanic, Prince Valiant, Bullet Is Waiting, The Swan, The Best of Everything, Lancelot & Guinevere, The Waltz King, Rosie.
AUTHOR: A Proper Job, A Dreadful Man.

AIMEE, ANOUK: Actress. b. April 27, 1934. Paris, France. r.n. Francoise Dreyfus. Studied, Bauer-Therond dramatic school, Paris. Debut: 1946. La Maison Sous La Mer.
FILMS INCLUDE: Les Amants De Verone. The Golden Salamander, Le Rideau Cramoisi, Nuit D'Orage, La Bergere et Le Ramoneur (Cartoon), Mauvais Rencontres, Nina, Stresemann, Tous Peuvent Me Tuer, Pot Bouille, Modigliani of Montparnasse, Le Tete Contres Les Murs, Les Dragueurs, La Dolce Vita, Le Farceur, Lola, Les Amours De Paris L'Imprevu, 8½, Sodom and Gomorrah, La Fuga, A Man and a Woman, Un Soir, Un Train, The Appointment, The Model Shop, Justine, The Mandarians.

AKINS, CLAUDE: Actor. b. May 25, Nelson, Ga. e. Northwestern Univ. Worked as salesman in Indiana before joining Barter Theatre for season. Came to New York; appeared in The Rose Tattoo. Spent severa seasons with touring cos.
PICTURES INCLUDE: From Here to Eternity, The Caine Mutiny, Skyjacked, Flap, Devil's Brigade, Inherit the Wind, The Great Bank Robbery.
TV: Police Story, Cannon, McCloud, Marcus Welby, M.D., Mannix, The Streets of San Francisco, Movin' On, The Rhinemann Exchange.

AKTER, MAHMOOD ALI KHAN: Producer-Distributor-Exhibitor: b. Hyderabad Deccan, India, Apr. 12, 1926. Graduate of Arts. Formerly with Columbia Films of Pakistan Ltd., gen. mgr., Hussein & Co. controlling film dist. of United Artists Corp., Selznick Intl., London Films, Pathe Overseas. Participant foreign leaders prog. of Intl. Educational Exchange Service of U.S.A. State Dept.; mgr., Paramount Films of Pakistan, Ltd., 1959–64; mgr. dir., General Film Distributors, Lahore. Chm., Motion Picture Importer's Group, Lahore, Mgr. dir., Pak Boor Corp.

ALBECK, ANDY: Executive. Pres. & chief exec. officer, United Artists Corp. b. U.S.S.R., Sept. 25, 1921. Industry career began in 1939 with Columbia Pictures Intl. Corp. 1947, Central Motion Picture Exchange. 1949, Eagle Lion Classics, Inc. Joined UA in 1951 in intl. dept., functioning in the area of operations. After filling a number of key posts, named asst. treas. in 1970. In 1972 became v.p. of UA and its subsidiary, UA Broadcasting, Inc. 1973, appt. pres. of UA Broadcasting and in 1976 named snr. v.p—operations. Named UA Corp. pres & chief exec. officer in 1978. Retired, 1981.

ALBERGHETTI, ANNA MARIA: Singer, actress. b. Pesaro, Italy, May 15, 1936; d. Daniele Alberghetti, cellist. Concert debut in 1948 in Pesaro, then toured Italy, Scandinavia, Spain; Am. debut Carnegie Hall, 1950, sang with N.Y. Philharmonic Society, Phila. Symphony, on television. Screen debut in The Medium, 1950; then Here Comes Groom, Stars Are Singing, Last Command, 10,000 Bedrooms, Cinderfella. Star: Carnival. Broadway stage. Winner Antoinette Perry (Tony) Award for best actress.
TV: Toast of the Town, Cavalcade of Stars, Arthur Murray Show, Bob Hope, Eddie Fisher, Red Skelton, Dinah Shore, Desilu Playhouse, G.E. Theatre, Chevy Show, Dupont Show, Voice of Firestone, Colgate Hour, Climax, Loretta Young, Ford Jubilee, Perry Como.

ALBERT, EDDIE: Actor. r.n. Eddie Albert Heimberger. b. Rock Island, Ill. Apr. 22, 1908, e. U. Minn., Radio NBC.
BROADWAY: Brother Rat, Say Darling, The Music Man, Room Service, The Boys from Syracuse, Seven Year Itch.
PICTURES INCLUDE: Brother Rat, Four Wives, Angel from Texas, Dispatch from Reuter's, Rendezvous With Annie, Perfect Marriage, Smash-Up, The Story of a Woman, Time Out of Mind, Hit Parade of 1947, Dude Goes West, You Gotta Stay Happy, Fuller Brush Girl, You're in the Navy Now, Carrie, Meet Me After the Show, Actors and Sin, Roman Holiday, Girl Rush, I'll Cry Tomorrow, Oklahoma Attack, Teahouse of the August Moon, Orders to Kill, Gun Runners, Roots of Heaven, The Young Doctors, The Longest Day, Captain Newman, The Heartbreak Kid, Miracle of the White Stallions, Seven Women, McQ, The Take, The Longest Yard, Escape to Witch Mountain, The Devil's Rain, Hustle, Birch Interval, Foolin' Around, Yesterday, The Concorde-Airport 79, The Border, Ladyfingers, How to Beat the High Cost of Living, Take This Job and Shove It.
TV: Green Acres, Switch!, The Yeagers, Benjamin Franklin, The Borrowers, Killer Bees, Nutcracker, Anything Goes, Crash, The Word, Evening in Byzantium, Pirates Key, Living in Paradise.

ALBERT, EDWARD: b. Los Angeles, Feb. 20, 1951. e. UCLA. Son of actor Eddie Albert and actress Margo. Was prod. asst. on Patton in Spain. Has appeared with father on radio and TV shows. Is photographer and has exhibited work in L.A.
PICTURES INCLUDE: Butterflies Are Free (debut), Forty Carats, Midway, The Domino Principle, The Greek Tycoon, When Time Ran Out.

ALBERTSON, JACK: Actor. b. Malden, Mass.
PICTURES INCLUDE: The George Raft Story, Convicts Four, The Harder They Fall, Never Steal Anything Small, Man of A Thousand Faces, Miracle on 34th Street, Teacher's Pet, Lover Come Back, Monkey on My Back, Kissin' Cousins, Days of Wine and Roses, Period of Adjustment, Roustabout, How to Murder Your Wife, The Subject Was Roses, Justine, Willie Wonka & the Chocolate Factory, Rabbit Run, Dead & Buried.
BROADWAY: Meet the People, Make Mine Manhattan, Tickets, Please, High Time, Top Banana, High Button Shoes, Strip For Action, Allah Be Praised, The Lady Says Yes, Showboat, The Red Mill, The Subject Was Roses (won Tony Award for performance).
OPERA: The Cradle Will Rock.
TV: Ensign O'Toole, The Thin Man, Jack Benny Show, Red Skelton Show, Playhouse 90, Dick Van Dyke, I Love Lucy, Joey Bishop, The Defenders, Chico and The Man, Grandpa Goes to Washington

ALBRIGHT, LOLA: Actress. b. Akron, Ohio, July 20, 1925; e. Studied piano 12 years. Switchboard operator and stenographer NBC; stenographer with WHAM and bit player; photographers' model. screen debut in "The Pirate" 1948.
PICTURES INCLUDE: Easter Parade, Girl from Jones Beach, Tulsa, Champion, Good Humor Man, Bodyhold, When You're Smiling, Sierra Passage, Killer That Stalked New York, Arctic Flight, Silver Whip, Treasure of Ruby Hills, Magnificent Matador, Beauty on Parade, Tender Trap, A Cold Wind in August, Kid Galahad, Lord Love a Duck, Where Were You When the Lights Went Out?, The Impossible Years, The Money Jungle.

ALDA, ALAN: Actor. b. New York City, Jan. 28, 1936. e. Fordham. f. Robert Alda. Studied at Cleveland Playhouse; performed with Second City, then on TV in That Was The Week That Was; Broadway credits include The Owl and The Pussycat, Purlie Victorious, Fair Game For Lovers (Theatre World Award), The Apple Tree (Tony Award nomination).
PICTURES INCLUDE: Gone Are The Days, The Moonshine War, Jenny, The Mephisto Waltz, Paper Lion, To Kill a Clown, Same Time Next Year, California Suite, The Seduction of Joe Tynan (also dir.), The Four Seasons (actor, dir., s.p.).
TV: M*A*S*H, The Glass House.

ALDA, ROBERT: Actor, r.n. Alphonso D'Abruzo; b. New York City, Feb. 26, 1914. e. New York U. Architectural draftsman, then in radio, dramatic stock. Signed by Warners 1943.
PICTURES INCLUDE: Cinderella Jones, Rhapsody in Blue, Cloak and Dagger, Man I Love, Nora Prentiss, Hollywood Varieties, Mr. Universe, Two Gals and A Guy.
N.Y. STAGE: in Guys and Dolls 1950–52; Harbor Lights 1955; What Makes Sammy Run? 1963–64; My Daughter, Your Son, 1969; Front Page, 1970.
OTHER FILMS: April Showers, Homicide, Beast With 5 Fingers, Force of Impulse, The Devil's Hand, Beautiful But Dangerous, Muskateers of the Sea, Revenge of the Barbarians, Cleopatra's Daughter, A Soldier and One Half, Imitation of Life, I Will . . . I Will . . . For Now, Bittersweet Love.

ALDON, MARI: Actress. b. Toronto, Canada, Nov. 17; e. Central High, Toronto. A.T.C.L. of Trinity College, London, England: With Canadian ballet six yrs.; nat'l co. of Streetcar Named Desire; sang on radio; m.p. debut in Distant Drums; since in The Tanks Are Coming, This Woman Is Dangerous, Tangier Incident, Barefoot Contessa, Race for Life, Summertime.

ALDRICH, ROBERT: Director. b. Cranston, R.I. 1918. e. U. of Va. Started as prod. clerk, RKO; then script clerk, asst. dir., prod. mgr., assoc. prod., dir. TV shows: The Doctor, China Smith.
PICTURES INCLUDE: Big Leaguer, World for Ransom, Apache, Vera Cruz, Kiss Me Deadly, Big Knife, Autumn Leaves, Attack!, Ten Seconds to Hell, Angry Hills, Last Sunset, What Ever Happened to Baby Jane?, 4 for Texas, Hush, Hush, Sweet Charlotte, Flight of the Phoenix, Dirty Dozen, The Legend of Lylah Clare, The Killing of Sister George, Too Late the Hero, The Grissom Gang, The Emperor of the North Hustle, The Longest Yard, Twilight's Last Gleaming, The Choirboys, The Frisco Kid, All the Marbles.

ALEXANDER, JANE: Actress. b. Boston, Oct. 28, 1939. e. Sarah Lawrence Coll., Univ. Edinburgh. Stage career includes appearances on Bdwy., at Arena Stage, Washington, and Shakespeare Festival at Stamford, Conn.

PICTURES: The Great White Hope, A Gunfight, The New Centurions, All the President's Men, The Betsy.
TELEVISION: Welcome Home, Johny Bristol, Miracle on 34th St., Death Be Not Proud, Eleanor and Franklin, Eleanor and Franklin: The White House Years.

ALEXANDER, JOHN: Actor. b. Newport, Ky. On N.Y. stage Arsenic and Old Lace (1943), Ondine (1954).
PICTURES INCLUDE: Jolson Story, New Orleans, Cass Timberlane, Fancy Pants, Sleeping City, Night Has Thousand Eyes, Summer Holiday, Winchester '73, Model and the Marriage Broker, Marrying Kind, Untamed Frontier, A Tree Grows in Brooklyn, Junior Miss Broadway: Kiss the Boys Goodbye, Born Yesterday, Never Too Late, Visit to a Small Planet, Teahouse of the August Moon.

ALGAR, JAMES: Prod.-writer-director. b. Modesto, Calif. June 11, 1912. e. Stanford U.; B.A., M.A. Journalism. Entire career since 1934. Walt Disney Prods. Animator: Snow White; Director: Fantasia, Bambi, Wind in the Willows, Ichabod & Mr. Toad.
DOCUMENTARIES: war. health. True-Life Adventures: Seal Island, The Living Desert, Vanishing Prairie, The African Lion, Secrets of Life.
FEATURE CREDITS: Ten Who Dared, The Incredible Journey, Gnome-Mobile, White Wilderness, Rascal, The Legend of Lobo, The Jungle Cat, Run, Cougar, Run.
TELEVISION CREDITS: Producer: Run Light Buck, Run, The Not So Lonely Lighthouse Keeper, One Day on Beetle Rock, Wild Heart, Along the Oregon Trail, The Best Doggoned Dog in the World, One Day at Teton Marsh, Solomon, the Sea Turtle, Manado the Wolverine, Wild Geese Calling, Two Against the Arctic, Bayou Bay, Secrets of the Pond, Boy Who Talked to Badgers, Big Sky Man. Wrote and coproduced Great Moments with Mr. Lincoln; New York World's Fair; Circarama, America the Beautiful, Circle Vision 1958 Brussels World's Fair, Disneyland, Hall of Presidents, Disney World, Florida.
Shares in nine Oscars.

ALIN, MORRIS: Editor, writer, publicist, lyricist. e. College of the City of New York came into the M.P. industry, auditor of Hunchback of Notre Dame, roadshow oper., 1924; asst. sls. prom. mgr. Universal, 1926–27; slsmn., Universal, 1927; assoc. editor. The Distributor, MGM publication, 1927; editor, 1928–33; writer, publicist, MGM Studio, 1933–34; writer, publicist, Hollywood, New York, 1935–38; rej. Universal. 1938, editor, Progress: (Univ. publication); Twice winner of award, International Competition on Industrial Journalism; Senior publicist and Progress editor, Universal, 1961–67; editor Enterprise Press, 1968; member, executive Enterprise Press 1973; American Guild of Authors and Composers, American Society of Composers, Authors and Publishers, National Academy of Popular Music, and Motion Picture Pioneers.

ALLAN, TED. b. Clifton, Ariz., Comm. artist, actor SAG 1945, cameraman; photo training at Technicolor, Consolidated, MGM labs.; portrait photog., MGM, 20th-Century Fox, Selznick; created own film studio, 1952, prod. M.P. features, TV series, & pilots; Dir. of photog. 1956; 2 yr. contract Sinatra Enter.; 9 yrs. head CBS Cam. Dept. Hollywood; 3 yrs. 20th Cen. Fox, Pub-Stills & Featurettes for Von Ryan's Express, Fantastic Voyage, Sand Pebbles, Dr. Doolittle, The Detective, Lady in Cement, Tora, Tora, Tora. Featurettes for Universal: Seven Per Cent Solution, Hindenburg, Rollercoaster, Two Minute Warning, Gray Lady Down, The Promise Dracula.

ALLAND, WILLIAM: Producer. b. Delmar, Del., March 4, 1916. e. Baltimore. Acted in semi-professional groups; with Orson Welles' Mercury Theatre as actor, stage mgr.; asst. prod. Mercury Theatre radio series; ac., dialogue dir., Citizen Kane; act., Macbeth; U.S. Air Force, World War II; then radio writer; prod., Universal, 1952.
PICTURES INCLUDE: The Raiders, Flesh and Fury, Stand At Apache River, It Came From Outer Space, Creature From The Black Lagoon, Johnny Dark, This Island Earth, Dawn At Socorro, Four Guns To The Border, Chief Crazy Horse, Revenge Of The Creature, Tarantula, Creature Walks Among Us, Gun for a Coward, Land Unknown, Deadly Mantis, Mole People, Raw Wind in Eden, The Lady Takes a Flyer, As Young As We Are; Paramount, The Party Crashers, Colsus of New York, The Space Children; Allied Artists, Look In Any Window, The Lively Set.

ALLEN, DAYTON: Performer. b. New York, N.Y., Sept. 24, 1919. e. Mt. Vernon H.S. Motion picture road shows, 1936–40; disc jockey, WINS, N.Y., 1940–41; writer, vaudeville comedy bits, 1941–45; then radio comic, puppeteer and voices; TV since 1948; film commercials; shows include Oaky Doky, Bonny Maid Varieties, Howdy Doody, Jack Barry's Winky Dink, Abe Lincoln in Illinois, The Circle of Chalk, The Steve Allen Show.

ALLEN, DEDE: Film Editor. b. 1924. Once a messenger at Columbia Pictures, moved to editing dept., then to commercials and features.
PICTURES INCLUDE: Odds Against Tomorrow (1959), America, America, Bonnie and Clyde, Rachel Rachel, Alice's Restaurant, Little Big Man, Serpico, Dog Day Afternoon.

ALLEN, HERB: Producer. b. San Francisco, Calif., Dec. 24, 1913. e. U. of San Francisco, Pasadena Playhouse. Radio actor, anncr., Los Angeles, San Francisco; m.c., packager. Meet the Champ, children's TV show; asso. prod., Bob Crosby Show; prod., Johnny Carson show. CBS-TV.

ALLEN, IRVING: Dir., Prod. b. Poland, Nov. 24, 1905; e. Georgetown U., Ph.B. 1926. Film ed. Universal, Paramount, Republic 1929–40; shorts dir. RKO & Warner 1941–42; Academy Award 1942 for ed. short subject "Forty Boys and a Song"; U.S. Signal Corps officer 1942–45; debut as dir. "Strange Voyage" 1945; Academy Award 1947, prod. "Climbing the Matterhorn," best two-reel short. Exec. prod., Warwick Films.
PICTURES INCLUDE: Avalanche, Strange Voyage, High Conquest, 16 Fathoms Deep, Man on the Eiffel Tower, New Mexico, Slaughter Trail, (prod.): Paratrooper, Hell Below Zero, The Black Knight, Prize of Gold, Cockleshell Heroes, Safari, Zarak, Pickup Alley, Fire Down Below, High Flight, No Time to Die, The Man Inside, Bandit of Zhobe, Idol on Parade, Adamson of Africa, Jazz Boat, In the Nick, Let's Get Married, The Trials of Oscar Wilde, The Hellions, The Long Ships, Ghengis Khan, The Silencers, Murderers Row, the Ambushers, Hammerhead, The Wrecking Crew, The Desperados, Run Wild, Run Free, Cromwell, Eyewitness, The Story of a Racehorse.

ALLEN, IRWIN: Producer. b. New York, N.Y. e. Columbia U., CCNY, journalism & adv. ed., Key Magazine, Hollywood; radio prod. dir., daily Hollywood program, 11 yrs.; radio & literary dir., own agency; then M.P. prod.
PICTURES INCLUDE: Double Dynamite, Girl in Every Port (co-prod.); assoc., prod., Where Danger Lives; prod., Dangerous Mission, The Animal World; prod., co-writer, director, The Story of Mankind; prod., writer, director, Sea Around Us (Academy Award), prod., co-writer, The Big Circus; prod., co-writer, director, The Lost World, Voyage to the Bottom of the Sea, Five Weeks in a Balloon; prod., director action sequences, The Poseidon Adventure, The Towering Inferno. 1978; The Swarm (prod.-dir.); Beyond the Poseidon Adventure (prod.-dir.); When Time Ran Out (prod.).
TV: Creator-prod., Voyage to the Bottom of the Sea, Lost in Space, Land of the Giant, The Time Tunnel, Adventures of the Queen, Swiss Family Robinson, Flood, Fire, The Return of Captain Nemo, Hanging by a Thread, Cave-In, Night the Bridge Fell Down, Memory of Eva Ryker. Producer: Code Red.

ALLEN, KAREN: Actress. b. Carrollton, Ill. e. George Washington Univ., Univ. of Maryland. Auditioned for theatrical company in Washington, D.C. and won a role in Saint, touring with it for 7 months. Spent several years with Washington Theatre Laboratory Co. Moved to N.Y., acting in student films at NYU and studying acting with Lee Strasberg at Theatre Institute. Major film debut in National Lampoon's Animal House in 1978.
PICTURES: Manhattan, The Wanderers, Cruising, A Small Circle of Friends, Raiders of the Lost Ark.
TELEVISION: Circle of Children, Part II.

ALLEN, LEWIS: Director. b. Shropshire, England, Dec. 25, 1905; e. Tettenhall Coll., Staffs, Eng. Exec. in chg. of N.Y. & London prod. for Gilbert Miller; actor & stage dir. N.Y. & England; dir. (N.Y.) Laburnum Grove, Amazing Doctor Clitterhouse, Ladies & Gentlemen; (London) The Women, studied m.p. prod. at Paramount 2 yrs.; apptd. dir. with debut of "The Uninvited" 1943.
PICTURES INCLUDE: Appointment With Danger, Suddenly, Another Time Another Place, The Lorelei.

ALLEN, MEL: TV commentator b. Birmingham, Ala., Feb. 14, 1913. e. U. Alabama, A.B. 1932, U. Alabama Law School LL.B. 1936. Started as sportscaster in Birmingham, while in law school; speech instructor in U. Ala. 1935–37; to N.Y. 1937, as staff announcer CBS to 1943; served in U.S. Army, World War II in infantry until the war ended, then before discharge was transferred to work on NBC Army Hour; sportscasting throughout U.S., joined N.Y. Yankees, 1946, concurrently narrating many shorts incl. How to Make a Yankee, appearing on radio & video and in Babe Ruth Story; sports commentator Fox Movietonews; voted best sportscaster in Motion Picture Daily-Fame radio. TV polls; Monitor, NBC; NCAA TV College Football, NBC; World Series (1938–1964), CBS-NBC; Rose Bowl (1951–1962), NBC; Sports Broadcasters Hall Of Fame.

ALLEN, NANCY: Actress. b. New York City, June 24.
PICTURES: The Last Victim, The Last Detail, Carrie, I Wanna Hold Your Hand, Home Movies, 1941, Dressed to Kill, Blow Out.

ALLEN, REX: Actor. b. Wilcox, Ariz., Dec. 31, 1922; e. Wilcox H.S. 1939. Vaudeville & radio actor; WLS, Chicago, 5 yrs.; was rodeo star appearing in shows through USA.
PICTURES INCLUDE: Arizona Cowboy, Hills of Oklahoma, Under Mexicali Stars, Thunder in God's Country,

Rodeo King & The Senorita, I Dream of Jeanie, Last Musketeer, South Pacific Trail, Old Overland Trail, Down Laredo Way, Phantom Stallion, For the Love of Mike, Tomboy and the Champ; narr., Walt Disney.

TV: Perry Como Special; commercials, Wonderful World of Color (NBC) Frontier Doctor.

RECORDINGS: Crying in The Chapel, Don't Go Near the Indians, Warner Bros. Records. 1966 Radio Man of the Year; 1966 Cowboy Hall of Fame.

ALLEN, ROBERT: Actor. e. Dartmouth Coll. Began film career in 1926.

PICTURES INCLUDE: Love Me Forever, Party Wire, Winter Carnival, Crime and Punishment, Pride of the Marines, Black Room Mystery, Death Flies East, Air Hawks, Bob Allen Ranger Series, Craig's Wife, Awful Truth, Lady of Secrets, Keep Smiling, Meet the Girls, Everybody's Baby, The City, Winner Take All, Fire-Away, Pie In The Sky.

PLAYS INCLUDE: Blessed Event, Popsy, Kiss Them for Me, Show Boat, I Killed the Count, Auntie Mame, A Few Wild Oats, Time Out for Ginger.

TV: Kraft, Armstrong Theatre, Philco, The Web, Suspense, Ethel & Albert, Danger, Lux Playhouse, Pulitzer Prize Playhouse, Lamp Unto My Feet.

ALLEN, STEVE: Performer. b. New York, N.Y., Dec. 26, 1921. U.S. Army 1942; radio shows, Los Angeles; TV, N.Y., 1950; shows include: What's My Line, Steve Allen Show, Tonight, I've Got a Secret, Meeting of Minds, Laughback.

PICTURES INCLUDE: Down Memory Lane, Warning Shot, I'll Get By, Benny Goodman Story, College Confidential.

COMPOSER: This Could Be the Start of Something Big, Pretend You Don't See Her, South Rampart St. Parade, Picnic, Houseboat, On the Beach, Sleeping Beauty, Bell, Book and Candle, Gravy Waltz, Impossible.

AUTHOR: Fourteen For Tonight, Steve Allen's Bop Fables, The Funny Men, Wry On the Rocks, The Girls on the Tenth Floor, The Question Man, Mark It and Strike It, Not All of Your Laughter, Not All of Your Tears, Letter to a Conservative, The Ground is Our Table, Bigger Than a Breadbox, A Flash of Swallows, The Wake, Princess Snip-Snip and the Puppykittens, Curses, Schmock-Schmock!, Meeting of Minds, Ripoff, Meeting of Minds-Second Series, Rip-off, Explaining China, The Funny People.

ALLEN, WOODY: Actor, Director, Writer. r.n. Allen Stewart Konigsberg. b. New York City, 1935. Began writing comedy at age 17, contributing to various magazine and top TV comedy shows.

STAGE: Author of Play It Again, Sam; Don't Drink The Water.

PICTURES INCLUDE: What's New Pussycat? (story, screenplay, actor); Casino Royale (actor); What's Up Tiger Lily? (dubbed scr., act.); Take the Money and Run (co-story, co-scr., dir., act.); Bananas (scr., dir., act.); Everything You Always Wanted to Know About Sex* But Were Afraid to Ask (scr., dir., act.); Sleeper, (scr., dir., act.); Love and Death (s.p., act.), The Front (act.), Annie Hall (co. s.p., dir., act.), Interiors (s.p.-dir.), Manhattan (co.-s.p., dir., act.), Stardust Memories (s.p., dir., act.)

ALLISON, FRAN: Performer. b. LaPorte City, Iowa; m. Archie Levington; e. Coe College, Cedar Rapids; sch. teacher, Iowa; started radio career as songstress Iowa Sta., 1934; starred in numerous programs, Chicago, 1937; Aunt Fanny of Breakfast Club; Fran of Kukla, Fran and Ollie on TV since 1947. Numerous commercials, daily TV variety show, 1958; Fac. mem., Careers Acad. Milwaukee.

ALLYSON, JUNE: Actress. r.n. Ella Geisman. b. Westchester, N.Y., Oct. 7, 1917. Started as chorus girl in school; m.p. debut in Best Foot Forward (1943). Voted one of ten best money-making stars in m.p. Herald-Fame poll, 1955.

BROADWAY INCLUDES: Best Foot Forward, 40 Carats, National Company, No No Nanette.

PICTURES INCLUDE: Girl Crazy, Thousands Cheer, Meet the People, Two Girls and a Sailor, Secret Heart, Reformer and Redhead, Right Cross, Too Young to Kiss, Girl in White, Battle Circus, Remains to be Seen, The Stratton Story, Little Women, Executive Suite, Glenn Miller Story, Woman's World, Strategic Air Command, The Strike, McConnell Story, You Can't Run Away From It, Opposite Sex, Interlude, My Man Godfrey, Stranger In My Arms, They Only Kill Their Masters, That's Entertainment!

TV: The June Allyson Show, Movie-of-the-Week: See the Man Run, 20th Century Follies.

ALMOND, PAUL: Producer-Director-Writer. b. Montreal, Canada, April 26, 1931; e. McGill U., Balliol Coll., Oxford. 1954–66 produced and directed over a hundred television dramas in Toronto, London, N.Y., and Hollywood; prod., dir., s.p., Isabel 1968, Act of the Heart 1970; Etrog (dir.; best Canadian feature); Journey 1972.

ALTERMAN, JOSEPH GEORGE: Executive. b. New Haven, Conn., Dec. 17, 1919. e. Wesleyan Univ., B.A., 1942; Institute for Organization Management, Yale Univ. 1957–59; exec. assist., SoundScriber Corp., 1945–48; district mgr., Industrial Luncheon Service, 1948–55; asst. secretary and admin. Secretary, Theatre Owners of America, 1955; Exec. dir. and vice pres., Natl. Assn. of Theatre Owners, 1966; Exec. v.p. COMPO., 1970.

ALTMAN, ROBERT: Director, writer, producer. b. Kansas City, Mo., Feb. 20; e. U. of Mo. Worked as Director for Dean Story; dir., Countdown, That Cold Day in the Park, M.A.S.H.; s.p. and dir., McCabe & Mrs. Miller, Brewster McCloud, Images, The Long Goodbye, Thieves Like Us, California Split, Nashville, Buffalo Bill and the Indians, The Late Show (prod.) Welcome to L.A. (prod.), Three Women (prod., dir., s.p.); A Wedding (prod., dir.); Remember My Name (prod.); Quintet (prod., dir.); A Perfect Couple (prod., dir.); Rich Kids (prod.); Health (prod., dir., s.p.), Popeye (dir.).

TV: Wrote, prod. and dir., Roaring Twenties, Bonanza, Bus Stop, Combat, Kraft Theatre and pilots for The Gallant Men and The Long Hot Summer.

ALTON, JOHN: Cinematographer. b. Oct. 5, 1901, Hungary. Laboratory technician, MGM 1924; Paramount cameraman 1928; sent to Paris, France, chg. camera dept., 1931; installed studios for Lumiton & Sono Film (South America); writer, photographer & dir. many Spanish pictures; returned to U.S. 1937, cinematographer Hollywood; served in U.S. Signal Corps as Capt., World War II. Books include: Painting With Light, Photography and Lighting in General; received Sintonia Medal for best photography in Argentina, 1937. Academy Award best color photog. American in Paris ballet, 1951.

PICTURES INCLUDE: Witness to Murder, Big Combo, Catered Affair. Member: ASC, AMPAS.

ALVIN, JOHN: Actor. r.n. John Alvin Hoffstadt; b. Chicago, Oct. 24, 1917; e. Pasadena (Calif.) Playhouse. On radio Chicago & Detroit; on N.Y. stage "Leaning on Letty," "Life of the Party," Screen debut 1944 in "Destination Tokyo."

PICTURES INCLUDE: Missing Women, Two Guys from Texas, Bold Frontiersman, Train to Alcatraz, Shanghai Chest, Carrie, April In Paris, Torpedo Alley.

ALWYN, WILLIAM: Composer. b. Northampton, England, 1905; e. Royal Academy of Music. Held Collard Fellowship in Music, Worshipful Company of Musicians, 1937–40. From 1936 author incidental music many British films.

PICTURES INCLUDE: Odd Man Out, Fallen Idol, I'll Never Forget You, Crimson Pirate, Magic Box, The Card (The Promoter), Mandy, Master of Ballantrae, Personal Affair, Million Pound Note (Man with a Million), Malta Story, The Seekers (Land of Fury), Bedevilled, Svengali, Cure for Love, Safari, Ship That Died of Shame, Manuela, I Accuse, Silent Enemy, Swiss Family Robinson, In Search of The Castaways, Life For Ruth, The Running Man.

AMATEAU, ROD: Director. b. New York, N.Y., Dec. 20, 1927. Staff writer, CBS radio; stage mgr., Lux Radio Theatre; U.S. Army, 1941; junior writer, 20th Century-Fox; then test dir., second unit dir.; TV dir., numerous shows including Schlitz Playhouse of Stars, Four Star Playhouse, General Electric Theatre, Private Secretary, Dennis Day Show, Lassie, Ray Milland Show, Bob Cummings Show, dir. & prod. Burns & Allen Show, Dobie Gillis.

PICTURES INCLUDE: The Statue, Where Does It Hurt?, The Wilby Conspiracy, Drive-In.

AMECHE, DON: Actor. b. Kenosha, Wis., 1908; e. Columbia Coll., U. Wisconsin. In stock; on radio; m.p. debut 1936. In television from 1951; on Broadway in Silk Stockings, 1955; International Show Time.

PICTURES INCLUDE: Sins of Man, Ladies in Love, Ramona, You Can't Have Everything, Alexander's Ragtime Band, Swanee River, Four Sons, That Night in Rio, Kiss the Boys Goodbye, So Goes My Love, That's My Man, Sleep My Love, Slightly French, A Fever in the Blood, Ring Around the World, Picture Mommy Dead, Suppose They Gave a War & Nobody Came, The Boatniks.

AMES, LEON: Actor. r.n. Leon Wycoff. b. Portland, Ind., Jan. 20, 1903. Aviator; with Champlin Players, Lansford, Pa., in stock and on stage; plays include: Tobacco Road, Male Animal, Land Is Bright; m.p. debut in "Murders in the Rue Morgue," 1932; owner & vice-pres. Studio City Ford Co.

PICTURES INCLUDE: Meet Me in St. Louis, Thirty Seconds Over Tokyo, Yolanda and the Thief, Merton of the Movies, On an Island With You, Velvet Touch, Date With Judy, Battleground, Ambush, Big Hangover, Little Women, Dial 1119, Happy Years, Alias a Gentleman, Watch the Birdie, Cattle Drive, On Moonlight Bay, It's a Big Country, Angel Face, By the Light of the Silvery Moon, Let's Do It Again, Sabre Jet, Peyton Place, From the Terrace, Absent Minded Professor, Son of Flubber, Misadventures of Merlin Jones, The Monkey's Uncle, On a Clear Day You Can See Forever, Tora! Tora! Tora!.

Under contract CBS TV; Life with Father, Father of the Bride, Mr. Ed.

AMES, LOUIS B.: Executive. b. St. Louis, Mo. Aug. 9, 1918; e. Washington U., St. Louis; m. Jetti Preminger; began as music consultant and staff director of musical programs for NBC; music dir. 1948, WPIX 1951 apptd. program mgr., WPIX; assoc. prod. Today, NBC TV, 1954; feature editor

Home, 1957; Adm.-prod. NBC Opera, 1958; dir. cultural prog. N.Y. World's Fair, 1960–63; dir. RCA Pavillion, N.Y. World's Fair, 1963–65; 1966 dir., Nighttime, TV; 1969, dir. of Programming N.W. Ayer & Sons, Inc. 1973 Mgr. Station Services, Television Information Office. NYC.

AMFITHEATROF, DANIELE: Composer, conductor. b. St. Petersburg, Russia, Oct. 29, 1901. e. Royal Conserv. Music, Rome (D. Mus.). Guest conductor symphony orchs. Paris, Rome, Budapest, Vienna, etc.; assoc. conductor Minneapolis Symphony, 1937–38; guest conductor Boston Symphony, 1938.
PICTURES INCLUDE: (Composer scores) Lassie, Come Home, Beginning or the End, Song of the South, O.S.S., I'll Be Seeing You, Ivy, Senator Was Indiscreet, Letter from an Unknown Woman, Another Part of the Forest, An Act of Murder, Somewhere in the City, Fan, House of Strangers, Sand, Storm Warning, Bird of Paradise, Tomorrow is Another Day, Angels in the Outfield, Desert Fox, Scandal at Scourie, Devil's Canyon, Naked Jungle, Human Desire, Day of Triumph, Trial, Last Hunt, The Mountain, Unholy Wife, Spanish Affair, From Hell to Texas, Heller in Pink Tights, Edge of Eternity, Major Dundee.

AMSTERDAM, MOREY: Prod., Writer, Composer, Musician, Actor. b. Chicago, Ill., Dec. 14, 1914. e. U. of Calif., Berkeley. Boy soprano. Radio KPO, 1922; night-club performer, Chicago, 1929; comedian, singer, musician. Rube Wolf Orchestra; comedian, Optimistic Doughnut Program, 1930; writer, vaudeville; comedian, writer, Al Pearce Gang, 1932; writer, MGM, 1937; co-writer, radio shows; m.p. Columbia, Universal; writer, performer, USO Shows, 1942–43; owner, performer, the Playgoers-club; v.p. American International Pictures.
TV: Stop Me If You've Heard This One, Broadway Open House, Morey Amsterdam Silver Swan Show, Dick Van Dyke Show. Exec. Star-prod. Can You Top This.
PICTURES INCLUDE: Don't Worry . . . We'll Think of a Title, Machine Gun Kelly, Murder, Inc.
SONGS: Rum and Coca Cola, Why Oh Why Did I Ever Leave Wyoming, Yak A Puk, etc.

AMYES, JULIAN: TV producer. b. Cambridge, England, Aug. 9, 1917. e. Cambridge Univ. Early career, stage-producer, actor, film-actor, 1946–51; ent. TV 1951 as drama producer.
PRODUCTIONS INCLUDE: Dial M for Murder, The Deep Blue Sea, The Liberators, The Small Victory, Fifty-Fifty, The Troubled Air; dir., A Hill in Korea, Miracle in Soho. TV series, The Third Man, Maigret; 1959 Granada TV; Head of Drama, Granada TV, 1961; hd. progs., Granada TV, 1963–68. Dir. Granada TV, 1968.

ANDERSON, DAPHNE: Actress. b. London, Eng., April 27, 1922; e. Kensington High School. Stage debut in pantomime chorus 1937; screen debut in Trottle True 1948.
PICTURES INCLUDE: Cloudburst, Beggar's opera, Hobson's Choice, A Kid for Two Farthings, The Prince and the Showgirl, No Time for Tears, Snowball, Stork Talk, Captain Clegg, The Launching, I Want What I Want.
TV: Silas Marner (Serial), Gideon's Way (Serial), Dr. Finlay's Casebook, Happy Family, The Imposter, The Harry Worth Show, The Whitehall Worrier, The Suede Jacket, Casanova, Haunted Series, Thirty Minute Theatre, Justice is a Woman, Paul Temple, Today, 2 Cars.

ANDERSON, EDDIE ("ROCHESTER"): Actor b. Oakland, Calif., Sept. 18, 1905. Vaud.; night clubs; featured Jack Benny radio program.
PICTURES INCLUDE: Transient Lady, Three Men on a Horse, Jezebel, You Can't Take It With You, Thanks for the Memory, Gone With the Wind, Tales of Manhattan, Cabin in the Sky, Broadway Rhythm, I Love a Bandleader, The Show-Off.

ANDERSON, GERRY: Hon. F.B.K.S., Producer, b. London, 1929. Ent. ind. 1946. Chmn./man. dir. Gerry Anderson Productions, Ltd. Over 320 pictures produced for TV worldwide. 1978 Prod. The Male Chauvinist Pig!

ANDERSON, HENRY: Executive. b. New York City; e. N.Y.U., B.S. 1910. In construction work; joined E.I. duPont de Nemours & Co. (Del.); mgr. insurance dept. Paramount 1924; chmn. of many comms. on theat. engineering, construction, operation, protection, projection; mem. Advisory comm. N.Y. State Labor Dept. 1944–48; chmn. insurance comm. MPAA; author of articles on insurance.

ANDERSON, JUDITH: Actress. b. Adelaide, Australia, Feb. 10, 1898. Stage debut (Sydney) in "Royal Divorce" 1915; on New York stage & in stock 1918; toured 1920; m.p. debut in "Blood Money" 1933. Entertained U.S. Armed Forces overseas, World War II; received Donaldson Award 1948, New York Critics Award 1948, American Academy of Arts & Sciences Award for Best Speech 1948. N.Y. stage, 1952, In the Summer House, 1953.
PLAYS INCLUDE: Dear Brutus, Cobra, Dove, Strange Interlude, Old Maid, Mourning Becomes Electra, Macbeth, Medea.
PICTURES INCLUDE: King's Row, Rebecca, Laura, Spector of the Rose, Diary of a Chambermaid, Strange Love of Martha Ivers, Red House, And Then There Were None, Tycoon, Pursued, Salome, Come of Age, Cinderfella, Cat on a Hot Tin Roof, A Man Called Horse.

ANDERSON, LINDSAY: director (cinema & theatre) and writer. b. Bangalore, India, 1923. e. Cheltenham College and Wadham College, Oxford (classical scholar). Ent. Ind. 1948 as writer-director in charge industrial film prod., also contributing editor independent film quarterly Sequence. Principal doc. films; O. Dreamland, Thursdays Children (collab. s.p. & dir.) Guy Brenton: Hollywood Short Subject Award, 1954, Three Installations, Pleasure Garden, Wakefield Express, Every Day Except Christmas (Venice Grand Prix, 1957), March to Aldermaston. Founder mem. Free Cinema group. National Film Theatre 1956/9. Recent Doc.: The Singing Lesson (Documentary Studio, Warsaw).
TV INCLUDES: Five episodes Robin Hood series, commercials, 1959–75. 1978 Dir. The Old Crowd.
PLAYS INCLUDE: The Long and the Short and the Tall, Billy Liar, The Fire Raisers, Diary of a Madman, Sergeant Musgraves Dance, The Cherry Orchard, Inadmissible Evidence (At Contemporary Theatre, Warsaw), In Celebration, The Contractor Home, The Changing Room, Artistic director, Royal Court theatre 1969–70. Associate, 1970–76. The Farm, Life Class, What the Butler Saw, The Sea Gull, The Bed Before Yesterday 1977 The Kingfisher, 1978 Alice's Boys, 1980, Early Days.
PICTURES INCLUDE: The Sporting Life, 1963; The White Bus, 1966; If . . . , 1968; O Lucky Man!, 1973, In Celebration, 1975.

ANDERSON, MELISSA SUE: Actress. b. Berkeley, CA, Sept. 26, 1962. Took up acting at suggestion of a dancing teacher. Did series of commercials; guest role in episode of Brady bunch; episode of Shaft. Gained fame as Mary Ingalls on Little House on the Prairie series.
TELEVISION: Very Good Friends, James at 15, The Loneliest Runner.

ANDERSON, MICHAEL: Director. b. London, England, 1920; e. France, Germany, Spain. Ent. m.p. industry as actor, 1936; asst. dir., unit manager on such films as In Which We Serve, Pygmalion, French Without Tears; Army service 1942–46; Co-dir., Private Angelo, 1949; since dir. Waterfront, Hell Is Sold Out, Night Was Our Friend, Will Any Gentleman?, The Dam Busters, 1984, Around the World in 80 Days, Yangtse Incident (Battle Hell), Chase A Crooked Shadow, Shake Hands with the Devil, The Wreck of the Mary Deare, All the Fine Young Cannibals, The Naked Edge, Flight From Ashiya, Monsieur Cognac, Operation Crossbow, The Quiller Memorandum, Shoes of the Fisherman, Pope Joan, Dr. Savage, Conduct Unbecoming, Logan's Run, Orca, Dominique, The Martian Chronicles.

ANDERSON, MICHAEL, JR.: Actor. b. London, 1943. son of dir. Michael Anderson. ent. films as child actor, 1954.
TV: Queen's Champion, Ivanhoe, The Monroes.
PICTURES INCLUDE: Tiger Bay, the Moonraker, The Sundowners, Reach For Glory, In Search of The Castaways, Play It Cool, Greatest Story Ever Told, Dear Heart, Major Dundee, The Glory Days, The Sons of Katie Elder, Logan's Run, The Martin Chronicles.

ANDERSON, RICHARD: Actor. b. Long Branch, N.J., Aug. 8, 1926. U.S. Army, World War II. Began acting career in summer theatre in Santa Barbara and Laguna Playhouse where spotted by MGM executives who signed him to six-yr. contract. Appeared in 26 films for MGM before leaving studio.
PICTURES INCLUDE: 12 O'Clock High, The People Against O'Hara, The Story of Three Loves, Escape from Fort Bravo, Forbidden Planet, The Long Hot Summer, Compulsion, Seconds, Paths of Glory, Tora! Tora! Tora!, Along Came a Spider, Macho Callahan, Doctors' Wives, Play It as It Lays, The Honkers, Stone, Redneck America.
TV: Bus Stop, Perry Mason, 12 O'Clock High (all regular), Ironside, The Big Valley, Mannix, My Friend Tony, The Mod Squad, Land of the Giants, The FBI, Gunsmoke. Series: The Six Million Dollar Man, The Bionic Woman.

ANDERSON, ROLAND: Art director. In 1933 collab. with Hans Dreier on "This Day and Age," Par. Since then collaborated in art direction on many productions, including Lives of a Bengal Lancer, Hands Across the Table, Give Us This Night, Till We Meet Again, The Buccaneer, Union Pacific, North West Mounted Police, Love Thy Neighbor, The Crystal Ball, Story of Dr. Wassel, Road to Utopia, Let's Dance, Branded, Carried, Son of Paleface, Just For You, Here Come the Girls, White Christmas, Country Girl, We're No Angels, Court Jester.

ANDERSON, SYLVIA: Producer/Writer. Sylvia Anderson Productions Ltd. (Pinewood Studios). b. London. e. London Univ. Entered m.p. ind. 1960.
TV: Thunderbirds; U.F.O.; Space: 1999.
PICTURES INCLUDE: Doppelganger; Thunderbird 6; Thunderbirds Are Go.

ANDERSON, WILLIAM H.: Producer; Member of Bd. of Dir. Walt Disney Productions. b. Utah, October 12, 1911. e.

Compton College. Firestone Rubber Co.; Universal Credit Co.; Walt Disney Prods. 35 years.

PICTURES INCLUDE: Old Yeller, Swiss Family Robinson, Happiest Millionaire, The Computer Wore Tennis Shoes, Barefoot Executive, $1,000,000 Duck, Superdad, Strongest Man In The World, Apple Dumpling Gang, Treasure of Matecumbe, Shaggy D.A.

TV: Zorro Series 1958–1959 (20 half-hour segments); 1959–1960 (38 half-hour segments), Wonderful World of Disney: Zorro, Texas John Slaughter, Daniel Boone, The Swamp Fox, Pop Warner Football (1960–1961), Johnny Shiloh, Mooncussers, 1962–1963; Bristle Face, The Scarecrow of Romney Marsh (1963–1964); The Legend of Young Dick Turpin (1965–1966); Willie and the Yank (1966–1967); A Boy Called Nuthin', The Young Loner (1967–1968); The Wacky Zoo of Morgan City (1970–1971); The Mystery of Dracula's Castle (1972–1973); The Bull From the Sky, co-producer; 1975–1976; Great Sleeping Bear Sled Dog Race, co-producer (1976–1977).

ANDERSSON, BIBI: Actress. b. Stockholm, 1935. e. Royal Dramatic Theatre School.

PICTURES INCLUDE: (for director, Ingmar Bergman) Smiles of a Summer Night, 1955; The Seventh Seal, Wild Strawberries, Brink of Life, The Magician, The Devil's Eye, Not to Mention All these Women, Persona; other films: My Sister, My Love Le Vio A Passion, Black Palm Trees, Story of a Woman, Duel at Diablo, The Kremlin Letter, The Passion of Anna, I Never Promised You a Rose Garden, Quintet.

ANDES, KEITH: Actor. b. Ocean City, N.J., July 12, 1920. e. Temple U., Oxford. Radio work; on Broadway in Winged Victory, Kiss Me Kate, Maggie.

PICTURES INCLUDE: Farmer's Daughter, Clash By Night, Blackbeard the Pirate, Split Second, Life at Stake, Second Greatest Sex, Away All Boats, Pillars of the Sky, Back from Eternity, Damn Citizen, Surrender, Hells Bloody Devils.

TV: This Man Dawson, Glynis.

BROADWAY: Wildcat.

ANDRESS, URSULA: Actress. b. Berne, Switzerland, 1936. M.p. Hollywood, 1955.

PICTURES INCLUDE: Dr. No, Four For Texas, Fun in Acapulco, Toys For Christmas, She, Tenth Victim, What's New Pussycat?, Chinese Adventures in China, Casino Royale, The Southern Star, Perfect Friday, The Red Sun, Scaramouche, Clash of the Titans.

ANDREWS, DANA: Actor, b. Collins, Miss., Jan. 1, 1912: e. Sam Houston Coll.; Pasadena Community Playhouse. Account with Gulf Refining Co., 1930; chief accountant Tobins, Inc., 1931; acted with Pasadena Playhouse 3 yrs.; m.p. debut in "The Westerner" 1939; Vice Pres., Screen Actor's Guild, 1962–63, Pres. 1964–66.

PICTURES INCLUDE: Best Years of Our Lives, Iron Curtain, Foolish Heart, Forbidden Street Bway play, Two for the See-Saw, Laura, Edge of Doom, Where Sidewalk Ends, Sealed Cargo, The Frogmen, I Want You, Assignment—Paris, Elephant Walk, Duel in the Jungle, Three Hours to Kill, Smoke Signal, Strange Lady in Town, While the City Sleeps, Commanche, Beyond a Reasonable Doubt, Spring Reunion, Night of the Demon, Zero Hour, Enchanted Island, The Fear Makers, The Crowded Sky, Madison Avenue, The Satan Bug, Crack in The World, In Harm's Way, Battle of the Bulge, Devil's Brigade, The Loved One, Innocent Bystander, The Last Tycoon, Airport, 1975, Born Again.

TV: Alas Babylon; G.E., The Playoff; Barbara Stanwyck; Dupont, Mutiny; 4 Star, Crazy Sunday; Checkmate, Alcoa, The Boy Who Wasn't Wanted; Twilight Zone, No Time Like The Past; Dick Powell, Last of the Big Spenders; Alcoa, The Town That Died, Bob Hope Chrysler, A Wind of Hurricane Force, Bright Promise (movie), Shadow in the Streets, One Small Step Forward (series), The Right Hand Man (Playhouse 90) Ike (ABC movie).

DINNER THEATRE: The Marriage-Go-Round; The Best of Friends, Angel Street, The Gang's All Here, Any Wednesday, A Man for All Seasons.

ANDREWS, EAMONN: C.B.E. Announcer and commentator. b. Dublin, Ireland, Dec. 19, 1922. e. Irish Christian Brothers (Dublin), Radio, vaudeville and legit. stage experience in Dublin; TV with BBC 1950; quiz master on several programs, sports com., columnist, author, playwright; Founder, chairman, Irish Television Authority; ABC-TV, 1964–67; Eamonn Andrews Show, and World of Sport, Thames TV's daily, Today, This Is Your Life, Time For Business.

ANDREWS, EDWARD: Actor. b. Griffin, Ga., Oct. 9. e. Univ. of Virginia. Broadway debut, How Beautiful With Shoes, 1935; So Proudly We Hail, Of Mice and Men, The Time of Your Life, They Knew What They Wanted, I Am a Camera, Come Back Little Sheba, A Visit to a Small Planet, The Gazebo.

TV: co-star, Broadside co-star, Mr. Terrific. Universal pilot; Thriller, U.S. Steel, Kraft Theatre, Studio One, Alfred Hitchcock, Gunsmoke, 11th Hour, Kildare, Bonanza, etc.

PICTURES INCLUDE: Birds Do It, The Thrill of It All, Advise and Consent, The Young Doctors, The Absent-Minded Professor, The Young Savages, Elmer Gantry, Those Wilder Years, The Tattered Dress, Tea and Sympathy, The Phoenix City Story, The Glass Bottom Boat, Charley and the Angel.

ANDREWS, HARRY: CBE, Actor. b. Tonbridge, Kent, England, 1911. First stage appearance, Liverpool, 1933. London stage 1935, "Worse Things Happen at Sea," to U.S., 1936, "Hamlet." Army Service 1939–45. Screen debut 1952, The Red Beret.

PICTURES INCLUDE: The Black Knight, The Man Who Loved Redheads, A Hill in Korea, Alexander the Great, Moby Dick, Ice Cold in Alex, Devil's Disciple, Solomon and Sheba, In the Nick, Circle of Deception, Cleopatra, The Best of Enemies, Reach For Glory, The Inspector, Barrabas, 55 Days at Peking, The Snout, The System, The Hill, The Agony and the Ecstasy, The Sands of the Kalahari, The Deadly Affair, A Dandy in Aspic, The Charge of the Light Brigade, The Night They Raided Minsky's, The Seagull, A Nice Girl Like Me, Too Late the Hero, The Gaunt Woman, Country Dance (in U.S. Brotherly Love), Entertaining Mr. Sloane, Wuthering Heights, I Want What I Want, Nicholas and Alexandra, The Nightcomers, The Ruling Class, Man of La Mancha, Theatre of Blood, The Mackintosh Man, Man at the Top, The Skyriders, The Bluebird, The Passover Plot, The Prince and The Pauper, Equus, Superman, The Medusa Touch, The Big Sleep, Death on the Nile, SOS Titanic.

TV: Affair of Honour, Clayhanger, Two Gentle People, Leo Tolstoy.

ANDREWS, JULIE: Actress, Singer. b. Walton-on-Thames. Eng., 1935 debut, Eng. Starlight Roof Revue; pantomime of Cinderella.

PLAYS: The Boy Friend, My Fair Lady, Camelot.

PICTURES INCLUDE: Mary Poppins, The Americanization of Emily, The Sound of Music, Hawaii, Torn Curtain, Thoroughly Modern Millie, Star!, Darling Lili, The Tamarind Seed, "10," "S.O.B.," Victor/Victoria.

TV: High Tor, Julie and Carol at Carnegie Hall, The Andy Williams Show, The Julie Andrews Show, An Evening with Julie Andrews and Harry Belafonte, The World of Walt Disney, Julie and Carol at Lincoln Center, Julie Andrews Hour (ABC Television series 1972–73); Julie on Sesame Street, 1973; Julie Andrews' Christmas Special, 1973; Julie and Dick in Covent Garden, 1974; Julie Andrews and Jackie Gleason Together, 1974; Julie Andrews—My Favorite Things, 1975.

AUTHOR: Mandy, Last of the Really Great Whangdoodles, 1973.

ANGEL, HEATHER: Actress. b. Oxford, England, Feb. 9, 1909; e. Wycombe Abbey School. On stage 1926–30; British m.p. debut in "City of Song" 1939; U.S. m.p. debut in "Pilgrimage" 1932.

PICTURES INCLUDE: Hound of the Baskervilles, After Office Hours, Berkeley Square, Charlie Chan's Greatest Case, Orient Express, Springtime for Henry, Informer, Three Musketeers, Last of the Mohicans, Bulldog Drummond series, Pride and Prejudice, That Hamilton Woman, Suspicion, Lifeboat, In the Meantime Darling, Saxon Charm, Alice in Wonderland, Peter Pan.

TV: Family Affair.

ANGERS, AVRIL: Actress, comedienne, singer. b. Liverpool, Eng., April 18; Stage debut at age of 14; screen debut in 1947 in Lucky Mascot.

PICTURES INCLUDE: Miss Pilgrim's Progress, Don't Blame the Stork, Women Without Men, Green Man, Devils of Darkness, Be My Guest, Three Bites of the Apple, The Family Way, Two a Penny, The Best House in London, Staircase, There's a Girl in My Soup, Forbush and the Penguins, Gollocks (1973), Confessions of a Driving Instructor, Dangerous Davies.

TV: How Do You View, Friends and Neighbors, Dear Dotty, Holiday Town, Charlie Fainsbarn Show, Arthur Askey Show, All Aboard, The Gold Hunter, Bob Monkhouse Show, Before The Fringe, Hudd, Coronation Street, Dick Emery Show, Dad's Army, Bright Boffins, The More We Are Together, The Millionairess, Liver Birds, Looks Familiar, No Appointment Necessary, The Songwriters, All Creatures Great and Small, Coronation Concert.

PLAYS INCLUDE: The Mating Game; Musical, Cockie; Murder at the Vicarage, Little Me, Norman, Is That You?, Blithe Spirit.

ANHALT, EDNA: Associate producer, writer. b. New York, N.Y., Apr. 10, 1914; e. Kathryn Gibbs Sch., 1932; U. doc. films; prod. mgr., Rockefeller Found. films.

PICTURES INCLUDE: (Orig. story, collab., assoc. prod., s.p.) Strange Voyage, Avalanche, Younger Brothers, Return of Frontiersman, Embraceable You, Sierra, The Sniper, My Six Convicts, Eight Iron Men, Member of the Wedding, Three Lives, The Big Kill, Not as a Stranger, Pride and the Passion, Girls, Girls, Girls, Decision at Delphi.

Academy Award, orig. story collab., Panic in the Streets. Member: Acad. of M.P. Arts & Sciences; S.W.G.

ANHALT, EDWARD: Writer. b. New York, N.Y. Academy Award. orig. story collab., Panic in the Streets.
PICTURES INCLUDE: The Young Savages, Academy Award, screenplay, Becket, In Enemy Country, Boeing-Boeing, The Young Lions, Day of the Guns, Not As a Stranger, Madwoman of Chaillot, Member of the Wedding, Boston Strangler, Jeremiah Johnson, Q B VII (TV), Luther, Escape to Athena.

ANKERS, EVELYN: Actress. b. Valparaiso, Chile (English parents); e. Latymer and Godolphyn School, Royal Academy Dramatic Art (London); wife of Richard Denning. In British pictures and stage, then on stage New York (in ("Ladies in Retirement") (West End, London) Bats in The Belfry.
PICTURES INCLUDE: Black Beauty, Queen of Burlesque, Wolf Man, Ghost of Frankenstein, Frozen Ghost, All By Myself, Texan Meets Calamity Jane, Parole, Inc., Tarzan's Magic Fountain, No Greater Love.
TV: TV shows, Mr. & Mrs. North, Ford Theatre, Screen Directors Playhouse, Star and the Story, 20th Century-Fox Hour, Cheyenne, etc.

ANN-MARGRET: Actress, singer. b. Stockholm, Sweden, April 28, 1941. e. grad., New Trier High School, Winnetka, Ill., attended Northwestern University. Radio shows, toured with band; worked with George Burns, Sahara, Las Vegas. TV debut, Jack Benny Show, April, 1961.
PICTURES INCLUDE: debut, Pocketful of Miracles, State Fair, Bye, Bye Birdie, Viva Las Vegas, Kitten With a Whip, Bus Riley's Back in Town, Once A Thief, Cincinnati Kid, The Pleasure Seekers, Made in Paris, The Tiger and the Pussycat, C. C. & Company, Carnal Knowledge, RPM, The Outside Man, The Train Robbers, Tommy, The Last Remake of Beau Geste, Joseph Andrews, The Cheap Detective, Magic, The Villain, Middle Age Crazy, Lookin' To Get Out.

ANNAKIN, KENNETH: Director, writer. b. Yorkshire, England. Graduated through Experimental Theatre and Documentary. Films incl: Across The Bridge, Swiss Family Robinson, Very Important Person, The Longest Day, The Fast Lady, The Informers, Those Magnificent Men In Their Flying Machines, Battle Of The Bulge, The Long Duel, Monte Carlo Or Bust, Call Of The Wild, Paper Tiger, The Fifth Musketeer, The Pirate.

ANSARA, MICHAEL: Actor. b. Lowell, Mass., April 15, 1927. e. Pasadena Playhouse. Served in U.S. Army; then summer stock, little theatre, road shows.
PICTURES INCLUDE: The Robe, Julius Caesar, Sign of the Pagan, Bengal Brigade, Soldiers Three, Only the Valiant, New Orleans Uncensored, Diane, Lone Ranger, Sol Madrid, Daring Game, The Bears And I, Mohammad, Messenger of God, The Manitou.
TV: Broken Arrow, Law of the Plainsman, Westerner.

ANSPACH, SUSAN: Actress. b. New York City. e. Catholic University (Washington, D.C.) After school returned to N.Y. and in 3 years had performed in 11 off-Bdwy. and moved to Los Angeles and entered films.
PICTURES INCLUDE: Five Easy Pieces, The Landlord, Play It Again Sam, Blume in Love, The Big Fix, Running, The Devil and Max Devlin, Gas.

ANTHONY, TONY: Actor-producer. b. Clarksburg, West Virginia. e. Carnegie Tech.
PICTURES INCLUDE: Force of Impulse, Pity Me Not, The Wounds of Hunger, A Stranger in Town, The Stranger Returns, A Stranger in Japan, Come Together, Blindman, Pete Pearl and the Pole, Let's Talk About Men, Get Mean, The Boy Who Owned a Melephant.

ANTONIONI, MICHELANGELO: Director. b. Ferrara, Italy, Sept. 29, 1913. e. Bologna University. Film critic on local paper, then script writer and asst. director. First films as a director were documentaries, including: Gente del Po (1947) and N.U. First feature Love Story, 1950; since The Lady Without Camelias, Love in a City, I Vinti, The Friend, The City, L'Avventura, The Night, The Eclipse, The Red Desert, Blow-Up, Zabriskie Point, The Passenger.

ANTONY, SCOTT: Actor. b. Gosforth, Newcastle-upon-Tyne, England. e. Leeds Univ. Rec'd. scholarship at Royal Academy of Dramatic Art. Ken Russell found him there and gave him role in Savage Messiah (1972). Has done much TV work in Britain.
PICTURES INCLUDE: Savage Messiah, The Mutations, Dead Cert.

ANZARUT, RAYMOND: Executive producer. b. Lebanon, 1912. e. French Lyceum, Alexandria, Egypt and Paris. Entered film industry 1941 on Jeannie. Subsequently worked on The Day Will Dawn, Talk About Jacqueline, Secret Mission, On Approval. Prod. asst. on They Met in the Dark. Prod. mgr. Flight from Folly. Prod. mgr. in Egypt on Caesar and Cleopatra in 1945. With Marcel Hellman as prod. mgr. on Meet Me at Dawn. Prod. mgr. Swiss location, Scott of the Antarctic; prod. mgr. of Wessex Film Productions Ltd. on Esther Waters, Once a Jolly Swagman, All Over the Town, Mr. Prohack, The Wooden Horse, Gen. prod. mgr. on The Grand Design (six technicolor two-reelers on European Recovery for E.C.A.). Appointed gen. mgr. of Wessex 1950. Appointed prod. & admin. head, E.C.A. Motion Picture Section, Paris, 1952. Then with Coronado (Sea Devils), Carol Reed 1-1 British Lion (Man Between), Watergate-British Lion (An Inspector Calls) 1953. British Lion-London Films (The Man Who Loved Redheads), David Lean-Lopert Films Inc. (Summer Madness) 1954, London Films (Storm Over the Nile) 1955, Sailor Beware 1955, Dry Rot 1956, Esther Costello 1956–57. Assoc. prod. on The Silent Enemy 1957, Room at the Top, 1958, Our Man in Havana, 1959, 1960; prod. sup. Mysterious Island (Charles Schneer 1-1 Columbia), 1961; Lolita, 1962; The War Lover, Lord Jim. Reconnaissance of South-East Asia, setting up and filming in Hong Kong and Cambodia. Decorated by Prince Sihanouk of Cambodia; prod. sup., Scene Nun, Take One. Assoc. prod. The Hill, 1966–67; The Dirty Dozen. Exec. prod. for Warner Bros. Seven Arts Prods. Ltd. London, May 1967–Sept. 1969 when Company was taken over by Kinney. Re-joined Ken Hyman in Nov., 1969 (Inter-Hemisphere Productions Ltd.) as prod. exec., 71/72 preparing projects for Universal Pictures Ltd. 1972 with Kenneth Hyman reverted to indep. prod. and helped prepare Emperor of the North Pole, in London. Organized and supervised all London filming for Gaumont International (Paris). Le Silencieux. 1973: preparing the Heights of Aervos for Inter-Hemisphere Productions. Organized and supervised Eng. loc. for La Gifle. Rejoined Inter-Hemisphere Prod., 1975–76.

APFEL, EDWIN R.: Executive. b. New York, N.Y., Jan. 2, 1934. e. Franklin and Marshall College, B.A., 1955. copywriter and account executive, Donahue and Coe, 1955–60; adv. publicity dept., Metro-Goldwyn-Mayer, 1960–62; dir. of adv. and creative services, MGM, Verve Records, 1962–63; dir. of adv., Embassy Pictures 1963–65; Advertising Writer, Consultant, 1965.

APPELBOOM, MAX: Producer. b. Amsterdam, Holland, March 7, 1924. e. art schools in Holland, Belgium and France. Started in the motion picture business in 1946 with J. Arthur Rank, then to RKO Radio Pictures. Worked from 1953–56 in Hollywood as writer, director and producer. Returned in 1956 to his home base, Amsterdam. Now pres. Max Appelboom Producties Inc. and Appletree Filmproductions Inc.
PICTURES INCLUDE: Floris, Monkey Business, Will O' The Wisp, Water For Waterland, The Bearboat, Gangrene, The Buggy Man and The New Adventures of Heidi.

APTED, MICHAEL: DIRECTOR. Broke into show business at Granada TV in England in early 1960s as trainee, researcher, and finally director. By 1965 was producer-director for local programs and current affairs; then staff drama dir. for TV series, plays and serials. In late 1960s left Granada to free-lance.
PICTURES: Triple Echo, Stardust, The Squeeze, Agatha, Coal Miner's Daughter, Continental Divide.
TELEVISION: Poor Girl, High Kampf, Highway Robbery, Kisses at 50, The Collection (won Intl. Emmy Award for last).

ARBEID, BEN: Producer. b. London, England, 1924. Ent. m.p. ind. 1947. Prod. man./executive some 20 features, TV series, 1961–1964 prod. The Barber of Stamford Hill, Private Potter, The Children of the Damned, Murder Most Foul, Murder Ahoy, The Alphabet Murders, 1966; The Jokers, 1967; Assignment K.; 1968: Exec. prod., Before Winter Comes; 1969, Hoffman; 1972, The Hireling, Grand Prix—Cannes 1973; 1977 The Water Babies; 1978 The Eagles Wing.

ARCHAINBAUD, GEORGE: Director. b. Paris, France, e. Paris U.; film director since 1915; In U.S. dir. World Films, later Selznick Pictures, RKO, Paramount, etc. Many Hopalong Cassidys, Gene Autrys for Col.; Lone Ranger, Gene Autry, Range Rider, Anne Oakley, Screen Gem, etc., film series for TV.

ARCHER, ANNE: Actress. Daughter of actress Marjorie Lord and actor John Archer. Married Terry Jastrow, actor and TV network sports producer-director.
PICTURES: The All-American Boy, Honkers, Cancel My Reservation, Trackdown, Lifeguard, Paradise Alley, Hero at Large, Green Ice.
TELEVISION: The Pirate (mini-series), The Blue Knight, Seventh Avenue.

ARCHER, JOHN: Actor. r.n. Ralph Bowman, b. Osceola, Neb., May 8, 1915; e. U. So. Calif. Won Jesse L. Lasky "Gateway to Hollywood" talent quest; m.p. debut in "Flaming Frontiers" 1938; on N.Y. stage also.
PICTURES INCLUDE: White Heat, After Nightfall, Destination Moon, Big Trees, Yank in Indo-China, Sound Off, Rodeo, Sea Tiger, The Stars Are Singing, Dragon's Gold, No Man's Woman, Emergency Hospital.

ARCHER, ROBERT V.: Executive. Joined Paramount Pictures in 1967 as adm. asst. to v.p. of facilities and TV prod. In 1975 named dir. of studio admin. 1977 promoted to dir. of studio operations and admin., responsible for prod. servicing and admin. depts. of studio.

ARCHERD, ARMY: Columnist-TV Commentator. rn. Armand Archerd, b. N.Y., Jan. 13. ed. UCLA Grad. '41, US Naval Academy Post Graduate School, 1943. Started as usher at Criterion Theatre, N.Y., while in high school. After grad. UCLA, worked at Paramount studios before entering Navy, Lt., joined AP Hollywood bureau 1945, Herald-Express, Daily Variety as columnist, 1953. M.C. Hollywood premieres and Academy Awards. President, founder Hollywood Press Club, awards from Masquers, L.A. Press Club, Hollywood Foreign Press Club, and "Newsman of the Year" award from Publicists Guild, 1970; "Movie Game" TV Series; People's Choice, co-host.

ARCHIBALD, JAMES: (M.B.E., OSr.J., J.P., M.A., Hon. F.T.C.L., F.R.S.A.). Producer, writer, b. 1920. e. Westminster School, Merten College, Oxford. With Rank Organization 1950–57, later J. Walter Thompson Co. Currently chairman James Archibald & Associates Ltd. Chmn. Nat. Mus. Counc. of G.B. Prod. writer credits incl.: Some People, The King's Breakfast, The Launching, The Scheme, Overture and Beginners, The More Man Understands, Breakthrough, Town Nurse Country Nurse, Soldier '67, Opus, Speaking of Britain, In Search of Opportunity, 16 shorts for British Pavilion Expo '67, Music! The Violin, The Threadmakers, Dialogue, Ambience, Island of Rainbows, For the Community, Enigma Variations, Shapes of The Future, The Engineer is a Woman, Magic Ride, The Science of Farming, Sweet Thames, 15 short color films for British Pavilion, Expo '70, Heuga, Images, International Company, Hovertrain, Talking of Science, What Did You Learn at School Today?, Richness and Complexity, Danger from Fire, 1000th Chance, Time for Thought?, Barriers, It's All Man Made, A Child's World, No Turning Back, Rehearsal, Cement, Birdstrike, A New Beginning, Sugar, Attitudes, Festival, Sea Symphony, Music in Scotland, The Single Team, To Sustain Life, Figures, Royal Doolton, Electronics of Film, Stock Exchange.

ARDEN, EVE: Actress. r.n. Eunice Quedens. b. Mill Valley, Calif., 1912. Appeared with Alcazar stock co., Bandbox Repertory Theatre; Broadway debut in "Ziegfeld Follies of 1936" also in "Ziegfeld Follies of 1938" and Let's Face It, Night and Day, Cover Girl, Voice of the Turtle, Goodbye, My Fancy, We're Not Married, Our Miss Brooks, Anatomy of a Murder, Dark at the Top of the Stairs, Sergeant Deadhead, The Strongest Man in the World, Grease.

TV: Our Miss Brooks, The Eve Arden Show, The Mothers-in-Law.

STAGE: Mame, Hello Dolly, Butterflies Are Free.

ARDEN, MARY: Actress. b. Litchfield, Mich., Dec. 9, 1910. e. Northwestern U., Anne Wolters School of Speech & Drama. Appeared in stock; radio, March of Time, Lux.

PICTURES INCLUDE: Jealousy, Youth Aflame, Missing Corps, Father Take a Wife, Best Years of Our Lives.

TV: Playhouse 90, Climax, Studio One.

ARKIN, ALAN: Actor-Director, b. New York City, March 26, 1934. e. Los Angeles City College, Los Angeles State College, Bennington (Vt.) College. Member folk singing group The Tarriers; off-Broadway, Second City, Man Out Loud, From the Second City; Broadway, Enter Laughing, Luv. Dir. Little Murders, White House Murders.

PICTURES INCLUDE: The Russians Are Coming, The Russians Are Coming, Wait Until Dark, The Heart is a Lonely Hunter, Popi, Catch 22, Deadhead Miles, Last of the Red Hot Lovers, Freebie and the Bean, Rafferty and the Gold Dust Twins. Director: Little Murders, Hearts of the West, 7½ Per Cent Solution, The In-Laws, Simon, Chu Chu and the Philly Flash.

ARKOFF, SAMUEL Z.: Producer, Motion Picture Executive. Chairman & President of the Samuel Z. Arkoff Company (formed 1980) and Arkoff Int'l. Pictures (formed 1981). b. Fort Dodge, Ia., June 12, 1918. e. U. of Colorado, U. of Iowa & Loyola U. Law School. m. Hilda Rusoff. USAF Cryptographer World War II. Co-founder American Releasing, 1954, and American International Pictures, 1955. President and Chairman of the Board American International Pictures until 1979. Named with partner James H. Nicholson as Producers of the Year in 1963 by Allied States Association of Motion Picture Theatre Owners and in 1964 as Master Showmen of the Decade by the Theatre Owners of America. Also named Producers of the Year at the Show-A-Rama VIII, and in 1966 and 1967 voted among top ten producers in exhibitor ratings by the independent theatre owners. In Rome in 1970 honored by President of the Republic of Italy with decree naming him a Commendatore of the Order of Merit. In 1971 he and Nicholson named Pioneers of the Year by the Foundation of the Motion Picture Pioneers, Inc. Since appointment in 1973, has served as International Ambassador of Variety Clubs International. Appointed Member of the Board of Trustees of Loyola Marymount Univ., Los Angeles, in 1979.

PRODUCTION CREDITS: On more than 200 pictures including Amityville Horror, Dressed to Kill, How to Beat the High Cost of Living, The Earthling, Love at First Bite, Meteor, California Dreaming, Force Ten from Navarone, The House of Usher, Pit and the Pendulum, Tales of Terror, Master of The World, Premature Burial, Panic in the Year Zero, The Raven, Beach Party, Haunted Palace, Comedy of Terrors, Bikini Beach, Masque of the Red Death, Muscle Beach Party, Pajama Party, Tomb of Ligeia, Wild Angels, Devil's Angels, The Trip, Three in the Attic, Wild in the Streets, The Oblong Box, Scream and Scream Again, Murders in the Rue Morgue, Cry of the Banshee, Bloody Mama, Wuthering Heights, Dr. Phibes, Frogs, Blacula, Dillinger, Heavy Traffic, Hennessy, Cooley High, Food of the Gods, Futureworld, Great Scout and Cathouse Thursday, Land that Time Forgot, People That Time Forgot, At The Earth's Core, Island of Dr. Moreau, and A Winning Season.

ARLEDGE, ROONE: Executive. b. Forest Hills, N.Y. e. Columbia Univ. Entered industry with Dumont Network in 1952; joined U.S. Army, 1953, serving at Aberdeen Proving Ground in Maryland, where produced and directed radio programs. Joined NBC in 1954 where held various production positions. Joined NBC in 1954 where held various production positions. In 1960 went to ABC TV; 1964, named v.p. in chg. of ABC Sports. Created ABC's Wide World of Sports in April, 1961. Named pres. of ABC News in 1968; pres. of ABC News and Sports, 1977. Holds four George Foster Peabody Awards for sports reporting; 19 Emmy awards.

ARLEN, HAROLD: Composer, b. Buffalo, N.Y., Feb. 15, 1905. Started as pianist, singer, arranger for popular orchestras; songs include: Stormy Weather, I've Got the World On a String, Let's Fall in Love Blues in the Night, Accentuate the Positive.

BROADWAY: You Said It, Earl Carroll's Vanities, Americana, Life Begins at 8:40, Bloomer Girl, Hooray for What, House of Flowers, Jamaica, etc.

PICTURES INCLUDE: Strike Me Pink, Let's Fall in Love, Wizard of Oz (Academy Award, 1939, for best song, Over the Rainbow), A Day at the Circus, Star-Spangled Rhythm, Love Affair, Sky's the Limit, Here Come the Waves, Cabin in the Sky, Country Girl, Gay Purr-ee, I Could Go On Singin

ARLING, ARTHUR E.: Cinematographer. b. Missouri, Sept. 2, 1906; e. N.Y. Institute of Photography. Entered m.p. Fox studio 1927 as asst. cameraman, 2nd cameraman 1931; operative cameraman on "Gone With the Wind" which won the Academy Award for technicolor photography 1939. Lt. Comdr. U.S.N.R., World War II; received Academy Award as dir. of Photography on "The Yearling" 1946. Member: Amer. Soc. of Cinematographers.

PICTURES INCLUDE: Homestretch, Captain from Castile, Mother Was a Freshman, You're My Everything, Wabash Avenue, My Blue Heaven, Call Me Mister, Belles on their Toes, The Glass Slipper, Three for the Show, Love Me or Leave Me, I'll Cry Tomorrow Ransom, Great American Pastime, Tammy & the Bachelor, Pay the Devil, Story of Ruth, Pillow Talk, Lover Come Back, Notorious Landlady, Boys Night Out, My Six Loves.

ARMOUR, REGINALD: Foreign dist., prod. b. Chicago, Ill., Nov. 28, 1905; e. Edinburgh, Scotland. Exec. RCA Victor Co. to 1933; estab. exch. India, Burma, Ceylon RKO Radio; Far East gen. mgr.; European gen. mgr. (Paris) 1937; exec. asst. to pres. RKO 1941; asst. gen. mgr. RKO studios 1942; British & European rep. Walt Disney Prod. 1942; financial field (N.Y.) 1943; exec. vice-pres. Richard Condon, Inc. 1944; with overseas br. OWI 1945; foreign rep. Columbia Internatl. Corp. (N.Y.) 1945–49; joined Republic Pictures Internatl. Corp. 1949; apptd. v.p., 1950; supvr., Europe & Near East for Republic 1952, exec. v.p. Republic Pictures International (Great Britain) and man. dir. Republic Prod. (Gt. Brit.) Ltd.; exec. v.p. Republic Pictures International Corp., 1955; pres., The Dorsey Corp., 1960; Vice Chmn., The Dorsey Corp., 1964; pres., SOS Photo-Cine-Optics, Inc., 1965; pres. F&B/Ceco of Calif. Inc., 1967; pres. Instant Protection Systems, Inc., 197-. Exec. V.P. and Treas. The Quedo Corporation and International Producers Services, Inc. 1973.; 1974, V.P. & Treas., Dyna-Soar, Inc.; 1977, exec. v.p. & treas., Two Feathers Prods.

ARMSTRONG, GORDON: Executive, b. East Orange, N.J., Nov. 26, 1937. e. Arizona State Univ., graduate studies at N.Y.U. Entered ind. as newspaper contact for MGM (1960–63); with Allan, Foster, Ingersoll and Weber 1963–70 as acct. exec.; joined 20th Century-Fox in 1970 as nat. pub. dir. In 1975 appt. dir. of adv.-pub.-promo. for Dino De Laurentiis Corp. In 1978, became vice pres., worldwide marketing for the company; 1980, named v.p., adv.-pub.-prom., Universal Pictures.

ARNALL, ELLIS GIBBS: Lawyer, Exec. b. Newnan, Ga. Mar. 20, 1907; e. Mercer U.; U. of South, A.B. 1928, D.C.L. 1947; U. of Georgia LL.B 1931; Atlanta Law School, LL.D. 1941; Piedmont Coll., LL.D. 1943; Bryant Coll., LL.D. 1948. Georgia state representative from Coweta County, 1936–38; asst. Attorney-General (Ga.) 1938–42; Attorney-General (Ga.) 1942–43; Governor of Ga. 1943–47; pres. Dixie Life Insurance Co.; pres., Columbus National Life Insurance Co. senior mem. law firm Arnall, Golden & Gregory; pres. Georgia State Jr. Chamber of Commerce 1939; trustee U.

of South; author of "The Shore Dimly Seen" 1946. "What The People Want" 1948; member U.S. Natl. Com. on UNESCO; member U.S. delegation to 4th annual conference UNESCO, Paris, France, 1949. Member: SIMPP (Pres. 1948); 1952 on leave as dir. Office of Price Stabilization, Washington; back to SIMPP, Sept. 1952; pres. Ind. Film Prod. Export Corp., 1953; member bd. of dir., exec. comm., COMPO, numerous honorary societies, mem. exec. com., U.S. Nat. Comm. for UNESCO, 1964–65, trustee, Mercer Univ. 1964–68; chair. bd. Coastal States Life Insurance Co., chair. bd. National Association of Life Companies. Member Academy Motion Picture Arts & Sciences; vice chm., Sun Life of America Group, Inc.

ARNAZ, DESI: Actor, b. Santiago, Cuba, Mar. 2, 1917. e. Colegio Dolores, Jesuit Prep School, Santiago; at the age of 17 was the vocalist with a band at Miami Beach, Fla.; later with own rhumba band; played in N.Y. night clubs; on stage in "Too Many Girls;" m.p. debut in "Too Many Girls" 1940; served as Staff Sgt. U.S. Army Medical Corps 3 years; musical dir. Bob Hope show 1946–47. Indep. prod.-owner, Corona Breeding Farm.

PICTURES INCLUDE: Father Takes a Wife, Four Jacks and a Jill, Navy Comes Through, Bataan, Cuban Pete, Long Long Trailer, Holiday in Havana; (actor-prod.) Forever Darling.

TV: (Actor-exec. prod.) I Love Lucy TV films, The Lucille Ball-Desi Arnaz Show; Pres. Desilu Productions, Inc., 1962; Pres. Desi Arnaz Prod. Inc., 1965; Prod. The Mothers-in-Law.

ARNAZ, JR., DESI: Actor-Singer, b. Jan. 19, 1953. Son of Lucille Ball and Desi Arnaz. Began appearing on I Love Lucy show on TV in infancy and on various Lucy shows since. Gained fame as rock singer and musician with the Dino, Desi and Billy group. Film debut in Red Sky at Morning in 1972.

PICTURES INCLUDE: Red Sky at Morning, Marco, She Lives (tv), Joyride, A Wedding.

ARNAUD, LEO: Composer, arranger, conductor. b. Lyons, France, July 24, 1904: e. Ormesson Music Acad., Ph.D. Faculty mem. & symph. cond., Ormesson Music Acad., 1922–27; mus. dir. & arr. in France & England, 1927–30; Fred Waring Orch., U.S., 1931–35; arr., vocal coach, comp., mus. dir., MGM, 1936–44; U.S. Armed Forces, 1945 dir. many orch. including; L.A. Concert Band, American Legion Band, Hollywood String Orch., Beverly Hills, Symph. Orch.; free-lance mus. dir., comp., arr. since 1945.

PICTURES INCLUDE: One Touch of Venus, Easter Parade, Date With Judy, Three Little Words, Lovely to Look At, Stars & Stripes Forever, Rose Marie, Seven Brides for Seven Brothers.

ARNE, PETER: Actor. b. British Malaya, 1922. Screen debut England 1955 in Timeslip. Under contract to Associated British Picture Corporation since 1956.

PICTURES INCLUDE: Murder Anonymous, Cockelshell Heroes, Tarzan and the Lost Safari, High Tide at Noon, The Moonraker, Intent to Kill, Danger Within, Conspiracy of Hearts, Scent of Mystery, Sands of the Desert, Hellfire Club, The Story of David, Treasure of Monte Cristo, Pirates of Blood River, Ice Cold in Alex, The Victors, Black Torment, The Girl in the Headlines, Strangers Meeting, The Sandwich Man, The Dambusters, The Purple Plain, Khartouri, Battle Beneath the Earth, Chitty Chitty Bang Bang, The Oblong Box, When Eight Bells Toll, House of Evil, Straw Dogs, Nobody Ordered Love, Antony and Cleopatra, Return of the Pink Panther, Providence, Agatha, Pentious Passage.

TV: The Saint, Danger Man, Crane, series. The Mask of Janus, The Spies, Baron, Man in the Suitcase, Gulliver's Travels, The Champion, Night Conspirators, Hereward The Wake, The Avengers, Father Tiger, The Third Man, This Way for Murder, Vendetta, Poor Charley, Departments, The Champions, The Survivors, The Trigaree Affair, All Out for Kangaroo Valley, Special Branch, The Shattered Eye, Take 3 Girls, The Stallion, Rivals of Sherlock Holmes, Softly Softly, Task Force, The Fox, A Place in Europe, The Ventures Quiller, The Expert, Tycoon.

ARNELL, PETER: Producer. b. Bridgeport, Conn., e. U. of Mich., A.B. Pub. dir., prog. dir., WJLS, Beckley, W.Va.; m.c. & disc jockey, WPEN, Philadelphia; actor, announcer, N.Y.; creator of Rate Your Mate, Name's the Same; creator & prod. of Wheel of Fortune, Balance Your Budget, I'll Buy That, Take a Guess, What's in a Word? prod. Chance of Romance, Take a Good Look, Celebrity Talent Scouts, Face the Facts, Talent Scouts, Take My Word, Ernie Kovacs' Take a Good Look.

ARNESS, JAMES: Actor. b. Minneapolis, Minn., May 26, 1923. e. Beloit College. Served in U.S. Army; worked in advertising, real estate; m.p. debut in Farmer's Daughter.

PICTURES INCLUDE: Battleground, Hell Gate, Man from Texas, People Against O'Hara, Iron Man, The Thing, Big Jim McLain, Horizon's West, Lone Hand, Island in the Sky, Veils of Bagdad, Hondo, Her Twelve Men, Them, Many Rivers to Cross, Flame of the Islands, Sea Chase.

TV SHOWS: Gunsmoke, How the West Was Won.

ARNOLD, DANNY: Writer, actor. r.n. Arnold Rothman; b. New York, N.Y., Jan. 23, 1925. Appeared in summer stock, night clubs, vaudeville; entered m.p. ind. as sound effects ed., Columbia, 1944–46; then legit., night clubs, vaudeville, 1946–50. Appeared in m.p., Breakthrough, Inside the Walls of Folsom Prison, Sailor Beware, Jumping Jacks, Scared Stiff, Stars Are Singing; featured on Martin and Lewis TV show 2 yrs., and wrote their Comedy Hours to 1953.

PICTURES INCLUDE: collab. s.p., story, The Caddy; collab. s.p., Desert Sands: story, s.p., Fort Yuma, story, s.p. Rebel in Town; s.p., Outside the Law. Writer-TV, Tennessee Ernie Ford Show.

ARNOLD, EDDY: Singer. Radio performer, Nashville, Tenn.; recording star since 1946; records include That's How Much I Love You, Anytime; star of Eddy Arnold Show, ABC-TV.

ARNOLD, JACK: Producer, Director. b. New Haven, Conn., Oct. 14, 1916; e. Ohio State U., Amer. Acad. of Dramatic Arts. Actor on B'way, Brit. m.p.; U.S. Air Force, 1942–45; prod. 25 documentary films for State Dept., Army & private industry including With These Hands; dir., 1952.

PICTURES INCLUDE: The Mouse That Roared, Bachelor in Paradise, Global Affair, The Incredible Shrinking Man, It Came From Outer Space, The Creature from the Black Lagoon, Girls in The Night, Revenge of The Creature, The Tattered Dress, Glass Web, Man in The Shadow, Man From Bitter Ridge; dir. story; Tarantula, Red Sundown, Decision at Durango, Outside The Law, Pay The Devil, No Name On The Bullet, The Lady Takes A Flyer, The Lively Set, High School Confidential. Producer, Universal 1956; dir., Black Eye, 1974, Swiss Conspiracy, 1975.

ARNOUL, FRANCOISE: Actress. r.n. Francoise Gautsch. b. Algeria, 1931. Studied acting in Paris; m.p. debut in 1949.

PICTURES INCLUDE: Companions of the Night, Sheep Has Five Legs, The Bed, Inside a Girl's Dormitory, Tempest in the Flesh, Only the French Can (French CanCan).

ARNOW, MAXWELL: Executive. b. New York; married; e. graduate New York University school of law. Manager New York office Leo Morrison, Inc.; then talent executive for Warner Brothers in east from Jan. 1, 1932. Coast casting director, August, 1932, July, 1937. Then executive capacity to 1938; production executive, Selznick-International, "Gone With the Wind"; production executive Walter Wanger Production, 1939–1940; talent exec. Columbia Pictures, 1941–56; v.p., Hecht-Lancaster Org., 1956; talent exec., Columbia Pictures, 1958–61; exec. v.p. Artists & Production Associates, Inc., 1962.

ARNOW, TED J.: Executive. b. Brooklyn, N.Y. e. St. Johns Univ. Washington and Lee Univ. Served as dir. of recreation for 262nd General Hospital in Panama. Veteran of over 40 yrs. in amusement industry. Is v.p. for adv., pub., & promo, for Loew's Theatres. Member: Motion Picture Pioneers, Variety Clubs, Will Rogers Hospital; former pres. of AMPA (Assoc. M.P. Advertisers).

ARTHUR, ART: Prod. writ. exec. b. Toronto, Can., July 24, 1911, e. Jarvis Collegiate Inst. M. Jessica Pepper. Broadway columnist, Brooklyn Daily Eagle; news mgr. Toronto Star News Service; newspaperman on various dailies; pub. rel. in 1936, entered m.p. ind. as writer 1937; U.S. Army 1943–46; wrote script for "Seeds of Destiny" which won 1947 Academy Award for documentary, & AAF films incl. "Air Pattern-Pacific" Member: Writers Guild of America (exec. bd. mem. 3 times), Prod. Guild of America, Nat. Academy of TV Arts & Science, AMPAS, Pub. Rel. Society of America, Pub. Guild, exec. staff C. B. DeMille, former exec. secy. Hollywood M.P. Ind. Council, exec. staff, Ivan Tors Films. Former member, exec. bd. COMPO; former Executive Coordinator, Film and Television Coordinating Committee.

PICTURES INCLUDE: Sun Valley Serenade; (collab. s.p.) Tight Shoes, Riding High, Everything Happens at Night, Fabulous Dorseys, Montana Mike, Northwest Stampede, Song of India, Clarence The Cross-Eyed Lion, Zebra in The Kitchen, Around the World Under the Sea, Birds Do It, The Daring Game, Hello Down There; co-prod., collab. story Battle Taxi, co-prod., writ. Rhino, Flipper's New Adventure; writer, TV., Sea Hunt, Ripcord, Beachcomber, Man and The Challenge, Flipper, Aquanauts, Daktari, Gentle Ben.

ARTHUR, BEATRICE: Actress. b. New York City, May 13. e. Franklin Institute of Sciences & Art. m. Gene Saks. Studied with Erwin Piscator at New School for Social Research; first stage role as Lysistrata; professional stage debut in Dog Beneath the Skin, 1947, followed by Gas, Yerma, No Exit, Six Characters in Search of an Author, The Owl and the Pussycat, etc.; stock appearances include Personal Appearance, Candlelight, Love or Money, The Voice of the Turtle; recent N.Y. stage includes Ullysses in Nighttown, Chic, Mame (Tony award), Fiddler on the Roof.

PICTURES INCLUDE: That Kind of Woman, Lovers and Other Strangers, Mame, History of the World—Part I.

TV: Numerous guest appearances; Maude.

ARTHUR, EDWARD B.: Executive. b. New York City, Sept. 30, 1915. e. San Jose State Coll.; Los Angeles Jr. Coll. Engaged in operation management, purchasing of films etc. in Calif., N.Y., Missouri, Illinois; served in armored div. World War II, 1st Lt., at close of war Special Service unit (Germany) in chg. of distribution of films to service units; resumed his post as gen. mgr. Arthur Enterprises, Inc. which he had joined in 1935; named pres. St. Louis Amusement Co., 1946. Member: Variety Club.

ARTHUR, JEAN: Actress b. New York City, Oct. 17, 1908. On stage; minor roles. m.p. debut as lead in "Warming Up" 1928.

PICTURES INCLUDE: Canary Murder Case, Mysterious Dr. Fu Manchu, Whirlpool, Mr. Deeds Goes to Town, You Can't Take It With You, Too Many Husbands, Talk of the Town, Lady Takes a Chance, Impatient Years, Foreign Affair, Shane.

N.Y. STAGE: Peter Pan, 1950.

TV: The Jean Arthur Show

ARTHUR, ROBERT: Actor, r.n. Robert Arthaud. b. Aberdeen, Wash., June 18; e. U. Washington (U.S. Navy training program). Won H.S. radio announcing contest; prof. announcer & disc jockey; m.p. debut in "Roughly Speaking" 1945.

PICTURES INCLUDE: Nora Prentiss, Sweetheart of Sigma Chi, Devil on Wheels, Mother Wore Tights, 12 O'Clock High, September Affair, Air Cadet, Big Carnival, On the Loose, Belles on Their Toes, Just For You, The Ring, Young Bess, Take the High Ground, Return from the Sea, Top of the World, Desperadoes Are in Town, The System, Green Grass of Wyoming, Switchblade, Hellcats of the Navy, Young and Wild, Yellow Sky, Mother is a Freshman.

ARTHUR, ROBERT: Producer, b. New York City, Nov. 1, 1909; e. Southwestern U., U. Southern Calif. Oil operator 1929–36; joined MGM as writer 1937. Served in the United States Army 1942–5.

PICTURES INCLUDE: (proc.) Buck Private Come Home, For the Love of Mary, Mexican Hayride, Abbott and Costello in Foreign Legion, Louisa, Wistful Widow of Wagon Gap, Abbott & Costello Meet Frankenstein, Francis, Golden Horde, Starlift, Story of Will Rogers, The Big Heat, The Black Shield of Falworth, Ricochet Romance, The Long Gray Line, Francis in the Haunted House, Day of Fury, Pillars of the Sky, Kelly and Me, Mr. Cory, Midnight Story, Man of a Thousand Faces, A Time to Love and a Time to Die, The Perfect Furlough, Operation Petticoat, The Great Impostor, Come September, Lover, Come Back, The Spiral Road, That Touch of Mink, For Love or Money, The Brass Bottle, Captain Newman, M.D., Bedtime Story, Father Goose, Shenandoah, Blindfold, A Man Could Get Killed, Hellfighters, Sweet Charity.

ASH, RENE: Publicist. b. Brussels, Belgium, March 14, 1939; e. U. of Omaha. Member of AMPA trustee since 1967; member of the Publicists Guild since 1968; Eastern Vice President of Pub Guild since 1973; Author of The Film Editor in Motion Pictures & Television; employed with I.A.T.S.E. 1968–1979, prior to which was assoc. editor, Greater Amusements; various articles published in foreign film magazines; editor-in-chief, Backstage; 1979–80; est. R.L. Ash, Inc., entertainment consultants.

ASHBY, HAL: Director. b. Ogden, Utah. e. Utah State Univ. Began career as film editor. Worked with Norman Jewison on The Cincinnati Kid, The Russians Are Coming, The Russians Are Coming, In the Heat of the Night (AA) and also as assoc. prod on The Thomas Crown Affair and Gaily, Gaily, Directing debut, The Landlord, 1970.

PICTURES INCLUDE: Harold and Maude, The Last Detail, Shampoo, Bound for Glory, Coming Home, Second Hand Hearts, Lookin' To Get Out.

ASHER, IRVING: Production executive. b. San Francisco, Sept. 16, 1903; m. Laura La Plante; began career in pictures in 1919; 1931–38 managing dir. of Warner Bros. First National Prod. in England, 1938; assoc. prod. London Films, exec. prod. Columbia British. In 1945 pres. Rainbow Prods. Exec. in charge TV prod., TCF Television Production, sales head, 1958.

PICTURES INCLUDE: Prison Without Bars, The Four Feathers, Billy the Kid, Blossoms in the Dust, Mr. and Mrs. North, Nazi Agent, The War Against Mrs. Hadley, Here Comes the Groom, Turning Point, The Stars Are Singing, Elephant Walk.

ASHER, JACK: Producer, dir. of photography. b. London, 1916. Ent. m.p. ind. 1930 as asst. cameraman for Gainsborough camera operator 1935; 1940 returned to Gainsborough as camera operator; dir., photography Gainsborough 1961. Many films since; 1961 Co. prod. with brother Robert, She'll Have to Go; dir. of photo., Scarlett Blade, Stitch in Time; dir., photo., The Intelligence Men, dir. photog., The Early Bird, 1965.

ASHER, JANE: Actress. Born London, England, 1946. She has acted in many films and TV plays. Films: Greengage Summer, Masque Of The Red Death, Alfie, Deep End, Henry The Eighth and His Six Wives, etc.

ASHERSON, RENEE: Actress. b. London, Eng., e. Gerrards Cross, Eng. On the stage in rep. Birmingham. London screen debut 1944 in Henry V.

PICTURES INCLUDE: Way to the Stars, Once a Jolly Swagman, Small Black Room, Cure for Love, Pool of London, Malta Story, Red Dress, The Day the Earth Caught Fire.

TV: The Red Dress The Man Who Could Find Things, At Home, Door of Opportunity. Balance of Her Mind, Life With Father, Black Limelight, The Long Memory, The Simple Truth, The Whole Truth, A Love Story, A Quiet Halfhour, Clayhanger, Jubilee, Quiet as a Nun, Disraeli, Flesh and Blood, Good-bye Dollins.

ASHLEY, ELIZABETH: Actress. b. 1939. Success on stage in 1960 with role in Take Her, She's Mine, winning Tony Award. Next starred in Barefoot in the Park.

PICTURES INCLUDE: The Carpetbaggers, Ship of Fools, The Third Day, The Marriage of a Young Stockbroker, Paperback Hero, Golden Needles, Rancho DeLuxe, 92 in the Shade, Great Scout and Cathouse Thursday, Coma, Windows, Captured.

ASHLEY, JOHN: Actor. r.n. John Atchley. b. Kansas City, Mo., Dec. 25, 1934; e. Oklahoma State U., BA 1956. Career started in Tulsa Little Theatre, 1956; screen debut, 1957, Dragstrip Girl; later: Motorcycle Gang, Suicide Battalion.

TV: Men of Annapolis, Sheriff of Cochise, Frontier Doctor; Matinee Theatre, Jefferson Drum.

ASHLEY, TED: executive b. Brooklyn, N.Y. Aug. 3, 1922. e. City College of N.Y. School of Business Administration. 1937–1939. With William Morris Agency, agent. 1939–1946; formed Ted Ashley Associates, 1946; pres. Ashley Famous Agency, 1954; director and chairman of executive committee of Warner Communications Inc., 1967 to 1974; chairman of bd. and chief exec. officer, Warner Bros. 1969–76.

ASKEY, ARTHUR: Actor. b. Liverpool, Eng., June 6, 1900. e. Liverpool Inst. Experience in concert parties, vaudeville, revues, musical comedy, pantomime; starred in several films including Charley's Aunt, Ghost Train, The Love Match, Ramsbottom Rides Again, and in BBC series. Many command performances.

PICTURES INCLUDE: Make Mine A Million, Friends and Neighbors.

ASNER, EDWARD: Actor. b. Nov. 15, Kansas City, Kans. e. Univ. of Chicago, where affilliated with campus acting group. Served two years with U.S. Army in France. Returned to Chicago to join Playwright's Theatre Club. Moved to N.Y.C.; Broadway debut in Face of a Hero. Appeared off-Broadway, in stock, and with American Shakespeare Festival. In 1961 moved to Hollywood to become active in films and TV.

PICTURES INCLUDE: Peter Gunn, The Slender Thread, The Satan Bug, Kid Galahad, The Wrestler, Fort Apache, The Bronx.

TELEVISION: Medical Center, Name of the Game, Mod Squad, Ironside, The Mary Tyler Moore Show (regular), Hey, I'm Alive (telefilm), Rich Man, Poor Man, Roots, Lou Grant.

ASSEYEV, TAMARA: Producer. Began career as asst. to Roger Corman, working on 8 films with him. In 1967 started to produce films independently. In 1972 met Alex (Alexandra) Rose and later began co-producing films with her, starting with Drive-In, released by Columbia.

PICTURES INCLUDE: The Wild Racers, Paddy, The Arousers, The History of Atlantic Records, Co-produced with Ms. Rose: Drive-In, I Wanna Hold Your Hand, Big Wednesday, Norma Rae.

ASTAIRE, FRED: Dancer, actor, b. Omaha, Neb. May 10, 1900; e. at home; sister: Adele Astaire, dancer. Brother-sister dance team formed 1907; vaudeville debut 1908; Broadway shows include, Over the Top, Passing Show of 1918, Apple Blossoms, For Goodness Sakes, Lady Be Good, Smiles, Funny Face, Gay Divorcee, Bandwagon; team dissolved 1932; m.p. debut in Dancing Lady 1933; voted Money-Making Star in Motion Picture Herald-Fame Polls 1935–36–37; served with U.S.O. overseas World War II.

PICTURES INCLUDE: Flying Down to Rio, Gay Divorcee, Roberta, Top Hat, Swing Time, Carefree, Story of Vernon and Irene Castle, Holiday Inn, Sky's the Limit, Ziegfeld Follies, Blue Skies, Easter Parade, Special Academy Award for "raising standards of all musicals," 1949. TV specials. Won 9 Emmy Awards, 1958 (An Evening with Fred Astaire) 1959, Another Eve, with Fred Astaire; 1960 Astaire Time; Alcoa Premiere, 1962. Barkleys of Broadway, Let's Dance, Three Little Words, Royal Wedding, Belle of New York, Band Wagon, Daddy Long Legs Funny Face, Silk Stockings, On the Beach, The Pleasure of His Company, Notorious Landlady, Finians Rainbow, 1967, Midas Run, 1968. The Towering Inferno, 1974, That's Entertainment, Part Two, 1976, The Purple Taxi, Ghost Story.

TV: Fred Astaire Show, 1968, It Takes A Thief, A Family Upside Down, Man in the Santa Claus Suit.

ASTIN, JOHN: Actor. b. Baltimore, Md., March 30, 1930. e. Washington and Jefferson College, Washington Drama School, Johns Hopkins U., grad. B.A., U. of Minnesota Graduate School; first prof. job., off broadway, Threepenny Opera; Broadway debut, Major Barbara; dir., co-prod., A Sleep of Prisoners, Phoenix Theatre; during 1955–59, did voices in cartoons, commercials; I'm Dickens . . . He's Fenster, as Harry Dickens, The Adams Family.
STAGE SHOWS INCLUDE: The Cave Dwellers, Ulysses in Nighttown, Tall Story.
PICTURES INCLUDE: West Side Story, That Touch of Mink, Candy, Viva, Max!, Bunny O'Hare, Get to Know Your Rabbi, Every Little Crook and Nanny, Freaky Friday.

ASTOR, MARY: Actress, r.n. Lucille V. Langhanke; b. May 3, 1906; e. Kenwood-Loring School (Chicago). Beauty contest winner 1920; m.p. debut in "Beggar's Maid" 1920; radio m.c. 2½ yrs.; on stage in: Among the Married, Tonight at 8:30, Male Animal.
PICTURES INCLUDE: (silent Don Juan, Bright Shawl, Beau Brummel; (sound) Lost Squadron, Jennie Gerhardt, Easy to Love, Man With Two Faces, Listen Darling, Dodsworth, Maltese Falcon, Across the Pacific, Meet Me in St. Louis, Fiesta, Cynthia, Cass Timberlane, Desert Fury, Act of Violence, Little Women, Any Number Can Play, Kiss Before Dying, Return to Peyton Place, Youngblood Hawke, Hush, Hush Sweet Charlotte.

ATHERTON, WILLIAM: b. New Haven, Conn., 1949. While in high school became youngest member of Long Wharf Theatre Co. Given scholarship to Pasadena Playhouse; then switched to Carnegie Tech School of Drama in 1965. In college years toured with USO prods. in Europe and in stock and industrial shows. Came to N.Y. where first prof. job was in nat'l. co. of Little Murders.
PICTURES INCLUDE: The New Centurions, The Sugarland Express, The Day of the Locust, The Hindenburg, Looking for Mr. Goodbar.
PLAYS: The House of Blue Leaves, The Basic Training of Pavlo Hummel, The Sign in Sidney Brustein's Window.

ATTENBOROUGH DAVID: Broadcaster. b. London, May 8, 1926; e. Wyggeston School, Leicester; Clare College, Cambridge, early career, editor in educational publishing house, ent. BBC-TC Sept. 1952. Prod. Zoo Quest series, Travellers Tales, Adventure and other prog., travel, Eastward with Attenborough, The Tribal Eye, Life on Earth, exploration natural History and anthropology. Controller BBC-2, 1965–68; Dir. of Prog. BBC-TV, 1969–72.

ATTENBOROUGH, RICHARD: Actor, producer, director. b. Cambridge, Eng., Aug. 29, 1923; m. 1945 Sheila Beryl Grant Sim. Wyggeston Grammar School, Leicester. Leverhulme Scholarship to Royal Academy of Dramatic Art, 1941 (Bancroft Medal). First stage appearance Ah Wilderness, 1941. London debut in Awake and Sing, 1942. First film appearance in In Which We Serve, 1942. Appeared in The Little Foxes, Piccadilly Theatre, 1942; Brighton Rock, Garrick Theatre 1943. Joined R.A.F. 1943; seconded to R.A.F Film Unit, 1944, and appeared in Journey Together; demobilised, 1946. Returned to stage, Jan. 1949, in The Way Back (Home of the Brave), To Dorothy, A Son, 1950; Sweet Madness, 1952; The Mousetrap, 1952–54; Double Image, 1956–1957; The Rape of the Belt, 1957–1958.
PICTURES INCLUDE: School for Secrets (Secret Flight), The Man Within (The Smugglers), Dancing with Crime, Brighton Rock (Young Scarface), London Belongs to Me (Dulcimer Street), The Guinea Pig, The Lost People, Boys in Brown, Morning Departure (Operation Disaster), Hell is Sold Out, The Magic Box, Gift Horse (Glory at Sea), Father's Doing Fine, Eight O'Clock Walk, The Ship That Died of Shame, Private's Progress, The Baby and the Battleship, Brothers-in-Law, The Scamp, Dunkirk, The Man Upstairs, Sea of Sand, Danger Within, I'm All Right Jack, Jetstorm, S.O.S. Pacific; 1959 formed Beaver Films with Bryan Forbes and appeared in and co-produced The Angry Silence; formed Allied Film Makers and appeared in their first film The League of Gentlemen; produced Whistle Down The Wind, appeared in Only Two Can Play, All Night Long, The Dock Brief (Trial & Error); produced The L-Shaped Room, appeared in The Great Escape, appeared in and produced Seance on a Wet Afternoon; appeared in The Third Secret, 1963; Guns at Batasi, 1964; The Flight of the Phoenix, 1965; The Sand Pebbles 1965–66; Dr. Dolittle, 1966; The Bliss of Mrs. Blossom, Only When I Larf, 1968. Prod. and dir.: Oh! What a Lovely War, 1968. Appeared in The Last Grenade, A Severed Head, Loot, and made guest appearances in The Magic Christian and David Copperfield, 1969, 10 Rillington Place, 1970, Young Winston, 1972 appeared in Rosebud, Ten Little Indians, Brannigan. Conduct Unbecoming 1974, directed A Bridge Too Far, 1976. Appeared in The Chess Players, 1977; dir. Magic, 1978, actor, The Human Factor, 1978.

ATTERBURY, MALCOLM: Actor. b. Philadelphia, Pa., Feb. 20, 1907, e. Hill School, Pa. Radio editor, Phila. Ledger; prog. mgr., gen. mgr., performer, Phila. Ledger radio station, WHAT, 7 yrs.; concert, vaudeville, Schubert musicals, 7 yrs.; Tamarack Playhouse, 5 yrs.; owned, oper., Albany Playhouse, 6 yrs.; conducted thea. seminars, Skidmore, R.P.I., College of St. Rose.
TV: Hallmark Hall of Fame, Lux Video Thea., Playhouse 90, Studio One, G.E. Theatre, Cheyenne, Perry Mason, Gunsmoke, 77 Sunset Strip, Alfred Hitchcock Presents, Route 66, The Defenders, The Untouchables, Dr. Kildare, Profiles In Courage, Fugitive, FBI, Hazel, Bonanza, Voyage to the Bottom of the Sea, Judd.
PICTURES INCLUDE: Dragnet, Storm Center, Crime in the Streets, Reprisal, Crime of Passion, Toward the Unknown, No Time for Sergeants, Rio Bravo, North by Northwest, Wild River, From the Terrace, Summer and Smoke, Advise and Consent, The Birds, The Chase, Hawaii.
BROADWAY: One Flew Over the Cuckoo's Nest.

ATTWOOLL, HUGH: Assoc. Producer. b. 1914. Ent. films 1928.
PICTURES INCLUDE: The Castaways, The Horse Without A Head, Three Lives or Thomasina, Dr. Syn, The Moonspinners, The Legend of Young Dick Turpin, The Fighting Prince of Donegal, Only When I Larf, Guns in the Heather, David Copperfield, Kidnapped, Diamonds on Wheels, Persecution, One of our Dinosaurs Is Missing, Escape From The Dark, Harness Fever, Candleshoe, The Spaceman and King Arthur, Watcher in the Woods, The London Connection.

ATWATER, GLADYS: Writer. Has written many m.p. scripts; member of writing team with Robert Bren since 1939; v.p. Bremex, Inc., writings include: (collab. orig.) Man Who Found Himself: (collab. orig. story) Argentine Nights, First Yank into Tokyo, in Old California; (collab. s.p.) Criminal Lawyer, Crashing Hollywood, This Marriage Business, Crime Ring; (collab. story) El Paso, Tag for Murder, Blood on the Shrine, Legacy in Nylons, Aztec Dagger, Female Menagerie. Collab. story, s.p., Great Sioux Uprising, Siege at Red River; collab. s.p. Overland Pacific; collab. stor. Naked Alibi, Treasure of Pancho Villa; TV: Lone Wolf in Mexico, Stolen Train, Casey Jones series, Winds of the Four Seasons, collab. s.p.; collab. s.p. orig. story The Hawaiians.

AUBREY, JAMES T., JR.: Executive. b. La Salle, Ill., Dec. 14, 1918. e. Princeton U., 1941, m. Phyllis Thaxter, U.S. Air Force, 1941–45; salesman, Street and Smith, Conde Nast Pub., 1946–47; account exec. KNX, Los Angeles, 1948; account exec., KNXT, 1951; sales mgr., then gen. mgr., KNXT and CTPN, 1952. Man. network prog., Hollywood CBS-TV, 1956; V.P. programs and talent, ABC-TV 1957; v.-p. CBS. 1958; exec. v.p. CBS-TV, 1959; pres. CBS-TV, 1959.
PICTURES INCLUDE: Futureworld (prod.).

AUDRAN, STEPHANE: Actress. b. France, 1938. Ex-wife of French star Jean-Louis Trintignant; now married to director Claude Chabrol.
PICTURES INCLUDE: Les Cousins (debut under direction of Chabrol), The Champagne Murders, Les Biches, La Femme Infidele, The Beast Must Die, The Lady in the Car, Le Boucher, Without Apparent Motive, Dead Pigeon on Beethoven Street, The Discreet Charm of the Bourgeoisie, The Devil's Advocate, The Blackbird (American film debut), Silver Bears, Blood Relatives.

AUERBACH, NORBERT T.: President & Chief Executive Officer, United Artists Corp. b. Vienna, 1923. Educated in U.S. and served with U.S. Army Intelligence in Europe during World War II. Joined m.p. business in 1946 after graduation from U.C.L.A. (business admin.). First asst. dir. at Service Studios in Calif. Moved to N.Y. to join domestic sales dept. of Film Classics. Left for position with Columbia Pictures in foreign dept. In 1950 assigned to Paris office, where remained for over decade, except for 18 mos. in Portugal as mgr. Returned to Paris in 1953 and filled number of exec. sis. positions for Columbia, ultimately rising to continental mgr. 1961, left Columbia to produce films in France. Resumed career in dist., this time at Paris office of United Artists, becoming continental mgr. In 1966 returned to production to make The Thief of Paris. 1967, joined Seven Arts Prods. heading theatrical and TV sis. operations in Paris. When Seven Arts acquired Warner Bros., he became continental sls. mgr. for Warners in Paris. 1968, set up European prod. and dist. org. for CBS Cinema Center Films, operating from London. 1972, moved to L.A. as v.p., foreign mgr. for CCF. Returned to London in 1973 to be consultant in prod. and dist. Rejoined UA in 1977 as sls. mgr. for Europe and the Middle East. Named snr. v.p. & foreign mgr. in 1978. Named pres. & COO, Jan. 1981; pres., CEO, Feb. 1981.

AUMONT, JEAN PIERRE: Actor. b. Paris, 1913. e. Conservatoire of Drama. Roles French stage and films. U.S. pictures: Assignment in Brittany, Cross of Lorraine. In 1943 enlisted Free French Army.
PICTURES INCLUDE: Heartbeat, Siren of Atlantis, Affairs of a Rogue, Wicked City, Lili, Life Begins Tomorrow, Gay Adventure, Charge of the Lancers, Hilda Crane, Enemy

General, The Devil at 4 O'Clock, Castle Keep, Cauldron of Blood, Day for Night, Mahogany, Catherine & Co.
STAGE: Tovarich, Incident at Vichy, Carnival, Camino Real.

AURELIUS, GEORGE M.: Executive. b. Grasston, Minn., Sept. 16, 1911; e. U. of Minnesota. Ent. m.p. ind. 1927 as usher Finkelstein & Ruben, St. Paul; asst. mgr. 1929–30; to Warner Theatres, New York 1931; man. Moss' Broadway; Minnesota Amusement Co. 1932–41; city mgr. Publix-Rickards-Nace. Paramount-Nace Theatres, Tucson, Ariz. 1941–46; city mgr. Paramount-Nace, Phoenix, Ariz. 1946–49; V.P. ABC Theas. of Arizona, Inc. 1949–67; pres. ABC North Central Theatres, Inc., 1967–1972; V.P. ABC Intermountain Theatres, Inc., V.P. ABC Theatres of California, Inc. 1972–1974; Mgmt. Consulting and ShoWest Convention & Trade Show since 1975, named exec. dir., 1979.

AURTHUR, ROBERT ALAN: Writer-producer. b. New York, N.Y., June 10, 1922. e. Univ. of Pennsylvania, BA, 1942. U.S. Marine Corps., 1942–46; author: History of the Third Marine Division, Inf. Journal Press, 1946; short story writer, The New Yorker, Harpers, Esquire, Sat. Eve. Post, etc., 1946–50; novel: The Glorification of Al Toolum; Rinehart, 1951; partner, Circle Records, 1947–50; freelance TV dramatist, 1951–58, vice-pres., Talent Associates-Paramount Ltd., 1961; pres., Edgewater Prods. Monthly column "Hanging Out" Esquire Magazine 1972, NYU School of Arts, Instructor screenplay writing, 1975; Delegate to Moscow Film Festival, International Writers Guild Symposium, Leningrad, 1975.
TV INCLUDES: script ed., Philco Playhouse, Mr. Peepers, assoc. prod., Philco Playhouse, then prod.; prod., Sunday Showcase, NBC, Years of Decision, pilots: Inside Danny Baker, The Laughmakers, ABC. Head of new prog. dev. United Artists-TV, 1963–4
PICTURES INCLUDE: screenwriter. Warlock, The Mountain Is Young; writer, co-prod., Edge of the City; s.p. Lilith, s.p. Grand Prix, For the Love of Ivy, s.p. and dir. The Lost Man, Indians (s.p.), Ending (s.p.).
BROADWAY: Author, A Very Special Baby, Kwamina, Carry Me Back to Morningside Heights.

AUSTIN, BUD: Executive. r.n. Harold M. b. New York, N.Y. e. Univ. of North Carolina, New York U., Pres. Austin Productions 1971; U.S. Army 1942–46; pres., Donna Hale Corp. 1948; pres., Lindley Corp. 1949; nat'l sis. mgr. Official Films, 1950–53; vice-pres., Telefilm Ent., 1954; vice-pres, NTA, 1955; pres. Austin Television Prods. Inc., 1956; exec. vice-pres., Goodson-Todman Prods., 1956; exec. vice-pres., Filmways, Inc., 1965; 1973, Vice President Creative Services and Marketing Paramount Pictures Corp. Executive Producer, Elaine May film "Mikey and Nicky," Executive Producer and Creator "Take Five" for CBS-TV; 1973–1976, exec. v.p., Paramount TV; pres., Bud Austin Productions. 1979, prod., Perfect Gentlemen; exec. prod., Whew (TV game show).

AUSTIN, EMERY M.: Executive. b. Dallas, Texas, Oct. 5, 1911. e. Central H.S., Tulsa, 1929. Usher with Ralph Talbot theatres, Tulsa, 1928; then doorman, chief usher, house mgr., mgr. & asst. gen. mgr.; MGM, New Orleans, 1941; Atlanta, 1942; New York 1953; asst. dir. of advtg.-publicity-exploitation. RXET. Feb., 1969. Now pres., Bali-Hai Marine Services, Inc., Miami, Fla.

AUSTIN, JOHN: Producer. b. Australia, 1923; Asst. Dir. of Entertainment for American Red Cross, 1942–45; assoc. West Coast rep. of Rodgers & Hammerstein, 1945–47; with Ministry of Info., Gov't of India, 1947–48; prod. TV films in Europe, 1949; shows include Intrigue, International Rendezvous, International Affair; prod. film, Lasca.

AUSTIN, RAY: Director-Writer-Producer. b. London, England, Dec. 5, 1932. Has written, produced and directed many TV series, specials and movies.
TELEVISION: Director of series: Avengers, The Champions, Department S, Randall & Hopkirk, Ugliest Girl in Town, Journey into the Unknown, etc. Writer: Randall & Hopkirk, Department S. Producer-Director: The Perfumed Garden. Director: It's the Only Way To Go, Virgin Witches, Fun and Games, Space 1999, New Avengers, Hawaii Five-0, Sword of Justice, Webb, Barnaby Jones, Hardy Boys, Wonder Woman, Salvage, B.G. and the Bears, Hart to Hart, The Yeagers, Man Called Sloane, From Here to Eternity, Bad Cast, Westworld. Director-Writer: Black Beauty.
PICTURES: House of the Living Dead.

AUSTRIAN, RALPH BROOKE: Executive. b. New York City, April 29, 1898. Asst. sound dir. Paramount-Publix Lasky Corp. 1928–29; in chg. retail dist. Westinghouse Elec. & Mfg. Co. 1930–32; gen. sales mgr. Koster Radio Corp. 1932; same Emerson Radio & Phono. Corp. 1933–34; asst. v.p. RCA Mfg. Co. 1935–42; WPB Planning Com. 1942–43; exec. v.p. RKO Television Corp. 1943–44, pres. 1945–47; v.p. Foote, Cone and Belding in chg. TV 1948; org. TV consultant service to adv. agencies, theatres, TV stations & m.p. cos. 1949; assoc. N. P. Rathvon & Co., exec. v.p. Multi-Deck Corp. of America; senior associate, Pereira & Luckman, architect-engineers; v.p. Paco, Inc.; west coast mgr., Allen B. DuMont Labs., Inc., SLS. Man. Ling Electronics, Inc.; Financial Consultant. Dir. Marketing of TV sta. KPLM. Member: SMPTE (fellow), m.p. Pioneers, IRE (senior mem.).

AUTANT-LARA, CLAUDE: Director. Began career as scenic designer for French films in early 1920's; then asst. director; first solo venture experimental film, 1923; in Hollywood, 1930–32. dir. Parlor Bedroom and Bath, Incomplete Athlete.
PICTURES INCLUDE: Devil in the Flesh, Seven Deadly Sins (seq.), Red Inn, Oh Ame ia, Game of Love, Red and the Black.

AUTEN, H. WILLIAM; Executive. b. Cardiff, Wales, Mar. 7, 1921; e. Neville House, Eastbourne; U. Coll. School of London. With sales dept., Warner Bros. (London) 1939; served as Major on Brit. Army Staff 1939–47, including custodianship of UFA, GMBH 1945–46; European & Near Eastern mgr. United Artists 1948; asst. foreign mgr. Eagle Lion Films, Inc. (N.Y.) 1948–49; founder & pres. Benagoss Productions, Inc. June, 1949, to produce and finance pictures; org. Ballentine Pict., 1951.

AUTRY, GENE: Actor. b. Tioga, Tex., Sept. 29, 1907. Railroad telegrapher at Sapulpa, Okla., 1925; became radio singer and recording artist (Columbia Records) 1928; screen debut 1934 at Mascot Pictures (later became Republic) as screen's first singing cowboy. Voted top money making Western star 1937–42, and in top Western stars 1936, 1946–54; first Western star to be in top ten money makers from 1938 to 1942. Served in U.S.A.A.F. as flight officer, 1942–45; on USO tour overseas 3 mos.; immediately thereafter resumed radio career with formed sponsor, the Wm. Wrigley Co., formed Gene Autry Productions, Inc.; star of Madison Square Garden Rodeo first in 1940; composed & recorded song "Here Comes Santa Claus"; owner radio & TV stations, and California Angels ball team.

AVAKIAN, ARAM: Director. b. N.Y.C. Was still photographer and editor before becoming director with The End of the Road.
PICTURES INCLUDE: Cops and Robbers, 11 Harrowhouse.

AVALON, FRANKIE: Entertainer. r.n. Francis Thomas Avalone. b. Phila., Sept. 18, 1940; e. South Phila. High. Trumpet Prodigy age 9 yrs.; Recording contract, Chancellor Records, Inc., 1957; Gold Record, Venus, 1959; Gold Album, Swingin' on a Rainbow, 1959; Night Clubs.
TV: Ed Sullivan, Perry Como, Pat Boone, Arthur Murray. Dick Clark Shows, Milton Berle, Golden Circle Spectacular; Dinah Shore Show, Steve Allen Show.
PICTURES INCLUDE: Guns of the Timberland, The Alamo, Voyage to the Bottom of the Sea, Sail a Crooked Ship, Panic in the Year Zero, Bikini Beach, Beach Blanket Bingo, Jet Set, I'll Take Sweden, Sgt. Deadhead, The Take, Grease.

AVEDON, DOE: Actress. b. Old Westbury, N.Y. Bookkeeper; then actress.
BROADWAY: Young and the Fair, My Name is Aquilon.
PICTURES INCLUDE: High and the Mighty, Deep In My Heart, The Boss.
TV: Big Town.

AVILDSEN, JOHN G.: Director-Cinematographer-Editor. b. Chicago. e. N.Y. Univ. After service in Army made film with friend, Greenwich Village Story. Then joined ad agency to write and produce film commercials. Entered m.p. industry as ass't. cameraman on Below the Hill, followed with prod. mgr. job on Italian film made in U.S. Then made first theatrical short, Smiles. Arthur Penn hired him as prod. mgr. on Mickey One. Produced, photographed and edited a short, Light, Sound, Diffuse. Returned to industry to make films for ad agencies before resuming theatrical career.
PICTURES INCLUDE: Turn on to Love (1st feature, dir., photo.), Out of It (dir., photo.), Sweet Dreams (dir., photo, editor), Guess What We Learned in School Today? (dir., photo editor), Joe (dir., photo.), Cry Uncle (dir., photo., editor), Save The Tiger (dir.), W.W. and the Dixie Dancekings (dir., editor), The President's Women, Rocky (dir.); Slow Dancing in the Big City (dir., editor, operator), The Formula (dir.), Neighbors (dir.).

AXELROD, GEORGE: Playwright writer, prod., dir. b. New York, N.Y., June 9, 1922, stage mgr., actor, summer stock, 1940–41; radio writer, 1941–52; writer, novels: Beggar's Choice, Blackmailer; co-writer, night club musical: All About Love, 1951.
BROADWAY: The Seven Year Itch, Will Success Spoil Rock Hunter?, Visit to a Small Planet, Once More With Feeling, Goodbye Charlie.
PICTURES INCLUDE: Phffft, The Seven Year Itch, Bus Stop, Breakfast at Tiffany's, The Manchurian Candidate, Paris When It Sizzles, How to Murder Your Wife, Lord Love a Duck, The Secret Life of an American Wife (prod., s.p.).

AXELROD, JONATHAN: Writer. b. N.Y.C. July 9, 1948. Stepson of writer George Axelrod. Started as on-set "gofer" before writing screen plays.
PICTURES INCLUDE: The Dirty Movie, Every Little Crook and Nanny.

AYRES, GERALD: Producer. e. Yale Univ., where had four plays produced. Became Broadway play doctor and then joined Columbia Pictures as freelance reader. Named story editor; exec. ass't. to v.p. Mike Frankovich; then v.p. in chg. creative affairs in Hollywood. Left in 1970 to become independent. Formed Acrobat Films.
PICTURES INCLUDE: Cisco Pike, The Last Detail.

AYRES, LEW: Actor. b. Minneapolis, Minn., Dec. 28, 1908. Toured Mexico with own orchestra; played with Henry Halstead's Orchestra; screen debut in The Sophomore, "The Kiss" 1929; served as medical corpsman & asst. chaplain World War II.
PICTURES INCLUDE: All Quiet on the Western Front, Common Clay, Doorway to Hell, Okay America, State Fair, Dr. Kildare series; (dir.) Hearts of Bondage, Dark Mirror, Unfaithful, Johnny Belinda, The Capture, New Mexico, No Escape, Donovan's Brain, The Carpetbaggers, Advise and Consent, Altars to the East (dir-nar. star-prod.), Last Generation, 1971, Biscuit Eater, The Man, Battle for the Planet of the Apes, Damien-Omen II, Battlestar Galactica.

AZZATO, ANTHONY: Executive; b. New York City, Oct. 7, 1912. e. Fordham Univ., City Col. of New York; production supervisor, film editor, director, documentary and comm. films, 1938–48; Sound Masters, Inc.; U.S. Army Signal Corps, 1942–46; film editor, Paramount, 1932–42; appt'd director of film programs, WPIX, 1948; also film program consultant, KLAC, Los Angeles; dir. of station relations, National Telefilm Assoc. 1956, v.p., Swan & Mason Adv., 1961; v.p., Sales, Teledynamics Corp.; v.p. Teledynamics Corp.

B

BABCOCK, DWIGHT V.: Writer. b. Ida Grove, Ia., Feb. 19, 1909. e. Modesto Jr. Coll. Author short stories, novelist. First Screenplay at Universal, 1943. 26 screen credits since at various studios. TV, free lance, includes 100 produced teleplays; co-author, Chautauqua.

BACALL, LAUREN: Actress. b. New York City, Sept. 16, 1924; e. American Acad. Dram. Arts. m. late Humphrey Bogart, actor. Fashion model; on stage, plays include: Johnny Two-by-Four, Franklin Street; m.p. debut in "To Have and Have Not" 1944.
BROADWAY: Cactus Flower, Goodbye Charlie, Applause (Tony Award, Best Actress in musical), Woman of the Year.
PICTURES INCLUDE: Big Sleep, Confidential Agent, Dark Passage, Key Largo, Young Man with a Horn, Bright Leaf, How to Marry a Millionaire, Woman's World, Cobweb, Blood Alley, Written on the Wind, Designing Woman, Flame Over India, Shock Treatment, Gift of Love, Sex and the Single Girl, Harper, Murder on the Orient Express, The Shootist, The Fan.
TV: Applause

BACH, STEVEN K.: Production Executive. b. Pocatello, Idaho, April 29, 1940. e. Northwestern Univ. (B.A., 1961; M.A., 1962); Univ. of So. Calif. (cinema, 1966); Universite de Paris (Sorbonne), 1960. Was public school teacher, Winnetka, Ill, 1961–65. Story editor for Mark Taper Forum, 1968; story editor, Katzka-Berne Productions, 1968–70. Joined Palomar Pictures as head of creative affairs in 1970; left to become v.p., Pantheon Pictures 1974–78. Sr. v.p., worldwide production, for United Artists Corp to May, 1981. Formed Lester (Richard)/Bach Prods., 1981.
PICTURES: Worked on Kelly's Heroes, Sleuth, The Heartbreak Kid, The Parallax View, The Taking of Pelham One-Two-Three. Producer of Mr. Billion and Butch and Sundance: The Early Days, both for 20th-Fox. Co-producer of Bdwy. play, Same Time Next Year.

BACK, LEON B.: Exhibiter. b. Philadelphia, Pa., Oct. 23, 1912; e. Johns Hopkins U., B.E., 1932; U. of Baltimore, LL.B., 1933. Entered m.p. ind. as mgr. for Rome Theatres, Inc., Baltimore, Md., 1934; booker, asst. buyer, 1936; asst. to gen. mgr., 1939; U.S. Navy, 1944–46; v.p., gen. mgr., Rome Theatres, 1946; Allied MPTO of Md. 1952–55; nat'l dir. Allied States, 1952–55; nat'l secy. 1954, Pres. NATO of Maryland 1969–80, Pres. USO Council, Greater Baltimore 1969–75; Chairman board of trustees, Employees Benefit Trust for Health & Welfare Council of Central Maryland, 1970–79.

BACKUS, JIM: Actor. b. Cleveland, Ohio, Feb. 25, 1913; e. Amer. Acad. of Dram. Arts. Began in stock & vaudeville; radio announcer; radio actor on Ilum On stage in Hitch Your Wagon; m.p. debut in Easy Living, Abner, Alan Young show & own shows
PICTURES INCLUDE: Half Angel, His Kind of Woman, Bright Victory, Hollywood Story, Man With a Cloak, Iron Man, Here Come the Nelsons, I'll See You in My Dreams, Deadline, U.S.A., Androcles and the Lion, Don't Bother to Knock, Pat and Mike, Above and Beyond, I Love Melvin, Angel Face, Geraldine, Human Jungle, Francis in the Navy, Rebel Without a Cause, Square Jungle, Meet Me in Las Vegas, The Wonderful World of the Brothers Grimm, Critics Choice, Six Loves, Boys' Night Out, The Horizontal Lieutenant, Zotz!, Hello Down There, Pete's Dragon.
TV: I Married Joan, Gilligan's Island.
RECORDINGS: Delicious, Cave Man.
BOOKS: Rocks On the Roof, Back to Backus, What Are You Doing After the Orgy.

BADDELEY, HERMIONE: Actress. b. Shropshire, Eng., Nov. 13, 1908; e. privately and at Margaret Morris School of Dancing. Stage debut in La Boite A Joujou at Court theatre, 1918; screen debut in Brighton Rock, 1947; many TV, radio, nightclub appearances.
BROADWAY: The Milk Train Doesn't Stop Here Anymore, Canterbury Tales.
PICTURES INCLUDE: Passport to Pimlico, Quartet, Christmas Carol, Pickwick Papers, Mr. Prohack, Belles of St. Trinians, Room at the Top, Jetstream, Expresso Bongo, Let's Get Married, Midnight Lace, Rag Doll, Mary Poppins, The Unsinkable Molly Brown.
TV: Richard of Bordeaux, Drink Doggie Drink, The Gambler, Airmail from Cyprus, The Castaway, Maude.

BADHAM, JOHN: Director. b. England. Raised in Alabama; schooled at Yale, where acquired 2 degrees. Landed first job in Universal Studio mailroom; later was Universal Tour guide and assoc. prod. to Steven Spielberg. Made film trailers. Started directing movies for TV which led to theatrical debut with The Bingo Long & Travelling All-Stars and Motor Kings.
PICTURES: Saturday Night Fever, Dracula, Whose Life Is It Anyway?
TELEVISION: The Law, Isn't It Shocking?, etc.

BAER, JOHN G.: Executive—Engineer. b. New York City, May 8, 1934. e. Univ. of Tenn. at Chattanooga (B.S., physics, 1955). Joined Bausch and Lomb, Inc., as optical and mechanical engineer, 1957–67; director of research and development, 20th Century-Fox, New York, 1967–72; sales/engineering, Century Projector Corp., named pres. & chief exec. officer, Jan. 1, 1975.
MEMBER: Optical Society of America; Society of American Magicians; Intl. Brotherhood of Magicians; Magic Circle (London); 25–30 Club (honorary). Fellow of SMPTE and dir. Theatre Equip. Assn. Chm., PH-22 (motion picture) American Natl. Standards Institute U.S. delegate to Intl. Standards Organization. Elected mayor West Windsor Township, N.J., 1977.

BAILEY, JOSEPH W.: Attorney, producer. b. Columbia City, Ind., Dec. 13, 1910; e. U. of Chicago, PHB, 1928–31, U. of Chicago Law Sch., J.D., 1931–34. Practicing atty. at law, Chicago, Oct., 1934–June, 1940; with Louis G. Cowan to Apr. 1942; U.S. Navy, 1942–46; v.-p. & treas., Louis G. Cowan, 1946–49; mgr., radio-TV dept., Grey Adv. Agency, 1949; exec. prod., Robert Montgomery Show & mgr., John Gibbs literary & talent agency, 1950–1957; Prod. v.p. Ziv-United Artists, Inc. 1957–61. Resigned June 1961 to become partner in firm, Bronstein, Van Veen & Bailey; in practice of law specializing in allied field of TV, motion pictures and the theatre.

BAILEY, PEARL: Entertainer. b. Newport News, Va., March 29, 1918; e. Philadelphia and Washington, D.C. Career started when 15 years old, winner of amateur stage contest; toured mining towns of Pennsylvania as dancer; dancer and singer in vaudeville with Noble Sissle's band; stage acting debut in St. Louis Woman (won Donaldson Award, best newcomer of 1946); many night club appearances and TV shows; screen debut 1947, Variety Girl.
PLAYS INCLUDE: Arms and the Girl, Bless You All, House of Flowers, Hello Dolly.
PICTURES INCLUDE: Carmen Jones, Isn't It Romantic, That Certain Feeling, St. Louis Blues, Porgy and Bess, All the Fine Young Cannibals, The Landlord, Last Generation, Norman, Is That You?

BAILEY, ROBIN: Actor. b. Hucknail (Nottingham) Eng., Oct. 5, 1919. e. Henry Mellish School, Nottingham.
STAGE: Barrets of Wimpole Street, Theatre Royal, Nottingham, 1938; Screen debut, School for Secrets, 1946.
PICTURES INCLUDE: Private Angelo, Portrait of Clare, His Excellency, Gift Horse, Folly to be Wise, Single Handed, Sailor of the King, The Young Lovers, For Better or Worse, Catch Us If You Can, The Whisperers, Spy With A Cold Nose, You Only Live Twice, The Eliminator, Blind Terror, Down by the Riverside, Nightmare Rally, The Four Feathers.
TV: Olive Latimer's Husband, Seven Deadly Sins, The Power Game, Public Eye, Person To Person, Troubleshooters, Armchair Theatre, Split Level, The Newcomers,

Discharge of Trooper Lusby. Brett, Owen M.D., Solidarity, General Hospital, Murder Must Advertise, Vienna 1900, Justice, The Pallisers, The Couch, Way of the World, Upstairs, Downstairs, Walk with Destiny, North and South, A Legacy, I Didn't Know You Cared, Punch Review, Play for Today, The Velvet Glove, Crown Court, Took and Co., The Good Companions

BAKER, CARROLL: Actress. b. Johnstown, Pa., May 28; e. schools there and St. Petersburg (Fla.) Junior Coll. Career started as dancer in night clubs. Actors' Studio N.Y. Stage debut: Escapade, Then, All Summer Long; screen debut; Easy to Love.

PICTURES INCLUDE: Giant. Starred in Baby Doll, But Not for Me, The Miracle, Bridge to the Sun, Something Wild, How the West Was Won, The Carpetbaggers, Station Six Sahara, Sylvia, Cheyenne Autumn, Mister Moses, Harlow, The Sweet Body of Deborah, Paranoia, Andy Warhol's Bad, Watcher in the Woods.

BAKER, DON: Theatre Executive. b. St. Louis, Mo., Dec. 16, 1931. e. St. Louis Univ. V.P., adv.-prom., Loews Theatres, New York. Member NATO nat'l. chm. adv. comm., 1972–present; mem. bd. of dir., Will Rogers Hospital, 1974.

BAKER, GEORGE: Actor. b. Varna, Bulgaria, April 1, 1931, e. Lancing College, Sussex. Stage debut Deal Repertory Theatre, 1946. Film debut The Intruder, 1953.

PICTURES INCLUDE: Dam Busters, Ship That Died Of Shame, Woman For Joe, Extra Day, Feminine Touch, A Hill In Korea, No Time for Tears, These Dangerous Years, Tread Softly Stranger, Lancelot and Guinivere, Curse of the Fly, Mister Ten Per Cent, Goodbye Mr. Chips, Justine, The Executioner, On Her Majesty's Secret Service, A Warm December, The Fire Fighters, The Spy Who Loved Me, Thirtynine Steps, A Nightingale Sang in Berkeley Square.

TV FILM: Fan Show, Ron Raudell's programme 1956, Guinea Pig, Death of a Salesman, The Last Troubadour, The Square Ring, Nick of the River, Mary Stuart, Probation Officers, Far Away Music, It Happened Like This, Boule de Suif, Maigret, Zero One, Rupert Henzau, Miss Memory, Any Other Business, The Navigators, Common Ground, Alice, The Queen and Jackson, The Big Man Coughed and Died, Up and Down, Call My Bluff, The Baron, St. Patrick, Love Life, Seven Deadly Virtues, The Prisoner, The Sex Games, Z Cars, Paul Temple, Candida, Fenn Street, Man Outside, The Persuaders, Main Chance, Ministry of Fear, Voyage in the Dark, Dial M for Murder, Zodiac, The Survivors, I, Claudius, Print Out, Goodbye, Darling.

BAKER, J. EDWIN: Executive. b. Detroit, Mich., June 27, 1914. e. St. John's Coll. m. Marjorie R. Baker, actress. In 1935 produced Asylum of Horrors, stage show featuring the Frankenstein Monster (in person), which is still playing dates after 40 years. Known professionally as Dr. Silkini. In 1960 organized Intl. Artists Pictures, of which he is pres., to produce films.

PICTURES INCLUDE: Phycoscope, Magic Land of Mother Goose, Santa Visits Mother Goose Land, Vampire's Coffin, Teenage Tramp, Aztec Mummy.

BAKER, JOE DON: Actor. b. Texas. Began career on N.Y. stage, impressing critics in Marathon 33 and Blues for Mr. Charlie. Film debut, Cool Hand Luke, 1967.

PICTURES INCLUDE: Guns of the Magnificent Seven, Adam at Six A.M., The Wild Rovers, Welcome Home, Soldier Boys, Junior Bonner, Walking Tall, Charley Varrick, The Outfit, Golden Needles, Mitchell, Framed.

BAKER, LESLIE F.: Executive. b. London 1903. dir. Michael Balcon Productions Ltd., Associated Talking Pictures Ltd., E.M.I. British Film Distributors Ltd., Santor Film Productions Ltd. Ent. film ind. with Gainsborough, 1924. Part. in Baker, Todman and Co., Chartered Accountants, 1930–1933; Ass. Gen. Mgr. of Prod. Gainsborough and Gaumont-British, 1933–1935; dir. sec. and gen. prod. mgr. New World Pictures, 1935–1937; dir. sec. gen. prod. mgr. and assoc. prod. Twentieth Century Productions Ltd., 1937–1943; in 1943 joined Ealing Studios Ltd. and assoc. co. as sec., later, as dir. 1961–77, chmn., man. dir., London Film Productions Ltd. Man. dir., Kew Ferry Management (Films) Ltd.

BAKER, ROBERT S.: Producer b. London, 1916; e. Tenterden Hall; Ent. m.p. ind., 1938 as assistant dir.; war service, later Army Film Unit as combat cameraman; dir., Tribune Productions, Bamore Film Productions.

PICTURES INCLUDE: Crossplot, Sea of Sand, Hellfire Club, Treasure of Monte Cristo, Jack The Ripper.

TV: The Persuaders series, The Saint series, Gideon C.I.D. "Return of the Saint."

BAKER, ROY: Prod.-Dir. b. London; e. Lycee Corneille, Rouen; City of London School. Asst. dir. with Gainsborough 1934–40; served in Army 1940–46.

PICTURES INCLUDE: Operation Disaster, Don't Bother to Knock, Inferno, One That Got Away, A Night to Remember, The Singer Not the Song, Flame in the Streets, Quartermass and the Pit, The Anniversary, Vampire Lovers, Dr. Jekyll and Mr. Hyde, Asylum (Paris Grand Prize), Seven Golden Vampires.

TV: The Human Jungle, The Saint, Gideon's Way, The Baron, The Avengers, The Champions, Department S., The Persuaders, Danger UXB, Minder.

BALABAN, A. J.: Theatre executive. b. Chicago, Apr. 18, 1889. In 1907 entered m.p. exhibition as operator of M.s. Nickelodeon Kedzie Theatre, Chicago; owned Movie Inn, restaurant with booths named for stars 1915; co-founder Balaban & Katz Corp. 1917; upon affiliation of B & K with Paramount named dir. of entertainment, then vice-pres.; abroad 9 yrs., returned to U.S. as circuit exhibitor 1935; gen. mgr. Roxy Theatre, N.Y., 1942; res. Jan 1, 1952, now consultant for 3 years; holds interest in Esquire, Carnegie theat., Chicago, with Harry and Elmer, brothers. Member: M.P. Pioneers, Jewish Theatrical Guild, French-American Club.

BALABAN, ELMER: President H. & E. Balaban Corp., Chicago, Ill. e. U of Pa., 1931. With Paramount Publix 1931; Balaban & Katz, 1933; then H. & E. Balaban Corp. Pres., Plains Television; interest in WTVO, Rockford, Ill.; interest in WHNB, Hartford, Conn. and numerous cable-TV systems.

BALABAN, HARRY: Executive. b. Chicago, Ill., July 23, 1903; e. U. of Illinois, 1925. With First Nat'l Pictures 1925–26; joined Balaban & Katz 1926–30; co-founder & Chairman of Board H. and E. Balaban Corp., 1932.

BALIN, INA: b. Bklyn, N.Y. e. Forest Hills High; Part time drama, psychology courses. N.Y. Univ. Comm. model, salesgirl, receptionist. Prof. debut, Perry Como Show. Off-Bway, summer stock, Compulsion, Majority of One. N.Y. Stage. Film debut, The Black Orchid. TV and Broadway. Toured Europe 1961. Int'l Star of Tomorrow. Hollywood Foreign Press Golden Globe Award.

PICTURES INCLUDE: From the Terrace, Young Doctors, The Comancheros, The Greatest Story Ever Told, The Patsy, Charro!, The Projectionist.

TV: American Heritage, Stoney Burke, Adventures in Paradise, Kraft Theatre, The Lieutenant, Bonanza, The Loner, Voyage to the Bottom of the Sea, Run for Your Life.

BALL, LUCILLE: Actress. b. Jamestown, N.Y., Aug. 6. e. Chautaqua Inst. Music; John Murray Anderson Dram. School, p. Desiree Ball, concert pianist. Hattie Carnegie model, stage debut in "Rio Rita"; on Broadway in "Hey Diddle, Diddle," "Dream Girl," "Wildcat;" m.p. debut in "Roman Scandals" 1933; entertained in Hollywood Canteen World War II; Voted most promising star in Motion Picture Daily's TV poll, 1951; best performer, 1952; best comedy team, 1954; best comedienne 1952, 54–55, 57; pres. Desilu Prod., Inc.; exec. prod., The Lucy Show; pres. Lucille Ball Prods., 1967.

PICTURES INCLUDE: Roberts, Stage Door, Big Street, Du Barry Was a Lady, Easy to Wed, Ziegfeld Follies, Lured, Sorrowful Jones, Easy Living, Miss Grant Takes Richmond, Fuller Brush Girl, Fancy Pants, Magic Carpet, Long Long Trailer, Forever Darling, Facts of Life, Critics Choice, Yours, Mine and Ours, Mame.

TV: I Love Lucy, The Lucy Show, Here's Lucy.

BALLARD, LUCIEN: Cinematographer. b. Miami, Okla., May 6, 1908; e. U. Okla., U. Pa. Started with Paramount as asst. cinematographer & cutter; since operative cinematographer.

PICTURES INCLUDE: Will Penny, The Party, How Sweet It Is!, The Wild Bunch, True Grit, The Ballad of Cable Hogue.

BALSAM, MARTIN: Actor. b. Nov. 4, 1919, N.Y.

PICTURES INCLUDE: On the Waterfront, 1954. Twelve Angry Men, Time Limit, Marjorie Morningstar, Al Capone, Middle of the Night, Psycho, Ada, Breakfast at Tiffany's, Cape Fear, Who's Been Sleeping in My Bed?, The Carpetbaggers, Youngblood Hawke, Seven Days in May, Harlow, The Bedford Incident, A Thousand Clowns, After the Fox, Hombre, 2001: A Space Odyssey, Trilogy, Catch 22, The Anderson Tapes, The Stone Killer, Summer Wishes, Winter Dreams, The Taking of Pelham 1,2,3, Murder on the Orient Express, Mitchell, All The President's Men, Two-Minute Warning, The Sentinel, Silver Bears, Cuba.

BALTER, ALLAN: Executive. b. Detroit, Mich., August 28, 1925. e. Univ. of Michigan, B.A. Asst. casting dept., 20th Century-Fox Film Corp. 1950–53; mgr., Encino theatre, 1953–54; asst. casting dept., MGM Studios, 1954–55; editor, Action mag., 20th Century-Fox house-organ 1955–56; trade planter, 20th Century-Fox publicity, 1956–57; unit publicist, Bryna Prods. (UA) 1957–58; unit publicist, 20th Century-Fox, 1958–62; dir. of pub., adv., Daystar Prods.

BANCROFT, ANNE: Actress. r.n. Anne Italiano; b. N.Y.C., Sep. 17, 1931; e. American Academy of Dramatic Arts. Acting debut on TV, Studio One; many TV shows; film debut, Don't Bother to Knock.

PICTURES INCLUDE: Tonight We Sing, Treasure of the Golden Condor, Kid from Left Field, Demitrius and the Gladiators, Gorilla at Large, The Raid, Life in the Balance,

New York Confidential, The Brass Ring, Naked Street, Last Frontier, Girl in the Black Stockings, Restless Breed, The Miracle Worker, The Pumpkin Eater (1964), Slender Thread, The Graduate (1967), Young Winston (1972), The Prisoner of Second Avenue, The Hindenburg, The Turning Point, Fatso (also dir. & s.p.), The Elephant Man.
PLAYS: Two For the Seasaw, The Miracle Worker, Mother Courage, The Devils, A Cry of Players.

BAND, ALBERT: Producer-director. b. Paris, France, May 7, 1924; e. Lyceum Louis le Grand, won French-English Literature Prize 1938; entered film industry as cutter Pathe Lab.; prod. ass't to John Huston at MGM; first screen credit adaptation Red Badge of Courage novel; first direction. The Young Guns, Allied Artists; formed Maxim Productions, Inc., Sept. 1956; prod.-dir. Killer on the Wall for United Artists. 1958 prod.-dir. Face of Fire; 1962 dir. The Avenger; 1963 dir. Grand Canyon Massacre; 1965 prod.-dir. The Tramplers; 1966 prod. The Hellbenders and then formed Albert Hand Prod. in Rome.
PICTURES INCLUDE: A Minute to Pray, A Second to Die, Little Cigars, Dracula's Dog, She Came to the Valley.

BANNEN, IAN: Actor. b. Airdrie, Scotland, 1928. Early career Shakespeare Memorial Theatre, Stratford-on-Avon.
STAGE: A View From the Bridge, The Iceman Cometh, Long Days Journey Into Night, Sergeant Musgrave's Dance. Royal Shakespeare Thea. Co. 1961–62. The Iceman Cometh, 1974. Hedda Gabler, 1977.
PICTURES INCLUDE: Carlton-Browne of the F.O., Man in Cocked Hat, Macbeth, Station Six Sahara, Mister Moses, The Hill, Rotten To the Core, Flight of the Phoenix, Sailor From Gibraltar, Penelope, Lock Up Your Daughters, Too Late the Hero, The Deserter, Jane Eyre, Fright, The Offence, The Macintosh Man, The Driver's Seat, The Voyage, Bite the Bullet, Sweeney, Darkness Into Darkness, Inglorious Bastards, Watcher in the Woods, Eye of the Needle.
TV: Johnny Belinda, Tinker, Tailor, Soldier, Spy, Cousin Phillis.

BANNER, BOB: Producer-director. b. Ennis, Tex., Aug. 15, 1921; e. Southern Methodist Univ., B.A., 1939–43, Northwestern Univ., M.A., 1946–48. U.S. Naval Reserve 1943–46; faculty, Northwestern Univ., 1948–50; staff dir., NBC-TV in Chicago, 1949–50; dir., Garroway at Large, 1949–50; prod. & dir., Fred Waring Show, 1950–53; dir. Omnibus. Metropolitan Opera Prod., 1953; Nothing But the Best (prod. dir.) NBC, 1953; Omnibus, CBS, 1953–54; Dave Garroway Show, NBC, 1953–54; (prod. dir); Dinah Shore Show, 1954–57; exec. prod. Garry Moore Show; exec. prod. Candid Camera TV show; exec. prod., Carnegie Hall Salutes Jack Benny, 1961; exec. prod., Julie & Carol at Carnegie Hall, 1962; exec. prod., Carol & Co., 1963; Jimmy Dean Show, 1963–66; Calamity Jane, Once Upon A Mattress, '64; The Entertainers, '65; Carol × 2, '66; Kraft Summer Music Hall, '66, 1966 Carol & Co., Ice Follies, Carol Burnett Show, 1969, Peggy Fleming at Madison Square Garden, 1967; John Davidson at Notre Dame, Here's Peggy Fleming, 1971; Peggy Fleming at Sun Valley, The American West of John Ford, 1971; Love! Love! Love!—Hallmark Hall of Fame, 1972; To Europe With Love, 1972.
PICTURES INCLUDE: Warning Shot (for TV), My Sweet Charlie, Mongo's Back in Town, 1971.
TELEVISION: Movies of the week: The Last Survivors, Journey From Darkness. Special: Peggy Fleming Visits the Soviet Union. Game Show: Almost Anything Goes. Specials: Perry Como's Lake Tahoe Holiday '75, Perry Como's Christmas In Mexico '75, Perry Como's Hawaiian Holiday '76, Perry Como's Spring In New Orleans '76. Daily Variety Series: Don Ho Show '76; Perry Como Las Vegas Style '76, Perry Como's Christmas in Austria '76, Jr. Almost Anything Goes '76, All-Star Anything Goes '77, Peggy Fleming and Holiday on Ice at Madison Square Garden '76; Julie Andrews, One Step Into Spring '78; Leapin Lizards It's Liberace '78; Perry Como's Easter By The Sea '78.

BANTAU, J. W.: Consulting electrical engineer and purchasing agent; b. Argyle, Tex. Jan. 23, 1904. e. Rice Institute; with public utilities, U.S. Navy and Paramount Public Corp. National General Corp., now theatre consultant.

BAR, JACQUES JEAN LOUIS: Executive. Producer. b. Chateauroux, France, Sept. 12, 1921; e. Lycees Lakanal and Saint Louis, France. Formed Cinema-Theatre d'Avallon, 1944; Cite-Films S.A., 1947; Spira in assoc. with MGM, 1961; first Hollywood film, co-prod. Scratch A Thief, 1964; prod. 43 pictures in France, Spain, Italy, Switzerland, Japan and Brazil, 1948–64.
PICTURES INCLUDE: Where the Hot Wind Blows, Bridge to the Sun, Riffifi in Tokyo, A Very Private Affair, Swordsmen of Siena, Monkey in Winter, The Turfist, Any Number Can Win, The Day and the Hour, Joy House, Guns for San Sebastian.

BARAGREY, JOHN: Actor. b. April 15, 1919; e. U. of Alabama, 1939.
PICTURES INCLUDE: Loves of Carmen, Shockproof, Pardners, The Fugitive Kind.

BROADWAY: The Enchanted, Royal Family, Richard III, The Devils, The Grass Harp, Murderous Angels.
TV: Studio One, Suspense, Television Playhouse, Danger, Kraft Theatre, You Are There, U.S. Steel Hour, Robert Montgomery Playhouse, Omnibus, Climax, Hitchcock Presents, G.E. Theatre, Suspicion, Schlitz Playhouse, Playhouse 90, Cimarron City, American Heritage, The Defenders.

BARBER, ARNOLD: Executive. b. London, England. Career in m.p. ind. started with RKO. Subsequently with AB-Pathe following war service until appointment with Warner-Pathe Dist. which led him through sales exec. positions until he headed company. Formed Sotia-Barber Distributors Ltd. in 1970. Now man. dir. Barber Intl. Films, Ltd.

BARBERA, JOSEPH R.: Executive. b. N.Y.C. e. N.Y. Univ., American Institute of Banking. After schooling joined Irving Trust Co. in N.Y.C.; started submitting cartoon drawings to leading magazines' selling one to Collier's. Left banking to seek career in cartooning. Joined Van Buren Associates as sketch artist, later going to work in animation dept. of MGM Studios. At MGM met William Hanna, who became his lifelong business associate. Made first animated short together in 1938, starting the famous Tom & Jerry series which they produced for 20 years. Left MGM in 1957 to form Hanna-Barbera Productions to make cartoons for TV. Series have included Yogi Bear, Huckleberry Hound, The Flintstones. Hanna-Barbera became a subsidiary of Taft Bdg. Co. in 1968 with both men operating studio under long-term agreements with Taft. Barbera is pres. until 1980. Company entered theatrical production with Charlotte's Web in 1973.

BARBOUR, ALAN G.: Writer, editor, publisher. b. Oakland, Calif., July 25, 1933. ed. e. Rutgers University, m. Catherine Jean Barbour, actress, teacher, American Academy of Dramatic Arts. U.S. Army, worked as computer programmer. Formed Screen Facts Press in 1963, Screen Facts Magazine. Compiled, edited: The Serials of Republic, The Serials of Columbia, Great Serial Ads, The B Western, Serial Showcase, Hit the Saddle, The Wonderful World of B-Films, Days of Thrills and Adventure, Serial Quarterly, Serial Pictorial, Karloff—A Pictorial History, Errol Flynn—A Pictorial Biography, A Pictorial History of the Serial, A Thousand and One Delights, Cliffhanger, The Old-Time Radio Quiz Book.

BARBOUR, MALCOLM: Executive. b. London, England, May 3, 1934; e. Radley College, Oxford, England, Columbia Coll. At N.B.C. was press info. asst., 1958–59; asst. magazine ed., 1959–62; assoc. mag. ed., 1962–64; sr. mag. ed., 1964–65; mgr. of magazine pub., National Broadcasting Company, 1965–67; pub. mgr., Buena Vista, 1967–68; Eastern story ed., Walt Disney Prod., 1968–69; dir. of adv. & pub. relations, Buena Vista, 1969.

BARDOT, BRIGITTE: Actress. b. Paris, France, 1934. Model.
PICTURES INCLUDE: Le Trou Normand, Un Acte D'Amour, Helen of Troy, Le Fils De Caroline Cherie, Les Grandes Moneuvres, Doctor at Sea, En Effeuillant La Marguerite, The Bride Is Much Too Beautiful, Moonlight Jewelers, La Femme et Le Pantin, Will You Dance With Me, Manina, La Fille Sans Voiles, Si Versailles M'Etait Conte, Trahi, Futures Vedettes, La Lumiere D'En Face, Une Sacre Gamine, And God Created Woman, Le Parisienne, En Cas De Malheur, Love Is My Profession, Babette Goes to War, La Verite Contempt, A Very Private Affair, Two Weeks in September, Viva Maria, Shalako, Spirits of the Dead, Les Femmes.

BARE, RICHARD L.: Producer, director. b. Turlock, Calif. Dir. or Warner: Smart Girls Don't Talk, Flaxy Martin, This Side of The Law, House Across The Street, Return of Frontiersman; SDG Best Dir. TV award, 1959; author, The Film Director (Macmillan), 1971.
TV: 77 Sunset Strip, Maverick, 1956, So This is Hollywood (1955-NBC), The Islanders (1960), Dangerous Robin (1961), This Rebel Breed, Twilight Zone, Bus Stop (1962), Adventures in Paradise, The Virginian (1963), Kraft Theatre, Run For Your Life, Green Acres Series (186 episodes-1969), Farraday and Son, 1974; Westwind (NBC) 1976.
PICTURES INCLUDE: Wrote, directed and produced Wicked, Wicked, MGM, 1973.

BAREN, HARVEY M.: Executive. b. New York City, Nov. 25, 1931. e. N.Y. State U. Served in U.S. Army, 1952–54; United Artists Corp., 1954–59 (contract dept., print dept., booker—N.Y. branch); asst. to general mgr., Magna Pictures Corp., 1959–61; road show mgr., national sales coordinator, 20th Century-Fox, 1961–71; asst. general sales manager, Allied Artists Pictures, 1971–79. Since 1980 pres. of Summit Film Distributors.

BARI, LYNN: Actress. r.n. Marjorie Bitzer; b. Roanoke, Va., 1917. First screen appearance as dancing girl in Dancing Lady (MGM). Became Fox contract player 1934.
PICTURES INCLUDE: Shock, Margie, Home Sweet Homicide, Nocturne, Amazing Mr. X, Kid from Cleveland, I'd Climb Highest Mountain, On the Loose, Sunny Side of

the Street, Has Anybody Seen My Gal, I Dream of Jeanie, Francis Joins the Wacs, A&C Meet the Keystone Kops, Women of Pitcairn Island, Damn Citizen, Trauma.

BARING, AUBREY: Producer. b. London, Eng., May 3, 1912. e. Eton & Cambridge. Ent. m.p. ind., 1935. Served in R.A.F. 1939–45, Squadron Leader, D.F.C.
PICTURES INCLUDE: Cairo Road, So Little Time. South of Algiers (Golden-Mask), Appointment in London. They Who Dare, Abominable Snowman, The Key, Cone of Silence, Wrong Arm of the Law.

BARKER, NORMAN: Executive. b. Toronto, Canada. Ent. m.p. industry as artist-copy writer, Famous Players Canadian Corp. Ltd., 1934; Presently advg. & pub. rel. Famous Players.

BARKER, ROBERT WILLIAM: TV Host. b. Darrington, Wash., Dec. 12. e. Springfield Central H.S., Drury Coll. News writer, announcer, disc jockey KTTS until 1949. News editor, staff announcer, Station WWPG; wife, Dorothy Jo. Emcee Truth or Consequences, 1956. Pres. Bob Barker Prod., Inc., M.C. Miss USA Pageant, CBS-TV, since 1967, M.C. Miss Universe Pageant, CBS-TV since 1967, M.C. Rose Parade, since 1970. Prod.-M. C. Pillsbury Bakeoff, since 1970 CBS. Prod. Lucky Pair, syndicated, M.C., Price Is Right-CBS, 1972; Narrator, 500 Festival Parade, Indianapolis since 1969.

BARKETT, STEVE: Actor-Director-Producer-Film Editor. b. Jan. 1, 1950, Oklahoma City, Okla. Exhibited and scored over 52 feature length classic silent films prior to coming to Los Angeles in 1970. Executive in several nontheatrical releasing companies including Independent Film Associates. Established the Nautilus Film Company in 1978.
PICTURES INCLUDE: Actor: The Egyptians are Coming, Corpsegrinders, Cruise Missile. Films as Actor, Director, Producer: The Movie People, Night Caller, The Aftermath.

BARNES, BILL: Motion Picture Executive b. Wilson, N.C., Jan. 28. e. Univ. of N.C., Chapel Hill, BA. Radio, TV, Motion Pictures. NBC-TV Stations Relations Dept., 1958–1959; prod. asst. to Broadway producers Bloomgarden and Shumlin, 1960–61; casting dir. and story ed. for Otto Preminger 1962–1969; motion picture agent with IFA in 1970 and merged into Intl. Creative Management; joined Producer Circle Co., Jan. 1979 as v.p., creative affairs.

BARNETT, GABRIEL: Graduate U.S.C. 1925 with B.L.L., publisher of theatre programs and commercial film advertising. Established in 1938. Creator of the Barnett theatre clock. Now engaged in producing intermission music for motion picture theatres.

BARR, ANTHONY: Producer, director, actor. r.n. Morris Yaffe. b. St. Louis, Mo., March 14, 1921. e. Washington Univ., B.S. 1942. Actor, asst. stage mgr., 1944–46; stage mgr., Katherine Dunham Dancers, 1946–47; teacher, actor, dir. in chg., Film Actors' Workshop, Professional Theatre Workshop, Hollywood; v.p. current prime time series, ABC-TV; director, current dramatic programming, CBS-TV.
BROADWAY: Jacobowsky and the Colonel, Winters' Tale, Embezzled Heaven.
PICTURES INCLUDE: (actor) People Against O'Hara, Border Incident, The Hollywood Story, The Mozart Story, (co-prod.) Dame With A Halo.
TV: (dir.) Art Linkletter's Houseparty, About Faces (ABC-TV), (assoc. dir.) Climax, Shower of Stars, (prod.) Climax, Summer Studio One, (assoc. prod. CBS-TV) Climax, Playhouse 90, Pursuit, G.E. Theatre, The Law and Mr. Jones, Four-Star.

BARR, PATRICK: Actor. b. Akola, India, Feb. 13, 1908. e. Radley and Trinity College, Oxford, as engineer; stage debut in Miss Black's Son, 1933; appeared in Young Madame Conti, Music Box, N.Y., 1937; entered m.p. 1939; winner "Daily Mail" award, outstanding TV actor, 1954 and 1957; Royal Shakespeare Company, 1970.
PICTURES INCLUDE: Brain Machine, Duel in the Jungle, Dam Busters, It's Never Too Late, The Valiant, The Longest Day, Ring of Spies.
TV: Flag Fall, Count-Down at Woomera, Quiet Game of Cards, Galileo, A Beast with Two Backs, The Vortex, The Pueblo Affair.

BARRETT, JAMES LEE: Screen writer, producer. b. Charlotte, N.C., Nov. 19, 1929. e. Anderson Jr. College, Furman Univ., Penn State Univ., Art Students League. U.S. Marine Corps.
PICTURES INCLUDE: The D.I. (Marine Corps Combat Correspondents Award), The Greatest Story Ever Told, The Truth About Spring, Shenandoah, The Green Berets, Bandolero, The Undefeated, tick . . . tick . . . tick, The Cheyenne Social Club, Fools' Parade, something big, Smokey and the Bandit.
STAGE: Shenandoah (Tony Award) best musical book, 1974–75; Theatre Club annual award.
TV: The Awakening Land, Parts I and II (Certificate of Commendation, American Women in Radio and Television), Stubby Pringles' Christmas (Humanitas Nomination), The Day Christ Died, Belle Starr, Angels Gates.

BARRETT, RONA: News Correspondent e. N.Y. Univ. (communications major) created the column, Rona Barrett's Young Hollywood, which led to featured column in 1960 in Motion Picture Magazine and a nationally syndicated column distributed to 125 newspapers by the North American Newspaper Alliance. Turned to TV; initial appearances on ABC Owned Stations in 5 cities, providing two-minute reports for local newscasts. This resulted in a network morning program, co-hosted by Joanna Barnes, called Dateline Hollywood. In 1969 she created first daily syndicated TV news segment for Metromedia. Now Art & Entertainment editor for Good Morning, America.

BARRIE, GEORGE: Executive. b. Brooklyn, N.Y. Feb. 9, 1918. e. N.Y.U. Left school to embark on business career, beginning with soda fountain concessions in drugstores. Played in dance band and later became booker for bands. Turned to beauty products, forming co. Caryl Richards. In 1963 acquired Faberge, merging it with own firm. Entered leisure time field with records, books, films and TV.
PICTURES INCLUDE: A Touch of Class, Night Watch, Welcome to Arrow Beach, Book of Numbers, Hang-Up, Miracles Still Happen, Hugo the Hippo, Whiffs, Thieves.

BARRIE, MONA: Actress. b. London, England, Dec. 18, 1909. On stage Australia, England (Hay Fever, Bitter Sweet Autumn Crocus, others). From 1933 on screen.
PICTURES INCLUDE: Devil's Mask, Just Before Dawn, Secret of the Whistler, I Cover Big Town, Cass Timberlane, When a Girl's Beautiful, The First Time, Strange Fascination.

BARRON, ARTHUR RAY: Executive. b. Mt. Lake, Minn., July 12, 1934. e. San Diego State College, 1956–60; B.S. Accounting. Certified public acc't, Calif., 1960. Lybrond, Rans Bros. and Montgomery, 1960–63; Desilu Productions, Inc., 1963–67; v.p. finance and administration, Paramount Television, 1967–70 v.p. Finance, Paramount Pictures Corp., 1970; senior v.p. finance and administration, 1971; exec. v.p., finance & admin., 1974; exec. v.p. 1980.

BARRON, WINSTON (Win): Sales executive, newsreel editor. b. Toronto, Ont., Canada; e. Toronto U. Started in radio; announcer, studio mgr., mgr. various stations; later freelance sports & newscaster; radio prod.; actor in Canadian m.p.; dir. Lux Radio Theatre (England) 1937–39; with Ortus Film Co. 1940; chg. Canadian sales prom. Paramount Pictures 1941; commentator & assoc. ed. Canadian Paramount News 1943; mgr. Canadian Public Relations Paramount 1944; ed. Canadian Paramount News since 1945; dir., pub. rel., sp. events, International Air Show; TV announcer and commentator. Variety Club, Toronto Press Club, Aviation Writers Ass'n; CIAS Scholarship Comm.; York Flying Club, chmn., Film Advertising Circle, Public Relations Society, Goodfellowship Club, Theatres Branch, Ontario Government; now dir. of promotion, O'Keefe Centre for the Performing Arts in Toronto, Canada. Received city of Toronto Award of Merit.

BARRY, DONALD (Red): Actor, r.n. Donald Barry de Acosta; b. Houston, Tex.; e. Tex. Sch. Mines. In 1936 in "Night Waitress" (RKO Radio), thereafter to 1939 in number of feature pictures, including "The Woman I Love," "Sinners in Paradise," "The Crowd Roars." From 1939 in Westerns, attaining wide popularity as Republic Western star. Entertained troops overseas. Voted one of first ten Money-Making Western Stars in Motion Picture Herald-Fame Polls 1942, 1943, 1944.
PICTURES INCLUDE: Jesse James' Women, Untamed Heiress, Twinkle in God's Eye, I'll Cry Tomorrow, Seven Men From Now, Shalako, The Shakiest Gun in the West, Orca, The Swarm.

BARRY, GENE: Actor. r.n. Eugene Klass; b. New York, N.Y., June 14, 1921; e. New Utrecht H.S., Bklyn. Appeared in little & summer theatre.
BROADWAY: Rosalinda, Catherine Was Great, Happy Is Larry, Bless You All.
PICTURES INCLUDE: Atomic City, Girls of Pleasure Island, War of the Worlds, Those Redheads from Seattle, Alaska Seas, Red Garters, Naked Alibi, Soldier of Fortune, Purple Mask, Houston Story, Back from Eternity, China Gato, 27th Day, Maroc 7.
TV: Bat Masterson, Burke's Law, The Name of the Game.

BARRY, JACK: Executive, performer. b. Lindenhurst, N.Y., March 20, 1918; e. U. of Pa. Prod. & mc. of Juvenile Jury, Life Begins at 80, Winky Dink & You, Oh! Baby, The Big Surprise, Twenty-one, Tic Tac Dough; pres., WGMA, Hollywood, Fla.; prod. Paramount TV Corp.

BARRY, JOHN: Composer, arranger, conductor. b. York, England, 1933. Artist and prod., CBS Records; TV: Elizabeth Taylor in London, Sophia Loren in Rome.
PICTURES INCLUDE: Beat Girl, Never Let Go, The Amorous Prawn, From Russia With Love, Seance On a Wet Afternoon, Zulu, Goldfinger, The Ipcress File, The Knack, King Rat, Born Free, Thunderball, The Chase, The Wrong Box, The Quiller Memorandum, The Whisperers, Deadfall, Petulia, You Only Live Twice, The Lion in Winter, Midnight

Cowboy, The Appointment, The Tamarind Seed, The Dove, The Day of the Locust, Robin and Marian, King Kong, The Deep, The Betsy, Moonraker, The Black Hole, Somewhere in Time.

BARRY, MICHAEL: Executive. b. May 15, 1910. Theatre & film dir., TV as BBC prod., 1938; during war in Royal Marines; rejoined TV as writer, prod., 1946; wrote and prod. Promise of Tomorrow, Passionate, Pilgrim, Shout Aloud Salvation; head of TV Drama Dept. June 1951. Awarded O.B.E. June, 1956. Programme Controller Telefis Eireann, Sept. 1961.

BARRYMORE, JOHN BLYTH, JR.; Actor. b. Beverly Hills, Calif., June 4, 1932. e. St. John's Military Acad., various public and private schools; p. late John Barrymore, Delores Costello (Mrs. John Vruwink). Debut in The Sundowners; many TV appearances.
PICTURES INCLUDE: High Lonesome, Big Night, Thunderbirds, While the City Sleeps, Shadow on the Window, War of The Zombies, Never Love A Stranger, The Night They Killed Rasputin, High School Confidential.

BART, PETER: Executive. e. Swarthmore College and The London School of Economics. Served eight years as correspondent for The New York Times and wrote for such magazines as Harper's, The Atlantic Saturday Review, etc. Joined Paramount Pictures in 1965. Named exec. ass't. to Robert Evans, exec. in charge of world-wide production. Appointed v.p. prod. Resigned post in 1973 to develop and produce own films for Para. Appointed pres. Lorimar Films, 1978. Resigned, 1979, to be independent producer
PICTURES INCLUDE: Islands in the Stream, Fun with Dick and Jane (both co-prod.).

BARTHOLOMEW, FREDDIE: Actor. b. London, Eng., Mar. 28, 1924. Stage debut 1927; m.p. debut U.S. in "David Copperfield" 1935.
PICTURES INCLUDE: Anna Karenina, Lloyds of London, Little Lord Fauntleroy, Devil Is a Sissy, Lord Jeff, Captains Courageous, Kidnapped, Swiss Family Robinson, Naval Academy, Yank at Eton, Junior Army, The Town Went Wild, St. Benny the Dip.

BARTLETT, HALL: Producer, Director, Writer. b. Kansas City, Mo., Nov. 27, 1922; e. Yale U., B.A. 1942. U.S. Naval Reserve 1942–47; formed Hall Bartlett Productions, 1952; prod. Navajo (winner of 27 nat'l awards & Festival of Brit. Award at Edinburgh); prod., s.p. Crazy-legs (winner of 9 nat'l awards, including Parents Mag. Gold Medal); prod., dir., s.p. Unchained (winner of Parents Mag. Gold Medal, Brotherhd. award of Nat. Con. of Christians and Jews); theme song, Unchained Melody, most pop. song of '55; prod., dir., s.p. Durango; prod. dir. s.p. Zero Hour, All the Young Men, Sol Madrid, photog., prod., Changes, prod. dir. Winner of the Sans Sebastian Festival. The Sandpit Generals, prod., dir., writer; winner of Grand Prize of VII International Film Festival of Moscow. Producer, director Jonathan Livingston Seagull; Producer, director, Cleo Laine Special, Zubin Mehta Special, and producer, director, The Children of Sanchez.

BARTON, CHARLES: Director, b. near San Francisco, Calif., May 25, 1902. In stock, vaudeville; then on screen in two comedies. Became "prop" boy for James Cruze, then William Wellman, including work on "Wings." Appt. asst. dir. Wellman Company; later dir. at Paramount; also directing TV films.
PICTURES INCLUDE: Men in Her Diary, Beautiful Cheat, Wagon Wheels, Vanishing Pioneer, Nevada Notorious Gentleman, Wistful Widow of Wagon Gap, Noose Hangs High, Abbott and Costello Meet Frankenstein, Mexican Hayride, Africa Screams, Abbott and Costello Meet the Killer, Free for All Double Crossbones, The Milkman, Ma and Pa Kettle at the Fair.
TV: McHales Navy, Hazel, Amos & Andy, Oh Susana, Dennis the Menace, Petti Coat Junction, Family Affair.

BARUCH, ANDRE: Announcer, commentator. b. Paris, France. e. Pratt Institute, Brooklyn, N.Y.; Columbia U.; Beaux Arts. Commercials for American Character, BC Headache Powder, Transogram Helene Curtis, Drajel, Lorillard; Prod. dir. for national sales conventions; TV, film prod. Helizer, Waring & Wayne, Inc.

BARUCH, RALPH M.: Executive. b. Paris, France, Aug. 5, 1923. e. The Sorbonne, Administrative aide, SESAC, Inc. 1944–48; account exec., DuMont Television Network, 1948–52; Eastern sales mgr., Consolidated Television Films, 1953–54; account exec., CBS Films, 1954; account supervisor, 1957; dir. international sales, 1959; v.p., CBS Enterprises, 1961–70; pres., Viacom Enterprises, 1971; pres., Viacom International, 1979; named chm. & chief exec. officer, Viacom International; gov., International Radio & Television Society; chmn., Rewrite Committee (Communications Act), NCTA; former member of the board of directors and former chmn., Pay Cable Committee, NCTA; gov. (New York), former chmn. of activities committee and former chmn. of forum program committee, International Council of the National Academy of Television Arts & Sciences.

BARWOOD, HAL: Writer-Producer. e. Univ. of S. Calif. School of Cinema. Writes scripts in collaboration with Matthew Robbins, Barwood branching out into producing with Corvette Summer in 1978.
PICTURES INCLUDE: Scripts, all with Robbins: The Sugarland Express, The Bingo Long Traveling All-Stars and Motor Kings, Corvette Summer (also prod.), Dragonslayer (also prod.).

BASCH, BUDDY: Publicist-promoter-producer-writer. b. South Orange, N.J., June 28, 1922. e. Columbia Univ. Began career as "Youngest Radio Editor in U.S." at 15, since written for national mags, syndicates, wire services, and newspapers. Edited and published "Top Hit Club News"-7 years. Joined Donahue and Coe 1940 on m.p. accounts, U.S. Army in Europe 1942–45. 1945 until 1967: own publicity and promotion office, working on m.p. company accounts and stars such as Burl Ives, Dinah Shore, Tony Martin, Danny Kaye, Peter Lorre, Tony Bennett, Gloria De Haven, McGuire Sisters, Rhonda Fleming, Sammy Davis, Jr., Anna Maria Alberghetti, Polly Bergen, Meyer Davis, The Beatles, Glenn Miller and Tommy Dorsey Orchestras. Produced many shows for radio, TV and stage in New York, Newark, Chicago, Hartford. Asst. to publisher, The Brooklyn Eagle 1962. Formed syndicate Buddy Basch Synd., 1965, 1966 formed Buddy Basch Feature Syndicate, formed BSB Creative Communications Unltd. in 1976, handling communications projects in all media, while continuing doing freelance writing on show bus, and travel subjects and general subject for N.Y. Daily News, A.P., Travel Holiday, Frontier Magazine, Kaleidoscope, True, United Features, Gannett Westchester-Rockland Newspapers, Bergen (N.J.) Record, Argosy, N.A.N.A., Womens' New Service, Today Magazine, Christian Science Monitor.

BASEHART, RICHARD: Actor. b. Zanesville, Ohio, Aug. 31, 1914. News reporter while in h.s.; surveying crew radio announcer, ward politician; with Hedgerow Theatre 5 yrs.; on N.Y. stage; won N.Y. Critics Award for "Hasty Heart"; plays include: Counter Attack, Land of Fame, Othello, etc.; m.p. debut in "Cry Wolf" 1945; summer stock.
PICTURES INCLUDE: Repeat Performance, He Walked By Night, Reign of Terror, Roseanna McCoy, Tension, House on Telegraph Hill, Fourteen Hours, Decision Before Dawn, Fixed Bayonets, Titanic, Good Die Young, Stranger's Hand, Canyon Crossroads, Moby Dick, La Strada, Five Branded Women, Hitler, Four Days In November, The Satan Bug, And Millions Must Die, The Island of Dr. Moreau, The Great Georgia Bank Hoax, Being There.

BASS, SAUL: Director-Producer. b. New York City May 8, 1920. e. Arts Students League. Pres. of, Saul Bass/Herb Yoger & Assoc. Has directed short films, m.p. titles/prologues/epilogues, TV show openings, TV commercials. Directorial feature debut in 1974 with Phase IV.
PICTURES INCLUDE: (shorts), The Searching Eye, From Here to There, Why Man Creates; Notes on the Popular Art, The Solar Film, Bass on Titles, (titles) Carmen Jones, The Man With The Golden Arm, Anatomy of a Murder, Vertigo, Psycho, Around the World in 80 Days, West Side Story, That's Entertainment II.

BATCHELOR, JOY: Director. b. Watford (Eng), May 12, 1914. e. Grammar School & Art School. Early career, illustrator & Designer. Entered m.p. in 1935 & is connected with the production of over 500 cartoon films. Co-director, co-prod., collab. s.p., Animal Farm, first British full length color feature cartoon; dir. Dam the Delta, The Colombo Plan, The Commonwealth, Ruddigore, The Five, Wotdot, Contact, Ass and the Stick. Christmas Feast, Carry in Milkmaids.

BATES, ALAN: Actor. b. Allestree, Derbyshire England, Feb. 17, 1934.
TV APPEARANCES: The Thug, A Memory of Two Mondays, The Jukebox, The Square Ring, The Wind and the Rain, Look Back in Anger, Three on a Gasring, Duel for Love, A Hero for Our Time, Plaintiff & Defendant, Two Sundays, The Collection, The Mayor of Casterbridge.
STAGE: On London and New York stage. The Caretaker, Long Days Journey Into Night, Poor Richard, In Celebration, Hamlet, Butley, The Taming of the Shrew, Life Class, Otherwise Engaged, The Seagull, Stage Struck.
PICTURES INCLUDE: The Entertainer, Whistle Down the Wind, A Kind of Loving, The Caretaker, Nothing But the Best, Zorba The Greek, Georgy Girl, King of Hearts, Three Sisters, Far From the Madding Crowd, The Fixer, Women in Love, Joe Egg, The Go Between, Impossible Object, Butley, In Celebration, Royal Flash, An Unmarried Woman, The Shout, Rose, Nijinsky, Quartet.

BATTY, PETER: Producer, director, writer. b. Sunderland, England, 1931. c. Bede Grammar School and Queen's College, Oxford. Feature-writer both sides Atlantic 1954–58. Joined BBC TV 1958 dir. short films. Edited Tonight programme 1963–4. Exec. prod. ATV 1964–68. Awarded Grand Prix for doc. at 1965 Venice and Leipzig festivals. Official entries 1970 and 1971 San Francisco and Melbourne festivals. Own company since 1968 prod. TV specials, series, commercials.

TV PRODUCTIONS INCLUDE: The Quiet Revolution, The Big Freeze, The Katanga Affair, Sons of the Navy Man, The Fall and Rise of the House of Krupp, The Road to Suez, The Suez Affair, Battle for the Desert, Vietnam Fly-In, The Plutocrats, The Aristocrats, Battle for Cassino, Battle for the Bulge, Birth of the Bomb, Search for the Super, Operation Barbarossa, Farouk: Last of the Pharaohs, Superspy, Spy Extraordinary, Sunderland's Pride and Passion, A Rothschild and His Red Gold, The World of Television, The Story of Wine, The Rise and Rise of Laura Ashley, The Gospel According to Saint Michael, Battle for Warsaw, Battle for Dien Bien Phu, Nuclear Nightmares. Contributed 6 episodes to Emmy-winning WORLD At WAR series.

BAUM, MARTIN: Executive. b. New York, N.Y. March 2, 1924; president, ABC Pictures; previously partner Baum & Newborn Theatrical Agency; head of West Coast office General Artists Corp., head of m.p. dept., Ashley Famous Agency; President of Martin Baum Agency; sr. exec. v.p. Creative Management Associations; president Optimus Productions, Inc., producing Bring Me the Head of Alfredo Garcia, The Wilby Conspiracy, and The Killer Elite. Partners with Michael Ovitz, Steven Roth, Michael Rosenfeld, Ron Meyer, Rowland Perkins, Bill Haber in Creative Artists Agency, Inc.

BAUM, MAURICE: Executive. b Chesterfield, Eng., April, 1912; e. Manchester Central High School. Ent. m.p. ind. in Manchester, 1925; to Ireland, 1936; pres. Theatre and Cinema Assn. 1945 & 1964–65 Nat'l Films Ltd., National Tele-Films Ltd., National TV; pres., Irish Cinema Trade Benevolent Fund; also man. dir. National Film Distributors, Ltd.

BAUMSTONE, HAROLD: Executive. b. New York City, June 27, 1911; e. U. No. Carolina, B.A. 1932; prod. & distrib. of short subjects; pres., Institutional Cinema Inc., National 16mm Film Exchange, now representing Columbia Pictures, Warner Bros., American International and ind. prod. Exclusive distributor for 16mm markets for foreign films.

BAXTER, ANNE: Actress. b. Michigan City, Ind., May 7, 1923; granddaughter of Frank Lloyd Wright, architect; e. Theodora Ervine School of Drama; with Mme. Maria Ouspenskaya 3 yrs. Stage debut in "Seen But Not Heard" 1935; in summer stock on Cape Cod; in N.Y. in "Madam Capet" 1938; film debut in Twenty Mule Team" 1940.
PICTURES INCLUDE: Great Profile, Charley's Aunt, Swamp Water, Magnificant Ambersons, North Star, The Razor's Edge, All About Eve, Follow the Sun, Outcasts of Poker Flat, O. Henry's Full House, My Wife's Best Friend, I Confess, Blue Gardenia, The Carnival Story, Bedeviled, One Desire, The Spoilers, The Come On, Three Violent People, Chase a Crooked Shadow, Summer of the Seventeenth Doll, Cimarron, The Ten Commandments, The Tall Women, The Late Liz.
STAGE: Applause (1972), Noel Coward in Two Keys (1974).
TV: The Moneychangers, East of Eden.
BOOK: Intermission, 1976.

BAXTER, BILLY: Executive. b. N.Y.C., Feb. 8, 1926. e. Holy Cross, 1948. Mgr., Ambassador Brokerage Group, Albany, 1957–58; Bill Doll & Co., 1959–63; organ., prod., radio show, Earl Wilson Celebrity Column, 1962; prod. Bwy show, Mandingo, with Franchot Tone, 1962; dir. of promotion, spec. events, Rumrill Ad Agency, 1963–64; dir. of promotion, exploitation, Landau Co., 1964–65; dir. of adv. and pub., Rizzoli Co., 1965; resigned, Rizzoli, 1966. Consultant on special events to the Philip Morris Corp. and American Express. Coproducer on movies Love and Anarchy, Daughters-Daughters, Outrageous, One Man, Dawn of the Dead. Produced Diary of the Cannes Film Festival with Rex Reed, 1980.

BAXTER, KEITH: Actor. b. April 29, 1933, Monmouthshire, South Wales. e. Wales, entered Royal Academy of Dramatic Arts in 1951. 1952–55 in national service; returned to RADA. Did years of repertory work in Dublin, Croydon, Chichester, London's West End, and New York. Biggest stage hit in Sleuth, both London and N.Y.
PICTURES INCLUDE: The Barretts of Wimpole Street, Peeping Tom, Chimese at Midnight, With Love in Mind, Ash Wednesday.
TELEVISION: For Tea on Sunday, Hold My Hand Soldier, Saint Joan.

BAXTER, STANLEY: Actor. b. Glasgow, Scotland, 1928. e. Hillhead High School, Glasgow. Principal comedian in Howard & Wyndham pantomimes. Summer revues. Televised regularly on BBC-TV, and also frequent broadcaster. M.P. debut 1955 in Geordie.
STAGE: On the Brighter Side, 1961, Chase Me Comrade, in Australia, 1965; Cinderella. 1966–68, What the Butler Saw; 1969; Phil The Fluter, 1969–70. Mother Goose Pantomime seasons 1970/71, 71/72, 72/73, 73/74. Jack & The Beanstalk, 75/76/77/78. Cinderella 77/80.
PICTURES INCLUDE: Very Important Person, Crooks Anonymous, The Fast Lady, Father Came Too Joey Boy.
TV: Baxter on (Series) 1964; The Confidence Course 1965, The World of Stanley Baxter, 1966; Stanley Baxter Show, 1967, Stanley Baxter Show, 1968, 1971, Time for Baxter, 1972, The Stanley Baxter Big Picture Show, 1973. The Stanley Baxter Moving Picture Show, 1974. Part III, 1975. "Stanley Baxter's Christmas Box," 1976. Bing Crosby's Merrie Olde Christmas, 1977. Stanley Baxter's Greatest Hits, 1977. Baxter on Television, 1979.

BEAL, JOHN: Actor, r.n J. Alexander Bliedung. b. Joplin, Mo., Aug. 13, 1909; e. Wharton School, Pa. U. On Broadway in: Another Language, She Loves Me Not, Petrified Forest, etc: m.p. debut in "Another Language" 1933; serves in U.S.A.A.F World War II.
PICTURES INCLUDE: Little Minister, Les Miserables, We Who Are About to Die, Man Who Found Himself, Arkansas Traveler, Cat and the Canary, Great Commandment, Doctors Don't Tell, Atlantic Convoy, Edge of Darkness; Stand By, All Networks, Key Witness, Madame X, Double Wedding, Beg, Borrow or Steal, Port of Seven Seas, Alimony, Song of Surrender, Chicago Deadline, My Six Convicts, Remains to Be Seen, The Vampire, The Sound and the Fury, That Night, Ten Who Dared.

BEATTY, NED: Actor. Worked at Barter Theatre in Virginia for seven years, appearing in over 70 plays. First N.Y. performance in The Great White Hope.
PICTURES: Deliverance (debut), Nashville, W.W. and the Dixie Dance Kings, Exorcist II The Heretic, Network, The Great Georgia Bank Hoax, Superman, Promises in the Dark, Hopscotch, The Incredible Shrinking Woman, Superman II.

BEATTY, ROBERT: Actor. b. Hamilton, Ont., Canada, Oct. 19, 1909. e. U. Toronto. London stage debut: Idiot's Delight, 1938. On screen first in Suspected Persons, 1942.
PICTURES INCLUDE: San Demetrio London, It Happened One Sunday, Appointment with Crime, Odd Man Out, Green Fingers, Against the Wind, Another Shore, Portrait from Life, Her Favorite Husband, Captain Horatio Hornblower, Calling Bulldog Drummond, Magic Box, Wings of Danger, The Gentle Gunman, The Oracle, The Net (Project M. 7), Man on the Tightrope, Broken Horseshoe, The Square Ring, Albert RN (Break to Freedom), Loves of Three Women, Out of the Clouds, Portrait of Alison (Postmark for Danger), Tarzan and the Lost Safari, Something of Value, Time Lock, The Shakedown, The Amorous Prawn, Stanley Kubrick's 2001, Where Eagles Dare, Sitting Target, Pope Joan, The Pink Panther Strikes Again, Golden Rendezvous.
TV: Weekly guide, Saturday Night Out, Dial 999 series, The Human Jungle (Series), Court Martial Series, Time Lock, Man at the Top, The Venturers, Thriller, Jesus of Nazareth, Blake's Seven, Park Ranger, The Martian Chronicles, Walk with Destiny, Suez, The Rose, Medallion.

BEATTY, WARREN: Actor. b. Richmond, Va., March 30, 1938. e. Northwestern U.: small roles on television; Compulsion, winter stock. North Jersey Playhouse; Broadway debut, A Loss of Roses.
PICTURES INCLUDE: Splendor in the Grass, The Roman Spring of Mrs. Stone, All Fall Down, Lilith, Mickey One, Promise Her Anything, Kaleidoscope; prod. & starred in Bonnie and Clyde, The Only Game in Town, McCabe and Mrs. Miller, Dollars, The Parallax View, Shampoo (producer, co-author, star), The Fortune, Heaven Can Wait (star, co-prod., co-dir., co-sp.), Reds (star, prod., dir.).

BEAUMONT, HUGH: Actor. b. Lawrence, Kan., Feb. 16, 1909. e. U. of Chattanooga; U.S.C.
PICTURES INCLUDE: Michael Shayne series, PRC; Blue Dahlia, Guilt of Janet Ames, Counterfeiters, Last Outpost, Lost Continent, Phone Call from a Stranger, Wild Stallion, Night Without Sleep, The Mississippi Gambler, Hell's Horizon.

BECK, ALEXANDER J.: Executive. b. Ung. Brod, Czechoslovakia, Nov. 5, 1922. e. Charles Univ. Prague, NYU. Owns many features and westerns for foreign distribution and library of 500 shorts. Importer and exporter; Pres., chairman of bd. Alexander Beck Films. 1955; formed Albex Films and A.B. Enterprises, 1959; formed & pres., Beckman Film Corp., 1960; formed Alexander Beck Productions, 1964. Sept. '69 formed Screencom Int'l Corp.

BECK, JACKSON: Actor-announcer-narrator. b. N.Y.C. TV and radio commercials. children's records, comm. & industrial films & shows; Narrator.

BECK, JOHN: Actor. b. Chicago. Acted with mid-western theatre groups; had basic training in scores of TV series. Theatrical debut in Three in the Attic.
PICTURES INCLUDE: Three in the Attic, Mrs. Pollifax, Spy, Lawman, Pat Garrett and Billy the Kid, Sleeper, Deadly Honeymoon, Rollerball, Paperback Hero, The Big Bus, Audrey Rose, The Other Side of Midnight.
TV: Guest star roles on Bonanza, Mannix, Hawaii Five-0, etc.

BECK, MICHAEL: Actor. b. Memphis, TN. e. Millsap College on football scholarship (quarterback). Became active in

college theatre. In 1971 went to London, England to attend Central School of Speech and Drama; studied 3 years, following which toured England with repertory companies for 2 years. Returned to U.S.; cast as lead in independent film, Madman (shot in Israel in 1977).
PICTURES: Madman, The Warriors, Xanadu.
TELEVISION: Holocaust, Mayflower: The Pilgrim Adventure, Alcatraz: The Rock and Clarence Carnes.

BECK, MYER P.: Sales, prod. rep. b. Meriden, Conn.; e. N.Y.U. School of Journalism. Newspaperman; publicist; publicity dir. United Artists; eastern mgr. Russell Birdwell Associates; formed own sls. prod. rep. office 1945, repr. Albert R. Broccoli; Jerome Hellman Productions, Stanley Kramer Pictures Corp., Harry Saltzman.

BECKER, VIOLA S.: Executive; r.n. Mrs. George J. Waas, b. New York, N.Y.; e. Columbia Univ.; organ V. S. Becker Productions, 1946, and V. S. Becker Adv. Svce.; AFRA International, Inc., Am. Women in International Radio & TV Society, Inc., Advertising Women of New York. Now free lance advertising consultant.

BEERY, NOAH, JR.: Actor. b. New York City, Aug. 10, 1916; son of Noah Beery, screen actor; e. Urban and Harvard Mil. Acad. Travelled with parents in stock company. Appeared as child in Mark of Zorro 1920.
PICTURES INCLUDE: Father and Son, Road Back, Only Angels Have Wings, Doolins of Oklahoma, Davy Crockett, Indian Scout, Last Outpost, Savage Horde, Rocketship XM, Two Flags West, Tropic Zone, Cimarron Kid, Wagons West, Story of Will Rogers, Wings of the Hawk, War Arrow, The Yellow Tomahawk, Black Dakotas, White Feather, Jubal, Fastest Gun Alive, Journey to Shiloh, Heaven With A Gun, Walking Tall, The Spikes Gang.

BEGELMAN, DAVID: President and Chief Operating Officer, Metro-Goldwyn-Mayer Film Co.; member of board of directors and member of executive committee. Independent Producer. President of Columbia Pictures; sr. exec. v.p., Columbia Pictures Industries, and member of board of directors, 1973 to 1978. Career in entertainment industry began with Music Corp. of America in N.Y. Resigned in 1960 as v.p. in charge of special projects to establish Creative Management Associates (talent agency) with Freddie Fields. CMA acquired General Artists Corp. in 1968 with Begelman as vice chairman of board of directors and member of exec. comm. Resigned to join Columbia.

BEICH, ALBERT: Writer. b. Bloomington, Ill., June 25, 1919. e. McGill U. Radio writer.
PICTURES INCLUDE: You Can't Beat the Law, Girls in Chains, Gangs of the Waterfront, Gay Blades, Bridge Goes Wild, Key to the City, Yellow Cab Man, The Milkman, Lieutenant Wore Skirts.

BEJTMAN, VINCENT: Executive. b. Poland. Started career producing entertainment films in native land. Came to U.S. during war. Named pres. of PIC Films, Inc., financed by Polish govt.-in-exile in London and the British Information Center. Company produced over 30 films depicting Poland participation in World War II. Formed Cine Classics Films, Inc., specializing in import of foreign films for U.S. market, including Rules of the Game, Zero De Conduit, etc. Produced documentaries (The Great Betrayal, Last Days of Warsaw.) Co-produced Naked Amazon. Now pres. of Vee Bee Productions, Inc. and involved in American-European co-production.

BELAFONTE, HARRY: Actor, singer. b. New York City, March 1, 1927. Professional debut; Royal Roost nightclub, N.Y., Village Vanguard, 1950. Broadway debut: John Murray Anderson's Almanac, 1953.
SHOWS INCLUDE: Three for Tonight, A Night With-Belafonte.
PICTURES INCLUDE: Bright Road, Carmen Jones, Island In The Sun, Odds Against Tomorrow, The World, The Flesh & The Devil, The Angel Levine, Buck and the Preacher, Uptown Saturday Night (prod., star).

BELFER, HAL B.: Exec. Producer-Director-Choreographer. b. Los Angeles, Feb. 16. e. Univ of So. Calif. (cinematography); Univ. of Calif. (writing). Head of choreography dept. at 20th Century-Fox. Head of choreography dept. at Universal Studios. Dir. of entertainment, in Las Vegas, Riviera and Flamingo Hotels. Prod., musical shows for Mexico City, Aruba, Puerto Rico, Montreal, Las Vegas. Dir., TV commercials and industrials. H. R. Pufnstuf TV series. Theatricals: Over 150 features. Producer-Director-Choreographer, Premore, Inc. Develop TV specials and sitcom, tape and film. Exec. prod., Once Upon a Tour and Dora's World, Rose on Broadway, Secret Sleuth, Inn by the Side of the Road, Imagine That!. Special staging "Tony The Pony" Series and prod., segment of What a Way to Run a Railroad; TV specials. Talent development programs, Universal Studios, 20th Century-Fox. Personal management and show packager.

BEL GEDDES, BARBARA: Actress. r.n. Barbara Geddes Lewis. b. New York City, Oct. 31, 1922; p. Norman Bel Geddes, scenic designer. Stage debut in "Out of the Frying Pan"; toured USO camps in "Junior Miss" 1941; screen debut in "The Long Night" 1946; Star of Tomorrow, 1949. N.Y. Stage: The Moon is Blue, Cat on a Hot Tin Roof, Mary, Mary, Everything in the Garden.
PICTURES INCLUDE: I Remember Mama, Blood on the Moon, Caught, Panic in Streets, Fourteen Hours, The Five Pennies, Five Branded Women, By Love Possessed, Summertree, The Todd Killings.

BELL, TOM: Actor. b. Liverpool, England, 1932. Early career in repertory and on West End stage. First TV appearance in Promenade.
TV: No Trams to Lime Street, Love On the Dole, A Night Out, Cul de Sac, The Seekers, Long Distance Blue, The Virginian. Film debut 1960 in Payroll.
PICTURES INCLUDE: The Kitchen, H.M.S. Defiant, Prize of Arms, L-Shaped Room, Rebels Against the Light, Ballad in Blue, He Who Rides a Tiger, In Enemy Country, The Long Days Dying, Lock Up Your Daughters.

BELLAMY, EARL: Prod. Dir. b. Minneapolis, Minn., March 11, 1917. e. L.A. City Coll. Universal Studios; latest m.p.: Fluffy, Gun Point, Munsters Go Home.
TV: Bachelor Father, Wells Fargo, Lone Ranger, Alcoa Premiere, Arrest and Trial, The Virginian, The Crusaders, Schlitz Playhouse, Heinz, Rawhide, The Donna Reed Show, Andy Griffith Show, Wagon Train, now president, The Bellamy Productions Co., Laramie, Laredo, I Spy, Mod Squad, Medical Center.

BELLAMY, RALPH: Actor. b. Chicago, Ill., June 17, 1905. Stock and repertory, 1922–30; pres. Actors Equity, 1952; council, Lambs, 1952; TV includes major live and film dramatic shows, voted many TV awards.
BROADWAY: Town Boy, Roadside, Tomorrow the World, State of the Union, Detective Story.
PICTURES INCLUDE: Great Impersonation, Lady on a Train, Guest in the House, Delightfully Dangerous, Court Martial of Billy Mitchell, Sunrise at Campobello, The Professionals, Rosemary's Baby, Oh, God!

BELLFORT, JOSEPH: b. New York. e. New York U., B'klyn Law Sch. Joined RKO Service Corp., Feb., 1930; trans. to RKO Radio Pictures, legal dept., May 1942; joined RKO Fgn. dept., Oct., 1944, handled Far Eastern division, Dec. 1946; then asst. to European gen. mgr.; gen. European mgr., 1949–1958; gen. sales mgr. National Screen Service, 1959; home office sup., Europe & Near East, 20th Century Fox, 1963; home office intl. mgr., 20th Century-Fox, 1966. Ass't v.p. and foreign mgr. 20th Cent.-Fox, 1967; vice president 20th Century-Fox, Int. Corp. & Inter-America, Inc. 1968; named sr. v.p., 1975. Resigned from Fox, 1977, to become v.p., m.p. Export Assn. of America in New York.

BELMONDO, JEAN-PAUL: Actor. b. Neuilly-sur-Seine, France, April 9, 1933. e. private drama school of Raymond Girard, and the Conservatoire d'Art Dramatique. Formed a theater group with Annie Girardot and Guy Bedos.
THEATRE INCLUDE: Jean Marais' production of Caesar and Cleopatra, Treasure Party. Oscar, 1958.
PICTURES INCLUDE: A Pied, A Cheval et En Voiture (By Foot, Horse and Car), Look Pretty and Shut Up, Drole de Dimanche, Les Tricheurs, Les Copains du Dimanche, Charlotte et Son Jules, A Double Tour, Breathless, Classe Tous Risques, Moderato Cantabile, La Francasie et l'Amour, Les Distractions, Mademoiselle Ange, La Novice, Two Women, La Viaccia, Une Femme Est Une Femme, Leon Morin, Pretre, Les Amours Celebres, Un Singe en Hiver, Le Doulos, L'Aine des Ferchaux, La Mer A Boire, Banana Peel, That Man From Rio, Cent Mille Dollars au Soleil, Echappement Libre, La Chasse a l'Homme, Dieu a Choisi Paris, Weekend a Zuydcocte, 100,000 Dollars au Soleil, Par Un Beau Matin d'Ete, Up to His Ears, Is Paris Burning?, The Thief of Paris, Tender Scoundrel, Pierrot le Fou, The Brain, Love Is a Funny Thing, Mississippi Mermaid, Borsalino, Inheritor, Stavisky, Fear Over the City.

BELUSHI, JOHN: Actor. b. Chicago, Jan. 24. Member of Second City improvisational group. Appeared in National Lampoon prod. of Lemmings on Broadway; directed one of the shows when it went on tour. Wrote, acted in and directed National Lampoon Radio Hour. Affiliation continued with The National Lampoon Show off-Bdwy., in which he performed and served as writer-director. Gained fame on NBC's Saturday Night Live.
PICTURES: Goin' South, Old Boyfriends, National Lampoon's Animal House, 1941, The Blues Brothers, Continental Divide, Neighbors.
TELEVISION: Richard Pryor Special, Beach Boys Special, All You Need is Cash, etc.

BENDHEIM, III, SAM: Executive. b. Richmond, Va., June 15, 1935. e. Washington & Lee Univ. Joined Neighborhood Theatre, Inc., 1958 as exec. trainee. Was dir. of adv. (1960–63); bd. member, ass't. treas. (1962); ass't. to gen. mgr. (1963–67) v.p. (1967). Bd. mem., ass't. sec., treas., Prince Edward Amusement Corp. Member of NATO of Va. bd.;

former vice-pres. NATO of Virginia, now pres.; Neighborhood Booking & Buying Service.

BENDICK, ROBERT: b. New York City, Feb. 8, 1917; e. New York U., White School Photography; U.S. Airforce, World War II. Documentary and still cameraman before joining CBS Television as cameraman and dir., 1940; rejoined CBS Television as dir. special events, 1946; promoted dir. news & special events; acting program dir. 1947; res. Oct. '51. Collab. with Jeanne Bendick on Making the Movies, Electronics for Young People, Television Works Like This, Filming Works Like This, 1971; Prod. Peabody Awardwinning U.N. show. The U.N. in Action; v.p., Cinerama Prod., co-prod. This Is Cinerama; co-dir., Cinerama Holiday; prod. Dave Garroway Show Today, prod., Wide Wide World 1955–56, NBC prod. dir. C. V. Whitney Pict., June, 1956; Merian C. Cooper Ent., 1957; prod NBC, 1958. Prod.; Garroway Today Show, Bob Hope 25 Yrs. of Life Show, 1961; Bell Telephone Threshold Science Series, Groucho Marx, Merrily We Roll Along, US Steel Opening New York World's Fair, 1964. Prod. First Look Series 1965 (Ohio St. Award); prod. & dir. American Sportsman, ABC; prod., pilot, Great American Dream Machine (NET) Emmy Award, 1971 and 1972; 1975, Co-exec. prod., Dick Cavett—Feeling Good. pres. Bendick Assoc. Inc.,; prod. of education audio-visual systems, Bd. of Governors, N.Y. Academy of TV Arts and Sciences. 1976, co-author with Jeanne Bendick, TV Reporting. Consultant, Warner Qube Cable Co.; 1978, produced/directed, Fight for Food (PBS). Program consultant to Times-Miror Cable Co., L.A. Produced segment ABC 20/20.

BENEDEK, LASLO: Director. b. Budapest, Hungary, Mar. 5, 1907; e. U. Vienna; Psycho-Analytic Clinic. Writer & photographer while in coll.; cameraman Budapest; first asst. cameraman several studios; cameraman UFA & Terra; film cutter; cutter & asst. prod. to Joe Pasternak; to Paris as cutter, then Engl dial. on "Antonia"; writer (England) 2 yrs.; montage dir. MGM 1937; writer (Mexico); assoc. prod. (Hollywood) on musicals; contract dir. MGM; Stanley Kramer Co., Col.; 1976–80, Chairman of Graduate Film Program, New York University.
PICTURES INCLUDE: Kissing Bandit, Port of New York, Death of a Salesman, Storm Over Tibet, Wild One, Bengal Brigade; Sons, Mothers and a General, Affair in Havana, Moment of Danger, Recours en Grace, Namu, The Killer Whale, A Daring Game, The Night Visitor.
TV: 4 Star Playhouse, Dupont Theatre, Stage 7, Loretta Young Show, Telephone Time, Perry Mason, Naked City, Untouchables, Outer Limits.
STAGE: Belial, Twelfth Night.

BENEDICT, DIRK: Actor. b. Helena, Mont. r.n. Dirk Niewoehner. e. Whitman College, Walla Walla, Wash. Enrolled in John Fernald Academy of Dramatic Arts, Rochester, Mich., after which had season with Seattle Repertory Theatre; other in summer stock at Ann Arbor, Mich. Broadway debut, 1970, Abelard and Heloise. Film debut, Georgia, Georgia, 1972.
PICTURES INCLUDE: Ssssss, W, Battlestar Galactica, Scavenger Hunt.
TELEVISION: Hawaii Five-O (one episode), Chopper One (series), Battlestar Galactica.

BENJAMIN, RICHARD: Actor. b. New York, N.Y., May 22, 1939; e. Northwestern U. Plays; Central Park productions of The Taming of the Shrew, As You Like It; toured in Tchin Tchin, A Thousand Clowns, Barefoot in the Park, The Odd Couple. Broadway debut in Star Spangled Girl, 1966. Directed London productions of Barefoot in the Park. TV series, He and She, with wife Paula Prentiss, 1967.
PICTURES INCLUDE: Goodbye, Columbus, Catch 22, Diary of a Mad Housewife, The Marriage of a Young Stockbroker, The Steagle, Portnoy's Complaint, The Last of Sheila, Westworld, The Sunshine Boys, Love at First Bite, Scavenger Hunt, The Last Married Couple in America, How to Beat the High Cost of Living, First Family.

BENNET, SPENCER G.: Director. b. Brooklyn, N.Y., Jan. 5, 1893. e. Manual Training H.S., Brooklyn, N.Y.; played stunt and thrill parts; asst. dir., Pathe; prod., dir., from 1925–64. Retired.
CREDITS INCLUDE: Play Ball (serial), The Green Archer (serial), Snowed in (serial), The Fighting Marine (serial) and many other serials and westerns. Submarine Seahawk, Atomic Submarine, Bounty Killer, Requiem For a Gunfighter.

BENNETT, BRUCE: Actor. r.n. Herman Brix. b. Tacoma, Wash.; e. U. Washington.
PICTURES INCLUDE: My Son is Guilty, Lone Wolf Keeps a Date, Atlantic Convoy, Sabotage, Underground Agent, More the Merrier, Sahara, Man I Love, Stolen Life, Nora Prentiss, Cheyenne, Treasure of Sierra Madre, Dark Passage, Smart Girls Don't Talk, Second Face, Great Missouri Raid, Last Outpost, Angels in the Outfield, Sudden Fear, Dream Wife, Dragonfly Squadron, Robbers Roost, Big Tip off, Hidden Guns, Bottom of the Bottle, Strategic Air Command.

BENNETT, CHARLES: Writer, director. b. Shoreham, England. British Army. Contract, Universal, 1937; wrote British Ministry of Information, WW II; dir. s.p. over 50 TV shows, including Cavalcade of America, Schlitz Playhouse, The Christophers, Four Star.
PLAYS INCLUDE: Blackmail, The Last Hour, Sensation, The Danger Line, Page From a Diary, The Return, After Midnight.
PICTURES INCLUDE: Blackmail, The 39 Steps, Secret Agent, The Man Who Knew Too Much (orig.), Sabotage, The Girl Was Young, Balalaika, The Young in Heart, Foreign Correspondent, Reap the Wild Wind, Joan of Paris, They Dare Not Love, The Story of Dr. Wassell, Unconquered, Ivy, Sign of the Ram, Kind Lady, The Green Glove, Dangerous Assignment, Madness of the Heart (dir.), The Big Circus, The Lost World, Voyage to the Bottom of the Sea, Five Weeks in a Balloon.

BENNETT, HARVE: Producer. r.n. Harve Fischman. b. Chicago, Ill., Aug. 17, 1930. e. U.C.L.A. Quiz Kids radio show, 5 yrs.; newspaper columnist, drama critic; freelance writer; Assoc. prod., CBS-TV; free-lance TV writer; prod. of special events. CBS-TV; dir., Television film commercials; program exec., ABC, vice pres., programs west coast—ABC-TV; prod.-writer, Mod Squad. Creator-writer, The Young Rebels, Screen Gems; exec. prod., Six Million Dollar Man, Bionic Woman, Rich Man, Poor Man. Universal Studios; Pres., Bennett-Katleman. Productions at Columbia Studios, and exec. prod. on American Girls series (CBS), From Here to Eternity (NBC), Salvage I (ABC).

BENNETT, JILL: Actress. b. Penang (Federated Malay Straits), Dec. 24, 1931. Stage debut at Stratford-on-Avon, 1948; screen debut 1952 in Moulin Rouge.
PICTURES INCLUDE: Hell Below Zero, The Criminal, Extra-Day, Lust For Life, The Skull, The Nanny, Charge of Light Brigade, Inadmissible Evidence, I Want What I Want, Quilp, Full Circle.
TV: The Heiress, Trilby, Jealousy, A Midsummer Night's Dream, The Three Sisters, Glimpse of the Sea, Return to the Sea, The Book Bag (Somerset Maugham Series), It's Only Us, The Parachute, Hotel in Amsterdam, Speaking of Murder, Intent to Murder, Jill and Jack, Almost A Vision, Hello Lola, The Old Crowd.

BENNETT, JOAN. Actress. b. Palisades, N.J., Feb. 27, 1910; e. St. Margaret's Boarding School, Waterbury, Conn.; L'Ermitage, Versailles, France; member of famous Bennett acting family; m. Walter Wanger, prod. On stage with father, Richard Bennett in "Jarnegan"; m.p. debut in "Bulldog Drummond" 1929; vice-pres. & treas. Diana Prods. 1945.
PICTURES INCLUDE: Three Live Ghosts, Disraeli, Moby Dick, Little Women, Pursuit of Happiness, Mississippi, Woman in the Window, Secret Beyond the Door, Man Who Reclaimed His Head, Hollow Triumph, Scar, Reckless Moment, For Heaven's Sake, Father of Bride, Father's Little Dividend, The Guy Who Came Back, Highway, Dragnet, We're No Angels, There's Always Tomorrow, Navy Wife, Desire in the Dust, Suspiria.
TV: Starring in ABC Dark Shadows.

BENSON, HUGH: Executive. b. New York. Sept. 7, 1917. e. New York public schools. Six yrs. infantry, then Air Force. Three yrs. European Theatre of War, As master sergeant, Air Force, two yrs., non com in chg. of soldier units under, prod.-dir., Joshua Logan. Joined Blaine-Thompson Advertising Agency as dir. of public relations; assist. to producers, Ed Sullivan and Marlo Lewis, Ed Sullivan Show, 1947–50; dir. television, public relations, Roy Dunstine Agency, 1951; hd. of radio, television promotion, publicity dept., Warner Bros., June, 1955; appointed assist. to Steve Trilling, vice-pres. and exec. associate of Jack L. Warner, 1956; exec. asst. to William T Orr, Warner v.p. in chg. of production, 1956; prod., United Artists, 1963; exec. asst. to William T. Orr, Warner Bros., 1964–65. Exec., Prod., Screen Gems, 1966, Prod., Cinema Center Films, 1969–71, prod., MGM, 1972; appointed exec. prod., MGM Television, 1972. 1975 indep. prod. On staff Col.-tv, 1975, pilots and long form.
PICTURES: Nightmare Honeymoon (prod.), Logan's Run (assoc. prod.), Billy Jack Goes to Washington (prod.).
TELEVISION: Contract On Cherry St. (prod.) Child Stealers, Goldie and the Boxer, A Fire in the Sky, Confessions of a Lady Cop, The Dream Merchants, Goldie and the Box Go to Hollywood, Goliath Waits.

BENSON, LEON: Producer-Director. b. Cincinnati, Ohio, Nov. 9. e. New York University, University of Cincinnati. Advertising Dept. Paramount, 1938 head of Trailer Dept., 1940. U.S. Army Air Corps, 1942–46, Major, First head of TV Dept. L.A. office of J. Walter Thompson Co., 1946–51. Head of Story Dept. Ziv TV, 1951; prod. and dir. West Point, 1955–56; Sea Hunt, 1956–60; Ripcord, 1960–62. Prod.-Dir. Flipper TV series, 1963–64. Under contract, Universal, 1965–66, prod.-dir. episodes, Kraft Suspense Theatre, The Virginian, dir. episodes, Chrysler Theatre, Laredo. Prod. Tarzan TV series, 1966. Under contract, NBC Prods., 1967–70, as staff dir. Bonanza, High Chaparral. Prod. theatrical feature, Chosen Survivors, 1973. Also dir. Owen Marshall, Counsellor at Law, Mission Impossible, The Eleventh Hour, Ben Casey, Wild Wild West, Rat Patrol, The Lieutenant, Outer Limits, The Loner, Empire.

BENSON, ROBBY: Actor. b. Dallas, 1957. Father, Jerry Segal, is writer, and mother, Ann Nelson, veteran of Dallas stage and natl. summer stock. Robby appeared on Broadway in The King and I at age five. Also worked 6 mos. in soap opera, Search for Tomorrow. Big break in feature films in Jeremy. Co-wrote screenplay of One on One (1977) with father and starred in film.
PICTURES: Jeremy, Jory, The Godfather—Part II, Ode to Billy Joe, Lucky Lady, One on One, The End, Ice Castles, Walk Proud, Die Laughing (also prod., co-s.p.), Tribute.
TV: The Death of Richie, Remember When, Virginia Hill, All the Kind Strangers, Death Be Not Proud.
STAGE: Zelda, The Rothschilds.

BENTLEY, JOHN: Actor. b. Warwickshire, Eng. Dec. 2, 1916. e. King Edward's Grammar Sch., Birmingham. Radio debut, 1933; Alexandra & Birmingham Rep. Cos. in 1934; stage debut in New Faces, London, 1940; m.p. debut in Hills of Donegal, 1946.
PICTURES INCLUDE: Calling Paul Temple, Paul Temple's Triumph, Happiest Days of Your Life, Torment, She Shall Have Murder, Salute the Toff, Hammer the Toff, Woman's Angle, Lost Hours, Tread Softly, Black Orchid, Paul Temple Returns, River Beat, Double Exposure, Scarlet Spear, Golden Ivory, The Flaw, Profile, Last Appointment, Confession, Deadliest Sin, Flight from Vienna, Strangehold, Blind Man's Bluff, Dial 999, Way Out, Escape in the Sun, Istanbul, Submarine Seahawk, The Singer Not the Song, Mary Had a Little, The Sinister Man, An Helligen Wassern, The Fur Collar, The Shadow of Treason.
TV: Strictly Personal, African Patrol, Crossroads, Armchair Theatre, Any Other Business, Misfire, Time Out of Mind.

BENTON, ROBERT: Writer-Director. b. Waxahachie, Texas, 1932. e. Univ. of Texas (B.A.); Was art director at Esquire Magazine where he met David Newman, a writer-editor, and formed writing partnership. All writing credits shared with Newman. Benton made directorial debut with Bad Company (1972).
PICTURES INCLUDE: Bonnie and Clyde, There Was a Crooked Man, What's Up, Doc?, Bad Company (also directed) The Late Show (written and directed solo), Superman (with Mario Puzo and Tom Mankiewicz), Money's Tight, Bad Company (also dir., solo); Superman (co.-s.p.); Kramer vs. Kramer (dir. s.p.); Stab (dir., s.p.).
STAGE: It's a Bird . . . It's a Plane . . . It's Superman (libretto), Oh! Calcutta (one sketch).

BERCUTT MAX: Executive. b. Omaha, Neb., Oct. 14, 1910. e. U. of Southern California, 1933. Pres., Paris-American Guide Services, Paris, France; publicist David O. Selznick Prod.; U.S. Navy officer World War II; Warner Bros. field rep.; assistant to exec. vice-pres.; director of publicity, Warner Bros. Seven Arts Studio; resigned, September, 1969. Formed Max Bercutt Co., Jan. 1970.

BERENGER, TOM: Actor. b. Chicago. e. Univ. of Missouri (drama dept.) Acted in regional theatres and off Bdwy. Plays include The Rose Tattoo, Electra, Streetcar Named Desire, End as a Man.
PICTURES: Looking for Mr. Goodbar, In Praise of Older Women, Butch and Sundance: The Early Days.
TELEVISION: One Life to Live (series), Johnny We Hardly Knew Ye.

BERG HARVEY M.: Executive. b. Portland, Oregon, June 9, 1943. e. BS and MBA from UCLA. 1969 Arthur Young & Co., leaving as audit manager/computer audit specialist. 1974 Warner Bros., Inc. as studio controller serving both Warner Bros. and The Burbank Studios. 1976 The Burbank Studio as exec. asst. to the pres. and member of exec. comm. 1977 dir., of studio services and secty. of exec. comm.

BERGEN, CANDICE: Actress. b. Beverly Hills, Calif., May 9, 1946. e. U. of Pennsylvania. Modeled during college; m.p. debut, The Group, 1965; free lance photo-journalist.
PICTURES INCLUDE: The Group, The Sand Pebbles, The Day the Fish Came Out, Live for Life, The Magus, The Adventurers, Getting Straight, Soldier Blue, Carnal Knowledge, The Hunting Party, T. R. Baskin, 11 Harrowhouse, Bite the Bullet, The Wind and the Lion, The Domino Principle, A Night Full of Rain, Olivers' Story, Starting Over, Rich and Famous.

BERGEN, POLLY: Singer, actress. b. Knoxville, Tenn., July 14, 1930. e. Compton Jr. Coll., Calif. Prof. debut radio at 14; in light opera, summer stock; sang with orchestra and appeared in night clubs; Columbia recording star; Emmy Award, best actress, 1957–58; pres. The Polly Bergen Co.; on stage, John Murray Anderson's Almanac, Champagne Complex, First Impressions.
TV: G.E. Theatre, Schlitz Playhouse, Playhouse 90, Studio One, Perry Como, Ed Sullivan Show, To Tell the Truth, Polly Bergen Show, Bob Hope Show, Bell Telephone, Wonderful World of Entertainment, Dinah Shore Show, NBC, CBS specials, Dean Martin Show, Andy Williams Show, Red Skelton Show, Mike Douglas Show.

BERGER, BENJAMIN N.: Executive. b. Poland, Mar. 5, 1896. Pres. Berger Amusement Co.; won decision in Fed. Court against payment to ASCAP for theatre seat tax.

BERGER, HELMUT: Actor. b. Salzburg, e. Feldkirk College and Univ. of Perugia. First film, small role in Luchino Visconti's The Witches (Le Streghe) in 1966.
PICTURES INCLUDE: The Young Tigers, The Damned, Do You Know What Stalin Did To Women?, The Garden of the Finzi-Continis, Dorian Gray, A Butterfly with Bloody Wings, The Greedy Ones, The Strange Love Affair, Ludwig, Ash Wednesday, Conversation Piece, The Romantic Englishwoman, Orders to Kill, Madame Kitty, Merry-Go-Round.

BERGER, RICHARD L.: Executive. e. UCLA. In 1964 joined acct. dept., 20th Century-Fox; promoted to exec. position in Fox-TV. Was dir. of programming, then v.p. of programs. Appt. asst. v.p., prod. 20th-Fox. Left in 1975 to join CBS-TV as v.p. dramatic development. Returned to 20th-Fox in 1977 as v.p., domestic prod., 20th Century-Fox Pictures.

BERGERAC, JACQUES: Actor. b. Biarritz, France. Career includes Five Minutes With Jacques Bergerac on radio; in the theatre, on tour in Once More With Feeling; on most major network TV shows.
PICTURES INCLUDE: Twist of Fate, The Time is Now, Strange Intruder, Come Away With Me, Les Girls, Gigi, Man and His Past, Thunder in the Sun, Hypnotic Eye, A Sunday in Summer, Fear No More, Achiles, A Global Affair, Taffy and the Jungle Hunter, The Emergency Operation, Lady Chaplin, The Last Party, One Plus One.

BERGMAN, INGMAR: Writer-director. b. Uppsala, Sweden, July 14, 1918. e. Stockhom U. Directed university play prods.; wrote & dir. Death of Punch, 1940; first theatrical success, dir., Macbeth, 1940; writer-director, Svensk Film-industri, 1942-present; first s.p., Frenzy, 1943; first directorial assignment, Crisis, 1946; chief prod., Civic Malmo, 1956–60.
PICTURES INCLUDE: Kris, 1946 Det Regnar Pa Var Karlek, Skepp Till Indialand, Night is My Future, Port of Call, The Devil's Wanton, Three Strange Loves, Till Gladje, Summer Interlude (Illicit Interlude), High Tension, Bris (soap ads), Secrets of Women, Monika, The Naked Light, A Lesson in Love, Dreams, Smiles of a Summer Night, Sista Paret Ut, The Seventh Seal, Wild Strawberries, Brink of Life, The Magician, The Virgin Spring, The Devil's Eye, Through a Glass Darkly, Winter Light, The Silence, All These Women, Persona, Hour of the Wolf, Shame, The Passion of Anna, Cries and Whispers, Scenes from a Marriage, Face to Face, The Serpent's Egg, Autumn Sonata, Life of the Marionettes.

BERGMAN, INGRID: Actress. b. Sweden, 1917; e. Lyceum for Flickor, Royal Dram. Theatre School. In Swedish m.p.; U.S. motion picture debut in Intermezzo, 1939; Academy Award best actress, Gaslight, 1944; voted one of ten best money making stars, Motion Picture Herald-Fame Poll, 1946–48; British TV, Hedda Gabler.
PICTURES INCLUDE: Adam Had Four Sons, Rage in Heaven, Dr. Jekyll and Mr. Hyde, En Enda Natt, Casablanca, For Whom the Bell Tolls, Notorious, Saratoga Trunk, Spellbound, Bells of St. Mary's, Arch of Triumph, Joan of Arc, Under Capricorn, Stromboli, Greatest Love, Strangers, Fear, Anastasia, Paris Does Strange Things, Indiscreet, The Inn of the Sixth Happiness, Goodbye Again, The Visit, The Yellow Rolls Royce, Cactus Flower, A Walk in the Spring Rain, From The Mixed-Up Files of Mrs. Basil E. Frankweiler, Murder on the Orient Express, A Matter of Time, Autumn Sonata.
STAGE: Captain Bassbound's Conversion (1972), More Stately Mansions.
TV: 24 Hours in the Life of a Woman.

BERGMANN, TED: Executive; b. Brooklyn, N.Y., Sept. 12, 1920. e. Gov. Dummer Acad. and Amherst Col.; joined NBC 1941; 1942–46 Captain, U.S. Army in pub. rel. div. of SHAEF; awarded Bronze Star Medal for combat reporting; returned to NBC 1946 international div.; 1947, acc't exec. DuMont TV Network; 1951 appt. dir. of sales of DuMont TV Network; named gen. mgr., 1953; managing dir. of broadcast div. DuMont, 1954–56; v.p. McCann-Erickson, 1956–57; pres. Parkson Adv. Agency, Inc., 1957–60; v.p. adv. Revlon Inc. 1960–62.
Member of Bd. Dir. NAB, 1954–56; V.P. Radio-TV Exec. Society; Bd. of Gov. 1957–62, Academy TV Arts & Sciences, 1959–61. President, Charter Producers Corp., 1962.

BERLE, MILTON: Actor. r.n. Milton Berlinger; b. New York City, July 12, 1908; e. Professional Children's Sch., N.Y. in vaudeville; on N.Y. stage (Ziegfeld Follies 1936, Life Begins at 8:40, etc.): night clubs; concurrently on radio & screen.
TV: Star of own NBC show; Texaco Star Theatre, 1948–54, Kraft Music Hall TV Show, 1958–59, Jackpot Bowling,

1960–61, Doyle Against the House, Dick Powell Show, 1961, Chrysler TV special, 1962, The Milton Berle Show, 1966.
PICTURES INCLUDE: Tall, Dark and Handsome, Sun Valley Serenade, Over My Dead Body, Margin for Error, Always Leave Them Laughing, Let's Make Love, It's a Mad Mad Mad World, The Happening, Who's Minding the Mint, Where Angels Go . . . Trouble Follows, Can Hieronymus Merkin Ever Forget Mercy Humppe and Find True Happiness?, Lepke.

BERLIN, IRVING: Composer. r.n. Israel Baline. b. Russia, May 11, 1888. To New York, 1893. As youth, singing waiter in N.Y.; early began writing songs. One of first successes, Alexander's Ragtime Band. Entered music pub.; member Waterson, Berlin & Snyder. W.W I, sgt. inf. U.S. Army. Composer many indiv. and stage musical songs. In 1937 wrote songs for screen prod. Top Hat (RKO); in 1942 prod. stage musical This Is the Army for U.S. Army Fund; in 1943 assoc. with Warner Bros. in prod. film version. Member: Lambs, Friars (N.Y.); pres. Irving Berlin Music Corp., music pub.
PICTURES INCLUDE: Follow the Fleet, On the Avenue, Carefree, Alexander's Ragtime Band, Second Fiddle, Holiday Inn, Blue Skies, Easter Parade, Annie Get Your Gun, Call me Madam, White Christmas, There's No Business Like Show Business.

BERLINGER, WARREN: Actor, b. Brooklyn, N.Y., Aug. 31, 1937. e. Col. Univ.
STAGE: Annie Get Your Gun, The Happy Time, Bernardine, Take A Giant Step, Anniversary Waltz, Roomful of Roses, Blue Denim, Come Blow Your Horn, How To Succeed in Business Without Really Trying, (London) Who's Happy Now?, California Suite, 77–78 tour.
PICTURES INCLUDE: Teenage Rebel, Three Brave Men, Blue Denim, Because They're Young, Platinum High School, The Wackiest Ship in the Army, All Hands on Deck, Billie, Spinout, Thunder Alley, Lepke, The Four Deuces, I Will . . . I Will . . . For Now, Harry and Walter Go to New York, The Shaggy D.A., The Magician of Dublin, The Cannonball Run.
TV: Secret Storm (serial), Alcoa, Goodyear, Armstrong, Matinee Theatre, Joey Bishop Show, The London Palladium, Kilroy, Billie, Bracken's World, The Funny Side (NBC 1971–72); Touch of Grace, (ABC-TV 1972–73); The Most Wanted Woman.

BERMAN, CHARLES. U.K. & European dir. of operations, the Saul Zaentz Production Co. Previously dep. man. dir. and dir., pub./adv./exploit., United Artists, London. Joined Odeon 1946 as trainee mgr.; later as publicity man with Hammer Theatrical Enterprises. Transferred to Exclusive Films pub. dept., before joining UA 1953. From pub. asst. promoted press officer. Later asst. pub. dir.; appt. dir. of pub./exploit. 1963; appt. to board, 1969.

BERMAN, MONTY: producer/cinematographer. b. London, England.
PICTURES INCLUDE: Voice of Merill, Tiger by the Tail, Blood of the Vampire, The Trollenberg Terror, Sea of Sand, Jack The Ripper, Flesh and the Fiends, The Siege of Sidney Street, Hellfire Club, Treasure of Monte Cristo, What a Carve Up.
TV: The Saint, Gideon's Way, The Baron, The Champions, Department S. Randall & Hopkirk (Deceased), Jason King, The Adventurer, Secret of Storm Mountain.

BERMAN, MONTY, M.: Film and theatrical costumier. b. London, England, 1912. Personally costumed films and shows since 1931. Squad Leader RAF 1940–45; Bomber Command, M.B.E. Since war has costumed major films and shows and numerous TV prod.
PICTURES INCLUDE: Doctor Zhivago, Tom Jones, Longest Day, My Fair Lady, Oliver, Cromwell, Patton, The Devils, Fiddler on the Roof, A Bridge Too Far, The Other Side of Midnight, Julia, The Seven Per Cent Solution, The Slipper And The Rose, A Little Night Music, Star Wars, Superman.

BERMAN, PANDRO S.: Producer. b. Pittsburgh, Pa. Mar. 28, 1905. Son of late Harry M. Berman, gen. mgr. Universal, FBO. Asst. dir. film ed., FBO; film & title ed. Columbia Studios; chief film ed. RKO, later asst. to William Le Baron & David Selznick; became prod. 1931 (RKO). A Champion of Champions Producers in Fame ratings. Joined MGM 1940.
PICTURES INCLUDE: What Price Hollywood, Symphony of Six Million, Bachelor Mother, The Gay Divorcee, Of Human Bondage, Morning Glory, Roberta, Alice Adams, Top Hat, Winterset, Stage Door, Vivacious Lady, Gunga Din, Hunchback of Notre Dame, Ziegfeld Girl, Honky Tonk, Seventh Cross, National Velvet, Dragon Seed, Portrait of Dorian Grey, Love Affair, Undercurrent, Sea of Grass, Three Musketeers, Madame Bovary, Father of the Bride, Father's Little Dividend, Prisoner of Zenda, Ivanhoe, All the Brothers Were Valiant, Knights of the Round Table, Long, Long Trailer, Blackboard Jungle, Bhowani Junction, Something of Value, Tea and Sympathy, Brothers Karamazov, Reluctant Debutante, Butterfield 8, Sweet Bird of Youth, The Prize, A Patch of Blue, Justine, Move.

BERNARD, JUDD: Writer-producer. b. Chicago, Ill., stage, The World of Carl Sandberg.
PICTURES INCLUDE: Double Trouble, Point Blank, Blue, Fade In, Negatives, The Man Who Had Power Over Women, Deep End, Glad All Over, and Now For Something Completely Different, The Marsailles Contract, Inside Out, The Class of Miss MacMichael.

BERNARD, MARVIN A.: Executive. b. New York City, Oct. 1, 1934. e. N.Y.U. Associated with Rapid Film Technique, Inc.; 1949–63: developed technological advances in film preservation, responsible for public underwriting; 1964–69; real estate sales & investments in Bahamas, then with Tishman Realty (commercial leasing div.); v.p. and operating head of International Film Treat 1970–1973; authored film damaged control chart, a critical analysis of film care and repair—1971; 1973-elected to and for inclusion in 1973–74 edition of Who's Who, because of his expertise in the field of motion picture film rejuvenation; for more than two decades; founded Filmlife Inc. with latest mechanical and technical advancement in field of film rejuvenation; 1973–75 bd. chrmn. and chief operating officer of Filmlife Inc., motion picture film rejuvenation company. Feb. 1975 elected President in addition to remaining bd. chairman.

BERNHARD, HARVEY: Producer. b. Seattle, Wash., March 5, 1924. e. Stanford Univ. In real estate in Seattle, 1947–50; started live lounge entertainment at the Last Frontier Hotel, Las Vegas, 1950. Partner with Sandy Howard, 1958–60; v.p. in chg. prod., David L. Wolper Prods., dividing time between TV and feature films, 1961–68; with MPC, v.p., chg. prod., 1968–70. Now pres. of Harvey Bernhard Ent., Inc.
PICTURES INCLUDE: The Mack (1973), The Omen (1976), Damien—Omen II (1978), The Final Conflict (1981), The Beast Within (1981).

BERNS, SEYMOUR: Director. Started as dir. CBS radio, 1946; radio shows include Art Linkletter's House Party, Double or Nothing, Hollywood Barn Dance, Free for All; joined CBS-TV network 1952; v.p. National General TV Productions, 1969; National pres. TV Academy, 1969–70; Chairman of the Board 1970–71; pres. The Valjon Prod. Co. 1971.
TV: Art Linkletter's House Party, Meet Millie, Red Skelton Show, Shower of Stars, My Friend Irma, Ford Star Jubilee, Edsel Show, Jack Benny Show, Gunsmoke, Lineup.

BERNSTEIN, Baron cr 1969 (Life Peer) of Leigh; Sidney Lewis Bernstein LLd; Chairman; Granada Group Limited in 1974–79. Pres. since 1979. (Granada Television, Granada Publishing, Granada Theatres, Granada TV Rental, Granada Motorway Services, Novello & Co). b. 30 Jan. 1899. A founder, Film Society, 1924. Films Adviser, Min. of Inf. 1940–45; Liason, British Embassy, Washington 1942; Chief, Film Section SHAEF 1943–45. Lecturer on Film and International Affairs, New York Univ. and Yale. Governor, Sevenoaks School. Member, Resources for Learning Committee, Nuffield Foundation. Address: 36 Golden Square, London, W1.

BERNSTEIN, BOB: Executive. Began public relations career 1952 at DuMont TV Network, followed by 2 yrs. as press agent for Liberace. With Billboard Magazine as review editor 3 yrs. Joined Westinghouse Bdg. Co. as p.r. director 1959. In 1963 named p.r. director for Triangle Publications, serving in various capacities to 1971. Joined Viacom Intl. as director of information services. In 1975 formed own co., March Five Inc., p.r. and promotion firm.

BERNSTEIN, ELMER: Composer-conductor. b. New York, N.Y. Apr. 4, 1922. Scholarship, Juilliard. e. Walden School, New York U., U.S. Army Air Force radio unit. After war 3 yrs. recitals, musical shows, United Nations radio dept; pres., Young Musicians Found.; 1st v.p. Academy of Motion Picture Arts & Sciences; co-chairman music branch. Music dir. Valley Symphony. Recording: United Artists. More than 90 major films. Pres. of Composers & Lyricists Guild of America past 5 years.
PICTURES INCLUDE: The Ten Commandments, Man With the Golden Arm, The Great Escape, To Kill a Mockingbird (received ten Academy Award nominations and Oscar, 1967), The Carpetbaggers, The Magnificent Seven, Birdman of Alcatraz, Walk on the Wild Side, Hallelujah Trail, The Reward, Seven Women, Cast A Giant Shadow, Hawaii, Thoroughly Modern Millie, True Grit, Cahill, U.S. Marshall, McQ., Gold, The Trial of Billy Jack, Report to the Commissioner, From Noon Till Three, Sarah, The Shootist, Bloodbrothers, Meatballs, Airplane!, Going Ape.
TV: Hollywood: The Golden Years (special), The Race for Space; Parts I & II (special), D-Day (special), The Making of the President—1960 (special), won Emmy Award), Hollywood and the Stars (special), Voyage of the Brigantine Yankee (special), Crucification of Jesus (special), Julia series, Owen Marshall series, Ellery Queen series, Serpico series.

BERNSTEIN, RICHARD: b. Rochester, N.Y., Aug. 8, 1922; e. Los Angeles City Coll. Radio prod., writer, dir. created shows Food for Fun, Mr. Celebrity. Press agent personalities and indep. co. 1951. M.P. prod. exec. 1951–54. Producing debut Walk The Dark Street, Prod. exec. Milner Bros. 1955. The Phantom from 10,000 Leagues. Co-authored, wrote screen play and ass't prod. From Hell It Came. Formed Viscount Films with Richard B. Duckett and George W. Waters, 1957; prod., co-author Tank Battalion; prod., co-author Speed Crazy; wrote Jail Break; prod. co-author, Why Must I Die?, prod. and story. Atlantic Attack, Woman By Night; prod., s.p. Terrified, prod., The Gun Hawk, 1968, co-author s.p., Treasure of Kota Twai, Max Orloff Productions; 1969 s.p., Laughter in Hell, Palisades Productions; 1970; s.p., The Wild Wild Wild West and Fast and Furious; presenter, the Challenges, 2nd place winner San Sebastian, Transworld Film Enterprises release; screen play, Backlash, William Productions; prod.; wrote story, screen play for The Dirt Gang; Creator TV Series, "Bureau of Missing Persons." Co-producer of Women of the World, Major Films Inc. Writer, producer, director, Ripoff; Original screen story, Win, Place and Steal, feature. Writer, Producer, TV series, My Mother . . . The Maid, Colebern Productions; pres., Pacific Global Film Enterprises.

BERNSTEIN, WALTER: Director-Writer. b. New York City. e. Dartmouth. Wrote for New Yorker Magazine; in World War II was roving correspondent for Yank Magazine. Returned to New Yorker after war. Wrote TV scripts; published book Keep Your Head Down (collection of articles).
PICTURES INCLUDE: Writer: The Molly McGuires, The Money Trap, Fail Safe, Paris Blues, Heller in Pink Tights, Semi-Tough, The Front, An Almost Perfect Affair, Yanks, Little Miss Marker. Director: Little Miss Marker (debut).

BERRY, JOHN: Director. b. N.Y.C., 1917. Directed films in Hollywood mid and late '40s; went abroad during McCarthy era in U.S. where worked in French film industry. Later went to London to do stage work, acting as well as directing. Returned to U.S. to do stage work; returned to Hollywood to do TV.
PICTURES INCLUDE: Cross My Heart, From This Day Forward, Miss Susie Slagle's, Casbah, Tension, He Ran All the Way, Ca Va Barder, The Great Lover, Je Suis un Sentimental, Tamango, On Que Mambo, Maya, Thieves.
TV: One Drink at a Time, Farewell Party, Mr. Broadway, etc.

BERTOLUCCI, BERNARDO: Director-Writer, b. Parma, Italy, March 16, 1941. e. Rome University. Son of Attilio Bertolucci poet and film critic. First worked in films at age of 20 as asst. dir. to Pier Paolo Pasolini on latter's first film, Accattone: in 1962 made debut film, The Grim Reaper, from script by Pasolini.
PICTURES INCLUDE: Before the Revolution, Partner, Love and Rage (one episode: Agony). The Spider's Stratagem, 1900, Luna.

BEST, BARBARA: Publicist. b. San Diego, Calif., Dec. 2, 1921. e. U.S.C., A.B., 1943. Pub., 20th Century-Fox, 1943–49; reporter, San Diego Journal, 1950 Stanley Kramer Co. 1950–53; own agency, Barbara Best & Associates, 1953–66; 1966 exec. v.p. Jay Bernstein Public rel.; Freeman and Best, 1967–74; Barbara Best Inc. current.

BEST, JAMES: Actor. b. Corydon, Ind., July 26, 1926. Magazine model; on stage; in European roadshow cast of "My Sister Eileen"; served as M.P. with U.S.A.A.F., World War II; m.p. debut in Comanche Territory, 1949.
PICTURES INCLUDE: Winchester 73, Air Cadet, Cimarron Kid, Steel Town, Ma & Pa Kettle At the Fair, Francis Goes to West Point, Seminole, The President's Lady, City of Bad Men, Column South, Riders to the Stars, The Raid, Caine Mutiny, Return from the Sea, The Left-Handed Gun, They Rode West, Seven Angry Men, Come Next Spring, Baby, The Rack, Sounder, Ode to Billy Joe.

BETHUNE, ZINA: Acress, dancer, singer. B. New York, 1950. B'way: Most Happy Fella, Nutcracker. New York City Ballet (Balanchine), Zina Bethune & Company Dance Theater.
PICTURES: Sunrise At Campobello, Who's That Knocking At My Door. National Tours: Sweet Charity, Carnival, Oklahoma, Damn Yankees, Member of the Wedding, The Owl & The Pussycat.
TV: The Nurses, Lancer, Cains Hundred, Naked City, Route 66, Little Women, Judy Garland Show, Jackie Gleason Show, Gunsmoke, Dr. Kildare, Emergency, Planet of The Apes, Police Story, Chips, Hardy Boys.

BETTGER, LYLE: Actor. b. Philadelphia, Pa., Feb. 13, 1915; m. Mary Rolfe, actress; e. Haverford School, Phila., Amer. Acad. Dramatic Art, N.Y.C. Started in summer stock; in road cos. of Brother Rat, Man Who Came to Dinner.
STAGE: John Loves Mary, Love Life, Eve of St. Mark.
PICTURES INCLUDE: No Man of Her Own, Union Station, First Legion, Greatest Show on Earth, Sea Chase, Gunfight at OK Corral, Johnny Reno, Nevada Smith, Return of The Gunfighter, The Hawaiians, Carnival Story, The Seven Minutes.
TV: Court of Last Resort, Grand Jury, Hawaii 5-0, Police Story.

BETTINSON, RALPH: Writer. b. London, Eng., Apr. 18, 1908; e. Radley Coll. Film ed. London Films, Korda (Paris), Brit. Paramount; prod. Sound City Films; writer Gaumont & Paramount Brit.; with Republic Pictures, Hollywood 1935; spec. supvr. Monogram Brit. Prod. 1937–42; prod. of training films & documentaries U.S. Army 1943–6; writer Eagle-Lion 1947–8; contract writer Columbia Pictures 1949; (orig. story) Rogues of Sherwood Forest; adapt. Mask of the Avenger; collab. s.p., asst. prod., 1948.

BEVILLE, HUGH M., Jr.: Executive; b. April 18, 1908; e. Syracuse U.; to NBC 1930 statistician, Chief Statistician; Research mgr., dir., research; U.S. Army 1942–46; dir. of research and planning for NBC, v.p., planning and research, 1956; v.p., Planning, 1964; Consultant, 1968; Professor Business Admin., Southampton College, 1968. Exec. dir., Broadcast Rating Council, 1971.

BEY, TURHAN: Actor. b. Vienna, Austria, March 30. Made U.S. m.p. debut in 1941.
PICTURES INCLUDE: Footsteps in the Dark, Burma Convoy, Bombay Clipper, Drums of the Congo, Destination Unknown, White Savage, Dragon Seed, Arabian Nights, Bowery to Broadway, Night in Paradise, Out of the Blue, Amazing Mr. X, Parole, Inc., Adventures of Casanova, Song of India, Prisoners of the Casbah, Stolen Identity.

BEYMER, RICHARD: Actor, r.n. George Richard Beymer, Jr., b. Avoca, Iowa, Feb. 21, 1939, e. N. Hollywood H. S., Actors Studio. Performer, KTLA, Sandy Dreams, Fantastic Studios, Inc., 1949, Playhouse 90.
PICTURES INCLUDE: Indiscretion of an American Wife, So Big, Johnny Tremaine, The Diary of Anne Frank, High Time, West Side Story, Bachelor Flat, Five Finger Exercise, Hemingway's Adventures of a Young Man, The Longest Day, The Stripper.

BIBAS, FRANK PERCY: Executive. b. New York City, 1917; e. Brown & Columbia Universities; Capt., Army Air Corps, World War II. Entered m.p. ind. adv. dept., Donahue & Coe; later pub. dept. American Display, thence to dist. dept., National Screen Service. In 1945 joined Casanave-Artlee Pictures Inc.; apptd. vice-pres. in charge of sales 1946 (dist. Selznick reissues in U.S.); vice-pres. Casanave Pictures, Inc. & Sixteen M.M. Pictures Inc., from 1949 mgr. m.p. dept., McCann-Erickson ad agency; joined Roland Reed TV Inc., v.p., prod., 1955. Prod-dir. with Hal Roach Studios, 1957. 1961 to present N.Y.C. partner, Bibas-Redford Inc. Production Co. 1962 won Academy Award for "Project Hope," documentary. Member: S.M.P.T.E., Acad. of TV Arts and Sciences, Directors Guild of America.

BICK, JERRY: Producer. b. New York City. e. Columbia Univ., Sorbonne. m. actress Louise Fletcher. Taught English at Univ. of Ga. and was radio sports announcer in Savannah, Ga., before entering firm industry in pub. dept. of MGM, N.Y. Opened own literary agency in Hollywood representing film, TV writers. Began career as producer in London; debut film, Michael Kohlhaas (1969), made in Czechoslovakia.
PICTURES INCLUDE: The Long Goodbye, Thieves Like Us, Russian Roulette.

BIEN, WALTER H.: Producer. b. San Francisco, April 16, 1923; e. UCLA, 1944. Entered m.p. ind. as film editor, MGM, 1943, Monogram, 1946; Eagle Lion, 1948; formed TV production company 1950, prod. live shows in L.A. area; Universal TV comm. dept., 1954; Warner Bros., to start comm. & ind. div., 1956; exec. prod. in chg. TV comm. & ind. films div., prod. Bell System Science Series; pres., SIB Productions, Inc., SIB-Tower 12.

BILBY, KENNETH W.: Executive. b. Salt Lake City, Utah, Oct. 7, 1918. e. Columbia U., U. of Ariz., B.A. With N.Y. Herald-Tribune, 47–50; author, New Star in the Near East, 1950; pub. rel. rep. to RCA Victor, Camden, New Jersey, 1950–54; exec. v.p. National Broadcasting Co., N.Y., 1954–60; v.p. public affairs, RCA, 1960–62, exec. v.p., 1962–75; exec. v.p., corporate affairs, 1976–present.

BILIMORIA, N. M.: Distributor. b. Oct. 7, 1922. Partner in M. B. Bilimoria & Son, Bombay; dir. Dominion Films (P) Ltd. Bombay; Modern Films Private Ltd., Bombay; Partner, Bilimoria & Lalji, Calcutta; Distributors Combine, Bombay; Bilimoria & Chhotubhai, Delhi; Moti Talkies, Delhi; Agent: Union Carbide India Ltd., Calcutta and Union Carbide International Co., N.Y.

BILL, TONY: Actor-Producer. b. San Diego, Ca. e. Notre Dame. Began career in films in 1963 acting in Come Blow Your Horn. In 1971 founded Bill/Phillips Prods. with Julia and Michael Phillips. Left in 1973 to be independent.
PICTURES INCLUDE: Actor: You're a Big Boy Now, Castle Keep, Shampoo, Heartbeat. Co-producer: Deadhead Miles, Steelyard Blues, The Sting. Prod.: Hearts of the West, Harry and Walter Go to New York, Boulevard Nights

(exec. prod.); Going in Style (co-prod.). Director: My Bodyguard.

BILLITTERI, SALVATORE: Executive. b. Belleville, N.J., March 21, 1921. e. Fordham U.: U.S. Air Force, 1943–45; ent. m.p. ind., doorman in N.Y. theatre, 1939; theatre mgr., 1947; film slsman, Casolaro Film Distributing Co., 1947–51; co-owner many foreign films; film editor, I.F.E., 1951; head of editing, prod. dept., Titra Sound Corp., 1956; head of east coast production, American International Pictures, 1961; dir. Post Prod., 1970; named v.p. 1974, sr. v.p. of Filmways Pictures, 1980.

BILSON, GEORGE: Prod., dir., writer. b. Leeds, Eng. e. C.C.N.Y. dir. advertising, W.B., 1927–32; West Coast adv. mgr. Warner Bros. Chg. of prod. of trailers 1932–37; prod., Universal 1938; exec. producer-writer short subjects RKO Studio, 1943–53. Freelance TV writer.

BINNS, EDWARD: Actor. b. Penn. Broadway debut in Command Decision, 1947. Has since appeared in TV, theatre, m.p. and commercial voice-overs.
PICTURES INCLUDE: Fail-Safe, Twelve Angry Men, Compulsion, Patton, The Americanization of Emily, Judgment at Nuremburg, Lovin' Molly, Night Moves, Oliver's Story.
TV: Kraft Theatre, Studio One, Danger, Brenner (series), The Defenders, The Nurses, It Takes a Thief (series).
STAGE: Detective Story Caligula, The Caine Mutiny Court Martial, A Touch of the Poet.

BIONDI, GUY: Executive. b. Newark, N.J. e. Oklahoma Univ. Entered m.p. career on roadshow of Henry V, UA; Universal pub. staff, 1952; Mike Todd Company, 1956; Harold Mirisch Co., o., 1960; v.p. Walter Reade-Sterling, 1964; AFT, 1972; v.p. adv. & pub. Gamma III.

BIRCH, WILLIAM: Cinematographer. b. Chicago, Ill., Mar. 1, 1918. Asst., Movietonews, 1937–40; army signal corp, 1940–44; head of midwestern bureau, did camera, NBC News & Document.; established own service; estab. William Birch & Assoc. Inc.
Won two Emmys, cameraman of the yr., Nat. Headliners Award.

BIRMINGHAM, PAUL A.: Executive. b. Burbank, Calif., Feb. 12, 1937. e. University of Calif., Univ. of South. Calif.; sr., v.p., studio operations and admin., Paramount Pictures.

BIRNEY, DAVID: Actor. b. Washington, D.C., April 23. e. Dartmouth, UCLA. m. actress Meredith Baxter Birney. Following college and the Army went to N.Y. where appeared in Lincoln Center prod. of Summertree. Appeared for two yrs. on TV daytime series, Love is a Many Splendored Thing, doing other stage roles in same period.
PICTURES INCLUDE: Trial by Combat, Oh, God, Book II.
TV: Bridget Loves Bernie (series), Police Story (segments), The Adams Chronicles, Serpico (series).

BIROC, JOSEPH F.: Cinematographer. b. New York, N.Y., Feb. 12, 1903. e. Emerson H.S. Union City, N.J.
PICTURES INCLUDE: Bwana Devil, Glass Wall, Tall Texan, Vice Squad, Donovan's Brain, Appointment in Honduras, Down Three Dark Streets, Lone Wolf, T-Men in Action, Man Behind the Badge, Dear Phoebe, Nightmare, Tension at Table Rock, Attack, Black Whip, Run of the Arrow, Ride Back, Garment Jungle, China Gate, Ice Palace, FBI Story, 13 Ghosts, Home Before Dark, Operation Eichmann, Devil at 4 O'clock, Gold of the Seven Saints, Reprieve, Opium Eaters, Hitler, Sail a Crooked Ship, Bye Bye Birdie, Toys in the Attic, Promises-Promises, Under the Yum-Yum Tree, Viva Las Vegas, Kitten With a Whip, Ride the Wild Surf, Renegade Posse, Gunfight at Commanche Pass, Hush . . . Hush Sweet Charlotte, I Saw What You Did, Flight of the Phoenix The Russians are Coming, The Russians are Coming, The Swinger, Warning Shot, Enter Laughing, Who is Minding the Mint, Garden of Cucumbers, The Killing of Sister George, What Ever Happened to Aunt Alice?, Too Late the Hero, Mrs. Polifax Spy, Escape from the Planet of the Apes, The Organization, The Grissom Gang, Cahill, U.S. Marshall, The Longest Yard, The Towering Inferno, Hustle, The Duchess And The Dirtwater Fox, The Choirboys, Beyond the Poseidon Adventure, Airplane!, Hammet, All the Marbles.
TV: Four Star Theatre, Readers Digest, Superman, Richard Diamond, Alcoa Theatre, Grindl, Solo, Ghost Breakers, Take Her She's Mine, Heaven Help Us, Hardy Boys, Brian's Song, Gidget Gets Married, Ghost Story, Thursdays Game, Lonely Hearts "555," Family Upside Down, S.S.T. Death Flight, Little Women, Scruples, The Gambler.
TV PILOTS: Wonder Woman, Honky Tonk, The Moneychangers, Washington, D.C., Clone Master

BISHOP, JULIE: Actress. b. Denver, Colo., Aug. 30, 1917. e. Westlake, Kenwood, Schools for Girls (Former screen name Jacqueline Wells); m.p. debut 1941.
PICTURES INCLUDE: Nurse's Secret, International Squadron, Steel Against the Sky, Lady Gangster, Busses Roar, Cinderella Jones, Murder in the Music Hall, Idea Girl, Strange Conquest, Last of the Redmen, High Tide, Deputy Marshal, Threat, Sands of Iwo Jima, Westward the Women, Sabre Jet, High and the Mighty, Headline Hunters.

BISSET, JACQUELINE: Actress. b. Weybridge, England, September 13, 1944. e. French Lycee, London. After photographic modeling made film debut in The Knack.
PICTURES INCLUDE: Two For The Road, The Sweet Ride, The Detective, Bullitt, The First Time, Airport, The Grasshopper, The Mephisto Waltz, Believe in Me, Stand Up and Be Counted, The Thief Who Came to Dinner, Murder on the Orient Express, End of the Game, The Spiral Staircase, St. Ives, Sunday Woman, The Deep, The Greek Tycoon, Who Is Killing the Great Chefs of Europe?, When Time Ran Out, Rich and Famous.

BIXBY, BILL: Actor. b. San Francisco, Calif., Jan. 22. e. Univ. of Calif., Berkeley. Worked in indust. films. Gen. Motors, Chrysler.
TV: Dobie Gillis, Danny Thomas Show, Joey Bishop Show, Andy Griffith Show, My Favorite Martian, Eddie's Father, The Hulk.
STAGE: Fantasticks (nat'l company), Under the Yum Yum Tree.
PICTURES INCLUDE: Lonely Are the Brave, Irma La Douce, Yum Yum Tree, Ride Beyond Vengeance, This Way Out Please.

BLACK, ALEXANDER F.: Publicist. b. New Rochelle, N.Y., Dec. 27, 1918. B.A. Brown U., 1940. Joined Universal 1941. U.S. Navy 1942–45, Lt. Sr. Grade Rejoined Universal 1946 serving in various capacities in Foreign Department, becoming Director of Foreign Publicity for Universal International Films, Inc. in 1967; 1974, named exec. in chg. intl. promotion for MCA-TV.

BLACK, KAREN: Actress. b. Park Ridge, Ill., 1943. e. Northwestern Univ. Left school for New York to join the Hecscher House, where she appeared in several Shakespearean plays. In 1965 starred in Playroom, which ran only 1 mth. but won her a nom. as best actress of season by N.Y. drama critics. Made m.p. debut in 1966 in You're a Big Boy Now.
PICTURES INCLUDE: Easy Rider, Five Easy Pieces, Drive, He Said, A Gunfight, Born To Win, Portnoy's Complaint, The Pyx, Rhinoceros, The Outfit, The Great Gatsby, Airport 1975, Law and Disorder, Day of the Locust, Nashville, Family Plot, Crime and Passion, Burnt Offerings, Capricorn One, Because He's My Friend, In Praise of Older Women, The Rip-off, Greed, Danny Travis.
STAGE: Happily Never After, Keep It in the Family.

BLACK, STANLEY: Composer, conductor, musical director. Resident Conductor, BBC, 1944–52. Musical Director 105 feature films and Pathe Newsreel music: Music dir. Associated British Film Studios 1958–64. Guest Conductor Royal Philharmonic Orchestra and London Symphony Orchestra; many overseas conducting engagements including (1977) Boston Pops and Winnipeg Symphony. Associated Conductor Osaka Philharmonic Orchestra. Exclusive recording contract with Decca Record Co. since 1944.
PICTURES INCLUDE: Crossplot, The Long, The Short and The Tall, Rattle of a Simple Man, The Young Ones, Hell is a City, Top Secret, etc., 1977 Valentino.

BLACK, THEODORE R. Attorney, b. New Jersey, Aug. 11, 1906; e. Harvard, B.A. 1927, LL.B., 1930 (Sigma Alpha Mu Frat.). Formerly General Counsel, bd. member, Republic Pictures Corp. Member: Nat'l Panel of Arbitrators, American Arbit. Assn., Bd. N.Y. Ethical Culture Society.

BLACKMAN, HONOR: Actress. b. London, Eng. Stage debut. The Gleam 1946. Screen debut, Fame Is the Spur.
TV: African Patrol, The Witness, Four Just Men, Probation Officer series, Top Secret, Ghost Squad, Invisible Man, The Saint, The Avengers series.
PICTURES INCLUDE: Quartet, Daughter of Darkness, A Boy A Girl and a Bike, Diamond City, Conspirator, So Long at the Fair, Set a Murderer, Green Grow the Rushes, Come Die My Love, Rainbow Jacket, Outsiders, Delavine Affair, Three Musketeers, Breakaway, Homecoming, Suspended Alibi, Dangerous Drugs, A Night to Remember, The Square Peg, A Matter of Who, Present Laughter, The Recount, Serena, Jason & The Golden Fleece, Goldfinger, The Secret of My Success, Moment to Moment, Life at the Top, A Twist of Sand, Shalako, Struggle for Rome, Twinky, The Last Grenade, The Virgin and the Gypsy, Fright, Something Big, Out Damned Spot, Ragtime, Summer, Cat and the Canary.

BLAINE, VIVIAN: Actress. r.n. Vivian S. Stapleton. b. Newark, N.J., Nov. 21, 1924; e. South Side H.S., Newark. Singer with various bands in New Jersey 1937–39, thereafter night clubs; 20th-Fox contract 1942. Personal appearance on Great Britain stage from 1947; co-starred Guys and Dolls N.Y. stage and London stage; Hatful of Rain, N.Y. stage; Guys and Dolls film 1955.

Member: Academy of M.P. Arts & Sciences, A.F.R.A., Equity, S.A.G., A.G.V.A.
PICTURES INCLUDE: He Married His Boss, Thru Different Eyes, Greenwich Village, Something for the Boys, Nob Hill, State Fair, Three Little Girls in Blue, If I'm Lucky, Skirts Ahoy, Public Pigeon No. 1.

BLAIR, BETSY: Actress. r.n. Betsy Boger. b. New York, N.Y., Dec. 11.
BROADWAY: Beautiful People, Richard II, actress in little theatre groups.
PICTURES INCLUDE: Another Part of the Forest, Kind Lady, Othello, Marty, Halliday Brand, A Delicate Balance.
TV: Appearances on U.S. Steel Hour, Ford Theatre, Philco, Kraft.

BLAIR, JANET: Actress. b. Blair, Pa., April 23, 1921. With Hal Kemp's Orchestra; on tour South Pacific play, 1950–52.
PICTURES INCLUDE: Three Girls About Town, Blondie Goes to College, Two Yanks in Trinidad, Burn Witch Burn, Broadway, My Sister Eileen, Something to Shout About, Once Upon a Time, Tars and Spars, Gallant Journey, Fabulous Dorseys, I Love Trouble, Black Arrow, Fuller Brush Man, Public Pigeon No. 1, Boys Night Out, The One and Only, Genuine, Original Family Band.

BLAIR, LARRY: Actor. r.n. Larry Bess. b. Newark, N.J., June 27, 1935; f. Herman Bess, NYC radio executive. e. public and private schools, Newark and Asbury Park, N.J. Worked as lifeguard. Began in showbusiness, age 13 as D.J. and actor in NYC; 1954–55, production assistant, WABC-TV, NYC; 1955–62, associate director and director, WNEW-TV, NYC; 1962–71, D.J., newscaster with WFIL, Phila., WIIC (Hartford), WHN, WMCA, NYC. 1971 to present, commercial spokesman, narrator and actor in educational and industrial films and videotapes.

BLAIR, LINDA: Actress. b. St. Louis, Jan. 22, 1959. Model and actress on TV commercials before going into films.
PICTURES: The Sporting Club, The Way We Live Now, The Exorcist, Airport '75, Sweet Hostage, Exorcist II: The Heretic.
TELEVISION: Born Innocent, Sarah T.—Portrait of a Teenage Alcoholic, Victory at Entebbe.

BLAKE, AMANDA: Actress. r.n. Beverly Louise Neill; b. Buffalo, N.Y., Feb. 20. Telephone operator; in m.p. 1950.
PICTURES INCLUDE: Battleground, Stars in My Crown, Counterspy Meets Scotland Yard, Scarlet Angel, Cattle Town, Lili, Sabre Jet, Miss Robin Crusoe, About Mrs. Leslie, Adventures of Hajji Baba, Star Is Born, Glass Slipper, High Society.
TV: Double Exposure; Schlitz Playhouse, General Electric Theater, Four Star Playhouse, Cavalcade of America, Lux Video Playhouse, My Favorite Husband, Professional Father, The Red Skelton Show, Climax, Gunsmoke.

BLAKE, DAVID M.: Producers' Representative. b. Trincomalee, Ceylon, April 19, 1948. Ent. m.p. ind. 1968 British Lion Films, London. Lion Int'l. Films, O'Seas Division. Appointed 1970 U.S. representative. New York. British Lion Films, Shepperton Studios, Lion TV.

BLAKE, ROBERT: Actor. b. Nutley, N.J., Sept. 18, 1938. m. actress Sondra Kerry. Started as Hollywood stunt man in Rumble on the Docks and The Tijuana Story. First acting job was in the Gallery Theater in Hatful of Rain.
PICTURES INCLUDE: Town Without Pity, Pt-109, The Greatest Story Ever Told, The Connection, This Property is Condemned, In Cold Blood, Tell Them Willie Boy is Here, Corky, Electra-Glide in Blue, Busting, Second Hand Hearts.
TV: Baretta.

BLAKELY, SUSAN: Actress. b. Germany, where father was stationed in Army. Studied at Univ. of Texas. m. writer Todd Merer. Became top magazine and TV commercial model in N.Y. Film acting debut in Savages.
PICTURES INCLUDE: The Lords of Flatbush, The Way We Were, The Towering Inferno, Report to the Commissioner, Shampoo, Capone, The Concorde—Airport '79.
TELEVISION: Rich Man, Poor Man, Secrets (film).

BLALOCK, RICHARD: Orchestra leader-composer. b. Delano, Calif.: e. Public Schools—Simi and Ventura, Calif., Composed score, Hobart Donavan feature, History in Blue, 1960. Wrote music for singers Nick Lucas, Geo. Wallace. Appearances with band: Highlands Inn, Carmel, Calif., Santa Barbara Biltmore Hotel, Los Posas Country Club, Saticoy Country Club, La Cumbre Country Club, Glendale Civic Auditorium, Mira Mar Hotel; Latest comp., The Flea.

BLANC, MEL: Executive, voice specialist. r.n. Melvin Jerome Blanc. b. San Francisco, Calif., May 30, 1908. e. Lincoln H.S., Portland, Ore., 1918–21. musician, NBC Radio, San Francisco, 1928; orchestra conductor, Orpheum theatre, Portland, Ore., 1931; free lance, violin, tuba, bass player, 1925–30; voices on Warner Bros. Cartoons, Bugs Bunny, Porky Pig, Daffy Duck, etc. since 1937; running characters, Jack Benny Show, Radio & TV since 1940.
In radio: Point Sublime, 1942–44; Judy Canova Show, 1940–48; Burns and Allen; Jack Carson Show; Abbott and Costello; Dagwood and Blondie; Al Pierce Show, 1938–50; Chairman of the Board, Blanc Communications Corp., 1960–63.
TV: The Bugs Bunny Show, The Porky Pig Show, The Munsters, Flintstones, Speed Buggy Animated Show (Hanna Barbera).

BLANCO, RICHARD M.: Executive, b. Brooklyn, N.Y., e. electrical engineering, Wentworth Institute. J.C., 1925–27; bus. admin., Univ. of Calif., 1939–40; U.S. Govt. College, 1942. Superv. Technicolor Corp., 1931–56; organ. and oper. Consumer Products, Kodachrome film process., Technicolor, 1956–62; dir. of MP Govt. and theatr. sales, N.Y. and Washington, D.C. 1963–65; gen. mgr. of Technicolor Florida photo. operations at Kennedy Space Center., prod. document and educ. films for NASA, 1965; VP of TV division, Technicolor Corp. of America; 1967 elected corporate v.p. Technicolor, Inc.; 1971 pres., Technicolor Graphic Services, Inc.; 1974, elected chm. of bd. of Technicolor Graphic Services; 1977, elected to board of directors of Technicolor Inc.

BLANE, RALPH: Composer. b. Broken Arrow, Okla., July 26, 1914. e. Northwestern U. Started as singer, then vocal arranger for Broadway shows; appeared on NBC radio. Formed partnership with Hugh Martin, wrote Best Foot Forward; m.p. composer since 1939.
PICTURES INCLUDE: Best Foot Forward, Meet Me in St. Louis, My Dream Is Yours, One Sunday Afternoon, My Blue Heaven, Friendly Islands, Skirts Ahoy, French Line, Athena, Girl Rush, The Girl Most Likely, Who is Sylvia?, Ziegfeld Follies, Broadway Rhythm, Abbott and Costello in Hollywood, Easy to Wed.
TV: The Great Quillow. 1961; same in color for NBC, 1963.
BROADWAY: Three Wishes for Janie, Tattered Tom, Something About Anne, Don't Flash Tonight.

BLANK, EDWARD L.: Critic. b. Pittsburgh, Pa. e. Duquesne University, Pittsburgh, Pa. Started working at Pittsburgh Press, 1967. Radio-TV critic, 1969–72. Became Drama Editor, 1972.

BLANK, MYRON: Circuit executive. b. Des Moines, Ia. Aug. 30, 1911; son of A. H. Blank, circuit operator; e. U. Mich. On leaving coll. joined father in operation Tri-States and Central States circuits. On leave 1943–46 in U.S. Navy, officer in charge visual educ. Now pres. Central States Theatre Corp.; pres. TOA, 1955; chmn. bd. TOA Inc. 1956–57; exec. chm. of NATO. Pres. of Greater Des Moines Comm.; treas., Iowa Methodist Medical Center; board, Iowa Des Moines Natl. Bank.; pres., Iowa Phoenix Corp., recipient of Brotherhood Award of National Conference of Christians & Jews; board, Simpson College.

BLANKFORT, MICHAEL: Writer. b. New York City, Dec. 10, 1907. e. U. Pa., B.A.; Princeton U., M.A. On faculty Bowdoin Coll., 1929–30; Princeton U., 1931–33; New Jersey State Prison psychiatrist, 1933. Writer, prod., dir., stage plays, 1933–36; novelist magazine contrib., U.S.M.C., 1942–45. Member exec. bd. Screen Writers Guild, 1939–40; 1962–66, secy.-treas.; 1963–65, president, Writers Guild 1967–69; Nat. Chm. 1971–1973; Gov. Acad. of Motion Picture Arts & Sciences, 1969; v.p. 1971–73, 1976–77, 1978–80.
PICTURES INCLUDE: Blind Alley, Texas, Adam Had Four Sons, An Act of Murder, Broken Arrow, Halls of Montezuma, Untamed, The Vintage, The Plainsman, See How They Run, The Other Man, Lydia Bailey, The Juggler, My Six Convicts, Tribute to a Bad Man, The Caine Mutiny, A Fire in the Sky.

BLATTNER, GERRY: Producer. b. Liverpool, 1913. Ent. industry 1928. Blattner Studios, Boreham Wood. Studies include Berlin Research Labs, 1930. Studio mgr., British National Studios, 1936. Gen. mgr., British Lion Studios, Isleworth and Shepperton studios. Joined board of Warner Bros. First National Prods., 1946–67. Was exec. in charge of Warner Bros. European Prods. Started own indep. prod. co. in Nov. 1967.
PICTURES INCLUDE: The Crimson Pirate, Helen of Troy, The Nun's Story, Indiscreet, The Sundowners.

BLATTY, WILLIAM PETER: Writer-Producer. Novelist who wrote John Goldfarb, Please Come Home (filmed) and Twinkle, Twinkle, Killer Kane before biggest success with The Exorcist, which became a 55-week best-selling book. He wrote s.p. and functioned as prod. on film version.
PICTURES INCLUDE: A Shot in the Dark (s.p.), What Did You Do in the War, Daddy? (s.p.), The Great Bank Robbery, Darling Lilli (co. s.p.), The Exorcist (s.p., prod.), The Ninth Configuration (s.p., prod. dir.).

BLAU, MARTIN: Executive. b. N.Y.C., June 6, 1924. e. Ohio U., 1948. Employed on newspapers in Ohio, Tex., W.Va., Pub. dept., Columbia Pictures, 1951; asst. pub. mgr. 1959; pub. mgr., Columbia Internat'l, 1961; admin. asst. to v.p. of adv. & pub. Columbia Pictures, 1966. Dir. adv. and publicity, Columbia Pictures International, 1970; v.p., 1971.

BLAUSTEIN, JULIAN: Producer. b. New York City, May 30, 1913. e. Harvard 1933. Ent. m.p. ind. as reader for Universal 1935; asst. story ed. 1935–36; story ed. 1936–38; in chg. story dept. Music Corp. of America 1938–39; story ed. Paramount 1939–41; Signal Corps Photo. Center 1941–46; edit. supervisor of Selznick 1946–48; to 20th-Fox as prod. 1949; apptd. exec. prod. 20th Cent.-Fox, 1951 to Dec. 1952.
PICTURES INCLUDE: Broken Arrow, Mister 880, Half Angel, Just One More Chance, Take Care of My Little Girl, Day the Earth Stood Still, Outcasts, Don't Bother to Knock, Desiree, The Racers, Storm Center, Cowboy, Bell Book and Candle, The Wreck of The Mary Deare, Two Loves, The Four Horsemen of the Apocalypse, Khartoum.

BLEES, ROBERT: Writer-Producer. Phi Beta Kappa, Dartmouth. Time and Life Magazines. Exec. boards of Writers Guild, Producers Guild.
PICTURES INCLUDE: Magnificent Obsession, Autumn Leaves, Dr. Phibes Rises Again, Glass Web.
TV: Producer of Combat!, Bonanza, Bus Stop, Kraft Theater. Writer also of Alfred Hitchcock, Cannon, Barnaby Jones, Harry O, Columbo, Flamingo Road, Magnum.

BLEIER, EDWARD: Executive. b. N.Y., N.Y., October 16, 1929. e. Syracuse University, 1951, C.C.N.Y. Graduate Courses. Reporter: Syracuse Herald Journal, Long Island Daily Press 1947–50. Sportscaster: WNEW, N.Y.; WSYR, Syracuse, 1947–50. American Broadcasting Company, 1952–57; 1959–68. Vice President in charge of Public Relations & Planning (Marketing, Advertising, Publicity), Broadcast Div. Vice President in chg. of daytime Programming & Sales, ABC-TV Network; vice pres./gen. mgr. sales, ABC-TV Network; Account Executive, ABC-TV Network & WABC-TV. Vice Pres. Radio-Television-Film, Tex McCrary, Inc. 1959; Program Service Manager, DuMont Televison Network, 1951; Trustee, New York Chapter, Natl. Academy of TV Arts & Sciences; Director, Int'l Academy TV Arts & Sciences; Steering Committee, Aspen Institute Broadcaster's Conference.
Board Member: IRTS; Overseas Press Club; exec. v.p., Warner Bros Television, New York (a Warner Communications company).

BLENDER, LEON PHILIP: Executive. b. Kansas City, Mo., Feb. 15, 1920; e. Kansas State Co., 1941. 20th Century-Fox Dist. Co., 1947–51; Kranz-Levin pictures, 1951–56. gen. sls. mgr. American International Film Dist. Corp., 1956–57; v.p. 1960; American Intl. Pictures, senior v.p., sales dist., later exec. v.p. sales and distribution; 1979, exec. v.p. sales & distribution, Film Ventures Intl

BLOCH, ROBERT: Writer. b. Chicago, Ill., April 5, 1917; novelist short-story writer; 47 published books, incl. Psycho, Yours Truly, Jack the Ripper, etc ; wrote radio series, Stay Tuned for Terror, adapting own stories; national pres., Mystery Writers of Amer. (1970–71); entered films, 1960.
PICTURES INCLUDE: The Couch, Cabinet of Caligari, Straitjacket, The Night-Walker, The Psychopath, The Deadly Bees, (collab.) Torture Garden, The House That Dripped Blood, Asylum 1. (Films adapted from published work). Psycho, The Skull.
TV: Features: The Cat Creature, The Dead Don't Die.

BLOCK, WILLARD: Executive. b. New York, N.Y., March 18, 1930. e. Columbia College, Columbia University Law School, 1952. Counter-Intelligence Corps., U.S. Army, 1952–54, account exec., Plus Marketing, Inc. 1954–55; joined sales staff, NBC Television Network, 1955–57; sales staff, CBS Enterprises, Inc., 1957; international sales mgr, 1960; dir., international sales, 1965; v.p., 1967; v.p., Viacom Enterprises, 1971; pres., 1972; v.p. MCA-TV, 1973; v.p., gen. mgr., Taft, H-B International, Inc.; pres., Willard Block, Ltd.; 1979, named pres., Viacom Enterprises.

BLOOM, CLAIRE: Actress. b. London, Eng., Feb. 15, 1931; e. Badminton School (U.S.A.). Stage debut at Playhouse, Oxford, Sept., 1947; Old Vic seasons, etc.; screen debut in Limelight, 1951.
PICTURES INCLUDE: Innocents in Paris, Limelight, The Man Between, Richard III, Alexander the Great, The Brothers Karamazov, Look Back in Anger, The Royal Game, The Wonderful World of the Brothers Grimm, The Chapman Report, The Haunting, Alta Infidelita, Il Maestro di Vigeuono, The Outrage, The Spy Who Came In From the Cold, Charley, The Illustrated Man, Three Into Two Won't Go, A Severed Head, Red Sky at Morning, A Doll's House, Islands In The Stream, Clash of the Titans.
TV: Misalliance, (Playhouse 90), Anna Karenina, Wuthering Heights, Ivanov, Wessex Tales, An Imaginative Woman, A Legacy, In Praise of Love, The Orestaia, Henry VIII, Brideshead Revisited.

BLOOM, VERNA: Actress. b. Lynn, Mass. e. Boston Univ. Studied drama at Uta Hagen-Herbert Berghof School. Performed with small theatre groups all over country; then started repertory theatre in Denver. Appeared on Broadway in Marat/Sade (played Charlotte Corday). Film debut, Medium Cool, 1969.
PICTURES INCLUDE: The Hired Hand, High Plains Drifter, Badge 373, National Lampoon's Animal House.

BLOOM, WILLIAM: Producer. b. New York, N.Y., Feb. 28, 1915. e. U. of Pa., BA, 1935. Story dept., Columbia, N.Y., 1936; U.S. Army Signal Corps, 1941–45; asst. to Ender Bohem, Columbia, Hollywood; then prod. and asst. to B. B. Kahane, 1947–51; asst. to Julian Blaustein, Fox, 1951; prod., 1952.
PICTURES INCLUDE: Glory Brigade, Inferno, On the Threshold of Space.

BLOWITZ, JOHN S.: President, Blowitz & Canton Co. Inc. Held publicist position with United Artists, Cinema Center Films, Universal Pictures. Handled publicity-promotion for Stanley Kramer Productions on Oklahoma Crude. Coordinated promotional activities on Mame for Warner Bros. In 1974 joined Columbia Pictures as worldwide director of promotion and publicity. Formed John Blowitz Public Relations in 1975. Merged with Arthur Canton to form Blowitz & Canton Co. Inc. in April, 1976. Member of Academy of Motion Picture Arts and Sciences and Film Information Council, exec. dir., Variety Clubs Intl.

BLUHDORN, CHARLES G.: Executive. e. CCNY in 1945. After U.S. Air Force resumed studies at CCNY and Columbia while employed in commodity and export-import business. Became pres. coffee co. in 1949; 1956 acquired interest in Michigan Bumper Co. which firm became nucleus of Gulf & Western Industries in 1958. Chairman & chief exec. officer, Gulf & Western Industries, Inc., chair. Famous Players, Ltd., char. exec. comm, Madison Square Garden Corp.

BLUM, HARRY N.: Executive. b. Cleveland, Oh., Oct. 3, 1932. e. Univ. of Mich. (BBA, LLB). Was attorney in Ohio, toy & hobby industry executive, management consultant, and venture capital and investment manager before entering industry. Now heads The Blum Group.
PICTURES: Executive Action (assoc. prod.); The Land That Time Forgot (assoc. prod.); At the Earth's Core (exec. prod.); Drive-In, (assoc. proc.) Diamonds (exec. prod.); The Bluebird (assoc. prod.); Obsession (co-prod.); Skateboard (prod.); The Magician of Lublin (exec. prod.).

BLUMENFELD, JOSEPH: Exhibitor. b. New York City, Jan. 14, 1901; owns and operates Blumenfeld Theatres, founded by father in 1917; hdqts., San Francisco, Calif., active real estate, shopping centers, bowling alleys, etc.

BLUMENSTOCK, SID: Ad. exec. New York City. Publicity dir. Warner Bros. Atlantic City theatres 1934–38; joined 20th Fox adv. dept. 1938; asst. exploitation mgr. 1944–49; advertising mgr. Paramount 1949; asst. nat'l. dir. adv. pub., exploit, 1951–58; MPH Coord 1959 Academy telecast; v.p. Embassy Pictures, 1959; v.p. Chas. Schlaifer Agency, Hollywood, October, 1959–62; v.p. Safranski Prod., Inc.; adv. dir., Embassy Pictures, 1964–67; Walter Reade Org., 1967–70; Levitt-Pickman Film Corp., adv. dir., 1971–72.

BLUMOFE, ROBERT F.: Producer. b. New York, N.Y. e. Columbia Coll., A.B., Columbia U. Sch. of Law. J.D. v.p., West Coast oper., U.A., 1953–66; independent prod., pres. RFB Enterprises, Inc; American Film Institute, director, AFI-West, Sept. 1, 1977 to date.

BLYTH, ANN: Actress. b. Mt. Kisco, N.Y., Aug. 16, 1928; e. New Wayburn's Dramatic School. On radio in childhood; with San Carlos Opera Co. 3 years; Broadway debut in "Watch on the Rhine"; on tour.
PICTURES INCLUDE: Chip Off the Old Block, Merry Monahans, Brute Force, Swell Guy, Mr. Peabody and the Mermaid, Woman s Vengeance, Mildred Pierce, Free for All, Top o' the Morning, Our Very Own, Great Caruso, Katie Did It, Thunder on the Hill, I'll Never Forget You, Golden Horde, One Minute to Zero, World in His Arms, Sally and Saint Anne, All the Brothers Were Valiant, Rose Marie, The Student Prince, King's Thief, Kismet, Buster Keaton Story, Jazz Age, Slander, The Helen Morgan Story.

BOASBERG, CHARLES: Sales executive. b. Buffalo, N.Y., Aug. 20, 1906; e. Cornell U., A.B. 1927. MGM special repr. 1927; RKO salesman, Buffalo 1930; br. mgr. Buffalo 1937; dist. mgr., Cleveland 1940; Metropolitan dist. mgr., N.Y.C. 1944; apptd. north south div. mgr. 1947. Named gen. sales mgr., dir., RKO Pict., Oct., 1952; v.p., gen. mgr. Dist. Corp. of America, 1954–55; joined Paramount, 1955, in chg. world wide sales of the Ten Commandments, gen. sls. mgr., pres. Warner Bros. Dist. Corp., Jan. 1958. Res., July, 1962; domestic gen. sls. mgr. & v.p., Paramount Film Dist. Corp., 1962; pres., Paramount Film Dist. Corp., 1963, and V.P. Paramount Pictures, Inc. 1967; v.p. National General Corporation, 1968; pres., Nat'l Gen'l Pictures Corp., 1969. In 1974 formed Boasberg/Goldstein, consultants to film producers, now Charles Boasberg, Inc.

BODE, RALF: Cinematographer. b. Berlin, Germany. Attended Yale where was actor with drama school and acquired degree in directing. Received on-job training teaching combat photography and making films for Army at Ft. Monmouth. First professional job in films was gaffer on Harry, followed by long association with director John G.

Avildsen, for whom served as gaffer and lighting designer on Guess What We Learned in School Today, Joe, and Cry Uncle. Later served as dir. of photography for Avildsen on Inaugural Ball and as East Coast dir. phot. for Rocky.
PICTURES: Saturday Night Fever, Slow Dancing in the Big City.
TELEVISION: PBS Theatre in America, working as lighting designer and dir. of photo. Also many TV commercials.

BOEHM, SYDNEY: Writer. b. Philadelphia, Pa., Apr. 4, 1908; Lehigh U., 1925–29. m. Ellen Kasputis. Reporter, N.Y. Journal-American & I.N.S., 1930–45.
PICTURES INCLUDE: Union Station, Big Heat, Atomic City, Six Bridges to Cross, The Raid, Rogue Cop, The Savage, Violent Saturday, Hell on Frisco Bay, Tall Men, Bottom of the Bottle, Revolt of Mamie Stover, Woman Obsessed, Seven Thieves, Sylvia, Rough Night in Jericho.

BOETTICHER, BUDD: Producer, director, writer. r.n. Oscar Boetticher, Jr. b. Chicago, Ill., July 29, 1918. e. Ohio State U. bull fighter; then technical dir., "Blood and Sand," 1941; asst. dir., Hal Roach studios and Columbia 1941–44; became feature director at Columbia in 1944; dir. Eagle Lion, 1946; dir., Universal; independ. prod., 1954.
PICTURES INCLUDE: Behind Locked Doors, Assigned to Danger, Black Midnight, Killer Shark, Wolf Hunters, Bullfighter and the Lady, Cimarron Kid, Bronco Busters, Red Ball Express, Horizons West, City Beneath the Sea, Seminole, Man from the Alamo, Wings of the Hawk, East of Sumatra, Magnificent Matador, Killer Is Loose, Seven Men From Now, Decision at Sundown, The Tall T, Buchanan Rides Alone, Ride Lonesome Westbound, The Rise and Fall of Legs Diamond, Comanche Station, The Carlos Arruza Story.

BOFFETY, JEAN: Cinematographer. Born in France.
PICTURES INCLUDE: Act of the Heart, Journey, Les Choses De La Vie, Cesar and Rosalie, Thieves Like Us.

BOGARDE, DIRK: Actor. b. Hampstead, London, Eng., March 28, 1921. e. Allen Glens College, Glasgow & University College, London. Started theatrical career with Amersham Repertory Co., then London stage; in Army in World War II; screen debut in Esther Waters Films. Top ten British star, 1953–54, 1956–64; number one British moneymaking star 1955, 1957, 1958, 1959; Variety Club Award—Best Performance 1961–64. British Academy Award, Darling, 1966.
PICTURES INCLUDE: Quartet, Once a Jolly Swagman, Mr. Prohack, Blue Lamp, Five Angles on Murder, Gentle Gunman, So Long at the Fair, Blackmailed, Woman in Question, Hunted (Stranger in Between), Penny Princess, Desperate Moment, They Who Dare, The Sleeping Tiger, Doctor in the House, For Better or Worse, The Sea Shall Not Have Them, Simba, Doctor at Sea, Appointment in London, Cast a Dark Shadow, Doctor At Large, Spanish Gardener, Ill Met by Moonlight, Campbell's Kingdom, A Tale of Two Cities, The Wind Cannot Read, The Doctor's Dilemma, Libel, The Angel Wore Red, Song Without End, The Singer Not The Song, Victim, H.M.S. Defiant, The Password Is Courage, I Could Go on Singing, The Mind Benders, The Servant (British Academy Award 1964), Hot Enough for June, Doctor Distress, The High Bright Sun, King and Country, Darling . . ., Modesty Blaise, Accident, Our Mother's House, Sebastion, The Fixer, Justine, Oh What a Lovely War, The Damned, Death in Venice, Le Serpent, The Night Porter, Permission to Kill, Providence, A Bridge Too Far, Despair.
TV: (USA) The Little Moon of Alban, Hallmark, 1964. Blythe Spirit, Hallmark, 1966, Upon This Rock.

BOGART, NEIL: Executive. Pres. of Casablanca Record and Film Works. Began as pop singer and moved into record prod. and promo. First associated with Cameo/Parkway Records Co. in New York; then pres. of Buddah Records, which he and partners formed in 1966. Formed Casablanca in 1974; company has had 29 Gold and 3 Platinum albums. Expanded into films in 1978 as exec. prod. of Thank God It's Friday.

BOGART, PAUL: Director. b. New York, N.Y. Nov. 21, 1919.
FILMS INCLUDE: Marlowe, Halls of Anger, Skin Game, Class of '44, Mr. Ricco.
TV: (Specials) Ages of Man, Mark Twain Tonight, The Final War of Ollie Winter, Dear Friends, Secrets, The House Without a Christmas Tree, Look Homeward Angel, The Country Girl, Double Solitaire, The Thanksgiving Treasure, Tell Me Where it Hurts, The Adams Chronicles.

BOGDANOVICH, PETER: Director, producer, writer, actor. b. Kingston, N.Y., July 30, 1939 e. Collegiate School, Stella Adler Theatre School, N.Y. 1954–58. Stage debut, N.Y. Shakespeare Festival, Amer. Shakespeare Festival, others, 1955–58. Off-Bway: dir.-co.prod. The Big Knife, 1959, Camino Real, Ten Little Indians, Rocket to the Moon, 1961, dir.-prod. Once in a Lifetime, 1964. Film critic and feature writer, general publications, 1958–79, Monographs for Museum of Modern Art Film Library on Orson Welles, Alfred Hitchcock, Howard Hawks, 1961–63. Books: John Ford, (U. of Calif. Press), Fritz Lang in America (Praeger), 1969; Allan Dwan—The Last Pioneer (Praeger), 1971; Pieces of Time (Arbor House), 1973. Second-unit dir.-writer, The Wild Angels (Roger Corman-Amer. International), 1966. Dir.-prod.-writer-actor Targets (Saticoy Prod.-Paramount), 1968. Dir.-writer, The Last Picture Show (BBS Prods. Col.) 1971, dir.-writer, Directed by John Ford (AFL-Calif. Arts Comm.), 1971; dir.-prod.-writer What's Up, Doc?, (Saticoy Prod.-WB), 1972; dir.-prod. Paper Moon (Saticoy Prod.-Directors Company-Para.), 1973; dir.-prod. Daisy Miller (Copa De Oro Prod. Directors Company-Para.), 1974; dir.-prod.-writer At Long Last Love (Copa De Oro Prod.-20th Century-Fox), 1975; dir.-writer Nickelodeon (Chartoff-Winkler Prod.-Col.), 1976; dir.-writer, Saint Jack (Copa de Oro-Orio-New World), 1979; dir., writer, They All Laughed, 1981.

BOHEM, ENDRE: Producer, writer. b. Hungary, e. U. Vienna. Exec. Asst. to E. Mannix, MGM and Harry Cohn, Columbia. Prod. shorts series: Passing Parade, Nostradamus, What do You Think, Crime Does Not Pay, Tell Tale Heart.
PICTURES INCLUDE: Night Has a 1000 Eyes, The Redhead and the Cowboy, Alias Nick Beal, Streets of Laredo, Postal Inspector, Wonder of Women, Thirst, House With a 1000 Candles, Two Wise Maids, Little Orphan Annie, Lord Jeff, Crime of the Century, Twin Stars, Bengazi.
TV: Revlon Mirror, Ford Theatre, Rawhide.

BOLGER, RAY: Actor, dancer. b. Dorchester, Mass., Jan. 10, 1906; e. O'Brien Dancing School, Boston. Bank clerk; vacuum salesman; accountant & student Russakoff's Ballet School; with insurance co.; stage debut with Bob Ott's Musical Comedy repertory touring New England; in dance team Sanford & Bolger; prod. two-reel M.P. Toured South Pacific with USO, World War II; m.p. debut in "Great Ziegfeld" 1936; Own TV show, ABC, 1954–55.
SHOWS INCLUDE: Passing Show of 1926, Heads Up, Scandals of 1931, Life Begins at 8:40, On Your Toes, By Jupiter; N.Y. stage, Three to Make Ready, Where's Charley?, All American, Come Summer.
PICTURES INCLUDE: Rosalie, Wizard of Oz, Sunny, Four Jacks and a Jill, Look for the Silver Lining, Make Mine Laughs, Where's Charley?, April in Paris, Babes in Toyland, The Runner Stumble.

BOLOGNA, JOSEPH: Actor-Writer. m. actress-writer Renee Taylor. b. Brooklyn, N.Y. e. Brown Univ., Providence, R.I. Service in Marine Corps and on discharge joined ad agency, becoming director-producer of TV commercials. Collaborated with wife on short film, 2, shown at 1966 N.Y. Film Festival. Together they wrote Lovers and Other Strangers, Bdwy. play, in which both also acted. Wrote s.p. for film version. Both wrote and starred in Made for Each Other, and created and wrote TV series, Calucci's Dept.
PICTURES INCLUDE: Lovers and Other Strangers (co.-s.p.), Made for Each Other (co.-s.p., star), Cops and Robbers (actor), Mixed Company (actor), The Big Bus (actor), Chapter Two (actor).
TV: Calucci's Dept. (co.-s.p.), Honor Thy Father (movie—actor).

BOLT, ROBERT: Writer, b. Sale, England, 1924. Ent. m.p. ind. 1961.
PICTURES INCLUDE: Lawrence of Arabia, Dr. Zhivago, A Man For All Seasons (play and film), Ryan's Daughter, Dir. own s.p. Lady Caroline Lamb.

BOND, ANSON: Producer, writer. b. Cleveland, Ohio, Mar. 21, 1914; p. Charles Anson Bond, founder, Bond Clothing Co. & Bond Stores, Inc. e. Yale Prep. & pre-legal, Stuyvesant Sch., Warrenton, Va. Washington corr., 30 Ohio newspapers, 1930–33; asst. to pres., C.W. Hord Co., N.Y. 1933–36; apprenticeship in m.p. prod. 1936–39; dir., prod., writer. U.S. armed forces, 1940–45; head pub. firm, Bond-Charteris, prod. radio program. The Saint; publ. Craig Rice Crime Digest & Movie Mystery Magazine.
PICTURES INCLUDE: The Judge, Not Wanted, Vicious Years, (co-prod.) Journey Into Light, Japanese War Bride, China Venture.

BOND, DEREK: Actor, Scriptwriter. b. Glasgow, Jan. 26, 1920; e. Haberdasher' Askes School, London. Stage debut in "As Husbands Go" 1937; served in Grenadier Guards H.M. Forces 1939–46, awarded Military Cross; m.p. debut in "Captive Heart" 1946; author of Unscheduled Stop, Two Young Samaritans, Ask Father Christmas, Packdrill, Double Strung, Order to Kill, The Riverdale Dam, Sentence Deferred, The Mavroletty Fund. Many TV appearances.
PICTURES INCLUDE: Nicholas Nickleby, Joanna Godden, Uncle Silas, Scott of the Antarctic, Marry Me, Poets Pub, Weaker Sex, Broken Journey, Christopher Columbus, Tony Draws a Horse, Quiet Woman, Hour of Thirteen, Distant Trumpet, Love's A Luxury, Trouble in Store, Svengali, High Terrace, Stormy Crossing, Rogues Yarn, Gideon's Day, The Hand, Saturday Night Out, Wonderful Life, Press For Time, When Eight Bells Toll, Intimate Reflections.

BONET, NAI: Actress—Producer. Worked in entertainment field since age of 13, including opera, films, TV, stage, night clubs and records.
PICTURES: Actress: The Soul Hustlers, The Seventh Veil, Fairy Tales, The Soul of Nigger Charlie, The Spy with the Cold Nose, John Goldfarb Please Come Home, etc. Wrote and starred in Nocturna and Hoodlums.
TELEVISION: Johnny Carson Show, Merv Griffin Show, Joe Franklyn Show, Beverly Hillbillies, Tom Snyder Show.

BONO, SONNY: Singer-Actor-Director-Writer. b. Detroit, Mich. Feb. 16. r.n. Salvatore Bono. Started writing songs at age 16; entered record business with Specialty Records as apprentice prod. Became ass't. to Phil Spector, rock music prod. and did background singing. Has recorded albums with former wife Cher and made two feature films, and formed nightclub act with her. CBS comedy-variety series began as summer show in 1971 and made regular later that year.
PICTURES INCLUDE: Good Times, Chastity (prod., s.p.), Escape to Athena.
TELEVISION: Sonny & Cher Comedy Hour, Sonny Comedy Revue.

BOOKMAN, ROBERT: Executive. b. Los Angeles. e. University of California, Yale Law School. Motion picture literary agent, IFA 1972–74, ICM 1974–79. 1979 joined ABC Motion Pictures as v.p., worldwide production.

BOONE, JR., ASHLEY A.: Executive. b. 1939. e. Brandeis Univ. Started career at United Artists in adv./pub.; later with Cinema Center Films; adm. asst., Motown Records; assoc. prod. for Sidney Poitier's E&R Productions. Joined 20th-Fox in 1972 in sls. dept.; advanced to sr. sls. & mktg. positions in feature film operation. In 1979 appt. pres. of 20th-Fox Distribution & Marketing.

BOONE, PAT: Singer. b. Jacksonville, Fla., June 1, 1934. e. David LipsComb College, No. Texas State Coll., grad. magna cum laude, Columbia U. Winner of Ted Mack's TV show; joined Arthur Godfrey TV show, 1955. m.p. debut in Bernadine. Most promising new male star, M.P.D.-Fame Poll 1957.
AUTHOR: Twixt Twelve and Twenty, and Between You & Me and the Gatepost. One of top ten moneymaking stars M.P. Herald-Fame Poll, 1957; wrote book, The Real Christmas.
RECORDINGS: Ain't That a Shame, I Almost Lost My Mind, Friendly Persuasion, Love Letters in the Sand, April Love, Tutti Frutti, Speedy Gonzalez, Days of Wine & Roses, Moody River.
PICTURES INCLUDE: Bernadine, April Love, Mardi Gras, Journey to the Center of the Earth, All Hands on Deck, State Fair, The Main Attraction, The Yellow Canary, The Horror of It All, The Perils of Pauline, The Cross and the Switchblade.

BOORMAN, JOHN: Producer, director. b. London, England. Wrote film criticism at age of 17 for British publications; served in National Service in Army; joined independent Television News as film editor; prod. documentaries for Southern Television; joined BBC, headed BBC Documentary Film Unit.
PICTURES INCLUDE: Catch Us If You Can, Point Blank, Hell In the Pacific, Leo The Last, Deliverance, Zardoz, Exorcist II: The Heretic, Excalibur (prod., dir., s.p.).

BOOTH, JAMES: Actor, Scriptwriter. b. Croydon, England, 1931. Early career with R.A.D.A., Old Vic and Theatre Workshop; on West End stage: Fings Aint Wot They Used To Be. Subsequently: King Lear, Comedy of Errors, A Thousand Clowns, The Fire Raisers, The Entertainer, The Recruiting Officer, Travesties, The Tempest.
PICTURES INCLUDE: Trials of Oscar Wilde, Sparrows Can't Sing, The Hellions, French Dressing, Zulu, 90 Degrees in the Shade, Secret of My Success, Bliss of Mrs. Blossom, Boomerang, The Man Who Had Power Over Women, Darker Than Amber (1969), Macho Callaghan (1970), Revenge, Percy's Progress, Brannigan, Robbery, Penny Gold, Airport 77.
TV: The Ruffians, The Great Gold Bullion Robbery, Stray Cats and Empty Bottles, Them.

BOOTH, SHIRLEY: Actress. b. New York, N.Y., Aug. 20, 1907. Joined Poli Stock Co., Hartford, Conn., at 12; Broadway debut in Hell's Bells, 1925; on radio in Duffy's Tavern; m.p. debut in Come Back Little Sheba (Academy Award, Best Actress, 1952).
PLAYS: After Such Pleasures, 3 Men on a Horse, Philadelphia Story, My Sister Eileen, Tomorrow the World, Goodbye My Fancy, Come Back Little Sheba, A Tree Grows in Brooklyn, Time of the Cuckoo, The Desk Set, Look to the Lillies, Miss Isobel, Juno, Colettes Second String, Glass Menagerie.
PICTURES INCLUDE: About Mrs. Leslie, Hot Spell, The Matchmaker.
TV: Perle Mesta Story, Hazel, A Touch of Grace.

BORCHERS, CORNELL: Actress. b. Heydekrug, Germany, March 16; e. Gottingen U. Berlin actors school, m. Dr. Anton Schelkopf.
PICTURES INCLUDE: (German) Martina, Das ewige Spiel, Haus des Lebens, Schule fur Ehegluck; (English) Big Lift, Divided Heart, Oasis, Never Say Goodbye, Istanbul, Alone Together.

BORGE, VICTOR: Comedian, pianist. b. Copenhagen, Denmark, Jan. 3. Child prodigy at age 10. Later became humorous concert artist. Wrote and starred in musical plays and films in Denmark. Fled Nazis in 1941, came to America.
Appeared on Bing Crosby radio show. Concert and nightclub tours. TV variety shows. One-man Broadway show. Comedy in Music, 1953, three-year run. Second edition in 1964. World tour. One-man TV show.

BORGNINE, ERNEST: Actor. b. Hamden, Conn., Jan. 24, 1918. e. Randall School of Dramatic Art, Hartford. Served in U.S. Navy; then little theatre work, stock companies; on Broadway in Harvey, Mrs. McThing; many TV appearances; m.p. debut in Whistle at Eaton Falls.
PICTURES INCLUDE: The Mob, China Corsair, From Here to Eternity, Demetrius & the Gladiators, Johnny Guitar, Vera Cruz, Bad Day at Black Rock, Marty (Acad. Award best actor 1955), Run for Cover, Violent Saturday, Last Command, Square Jungle, Catered Affair, Jubal, Best Things in Life are Free, Three Brave Men, Pay or Die, Go Naked in the World, Rabbit Trap, 10 Years a Counterspy, Summer of the Seventeenth Doll, Barabbas, Chuka, The Dirty Dozen, The Wild Bunch, The Adventurers, Suppose They Gave a War and Nobody Came?, A Bullet for Sandoval, Bunny, O'Hare, Hannie Caulder, The Revengers, Legend of Lylah Clare, The Poseidon Adventure, The Emperor of the North, Law and Disorder, The Devil's Rain, Hustle, Shoot, The Greatest, Crossed Swords, Convoy, The Black Hole, When Time Ran Out, Escape from New York, Deadly Blessing.
TV: Philco Playhouse, General Electric Theater, Wagon Train, Laramie, Zane Grey Theater, Alcoa Premiere, McHale's Navy.

BORNSTEIN, ALLEN: Executive. b. Bronx, N.Y. e. CCNY. Entered m.p. career 1957 as asst. branch cashier, 20th Century Fox; booker, Buena Vista, 1962; N.Y. branch mgr., Cinerama, 1968; natl. dir. branch operations, Gamma III.

BORODINSKY, SAMUEL: Executive. b. Brooklyn, NY, Oct. 25, 1941. e. Industrial School of Arts & Photography. Expert in film care and rejuvenation. Now exec. v.p., Filmtreat International Corp. Previously with Modern Film Corp. (technician) and Comprehensive Filmtreat, Inc. & International Filmtreat (service manager).

BOSLEY, TOM: Actor. b. Chicago, Oct. 1, 1927. e. DePaul Univ. Had roles on radio in Chicago and in stock productions before moving to New York. Appeared off-Bdwy and on road before signed to play lead in Fiorello! for George Abbott on Bdwy. First actor to win Tony, Drama Critics, ANTA and Newspaper Guild awards in one season for that role. Has since done many plays, movies and TV appearances.
PICTURES: Love with the Proper Stranger, The World of Henry Orient, Divorce, American Style, Yours, Mine & Ours, The Secret War of Harry Frigg, Mixed Company, Gus.
TELEVISION Series: Debbie Reynolds Show, Sandy Duncan Show, Happy Days, Movies: The Girl Who Came Gift Wrapped, Death Cruise, The Night the Martians Landed, Love Boat. Specials: Testimony of Two Men, The Bastard.

BOSTICK, ROBERT L.: b. Waynesboro, Georgia, Oct. 25, 1909. e. Georgia Tech, Atlanta (ME, Eng.) 1932. Started with National Theatre Supply, Atlanta, salesman, Memphis, 1933; br. mgr. 1937; br. mgr. Dallas, 1942; Vice Pres., Southern Division Mgr., 1952; retired 1968. Chief Barker Variety Club Tent 20, Memphis 1950–51. Since 1957, has served as International Rep., International Ambassador-at-Large, and International Vice Pres. for Variety Clubs International. Since 1968, owner & operator of theatres in Memphis and Charlotte areas.

BOSUSTOW, NICK: Producer. b. Los Angeles, Calif., March 28, 1940. e. Menlo College, Calif. Administration. MCA, International sales, 1963. Pres., Stephen Bosustow Productions, 1967; pres., ASIFA-West; Academy Award '70 best short, Is it always right to be right?; 1973 Academy Award nomination, The Legend of John Henry. The Incredible Book Escape, Misunderstood Monsters, A Tale of Four Wishes (TV specials). 1977, pres., Bosustow Entertainment, Inc.

BOSUSTOW, TEE: Producer-Director-Editor. b. Hollywood, Calif., Feb. 18, 1938. e. UCLA Cinema in Westwood, La Sorbonne, Paris.
PICTURES INCLUDE: Beware of Thin Ice, Big Boys Don't Cry, Avati and the Mezzotint. Edited short, Is It Always Right to Be Right? (AA, 197-).

TELEVISION: About a Week (community affairs series; Emmy award).

BOTTOMS, JOSEPH: Actor. b. Santa Barbara, Calif. April 22, 1954. Did plays in jr. high school in Santa Barbara and then with community theatre. Made m.p. debut in The Dove (1974).
PICTURES: Crime and Passion, King of the Mountain.
TV: Owen Marshall, Winesburg, Ohio.

BOTTOMS, TIMOTHY: Actor. b. Santa Barbara, Calif., Aug. 30, 1951. Early interest in acting; was member of S.B. Madrigal Society, touring Europe in 1967. Sang and danced in West Side Story local prod.
PICTURES INCLUDE: Johnny Got His Gun (debut), The Last Picture Show, Love and Pain, The Paper Chase, The White Dawn, The Crazy World of Julius Vrooch, Seven Men at Daybreak, A Small Town in Texas, Rollercoaster, The Other Side of the Mountain: Part II, Hurricane.

BOUDOURIS, A: Executive. b. Toledo, Ohio, Jan. 31, 1918. e. U. of Toledo, 1945–47. USN, radio-radar, 1941–45; carrier pilot. Service, installation, projection equip., 1934–36; Strong Electric, 1936–41; Pres., Theatre Equip. Co., 1945–65; Theatre Operating Co., since 1947; pres. Eprad, Inc., 1949, bd. chmn., 1980. Member (ex-chm.) of NATO's Technical Advisory Comm., Pres. NATO of Ohio, NATO representative on American National Standards Committee; now Vice Pres. NATO. Chairman, NATO Statistical Committee; Member, Projection Practices and Sound Comm. of SMPTE; NATO Representative International Standards Organization (Geneva), Member National Electrical Code Committee Panel #15. Recipient, SMPTE Fellowship Award, 1980.

BOULTING, JOHN: Producer, director, writer. b. Bray, Buckinghamshire, England, Nov. 21, 1913. With brother, Roy, formed Charter Film Productions in 1937. Produced 1938/40; Consider Your Verdict, Trunk Crime, Inquest, Pastor Hall. Went into RAF 1940. Seconded RAF Film Unit, directed several documentaries and Journey Together feature. Post-war: Prod. Fame is the Spur; The Guinea Pig; Seagulls over Sorrento; Josephine and Men; Brothers in Law; Dir. Brighton Rock, Seven Days to Noon, The Magic Box, Lucky Jim (Edinburgh Festival 1957); dir. co-author s.p. Private's Progress, 1955; prod. Carlton-Browne of the FO, 1958; dir. co-author s.p. I'm All Right Jack, 1959; produced The Risk, and The French Mistress, 1960; co-author s.p. dir. Heavens Above, 1962; prod. The Family Way, 1966; There's a Girl in My Soup, 1970; Soft Beds, Hard Battles, 1974. Dir. 1958–72, Man. Dir. 1966–72, Consultant, 1972, British Lion Films Ltd. Former Chm., Local Radio Assoc.

BOULTING, ROY: Producer, director. b. Bray, Buckinghamshire, England, Nov. 21, 1913. e. McGill U., Montreal. Capt., Brit. Army, World War II. Dir. Charter Film, Charter Film Prod. Ltd., London; dir. British Lion Films, Ltd., 1958.
PICTURES INCLUDE: Inquest, Trunk Crime, Pastor Hall, Thunder Rock, Desert Victory, Burma Victory, Fame is the Spur, Brighton Rock, Guinea Pig, Seven Days to Noon, Lucky Jim, High Treason, Singlehanded (Sailor of the King), Seagulls Over Sorrento (Crest of the Wave), Josephine and Men, Private's Progress, Run for the Sun, Brothers in Law, Happy Is the Bride, Carlton-Browne of the F.O., I'm All Right Jack, The Risk, The French Mistress, Heavens Above! Rotten to The Core, The Family Way, Twisted Nerve, There's a Girl in My Soup, Soft Beds, Hard Battles, 1977, co-author with Leo Marks of play, Favourites, Danny Travis, 1978; The Last Word, 1979.

BOWER, DALLAS: Producer, director. b. London, 1907. Ent. film prod. 1927; film ed., writer, dir., prod. with BBC-TV 1936. Commissioned in Royal Corps of Signals, 1939; supvr. film prod. Ministry of Inf., Films Div. 1942; prod. official and comm. documentaries; author, Plan for Cinema, 1936.
PICTURES AND TV INCLUDE: Aida, Tristan & Isolde, Master Peter's Puppet Show, Cinderella, Julius Caesar, The Tempest, The Taming of the Shrew, The Silver Box, The Mock Emperor, The Emperor Jones, Rope, Path of Glory, Victory Over Space, Henry V. Alice in Wonderland, Second Mrs. Tanquery, Fire One, Doorway to Suspicion, Adventures of Sir Lancelot. In prep. for T.V., The Ring, The Trojans.

BOWERS, WILLIAM: Writer, producer. b. Las Cruces, New Mexico, Jan. 17, 1916; e. U. Mo. Playwright (Where Do We Go From Here, Back to Eden). Reporter, feature writer, N.Y. World-Telegram: then N.E.A. Hollywood corresp.; later radio scripts (Helen Hayes Theat., etc.). Began screen writing 1942 with collab. s.p. My Favorite Spy.
PICTURES INCLUDE: Alias Jesse James, Imitation General, The Law and Jake Wade, The Sheepman, My Man Godfrey, The Best Things in Life, Tight Spot, Split Second, The Gunfighter, The Web, Support Your Local Sheriff.

BOX, BETTY: OBE. Producer. b. Beckenham, Kent. Assisted Sydney Box in prod. 200 propaganda films in World War II. Assoc. prod. Upturned Glass.
PRODUCTIONS INCLUDE: Dear Murderer, When the Bough Breaks, Miranda, Blind Goddess, Huggett Family series. It's Not Cricket, Marry Me, Don't Ever Leave Me, So Long At the Fair, The Clouded Yellow, Appointment With Venus (Island Rescue). Venetian Bird (The Assassin), A Day to Remember, Doctor in the House, Mad About Men, Doctor at Sea, The Iron Petticoat, Checkpoint, Doctor at Large, Campbell's Kingdom, A Tale of Two Cities, The Wind Cannot Read, The 39 Steps, Upstairs and Down-Stairs, Conspiracy of Hearts, Doctor in Love, No Love for Johnnie, No, My Darling Daughter, A Pair of Briefs, The Wild and the Willing, Doctor in Distress, Hot Enough for June (Agent 8¾), The High Bright Sun, (McGuire Go Home), Doctor in Clover, Deadlier Than the Male, Nobody Runs Forever, The High Commissioner, Some Girls Do, Doctor in Trouble, Percy, The Love Ban, Percy's Progress.

BOYARS, ALBERT: Executive. b. N.Y.C. Aug. 11, 1924: e. N.Y.U.: U.S. Navy, 1941–45. David O. Alber Assoc., 1945–51: Greater N.Y. Fund, Robert S. Taplinger Assoc., Michael Myerberg Prod., 1951–54: pub. rel. dir., Transfilm-Caravel Inc., and parent co. Buckeye Corp., 1954–63: director spec. projects in adv-pub-exploit. M-G-M, 1963–64; dir. of adv. & pub. Trans-Lux Corp., 1964, v.p. of adv. and pub. rel., Trans Lux, 1972, v.p. special projects, Trans-Lux, 1974; v.p. chg. Trans-Lux Multimedia Corp., 1976. Under his aegis are the attractions New York Experience & Space Works.

BOYER, PHIL: TV Executive. b. Portland, Oreg. e. Sacramento State Univ. Began broadcasting career as 12-year-old in Portland, establishing nation's first youth radio facility—a 5-watt facility in the basement of his home. At 16 began working at KPDQ, Portland; two years later joined KPTV, Portland, as announcer. In 1960 joined KEZI-TV, Eugene, Oreg., heading prod. and prog. depts. In 1965 named staff prod.-dir. for KCRA, TV, Sacramento, Calif., becoming prod. mgr. in 1967 and prog. mgr. in 1969. In 1972 joined KNBC-TV, Los Angeles, as prog. dir. In 1974 named v.p., programming, of ABC Owned Television Stations; 1977, v.p.-gen. mgr., WLS-TV, Chicago; 1979, v.p.-gen. mgr. of WABC-TV, New York.

BOYETT, ROBERT: Producer. e. Duke University (B.A.) and Col. Univ. (Master's in mkt.). Began career in media and mkt. research at Grey Advertising, Inc. Was program development consultant for PBS. In 1973 joined ABC as dir. of prime time series TV, East Coast. In 1975 named ABC TV v.p. & asst. to v.p. programs for West Coast. In 1977 joined Paramount Pictures in newly created position of v.p., exec. asst. to pres. & chief operating officer. 1979, formed Miller-Milkis—Boyett Productions to produce for Paramount Television.

BOYLE PETER: Actor. b. Philadelphia, e. LaSalle College. Was monk in Christian Bros. order before leaving in early 60's to come to N.Y. Acted in off-Broadway shows and joined The Second City in Chicago. Also did TV commercials.
PICTURES INCLUDE: Joe (debut), T.R. Baskin, The Candidate, Steelyard Blues, Slither, The Friends of Eddie Coyle, Kid Blue, Crazy Joe, Young Frankenstein, Swashbuckler, F.I.S.T., Brinks' Job, Hardcore, Beyond the Poseidon Adventure, Where the Buffalo Roam, In God We Trust, Outland.

BOYLE, WILLIAM NEAL (Billy): Producer. b. London, England; e. Rothschild Grammar School. Ent. m.p. 1927 with British International Pictures on set design & construction. Subsequently con. manager, first asst. Dir., Pro. & Loc.; several American and British Cos.; Produced 17 pictures for Republic Pictures International; Productions Controller, Dudley Pictures International-Hollywood; United Artists. Paramount Pictures. dir., W.N.B. Prods., 1966. Incorporated Films and W.N.B. Prods., 1970, W.N.B. World Locations, 1971, TV series.
PICTURES INCLUDE: Demi Paradise, Tawny Pipit, Good Time Girl, Man Within, Quartet, Trio, Encore, So Long at The Fair. Highly Dangerous, Holiday Camp, Easy Money, Lady of Vengeance, High Hell, African Story (World locations, 1978).

BRABOURNE, THE LORD: Producer. b. London, Nov. 9, 1924.
PRODUCTIONS INCLUDE: Harry Black, Sink the Bismarck, H.M.S. Defiant, Othello, The Mikado, Up the Junction, Romeo and Juliet, Dance of Death, Tales of Beatrix Potter, Murder on the Orient Express, Death On The Nile, Stories from a Flying Trunk, The Mirror Crack'd.

BRACKEN, EDDIE: Actor. b. New York City, Feb. 7, 1920: e. Prof. Children's School for Actors, N.Y.C.: m. Connie Nickerson, actress. Vaudeville & night club singer: stage debut in "Lottery" 1930; m.p. debut in "Life with Henry" 1940.
PLAYS: Lady Refuses, Iron Men, So Proudly We Hail, Brother Rat, What A Life, Too Many Girls, Seven Year Itch, Shinbone Alley, Teahouse of the August Moon, You Know I Can't Hear You when The Water's Running, The Odd Couple, Never Too Late, Sunshine Boys, Hotline to Heaven, Hello, Dolly.
PICTURES INCLUDE: Fleet's In, Happy Go Lucky, Sweater Girl, Star Spangled Rhythm, Young and Willing, Hail the Conquering Hero, Miracle of Morgan's Creek, Girl

From Jones Beach, Summer Stock, Two Tickets to Broadway, About Face, We're Not Married, Slight Case of Larceny.
TV: Masquerade Party, NBC.

BRADEN, WILLIAM: Executive/Producer. b. Alberta, Canada, 1939. e. Vancouver, British Columbia. Began career as stuntman in Hollywood, later working in all aspects of industry for 22 years. Worked for Elliott Kastner as prod. executive with offices at Paramount and with Jeffrey Bloom, of Feature Films, Inc., as prod. and v.p. in chg. of prod. before forming own co., Dunatai Corp., for film and t.v. prod.
PICTURES: Pyramid (assoc. prod., prod. supv.), Russian Roulette (prod. exec.), 92 in the Shade (prod. exec.), Breakheart Pass (prod. exec.), Dogpound Shuffle (asst. dir., prod. supv.); He Wants Her Back (prod.); Goldengirl (prod. exec.); Running Scared (prod.); Death Valley (asst. dir.); The Seduction (prod. exec.).
TELEVISION: Requiem for a Planet (series, prod./creator); Nothing Great is Easy (special, exec. prod.); King of the Channel (special, exec. prod.); I Believe (special, prod.); If My People . . . (special, prod.), Jude (pilot, prod.). America: Life in the Family (dir./prod.) Also various Movies of the Week for networks.

BRADFORD, PETER: Producer and Director. b. Bradford Abbas, 1919. e. Bryanston School, Reimann School of Art and Cinematography, 1941–44 Head of Photo. Dept. Naval Intelligence, Oxford, and concurrently lecturer in photography and cinematography at School of Art and Technology, Oxford. 1944–48 Film Director with Paul Rotha Productions and Films of Fact Ltd. 1948–57 Associated British-Pathe. 1957–59 Film Prod., with Ogilvy & Mather advertising agents; prod., World Wide Pictures, 1960–81. Now freelance.

BRADLEY, BILL: Performer, r.n. William M. Silbert. b. Jan. 1, 1921, Detroit, Mich.; e. Univ. of Detroit, So. Methodist; Disc jockey, m.c., many radio-TV shows, Detroit; panelist, Songs for Sale, CBS-TV, 1952; emcee, Bill Silbert Show, Let's Go Bowling WABD-DuMont TV, 1952–53; Bill Silbert Show. WMGM radio; announcer, Red Buttons Show, NBC-TV; Philco Phonorama Time, Mutual; m.c., National Radio Fan Club, NBC; Magazine of the Air, Mutual, KLAC Hollywood, Bill Bradley Show KTLA, Hollywood, M.C. Crime Story, Greet the People, Ad Lib, Hollywood Diary. Sales mgr., KLOS, Los Angeles.
PICTURES INCLUDE: Bundle of Joy, Thunderjets, The Alligator People, Young Jesse James, Lost Missile, Breakfast at Tiffanys, Return to Peyton Place, Looking for Love.
TV: Bronco, 77 Sunset Strip, Hawaiian Eye, Sugarfoot, Combat, Adventures in Paradise, Police Station, Michael Shayne, Roaring 20's, The Outlaws, Breaking Point, The Fugitive, Bill Dana Show, My Living Doll, Joey Bishop Show, Ben Casey, Bing Crosby Show. Many commercials, Mannix, Wild Wild West, Name of the Game.

BRADY, FRANK L.: Theatre Executive. b. Columbus, Ga., Jan. 11, 1914. Former pres., Martin Theatres.

BRADY, SCOTT: Actor. r.n. Jerry Tierney; b. Brooklyn, N.Y., Sept. 13, 1924; b. Lawrence Tierney, actor; e. Bliss-Hayden Dram. School, Beverly Hills, Calif. Served in U.S. Navy, won light-heavyweight boxing championship of naval base, discharged 1945. First starring role Born to Fight, Star of Tomorrow, 1953.
PICTURES INCLUDE: Canon City, He Walked By Night, In This Corner, Port of New York, Undertow, Kansas Raiders, Undercover Girl, Model and the Marriage Broker, Bronco Buster, Montana Belle, Untamed Frontier, Yankee Buccaneer, Bloodhounds of Broadway, Perilous Journey, El Alamein, White Fire, Johnny Guitar, Law vs. Billy the Kid, Gentlemen Marry Brunettes, Vanishing American, They Were So Young, Terror at Midnight, Mohawk, Maverick Queen, Fort Utah, Arizona Bushwackers, Dollars, The Loners, Wicked, Wicked. The China Syndrome.

BRAMBELL, WILFRID: Actor, b. Dublin, Ireland, 1912. Early career as reporter. Gate and Abbey Theatres, Dublin. Repertory, variety, pantomime, tours, etc.
STAGE: Blind Man's Bluff (1953), Stop It, Whoever You Are (1961), first Steptoe and Son.
TV: The Importance of Being Earnest, Canterbury Tales.
PICTURES INCLUDE: Dry Rot, Serious Charge, The Long Hot Summer, What a Whopper, The Castaways, Crooks in Cloisters, The Small World of Sammy Lee, Fast Lady, Thomasina, Hard Day's Night, Call Me a Spy, Steptoe and Son.

BRAND, NEVILLE: Actor. b. Kewanee, Ill., Aug. 13, 1921. e. high sch., Kewanee. U.S. Army, 10 yrs to 1946; studied acting in N.Y.: M.P. debut in D.O.A.
PICTURES INCLUDE: Halls of Montezuma, Only the Valiant, The Mob, Flame of Araby, Stalag 17, Charge at Feather River, Man Crazy, Gun Fury, Riot in Cell Block 11, Long Gun, Prince Valiant, Return from the Sea, Fury at Gunsight Pass, The Prodigal, Return of Jack Slade, Bobby Ware Is Missing, Mohawk, Raw Edge, The Adventures of Huckleberry Finn, The Desperadoes, Three Guns for Texas, Birdman of Alcatraz, That Darn Cat, Scalawag, Cahill, U.S. Marshall, The Deadly Trackers, Psychic Killer.
TV: Laredo, The Captain and the Kings.

BRANDO, JOCELYN: Actress. b. San Francisco, Calif. e. Lake Forest Coll.; m.p. debut in The Big Heat; on B'way in Mr. Roberts, Desire Under the Elms, Golden State.
PICTURES INCLUDE: China Venture, Ten Wanted Men.

BRANDO, MARLON: Actor. b. Omaha, Neb., April 3, 1924. e. Shattuck Military Acad., Faribault, Minn. Studied Dramatic Workshop, New York; played stock Sayville, Long Island, m.p. screen debut. The New state debut: I Remember Mama, then Truckline Cafe, Candida, A Flag is Born, Street Car Named Desire.
PICTURES INCLUDE: Streetcar Named Desire, Viva Zapata, Julius Caesar, Wild One, On the Waterfront (Acad. Award best actor, 1954). One of top ten Money-Making Stars, M.P. Herald-Fame poll, 1954–55. Guys and Dolls, Teahouse of the August Moon, Sayonara, The Young Lions, Mutiny on the Bounty, One-Eyed Jacks, The Ugly American, Bedtime Story, The Saboteur, Code Name, Morituri, The Fugitive Kind, The Chase, Appaloosa, The Countess From Hong Kong, Reflections in A Golden Eye, Night of the Following Day, Candy, Burn!, The Nightcomers, The Godfather, Last Tango in Paris, The Missouri Breaks, Superman, The Formula.

BRANDT, JANET: Actress. b. New York City. Acting debut at 6 years old with Yiddish Theater. Over 100 stage appearances in New York where she studied with Martha Graham and with Michael Chekov. Lectured on Shakespeare at the Strasberg Institute in L.A.
PICTURES INCLUDE: Actress: Cold Wind in August, Kotch, Mad Adventures of Rabbi Jacob, Sheila Levine, Semi-Tough, Dialogue Director: Battle of the Bulge, El Cid, Phaedre, View from the Bridge, King of Kings, etc.
TV: Actress: Odd Couple, Mannix, Lou Grant, etc.

BRANDT, RICHARD PAUL: Executive. b. New York, N.Y., Dec. 6, 1927; e. Yale U., chief exec. off. & bd. chm. Trans Lux Corp.; bd. chm., Brandt Theatres; dir., American Book-Stratford Press, Inc.; dir., Harris Creatline Corp.; member President's Advisory Council, National Association of Theatre Owners; member, bd. of trustees & chm. of exec. committee, American Film Institute; member, bd. of dirs., Presidential Realty Corp.; Bd of Dir.: American Theatre Wing; trustee, Museum of Holography.

BRANDY, HOWARD: Executive. b. Brooklyn, N.Y., 1929. e. Emory Univ. Served with Marine Corps in Korean War. Worked as N.Y. press agent; in 1956 went with NBC. Was columnist for Hollywood Reporter in Las Vegas then went back into publicity as acct. exec. for Rogers and Cowan. Started own business; clients included Frankie Avalon, Fabian, Monkees, Fifth Dimension, Johnny Cash, etc. Joined Commonwealth United in London working for co. 5 yrs. before it folded. Entered film prod.
PICTURES INCLUDE: Blood from the Mummy's Tomb, The Take.

BRANNUM, HUGH: Performer. b. Sandwich, Ill., 1910. Musician, photographer, TV writer, artist; with Fred Waring's Pennsylvanians, 1940–55; writer, Waring TV show, 3 yrs.; children's songs; TV show, Captain Kangaroo.

BRANSON, WALTER E.: Western division sales manager, RKO Pictures Corp. e. U. of Nebraska. Booker for Pathe, Omaha, 1922–24; then salesman in South Dakota; city salesman, br. mgr., Omaha, 1925–27; br. mgr.; Des Moines, 1927–28; St. Louis, 1928–29; to Pathe h.o. as western sales mgr.; br. mgr. Chicago, 1931; midwest dist. mgr., 1932; Midwest dist. mgr., 1932–41; Western sales mgr., 1941–51, asst. gen. sales mgr., 1952; gen. mgr. foreign operations, RKO-Radio Pictures, February, 1954; gen. sales mgr., Aug., 1954; v.p. in charge of world-wide distribution 1955; exec. asst. to pres., National Screen Service, 1960; v.p. distrib., National Screen Service, 1963; v.p. in charge of prod., National Screen Service, 1966.

BRASLOFF, STANLEY H.: Producer, director. b. Philadelphia, Pa., July 23, 1930. In U.S. Army was entertainment dir., Heidelberg military post, 1948–50; prod., dir. & acted for various little theatre & summer stock groups, 1950–60; created, wrote, co-prod., dir. & starred in "Time For Teens" TV show (Chicago); formed VCS Pictures Ltd., prod. & dir. "Two Girls for a Madman," 1967; formed SHB Productions, Inc., prod. "Nightmare at Ghost Beach," 1968; wrote, co-prod, & dir. "Toys Are Not for Children," 1971.

BRAUDY, SUSAN: Executive. e. Bryn Mawr College; taught writing at Brooklyn College (1974). 1970–72, editor and writer at Newsweek; same posts at Ms. Magazine, 1973–78. Worked in 1978 for producer Alberto Grimaldi. Has published numerous articles in New York Magazine, New Journal (Yale), The Atlantic, Village Voice, etc. In 1981 appt. Warner Bros. v.p., east coast production. 1981, novel, Who Killed Sal Mineo?, published.

BRAUER, LE ROY: Executive. b. Cincinnati, Ohio, Sept. 30, 1902, e. Col. U. Salesman, Cincinnati Standard Film Service, 1919. Universal, 1921–46, as salesman, branch mgr., Supvsr., China, Japan; br. mgr. New Orleans, other cities, then sales mgr., Europe, Middle East and India. Columbia Pictures, Intl. Corp. 1949–1951. Man. dir. Australia, 1951; chmn. bd., Exotic TV Film Distribs. Pty. Ltd.

BRAUNAGEL, JACK D.: Exhibitor. b. San Francisco, Calif., Jan. 4, 1912; e. U. of Utah, 1929. Entered m.p. ind. as usher, Paramount-Publix, San Francisco, 1927; Des Moines, Ia., 1929; asst. mgr. in Ogden and Salt Lake City, Utah 1930; mgr., Ogden, 1932; mgr., Boise, Ida., 1935; city mgr., Logan, Utah, 1936; city mgr., Provo, Utah, 1937; dist. mgr., Intermountain Theatres, 1939–43; U.S. Army, 1943–46; supervisor, Northio Theatres, Cincinnati, 1946; film buyer, supervisor, Durwood Theatres, Kansas City, 1947; gen. mgr. drive-ins, Commonwealth Theatres, 1948–54; asst. to pres., in charge of theatre oper., United Theatres, No. Little Rock, 1955; exec. bd., TOA, 1953–55; formed Jay D. Bee Circuit, No. Little Rock, 1958; mem. bd. Motion Picture Investors, Inc.; pres., Money Investors, Inc., 1964; pres., Helene Theatres, Inc.

BRAUNSTEIN, GEORGE GREGORY: Producer. b. New York City, May 23, 1947. e. Culver Military Academy, Univ. of CA (b.a. biology, chemistry—1970). Father is Jacques Braunstein (Screen Televideo Prods. At War with the Army, Young Lions, etc.). 1971, co-founder of The Crystal Jukebox—record productions, music management, music publishing co. Produced 12 hit albums for Decca Records of England and London Records, U.S.A. Entered m.p. production in 1975–76 producing Train Ride to Hollywood for Taylor-Laughlin Co. 1979–80, produced Fade to Black for American Cinema Corp. Now pres. of Solar Survival Film Corp.

BRAVERMAN, CHARLES: Producer-Director. b. Los Angeles, Calif., March 3, 1944; e. L.A. City Coll., U. of S. Cal. (BA); m. Kendall Carly Browne, actress. Child actor, 1950–57. Two time Emmy winner.

TV: An American Time Capsule, The Smothers Brothers Racing Team Special, How to Stay Alive, David Hartman . . . Birth and Babies, Breathe a Sigh of Relief, The Television Newsman, Getting Married, The Making of a Live TV Show, Televisionland, Nixon: Checkers to Watergate, Braverman's Condensed Cream of Beatles, Two Cops, Peanuts to the Presidency: The Jimmy Carter Campaign; The Making of Beatlemania, Willie Nelson Plays Lake Tahoe, Tony Bennett Sings, What's Up, America?, The Big Laff Off, Engelbert at the MGM Grand, Oscar's First 50 Years, Frankie Valli Show, The Sixties.

PICTURES: Dillinger, Soylent Green, Same Time Next Year (all montages, titles), Can't Stop the Music (titles).

BRAVERMAN, MILLICENT: Adv. agency-pub. relations agency president (Braverman-Mirisch Inc.) KFAC AM & FM, syndicated radio Los Angeles commentator on books, literary critic, publishing specialist, b. New York City, B-M. Inc. founded 1963 to date. West coast consultant for number of NYC publisher accounts, general advertising, print and broadcast media nationally.

BRAY, PAUL, JR.: e. Rensselaer Polytechnic Institute. Experienced airline transport pilot and instrument flight instructor. Writer and director specializing in motion picture scripts and productions on all phases aviation, engineering and electronic from theoretical to practical industrial personnel training, and for Armed Forces and Vocational-Technical school instruction.

BRAZZI, ROSSANO: Actor. b. Bologna, Italy, 1916. e. U. of Florence. Started career on Italian stage; has appeared in numerous Italian pictures.

PICTURES INCLUDE: (U.S.) Little Women, Three Coins in the Fountain, Barefoot Contessa, Summertime, South Pacific, Light in the Piazza, The Battle of Villa Fiorita, Rome Adventure, Dark Purpose, The Christmas That Almost Wasn't, The Bobo, Woman Times Seven, Krakatoa, East of Java, The Adventurers, Psychout for Murder, The Great Waltz, The Final Conflict.

BREACHER, HAROLD: Executive. b. June 17, 1910. e. University of Calif. at L.A., B.A.; Univ. of Southern Calif., LL.B.; Univ. of South Calif., JD. Attorney at Law, 7 years. Bergerman-Breacher Motion Picture Agency, part owner. William Morris Agency, exec. agent; Famous Arts Agency, head Television Dept.; Ashley-Famous Agency, exec. packaging agent; United Artists Pictures Corp., head of sales & distribution United Kingdom, Europe, Middle East and Africa (TV); director, sales, U.K., Europe, Middle East & South Africa, Warner Bros. Television.

BRECHER, IRVING: Writer. b. New York City, Jan. 17, 1914. Yonkers Herald reporter; network programs writer for Milton Berle, Willie Howard, Al Jolson, etc., m.p. writer since 1937.

PICTURES INCLUDE: (collab. s.p.) Shadow of the Thin Man, Best Foot Forward, Meet Me in St. Louis; (collab, adapt.) Summer Holiday: (s.p.), Go West, At the Circus, Du Barry Was a Lady, Yolanda and the Thief, Life of Riley, Somebody Loves Me, Cry for Happy, Bye Bye Birdie, Sail a Crooked Ship.

TV: The Peoples Choice, The Life of Riley.

BREGMAN, MARTIN: Motion Picture Producer.

PICTURES INCLUDE: Serpico, Dog Day Afternoon, The Next Man, The Seduction of Joe Tynan, S*H*E, Simon, The Four Seasons.

BREN, ROBERT J.: Producer, writer. b. Guanajuato, Mex. Film writing debut with "Looking for Trouble" 1934; in U.S. Signal Corps, World War II, also co-ordinator Inter-American Affairs as m.p. program office head, N.Y.C. & Wash. D.C.

PICTURES INCLUDE: This Marriage Business, Crime Ring, Parents on Trial, American Empire, Underground Agent, Charter Pilot, In Old California, Gay Senorita, Five Little Peppers and How They Grew, prod. First Yank into Tokyo, assoc. prod. El Paso, Overland Pacific; Naked Alibi, Treasure of Pancho Villa, The Great Sioux Uprising, Siege at Red River, Without Orders, Maltrata (O.S., S.P., producer) for Bremex Corp., Mexico; Cajeme (O.S., S.P., producer) for Bremex, S.A., Mexico.

TV: Lone Wolf in Mexico, Stolen Train, Casey Jones series. Formed Bremex, indep. picture co., 1963; collab. s.p., Winds of the Four Seasons; collab. s.p., orig. story, The Hawaiians, 1967. Wrote & collaborated on Blood On the Shrine, Legacy in Nylons, Aztec Dagger, Female Menagerie, Tag for Murder.

BRENNAN, EILEEN: Actress. b. Los Angeles. Daughter of silent film actress Jean Manahan. e. Georgetown Univ. American Academy of Dramatic Arts (N.Y.). Big break came with lead in off-Bdwy. musical, Little Mary Sunshine.

PICTURES INCLUDE: Divorce, American Style, The Last Picture Show, Scarecrow, The Sting, Daisy Miller, At Long Last Love, Hustle, Murder by Death, The Cheap Detective, Private Benjamin.

STAGE: The Miracle Worker (tour), Hello, Dolly! (Bdwy.), and revivals of The King and I, Guys and Dolls, Camelot, Bells Are Ringing, etc. Did one-woman revue, An Evening with Eileen Brennan.

BRENNER, JOSEPH: Executive. b. Brooklyn, N.Y., Oct. 27, 1918; e. Brooklyn Coll. Started as usher, 1935, becoming mgr., Rogers Theatre, 1936; salesman, Eagle Lion Films, 1946; Screen Guild Prods., 1948; sales mgr., Ellis Films, 1949; formed Joseph Brenner Associates, 1953.

BREWER, ROY M.: b. Cairo, Neb., Aug. 9, 1909. Was projectionist 1925; chief projectionist, Capitol Theatre, Grand Island, Neb. 1927–39; pres. Nebraska State Fed. of Labor 1933; active in Nebraska labor movement in this period; NRA labor compliance officer for Nebraska 1934–35; campaigned for U.S. Senator George W. Norris in chg. campaign hdqts. 1936; re-elected pres. Nebraska State Federation of Labor 1937; covered legislature for Federation 1939, 1941, 1943; chief, plant & community facilities service War Prod. Bd. Washington 1943–45; apptd. specl. internl. rep. LATSE 1945; then made internl. rep. in Hollywood, rep. Richard F. Walsh, LATSE pres.; exec. ass't Steve Broidy, Allied Artists president, October, 1953; mgr. of br. oper., 1955; apptd. admin. sls. asst. to Edward Morey, vice-pres., Allied Artists, 1962; mgr. prod. dev., 1965. Member: Hollywood AFL Film Council, M.P. Ind. Council, M.P. Alliance, Pres, Pern Char. Committee 1965 Studio mgr. and prod. rep. for West Coast.

BRIAN, DAVID: Actor. b. New York City, Aug. 5, 1914; e. C.C.N.Y. Doorman; chorus boy stage debut in New Moon.

STAGE: You Said It, Bittersweet, Princess Charming, Let 'Em Eat Cake, Crazy Quilt, Beat the Band, Candle in the Wind. By Jupiter and night club singer & dancer; in vandeville; m.c. in South American night club; served as instructor in U.S. Coast Guard 1942–45; m.p. debut 1949, in Flamingo Road.

PICTURES INCLUDE: Intruder in the Dust, Beyond the Forest, Breakthrough, Damned Don't Cry, Inside Straight, Forth Worth, Inside Walls of Folsom Prison, This Woman is Dangerous, Springfield Rifle, Million Dollar Mermaid, Ambush at Tomahawk Gap, Perilous Journey, Dawn at Socorro, High and the Mighty, Timberjack, Fury at Gunsight Pass, The Rabbit Trap, The Seven Minutes.

TV: Mr. District Attorney.

BRICKNER, ROY: Executive. b. New York City, January 3, 1904. Advertising copywriting, sales promotion New York, Baltimore. Started motion picture career 1931 MGM film editing dept.; transferred advertising & trailer production, 1938; 1945 joined Warner Bros.; in 1955 appointed head Warner trailer and main title dept.

BRIDGES, BEAU: Actor. b. Hollywood, Calif., Dec. 19, 1941. f. Lloyd Bridges; e. U.C.L.A., U. of Hawaii. After several feature film bit parts and more than 80 TV credits, made screen feature debut in The Incident.

PICTURES INCLUDE: For Love of Ivy, Gaily, Gaily, The Landlord, Adam's Woman, The Christian Licorice Store, Hammersmith Is Out, Child's Play, Your Three Minutes

Are Up, Lovin' Molly, The Other Side of the Mountain, Dragonfly, Swashbuckler, Two Minute Warning, Greased Lightning, Norma Rae, The Runner Stumbles, Honky Tonk Freeway.

TV: Ensign O'Toole, Sea Hunt, Ben Casey, Dr. Kildare, Mr. Novak, Combat, Eleventh Hour.

BRIDGES, JAMES: Actor-Writer-Director. b. Paris, Ark., Appeared as an actor in 50 TV shows and five feature films. Has written 16 plays and is published in New Theatre for Now, 18 hour Hitchcock Shows and one Great Adventure, Go Down Moses. Worked as writer on 14 features. As director, worked in New York, Edinburgh Festival, Mark Taper and Ahmanson in LA.

PICTURES INCLUDE: The Appaloosa (A.P.), Forbin Project, (s.p.), The Baby Maker (s.p., dir.), The Paper Chase (s.p., dir.); September 30, 1955 (s.p., dir.); The China Syndrome (dir., co-s.p.); Urban Cowboy (co-s.p., dir.).

BRIDGES, JEFF: Actor. b. Los Angeles, 1951. Made acting debut at eight in the TV Sea Hunt series of his father, Lloyd Bridges. Acted also in Lloyd Bridges Show. At 14 toured New England with father in stock production Anniversary Waltz. Studied acting with Uta Hagen in New York. Brother of actor Beau Bridges.

PICTURES: The Last Picture Show, Fat City, Bad Company, Lolly Madonna (XXX), The Last American Hero, The Iceman Cometh, Thunderbolt and Lightfoot, Rancho Deluxe, Hearts of the West, Stay Hungry, King Kong, Somebody Killed Her Husband, Winter Kills, Heaven's Gate, Cutter and Bone.

BRIDGES, LLOYD: Actor. b. San Leandro, California, January 15, 1913. Went into stock from college dramatics.

PICTURES INCLUDE: Miss Susie Slagle's, Abilene Town, Canyon Passage, Ramrod, Trouble with Women, Hideout, Calamity Jane and Sam Bass, Trapped, Rocketship XM, Try and Get Me, Colt .45, Three Steps North, Whistle at Eaton Falls, Last of the Comanches, High Noon, Plymouth Adventure, Tall Texan, Kid from Left Field, City of Bad Men, Limping Man, Pride of the Blue Grass, Deadly Game, Apache Woman, Wichita, Wetbacks, The Rainmaker, Daring Game. The Goddess, The Happy Ending, Silent Night, Lonely Night, Airplane!

TELEVISION: Sea Hunt, The Loner, Movieola.

BRIEN, LIGE: Executive, b. Pittsburgh, Pa., Mar. 23; e. U. Pittsburgh, B.A. Asst. theatre mgr. 1937; theatre mgr. (Pa.); winner of many prizes for publicity, exploitation, grosses; winner of M.P. Herald Bronze Grand Award 1939, Silver Award 1944 & War Showmanship Award; press-book ed. PRC Pictures 1945, appt'd exploitation mgr., 1946; since exploit. mgr. Eagle Lion Classics; to U.A. as dir., spcl. events, April, 1951. Pres. AMPA, 153–54; asst. exploitation mgr., 20th Fox, 1960; nat'l dir. mdsg., Paramount Pictures, 1961–65; Expl. Dir., Seven Arts Pictures. (Warner Bros.). Now exploitation liaison between Hollywood studio and New York offices.

BRIGHT, RICHARD S.: Executive. b. New Rochelle, N.Y., Feb. 28, 1936. e. Hotchkiss School (1953–54); Wharton School of Finance, Univ. of Penna. (1954–58). With U.S. Army Finance Corp., 1959–60. Was corporate exec. prior to founding in 1973 Persky-Bright Organization, private investment group to finance films. Now bd. chm., Persky-Bright Productions, Inc.; adjunct professor at Columbia Univ. School of Fine Arts.

PICTURES: Last Detail, Golden Voyage of Sinbad, For Pete's Sake, California Split, The Man Who Would Be King, Funny Lady, The Front, and Equus. Financing/production services for: Hard Times, Taxi Driver, Missouri Breaks, Bound for Glory, Sinbad and the Eye of the Tiger, Hair.

TELEVISION: The President's Mistress (co-prod.), a CBS TV feature.

STAGE: A History of the American Film (1978).

BRIGHTMAN, HOMER H.: Writer b. Port Townsend, Washington. e. at the Washington State Nautical School-ship. Apprentice at 14, spent 10 years at sea as 2nd and 3rd officer; claims dept., Dollar Steamship Lines Shanghai, Hong Kong, Singapore: Academy Award for short Lend a Paw 1939; assoc. Walt Disney Prod. 13 years; created It's A Dog's Life, TV show; animated cartoon shorts for MGM, Walter Lantz, U.P.A.

PICTURES INCLUDE: Saludos Amigos, Three Caballeros, Make Mine Music, Fun and Fancy Free, Melody Time, Cinderella. Writer (collab.) TV shorts.

BRINKLEY, DAVID: TV News Correspondent. b. Wilmington, N.C., July 10, 1920. Started writing for hometown newspaper; attended Univ. of N.C., Vanderbilt University. Joined United Press before entering Army, World War II. After discharge in 1943 jointed NBC News in Washington as White House corr. Co-chm. for many years with late Chet Huntley on NBC Nightly News. Now delivers personal reports (David Brinkley's Journal) on same program.

BRISKIN, MORT: Producer-Writer. b. Oak Park, Ill. e. Univ. of So. Calif. Attended Harvard and Northwestern law schools, being admitted to the Bar at 20. Practiced law before entering m.p. industry in management with such stars as Mickey Rooney. Turned to production and also wrote screenplays for 16 of his 29 films. Created nine TV series and was prod. or exec. prod. of some 1,350 TV segments, of which he wrote more than 300.

PICTURES INCLUDE: The River, The Magic Face, No Time for Flowers, The Second Woman, Quicksand, The Big Wheel, The Jackie Robinson Story, Ben, Williard, Walking Tall, Framed.

TV: Sheriff of Cochise, U.S. Marshal, The Texan, Grand Jury, The Walter Winchell File, Official Detective, Whirlybirds.

BRISSON, FREDERICK: producer; b. Copenhagen, Denmark, Mar. 17, 1913; s. of Carl Brisson, the stage and film actor, and his wife Cleo; e. Rossall College, England; m. the late Rosalind Russell. Formerly a successful actor's agent with offices in London, Paris and Hollywood; was asst. mgr. at London Hippodrome for The Merry Widow, Adelphi, London, Oct. 1937, co-produced Transatlantic Rhythm; came to America in 1939 and engaged in artists' representation and film production; first Broadway production was at the St. James, May, 1954, when he co-produced The Pajama Game; has since produced or co-produced Damn Yankees, 1955; New Girl in Town, 1957; The Pleasure of His Company, The Gazebo, 1958; Five Finger Exercise, 1959; Under the Yum-Yum Tree, 1960; The Caretaker, First Love, 1961; The Time of the Barracudas, 1963; Alfie, 1964; Generation, 1965; The Flip Side, 1968; Coco, 1969; Twigs, 1971; Jumpers, 1974; So Long, 174th Street, 1976; Mixed Couples, 1980; entered film production in England in 1937 with Two Hearts in Three-Quarters Time, and subsequently co-produced Prince of Arcadia and, 1938, produced Moonlight Sonata; entered Hollywood Film production in 1948 when he founded Independent Artists Pictures and produced The Velvet Touch; has subsequently produced Never Wave at a Wac, The Pajama Game, Damn Yankees, Five Finger Exercise, Under the Yum-Yum Tree, Generation, Mrs. Pollifax—Spy, etc.; served in the Air Force during the war as a Lt. Col. and was awarded the US Legion of Merit and the King Christian X Medal of Denmark. Since 1979 co-chm., Rosalind Russell Medical Research Center for Arthritis.

BRITTON, TONY: Actor, b. Birmingham, England, 1924; e. Thornbury Grammar School, Glos. Early career as clerk and in repertory; TV debut, 1952, The Six Proud Walkers (serial), m.p. debut 1955 in Loser Takes All.

PLAYS INCLUDE: The Guv'nor, Romeo and Juliet, The Scarlet Pimpernel, The Other Man, The Dashing White Sergeant, Importance of Being Earnest, An Ideal Husband, School for Scandal, A Dream of Treason, That Lady, The Private Lives of Edward Whiteley, Affairs of State, The Night of The Ball, Gigi, The Seagull, Henry IV Part 1, Kill Two Birds, Cactus Flower, A Woman of No Importance, The Boston Story, Lady Frederick, My Fair Lady, Move Over Mrs. Markham, No No Nanette, Dame of Sark, The Chairman.

PICTURES INCLUDE: Birthday Present, Behind the Mask, Operation Amsterdam, The Heart of a Man, The Rough and the Smooth, The Risk, The Horsemasters, Stork Talk, The Break, There's a Girl in My Soup, Forbush and The Penguins, Sunday Bloody Sunday, Night Watch, The Day of the Jackal.

TV: Melissa series, Father Dear Father series, The Man Who Understood Women, Och La La, Call My Bluff, The Nearly Man.

BROCCOLI, ALBERT: Producer. b. New York City, April 5, 1909, e. C.C.N.Y. Agricuturist in early years; entered m.p. ind. as asst. director, 20th Century-Fox, 1938; exec. prod., Warwick Films.

PRODUCTIONS INCLUDE: Red Beret (Paratrooper), Hell Below Zero, Black Knight, Prize of Gold, Cockleshell Heroes, Safari, Zarak, Pickup Alley, Fire Down Below, Odongo, High Flight, No Time to Die, The Man Inside, Idle on Parade, Adamson of Africa, Bandit of Zhobe, Jazz Boat, In the Nick, Let's Get Married, The Trials of Oscar Wilde, Johnny Nobody, Carolina, Dr. No, Call Me Bwana, From Russia With Love, Goldfinger, Thunderball, You Only Live Twice, Chitty Chitty Bang Bang, On Her Majesty's Secret Service, Diamonds Are Forever, Live and Let Die, The Man With The Golden Gun, The Spy Who Loved Me.

BRODERICK, JAMES: Actor. b. Charlestown, N.Y., March 7. Bdwy. debut after years of study and stock in musical, Maggie. Has appeared in many plays, both on Bdwy. and the road.

PICTURES: Alice's Restaurant, The Group, The Taking of Pelham One Two Three, Dog Day Afternoon.

TELEVISION: Series; Brenner, Family. Specials: The Iceman Cometh, The Children of Innocence, John Brown.

BRODKIN, HERBERT: Producer. b. New York, N.Y., Nov. 9, 1912. e. Michigan U., Yale Drama School, U.S. Army, World War II; then scenery designer, prod. mgr., dir., summer stock; designer, dir., prod. mgr., Broadway theatres; prod.

exec., Paramount, Universal; prod., designer, Charlie Wild Private Detective, CBS-TV; prod., ABC, 1951.
TV: Album, TV Hour, Motorola TV Hour, Center Stage, Elgin Hour, Goodyear Playhouse, Alcoa Hour, Sebastian.

BRODNEY, OSCAR: Writer, Producer. b. Boston, Mass., e. Boston U., LLB, 1927; Harvard, LLM, 1928. Atty., Mass. Bar, 1928–35.
PICTURES INCLUDE: She Wrote the Book, If You Knew Susie, Are You With It?, For the Love of Mary, Mexican Hayride, Arctic Manhunt, Yes Sire, That's My Baby, Double Crossbone, Gal Who Took the West, South Sea Sinner, Comanche Territory, Harvey; story, Frenchie, Francis Goes to the Races, Little Egypt, Francis Covers the Big Town, Willie & Joe Back at the Front, Scarlet Angel, Francis Goes to West Point, Walking My Baby Back Home, Sign of the Pagan, Glenn Miller Story, Black Shield of Falworth, Captain Lightfoot, The Spoilers, Purple Mask, Lady Godiva, Day of Fury, Star in the Dust, Tammy and the Bachelor, When Hell Broke Loose, Bobbikins, Tammy Tell Me True, The Right Approach, All Hands on Deck, Tammy and the Doctor, The Brass Bottle, I'd Rather Be Rich.

BRODSHAUG, MELVIN: Producer. b. Davenport, N.D., May 22, 1900; e. No. Dakota State University, B.S., 1923; Litt.D. 1958 Chicago U., M.A. 1927, Columbia U., Ph.D. 1931; Supt. of Schools, Arnegard, N.D. 1923–28; research assoc. Erpi Classroom Films 1930; apptd. dir. research Encyclopaedia Britannica Films, Inc., apptd. vice-pres. 1945, in chg. of product development; apptd. member bd. dir. 1951, Dean of Boston U. Sch. of Public Relations & Communications, 1954–62, dir., Boston U film library, 1962–65. Prof. and consultant, Norfolk State Coll., Va., 1965–70; independent educational film producer 1970.

BRODSKY, JACK: Producer. b. Brooklyn, N.Y. e. George Washington High School. Writer for N.Y. Times; joined 20th-Fox publicity in N.Y. in 1956. Left in 1961 to head national ad-pub for Filmways. Joined Rastar Productions to work on "Funny Girl"; later named v.p. in charge of prod. In 1976 named v.p. in chg. film prod. prom., Rogers & Cowan; 1978, Columbia Pictures v.p., adv., pub., promo.; 1979, named exec. v.p. of Michael Douglas' Big Stick Productions.
PICTURES INCLUDE: Little Murders, Everything You Always Wanted To Know About Sex (exec. prod.), Summer Wishes, Winter Dreams.
AUTHOR: The Cleopatra Papers, with Nat Weiss.

BROIDY, STEVE: Executive. b. Malden, Mass., June 14, 1905. e. Boston U. Sales exec. Franklin Film Co., N.Y., 1935, Universal, 1926–1930; Warner Bros. 1931–1933; joined Monogram (original corporate name of Allied Artists) 1933 as sales mgr., in 1940 elected to bd. of dir., in 1945, elected v.p. in charge of oper.; Nov. 1945 pres. to 1964; pres., Motion Pictures Intl., 1965. MPI name changed to Associated Film Enterprises.

BROKAW, NORMAN R.: Executive. b. April 21, 1927, New York City. Joined William Morris Agency as trainee, in 1943, junior agent, 1948; sr. agent, company exec. in m.p. and TV, 1951; 1974, v.p., William Morris Agency, World Wide all areas. 1981, named exec. v.p. & member of board, William Morris Agency, worldwide. Member Academy of TV Arts & Sciences, Academy M.P. Arts & Sciences. Member bd. of dir. of Cedars-Sinai Medical Center, Los Angeles; pres., The Betty Ford Cancer Center. Clients include President and Mrs. Gerald R. Ford.

BROKAW, TOM: TV Host—Anchorman. e. Univ. of So. Dakota. Joined KMTV, NBC affiliate in Omaha, in 1962; 1965, joined WSB-TV, Atlanta. Worked in L.A. bureau of NBC News, anchored local news shows for KNBC, NBC station (1966–73). In 1973 named NBC News' White House correspondent; was anchorman of NBC Saturday Night News. Named host of Today Show in August, 1976.

BROLIN, JAMES: Actor. b. Los Angeles, July 10, 1942. e. UCLA. Debut in Bus Stop (TV series); named Most Promising Actor of 1970 by Fame and Photoplay magazines.
PICTURES INCLUDE: Take Her She's Mine, Goodbye, Charlie, Von Ryan's Express, Morituri, Our Man Flint, The Boston Strangler, Skyjacked, West World, Gable and Lombard, The Car, Capricorn One, The Amityville Horror.
TV: The Monroes, Marcus Welby, M.D. (Emmy award), Short Walk to Daylight, Class of '63.

BROMFIELD, JOHN: Actor. r.n. Farron Bromfield; b. South Bend, Ind., June 11, 1922; e. St. Mary's College. College football star. Served in U. S. Navy, medical discharge; worked as tuna fisherman. Film debut in Harpoon, 1948; under contract to Hal Wallis.
PICTURES INCLUDE: Sorry Wrong Number, Harpoon, Rope of Sand, Hold That Line, Flat Top, Easy to Love, Ring of Fear, Black Dakotas, Revenge of the Creature, Big Bluff, Three Bad Sisters, Manfish, Crime Against Joe, Quincannon Frontier Scout, Hot Cars.

BROMHEAD, MICHAEL: Executive. b. London, England, May 30, 1942. e. Charterhouse and St. John's College, Cambridge, RAF pilot during war, Ent. m.p. ind. 1947, with Eagle-Lion; resigned, 1950. Appt. overseas man. Independent Film Dist.; dir. One World Film Dist. 1953; dir. Lion International Films, 1955; New York rep. Lion International, 1958; pres., Lion International Films, Inc., 1959; general manager Lion International Films, 1962; dir. London Independent Producers, Ltd.; man. dir., Alliance Intl. Film Distribs., Ltd., 1966. Dir. EMI Film Dist., 1970. Dir., gen. mgr., 1971; appt. dir. overseas dist., 1975; dir., EMI Film Theatre Corp., man. dir., EMI Film Dist. Overseas Ltd., 1979.

BROMHEAD, RALPH S. (M.V.O., F.C.A.): b. Ealing, Dec. 22, 1906; e. St. Bees, Cumberland, 1927–29; theatre controller Gaumont-Brit. & P.C.T. Ltd.; 1929–34, asst. gen. mgr. Associated British Cinemas Ltd., Managing dir. Regent Circuit Ltd.; 1934–39, dir. and gen. mgr. County Cinemas Ltd. & Entertainments Investment Corporation; 1928, Chartered Accountant; 1939–45, Royal Artillery; 1945–50, dir. & controller of export Eagle Lion; 1950 to date, dir. Romulus Films, Remus Films, etc. and later British & American Film Holdings; 1950 dir. of administration Better Business Drive; 1934, Council C.T.B.F.; 1956–69, Chairman C.T.B.F.; 1969 to date Trustee C.T.B.F. 1971, Appointed M.V.O.

BRON, ELEANOR: Actress. Started career in Establishment Club, London, and on American tour. Leading lady on British TV show Not So Much A Programme, More a Way of Life.
PICTURES INCLUDE: Help, Alfie, Two For The Road, Bedazzled, Women in Love, The Millstone.
THEATRE: The Doctor's Dilemma, Howard's End, The Prime of Miss Jean Brodie.

BRONDFIELD, JEROME: b. Cleveland, O., Dec. 9, 1913; e. Ohio State U., 1936. Reporter, ed. on Columbus Dispatch, Cleveland News, Scripps-Howard NEA Svce., I.N.S., A.P.; story ed., script head, RKO Pathe, Oct., 1944; writer, dir. & supvr. of many doc. shorts incl. This Is America series; TV writer; short story writer; collab. s.p., Below the Sahara; s.p., Louisiana Territory; doc. film writer; now Sports Editor, Scholastic Magazines, Inc. Author, Woody Hayes, The 100-Yard War, Knute Rockne, The Man & the Legend.

BRONSON, CHARLES: Actor. b. Penn., 1920. m. Jill Ireland, actress. Worked as a coal miner. Starred on his own TV series, Meet McGraw.
PICTURES INCLUDE: Red Skies of Montana, Pat and Mike, House of Wax, Drumbeat, Vera Cruz, Jubel, Lonely Are the Brave, Machine Gun Kelly, Never So Few, A Thunder of Drums, The Magnificent Seven, The Great Escape, The Sandpiper, The Battle of the Bulge, Pancho Villa, The Dirty Dozen, This Property is Condemned, Once Upon a Time in the West, Goodbye, Friend, Rider in the Rain, You Can't Win 'Em All, Someone Behind the Door, Red Sun, Chato's Land, The Mechanic, The Valachi Papers, The Stone Killer, Mr. Majestyk, Death Wish, Breakout, Hard Times, Breakheart Pass, From Noon Till Three, St. Ives, The White Buffalo, Telefon, Borderline, Caboblanco, Death Hunt.

BRONSTON, SAMUEL: Producer. b. Russia; e. Sorbonne, Paris. Film distributor, Paris; prod. exec. Columbia Studios, Hollywood; Martin Eden, City Without Men, Producer, Columbia Pictures; resigned to form Samuel Bronston Pict., Inc.; exec. prod. Jack London, A Walk in the Sun. Color documentaries produced first time in Vatican; prod. John Paul Jones 1959; Received U.S. Navy Meritorious Pub. Serv. Citation; pres. Samuel Bronston Productions; 1960 prod. King of Kings, El Cid; 1962, 55 Days at Peking; 1963. The Fall of the Roman Empire; prod. Circus World, 1964, Condor Award, Society for Pan American Culture—for El Cid, 1962 (Award shared with Stanford University), The Hollywood Foreign Press Association Golden Globe for the achievement of his outstanding production of El Cid, 1962, Italian Order of Merit with Medal of Commendatore. Grand Cross of Merit by the Equestrian Order of the Knights of the Holy Sepulchre. (The highest honor of the Catholic Church), The Encomienda of the Order of the Great Cross of Isabel la Catolica. President Samuel Bronston Studios, Madrid, Spain.

BROOKE, PETER R.: Writer. b. Berlin, Germany, April 12, 1929. e. Germany, Switzerland, England. With Warner Bros., 1944–46, mail room, pub. dept., then asst. casting dir.; contract writer, Robert Riskin Prod., Paramount; then free lance; radio writer many shows; TV writer for Teevee Prod., Celebrity Playhouse, Science Fiction Theatre; staff writer, Universal Pictures; left Universal to join Excelsior under new banner of Warbrook Inc. as writer-producer.
PICTURES INCLUDE: Basketball Fix, Sweet Violence orig., Outside the Law.
TV: Sea Hunt, 77 Sunset Strip, Sugar Foot, The Fugitive, Six Million Dollar Man.

BROOKING, DOROTHEA: TV producer. b. Slough, Bucks, Eng., Dec. 7, 1916. e. St. George's School, Switzerland. Stage trained at Old Vic Theatre; TV as prod. BBC children's programmes 1950; from July 1972 freelance producer, actress.

PRODUCTIONS INCLUDE: Railway Children, Secret Garden, Great Expectations, The Viaduct, Thursday's Child, Kizzy, The Phoenix and the Carpet.

BROOKS, DICK: Executive. b. New York, N.Y., e. U. of Georgia, grad. 1951; reporter, Atlanta (Ga.) Journal, 1951; sports ed., Gainesville Times, 1952; war correspondent, Pacific Stars & Stripes, 1952–54; entered m.p. industry, 1957 as staff writer, 20th Century-Fox pub. dept; N.Y. press rep., Embassy Picts., 1959; N.Y. press rep., 20th Century Fox, 1960; nat'l publ. dir., Seven Arts Productions, 1962; Paramount pub. mgr., 1965; nat'l pub. dir., 20th-Fox, 1967; adv.-pub. dir., Rastar Prods., 1970; formed Dick Brooks Organization, independent p.r. and prod. co., 1972. In 1977 joined Rogers & Cowan as dir. of mktg., film division. In 1978 named Warner Bros. pub. dir.; 1980, formed Dick Brooks Unlimited, p.r. & m.p. prod. co.

BROOKS, JOSEPH: Producer-Director-Writer-Composer-Conductor. Well-known for composing music for TV commercials before turning to producing, directing, writing and scoring theatrical feature, You Light Up My Life, in 1977. Winner of 21 Clio Awards (made by adv. industry); created music for 100 commercials. Has also composed for theatrical films.

PICTURES INCLUDE: Scores: The Garden of the Finzi Continis, Marjoe, Jeremy, The Lords of Flatbush. Produced, directed, wrote, and composed, arranged and conducted music for You Light Up My Life. Produced, directed, wrote, starred in and composed music for If Ever I See You Again.

BROOKS, MEL: Writer-Director-Actor. b. Brooklyn, 1926. m. actress Anne Bancroft. As child did impressions and was amateur drummer and pianist. First appearance as actor in Golden Boy in Redbank, N.J. Was also director and social director in the Catskills. Became writer for Sid Caesar on TV's Broadway Review and Your Show of Shows, writing for latter for decade. Teamed with Carl Reiner for comedy record album, 2000 Year Old Man and The 2000 and 13 Year Old Man.

PICTURES INCLUDE: The Critic (short, s.p.-narrator), The Producers, The Twelve Chairs, Blazing Saddles, Young Frankenstein, Silent Movie, High Anxiety (prod., dir., s.p.). History of the World—Part I (prod., dir., s.p., actor, composer).

TV: Get Smart (writer), When Things Were Rotten (creator).

STAGE: Shinbone Alley, All-American (books for musicals).

BROOKS, RICHARD: Writer, director. b. Philadelphia, May 18, 1912; e. Temple Univ.; Radio writer, narrator and commentator, NBC. Author of several short stories. Novels, Brick Fox Hole, Boiling Point, The Producer.

PICTURES INCLUDE: Swell Guy, Brute Force, To the Victor, Crossfire, Key Largo, Mystery Street, Storm Warning, Deadline U.S.A., Battle Circus, Take The High Ground, Flame and the Flesh, Last Time I Saw Paris, Blackboard Jungle, Last Hunt, Catered Affair, Something of Value, The Brothers Karamazov, Cat on A Hot Tin Roof, Elmer Gantry (Academy Award for s.p. 1961); Sweet Bird of Youth, writer, Lord Jim, The Professionals, In Cold Blood, The Happy Ending, Dollars, Bite the Bullet, Looking for Mr. Goodbar.

BROOKS, THOR L.: Director. b. Stockholm, Sweden. Studied m.p. photographer at Reinman School, Berlin, dramatics & dir. in Paris & Stockholm; dir. 36 films in Europe, 1936–41; U.S. Air Force, 1942–46; dir., commercial films, 1947–49; staff dir., KLAC-TV; dir., assoc. prod., TV film series supervisor, I Search for Adventure, 1955; dir. Medi; prod. docum. feature, Kwaheri.

PICTURES INCLUDE: Legion of the Doomed, Arson for Hire.

TV: 7 League Boots, Sweet Success; dir., Devils Disciple.

BROUMAS, JOHN G.: Exec. b. Youngstown, O. Oct. 12, 1917, e. Youngstown, O. Usher, Altoona Publix Theatres, 1933, usher to asst. mgr., Warner Thea. 1934–39; mgr. Grand 1939–40; mgr. Orpheum 1940–41. WW-II active, Officer Chemical Corps, past Commander 453rd Chem. Grp. (Reserve); Member Reserve Officers Assoc.; Gen. mgr. Pitts & Roth Theatres 1946–54; President, Broumas Theatres; bd. govnrs., Variety Club, Tent No. 11, 1959–73; Wash., first asst. chief barker, 1964 & 1971; chief barker, 1965–1966, 1972 V.P.T.O.A. 1965; V.P. NATO, 1969; past bd. chmn. Maryland Theatre Owners; bd. of directors of NATO of Va., Md., D.C.; pres., Broumas Theatre Service; pres., Showcase Theatres; pres., Capital Plaza Cor.; mem. bd. Foundation of Religious Action (FRASCO). Mem. Nat'l Immigration Comm. Founder of John G. Broumas Scholarship for Drama at Youngstown State Univ.; past pres. & Board Chairman Maryland Theatre Owners; v.p. & bd. of dir.—Virginia Theatre Owners; bd. of dir. NATO of D.C.; pres. B.C. Theatres; Sponsor of Andre G. Broumas Memorial Award—West Point; pres., McLean Theatre Corp.; member, Motion Picture Pioneers; sponsor, Broumas Scholastic and Athletic Scholarship for AHEPA Wash. D.C. area; director, McLean Bank McLean, Va.; honorary member, West Point Class 1954; Director, Exchange, Ltd., Wash., D.C.; President, University III Cinema Corp.; Secretary-Treasurer, Tenley Circle Twin Theatre Corp.; President, Cinema Mgt. Corp; Director: Motion Picture Pioneers; Advisory Council: Will Rogers Memorial Hospital; Chief Barker: Washington, D. C. Variety Club, Tent #11, 1978–79 and bd. chmn., 1980.

BROWN, CLARENCE: Director. b. Clinton, Mass., May 10, 1890; e. U. Tenn. Aviator, U.S. Army, World War I. Engineer, automobile industry; formed Brown Motor Co., Birmingham, Ala. To Hollywood 1915 as asst. to dir. Maurice Tourneur for whom he directed his first picutre. The Great Redeemer. Joined MGM 1924. Launched Greta Garbo as star in Flesh and the Devil; other pictures with Garbo are Woman of Affairs, Anna Christie, Romance, Inspiration, Anna Karenina, Conquest.

PICTURES INCLUDE: The Eagle, Kiki, Night Flight, Sadie McKee, Ah, Wilderness!, Wife vs. Secretary, The Gorgeous Hussy, Of Human Hearts, Idiot's Delight, Edison the Man, The Human Comedy, White Cliffs of Dover, National Velvet, Letty Lynton, Possessed, Trail of '98, The Rains Came, The Human Comedy, Song of Love, Intruder in the Dust, To Please a Lady, Angels in the Outfield, When in Rome, The Yearling, Plymouth Adventure, The Secret Garden, Never Let Me Go.

BROWN, DAVID: Executive and Producer. b. New York, N.Y., July 28, 1916; e. Stanford U., A.B., 1936; Columbia U. (school of Journalism), M.S., 1937. Apprentice reporter, copy-editing, San Francisco News & Wall St. Journal, 1936; night ed. asst. drama critic, Fairchild Publications, N.Y., 1937–39; edit. dir. Milk Research Council, N.Y., 1939–40; assoc. ed., Street & Smith Publ., N.Y., 1940–43; assoc. ed., exec. ed., then ed.-in-chief, Liberty Mag., N.Y., 1943–49; edit. dir., nat'l education campaign, Amer. Medical Assn., 1949; assoc. ed., man. ed., Cosmopolitan Mag., N.Y., 1949–52; contrib. stories & articles to many nat'l mags.; man. ed., story dept., 20th-Fox, L.A., Jan., 1952; story ed. & head of scenario dept., May, 1953–56; appt'd. member of exec. staff of Darryl F. Zanuck, 1956; mem. of exec. staff, 20th-Fox studios, and exec. studio story editor, 1956–1960; Prod. 20th-Fox Studios, Sept. 1960–62; Editorial v.p. New American Library of World Literature, Inc., 1963–64; exec. story opers., 20th Century-Fox, 1964–67; v.p. dir. of story operations, 1967; exec. vice-pres., creative operations and members, Board of Directors, 1969–71. Exec. v.p., member board of directors Warner Bros. 1971–1972; partner and director, The Zanuck/Brown Co., 1972; mbr., bd. of trustees, American Film Institute, 1972–80.

FILMS INCLUDE: The Sugarland Express, The Black Windmill, Sssssss, The Eiger Sanction, The Sting, MacArthur, Jaws, Jaws 2, The Island, Neighbors.

BROWN, GEORGE: Producer. b. London, Eng., e. Latymer Upper Sch., Colegio Franco, Barcelona. C.O., RAF Film Unit (overseas), World War II. Asst. dir., Ealing and others.

PICTURES INCLUDE: Fire Over England, Farewell Again, Vessel of Wrath, Jamaica Inn, 49th Parallel, Journey Together, School for Secrets, Vice Versa, Sleeping Car to Trieste, Chiltern Hundreds, Hotel Sahara, Made in Heaven, Desperate Moment, The Seekers (Land of Fury), Jacqueline, Dangerous Exile, Rooney, Tommy the Toreador, The Boy Who Stole A Million, Double Bunk, Murder at The Gallop, Ladies Who Do, Murder She Said, Village of Daughters, Kill or Cure, Guns of Batasi, The Trap, Finders Keepers, Ghost of a Chance, Up in the Air, Fanfare Films Ltd., Eyeline Films, Parallel Prods. Hoverbug, Assault, Revenge, All Coppers Are . . . , Innocent Bystanders, Penny Gold, Open Season.

BROWN, HAROLD: Executive. Pres., International Picture Show Co. since Dec., 1980. Previously pres., American International TV; v.p. & bd. mbr., American International Pictures (now Filmways).

BROWN, HARRY: Writer. b. Portland, Me., April 30, 1917. e. Harvard. With The New Yorker 1940; U.S. Army 1941–45.

PICTURES INCLUDE: (collab.) True Glory 1944, Walk in the Sun 1945; The Other Love, Arch of Triumph 1947; Wake of Red Witch, Man on Eiffel Tower, Sands of Iwo Jima (s.p.), Kiss Tomorrow Goodbye, Only the Valiant, Bugles in Afternoon, Place in the Sun (orig. story;) Academy Award collab., Apache Drums, The Sniper, Eight Iron Men, All the Brothers Were Valiant, Many Rivers to Cross, Virgin Queen, D-Day the Sixth of June, Between Heaven and Hell, Ocean's 11, El Dorado.

BROWN, HENRY: Executive. b. New York, N.Y., Feb. 18, 1899; e. C.C.N.Y., 1920; Columbia University, 1923. pres., Lakewood Amusement Corp.; pres., Atlas TV Corp.; pres. Capt. Zero Prods., Inc.

BROWN, HIMAN: M.P. Producer & director, b. N.Y.C. July 21, 1910; e. C.C.N.Y., St. Lawrence Univ. Radio & TV package prod. since 1927 include: Inner Sanctum, Thin Man, Bulldog Drummond, Dick Tracy, Terry and the Pirates, Joyce Jordan MD, Grand Central Station, CBS Radio Mystery Theatre, pres. Production Center, Inc.

PICTURES INCLUDE: That Night, Violators, The Stars Salute, The Price of Silence, The Road Ahead.

BROWN, HOWARD C.: Executive. b. Newport, N.H., Aug. 15, 1901. e. Boston U. Assoc. with Technicolor Motion Picture Corp., before org. Colorart Pictures, 1926. Merged with Synchrotone Pictures & Kennedy Pictures, 1929, forming Colorart Synchrotone Corp. In 1930 org. Brown-Nagel Productions. Since 1933 has produced numerous shorts. Sales mgr., Cinecolor, 1936–37; pres. Howard C. Brown Co. 1937–1938. Sales mgr., Trimble Laboratories, Inc. 1947; formed Howard C. Brown Productions 1948, to release for United Artists. Prod. Kangaroo Kid; prod. Mating Urge, inter-nations, documentary; 1969, Film Exporter.

BROWN, JIM: Actor. b. St. Simons Island, Feb. 17, 1936. e. Manhasset H.S., Syracuse U. For nine years played football with the Cleveland Browns; in 1964 won Hickock Belt as Professional Athlete of the year; founder, Black Economic Union.

PICTURES INCLUDE: Rio Conchos, The Dirty Dozen, Ice Station Zebra, The Split, Riot, Dark Of The Sun, 100 Rifles, Kenner, Tick . . . Tick . . . Tick, The Grasshopper, Slaughter, Slaughter's Big Rip-Off, I Escaped from Devil's Island, The Slams, Three the Hard Way, Take a Hard Ride.

BROWN, TOM: Actor. b. New York City, Jan. 6, 1913; son of Harry Brown, actor-producer, and Marie Francis, actress; e. N.Y. Prof. Children's School. On radio, then dram. stage (Neighbors, Many a Slip, Pardon My Glove, etc.). Screen debut in A Lady Lies, 1929.

PICTURES INCLUDE: Buck Privates Come Home, Duke of Chicago, Operation Haylift, Fireman Save My Child.

TV: General Hospital.

BROWN, WILLIAM: Executive. b. Ayr, Scotland, June 24, 1929. e. Ayr Academy, University of Edinburgh, where graduated Bachelor of Commerce 1950. Served to Lieutenant, Royal Artillery, 1950–52. Sales mgr. for Scotland Television Ltd. in London, England, 1958 to 1961, sales director 1961 to 1963. Deputy mng. dir. of Scottish Television Ltd. at Glasgow 1963–66, mng. dir. in 1966. Deputy chm. and mng.-dir. of STV from 1974. Director, independent TV Publications, London, England, 1968. On board of directors for Scottish National Orchestra, 1973. Member of Independent TV Companies Association, and chairman of same 1978–80. Holds C.B.E. honor from British Royalty.

BROWN, WILLIAM H., JR: Director, concert mgr., Juilliard School of Music, 1946; TV prod., Young and Rubicam, 1948; dir., Life with Father, Climax, CBS-TV, 1954, Studio One 1955; prod. Paramount Television, 1958; Exec. prod., Shirley Temple Show, NBC, 1960; TV prod., Norman, Craig and Kummel, 1962; Compton, 1963; Post-Keyes-Gardner, 1965; Free lance, 1967.

BROWN, WILLET HENRY: Executive. Assist. gen. mgr., Don Lee, Inc., 1932–33; exec. v.p. Don Lee Broadcasting System, Hollywood, 1938–48; pres. Don Lee Broadcasting System, 1949–58; consultant, R.K.O. General Inc.: pres. Don Lee, Inc. pres. Laurie Leasing Corp.: pres. Pacific States Invest. Corp.: pres. Hillcrest Motor Company: pres., KGB INC, KGB-AM and KGB-FM, San Diego, Calif.: pres. bd. and dir. of KXOA, Inc., KXOA, AM and KXOA-FM, Sacramento, Calif.

BROWNE, ROSCOE LEE: Actor, director, writer. b. Woodbury, N.J. e. Lincoln University, Pa. Post graduate studies in comparative literature and French at Middlebury College, Vermont, and Columbia U., N.Y. Taught French and lit. at Lincoln U. until 1952. National sales rep. for Schenley Import Corp. 1946–56; he was one of the United States' international track stars and a member of ten A.A.U. teams. He was twice American champion in the 1000-yard indoor competition, twice all-American and, in 1951 in Paris, ran the fastest 800 meters in the world for that year. Professional acting debut, 1956, in Julius Caesar at the N.Y. Shakespeare Festival; published poet and short story writer.

BROADWAY: The Ballad of the Sad Cafe, The Cool World, General Seeger, Tiger, Tiger Burning Bright!, The Old Glory. Star and dir. A Hand is On The Gate; off Broadway; The Blacks, Aria da Capo.

PICTURES INCLUDE: The Comedians, Uptight, The Liberation of L.B. Jones, The Cowboys, World's Greatest Athlete, Superfly T.N.T., Logan's Run, Twilight's Last Gleaming, The Fifth Door.

BROWNING, KIRK: TV Director; b. N.Y.C., Mar. 28, 1921; e. Brooks School, Andover, Mass., Avon Old Farms, Avon, Conn., and Cornell U.; 1940, reporter to News-Tribune in Waco, Texas; with American Field Service, 1942–45; adv. copywriter for Franklin Spier, 1945–48; became floor manager NBC-TV 1949; app'ty asst. dir. NBC-TV Opera Theatre in 1951 directing NBC Opera Theatre, TV Recital Hall, and Toscanini Simulcasts.

BRUCE, BRENDA: Actress. b. Manchester, England, 1922, e. privately. London stage debut, 1066 And All That; On screen 1944; Millions Like Us; TV Best Actress Award 1962.

PICTURES INCLUDE: Night Boat to Dublin, I see A Dark Stranger, They Came To A City, Carnival, Piccadilly Incident, While The Sun Shines, When The Bough Breaks, My Brother's Keeper, Don't Ever Leave Me, The Final Test, Law and Disorder, Behind the Mask, Peeping Tom, Nightmare, The Uncle.

BROADWAY: Gently Does it (1953), This Year Next Year, Happy Days, Woman In A Dressing Gown, Victor Eh!, Merry Wives of Windsor, The Revenger's Tragedy, Little Murders, Winters Tale, Pericles, Twelfth Night, Hamlet.

TV Appearances: Mary Britton series, Nearer to Heaven, Wrong Side of the Park, The Lodger, The Monkey And the Mohawk, Love Story, A Piece of Resistance, Give the Clown His Supper, Knock on Any Door, The Browning Version, Death of a Teddy Bear, Softly, Softly, The Girl, Happy, Family at War, Budgie.

BRUCE, GEORGE: Writer. b. Danville, Pa., Sept. 15, 1898; e. priv. Aviator, 1919–27; then free lance writer; mag. pub. From 1937 sole author or collab. s.p.

PICTURES INCLUDE: She's No Lady, The Crowd Roars, Navy Blue and Gold, Duke of West Point, King of the Turf, South of Pago Pago, Man in Iron Mask, Son of Monte Cristo, Corsican Brothers, Two Years Before the Mast, Miss Annie Rooney, Fiesta, Killer McCoy, Walk A Crooked Mile, Rogues of Sherwood Forest, Lorna Doone, Valentino, The Brigand, Kansas City, Confidential, Fury in Paradise, Solomon and Sheba.

BRUCKHEIMER, JERRY: Producer. Was adv. agency exec. in TV commercials before becoming producer of films.

PICTURES: Assoc. Prod.: Culpepper Cattle Company, Rafferty and the Gold Dust Twins. Producer: Farewell My Lovely, March or Die, Defiance, American Gigolo, Thief.

BRUCKNER, SIDNEY THOMAS: Executive. b. New York City, May 28, 1914; e. C.C.N.Y., B.B.A. 1931; Columbia U., post-graduate study 1931–32. Textile Buyer (partner) 1931–35; joined Columbia Pictures Intern. Corp. as home office rep., covered all countries south of U.S. 1935–38 and 1940–42, covered Europe with hdqtrs in England and Paris 1938–40; apptd. prod. supervisor (Mexican prod.) 1943; res. 1950; ind. prod., 1950, of Mexican films, American TV films; pres. Prod. Espada S.A.; technical advisor, Garden of Evil, Tall Men; prod. 19 TV shows, Capt. David Grief. Spanish/English dubbing of features, TV films, commercials. Mem. Prod. Guild (Mexico), Variety Club (Mexico), American Club, Am. Soc.

BRUNING, RICHARD R.: Executive. b. Kansas City, MO. e. Yale University, Harvard Business School. Asst. treas. Transamerica Corp., parent corp. of United Artists, from 1972. In 1979 named treas. of UA. 1980, named v.p. and treas. of UA Corp.

BRUSKI, NATALIO: Writer, Journalist, Buenos Aires. b. Buenos Aires, B.A. Argentina, Aug. 9, 1906; p. non-professionals; e. high school graduate; married; Had commercial career before entering m.p. ind. In 1932, mgr. Coliseo Theatre. In 1933, mgr. Govena Theatre. In 1934, film critic. La Pelicula, show business professional organ. In 1939; founded own trade paper, Cine-Prensa, since 1951, publicity mgr. for Columbia Pictures. Buenos Aires corresp., M.P. Herald since 1931.

BRUSSEAU, WILLIAM E.: Executive. b. Detroit, Mich. June 6, 1926. e. Engineering, Lawrence Institute of Technology, 1942–44; theatre arts UCLA, 1948; m.p. prod., USC, 1948–50. Joined staff engaged in producing documentaries; prod. films, science, industry, medical profession and the military. gen. doc., TV in Europe, Africa, Middle East, Far East, Latin America, U.S.; formed & pres. Westminster Films, 1950.

PICTURES INCLUDE: Mr. Texas, orig. story, A Cry in the Night, Skidrow, Stop Gap. The Glass Mountain, Dream of Paradise, Flamingo Feather, Shackleton.

BRYAN, DORA: Actress. b. Southport (Lancashire), Eng. e. Council School. Stage debut 1935. Screen debut Fallen Idol, 1948.

PICTURES INCLUDE: No Room at the Inn, Once Upon a Dream, Blue Lamp, Cure for Love, Now Barabas, The Ringer, Women of Twilight, The Quiet Woman, The Intruder, You Know What Sailors Are, Mad About Men, See How They Run, Cockleshell Heroes, Child in the House, Green Man, Carry on Sergeant, Operation Bullshine, Desert Mice, The Night We Got the Bird, A Taste of Honey.

TV: Dora Bryan Show.

STAGE: Hello Dolly, 1966.

BRYNNER, YUL: Actor. b. Sakhalin Island, July 11, Started singing in night clubs in Paris with gypsy groups at age of 13; then joined Cirque d'Hiver as trapeze artist; joined repertory company of Theatre des Maturins, Paris; made American stage debut in Twelfth Night; TV director for CBS.

BROADWAY: Lute Song, Dark Eyes. The King and I (Donaldson award, Antoinette Perry Award, Critics Circle Award.)

PICTURES INCLUDE: Port of New York, King and I (Academy award, 1956, best actor), Ten Commandments,

Anastasia, The Journey, Solomon and Sheba, Once More, With Feeling, The Magnificent Seven, Taras Bulba, Escape to Zahrain, Kings of the Sun, The Saboteur, Code Name, Morituri, Invitation To A Gunfighter, Return of the Seven, Triple Cross, The Double Man, The Long Duel, The Battle of Neretva, The Magic Christian, File of the Golden Goose, Madwoman of Chaillot, Romance of a Horsethief, Light at the Edge of the World, Fuzz, Westworld, Futureworld.

TV: Anna and the King of Siam.

BUCHHOLZ, HORST: Actor. b. 1933. e. High School. In radio and stage plays. Film debut. Marianne (French) 1955.

PICTURES INCLUDE: Himmel Ohne Sterne (No Star in the Sky), Robinson Must Not Die, Mon Petit, The Confessions of Felix Krull, Tiger Bay, Fanny One, Two, Three, Nine Hours to Rama, The Empty Canvas, The Great Waltz, Cervantes.

TV: Raid on Entebbe.

BUCK, JULES: Producer. b. St. Louis, Mo., July 30, 1917. Asst. to prod., The Killers; assoc. prod., Brute Force, Naked City, We Were Strangers; prod., Love Nest, Fixed Bayonets, Treasure of the Golden Condor, O.S.S., TV series; prod., The Day They Robbed the Bank of England, Great Catherine; formed Keep Films with Peter O'Toole. Co-prod., Under Milkwood. Prod., The Ruling Class. Exec. Prod., Man Friday. Prod., The Great Scout and Cathouse Thursday.

BUCKNER, ROBERT H.: Producer, writer b. Crewe Va., May 28, 1906; e. U. Va.: Edinburgh (Scotland) U. Newspaper corresp. New York World. 1926–27 in England: instructor, Belgian Mil. Acad. 1927–28: with Alfred A. Knopf, Inc., Doubleday, Doran, N.Y., pub., 1928–33; corresp. in Russia, Scandinavia, 1933–35; contrib. fiction, Amer. & Brit. mags., 1926–36. Author & collab. many s.p.

PICTURES INCLUDE: Gold Is Where You Find It, Jezebel, Dodge City, Virginia City, Knute Rockne, Santa Fe Trail, Dive Bomber, Yankee Doodle Dandy, Desert Song. In 1943 prod. Gentleman Jim, Mission to Moscow, Desert Song. A Champion of Champion Producer in Fame ratings. (See Index). The Gang's All Here, Cheyenne, Life With Father, prod. Rogue's Regiment. Sword in the Desert; Free For All, Deported, Bright Victory, The Man Behind the Gun, Safari.

BUJOLD, GENEVIEVE: Actress. b. Montreal, July 1, 1942, e. Montreal Conservatory of Drama. m. TV Director Paul Almond. Worked in a Montreal cinema as an usherette; American TV debut: St. Joan.

STAGE: The Barber of Seville, A Midsummer Night's Dream, A House . . . A Day.

PICTURES INCLUDE: La Guerre est Finie, La Fleur de L'Age, Entre La Mer et L'eau Douce, King of Hearts, The Thief, Isabel, Anne of The Thousand Days, The Act of the Heart, The Journey, Earthquake, Swashbuckler, Obsession, Alex and the Gypsy, Another Man, Another Chance, Coma, Murder by Decree, The Last Flight of Noah's Ark.

TV: Anthony and Cleopatra.

BUNDY, FRANK: Producer. b. Hanwell, England. Jan. 12, 1908. Entered m.p. industry with Herbert Wilcox Productions, London: from 1925–29 successively asst. Brit. Natl. Pictures. Ltd.; Brit. Int. Pictures, Ltd.; Brit. Instruct. Films, Ltd.; G.B. Pictures Corp.; Imperator Films Ltd.: joined Brit. Council as prod. on propaganda films 1939: apptd. assoc. prod. Sydney Box Productions, Ltd., Jan. 1946; apptd. personal asst. Sydney Box Aug. 1946; apptd., prod. Gainsborough Pictures 1947; Dir. Screen Audiences Ltd., 1955; Man. dir., Wallace Prods Mercurio D'Oro Award Venice, 1960.

PICTURES INCLUDE: Steel, Easy Money, Christopher Columbus, Diamond City, Daybreak, Stone Into Steel, Fair Oriana, Caught in the Net, Master of Venus, Count Down to Danger, for children's Film Foundation.

BUNUEL, LUIS: Director. b. 1900, Calanda, Spain. Jesuit education. Dir. first performance of Falla's puppet opera, El Retablo de maese Pedro. Founded one of the first European film clubs in 1920. After taking a degree in philosophy and letters, worked as assistant dir. to Jean Epstein on Mauprat and The Fall of the House of Usher. 1927, Assist. dir. La Sirene des Tropiques; 1928, Un Chien Andalou. 1930, L'Age D'or. Spanish films: 1932, Land Without Bread, Exec. prod.; Don Quentin el Amargao, La Hija de Juan Simon, Quien me Quiere a Mi?, Centinela!, Centinela! Alerta! 1939 & 42 worked on American propaganda films. Dir. in Mexico: 1947 Gran Casino; 1949, El Gran Calavera; 1950, Los Olvidalos (The Young and the Damned), Susana; 1951, Daughter of Deceit, Woman Without Love, Ascent to Heaven; 1952, El Bruto, The Adventures of Robinson Crusoe, El; 1953, Wuthering Heights, Illusion Travels By Streetcar; 1954, El Rio Y La Muerte; 1955. The Criminal Life of Archibaldo De La Cruz; 1956, La Mort en Ce Jerdin; 1958, Nazarin; 1959, La fievre monte a El Pao; 1960, The Young One; 1961, Viridana; 1962, The Exterminating Angel; 1965, Diary of Chambermaid, Simon of the Desert; 1967. Belle de Jour; 1969, The Milky Way, Tristana, The Discreet Charm of the Bourgeoisie, Le Fantome de la Liberte.

BUONO, VICTOR: Actor. b. San Diego, Calif. 1939. Studied English and drama at Vallanova and acted at Old Globe Theatre, San Diego.

PICTURES INCLUDE: Whatever Happened to Baby Jane?, Four for Texas, The Strangler, Robin and the Seven Hoods, Hush, Hush . . . Sweet Charlotte, The Greatest Story Ever Told, Young Dillinger, The Silencers, Who's Minding the Mint?, The Wrath of God, Arnold.

BURGER, HENRY: Executive. b. Pittsburgh, Pa., Mar. 26, 1915. Asst. mgr. and mgr., various theatres in Pittsburgh and W.Va.; asst. ad. pub. mgr., Pittsburgh Zone Warner Bros. Theatres, 1942; ad., pub. mgr., 1948; district mgr., Stanley Warner Theatres, Erie district 1951; district mgr., Pittsburgh City Theaters, zone ad. pub. mgr., 1953; Zone mgr., Stanley-Warner Midwest-Zone Theatres, 1962; nat. adv. pub. dir., Stanley Warner Theatres, 1965; zone manager, 1967.

BURKE, ALFRED, Actor. b. London, England, 1918. Ent. films 154.

PICTURES INCLUDE: The Angry Silence, Moment of Danger, The Man Inside, The Man Upstairs, No Time To Die, Law and Disorder, Yangtse Incident, Interpol, Bitter Victory.

TV: The Crucible, Mock Auction, Parole, No Gun, No Guilt, The Big Knife, Parnell, The Strong Are Lonely, Home Of The Brave, The Birthday Party, The Watching Eye, Public Eye series.

BURKE, PAUL: Actor. b. New Orleans, La., July 21, 1926, e. Prep Schools, N.O., Pasadena Playhouse.

TV: Five Fingers, Harbor Master, Noah's Ark, Naked City, Playhouse 90, Studio One, Medic, Frontier, Men in Space, Man and the Challenge, Target, M-Squad, Black Saddle, Philip Marlowe, Martin Kane, Line Up, Dragnet, Man Without a Gun, Tightrope, Panic, Highway Patrol, Men of Annapolis, Flight, Naked City, 12 O'Clock High.

PICTURES INCLUDE: Once You Kiss a Stranger, Daddy's Gone A-Hunting, Valley of the Dolls.

BURKS, J. COOPER: Executive. b Pilot Point, Texas, April 26, 1919. e. Oklahoma City University Oklahoma City Law School, American Institute of Banking (Special Finance Courses), U.S. Navy Schools at Wesleyan College, Middletown, Conn. University of North Carolina, Chapel Hill, N.C., Virginia Polytechnic Institute, Blacksburg, Va. Military: Commander USNR-Ret. WWII PT Boats, North Atlantic then as Naval Aviator—latter part WWII and Korean War. Business experience: Prior to theatre business, was in banking & construction business. Formed present company March, 1970. Now president at chairman of the board of AATI (American Automated Theatres, Inc.) in Oklahoma City. Military Affairs Committee, Chamber of Commerce, Life Member-Navy League.

BURLINSON, JOHN J. JR.: Executive. b. New York, N.Y., Dec. 26, 1930. e. Collegiate Sch. N.Y., Fordham U.; dir. promotion and adv., Henry Regnery Co. (book publishers). 1960; sls. mgr. Regnery, 1962; N.Y. rep., U.S. Press Association, 1963; v.p., Quigley Publishing Co., N.Y., 1964. Ad Dir. Better Theatres Theatre vending; adv. mgr. Motion Picture Daily, 1966; National Screen Service Corp. 1970, dir. inter-corporate promotion 1972; dir. of admin., National Theatre Supply; 1974, gen. mgr., NTS; 1977, v.p. & gen. mgr., EPRAD, Inc.; 1980, named pres., EPRAD: Elected exec. dir. & treasurer of TESMA in 1967. Co-director NATO/NAC/TESMA Trade Show; sec. & dir., M.P. Pioneers, 1970; hon. dir. TESMA, 1971; member SMPTE, Variety Club. 1972; Vice President Equipment Assoc.; 1974; bd. mem. TEA and Conference chm.; Chm. annual NY Variety Golf Tournament 1972–75; vice chairman, Motion Picture Div., Catholic Charities of NY, 1972–75.

BURNETT, CAROL: Actress-comedienne-singer; b. San Antonio, Apr. 26, 1936; Jody and Louise (Creighton) Burnett; e. Hollywood H.S., UCLA. m. Joseph Hamilton, 1963; children-Carrie Louise, Judy Ann, Erin Kate. Introduced comedy song, I Made a Fool of Myself Over John Foster Dulles, 1957; Broadway debut in Once Upon a Mattress, 1959; regular performer Garry Moore TV show, 1959–62; appeared several CBS-TV spls., 1962–63, Broadway play Fade Out-Fade In, 1964, play Plaza Suite, 1970, film Pete 'n Tillie, 1972, musical play I Do, I Do, 1973; Same Time Next Year, 1977; Film Front Page, club engagements Harrah's Club, The Sands, Caesar's Palace; star Carol Burnett Show, CBS-TV. CBS-TV Spec. "6 Rooms Riv Vu" 74; "Twigs" The Grass Is Always Greener, Friendly Fire, The Tenth Month. 75. Films: The Front Page, A Wedding (best actress, San Sebastian Film Festival), Health, The Four Seasons, Chu Chu and the Philly Flash. Recipient outstanding comedienne award Am. Guild Variety Artists, 5 times; 5 Emmy awards for outstanding variety performance Acad. TV Arts and Scis., TV Guide award for outstanding female performer 1961, 62, 63; Peabody award, 1963; Golden Globe award for outstanding comedienne of year Fgn. Press Assn.; Woman of Year award Acad. TV Arts and Scis. One of the world's 20 most admired women voted in poll conducted by George Gallup (1977). First Annual National

Television Critics Award for Outstanding Performance, 1977.

BURNETT, W. R.: Writer. b. Springfield, Ohio, Nov. 25, 1899; e. Miami Mil. Inst., Ohio State U. Novelist (Little Caesar, Dark Hazard, Iron Man). Author of collab. s.p. including Scarface, Heart of a City, Whole Town's Talking, Dr. Socrates, High Sierra, Wake Island, This Gun for Hire, Yellow Sky, Belle Starr's Daughter, Asphalt Jungle, Vendetta, Iron Man, I Died a Thousand Times, The Racket, Dangerous Mission, Captain Lightfoot, Illegal.

BURNS, GEORGE: Actor. r.n. Nathan Birnbaum; b. New York City. In vaudeville as singer in children's quartet, later as roller skater, then comedian; formed team Burns & (Gracie) Allen 1925; m. partner 1926. Team many years on Keith and Orpheum vaudeville circuits, then on screen in Paramount short subjects, on radio in England; in 1930 began long career Amer. radio. Feature picture debut 1932 in The Big Broadcast.

PICTURES INCLUDE: International House, Love in Bloom, College Swing, Honolulu, Two Girls and A Sailor, The Sunshine Boys, Oh, God, Just You and Me, Kid, Oh, God! Book II.

TV: Co-star Burns & Allen show, CBS-TV.

BURR, RAYMOND: Actor. b. New Westminister, B.C., Canada, May 21, 1917; e. Stanford U., U of Calif., Columbia U, U of Chungking. Forestry svce.: appeared on stage in many countries in Night Must Fall, Mandarin, Crazy with the Heat, Duke in Darkness; dir., Pasdena Community Playhouse, 1943; on radio.

PICTURES INCLUDE: Pitfall, Raw Deal, Key to the City, His Kind of Woman, Place in the Sun, New Mexico, Meet Danny Wilson, Mara Maru, Horizons West, Blue Gardenia, Fort Algiers, Casanova's Big Night, Gorilla at Large, Khyber Patrol, Rear Window, Passion, They Were So Young, You're Never Too Young, A Man Alone, Count Three and Pray, Please Murder Me, Godzilla King of the Monsters, Great Day in the Morning, Secret of Treasure Mountain, Cry in the Night, P.J.

TV: Perry Mason series, CBS-TV, 1957. Ironside series, CBS, 1970, Centennial (1978).

BURRELL-DAVIS, Derek: Freelance producer, director with own company "B.D. Productions": b. York 1918 entered m.p 1946 with J. Arthur Rank Org. 1950 joined B.B.C. T.V. as stage manager then producer. 1951 producer North Region. 1955 returned London, senior producer then executive producer. 1970 editor in charge B.B.C. T.V. O.B. entertainment progs. 1971 Head of Network Production Centre Manchester. 1977 left B.B.C. to freelance. Reponsible for wide range of television productions covering sport, features and entertainment in U.K. and Europe.

PRODUCTIONS INCLUDE: Continental, Russian and British Circuses, State Dance Groups, The Bolshoi Ballet. Satellite Work Hook-Up programmes. The Beatles, It's a Knock-Out, Gala Variety, etc. Currently producer "3-2-1" (network) series for Y.T.V.

BURRILL, TIMOTHY: Producer, executive. b. North Wales, 1931. e. Eton College, Sorbonne Univ., Paris, Grenadier Guards 2 yrs., then London Shipping Co. Ent. m.p. ind. as resident prod. mgr. Samaritan Films Ltd., working on shorts, commercials, documentaries, 1954. Ass't dir. on feature films: The Criminal, The Valiant Years (TV series), On the Fiddle, Reach for Glory, War Lover. Prod. mgr.: The Cracksman, Night Must Fall, Lord Jim, Yellow Rolls Royce, The Heroes of Telemark. Resident producer with World Film Services, 1968–1969: Privilege, Oedipus The King, 1969–1970: A Severed Head, Three Sisters. 1970 produced two films on pop music for Anglo-EMI. 1971 assoc. prod., Macbeth. 1972 first production administrator National Film School in U.K. 1973 Prod. Alpha Beta. 1974 Post production administrator The Three Musketeers. 1974: Prod. TV Special The Canterville Ghost; 1975: asst. prod., That Lucky Touch; 1976: UK Administrator, The Prince and the Pauper; 1977: North American production controller, Superman; 1974/80: Council Member of BAFTA. Produced Polanski's Tess. 1979–80, man. dir., Allied Stars Council mbr., BAFTA, 1974–80. V. chmn. film, 79–80.

BURROWS, JOHN, H.: Producer. b. Brooklyn, N.Y., September 4, 1924. e. Yale, New York U. Law with interruption for service. U.S. Navy, World War II, lieut., son of George D. Burrows, vice-president Allied Artists, started film career as salesman for AA later joining studio; ass. prod. Lindsley Parsons Productions.

BURRUD, BILL: Exec. b. Los Angeles, Calif., Jan. 12, 1925. e. U. of Southern Calif., B.S.; Harvard Bus. School, ent. ind. as child actor; U.S. Navy, W.W. II; formed Bill Burrud Prods., 1952.

TV: Animal World, World of the Sea, Safari to Adventure, Challenging Sea, Secret World of Reptiles, Vanishing Africa, The Great American Wilderness, Predators of the Sea, Creatures of the Amazon, The Amazing Apes, The Amazing World.

BURSTYN, ELLEN: Actress. b. Detroit, 1932. Majored in art; was fashion model in Texas when 18. Moved to Montreal as dancer; then N.Y. to do TV commercials, appearing for a year on the Jackie Gleason show. In 1957 turned to dramatics and won lead in show, Fair Game. Then went to Hollywood to do TV and films. Returned to N.Y. to study acting with Lee Strasberg; worked in TV serial, The Doctors.

PICTURES INCLUDE: For Those Who Think Young, Goodbye, Charlie, Tropic of Cancer, Alex in Wonderland, The Last Picture Show, The King of Marvin Gardens, The Exorcist, Harry and Tonto, Alice Doesn't Live Here Anymore, Providence, A Dream of Passion, Same Time Next Year, Resurrection, Silence of the North.

BURTON, BERNARD: Producer. Started as film ed.; prod. "Vagabond Lady" 1935.

PICTURES INCLUDE: She Gets Her Man, Fighting Youth, Invisible Ray, Showboat, When Love is Young, One Hundred Men and a Girl, You're a Sweetheart, Midnight Intruder, Little Tough Guy, Three Smart Girls Grow Up, Spring Parade, Moonlight in Havana, Get Help to Love, Gentleman Joe Palooka, Fighting Mad, Smart Woman, Underworld Story, Cry Danger, The Highwayman, Triple Cross, Beast from 20,000 Fathoms.

BURTON, LeVAR: Actor. b. Landstuhl, Germany, Feb. 16, 1957. e. Univ. of So. Calif. Signed to play role of Kunta Kinte in TV series, Roots, while still in school. Appears frequently on public TV, hosts children's show, Bebop, (PBC).

PICTURES: Looking for Mr. Goodbar, The Hunter.

TELEVISION: Roots, Billy: Portrait of a Street Kid, Battered, The Ron LeFlore Story, Dummy.

BURTON, MARGARET: Actress, opera singer. b. Keighley, Eng., March 18, 1924. Leading opera roles at Sadler's Wells, London, in La Belle Helene, La Vie Parisienne, and Bluebeard. M.p. debut, 1962, in The Comedy Man. Numerous broadcasts and TV appearances in England and the U.S. London stage debut in Mother Goose, Palladium, 1954–55. Broadway debut in Time Gentlemen Please, 1961–62.

STAGE: (London) Twenty Minutes South, Romance in Candlelight, The Three Caskets, Annie, Mandrake, Oliver! (1978–80).

TV: The Gold Old Days, The Quick One, Harpers West One, The Plane Makers, Z-Cars, A Variety of Reg Varney, The Harry Secombe Show, Free As a Bird, Orson Welles Great Mysteries, Pygmalion Smith, South Riding, The Tomorrow People, The Last of the Summer Wine, Angels, Send in the Girls (1978), Sally Ann (1979), Coronation Street (1980).

BURTON, RICHARD: Actor. r.n. Richard Jenkins; b. Pontrhydyfen, So. Wales, Nov. 10, 1925. e. Oxford. Actor, legit. stage, England, Lady Is Not for Burning, Monserrat; actor, Bway, legit. stage, The Legend of Lovers; 8 yrs. on radio BBC, London.

PICTURES INCLUDE: The Last Days of Dolwyn, Now Barrabas, Waterfront, The Woman with No Name; with Old Vic; appeared in Henry V, Othello, Troilus and Cressida, Iago; Time Remembered, 1957–8, New York; Camelot, 1960–61, N.Y.U.S. film debut, My Cousin Rachel, Star of Tomorrow, The Desert Rats, The Robe, Prince of Players, Alexander The Great, Rains of Ranchipur, Seawife, Bitter Victory, Look Back in Anger, The Bramble Bush, Cleopatra, The V.I.P.'s, Becket, Night of the Iguana, The Sandpiper, The Spy Who Came in From the Cold, Who's Afraid of Virginia Woolf?; The Taming of The Shrew, Dr. Faustus, The Comedians, Candy, Boom, Where Eagles Dare, Staircase, Anne of the Thousand Days, Hammersmith is Out, The Assassination of Trotsky, Bluebeard, Massacre in Rome, The Klansman, The Voyage, Exorcist II: The Heretic, Equus, The Medusa Touch, The Wild Geese, Absolution, Sgt. Steiner II, Tristan & Iseult, Circle of Two.

TV: His and Hers (movie), Brief Encounter.

PLAYS: Camelot, Hamlet, Othello, The Tempest, Henry V, Equus

BUSCH, NIVEN: Writer. b. New York City, N.Y., Apr. 26, 1903; e. Princeton U.; p. Briton Niven Busch, former v-p Lewis J. Selznick Enterprises; Exec. prod., Samuel Goldwyn Productions; m. Phyllis Cooper, 1936; children: Peter, Briton, Terence, Mary Kelly, Joseph, Nicholas, and Eliza. m. Carmencita Baker, 1956. div. 1969, m. Suzanne de Sanz 1973. Assoc. ed. Time mag., 1923–31; assoc. ed. & contrib. New Yorker, 1927–31; numerous articles, many national magazines. Regents professor, Univ. of Calif. 1971–78. Special lectures on film.

AUTHOR: Books: Twenty One Americans, Carrington Incident, Duel in the Sun, They Dream of Home, Day of the Conquerors, The Furies, The Hate Merchant, The Actor, California Street, The San Franciscans, The Gentlemen from California, The Takeover, No Place for a Hero, Continent's Edge.

PICTURES INCLUDE: The Crowd Roars, In Old Chicago, The Westerner, The Postman Always Rings Twice, Duel in the Sun, Till the End of Time, Pursued, Moss Rose; prod., The Capture, Distant Drums, Man from the Alamo, Moonlighter, Treasure of Pancho Villa, Moss Rose, He

Was Her Man, The Angels Wash Their Face, Big Shakedown, Little Miss Pinkerton, Regents Professor, Univ. of Cal., Irvine, 1971–75. San Diego, 1974; Lecturer Berkeley, 1977.

BUTLER, DAWS: Actor (Voice), r.n. Charles Dawson Butler. b. Toledo, Nov. 16, 1916. With Time for Beany, 1948–53. Host (voice) of Off to See the Wizard TV program. From 1953 to present, various cartoon voices, including: Yogi Bear, Huckleberry Hound, Quick Draw McGraw, Loopie DeLoop, Chilly Willy, Peter Potamus, Mr. Jinx.

BUTTERFIELD, ELIZABETH: Production Manager. b. London, England e. London. Extensive theatrical experience as stage manager for The Old Vic Company (Michael Benthall, Tyrone Guthrie) and Donald Wolfit Company. Emigrated to Canada in 1957. Television experience includes: Quentin Durgens M.P. and Whiteoaks Of Jaina for C.B.C., Seaway, Ind.

PICTURES INCLUDE: A Fan's Notes, Warner Bros. 1971; Sunday In The Country, Quadrant, 1973; It Seemed Like A Good Idea At The Time, Quadrant, 1975; Find the Lady, Quadrant, 1976; Running, Buena Vista, 1978; Middle Age Crazy, Torment, 1979.

BUTTOLPH, DAVID: Music director.

PICTURES INCLUDE: Phantom of Rue Morgue, Secret of the Incas, Bounty Hunter, Long John Silver, Jump Into Hell, Target Zero, Lone Ranger, Steel Jungle, Santiago, Burning Hills, Cry in the Night.

BUTTONS, RED: Performer. r.n. Aaron Chwatt; b. N.Y.C., Feb. 5, 1919. Singer at the age of 13; comic, Minsky's. U.S. Army, 1943; in Army and film version of Winged Victory.

TV: Vickie, Wine, Women and Song, Barefoot Boy With Cheek, Hold It. Many TV appearances; star of Red Buttons Show CBS-TV, 1953; Best Comedian award 1953; NBC-TV, 1954; on screen in Sayonara (Academy Award best actor in supporting role 1957).

PICTURES INCLUDE: Imitation General, The Big Circus, Five Weeks in a Balloon, Harlow, They Shoot Horses Don't They, Who Killed Mary What's 'er name?, The Poseidon Adventure, Gable and Lombard, Viva, Knievel!, Movie Movie, When Time Ran Out.

BUTTS, R. DALE: Composer. b. Lamasco, Ky., March 12, 1910; e. Louisville Conservatory of Music. Started in vaudeville at 16, later playing, arranging for dance bands; staff arranger, pianist, radio stations; to Republic Pictures, 1944.

PICTURES INCLUDE: The Plunderers, Too Late for Tears, House by the River, Sea Hornet, The Outcast, Geraldine, Shanghai Story, Hell's Outpost, Santa Fe Passage, Fighting Chance, No Man's Woman, Headline Hunter's, Lay That Rifle Down, Double Jeopardy, City of Shadows, Terror at Midnight, Stranger at My Door, Dakota Incident.

BUXBAUM, JAMES M.: Executive, producer, writer. b. Jamaica N.Y., March 8. e. Harvard, B.A., 1949, Columbia Law School, LL.B., 1955, spec. studies, The Hague Acad., Netherlands, 1953. Story ed. ZIV, Ivan Tors' Sea Hunt series, then assoc. prod., ZIV-CBS, Ivan Tors' Aquanauts series, story ed., 1957–60. Attorney, Wm. Morris Agency, 1962–63. Joined Ivan Tors Films, Inc. as v.p., prod. and wrote many episodes of Flipper series, 1964; exec v.p. Ivan Tors Miami Studios, 1967–69; exec. v.p. Ivan Tors Films, Inc. 1968–69. Gen. mgr. Amer. Film Institute Center, Beverly Hills, 1969. Member: State Bar of Calif., Amer. Bar Assoc. Exec. in charge of prod. for Island of the Lost.

BUZZELL, EDWARD: Actor, director. b. Brooklyn, N.Y., Nov. 13, 1907. Musical comedy star prior to directorial debut 1932.

PICTURES INCLUDE: Best Foot Forward, Youngest Profession, Keep Your Powder Dry, Three Wise Fools, Song of Thin Man, Neptune's Daughter, At the Circus, Woman of Distinction, Emergency Wedding, Easy to Wed, Confidentially Connie, Ain't Misbehavin', My Favorite Husband.

BUZZI, RUTH: Actress. b. Wequetequak, CT, July 24, 1939. e. Pasadena Playhouse. Launched TV career on Garry Moore Show as comedy partner of Dom DeLuise. Created character Gladys Ormphby with Artie Johnson on TV's Laugh-In.

PICTURES: Freaky Friday, The Apple Dumpling Gang Rides Again, The Villain, Chu Chu and the Philly Flash, Easter Sunday.

TELEVISION: The Entertainers, The Donny & Marie Show (semi-regular), Rowan & Martin's Laugh-In, The Dean Martin Show, The Dean Martin Roasts, The Tonight Show, Kroft's Superstars, Rhoda (recurring role), Gladys and Tyrone (cartoon show, partial creator and voice of character) etc.

STAGE: Sweet Charity (Broadway), The Ruth Buzzi Show (Las Vegas), Pasadena Playhouse Theatre, and summer stock.

AWARDS: 4 Grammy nominations; Golden Globe winner, AGVA Variety Artist of the Year, 1977, Rhode Island Hall of Fame, Presidential Commendation for outstanding artist in the field of entertainment, 1980.

BYGRAVES, MAX: Comedian, actor. b. London, Eng., October 16, 1922. e. St. Joseph's R.C. School, Rotherhithe. After RAF service, touring revues and London stage, M.P. debut 1949 in Skimpy in The Navy. TV debut in 1953, with own show.

TV: Roamin Holiday series.

PICTURES INCLUDE: Tom Brown's Schooldays, Charlie Moon, A Cry from the Streets, Bobikins, Spare the Rod, Autobiograph "I Wanna Tell you A Story" Pub. 1976.

NOVEL: The Milkman's on His Way, pub. 1977.

BYRD, CARUTH C.: Production Executive. b. Dallas, Texas, March 25, 1942. Multi-millionaire businessman, chm. of Caruth C. Byrd Enterprises, Inc., who entered entertainment industry forming Communications Network Inc. in 1972, producer of TV commercials. Was principal investor in film Santee (1972) and in 1973 formed Caruth C. Byrd Prods. to make theatrical features.

PICTURES INCLUDE: Allan and Murph the Surf, The Monkeys of Bandapur (both exec. prod.).

BYRD, JOHN: Writer, director, producer. Entered m.p. industry 1934 via scenario dept. Rock Studios, Elstree; visited Hollywood to study requirements of U.S. market, 1939; 1940–45 B.B.C. War Correspondent, 1945–46 Riverside Studios, 1947–8 British National Studios.

PICTURES INCLUDE: Murder in Reverse, World Owes Me a Living, Waltz Time. 1949 formed own org. prod. bus. TV films. Prod. films in South Africa, Uganda, Kenya, Portugal and Mexico. Filming Durgapur Steelworks project, India, 1961; Steel Strides Ahead. Concerning Colour, Design for Tall Building, Jungle to Steel Town, Steelworks in Action, 1962–4; Dryer Felt Loom, Bridging the Highways, 1965; Filming in Afghanistan, 1966; Filming in South Africa, 1967; Built from Top, 1968; Building With Steel and Plastics, 1969; The Desert Will Yield, 1970; Long Sea Outfall, One Steelworks, One Contract, Highweld Process. 1971: Go-Con Process, Mechanical Handling, 1972. Undersea Tunnel to Hong Kong. 1973: Contract 1306—Zambia, Bridge Across the Avon; 1974 filming for British Steel Corp.

OTHER PICTURES: Murder in Reverse, World Owes Me a Living, Waltz Time. 1974–5 Anchor Project for British Steel Corporation. 1976–77 Filming in Mexico AHMSA Steelworks project. 1977, More Steel for Mexico; 1978, Filming in South Africa; 1979, Filming in South Africa; 1980, Filming in France & Spain.

BYRNES, EDWARD: Actor. b. New York, N.Y., July 30, 1933. e. Harren H.S.; prof. debut, Joe E. Brown's Circus Show; appeared on stage in Tea and Sympathy.

PICTURES INCLUDE: Periscope, Marjorie Morningstar, Secret Invasion, Wicked, Wicked.

TV: Matinee Theatre, Crossroads, Jim Bowie, Wire Service, Navy Log, Oh Susanna!, 77 Sunset Strip.

BRYON, CHARLES ANTHONY: Producer. b. Constantinople, December 15, 1919. e. University of Paris, Harvard; Army Air Force 1942–1947. With UNESCO in Paris; NBC-International produced and wrote foreign programs for Dept. of State; public relations ECA mission to France. Film industry 1952 as director of public relations, research, Louis de Rochemont Associates European prod. Cinerama Holiday; Dragon Prod., 1959, pres., exec. prod. of: The Silken Affair, Chase a Crooked Shadow, etc.

BYRON, KATHLEEN: Actress. b. London, England, Jan. 11, 1922; e. London U. Old Vic. Co. student 1942. Screen debut in Young Mr. Pitt 1943.

PICTURES INCLUDE: Silver Fleet, Black Narcissus, Matter of Life and Death, Small Back Room, Madness of the Heart, Reluctant Widow, Prelude to Fame, Scarlet Thread, Tom Brown's Schooldays, Four Days, Hell Is Sold Out, I'll Never Forget You, Gambler and the Lady, Young Bess, Night of the Silvery Moon, Profile, Secret Venture, Hand in Hand, Night of the Eagle, Hammerhead, Wolfshead, Private Road, Twins of Evil, Craze, Abdication, One of Our Dinosaurs Is Missing.

TV: The Lonely World of Harry Braintree, All My Own Work, Emergency Ward 10, Probation Officer, Design for Murder, Sergeant Cork, Oxbridge 2000, The Navigators, The Worker; Hereward the Wake, Breaking Point, Vendetta, Play To Win, Who is Sylvia, Portrait of a Lady, Callan, You're Wrecking My Marriage, Take Three Girls, The Confession of Mariona Evans, Paul Temple, The Worker, The Moonstone, The Challengers, The Golden Bowl, The Edwardians, The New Life, Menace, The Rivals of Sherlock Holmes, The Brontes, On Call Edward VII, Sutherland's Law, Crown Court, Anne of Avonlea, Heidi, Notorious Woman, General Hospital, North & South, Angelo, Within these Walls, Jubilee, Z Cars, Tales from the Supernatural, Secret Army, An Englishman's Castle, The Professionals, Forty Weeks, Emmerdale Farm, Blake Seven.

BYRON, WARD: TV Prod.-writer-dir. b. N.Y.C., June 2, 1910. e. New York. Orchestra leader, 1929–32; radio prod., writer, 1932–39. Ent. TV as asst. to prog. dir. of ABC; writer & producer of Paul Whiteman show in 1950; app'td exec. prod. of ABC-TV in 1950. Exec. prod., Canadian TV, Young & Rubicam, 1955–57. Freelance writer prod., TV Specials

& film series, v.p. gen, mgr. Trans-Nat'l Communications O'way Recording Studios; 1970, v.p. & sec. International Coproductions, motion picture distributors.

BYRUM, JOHN: Writer—Director. b. Winnetka, Ill., March 14, 1947. e. N.Y. Univ. Film School. First job as "go-fer" on industrial films and cutting dailies for underground filmmakers. Went to England where wrote 1st s.p., Comeback. From 1970–73 was in N.Y.C. writing and re-writing scripts for low-budget films.
PICTURES INCLUDE: Mahogany (s.p.), Inserts (s.p., dir.), Harry and Walter Go to New York (story, s.p.), Heart Beat (s.p. dir.), Sphinx (s.p.).

C

CAAN, JAMES: Actor. b. New York, N.Y., March 26, 1939; e. Hofstra Coll.: Appeared off-Broadway in La Ronde, 1961.
PICTURES INCLUDE: Lady in a Cage, The Glory Guys, Red Line 7000, Eldorado, Games, Journey to Shiloh, Submarine XQ/⅛, Rain People, Rabbit, Run, Brian's Song (TV), The Godfather, Slither, Cinderella Liberty, Freebie and the Bean, The Gambler, Godfather II, Funny Lady, Rollerball, The Killer Elite, Harry and Walter Go To New York, Silent Movie, A Bridge Too Far, Another Man, Another Chance, Comes a Horseman, Chapter Two, Hide in Plain Sight (also dir.), Thief.

CACOYANNIS, MICHAEL: Producer, director, writer. b. Cyprus, June 11, 1922. Studied law in London, admitted to bar at age 21. Became a producer of BBC's wartime Greek programs while attending dramatic school. After acting on the stage in England, left in 1952 for Greece, where he made his first film, Windfall in Athens, with his own original script. While directing Greek classical plays, he continued making films.
PICTURES INCLUDE: Stella, The Girl in Black, A Matter of Dignity, Our Last Spring, The Wastrel, Electra, Zorba the Greek, The Day the Fish Came Out, The Trojan Women.

CADELL, JEAN: Actress. b. Edinburgh, Scotland, Sept. 13, 1884; e. Edinburgh & Paris. Early prof. stage appearances on tour in Gt. Brit., beginning in London 1906 (The Inspector General); with Sir George Alexander's Co. in Eng.; later numerous London, N.Y. prods.
PICTURES INCLUDE: Madeleine, Reluctant Widow, No Place for Jennifer, Obsessed, Whisky Galore, Meet Mr. Lucifer, Little Hut, The Surgeon's Knife, Rockets Galore, Upstairs and Downstairs, A Taste of Money, V.I.P.
TV: Whiteoakes, Arsenic and Old Lace, Doomsday for Dyson, Mary Rose, Eustace Diamonds (serial).

CAESAR, IRVING: Author, composer, publisher. b. New York, N.Y., July 4, 1895, e. C.C.N.Y. Abroad with Henry Ford on Peace Ship, World War I; songwriter since then, several songs with George Gershwin, songwriter for stage, screen and radio, including Swanee, Tea for Two, Sometimes I'm Happy, I want to Be Happy, Lady Play Your Mandolin, Songs of Safety, Songs of Friendship, Songs of Health and Pledge of Allegiance to the Flag.

CAESAR, SID: Performer. b. Yonkers, N.Y. Sept. 8, 1922. Musician. Joined Coast Guard in 1942. Appeared in "Tars and Spars." Signed for Broadway Revue by Max Liebman, 1949. Star of Your Show of Shows on NBC-TV 1950–54; star of Caesar's Hour, 1954–1957; voted best comedian (tied) in M.P. Daily's TV poll, 1951. Best Comedian, 1952; Best Comedy Team (with Imogene Coca), 1953; 5 Emmy Awards 1957; Sylvania Award, 1958. Sid Caesar Invites You, 1958; Formed Shelbrick Corp., TV, 1959; TV series, As Caesar Sees It, 1962; Bway Play: Little Me, 1962–64; The Sid Caesar Show, 1963–64.
PICTURES INCLUDE: It's a Mad, Mad, Mad, Mad World, The Spirit is Willing, The Busy Body, Guide for the Married Man, Ten from Your Show of Shows, Airport 1975, Silent Movie, Grease, The Cheap Detective, History of the World—Part I.

CAGNEY, JAMES: Actor. b. New York City, July 1, 1904; e. Columbia U. In vaudeville 1924 and later; on dramatic stage ("Outside Looking In," "Women Go On Forever," etc.). On screen from 1931. Pres. Screen Actors Guild 1942–43. Voted one of best ten Money-Making Stars in Motion Pictures Herald-Fame Poll 1935, 39, 40, 41, 42, 43. Academy Award, 1942 (Yankee Doodle Dandy).
PICTURES INCLUDE: Doorway to Hell, Steel Highway, Crowd Roars, Jimmy the Gent, A Midsummer Night's Dream, Angels With Dirty Faces, Each Dawn I Die, The Strawberry Blonde, Yankee Doodle Dandy, Johnny Come Lately, Blood on the Sun, 13 Rue Madeleine, Time of Your Life, White Heat, The West Point Story, Come Fill the Cup, Starlift, What Price Glory? A Lion is in the Streets, Run for Cover, Love Me Or Leave Me, Mr. Roberts, Seven Little Foys, Tribute to a Bad Man, These Wilder Years, Man of a Thousand Faces; dir., Short Cut to Hell, Shake Hands with the Devil, The Gallant Hours, One, Two, Three, Ragtime.

CAGNEY, JEANNE: Actress. b. New York City, Mar. 25, e. Hunter College, sister of James Cagney. m. Jack Morrison.
THEATRE: Broadway: I'll Take The High Road, A Place of Our Own, Streets Are Guarded, The Iceman Cometh, La Jolla, Marriage Go Round, World of Carl Sandburg, The Just.
PICTURES INCLUDE: All Women Have Secrets, Golden Gloves, Queen of the Mob, Yankee Doodle Dandy, Time of Your Life, Don't Bother to Knock, Quicksand, A Lion is in the Streets, Kentucky Rifle, Man of A Thousand Faces, Town Tamer.
TV FILMS: Big Hello, Legal Tender, Hill, Number One, Trial at Tara, Mr. and Mrs. North, Big Town, TV Readers Digest, Wild Bill Hickok.
LIVE TV: Jack Benny Show, This Is Your Life, Lux Video, Stairway to Stardom, Pantomime Quiz, Wild Bill Hickok, Pstter Potter Platter Panel, A Capture, Red Skelton Show, Fashion Commentator on Queen for a Day.

CAGNEY, WILLIAM J.: Producer. b. New York; brother of James Cagney, actor. e. City Coll., N.Y. (law student). Adv. rep., Fairchild Publications. From 1933 on screen in Ace of Aces, Lost in the Stratosphere, Stolen Harmony, etc. Became talent agent, Hollywood, then in 1940 joined Warner Bros. as assoc. prod. (The Strawberry Blonde), The Bride Came C.O.D., Captains of the Clouds, Yankee Doodle Dandy, etc.). In 1942 org. (with James C.) Cagney Prods.: pres. Cagney Prod.
PICTURES INCLUDE: Blood on the Sun, Johnny Come Lately, Time of Your Life, Kiss Tomorrow Goodbye, Only the Valiant, Bugles in the Afternoon, A Lion Is in the Streets.

CAHN, SAMMY: Lyricist, Producer. b. New York, N.Y. June 18, 1913; e. Seward Park H.S. Org. dance band with Saul Chaplin; collab. song writer for shows. m.p.
PLAYS INCLUDE: High Button Shoes, Skyscraper, Walking Happy, many songs, 30 Acad. Award Nominations, Four Oscars for 3 Coins in Fountain, All The Way, High Hopes, and Call Me Irresponsible! Only TV Emmy ever given a song "Love & Marriage" From the TV award winning "Our Town".
PICTURES INCLUDE: Anchors Aweigh, 3 Coins in Fountain, Romance on High Seas, Some Came Running, Robin & Seven Hoods, The Tender Trap, Pocketfull of Miracles, Thoroughly Modern Millie. 1972, inducted into Songwriters' Hall of Fame.

CAIN, SUGAR: Actress. r.n. Constance McCain. b. Jerome, Ariz., Mar. 13. e. Univ. Washington, B.A., radio-drama; Colo. Woman's Col., A.A., journalism; Columbia Univ., grad. TV. MP. Reg. on radio serials. Hopalong Cassidy, Clyde Beatty Show, Night club act, Mexico; appeared TV, Hollywood & N.Y. Plays include: nat. tours, Pajama Tops, In One Bed, Idiots Delight, Streetcar Named Desire, The Rainmaker, Dead End, Druid Circle.
PICTURES INCLUDE: Breakstone Affair, Life with Edmer, I've Been Here Before, Monster from 20,000 Fathoms, Dalton Women, I Shot Billy the Kid, Hostle Country, Dalton Gang, Africa Screams; Sporting Club, Apple Man, Klute, Panic in Needle Park, The French Connection, Effect of Gamma Rays, Summer Wishes, Winter Dreams; Dir.-writer, m.p. short, Rama.

CAINE, MICHAEL: Actor. b. London, Eng., 1933.
TV: The Compartment, The Playmates, Hobson's Choice, Funny Noises With Their Mouths, The Way With Reggie, Luck of the Draw, Hamlet, The Other Man.
PICTURES INCLUDE: Zulu, Ipcress File, Alfie, The Wrong Box, Gambit, Funeral in Berlin, Hurry Sundown, Billion Dollar Brain, Deadfall, The Magus, Play Dirty, Italian Job, The Battle of Britain, Too Late the Hero, Get Carter, Kidnapped, Zee and Company, Pulp, Sleuth, The Black Windmill, The Destructors, The Wilby Conspiracy, Peepers, The Romantic Englishwoman, The Man Who Would Be King, Harry and Walter Go to New York, The Eagle Has Landed, A Bridge Too Far, The Silver Bears, The Swarm, Ashanti, California Suite, Beyond the Poseidon Adventure, The Island, Dressed to Kill, The Hand, Victory.

CALHOUN, RORY: Actor. r.n. Francis Timothy Durgin. b. Los Angeles, Calif., Aug. 8, 1923, e. Santa Cruz H.S. Worked as logger, miner, cowpuncher, for firefighter; m.p. debut in Something for the Boys, 1944.
PICTURES INCLUDE: Sunday Dinner for a Soldier, Nob Hill, Great John L, Red House, Adventure Island, That Hagen Girl, Sand, Massacre River, Ticket to Tomahawk, County Fair, I'd Climb the Highest Mountain, Rogue River, Meet Me After the Show, With a Song in My Heart, Way of a Gaucho, The Silver Whip, Powder River, How to Marry a Millionaire, River of No Return, Yellow Tomahawk, Bullet Is Waiting, Dawn at Socorro, Four Guns to the Border, The Looters, Ain't Misbehavin', Shotgun, Domino and Actor; actor, Treasure of Pancho Villa, The Spoilers, Red Sundown, Raw Edge, Flight to Hongkong, Utah Blaine,

Big Caper, Finger On the Trigger, Young Fury, Black Spurs, Adventures of Marco Polo, Colossus of Rhodes.
TV: The Texans.

CALL, CHARLES W.: Exec. b. New York, N.Y., Century Theatres, 1933; chief usher; asst. mgr.; student mgr.; mgr., 1937–42; U.S. Army Air Force, 1942–46; mgr., various theatres, Century Circuit, 1946–50; field rep., L.I. theatres; adv. & publ. exec., home office, 1957–63; asst. v.p., 1963–78; Joined United Artists as mgr. co-op adv., 1979.

CALLAN, MICHAEL: Actor, singer, dancer. b. Philadelphia, Pa. Singer, dancer, Philadelphia night clubs; to New York in musicals, including The Boy Friend and West Side Story; dancer Copacabana; in short-run plays, Las Vegas: 5 month run starred in That Certain Girl. Guest starred on major dramatic TV shows and Occasional Wife (series, star).
PICTURES INCLUDE: debut, They Came to Cordura, The Flying Fontaines, Because They're Young, Pepe, Mysterious Island, Gidget Goes Hawaiian, 13 West Street, Bon Voyage, The Interns, The Victors, The "New" Interns, Cat Ballou, Frasier, The Sensuous Lion, Lepke, The Photographer, The Cat and The Canary (1977), Record City (1977); The Donner Party (1978).
TELEVISION: Blind Ambition, Scruples.

CALLEY, JOHN: executive. b. New Jersey, 1930. Dir. of night time programming, dir. programming sales, NBC, 1951–57; prod. exec. and TV prod. Henry Jaffe Enterprises, 1957; v.p. in charge of Radio and Television, Ted Bates Advertising Agency, 1958; joined Filmways, Inc., 1960; exec. v.p. and prod. to 1969; exec. v.p. chg. world-wide prod. Warner Bros.; Pres., W.B. from Jan. 6, 1975; vice chm. bd., 1977.
PICTURES INCLUDE: Wheeler Dealer, The Americanization of Emily, Topkapi, The Cincinnati Kid, Loved One, Don't Make Waves, Ice Station Zebra, Catch-22.

CALLOW, EVERETT C.: Publicist. b. Perth Amboy, N.J. e. St. John's Coll. On ed. staff N.Y. World; service staff chief Paramount Thea., N.Y., 1926; mgr. Keith & Albee Theas., Phila., 1928; mgr. Stanley Theatre, Phila., 1929–31; dist. theatre mgr. various dists. in N.J., Pa. & Del.; apptd. dir. of adv. & publicity Warner Bros., Phila. zone, 1939; winner Quigley Silver Award, 1940; Lt. Col. U.S.M.C., 1943; returned to Warner post after war; 1950–51, internat'l dir. adv., pub., Cinerama, Feb., 1954. Member: American Legion Variety Post 713 (past commdr.); Military Order of Foreign Wars of U.S.; Adv. pub. dir. National Screen Service, 1965; VP in charge of adv. and pub. sales. TV, Foreign; pres. United Screen Arts, Inc. 1966; world wide dist. and adv. and production. Presently consultant.

CALVERT, PHYLLIS: Actress. b. London, England, Feb. 18, 1917. e. Margaret Morris Sch. Dancing, London; m. Peter Murray Hill, actor. On stage at 10 as dancer; later in provincial repertory; London debut 1939 in "A Woman's Privilege." Screen debut 1939 in "They Came By Night"; voted one of Britain's ten best money-making stars in Motion Picture Herald-Fame Poll 1945–46.
PICTURES INCLUDE: Kipps, Young Mr. Pitt, Uncensored, Fanny by Gaslight, 2,000 Women, Madonna of the Seven Moons, They Were Sisters, Men of Two Worlds, Magic Bow, Root of All Evil, Time Out of Mind, Broken Journey, My Own True Love, Golden Madonna, Man of Evil, Appointment With Danger, Woman With No Name, Mr. Denning Drives North, Mandy, The Net, Its Never Too Late, Child in the House, Indiscreet, Twisted Nerve, Oh! What a Lovely War, The Walking Stick.
TV: Little Women (series), Good Wives, Mrs. Dot, Parnell, You'll Know Me By The Stars in My Eyes, Love Story, The Kindness of Mrs. Radcliffe, Kate (series), Owen M.D.

CALVET, CORINNE: Actress. r.n. Corinne Dibos. b. Paris, France, April 30; e. U. of Paris School of Fine Arts; French stage and radio. Screen debut in France; La Part de L'Ombre, Nous Ne Sommes Pas Maries, Petrus. American screen debut: Rope of Sand.
PICTURES INCLUDE: When Willie Comes Marching Home, My Friend Irma Goes West, Quebec, On the Riviera, Peking Express, Thunder in the East, Sailor Beware, What Price Glory? Powder River, Flight to Tangier, The Far Country, So This Is Paris, Bluebeard's Ten Honeymoons, Adventures of a Young Man, Pound.

CAMERON, JOANNA: Actress—Director. b. Sept. 20, Aspen, Colo. r.n. Patricia Cameron, e. Univ. of Calif., Sorbonne. Pasadena Playhouse, 1968. Guinness Record: Most network programmed TV commercials. Prod., dir., documentary, Razor Sharp, 1981.
PICTURES INCLUDE: How To Commit Marriage (debut), P.S., I Love You, Pretty Maids All in a Row.
TV: Features: The All-American Beauty Contest, It Couldn't Happen to a Nicer Guy, High Risk. Other: The Survivors, Love American Style, Danial Boone, Mission Impossible, The Partners, Search, Medical Center, Isis, Name of the Game, The Bold Ones, Marcus Welby, Petrocelli, Columbo, Switch, Bob Hope Special, Bob Hope 25th NBC Anniversary Special, Westwind, MacMillian, Spiderman, Swan Song, Johnny Carson, numerous commercials. Director: Various commercials, CBS Preview Special, closed circuit program host U.S N., all TV equipped ships—actress and director.

CAMERON, ROD: Actor, r.n. Rod Cox; b. Calgary, Alberta, Canada, Dec. 7, 1912.
PICTURES INCLUDE: Christmas in July, Quarterback, North West Mounted Police, Monster and the Girl, Henry Aldrich for President, True to the Army, Wake Island; Salome Where She Danced; Swing Out Sister; Panhandle, Belle Starr's Daughter, Plunderers, Strike It Rich, Brimstone, Dakota Lil, Cavalry Scout, Short Grass, Stage to Tucson, O! Susanna, Fort Osage, Sea Hornet, Wagons West, Woman of the North Country, The Jungle, Ride the Man Down, San Antone, Steel Lady, Southwest Passage, Hell's Outpost, Santa Fe Passage, Fighting Chance, Headline Hunters, Double Jeopardy, Zex, The Bounty Killer, Evel Knievel, Psychic Killer.

CAMPAGNOLA, GINO: Executive. Joined Paramount Pictures in 1975; held various sales positions, including v.p., sls. admin. In 1980 named v.p., asst. sls. mgr.

CAMPBELL, GLEN: Actor, singer. b. Delight, Ark., April 22. After forming local band became guitarist in Hollywood; records won two Grammy awards, 1967.
PICTURES INCLUDE: True Grit, Norwood.

CAMPBELL, MAE E.: Actress. b. March 19, St. Louis, Mo. Studied with the Actors Studio and Lee Strasberg.
PICTURES INCLUDE: The Old Boyfriends; two industrial films for the Veterans Administration Hospital.

CAMPBELL, ROBERT MAURICE: Producer, director. b. Detroit, Mich., Feb. 16, 1922 e. Mich. Conservatory of Music. Detroit Inst. of Musical Arts, American Theatre Wing. Featured player. assoc. dir., Detroit stage prod.; numerous radio appearances; pub. relations, U.S. Coast Guard, World War II; since 1946 dir., off-Broadway shows; dir.-prod., Springer Pictures, Inc.; TV show: All-Star News; exec., Visual Transcriptions, Inc.; prod. Jazz Dance (Robert J. Flaherty Award, Edinburgh Film Festival Merit Award); co-author, prod. designer, See No Evil.

CAMPBELL, WILLIAM: Actor. b. Newark, N.J.; e. Feagin School of Drama Appeared in summer stock. B'way plays; m.p. debut in The Breaking Point; then in People Against O'Hara, Holiday for Sinners, Battle Circus, Code Two, Big Leaguer, Escape from Fort Bravo, High and the Mighty, Man Without a Star, Cell 2455 Death Row, Battle Cry, Running Wild, Backlash, Pretty Maids All in a Row, Black Gunn.

CANALE, GIANNA MARIA: Actress. b. Reggio Calabria, Italy, Sept. 12, e. Florence. Debut in Mysterious Knight.
PICTURES INCLUDE: Kiss of a Dead Woman, Count Orgolino, Son of D'Artagnan, Go for Broke, Adventure in Algiers, Eternal Chain, Theodora Slave Empress, Madame DuBarry, The Whole Truth.

CANBY, VINCENT: Journalist-Critic. b. Chicago, July 27, 1924. e. Dartmouth Coll. Navy officer during World War II. Worked on newspapers in Paris and Chicago. Joined Quigley Publications in 1951 in editorial posts on Motion Picture Herald. Reporter for Weekly Variety 1959–1965. Joined New York Times film news staff, 1965; named film critic, 1969.

CANFIELD, ALYCE: Writer. b. Los Angeles. e. U. of Calif. Magazine writer; author Mervyn LeRoy book It Takes More Than Talent; in collab. with Howard Duff original story for Models, Inc.; story Journey Into Fire, Assignment for Murder; s.p. for latter in collab. with Jerry Jerome; creator, exec. prod. Ziv TV series, Underground, USA.

CANNON, DYAN: Actress. b. Tacoma, Wash. January 4, 1937. e. U. of Washington. Studied with Sanford Meisner. Modelled.
TV: Playhouse 90, Diane's Adventure.
BROADWAY: The Fun Couple, Ninety-Day Mistress.
ROAD TOUR: How to Succeed in Business Without Really Trying.
PICTURES INCLUDE: Bob and Carol and Ted and Alice (Academy Award nomination), Doctors' Wives, The Anderson Tapes; The Love Machine; The Burglars, Such Good Friends, Shamus, The Last of Sheila, Child Under a Leaf, Heaven Can Wait, Revenge of the Pink Panther, Coast To Coast, Honeysuckle Rose, Deathtrap.

CANNON, JUDY: Actress. b. Santa Fe, N.M., June 16, 1938. e. Santa Monica City College, 1957; teenage Adrian model, 1954–56; toured as band singer, Orlo Wagner Orchestra, 1957; appeared in: The Little Hut, Girls of Summer, Made in Japan, Los Angeles Little Theatre, 1959–61; midwest musical comedy tour group, 1962; m.p. debut, Lullaby, 1963; Critics Choice, '63; A Comedy Tale of Fanny Hill, 1964.
TV: Smothers Bros. Show, Man From U.N.C.L.E., Chrysler Theatre, Run For Your Life, The Road West, My Three Sons, Death Valley Days.
PICTURES INCLUDE: More, 1970.

CANNON, WILLIAM: Writer, Producer, Director. b. Toledo, Ohio, Feb. 11, 1937. e. Columbia Coll., B.A., 1959, M.B.A., 1962. Dir. Off-B'way, Death of a Salesman, Pirates of Penzance, 1960. Wrote, prod., dir., Square Root of Zero, Locarno and San Francisco Film Festivals, 1963–65; Distrib., Doran Enterprises, Ltd; author, Skidoo, (Par-Otto Preminger), 1968, Brewster McCloud, (MGM), 1970; HEX (20th Century Fox) 1974; Author, Novel, The Veteran (E.P. Dutton) 1974.

CANOVA, JUDY: Actress, singer. b. Jacksonville, Fla., Nov. 20, 1916; p. Joe and Henrietta Canova. In Ziegfield Follies, "Calling All Stars," and with Paul Whiteman on the air; 10 yrs., Judy Canova Show, NBC Radio.
PICTURES INCLUDE: Going Highbrow, Artists and Models, Thrill of a Lifetime, Scatterbrain, Singin' In the Corn, Honeychile, Oklahoma Annie, Wac from Walla Walla, Untamed Heiress, Carolina Cannonball, Lay That Rifle Down, Cannonball.
TV: The Adventures of Huckleberry Finn, Party Line, Alfred Hitchcock, Pistols 'n Petticoats, Love American Style, Paramount TV, 1973; Joy Commercial, Proctor, Gamble, 1974, Policewoman, 1974. Recent: Dinah Shore Show, Mike Douglas Show, I'm a Big Girl Now (1980 series), Love Boat, Mother's Day Special (1981).

CANTON, ARTHUR H.: Executive, Blowitz & Canton Co. Inc. b. New York. e. N.Y. University, Columbia University, Capt. USAF. Pres., Canton-Weiner Films, indep. foreign films importers, 1947; MGM Pictures, eastern div. publicity manager, executive liaison, advertising-publicity, Independent Productions; public relations executive, v.p.; pres., Blowitz, Thomas & Canton Inc., 1964; pres., Arthur H. Canton Co. Inc.; prod. exec., Warner Bros., 1968–70; advertising-publicity v.p., Columbia Pictures, 1971; executive v.p. of advertising and publicity, Billy Jack Productions, 1974–76. Co-founder of Blowitz & Canton Co. Inc., 1976, chairman of the board. Member Academy of Motion Picture Arts and Sciences, Variety Clubs International, Film Information Council; Friars Club.

CANUTT, YAKIMA: Director. b. Colfax, Wash., Nov. 29, 1895. World's Champion All-around Rodeo, 1917–24. From 1924 in pictures; stunt man, double, asst. dir.; dir. Republic Pictures, Hollywood, 1944.
PICTURES INCLUDE: Angel and the Badmen, Oklahoma Badlands, Carson City Raiders, Sons of Adventure, G-Men Never Forget, Dangers of the Canadian Mounted, Adventures of Frank and Jesse James, Lawless Rider.

CAPLAN, HARRY: Executive. b. New Haven, Conn., June 2, 1908. e. UCLA, Southwestern U.; mem. King Brothers dance team on stage, Orpheum and Loew's Circuit; 1930, joined Paramount as property master, became unit production mgr., did directorial assignments; 1960, assoc. producer UA Love is a Ball; 1962, joined Filmways as exec. prod. mgr.; 1964–7, Unit prod. mgr. 20th Century-Fox; 1967, joined National General Productions as exec. prod. mgr.

CAPRA, FRANK: Producer, Director. b. Palermo, Sicily, Italy, May 19, 1897; e. Manual Arts H.S.; Calif. Inst. of Tech., B.Sc., 1918 dir., The Strong Man, Long Pants, Platinum Blonde, American Madness, Bitter Tea of General Yen, Lady for a Day, Broadway Bill, It Happened One Night, 1934; Mr. Deeds Goes to Town, 1936; You Can't Take It With You, 1938, Lost Horizon, Mr. Smith Goes to Washington, Meet John Doe, Arsenic and Old Lace, It's a Wonderful Life, State of the Union, Riding High, Here Comes the Groom, A Hole in the Head, Pocketful of Miracles; prod., dir., Why We Fight Series, U.S. Army; pres. Academy of Motion Picture Arts and Sciences, 1935–39; pres., Directors Guild of America, 1938–40, 1959. 1964 Prod. & Dir. Rendezvous in Space.

CAPRA, JR., FRANK: Executive. Son of famed director Frank Capra. Served in various creative capacities on TV series (Zane Grey Theatre, Gunsmoke, The Rifleman, etc.). Associate producer on theatrical films (Planet of the Apes, Play It Again Sam, Marooned, etc). Joined Avco Embassy Pictures in March, 1981, as v.p., worldwide production. In July, 1981, became pres. of A-E.
PICTURES: Producer: Born Again, The Black Marble, An Eye for an Eye.

CAPUCINE: Actress. r.n. Germaine Lefebvre. b. Toulon, France, January 6. e. attended schools at Saumur, France, B.A. Photographer's model, Paris. Came to N.Y. to model. Signed contract, Famous Artists.
PICTURES INCLUDE: debut, Song Without End, North To Alaska, Walk On the Wild Side, The Lion, The 7th Dawn, What's New Pussycat?, The Honey Pot, Fraulein Doktor, Fellini's Satyricon, Red Sun, Arabian Adventure.

CAPUTO, GEORGE: Executive. b. Genoa, Italy, Oct. 24, 1916. e. Lycee de Nice (France) 1930, Sorbonne, Paris, 1934. Partner, gen. mgr., Caputo & Co., Colombia, S.A.; pres., Master Films Dist., Inc., 1950; pres., Fiesta Tele-Cinema, S.A.; pres. International Sound Studios Inc.; pres., Fall River Investments S.A.; acquired Latin America and South America rights in perpetuity to full RKO package feature films library, 1961.

CARDIFF, JACK: Director. b. Yarmouth, England, September 18, 1914. Early career as child actor, later cinematographer on Stairway to Heaven, Black Narcissus, Red Shoes, Scott of the Antarctic, Black Rose, Under Capricorn, Pandora and the Flying Dutchman, African Queen, Magic Box, The Brave One, War and Peace, The Vikings; started as dir. 1958.
PICTURES INCLUDE: Intent to Kill, Beyond This Place, Scent of Mystery, Sons and Lovers, Fanny, My Geisha, The Lion, The Long Ships, Young Cassidy, The Liquidator, Dark of the Sun, The Girl on the Motorcycle, Penny Gold, The Mutations, Ride A Wild Pony, The Prince and The Pauper, Behind the Iron Mask, Death on the Nile, The Dogs of War (Cinematographer).

CARDINALE, CLAUDIA: Actress. b. Italy, 1939.
PICTURES INCLUDE: Persons Unknown, 1958; Upstairs and Downstairs, Il Bell' Antonio, Rocco and His Brothers, Cartouche, The Leopard, Eight and a Half, The Pink Panther, Circus World, Of A Thousand Delights, Blindfold, Last Command, The Professionals, Don't Make Waves, The Queens, The Red Tent, Conversation Piece, Escape to Athena, Immortal Bachelor.

CAREY, HARRY JR.: Actor. b. Saugus, Calif., May 16, e. Newhall, Cal., public school, Black Fox Military Acad., Hollywood; m. Marilyn Fix. two children, Steven and Melinda; Summer stock, Skowhegan, Me., with father; page boy, NBC, New York; U.S. Navy 1941–46; Screen debut: Pursued, then Red River, Moonrise.
PICTURES INCLUDE: Three Godfathers, She Wore a Yellow Ribbon, Wagonmaster, Rio Grande, Copper Canyon, Warpath, Wild Blue Yonder, Monkey Business, San Antone, Island in the Sky, Gentlemen Prefer Blondes, Beneath the 12-Mile Reef, Silver Lode, The Outcast, Long Gray Line, Mister Roberts, House of Bamboo, Great Locomotive Chase, The Undefeated Big Jake, Something Big, One More Train To Rob.

CAREY, JOYCE: Actress. b. London, England, p. Lillian Braithwaite, actress. e. Florence Ettinger Dramatic School, London. Stage debut in Mr. Wu 1916; in many plays, England & U.S.; writings include: Sweet Aloes, Thing Apart, under pseud. Jay Mallory; m.p. debut in In Which We Serve.
PICTURES INCLUDE: Brief Encounter, Blithe Spirit, October Man, Astonished Heart, Dulcimer Street, The Chiltern Hundreds, Cry the Beloved Country, Street Corner (Both Sides of the Law), End of the Affair, Loser Takes All.
TV PLAYS: Full House, The Farmer's Wife, The Family First, Before the Party, Not Many Mansions, Silas Marner, Sergeant Cork, Sally, Tarnish On a Golden Boy, Wings of The Dove, The Secret Agent, The Chink in the Wall.
TV: Mr. Samson, A Question of Fact, Frolic Wind, No Escape, The Carrington Case, The Bristol Case, Greyfriars Bobby, Barchester Towers, Maggie, No Hiding Place, The Flemish Shop (Maigret), Take the Plunge, The Scissors, The Case of the Self-Made Man, Quite Nice, Gideon's Way, Danger Man, Eyes of Annie Jones, A Family and a Fortune, The Cocktail Party, The Ideal Husband, Howard's End, A Nice Girl Like Me, The Avengers, Father Dear Father, The Shopper, The Last Witness, At the Villa Pandora, The Coffee Lace, Barbara's I, The Cedar Tree, Wedding, Only a Scream Away, Edward VI.
STAGE: Heirs and Graces, Maigret and the Lady, Sergeant Cork, Dear Octopus, The Old Ladies, Trelawny, A Ghost on Tiptoe.

CAREY, MACDONALD: Actor. b. Sioux City, Iowa, March 15, 1913; e. Phillip Exeter Acad., U. Wis., U. Iowa. On stage (stock); in radio serials. On B'way in Anniversary Waltz; m.p. debut in 1942.
PICTURES INCLUDE: Dr. Broadway, Take a Letter Darling, Wake Island, Suddenly It's Spring, Variety Girl, Dream Girl, Hazzard, Streets of Laredo, Song of Surrender, South Sea Sinner, Copper Canyon, Great Missouri Raid, Mystery Submarine, Excuse My Dust, Meet Me After the Show, Let's Make It Legal, Cave of the Outlaws, My Wife's Best Friend, Count the Hours, Outlaw Territory, Fire Over Africa, Stranger At My Door, American Gigolo.
TV: Days of our Lives, Roots.

CAREY, PHIL: Actor. b. Hackensack, N.J., July 15, 1925; e. Mohawk U., U. of Miami, U.S. Marines; New England stock; m.p. debut in Operation Pacific.
PICTURES INCLUDE: Inside the Walls of Folsom Prison, This Woman Is Dangerous, Springfield Rifle, Calamity Jane, Gun Fury, The Nebraskan, Massacre Canyon, Outlaw Stallion, They Rode West, Pushover, The Long Gray Line, Wyoming Renegades, Mister Roberts, Count Three and Pray, Three Stripes in the Sun, The Time Traveler, Once You Kiss a Stranger, Three Guns for Texas, The Seven Minutes. TV: Laredo.

CARLIDGE, WILLIAM: Executive. b. Fenton, Eng., May 2, 1910. e. Grafton Boys' School. Entered m.p. ind. in 1930 as mgr. Ilford Hippodrome; circuit mgr. Union Cinemas, 1935; supervisor for Northern Ireland. Associated British Cinemas Ltd., 1947; first regional controller for Northern Region, personal asst. to man. dir. 1949; gen. man. Associated British Cinemas Ltd. 1951; dir. 1955; asst. man. dir. 1959; Board of Associated British Picture Corp., 1964, now known as EMI Film and Theatre Corp.

CARLILE, TOM: Publicist-Writer, b. Lindsay, Okla., June 9. e. Los Angeles City College (A.A., Journ.), Fresno State College (B.A.) U.C.L.A. (grad work); So. Calif. PR rep. Time, Inc., 1941; correspondent, Life Mag., 1942–43; movie editor, Life Mag., 1944–45; pres. Picture Surveys, Inc., 1945–47; western ed. director, Dell Publishing Co., 1947–49; western ed. PIC Magazine and freelance writer, 1950–54; Pub., Public Affairs Dept., CBS Television, 1954–55; ed. and publisher, Weekend Outdoor News, 1955–56; Sr. pub., Warner Bros. (Giant, The D.I. and Old Man and the Sea), 1956–1957; manager, Advertising Associates, Inc., San Diego, 1958–59; unit. pub., The Alamo, 1959; studio pub. manager, Samuel Bronston Productions, Madrid, (King of Kings and El Cid), 1960–61; pub. dir., Eon Productions Ltd., London (Call Me Bwana, From Russia With Love, Goldfinger, Thunderball, You Only Live Twice) 1962–67; Creative supervisor, European pub., Paramount Pictures, 1967–68; pub. dir., The Adventurers, 1968; ad-pub. dir., Dino De Laurentiis Studios, Rome, 1968–1971; pub. supervisor, Man of La Mancha, 1971–72; U.S. pub., Diamonds Are Forever and Live and Let Die, 1973–74; pub. dir., Algonquin Productions Ltd., London (Rollerball) 1975–76; U.S. pub., A Bridge Too Far, 1977; spec. projects, United Artists, 1977–78; pub. supervisor, Bloodline, Paramount 1978–79. Has written numerous magazine articles and fact books on motion picture subjects.

CARLINO, LEWIS JOHN: Writer. b. N.Y.C., Jan. 1, 1932. e. USC. Early interest in theatre, specializing in writing 1-act plays. Winner of Obie award (off-Bdwy. play). Won Rockefeller Grant for Theatre, the Int'l. Playwriting Competition from British Drama League, Huntington Hartford Fellowship.

PICTURES INCLUDE: Seconds, The Brotherhood, The Fox (co-s.p.), The Mechanic, The Sailor Who Fell from the Sea (s.p., dir.), I Never Promised You a Rose Garden (co. s.p.), The Ace (s.p., dir.), Resurrection (s.p.).

PLAYS: Cages, Telemachus Clay, The Exercise, Double Talk, Objective Case, Used Car for Sale, Junk Yard.

CARLISLE, ROBERT: Producer. b. Los Angeles, Calif., Sept. 19, 1906. Cutter, Metro; Universal; John Stahl; Columbia; film ed. in chf. Columbia; prod. short subjects with Jerry Fairbanks including: Popular Science, Unusual Occupations, Speaking of Animals (two Academy of M.P. Arts & Sciences awards); prod. industrials for Dupont, Union Pacific RR; trophy at Italy's Cortina Sports Film Festival for ski film, 1957; prod. and dir. films on world affairs for Dept. of Defense; prod. Wondsel, Carlisle & Dunphy, Inc., 1957.

CARLTON, RICHARD: Executive. b. New York City, Feb. 9, 1919. e. Columbia U., Pace Inst. Columbia Pictures 1935–41; U.S. Army 1941–45; National Screen Serv. 1945–51, head of agency sales div. of Sterling Television 1951–53; v.p. of Sterling Television, 1954; eastern sales mgr., U.M. & M. TV Corp. 1955; v.p. in charge of sales, Trans-Lux Television Corp., 1956; exec. vice-pres., Television Affiliates Corp., 1961; v.p., Trans-Lux TV Corp.; exec. v.p. Trans-Lux Television Corp.; v.p. Entertainment Div. Trans-Lux Corp., 1966. Pres., Schnur Appel, TV, Inc. 1970; Deputy Director, American Film Institute, 1973.

CARMICHAEL, HOAGY: Composer, actor. b. Bloomington, Ind., Nov. 22, 1899; e. Indiana U., LL.B., 1926. Composer song hits including: Star Dust. Member: Bel Air Club, Kappa Sigma.

PICTURES INCLUDE: To Have and Have Not, Johnny Angel, Canyon Passage, Best Years of Our Lives, Night Song, Young Man With a Horn, Johnny Holiday, Belles on Their Toes, Las Vegas Story, Timberjack. Academy Award, best song in Cool, Cool, Cool Evening from Here Comes The Groom, 1951.

TV: Laramie, also commercials.

CARMICHAEL, IAN: Actor. b. Hull, England, June 18, 1920. e. Scarborough College, Bromsgrove School, Stage debut; R.U.R. 1939. Film debut: Bond Street, one of the top ten British money making stars M.P.H.—Fame Poll 1957, 1958.

TV: New Faces, Twice Upon a Time, Passing Show, Tell Her The Truth, Lady Luck, Give My Regards to Leicester Square, Jill Darling, Don't Look Now, Regency Room, Globe Revue, Off the Record, Here and Now, The Girl at the Next Table, Gilt and Gingerbread, The Importance of being Earnest, Simon and Laura, 90 Years On, The World of Wooster, The Last of the Big Spenders, The Coward Revue, Odd Man In, Bachelor Father, Lord Peter Wimsey, Alma Master, Comedy Tonight, Song by Song, Country Calendar.

PICTURES INCLUDE: Trottie True, Mr. Prohack, Time Gentlemen Please, Meet Mr. Lucifer, Betrayed, Coldits Story, Storm Over the Nile, Simon and Laura, Private's Progress, Brothers in Law, Lucky Jim, Happy Is the Bride, The Big Money, Right, Left and Center, I'm Alright Jack, School for Scoundrels, Light Up the Sky, Double Bunk, The Amorous Prawn, Hide and Seek, Heavens Above, Smashing Time, The Magnificent Seven, Deadly Sins, From Beyond the Grave, The Lady Vanishes.

CARNEY, ART: Performer. b. Mt. Vernon, N.Y., Nov. 4, 1918. Many radio shows. U.S. Army, 1944–45; performer, Jackie Gleason's Honeymooners.

TV: Studio One, Kraft, Omnibus, Chevy Show, Playhouse 90, Dupont Show of the Month—"Harvey," Sid Caesar Show, Alfred Hitchcock Presents—"Safety for the Witness," Art Carney Meets Peter and the Wolf," Playhouse 90—"Fabulous Irishman" "Charley s Aunt," "Velvet Alley," "Art Carney Meets the Sorcerer's Apprentice," The Sid Caesar-Art Carney Show, Our Town, Very Important People, Man in the Dog Suit, Call Me Back, Batman, Carol Burnett Show, Jonathan Winters Show, Tonight Show.

PICTURES INCLUDE: Harry and Tonto, W. W. and the Dixie Dancekings, Won Ton Ton, The Late Show, Take This Job and Shove It.

STAGE: The Rope Dancers, Take Her She's Mine, (Bdwy.), The Odd Couple (Bdwy.), Lovers (Bdwy.), The Prisoner of Second Avenue (Bdwy.), Sunburn, Going in Style, Defiance, Roadie.

CARNEY, FRED: Producer, director. b. Brooklyn, N.Y., June 10, 1914; e. Mt. Vernon H.S., 1932. Actor on B'way & summer stock; prod. mgr. for radio show, Truth or Consequences; asst. to prod.-dir of Kraft TV Theatre, 3 yrs.; dir., Kraft, Pond's Show; creator-prod., Medical Horizons; dir., Lux Video Theatre; prod. commercials at Cunningham & Walsh. Assoc. Prod. Everybody's Talking for ABC-TV; Assoc. Prod. 40th Acad. Award show, ABC-TV. Ass't. exec. dir., Hollywood chpt., Nat'l. Acad. TV Arts & Sciences; currently acting in feature films and TV.

CARNOW, HOWARD N.: Executive. b. Brooklyn, N.Y., Sept. 25, 1912. e. Brooklyn College, 1929–35. Pub., Columbia Pictures, 1929–33; ad mgr., Radio Attractions Inc., 1935–36; ad. & pub mgr., Brandt Theatres, Subway Circuit, 1937–38; Warner Bros., 1939–44; U.S. Navy, 1944–46; Warner Bros., 1946; ad. mgr., United Artists, 1966. Retired 1977.

CARON, LESLIE: Dancer, actress. b. Paris, France. July 1, 1931. e. Covent of Assumption, Paris; Nat'l Conservatory of Dance, Paris. Did solo bit parts, then starred joined Ballet des Champs Elysees. In 1950, signed MGM lead film role in An American in Paris; since in Man with a Cloak, Glory Alley, Story of Three Loves, Lili, Glass Slipper, Daddy Long Legs, Gaby, Gigi, The Doctor's Dilemma, The Man Who Understood Women, The Subterraneans, Fanny, Guns of Darkness, The L Shaped Room, Father Goose, Promise Her Anything, A Very Special Favor, Is Paris Burning, Head of the Family, The Beginners, Madron, Chandler, Purple Night, Valentino.

CARPENTER, CARLETON: Actor. b. Bennington, Vt., July 10; e. Bennington H.S.; Northwestern U. (summer scholarship). Began career with magic act, clubs, camps, hospitals, New Eng.; then toured with carnival; first N.Y. stage appearance in Bright Boy.

STAGE: Career Angel, Three To Make Ready, The Magic Touch, The Big People, Art of Dust, John Murray Anderson's Almanac. Appeared night clubs; radio; as model magazines, TV debut, Campus HoopLa show. Screen debut Lost Boundaries (also wrote song for film, I Wouldn't Mind).

PICTURES INCLUDE: Summer Stock, Father of the Bride, Two Weeks With Love, Whistle at Eaton Falls, Fearless Fagan, Sky Full of Moon, Take the High Ground, Some of My Best Friends Are . . .

CARPENTER, JOHN: Director-Writer. b. Carthage, NY, Jan. 16, 1948. e. USC. Married to Adrienne Barbeau. At USC became involved in film short, Resurrection of Bronco Billy, which won Oscar as best live-action short of 1970. Also at USC began directing what ultimately became Dark Star, science fiction film that launched his career.

PICTURES: Assault on Precinct 13, Halloween (also music), Eyes of Laura Mars (s.p. only), The Fog (also music), Escape from New York (also music).

CARPENTER, ROBERT L.: Executive. b. March 20, 1927, Memphis, Tenn. Joined Universal Pictures in 1949 as booker in Memphis exchange; promoted to salesman there in 1952 and branch mgr. in 1958. In 1963 named Los Angeles branch manager. In Dec., 1971 moved to New York to become asst. to general sales mgr. Named gen. sls. mgr. on April 16, 1973, replacing Henry H. Martin when latter became pres. of Universal.

CARR, ARNOLD: Publicist. b. Chicago, Ill., Aug. 9, 1931. e. Univ. of Missouri. Began public relations handling tour with Dimitri Mitropoulos, U.S. State Depart.; Columbia Broadcasting System; pub. dir., KABC-TV; Screen Gems; pub. rel. dir., Hanna-Barbera Prod.; formed own firm, 1962; partnered with Richard Carter 1964; resumed business under own name, Arnold Carr Public Relations.

CARR, ALAN: Producer—Personal Manager. b. Highland Park, Ill. e. Lake Forest College, Northwestern Univ. First venture in show business as one of creators of Playboy Penthouse TV series. Produced plays at Civic Theatre, Chicago. Asst.

to Nicholas Ray on King of Kings shot in Madrid. Became talent scout, launching Marlo Thomas in play, Sunday in New York, on West Coast. Other clients have included Ann-Margret, Peter Sellers, Nancy Walker, Marvin Hamlisch, Stockard Channing, Melina Mercouri, Franke Valli, etc. Served as creative consultant to Robert Stigwood Organisation in marketing and promotion of movie, Tommy. Presented Survive! with Stigwood.

PICTURES: Producer: The First Time, C.C. and Company, Grease (co-prod.), Can't Stop the Music (co-prod., co. s.p.).

CARR, MARTIN: Producer-Director-Writer b. N.Y.C. Jan 20, 1932. e. Williams Coll. Worked for all three networks.

TELEVISION: For CBS produced, wrote and directed CBS Reports: Hunger in America, The Search for Ulysses, Gauguin in Tahiti, Five Faces of Tokyo, Dublin Through Different Eyes. For NBC produced, wrote and directed NBC White Paper: Migrant, NBC White Paper: This Child Is Rated X. Also directed drama, dance, music, opera specials and daytime serial for CBS-TV. ABC Close-Up. The Culture Thieves. PRS Global Paper: Waging Peace, ABC News 20/20.

AWARDS: Winner of 4 Emmys; 3 Peabody awards; 2 Du-Pont Col. Journalism awards; Robert F. Kennedy award; Sidney Hillman award; Writers Guild Award.

CARR, THOMAS: Director; b. Philadelphia, Pa., July 4, 1907. On screen at 2½ yrs. for Lubin Co., Phil.; starred in Little Britches at 5 yrs., actor on stage, radio & screen until 1937; then became dialogue clerk, Republic Studios, advancing to script; asst. assoc. prod.; assoc. prod. 1944; dir. 1945. Pictures include many westerns, Bobby Ware is Missing.

CARRADINE, DAVID: Actor. b. Hollywood, Oct. 8, 1940. f. John Carradine. e. San Francisco State. Began career in local repetory; first TV on Armstrong Circle Theatre and East Side, West Side; later TV includes Shane series and Kung Fu; N.Y. stage in The Deputy, Royal Hunt of The Sun.

PICTURES INCLUDE: Taggart, Bus Riley's Back in Town, Too Many Thieves, The Violent Ones, Heaven With a Gun, Young Billy Young, The Good Guys and the Bad Guys, Gallery of Horrors, The McMasters, Macho Callahan, A Gunfight, McCabe & Mrs. Miller, Boxcar Bertha, Death Race 2000, Bound for Glory, The Serpent's Egg, Gray Lady Down, The Long Riders, Cloud Dancer.

CARRADINE, JOHN: Actor. b. New York City., Feb. 5, 1908. On stage in Shakespearian roles. A Funny Thing Happened on My Way to the Forum. On screen from 1936.

PICTURES INCLUDE: Fallen Angel, House of Dracula, Captain Kidd, Face of Marble, House of Frankenstein, It's in the Bag, Private Affairs of Bel Ami, C-Man, Cassanova's Big Night, Thunder Pass, Johnny Guitar, The Egyptian, Stranger on Horseback, Hidden Guns, Court Jester, Desert Sands, The Kentuckian, Dark Venture, Black Sheep, The Good Guys and the Bad Guys, Boxcar Bertha, Everything You Always Wanted to Know About Sex*, The Shootist, The Last Tycoon, The Sentinel, Zorro, The Gay Blade.

CARRADINE, KEITH: Actor. b. San Mateo, Calif. Aug. 8 Son of actor John Carradine, brother of David. e. Colorado State Univ. Got first break in rock opera "Hair." Screen debut in A Gunfight (1971).

PICTURES INCLUDE: McCabe and Mrs. Miller, Idaho Transfer, Emperor of the North, Thieves Like Us, Nashville, Lumiere, Welcome to L.A., The Duellists, Pretty Baby, Old Boyfriends, An Almost Perfect Affair, The Long Riders, Southern Comfort.

CARRERAS, SIR JAMES M.B.E.: Chairman and chief executive. Hammer Film Productions Ltd. 1946–1973. Special advisor to the EMI group of companies; director, Studio Film Laboratories Ltd.; chairman, Variety International Executive board; deputy chairman, Royal Naval Film Corporation; deputy chairman, Police Dependents Trust; cochairman, Friends of the Duke of Edinburgh's Award Scheme; director, National Screen Service; trustee, Young Volunteer Force Foundation; chairman, Sobell Variety Islington Trust; vice-president, London Federation of Boys' Clubs; member of the board, Services Kinema Corporation; member of the council, Cinema and Television Benevolent Fund.

CARROLL, CARROLL: Writer, producer, b. New York, N.Y. m.p. critic, feature writer, N.Y. Sunday World; free lance writer, nat'l mags: head writer & edit. supvr., J. Walter Thompson, N.Y. 1932–36; Coast, 1936–46; v.p., writer, prod., Ward Wheelock Co., West Coast, 1946–53. CBS, NBC, JWT, East Coast 1957–67. Now based in Hollywood.

SHOWS INCLUDE: (radio) Bing Crosby, Al Johnson, Rudy Vallee, Eddie Cantor, Burns & Allen, Joe Penner, Kraft Music Hall, Edgar Bergen, Frank Sinatra, Bob Crosby, Corliss Archer, Double or Nothing. Club 15; (TV) head writer, Bob Crosby Show; CBS-TV creative staff, 1953–55; General Electric Hour; Young & Rubicam TV, 1955; NBC-TV creative staff, 1956–57; ed. dept., J. Walter Thompson Co., 1957–67. Writer, prod., Chase & Sanborn Anniv. Radio Show with Edgar Bergen, 1964; C & S 101. Anniv. Show with Fred Allen, 1965; 1966 NBC 40th Anniv. Show. Free-lance writer and adv. consultant, 1967–77. Book Reviewer for Variety and Columnist, "And Now a Word from . . ." in Variety.

AUTHOR: "None of Your Business or My Life with J. Walter Thompson," 1970. "My Life With . . ." 1978. Co-author with Henny Youngman, "Take My Wife . . . Please!" Editorial consultant "Liberace." Editorial Consultant "Mike Douglas, My Story," Co-author with Ed McMahon, "Here's Ed." Co-author with Bob Hope: "I Never Left Home", "So This is Peace," Author, Carroll's First Book of Proverbs.

CARROLL, GORDON: Producer. b. Balt. Md., Feb. 2, 1928. e. Princeton U. Advtg. exec., Foote, Cone & Belding, 1954–58; Ent. industry, Seven Arts Prods., 1958–61; Staff Prod., Columbia Pictures, 1962; V.P., Jalem Prods., 1966–1969; Independent Producer to present.

PICTURES INCLUDE: How to Murder Your Wife, Luv, Cool Hand Luke, The April Fools, Pat Garrett and Billy the Kid, Alien.

CARROLL, LAURIE: Singer, actress. b. Cambridge, Mass., April 15, 1935. e. U.S.C. Actress in little theatre group, Los Angeles; on Broadway in Plain and Fancy.

TV: You Are There, Damon Runyon Theatre, Shower of Stars; feature singer, Johnny Carson Show, CBS-TV.

CARROLL, MADELEINE: Actress. b. West Bromwich, England, Feb. 26, 1906; e. Birmingham U., B.A. Taught school. Given leading role in Brit. "Guns of Loos" 1929; thereafter in many Brit. & U.S. (from 1934).

PICTURES INCLUDE: Young Woodley, Madame Guillotine, I Was a Spy, Thirty-Nine Steps, Case Against Mrs. Ames, General Died at Dawn, Lloyds of London, Prisoner of Zenda, Cafe Society, Convoy, Lady in Distress, My Favorite Blonde. From 1942, prominent in USO Fund and War Bond drives, soldier shows.

CARROLL PAT: Performer. b. Shreveport, La., May 5, 1927. e. Immaculate Heart College, LA.; Catholic University, Wash., D.C. Joined U.S. Army in capacity of "civilian actress technician." Night club entertainer in N.Y., 1949.

TV: Red Buttons, George Gobel, Jimmy Durante and Mickey Rooney; Caesar's Hour, 1956. Bdwy.

STAGE: Catch a Star.

CARSON, JEANNIE: Actress. b. Yorkshire, England. Amer. Citizen, 1966, in musicals Ace of Clubs, Love from Judy; Ent. Motion Pictures in 1954 in As Long as They're Happy; Alligator Named Daisy, Mad Little Island.

TV: Best Foot Forward, Little Women, Berkeley Square, The Rivals, Frank Sinatra Show, Own TV Series, Hey Jeannie and Jeannie Carson Show. Toured in Sound of Music, Camelot, 110 in the Shade. Under contract to Decca.

ARSON, JOHNNY: Comedian. b. Corning, Ia., Oct. 23, 1925. e. U. of Neb. U.S. Navy service; with station KFAB, Lincoln, Neb.; WOW radio-TV, Omaha, 1948; announcer, KNXT, Los Angeles, 1950; then program, Carson's Cellar; quizmaster, Earn Your Vacation, 1954; writer for Red Skelton; star of Johnny Carson Show, CBS-TV; Who Do You Trust, ABC-TV; The Tonight Show Starring Johnny Carson, NBC-TV.

CARSON, ROBERT: Writer. Served in U.S. Armed Forces 1942–45.

WRITINGS: Western Union, Tuttles of Tahiti, Desperadoes, Bedside Manner, Perilous Holiday, Reformer and the Redhead, The Groom Wore Spurs, Just For You, Men With Wings, Light That Failed, Beau Geste, Bundle of Joy, Action of the Tiger, You Gotta Stay Happy.

CARSTENSEN, VERN: Exec. b. Clinton, Ia., May 24, 1914. e. U. of Iowa, 1933–38.

RADIO: announcer, 1936–38, WSUI, Iowa City, WOC, Davenport, WHBF, Rock Island, CBS, Chicago; prog. dir., sls. prom., KROS, Clinton, 1939–42; U.S. Army, 1942–46, Armed Forces Radio Network, No. Africa, orig. Mobile Radio Station, Italy, Armed Forces Radio Service Hdqtrs., Hollywood, OIC overseas liaison; prod. mgr., Mayfair Transcription Co., 1946–51; exec. asst., Alan Ladd Enterprises Inc., 1951–58; assoc. prod., exec. coord., Jaquar Prods., 1958–61; v.p., sec., Eagle Animation Corp.; sec.-treas., Juggernaut Prods. Inc., 1962–64; exec. v.p. Dale Robertson Assoc., 1963; exec. v.p., United Screen Arts Inc. 1964–66; exec. v.p. Juggernaut, Inc. 1966–68, v.p. Entertainment Associates, Inc. 1968–70; v.p. Tommy Walker Productions 1970–1973; pres. Center Features, Inc. 1973–1976; 1977–1978, exec. v.p., Justin & Assoc. Inc.; 1979–81, v.p. Great Rollerworks Corp.

CARTER, JACK: Actor, r.n., Jack Chakrin. b. New York, June 24, 1923. e. Brooklyn College, Feagin Sch. of Dram. Arts. Worked as comm. artist for adv. agencies. Debut Broadway in Call Me Mister, 1947; starred in TV Jack Carter Show. Seen on most major variety, dram. programs, incl. Ed Sullivan Show. Emmy nom. 1962 for Dr. Kildare seg. Played most major nightclubs. App. Broadway in Top Banana; Mr. Wonderful.

PICTURES INCLUDE: The Horizontal Lieutenant, Viva

Las Vegas, The Extraordinary Seaman, The Resurrection of Zachary Wheeler.

CARTER, LYNDA: Actress. b. Phoenix, Ariz., July 24. e. Arizona State Univ. Wrote songs and sang professionally in Arizona from age of 15; later toured for 4 yrs. with rock 'n' roll band. Won beauty contests in Ariz. and became Miss World-USA in 1973. Dramatic training with Stella Adler and Charles Conrad.
TELEVISION: The New Adventures of Wonder Woman.

CARTER, MAURICE: Art director/designer. b. London, England, 1913. Grad. of The Royal Society of Art Masters. Early career as interior decoration designer. Ent. m.p. ind. 1934/5 Islington Studios, became art dir. 1938. Involved in development back projection. Has art dir. or designed over eighty features includ. Man in Grey, Becket, The Battle of Britain, Anne of the Thousand Days, Innocent Bystanders, The Land that Time Forgot, At The Earth's Core, The People That Time Forgot, The Great Train Robbery. Specializes in period subject and special effect. Three times nominated for American Academy awards and twice for British Film Academy awards. Founder of Guild of Film Art Directors.

CARTER, TRACY: Actress. r.n. Tracy Olsen. b. Los Angeles Calif., Feb. 16, 1940. e. Hollywood Sch. of Drama, 1957–58; Pasadena Playhouse, 1958–61. Pianist, 15 years, Los Angeles Conservatory.
TV: Have Gun Will Travel, Ripcord, 77 Sunset Strip, Fred McMurray Show, The Deputy June Allyson Show, Michael Shayne, Betty Hutton Show, Divorce Court, Kraft Suspense Theatre, Bonanza, Beauty Fair.
PLAYS: Desperate Hours, Look Back in Anger, Puss 'n Boots, The Thousand Dollar Mule, Taming of the Shrew, Richard III.
PICTURES INCLUDE: The Couch, Boy on the Run, Terrified, Dust Is My Tomorrow, Sweet Hell, I'll Love You Forever, Johnny Reno, Waco, Red Tomahawk, Journey to the Center of Time.

CARTIER, RUDOLPH: TV producer, director. b. Vienna, Austria, April 17, 1908. Stage designer, journalist, writer, film dir. and prod.; entered m.p. industry 1929. Joined BBC-TV drama dept., 1952; senior producer-director, 1953.
PICTURES INCLUDE: L'Aiglon, Quartermass Experiment, Wuthering Heights, That Lady, Such Men Are Dangerous, Rebecca, Liebelei, Captain Banner, Nineteen Eighty-Four, The Creature, Moment of Truth, Midsummer Fire, Thunder Rock, Quatermass II, Devil's General, White Falcon, Mayerling Affair, Arrow to the Heart, Fugitive, Cold Light, Saint of Bleecker Street, Dark Victory, Clive of India, Queen and the Rebels, Salome, Mother Courage, A Tale of Two Cities, Midsummer Night's Dream, Quatermass and the Pit, Tobias and the Angel, Roshomon, Adventure Story, Anna Karenina, Cross of Iron, Golden Fleece, The Liars, Carmen, Dr. Korczak and the Children, Sword of Vengeance, Aspern Papers, Anna Christie, Stalingrad, Respectable Prostitute, Lady of the Camellias, Midnight Men, July Plot, Wings of the Dove, Ironhand, Joel Brand Story, Gordon of Khartoum, Lee Oswald-Assassin, The Survivors, Level Seven, Firebrand, The Burning Bush, The Fanatics, The Rebel, Conversation at Night, An Ideal Husband, Rembrandt, The Bear, The Year of the Crow, The Proposal, Lady Windermere's Fan (1972), The Deep Blue Sea, Fall of Eagles (1974), Loyalties (1976).
TV: Winslow Boy, Gaslight, Sacrifice to the Wind, The Patriot, A Night Out, Death Has Many Faces, Attorney-General, The Consul, Aspern Papers, That Lady, Corinth House, The Savages, Alibi for James, Mr. Pickwick, The Spanish Civil War, In the Name of Conscience.

CASE, NELSON: Announcer. b. Long Beach, Calif. e. William and Mary Coll. Joined NBC, San Francisco, 1931, announcer, NBC, New York, 1934; freelance 1939; U.S. Navy, 1942–45; Lowell Thomas Show, 1946; TV announcer-host.
SHOWS: Pulitzer Prize Playhouse, Robert Montgomery Presents, Jackie Gleason Show, Omnibus, Red Buttons Show, Masquerade Party, Wide Wide World.

CASELLI, LAMAR: Director. b. San Francisco, Calif., June 16, 1921. e. U. of Wash.; U. of Calif. at L.A.; B.A., 1946; M.A. Telecommunications UCLA 1971; Pilot-Air Corps, 1943–45, legit actor & stage mgr., 1947–50; stg. mgr., assoc. dir. and dir., CBS-TV and NBC-TV, 1950–56; C.C. medical and industrial TV, 1955–63; staff dir., CBS-TV, The Last Word, Lamp Unto My Feet, Look Up and Live, 1957–58; F.L. dir., G.E. College Bowl, CBS-TV. NBC-TV, 1959–64; dir. Alumni Fun CBS-TV, 1964; 1965 Insight CBS-TV; Morning Star, serial NBC-TV; 1966 Young Marrieds, serial ABC-TV; 1967-present, prof., Loyola Univ., 1969, dir. film TV, Death Valley Days; 1970–71, also 1976–77 dir., General Hospital (ABC-TV serial), dir., This is the Life, synd. series, 1975–79.

CASSAVETES, JOHN: Director, Actor. b. New York, N.Y., Dec. 9, 1929. e. Colgate College, N.Y. Acad. of Dramatic Arts. m. Gena Rowlands. Actor in stock co.; asst. stage mgr. for Broadway play Fifth Season; many TV credits including Omnibus, Elgin Playhouse.
PICTURES INCLUDE: Taxi, Night Holds Terror, Crime in the Streets, Edge of the City, Fever Tree, Shadows (dir.), Too Late Blues (dir., s.p.), A Child is Waiting (dir.), Faces, Husbands (actor, s.p., dir.), Minnie & Moskowitz (s.p., dir.), Devil's Angels, The Dirty Dozen, Rosemary's Baby, Machine Gun McCann, Woman Under The Influence, (dir.), Mikey and Nicky, Killing of a Chinese Bookie (dir., s.p.); Two Minute Warning (actor); Opening Night, 1977 (s.p., dir.), The Fury (actor) 1978. Brass target (actor) 1978, Gloria (s.p., dir.), Whose Life Is It Anyway? (actor).

CASSEL, JEAN-PIERRE: Actor. b. Paris, Oct. 27, 1932. Began as dancer, attracting attention of Gene Kelly in Left Bank nightspot, who gave him film work. Appeared in plays before becoming established as leading French screen star.
PICTURES INCLUDE: Games of Love, The Gay Deceiver, Five Day Lover, The Vanishing Corporal, The Male Companion, A Woman Passed By, Is Paris Burning?, Those Magnificent Men in Their Flying Machines, The Killing Game, Oh! What a Lovely War, The Bear and The Doll, The Rupture, The Boat on the Grass, Baxter!, The Discreet Charm of the Bourgeoisie, The Three Musketeers, Le Mouton Enrage, Murder on the Orient Express, Who Is Killing the Great Chefs of Europe?

CASSIDY, SHAUN: Actor/Singer/Composer. b. Los Angeles, Sept. 27. One of 3 sons of Shirley Jones and late Jack Cassidy. e. Bev. Hills high school. Began recording in 1976 and toured Europe and Australia, appearing on numerous TV shows. Has had several hit records.
TELEVISION: Hardy Boys Mysteries, numerous specials, talk shows.
PICTURES: Born of Water (debut for Ameri. Film Inst.).

CASTELL, ANTONIO: Executive. b. Havana, Cuba, Dec. 9, 1911; e. U. of Havana; publ city dir., Cuba, 1937; motion picture exhibitor, 1937–48; distributor, Cooperative Cinematografica; owner dir., Alex and Atlas Films, up to 1961; distributor, Spanish pictures for the Americas.

CASTELLANO, RICHARD S.: Actor Producer. b. New York City, September 4, 1933. e. Commerce H.S., Columbia U. Before acting career was owner of construction company and employed as construction consultant.
THEATRE: 1963: New Yiddish Theatre; 1965–66: starred off-Broadway in A View From The Bridge; 1966: on Broadway in The Investigation and That Summer That Fall; 1967: Stratford Shakespeare Company Festival; 1968: Milk Downstairs, Why I Went Crazy. Also on Broadway in Sheep On The Runway and Lovers and Other Strangers (Tony nomination).
PICTURES: A Fine Madness, Lovers and Other Strangers (Academy nomination, The Godfather.)
TELEVISION: NYPD, The Super, Incident on a Dark Street, Honor Thy Father.

CASTLE, WALTER H.: Cinematographer. b. Santa Ana, Calif., May 19, 1905, Engaged continuously in photographic work for 27 years with Fox Film Corp. and 20th Century-Fox Film Corp., as director of photography; Walt Disney Prod., TV, Hardy Boys, American Dairy Story, Spin, Mickey Mouse Club, Our Friend the Atom, Saga of Andy Burnett; spec. effects for Trapeze.

CATES, GILBERT: Director-Producer. b. New York City, June 6, 1934. e. Syracuse Univ. Began TV career as guide at NBC studios in N.Y., working way up to prod. and dir. of game shows (Camouflage, Haggis Baggis, Mother's Day, etc.). Created Hootenanny and packaged and directed many TV specials.
PICTURES INCLUDE: The Painting (short), Rings Around the World, I Never Sang for My Father, Summer Wishes, Winter Dreams (dir. only); One Summer Love, Dragonfly (prod.-dir.), 1975; The Promise, 1978, The Last Married Couple in America, 1980; Oh, God! Oh, God!—Book II, 1980.
TELEVISION: International Showtime (1963–65 exec. prod.-dir.), Electric Showcase Specials ('65, dir.-prod.), To All My Friends on Shore ('72-dir.-prod.), The Affair ('73-dir.), After the Fall ('74-dir.-prod.), Johnny, We Hardly Knew Ye, (1976-prod.-dir.), The Kid from Nowhere (1981-prod.).
STAGE: Director: Tricks of the Trade, Voices, The Price (Long Wharf Theatre). Producer: Solitaire/Double Solitaire, The Chinese and Mr. Fish, I Never Sang for My Father, You Know I Can't Hear You When the Water's Running.

CAULFIELD, JOAN: Actress. b. Orange, N.J., June 1; e. Columbia U. Harry Conover model 1942–3; on stage 1943–4; screen debut in "Miss Susie Slagle's" 1945.
PICTURES INCLUDE: Dear Ruth, Variety Girl, Unsuspected, Sainted Sisters, Larceny, Pretty Girl, Lady Said No, Rains of Ranchipur, Buckskin.
TV: Sally, My Favorite Husband, CBS.

CAVALCANTI, ALBERTO: Director. b. Rio de Janeiro, Brazil. Feb. 6, 1897; e. Geneva Fine Arts School (architecture). Began in m.p. as art dir., Paris; then dir., later prod. many pictures Paris. To London as head of GPO Film Unit (later Crown Unit).
PICTURES INCLUDE: North Sea, First Days, Squadron

992, Men of the Lightship, Foreman Went to France, Halfway House, Went the Day Well, Champagne Charlie, Dead of Night, Nicholas Nickleby, They Made Me a Fugitive, First Gentleman, For Them That Trespass, & features in Brazil, One Man and The Cinema, Parts I and II.
STAGE: Blood Wedding, Fuente-Ovejuna.
TV: Thus Spake Theodor Herzl, Les Empailles, La Visite de la Vieille Dame.

CAVETT, DICK: Actor-Writer. b. Kearny, Nebr., 1937. e. Yale Univ. Acted in TV dramas and Army training films. Was writer for Jack Paar, and his successors on the Tonight Show and had comedy writing assignments with Merv Griffin, Jerry Lewis, Johnny Carson. In 1967 wrote for self and appeared in night clubs. On TV starred in Where It's At (special on ABC Stage 67) and What's In (special). Began daytime series for ABC-TV in 1968, three-weekly series summer of 1969. Current program: The Dick Cavett Show.

CHABROL, CLAUDE: Director. b. 1930. One of the first directors involved in the French New Wave.
PICTURES INCLUDE: Le Beau Serge, The Cousins, Leda, Les Bonnes Femmes, Les Godelureaux, The Third Lover, Seven Capital Sins, Ophelia, Landru, Le Tigre Aime la Chair Fraiche, Marie-Chantal Contre le Docteur Kah, Le Tigre Se Parfume a la Dunamite, Paris vu par . . . Chabrol, La Ligne de Demarcation, The Champagne Murders, The Route to Corinth, Les Biches, Le Femme Infidele, This Man Must Die, Le Boucher, La Rapture, Ten Days' Wonder, Just Before Nightfall, Dr. Popaul, Les Noces Rouges, Nada.

CHAFFEY, DON: Director. b. Hastings, England, 1917. Trained as artist; secured work in art dept. of Gainsborough Studios in early 40s. Eventually moved to Gaumont British International where directed first film, a documentary on dogfish.
PICTURES INCLUDE: Time Is My Enemy, The Girl in the Picture, The Flesh is Weak, A Question of Adultery, The Man Upstairs, Danger Within, Dentist in the Chair, Greyfriars Bobby, Nearly a Nasty Accident, A Matter of Who, The Prince and the Pauper, Jason and the Argonauts, A Jolly Bad Fellow, One Million Years B.C., The Viking Queen, A Twist of Sand, Creature the World Forgot, The Three Lives of Thomasina, The Horse Without a Head, Ride a Wild Pony.

CHAIKIN, WILLIAM E.: Executive. b. Cleveland, Ohio, April 7, 1919. e. Ohio State Univ. (BS, journalism; MS political science). Was newspaper reporter and columnist for paper in Ohio, Indiana and Florida before joining 20th-Fox pub. dept. in 1945. And in pub. depts. of Republic and Eagle Lion Films. Later pres., Chaikin-Perret, p.r. firm. Became v.p. and treas. of Standard Capital, investment banking firm, where financed and supervised prod. of over 60 films. In 1963 named pres. and bd. chm. of Charter Title Insurance Co. in Los Angeles. Is member of the Board of Mariners Savings and Loan. Is mem. of bd. of Mariners Financial Corp., financial holding co. diversified building firm. From 1968–74 was v.p. in chg. of West Coast operations of Avco Embassy Pictures Corp. and asst. to James R. Kerr, pres. of Avco Corp. in July, 1974, assumed presidency of Avco Embassy Pictures Corp to 1980.

CHAKERES, MICHAEL H.: Executive b. Ohio, Ky. Graduate of Wittenberg University; pres. and chm. of bd. of Chakeres Theatres of Ohio and Kentucky. U.S. Army AF 1942–45. Bd. of Dir.: National NATO, NATO of Ohio, Will Rogers Memorial Fund, Motion Picture Pioneers, Society National Bank, Wittenberg University, Mercy Medical Center. Member: Variety Club, Masonic Lodge, I.O.O.F. Order of AHEPA, Ohio Heart Ass'n.

CHAKIRIS, GEORGE: Actor. b. Norwood, Ohio, Sept. 16, 1934. Entered m.p. industry as a dancer in There's No Business Like Show Business and Brigadoon; academy award for West Side Story. Starred in the Young Girls of Rocheford, The Big Cube.

CHALMERS, PATRICK: b. Aberdeenshire, Scotland, 1940. e. Fettes College, Edinburgh, Aberdeen School of Agriculture, and Durham University. Joined BBC in 1973 as radio producer, and became television producer, 1975. Senior producer at Aberdeen 1970, helping with launch of Radio Aberdeen. Has won awards for films at Berlin Film Festival. Head of Television, BBC Scotland, 1978.

CHAMBERLAIN, RICHARD: Actor. b. Los Angeles, 1935. Became TV star in Dr. Kildare series (1961–65). Theatrical film debut in 1962. A Thunder of Drums.
PICTURES INCLUDE: Twilight of Honor, Joy in the Morning, Petulia, The Madwoman of Chaillot, Julius Caesar, The Music Lovers, Lady Caroline Lamb, The Three Musketeers, The Towering Inferno, The Slipper and the Rose, The Last Wave, The Swarm.
TV: Hamlet (1970), F. Scott Fitzgerald and the Last of the Belles, The Count of Monte Cristo, Centennial, Shogun.

CHAMBERS, EVERETT: Producer, director. b. Montrose, Calif.; Aug. 19, 1926. e. New School For Soc. Res., Dramatic Workshop, NYC. Entered industry as actor; worked with Fred Coe as casting dir. and dir., NBC, 1952–57; prod., dir. Run Across the River, indep. feature and The Kiss, short nominated for Acad. Award, 1958; prod. Johnny Staccato, MCA, 1959; prod., Tess of the Storm Country, 20th-Fox.
TV: Target: The Corrupters, Dick Powell Theater, Saints and Sinners (pilot), The New Lloyd Bridges Show, numerous pilots; prod., dir., co-author with Don Gordon, The Lollipop Cover, Chicago Film Fest. awards, Best Film, Best Actor, 365 episodes Peyton Place, prod. writer for Nightslaves-ABC Movie of the Week, exec. prod. Monty Nash, prod. staff producer ABC TV; produced movies of the week, Moon of the Wolf, Trouble Comes to Town, The Great American Beauty Contest, Can Ellen Be Saved?, The Girl Must Live, They Only Come Out at Night. Currently prod. of Columbo.

CHAMPION, JOHN C.: Director, producer, writer. b. Denver, Col., Oct. 13, 1923; e. Stanford U., Wittenberg Coll.; p. Lee R. Champion, Supreme Court Judge. Col. Entered m.p. in Fiesta; did some radio work; in stock at MGM briefly; co-pilot Western Air Lines, Inc., 1943; served in U.S. Army Air Forces, air transport command pilot 1943–45; public relations officer AAF; writer & prod. for Allied Artists; v.p. prod. Commander Films Corp.: press. Champion Pictures, Inc.; prod., MGM, Warner, Para., Universal, Member: SAG, SWG, SIMPP, SPG; TV Academy, Prod. Writer, MIRISCH-U.A.; prod. TV Laramie series; created McHales Navy; author, novel, The Hawks of Noon, 1965; National Cowboy, Hall of Fame Award, 1976.
PICTURES INCLUDE: Panhandle, Stampede, Hellgate, Dragonfly Squadron, Shotgun, Allied Artists; Zero Hour, Paramount; The Texican, Columbia; Attack on the Iron Coast, Submarine X-1, The Last Escape, Mirisch-United Artists; Brother of the Wind, Sun International, dir-prod-writer, Mustang Country, Universal.

CHAMPION, MARGE: Dancer, actress. b. Los Angeles, Calif., Sept. 2. e. Los Angeles public schools; p. Ernest Belcher, ballet master; m. Gower Champion, dancer, actor. In Blossom Time, Student Prince for Los Angeles Civic Opera; Dark of the Moon. Beggar's Holiday in N.Y. Made debut with husband as dancing team, played many night clubs; m.p. Debut in Mr. Music; then signed by MGM; Fame Star of Tomorrow, 1952; on B'way in 3 for Tonight, 1955; toured, invitation to A March 1962.
PICTURES INCLUDE: Show Boat, Lovely to Look At, Everything I Have Is Yours, Give a Girl a Break, Three for the Show, Jupiter's Darling, The Swimmer, The Party, The Cockeyed Cowboys of Calico County.
TV: GE Theatre, Chevy Show, Telephone Hour, etc.

CHANCELLOR, JOHN: TV Anchorman and News Reporter. b. Chicago, 1927. e. Univ. of Ill. Following military service joined Chicago Sun-Times (1948) and after two years moved to NBC News as Midwest corr. In 1948 assigned to Vienna bureau. Subsequently reported from London; was chief of Moscow bureau before appt. as host of Today program for one year (1961). Left NBC 1965–67 to become dir. of Voice of America. In recent yrs. anchorman for special coverage of moon landings, political conventions, inaugurations etc. Now anchorman, principal reporter on NBC Nightly News and writes and delivers daily series on NBC Radio, The Chancellor Report.

CHANDLER, CHICK: Actor. b. Kingston, N.Y.
STAGE: Great Magoo, Best Man, Harvey, Our Town, Dinner at Eight, Star Dust, Show Boat, Maytime, Roberta, Firefly.
MUSICAL COMEDY: Pajama Game, Damn Yankees, Brigadoon. On screen from 1934.
PICTURES INCLUDE: Blood Money, Big Shot, Alexander's Ragtime Band, Mother Wore Tights, Battle Cry, Mad Mad Mad World, Lost Continent and over 100 more.
TV: Soldiers of Fortune, One Happy Family, Freelance major TV shows.

CHANNING, CAROL: Actress. b. Seattle, Jan. 31, 1923. e. Bennington Col. Long career on Broadway and road; most notably in Gentlemen Prefer Blondes, Lend an Ear, and Hello, Dolly!
PICTURES: First Traveling Saleslady, Thoroughly Modern Millie, Skidoo.
TELEVISION: Svengali and the Blonde, Three Men on a Horse, Crescendo; many guest appearances.

CHANNING, STOCKARD: Actress. b. New York City. Has performed in some 25 plays as well as small roles in two Hollywood films, leading to the starring role in the ABC Movie-of-the-Week, The Girl Most Likely To.
PICTURES INCLUDE: The Fortune, The Big Bus, Sweet Revenge, Grease, The Cheap Detective, The Fish That Saved Pittsburgh.

CHAPIN, SLOCUM: Executive. b. Quincy, Mass. May 12, 1913; e. Dartmouth College, 1936. With Broadcast Builders, Hanover, N.H.; sales prom., World Broadcasting System; sales exec., WOC, Davenport, Ia.; sales mgr., WKBN, Youngstown, O.; gen. mgr., WSTC, Stamford, Conn.; network sales staff, ABC, 1942–48; east. TV sales mgr., ABC, 1948–51; v.p. in charge of owned TV stations, ABC, 1951; sales vice pres., ABC-TV, 1954; V.P. client relations, 1957; v.p. sls., Western division 1959; v.p. sls. National division, 1963; v.p. exec. relations, 1964.

CHAPLIN, CHARLES S.: Executive. b. Toronto, Ont., Canada, June 24, 1911. Studied Law. Entered m.p. ind. 1930 as office boy with United Artists; then office mgr. booker, St. John, N.B., 1933; br. mgr. 1935; to Montreal in same capacity, 1941; named Canadian gen. mgr., June 11, 1945; resigned January 1962. Vice-pres. Canadian sls. mgr., Seven Arts Prod., 1962. Member: B'nai Brith, Toronto Bd. of Trade, various charitable org., many trade assns., pres. Canadian M.P. Dist. Assn., Chmn. m.p. section Com. Chest, chmn. publ rel comm. & past-chmn., M.P. Industry Council.

CHAPLIN, GERALDINE: Actress. b. Santa Monica, CA., 1944. Daughter of Charles Chaplin. Starred in over 20 European productions, including seven with Spain's leading filmmaker, Carlos Saura.
PICTURES INCLUDE: Doctor Zhivago, Stranger in the House, I Killed Rasputin, The Hawaiians, Zero Population Growth, Innocent Bystanders, The Three Musketeers, Nashville, Buffalo Bill and the Indians, Welcome to L.A., Roseland, Remember My Name, The Mirror Crack'd.

CHAPLIN, SAUL: Musical director and producer. b. Brooklyn, N.Y., Feb. 19, 1912. e. NYU 1929–34. Wrote vaude. material, 1933–36; songwriter Vitaphone Corp.; other, 1934–40; Columbia, 1940–48; MGM, from 1948; songs include: Bei Mir Bist Du Schoen, Shoe Shine Boy, Anniversary Song.
PICTURES INCLUDE: Acad. Award, collab. best scoring of mus., American in Paris, 1951, 7 Brides for 7 Brothers, 1954; West Side Story, 1961; mus. dir., Lovely to Look At, Give A Girl a Break, Kiss Me Kate, Jupiter's Darling; mus. supvr. Interrupted Melody, High Society; assoc. prod. Les Girls; music assoc. prod. Merry Andrew; assoc. prod. Can-Can, West Side Story, The Sound of Music; prod. Star, 1968; assoc. prod., The Man of La Mancha, 1972; co-prod. That's Entertainment, Part Two.

CHAPMAN, MICHAEL: Cinematographer. b. New York. Early career in N.Y. area working on documentatries before becoming camera operator for cinematographer Gordon Willis on The Godfather.
PICTURES: The Last Detail, White Dawn, Taxi Driver, The Front, The Next Man, Fingers, The Last Waltz, Invasion of the Body Snatchers.

CHAPMAN, TEDWELL: Playwright, journalist. b. Chicago, Ill., April 12, 1917. e. De Paul University, Layola University, Columnist, Hollywood Feature Syndicate, editor, Tailwagger Magazine, Hoja de Lunes (weekly, Malaga, Spain).
CREDITS: Abroad With Two Yanks (collab. orig. story & s.p.), The Fabulous Suzanne (collab. orig. s.p), The Flanagan Boy (collab. s.p.), Drop In Anytime (orig. s.p.), Forever Is a Long Time (TV, orig. story & s.p.), The Crowded Bed (TV, orig. s.p.), Twice Upon a Time (Stage), The Intriguing Stranger (Stage), Bedlam (Stage), The Shame of Cain (Stage, book & lyrics), South of Granada (Stage, book & lyrics).

CHARI, V.K.N.: Public relations counsel & Publicity advisor. b. Salem, 1913. Till 1956 publicity & pub. relations officer, Gemini Studios, Madras. Presently film publicity adviser to film studios and film producers. Editor: Advertiser's Vademecum, Indian Advertising Year Book, An Economic Guide to India, The Languages of India. Refer-india, etc. Hon. Genl. Secretary. The Indian Council of Public Affairs, Madras. Journalist & news correspondent, B44 First Mainroad Sastrinagar, Madras 20.

CHARISSE, CYD: Dancer, actress. r.n. Tula Ellice Finklea. b. Amarillo, Tex., Mar. 8, 1923. m. Tony Martin, singer; e. Hollywood Prof. School. Toured U.S. & Europe with Ballet Russe; screen debut in "Something to Shout About," 1943. Elected Star of Tomorrow 1948.
PICTURES INCLUDE: Mission to Moscow, Three Wise Fools, Till the Clouds Roll By, On an Island with You, Words and Music, Kissing Bandit, Tension, East Side, West Side, Mark of the Renegade, Wild North, Singin' in the Rain, Sombrero, Band Wagon, Easy to Love, Brigadoon, Deep in My Heart, It's Always Fair Weather, Meet Me in Las Vegas, Silk Stockings, Black Tights, Two Weeks in Another Town, The Silencers, Maroc 7.

CHARLES, MARIA: Actress. b. London, England, Sept. 22, 1929. Trained at Royal Academy of Dramatic Art. London Stage Debut 1946 in Pick Up Girl. Subseq.
STAGE (London): Women of Twilight, The Boy Friend, Divorce Me, Darling!, Enter A Free Man, They Don't Grow on Trees, Winnie the Pooh, Jack the Ripper, The Matchmaker, Measure for Measure, Annie (1979–80).
PICTURES INCLUDE: Folly To Be Wise, The Deadly Affair, Eye of the Devil, Great Expectations, The Return of the Pink Panther, Cuba.
TELEVISION: The Likes of 'Er, The Moon and the Yellow River, Down Our Street, Easter Passion, Nicholas Nickleby, The Voice of the Turtle, The Fourth Wall, The Good Old Days, Turn Out the Lights, Angel Pavement, The Ugliest Girl in Town, Other Peoples Houses, Rogues Gallery, The Prince and the Pauper, Crown Court, Barmitzvah Boy, Secret Army, Agony.

CHARPENTIER, GASTON L. G.: Producer, director. b. London 1912. Early career: dramatic school, theatre and film actor, author short stories, asst. dir. Ent. m.p. industry as prod. man., writer, dir. scriptwriter, 1932; prod. for Strand Films Ltd., National Studios, ABPC., Denham; with Rank Organ. 4½ years; exec. prod. and man. dir., Impact Production Ltd., 1957. At present MGM Studios, Boreham Woods.

CHARTERIS, LESLIE: Writer, producer. b. Singapore, 1907; e. Cambridge, Eng. Novelist; creator The Saint, protagonist of mystery novels and series of pictures based on them prod. by RKO from 1938 and by Julian Lesser from 1952.

CHARTOFF, ROBERT: Producer. b. N.Y.C. e. Union College (A.B.), Col. Univ. (LLB). Met Irwin Winkler through mutual client at William Morris Agency (N.M.) and established Chartoff-Winkler Prods. All films co-produced with Winkler.
PICTURES INCLUDE: Double Trouble, Point Blank, The Split, They Shoot Horses, Don't They?, The Strawberry Statement, Leo The Last, Believe in Me, The Gang That Couldn't Shoot Straight, The New Centurions, Up the Sandbox, The Mechanic, Busting, The Gambler, SPYs, Breakout, Nickelodeon, New York, New York, Valentino, Rocky, Comes a Horseman, Uncle Joe Shannon, Rocky II, Raging Bull, True Confessions.

CHASE, BRANDON: Producer-Director. President MPA Feature Films, Inc.; newscaster-news director NBC-TV 1952–57. Executive director Mardi Gras Productions, Inc. and member of Board of Directors. Now pres., Group I Films, Ltd., and V.I. Prods., Ltd.
PICTURES INCLUDE: The Dead One, The Sinner and the Slave Girl, Bourbon Street Shadows, Verdict Homicide, Face of Fire, Four for the Morgue, Mission To Hell, The Wanton, Harlow, Girl In Trouble, Threesome, Wild Cargo, The Depraved, Alice in Wonderland, The Models, Sensuous 3, The Four of Us, Against All Odds, The Giant Spider Invasion, House of 1,000 Pleasures, The Rogue, Eyes of Dr. Chaney, Alligator, Crash!, Take All of Me, The Teasers, The Psychic, UFOs Are Real, The Actresses.
TELEVISION: Wild Cargo (series prod.-dir.); This Strange and Wondrous World (prod.-dir.).

CHASE, CHEVY: Actor. b. New York City, Oct. 8. e. Bard College. Teamed with Kenny Shapiro and Lane Sarasohn while still in school to collaborate on material for underground TV, which ultimately became off-Broadway show and later movie called Groove Tube. Co-wrote and starred in Saturday Night Live on TV, winning 2 Emmys for outstanding continuing or single performance by a supporting actor in variety and music and as writer for show. Has exclusive contract as performer with NBC. Feature film debut in Foul Play in 1978.
PICTURES: Caddyshack, Seems Like Old Times, Under the Rainbow.

CHASE, STANLEY: Producer. b. Brooklyn, N.Y., May 3. e. NYU (B.A.); Columbia (post-grad.). M. actress/artist Dorothy Rice. Began career as assoc. prod. of TV show Star Time; story dept., CBS-TV; then produced plays off-Bdwy. and on Bdwy., winner Tony and Obie awards for Threepenny Opera. Joined ABC-TV as dir. in chg. programming; exec. consultant, Metromedia Producers Org.; prod. & exec., Alan Landsburg Productions. Formed Stanley Chase Productions, Inc. in 1975, which heads as pres.
PICTURES INCLUDE: To Hell with Heroes, Colossus: The Forbin Project; High-Ballin' (exec. prod.); Fish Hawk (exec. prod.).
TELEVISION: Inside Danny Baker (pilot); Al Capp special (prod., writer); Happily Ever After (pilot; prod., writer); Bob Hope Presents the Chrysler Theatre series; Jigsaw (pilot); Fear on Trial (Emmy nomination); Courage of Kavik (exec. prod.); An American Christmas Carol.
STAGE: Producer of following plays: Threepenny Opera, The Potting Shed, The Cave Dwellers, A Moon for the Misbegotten, Free and Easy

CHASEMAN, JOEL: Executive. b. Albany, N.Y., Feb. 18, 1926. e. Cornell. Entered broadcasting in 1943 as engineer and announcer at New York State radio stations. Worked at WAAM—TV, Baltimore, from 1948 to 1955 as announcer, director, news director and public service producer; later as advertising and promotion mgr. 1955–57, returned to radio as air personality for WITH, Baltimore; returned to WAAM as asst. gen. mgr. & dir. of natl sls. Named program mgr. in 1957. 1960, moved to N.Y. for Group W as natl. radio program mgr. 1961 named v.p. & gen. mgr. Moved to Los Angeles in 1962 as exec. prod. of The Steve Allen Show; returned to Group W, 1964 as v.p. & gen. mgr. of WINS. Named pres. of Group W Radio, 1967; 1970, made snr. v.p., programming & production. Later named pres. of Group W Television. 1973, moved to Washington to assume presidency of Post-Newsweek Stations, Inc.

CHASMAN, DAVID: Executive. b. N.Y., Sept. 28, 1925. e. School of Industrial Art, 1940–43; Academie De La Grande-Chaumiere, 1949–50. Monroe Greenthal Co., Inc. 1950–53; Grey Advertising Agency, Inc., 1953–60. Free lance consultant to industry 1950–60; worked on pictures for UA, 20th-Fox, Columbia, Samuel Goldwyn, City Film; Adv. mgr. United Artists, 1960; exec. dir. adv., United Artists, 1962; exec. production, United Artists, London, 1964; v.p. in prod. United Artists, 1969; v.p. of west coast operations, U.A. 1970; Senior Vice President in charge of Production, U.A. 1972; President, Convivium Productions Inc., 1974.

CHATELAIN, DIDIER: Production Executive. b. Jan. 24, Paris, France. Educated Paris and Polytechnic, London. Degree in Business Administration, Marketing & Advertising. Entered industry in 1970 joining Herman Cohen Productions in London and Hollywood, starting with the advertising & promotion of Crooks and Coronets, 1971; Trog, 1972, Today We Kill . . . Tomorrow We Die!, 1973; The Strangers Gundown, 1974. Assistant Producer, Craze, 1974.

CHAUDHRI, AMIN QAMAR: Director, Cinematographer-Editor. b. Punjab, India, April 18, 1938; e. Hampstead Polytechnic, London, City U., N.Y. President, Artscope Ltd., 1967.
PICTURES INCLUDE: Direction, Kashish, Sweet Vengeance, Khajuraho, Eternal, Vice Girl, Urvasi, Konarak, Black Rodeo, The Land of Buddha, Love Generation, Woman's Urge.
CINEMATOGRAPHY: Right On, Sweet Vengeance, The Hopefuls, The Wicked Die Slow, The Love Statute, Vice Girls, Ltd., Who Says I Can't Ride a Rainbow, Black Rodeo, Medium is the Message, Death of a Dunbar Girl, Kashish.

CHER: Singer-Actress. b. El Centro, Calif. May 20, 1946. r.n. Cherilyn Sarkisian. Began singing with then husband Sonny Bono in 1965; first hit record, I Got You Babe, sold 3 million. Made two films and then debuted nightclub musical-comedy act in 1969. CBS comedy-variety series started as summer show in 1971; became regular following December.
PICTURES INCLUDE: Good Times, Chastity.
TELEVISION: Sonny & Cher Comedy Hour, Cher.

CHERMAK, CY: Producer-Writer. b. Bayonne, N.J., Sept. 20, 1929. e. Brooklyn College; Ithaca College.
TELEVISION: Produced Ironsides, The Virginian, The New Doctors, Amy Prentiss, Night Stalkers, Barbary Coast, CHIPS. TV movie: Murder at the World Series.

CHERTOK, HARVEY: Executive. b. New York, N.Y., Oct. 29, 1932. e. N.Y.U., grad. school, bus. admin. Merchandising mgr., National Telefilm Assocs., Jan. 1956–Aug. 1959; supv., adv., United Artists Associated, Aug. 1959–March 1961; dir., adv., sls. prom., publicity, Seven Arts Associated Corp., 1961–1968; prod., 7 Surprizes, 1964; 1968 TV v.p. advertising-publicity, Warner-7 Arts, co.-prod., The Great Charlie Chan (TV spec.) 1968, co-author Quotations from Charlie Chan (Golden Press) 1968; Pres. The Children's Movie of the Month, Inc., 1969–71; marketing consultant, United Artists Television, 1972–73; V.P.-special-projects, The American Film Theatre, Inc. 1973–75; v.p. adv., sls. prom., pub. for Time-Life Television, 1975; named v.p., adv./promo, Time Life-tv, 1978–81.

CHERTOK, JACK: Producer. b. Atlanta, Ga., July 13. Began career as script clerk, MGM; later asst. cameraman, asst. dir., head of music dept., short subjects prod. Feature prod. MGM 1939–42 ("The Penalty," "Joe Smith, American," "Kid Glove Killer," "The Omaha Trail," "Eyes in the Night," etc.). In July, 1942, apptd. Hollywood prod. chief, Co-Ord. Inter-Amer. Affairs, serving concurrently with regular studio work. Left MGM 1942 and prod. for Warner Bros. to late 1944; pres.: Produced The Corn is Green and Northern Pursuit for Warner Bros. Pres. Jack Chertok TV, Inc.

CHESTER, HAL E.: Producer. b. Brooklyn, N.Y., Mar. 6, 1921. On N.Y. stage; film debut in "Crime School" 1937, followed by "Little Tough Guy" series. "Juvenile Court"; personal appearance tour of U.S and Canada 1941–44; entered m.p. production 1944, doing a series of musical shorts. Member: IMPPA. Prod. Joe Palooka series beginning with "Joe Palooka Champ" 1946, released via Monogram, Smart Woman; v.p. in chge. prod., Mutual Films Corp., 1951.
PICTURES INCLUDE: Joe Palooka in The Squared Circle, Underworld Story, The Highwayman, Triple Cross, Models, Beast from 20,000 Fathoms, Crashout, Bold and The Brave, The Weapon, The Haunted, School for Scoundrels, Two-Headed Spy, His and Hers, Hide and Seek, The Comedy Man, The Double Man, The Secret War of Harry Frigg.

CHETWYND, LIONEL. Executive, writer. b. London, England, 1940. Emigrated to Canada, 1948. B.A. (Economics, Sir George Williams Univ. Montreal), B.C.L.-McGill University, Montreal. Graduate Work-Law—Trinity College, Oxford. Admitted to Bar—Province of Quebec, 1968. C.B.C.—TV-Public Affairs and Talks, 1961–1965. CTV Network 1965–67. Controller commercial TV and film rights, Expo '67. Freelance writer and consultant 1961–68. Joined Columbia Pictures Intern. New York July 1968, transferred to Columbia Pictures Corp. Ltd., London, Sept. 1968. Asst. man. dir. January 1969. Asst. man. dir. Columbia-Warner UK, 1971. Story and book for musical Maybe That's Your Problem, 1971–1973: s.p., Apprenticeship of Duddy Kravitz; The Guest Room (TV).

CHIKADA, TADASHI: Executive. b. Japan Nov. 20, 1919. e. Tokyo Foreign Language Univ. 1940–6 with Showa Tsusho in Japan and China; 1947–50 Izumi Sangyo, transferring to Nihon Eiga Shizai (Japan Movie Equipment Co.) which later became Toshiba Photo Phone Co.; appt. director and gen. mgr. of Trade Div. in 1972, and in 1974 director in charge of trade.

CHILTON, T.E.: Executive. b. Kent, England, 1929. Ent. m.p. ind. 1951 with J.F. Brockliss. 1953 Simplex Apro. 1955. Rank Precision Industries, then various appointments in equipment industry in Rank Organisation becoming Asst. Man. Dir. 1968, Man. Dir. 1971. Appt. man. dir. Rank Leisure Service 1973. Appt. chairman Rank Leisure Services, Rank Hotels, Rank Advertising Films, Rank Film Distributors, Pinewood Studios, 1976, dir. Rank Organisation 1976.

CHIN, ELIZABETH MAE: Executive. b. New York, N.Y. Nov. 2, 1938. e. George Washington H.S. 1956; C.C.N.Y., Business Administration 1960; Internal Auditor 1960; International trade and entrepreneur of international dairy franchise 1968. Entered motion picture industry in 1974 as agent for independent film distribution in theatrical and television. Presently, executive of Far East Enterprises, and Emcee World Associates, Inc. and Bergen Intl. Corp.

CHING, WILLIAM: Actor. b. St. Louis, Mo., Oct. 2, 1913. e. Spring Hills Coll. Radio Singer; on Broadway in Allegro.
PICTURES INCLUDE: Pat and Mike, Never Wave at a Wac, Give a Girl a Break, Scared Stiff, The Moonlighter, Tall Man Riding, Magnificent Matador.
TV: featured actor, Our Miss Brooks; NBC-TV.

CHINICH, JESSE: Executive. b. Hoboken, N.J., Dec. 17, 1921; e. N.Y.U. Law S., L.L.B. 1938. Lawyer with Hovell, Clarkson and Klupt, 1938–41; U.S. Army, 1941–46; joined Paramount Theatres as film buyer, 1946–51; circuit supervisor, Rugoff & Becker, NYC, 1951–53 western div. sales mgr., Buena Vista Film Dist., Co., 1953–61; Ass't gen. sales mgr. co-ordinator Cinema V Dist.; ass't sales mgr., Allied Artists; sales executive, special projects, Warner Bros. Dist. Corp. Now Publisher of weekly newsletters.

CHISSELL, NOBLE (KID): Actor. b. Indianapolis, Ind., Feb. 16, 1910; e. City Coll. of L.A., U.S. Merchant Marine, prof. boxer, prison officer, U.S. Navy Middleweight Boxing Champ 1932, World Marathon Dance Champion 1928, Champ Walkathoner 1934–35. Chairman/m.c. world famed Cauliflower Alley Club, co-ed boxing/wrestling group, since 1968. Middleweight Senior Olympic Skill Boxing Champ 1974–78 m.c. Burbank Easter Parade 1979–80. Nostalgic Movie/Boxing columnist, Hollywood Independent; Boxing columnist, Official Boxing Record; Master of Ceremonies, Film Welfare League benefit Show, 1980.
STAGE: Cafe Frisco, All the Kings Men, J. Silverheel's Indian Pageant, Love in Upper Sandusky, Lower Depths, Meet the World, Twelfth Night.
PICTURES INCLUDE: Abe Lincoln in Illinois, Home in Indiana, Dillinger, Grand Canyon, Young At Heart, Guys and Dolls, Big Chase, Friendly Persuasion, Bonnie and Clyde, More Dead Than Alive, Hell Cats, Big Foot, They Shoot Horses, Don't They?, Mule Feathers, The Clones, Ex-Champ, Jealousy, The Leather-pusher, Song of Arizona, The Polygamist, The Traveling Executioner, Three in The Saddle, Arizona Mission, El Salvaje, Pistol Packing Mama, The Rebel Breed, etc.
TV: Marshall of Gunsight Pass, Colgate Comedy Hour, Life of Riley, It's a Great Life, Dragnet, Playhouse 90, Climax, People's Court, Gunsmoke, Gallagher-Disney Series, High Chaparral, To Tell the Truth, Dating Game, 20 Bob Hope Comedy Specials. Also, Minsky Follies 1975, Hollywood.

CHIVERS, ALAN: Executive. b. 1918. Joined BBC (Radio) 1946 transferred to Television Outside Broadcasts 1948 as prod.; moved to Independent Television 1959. Returned to BBC as Exec. Prod. special projects television outside broadcasts from 1969.

CHODOROV, EDWARD: Writer, dir. b. New York City, April 17, 1914. e. Brown U. Writer, dir., plays; Writer, prod., films.
PLAYS INCLUDE: Wonder Boy, Kind Lady, Cue for Passion, Those Endearing Young Charms, Decision, Common Ground, The Spa, Signor Chicago, Oh Men Oh Women, M. Lautrec, Erskine, Listen to the Mocking Bird, etc. Writer-prod. Warners, 20th, Col., MGM. Wrote Billy Rose Show ABC; writer dir., prod. Sam Spiegel, Horizon Pictures, 1961–63; Writer-prod.-dir. A.C.E. Prods. (London) 1964; Columbia, 1965; 20th-Fox, 1966–67; writer-producer, Warner-7 Arts, 1968, 1969; writer-dir., Dowling, Whitehead, Stevens, 1971–72; Cheryl Crawford Prods., 1973, Universal, 1974–75; Theatre Guild 1976; The Lecturer, California State University 1977–78; the Sorbonne 1979. Play commissioned

by Roger Stevens-Robert Whitehead, A Very Important Man Is Dead.

CHOMSKY, MARVIN J.: Director. b. Bronx, N.Y. Started in theatre business at early age as art dir. with such TV credits as U.S. Steel Hour, Playhouse 90, Studio One, etc. Later worked with Herb Brodkin who advanced him to assoc. prod. with such TV shows as The Doctors and The Nurses. Brought to Hollywood in 1965 as assoc. prod. for Talent Associates, producing series of TV pilots. First theatrical film, Maya, for King Bros.
PICTURES INCLUDE: Evel Knievel, Murph the Surf, Mackintosh and T.J.
TV: The Wild, Wild West, Gunsmoke, Star Trek, Then Came Bronson, and many Movies of the Week, Victory at Entebbe, Roots.

CHONG, TOMMY: Singer-Actor-Writer. b. Edmonton, Atla., Canada. Was guitar player with various Canadian rhythm and blues combinations, teamed with Richard Marin (Cheech) in improvisational group. Has made comedy recordings.
PICTURES: Cheech and Chong's Up in Smoke, Cheech and Chong's Next Movie, Cheech and Chong's Nice Dreams.

CHOOLUCK, LEON: Producer-Director. b. New York, March 19, 1920; e. CCNY, 1938. Production, distribution, editing with Consolidated Film Industries 1936–40; Army Pictorial Service as news photographer 1941–45; Indep. as ass. dir. and prod. mgr. 1946–56; prod. for Regal Films and Orbit Pro., 1957–58; dir. Highway Patrol, 1958. Various occupations on stage 1947–58; prod. mgr., Daystar Prods., ABC-TV series, Stoney Burke; prod. mgr., assoc. prod., Daystar Prods., 1962–63; prod. supv., Encyclopedia Britannica Films, in Spain, 1964; prod. supv., U.S. Pictures, Battle of the Bulge; loc. mgr., Three F Prods., assoc. prod. I Spy, TV Series, 1965–67. Vice Pres. Fouad Said Cinemobile Systems, 1969–70. Pres., The Society (film prod. org.).
PICTURES INCLUDE: Hell on Devil's Island, Plunder Road, Murder by Contract, City of Fear, The Fearmakers, Day of the Outlaw, Bramble Bush, Rise and Fall of Legs Diamond, Studs Lonigan, Three Blonds in His Life, El Cid, The Outer Limits, Midas Run, The Grissom Gang, Kotch, Payday, 1971; Three the Hard Way-'73–74; Take a Ride-'74–75, Apocalypse Now, Loving Couples.
TELEVISION: Prod. supvr. 1974–76; NBC (2 hr.) Specials, Strange Homecoming, Michener's Dynasty, Judge Horton and the Scottsboro Boys, Pearl, A Rumor of War, Murder in Texas.

CHOW, RAYMOND: Producer. b. Hong Kong. e. St. John's University, Shanghai. Worked as journalist for Hong Kong Standard; then joined the U.S. Information Agency, establishing its Chinese radio station which broadcast to the Chinese mainland. In late 1950s joined Shaw Brothers as head of their publicity operations. In next 12 years became right-hand man of Run Run Shaw and head of production. In 1971 formed The Golden Harvest Group to produce Chinese-language films in Hong Kong. Two kung-fu films featuring Bruce Lee put Harvest into intl. market. Started English-language films in 1977, beginning with The Amsterdam Kill and The Boys in Company C.

CHRISMAN, H. ED: Executive. b. Nashville, Tenn., Sept. 6, 1914. e. Vanderbilt Univ., 1932. Sales, br. mgr., Dun and Bradstreet, 1935–50; v.p. and sales mgr., Cretors and Co., Popcorn Machine manufacturer, Chicago and Nashville, 1950–66; purchased Shoemaker Popcorn Co., Murray, Ky.; incorporated bus. as Chrisman Popcorn Co., processors and packers of Rocket popcorn; regional v.p. & comm. chmn, National Assc. of Concessionaires.

CHRISTIAN, LINDA: Actress. r.n. Blanca Rosa Welter. b. Tampico, Mex., Nov. 13, 1924, e. Mexico, Venezuela, Palestine, So. Africa, Holland, Italy; attended medical school in Palestine; Worked for British Censorship Bureau in Palestine; asst. to plastic surgeon; screen debut in Holiday in Mexico.
PICTURES INCLUDE: Green Dolphin Street, Tarzan and the Mermaids, The Happy Time, Battle Zone, Slave of Babylon, Athena, Thunderstorm.

CHRISTIANSEN, ROBERT W.: Producer. B. Porterville, CA. e. Bakersfield College. Spent 3 years in Marine Corps. Worked on Hollywood Reporter in circulation and advertising. Joined Cinema Center Films; prod. asst. on Monte Walsh and Hail Hero. Co-produced first feature in 1970, Adam at Six A.M., with Rick Rosenberg, with whom co-produced all credits listed.
PICTURES INCLUDE: Adam at Six A.M., Hide in Plain Sight.
TV: Features: Suddenly Single, The Glass House, A Brand New Life, The Man Who Could Talk to Kids, The Autobiography of Miss Jane Pittman, I Love You . . . Goodbye, Queen of the Stardust Ballroom, Born Innocent, A Death in Canaan, Strangers.

CHRISTIE, HOWARD J: Producer. b. San Francisco, Calif., Sept. 16, 1912. e. Berkeley, U. of Calif. Entered m.p. ind. following graduation from coll. in 1934, as actor; asst. dir. 1936–40; assoc. prod., 1942–44, Deanna Durbin pictures. Member: Screen Directors Guild, Screen Producers Guild, Delta Tau Delta, vice pres., Reeve M.C.A. Inc., 1961.
PICTURES INCLUDE: Lady on a Train, Because of Him I'll Be Yours, A & C Meet the Invisible Man, Golden Horde, Comin' Round the Mountain, Lost in Alaska, Against All Flags, Yankee Buccaneer, Seminole, Lone Hand, A & C Go To Mars, A & C Meet Dr. Jekyll and Mr. Hyde, Back to God's Country, Yankee Pasha, A & C Meet Keystone Kops, Smoke Signal, The Looters, Purple Mask, A & C Meet the Mummy, Price of Fear, Congi Crossing, Showdown at Abilene, Toy Tiger, Away All Boats, I've Lived Before, Wagon Train, The Raiders, Sword of Ali Baba, Senes, Laredo, Ride to Hangman's Tree, Journey to Shiloh, Nobody's Perfect, A Man Called Gannon.
TV: The Virginian.

CHRISTIE, JULIE: Actress. b. Chukua, Assam, India, July 14, 1941. e. Year in France followed by Drama School in England. TV debut in A for Andromeda. Birmingham Rep.; Royal Shakespeare Co.; East European and American tour, Starred (NY) in Uncle Vanya.
PICTURES INCLUDE: Fast Lady, Billy Liar, Young Cassidy, Darling, Dr. Zhivago, Farenheit 451, Far From the Madding Crowd, Petulia, In Search of Gregory, The Go-Between, McCabe and Mrs. Miller, Don't Look Now, Shampoo, The Demon Seed, Heaven Can Wait.

CHRISTINE, VIRGINIA: Actress. b. Stanton, Iowa, March 5. e. UCLA Los Angeles, Calif.; has appeared in more than 400 motion pictures and television productions.
PICTURES INCLUDE: Mission to Moscow, Counter Attack, The Killers, Cover Up, The Men, Cyrano De Bergerac, Cobweb, High Noon, Not as a Stranger, Spirit of St. Louis, Three Brave Men, Judgment in Nuremberg, The Prize, Four For Texas, A Rage to Live, Guess Who's Coming to Dinner, Hail Hero, Daughter of the Mind.
TV: Dragnet, Abbott & Costello, Dangerous Assignment, Racket Squad, Superman, Schlitz Playhouse, Four Star Playhouse, The Whistler, Code 3, Ford Theatre, You Are There, Stage 7, Passport to Danger, Soldier of Fortune, Heinz Show, Anthology, Four Star Playhouse, Cavalcade of America, Alfred Hitchcock Presents, Father Knows Best, Crusader, Kellogg, Front Row Center, The Twisted Road, Matinee, Private Secretary, Fort Laramie, Big Town, Science Fiction, Frontier Detective, Lone Ranger, Jim Bowie, Wire Service, Whirlybirds, Gunsmoke, Las Vegas Story, Trackdown, The Thin Man, San Francisco Beat, Stranger In Town, T Men, Casey Jones, Climax, Mickey Spillane, The Millionaire, Behind Closed Doors, Target, Peter Gunn, Zane Grey Theatre, Wyatt Earp, Secret Mission, The Donna Reed Show, Wanted Dead Or Alive, Buckskin, Loretta Young Show, Starperformance Rescue 8, Steve Canyon, State Trooper, Twilight Zone, General Electric Theatre, How to Marry A Millionaire, Rifle Man, June Allyson Show, Rawhide, M Squad, Coronado 9, Man From Black Hawk, Grand Jury, Riverboat, 77 Sunset Strip, Happy, The Thriller, Lawless Years, The Untouchables, Death Valley Days, Verdict Is Yours, Deputy, Perry Mason, Wagon Train, Mr. Ed, Maverick, The Shirley Temple Show, Asphalt Jungle, Harrigan and Son, Tales of Wells Fargo, Bronco, Line Of Duty, The New Breed, Going My Way, Stoney Burke, Bonanza, The Eleventh Hour, Ben Casey, The Virginian, The Fugitive, Day In Court, Hazel, Singing Nun, Billy The Kid, The Big Valley, The FBI, Laredo, A Man Called Shenandoah, Jericho, Judd for the Defense, The Invaders, Lancer, Nanny And The Professor.
STAGE: Hedda Gabler, Mary, Queen of Scots, Miss Julie, Desdemona.

CHRISTOPHER, JORDAN: Actor, musician. b. Youngstown, Ohio, 1941. e. Kent State U. Led rock 'n' roll group, "The Wild Ones." Broadway debut, Black Comedy, 1967.
PICTURES INCLUDE: Return of the Seven, The Fat Spy, The Tree, Pigeons.

CHURCHILL, SARAH: Actress. b. Oct. 7, 1916, Admiralty House, London. p. Winston Churchill. e. England and France. At 17 attended ballet school. Made first appearance in chorus of C.B. Cochrane's Follow the Sun, in London; and continued in theatre. Women's Auxiliary Air Force, World War II. In 1945, returned to professional life: two Italian films, When In Rome and Daniele Cortis. Also on stage, London, The Barretts of Wimpole Street. A personal appearance in America with J. Arthur Rank's All Over the Town brought her a Theatre Guild contract for touring with The Philadelphia Story, after which she danced for MGM in The Royal Wedding. Has appeared on Broadway in Gramercy Ghost, and for two years actress, hostess Hallmark's Hall of Fame, TV.

CIANNELLI, LEWIS E: Associate Producer. b. New York, Jan, 12, 1923 Univ. of Calif., 1941, Leland Stanford Univ.-AB 1947. Meschini Institute, Rome, 1950. World War II 1942–46. Acct. exec. adv. firm 1947–48. Actor: TV and motion pictures, 1948–50. Exec. asst. Industrie Cinematografiche Sociali, 1950, Director of Foreign Sls. I.C.S. and on board of A.N.I.C.A. 1953. Production exec. Lux Film, Rome 1953.

Head of Production Republic Pictures, Italy 1954–55; assoc. prod. The Best of the Post; assoc. prod., exec. asst. Ponti & Girosi, That Kind of Woman; Heller in Pink Tights, assoc. prod. NBC-TV; Ford Startime, Chevy Special Mexico, Rivak The Barbarian, The Americans, Outlaws.
PLAYS: OK Nero! Twin Trouble, The Bed.

CILENTO, DIANE: Actress. b. Queensland, Australia, 1934. e. Toowoomba. Went to New York and finished schooling and then American Acad. of Dramatic Art; first theatre job at 16; toured U.S. with Barter Co.; returned to London and joined Royal Acad. of Dramatic Art; several small parts and later repertory at Manchester's Library Theatre.
PICTURES INCLUDE: Angel Who Pawned Her Harp, Passing Stranger, Passage Home, Woman for Joe, Admirable Crichton, Truth About Women, Jet Storm, The Full Treatment, The Naked Edge, The Breaking Point, Once Upon A Tractor for UN, Hombre.
THEATRE: N.Y. stage: Tiger at the Gates, I Thank a Fool, The Third Secret, Tom Jones, The Rattle of a Simple Man, The Agony and the Ectasy, Hombre, Once Upon a Tractor, Negatives, The Four Seasons, The Bonne Soup, Heartbreak House. (NY), the Big Knife, Orpheus, Altona, Castle in Sweden, Naked, Marys, I've Seen You Cut Lemons.
TV: La Bell France (series), Court Martial, Blackmail, Dial M for Murder, Rogues Gallery, Rain, Lysistrata.

CIMINO, MICHAEL: Writer-Director. b. 1943. Protege of Clint Eastwood, who signed him to do s.p. for Magnum Force and then to write and direct Thunderbolt and Lightfoot, in both of which Eastwood starred.
PICTURES: The Deer Hunter, Heavens' Gate.

CIPES, ARIANNE ULMER: Executive. b. New York City, July 25, 1937. Daughter of film director Edgar G. Ulmer. e. Royal Academy of Dramatic Arts, London, University of London; actress, then production and dubbing, Paris; CDC, Rome; Titra, New York; 1975–77, v.p., Best International Films (international film distributor), Los Angeles; 1977 co-founder and v.p./sales & services of Producers Sales Organization.

CIPES, J.H.: Executive. b. N.Y., Dec. 14. e. Cornell U. 1960–66, Independent Producer-Packager-Distributor European features for U.S. TV sales; 1967, Producer Columbia Pictures Hollywood; 1968, Producer 20th Century-Fox TV; 1970, Producer Four Star TV; 1971, Marketing Executive, Technicolor, Inc.; 1973, Vice President, Marketing Technicolor, Inc.

CLAGETT, MANNING (TIM): Executive. b. Washington, D.C., Aug. 22, 1913. e. U. of Southern Calif.; U. of Calif., Los Angeles; Geo. Wash. U. Correspondent for Washington Herald, Washington Times-Herald 1937–41; Washington Star, 1941–48. Joined U.S. Merchant Marine 1943–44; corr., Washington Post 1945–46; own news bureau 1946–48. Assoc. Director of information, Motion Picture Assoc. of America, Washington, 1948–1958; dir. pub. rel. MPAA, 1953; dir. State Leg. Service, MPAA, 1956; v.p., 1961; special consultant, 1970.

CLARK, CANDY: Actress. b. Oklahoma. Was successful model in N.Y. before landing role in Fat City, 1972. Continues to model between films.
PICTURES INCLUDE: Fat City, American Graffiti, The Man Who Fell To Earth, Citizens Band, The Big Steep, When You Comin' Back, Red Ryder, More American Graffiti.

CLARK, DANE: Actor. b. New York City, Feb. 18, 1915; e. Cornell U., Johns Hopkins U. In radio series 2 yrs.; on N.Y. stage (Of Mice and Men, Dead End, etc.).
PICTURES INCLUDE: Glass Key, Sunday Punch, Pride of the Yankees, Tennessee Johnson, Action in the North Atlantic, Destination Tokyo, Very Thought of You, Hollywood Canteen, God Is My Co-Pilot, Her Kind of Man, That Way With Women, Deep Valley, Embraceable You, Moonrise, Whiplash, Without Honor, Backfire, Barricade, Never Trust a Gambler, Fort Defiance, Highly Dangerous, Gambler and the Lady, Go Man Go, Blackout, Paid to Kill, Thunder Pass, Port of Hell, Toughest Man Alive.
TV: No Exit, The Closing Door, Wire Service, Bold Venture (series).

CLARK, DICK: Performer. b. Mt. Vernon, N.Y. Nov. 30, 1929. Syracuse Univ. graduated 1951, summer announcer WRUN, Utica 1949, staff announcer WOLF, Syracuse 1950. After graduation 1951, took regular job with WOLF. Rejoined WRUN, Utica, then joined WKTV, Utica. Announcer WFIL Philadelphia 1952. Host of American Bandstand ABC-TV nationwide, the Dick Clark Beechnut Show, Dick Clark's World of Talent, The Object Is, Missing Links, Record Years, Years of Rock. Producer of TV shows, Where The Action Is, Swinging Country, Happening, Get It Together, Shebang, Record Years, Years of Rock. Author of Your Happiest Years, Random House . . . 1959, Author of Rock, Roll & Remember, T.Y. Crowell, 1976 and To Goof or Not to Goof, Bernard Geis, 1963. Formed Dick Clark Productions 1956 . . . Business engaged in motion picture production: Because They're Young, The Young Doctors, Psychout, The Savage Seven, Killers Three, Elvis, the Movie. In-person concert division, marketing and communications division, cable TV programing department and radio division. Executive Producer of Dick Clark's Rockin' New Year's Eve, American Music Awards, Academy of Country Music Award, and other TV specials. 1978: Dick Clark's Live Wednesday. Produced TV Movies: Elvis, Man in the Santa Claus Suit.

CLARK, GREYDON: Producer-Director-Writer. b. Niles, MI, Feb. 7, 1943. e. Western Michigan Univ. (B.A., theatre arts, 1963). Heads own company, World Amusement Corp., Sherman Oaks, CA.
PICTURES: Writer: Satan's Sadists, Psychic Killer. Director-Writer: Mothers, Fathers, & Lovers, Bad Bunch. Producer-Writer-Director: Hi-Riders, Angel's Brigade, The Warning.

CLARK, JOHN L.: Executive. b. Owenton, Ky., Sep. 13, 1907; e. Univ. of Tenn. Talent scout, Brunswick Phono. Co.; mgr., WLW, Cincinnati; formed Transamerican Broad. & Tele. Corp. with Warner Bros.; prod. 50 TV films, 1941–44; created & prod. many radio & TV programs; pres. & sole owner, Transamerican.

CLARK, JOHN R., JR.: Executive. b. Owensburg, Ind., Aug. 18, 1915. e. Columbia U., 1930–34; Columbia Teachers Coll., 1934–35; U. of Calif., 1935–36. Employed in manufacturing plant, Technicolor Corp., 1936–38; research dept., administrative asst., 1938–46; civilian, Office of Scientific Research and Development, 1944; asst. to pres., Technicolor Corp., 1946; exec. v.p., 1955; pres., gen. mgr. Technicolor companies, 1960; dir., Technicolor, Inc., and Technicolor Corp. Fellow member: SMPTE.

CLARK, KENNETH: Journalist, executive. b. Dekalb, Ill., Aug. 1, 1899; e. U. Ill. Washington & foreign corresp. United Press, Internatl. News Service, Universal Service. Joined M.P. Prod. & Dist. Amer., 1936. Leave of absence 1942 to join U.S. Army; Maj., Lt. Col. and Col.; Returned to MPAA, Nov., 1945; v.p., 1953; exec. v.p., MPAA, 1962; v.p., MPEAA, 1953.

CLARK, PETULA: Actress and vocalist. b. Ewell, Surry, England, Nov. 15, 1932.
PICTURES INCLUDE: The Huggets, Dance Hall, White Corridors, The Card, Made in Heaven, Gay Dog, Runaway Bus, That Man Opposite, Finian's Rainbow, Goodbye Mr. Chips.

CLARK, ROBERT: Executive. b. Paisley, Renfrewshire, Scotland, July 26, 1905; e. Glasgow Univ. M.A., LL.B. Member of English bar 1929; joined m.p. industry at Elstree Studios, 1929; mem. bd. dir. Associated British Picture Corp., Aug., 1945; exec. dir. of prod. at Elstree Studios, 1948. Pres. Brit. Film Prod. Assn., 1953–55, 62–65; chmn. Nat. Film Thea. Appeal Comm., 1957; deputy chmn., Associated British Picture Corp. Ltd., Nov. 1964 & chief exec. ABP Ltd. Oct. 1966.

CLARK, SAMUEL H.: Exhibitor. b. Boston, Mass., Jan. 10, 1914. Owner, Music Suppliers, Boston, Mass.; founded Music Suppliers of New England, 1945; pres., ABC-Paramount Records Inc., 1955–65; v.p., head thea. operations, American Broadcasting-Paramount Theatres, Inc., Jan. 1965; dir., May, 1965.

CLARK, SUSAN: Actress. b. Canada. Trained at Royal Academy of Dramatic Arts in London. Made m.p. debut in Banning, 1967.
PICTURES INCLUDE: Coogan's Bluff, Madigan, Colossus: The Forbin Project, Tell Them Willie Boy Is Here, Skullduggery, Skin Game, Valdez Is Coming, Showdown, The Midnight Man, Night Moves, Airport 1975, The Apple Dumpling Gang, The North Avenue Irregulars, Murder by Decree, City on Fire, Promises in the Dark.

CLARKE, CHARLES G.: Cinematographer. b. Potter Valley, Cal., Mar. 19, 1899. Started in film laboratory for D. W. Griffith and others, then on camera for Natl. Film Co. in 1919. Cinematographer ever since for Famous Players, Wm. Fox, MGM, under contract to 20th Cent.-Fox, 1937–61.
PICTURES INCLUDE: Barbarian and Geisha, The Hunters, Sound and Fury, These Thousand Hills, A Private Affair, Holiday for Lovers, Return to Peyton Place.
TV: Arnie.

CLAYBURGH, JILL: Actress. Appeared on Broadway in Jumpers, Pippin, The Rothschilds, etc.
PICTURES INCLUDE: Portnoy's Complaint, The Thief Who Came To Dinner, Terminal Man, Gable and Lombard, Silver Streak, Semi-Tough, An Unmarried Woman, Starting Over, Luna, It's My Turn, First Monday in October.
TV: Griffin and Phoenix.

CLAYTON, JACK: Producer, director. b. 1921. Ent. m.p. 1935 as asst. dir. for London Films, Fox, Warner. Served H.M. Forces 1940–46. prod. man. Ideal Husband; assoc. prod. Queen of Spades, Flesh and Blood, Moulin Rouge, Beat the Devil, The Good Die Young, I am a Camera; prod. dir. Bespoke Overcoat, 1955, Venice Festival prize-winning film; prod. Sailors Beware, Dry Rot, Three Men in a Boat; dir.

Room at the Top, The Innocents, The Pumpkin Eater; prod.-dir., Our Mother's House, 1967; The Great Gatsby, 1974.

CLAYTON, JAN: Actress, singer. b Tularosa, N.M. Numerous motion picture roles; on Broadway in Showboat, Carousel; TV series, Lassie, CBS.

CLEARY, CHIP: Pub. Relations. b. New Bedford, Mass., Jan. 20, 1910; e. Amherst, 1929–31. Reporter, New Bedford Times, Standard-Times; legis. reporter, U.P., Pa., pub. rel., Wash., D.C.; feat. writer, I.N.S.; ed. writer, Phoenix Gazette; man. ed., Hollywood Reporter, 1936–40; Hollywood ed. consultant, Time, Inc., 1938–42; U.S. Army, 1942–45; ad., pub. dir., Mark Hellinger Prod., 1945–46; asst. ad., pub. dir., Liberty Films, 1946–47; mag. dept. head, Columbia Pictures, 1947; org. pub. rel. firm, Cleary-Strauss & Irwin, 1949; pres. since 1959, Chip Cleary Co.

CLEMENS, BRIAN. Writer, producer, director. b. Croydon, England, 1931. Early career in advertising then wrote BBC TV play. Later TV filmed series as writer, script editor and features. Script editor "Danger Man"; Won Edgar Allen Poe Award for Best TV Thriller of 1962 (Scene of the Crime for U.S. Steel Hour). Various plays for Armchair Theatre; ATV Drama 70; Love Story. Writer and Producer of The Avengers (two Emmy nominations for Best Production 1967 and 1968), Winner two Edgar Allen Poe Awards, Cinema Fantastique Award for best Screenplay.

PICTURES INCLUDE: The Tell-Tale Heart, Station Six-Sahara, The Peking Medallion, And Soon The Darkness, The Major, and, When The Wind Blows, See No Evil, Dr. Jekyll and Sister Hyde, Golden Voyage of Sinbad, Watcher in the Woods, Stiff.

TV: Wrote and prod.: The New Avengers, The Professionals, Escapade (in US for Quinn Martin).

CLIFT, DENISON: Author and director. b. May 2, 1893; e. Stanford U. Editor, Overland Monthly and other publications; publicity expert, San Francisco, 1917. Scenario writer and dir. for Fox, Paramount, MGM; dir. Honor Among Men, Ports of Call, This Freedom, A Bill of Divorcement, Paradise, The Mystery of the Mary Celeste, etc. Dir. for British International, Ideal Films, Gainsborough Pictures and Hammer Prod., London. Author of novels, Guns of Galt, Man About Town, The Mighty Thing, I Belong to You. Author of the New York plays, The Woman Disputed, Scotland Yard; and London plays, The Moon Is Red and Gentle Rain.

CLORE, LEON: Producer. Co-prod. Conquest Of Everest. Virgin Island. exec. prod. Time Without Pity; Prod. Sunday By The Sea, Bow Bells, Every Day Except Christmas, We Are the Lambeth Boys, Dispute, I Want to Go to School, Morgan, A Suitable Case for Treatment, All Neat in Black Stockings.

CLORK, HARRY: Writer. r.n. Clarke; b. Galveston, Tex. e. St. Paul's Academy, Garden City, Long Island, N.Y. Was husband of late Nora Bayes, writer of her stage material; author several stage plays (Smiling Faces, The Milky Way, See My Lawyer) (collab., filmed in 1936 by Paramount); adapt. many plays for Shubert stage enterprises; radio writer. From 1935 writer many s.p. & adapt.

PICTURES INCLUDE: Thrill of Brazil, Mighty McGurk, Sainted Sisters, Painting the Clouds with Sunshine, Tea for Two, Ma and Pa Kettle at Waikiki, The Prisoner.

TV: Beulah, Life of Riley.

CLOVER, DAVID: Actor. b. November 10.

TELEVISION INCLUDES: Policewoman, Police Story, McLaren's Raiders; Kate McShane; One Day At A Time; Delphi Bureau; On The Rocks; The Dick Cavett Special.

CLUG, A. STEPHEN: Executive. b. New York City, Apr. 2, 1929. e. Colgate Univ., Columbia Law School. Was special agent, U.S. Army Counter Intelligence Corps, Far East Command, 1951–4. Started with industry as branch mgr., Paramount Films of Indonesia, 1954; mgr. dir., Paramount Films of India, Ltd., supervising India, Burma, Pakistan, Persian Gulf, 1955–57; Far East supvr. for special handling, The Ten Commandments, 1957–8; co-ordinator world-wide program theatrical re-releases, Paramount Intl., 1958–9; mgr., dir., Paramount Films of Brasil, 1959–65; v.p. & exec. prod., Gold Dolphin Prods., Inc., N.Y., 1965–69; gen. mgr. & supvr. Central America/Caribbean areas, MGM Intl. Inc., Panama, 1969–71; gen. mgr., MGM Puerto Rico/supvr. Caribbean area/MGM Cinemas, 1971–3; dir. sls. & asst. continental supvr. for Europe, UK, Middle East, MGM Intl., Inc. 1973–79; now v.p./sls. mgr., MGM Intl., headquartered in Paris.

CLYDE, THOMAS: Producer. b. Battle, England, Ent. m.p. ind. 1949. After writ. dir. documentaries, with Ivan Foxwell as prod. mgr. The Intruder, Colditz Story; 1956 to Douglas Fairbanks, Ltd., prod. The Hostage, 1957 for own Company, Cavalcade Films, Ltd. Chase a Crooked Shadow; Moment of Danger, Follow That Horse, Guns of Darkness, Work Is a Four Letter Word.

CLYNE, LIONEL: Executive. b. London, Eng., 1908. Entered m.p. ind. 1934 as sales representative theatre catering. 1942 Granada Circuit. Joined Pathe 1943. 1950: gen. sales mgr., Republic International (G.B.), Inc.; dir. Rep. Prod. (G.B.) Ltd., 1953; joined Columbia, 1957; man. dir. Regent Film & TV Prod. Ltd.; Sherwood Picture House Co. Ltd., 1959.

COBE, SANDY: Executive, Producer-Distributor. b. New York City, N.Y., Nov. 30, 1928. e. Tulane Univ. (B. of A., fine arts). U.S. Army WW II & Korea, combat photographer; Loews Theatre Corp., theatre mgr., N.Y. Daily News, photographer; WPIX-TV, cinematographer; Artmark Pictures, N.Y., produced 11 features. Pepsicola, v.p., Pepco Special events, in charge of N.Y.'s World's Fair operations and promotion. General Studios, exec. v.p., distribution; First Cinema Releasing Corp., pres. Formed Sandy Cobe Productions, Inc., producer, packager, European features for U.S. theatrical & television. Pres., Intercontinental Releasing Corporation, domestic and foreign distribution of theatrical features. Golden Image Motion Picture Corp., bd. chm.

MEMBER: Masons (Scottish-Rite 32nd), Shrine, Variety Clubs Tent 25, Academy of Television Arts and Sciences. Special commendations from: California State Senate, Los Angeles City Counsel, California State Assembly and County of Los Angeles.

COBE, SHARYON REIS: Executive. b. Honolulu, Hawaii. e. Univ. of Hawaii. Dancer Fitzgerald, & Sample, N.Y. United Air Lines, N.Y.; v.p., story editor, Gotham Publishing N.Y.; v.p., distribution-foreign sales, World Wide Film Distributors, L.A.; current v.p., foreign sales-finance, Intercontinental Releasing Corp., L.A.

COBURN, JAMES: Actor. b. Laurel, Neb., Aug. 31, 1928. e. Los Angeles City College where he studied drama. Served in U.S. Army. First acting role in coast production of Billy Budd. Later to New York, where he worked on TV commercials, then in teleplays on Studio One, GE Theatre, Robert Montgomery Presents. Summer stock in Detroit before returning to Hollywood. First film, 1959, was Ride Lonesome.

PICTURES INCLUDE: Foes of the Fugitive, The Magnificent Seven, Hell Is for Heroes, The Great Escape, Charade, The Americanization of Emily, The Loved One, Major Dundee, A High Wind in Jamaica, Our Man Flint, What Did You Do in the War, Daddy?, Dead Heat on a Merry-Go-Round, In Like Flint, Waterhole No. 3, Candy, Hard Contract, The Last of the Mobile Hot Shots, The Carey Treatment, The Honkers, Duck, You Sucker, The Last of Sheila, Harry in Your Pocket, A Reason to Live, A Reason to Die, The Internecine Project, Bite the Bullet, Hard Times, Sky Riders, The Last Hard Men, Midway, Cross of Iron, Firepower, The Big Bullet, Loving Couples, Looker.

COCA, IMOGENE: Actress. b. Philadelphia; p. the late Joe Coca, orchestra leader, and Sadie Brady, retired vaudevillian. At 11, debut tap dancer in New York vaudeville; solo dancer Broadway musicals; as comedienne, in New Faces of 1934; with husband, Bob Burton, in Straw Hat Revue in 1939, and others through 1942. New York night clubs, Cafe Society and Le Ruban Bleu, New York; Palmer House, Chicago; Park Plaza, St. Louis. 1949 to TV via Broadway Revue, evolved routines with Sid Caesar, Your Show of Shows, 1950–54. NBC-TV; Imogene Coca Show, 1954, 1955; Sid Caesar invites you, 1958, Glinda.

PICTURES INCLUDE: Under the Yum Yum Tree.

COCCHI, JOHN: Writer-Critic. b. Brooklyn, N.Y., June 19, 1939. e. Fort Hamilton High School 1957; Brooklyn College, AAS, 1961. U.S. Army 1963–65. Puritan Film Labs, manager, 1967–9. Independent-International Pictures, biographer-researcher, 1969. Boxoffice Magazine, critic, reporter, columnist, 1970. Co-author: The American Movies Reference Book (Prentice-Hall). Contributor: Screen Facts, Film Fan Monthly. Actor in: The Diabolical Dr. Ongo, Thick as Thieves, Captain Celluloid VS. the Film Pirates. Worked on dubbing: Dirtymouth, 1970 Author, The Westerns, a Movie Quiz Book. Contributor to various books on films.

COCHRAN, RON: Producer, writer, narrator. b. Saskatchewan, Canada, Sept. 20, 1912. e. Parsons College U. of Fla. Newscaster for radio stations in Rock Island, Ill., Des Moines, St. Louis; special agent, FBI 1942–45; then returned to radio as news dir., dir. of news WCOP Boston 1945–51; joined CBS-TV news staff, 1951, in Washington; newscaster, Late News Show; news roundups, CBS radio; moderator, Capitol Cloakroom, Where Do You Stand, Man of the Week; CBS news staff, N.Y., 1954; newscaster. The Late News; moderator, Youth Takes a Stand, One O'clock Report, CBS-TV; In Person, CBS-Radio; News, ABC-TV, 1962; Ron Cochran With the News 1967; Pres., Ron Cochran Enterprises Inc., producer and syndicator or radio and television.

COCO, JAMES: b. Bronx, N.Y., Mar. 21, 1929. Became stage mgr. of touring children's theatre while still in teens. Was character actor in summer stock and off-Bdwy. prods. before appearing on Bdwy. in Man of La Mancha and The Last of the Red Hot Lovers.

PICTURES INCLUDE: A New Leaf, The Strawberry

Statement, Tell Me That You Love Me, Junie Moon, Such Good Friends, Man of La Mancha, The Wild Party, Murder by Death, The Cheap Detective, Scavenger Hunt, Wholly Moses, Only When I Laugh.

TELEVISION: Marcus Welby, M.D, The Flip Wilson Show, The Trouble With People, VD Blues, Calucci's Dept.

COE, RICHARD L.: Newspaper editor and critic. b. New York, N.Y., Nov. 8, 1916; e. George Washington U. Joined Washington Post as asst. dramatic ed. & radio ed., 1937; Stars & Stripes, 1942–46; dramatic & film ed., Washington Post and Times Herald, 1946–69. Drama Critic, The Washington Post 1946. Film and theater commentator, NBC-WRC-TV 1969–75.

COEN, GUIDO: Producer and Executive. Prod. series pictures for Fortress Films and Kenilworth Films.

PICTURES INCLUDE: One Jump Ahead, Golden Link, The Hornet's Nest, Behind the Headlines, Murder Reported, There's Always a Thursday, Date with Disaster, The End of the Line, The Man Without A Body, Woman Eater, Kill Her Gently, Naked Fury, Operation Cupid, Strictly Confidential, Dangerous Afternoon, Jungle Street, Strongroom, Penthouse, Baby Love, One Brief Summer, Burke and Hare, Au Pair Girls, Intimate Games, Twickenham Studios.

COHEN, CHARLES: Executive. b. Brooklyn, N.Y., Sept. 15, 1912; e. Brooklyn Coll. Pub. dept., MGM, 1931; U-I, 1948–50; 20th-Fox, 1950; assoc. adv. mgr.; eastern asst. adv & pub. mgr. Allied Artists, 1955; home off. pub. mgr. Warner Bros., 1957. Eastern adv. mgr. 1958; eastern adv., pub. mgr., 1960; Exec. ass't to adv. pub. director, 1960; expl. dir., Embassy Pictures, 1961; nat. adv. & pub. dir., Cinema V Dist., 1964; Adv. & Pub. Dir., Sigma III, 1968; adv., pub. mgr., The Cannon Group, 1970, v.p., adv.-pub. rel. 1971; Adv.-Pub. Dir., U.A. Eastern Theatres, 1973–77.

COHEN, ELLIS A.: Executive, Producer-Writer. b. Baltimore, Md. Sept. 15, 1945. e. Balt. Jr. Coll. (A.A.), Univ. of So. Carolina (B.S.). 1980, pres. Ellis A. Cohen Prods., 1978, Director of TV Network Boxing Events, Don King Productions, Inc., NYC 1978, Account Exec., Solters & Roskin P.R. L.A. 1977, Exec. Prod., Henry Jaffe Enterprises, Inc., 1974–76 dir., worldwide pub./adv., William Morris Agency. 1963, talent coordinator, Cerebral Palsy Telethon, WBAL-TV, Baltimore; 1964, p.r. asst. Campbell-Ewald Adv. Agency, L.A.; 1966, and mgr., Hochschild Kohn Dept. Stores, Baltimore; 1968–69, asst. dir., p.r. entertainment, talent booking for Club Venus Night Club, Baltimore; 1968, created-edited The Forum Oracle, national entertainment mag. 1969–72, dir., p.r., Jewish Community Center of Baltimore; 1970, leave of absence to work as corr. in Israel, Denmark & London; 1972, dir., p.r. & adv., The Camera Mart, New York; 1972–74 creator & editor-in-chief, TV/New York Magazine, nationwide TV mag.; Producer, New York Area Emmy Awards Telecast (1973 & 1974), WOR-TV (Co-Prod.), chm., exec. prod. of TV Academy Celebrity drop-in luncheon series; 1972, talent coordinator Bob Hope's Celebrity Flood Relief Telethon. Member: Writers Guild of America, Producers Guild of America, Friars Club, Amer. Newspaper Guild, Intl. Press Corp., Israeli Press Corp., National Academy of TV Arts & Sciences, Academy of Television Arts & Sciences, Screen Actors Guild. Mayor Beame's Committee In The Public Interest for NYC; executive producer, 1976 Democratic National Convention Gala.

TV: Aunt Mary (Dec. 1979), a Hallmark Hall of Fame, 2-hr. special, CBS-TV starring Jean Stapleton (Exec Prod-Story By); New York Area Emmy Awards. Telecast; January 1973 WOR-TV (co.-prod.) and January 1974, WOR-TV (co.-prod.).

FILM: Slob Sisters (prod.).

COHEN, HERMAN: Producer, writer. b. Detroit, Mich. Age 12, usher, Detroit, Dexter theatre; chief of service, Detroit, Fox theatre; mgr., booker, Dexter Theatres; military service; house mgr., Fox theatre; sales mgr. Columbia Pictures, Detroit; prod., Realart, Jack Broder Productions, Hollywood; formed own prod. Co., 1953; owner, Fox and Cinderella Theatres, Detroit; prod. pictures for Columbia and Warner Bros.

PICTURES INCLUDE: River Beat, Magnificent Roughnecks, The Brass Legend, Target Earth, Crime of Passion, Horrors of the Black Museum, I was a Teenage Werewolf, I Was a Teenage Frankenstein, Blood of Dracula, How to Make a Monster, Konga, Headless Ghost, Black Zoo, A Study in Terror, Berserk, Crooks and Coronets, TROG, Craze.

COHEN, IRWIN R.: Exhibition Executive. b. Baltimore, Md., Sept. 4, 1924. Attended Massanutten Academy, Woodstock, Va., Member of alumni board of directors; attended University of Baltimore, bachelor of laws, 1948. Admitted to Maryland bar, 1948, partner law firm Cohen & Dackman since 1952. In exhibition 1938 to date. Founded R/C Theatres 1966, chief exec. officer 1966 to date. Member of Variety Club, Tent #11, Member of Board of Directors NATO of Maryland, past pres. NATO of Virginia 1975–77, past chmn. board NATO of Virginia 1977–79. Member of exec. comm. and member of board of directors National NATO. One of the founders, co-atny., and vice-chmn. of Key Federal Savings & Loan Association 1961 to date.

COHEN, MILTON E.: Distributor. b. Chicago, Ill., e. University of Illinois. Entered motion picture industry as salesman. United Artists, Chicago, 1925–29; Columbia, Chicago, 1929–31; salesman, Detroit, 1931–38; sales mgr., RKO, Detroit, 1938; branch mgr., 1943; Eastern-Central dist. mgr., 1946; Eastern sales mgr., Eagle Lion Classics, 1948; genl. sales mgr., 1951; western & southern sales mgr., United Artist, 1951; Eastern & Southern sales mgr., 1952; supervisor, sales, Around The World In 80 Days, 1958; Eastern Canadian sales mgr., 1959; national dir., Roadshow Sales, 1961; retired. United Artists, 1969; exec. v.p. in charge of theatrical distrib. for Trans Globe Films, Inc., 1971.

COHEN, NATHAN: Executive. b. London, Eng., Dec. 23, 1905. Entered m.p. ind. as exhibitor 1930; into production and distribution 1942. Director: EMI Films Ltd.

COHEN, ROB: Producer. Formerly exec. v.p. in chg of m.p. and TV for Motown. Started as dir. of m.p. for TV at 20th Century-Fox. Joined Motown at age of 24 to produce films. Now heads own production co.

PICTURES INCLUDE: Mahogany, Bingo, Long and the Traveling All-Stars, Scott Joplin, Almost Summer, Thank God It's Friday, The Wiz, A Small Circle of Friends (dir.).

COHN, ROBERT: Producer. b. Avon, N.J, Sept. 6, 1920; e. U. Michigan, B.A., '41; p. Jack Cohn. Joined Columbia as asst. dir. In World War II, as Signal Corps film cutter. Air Corps Training Lab. unit mgr., combat aerial m.p. camera man with 13th A.A.F. Awarded: DFC, Air Medal & 3 clusters, Purple Heart. Assoc. prod. "Lone Wolf In London" 1947; prod. "Adventures in Silverado" 1948, all Col. Headed Robert Cohn prod. unit at Columbia, pres. International Cinema Guild. Columbia European prod.: exec. Columbia Studios. Hollywood: formed Robert Cohn Prod.

PICTURES INCLUDE: Black Eagle, Rusty Leads the Way, Palomino, Kazan, Killer That Stalked New York, The Barefoot Mailman, Mission Over Korea, The Interns, The New Interns, The Young Americans.

COLBERT, CLAUDETTE: Actress: r.n. Claudette Chauchoin. b. Paris, Sept. 13, 1907. e. public schools, Paris, New York; Art Students League (N.Y.). On N.Y. stage (debut, Wild Wescotts; following Marionette Man, We've Got to Have Money, Cat Came Back, Kiss in a Taxi, Ghost Train, The Barker, Dynamo, etc.). First screen role in Love O' Mike (silent); thereafter featured and starred in many prod. Academy Award best actress, 1934 (It Happened One Night, Columbia); voted one of ten best Money Making Stars in Fame Poll, 1935, 36, 47.

PICTURES INCLUDE: Cleopatra, I Cover the Waterfront, No Time for Love, Practically Yours, Since You Went Away, So Proudly We Hail, Guest Wife, Tomorrow is Forever, The Egg and I, Sleep My Love, Family Honeymoon, Bride for Sale, Three Came Home, The Secret Fury, Thunder on the Hill, Let's Make It Legal, Outpost in Malaya, Daughters of Destiny, Texas Lady, Parrish.

BROADWAY: Marriage Go Round, Irregular Verb to Love.

COLBY, ETHEL: Drama critic and m.p. amuse. ed. b. N.Y.C., Sept. 15, 1908; e. N.Y.C. Vaudeville, operettas.

RADIO: Programs include B'way Busybody, Miss Hollywood.

TV: Programs include Personality Spotlight, Here and There in N.Y., Curtain Call, Mad About Music, B'way Matinee.

BROADCAST: for Army & Navy during war; drama critic and film editor-critic, Journal of Commerce; treas., New York Drama Critics Circle, N.Y. Drama Desk. Member: Women's Broadcasters of America, Newspaper Womens Club.

COLE, GEORGE: Actor. b. London, Eng., Apr. 22, 1925. e. Secondary Sch. Surrey. Stage debut in White Horse Inn, 1939; m.p. debut in Cottage to Let, 1941.

PICTURES INCLUDE: Henry V, Quartet, My Brother's Keeper, Laughter in Paradise, Scrooge, Lady Godiva Rides Again, Who Goes There (Passionate Sentry), Morning Departure (Operation Disaster), Top Secret (Mr. Potts Goes to Moscow), Happy Family, Will Any Gentleman, Apes of the Rock, The Intruder, Happy Ever After (Tonight's the Night), Our Girl Friday (Adventures of Sadie), Belles of St. Trinian's, Prize of Gold, Where There's a Will, Constant Husband, Quentin Durward, The Weapon, It's a Wonderful Life, Green Man, Bridal Path, Too Many Crooks, Blue Murder at St. Trinians, Don't Panic Chaps, Dr. Syn, One Way Pendulum, Legend of Young Dick Turpin, The Great St. Trinian's Train Robbery, Cleopatra, The Green Shoes, Vampire Lovers, Fright, The Bluebird.

TV: Life of Bliss, A Man of Our Times, Don't Forget To Write, The Good Life, Minder.

COLE, SIDNEY: Prod., dir., writer. b. London, England, 1908; e. Westminster City Sch.; London U., BSc. Began at Stoll Studios 1930; then various feature and documentary studios. Dir. Roads Across Britain. Spanish ABC. Behind the Spanish Lines. Ed. Gaslight. Mr. Pimpernel Smith, San

Demetrio, London, Nine Men, etc.; tech. adviser First of the Few; writer and assoc. prod. They Came to a City, Return of the Vikings; assoc. prod. Dead of Night, Loves of Joanna Godden, Against the Wind, Scott of the Antarctic, The Magnet, Man in the White Suit, Train of Events (co-dir.); Operation Swallow (Eng. adapt.); The Angel Who Pawned Her Harp (writer-prod.); s.p., North Sea Bus. Assoc. prod. Escapade; prod. Sword of Sherwood Forest; prod. s.p. The Kitchen; prod. We Are The Engineers, One in Five.

TV: prod. Adventures of Robin Hood, The Buccaneers, Sword of Freedom, The Four Just Men, Danger Man (Secret Agent), Man In A Suitcase, Adventures of Black Beauty, Dick Turpin.

COLEMAN, NANCY: Actress. b. Everett, Wash. e. U. Wash. In radio serials; on New York stage (with Gertrude Lawrence, in Theatre Guild productions, etc.). Desperate Hours, 1955; m.p. debut, 1941, Rep. Co tour of Europe and So. Am.

PICTURES INCLUDE: Kings Row, Dangerously They Live, Gay Sisters, Desperate Journey, Edge of Darkness, In Our Time, Devotion, Violence, Mourning Becomes Electra, That Man from Tangier, Slaves.

TV: Valiant Lady Producers Showcase. Am. Theatre Guild.

COLER, JOEL H.: Executive. b Bronx, N.Y., 1931. e. Syracuse University (B.A., journalism). Married and has two daughters. Worked as adv. asst. NBC; acct. exec. Grey advertising. Joined 20th Century-Fox 1964 as adv. coordinator Fox Intl.; 1967, named intl. adv./pub. mgr. 1974, named dir., intl. adv./pub.

COLLERAN, BILL: Producer-Director. b. April 16; story department 20th Century-Fox 1945–46; Director Louis de Rochemont 1946–50; stage mgr. NBC 1951; Assoc. Dir. The Hit Parade 1952–53; Dir. The Hit Parade, various TV specs. 1954–56; Dir. Cinerama Windjammer 1956; TV Specs. Bing Crosby, Frank Sinatra, Debbie Reynolds 1957–60; various TV Specs. 1960–65, Exec. Prod. Judy Garland Show, Dean Martin Show, 1965–66; Dir. Richard Burton's "Hamlet" film, Prod. "Popendipity" ABC-TV spec. various other TV specs. 1967–70 1971–77 Various TV specials and series; 1978–79, prod., dir., writer for Hill-Eubanks Group.

COLLIN, REGINALD: Producer/director. Trained at the Old Vic Theatre School as actor. Directed in repertory, also pantomimes and summer shows. Entered television 1958. Producer ITV's First Arts Programme Tempo. Producer of Callan, Mystery And Imagination, Special Branch, Napoleon And Love. 1975: Producer/director two Armchair Cinemas, also documentary on the Royal Shakespeare Theatre. Board member IBA's Repertory Theatre Trainee Director Scheme. 1976; Vice-Chairman of The British Academy of Film and Television Arts. Chairman of the Awards Committee. 1977: Appt. first dir. of BAFTA.

COLLINS, HAL: Writer, assoc prod. b. New York, N.Y., Nov. 1, 1920. e. Bklyn. Technical H.S., James Monroe H.S., 1932–35.

PICTURES: Monogram, Columbia, 1945–48.

TV: Texaco Star Theatre, Red Buttons Shows, Celeste Holmes Shows, Milton Berle Shows, Kraft Shows, Jackpot Bowling, Share Boomtown Shows, Andy Williams Shows, Bob Hope Shows, Ed Sullivan Shows.

RADIO: Hearns Kiddie Radio Revue, Eddie Cantor Shows, Milton Berle Shows, Rudy Vallee Shows, Duffy's Tavern, Take It or Leave It, Zeke Manners Radio Shows.

COLLINS, JOAN: Actress. b. London, Eng., May 23. e. Francis Holland School (London). Made stage debut in A Dolls House, Arts Theatre 1946. Screen debut in I Believe in You, 1952. TV appearances.

PICTURES INCLUDE: I Believe in You, Judgment Deferred, Decameron Nights, Cosh Boy, The Square Ring, Turn the Key Softly, Our Girl Friday (Adventures of Sadie), The Good Die Young, Land of the Pharaohs, Virgin Queen, Girl in the Red Velvet Swing, Opposite Sex, Sea Wife, Island in the Sun, Wayward Bus, Stopover Tokyo, The Bravados, Rally Round the Flag Boys, Seven Thieves, Esther and the King, Warning Shot, Can Hieronymus Merkin Ever Forget Mercy Humpe and Find True Happiness?, The Executioner, Tales from the Crypt, The Bawdy Adventures of Tom Jones, Empire of the Ants, The Big Sleep, Sunburn.

COLLINS, NORMAN: Executive. b. Beaconsfield, Bucks, Eng., Oct. 3, 1907. e. William Ellis Sch., Hampstead, London. Oxford University Press, 1926–29; asst. literary ed. London News Chronicle, 1929–33; deputy-chmn. Victor Gollancz, publishers, 1939–41; joined BBC as controller Light Program, 1946 & later dir. BBC Overseas Services; TV as controller BBC service, 1947; Dep. chmn. Associated-Tele-Vision Corp. Ltd., Dir., Independent Television News Ltd., 1971–73. Chmn. Independent TV Companies Assoc. Ltd. 1960, Chmn., Central School of Speech and Drama. Gen. commiss. of taxes 1967. Dir. English Stage Soc. Governor, Sadler's Wells Foundation. Author numerous works.

COLLINS, PAT: Performer. r.n. Patricia Colinaka Allan. b. Detroit, Mich., May 7, 1935. e. U. of Ill., 1953. District mgr., Yardley Created Prods., Chicago; played in various nightclubs, Chicago, Miami, Las Vegas, etc.

PICTURE: Experiment in Nightmares.

TV (1963–65): Steven Allen Bill Dana Show, No Time for Sergeants, Mike Douglas, Art Linkletter, Honey West.

COLLINS, STEPHEN: Actor. b. Hastings-on-Hudson, N.Y., 1948. Appeared off-Bdwy. in Joseph Papp production before acquiring role on Bdwy. in The Ritz.

PICTURES: All the President's Men, Between the Lines, The Promise, Star Trek, Fedora, Loving Couples.

TELEVISION: The Waltons, Barnaby Jones.

COLVILLE, JOHN: Producer, Director. Started in theatre. Worked in all departments before becoming Film Editor on Features, Documentaries, and Television Series for both BBC/ITV. Extensive experience in Greece, Singapore, Spain, Switzerland—now establishing new Unit, Studios, and Producing films in Nigeria.

COMDEN, BETTY: Writer. b. Brooklyn, N.Y., May 3; e. NYU. Nightclub performer, 1939–44; all writing in collab. with Adolph Green; writer book, sketches & lyrics for many Bway. shows including: On the Town, Billion Dollar Baby, Two On the Aisle, Wonderful Town, Peter Pan, Fade Out . . . Fade In.

PICTURES INCLUDE: Good News, On the Town, Barkleys of Broadway, Singin' in the Rain, Band Wagon, It's Always Fair Weather, What a Way to Go, Bells Are Ringing, Auntie Mame, Say Darling.

BROADWAY: Lyrics with Adolph Green—Do Re Mi. Collab. writing, Adolph Green, A Party; worked on Bells Are Ringing; book, lyrics, Subways Are For Sleeping; Show, Hallelujah, Baby! Book for show Applause. Tony Awards for: Wonderful Town, Hallelujah Baby, Applause.

COMO, PERRY: Singer. b. Canonsburg, Penn., May 18; p. Pietro and Lucille Travaglini Como; e. Canonsburg local schools. Barber at 15; joined Carlone Band, then Ted Weems in 1936; CBS sustaining show; played many night clubs, records for Victor. Screen debut, Something for the Boys, 20th Century Fox.

PICTURES INCLUDE: Doll Face, If I'm Lucky.

TV: Perry Como show. NBC-TV. Best Male vocalist M.P. Daily, TV poll, 1952–55; radio poll, 1954. Best Male vocalist M.P. Daily, TV Poll, 1956: best TV performer M.P.D.—Fame poll 1957. Interfaith Award, 1953; Emmy, Peabody, Christopher Awards, 1955–56; Knight Commander and Lady Com. (Mrs. Como) of Equestrian Order of Holy Sepulchre of Jerusalem; personality of the yr., Variety Club, 1956. Perry Como Kraft Music Hall, NBC-TV.

COMPTON, JOYCE: Actress. b. Lexington, Ky.; e. Tulsa U. Screen debut in Ankles Preferred.

PICTURES INCLUDE: If I Had a Million, Luxury Liner, Only Yesterday, Affairs of a Gentleman, Christmas in Connecticut, Danger Signal, Hitchhike to Happiness, Charlie Chan of Alcatraz, Exposed, Incident, Grand Canyon, Jet Pilot.

CONAWAY, JEFF: Actor. b. New York City, Oct. 5, 1950. Started in show business at the age of 10 when he appeared in Bdwy. production, All the Way Home. Later toured in Critics Choice before turning to fashion modeling. Toured with musical group, 3½, as lead singer and guitarist. Entered theatre arts program at N.Y. Univ. Film debut at 19 in Jennifer on My Mind.

PICTURES: The Eagle Has Landed, Pete's Dragon, I Never Promised You a Rose Garden, Grease.

TELEVISION: The Mary Tyler Moore Show, Happy Days, Movin' On, Barnaby Jones, Kojak. TV movies: Having Babies and Delta County, U.S.A.

SERIES: Taxi.

CONDON, CHRIS J.: Producer-Director-Motion Equipment Designer. b. Chicago, Dec. 7, 1922. e. Davidson Institute, USC, U.S. Air Force 1943–46 Founded Century Precision Optics, 1948. Designed Athenar telephoto lenses, Century Super wide-angle lenses and Duplikins. Co-founded StereoVision International, Inc. 1969 specializing in films produced in new 3-D process.

FILMS PRODUCED: The Wild Ride, The Surfer, Girls, Airline, The New Dimensions. Member SMPTE. Lecturer and consultant on motion picture optics and 3-D motoion picture technology.

CONLEY, DICK: Executive. r.n. Richard Glen. b. La Cygne, Kansas, Feb. 7, 1920. e. Lamar Union H.S., Lamar, Colorado. Atlas Theatre Corp., Denver, 1938–43; merchandising specialist, National General Corp., 1944–62; district mgr., Kansas City-St. Louis area, 1962–68; area superv. Kansas City-Omaha-Des Moines-St. Louis area; pres., Petite Amusement Corp., Kansas City.

CONN, ROBERT A.: Executive. b. Philadelphia, Pa., Jan. 16, 1926; e. Lehigh U. 1944; U. of Pa., 1948. 1st Lt. Army Security Agency, 1944–46, 1951–52; furniture dist., Philadelphia, 1948–51; band & act. dept., MCA, 1952–53; dir.

of adv. & prom. Official Films N.Y. 1954; head of Official Films Philadelphia sales office serving Pennsylvania, Baltimore, Washington, Cleveland and Detroit, 1956. Eastern Reg. Sls. Mgr. Flamingo Films, 1957; acct. exec. Dunnan and Jeffrey, Inc., June, 1961; v.p., Dunnan and Jeffrey, Jan. 1962; pres., adv. mgr., Suburban Knitwear Co., 1963; exec. v.p. Rogal Travel Service, 1964–68. 1968–78, pres. RAC Travel, Inc., Jenkintown, Pa. and pres. Royal Palm Travel, Inc. Palm Beach, Florida, 1978; Rosenbluth Travel Service, 1979.

CONNELLY, PAUL V.: Executive. b. Boston, Mass., June 11, 1923. e. Boston Coll., MA, 1951, B.S.B.A., 1949; Fordham Univ., 1951–54; Asst. professor of Economics, Manhattan Coll., 1950–54; treas., America Corp. (formerly Chespeake Industries), 1957–59; treas., dir., Pathe-America Dist. Co. Inc., Sutton Pictures Co.; vice-pres., dir., Pathe Labs., Inc.; pres., dir., Pathe-DeLuxe of Canada, Ltd. 1959–65; pres. International Business Relations, 1965–67; v.p., treas., dir. Movielab, Inc., 1968; v.p.—finance, Tele-Tape Corp., 1970.

CONNERY, SEAN: Actor. b. Edinburgh, Scotland, Aug. 25, 1930. Stage debut South Pacific, 1953.

TV: Requiem for a Heavyweight, Anna Christie, Boy With the Meataxe, Women in Love, The Crucible, Riders to the Sea, Colombe, Adventure Story, Anna Karenina, Macbeth (Canadian TV).

PICTURES INCLUDE: The Longest Day, Dr. No, From Russia With Love, Marnie, Woman of Straw, Goldfinger, The Hill, Thunderball, A Fine Madness, You Only Live Twice, Shalako, The Molly Maguires, The Red Tent, The Anderson Tapes, Diamonds Are Forever, The Offence, Zardoz, Ransom, Murder on the Orient Express. The Wind and The Lion, The Man Who Would Be King, Robin and Marian. The Next Man, A Bridge Too Far, The Great Train Robbery, Meteor, Cuba, Outland.

Prod. dir., The Bowler and the Bonnet (film doc.), I've Seen You Cut Lemons (London stage). Director of Tantallon Films Ltd. (First production: Something Like the Truth).

CONNOR, KENNETH: Actor. b. London, England. Ent. m.p. industry 1949 in The Lady Killers.

TV: Ted Ray Show, Show Called Fred, Charlie Farnabarn's Show, Alfred Marks Time, As You Like It, Dickie Valentine Show, Black and White Minstrel, Anne Shelton, Hi Summer, series Don't Say A Word, Room at the Bottom, On the Houses, Frankie Howard Reveals All.

PICTURES INCLUDE: Carry on Sergeant, Carry on Nurse, Carry on Constable, Watch Your Stern, Carry on Regardless, Nearly a Nasty Accident, What a Carve Up, Call Me a Cab, Carry on Cleo, Captain Nemo, Carry On Up The Jungle, Carry On Matron, Carry On Abroad, Carry on England, Carry on Emanuelle.

CONNORS, CHUCK: Actor. r.n. Kevin Joseph Connors. b. Brooklyn, N.Y., April 10, 1924; e. Adelphi Acad., Seton Hall Coll., U.S. Army 3 yrs., then prof. baseball player.

TV: Dennis Day Show, Four Star Playhouse, Star and the Story, Reader's Digest, Gunsmoke, West Point, Rifleman series, Roots.

PICTURES INCLUDE: Pat and Mike, Code Two, Trouble Along the Way, South Sea Women, Dragonfly Squadron, Naked Alibi, Human Jungle, Target Zero, Three Stripes in the Sun, Good Morning Miss Dove, Hold Back the Night, Hot Rod Girl, Tomahawk Trail, Walk the Dark Street, Geronimo, Kill Them All and Come Back Alive, Capt. Nemo and the Underwater City, Profane Comedy, The Deserter, Support Your Local Gunfighter, Soylent Green, 99 and 44/100ths½ Dead.

CONNORS, MIKE: Actor. b. Fresno, Cal., Aug. 15, 1925. r.n. Krekor Ohanian. e. UCLA. Film debut in Sudden Fear; other pictures include Sky Commando, Day of Triumph, Flesh and Spur, Seed of Violence, Harlow, Good Neighbor Sam, Where Love Has Gone.

TV: Tightrope, Mannix (Golden Globe award).

CONRAD, ROBERT: Actor, singer. r.n. Conrad Robert Falk; b. Chicago, Ill., March 1, 1935; e. public schools, Northwestern U.; prof. debut, nightclub singer. Formed Robert Conrad Production, 1966.

TV: Lawman, Maverick, 77 Sunset Strip, Hawaiian Eye (series), Wild Wild West (series), Baa, Baa, Black Sheep (series), Centennial (mini-series), A Man Called Sloane (series).

PICTURES: Thundering Jets, Palm Springs Weekend, Young Dillinger, Murph the Surf, Sudden Death, The Women in Red.

CONRAD, WILLIAM: Actor, Producer, Director. b. Louisville, Ky., Sept. 27. e. Fullerton College. Announcer-writer-director for L.A. radio station KMPC before becoming WW II fighter pilot in 1943; returned to radio drama as original Matt Dillon of Gunsmoke Series; m.p. debut in The Killers, 1946; later acting credits include 30, Body and Soul, Sorry, Wrong Number, East Side, West Side, The Naked Jungle; TV credits include This Man Dawson, Klondike (produced and directed), 77 Sunset Strip (producer), True (directed 35 episodes); as m.p. producer for Warner Bros. credits include: Two On a Guillotine, Brainstorm, An American Dream, A Covenant With Death, First To Fight, The Cool Ones, The Assignment.

RECENT TV: Acting credits include The Brotherhood of the Bell, The D.A., Conspiracy to Kill, O'Hara, U.S. Treasury, Cannon (series).

CONROY, THOMAS: Exec. r.n. Coleman Thomas Conroy. b. Camden, N.J., Nov. 3, 1924; e. Yale U., hd. prof. camera dept., Bell and Howell Inc., New York, 1951; hd. camera dept., Cinerama Inc., 1952–55; National Theatres Inc., 1955; hd. camera dept., Cinemiracle; v.p. prod., prod. supv., Cinerama, Inc. 1960. Developed Cinerama single-lens system; v.p., production, Shannon Prod., Inc., 1967.

PICTURES INCLUDE: This Is Cinerama, Cinerama Holiday, Seven Wonders of the World, Windjammer, The Wonderful World of The Brothers Grimm, How the West Was Won, The Best of Cinerama, The Golden Head.

CONTE, JOHN: Actor, singer. b. Palmer, Mass.; e. Lincoln h.s., Los Angeles. Actor, Pasadena Playhouse; radio anncr., m.c.; Armed Forces World War II.

THEATRE: On Broadway in Windy City, Allegro, Carousel, Arms and the Girl.

TV: John Conte's Little Show, Max Liebman Spectaculars and dramatic shows, host and star of NBC Matinee Theatre; TV Hour of Stars; Mantovani Welcomes You.

PICTURES INCLUDE: Debut in Man With the Golden Arm, The Carpetbaggers.

Pres. KMIR-TV, Channel 36, Desert Empire Television Corp., Palm Springs.

CONTI, BILL: Composer. b. Providence, Rhode Island, 1943. Studied piano at age 7, forming first band at age 15. e. Juilliard School of Music. Toured Italy with jazz trio where scored first film, Candidate for a Killing. Returned to U.S. to be music supvr. on Blume in Love. for Paul Mazursky.

PICTURES: Harry and Tonto, Rocky, Handle with Care, F.I.S.T., Slow Dancing in the Big City, The Big Fix, Uncle Joe Shannon, An Unmarried Woman, Hurricane, Paradise Alley, Rocky II, Golden Girl, A Man, A Woman, and A Bank, Gloria, Private Benjamin, For Your Eyes Only, Carbon Copy.

TELEVISION: Kill Me If You Can.

CONVERSE, FRANK: Actor. b. St. Louis, Mo., May 22, 1938. e. Carnegie Tech. Early training on stage in New York. Active in repertory theatres. Two seasons with Amer. Shakespeare Fest.

PICTURES INCLUDE: Hurry Sundown, Hour of the Gun.

TV: Mod Squad, Medical Center, The Bold Ones, N.Y.P.D. (series) Coronet Blue (series), The Guest Room, Movin' On, The Widowing of Mrs. Holyrod. Movies: The Rowdyman, Shadow of a Gunman, Tattered Web, Dr. Cook's Garden.

STAGE: The Seagull, Death of a Salesman, Night of the Iguana, A Man for All Seasons, The House of Blue Leaves, First One Asleep Whistle, Arturo Ui.

CONWAY, GARY: Actor. b. Boston, 1938. r.n. Gareth Carmody, e. UCLA. As college senior was chosen for title role in Teen-Age Frankenstein. After graduating served in military at Ford Ord, CA. In 1960 began contract with Warner Bros., doing bits for films and TV. Has also appeared on stage.

PICTURES INCLUDE: The Guns of Texas, Once is not Enough, The Farmer (also prod.).

TV: Burke's Law.

CONWAY, KEVIN: Actor. Gained reputation on N.Y. stage, winning Obie for performance in When You Comin Back, Red Ryder? Also on and off Broadway in such plays as One Flew Over the Cuckoo's Nest, Of Mice and Men, Moonchildren, The Plough and the Stars, etc.

PICTURES INCLUDE: Slaughterhouse Five, Shamus, Portnoy's Complaint, F.I.S.T.

TV: Johnny We Hardly Knew You, The Deadliest Season.

CONWAY, TIM: Actor. b. Dec. 15, 1933, Willoughby, Ohio. e. Bowling Green State Univ. After 2 yrs. Army service joined KWY-TV in Cleveland as writer-director and occasional performer. Comedienne Rose Marie discovered him and arranged audition for The Steve Allen Show on which he became regular. In 1962 signed for McHale's Navy, series. Also has done night club appearances.

PICTURES INCLUDE: McHale's Navy, The World's Greatest Athlete, The Apple Dumpling Gang, Gus, The Shaggy D.A.

TV: The Steve Allen Show, The Gary Moore Show, McHale's Navy, guest appearances on Hollywood Palace and shows starring Carol Burnett, Red Skelton, Danny Kaye, Dean Martin, Cher, Doris Day, etc.

CONWAY, SHIRL: Actress. r.n., Shirl Conway Larson; b. Franklinville, New York, June 13, 1916; e. U. of Michigan, 1934–38. Singer, comedian, nightclubs.

PLAYS: Gentlemen Prefer Blondes, Plain and Fancy.

TV: The Doctors and The Nurses.

COOGAN, JACKIE: Actor. b. Los Angeles, Calif., Oct. 26, 1914; e. Villanova Coll.

PICTURES INCLUDE: Debut in The Kid at 4. Peck's Bad

Boy, My Boy Trouble, Daddy, Oliver Twist, Circus Days, Long Live the King, Boy of Flanders, Rag Man, Old Clothes, Johnny Get Your Gun, Bugle Call, Buttons, Tom Sawyer, Huckleberry Finn, Home on the Range, College Swing, Million Dollar Legs, Sky Patrol, Kilroy Was Here, French Leave, Skipalong Rosenbloom, Outlaw Women, Lost Women, Proud Ones, Marlowe.

COOK, ELISHA, JR.: Actor. b. San Francisco, December 26, 1907; e. St. Albans, Chicago boarding school. Joined Frank Bacon in "Lightnin'," at the age of 14.
THEATRE: Appeared with Ethel Barrymore in "Kingdom of God," in "Henry, Behave," "Many a Slip," "Three Cornered Moon," in London in "Coquette." Played in vaudeville and summer stock companies. Appeared in "Chrysalis," then "Ah, Wilderness," Theatre Guild success. Signed by Par. in 1936: Two in a Crowd, U; Pigskin Parade, etc.
PICTURES INCLUDE: Up in Arms, Cassanova Brown, Cinderella Jones, Dillinger, The Big Sleep, The Long Night, Don't Bother to Knock, I the Jury, Shane, Thunder Over the Plains, Drum Beat, Outlaw's Daughter, Timberjack, Indian Fighter, The Killing, Voodoo Island, Rosemary's Baby, Welcome to Hard Times, Blacula, The Great Northfield, Minnesota Raid, Emperor of the North, Electra-Glide in Blue, the Outfit, The Black Bird, The Champ, Parny.

COOK, FIELDER: Prod., dir. b Atlanta, Ga., Mar. 9, 1923. e. Washington & Lee U., BA, 1947; U. of Birmingham, Eng., post grad., 1948. Doctor of Fine Arts (Hon) (1973) (W&L). USNR, 1944; 7th Amphibious Force, 1944–45, J. Walter Thompson Co., 1947–48.
TV: Asst. prod-dir., Kraft Television Theatre; dir., Believe It or Not; dir., Lux Video Theatre; prod.-dir. (owner, Unit Four Prod. Co.). Kaiser Aluminum Hour, 1956–57; freelance, 1952–56; Alcoa-Goodyear Hour, Studio One, Theatre Guild on The Air (U.S. Steel), Playhouse 90; prod.-dir., Amer. Jewish Comm.'s Destiny Tot; prod.-dir., The Fifty Minute Hour, Focus, 1962; formed Eden Prods. to prod.: DuPont Show of the Week, The Directors Company dramas; TV pilots; Ben Casey, The Eleventh Hour, Going My Way, Mr. Roberts, The Waltons, Beacon Hill.
PICTURES INCLUDE: Patterns, Home Is the Hero, A Big Hand for the Little Lady, How to Save a Marriage and Ruin Your Life, Prudence and the Pill, Eagle In a Cage, From the Mixed Up Files of Mrs. Basil E. Frankweiler.
PLAY: A Cook For Mr. General.
AWARDS: Various for (after 1956): Patterns, Project Immortality; Big Deal in Lareco; Teacher, Teacher; Ben Casey Pilot; The Price; Brigadoon; Judge Horton And The Scottsboro Boys; The Homecoming.

COOK, KWENAM DAVID: Exec. b. Seoul, Korea, Mar. 16, 1922. e. Waseda U., Tokyo, Japan. ROK national assemblyman (Congress) 1959–61. Pres. Korean Gymnastic Assoc. 1967–70. Chairman: Century Co., Ltd., Seoul, producers, distributors, exhibitors, and importers of films. Honorary consul general for Peru.

COOK, ZALE EDWARD: Executive. b. Corpus Christi, TX., June 29, 1942. e. Del Mar Jr. College, Corpus Christi; Texas A&L; Univ. of Maryland; Univ. of Hawaii. Partner in Corpus Christi Egg Co., 1961–72. U.S. Army, 1966–70. Formed Cinema Southwest, Inc. in April, 1972, acquiring 1st theatre in April, 1973, in Murray, Ky. Opened 1st theatre in Texas in Feb., 1974. Cinema Southwest currently owns and manages 28 screens, all in Texas; has 6 screens under construction (3 in New Mexico). Theatres previously owned in Kentucky, Indiana and Arkansas have been sold.

COOKE, ALISTAIR: Journalist and Broadcaster. b. Eng., Nov. 20, 1908; e. Jesus Coll., Cambridge; Yale; Harvard. Film crit. of BBC 1934–37. London corr. NBC 1936–37. BBC commentator in U.S. since 1937. Chief Amer. corr., Manchester Guardian, 1948–72; English narrator, The March of Time, 1938–39; v.o. narrator, Sorrowful Jones, 1948; narrator, Three Faces of Eve, 1957; narrator, Hitler, 1973; Peabody award winner for International reporting, 1952; author, Douglas Fairbanks, Garbo & The Night Watchmen. A Generation on Trial, One Man's America, Christmas Eve, The Vintage Mencken, etc.; m.c. Omnibus, TV show, 1952–61; narrator, Three Faces of Eve, 1957, m.c. prod. U.N.'s International Zone, Emmy Award, 1958, Talk About America; writer & narrator, America: A Personal History of The United States, TV series BBC, NBC, PBS, for which 5 Emmy Awards, 1973; Franklin Medal, Royal Society of Arts, 1973; Knighted, KBE, 1973; America (book) 1973; Six Men, 1977, The Americans, 1979; Above London, 1980.

COOPER, BEN: Actor. b. Hartford, Conn., Sept. 30. e. Columbia U. On stage in Life with Father, Theatre; numerous radio, TV appearances.
PICTURES INCLUDE: Woman They Almost Lynched, A Perilous Journey, Sea of Lost Ships, Flight Nurse, Fortune Hunter, Johnny Guitar, Hell's Outpost, Eternal Sea, Last Command, Fighting Chance, Headline Hunters, Rose Tattoo, Rebel in Town, Chartroose Caboose, Red Tomahawk, The Fastest Gun Alive, One More Train to Rob, Support Your Local Gunfighter.

COOPER, EDWARD: Exec. b. New York, Oct. 21, 1903; e. Montana State Univ., 1928; Washington Coll. of Law, 1936. Reporter, city ed. papers in Montana, Colorado, Wash., D.C.; Chief of Staff, Senate Interstate Commerce Com., investigating communications, domestic and intern'l, 1939–41; U.S. Navy officer, 1942–45. Returned to previous post, handling radio broadcast, TV, m.p. matters, 1946–50; staff dir., Senate Democratic Policy com., 1951–52.
Member: Masons (Scottish Rite-32nd), Army & Navy Club, Armed Forces Communications Assn.; Joined MPAA 1952 as director national legislative dept., v.p. since 1967. Member advisory council of Endowment and Res. Found., Montana State U., 1966.

COOPER, HAL: Director and Performer. b. Feb. 22, 1923; N.Y.C.; e. U. of Mich.; m. Marta Salcido; child actor in various radio prog. since 1932; featured Bob Emery's Rainbow House, Mutual, 1936–46; asst. dir. Dock St. Theatre, Charleston, S.C., 1946–48.
TV: Your Sch. Reporter, TV Baby Sitter, The Magic Cottage as writer and producer; dir., Valiant Lady, Search for Tomorrow, Portia Faces Life, CBS-TV; dir., assoc. prod. Kitty Foyle, NBC-TV; prod. dir. Indictment, Para. TV; assoc. prod. dir. The Happy Time, Para. TV; prod. dir. For Better or Worse, CBS-TV; dir., The Clear Horizon, CBS-TV; Assoc., prod., dir., Surprise Package, CBS-TV; dir., Dick Van Dyke Show; prod., dir., The Art Linkletter Show, The Object Is. Dir.: Death Valley Days, I Dream of Jeannie, That Girl, I Spy, Hazel, Gidget, Gilligan's Island, NYPD, Mayberry, Courtship of Eddie's Father, My World and Welcome to It, The Odd Couple, Mary Tyler Moore, All in the Family. Exec. prod., dir. Maude, 1972–78.

COOPER, JACKIE: Actor-Director-Producer. b. Los Angeles, Sept. 15, 1922. Began theatrical career at age of 3 as m.p. actor; was member of Our Gang comedies. First starring role in 1930 in Skippy. Worked at every major studio, always with star billing. At 20 enlisted in Navy. After three-yr. tour of duty went to N.Y. to work in live TV. Appeared on Broadway stage in Mr. Roberts and on natl. tour and in London. Directed as well as acted in live and filmed TV. Served as v.p. in chg. of TV prod., for Screen Gems, from 1964 to 1969, when resigned to return to acting, directing, producing.
PICTURES INCLUDE: Actor—Movietone Follies, Sunny Side Up, Skippy, Sooky, The Champ, When a Fellow Needs a Friend, Lumpy, Lost, The Bowery, Treasure Island, O'Shaughnessy's Boy, The Devil Is a Sissy, Gangster's Boy, Seventeen, Gallant Sons, Her First Beau, Syncopation, Stork Bites Man, Kilroy Was Here, Everything's Ducky, The Love Machine, Chosen Survivors, Superman, Superman II. Director: Stand Up And Be Counted.
TELEVISION: People's Choice (directed 71 segments; also starred), Hennesy (dir. 91 segments; also starred) 1975 series: Mobile Two (star).

COOPER, TOMMY: Comedian. b. Caerphilly, Wales, 1921. TV debut, 1947; stage debut, 1948. First starring TV series, It's Magic, 1952; London Palladium debut, 1952; U.S.A. cabaret debut, Las Vegas, 1954; Royal Variety Performances.
TV: Coopers Capers, Life With Tommy, all principal London cabarets; Ed Sullivan, USA; Sunday Night at the London Palladium. Cooperama series, 1966; Life with Cooper, 1967–68 and 1969–70.
STAGE: Paris By Night, Blue Magic, and pantomimes, London Palladium, 1971.
PICTURES INCLUDE: The Cool Mikado, The Plank.

COOPERMAN, ALVIN: Producer. b. Brooklyn, N.Y., July 24, 1923; e. NYU, 1939–43. Started career with Lee & J. J. Shubert, 1939–51; color team, dev. color TV for FCC approval, 1953; prod. first color TV shows with mobile unit, 1954; developed & prod. first Wide Wide World, June 1955; mgr. program sls., NBC Network, 1955; exec. prod. Producers Showcase, Dodsworth, Rosalinda, 1956; prod. Jack and the Beanstalk, Festival of Music, 1957; dir. prog. NBC-TV, Apr. 1957; joined HJ Enterprises as prod. NBC-TV, The Shirley Temple Storybook, 1957; exec. prod. Screen Gems, 1958; prod. Du Pont Show, 1959; exec. prod. Roncom Prod. 1960; Prod., Untouchables, 1961–63; exec. dir., Shubert Thea. Ent. 1963; v.p., special programs, NBC, 1967–68; exec. v.p., Madison Square Garden Center, 1968–1972 Exec. Vice President Madison Square Garden Center, Inc.; Present-Chairman of the Board Athena Communications Corporation.

COOTE, BERNARD: Producer, exec., ent. m.p. industry 1945. Denham Studios; 1951, joined Group 3, Ltd.; 1953, studio manager, Beaconsfield Studios; 1955, gen. man.; 1957, dir. prod. Ivanhoe TV series; 1958, prod. Violent Moment, dir. Shepperton TV Productions, Ltd.; Assoc. prod. TV series, The Third Man, Samson of the Circus, Voyage of Discovery; 1960–63, mgr. Shell Intntl. documentary film unit; 1963–65, dir. documentary films in Africa; 1965–67, freelance feature prod., Europe and Middle East; 1967–present, U.K. prod. supervisor, United Artists.

COPPOLA, FRANCIS FORD: Writer-Producer-Director. b. Detroit, April 7, 1939. e. Hofstra Univ. (BA 1958); UCLA (1958–68, masters cinema). Heads own production facility, American Zoetrope, San Francisco.
PICTURES INCLUDE: Director: Dementia 13, You're a Big Boy Now, Finian's Rainbow, The Rain People. Writer: Patton, This Property is Condemned, Reflections in a Golden Eye, The Rain People, Is Paris Burning?, The Great Gatsby. Writer-Producer-Director: The Conversation, The Godfather, The Godfather, Part II; Apocalypse, Now. Producer: American Graffiti, THX 1138. Executive Producer: The Black Stallion, Hammett. Director-co-writer: One from the Heart.
TV: The People (movie of the week).

CORBETT, HARRY H.: Actor. b. Rangoon, 1925. Ent. m.p. ind. 1955.
PICTURES INCLUDE: Floods of Fear, The Big Day, Cover Girl Killer, The Unstoppable Man, Marriage of Convenience, Shakedown, Sammy Going South, What A Crazy World, Ladies Who Do, The Barqee, Rattle of a Simple Man, Joey Boy, Carry on Screaming, Crooks and Coronets.
TV: Song in a Strange Land, Thunder on the Snow, Steptoe and Son, Mr. Aitch.

CORD, ALEX: Actor. b. Floral Park, L.I., Aug. 3, 1931. r.n. Alexander Viespi. Early career in rodeo; left to become actor. Studied at Shakespeare Academy (Stratford, Conn.) and Actor's Studio (N.Y.). Spent two yrs. in summer stock; in 1961 went on tour with Stratford Shakespeare Co. Made m.p. debut in Synanon, 1965.
PICTURES INCLUDE: Stagecoach, The Scorpio Letters, The Prodigal Gun, The Brotherhood, Stiletto, The Dead Are Alive, Chosen Survivors, Sidewinder One, Grayeagle.
TELEVISION: Guest in many series; Genesis II (tf).

COREY, JEFF: Actor. b. New York City, Aug. 10, 1914; e. Feogin Sch. of Dram. Art. On stage in Leslie Howard prod. of Hamlet, 1936; Life and Death of an American, 1938. Screen debut 1940 in All That Money Can Buy, In The Matter of J. Robert Oppenheimer and Hamlet-Mark Taper Forum, L.A. King Lear, Beverly Music Center '73.
PICTURES INCLUDE: Syncopation, The Killers, Ramrod, Joan of Arc, Roughshod, Black Shadows, Bagdad, Outriders, Singing Guns, Seconds, In Cold Blood, Golden Bullet, Boston Strangler, True Grit, Butch Cassidy and The Sundance Kid, Beneath the Planet of the Apes, Getting Straight, Little Big Man, They Call Me Mister Tibbs, Clear and Present Danger, High Flying Lowe, Catlow, Something Evil, Premonition, Shine, Rooster, Oh, God! Butch and Sundance: The Early Days, Up River.
TV: The Untouchables, The Beachcomber, The Balcony, Yellow Canary, Lady in A Cage, Outer Limits, Channing, The Doctors and The Nurses, Perry Mason, Gomer Pyle, Wild, Wild West, Run for Your Life, Bonanza, Iron Horse, Judd for Defense, Garrisons Gorillas, Gunsmoke, Hawaii Five O, Star Trek, dir. The Psychiatrist, Night Gallery, Alias Smith and Jones, Sixth Sense, Hawkins, Owen Marshall, Police Story, Bob Newhart Show, Six Million Dollar Man, Doctors Hospital, Starsky and Hutch, Land of the Free (film), Kojak, McCloud, Captains Courageous (Bell Tel. Hr.), Bionic Woman, Barney Miller, One Day At a Time, The Pirate, Lou Grant.

CORMAN, GENE: Producer. b. Detroit, Mich. e9 Stanford Univ. Went to work for MCA as agent; left to produce his first feature film, Hot Car Girl. Partner with brother Roger in Corman Company and New World Distributors.
PICTURES INCLUDE: Attack of the Giant Leaches, Not of This Earth, Blood and Steel, Valley of the Redwoods, Purple Reef, Beast from Haunted Cave, Cat Burglar, Tobruk, You Can't Win 'Em All, The Intruder, Von Richthofen and Brown, I Escaped from Devil's Island, Secret Invasion, Vigilante Force, F.I.S.T. (exec. prod.), The Big Red One.

CORMAN, ROGER WILLIAM: Executive-Director, Producer, Writer, Distributor. b. Detroit, Michigan, April 5, 1926. e. Stanford U. 1947; Oxford U., England 1950; U.S. Navy 1944; 20th Century-Fox, production department, 1948, story analyst 1948–49; Literary agent, 1951–52; story, s.p., assoc.-prod. Highway Dragnet. Formed Roger Corman Prod. and Filmgroup. Produced over 200 feature films and directed over 60 of them. Formed production-releasing company, organization, New World Pictures, Inc., 1970.
PICTURES INCLUDE: Five Guns West, 1953, House of Usher, Little Shop of Horrors, 1961; Pit and the Pendulum, The Intruder, Masque of the Red Death, Tomb of Ligeia, The Secret Invasion, The Wild Angels, The Trip, Bloody Mama, Von Richtofen and Brown, Gasss, St. Valentine's Day Massacre, Box Car Bertha, Big Bad Mama, Death Race 2000, Eat My Dust, Capone, Jackson County Jail, Fighting Mad, Thunder & Lightning, Grand Theft Auto, I Never Promised You A Rosegarden, Deathsport, Avalanche.

CORNFIELD, HUBERT: Director-Writer-Producer. b. Istanbul, Turkey. e. University of Pennsylvania. Actors' Studio, 1952–54. Story analyst Allied Artists 1954. Directed first picture 1955. Lives in Paris since 1964.
PICTURES INCLUDE: Sudden Danger, Lure of the Swamp, Plunder Road, The Third Voice, Pressure Point, The Night of the Following Day.

CORRADINE, TOM J.: Producer, distributor. b. Chicago, Ill., Nov. 14, 1924. e. parochial schools; theatre mgr., booker, 1944; ent. TV ind. as film dir., CBS, west coast, 1949; formed Associate Co., 1952; assoc. prod., TV, Wild Bill Hickock, Judge Roy Bean, Rendezvous with Adventure; formed Artists Productions, Inc., 1963. Dir. west coast operations Republic Film Corp., 1965. Reactivated co. of artists to prod. TV commercials and specials, 1968. V.P. charge of production and V.P. national sales, TVNATIONAL Releasing Corp., 1977; 1978, formed Videobrary Inc. to produce and distribute pre-recorded video tapes for the home consumer.
PICTURES INCLUDE: The Big Tipoff, Night Freight, Port of Hell, Highway Dragnet, Kroma.

CORRI, ADRIENNE: Actress. b. Glasgow, Scotland, Nov. 13, 1933. Ent. Royal Acad. of Dramatic Art at 13; parts in several stageplays including The Human Touch; m.p. debut in The River.
PICTURES INCLUDE: Quo Vadis, The Kidnappers, A Clockwork Orange, Lease of Life, Make Me An Offer, Feminine Touch, The Big Chance, Corridors of Blood, Doctor of Seven Dials, The Rough and the Smooth, Hellfire Club, The Tell-Tale Heart, Dynamite Jack, Doctor Zhivago, Epilogue to Capricorn, Rosebud. Numerous TV appearances.

CORT, ROBERT W.: Executive. e. Univ. of Pennsylvania (Phi Beta Kappa). Has worked primarily in marketing/advertising. Joined Columbia Pictures as v.p., 1976; elevated to v.p., adv./pub./promo. Named exec. v.p. of mktg. for 20th-Fox, 1980.

CORTESA, VALENTINA: Actress. b. Milan, Italy, Jan. 1, 1925. Started career at 15 in Rome. Screen debut: La Cens Delle Beffe, 1941; brought to Hollywood by 20th Century-Fox, following picture, A Yank in Rome; experience on dramatic stage in variety of roles inc. Shakespeare, O'Neill, Shaw.
PICTURES INCLUDE: Cagliostro, Glass Mountain, House on Telegraph Hill, Thieves' Highway, Les Miserables, Secret People, Barefoot Contessa, Shadow of the Eagle, Magic Fire, The Legend of Lylah Clare, Juliet of the Spirits, Day for Night, When Time Ran Out.

CORTEZ, STANLEY: Dir. Photography. b. New York City, br. Ricardo Cortez, actor; e. N.Y.U. Began working with portrait photographers (Steichen, Pirie MacDonald, Bachrach, etc.), N.Y. Entered film industry with Paramount Pictures; to Hollywood as camera asst. and later second cameraman, various studios; pioneer in use of montage; Signal Corps World War II, Yalta, Quebec, etc. Received Film Critics of Amer. award for work on Magnificent Ambersons. Under personal contract to David O. Selznick, Orson Welles, Walter Wanger, David Wolper. Contributor Encyclopedia Britannica.
PICTURES INCLUDE: Man on the Eiffel Tower, Shark River, (A.A. nominee) Since You Went Away, (A.A. nominee), Bad Lands of Dakota, Magnificent Ambersons, Eagle Squadron, Powers Girl, Since You Went Away, Smash Up, Flesh and Fantasy, Captain Kidd, Secret Beyond the Door, Fort Defiance, Riders to the Stars, Black Tuesday, Night of the Hunter, Man from Del Rio, Three Faces of Eve, Top Secret Affair, Angry Red Planet, Dinosaurus, Back Street, Shock Corridor, Nightmare in the Sun, The Naked Kiss, Blue, The Bridge of Remagen, The Date, Another Man, Another Chance. Special sequences on Damien, Omen II, Day the World Ended, Le Bon Vivant.

CORWIN, BRUCE CONRAD: Exhibitor. b. June 11, 1940, Los Angeles, Calif.; e. Wesleyan U.; Pres., Metropolitan Theatres Corp.; chmn., Will Rogers Hospital area ind. campaigns; exec. v.p., Variety Boys Club; chmn., Education Task Force; bd. member, Coro Foundation; Board of Trustees American Film Institute; Board of Trustees U.C.L.A. Foundation.

CORWIN, NORMAN: Writer, producer, director. b. Boston, Mass., May 3, 1910. Sports ed. Greenfield, Mass. Daily Recorder, 1926–29; radio ed., news commentator, Springfield Republican & Daily News, 1929–36; prog. dir., CBS, 1938. Bok Medal "for distinguished services to radio," 1942; Peabody Medal, 1942; awarded grant by Amer. Acad. of Arts & Letters, 1942; Page One Award, 1944; Distinguished Merit Award, Nat'l Conf. of Christians & Jews, 1945; Unity Award, Interracial Film & Radio Guild, 1945; Wendell Willkie One World Flight Award, 1946; Met. Opera Award for modern opera in collab. Bernard Rogers, 1946; first award, Res. Comm. of U.N., 1950; Radio & TV first award, Nat'l Conf. of Christians & Jews, 1951; Honor Medal of Freedom Foundation for TV show, Between Americans, 1951; ent., Radio Hall of Fame, 1962. Hon. doctorate Columbia Col. of Comms., 1967; Valentine Davies Award, WGA, 1972; author of Thirteen by Corwin, More by Corwin,

Untitled & Other Plays; The Plot to Overthrow Christmas, Dog in the Sky, Overkill and Megalove, Prayer for the 70s. Holes in a Stained Glass Window; lectured at various colleges; taught courses U.C.L.A., U.S.C., San Diego State U., regents lecturer, UC at Santa Barbara; Chairman, Creative Writing, USC-Isomata; U. of Alberta, U. of So. Col., Witswatersrand U., Rand Afrikaans U., So. Africa, Cantatas, The Golden Door, 55; Yes Speak Out Yes (commissioned by UN, 1968)

STAGE PLAYS: The Rivalry, The World of Carl Sandburg, The Hyphen, Overkill and Megalove, Cervantes. Together Tonight: Jefferson, Hamilton and Burr.

PICTURES INCLUDE: Once Upon a Time, Blue Veil, The Grand Design, Scandal in Scourie, Lust for Life, The Story of Ruth.

TV: Inside the Movie Kingdom, The FDR series, The Plot to Overthrow Christmas, Norman Corwin Presents (for Westinghouse), The Court Martial of General Yamashita, Network at 50 (CBS). Critic, monthly media column, Westways Magazine, Corwin on Media.

Chmn. Doc. Award Com., Motion Picture Academy 1965–80; elected to bd. of governors, 1980. Chm., writers' exec. comm., M.P. Academy; co-chmn. scholarship com., m.p. Academy; member, Film Advisory Bd.; L.A. County Museum; Norman Corwin Presents series; member, bd. of trustees, Advisory Board, Filmex, member, board of directors, WGA.

COSBY, BILL: Actor, comedian. b. Philadelphia, Pa., July 12, 1938. e. Temple U.; served in United States Navy Medical Corps. Started as night club entertainer. Has appeared many times on TV variety shows, in numerous one-nighters across the country, and concert tours.

TV: Emmy Award 1966, 1967, 1968: Best Actor in Dramatic Series—"I Spy," NBC-TV. Emmy Award, 1969; The First Bill Cosby Special, I Spy, The Bill Cosby Special, The Second Bill Cosby Special, The Bill Cosby Show, Fat Albert, Fat Albert and the Cosby Kids.

COMEDY ALBUMS: Bill Cosby Is A Very Funny Fellow . . . Right!; I Started Out As a Child; Why Is There Air?; Wonderfulness; Revenge; To Russell, My Brother, Whom I Slept With; 200 MPH; It's True, It's True; 8:15, 12:15.

SINGING ALBUMS: Silverthroat, Hooray for the Salvation Army Band. Grammy Award, 1964, 1965, 1966, 1967, 1969: Best Comedy Album.

RADIO: The Bill Cosby Radio Program.

PICTURES INCLUDE: Man and Boy, Uptown Saturday Night, Let's Do It Again, Mother, Jugs, and Speed, A Piece of the Action, California Suite, The Devil and Max Devlin.

COSLOW, SAM: Producer, song writer. b. New York City, Dec. 27, 1905. Writer popular songs, many for screen prod. (Cocktails for Two, Just One More Chance, Sing You Sinners, etc.). In 1941 formed RCM Prod., Inc., with James Roosevelt. Apptd. Paramount prod. 1944. Academy Award (with Jerry Bresler), best two-reel short, 1943 (Heavenly Music). In 1945 joined Mary Pickford Prods. as prod.

PICTURES INCLUDE: Out of This World, Copacabana, Dreaming Out Loud.

COSELL, HOWARD: Sports Commentator. b. Winston-Salem, N.C., 1920. Studied law and practiced. Broadcasting career began in 1953 when hired to host program on which N.Y. area Little Leaguers were introduced to baseball stars. In 1956 ABC hired him for ten five-minute sports shows on weekends. He dropped legal work to concentrate on sports reporting. Has had wide variety of roles in TV: commentator on ABC Monday Night Football, Monday Night Baseball, and various sports specials. Has hosted the Howard Cosell Sports Magazine for 4 yrs. and hosts 14 shows each week on American Contemporary Radio Network. Has guested on numerous prime-time TV shows.

PICTURES: Bananas, Sleeper, The World's Greatest Athlete.

COSMATOS, GEORGE P.: Director-Producer-Writer. b. Jan. 4, 1941. e. London Univ. Asst. dir., Exodus and Zorba the Greek.

PICTURES: Restless (co.-prod., s.p., dir.); Massacre in Rome (co-s.p., dir.); The Cassandra Crossing (co.-s.p., dir.); Escape to Athena (co-s.p., dir.).

COSTA-GAVRAS, HENRI: Director. b. Athens, Greece, 1933. e. Studied at the Sorbonne. Worked as second, then first assistant to Marcel Ophuls, Rene Calir, and Rene Clement, on Banana Peel, All the Gold in the World, The Thieves, 1964; dir. and wrote s.p. The Sleeping Car Murders. Second film Un Homme De Trop won prize at the Moscow Festival, 1966. Dir. and collab. s.p. Z (won Acad. Award, Best Foreign Lang. Film). L'Aveu (The Vow), State of Siege, Special Section.

COTES, PETER: F.R.S.A. Producer, dir., writer. Began early career as actor; stage debut Henry V, London 1926; debut as stage producer, 1942; founded Peter Cotes prods. 1946 and Peter Cotes Players 1948, prod., dir. numerous plays; prod. dir. for BBC: Same Sky, The Mask, Shadow of the Vine, Infinite Shoeblack, The Road, What the Public Wants; exec. prod., dir. Associated-Rediffusion, 1956; Sup. prod. TV series, London Playhouse. Dir. Woman in a Dressing Gown, Ashes in the Wind, Hungry God. Co. dir., Waterfront; dir., Young and the Guilty, The Right Person Publications; No, Star Nonsense, Little Fellow (Charles Chaplin), A Handbook for the Amateur Theatre, George Robey. Head prod. Channel 7, Melbourne, Australia 1961. dir. doc., Australian Snowy Mountains; setting, dir., plays London, Broadway, 1962–63; TV consultant, McCann-Erickson (London); prod., dir.

TV: The Offence, Winter in Ischia, Also stage prods. in London. Assoc. prod., The World of Wooster, TV series for BBC, Prod. & Dir. West End Theatre, among others: Impossible Years, Staring At the Sun, Janie Jackson, West End Revival, The Old Ladies; wrote TV Omnibus program, gave R.S.A. theatre lectures. Devised-directed, Down With All Parties, Look, No Hands, My Brother, Willie, Lecturing, writing series of radio portraits; working on auto-biography. Publications: The Trial of Elvira Barney, Circus, The Man Called Mitch. 1978: Music Hall Tragedies, Charlie Chaplin Merryn Peake (radio), The Diminutive One.

COTTEN, JOSEPH: Actor. b. Petersburg, Va., 1905. In stock and on New York Stage, 1930–40, including Accent on Youth, Philadelphia Story; also Orson Welles' Federal Theatres productions and Mercury Theatre of the Air.

STAGE: Sabrina Fair, Once More With Feeling, Prescription: Murder, Calculated Risk.

TV: narrator, Hollywood and the Stars, own show.

PICTURES INCLUDE: Citizen Kane, Lydia, Magnificent Ambersons, Journey Into Fear Shadow of a Doubt, Hers to Hold, Temptation, The Last Sunset, Two Flags West, September Affair, Walk Softly, Stranger, Half Angel, Peking Express, Man With a Cloak, The Untamed Frontier, The Steel Trap, Niagara, Blueprint for Murder, Special Delivery, Bottom of the Bottle, Killer Is Loose, Halliday Brand, From the Earth to the Moon, The Angel Wore Red, The Great Sioux Massacre, The Oscar, The Hellbenders, The Tramplers, The Money Trap, They Also Killed, The White Comanche, Petulia, The Grasshopper, The Abominable Dr. Phibes, Soylent Green, A Delicate Balance, Twilight's Last Gleaming, Airport '77, Caravans, Heaven's Gate.

COTY, HENRI ROLAND: Executive. b. Neuilly-sur-Seine, France, Feb. 13, 1922. e. Lycee Descartes, Tours, France. Baccalaureat, 1939–40, St. Jean DePassy, WWII Buchmaster Underground network; arrested 1942, deported to Dachau, Dora, Manthauson, Bergen-Belsen concentration camps; liberated, 1945. Sales mgr., Coty Perfume Co., 1946–55; found own company, s.a. Coty & Co., promotion, publicity, public relations, printing, now TV program prod.

COURTENAY, TOM: Actor. b. Hull, England, 1937; e. University Coll., London, Royal Academy of Dramatic Art, 1960–61; Old Vic. Ent. TV 1961 in Private Potter and The Lads; I Heard the Owl Call My Name (U.S.).

STAGE: Billy Liar, Andorra, Hamlet, She Stoops to Conquer, Otherwise Engaged (N.Y. debut).

PICTURES INCLUDE: Loneliness of the Long Distance Runner, Private Potter, Billy Liar, King and Country, Operation Crossbow, King Rat, Dr. Zhivago, Night of the Generals, The Day the Fish Came Out, A Dandy in Aspic, Otley, One Day in the Life of Ivan Denisovich.

COURTLAND, JEROME: Actor-Producer-Director. Began career in 40s as actor, then turned to directing and producing.

PICTURES INCLUDE: Actor: Kiss and Tell, Man from Colorado, Battleground, The Barefoot Mailman, The Bamboo Prison, Tonka, Black Spurs. Director: Run, Cougar, Run, Diamonds on Wheels. Producer: Escape to Witch Mountain, Ride a Wild Pony, Return from Witch Mountain.

TV: Actor: The Saga of Andy Burnett, Tonka. Director: Hog Wild (also co-prod.), Harness Fever.

COWAN, THEODORE: Executive. b. Letchworth, Eng. e. Parmiters School. Entered m.p. ind., J. Arthur Rank Productions, pub. div., asst. dir. pub. & adv., 1960. Formed own pr. & publ. Co., 1962.

COWAN, WARREN J.: Publicist. b. New York, Mar. 13. e. Townsend Harris H.S., U.C.L.A., graduated 1941. Entered public relations, 1941 with Alan Gordon & Associates; three yrs. Air Force; to Henry C. Rogers office; became partner, 1949 and changed name to Rogers & Cowan, Public Relations Advisor, Rogers & Cowan, Inc., 1960; pres., Rogers & Cowan, Inc., 1964-present.

CRABBE, (LARRY) BUSTER: Actor. r.n. Clarence Linden C.; b. Oakland, Calif.; e. U. So. Calif. Spent boyhood Hawaii; expert swimmer. In 1933 on screen in King of the Jungle; thereafter in many adventure pictures, serials and others.

PICTURES INCLUDE: The Sweetheart of Sigma Chi, Wanderer of the Wasteland, Desert Gold, various Flash Gordon & Billy the Kid, Gunfighters of Abilene.

TV: The Foreign Legion.

CRAIG, MICHAEL: Actor. b. India, 1929. At 16 joined Merchant Navy. 1949 returned to England and made stage debut in repertory. M.P. debut as crowd artist 1950.

PICTURES INCLUDE: Malta Story, The Love Lottery, Passage Home, The Black Tent, Yield to the Night, Eyewitness, House of Secrets, High Tide At Noon, Sea of Sand, Sapphire, Upstairs and Downstairs, The Angry Silence, Cone of Silence, Doctor In Love, The Mysterious Island, Payroll, No My Darling Daughter, A Pair of Briefs. Under contract Rank Organisation. A Life for Ruth, The Iron Maiden, Captive City, Summer Flight, Of a Thousand Delights, Life at the Top, Modesty Blaise, Star, A Town Called Hell, Ride a Wild Pony.

CRAIG, STANLEY HERBERT: Executive. Began career with Co-operative Films, Melbourne, Australia, 1914; Paramount, 1917; asst. mgr. Melbourne, 1921; br. mgr., Adelaide, 1923; gen. mgr., New Zealand, 1928–67; Pres. M.P. Distributors Assn. of New Zealand, 1938–55; chmn., 1963–64; mem. Cinematograph Films Licensing Authority, 1968–70.

CRAIN, JEANNE: Actress. b. Barstow, Calif., May 25, 1925. Model; crowned Miss Long Beach of 1941: Camera Girl of 1942.
PICTURES INCLUDE: Home in Indiana, In the Meantime Darling, Winged Victory, State Fair, Leave Her to Heaven, Margie, Centennial Summer, You Were Meant for Me, Apartment for Peggy, Letter to Three Wives, Pinky, Cheaper by the Dozen, Take Care of My Little Girl, People Will Talk, Model and the Marriage Broker, Belles on Their Toes, O. Henry's Full House, City of Bad Men, Dangerous Crossing, Vicki, Duel in the Jungle, Man Without a Star, Second Greatest Sex, Gentlemen Marry Brunettes, Fastest Gun Alive, Tattered Dress, The Joker, Guns of the Timberland, Skyjacked.

CRAMER, WM. DOUGLAS: Executive. b. Aug. 22, 1931. e. Northwestern U., Sorbonne, Cincinnati U., B.A.; Columbia U.M.F.A., M. Joyce Haber, columnist. Taught at Carnegie Inst. of Tech., 1954–55; Production asst. Radio City Music Hall 1950–51; MGM Script Dept. 1952; Manag. Dir. Cincinnati Summer Playhouse 1953–54. TV supvr. Procter and Gamble 1956–59; Broadcast supvr. Ogilvy, Benson and Mather, Adv. 1959–62; V-P Program Dev. ABC-TV 1962–66; V-P Program Dev. 20 Cent.-Fox TV 1966; Exec. v.p. in chg. of production, Paramount Television, 1968.
PLAYS: Love is a Smoke, Whose Baby Are You, Call of Duty.

CRAWFORD, BRODERICK: Actor. b. Philadelphia, Dec. 9, 1911; p. Helen Broderick and Lester Crawford, prof.
PICTURES INCLUDE: Submarine D-1, Ambush, Undercover Doctor, Real Glory, Eternally Yours, I Can't Give You Anything But Love Baby, Slightly Honorable, When the Daltons Rode, Butch Minds the Baby, Broadway, Sin Town, Night Unto Night, Flame, Anna Lucasta, All the King's Men, Born Yesterday, The Mob, Lone Star, Scandal Sheet, Last of the Comanches, Stop You're Killing Me, Last Posse, Night People, Human Desire, Down 3 Dark Streets, New York Confidential, Big House, Not As a Stranger, Fastest Gun Alive, Between Heaven and Hell, The Oscar, Gregorio and His Angels, The Texican, Hell's Bloody Devils, The Private Files of J. Edgar Hoover.
TV: Highway Patrol (series), The Interns (CBS).

CRAWFORD, MICHAEL: Actor. b. Salisbury, England, 1942. Early career as boy actor in children's films and sound radio.
TV: Still Life, Destiny, Byron, Move After Checkmate, Three Barrelled Shotgun, Home Sweet Honeycomb, Some mothers do 'ave 'em, Chalk and Cheese.
PICTURES INCLUDE: Two Left Feet, The War Lover, Two Living, One Dead, The Knack, A Funny Thing Happened On The Way to the Forum, The Jokers, How I Won the War, Hello, Dolly!, The Games, Hello and Goodbye, The Adventures of Alice in Wonderland.
STAGE: Come Blow Your Horn, Traveling Light, The Anniversary, Black Comedy (N.Y.), No Sex Please We're British, Billy, Same Time Next Year, Flowers for Algernon.

CRENNA, RICHARD: Actor. b. L.A., Calif., Nov. 30, 1927. e. Belmont H.S., U. of Southern Calif.
RADIO: Boy Scout Jambouree, A Date With Judy, The Great Gildersleeve, Johnny Dollar, Our Miss Brooks.
TV: Our Miss Brooks, The Real McCoys, Slattery's People, All's Fair.
PICTURES INCLUDE: Pride of St. Louis, It Grows on Trees, Red Skies Over Montana, John Goldfarb, Please Come Home, Pendick Enterprises, The Sand Pebbles, Star, Marooned, The Deserter, Doctor's Wives, Red Sky at Morning, A Man Called Noon, Catlow, Dirty Money, Death Ship.

CRICHTON, CHARLES: Director. b. Wallasey, Aug. 6, 1910; e. Oundle & Oxford. Collab. dir. Dead of Night.
PICTURES INCLUDE: Painted Boats, Hue and Cry, Against the Wind, Another Shore, Dance Hall, Lavender Hill Mob, Hunted (Stranger in Between), Titfield Thunderbolt, The Love Lottery, Divided Heart, Man in the Sky, Floods of Fear, Battle of the Sexes, The Third Secret, He Who Rides a Tiger.
TV: The Wild Duck, Danger Man, The Avengers, Man in a Suitcase, The Strange Report, Shirley's World, Black Beauty, The Protectors, Space 1999, Return of the Saint, Dick Turpin 1 & 2 Series.

CRICHTON, MICHAEL: Writer-Director. Entered Harvard Medical School in 1965; while there completed first novel, Easy God, under pseudonym, John Lange. Has written 15 books under four different names, including A Case of Need (filmed as The Carey Treatment) and The Andromeda Strain (first book to appear under his own name). Did post-graduate work at Salk Institute in La Jolla. Also wrote novels Dealing and The Terminal Man, both filmed.
PICTURES INCLUDE: Westworld (s.p., dir.), Coma (s.p., dir.), The Great Train Robbery (s.p., dir.), Looker (s.p., dir.).
TELEVISION: Pursuit (Movie of Week), dir.; script by Robert Dozier based on Crichton's book, Binary.

CRIST, JUDITH: Journalist-Critic. b. N.Y.C. e. Hunter College, Columbia Univ. (School of Journalism). Joined New York Herald Tribune, serving as reporter, arts editor, assoc. drama critic, film critic. Continued as film critic for New York World Journal Tribune, NBC-TV Today Show, New York, Saturday Review, New York Post. Now regular film critic for TV Guide, Saturday Review. 50 Plus, Medical/Mrs. Teaches at Col. Grad. School of Journ.
AUTHOR: The Private Eye, the Cowboy and the Very Naked Girl, Judith Crist's TV Guide to the Movies.

CRISTALDI, FRANCO: Producer. b. Turin, Italy, Oct. 3, 1924. Owner, prod. Vides Cinematografica; President of Italian Producer's Union.
PICTURES INCLUDE: White Nights, The Strawman, The Challenge, Big Deal On Madonna Street, Kapo, The Dauphins, Salvatore Giuliano, The Assassin, Divorce Italian Style, The Organizer, Bebo's Girl, Seduced and Abandoned, Time of Indifference, Sandra, A Rose for Every-One, China Is Near, A Quiet Couple, The Red Tent.

CRITCHFIELD, EDWARD: Producer. B. Cleveland, Ohio, Jan. 9, 1919. e. Glenville H.S., 1937. Gen. insurance broker; real estate mortgage financing; motion picture financier, Delta Diamond Productions; then joined Bernfield Publications Inc., as vice-pres.; assoc. prod. Terrified.

CRON, JOHN B.: Executive. b. Mt. Vernon, New York, June 11, 1923. e. Princeton U.: Exec. vice-pres., Robert Lawrence Productions; managing dir., Screen Gems (Europe); sales dir., N.B.C. Film Division; vice-pres., sales dir. SIB Productions.

CRONKITE, WALTER: Correspondent. b. St. Joseph, Mo., Nov. 4, 1916; e. Univ. of Texas. Reporter: radio reporter; U.P. Correspondent. Joined CBS as Washington news correspondent, 1950; many TV shows including You Are There, Twentieth Century, Eyewitness to History; CBS Reports: 21st Century, past nat'l pres. & mem. bd. Trustees, Acad. TV Arts & Sciences. Managing Editor of CBS Evening News since 1963.

CRONYN, HUME: Actor, writer, director. b. London, Ontario, Canada, July 18, 1911; e. Ridley Coll.; U. of McGill; Amer. Acad. of Dramatic Art; m. Jessica Tandy, actress. m.p. acting debut in Shadow of a Doubt.
STAGE: (Actor N.Y. plays) High Tor, Escape This Night, Three Men on a Horse, Boy Meets Girl, Three Sisters, Mr. Big, The Survivors; dir. plays Now I Lay Me Down to Sleep, Hilda Crane, Madam Will You Walk. N.Y. stage, The Fourposter, 1951–53; Madam Will You Walk, 1953–54; The Honeys, A Day by the Sea, 1955; The Man in The Dog Suit, 1957; Directed The Egghead, 1957; Directed and toured with his wife, Miss Tandy, in Triple Play, 1958–59; Big Fish, Little Fish, 1961; also in London, 1962; The Miser, 1963; The Three Sisters, 1963; Hamlet, 1964; The Physicists, 1964; Produced Slow Dance on The Killing Ground, 1964; Appeared at the White House; Hear America Speaking 1965; Richard III, 1965; Revival The Miser, 1965; A Delicate Balance, 1966 and tour, 1967; The Miser, 1968; Hadrian VII tour, 1969–70; Caine Mutiny Court Martial, 1971; Promenade All 1972–73; Krapp's Last Tape, Happy Days, Act Without Words I, 1972; Coward In Two Keys, 1974, Concert recital Many Faces Of Love 1974; National tour, Noel Coward in Two Keys, 1975, Merchant of Venice and A Midsummer Night's Dream, at Stratford Festival Theatre, Canada 1976. 1976; Many Faces of Love limited tour 1975–76 and at Theatre London, Ontario, Canada '76 as well as for CBC, Canada, 1977; Appeared in The Gin Game, Long Wharf Thea., New Haven, Conn. June, 1977. Co-produced with Mike Nichols and also appeared in The Gin Game, The Golden Theatre, N.Y.C. 1977–78. Toured U.S., Toronto, London, U.S.S.R., 1978
TV: The Marriage, and other network dramatic shows, 1957.
PICTURES INCLUDE: Cross of Lorraine, Lifeboat, Seventh Cross, Main Street, After Dark, The Sailor Takes a Wife, A Letter for Evie, The Green Years, Brute Force, Bride Goes Wild, Postman Always Rings Twice, Top O' the Morning, People Will Talk, Crowded Paradise, Sunrise at Campobello, Cleopatra, Gaily, Gaily, The Arrangement,

There Was a Crooked Man, Conrack, Parallax View, Honky Tonk Freeway, Roll-over, Garp.

CROSBY, BOB: Band leader, actor; r.n. George Robert C.; b. Spokane, Wash., Aug. 23, 1913; e. Gonzaga U. Began as singer; later featured vocalist Jimmie & Tommy Dorsey band. Org. own band (Bobcats); appeared with orch. on screen in Let's Make Music, 1940.
PICTURES INCLUDE: Sis Hopkins, Reveille with Beverly, Thousands Cheer, Presenting Lily Mars, See Here Private Hargrove; Meet Miss Bobby Socks, Kansas City Kitty, My Gal Loves Music, Pardon My Rhythm, Singing Sheriff, Two Tickets to Broadway.
TV: Bob Crosby Show.

CROSBY, KATHRYN: Actress. r.n. Kathryn Grandstaff. b. Houston, Tex., Nov. 25, 1933. e. University of Texas, U.C.L.A. m. Bing Crosby.
CREDITS: Forever Female, Rear Window, Living It Up, Arrowhead, Cassanova's Big Night, Unchained, Cell 2455 Death Row, Tight Spot, Five Against the House, Reprisal, Guns of Fort Petticoat, Phoenix City Story, Wild Party, Mister Cory, Night the World Exploded, Brothers Rico, Operation Mad Ball, The Big Circus, Bob Hope Chrysler Theatre (TV), Bing Crosby Christmas Specials (TV), Suspense Theatre (TV), Ben Casey (TV), The Kathryn Crosby Show KPIX-TV-San Francisco.

CROSS, PERRY: Performer. b. Brooklyn, N.Y., Feb. 26. e. R.I. State Coll. At various clubs; to NBC, 1947. Guest relations staff, then TV production supervisor, coordinator, unit mgr.
TV: Ernie Kovacs Show, Henie Ice Spectacular, Ruggles of Red Gap, Tonight, Jack Paar Show, Today, TV network programs, night time.

CROUSE, LINDSAY: Actress. b. New York City. Daughter of playwright Russel Crouse. e. Radcliffe. Began career as modern and jazz dancer; is also flutist and pianist. Has worked in films, theatre and TV.
PICTURES INCLUDE: All the President's Men, Slap Shot, Between the Lines.
TV: Eleanor and Franklin, The Tenth Level.

CROWN, ALFRED W.: Executive. b. New York City, Sept. 30, 1913. e. Columbia U. To South America for MGM 1933–35; foreign mgr. Grand Natl. Pictures, 1935–38; coordinator of Inter-American Affairs, 1939–40; served in U.S. Army, 1940–42; with Chas. Feldman, Hollywood 1942–43; Film Classics, 1943–45; world-wide sales mgr. for Goldwyn; vice-pres., Samuel Goldwyn Prod., Inc., 1950; v.p., foreign sales mgr., RKO Pictures, Oct. 1952; pres. Moulin Prod., Inc., Jan. 1954; v.p., Allied Artists; indep. prod. Motion Pictures; TV packager; prod., B'way play: The Deputy; prod., electronovision, Hamlet with Richard Burton. Film Prod. Midsummer Night's Dream, Columbia; Prod. Last Summer, Allied Artists; Taking Off, Universal, 1975—senior vice pres., Penthouse Films.

CROYDON, JOHN: Consult. & Rep. Film Finances, Ltd. b. London, England, Nov. 3, 1907. Entered industry 1931.
CREDITS INCLUDE: Foreman Went to France, Goose Steps Out, Champagne Charlie, Dead of Night, Nicholas Nickleby, One Wild Oat, White Corridors, Tarzan and the Lost Safari, The Entertainer, First Man Into Space, The Projected Man, High Wind in Jamaica.

CRUEA, EDMOND D.: Executive. b. Jersey City, N.J., June 3, 1918. Joined Grand National Pictures, Los Angeles, 1935; Monogram Pictures, 1938–41 (L.A. & Seattle); U.S. Army Signal Corp., 1942–46; Monogram Pictures, Seattle, 1946–48; branch mgr., 1948–49; branch mgr. Allied Artists, 1950–65 (Seattle, Portland, San Francisco); Western Division Mgr., 1965–66; v.p., sales, 1970; v.p., gen. sls. mgr., 1970–71; v.p., Western Division, 1971; dir. of distribution for Abkco Films div. of Abkco Industries, Inc., 1971–73; pres. of Royal Dist. Corp., 1974 pres., Esco Film Corp., 1975; joined Film Ventures Intl. in 1976 as exec. v.p., succeeding to pres. and chief operating officer in August, 1976. Resigned Sept. 1977 to form Fil-Mark Inc.

CRYSTAL, BILLY: Actor. b. Long Island, NY, Mar. 14. Father, Jack, produced jazz concerts; family owned Commodore jazz record label. e. Marshall Univ., New York Univ. Worked with Alumni Theatre Group at Nassau Community College. Later teamed with two friends (billed as 3's Company) and toured coffee houses and colleges. Became stand-up comedian on own, appearing on TV.
PICTURES INCLUDE: Rabbit Test.
TELEVISION: Tonight Show, Dinah, Mike Douglas Show, That Was the Year That Was, All in the Family, Love Boat, Soap. Movies: Death Flight, Breaking Up Is Hard to Do.

CUKOR, GEORGE: Director. b. N.Y.C. Began on Broadway stage. Some noted plays were The Constant Wife, The Cardboard Lover, The Furies, Great Gatsby. Stage dir. & mgr. Charles Frohman Co.: with Lyceum Theatre Co. 1920–8. Rochester, N.Y.
PICTURES INCLUDE: dial. All Quiet on the Western Front, River of Romance, Grumpy, Virtuous Sin, Royal Family, Bill of Divorcement, Sylvia Scarlett, Tarnished Lady, Girls About Town, One Hour with You, Little Women, Dinner at Eight, David Copperfield, Romeo and Juliet, Holiday, Philadelphia Story, Two-Faced Woman, Keeper of the Flame, Winged Victory, Camille, Gaslight, Double Life, Edward, My Son; Adam's Rib, A Life of Her Own, Born Yesterday, Model and Marriage Broker, The Marrying Kind, Pat and Mike, The Actress, It Should Happen to You, A Star Is Born, Bhowani Junction, Wild is the Wind, Les Girls, Heller in Pink Tights, Song Without End, Let's Make Love, The Chapman Report, My Fair Lady Justine, Travels with My Aunt, The Blue Bird, Rich and Famous.
TV: Love Among the Ruins, The Corn Is Green.

CULLEN, BILL: Performer. b. Pittsburgh, Pa., Feb. 18, 1920; U. of Pittsburgh, B.A. Asst. disc jockey; announcer, KDKA, Pittsburgh; sportcaster; staff announcer. CBS radio, 1943; then m.c., panel member; shows include: m.c. Three on a Match, To Tell the Truth, Winning Streak, 25,000 Pyramid; Love Experts.

CULLEN, JAMES V.: Executive. b. San Francisco, Dec. 29, 1938; e. City College of S.F. Sales mgr., M.P.I. Toys, Los Angeles, 1963–65; entered film industry 1965 with 20th-Fox in S.F. exchange as ass't. field man; Named Southwest ad-pub. mgr. for Fox 1966 and Western Division ad-pub mgr. 1968. Appointed Fox director of exploitation 1971; then nat'l. dir. field adv. & pro.; Joined Sandy Howard Prods., v.p. creative affairs, 1974. Produced The Devils' Rain, 1975.

CULP, ROBERT: Actor, Writer, Director. b. Berkeley, Cal., Aug. 16, 1930. e. Stockton, College of the Pacific, Washington U., San Francisco State; to N.Y. to study with Herbert Berghof; starred in off B'way prod. He Who Gets Slapped. Best Actor of the Year in an off B'way Play; motion picture debut, 1962; P.T. 109; television guest appearances in Rawhide, Wagon Train, Bob Hope Presents the Chrysler Theatre; wrote and acted in Rifleman, Cain's Hundred, The Dick Powell Show.
PLAYS: The Prescott Proposals, He Who Gets Slapped, A Clearing in the Woods.
TV: Trackdown, I Spy.
PICTURES INCLUDE: PT 109, Sammy, The Way Out Seal, The Raiders, Sunday in New York, Rhino, The Hanged Man, Bob & Carol & Ted & Alice, The Grove, Hannie Caulder, Hickey and Boggs, Sky Riders, Great Scout and Cathouse, Thursday.

CULVER, ROLAND: Actor. b. London, England, Aug. 31, 1900; e. Highgate Coll.; Royal Acad. of Dram. Arts, London. On London stage (Old Bachelor, Average Man, Fortune, French Without Tears, Five Finger Exercise, Hay Fever.
TV: Spread of the Eagle, The Caesars, The Letter, The Yellow Rolls-Royce
PICTURES INCLUDE: 77 Park Lane, Love on Wheels, Accused, French Without Tears, Folly to Be Wise, Holly & the Ivy, True and the Brave (Betrayed), Man Who Loved Redheads, Teckman Mystery, Ship that Died of Shame, An Alligator Named Daisy, Touch and Go, Safari, The Hypnotist, My Friend Charles, The Rise and Rise of Michael Rimmer, Fragment of Fear, The Magic Christian, Thunderball, A Man Could Get Killed, To Each His Own, Down to Earth, Emperor Waltz, Singapore, Isn't it Romantic, The Great Lovers.

CUMMINGS, BOB: Godson of Orville Wright. Actor, dir. b. Joplin, Mo., June 9, 1910; e. Drury Coll., Carnegie Inst. Tech.; American Academy of Dramatic Arts: on dram. & musical stage, also radio.
TV: Bob Cummings Show; My Hero, Twelve Angry Men (Emmy, 1954).
PICTURES INCLUDE: Three Smart Girls Grow Up, Spring Parade, Saboteur, Heaven Only Knows, So Red the Rose, Kings Row, Flesh and Fantasy, Princess O'Rourke, You Came Along, Reign of Terror The Accused, Free for All, Tell It to the Judge, Paid in Full, Petty Girl, For Heaven's Sake, Barefoot Mailman, The Lost Moment, Free and Easy, First Time, Marry Me Again, Lucky Me, Dial M for Murder, How to Be Very Very Popular, My Geisha, Beach Party, The Carpetbaggers, What A Way to Go, Stage Coach, Promise Her Anything, Gidget Grows Up, The Great American Beauty Contest, (TV). Has toured Dinner Theater Circuit since '73 and established new attendence records in 16 theaters playing Never Too Late, Marriage Go Round, No Hard Feelings, Harvey, Fun and Games.

CUMMINGS, CONSTANCE: Actress. b. Seattle, Wash., May 15, 1910; p. D.V. Halverstadt, attorney, and Kate Cummings, concert soprano; m. Benn Levy, English playwright. Was chorus girl in The Little Show and also appeared in June Moon.
PICTURES INCLUDE: Behind the Mask, Washington Merry-Go-Round, Broadway Through a Keyhole, Channel Crossing, Glamour, Doomed Cargo, Busman's Honeymoon, This England, The Outsider, The Foreman Went to France, Somewhere in France, Blithe Spirit, Into the Blue, The Scream, John and Julie, The Intimate Stranger, Battle of the Sexes, A Boy 10 Feet Tall.

TV: Touch of the Sun, Clutterbuck, The Last Tycoon, Ruth, Late Summer.

CUMMINGS, IRVING, JR.: Writer, producer.
PICTURES INCLUDE: Yesterday's Heroes, Lone Star Ranger, He Hired the Boss, Sign of the Ram, It's Only Money, Where Danger Lives, Double Dynamite, Girl in Every Port, Edward Small Productions, 1953–54, Television Programs of America, Fury.

CUMMINGS, JACK: Producer. b. New Brunswick, Canada. Started at MGM as office boy. Later worked as script boy, assistant director, director, producer of short subjects. Became producer of feature pictures. In 1934 produced The Winning Ticket.
PICTURES INCLUDE: Bathing Beauty, I Dood It, Broadway Melody, Girl from Rectors, Romance of Rosy Ridge, Neptunes Daughter, Two Weeks with Love, Three Little Words, Excuse My Dust, Texas Carnival, Lovely to Look At, Fiesta, Stratton Story, Sombrero, Give a Girl a Break, Kiss Me Kate, Seven Brides for Seven Brothers, Last Time I Saw Paris, Many Rivers to Cross, Interrupted Melody, Teahouse of the August Moon, Can Can, Bachelor Flat, Viva, Las Vegas.

CUMMINGS, SANDY: Executive. r.n. Sanford Cummings. b. Oct. 31, 1913. e. University of Southern Calif. Child actor; Broadway. Summer stock, 1935; asst. prod., Paramount; assist. to Walter Wanger, 1936; started Hollywood office of Benton & Bowles, 1937; major, army, 1939–46; Head American Forces Network, Germany; prod. Columbia, 1946; ABC Coordinator, Walt Disney Studios, 1954; manager, TV Net Programs, Western Div. ABC, 1957; dir. TV Net Prog., 1958; V.P. & Dir. of pgms., Western Div., 1958–62; prod. specials, STV, 1964; prod., 20th Fox Television, 1965–66; NBC Broadcast Standards; 1969 mgr., live night-time and special programs, NBC.

CUMMINS, DWIGHT W.: Writer. b. San Francisco, Cal., Feb. 20, 1901; e. U. Calif., L.A., B.A. '26. m. Dorothy Yost, writer. Tech. dir. & prod. asst. Film Booking Office studio, 1926–27. Began writing scripts 1927. Ed. & story asst. various major producers 1932–38. Instructor in Cinematog., U. So. Calif., 1938–42. Member Phi Gamma Delta.
PICTURES INCLUDE: Naughty but Nice, None but the Brave, The River, Thunderhead, Smoky, The Strawberry Roan, The Saginaw Trail.
TV: Gene Autry, Range Rider; TV writer for Roy Rogers, 1955–60; TV free lance.

CUMMINS, PEGGY: Actress. b. Prestatyn, North Wales, Dec. 18, 1926. e. Alexandra School, Dublin, Gate Theatre, Dublin. Starred in Let's Pretend on London Stage 1938. Made Dr. O'Dowd WB, Eng., 1939. In 1942; Salute John Citizen; Welcome Mr. Washington; On London stage in Junior Miss, Alice in Wonderland, Peter Pan. From 1946 Hollywood, starred in Late George Apley.
PICTURES INCLUDE: Moss Rose, Green Grass of Wyoming, Escape, That Dangerous Age, Gun Crazy, My Daughter Joy, Returned to Eng. 1950. Who Goes There (Passionate Sentry), Street Corner (Both Sides of the Law), Meet Mr. Lucifer, Always a Bride, Love Lottery, To Dorothy a Son, Cash on Delivery, March Hare, Carry on Admiral, Night of the Demon, Hell Drivers, The Captain's Table, Your Money or Your Wife, Dentist in the Chair, In the Doghouse.
TV: The Human Jungle, Looks Familiar.

CUNNINGHAM, SEAN S.: Producer-Director. b. New York City, 1949. e. B.A. Franklin & Marshall, MFA Stanford Univ. Worked briefly as actor, moving into stage-managing. Became producer of Mineola Theatre (Long Island, NY) and took several productions to Broadway. Formed Sean S. Cunningham Films, Ltd., 1971. Produced commercials, industrial film, documentaries, features.
PICTURES: Together (prod.-dir.), Last House on the Left (prod.); The Case of the Full Moon Murders (prod.); Here Come the Tigers (prod.-dir.); Kick (prod.-dir.); Friday the 13th (prod.-dir.).

CURTIN, JOHN P.: Distributor. b. Pittsfield, Mass., Nov. 24, 1915; e. Williams Coll., 1938; B.A. Columbia U. Graduate School, 1939. Eastern div. man., Republic; v.p., Color Service Corp.; sls. mgr. Guffanti Film Lab., Inc.
MEMBER: Montclair Golf Club, Alpha Delta Phi Fraternity, Dir., Seaboard Corp.

CURTIS, DAN: Producer-Director. b. Bridgeport, Conn., Aug. 12. e. Univ. of Bridgeport; Syracuse Univ. (BA, 1950). Was sales exec. for NBC and MCA before forming own company, Dan Curitis Productions, which he now heads. Producer/owner of CBS Golf Classic (1963–73).
PICTURES INCLUDE: House of Dark Shadows, Night of Dark Shadows (features for MGM, prod.-dir.), Burnt Offerings (prod., dir., co-s.p.).
TV: Producer: Dark Shadows (ABC serial, 1966–71), and following movies: The Night Stalker, Frankenstein, The Picture of Dorian Gray. Producer-Director of movies: The Night Strangler, The Norliss Tapes, The Turn of the Screw, Dracula, The Scream of the Wolf, Purvis.

CURTIS, KEN: Actor. b. Lamar, Colo., July 2, 1916. e. Colorado Coll. Success as songwriter for college prod. steered him to musical career in Hollywood. Sang with several groups, including Sons of the Pioneers, Tommy Dorsey Orchestra, Shep Fields band. In infantry and anti-aircraft unit in World War II. Signed by Columbia Pictures for series of westerns with "Big Boy" Williams in 1945.
PICTURES INCLUDE: The Searchers, The Alamo, The Quiet Man, Cheyenne Autumn, How the West Was Won, The Killer Shrews, The Giant Gila Monster, My Dog, Buddy.
TV: Ripcord, Perry Mason, Rawhide, Have Gun, Will Travel, Gunsmoke (joined later series as Festus Haggin in 1963).

CURTIS, TONY: Actor. r.n. Bernard Schwartz. b. New York, N.Y., June 3, 1925; e. Seward Park H. Sch. In U.S. Navy, amateur dramatics, NYC, started Empire Players Theatre, Newark, N.J., with Dramatic Workshop, Cherry Lane Theatre, Jr. Drama Workshop of Walt Whitman School; first prod. work with Stanley Woolf Players; m.p. debut in Criss-Cross; signed with U-I Star of Tomorrow, 1953.
PICTURES INCLUDE: City Across the River, Johnny Stool Pigeon, Francis, Sierra, I Was a Shoplifter, Winchester 73, Kansas Raiders, Prince Who Was a Thief, Flesh and Fury, Son of Ali Baba, No Room for the Groom, Houdini, All American, Forbidden, Beachhead, Johnny Dark, Black Shield of Falworth, 6 Bridges to Cross, So This is Paris, Purple Mask, Square Jungle, Rawhide Years, Trapeze, Mister Cory, Midnight Story, Sweet Smell of Success, Some Like It Hot, Spartacus, Operation Petticoat, Who Was That Lady?, The Rat Race, The Defiant Ones, Perfect Furlough, The Great Impostor, The Outsider, Taras Bulba, 40 Pounds of Trouble, Paris When It Sizzles, The List of Adrian Messenger, Captain Newman, M.D., Wild and Wonderful, Sex and the Single Girl, Goodbye, Charlie, The Great Race, Boeing-Boeing, Arrivederci, Baby!, Not With My Wife, You Don't!, Don't Make Waves, On My Way to The Crusades I Met a Girl Who—; The Boston Strangler; Those Daring Young Men in Their Jaunty Jalopies, Suppose They Gave a War and Nobody Came; You Can't Win 'Em All, Lepke, The Last Tycoon, The Manitou, Bad News Bears Go To Japan, The Mirror Crack'd.
TV: The Persuaders, The Second Girl on the Right (movie), Vegas (series).

CUSACK, CYRIL: Actor. b. Durban, S. Africa, Nov. 26, 1910. e. Newbridge, Co. Kildare; University College, Dublin, Eire. LL.D (Honoris Causa-National University of Ireland). Stage debut: Candida, Abbey Theatre, 1932. Screen debut: Odd Man Out, 1945.
PICTURES INCLUDE: Esther Waters, Escape, The Blue Lagoon, Once a Jolly Swagman, All Over the Town, Small Back Room, The Elusive Pimpernel, Soldiers Three, Blue Veil, Secret of Convict Lake, Gone to Earth (Wild Heart), Saadia, Passage Home, Man in the Road, Man Who Never Was, March Hare, Jacqueline, Spanish Gardner, Ill Met By Moonlight, Rising of the Moon, Miracle in Soho, Shake Hands with the Devil, Floods of Fear, Gideon's Day, A Terrible Beauty, Johnny Nobody, The Waltz of the Toreadors, I Thank a Fool, 80,000 Suspects, Passport to Oblivion, The Spy Who Came In From The Cold, Fahrenheit 451, Taming of the Shrew, I Was Happy Here, Oedipus Rex, Galileo, King Lear, Country Dance, David Copperfield, Harold and Maude, Sacco Vanzetti, La La Polizia Ringrazia, The Day of the Jackal, Juggernaut, Homecoming, Gallileo, Tristan and Iseult, True Confessions.
TV: The Dumms, The Moon and Sixpence, What Every Woman Knows, The Enchanted, The Power and The Glory, The Chairs, Don Juan in Hell, The Lotus Eater, Krapp's Last Tape, Murder in the Cathedral, Six Characters in Search of An Author, The Big Toe, Workhouse Ward, In the Train, Purgatory, The Moon in the Yellow River, Passage to India, Deirdre, The Tower, Dial M for Murder, St. Francis, The Physicists, Trial of Marshal Petain, In the Bosom of the Country, Uncle Vanya, A Time of Wolves and Tigers, Them, Clochemerle, The Golden Bowl, The Reunion, I Stand Well With All Parties, Catholics, Crystal & Fox, Jesus of Nazareth, The Plough and The Stars, You Never Can Tell, The Hitchhiker.

CUSHING, PETER: Actor. b. Kenley, Surrey, Eng., May 26, 1913. e. Purley Secondary Sch. Stage debut with Worthington Repertory Co.
TV: Asmodee, Anastasia, 1984, Gaslight, Home at Seven, Tovarich, Beau Brummell, Epitaph for a Spy, Pride and Prejudice, The Moment of Truth, Uncle Harry, Eden End, Rookery Nook, The Creature, The Browning Version, Winslow Boy, Peace With Terror, Julius Caesar (Cassius), Monica. Daily Mail TV award actor, 1953–54; Guild of TV award, 1955; News Chronicle T.V. Top Ten award, 1956, The Plan, Caves of Steel, Sherlock Holmes series, 1968, Morecambe & Wise Show, Wild-life Spectacular, The Zoo Gang, Orson Welles Great Mysteries, Space 1999, The New Avengers, The Great Houdini, A Land Looking West.
PICTURES INCLUDE: Vigil in the Night, Moulin Rouge, Hamlet, Black Knight, End of the Affair, Alexander the Great, Magic Fire, Time Without Pity, Curse of Frankenstein,

Abominable Snowman, Dracula, John Paul Jones, The Hound of the Baskervilles, Violent Playground, The Mummy, Suspect, The Flesh and the Friends, The Revenge of Frankenstein, Cone of Silence, Brides of Dracula, Sword of Sherwood Forest, The Naked Edge, Cash on Demand, The Devil's Agent, Captain Clegg, Fury at Smuggler's Bay, Hell-Fire Club, The Man Who Finally Died, The Evil of Frankenstein, The Gorgon, Dr. Terror's House of Horrors, She, The Skull, Dr. Who and the Daleks, The Frightened Island, Daleks Invade Earth, Frankenstein Created Woman, Torture Garden, Some May Live, The Night of the Big Heat, Corruption, Death's Head Moth, Frankenstein Must Be Destroyed, Doctors Wear Scarlet, The Vampire Lovers, Scream and Scream Again, House That Dripped Blood, I Monster, Twins of Evil, Tales from the Crypt, Dracula Today, Fear in the Night, Horror Express, The Creeping Flesh, Asylum, Nothing But the Night, Bride of Fengriffen, Frankenstein and the Monster from Hell, The Satanic Rites of Dracula, The Revenge of Dr. Death, From Beyond the Grave, The Beast Must Die, Dracula and the Legend of the Seven Golden Vampires, Shatter, Tender Dracula, The Ghoul, Legend of the Werewolf, The Devil's People, Death Corps, Trial by Combat, At The Earth's Core, Star Wars, Battleflag, The Uncanny, Hitler's Son, Touch of the Sun, Arabian Adventure.

D

DAFF, ALFRED EDWARD: Executive. b. Melbourne, Australia, Aug. 18, 1902. e. McDonald's Private Coll., Melbourne. Office boy Progressive Films, night asst. projectionist Moonee Ponds Theatre 6 mos., then in chg. of dispatch, adv. & accessory depts., booker & in chg. of shipping, repairing, adv. & accessory depts. Cooperative Films; joined Universal Mar., 1920 booker & suburban salesman 2 yrs., asst. mgr., then mgr. of Victoria & Tasmania; apptd. managing dir. of Japanese subsidiaries, 1935; Far Eastern supvr., 1938; also of Far and Middle East, 1940; to N.Y.C., 1942; apptd. foreign supvr. 1943; vice-pres. Universal Internatl Films, Inc., 1944 pres. same co., 1950; v.p. dir. world sales, Universal Pictures Co., since 1951. Named dir. Universal, May, 1952 exec. v.p., July, 1952; res. 1958; consultant motion pictures, TV.

DAHL, ARLENE: b. Aug. 11, 1928, Mpls., Minn., USA Actress-writer; Designer. e. Mpls. Bus. Coll.; Univ. of Minn., summers 1941–44; Mpls. Coll. of Music. m. (1) Fernando Lamas, 1954, 1 s, Lorenzo Lamas b. 1958, (2) Christian R. Holmes III, 1960, 1 d, Carole Christine b. 1961 (3) Rounsevelle W. Schaum, 1969, 1 s. Sonny b. 1970. Acting career begun at age 8, playing heroine of children's adventure serials on radio, Broadway stage debut in Mr. Strauss Goes to Boston, 1946; other Broadway starring roles include: Cyrano de Bergerac; Applause; other starring roles onstage in major US cities include: The King and I, One Touch of Venus, Pal Joey, Bell Book and Candle, The Camel Bell, Lilliom, Marriage Go Round, Blythe Spirit; 28 film appearances include: debut in My Wild Irish Rose (Warner Bros., (1947), Three Little Words, Desert Legion, Here Come the Girls, Sangaree, Woman's World, Journey to the Center of the Earth, Kisses for My President. Num. TV appearances include: Series of 65 films for ABC-TV—Arlene Dahl's Beauty Spot, 1965; Hostess, CBS Model of the Year Show, 1969, Arlene Dahl's Starscope, 1980. Appts: Internationally syndicated beauty columnist, Chgo. Tribune-N.Y. News Syndicate, 1951–71; Sleepwear Designer, A.N. Saab & Co., 1952–57; Nat. Beauty Advisor, Sears Roebuck & Co., 1970–75; v.p. Kenyon & Eckhart, pres., Women's World Div., Kenyon-Eckhart, 1967–72, Fashion Consultant, O.M.A. 1975–1978, Int'l. Director of S.M.E.I, 1973–76, Pres., Dahlia Parfums Inc., 1975, pres., Dahlia Prods., 1978–, Publs: Always Ask a Man, 1965 (6th ed.), Your Beautyscope (series of 12), 1969, Secrets of Hair Care, 1971, Your Beautyscope (series of 12) 1977–1978. Profl. Assns. include: Screen Actors Guild, Actors Equity, Am. Fedn. of TV & Radio Artists, International Platform, Inc., Acad. of Motion Picture Arts and Sciences. Honrs. include: 8 Motion Picture Laurel Awards, 1948–63; Hds. of Fame Award, 1971, Woman of Year, N.Y. Adv. Council 1969.

DALES, L. JOHN: Executive. b. Santa Monica, Calif., Feb. 24, 1907; e. Stanford U., B.A., 1929; J.D., 1932. Practiced law 1932–37; v.p. Calif. Labor Federation, Nat'l exec. sec. Screen Actor's Guild, 1937–1972. Retired.

DALEY, ROBERT: Producer. Began career in TV; later joined Clint Eastwood's Maplasco Co.
PICTURES INCLUDE: Dirty Harry, Magnum Force (both exec. prod.), Thunderbolt and Lightfoot (prod.), The Enforcer (prod.), The Gauntlet (prod.), Every Which Way But Loose (prod.), Escape from Alcatrez (exec. prod.)., Any Which Way You Can (exec. prod.).
TV: The Untouchables, Ben Casey, The FBI, 12 O'Clock High, The Invaders, etc.

DALY, JOHN: Chm., Hemdale Group Ltd. b. London, England, 1937. After working in journalism joined Royal Navy. On leaving Service after three years, trained as underwriter with an Assurance Company. In 1966 became David Hemmings manager and in 1967 formed the Hemdale Company. Subsequently, in 1968, Hemdale obtained public status and produced features including The Passage, Sunburn, Cattle Annie and Little Britches, High Risk, Going Ape and Carbon Copy (exec. prod.).

DAMON, MARK: Executive. b. Chicago, April 22, 1933. e. University of California at Los Angeles, B.A. Literature, M.A. Business Administration. Actor: 1958 under contract to 20th Century Fox. 1960 winner Golden Globe Award—Newcomer of the Year; early career includes The Fall of The House of Usher, The Longest Day; 1961 moved to Italy, stayed 16 years appearing in leading roles in 50 films; 1973 co-producer of The Arena with Roger Corman; 1974 head of foreign department for PAC, a leading film distributor in Italy; 1976 returned to the U.S.A. as executive producer of The Choirboys and in charge of its foreign distribution; 1977 founder and pres. of Producers Sales Organization, international distribution organization.

DAMONE, VIC: Singer, actor. r.n. Vito Farinola. b. Brooklyn, N.Y., June 12. e. Lafayette H Sch., B'klyn. Winner Arthur Godfrey talent show, 1945; then night clubs, radio, theatres, hotels. m.p. debut in Rich, Young and Pretty; U.S. Army, 1951–53.
PICTURES INCLUDE: The Strip, Athena, Deep in My Heart, Hit the Deck, Kismet, Hell to Eternity.
TV: Vic Damone Show, 1958. Lively Ones, 1962.

D'ANGELO, BEVERLY: Actress. b. Columbia, Ohio. Studied visual arts and was exchange student in Italy before working as cartoonist for Hanna-Barbera Studios in Hollywood. Toured Canada's coffeehouse circuit as singer and appeared with rock band called Elephant. Joined Charlotte Town Festival Company. Bdwy. debut in rock musical, Rockabye Hamlet. Film debut in The Sentinel.
PICTURES: Annie Hall, First Love, Every Which Way But Loose, Hair, Coal Miner's Daughter.

DANHEISER, MELVIN B.: Executive. Held positions at RKO Radio Pictures and National Telefilm Associates before joining United Artists in 196[illegible] in intl. div.'s sls. dept. Since served in various capacities—asst. intl. sls. mgr. for Europe, Middle East, and English-speaking mkts. overseas. 1978 promoted to exec. asst. to foreign mgr. Appt. v.p. of UA in 1979; 1981, named exec. asst. to Norbert T. Auerbach, pres. & chief operating officer for UA.

DANIEL, SEAN: Executive. b. 1952. e. California Inst. of Arts film school. Was journalist for Village Voice before starting m.p. career as documentary filmmaker and asst. dir. for New World Pictures. In 1976 joined Universal Pictures as prod. exec.; 1979, named v.p.

DANIELEWSKI, TAD: Director-Writer. b. Poland. Studied Royal Academy of Dramatic Art, London; Ohio University; State Univ. of Iowa. At Johns Hopkins Univ. won national award for research in prod. and dir. methods for TV. Produced and directed a series of dramatic shows on WAAM-TV in cooperation with NBC-TV, 1953–54.
PICTURES INCLUDE: Imperial Woman (co-writer), No Exit (dir.), The Guide (writer-dir.), Spain (writer-dir.), Copernicus (writer-dir. of English version).
TELEVISION: Omnibus, Wide Wide World, Eddie Fisher Show, Matinee Theatre, Robert Montgomery Presents, etc.
STAGE: Artistic director New York Repertory Theatre; directed Man with a Load of Mischief (London), A Desert Incident (Bdwy.) Brouhaha (off-Bdwy.), etc.

DANIELS, HAROLD: Director, producer, writer, b. Buffalo, N.Y. e. Carnegie Tech Drama Dept. B.A., U. of Pittsburgh, PHG.
STAGE: Director in Pittsburgh for Prof. Stage Guild Co., N.Y., Rhode Island and Boston Repertoire. Directed over 50 plays.
PICTURES INCLUDE: Joined MGM in 1940, directed shorts and won award from M.P. Council; joined David Selznick as director 1943–45; produced, directed and wrote Prince of Peace, won spec. award for dir., directed Woman from Tangier, Sword of Venus, Port Sinister, Roadblock, Daughter of the West Classics, Terror in the Haunted House, Date With Death, My World Dies Screaming, Bayou, Poor White Trash, Ten Girls Ago; directed, Night of the Beast, House of Black Death, Annabelle Lee, 1968–69, Moonfire, 1969–70. Pigmy, 1971.
TV: directed over 200 half-hour and hour films including My Hero, Readers Digest, Fury, Colt 45, Ellery Queen, Jim Backus Theatre G.E. Theatre, etc. Wrote many original screenplays for both films and TV. The Phantom, On Guard, Death Valley Days, Hannibal Cobb.

DANIELS, WILLIAM: Actor. b. Brooklyn. e. Northwestern University. Stage debut in Life with Father. Brought to national

attention by role in A Thousand Clowns in original Bdwy. play and film version.
PICTURES: Marlowe, The Parallax View, Two for the Road, Black Sunday, Oh God, Sunburn, The Blue Lagoon.
TELEVISION: The Adams Chronicle, A Case of Rape, Blind Ambition, The Nancy Walker Show, Captain Nice.
STAGE: On a Clear Day You Can See Forever, 1776, Dear Me, the Sky Is Falling, A Little Night Music, etc.

DANIELSON, DALE H.: Exhibitor, Operator, Dream, Sky-Vu Drive-In theatres, Russell, Kan.; past pres., Kansas-Missouri Theatre Assn.

DANISCHEWSKY, J.: Producer-Director. b. London, England, 1940. Chelsea Art School, 1957. Ent. m.p. ind. 1959 as asst. dir. 1967; Prod. dir. Avalanche, 1968. Prod., Run Wild, Run Free.

DANNER, BLYTHE: Actress. b. Philadelphia. e. Bard College. m. writer-producer Bruce Paltrow. Appeared in repertory cos. in U.S. before Lincoln Center (N.Y.) productions of Cyrano de Bergerac, Summertree, and The Miser (Theatre World Award for last).
PICTURES INCLUDE: 1776, To Kill a Clown, Lovin' Molly, Hearts of the West, Futureworld, The Ace.
TV: Dr. Cook's Garden, To Confuse the Angel, George M, To Be Young, Gifted and Black, The Scarecrow, Adam's Rib (series), F. Scott Fitzgerald and The Last of the Belles, Eccentricities of a Nightingale.

DANSON, HAROLD L.: Executive. b. New York, N.Y., Nov. 15, 1905. e. Brooklyn Polytechnic Institute, 1926. Feature ed., Brooklyn Eagle, 1927; press rep., Shubert Theatrical Co., asst. city ed., Newark Star-Eagle, 1928; dir. of adv. pub., RCA Photophone, 1929–30; asst. nat'l adv. mgr., RKO, 1930–32; exploit. mgr., studio adv. dir., asst. nat'l adv. dir., head of trailer prod., Paramount, 1932–44; account exec., 20th Century-Fox, 1944–46; nat. adv. dir., Eagle Lion, 1946–49; v.p., gen. sales mgr., Adler Communications Labs., 1951–54; dir. of adv., pub., TV National Screen Service Corp., 1955; acct. exec. ZIV TV Programs, 1956; Nat'l Telefilm Assoc. program sales, 1959; nat'l sales mgr., syndication Independent Television Corp., 1961–68; consultant 1969.

DANTINE, HELMUT: Actor, Director, Executive b. Vienna, e. Europe; UCLA, Calif. Member: Pasadena Community Players, then on screen in International Squadron, 1941. V.p. Joseph M. Schenck Ent. 1959; pres., Hand Enterprises, Inc., 1963; pres., N.M. Schenck Enterprises, Inc., 1970.
PICTURES INCLUDE: Mrs. Miniver, Casablanca, Edge of Darkness, Northern Pursuit, Mission to Moscow, Passage to Marseille, Hotel Berlin, Escape in the Desert, Shadow of a Woman, Whispering City, Guerilla Girl, Call Me Madam, Alexander the Great, War and Peace, The Story of Mankind, Hell on Devil's Island, Fraulein, Tempest, Operation Crossbow, Dir. Thundering Jets; 1974: Bring Me the Head of Alfredo Garcia (exec. prod., actor); 1975; The Wilby Conspiracy (exec. prod., actor); 1975: The Killer Elite (exec. prod., actor); Tarzan (exec. prod.).

DANTON, RAY: Actor. b. New York, N.Y., Sept. 19, 1931. e. Horace Mann School, Carnegie Tech. m. Julie Adams. Radio actor on many programs; summer stock; London prod. of Mr. Roberts; U.S. Army, 1951–53; then TV actor numerous programs. Became dir. in 1976 with Psychic Killer (also co-s.p.).
PICTURES INCLUDE: Chief Crazy Horse, The Looters, The Spoilers, I'll Cry Tomorrow, Outside the Law, The Night Runner, Onionhead, Too Much Too Soon, Ice Palace, Legs Diamond, Tarawa Beachhead, Majority of One, The George Raft Story, Fever In The Blood, Portrait of a Mobster, The Chapman Report, The Longest Day, Sandokan, FBI Code 98, New York Calling Superdrago.

D'ANTONI, PHILIP: Producer-Director. b. NYC, Feb. 19, 1929. e. Fordham Univ. Has degree in business admn. Joined CBS in mailroom, advanced into prod., sales development, prog. analysis, mkt. rsrch. Became indep. radio-TV repr. in 1954 for two years; then joined Mutual Broadcasting as sales manager; later, exec. v.p. Resigned in 1962 to form own prod. co. Made theatrical film debut with Bullitt as producer; directing debut with The Seven Ups. Heads Phil D'Antoni Prods.
PICTURES INCLUDE: Producer: Bullitt, The French Connection. Prod.-Dir.: The Seven Ups.
TELEVISION: Elizabeth Taylor in London, Sophia Loren in Rome, Melina Mercouri in Greece, Jack Jones Special, This Proud Land, and two movies: Mr. Inside/Mr. Outside, The Connection.

DANZ, FREDERIC A.: Executive. b. Seattle, Wash., Feb. 28, 1918. Is president of Sterling Recreation Organization Co., Seattle; member, Foundation of M.P. Pioneers; v.p., Variety Club Intl.

DANZA, TONY: Actor. b. Brooklyn, NY, Apr. 21. e. Univ. of Dubuque. Professional boxer before tested for role in TV pilot (Fast Lane Blues) which he won. Back to New York and fighting until called to coast to appear as Tony Banta in Taxi series.
PICTURES: Hollywood Knights.

DARBY, KIM: Actress. r.n. Derby Zerby. b. July 8, 1948, Hollywood, Calif. e. Swanson's Ranch School, Van Nuys High School. Studied at the Desilu Workshop in Hollywood. Professional debut on the Mr. Novak TV series.
TV: Eleventh Hour, Gunsmoke, Flesh and Blood (special).
PICTURES INCLUDE: Bus Riley's Back in Town, True Grit, Generation, Norwood, The Strawberry Statement, The Grissom Gang, The One and Only

DARK, JOHN: Producer. Films include, as associate: Light Up the Sky, Lost Innocence, The 7th Dawn, Casino Royale. Production exec., Paramount Pictures, exec. producer: Half a Sixpence, There's a Girl in My Soup. Produced: Wind of Change, Bachelor of Arts, Land That Time Forgot, At the Earth's Core, People That Time Forgot, Warlords of Atlantis, Arabian Adventure.

DARLEY, DICK: Director, Producer network TV series and specials. Over 180 TV film shows; over 1,370 TV live/tape shows. Numerous pilots and commercials. Credits in U.S.A. and 27 foreign countries include drama, musical-variety, comedy, sports and documentary.

DARNBOROUGH, ANTONY: Producer, director. b. London, Eng., ent. m.p. ind. 1944.
PICTURES INCLUDE: Seventh Veil, Years Between, Daybreak, Girl in a Million, Upturned Glass, When the Bough Breaks, Dear Murderer, The Calendar, My Brother's Keeper, The Girl in the Painting, Helter Skelter, Once Upon a Dream, Traveller's Joy, Boys in Brown, The Astonished Heart, So Long at the Fair, Quartet, Trio, Highly Dangerous, Encore, The Net (Project M. 7), Personal Affair, To Paris With Love, Baby and the Battleship.
TV: Also made many TV and documentary films for own co. and in assoc. with NSS.

DARREN, JAMES: Actor. b. Philadelphia, Pa., June 8, 1936. Studied with Stella Adler group. Hollywood debut: Rumble on the Docks.
PICTURES INCLUDE: The Brothers Rico, Guns of Navarone, Gidget Goes Hawaiian, The Lively Set, 633 Squadron, Venus in Furs.

DARRIEUX, DANIELLE: Actress. b. Bordeaux, France, May 1, 1917. e. Lycee LaTour; Conservatoire de Musique.
PICTURES INCLUDE: La Crise Est Finis, Mayerling, Club des Femmes, Abus de Confiance, Counsel for Romance, Mademoiselle Ma Mere, Katia, Orders from Tokyo, Oh Amelia, Rage of Paris, Rich Young and Pretty, 5 Fingers, La Ronde, Le Plaisir, Earrings of Madame De, Alexander the Great, Adorable Creatures, A Friend of the Family, Loss of Innocence.
THEATRE: Coco.

DARTIGUE, JOHN: Executive. b. Port-au-Prince, Haiti, Sept. 12, 1940. e. Brandeis Univ. (B.A., 1961), Columbia Univ. (M.A., 1965). United Artists: named director of worldwide pub. in 1975; appointed executive assistant to the senior vice president for advertising and publicity in 1977; named v.p. adv.-pub., 1978. Joined Warner Bros. in 1978 as project executive. Named sr. exec. asst. to the exec. v.p. of worldwide adv./pub., Jan., 1980; v.p. & exec. asst., July 1980.

DARTNALL, GARY: Exec. b. Whitchurch, Eng., May 9, 1937. e. Kings Coll., Taunton. Overseas Div., Associated British Pathe; Eur. rep., 1958–60; Middle & Far East rep., Lion International Films; Amer. rep., 1962; pres., Lion International Inc., 1963; Amer. rep., Alliance International Films Distributors, Ltd., and London Independent Producers, Ltd.; pres., Alliance International Films Corp. & Dartnall Films Ltd., 1966; managing dir., Overseas Division, Walter Reade Organization, 1969. pres., EMI Film Distributors, Inc., 1971; vice chairman, EMI Television Programs, Inc., 1976; pres., EMI Videograms, Inc., 1979; pres., VHD Programs, Inc. & VHD Disc Mfg. Co., 1980.

DA SILVA, HOWARD: Actor. b. Cleveland, Ohio, May 4, 1909; e. Carnegie Inst. Tech. With Civic Repertory Theatre, Theatre Union, Group Theatre & Mercury Theatre.
PLAYS: Master Builder, Alice in Wonderland, Three Sisters, Alison's House, Liliom, Doll's House, Sailors of Cattaro, Waiting for Lefty, Golden Boy, Cradle Will Rock, Abe Lincoln in Illinois, Oklahoma, 1776; m. p. debut in "I'm Still Alive," 1940.
PICTURES INCLUDE: Abe Lincoln in Illinois, Sea Wolf, Nine Lives Are Not Enough, Big Shot, Omaha Trail, Tonight We Raid Calais, Lost Weekend, Duffy's Tavern, Unconquered, Blaze of Noon, They Live By Night. Tripoli, Underworld Story, Three Husbands, Wyoming Mail, Fourteen Hours, M. David and Lisa, 1776, The Great Gatsby.
BROADWAY STAGE: 1776.

DASILVA, RAUL: Creative consultant, writer, director, producer, b. New York City. June 12, 1933. e. Adelphi U., B.A. 1958, elected to Academy of Distinction Adelphi Alumni Assoc. 1978. Specializes in script evaluation, rework of script, art and graphic look. Exp. Corporate AV dir., exec. prod. adv.

agencies; indep. writer, producer, dir., univ. lecturer (business aspects of film prod. writing).
PUBLICATIONS: The World of Animation (Eastman Kodak Co. Rochester, 1979); The Business of Filmmaking (Eastman Kodak Co., Rochester, 1978); SOUND; Magnetic recording for Motion Pictures (Eastman Kodak Co., Rochester, 1977); Wrote articles for American Cinematographer, Filmmakers Monthly, Today's Film Maker Millimeter. ProFilm; Filmmaker's Monthly.
PICTURES: Fear No Evil (creative consultant)
TV: Nat Hurst, M.D.; The Strangest Voyage (dir., prod)
AWARDS: Numerous national & international for films, 6 in 1975–76 for The Strangest Voyage, (Coleridge and his masterpiece). The Ancient Mariner with Sir Michael Redgrave.

DA SILVA, RAY: Director and designer of animated films, animator, illustrator, character designer. b. New York City, July 13, 1934. e. School of Visual Arts, NYC, also instructor there. Specializes in animation direction, character design. Numerous national and international TV spots for the advertising industry.
PICTURES: Raggedy Ann & Andy; Heavy Metal.
TELEVISION: The Strangest Voyage, Noah's Animals, The Little Brown Burro, Ichabod Crane.

DASSIN, JULES: Director. b. Middletown, Conn. Actor on dramatic stage several years; radio writer. Joined MGM, 1940, as dir. short subjects; later dir. features.
PICTURES INCLUDE: Canterville Ghost, Brute Force, Naked City, Thieves' Highway, Night and the City, Rififi, He Who Must Die, Never On Sunday, Phaedra, 10:30 p.m. Summer, Promise At Dawn, Uptight, A Dream of Passion.
PLAYS: Ilya, Darling, 1967.

DAUPHIN, CLAUDE: Actor. b. Corbeil, France, Aug. 19. e. Lycee de Grand, Paris. In French & Allied armies, 1940–45; org. French stock co.; screen debut in Deported; B'way stage debut in Happy Time.
PICTURES INCLUDE: Casque D'or, April in Paris, L'enentaile (Naughty Martine), Little Boy Lost, Phantom of the Rue Morgue, Le Plaisir, Innocents in Paris, Two for the Road, Hard Contract, Madwoman of Chaillot, Rosebud, Madame Rosa.

DAVEE, LAWRENCE W.: Engineer. b. Foxcroft, Me., March 28, 1900; e. U. Me., B.S. (elec. eng.). Research eng. Bell Telephone Lab.; Fox Case Corp.; studio mgr. Fox Hearst Corp.; Bronx Studio. Elec. Research Prods., Inc.; Century Projector Corp., N.Y.; engineer & sales mgr.; pres. 1959. Member: 25–30 Club (Honorary). Lifemember, N.Y. State Projectionists.

DAVENPORT, NIGEL: Actor. b. Cambridge, England, May 23, 1928. e. Trinity Coll., Oxford. Began acting after stint in British military at 18 years. First 10 years of professional career in theatre. Majority of screen work in British films in 1950s and 60s.
PICTURES INCLUDE: Peeping Tom, A High Wind in Jamaica, A Man for All Seasons, Royal Hunt of the Sun, No Blade of Grass, Villain, Mary, Queen of Scots, Island of Dr. Moreau, Zulu Dawn, Nighthawks.

DAVID, SAUL: Producer. b. Springfield, Mass., June 27, 1921. e. Classical High School, Springfield, Mass., Rhode Island School of Design. Started in radio, newspaper work and as editorial director for Bantam Books. Worked for Columbia Pictures, 1960–62; Warner Bros., 1962–63; 20th Century-Fox, 1963–67, and Universal, 1968–69; Executive story editor at MGM, 1972.
PICTURES INCLUDE: Produced Von Ryan's Express, Our Man Flint, Fantastic Voyage, In Like Flint, Skullduggery, Logan's Run.

DAVIDSON, JOHN: Actor-Singer. b. Pittsburgh, Penna., Dec. 13, 1941. e. Denison Univ. In numerous school stage prods. before coming to N.Y. where won co-star role with Bert Lahr in Bdwy. show, Foxy. Signed as regular on The Entertainers with Carol Burnett.
PICTURES INCLUDE: The Happiest Millionaire, The Concorde—Airport '79.
TELEVISION: The Fantasticks, USA, The FBI, The Interns, Owen Marshall, Kraft Summer Music Hall (own variety series), The Tonight Show, The Girl with Something Extra, The John Davidson Talk Show (1980).

DAVIDSON, MARTIN: Director. Acted in and directed off-Bdwy. plays, wrote material for Sid Caesar, worked as agent with N.Y. offices of Int'l. Famous Artists.
PICTURES: The Lords of Flatbush, Almost Summer, Hero at Large.

DAVIES, JACK: Screen writer. b. London 1913. Ent. m.p. ind. 1932 as screen writer with B.I.P. Collab. number of films. Also coll. on Convict 99 before joining Daily Sketch as film critic 1938. Served with R.A.F. 1940–5. Became columnist Daily Graphic and film critic Sunday Graphic 1948. Wrote orig. story and s.p. Trouble in the Air, Helter-Skelter.
PICTURES INCLUDE: Dance Band, Heart's Desire, Radio Parade, One in a Million, Mimi, The Tenth Man. Laughter in Paradise, Top Secret, Curtain Up, Happy Ever After, Doctor at Sea, An Alligator Named Daisy, High Flight, Up in the World, True as a Turtle, The Square Peg, Don't Panic Chaps, Follow a Star, It Started in Naples, The Bulldog Breed, Very Important Person, Crooks Anonymous, On The Beat, The Fast Lady, A Stitch In Time, Father Come Too, Those Magnificent Men In Their Flying Machines, The Early Bird, Doctor in Clover. Gambit, Doctor in Trouble, Monte Carlo or Bust, Paper Tiger, Le Hold-Up, ffolkes.

DAVIES, JOHN W. (O.B.E.): Executive. b. London, Eng., 1908. Originally shipbroker; entered m.p. ind., 1931; man. dir., Davies Cinemas, Ltd. and British Cinematograph Theatres, Ltd.; R.A.F. 1939–45, awarded U.S. Legion of Merit; chmn. London & Home Counties Branch, C.E.A. 1948; apptd. mem., Cinematograph Films Council, Dept. of Trade, 1951–76; v.p. C.E.A., 1952; pres. 1953; dir., FIDO, 1962.

DAVIES, PAUL: Executive. b. Toronto, Canada, M.A. Columbia Univ. Service with RCAF. Prod. Your Man Everywhere, many documentaries; Isle of Rainbow, Kansanshe, African Heritage, etc. Originator and pilot prod. Jazz in Europe, Man About the House. Manager of Malachite Films, NBC reporter and cameraman, bureau chief for Visnews, Israel 1980: Jerusalem Capital Studios.

DAVIES, WILLIAM C.: Director. b. Auburn, N.Y., June 17, 1932. e. Auburn Community Coll., School of Radio Technique. Radio & T.V. announcer & narrator, then cameraman & editor. Formed Virgo Productions, 1965.
PICTURES INCLUDE: Legend of Horror, Orgy of the Dead, The Seekers, Day the Adults Died, Pink Garter Gang, Night at the Feast, Laughter in Hell, The Crimson Cult, Submarine X-1.

DAVIS, ALFRED: Exhibitor. b. Brighton, England, Aug. 19, 1899. e. University College Sch., London. In Brit. Army 1914 before entering m.p. ind. as dir. & booking mgr. Davis Pavillion circuit; later same position GaumontBritish, New Era Pictures. Chr. London & Home Counties Branch, C.E.A., 1937, 1953, 1955. Dep. asst. dir., War Office Army Kinematograph, World War II; v.p. CEA; 1959–60; pres. CEA, 1960–61.

DAVIS, ARTHUR: Executive. b. New York, N.Y., Dec. 28, 1927. e. N.Y. Univ. Began as exhibitor in U.S.; founded Arthur Davis Organization in Tokyo, Japan, in 1963, representing European and American films for sales to Japanese TV and motion picture industry. 1979, released two documentaries, Brutes and Savages, The Art of Killing.

DAVIS, BETTE: Actress. b. Lowell, Mass., April 5, 1908; e. Cushing Academy; Mariarden School of Dancing; John Murray Anderson Dram. School. Stage debut in "Broadway" (Rochester, N.Y.); with Provincetown & Cape Players; on Broadway in: Broken Dishes, Solid South; m.p. debut in "Man Who played God" 1932; won Academy Award twice: best actress 1935, in Dangerous and 1938, in Jezebel; voted one of the 10 best Money-Making Stars in Motion Picture Herald-Fame Poll 1939; 40, 41, 44; founder & pres. Hollywood Canteen 1942; pres. AMPAS 1940. Received American Film Institute Life Achievement Award, 1977; 1979, Emmy for performance in Strangers.
PICTURES INCLUDE: Of Human Bondage, Fog Over Frisco, Sisters, Dark Victory, Juarez, Old Maid, Private Lives of Elizabeth and Essex, All This and Heaven Too, Letter, Now Voyager, Corn Is Green, Little Foxes, Deception, Winter Meeting, N.Y. stage, Night of the Iguana, Sandburg, June Bride, Beyond the Forest, All About Eve, Payment on Demand, Another Man's Poison, Phone Call From a Stranger, The Star, Virgin Queen, Catered Affair, Storm Center, John Paul Jones, The Scapegoat, A Pocketful of Miracles, What Ever Happened To Baby Jane, Dead Ringer, Empty Canvas, Where Love Has Gone, Hush . . . Hush Sweet Charlotte, The Nanny, The Anniversary, Connecting Rooms, Bunny O'Hare, Madame Sin (TV in U.S.), Scientific Card Game, Burnt Offerings, Return from Witch Mountain, Death on the Nile, Watcher in the Woods.
TV: The Disappearance of Aimee, Laugh-In, The Dark Secret of Harvest Home, Strangers, The Story of a Mother and Daughter, White Mama, Skyward, Family Reunion.

DAVIS, BRAD: Actor. b. Florida. Won music talent contest at 17; moved to Atlanta and acted in theatres before settling in New York, N.Y. stage debut in Crystal and Fox; did several off-Bdwy plays. Appeared in TV soap opera, How to Survive a Marriage, for 10 months.
PICTURES: Midnight Express, A Small Circle of Friends.
TELEVISION: Walt Whitman, Sybil, Roots, A Rumor of War, The Greatest Man in the World.

DAVIS, FRANK I.: Executive. b. Poolesville, Md. Feb. 18, 1919. e. U. of Md., AB, 1941; Harvard Law School, LL.B., 1948; law firm, Donovan, Leisure Newton, Lombard and Irvine, 1948–50; v.p., gen. counsel, Vanguard Films, 1951; v.p., gen. counsel, Selznick Releasing Org., 1951–53; pres., The Selznick Company, 1953–55; v.p., Famous Artists Corp., 1956–62; v.p. George Stevens Productions Inc., 1962–65; exec. prod., The Greatest Story Ever Told; v.p. in charge of m.p. affairs, Seven Arts, 1966; exec. in chg. talent and exec. asst. to v.p. in chg. prod., MGM, 1967; dir. m.p. business affairs, MGM, 1970; v.p., business affairs, MGM, 1972.

DAVIS, GEORGE W.: Art director, b. Kokomo, Ind., Apr. 17, 1914. e. U. Southern California.
PICTURES INCLUDE: The Robe, The Diary of Anne Frank, Love Is A Many Splendored Thing, All About Eve, David and Bathsheba, Americanization of Emily, Unsinkable Molly Brown, Funny Face, Cimarron, Period of Adjustment, Mutiny on the Bounty, Twilight of Honor, How The West Was Won, Patch of Blue, The Wonderful World of the Bros. Grimm, Mr. Buddwing, The Shoes of the Fisherman, etc. Including 200 feature films and 2000 TV segments.

DAVIS, JORDAN P.: Executive. b. N.Y., N.Y., October 29, 1933. e. Lehigh University, 1954; Columbia University Law School, 1958. Attorney for five years. Director of Business Affairs, ABC TV Network, 1963–67. NBC TV Network 1967–69. Director of Talent and Program Admin. v.p. Warner Bros. Television, 1969–1972. Executive, The Sy Fischer Company, 1972–1973. Executive in charge of production, Four D. Productions, Inc.; The Mimus Corp., The Triseme Corporation to present.

DAVIS, LUTHER. Writer, producer. b. N.Y., Aug. 29, 1921. e. Yale Univ., WWII major U.S. Air Corps.; B'way play 1945. Kiss Them for Me. Ent. m.p. 1946. Solo s.p., The Hucksters, B.F.'s Daughter, Black Hand, A Lion Is in the Streets, The Gift of Love, Holiday for Lovers, The Wonders of Aladdin; Across 110th St.; collab. book B'way musical Kismet, collab. s.p.; author, prod., Lady In A Cage; s.p. Across 110th St. Prod. and wrote book for Bdwy musical, Timbuktu!, 1978–79.
TV: Wrote, prod., Kraft Suspense Theatre and several pilots for series (Run for Your Life, The Silent Force, etc.). Wrote, prod. Arsenic and Old Lace (TV special). Wrote teleplays for MOW's Daughter of the Mind, The Old Man Who Cried Wolf, Colombo, etc.

DAVIS, MAC: Singer-Songwriter-Actor. b. Lubbock, TX, Jan. 21, 1942. e. Emory Univ., Ga. State Coll. Employed as ditch digger, service station attendant, laborer, probation officer and record company salesman before gaining fame as entertainer-singer in 1969. Recording artist and composer of many popular songs.
PICTURES: North Dallas Forty, Cheaper To Keep Her.

DAVIS, MARTIN S.: Executive. b. New York City. e. City Coll., N.Y.; New York Univ. U.S. Army, 1943–46; joined Saml. Goldwyn Prod., Inc., 1946; Allied Artists, 1955; Paramount Pictures, 1958. Dir. adv., pub. expl. 1960; vice-pres., 1963; exec. v.p., 1966; exec. comm. & bd. of dir. Member of Bd., Gulf & Western, 1967. Exec. v.p. Gulf and Western, member exec. comm.

DAVIS, SIR JOHN: Executive. b. London, Eng., 1906; e. City of London Sch.; fellow Inst. of Chartered Secretaries and Administrators. Secy. to various ind. gps.; entered m.p. ind. as acct.-secy. to Odeon Theatres, Ltd., 1938; joint man. dir., 1941; man. dir., 1948; deputy chmn. 1951–62; chmn. and chief exec. The Rank Organisation Ltd. 1962–74; chmn. 1974–77; pres. 1977– . Dir. of Eagle Star Insurance Co. Ltd., Jt.-Pres. Rank Xerox Ltd., Chairman, Children's Film Foundation, National Centre of Films for Children, Trustee, Westminster Abbey Trust and Chairman, fund raising committee, Westminster Abbey Appeal 1973. Knighted June, 1971; Hon. D. Tech. (Loughborough) 1975; Knight Commander of the Order of St. John. Commandeur de l'Ordre de la Couronne (Belgium) 1974.

DAVIS, OSSIE: Actor-Writer-Director. b. Cogdell, Ga., Dec. 18, 1917. e. Howard Univ., Washington, D.C. m. actress Ruby Dee. Studied acting in N.Y. with Rose McLendon Players, leading to Broadway debut in 1946 in Jeb. For 11 years thereafter was one of most well-known blacks on Broadway stage (Anna Lucasta, Jamaica, A Raisin in the Sun, etc.) Wrote and starred in Purlie Victorious, repeating role for film version.
PICTURES INCLUDE: Actor: The Joe Louis Story, Fourteen Hours, Shock Treatment, No Way Out, The Cardinal, The Hill, Man Called Adam, The Scalphunters, Sam Whiskey, Let's Do It Again. Director: Cotton Comes to Harlem, Black Girl, Gordon's War, Countdown at Kusini (also star).
TELEVISION: Author: East Side/West Side, The Eleventh Hour. Acted in many top dramatic series.

DAVIS, ROGER H.: Attorney. b. June 19, 1923, Chicago. e. U. Calif.-Berkeley, 1951, A.B., LL.B. Private law practice, partner, Loeb and Loeb, Los Angeles, 1951–1961; atty., William Morris Agency, head of west coast legal, literary and M.P. depts., exec./admin., asst. sec. WMA, Inc., 1969-present, made v.p., 1974. exec. v.p. & bd. member, 1980.

DAVIS, SAMMY, JR.: Entertainer. b. New York City, Dec. 8, 1925. Father: Sammy Davis, Sr., entertainer. Professional debut at age 2, Gibson Theatre, in act with father and uncle, Will Mastin; singer, dancer, comedian, vocal and acting impressions; song writer; recording artist; starred in Mr. Wonderful, & Golden Boy on B'way; in many major TV shows.
PICTURES INCLUDE: Anna Lucasta, Porgy & Bess, Ocean's 11, Robin and the Seven Hoods, Salt and Pepper, A Man Called Adam, Sweet Charity, One More Time, The Cannonball Run.

DAVIDSON, BRUCE: Actor. b. Philadelphia. e. Penn State, New York University. debut, Lincoln Center Repertory.
PICTURES INCLUDE: Last Summer, The Strawberry Statement, Willard, Been Down So Long It Looks Like Up To Me, The Jerusalem File, Ulzana's Raid, Mame, Mother, Jugs, and Speed, Short Eyes, Brass Target.

DAWBER, PAM: Actress-Singer. b. Detroit, Mich. e. Farmington High School, Oakland Community College. Travelled nationally to trade shows to introduce new auto model; also began working in commercials. In 1971 came to N.Y.C.; hired as model, also doing commercials. First professional performance as singer in Sweet Adeleine at Goodspeed Opera House, East Haddam, Conn. Signed by Robert Altman to appear in theatrical film, A Wedding. On TV in comedy special, Sister Terri. Plays Mindy in Mork and Mindy.

DAWSON, ANTHONY: Actor-producer. b. Edinburgh, Scot., Oct. 18, 1916. e. Bryanstone School, Dorset, Eng., Royal Academy of Dramatic Art. Jockey, trainer, 1939–43; British Army.
PICTURES INCLUDE: Queen of Spades, Way to Stars, Beware of Pity, School for Secrets, Long Dark Hall, Valley of the Eagles, Haunted Strangler, Curse of the Werewolf, That Lady, Action of the Tiger, Tiger Bay, Dial M for Murder, Death Rides a Horse, Dr. No, Midnight Lace, Red Sun, Battle of Neretva.
TV: USA. '51–'62. US Steel Hour, Hitchcock Presents, Philco TV Playhouse, Lux, Montgomery Presents, Studio One, Hallmark.
WRITER: The Deserter, The Snorkel, The Papoose Sitter, Oil, Break Point, 1976. Formed Arengo Prod. (Italy).

DAY, DORIS: Singer, actress. r. n. Doris Kappelhoff. b. Cincinnati, Ohio, Apr. 3, 1924. e. dancing, singing. Toured as dancer; radio and band singer; screen debut in Romance on the High Seas, 1948. Voted one of Top Ten Money-Making Stars in Motion Picture Herald-Fame poll, 1951–52. Best female vocalist. M. P. Daily radio poll, 1952.
PICTURES INCLUDE: Young Man With a Horn, Tea for Two, Storm Warning, West Point Story, Lullaby of Broadway, On Moonlight Bay, I'll See You in My Dreams, Starlift, Winning Team, April in Paris, By The Light of the Silvery Moon, Calamity Jane, Lucky Me, Young at Heart, Love Me or Leave Me, Man Who Knew Too Much, Julie, Pajama Game, Teacher's Pet, Pillow Talk, Please Don't Eat the Daisies, Midnight Lace, Lover, Come Back, That Touch of Mink, Jumbo, Thrill of It All, Move Over, Darling, Do Not Disturb, Send Me No Flowers, Glass Bottom Boat, Caprice, The Ballad of Josie, Where Were You When the Lights Went Out, With Six You Get Eggrolls.
TV: The Doris Day Show 1968–73.

DAY, LARAINE: Actress. r.n. Laraine Johnson. b. Roosevelt, Utah, Oct. 13, 1920. In school dramatics; with Players Guild, Long Beach, Calif.; toured in church prod. Conflict; Professionally on stage in Lost Horizon, The Women, Time of the Cukoo, Angel Street; m.p. debut in Border G-Men 1938.
PICTURES INCLUDE: Story of Dr. Wassell, Those Endearing Young Charms, Locket, Tycoon, My Son, My Son, Foreign Correspondent, Woman on Pier 13, Without Honor, High and the Mighty, Mr. Lucky, Toy Tiger, Three for Jamie Dawn, The Third Voice, Yank on The Burma Road, The Bad Man, Fingers at The Window, Bride By Mistake, My Dear Secretary.
RECENT ACTIVITIES: mem bd., Te Arohanui Maori Co., New Zealand, Share, Inc.
TV: Climax, Playhouse 90, Alfred Hitchcock, Wagon Train, Let Freedom Ring, Name of the Game, FBI, Sixth Sense, Medical Center, Murder on Flight #504 (ABC Movie), Fantasy Island, Love Boat, Lou Grant.

DAY, ROBERT: Director. b. London, England, September 11, 1922. Ent. m.p. ind. 1938 in camera dept. Warners' Teddington Studios.
PICTURES INCLUDE: The Green Man, Grip of the Strangler, First Man into Space, Life in Emergency War 10, Bobbikins, Two-Way Stretch, Tarzan The Magnificent, The Rebel, Operation Snatch, Tarzan's Three Challenges, She, Tarzan, 1966, Tarzan and the Great River.
TV: Robin Hood, O.S.S., Rendezvous, Danger Man, The Avengers.

DAYTON, LYMAN D.: Producer. b. Salt Lake City, Aug. 17, 1941. m. Elizabeth Doty Dayton. e. Brigham Young Univ. After college worked in film lab at MGM, 1967–68; joined Screen Gems and General DeLuxe, 1968–69; became indep. prod. 1969. Heads Doty-Dayton Productions.
PICTURES INCLUDE: Where the Red Fern Grows, Seven Alone, Against A Crooked Sky, Pony Express Rider, Baker's Hawk, Young Rivals, Powder Keg.

DEAN, EDDIE: Actor. r.n. Edgar D. Glosup. b. Posey, Texas; married. From 1930–33 in radio throughout middle west; 1934 National Barn Dance, Station WLS; 1935 on CBS & NBC with same program. Featured male singer on TV KTLA Western Varieties 1944–55 Came to Hollywood in 1936; since then has starred in many westerns. Featured performer in western series for PRC in 1945. Voted one of the ten best money making Western Stars in Motion Picture Herald-Fame Poll 1936–47; recording artists, personal appearances, rodeos, fairs, etc.; 1966 V.P. Academy of Country & Western Music; 1967–68 on Bd. of Dir. of Academy of Western Music, Calif. Winner, Pioneer Award of Academy of Country Music, 1978.

DEAN, JIMMY: Performer. b. Plainview, Tex., Aug. 10, 1928. Joined armed forces, 1946–49; first appeared in various clubs in Wash., 1949; then app. on Town and Country Jamboree; toured Caribbean and Europe with his troupe.
SONGS: Comp. of following songs: Bummin' Around, Freight Train Blues, Glad Rags, Hello, Mr. Blues, I Found Out, Big Bad John, PT-109, Little Black Book, This Ole House.
TV: Sunday Night at the Palladium (London), The Jimmy Dean Show.

DEAN, MERRILL C.: Executive. b. May 24, 1941. e. UCLA (bus. admin.). 1967–68, involved with Walt Disney Productions, starting in m.p. distribution internationally with Buena Vista Intl., concentrating on Latin America. Later involved with world-wide publications and gen. mgr. of direct mktg. Was v.p., Walt Disney Records Music Division, four years. 1978–present, pres. of Televisa Intl. Marketing Group, Inc., and pres. of Televicine Intl. Distribution Corp., whose activities are principally involved with dist. of films to the Hispanic mkt. in U.S. & Puerto Rico

DE BROCA, PHILIPPE: Director. b. 1933. French director.
PICTURES INCLUDE: Les Jeux de l'Amour, L'Amant, de Cinq Jours, Cartouche, That Man from Rio, Un Monsieur de Compagnie, Tribulations, Chinoise en Chine, King of Hearts, Devil By the Tail, Give Her the Moon, Louise.

DECAE, HENRI: Cinematographer b. France, 1915. Career spans 30 years and 60 films.
PICTURES INCLUDE: Le Silence de la Mer, Les Enfants Terribles, Creve-Coeur, Bob le Flambeur, Lift to the Scaffold, Le Beau Serge, A Double Tour, Les Quatre Cents Coup, Les Cousins, Plein Soleil, Les Bonnes Femmes, Leon Morin, Priest, Sundays and Cybele, Dragees au Poivre, Viva Maria, Weekend at Dunkirk, Night of the Generals, La Voleur, The Comedians, Castle Keep, The Sicilian Clan, The Light at the Edge of the World, Bobby Deerfield, The Island.

De CAESAR, GABRIEL: Producer. b. New York, Nov. 15, 1928. e. L.A. City College, 1952 Dir.-mgr. Little Theatre, L.A. Group, 3 yrs.; actor, 6 years; acting member Catholic Coll. Thea., Santa Monica Thea. Group, Pilgrimage Play Thea. Group, 3 yrs., producer: Three Wishes for Jamie, Bullfight; assoc. prod., asst. prod., Viscount Films, Inc. 1957–59; dir., N.A.B.E.T.; acted: Unexpected, TV series; ABC-TV series; producer, TV pilots, 1958; asst. prod., pictures: Tank Battalian, Speed Crazy; pres., exec. prod. Marqhis Productions, Inc., 1959; assoc. prod., Why Must I Die?

DE CAMP ROSEMARY: Actress. b. Prescott Ariz., 1913.
TV: Robert Cummings Show, That Girl, Life of Riley (with Jackie Gleason); Death Valley Days; Partridge Family; Love American Style, Police Story, Rockford Files, Days of Our Lives, Misadventures of Sheriff Lobo, Love Boat, Blind Ambition, B.J. & the Bear.
PICTURES INCLUDE: Cheers for Miss Bishop, Hold Back the Dawn, Jungle Book, Yankee Doodle Dandy, Eyes in the Night, Pride of the Marines, Look for the Silver Lining, Story of Seabiscuit, Big Hangover, Scandal Sheet, On Moonlight Bay, Treasure of Los Canyon, By the Light of the Silvery Moon, Main Street to Broadway, So This Is Love, Many Rivers to Cross, Strategic Air Command, Saturday the 14th.

DE CAPRIO, AL: Producer-director. e. Brooklyn Tech., NYU. Started as radio engineer, cameraman, tech. dir., prod. & dir. CBS; dir. Sgt. Bilko, Car 54 Where Are You?, Musical specials for ABC, CBS, NBC; v.p. exec. prod. dir., MPO Videotronics, Pres. World Wide Videotape.

DE CARLO, YVONNE: Actress. b. Vancouver, B.C., Sept. 1, 1924; e. Vancouver School of Drama, Fauchon & Marco, Hollywood; specialty dancing at Florentine Gardens, Earl Carroll's; m.p. debut in This Gun for Hire 1942.
PICTURES INCLUDE: Harvard Here I Come, Youth on Parade, Road to Morocco, Story of Dr. Wassell, Salome, Where She Danced; Frontier Gal, Brute Force, Song of Scheherazade, Black Bart, Casbah, River Lady, Criss Cross, Gal Who Took the West, Calamity Jane and Sam Bass, Buccaneer's Girl, Tomahawk, Hotel Sahara, Silver City, Scarlet Angel, San Francisco Story. Hurricane Smith, Sombrero, Sea Devils, Fort Algiers, Border River, Captain's Paradise, Passion, Tonight's the Night, Shotgun, Magic Fire, Flame of the Islands, Ten Commandments, Raw Edge, Death of a Scoundrel, Band of Angels, Timbuktu, Law of the Lawless, Hostile Guns, The Power, McClintock.
BROADWAY STAGE: Follies.

DE CORDOVA, FREDERICK: Director. b. New York City, Oct. 27, 1910; e. Northwestern U. B.S. '31. Gen. stage dir. Shubert enterprises, N.Y., 1938–41; same for Alfred Bloomingdale Prods., N.Y., and prod. Louisville (Ky.) Amphitheatre 1942–43 m.p. dir.
PICTURES INCLUDE: (dial. dir.) San Antonio, Janie, Between Two Worlds; (dir.) Too Young to Know, Her Kind of Man, That Way with Women, Always Together, Wallflower, For the Love of Mary, Countess of Monte Cristo, Illegal Entry, Girl Who Took the West, Buccaneer's Girl, Peggy, Desert Hawk, Bedtime for Bonzo, Katie Did It, Little Egypt, Finders Keepers, Here Come the Nelsons, Yankee Buccaneer, Bonzo Goes to College, Column South.
Under contract CBS-TV, prod., dir. Burns and Allen, 1955–56; prod., December Bride, 1954–55; prod. and dir., Mr. Adams and Eve, prod. dir. December Bride; prod. dir. George Gobel Show; prod. dir., Jack Benny Program, 1960–63; dir., program planning, Screen Gems, 1964; prod. dir., Smothers Bros. Show, 1965–66; 1965 dir. I'll Take Sweden and Frankie & Johnny; 1966–70 dir. My Three Sons; 1971–1981, prod., Tonight Show.

DECUIR, JOHN: Art Dir. Prod. Des. b. San Francisco Universal 1938–49; Mark Hellinger Prods., 1946. 20th Fox 1949. Nominated for Acad. Award 11 times.
PICTURES INCLUDE: The Naked City, Snows of Kilimanjaro, My Cousin Rachel, Call Me Madam, Three Coins in the Fountain, There's No Business Like Show Business, Daddy Long Legs, The King and I (A.A., Art Dir.), Island in the Sun, South Pacific, The Big Fisherman, A Certain Smile, Cleopatra (A.A.: Prod. Des.), The Agony and the Ecstasy, The Honey Pot, Zeffirelli's Taming of The Shrew, Dr. Faustus, Hello, Dolly! (A.A.: Prod. Des.), On A Clear Day You Can See Forever, The Great White Hope, Once Is Not Enough, That's Entertainment, Too!, The Other Side of Midnight, Ziegfeld: The Man and his Women, Love and Bullets, Charlie.

DEE, RUBY: Actress. b. Cleveland, Oct. 27, 1924. m., Ossie Davis. e. Hunter Coll. Stage appearances include Jeb, Raisin in the Sun, Purlie Victorious, Wedding Band, etc.
PICTURES: Jackie Robinson Story, Take a Giant Step, St. Louis Blues, Raisin in the Sun, Purlie Victorious, Buck and the Preacher.

DEE, SANDRA: Actress. b. Bayonne, N.J., April 23, 1942. Modeled, Harry Conover and Huntington Hartford Agencies, N.Y., 1954–56; signed long term exclusive contract, U-I, 1957.
PICTURES INCLUDE: Until They Sail, The Restless Years, Stranger in My Arms, The Reluctant Debutante, Gidget, The Wild and the Innocent, Imitation of Life, Summer Place, Portrait in Black, Romanoff and Juliet, Come September, Tammy Tell Me True, If a Man Answers, Tammy and the Doctor, Take Her, She Is Mine, I'd Rather Be Rich, That Funny Feeling, Doctor You've Got to Be Kidding!, Rosie, The Dunwich Horror.

DEELEY, MICHAEL: Chief Exec. EMI Film Inc. and Int. Head of Worldwide EMI Film Prod. and Dist. 1978. b. London, England, 1932. Ent. m.p. ind. 1951 and TV, 1967, as alt. dir. Harlech Television Ltd. Film editor, 1951–58. MCA-TV 1958–61, later with Woodfall as prod. and assoc. prod. Proc. One Way Pendulum. Assoc. prod. The Knack, The White Bus, Ride of the Valkyrie, Prod. Robbery, The Italian Job. Executive proc. Long Days Dying, Where's Jack, Sleep Is Lovely, Murphy's War, Convoy, The Deer Hunter (AA), Great Western Investments Ltd.; 1972; Great Western Festivals Ltd.; 1973, man. dir. British Lion Films Ltd. 1975, purchased BLF, Ltd. Appt. Jnt. man. dir. EMI Films Ltd., 1977; pres., EMI Films, 1978.

DE GRUNWALD, DIMITRI: Producer. Managing director Script Development Ltd. de Grunwald Films Ltd. and Films include: That Lucky Touch, The Millionairess, Shalako, Stranger in the House, The McMasters, Murphy's War, The Virgin and the Gypsy, Time for Loving, Perfect Friday.

DE HAVILLAND, OLIVIA: Actress b. Tokyo, Japan, July 1, 1916. e. California schools and Notre Dame Convent, Belmont. Acting debut, Max Reinhardt's stage prod., Midsummer Night's Dream; film debut in m.p. version, 1935; won Academy Award twice, best actress performance in To Each His Own, 1946, and The Heiress, 1949; N.Y. Film Critics award twice, 1948–49; Women's Natl. Press Club Award, 1949; Look award, best perf. 3 times, 1941-48-49.
PICTURES INCLUDE: Anthony Adverse, Gone With the Wind, Strawberry Blonde, Hold Back the Dawn, Princess O'Rourke, Dark Mirror, Snake Pit, My Cousin Rachel, Not as a Stranger, Ambassador's Daughter, Proud Rebel, Libel, Light in the Piazza, The Adventurers, Pope Joan, Airport '77, The Swarm.
STAGE: Romeo and Juliet, 1951; U.S. tour Candida, 1951–52, N.Y., 1952; A Gift of Time, Bdwy. 1962.
TV: Noon Wine for Stage 67.

DEIN, EDWARD: Director-Writer. b. Chicago, Ill.
PICTURES INCLUDE: dir., s.p. Shack Out on 101, Seven Guns to Sin, Curse of the Undead, Calypso Joe, I Trust My Wife, Hard Yellow, Come Die My Love (Spain), The Naked Sword (Mexico), Gaby (Brazil); story s.p. The Fighting Guardsman, Calling Dr. Death, The Gallant Blade, Leopard Man, All this and More, 13 French Street, Hang Me High; dir., prod., wrote, Capito, in Italy.
TV: Lawless Years, Hawaiian Eye, Bronco, Philip Marlowe, Not for Hire, Wild, Wild West series.

DE LANE LEA, JACQUES: Executive. b. Nice, France, 1931. Early career as prod. shorts, second features; later prod. sup. features. Ent. m.p. ind. 1952, prod. own films. Became man. dir. De Lane Lea Ltd., after death his father 1964. Also dir. Delmore Film Productions Ltd., Mole-Richardson (England) Ltd., Humphries Holdings, Ltd., Int'l Library Service Ltd., CTS Ltd.

DELANNOY, JEAN: Director. b. Noisy-le-Dec, France, Jan. 12, 1908; e. Universite de Paris, Lycee Montaigne, Lycee Louis le Grand, La Sorbonne; actor, 1927–29; film edit., 1929–37; dir., 1937–52.
PICTURES INCLUDE: Black Diamond, The Eternal Return, Symphonie Pastorale, The Chips Are Down, Souvenir, The Secret of Mayerling, Savage Triangle, Moment of Truth, Daughter of Destiny, The Bed.

DE LAURENTIIS, DINO: Producer. b. Torre Annunziata, Italy, Aug. 8, 1919. Took part in Rome Experimental Film Center; dir., prod.
PICTURES INCLUDE: The Bandit Mussolino, Bitter Rice, Girls Marked Danger, Side Street Story, She Wolf (La Lupa), Lure of the Sila, Attila Scourge of God, Europe 51, Anna, War and Peace, Gold of Naples, Ulysses, Mambo, The Great War, La Strada, Nights of Cabiria, The Tempest, Barabbas, The Best of Enemies, Il Mafioso, The Trial of Verona, The Bible, The Hills Run Red, The Stranger, Barbarella, L'Odissea, Anzio, The Valachi Papers, The Stone Killer, Crazy Joe, Serpico, Death Wish, Mandingo, Three Days of the Condor, Lipstick, Face to Face, Buffalo Bill & The Indians, The Shootist, Drum, King Kong, Orca, The Serpent's Egg, The Brinks Job, King of the Gypsies, Hurricane, Flash Gordon.

DE LA VARRE, ANDRE, JR.: Producer-Director. b. Vienna, Austria, Oct. 26, 1934. Prod. "Grand Tour" Travelogues; producer of promotion films for KLM, Swissair, tourist offices, recent productions: Bicentennial films for State of Virginia, City of Charleston, NY State; winner, Atlanta Film Festival, Sunset Travel Film Festival; Burton Holmes Travelogue subjects; V-P-R Educational Films; producer, director, lecturer, narrator.

DEL VALLE, JOHN: Publicist. b. San Francisco, Calif., Mar. 23, 1904; e. U. California. Adv., edit. staff various newspapers including asst. drama ed. S.F. Call-Bulletin, L.A. Mirror; adv.-publicity dir. San Francisco Fox Theatre 1933–36; publicist, Paramount Studio, 1936–42; dir. pub., adv. Arnold Prod. 1946; Chaplin Studios, 1947; Nat Holt Prod., 1948–52; Editor, TV Family mag., N.Y., 1952–53; adv. pub. dir. Century Films, 1954; pub. rel. Academy M.P. Arts & Sciences, 1965; publicist, various U.A. indep. film prod., 1955–56; unit publicist, Para., 1956; TC-F 1957–62, Para., 1962–63; Universal, 1964–65; Mirisch Corp.—UA Filming, "Hawaii," 1965; pub. rel. and editor, Atomics Int'l div. North American Rockwell, 1966–71; present, free-lance writer; res. Santa Barbara.

DEL BELSO, RICHARD: Executive. e. New York Univ. (1965). Began career in adv./research dept. at Benton & Bowles. Served as research dept. group head for Kenyon and Eckhart; group head for Gruden/Appell/Haley Research Co. (now known as A/H/F Marketing Research, Inc.). Two years as assoc. dir. of mktg., research for Grey Advertising (N.Y.). Joined MCA/Universal in 1977 as assoc. dir., mktg. research. In 1980 named v.p. & dir. of mktg. research for Warner Bros.

DELERUE, GEORGES: Composer-Conductor. b. Roubaix, France, 1924. Studied with Busser and Milhaud at Paris Cons. Was conductor for RTF (Radio-TV Francais).
PICTURES INCLUDE: Hiroshima Mon Amour, King of Hearts, Shoot the Pianist, Jules and Jim, Silken Skin, The Pumpkin Eater, Viva Maria, The 25th Hour, Interlude, The Day of the Jackal, The Day of the Dolphin, A Little Romance.

DELFONT, BERNARD LORD: Chief Executive EMI, Division of Tharn-EMI, Ltd. b. Tokmak, Russia, September 5, 1909. e. London, England. Dir. more than 30 companies incl. Forte's Trust Houses, chairman Grade and Delfont Organization. Presents entertainment at London's Talk of the Town. Ent. theatrical management 1941. Since presented over 250 shows in London; pantomimes in provinces and seaside resorts; presented annual Royal Variety Performance. Controls New London and Prince of Wales Theatres in London's West End. 1969: Chief Barker (Pres.) Variety Club of Great Britain; Companion of the Grand Order of Water Rats, President of Entertainment. Artistes' Benevolent Fund, member of Saints and Sinners organisation.

DELMAINE, BARRY: Writer, director. b. London. Began motion picture career as extra. Later roles in comedies; then asst. dir. for Gloria Swanson prod. in England. From 1936–39 wrote sketches, lyrics London stage musicals; radio scripts. In 1939 apptd. chief scenarist Gaumont-British Instructional. In 1941 dir. films for Brit. Ministry of Information; 1942–43 scenarist, Brit. War Office mi. training films. In 1944 author & dir. Here We Come Gathering, J. Arthur Rank experimental film for child audiences. In 1946 joined Film Producers Guild, writer, dir.; 1949–50, dir. Concord Productions, Ltd., freelance writer films, radio, TV; s.p. Heir Conditioned; dir. ind. films, Concord Prod. Ltd., ass't dir. Advance Films; The Great Game, Les Parent Terrible (Eng. version), Is Your Honeymoon Really Necessary? Unit dir. Don't Blame the Stork; Unit dir., loc. mgr. Douglas Fairbanks Ltd.; prod. mgr. 1955; prod. man. Chase a Crooked Shadow, Ice Cold in Alex, Jetstream; prod. sup. The Night We Dropped a Clanger. Prod. Sup., Nothing Barred; prod. man., The Hellions; prod. The Golden Rabbit; prog. sup., Man of The World, Sentimental Agent, assoc. prod., Danger Man (secret agent) series. Assoc. prod. Man in the Suitcase series; Assoc. prod., Strange Report series; assoc. prod. Mr. Jerico feature. TV series: assoc. prod. The Firechasers; prod. Shirley's World; assoc. prod. The Adventurers. Assoc. Prod. Three For All. Currently: Free lance Writer and Prod. Consultant.

DELON, ALAIN: Actor. b. Sceaux, France, Nov. 8, 1935. Discovered by Yves Allgret. Served in French Navy as a Marine. Worked as cafe waiter, heavy-load carrier.
PICTURES INCLUDE: When a Woman Gets Involved, Be Beautiful, and Keep Quiet, Plein Soleil and Quelle Joie de Vivre, Rocco and His Brothers, L'Eclipse, The Leopard, The Big Crab, The Black Tulip, The Felines, The Yellow Rolls Royce, The Killers of San Francisco, The Centurions, Is Paris Burning?, Texas, Here We Come, The Adventures, Extraordinary Stories, Le Samourai, Diaboliquement Votre, The Motorcycle, Goodbye Friend, The Swimming Pool, Jeff, The Sicilian Clan, Borsalino, Red Sun, Assassination of Trotsky, Scorpio, Dirty Money, Mr. Klein, The Concord-Airport '79.

DEL RIO, DOLORES: Actress. r.n. Dolores Ansunsolo; b. Durango, Mexico, Aug. 3, 1905; e. St. Joseph's Conv., Mexico, D.F.; studied voice Madrid & Paris. Screen debut in Joanna, 1925.
PICTURES INCLUDE: What Price Glory?, Resurrection, Ramona, Evangeline, Flying Down to Rio, Wonder Bar, Madame DuBarry, Accused, The Devil's Playground, Journey Into Fear, Portrait of Maria, The Fugitive, More Than a Miracle, Cheyenne Autumn.

DeLUISE, DOM: Comedian-Actor. b. Brooklyn, Aug. 1. e. Tufts Coll. Spent two seasons with Cleveland Playhouse. Launched TV career on The Garry Moore Show with character, Dominick the Great, a bumbling magician.
PICTURES INCLUDE: Fail Safe, The Busybody, The Glass Bottom Boat, 12 Chairs, Who Is Harry Kellerman?, Every Little Crook and Nanny, The Adventures of Sherlock Holmes' Smarter Brother, Silent Movie, The World's Greatest Lover, The End, The Cheap Detective, The Last Married Couple in America, Fatso, Wholly Moses, Smokey and the Bandit II, History of the World—Part I, The Cannonball Run.
TV: The Entertainers, The Dean Martin Summer Show, Dom DeLuise Variety Show, The Barrum-Bump Show, The Dean Martin Show, The Tonight Show, Lotsa Luck.
STAGE: The Student Gypsy, Last of the Red Hot Lovers, Here's Love.

DEL VALLE, JAIME: Producer. b. Los Angeles, Calif., Jan. 19, 1910. e. Stanford U. Motion Pictures 1932 as writer. Since then writer, director, producer in motion pictures, radio and television with exception of four years USAAF. Currently Metro-Goldwyn-Mayer.

DEMAREST, WILLIAM: Actor. b. St. Paul, Minn., Feb. 27, 1892. Began theat. career 1897 with brothers Reuben & George in vaudeville; later also stock, carnival acts, occasionally professional boxer; played Palace, N.Y., 1917; on Broadway in: Sketch Book, Vanities; talent agent one yr.; m.p. debut in Fingerprints, 1926. Starred in first released talking picture (a short), 1927, A Night in Coffee Dans.
PICTURES INCLUDE: Diamond Jim, Wedding Present, Love on the Run, Mr. Smith Goes to Washington, Pardon My Sarong, Hail the Conquering Hero, Miracle of Morgan's Creek, Duffy's Tavern, Perils of Pauline, Jolson Story, Miracles Can Happen, Red, Hot and Blue, The First Legion, When Willie Comes Marching Home, Riding High, Jolson Sings Again, Never a Dull Moment, Excuse My Dust, The Strip, Blazing Forest, The Lady Wants Mink, Dangerous When Wet, Here Come the Girls, Escape from Fort Bravo, Yellow Mountain, Far Horizons, Jupiter's Darling, Private War of Major Benson, Hell on Frisco Bay, Lucy Gallant, Rawhide Years, Son of Flubber, It's a Mad, Mad, Mad, World.
TV: My 3 Sons.

DEMBY, EMANUEL H.: Producer, writer. b. New York, N.Y., Dec. 12, 1919. e. C.C.N.Y., New School. Chung Ang University (Ph.D.). Pioneered new programs for radio, 1936–47; story dept., Universal Pictures, 1937; writer, Gangbusters, Crime of the Week, Thrill of the Week (NBC, CBS); TV shows; What's Playing, 1950–52; Hollywood to Broadway, currently in TV syndication, How To Be a Cowboy; The Shadow; prod., theatrical features, filmed commercials; pub. rel. consultant. Author: My Forty Year Fight for Korea; Indonesia; King of The Hill; Hot Tip on Wall St.; prod. Cavalcade of Music (TV); The World in Space (TV, theatrical); The Edge of Violence (stage); Man Into Space; Year III-Space Age (TV); The Communications Gap, The Creative Consumer, (P.R. films). Book: Who's Alienated, Youth or Society? Research Consultant, NBC TV News; consultant, Radio Advertising Bureau, Smash, Crash, Pow! (feature).

DEMME, JONATHAN: Director-Writer. b. Rockville Center, N.Y., 1944. e. Univ. of Florida. First job in industry as usher; was film critic for college paper, The Florida Alligator. Did publicity work for United Artists, Avco Embassy, Pathe Contemporary Films; wrote for trade publication, Film Daily, 1966–68. Moved to England in 1969; served as musical co-ordinator on Irving Allen's EyeWitness in 1970. In 1972 produced and co-wrote first film, Angels, Hard As they Come.
PICTURES: Hot Box (prod., co-writer); Caged Heat (writer-dir.); Crazy Mama (writer-dir.); Fighting Mad (writer-dir.), Citizen's Band (dir.), Last Embrace (dir.), Melvin and Howard (dir.).

DEMOSS, LYLE: Executive. b. Anthony, Kansas, Sept. 14, 1907; e. York Coll. and U. of Nebraska School of Music; 1925. KGBZ in York, Nebr.; 1928. KFAB in Lincoln, Nebr.; in 1932 became Prog. Mgr. of KFAB-KFOR; held same position for WOW from 1936 to 1946; asst. gen. mgr. Meredith WOW, 1946–59. Allen & Reynolds v.p. radio-TV dept. Omaha Advertising Club, Variety Club Canvasman, past pres. Omaha Ad. Club. Radio KLNG, 1965–67; freelance TV; Elected to Nebr. Broadcasters Hall of Fame—1972; KYNN-community relations director.

DENCH, JUDI: O.B.E. Actress. b. York, England, 1934. Studied for stage at Central School of Speech and Drama. Theatre debut Old Vic, 1961.
TELEVISION: Major Barbara, Pink String and Sealing Wax, Talking to a Stranger, The Funambulists, Age of Kings, Jackanory, Hilda Lessways, Luther, Neighbours, Parade's End, Marching Song, On Approval, Days to Come, Emilie, The Comedy of Errors, Macbeth (both RSC productions), Langrishe Go Down, On Giant's Shoulders, Love in a Cold Climate, Village Wooing.
PICTURES: He Who Rides a Tiger, Study in Terror, Four in the Morning, A Midsummer Night's Dream (RSC Production), The Third Secret, Dead Cert.

DENEAU, SIDNEY G.: Sales executive. Head film buyer Fabian Theatres; U.S. Army 1942–46; gen. mgr. Schine Theatres 1947; v.p., gen. sales mgr., Selznick Releasing Orgn., 1949; 1956; v.p. asst. gen. sls. mgr., Para. Film Dist., 1958; exec. v.p., Rugoff Theatres, 1964. Resigned, September, 1969 to engage in own theatre consultant business.

DENEUVE, CATHERINE: Actress. b. 1945. r.n. Catherine Dorleac. Sister was the late Francoise Dorleac.
PICTURES INCLUDE: The Doors Slam, Vice and Virtue, Satan Leads the Dance, Umbrellas of Cherbourg, (received the French Film Academy's best actress award); Repulsion, Le Chant du Monde, La Vie de Chateau, Les Creatures, The Young Girls of Rochefort, Belle de Jour, Benjamin, Manon 70, Mayerling, La Chamade, April Fools, Mississippi Mermaid, Tristana, It Only Happens to Others, Dirty Money, Hustle, March or Die.

DENHAM, MAURICE: Actor. b. Beckenham, Kent, Eng., Dec. 23, 1909. e. Tonbridge School. Started theatrical career with repertory com. 1934. Served in World War II and since demobilization played in numerous plays, films & radio shows.
PICTURES INCLUDE: Blanche Fury, London Belongs To Me, It's Not Cricket, Traveller's Joy, Landfall, Spider and the Fly, No Highway in the Sky, The Net, Time Bomb, Street Corner (Both Sides of the Law), Million Pound Note (Man With a Million), Eight O'Clock Walk, Purple Plain, Simon and Laura, 23 Paces to Baker Street, Checkpoint, Carrington V.C. (Court Martial), Doctor at Sea, Night of the Demon, Man With a Dog, Barnacle Bill, The Captain's Table, Our Man in Havana, Sink the Bismark, Two-Way Stretch, Greengage Summer, Invasion, Quartette, The Mark, HMS Defiant.
TV: appearances include: Uncle Harry, Day of the Monkey, Miss Mabel, Angel Pavement, The Paraguayan Harp, The Wild Bird, Soldier Soldier, Changing Values, Maigret, The Assassins, Saturday Spectacular, Vanishing Act, A Chance in Life, Virtue, Somerset Maugham, Three of A Kind, Sapper, Pig In The Middle, Their Obedient Servants, Long Past Glory, Devil in The Wind, Any Other Business, The Retired Colourman, Sherlock Holmes (series), Blackmail, Knock on Any Door, Danger Man, Dr. Finley's Casebook, How To Get Rid of Your Husband, Talking To a Stranger, A Slight Ache, From Chekhov With Love, Home Sweet Honeycomb, St. Joan, Julius Caesar, Golden Days, Marshall Petain, The Lotus Eaters, Fall of Eagles, Carnforth Practice. The Unofficial Rose, Omnibus, Balzac, Loves Labour Lost, Angels, Huggy Bear, The Portrait, The Crumbles Murder, A Chink In The Wall, Porridge, For God's Sake, Bosch, Marie Curie, Upchat Line, Secret Army, My Son, My Son, Edward and Mrs. Simpson, Gate of Eden, Potting Shed, Double Dealer.
OTHER PICTURES: The Very Edge, Paranoiac, The Set Up, Penang, The Kings Breakfast, Downfall, Hysteria, The Uncle, Operation Crossbow, Legend of Dick Turpin, The Alphabet Murders, The Night Callers, The Nanny, Those Magnificent Men in Their Flying Machines, Heroes of Telemark, After the Fox, The Torture Garden, The Long Duel, The Eliminator, Danger Route, Attack on the Iron Coast, The Best House in London, Negatives, The Midas Run, Some Girls Do, The Touch of Love, The Virgin and the Gypsy, Bloody Sunday, Countess Dracula, Nicholas and Alexandra, The Day of the Jackal, Luther, Shout at the Devil, Julia, The Recluse.

DE NIRO, ROBERT: Actor. b. New York City, 1945. Studied acting with Stella Adler and Lee Strasberg. In off-Broadway productions and touring shows.
PICTURES INCLUDE: The Wedding Party, Jennifer on My Mind, Bloody Mama, Born To Win, The Gang That Couldn't Shoot Straight, Bang the Drum Slowly, Mean Streets, Godfather II, Taxi Driver, 1900, The Last Tycoon, New York, New York, The Deer Hunter, Raging Bull, True Confessions.

DENISON, MICHAEL: Actor. b. Doncaster, York, Eng., Nov. 1, 1915; e. Harrow & Magdalen Coll., Oxford and Webber Douglas Sch.; m. Dulcie Gray, prof. Served overseas, Capt. Intelligence Corps, 1940–46. On stage first, 1938, Charlie's Aunt. Screen debut 1940, Tilly of Bloomsbury.
THEATRE: Ever Since Paradise, Rain on the Just, Queen Elizabeth Slept Here, Fourposter, Dragon's Mouth, Bad Samaritan; Shakespeare Season Stratford-on-Avon; Edinburgh Festival. Meet Me By Moonlight, Let Them Eat Cake, Candida, Heartbreak House, My Fair Lady (Australia), Where Angels Fear to Tread, Hostile Witness, An Ideal Husband, On Approval, Happy Family, No. 10, Out of the Question, Trio, The Wild Duck, The Clandestine Marriage, The Dragon Variation, At the End of the Day, The Sack Race, Peter Pan, The Black Mikado, The First Mrs. Fraser, The Earl and the Pussycat, Robert and Elizabeth, The Cabinet Minister, Old Vic Season: Twelfth Night, Lady's Not for Burning, Ivanov, Bedroom Farce.
PICTURES INCLUDE: Hungry Hill, My Brother Jonathan, The Blind Goddess, The Glass Mountain, Landfall, The Franchise Affair, Angels One Five, Tall Headlines, Importance of Being Earnest, There Was a Young Lady, Contraband Spain, The Truth About Women, Faces in the Dark.
TV: Boyd QC Series, Funeral Games, Unexpectedly Vacant, Tale of Piccadilly. 1973: The Twelve Pound Look, The Provincial Lady, Subject: This Is Your Life, 1977, Bedroom Farce.

DENNIS, SANDY: Actress. b. Hastings, Nebraska, April 27. m. Gerry Mulligan. Joined the Lincoln Community Theatre group. Made her professional acting debut in summer stock at London, N.H. Studied with Herbert Berghof, N.Y. Understudy for N.Y. prod. Dark at the Top of the Stairs. Broadway debut in Any Wednesday (Antoinette Perry Award), then A Thousand Clowns (Tony Award for best supporting actress).
PICTURES INCLUDE: Who's Afraid of Virginia Wolf? (won an Academy Award as best supporting actress), Up the Down Staircase, Sweet November, The Fox, Thank You All Very Much, That Cold Day in the Park, The Out-of-Towners, Nasty Habits, The Four Seasons.

DE NORMAND, GEORGE: Actor, director. b. New York, N.Y., Sept. 22. Actor many westerns & TV series; dir., TV series, Man of Tomorrow. Member: SAG, SEG, NSTP.

DE PALMA, BRIAN: Director. b. Newark, N.J., 1940. e. Columbia Univ. (B.A.) and Sarah Lawrence (M.A.). While in college made series of shorts, including Wotan's Wake, winner of Rosenthal Foundation Award for best film made by American under 25. Also judged most popular film of Midwest Film Festival (1963); later shown at San Francisco Film Festival.
PICTURES INCLUDE: The Wedding Party, Murder A La Mod, Greetings, Hi Mom, Dionysus in '69, Get To Know Your Rabbit, Sisters, Phantom of the Paradise, Obsession, Carrie, The Fury, Home Movies, Dressed to Kill, Blow Out.

DEPARDIEU, GERARD: Actor. b. Chateuaroux, France, Dec. 27, 1948. Studied acting at Theatre National Populaire in Paris. Made film debut at 16 in short by Roger Leenhardt (Le Beatnik et Le Minet). Acted in feature film by Agnes Varda which was never finished.
PICTURES INCLUDE: Going Places, Stavisky, 1900, Get

Your Handkerchiefs, LouLou, The Last Metro, Mon Oncle d'Amerique.

DEPEW, RICHARD H.: Executive. b. New York, N.Y., Jan. 16, 1925. e. Princeton U. U.S. Navy; American Broadcasting Co., 1947; television director, 1950; assistant, eastern TV network program director, 1953; mgr. of TV network program oper., 1954–57; Cunningham & Walsh; Radio & TV acct. supv. & programming coordinator, 1961–65 v.p. & dir. of TV programming; 1965 Broadcast supr., Ogilvy and Mather; 1967, v.p. dir. TV program Fuller & Smith & Ross. V.P. Media and Programming 1969, FS&R.; 1973, Director of Corporate Underwriting Department WNET/13, Educational Broadcasting Corp. 1977 Management Supervisor, J. Walter Thompson, N.Y.C.; 1978, v.p., Account Director, 1980, Marstella, Inc., mgt. supvr.

DEREK, BO: Actress. r.n. Cathleen Collins. Discovered by John Derek, actor turned film producer, who changed her first name to Bo, now married to him. Film debut in "10" in 1979.
PICTURES: Change of Seasons, Tarzan, the Ape Man.

DERN, BRUCE: Actor. b. Chicago, Ill., 1937. e. University of Pennsylvania. Studied acting with Gordon Phillips, member, Actor's Studio, 1959 after N.Y. debut in Shadow of a Gunman.
CREDITS: The Trip, The Wild Angels, Hush, Hush, Sweet Charlotte, Support Your Local Sheriff, Thumbtripping, Drive, He Said; They Shoot Horses, Don't They?; The Cowboys, Silent Running, Sweet Bird of Youth (Stage), Orpheus Descending (Stage), The King of Marvin Gardens, The Laughing Policeman, The Great Gatsby, Smile, Posse, Family Plot, Won Ton Ton, Black Sunday, Coming Home, The Driver, Middle Age Crazy.

DE ROCHEMONT, RICHARD: Producer. b. Chelsea, Mass.; e. Cambridge (Mass.) Latin Sch.; Williams Coll.; Harvard U., B.A. On ed. staffs Boston Daily Advertiser, Pathe News (newsreel), N.Y. American, N.Y. Sun. Joined Fox Movietone News; ed. Paris edition 1931–34. Joined March of Time 1934; European mgr., man. dir. March of Time, Ltd., England, to 1940 (1939–40, 1944–45 accredited Amer. war corresp., France). In 1941 prod. Story of the Vatican for RKO Radio; in 1942 man. ed. March of Time; prod. 1943–51; prod. A Chance to Live, March of Time short (Academy Award, 1949;) prod. The Golden Twenties, 1950; Modern Arms and Free Men, 1951; World Assembly of Youth, 1952, feature documentaries. Prod. TV series, Mr. Lincoln for Ford Foundation; v.p., TV prod., J. Walter Thompson ad agency, New York, 1953–55; pres., Vavin Inc., 1955. Contrib. to Time, Life, Readers Digest, American Mercury, etc. Clubs: Harvard, Williams, Coffee House, Sky, Explorers, Overseas Press (bd. gov. 1943–55, 1961–1963). France Forever (pres. 1943–44), Paris-American (pres. 1963–66). Decorations: Chevalier du Merite Agricole (1938); officer French Legion of Honor (1945); officer, Order of Merite Touristique (1959); officer, Merite Agricole (1959).; order of National Merit (commander) France, 1979.

DE SANTIS, GIUSEPPE: Director. b. Italy, 1917. Film publicists; asst. dir.
PICTURES INCLUDE: Tragic Hunt, Bitter Rice, Under the Olive Tree, Rome 11 O'Clock, A Husband for Anna Zaccheo, Days of Love.

DE SCHAAP, PHILIP: Dutch correspondent, Quigley Publications, N.Y. b. Amsterdam, Holland, May 26, 1911. Entered m.p. ind. 1934 as publicity mgr., MGM br. office, Amsterdam; in 1936 corresp. for Quigley Publications. Also in 1937 exploitation mgr., RKO Radio Pictures, Amsterdam; to Paris 1938 as spec. publicity repr. in Paris and Brussels, RKO: in 1939 org. own co. for exploitation and publicity; non-active 1940–45; on Oct. 1, 1945 estab. Phidesa Publiciteitsbureau voor de Filmbranche.

DESMOND, JOHNNY: Performer; r.n. Giovanni Alfredo De Simone; b. Nov. 14, 1923, Detroit, Mich.; began singing in Detroit, 1932; organ. Bob-O-Links which were hired by Bob Crosby, 1939; enlisted, Army Air Corps and transferred to Glenn Miller Air Corps band; sang over Armed Forces network; TV singer, Don McNeill's Breakfast Club Show, 1949–51. Songs for Sale, 1951–52, Jackie Gleason Show, 1952; Ed Sullivan, 1952; Coral Records, TV shows, radio.
BROADWAY: Funny Girl.

DE TITTA, ARTHUR A.: Newsreel editor. b. North Bergen, N.J., July 9, 1904. Began work 1916 in employ newspaper in Union City, N.J.; joined Fox Film Corp., N.Y. office & lab. staff; later became asst. & second cameraman, performing in that capacity numerous pictures Fox and other produs. (The Mountain Woman, When Knighthood Was in Flower, The Royale Girl, East Side, West Side, Untamed, etc.). In 1928 joined Movietonews, Inc.; Wash. (D.C.) suprv. to 1938; asst. Europ. supvr. to 1940; Pacific Coast supvr. to 1943. Commis. Lt. USNR, assigned to M.P. Sect., 1943; ret. 1944. Appt. news ed. Movietonews; west coast superv., 1951–63; pub. rel., prod., documentaries; West Coast Bureau mgr. for foreign editions of Movietone News, now retired.

DE TOTH, ANDRE: Writer-Director-Producer. b. Hungary. Dr.-writer European films, 1931–39; U.S. assoc. Alexander Korda prod., 1940; dir. Columbia, 1943; assoc. David Selznick, 1943; assoc. Hunt Stromberg-UA, 1944–45; staff dir., Enterprise 1946–47; dir., 20th-Fox, 1948–49; collab. story, The Gunfighter; dir., Columbia & Warner Bros., 1951; contract dir., Warner Bros., 1952; U.A. Columbia, W.B., 1953–55; Palladiums U.A., Denmark, 1956; Col. U.A. 1957; Columbia, 1960; assoc., Sam Spiegel, Horizon Pictures, Columbia, 1962; Harry Saltzman, Lowndes Prod., U.A., 1966–68; National General, 1969–70.
PICTURES INCLUDE: Passport to Suez, None Shall Escape, Since You Went Away, Pitfall, Springfield Rifle, Thunder Over the Plains, House of Wax, Bounty Hunter, Tanganyika, Indian Fighter, Monkey on My Back, Two-Headed Spy, Man on a String, Morgan, The Pirate, The Mongols, Gold for the Caesars, Billion Dollar Brain, Play Dirty, El Condor.

DEUTSCH, ARMAND S.: Producer. b. Chicago, Ill., Jan. 25, 1913. e. U. of Chicago, 1935; B.A. Producer, MGM.
PICTURES INCLUDE: Ambush, Right Cross, Magnificent Yankee, Three Guys Named Mike, Kind Lady, Girl in White, Carbine Williams, Girl Who Had Everything, Green Fire, Slander.

DEUTSCH, DAVID: Producer-Writer. b. Birmingham, England, January 10, 1926. e. in Birmingham and Queen Mary College, London Univ. Ent. m.p. ind. apprentice at Gainsborough Pictures, Ltd., 1949; asst. Sydney Box, 1954; J. Arthur Rank asst. to prod. 1955 Lost; assoc. prod. High Tide at Noon, The One That Got Away, Floods of Fear; prod. Blind Date; Anglo Amalgamated Film Dist. Ltd., as exec. in chrg. prod., 1960. Prod., Play It Cool, 1962, Nothing But the Best. 1963: Domino Prods. Ltd., Catch Us If You Can, 1965; Interlude, 1967; Lock Up Your Daughter, 1968. A Day in the Death of Joe Egg, 1970. Co-prod. The Day of the Jackal, 1972; s.p., The Blue Train, The Jacaranda Tree.

DEUTSCH, HELEN: Writer. b. New York, N.Y. e. Barnard Coll., B.A. Author: The Provincetown.
PICTURES INCLUDE: Seventh Cross, National Velvet, King Solomon's Mines, Golden Earrings, Kim, Plymouth Adventure, Lili, The Glass Slipper, I'll Cry Tomorrow, Forever Darling.
TV: Jack and the Beanstalk, NBC Gen. Motors 50th Anniv. Show, 1957; book of Broadway musical, Carnival, 1961.

DEVANE, WILLIAM; Actor. b. New York City. Appeared in some 15 productions with N.Y. Shakespeare Festival and off-Bdwy. shows before heading for California and films and TV. Has own theatre, Actors for Themselves, in Los Angeles. Has also directed plays.
PICTURES: Rolling Thunder, Marathon Man, Yanks, Honky Tonk Freeway.
TELEVISION: The Missiles of October, Fear on Trial, From Here to Eternity (series).

DE VITO, DANNY: Actor. b. Neptune, NJ, Nov. 17. e. Academy of Dramatic Arts, Manhattan. Produced two short motion pictures (with associate Rhea Perlman), The Sound Sleeper and Minestrone (latter on grant from American Film Institute). Acted off Broadway in The Man with the Flower in His Mouth, Shrinking Bride, etc.
PICTURES: One Flew Over the Cuckoo's Nest, Goin' South.
TELEVISION: Taxi (series), Valentine (movie).

DEWHURST, COLLEEN; Actress. b. Montreal, Canada, June 3, 1926. Gained fame as stage actress, appearing in, among others, Mourning Becomes Electra, All Over, A Moon for the Misbegotten, More Stately Mansions, All the Way Home, The Ballad of the Sad Cafe, etc.
PICTURES: The Nun's Story, A Fine Madness, McQ, The Cowboys, The Last Run, Annie Hall, Ice Castles, When a Stranger Calls, Tribute.
TELEVISION: The Price, The Crucible, The Hands of Cormac Joyce, Jacob and Joseph, Studs Lonigan, The Kitty O'Neill Story, And Baby Makes Six, etc.

DEWITT, JOYCE: Actress. b. Wheeling, W. Va., April 23. e. Ball State Univ. (B.A.), theatre. Performed in summer stock musical comedies (Damn Yankee, Li'l Abner, etc.). Worked at Indianapolis dinner theatres, directing, staging and starring in musical revues.
TELEVISION: Baretta, The Tony Randall Show, Most Wanted, Risko, Three's Company. Movie: With This Ring.

DEY, SUSAN: Actress. b. Pekin, Ill. Began modeling at 15 for magazines. Has been frequent guest star on TV series: The Rookies, S.W.A.T., Hawaii Five-O, Born Free, Petrocelli, etc.
PICTURES INCLUDE: First Love, Looker.
TV: Movies: Terror on the Beach, Cage without a Key.

DEZEL, ALBERT: Exhibitor. b. Russia, May 21, 1900; e. Chicago Acad. Fine Arts; also Art Inst. & Crane Coll., Chicago. Art work, theatre publicity, exploitation & Middle West. Org. Albert Dezel Roadshows, exploitation films, 1926; in 1945 owner, operator, Film Classics Franchise, Detroit, Cleveland, Cincinnati; in 1946 acquired rights in same territory for Screen Guild Products; operator, operator, Avon Art Theatre, South Bend; Cinema, Hamilton Theatre, Indianapolis; Guild Theatres, Detroit.

DIAMANT, LINCOLN: Executive. b. New York City, Jan. 25, 1923. e. Columbia College, AB cum Laude 1943; cofounder, Columbia University Radio Station. Advertising mgr., World Publ. Co.; writer-prod, CBS; prod.-dir., Blue Network (NBC); prom. dir., Book Find Club; creative dir., Ben Sackheim; group hd., McCann-Erickson; copy chief, DKG; Sr. TV prod., Ogilvy & Mather; TV prod. super., Grey Adv; founder, pres., Spots Alive, Inc., broadcast advertising consultants, 1969; author, The Broadcast Communications Dictionary, Anatomy of a Television Commercial, Television's Classic Commercials, contrib., to Effective Advertising, to Messages and Meaning; member, Broadcast Pioneers, Acad. TV Arts & Sciences; v.p. Broadcast Advertising Producer's Society of America. Adjunct faculty member, Pace College, Hofstra University. Fellow, Royal Society of Arts. President, Teatown Lake Reservation of Brooklyn Botanic Garden.

DIAMOND, BERNARD: Theatre Executive. b. Chicago, Jan. 24, 1918. e. Univ. of Indiana, Univ. of Minnesota. Except for military service was with Schine Theatre chain from 1940 to 1963, working up from ass't. mgr., booker, buyer, dir. of personnel to gen. mgr. Then joined Loews Theatres; now exec. v.p.

DIAMOND, I.A.L.: Writer. b. Unghani, Roumania, June 27, 1920. r.n. Isidore Diamond. e. Col Univ. Came to US when 9 yrs. old. Hired as Junior writer by Paramount in 1941. Usually collaborates on scripts with other writers, particularly in recent years with Billy Wilder.

PICTURES INCLUDE: (All co-written except where indicated by (s) for solo) Murder in the Blue Room, Never Say Goodbye, Two Guys from Milwaukee, Two Guys from Texas, Always Together, The Girl from Jones Beach (s), It's a Great Feeling (s), Love Nest (s), Let's Make It Legal, Monkey Business, Something for the Birds, That Certain Feeling, Love in the Afternoon, Merry Andrew, Some Like It Hot, The Apartment, One, Two, Three, Irma La Douce, Kiss Me, Stupid, The Fortune Cookie, Cactus Flower (s), The Private Life of Sherlock Holmes, Forty Carats (s), Avanti!, The Front Page, Fedora.

DICKINSON, ANGIE: Actress. b. Kulm, N.D. Sept. 30, 1936. e. Parochial Schools, Heart Coll., Glendale Coll., secretarial course. m. Burt Bacharach. Won a beauty contest.

PICTURES INCLUDE: Sins of Rachel Cade, Rio Bravo, Bramble Bush, Ocean's 11, Jessica, Capt. Newman, M.D., The Killers, The Art of Love, The Chase, The Last Challenge, Point Blank, Sam Whiskey, Young Billy Young, Pretty Maids All in a Row, The Outside Man, Dressed to Kill, Charlie Chan and the Curse of the Dragon Queen, Death Hunt.

DICKINSON, DESMOND: Dir., photog. b. Surrey, May 25, 1902. e. privately. Entered m.p. indust. old Clarendon Film Co., 1919; 1922 camera asst. Stoll. Film Co.; with various motion picture cos. since.

PICTURES INCLUDE: Hamlet, Rocking Horse Winner, The Woman in Question, The Browning Version, Encore, Importance of Being Earnest, The Net, Man Between, Carrington V.C., Gentlemen Marry Brunettes, Black Tent, Fire Down Below, Action of the Tiger, Orders to Kill, Intent to Kill, Horrors of the Black Museum, Moment of Danger, City of Death, Foxhole in Cairo, Kongo, Hands of Orlac, Mary Had a Little, The Frightened City, The Devils Daffodil, Two & Two Make Six, Cairo, Sparrows Can't Sing, Murder Most Foul, Murder Ahoy, Alphabet Murders, A Study in Terror, Berserk, Decline and Fall, Baby Love, Crooks and Coronets, Trog, Incense for the Damned, The Fiend, Burke and Hore, Nobody Ordered Love, Who Slew Auntie Roo?, Tower of Evil, The Man From Nowhere.

DICKINSON, GLEN JR.: Executive. b. Brookfield, Mo., May 29, 1914. e. Pembroke County Day, Kansas U.; officer, U.S. Navy; chm., Dickinson Operating Co., Inc.; chm., Dickinson Inc.; chm., Glenwood Manor, Inc.

DICKINSON, ROBERT PRESTON: Actor. b. Newhall, W. Va.; Dec. 24, 1924. e. Va. Polytechnic Inst. & State University, B.S., 1947; N.Y. Univ., U. Calif., American Academy of Dramatic Arts, Pasadena, Ca. Private schools. U.S. Army Signal, Infantry Corp, W.W. II U.S. chemical-aerospace firms, 10 yrs., prof. chemical engineer-writer; legit. stage plays include: Murder in the Red Barn; Career; The Ponder Heart. M.p. include: The Brain Eaters; Her Conversion. Elected member Int'l. Platform Ass'n.

TV: The Rank & File (Playhouse 90 series, CBS).

DICKINSON, THOROLD: C.B.E. 1973. Prof. Emeritus of Film. b. Bristol, England, Nov. 16, 1903. e. Clifton Coll., Bristol; Oxford U. Entered m.p. ind. 1925 as asst. dir.; from 1927–36 film ed., prod. supvr. Midshipman Easy; from 1936–41 dir. (High Command, Spanish A.B.C., Arsenal Stadium Mystery, Gaslight, Prime Minister, Next of Kin, M.O.I. short subjects, etc.). From 1941–45 Army Kine Service. Since 1945 dir. Men of Two Worlds, Queen of Spades; A.B.P.C. 1949–50; dir., collab. s.p., Secret People; 1952–53 chmn. British Film Acad.; dir., collab. story & s.p., Hill 24 Doesn't Answer; 1956–60 in chge. U.N. Film Services, N.Y.; Slade School of Fine Arts, Univ. Coll., London, Eng.: Film Dept. 1967–71; Prof. of Film U. of London. Ph.D. (London U.). 1973–6 Hon. Consultant. Council of National Academic Awards. 1976 D. Univ (Surrey). Visiting Prof. of Film (Surrey U.) until 1977.

PUBLICATIONS: A Discovery of Cinema, Soviet Cinema (with Catherine de la Roche).

DIEBOLD, JEROME C.: Executive. b Madison, Wis., May 30, 1909. e. U. of Wis. in economics and law; photographer and teacher, Univ. of Wisc.; visual education, 1933–43; served in U.S. Navy as Lt. Commdr., 1943–46; producer and director of Navy training motion pictures; head, Training Film Branch 1945–46; asst. prod. mgr., Wilding Inc., 1946–48; assist. to the pres. Wilding Inc. 1948–53; executive producer, 1954–63; dir. of prod., Enclop. Brit. Films, 1963–65; pres., Diebold Film Prod., 1965, Life Fellow SMPTE.

DIEHL, WALTER F.: Executive. International President, I.A.T.S.E. b. Revere, Mass., April 13, 1907. e. Northeastern Univ., Boston Catholic Labor Guild School. Became projectionist in Boston in 1927 and member of Boston Moving Picture Machine Operators Local 182 in 1933. After several years of service on the executive board became business agent in 1946 and held that post until 1953, when he resigned to accept an appointment as an International Representative, handling assignments throughout New England until 1957, when was appointed assistant international president. In 1952 made a member of the Minimum Wage Commission for the amusement industry in the state of Massachusetts. Also served several years on the executive board of the Boston Central Labor Union and as Labor Representative on the Suffolk County March of Dimes. IATSE pres. since March, 1974.

DIENER, JOAN: Actress. b. Cleveland, Ohio, Feb. 24. e. Sarah Lawrence College.

THEATRE: Plays: H.M.S. Pinafore, Cleveland Playhouse 1947; Broadway debut, Small Wonder, 1948; Season in the Sun, 1950; Kismet 1953, also London production, 1955. Road tours, The Ziegfeld Follies, 1956; At the Grand, 1958; La Belle, 1962. Man of the La Mancha, 1965. New York, Los Angeles and London productions, Cry For Us All.

DIETRICH, MARLENE: Actress. r.n. Maria Magdalene von Losch. b. Berlin, Germany, Dec. 27, 1904; e. pvt. sch., Weimar; Musical Acad., Berlin; m. Rudolph Sieber; d. Maria Riva, actress. Stage training, Max Reinhardt's Sch. of Drama; debut in Viennese version of play, Broadway; then mus. com.

PICTURES INCLUDE: In German film, Blue Angel; in Hollywood: Morocco, Dishonored, Blonde Venus, Shanghai Express, Scarlet Empress, Song of Songs, Desire, Angel, Garden of Allah, Knight Without Armour, Destry Rides Again, Seven Sinners, Flame of New Orleans, Manpower, Pittsburgh, The Spoilers; with Army overseas (USO), 1942–45. Golden Earrings, Martin Roumagnac (French), Foreign Affair, Stage Fright, No Highway in the Sky, Rancho Notorious, Witness for the Prosecution, Judgment at Nuremberg.

DIFFRING, ANTON: Actor. b. Koblenz, Germany. Early career Academy of Drama, Berlin. Acted in Canada and U.S. before ent. m.p. ind. in Britain, 1951.

PICTURES INCLUDE: I Am a Camera, Seven Thunders, House of Secrets, Albert RN, Circus of Horrors, The Man Who Could Cheat Death, A Question of Adultery, Lady of Vengeance, The Traitor, The Crooked Sky, The Sea Shall Not Have Them, The Colditz Story, The Black Tent, Enter Inspector Duval, Incident at Midnight. 1963/65 working in Germany, The Heroes of Telemark, The Blue Max, Farenheit 451, The Double Man, Counterpoint, Where Eagles Dare, Michael Kohlhaas, Horse Dealer, Uccidete Rommel, Piggies, Zeppelin, L'Iguana della lingua del Fuoco, Sutjeska, Corringa, Assassination of Little Sister, Dead Pigeon on Beethoven St., Der Stoff aus dem die Traume sind, The Day the Clown Cried, Das Schicksal ist der Jager, Tony Arzenta, The Beast Must Die, Shatter, Swiss Conspiracy, Vanessa, Operation Daybreak, Potato Fritz, Valentino, Les Indiens sont encore loin, L'Imprecateur, Waldrausch, Anna Ferroli, Das Einhorn, Return to Munich, Arsene Lupen, Escape to Victory, Tristan und Isolde.

TV: The Last Hours, A Small Revolution, Wedding Day, The One Who Came Back, The Man in Half Moon Street, The Queen and the Rebels, The Cold Light, The Magnificent Egoist, The One Who Got Away, Autumn Crocus, The Fourposter, Dear Charles, Cross of Iron, One Step Beyond, Ghost Squad, Dr. Korczak and His Children, Dixon of Dock Green, The Joel Brandt Story, Firebrand, Scobie in September, The Trouble Shooters The Million Pound Note, Strange Report, Kommissar, A Place in the Sun, Die Krise, Assignment Vienna, Liebe 74, Yoster, Kiss Me and Die, Die Himmlischen Tochter, Car along the Pass, MM Le Mutant, Flambards, Plutoniumaler Alte.

DILLER, BARRY: Executive. Joined ABC in April, 1966, as asst. to v.p. in chg. programming. In March, 1968, made exec. asst. to v.p. in chg. programming and dir. of feature films. In May, 1969, named v.p., feature films and program development, east coast. In Feb., 1971, made v.p., Feature Films and Circle Entertainment, a unit of ABC Entertainment, responsible for selecting, producing and scheduling The Tuesday Movie of the Week, The Wednesday Movie of the Week, and Circle Film original features for airing on ABC-TV, as well as for acquisition and scheduling of theatrical features for telecasting on ABC Sunday Night Movie and ABC Monday Night Movie. On May 30, 1973, named v.p. in chg. of prime time TV for ABC Entertainment. In 1974 joined Paramount Pictures as bd. chm. and chief exec. officer.

DILLER, PHYLLIS: Comedienne. b. Lima, O., July 17, 1917. e. Sherwood Music School, 1935–37; Bluffton College, O., 1938–39.

PICTURES INCLUDE: Boy, Did I Get A Wrong Number! 1966; Mad Monster Party, 1967; Eight on the Lam, 1967; Did You Hear the One About the Traveling Saleslady, 1967; The Private Navy of Serg. O'Farrell, 1967; The Adding Machine.

TV: Series, The Phyllis Diller Show, ABC 1966–67; The Beautiful Phyllis Diller Show (NBC) 1968. Guest spots on major shows here and abroad.

BOOKS: Phyllis Diller's Housekeeping Hints, 1968, Phyllis Diller's Marriage Manual, Phyllis Diller's The Complete Mother.

RECORDS: Verve and Columbia.

BROADWAY: Hello, Dolly!

CONCERTS: A dozen annual piano concerts with symphony orchestras.

DILLMAN, BRADFORD: Actor. b. San Francisco, Calif., April 14, 1930. e. Yale, 1951.

PICTURES INCLUDE: A Certain Smile, In Love and War, Compulsion, Crack in the Mirror, Circle of Deception, Sanctuary, Francis of Assisi, A Rage to Live, Sergeant Ryker, Bridge at Remagen, Jigsaw, Suppose They Gave a War and Nobody Came, Mastermind, Brother John, The Mephisto Waltz, Escape from the Planet of the Apes, The Way We Were, The Iceman Cometh, Chosen Survivors, 99 and 44/100ths ½ Dead, Gold Bug, A Black Ribbon for Deborah, One Away, The Enforcer, Amsterdam Kill, The Lincoln Conspiracy, The Swarm, Love and Bullets, Piranha, Guyana-Crime of the Century, Running Scared.

DILLON, MELINDA: Actress. Launched career on Broadway in original company of Who's Afraid of Virginia Woolf? Screen debut with bit in The April Fools (1969).

PICTURES INCLUDE: Bound for Glory, Slap Shot, Close Encounters of the Third Kind, F.I.S.T.

DILLOW, JEAN CARMEN: Producer, writer, actress, director. Began as baby star at age 5 on stage, screen. Film work in England, Germany, Switzerland, Mexico, Italy. Playwriting with George S. Kaufman; screenplay-writing with Andrew Solt. Wrote and produced with John Croydon, London producer. Wrote produced, directed and starred in The Pawn, filmed in Italy, Vienna, Warner Bros. TV feature. Author, Morning Angel, Boy Doll; Latest play to music for London, in which she stars, There is No Other Prince But Aly. Completed scripts Spirit-Doll. The Resurrection, for Pinewood Studios, London. Features in production, The Lavender Mule From Leningrad and Little Yakov's Visions, Norway. See Carmen Productions under Corporations. 1979, non-fiction books on phenomena: Do You Hear the Voices? and The Kidnapping of Aldo Moro. Wrote TV show, Portrait of Zsa Zsa.

DIMMOCK, PETER: CVO, OBE. Vice President-managing Director ABC Sports Worldwide Enterprises Ltd. b. Dec. 6, 1920. e. Dulwich Coll. & in France, R.A.F. pilot & Staff Officer during war. TV as prod. & commentator BBC Outside Broadcasts Dept., 1946. Prod. more than 500 TV relays including telecast from Windsor Castle of King George VI's funeral, 1952. Coronation telecast Westminster Abbey; State Opening Parliament. Commentator, introduced BBC-TV weekly Sportsview, 1954–64. Head of BBC-TV Outside Broadcasts 1954–72. Head of BBC Enterprises 1973, v.p. ABC Cos., Inc., 1977. Fellow Royal TV Society.

DIPSON, WILLIAM D.: Circuit executive. b. U.S.A., July 24, 1916; e. Phillips Exeter Acad., 1933; Dartmouth Coll., 1937, A.B.; Harvard Law School. Currently pres., Dipson Theatres, Inc.; Army 1940–45. Member: Variety Club of Buffalo, chief barker, 1948.

DISNEY, ROY E. Prod.-dir.-writer, cameraman, film editor. b. January 10, 1930, Los Angeles; e. Pomona College, Calif. 1951 started as page, NBC-TV. Asst. film editor "Dragnet" TV series. 1952–78, Walt Disney Prods., Burbank, Calif., various capacities; now pres. Roy E. Disney Prods.; bd. chm., Shamrock Bldg. Co., bd. dir., Walt Disney.

PICTURES INCLUDE: Perri, Mysteries of the Deep, Pacific High.

TV: Walt Disney's Wonderful World of Color; The Hound That Thought He Was A Raccoon, Sancho, The Homing Steer, The Silver Fox and Sam Davenport, Wonders of the Water World, Legend of Two Gypsy Dogs, Adventure in Wildwood Heart. Also, The Postponed Wedding, (Zorro series), (Wonder World of Color); An Otter in the Family, My Family Is A Menagerie, Legend of El Blanco, Pancho, The Fastest Paw In The West, The Owl That Didn't Give A Hoot, Varda the Peregrine Falcon, Cristobalito, The Calypso Colt, Three Without Fear, Hamad and the Pirates, Chango, Guardian of the Mayan Treasure, Nosey, the Sweetest Skunk in the World, Mustang!, Call It Courage, Ringo, the Refugee Raccoon, Shokee, the Everglades Panther, Deacon, the High-Noon Dog, Wise One, Whale's Tooth, Track of African Bongo, Dorsey, the Mail-Carrying Dog.

DMYTRYK, EDWARD: Director. b. Grand Forks, B.C., Canada, Sept. 4, 1908. Entered employ Paramount 1923, working after school. Film editor 1930–1939; dir. from 1939.

PICTURES INCLUDE: Murder My Sweet, Back to Bataan, Cornered Crossfire, So Well Remembered, Till the End of Time, Give Us This Day, Mutiny, The Sniper, Eight Iron Men, The Juggler, Caine Mutiny, Broken Lance, End of the Affair, Left Hand of God, Soldier of Fortune, The Young Lions, Where Love Has Gone, Mirage, Alvarez Kelly, Anzio, Bluebeard.

DODD, WILFRID E.: Executive. b. Hamburg, Germany, March 19, 1923. e. Institut Fisher, Montreux (Switzerland), Downside College, Bath (England), Royal Military College, Sandhurst (England). In the '50s formed the Canadian Investing Banking firm: Oswald Drinkwater & Graham Ltd., Vice Pres. and partner. Pres. Allied Artists International Corp., 1964. President of Allied Artists Productions, Inc., 1969. Formed Cinepix Inc., Canadian dist. firm, 1963.

DOLLINGER, IRVING: Exhibitor. Columbia Amusement Co. b. N.Y.C., Sept. 20, 1905; e. U. of Pennsylvania. Stanley-Fabian mgr., 1926. Then with Warner Theatres in New Jersey. Owner and operator of theatre since 1929. Past pres. Allied Theat. Owners of N.J. Pres., Assoc. Theats. of N.J., booking org. 1938–44; v.p. & buyer, Independent Theatre Service Eastern regional v.p. Allied States Assoc., 1949–54; treas., Nat'l Allied, 1955–56; partner, Triangle Theatre Service, 1957; chief barker, N.Y. Tent 35, Variety Club, 1966.

DONAHUE, PHIL: Television Personality. b. 1936. Started daytime talk show in Dayton, Ohio, 1967. Originally called The Phil Donahue Show. Syndicated show, Host, Donahue, now in 165 outlets in U.S. Home base: WGN—TV, Chicago, an independent station. In 1979 a mini-version of show became 3-times-a-week segment on NBC's Today Show. Winner of 2 Emmys.

DONAHUE, TROY: Actor. r.n. Merle Johnson, Jr. b. New York City, Jan. 17, 1937. e. Bayport H.S., N.Y. Military Acad. Directed, wrote, acted in school plays. Journalism, Columbia Univ.; Summer stock, Bucks County Playhouse, Sayville Playhouse; contract, Warner Brothers, 1959.

PICTURES INCLUDE: A Summer Place, The Crowded Sky, Parrish, Susan Slade, Palm Springs Weekend, Rome Adventure, A Distant Trumpet, My Blood Runs Cold, Come Spy With Me, Sweet Savior, Godfather, Part II.

TV: Surfside 6, Hawaiian Eye.

DONALD, JAMES: Actor. b. Aberdeen, Scotland, May 18, 1917; e. Rossall School, McGill U., Edinburgh U. Stage debut: The Scarlet Pimpernel, 1936; screen debut: The Missing Million, 1940.

PICTURES INCLUDE: In Which We Serve, The Way Ahead, Broken Journey, The Small Voice, Trottie True (The Gay Lady), White Corridors, Brandy for the Parson, The Gift Horse, The Pickwick Papers, Lust for Life, Bridge on the River Kwai, The Vikings, Third Man on the Mountain, The Great Escape, King Rat, Cast a Giant Shadow, The Jokers, Epitaph For a Spy, The Royal Hunt of the Sun, David Copperfield, Conduct Unbecoming, The Big Sleep.

THEATRE & TV: Numerous U.S. and Br. TV and stage appearances.

DONCASTER, CARYL: TV writer, prod. b. Liverpool, Eng., Sept. 28, 1923. e. London U. (B.A. Honors). Educational film & film strip prod.; TV as BBC Talks producer in 1948; docum. writer & prod., 1950.

PRODUCTIONS: Made by Hand series, Report on Gambling, Report on Women, Born with the Century, The Rising Twenties, The Declining Years, Make Me an Offer, Pattern of Marriage, Children in Trust, Return to Living, New Canadians, Those Who Dare. Writer, dir., Canadian Venture, 1955; joined Associated Rediffusion, prod., Big City, This Week, People Are Talking, Secret Mission, Mantrap, Boyd Q.C., Outlook, Look In on London, Northern Journey, Members Mail, The Unwanted; head, Assoc. Rediffusion features; free lance TV writer-prod.; writ., dir., School at Sea, Supermarket; Lecturer in prod., Thomson TV; Four Men of India. Prod. BBC TV, Report on Immigrant School Children, on Nurses, 1966; TV experiments in Sierra Leone, 1966. 1968 Nigerian TV; 1969 Documentaries in Mexico, Peru, Java, Zambia, Baffin Island, India.

DONEN, STANLEY: Prod dir. b. Columbia, S.C., April 13; p. Mortie and Helen Donen; e. local public, high schools, U. of South Carolina.
PICTURES INCLUDE: Royal Wedding; Singin' in the Rain; It's Always Fair Weather, Deep in My Heart, Seven Brides for Seven Brothers, On the Town, Funny Face, Pajama Game, Kiss Them for Me, Indiscreet, Damn Yankees, Once More With Feeling, The Grass Is Greener, Charade, Arabesque, Two for the Road, Bedazzled, Staircase, The Little Prince, Lucky Lady, Movie, Movie.

DONIGER, WALTER: Writer, dir. prod. b. New York City. e. Valley Forge Military Academy, Duke University, Harvard Graduate Business School. Entered m.p. business as writer later writer-prod-dir. Wrote documentaries in Army Air Forces M.P. Unit in World War II.
PICTURES INCLUDE: Rope of Sand; Desperate Search, Cease Fire, Safe At Home, House of Women, Duffy of San Quentin, Along the Great Divide, Tokyo Joe, Alaska Seas, Steel Cage, Steel Jungle, Hold Back the Night, Guns of Fort Petticoat, Shannon (pilot) and Movies of Week.
TELEVISION: Delvecchio, Mad Bull, Switch, Moving On, Baa, Baa, Blacksheep, McCloud, The Man and the City, Sarge, Owen Marshall, Peyton Place (200 episodes), Mr. Novak, The Greatest Show On Earth, Travels of Jaimie McPheeters, Outlaws, Hong Kong, Checkmate, Bat Masterson, The Web, Bold Venture, Tombstone Territory, Maverick, Rough Riders, Captain Grief, Lockup, Dick Powell, The Survivors. Bracken's World, Bold Ones, Kung Fu, Barnaby Jones, Marcus Welby, Lucas Tanner.

DONLAN, YOLANDE: Actress, b. Jersey City, June 2; e. Immaculate Heart Convent, Hollywood, Calif. English stage debut in Born Yesterday at the Garrick Theatre (London) 1947; screen debut in Miss Pilgrim's Progress 1950.
PICTURES INCLUDE: To Dorothy a Son, The Body Said No, Traveller's Joy, Mr. Drake's Duck, Penny Princess, They Can't Hang Me, Expresso Bongo, Jigsaw, 80,000 Suspects, The Adventurers, Seven Nights in Japan.

DONNELL, JEFF: Actress. b. South Windham, Me., July 10, 1921. r.n. Jean Donnell, e. Leland Powers Sch. of Theat., Yale Sch. Dram. On stage Farragut Playhouse, Rye Beach, N.H., before screen debut 1942. Summer stock, N.Y., Fla., Ohio, Wash., Calif.
PICTURES INCLUDE: Sweet Smell of Success, My Man Godfrey, Gidget Goes Hawaiian. Gidget Goes to Rome, Fran Maidin, Stand Up and Be Counted.
TV: George Gobel Show. U.S. Steel Hour, Play of the Week, June Allyson, Ann Southern Shows.

DONNELLY, RALPH E.: Executive. b. Lynbrook, New York, Jan. 20, 1932. e. Bellmore (L.I.) Public School; W.C. Mepman High School (1949). Worked for Variety (publication) as writer, 1950; Long Island Press as daily columnist, 1951; joined Associated Independent Theatres, 1953, as gen. mgr.; later film buyer; in 1973 left to become independent buyer and booker for Creative Films; film buyer and v.p., RKO/Stanley Warner Theatres, 1976–79; now v.p. & gen. mgr. for Cinema Ltd. circuit in New York.

DONNENFELD, BERNARD: Exec. b. New York, N.Y. Oct. 28, 1926. e. New York U., LL.B. Admitted to New York, Cal. Bar. Spvsr., corp. affrs., legal dept., Paramount Pictures Corp., New York, 1957–61; exec. asst., asst. secy. corp., Hollywood Studio, 1961–64; asst. to pres., asst. secy. corp., New York, 1964–65; assoc. hd. stud. activities, asst. secy. corp., Hollywood Studio, 1965, v.p. in charge of World Wide Production, Administration, Paramount, 1965–69. Pres. The Filmakers Group, 1970.

DONNER, CLIVE: Director. b. London, Eng. Ent. m.p. ind. 1942.
PICTURES INCLUDE: The Card, Meet Me Tonight, Genevieve, Million Pound Note, Purple Plain, I Am a Camera, The Secret Place, Heart of a Child, Marriage of Convenience, Some People, The Caretaker, Nothing But the Best, What's New Pussy Cat, Luv, Here We Go Round The Mulberry Bush, Alfred the Great, Vampira, Spectre, Rogue Male, Three Hostages, She Fell Among Thieves, The Thief of Baghdad, The Nude Bomb, Carlie Chan and the Curse of the Dragon Queen.
TV: Danger Man, Sir Francis Drake, Mighty and Mystical, British Institutions, Tempo.

DONNER, RICHARD: Director. b. New York City. Began career as actor off-Bdwy. Worked with director Martin Ritt on TV production of Maugham's Of Human Bondage. Moved to California 1958, directing commercials, industrial films and documentaries. First TV drama: Wanted: Dead or Alive (Steve McQueen series).
TELEVISION: Many Movies of the Week; episodes of Kojak, Bronk, etc.; Portrait of a Teen-Age Alcoholic.
PICTURES: The Omen, Superman, The Final Conflict, exec. prod.

DONOHUE, JACK: Director. b. New York City, Nov. 3, 1914. Entered show business as dancer in Ziegfeld Follies; later danced in and directed dances for numerous productions on Broadway and London stage. Dance dir. Fox, Pana., MGM.
PICTURES INCLUDE: Close Up Yellow Cab Man, Watch the Birdie, Calamity Jane, Lucky Me, Babes in Toyland, Marriage on the Rocks, Assault on a Queen.
THEATRE: Top Banana, Mr. Wonderful.
TV: Frank Sinatra, Dean Martin, Red Skelton, George Gobel, Ed Wynn shows, The Lucy Show, The Jim Nabors Show, Chico and the Man.

DONOVAN, HENRY B.: Executive. producer. b. Boston, Mass. Entered m.p. ind. for RKO Pathe Studios 1932; asst. cameraman, head of property dept., special effects dept., unit mgr., asst. dir., prod. mgr.; worked on over 300 pictures for many studios; dept. head Sherman Prod. 10 yrs., U.S. Army Signal Corps, as head of dept. of California studios prod. training m.p.; pres. Telemount Pictures, Inc. Prod., dir., writer for Cowboy G Men (TV series).
PICTURES INCLUDE: Hopalong Cassidy Features, Gone with the Wind, Becky Sharp, dir. Our Flag; Magic Lady (live-action TV show); Magic Lady (13 one-reel featurettes), others. Prod. writer, Cowboy G. Men (39 films), TV, 1952–54; Law in Action; Frontier Lawyer; Safe Crackers.
TV: programming, financing, distribution. Global Scope; International TV; Dist., Financing, programming; sls. consultant, Internat'l TV & motion pictures. Cable TV & distribution & program development, collector of movie memorabilia.

DOOLEY, PAUL: Actor. Began career on N.Y. stage in Threepenny Opera. Later member of Second City. Bdwy. credits include The Odd Couple, Adaptation/Next, The White House Murder Case, Hold Me, etc. Co-creator and writer for The Electric Company on PBS. Owns co. called All Over Creation, which produces original industrial films and shows and has created over 1,000 radio commercials.
PICTURES: What's So Bad About Feeling Good?, Slap Shot, Gravy Train, Death Wish, The Out-of-Towners, A Wedding, A Perfect Couple, Rich Kids, Breaking Away, Popeye.

DORAISWAMY, V.: Executive, publicist. b. Trichur, India, 1912. Nat'l Council of YMCA's for India, Burma & Ceylon, 1928; dist. secy., Rotary Inter'l, India, Burma & Ceylon, 1935–42; secy., RCA Photophone Equip., Bombay, 1942; later sales admin. Ed. & Compiler, Asian Film Directory & Who's Who; Film Technician and Theatreman's Diary. India-Ceylon representative, Motion Picture Herald & Motion Picture Daily; Secretary, Cinematographic Suppliers Association of India, Post Box No. 311A, Bombay-1.

DORFMAN, IRVIN S.: Executive. b. New York, Sept. 3, 1924. e. Yale University, 1945 B.A. Following three years in U.S. Navy, joined Dorfman Associates, press rep. organ. handling film, and TV personalities. Handled pub. and promotion for more than 100 B'way shows and worked for 20th Cen.-Fox and UA. Also produced off-B'way "The Lion in Love." Top ranking amateur tennis star, 1950; pres., Surrogate Releasing and Dancrea Releasing.

DORFMAN, ROBERT S.: Executive. b. New York, N.Y. March 23, 1930. e. CCNY. Newspaper and magazine contact, exploit. mgr. publicity and asst. adv. and publ. dir., Walt Disney Prod. and Buena Vista 1948–64. Joined Seven Arts in 1964 as national promotion dir.; appointed adv. dir. 1966; app'ted nat'l adv. dir. Warner Bros. 1967; Eastern adv.-pub. dir., 1970. Joined 20th Century-Fox as eastern pub. dir., 1980.

DORS, DIANA: Actress. b. Swindon, Wilshire, Oct. 23, 1931; e. Colville House, Swindon; studied at London Acad. of Music and Dramatic Art; first screen role in The Shop at Sly Corner at 14; contract work with J. Arthur Rank Organization at 15.
PICTURES INCLUDE: Lady Godiva Rides Again, The Weak and the Wicked, Is Your Honeymoon Really Necessary, A Kid for Two Farthings, Value for Money, An Alligator Called Daisy, As Long as They're Happy, Yield to the Night (Blonde Sinner), I Married a Woman, Unholy Wife, Long Haul, Tread Softly Stranger, Passport to Shame, Room 43, Love Specialist, Ladies May, On the Double, Mrs. Gibbon's Boys, West Eleven, Allez France, Circus of Blood, Baby Love.
TV: The Innocent.

DORTORT, DAVID: Executive Producer. b. New York, Oct. 23, 1916. e. City Coll. of N.Y. Served U.S. Army, 1943–46. Novelist and short story writer, 1943–49. Also TV writer. Now pres. of Xanadu Prods. and Aurora Enterprises, Inc., Pres. TV branch, WGA, West, 1954–55; TV-radio branch, 1955–57; v.p. PGA, 1967; pres. 1968. Chairman of The Caucus for Producers, Writers and Directors, 1973–75. Pres., PGA, 1980; campaign director, Permanent Charities Comm., 1980–81; chm., Interguild Council 1980–81.
AUTHOR: Novels include Burial of the Fruit and The Post of Honor.
PICTURES INCLUDE: The Lusty Men, Reprisal, The Big Land, Cry in the Night, Clash by Night.
TV: Creator, of Bonanza TV series; High Chaparral, The Chisholms.

DOUGHERTY, MARION: Executive. Gained fame as casting director. Acted as co-executive producer on Smile (1975). In 1977 named v.p. in chg. talent for Paramount Pictures. In 1979 joined Warner Bros. v.p. in chg. talent to work with production dept. and producers and directors.

DOUGLAS, GORDON: Director, b. New York, New York. Actor. Hal Roach stock company; writer; callab. Topper series, Housekeeper's Daughter; dir., 30 Our Gang shorts.
PICTURES INCLUDE: Saps at Sea, Broadway Limited, Devil with Hitler, First Yank into Tokyo, San Quentin, If You Knew Suzie, Black Arrow, Walk a Crooked Mile, Doolins of Oklahoma, Mr. Soft Touch, The Nevadan, Between Midnight and Dawn, Kiss Tomorrow Goodbye, Great Missouri Raid, Only the Valient, I Was a Communist for the FBI, Come Fill the Cup, Mara Maru, Iron Mistress, She's Back on Broadway, So This Is Love, The Charge at Feather River, Them, Young at Heart, McConnell Story, Sincerely Yours, Santiago, The Big Land, Bombers B-52, Fort Dobbs, Yellowstone Kelly, Rachel Cade, Gold of 7 Saints, Follow That Dream, Call Me Bwana, Rio Conchos, Robin and the Seven Hoods, Sylvia, Harlow, Stagecoach, Way Way Out, In Like Flint, Chuka, Tony Rome, The Detective, Lady in Cement, Barquero, They Call Me Mr. Tibbs, Slaughter's Big Rip Off, Nevada Smith (MOW), Viva Knieval.

DOUGLAS, KIRK: Actor, Producer, Director. b. Amsterdam, N.Y. Dec. 9, 1918; e. St. Lawrence U; B.A. Stage debut in New York: Spring Again; others: The Sisters, Kiss and Tell; U.S. Navy during World War II; resumed stage work in Trio; did radio soap operas. Signed by Hal B. Wallis from The Wind is Ninety, B'way show, 1945. Screen debut: The Strange Love of Martha Ivers. *President of The Bryna Company. Recipient of U.S. Freedom Medal, 1981.
TV: Mousey, 1974; The Money Changers, 1976.
PICTURES INCLUDE: Out of the Past, I Walk Alone, Mourning Becomes Electra, The Walls of Jericho, Letter to Three Wives, Champion, Young Man With a Horn, The Glass Managerie, Big Carnival, Along the Great Divide, Detective Story, Big Trees, Big Sky, Bad and the Beautiful, Story of Three Loves, The Juggler, Art of Love, Ulysses, Man Without a Star, 20,000 Leagues Under the Sea, The Racers, Indian Fighter, Lust for Life, Top Secret Affair, Gunfight at the OK Corral, Paths of Glory, The Vikings, Last Train from Gun Hill, The Devil's Disciple, Spartacus, Town Without Pity, Lonely Are the Brave, Two Weeks in Another Town, List of Adrian Messenger, For Love or Money, Seven Days in May, In Harms Way, Cast a Giant Shadow, The Way West, The War Wagon, Heroes of Telemark, A Lovely Way to Die, The Brotherhood, The Arrangement, There Was a Crooked Man, A Gunfight, Light at the Edge of the World, Catch Me a Spy, A Man to Respect. Producer: Indian Fighter, The Vikings, Spartacus, Lonely Are The Brave, List of Adrian Messenger, Seven Days in May, The Brotherhood, Summertree, (prod.) Scalawag; (prod., dir., act.) Once Is Not Enough, Posse (prod., dir., act.) The Chosen, The Fury, Home Movies, The Villain, Saturn III, The Final Countdown, The Man from Snowy River.

DOUGLAS, MICHAEL: Actor. b. N. Brunswick, N.J., Sept. 25. f. Kirk Douglas. e. Black Fox Military Academy, Choate, U. of Cal. Worked as asst. director on Lonely Are The Brave, Heroes of Telemark, Cast A Giant Shadow; after TV debut in The Experiment, appeared off-Broadway in City Scene, Pinkville; m.p. debut in Hail Hero, followed by Adam at 6 A.M., Summertree, Napoleon and Samantha, Coma, China Syndrome (also prod.). Co-prod., One Flew Over the Cuckoo's Nest; Running (actor). It's My Turn (actor).
TV: The FBI, Medical Center, When Michael Calls, Streets of San Francisco (series).

DOUGLAS, MIKE: TV host, commentator. r.n. Michael Delaney Dowd, Jr. b. Chicago, 1925. Started career singing with bands in and around Chicago. In 1950 became featured singer with Kay Kyser's band, staying four years. In 1953 became host of WGN-TV's Hi Ladies in Chicago; also featured on WMAQ-TV, Chicago, as singer and host. Moved to Hollywood in late '50s, working as piano bar singer. In 1961 hired as host for new show on station KYW-TV in Cleveland, owned by Westinghouse Bdg. Co., featuring celebrity guests. This became the Mike Douglas Show which was later nationally syndicated and moved base of operations to Philadelphia. Presently is seen in about 130 U.S. outlets and 15 in Canada.

DOUGLAS, VALERIE: Executive. b. Hollywood, CA, Dec. 3. e. UCLA (Journalism). Uncle was actor Elmo Lincoln (first Tarzan). 1945–46, publicist, Vic Shapiro Public Relations; 1946–49, sub-agent, Manning O'Connor Agency; 1949–51, TV coordinator, Bing Crosby Enterprises, Fireside Theatre Series; 1951–52, publicist, RKO Studios; 1952–59, personal mgr., Richard Burton, v.p. & dir. Denham Films, Ltd., London; 1959–61, pub. coordinator, Hecht-Lancaster-Hill; 1961–64, dir. of pub. relations, IPAR Productions, France; 1964–67, v.p., Illustra Films, West Coast branch; 1967–75, asst. dir. pub., United Artists Corp., West Coast; 1975–78, exec. v.p., Guttman & Pam Public Relations; 1978, formed Suvarie, Inc., m.p. representation, of which is pres., Personal manager, Richard Burton.

DOURIF, BRAD: Actor. b. Huntington, WV. Stage actor before films and TV.
PICTURES INCLUDE: Split, W.W. and the Dixie Dancekings, One Flew Over the Cuckoo's Nest, Group Portrait with Lady, Eyes of Laura Mars, Wise Blood, Heaven's Gate, Ragtime.
TELEVISION: Mound Builders, The Gardeners's Son, Studs Lonigan, Guyana Tragedy—The Story of Jim Jones.

DOWN, LESLEY-ANN: Actress. b. London, England, March 17, 1954. At age of 10 modeled for TV and film commercials, leading to roles in features. At 14 appeared in All the Right Noises (debut).
PICTURES: Pope Joan, Scalawag, Brannigan, The Pink Panther Strikes Again, The Betsy, A Little Night Music, The Great Train Robbery, Hanover Street, Sphinx.
TELEVISION: Upstairs, Downstairs (series), Heartbreak House, The One and Only Phyllis Dixey (film).
STAGE: Great Expectations, Hamlet, etc.

DOWNS, HUGH: Broadcaster. b. Akron, Ohio, Feb. 14, 1921. e. Bluffton Coll., 1938. Wayne Univ., 1941. Col. Univ., NYC, 1955; Supervisor of Science Programming, NBC's Science Dept. one yr.; science consultant for Westinghouse Labs., Ford Foundation, etc.; chm. of bd., Raylin Prods., Inc. Today, Chairman, U.S. Committee for UNICEF. Pres. National Space Institute.
TV: Hawkins Falls, Kukla Fran & Ollie, Short Story Playhouse, American Inventory, Home, Sid Caesar Show, Tonight (Jack Paar Show), Concentration, Today. Host: ABC-TV 20/20 Program, Host: Over-Easy-PBS Network.
RADIO: NBC's Monitor.

DOWSON, GRAHAM R.: Dep. Chief Exec. Rank Organization, 1972–75. Chrm. Erskine House Investments Ltd., Cambridge Communication Ltd., Nimslo Ltd., Pincus Vidler-Arthur Fitzgerald Ltd., 1979. Dir. Carron Company (Holdings) Ltd. b. Southend England, 1923. e. City of London School and Ecole Alpina, Switzerland. 1941; commissioned in RAF serving as pilot with RAF and US Navy. 1946; joined United States Steel Corp. as salesman. 1952; returned to England joining A.C. Nielsen Co., appt. dir. 1954. Council Mem. Incorp. Soc. British Advertisers; Cinema and Television Benevolent Fund. Fellow Inst. Directors and Inst. of Marketing. Mem. Industrial Advertising Assoc. and European Soc. for Opinion Survey and Market Survey. Chrm. British Section European League Economic Co-operation, June 1972. Past pres. of Appeals for National Playing Fields Assoc. Member Carlton Club St. James's, United Service and Royal Aero, vice commodore Royal London Yacht Club, Royal Cork Yacht Club and Royal Air Force Yacht Club. Royal Southern Yacht Club.

DOYLE, KEVIN: Executive. b. Sydney, Aust., June 21, 1933. e. N. Sydney Tech. HS, Aust. Jr. exec., asst. adv. & pub. div., 20th Century-Fox, Aust., 1947–59; adv. & pub. dir., Columbia Pictures Aust., 1960–66; international ad/pub. mgr.; Columbia Pictures Int'l, N.Y. 1966; intl. pub./promo. mgr., 1980.

DRAKE, CHARLES: Actor. r.n. Charles Ruppert; b. New York City, Oct. 2, 1914; e. Nicholas Coll. 1937. With Electric Boat Co., Groton, Conn.; adv. salesman; in little theatres; in m.p. as an actor.
PICTURES INCLUDE: I Wanted Wings, Man Who Came to Dinner, Now, Voyager, Air Force, Mr. Skefflington, Whistle Stop, Pretender, You Came Along, Tender Years, Bowie Knife, Comanche Territory, Air Cadet, Winchester '73, Harvey, Little Egypt, You Never Can Tell, Treasure of Lost Canyon, Red Ball Express, Bonzo Goes to College, Gunsmoke, Lone Hand, It Came from Outer Space, War Arrow, Glenn Miller Story, Tobor the Great, Four Guns to the Border, Female on the Beach, All That Heaven Allows, Price of Fear, The Arrangement.

DRAKE, CHARLIE: Actor, comedian, scriptwriter. b. London, England. Ent. TV in Nuffield Centre Show, 1950. Two years children's television followed by Askey Show, Star Time, Benny Hill Show, Fast and Loose. Two-and-a-half years children's television programme, Mick and Montmorency. Own BBC show 1958–9; signed film contract with Associated British studios, 1959–61.
TV: Charlie Drake Show, BBC TV, 1965; TV show, The Worker, Meet Peters & Lee 1976.
PICTURES INCLUDE: Sands of the Desert, Petticoat Pirates, The Cracksman, Mr. Ten Per Cent, Professor Poppers Problems 1975.

DRAKE, DONA: Actress. r.n. Rita Nevella; b. Mexico City.
PICTURES INCLUDE: Louisiana Purchase, Road to Morocco, Star Spangled Rhythm, So This is New York, Another Part of the Forest, Beyond the Forest, Valentino, Bandits of Corsica, Kansas City Confidential, Son of Belle Starr, Princess of the Nile.

DREIFUSS, ARTHUR: Producer-director-writer. b. Frankfurt on Main. e. Univ. of Frankfurt on Main, Germany, Conservatory of Music, Columbia University; choreographer, producer; many U. lectures; legit. producer: Allure, Baby Pompadour; producer many night club shows; associate producer Fanchon & Marco, Hollywood; dir. over 50 features, many musical shorts for Columbia, RKO, Universal, Allied Artists, Ind. to 1952; director Paul Muni debut Screen Gems-Ford Theatre, TV 1953; prod-director Secret File, U.S.A. Official Films; TV 1954–54; prod-dir. features Secret File and Assignment Abroad, Anglo-Amalgamated, London, 1956; dir: The Last Blitzkrieg; Life Begins at Seventeen; Juke Box Rythm, Columbia 1958–59; writer-director Brendan Behan's The Quare Fello British-Lion-Bryanston, London, 1961–62; director: Riot on Sunset Strip, AIP; The Love-Ins, For Singles Only, Columbia; The Young Runaways, A Time to Sing, MGM, 1963 to 1968; assoc. producer Angel, Angel, Down We Go, AIP, 1969; executive producer New Age Productions, 1970 to 1972; executive producer The Peter Hurkos Show, CBS; writer Owl Hill; Dolls behind Walls, 1973, director, Wildlife in Crisis, Viacom TV series 1975. Assoc. prod. Chennault China Tiger (Genson Prods.), 1978; assoc. video editor Creative Editing, Inc., 1979–80.

DREW, ELLEN: Actress. Previously known as Terry Ray. b. Kansas City, Mo. Nov. 23, 1915. Played on stage. Appeared in many pictures for major companies. In 1940: Women Without Names, Buck Benny Rides Again, Christmas in July, Texas Rangers Ride Again, Par. Many others.

DREXLER, ELIAS J.: Executive. b. New York City, Dec. 24, 1911. e. Columbia University, Columbia Law School, L.L.B., 1936. Special agent. Div. of Investigations, U.S. Dept. of Interior, 1937–40; dir., Preclusive operations, 1940–45; dir., Surlluettgterty Div., Foreign Economic Admin; exec., National Screen Service, 1945–56; general counsel and nat. sales mgr. AGFA-Gavaert; 1956–65; nat. sales mgr. and exclusive U.S. and Mexican dist. of motion picture products manufactured by Fuji Photo Film Co., Ltd., Tokyo, Japan, 1965-present.

DREYFUSS, RICHARD: Actor. b. Brooklyn, N.Y., 1949. e. Beverly Hills High. Prof. career began at Gallery Theatre (L.A.) in In Mama's House. Has also acted off-Bdwy. and on Bdwy.

PICTURES INCLUDE: Hello Down There, The Young Runaways, American Graffiti, The Apprenticeship of Duddy Kravitz, Jaws, Inserts, Close Encounters of the Third Kind, The Goodbye Girl, The Big Fix (also co-prod.) The Competition, Whose Life Is It Anyway?

TV: The Big Valley, Room 222, Judd For the Defense, Mod Squad, The Bold Ones.

STAGE: Journey to the Day, Incident at Vichy, People Need People, Enemy, Enemy, Line, Whose Little Boy Are You, But Seriously, Major Barbara, The Time of Your Life.

DRU, JOANNE: Actress. r.n. Joanne La Cock; b. Logan, W. Virginia, Jan. 31, 1923. John Robert Powers model: on stage as showgirl in Hold on to Your Hats; a Samba Siren at Ritz Carlton & Paramount; with theatrical group under Batami Schneider. Hollywood: m.p. debut in Abie's Irish Rose, 1946.

PICTURES INCLUDE: Red River, She Wore a Yellow Ribbon, All the King's Men, Wagonmaster, Vengeance Valley, 711 Ocean Drive, Mr. Belvedere Rings the Bell, Siege at Red River, Outlaw Territory, Southwest Passage, Three Ring Circus, Day of Triumph, Hell on Frisco Bay, Sincerely Yours, The Warriors.

DRURY, JAMES: Actor. b. N.Y.C. ed. N.Y. Univ. Acting debut at age 8 in Biblical play for children at Greenwich Settlement Playhouse. Performed on stage while youngster. Signed by MGM in 1955, working one day in each of seven movies that year, including Blackboard Jungle. Then got two-year contract at 20th-Fox. Gained fame as hero of TV series, The Virginian, which had nine-year run.

PICTURES INCLUDE: Love Me Tender, Bernardine.

TELEVISION: The Virginian, The Devil and Miss Sarah (movie of week), Firehouse (series).

DRURY, WESTON, JR.: Freelance casting director. b. Leeds, England, April 1916. Ent. m.p. ind. 1933. Rank Organization 1946–71, I.T.C. Ltd. 74–76. TV series include Return of the Saint, TV series 1977–78, Danger U. X. B. Minder; Masada, The Acts of Peter and Paul.

DRUXMAN, MICHAEL B.: Publicist, producer. b. Seattle, Wash., Feb. 23, 1941. e. Univ. of Washington. exec. v.p., Seattle Civic Playhouse, 1962–63; founder, exec. dir., Actor's Theatre, Seattle; formed public relations firm, Michael B. Druxman and Associates.

CREDITS: Genesis (wrote, prod., dir.); produced and directed plays in the Los Angeles area—A Shot in the Dark, A Thousand Clowns.

DUBBINS, DON: Actor. b. Brooklyn, N.Y., June 28. Appeared in stock company of Mr. Roberts, also Tea and Sympathy. Many television shows including Ed Sullivan show.

PICTURES INCLUDE: From Here to Eternity, Caine Mutiny, Tribute to a Bad Man, These Wilder Years, The D.I., Gunfight in Abilene, The Illustrated Man.

DUBENS, STANLEY: Writer-Producer. b. London, England, 1920. e. St. Paul's School, s. of Harry Dubens, Royal Corps Signals 1939–1945. From 1946, agent and theatrical productions. Ent. film production 1963. Currently making commercials, Doc., Film and video.

THEATRE: prod. incl. Sweet and Low, And No Birds Sing, The Train for Venice, The Man Who Let it Rain, The Hands of Eurydice, I Want to See Mioussov, Sign Here Please.

PICTURE INCLUDE: Tperaticn Snatch, Live Now Pay Later, The World Ten Times Over, Modesty Blaise, Marble Heroes, Matchgirls, Archie's Caper. 1975 formed with Roger Moss, The Original Electric Picture Company, making commercials, doc., video, AV presentations.

DUBIN, JOSEPH S.: Attorney, Los Angeles, Calif.: Chief Studio Counsel, Universal Pictures, Calif., 1950–1970; Member, U.S. State Department International Copyright Panel, Panel of Consultants on General Revision of Copyright Law, Panel of Neighboring Rights; Chairman, Copyright Subsection, ABA (1956); Member of Council, Patent Trademark and COLTIGT ABA (1957–61); Past Chairman, Committee on Related Rights, ABA; Past Chairman, Committee on Program for Revision of Copyright Law, ABA (1961–1969); Past Chairman, Committee on International Copyrights, AIPPI; Past Chairman, Committee on Copyright Law, APLA (1964–1969); Member, Board of Managers, APLA (1969–1972); President, Los Angeles Copyright Society (1964–65).

AUTHOR: Copyright Relations Between Nations; Copyright—Duration of Protection. Copyright—Formalities; Moral Rights and Protection for U.S. Works; Copyright Aspects of Sound Recordings; Universal Copyright Convention; Motion Picture Recordings; Universal Copyright Convention; Motion Picture Rights—U.S. and International; An Exclusive Federal System for All Works; Neighboring Rights; The New French Copyright Law of 1957; Wartime Extensions of Copyright in Europe; International Developments in Copyrights; Copyright Relations with the Soviet Bloc and Less Industrialized Nations; The New German Copyright Law of 1965; Copyright Duration. Waiver of the exercise of the Droit Moral.

DUDELHEIM, HANS R.: Producer—Director—Editor. Began film career in Germany after World War II. Hired by U.S. Army Signal Corps as photographer, being sent to Far East Command. Came to New York, where joined ABC-TV as film editor. Beginning with hard news, moved to documentaries, editing many films. In 1981 established Cinema Arts Associates, Inc. as film society; 1966 expanded into indep. m.p. prod. co., of which he is pres.

DUDELSON, STANLEY E.: Executive. b. Cleveland, O., July 12, 1924. Regnl. sales mgr., Hygo-Unity; natl. sales mgr. Screen Gems; producer's rep., Selma Ent., Italian Intl. Uni-Export Films of Rome, AIP; first vice pres., American Intl. Television, Inc. Now with New Line Dist. Corp., pres., New Line Intl. Releasing.

DUFF, HOWARD: Actor Bremerton, Wash. Nov. 24, 1917: e. Reportory Playhouse, Seattle. With KOMO radio station 1935; served with U.S. Army 1941–5; entered m.p. 1947. Radio's original Sam Spade.

PICTURES INCLUDE: Brute Force, Naked City, All My Sons, Calamity Jane and Sam Bass, Illegal Entry, Johnny Stoolpigeon, Woman in Hiding, Shakedown, Lady From Texas, Models Inc., Spaceways, Roar of the Crowd, Jennifer, Tanganyika, Private Hell 36, Women's Prison, Yellow Mountain, Flame of the Islands, Broken Star, Sierra Stranger, Blackjack Ketchum Desperado, While the City Sleeps, Boy's Night Out, Syria Against Babylon, Panic in the City, The Late Show, A Wedding, Kramer vs. Kramer, Oh, God! Book II.

TV: Mr. Adams and Eve, Dante, Felony Squad; Movies of the Week: The D.A. and The Heist.

DUFFY, JAMES E.: Executive. b. Decatur, Ill. e. Beloit College. Radio announcer, then reporter; joined publicity dept., ABC in 1949; named director of adv. & Promo., then account exec. for Central division of ABC Radio Network; director of sales ABC Radio, 1957; central division account exec., ABC Television Network, 1955; natl. director of Sales, ABC Radio central division, 1960; vice president, ABC Radio Network, 1961; exec. v.p. & natl. director of sales, 1962; vice president in charge of sales, ABC Television Network, 1963; president, ABC Television Network, 1970.

DUFFY, PATRICK: Actor. b. Townsend, Mont., March 17. e. Univ. of Washington. Became actor-in-residence in state of Washington, where performed with various statefunded groups. Acted off-Bdwy. Taught mime and movement classes in summer camp in Seattle. Moved to L.A. and began TV acting career.

TELEVISION: Movies: The Stranger Who Looks Like Me, Hurricane. Series: Man from Atlantis (star), Switch (guest), Dallas (star).

DUKE, PATTY: Actress. b. New York, N.Y., Dec. 14, 1947. e. Quintano School for Young Professionals.

THEATRE: The Miracle Worker, Isle of Children.
PICTURES INCLUDE: The Miracle Worker, Valley of the Dolls, Billie, My Sweet Charlie, Me, Natalie, The Swarm.
TV: Armstrong Circle Theatre, The Prince and the Pauper, Wuthering Heights, United States Steel Hour, Meet Me in St. Louis, Swiss Family Robinson, The Power and the Glory; now star, The Patty Duke Show.

DULLEA, KEIR: Actor. b. Cleveland, Ohio, May 30, 1936. e. San Francisco State College, Sanford Meisner's Neighborhood Playhouse. Worked as ice cream vendor, carpenter with a construction co. Acted as resident juvenile at the Totem Pole Playhouse in Pa. N.Y. theatre debut in the revue Sticks and Stones, 1956; appeared in stock co. prods. at the Berkshire Playhouse and Philadelphia's Hedgerow Theatre, 1959, off-Broadway debut in Season of Choice, 1969, Butterflies Are Free.
TV: All Summer Long.
PICTURES INCLUDE: The Hoodlum Priest, David and Lisa (Best Actor Award, San Francisco Intern'l Film Festival), The Thin Red Line, Mail Order Bride, The Naked Hours, Madame X, Bunny Lake Is Missing, The Fox, 2001: A Space Odyssey, de Sade, Pope Joan, Paul and Michelle, The Paperback Hero, Silent Night, Evil Night, Leopard in the Snow.

DUNAWAY, FAYE: Actress. b. Bascom, Fla., Jan. 14, 1941. e. Texas, Arkansas, Utah, Germany, University of Fla., Awarded a Fulbright scholarship in Theatre. Boston U. of Fine Applied Arts. Appeared on N.Y. stage in: A Man For All Seasons, After the Fall (with Lincoln Center Repertory Co., three years), Hogan's Goat.
PICTURES INCLUDE: The Happening, Bonnie and Clyde, Hurry Sundown, The Thomas Crown Affair, The Extraordinary Seaman, A Place for Lovers, The Arrangement, Puzzle of a Downfall Child, Little Big Man, Doc, Oklahoma Crude, The Three Musketeers, Chinatown, The Towering Inferno, The Four Musketeers, Three Days of the Condor, Network, Voyage of the Damned, Eyes of Laura Mars, The Champ, The First Deadly Sin, Mommie Dearest.
TV: The Disappearance of Aimee, Evita.

DUNCAN, SANDY: Actress. b. Henderson, Tex., Feb. 20, 1946, e. Len Morris Col.
STAGE: Ceremony of Innocence, Your Own Thing, Canterbury Tales, etc.
PICTURES INCLUDE: $1,000,000 Duck, Star Spangled Girl, The Cat From Outer Space.
TV: The Sandy Duncan Show, Roots.

DUNING, GEORGE: Composer, Conductor, Arranger. b. Richmond, Ind., Feb. 25; e. Cincinnati Conservatory of Music; University of Cincinnati. Musical director; music scores for many m.p. including: Jolson Sings Again, Eddy Duchin Story, From Here to Eternity, Picnic, World of Susie Wong, Devil at 4 O'Clock, Toys in the Attic, Any Wednesday, The Man with Bogart's Face. TV: No Time for Sergeants, Wendy and Me, The Farmer's Daughter, Big Valley, The Long Hot Summer, The Second Hundred Years, Star Trek, Mannix, Then Came Bronson; music dir. Aaron Spelling Prods., 1970–71, Bobby Sherman Show, "Movies" of Week—NBC, CBS, ABC. OPT. Board of directors, ASCAP.

DUNLAP, RICHARD D.: Producer, director. b. Pomona, Calif., Jan. 30, 1923; e. Yale U., B.A., 1944; M.F.A., 1948. U.S. Navy 1943–46; Instructor, English dept., Yale U., 1947–48; Prod.-dir., Kraft TV Theatre, 3 years; Dir., Assoc. Prod., Omnibus, 3 seasons; Dir., 25 half-hr. Dramatic Film Shows. Frank Sinatra Specials, Prod.-Dir., 10 Academy Award Shows, 4 Emmy Award Shows.

DUNNE, DOMINICK: Producer. Began career as stage manager at NBC-TV; then produced shows for CBS Studio One. Later exec. prod. at 20th-Fox TV, v.p. at Four Star.
PICTURES INCLUDE: Boys in the Band (exec. prod.), The Panic in Needle Park, Play It As It Lays, Ash Wednesday.

DUNNE, IRENE: Actress. b. Louisville, Ky., Dec. 20. e. Loretta Academy in St. Louis, Mo.; Chicago Coll. of Music (D.M., 1945), N.Y.; one child, Mary Frances. Screen debut 1931.
THEATRE: Stage prods. include Sweetheart Time, The City Chap, Show Boat.
PICTURES INCLUDE: Cimarron, Consolation Marriage, Back Street, Silver Cord, Ann Vickers, The Age of Innocence, Sweet Adeline, Roberta, The Magnificent Obsession, Showboat, The Awful Truth, Invitation to Happiness, My Favorite Wife, Penny Serenade, Lady in a Jam, A Guy Named Joe, The White Cliffs, Together Again, Over 21, Anna and the King of Siam, Life With Father, I Remember Mama, Never A Dull Moment, The Mudlark, It Grows on Trees, Love Affair, Theodora Goes Wild.

DUNNE, PHILIP: Writer-prod. dir. b. N.Y.C., Feb. 11, 1908; p. Finley P. and Margaret Abbott Dunne; e. Harvard Univ.; m. Amanda Duff.
PICTURES INCLUDE: How Green Was My Valley, Stanley and Livingstone, The Rains Came, Johnnie Apollo, Son of Fury, Suez, Lancer Spy, The Last of the Mohicans, The Count of Monte Cristo, The Late George Apley, Forever Amber, The Ghost and Mrs. Muir, Escape, Luck of the Irish, Pinky, David and Bathsheba, Anne of the Indies, Lydia Bailey, Way of a Gaucho, Demetrius and the Gladiators, The Egyptian, Prince of Players, View from Pompey's Head, Hilda Crane, Three Brave Men, Ten North Frederick, In Love and War, Blue Denim, Wild In The Country, Lisa, Blindfold, The Agony and the Ecstasy.
PUBLICATIONS: Mr. Dooley Remembers, Atlantic-Little Brown. Take Two (McGraw Hill, 1980).

DUNNING, JOHN: Film editor. b. Los Angeles, Calif., May 5, 1916; e. U.C.L.A., A.B., 1939. With MGM since 1935; U.S. Armed Forces, 1942–45.
PICTURES INCLUDE: Cass Timberlane, Homecoming, Julia Misbehaves, Battleground, Show Boat, Julius Caesar, Take the High Ground, Rhapsody, Last Time I Saw Paris, Tender Trap, Interrupted Melody, The Swan, Brothers Karamozov, Raintree County, Cimarron, Ben-Hur (Academy Award).
Post prod. supervisor, MGM, Inc.

DUNNOCK, MILDRED: Actress. b. Baltimore, Md., Jan. 25,; e. Goucher Co., Baltimore. Teacher, Brearly Sch., N.Y.; acted in summer stock.
THEATRE: Prof. B'way debut, Corn Is Green; since in Richard III, Vicki, Lute Song, Another Part of the Forest, Cat on a Hot Tin Roof, The Chinese, Colette.
PICTURES INCLUDE: Corn Is Green, Kiss of Death, I Want You, Death of a Salesman, Viva Zapata, Girl in White, The Jazz Singer, Bad for Each Other, Hansel & Gretel (voice), Trouble with Harry, Love Me Tender, Sweet Bird of Youth, Whatever Happened to Aunt Alice?, Dragonfly.
TV: Death of a Salesman.

DUPONT, ADLEY: Actress. r.n. Adley Stoltz, b. New York City, Aug. 7, 1946. e. Geller School of Theatre, Film Industry Workshop, Desilu School.
PICTURES INCLUDE: Movie Star, The Golden Poodle.

DURBRIDGE, FRANCIS: Writer. b. Hull, England, Nov. 25, 1912. e. Birmingham University. Early career, radio playwright and novelist; entered TV 1946.
TV PLAYS AND SERIALS INCLUDE: Broken Horseshoe, Operation Diplomat, Teckman Biography, Portrait of Alison, My Friend Charles, A Time of Day, The Scarf, The Other Man, The Desperate People, Bat Out of Hell, The Passenger, The Dool.
TV: The World of Tim Frazier, Melissa.
PICTURES INCLUDE: The Teckman Mystery, The Vicious Circle.
THEATRE: London stage play: Suddenly At Home, The Gentle Hook, Murder With Love.
RADIO PLAYS: Paul Temple plays, La Boutique.

DURNING, CHARLES: Actor. Many stage credits on Broadway (That Championship Season, The Happiness Cage, etc.).
PICTURES: Harry and Walter Go to New York, Twilight's Last Gleaming, The Choirboys, Breakheart Pass, The Hindenburg, Enemy of the People, The Sting, Dog Day Afternoon, The Fury, The Greek Tycoon, Tilt, The Muppet Movie, North Dallas Forty, Starting Over, When a Stranger Calls, The Final Countdown.
TELEVISION: Captains and the Kings, The Rivalry, The Dancing Bear, The Copy and the Kid (1975 series), Queen of the Stardust Ballroom, Studs Lonigan.

DURSTON, DAVID E.: Writer-Director. b. Newcastle, Pa., Sept. 10, 1925. e. Evanston Township H.S. Served as TV-Radio director, Lynn Baker Adv. Agency, 1952–57; assoc. producer, Your Hit Parade, 1957–58. Acting credits include Winged Victory (B'way, film), Young Man's Fancy (B'way); TV writing credits include Tales of Tomorrow, Navy Log.
PICTURES: Felicia, Love Statue, Reflections, Blue Sextet (also edited), I Drink Your Blood, Stigma.

DURWOOD, RICHARD M.: Executive. b. Kansas City, MO., Aug. 18, 1929. e. Brown Univ. (A.B.). Pres. Crown Cinema Corp. Member: Motion Picture Assn. of Kansas City (pres.); United Motion Picture Assn. (pres. 1972–73); Young NATO (chm., 1968–69); National NATO bd., Past Chief Barker, Tent #8; exec. comm., National NATO.

DURWOOD, STANLEY H.: President, American Multi-Cinema, Inc.; BS Harvard College (Varsity Football, Varsity Wrestling); Air Force navigator 3 years; six children. Member: Harvard Club of Kansas City; Harvard Club of New York; Rotary, Chamber of Commerce (board of directors); Downtown, Inc. (board of directors), United Motion Picture Assoc. of America, Variety Club International. Board of Directors and major stockholder First National Charter Corp. (23 bank holding company), Kansas City, Mo.

DUTFIELD, RAY: Executive. Early career as chief accountant to in Hawker Siddeley Group. Ent. m.p. ind. 1957, as exec. accountant with Rank Film Laboratories, gen. man., 1967; Appt. to board of dir., 1969; Appt. man. dir., 1970. Appt. man. dir., Rank Leisure Services, 1976 and vice-chmn., Rank Film Labortories. 1979 January appt. chmn. and chief executive officer, Technicolor Limited, and v.p., Technicolor Cinema Systems, Hollywood, Fellow Member of Inst.

Chartered Accounts, SMPTE, BKSTS, Member Institute of Directors, Institute of Marketing.

DUVALL, ROBERT: Actor. b. San Diego, Calif., 1931.
PICTURES INCLUDE: To Kill a Mockingbird, Captain Newman, M.D., The Chase, Countdown, The Detective, Bullitt, True Grit, The Rain People, M*A*S*H*, The Revolutionary, THX-1138, The Godfather, Tomorrow, The Great Northfield, Minn. Raid, Joe Kidd, Lady Ice, Badge 373, The Outfit, The Conversation, Godfather, Part II, Breakout, The Killer Elite, 7½ Per Cent Solution, Network, The Eagle Has Landed, The Greatest, The Betsy, True Confessions, Pursuit.

DUVALL, SHELLEY: Actress. b. Houston, TX, 1949.
PICTURES INCLUDE: Brewster McCloud, McCabe and Mrs. Miller, Thieves Like Us, Nashville, Three Women, Annie Hall, The Shining, Popeye.

DWAN, ALLAN: Director. b. Toronto, Ont., Canada, April 3, 1885; e. Notre Dame, Ind.; m. First film work writing and selling a scenario to Essanay. With American Film Co., San Diego, Cal., as scenarist, scenario ed. and dir. for three years. Dir. for Selznick, Goldwyn, Triangle, C. K. Young Prod., Louise Glaum Prod., then organized own co. Later connected with Associated Exhibitors, American Releasing Corp., UA, Douglas Fairbanks' pictures, Robin Hood, and Par., switching in 1926 to Fox, and later to FN. In 1941; prod.-dir. Look Who's Laughing, RKO; director Rise and Shine, 20th-Fox. In 1942; Prod. and dir. Here We Go Again; dir. Friendly Enemies, Up in Mabel's Room, Around the World, Abroad with Two Yanks, Brewster's Millions.
PICTURES INCLUDE: Northwest Outpost, Driftwood, Montana Belle, Calendar Girl, Inside Story, Sands of Iwo Jima, Surrender, Belle Le Grand, Wild Blue Yonder, I Dream of Jeanie, Woman They Almost Lynched, Flight Nurse, Silver Lode, Passion, Cattle Queen of Montana, Escape to Burma, Pearl of the South Pacific, Tennessee's Partner, Slightly Scarlet, Hold Back the Night, Restless Breed, River's Edge.

E

EADES, WILFRID: Director, producer, writer. b. England, 1920. Early career studying civil engineering. Ent. m.p. ind. as camera asst. 1939; asst. dir. prod. man., 2nd unit dir. since 1950.
PICTURES INCLUDE: No Place for Jennifer, The Last Holiday, Yellow Balloon, Moby Dick, Tarzan and the Lost Safari, You Can't Escape, The Small Hotel, The Moonraker, Gorgo.
TV: The Man in Half Moon Street, Portrait in Black, Wolf Pack, The Franchise Affair, I Can Destroy the Sun, Please Murder Me, Soundings, The Dangerous Word, To Ride a Tiger, The Angry Flower, Other People's Houses, The Upstart, Each Wind That Blows, The Intruder, Harriet, Then We Fail, Exits and Entrances, A Sickening Thud, Deadline Midnight, Echo Two Four, Tales of Mystery, No Hiding Place, The Pursuers, The Danny Thomas Show in Europe, TV Script adaptations: The Man in Half Moon Street, Portrait In Black, Wolf Pack, The Franchise Affair, Each Wind That Blows.
1963: exec. prod., Zenith Film Prod. Ltd.; 1964: prod. sup., Light Out of Darkness, 2nd unit dir., Operation Crossbow. 1965–68; gen. man., Ardmore Studios, Ireland. 1966: assoc. prod. Ulysses. 1968–69: Prod., The Violent Enemy; 1969 2nd unit dir. The Kremlin Letter. 1970–71: Exec. in chg. of prod., Interservice Mediterranean Ltd., Britannia Studios, Inc., Ltd. 1972: prod. controller. Pictures at an Exhibition for CIP Ltd. 1973–75 Story and Prod. Adviser, Star Services (Isle of Man) Ltd., 1976 Chief Exec. Eurofilm Advisers Ltd.

EASTWOOD, CLINT: b. San Francisco, Calif., May 31. e. Oakland Technical High School, Los Angeles City College. Worked as a lumberjack in Oregon before being drafted into the Army. Starred in TV series Rawhide, for seven and a half years. Formed Malpaso Productions, 1969.
PICTURES INCLUDE: A Fistful of Dollars, For a Few Dollars More, The Witches, The Good The Bad and The Ugly, Paint Your Wagon, Hang 'Em High, Coogan's Bluff, Where Eagles Dare, Kelly's Heroes, Two Mules For Sister Sara, Beguiled, Play Misty For Me (dir. and star); Magnum Force; Breezy (dir.); Thunderbolt & Lightfoot; The Eiger Sanction, The Outlaw Josey Wales, The Enforcer, The Gauntlet (dir., star), Every Which Way But Loose, Escape from Alcatraz, Bronco Billy (dir., star), Any Which Way You Can.

EBERSON, DREW: Architect. Stamford, Conn; b. Hamilton, Ohio, Feb. 29, 1904; e. Northwestern, Military and Naval Academy; University of Pennsylvania, Architect U.S.A. and England. WB production, Burbank. In 1931–38: Director, Coronet Pictures. Then theatre architect practicing throughout U.S.A. in 1942: Colonel, Corps of Engineers.

EBSEN, BUDDY: Actor. b. Belleville, Ill., April 2, 1908. r.n. Christian Ebsen, Jr. e. Univ. of Fla., Rollins College. Won first Broadway role as dancer in Ziegfeld's Whoopee in 1928. Sister, Vilma, became dancing partner and they played nightclubs and did road tours. Went to Hollywood and appeared in many musicals as single. Later became dramatic actor and appeared on TV.
PICTURES INCLUDE: Broadway Melody of 1936, Captain January, Banjo on My Knee, Four Girls in White, My Lucky Star, Thunder in God's Country, Red Garters, Davy Crockett, Attack, Breakfast at Tiffany's, Mail Order Bride, The Family Band.
TELEVISION: Hawaii Five-O, Gunsmoke, The Beverly Hillbillies, Davy Crockett, Barnaby Jones.
STAGE: Take Her, She's Mine, Our Town.

EBY, GEORGE W.: Executive. b. Pittsburgh, Pa., Jan. 2, 1914. e. Carnegie Tech.: Penna. State U., BA 1934. pres., Ice Capades, Inc., 1963–78, chm., Jan. 1, 1979; Int. Chief Barker Variety Clubs, 1958–60.

ECKERT, JOHN M.: Co-Producer, Assoc. Producer. b. Chatham, Ontario, Canada, 1948.
PICTURES INCLUDE: Middle Age Crazy, 1979 (co-prod.), Running, 1978 (co-prod.), Power Play, 1977 (assoc. prod.), The Clown Murders, 1975 (Associate Prod.) As Prod. Man/Assist. Dir.; Second Wind, 1975, Sudden Fury, 1974. First assist. dir. Rituals, Find the Lady, It Seemed Like a Good Idea at the Time, Lions for Breakfast. Second unit asst. dir., Silver Streak. Second asst. dir., Equus, Black Christmas. Various TV dramas and series.

EDELE, DURAND (BUD) J.: Executive e. St. Louis U., Washington U. Reportorial staff, St. Louis Globe Democrat, 1934–35; booking dept., Paramount, 1935; booker & salesman, Warner Bros., 1936–41; branch & dist. mgr., Film Classics, 1946; branch mgr., United Artists, 1952–60; managed various divisions, United Artists, 1960–63; gen. sales mgr., v.p., Avco Embassy Pictures Corp., 1964, gen. sales mgr., Venture Dist. Inc., 1975; Consultant, RCMC 1976–78.

EDWARDS, BLAKE: Writer-director-producer. b. Tulsa, Okla., July 26, 1922. e. Beverly Hills High. Coast Guard during war.
RADIO: Johnny Dollar. Line-Up; writer-creator: Richard Diamond.
TV: creator: Dante's Inferno, Peter Gunn, Mr. Lucky.
PICTURES INCLUDE: Panhandle, Stampede, Sound Off, Bring Your Smile Along, All Ashore, Crusin' Down the River, Rainbow Round My Shoulder, He Laughed Last, Drive a Crooked Road, My Sister Eileen, Mr. Cory, This Happy Feeling, The Perfect Furlough, Operation Mad Ball, Notorious Landlady, Operation Petticoat, High Time, Breakfast at Tiffany's, Experiment in Terror, Days of Wine and Roses, Soldier in the Rain, The Pink Panther, Shot in the Dark, The Great Race, What Did You Do in the War, Daddy, Gunn, Darling Lili, The Party, The Wild Rovers, Carey Treatment, The Tamarind Seed (dir. s.p.) The Return of the Pink Panther (prod., dir., co-s.p.), The Pink Panther Strikes Again (prod., dir., co-s.p.), Revenge of the Pink Panther (prod., dir., co-s.p.); "10" (co-prod., dir., s.p.), "S.O.B." (co-prod., dir., Victor/Victoria (co-prod., dir., s.p.).

EDWARDS, DOUGLAS: News Correspondent. b. Ada, Okla, July 14, 1917. Became radio reporter in Troy, Ala., at 15 yrs. e. Univ. of Ala., Emory Univ., Univ. of Ga. News reporter for WAGF, Dothan, Ala.; ass't. news editor for Atlanta Journal and its station, WSB. Transferred to WXYZ, Detroit, returning to WSB as ass't. news editor 2 yrs. later. Joined CBS Radio News staff, 1942, appearing on Report to the Nation, and The World Today. Chief of CBS News Paris Bureau. First major newsman to make transition to TV, 1947. Anchorman on CBS Afternoon News and Douglas Edwards with the News for 15 years on CBS-TV. Presently on CBS Mid-Day News with Douglas Edwards.

EDWARDS, JAMES H.: Executive. President, Storey Theatres, Inc. b. Cedartown, GA, Aug. 14, 1927. e. Georgia State. U.S. Navy, 1948–50. With Ga. Theatre Co., 1950–1952; Storey Theatres, 1952–present. Pres., NATO of Ga; formerly pres., Variety Club of Atlanta. Director at large, Nat'l. NATO. Director, numerous theatre cos.

EDWARDS, VINCE: Actor. b. New York City, July 9, 1928. e. Ohio State U., Univ. of Hawaii, Amer. Acad. of Dram. Arts. N.Y. stage. High Button Shoes.
TV APPEARANCES INCLUDE: Studio One, Philco, Kraft, Ben Casey, The Untouchables, General Electric Theatre, Hitchcock, The Deputy, Firehouse (movie), The Rhinemann Exchange.
PICTURES INCLUDE: Sailor Beware, Hiawatha, Rogue Cop, Night Holds Terror, Serenade, The Killing, Hit and Run, The Scavengers, The Three Faces of Eve, City of Fear, Murder by Contract, The Victors, Devil's Brigade, The Desperados, Las Vegas, Los Angeles.

EGAN, RICHARD: Actor. b. San Francisco, Calif., July 29, 1923. e. U. of San Francisco, B.A.; Stanford U., M.A.; U.S. Army, 1942–46; then taught public speaking; several TV appearances.
PICTURES INCLUDE: Damned Don't Cry, Undercover Girl, Bright Victory, One Minute to Zero, Devil Makes Three, Blackbeard the Pirate, Split Second, Demetrius and the Gladiators, Wicked Woman, Gog, Khyber Pass, Underwater, Untamed, Violent Saturday, View from Pompey's Head, Seven Cities of Gold, Revolt of Mamie Stover, Love Me Tender, Tension At Table Rock, These Thousand Hills, A Summer Place, Pollyanna, Esther and the King, The Lion of Sparta, Valley of Mystery, The Big Cube.

EGGAR, SAMANTHA: Actress. b. London, Eng., 1940.
PICTURES INCLUDE: The Wild and the Willing, Dr. Crippen, Doctor in Distress, Psyche 59, The Collector, Walk, Don't Run, Return From the Ashes, Doctor Dolittle, Molly Maguires, The Lady in the Car, Walking Stick, The Grove, Light at the Edge of the World, 7½ Solution, Why Shoot the Teacher, Blood City, The Uncanny.
TV INCLUDES: Anna and the King (CBS), Man of Destiny, Double Indemnity, The Hope Diamond, Columbo, Baretta, The Hemingway Play, Love Story, Kojak, McMillan & Wife, Streets of San Francisco, Starsky and Hutch.

EICHHORN, LISA: Actress. b. Reading, Pa. Went to England for literature studies at Oxford. Studied at RADA. Made film debut in Yanks (1979).
PICTURES: The Europeans, Why Would I Lie?, Cutter and Bone.

EISNER, MICHAEL D.: Executive. b. New York City, 1942. e. Denison Univ., (B.A.). Started career with programming dept. of CBS TV network. Joined ABC in 1966 as asst. to v.p. & natl program director. Two years later named manager of specials and talent. Dec., 1968 became dir. of program development—east coast. In March, 1971 named v.p., daytime programming, ABC-TV. In June, 1975 made v.p., program planning and development. In May, 1976 named sr. v.p., prime time production and development, ABC Entertainment. In Nov., 1976, left ABC to join Paramount Pictures as pres. & chief operating officer.

EKBERG, ANITA: Actress. b. Malmoe, Sweden, Sept. 29, 1931. Started career as a model.
PICTURES INCLUDE: Man in the Vault, Blood Alley, Artists and Models, War and Peace, Back from Eternity, Zarak, Pickup Alley, Sheba and the Gladiator, Sign of the Gladiator, La Dolce Vita, Boccaccio '70, The Alphabet Murder, The Cobra, Fellini's Clowns.

EKLAND, BRITT: Actress. b. Stockholm, Sweden, 1942.
TV: England: A Cold Peace. U.S.A.: Trials of O'Brien, McCloud, Six Million Dollar Man.
PICTURES INCLUDE: After the Fox, Double Man, Bobo, Night They Raided Minsky's, At Any Price, Stiletto, Cannibals, Tintomara, Percy, Carter, Night Hair Child, Endless Night, Baxter, Asylum, Wickerman, Ultimate Thrill, Man With Golden Gun, Royal Flash, Cassanova and Company, Slaves, King Soloman's Treasures.

ELAM, JACK: Actor. b. Miami, AZ, Nov. 13, 1916. e. Santa Monica Jr. College, Modesto Jr. College. Worked in Los Angeles as bookkeeper and theatre mgr.; served in Navy in World War II. Introduction to show business was as bookkeeper for Sam Goldwyn. Later worked as controller for other film producers. Given first acting job by producer George Templeton in 1948; has since appeared in over 100 films.
PICTURES: Rawhide, Kansas City Confidential, The Moonlighter, Vera Cruz, Moonfleet, Kiss Me Deadly, Gunfight at the OK Corral, Baby Face Nelson, Edge of Eternity, The Comancheros, The Rare Breed, The Way West, Firecreek, Once Upon a Time in the West, Support Your Local Sheriff, Rio Lobo, Support Your Local Gunfighter, Dirty Dingus Magee, The Cannonball Run, etc.
TELEVISION: The Texas Wheelers, The Dakotas, Temple Huston, Gunsmoke, Struck by Lightning, etc.

ELFAND, MARTIN: Executive. Was talent agent for ten years with top agencies; joined Artists Entertainment Complex in 1972. First film project as producer: Kansas City Bomber, first venture of AEC, of which he was snr. v.p. In 1977 joined Warner Bros. as production chief. 1980, produced It's My Turn

ELG, TAINA: Actress, dancer. b. Helsinki, Finland, March 9. e. Helsinki. Sadler's Wells Ballet. Toured with Marquis de Cuevas Ballet; m.p. debut in The Prodigal.
PICTURES INCLUDE: Diane, Gaby, Les Girls, Imitation General, The 39 Steps.

ELIAS, HAL: Executive. b. Bklyn., N.Y. Dec. 23. Publicity dir., State Theatre, Denver; western exploitation mgr., MGM; adv. dept., pub. dept., MGM, Culver City studios; Head, MGM cartoon studio (Tom and Jerry); UPA Pictures, Inc., vice-pres. studio mgr.: Hollywood Museum; bd. dir., Academy of Motion Picture Arts & Sciences, 30 years; treasurer, AMPAS 1976–1979.

ELIZONDO, HECTOR: Actor. b. N.Y.C. m. actress Carolee Campbell. Studied with Ballet Arts Co. of Carnegie Hall and Actors Studio. Many stage credits in N.Y. and Boston.
PICTURES INCLUDE: Pocket Money, Born to Win, Deadhead Miles, Stand Up and Be Counted, One Across, Two Down, The Taking of Pelham One Two Three, Report to the Commissioner.
STAGE: Mr. Roberts, Island in Infinity, Madonna of the Orchard, Drums in the Night, The Prisoner of Second Avenue, Dance of Death, Steambath, The Great White Hope.
TV: The Impatient Heart (tf), Kojack, the Jackie Gleason Show, All in the Family.

ELKINS, SAUL: Producer. b. New York City, June 22, 1907; e. C.C.N.Y., B.S. 1927. Radio writer, dir., prod. 1930–2; dir., prod. stock co. touring Latin America 1932–4; writer Fox Films, 20th Century-Fox; writer RKO, Columbia 1937–42; writer, dial-dir., dir. Warner Bros. 1943–7; prod. Warner Bros. since 1947. Member: AMPAS, Screen Writer's Guild. Presently: exec. prod., Comprenetics, Inc.
PICTURES INCLUDE: Younger Brothers, One Last Fling, Homicide, House Across the Street, Flaxy Martin, Barricade, Return of the Frontiersmen, This Side of the Law, Colt 45, Sugarfoot, Raton Pass, The Big Punch, Smart Girls Don't Talk, Embraceable You.

ELLIOTT, DENHOLM: Actor. b. London, Eng., May 31, 1922; e. Malvern College. Screen debut in Dear Mr. Proback, 1948; films include: The Sound Barrier, The Holly and the Ivy, The Ringer, The Cruel Sea, Heart of the Matter, They Who Dare, Man Who Loved Redheads, Lease of Life, Night My Number Came Up, Pacific Destiny, Scent of Mystery; TV appearances; Stage: Write Me A Murder, N.Y., Seagull, N.Y., 1964.
PICTURES INCLUDE: Nothing But the Best, Station Six Sahara, King Rat, You Must Be Joking, The High Bright Sun, Alfie, Here We Go Round The Mulberry Bush, The Night They Raided Minskys, The Seagull, Too Late the Hero, The Rise and Rise of Michael Rimmer, A Doll's House, The Apprenticeship of Duddy Kravitz, Robin & Marion, To The Devil A Daughter, Russian Roulette, Voyage of the Damned, The Boys from Brazil, St. Jacques, Zulu Dawn, A Game for Vultures, Cuba, Illusions, Saint Jack, Bad Timing, Sunday Lovers.

ELLIOTT, LANG: Producer-Director. b. Los Angeles, CA, Oct. 18, 1949. Given first job in films by his uncle, the late actor William Elliott (known as Wild Bill Elliott). Worked as actor from early years; employed by, among others, the McGowan Brothers. Turned to film production; co-founded distribution co., The International Picture Show Company, serving as exec. v.p. in chg. of financing, production & distribution. Under banner of TriStar Pictures, Inc. also finances productions.
PICTURES: Produced Ride the Hot Wind, Where Time Began, The Farmer, The Billion Dollar Hobo, They Went That-a-Way & That-a-Way, The Prize Fighter. Produced and directed the Private Eyes.
TELEVISION: Produced Experiment in Love (special).

ELLSWORTH, JAMES: Producer. b. Deltaville, Virginia, March 12, 1927. e. U.C.L.A., 1949. Pro. baseball player.
TV SERIES: Champions of Sports.
PICTURES INCLUDE: Naked Fury, Marine, Life Story of Lt. General Lewis B. (Chesty) Fuller, Door to Door Maniac (Five Minutes to Live.)

ELMES, GUY: Writer. b. London, Eng., July 22, 1920; e. Aberdeen U. Painter and writer for radio after War Service (1939–45) in Royal Navy.
PICTURES INCLUDE: The Planter's Wife, The Flanagan Boy, Counterspy, Bang! You're Dead, Across the Bridge, Nor the Moon by Night, Wheel of Fate, Stranger's Hand, Serious Charge, Hello London, The Lady Takes a Gun, Joseph and His Brothers, Hannibal, The Blue Countess, Pontius Pilate, A Face in the Rain, The Captive City, Stranglehold, Flames Over India, El Greco, Submarine X, The Heroes, The Night Visitor, s.p.

ELSON, NORMAN: Executive. b. New York City, July 5, 1907; e. U. of Maryland. Contact clerk for Columbia; accessory mgr. N.Y. Columbia Exchange; district mgr. film buyer & gen. mgr. Brandt Theatres; gen. mgr. Trans-Lux Theatre Circuit, then apptd. vice-pres. 1950. Guild Enterprises, Inc., circuit pres. operating nine theatres in New York City.

EMANUEL, DAVID: Executive. b. New York, N.Y., Dec. 14, 1910; e. grad. Coll. 1932; ent. m.p. ind., slsman, Classic Pictures, 1948; formed, pres., sls. mgr., Phoenix Films, 1951; formed, pres., gen. sls. mgr., Governor Films Inc.; Pres. Gen. sls. mgr. Phase One Films, Inc.

ENDERS, ROBERT: Producer, Writer. Began in television, being responsible for 64 hrs. of live programming weekly for industry and govt. as pres. of Robert J. Enders, Inc. In 1961 turned to theatrical prod. and writing.
PICTURES INCLUDE: A Thunder of Drums (prod.), The Maltese Bippy (prod.), How Do I Love Thee (prod.), Zig Zag (story), Voices (prod., s.p.), The Maids, (s.p.), Winter

Rates (exec. prod.), Conduct Unbecoming (s.p.), Hedda (prod.), Nasty Habits (prod., s.p.).
TV: The Best of the Post (prod. of series), Ben Franklin, High Noon (prod., special). Co-prod. of Acad. Award Show, 1968.

ENDFIELD, CY: Producer, director, writer. b. Nov. 1914. Ent. m.p. ind. 1942; credits include: Joe Palooka series. Argyle Secrets, Underworld Story, Sound of Fury, Limping Man, Impulse, The Secret Child in the House, shorts for MGM; dir., Hell Drivers, Sea Fury.
PICTURES INCLUDE: dir., Jet Storm, Mysterious Island, Hide and Seek; dir., co-prod., co-scripted, Zulu, Sands of the Kalahari. Dir. De Sade.

ENGELBERG, MORT: Producer. b. Memphis, TN. e. Univ. of Illinois; Univ. of Missouri. Taught journalism; worked as reporter for UPI, AP. Worked for US government, including USIA, Peace Corps., Office of Economic Opportunity; President's Task Force on War on Poverty. Left government service in 1967 to become film unit publicist, working on three films in Europe: Dirty Dozen, Far From the Madding Crowd, The Comedians. Returned to US; appt. pub. mgr. for United Artists. Sent to Hollywood as asst. to Herb Jaffe, UA head of west coast production, which post he assumed when Jaffe left. Left to join with independent producer, Ray Stark.
PICTURES: Smokey and the Bandit, Hot Stuff, The Villain, The Hunter.

ENGLUND, GEORGE H.: Producer-Director. b. Washington, D.C., June 22, 1926. Producer, Paramount, Pennebaker Prods.; prod., MGM; prod.-dir., Universal.
PICTURES INCLUDE: The World the Flesh and the Devil, The Ugly American, Signpost to Murder, Snow Job.

ENGLUND, KEN: Writer. b. Chicago, Ill., May 6, 1914. e. Lane Tech., Chicago. Started career as magazine writer; then vaudeville routines, comedy radio shows. B'way revues and musicals. Twice elec. pres. screen branch Writers Guild of America, West, then pres. of WGAW. Member national council.
TV SHOWS INCLUDE: Jackie Gleason Show, Dear Phoebe, Ray Milland Show, Loretta Young Show, several spectaculars. Prod.-writer CBS-TV staff 1957; Sonja Henie Spect. London, 1958; 20th Fox and Warner Bros. TV Films, My Three Sons; sev. teleplays, Bewitched, 1965, That Girl!, Dr. Joyce Brothers (head writer for series).
PICTURES INCLUDE: Big Broadcast of 1938, Artists and Models Abroad, Good Girls Go to Paris, Slightly Honorable, Doctor Takes a Wife, No No Nanette, This Thing Called Love, There's That Woman Again, Nothing But the Truth, Springtime in the Rockies, Sweet Rosie O'Grady, Here Comes the Waves, Secret Life of Walter Mitty, Androcles and the Lion, Good Sam, A Millionaire for Christy, The Caddy, Never Wave at a Wac, Vagabond King, The Wicked Dreams of Paula Schultz, Surviving the Savage Sea.
AUTHOR: Tour D'Amour, Larks in a Casserole; B'way musical in work, and literary projects. Co-author (with George Marshall) The Ghosts in Emily's Trunk.

EPSTEIN, MEL: Producing. b. Dayton, O., Mar. 25, 1910, e. Ohio State U. Adv. & edit. depts. on newspapers; entered m.p. ind. as player in 1931; then asst. dir., unit prod. mgr., second unit & shorts dir.; U.S. Army Signal Corps (1st Lt.); apptd Para. prod., 1946. Now retired.
PICTURES INCLUDE: Whispering Smith, Hazard, Copper Canyon, Dear Brat, Branded, The Savage, Alaska Seas, Secret of the Incas.
TV: Broken Arrow, Men into Space, The Islanders, Asphalt Jungle, Rawhide, Long Hot Summer, The Monroes, Custer TV, Lancer TV Pilot, unit mgr. Lancer TV series, Medical Center, TV series.

ERDMAN, RICHARD: Actor-director. b. Enid, Okla., June 1, 1925, e. Hollywood High Sch.
PICTURES INCLUDE: Janie, Objective Burma, Time of Your Life, Four Days Leave, The Men, Cry Danger, Jumping Jacks, Happy Time, The Stooge, Stalag 17, The Power and the Prize, Saddle the Wind, Namu, The Killer Whale. Features directed: Bleep, The Brothers O'Toole.
TV: Ray Bolger Show, Perry Mason, Police Story, Tab Hunter Show, Alice, Bionic Woman, One Day at a Time, Playhouse of Stars, Twilight Zone, The Lucy Show, Lou Grant.
DIRECTED: The Dick Van Dyke Show, Mooch (special).

ERICKSON, CORYDON: Actor-Singer. b. New York, Dec. 26, 1944.
TV SHOWS: Robert Montgomery Presents, Goodyear, Lux Video Theater, Starstage, Omnibus, U.S. Steel, The Naked City, Phil Silvers Show, Ingrid Bergman Spectacular, True Story, The Defenders, Model.
Special for the United Nations, Declaration of Human Rights, The Children of Strangers, V.O. Paramount Pictures, Slide Films, Transcriptions, Records, Radio comm., comm. live and on film; Girls in the Night, It Should Happen To You, Man Against Crime, Norby, Radio, The Bible, etc.

ERIKSON, LAUNCELOT: Actor-singer. b. New York, Sept. 6, 1949. Industrial films, Stanley Neal Prod., Faith for Today, Gifts, Take My Name, Odds Against Tomorrow, dubbing for Carosello Napolitano; Commercials; Transcriptions, Radio comm. & dramas. Model.
TV LIVE: Studio One, Omnibus, U.S. Steel, Du Pont Show of the Month.

ERIKSON, LEIF: Actor b. Alameda, Calif., Oct. 27, 1911; e. Glenn Taylor Mil. Acad., Calif. U., L.A., studied singing. With Ted Fio Rito orch., San Francisco, then Max Reinhardt's stage prod., A Midsummer Night's Dream; next joined Olsen & Johnson, comedy team. Screen debut in Wanderer of the Wasteland, 1935. On Bway, in Tea & Sympathy, 1953–55.
PICTURES INCLUDE: Sorry, Wrong Number, Miss Tatlock's Millions, Joan of Arc, Snake Pit, Stella, Three Secrets, Dallas, Tall Target, Show Boat, Reunion in Reno, Cimarron Kid, With A Song in My Heart, Sailor Beware, Carbine Williams, My Wife's Best Friend, A & C Meet Captain Kidd, Never Wave at a Wac, Trouble Along the Way, Perilous Journey, Paris Model, On the Waterfront, Star in the Dust, Fastest Gun Alive, Tea and Sympathy, The Vintage, Man and Boy, Twilight's Last Gleaming.
TV: High Chapparal.

ERICSON, JOHN: Actor. b. Dusseldorf, Germany, Sept. 25, 1926. e. American Acad. of Dramatic Arts. Appeared in summer stock; m.p. debut, Teresa, 1951, then Stalag 17 on Broadway.
PICTURES INCLUDE: Rhapsody, Student Prince, Green Fire, Bad Day at Black Rock, Return of Jack Slade, Pretty Boy Floyd, Bedknobs and Broomsticks.
TV: Honey West.

ESBIN, JERRY: Executive. Joined Paramount Pictures in 1975 as mgr. of branch operations; later named v.p., asst. sls. mgr. In 1980 named v.p., gen. sls. mgr. 1981, snr. v.p., domestic sls. & mktg.

ESMOND, CARL: Actor. b. Vienna, Austria, June 14, 1906; e. U. Vienna. On stage Vienna, Berlin, London, etc. (Shakespeare, Shaw, German modern classicals. On screen in Brit. prod. including Blossom Time, Even Song, Invitation to the Waltz. To U.S. in 1938.
PICTURES INCLUDE: The Catman of Paris, Smash-Up, Story of a Woman, Slave Girl, Walk a Crooked Mile, Mystery Submarine, World In His Arms, Agent for H.A.R.M., Morituri.

ESSENFELD, BARRY: Executive. b. Bronx, N.Y., Feb. 16, 1936. e. De Witt Clinton, N.Y.U. Acct., American Broadcasting Company, 1957–62; division controller, International Paper Co., 1962–67; group controller & division exec., New England Industries, 1968–70; Controller & asst. secty., Allied Artists Pictures Corp., 1970–73; named asst. treas. & dir. admin., Sept. 1973; gen. sls mgr. 1979.

ESSERT, GARY: Executive. b. Oakland, Calif., Oct. 15, 1938. e. UCLA. Entered the entertainment field through motion picture exhibition; managed theatres in San Francisco and Los Angeles has created advertising art campaigns for major motion picture studios and main titles for feature films; acts regularly as technical consultant for professional motion picture presentation installations, including Kennedy Center for the Performing Arts, Lincoln Center film theatre, Juilliard School of Music, American Zoetrope; Coordinated and supervised the design and planning of UCLA Motion Picture Center (Melnitz Hall), 1964–1967. Co-owned and operated a multi-media dance/concert hall in Hollywood known as The Kaleidoscope; technical coordinator for the American Film Institute's Center for Advanced Film Studies (Greystone) in Beverly Hills, 1968–1970, designing and outfitting complete facilities. Produced The Movies, a two-part, four-hour compilation documentary on the history of American motion pictures, telecast nationally on ABC-TV in March, 1975. Co-founded (1971) and currently director of the Los Angeles International Film Exposition (Filmex).

ESSEX, DAVID: Actor-singer. B. Plaistow, London, Eng. July 23, 1947. e. Shipman Sch. Custom House. Started as a singer-drummer in East London band. 1967: Joined touring Repertory Co. in "The Fantasticks," "Oh, Kay," etc. 1970: West End debut in "Ten Years Hard," 1972: Jesus Christ in Godspell, Che in Evita; Phonogram recording artist.
TV: appearances include: Top of the Pops, Own Specials, BBC series. Appearances on TV: France, Japan, Germany, Spain, Denmark Australia.
PICTURES INCLUDE: Assault, All Coppers Are . . . , Recent pictures: That'll Be the Day, Stardust, Silver Dreaam Racer.
U.S. TV: Merv Griffin, Johnny Carson, Dinah Shore, American Bandstand, Midnight Special, Grammy Awards, Salute To The Beetles, Don Kirshners Rock Concert, A.M. America, Phil Everle in Session.

ESSEX, FRANCIS: Prod. b. Essex, Eng., 1929. Began career in films and theatre, including The Bells of St. Martin's, London, 1952; prod. BBC first anniversary program, 1954–60; prod. Associated TeleVision Ltd., London, 1960; prod.

Six of One (stage musical), London, 1964; controller programs, Scottish Television Ltd., Glasgow, 1955.
TV: Saturday Spectacular, Sunday Night at London Palladium, The Eartha Kitt Show, Anthony Newley Show, Mainly Millicent (tv series). 1969 Rejoined ATV as Prod. Controller, elected to Board of Dir. 1974. Appt. Dir. Prod. 1976.

ESSEX, HARRY J.: Writer. b. New York City, N.Y. e. St. John's Univ., Brooklyn, B.A. With Dept. Welfare, author play Something for Nothing, on New York stage, followed by Stronger than Brass. Wrote orig. story, Man Made Monster, for Universal. During World War II in U.S. Army Signal Corps; scenarist, training films on combat methods, censorship. Since release has written orig. s.p. Boston Blackie and the Law, Dangerous Business, (s.p.) Desperate Bodyguard, He Walked by Night, Dragnet, Killer That Stalked New York, Wyoming Mail, The Fat Man, Undercover Girl, Las Vegas Story, Models, Inc., Kansas City Confidential, The 49th Man, It Came From Outer Space; s.p., dir., I the Jury; s.p. Creature from the Black Lagoon; collab. s.p., story, Southwest Passage; adapt., Devil's Canyon; s.p., dir., Mad at the World; collab. s.p., Teen-age Crime Wave, Raw Edge; story, s.p. Lonely Man; play. One for the Dame; collab., Sons of Katie Elder, Man and Boy, s.p. & dir. Octman, s.p., dir., prod. The Cremators, The Amigos, story and SP Metro, release 1973. Collaboration with Oscar Saul; Chrysalis, in preparation from a story by Ray Bradbury.
TV: Untouchables (various series). The Racers, Alcoa Hour, Westinghouse, Desilu; story consultant and head writer: Target, The Corruptors, The Dick Powell Show, Bewitched, I Dream of Jeannie, Kraft Suspense Theatres.
NOVELIST: I Put My Right Foot In, Little Brown. Stage prod., Fatty, B'way, Sept., 1976; Man and Boy, (Dell), 1971, Marina, Playboy Press, 1981.
PLAYS: Something for Nothing; Neighborhood Affair; One for the Dame; Fatty; Current production, stage: Twilight, When the Bough Breaks.

ETKES, RAPHAEL: Executive. Joined MCA in 1961; named v.p. of Universal Pictures in 1973. Appt. v.p. of MCA, Inc. in 1978; named sr. v.p., Universal Pictures, 1979; 1980, named pres. chief exec. off., AIP; 1981, joined United Artists as snr. v.p.—Worldwide Prod.

ETTINGER, EDWIN D.: Publicist. b. New York City. 1921. Entered m.p. ind. as office boy, MGM; pub. rel. and publ. for industrial, comm. clients, 1946–52; joined Ettinger Co., pub. rel., 1952; pub. rel. dir., Disneyland Inc., 1955; marketing dir., Disneyland, 1955–65; v.p., M.C.A. Enterprises, Inc., 1955.

ETTLINGER, JOHN A.: Prod.-dir., distributor. b. Chicago, Ill., Oct. 14, 1924. e. Peddie Inst., Cheshire Acad., Signal Corps Photog. Center, 1942–45; with Paramount Theatres Corp., 1945–47; dir., KTLA, Paramount TV Prod., Inc., Los Angeles, 1948–50; radio-TV dir., Nat C. Goldstone Agency, 1950–53; pres. Medallion TV Enterprises, Inc.; TV prod., View the Clue, Greenwich Village, High Road to Danger, Medallion cartoons; Sur Demande, Star Route, Las Vegas Fights, Celebrity Billiards; Pres., KUDO-FM, Las Vegas.

EVANS, BARRY: Director. b. Guildford, England, 1943. Trained Central School. Repertory: Barrow, Nottingham, Chester, Royal Court, Nat. Theatre, Hampstead Th. Club, Chips with Everything, London and B'way. Young Vic. Theatre Clwyd Mold.
TV: Redcap, Undermined, The Baron, The Class, Armchair Theatre, Love Story, Doctor at Large, Short Story, Crossroads, Mind Your Language.
PICTURES INCLUDE: The White Bus, Here We Go 'Round the Mulberry Bush, Alfred the Great, Die Screaming, Marriane, The Adventures of a Taxi-Driver, Under the Doctor.

EVANS, CLIFFORD: Actor, director, writer, b. Cardiff, February 17, 1912; e. Royal Acad. Dram. Art, London.
STAGE: London debut 1930 in The Witch; later also on tour; N.Y. debut 1934 in The Distaff Side; other appearances include many Shakespearean prods., Gallows Glorious, Cul-de-Sac, Ghosts, The Emperor of the World, The Doctor's Dilemma, The Judge's Story, The Marquise.
PICTURES INCLUDE: The Mutiny of the Elsinore, Mademoiselle Docteur, The Proud Valley, Fingers, Love on the Dole, The Foreman Went to France, The Flemish Farm. Served H.M. Forces during World War II, The Silver Darlings, While I Live, Twenty Question Murder Mystery, A Run for Your Money, Escape Route, Valley of Song, Stryker of the Yard, Passport to Treason, Violent Playground, At the Stroke of Nine, SOS Pacific, The Curse of the Werewolf, Kiss of the Vampire, The Long Ships, Twist of Sand, One Brief Summer.
TV: The Accused, The Red Dress, White Hunter series, The Hour of the Rat, Treason, The Quiet Man, Roman Gestures, Man With a Feather In His Hat, Madras House, Traitor At the Gate, 12th Hour, Dr. Finlay's Casebook, Woman Beware Women, The Local Boy, Old Man's Fancy, Lime Street, Power Game series, The Avengers, Treason, War and Peace, The Prisoners, Who's For Tennis?, The Saint, Randall and Hopkirk, Codename. 1973: Co-producer series The Inheritors, Memories of Xmas (ABC Am.), Strife, Land of Ice Cream, Childs Play, 1974. The Kilvert Diaries, 1977; Dylan, 1978; Coal's Prayer, 1979.

EVANS, GENE: Actor. b. Holbrook, Ariz., July 11, 1924; e. Colton H.S. Started career in summer stock, Penthouse Theatre, Altadena, Calif. Screen debut: Under Colorado Skies, 1947.
PICTURES INCLUDE: Crisscross, Larceny, Berlin Express, Assigned to Danger, Mother Was a Freshman, Sugarfoot, Armored Car Robbery, Steel Helmet, I was an American Spy, Force of Arms, Jet Pilot, Fixed Bayonets, Mutiny, Park Row, Thunderbirds, Donovan's Brain, Golden Blade, Hell & High Water, Long Wait, Cattle Queen of Montana, Wyoming Renegades, Crashout, Helen Morgan Story, Bravados, Sad Sack, The Hangman, Operation Petticoat, Support Your Local Sheriff, War Wagon, Nevada Smith, Young and Wild, Ballad of Cable Hogue, There Was a Crooked Man, Support Your Local Gunfighter, Camper John, Walking Tall, People Toys, Pat Garrett & Billy The Kid, Magic of Lassie.
TELEVISION: Matt Helm (series), Spencer's Pilots (series); Kate Bliss & Ticker Tape Kid Fire.

EVANS, JEROME M.: Executive. b. N.Y., Jan. 6, 1923. e. Univ. of Pa., B.S.; N.Y. Univ., M.B.A.; Ph.D.; Exec. in Charge Eastern Advertising & Publicity, Universal Pictures; Lecturer in Economics and Marketing, Bernard Baruch College; Graduate School of Business, City University of New York; Nat'l. Panel American Arbitration Assn.

EVANS, JULIUS: Writer, producer. e. N.Y.U., B.S.
THEATRE: Dir. B'way prods., Searching for the Sun, Dream Child; prod. and dir. summer stock. Later Assoc. Administrator So. Calif. Fed Theatre Project; organized Caravan Theatre Circuit, N.Y. Eastern play editor, Samuel Goldwyn, 1937; Test dir. in charge of new talent at RKO, 1938–40; assoc. prod. to Jesse L. Lasky on Adventures of Mark Twain, Sergeant York, Rhapsody in Blue. Signal Corps, U.S. Army, became chief of training films branch, 1942–46. Story ed., Sol Lesser Prod.; Ziv-TV Prod.; wrote screenplay for Sword of the Avenger; wrote for Mayor of the Town Series and The Phantom. Wrote and prod. ind. doc. films, including Achievement in Hong Kong (Edinburgh Film Festival); TV journalist with wife in personal interview series Topic for British Foreign Office. Partner in Chelsea International Films (Goldwyn Studios), s.p. & Assoc. prod.: Rain for a Dusty Summer, v.p., World View Prod. Co-Prod. The Challenge (A Tribute to Modern Art).

EVANS, MAURICE: Actor. b. Dorchester, Eng., June 3, 1901. On the stage since 1926.
THEATRE: London theatrical, career included Justice, Loyalties, Diversion, Journey's End, The Man I Killed, The Queen Bee. The Swan, See Ourselves, The Heart Line and starred for one season at the Old Vic; to U.S. 1935 and appeared in Romeo and Juliet, St. Helena, Richard II, Hamlet (Entirety), Henry IV (Part I), Twelfth Night, Macbeth, 1942; after War returned to Broadway, 1945 and appeared in G.I., Hamlet, Man and Superman, The Devil's Disciple, The Wild Duck, Dial 'M' For Murder, The Apple Cart, Heartbreak House, Tenderloin, The Aspern Papers, Shakespeare Revisited, A Program for Two Players.
PICTURES INCLUDE: Kind Lady, Androcles and the Lion, Gilbert and Sullivan, Macbeth, Warlord, Planet of The Apes, Rosemary's Baby, The Body Stealers.
Prod. Teahouse of the August Moon. No Time for Sergeants, N.Y. stage.
TV SHOWS: Hamlet, Richard II, Macbeth, Devil's Disciple, Taming of the Shrew, Man and Superman, The Tempest, St. John, series of 7 plays for Hallmark, NBC; G.E. Theatre (Caesar & Cleopatra), Theatre Guild, etc., Gen. Motors Motorama, Bewitched.

EVANS, RAY: Songwriter. b. Salamanca, N.Y., Feb. 4, 1915. e. Wharton School of Pa. U. Musician on cruise ships, radio writer spec. material. Hellzapoppin', Sons o' Fun.
SONGS INCLUDE: To Each His Own, Golden Earrings, Buttons and Bows (Academy Award, 1948), Mona Lisa (Academy Award, 1950), Whatever Will Be Will Be (Academy Award, 1956), A Thousand Violins, I'll Always Love You, Dreamsville, Love Song From Houseboat, Tammy, Silver Bells, Dear Heart, Angel, Never Let Me Go, Almost In Your Arms, As I Love You, In the Arms of Love, Wish Me a Rainbow.
PICTURES INCLUDE: Paleface, Sorrowful Jones, Fancy Pants, My Friend Irma, Capt. Carey U.S.A., Off Limits, Here Come the Girls, Red Garters, Man Who Knew Too Much, Stars Are Singing, Tammy, Houseboat, Blue Angel, A Private's Affair, All Hands on Deck, Dear Heart, The Third Day, What Did You Do In The War Daddy?, This Property Is Condemned.
BROADWAY MUSICAL: Oh Captain! Let It Ride!, Sugar Babies.
TV THEMES: Bonanza, Mr. Ed, Mr. Lucky, To Rome With Love.

EVANS, ROBERT: Producer. b. N.Y.C., June 29, 1930. Made appearance on radio as actor at age 11; went on to appear in more than 300 radio programs on major networks. Also appeared on TV in its early days. At 20 joined brother, Charles, and Joseph Picone, as partner in firm of Evans-Picone, Inc. In 1957 signed by Universal to play Irving Thalberg in The Man of a Thousand Faces after recommendation by Norma Shearer, Thalberg's widow. (See credits below for other acting assignments.) Turned to production and signed as independent by 20th Century-Fox. In August, 1966, joined Paramount Pictures as head of prod., a post he held ten years. Resigned to become indep. prod. again; has exclusive contract with Para.

PICTURES INCLUDE: Actor: The Sun Also Rises, The Fiend Who Walked the West, The Best of Everything. Independent Producer: Chinatown, Marathon Man, Black Sunday, Players, Urban Cowboy, Popeye.

EVANS, SY: Exec. r.n. Seymour H. Evans. b. New York, N.Y., March 14. e. Columbia U., B.A., 1941. U.S.A.F., 1941–46. Field prom. rep., Universal Pictures, 1946; asst. dir. adv., pub. & exploit., Schine Circuit, 1951; hd. dept., 1960; gen. mgr., 1963–65; dir. adv., pub. & prom., General Cinema Corp., 1965; named v.p., pub. and p.r., 1976.

EVERETT, CHAD: Actor. b. South Bend, Ind., June 11, 1936. r.n. Ray Canton. e. Wayne State Univ., Detroit. Signed by William T. Orr, head of TV prod. for Warner Bros. to 3-year contract. Appeared in many TV series as well as films. Next became contract player at MGM.

PICTURES INCLUDE: Claudelle Inglish, The Chapman Report, The Singing Nun, The Last Challenge, Made in Paris, Johnny Tiger, Return of the Gunfighter, The Impossible Years.

TELEVISION: Hawaiian Eye, 77 Sunset Strip, Surfside Six, Lawman, Bronco, The Dakotas, The Lieutenant, Redigo, Route 66, Ironside, Medical Center.

EVERSON, WILLIAM K.: Writer b. Yeovil, Eng., April 8, 1929. Pub. dir., Renown Pictures Corp., Ltd., London, 1944; film critic; m.p. journalist; in armed forces, 1947–49; thea. mgr., pub. & booking consultant, Monseigneur News Theatres, London, 1949; pub. dir., Allied Artists Inc. Corp., 1951; prod., writer Paul Killiam Org., 1956. Writer-editor-researcher on TV series "Movie Museum" and "Silents Please," also on TV specials and theatrical features "Hollywood the Golden Years," "The Valentino Legend," "The Love Goddesses" and "The Great Director."

AUTHOR: Several books on movie history, including "The Western," "The Bad Guys," "The American Movie," "The Films of Laurel & Hardy," "The Art of W. C. Fields," "Hal Roach," "The Detective in Film," "Classics of the Horror Film," "Claudette Colbert."

Lecturer, archival consultant, American Film Institute representative. Film History instructor at New York University, The New School and The School of Visual Arts, all in New York. Also, Harvard Univ.

EWELL, TOM: Actor. r.n. Yewell Tompkins. b. Owensboro, Ky., April 29. e. U. of Wisconsin. Active in coll. dramatics; salesman at Macy's. Stage debut, New York: They Shall Not Die, 1934; thereafter many unsuccessful plays. U.S. Navy 1942–46.

THEATRE: Returned to stage in John Loves Mary; Small Wonder; on road in Stage Door, Tobacco Road, Roberta, Key Largo; on Broadway in Seven-Year Itch, Tunnel of Love.

RADIO: Many programs.

PICTURES INCLUDE: Adam's Rib, Mr. Music, A Life of Her Own, American Guerilla in the Philippines, Up Front, Finders Keepers, Lost in Alaska, Willie & Joe Back at the Front, Seven-Year Itch, Lieutenant Wore Skirts, Girl Can't Help It, Tender Is the Night, State Fair, Suppose They Gave A War and Nobody Came?, To Find a Man, They Only Kill Their Masters, The Great Gatsby.

F

FABRAY, NANETTE: Actress. b. San Diego, Calif., Oct. 27, 1920. d. Raoul Fabares and Lillian (McGovern) Fabares; student Los Angeles City College. m. David Tebet 1947 (div. July, 1951); m. 2d Ranald MacDougall 1957, deceased Dec. 11, 1973. One son, Jamie. First prof. stage appearance at age of three and a half in vaudeville as Baby Nan. Leading lady in Charlie Chan radio series. Member of the cast, radio, Showboat. 1938, won two-year scholarship to Max Reinhardt school in Hollywood, and starred in his Calif. productions of The Miracle, Six Characters in Search of an Author, and Servant With Two Masters.

PICTURES INCLUDE: Elizabeth and Essex, A Child is Born, The Happy Ending, Cockeyed Cowboys, The Bandwagon.

TV: Own TV Series, Yes, Yes Nanette 1961–62. Also starred on shows with every major TV star. Two years on Caesar's Hour.

BROADWAY SHOWS: Meet the People, By Jupiter, Jackpot, May Dear Public, Let's Face It, Bloomer Girl, Arms and the Girl, High Button Shoes, Make A Wish, Love Life, Mr. President, No Hard Feelings.

AWARDS: High Button Shoes 1948, two Donaldson Awards for best performance in a musical and best supporting performance; Love Life 1947, the Tony Awards best musical actress of the year; Three Emmy Awards for Caesars Hour as best comedienne, 1955, 1956, best supporting actress 1955; Woman of the Year, 1955, Radio and TV Editors of America; Hollywood Women's Press Club, 1960; Honorary Mayor of Pacific Palisades 1967–68; One of Ten Best Dressed Women in America, Fashion Academy Award, 1950.

FABRIZI, ALDO: Actor. b. Rome, Italy. Stage actor to 1942.

PICTURES INCLUDE: Go Ahead, Passengers, Square of Rome, Open City, My Son, the Professor, To Live in Peace, Christmas at Camp 119, Emigrants, Welcome, Reverend, First Communion (Father's Dilemma), Flowers of St. Francis, Mishappy Family, Thieves and Cops, Times Gone By, Lucky Five, Too Young for Love, Voice of Silence.

FACTOR, ALAN JAY: Producer-Director, b. Chicago, Ill. Dec. 25, 1925; e. Northwestern Univ. & Goodman Theatre. Former actor under name Alan Frost. Broadway; Richard III, Pick Up Girl, E-MC2, Battle For Heaven (ANTA).

PICTURES INCLUDE: Portrait of Jennie, All Hands On Deck, Double Lift, The Braineaters, The Right Approach, Female Jungle, Producer, Screen Gems, Metromedia, Columbia Broadcasting System. Currently President, Bedford Productions. Recent TV credits: Bewitched, (Director), Something Evil, Let Me Live Unseen

FADIMAN, CLIFTON: Performer. e. Columbia Univ. Contributor to magazines since 1924. Asst. ed. Simon & Schuster, 1927–29, ed. 1929–35. Book ed. The New Yorker, 1933–43. Mc. on Information Please radio program 1938–48; TV, 1952; Judge, Book of the Month Club, since 1945; Mc. This Is Show Business, TV; author, Party of One; Any Number Can Play; Enter Conversing; The Lifetime Reading Plan; editor, American Treasury, 50 years; m.c., Conversation, 1954; m.c., Quiz Kids, 1956; Metropolitan Opera Roving Reporter, 1955–60; m.c., Alumni Fun, 1964; bd. of editors, Encyclopaedia Britannica; Edit. consultant Encyclopaedia Britannica Educational Corp.; Editorial Board, CRICKET: The Magazine for Children, Co-author, The Joys of Wine, 1975. Co-author, The Wine Buyers Guide, 1977; Co-author, Empty Pages: A Search for Writing Competence in School and Society, 1979.

FAICHNEY, JAMES B.: Producer-Director-Writer-Consultant. b. Yonkers, N.Y. Pathe News; RKO Pathe, film editor, writer, director; U.S. Army WW II, chief motion picture security censorship; Soundmasters; James B. Faichney Prod.; U.S. Dept. of State—USIA; Freelance producer, director, writer for CBS Films, ABC-CBS-NBC News, U.S. Army, National Foundation, Communications Four, Cul-Crieff Associates. Consultant to the U.S. Atomic Energy Commission, U.S. Air Force, Committee for Free Asia, etc.

PICTURES INCLUDE: The Most Rewarding Law, The Third Pillar, Air Power, Dead Man's Voice, Dawn of a New Day, Kampong Sentesa.

TV: Air Power, Twentieth Century, 21st Century, David Brinkley's Journal, NBC White Paper, ABC Close-up, The Great Conventions (the Republicans, the Democrats), Viet Cong, The Green Scene, The Revolutionaries, The Will for Peace, A Better World, 1945, The Diplomat, The General.

AWARDS: George Foster Peabody, Sylvania. Billboard and United States Air Force exceptional service awards; two ATVAS nominations for best documentary program and series.

FAIN, SAMMY: Composer. b. New York. e. Public and high schools. Self taught on piano. Commenced career in Tin Pan Alley as teenage staff pianist-composer for leading music publishers, then in vaudeville. Recording contract at Harmony Records (Columbia); featured solo performer on many New York radio stations. Joined ASCAP, 1926. Wrote first song hit "Let a Smile Be Your Umbrella." Lyrics by Irving Kahal, 1927.

STAGE: Scores for Hellzapoppin, Son's of Fun, George White's Scandals, Ed Wynn's Boys & Girls Together, Flahooley, Christine & others.

PICTURES INCLUDE: Scores & Songs—Big Pond, Young Man of Manhattan, Footlight Parade, Sweet Music, Harold Teen, New Faces of '37, Geo. White's Scandals, Call Me Mister, Alice in Wonderland, Peter Pan, Jazz Singer, Weekend at the Waldorf, Anchors Aweigh, Imitation of Life, Mardi Gras, Calamity Jane, April Love, Marjorie Morningstar, Certain Smile, Love is a Many Splendored Thing, Tender is the Night, Made in Paris, The Stepmother and many others.

SONGS: Let a Smile Be Your Umbrella, Wedding Bells Are Breaking Up That Old Gang of Mine, When I Take My Sugar to Tea, You Brought a New Kind of Love to Me, Was That the Human Thing to Do, By A Waterfall, Are You Having Any fun, I'm Late, I Can Dream Can't I, That Old Feeling, I'll Be Seeing You, Dear Hearts and Gentle People, April Love, A Certain Smile, A Very Precious Love, Tender

Is the Night, Secret Love, Love Is a Many Splendored Thing, many others. Elected to Songwriters' Hall of Fame.

CITATIONS: Oscar awards for Secret Love and Love is a Many Splendored Thing, plus eight Acad. Oscar nominations: International awards—Diploma Di Benemerenza most coveted honor by the Hall of Artists in Nice, France, and Augasto Messinesse Gold Award from Messina, Italy. Also two Laurel Awards.

FAIRBANKS, DOUGLAS, JR.: Actor, prod. Bus. exec. b. N.Y. Dec. 9, 1909; son of late Douglas Fairbanks. e. Pasadena (Calif.) Polytech. Sch.; Harvard Mil. Acad., Los Angeles; Bovee and Collegiate Sch., N.Y.; also Paris, London. Began as screen actor 1923 in Stephen Steps Out; thereafter in more than 75 pictures. Stage from 1927 in U.S., U.K. (Young Woodley, Saturday's Children, etc.); 1934 in Manchester, England (The Winding Journey), London (Moonlight in Silver), 1968–1977 My Fair Lady, The Secretary Bird, Present Laughter (U.S.), The Pleasure of His Company (U.S., U.K., Ireland, Australia and Hong Kong), Out on a Limb. Since sound in both British & U.S. prods.; formed own co. 1935, prod. & starring, The Amateur Gentleman, etc. U.S. Navy, 1941; Silver Star Combat Legion of Merit; K.B.E., D.S.C., Legion of Honor, Croix de Guerre, etc. 1949. Entered TV film prod., 1952.

PICTURES INCLUDE: (since sound): Dawn Patrol, Little Caesar, Outward Bound, Union Depot, Captured, Catherine the Great, Accused, The Prisoner of Zenda, Having Wonderful Time, Gunga Din, Morning Glory, Angels Over Broadway, The Corsican Brothers, The Exile, Sinbad the Sailor, That Lady in Ermine, Joy of Living, Fighting O'Flynn, State Secret; prod. Another Man's Poison; Chase Crooked Shadow, Ghost Story.

TV: prod. Douglas Fairbanks Presents, etc.

FAIRBANKS, JERRY: Exec. prod. b. San Francisco, Calif.; cameraman, 1924–29; prod., shorts, Universal, 1929–34; prod., Popular Science, Unusual Occupations, Speaking of Animals Series, Para., 1935–49; Winner two Acad. Awards; set up film div., NBC, 1948; formed, NBC Newsreel, 1949; devel. Zoomar Lens; formed Jerry Fairbanks Prods., 1950.

TV: Front Page Detective, Crusader Rabbit, Hollywood Theatre; owner, Jerry Fairbanks Productions.

PICTURES INCLUDE: Lost Wilderness, Down Liberty Road, With This Ring, Counterattack, Collision Course, Land of the Sea, Brink of Disaster, The Legend of Amaluk, North of The Yukon, Damage Report, Seven Lies South, The Boundless Seas.

FAIRCHILD, WILLIAM: Director, screenwriter, playwright. b. Cornwall, England. e. Royal Naval College, Dartmouth. Early career Royal Navy.

PICTURES INCLUDE: Screenplays: Morning Departure, Outcast of the Islands, The Gift Horse, The Net, Newspaper Story, Malta Story, The Seekers, Passage Home, Value For Money, Star!, Embassy, The Darwin Adventure, Dir.: John and Julie, The Extra Day, The Silent Enemy.

SCREENPLAYS AND TV PLAYS INCLUDE: The Man With the Gun, No Man's Land, The Signal, Four Just Men, Some Other Love, Cunningham 5101, The Break, The Zoo Gang, Lady with a Past.

STAGE: Sound of Murder, Breaking Point, Poor Horace, The Pay-Off, The Flight of The Bumble B, The Paternal Optimist.

BOOKS: A Matter of Duty, The Swiss Arrangement.

FALK, PETER: Actor. b. New York City, Sept. 16, 1927.

THEATRE: Off-Broadway: The Iceman Cometh, Comic Strip, Purple Dust, Bonds of Interest, The Lady's Not For Burning, Saint Joan, Diary of a Scoundrel. On Broadway: Saint Joan, The Passion of Josef D., The Prisoner of Second Avenue.

TV: Studio One, Kraft Theatre, Alcoa Theatre, N.T.A. Play of the Week, Armstrong Circle Theatre, Omnibus, Robert Montgomery Presents, Brenner, Deadline, Kraft Mystery Theatre, Rendezvous, Sunday Showcase, The Untouchables, N.B.C.'s Sacco—Vanzetti Story, Dick Powell Show—The Price of Tomatoes, Danny Kay Show, Edie Adams Show, Bob Hope Chrysler Hour, Columbo (1971).

PICTURES INCLUDE: Murder, Pocketful of Miracles, The Balcony, Robin and the Seven Hoods, Mad, Mad, Mad World, Italiano Brava Gente, Luv, Castle Keep, Anzio, Husbands, Machine Gun McCann, Mikey and Nicky, Murder by Death, The Cheap Detective, Brinks' Job, The In-Laws, All the Marbles.

FANTOZZI, TONY: Theatrical Agent. b. May 1, 1933. New Britain, Conn. William Morris Agency.

FARBER, BART: Executive. Joined United Artists Corp. in early 1960s when UA acquired ZIV TV Programs. Served as v.p. United Artists Television and United Artists Broadcasting. In 1971 named v.p. in charge of legal affairs of the cos. In January 1978, named snr. v.p.—TV, video and special markets; now indep. consultant, TV, Pay TV, home video.

FARENTINO, JAMES: b. Brooklyn, N.Y., Feb. 24, 1938. m. Michele Lee, actress, singer. e. American Academy of Dramatic Arts.

THEATRE: Appeared in Broadway prod. of Night of the Iguana. A Streetcar Named Desire (revival, 1973).

TV INCLUDES: Naked City, daytime soap operas, Laredo, Route 66, The Alfred Hitchcock Hour, Ben Casey, Twelve O'Clock High, Death of a Salesman.

PICTURES INCLUDE: The Pad (And How to Use It) (Golden Globe Award, 1966), The Ride of Hangman's Tree, Banning Rosie!, The Story of A Woman, The Final Countdown, Dead & Buried.

FARR, DEREK: Actor. b. London, England, Feb. 7, 1912; e. St. Michael's, Uckfield & Cranbrooke, Lieut. Royal Artillery World War II. Stage debut, "Gate Revue." On screen 1940.

PICTURES INCLUDE: Outsider, Spellbound, Bond Street, Murder Without Crime, Young Wives' Tale, Reluctant Heroes, Double Confession, Bang! You're Dead, Dam Busters, Front Page Story, Eight O'Clock Walk, Man in the Road, Town on Trial, Doctor At Large, Vicious Circle, The Truth About Women, Attempt to Kill, 30 Is a Dangerous Age, Cynthia, Pope Joan.

TV: Murder Mistaken, Escapade, Two Dozen Red Roses, Deadline Midnight, Faraway Music, Zero One, The Saint, The Human Jungle, series. 1963–64; acting in Australia; The Projected Man, in England, 1956–66. 1966: Dixon of Dock Green 1967; Inspector Rose; Adam Adamant, Owen MD, The Visitors, Crossroads. Garrick Theatre, 1967–68. Strand Theatre, 1969. Treasure Island Tour for Mermaid, 1970–71.

FARR, FELICIA: Actress. b. Westchester, N.Y., Oct. 4. e. Penn State College. m. Jack Lemmon. Stage debut: Picnic (Players Ring Theatre).

PICTURES INCLUDE: Timetable, Jubal, Reprisal, First Texan, Last Wagon, 3:10 to Yuma, Hell Bent For Leather, Kiss Me Stupid, Kotch, Charley Varrick.

FARRAR, DAVID: Actor. b. Forest Gate, Essex, England.

PICTURES INCLUDE: Black Narcissus, Frieda, Mr. Perrin & Mr. Traill, Small Back Room, Diamond City, Gone to Earth. One of top ten British money makers Motion Picture Herald-Fame poll, Cage of Gold, Night Without Stars, The Golden Horde, Obsessed, Wild Heart, Duel in the Jungle, Black Shield of Falworth, Lilacs in the Spring, Escape to Burma, Sea Chase, Pearl of the South Pacific, Lost, Woman and the Hunter, I Accuse, Son of Robin Hood, John Paul Jones, Watusi, Solomon and Sheba, Best Girl, Middle of Nowhere, The 300 Spartans, Beat Girl.

TV: Dick Powell Show.

FARRELL, CHARLES: Actor. b. Dublin, Eire, Aug. 6, 1906. m. Babbie McManus. First stage appearance as a child with Detroit (Mich.) stock co., 1912; London debut at the Coliseum, May, 1921, in The Crown Diamond. Since in many Brit. stage prods.

THEATRE: (Brit. stage) The Nervous Wreck, Derby Day, The Bowery Touch, Get a Load of This, My Sister Eileen, Carissima, Detective Story, Wise Guys, Remains to Be Seen, Never Too Late, Harvey with James Stewart 1975.

PICTURES: Began screen career 1912, Vitagraph. Later in many British films including Jack's the Boy, Meet Sexton Blake, The Way to the Stars, Don Chicago, This Man Is Mine, Black Michael, They Made Me a Fugitive, Boys in Brown, Night and the City, The Turners of Prospect Road, The Crimson Pirate, Madame Louise, There Was a Lady, Miss Robin Hood, Far and Wide, See How They Run, Hornet's Nest, Round the Bend, My Infallible Uncle, Morning Call, Sheriff of Fractured Jaw, Best of All, Operation Cupid, No Hiding Place, Too Young to Love, Too Hot to Handle, Humanite, Moonlight, Girl Hunters, The Unkind Philanthropist, Chimes at Midnight, Man of Atlas, Golden Silence, Mr. Sebastian; Generalissimo, 1966–67. Crossplot, Oh, What a Lovely War, From a Bird's Eye View, Vampire Lovers, Doctor Phibes, Countess Dracula, The Common Error, Scoop, Rudolph Valentino, 1976.

TV: Many appearances, including BBC-TV series, Hans Christian Andersen series in Copenhagen; Look Out series, TV films. Dir. English Dialogue 13 TV Films in Italy; Hon.-treas. British Actors Equity 1949–75.

FARRELL, CHARLES: Actor. b. Onset Bay, Mass., Aug. 9, 1901; e. Boston U. Entered m.p. ind. in silent films.

PICTURES INCLUDE: Seventh Heaven, Street Angel, Lucky Star, Sunny Side Up, Wild Girl, Big Shakedown, Moonlight Sonata, Just Around the Corner, Tailspin.

U.S.Navy, 1942–45; mayor of Palm Springs since 1947.

TV: My Little Margie; The Charles Farrell Show.

FARRELL, HENRY: Author of novels and screenplays

SCREENPLAYS INCLUDE: Whatever Happened to Baby Jane? Hush . . . Hush, Sweet Charlotte, and What's the Matter with Helen?

TV: How Awful About Allan, The House That Would Not Die.

FARREN, JACK: Prod.-Writer. b. New York City; e. B.A., NYU. Theatre: prod., Take a Giant Step, Lorenzo 1967, Robert

Anderson's You Know I Can't Hear You When the Water's Running; 1968, The Education of Hyman Kaplan.
PICTURES INCLUDE: Silent Night, Lonely Night; Fuzz.
TV: Goodson-Todman Productions; NBC, CBS.

FARROW, MIA: Actress. b. Los Angeles, Feb. 9. e. Marymount, Los Angeles, Cygnet House, London. Bway. debut. The Importance of Being Earnest, 1963. Screen debut: Guns at Batasi.
PICTURES INCLUDE: Guns at Batasi, A Dandy in Aspic, Rosemary's Baby, John and Mary, Secret Ceremony, See No Evil, The Public Eye. The Great Gatsby, Full Circle, A Wedding, Death on the Nile, Hurricane.
TV: Peyton Place, Johnny Belinda.
THEATRE: Mary Rose (London).

FASS, M. MONROE: Theatre Broker. b. New York City, Feb. 26, 1901. e. City College of N.Y. (M.E. degree, engineering). Entered real estate business in 1925, making first deal with Marcus Loew for land on which Paradise Theatre, Bronx, N.Y., was built. Thereafter made theatre deals (sale or lease of land or building) in most major cities in the U.S. and in major shopping centers. Member: Real Estate Board of N.Y.; Natl. Institute of Real Estate Brokers, Natl. Assoc. of Real Estate Board; Amer. Society of Real Estate Appraisers, Natl. Assoc. of Theatre Owners.

FAWCETT, FARRAH: Actress. b. Corpus Christi, Texas, Feb. 2. e. Univ. of Texas. m. Actor Lee Majors. Picked as one of the Ten Most Beautiful Girls while a freshman; went to Hollywood and signed by Screen Gems. Did films, TV shows, and made over 100 TV commercials.
PICTURES INCLUDE: Love Is a Funny Thing. Myra Breckenridge, Logan's Run, Somebody Killed Her Husband, Sunburn, Saturn III, The Cannonball Run.
TV: Owen Marshall, Counselor at Law, The Six Million Dollar Man. The Feminist and the Fuzz (movie), Charlie's Angels (regular).

FAY, PATRICK J.: Director, producer. b. June 7. e. Carnegie Tech; Dumont TV Network 10 years. Director of over 100 Army training films.
TV: Bishop Sheen, Broadway to Hollywood, Manhattan Spotlight, Life is Worth Living, Front Row Center, Ilona Massey Show, Alec Templeton Show, Maggi McNellis Show, Key to Missing Persons, Kids and Company; co-prod., dir., Confession; dir., TV film series, Confession; dir. IBM Industrials, IBM World Trade, Europe Industrial, The Big Picture.
AUTHOR: Melba, The Toast of Pithole, The Last Family Portrait in Oil, Coal Oil Johnny, French Kate, The Water's Fine, No Pardon in Heaven.
FILMS: Director for RCA, General Electric H.G. Peters Company, Bransby Films.

FAYLEN, FRANK: Actor.
PICTURES INCLUDE: Address Unknown, The Angels Sing, The Lost Weekend, You Came Along, Road to Rio, The Centerville Ghost, Cooper Canyon, Whispering Smith, Detective Story, Blood on the Moon, Sniper, The Lusty Men, 99 River Street, Red Garters, Lone Gun, Riot in Cell-block 11, The Looters, McConnell Story, Away All Boats, Return of Custer, Gunfight at the OK Corral, Death of a Nobody, Fluffy, The Monkey's Uncle, Funny Girl.
TV: The Dobie Gillis Show, That Girl, Tell Tale Gun, Joe & Lillian, Quincy.

FAZALBHOY, M.A.: Executive. b. Bombay, India, 1902. equip. mfg., mgr., dir., Photophone Ltd.; dir. several business houses.
Member of Rotary. International Business Men's Convention in America, 1944. Development Panel of Manufacture of Cinema & Studio Equipment of Government of India.

FEHR, RUDI: Executive. b. Berlin, Germany; m. Maris Wrixon actress. Started career with Tobis-Klangfilm, Berlin. Joined Warner Bros. editorial department, 1936. Became producer, 1952; promoted to executive, 1956; Post Production Exec. Warner Bros.; WB title changed to dir. of editorial & post-prod. operations. Now retired; is consultant to industry.

FEINBERG, MILTON: Executive. b. Milwaukee, Wis. e. Univ. of Minn., school of Bus. Admin. Salesman, 20th Century-Fox, Des Moines, 1942–44; branch mgr., National Screen Service, Des Moines, 1945–57; branch mgr., in Chicago, superv. Chicago, Milwaukee, Des Moines exchange territories, National Screen Service, 1958–65; gen. sales mgr., NSS, 1966; v.p. & gen. sls. mgr. 1967; May, 1974, sr. v.p., N.S.S.

FEINSTEIN, HARRY: Exhibitor. b. Pittsburgh, Pa., Aug. 22, 1906; e. U. of Pittsburgh. Dept. store exec. prior to entering m.p. ind. 1930, as asst. mgr. Embassy Theatre, East Orange, N.J.; mgr. Warner theatres, Newark & Pittsburgh, asst. to various zone mgrs.; apptd. zone mgr. Warner New Eng. Theatres, hdqts. New Haven, Conn., 1951. Apptd. zone mgr., Stanley Warner, Pittsburgh, May, 1960.

FEITSHANS, BUZZ: Executive. Worked for 10 years at American-International as supvr. of prod., jobs including film editing and supervision of editing and dubbing. Bowed as producer on Dillinger (1973). In 1975 formed A-Team Productions with John Milius.
PICTURES: Big Wednesday (prod.), Hardcore (prod.), 1941 (prod.), Extreme Prejudice (exec. prod.).

FELD, FRITZ: Actor. b. Berlin, Germany, Oct. 15, 1900. e. U. Berlin; Max Reinhardt Sch. of Drama, Berlin. Actor, prod. asst. dir. for Reinhardt in Berlin 7 yrs.; on screen 1918 in The Golem, Ufa. Since found variously assoc. U.S. m.p. prod., as writer, director, actor: has appeared in more than 410 pictures. 1971; nat'l bd. dir., Screen Actors Guild. 1968 chairman of the "American National Committee," Southern California Chapter. 1976 chairman Hollywood Museum Project committee—Screen Actors Guild, 20th Century-Fox Film Corp. staged A Tribute To Fritz Feld-60 Years In The Movies and the Los Angeles City named a theater in Brentwood The Fritz Feld Community Theatre. In April 1979 Feld was elected honorary Mayor of Brentwood by the Chamber of Commerce.
TV: Dangerous Assignment, Racket Squad, Mr. & Mrs. North, Jimmy Durante Show, Jack Paar, Thin Man, Chevy Show, Red Skelton, Milton Berle, Colonel Flack, Accused, Peter Gunn, General Electric Thea., Kraft Music Hall, Danny Thomas, Bachelor Father, Commercials. Adventures in Paradise, Follow the Sun, The Donna Reed Show, Valentine Day, No Time for Sergeants, The Farmer's Daughter, The Bing Crosby Show, Batman, Lost in Space, The Man From U.N.C.L.E., Laredo, Please Don't Eat the Daisies, Girl From U.N.C.L.E., The Smothers Bros., The Wild, Wild, West, Bewitched, Donald O'Conner Show, The Beverly Hillbillies, Land of the Giants, Arnie, Love, American Style, The Merv Griffin Show, Movie of the Week, The New Bill Cosby Show, The Julie Andrews Hour, Fire House, The Odd Couple, The Night Couple, The Night Stalker, Only With Married Men, The Mike Douglas Show, The Tonight Show, Tabitha, The Hardy Boys, Flying High, Over Easy, Hizzonner, NBC's Sunday Show, General Hospital, Heidi, Supertrain.
STAGE: The Miracle, Once More With Feeling, Would Be Gentleman, Midsummer Night's Dream, Arsenic and Old Lace.
PICTURES INCLUDE: Wives and Lovers, Promises, Promises, Who's Minding the Store?, Four for Texas, The Patsy, Harlow, Made in Paris, Three on a Couch, Way . . . Way Out, The Comic, Hello, Dolly!, The Computer Wore Tennis Shoes, The Phynx, Which Way to the Front? Hebie Rides Again, The Strongest Man in the World, Only With Maried Men, Hoyd Axton's Country Western Rock 'Roll Show, The Sunshine Boys, Won Ton Ton, The Dog Who Saved Hollywood, Broadway Rose, Mel Brooks' Silent Movie, Pennsylvania Lynch, Freaky Friday, The World's Greatest Lover, Herbie Goes Bananas, Mel Brook's History of the World.

FELDKAMP, FRED: Producer. b. Newark, N.J., Mar. 2, 1914. Reporter, writer, editor many publications; writer, chair., edit. bd. March of Time, motion picture release, 1947–51; adapted Crusade in Europe for TV: Crusade in the Pacific for TV: produced Operation Manhunt, Silken Affair. Executive prod. Triple Cross.
Co-author with wife, Phyllis—The Good Life—Or What's Left of It. New book: Not Everybody's Europe.

FELDMAN, EDWARD S.: Producer b. New York, Sept. 5, 1929. e. Michigan State Univ. Trade press contact, newspaper and mag. contact, 20th Century Fox, 1950; dir. info. services, Dover Air Force Base. 1954–56; publicity coordinator, The World of Suzie Wong, Para., 1959; joined Embassy, dir. of publicity, 1969; vice pres. in chg., adv. & pub., Seven Arts Prods., 1962; v.p. exec. asst. to head prod. Warner-7 Arts Prods., 1967, pres., m.p. dept., Filmways, 1970; exec. prod., What's the Matter With Helen?, 1971, exec. prod. Fuzz, 1972, Save the Tiger, 1972; prod. The Other Side of the Mountain, 1974 prod. Two-Minute Warning, 1976; exec. prod. 21 Hours At Munich, 1976; prod., King, 1977. Formed Edward S. Feldman Co., 1978. Co-prod. The Last Married Couple in America; Valentine (exec. prod.), 1980, 300 Miles for Stephanie (exec. prod.).

FELDMAN, MARTY: Actor-Writer-Director. b. England, 1933. Started as comic on TV in Britain (Every Home Should Have One, Marty, etc.). Appeared in theatrical films for Mel Brooks before co-writing and directing and starring in The Last Remake of Beau Geste in 1976.
PICTURES INCLUDE: The Bed-Sitting Room, Young Frankenstein, The Adventures of Sherlock Holmes' Smarter Brother, Silent Movie, The Last Remake of Beau Geste, In God We Trust (co. s.p., dir., act.).

FELDMAN, PHIL: Producer. b. Boston, Mass. Jan. 22, 1922. e. Harvard, 1943; Georgetown U., 1946; Harvard Law School, 1949, Harvard Business School, 1947. First Lieutenant, U.S. Army, 1943–46. Owner, Wholesale and retail dry goods firm, 1946–49. Law practice, 1950–51. Legal counsel Famous Artists Corp. 1951–53. Assoc. Dir. of Business Affairs CBS, 1953; Dir. of Business Affairs CBS, 1954–57. V.P. Talent & Contract Properties CBS, 1957–60. Exec. v.p. Broadcast Management, 1960–62. Head Business Relations

20th-Century Fox, 1962. V.P. 7 Arts Assoc. Corp. 1962–66 Pres., Phil Feldman Prods., Inc., 1967. Pres., First Artists, 1969; 1980, joined Rastar Films as exec. v.p.
PICTURES INCLUDE: You're a Big Boy Now, The Wild Bunch, The Ballad of Cable Hogue.

FELL, NORMAN: Actor. b. Philadelphia, March 24. e. Temple Univ. Studied acting with Stella Adler. Member, Actors Studio. Professional debut at Circle-in-the-Square Theatre in N.Y. in Bonds of Interest.
Summer stock; appearances on TV; moved to Hollywood in 1958 to begin theatrical film career.
PICTURES: The Graduate, Pork Chop Hill, Oceans 11, Rat Race, Inherit the Wind, Bullitt, If It's Tuesday, This Must Be Belgium, Catch 22, The End.
TELEVISION: Over 150 live plays from New York and some 200 shows filmed in Hollywood. Series: Dan August, 87th Precinct, Joe and Mable, Tom Ewell Show, Needles and Pins, Three's Company, The Ropers. Mini-Series: Rich Man, Poor Man, Roots: The Next Generation.

FELLERMAN, MAX: Theatre executive. b. New York City, June 5, 1899. Entered m.p. ind. with Pathe Films in 1916, becoming salesman in 1918; resigned from RKO Distributing Co. to join RKO Theatre Co. 1932; joined United Paramount Theatres 1944 in theatre administration: to Lopert Films; v.p., gen. mgr., March, 1954–59; MGM, 1959–61, v.p. dir. film industry affairs, Inflight Motion Pictures, 1961–74. Now m.p. consultant.

FELLINI, FEDERICO: Director. b. Rimini, Italy. Jan. 8, 1920. Cartoonist, caricaturist; then m.p. dir., writer, actor; s.p., actor.
PICTURES INCLUDE: The Miracle, Open City, Paisa, Variety Lights, White Sheik, I Vitelloni, The Matrimonial Agency (Love in the City), Il Bidone, The Wastrels, La Strada, The Swindlers, La Dolce Vita, Boccaccio '70, "8½", The Nights of Cabiria, Juliet of the Spirits, Spirits of the Dead, Fellini Satyricon, The Clowns, Fellini's Rome, Amarcord, Casanova, City of Women.

FELLMAN, DANIEL R.: Executive. b. Cleveland, Ohio, March 14, 1943. e. Rider College (BS, 1964). Joined Paramount N.Y. 1964; later sales mgr. Washington DC, Dallas. Next branch mgr. Cleveland; then Chicago. In 1969 joined Loews Theatres as film buyer. In 1971 joined Cinema National Theatres, division of Carrols Development Corp., as v.p./chief film buyer. In 1973 named v.p./dir., Cinema National Theatres. Joined Warner Bros. in 1978 as eastern sales mgr. 1979 named v.p./ass't. gen. sales mgr. Elected to: board directors NATO, 1975–76. Assistant Chief Barker Variety Club, Tent 35 NY 1976. President Variety Club Tent 35, 1977–78. 1980 Chairman Will Rogers Raffle Committee. v.p., bd. member, Variety Club Tent 25.

FELLMAN, NAT D.: Executive. b. New York, N.Y., Feb. 19, 1910. 1928. Started as office boy, Warner Bros. Pictures, 1928; transferred to Warner Bros. Theatres, asst. to chief booker; handled pool, partnership operations; head buyer, booker for Ohio zone, 1941; asst. to chief film buyer in New York, 1943; apptd. chief film buyer, 1952; exec. asst. to v.-p. and gen. mgr., Stanley Warner Theatres, 1955; asst. gen. mgr., Stanley Warner Theatres, 1962; acting gen. mgr., Stanley Warner Theatres, July, 1964; Stanley Warner Theatres, v.p. and gen. mgr., 1965; v.p., NGC Theatre Corp. and division mgr. Fox Eastern Theatres, 1968; v.p. National General Corp., and pres., National General Theatres, 1969; 1974, formed Exhibitor Relations Co., operations consultant; vice pres., Variety Clubs International; NATO, Chrm., presidents' advisory comm.

FENADY, ANDREW J.: Producer-writer. b. Toledo, Ohio, Oct. 4, 1928. e. Univ. of Toledo, 1946–50. Radio-prod.-actor-writer, WTOD; Actor, Summer Stock, 1948–51; National Classic Theatre, 1950–51; writer-prod., Confidential File, TV, 1954–56; joined Screen Gems as indep. prod. with own company, Fenady Associates, 1962.
PICTURES INCLUDE: Stakeout on Dope Street, The Young Captives, Ride Beyond Vengeance, Chisum, Terror in the Wax Museum, Arnold.
TV: The Rebel, Branded series, Hondo, Mission: Impossible.

FENNELLY, VINCENT M.: Producer. b. Brooklyn, July 6, 1920. e. Holy Cross, 1938–42. U.S. Navy, 1942–46; salesman, Monogram, Des Moines, 1947; entered prod. field, 1949; indep. prod. for Monogram, 1950; Ent. TV field, 1957; prod. Transwestern Films, Inc., Frontier Pictures, Silvermine Productions Co., Allied Artists; Malcolm Enterprises, Hilgarde Enter.
PICTURES INCLUDE: Kansas Territory, Wagons West, Fargo, Marksman, Star of Texas, Topeka, Texas Bad Man, Bitter Creek, The Desperado, Seven Angry Men, Dial Red O, Bobby Ware Is Missing, At Gunpoint, Crime in the Streets, Last of the Badmen.
TV: Four Star Films, Alcoa-Goodyear, Trackdown, Wanted Dead or Alive, David Niven Show, Richard Diamond, Stagecoach West, The Dick Powell Theatre, Target, The Corruptors, Rawhide; A Man Called Shenandoah, MGM.

FENNEMAN, GEORGE: M.C., Announcer. b. Peking, China, Nov. 10.; e. San Francisco State Univ.
CREDITS INCLUDE: Groucho Marx Show You Bet Your Life, M.C. Host: Surprise Package, Funny Funny Films, Talk About Pictures, On Campus, Donny & Marie, Spokesman for Home Savings & Loan Assoc.

FERGUSON, HELEN: Public rel., career counsel. b. Decatur, Ill., July 23. e. Chicago Academy of Fine Arts. Film debut at 13, Essanay Studio, with Vitagraph, Lasky, Goldwyn, etc.; stage debut in Alias the Deacon, 1927.
THEATRE: New Brooms, Lomardi Ltd., This Thing Called Love, Intimate Strangers.
Freelance film work for many companies; retired from acting, 1930; entered pub. & pub. rel., 1933.
TV Panelist, RSVP, 1949–50.
Author: Loretta Young's The Things I Had to Learn, 1961. Bobbs-Merrill, other professional biographies.

FERGUSON, ROBERT S.: Executive. b. New York, N.Y., May 8, 1915. e. N.Y.U., BS, 1936. Started in newspaper field, Scripps Howard, 1936–37; m.p. ind., 1938 with Warner Bros. Pictures; Columbia Pictures, 1940; asst. dir. of adv. & pub., 1950; dir. adv., exploitation & pub., 1957; pres. Screen Publicists Guild, 1946; V.P. World-Wide Adv. & Pub. Columbia, 1963; V.P. Corporate Relations, Columbia Pictures Industries, 1974; V.P. Marketing—The American Film Theatre, 1974; V.P. and mgr., Entertainment Division—Rosenfeld, Sirowitz and Lawson Advertising, 1975; Director Marketing—Horizon Pictures, 1976; pres. Cinema Think Tank 1977–; instructor, Adelphi University, 1978. Rejoined Columbia Pictures Industries as consultant to pres., 1979.

FERRER, JOSE: Producer, director, actor. b. Santurce, Puerto Rico, Jan. 8, 1912; e. Princeton U.; Director and actor, New York stage, prior to screen career; dir. stage play, Strange Fruit; New York stage; Twentieth Century, 1951; dir., prod. Stalag 17, N.Y. stage, 1951; prod. and dir., The Chase; prod. dir., act., The Shrike, N.Y. stage, 1952; dir. My Three Angels.
PICTURES INCLUDE: Joan of Arc, Whirlpool, Crisis, Cyrano de Bergerac, Anything Can Happen, Moulin Rouge, Miss Sadie Thompson, Caine Mutiny, Deep in My Heart, The Shrike, Cockleshell Heroes, Great Man, I Accuse, State Fair; Lawrence of Arabia, Stop Train 349 From Berlin, Enter Laughing, Voyage of the Damned, The Sentinel, The Swarm, Fedora.

FERRER, MEL: Producer, director, actor. b. Elberon, N.J., Aug. 25, 1917; e. Princeton U. During coll. and early career spent summers at Cape Cod Playhouse, Dennis, Mass.; then writer in Mexico, authored juvenile book, Tito's Hats; later ed. Stephen Daye Press, Vt. Left publishing upon reaching leading-man status at Dennis; first two Broadway appearances, dancing roles, You'll Never Know, Everywhere I Roam; others. Kind Lady, Cue For Passion; then to radio, serving apprenticeship in small towns; prod.-dir., NBC; dir. Land of the Free, The Hit Parade, and Hildegarde program. Entered m.p. ind., 1945, when signed by Columbia, dial. dir.: The Girl of the Limberlost; later, returned to Broadway, leading role, Strange Fruit; signed by David Selznick as producer-actor, on loan to John Ford as prod. asst. on The Fugitive; then to RKO for Vendetta. Acted in Lost Boundaries, Born to be Bad, Ondine on Bway., 1954.
PICTURES INCLUDE: The Secret Fury, The Brave Bulls, Rancho Notorious, Scaramouche, Lili, Knights of the Round Table, Saada, War and Peace, Paris Does Strange Things, The Vintage, The World, The Flesh, and the Devil, Blood and Roses, El Greco; prod., Wait Until Dark, "W."

FERRI, ROGER: Executive. b. Providence, R.I. Sports ed. Providence News, Springfield, Mass. Union; others; feature writer, Boston Post (co-author, Life of Gen. Pershing); Detroit Free Press; New York World, Press repr., Shubert Attractions, Comstock, Elliott & Gest, John Cort; agent, Sells-Floto and Forepaugh circuses; adv. pub. dir. Emery Circuit, R.I.; Nathanson Amusements, New England; on staff, Exhibitors' Trade Review, M.P. World, Variety, M.P. News; adv. pub. mgr., Chadwick Pictures; 1924, dist. expl. rep., Fox Film Corp., Cleveland, Cincinnati, Detroit, Indianapolis, Chicago; 1925, Home Office executive, 1926–62, sales promotions director, publisher-editor 20th Century-Fox Dynamo, 1963, adv. pub. dir., National Screen Service Corp.

FETZER, JOHN E.: Executive. b. Decatur, Ind., March 25, 1901. e. Purdue Univ., U. of Mich. A.B., Andrews U. Hon. LL.D., West. Mich. Univ. 1958; chairman, owner Fetzer Broadcasting Co., Kalamazoo, Mich.; chmn., owner Detroit Baseball Club; pres., owner Fetzer Communications, Inc., Kalamazoo; chairman, dir., Cornhusker TV Corp.; Ch. Wolverine Cablevision; Mem., former bd. dir., Amer. Nat'l Bank and Trust Co., former mem. Bd. of Trustees, Kalamazoo College. Radio research Europe, 1925; asst. dir., U.S. Censorship in charge of radio. 1944–46; served as war corr. in ETO, 1945; spl. assignment radio, TV, newspaper mission to Europe and Middle East, 1952. chmn., TV code review bd., NARTB, 1952–55. Fellow, Royal Society of Arts, London; mem., Acad. of Polit. Science. Clubs: Park, Kalamazoo Country, (Kalamazoo); Radio and Television Execs. Soc. (Assoc.); Broadcast Pioneers, N.Y.C.

FIELD, DAVID M.: Executive. b. Kansas City, MO. e. Princeton University. Worked as reporter on city desk at Hartford (Conn.) Courant. In 1968 with NBC News in N.Y.C. and Washington, D.C. Entered film school at U.S.C. (L.A.) after which joined Columbia Pictures as west coast story editor. In 1973 went to ABC-TV Network as mgr., movies of the week. 1975, moved to 20th-Fox as v.p., creative afffairs. Joined United Artists in 1978; named snr. v.p.—west coast production. Left in 1980 to become 20th-Fox exec. v.p. in chg. of worldwide production.

FIELD, SALLY: Actress. b. Pasadena, Calif., 1946. Gained stardom lead of TV series, Gidget, 1965, followed by The Flying Nun, 1967–68. Theatrical film debut in The Way West (1967).
PICTURES INCLUDE: Stay Hungry, Smokey and the Bandit, Heroes, The End, Hooper, Norma Rae, Beyond the Poseidon Adventure, Smokey and the Bandit II, Back Roads, Absence of Malice.
TV: Home for the Holidays, The Girl with Something Extra, Sybil.

FIELD, SHIRLEY ANNE: Actress. b. London. Ent. films after repertory experience 1955. Under contract to Ealing-M.G.M. 1958.
THEATRE: The Lily White Boys, Kennedys Children, Fire.
PICTURES INCLUDE: Saturday Night and Sunday Morning, The Man in the Moon, War Lover, Lunch Hour, Kings of the Sun, Doctor in Clover, Alfie, Hell is Empty.
TV: Risking It.

FIELD, VIRGINIA: Actress. r.n. Margaret Cynthia Field. b. London; e. England, Paris, Vienna. On N.Y. stage in Victoria Regina & others. On screen from 1934.
PICTURES INCLUDE: The Lady is Willing, Little Lord Fauntleroy, Ladies in Love, Lloyds of London, Singapore Woman, Atlantic Convoy, The Crystal Ball, Christmas Eve, Repeat Performance, Ladies Man, Dial 1119, Weekend With Father, Veils of Bagdad.
THEATRE: The Doughg rls.

FIELDS, FREDDIE: Executive. b. Ferndale, N.Y. July 12, 1923. Vice-pres., member of bd. of directors, MCA-TV, MCA Canada Ltd., MCA Corp.; mem., Pres. Club, Wash., D.C.; pres., Freddie Fields Associates Ltd., 1961 Pres., Kam Ent., Ltd.; v.p. F/B Prod., Inc., pres., chief exec. officer Creative Management Assoc. Ltd. Agency, Chicago, Las Vegas, Miami, Paris, Los Angeles, N.Y., London, Rome, 1962. Exclusive agent of Henry Fonda, Phil Silvers, Joanne Woodward, Paul Newman, Peter Sellers, Barbara Streisand, Steve McQueen, Ali MacGraw, Woody Allen, Robert Redford, Ryan O'Neal, Liza Minnelli and others. Dir. Chelsea Nat'l Bank, N.Y., Polly Bergen Co., First Consolidated Corp. In 1975 resigned post with CMA (now International Creative Mgt.) but continues as consultant. Presently producing for Paramount Pictures. 1977: Looking for Mr. Goodbar. 1980: Wholly Moses. 1981: Victory.

FINBERG, JACK GERALD: Writer technician. b. London, England. Asst. to exec. prod., story dept., British National Studios; Offic. shorthand writer, Royal Courts of Justice and Central Criminal court; hon. sec., Film Studios Club; personal asst. to Sam Spiegel; verbatim report., nuclear weapons, testing and disarmament conferences, UN; asst. prod., ABC TV; two feature writing credits; six major, 18 second feature film adaptations; Film/TV prod., Australia.

FINCH, JON: Actor. b. England, 1941. Came to acting via backstage activities, working for five years as stage manager in repertory theatres all over Britain.
PICTURES: MacBeth, Sunday, Bloody Sunday, Frenzy, Lady Caroline Lamb, The Final Program, Une Femme Fidele, The Man of the Green Cross, Battleflag, Death on the Nile.

FINE, HARRY: Producer. e. St. Andrew's College and Dublin Univ.
FILMS INCLUDE: Up The Junction, The Liquidator, The Penthouse, Long Days Dying, Vampire Lovers, Fright, Twins of Evil, Too Far To Go, Quadrophenia, McVicar.
T.V. FILM SERIES: Sir Francis Drake, Man of The World, The Sentimental Agent.

FINESHRIBER, WILLIAM H., JR.: Executive. b. Davenport, Ia. Nov. 4, 1909; e. Princeton U., B.A., 1931. Pub., CBS, 1931–34; mgr. Carnegie Hall, N.Y., 1934–37; script writer, dir., music comm., dir. of music dept., CBS, 1937–40; dir. of short wave programs, CBS, 1940–43; gen. mgr. CBS program dept. 1943–49; v.p. in charge of programs MBS, 1949–51; exec. v.p. & dir., MBS, 1951–53; v.p. & gen. mgr. of networks, NBC, 1953–54; v.p. in charge of Radio Network, NBC, 1955; v.p., Television Programs of America, 1956; director International operations, Screen Gems, 1957; v.p., Motion Picture Assoc. of America and Motion Picture Export Assoc. of America, 1960; bd. of dir., NARTB; exec. comm., bd. of dir., R.A.B.; v.p. Radio Pioneers.

FINKELSTEIN, HERMAN: Attorney. b. Torrington, Conn., Jan. 9, 1903; e. Clark U., A.B.; Yale Law School, LL.B., member of Conn., Pa. & N.Y., U.S. Sup. Ct. bar; American Law Institute; director, ASCAP Nathan Burkan Memorial Competition; Mem.; State Dept. Panel on intl. Copyright; U.S. National Commission for UNESCO 1957–62; Consultant to U.S. Delegation, Inter-Amer. Conf. of Copyright Exports, Wash., D.C., 1946; Mem. U.S. Delegation, Intergovernmental Conference on Universal Copyright Convention Geneva (1952), Berne Convention-Stockholm Revision Conf. 1967, Paris Revision Conf. 1971, Communication Satellite Conf. 1973, past pres., Copyright Soc. U.S.A., past pres., Exec. Bureau, CISAC; past pres., SCRIBES.
AUTHOR: Numerous articles on copyright.

FINNEY, ALBERT: Actor. b. Salford, England, May 9, 1936. Studied for stage at R.A.D.A. making his West End debut 1958 in The Party. Appeared at Stratford-on-Avon 1959 season playing title role in Coriolanus, etc. 1960; On West End stage in The Lily White Boys and Billy Liar, Luther; New York, 1963, Luther. National Theatre 1965; Much Ado About Nothing, Armstrong's Last Goodnight, Love for Love, Miss Julie, Black Comedy, A Flea in Her Ear; Broadway, Joe Egg; London Stage, Alpha Beta, Krapp's Last Tape, Cromwell, Chez Nous, Hamlet Tamburlaine, Uncle Vanya, Present Laughter. National Theatre: The Country Wife, The Cherry Orchard, Macbeth 1978.
PICTURES INCLUDE: The Entertainer, Saturday Night and Sunday Morning, Tom Jones, Night Must Fall, Two for the Road, Charlie Bubbles, Scrooge, Gumshoe, Alpha Beta, Murder on the Orient Express, Wolfen, Loophole, Looker.

FINEY, EDWARD: Producer. b. New York City; e. C.C.N.Y. Began as engineer, Western Electric. Then property man, Johnny Hines, C.C. Burr comedies. To MGM as press sheet ed.; sales promotion mgr., Pathe; asst. adv. dir. UA; story ed. and adv. dir., Mon-Rep., prod. & adv. dir., Grand Natl.; then prod. Monogram, many westerns. In 1941 org. Edward Finney Prod. Discovered Tex Ritter and made his first 40 features. Prod. & dir. Silver Stallion, Code of the Redman, Queen of the Amazons, etc. Clubs: Masquers. ex. pres. Catholic Press Council, Board Member Permanent Charities, M.P. Acad. of Arts and Sciences Dramatic Guild; pres., Indep. M.P. Prod. Assn. Film chrm. Catholic Press Council.
PICTURES INCLUDE: Corregidor, Golden Stallion, Call of the Forest, Hi Diddle Diddle, Primitive Passion, Mormon Battalion, This Is Pippin (TV); Seven Wonders, Baron Munchausen, Enchanted Years, Buffalo Bill in Tomahawk Territory, Slash of the Knife, Gun Girls, Journey to Freedom, London After Dark, Dark Road, Halfway to Hell, The Prairie, Secret Tower, Barrels Away, Things That Make America Great, (TV); Spring Affair, Dragons Across Asia, The Madcaps, The Happy Clown, The Great Dogtown Robbery.
TV: Pilots for Law and the People and the Circle Kids. Now preparing Small Crossing, Dog House Blues and The Italian.

FIRTH, PETER: Actor. b. Bradford, Yorkshire, Oct. 27, 1953. Selected to appear in local TV children's show where casting director spotted him and got him role in series, The Flaxton Boys. Moved to London and worked in TV, first in children's show, later on dramas for BBC. Big break came in 1973 when signed for Equus at National Theatre, London, which role repeated for film.
PICTURES: Joseph Andrews, Aces High, Equus, When You Comin Back, Red Ryder.
STAGE: Equus, Romeo and Juliet, Spring Awakening.

FISCHER, JOSEPH A. Executive. e. Syracuse Univ.; As CPA was mgr. with Price Waterhouse & Co. Joined Columbia Pictures in 1967 as comptroller; In 1972 made v.p. In 1973 named financial v.p.-treas. of Col. Pictures Industries; 1976, sr. v.p. & chief financial officer; 1979, exec. v.p. Mem. American Inst. of CPA's and N.Y Society of CPA's.

FISHER, AL: Executive. b. Brooklyn, N.Y. Entered m.p. industry as office boy, Fox Metropolitan Theatres; U.S. Army Provost Marshal General's Office, 1942–46; Universal Pictures, mgr., Park Avenue Theatre, N.Y & Copley Plaza Theatre, Boston, 1946; Eagle Lion Film Co., mgr., Red Shoe's Bijou Theatre, N.Y., 1947; Stanley Kramer Prods., exploitation, Cyrano de Bergerac, 1951; press agent, 1951; prod., Bway show, Daphine, 1952; joined United Artists Corporation, 1952, named dir. of exploitation; now freelancing as producer's repr.

FISHER, CARRIE: b. Beverly Hills, CA, Oct. 21, 1956. e. London Central School of Speech & Drama. Daughter of Debbie Reynolds and Eddie Fisher. On Broadway in revival of Irene and Censored Scenes from King Kong.
PICTURES: Shampoo, Star Wars, Mr. Mike's Mondo Video, The Empire STrikes Back, The Blues Brothers, Under the Rainbow.

FISHER, EDDIE: Singer. b. Philadelphia, Pa., Aug. 10, 1928. Band, nightclub, hotel singer; discovered by Eddie Cantor, 1949; U.S. Army, 1951–53; many hit records include: Wish You Were Here, Lady of Spain; radio & TV shows, NBC.
PICTURES INCLUDE: Butterfield 8.

FISHER, FRANK H.: Executive. b. Toronto, Ont., Canada, May 16, 1907. e. Hillhurst public sch., Calgary, Alberta, 1912–19; Crescent Hghts. Collegiate Institute, Calgary, 1919–23. Bank of Toronto, 1923–26. Cashier, booker, Educational Films, 1926–32; br. mgr., Empire Universal Films, Calgary, 1932–34; office mgr., Toronto, 1935; br. mgr., Montreal, 1938; br. mgr. Toronto, 1939; asst. gen. mgr. 1939–45; chief buyer, then western div. mgr., Odeon Theatres, 1945–47; dir., vice-pres. gen. mgr., Rank Film Distr., 1947–54; apptd., vice-pres. & gen. mgr., Odeon Theatres of Canada, 1954. In 1973 joined Canadian Theatre Group Ltd. as Toronto rep. and v.p.-asst. to pres. Past Pres., Canadian Picture Pioneers; past pres., Canadian Motion Picture Industry Council; past pres., Rotary Club of Toronto; past pres., Convention and Tourist Bur. of Metro Toronto; dir. Health League of Canada and Ontario Crippled Children's Society; retired May, 1972; chairman of board of Governors O'Keefe Centre for the performing arts Toronto; director of Y.E.S. (Youth Employment Service.)

FISZ, BENJAMIN: Producer. b. Warsaw, Poland, Nov. 4, 1922. Formed in assoc. Golden Era Films, 1954. Later Aqua Films Ltd., 1957 as ind. prod. for Rank Org. and formed S. Benjamin Fisz Prod. Ltd., 1963 formed Benton Film Prod. Ltd. in assoc. with Anthony Mann. 1967 Spitfire Prods. with Harry Saltzman. 1969 Benmar Prods. Incorporated into Scotia Investments Ltd. Formed Scotia Intl. under name group for dist. Dir., Scotia-Barber. 1972: Formed Scotia Int. (Film Sales.) Jointly formed Scotia Deutschland and International Film Theatre Ltd.
PICTURES INCLUDE: Capri, The Secret, Child in the House, Hell Drivers, Sea Fury, On the Fiddle, The Heroes of Telemark, The Battle of Britain, A Town Called Bastard, Aces High, A Nightingale Sang in Berkeley Square, Philby.

FITELSON, H. WILLIAM: Attorney, counsel to Fitelson, Lasky and Aslan, law firm, New York City, specializing in the field of communications (publishing, motion pictures, theatre, television, radio); born New York City Jan. 21, 1905; educated Columbia University (extension), New York Law School; newspaper and editorial work, librarian and tutor New York Law School, Story Department Tiffany-Stahl Productions 1929, counsel Tiffany-Stahl Productions 1929–1932, United States counsel to British and foreign motion picture and theatrical producers, counsel to The Theatre Guild and numerous members motion picture industry, theatre industry, newspaper and publishing, television and radio interests, managing director Theatre Guild Television and radio divisions, including U.S. Steel Hour. Non-legal consultant on communications projects to Allen & Co., and Columbia Pictures Industries.

FITZGERALD, GERALDINE: Actress. b. Dublin, Eire, Nov. 24, 1914; e. Dublin Art Sch. On stage Gate theat., Dublin; then in number of Brit. screen prod. including Turn of the Tide, Mill on the Floss. On N.Y. stage in Heartbreak House. In U.S. screen prod. from 1939 Wuthering Heights. Pictures (U.S.) include Dark Victory, Till We Meet Again, Flight from Destiny, Gay Sisters, Ladies Courageous, Watch on the Rhine, Wilson, The Strange Affair of Uncle Harry, Three Strangers, Nobody Lives Forever, So Evil My Love, Obsessed, Ten North Frederick, The Pawnbroker, Rachel, Rachel, Believe in Me, The Last American Hero, Harry and Tonto, Echoes of a Summer, Arthur.
STAGE: A Long Day's Journey Into Night (1971).
TV: The Best of Everything.

FITZGERALD, PEGEEN: Commentator, executive. b. Norcatur, Kans., Nov. 24. e. Coll. of St. Theresa, Winona, Minn.; LL.D. (hon.) Seton Hall U. m. Ed Fitzgerald. Sales prom. adv. in retail field; magazine editor; originator (with husband) of husband and wife radio-TV conversation; radio-TV commentator, WOR, N.Y.; pres. Vivisection Investigation League; founder, Crusade for the Abolition of Cruelty; president Millennium Guild; co-author of Cast of Characters, author of Meatless Cooking pub. by Prentice-Hall.

FITZGERALD, ROBERT: Actor-writer. b. San Francisco, Calif., Nov. 28, 1925. e. Hollywood H.S., 1943; Univ. of Calif, 1949. John Golden Award, 1945; Erskine Johnson Award, 1948; pioneered Late Show on TV Midnite Shack, CBS, 1951; prod., Peter Lind Hayes Review, 1953; Actor: Verdict is Yours, CBS-TV, 1960; prod., Japanese Plan, Hotel Bunker, The Big Trip (Docu.), In The Looking Glass (Short), Spook Tours Europe (Docu.), Funnyman (Short), The Welt Story (Docu.); prod. dir., Whoever He Was?, short; The Southern Gentleman, short; The World is Three Drinks Behind, docu.; Bradley For President, (doc.), Gunga Din Massey (short); 1967 prod. The Bogart Years; 1968 London Girl, Bogie Revisited. Prod. I Love You Rick Leland, 1968; John Garfield Story.

FITZPATRICK, JAMES A.: Producer. b. Shelton, Conn. February 26, 1902; e. Yale and American Academy of Dramatic Arts; m. Lesley Champlin. Newspaper field and stage. Writing, directing and producing since 1925. Pioneer producer and originator of off-stage voice for FitzPatrick Traveltalks and others through his own company. Distributes through MGM, Paramount, Warner Bros.; in 1944 org. own prod. co., Fitz-Patrick Pictures, Hollywood, Calif.

FIX, PAUL: Actor. b. Dobbs Ferry, N.Y., March 13, 1902. Toured with stock companies in the eastern states, also in numerous stage productions in the twenties and thirties on the west coast. In motion pictures since 1922.
PICTURES INCLUDE: High and the Mighty, Giant, Bad Seed, To Kill a Mockingbird, Pat Garret and Billy the Kid, 1973.
TV: Numerous prods., five yrs. as Marshall Micah on The Rifleman, Zabreskie Point.

FLAKS, STEPHEN: Executive. b. Jan. 1, 1941, New York. e. City College of New York, B.B.A. Early career as an investment banker & stock broker for various companies, including First Hanover, First Devonshire, 1962–69. Founded Flaks-Zaslow in 1969. Entered motion picture industry in 1974 in capacity of production financing. Established VideoVision, Inc., in 1978.

FLATTERY, THOMAS L.: Executive-Lawyer b. Detroit, Mich., Nov. 14, 1922. e. U.S.M.A., West Point, B.S., 1944–47; U.C.L.A., LL.B., 1952–55; U.S.C., LL.M, 1955–65. Gen'l counsel and asst. sec'y, McCulloch Corp., Calif., 1957–64; sec'y and corporate counsel, Technicolor, Inc., 1964–70; Vice President, Secretary & General Counsel, Amcord, Inc (formerly American Cement Corporation) 1970–72; Vice President, Secretary & General Counsel, Schick Incorporated, 1972–1975; counsel asst. secretary, C.F. Braun & Co., 1975–1976; sr. vice pres., secretary & general counsel Automation Industries, Inc. since 1976 to present.

FLAXMAN, JOHN P.: Producer. b. New York City, March 3, 1934. e. Dartmouth Univ. B.A. 1956; 1st Lt. U.S. Army, 1956–58. Ent. m.p. industry in executive training program, Columbia Pictures Corp., 1958–1963; exec. story consultant, Profiles in Courage, 1964–65; head of Eastern Literary Dept., Universal Pictures, 1965; writer's agent, William Morris Agency, 1966; partner with Harold Prince in Media Productions, Inc. 1967; pro. m.p. Something for Everyone. Founded Flaxman Film Corp., 1975. Prod. Jacob Two-Two Meets the Hooded Fang, 1976. President-Tricorn Productions 1977; pres. Filmworks Capital Corp., 1979–present: Becker/Flaxman & Associates, 1979–present.

FLEISCHER, RICHARD O.: Director. b. Brooklyn, N.Y., Dec. 8, 1916; e. Brown U., B.A.; Yale U., M.F.A. Stage dir.; joined RKO Pathe 1942.
PICTURES INCLUDE: Flicker Flashbacks, This is America, Design for Death, Child of Divorce, Banjo, So This Is New York, Bodyguard, Follow Me Quietly, The Clay Pigeon, Narrow Margin, The Happy Time, Arena, 20,000 Leagues Under the Sea, Violent Saturday, Girl in the Red Velvet Swing, Bandido, Between Heaven and Hell, The Vikings, These Thousand Hills, Compulsion, Crack in the Mirror, The Big Gamble, Barabbas, Fantastic Voyage, Doctor Dolittle, Boston Strangler, Tora! Tora! Tora!, Ten Rillington Place, The Last Run, See No Evil, The New Centurions, Soylent Green, The Don is Dead, The Spikes Gang, Mister Majestyk, Mandingo, The Incredible Sarah, Crossed Swords, Ashante, The Jazz Singer, Tough Enough.

FLEMING, JANET BLAIR: Executive. b. Ottawa, Canada, November 29, 1944. e. Bachelor Arts Degree, Carlton University, Ottawa, Canada; Special asst. to Canada's Federal Minister of Transport 1967–1972; 1973–77, asst. to Sandy Howard—business affairs; 1977, co-founder and v.p./sales & admin. of Producers Sales organization.

FLEMING, RHONDA: Actress. r.n. Marilyn Lewis. b. Los Angeles, Aug. 10. p. Harold and Effie Graham Lewis, prof.; e. Beverly Hills High School. Calif. Screen debut: Spellbound. L.A. Civic Light Opera in Kismet revival; 1976. Broadway debut in 1973: Revival of The Women.
PICTURES INCLUDE: Spiral Staircase, Adventure Island, Out of the Past; singing debut: A Connecticut Yankee in King Arthur's Court, The Last Outpost, Cry Danger, The Redhead and the Cowboy, The Great Lover, Crosswinds, Little Egypt, Hong Kong, Golden Hawk, Tropic Zone, Pony Express, Serpent of the Nile, Inferno, Those Redheads from Seattle, Jivaro, Yankee Pasha, Tennessee's Partner, While the City Sleeps, Killer Is Loose, Slightly Scarlet, Odongo, Queen of Babylon, Gunfight at the OK Corral, Buster Keaton Story, Gun Glory, Home Before Dark, Alias Jesse James, The Big Circus, The Crowded Sky, Instant Love, The American Wife, Won Ton Ton, The Dog Who Saved Hollywood, The Nude Bomb.

FLEMYNG, ROBERT: Actor. b. Jan. 3, 1912, Liverpool, Eng. e. Halleybury Coll. Stage debut: Rope 1931. Screen debut; Head Over Heels 1937.
PICTURES INCLUDE: Bond Street, The Guinea Pig, The Conspirators, The Blue Lamp, Blackmailed, The Magic

Box, The Holly and the Ivy, Cast a Dark Shadow, Man Who Never Was, Funny Face, Let's Be Happy, Windom's Way, Blind Date, A Touch of Larceny, Radtus (Italian), The King's Breakfast, The Deadly Affair, The Spy with the Cold Nose, The Quiller Memorandum, Deathhead Avenger, OH! What a Lovely War, Battle of Britain, Cause for Alarm, Young Winston, The Darwin Adventure, Travels with my Aunt, Golden Rendezvous, The Medosa Touch, The Four Feathers, The Thirtynine Steps.

TV: appearances in England, U.S. inc.: Rainy Day, Playhouse 90, Wuthering Heights, Browning Version, After the Party, Boyd Q.C., They Made History, Somerset Maugham Show, Woman in White, The Datchet Diamonds, Probation Officer, Family Solicitor (series), Man of the World, Zero One, Compact, (serial), Day By the Sea, The Living Room, Hawks and Doves, Vanity Fair, The Inside Man, The Doctor's Dilemma, The Persuaders, Major Lavender, Public Eye, Florence Nightingale, Edward VIII, Spy Trap, The Venturers' Loyalties, The Avengers, Crown Court, Enemy at the Door, Rebecca, Edward and Mrs. Simpson.

FLETCHER, LOUISE: Actress. Came to Hollywood at age 21; studied with Jeff Corey. Worked on TV shows (including Playhouse 90). Gave up career to be housewife for 10 yrs.; returned to work in 1973.

PICTURES INCLUDE: Thieves Like Us, One Flew Over the Cuckoo's Nest, Exorcist II: The Heretic, The Cheap Detective, Natural Enemies.

FLEXER, DAVID: Executive. b. Mt. Plasant, Tenn., June 29, 1909; e. Wallace U. School, 1926; Vanderbilt U., 1930; m. Eleanor Handmacher, concert pianist. Was adv. rep. on national magazines, 1930–32; joined United Artists 1932–33; entered exhib. 1933, head of Flexer Theatres, operating in Tenn. and Miss.; The Albany Corp.; Theatre Real Estate Corp.; v.p. Impossible Pictures, Inc., producers of cartoons currently distrib. by Republic; constructed first drive-in theatres in Minneapolis, and St. Louis, 1947; pres., Flexer Theatres of Tenn., Inc., Flexer Theatres of Miss., Inc., The Albany Corp. Pres., Inflight Motion Pictures.

FLINN, JOHN C.: Publicist. b. Yonkers, N.Y. May 4, 1917; p. late John C. Flinn, pioneer m.p. executive; e. U. of Calif. In pub. dept. David O. Selznick, 1936–39; unit publicist, then head planner, Warner, 1936–46; joined Monogram as asst. to nat'l adv. & pub. head & pub. mgr. Aug. 5, 1946; apptd. nat'l dir. of pub. & adv. of Allied Artists Pictures Corp., March, 1951, appt'd studio dir. adv. & pub., Columbia, March, 1959; v.p., Jim Mahoney & Associates (pub. rel. firm) in 1971. Joined MGM West Coast publicity department as publicity coordinator, January, 1973; Rejoined Columbia Pictures in Feb., 1974 as studio pub. dir.; 1979, promoted to dir. industry relations.

FLOREA, JOHN: Producer-Director. b. Alliance, O., May 28, 1916. Served as photo journalist with Life magazine, 1940–50; assoc. editor Colliers magazine, 1950–53. Prod.-dir.

TV: Sea Hunt TV series, 1957–60; also dir. TV shows Bonanza, The Virginian, Honey West, Daktari, Gentle Ben, Cowboy in Africa. Prod.-dir. of film Islands of the Lost. With Ivon Tors Films. Nominated as one of the top 10 directors in America by DGA for 1968 Mission Impossible episode. Dir. several Ironsides episodes.

PICTURES INCLUDE: A Time to Every Purpose.

FLYNN, JOHN: Director-Writer. Began career as trainee script supvr. for dir. Robert Wise. Soon working as ass't. dir. on MGM-TV shows. Made dir. debut with The Sergeant, 1969.

PICTURES INCLUDE: The Jerusalem File (dir.), The Outfit (s.p., dir.), Rolling Thunder (dir.); Defiance (dir.).

FOCH, NINA: Actress. b. Leyden, Holland, April 20, 1924; daughter of Consuelo Flowerton, actress, & Dirk Foch, symphony orch. conductor. Many stage, radio, television appearances. Prof. Univ. South. Calif, 1966–67; 1978–80, senior faculty, American Film Institute, 1974–77; Board of Governors, Hollywood Acad. Television Arts & Sciences, 1976–77; Exec. Comm. Foreign Film Award, Acad. Motion Picture Arts & Sciences, 1970–.

PICTURES INCLUDE: Nine Girls, Song to Remember, My Name is Julia Ross, An American in Paris, Scaramouche, Young Man in Hurry, Undercover Man, Sombrero, Fast Company, Executive Suite Four Guns to the Border, Ten Commandments, Illegal, You're Never Too Young, Cry of the Werewolf, Escape in the Fog, I Love A Mystery, Johnny Allegro, Johnny O'Clock, Prison Ship, Return of the Vampire, Shadows In The Night, The Dark Past, The Guilt of Janet Ames, Three Brave Men, Spartacus, Cash McCall, Such Good Friends, Salty, Mahogany, Jennifer.

FOGARTY, JACK V.: Executive-Writer. b. Los Angeles, Calif. e. UCLA. hd. travel dept., MGM, 1960–62; exec. prod. mgr., Cinerama, Inc., 1962–64; assoc. prod., The Best of Cinerama, 1963; co-creator, The Boilermakers, CBS-TV, 1964; own p.r. firm, 1965; pres., AstroScope, Inc., 1969; named chm. bd., chief exec. off., 1974.

TV: (writer) The Rookies, SWAT, D'Angelo Superstar, Charlie's Angels, Most Wanted, Barnaby Jones, A Man Called Sloane, others. Story Editor, Charlie's Angels; exec. story editor, A Man Called Sloane exec. story editor, Most Wanted, exec. story consultant, Sheriff Lobo.

FOGELSON, ANDREW: Executive. b. New Rochelle, N.Y., August 5, 1942. e. Union College (1960–64). First entered m.p. industry in 1968 with Warner Bros., starting as copywriter and soon made exec. asst. to v.p. of adv.-pub. In 1973 appt. v.p., marketing services. Joined Columbia Pictures in Dec. 1973, as v.p. in chg. world-wide adv.-pub. In 1977 went to Warner Bros. as exec. v.p. in chg. worldwide adv.-pub, 1979, left to become pres. of Rastor Prods. Resigned 1981 to become indep. producer.

FOGELSON, DAVID: b. Netcong, N.J., Mar. 15, 1903; e. New York U. (Sch. of Commerce), B.C.S., '23; Fordham U., LL.B. '26. Trustee Joe & Emily Lowe Fund, CPA, (N.Y.).

FOLEY, GEORGE F.: TV. m.p. prod., lawyer. b. Grantwood, N.J., March 26, 1919; e. Fordham U., 1941. Attorney, N.Y.; dir. of pub. rel., Newell-Emmett Co., 1946–48; dir. radio & TV Newell-Emmett Co., 1948–49 Cecil & Presbrey, 1949–50; prod. package shows including Tales of Tomorrow, Gallup Poll, Herman Hickman, etc.

PICTURES INCLUDE: Public Enemy No. 1, Frankenstein's Daughter, Missile to the Moon, Leave Me Alone, The Vulture, Importer of films: La Dolce Vita, Last Year at Marienbad, Rocco and His Brothers.

Counsel, CBS-EVR, Believe It Or Not, Inc., Crosby Enterprizes.

FOLSEY, GEORGE: Cameraman.

PICTURES INCLUDE: Reunion in Vienna, Storm at Daybreak, Stage Mother, Going Hollywood, MGM, Men in White, Operator 13, MGM. In 1935, Reckless, I Live My Life, The Great Ziegfeld. The Gorgeous Hussy, MGM; Hearts Divided, Meet Me in St. Louis, Mr. Imperium, Law and Lady, Lovely to Look At, Million Dollar Mermaid, All The Brothers Were Valiant, Executive Suite, Seven Brides for Seven Brothers, Deep in My Heart, The Cobweb, Fastest Gun Alive, House of Numbers, Imitation General, Cash McCall, Glass Houses.

FONDA, HENRY: Actor. b. Grand Island, Neb., May 16, 1905; e. Minn. U. With Community Playhouse (Omaha), then Cape Playhouse (Dennis, Mass.), also other theat. groups & tourist cos.; later with N.Y. Theatre Guild as extra & understudy.

STAGE: New Faces, The Swan, The Farmer Takes a Wife; extended N.Y. run in Mr. Roberts, 1949–50; Point of No Return, 1952; Caine Mutiny, Court Martial, 1953, Gift of Time, Generation, Our Town, The Time of Your Life, Two For The Seasaw, Silent Night, Lonely Night, Critics Choice. U.S. Navy, WW II. Screen debut in Farmer Takes a Wife, 1935; Clarence Darrow, 1974–75, First Monday in October, The Oldest Living Graduate (1980), Showdown at the Adobe Hotel (1981).

PICTURES INCLUDE: Jesse James, Young Mr. Lincoln, Drums Along the Mohawk, Grapes of Wrath, Male Animal, Immortal Sergeant, Ox-Bow Incident, My Darling Clementine, Daisy Kenyon, Long Night, The Fugitive, Fort Apache, On Our Merry Way, Mister Roberts, War and Peace, Wrong Man, 12 Angry Men, Warlock, Advise and Consent, Spencer's Mountain, Fail-Safe, The Rounders, The Best Man, Battle of the Bulge, A Big Hand for the Little Lady, Welcome to Hard Times, Firecreek, Yours, Mine, and Ours, Madigan, Once Upon a Time in the West, Boston Strangler, Too Late the Hero, There Was a Crooked Man, Cheyenne Social Club, Sometimes a Great Notion, Ash Wednesday, The Serpent, The Last Days of Mussolini, My Name is Nobody, Midway, Great Smokey Roadblock, Rollercoaster, Tentacles, The Swarm, Fedora, Meteor, Battle of Mareth, Wanda Nevada, City on Fire, On Golden Pond.

TV: Films: The Red Pony, Clarence Darrow, Home to Stay, AFI Henry Fonda Life Achievement Award; Roots 2, Gideon's Trumpet, Oldest Living Graduate.

FONDA, JANE: Actress. b. New York City, Dec. 21, 1937. p. Henry Fonda. e. Emma Willard School, Troy, N.Y. active in dramatics. Vassar. Appeared with father. Summer stock production. The Country Girl, Omaha, Neb. Studied painting, languages, Paris. Art Students League, N.Y. Appeared in The Male Animal, Dennis, Mass. Modeled, appeared on covers, Esquire, Vogue, The Ladies Home Journal, Glamour, and McCall's, 1959.

PICTURES INCLUDE: Tall Story (debut), Walk on the Wild Side, The Chapman Report, Period of Adjustment, In the Cool of The Day, Sunday in New York, The Love Cage, La Ronde, Cat Ballou, The Chase, La Curee, Any Wednesday, The Game Is Over, Hurry Sundown, Barefoot in the Park, Barbarella, Spirits of The Dead, They Shoot Horses, Don't They? (Acad. Award nom.), Klute, Steelyard Blues, F.T.A., A Doll's House, The Bluebird, Fun with Dick and Jane, Julia, Coming Home, Comes a Horseman, California Suite, The China Syndrome, The Electric Horseman, Nine To Five, Roll-Over, On Golden Pond.

STAGE: There Was A Little Girl, Invitation to a March, The Fun Couple, Strange Interlude.

FONDA, PETER: Actor. b. 1939. Son of Henry Fonda. Brother of Jane Fonda.
PICTURES INCLUDE: Tammy and the Doctor, The Victors, Lilith, The Wild Angels, Easy Rider, The Last Movie, The Hired Hand (also dir.), Two People, Dirty Mary Crazy Larry, Open Season, Race with the Devil, 91 In The Shade, Killer Force, Futureworld, Outlaw Blues, High Ballin!, UWanda Nevada (also dir.).

FONTAINE, JOAN: Actress. b. Tokyo, Oct. 22, 1917; e. art, privately; sister of Olivia de Haviland, actress; on B'way in Tea and Sympathy; m.p. debut in "Quality Street" 1937; won Academy Award best performance, 1941, for "Suspicion."
PICTURES INCLUDE: Women, Rebecca, Man Who Found Himself, You Can't Beat Love, Music for Madame, Blond Cheat, Gunga Din, This Above All, Jane Eyre, Frenchman's Creek, Affairs of Susan, Emperor Waltz, Kiss the Blood Off My Hands, You Gotta Stay Happy, Letter from an Unknown Woman, Duke of West Point, Ivy, Born to Be Bad, Something to Live For, September Affair, Darling How Could You?, Ivanhoe, Decameron Nights, Flight to Tangier, The Bigamist, Casanova's Big Night, Island in the Sun, Until We Sail, A Certain Smile, Tender Is the Night, The Devil's Own.

FORBES, BRYAN: Actor, writer, producer, director. Former head of prod., man. dir., Associated British Prods. (EMI). b. Stratford (London) July 22, 1926. Stage debut, The Corn is Green (London), 1942; screen debut, The Small Back Room 1948.
PICTURES INCLUDE: All Over the Town, The Wooden Horse, Dear Mr. Prohack, Appointment in London, Sea Devils, Wheel of Fate, Million Pound Note (Man with a Million), An Inspector Calls, Colditz Story, Passage Home, Extra Day, It's Great to be Young, Satellite in the Sky, Quartermass 11, Cockleshell Heroes, Black Tent, House of Secrets, Baby and the Battleship, I was Monty's Double, A Piece of Cake, Man in the Moon, Whistle Down The Wind, League of Gentlemen, Only Two Can Play, The L-Shaped Room, Of Human Bondage, Station Six Sahara, Seance on a Wet Afternoon, King Rat, The Wrong Box, The Whisperers, Deadfall, The Madwoman of Chaillot, The Go-Between (executive producer), The Railway Children (exec. prod.) Tales of Beatrix Potter (exec. prod.) The Raging Moon (US title; Long Ago Tomorrow). Also: I Caught Acting Like the Measles (produced/directed for television), Goodbye Norma Jean and Other Things, The Stepford Wives (dir.), The Slipper & The Rose, International Velvet, Sunday Lovers, Jessie, Hopscotch (s.p.).
TV: The Breadwinner, French Without Tears, Johnnie was a Hero, The Gift, The Road, The Heiress. Theatre: Flarepath, Fighters Calling, Gathering Storm, September Tide, The Holly and the Ivy, Tobias and the Angel, A Touch of Fear, Three Way Switch.
PUBLICATIONS: Truth Lies Sleeping (short stories) 1951; Distant Laughter (novel) 1972; Notes For a Life (Autobiog.) 1973/4. The Slipper & The Rose (Book of Film) 1975/6. Ned's Girl, 1977, International Velvet 1978 (novel); *Famililar Strangers* (novel), 1979 (US Title "Stranger"), That Despicable Race, 1980.

FORBES, GORDON B.: Executive. b. New York, New York. e. Yale University Drama Dept. Republic Studio-1945 Budget and Cost Control Dept.; 1950 asst. to production mgr.; 1952 head of production operations; 1954–1958 UPM Batjack Productions, production manager RKO-El Dorado Pictures (Cuba); 1959–1976 v.p. Universal TV, production operations mgr.; 1977 executive assistant to the pres. of The Burbank Studios and member of the executive committee. In 1979 named v.p. of studio productions for Paramount Pictures.

FORD, GLENN: Actor. r.n. Gwylin Ford; b. Quebec, Canada, May 1, 1916. m. Eleanor Powell, actress. On stage in western co. "Children's Hour" 1935; on Broadway in: Broom for a Bride, Soliloquy; served in U.S. Marine Corps 1942–45; m.p. debut in Heaven With a Barbed Wire Fence 1940.
PICTURES INCLUDE: Men Without Souls, Lady in Question, So Ends Our Night, Desperadoes, Stolen Life, Gilda, Framed, Mating of Millie, Return of October, Loves of Carmen, Mr. Soft Touch, Man from Colorado, Undercover Man, Lust for Gold, Doctor and the Girl, Flying Missile, The Redhead and the Cowboy, Follow the Sun, The Secret of Convict Lake, Green Glove, Young Man With Ideas, Affair in Trinidad, Man from the Alamo, Terror on a Train, Plunder of the Sun, Big Heat, Appointment in Honduras, Human Desire, The Americano, Violent Men, Blackboard Jungle, Interrupted Melody, Trial, Ransom, Fastest Gun Alive, Jubal, Teahouse of the August Moon, Don't Go Near the Water, Cowboy, The Sheepman, Imitation General, Torpedo Run, It Started with a Kiss, The Gazebo, Cimarron, Cry for Happy, The Four Horsemen of the Apocalypse, Experiment in Terror, Love Is a Ball, The Courtship of Eddie's Father, The Rounders, The Money Trap, Fate Is the Hunter, Dear Heart, The Last Challenge, Heaven With a Gun, Smith! The Day of the Evil Gun, A Time for Killing, Midway, Superman, Happy Birthday To Me.
TV: The Brotherhood of Ty Bell (movie); Cade's Country (series); Jarret; The Holvak Family.

FORD, HARRISON: Actor. b. Chicago, July 13, 1942. e. Ripon Coll.
PICTURES: Dead Heat on a Merry-Go-Round, Luv, Getting Straight, The Long Ride Home, Journey to Shiloh, Zabriskie Point, The Conversation, American Graffiti, Star Wars, Heroes, Force 10 from Navarone, Hanover Street, Apocalypse Now, The Frisco Kid, The Empire Strikes Back, Raiders of the Lost Ark.
TELEVISION: Dynasty, Trial of Lt. Calley, The Possessed.

FORD, TENNESSEE ERNIE: Singer. r.n. Ernest J. Ford, b. Bristol, Tenn., Feb. 13, 1919. e. Cincinnati Conserv. of Music. Radio anncr., 1939–41; U.S. Air Force, 1942–45; radio anncr., San Bernardino, Calif.; hillbilly disc jockey, Pasadena; Capital Recording Star.
TV: own TV shows, NBC; guest appearances on I Love Lucy, Red Skelton Show, Perry Como Show, George Gobel Show; own show, Tennessee Ford Show, ABC.
RECORDS INCLUDE: 16 Tons, etc.

FORD, TONY: Executive. b. August 6, 1925, New York City. e. St. Johns Univ. In U.S. Navy World War II; in life insurance business, 1946–48. Agent with MCA 1949–53. TV indep. prod., specials for Timex, Ringling Bros. Circus, Pontiac Victor Borge. Re-entered agency business, General Artists Corp., v.p., TV dept. Headed own agency, Tony Ford Mgt., Inc., which was acquired by William Morris Agency. Headed creative services division WMA representing producers, directors, writers specializing in the packaging of TV series and specials, left Morris Agency Jan. 1, 1977; now pres. of Tony Ford Productions, affiliated with Universal Studios.

FOREMAN, CARL (C.B.E.): Writer, prod., dir., b. Chicago, July 23, 1914. e. U. of Ill., Northwestern U., Crane Coll., John Marshall Law Sch.; m. In circus & carnival exploit.; then freelance writer, newspaper pub. rel., radio U.S. Army Signal Corps. World War II; former Executive producer, Managing director, Open Road Films, Ltd., England. Pres., Writer's Guild of Great Britain, 1968–75.
PICTURES INCLUDE: So This Is New York, Champion, Young Man With a Horn, Home of the Brave, The Men (Robert Metzger Award Screen Writers Guild), Cyrano de Bergerac, High Noon, The Key, The Bridge over the River Kwai, The Mouse that Roared, The Guns of Navarone, The Victors, Born Free, MacKenna's Gold, Otley, The Virgin Soldiers, Living Free, Young Winston, When Time Ran Out.

FOREMAN, JOHN: Producer. b. Idaho Falls, Idaho. Cofounder of CMA Agency. Resigned Jan., 1968, to form production co. with Paul Newman: Newman-Foreman Company.
PICTURES INCLUDE: Winning, Butch Cassidy and the Sundance Kid, WUSA (co-prod.), Puzzle of a Downfall Child, They Might Be Giants, Sometimes a Great Notion, Pocket Money, The Effect of Gamma Rays on Man-in-the-Moon Marigolds (exec. prod.), The Mackintosh Man, The Man Who Would Be King, Bobby Deerfield, The Great Train Robbery.

FORMAN, SIR DENIS, O.B.E., M.A.: Executive. b. Moffat, Dumfriesshire, Scot., Oct. 13, 1917. e. Loretto School, Musselburgh, Pembroke College, Cambridge. Served in Argyll & Sutherland Highlanders, World War II. Entered film business May, 1946, production staff Central Office of Information, 1947; Chief Production Officer C.O.I. 1948; appointed dir. of the British Film Institute, 1949; joined Granada Television Ltd., 1955. Jnt. Mng. Dir., 1965 chm., British Film Institute, Bd. of Gov., 1971–73. Chmn. Granada T.V. 1975. Chmn. Novello & Co. 1972. Fellow, British Acad. Film & TV Arts, 1976.

FORMAN, MILOS: Director. b. Czechoslovakia, 1932. Trained as writer at Czech Film School and as director at Laterna Magika. Won Int'l. attention with first film, Black Peter, 1963. Emigrated to U.S. after collapse of Dubcek Govt. in Czech.
PICTURES INCLUDE: The Loves of a Blonde, Fireman's Ball, Taking Off, Visions of Eight (segment), One Flew Over the Cuckoo's Nest, Hair, Ragtime.

FORREST, FREDERIC: Actor. Began career off-Broadway in Futz, Massachusetts Trust and Tom Paine, all with La Mama Troupe under direction of Tom O'Horgan. Moved to Hollywood in 1970. Made film debut in When The Legends Die in 1972.
PICTURES INCLUDE: The Don is Dead, The Conversation, The Gravy Train, Permission to Kill, The Missouri Breaks, It Lives Again, Apocalypse Now, The Rose, One from the Heart.

FORREST, STEVE: Actor. b. Huntsville, Texas, Sept. 29. e. UCLA, 1950. Acted at La Jolla Playhouse; appeared on radio, TV; m.p. debut in Geisha Girl.

PICTURES INCLUDE: Bad and the Beautiful, Battle Circus, The Clown, Band Wagon, Dream Wife, So Big, Take the High Ground, Phantom of the Rue Morgue, Prisoner of War, Rogue Cop, Bedeviled, It Happened to Jane, Heller in Pink Tights, Five Branded Women, Flaming Star, The Second Time Around, Rascal, The Wild Country, The Late Liz, North Dallas Forty.

FORRESTER, LARRY: Writer. b. Glasgow, Scotland, 1924. Ent. TV industry with his wife Pauline as Special investigators on audience participation shows. Later scripting for TV.
SERIES INCLUDE: Ivanhoe, William Tell, Flying Doctor, International Detective, Interpol Calling, Knight Errant, No Hiding Place, The Golden Girl, Sir Francis Drake, Top Secret, Ghost Squad, This Man Craig, Danger Man, Espionage, The Saint.
PICTURES INCLUDE: Hercules and the Princess of Troy; Tora! Tora! Tora!
BOOKS INCLUDE: Battle of the April Storm, A Girl Called Fathom, Fly for Your Life, Skymen, Diamond Beach.

FORSTER, ROBERT: Actor. b. Rochester, N.Y., July 13, 1941. e. Heidelberg College, Alfred University, Rochester U., B.S.
THEATRE: Plays, Mrs. Dally, 1965. Summer tour, A Streetcar Named Desire, 1967.
PICTURES INCLUDE: Reflections in a Golden Eye, Medium Cool, Justine, Journey Through Rosebud, The Don is Dead, The Black Hole.

FORSYTHE, JOHN: Actor. b. Penn's Grove, N.J., Jan. 29, 1918.
STAGE CREDITS: Mr. Roberts, All My Sons, Yellow Jack, Teahouse of the August Moon, and others.
TV: Started in 1947; appeared on Studio One, Kraft Theatre, Robert Montgomery Presents, and others. Bachelor Father series; To Rome with Love series.
PICTURES INCLUDE: Captive City, It Happens Every Thursday, The Glass Web, Escape from Fort Bravo, Trouble With Harry, Ambassador's Daughter, Everything But the Truth, Kitten With a Whip, Madame X, in Cold Blood, The Happy Ending, Topaz, And Justice for All.

FORTE, FABIAN: Singer, Actor. b. Philadelphia, Feb. 6, 1940. e. South Philadelphia high sch. At 14, discovered by Marcucci and DeAngelis, voice teacher. Chancellor Records; signed contract. Studied with Carlo Menotti.
RECORDS INCLUDE: Turn Me Loose, This Friendly World, The Fabulous Fabian (Gold album), Foggy Foggy Dew (High Time).
PICTURES INCLUDE: Hound Dog Man, High Tim, North to Alaska, Love in a Goldfish Bowl, Mr. Hobbs Takes a Vacation, Ride the Wild Surf, Dear Brigette, Longest Day, 10 Little Indians, Fireball 500, Thunder Alley, Five Weeks in a Balloon, Devil's Eight, The Longest Day, A Bullet for Pretty Boy, Little Laura and Big John.

FOSSE, BOB: Director, Choreographer, Actor. b. Chicago, 1927. Began dancing at age of 9 and made first professional appearance at 13. After service with U.S. Navy in World War II toured with Mary Ann Miles in Call Me Mister. Signed by MGM in 1953 and acted in three films. First big Broadway success as choreographer of Pajama Game in 1955.
PICTURES INCLUDE: Actor: Give a Girl A Break, The Affairs of Dobie Gillis, Kiss Me Kate, The Little Prince. Choreographer: My Sister Eileen. Director-Choreographer: Sweet Charity, Cabaret. Director: Lenny. All That Jazz (also co. s.p.).
BROADWAY: Choreographer: Pajama Game, Damn Yankees, Redhead, Little Me, New Girl in Town, How To Succeed in Business Without Really Trying, Sweet Charity, Pippin, Chicago. Director: Redhead, Little Me, Sweet Charity, Pippin, Chicago, Dancin!.

FOSTER, DAVID: Producer. b. New York City, Nov. 25, 1929. e. Dorsey H.S., U. of Southern California (School of Journalism). U.S. Army, 1952–54; entered public relations field in 1952 with Rogers, Cowan & Brenner; Jim Mahoney, 1956; Allen, Foster, Ingersoll & Weber, 1958; left field in 1968 to enter independent m.p. production.
PICTURES INCLUDES: Co-produced (with Mitchell Brower) McCabe and Mrs. Miller, The Getaway, The Nickel Ride (exec. prod.), The Drowning Pool (co-prod.), The Legacy.

FOSTER, JODIE: Actress. b. Los Angeles. Acting debut on Mayberry, R.F.D. TV series. Followed with many TV appearances, from series to Movies-of-the-Week.
PICTURES INCLUDE: Napoleon and Samantha, Menace of the Mountain, One Little Indian, Tom Sawyer, Kansas City Bomber, Taxi Driver, Echoes of a Summer, Bugsy Malone, Freaky Friday, The Little Girl Who Lives Down the Lane, Candleshoe, Foxes, Carny.
TV: The Courtship of Eddie's Father (regular), My Three Sons (regular), Paper Moon (regular), Smile, Jenny, You're Dead (movie), Rookie of the Year (movie).

FOSTER, JULIA: Actress. b. 1944, in Lewes, Sussex, England. First acted with the Brighton Repertory Company, then two years with the Worthing, Harrogate and Richmond companies. 1956, TV debut as Ann Carson in Emergency Ward 10.
PICTURES INCLUDE: Term of Trial, The Loneliness of the Long Distance Runner, The Small World of Sammy Lee, The System, The Bargee, One Way Pendulum, Alfie, Half A Sixpence.
TV: A Cosy Little Arrangement, The Planemakers, Love Story, Taxi, Consequences, They Throw It At You, Crime & Punishment, The Image.
STAGE: No. 1 tour of The Country Wife; 1969, What the Butler Saw.

FOSTER KEMP, CECIL R.: (M.B.E.) Assoc. prod. and prod. supervisor. b. Shanghai. e. Brighton College.
PICTURES INCLUDE: Paris Holiday, Interpol, Tarzan and the Lost Safari. Around the World in Eighty Days. Front Page Story, Appointment in London, South of Algiers, Under Capricorn, Scott of the Antarctic, Untamed, John Paul Jones, King of Kings, Offbeat, El Cid, Fifty Five Days at Peking, The Fall of the Roman Empire. William The Conqueror. Girl on the Motorcycle, Arthur! Arthur!, I Saw Him Die, Ferdinand and Isabella, Wild Violets.

FOSTER, MAURICE DAVID: Prod.-Wrt. e. St. Paul's Ent. m.p. ind. 1943 as asst. to gen. mar. and prod. sup. Easling Studios. From 1950–61, prod. sup. and dir. Film Finances, 1961–63, dir. and gen. man. MGM British Studios.
PICTURES INCLUDE: The Jokers (Universal), Assignment K (Columbia).

FOWLER, HARRY: Actor. b. Lambeth Walk (London). Stage debut, Nothing Up My Sleeve (London) 1950; Screen debut, 1941.
PICTURES INCLUDE: Demi-Paradise, Don't Take It to Heart, Champaigne Charlie, Painted Boats, Hue and Cry, Now Barabbas, The Dark Man, She Shall Have Murder, The Scarlet Thread, High Treason, The Last Page, I Believe in You, Pickwick Papers, Top of the Form, Angels One Five, Conflict of Wings (Fuss Over Feathers), A Day to Remember, Blue Peter, Home and Away, Booby Trap, Town on Trial, Lucky Jim, Birthday Present, Idle on Parade, Don't Panic Chaps, Heart of a Man, Crooks Anonymous, The Longest Day, Lawrence of Arabia, Flight From Singapore, The Golliwog, Ladies Who Do, Clash By Night, The Nanny, Life At the Top, Start the Revolution Without Me, The Prince & The Pauper.
TV: Stalingrad, I Remember the Battle, Gideon's Way, That's For Me, Our Man At St. Mark's, Dixon of Dock Green, Dr. Finlay's Case Book, I Was There, Cruffs Dog Show, The Londoners, Jackanory, Get This, Movie Quiz, Get This series, Going a Bundle. Ask a Silly Answer, London Scene, Flockton Flyer, Sun Trap, The Little World of Don Camillo.

FOWLEY, DOUGLAS: Actor. b. New York City, May 30, 1911; e. St. Francis Xavier's Mil. Acad., N.Y. In stock; operated dramatic sch. N.Y.; on screen in bit parts. From 1934 in regular roles. Has appeared in many pictures.
PICTURES INCLUDE: Battleground, Just This Once, This Woman Is Dangerous, Singin' in the Rain, Man Behind the Gun, Slight Case of Larceny, Naked Jungle, Casanova's Big Night, Lone Gun, High & Mighty, Three Ring Circus, Texas Lady, Broken Star, Girl Rush, Bandido, Nightmare in the Sun.

FOX, EDWARD: Actor. b. England, 1937. Comes from theatrical family; father was agent for leading London actors; brother is actor James Fox.
PICTURES INCLUDE: The Mind Benders, The Long Duel, Morgan, The Naked Runner, The Jokers, I'll Never Forget What's 'Is Name, The Battle of Britain, Oh! What a Lovely War, Skullduggery, The Breaking of Bumbo, The Day of the Jackal, A Doll's House, Galileo, The Squeeze, A Bridge Too Far, The Duellists, The Big Sleep, Force 10 from Navarone, The Mirror Crack'd.

FOX, JAMES: Actor. b. London, England, 1939. Ent. films as child actor in 1950 in The Magnet and Mrs. Minniver.
TV APPEARANCES INCLUDE: The Door, Espionage.
PICTURES INCLUDE: The Loneliness of the Long-Distance Runner, Tamarine, The Servant, Those Magnificent Men in Their Flying Machines, King Rat, The Chase, Thoroughly Modern Millie, Arabella, Duffy, Loves of Isadora, Performance.

FOXWELL, IVAN: Producer. b. Feb 22, 1914, London, Eng. Entered m.p. ind. 1933 as technician with British & Dominions Film Corp., subsequently with Paramount British & London Films; Assoc. with Curtis Bernhardt in Paris 1937 becoming producer & collaborating on story, s.p. of Carefour, Le Train pur Venise, Sarajevo, others. In World War II with BEF and AEF 1939–46. Returned to British films 1947 as producer, co-author screen adapt., No Room at the Inn; prod., collab. s.p., Guilt Is My Shadow; prod., Twenty-Four Hours of a Woman's Life; co-author s.p. and prod. The Intruder, The Colditz Story (TV series adapt. 1972), Manuela, A Touch of Larceny, Tiara Tahiti, The Quiller Memorandum; s.p. and prod. Decline and Fall. Director, Foxwell Film Prods. Ltd.

FOXWORTH, ROBERT: Actor. b. Houston, Tx. e. Mellon Univ. Began acting at age 10 at Houston Alley Theatre and stayed with stage part-time while completing formal education.

Returned to theatre on full-time basis after graduation from Mellon. Made TV debut in Sadbird, 1969.
TELEVISION: The Storefront Lawyers (series), Mrs. Sundance, Hogan's Goat.
PICTURES: Treasure of Matecumbe (debut), The Astral Factor, Airport '77, Damien-Omen II, The Black Marble.
STAGE: P.S., Your Cat Is Dead.

FOXX, RED: Actor, Performer. b. St. Louis, Mo., Dec. 9, 1922. Began career by running away from home to join a washboard band at age 13; first nightclub appearances as a single, then with Slappy White from 1951 to 1956; 50 record albums in last 17 years.
TV: Sanford and Son.
PICTURE: Norman, Is That You?

FRAKER, WILLIAM A.: Cinematographer—Director. b. Los Angeles, 1923. Graduate USC Film School. Worked as operator with Conrad Hall; moved to TV before feature films.
PICTURES: Cinematographer: Games, The Fox, President's Analyst, Fade In, Rosemary's Baby, Bullitt, Paint Your Wagon, Dusty and Sweets McGee, Day of the Dolphin, Rancho Deluxe, The Killer Inside Me, Aloha, Bobby and Rose, Gator, Close Encounters of the Third Kind, Looking for Mr. Goodbar, Heaven Can Wait, Odd Boyfriends, 1941. Director: Monte Walsh, Reflection of Fear, Bette Midler's Divine Madness, The Legend of the Lone Ranger.
TELEVISION: Stony Burke, Outer Limits, Ozzie and Harriet, Daktari.

FRAMER, WALT: TV Producer. b. Pittsburgh, Pa., Apr. 7, 1908; e. Univ. of Pittsburgh. Former theatre magr.; news commentator, KDKA, Pitts.; sidewalk interviewer of Radio Swaps; creator-prod. Strike It Rich, The Big Payoff, Million Dollar Family, Futures Unlimited, Penny-a-Mile, Lady Luck, For Love or Money, Man on the Beat, It's in the Cards, Meet the People, Make It Big, The Fashion News Reel, Make Your Move, Holiday for Two, Take A Letter. Your Great Escape, Run For the Money, Heartline U.S.A., Look For Tomorrow, What's It Worth?

FRANCIOSA, ANTHONY: Actor. b. New York City, Oct. 25, 1928. e. High School there. First stage part in Y.W.C.A. play; joined Off-Broadway stage group; stock at Lake Tahoe, Calif., Chicago and Boston.
THEATRE: N.Y. stage in End As a Man, The Wedding Breakfast; lead in A Hatful of Rain.
PICTURES INCLUDE: A Face in the Crowd, This Could Be The Night, A Hatful of Rain, Wild Is The Wind, The Long Hot Summer, The Naked Maja, Career Story, On Page One, Go Naked in the World, Period of Adjustment, The Swinger, Fathom, A Man Called Gannon, The Sweet Ride, Rio Conchos, In Enemy Country, Across 110th Street, The Drowning Pool, Firepower.
TV: Valentine's Day, The Name of the Game, Search, Matt Helm.

FRANCIS, ANNE: Actress b. Ossining, N.Y., Sept. 16. Child model; radio TV shows as child & adult; on B'way in Lady in the Dark.
PICTURES INCLUDE: Summer Holiday, So Young So Bad, Whistle at Eaton Falls, Lydia Bailey, Elopement, Dream Boat, Lion is in the Streets, Rocket Man, Susan Slept Here, Rogue Cop, Bad Day at Black Rock, Battle Cry, Blackboard Jungle, Scarlet Coat, Forbidden Planet, The Rack, Great American Pastime, Don't Go Near the Water, Crowded Sky, Girl of the Night, Satan Bug, Brainstorm, Hook, Line, and Sinker, More Dead Than Alive, The Impasse, The Love God, Funny Girl, Born Again.

FRANCIS, ARLENE: Actress. r.n. Arlene Kazanjian; b. Boston, Mass. e. Convent of Mount St. Vincent Academy, Riverdale, N.Y.; Finch Finishing School, Theatre Guild School, N.Y.; m. Martin Gabel, actor.
STAGE: All That Glitters, Doughgirls, The Overtons, Once More With Feeling, Tchin-Tchin, Beekman Place, Mrs. Dally, Dinner at Eight, Kind Sin, Lion in Winter, Pal Joey, Who Killed Santa Claus?, Gigi.
TV: Regular panelist What's My Line, WCBS-TV; Home, Arlene Francis Show, NBC-TV: Talent Patrol, etc.
RADIO: Arlene Francis Show, Emphasis, Monitor.
PICTURES INCLUDE: All My Sons, One, Two, Three, The Thrill of It All.

FRANCIS, CONNIE: Singer. r.n. Constance Franconero. b. Newark, N.J., Dec. 12, 1938. Appeared, Star Time since 12 years old; won Arthur Godfrey's Talent Scout Show, 12 years old.
GOLD RECORDS: Who's Sorry Now, My Happiness. Numerous vocalist awards: All Major TV shows.
PICTURES INCLUDE: Where the Boys Are, Follow the Boys, Looking For Love.

FRANCIS, FREDDIE: Producer-Director-Cameraman. b. London, 1917. Joined Gaumont British Studios as apprentice to stills photographer; then clapper boy at B.I.P. Studios, Elstree; camera asst. at British Dominion. After World War II returned to Shepperton Studios to work for Korda and with Powell and Pressburger as cameraman.
PICTURES INCLUDE: Cameraman: Moby Dick (second unit photo., special effects), Room at the Top, Saturday Night and Sunday Morning, Sons and Lovers (Oscar), The Innocents, The Elephant Man. Director: Paranoiac, Vengeance, Evil of Frankenstein, Nightmare, Hysteria, Dr. Terror's House of Horrors, The Skull, Traitor's Gate, The Psychopath, The Deadly Bees, They Came from Beyond Space, Torture Garden, Dracula Has Risen from the Grave, Girly, Trog, Tales from the Crypt, The Creeping Flesh, Tales That Witness Madness, The Ghoul, Legend of The Werewolf.

FRANCIS, KEVIN: Producer. b. London, England, 1947, 1967–1970 Production mgr., assoc. producer 1970–1972. Produced It's Life, Passport, Trouble With Canada: 1972 Founder Tyburn Productions Limited: 1973 Produced Persecution: 1974 Produced The Ghoul and Legend of the Werewolf, 1975 Founder Tyburn Prods. Inc. 1976 Prod. Film Techniques Educ. Course for BFI. 1977–81 Exec. Prod. the Satanists, Master of the Shell, Courier.

FRANCISCUS, JAMES: Actor. b. Clayton, Mo., 1934; e. Taft Prep, Yale U.
PICTURES INCLUDE: Four Boys And A Gun, I Passed For White, The Outsiders, The Miracle of the White Stallions, Youngblood Hawke, The Valley of Gwangi, Marooned, Beneath the Planet of the Apes, The Cat O' Nine Tails, The Greek Tycoon, City on Fire, When Time Ran Out.
TV: Naked City, Mr. Novak, Longstreet (TV film feature), The 500 Pound Jerk (TV feature).

FRANK, MELVIN: Writer, producer, director. In 1942: collaborated on story, My Favorite Blonde, Par. In June, 1942: appointed consultant in the radio branch of the War Department, working under Lt. Col. E. M. Kirby. In 1942 collab. s.p. Happy-Go-Lucky, Thank Your Lucky Stars; collab. Star Spangled Rhythm, (collab. s.p.) And the Angels Sing, Monsieur Beaucaire, Road to Utopia, Duffy's Tavern, The Return of October, It Had To Be You.
PICTURES INCLUDE: Mr. Blandings Builds His Dream House, Southern Yankee, The Redhead and the Reformer, Strictly Dishonorable, Callaway Went Thataway, Above and Beyond, Knock on Wood, White Christmas, Court Jester, That Certain Feeling, Li'l Abner, A Funny Thing Happened on the Way to the Forum, The Facts of Life, The Road to Hong Kong, Strange Bedfellows, Buona Sera, Mrs. Campbell, A Touch of Class, The Prisoner of Second Avenue, The Duchess and the Dirtwater Fox, Lost and Found.

FRANKEL, DANIEL: Executive. b. New York City, Aug. 21, 1903; e. U. Mich., U. Berlin (Germany). Gen. mgr., Films Erka, Paris, 1925–1930. Ran Theatre des Champs Elysees, 1928–1929. Produced two French pictures. Returned to New York and joined Pathe 1930; dir. foreign dept., later of all sales to 1941; resigned to engage in prod. unit org. Vice-pres. & exec. dir., Four Continents Films, 1945; with U.A. in Europe 1953–56; ret. U.S. & org., pres., Zenith Int'l Film Corp., 1956, distr., Lovers, 400 Blows, Hiroshima, Mon Amour, etc. Governor IFIDA.

FRANKENHEIMER, JOHN: Director. B. Malba, N.Y. e. Williams College. Actor, dir., summer stock: radio-TV actor, dir., Washington, D.C.; then joined CBS network.
TV: Mama, You Are There, Danger, Climax, Studio One, Playhouse 90, Du Pont Show of the Month; Ford Startime, Sunday Showcase.
PICTURES INCLUDE: The Young Stranger, The Young Savages, Birdman of Alcatraz, All Fall Down, The Manchurian Candidate, Seven Days in May, The Train, Seconds, Grand Prix, The Extraordinary Seaman, The Fixer, The Gypsy Moths, I Walk the Line, Horsemen, The Impossible Object, The Iceman Cometh, 99 and 44/100½ Dead, French Connection II, Black Sunday.

FRANKLIN, MICHAEL HAROLD: Executive. b. Los Angeles, Dec. 25, 1923. e. Univ. of Cal. (A.B.), Univ. So. Cal. (LL.B.). Admitted to Cal. bar, 1951; pvt. practice in L.A. 1951–52; atty. CBS, 1952–54; atty. Paramount Pictures, 1954–58; exec. dir. Writers Guild Am. West, Inc. 1958–1978; national exec. dir., Directors Guild of America 1978– . Mem. Am. Civil Liberties Union, Los Angeles Copywright Soc., Order of Coif.

FRANKLIN, PAMELA: Actress. b. Tokyo, Japan. Feb. 4, 1950. Attended, Elmshurst Ballet School, Camberley, Surrey.
PICTURES INCLUDE: The Innocents, The Lion, Flipper's New Adventure, The Prime of Miss Jean Brodie, The Night of the Following Day, And Soon the Darkness, The Legend of Hell House, Food of the Gods.

FRANKLYN, ARTHUR R.; b. N.Y.C. July 19, 1928. Child actor with Life With Father, Farewell Tour and Just Because. New York stage 1941–44. Featured skater, Ice Frolic Latin American tour, 1948–49; star and prod. ice revues at Havana Coliseum, 1949–50; served with U.S. Marines, Korea 1950–51; prod. dir., Los Angeles Ice Palace, 1951–56; exploitation and special events, Fox West Coast Theatres, 1956–59; Lippert Theatres, 1960; purchased Amador Chain of Central California Theatres. Pres. U.S. Bowling Photographers, Inc.; 1960–65 Dist. Mgr. Theatre Amuse Co., Tarzana, Calif.; Mgr. Florida State Theas., Dist. mgr. Arlo Enterprises.

FRANKOVICH M. J.: Executive. b. Bisbee, Arizona, Sept. 29, 1910. e. U.C.L.A., B.A. Ent. Radio 1934 as producer, commentator; began writing screenplays for Universal, 1938; Republic Pictures, 1940–49 (except for period war service U.S. Army); to Europe 1949 to make Fugitive Lady, Lucky Nick Kane, Thief of Venice etc.; England 1952. prod., Decameron Nights, Malaga, Footsteps in the Fog, Joe Macbeth; apptd. man. dir. Columbia Pictures Corp. Ltd., in U.K. and Eire Aug. 1955; elected vice-pres., Columbia Pictures International Corp., Dec. 1955; head Col. Pictures Int. Prod., 1958; vice-pres., Col. Pics. Corp., 1959. Appt. Chrmn., Columbia Pics. Corp. Ltd., 1959. Variety Club Crew Member; chief barker, Tent 36 Variety Club, 1957. dir., BLC Films; chmn., Screen Gems Ltd.; first v.p. chge world prod., Columbia; resigned July, 1967 to return to independent production. Long Term Deal, 1969, rel. through Columbia.
PICTURES INCLUDE: Bob & Carol & Ted & Alice, Marooned, Cactus Flower, The Looking Glass War, Doctors' Wives, There's A Girl in My Soup, The Love Machine, Dollars, Butterflies Are Free, Stand Up and Be Counted, 40 Carats, A Report to the Commissioner, From Noon Till Three, The Shootist.

FRANZ, ARTHUR: Actor. b. Perth Amboy, N.J. Feb. 29, 1920; e. Blue Ridge Coll., Md. U.S. Air Force. Radio, TV shows.
THEATRE: Streetcar Named Desire, Second Threshold.
PICTURES INCLUDE: Jungle Patrol, Roseanna McCoy, Red Light, Doctor and the Girl, Sands of Iwo Jima, Strictly Dishonorable, Submarine Command, Member of the Wedding, Flight Nurse, Bad for Each Other, Eddie Cantor Story, Caine Mutiny, Steel Cage, Battle Taxi, New Orleans Uncensored, Bobby Ware Is Missing, Atomic Submarine, The Human Factor.

FRASER, SIR ROBERT: TV executive. b. Adelaide, South Australia, Sept. 26, 1904; e. Melbourne and London Universities; OBE, 1944 controller of production, Ministry of Information, 1945; director-general, Central Office of Information, 1946–54; director-general, Independent Television Authority, 1954–70; Hon. Fellow, London School of Economics, 1965; chmn., Independent TV news, 1971–74.

FRASER, RONALD: Actor. b. Ashton-under-Lyne, England, 1930. Ent. films and TV, 1954.
TV: Lonesome Road, Sealed With A Loving Kiss, Stray Cats and Empty Bottles, John Bulls Other Island, Captain Rubian, And the Walls Came Tumbling Down, Sword of Honour, The Corn Is Green, The Brahmin Widow.
PICTURES INCLUDE: There Was a Crooked Man, The Long & the Short and the Tall, The Best of Enemies, The Hellions, The Pot Carriers, The Punch and Judy Men, Private Potter, Crooks in Cloisters, Table Bay, The Beauty Jungle, Code 7, Victim 5, Flight of the Phoenix, The Whisperers, Fathom, Sinful Davey.

FRAWLEY, JAMES: Director. b. Houston, Texas. Studied drama at Carnegie Tech and Actors Studio, where later taught and ran the directors unit. Was charter member of comedy group, The Premise; has acted in plays on and off Bdwy. Won Emmy Award for Monkees series, where staged two musical numbers a week for two seasons.
PICTURES: Kid Blue, The Big Bus, The Muppet Movie.
TELEVISION: Columbo, Delancy Street, Capra, etc.

FRAZER, AUSTIN: Prod. consultant. b. London, England. Early career, designer and writer. Fellow SIAD. Writer's Guild Award 1967, Dante's Inferno.

FRAZIER, CLIFF: Director, actor. b. Detroit, Mich., Aug. 27, 1934; e. Wayne State U., Harold Clurman. Exec. dir., The Community Film Workshop Council; was co-founder & artistic dir.: Stables Theatre, Detroit, 1960–63; Concept East Theatre, Detroit, 1962–64; co-produced and acted in Study in Color, 1964–65; assoc. dir., Theatre of Latin America, Inc., N.Y.C., 1968; dir. of drama, Mobilization for Youth, N.Y.C., 1966–68; East Coast dir. of rsch., Brooks Foundation, Sta. Barbara, 1966–68.
TV ACTING INCLUDES: Today, Tell It Like It Is, NYPD. The Nurses, The Negro Experimental Theatre, Othello.
THEATRE: Stage repertories: Washington Theatre Club, Irish Hills Rep., Will-O-Way Playhouse.
Co-author, Discovery in Drama and book critic, Catholic Reporter, 1969.

FRAZIER, SHEILA E.: Actress. b. Bronx, N.Y., Nov. 13, 1948. e. Englewood, N.J. Was exec. sect'y. and high-fashion model. Steered to acting career by friend Richard Roundtree. Studied drama with N.Y. Negro Ensemble Co. and New Federal Theatre, N.Y.
PICTURES INCLUDE: Super Fly (debut), Superfly T.N.T., The Super Cops.

FREDERIC, MARC: Producer. b. New York, Apr. 25, 1916. e. Washington & Lee Univ. Capt. U.S. Air Force. Prod., dist. "Little Theatre" 1948, 104 episodes. First to make deal through AFTRA for kinescope release of live shows for dist. Tales of Tomorrow series. Org., Marc Frederic Prods., Inc.
PICTURES INCLUDE: Giant of the Unknown, She Demon, Missile to the Moon, Frankenstein's Daughter, Girl in Room 13, Career Girl, Festival Girl.

FREDERICKSON, H. GRAY, JR.: Producer. b. Oklahoma City, Okla., July 31, 1937. e. U. of Lausanne, Switzerland, 1958–9; U. of Okla. B.A., 1960. m. Victoria Schmidlapp. Worked one yr, with Panero, Weidlinger & Salvatori Engineering Co., Rome, Italy. In 1978 named v.p. for prod. of Lorimar Films.
PICTURES INCLUDE: Candy, Inspector Sterling, Gospel 70, An Italian in America, The Man Who Wouldn't Die, The Good, Bad & Ugly, Intrigue in Suez, How to Learn to Love Women, God's Own Country, Wedding March, An American Wife, Natika, Echo in the Village, Little Fauss & Big Halsey, Making It, The Godfather, The Godfather, Part II, Hit (exec. prod.), Apocalypse Now (co.-prod.).

FREEMAN, EVERRETT: Writer. b. New York City. Contrib. Sat. Eve. Post. From 1942 writer for screen.
PICTURES INCLUDE: George Washington Slept Here, Princess and the Pirate, It Happened on Fifth Ave., Secret Life of Walter Mitty, Too Young to Kiss, Lady Takes a Sailor, Jim Thorpe—All American, Pretty Baby, Million Dollar Mermaid, Destination Gobi, Kelly and Me, My Man Godfrey, Marjorie Morningstar, Sunday In New York, The Glass Bottom Boat, Where Were You When the Lights Went Out, The Maltese Bippy, How Do I Love Thee, ZigZag.
TV: Prod. Bachelor Father series.

FREEMAN, JOEL: Producer. b. Newark, N.J., June 12, 1922. e. Upsala Coll. Obtained job as messenger at MGM studios in 1941; promoted to short subjects dep., then prod. planning dept. In Air Force 1942–46. Became assist. dir. at RKO and David O. Selznick Organization. In 1948 returned to MGM as asst. dir; later assoc. prod. In 1956 entered indep. field as prod. supvr. on films and TV series. In 1960 to Warner Bros., working on Sunrise at Campobello and The Music Man. Worked in N.Y. for CBS, returning to Calif. in 1965 to be assoc. prod. at Warners. Named studio prod. exec; assigned to prod. contract when Seven Arts bought co. Also produced for 20th-Fox and MGM. Now heads Joel Freeman Productions. Most recent credit: It's a Wonderful Life, exec. prod. for Universal-ABC.

FREEMAN, SIR N. BERNARD: (C.B.E.) Executive. b. Sydney, Sept. 1, 1896; e. public schools, Xavier College, Melbourne; established MGM throughout Australia, New Zealand, man. dir., chmn., retired 1967. European War, 1914–18.
Member: Admin. Comm. of Brit. Centre, 1944–46. Chairman: Anzac House Appeal 1945, Anzac House Trust 1946, Miss Australia Quest, 1948–49. N.S.W. Comm. for World Refugee Year, 1960. Nat'l Chmn., UNICEF, 1953; pres. Rotary Club of Sydney, 1960–61; Created C.B.E. (Civil) 1946; chmn., Motion Picture Distributors' Association of Australia, 1939–41, '63 Deputy Chairman, Council of International House University of Sydney.

FREGONESE, HUGO: Director. b. Argentina. e. Buenos Aires College, Buenos Aires U. Prior to entering m.p. ind., newspaperman, Buenos Aires; ed. All Sports, in cattle ind., Azul; arrived New York City, for study at Columbia U.; to Los Angeles, as tech. adv.; back to Buenos Aires, 1938; made shorts in Argentina, cutter; asst. dir.; co-dir., Pampa Barbara; dir. Where Words Fail; signed by Hollywood, when MGM bought releasing rights; dir. Hardly A Criminal, Argentina; to Universal-International, 1949. Stanley Kramer Prod., 1951.
PICTURES INCLUDE: One Way Street, Saddle Tramp, Apache Drums, Mark of the Renegade, My Six Convicts, Untamed Frontier, Blowing Wild, Decameron Nights, Man in the Attic, The Raid, Black Tuesday, Thunders, Harry Black and the Tiger, I Girovaghi, Marco Polo, Moneta, Sperzzata (Italy), Shatterhand (Germany), The Tramps, Broken Coin, Savage Pampas (Spain).

FRENCH, HAROLD: Film and stage director. b. London, England, April 23, 1900. Popular London player in revue musical comedy, etc., 1930–35. Started directing 1936. Cavalier of the Streets, at Pinewood. Since: The House of the Arrow, Dead Men Are Dangerous, Jeannie, Unpublished Story, The Day Will Dawn. Directed stage plays: Youth at the Helm, Design for Living, No Time for Comedy, French Without Tears. In 1942: Secret Mission, Helm-General Film: The Avengers, Par. In 1944 dir. Mr. Emmanuel.
PICTURES INCLUDE: English Without Tears, Quiet Weekend, My Brother Jonathan, The Blind Goddess, Quartet, Adam & Evelyn, The Dancing Years, Trio, Hour of Thirteen, Encore, Isn't Life Wonderful, Rob Roy, Man Who Watched Trains Go By (Paris Express), Man Who Loved Redheads.

FRENKE, EUGENE: Producer. m. Anna Sten.
PICTURES INCLUDE: Life Returns, Two Who Dared, Three Russians Girls, Let's Live a Little, Lady in Iron Mask, Miss Robin Crusoe, Heaven Knows, Mr. Allison, Barbarian and the Geisha, The Last Sunset, Royal Hunt of the Sun.

FRESCO, ROBERT M.: Writer. b. Burbank, Calif., Oct. 18, 1928. e. Los Angeles City Coll. Newspaperman. Los Angeles, 1946–47; U.S. Army, 1948–49; staff writer, Hakim Prod., 1950–51; various screenplays, 1951–56.
PICTURES INCLUDE: Tarantula, They Came to Destroy the Earth, Monolith.
TV: Scripts for Science Fiction Theatre, Highway Patrol.

FREULICH, HENRY: Cinematographer. b. New York City, April 14, 1906. Four years with Universal as assistant cameraman and still man. Seven years with First National as still man and second cameraman. Five years with Columbia as second cameraman; then with Columbia as director of photography.

FRIEDKIN, JOHN: Executive. Entered industry in New York as publicist for Columbia Pictures; spent eight years at Young & Rubicam adv. agency. Formed Sumner & Friedkin with Gabe Sumner as partner; left to join Rogers & Cowan, where named v.p. In 1967 resigned to join 20th-Fox, moving to California in 1972 when home offices were transferred. Appointed Fox v.p. worldwide publ. & promo. In 1979 joined Warner Bros. as v.p.—adv. pub. for intl. div.

FRIEDKIN, WILLIAM: Director. b. Chicago, Ill., 1939. Joined NBC-TV, 1957, worked for National Education TV, did TV documentaries before feature films.
PICTURES INCLUDE: Good Times, The Night They Raided Minsky's, The Birthday Party, The Boys in the Band, The French Connection. (Academy Award, best picture, dir. 1971), The Exorcist, Sorcerer (prod.-dir.), Cruising (dir.-s.p.).

FRIEDMAN, ARNOLD J.: Executive b. New York, N.Y. Entered m.p. industry as exploiteer for Columbia Pictures Corp. 1957–60; advng dept. United Artists Corp. 1960–62; adv. mgr., Embassy Pictures Corp. and later director of adv., publ., prom. for Embassy Pictures Television, 1962–67; pres. Arnold Friedman's Company, 1967–70; dir. of ad./pub./promo for Metromedia Producers Corp., 1970–72; creative dir., ITC, 1973–76; vice pres. of ad./pub./promo for Cinema Shares International Television. Ltd., 1976, present.

FRIEDMAN, DAVID F.: Executive. b. Birmingham, Ala., Dec. 24, 1923. e. Cornell Univ. With U.S. Army Signal Corps, 1944–46. With Paramount Pictures as booker office manager 1946–50; circus press agent, 1950–51; Para. publ. agent, 1952–55; partner, Modern Film Distributors, 1956–60; partner, Sonney Amusement, 1960–64. Formed Entertainment Ventures, Inc. in 1965, of which he is pres. Produced a total of 53 features between 1958 and 1981.

FRIEDMAN, JOSEPH: Executive. b. New York, N.Y. e. City Coll. of N.Y., 1940–42, N.Y. Univ., 1946–47. U.S. Navy 3 yrs. Asst. to nat'l dir. field exploitation, Warner Bros. Pictures, 1946–58; nat'l exploitation mgr., Paramount Pictures, 1958–60; exec. asst. to dir. of adv., publicity & exploitation, Para., 1961; dir. adv. & pub., Paramount 1964; a v.p., Para., 1966; v.p. in charge of Marketing, 1968; v.p., adv., and p.r., Avco Embassy Pictures, 1969; v.p., p.r. American Film Theatre, 1973; v.p., adv. and p.r., ITC, motion picture div., 1976, pres., Joseph Friedman Marketing and Advertising, Inc., 1977. Exec. dir. New Jersey M.P. & T.V. Commission, 1978; v.p. worldwide adv./pub./promo., Edie & Ely Landau, Inc., 1980.

FRIEDMAN, MARTIN: Executive. b. New York City, Oct. 9. Warner 1928 sls. and contract dept.; booking sales, Cleveland 1939–42; U.S. Marine Corps, production unit; 1945 joined A. W. Schwalberg at Artists Producers, when company was dissolved in 1946 then joined Liberty Films as ass't. to E. K. (Ted) O'Shea. Liberty was taken over by Paramount Pictures; remained with Paramount in their sales division until 1957 at which time joined American Broadcasting-Paramount Theatres as their producers representative. Production was stopped, then Friedman assigned to the theatre division in 1960; v.p., gen. mgr., ABC Theas. of Penn. Inc., ABC Theas. of N.Y., Inc., New England Theas., Inc.; pres., Countrywide Theaters, 1970. In 1973 formed own corp., Martin Films, Inc., of which is pres. Co. buys and distributes indep. product.

FRIEDMAN, ROBERT L.: Executive. b. Bronx, N.Y., March 1, 1930. e. DeWitt Clinton High School. Started as radio announcer and disc jockey with Armed Forces Radio Service in Europe and U.S. v.p., marketing, United Artists Corp.; now pres. domestic distribution, Columbia Pictures.
MEMBER: M.P. Associated Foundation, Phila., pres. 2 yrs.; Variety Club (on board) M.P. Pioneers; (on board) area chm. and N.Y. participation in Will Rogers Hospital Foundation, American Film Institute, Academy of M.P. Arts & Sciences.

FRIEDMAN, SEYMOUR MARK: Director. b. Detroit, Mich., Aug. 17, 1917; e. Magdalene Coll., Cambridge, B.S. 1936; St. Mary's Hospital Medical School, London. Entered m.p. ind. as asst. film ed. 1937; 2nd asst. dir. 1938; 1st asst. dir. 1939, on budget pictures; entered U.S. Army 1942; returned to ind. 1946; dir. Columbia Pictures 1947. Member: Screen Directors Guild; Vice president & executive production for Columbia Pictures Television, division of Columbia Pictures Industries, since 1955.
PICTURES INCLUDE: To the Ends of the Earth, Rusty's Birthday, Prison Warden, Her First Romance, Rookie Fireman, Son of Dr. Jekyll, Loan Shark, Flame of Calcutta, I'll Get You, Saint's Girl Friday, Hyber Patrol, African Manhunt, Secret of Treasure Mountain.

FRIEDMAN, STEPHEN: Writer-Producer. e. Univ. of Penn., Harvard Law School. Worked as lawyer for Columbia Pictures and Ashley-Famous Agency. Formed and heads Kings Road Productions.
PICTURES INCLUDE: The Last Picture Show (prod.), Lovin' Molly (prod.-s.p.), Slap Shot (prod.), Bloodbrothers (prod.) Fast Break (prod.), Hero at Large (prod.); Little Darlings (prod.).

FRIEND, PHILIP: Actor. b. Horsham, Sussex, England, Feb. 20, 1915; e. Bradfield Coll.; m. Eileen Erskine, prof.
THEATRE: With musical comedy tour 1935; at Drury Lane Theatre 1936; Broadway debut in "French Wihout Tears" 1937; plays include: Pink Strings and Sealing Wax, First Gentlemen, Hide Out, 1969; Dictator's Slippers, The Ladder, 1970. Blindsight, Forgotten Factor, 1970.
Brit. m.p. debut in "Midas Touch" 1939; on Brit. radio; served in Brit. Army 3 yrs. World War II; made training m.p. for Brit. govt.
PICTURES INCLUDE: My Own True Love, In Which We Serve, Next of Kin, Great Day, Enchantment, Sword in the Desert, Buccaneer's Girl, The Highwayman, Thunder on the Hill, Desperate Moment, Background (Edge of Divorce), The Diamond, The Betrayal, Son of Robin Hood, Web of Suspicion, The Solitary Child, Stranglehold, The Fur Collar, Manutara.
TV: Rendezvous, Invisible Man series, The Third Man Series, Z Cars, Freewheelers, Short Story, Dial M for Murder, Moody & Peg, Suez.

FRIES, CHARLES W.: Executive. b. Cincinnati, Ohio. Ohio State U., B.S.; exec.-prod., Ziv Television; v.p., prod., Screen Gems; v.p., prod., Columbia Pictures; exec. v.p., prod. and exec. prod., Metromedia Prod. Corp., 1970–74; pres., exec. prod., Alpine Prods. and Charles Fries Prods. 1974 to date. Nat'l treas., TV Academy; pres., Alliance TV Film Producers; exec. comm., MPPA. Chm., Comm. of Producers, Writers and Directors, board of governors and exec. comm. of Academy of TV Arts and Sciences.

FRISCH, LARRY: Producer, Director, Writer. b. Indianapolis, Ind., Dec. 27, 1929; e. Univ. of So. California, 1947–49, Columbia Univ., 1951–54. Prof. child actor, Radio 1944–47; writer/dir. with Caravel Films, NYC, 1950; Assist. dir., Exodus, El Cid (Spain); Pro., Dir.; Writer, Tel Aviv Taxi, 1956 (Israel), Pillar of Fire 1962 (Israel), Casablan, 1963 (Greece).
TV: Dir., Story of a Teenage Drug Addict, Power of Pot Roast, Beyond Three Doors, Destination Vietnam, Bus to Sinai (NBC-TV), Biafra Eye-Witness (UPI-TV); Metromedia TV News, overseas correspondent 1969–71; prod., Miracle of Survival (TV special), An American Family In China, 1971.

FROBE, GERT: Actor. b. Zwickau, Germany, 1912. Began acting career in 1935 in German films and theatrical productions. In post-war period there became street comedian; later worked in more than 70 German and French films, usually as villain.
PICTURES: Those Magnificent Men in Their Flying Machines, Is Paris Burning?, The Three Penny Opera, He Who Must Die, Those Daring Young Men in Their Jaunty Jalopies, Chitty Chitty Bang-Bang, Goldfinger, Bloodline.

FRONTIERE, DOMINIC: Executive. b. New Haven, Conn., June 17, 1931. e. Yale School of Music. Studied composing, arranging and conducting; concert accordionist, World's Champion Accordionist, 1943; An Hour with Dominic Frontiere, WNHC-TV, New Haven, 3 years, 1947; exec. vice-pres., musical dir., Daystar Prods. Composer or arranger over 75 films.
PICTURES INCLUDE: Giant, Gentlemen Prefer Blondes, Let's Make Love, High Noon, Meet Me in Las Vegas, 10,000 Bedrooms, Hit the Deck; composer-conductor; Marriage-Go-Round, The Right Approach, One Foot in Hell, Hero's Island, Popi, Barquero, Chisum, A For Alpha.
TV: Composer-conductor: The New Breed, Stoney Burke, Bankamericard commercials (Venice Film Festival Award Best Use of Original Classical Music for filmed TV commercials), Outerlimits, Branded, Iron Horse, Rat Patrol, Flying Nun, The Invaders, Name of the Game, That Girl, Twelve O'Clock High, Zig Zag, The Young Rebel, The Immortal, Jean C. Killy, Fugitive, The Love War.

FROST, DAVID: Performer. b. Tenterdon, England, April 7, 1939; e. Cambridge U. Writer, co-creator, co-star, That Was The Week That Was, 1962–63; star of The David Frost Show, nationally syndicated; honors include 2 Emmy Awards, Golden Rose of Montreux Award, TV Personality of the Year, Guild of TV Producers Award, Order of the British Empire.
AUTHOR: The Americans; co-author, The English.

FRUCHTMAN, MILTON A.: Producer, director. b. New York, N.Y. e. Columbia U.
TV: High Adventure Series, CBS, Every Man's Dream, ABC, Verdict for Tomorrow, NBC Spectaculars, Assignment Southeast Asia, It Happened in Naples; Son of Sulan, worked for Columbia and independent producers in various production capacities, set up first worldwide TV network, Eichmann Trial, Capital Cities Broadcasting, The Makebelievers, Odyssey Prods.; received Peabody Emmy, and Gabriel awards.

FRYE, WILLIAM: Producer. Was agency exec. before beginning prod. career as associate prod. of Four Star Playhouse in assoc. with late Dick Powell, David Niven and Charles Boyer. Later joined Revue Prods., which became Universal TV; he produced General Electric Theatre and other series. Has produced many Movie of the Week entries for ABC-TV.
PICTURES INCLUDE: The Trouble with Angels, Where Angels Go, Trouble Follows, Airport 1975, Airport 1977.

FUCHS, LEO L.: Indep. prod. b. Vienna, June 14, 1929; moved to U.S., 1939; ed. Vienna and New York; U.S. Army cameraman 1951–3; int'l mag. photographer until entered motion pictures as producer with Universal in Hollywood in 1961.
PICTURES INCLUDE: Gambit, A Fine Pair; Jo (French version of The Gazebo).

FUEST, ROBERT: Director. Early career as painter, graphic designer. Ent. TV industry as designer with ABC-TV, 1958. 1962: directing doc., commercials. 1966: Wrote and dir. Just Like A Woman, 1967–68; dir. 7 episodes of The Avengers, 1969: dir. And Soon The Darkness, 1970: dir. Wuthering Heights, Doctor Phibes. 1972: wrote and dir. Doctor Phibes Rises Again. 1973: wrote, designed and dir. The Final Programme. 1975: The Devils Rain.

FULLER, JACK DUBOSE: Exhib. b. Columbia, S.C. March 7, 1921. U. of So. Carolina, B.S. Palmetto Theatre Co. 1936–1942; U.S. Navy, 1942–45; theatre mgr. Wilby-Kincey Service Corp., 1945–47; booker, 1947–1951; partner Irvin-Fuller Theatres, operate 24 theatres No. & So. Carolina, since 1951; Pres., Theatre Owners of N.C. and S.C. 1957; dir. NATO since 1956.

FULLER, SAMUEL: Director, writer, producer. Copy boy, N.Y. Journal; reporter, rewrite man, N.Y. Graphic, N.Y. Journal, San Diego Sun; journeyman reporter many papers. Author of novel The Dark Page; many orig. s.p.; in U.S. Army, 16th Inf. 1st U.S. Inf. Div. 1942–45; s.p., dir., I Shot Jesse James, Baron of Arizona; prod., dir., also screen play author of The Steel Helmet; dir., Fixed Bayonets; s.p., dir., prod., Park Row; s.p., dir., Pickup On South Street; s.p., dir., Hell and High Water, House of Bamboo; prod. dir., s.p. Run of the Arrow, China Gate, Forty Guns, Verboten, The Crimson Kimono, Underworld U.S.A., Merril's Marauders, (collab. s.p., dir.), s.p., dir., prod. Shock Corridor, s.p. prod., dir., The Naked Kiss; s.p. dir., Dead Pigeon on Beethoven Street. s.p. dir., The Big Red One; co-s.p., dir., White Dog.
AUTHOR: Novel, Crown of India. Novel; 144 Piccadilly Street; Dead Pigeon on Beethoven Street; The Rifle; The Big Red One.

FUNT, ALLEN: Producer—Performer. b. New York City, 1914. Best known as producer and creator of Candid Camera series which originated on radio in 1947 as Candid Microphone which inspired theatrical film shorts. TV version began in 1948 as Candic Mike, changed in 1949 to Candid Camera which played off and on until 1960 when became regular series on CBS, lasting until 1967. Revived briefly in early '70s in new format; then syndicated as The New Candid Camera. Funt produced and starred in film, What Do You Say to a Naked Lady?

FURBER, PERCIVAL E.: b. Paris, France, May 16, 1906; e. Hotchkiss School, Princeton '27; chm. emeritus div., Trans-Lux Corp.; dir., Tampax, Inc., Fedders Corp. & Fedders Fin. Corp.

FURIE, SIDNEY J.: Director, writer, producer. b. Toronto, Canada, 1933. Ent. TV and films 1954. Canadian features include: Dangerous Age, A Cool Sound from Hell. Also dir. many Hudson Bay TV series. To England 1960. Films since include Dr. Blood's Coffin, During One Night, Brewster's Millions, The Young Ones, The Boys. 1961 appt. exec. dir. Galaworldfilm Productions, Ltd.
PICTURES INCLUDE: The Leather Boys, Wonderful Life, The Incress File, The Appaloosa, The Naked Runner, The Lawyer, Little Fauss and Big Halsy, Lady Sings the Blues, Hit!, Sheila Levine, Gable and Lombard, The Boys in Company C.

FURNEAUX, YVONNE: Actress. English. b. France. e. St. Paul's London and Oxford Univ. Student at R.A.D.A.
PICTURES INCLUDE: (In England:) Master of Ballantrae, Beggar's Opera, Dark Avenger, House of the Arrow, Lisbon, Repulsion (1965). (Films in Italy:) Le Amiche, Dolce Vita, Via Margutta, A Noi Place Freddo, I Lancieri Neri, Semiramide, Helen of Troy, Le Ultime Sette Ore, I Quattro Tassisti, Caccia all'Uomo, The Man Who Never Was. In France: Le Comte de Monte Cristo, Le Meurtrier. In Germany: Dr. Mabuse, The Champagne Murder.
TV: Children's hour for Rex Tucker. Tigress on the Hearth, Hereward the Wake. TV Films: Danger Man, The Baron. The Survivors (TV series).
STAGE: Bristol Old Vic.

FURNESS, BETTY: Actress—TV Correspondent. b. New York City, 1916. Stage and screen actress in the '40s; TV commercial spokeswoman in '50s' leading consumer advocate in '70s. Host on both local TV and radio shows in New York (At Your Beck and Call, Ask Betty Furness, etc.). Named President Johnson's asst. for consumer affairs in 1967. 1970, head of New York State's consumer protection board. Joined WNBC-TV in 1970 as consumer reporter and occasional contributor to network news specials and Today Shows.

FURSE, RUSSELL L.: Executive. b. Humansville, Mo., Jan. 27, 1908. e. Colorado Teachers College, Greely, Colo. Film. dir., prod. Warner Bros., Hal Roach: Prod. vice pres. Televisions Programs, Inc. Executive prod., ABC Television, Hollywood. Gen. mgr. network prog. dept., CBS Television: dir; prog. oper.; KEY T Santa Barbara. Gen. mgr. CHEK TV Victoria B.C. Santa Barbara; exec. v.p. Santa Maria Telecasting Corp. Co-founder Academy of Televisions Arts & Sciences. Chmn, of board, Communications Foundation, pres. Tri-County Public Service Corporation, pres. KBBY-KBBQ Radio Oxnard, member Affiliated Producers Work Shop, Board of Directors Affiliates University of California Santa Barbara, Managing director, Riviera Park Research & Communications Center; pres. & chief exec. off., California Video Communications.

G

GABOR, ZSA ZSA: Actress. r.n. Sari Gabor; b. Hungary, Feb. 6, 1923. e. Budapest, Lausanne, Switzerland; Stage debut in Europe.
PICTURE INCLUDE: Lovely to Look At, We're Not Married, The Story of Three Loves, Lili, Moulin Rouge, Three Ring Circus, Death of a Scoundrel, Girl in the Kremlin, For The First Time, Boys' Night Out, Picture Mommy Dead, Jack of Diamonds.

GAFFNEY, ROBERT: Prod.-dir. b. New York, Oct. 8, 1931. e. Iona College, 1949–51. Staff, Louis de Rochemont mer; prod., Rooftops of New York, in assoc. with Robert Associates, prod. staff, Cinerama Holiday, Man on a String; camera technical consultant, Cinemiracle prod., Windjam-McCarty, 1960; prod. industrial films, Seneca Prods., prod., Light Fantastic; prod., Troublemaker; assoc. prod., Harvey Middleman Fireman dir., Frankenstein Meets the Space Monster.

GAGE, EDWIN: Executive. b. Deal, N.J., July 20, 1915; e. U. of Pennsylvania, 1938; Admin. v.p. & treas., secy., Walter Reade-Sterling, Inc., resigned, 1966; TV dist. & real estate oper. in Oakhurst, N.J.

GAINES, JAMES M.: Executive. b. Saxe, Va., May 8, 1911; m. Associated with Major Bowes Amateur Hour; former manager of WNBC-WNBT, New York; vice-president of NBC and director of NBC owned and operated stations; Pres., gen. mgr. Southland Industries, Inc. (WOAI, WOAI-TV), v.p. Blanco Oil Co.

GAITHER, GANT: Producer, writer. b. Hopkinsville, Kentucky, Aug. 1. e. Royal Academy, London, Yale U.; build Miami Beach Playhouse, 1940–42; stage mgr., The Army Air Forces, 1942–43; asst. dir. George Cukor, 1944; Pacific theatre operations, W.W. II, Winged Pigeons' assoc. prod., Beverly Hills, Broadway; prod., Craig's Wife, The First Mrs. Fraser, The Shop At Sly Corner, Gayden, Alexander, Seventh Heaven; wrote: The Long Street, The Swan, Princess of Monaco; exec. prod., Paramount Pictures, 1960–61.
PICTURES INCLUDE: My Six Loves.
TV: Simon Says, for Qualis Productions.

GALANTE, M. CHRISTINA: Executive. b. Tucson, AZ, Aug. 30, 1942. e. U. of Arizona, New School for Social Research (NYC). Started career as child model in 1948. Entered film career as stunt double in 1955 working in dozens of major films made in Arizona. 1962–66, feature vocalist with big bands for USO tours and major night clubs. 1967–70, spokeswoman and actress in TV commercials and feature films. 1970–74, television commercials producer. 1974–77, film coordinator for mayor's office, City of New York, coordinating and supervising all m.p. and television filming in City. 1977 named special consultant to California Motion Picture Council. 1980, named to Council by Governor Edmund G. Brown and elected it's Chairman. 1980, named to Los Angeles Film Commission by Mayor Tom Bradley, currently on executive board. Professional photographer and writer, former President of Women in Film, Sponsor, Women in the Performing Arts Coalition, Member of the Board, Inter Agency Council on Child Abuse and Neglect.

GALE, BOB: Writer-Producer. b. St. Louis, 1952. e. USC School of Cinema. Joined with friend Robert Zemeckis to write screenplays, starting with episode for TV series, McCloud. Also co-wrote s.p. for The Nightstalker series. Turned to feature films, co-writing with Zemeckis script for I Wanna Hold Your Hand, on which Gale also acted as associate producer.
PICTURES: 1941 (s.p.).

GALE, GEORGE: Executive, e. Sorbonne U., Paris, France. Feature editor, Budapest Ed., U.S. Army Pictorial Service, Astoria, N.Y., 1944–45; shorts editor, MGM; Hal Roach Studios, as super, film ed.; ed. Oriental-Internat'l films; ed. The River, winner of Int. Award in Venice, 1951; free lance features editor, 1951–54; ed. Disney Studios, films and TV, Disneyland, 1955–65 asst. prod. to Ivan Tors; assoc. prod. Daktari; prod. exec. Cowboy in Africa. assoc. prod. Jambo; assoc. prod., From Africa With Love; post prod. exec. "Primus" Metromedia. 1971–1976, American National Enterprises, Inc., vice-president in charge of production. Producer and director. Supervised the production of approximately 30 features for television syndication. Producer-director of: UFO: Fact or Fiction, Animals Are the Craziest People, Mysteries From Beyond Earth (theatrical feature). Formed George Gale Productions, Inc. in 1976. Wrote story and produced feature, Escape from Angola, in South Africa; prod. for American National Ent., Mysteries from Beyond the Triangle; The Force on Thunder Mountain, Mysteries of the Mind, Anne of the Seven Seas, Lure of the Blue Planet, Quest of Courage. Produced for own co.: Neptune's Squad (TV pilot).

GANIS, SIDNEY M.: Publicist. b. New York, Jan. 8, 1940. e. Brooklyn College. Staff writer, newspaper and wire service contact, 20th Century-Fox 1961–62; radio, TV contact and special projects, Columbia Pictures 1963–64. Joined Seven Arts Prod. 1965 as publicity mgr; 1967, appt. prod. publicity mgr. Warner-7 Arts, Ass't prod. There Was a Crooked Man, 1969. Studio Publicity dir., Cinema Center Films, 1970. Director of Ad-Pub for "Mame", Warner Bros., 1973; Director of Advertising, Warner Bros., 1974; named WB v.p., worldwide adv. & pub., 1977; 1979, exec. v.p., Lucasfilm, Ltd..

GARBO, GRETA: Actress. r.n. Greta Gustafson. b. Stockholm, Sept. 18, 1906; e. Stockholm. Stage career as a dancer in Sweden. Hollywood screen career started 1926 with Torrent. Voted one of the ten best Money-Making Stars in Motion Picture Herald-Fame Poll 1932. Spec. Academy Award, 1954.
PICTURES INCLUDE: Temptress, Flesh and the Devil, Love, Divine Woman, Mysterious Lady, Single Standard, Wild Orchids, Woman of Affairs, Kiss, Anna Christie, Susan Lennox, Her Fall and Rise, Romance, Mata Hari, Grand Hotel, Queen Christina, Anna Karenina, Camille, Conquest, Ninotchka, Two-Faced Woman.

GARDENIA, VINCENT: Actor. b. Naples, Italy, Jan. 7, 1923. f. was actor and singer who brought Vincent to U.S. at age 2, formed theatrical co. in which Vincent took part. Vincent served in U.S. Army, World War II, and after landed first English-speaking role in summer stock prod. of Burlesque. Made off-Bdwy. debut in The Man with the Golden Arm; first Bdwy. role in The Visit (1958) with the Lunts.
PICTURES INCLUDE: Cop Haters, Cold Turkey, Little Murders, Hickey and Boggs, Bang the Drum Slowly, Luciano, The Manchu Eagle, Death Wish, The Front Page, Heaven Can Wait, Firepower, Home Movies, The Last Flight of Noah's Ark.
TV: All in the Family.
STAGE: Machinal, Passing Through from Exotic Places, The Prisoner of Second Avenue.

GARDNER, ARTHUR: Producer, b. Marinette, Wis.
PICTURES INCLUDE: Without Warning, Vice Squad, Down Three Dark Streets, Geronimo, The Glory Guys, Clambake, Scalphunters, Sam Whiskey, Underground, McKenzie Break, The Honkers, Hunting Party, Kansas City Bomber, White Lighting, McQ, Brannigan, Gator, Rally.
TV: Rifleman, Robert Taylor's Detectives, Law of The Plainsman, The Big Valley.

GARDNER, AVA: Actress. b. Smithfield, N.C., Dec. 24, 1922. e. Atlantic Christian Coll.
PICTURES INCLUDE: We Were Dancing, Joe Smith, American, Lost Angel, Three Men in White, Maisie Goes to Reno, Whistle Stop, The Killers, The Hucksters, Singapore, One Touch of Venus, The Bribe, Great Sinner, East Side, West Side, Show Boat, Pandora and the Flying Dutchman, Lone Star, Snows of Kilimanjaro, Ride Vaquero, Mogambo, Knights of the Round Table, Barefoot Contessa, Bhowani Junction, Little Hut, The Naked Maja, On The Beach, The Fair Bride, 55 Days at Peking, Night of the Iguana, The Bible, Mayerling, Life and Times of Judge Roy Bean, The Devil's Widow, Earthquake, The Bluebird, Permission to Kill, The Cassandra Crossing, The Sentinel, City on Fire.

GARFIELD, WARREN: Writer, publicist, b. Nov. 18, 1936. e. UCLA, Loyola Of Los Angeles School of Law. Story analyst, assistant to dir., Hecht-Hill-Lancaster, 1948–58; casting dir. Assist., assist. film editor, Columbia Pictures, films, featurettes, trailers, TV spots, Paramount Pictures, 1966–68; creative film services, Walt Disney Productions, 1968–present. Member: Academy of Motion Picture, Arts & Sciences.
CREDITS: A Stranger in Town (wrote s.p.), Cat Ballou, Wild and Wonderful, Flight From Ashiya, Taras Bulba (as assist. to Harold Hecht); The High Chaparral (wrote, TV).

GARFINKLE, LOUIS: Writer-Director-Producer. b. Seattle, Wash., February 11, 1928. e. U. of Calif., U. of Washington, U. of So. Calif. Writer KOMO, Seattle, 1945; Executive Research, Inc., 1948; Writer, educ. doc. screenplays, Emerson Films, EBF. 1948–50; s.p. You Can Beat the A-bomb (RKO), 1950; Writer-dir. training films, info. films, Signal Photo, 1950–53; Copy, Weinberg Adv., 1953; Head of Doc. Research in TV, U. of Calif., Berkeley, 1954–55; staff, Sheilah Graham Show, 1955; story and s.p. The Young Guns (AA), formed Maxim Prod. Inc. with Albert Band, 1956; story, s.p. and co-producer I Bury the Living (UA), 1957 (Killer on the Wall); s.p. and co-producer Face of Fire (AA), 1958; Writer, 712 teleplays for Day in Court, Morning Court, Accused for Selmur—ABC-TV, 1959–66; Co-writer-creator Direct Line pilot, Selmur, 1960; Story and t.p. June Allyson Show, Threat of Evil, 1960; s.p. in collab. The Hellbenders (Embassy) 1967; story and t.p. Death Valley Days, Crullers At Sundown, 1967; story, Death Valley Days, Captain Dick Mine, 1967; s.p. in collab. A Minute to Pray A Second to Die (ABC-Cinerama) 1968; story, s.p. and co-producer The Love Doctors (Sigma III) 1969; story, s.p. and director Beautiful People, (I.P.C.) 1970; s.p. (collab.) The Models, 1971; head writer, No. 3 Peanut Place (pilot), 1972; story & s.p. (collab.) The Doberman Gang, (Dimension) 1971; story & s.p. (collab.) Little Cigars (AIP) 1973; The Deerhunter (U-EMI) 1978, (story collab.).

GARFUNKEL, ART: Actor-Singer-Composer. b. New York City, 1942. e. Columbia Univ. Began singing at age 4. Long partnership with Paul Simon began in grade school at 13 (Queens); first big success in 1965 with hit single, Sound of Silence. Partnership dissolved in 1970. Film debut in Catch 22 (1969).
PICTURES: Carnal Knowledge, Bad Timing/A Sensual Obsession.

GARLAND, BEVERLY: Actress. b. Santa Cruz, Calif., Oct. 17, 1930. e. Glendale College, 1945–47.
TV: starred, Decoy, Bing Crosby; Twilight Zone, Dr. Kildare, Medic, Emmy Nomination, 1954; My Three Sons; M.P. The Mad Room, Where the Red Fern Grows, Airport, 1975, Roller Boogie, It's My Turn.

GARLAND, P. HARVEY: Theatre Executive. b. Jacksonville, Fla., Apr. 4, 1919. e. Univ. of Fla. Entire career with same co., joining Fla State Theatres in 1935 as usher, advancing to doorman, ass't., mgr., mgr., film booker, buyer, v.p. and then president. Named v.p. ABC Theatres; now pres.

GARNER, JAMES: Actor. b. Norman, Okla., April 7, 1928; e. High School there, then joined Merchant Marine, U.S. Army ser. in Korean War. Prod. Paul Gregory suggested acting career. Studied drama at N.Y. Berghof School. Toured with road companies; Warner Bros. studio contract followed with screen debut in Toward the Unknown.
PICTURES INCLUDE: Shoot-out at Medicine Bend, Darby's Rangers, Sayonara, Up Periscope, Cash McCall, The Children's Hour, The Great Escape. Thrill of It All, Move Over Darling, The Americanization of Emily, 36 Hours, The Art of Love, Mister Buddwing, Duel at Diablo, Grand Prix, Hour of the Gun, Support Your Local Sheriff, Marlowe, Support Your Local Gunfighter, Skin Game, They Only Kill Their Masters, One Little Indian, Hawaiian Cowboy, Health, The Fan, Victor/Victoria.
TV: Cheyenne, Maverick, Rockford Files.

GARNER, PEGGY ANN: Actress. b. Canton, Ohio, Feb. 8, 1932. John Powers child model. Screen debut 1938 in Little Miss

Thoroughbred; Acad. Award most promising juvenile of the yr. 1945.
PICTURES INCLUDE: In Name Only, Blondie Brings Up Baby, Abe Lincoln in Illinois, Eagle Squadron, The Pied Piper, Jane Eyre, Keys of the Kingdom, Tree Grows in Brooklyn, Nob Hill, Junior Miss, Home Sweet Homicide, Daisy Kenyon, Thunder in the Valley, Sign of the Ram, Bomba, The Jungle Boy, Teresa, Black Widow, The Cat.
TV: Studio One, Lux Playhouse, Schlitz Playhouse, Best of Broadway, etc.

GARRETT, BETTY: Singer, actress. b. St. Joseph, Mo., May 23. e. scholarships; Annie Wright Seminary, Tacoma, Wash., Neighborhood Playhouse, N.Y.C. Sang in night clubs, hotels, Broadway shows (1942–46); won Donaldson award for best musical comedy performance of 1946, Spoon River Anthology.
PICTURES INCLUDE: (Screen debut) The Big City 1947, Star of Tomorrow, '49; Words and Music, Neptune's Daughter, On The Town, My Sister Eileen, Shadow on the Window.
TV: Laverne and Shirley.

GARRISON, PAUL: Writer, producer, director. Adv. photog, in N.Y., Hollywood; production of TV films since 1948; dir. first series of half hour TV pictures made, Cases of Eddie Drake; dir., writer for more than 100 TV film shows; currently in chge. of TV Workshop, Hollywood.

GARROWAY, DAVE: Performer. b. Schenectady, N.Y., July 13. e. Washington Univ., St. Louis. U.S. Navy, 1942–45; star of Garroway at Large, NBC-TV, 1948–51 and Today Show, 1952–61; Dave Garroway Show, NBC-TV, Oct. 1953–54; NBC radio; Friday with Garroway, 1954–56; Monitor, 1955–61; NBC-TV, Wide Wide World, 1955–58; Exploring the Universe, NET, 1962–63; Bd. of Gov. N.A.T.A.S., 1968–69. Dave Garroway Show (syndication) Boston, 1969–70; Los Angeles, 1970–71.

GARSON, GREER: Actress. b. County Down, Northern Ireland; e. London U., B.A. cum laude; Grenoble U.
THEATRE: Stage debut Birmingham (England) Rep. theat. 1932 in Street Scene; London debut 1935 in Golden Arrow; continued London stage to 1938 (Vintage Wine, Mademoiselle, Accent on Youth, Page From a Diary, Old Music, etc.).
Screen debut 1939 in Goodbye, Mr. Chips. Academy Award best performance by actress, 1942 (Mrs. Miniver, MGM). Voted one of the ten best Money-Making Stars in Motion Picture Herald-Fame Poll 1942–46 inclusive. Photoplay Mag. Gold Medal 1944–45.
PICTURES INCLUDE: Pride and Prejudice, Blossoms in the Dust, When Ladies Meet, Mrs. Miniver, Random Harvest, Mme. Curie, Mrs. Parkington, Valley of Decision, Adventure, Desire Me, Julia Misbehaves, That Forsythe Woman, The Miniver Story, The Law and the Lady, Scandal at Scourie, Julius Caesar, Her Twelve Men, Strange Lady in Town, Sunrise at Campobella, The Singing Nun, The Happiest Millionaire.

GASSMAN, VITTORIO: Actor. b. Genoa, Italy, Sept. 1, 1922; e. Academy of Dramatic Art, Rome. Stage actor, 1943; m.p. debut, 1946.
PICTURES INCLUDE: Daniele Cortis, Mysterious Rider, Bitter Rice, Lure of Sila, The Outlaws, Anna, Streets of Sorrow; to U.S., Cry of the Hunted, Sombrero, The Glass Wall, Rhapsody, Girls Marked Danger, Mambo, War and Peace, World's Most Beautiful Woman, Tempest, The Love Specialist, The Great War, Let's Talk About Women, Il Successo, The Tiger, Woman Times Seven, Ghosts—Italian Style, Scent of a Woman, Viva Italia!, A Wedding, Quintet, Immortal Bachelor, The Nude Bomb.

GAVIN, JOHN: Actor. b. Los Angeles, Calif., April 8. e. St. John's Military Acad., Villanova Prep at Ojai, Stanford Univ., Naval service: air intelligence officer. Married to actress Constance Towers, 1974. Four children. pres. Screen Actors Guild. Broadway Stage debut: Seesaw, 1973. Named U.S. Ambassador to Mexico, 1981.
PICTURES INCLUDE: A Time to Love and a Time to Die, Imitation of Life, Spartacus, A Breath of Scandal, Romanoff and Juliet, Tammy, Tell Me True, Back Street, Thoroughly Modern Millie, Mad Woman of Chaillot, Psycho, Midnight Lace.

GAYLARD, JAMES W., III: Exhibitor b. Knoxville, Tenn., Sept. 18, 1943. e. Henderson H.S., Troy, Ala., 1961. Troy State Univ. 1965 BS Mathematics Ent. industry, managed theatres, Troy Drive-In Theatres, Inc. 1966.

GAYLE, PETER: Producer, writer, actor. b. New York City. American Academy of Dramatic Arts; New York School of Theatre Arts. Formed Peter Gayle Productions; v.p. television, Screen Gems–Columbia Pictures v.p., MGM Studios; v.p., m.p. development, United Artists.
PICTURES INCLUDE: Without Each Other (winner of Cannes Film Festival), The Gang; Never Too Young To Rock; Black Cobra; Shoot; The Boy Who Owned A Melephant (winner of the Venice Film Festival); Intercept, Scenes from a Murder.
TV: (Specials) Producer, writer, The World of Brigette Bardot, The Goldie Hawn Special The Rolling Stones in Concert, The Janis Joplin Special Hollywood: The Magic Kingdom, On Location With.
TV: (Series) The Young Revels (ABC-TV), The Protectors (ABC-TV), Spindletop (Universal Studios).

GAYNOR, MITZI: Actress. b. Chicago, Ill., e. Powers Professional H.S., Hollywood. Studied ballet since age four; was in L.A. Light Opera prod. Roberta.
OPERA: Fortune Teller, Song of Norway, Louisiana Purchase, Naughty Marietta, Great Waltz.
PICTURES INCLUDE: My Blue Heaven, Take Care of My Little Girl, Golden Girl, The I Don't Care Girl, We're Not Married, Bloodhounds of Broadway, There's No Business Like Show Business, Anything Goes, Birds and the Bees, The Joker, Les Girls, South Pacific, Surprise Package, Happy Anniversary, For Love or Money.

GAZZARA, BEN: Actor. b. N.Y.C., 1930. Won scholarship to study with Erwin Piscator; joined Actor's Studio, where students improvised a play, End As a Man, which then was performed on Broadway with him in lead. Screen debut (1957) in film version of that play retitled The Strange One.
PICTURES INCLUDE: Anatomy of a Murder, The Passionate Thief, The Young Doctors, Convicts Four, Conquered City, A Rage To Live, The Bridge at Remagen, Husbands, Capone, Killing of a Chinese Bookie, Voyage of the Damned, Bloodline, They All Laughed.
TV: Arrest and Trial, Run for Your Life, etc.
STAGE: Cat on a Hot Tin Roof, A Hatful of Rain, Hughie, etc.

GEBHARDT, FRED: Prod.-writer, exhibitor. b. Vienna, Austria, Mar. 16, 1925. e. Schotten Gymnasium, Vienna. U.C.L.A., 1939. Usher Boyd Theatre, Bethlehem, Pa.; Mgr., Rivoli Thea. L.A., 1944; 18 yrs. mgr. many theatres. Fox West Coast, then Fine Arts Theatre. Writer, prod.: 12 To the Moon, The Phantom Planet; prod., Assignment Outer Space, Operation M; s.p., 12 to The Moon, All But Glory, The Starmaker, Phantom Planet Shed No Blood, Fortress in Heaven, Eternal Woman.
BOOKS: Mental Disarmament All But Glory, Starmaker, Shed No Blood, The Last of the Templars.
Pres., Four Crown Prods., Inc.; recipient of Medal of Americanism, D.A.R., 1963; Honorary Lifetime Member, P.T.A., Young Man of The Year Award, 1956, 24 Showmanship Awards; Mem. Acad. M.P. Arts and Sciences, Ind. M.P. Prod. Assoc.

GEDDES, HENRY: Executive producer and chief executive Children's Film Foundation since 1964; President, International Centre Films for Children, Mem. Governing Body Prix Jeunesse. b. Dover, England, 1912; e. Dulwich College, Polytechnic School of Kinematography. Early career Hudson's Bay Company, N. Canada 1928–32. 1934–39 freelance asst. dir., prod. man. 1939–46. Served in Royal Artillery. 1946–52 Gen. Prod. Man. Crown Film Unit. 1953–63 freelance prod./dir.
ORIG. S. P. INCLUDE: Seal Island, Ali and the Camel, Last Rhino, Magnificent 6½.

GEESON, JUDY: Actress. b. Arundel, Sussex, England, Sept. 10, 1948. e. Corona Stage School. Began professional career on British TV, 1960.
PICTURES INCLUDE: To Sir With Love, Circus of Blood, Here We Go Round the Mulberry Bush, Hammerhead, Three into Two Won't Go; Two Gentlemen Sharing, The Executioner. 10 Rillington Place, Brannigan, The Eagle Has Landed.
TV: Dance of Death, Lady Windermere's Fan, Room with a View, The Skin Game, Star Maidens, Poldark, She.
THEATRE: Othello, Titus Andronicus, Two Gentlemen of Verona, Section Nine, An Ideal Husband.

GENOCK, TED: Newsreel Editor. r.n. Edouard Paul Genock. b. Paris, France, Mar. 7, 1907; e. Chelsea Polytechnic London U. (BS), Columbia (MA). Initiated prod. and pictorial news, Egypt. From 1930–39 covered Ethiopia, Iraq, Iran, India, Malaya, Netherlands Indies, Australia as newsreel war correspondent. British Paramount News. Apptd. assignment editor, Paramount News (N.Y.C.), 1942. Tech. dir., Neighbor to North (1948), Cassino to Korea (1950) for Parat.; dev. 22-second high-speed processing machine for TV film recording, now inc. in Parat. theatre TV system. Ed. in chief, Telenews Prod., Inc., 1952; Television, Eastman Kodak adv. dept., 1954. Fellow: Soc. M.P. & TV Engineers. Member, Dir. American TV Commercials Festival, 1962; dir. TV adv. Eastman-Kodak; chmn. Assn. Nat'l Advertisers Television Committee, 1969; Director Media Development Film & TV, Mercer County College, N.J., 1972.

GEORGE, GEORGE LOUIS: Director. b. Moscow, Russia, July 31, 1907. e. U. Paris (law). Began as cutter, later asst. dir., French prods. To Hollywood 1935 as corresp. French pubs. Re-entered prod. 1939 as asst. dir. & prod. French version of The Four Hundred Million; in similar capacity numerous French versions of Hollywood prods. thereafter. In

1942 asst. prod. Our Russian Front. Became assoc. Natl. Film Board of Canada in prod. war information films; also dir. The Labor Front for World in Action series. In 1944 dir. University Town for Co-ord. Inter-Amer. Affairs; 1946–49 dir. 16 films for Signal Corps Photo. Center, Astoria, L.I., incl. Acad. Award Winner Toward Independence. In 1949 dir. in TV, also Women of Tomorrow, WB. 1950, dir. in Israel for Palestine Films; dir., prod., TV & documentary films, 1950–55; exec. secr., Screen Directors International Guild, 1956–63; admin., SDIG Trust Fund, 1963–67; member, Natl. Bd. Directors Guild of America 1966–69; since 1966, member Eastern Directors Council, DGA; pres. Film & TV Book Club, Inc. 1967–69. Currently free-lancing.

GEORGE, GEORGE W.: Writer/Producer. b. New York, N.Y., Feb. 8, 1920. e. Williams College; U.S. Navy, 1941–44; screen writer since 1948. President, SAGA PRODUCTIONS. N.Y.C. based, Industrial Video/film company.
PICTURES INCLUDE: Bodyguard, The Nevadan, Woman on Pier 13, Peggy Mystery Submarine, Red Mountain Experiment, Alcatraz, Fight Town, Smoke Signal, Desert Sands, Uranium Boom, Halliday Brand, Doc, The James Story, The Two Little Bears.
PRODUCER STAGE: Dylan, Any Wednesday, Ben Franklin in Paris, The Great Indoors, Happily Never After, Night Watch, Via Galaetica, Bedroom Force.
PRODUCER PICTURES: A Matter of Innocence, Twisted Nerve, Hello-Goodbye, Night Watch, Rich Kids, My Dinner with Andre.
TV: Climax, Screen Gems, Loretta Young Show, The Rifleman, Peter Gunn, The Real McCoys, Adventures in Paradise, Hong Kong, Follow The Sun.

GEORGE, LOUIS: Executive. Pres. & chief operating officer, Arista Films, Inc. b. Karavas, Kyrenia, Cyprus, June 7, 1935. e. Kyrenia Business Academy, Cyprus (honored 1951). Emigrated to U.S. in 1952. After brief stint in Foreign Exchange Dept. of City National Bank New York, served in U.S. Army, 1953–55. Entered industry in 1956 as theatre manager with Loew's Theatres in N.Y. metro area, managing Metropolitan, Triboro, New Rochelle, between 1958–66. In 1966 joined MGM as dir. of intl. theatre dept. In 1969 promoted to dir. of world-wide non-theatrical sales. From 1972 to 1974 served as regional dir. of MGM Far East operations. In 1974 left MGM to establish Arista Films, Inc., handling foreign sales of independently produced U.S. films.

GEORGE, SUSAN: Actress. b. England. Studied at London's Corona Stage School, appearing in The Sound of Music at 12 on London stage. Has appeared in many TV programs.
PICTURES INCLUDE: Billion Dollar Brain, Up the Junction, Spring and Port Wine, The Strange Affair, Lola, Straw Dogs, The Looking Glass War, Sunny and Jed, Dirty Mary Crazy Larry, Mandingo, A Small Town in Texas.

GERAGHTY, MAURICE: Writer, director, producer. b. Rushville, Ind., Sept. 29, 1908; e. Princeton. Western story suprv. Republic 3 yrs; prod Falcon series, China Sky, Action In Arabia at RKO; Flash Gordon (feature and serial).
PICTURES INCLUDE: Red Canyon, Dakota Lil, Calamity Jane and Sam Bass, Tomahawk, Sword of Monte Cristo, Rose of Cimarron, Mohawk, Love Me Tender (Elvis Presley), Tall In The Saddle (John Wayne), Gene Autry and Roy Rogers series, Hopalong Cassidy series, Mesquiteers series (John Wayne) Darby O'Gill And The Little People.
TV: Created Cheyenne series, Warner Bros, Virginian, Bonanaza, Daniel Boone, 87th Precinct, Thriller, 21 Beacon Street, Cavalcade of America, Panic, Whiplash (Australia) The Addams Family, Annie Oakley, Gene Autry series, Champion series, No Warning, Flight, Jane Wyman Theater, Laramie, Lassie (Emmy and Peabody Awards).

GERALD, HELEN: Actress. b. New York City, Aug. 13. e. U. of Southern Calif. 1948. Stage: Italian Teatro D'Arte, Les Miserables, The Civil Death, Feudalism.
PICTURES INCLUDE: The Gay Cavalier, The Trap, Tarzan nd the Leopard Woman, Cigarette Girl, Meet Miss Bobby Socks, G.I. War Brides, Gentleman's Agreement, A Bell For Adano, Tomorrow is Forever, Janie, Grand Prix, The Sandpiper, Make Mine Mink, Best of Everything.
TV: Robert Montgomery Presents, Frontiers of Faith, Valiant Lady, Kraft Theatre, Gangbusters, Adventures of the Falcon, Schlitz Playhouse of Stars, This Is the Answer, Man from UNCLE, Run for Your Life, Perry Mason.

GERARD, LILLIAN: Publicist-Writer, Publicity, Rialto Theatre, 1936; publicity-adv. Filmarte Theatre, 1938, Gerard Associates, 1938–47; V.P. and managing dir. of Paris Theatre, 1948–62; publicity-adv. dir., Rugoff Theatres, 1962. Film consultant to Times Films, Lopert Films, Landau Co., 1962–65. Exec. secy. to the National Soc. of Film Critics, 1966–68. Adjunct Professor, Film, 1968–1970, Columbia Univ., School of the Arts, Special Projects Co-ordinator, Museum of Modern Art, 1968–80. Contributor to American Film. Published by American Film Institute. Now associated with Philip Gerard in Gerard Associates.

GERARD, PHILIP R.: Executive. b. New York City, Aug. 23, 1913. e. C.C.N.Y., B.B.A. 1935; Columbia U.; Publicity dir. Mayer-Burstyn 1936–9; Gerard Associates, 1939–41; in public relations U.S. War Dept. 1942–4; with MGM 1944–8; with Universal Pictures since 1948; Eastern pub. mgr., 1950–59; Eastern ad. and pub. dir., Dec. 1959; N.Y. Production Exec., 1968–76. As of Jan. 1, 1977 formed Gerard Associates, film consultants on marketing, production and acquisitions; Film Consultant to The Jewish Film Archives, Jerusalem, International Board of American Friends of Hebrew University; and American Arbitration Association.

GERAY, STEVEN: Actor. b. Uzhored, Czechoslovakia, Nov. 10, 1904. Member Hungarian Natl. Theat.
THEATRE: Stage roles (abroad) include Tulip Time, Let's Go Gay, Silver Swan, etc.
PICTURES INCLUDE: Appeared in 40 European films. Hollywood career started 1941 with Man at Large; Cornered, The Unfaithful, I Love Trouble, Mr. District Attorney, House on Telegraph Hill, Little Egypt, Lady Possessed, Big Sky, Affair in Trinidad, O. Henry's Full House, Night Without Sleep, Tonight We Sing, Call Me Madam, Gentlemen Prefer Blondes, French Line, Knock on Wood, Paris Playboys, Tobor the Great, New York Confidential, A Bullet for Joey, Daddy Long Legs, Artista and Models, Birds and the Bees, Attack, Ship of Fools, Jesse James Meets Frankenstein's Daughter, The Swinger.

GERBER, DAVID: Executive. b. Brooklyn, N.Y. e. Univ. of Pacific. m actress Laraine Stephens. Joined Batten, Barton, Durstine and Osborn ad agency in N.Y. as TV supvr. Left to become senior v.p. of TV at General Artists Corp. In Jan., 1956, named v.p. in chg. sales at 20th-Fox TV where sold and packaged over 50 prime-time series and specials. Entered indep. prod. with The Ghost and Mrs. Muir, followed by Nanny and the Professor. In 1970 was exec. prod. of The Double Deckers, children's series made in England. In 1971 was exec. prod. of Cade's County (CBS). In 1972 he joined Columbia Pictures Television as an independent producer and in 1974 was named executive vice president worldwide production for CPT. In 1976 he returned to independent production and has been the executive producer on such programs as Police Story, Police Woman, The Lindbergh Kidnapping Case, Joe Forrester, The Quest and Gibbsville. For Police Story he received an Emmy award for best dramatic series.

GERBER, MICHAEL H.: Executive. b. New York City, Feb. 6, 1944. e. St. Johns Univ. (BA, '69); St. Johns Univ. School of Law (Juris Doctor, '69). Atty, for Screen Gems, 1969–71; asst. secy. & asst. to gen. counsel, Columbia Pictures Industries, 1971–74; corporate counsel and secretary, Allied Artists Pictures, 1974, v.p. Corporate affairs, Allied Artists, 1978; Parther, Talmodge, Pritzler, & Gerber.

GERE, RICHARD: Actor. e. Univ. of Massachusetts. Started acting in college; later joined Provincetown Playhouse and Seattle Repertory Theatre. Composed music for productions of these groups. Appeared on Broadway in Grease, Soon, Habeas Corpus, and Bent and in London in Taming of the Shrew with Young Vic.
PICTURES INCLUDE: Report to the Commissioner, Baby Blue Marine, Looking for Mr. Goodbar, Days of Heaven, Bloodbrothers, Yanks, American Gigolo.
TV: Kojak, Strike (movie).

GERETY, T. MICHAEL: Executive. b. Rockville Center, N.Y., Oct. 30, 1942. U.S., Navy, 1961–64. Started industry career in Chicago with MGM as regional adv/pub. repr.; transferred to Atlanta in same capacity, 1966 and then to Dallas, 1967–70. Joined Durwood Theatres in Dallas in 1971 as div. operations mgr. In 1972 returned to MGM as south west div. dir. adv./pub. in Dallas. Moved to MGM Studios in 1973 as asst. natl. adv. coordinator. In 1974 joined American International as dir. of cooperative adv./exploit. In 1976 named exec. dir. of adv./pub. for AIP (now Filmways Pictures). In 1980 joined the Milton I. Moritz Co., film marketing firm, as v.p. Member: Variety Clubs Int'l., Academy of MP. Arts & Sciences.

GERSHENSON, JOSEPH: Musician, b. Russia, Jan. 12, 1904; e. N.Y. public sch.; Pace Inst., N.Y.; studied music priv. Conductor orch. B.F. Keith Theats. 1920–28; asst. genl. music dir. RKO Theats., 1928–33. Joined Universal Pictures 1933; assoc. prod. & music dir. of shorts in 1939; assoc. prod. (under name of Joseph G. Sanford) of features from 1940; in 1941 apptd. exec. prod.; head of music dept. Universal International.
PICTURES INCLUDE: Glenn Miller Story, Magnificent Obsession, So This Is Paris, Man Without a Star, Foxfire, Benny Goodman Story, All That Heaven Allows. There's Always Tomorrow, Spoilers, Never Say Goodbye, Battle Hymn, Away All Boats, Written on the Wind, My Man Godfrey, Man of a Thousand Faces, Never Steal Anything Small, Imitation of Life, This Earth Is Mine, Pillow Talk, Operation Petticoat, Back Street, Come September, Lover Come Back, Spiral Road, Freud, Captain Newman, Father Goose, Send Me No Flowers, I'd Rather Be Rich, Warlord, Madame X,

Shenandoah, Blindfold, Thoroughly Modern Millie, Sweet Charity.

GERSHWIN, IRA: Music lyricist. b. New York City, Dec. 6, 1896. e. Coll. City of New York; Pulitzer Prize, 1932; brother of late George Gershwin. Began writing songs when 22.

FILMS INCLUDE: 1931 Delicious, Fox; 1937 Shall We Dance, RKO; 1937 Damsel in Distress, RKO; 1938 The Goldwyn Follies, Goldwyn; 1943 Rhapsody in Blue, Warner Bros.; Cover Girl, Where Do We Go From Here? The Shocking Miss Pilgrim, The Barkleys of Broadway, An American in Paris, Give the Girl A Break, A Star is Born, The Country Girl, Kiss Me Stupid.

AUTHOR: Lyrics on Several Occasions, 1959.

GERTNER, RICHARD: Editor. b. Tampa, Florida, Nov. 1; e. Rollins College, B.A., 1949, Columbia U., M.A., 1950; editor, Hoffman Publications, 1950–52; editorial asst., Better Theatres, 1952; news editor Motion Picture Daily, 1956; managing editor, Motion Picture Daily, 1962; managing editor, Motion Picture Herald, 1963; editor, Better Theatres, 1963; executive editor, Motion Picture Herald, 1964; editor, Motion Picture Herald, 1966; editor, Motion Picture Almanac and TV Almanac, 1970. Managing editor Motion Picture Daily, Oct. 1970; Editor, Motion Picture Daily, 1971; Editor-in-chief, Quigley Publications, Nov., 1971.

GERTZ, IRVING: Composer, musical director. b. Providence, R.I., May 19, 1915; e. Providence Coll. of Music, 1934–37. Assoc. with Providence Symph. Orch., comp. choral works for Catholic Choral Soc.; music dept., Columbia, 1939–41; U.S. Army, 1941–46; then comp. arranger, mus. dir. for many cos. incl. Columbia, U-I, NBC.

PICTURES INCLUDE: Bandits of Corsica, Gun Belt, Long Wait, The Fiercest Heart, First Travelling Saleslady, Fluffy, Nobody's Perfect, Marines, Let's Go! It Came From Outer Space, The Man From Bitter Ridge, Posse From Hell, The Creature Walks Among Us, The Incredible Shrinking Man, Hell Bent For Leather, Seven Ways From Sundown, Francis Joins the WACS, Raw Edge, East of Sumatra, A Day of Fury, To Hell and Back, Cult of the Cobra, Plunder Road, Top Gun, Tombstone Express, The Alligator People, Khyber Patrol, The Wizard of Baghdad.

Record album (Dot Records) Leaves of Grass; published works for mixed voices; Fluffy, feature, Universal. Marines, Let's Go! feature, 20th Century Fox Serenata for String Quartet, Divertimento for String Orchestra, Tableau for Orchestra.

TV: Orig. theme & scores for TV: America, The Golden Voyage, Across the Seven Seas, The Legend of Jesse James, Daniel Boone, Voyage to the Bottom of the Sea, Peyton Place, Land of the Giants, Lancer, Medical Center.

GETTY, J. RONALD: Producer. b. -Berlin, Germany, Dec. 19, 1929. e. Zurich U., Heidelberg U., U. of Southern California. Prior to entering m.p. industry was oil executive, Tidewater Oil Co., mgn. dir. Veedol GmbH, 1955; Getty Oil, Hamburg, 1954–61; pres. & chm. of bd., Veedol Petroleum Int'l, Switzerland, 1961–69; dir., Huiles Veedol France, 1961–69. President & Chm. of bd., Getty Picture Corp.; pres. & chm., Getty Labs, Inc.; dir., Home Theatre Network.

PICTURES: Flare Up, Zeppelin, Shelia.

GIANNINI, GIANCARLO: Actor. b. Spezia, Italy, Aug. 1, 1942. Acquired degree in electronics but immediately after school enrolled at Academy for Drama in Rome. Cast by Franco Zeffirelli as Romeo at age of 20. Subsequently appeared in a play also directed by Zeffirelli, Two Plus Two No Longer Make Four, written by Lina Wertmuller.

PICTURES INCLUDE: Love and Anarchy, The Seduction of Mimi, Swept Away by an Unusual Destiny in the Blue Sea of August, Seven Beauties . . . That's What They Call Him, How Funny Can Sex Be?, A Night Full of Rain, The Innocent, Immortal Bachelor.

GIBSON, HENRY: Actor. b. Germantown, PA, Sept. 21, 1935. e. Catholic Univ. Amer. Appeared as child actor with stock companies, 1943–57; Bdwy. debut in My Mother, My Father and Me, 1962. Many appearances on stage since and in films and TV.

PICTURES: The Long Goodbye, Charlotte's Web, Nashville, The Last Remake of Beau Geste.

TELEVISION: Laugh-In co-star.

GIELGUD, SIR JOHN: Actor. b. London, England, Apr. 14, 1904; e. Westminster Sch., Lady Benson's Sch. (dram.), London; Royal Acad. Dram. Art. Knighted, 1953.

THEATRE: Began stage career in Shakespearean roles; on London stage also in the Constant Nymph, The Good Companions, Dear Octopus, The Importance of Being Earnest, Dear Brutus, etc., various Shakespearean seasons, London & N.Y.

PICTURES INCLUDE: On screen 1932 in Brit. prod. Insult; later (Brit.) films The Good Companions, Secret Agent (Warner-Brit.); The Prime Minister, Other films: Julius Caesar, Richard III. Around the World in 80 Days, Barretts of Wimpole Street, Becket. The Loved One, Chimes At Night, St. Joan, Sebastian, The Assignment, Charge of the Light Brigade, The Shoes of the Fisherman, Oh, What a Lovely War, Eagle in a Cage, Lost Horizon, 11 Harrowhouse, Gold, Murder on the Orient Express. Providence, Portrait of the Artists as a Young Man, Joseph Andrews, Murder by Decree, The Human Factor, The Elephant Man, The Formula, Sphinx, Lion of the Desert, Arthur.

TV: A Day by the Sea, The Browning Version (U.S.A.), The Rehearsal, Great Acting, Ages of Man, Mayfly and the Frog, Cherry Orchard, Ivanov, From Chekhov With Love, St. Joan, Good King Charles' Golden Days, Conversation at Night, Hassan, Deliver Us From Evil, Heartbreak House.

GIFFORD, ALAN: Actor. b. Boston, Mass., March 11, 1911. Early career in Group Theatres and summer stock. Served with SHAEF as Col. U.S. Army during World War II. To England 1951 for Love of Four Colonels. Since extensive appearances films and television and stage roles.

PICTURES INCLUDE: The Iron Petticoat, Time Lock, The Flying Scot, Paris Holiday, Screaming Mimi, Onionhead, Table in the Corner, Mouse That Roared, Too Young to Love, I Aim at the Stars, The Royal Game, No Kidding, The Road to Hong Kong, Town Without Pity, Devil Poll, One Spy Too Many, Drop Dead Darling, Only When I Larf, 2001, A Space Odyssey, Isadora, Legend of Nigger Charlie, Phase IV.

TV PLAYS INCLUDE: High Tension, No Deadly Medicine, Ernie Barger is 50, My Lost Saints, Philadelphia Story, Flight For Martin Wheeler, Man on the Mountain Top, Woman of Fifty, The Executioners, Means To an End, A Quiet Game of Cards, Winter Journey, The Velvet Cage, Marching Song, Saint Series, Sentimental Agent, The Monster, Espionage Series; stage: License to Murder, Danger Man series, The Ambassadors, The Champions, Troubleshooters, Champion House, Randall and Hopkirk (Deceased), The Male Animal, Softly, Softly, The Name of the Game, As the World Turns, Edge of Night, Somerset, Duchess of Duke Street, Crossroads series.

GIL, DAVID: Producer. b. Tel Aviv, Israel, Jan. 24, 1930; e. U. of Jerusalem; m. Joan Andre. After commission in Israeli Army worked for Israeli Embassy, Paris, 1950–52; Israeli film ind., 1953–55; prod. educational films, 1955–61; headed Gilart Productions, 1962–68; foreign sales dir., Commonwealth United, 1968; prod. Guess What We Learned in School Today, Joe, A Journey Through Rosebud, A Change in the Wind.

GILBERT, ARTHUR N.: Producer. b. Oct. 17, 1920, Detroit; Lt., U.S.M.C., 1941–45; ed. Univ. of Chicago, 1946; Special Agent, FBI, 1946–53; world sales dir., Gen. Motors, Cadillac Div., 1953–59; investments in mot. pictures and hotel chains, 1959–1964; exec. prod., Mondo Hollywood, 1965; exec. prod. Jeannie-Wife Child, 1966; assoc. prod., The Golden Breed, 1967; commissioned rank of Colonel U.S.M.C., 1968.

PICTURES INCLUDE: The Glory Stompers, Fire Grass, Cycle Savages, Bigfoot, Incredible Transplant.

GILBERT, BRUCE: Producer. e. University of Calif. Pursued film interests at Berkeley's Pacific Film Archive; in summer involved in production in film dept. of San Francisco State Univ. Founded progressive pre-school in Bay Area. Became story editor in feature film division of Cine-Artists; involved in several projects, including Aloha, Bobby and Rose. Partnered with Jane Fonda and IPC Films, Inc.

PICTURES: Coming Home (assoc. prod.); The China Syndrome (exec. prod.).

GILBERT, LEWIS: Writer, director, actor. b. London, England, Mar. 6, 1920. In RAF, World War II. Screen debut, 1932; asst. dir. (1930–39) with London Films, Assoc. British, Mayflower, RKO-Radio; from 1939–44 attached U.S. Air Corps Film Unit (asst. dir., "Target for Today"). In 1944 joined G.B.I. as writer and dir. In 1948, Gainsborough Pictures as writer, dir., 1949; Argyle Prod. 1950; under contract Nettlefold Films, Ltd. as dir.

PICTURES INCLUDE: Under One Roof, I Want to Get Married, Haunting Melody, Once a Sinner, Scarlet Thread, There Is Another Sun, Time Gentlemen Please, Emergency Call, Cosh Boy, Johnny on the Run, Albert R.N., The Good Die Young, The Sea Shall Not Have Them, Reach for the Sky, Cast a Dark Shadow, The Admirable Crichton, Carve Her Name with Pride, A Cry from the Street, Ferry to Hong Kong, Sink the Bismarck, Light Up the Sky, The Greengage Summer, H.M.S. Defiant, The Patriots, Spare the Rod, The Seventh Dawn You Only Live Twice, The Adventurers, Friends, Paul & Michelle, Operation Daybreak, Seven Nights in Japan, The Spy Who Loved Me. 1978–79: Dir. Moonraker.

GILBERT, MELISSA: Actress. b. Los Angeles, May 8, 1964. Made debut at age of 3 in TV commercial. Comes from show business family: father, late comedian Paul Gilbert; mother, former dancer-actress Barbara Crane.

TELEVISION: Gunsmoke, Emergency, Tenafly, The Hanna-Barbera Happy Hour, Christmas Miracle in Caufield U.S.A., Love Boat, Little House on the Prairie (regular).

GILFORD, JACK: Actor. Made first stage appearances in amateur night performances as stand-up comic. Wrote own material, specializing in imitations. First theatrical role in 1952 in The World of Sholem Aleichem, which he repeated in TV special based on play.
PICTURES INCLUDE: They Might Be Giants, Catch-22, Save the Tiger, Harry and Walter Go to New York, Wholly Moses.
STAGE: The Diary of Anne Frank, Romanoff and Juliet, Once Upon a Mattress, Cabaret, No, No, Nanette.

GILKISON, ANTHONY: Producer. b. York, England, June 3, 1913. e. Stowe School. Assistant dir. and prod. mgr. European and British companies, 1932–38; film director Rayant Pictures, 1938; executive producer, managing director Rayant Pictures, 1945; chmn. Rayant Television, Ltd.; founded Anthony Gilkison Associates, International Motion Picture and Television Consultants, 1963. 1976, Chmn. Anthony Gilkison Assoc. (Holdings) Ltd.

GILLASPY, RICHARD M.: Producer, director. b. St. Louis, Mo., Dec. 10, 1927; joined NBC in New York as exec. trainee, 1947; headed Marine Corps Radio, 1952–54; rejoined NBC-TV as stage mgr., became prod. dir.; won Emmy Award, 1960, for Nixon-Khrushchev debate; pres., Seven League Productions, 1961–63, owner, Radio Station WIII, Homestead, Fla., 1963–67; freelance prod.-dir., 1967–70; v.p. of Ivan Tors Studios, N. Miami, 1970–1972; pres. RMG Productions, 1972.

GILLIAT, LESLIE: Producer. b. New Malden, England, 1917. e. Epson College. Ent. m.p. ind. 1935.
PICTURES INCLUDE: Only Two Can Play, The Amorous Prawn, Joey Boy, The Great St. Trinians Train Robbery, A Dandy in Aspic, The Virgin Soldiers, The Buttercup Chain, Endless Night, Priest of Love (prod. supvr.).

GILLIAT, SIDNEY: Writer, producer, director of many British films. b. Cheshire, Eng., 1968.

GILLIN, DONALD T.: Executive. b. Council Bluffs, Iowa, June 17, 1914. e. Univ. of Minn., bus. adm., 1929–32; sls., Warner Bros., 1932–39; World War II; home office sup., Warner Bros., 1946–48; sls. exec., Universal, 1948–50; v.p., gen. sls. mgr., mgr., Sol Lesser, 1950–59; establ., Donald T. Gillin Inc., producers' rep., 1960. Muchnic-Gillin Internat'l, Inc., 1961; Producers' Representatives, Inc., 1967.

GILLING, JOHN: Writer, producer, director. b. 1912. Started in show business in U.S., 1930. Ent. Brit. m.p. ind. 1933 as asst. dir. B.I.P., Gainsborough, Gaumont. Served in R.N.V.R. during war. Has written 56 s.p. inc. 29 orig., all prod. since 1946; dir. 26 pictures.
PICTURES INCLUDE: Interpol, The Man Inside, Odongo, High Flight, The Flesh and the Fields, The Challenge, Fury at Smugglers Bay, The Pirates of Blood River, The Shadow of the Cat, Blood River, Scarlet Blade, Brigand of Kandahar, Plague of the Zombies, The Reptile, The Night Caller.
TV: Written and dir. TV films.

GILLIS, ANNE: Actress. r.n. Alma Mable O'Connor. b. Little Rock, Arkansas, Feb. 12, 1927. Screen debut 1936 in The Garden of Allah; thereafter in number child roles.
PICTURES INCLUDE: King of Hockey, Off to the Races, The Californian, The Adventures of Tom Sawyer, Peck's Bad Boy with the Circus, Little Orphan Annie, Edison the Man, All This and Heaven Too, Janie; A Wave, a Wac, a Marine; The Cheaters, Gay Blades, Big Town After Dark.

GIMBEL, ROGER: Producer. Began television production career as copy and creative chief of RCA Victor TV, then became assoc. prod. of the Tonight Show for NBC. Subsequently named head of program development of NBC daytime programming; later became producer of the 90-minute NBC Tonight Specials, including The Jack Paar Show and the Ernie Kovacs Show. Became prod. and co-packager of the Glen Campbell Goodtime Hour for CBS, 1969; vice pres. in charge of production for Tomorrow Entertainment, 1971. Produced 15 major movies for television including The Autobiography of Miss Jane Pittman, Born Innocent, Birds of Prey, Brand New Life, Gargoyles, Glass House, In This House of Brede (General Electric Theatre), I heard the Owl Call My Name (G.E. Theatre), I Love You, Good-Bye, Larry (G.E. Theatre), Miles to Go (G.E. Theatre), Queen of the Stardust Ballroom, Tell Me Where It Hurts (G.E. Theatre) Things in Their Season (G.E. Theatre), War of the Children (Emmy award, Outstanding Single Program, Drama or Comedy). Formed his own production company, Roger Gimbel's Tomorrow Enterprises, Inc., 1975; produced two-hour CBS special, Minstrel Man. Became U.S. pres. of EMI-TV, 1976. Received special personal Emmy as exec. prod. of War of the Children, 1975.

GINGOLD, DAN: Freelance Executive Producer, Producer, Director. Credits include all types live and film prod. Specialist in Special Events and Documentary. Awards include Emmy, Ohio State, Northwestern, Ed Murrow, Assoc. Press, San Francisco State.

GINGOLD, HERMIONE: Actress. b. London, England, 1897. Appeared with Old Vic and at Shakespeare at Stratford on Avon and in Paris before coming to America and Bdwy.
PICTURES: Gigi, Naked Edge, Music Man, Gay Paree, Harvey Middleman—Fireman, I'd Rather Be Rich, Promise Her Anything, Rocket to the Moon.
TELEVISION: Ed Sullivan Show, Omnibus, Matinee Theatre, Jack Paar, Alfred Hitchcock Presents, Merv Griffin Show, Mike Douglas Show, etc.

GINNA, ROBERT EMMETT, JR.: Producer, writer. b. New York City, Dec. 3, 1925. e. U. of Rochester, Harvard, M.A. Lt. j.g. U.S. Navy Amphibious Forces, W.W. II. Journalist, editor, Life, Scientific American, Horizon, 1950–61; contributor many other magazines, particularly on stage and screen. Worked as writer, director, producer or associate producer, Public Affairs TV, Wisdom series, FDR series, Specials. 1955–58, 1961–63; v.p., Sextant, Inc., 1961–64; dir. Sextant Films, Ltd.; co-prod., Young Cassidy, 1964; founder and pres., Windward Productions, Inc., 1965; founder and mgr. dir., Windward Film Productions, Ltd., 1965.

GINSBERG, SIDNEY: Executive. b. New York City, Oct. 26, 1920; e. City Coll., 1938; entered m.p. ind., as asst. mgr., Loew's Theatres; joined Trans-Lux, 1943, as thea. mgr.; film booker; helped form Trans-Lux Distributing Corp., 1956; asst. to pres., Trans-Lux Dist. Corp.; asst. vice-pres., Trans-Lux Picture, Distributing and TV Corp., 1961, V.P. Trans-Lux Dist. Corp., 1967, V.P. in charge of world wide sales, 1969. Haven International Pictures, Inc., Haven Intern'l 1970; IFIDA gov., 1970, v.p. sales, Scotia International Films, Inc., 1971; exec. v.p., Scotia American Prods; 1977, pres., Rob-Rich Films Inc.; 1979, exec. v.p., A Major Studio, Inc.; 1980, exec. v.p., The Health and Entertainment Corp. of America.

GINSBURG, LEWIS S.: Distrib., Import., prob. b. New York, May 16, 1914. e. City Coll., N.Y., 1931–32. Columbia Univ., 1932–33. Ent. film industry, tabulating dept., United Artists, Sept. 1932; sls. contract dept. 1933; asst. to eastern district mgr., 1938; slsmn., New Haven exch., 1939. Army, 1943. Ret. to U.S., then formed first buying & booking service in Connecticut, 1945–55; in chg., New England Screen Guild Exchanges, 1955; TV film distr., 1955; Formed & org. International Film Assoc., Vid-EX Film Distr. Corp., 1961. Prod., TV half-hour series; vice-pres. in chg., dist., Desilu Film Dist. C., 1962; organized Carl Releasing Co., x963; Walter Reade-Sterling Inc., 1964–65; formed L.G. Films Corp.; contract and playdate mgr., 20th Fox, 1965–68. Cinerama Releasing Corp. Adm. Ass't to sales mgr., 1968–69; 20th Cent.-Fox. Nat'l sales coordinator, 1969–present. 1970, 20th Century-Fox, Asst. to the Sales Mgr. 1971, Transnational Pictures Corp., v.p. in chg. of dist., pres., Stellar IV Film Corp., 1972.

GIRARDOT, ANNIE: Actress. Studied nursing. Studied acting at the Paris Conservatory, made her acting debut with the Comedie Francaise. Has acted on the French stage and in reviews in the Latin Quarter.
PICTURES INCLUDE: Rocco and His Brothers, The Organizer, Les Galoise Bleues, Live For Life, Trois Chambres A Manhattan (Best Actress Award at the Venice Film Festival), Story of A Woman, Love Is A Funny Thing, The Slap, No Time for Breakfast.

GISH, LILLIAN: Actress. b. Springfield, Ohio, Oct. 14, 1899; sister of Dorothy G., actress. At 5 appeared in In Convict's Stripes, at Rising Sun, Ohio; following year danced in Sarah Bernhardt prod. in N.Y. In 1913 appeared with Mary Pickford in A Good Little Devil, N.Y. Began screen career 1912 with Biograph, beginning there assn. with D. W. Griffith, dir., which culminated in such films as The Birth of a Nation, Intolerance, Hearts of the World, The Great Love, Broken Blossoms, Way Down East and Orphans of the Storm. Continued in films: White Sister, Romola, The Wind, La Boheme, Scarlet Letter. From 1930 on N.Y. stage in number orig. prods. & classics Uncle Vanya, Camille, 9 Pine Street, Within the Gates, Hamlet, Star Wagon, Old Maid, Dear Octopus, Life With Father, Mr. Sycamore, The Marquise, Legend of Leonora, Crime and Punishment, Miss Mable, Curious Savage, A Passage to India, Too True to be Good; Romeo and Juliet, Stratford Shakespeare Theatre, 1965. 1969–71, One woman int'l. concert tour, Lillian Gish and the Movies. Received honorary Oscar, 1971. International touring with illustrated lecture on the art of film and TV, 1974. Lecture tour on Queen Elizabeth 2, 1975.
PICTURES INCLUDE: One Romantic Night, His Double Life, Commandos Strike at Dawn, Duel in the Sun, Miss Susie Slagle's, Portrait of Jenny, Follow Me Boys!, Night of the Hunter, 1954; The Cobweb, 1955; The Unforgiven, Orders to Kill; Warning Shot, The Comedians, A Wedding (her 100th film).
PLAYS: Theatre Guild play The Trip to Bountiful, 1954; Chalk Garden, 1956; Phoenix Theatre, N.Y., The Family Reunion; in Berlin, Ger., Portrait of a Madonna, Wreck of the 5:25, Uncle Vanya, 1973. All The Way Home, 1960;

Anya, 1965; I Never Sang For My Father, 1968; A Musical Jubilee, 1976.

TV: I Mrs. Bibbs, Sound and the Fury, Day Lincoln Was Shot, Mornings at Seven, The Grass Harp, Grandma T.N.T., Mr. Novak, Alfred Hitchcock Hour, Breaking Point, The Defenders, The Silent Years (hostess). Twin Detectives—(ABC Movie-of-the-Week), 1976.

AUTHOR: The Movies, Mr. Griffith and Me, published by Prentice-Hall, 1969; Dorothy and Lillian Gish published by Charles Scribner, October, 1973.

GITHENS, W. FRENCH: Executive. b. New York City, April 30, 1906; e. Columbia U. in 1928 secy. Amer. embassy Madrid. Entered m.p. ind. 1929 as member ed. staff Movietonews, editing program Embassy Newsreel theat., N.Y.; asst. ed. Pathe News 1933. On 1934 org. Newsreel Theats. (Embassy theat., Broadway, N.Y. & others), becoming pres. & dir. Org. Sound Masters 1937 to prod. industrial & educational films. Lt. Comdr., U.S.N.R. Bureau Aeronautics World War II; com. U.S.N.R.; pres. National Educational Films, Inc.; on bd. of Viterama & Cinerama Corps. developers of 3-dimensional projection system, 1950. Mayor of Bernardsville, N.J., 2 terms; prod., Movietime Digest; prod., V.P. and dir. Steel Properties, N.Y.C.

GLASER, PAUL MICHAEL: Actor. b. Cambridge, Mass., March 25. e. Tulane Univ., Boston Univ. (M.A.). Did five seasons in summer stock before starting career in New York City, making stage debut in rock version of Hamlet in 1968. Appeared in numerous off-Bdwy. plays and got early TV training as regular in daytime series, Love of Life and Love Is a Many Splendored Thing.

PICTURES INCLUDE: Fiddler on the Roof, Butterflies Are Free.

TV: Kojak, Toma, The Streets of San Francisco, The Rockford Files, The Sixth Sense, The Waltons. Movies: Trapped Beneath the Sea, The Great Houdini. Series: Starsky & Hutch.

GLASER, SIDNEY: Exec. adv., prod. b. New York, N.Y., July 12, 1912. e. CCNY, b. Eng., 1936; NYU, m. Eng., 1942; office boy, adv. dept., Metro-Goldwyn-Mayer, 1929; prod. asst. 1934; prod. mgr., 1957; adv. prod. mgr., 1972; gen. adv. exec.

GLASS, GEORGE: Producer, publicist. b. Los Angeles, Calif., Aug. 19, 1910. e. L.A. schools. Reporter, sports writer L.A. Herald, Record; news commentator, special events broadcaster L.A. Examiner; joined Selznick publ. dept., 1936; West Coast dir. adv. for United Artists 1941; dir. pub. and adv. Samuel Goldwyn 1944. In 1945 special consultant to ind. prod. cos. In 1947 joined Stanley Kramer as partner and v.p. Stanley Kramer Co.; assoc. prod. The Men, Cyrano de Bergerac, Death of a Salesman. In 1954 dissolved Kramer partnership, exec. staff, Hecht-Lancaster, 1956; org. with Walter Seltzer, Glass-Seltzer Co.; joined Pennebaker, Inc. as exec. prod., v.p.; co-prod. Shake Hands With the Devil; exec. prod. One-Eyed Jacks. Co-prod., The Naked Edge, Paris Blues, 1960–61. Prod., Universal-International, 1963; assoc. prod., The Day Custer Fell, 20th Century-Fox, 1964. Rejoined Stanley Kramer as v.p.-assoc. prod. 1967. Assoc. Prod. Guess Who's Coming to Dinner, 1968. Assoc. prod. The Secret of Santa Vittoria, 1969; Assoc. prod. R.P.M., 1970; assoc. prod., Bless the Beasts and Children, 1971. Now consultant for production and p.r.

GLAZIER, SIDNEY: Producer. b. Philadelphia, May 29, 1918. Managed movie and legitimate theatres in Pennsylvania and Ohio. Air Force captain during World War II. Became v.p. of Washington Federal Savings and Loan Bank in Miami. Org. and became exec. dir. of Eleanor Roosevelt Cancer Foundation. Prod. network TV public service programs. First m.p. The Eleanor Roosevelt Story won Academy Award, 1966. Pres. of U-M Film Distributors, 1969.

PICTURES INCLUDE: The Producers, Take the Money and Run, The Gamblers, Quackser Fortune Has a Cousin in the Bronx, The 12 Chairs, Glen and Panda, The Night Visitor, The Only Way.

GLEASON, JACKIE: Performer. b. Brooklyn, N.Y., Feb. 26, 1916. Started as amateur entertainer, clubs, WB pact, appeared in Navy Blues, All Through the Night, Larceny, Inc., Orchestra Wives, Springtime in the Rockies; returned to Broadway appearing on the stage and summer stock.

BWY. STAGE: Take Me Along, 1959–60 (Tony Award).

TV: Started as lead in The Life of Riley; then star and m.c. Cavalcade of Stars on the Du Mont network; The Honeymooners; Jackie Gleason Show, CBS-TV network. Best comedian, Fame poll, 1953; best comedy show, 1954.

PICTURES INCLUDE: The Hustler, Requiem For a Heavyweight, Gigot, Papa's Delicate Condition, Soldier in the Rain, Ski-doo; How to Commit a Marriage, Don't Drink the Water, How Do I Love Thee?, The Late Show, Smokey and the Bandit, Smokey and the Bandit II.

GLEASON, KEOGH: Set decorator. b. Minneapolis, Minn., April 14, e. U. of Minn. Vaudeville, 1929–30; interior decorator, portrait painter, 1932–34; set decorator, MGM, 1936; Acad. Award, An American in Paris, 1951; Bad and the Beautiful, 1952. Somebody Up There Likes Me, 1956; Gigi, 1958; Nominations, Story of Three Loves, 1953; Brigadoon, 1954; Lust for Life, 1956.

PICTURES INCLUDE: Father of the Bride, Fathers Little Dividend, The Trailer, Cob Web, Kismet, Tea and Sympathy, For Me and My Gal, Bells are Ringing, Four Horsemen of the Apocalypse, Two Weeks in Another Town, Goodbye Charlie, Courtship of Eddie's Father, The Sandpiper, Point Blank, Legend of Lylah Clare, There was a Crooked Man.

TV: Richard Boone Rep. Theatre, Twilight Zone.

GLENN, CHARLES OWEN: Executive. b. Binghamton, N.Y., March 27, 1938; e. Syracuse U. (BA), U. of Pa. Asst. to dir. of adv., 20th Cent. Fox, 1966–67; asst. adv. mgr., Paramount, 1967–68; acct. spvsr. & exec., MGM record & m.p. div., 1968–69; nat'l adv. mgr., Paramount, 1969–70; nat'l. dir. of adv., Paramount, 1970–71; v.p. adv.-pub.-prom., 1971–73; v.p. marketing, 1974; v.p. prod. mktg., 1975; joined American Intl. Pictures as v.p. in chg. of adv./creative affairs, 1979. 1980, when Filmways took AIP over he was named their v.p. in chg. worldwide adv./pub./promo.

GLICK, HYMAN J.: Executive. b. Russia, Dec. 15, 1904; e. N.Y.U., B.C.S., '26, C.P.A. (N.Y.); with public accounting firm, N.Y., 1923–29; own public accounting Business, 1929–32. Became assoc. m.p. ind. as member comptrollers' com., repr. Republic; 1932–36, tax & financial counsel Mascot Pictures Corp. Joined Republic 1936 as comptroller; apptd. asst. secy.-asst. treas. Jan., 1945.

Member: B'nai Brith Lodge 1325; YMCA, State Soc. of Certified Public Accountants, Am. Inst. of Accountants. Resigned Republic, 1959; CPA (Calif.), member, Calif. Soc. of CPA; sec'y-treas., Own Accounting and Tax Practice.

GLICKMAN, JOEL: Producer. b. Los Angeles, July 29, 1930. e. UCLA (film dep't.). Was actor and director for LA little theatre groups and writer-director for industrial films. Early TV work on series, documentaries and commercials. Directed videotaping of off-Broadway shows. Was production assoc. on film, Wedding and Babies, 1958. Was assoc. prod. and prod. mgr. on films (Terror in the City, The Balcony, All the Way Home, For Love of Ivy, Hamlet (Richard Burton, assoc. director), Dion Brothers, Last Summer and TV (East Side, West Side, Mr. Broadway), N.Y.P.D., Love Song of Barney Kempinski, Among the Paths To Eden. Member Motion Picture Academy & D.G.A. Has own co.: Selznick/Glickman Productions.

PICTURES INCLUDE: Brother John, Buck and the Preacher, Trial of the Catonsville Nine (exec. prod.), Night Terror (NBC-TV movie), Angel on Horseback (CBS-TV movie), Kennedy-Hoffa War (NBC-TV miniseries).

GLUCK, NORMAN E.: Executive. b. Racine, Wisc., Apr. 17, 1914. Worked for RKO Theaters, Albany dist., 1932–34; Skouras Theatres, 1935–46; U.S. Army; oper. Park Ave. Theatre, N.Y., 1947–48; United World Films (Univ. subsid.), 1949; v.p. and dir., head Universal Pictures TV dept.; dir. corp. oper., Screen Gems, 1959–60; Universal Pictures, 1960; Asst. to gen. sales mgr. Now retired.

GLUCKSMAN, ERNEST D.: Producer, director, writer. b. Vienna, Austria, March 21. Broadway musicals, summer stock, vaudeville, night clubs. Legit: Odd Couple. Consultant to Commissioner of Cultural Affairs, City of New York.

PICTURES INCLUDE: The Nutty Professor, The Errand Boy, The Patsy, Rock-A-Bye-Baby, Bell Boy, Cinderfella, Geisha Boy.

TV: Here Come the Stars, Can You Top This, Colgate Comedy Hour, Saturday Night Revue. Specials: Timex All-Star Swing Festival Dr. Jekyll and Mr. Hyde, Bobby Sherman, Fifth Dimension, Bobby Darin, Sonny & Cher, George Kirby, Kenny Rogers and The First Edition, Charles Aznavour & Liza Minnelli; game show What's the Good Word?; series Story Theatre.

GOATMAN, ALAN H.: Consultant, Entered m.p. industry 1934. Officer of the Venerable Order of St. John of Jerusalem. Fellow Society of Company and Commercial Accountants. Memb. Brit. Computer Soc. Fellow Brit. Inst. Management; member Inst. Data Processing; Management Dir. Gen. Man. E.M.I. Films Ltd. and E.M.I. Film Distributors Overseas Ltd. Dir. E.M.I. Pathe Ltd., Anglo-E.M.I. Productions Ltd. Databank Computing Services Ltd., Columbia-EMI-Warner Dist. Ltd.

GODARD, JEAN-LUC: Writer-Director. b. Paris, France, Dec. 3, 1930.

PICTURES INCLUDE: A Bout de Souffle, Une Femme Est Une Femme, Vivre Sa Vie, Le Petie Soldat, Les Carabiniera, Bande a Part, Une Femme Marlee, A Little Godard; Pierrot Le Fou, Made in USA, Weekend, Tout va Bien.

GODBOLD, GEOFF: Executive. b. London, England, 1935. Ent. ind. 1959. Specialized in supply of location requirements for film and TV prods. Formed Prop Workshops Ltd., copromoted Television Recordings Ltd. Man. dir. Facilities

(Screen & Television) Ltd. Dir. TV Recordings, Investments, Ltd.; Centrepoint Screen Prod. Ltd.; Tape Commercials, Ltd. Council Mem. Film and TV Contractors Assoc. Dir. Lancair Export Services Ltd.; 1968 Freelance production buyer, Screen Gems and Tigon. 1969. Feature Prod. Rep. Film Div. N.A.T.T.K.E.

PICTURES INCLUDE: Dubious Patriot, Every Home Should Have One, Melody, Up Pompeii. Morocco location, Young Winston. The Asphyx, Our Miss Fred, Death of a Snow Queen, Man in the Iron Mask.

GODDARD, PAULETTE: Actress. b. Great Neck, N.Y., June 3, 1911. On N.Y. stage in Rio Rita; then member of Hal Roach Studios stock co. In 1936 opposite Charles Chaplin in Modern Times.

PICTURES INCLUDE: The Young in Heart, The Women, North West Mounted Police, The Great Dictator, Hold Back the Dawn, Reap the Wild Wind, Star Spangled Rhythm, I Love a Soldier, So Proudly We Hail, Standing Room Only, Duffy's Tavern, Kitty, Unconquered, An Ideal Husband, Hazard, On Our Merry Way, Bride of Vengeance, Anna Lucasta, Babes in Bagdad, Vice Squad, Paris Model, Sins of Jezebel, Charge of the Lancers, Unholy Four, Time of Indifference.

GODFREY, ARTHUR: Performer. b. New York City, Aug. 31, now 54th year in show business—actor, writer, narrator, entertainer.

CLUBS: New York Yacht, Explorers, Quiet Birdmen. Calif. Yacht, Sky, Illustrators.

PICTURES INCLUDE: The Glass Bottom Boat, Where Angels Go . . . Trouble Follows, Four for Texas, The Great Georgia Bank Hoax.

GODWIN, FRANK: Producer, Director. Joined Gainsborough Pictures 1943, subsequently exec. with Rank Organization; asst. to exec. prod. Earl St. John.

PICTURES INCLUDE: Woman in a Dressing Gown, No Trees in the Street, Operation Bullshine, Don't Bother to Knock, The Small World of Sammy Lee, Danny the Dragon, Headline Hunters, Demons of the Mind, The Boy With Two Heads, The Firefighters, Sky Pirates, Sammy's Super T-Shirt, Electric Eskimo, The Boy Who Never Was.

GOLD, MELVIN: Executive. b. Chicago, Ill., Sept. 3, 1909. In 1930 joined MGM, Chicago, as assistant office manager. To Reinheimer Circuit 1932 to operate theatres Hammond, Indiana; in 1940 organized own advertising agency, Sales, Inc., mgr.; mgr. Vogue Theatre, Hollywood, Calif., 1943. Joined National Screen Service 1943 as editor Mister Showman; Feb., 1945 named director of advtsng. and publicity; assumed east coast film prod. and TV, Sept., 1948 to May, 1954. Formed Mel Gold Productions, Inc., June 1954.

Member Publicity Club of N.Y.; Associated Motion Picture Advertisers; founded National Television Film Council in 1948, elected Honorary Lifetime President, 1955. Partner Melmon Productions; Melvin L. Gold Enterprises, 1958; pres., Mesal Prods., Inc., 1961; wrote s.p., Not For Love, 1961; pres., Associated Motion Picture Advertisers, 1963–64–65–66; gen. sls. mgr., National Screen Service, 1963. Pres., Melvin L. Gold Enterprises, Inc., 1966. Pres., Manhattan Sound Studios, 1967, v.p. in chg. m.p. div., National Showmanship Services, 1968; 1971, Pres. East Side Productions; Pres. Melvin L. Gold Enterprises. Motion picture consultant, Philip Morris, Inc., 1976–present. Operations director, Benson & Hedges 100, Film classics, 1976–79. Wrote s.p., The Sheriff is a Lady, 1979.

GOLDBERG, BERNARD: Executive. b. Bronx, N.Y., Aug. 25, 1932. e. Queens College (B.A.). Co-owner, Golden Theatre Mgt. Corp.; named vice pres. 1973.

GOLDBERG, FRED: Publicist. b. N.Y.C., Aug. 26, 1921. e. Pace College, School of Marketing and Advertising. expl., Paramount, 1946; asst. expl. mgr., trade paper contact, syndicate contact, N.Y. newspaper contact promotion mgr., 1946–52; ass't publ. mgr., RKO, 1952; national publ. mgr., IFE, 1953; v.p. Norton and Condon, pub., 1953; returned to IFE Sept., 1954, as nat'l pub. mgr.; head of N.Y. office, Arthur Jacobs, then Blowitz-Maskel, 1956; exec. asst. to dir. pub., adv., United Artists Corp., 1958; exec. dir., adv., pub., exploitation, United Artists Corp., 1961; named vice pres., 1962, senior vice president, 1972, Senior Vice President, Director of Marketing, 1977. Left in 1978 to be consultant with Piener, Hauser & Bates Agency. In 1979 joined Columbia Pictures as snr. v.p. in chg. adv./pub. Left in 1981 to form new company.

GOLDBERG, LEON: Executive. b. Providence, R.I., Aug. 29, 1900; e. A.B., Brown U., 1921; M.B.A. Harvard Graduate School of Business Administration, 1923. Assoc. with Goldman, Sachs, investment bankers, from September, 1923 to April, 1934. In April, 1934 became treasurer of Keith-Albee-Orpheum Corp. In December, 1941, became treasurer of RKO and RKO Radio Pictures, Inc. Resigned October, 1943, to become studio mgr., RKO Studios, Hollywood; elected vice-pres., 1946; elected fin'l. v.p., treas., Universal-International 1950; res. Oct., 1953; to United Artists as financial v.p., treas., November 1953–67, dir., National Screen Service Corp. 1969.

GOLDBERG, LEONARD: Executive. b. Brooklyn. e. Wharton School, Univ. of PA. Began career in ABC-TV research dept.; moved to NBC-TV research div.; 1961 joined Batten. Barton, Durstine & Osborn ad agency in chg. of daytime TV shows and overall bdcst. coordinator. In 1963 rejoined ABC-TV as mgr. of program devel. In 1966 named VP in chg of network TV programming. Resigned in 1969 to join Screen Gems as VP in chg. of prod. Left for partnership with Aaron Spelling in Spelling/Goldberg Prods. Goldberg is now producing TV and theatrical films under own banner, Mandy Prods. 1981: co-prod., All Night Long.

GOLDEN, HERBERT L.: b. Philadelphia, Feb. 12. e. Temple U., 1936, B.S. Reporter, rewrite man. asst. city ed., Philadelphia Record, 1933–38; joined Variety, 1938; on leave of absence, 1942–43, when asst. to John Hay Whitney and Francis Alstock, directors, M.P. Division, Coordinator of Inter-American Affairs (U.S.); commissioned in U.S. Navy, 1943, served on destroyer to 1946; then returned to Variety. M.p. ed. Consultant on motion pictures, Good Housekeeping magazine, McGraw-Hill Publications, American Yearbook. Ent. Ind. Div. Bankers Trust Co., N.Y.C., 1952; named v.p. 1954–56; treas., Children's Asthma Research Institute, 1956; v.p. & mem. of bd. United Artists Corp., 1958; member of board, MPAA, 1959; pres., Lexington Int., Inc. investments, 1962; mem. bd., chmn. exec. com., Perfect Photo Inc., 1962; 1965 sect. & mem. bd. Century Broadcasting Group; chmn. G & G Thea. Corp.; pres. Diversifax Corp., 1966; consult. Pathe Lab, 1967; Mem. bd. Childhood Prod. Inc., 1967. Member bd. Music Makers Group, Inc., 1962. Mem. bd. Cinecom Corp., 1968; pres., Vere/Swiss Corp., 1977; mem. bd., Coral Reef Publications, Inc., 1977. Returned to Bankers Trust, 1979, to head its Media Group (service to film and TV industries).

GOLDEN, JEROME B.: Executive. b. New York, N.Y., Nov. 26, 1917. e. St. Lawrence Univ., LL.B., 1942. Member legal dept., Paramount Pictures, Inc., 1942–50; United Paramount Theatres, Inc., 1950–53; American Broadcasting Companies, Inc., 1953; secy., ABC, 1958; vice-pres., ABC, 1959.

GOLDEN, NATHAN D.: Director, Scientific, M.P. Photographic Products Division, Business-Defense Services Adm., Dept. of Commerce, Washington. b. Bellaire, O., July 4, 1895. e. Emerson Inst. (Washington, D.C.); Columbus U. & Washington Coll. of Law, LL.B., J.D. American University. Wounded during World War I, serving in the U.S. Infantry. In 1926 entered Bureau of Foreign & Domestic Commerce as asst. chief; chief of section, 1933; chief, m.p. div., 1937. Has written numerous articles on world m.p. market developments. Motion Picture Consultant to the Golden Gate International Exposition, 1939. Awarded first Gold Medal for outstanding contribution field of m.p. projection, by Projection Advisory Council, 1930. Fellow, Soc. of M.P. Engineers; Chief Barker, 1947, Variety Club of Washington, D.C.; Int'l Heart Chmn., Variety Clubs International. Headed mission to Europe for the Joint Chiefs of Staff 1945 to obtain technical data on Agfa Color Film for which rec. War Dept. Certificate of Appreciation. Dept. of Commerce Silver Medal Award for meritorious service for the m.p. ind. Feb., 1954; Chevalier French Legion of Honor, 1956. Awarded Order of Merit, First Class, from Federal Republic of Germany, Oct. 2, 1961; presented with Medaille de Vermeil by Society for the Encouragement of Progress, Paris, France, Nov. 19, 1961; Decorated by French Photographic Industry with their "Gold Medallion" at Photokima in September 1976; retired from Dept. of Commerce, 1963; practicing Law, Washington, D.C.

GOLDENSON, LEONARD H.: Executive. b. Scottsdale, Pennsylvania, December 7, 1905; e. Harvard College, B.A., Harvard Law School, LL.B. Practiced law, New York; counsel in reorg. Paramount theats. in New England, 1933–37; in 1937 apptd. asst. to vice-pres. Paramount in charge theat. operations; became head of theat. operations, 1938; elected pres. Paramount Theat. Service Corp., vice-pres. Paramount Pictures, Inc., 1938; dir. Paramount Pictures, 1942 (also pres. various Paramount theat. subsids., see Paramount circuits in Theatre Circuits section). Pres., chief exec. off. and director United Paramount Theatres, Inc., 1950, and of American Broadcasting-Paramount Theatres, Inc., 1953, result of merger of ABC and United Paramount Theatres, Inc.; name changed to American Broadcasting Companies, Inc. 1965; Chairman of the Board and Chief Executive Officer; of American Broadcasting Companies, Inc. since January 17, 1972; mem., board chm. of United Cerebral Palsy Assns.; trustee, John F. Kennedy Center for the Performing Arts; dir., Daughters of Jacob Geriatric Center; mem., International Radio and Television Society; Founder Member of Hollywood Museum; Trustee of Children's Cancer Research Foundation of the Children's Medical Center, Boston, Mass.; Director of Allied Stores Corporation; Trustee of Highway Users Federation for Safety and Mobility; Member of National Academy of Television Arts and Sciences; Member of Uptown Advisory Committee of Bankers Trust Company; Graduate Director of The Advertising Council, Inc.; Associate Trustee and Member of Advisory Council for the Performing Arts of University of Pennsylvania; Member of Broadcast Pioneers; Member of Inter Lochen Arts Academy-National Advisory Board; Member of Motion Picture Pioneers; Member of National Citizens' Advisory Committee on Vocational Rehabilitation; Member of United Negro College Fund-National Corporations Committee; Director of World Rehabilitation Fund, Inc.

GOLDFARB, HOWARD: Executive. b. New York City, Sept. 19, 1941. e. Wharton School of Business. Joined Columbia Pictures as exec. trainee; 1964–69 mgn. dir., U.A., central America; 1969–70, supervisor of sales, Buena Vista Intl.; 1972–75, pres. H.C. Entertainment Ltd.; 1975–76, dir. of dist., CIC, London; v.p., intl. operations, NTA; 1978–79, v.p., intl. sls., Dino de Laurentis pres., Goldfarb Distributors, since 1979.

GOLDIN, BERNARD: Executive. Began career with Universal Pictures as branch mgr. in number of territories. Named Philadelphia mgr. for United Artists; 1977–81, Midwestern div. mgr. for Columbia Pictures. Returned to UA, 1981, as v.p. & asst. gen. sls. mgr.

GOLDING, DAVID: Executive. b. New York City, U. of Wisc., B.A. Now operating public relations consultancy in London, David Golding and Associates Ltd. Formerly Universal advertising and publicity representative in UK. During war, mg. ed. Mediterranean edition, The Stars and Stripes. Also pub. dir. Samuel Goldwyn, 20th-Fox, Hecht-Hill-Lancaster, Otto Preminger.

GOLDMAN, EDMUND: Executive-Producer. b. Shanghai, China, Nov. 12, 1906. e. in Shanghai and San Francisco. Entered ind. as asst. mgr., for Universal in Shanghai, 1935–36; named mgr. Columbia Pictures' Philippine office, 1937. In 1951 named Far East. supvr. for Columbia, headquartering in Tokyo. From 1953 to present indep. m.p. dist., specializing in foreign marketing, representing indep. producers and distributors. Now chm. of bd. for Manson International.

PICTURES: Surrender Hell (prod.), The Quick and the Dead (exec. prod.).

GOLDMAN, LES: Executive. b. New York, N.Y., July 2, 1913. e. N.Y. Univ., Magna Cum Laude, 1933. Prod. supv., MGM, Tom & Jerry cartoons; prod. supv., animated commercials, industrials, theatrical & TV films; produced educational & public affairs films, Information Films, Inc. Commercial prod., Tempo, Academy and Transfilm, Inc.; prod., combat & staff reports, Army Pictorial Service; org. teacher-training & instruction curriculum, APS School, London; cinema dept., USC est. animation curriculum, 1960 & taught courses first yrs., USC prog.; officer, Storyboard, Inc., 1955–56; org., Quartet Films, Inc., 1956–63; prod.: The Hangman; prod. mgr., SIB-TOWER 12, Animation Studio Hollywood; hd., prod., MGM Animation-Visual Arts Div.: Tom and Jerry Cartoons, animated short subjects; co-prod., The Dot and the Line, MGM; Prod. The Phantom Tollbooth, MGM: Prod. exec. Dr. Seuss' How the Grinch Stole Christmas; Off to See the Wizard; exec. prod.: FILM/SENSE, live action and animation film for industry, TV, and special applications; prod. Breakthrough for IBM Corp. Pres. & exec. prod. Film Sense, Inc. (a Media Creations, Ltd. Co.), prod. Mr. and Mrs. for IBM Corp.; prod. Apres le Silence. prod. On Your Mark, dir. of prod. Bonanza Films: The Cosmic Bicycle, Director/Producer Cinetics: The Curious Fish, The Adventures of Menachem/Mendl, The Perfectly Fair, The Romantic Revolutionaries. President: International Tournee of Animation. In production: Beyond the Milky Way.

GOLDMAN, MARVIN: Executive. b. New York City. e. Fordham & Univ. of Miami. Served in U.S. Navy in WW II, leaving with rank of Lt. commander. In 1947 went to Washington, D.C. to work for K-B Theatres; with present partner, Fred Burka, purchased the chain of five theatres. Circuit now operates 28 theatres in the D.C., Maryland, Virginia area. (Partnership now includes Goldman's son, Ronald, and Burka's son, David.) Goldman also involved with son in indep. film prod. co., which has completed 6 features and planning more. Active in theatre organizations for many years; named pres. of National NATO in 1976. Former pres. of Metropolitan, Washington, D.C. area chapter of NATO and now bd. chmn. Past chief barker of Variety Club of Washington. Named exhibitor of year by IFIDA in 1966.

GOLDMAN, MICHAEL F.: Executive. b. Manila, Philippines, Sept. 28, 1939. e. UCLA (B.S. in acct., 1962). C.P.A. certificate issued June, 1972. Owner and sole proprietor, Taurus Film Co. of Hollywood. In 1962 incorporated Manson Distributing Corp. as specialist in exporting indep. American feature films. Presently owner and pres. of Manson International, succeeding company to Manson Distributing Corp. Co-founder and chief financial officer of American Film Marketing Association. Sponsor of First American Film Market in Los Angeles 1981. Produced Jessi's Girls, 1975.

GOLDMAN, WILLIAM: Writer. b. Chicago, 1931. e. Oberlin College (B.A.), Columbia Univ. (M.A.). Novels include The Temple of Gold, Soldier in the Rain (filmed), No Way to Treat a Lady (filmed), Father's Day, The Princess Bride, Marathon Man (filmed), Magic (filmed), Tinsel, etc.

PICTURES INCLUDE: Harper, Butch Cassidy and the Sundance Kid, The Great Waldo Pepper, Marathon Man (based on own book), All the President's Men, A Bridge Too Far, Magic (based on own book).

GOLDSMITH, JERRY: Composer. b. Los Angeles, 1930. e. L.A. City College. Studied piano with Jacob Gimpel and music composition, harmony, theory with Mario Casteinuovo Tedesco. Taught music; went with CBS radio first with own show (Romance) and then moved on to others (Suspense). Began scoring for TV, including Climax, Playhouse 90, Studio One, Gunsmoke, etc.

PICTURES INCLUDE: Black Patch (debut), Freud, Lilies of the Field, The Stripper, The Prize, Seven Days in May, In Harm's Way, Von Ryan's Express, A Patch of Blue, Our Man Flint, Stagecoach, The Blue Max, Seconds, Sand Pebbles, In Like Flint, Patton, Papillon, The Reincarnation of Peter Proud, The Cassandra Crossing, Islands in the Stream, MacArthur, Coma, Damien-Omen II, The Boys from Brazil, The Great Train Robbery, Alien, Players, Star Trek, The Final Conflict, Outland.

GOLDSMITH, MARTIN M.: Writer. b. New York, N.Y., Nov. 6, 1913. Bush pilot.

AUTHOR: Novels include: Double Jeopardy, Detour, Shadows at Noon, Miraculous Fish of Domingo Gonzales.

PICTURES INCLUDE: Detour, Blind Spot, Narrow Margin, Mission Over Korea, Overland Pacific, Hell's Island.

GOLDSTEIN, DAVID M.: Producer's rep. b. Cairo, Egypt, 1973. Ent. industry 1958 with Eros Films Ltd. Then to Sweden as owner/director Swedish International Motion Pictures. Opened up Savant Films, 1965, conducting all import and export from London specializing Scandinavian, Benelux Distribution, 1969; vice pres. in charge world sales, Counselor Films Ltd.

GOLDSTEIN, MILTON: Executive. b. N.Y.C., Aug. 1, 1926. e. N.Y.U., 1949. In exec. capac., Paramount; foreign sales coord., "The Ten Commandments," "Psycho"; v.p. foreign sales, Samuel Bronston org.; asst. to Pres., Paramount Int'l., special prods., 1964; Foreign sales mgr., 1966; v.p., world wide sales, 1967, Cinerama; Sr. v.p. Cinema Center Films, 1969; pres., Cinema Center Films, 1971; v.p. Theatrical Mktg. & Sales, Metromedia Producers Corp., 1973; in March, 1974, formed Boasberg-Goldstein, Inc., consultants in prod. and dist. of m.p.; 1975, named exec. vice pres., Avco Embassy Pictures; 1978, named exec. v.p. & chief operating officer, Melvin Simon Prods. 1980, named pres.

GOLDSTONE, JAMES: Director. b. Los Angeles, Calif. June 8, 1931. e. Dartmouth College, BA, Bennington College, MA. Film Editor from 1950. Writer, story editor from 1957. Director, TV, from 1958, including pilots of Star Trek, Ironside, The Senator, etc., A Clear and Present Danger, (Emmy nomination); Eric, Virgin Islands Int'l. Film Festival Gold Medal, Journey From Darkness, Christopher Award. Feature director since 1967, Studs Lonigan, 6 hr. miniseries, 1978, Kent State, 1980.

PICTURES INCLUDE: Jigsaw, Man Called Gannon, Winning, Brother John, Red Sky at Morning, The Gang That Couldn't Shoot Straight, They Only Kill Their Masters, Swashbuckler, Rollercoaster, When Time Ran Out.

GOLDSTONE, RICHARD: Producer. b. New York City, July 14, 1912. e. University of Cal., Los Angeles, B.A., '33. Capt., Army Air Forces, office of Motion Picture Services, World War II. Adv. exec. Entered m.p. ind. 1934 as writer, MGM; in 1935 asst. head short subjects dept.; in 1939 assoc. prod. short subjects; apptd. prod. mgr., MGM short subjects dept., co-holder 3 Academy Awards short subjects, 1942. Prod. The Set-Up, (RKO), 1948, won Int'l Critics Grand Prix, Cannes, 1949. v.p. Dudley Pict. Corp.; v.p., Goldcoast Productions, Inc.

PICTURES INCLUDE: The Outriders, Yellow Cab Man, Dial 1119, Inside Straight, The Tall Target, Talk About a Stranger, Devil Makes Three, Terror on the Train, Tabor the Great, The Big Search, East of Kilimanjaro, South Seas Adventures; No Man Is an Island, Rage, The Sergeant, The Babymaker.

TV: Prod. Adventures in Paradise, Combat, Peyton Place, We Ask, Why Not? (doc.)

GOLDWURM, JEAN: Executive. b. Bucharest, Rumania, Feb. 21, 1893. e. Univ. of Vienna. Owner, Little Carnegie and World theatres, New York; pres. Times Film Corp. officer French Legion of Honor; commander Italian Order of Merit, Commander French Order of Arts and Letters.

GOLDWYN, SAMUEL, JR.: Producer director. b. Los Angeles, Calif., Sept. 7, 1926; e. U. of Va. U.S. Army, 1944; following

war writer, assoc. prod., J. Arthur Rank Org.; prod. Gathering Storm on London stage; returned to U.S., 1948; assoc. prod., Universal; recalled to Army service, 1951; prod., dir., Army documentary films including Alliance for Peace (Edinburgh Film Festival prize); prod. TV shows, Adventure series for CBS, 1952–53; prod. TV series, The Unexpected, 1954; pres., The Samuel Goldwyn Company.
PICTURES INCLUDE: Man With the Gun, The Sharkfighters, The Proud Rebel, The Adventures of Huckleberry Finn, The Young Lovers, Cotton Comes to Harlem, Come Back Charleston Blue.

GOLLINGS, FRANKLIN: Producer, writer, director. b. Llandudno, Wales. Ent. m.p. ind. 1936. Film prod. adviser to the Admiralty, 1945–8. Shorts pro. Rayant Pictures, 1948–52. Unit man. Mogambo. First asst. The Maggie. S.p. The Sitting Duck. Prod. man. Bhowani Junction. Exec. prod. Associated-Rediffusion. Writ. 2nd unit dir. assoc. prod. Yangtse Incident, 1960, Writ. Tempt Not the Stars, The Stirling Moss Story, 1961; Writ. Force 136 and The Stone; writer, dir., The Changing Face, 1963; dir., Woman; res., writer, The Story of Yang-Kwei-Fei; writer, prod., Jason and the Argonauts, 1964; writer, dir., Mystery Horse, A Price for Dorothy, The Big Design, 1965; Writer Dictionary of 1001 Famous People, Great Moments in Amer. History. S.P. dir. prod. Connecting rooms. Formed Four Cities Int. Corp., 1968; program of Internat'l co-prods., 1969. Research and Script High Risk (based on The Junkie Priest), 1970. Prod., dir. High Risk. Dir. The Todd Dossier.

GONZALEZ-GONZALEZ, PEDRO: Actor. b. Aguilares, Tex., Dec. 21, 1926. Comedian in San Antonio Mexican theatres.
PICTURES INCLUDE: Wing of the Hawk, Ring of Fear, Ricochet Romance, High and the Mighty, Strange Lady in Town, Bengazi, I Died a Thousand Times, Bottom of the Bottle, Gun the Man Down, Wetbacks, The Love Bug, Hellfighters, Support Your Local Gunfighter.
TV: O'Henry Stories, Felix, the Fourth, Hostile Guns.

GOOD, CHUCK: Executive. b. 1922. Joined Buena Vista in 1957 in Chicago office; progressed from salesman to branch manager and then district manager. Later moved to Burbank as domestic sales manager in 1975; 1978, named v.p. & general sales manager; 1980, appointed pres., BV Distribution Co.

GOODFRIED, ROBERT: Publicity executive. b. New York City, Apr. 8. Entered motion picture industry with Skouras Theats. Corp.; management & adv.; in 1943 joined United Artists as asst. exploitation dir. In July 1945 named pub. dir., PRC Studios, Hollywood. Then mem. pub. staff, Eagle-Lion Studios; adv., pub. rel. dir. Metropolitan Theatres, Los Angeles; rejoined UA field force, 1951; chge. of expl., prom., Paramount Studios, 1952; pub. mgr., 1958; pub. mgr. Columbia Studio, 1958; studio publicity div., Paramount Studio, 1963. Appt. v.p. in chg. of studio publicity, 1971; exec. consultant to West coast pub. dept., 1980.

GOODMAN, ABE I.: Advertising executive. b. New York City. Entered Motion Picture Industry with Universal Pictures, N.Y.; later in foreign dept. Engaged in theat. circuit exploitation then joined Fox Film Corp. as adv. dir.

GOODMAN, GENE: Executive. Joined United Artists as salesman in 1954; promoted to branch mgr., Atlanta (1958–61). Served successively as branch mgr., New Orleans (1961–68); southern regional mgr. (1968–70); and southern div. mgr., (1970–78), New Orleans. Moved to N.Y. in 1978 to become asst. gen. sls. mgr. Promoted to v.p. & gen. sls. mgr. in 1980. Later in year named snr. v.p. for domestic sls.

GOODMAN, JOHN B.: Art director. b. Denver, Colo., Aug. 15, 1901. Studied architecture. Joined Famous Players Lasky 1920–27; later Pathe, RKO Radio, Warner; Paramount, 1932–39; Frank Lloyd Prods., 1940–41; Universal as superv. art dir. Academy Award (with Alexander Golitzen), art direction, color picture, 1943 (Phantom of the Opera, Universal), Rainbow Prod., Nuys Theatre Corp., Arfan Prods., 1949–50; Nat Holt Pictures, 1950; Paramount 1953–58; MGM, 1959–62; Republic Prods., 1962; CBS Studio Center, 1963–66; M-G-M, 1967. Semi-retired, 1968. Retired, 1969.

GOODMAN, JULIAN: Executive. b. Glasgow, Ky., May 1, 1922. e. B.A. Western Kentucky Univ; B.A. Geo. Washington Univ; Office mgr., Comb. Prod. & Resources, 1943–1945. History at NBC: news writer, WRC, NBC owned station in Washington, D.C., 1945; Washington ed., News of the World, NBC Radio network; mgr. of news and special events, NBC-TV, 1951; dir. of news and public affairs, NBC News Div., 1959; v.p., NBC News, 1961; exec. v.p., 1965; sr. exec. v.p., Operations, Jan. 1, 1966; pres., NBC, April 1, 1966. Mem. bd. of dirs., NBC, 1966; chief executive officer, Jan. 1, 1970; elected director of RCA Corporation, Jan. 1, 1972; chmn., chief exec. off., NBC, April, 1, 1974; chmn., Jan. 5, 1977.

GOODMAN, MORT: Advertising-public relations executive. b. Cleveland, O., Oct. 17, 1910. e. Western Reserve U. With sports dept. Cleveland News 1928–9; joined pub. dept. Warner Bros. Theatres (Ohio), 1930; transferred to Hollywood as pub. dir. for Pacific Coast zone, 1937; pub. dir. Republic Studios, 1946; res. 1952, apptd. v.-p. Stodel Adv. Agcy., formed Mort Goodman Adv., 1953; formed Goodman Org., advertising agency, pub. rel., 1955. Currently pres., Goodman Nemoy and Partners, adv.-p.r. firm started 1977. In 1980 became adv. & p.r. consultant for m.p. industry.

GOODSON, MARK: TV producer; b. Sacramento, Jan. 24, 1915; s. Abraham Ellis and Fannie (Gross) G.; A.B., U. Cal., 1937. Announcer, newscaster, dir. Radio Sta. KFRC, San Francisco, 1938–41; radio announcer, dir., N.Y.C. 1941–43; producer—Appointment with Life, ABC, 1943; dir. Portia Faces Life, Young & Rubicam, advt. agy., 1944; radio dir. U.S. Treasury War Bond Drive, 1944–45; formed Goodson-Todman Prodns., 1946, originated radio shows Winner Take All, 1946, Stop the Music, 1947, Hit the Jackpot, 1947–49. Creator of TV game programs What's My Line, It's News To Me, The Name's the Same, I've Got a Secret, Two For The Money, To Tell the Truth, The Price Is Right, Password, Match Game, Family Feud, and others; TV film series, The Web, The Rebel, Richard Boone Theater, Branded. 1st v.p. Capitol City Pub. Co.; v.p. New Eng. Newspapers, Inc.; dir. City Center Music and Drama, Mem. American Film Institute. Recipient nat. television award Great Britain, 1951; Emmy award Acad. TV Arts and Scis., 1951, 52; Sylvania award. Pres. N.Y. Acad. TV Arts and Sci. 1957–58. Phi Beta Kappa.

GOODWIN, BERNARD: Executive and lawyer. b. New York, N.Y., Dec. 19, 1907. e. N.Y.U., Harvard Law School; with Paramount Pictures Corporation 1934–1957, vice president, general manager, director, Paramount music publishing subsidiaries, Famous Music Corporation and Paramount Music Corporation, executive producer Paramount short subjects, vice-president, director Paramount Television Productions, Inc.; secretary, director Allen B. DuMont Laboratories, Inc., 1938–1955; president, director Metropolitan Broadcasting Corporation, 1955–1959; director, ASCAP, since 1951.

GOODWIN, RICHARD: Producer. b. Bombay, India, Sept. 13, 1934. e. Rugby. Entered film world by chance: while waiting to go to Cambridge Univ. took temporary job as tea boy at studio which led to 20-year-long association with producer Lord Brabourne. First worked as asst. on Seven Thunders, of which Brabourne was assoc. prod. Named location manager for Brabourne's company, Mershal Productions; promoted to prod. mgr. Named assoc. prod. for Brabourne's British Home Entertainment Productions in 1960 (now defunct).
PICTURES: Prod. Mgr.: The Sheriff of Fractured Jaw, Carve Her Name with Pride, The Grass Is Greener, Sink the Bismarck, HMS Defiant. Prod.: The Tales of Beatrix Potter. Co-Prod.: Murder on the Orient Express, Death on the Nile, The Mirror Crack'd.

GOODWIN, RONALD: Composer, arranger, conductor. b. Plymouth, England. Early career: arranger for BBC dance orchestra; mus. dir., Parlophone Records; orchestra leader for radio, TV and records. Fut. m.p. ind., 1958. Many major film scores. Guest cond. R.P.O., B.S.O., Toronto Symph. Orch. New Zealand Symphony Orch., Sydney Symphony Orch.
PICTURES INCLUDE: The Trials of Oscar Wilde, 633 Squadron, Those Magnificent Men in Their Flying Machines, The Trap, Where Eagles Dare, Monte Carlo or Bust, Battle of Britain, The Executioner, The Selfish Giant, Frenzy (1972), Diamonds on Wheels, The Little Mermaid, The Happy Prince, One of Our Dinosaurs Is Missing, Escape From the Dark, Born to Run, Beauty and the Beast, Candleshoe, Force Ten from Navarone, Spaceman and King Arthur.

GORDON, ALEX: Producer. b. London, Eng., Sept. 8, 1922; ed. Canford Coll., Dorset, 1939. Writer, m.p. fan magazines, 1939–41; British Army, 1942–45; pub. dir. Renown Pictures Corp., 1946–47; P.R. and pub. rep. for Gene Autry, 1948–1953; v.p. and prod. Golden State Productions, 1954–58; prod. Alex Gordon Prods., 1958–66; producer Twentieth Century-Fox Television.
PICTURES INCLUDE: Lawless Rider, Bride of the Monster, Apache Woman, Day the World Ended, Oklahoma Woman, Girls in Prison, The She-Creature, Runaway Daughters, Shake Rattle and Rock, Flesh and the Spur, Voodoo Woman, Dragstrip Girl, Motorcycle Gang, Jet Attack, Submarine Seahawk, Atomic Submarine, The Underwater City, The Bounty Killer, Requiem for a Gunfighter.
TV: Movie of the Year, Golden Century, Great Moments in Motion Pictures.

GORDON, BERT I: Producer-director-writer. b. Kenosha, Wis., U. of Wisconsin, grad.
PICTURES INCLUDE: Beginning of the End, The Amaz-

ing Colossal Man, the Fantastic Puppet People, The Colossal Beast, The Cyclops, The Spider, Tormented, Boy and The Pirates, The Magic Sword, Village of the Giants, Picture Mommy Dead, How to Succeed With the Opposite Sex, Necromancy, Geronimo, The Mad Bomber, The Police Connection, The Food of the Gods, Empire of the Ants, The Coming.

GORDON, GALE: Actor. r.n. Charles T. Aldrich, Jr. b. New York City, Feb. 2, 1906. Stage debut in the Dancers; m.p. debut in The Pilgrimage Play, 1929.
PICTURES INCLUDE: Rally Round the Flag Boys, All in a Night's Work, Don't Give Up the Ship, Visit to a Small Planet, All Hands on Deck, Speedway.
TV: My Favorite Husband, Our Miss Brooks, The Brothers, Dennis The Menace, The Lucy Show, Here's Lucy.

GORDON, JACK: Executive. b. Brooklyn, N.Y., Mar. 13, 1929. e. UCLA. Father is Mack Gordon, songwriter. Started in industry as TV production asst. in 1949. Served in U.S. Infantry, Korea, 1951–52. Joined MGM in 1953; named dir. of non-threatrical div., 1956; v.p., MGM Intl., 1972. Appointed exec. v.p., MGM, Intl. 1979.
Member: ASCAP, Motion Picture Pioneers, Academy of Motion Picture Arts & Sciences.

GORDON, LAWRENCE: Producer. b. Yazoo, Miss., March 25, 1936. e. Tulane Univ. First show business venture with TV producer Aaron Spelling on Burke's Law series, 1964. Joined ABC-TV as head of west coast talent development 1965, left to become v.p. of Screen Gems (now Columbia Pictures TV) where worked on developing new projects. Joined American Intl. Pictures as v.p. in chg. of worldwide production. Resigned to form his own company, Lawrence Gordon Productions, Inc.
PICTURES: Hard Times Rolling Thunder, The Driver, The End, Hooper, Warriors, Xanadu, Paternity.

GORDON, MICHAEL: Director. b. Balimore, Md., Sept., 1909; e. Johns Hopkins Univ., B.A.; Yale, M.F.A. Stage experience as technician, stage mgr., actor, stage dir. (Stevedore, Golden Boy, Home of the Brave, Anna Christie, One Bright Day); m.p. debut: dial. dir. Columbia on over 20 films; dir. in 1942. Professor of Theater Arts, UCLA.
PICTURES INCLUDE: Boston Blackie Goes Hollywood, Underground Agent, One Dangerous Night, The Crime Doctor, The Web, Another Part of the Forest, An Act of Murder, The Lady Gambles, Woman in Hiding, Cyrano de Bergerac, I Can Get It for You Wholesale, The Secret of Convict Lake, Any Way the Wind Blows, Pillow Talk, Portrait in Black, Boy's Night Out, For Love or Money, Move Over Darling, Texas Across the River, A Very Special Favor, The Impossible Years, How Do I Love Thee?
PLAYS: Deadfall, The Lovers, Tender Trap, Male Animal, His & Hers, Champagne Complex, Home of the Brave, etc.

GORDON, RICHARD: Producer. b. London, Eng., Dec. 31, 1925; e. U. of London, 1943. Served in Brit. Royal Navy, 1944–46; ed. & writer on fan magazines & repr. independent American cos. 1946, with publicity dept. Assoc. Brit. Pathe 1947; org. export-import business for independent, British and American product; formed Gordon Films, Inc., 1949; formed Amalgamated prod., 1956; formed Grenadier Films, Ltd. 1971.
PICTURES INCLUDE: The Counterfeit Plan, The Haunted Strangler, Fiend Without a Face, The Secret Man, First Man into Space, Corridors of Blood, Devil Doll, Curse of Simba, The Projected Man, Naked Evil, Island of Terror; Tales of the Bizarre, Tower of Evil, Horror Hospital, The Cat and the Canary, Inseminoid.

GORDON, RUTH: Actress. b. Wollaston, Mass., Oct. 30, 1896; e. Amer. Acad. Dram. Art; m. Garson Kanin, producer, director. First stage appearance with Maude Adams in Peter Pan, 1915. Mem. Screen Actor's Guild, Academy of Motion Picture Arts and Sciences. Author of Book, Myself Among Others, 1971; My Side, 1976 (autobiography).
STAGE: The Doll's House, Serena Blandish, Ethan Frome, Here Today, Seventeen, Clarence, The First Year, Three Cornered Moon, The Violet, Hotel Universe, Saturday's Children, The Three Sisters, Over Twenty-One, The Leading Lady, The Matchmaker (London, Berlin), The Smile of the World, The Good Soup, A Time to Laugh (London), My Mother, My Father and Me, A Very Rich Woman (playwright and star), Fair and Warmer, The Loves of Cass McGuire, Dreyfus in Rehearsal, Mrs. Warren's Profession.
PICTURES INCLUDE: Abe Lincoln in Illinois, Two Faced Woman, Dr. Ehrlich's Magic Bullet, Action in the North Atlantic, Edge of Darkness, Inside Daisy-Clover, Lord Love a Duck, Whatever Happened to Aunt Alice?, Rosemary's Baby (Academy Award for Best Supporting Actress), Where's Poppa?, Harold and Maude, The Big Bus, Every Which Way But Loose, Boardwalk, Scavenger Hunt, My Bodyguard, Any Which Way You Can, Jimmy the Kid.
TV: The American Dream, Blithe Spirit, The Prince of Central Park.
AUTHOR: Plays: Over Twenty-One, Years Ago, The Leading Lady. Screenplays: A Double Life (with Garson Kanin), Adam's Rib, with Garson Kanin, The Marrying Kind, Pat and Mike Contbr. to Reader's Digest, Forum Magazine, Atlantic Monthly, McCalls, Ladies Home Journal.

GORDY, BERRY: Executive. Was working on auto assembly line in Detroit when decided to launch record co., Motown. In 1961 wrote song, Shop Around; recording by Smokey Robinson made it his first million dollar record. Expanded into music publishing, personal mgt., recording studios, film and TV, also backing stage shows. Now bd. chm., Motown Industries.
PICTURES INCLUDE: Lady Sings the Blues (prod.), Mahogany (dir.).

GORING, MARIUS: Actor. b. Newport, Isle of Wight. e. Cambridge & Universities of Frankfurt-on-Main, Munich, Vienna, Paris. Early career with Old Vic; stage debut 1927, Jean Sterling Rackinlay's Children's Matinees.
PICTURES INCLUDE: Rembrandt, UA-London Films. Dead Men Tell No Tales, Flying 55 Admiral Films, Consider Your Verdict, Spy in Black, Pastor Hall, The Case of the Fightened Lady Br. Lion. 1940–46 serving with H.M. Forces & Foreign Office. In 1942, The Big Blockade, The Night Raider. In 1944, Lilli Marlene. Others: Stairway to Heaven, Night Boat to Dublin, Take My Life, Red Shoes, Mr. Perrin and Mr. Traill, Odette, Pandora and the Flying Dutchman, Circle of Danger, Highly Dangerous, So Little Time, The Man Who Watched Trains Go By, Rough Shoot, The Barefoot Contessa, Break in the Circle, Quentin Durward, Ill Met By Moonlight, The Moonraker, Family Doctor, Angry Hills, Whirlpool, Treasure of St. Teresa, Monty's Double, Beyond the Curtain, Desert Mice, The Inspector, Girl On a Motorcycle, Subterfuge.
TV: Numerous appearances, Sleeping Dog, Man In a Suitcase, Scarlet Pimpernel, The Expert.

GORMAN, KENNETH F.: TV Executive. Career in communications started at NBC on corporate staff. Joined NBC radio division then NBC Enterprises, program syndication arm. Joined CBS radio division, 1967; later CBS Broadcast Group. Career with Viacom International began at spin-off in 1971; named controller, 1974; v.p., finance, 1978; snr. v.p., January, 1979. In April 1979 named pres. of Viacom Entertainment Group, consisting of Viacom Enterprises and Showtime.

GOROG, LASZLO: Writer. b. Hungary, Sept. 30, 1903. e. U. of Sciences, Budapest. Playwright, short story writer, asst. editor, Budapest, 1928–39.
PICTURES INCLUDE: Tales of Manhattan, The Affairs of Susan, She Wouldn't Say Yes, The Land Unknown, for U-I.
TV: 4 Star, Dupont, The Roaring Twenties, 77 Sunset Strip, Maverick, etc.

GORTNER, MARJOE: Actor—Producer. Was child evangelist, whose career as such was basis for documentary film, Marjoe (AA). Acted in films and TV; turned producer in 1978 for When You Comin' Back, Red Ryder.
PICTURES: Bobbie Joe and the Outlaw, The Food of the Gods, Viva Knievel, Sidewinder One, Earthquake, When You Comin' Back, Red Ryder.
TELEVISON: Films: The Marcus-Nelson Murders, The Gun and the Pulpit, Pray for the Wildcats. Guest appearances on Police Story, Maycay: 40,900 Feet, etc.

GOTTLIEB, ALEX: Producer, writer. b. Dec. 21, 1906; e. U. of Wisconsin. Pub. Dir. Paramount theatre, New York; adv. mgr. United Artists, Columbia 1930–37; prod., W.B., Universal, RKO, 1941–50; then ind. prod. Member, ASCAP, WGA, Dramatists Guild, Academy of Motion Picture Arts and Sciences, Sigma Delta Chi, Pi Lambda Phi.
TV: writer, prod., Dear Phoebe, The Gale Storm Show, The Tab Hunter Show, Bob Hope Chrysler Theatre, Donna Reed Show, The Smothers Brothers Show, ABC Movie of the Week.
PICTURES INCLUDE: I'll Take Sweden, Frankie and Johnny, Arizona Ranger, The Pigeon, Blue Gardenia, Macao, Susan Slept Here, Hollywood Canteen.
AUTHOR OF B'WAY PLAYS: Wake Up Darling, Separate Rooms, Susan Slept Here, Stud, Your Place or Mine?, Divorce Me, Darling; Marcus, Farkus and O'Brien.
ONE-ACT PLAYS: Published in Best Short Plays of 1969, Best Short Plays of 1976.

GOTTLIEB, MARTIN Executive. b. New York, April 16, 1903. Entered m.p. ind. 1919 with Louis Meyer, m.p. Title Studio, New York. Nine years with Consolidated Film Industry (Craftmen Lab.). Chief cameraman and business manager of 19th St. plant. Leave of absence, 1921 and 1925. Photog. m. p. in Mexico and Hollywood. Helped organize Meyer-Rieger Lab. Became secretary 1930. In 1932 organized Welgot Trailer Service. Partnership with C.L. Welsh, liquidated, 1959; C&G Film Effects, Inc. 1959–65, partner, secy., treas. with Hugo A. Casolaro; secy., treas., dir., Colodzin Productions Inc., 1965; pres., Celebrity In'l Films; v.p. Zodiak International Prod. Inc. of Canada; Pres. Wellmax Film Corp.; Pres. Celebrity Holdings, Inc. Semi-retired since 1965. Associated with Globe Picture Corp., 1968 to present.

GOTTLIEB, STAN: Actor, Film Distributor. b. N.Y.C. April 22, 1917. e. N.Y.U., Col. Univ. Has acted in films, theatrical, stage and TV since 1970. Gen. mgr. for Impact Films, Inc. 1966–72. Gen. Mgr., Mammoth Films (1973–76).
PICTURES INCLUDE: Actor: Putney Swope, You Gotta Walk It Like You Talk It, The Owl and the Pussycat, Pound, The Anderson Tapes, Cold Turkey, Slaughterhouse Five, Black Fantasy, Greaser's Palace, Compliments to the World.
TV: Actor: Sticks & Bones, Hot L Baltimore, The Jeffersons.
STAGE: Steambath.

GOTTSCHALK, ROBERT: Executive. b. Chicago, Ill., March 12, 1918. e. Carleton College. B.A., 1939. Short subject prod., 1940–49; org. R.E.G., Inc., camera equipment, 1949; pres., 1949–53; pres., Panavision, Inc., 1953.

GOULD, ELLIOTT: Actor. b. Brooklyn, N.Y., August 29, 1938. Broadway debut in Rumple.
STAGE: Say Darling, Irma La Douce, I Can Get It For You Wholesale, On the Town (London), Fantasticks (tour), Drat the Cat, Little Murders, Luv (tour), A Way of Life.
TV: Once Upon A Mattress. Formed, with prod. Jack Brodsky, independent m.p. co.
PICTURES INCLUDE: The Night They Raided Minsky's, Bob & Carol & Ted & Alice, M*A*S*H*, Getting Straight, Move, Little Murders, The Touch, I Love My Wife, The Long Goodbye, Busting, S*P*Y*S!*, California Split, Who?, Nashville, (guest), Whiffs, I Will . . . I Will . . . For Now, Harry and Walter Go to New York, A Bridge Too Far, Capricorn One, Matilda, Escape to Athena, The Last Flight of Noah's Ark, The Devil and Max Devlin, Dirty Tricks.

GOULD, JAMES F.: Exec. Radio City Music Hall. b. Plymouth, Eng., Sept. 30, 1908; e. City of London Coll., C.P.A., N.Y.; asst. cashier, Dakin Bros. Ltd., London; came to USA, joined Radio Keith Orpheum Corp., acct. dept., 1930; joined opening staff, Radio City Music Hall, Nov. 1932; asst. treas., 1942; treas., 1948; vice-pres., 1957; exec. v.p., 1964; pres., 1966, member Board of Directors Rockefeller Center Inc. Retired Feb., 1973.

GOULD, JON J.: Executive. e. New England College, Harvard Graduate Program in Publishing. In 1977 joined Straight Arrow Publishers as east coast sales mgr. for publications, Rolling Stone and Outside. In 1978 left for post at Paramount Pictures in m.p. div. as dir. of mktg. admin. for mktg. group. 1979, appt. exec. asst. to Frank G. Mancuso, then exec. v.p., dist. & mktg. In 1980 named v.p., corporate communications, for Paramount.

GOULD, JOSEPH: Executive. b. New York, N.Y. Jan. 30, 1915. e. B.A., New York U., 1935; M.S., Pulitzer School of Journ., Columbia U., 1936. Adv. dept., United Artists, 1939–46; Universal, 1947–48; asst. adv. mgr. 20th Century-Fox. 1949–52; adv. consultant, Joseph Burstyn. Louis De Rochemont Assoc., I.F.E. Releasing Corp. 1953–55; adv. mgr. United Artists, 1955–60; adv. mgr. Paramount Pictures, 1960–62; David Singer Associates, 1962–64; partner, Konheim Gould & Ackerman, 1968–1974; principal, Joseph Gould Associates, 1974–1977; creative services director, WIXT-TV, Syracuse, 1978.

GOULET, ROBERT: Singer, actor. b. Lawrence, Mass., Nov. 26, 1933. e. attended school, Edmonton; scholarship, Royal Conservatory of Music. Sang in choirs, appeared with numerous orchestras; disk jockey, CKUA, Edmonton; pub. rel., The Merrick Co.
STAGE: Camelot, The Happy Time.
TV: The Ed Sullivan Show, Garry Moore, The Enchanted Nutcracker, Omnibus, The Broadway of Lerner and Loewe, Rainbow of Stars, Judy Garland Show, Bob Hope Show, The Bell Telephone Hour; England: Granada—TV special, Jack Benny; Dean Martin; Andy Williams; Jack Paar; Red Skelton; Hollywood Palace; Patty Duke Show, Star of Robert Goulet Show, Blue Light series for ABC-TV, Brigadoon & Carousel for ABC-TV. Kiss Me Kate.
PICTURES INCLUDE: Honeymoon Hotel, I'd Rather Be Rich, I Deal in Danger, Underground.
RECORDS: Always You, Two of Us, Sincerely Yours, The Wonderful World of Love, Robert Goulet in Person, This Christmas I Spend With You, Manhattan Tower, Without You, My Love Forgive Me, Travalling On, Robert Goulet on Tour, Robert Goulet on Broadway, Robert Goulet on Broadway II, Camelot, Happy Time.

GOWDY, CURT: Sportscaster. b. Green River, Wyoming, 1919. Basketball star at U. of Wyoming. All-Conference member; graduated U. of Wyoming. 1942; officer in U.S. Air Force WWII, then became sportscaster. Voted Sportscaster of the Year, 1967, Nat'l Assn. of Sportwriters Broadcasters. Best Sportscaster, FAME, 1967. Sports staff at NBC; American Sportsman, ABC. Won Peabody Award, 1975. Host of The Way It Was on PSS.

GRADE, LORD: Chmn & Man. dir., Incorporated Television Co. Ltd. Chmn. Associated TeleVision Corporation Ltd. Created Life Peer 1976.

GRADUS, BEN: Producer-Director-Writer. b. N.Y.C. e. Brooklyn Coll. Is principal in Directors Group Motion Pictures, Inc. Positions held include v.p. Filmways, Inc.; prod.-dir. for Screen Gems. Specialist in educational and doc. films; also has done feature, children's comedy TV shows and series, Decision: The Conflicts of Harry S. Truman. Author, Directing the Television Commercial.
CREDITS INCLUDE: Dawn over Ecuador, Gentlemen of the Jury, To Save Your Life, Lifewatch Six, Ford Around the World, Art Heritage, A Girl from Puerto Rico, Span of Life, Crowded Paradise.

GRADY, BILLY, Jr.: Producer. b. Brooklyn, July 4, 1917. Entered m.p. ind. as asst. film ed., MGM, 1934; film ed. MGM, 1938; asst. to Arthur Freed, 1940; U.S. Air Force four years; then joined Frank Capra as asst.; thereafter to U-I as assoc. prod., Ma and Pa Kettle series; Arctic Manhunt, Hollywood Story; then to MGM prod., Cry of the Hunted, Code Two, Gypsy Colt; CBS-TV prod. That's My Boy.

GRAF, BILLY: Executive. b. 1945. Entered industry 1965. Was asst. dir.; unit/location mgr./production mgr. for American, British and European compies. Now v.p., Legion Films, Inc., Beverly Hills.
PICTURES: Alfie, A Man for All Seasons, Billion Dollar Brain, Scrooge, Run Wild Run Free, The Best House in London, Chitty Chitty Bang Bang, Galaxina, Khartoum, You Only Live Twice, Women in Love, Song of Norway, Prudence and the Pill, Hammerhead, Underground Aces.
TELEVISION: Secret Agent, The Saint, "Q" Branch, Private Eye Public Ear, The Avengers, Department "S", The Dave Cash Radio Show.

GRAF, WILLIAN N.: Executive. b. New York City. Entered industry in 1934. 1937–42, exec. secty./asst. to Mark Hellinger at Warner Bros. & 20th-Fox. 1942–45, writer of armed forces training films. 1946–50, exec. secty./asst. to Harry Cohn, pres., Columbia Pictures; 1951, prod. asst. to Jack Fier, prod. head, Columbia. 1952–65, Amer. repr. for British productions, Columbia Pictures; 1965–66, exec. asst. to J.J. Frankovich, Columbia Pictures, London; 1965–66, v.p., Columbia Pictures Intl., London; 1969–70, indep. prod., Cinema Center Films; 1980–present, v.p. in chg. of prod., Legion Films, Inc., Beverly Hills
PICTURES: (Producer): The Red Beret, A Man for All Seasons, Sinful Davey, The African Elephant.

GRAFF, RICHARD: Executive. b. Milwaukee, Wisc., Nov. 9, 1924; e. U. of Illinois; served U.S. Air Force, 1942–46; 1946 joined Universal Pictures in Chicago as film booker; later salesman, then sales mgr., in 1956 became Universal branch mgr. in Detroit, served until return to Chicago as branch mgr. in 1961. 1963 became asst. to gen. sales mgr. in New York for Universal Pictures. 1964 went to National General Corp. in Los Angeles as asst. to exec. v.p. Became asst. v.p. of National General Corp. in 1965. 1967 became v.p. and gen. sales mgr. of National General Pictures and formed company. Became exec. v.p. of National General Pictures 1968 in charge of worldwide sales and marketing. 1968 became v.p. of parent National General Corp. v.p. and gen. Sales Mgr., American International Pictures, December, 1971; pres., Cine Artists Pictures, 1975; pres., The Richard Graff Co., 1977.

GRAFF, ROBERT D.: Producer. b. New York City 1919. Early career 1951–60 with National Broadcasting Co. and consultant to m.p. and t.v. orgs. in Iran, Venezuela, U.S.A. Formed Sextant Inc., 1960. Sextant Films Ltd., 1963.
PICTURES INCLUDE: Young Cassidy.

GRAHAM, RODERICK: b. Edinburgh, Scotland, 1934. e. Royal High School, Edinburgh, and Edinburgh University. Directed TV drama in Scotland and London, and was senior producer of Z Cars series. Won Emmy awards as producer of The Sextet and Elizabeth R. Deputised for Head of Serials, Drama, at BBC Television Centre, London. Since 1959 actively associated with BBC work in drama, both radio and television. Appointed head of drama, television, Scotland, June 1976.

GRAHAME, GLORIA: Actress. r.n. Gloria Grahame Hallward. b. Los Angeles, Calif., Nov. 28, 1929. In numerous stage plays; screen debut in Blonde Fever, 1944, Academy Award, best supporting actress, The Bad and the Beautiful, 1952; Star of Tomorrow, 1953.
PICTURES INCLUDE: It Happened in Brooklyn, Crossfire, Merton of the Movies, Roughshod, Woman's Secret, The Greatest Show on Earth, Sudden Fear, The Glass Wall, Man On a Tight Rope, The Big Heat, The Good Die Young, Human Desire, Naked Alibi, Not as a Stranger, The Cobweb, Man Who Never Was, Oklahoma!, Blood and Lace, Ride Out for Revenge, The Todd Killings, Head Over Heels.

GRANET, BERT: Producer, writer. b. New York City, July 10, 1910; e. Yale U., Sch. Fine Arts (47 Workshop). From 1936

author s.p. orig. & adapt. numerous pictures. Exec. prod., Universal, 1967–69.
PICTURES INCLUDE: Quick Money, The Affairs of Annabel, Mr. Doodle Kicks Off, Laddie, A Girl a Guy and a Gob, My Favorite Wife, Bride by Mistake, Sing Your Way Home, Those Endearing Young Charms, Do You Love Me?, The Marrying Kind, Berlin Express, The Torch, Scarface.
TV: Desilu (1957–1961), Twilight Zone, The Mob, The Untouchables Pilot, Scarface Mob; Loretta Young Show (1955–56), Walter Winchell File 1956–57, Lucille Ball-Desi Arnaz Show 1957–60, Westinghouse Desilu Playhouse, The Great Adventure.

GRANGER, FARLEY: Actor. b. San Jose, Calif., July 1, 1925; e. Hollywood. School prior to m.p. career; screen debut in North Star 1943; in U.S. Armed Forces 1944–6.
PICTURES INCLUDE: They Live By Night, Rope, Enchantment, Roseanna McCoy, Side Street, Our Very Own, Edge of Doom, Strangers on a Train, Behave Yourself, I Want You, O. Henry's Full House, Hans Christian Andersen, Story of Three Loves, Small Town Girl, Summer Hurricane, Brass Ring, Naked Street, Girl in the Red Velvet Swing, Arrow Smith, The Heiress, The Prisoner of Zenda, Senso, The Serpent, A Man Called Noon, Those Days in the Sun, The Chief of Homicide, The Painter and the Red Head, Call Me Trinity, The Syndicate, Arnold, A Crime for a Crime.
TV: Playhouse of Stars, U.S. Steel Hour, Producer's Showcase, Climax, Ford Theatre, Playhouse 90, 20th Century Fox Hour, Robert Montgomery Presents, Arthur Murray Dance Party, Wagon Train, Masquerade Party, Kojak, 6 Million Dollar Man, Ellery Queen, 9 Lives of Jenny Dolan, Widow, National Repeitory Co., One Life To Live.

GRANGER, STEWART: Actor. r.n. James Stewart. b. May 6, 1913; e. medicine; Webber-Douglas Sch. of Acting, London. m. Jean Simmons, actress. In Brit. Army, World War II. On stage from 1935, Hull Repertory theat.; Birmingham Repertory; Malvery Festivals (1936–37); Old Vic Co. Screen debut 1940 in So This Is Voted one of Brit. top ten money-making stars in M.P. Herald-Fame Poll, 1945, 1946, 1947, 1949.
PICTURES INCLUDE: Convoy, Secret Mission, Thursday's Child, Man in Grey, The Lamp Still Burns, Fanny by Gaslight, Love Story, Waterloo-Road, Madonna of the Seven Moons, Caesar and Cleopatra, Caravan, Magic Bow, Captain Boycott, Blanche Fury, Saraband for Dead Lovers, Woman Hater, Adam and Evalyn, King Solomon's Mines, Soldiers Three, Light Touch, Wild North, Scaramouche, Prisoner of Zenda, Salome, Young Bess, All the Brothers Were Valiant, Beau Brummell, Green Fire, Moonfleet, Deadlock, Bhowani Junction, Last Hunt, Footsteps in the Fog, The Little Hut, Gun Glory, The Secret Invasion, Flaming Frontier, The Trygon Factor, The Last Safari, The Wild Geese.

GRANT, CARY: Actor. r.n. Archibald Alexander Leach. b. Bristol, England, Jan. 18, 1904. On dram. stage in England, then to U.S. and appeared with St. Louis (Mo.) Municipal Opera Co. a season. Screen debut 1932 in Paramount prod. This is the Night. Voted one of the ten best Money-Making Stars in M.P. Herald-Fame Poll 1944, 1949.
PICTURES INCLUDE: Gambling Ship, Alice in Wonderland, Sylvia Scarlett, Suzy, Holiday, Gunga Din, The Philadelphia Story, The Howards of Virginia, Penny Serenade, Suspicion, Once Upon a Honeymoon, Mr. Lucky, Destination Tokyo, Once Upon a Time, None But the Lonely Heart, Arsenic and Old Lace, Bachelor and the Bobby Soxer, The Bishop's Wife, Every Girl Should Be Married, Mr. Blandings Builds His Dream House, I Was a Male War Bride, Crisis, People Will Talk, Room for One More, Monkey Business, Dream Wife, To Catch A Thief, The Pride and the Passion, An Affair to Remember, Kiss Them for Me, Houseboat, North by Northwest, Operation Petticoat, The Grass is Greener, That Touch of Mink, Charade, Father Goose, Walk Don't Run.

GRANT, LEE: Actress. At 4 was member of Metropolitan Opera Company; played princess in L'Orocolo. Member of the American Ballet at 11. e. Juilliard School of Music, studied voice, violin and dance. At 18 with road co. Oklahoma as understudy. Acting debut: Joy To The World.
THEATRE: acted in a series of one-acters at ANTA with Henry Fonda. 1949, played shoplifter in The Detective Story (won Critics Circle Award); Lo and Behold, A Hole in he Head, Wedding Breakfast; road co. Two for the Seesaw, The Captains and the Kings; toured with Electra, Silk Stockings, St. Joan, Arms and the Man, The Maids (Obie Award), Prisoner of Second Avenue.
TV: Studio One, The Kraft Theatre, Slattery's People, The Fugitive, Ben Casey, The Nurses, The Defenders, East Side/West Side, Peyton Place (Emmy Award, Best Supporting Player, 1966), Bob Hope Show (Emmy nom.), The Love Song of Bernard Kempenski, BBC's The Respectful Prostitute, The Neon Ceilings (Emmy Award), Fay (series), The Spell (movie).
PICTURES INCLUDE: The Detective Story, Terror in the Streets, Affair of the Skin, The Balcony, Divorce American Style, Valley of the Dolls, In the Heat of the Night, Buona Sera, Mrs. Campbell, The Big Bounce, Marooned, The Landlord, There Was a Crooked Man, Plaza Suite, Portnoy's Complaint, Shampoo, Voyage of the Damned, Airport '77, When You Comin' Back, Red Ryder, Charlie Chan and the Curse of the Dragon Queen.

GRASGREEN, MARTIN: Executive. B. New York City, July 1, 1925. Entered m.p. ind. Jan., 1944, Columbia Pictures h.o. in contract dept. Promoted to travelling auditor April, 1946. Appt. office mgr. Omaha branch Dec., 1948; salesman Omaha, Dec., 1950. Transferred to Indianapolis, 1952, as city salesman; transfered to Cleveland as sales mgr., 1953. Left Columbia in 1960 to become 20th-Fox branch mgr. in Cleveland. Transferred to Philadelphia in 1965 as branch mgr.; transferred to N.Y. in 1967 as Eastern dist. mgr. Resigned in 1970 to form Paragon Pictures, prod.-dist. co. In Jan., 1975, formed Lanir Releasing Corp., representing producers for U.S. and Latin American sales and dist. of films in U.S., Canada.

GRASSHOFF, ALEX: Writer, producer, director. b. Boston, Mass. e. Tufts Coll., and U. of Southern Calif., 3 Academy Award nominations for feature documentaries; 1974, Journey to the Outer Limits; 1966, Really Big Family; 1968, Young Americans won Oscar; Future Shock: 1973 Cannes Film Festival. T.V. Series (Dir.): The Rockford Files, Chips, Foul Play, Harper Valley, Toma. Worked in Berlin, Paris, Rome.

GRAVES, PETER: Actor. b. London, Oct. 21, 1911: e. Harrow. With Knight, Frank & Rutley, then Lloyds prior to theat. career. First stage appearance 1934 in Charles B. Cochran's Streamline.
THEATRE: Novello musicals at Drury Lane; repertory at Windsor, Old Chelsea, The Merry Widow, The Sound of Music, Private Lives. Recent: The Reluctant Peer, The Last of Mrs. Cheyney, Dear Charles, An Ideal Husband (S. Africa), A Boston Story (tour), His, Hers, and Theirs, The Great Waltz, No Sex Please We're British.
PICTURES INCLUDE: Mrs. Fitzherbert, Spring in Park Lane, Maytime in Mayfair, Lady With a Lamp, Encore, Derby Day (Four Against Fate), Lilacs in the Spring (Let's Make Up), Admirable Crichton, Alfie, The Wrong Box, The Jokers, I'll Never Forget What's Is Name, How I Won the War, Assassination Bureau, The Adventurers, The Slipper and the Rose, Airplane!.
TV: Those Wonderful Snows, Chelsea at 9, One O'Clock Show, Lunch Box, 2 Cars, Dickie Henderson Show, Ivor Novello Series, East Lynne, Ninety Years On, The Sleeping Doe, The Frobisher Game, The Jazz Age series, Kate series, Crown Court, Softly, Softly, 10 from the 20s. Quiller, Duchess of Duke Street.

GRAVES, PETER: Actor, b. Minneapolis, Minn., March 18, 1936, brother, James Arness, e. U. of Minn. Played with bands, radio announcer, while at school; U.S. Air Force 2 yrs.; summer stock appearances; m.p. debut in Rogue River.
PICTURES INCLUDE: Fort Defiance, Stalag 17, East of Sumatra, Beneath the 12-Mile Reef, The Raid, Black Tuesday, Wichita, Long Gray Line, Night of the Hunter, Naked Street, Fort Yuma, Court Martial of Billy Mitchell, The Ballad of Josie, Sergeant Ryker, The Five Man Army, Sidecar Racers, Airplane!.
TV: Fury, 1955–58, Mission Impossible, 1967–1973.

GRAY, BARRY: Composer-arranger, Musical Dir. B. Blackburn, England. 1949–59, Accompaniest-arranger for Dame Vera Lynn 1956–75, musical dir. for AP Films Ltd. Century 21 ATV productions, and composed, arranged and directed original music for TV series: Four Feather Falls, Supercar, Fireball XL5, Stingray, Thunderbirds, Joe 90, Captain Scarlet, The Secret Service, UFO, Space 1999, and feature films: Thunderbirds are Go!, Thunderbird Six, Doppelganger, Journey to the Far Side of the Sun.

GRAY, COLLEEN: Actress. r.n. Doris Jensen. b. Staplehurst, Neb., Oct. 23, 1922; e. Hamline U., B.A. summa cum laude, 1943, Actor's Lab.; m. William C. Bidlack. Member: Nat'l Collegiate Players, Kappa Phi, a Capella choir, little theatres, 1943–4; screen debut State Fair, 1945.
PICTURES INCLUDE: Kiss of Death, Nightmare Alley, Fury at Furnace Creek, Red River, Sleeping City, Riding High, Father Is a Bachelor, Models Inc., Kansas City Confidential, Sabre Jet, Arrow in the Dust, The Fake, The Vanquished, Las Vegas Shakedown, Twinkle in God's Eye, Tennessee's Partner, The Killing, Wild Dakotas, Death of a Scoundrel, Frontier Gambler, Black Whip, Star in the Dust, The Vampires Hell's Five Hours, Copper Sky, Johnny Rocco, The Leech Woman, The Phantom Planet, Town Tamer, P.J., The Late Liz.
TV: Day of Our Lives, NBC-TV, 1966–67. Family Affair, Ironside, Bonanza, Judd for the Defense, Name of the Game, Bright Promise, The FBI, The Bold Ones, World Premiere, Mannix, Sixth Sense, McCloud, The Best Place To Be.

GRAY, DULCIE: Actress b. Malaya, Nov. 20; e. Webber Douglas Sch., London Stage debut 1939, Aberdeen, Hay Fever,

Author: Love Affair (play), 18 detective novels, book of short stories. 8 radio plays; co-author with husband Michael Denison, An Actor and His World.

STAGE: 37 plays including Little Foxes, Brighton Rock, Dear Ruth, Rain on the Just, Candida, An Ideal Husband, Where Angels Fear to Tread, Heartbreak House, On Approval, Happy Family, No. 10, Out of the Question, Village Wooing, Wild Duck, At The End of the Day, The Pay Off, A Murder Has Been Announced, Bedroom Farce.

PICTURES INCLUDE: Two Thousand Women, A Man About the House, Mine Own Executioner, My Brother Jonathan, The Glass Mountain, They Were Sisters Wanted for Murder, The Franchise Affair, Angels One Five, There Was a Young Lady, A Man Could Get Killed.

TV: Milestones, The Will, Crime Passionel, Art and Opportunity, Fish in the Family, The Governess, What the Public Wants, Lesson in Love, The Happy McBaines, Winter Cruise, The Letter, Tribute to Maugham, Virtue, Beautiful Forever, East Lynne, Unexpectedly Vacant, The Importance of Being Earnest, This Is Your Life, Crown Court, Making Faces, Read all about it, The Voysey Inheritance.

GRAY, GORDON: Broadcast Management Consultant. b. Albert Lea, Minn., Nov. 16, 1905; e. U. of Missouri. Entered broadcasting ind. 1932, v.p., gen. mgr., WOR, WOR TV; pres. WKTV, Utica, N.Y., KAUZ, Wichita Falls, Tex. Founded Central N.Y. Cable, Utica, N.Y. Board of Governors, WFTV, Orlando, Fla. and management consultant, presently.

GRAY, VERNON: Actor b. Saskatoon, Canada, e. Elk Point. Early career engineering plant; repertory Canada and America; studied Irving School of the Theatre; in Britain for more repertory; London stage, Stalag 17; screen debut: A Day to Remember.

PICTURES INCLUDE: To Paris With Love, Now and Forever, Barretts of Wimpole Street.

GRAYSON, KATHRYN: Actress. r.n. Zelma Hedrick; b. Winston-Salem, N.C. Feb. 9, 1923. e. St. Louis schools. Singer. Screen debut; Andy Hardy's Private Secretary, 1940; The Vanishing Virginian, Rio Rita, Seven Sweethearts, Thousands Cheer; achieved stardom in Anchors Aweigh, 1945.

PICTURES INCLUDE: Ziegfeld Follies of 1946, Two Sisters from Boston, Showboat, That Midnight Kiss, Grounds For Marriage, The Toast of New Orleans, The Kissing Bandit, Lovely To Look At, Desert Song, So This Is Love, Kiss Me Kate, Vagabond King.

TV: GE Theatre, Emmy Nomination, 1960; Die Fledermaus, ABC, '66.

STAGE: Debut in N.Y. and tour, Camelot, 1963; Rosalinda, Merry Widow, Kiss Me Kate, Showboat, (N.Y. and U.S. tour).

GREEN, ADOLPH: Writer. b. New York, N.Y., Dec. 2. Wrote book, sketches & lyrics for many Broadway shows including: On the Town, Billion Dollar Baby, Two on the Aisle, Wonderful Town, Peter Pan (Mary Martin), Bells Are Ringing, Applause (Tony Award).

PICTURES INCLUDE: Good News, On the Town, Barkleys of Broadway, Take Me Out to the Ball Game, Singin' in the Rain, Band Wagon, It's Always Fair Weather, Auntie Mame; all in collab. with Betty Comden.

GREEN, GUY: Director. b. Somerset, Eng. 1913. Joined Film Advertising Co. as projectionist & camera asst. 1933; camera asst., Elstree Studios (BIP) 1935; started as camera operator on films including One of Our Aircraft Is Missing, In Which We Serve, This Happy Breed. 1944: Director of Photography; Dir of Allied Film Makers Ltd.

PICTURES INCLUDE: The Way Ahead, Great Expectations, Oliver Twist, Captain Horatio Hornblower, I Am A Camera, River Beat, Tears For Simon, House of Secrets, Sea of Sand, The Angry Silence, The Mark, Light In The Piazza, Diamond Head, A Patch of Blue, Pretty Polly, A Matter of Innocence, The Magus, A Walk in the Spring Rain, Luther, Once Is Not Enough, The Devil's Advocate.

GREEN, JANET: Actress, screen writer. b. Hertfordshire, England, 1914. Early career as actress; Wrote first play, 1945; Ent. m.p. ind. 1947, wrote orig. s.p., The Clouded Yellow; further orig. s.p. incl. The Good Beginning, Lost Eye Witness, The Long Arm (in coll.), The Fever Tree, Saphire, Cast a Dark Shadow, Midnight Lace, Gathering of Eagles, Ashenden, Victim, Life for Ruth, Christobel; Show Me A Murderer, Walk in the Shadow, Seven Women, The Golden Keel, It Happened in Havana, My Turn Now, Gutsy, The Sweetest Voice in the World (USA).

GREEN, JOHN: Music director-conductor-composer-pianist. b. New York City, Oct. 10, 1908; e. Horace Mann School, Harvard U., 1928, A.B. Entered m.p. industry as rehearsal pianist. Paramount Astoria, N.Y.; composer-conductor 1930–32. Conductor, Paramount-Publix Theatres, 1930–33. Name band leader, 1933–40. Made numerous shorts, appeared on many TC commercial radio programs. Composer of songs: Coquette, I'm Yours, Out of Nowhere, I Cover the Waterfront, Song of Raintree County, Body and Soul, Easy Come, Easy Go, I Wanna Be Loved etc.; also symphonic works performed by major orchestra U.S. and abroad; Symphony, Mine Eyes Have Seen, 1977 Joined MGM, composer-conductor, 1942–46; Univ. Internat'l. 1947; returned to MGM for Easter Parade (Acad. Award, scoring of musical); Warner 1948–49. Inspector General (Golden Globe Award 1949, best score of year). Apptd. gen. music dir. and exec-in-chg-of-music MGM 1949; produced shorts series MGM Concert Hall (Acad. Award, best one real short 1953); Acad. Award, best scoring of musical, American in Paris 1951; composed score Raintree County (Acad. Nom. 1957): left MGM to freelance 1958. Prod. Desilu 1958–59. Comp. many original scores. V.p., Academy M.P. Arts and Sciences, 1953–55; chmn. Music Br. several terms: Acad. Life Member; prod. Acad.'s first Award Show TV, 1953. Acad. Award, best scoring of musical, West Side Story, 1961, nominated for Acad. Award, 14 times. Other films incl. High Society, Bye Bye Birdie, The Great Caruso, etc. Music dir. film, Oliver, won Acad. Award, 1968. Assoc. prod. They Shoot Horses, Don't They? (Acad. Award nomination, best music score). Pres. Screen Comp. Assoc., U.S.A. 1966–1972; Member Bd of Governors Performing Arts Council L.A. Music Center, 1965–1971; Chmn. bd. Emeritus, Young Musicians Foundation since 1965. Pres., Young Musicians Foundation 1973–1977; Since 1949, guest conductor, many U.S. orchestras; 1980–24th season Hollywood Bowl; co-chairman, music branch and v.p., Academy of Motion Picture Arts and Sciences, 1973–77.

GREEN, JOSEPH: Executive. b. Warsaw, Poland, Apr. 23, 1905. e. high school, prep school. Industry, legitimate theatre prod. Foreign film dist. since 1933; headed Green Film Co., Warsaw, Poland; head of Sphinx Film Dist. Co., N.Y., until 1940. Co-owner, Art Theatre Circuit, N.Y., 1940–52; pres., Globe Dist. Co. of Foreign Films; formed President Films, Inc., 1954, now Globe Pictures, Inc.

GREEN, MICHAEL L.: Executive. Eng. m.p. ind. United Artists 1934 terminating 1948, as Sales sup. Cont. Europe and Near East. Man. dir. Michael Green Enterprises Ltd., Nitrev Entertainments, Ltd. Chm., Entertainment Film Dist. Ltd.

GREEN, NATHANIEL CHARLES: Executive (Retired) b. Spokane, Wash. Feb. 16, 1903. e. Pleasant Prairie School until 14 yrs. of age; ent. banking bus., Spokane & Eastern Trust Co., until 1921; came to Calif., became associated Bank of America. Br. bank mgr., 26 yrs. of age, youngest in the U.S. at the time; became vice-pres., 1941; Jan. 1960 became v.p., and senior pub rel. officer between bank and m.p. and TV industry.

GREENBERG, BERRY: Executive. b. Shanghai, China, June 1, 1912. e. Shanghai Public School London Chamber of Commerce. MGM International Dept., Gen. Mgr. 15 yrs. from 1933, in Far East Branches. Joint rep. and dir. for Far East sales for Walt Disney Productions, Samuel Goldwyn Prods. Warner Bros. Pictures Int'l, div. sales mgr. for Far East, Australia and New Zealand, theatrical and TV. Vice pres. in chg. of all foreign TV sales and admin. and supervisor theatrical sales Australia and New Zealand, Warner Bros. Int'l Corp. (Sunset Int'l). Vice pres. foreign TV admin. and sales, Warner Bros-7 Arts. Vice pres. International sales and admin., Warner Bros. Television; vice-pres. International sales and admin., 1970 to June 30, 1977. Retired 1977.

GREENE, CLARENCE: Producer, writer. b. New York City. e. St. John's Univ. (L.L.B.); author of play Need a Lawyer. Formed Greene-Rouse Prods. with Russell Rouse; Acad. Oscar co-orig. story Pillow Talk. Acad. award nom. co-orig. s.p. The Well. Two Writers Guild nominations. Writers Guild award outstanding teleplay, One Day in the Life of Ivan Denisovitch.

PICTURES INCLUDE: Prod., collab. s.p. The Town Went Wild, D.O.A., The Well, The Thief, Wicked Woman, New York Confidential, A House Is Not a Home, The Oscar. Prod. Unidentified Flying Objects, The Gun Runners, Fastest Gun Alive, Thunder in the Sun, The Caper of the Golden Bulls. Collab. prod., writer TV series Tightrope.

GREENE, DAVID: Director, Writer. Early career as actor. To U.S. with Shakespeare company early 1950's; remained direct TV in Canada, New York and Hollywood.

TV: The Defenders. Emmy Awds. The People Next Door, Rich Man, Poor Man, Roots. Recent: Friendly Five.

PICTURES INCLUDE: The Shuttered Room, Sebastian, The Strange Affair, I Start Counting, Madame Sin, Godspell, Count of Monte Cristo, Gray Lady Down, Hard Country.

GREENE, LORNE: Actor. b. Ottawa, Can., Feb. 12, 1915. e. Queene's University, where began acting and directing plays. Entered radio and became chief announcer for Canadian Broadcasting Corp. and chief news broadcaster. Following World War II service returned to Canadian radio. Came to N.Y. and appeared in TV production and Broadway plays. Went to Hollywood for films and gained biggest fame as Ben Cartwright in Bonanza series started in Sept., 1959.

PICTURES INCLUDE: The Silver Chalice, Autumn Leaves,

The Gift of Love, The Buccaneer, etc. 1974: Earthquake. 1975: Tidal Wave.

TV: Destiny of a Spy, The Harness (films), Lorne Greene's American West, Christmas Special with UNs Children's Chorus, Lewis and Clark Expedition, Big Cats, Little Cats, Wonderful World of Horses, Swing Out Sweet Land, Andy Williams Show, Johnny Cash Show, Sonny and Cher Show, Celebration, The Barnum & Bailey NBC-TV Special, Roots, Battlestar Galactica, The Bastard.

GREENFIELD, IRVING H.: b. Nov. 15, 1902; e. Brooklyn Law School. Formerly president, Cinema Lodge, B'nai B'rith. Former sec'y, attorney, Metro-Goldwyn-Mayer, Inc. and Key chairman M.P. Amusement Div. of U.J.A. Federation. Engaged in law practice: Greenfield, Lipsky, and Bress.

GREENFIELD, LEO: Executive, b. N. Y. C., Education: St. John's Univ, college Arts & Sciences; v.p., gen. sales mgr. Buena Vista, 1965; Columbia road show sales mgr. 1966; v.p.-gen. sales mgr., Cinerama Rel Corp. 1966; v.p.-gen. sales mgr., Warners, 1969; sr. v.p. worldwide distribution, MGM 1975; v.p. distribution & marketing, Marble Arch Productions, 1978; sr. v.p., mkt. & dist. (domestic) Associated Film Distribution, 1979.

GREENWOOD, JACK: Producer. b. 1919.

PICTURES INCLUDE: Horrors of the Black Museum, Concrete Jungle, Invasion, Act of Murder, We Shall See, Face of A Stranger.

Also prod. Scotland Yard, The Scales of Justice and the Edgar Wallace series, Avengers series, 1967–68.

TV: From a Bird's Eye View, TV series with Sheldon Leonard 1969.

GREENWOOD, JOAN: Actress. b. Chelsea, London, Eng. e. St. Catherines, Bramley, Surrey, Royal Academy Dramatic Art. Stage debut: Le Malade Imaginaire, London; in other London plays & repertory.

PICTURES INCLUDE: The Gentle Sex, They Knew Mr. Knight, Latin Quarter, Girl In a Million, The Man Within, The October Man, The White Unicorn, Saraband for Dead Lovers, Bad Lord Byron, Whiskey Galore, Kind Hearts & Coronets, Train of Events, Flesh & Blood, Young Wives' Tale, Mr. Peek-A-Boo, Man in the White Suit, Importance of Being Earnest, Father Brown (The Detective), Knave of Hearts (Lovers, Happy Lovers), Moonfleet, Stage Struck, Mysterious Island Pawn, The Amorous Pawn, Tom Jones, The Moonshiners, Boy Stroke Girl, Water Babies, The Uneasy, Hound of the Baskervilles.

THEATRE: N.Y. Stage, Confidential Clerk 1954, Those That Play the Clowns. London Stage, Moon & the Chimney, Bell Book and Candle, Doll's House, Grass is Greener, Hedda Gabler, Irregular Verb to Love, Au Pair, Chalk Garden, Eden End, In Praise of Love.

TV: Doll's House, Confidential Clerk, King and Mrs. Candle, Man and Superman, Fat of the Land, Good King Charles, Golden Days, Love is the Flame, Love Among the Artists, Wainwright's Law.

GREER, JANE: Actress. b. Washington, D.C., Sept. 9, 1924. Orchestra singer; photograph as WAC on Life Magazine cover won screen debut in Pan-Americana 1945.

PICTURES INCLUDE: They Won't Believe Me, Out of the Past, Station West, Big Steal, You're in the Navy Now, The Company She Keeps, You For Me, Prisoner of Zenda, Desperate Search, The Clown, Down Among the Sheltering Palms, Run for the Sun, Man of a Thousand Faces, Where Love Has Gone, Billie, The Outfit.

GREFE, WILLIAM: Director, Producer, Writer. b. Miami, Fla., May 17. Pres., Film Artists Corp., Miami.

PICTURES INCLUDE: Naked Zoo, The Hooked Generation, Racing Fever, The Checkered Flag, Wild Rebels, Devil's Sisters, Death Curse of Tartu, Sting of Death, Beatrice, Caribbean Adventure. Electric Shades of Grey, V, Stanley, The Godmothers. Impulse, Live and Let Die, Mako, Whiskey Mountain.

AUTHOR: Writer, Daytona, Jo-Ann, Film Financing.

GREGORY, DAVID: Writer. b. Nashville, Tenn. e. U. of Pittsburgh, Carnegie Tech.

Writer of sketches and lyrics for Broadway shows including Hot Mikado, Ziegfeld Follies, Fools Rush In, Star and Garter, Make Mine Manhattan; night club and special material for Andrews Sisters, Bob Crosby, Betty Garrett, others.

RADIO: Writer for shows including Bob Crosby, Kraft Music Hall, Fred Allen, Doris Day, Double or Nothing, Corliss Archer.

TV: Colgate Comedy Hour, Saturday Night Revue, Hollywood Music Hall, Milton Berle Show, Frank Sinatra Show, Frankie Laine Show, Jackie Gleason's Honeymooners, Johnny Carson Show.

GREGORY, JOHN R.: Executive, producer, writer. b. Bklyn, N.Y., Nov. 19, 1918. e. Grover Cleveland H.S., 1935, New Institute of M.P. & Telev; Sls., adv. dept. Fotoshop, Inc., N.Y., 1938–42; Spec. Serv., Photo. instructor, chief projectionist, supv., war dept. theatres, U.S. Army, 1942–46; sls. mgr., J.L. Galef & Son, N.Y. 1948–49; gen. mgr., Camera Corner Co., 1949–58; pres., City Film Center, Inc. 1957; exec. v.p., Talent Guild of New York, Inc., 1958; pres., Teleview Prods., Inc., 1961; executive producer, City Film Productions, 1970. Executive post-production supervisor, Jerry Liotta Films, 1977.

AUTHOR: many articles in nat'l publications dealing with m.p. practices and techniques; tech. editor, "Better Movie-Making" Magazine, 1962; editor, pub., National Directory of Movie-Making Information, 1963; assoc. ed., Photographic Product News, 1964; contrib. editor, U.S. Camera. M.P. columnist, contributing ed. Travel and Camera magazine, 1969; Advisory panelist, Photo-methods (NY), 1975. Consultant, Photographic Guidance Council, 1957, assoc. Society of M.P. & Television-Engineers, 1952.

GRESHLER, ABNER J.: Producer. b. New York City. e. Fordham U., St. John's School of Law. Prior to entering m.p. ind., prod. shows for resort hotels; mgd., booked artists for vaudeville, hotels, cafes. Now pres., York Pictures Corp.; Abner J. Greshler Prod., Inc. Executive prod. At War with the Army, 1951; Hundred Hour Hunt, Yesterday and Today, 1953; Johnny and the Gaucho, 1955; prog. consultant, NBC; coord. of prog. development dept., NTA; pres., R.G. Prod., Ltd. Pres. Diamond Artists Ltd.; dir. Yesterday and Today, The Fugitive, Odd Couple; v.p., Astron Prod Ltd.

GREY, JOEL: Actor-Singer. b. Cleveland, 1932. e. Alexander Hamilton High, L.A. Acting debut at 10 years in On Borrowed Time at Cleveland Playhouse. Extensive nightclub appearances before returning to theatres and TV.

PICTURES INCLUDE: About Face, Calypso Heat Wave, Come September, Cabaret, Man on a Swing, Buffalo Bill and the Indians, 7 percent Solution.

TV: Maverick, December Bride, Jack and the Beanstalk (special), Ironside, Night Gallery, The Burt Bacharach Show, The Tom Jones Show, The Englebert Humperdinck Show, George M! (special), The Carol Burnett Show, The Julie Andrews Hour.

STAGE: Come Blow Your Horn. Stop the World—I Want to Get Off, Half a Sixpence, Harry, Noon and Night, West Side Story, Finian's Rainbow, Tom Sawyer, Mardi Gras, Littlest Revue, Cabaret, George M!

GREY, VIRGINIA: Actress. b. Los Angeles, Calif., March 22, 1923. Screen career started 1935 with She Gets Her Man.

PICTURES INCLUDE: Who Killed Doc Robbin, Bullfighter and the Lady, Highway 301, Slaughter Trail, Desert Pursuit, Perilous Journey, Forty-Niners, Target Earth, Eternal Sea, Last Command, Rose Tattoo, All That Heaven Allows, Tammy Tell Me True, Bachelor In Paradise, Back Street, Madame X, Rosie, Airport.

GRIFFIN, MERV: Singer, M.C. b. San Mateo, Calif., July 6, 1925. e. Univ. of San Francisco and Stanford. w. Julann Wright. The Merv Griffin Show KFRC-Radio, 1945–48; vocalist, Freddy Martin's orch., 1948–52; Contract, Warner Bros., 1952–54. co-starred, So This Is Love, The Boy From Oklahoma; CBS-TV, toured Niteclubs, 1954–55; Prod. Finians Rainbow, City Center, N.Y.C., 1955; Vocalist, CBS-TV, Morning Show, and Robt. Q. Lewis Show, 1956; M.C., Look Up And Live, CBS-TV, Going Places, ABC-TV, 1957, Merv. Griffin Show, 1958, Play Your Hunch, 1960; M.C., Keep Talking, ABC-TV, 1960; Merv. Griffin Show, NBC-TV, 1962; Word for Word, NBC-TV, 1963; Merv. Griffin Show, Westinghouse Broadcasting, 1965; pres., Griffin Prods.

GRIFFITH, ANDY: Actor. b. Mount Airy, N.C., June 1, 1926. e. U. of North Carolina.

THEATRE: Broadway stage: No Time for Sergeants, Destry Rides Again.

TV: The Andy Griffith Show, Andy of Mayberry, No Time for Sergeants, The Headmaster (series), Go Ask Alice (movie). Salvage I.

PICTURES INCLUDE: A Face in the Crowd, No Time for Sergeants, Onionhead, The Second Time Around, Angel in My Pocket, Hearts of the West.

GRILLO, BASIL F.: Executive. b. Angel's Camp, Calif., Oct. 8, 1910, e. A.B., Univ. of Calif., Berkeley. Certified public accountant, exec. vice-pres., dir., Bing Crosby Ent., Inc., 1948–57; bus. mgr., Bing Crosby, 1945; co-organizer, dir., 3rd pres., & treas., Alliance of T.V. Film Producers, 1950–54; exec. prod., BCE. Inc., shows incl. Fireside Thea., Rebound, Royal Playhouse, The Chimps; dir., KCOP, Inc., 1957–60; dir. KFOX, Inc., 1958–62; pres., dir., Bing Crosby Prods., 1955–69; dir., Seven Leagues Ent., Inc., 1958; dir. Electrovision Prods., 1970, chief exec. off., Bing Crosby Enterprises.

GRIMALDI, ALBERTO: Producer. b. Naples, 1927. Studied law, serving as counsel for Italian film companies, before turning to production with Italian westerns starring Clint Eastwood and Lee Van Cleef. Is pres. of P.E.A. (Produzioni Europee Associate, S.A.S).

PICTURES INCLUDE: For a Few Dollars More, The Good, the Bad and the Ugly, Three Steps in Delerium, Satyricon, Burn!, The Decameron, The Canterbury Tales, 1001 Nights,

Salo, or the 100 Days of Sodom, Bawdy Tales, Man of La Mancha, Last Tango in Paris, Avanti, Fellini's Casanova, 1900, The True Story of General Custer.

GRIMES, GARY: Actor. b. San Francisco, 1955. Family moved to L.A. when he was nine. Made film debut at 15 yrs. in Summer of '42, 1971. Voted Star of Tomorrow in QP poll, 1971.
PICTURES INCLUDE: The Culpepper Cattle Company, Cahill, United States Marshall, Class of '44, The Spikes Gang, Gus.

GRINBERG, SHERMAN: Executive. b. Los Angeles, June 29, 1927; e. grad., Woodbury School of Business, 1950; accountant, 20th Century-Fox Studios, 1950; agent, 20th Century-Fox, rep. stock footage film libraries, 1953; formed Sherman Grinberg Film Libraries, Inc., 1958; agent, Columbia Studios, 1959; ABC News, 1963; MBM, 1970; 20th Century-Fox, 1971; pres., chmn. of bd., Sherman Grinberg Film Libraries, Inc.; co-prod., with David L. Wolper, Biography, 1961; pres., chmn. of bd., Filmrite Associates, Inc., 1963–64; pres., chmn of bd., American Color Laboratories, Inc.; pres., Brilliant Music, Inc.; TV, prod., Battle Line series; prod., Survival series. Partner, Flaum-Grinberg Prod. Bogart Special. Exec., prod. Untamed World, 1969. with exec. prod., Whatever Happened To, Explorers, 1971; producer of Adventures in Life Pilot, 1972; 1975, created feature film digests for airline usage; pres., Classic Film Digest, Inc. Distributor, Bogart & Hollywood—the Selznick Years. 1976, Exec. prod. for KABC-TV & synd; 1980, prod. for Landsbury Prods., ABC-TV, No Safe Place.

GRIPPO, JAN: Producer. b. Beacon, N.Y., Dec. 15, 1906; e. Eastman Coll., 1925; N.Y. Sch. of Design; m. Flo Browne. From 1926–28 cartoonist New York Herald-Tribune Syndicate; com. artist, caricaturist for Shubert shows, cartoonist national magazines; agent in Hollywood, 1940–49; prod. since 1945. Produced 23 Bowery Boys features for Monogram; assoc. producer Edward Small on Valentino for Columbia, 1950; prod. Bowery Battalion, Ghost Chasers, Let's Go Navy for Monogram, 1951.
MEMBER: Soc. American Magicians.

GRIZZARD, GEORGE: Actor. b. Roanoke Rapids, N.C., 1928. Debut on Broadway in The Desperate Hours. Has been member of APA repertory company and Tyrone Guthrie resident company in Minneapolis. Broadway plays include The Disenchanted, The Country Girl, The Royal Family, Who's Afraid of Virginia Woolf, California Suite, etc.
PICTURES: From the Terrace, Advise and Consent, Warning Shot, Happy Birthday, Wanda June, Comes a Horseman.

GRODIN, CHARLES: Actor-Director. e. Univ. of Miami. Studied acting with Uta Hagen and Lee Strasberg; began directing career in New York 1965 in New York as asst. to Gene Saks. Has appeared in some 75 plays all over country and has been quest of major network TV shows. Has also written scripts produced plays.
PICTURES INCLUDE: Actor: Rosemary's Baby, Catch-22, The Heartbreak Kid, 11 Harrowhouse, Thieves, King Kong, Heaven Can Wait, Real Life, Sunburn, It's My Turn, Seems Like Old Times, The Incredible Shrinking Woman, Great Muppet Caper.
TV: Candid Camera (writer, dir.); Simon and Garfunkel Special (writer, dir.); Acts of Love and Other Comedies (dir.); Paradise (prod., dir.).

GROESSE, PAUL: Art director, MGM Studios, Culver City, Calif. b. Hungary, Feb 28, 1906; e. Northwestern U., 1928; Yale U. (BFA), 1930. Architectural practice, Chicago, 1930–31; designer, Century of Progress exposition, Chicago, 1931–34.
PICTURES INCLUDE: The Firefly, The Great Waltz, Pride and Prejudice, Madame Curie, Human Comedy, Thirty Seconds Over Tokyo, The Yearling, Little Women, Annie Get Your Gun, King Solomon's Mines, The Merry Widow, Rhapsody, Catered Affair.
MEMBER: M.P. Ind. Council; Soc. of M.P. Art Directors.

GROSS, JERRY: Executive, producer, writer, director. e. N.Y. pub. schools, night sch.
PICTURES INCLUDE: Girl on a Chain Gang, Teen Age Mother and Female Animal.
Pres., Jerry Gross Productions, 1962; Pres., Cinemation Industries, Inc., 1964–75. Pres., Jerry Gross Org., 1979.

GROSS, KENNETH H.: Executive. e. New School for Social Research, Univ. of London. Conducted film seminars at New School and active in several indep. film projects. Published film criticism in various journals and magazines. Joined ABC Entertainment Oct. 1971. Named supvr. of feature films for ABC-TV. Appt. mgr. of feature films, Jan. 1974. Promoted in Nov, 1975 to program executive, ABC Entertainment Prime Time/West Coast. Moved to L.A. offices, Nov. 1976, promoted to exec. producer, movies for TV, ABC Ent.; 1978, with literary agency F.C.A. as partner in Los Angeles; 1979 producer for Lorimar; currently with Intl. Creative Mgt.

GROSSBERG, JACK: Producer/Executive. b. Brooklyn, N.Y., June 5, 1927.
FEATURES PRODUCED: Requiem for a Heavyweight, Pretty Poison, The Producers, Take The Money and Run, Don't Drink the Water, They Might Be Giants, Bananas, Everything You Always Wanted to Know About Sex.

GROSSMAN, ERNIE: Exec. b. N.Y.C., Sept. 19, 1924. Still dept., pressbook edit., asst. field mgr., Warner Bros., 1940–58; Studio publicist, 1958–60; exploitation, promo. mgr. field dept., 1960–64; nat'l mgr., pub., exploit., promo.; 1964–67 exec. co-ord. advt., pub. & promo., Warner-7 Arts, 1967; WB nat'l supv. ad.-pub., 1970. exec. assist. to Richard Lederer, 1971–72; 1973 nat'l dir. of Pub. & Promotion, Warner Bros. Inc.; 1977, natl. dir. of adv.-pub.; 1980, natl. dir. promo.

GROSZ, PAUL: Art Director. b. New York, N.Y. e. Cooper Union, 1928–30; Pratt, 1930–32; National Academy. Ent. m.p. ind. as artist-apprentice, Universal Pictures, 1929; art, adv., Warner Bros., Paramount, Buchanan Ad Agency, Monroe Greenthal Agency, Roxy Theatre; 20th Century-Fox, 1948–62; art dir. United Artists, 1962 to date, art director.

GRUEN, ROBERT: Executive. b. New York, N.Y., Apr. 2, 1913. e. Carnegie Mellon University, B.A. Stage designer, 1934–35; designer, 20th-Fox, 1936; prod. exec., National Screen Service Corp., 1936; head, Robert Gruen Associates, ind. design org., 1940; nat. pres. Industrial Designers Institute, 1954–55; dir. and v.p., National Screen Service Corp. since 1951; senior v.p. 1975–1978; mkt. consultant, 1978–; dir., 1978–; design & mkt. consultant, 1978–.

GRUENBERG, AXEL: Producer, director. b. Riga, Latvia, Aug. 17, 1902. e. U. of Ill., A.B., 1925. High sch. teacher 1930–36; dir. WWJ Detroit 1934–39; prod. dir., NBC Chicago, 1949–40; freelance dir. many radio shows, N.Y., 1940–43; dir., writer, 1944–50, incl. This Is Your Life; writer, dir., TV films: Stars Over Hollywood, General Electric, etc.; prod., This Is Your Life, TV, 1952. Now freelance writer.

GRUENBERG, JERRY: Executive. b. Minneapolis, Minn., June 7, 1927. e. U. of Minn. United Artists, Minneapolis, 1952; m.p. exhibitor 1952–67; v.p. Sigma III Corp. 1965–67. division mgr., 20th Century-Fox, 1967–71; Senior V.P., general sales mgr., Allied Artists Pictures, 1971–79. Now pres., Blossom Pictures, Inc. Member of Variety Club, Friars Club. Director of Will Rogers Memorial Fund.

GRUENBERG, LEONARD S.: Executive. b. Minneapolis, Minn., Sept. 10, 1913. e. Minnesota U. Began as salesman Republic pictures (Minneapolis), 1935; with RKO in same capacity, 1936; promoted to city sales mgr., (St. Louis), 1937, then branch mgr. (Salt Lake City), 1941; later that year apptd. Rocky Mt. Dist. Mgr. (hqts., Denver, Colo.); 1946 Metropolitan, div. mgr., v.p. NTA, v.p. Cinemiracle Prods.; Pres., Chmn. of bd., Sigma III Corp., 1962. Chmn. of bd., Filmways, 1967. Chmn. of bd. Gamma III Dist. Co. & Chmn of bd. and Pres. Gamma III Group Ltd., 1976. Member variety Club, Sigma Alpha Mu Fraternity; Lieut. Civil Air Patrol, Lieut. Comdr., U.S.N.R.

GRUSIN, DAVID: Composer-Conductor-Performer. Directed music for the Andy Williams Show on TV for 7 yrs in the 1960s, where met Norman Lear and Bud Yorkin, producers of the series. They signed him to score their first feature film, Divorce, American Style.
PICTURES INCLUDE: Winning, The Heart Is a Lonely Hunter, Tell Them Willie Boy Is Here, The Front, Murder by Death, The Yakuza, Three Days of the Condor, Bobby Deerfield, And Justice for All, My Bodyguard.

GUARDINO, HARRY: Actor. b. Brooklyn, N.Y., Dec. 23, 1925. e. Haaren H.S.
THEATRE: B'way: End As A Man, A Hatful of Rain, Any-one Can Whistle, One More River, Natural Affection, The Rose Tattoo (revival), Seven Descents of Myrtle.
TV: Studio One, Playhouse 90, The Alcoa Theatre, Naked City, Dr. Kildare, The Untouchables, The Dick Powell Show, The Reporter, The New Perry Mason.
PICTURES INCLUDE: Pork Chop Hill, The Five Pennies, Houseboat, The Pigeon That Took Rome, Treasure of San Grennaro, Madigan, Lovers and Other Strangers, Red Sky at Morning, Dirty Harry, They Only Kill Their Masters, St. Ives, The Enforcer, Rollercoaster, Matilda, Any Which Way You Can.

GUARINI, ALFREDO: Producer. b. Genoa, Italy, May 23, 1901; m. Isa Miranda, prof.; p. Maria and Ettore Guarini. e. U. of Commerce, Genoa. In 1930, executive of various film companies in Italy, France and Germany; represented in Italy, the Consortium Emelka and Bavaria Films Co. In 1939; writer, Adventure in Diamonds, Par. in 1940: prod. Without Heaven (Rome, Italy) Academy Award for Walls of Malapaga, best foreign film, 1950.

GUBER, PETER: Producer. e. Syracuse Univ. (B.A.); Univ. at Florence (Italy), (S.S.P.); N.Y.U. School of Law (Juris Doctor, L.L.M.). Recruited by Columbia Pictures as exec. asst. in 1968 while pursuing MBA degree at N.Y.U. Graduate School of Business Adm. With Col. seven yrs. in key prod.

exec. capacities, serving last three as exec. v.p. in chg. of worldwide prod. Formed own company, Peter Guber's Filmworks, which in 1976 was merged with Casablanca Records to become Casablanca Record and Filmworks. He was chm of bd. of co. now called PolyGram, of which he is co.-mgr. dir.

PICTURES INCLUDE: The Deep (first under own banner), Midnight Express (6 Golden Globes, 2 Oscars).

GUEST, VAL: Writer, dir., prod. b. London, England, 1911; e. England and America. Journalist with Hollywood Reporter, Zit's Los Angeles Examiner and Walter Winchell.

PICTURES INCLUDE: Murder at the Windmill, Miss Pilgrim's Progress, The Body Said No, Mr. Drake's Duck, Happy Go Lovely, Another Man's Poison, Penny Princess, The Runaway Bus, Life With the Lyons, Dance Little Lady, Men of Sherwood Forest, Lyons in Paris, Break in the Circle, It's A Great Life, Quatermass Experiment, They Can't Hang Me, The Weapon, The Abominable Snowman, Carry on Admiral, It's a Wonderful World, Camp on Blood Island, Up the Creek, Further Up the Creek, Life is a Circus, Yesterday's Enemy, Expresso Bongo, Hell Is a City, Full Treatment, The Day the Earth Caught Fire, Jigsaw, 80,000 Suspects, The Beauty Jungle, Where the Spies Are, Casino Royale, Assignment K, When Dinosaurs Ruled the Earth, Tomorrow, The Persuaders, Au Pair, Girls, The Adventurers, Confessions of a Window Cleaner, Killer Force, Diamond Mercenaries. TV. Space 1999, The Shillingbury Blowers, The Band Played On, Sherlock Holmes & Dr. Watson.

GUFFEY, BURNETT: Cinematographer. b. Del Rio, Tenn., May 26, 1905. e. Etowah, Tenn., H.S. Asst. cameraman, William Fox Co., 1923–27; operative cameraman, 1928; dir. of photography, Columbia, 1943.

PICTURES INCLUDE: The Informer, Foreign Correspondent, Cover Girl, Gallant Journey, All the King's Men, From Here to Eternity (Acad. Award, best black and white photography, 1953), Human Desire, Violent Men, Harder They Fall, Good Neighbor Sam, Mad Woman of Chaillot, Halls of Anger, Bonnie and Clyde (AA, 1967), The Steogle, The Great White Hope.

GUILLAUME, ROBERT: Actor. b. St. Louis, Nov. 30. e. St. Louis Univ., Washington, Univ. Studied voice; acting career debut in Carousel. Has appeared in many Bdwy. plays and musicals.

TELEVISION: Dinah, Mel and Susan Together, Rich Little's Washington Follies, Jim Nabors, All in the Family, Sanford and Son, The Jeffersons, Marcus Welby, M.D., Soap, Benson.

GUILLERMIN, JOHN: Director. b. London, England, Nov. 11, 1925. e. City of London School, Cambridge Univ. RAF pilot prior to entering film industry.

PICTURES INCLUDE: Waltz of the Toreadors, Guns at Batasi, Rapture, The Blue Max, P.J., House of Cards, Bridge at Remagen, El Condor, Skyjacked, Shaft in Africa, The Towering Inferno, King Kong, Death on the Nile.

GUINNESS, SIR ALEC: Actor. b. London, Eng., April 2, 1914. e. Pembroke Lodge, Southbourne & Roborough School, Eastbourne. created C.B.E. 1955; Knighted 1959. Stage debut: London, 1934. Theatre appearances in London, New York & Continent. Special AA (1979) foir services to film.

PICTURES INCLUDE: Kind Hearts & Coronets, Run For Your Money, Last Holiday, The Mudlark, Lavender Hill Mob (Academy Nomination), The Man in the White Suit, The Card (The Promoter), Malta Story, Captain's Paradise, Father Brown (The Detective). To Paris With Love, The Prisoner, The Ladykillers, The Swan, The Bridge On the River Kwai (Academy Award 1957, best actor); The Horse's Mouth (Academy Nomination for s.p.), The Scapegoat, Our Man in Havana, Tunes of Glory, A Majority of One, H.M.S. Defiant, Lawrence of Arabia, Dr. Zhivago, The Comedians, Cromwell, Scrooge, Brother Sun and Sister Moon, Hitler: The Last Ten Days, Murder by Death, Star Wars (Academy Nomination), The Empire Strikes Back.

TV: The Wicked Scheme of Jehel Deeks (National Academy Nomination), Twelfth Night, Conversation at Night, Solo, E.E. Cummings, Little Gidding, The Gift of Friendship, Caesar & Cleopatra, Tinker, Tailor, Soldier, Spy (7-part series).

GULAGER, CLU: Actor. Father John Gualger, cowboy entertainer. e. Baylor Univ. Starred at school in original play, A Different Drummer, where spotted by prod. of TV's Omnibus; invited to New York to recreate role on TV.

PICTURES INCLUDE: The Last Picture Show, Winning, The Other Side of Midnight, Touched by Love, etc.

TELEVISION: The Virginian, San Francisco International, Glass House, Ski Lift to Death, etc.

GUNN, MOSES: Actor. b. St. Louis. e. Tenn. State Univ. Taught speech and drama at Grambling College. Came to N.Y. and first cast in off-Bdwy prod. of Genet's The Blacks. Later joined Negro Ensemble Company.

PICTURES INCLUDE: WUSA, The Great White Hope, The Wild Rovers, Shaft, Shaft's Big Score, Eagle in a Cake, Hot Rock, Amazing Grace, The Iceman Cometh, Rollerball, Aaron Loves Angela, Ragtime.

STAGE: In White America, Day of Absence, Song of the Lusitanian Bogey, Summer of the 17th Doll, Daddy Goodness, Harvest, Titus Andronicus, Measure for Measure, Romeo and Juliet, The Tempest, As You Like It, Macbeth, Othello, A Hand Is on the Gate.

TELEVISION: Mr. Carter's Army, The Borgia Stick, Of Mice and Men, Haunts of the Very Rich, Hawaii Five-O, The FBI, Kung Fu, If You Give a Dance, You Gotta Pay the Band, The Cowboys, Roots.

GUNSBERG, SHELDON: Executive. b. Jersey City, N.J., Aug. 10, 1920; e. St. Peters Coll., N.J. State Normal, N.Y. Univ. With Night of Stars, Madison Sq. Garden, 1942; for. pub., 20th-Fox 1942; United Artists, 1945–47; Universal, roadshows. Rank product, asst. adv., pub. dir., 1947–54; v. pres., Walter Reade Theatres; exec. v.p. & dir., Walter Reade Org. 1962; Made chief operating officer, 1971; president, and Chief Executive Officer, 1973.

GUNZBURG, M.L.: President, Natural Vision Corporation. b. Denver, Colo. e. UCLA; Columbia U. Newspaper man, columnist, Los Angeles Times, New York Times; radio writer, screen writer many m.p. cos.

Developed 3-D process, Natural Vision; contract Polaroid Corp. for 3-D, 1953.

GURIAN, PAUL R.: Executive. b. New Haven, CT., Oct. 18, 1946. e. Lake Forest Coll., Univ. of Vienna, N.Y. Univ. Started producing films in 1971 with Cats and Dogs, a dramatic short which won prizes at Chicago Int. Film Fest and Edinburgh Fest. In 1977 formed Gurian Entertainment Corp., to acquire film properties for production.

PICTURES INCLUDE: The Garden Party (PBS program), Profile Ricardo Alegria (short) Bernice Bobs Her Hair (shown at 1977 N.Y. Film Festival); Cutter and Bone (United Artists, distributor).

GUYETT, HAROLD P.: Executive. b. Woodbine, Iowa, Nov. 21, 1920. e. K.C. School of Watchmaking and Jewelry; 1938–41; ass't mgr. Golden State Theaters, San Francisco; 1946–9; ass't mgr. Fox Midwest Theatres, Kansas City, Mo.; 1950–64, managed var. Fox Midwest Theatres, Kansas City, Mo.; 1964–5, Booker Fox Midwest Theatres, Kansas City, Mo.; 1965–6, Film buyer and booker Fox Midwest Theatres, St. Louis, Mo.; 1966, transferred to Fox Eastern Theatres, NYC; 1966–67, ass't div. mgr. Fox Eastern Theatres; 1968–1972; dist. mgr. Fox Eastern Theatres, div. National General Corp.; 1974, resigned and moved to Des Moines as film buyer and booker Davis Theatres, Inc., and heading B&I Booking Agency, Des Moines; Inc., and heading B&I Booking Agency, Des Moines; 1975, resigned from Davis Theatres. At present president and film buyer Guy-Con Enterprises, Shawnee Mission, Kansas.

GWYNNE, ANNE: Actress. r.n. Marguerite Gwynne Trice. b. Waco, Tex., Dec. 10, 1918. e. San Antonio High School, Stephens Coll.

THEATRE: Stage Door, Inside Story, The Colonel's Lady.

PICTURES INCLUDE: Charlie McCarthy, Detective; Framed, Unexpected Father, Honeymoon Deferred, Fear, Ghost Goes Wild, Glass Alibi Dick Tracy Meets Gruesome, Panhandle, Enchanted Valley, Blazing Sun, Call of Klondike, Breakdown, Teenage Monster.

H

HABEEB, TONY G.: Executive. b. San Francisco, Calif., Oct. 26, 1927. e. San Francisco City College, 1945–47, Univ. of Calif., 1947–49, Lincoln Univ. Law School, LLB, 1949–53. TV Editor, San Francisco Chronicle, 1949–54; Information Specialist, U.S. Army, 1954–56; managing editor; Torrance Press, 1956–57; CBS-TV Network Press information Division, 1957–61; ass't. dir. promo. and pub., Screen Gems, 1961–63; dir. promo. and pub., Screen Gems, 1963–67; world-wide dir. advertising, pub. and promo., Paramount Television; 1967–70. v.p., adv. and publicity, Metromedia Producers Corp. 1970–1974. v p. adv. and publicity, Irwin Allen Productions, 1974–present. President, Publicist Guild of America.

HABER, JOYCE: Writer, syndicated Hollywood columnist. r.n. Mrs. Joyce Haber Cramer. b. Dec. 28, 1932, New York, N.Y. e. Brearley School, N.Y.C., class of '48; Bryn Mawr College, 1949–50, Cum Laude List; Barnard College, B.A., 1953.

Researcher, Time magazine, 1953–63; Hollywood Reporter, L.A. Bureau Time, 1963–66; Columnist, Los Angeles Times, 1966 to 1975; contributing editor, Los Angeles Magazine, 1977– .

Published "Caroline's Doll Book," illus. by R. Taylor, 1962. freelance writing: Esquire, Herald Tribune's New

York Magazine, Harper's Bazaar, New York Magazine, Town and Country. Published "The Users," a novel, 1976.

HACK, SHELLEY: Actress. b. Connecticut, July 6. e. Smith College. Made modeling debut at 14 on cover of Glamour Magazine. Gained fame as Revlon's Charlie Girl on TV commercials.
PICTURES: Annie Hall, If Ever I See You Again.
TELEVISION: Death Car, Charlie's Angels.

HACKER, CHARLES R.: Executive. b. Milwaukee, Oct. 8. e. Univ. Wisconsin. Thea. mgr., Fox Wisc. Amuse. Corp., 1940; served in U.S.A.F., 1943–5; rejoined Fox Wisconsin Amusement Corp.; joined Standard Theatres Management Corp. 1947, on special assignments; apptd. district mgr. of Milwaukee & Waukesha theatres 1948; joined Radio City Music Hall Corp. as administrative asst. July, 1948; mgr. of oper., 1952; asst. to the pres., Feb. 1957. Award: Quigley Silver Grand Award for Showmanship, 1947; v.p., Radio City Music Hall Corp., 1964; appointed executive vice president and chief operating officer, February 1, 1973.

HACKETT, BUDDY: Actor, comedian. b. Brooklyn. Prof. debut, Borscht Circuit.
TV: Stanley series, Bud and Lou (movie).
THEATRE: B'way, Lunatics and Lovers, I Had A Ball.
PICTURES INCLUDE: God's Little Acre, All Hands on Deck, The Music Man, Everything's Ducky, It's a Mad, Mad, Mad, Mad World, Muscle Beach Party.

HACKETT, JOAN: Actress. b. N.Y.C. 1942. Made stage debut on Bdwy. in A Clearing the Woods which lead to other N.Y. plays. TV debut on Ellery Queen, followed by many other shows. Film debut in The Group (1966).
PICTURES INCLUDE: Will Penny, Support Your Local Sheriff, Assignment to Kill, The Rivals, The Last of Sheila, Mackintosh and T. J., Treasure of Matecumbe, One Trick Pony, Only When I Laugh.
THEATRE: Laurette, She Didn't Say Yes, etc.
TV: Young Doctor Malone, The Defenders, Alfred Hitchcock Show, Ben Casey, The Nurses, Dr. Kildare, The Twilight Zone, The Young Country (movie), etc.

HACKMAN, GENE: Actor. b. San Bernardino, Calif., 1930. First major broadway role in Any Wednesday. Other stage productions include: Poor Richard, Children from their Games, A Rainy Day in Newark, The Natural Look. Formed own production co., Chelly Ltd.
TV: CBS Playhouse's My Father, My Mother, The F.B.I., The Invaders, The Iron Horse, etc.
PICTURES INCLUDE: Lilith, Bonnie and Clyde (Acad. Award nom. for Best Supporting Actor), First to Fight, Out by the Country Club, Hawaii, Riot, The Split, The Gypsy Moths, Downhill Racer, Marooned, I Never Sang for My Father, Doctor's Wives, Cisco Pike, The French Connection (Acad. Award best actor, 1971), Prime Cut, The Poseidon Adventure, Scarecrow, The Conversation, Zandy's Bride, Young Frankenstein, Night Moves, Bite the Bullet, French Connection II, Lucky Lady, The Domino Principle, A Bridge Too Far, March or Die, Superman, All Night Long, Superman II.

HADLOCK, CHANNING M.: Mktg. TV exec. b. Mason City, Iowa. e. Duke Univ., Univ. of North Carolina. Newspaperman, Durham, N.C. Herald, war corr., Yank; NBC, Hollywood; television prod.-writer, Cunningham & Walsh Adv.; v.p. account supr. Chirug & Cairns Adv.; v.p. Marketing Innovations; dir. mktg. Paramount Pictures; mktg. svcs, Ogilvy & Mather; mktg, Time Life Books.

HAGEL, ROBERT K.: Executive. b. Salmon, Idaho, Aug. 7, 1940. e. Univ. of Idaho, UCLA. Became TV production exec. for The Johnny Cash Show, The Ugliest Girl in Town and Three for Tahiti. Later joined Screen Gems as dir. of budgets, controls; promoted to ass't. prod. mgr. Named v.p. of Columbia Pictures in facilities and operations div. Appointed gen. mgr., chief operations exec. of newly-formed corp., The Burbank Studios, Inc., Dec. 1971.

HAGMAN, LARRY: Actor. b. Fort Worth, TX, Sept. 21. Son of actress Mary Martin. e. Bard College. First stage experience with Marg Jones Theatre in the Round in Dallas. Appeared in N.Y. in Taming of the Shrew; two years with London production of South Pacific. 1952–56 was in London with US Air Force where produced and directed show for servicemen. Returned to N.Y. for plays on and off Bdwy. Starred in daytime serial, Edge of Night, for over 2 years.
PICTURES: Fail Safe, Ensign Pulver and the Captain, The Cavern, Stardust, 3 in the Cellar, Mother, Jugs and Speed, Harry and Tonto, The Eagle Has Landed, Superman, etc.
TELEVISION: Series: I Dream of Jeannie, The Good Life, Here We Go Again, Dallas. Movies: The President's Mistress, Last of the Good Guys, Battered.

HAHN, HERBERT R.: Executive. b. Boston, Mass., June 12, 1924. e. Harvard Coll, 1941–43, 1946–47, BA, Harvard Bus. School, 1947–49, M.B.A. Theatre exec., United Paramount Theatres, Inc., 1949–53; dir. of public & stockholder relations, American Broadcasting Companies, Inc., vice-pres., investor relations, 1959.

HALAS, JOHN: Director. b. Budapest, Apr. 16, 1912; e. Hungary & Paris. Entered m.p. ind. 1928; prod. over 500 documentary, educational shorts & cartoons.
PICTURES INCLUDE: Animal Farm, The Owl and the Pussycat (3-D), History of the Cinema, Animal Vegetable and Mineral, The Candlemaker; anim. seq., The First 99, The Energy Picture, Habatales, A Christmas Visitor, Hamilton Cartoon series, Automania 2000, Is There Intelligent Life On Earth, Midsummer Nightmare, Hoffnung Cartoon Series, Dodo, The Kid From Outer Space (series), Ruddigore, The Question, Children and Cars, Parkinson's Law, Tomfoolery Show, Max and Moritz series, Ten for Survival, Autobahn.

HALE, ALAN, JR.: Actor. Son late Alan Hale, actor.
PICTURES INCLUDE: Short Grass, The Gunfighter, West Point Story, At Sword's Point, Wait Till the Sun Shines Nellie, Big Trees, Lady in the Iron Mask, Springfield Rifle, Man Behind the Gun, Capt. John Smith & Pocahontas, Iron Glove, Silver Lode, Rogue Cop, Young at Heart, Many Rivers to Cross, Destry, A Man Alone, Sea Chase, Indian Fighter, Killer Is Loose, Up Periscope, Advance to the Rear, Hang 'Em High, The True Story of Jesse James.
TV: The Lucy Show, Hazel, Wagon Train, Cheyenne, Maverick, Route 66, Jack Benny Show, Biff Baker, U.S.A. (series), Casey Jones (series), Gilligan's Island (series).

HALE, BARBARA: Actress. b. DeKalb, Ill., April 18, 1922; e. Chicago Acad. Fine Arts. Beauty contest winner, Little Theatre actress. Screen debut, 1943: Higher and Higher.
PICTURES INCLUDE: Belle of the Yukon, Goin' to Town, Boy with Green Hair, Window, Jolson Sings Again, And Baby Makes Three, Emergency Wedding, Jackpot, Lorna Doone, First Time, Last of the Comanches, Seminole, Lone Hand, Lion Is in the Streets, Unchained, Far Horizons, Houston Story, Buckskin.

HALE, MONTE: Actor, b. San Angelo, Texas, June 8, 1919. Night clubs and theatres, radio singer. Under contract to Republic Pictures, 1942–50.
PICTURES INCLUDE: Home on the Range, Timber Trail, Prince of the Plains, Bandits of the Badlands, Pioneer Marshal, The Old Frontier, Giant, The Chase.
TV: Gunsmoke, Honey West, Circus Boy, Wild Bill Hickok.

HALEY, JR., JACK: Executive. Sr. v.p. at Wolper Productions before joining MGM. Named dir. of creative affairs. Produced, wrote and directed That's Entertainment! (1974). Left Nov., 1974, to join 20th Century-Fox as pres. of TV Div. and v.p., TV for 20th-Fox Film Corp. Winner of 2 Peabody Awards, best prod. at Int'l. TV Festival at Monte Carlo and 3 Silver Lion Awards at Venice Film Festival. Won Emmy for best dir. in music or variety shows for Movin' on with Nancy. Directed M.P. Academy Awards Show in 1970; prod. it in 1974.

HALL, CONRAD: Cinematographer. b. Tahiti, 1926. Worked as camera operator with Robert Surtees, Ted McCord, Ernest Haller; moved to TV as director of photography before feature films.
PICTURES INCLUDE: Wild Seed, The Sabateur—Code Name Morituri, Harper, The Professionals, Rogue's Gallery, Incubus, Divorce, American Style, In Cold Blood, Cool Hand Luke, Hell in the Pacific, Butch Cassidy and the Sundance Kid, Tell Them Willie Boy Is Here, The Happy Ending, Fat City, Electra-Glide in Blue, The Day of the Locust, Smile, Marathon Man.

HALL, HUNTZ: Actor. In 1937 appeared in stage and screen production Dead End.
PICTURES INCLUDE: Crime School, Angels with Dirty Faces, They Made Me a Criminal, Hell's Kitchen, Muggs Rides Again, Live Wires, A Walk in the Sun, Jinx Money, Smuggler's Cove, Fighting Fools, Blues Busters, Bowery Battalion, Ghost Chasers, Crazy Over Horses, Let's Go Navy, Here Come the Marines, Hold That Line, Feudin' Fools, No Holds Barred, Private Eyes, Paris Playboys, Bowery Boys Meet the Monsters, Clipped Wings, Jungle Gents, Bowery to Bagdad, High Society, Spy Chasers, Jail Busters, Dig That Uranium, Up in Smoke, Second Fiddle to a Steel Guitar, Gentle Giant.
TV: The Teddy Bears (series).

HALL, KEN G.: Producer, director. b. Sydney, Australia, Feb. 22, 1901. Reporter. Joined pub. dept., Union Theats., Sydney; adv. mgr. 1920. In 1922 adv. mgr. FN (Asia) Ltd. Became adv. mgr. & pub. dir., State theat., Sydney, 1928; then asst. to managing dir., Union Theats. In 1931 gen. mgr. & dir., Cinesound Prods., Ltd.
PICTURES INCLUDE: (dir.) On Our Selection, Squatter's Daughter, Tall Timbers, Lovers and Luggers, Mr. Chedworth Steps Out, Vengeance of the Deep, Pacific Adventure.

HALPERN, NATHAN L.: Executive. b. Sioux City, Ia., Oct. 22, 1914; e. U. of South. Calif. 1936, B.A.; Harvard Law School 1939, L.L.B. With general counsel's office, U.S. Securities

& Exchange Commission, 1939–41; exec. asst. to dir., Civilian Supply, U.S. War Prod. Board, 1941–43; naval officer, Psychological Warfare Div., Supreme Headquarters, Allied Expeditionary Force, 1943–45; exec. asst. to dir., U.S. Information Service, Paris, 1945; ast. to pres., Columbia Broadcasting System, 1945–49; 1949–present; pres., TNT Communications, Inc.

HAMILL, MARK: Actor. Started in TV, including General Hospital (serial) and the Texas Wheelers (series). Film debut in Star Wars.
PICTURES INCLUDE: Star Wars, Corvette Summer, The Empire Strikes Back, The Night the Lights Went Out in Georgia.
TV: Sarah T: Portrait of A Teenage Alcoholic, The F.B.I., Owen Marshall, Room 222, The Partridge Family.

HAMILL, PETE: Journalist-Writer. b. Brooklyn. Worked as ad designer, NBC page boy and sheetmetal worker before joining staff of New York Post. In 1962 won Mike Berger Award of Col. Univ. Graduate School of Journalism for N.Y.'s worst slum. Received citation from Newspaper Reporters' Assn. for series on NY Police Dept. Made s.p. writing debut with Doc, 1971.
PICTURES INCLUDE: Doc, Death at an Early Age, Badge 373, Report from Engine Co. 82.
BOOKS: A Killing for Christ (novel), Irrational Ravings (collection of N.Y. Post columns), The Seventeenth Christmas (novel).

HAMILTON, GEORGE: Actor. b. Memphis, Tenn. Aug. 12, 1939. e. grammar, Hawthorne, Calf.; military sch., Gulfport, Miss., N.Y. Hackley Prep Sch., Florida, Palm Beach H.S. Won Best Actor Award for the State of Florida, high sch. contest.
TV: Rin Tin Tin, The Donna Reed Show, prod., The Veil, Roots.
PICTURES INCLUDE: Crime and Punishment, USA (debut), Home from the Hill, All The Fine Young Cannibals, Angel Baby, Where the Boys Are, By Love Possessed, A Thunder of Drums, Light in the Piazza, Two Weeks in Another Town, The Victors, Your Cheatin' Heart, Viva Maria, That Man George, Doctor, You've Got to Be Kidding!, The Long Ride Home, Jack of Diamonds, A Time for Killing, The Power, Evel Knievel, The Man Who Loved Cat Dancing, Once is Not Enough, Love at First Bite (also exec. prod.), Zorro, the Gay Blade (also co-prod.).

HAMILTON, GUY: Director. b. Paris, Sept. 1922. Ent. m.p. industry 1939 as apprentice at Victorine Studio, Nice; Royal Navy, 1940–45, in England asst. dir., Fallen Idol, Third Man, Outcast of the Islands, African Queen.
PICTURES INCLUDE: The Ringer, The Intruder, An Inspector Calls, Colditz Story, Manuela, The Devil's Disciple, A Touch of Larceny, The Best of Enemies, The Party's Over, Man In the Middle, Goldfinger, Funeral in Berlin, Battle of Britain, Diamonds Are Forever, Live and Let Die, The Man with the Golden Gun, Force Ten From Navarone, The Mirror Crack'd.

HAMILTON, MARGARET: Actress. b. Dec. 9, 1902. Kindergarten teacher; joined Cleveland Playhouse group. Played in Another Language on Broadway in 1932 and same role in screen version (1933), Wizard of Oz.
PICTURES INCLUDE: Driftwood, Bungalow 13, Sun Comes Up, Beautiful Blond from Bashful Bend, Riding High, Wabash Avenue, Great Plane Robbery, Brewster McCloud, Anderson Tapes.

HAMLIN, HARRY: Actor. b. California. e. Univ. of Calif., Yale (graduate 1974 in theatre, psychology). Joined American Conservatory Theatre (San Francisco) for two years' study before joining as professional actor. Screen debut in Movie, Movie (1979).
PICTURES: King of the Mountain, Clash of the Titans, Making Love.
TELEVISION: Studs Lonigan (mini-series).

HAMLISCH, MARVIN: Composer. b. New York City. Winner of Tony Award for score of Broadway show, A Chorus Line.
PICTURES INCLUDE: The Way We Were (Oscars for original score and song), The Sting (Oscar for adaptation), The Spy Who Loved Me, Starting Over, Ordinary People, The Fan.

HAMMOND, PETER: Actor, writer, dir. b. London, Eng., Nov. 15, 1923. e. Harrow School of Art. Stage debut: Landslide, Westminster Theatre. Screen debut: Holiday Camp.
PICTURES INCLUDE: The Huggetts, Helter Skelter, Fools Rush in, The Reluctant Widow, Fly Away Peter, The Adventurers, Operation Disaster; Come Back, Peter; Little Lambs Eat Ivy, Its Never Too Late, The Unknown, Morning Departure, Confession, dir.: Spring and Port Wine.
TV: William Tell, Robin Hood, The Buccaneers series. 1959–61. writ., dir. TV plays. TV dir. credits: Avengers, 4 Armchair Theatres, Theatre 625, BBC classic serials Count of Monte Cristo, Three Musketeers, Hereward the Wake, Treasure Island, Lord Rango, Cold Comfort Farm, The White Rabbit, Out of the Unknown, Follyfoot; Lukes Kingdom, Time to Think, Franklin's Farm, Sea Song, Shades of Greene, Our Mutual Friend, The House that Jack Built, The King of the Castle, The Black Knight, Kilvert's Army, Turgenev's Liza, Wuthering Heights, Funnyman, Little World of Don Camillo.

HAMPSHIRE, SUSAN: Actress. b. London, Eng., 1941.
STAGE: Expresso Bongo, Follow That Girl, Fairy Tales of New York, Ginger Man, Past Imperfect, She Stoops to Conquer, On Approval, The Sleeping Prince, A Doll's House, Taming of the Shrew, Peter Pan, Romeo & Jeanette, As You Like it, Miss Julie, The Circle, Arms and the Man, Man and Superman, Tribades, An Audience Called Edward, The Crucifer of Blood, Night and Day.
TV: Andromeda, The Forsyte Saga, Vanity Fair, Katy, The First Churchills; An Ideal Husband, The Lady is a Liar, The Improbable Mr. Clayville, musical version of Dr. Jekyll and Mr. Hyde, The Pallisers.
PICTURES INCLUDE: The Three Lives of Thomasina, Night Must Fall, Wonderful Life, Paris Au Mois d'Aout, The Fighting Prince of Donegal, The Trygon Factor, Monte Carlo or Bust, Rogan, David Copperfield, A Room in Paris, Living Free, Time for Loving, Malpertius, Baffled, Neither the Sea nor the Sand, Roses and Green Peppers, David the King, Bang.

HANCOCK, JOHN: Director. b. Kansas City, Mo., Feb. 12, 1939. e. Harvard. Was musician and theatre director before turning to films. Nominated for AA for short, Sticky My Fingers, Fleet My Feet.
PICTURES INCLUDE: Let's Scare Jessica to Death, Bang the Drum Slowly, Baby Blue Marine, California Dreaming.

HAND, BETHLYN J.: Executive. b. Alton, Illinois. e. University of Texas. Entered motion picture industry in 1966 as administrative assistant to president of Motion Picture Association of America, Inc. In 1975 became associate director of Code for advertising of MPAA. In 1976 became director of Code for advertising; v.p.—west coast activities, board of directors, SPCA, Los Angeles.

HANDEL, LEO A.: Producer. Organized a film research bureau, New York, 1941; dir. audience research, MGM, 1942–51; organized Meteor Prod., 1951; organized Leo A. Handel Prod., for TV films, 1953; author, Hollywood Looks at Its Audience, also TV plays; pres., Handel Film Corp.; prod. TV series including Everyday Adventures, Magic of the Atom. Exec. prod. & v.p., Four Crown Prods., Inc., exec. prod., Phantom Planet, 1961; prod.-writer-dir., feature film, The Case of Patty Smith, 1961; Americana Series, 1963; book, A Dog Named Duke, 1965. TV specials, Age of the Atom, 1966; Sweden-Vikings Now Style, 1968, Benjamin Franklin, 1969; The Mexican American Heritage and Destiny, 1970, The American Indian, 1971, Police Dog (TV special), 1972; Art in America, 10 half-hour films, 1981.

HANNA, LEE: Executive. e. Univ. of Colo., N.Y. City Coll., Long Island Univ. Joined CBS as news writer, 1954, later becoming producer-writer. Left in 1958 to write speeches and make documentaries for Gov. Averell Harriman, Gov. George Leader, Se. Harrison Williams. Ass't. dir., WNEW News, Dec. 1959; later dir. of news to 1964 when rejoined CBS News ass't. gen. mgr. Named dir. of news, WCBS-TV, N.Y., 1966. Joined NBC News, Aug., 1972, as gen. mgr., news, NBC TV stations. Appt. v.p., news, NBC TV, March, 1973.

HANNA, WILLIAM: Executive. b. Melrose, N.M. Studied engineering and journalism. Joined firm in Calif. as structural engineer; turned to cartooning with Leon Schlessinger's company in Hollywood. In 1937 hired by MGM as director and story man in cartoon dept. There met Joseph R. Barbera and created famous cartoon series Tom & Jerry, continuing to produce it from 1938 to 1957. Left MGM in 1957 to form Hanna-Barbera Productions to make cartoons for TV. Series have included Yogi Bear, Huckleberry Hound, The Flintstones, Hanna-Barbera became a subsidiary of Taft Bdg. Co. in 1968 with both men operating studio under long-term agreements with Taft. Hanna is senior v.p. of Hanna-Barbera Productions. Company entered theatrical production with Charlotte's Web in 1973.

HANNEMANN, WALTER A.: Film editor. b. Atlanta, G., May 2; U.S.C., the Class of 1935. Editorial training, RKO 1936–40; edit. supvr., Universal, 1941–42; consultant 1970–75 national educational media.
PICTURES INCLUDE: Interval, The Revengers, Dream of Kings, Guns of the Magnificent Seven, East of Java, Pay or Die, Al Capone, Hell's Five Hours, Armoured Command, Only the Valiant, Time of Your Life, Kiss Tomorrow Goodbye, Blood on the Sun, Guest in the House, Texas Masquerade, Cannon for Cardoba, El Condor, Maurie, Lost in the Stars, Mad Mad Movie Making, Big Mo, Two Minute Warning, (Oscar nominee) The Peter Hill Puzzle, Smokey and the Bandit, Other Side of the Mountain—Part II, The Visitor, The Villain, Return of Maxwell Smart.
TV: Death Valley Days, Reader's Digest, Rosemary Clooney Show, The New Breed, The Fugitive, Twelve O'Clock High, The Invaders, Hawaii Five-O, Streets of San Francisco, Cannon, Barnaby Jones, Caribe.

HANSON, JOHN: Independent Film-maker. Aegis Productions Ltd. b. St. Leonards-on-Sea, England. Early career: studio porter, designer, ecologist and actor. Docs incl: Days of Our Youth (John Irwin dir.) won 1st prize Mexico Olympics film festival.
PICTURES INCLUDE: Secrets, Shh!, The Dilessi Affair.

HANSON, PETER: Actor. b. Oakland, Calif.; p. Sydney and Lee Hansen; e. Cranbrook Acad. of Arts, Mich.; U. of Mich., 1940–41. Studied acting at Pasadena Playhouse, Calif., 1946, following war service as marine fighter pilot. Campagne, Arsenic and Old Lace, This Happy Breed; m.p. debut in Branded.
PICTURES INCLUDE: Molly, Something to Live For, When Worlds Collide, Passage West, Darling How Could You, The Savage, Violent Men, Proud and Profane, Harlow.

HARARI, ROBERT: Writer. Served in U.S. Armed Forces, 1942–45.
PICTURES INCLUDE: Music for Madame, Hitting a New High, Daytime Wife, Ice Capades, Everything Happens at Night, Sun Valley Serenade, Larceny With Music, Foreign Affair, Millionaire for Christy, story, Three Steps North.
TV: Haunted Honeymoon, Death Takes A Pass, It Happened in a Pawnshop.

HARBACH, WILLIAM O.: Producer. b. Yonkers, N.Y., Oct. 12, 1919; p. Otto Harbach, lyricist, author; e. Brown U. U.S. Coast Guard, 1940–45; actor, MGM, 1945–47; broadcast co-ordinator. NBC, 1947–49; stage mgr., 1949–50; dir., NBC, 1950–53; prod., Tonight, 1954; prod. Steve Allen Show, 1960–61 prod., dir., Bing Crosby shows; prod., Milton Berle Special, 1962; prod., Hollywood Palace Shows, 1963–69.; co-produced Julie Andrews Show, 1972–73. Emmy for Shirley MacLaine's Gypsy In My Soul, 1976.

HARDIMAN, JAMES W: Publicist. b. Brighton, England 1926. Director Advtg. & Pub., Rank Organization of Canada & Odeon Canada, 1947–56. Asst. dir. ad & pub, National Theatres, 1956–58. Dir. radio & TV promotion, Walt Disney, 1959–60. Dir. pub, Screen Gems Hollywood, 1960–64. Dir. press infmtn, CBS TV Network, Hollywood, 1964–67. Studio dir. promtn. & pub, Screen Gems, Hollywood, 1967–70. VP Yuni public rels, Tokyo and v.p. and resident director, Sijohn o Enterprises, Japan, 1970–75. Dir. pub, (entertainment) Rogers & Cowan, Beverly Hills, 1975–77. President, Suhosky & Hardiman Public Rels, 1977 to date. 1968 named Hollywood Showman of the Year by the Publicists Guild.

HARDING, ANN: Actress. r.n. Dorothy Walton Gatley. b. Fort Sam Houston, Tex., August. 17, 1904. Provincetown Players, stock, Broadway Stage: Trial of Mary Dugan, Screen: Paris Bound, Holiday, Animal Kingdom, Peter Ibbetson, others.
PICTURES INCLUDE: Mission to Moscow, The North Star, Cinderella Jones, Nine Girls, Janie, Those Endearing Young Charms, It Happened on Fifth Avenue, Christmas Eve, The Magnificent Yankee, Unknown Man, Man in the Gray Flannel Suit.

HARGREAVES, JOHN: Executive. Joined Gainsborough Pictures 1945. Transferred to Denham Studios 1946 and later Pinewood Studios. Joined Allied Film Makers 1960, then Salamander Film Productions as Bryan Forbes' financial controller and Asst. Prod. 1965. Joined EMI Film Prods. Ltd. as Asst. Man. Dir. and prod. controller May 1969–May 1972. Produced Don Quixote (with Rudolf Nureyer, in Australia), 1973. Asst. Prod. The Slipper & The Rose 1975, Man. Dir; Cinderella Promotions Ltd, 1978: assoc. prod., International Velvet. Orion repr., 1979, The Awakening; post prod. exec., Fiendish Plot of Dr. Fu Manchu, 1980.

HARINGTON, JOY: Producer. b. London, Eng., Feb. 22, 1914. e. private school, Kate Rorke Dram. School. 6 years in English theatre; New York theatre and radio; Hollywood Studios. Since 1971, freelance writer, producer and actress.
PICTURES INCLUDE: Our Hearts Were Young and Gay, Gaslight, National Velvet, Salome Where She Danced, Tony Draws A Horse.
TV: Treasure Island, Holly and the Ivy, Bridesman Danger, Midsummer Night's Dream, Beauty andThe Beast, Billy Bunter of Greyfriars, Heidi, Heidi Grows Up, Clementine, Vice Versa, Robin Hood, Kidnapped, Jesus of Nazareth, Little Women, Jo's Boys, Paul of Tarsus, Hurricane, The Chen-Lab. Mystery serials. Prod. Fothergale Co. Ltd., Law and Life, Make the Best of Yourself.
AUTHOR: Jesus of Nazareth, Paul of Tarsus, TV Script Pollyanna, Radio Play, Play It By Ear.

HARMON, MARK: Actor. b. Burbank, CA, Sept. 2. Son of actress Elyse Knox and football star Tom Harmon.
PICTURES: Beyond the Poseidon Adventure; Comes a Horseman.
TELEVISION: Series: Sam, Laverene & Shirley, Nancy Drew, Police Story, Adam-12, 240-Robert. Movies: Eleanor and Franklin: The White House Years, Little Moe, Getting Married.

HARMON, TOM: Performer. b. Rensselaer, Ind., Sept. 28, 1919; e. U. of Mich., B.S., 1941; m. Elyse Knox, actress. All American football player, 1939–40; U.S. Air Corps, 1941–46; sports dir., WJR, Detroit, 1941; KFI, Los Angeles, 1947; broadcaster, many football, baseball games; sports dir., Columbia Pacific Radio Network, 1948–61; Tom Harmon Sports Show (ABC) 1961–70; Golden West Broadcasters (Channel 5, Hollywood), 1970; 1974—Hughes Television Network—sports dir.; 1976—Editor-Publisher—Tom Harmon's Football Today. (weekly national football paper).

HARNELL, STEWART D.: Executive. b. New York City, Aug. 18, 1938. e. Univ. of Miami, UCLA, New School of Social Research, Entertainer with Youth on Parade in Coral Gables, 1948–55, performing for handicapped children, Variety Club, etc. Singer, dancer, musician. Had own bands, Teen Aces & Rhythm Rascals, 1950–56; performed on Cactus Jim TV show and Wood & Ivory, 1959–61, WTVJ. Miami. Catskills, Sand Lake, N.Y., 1954–55. Joined National Screen Service as exec. trainee in 1960 in Chicago; worked as booker & salesman. Transferred to N.Y. home office, 1963; worked in special trailer production. Promoted to asst. gen. sls. mgr., 1964–66; New Orleans branch mgr., 1966–67; Atlanta division mgr., 1967–80. Formed own independent distribution co., 1970–Harnell Independent Productions. Resumed post as gen. sls. mgr. of NSS, New York, 1977–78; resigned to become pres. of Cinema Concepts, Inc. in 1978 to present. Pres. of Variety Club of Atlanta, Tent 21, 1972, 1976, 1979.

HARPER, JOE: Theatre Executive. b. Dallas, Texas, Aug. 11, 1941. e. El Centro Coll., Dallas; Ga. State Univ. Was in public relations, N.Y.C., 1961–65 when joined Academy Theatres, Dallas. In 1968–69, with American Multi-Cinema, Dallas. In 1969 joined R. C. Cobb Theatre, Atlanta. Now v.p. in chg. booking, buying, Member: Democratic Executive Committee, Officer Ruritan International Active in Scouting and Little League sports.

HARPER, KENNETH: Producer. Formerly an agent with MCA, became an indep. prod. 1955.
PICTURES INCLUDE: Yield to the Night, Action of the Tiger, The Passionate Summer. Formed Elstree Film Distributors, 1961. Since prod. Go to Blazes, The Young Ones, Summer Holiday, French Dressing, Wonderful Life, Prudence and the Pill, 1966; Virgin and the Gypsy, 1970.

HARPER, RICHARD A.: TV Executive, b. Johnstown, Pa. e. Univ. of So. Calif. 1939; Columbia Univ. 1943 (MA); Harvard Grad. School of Bus. Admin. 1945; National City Bank of New York in Japan, China, Malaya, India and Burma 1939–42; Lt. Cmd. USNR 1942–45; MGM-New York, ass't to gen. sales mgr., circuit sales mgr. 1945–1956; director world-wide syndication MGM-TV 1956–63; gen. sales mgr. Samuel Goldwyn Prod's. 1963–64; Warner Bros.-Seven Arts v.p.-administration 1964–71; Twentieth Century-Fox Television, v.p. World-Wide syndication 1971; sr. v.p., 20th-Fox Intl. TV, 1980.

HARPER, VALERIE: Actress. b. Aug. 22, Suffern, N.Y. m. actor-writer Dick Schaal. Started as dancer in stage shows at Radio City Music Hall. First professional acting in summer stock in Conn. Appeared on Bdwy. in Take Me Along, Wildcat, Subways Are for Sleeping. Won 3 Emmys for best performance in supporting role in comedy for portrayal of Rhoda on The Mary Tyler Moore Show.
PICTURES INCLUDE: Freebie and the Bean, Chapter Two, The Last Married Couple in America.
TV: The Mary Tyler Moore Show, Rhoda.

HARPER, WILLIAM A.: Producer, b. Port Jervis, N.Y., Sept. 3, 1915; e. Univ. So. Calif. (BS), 1936. Founder member Delta Kappa Alpha frat (cinematography), U.S.C.; started in m.p. ind. in pub.; later asst. dir., prod. mgr., director; In WW II Major, USMC. Organized Marine Corps Photo Service, then in chg. Marine Corps Photo activities Pacific Ocean Areas. End of war organized Reliance Film Co., Inc., eng. in prod., dist., early TV prod. Recalled active duty for Korean War, headed USMC Photo Service. Co-prod. with Navy (Adm. John Ford) feat. documentary, This is Korea; govt. liaison with March of Time, Crusade in Pacific and with NBC US Navy Victory at Sea; prod. dir., writer, Loucks & Norling Studios, 1952; free lance writer-director 1953–55; with Fred Feldkamp Prods. 1955–56; in Europe 1956–58. Assoc. prod. The Silken Affair (RKO) 1956; Producer, The Stranding in Holland 1957; prod. advisor The Last Blitzkreig (Col.), Europe, (1958); producer, The Stone, 1962; Managing Director, St. James Productions Ltd., (England) and Pres. American-European Film Service (Paris).

HARRINGTON, CURTIS: Director, writer. b. Los Angeles, Calif., Sept. 17; e. U. Southern Calif., B.A. Exec. assts. to Jerry Wald, 1955–61 Associate Producer at 20th Cent. Fox.
PICTURES INCLUDE: Hound Dog Man, Return to Peyton Place, The Stripper, Night Tide and Queen of Blood, released by A.I.P. For Universal, What's the Matter with Helen? (UA) 1971, Gingerbread House, The Killing Kind.

HARRINGTON, PAT (JR): Actor. b. e. Fordham U. Served USAF as 1st Lt., 1952–54. Time salesman for NBC, 1954–58. Some

150 TV appearances on Jack Paar, Steve Allen and Danny Thomas TV shows, 1958–61; nightclub appearances, 1960–63. TV and films, 1963 to present. Television Series include: Mr. Deeds Goes to Town, Owen Marshall, and One Day at a Time.
PICTURES INCLUDE: The Wheeler Dealers, Move Over Darling, Easy Come, Easy Go, The President's Analyst, 2000 Years Later, The Candidate.

HARRIS, BARBARA: Actress. b. Evanston, Ill.; e. Wright Junior College (Chicago), Goodman School of the Theatre and U. of Chicago. Joined acting troop, The Compass, and came to N.Y. where first role was in Oh, Dad, Poor Dad, Mamma's Hung You in the Closet and I'm Feeling So Sad, repeating in m.p.
PLAYS: On a Clear Day You Can See Forever, The Apple Tree.
PICTURES INCLUDE: Thousand Clowns, Who Is Harry Kellerman?, The War Between Men and Women, Mixed Company, Nashville, Family Plot, Movie Movie, The North Avenue Irregulars, The Seduction of Joe Tynan.

HARRIS, EDWARD M.: Exhibitor. b. Holdenville, Okla., Aug. 21, 1916. e. Tulsa U. Theatre mgr. in Neosho, Mo., since 1933; now mgr., co-owner, Gardner Theatres there. Member: Kans.-Mo. Theatre Assn. (pres. 1954–56).

HARRIS, JAMES B.: Prod. Dir. b. New York City, Aug. 3, 1928; e. Julliard School. U.S. film export, 1947; Realart Pictures, 1948; formed Flamingo Films, 1949; formed Harris-Kubrick Productions, 1954. Producer of The Killing, Paths of Glory, Lolita; formed James B. Harris Prods., Inc., 1963; prod., dir., The Bedford Incident, 1965; produced, directed & wrote screenplay, Some Call It Loving, 1973; prod., Telefon, 1977; Fast-Walking, 1981 (prod., dir., s.p.).

HARRIS, JOHN H.: Circuit executive. b. Pittsburgh, Pa., July 9, 1898; fr. John Harris, founder of first all-film theatre, coined name Nickelodeon; e. Georgetown U., LL.B. Born in show business, managed theatres in summer as relief manager while going to school; 1917 on first definite assignment, 1922 as mgr. Strand Theat., Youngstown, Ohio; in 1923, mgr. Harris theat., McKeesport, Pa. On death of father, Senator J. P. Harris, 1926, became gen. mgr. of all Harris Theats. in Pittsburgh office. In 1930 sold Harris Theats. to RKO and Warner Bros., becoming asst. to gen. mgr., Warner Bros. Theats., N.Y., then zone mgr., Pittsburgh. In 1935 on expiration of Warner contract, assumed control of new theats., comprising Harris circuit; later pres. and gen. mgr. Harris Amusement Cos., Pittsburgh. Also pres., Arena Managers, Inc.; Pres. Prod. Ice Capades, Inc.; Ice Cycles, Int.; H. Harris Assoc.; Penna. Promotions; Arena Promotions; John Harris Assoc.; Harris & Harris (insurance); F. J. Harris Real Estate; Founder and international Big Boss of Variety Clubs International for Life.

HARRIS, JULIE: Designer. b. London, England. e. Chelsea Arts School. Entered industry in 1945 designing for Gainsborough Studios. First film, Holiday Camp.
PICTURES INCLUDE: Greengage Summer, Naked Edge, The War Lover, Fast Lady, Chalk Garden, Psyche 59, A Hard Day's Night, Darling, Help, The Wrong Box, Casino Royale, Deadfall, Prudence and the Pill, Decline and Fall, Goodbye Mr. Chips, Sherlock Holmes, Follow Me!, Live and Let Die, Rollerball, Slipper and The Rose, Dracula.
TV: Laura (with Lee Radziwill), Candleshoe, The Sailor's Return, Lost and Found.

HARRIS, JULIE: Actress. b. Grosse Pointe, Mich., Dec. 2, 1925; e. Yale Drama Sch. m.p. debut in Member of the Wedding.
PICTURES INCLUDE: East of Eden, I Am a Camera, The Trouble with Women, The Haunting, Harper, Reflection in a Golden Eye, The Poacher's Daughter, You're a Big Boy Now, The Split, The People Next Door, The Hiding Place, Voyage of the Damned, The Bell Jar.
B'WAY PLAYS: Sundown Beach, Playboy of the Western World, Macbeth, Young and the Fair, Magnolia Alley, Monserrat, Member of the Wedding, I Am a Camera, Colombe, The Lark, And Miss Reardon Drinks a Little, The Last of Mrs. Lincoln (Tony Award).
TV: Little Moon of Alban, Johnny Belinda, A Doll's House, Ethan Frome, The Good Fairy, The Lark, He Who Gets Slapped, The Heiress, Victoria Regina, Pygmalion, Anastasia, The Holy Terror, The Power and The Glory.

HARRIS, PHIL: Orchestra leader. b. Linton, Ind., June 24, 1906; husband of Alice Faye, actress. In 1933: with orchestra in Melody Cruise. In 1936: Vitaphone short prod. In 1937: Turn Off the Moon. In 1939: Man About Town. In 1940: Buck Benny Rides Again, Dreaming Out Loud.
PICTURES INCLUDE: I Love a Bandleader, Wabash Avenue, Wild Blue Yonder, Starlift, High and the Mighty, Anything Goes, Good-Bye My Lady. Co-starred with Alice Faye in weekly radio show; many TV appearances.

HARRIS, RICHARD: Actor. b. Limerick, Ireland, Oct. 1, 1930.
PICTURES INCLUDE: Alive and Kicking, Shake Hands With the Devil, The Long, The Short and The Tall, Guns of Navarone, Mutiny on the Bounty, This Sporting Life, The Red Desert (Italy), Major Dundee, The Heroes of Telemark, The Bible, Hawaii, Camelot, The Molly Maguires, A Man Called Horse, Cromwell, Bloomfield (actor & dir.), Man in the Wilderness, The Snow Goose, The Deadly Trackers, Gulliver, 99 and 44/100% Dead, Juggernaut, Echoes of a Summer, Robin and Marian, Return of the Man Called Horse, The Cassandra Crossing, Orca, Golden Rendezvous, The Wild Geese, The Ravagers, The Number, Game for Vultures, High Point, Your Ticket is No Longer Valid.

HARRIS, ROBERT: Executive. Senior v.p., Universal TV, with responsibility of supervising Universal programming on CBS network. In 1981 named pres. of Universal TV.

HARRIS, ROSEMARY: Actress. b. Ashby, Suffolk, 1930; e. India and England. Early career, nursing; studied Royal Acad. of Dramatic Art, 1951–52.
PLAYS INCLUDE: Climate of Eden (New York), Seven Year Itch, Confidential Clerk (Paris Festival), and with Bristol Old Vic, The Crucible, Much Ado About Nothing, Merchant of Venice.
Screen debut in Beau Brummell, The Shiralee, TV, Cradle of Willow, Othello, The Prince and the Pauper, The Tale of Two Cities, Dial M for Murder, etc. On stage, at Old Vic, 1955–56; U.S. tour, 1956–57; U.S. stage, 1958–63.
TV: Profiles in Courage, A Dickens Chronicle, Athens Where the Theatre Began; Blithe Spirit.
Chichester Festivals 1962 and 63; Nat'l Theatre 1963–64; You Can't Take It With You, 1965; The Lion in Winter, 1966, (U.S. Stage). 1967, APA Repertory Co., A Lion in Winter.
PICTURES INCLUDE: Camelot, The Boys from Brazil.

HARRISON, JOAN: Writer, producer. b. Guildford, Surrey, England; e. U. Sorbonne; Oxford U., B. A. Began screen career in England as asst. & writer with Alfred Hitchcock, dir. To U.S. 1939 to write s.p. Rebecca.
PICTURES INCLUDE: Foreign Correspondent, Suspicion, Saboteur, Phantom Lady, Dark Waters, The Strange Affair of Uncle Harry, Ride A Pink Horse, Circle of Danger, Eye Witness.
Prod. Alfred Hitchcock Presents, TV.

HARRISON, KATHLEEN: Actress. b. Lancashire, Eng., Trained at R.A.D.A.
PLAYS: Night Must Fall, Corn Is Green, Flare Path, Winslow Boy, Flowers for the Living, The Silver Box, Waters of the Moon, All for Mary, Nude With Violin, How Say You, Watch It Sailor, The Chances, Norman, Collapse of Stout Party, Mrs. Puffin, Harvey, She Stoops to Conquer (1972).
PICTURES INCLUDE: Bank Holiday, A Girl Must Live, Night Must Fall, In Which We Serve, Major Barbara, Holiday Camp, Bond Street, Oliver Twist, Winslow Boy, Trio, Magic Box, Waterfront, Here Comes the Huggetts, Vote for Huggett, Huggetts Abroad, Scrooge, The Happy Family, Pickwick Papers, Turn the Key Softly, Lilacs in the Spring, Where There's a Will, Cast a Dark Shadow, All for Mary, It's a Great Life, Big Money, Home and Away, Seven Thunders, A Cry From the Streets, Alive and Kicking, Mrs. Gibbons Boys, Fast Lady, West Eleven, Lock Up Your Daughters, Our Mutual Friend.
TV: Mrs. Thursday, series, Waters of the Moon, High Finance, Dixon of Dock Green, Spring & Autumn, VXB.

HARRISON, REX: Actor. b. Derry House, Huyton, Lancashire, England, Mar. 5, 1908; m. Rachel Roberts. In the RAF, World War II. Stage debut 1924 in "Thirty Minutes in a Street," Liverpool Repertory theatre, England; later on British tour ("Charley's Aunt," "Alibi," etc.) London debut 1930 in "Getting George Married." To New York, 1936, in "Sweet Aloes."
PLAYS INCLUDE: London, repertory, tour. New York stage: Anne of a Thousand Days, Henry VIII, 1948; Cocktail Party, London, 1950; Bell, Book & Candle, New York, 1951; Venus Observed, Love of 4 Colonels, N.Y.; Bell, Book & Candle, London.
PICTURES INCLUDE: Began screen career, 1929. Films include Men Are Not Gods, Storm in a Teacup, School for Husbands, Over the Moon, The Citadel, Ten Days in Paris, Sidewalks of London, Night Train, Major Barbara, Blithe Spirit. Voted one of top ten British money-making stars in Motion Picture Herald-Fame Poll, 1945–46. To Hollywood, 1946. I Live in Grosvenor Square, The Rake's Progress, Anna and the King of Siam, The Ghost and Mrs. Muir, Foxes of Harrow, Escape, Unfaithfully Yours, The Long Dark Hall, The Four Poster, King Richard & the Crusaders, Constant Husband, The Reluctant Debutante, Midnight Lace, Once a Thief, Cleopatra, My Fair Lady, The Yellow Rolls-Royce, The Agony and the Ecstasy, The Honey Pot, Dr. Doolittle, A Flea In Her Ear, Staircase, The Prince and the Pauper.
BROADWAY STAGE: in My Fair Lady, 1956–57; B'way stage in The Fighting Cock, 1959–60; London, My Fair Lady, 1958–59; Platinov, London, 1960; 1961, August for The People. Both Seasons at Royal Court Theatre, London. nom. best actor for Cleopatra; acad. award, 1965, My Fair Lady; New York Film Critics Award, 1965; Golden

Globe, 1965; David di Donatello award, 1965; Top ten world box office star, 1966.

HARROLD, KATHRYN: Actress. b. Tazewell, VA. e. Mills College. Studied acting at Neighborhood Playhouse in N.Y.: also with Uta Hagen. Appeared in off-off-Bdwy. plays for year; then joined experimental theatre group, Section Ten, touring East, performing and teaching at Connecticut College and New York univ. Cast in TV daytime serial, The Doctors.
PICTURES: Nightwing (debut), The Hunter, Modern Romance.
TELEVISION: Movies: Son-Rise, a Miracle of Love, Vampire, Bogie. Series: Starsky and Hutch.

HARRYHAUSEN, RAY: Producer, Writer, Special Effects Expert. b. L.A. e. L.A. City Coll. While at coll. made 16mm animated film, Evolution, which got him job as model animator for George Pal's Puppetoons in early '40s. Served in U.S. Signal Corps; then made series of filmed fairy tales with animated puppets for schools and churches. In 1946 worked on Mighty Joe Young as ass't. to Willis O'Brien. Designed and created special visual effects for The Beast from 20,000 Fathoms; then began evolving own model animation system called Dynarama. In 1952 joined forces with prod. Charles H. Schneer, using new process for first time in It Came from Beneath the Sea. Subsequently made many films with Schneer in Dynarama.
PICTURES INCLUDE: Twenty Million Miles to Earth, The Three Worlds of Gulliver, Jason and the Argonauts, The First Men in the Moon, One Million Years B.C., The Valley of Gwangi, The Golden Voyage of Sinbad, Sinbad and the Eye of the Tiger, Clash of the Titans (co. prod., special effects).

HART, HARVEY: Director. b. Canada, 1928. Began career on TV in native country then went to Hollywood.
PICTURES INCLUDE: Bus Riley's Back in Town, Sullivan's Empire, The Sweet Ride, Fortune and Men's Eyes.

HARTFORD, K.: Executive, b. New York, July 5, 1922. Graduate Univ. Cincinnati, Los Angeles City college. Active in Film Financing, co-production, packaging. Executive of Hartford Industries, Latin American Development & Investment Co., Western International. Has offices in 8 countries.

HARTMAN, DAVID: Actor. b. Pawtucket, R.I., May 19. e. Duke Univ. Was 2nd lt. in Air Force; entered Amer. Acad. of Dramatic Arts in N.Y. Appeared in off-Bdwy. musicals and summer stock; Bdwy. debut in Hello, Dolly!
PICTURES INCLUDE: The Island at the Top of the World.
TV: World Premiere: I Love a Mystery, Nobody's Perfect, Ballad of Josie. Series: The Virginian, The Bold Ones, Lucas Tanner, Good Morning, America.

HARTMAN, ELIZABETH: Actress. b. Youngstown, O., December, 1943. e. Carnegie Tech.
PICTURES INCLUDE: Patch of Blue, The Group, You're a Big Boy Now, The Fixer, The Beguiled, Walking Tall.

HARTMAN, PAUL: Actor, dancer, comedian. e. U. of Calif.; pr. Ferris Hartman. In show business since a child; toured West Coast with own vaude. act; toured Asia & Europe; worked in supper clubs.
BROADWAY: Red Hot & Blue, You Never Know, Of Thee I Sing, Angel in the Wings, Tickets Please, Ballyhoo, Topnotchers, Keep 'Em Laughing, All for Love, Showboat, The Pajama Game, Drink to Me Only, What A Killing, A Funny Thing Happened on the Way to the Forum.
PICTURES INCLUDE: 45 Fathers, Sunny, Higher & Higher, Man on a Tightrope, Inherit the Wind, The Thrill of It All, Soldiers in the Rain, Those Calloways, How To Succeed in Business Without Really Trying, The Reluctant Astronaut, Luv.
TV: Our Town, Petrified Forest, 12 Angry Men, Good Old Charlie Faye, Alfred Hitchcock, Ben Casey, The Defenders, Have Gun, Will Travel, Mary Martin Easter Show, Hazel, Ozzie & Harriet, Chrysler Thea., Pride of the Family, Andy Griffith Show, Petticoat Junction, Mayberry RFD, The Hartmans At Home.
AWARDS: Tony Award, Comedian of the Year; Donaldson Award, Best Comedian on Broadway; Newspaper Guild Page One Award, Best Night Club Act; Lambs Club Award, contribution to Amer. Theatre; Daily Mirror Night Club Award; Critics Award for Best Comedian in Musical; Variety Poll Award, 1947–48.

HARTZ, JIM: TV Newsman and Panelist. b. Feb. 3, 1940, Tulsa, Okla. Pre-med student at University of Tulsa, where worked in spare time as reporter for radio station KRMG. In 1963 left studies for career as newsman and joined KOTV in Tulsa. In 1964 moved to NBC News in New York, acting as reporter and anchorman. In 1974 became co-host of Today Show, joined Barbara Walter.

HARVEY, ANTHONY: Director. b. London, Eng., June 3, 1931. Royal Academy of Dramatic Arts; two yrs. as actor. Ent. m.p. ind. 1949 with Crown Film Unit.
PICTURES INCLUDE: Private's Progress, Brothers-in-Law, Man in the Cocked Hat, Carlton Brown of the F.O., I'm Alright Jack, The Angry Silence, The Millionairess, Lolita, The L-Shaped Room, Dr. Strangelove, Spy Who Came In From the Cold, The Whisperers, Directed Dutchman; Dir. The Lion in Winter, They Might Be Giants, Eagles' Wing, Players, The Abdication, The Glass Menagerie, Richard's Things.
TV: The Disappearance of Aimee.

HARVEY, WALTER F.W.: Cinematographer, director. b. London, England, Feb. 9, 1903; e. Germany. In 1921 with Anfa Laboratory, Germany; then to Decla-Bioscope, Neubabelsberg in 1922; until 1928 with UFA and subsidiaries as asst. to such photographers as Carl Hoffman, Carl Freund, Franz Planer, Carl Hoesch, F.A. Wagner, etc. To England as asst. to Charles Rosher, Dupont Prods., Elstree, 1928; m.p. photographer 1930; dir.-photog. G.B. on Ministry of Inf. War Office & Admiralty Training m.p., 1941; children's m.p., 1946; dir. of photography Exclusive Films Ltd.
PICTURES INCLUDE: Crime at Blossoms, Badgers Green, Dark Street, Cloudbust, Case for P.C. 49, Death of An Angel, Whispering Smith Hits London, Last Page, Wings of Danger, Stolen Face, Lady in the Fog, Gambler and the Lady, Flannagan Boy, Noose for a Lady, Saint Returns, Blood Orange, Thirty-six Hours, House on the Lake, Face the Music, Murder by Proxy, Life with the Lyons, Five Days, The Stranger, Third Party Risk, Mask of Dust, Men of Sherwood Forest, Lyons in Paris, Glass Cage, Break in the Circle, Quartermass Experiment, Prison Story, You Pay Your Money, West of Suez, Morning Call, Kill Her Gently, Spy in the Sky, Broth of a Boy, Naked Fury, Trouble with Eve, The Hand, Jackpot, The Man Who Couldn't Walk, Operation Cupid, Strictly Confidentially, Transatlantic, Stranger in Town, Part Time Wife, The Twisted Tape, Jungle Street, Out of The Fog, Murder In Eden, Three Spare Wives, Spanish Sword, Danger By My Side, Night of The Prowler, Echo of Diana, Shadow of Fear, The Hi-Jackers, Beware of the Dog, Boy With a Flute, The Runaways Delayed Flight, Scales of Justice series.
TV: The Cheaters, Richard the Lionheart, The Saint, Gideon's Way, The Baron, The Champions, The Saint, The Avengers, commercials and documentaries.

HARWOOD, RONALD: Writer. b. Cape Town, South Africa, 1934. To England to study acting at RADA, 1951.
TV: The Barber of Stamford Hill, Private Potter, Take a Fellow Like Me, The Lads, Convalescence, Guests of Honor, The Paris Trip, New Assistant, The Guests. Adapted several of the Tales of the Unexpected.
PICTURES INCLUDE: Barber of Stamford Hill, Private Potter (written with Casper Wrede), subsequently High Wind in Jamaica, Eye Witness, One Day in the Life of Ivan Denisovich, Operation Daybreak.

HASKIN, BYRON: Director. e. U. Calif., Berkeley. Cartoonist, San Francisco Daily News; later adv. man., traveling & photog. industrial pictures. Entered m.p. ind. as cameraman Pathe; Internatl. Newsreel; then asst. dir., Selznick, 1920; later asst. cameraman, Marshall Neilan, Sidney Franklin, Raoul Walsh; in 1923 first cameraman, Holubar; then made own prods. To Warner, filming John Barrymore pictures.
PICTURES INCLUDE: Ginsborg the Great, The Siren. Dir.: Too Late for Tears, Man-Haters of Kumson, Treasure Island, Warpath, Tarzan's Peril, Silver City, Denver and Rio Grande, War of the Worlds, His Majesty O'Keefe, Naked Jungle, Long John Silver, Conquest of Space, The Power.

HASSANEIN, SALAH M.: Executive. b. Suez, Egypt, e. British School, Alexandria, Egypt, Nat'l Bank of Egypt, Cairo, 1939–42. Asst. division mgr. Middle East, 20th-Fox, Cairo, Egypt, 1942–44: U.S. armed forces, 1945–47; usher, asst. mgr., Rivoli Theatre, N.Y., 1947–48. Film buyer, booker, oper. v.p. U.A. Eastern Theas., 1948–59; pres. 1960; exec. v.p. U.A. Theatre Circuit, Inc. 1960; pres. Todd-AO Corp., 1963.

HASTINGS, DON: Performer; b. Apr. 1, 1934, Bklyn., N.Y., e. Professional Children's Sch.; on B'way in Life With Father, I Remember Mama, Summer and Smoke, etc.; on various radio shows, video ranger on Capt. Video since 1949, also occasional TV guest appearances.

HATFIELD, BOBBY: Performer. b. Beaver Dam, Wis., Aug. 10, 1940. e. Long Beach State Coll. Mem., Righteous Bros. recording team.
TV: Shindig, The Danny Kaye Show, Ed Sullivan Show.
PICTURES INCLUDE: Beach Ball, Swingin' Summer.

HATFIELD, HURD: Actor. b. New York City; e. Morristown prep, Horace Mann; Riverdale Acad., Columbia U., Chekhov Drama School, Devonshire, Eng., on dramatic stage, Lower Depths, Twelfth Night, Cricket on the Hearth, King Lear, then Screen debut Dragon Seed, The Picture of Dorian Gray, 1943–44.
PICTURES INCLUDE: The Unsuspected, Joan of Arc, Destination Murder, Tarzan & Slave Girl, Left-Handed Gun, Mickey One, Harlow, Von Richtofen and Brown.
N.Y. stage 1952, Venus Observed. TV appearances.

HATFIELD, TED: Executive. b. Wilton Junction, Iowa, Aug. 26, 1936. e. Hot Springs, Ark., U.S. Army-NCO Academy (1954). Entered industry as usher, ABC Paramount Theatre, Hot Springs, Ark (1949) continuously employed by ABC Theatres as doorman, ass't manager, manager, city manager (1949–1967). Joined MGM Studios as assistant exploitation director (1967) assistant national advertising coordinator (1970) national advertising coordinator (1971).

MEMBER: Variety Club; Motion Picture Pioneers; past v.p., Louisiana Jr. Chamber of Commerce; Past pres. Ark.-La.-Miss. Advertising Club; Great Western Boy Scout Council of L.A.

HATHAWAY, HENRY: Director. b. Sacramento, Calif., Mar. 13, 1898. Child star with American Film Co., 1908; property boy Universal, 1914–17; juvenile roles; gunnery instructor in U.S. Army, 1918–19; with Morris Audit Co.; property man for Frank Lloyd, 1921; to India with Paul Bern; dir. Westerns for Paramount, then features.

PICTURES INCLUDE: Come On, Marines, Trail of the Lonesome Pine; Go West, Young Man; Souls at Sea, Spawn of the North, Johnny Apollo, Ten Gentlemen from West Point, House on 92nd Street, Nob Hill, The Dark Corner, 13 Rue Madeleine, Kiss of Death, Call Northside 777, Down to Sea in Ships, The Black Rose, You're in the Navy Now, Fourteen Hours, Rawhide, Desert Fox, Diplomatic Courier; seq., O. Henry's Full House; Niagara, White Witch Doctor, Prince Valiant, Garden of Evil, The Racers, Bottom of the Bottle, 23 Paces to Baker Street, Legend of the Lost, Woman Obsessed, North to Alaska, Circus World, The Sons of Katie Elder; prod., dir., Nevada Smith, True Grit, Raid on Romel, Hangup.

HAUER, RUTGER: Actor. b. Breukelen, Netherlands, Jan. 23, 1944. Stage actor in Amsterdam for six years. Motion picture debut in Turkish Delight, 1975.

PICTURES INCLUDE: The Wilby Conspiracy, Keetje Tippl'e, Max Havelaar, Solcier of Orange, Pastorale 1943, Femme Entre Chien et Loup, Mysteries, Nighthawks.

HAVELOCK-ALLAN, SIR ANTHONY: Baronet; film producer, born Durham County, England; e. Charterhouse, Switzerland; 1924–1929 Artists and Recording Manager, Brunswick Grammophone Co. London & Vox AG. Berlin; 1930 Advertising Dept. "Evening Standard"; 1930–1933 presented Cabaret-Ciro's Club, London; 1933 entered Films as Casting Director and Producer's Assistant; 1935–1937 produced 23 Quota Films for Paramount; 1938–1940 produced for Pinebrook Ltd. and Two Cities Films; 1941 Associate producer to Noel Coward; 1942 with David Lean and Ronald Neame formed Cineguild; 1942–1947 Producer, associate Producer or in charge of production for Cineguild; 1949 formed Constellation Films, independent company producing for Rank Org. and British Lion; from 1948–1951 Member of the Cinematographic Films Council and the National Film Production Council; also Member of Home Department's Committee on Employment of Children in Entertainment; 1952 Chairman of the British Film Academy; 1958 formed with John Brabourne and Daniel Angel British Home Entertainment; 1962 and 1963 Chairman of the Council of the Society of Film and Television Arts (B.A.F.T.A.); Member of National Film Archive Committee; from 1958–1965 a Governor of the British Film Institute and Member of the Institute's Production Committee; 1970 Member of the U.S. Academy of Motion Pictures Arts & Sciences.

PICTURES INCLUDE: This Man Is News, This Man in Paris, Lambeth Walk, Unpublished Story, From the Four Corners (documentary produced and directed), Brief Encounter (shared Academy script nomination), Great Expectations (shared Academy script nomination), Take My Life, Blanche Fury, Shadow of the Eagle, Never Take No for an Answer, Young Lovers, (director Anthony Asquith), Orders to Kill, (director Anthony Asquith), Brendon Behan's The Quare Fellow, An Evening with The Royal Ballet (directed two ballets); (for Television only): National Theatre's "Uncle Vanya"; Olivier's "Othello," Zeffirelli's "Romeo and Juliet," David Lean's "Ryan's Daughter."

HAVOC, JUNE: Actress. r.n. Hovick. b. Seattle, Washington. Sister of Gypsy Rose Lee, actress. Made film bow when two in Hal Roach prod. Danced with Anna Pavlova troupe, then entered vaudeville in own act. Later, joined Municipal Opera Company, St. Louis, and appeared in Shubert shows. To Hollywood, 1942.

PICTURES INCLUDE: Hello Frisco, Hello; No Time for Love, Sweet and Low Down, Brewster's Millions, Intrigue, Gentleman's Agreement; Red, Hot and Blue; Chicago Deadline, Once a Thief, Follow the Sun, Lady Possessed, Can't Stop the Music.

PLAYS: Pal Joey, Sadie Thompson, Mexican Hayride, Dunnigan's Daughter, Dream Girl, Affairs of State, The Ryan Girl, The Infernal Machine, The Beaux Strategem.

American Shakespeare Festival, Stratford, 1958; The Skin of Our Teeth, Tour for U.S. Dept. of St., 1961; wrote Marathon 33.

TV: Anna Christie, The Bear, Cakes and Ale, Daisy Mayme, The Untouchables; co-owner, Willy.

HAWN, GOLDIE: Actress. b. Washington, D.C., November 21, 1945; was a professional dancer and made TV debut dancing on an Andy Griffith Special; TV acting debut, "Good Morning World" followed by "Laugh-In." Screen debut: Catus Flower.

PICTURES INCLUDE: There's A Girl In My Soup, Dollars, Butterflies Are Free, The Sugarland Express, The Girl from Petrovka, Shampoo, The Duchess and the Dirtwater Fox, Foul Play, Private Benjamin (also prod.), Seems Like Old Times.

HAYDEN, STERLING: Actor. b. Montclair, N.J.; e. Browne and Nichols School, Cambridge, Mass. In 1937: cruised around world as mate. In 1939: captain of the "Aries" owned by Warner Buxton.

PICTURES INCLUDE: Virginia, Bahama Passage, Blaze of Noon, El Paso, Manhandled, The Asphalt Jungle, Flaming Feather, Journey Into Light, Denver and Rio Grande, Golden Hawk, Hellgate, Flat Top, The Star, Take Me to Town, So Big, Crime Wave, Arrow in the Dust, Prince Valiant, Johnny Guitar, Naked Alibi, Suddenly, Battle Taxi, Timberjack, Eternal Sea, Last Command, Shotgun, Top Gun, The Killing, Hard Contract, The Godfather, The Long Goodbye, King of the Gypsies, Winter Kills, Nine to Five, Gas.

HAYDN, RICHARD: Director, actor. In 1941: Charley's Aunt. In 1942: Are Husbands Necessary?, Thunder Birds. In 1943: Forever and a Day.

PICTURES INCLUDE: The Beginning or the End, Ball of Fire, Sitting Pretty, The Lost World, Five Weeks in a Balloon, Clarence the Crosseyed Lion, Bullwhip Griffin, And Then There Were None, Adventure, The Green Years, Cluny Brown, The Late George Apley, Foxes of Harrow, Singapore, Forever Amber, (dir. actor) Miss Tatlock's Millions; (dir.) Mr. Music, Dear Wife, Merry Widow, You Belong to Me, Never Let Me Go, Money from Home, Her Twelve Men, Jupiter's Darling, Toy Tiger, Please Don't Eat the Daisies, Mutiny On The Bounty, Sound of Music, retired in 1965.

HAYES, HELEN: Actress, r.n. Helen H. Brown. b. Wash., D.C., Oct. 10, 1901; e. Sacred Heart Convent, Wash.; wife of Charles MacArthur, writer. Started film career in 1931.

PLAYS INCLUDE: What Every Woman Knows, Coquette, Petticoat Influence, The Good Fairy, Mary of Scotland, Victoria Regina, Harriet Happy Birthday, Wisteria Trees, Mrs. McThing, Skin of Our Teeth, Glass Menagerie, The Show Off, Front Page (revivals).

PICTURES INCLUDE: Arrowsmith, The White Sister, Another Language Night Flight, A Farewell to Arms, Academy Award 1931–32 (The Sin of Madelon Claudet), My Son, John, Main Street to Broadway, Anastasia, Airport (AA, 1970), Herbie Rides Again, One of Our Dinosaurs Is Missing, Candleshoe.

TV: Twelve Pound Look, Mary of Scotland, Dear Brutus, Skin of Our Teeth, Christmas Tie, Drugstore on a Sunday Afternoon, Omnibus.

HAYES, JOHN MICHAEL: Writer. b Worcester, Mass., May 11, 1919; e. University of Massachusetts, 1941.

PICTURES INCLUDE: Red Ball Express, Thunder Bay, Torch Song, War Arrow, Rear Window, To Catch a Thief, Trouble with Harry, It's A Dog's Life, Man Who Knew Too Much, The Matchmaker, Peyton Place, But Not for Me, Butterfield 8, The Children's Hour, Where Love Has Gone, The Chalk Garden, Judith, Nevada Smith.

HAYES, PETER LIND: Actor: m. Mary Healy, prof.

PICTURES INCLUDE: Million Dollar Legs, All Women Have Secrets, These Glamour Girls, Seventeen, Dancing on a Dime, Playmates, Seven Days Leave.

Was radio singer, actor, vaudeville, night clubs. Producer, Grace Hayes Lodge Review: on TV show with Mary Healy; 5000 Fingers of Dr. T. Recent films: Once You Kiss A Stranger.

HAYES, RICHARD: Film censor. b. Bruree, Co. Limerick, Ireland; e. Catholic U. Medical Sch., Dublin, 1908; D. Litt. (Honoris Causa), Natl. U. Ireland, 1940. Practised prof. of medicine & surgery until 1940. In Irish Insurrection, 1916; member of Dail Eireann (Irish Parliament) 1918–23 Dir. Abbey Theat., Dublin. M.R.I.A. Author various books (Ireland and Irishmen in the French Revolution, Irish Swordsmen of France, The Last Invasion of Ireland, Old Irish Links with France). In 1940 apptd. Film Censor for Eire.

HAYS, ROBERT: Actor. b. Bethesda, MD, July 24. e. Grossmont College, San Diego State. Left school to join Old Globe Theatre for five years. Cast in ABC series, Harry O.

PICTURES: Airplane, Take This Job and Shove It!

TELEVISION: Series: Love Boat, Laverne & Shirley, Most Wanted, Wonder Woman, Angie. Movie: California Gold Rush. Mini-series: Will Rogers: Champion of the People.

HAYTER, JAMES: Actor. b. Lanoula, India, April 23, 1907; e. Dollar Acad., Scotland; Royal Acad. Dram. Art, London. In Royal Armored Corps, World War II. On prof. stage from 1925, tour. repertory; London debut 1936 in The Composite Man. Screen debut, 1936.

PICTURES INCLUDE: Morning Departure (Operation Disaster), Trio, Robin Hood, Crimson Pirate, Pickwick Papers, Great Game, Will Any Gentleman, Always a Bride, A Day to Remember, Beau Brummell, For Better For Worse, Land of the Pharaohs, Touch and Go, Port Afrique, Keep It Clean, Seven Waves Away (Abandon Ship), Sail Into Danger. The Heart Within, The Captain's Table, The Boy on the Bridge, The 39 Steps, Carry On Admiral, Go To Blazes, Out of The Fog, Stranger in the House, Oliver, A Challenge to Robin Hood, Song of Norway, David Copperfield, Cause for Alarm, Horror of Frankenstein, Burke & Hane, The Loving Game.

STAGE: My Fair Lady 1959–66, A Woman of No Importance, Oh Clarence, Yellow Sands, Pride and Prejudice, Pygmalion, 1966; A Woman of No Importance, 1967; The Duel, 1968; Forget Me Not Lane, 1971; The Grouse Moor, 1973; There Goes The Bride, 1976; Babes In The Wood, 1974; Gaslight, Rain, She Stoops To Conquer, 1977; Crime & Punishment, 1978.

T.V. SERIES: Huntingtower, The Moonstone, Last Chronicles of Barsett, The Flaxman Boys, The Onedin Line, For the Love of Ada, Doctor at Large, Dr. Finlay's Casebook, Old Newsom, Under The Same Sun, Are You Being Served.

HAYWARD, LOUIS: Actor. b. Johannesburg, So. Africa, March 19. On English stage in Dracula, Vinegar Tree, Another Language, Conversation Piece. The Ringer; to N.Y., 1935; won N.Y. Critics Award in first show, Point Valaine; m.p. debut in Self-Made Lady. U.S. Marines, W.W. II; owner Assoc. Film Artists.

PICTURES INCLUDE: Flame Within, Anthony Adverse, Duke of West Point, Man in the Iron Mask, My Son, My Son, Son of Monte Cristo, And Then There Were None, Ruthless, Black Arrow, Walk a Crooked Mile; Road Show, Camelot, Pirates of Capri, Fortunes of Captain Blood, Lady and the Bandit, Son of Dr. Jekyll, Lady in the Iron Mask; Captain Pirate, The Saint's Girl Friday, Duffy of San Quentin, Royal African Rifles, Search for Bridey Murphy, The Christmas Kid, Chuka.

TV: The Lone Wolf series, Climax, Studio One, Schlitz Playhouse, Pursurers, The Survivors.

HAYWORTH, RITA: Actress. r.n. Margarita Cansino; b. New York City, Oct. 17, 1919; e. private schools. Danced at Foreign Club in Tiajuana, Mexico, then at Agua Caliente. Started screen career 1935 with Dante's Inferno. Formed prod. co. with Dick Haymes, Crystal Bay Prods., 1955.

PICTURES INCLUDE: Meet Nero Wolfe, Who Killed Gail Preston?, There's Always a Woman, Convicted, The Lone Wolf, Spy Hunt, Gilda, Cover Girl, Tonight and Every Night, Down to Earth, Lady from Shanghai, Loves of Carmen, Affair in Trinidad, Salome, Miss Sadie Thompson, Fire Down Below, Pal Joey, Separate Tables, They Came to Cordura, Story on Page One, The Happy Thieves, Circus World, Money Trap, The Poppy Is Also a Flower, The Naked Zoo, Road to Salina, The Wrath of God.

HAZEN, JOSEPH H.: Attorney b. Kingston, N.Y., May 23, 1898; e. George Washington U. & Law School. Formerly assoc. as member of firm, Thomas and Friedman. In 1939 member ind. com. in discussions with Commerce Dept. Member com. which negotiated consent decree. Member ind. Committee of Six, lawyers to study and reorganize ind. activities. Mem. War Activities Committee. In 1944 resigned as v.p. and dir., Warner Bros. to join Hal Wallis Prods. as pres., 1944–48; pres. Wallis-Hazen, Inc., Tillco. dissolved, June 1953; photoplay prod., Hal B. Wallis, Paramount Pictures Corp., since 1953. Dir. USO Camps, Inc., M.P.A.A.

HEAD, EDITH: Chief designer. Universal. b. Los Angeles, Calif., October 28. e. Univ. of Calif., A.B.; Stanford Univ., M.A.; Otis Art School and Chouinard Art Sch., L.A. Taught French, Spanish and Art, Hollywood Sch. for Girls and Bishops Sch. at La Jolla, Calif.

PICTURES INCLUDE: The Heiress (Acad. Award B&W, 1949), All About Eve (Acad. Award B&W, 1950), Samson and Delilah (Acad. Award Tech., 1950), A Place In The Sun (Acad. Award B&W, 1951), The Greatest Show On Earth, Carrie, Roman Holiday (Acad. Award B&W, 1953), Sabrina (Acad. Award B&W, 1954), The Rose Tattoo, To Catch A Thief, The Ten Commandments, The Proud and Profane, Funny Face, The Facts of Life (Acad. Award B&W, 1969), Pepe, Pocketful of Miracles, Tell Them Willie Boy Is Here, Airport, Butch Cassidy and The Sundance Kid, The Sting. (Acad. Award 1973) Author of The Dress Doctor and How to Dress for Success; 1967, with Universal Studios.

HEALEY, MYRON: Actor, writer, dialogue director. b. Petaluma, Calif., e. East Central State Teachers Coll., Okla.; bombardier, navigator, World War II.

PICTURES INCLUDE: Monsoon, Son of Belle Starr, The Longhorn, African Manhunt, Gang Busters, Man from Bitter Ridge, Cattle Queen of Montana, Rage at Dawn, Shoot Out at Medicine Bend.

Writer on several s.p.; TV shows. Stars in numerous live and filmed TV shows.

HEALY, JOHN T.: Executive. e. Brooklyn Coll. Taught economics and was associated with Lehigh Valley Industries and General Food Corp. before joining ABC, Inc. in 1970 as assoc. dir. of corp. planning. Named dir. of planning and develop. June, 1972. Elected v.p., planning and admin. of ABC Leisure Group, March, 1974; elected vice pres. of corporate planning, Feb., 1976.

HEARD, JOHN: Actor. b. 1947. Married, actress Margot Kidder. Career began at Organic Theatre, starring in Chicago & N.Y. productions of Warp. Other stage roles include Streamers, G.R. Point, Othello.

PICTURES: Between the Lines, On the Yard, Head Over Heels, Heartbeat, Cutter and Bone.

TELEVISION: The Scarlet Letter.

HEARD, PAUL F.: Producer. b. Olivia, Minn., Oct. 14, 1913; e. Lawrence Coll., 1929–31; U. of Minn., 1933–35. Production supvr., Visual Ed. Dept., U. of Minn., 1938–40; dir. of films, Methodist Church Hdqts., N.Y., 1940–42; Orientation Film Officer, Bureau of Personnel, U.S. Navy, Washington, 1942–45; nat'l exec. dir. Protestant Film Comm., N.Y., 1945–50; dir. of films. Broadcasting & Film Comm., Nat'l Council of Churches of Christ in U.S., 1950–52; pres., Paul F. Heard, Inc., m.p. & TV prod. co., 1952–63.

PICTURES INCLUDE: Kenji Comes Home, The Broken Mask, For Every Child, Hong Kong Affair, What's Your Trouble.

HECHT, HAROLD: Producer. b. New York City, June 1, 1907. New York stage, asst. late Richard Boleslavsky, dance dir. Screen: dance dir. Horsefeathers, She Done Him Wrong, Bottoms Up, many others; literary agt. Org. Norma Prod. with Burt Lancaster, 1947, and later other prod. companies now known as Hecht-Hill-Lancaster Companies.

PICTURES INCLUDE: The Flame and the Arrow, Ten Tall Men, First Time, Crimson Pirate, His Majesty O'Keefe, Apache, Vera Cruz, Marty (Academy Award, best picture, 1955), The Kentuckian, Trapeze, The Bachelor Party, Sweet Smell of Success, Run Silent Run Deep, Separate Tables, The Devil's Disciple, The Unforgiven, The Young Savages, Bird Man of Alcatraz, Taras Bulba, Wild and Wonderful, Flight From Ashiya, Cat Ballou, The Way West.

HECKART, EILEEN: Actress. b. Columbia, Ohio. e. Ohio State U., American Theatre Wing. m. Jack Yankee.

BROADWAY: Bad Seed, View from the Bridge.

TV: Kraft, Suspense, Philco Playhouse, The Web (CBS Playhouse). m.p. debut in Miracle in the Rain.

AWARDS: Daniel Blum and Outer Circle (Picnic), Foreign Press, and Donaldson, Oscar nom. and Film Daily Citation, (Bad Seed), TV Sylvania for the Haven, Variety Poll of N.Y. and Drama Critics (Dark at The Top of the Stairs); Emmy, (Save Me a Place at Forest Lawn).

PICTURES INCLUDE: Somebody Up There Likes Me, Bus Stop, Hot Spell, Heller in Pink Tights, My Six Loves, Up the Down Staircase, No Way to Treat a Lady, Butterflies Are Free, Zandy's Bride, The Hiding Place, Burnt Offerings.

STAGE: Bad Seed, Family Affair, Pal Joey, Invitation to a March, Everybody Loves Opal, Dark at the Top of the Stairs, And Things That Go Bump in the Night, You Know I Can't Hear You When the Water's Running, Too True to Be Good, Butterflies Are Free.

HEDRICK, A. EARL: Art director. b. Los Angeles, Calif., March 2, 1896; e. U. So. Calif. (Sch. Archit.), Beaux Art Sch. Design. Began as architect. Joined Paramount 1930, as art dir., 1933.

PICTURES INCLUDE: So Proudly We Hail, Wake Island, Lost Weekend, Cocoanut Grove, The Forest Rangers, Road to Rio, Fancy Pants, Mr. Music, Union Station, Great Lover, Here Comes the Groom, Detective Story, Somebody Loves Me, About Mrs. Leslie, Strategic Air Command, Far Horizons, You're Never Too Young, Proud and Profane, Teacher's Pet, But Not for Me, Michael O'Hare IV, Jory

TV: Bonanza series, The High Chaparral.

HEDWIG, GORDON W.: President, Nu-Art Films, Inc. b. Jersey City, N.J., April 27, 1913; p. William K. and Elsa Hedwig; e. Newtown high school, Rensselaer Polytechnic Inst., Lt., Signal Corp. World War II. Part-time work film lab., 1927–33; salesman, 1933–35. Started 16 mm. distribution in 1935 as Nu-Art Film Co., an individual enterprise and incorporated as Nu-Art Films, Inc., 1937. President NuArt Films, Inc., org. Telecast Films, Inc., org. in 1947 for TV film distribution. Sec. Theatre-on-films, Inc. Now active in all corporations, in TV & non theatrical.

HEFFNER, RICHARD D.: Executive. b. Aug. 5, 1925. e. Columbia Univ. Instrumental in acquisition of Channel 13 (WNET) as New York's educational tv station; served as its first general manager. Previously had produced and moderated Man of the Year, The Open Mind, etc. for commercial tv. Served as dir. of public affairs programs for WNBC-TV in N.Y. Was also dir. of special projects for CBS TV Network and editorial consultant to CBS, Inc. Editorial Board. Was radio newsman for ABC. Taught history at Univ. of Calif. at Berkeley, Sarah Lawrence College, Columbia Univ.

and New School for Socia Research, N.Y. Served as American specialist in communications for U.S. Dept. of State in Japan, Soviet Union, Germany, Yugoslavia, Israel, etc. Now produces and moderates The Open Mind on Channel 11 (WPIX) in New York. Is Univ. Prof. of Communications and Public Policy at Rutgers University. In July, 1974 appt. chm. of MPAA Rating Board.

HEFFRON, RICHARD T.: Director.
PICTURES INCLUDE: Fillmore, Newman's Law, Trackdown, Futureworld, Outlaw Blues.
TV: The Morning After, Dick Van Dyke Special, I Will Fight No More Forever, Toma (pilot), Rockford Files (pilot).

HEIDER, FREDERICK: Producer. b. Milwaukee, Wis., April 9, 1917. e. Notre Dame U., Goodman Theatre, Chicago. Actor in Globe Theatre, Orson Welles' Mercury Theatre.
TV: Chesterfield Supper Club, Sammy Kaye's So You Want to Lead a Band, Frankie Carle Show, Jo Stafford Show, prod., writer, Paul Whiteman Goodyear Revue, Billy Daniels Show, Martha Wright Show, Earl Wrightson Show, Club Seven, Mindy Carson Show; prod., ABC, Ted Mack Family Hour, Dr. I.Q., Miss America Pageant, Bishop Sheen's Life Is Worth Living, Voice of Firestone, Music for a Summer Night. Music for a Spring Night, The Bell Telephone Hour. (Warner Bros. TV) publisher, Television Quarterly, National Academy of Television Arts and Sciences. Presently syndicated columnist, Desert Post, Palm Desert, CA.

HEINECKE, RUTH C.: executive. b. New York, N.Y. Treas., Breitkopf & Hartel, Inc., N.Y., and treas., Breitkopf Pub., Inc. N.Y., 1914–26; v.p., Assoc. Music Pub., Inc., N.Y., 1927–29; treas., SESAC Inc. 1930.

HELLER, FRANKLIN: Producer-director. b. Dover, N.J.; Carnegie Inst. of Tech., B.A., 1934. Actor, 1934–36; stage mgr., Sam Harris-Max Gordon Prods., 1936–44; exec. prod., USO shows N.Y., 1944–45; prod. & dir., Paramount, 1945–47; dir., summer stock, 1947–48; prod. & dir., CBS TV, 1949–54; exec., prod. and dir. Goodson-Todman Prods., 1954–69; exec. prod. Protocol Prods., 1969–72 Literary Representative 1972. Dirs. Guild of America, Nat'l bd. 1965–77; Treas. 1965–69; Sec. 1970–73; Chr. Publications 1966–76.
TV SHOWS: What's My Line?, Beat the Clock, The Front Page, The Web, Danger, To Tell the Truth, I've Got a Secret.

HELLER, HUGH: Executive. b. Fargo, N.D. e. San Jose State Coll. Prod.-agent, for MCA; asst. exec. prod., George Gobel Show; prog. consultant to CBS, other networks; winner of IBA and AAW awards since 1963; clients incl.: Hamm's Beer, Folger's Coffee, Dial Soap, Signal Oil, Bargermeister Beer; pres., Heller Ferguson and Heller Corp.

HELLER, PAUL M.: Producer. b. N.Y.C. Sept. 25, 1927. e. Hunter Coll., Drexel Inst. of Technology. President, Sequoia Pictures, Inc. Studied engineering until entry into U.S. Army as member of security agency, special branch of signal corps. Left Army and went into fine arts and theatre. Worked as set designer (Westport, East Hampton, Palm Beach) and in live TV and then on theatrical films. Debut as film producer, David and Lisa (1963). From 1964 to 1969 was president of MPO Pictures Inc. Joined Warner Bros. as prod. exec., 1970. In 1972 founded Sequoia Pictures, Inc. with Fred Weintraub. Now pres. of Paul Heller Prods. Inc.
PICTURES INCLUDE: David & Lisa, The Eavesdropper, Secret Ceremony, Enter the Dragon, Truck Turner, Golden Needles, Dirty Knight's Work, Outlaw Blues, The Pack, The Promise, First Monday in October.

HELLER, WILSON B.: Publicist. b. Omaha, Neb., Aug. 1, 1893; e. U. of Mo., 1914. Free-lance pub. 1922 to 1945 maintaining own office, L.A.; clients, mainly m.p. and radio players, some prod. cos. Was long sec'y & treas. of Wampas; permanent secy., Pioneer Publicists; mgr. owner of College Survey Bureau since 1912; club and college lectures and Commencement addressing; publ., author, Annual, College Fraternities & Sororities Comparisons office; 1574 S. Hayworth, L.A. Calif. 90035. Retired 1970.

HELLMAN, JEROME: Producer. b. New York City. e. N.Y.U. Joined ad dept. of New York Times then went to William Morris Agency as apprentice. Made Asst. in TV dept. Worked as agent for Jaffe Agency. After hiatus in Europe joined Ashley-Steiner Agency (later IFA) where clients included Franklin Schaffner, Sidney Lumet, George Roy Hill, John Frankenheimer. Functioned as TV prod., including Kaiser Aluminum Hour. Left to form own agency, Ziegler, Hellman and Ross. Switched to feature prod. with The World of Henry Orient in 1964.
PICTURES INCLUDE: A Fine Madness, Midnight Cowboy (AA), The Day of the Locust, Coming Home, Promises in the Dark (also dir.).

HELLMAN, LILLIAN: Playwright. b. New Orleans, La.; e. New York U., Columbia. Hon. Degrees: M.A., Tufts College, 1950; LL.D., Wheaton College, 1962; LL.D., Rutgers, 1963; Brandeis, 1965; Mt. Holyoke, 1966; L.L.D. Yale University, 1974; Smith College, 1974; New York University, 1974; Franklin and Marshall College, 1975; Columbia Univ., 1976. With Horace Liveright, Inc. (N.Y. pubs.) 1924–25; theat. playreader. 1927–30; book reviewer, 1925–28. Author: The Children's Hour, The Little Foxes, Watch on the Rhine, The Searching Wind, Another Part of the Forest, Autumn Garden, Toys In the Attic; adapt., The Lark, Montserrat, Candide, My Mother, My Father and Me. Member: Dramatists Guild, Academy of Arts & Letters, Am. Acad. of Arts & Sciences. Started film career 1935 (collab.) The Dark Angel. American Academy of Arts & Letters, Awarded Gold Medal for Drama from Institute of Arts & Letters, 1964; Awarded Brandeis U. Medal of Achievement, 1961. An Unfinished Woman, memoir, 1969, Winner of National Book Award, "The Collected Plays," 1972; "Pentimento," 1973; Scoundrel Time, 1976.
PICTURES INCLUDE: (s.p. based on The Children's Hour), These Three; (s.p.) Dead End, The Little Foxes, The North Star, The Searching Wind The Children's Hour, The Chase (s.p.).

HELLMAN, MARCEL: Producer. b. Bucharest, Rumania. e. Commercial U., Berlin. Entered m.p. Ind. 1924 as managing dir. & prod., Nero Film A.G., Berlin (And Let It Be Light with Conrad Veidt). Later to England as prod. various Brit. cos.
PICTURES INCLUDE: Since 1936, The Amateur Gentleman, Accused, When Thief Meets Thief, Crime Over London, Double Crime in the Maginot Line (English version), Jeannie, Talk About Jacqueline Secret Mission, They Met in the Dark, Wanted for Murder, Meet Me At Dawn, This Was a Woman, Happy Go Lovely, Duel in the Jungle, Let's Be Happy, North West Frontier, The Amorous Adventures of Moll Flanders. Exec. prod. Eng. version Mayerling.

HELPMANN, ROBERT SIR: CBE., Actor-dancer, producer, choreographer. b. Mount Gambier, Australia, April 9, 1909; e. Prince Alfred's Coll., Adelaide, Aus.; studied dancing under Laurent Novikoff. Solo dancer, 1923, Theatre Royal, Adelaide. London debut, 1931, at Sadler's Wells. Appeared on screen, 1942 in One of Our Aircraft Is Missing.
PICTURES INCLUDE: Henry V, Caravan, The Red Shoes, Tales of Hoffmann, Iron Petticoat, Big Money, The Quiller Memorandum, 55 Days at Peking, Chitty Chitty Bang Bang, Alice's Adventures in Wonderland, Don Quixote, The Mango Tree, Patrick.
London & N.Y. stage in Caesar and Cleopatra, Anthony and Cleopatra, The Millionaires, TV, U.S. and England; Old Vic Australian tour, 1955; Old Vic London, 1956–57, Nude with Violin, Royal Ballet 1958–59; Australian tour. Prod.: The Sleeping Beauty, Paris 1960; The Ghost Sonata, BBC TV play, 1962; awarded CBE 1965; Australian of the Year, 1966; co-artistic dir. of Australian Ballet, 1965–1976. Knighted, 1968.

HELPRIN, MORRIS: Executive. b. Paterson, N.J., Dec. 20, 1904; p. Henrietta Helprin; e. Columbia. Assistant film critic, New York Times. General newspaper experience. Publicity, Par. L.I. Studio. Special writer, Samuel Goldwyn, Hollywood. Publicity, Fox Films. Publicity manager, UA. Joined Alexander Korda. In July, 1942 apptd. by Samuel Goldwyn as adv. & pub. repr. in the east. Asst. Chief Overseas M.P. Div. OWI; 1943, Spec. Film Mission OWI. In 1946 rejoined Alexander Korda; pres. London Film Prod. Inc. (America); v.p. Tricolore Films, Inc.; g.m. British Lion Film Corp. Ltd. U.S. rep. all Korda film interests.

HEMMINGS, DAVID. Actor, director. b. Guildford, England, 1941. Early career in opera. Ent. m.p. ind. 1956.
THEATRE: Dylan Thomas in Adventures in the Skin Trade.
TV: Auto Stop, The Big Toe, Out of the Unknown.
PICTURES INCLUDE: Some People, Two Left Feet, The System, The Eye of the Devil, Blow Up, Camelot, Charge of the Light Brigade, Only When I Larf, Barbarella, The Best House in London, Alfred The Great, The Walking Stick, Fragment of Fear, Unman Wittering and Zigo, Voices, Juggernaut, Crossed Swords. Directed: Running Scared, The 14, Quilp, Profundo Rosso, Islands In The Stream, The Squeeze, Murder by Decree.

HENDERSON, DICKIE: Entertainer, actor. Son of late Dick Henderson. Began career in 1933 as child actor in film Cavalcade. Since starred in London West End plays and musicals; variety and revue London Palladium; seven appearances in The Royal Variety Show; cabaret seasons at Savoy, London: and Talk of the Town, as well as New York, Las Vegas, Monaco, Hong Kong, Australia, Kenya, South Africa, Toronto. On TV starred in his own show for 12 years doing 120 shows; guested in every major British show; has done 50 TV shows in USA for Ed Sullivan, Jack Paar etc. London, Palladium, one-man show in UK. 1978: Series for ATV, 'I'm Dickie—That's Show Business.'

HENDERSON, DONALD A.: Executive. b. Melrose, Mass., March 19, 1903. e. Mass. Inst. Technology (1925). From 1925–35 in buying dept. Halsey-Stuart Co.; 1935–45, industrial dept., Lehman Bros. Joined 20th Century-Fox as treasurer and financial dir., later named treasurer and secretary. 1962 elected financial vice-pres., dir., 1968–71 Financial Consultant and Dir. former dir. Gen. Public Utilities Corp. and P. Lorillard Co.; trustee Athens College, Greece.

HENDERSON, MARCIA: Actress. b. Seattle, Wash., Dec. 11, 1934. e. Univ. of Arizona. Appeared in stock in west; m.p. debut in Delicate Delinquent, 1957.
PICTURES INCLUDE: Short Cut to Hell, I Married a Monster from Outer Space, G.I. Blues, Guns of Wyoming, Town Tamers, Dear Phoebe.
Many TV appearances.

HENDERSON, SKITCH: Mus. dir. r.n. Lyle Henderson, b. Birmingham, England. e. U. of Calif.
TV: Steve Allen Show, Tonight Show.

HENDRICKS, BILL L.: Owner/Director Bill Hendricks Films; writer, prod. dir. of documentaries & shorts and TV commercials. b. Grand Prairie, Texas, May 3; e. St. John's College. Formerly Warner Bros. Studio Publicity Director; special asst. to Jack Warner; and Director WB Cartoon Studios. Winner first Quigley Grand Award; mgr., Friendship Train & Merci Train. Author "Encyclopedia of Exploitation" with Howard Waugh, "Showmanship in Advertising" with Montgomery Orr; writer, prod., A Force in Readiness, (Academy Award). The John Glenn Story; The FBI, Seapower, The Land We Love, A Free People, Top Story, A World of Pleasure, Red Nightmare, Star Spangled Revue, This is Eucom, An American Legend, Scenes to Remember, Freedom and You, Jobs, Wonderful World of Warner, Global Marine; Football Safety; An American Partnership, Free Enterprise, Today's Demand, Tomorrow's Challenge, That's Us in the USA, A Special Day.

HENDRY, IAN: Actor. b. Ipswich, Eng., 1931.
TV: Probation Officer, Beyond the Horizon, Afternoon of a Nymph, Ben Spray, A Suitable Case for Treatment, The Avengers, Joy, Police Surgeon, Lotus Eaters.
PICTURES INCLUDE: In the Nick, Live Now, Pay Later, The Girl in The Headlines, This is my Street, The Beauty Jungle, The Hill, The Southern Star, The Internecine File, Get Carter, The Assassin, The Birch, Theatre of Blood.

HENKIN, HOWARD, H.: Executive writer, producer, director. b. New York, Sept. 13, 1926. e. U. of Delaware, 1944. U.S. Army, 1944–46, TV dept., Newell Emmett Agency, 1947–48; gen. mgr., TelePrompter, 1950–54; eastern sales mgr., Shamus Culhane Prod. 1955–57; Academy Pictures 1957–58; pres. HFH Productions, 1958; pres., Henkin Prods. Inc. & Henkin-Faillace Prods. Inc., 1962–1968; ch. of bd., Trio Prods., Inc., 1968; pres., Dot Enterprises, Ltd., 1981.

HENNING, LINDAY KAY: Actress, singer. b. Toluca Lake, Calif. Sept. 16, 1944. e. San Fernando Valley State College, Stage Theatre work in Carousel, Brigadoon, Best Foot Forward, Gidget, Rebel Without a Cause, Bus Stop, Sound of Music, Generation, Enter Laughing, High Button Shoes.
PICTURES INCLUDE: Bye Bye Birdie.
TV: Mr. Ed, Dobie Gillis, P.J., Petticoat Junction

HENNING, PAUL: TV producer-writer. b. Independence, Mo., Sept. 16, 1911. e. Kansas City School of Law. grad. 1932. Radio singer and disc jockey. Also acted, ran sound effects, sang, wrote scripts. To Chicago 1937–38, to write Fibber McGee and Molly. w. Don Quinn. To Hollywood as writer for Rudy Vallee, 1939. Wrote scripts for Burns and Allen 10 years, including transition radio to TV. In 1953 wrote, produced live and film shows for Dennis Day. Created, wrote, produced Bob Cummings Show, 1954–59. Wrote Beverly Hillbillies, 1962 to date. Created, prod. wrote Petticoat Junction. Exec. prod. Green Acres series. Wrote motion pictures, Lover Come Back, Bedtime Story.

HENREID, PAUL: Actor, director. b. Trieste, Italy, 1908; son of Baron Carl von Henreid. e. Maria Therese Acad. Inst. Graphic Arts, Vienna. With book pub. taking night courses at Konservatorium, dramatic arts school in Vienna. In Max Reinhardt's Vienna Theat., then in Austrian films. On dramatic stage and in films among them Jersey Lily and Victoria Regina, London. Arrived in N.Y. 1940 for stage play Flight to the West. In Brit. film Night Train. In 1980 awarded Austrian Cross of Honor, 1st Class, for Science and Arts.
PICTURES INCLUDE: Joan of Paris, Now Voyager, Casablanca, Of Human Bondage, Song of Love, In Our Time, Devotion, Deception, Between Two Worlds, The Spanish Main, Conspirators, Rope of Sand, Last of the Buccaneers, Pardon My French, So Young—So Bad, Stolen face, Man in Hiding, Pirates of Tripoli; actor, dir., prod. For Men Only. Actor, dir., Acapulco, Ten Thousand Bedrooms, etc. Dir. of TV films for Revue Productions, Warner, Desilu, 4-Star, CBS, Screen Gems, Universal, 20th Century-Fox Studios, Actor; Holiday for Lovers, Never So Few, The Four Horsemen of The Apocalypse, The Great Spy Mission; dir., Live Fast Die Young, Take Five From Five, Deadringer, Blues For Lovers, actor. The Madwoman of Chaillot, Colors of Love, Exorcist II: The Heretic; dir., Bracken's World, dir., The Man and the City, 1972–73, "Don Juan in Hell" on major theatrical tour in the U.S. and Canada for 6 months.
TV: Film: Any Number Can Kill.

HENRY, BUCK: Actor, writer. e. Dartmouth College. Acted in Life with Father, tour, 1948; Fortress of Glass, Circle in the Square; Bernardine, Bwy; 1952–54, U.S. Army; No Time for Sergeants, Nat'l. Co.; The Premise, improvis. theatre, off-Bwy; TV's The Steve Allen Show, writer, performer; That Was the Week That Was, writer, performer.
TV: Garry Moore Show; Steve Allen Show; Get Smart, co-creator with Mel Brooks, story editor; producer of Captain Nice series, pilot for NBC-TV.
PICTURES INCLUDE: The Troublemaker; The Graduate, 1966 for Laurance Turman, (producer); Catch 22 (author & actor); Taking Off, Candy, The Owl and the Pussycat, What's Up, Doc?, The Day of the Dolphin (s.p.), The Man Who Fell to Earth, Heaven Can Wait (actor, co-s.p.), Gloria, First Family (s.p., dir., act.).

HENRY, FRED: Producer. b. L.A., Calif., May 30, USC Journalism and business administration; CBS-LA radio station KNX newscaster, writer; news, spec. events dir., KMPC-LA. Co-News commentator, CBS, Let's Talk Over the News; Served, U.S.N. Office of Naval Intelligence, Radio Relations Officer on Staff of Supreme Allied Commander, C-B-I Theatre, Admiral Lord Louis Mountbatten; officer, chg. short-wave broadcasting, AFRS, S.F., 1945; appointed news, spec. events dir., KLAC, Don Fedderson; prg. dir. asst. gen. mgr., prg. dir. combined radio-tv operations, KLAC; res. KLAC, Jan. 1954, as vice-pres., exec. prod., Don Fedderson Prod.; supv. prod.: The Millionaire, Do You Trust Your Wife, Betty White Show, Date With Angels, Liberace, Life with Elizabeth, Who Do You Trust; My Three Sons, 1960; Family Affair, 1966; To Rome With Love, 1969; Henry Fonda Show, 1970.

HENSON, JIM: Producer-Director-Writer. b. Greenville, Miss., Sept. 24, 1936. e. Univ. of Maryland. Early TV work with Washington station, appearing with puppets he built called The Muppets. Did commercials in Washington area, followed by bookings on Today, Tonight shows and Ed Sullivan Show. Followed with Sesame Street, The Muppet Show, and series of TV specials he produced and/or directed. Heads own production company, Henson Associates. First theatrical feature: The Muppet Movie, 1979. 1981: Great Muppet Capor.

HEPBURN, AUDREY: Actress. b. Brussels, May 4, 1929. Appeared in London plays; screen debut in Laughter in Paradise; on B'way in Gigi, Ondine; Star of Tomorrow, 1954.
PICTURES INCLUDE: One Wild Oat, Young Wives' Tales, Lavender Hill Mob, Secret People, Roman Holiday (Academy Award, best actress, 1953), Sabrina, War and Peace, Funny Face, Love in the Afternoon, Green Mansions, The Nun's Story, The Unforgiven, Breakfast at Tiffanys, Paris When it Sizzles, Charade, My Fair Lady; Two for the Road, Wait Until Dark, Robin and Marian, Bloodline, They All Laughed.
TV: Producers Showcase, Mayerling, 1957.

HEPBURN, KATHARINE: Actress. b. Hartford, Conn., Nov. 8, 1909. On stage in Death Takes a Holiday, Warrior's Husband, The Lake, Philadelphia Story, (film version 1940), As You Like It, Millionairess. Started screen career 1933 with A Bill of Divorcement; then Christopher Strong and won Academy Award same year for Morning Glory. On stage 1956 in Merchant of Venice, Taming of the Shrew, Measure for Measure, Coco.
PICTURES INCLUDE: Little Women, Spitfire, The Little Minister, Alice Adams, Sylvia Scarlett, Mary of Scotland, A Woman Rebels, Quality Street, Stage Door, Bringing Up Baby, Holiday, Woman of the Year, Keeper of the Flame, Dragon Seed, Without Love, Song of Love, State of the Union, Adam's Rib, Sea of Grass, African Queen, Pat and Mike, Summertime, Iron Petticoat, The Rainmaker, Desk Set, Suddenly Last Summer, Long Day's Journey Into Night, Guess Who's Coming to Dinner, Lion in Winter, The Madwoman of Chaillot, The Trojan Women, A Delicate Balance, Rooster Cogburn, Olly, Olly, Oxen Free, On Golden Pond.
TV: The Glass Menagerie, Love Among the Ruins.

HERALD, PETER: Executive. UCLA, BA. Supervisory production mgr., Columbia Pictures.
PICTURES INCLUDE: assoc. prod. Almost Angels, Magnificent Rebel, Miracle of the White Stallions, Waltz King, Emil and the Detectives, The Young Loner, prod. supr. There Was a Crooked Man, Soldier Blue; co-prod., Ballerina, assoc. prod. The Great Waltz, Assignment: Vienna, Crazy World of Julius Vrooder, Foul Play, Nightwing. Also: W.W. and the Dixie Dancekings, Mandingo, W.C. Fields and Me, Alex and the Gypsy, Silver Streak, Fire Sale, Star Wars.

HERBERT, HOLMES: Actor. b. Mansfield, Nottinghamshire, England, July 3, 1882; e. Rugby, England, First appearance on stage 1890, London; later with John Sanger circus, Robys Minstrels. F.R. Benson Co., in English provinces; also in many London prods.; on Amer. stage with Charles Frohman Co. Began m.p. career in The Terror 1928.
PICTURES INCLUDE: Gentlemen Prefer Blondes, Sporting Age, Madame X, The Kiss, Untamed, Dr. Jekyll

and Mr. Hyde (1931), The House of Rothschild, Captain Blood, Lloyds of London, Lancer Spy, British Intelligence, This Above All, Sherlock Holmes in Washington, The Verdict, The Swordsman, Singapore, The Brigand.

HERBUVEAUX, JULES: Executive. b. Utica, N.Y., October 2, 1897; e., Culver Military Academy, McGill U., Montreal; m. Jeannette McNulty. Professional musician, 1919–21; orchestra dir., 1921–32; orchestra dir. NBC, 1927–32. asst. prod. mgr., 1932–38; program mgr., central div., 1938–47; mgr. television, 1948–53; ass't gen. man., 1953, gen. mgr., 1954; v.p. and gen. mgr., 1955, v.p. Central Div.; Merchants and Manufacturers (Chicago); retired, NBC, 1962; communications consultant for Zenith Radio Corp., Chicago; retired Jan., 1972.

HERDER, W. ED: Producer. b. New York City. Univ. of Miami (BBA). Pres. and chm. of bd., Worldfilm Corp. Financial investor in many m.p. and t.v. prods.
PICTURES: Force of Impulse, Without Each Other, Black Cobra, Shoot, Never Too Young to Rock, The Liberian Move.
TELEVISION: Hollywood, The Magic Kingdom.

HERFEL, CHRIS.: Executive. b. Bklyn, N.Y., Sept. 5, 1927. e. Univ. of Delaware, 1944–45; N.Y. Univ., 1949. Studio mgr., prod., exec. prod., Transfilm-Carnvel, Inc., 1957–61; Asst. dir., MPO-TV, 1957; studio mgr., assoc. prod., ATV-Films, 1956–57; Asst. dir., dir., Army Pictorial Centre, 1950–56; v.p. chg. of prod. Wylde Films, Inc.

HERMAN, LEONARD WOOD: Editor, assoc. prod. b. Springfield, Ohio, Jan. 16, 1913. Managed theat. circuits, Iowa, Nebraska, 1931–36. With Universal in mailing, casting, projection, library cutting depts., 1938–42; in 1943 film ed. In 1945 to Monogram in same capacity.
PICTURES INCLUDE: Great Alaskan Mystery, Raiders of Ghost City, Mystery of the River Boat, Jungle Queen, The Master Key, Shanghai Cobra, Suspense, Swing Parade, Charlie Chan in Mexico, Vacation Days, and Louisiana (edit.), Ginger, Rocky (assoc. prod.); (ed.) Big Timber, Jack Slade.

HERMAN, NORMAN: Producer-Director. b. Newark, N.J. e. Rutgers Univ., N.Y. Univ. Was accountant in California; in 1955 switched to film ind., joining American Int'l Pictures. Headed AIP prod. dept. 4 years, incl. prod., post-prod., labor negotiations, supervising story dept., etc. Now pres. of Zide-Herman Co., Inc.
PICTURES INCLUDE: All producer in some capacity except where noted as also directing and/or writing: Sierra Stranger, Tokyo After Dark (dir.), Hot Rod Girl, Hot Rod Rumble, Crime Beneath Seas, Look in any Window, Mondy Teeno (dir., co-s.p.), Glory Stompers, Three in the Attic, Bloody Mama, Pretty Boy Floyd, Dunwich, Three in the Cellar, Angle Unchained, Bunny O'Hare, Psych-Out, Killers Three, Frogs, Blacula, Legend of Hell House, Dirty Mary Crazy Larry.
TELEVISION: Writer: Robert Taylor Detectives, Iron Horse, Invaders, Adam 12, Lancer. Director-Producer: Hannibal Cobb, You Are the Judge.

HERMAN, PINKY: Journalist, songwriter. b. N.Y.C., Dec. 23, 1905; e. N.Y.U. Song writer; member, ASCAP. Counsel member, S.P.A.; writer, M.P. News, 1930; 1934; charter member, Songwriters Protective Assoc.; writer, M.P. Daily, 1935–43; columnist, Radio Daily, 1943–50; TV columnist for M.P. Daily. Retired, 1973.
SONGS INCLUDE: (collab.) Face the Sun, All Around the Town, Boom Ta Ra Ra, It Must Be LUV, Piano Teacher Song, Manhattan Merry Go Round, Lucky, I'm Still in Love With You, Havin' A Wonderful Time, Where Can You Be, Seven Days a Week, Texas Lullaby, Lighthouse in the Harbor, I'm Cuckoo Over You, It's a Coincidence, If I Had a Million Dollars, Masquerade of Love, Mademoiselle Hortensia, Come Back to Me My Love, Someday When Shadows Fall, Myrtle the Turtle & Flip the Frog, Cowboy Isn't Speaking to His Horse, Gotta Horse That Loves to Rhumba, Girl on My Grandfather's Shaving Mug, Definition Song, Got a Gal in a Town Way Out West, Sing Something Irish, Little Sweetheart of the Mountains, Make a Million, Music Sets Me Free, She's the Daughter of the Sweetheart of Sigma Chi, Who Wants Love, Good-Lookin' It's Good Lookin' at You, Carib Gold, What Makes the Rainbow, Right Across de Ribber, Shadows in the Moonlight, Bible My Mother Left to Me, Poor Little Doll, Little Bit O'Rhythmn in the Best of Us, Lovely Lady, When a Girl's In Love, Never Leave A Lady When She Loves You, Without You, That's The Way to Live, It's Time to Sing, Acapulco (By The Sea), Councilman; The Lambs.

HERSHEY, BARBARA: Actress. e. Hollywood H.S.; Screen debut in With Six You Get Eggroll.
PICTURES INCLUDE: Last Summer, The Liberation of Lord Byron Jones, The Baby Maker, The Pursuit of Happiness, Boxcar Bertha, Vrooder's Hooch, Diamonds, The Last Hard Men, The Stunt Man, Take This Job and Shove It.
TV: Gidget, The Monroes, The Farmer's Daughter, Run For Your Life, The Invaders, Daniel Boone, CBS Playhouse, Chrysler Theatre.

HERSKOVITZ, ARTHUR M.: Executive. b. Mukden, China. Nov. 28, 1920. e. CCNY, 1939. Joined RKO, scenario dept., 1939. U.S. Army, 1942–46; appt. mgr. RKO Radio Pictures of Peru, 1955; RKO Pict. Peru, 1958; Warner Bros., Peru, 1958–64; MGM, Panama, 1965–67; MGM rep. in Japan, 1968; Far East supvr., 1970; Joined National General Pictures, 1973 as foreign sls. mgr. In 1974 appt. dir. of sls., JAD Films Int'l.; named v.p., 1978; pres., 1979.

HERTZ, WILLIAM: Executive. Began theatre career in 1939; 1946 joined Fox West Coast Theatres; theatre mgr., booking dept.; 1965 appointed Los Angeles first-run district mgr.; promoted to Pacific Coast Division Mgr., National General Corp., 1967; v.p. Southern Pacific Div. Mgr., National General Theatres, Inc. 1971. Now with Mann Theatres as dir. of theatre operations.

HESSLER, GORDON: Producer, director. b. Berlin, Germany. e. Reading U., England; dir., vice pres., Fordel Films, Inc., 1950–58; dir., St. John's Story (Edinborough Film Festival), March of Medicine Series, Dr. Albert Lasker Award; story edit., Alfred Hitchcock Presents 1960–62; assoc. prod., dir., Alfred Hitchcock Hour, 1962; prod., Alfred Hitchcock Hour; prod., dir., Universal TV 1964–66.
PICTURES INCLUDE: The Woman Who Wouldn't Die, The Last Shot You Hear, The Oblong Box, Scream and Scream Again, Cry of the Banshee, Murders of the Rue Morgue, Sinbad's Golden Voyage, Medusa, Next Week Rio, Puzzle, Starlock.
TV: Alfred Hitchcock Presents, 1960–62; Alfred Hitchcock Hour, 1962–65; Run For Your Life, Convoy, Bob Hope Chrysler Show, 1964–66; ABC Suspense Movies of the Week, ABC Movies of the Week, 1973. Lucas Tanner, Night Stalker, Amy Prentis Switch, 1974; Kung Fu, Sara, Hawaii Five O, 1975–76; Blue Knight; Wonder Woman.
Pilots: Tender Warriors, Chips, Thief.

HESTON, CHARLTON: Actor. b. Evanston, Ill., Oct. 4, 1924; e. Northwestern U. School of Speech, Radio, stage, TV experience. Following coll. served 8 yrs. 11th Air Force, Aleutians. After war, co-acted (leads) and dir. with wife, Thomas Wolfe Memorial Theatre, Asheville, N.C. in State of the Union, Glass Menagerie; member, Katharine Cornell's Co., during first year on Broadway; Anthony and Cleopatra, other Bway. plays, Leaf and Bogh, Cockadoodle Doo; Studio One (TV): Macbeth, Taming of the Shrew, Of Human Bondage, Julius Caesar, etc.
PICTURES INCLUDE: Dark City, Greatest Show on Earth, The Savage, Ruby Gentry, President's Lady, Pony Express, Arrowhead, Ead for Each Other, Naked Jungle, Secret of the Incas, Far Horizons Lucy Gallant, Private War of Major Benson, Ten Commandments, The Maverick, Ben-Hur, Wreck of the Mary Deare, El Cid, The Pigeon That Took Rome, 55 Days at Peking Major Dundee, The Agony & The Ecstasy, The War Lord, The Greatest Story Ever Told, Khartoum, The Battle Horns, Planet of the Apes, Beneath the Planet of the Apes The Omega Man, Anthony and Cleopatra, (star & dir.), Skyjacked, Soylent Green, The Three Musketeers, Airport 1975, Earthquake, The Four Musketeers, The Last Hard Men, Midway, Two Minute Warning, The Prince and the Pauper (Crossed Swords), Gray Lady Down, The Awakening.

HETZEL, RALPH D.: Executive. b. Corvallis, Ore., August 18, 1912; e. Penn. State Univ. (A.B.), 1933; University of London, 1935–36; Private secy. to Gov. Pinchot of Pa. 1933–35; did research & study, 1936–39; exec. secty. natl. hdqts., CIO, 1937–40; economic dir., 1938–40; in service, 1942–45. Consultant on labor, Nat. Selective Service hdqts., 1942; manpower consultant War Prod. Bd., 1942–43; dept. vice chmn., manpower requirements, W.P.B., 1943–45; acting vice-chmn., 1945; dir. Office of Labor Requirements, Civilian Prod. Admin., 1945–46; asst. to Secy. of Commerce, U.S. Dept. of Comm., 1946–48; asst. to secty. & dir. Office of Program Planning, 1948–51; asst. admin., Economic Stabilization Agency, 1951; exec. v.p. Motion Picture Association, MP Export Assn., 1951–71, Past president, International Federation of Film Producers Assns.; member of Board of Trustees, California Institute of the Arts and of Penn State University; member of Film Advisory Committee of Museum of Modern Art, and of Advisory Council of Edward R. Murrow Center of Public Diplomacy at Tufts University. Dean, College of Fine and Professional Arts, Professor of Art, Kent State University, 1971–76; provost and vice president, academic affairs, California Institute of the Arts 1976–80; faculty member, 1976–.

HEYER, JOHN: Producer, director. b. Devenport, Australia, 1916. e. Scotch College, Melbourne. Dir., s.p., Native Earth, Journey of A Nation, The Valley Is Ours, Back of Beyond (Venice Grand Prix, 1954) Forerunner. Prod. for Allied Works Council, Film advisor to Prime Minister's Propaganda Comm., World War II; collab. s.p., dir., The Overlanders, 1945; prod., Australian Nat'l. Film Board, 1945–48; prod., Shell Film Unit. Aust., 1948–1956; Appt'd. by Government, exec. of New South Wales Film Council, several yrs. pres., Federation of Film Societies, Sydney Film Society; v.p., Scientific Film Soc., Governor, Australian

Film Institute; many Int. film awards; exec. prod. films & TV Shell Inter'l., London, 1957–67.

HEYWARD, LOUIS M.: Writer, prod., exec. b. New York, N.Y., June 29, 1920. e. NYU, Brooklyn Law School. Sr. ed., Press Assoc. Div., AP; formed Radio Transcript Div.; Radio writer: CBS Garry Moore Show, Ernie Kovacs Show; co-owner, Heyward-Wilkes Industrial Film Co.; prod., Warner Bros.; s.p., Pajama Party. TV: Dick Clark Show, Pete and Gladys Show; dir. TV prog. dev., 20th Century Fox; dev. of live progs., MCA-TV Ltd.; dir. m.p. & TV dev., American International Pictures.

HEYWOOD, ANNE: Actress. m. Raymond Stross.
PICTURES INCLUDE: Checkpoint, Doctor at Large, Dangerous Exile, The Depraved, Violent Playground, Floods of Fear, Heart of a Man, Upstairs and Downstairs, A Terrible Beauty, Carthage in Flames, Petticoat Pirates, Stork Talk, Over My Dead Body, The Very Edge, 90 Degrees In The Shade, The Fox, A Run On Gold, The Most Dangerous Man in the World, The Nun of Monza, I Want What I Want, And Presumed Dead.

HICKMAN, DARRYL: Actor. b. Hollywood, July 28, 1933. Started screen career 1938 with The Starmaker. Now with CBS as exec. prod., daytime programming.
PICTURES INCLUDE: Grapes of Wrath, Young People, Jackass Mail, Northwest Rangers, Keeper of the Flame, And Now Tomorrow, Salty O'Rourke, Captain Eddie, Kiss and Tell, Leave Her to Heaven, Black Gold, Happy Years, Submarine Command, Destination Gobi, Island in the Sky, Sea of Lost Ships, Southwest Passage, The Human Comedy, Men of Boystown, Fighting Father Dunn, Tea and Sympathy.

HIFT, FRED: Executive. b. Vienna, 1924. e. Vienna, London and Chicago. Early career reporter Chicago Sun and radio work with CBS News, New York. 1946 joined Boxoffice magazine; 1947 Quigley Publications; 1950 Variety. 1960 began career as publicist on Exodus. 1961 dir. pub., The Longest Day for Darryl Zanuck. 1962 joined Fox in Paris as ad-pub. dir. for Europe. 1964 became dir. European prod. pub. with headquarters London. Formed own pub., p.r. Co., Fred Hift Associates, 1970. 1979, joined Columbia Pictures as dir. of eastern ad-pub operations in N.Y.; 1980, to United Artists as intl. adv./pub. v.p.

HIGGINS, COLIN: Writer-Director. e. Sanford Univ., Sorbonne, UCLA (masters in film). Co-wrote The Ik for Peter Brook's International Center of Theatrical Research; brought to U.S. in 1977 where toured college campuses. Wrote s.p. for Harold and Maude and Silver Streak before turning to directing with Foul Play (also s.p.).
PICTURES: Nine to Five (co. s.p., dir.).

HILFORD, LAWRENCE B.: Executive. b. New York City. e. Yale University; Harvard Business School. Held exec. positions with Columbia Pictures Industries, as v.p. & gen. mgr. of Screen Gems Intl.; named exec. v.p., SGI; v.p. & gen. mgr., Cassette Division of Columbia Pictures. Pres. of Cartridge Rental Network. Pres. Viacom Enterprises Division. Served as exec. v.p. & dir. of Viacom Intl. Inc. In 1979 named snr. v.p. of Columbia Pictures Industries to develop new markets in area of video cassettes and discs.

HILL, ARTHUR: Actor. b. Saskatchewan, Canada, 1922. e. U. of British Columbia. Moved to England in 1948, spending ten years in varied stage & screen pursuits; starred on Broadway in Who's Afraid of Virginia Woolf? (Tony Award), later in More Stately Mansions; film debut in Miss Pilgrim's Progress; other British work includes The Body Said No, Raising A Riot, The Deep Blue Sea.
PICTURES INCLUDE: The Young Doctors, The Ugly American, In the Cool of the Day, Moment to Moment, Harper, Petulia, Don't Let the Angels Fall, The Chairman, The Pursuit of Happiness, Rabbit Run, The Andromeda Strain, The Killer Elite, Future World, A Bridge Too Far, A Little Romance, The Champ, Dirty Tricks, Making Love.
TV: Owen Marshall, Counselor-At-Law, Death Be Not Proud, Churchill and the Generals.

HILL, BENJAMIN (BENNY): TV performer. b. Southampton, England, Jan. 21, 1925. TV debut 1952; since many TV appearances including Showcase, Benny Hill Show; Midsummer Night's Dream, 1964; winner Daily Mail Award TV personality of the year 1954; m.p. debut, Who Done It, 1955; Light Up the Sky, 1960; Those Magnificent Men in Their Flying Machines, 1965; Chitty Chitty Bang Bang.

HILL, GEORGE ROY: Director. b. Minneapolis. e. Yale and Trinity College in Dublin. Started as actor, Irish theatres and U.S. Shakespeare repertory company, also off-Broadway. Served as Marine pilot in World War II and again in Korean War. Wrote TV play, My Brother's Keeper, for Kraft Theatre, later rose to director with show. TV assignments included A Night to Remember, The Helen Morgan Story, Judgment at Nuremberg. Directed first Broadway play in 1957, Look Homeward Angel, followed by The Gang's All Here, Greenwillow. First film, Period of Adjustment, 1962.
PICTURES INCLUDE: Toys in the Attic, The World of Henry Orient, Hawaii, Thoroughly Modern Millie, Butch Cassidy and the Sundance Kid, Slaughterhouse Five, The Sting, The Great Waldo Pepper, Slap Shot, A Little Romance.

HILL, TERENCE: Actor. b. Venice. r.n. Mario Girotti. First attracted attention as actor in Visconti's The Leopard (1963). Gained fame in European-made westerns.
PICTURES INCLUDE: God Forgives, I Don't, Boot Hill, Ace High, Barbagia, Anger of the Wind, They Call Me Trinity, Trinity Is Still My Name, My Name Is Nobody, Mr. Billion, March or Die.

HILL, WALTER: Writer. b. Long Beach, Calif. Jan. 10, 1942. e. Michigan State Univ. Was construction and oilfield worker before writing screenplays.
PICTURES INCLUDE: Hickey & Boggs, The Getaway, The Thief Who Came to Dinner, The Mackintosh Man, The Drowning Pool, Hard Times (also dir.). The Driver (s.p., dir.). The Warriors (s.p., dir.), Alien (co-prod.); The Long Riders (dir.); Southern Comfort (dir., co-s.p.

HILLER, ARTHUR: Director. b. Edmonton, Alberta, Can., Nov. 22, 1923. e. Univs. of Alberta, Toronto, Brit. Columbia.
TV: Matinee Theatre, Playhouse 90, Climax, Alfred Hitchcock Presents, Gunsmoke, Ben Casey, Rte. 66, Naked City, The Dick Powell Show.
PICTURES INCLUDE: The Careless Years, Bryne Prods., Americanization of Emily, Tobruk, The Tiger Makes Out, Popi, The Out-of-Towners, Love Story, Plaza Suite, The Hospital, Man of La Mancha, The Man in the Glass Booth, W. C. Fields and Me, Silver Streak, Nightwing, Making Love.

HILLER, WENDY: (DBE, 1975, OBE, 1971). Actress. b. Bramhall, Cheshire, England, Aug. 15, 1912. e. Winceby House Sch., Bexhill. On stage 1930, Manchester Repertory Theatre, England; then on British tour. London debut 1935 in "Love On the Dole"; to N.Y., same role 1936. m.p. debut in "Lancashire Luck," 1937.
PLAYS: First Gentleman, Cradle Song, Tess of the D'Urbervilles, Heiress, Ann Veronica, Waters of the Moon, Night of the Ball, Old Vic Theatre, Wings of the Dove, Sacred Flame, Battle of Shrivings, Crown Matrimonial, John Gabriel Borkman, Waters of the Moon (revival).
PICTURES INCLUDE: Pygmalion, Major Barbara, I Know Where I'm Going, Outcast of the Islands, Single Handed (Sailor of the King), Something of Value, Uncle George, Separate Tables, Sons and Lovers, Toys in the Attic, Man For All Seasons, David Copperfield, Murder on the Orient Express, Voyage of The Damned, The Elephant Man.

HILLIER, ERWIN: Director of Photography-Producer. e. Academia of Arts, Berlin.
PICTURES INCLUDE: I Know Where I am Going, Private Angelo, Interrupted Journey, October Man, Canterbury Tale, Shadow of the Eagle, Chase A Crooked Shadow, Happy Go Lovely, Where's Charley, Will Any Gentleman, House of the Arrow, Isn't Life Wonderful, Duel in the Jungle, Dam Busters, Now and Forever, The Mark of the Hawk, Let's Be Happy, Naked Earth, Shake Hands with the Devil, The Long and the Short and the Tall, School for Scoundrels, The Naked Edge, A Matter of Who, Sammy Going South, The Great Spy Mission, Sands of Kalahari, The Eyes of the Devil, Berlin Memorandum, The Valley Time Forgot, The Shoes of the Fisherman; preparing for production: Anna Pavlova, Eliza Lynch, The Lost City, Shaka Zulu, African Adventure, El Dorado by Alistair Maclean, Los Vagabundos, Uganda Queen.

HILTON, MARS: Actor. b. New York City, Nov. 6, 1951. Since age 3 has appeared on TV and five films in Europe and U.S. Title role, The Pawn, for Seven Arts—Frank Sollento. Speaks Italian; dubbing Pawn and other films at Fono-Roma Studios, Rome.

HINDIN, PHILIP: Executive. b. London, England, 1916; prod., stage dir., m.p. casting, 1937–39; Wielands Ltd. prod., revues, Far East; mgr. dir., Maurice Winnick Org.
TV: Password, Play Your Hunch, Call My Bluff, 20 Questions, He Said; She Said, Watchword, G.O. Water Rats, Goodson-Todman Prods., Chuchin Chow, Hit the Deck, Gentlemen Prefer Blondes, The Hanging Man, Funny Girl, Rumanian State Dance Co., Where Did That Come From, Beat the Clock, The Merry Widow, Bayerische Rundfunk Munich, Belgische Radio en Televisie, Radio Telefis Eireann, Dansk Rad. and Television, What's My Line, Music Bingo, Was Bin Ich, The Sunshine Boys, What's The Origin?, 1000 Pd. Pyramid Show: Take My Word.

HINGLE, PAT: Actor. b. 1923.
THEATRE: Man for All Seasons (Oherbein College, Ohio); The Price, Macbeth (American Shakespeare Festival).
RADIO: Voice of America.
PICTURES INCLUDE: On the Waterfront, The Strange One, No Down Payment, Splendor in the Grass, The Ugly American, Sol Madrid, Hang Em High, Jigsaw, Norwood, Bloody Mama, WUSA, Corporal Crocker, The Carey Treatment, One Little Indian, The Super Cops, The Gauntlet, When You Comin' Back, Red Ryder, Norma Rae.

HINKLE, ROBERT: Actor-Prod.-Dir. b. Brownfield, Texas, July 25, 1930; e. Brownfield High School. Joined Rodeo Cow-

boys Association, 1948 and rodeoed professionally until 1953 when began acting career in Outlaw Treasure.
PICTURES INCLUDE: The First Texan, Dakota Incident, Gun the Man Down, The Oklahoman, First Traveling Saleslady, No Place to Land, Under Fire, Speed Crazy, The Gunfight at Dodge City, Broken Land, Law in Silver City, Ole Rex (award for Family Entertainment), Born Hunter, Trauma, Something Can Be Done, Mr. Chat, Stuntman, Jumping Frog Jubilee, Mr. Chat-Mexico Safari, Oregon Today, California Today, Alabama Today, Ride Trail, Virginia City Cent., Texas Today, Texas Long Horns, Kentucky Thoroughbred Racing, Produced & Directed Country Music, Guns of a Stranger.
TV: Test Pilot, Dial 111, Juvenile Squad, X13 Vertijet, Giant, Opposite Sex, All the Fine Young Cannibals, Hollywood Jamboree, Color Me Lucky, Cellist Extraordinary.
Pres. Hinkle Pictures, Inc.

HIRD, THORA: Actress. b. Morecambe (Lancs), Eng., May 28, 1914. e. The Nelson School, Morecambe.
PICTURES INCLUDE: (Screen debut, 1940) The Black Sheep of Whitehall; Street Corner, Turn the Key Softly, Personal Affair, The Great Game, Storks Don't Talk, Shop Soiled, For Better or Worse; Love Match, One Good Turn, Quatermass Experiment, Simon and Laura, Lost, Sailor Beware, Home and Away, Good Companions, The Entertainer, A Kind of Loving, Term of Trial, Bitter Harvest, Rattle of A Simple Man, Some Will, Some Won't, The Nightcomers.
TV: The Winslow Boy, The Bachelor, What Happens to Love, The Witching Hour, So Many Children, The Queen Came By, Albert Hope, All Things Bright and Beautiful, Say Nothing, Meet the Wife, Who's A Good Boy Then? I AM! Dixon of Dock Green, Romeo & Juliet, The First Lady, Ours Is a Nice House, The Foxtrot, Seasons, She Stoops to Conquer, Villa Maroc, When We Are Married, The Loving Memory, Flesh and Blood, Your Songs of Praise Choice.

HIRSCH, JUDD: Actor. b. New York City, March 15. e. City College of N.Y. Studied physics but turned to acting; first job in 1962 when performed with stock co. in Colorado. Returned to N.Y. to work on stage and since has also done films and TV.
PICTURES INCLUDE: Serpico, Ordinary People.
TV: Medical Story, Delvecchio. Movies: Fear on Trial, Valentino, The Law, The Keegans, Taxi.

HIRSCHFIELD, ALAN J.: Executive. e., University of Oklahoma (BA); Harvard Business School (MBP). Presently vice-chm., chief operating officer, 20th Century-Fox. V.P., Allen & Co., 1959–66; Financial v.p. & dir. Warner/7 Arts, 1967–68; v.p. & dir., American Diversified Enterprises, 1969–73; pres. & chief exec. officer, Columbia Pictures Industries, 1973–78; consultant, Warner Communications, 1979.
Trustee: Cancer Research Institute. Director: Film Society of Lincoln Center, Motion Picture Association of America, N.Y. State Motion Picture and Television Advisory Board, Straight Arrow Publishing Co., John B. Coleman Co.

HIRSCHMAN, HERBERT: Prod.-dir. b. N.Y.C.; e. Univ. of Michigan, B.A., Yale Univ. Drama School, M.F.A. Story dept. RKO & Paramount; Prod. & dir. summer stock; B'way stage mgr., production manager, TV since 1948. Awards: Peabody (dir., Mr. I-Magination), Robert E. Sherwood (dir., PH 90, "Made in Japan"), TV Guide (prod. Perry Mason); 2 Christophers, dir., London Stage. A Thousand Clowns; prod., m.p. features, Scalplock, Halls of Anger, They Call Me Mister Tibbs.
TV: Dir. Omnibus, Starlight Theatre, Mr. I-Magination, Steve Allen Show, Celebrity Time, What's My Line, Beat the Clock, Inside USA, Les Paul & Mary Ford, The Nurses, The Defenders, For the People, Iron Horse, Felony Squad, Hawaii 5-O, Waltons, prod. and dir., The Web, Alcoa Hour, Goodyear Playhouse, Studio One, Playhouse 90, Perry Mason, Hong Kong, Dr. Kildare, Twilight Zone, Espionage, Wackiest Ship in the Army; Men from Shiloh, Young Lawyers, Bold Ones (The Doctors), GE Theatre: Tell Me Where It Hurts, Larry, Things In Their Season, Miles to Go Before I Sleep; MOW's: Returning Home, Eric, The Zoo Gang, The Amazing Howard Hughes, The Greatest Thing that Almost Happened, The President's Mistress, Human Feelings, Flesh and Blood, And Baby Makes Six: Exec. prod. Iron Horse, Planet of the Apes, The Scarlet Letter, Sophia.

HIRSHAN, LEONARD: Theatrical Agent. b. Dec. 27, 1927, New York City. e. NYU. Joined William Morris Agency as agent trainee, New York, 1951. Agent legit theatre & TV dept. 1952–1954. Sr. exec. agent M.P. dept., California office, 1955.

HIRSHBERG, JACK: Publicist. b. Montreal, Canada, July 17, 1917; e. Sir George Williams Univ., Montreal. Did pub., Montreal Symph. Orch. Canadian Broad. Co.; Hollywood staff rep. Toronto Star, Montreal Star, 1939. Paramount Pictures pub., 1940; in U.S. Navy, 1941–45. Returned to Paramount; pub., expl. rep. Perlberg-Seaton Prod., Frank Ross, Mirisch Co., publicist for C. B. DeMille; 1963 p.b. dir. for Geo. Stevens, Greatest Story Ever Told, 1964; in-dep. prod., corp. pub. rel. publicist, APJAC Prod. Dr. Doolittle, Planet of the Apes, 1966–67; dir. of publicity, APJAC Prod., 1969; v.p., APJAC Prod., 1971.

HIRT, AL: Musician. b. New Orleans, La., Nov. 7, 1922. e. Cincinnati Conservatory of Music. Military service four years. Played with Tommy and Jimmy Dorsey bands, Ray McKinley and Horace Heidt; appeared, Dunes Hotel, Harrah's Club, Basin Street East, Cloisters, Palmer House, Eden Roc Hotel, Greek Theatre. European tour concerts. Inaugural Ball, President John F. Kennedy, Jan., 1961.
TV: Dinah Shore Show, Jack Paar Show, Ed Sullivan, NBC Special Home For the Holidays, Bell Telephone Hour Rainbow of Stars, Andy Williams Show, Tonight Show, Today Show, Perry Como Show, Lively Ones, Jerry Lewis Show.
PICTURES INCLUDE: World By Night, Rome Adventure.
RECORDINGS: RCA: The Greatest Horn in the World, Al—He's the King—Hit, Horn-A-Plenty, Al Hirt at the Mardi Gras, Trumpet & Strings, Our Man in New Orleans, Honey In The Horn, Beauty And The Beard, Pop Goes the Trumpet, Boston Pops.

HITT, J. T.: Executive. b. Spearman, Texas, Dec. 16, 1922. e. Grad., Gentry, Ark. H.S., 1940. Farming and ranching. Ent. motion picture industry 1947; Pres., Independent Theatre Owners of Arkansas, 1961–63; chmn. bd., 1964; sec.-treas., Tri-States Theatre Owners, mem. bd., Theatre Owners of America; pres. Tri-State T.O. Ark., Tenn. & Miss., 1965–66.

HIX, AL: Publicist. b. N.Y., Feb. 19, 1918. e. U.S.C., B.A., Phi Beta Kappa, 1949. Pubrel, Ephebian Soc. L.A., 1937–40; U.S. Army Signal Corps, 1941–45; edit.-prod. office, Time mag., & editor Underwriters Report, L.A., 1948–52; unit publicist Europe, & Paris col., Hlywd Rpt, 1952–54; senior publicist, 20th-Fox, 1954–56; since then film publicist in Europe, Asia, Africa & Mid-East 60-plus films) except for publ. "The Rain People," WB, 1968, and publicity for Academy Awards in L.A., 1971.

HOARE, VICTOR J.: Executive. Entered m.p. ind. with United Artists 1931; Dir., British Lion Films, 1961; man. dir., BLC Films, 1962; dep. chmn., BLC Films, 1965; man. dir. British Lion Films, 1965; v.p. in chg. of Continental Europe, Columbia International, in Paris, 1966; v.p., theatres, C.I.C., 1974.

HOBBES, HALLIWELL: Actor. b. Stratford-on-Avon, England, November 16, 1877. First appearance on stage in Glasgow, 1898. First sound picture with Jeanne Eagels in Jealousy, then Grumpy.
PICTURES INCLUDE: Dr. Jekyll and Mr. Hyde, Should Ladies Behave? Captured, Lady for a Day, Mandalay, The Key, Sherlock Holmes Faces Death, The Invisible Man's Revenge, Gaslight, Mr. Skeffington, Casanova Brown, If Winter Comes, You Gotta Stay Happy, You Can't Take It With You, Here Comes Mr. Jordan, Black Arrow, That Forsythe Woman, Life of Pasteur, Miracle in the Rain.

HOBIN, BILL: TV prod. dir.; r.n. Charles W. Hobin; b. Nov. 12, 1923, Evanston, Ill.; e. U. of So. Calif.; prod. mgr., Coronet Instructional Films, Glenview, Ill.; dir., Garroway at Large, Wayne King Show from Chic., 1949–51; dir., Assignment Manhunt, 1952; Nothing But the Best, 1953; Les Paul and Mary Ford, 1953; Your Show of Shows, 1951–54; assoc. prod.-dir., Max Liebman Presents (Spectaculars), prod. dir. Fred Waring Show, Andy Williams Show, Pat Boone, Timex All-Star Jazz Show. dir. Your Hit Parade. Prod.-dir. The Golden Circle, 1959; The Bell Telephone Hour, 1959–60; The American Cowboy, 1960; Sing Along with Mitch, 1960–63; dir., Judy Garland Show, 1963; prod., dir., Victor Borge at Carnegie Hall, 1964; dir., m.p. Chrysler Show-Go-Round, N.Y. World's Fair. 1964. Dir.: Meredith Wilson Special, 1964; Jack Jones on the Move Special, 1966; Red Skelton Hour, 1964–68; prod.; Red Skelton Hour, 1968–70; prod. and dir. The Bill Cosby Special I, 1968; prod., dir., The Tim Conway Comedy Hour, 1970–71; prod., dir. The CBS Newcomers series, 1972; dir. An Evening With My Three Sons special, prod., dir. Michel Legrand Special; prod., dir. Fred Astaire special; director, Maude, 1972; director, A Touch of Grace, 1972–73; 1974, prod.-dir. credits: Dinah, Won t You Please Come Home! (NBC Special); Bobby Goldsboro Show (syn); Your Hit Parade (CBS), prod.-dir. Flip Wilson Special, Dionne Warwick Special, Wayne Newton Special, all 1975; 1976—dir., Welcome (Back), Kotter, Prod.-dir., Monty Hall's Variety Special. Dir., Bert Convy Show; 1976 Director, George Burns Special; McLean Stevenson Show, 1976–78, Director, Three's Company. Producer-director Bob Hope Specials, (1978–79) (NBC); director Celebrity Challenge of the Sexes, 1979 (CBS).

HOBLEY, McDONALD: b. Port Stanley, Falkland Islands; e. South America; Brighton Col. Early career in repertory cos. tours, etc.; entered TV in 1946 as TV announcer; won Nat'l TV Award for Personality of the Year, 1953–54; resigned from BBC, 1956; now free-lance.
PICTURES INCLUDE: Checkpoint, The Entertainer.

TV: Holiday Town Parade, The Dave King Show, Tell the Truth, The Tommy Steele Show, Close-up.

HOCK, MORT: Executive. b. New York, N.Y., June 24, 1929. United States Army, 1951–53. Blaine-Thompson Agency; A. E. Warner Bros., 1945; David Merrick B'way Prod., 1958; asst. adv. mgr., Paramount Pictures Corp., 1960; adv. mgr., United Artists Corp., 1962; dir. adv., United Artists Corp., July 1964; adv. dir., Para., 1965; v.p. advertising and public relations, Paramount Pict. Corp., 1968–71; pres., Atina Enterprises, Inc. 1971; v.p., marketing, Rastar Prods., 1973; exec. v.p., Charles Schlaifer & Co., 1974.

HODDER-WILLIAMS, CHRISTOPHER: Writer. b. 1927. e. Eton. Military Service, 1944–48; early career as entertainer, property man, songwriter, prof. novelist, writer, 1957.
AUTHOR: The Cummings Report, Chain Reaction, Final Approach, Turbulence, The Higher They Fly. Novels: 1964, The Main Experiment; 1967, The Egg-Shaped Thing; 1968, Fistful of Digits, 98.4, Panic O'Clock, Coward's Paradise, The Prayer Machine.
TV: (Writer) The Ship That Couldn't Stop, The Higher They Fly.

HOFF, J. ROBERT: Executive. b. Chicago, Ill., Mar. 6, 1909; e. U. of Illinois, A.B. 1930; John Marshall Law School, J.D. 1934. With law firm of Hoff & Collis, Chic. 1934–41; Lt. Comdr. U.S.N.R. 1941–45; sales mgr. Ballantyne Co. 1945–55; pres., 1957–61; exec. v.p., Ballantyne Instr. and Elec. Inc., subsid. ABC Consol., 1962; ABC Consol. Sub., Ogden Corp., 1967 (repurchased and renamed Ballantyne of Omaha, Inc.); bd. chm., Ballantyne of Omaha, 1970–1978, retired. Now protecting law in Omaha.
MEMBER: TESMA (dir. 1947–49; vice-pres., 1949–1951; pres. 1951–53); SMPTE, Variety Club, Life member, Variety Int'l., 1973. (Chief Barker, 1954–56; Internat'l repr., 1955–57–59). Elected board member Theatre Equip. & Supply Mfrs. Assn., member board dir. Will Rogers Memorial Hospital; honorary life member TEA, 1975.

HOFFMAN, DUSTIN: b. Los Angeles, Calif., Aug. 8, 1937. m. Anne Byrne. e. Los Angeles Conservatory of Music, Santa Monica College, Pasadena Playhouse, 1958. Worked as an attendant at a psychiatric institution, a demonstrator in Macy's toy dept., and a waiter. First stage role in Yes Is For A Very Young Man, at Sarah Lawrence College. Acted in summer stock, television and dir. at community theatre. Broadway and Off Broadway plays include: 1964, Harry, Noon and Night (American Place Theatre); 1966, Journey of the Fifth Horse (APT, won Obie Award); Eh? (won Vernon Rice & Theatre World Award); Jimmy Shine. Asst. Dir.: A View From The Bridge. Dir. (1974): All Over Town
TV: Journey of the Fifth Horse, The Star Wagons.
PICTURES INCLUDE: Madigan's Millions, The Graduate (Acad. Award nom.); Midnight Cowboy (Acad. Award nom.); John and Mary, Little Big Man, Who Is Harry Kellerman?, Straw Dogs, Alfredo, Papillon, Lenny (Acad. Award nom), All the President's Men, Marathon Man, Straight Time, Agatha, Kramer vs. Kramer (AA).

HOFFMAN, HERMAN: Wrt. Dir. b. Montgomery, Ala., e. Baltimore City Coll., U.S. Navy.
FEATURE DIRECTING: The Bar Sinister, Great American Pastime, The Invisible Boy, The Battle of Gettysburg.
FEATURE WRITING: Attack on the Iron Coast, Guns of the Magnificent Seven, Escapeline, Reluctant Heroes (Movie-of-Week), The Icemen.
TV DIRECTING: Alcoa Premiere, Untouchables, Bonanza, Virginians, M-Squad, The Nurses, Room 222, Run Joe Run, Monster Squad, The Phantom Rebel.

HOFFMAN, JOSEPH: Writer. b. New York City, Newspaperman, radio writer, magazine writer. TV prod. for Screen Gems, Warner Bros., Four Star. Now TV and screen free lance writer.
PICTURES INCLUDE: China Sky, Don't Trust Your Husband, Gung-Ho, And Baby Makes Three, Weekend with Father, Duel at Silver Creek, At Sword's Point, Has Anybody Seen My Gal?, Against All Flags, No Room For the Groom, Lone Hand, Yankee Pasha, Rails into Laramie, Tall Man Riding, Chicago Syndicate, Live A Little, How to Make Love and Like It, Sex and the Single Girl.

HOGARTH, JOHN M.: Executive. b. Hampstead, London, England, 1931. Ent. m.p. ind. 1947 in domestic distri. div. British Lion; appt. indep. circuits man., 1958. Appt. prod. rep. in charge UK dist. Britannia Film. Bryanston Films, Pax Films, 1962. Joined Sydney Box group, 1964 as gen. man. and dir. Monarch Film Corp. Exec. prod. various films. Formed Crispin Film Distribs. 1969. Apptd., man. dir. London Screen Dists. Ltd. 1970. Dir. of U.K. Dist. EMI Film Dist. 1975–77. Man. Dir. Enterprise Pictures Ltd. 1977.

HOHL, ARTHUR: Actor. b. Pittsburgh, Pa., May 21, 1889. e. Leland Stanford U., Wsh. Sq. Players (later Theatre Guild). On dramatic stage in White Cargo, Mary Dugan, etc. Made screen debut 1933. More than 100 films.
PICTURES INCLUDE: The Whole Town's Talking, Eight Bells, The Unknown Woman, Woman on the Town, The Eve of St. Mark, Salome, Where She Danced, Love Letters, It Happened on Fifth Avenue, Monsieur Verdoux, Down to Sea in Ships.

HOLBROOK, HAL: Actor. b. Cleveland, O., Feb. 17, 1925. Gained fame as impersonator of Mark Twain on stage.
PICTURES INCLUDE: The Group, Wild in the Streets, They Only Kill Their Masters, Magnum Force, The Girl rom Petrovka. All the President's Men, Midway, Julia, Capricorn One, Natural Enemies, The Fog.

HOLDEN, GLORIA: Actress. b. London, Eng. Sept. 5, 1911. e. Amer. Acad. of Dramatic Arts. Church soloist and singer, artists' & clothes model; mgr. Elizabeth Arden salon, Southhampton, L.I.; stage and radio roles, lead in As Husbands Go, Theatre Guild road tour. Began screen career 1937.
PICTURES INCLUDE: Life of Emile Zola, The Girl of Limberlost, Dracula's Daughter, Test Pilot, Girls' School, Behind the Rising Sun, The Hucksters, Kiss for Corliss, Dream Wife, Music by Duchin.
TV: Lux Theatre, Hallmark Theatre.

HOLDEN, WILLIAM: Actor. b. O'Fallon, Ill., April 17, 1918; e. South Pasadena junior coll.; husband of Brenda Marshall, actress. Began screen career 1939. Voted one of ten best Money-Making Stars, M.P. Herald-Fame poll, 1954, 1955, 1957.
PICTURES INCLUDE: Golden Boy, Invisible Stripes, Those Were the Days, Our Town, Arizona, I Wanted Wings, Texas, The Remarkable Andrew, The Fleet's In, Meet the Stewarts, Young and Willing, Dear Ruth, Rachel and the Stranger, Apartment for Peggy, The Dark Past, Dear Wife, Miss Grant Takes Richmond, Union Station, Sunset Boulevard, Born Yesterday, Submarine Command, Force of Arms, Turning Point, Boots Malone, Stalag 17 (Academy Award, best actor, 1953), Moon Is Blue, Forever Female, Escape from Fort Bravo, Executive Suite, Sabrina, Bridges at Toko-Ri, Country Girl, Love Is a Many-Splendored Thing, Picnic, Proud and Profane, Toward the Unknown, Bridge on the River Kwai, The Horse Soldiers, World of Susie Wong, Counterfeit Traitor, Satan Never Sleeps, The Lion, The Longest Day, Paris When It Sizzles, The 7th Dawn, Alvarez Kelley, Casino Royale, Devil's Brigade, The Wild Bunch, The Christmas Tree, The Wild Rovers, The Revengers, Breezy, Open Season, The Towering Inferno, Network, Damien-Omen II, Fedora, When Time Ran Out, The Earthling.

HOLDSWORTH, GERARD (D.S.O., O.B.E.): b. Stourbridge (Worcester), Eng., Dec. 27, 1904; e. Tettenhall College (Staffordshire). Began career as rubber planter; ent. m.p. ind. 1930 with Publicity Films Ltd.; joined J. Walter Thompson Co. 1933; Royal Navy 1939–45; man. dir. Signal Films Ltd. 1946; dir. Wallace Productions Ltd. 1947; formed Gerard Holdsworth Prod. 1950 & Screen Audiences Ltd., 1951; man. dir., Rank Screen Services Ltd., 1954; res. 1959 to re-form Gerard Holdsworth, Ltd., and Gerard Holdsworth Prods. Ltd. 1971: Involvement with video-cassettes and programming for new outlets.

HOLLENDER, ALFRED L.: Advertising executive. b. Chicago, Ill.; e. U. of Illinois. Was associated with Radio Stations WIND-WJJD, Chicago as continuity ed., program dir. & asst. to pres.; entered military service in 1943; exec. v.p. & partner of Louis G. Cowan, Inc., exec. v.p. and dir. radio-TV dept., Grey Adv. Agency; pres., Grey Int'l.

HOLLIMAN, EARL: Actor. b. Tennasas Swamp, Delhi, La., Sept. 11. e. Louisiana State U. U.S.L.A.
PICTURES INCLUDE: Girls of Pleasure Island, Destination Gobi, East of Sumatra, Devil's Canyon, Tennessee Champ, Bridge at Toko-Ri, Broken Lance, Big Combo, I Died a Thousand Times, Forbidden Planet, Giant, Burning Hills, Gunfight at the OK Corral, The Rainmaker, Hot Spell, Visit to a Small Planet, The Power.

HOLLINGER, HY: Executive. b. New York, N.Y. e. CCNY, Columbia Sch. of Journalism; reporter, Upper Darby Review, The Morning Telegraph; Warner Bros.; reporter on Variety, Paramount Pictures Corp.; publ. mgr.; Dir., European prod. publicity, London; Paramount Pictures Corp. Dir. adv. & publicity; assoc. editor, Variety.

HOLLOWAY, STANLEY: Actor. b. London, Eng.; London stage debut 1919 in Kissing Time. Screen debut 1921 in The Rotters; from 1933 in many Brit. features. On Broadway & London in My Fair Lady, 1956–59. Toured Australia in The Pleasure of His Company, 1977.
PICTURES INCLUDE: One Night With You, Hamlet, Snowbound, Noose, The Winslow Boy, Another Shore, Passport to Pimlico, The Perfect Woman, Old Wild Oat, Lavender Hill Mob, The Magic Box, The Happy Family, Meet Me Tonight, Titfield Thunderbolt, Beggar's Opera, Meet Mr. Lucifer, A Day to Remember, Fast & Loose, An Alligator Named Daisy, Jumping for Joy, No Trees in the Street, Alive and Kicking, No Love for Johnnie, My Fair Lady, In Harms Way, Ten Little Indians, The Sandwich Man, Mrs. Brown You Have a Lovely Daughter, 1969; The Private Life of Sherlock Holmes, Run a Crooked Mile, Flight

of The Doves, Up the Front! (1972), Journey into Fear, Dangerous Journey.

TV: Sunday Night at London Palladium, The Mikado, Amer. TV, Lerner and Loewe of Broadway, Our Man Higgins series, The Fantastiks, Perry Como Show, Bell Telephone Hour, Dean Martin Show, Red Skelton Show, Bell Telephone Hour, Tribute to Alan Lerner, Danny Kaye Show, Dean Martin Show. Blanding Castle, What's In It For Harry, Dr. Jekyll and Mr. Hyde.

HOLLOWAY, STERLING: Actor. b. Cedartown, Ga.; e. Ga. Military Acad., Atlanta, Amer. Acad. of Dramatic Art.

THEATRE: N.Y. Plays include: Shepherd of the Hills, The Failures, Garrick Gaieties (4 editions), Donna Magana; vaude.; night clubs; radio; U.S. armed forces, 1943.

PICTURES INCLUDE: Walk in the Sun, Death Valley, Beautiful Blonde from Bashful Bend, Her Wonderful Lie, Alice in Wonderland, Live a Little, Love a Little, The Jungle Book (voice).

HOLM, CELESTE: Actress. b. New York City, Apr. 29. p. Theodor Holm and Jean Parke Holm. Ed. Univ. School for Girls, Chicago, Francis W. Parker, Chicago, Lycée Victor Durui (Paris), Chicago U & UCLA.

THEATRE: On Bdwy inc. Time of Your Life, Return of the Vagabond, Papa Is All, Oklahoma, Bloomer Girl, Affairs of State, Anna Christie, The King and I, Third Best Sport, Invitation to a March, A Month in the Country; Theatre-in-Concert for the US State Department in 8 countries May–July 1966. Appeared on B'way 1975–76, Habeas Corpus.

PICTURES INCLUDE: The Little Girls in Blue, won Academy Award 1947, Gentlemen's Agreement, Come to the Stable, All About Eve, Snake Pit, Road House, Chicken Every Sunday, Everybody Does It, Champagne for Caesar, Tender Trap, High Society, Bachelor Flat, Doctor, You've Got To Be Kidding, Tom Sawyer, Bittersweet Love.

TV: A Clearing House in the Wood, Play of the Week, Cinderella.

RADIO: People at the U.N.

HOLT, DENIS: Associate Producer. 30 years in the industry working on major feature pictures. Currently a director of Winkast Programming Ltd., based at Pinewood Studios, England.

HOLT, PATRICK: Actor. b. Cheltenham, England, 1912; e. Christ's Hospital. From 1936–39 played in repertory. London stage debut, "Saloon Bar," 1939. Enlisted in Royal Artillery, 1940; served in 1st Punjab Reg., Indian Army, 1942–46. On screen first, 1938, "Return of the Frog."

PICTURES INCLUDE: Sword of Honor, Convoy, Hungry Hill, Frieda, October Man, When the Bough Breaks, Mark of Cain, My Sister and I, Fly Away Peter, Portrait from Life, Master of Bankdam, A Boy A Girl and A Byke, Marry Me, Boys in Brown, Guilt Is My Shadow, Golden Link, Men of Sherwood Forest, Miss Tulip Stays the Night, The Dark Avenger, Alias John Preston, Stolen Assignment, Gelignite Gang (The Dynamiters), Suspended Alibi, Count of Twelve, Fortune Is a Woman, Monty's Double, Further Up the Creek, Too Hot To Handle, The Challenge, Guns at Batasl., 13 East Street, Ivanhoe, Serena, Ghengis Khan, Thunderball, Murderers Row, Hammerhead, The Desperadoes, When Dinosaurs Ruled the Earth, Cromwell, No Blades of Grass, Young Winston, Psychomania, Legend of the Werewolfe, The Wild Geese, Sea Wolves.

TV: King Richard the Lionheart (series), African Patrol, Rendez-vous, Flying Doctor Series, The Avengers, The Liars, The Saint series, The Commander, Mr. Rose, No Hiding Place, Randall and Hopkirk, Man Charged, Whom God Hath Joined, War and Peace, Father Dear Father, Emmerdale Farm, Shabby Tiger, Howerd's History of England, Crown Court, Warship, Holding On, Mother Knows Best, Poldark, The Survivors, Rob Roy, Whodunnit, Philby, Burgess & Maclean, Eagle of the IXth, Lillie, Family Affair.

HOMEIER, SKIP: Actor. r.n. George Vincent Homeier. b. Chicago, Oct. 5, 1930. e. U.C.L.A. Started in radio, 1936–43; on B'way stage, Tomorrow the World, 1943–44.

PICTURES INCLUDE: Tomorrow the World, Boy's Ranch, Mickey, Arthur Takes Over, The Big Cat, The Gunfighter, Halls of Montezuma, The Black Widow, Cry Vengeance, Dakota Incident, The Captives, No Road Back, Decision at Durango, Showdown.

TV: Playhouse 90, Alcoa Hour, Kraft Theatre, Studio 1, Armstrong Circle Theatre, Alfred Hitchcock, etc.

HOMER, RAYMOND R.: Producer /director: Queen of Diamonds. Executive producer: Dream City, Double Identity, Left Hand of the Law, Swiss Conspiracy, The Inheritance. Producer: Shadow of a Killer, The Pawn, The Kids, Sharpies, For Love of Anna.

HONG, WILSON S.: Cinematographer. b. Butte, Mont., Dec. 18, 1934; ed. Montana State University, Brooks Institute of Photography; 1965 Free lance photographer for national magazine and world-wide newspaper syndication; 1966, Photographic Director of U.S. Forest Service Fire & Research Division; 1967, first cameraman on various documentaries, industrials, commercials, sports specials.

PICTURES INCLUDE: Bigfoot, Operation North Slope, John Wayne's 'No Substitute for Victory', The Hellcats, Dear Dead Delilah, Zodiac Killer, Velvet Vampire, Mrs. McGrudy, The Day The Adults Died, Sundown in Watts, Parallax View, 1776, Don't Go West, Drum, White Buffalo, MacArthur, An Enemy of the People, Sergeant Pepper's Lonely Hearts Club Band, Mulefeathers, Winterkills, Mission to Glory, The Unfinished, They Only Kill Their Masters, The Fearless Five.

TELEVISION: Snowmobile Grand Prix (ABC), Indianapolis International Drag Races (ABC), The Great Outdoors (23 episodes, ABC), Thank you America (spec.), The Unser Story (ABC), Gun Hawks (pilot), Keep it Up (pilot), Where are they Now? (pilot), Hunting and Fishing the North American Continent (spec.), The Blue Knight, A Dream for Christmas, The Toy Game, Moose (pilot), The Jerry Show, Apple's Way, The Moneychangers, Young Maverick, Rudi Gernreich/Future, Ours, Max Factor and Pepsi-Cola Commercials, The Groovy Seven (pilot).

HOPE, BOB: Actor. b. Eltham, England., May 29, 1903. Started in vaude.; plays include: Roberta, Ziegfeld Follies, Red, Hot & Blue; author, They Got Me Covered, I Never Left Home, So This Is Peace. Voted one of ten best Money-Making Stars in M.P. Herald-Fame Poll, 1941–47, 49–53. On radio and TV.

PICTURES INCLUDE: Big Broadcast of 1938, College Swing, Give Me a Sailor, Thanks for the Memory, Never Say Die, Some Like It Hot, Cat and Canary, Road to Singapore, Ghost Breakers, Road to Zanzibar, Caught in the Draft, Louisiana Purchase, My Favorite Blonde, Road to Morocco, Nothing But the Truth, They Got Me Covered, Star Spangled Rhythm, Let's Face It, Road to Utopia, Princess and Pirate, Monsieur Beaucaire, My Favorite Brunette, Where There's Life, Road to Rio, Paleface, Sorrowful Jones, Great Lover, Fancy Pants, Lemon Drop Kid, My Favorite Spy, Son of Paleface, Road to Bali, Off Limits, Here Come the Girls, Casanova's Big Night, Seven Little Foys, That Certain Feeling, Iron Petticoat, Beau James, Paris Holiday, Alias Jesse James, Facts of Life, Bachelor In Paradise, Road to Hong Kong, Call Me Bwana, A Global Affair, I'll Take Sweden, Boy Did I Get a Wrong Number, Eight on the Lam, The Private Navy of Sgt. O'Farrell, How to Commit Marriage, Cancel My Reservation.

HOPE, HARRY: Producer, director, writer. e. UCLA, Etudes Universitaires Internationales, Ph.D. Entered m.p. industry as special effect man, Republic Studios, 1944; assoc. prod., Star Productions, 1945; prod. for Exploitation Films, 1951; formed Blue Bird Film Co., Taiwan; prod., dir., wrote 22 features for Flame Films, Hong Kong, including Like The Gull, 1967, which won creative classical film award at Asian Film Festival. In 1967 co-founder, gen. mgr. and prod., Western International Motion Pictures, Inc., which merged with Inter-Associates to form First Leisure Corp., of which Hope was prod. and v.p. chg. of foreign operations. In 1973 withdrew Inter-Associates from FLC and now operating it as indep. prod.-dist. co.

HOPKINS, ANTHONY: Actor. b. Port Talbot, 1941. Trained at Royal Academy of Dramatic Arts. Joined National Theatre, gaining fame on stage in England, then TV and films.

PICTURES INCLUDE: White Bus, Lion in Winter, Hamlet, The Looking Glass War, When Eight Bells Toll, Young Winston, The Girl from Petrovka, Audrey Rose, A Bridge Too Far, International Velvet, Magic, Change of Seasons, The Elephant Man.

TV: QB VII, All Creatures Great and Small, The Bunker, Peter and Paul.

HOPKINS, BO: Actor. B. Greenwood, S.C. Studied with Uta Hagen in N.Y. then went to Calif. with Desilu Playhouse training school in Hollywood. Parts in several prods. for that group won him an agent, an audition with director Sam Peckinpah and his first role in latter's The Wild Bunch.

PICTURES INCLUDE: Monte Walsh, The Culpepper Cattle Co., The Moonshine War, White Lightning, The Getaway, The Man Who Loved Cat Dancing, American Graffiti, The Nickel Ride, The Day of the Locust, Posse, The Killer Elite, A Small Town in Texas, More American Graffiti, The Fifth Floor.

TV: Doc Elliott (series).

HOPPER, DENNIS: Actor. b. Dodge City, Kans., May 17, 1936. e. San Diego, Calif., public schools.

TV: Medic, Loretta Young Show.

PICTURES INCLUDE: Jagged Edge, Rebel Without a Cause, Giant, Wrote, dir. starred in Easy Rider, The Last Movie, Hex, The American Friend, King of the Mountain.

HOPPER, JERRY: Director. b. Guthrie, Okla., July 29, 1907. Wrote radio scripts; asst. dir. & actor. Universal, MGM; casting office, Paramount; Combat/photog—U.S. Army, m.p. div.; asst. edit. dept. head, Paramount; dir., Paramount.

PICTURES INCLUDE: The Atomic City, Hurricane Smith, Pony Express, Alaska Seas, Secret of the Incas, Naked Alibi, Smoke Signal, Private War of Major Benson, One Desire, Square Jungle, Never Say Goodbye, Toy Tiger,

Shark Fighters, Missouri Traveler, Blueprint for Robbery, Maharlica, Madrone.

TV: pilots—Bachelor Father, Leave it to Beaver, Naked City, Wagon Train, Witchita Town. Many television productions.

HORDERN, MICHAEL: Actor. b. Berkhampstead, England, 1911. e. Brighton College. early career in business before stage appearance, 1937. M.P. debut 1939. TV debut 1946. 1939–45 Naval Service.

PICTURES INCLUDE: Passport to Pimlico, The Constant Husband, Alexander the Great, the Black Prince, Storm Over The Nile, Pacific Destiny, The Baby and the Battleship, The Spanish Gardener, No Time for Tears, Windom's Way, Monty's Double, Girls at Sea, Moment of Danger, Sink the Bismarck, Man in the Moon, El Cid, Cleopatra, V.I.P.s, Dr. Syn, The Yellow Rolls Royce, Ghengis Khan, The Spy Who Came in From the Cold, Khartoum, Cast a Giant Shadow, A Funny Thing Happened on the Way to the Forum, Taming of the Shrew, The Jokers, How I Won the War, I'll Never Forget What's 'Is Name, Prudence and the Pill, The Bed-sitting Room, Where Eagles Dare, Anne of the Thousand Days, Futtocks End, Some Will, Some Won't, Up Pompeii, The Possession of Joel Delaney (USA), Pied Piper, Blood Will Have Blood, England Made Me, Girl Stroke Boy, Alice's Adventures in Wonderland, Theatre of Blood, The Mackintosh Man, Quilp, Barry Lyndon (narrator), The Slipper and the Rose, Joseph Andrews, The Medusa Touch, Wildcats of St. Trinians.

TV: Doctor's Dilemma, The Great Adventure, The Witness, The Indifferent Shepherd, Dock Brief, Mr. Kettle & Mrs. Moon, Guinea Pig, The Gathering Dusk, Farewell My City, Flowering Cherry, I Have Been Here Before, Without the Grail, The Outstation, The Square, Any Other Business, The Stone Dance, The Quails, A Waltz on the Water, August for the People, Land of My Dreams, Condemned to Acquittal, Nelson, The Browning Version, Whistle and I'll Come to You, The Man Who Murdered in Public, A Crack in the Ice, Sir Jocelyn the Minister Would Like A Word, Six Dates With Barker, Don Juan in Hell, Tartoffe, Tall Stories, The Magistrate, Edward VII, King Lear, Cakes and Ale, Chester Mystery Cycle, Paddington Bear (story teller), The Saints Go Marching In, Mrs. Bixby and the Colonel's Coat, Romeo and Juliet, The Tempest.

HORNBECK, WILLIAM W.: Prod. executive. b. Los Angeles, Calif., Aug. 23, 1901. Lt. Col. in Signal Corps, U.S. Army, World War II (Legion of Merit). Entered m.p. ind. 1916 in film lab., Keystone Comedies; to cutting room; then editor supvr., Mack Sennett through 1933, 1934 to England with Alexander Korda and London Films; ed. The Scarlet Pimpernel; in 1935 supervising ed. for London Films: The Ghost Goes West, Things to Come, Drums, Sanders of the River, Four Feathers, among others. In 1946 apptd. asst. to vice-pres. in charge of studio operations, Republic Pictures. Later, supervising editor, Liberty Films: It's a Wonderful Life, State of the Union. Then Paramount, edited The Heiress, Shane. Academy Award for film ed., Place in the Sun, 1951. Edited Act of Love in France 1953, The Barefoot Contessa in Italy 1954. Then Warner Bros. for Giant in 1955 and The Quiet American for United Artists in Italy 1957. I Want to Live, 1958. A Hole in the Head, and to England for Suddenly Last Summer 1959. Appointed supervisor of editorial operations Universal Studio, 1960. Vice pres. Universal Pictures, 1966. Nominated for Academy Award, 4 times.

HORNE, DAVID: Actor. b. Balcome, Sussex, July 14, 1898; e. Eton & Sandhurst. Began screen career 1931.

PICTURES INCLUDE: Doomed Cargo, The Mill on the Floss, The Wrecker, Four Dark Hours, The First and the Last, Inspector Hornleigh Goes To It, Chamber of Horrors, Spitfire, Adventure in Blackmail, Rake's Progress, Seventh Veil, Men of Two Worlds, Gaiety George, The Magic Bow, Saraband for Dead Lovers, Easy Money, Martin Luther, Spaceways, Five Pound Note, Policewoman, Wedding Gift, The Intruder, 3 Cases of Murder, Beau Brummell.

HORNE, LENA: Vocalist, actress. b. Brooklyn, N.Y.

THEATRE: Blackbirds, Dance With Your Gods.

Radio with Noble Sissle, Charlie Barnet, other brands. Floor shows at Cotton Club, Cafe Society, Little Troc, etc. Started screen career 1942.

PICTURES INCLUDE: Panama Hattie, Cabin in the Sky, Stormy Weather, I Dood It, Thousands Cheer, Broadway Rhythm, Swing Fever, Two Girls and a Sailor, Ziegfeld Follies, Till the Clouds Roll By, Words and Music, Meet Me in Las Vegas, Death of a Gunfighter, The Wiz.

HORNER, HARRY: Director. b. Holic, Czechoslovakia, July 24, 1910; e. U. of Vienna, Dept. of Architecture, 1928–33; Acad. of the Theatre, Vienna, 1930–32, Max Reinhardt's Seminary. Joined Max Reinhardt Thea. Co., Vienna and Salzburg Festivals; to U.S. as asst. to Reinhardt on pageant, The Eternal Road. Stage designer, N.Y. theatre 10 yrs. (Lady in The Dark, Family Portrait, others). First m.p. as prod. designer, 1938, Our Town (co-credit, Wm. Cameron Menzies). later Little Foxes, Stage Door Canteen. Army service, 1942–45; designed Winged Victory for Air Force. In 1949, won Acad. award black and white art dir., The Heiress. Other films designed: Born Yesterday, Separate Tables, They Shoot Horses; Acad. nomination, The Hustler; (Acad. award). Since 1959 prod., dir., TV series, The Royal Canadian Mounted Police; exec. prod., dir., Enterprise Films of Can., Anglo Ent. Films, London; co-prod., Fahrenheit 451, 1966, They Shoot Horses, Don't They? (Acad. nom.), Who Is Harry Kellerman?, Up the Sandbox (Streisand)

FURTHER PICTURES INCLUDE: Born Yesterday, He Ran All the Way, Androcles and the Lion, Outrage. Films Directed: Beware My Lovely, Red Planet Mars, Vicki, New Faces, Life in the Balance, Step Down to Terror, Lonesome Gun, Wild Party Man From Del Rio, Winner of 2 acad. awards, 6 acad. nominations for Art Dir. 1975: The Black Bird, Harry and Walter Go To New York, Audrey Rose, The Driver, Moment by Moment, The Jazz Singer.

TV: Omnibus, Cavalcade, Reader's Digest, Author's Playhouse, Four Star Theatre, Gunsmoke, Revue Productions, Dupont Theatre.

OPERA: designer & director at San Francisco Opera, Metrop. Opera N.Y. Vancouver Festivals Hollywood Bowl Operas designed & directed: David, Joan at the Seake, Magic Flute. Amer. Premiere of New Opera, Midsummers Night Dream. Designed Idiots Delight, Ahmanson Theatre, 1970, winner, L.A. Drama Critics Award, best stage design. Designed Time of the Cuckoo, Ahmanson Theatre 1974.

HORTON, ROBERT: Actor. b. Los Angeles, July 29, 1924. e. Miami Univ., U.C.L.A. U.S. Coast Guard; many legit. plays; many radio & TV appearances; co-star, Wagon Train; screen debut in The Tanks Are Coming.

PICTURES INCLUDE: Return of the Texan, Pony Soldier, Apache War Smoke, Bright Road, The Story of Three Loves, Code Two, Arena, Prisoner of War, Men of the Fighting Lady, The Green Slime.

HORWICH, FRANCES R.: b. Ottawa, Ohio. e. U. of Chicago, Ph.B., 1929; Columbia U., M.A., 1933; Northwestern U., Ph.D., 1942; hon. deg., D. Ped., Bowling Green (Ohio) State U., 1954; Grade-sch. teacher, Evanston, Ill., 1929–32; supvr., nursery schools, Chicago, 1932–35; dir., jr. kindergarten, Winnetka, Ill., 1935–38; dean of educ., Pestalozzi Froebel Teachers Coll., Chicago, 1938–40; counsellor of student teachers, Chicago Teachers Coll., 1940–42; dir., Hessian Hills Sch., N.Y., 1943–45; prof., U. of N.C., 1945–46; Roosevelt U., 1947–52; supvr. of children's programs, NBC, 1952–56; Miss Frances on TV show, Ding Dong School, NBC, 1955–56; WGN-TV, 1957–58; I.T.C. syndication, 1959; ed. dir., The Curtis Publishing Co., 1962–66; dir. of children's programs at WFLD-TV Chicago, 1966–68.

TV PROGRAMS: Time For Children with Miss Francis, Time for Parents with Miss Francis, 1966–67; Sunday in Chicago segment, WMAQ-TV, 1969.

HORWITZ, LEWIS M.: Exhibitor, Producer. b. Cleveland, Ohio, Sept. 14, 1931. e. Miami University, Ohio State University, Western Reserve University. 1943, popcorn boy, Haltnorth Theatre, Cleveland, Ohio; 1950, assistant manager, State Theatre, Cuyaghoga Falls, Ohio; 1953, booker, Washington Theatre Circuit, Cleveland; 1956, film buyer, Washington Theatre Circuit; 1959, Head Film Buyer, Washington Theatre Circuit; 1960, partner and chief executive officer, Washington Theatre Circuit; 1965, secretary-treasurer, Global Screen Associates, Inc., Los Angeles; 1966, indep. producer, A Man Called Dagger; 1967, formed independent production company—LMH Productions; 1967–70, independent producer, Metro-Goldwyn-Mayer Studios, Culver City; 1971, independent producer, Woodshed Films; 1972–73, exec., General Cinema Corporation, Boston; 1973–74, producer of The Prisoners, which was invited Berlin, Moscow and Belgrade International Film Festivals. 1974–75, head film buyer, Royal Cinemas, Inc.

HOUGH, JOHN: Director, Worked in British film prod. in various capacities; impressed execs. at EMI-MGM Studios, Elstree, London, so was given chance to direct The Avengers series for TV. Began theatrical films with Sudden Terror for prod. Irving Allen, 1971.

PICTURES INCLUDE: Twins of Evil, Treasure Island, The Legend of Hell House, Dirty Mary Crazy Larry, Escape to Witch Mountain, Return to Witch Mountain, Brass Target, Watcher in the Woods.

HOUSEMAN, JOHN: Producer, director, writer. r.n. J. Haussmann. b. Rumania, Sept. 22, 1902. e. Clifton Coll., Eng. Moved to Argentina 1923 to conduct wheat export business. To U.S., Canada, 1925; contrib. to magazines; adapt. plays from German, French. In 1932 prod. Four Saints in Three Acts; later dir. Valley Force, Panic, for Theatre Guild. Assoc. with Orson Welles in founding Mercury Theat., 1937; apptd. assoc. prof. English at Vassar same yr. Prod. Julius Caesar, Native Son for stage. Co-author Woodrow Wilson (prod. on stage as In Time to Come), bought by 20th Century-Fox. Wrote radio scripts for Helen Hayes. Exec. Selznick Prod., 1941–42. Chief of overseas radio div. OWI, 1942–43. In 1943–44 (joint s.p.) Jane Eyre; New York stage; dir. Lute Song, 1948; King Lear, 1950; Coriolanis, 1953.

PICTURES INCLUDE: Miss Susie Slagle's, The Unseen, Blue Dahlia, Letter from an Unknown Woman, They live By Night, The Company She Keeps; prod. On Dangerous Ground, Holiday For Sinners, The Bad and The Beautiful, Julius Caesar, Executive Suite, Her Twelve Men, Moonfleet, The Cobweb, Lust For Life, All Fall Down, Two Weeks in Another Town, In The Cool of the Day. Actor: The Paper Chase, 1973, Rollerball, 1975, Three Days of the Condor, St. Ives, The Cheap Detective, The Fog, Wholly Moses.

TV: Paper Chase (series).

HOVEY, TIM: Actor. b. Los Angeles, Calif., June 19, 1945. Child model; on TV in Lassie.

PICTURES INCLUDE: Queen Bee, Private War of Major Benson, Toy Tiger, Everything But the Truth, Man Afraid, Slim Carter. Money Women and Guns.

TV: Playhouse 90, Lux Theatre, Kraft Theatre, Matinee Playhouse, Schiltz Playhouse, G.E. Theatre, Cimarron City.

HOWARD, CY: Producer, director. b. Milwaukee, Wis., Sept. 27, 1915. e. U. of Minnesota, U. of Wisconsin, Entered radio Station KTRH, Houston, Tex. as writer, prod., actor; served 1 yr. Army Air Corps; to WBBM, Chicago, as writer, prod., actor, 1942; to Jack Benny's radio writing staff; actor, Storm Operation (stage), 1943; to ABC as writer, comedian on What's New program; to NBC where orig. wrote Palmolive Party; radio writer for Milton Berle, Danny Thomas, Bert Lahr, Jerry Lewis; orig. My Friend Irma, Life with Luigi, radio and TV shows; to Hall B. Wallis Prod. as writer and assoc. prod. My Friend Irma; exec. prod., Desilu Studios, created and produced Harrigan & Son, Westward Ho, Fair Exchange, My Friend Irma Goes West (films), That's My Boy; writer, Marriage on the Rocks; co-writer: Won Tonton, The Dog Who Saved Hollywood; director, Lovers & Other Strangers and Every Little Crook and Nanny.

HOWARD, CYRIL: Managing Director of Rank's Pinewood Studios, England. Formerly Secretary and general manager of the studios. Joined Rank Organization 1942.

HOWARD, KEN: Actor. b. El Centro, CA, March 28. e. Yale Drama School. Left studies to do walk-on in Bdwy. musical, Promises, Promises. Starred at Thomas Jefferson in 1776 on Bdwy. (Theatre World Award).

PICTURES: Tell Me That You Love Me, Junie Moon (debut), Such Good Friends, 1776.

TELEVISION: Series: Manhunter, Bonanza, Medical Center, Adam's Rib, The White Shadow. Movie: The Trial of George Armstrong Custer.

STAGE: Child's Play, Seesaw, 1600 Pennsylvania Avenue, The Norman Conquests, Equus.

HOWARD, ROBERT T.: Executive. b. Red Bank, N.J. e. Univ. of Virginia, Columbia Univ. (N.Y.). Began prof. career as NBC page, joining guest relations staff in N.Y., 1947. Moved up to NBC TV research dept., where worked in program testing and audience measurement. Became head of research for NBC Radio Spot Sales in 1953; 1955, promoted to acct. exec. in same dept. Joined TV Spot Sales in 1959 and named natl. sales mgr. of WNBC-TV, New York in 1963 named gen. mgr. of KNBC, Los Angeles. Elected v.p. of NBC a month later. Returned to N.Y. and named pres. of NBC TV Network April 1, 1974. Elected to NBC bd. and named exec. v.p.; 1977, v.p. & gen. mgr. WNBC TV, New York; 1980, pres. & chm., Citicom Communications Co.

HOWARD, RON: Actor. b. Duncan, Okla., March 1, 1954. e. Univ. of South Calif. Acting debut at age of two with parents, Rance and Jean Howard, in The Seven Year Itch at Baltimore's Hilltop Theatre. Two years later travelled to Vienna for his first film, The Journey. Many TV appearances over years. Is brother of Clint Howard, also actor from childhood.

PICTURES INCLUDE: The Journey, Wild Country, The Music Man, Courtship of Eddie's Father, Happy Mother's Day . . . Love, George, American Graffiti, The Spikes Gang, The Shootist, More American Graffiti.

TELEVISION: Red Skelton Hour, Playhouse 90, Dennis, the Menace, Many Loves of Dobie Gillis, Five Fingers, Twilight Zone, Dinah Shore Show, Andy Griffith Show (regular), The Fugitive, Dr. Kildare, Big Valley, I Spy, Danny Kaye Show, Gomer Pyle, USMC, The Monroes, The FBI, Judd for the Defense, Daniel Boone, Lancer, Land of the Giants, Gentle Ben, Gunsmoke, Disney TV films (A Boy Called Nuthin', Smoke), Happy Days (regular).

HOWARD RONALD: Actor. b. Norwood, England; e. Tonbridge School and Jesus Coll.; s. the late Leslie Howard. m. Jean Millar. Reporter, Allied Newspapers. In British armed forces.

PICTURES INCLUDE: While the Sun Shines, 1946; Night Beat, My Brother Jonathan, Bond Street, Queen of Spades, Now Barabbas, Portrait of Eve, Double Confession, Assassin for Hire, The Browning Version, Tom Brown's School Days, Street Corner (Both Sides of the Law), Rustle of Silk, Drango; No Trees in the Street, Babette Goes to War, Come September (Italy), Monster of Highgate, The Naked Edge, Bomb in the High Street, The Curse of the Mummy's Tomb.

TV: Sherlock Holmes, Robin Hood, The Man Who Was Two, Mary Britten, M.D., Four Just Men.

HOWARD, SANDY: Producer. b. 1927. Ent. m.p. ind. 1946.

PICTURES INCLUDE: Perils of the Deep, One Step to Hell, Jack of Diamonds, Tarzan and the Trappers, A Man Called Horse, Man in the Wilderness, Together Brothers, Neptune Factor, The Devil's Rain, Sky Riders, The Last Castle, Embryo, Magna I—Beyond the Barrier Reef, The Battle, Island of Dr. Moreau, City on Fire, Death Ship (exec. prod.).

TV: Over 50 TV series.

HOWARD, TREVOR: Actor. b. Kent, England, Sept. 29, 1916; e. Clifton Coll., Bristol, Royal Acad. of Dramatic Art. Stage debut: Revolt in a Reformatory, 1933; played Petruchio in Old Vic production of The Taming of the Shrew and in James Bridie's The Anatomist. Screen debut: The Way Ahead, 1943. In H. M. Forces in World War II.

PICTURES INCLUDE: Brief Encounter, Way to the Stars, Green for Danger, So Well Remembered, Passionate Friends, Third Man, Golden Salamander, Odette, Outcast of the Islands, Lovers of Lisbon (French), The Clouded Yellow, The Gift Horse, Heart of the Matter, Stranger's Hand, Cockelshell Heroes, Run for the Sun, Around the World in 80 Days, Interpol, Manuela, The Key, Roots of Heaven, Moment of Danger, Sons and Lovers, Mutiny On the Bounty, Father Goose, The Great Spy Mission, Von Ryan's Express, The Sabateur Code Name-Morituri, The Lion, Man in the Middle, Pretty Polly, A Matter of Innocence, The Battle of Britain, Ryan's Daughter, Pope Joan, Ludwig, 11 Harrowhouse, Who?, Hennessy, Conduct Unbecoming, The Bawdy Adventures of Tom Jones, The Last Remake of Beau Geste, Superman, Hurricane, Meteor, The Sea Wolves.

TELEVISION: Hedda Gabler, Mr. Disraeli, Napoleon, Catholics, Night Flight.

HOWELLS, URSULA: Actress. b. Sept. 17, 1922. e. St. Paul's Sch., London. Stage debut, 1939, at Dundee Repertory with Bird in Hand followed by several plays inc. Springtime for Henry in N.Y., 1951; m.p. debut in Flesh and Blood, 1950; TV debut in Case of the Frightened Lady for BBC, 1948.

TV: Many appearances including The Small Back Room, A Woman Comes Home, For Services Rendered, Mine Own Executioner, The Cocktail Party.

PICTURES INCLUDE: The Oracle (Horse's Mouth), Track the Man Down, They Can't Hang Me, Keep It Clean, Long Arm (Third Key), Death and The Sky Above, Mumsy, Nanny, Sonny, and Girly.

HOWERD, FRANKIE: Performer. b. York, Eng., e. Shooters Hill, London, Sch. Stage debut, Sheffield Empire, 1946. Film: Runaway Bus; BBC radio & TV appearances include Tons of Money, Frankie Howerd Show, Howerd Crowd.

PICTURES INCLUDE: Jumping for Joy, An Alligator Names Daisy, The Lady Killers, A Touch of the Sun, Further Up the Creek, Cool Mikado, Fast Lady, The Great St. Trinian's Train Robbery, Carry On Doctor, Up Pompeii, Up the Chastity Belt, Up The Front, The House in Nightmare Park, Sgt. Pepper's Lonely Hearts Club Band.

TV: BBC series. Comedy Playhouse, Frankie Howerd Show, Up The Convicts—Series (4) (Australia 1975), The Frankie Howerd Show—Series (13) (Canada 1976), Up Pompei.

STAGE: Old Vic Midsummers Night's Dream, A Funny Thing Happened on the Way to the Forum, Way Out in Piccadilly, Wind in the Sassafras Trees.

HUBBARD, GORDON: Executive. b. Los Angeles, Calif., Aug. 11, 1921. e. Santa Monica City Coll. Air Force, 1940–45; sls. office, Maidenform Bra Co., 1945; ret., Air Force, 1950; special training program for sls. contracts and representation, Pacific Title, 1953; worked two yrs. in plant, in laboratory and optical depts., then represented the company in the field, acct. exec., Pacific Title; pres., Pacific Title.

HUDGINS, TORRENCE B.: Film Buyer. b. Merkel, Tex. e. SMU. Interstate Theatres, Inc. 1923–52; Trans-Texas Theatres, 1951–54, advertising; Cinema Art Theatres Inc., 1954–1974, film buyer and advertising. 1974–78 Torrence Hudgins Buying and Booking Service.

HUDIS, NORMAN: Writer. b. London, England. Now based in L.A.

PICTURES INCLUDE: 6 Carry On comedies, 2 Man From U.N.C.L.E. features.

TV: Many series include own series, Our House.

HUDSON, ROCK: Actor. r.n. Roy Fitzgerald. b. Winnetka, Ill., Nov. 17, 1925; e. New Trier H.S. Winnetka. U.S. Navy, 1944–46; m.p. debut in Fighter Squadron; No. 1 top money-making star M.P. Herald-Fame poll, 1957–65.

PICTURES INCLUDE: Undertow, I was a Shoplifter, One-Way Street, Peggy, Winchester 73, Tomahawk, Fat Man, Air Cadet, Iron Man, Bright Victory, Desert Hawk, Here Come the Nelsons, Bend of the River, Has Anybody Seen My Gal, Scarlet Angel, Horizons West, Lawless Breed,

Seminole, Sea Devils, Golden Blade, Back to God's Country, Gun Fury, Taza Son of Cochise, Magnificent Obsession, Bengal Brigade, Captain Lightfoot, One Desire, All that Heaven Allows, Never Say Goodbye, Giant, Written on the Wind, Battle Hymn, Something of Value, Pylon, Farewell to Arms, This Earth Is Mine, Pillow Talk, The Last Sunset, Come September, Lover, Come Back, The Spiral Road, A Gathering of Eagles, Mans Favorite Sport, Strange Bed Fellows, Send Me No Flowers, A Very Special Favor, Blindfold, Seconds, Tobruk, Ice Station Zebra, A Fine Pair, Darling Lilli, The Hornet's Nest, Pretty Maids All in a Row, Showdown, Embryo, Avalanche, The Mirror Crack'd.

TV: MacMillan and Wife series.

HUGGINS, ROY: Writer, director. b. Litelle, Wash., July 18, 1914; e. U. of Calif. 1935–39; U. of Calif. Grad. Sch., 1939–41; m. Adele Mara, actress. Spec. rep., U.S. Civil Service Comm., 1941–43; industrial eng., 1943–46; writer 3 novels and many stories for Sat. Eve. Post; pres., Public Arts, Inc., 1968. V.P., 20th Century-Fox TV, 1961.

PICTURES INCLUDE: I Love Trouble, Too Late for Tears, Lady Gambles; story, Fuller Brush Man, Good Humor Man; adap., Woman in Hiding; collab., s.p., Sealed Cargo; Sp., dir., Hangman's Knot; collab. s.p., Gun Fury, Three Hours to Kill; s.p., Pushover; Prod. motion picture A Fever in the Blood for Warner Bros.

TV: Prod. Warner Bors. Presents Cheyenne series, anthologies, 1955–56; Conflict (series); Produced Pilots of Colt .45, 77 Sunset Strip, Maverick, Prod. Maverick, 1957–58; won Emmy Award 1958. Created, The Fugitive, 1962; v.p., MCA Revue, 1963; exec.-prod. Run For Your Life, 1965; exec. prod., The Outsiders, The Bold Ones, Alias Smith and Jones, Toma, 1968–74. ABC Movie of Week, Pretty Boy Floyd. 1974: Co-creator of The Rockford Files pilot and series for NBC; 1975: Co-creator of City of Angels for NBC; 1976: executive producer of Captains and the Kings, a segment of Best Sellers for NBC. 1977: exec. prod. of Aspen, a segment of best sellers for NBC; 1978: exec. prod. of Wheels, a segment of The Big Event for NBC.

HUGHES, KATHLEEN: Actress. r.n. Betty von Gerkan; b. Hollywood, Calif., Nov. 14, 1928; e. L.A. City Coll., UCLA. m. Stanley Rubin, producer, mother of 4, Michael played Baby Matthew on Peyton Place. Studied drama; under contract, 20th-Fox, 1948–51; starred in 7 Year Itch 1954, La Jolla Playhouse; parts m.p.; signed by UI, 1952.

PICTURES INCLUDE: For Men Only, Sally and Saint Anne, Golden Blade, It Came From Outer Space, Thy Neighbor's Wife, Glass Web, Dawn at Socorro, Cult of the Cobra, Three Bad Sisters, Promise Her Anything, The President's Analyst, The Late Liz, The Take.

TV: Bob Cummings Show, Hitchcock, 77 Sunset Strip, G.E. Theatre, Bachelor Father, The Tall Man, Dante, Tightrope, Markham, I Dream of Jeannie, Peyton Place, Gomer Pyle, Kismet, Ghost and Mrs. Muir, Bracken's World, The Survivors, Julia, Here's Lucy, To Rome with Love, The Interns, The Man and the City, Mission Impossible, The Bold Ones, Lucas Tanner, Marcus Welby, Barnaby Jones, Medical Center, Babe (mov.).

HUGHES, KEN: Writer, dir. b. Liverpool, Eng. Ent. ind. as sound engineer, BBC, 1940; dir., doc. films. Novels: High Wray, The Long Echo.

TV: Script of serials Solo for Canary, An Enemy of the State; Sammy (Eddie in USA); The Haunting, Lenin in 1917 (Fall of Eagles); The Voice (Dial M For Murder).

PICTURES INCLUDE: Joe Macbeth, Confession, The Trials of Oscar Wilde, The Small World of Sammy Lee, Arrivederci Baby, Casino Royale, Chitty Chitty Bang, Bang, Cromwell, The Internecine Project, Sextette.

AWARDS INCLUDE: Golden Globe, Emmy, Script Writer of Year, W.G. Merit Award.

HUGO, LAURENCE: Actor. b. Berkeley, Calif. e. U. of Calif.

THEATRE: On Broadway in Skin of Our Teeth, Decision, Born Yesterday, Stalag 17.

TV: Studio One, Omnibus, Danger, The Web, Mama, over the CBS-TV network.

HUKE, BOB, B.S.C.: Cinematographer. Ent. m.p. ind. 1937, Asst. cameraman Pygmalion, French Without Tears, etc.; 1939–44 Royal Navy. 1945–9 camera operator Great Expectations, Uncle Silas, Seven Days To Noon and others. 1950–56 Brazil. Contract dir. of photo. for Cia Cinematographica Vera Cruz, 1957–59 dir. own company, Zenith, 1960. dir. photo, 3 Dinah Shore Shows, Spain, Paris, Copenhagen. NBC 1961. Dir. photo. Reach For Glory, The War Lover, The Very Edge, 1962; The Brain, 8 Danny Thomas Shows in Europe. 1963 Sandres of the River. 1964, Ballad in Blue, License to Share. TV & Cinema Commericals, 1955. 1966, 2nd Unit You Only Live Twice; 1968, 2nd Unit, Battle of Britain, 1969; dir. photo, Virgin and Gypsy. 1971; Under Milk Wood.

HUNNICUT, GAYLE: Actress. b. Fort Worth, Texas, February 6, 1943. e. BA (with honors) University of California, Los Angeles, (UCLA) Theater Arts & English major. Early career, community theatres in Los Angeles. Ent. m.p. ind. 1967.

PICTURES INCLUDE: PJ (New Face in Hell), Eye of the Cat, Marlowe, Fragment of Fear, The Freelance, Voices, Running Scared, New Face in Hell, Scorpio, L'Homme Sans Visage, The Spiral Staircase, The Sell Out, Tony Siatta, Once in Paris, One Take Two, Fantomas.

TV: Man and Boy, The Golden Bowl, The Ripening Seed, Fall of Eagles, The Switch, Humboldts Gift, The Life and Death of Dylan Thomas, Return of the Saint, Martian Chronicles, A Man Called Intrepid.

THEATRE: The Ride Across Lake Constance, Twelfth Night, The Tempest, Dog Days, The Admirable Crichton, A Woman of No Importance.

HUNT, BILLY H.: Executive, attorney. b. Alton, Ill., Aug. 19, 1926. e. University of Southern California, University of California, Berkley, Juris Doctor, 1957. Attorney, Morrison, Forster, Holloway, Schuman & Clark, 1957–59; attorney for Mitchell, Silberberg & Knupp, working in legal dept. at Columbia Pictures, 1959–63; entertainment, corporate, labor law for entertainment companies at Mitchell, Silberberg & Knupp, was exec. v.p. of Association of Motion Picture and Television Producers, Inc., 1972–75; pres. of Central Casting Corporation, 1972–75; named exec. v.p. of Motion Picture Association of America, 1973–75.

HUNT, G. CARLETON: Executive. b. Bridgeport, Conn., Jan. 5, 1908. With Warner Bros. Theatres, 1929–35; with New England Theatres, 1935–38; with stage groups, 1938–41; with RKO Studios, 1941–51; secy., gen. mgr., Unicorn Theatres, 1952–53; pres., General Film Labs., 1953–62; Chairman of the Board; Pacific Industries, Inc. 1962–64, pres., Glen Glenn Sound Co., now bd. chm.; Pres. DeLuxe General Inc., 1964–1972. Retired 1973.

MEMBER: Acad. of M.P. Arts & Sciences, Acad. of TV Arts & Sciences; Assoc. mem. ASC; past pres. SMPTE, v.p. UNIATEC.

HUNT, MARSHA: Actress. b. Chicago, Ill., Oct. 17, 1917. Started screen 1935.

B'WAY PLAYS INCLUDE: Devils Disciple, Joy to the World, Legend of Sarah, Born in Texas, Tunnel of Love, The Paisley Convertible.

PICTURES INCLUDE: Virginia Judge, College Holiday, Easy to Take, Blossoms in the Dust, Panama Hattie, Joe Smith American, These Glamour Girls, Winter Carnival, Irene, Pride and Prejudice, Flight Command, Cheers for Miss Bishop, Trial of Mary Dugan, Thousands Cheer, None Shall Escape, Lost Angel, Cry Havoc, Bride by Mistake, Music for Millions, Valley of Decision, A Letter for Evie, Smash-Up, Carnegie Hall, Raw Deal, Take One False Step, Actors and Sin, Happy Time, No Place To Hide, Bombers B-52, Blue Denim, Johnny Got His Gun.

TV: Philco, Studio One, Ford Theatre, Show of Shows, G.E. Theatre, Climax, Hitchcock, Peck's Bad Girl Series, The Defenders, Twilight Zone, Cains Hundred, Gunsmoke, The Breaking Point, Outer Limits, Profiles in Courage, Ben Casey, Accidental Family, Run For Your Life, My Three Sons, The Outsiders, Name of the Game, Univ.'s 120, Ironside, Marcus Welby, M.D., Police Story, The Young Lawyers.

HUNT, PETER: Director—Editor. b. London, England. 1928. e. Romford England—Rome, Italy. London School of Music. Actor English Rep. Entered film as camera asst. Documentary, later asst editor documentary, then asst editor features. London Films then editor—credits incl. Hill in Korea, Admirable Crichton, Scripting films. Editor: Cry From the Streets, Greengage Summer (Loss of Innocence in USA), Ferry To Hong Kong, H.MS Defiant (Damn the Defiant in USA), Supervising editor/2nd Unit Director, Dr. No, Call Me Bwana, From Russia With Love, Goldfinger, Ibcress File, Thunderball, You Only Live Twice, . . . associate producer, Chitty Chitty Bang Bang. . . . Director, On Her Majesty's Secret Service, Gullivers Travels (Live & Animation) Gold, Shout At The Devil, at present filming for Golden Harvest Group (USA), Death Hunt.

HUNT, PETER H.: Director. b. Pasadena, Calif., Dec. 16, 1938. e. Hotchkiss, Yale, Yale Drama School. m. actress Barbette Tweed. Director for Williamston Theatre since 1957. Lighting designer on Bdwy. (1963–69).

PICTURES INCLUDE: 1776, Give 'Em Hell, Harry, Bully, A New Start.

TV: Adams Rib (pilot and series), Hello Mother Goodbye (pilot), Ivan the Terrible (pilot), Quark (pilot), Mixed Nuts (pilot), Flying High (pilot), Wilder and Wilder (pilot), Rendezvous Hotel (pilot).

STAGE: 1776 (London & Bdwy), Georgy (Bdwy.), Scratch (Bdwy.), Goodtime Charley (Bdwy.), Give 'Em Hell Harry (Tour), Magnificent Yankee (Kennedy Center), Bully (Tour).

HUNT, WILLIE: Executive. b. Van Nuys, Ca., Oct. 1, 1941. e. Utah State Univ. (BA, 1963). Married to actor Tim Considine. Started in industry as secretary at Warner Bros., 1965; named exec. secty. to Ted Ashley, WB, 1969; story analyst, WB, 1974; story editor, WB, 1975; named West Coast story editor for WB, 1978; joined MGM in 1979 as v.p., motion picture development.

HUNTER, KIM: Actress. r.n. Janet Cole; b. Detroit, Mich., Nov. 12, 1922. e. Miami Beach, Fla. schools; m. Robert Emmett,

Dec., 1951; daughter and son. Started career in summer stock, Gant Gaither Theatre, Fla.; Theatre of the 15, Coral Gables; Hendersonville, N.C.; Baltimore, Md.; Pasadena, Calif. Community Playhouse appeared in Arsenic and Old Lace; others. Screen debut; in leading role, The Seventh Victim, David O. Selznick; other films, Tender Comrade, When Strangers Marry, You Came Along; then, to England to co-star in Stairway To Heaven, A Canterbury Tale; then N.Y. stage in A Streetcar Named Desire (Pulitzer Prize play); signed by Elia Kazan to play orig. role in screen version for which Academy Award as best supporting actress, 1951. N.Y. stage; Darkness at Noon, The Chase, Children's Hour, Tender Trap, Write Me A Murder, bdway., 1961–62; films, Anything Can Happen; Deadline U.S.A., Storm Center, Bermuda Affair, Young Stranger, Money, Women & Guns, Lilith; 1961 American Shakespeare Festival, Stratford, Conn., As You Like It, Troilus and Cressida, Macbeth.

TV APPEARANCES: Climax, Lux Video, G.E. Theatre, U.S. Steel Hour, Omnibus, Playhouse 90, The Play of the Week, Hallmark Hall of Fame, Hitchcock, Arrest and Trial, TWTWTW, Jackie Gleason Show, Hawk, Mannix, Disney's Wonderful World of Color, Bonanza, CBS Playhouse, NET, 1970: Eternal Light, David Frost Show, Young Lawyers, Bracken's World. 1971: Bold Ones, In Search of America (Special), Medical Center, Gunsmoke, Cannon, Columbo. 1972: Night Gallery, Owen Marshall, Love, American Style, Young Dr. Kildare, The Evil Touch. 1973: Mission: Impossible, NBC Mystery Movie, Marcus Welby, M.D., Hec Ramsey, Griff, The Police Story, "Unwed Father"—ABC Movie-of-the-Week, Ironside; This Is The Life, Insight, Lucas Tanner, Wide World of Mystery, and Movies of the Week: Born Innocent, Bad Ronald, Ellery Queen, This Side of Innocence, Once An Eagle, Gibbsville, Hunter, Baretta, Crazy Annie received Emmy nomination; Oregon Trail, and Backstairs at the White House; Stubby Pringle's Christmas, Project UFO, 1978; The Rockford Files, Specter on the Bridge (movie), 1979; Edge of Night (1978–80—Emmy nomination; FDR, The Last Year, 1980; Skokie, 1981.

BROADWAY: Weekend, 1968. The Penny Wars, 1969. The Women, 1973; Nat'l Tour: And Miss Reardon Drinks a Little, 1971–72; In Praise of Love (1975 summer tour), The Cherry Orchard (1976, off B'way), Elizabeth The Queen (Buffalo, N.Y.), Semmelweiss (1977, Buffalo, N.Y.), The Bell of Amherst (1978, New Jersey); The Little Foxes (1980, Berkshire Theatre Festival), To Grandmother's House We Go, 1981.

PICTURES INCLUDE: Planet of the Apes, The Swimmer, Beneath the Planet of the Apes, Escape from the Planet of the Apes, Dark August.

AUTHOR: Kim Hunter—Loose in the Kitchen (Domina Books).

HUNTER, ROSS: Producer. r.n. Martin Fuss. b. Cleveland, Ohio, May 6; e. Western Reserve U., M.A. School teacher, 1938–43; actor, Columbia Pictures, 1944–46; returned to school teaching; stage prod. & dir.; m.p. dialogue dir.; assoc. prod. U-I, 1950–51; prod., U-I, 1951. Moved production Co. from Universal to Columbia, April 1, 1971. Moved to Paramount, Oct. 1, 1974.

PICTURES INCLUDE: Louisiana Hayride, Ever Since Venus, Bandit of Sherwood Forest, Groom Wore Spurs, Take Me to Town, All I Desire, Tumbleweed, Taza Son of Cochise, Magnificent Obsession, Naked Alibi, Yellow Mountain, Captain Lightfoot, One Desire, The Spoilers, All That Heaven Allows, There's Always Tomorrow, Battle Hymn, Tammy and the Bachelor, Interlude, My Man Godfrey, The Wonderful Years, Stranger in My Arms, Imitation of Life, Pillow Talk, Portrait in Black, Midnight Lace, Back Street, Flower Drum Song, Tammy and the Doctor, The Thrill of It All, The Chalk Garden, I'd Rather Be Rich, The Art of Love, Madame X, The Pad, Thoroughly Modern Millie, Rosie, Airport, Lost Horizon.

HUNTER, TAB: Actor. b. New York, N.Y., July 11, 1931. U.S. Coast Guard; odd jobs. discovered in 1948.

PICTURES INCLUDE: The Lawless, Island of Desire, Gun Belt, Steel Lady, Return to Treasure Island, Track of the Cat, Battle Cry, Sea Chase, Burning Hills, Girl He Left Behind, Lafayette Escadrille, Gunman's Walk, Damn Yankees, That Kind of Woman, Pleasure of His Company, The Golden Arrow, War Gods of the Deep, Ride the Wild Surf, Hostile Guns, Life and Times of Judge Roy Bean.

HUNTLEY, RAYMOND: Actor. b. Birmingham, England, Apr. 23, 1904. On stage from 1922 repertory, tour; London debut in "Back to Methuselah" 1924; N.Y. debut in "Venetian Glass Nephew" 1931; m.p. debut 1934; many TV appearances.

PICTURES INCLUDE: Rembrandt, Knight Without Armour, Night Train to Munich, They Came to a City, Pimpernel Smith, Freedom Radio, Way Ahead, I See a Dark Stranger, School for Secrets, Broken Journey, So Evil My Love, Mr. Perrin and Mr. Traill, It's Hard to Be Good, Passport to Pimlico, Trio, I'll Never Forget You, The Last Page, Laxdale Hall, Meet Mr. Lucifer, Mr. Denning Drives North, Hobson's Choice, Orders Are Orders, Constant Husband, Geordie, The Prisoner, Doctor at Sea, Dam Busters, Teckman Mystery, Green Man, Brothers in Law, Room at the Top, Carlton-Browne of the F.O., Suspect, A French Mistress, Pure Hell of St. Trinians, Only Two Can Play, Waltz of The Toreadors, Crooks Anonymous, On the Beat, Carry On Nurse, The Great St. Trinian's Train Robbery, Hostile Witness, The Gaunt Woman, Arthur! Arthur!, Young Winston, That's Your Funeral, Symptoms.

HURLOCK, ROGER W.: Pres. Hurlock Cine-World. b. Cambridge, Md., May 30, 1912. e. Baltimore City College; ent. m.p. ind. as publicist, Hippodrome Theatre, Balt.; asst. mgr., Lessor-operator Imperial and Majestic Theatres, Balt., 1931–35; real estate, bldg., farming, Md. and Alaska, 1936–58; elected bd. mem., Allied Artists, 1958; asst. to pres., 1961–63; chmn. budget comm., 1963; chmn. policy comm., 1964; c.p. exec. comm. member, 1964; v.p., chf. operating officer 1965; chmn. exec. comm., 1966; pres., 1967. pres., Hurlock Cine-World, 1969.

HURT, JOHN: Actor. b. Shirebrook, Derbyshire, Jan. 22, 1940. e. St. Martin's School for Art, London. Professional debut in 1962 in British film, The Wild and the Willing.

PICTURES: A Man for All Seasons, Before Winter Comes, Sinful Davey, In Search of Gregory, 10 Rillington Place, Forbush and the Penguins, East of Elephant Rock, Disappearance, Spectre Pied Piper of Hamelin, The Ghoul, Little Malcolm, The Shout, Midnight Express, Alien, Heaven's Gate, The Elephant Man, Partners.

TELEVISION: Playboy of the Western World, A Tragedy of Two Ambitions, Green Julia, Nijinsky, Shades of Green, Ten from the Twenties, The Pedlar, The Naked Civil Servant, I, Claudius, Crime and Punishment.

STAGE: The Dwarfs, Little Malcolm and His Struggle Against the Eunuchs, Man and Superman, Belcher's Luck, Ride a Cock Horse, The Caretaker, Romeo and Juliet, Ruffian on the Streets, The Dumb Waiter, Travesties, The Arrest.

HURT, MARY BETH: Actress. e. NYU School of Arts. Stage debut in 1973 in N.Y. Shakespeare Festival prod., More Than You Deserve. Other stage work includes Trelawny of the Wells, Father's Day.

PICTURES: Interiors, Head Over Heels, Change of Seasons.

TELEVISION: Secret Service (NET Theatre), Kojak, etc.

HURT, R.N.: Circuit executive. b. Veedersburg, Ind., Jan. 30, 1902; e. Brown's Business Coll., Danville, Ill.; m. Entered m.p. ind. as operator at Veedersburg for H. C. Whisler, 1918–22; owned, oper. Lyric Edinburgh, Ind., 1920; for six months, operator and later booker Lincoln Paris, Ill., for Mrs. L. Jarodsky. To Crown Point, Ind., Palace Theatre, manager four years; State, Princeton, in 1930 as mgr., later owner. Sold out to E. E. Alger, 1932 and became mgr. Apollo, Princeton. Transferred to main office, Alger Circuit as booker in 1934, then gen. mgr. oper. in 10 towns controlling 14 theatres. Left 1945; Dixon Theatre Co., 1945–46; Isis Theatre, Toluca, Ill., 1946–48; returned to Alger Sept. 1948–58; booker, Kerasotes Theatres, hdq. Springfield, Ill., 1958–71. Mgr., Times Danville, Ill., Oct. 1, 1971. Retired.

HURT, WILLIAM: Actor. b. Washington, D.C. 1951. e. Tufts, Julliard. Leading actor with New York's Civic Repertory Company since 1976, appearing in among other plays, The Fifth of July, My Life, Ulysses in Traction, The Runner Stumbles, Hamlet, Childe Byron.

FILMS: Altered States (debut), Eyewitness, Body Heat.
TELEVISION: Verna: USO Girl.

HUSSEIN, WARIS: Director. b. India, 1938. TV incl: Sleeping Dog, Death of a Teddy Bear, Toggle, Spoiled, Days In the Trees, A Passage to India, Girls in Uniform, St. Joan, A Casual Affair, Divorce His, Divorce Hers, Shoulder To Shoulder, Gorges Sand, Chips With Everything, The Glittering Prizes, Love Letters on Blue Paper, Sarah Bernhardt, Blind Love, Romance, Daphne Laureola, Waiting for Sheila, Armchair Thriller, Edward and Mrs. Simpson, Death Penalty, And Baby Makes Six.

PICTURES INCLUDE: A Touch of Love, Quackser Fortune, Melody, The Possession of Joel Delaney, Henry VIII and His Six Wives.

HUSTON, JOHN: Director, writer. b. Aug. 5, 1906, Nevada. Mo. p. Walter Huston. In 1938: collab. s.p., Amazing Dr. Clitterhouse, WB. In 1939: Juarez. In 1940: Dr. Ehrlich's Magic Bullet, WB. In 1941: High Sierra, Sergeant York, WB.; dir., s.p., The Maltese Falcon, WB, in 1942: dir. In This Our Life, Across the Pacific, WB. U.S. documentaries (dir.).

PICTURES INCLUDE: Three Strangers, Treasure of the Sierra Madre, Key Largo, WB. 1949 founded Horizon Films with S.P. Eagle, Were Strangers, The Asphalt Jungle, The Red Badge of Courage, The African Queen, Moulin Rouge, Beat the Devil, Moby Dick, The Misfits, Freud, List of Adrian Messenger, The Night of the Iguana, The Bible, Reflections in a Golden Eye, Candy, acted in De Sade, Walk with Life and Death, The Kremlin Letter, The Deserter, acted in Man in the Wilderness, Fat City, The Life and Times of Judge Roy Bean (dir. & acted), Battle for the Planet of the Apes (actor), The McIntosh Man, Chinatown

(actor), The Wind and The Lion (actor), The Man Who Would Be King (dir.), Winter Kills (actor), Wise Blood (dir.), Victory (dir.).
TV: The Rheinmann Exchange.

HUSTON, VIRGINIA: Actress. b. Omaha, Neb., Apr. 24, 1925. Did radio and stage work while in school; acted at Omaha Community playhouse; started in m.p. with RKO 1945.
PICTURES INCLUDE: Nocturne, Out of the Past, Tarzan's Peril, The Highwayman, Racket, Flight to Mars, Night Stage to Galveston, Sudden Fear, Knock on Wood.

HUTNER, MEYER MICHAEL: Executive. b. Brooklyn, N.Y., July 11. e. St. John's U., Brooklyn; m. Roberta Bogart. Reporter, ed., N.Y. Post, Journal American; pub. dir., Billy Rose Enterprises, with 20th Century-Fox as newspaper contact, pub. dir., 1948–55; adv. pub. dir., Samuel Goldwyn Prod., 1956; nat. pub. mgr. Warner Bros., 1956–60; v.p., dir. & pub., Astor Pictures, 1961; apptd. pub. mgr., United Artists Corp., July 1962; pub. dir. 1965; exec. ass't in chg. publicity Paramount Pictures, 1966. v.p. chg. adv. publicity, Filmways, 1967. World-Wide pub., dir. United Artists, 1969. Eastern Co-ordinator, adv.-pub., Columbia Pictures, 1975; joined 20th Century Fox as natl. pub. director, Oct., 1976. Appt. Warner Bros., dir. special projects, 1980.

HUTTON, BETTY: Actress. b. Battle Creek, Mich., Feb. 26, 1921. Made screen debut 1942.
PICTURES INCLUDE: The Fleet's In, Star Spangled Rhythm, Happy Go Lucky, Miracle of Morgan's Creek, Incendiary Blonde, And the Angels Sing, Here Come the Waves, Duffy's Tavern, The Stork Club, Perils of Pauline, Annie Get Your Gun, Let's Dance, The Greatest Show on Earth, Somebody Loves Me, Spring Reunion.
TV: Goldie.

HUTTON, BRIAN, G.: Director. b. N.Y., N.Y.
PICTURES INCLUDE: The Wild Seed (Universal); The Pad (Universal); Sol Madrid (MGM); Where Eagles Dare (MGM); Kelly's Heroes (MGM); X, Y, and Zee, Night Watch, The First Deadly Sin.

HUTTON, LAUREN: Actress. b. Charleston, S.C. e. Univ. of So. Florida, Sophie Newcombe College. As model featured on more covers than any other American.
PICTURES: Paper Lion (debut, 1968), Little Fauss and Big Halsey, The Gambler, Gator, Welcome to L.A., A Wedding, American Gigolo, Zorro, the Gay Blade, Paternity.

HUTTON, ROBERT: Actor. r.n. R. Winne. b. Kingston, N.Y., June 11, 1920. e. Blair Acad., N.J. In summer stock prior to screen career, 1943.
PICTURES INCLUDE: Destination Tokyo, Janie, Roughly Speaking, Hollywood Canteen, Too Young to Know, Love and Learn, Always Together, Steel Helmet, New Mexico, Racket, Slaughter Trail, Casanova's Big Night, Cinderella; co-prod., dir., star, The Slime People; asso. prod., Now It Can Be Told. The Vulture, You Only Live Twice, They Came From Beyond Space, Torture Garden, Tales From the Crypt.

HUTTON, TIMOTHY: Actor. b. Malibu, CA, 1961. Father, late actor Jim Hutton. In plays in high school; toured with father in Harvey during vacation. Theatrical film debut in Ordinary People (1980).
TELEVISION: Movies: Zuma Beach, Best Place To Be, Baby Makes Six, Sultan and the Rock Star, Young Love, First Love, Friendly Fire.
PICTURES: Ordinary People, Taps.

HUYCK, WILLARD: Writer-Director. e. USC. Went to work as reader for Larry Gordon, executive at American-International Pictures; named Gordon's asst., working on scene rewrites for AIP films. First screen credit on The Devil's Eight as co-writer with John Milius, also USC graduate. Left AIP to write original scripts, joining with Gloria Katz. Both signed by Francis Ford Coppola to write and direct films for his America Zoetrope but projects never materialized. Co-wrote American Graffiti with Katz (1973) and Lucky Lady (1975). Huyck made directorial debut in 1979 with French Postcards, co-written with Katz, who also produced.

HYAMS, JEROME: Executive. b. July 19, 1915, N.Y.C. e. New York Univ.; with Guaranteed Pictures Co., 1937; mgr., Commonwealth Pictures Corp., 1938; U.S.N., 1941–45; returned to Commonwealth 1945; sales mgr., Commonwealth Film & TV, Inc., 1948; apptd. v.p. chg. sales, Commonwealth; organ. Hygo TV Films, Inc., 1951; dir. of sync. dist. for Screen Gems, 1957; exec. v.p., gen. mgr., Screen Gems Inc. Pres. Screen Gems, 1968; senior exec. v.p. Columbia Pictures Industries, Inc., 1973, created Chief Operating Officer.

HYAMS, JOSEPH: Adv. & pub. exec. b. N.Y., Sep. 21, 1927. e. N.Y. Univ.; ent. industry, 1947. Various publicity posts, 20th Century-Fox, Columbia Pictures, 1947–55; eastern pub. mgr., Figaro Prods., 1955–56; West Coast pub. mgr., Hecht-Hill-Lancaster, 1955–58; pub. adv. dir., Batjac Prods. 1959–60 national adv. & pub. dir., Warner Bros., Seven Arts, 1960. v.p., Warner Bros., Inc., 1970.

HYAMS, PETER: Director-Writer. b. N.Y.C., July 26, 1943. e. Hunter College, Syracuse Univ. Joined CBS news staff N.Y. and made anchor man. Filmed documentary on Vietnam in 1966. Left CBS in 1970 and joined Paramount in Hollywood as writer. Hired by ABC to direct TV features.
PICTURES INCLUDE: Busting (s.p., dir.), Our Time (dir.), Peeper, Telefon (co-s.p.), Capricorn One (s.p., dir.), The Hunter (co. s.p.), Outland (dir., s.p.).
TV: The Rolling Man, Goodnight My Love (both s.p., dir.).

HYDE, JOHN W.: Executive. b. Jackson, MI. e. New York University, B.A. 1963. Joined ABC upon graduation from N.Y.U. 1963 hired by MCA-Universal as executive assitant to then-v.p. Ned Tanen. As Tanen's assistant, became assoc. prod. on several MCA features. Also wrote, produced and directed musical shorts for MCA Records. 1969 joined Filmways, Inc. as v.p. and exec. asst. to then-pres. Richard R. St. Johns. 1972 formed own company, Acmelab, Ltd., multi-faceted special effects, commercial and post-production facility, which included Cinefex, Videoconversion and Acme Film Laboratories. 1976 Hyde sold Acmelab to produce The Ravagers for Columbia Pictures. During production of that film, Hyde rejoined Richard St. Johns as v.p. of the Guinness Film Group. 1977 Hyde became vice president in charge of production for all Guiness films.
TV INCLUDES: The Andy Williams Show, The Lloyd Thaxton Show.
PICTURES INCLUDE: Games, Midnight Patient, Skulduggery, The Ravagers, Death Hunt, Rituals, The Uncanny, Silent Flute (Circle of Iron), Matilda, The Wanders, Nightwing, The Mountain Men, The Final Countdown, A Change of Seasons, Dead & Buried, American Pop and Venom.

HYDE, TOMMY: Executive. r.n. Thomas L. b. Meridian, Miss., June 29, 1916. e. Lakeland H.S., grad., 1935. Worked, E.J. Sparks Theatres, 1932–41; Florida State Theatres, 1941–42. U.S. Navy, 1942–46. Florida State Theatres, 1946–47; city mgr. (Tallahassee); Talgar Theatres, 1947–58; v.p. and gen. mgr., Kent Theatres, 1958–81; vice-pres., Motion Picture Films, Inc.; pres., NATO of Florida, 1961–62; chmn. bd. 1963–70.

HYER, MARTHA: Actress. b. Fort Worth, Tex., Aug. 10, 1930; e. Northwestern U., Pasadena Playhouse.
PICTURES INCLUDE: Thunder Mountain, Indian Summer, Roughshod, Velvet Touch, The Lawless, Outcast of Black Mesa, Salt Lake Raiders, Frisco Tornado, Abbott and Costello Go to Mars, Scarlet Spear, So Big, Sabrina, Kiss of Fire, Paris Follies of 1956, Francis in the Navy, Red Sundown, Showdown at Abilene, Kelly and Me, Battle Hymn, Mister Cory, My Man Godfrey, Paris Holiday, Once Upon a Horse, Houseboat, Some Came Running, Big Fisherman, Best of Everything, Ice Palace, Desire In the Dust, The Right Approach, The Last Time I Saw Archie, Girl Named Tamiko, Man From the Diner's Club, Wives and Lovers, Pyro, The Carpetbaggers, First Man in the Moon, Blood On the Arrow, Bikini Beach, Sons of Katie Elder, The Chase, Night of the Grizzly, Picture Mommy Dead, The Happening, Some May Live, House of a Thousand Dolls, Once You Kiss A Stranger, Crossplot.

HYLAND, DICK IRVING: Writer-producer-management. b. N.Y.C., Aug. 22, 1906. e. Syracuse U. Reporter N.Y. Evening Journal, ed., trade pubs., adv., publicity dir. for Mills Artists, Inc., publicity dir. Warner Bros., Pres., Cinemasters, Inc., film prod. co. Made over 100 short subjects in conjunction with James Roosevelt-Mills Co., also commercial films. In Hollywood, 20 feature screen credits as writer-producer at Universal, R.K.O., Col. Became V.P. Frank Cooper Associates, Inc., artists management, 6 years. V.P. The Jaffe Agency, Inc. Currently Pres., Dick Irving Hyland Associates, literary agents, packagers. Guest lecturer U. So. Calif., Amer. Nat. Theater Academy. Pac. Coast Writers Conf. Recipient Wisdom Award of Honor for advancement of knowledge, learning and research in education.

I

IANNUCCI, SALVATORE J.: Executive. b. Brooklyn, N.Y., Sept. 24, 1927. e. NYU 1949 B.A., Harvard Law School 1952 L.L.D. 2 yrs. legal departments RCA and American Broadcasting Company, Inc.; 14 yrs. with CBS Television Network: asst. dir. of bus. affairs, dir. of bus. affairs, v.p. of bus. affairs; 2 yrs. v.p. admin. National General Corp.; 2' yrs. pres. of Capital Records; 4½ yrs. Corpor. Vice pres. and dir. of Entertainment Div. of Playboy Enterprises, Inc.; 4 yrs. partner with Jones, Day, Reavis & Pogue in Los Angeles office, handling entertainment legal work; Presently, pres. Filmways Entertainment, Inc. and sr. v.p., Filmways, Inc.

IMMERMAN, WILLIAM J.: Producer. b. 1937. Joined 20th Century-Fox in 1972 as v.p., business affairs. Promoted 1975 to senior vice pres., administration and worldwide business affairs of the Feature Film Division. Previously was with American International Pictures for 6 yrs. as v.p. business affairs & assoc. counsel. From 1963–65 was deputy dist. atty. for Los Angeles County. Presently bd. chm. of Cinema Group, Inc. Exec. prod., Highpoint.

INGALLS, DON: Producer, writer. b. Humboldt, Nebr. e. George Washington U. Columnist, Washington Post; producer-writer, ATV England and Australia; writer-producer, Have Gun Will Travel, CBS, also produced for TV: The Travels of Jamie McPheeters, The Virginian, Honey West, Serpico, Kingston: Confidential. Exec. story consultant The Sixth Sense, ABC.
WRITER: Gunsmoke, Have Gun Will Travel, The Bold Ones, Marcus Welby M.D., Mod Squad, Star Trek, Honey West, Bonanza, The Sixth Sense, Then Came Bronson, Police Story, World Premier Movie, Shamus, Flood, Capt. America, The Initiation of Sarah, others. Currently prod.-writer for TV's Fantasy Island.
FEATURE FILMS: Airport-1975, Story and Screenplay. Who's Got the Body?

INGELS, MARTY: Actor. b. Brooklyn, N.Y., Mar. 9, 1936. U.S. Infantry 1954–58. Ent. show business representing Army, Name That Tune. Stage: Sketchbook revue, Las Vegas.
TV: Phil Silvers Show, Steve Allen, Jack Paar, Playboy Penthouse, Bell Telephone Hour, Manhunt, Ann Southern Show, Peter Loves Mary, The Detectives, Joey Bishop Show, Hennessey, Dick Van Dyke Show, I'm Dickens . . . He's Fenster, Burke's Law, Hollywood Palace.
PICTURES INCLUDE: Ladies Man, Armored Command, Horizontal Lieutenant, Wild & Wonderful.

INGIS, ROBERT L.: Executive. b. Brooklyn, N.Y., Feb. 19, 1935. e. N.Y.U. Accounting staff, S. D. Leidesdorf & Co., 1956–61; officer of Kalvex, Inc., 1961 to present.

INGSTER, BORIS: Writer, director. In 1935; collaborated on adaptation, The Last Days of Pompeii, RKO. In 1936: Dancing Pirate, RKO. In 1937: collaborated on screen play Thin Ice, 20th-Fox. In 1938: Happy Landing.
PICTURES INCLUDE: Judge Steps Out, Southside 1–1000; Something for the Birds, Abdullah's Harem, California-story, Cloak & Dagger, The Amazing Mrs. Holliday.
TV: Wagon Train, The Alaskans, The Roaring 20's, Travel of Jaimie McFeathers, The Man From U.N.C.L.E.

IRELAND, JILL: Actress. b. London, England, 1936. Began career in music halls of England at age of 12; went on to sing, dance and entertain at London's Palladium, in cabarets and a tour of the continent in ballet. Began acting in West End repertory; then signed to major film studio contract at 16 by J. Arthur Rank. Screen debut as ballet dancer in Oh, Rosalinda, first of 16 feature films for Rank.
PICTURES INCLUDE: Three Men in a Boat, Hell Drivers, Robbery Under Arms, Carry On, Nurse, Raising the Wind, Twice Round the Daffodils, Villa Rides, Rider on the Rain, Cold Sweat, The Family, Someone Behind the Door, The Mechanic, The Valdez Horses, The Valachi Papers, Breakout, Hard Times, Breakheart Pass, From Noon Till Three.
TV: Shane (series), The Man from U.N.C.L.E., Ben Casey, Night Gallery, Daniel Boone, Mannix, Star Trek (series).

IRELAND, JOHN: Actor. b. Vancouver, B.C., Jan. 30, 1915; m. Stock 5 yrs. To Hollywood, 1945.
THEATRE: On N.Y. stage: Macbeth, Moon Is Down, Native Son, Counter Attack, etc.
PICTURES INCLUDE: A Walk in the Sun, Wake Up and Dream, Red River, The Gangsters, Roughshod, All The King's Men, Return of Jesse James, Vengeance Valley, Red Mountain, Basketball Fix, Bushwackers, Hurricane Smith, 49th Man, Combat Squad, Southwest Passage, Security Risk, Steel Cage, Outlaw Territory, Fast and Furious, Good Die Young, Queen Bee, Hell's Horizon, Faces in the Dark, I Saw What You Did, Fort Utah, Once Upon a Time in the West, The Adventurers, The Dirty Heroes, Escape to the Sun, Farewell, My Lovely.

IRETON, GLENN F.: Executive. b. Hammond, N.Y., July 11, 1906. e. Colgate Univ. Actor, stock in Balt., Md., Englewood, N.J.; vaude.; juvenile lead in Lew Cantor's, The Skull; musician, Frank Silver's orch., Loew's Theatres, NYC, publicity, Montclair, Commodore Hotels; 24-hour man. carnival, Endy Bros. Railroad shows; Public relations dir. Warner Bros., Canada, 1940–47. Free-lance publicist & promoter, 1947–50. Occupation film post, Japan, 1951. Editor & publisher, and co-founder with wife (Kikuko M. Ireton) of Far East Film News in 1953, renamed Movie Marketing in 1961, Movie/TV Marketing in 1966.

IRETON, KIKUKO MONICA: Managing Editor. Movie/TV Marketing; Tokyo, Japan. b. May 22, 1929. e. Seijo Gaguen, Tokyo. Daughter of late Shoichiro Kobayashi, manager of Bank of Japan in New York (1937–41) and later director. Co-founder with husband (Glenn F. Ireton) of Far East Film News in 1953, renamed Movie Marketing in 1961, Movie/TV Marketing in 1966.

IRVING, GORDON: Film, stage, TV Journalist and Broadcaster. b. Annan, Scotland, Dec. 4, 1918. e. Dunfries Acad., Scotland, Edinburgh Univ., grad. M.A. Hons. Classics. Writer on show business since 1939. Columnist for 18 years with Daily Record and Sunday Mail, Glasgow. Contributor to "The Performer," former weekly of variety Artistes' Federation and The Stage, London.
AUTHOR: The Devil on Wheels, Annie Laurie Story, Great Scot! (Harry Lauder biography), The Good Auld Days (Story of Scotland's Music-Hall), Brush Up Your Scotland!, The Wit of the Scots, The Wit of Robert Burns, Take No Notice, etc. Contributor to trade and fan magazines and journals in Britain, America and South Africa. Free lance writer for worldwide publications. Official correspondent of Stage, London "Variety," New York "Daily Variety," Hollywood, Argus South African Newspapers in Johannesburg, Durban, Cape Town, etc.

IRWIN, CHRISTOPHER: Producer-Executive, b. England 1948. e. University of Sussex, Eng., from where he holds a B.A. degree in social studies. From 1967 to 1969 worked as freelance producer-presenter for BBC Radio Brighton. Was with the Federal Trust (the institute concerned with European affairs) 1969 to 1975, during which time worked for the Secretariat of the North Atlantic Assembly. Joined the BBC's External Services as talks producer in 1975, moving in 1977 to the BBC's central Secretariat. From 1977 to 1978 was seconded to the International Institute for Strategic Studies. In 1978 went to Scotland as Secretary, BBC Scotland, and was closely associated with early stages of Radio Scotland. Is a member of BBC's Future Policy Group, the "think tank" of the Corporation. Appointed Head of Radio Scotland May 1980.

ISAACS, PHIL: Executive. b. N.Y.C., May 20, 1922. e. City College of N.Y., 1938–42. In U.S. Navy, 1943–46. Joined Paramount Pictures in 1946 as bookers asst., N.Y. exch.; following year was booker. In 1951 branch mgr. in Washington. In 1955 named mgr. of Rocky Mountain Div. and in 1957 made asst. east sales mgr. In 1966 was Eastern-Southern sls. mgr In 1967 joined Cinema Center Films as v.p. domestic dist. In 1972 named v.p., marketing, for Tomorrow Entertainment; Joined Avco-Embassy 1975 as v.p., gen. sls. mgr., named exec. v.p., 1977. 1978 joined General Cinema Corp. as v.p. Now v.p., gen. sls. mgr., Orion Pictures.

IVANO, PAUL: Director of Photography. r.n. Paul Ivanichevitch. b. Nice, France, May 13, 1900. e. Lycee de Nice. First director of photography to start air-shots from a helicopter (They Live by Night, Johnny Belinda, Criss Cross, My Four Sons, Japanese War Bride, etc.)
PICTURES INCLUDE: Chubasco, Lizzie, Three and a Day, Champagne for Caesar, The Seagull, Shanghai Gesture, Blond Venus, The Suspect, The Impostor, Flesh and Fantasy, Queen Kelly, For Men Only, Hold Back Tomorrow, Johnny Belinda (exteriors & helicopter shots), The Gangster, Black Angel, Frozen Ghost, Dead Man's Eyes, Black Dawn.
TV: The Jim Backus Show, The Lawless Years, Westinghouse Playhouse, Screen Director's Playhouse, The Texan, Telephone Time, Please Don't Eat the Daisies, Daktari, Man From Uncle, Family Affair.

IVERS, IRVING N.: Executive. Worked for 10 years in radio and TV in variety of executive capacities in station management before entering film business. Joined Columbia Pictures in 1973, serving as director of mktg. and dir. of adv. 1973–77; named Canadian sls. mgr. 1977–78; v.p. of adv./pub. 1978–80. Now with 20th Century-Fox as snr. v.p. of adv./pub./promo.

IVES, BURL: Ballad singer. b. Hunt Township, Ill., June 14, 1909. e. Teacher's College, Charleston, Ill. Professional football player, itinerant worker, radio singer, specializing ballads.
STAGE: On B'way in Cat on a Hot Tin Roof.
PICTURES INCLUDE: Smoky, Green Grass of Wyoming, Station West, So Dear to My Heart, Sierra, East of Eden, The Big Country, Cat on a Hot Tin Roof, Day of the Outlaw, Our Man in Havana, Robin and the Seven Hoods, Just You and Me, Kid.

J

JABLONOW, SCOTT: Exhibitor Executive. b. St. Louis, Mo., Feb. 27, 1950. e. Arizona State Univ. Joined Mid-America Theatres as film buyer in 1971, named president in February, 1979. Member of board of National Endowment for Arts; served on Short Film Showcase.

JACKS, ROBERT L.: Producer. b Oxnard, Calif., June 14; e. Santa Monica City Coll.; U.S.C. U.S. Navy; expl. dept., 20th-Fox, 1949; then story dept.; admin. asst. to prod. Julian Blaustein & Samuel G. Engel; prod., 1952.
PICTURES INCLUDE: Lure of the Wilderness, Man on a Tightrope, Desert Rats, Prince Valiant. Prod (with late Leonard Goldstein) as Panoramic Prod. for 20th-Fox: The Gorilla, Gambler from Natchez, Man in the Attic, Princess of the Nile; The Raid, Siege at Red River, Three Young

Texans, White Feather, A Life in the Balance, Robt. Jacks Prods.: Killer Is Loose, A Kiss Before Dying, Proud Ones, Bandido, Man from Del Rio, Roots of Heaven, Darryl Zanuck Prods.; prod. rep. 20th Century-Fox, 1963; Man in the Middle, The Third Secret, Guns at Batasi, Zorba the Greek; prod. exec. 20th Century-Fox TV, 1964. Prod.; Victor Borge Show (pilot), 1964; Three Coins in the Fountain (pilot), 1964; The Man Who Never Was, 1965; Custer (pilot), 1967; Honeymoon With a Stranger (ABC Movie of Week), 1969; Arnie (pilot), 1970; Bandolero (film), The Undefeated (film).

TELEVISION: Eight Is Enough (series), 1977–78; State Fair (1 hour pilot); Crunch (2 hour pilot); The Waltons (series), 1972–1976; Lovejoy (pilot); Pomroy's People (pilot/co-creator); The Girls of Huntington House (MOW); Binary (MOW); The Homecoming (CBS 2 hour special); Do Not Fold, Spindle or Mutilate (MOW); A Girl and a Boy (MOW); Wild, Wild West Revisited (special), Mr. Horn (mini-series); Young Pioneers (mini-series); Young Love, First Love (MOW); Murder Can Hurt You (MOW); More Wild Wild West (MOW); A Matter of Death & Life (MOW); The Day the Loving Stopped (MOW), A Few Days in Weasle Creek (MOW).

JACKSON, BRIAN: Actor. film/stage prod. b. Bolton, England, 1931. Early career in photography then numerous stage performances incl. Old Vic, Royal Shakespeare. Ent. film/TV industry 1958. Formed Quintus Plays, 1965; formed Brian Jackson Productions 1966; formed Hampden Gurney Studios Ltd. 1970. Co-produced The Others 1967; presented The Button, 1969; co-produced the documentary film Village in Mayfair, 1970; 1971: Formed Brian Jackson Films Ltd.; produced Yesterday, The Red Deer; 1972: produced The Story of Tutankhamen.

TV INCLUDES: Moon Fleet, Private Investigator, Life of Lord Lister, Z Cars, Vendetta, Sherlock Holmes, Mr. Rose, Hardy Heating International, Nearest and Dearest, The Persuaders, The Paradise Makers, The New Avengers, Smugglers Bay, The Tomorrow People, Secret Army, Last Visitor for Hugh Peters.

PICTURES INCLUDE: Incident in Karandi, Carry On Sergeant, Gorgo, Jack the Ripper, Taste of Fear, Heroes of Telemark, Only the Lonely, The Deadly Females, The Revenge of the Pink Panther.

STAGE INCLUDES: Mame, Drury Lane, Fallen Angels, South African Tour, Seasons with Old Vic and Royal Shakespeare Co., In Praise of Love.

JACKSON, FREDA: Actress. b. Nottingham, Eng., Dec. 29, 1909; e. U. Coll., Nottingham. Stage debut 1933, Northampton Repertory Theatre; London debut 1936, Old Vic; screen debut in Canterbury Tale, 1942.

PICTURES INCLUDE: Henry V, Beware of Pity, Great Expectations, No Room at the Inn, Flesh and Blood, Women of Twilight, The Good Die Young, The Crowded Day, The Enchanted Doll, Bhowani Junction, Last Man to Hang, The Flesh Is Weak, Brides of Dracula, Greyfriar's Bobbie, Shadow of the Cat, Attempt to Kill, West Eleven, Monster of Terror, The Third Secret, Gwangi, Tom Jones, Clash of the Titans.

TV: Macadam and Eve, Sorry Wrong Number, Trial of Marie Lafarge, Release, Colombe, Maigret in Montmartre, Sergeant Musgrave's Dance, Dr. Finlay's Casebook, Sunset, Knock On Any Door, The Spies, Adam Adamant, Midland Profile, Owen Md., The Kilvert Diaries, She Fell Among Thieves, Randall and Hopkirk, The Old Curiosity Shop.

JACKSON, GLENDA: Actress. b. Birkenhead, England, 1938. Ent. m.p. ind. 1955.

PICTURES INCLUDE: Marat-Sade, Negatives, Women In Love, The Music Lovers, Sunday, Bloody Sunday, Mary Queen of Scots, Triple Echo, The Nelson Affair, A Touch of Class, The Maids, The Tempter, The Romantic Englishwoman, Hedda, Sarah, Nasty Habits, House Calls, Stevie, The Class of Miss McMichael, Lost and Found.

TV: Queen Elizabeth.

JACKSON, GORDON: Actor. b. Glasgow, Scotland, Dec. 19, 1923; e. Glasgow. On radio since 1939; screen debut in "Foreman Went to France" 1940.

PICTURES INCLUDE: Nine Men, San Demetrio, Millions Like Us, Pink String & Sealing Wax, Captive Heart, Against the Wind, Eureka Stockade, Whiskey Galore, Floodtide, Stop Press Girl, Bitter Springs, Happy Go Lovely, The Lady with the Lamp, Meet Mr. Lucifer, Malta Story, Castle in the Air, Quartermass Experiment, Pacific Destiny, Baby and the Battleship, Sailor Beware, Seven Waves Away (Abandon Ship), As Long As You're Happy, Hell Drivers, Rockets Galore, Bridal Path, Yesterday's Enemy, Blind Date, Cone of Silence, Tunes of Glory, Greyfriar's Bobbie, Mutiny on the Bounty, The Great Escape, The Long Ships, Those Magnificent Men in Their Flying Machines, The Great Spy Mission, The Ipcress File, Cast A Giant Shadow, Fighting Prince of Donegal, Night of the Generals, Triple Cross, The Eliminator, Prime of Miss Jean Brodie, Run Wild, Run Free, Hamlet, Scrooge, Kidnapped, Madame Sin, Russian Roulette, Spectre, Golden Rendezvous, Medusa Touch, Raising Daisy Rothschild.

TV: Numerous TV appearances incl: Upstairs, Downstairs, The Professionals.

JACKSON, JAY: Performer. b. Stockdale, Ohio. e. Miami Univ. of Ohio, Ohio State Univ. Actor & announcer, many radio shows; U.S. Army; m.c., Twenty Questions TV show; Masquerade Party, Perry Como Show, Father Knows Best.

JACKSON, JOHN HENRY: Executive. b. New York, April 27, 1916. e. Holy Cross Academy, 1930; Professional Children's School, 1934; Georgia Tech, 1936. Performed as vaudeville artist touring Europe, 1928–29; Fair dates and indoor circus, 1929–35; Texas Centennial, 1936; George Abbott's Too Many Girls, 1938. Joined Radio City Music Hall, 1943 with Glee Club; stage manager, 1944–51; director of stage operations, 1958; v.p., 1970; prod., 1971; pres., Tri-Marquee Productions, Ltd., 1979.

JACKSON, KATE: Actress. b. Birmingham, Ala., Oct. 29. e. Univ. of Mississippi, Birmingham Southern Univ. Did summer stock before going to N.Y. to enter American Academy of Dramatic Arts, appearing in Night Must Fall, Constant Wife, Little Moon of Alban, etc. Worked as model and became tour guide at NBC. First role on TV in Dark Shadows (series).

PICTURES INCLUDE: Thunder and Lightning, Dirty Tricks, Making Love.

TV: The Jimmy Stewart Show (pilot). Movies: Killer Bees, The Shrine of Lorna Love, The Jenny Storm Homicide. Series: Charlie's Angels (regular).

JACKTER, NORMAN: Executive. b. New York City, May 18, 1922. e. N.Y. City Coll. Entered industry directly from school. Spent 35 years with Columbia Pictures, holding all jobs from booker thru division mgr. Was v.p. dist. from 1964–73. Also v.p., marketing, for Radnitz/Mattel Prods. Inc. Now indep. prod. and dist. consultant.

JACKTER, RUBE: Sales executive. b. New York City, Dec. 12, 1900. Entered m.p. ind. 1913 with Jesse L. Lasky Feature Play Co. in sales dept., joined Samuel Goldwyn upon org. Goldwyn Pictures Corp. as spcl. rep.; to Fox Film Corp. as Fox News staff cameraman, also a special rep. in charge of Fox News sales. Resigned to go with Universal and in 1924 left to join Columbia as first special field rep. and assisted in nationalizing Columbia Film Exchanges; then gen. sales mgr.; v.p., 1958.

JACOBS, BARRY: Executive. b. London, England, 1924. Ent. m.p. ind. 1938. Served in RAF 1943–46. Circuit rep. Warner Bros. 1938–59. Overseas sales rep. independent producers 1960–62. Formed Eagle Films Ltd. dist. organization UK 1962. Entered prod. 1969. Exec. prod. The Wife Swappers, Groupie Girl, Bread, Naughty, The Love Box, Sex and The Other Woman, On the Game, Eskimo Nell.

JACOBY, FRANK DAVID: Director/Producer. b. New York, N.Y., July 15, 1925. Hunter College, Brooklyn College. m. Doris Storm, producer/director educational films, actress. 1949–52: NBC network TV director; 1952–56: B.B.D.O., Biow Co., TV producer/director; 1956–58 Metropolitan Educational TV Association, Director of Production; 1958–65: United Nation, film producer/director; 1965 to present: President, Jacoby/Storm Productions, Inc., Westport, Conn., documentary, industrial, educational films and filmstrips. Clients include Burlington Industries, United Federation of Teachers, USIA, Time/Life, Inc., Xerox Corp., Random House, Publ., Lippincott Co., Globe Publ., Harper & Row, IBM, Heublein, G.E., and Pitney Bowes. Winner, Sherwood Award, Peabody Award. Member, Director's Guild of America; winner, Int'l TV & Film Festival, National Educational Film Festival, American Film Festival.

JACOBY, JOSEPH: Producer-Director-Writer. b. Brooklyn, N.Y., 1942. e. N.Y. Univ. School of Arts and Sciences, majoring in m.p. As undergraduate worked part-time as prod. asst. on daytime network TV shows and as puppeteer for Bunin Puppets. In 1963 joined Bill Baird Marionettes as full-time puppeteer, working also on Baird film commercials. Made feature m.p. debut as prod.-dir of Shame, Shame, 1968.

PICTURES INCLUDE: Hurry Up, or I'll Be 30, The Great Georgia Bank Hoax (co.-prod., dir., s.p.).

JACON, BERNARD: Executive. b. Louisiana. Manager, promotion, Small & Strausberg Theatres, New York; buyer & gen. mgr., Mantell Theatres N.Y., gen. mgr. & assoc., Rockaway Beach Theatres, Universal Pictures, home office & field, br. operations & sales. Mgr., sales & dist. (Continental Films) Superfilms Distribution Corp., New York 1946–49; vice-pres. in chg. of sales & dist., Lux Film Distributing Corp., N.Y. 1949–52; v.p. sales, dist. IFE Releasing, 1952–55; org. nat'l distrib. co., Jacon Film Distributors, Inc., as pres., 1956; nat'l consultant to independent distributors, producers & exhibitors, member of the Pioneers & Variety; 1979, formed Bernie Jacon, Inc., natl. co-ordinator for producers and distributors.

JACQUES, ROBERT C.: Film editor. b. Cincinnati, Ohio, Feb. 24, 1919. e. U. of Mich., U.S.C. Asst. film ed., RKO Radio 1939–42; film ed., Pathe News, 1942–44; chief film ed., NBC Television, 1944–48; prod. chief, Ziv TV film dept. 1948–51; free lance m.p. film ed., 1951–52; supervising film ed., prod. mgr., American Film Prod., 1952–53; prod. supvr., ed., Telenews and Screen Gems 1953; film ed.,

RKO Radio, 1953–55; supervising film ed., George Blake Enterprises, 1955–57; v.p. Peter Elgar Productions 1957–60; supv. film editor, Transfilm-Caravel Inc. 1960–62; v.p. in chg. of completion, Filmex, Inc., 1962. V.P. in charge of completion, Filmex Inc. 1962–69. Free lance film editor, May–Nov. 1969. V.P. of own co., Double Image Inc., film and videotape editorial service, Nov. 1969 to date.

JAECKEL, RICHARD HANLEY: Actor. b. Long Beach, N.Y., Oct. 10, 1926; e. Hollywood High School, 1943. Performed odd jobs upon graduation, with plans toward entering Navy when of age; worked as delivery boy in mail room, 20th Century-Fox; signed for role of Baby Marine, in Guadalcanal Diary.

PICTURES INCLUDE: The Gunfighter, Sea Hornet, Hoodlum Empire, My Son John, Come Back Little Sheba, Big Leaguer, Sea of Lost Ships, Shanghai Story, Violent Men, Apache Ambush, Sands of Iwo Jima, Fragile Fox, The Dirty Dozen, Sometimes a Great Notion, Ulzanas Raid, Pat Garrett and Billy the Kid, The Outfit, Chosen Survivors, The Drowning Pool, Part II—Walking Tall, Twilights Last Gleaming.

TV: U.S. Steel Hour, Elgin Hour, Goodyear Playhouse, Kraft, Firehouse (movie).

JAEGER, ANDREW P.: Executive. b. New York City, Oct. 15, 1917. e. Fordham, Columbia U. American Tobacco Co., 1937–41 as asst. to sales & adv. division head; trainee through managing director, Puerto Rican office of 20th Century-Fox Int'l, 1946–49; network dir., film programming, Dumont TV Network, 1949–53; v.p. sales & program devel., Procker Syndications Int'l, 1953–57; network salesman, division mgr., Ziv Television Programs (later United Artists TV), 1957–63; Latin American supervisor, 1963–66; account exec., Telcom Inc., and program consultant to film buyers, 1966–69; sales mgr., then V.P. and currently president, Allied Artists Television, 1969 to present.

JAFFE, HERB: Executive-Producer. b. New York, N.Y. e. Brooklyn Coll., Columbia U. Press agent; then talent agent, MCA personal appearance div.; sales exec., MCA-TV, Ltd., syndic. film div.; eastern sales mgr., Motion Pictures for Television; joined Official Films as v.p., sales mgr.; Herb Jaffe Assoc. 1957; sold Herb Jaffe Assoc. to Ashley-Steiner-Famous Artists, Inc.; v.p. in prod., United Artists, 1965; v.p. of West Coast Operations, 1966. 1970, v.p. charge of World-Wide prod, 1973 entered field of independent motion picture production.

PICTURES: The Wind and the Lion, Demon Seed, Who'll Stop the Rain, Time After Time, Those Lips, Those Eyes, Motel Hell, Jinxed.

JAFFE, LEO: Executive. b. April 23, 1909. e. New York U. Columbia, 1930; v.p., Columbia Pictures, January, 1954; 1st v.p., treas., member of board, 1956; v.p. & treas., 1958; exec. v.p., Columbia Pictures, 1962; Pres. Columbia Pictures, 1968; pres., Columbia Pictures Industries, Inc, 1970, president & chief executive officer Columbia Pictures Industries, Inc., 1973; chm., August, 1973, Chairman of Board of directors since Aug., 1973.

JAFFE, SAM: Talent agent. b. New York City, May 21, 1901. In 1920 joined Mayer-Schulberg prod. org. as studio mgr.; later prod. mgr., prod. B.P. Schulberg Prod. 1922; 1923–29 prod. mgr. & exec. mgr. Paramount Studios; 1930–32 prod. mgr. & asst. to David O. Selznick, RKO Radio; 1933 asst. to Sam Briskin, Columbia. In 1934 became partner Schulberg-Feldman-Jaffe, talent agents, Hollywood; 1935 org. Sam Jaffe, Inc., Sam Jaffe, Ltd.; later dissolved; then headed Jaffe Agency; Ind. prod., head Sam Jaffe Prod., 1959; prod., The Sullivans, Damon & Pythias, co-prod., Born Free; executive producer, Theatre of Blood.

JAFFE, SAM: Actor. b. New York City, Mar 8, 1896. Went on stage 1915, member Washington Square Players, N.Y.; toured with Shakespearean Company; later in many N.Y. stage prods. (Samson and Delilah, The God of Vengeance, The Jazz Singer, Grand Hotel, A Doll's House, etc.). On screen 1933.

PICTURES INCLUDE: The Scarlet Empress, Lost Horizon, Gunga Din, 13 Rue Madeleine, Gentleman's Agreement, The Barbarian and the Geisha, Ben Hur, Les Espions, Rope of Sand, The Asphalt Jungle (Venice Int. Film Festival Award, Best Male Performance), Under the Gun, I Can Get It for You Wholesale, Day the Earth Stood Still, Guns for San Sebastian, The Great Bank Robbery, Dunwich Horror, Bedknobs and Broomsticks.

TV: Ben Casey series, Bonanza, Daniel Boone, Play of he Week, The Lovers (ABC movie of the week), Quarantine (World Premiere), Night Gallery, The Old Man Who Cried Wolf, Enemies (Hollywood Television Theatre), Ghost Story, Love, American Style, Owen Marshall, QB VII, Streets of San Francisco, S.W.A.T, Harry O, Medical Story.

JAFFE, STANLEY R.: Producer. b. New Rochelle, N.Y., July, 1940. Graduate of University of Pennsylvania's Wharton School of Finance. Joined Seven Arts Associates, 1962; named exec. ass't to president, 1964; later, head of East Coast TV programming. Produced Goodbye, Columbus, in 1968 for Paramount; then joined that company as exec. v.p., 1969. Produced A New Leaf, 1970. Named pres. of Paramount in 1970; resigned June 10, 1971 to form own prod. unit. First film: Bad Company (1972); Man on the Swing, Bad News Bears. Joined Columbia as v.p. of global prod. in 1976, but resigned to be independent producer. 1979: Kramer vs. Kramer.

JAFFE, STEVEN-CHARLES: Prducer. b. 1954. e. Univ. of So. Calif. (cinema). Worked on production in Holland and Switzerland; served as prod. asst. on The Wind and the Lion in Spain. Assoc. prod. on Demon Seed (written by brother Robert); served as location mgr. on Who'll Stop the Rain; assoc. prod. on Time After Time. Full producer on Those Lips, Those Eyes.

JAFFE, WILLIAM B.: Attorney. b. New York City, Mar. 11, 1904; e. Union Coll. 1926, A.B.; Columbia U. Law School 1929, LL.B.; m. Evelyn Annenberg. Assoc. Nathan Burkan 1929–35; gen. counsel Allied Artists; 1933–69 spcl. counsel to interests financing independent prod.; to leading personalities m.p. ind.; dir. MPPDA rep. Columbia; gen. counsel & dir. Allied Artists Internat'l 1933–69; chmn. Manning Plan, State of N.Y.; counsel, War Manpower Commission, N.Y. Region Spec. counsel to Revlon, Inc., in TV field; s.p. counsel Triangle Pub.; chmn., Art Advisory Committee Dartmouth College; trustee, Union Coll., Schenectady, N.Y. Vice Chr. & Bd. of Gov Jewish Museum, N.Y.C.; Benefactor Metropolitan Museum of Art; Patron, Museum of Modern Art, Group Chairm. United Hospital Fund 1964–69; Trustee, Phi Epsilon Pi Foundation. Recipient Phi Epsilon Pi Achievement Award, 1964; Doctor Humane Letters, Dartmouth College 1964; Trustee Mt. Sinai Hospital, N.Y., N.Y. Eye & Ear Infirmary. Partner law firm, Shea, Gallop, Climenko & Gould.

JAFFEY, HERBERT: Executive. b. Somerville, N.J., Cap't. U.S. Army, pressbook writer, 20th Century-Fox; theatre management, Loew's; field exploitation, United Artists, Ad. pub. dir. 20th Century-Fox Int and Inter-Amer. Corp.; dir., adv. & publicity, Rugoff Theatres; Independent Film Prod.; Paramount Foreign Advertising & Publicity Manager, screenplay writer.

JAGGER, DEAN: Actor. b. Lima, Ohio, Nov. 7, 1903; e. Wabash Coll. (Crawfordsville, Ind.). On N.Y. stage, stock, vaudeville. Screen debut 1929 in Woman from Hell. Academy Award sup. role 1949 for 12 O'Clock High.

PICTURES INCLUDE: College Rhythm, Home on the Range, Car 99, Woman Trap, 13 Hours By Air, Woman in Distress, Escape by Night, Brigham Young, Western Union, Valley of the Sun, I Escaped from the Gestapo, North Star, Alaska, When Strangers Marry, Pursued, Driftwood, Dark City, Rawhide, Warpath, Denver and Rio Grande, My Son John, It grows on Trees, The Robe, Executive Suite, Private Hell 36, White Christmas, Bad Day at Black Rock, The Eternal Sea, On the Threshold of Space, It's a Dog's Life, Red Sundown, Great Man Three Brave Men, The Nun's Story, Cash McCall, Elmer Gantry, Parrish, Jumbo, Stay way Joe, Firecreek, Smith, The Kremlin Letter, Vanishing Point, Tiger by the Tail.

TV: Glass House, Mr Novak (series), Brotherhood of the Bell, The Lie.

JAGLOM, HENRY: Director-Writer-Editor. b. New York. Studied acting, writing and directing at Univ. of Pennsylvania and with Actors Studio. Did off-Bdwy. shows; went to West Coast where appeared in TV series (Gidget, The Flying Nun, etc.). Started shooting documentary film in Israel during Six Day War; turned it into 3-hr. silent film. Hired to edit Easy Rider by producer Bert Schneider. Acted in Drive, He Said and Last Movie. Directed first feature, A Safe Place, in 1971.

PICTURES; Tracks (1977); Sitting Ducks (1980).

JAHNCKE, ERNEST LEE, JR.: Executive. b. New Orleans, La., Aug. 8, 1912. e. U.S. Naval Acad. 1933. Officer, U.S. Navy; traffic clerk, NBC, 1937–39; station rel. dept., 1939–41; active duty, U.S. Naval Reserve 1941–46; joined station rel. dept., American Broadcasting Co., 1946; mgr., stat. rel., 1948–49; v.p., stat. rel., 1949–51; v.p. in charge of radio network, 1951–52; v.p., asst. to pres., ABC 1952–56; v.p., asst. to bd. chrm., Edward Petry & Co., Inc.; 1957–59; div., Standards & Practices, NBC, 1960–61; Vice pres., 1962–69; East. Dir., American Humane Ass'n, 1970.

JAMES, CLIFTON: Actor. b. N.Y.C, 1922. e. Oregon Univ. Studied at Actors Studio. Made numerous appearances on stage and TV, as well as theatrical films.

PICTURES INCLUDE: On The Waterfront, The New Centurions, Live and Let Die, The Last Detail, Bank Shot, Juggernaut, The Man with the Golden Gun, Rancho DeLuxe.

JAMES, DENNIS: Performer. b. Jersey City, N.J. Aug. 24, 1917. e. St. Peter's Coll., Jersey City. TV personality for over 30 years. Formerly M.C. actor, sports commentator in Radio; award winning sports commentator for wrestling, 25 TV first to credit; currently pres., Dennis James Prod.

TV: Chance of A Lifetime, High Finance, First Impressions, What's My Line, PDQ, Your All-American College Show; host new Price Is Right, Nightime; Host, "Name That Tune," Daytime.

JAMES, HARRY: Band leader. b. Albany, Ga. Mar. 15, 1916. With Mighty Haag Circus as child; contortionist until 1922; trumpet player until 1931; with Benny Goodman band; formed own orchestra 1939-till present. Dot records, 1966 Capitol, MGM, Columbia, London Records, Longines, RCA Readers Digest Albums & Sheffield Lab.
PICTURES INCLUDE: Syncopation, Private Buckaroo, Springtime in the Rockies, Best Foot Forward, Two Girls and a Sailor, Bathing Beauty, Do You Love Me, If I'm Lucky, Carnegie Hall, mus. dir., Young Man With a Horn, To Catch a Thief, Anything Goes.

JAMES, LAURA: Actress. b. Hazelton, Penn., 1933. e. The Theatre Wing, The American Academy of Dramatic Arts, 1948–50; worked Off-Broadway, Summer Stock, TV; starred in West Coast premieres of The Second Man, Champagne Complex.
PICTURES INCLUDE: Four Girls in Town, St. Francis of Assisi.
TV: Matinees Theatre, The Big Story, Day in Court, Divorce Court, Official Detective, Congressional Investigator, Sheriff of Cochise; now working overseas.

JAMES, POLLY: Writer. b. Ancon, Canal Zone; e. Smith Coll. Newspaper work, Panama; with trade mag., N.Y.; screen writer since 1942.
PICTURES INCLUDE: Mrs. Parkington, The Raiders, Redhead from Wyoming, Quantrill's Raiders.
TV: Several shows.

JANIS, HAROLD E.: Producer. b. July 8, 1906. e. Peddie, Columbia School of Journalism. AP Latin American reporter; INS feature writer; War ser. U.S.A. 1941–42, major; U.S. Dept. of State information ser., 1945–47; Radio Stat. WMCA 1934–40. Joined NBC 1950 as program exec., assigned to Show of Shows; exec. prod. Martha Raye Show, Caesar's Hour, Bob Hope Show, Max Liebman presents, others inc. Jackie Gleason, Martin & Lewis; joined Sid Caesar as prod. Sid Caesar Invites You; exec. prod. Shellrich Corp.

JANNI, JOSEPH: Producer. b. Milan, Italy, May 21, 1916. e. Milan University; Rome Film School. Entered m.p. industry 1941. Assistant producer. Founded Vic Films Ltd.
PICTURES INCLUDE: Own productions: The Glass Mountain, White Corridors, Something Money Can't Buy, Romeo and Juliet (co-prod.), A Town Like Alice (prod.); The Captain's Table, Savage Innocents, A Kind of Loving, Billy Liar, Darling, Modesty Blaise, Far From The Madding Crowd, Poor Cow, In Search of Gregory, Sunday, Bloody Sunday, Made, Yanks (co-prod.).

JARMAN, CLAUDE, JR.: Actor. b. Nashville, Tenn., Sept. 27, 1934; e. MGM School. Film debut in "Yearling," 1946.
PICTURES INCLUDE: High Barbaree, Sun Comes Up, Intruder in the Dust, Roughshod, Outriders, Inside Straight, Rio Grande, Hangman's Knot, Fair Wind to Java, Great Locomotive Chase.

JARRATT, ALFRED: Executive. b. London, England. Ent. m.p. ind. 1928 with W&F Film Services as sales rep. Joined Gaumont British Distributors as circuit rep. 1933–37; Grand National Pictures 1937–39; Army Career 1939–46; gen. sales man. Walt Disney Productions in London prior to going to Paramount Pictures, Australia, 1968. Appt. man. dir. Paramount Pictures (UK) Ltd., 1972. From 1973 man. dir., C.I.C. (UK).

JARRE, MAURICE: Composer. b. Lyon, France, 1924. Studied at Paris Cons. Was orchestra conductor for Jean Louis Barrault's theatre company four years. In 1951 joined Jean Vilar's nat'l theatre co., composing for plays by Shakespeare, Moliere, O'Neill, Eliot, and Victor Hugo. Started writing music for films in 1952.
PICTURES INCLUDE: Hotel des Invalides, La Tete contre les Murs, Eyes without a Face, Crack in the Mirror, Sundays and Cybele, The Longest Day, Lawrence of Arabia (AA, 1962), The Collector, Is Paris Burning?, Weekend at Dunkirk, Dr. Zhivago (AA, 1965), Night of the Generals, The Professionals, Grand Prix, Five Card Stud, Isadora, The Damned, Ash Wednesday, The Life and Times of Judge Roy Bean, The Mackintosh Man, The Effect of Gamma Rays on Man-in-the-Moon Marigolds, Island at the Top of the World, Mandingo, Posse, Winter Kills, The Black Marble.

JARRICO, PAUL. Writer, Producer. b. Los Angeles, Jan. 12, 1915.
PICTURES INCLUDE: Salt of the Earth, Tom, Dick & Harry (Academy nomination), Thousands Cheer, The Search (Academy Award), The White Tower, Not Wanted, The Girl Most Likely.

JARROTT, CHARLES: Director. b. London, England June 16, 1927. Son of British businessman and former singer-dancer at Gaiety Theatre. Joined British Navy; wartime service in Far East. After military service turned to theatre as asst. stage mgr. with Arts Council touring co. In 1949 joined Nottingham Repertory Theatre as stage dir. and juvenile acting lead. In 1953 joined new company formed to tour Canada; was leading man and became resident leading actor for Ottawa Theatre. In 1955 moved to Toronto and made TV acting debut opposite Katharine Blake whom he later wed. In 1957 made directing debut in TV for Canadian Bdcstg. Co. Became CBC resident dir. Moved to London to direct for Armchair Theatre for ABC-TV. Then became free-lance dir., doing stage work, films, TV.
PICTURES INCLUDE: Anne of the Thousand Days, Mary, Queen of Scots, Lost Horizon, The Dove, The Littlest Horse Thieves, The Other Side of Midnight, The Last Flight of Noah's Ark.
TELEVISION: The Hot Potato Boys, Roll On, Bloomin' Death, Girl in a Birdcage, The Picture of Dorian Gray, Rain, The Rose Affair, Roman Gesture, Silent Song, The Male of the Species, The Young Elizabeth, A Case of Libel, Dr. Jekyll and Mr. Hyde.
STAGE: The Duel, Galileo, The Basement, Tea Party, The Dutchman, etc.

JASON, RICK: Actor. b. New York City, May 21, 1926. e. American Academy of Dramatic Arts.
THEATRE: Broadway debut: Now I Lay Me Down To Sleep (Theatre World award).
PICTURES INCLUDE: Sombrero, Saracen Blade, This Is My Love, Lieutenant Wore Skirts, Wayward Bus.
TV: The Case of the Dangerous Robin, Combat.

JAY, MORTY: Exec.-prod. r.n., Morton Saroff. b. Brooklyn, N.Y., Oct. 15, 1924. e. Brooklyn Coll. Pianist arranger, Sammy Kaye, 1953–54; mus. dir., Crew Cuts, 1955–61; v.p., Japsina Enterprises Inc., since 1961; v.p., Motivation Concepts of America, since 1962; v.p., International Concepts Inc., since 1963; indep. mus. arranger, since 1963; prod., commercials.
TV: Ed Sullivan Show, Masquerade Party, Stage Show.

JEFFEE, SAUL: Executive. b. Elizabeth, N.J., March 30, 1918. Established the Movielab, Inc., 1936, pres., bd. chmn., & chief exec. officer. Fellow, SMPTE, Treas.; Pres. & chm., Movielab Theatre Service; Pres. Assoc. of Cinema Labs, 1963; American Tech. Rep. & Deleg. in the U.S.-U.S.S.R. Cult. Exch. 1965; Ch. Enter. & Commu. Div., U.J.A., 1967–68; Life member, Motion Picture Pioneers, Friars, City Athletic Club; Author of Narcotics—An American Plan, chmn. Bd. of Trustees, Lorge School; Chm., Movielab-Hollywood, Inc., vice chrm., Film Society of Lincoln Center, Patron, Lincoln Center for The Performing Arts, Mem. Bd. of Gov. U.J.A., life member, chm., Life Patron Program, Variety Clubs International Patentee, Professional Motion Picture Equipment, Trustee, Federation of Jewish Philanthropies; Trustee, Will Rogers Hospital & Research Center; Mason (Shriner, 32 deg); Life Member, The Jewish Chautaqui Society. Member, Boy Scouts Advisory Committee on Scouting For The Handicapped; Fairview Country Club, Greenwich, Connecticut; Advisory Board, Cinema Lodge No. 1366 B'nai B'rith. Academy of Motion Picture Arts & Sciences; Mayor of N.Y.C. Advisory Council on Motion Pictures. Member, Natl. Academy of TV Arts & Sciences.

JEFFREYS, ANNE: Actress. b. Goldsboro, N.C.
PICTURES INCLUDE: I Married an Angel, Step Lively, Dillinger, Sing Your Way Home, Trail Street, Riffraff, Return of the Bad Men, Boys' Night Out.
THEATRE: On Bway. in Street Scene, Kiss Me Kate, Romance, Three Wishes for Jamie, Kismet.
TV: Topper, Love That Jill, many guest appearances, Delphi Bureau.
STOCK: Camelot, King & I, Kismet, Song of Norway, Bells Are Ringing, Marriage Go Round, No Sex Please, We're British, Take Me Along, Carousel, Anniversary Waltz, Do I Hear a Waltz, Ninotchka, Pal Joey, Name of the Game, Destry Rides Again, The Merry Widow, Bitter Sweet, Desert Song, High Button Shoes, Sound of Music.

JEFFRIES, LIONEL: Actor. b. Forest Hill, London, England. Ent. m.p. ind. 1952.
PICTURES INCLUDE: The Nun's Story, Two-Way Stretch, The Trials of Oscar Wilde, Fanny, The Notorious Landlady, The Wrong Arm of the Law, First Men in the Moon, Call Me Bwana, The Truth About Spring, You Must Be Joking, The Crimson Blade, Arrividerci Baby, Spy with the Cold Nose, Journey to the Moon, Camelot, Chitty Chitty Bang Bang, Eye Witness, The Prisoner of Zenda. Dir. s.p., Railway Children, 1970. Gingerbread House, Dir. Baxter, Dir. s.p., The Amazing Mr. Blunden, The Water Babies, Cream in My Coffee.

JENKINS, ALLEN: Actor. r.n. Al McConegal. b. New York, Apr. 9, 1900; e. Amer. Acad. Dramatic Arts, N.Y.
THEATRE: On stage: What Price Glory, Rain, The Last Mile, The Front Page, Five Star Final. In play & m.p. Blessed Event.
PICTURES INCLUDE: Three on a Match, I am a Fugitive from a Chain Gang, 42nd Street, Bureau of Missing Persons, Twenty Million Sweethearts, Big Shakedown, Whirlpool, Footsteps in the Dark, Ball of Fire, Tortilla Flat, Falcon Takes Over, Maisie Gets Her Man, They All Kissed the Bride, Wonder Man, Hat Box Mystery, Senator Was Indiscreet, Inside Story, Behave Yourself, Crazy Over Horses, Let's Go Navy, Oklahoma Annie, Wac from Walla Walla, Doctor, You've Got to Be Kidding.

JENKINS, CHARLES: Animation producer. b. Yorkshire, England, 1941. Ent. m.p. ind. 1957. Joined T.V.C. as gen. ass't 1958. Animated and prod. various prods. Joined Dick Williams, pioneered use of Oxberry Camera for animation, 1966. Esta. Trickfilm Ltd., prod. optical effects for Yellow Submarine, 1967.

JENKINS, DAN: Public Relations Executive. b. Montclair, N.J., Dec. 5, 1916. e. Univ. of Va., 1938. U.S. Army, 1940–45; major, infantry. P.R. officer, Hq. Eighth Army. Mng. ed., Motion Picture Magazine, 1946–48; editor, Tele-Views Magazine, 1949–50; TV editor, columnist, Hollywood Reporter, 1950–53; Hollywood bureau chief, TV Guide, 1953–63; v.p., exec. dir., TV dept. Rogers, Cowan & Brenner, Inc., 1963–71. Formed Dan Jenkins Public Relations, Inc., 1971. Joined Charles A. Pomerantz Public Relations, Ltd. as v.p., 1975, while retaining own firm. Mem. bd. trustees, Natl. Academy of TV Arts & Sciences; bd. gov., Hollywood chapter, Natl. Academy of TV Arts & Sciences, 1967–71. Rejoined Rogers & Cowan, 1979. Reactivated own firm, 1981.

JENKINS, GEORGE: Art dir. b. Baltimore, Md.; e. Univ. of Pennsylvania. Holly-N.Y. art dir. since 1946; TV pictures for Four Star Playhouse and revue productions; NBC-TV opera, Carmen; color dir., CBS-TV, 1954; NBC color spec. Annie Get Your Gun, 1957; TV music with Mary Martin, 1959.

STAGE: Mexican Hayride, I Remember Mama, Dark of the Moon, Lost in the Stars, Bell, Book & Candle, Bad Seed, Happiest Millionaire, Two for the Seesaw, Icecapades. Jones Beach spec., Song of Norway, Paradise Island, Around the World in 80 Days, Mardi Gras, Miracle Worker, Critics Choice, A Thousand Clowns, Jennie, Generation, Wait Until Dark, Only Game in Town, Night Watch, Sly Fox.

PICTURES INCLUDE: Best Years of Our Life, Secret Life of Walter Mitty, Miracle Worker, Mickey One, Up The Down Staircase, Wait Until Dark, Subject Was Roses Klute, 1776, Paper Chase, Parallax View, Night Moves, Funny Lady. Won Oscar 1977 for All the President's Men, Comes A Horseman, China Syndrome, Starting Over, The Postman Always Rings Twice, Rollover.

JENNINGS, TALBOT: Writer. b. Shoshone, Ida.; e. Harvard, M.A.; Yale Drama Sch.; playwright, N.Y. & London.

PICTURES INCLUDE: Mutiny on the Bounty, Good Earth, Romeo and Juliet; Rulers of the Sea, Spawn of the North, Northwest Passage, Edison the Man, So Ends Our Night, Landfall, Frenchman's Creek, Anna and the King of Siam, Black Rose, Across the Wide Missouri, Landfall, Knights of the Round Table, Escape to Burma, Pearl of the South Pacific, Untamed, The Naked Maja, Gunsight Ridge, The Sons of Katy Elder.

JEPHCOTT, SAMUEL C.: Executive. b. 1944, Southampton, England. e. Arts Educational School, London. Entered industry as child actor The Grove Family BBC TV, 1956. Eight years in advertising producing TV Commercials. Emigrated to Toronto, Canada in 1968. Exec. Secty., Directors Guild of Canada, 1969–72. Film production management, 1972–75. Joined Compass Film Sales Ltd., a division of Quadrant Films Ltd. 1975–77. Joined Nielsen-Ferns Intl. Ltd. in chg. of distribution. Is secretary of the Canadian Association of Motion Picture Producers. Now supvr. of prod., Nielsen-Ferns Intl. Ltd.

PICTURES INCLUDE: The Hard Part Begins 1973, Odyssey Films (prod. Manager); Sunday In The Country, 1973, Quadrant (2nd A.D.); Me, 1974, Muddy York Prods. (prod. Manager); Lions For Breakfast, 1974, Burg Prod. (prod. Manager), It Seemed Like A Good Idea At The Time, 1975, Quadrant (2nd A.D.); Love at First Sight, 1976 (prod. accountant) Quadrant; Find The Lady, 1976 (stills) Quadrant; The New Avengers, TV, 1977 (prod. manager) Neilson-Ferns Inc.

JERGENS, ADELE: Actress. b. Brooklyn, N.Y., Nov. 28, 1922. Began career in musical shows during summer vacation at 15; won contest, New York's World Fair, as model; appeared on New York stage, night clubs, U.S. and abroad.

PICTURES INCLUDE: Edge of Doom, Side Street, A&C Meet the Invisible Man, Sugarfoot, Try and Get Me, Show Boat, Somebody Loves Me, Aaron Slick From Punkin' Crick, Overland Pacific, Miami Story, Fireman Save My Child, Big Chase, Strange Lady in Town, The Cobweb.

JESSEL, IAN: Executive. b. London, England, 1939. e. Oxford. Joined Rank Organisation in 1962 as graduate trainee and became responsible for acquisition of indep. product. Formed Target International Pictures in 1968. In 1970 joined World Film Sales becoming man. dir. 1972. World Film Sales merged with ITC in 1974. Elected to board 1975. Appointed dir. Classic Cinemas and ITC Film Distributors (UK), 1979. Appt. man. dir. ITC Films Int., 1980.

JEWISON, NORMAN P.: Producer-director. b. Canada. e. Malvern Collegiate Institute, Toronto, 1940–44; Victoria College, Univ. of Toronto, Canada, 1946–50 (BA).

TV: Exec. prod. of 8 Judy Garland shows; prod.-dir., Judy Garland specials.

PICTURES INCLUDE: 40 Pounds of Trouble, The Thrill of It All, Send Me No Flowers, Art of Love, The Cincinnati Kid, The Russians Are Coming, The Russians Are Coming. Dir. In The Heat of the Night. Prod.-dir. Gaily, Gaily, Prod.: The Landlord, Fiddler on the Roof, Prod./dir.: Jesus Christ Superstar; Prod.: Billy Two Hats; Prod.-dir.: Rollerball. Prod.-dir.: F.I.S.T.; co-prod.-dir.: And Justice For All.

JOELS, MERRILL E.: Performer. b. Hartford, Conn.; began career, Hartford, 1930; organ. Mark Twain Masquers Thea.; radio, New York since 1945; TV since 1946; actor, announcer, narrator; over 500 TV shows; feature, Army, Navy and industrial films; narr., children's records, dir., AFTRA; member, The Players; Credits also Off-Broadway, Summer Stock, Daytime TV Drama. Announcer and Co-host WEVD Home Show 27 years. Announcer-engineer, WEVD.

AUTHOR: Acting is a Business, How to Get Into Show Business, Touring in Theatre of Peretz and Three by Chayefsky; Talking Books for Library of Congress.

JOFFE, CHARLES H.: Executive. b. Brooklyn, N.Y. e. Syracuse Univ. Joined with Jack Rollins to set up management-production org., clients including Woody Allen, Dick Cavett, Ted Bessell, Billy Crystal, Robert Klein, David Letterman, Martin Mull, Tom Poston, Robin Williams.

PICTURES INCLUDE: Produced: Don't Drink the Water, Take the Money and Run, Everything You Always Wanted to Know About Sex but Were Afraid To Ask, Love and Death, Annie Hall, House of God, Arthur. Exec. Prod.: Play It Again, Sam, Bananas, Sleeper, Arthur.

TV: Woody Allen specials.

JOFFE, EDWARD: producer, director, writer. b. Johannesburg, S. Africa. e. Marist Brothers College and Univ. of the Witwatersrand. Worked in m.p., theatre, commercial radio and as Journalist before ent. TV ind. in Britain 1957 as writer/prod with ATV. Has prod. & dir. some 3000 progs. 1959–61 staff prod. Granada TV. 1962 prod./dir. Grampian TV. 1966–7 dir. film The Price of a Record-r.u. Int. Emmy Award, Special Mention XXI Salerno Film Fest. Prod. dir. Columbia's Folk and So Many Partings, successive ITV entries Golden Harp TV Fest. 1968 prod. dir. Tony Hancock series in Australia; prod. dir. ind. film Will Ye No' Come Back Again for ITV; prod. dir. film Up At The Cross. 1969 ATV Network dir. This Is . . . Tom Jones. 1970 prod. dir. The Golden Shot. 1971–2 Senior Production Lecturer Thomson TV College. 1972 dir. for London Television Service, and Thames Television, co-prod., dir. Sound Scene. 1972–73 dir. Opportunity Knocks! Today, The David Nixon Show, Magpie. Contract Prod/Dir. Thames TV. 1976–80. Productions include Problems (1st UK. Adult Sex Series), Finding Out, Writer's Workshop, About Books series.

JOHNS, GLYNIS: Actress. b. Durban, So. Africa, Oct. 5, 1923; daughter of Mervyn Johns, actor, and Alys Steele, pianist; e. In England. On London stage from 1935 (Buckie's Bears, The Children's Hour, A Kiss for Cinderella, Quiet Week-End; Gertie, N.Y. stage, 1952; Major Barbara, N.Y., 1956–57.) On screen 1936: South Riding. Voted one of top ten British Money-making stars in Motion Picture Herald-Pathe poll, 1951–54.

STAGE: N.Y. 1964 Too Good to Be True, London, 1967: The King's Mare; 1970, Come As You Are. 1971–2 Tour of Britain, Canada and USA in The Marquise, A Little Night Music, NY 1973 (Emmy Award.)

PICTURES INCLUDE: Murder in Family, Prison Without Bars, Mr. Brigg's Family, 49th Parallel, Adventures of Tartu, Half-Way House, Perfect Strangers, This Man Is Mine, Frieda, An Ideal Husband, Miranda, Third Time, Lucky Mr. Proback, The Great Manhunt, Flesh and Blood, No Highway in the Sky, Appointment With Venus (Island Rescue), Encore, The Card (The Promoter), The Sword and the Rose, Rob Roy, Personal Affair, The Weak and the Wicked, The Seekers (Land of Fury) The Beachcomber, Mad About Men, Court Jester, Josephine and Men, Loser Takes All, Day They Gave Babies Away, Another Time Another Place, Shake Hands With the Devil, The Sundowners, The Spider's Web, The Chapman Report, Mary Poppins, Dear Brigette, Don't Just Stand There, Lock Up Your Daughters, Vault of Horror.

TV: Own series (US): Glynis. Guest star TV includes: Dr. Kildare, Roaring Twenties, Naked City, The Defenders, Danny Kaye Show, Noel Coward's Star Quality, Mrs. Amworth, All You Need Is Love, Across A Crowded Room.

JOHNS, MERVYN: Actor. b. Pembroke, Wales, Feb. 18, 1899; married; p. non-prof.; e. Llandovery College. Abandoned study of dentistry for stage.

PICTURES INCLUDE: Lady in Danger, Guynor, Foreign Affairs, Pot Luck, Everything Is Thunder, Dishonour Bright, Storm in a Teacup, Jamaica Inn, Midas Touch, Girl in the News, Next of Kin; San Demetrio, London: Half-Way House, Twilight Hour, Dead of Night, Pink String and Sealing Wax, Captive Heart, Captain Boycott, Counterblast, Edward My Son, Quartet, Helter Skelter, Diamond City, Tony Draws a Horse, The Magic Box, A Christmas Carol, Tall Headlines, The Oracle, Master of Ballantrae, The Valley of Song, Romeo & Juliet, The Blue Peter, Moby Dick, Shield of Faith, 1984, Intimate Stranger, The Tunnel, Counterfeit Plan,

Doctor at Large, My Friend Charles, Dangerous Drugs, The Gypsy and the Gentleman, The Devil's Disciple, Once More With Feeling, Moment of Truth, No Love for Johnnie, Francis of Assisi, The Day of the Triffids, The Old Dark House, The Victors, 80,000 Suspects, Fifty-five Days at Peking, For He's a Jolly Bad Fellow, The Heroes of Telemark, The National Health, Q.B. VII, The Confessional.

TV: Many appearances, including Danger Man, The Avengers, The Sullivan Brothers, The Strauss Family, The Adventurer, Crown Court, Beryl's Lot, A Thinking Man As Hero, Kilvert's Diaries, The New Avengers, Shoestring.

JOHNSON, BEN: Actor. b. Pawhuska, Okla. Stunt rider & performer in rodeos, touring country; did stunt work in War Party; m.p. debut in Mighty Joe Young.

PICTURES INCLUDE: Three Godfathers, She Wore a Yellow Ribbon, Wagonmaster, Rio Grande, Wild Stallion, Fort Defiance, Shane, Rebel in Town, The Wild Bunch, The Undefeated, Chisum, The Last Picture Show (Academy Award, Best Supporting Actor, 1971), Junior Bonner, The Getaway, Dillinger, The Train Robbers, Kid Blue, The Sugarland Express, Bite The Bullet, Hustle, The Town That Dreaded Sundown, The Greatest, Grayeagle, The Swarm, The Hunter, Terror Train.

JOHNSON, CELIA: Actress. b. Richmond, Surrey, England, Dec. 18, 1908. e. Royal Acad. Dram. Art, London; m. Peter Fleming. On stage from 1928. London debut 1929 in A Hundred Years Old; N.Y. debut as Ophelia in Hamlet, 1931; screen debut 1942 in In Which We Serve.

THEATRE: Appearances include The Artist and the Shadow, The Circle, Death Takes a Holiday, The Wind and the Rain, Pride and Prejudice, Old Music, Sixth Floor Brit.

PICTURES INCLUDE: This Happy Breed, Brief Encounter, Dear Octopus. Received N.Y. Critics Award best perf. 1948 in Brief Encounter; 1947, This Happy Breed; pictures since include: Astonished Heart, I Believe In You, Holly & the Ivy, Captain's Paradise, Kid for Two Farthings, Good Companions, The Prime of Miss Jean Brodie.

JOHNSON, G. GRIFFITH: Executive. b. New York, Aug. 15, 1912; e. Harvard U., 1934, (AM), 1936, (Ph.D.), 1938; U.S. Treasury Dept., 1936–39; Dept. of Comm., 1939–40; O.P.A. & predecessor agencies, 1940–46; consulting economist, 1946–47; dir., Econ. Stab. Div., Nat'l. Security Resources Bd., 1948–49; chief econ., U.S. Bur. of Budget, 1949–50; econ. advisor to Econ. Stab. Admin. 1950–52; Exec. v.p. MPEAA, 1965, MPAA, 1971; Asst. Sec'y of State for Economic Affairs, 1962-65; v.p. MPAA, 1953–62; Author of several books & articles.

JOHNSON J. BOND: Producer, Exec. b. Fort Worth, Texas, June 18, 1926. e. Texas Wesleyan Coll., B.S., 1947; Texas Christian Univ. M. Ed., 1948; Southern Methodist Univ., B.D., 1952; Ph.D., Univ. of Southern Calif., 1967. Army Air Forces, WW II; public information officer, captain, U.S. Marine Corps. Korean War. Formerly member Marine Corps Reserve, Motion Picture Production Unit, Hollywood. Now Colonel, U.S. Army Reserve. Newspaper reporter, Fort Worth Star-Telegram, 1942–48; pres., West Coast News Service, 1960; pres., exec. prod., Bonjo Prods., Inc., 1960, President, chief executive officer, Cine-Media International, 1975 managing partner, Capra-Johnson Productions, Ltd., 1978.

PICTURES INCLUDE: Sands of Iwo Jima, Retreat Hell, Flying Leathernecks; photographed aerial portions, Jamboree 53, Norfleet, Devil at My Heels, Kingdom of the Spiders, Ordeal at Donner Pass, Place of the Dawn, Lies I Told Myself, Backstretch, Airs Above The Ground, The Jerusalem Concert, The Berkshire Terror, The Seventh Gate.

TV: Creator, story consultant, tech. advisor, Whirlpool, 1962, one hour dramatic series for Screen Gems/CBS-TV, Executive Producer, Creator, On The Go. (½ hour TV News-Sports series), 1977, Coasties (1 hour TV series), 1978, Desert Rangers (1 hour series) 1979. Producer, Fandango (series), 1981.

JOHNSON, LAMONT: Director, producer, actor. b. Stockton, Calif.; e. UCLA.

TV: The Defenders, Profiles in Courage, Twilight Zone, That Certain Summer, My Sweet Charlie, The Execution of Pvt. Slovik, Fear On Trial.

PLAYS: The Egg, Yes is For a Very Young Man; dir., two operas, L.A. Philharmonic, 1964; founder, dir., UCLA Professional Theatre Group.

THEATRICAL PICTURES INCLUDE: McKenzie Break, A Gunfight, The Grundstar Conspiracy, You'll Like My Mother, The Last American Hero, Lipstick, One on One, Somebody Killed Her Husband, Cattle Annie and Little Britches.

JOHNSON, LAURIE: Composer, mus.; dir. b. 1927. Studied Royal College of Music.

STAGE: Lock Up Your Daughters, Pieces of Eight, The Four Musketeers.

TV: Themes for No Hiding Place, Echo Four-Two, Score, When the Kissing Had to Stop, The Avengers, Jason King, Shirley's World, The Avengers, Thriller, The New Avengers, The Professionals, The Botanic Man, The Adventure.

PICTURES INCLUDE: Good Companions, Moonraker, Girls at Sea, Operation Bullshine, Tiger Bay, I Aim at the Stars, Spare The Road, What a Whopper, Bitter Harvest, Seige of the Saxons, Dr. Strangelove, The First Men in The Moon, Beauty Jungle, East of Sudan, Hot Millions, And Soon the Darkness, Mister Jerico, Cause for Alarm, The Beltstone Fox, Hedda, It Lives Again, All Things Bright and Beautiful.

JOHNSON, RICHARD: Actor. b. Essex, England, 1927. Studied at RADA. First stage appearance Opera House, Manchester, then played in John Gielgud's repertory season, 1944. Served in Royal Navy 1945–48. Subsequent stage appearances incld. The Mad Woman of Chaillot, The Lark. Visited Moscow with Peter Brook's production of Hamlet. Royal Shakespeare Thea.: Stratford & London, 1957–62. Royal Shakespeare Company 1972–73. National Theatre, 1976–77.

PICTURES INCLUDE: The Haunting, 80,000 Suspects, Moll Flanders, Operation Crossbow, Khartoum, La Strega in Amore, Deadlier than the Male, The Rover, Danger Route, Twist of Sand, Oedipus The King, Trajan's Column, Lady Hamilton, Some Girls Do, Julius Caesar, The Deserters, The Beloved, Hennessy, Aces High, The Four Feathers.

JOHNSON, RUSSELL: Actor. b. Ashley, Pa. e. Girard College, Actors Laboratory, L.A.: W.W. II, Army Air Corps.

PICTURES INCLUDE: For Men Only, The Greatest Story Ever Told.

TV: Black Saddle, Climax, You Are There, Rawhide, Twilight Zone, Gilligan's Island.

JOHNSON, VAN: Actor. b. Newport, R.I., Aug. 28, 1916. Began in vaudeville; then on N.Y. stage New Faces of 1937, Eight Men of Manhattan, Too Many Girls, Pal Joey. On screen 1941 in Murder in Big House. Voted one of the top ten Money Making Stars in Motion Picture Herald-Fame Poll 1945–46.

PICTURES INCLUDE: War Against Mrs. Hadley, Dr. Gillespie's New Assistant, Pilot No. 5, Dr. Gillespie's Criminal Case, Guy Named Joe, Three Men in White, Two Girls and a Sailor, Thirty Seconds Over Tokyo, Ziegfeld Follies, Between Two Women, Thrill of Romance, Week-End at the Waldorf, Romance of Rosy Ridge, Bride Goes Wild, State of the Union, Command Decision, In the Good Old Summertime, Scene of the Crime, Battleground, Big Hangover, Three Guys Named Mike, Grounds for Marriage, Go For Broke, Too Young To Kiss, It's A Big Country, Invitation, When In Rome, Washington Story, Plymouth Adventure, Confidentially Connie, Remains to Be Seen, Easy to Love, Caine Mutiny, Siege at Red River, Men of the Fighting Lady, Brigadoon, Last Time I Saw Paris, End of the Affair, Bottom of the Bottle, Miracle in the Rain, 23 Paces to Baker Street, Slander, Kelly and Me, The Lost Blitzkreig, Beyond This Place, Subway in the Sky, Web of Evidence, Enemy General, Wives and Lovers, Where Angels Go . . . Trouble Follows, Eagles Over London.

JOHNSTON, MARGARET: Actress. e. Sydney Univ., Australia. London stage debut: Murder Without Crime. Screen debut: Rake's Progress, 1945.

TV: Always Juliet, Taming of the Shrew, Man With a Load of Mischief, Light of Heart, Autumn Crocus, Androcles and the Lion, Sulky Five, Windmill Near a Frontier, The Shrike, The Out of Towners, Looking for Garrow, The Typewriter, The Glass Menagerie, That's Where The Town's Going, The Vortex.

PICTURES INCLUDE: A Man About the House, Portrait of Clare, The Magic Box, Knave of Hearts, Touch and Go, Night of the Eagle, The Nose on My Face, Life at the Top, Schizo, Mr. Sebastian.

THEATRE: Ring of Truth, The Masterpiece, Lady Macbeth, Merchant of Venice, Measure for Measure, Othello.

JOHNSTONE, DAVID: b. Kilmarnock, Scotland, July 4, 1926. e. Ayr Academy. Journalist: Ayrshire Post; Glasgow Herald. Night News Editor Scottish Daily Mail; Scottish Correspondent News Chronicle. Joined Scottish Television May, 1958 as News Editor. Later producer, director. Originated daily magazine, Here and Now. Producer ITV World Cup (part) 1976. Head of News, Current Affairs, Documentaries and Sport. Assistant controller of programmes; controller (now director) of programmes 1977.

JOLLEY, STAN: Producer-Director-Production Designer-Art Director. b. New York City, May 17, 1926. e. Univ. of S. Calif., col. of architecture, son of actor I. Stanford Jolley. In Navy in World War II. Has acted in capacities listed for many feature films and TV series.

PICTURES INCLUDE: Producer and Production Designer: Knife For the Ladies. Assoc. Producer and Prod. Designer, The Good Guys & The Bad Guys, 2nd Unit director, Superman. Production Designer: Weeds, Caddyshack, Cattle Annie and Little Britches, Americathon (also second unit director), Swarm, Drum, Framed, Dion Brothers, Mixed Company, Walking Tall, Terror In The Wax Museum, Night of The Lepus (also second unit director), War Between Men & Women, Law Man, The Phynx. Art

Director: Young Billy Young, Ride Beyond Vengeance, Broken Saber, The Restless Ones, Mail Order Bride, Toby Tyler, Nine Lives of Elfego Baca. Assoc. producer & prod. designer & 2nd unit director, Happily Ever After.
TV FEATURES: 2nd Unit Director and Production Designer: Swiss Family Robinson, Voyage of The Queen, Woman Hunter, Production Designer: Like Normal People, Howard Hughes, Flood, Voyage of The Yes, The Stranger, Punch & Jody.
TV SERIES: Art Director: Walt Disney Presents, Pete & Gladys, Gunsmoke, Mr. Ed., Branded, Voyage to The Bottom of The Sea, Land of The Giants, O'Hara, etc.

JONES, BARRY: Actor. b. Guernsey (Channel Islands) March 6, 1893. Stage debut in 1921 acting extensively in London and United States. Film debut in Arms and the Man, 1931. During World War II, in Royal Navy.
PICTURES INCLUDE: Dancing with Crime, Frieda, Uneasy Terms, Seven Days to Noon, Clouded Yellow, White Corridors, Plymouth Adventure, Return to Paradise, Prince Valiant, Brigadoon, Demetrius and the Gladiators, Glass Slipper, Alexander the Great.
THEATRE: Starred B'way in Mis-Alliance.

JONES, CAROLYN: Actress. b. Amarillo, Texas, 1933.
STAGE: Summer and Smoke, Live Wire.
TV: Colgate Comedy Hour, Schlitz Playhouse of Stars, Dragnet, Burke's Law, The Addams Family, Roots.
PICTURES INCLUDE: Turning Point, Road to Bali, Military Policeman, Desiree, House of Wax, Invasion of the Body Snatchers, Seven Year Itch, Big Heat, Man Who Knew Too Much, Tender Trap, Opposite Sex, Bachelor Party, King Creole, Last Train from Gun Hill, Man in the Net, A Hole in the Head, Ice Palace, Sail a Crooked Ship, A Ticklish Affair, Heaven with a Gun.

JONES, CHUCK: Producer, Director, Writer. b. Spokane, Wash., Sept. 21, 1912. e. Chouinard Art Institute. Dir., Warner Bros. Animation until 1962 where he created and directed Roadrunner & Coyote, Pepe le Pew; directed and helped create Bugs Bunny, Porky Pig, Daffy Duck etc. Created Snafu character, U.S. Armed Services. Co-prod., wrote, dir., Bugs Bunny Show, ABC-TV. Headed MGM Animation Dept. dir. How the Grinch Stole Christmas, Horton Hears a Who, The Dot and the Line, Pogo, The Phantom Tollbooth. Lecturer and teacher at many universities. Currently independent, Chuck Jones Enterprises, Producer, Director, Writer (for ABC-TV) The Cricket in Times Square; A Very Merry Cricket; Yankee Doodle Cricket and (for CBS-TV) Rudyard Kipling's Rikki-Tikki-Tavi; The White Seal; Mowgli's Brothers, Saint-Saëns' The Carnival of the Animals, Ogden Nash lyrics, with Daffy Duck & Bugs Bunny; and most recently, A Connecticut Rabbit in King Arthur's Court, based on Mark Twain's original story, with Bugs Bunny, Daffy Duck, Porky Pig, Elmer Fudd, etc., two specials featuring Raggedy Ann and Andy in The Great Santa Claus Caper and The Pumpkin Who Couldn't Smile; plus a feature compilation of past work: Chuck Jones' Bugs Bunny Road Runner Movie. Most recently: Daffy Duck's Thanks-for-Giving Special and Bugs Bunny's Bustin' Out All Over. Recipient Academy Award for best animated cartoons for Scenti-Mental Reasons-1950, The Dot and The Line-1965; best documentary short subject for So Much for So Little, 1950. Honored with retrospectives at Deauville Film Fest (1978) British Film Institute (1979).

JONES, CHUCK: Public Relations Executive. b. Detroit, Mich., Dec. 6, 1942. e. Michigan State University (B.A. Advertising) U.S. Marine Corps 1964–66. Staff writer, Pacific Stars & Stripes, Tokyo, Japan and DaNang, Vietnam. Contributing editor, Leatherneck Magazine, Naval Aviation News, Navy Times. Entered motion picture industry 1969. Critic/writer Motion Picture Daily, 1969. Publicist, Harold Rand & Co., Public Relations, 1970–71. Publicist, American International Pictures, 1971. Radio/TV, newspaper, magazine & syndicate contact, United Artists Corp., 1972–73. Established Chuck Jones Public Relations in 1973 and has been pres. since then.

JONES, CLARK R.: Prod. dir. b. Clearfield, Pa., April 10, 1920. e. Northwestern U.
TV SHOWS INCLUDE: The Tony Awards, Peter Pan, Sleeping Beauty Ballet, Ford 50th Anniversary Show, Caesar's Hour, Your Hit Parade, Jack & the Beanstalk; Romeo & Juliet, Perry Como, Bell Telephone Hour, Carol Burnett series; Carol Channing and Pearl Bailey on Broadway, Tony Awards, 6 Rms Riv Vu, Twigs, CBS 50th Anniversary, Emmies, Miss USA, Miss Universe.

JONES, DAVID: CBE, publicity director, EMI Film and Theatre Corporation, Ltd. Director of the Corpn. and also of EMI Cinemas. b. London, Eng., 1913. Entered industry through the CEA Organ, "Cinematograph Times"; editorial assistant RKO-Radio, 1930–36 served in advertising and editorial sections, London "Hollywood Reporter"; rejoined RKO-Radio as exploiteer, appointed press representative, 1938; became director of publicity, 1942 until 1958, chairman, FIPC, 1951–52, 1955–62; president, FIPC, 1952–54, 1960–63; Press Guy, Variety Club of Gt. Britain, 1949–50, 1953–63, Chief Barker, Variety Club, 1964, International Press Guy Variety Club of Great Britain, 1965–66, 1968–71. European Press Guy, 1967–76. Int. Press Guy, 1977–80.

JONES, DEAN: Actor. b. Morgan County, Ala., Jan. 25, 1936; e. Asbury College, Wilmore, Ky.; prof. debut as blues singer, New Orleans; U.S. Navy, 1953.
TV: Ensign O'Toole The Teddy Bears.
STAGE: There Was a Little Girl, Company.
PICTURES INCLUDE: Handle With Care, Never So Few, Under the Yum-Yum Tree, The New Interns, That Darn Cat, Two On A Guillotine, The Ugly Daschshund, Monkeys, Go Home, Blackbeard's Ghost, The Love Bug, The $1,000,000 Duck, Snowball Express, Mr. Super Invisible, The Shaggy D.A., Herbie Goes to Monte Carlo, Born Again.

JONES, GRIFFITH: Actor. b. London, England. e. University College, London; Royal Academy Dramatic Art (gold medal 1932). In H.M. Forces, World War II. Stage debut, London, 1930, in Carpet Slippers; N.Y. debut 1935 in Escape Me Never. Many stage successes. In many Brit. pictures from 1932.
PICTURES INCLUDE: Escape Me Never, Four Just Men, Atlantic Ferry, This Was Paris, The Day Will Dawn, Uncensored, Henry V. Fake's Progress, Wicked Lady, They Made Me a Fugitive, Good Time Girl, Miranda, Look Before You Love, Once Upon a Dream, Honeymoon Deferred, Star of My Night, Scarlet Web, The Sea Shall Not Have Them, Face in the Night, Wanted on Voyage, The High Wall, Hidden Homicide, Strangler's Web, Decline and Fall.
PLAYS: The Moonraker, Quadrille, Alice Thro' the Looking Glass, Love Machine, Dead on Nine, The Entertainer, Expresso, The Sound of Murder, Treasure Island, Two Accounts Rendered, The Cavern, The Doctor's Dilemma, Jockey Club Stakes. 1973, Nottingham P'House, 1974 Crucible, Sheffield. Member of Royal S'Peare Co. 1975–80.
TV: The Breaking Point, The Ware Case, When in Rome, A Moment in the Sun, Hell Hath no Fury, Margret, No Hiding Place, The Collection, By Invitation Only, A Woman of No Importance, Freedom in September, Blythe Spirit, Treasure Island, The Three Sisters, Emergency Ward, Vendetta, The Cabinet Papers, Man in a Suitcase, Boy Meets Girl, Troubleshooters, Strange Report, Avengers, Inside Man, A Matter of Principle, Doom Watch, Warm Feet, Warm Heart, Paul Temple The Persuaders, The Lotus Eaters, The Black Arrow, Arrow, Spy Trap, Crown Court, Fallen Eagles, The Apple Cart, Comedy of Errors, Macbeth.

JONES, JENNIFER: Actress. r.n. Phyllis Isley. b. Tulsa, Okla., e. Northwestern U., Amer. Acad. Dram. Arts; daughter of Phil R., Flora Mae (Suber) Isley exhib.; m. David O. Selznick prod. Toured with parents stock company as child; in summer stock in East; little theat. East & West. Began screen career (as Phyllis Isley) in several Republic Westerns, first major role 1943 as Marie Soubirous in The Song of Bernadette, Winning Academy Award 1943 (best performance by an actress).
PICTURES INCLUDE: Since You Went Away, Love Letters, Duel In the Sun, Cluny Brown, Portrait of Jennie, We Were Strangers, Madame Bovary, Carrie, Wild Heart, Ruby Gentry, Indiscretion of an American Wife, Beat the Devil, Love Is a Many-Splendored Thing, Good Morning Miss Dove, Man in the Gray Flannel Suit, Barretts of Wimpole Street, Farewell to Arms, Tender Is the Night, The Idol, Angel, Down We Go, The Towering Inferno.

JONES, QUINCY: Composer. b. 1935. Formerly in trumpet section of Lionel Hampton's band has played with Dizzy Gillespie, Count Basie before scoring films.
PICTURES INCLUDE: In the Heat of the Night, In Cold Blood, (Acad. Award nom.), The Pawnbroker, Mirage, The Slender Thread, Made In Paris, Walk, Don't Run, Banning, The Deadly Affair, Enter Laughing, A Dandy in Aspic, The Counterfeit Killer, For Love of Ivy, The Split, Bob & Carol & Ted & Alice, The Lost Man, Cactus Flower, John and Mary, The Last of the Mobile Hotshots, The Out-of-Towners, They Call Me Mister Tibbs, Brother John, Dollars, The Anderson Tapes, The Hot Rock, The New Centurions, The Getaway.

JONES, SHIRLEY: Actress. b. Smithon, Pa., March 31.
THEATRE: Appeared with Pittsburgh Civic Light Opera in Lady in The Dark, Call Me Madam. Broadway: South Pacific, Me and Juliet, Maggie Flynn.
PICTURES INCLUDE: Oklahoma, Carousel, Bobbikins, Elmer Gantry, Pepe, A Ticklish Affair, The Secret of My Success, Fluffy, The Happy Ending, Beyond the Posiedon Adventure.
TV: Silent Night Lonely Night, The Partridge Family (series).

JONES, TOMMY LEE: Actor. b. Texas. Worked in oil fields; studied acting at Harvard, where earned a degree in English. Broadway debut in A Patriot for Me; appeared in Four in a Garden, Ulysses in Nighttown, Fortune and Men's Eyes. Film debut in Love Story (1970).

PICTURES: Jackson County Jail, Rolling Thunder, The Betsy, Eyes of Laura Mars, Coal Miner's Daughter, Back Roads.
TELEVISION: The Amazing Howard Hughes.

JORDAN, HENRIETTA: Exec. b. New York, N.Y., Feb. 26. Ent. m.p. ind. as ass't to exec. v.p., UPA Pictures, 1950; v.p. in charge of sales, Format Prods, Inc. v.p. and assoc. prod., 1962–71. Assoc. prod. Levitow-Hanson Films, Inc., 1972–74; Producer, Image West, Ltd., 1975.

JORDAN, MARION F.: Executive. b. Hungary of Polish parents. e. grammar school in Yugoslavia, military academies in Austria and Poland, and Trade Academy of Warsaw. Entered m.p. ind. in 1928 as pub. in Paramount's Warsaw office; named mgr. of branch in Katowice. In 1930 named Para. mgr. in Yugoslavia; 1938, gen. mgr., Poland. After German invasion of Poland went with wife to India, where, from 1940 to 1946, he again served Para., the last four years as gen. mgr. of the territory. In 1946 transferred to H.O. (N.Y.); in 1947 was loaned to the Motion Picture Export Assn., which he represented until 1950 as gen. mgr., West Germany. For brief period turned exhib. in Providence, R.I.; re-entered dist. as Univ.-Intl.'s Latin-Amer. supvr. in 1959. Moved following year to Paris as Cont. sales mgr. In 1958 named U-I's Cont. mgr.; in 1960 joined Columbia Pictures as v.p. and Cont. mgr., headquartered in Paris. In 1966 transferred to N.Y. as foreign mgr.; same year was promoted to exec. v.p. of Col. Intl. In 1974 named special repr. for product acquisition, headquartered in Europe.

JORY, VICTOR: Actor. b. Dawson City, Canada, Nov. 28, 1902. e. Calif. U. Won National Guard wrestling and boxing championship.
THEATRE: On stage: Berkeley Square, Tonight or Never, What Every Woman Knows, The Truth Game, etc. Wrote New York stage production, Five Who Were Mad.
PICTURES INCLUDE: (Screen debut, 1932) Sailor's Luck; Hoppy Serves a Writ, Leather Burners, Unknown Guest, Gallant Blade, Cariboo Trail, The Highwayman, Flaming Feather, Cave of Outlaws, Son of Ali Baba, Toughest Man in Arizona, Man from the Alamo, Valley of the Kings, Cat-Women of the Moon, Sabaka, Mandsu, Blackjack Ketchum Desperado, Cheyenne Autumn, The Miracle Worker, A Time for Dying, Papillon.

JOSEPH, KENNETH. Executive. b. New York City, July 15, 1922. e. Syracuse U. Entered industry as announcer and program director of various NY radio stations, including WNYC, 1946–53; exec. in various capacities of United Artists Television and predecessor companies, 1953–68; exec. v.p., syndication, Four Star Entertainment Corp., 1968–70; exec. v.p., world-wide syndication, Metromedia Producers Corp., 1970 to present.

JOURDAN, LOUIS: Actor, b. Marseille, France, June 19, 1921. Stage actor, prior to m.p.
PICTURES INCLUDE: Her First Affair, La Boheme L'Arlesienne, La Belle, Adventure, Felicie Nanteuil, The Paradine Case, Letter from an Unknown Woman, Madame Bovary, Bird of Paradise, Anne of the Indies, The Happy Time, Three Coins in the Fountain, No Minor Voices, Decameron Nights, The Swan, Julie, Gigi, Streets of Montmartre, Story of the Count of Monte Cristo, VIP's, Made in Paris, A Flea in Her Ear, To Commit a Murder, Silver Bears.

JUDD, EDWARD: Actor. b. Shanghai, 1934. e. Far East. Stage; The Long and the Short and the Tall, The Tinker. Numerous TV appearances.
PICTURES INCLUDE: The Day the Earth Caught Fire, Stolen Hours, The World Ten Times Over, Mystery Submarine, The Long Ships, First Men On The Moon, Strange Bedfellows, Invasion, Island of Terror, The Vengeance of Shee, Shakedown.

JUDELSON, DAVID N.: Executive. e. Graduated NYU Coll. of Engineering. Started business career in family's manufacturing business in N.J. Following military service with U.S.A. Engineers Corp. joined Gulf & Western and in 1959 was elected a director. In 1965 was appointed chairman of the exec. comm. and the following year became exec. v.p. of Gulf & Western; elected pres., 1967.

JUNKIN, RAYMOND: Executive. b. New York, N.Y., Nov. 4, 1918. e. Univ. of Alabama, Univ. of Penn. Air Force, World War II. Eastern sls. mgr., Robert H. Clark Co., southwestern sls. mgr., then ass't to president; v.p., director of sales, Official Films, Inc., 1951–58; self-employed as a film and program consultant, 1959–60; pres. Program Sales, Inc.; gen. sls. mgr., Programs For Television, Inc., since 1961; v.p. & gen. mgr., Screen Gems (Canada), Ltd.; v.p. Domestic Sales Mgr. Screen Gems Internat'l; 1967 gen. sls. mgr., Trans-Lux TV Corp., v.p. & gen. sls. mgr., 1968 to date. CTV Television Network Ltd.

JURADO, KATY: Actress. r.n. Maria Christina Jurado Garcia. b. Guadalajara, Mexico, 1927. Numerous Mexican films m.p. columnist for Mexican publications; American m.p. debut in Bullfighter and the Lady.
PICTURES INCLUDE: High Noon, San Antone, Arrowhead, Broken Lance, The Racers, Trial, Trapeze, Man from Del Rio, The Badlanders, One Eyed Jacks, Barabbas, A Convenant with Death, Pat Garrett and Billy the Kid.

JURGENS, CURT: Actor. b. Munich, Germany, Dec. 12, 1915. Leading man on German stage, turning to films in 1939.
PICTURES INCLUDE: Les Heros Sont Fatigues, An Eye for an Eye, Without You It Is Night (also dir.), And Woman Was Created, Me and the Colonel, The Enemy Below, The Devil's General, Inn of the Sixth Happiness, The Blue Angel, Ferry to Hong Kong, I Aim at the Stars, Tamango, Lord Jim, The Threepenny Opera, Das Liebeskarussel, The Assassination Bureau, The Battle of Neretva, Nicholas and Alexandra, Vault of Horror, The Spy Who Loved Me.

JUROW, MARTIN: Producer. b. New York, N.Y. e. William and Mary, Harvard Law School. Associated with MCA, William Morris, pres., Famous Artists; prod., G & E Productions; co-prod. with Richard Shepherd Jurow-Shepherd Productions.
PICTURES INCLUDE: The Hanging Tree, The Fugitive Kind, Love In a Goldfish Bowl, Breakfast At Tiffany's, The Great Race, Highroad Films.

JUSTIN, GEORGE: Executive. Major credits include producer on Middle of the Night and The Tiger Makes Out; assoc. prod. for Marathon Man and On the Waterfront; prod. exec. on The Goddess and Twelve Angry Men. V.p. & exec. prod. mgr. at Paramount Pictures; four years; left to join Columbia as prod. exec, where was assoc. prod. for The Deep and prod. exec. on The Eyes of Laura Mars. In April, 1979 joined Orion Pictures; named exec. prod. mgr. Promoted to v.p./exec. prod. mgr. in April, 1980.

JUSTIN, JOHN: Actor. b. London, Nov. 23, 1917; e. Bryanston (Dorset). Stage debut, repertory at Plymouth 1933; entered m.p. ind. in 1939. Many TV appearances.
PICTURES INCLUDE: Thief of Bagdad, Journey Together, The Sound Barrier, The Village, Melba, Crest of the Wave, King of the Khyber Rifles, The Man Who Loved Redheads, The Teckman Mystery, Untamed, Safari, Crime Passionelle, Guilty?, Island in the Sun, Spider's Web, Les Hommes Veulent, Vivre, The Savage Messiah, Barcelona Kill, Lisztomania, Valentino, The Big Sleep.

K

KAEL, PAULINE: Critic. b. June 19, 1919, Sonoma County, CA. e. Univer of Calif. at Berkeley (1936–40), majoring in philosophy. Managed two art theaters in Calif. for which she wrote program notes. Broadcast weekly about films on Pacifica network. Made experimental shorts. Has written on films for many magazines. Since 1968 movie critic for The New Yorker
AUTHOR: I Lost it at the Movies, Kiss Kiss Bang Bang, Going Steady, The Citizen Kane Book, Deeper into Movies, Reeling, When the Lights Go Down.
AWARDS: Guggenheim Fellow, 1964; George Polk Memorial Award for Criticism, 1970; The National Book Award, 1974, for Deeper into Movies; Front Page Award for best magazine column in 1974 from Newswomen's Club of N.Y.

KAHN, MADELINE: Actress-Singer. Broadway bow in New Faces of '68. Featured in off-Broadway musical Promenade. Trained as opera singer and appeared in La Boheme, Showboat, Two by Two, Candide.
PICTURES INCLUDE: What's Up Doc? (debut), Paper Moon, From the Mixed-Up Files of Mrs. Basil E. Frankweiler, Blazing Saddles, At Long Last Love, Young Frankenstein, The Adventures of Sherlock Holmes' Smarter Brother, Won Ton Ton, High Anxiety, The Cheap Detective, Simon, Happy Birthday, Gemini, Wholly Moses, First Family, History of the World—Part I.
RECENT THEATRE: In the Boom Boom Room, On the Twentieth Century.

KAHN, MILTON: Publicist. b. Brooklyn, N.Y. May 3, 1934, e. Syracuse and Ohio University; BSJ 1957, formed Milton Kahn Associates, Inc. in 1958.

KAHN, RICHARD: Executive. b. New Rochelle. N.Y., Aug. 29, 1929; e. Wharton School, U. of Penn., B.S., 1951; U.S. Navy, 3 yrs.; joined Buchanan & Co., 1954; ent. m.p. ind., pressbook writer, Columbia Pictures, 1955; exploitation mgr., 1958; adv. and pub., 1963; natl. dir. of adv., pub. and exploitation, 1963; v.p., 1969; 1974 v.p. in chg. of special marketing projects; 1975; moved to MGM as v.p. in chg. of worldwide advertising, publicity and exploitation; 1978, named sr. v.p. in chg. worldwide mktg. & pres., MGM Intl. 1980, elected bd. of governors, m.p. Academy of Arts & Sciences.

KALB, MARVIN: TV news reporter. e. CCNY: Harvard, M.A., 1953; Russian Language School, Middlebury College. Worked for US State Dept., American Embassy, Moscow; CBS News, 1957; writer, reporter-researcher. Where We Stand: reporter-assignment editor; Moscow Bureau Chief, 1960–63; first dip. corres., Washington Bureau, 1963.

BOOKS: Eastern Exposure, Dragon in the Kremlin, A Day in the Life of Ivan Denisovich, Roots of Involvement, The U.S. in Asia, 1784–1971.

KALISCH, BERTRAM: Writer, supervising editor. TV Consultant, film documentaries. b. New York City, June 5, 1902; e. U.S. Naval Acad., '24, Began as reporter; Wash. corresp. 1924–29, man. ed. Army & Navy Journal 1926–29; asst. ed. RKO Pathe News, 1934–37; chief script writer, MGM News of the Day. Also author & supvr. many short subjects. Army from June, 1941; exec. prod., A.A.F. training films, TFPL, Wright Field. 1941–43; pub. rel. pictorial officer, 12th Army Group 1944; G.H.Q. Photo Officer, Pacific. 1945; Army Pictorial Center, 1948–1950, 1954–1956; Office Sec. of Defense 1950–64; Chief, Audio-Visual Div., Executive Producer Sec. Defense Film Reports. Member: Broadcast Pioneers, Academy TV Arts & Science, Overseas Press Club, Nat'l Headliner's Club.

KALISH, EDDIE: Executive. Joined Paramount Pictures in 1975 as dir. of intl. mktg.; later named v.p., worldwide pub. & promo. In 1980 appt. senior v.p., mktg. 1981, named United Artists v.p.—domestic mktg.

KALSER, KONSTANTIN: Executive, b. Munich, Germany, Sept. 4, 1920; e. Switzerland, and UCLA in United States. Color photographer, newsreel cameraman, founded Marathon International Prods., Inc., pres., executive producer; Crashing the Water Barrier, 1956 (Acad. Award); Give and Take (Venice Award), The Carmakers, Chris Award. Director, The One for the Road, 1973 Gold Award, Internatl. Film & TV Festival of N.Y. Director, We Did It! 1978 Gold Medal Intl. Film & TV Festival of N.Y.; Production Executive, The Unknown War 1978 Grand Award, Intl. Film & TV Festival of N.Y.

KAMBER, BERNARD M.: Executive. e. U. of Pa. New England exploitation rep. U.A. 1940; Army service 1941–43; dir. special events dept. U.A., 1943; asst. to Gradwell L. Sears, nat'l distrib. chmn. 6th War Loan Drive; dir. pub. 7th War Loan Drive, 1943–47; dir. pub. & prom. Eagle Lion Classics, 1951; org. Kamber Org., pub. rel. rep. for ind. prod. v.p. sales, adv. pub. Ivan Tors Prod. Greene-Rouse Prods.; June 1953; exec. asst. Hecht-Hill-Lancaster, chg. of N.Y. off., 1957; v.p. Hecht-Hill-Lancaster Companies, 1958; formed Cinex Distr. Corp., 1962; Pres. Cinex and Posfilm, Inc.; 1967 v.p. in chg. sls. Du Art Film Lab. Inc; 1975 joined Technicolor, Inc.

KAMEN, STANLEY A.: Executive. b. Jan. 13, 1928, New York City. e. Washington and Lee U. grad. 1948; Law School, grad. 1950; member, N.Y. Bar; atty., New York office, William Morris Agency; agency representative on West Coast, 1954. U.S. Navy, 1944–46.

KAMEY, PAUL: Executive. b. New York, N.Y., Aug. 25, 1912. Worked on newspapers including NY Journal American. Ent. m.p. industry 1938; worked for MGM and 20th Century Fox; during war, writer, Office of War information; joined Universal, 1949; eastern pub. mgr., Universal Pictures. 1968, Freelance publicist.

KAMINS, BERNIE: Publicist. b. Cambridge, Mass., Jan. 2, 1915; e. Harvard U., B.A. Harvard U. corresp. various newspapers and press services while in coll.; joined Russell Birdwell & Assoc., publicists unit man. Paramount studios. Author screen story Forty Thieves; co-author Sky Princess. Dir. adv. & publ. Harry Sherman Prod., then Charles R. Rogers Prod.; apptd. same Principal Artists Prods., April, 1944; later same yr. joined Chas. R. Rogers Prods.; identical posts Cagney Prod. Inter-American Prod., Jesse L. Lasky Prod. Author, The Copy Rider; pub. relations, Hollywood Actors Council; 1953. instructor, propaganda, UCLA; 1953–54 pub. Peter Goelet Ent. Nat'l Bd. of Review of M.P., Inc.; 1954–57. p.r. and exec. v.p. national Audience Board, Inc.; p.d. Clarion Enterprises, 1956; p.d. Romson Productions, 1957; p.r. Guy Madison Prod., 1958; Julian Lesser Prod., Principal Securities, 1959. Pacific Title and Art Studio p.r., Counsel, 1960; elected pres. Opinion Builders, Inc., 1961. Exec. hd. of United Republican Finance Committee of Los Angeles County, 1962; Buddy Ebsen Enterprises (Turquoise Prods.), 1963; finance pub. rel. counsel. Herts-Lion International Corp., 1963–64; acting P.R. chmn., Republican State finance committee; P.R. chmn., Western US Amer. Jewish Comm., 1964; Publicists Assoc. Awards Comm. 1966–67. V.P. Commu. Rel. South. Calif. Industry-Educ. Council, 1966–68; Co-ord. Harvard Med. School Alumni Assoc. 1967; Exec. Sect. Insurance Research Foundation, 1968. Statewide education comm. of Calif. State Chamber of Commerce. National advisory comm. for UCLA Allied Health Projects Div.; staff, UCLA Alumni & Development; board member, Foundation for California Community Colleges; Advisory board, L.A. City Board-of-Education on Unified School District Proposal for Utilization of UHF-TV Channel 58; Exec. Secty., American Consumers Council, 1966–; Member, general advisory board, Santa Monica College; 1972–; Exec. Secty., Foundation of Oral and Maxillofacial Surgery Inc., 1976–, 1977 appointed to Los Angeles County Youth Commission; 1977—UCLA Chancellor's Associates member. 1978—lecturer on P.R. for Community Organizations, West Los Angeles College. 1979—lecturer, same, Santa Monica College. 1979—Appointed to Board of Directors of International Students Center, UCLA. 1979—Coordinator and public relations consultant, Foundation for Effective Government Inc.

KANE, CAROL: Actress. b. Cleveland, Ohio, 1952. Began professional acting career at age 14, touring in The Prime of Miss Jean Brodie. Other stage credits include, Arturo Ui, The Enchanted, The Tempest, Macbeth, Tales of the Vienna Woods, etc.

PICTURES: Wedding in White, The Last Detail, Carnal Knowledge, Hester Street, Dog Day Afternoon, Annie Hall, The World's Greatest Lover, The Muppet Movie, When a Stranger Calls.

KANE, STANLEY D.: Judge. b. Minneapolis, Minn., Dec. 21, 1907; e. U. of Minn., B.A. (magna cum laude, 1930), M.A., 1931; Minn. Coll. of Law, LL.B., 1940. Instructor, U. of Minn., 1930–33; exec. sec. Allied Theatre Owners of the Northwest, 1933–37; city attorney, Golden Valley, Minn., 1940–63; City Atty., Champlin, Minn., 1955–60; on faculty, Minn. Coll. of Law, 1940–44; trial attorney & trial examiner, NLRB, Minneapolis, New Orleans, N.Y., 1943–46; special, gen. counsel, Puerto Rico Labor Relations Bd., 1946; exec. vice-pres. & gen. counsel, North Central Allied Independent Theatre Owners, 1946–63; recording sec. Allied States Assn., 1947 to 1956; Dist. Court Judge, Hennepin County, 1963; elected to 6-yr. term, 1964; re-elected, 1970.

KANIN, FAY: Writer. b. New York, N.Y. e. Elmira Coll., U.S.C., m. Michael Kanin, writer. Contrib. fiction to mags.; mem. and officer of WGA. AMPAS. also bd. mem. of latter.

PICTURES INCLUDE: Blondie for Victory, Sunday Punch, My Pal Gus, Rhapsody, The Opposite Sex, Teacher's Pet, Swordsman of Siena, The Right Approach.

BROADWAY: Goodbye My Fancy, His and Hers, Rashomon, The Gay Life.

TELEVISION: Heat of Anger, CBS Friday Night Movie 1972; Tell Me Where It Hurts, CBS General Electric Theater, 1974.

KANIN, GARSON: Director, writer. b. Rochester, N.Y., Nov. 24, 1912; e. American Acad. Dram Arts; m. Ruth Gordon, actress. Musician actor, appearing in Spring Song, Little Ol' Boy, and others. Prod. assist George Abbott on Three Men on a Horse, Brother Rat, Room Service; dir. Hitch Your Wagon, Too Many Heroes, Broadway plays; In June, 1937, Samuel Goldwyn's prod. staff, 1938, joined RKO, prod.-dir. contract. In 1942: prod. for U.S. Office of Emergency Management. Joined armed forces, World War II; co-dir. True Glory.

PICTURES INCLUDE: A Double Life, Adam's Rib, Born Yesterday, Marrying Kind, Pat and Mike, It Should Happen to You, Next Time Marry, Man to Remember, Great Man Votes, Bachelor Mother, My Favorite Wife, They Knew What They Wanted, Tom, Dick and Harry, From This Day Forward, The Girl Can't Help It, The Rat Race, High Time, The Right Approach, Where It's At, Some Kind of a Nut, Woman of the Year, The More the Merrier.

THEATRE: Born Yesterday, The Smile of the World, The Rat Race, The Live Wire, A Gift of Time, Do Re Mi, Come on Strong, The Amazing Adele, The Good Soup, Dreyfus in Rehearsal. (Writer &/or Dir.) Dir.: The Rugged Path; Years Ago; How I Wonder; The Leading Lady; The Diary of Anne Frank; Into Thin Air Small War on Murray Hill; Hole in the Head; Sunday in New York; Funny Girl; I Was Dancing; A Very Rich Woman; We Have Always Lived in a Castle; Idiot's Delight; Ho! Ho! Ho!

BOOKS: Remembering Mr Maugham, Cast of Characters, Tracy and Hepburn; Hollywood; Blow Up A Storm; The Rat Race; A Thousand Summers; One Hell of an Actor; It Takes a Long Time to Become Young; Moviola, Smash, Together Again!

KENIN, MICHAEL: Writer. b. Rochester, N.Y., Feb. 1, 1910; m. Fay Mitchell. Member SWG, officer; WGA, AMPAS. Commercial artist, musician, N.Y. prior to m.p. career; contrib. fiction mags.; s.p. Panama Lady, They Made Her a Spy, 1939.

PICTURES INCLUDE: Anne of Windy Poplars, Woman of the Year, Sunday Punch, The Cross of Lorraine, Centennial Summer Honeymoon, A Double Life, My Pal Gus, When I Grow Up, Rhapsody, The Opposite Sex, Teacher's Pet, The Right Approach, The Swordsman of Siena, The Outrage, How to Commit Marriage.

BROADWAY: Goodbye My Fancy, Seidman and Son, His and Hers, Rashomon, The Gay Life.

KANTER, HAL: Writer, director, producer. b. Savannah, Ga., Dec. 18, 1918. Writer, Danny Kaye Show, Amos 'n Andy, Bing Crosby Show, Ed Wynn TV Show, 1949; creator, writer, prod., stager, George Gobel Show; writer, Paramount, 1951–54; dir., RKO Radio, 1956; Prod., dir., writer, Kraft Music Hall, 1958–59; exec. prod., TCF-TV; Valentine's Day; writer, prod. dir. Chrysler Theatre 1966–67. Creator W.D.P., Julia, NBC, 1968–71; creator W.D.P., Jimmy Stewart (NBC), 1971. W.P. Many TV specials. 1975–76 exec. prod. All In The Family (CBS). Sup. Prod. Chico & The Man (NBC),

1976–77; exec. prod., WB-TV; 1978; prod./writer, Lucille Ball Prod., 1979–80; dir./writer, ABC TV Movies, 1980; Walt Disney Prods., 1981.

PICTURES INCLUDE: My Favorite Spy, Off Limits, Road to Bali, Casanova's Big Night, About Mrs. Leslie, Money from Home, Artists and Models, Mardi Gras, Rose Tattoo, Married a Woman, Loving You, Once Upon a Horse, Blue Hawaii, Pocketful of Miracles, Bachelor in Paradise, Move Over, Darling, Dear Brigitte.

KANTER, JAY: Executive. b. 1927. Entered industry with MCA, Inc., where was v.p. Left after more than 20 yrs. to become pres. of First Artists Production Co., Ltd. In 1975 joined 20th-Fox as v.p. prod.; 1976, named sr. v.p., worldwide prod. Named v.p., The Ladd Co., 1979.

KANTOR, IGO: Producer, Film Editor. b. Vienna Austria, Aug. 18, 1930. e. UCLA, AA. 1950. BS, 1952. MS, 1954. Foreign corres., Portugal magazine, FLAMA, 1949–57, music supvr., Screen Gems, Columbia 1954–63; post-prod. supvr., film ed., features, TV; assoc. prod., 1963–64; prod., exec., International Entertainment Corp., 1965. pres., Synchrofilm, Inc., post-production co. and Duque Films, Inc., production co. 1968–74. 1975—present, produced and edited films.

PICTURES INCLUDE: Bye Bye Birdie, Under the Yum Yum Tree, Gidget Goes to Rome, A House Is Not a Home, Pattern for Murder, Willy. Co.-prod., editor: Assault on Agathon (1975); assoc., prod., editor, Dixie Dynamite (1976); prod., editor, music supvr., Kingdom of the Spiders; assoc. prod., The Dark (1977); prod. supvr., Good Luck Miss Wyckoff, (1978); prod., Hardly Working,: (1980); Kill and Kill Again (1981).

KAPER, BRONISLAU: Composer. b. Warsaw, Poland Feb. 5, 1902; e. Warsaw Cons. of Music. Wrote concert music, songs and music for films in Warsaw, Berlin, Vienna, London and Paris.

PICTURES INCLUDE: Gaslight, Without Love, Mrs. Parkington, Green Dolphin Street, That Forsyte Woman, Naked Spur, Lili (Acad. Award, 1953), Saadia, Glass Slipper, Quentin Durward, Forever Darling, The Prodigal, The Swan, Brothers Karamazov, Don't Go Near the Water, Auntie Mame, Green Mansions, San Francisco, Butterfield 8, Night at the Opera, Red Badge of Courage, Day at The Races, Mutiny on The Bounty, Lord Jim.

KAPLAN, BORIS: Executive. b. New York City, Sept. 23, 1897; m.; e. Coll. of City of New York. General mgr. of Selwyn Theatrical Enterprises, 1925–32. Prod. Plays for legitimate stage, Broadway; joined Paramount Pictures Corp., 1933; head of eastern casting and talent dept., Paramount 1936–62; independent casting dir.; m.p. consultant.

KAPLAN, GABRIEL: Actor/Comedian. b. Brooklyn, N.Y., March 31. After high school worked as bellboy at Lakewood, N.J. hotel, spending free time studying comedians doing routines. Put together a comedy act, landing engagements in small clubs and coffee houses all over U.S. Made several appearances on Tonight Show, Merv Griffin Show, Mike Douglas Show, etc. Has played Las Vegas clubs.

TV: Welcome Back, Kotter.

PICTURES INCLUDE: Fast Break, Tulips, Nobody's Perfect.

KAPLAN, MURRAY M., Executive. In motion picture distribution since 1936. Distributor for Warner Bros., Paramount, Monogram, Eagle-Lion, United Artists, Allied Artists, Cinemation Industries. Became independent distributor in 1950, independent producer in 1966. Co-founder of PRO-International Films, 1975; 1978, formed M & M Films, to produce and distribute worldwide. Member: Motion Picture Pioneers of America.

KARLIN, FRED: Composer. Won Academy Award for Best Song for For All We Know (from Lovers and Other Strangers) and Emmy for original music in The Autobiography of Jane Pittman on TV. Adapted Huddie Ledbetter melodies for film Leadbelly.

PICTURES INCLUDE: Up the Down Staircase, Yours, Mine and Ours, The Sterile Cuckoo, Westworld, Gravy Train, Mixed Company, Leadbelly, Loving Couples.

KARLIN, MYRON D.: Executive. b. Revere, Mass., Sept. 21, 1918. e. UCLA. Joined m.p. business in 1946 as gen. mgr. for MGM in Ecuador. Two yrs. later assigned same spot for MGM in Venezuela. In 1952–53 was gen. sales mgr. for MGM in Germany, after which held same post in Argentina, returning to Germany as mgr. dir. in 1956. Named mgn. dir. for United Artists in Italy. 1960–68 was pres. of Brunswick Int'l., while also serving as advisor to World Health Organization and UNESCO. In 1969 was European mgr. for MGM and mgn. dir. in Italy. Joined Warner Bros. Int'l. in May, 1970 as v.p. of European dist. In March, 1972 appt. v.p. in chg. of int'l. operations for WB; 1977, appt. pres., WB Intl. & exec. v.p., Warner Bros., Inc.

KARLSON, PHIL: Director. b. Chicago, Ill., 1908. r.n. Philip N. Karlstein.

PICTURES INCLUDE: The Phoenix City Story, Gunman's Walk, Hell to Eternity, Key Witness, The Secret Ways, The Young Doctors, Kid Gallahad, Rampage, The Silencers, A Time for Killing, The Long Ride Home, The Wrecking Crew, Hornet's Nest, Ben, Walking Tall, Framed.

KASTNER, ELLIOTT: Executive. b. N.Y.C. Was agent before becoming indep. prod., financing and personally producing 23 feature films in 9 yrs. Based in London.

PICTURES INCLUDE: Harper, Kaleidoscope, Sweet November, Laughter in the Dark, Where Eagles Dare, When Eight Bells Toll, X, Y, & Zee, The Nightcomers, Fear Is The Key, The Long Goodbye, Cops and Robbers, Jeremy, 11 Harrowhouse, Rancho Deluxe, 92 in the Shade (exec. prod.,) The Missouri Breaks (co-prod.), Equus (co-prod.), The Big Sleep (co-prod.); ffolkes (prod.).

KATLEMAN, HARRIS L.: Executive. b. Omaha, Neb., Aug. 19, 1928. e. U.C.L.A. (B.A. in Admin., 1949). Joined MCA in 1949; in 1952 transferred to N.Y. as head of TV Packaging Dept. Left to join Goodson-Todman Prods. in 1955, where named v.p., 1956; exec. v.p., 1958; sr. exec. v.p., 1968. Was directly responsible for all film prod. in L.A., including such shows as The Rebel, Branded, The Richard Boone Show (Emmy nominations, Fame Award of Year), and Don Rickles Show, on which was exec. prod. Joined Metro-Goldwyn-Mayer in 1972 as v.p. of MGM-TV; promoted following year to pres., MGM-TV and sr. v.p. of MGM, Inc. Resigned as pres. of MGM-TV September, 1977. Formed Bennett/Katleman Productions under contract to Columbia Pictures. Exec. prod.: From Here to Eternity, Salvage 1; 1980, named bd. chm. 20th-Fox Television.

KATSELAS, MILTON: Director. b. Pittsburgh. e. drama dept. Carnegie Tech (now Carnegie Mellon Univ.). Has directed more than 30 stage prods., including in New York, The Rose Tattoo and Camino Real (both revivals), The Zoo Story, Butterflies Are Free, etc.

PICTURES INCLUDE: Butterflies Are Free (debut), 40 Carats, Report to the Commissioner, When You Comin' Back, Red-Ryder.

KATT, WILLIAM: Actor. Son of Barbara Hale and Bill Williams. e. Orange Coast College. Majored in music, playing piano and guitar. Acted with South Coast Repertory Theatre, later working in production at the Abmanson and Mark Taper Theatres in L.A. Made theatrical film debut in Carrie.

PICTURES INCLUDE: Carrie, First Love, Big Wednesday, Butch and Sundance: The Early Days.

KATZ, GLORIA: Producer-Writer. e. UCLA Film School. Joined Universal Pictures as editor, cutting educational films. Later joined forces with Willard Huyck, whom she had met at UCLA. Pair signed by Francis Ford Coppola to write and direct for his newly created company, American Zoetrope. Projects didn't materialize but Katz and Huyck teamed to write script for America Graffiti for director Paul Lucas. Wrote Lucky Lady (1975) together. Katz made debut as producer with French Postcards in 1979 which co-wrote with Huyck, who directed.

KATZ, MARTY: Producer-Writer. b. Landsburg, West Germany, Sept. 2, 1947. e. UCLA, Univ. of Maryland. Combat pictorial unit commander, US Army; 1st lt. US Army Southeast Asia Pictorial Unit, Vietnam, 1969; awarded Bronze Star. Director, film production operations, ABC 1971–76; exec. v.p., prod., Quinn Martin Productions, 1976–77; prod., movies for TV, QM, 1977–78; producer for Paramount Pictures, 1979–80; formed Marty Katz Productions in 1980.

TELEVISION: Supervised production for Love Among the Ruins, Eleanor and Franklin, Eleanor and Franklin: The White House Years. Produced for Paramount: 11th Victim, Catalina C-Lab.

PICTURES: The Man, Nashville, Sounder II.

KATZ, NORMAN B.: Executive. b. Scranton, Penna., Aug. 23, 1919. e. Columbia Univ. In U.S. Army 1941–46 as intelligence officer, airborne forces. Entered m.p. industry in 1947 with Discina Films, Paris, France, as prod. asst. Named exec. asst. to head of prod. in 1948. In 1950 named v.p. Discina Int'l. Films and in 1952 exec. v.p. In 1954 joined Associated Artists Prods. as foreign mgr.; named dir. of foreign operations in 1958. In 1959 became dir. of foreign operations for United Artists Associated. 1961 joined Seven Arts Associated Corp. of v.p. in chg. of foreign operations; named exec. v.p., Seven Arts Prods. Int'l. in 1964. Named exec. v.p. Warner Bros.—Seven Arts Int'l. in 1967. In 1969 appt. exec. v.p. and chief exec. off. Warner Bros. International and bd. mem. of Warner Bros. Inc. In 1974 named snr. v.p. int'l. div. of American Film Theatre. Pres. of Cinema Arts Associated Corp. 1979, exec. v.p. and bd. member, American Communications Industries and pres., chief exec. off. of ACI subsidiary, American Cinema.

KAUFMAN, HAL: Creative director, TV writer-producer. b. New York, N.Y., Dec. 16, 1924; e. U. of Texas. 1943–44; U. of Michigan. 1944–47. Started career as petroleum geologist, Western Geophysical Co., 1947–48; TV writer-prod-dir., KDYL-TV, Salt Lake City, 1948–49; prog. dir., WLAV-TV, Grand Rapids, 1949–51; prod. mgr., WOOD-TV, Grand

Rapids, 1951–54; Radio-TV dir., Webber Advertising Agency, Grand Rapids, 1951–54; TV writer-prod., Leo Burnett Company, Chicago, 1954–56; TV writer-prod., Gordon Best Company, Chicago, 1956–59; senior writer, TV/Radio creative dept., Needham, Louis & Brorby, Inc., 1959; vice-pres., asst. copy dir., Needham, Louis & Brorby, Inc., 1962; dir., TV, Radio prod., Needham, Louis & Brorby, Inc., 1963; dir., broadcast design, production, Needham, Louis & Brorby, Inc., 1964; assoc. creat. dir., asst. Exec. v.p., Needham, Harper & Steers, Inc., 1965; Creat. dir. L.A., 1966; Sr. v.p. and Member Bd. of Dir., 1966. Aug. 1969, creative & marketing consultant in Beverly Hills. Aug., 1970, v.p., principle, Kaufman, Vinson, Lansky, Inc., Beverly Hills and San Diego; Aug. '72, Exec. V.P., principle, Kaufman, Lansky Advertising, Inc., Los Angeles and San Diego, Member, Directors Guild of America, SAG, AFTRA. 1979, program dir., Z Channel, Theata Cable TV. 1979 named snr. v.p./adv. & p.r. & asst. to pres. & bd. chmn., World Airways, Inc.

KAUFMAN, J. L. (Les): Publicist. b. Chicago, Ill., June 3: e. Morgan Park (Ill.) Military Acad. Police reporter, City News Bureau, Chicago. In 1926 adv. dept., Balaban & Katz Theats. In 1929 pub. Paramount Public Theats. To Columbia, 1933 asst. exploit, dir. In 1938 adv. & pub. dir., Fanchon & Marco Serv. Corp., St. Louis. Adv.-pub. dir., Republic Studios, 1944–46; Nat'l. adv. pub. dir. International Pictures Corp. In 1947 named studio pub. dir., Universal-International; sales prom. counsel, Kaiser-Frazer Corp., 1950; v.p. in chge. West Coast, Ettinger Co., 1950; v.p. Grant Adv. Inc., Hollywood, Detroit, 1952–55; v.p., adv. dir., UPA Pictures, Inc., 1956; P.R. dir., Fedderson Productions & Lawrence Welk, 1957; My Three Sons, Hollywood Palladium, 1961, Family Affair, To Rome with Love. The Smith Family; pres., Hollywood Press Club, 1964–65, 1967–68. Bd. governors—NATVAS, Hollywood chapter.

KAUFMAN, LEONARD: Executive, attorney. b. New York, N.Y., April 30, 1913. e. C.C.N.Y., Fordham U. School of Law, 1933–36. LL.B. admitted New York Bar, 1937. U.S. Supreme Ct.; Law offices of Nathan Burkan, 1933–36; law offices of Schwartz & Frohlich. 1936–49, Member legal staff Paramount Pictures Corp. 1949–67, gen. counsel, 1964–67; Mem. law firm, Kaufman & Kaufman, 1969–75; Kommel, Rogers, Kaufman, Lorber & Shenkman 1975–78; Assoc. Bar City New York, American Bar Assoc., New York State Bar Assoc., President Cinema, Radio, TV Unit, B'nai B'rith, Motion Picture Pioneers, Variety Clubs Intl. Gen. Counsel, Variety Club, N.Y.

KAUFMAN, LEONARD B.: Producer, writer, director. b. Newark, N.J., Aug. 31. e. N.Y.U. In World War II served with Army Special Services writing and directing camp shows. Nat'l magazine writer, 1945–48; radio writer, including Errol Flynn Show, 1948–50; radio and TV writer, 1950–52. Headed own public relations firm: Kaufman, Schwartz, and Associates, 1952–64. Joined Ivan Tors Films as writer-prod., 1964. Films Corp., 1958.
PICTURES INCLUDE: Clarence, the Cross-eyed Lion, Birds Do It, Story.
TV: Daktari, Ivan Tors' Jambo, O'Hara, U.S. Treasury (CBS) pilot feature and series for Mark VII Prods.

KAUFMAN, PHILIP: Writer-Director. b. Chicago, Oct. 23, 1936. e. Univ. of Chicago, Harvard Law School. Was teacher in Europe before turning to film medium.
PICTURES INCLUDE: Goldstein, Fearless Frank, The Great Northfield, Minnesota Raid, The White Dawn, Invasion of the Body Snatchers, The Wanderers.

KAUFMAN, SIDNEY: Producer, Completion Guarantor. b. N.Y. Mar. 28, 1910. e. Cornell, N.Y.U., C.C.N.Y., L.I.U., Member faculty New School for Social Research, 1938–59. Newspaper editor; film critic, WQXR, N.Y., 1936–40. Production staff Walter Wanger, MGM. America associate—Pascal, Del Guidice, etc.; producer-writer-director numerous documentaries. Writer-prod. The Sorcerer's Village (1958—Continental); Behind the Great Wall (writer-adapter) 1959; man. dir. Grand Prize Films Ltd., London, 1960 release Macbeth (exec. prod.); since 1952, pres., Grand Prize Films, Inc., importer, dist. over 400 foreign features, since 1958, providing completion bonds, guarantor and prod. exec. numerous dom. and foreign features; Pres., Performance Guarantees, Inc.; Pres., Jermyn Venture Capitol Corp.

KAUFMANN, CHRISTINE: Actress. b. Lansdorf, Graz, Austria, Jan. 11, 1945. e. attended school in Munich, Germany. Film debut as a dancer. Salto Mortale at 7 yrs of age.
PICTURES INCLUDE: Rosenrosli (Little Rosie), Schweigende Engel (Silent Angel), Maedchen in Uniform, Winter Vacation, The Last Days of Pompeii, Red Lips, Town Without Pity, Taras Bulba (first American-made film), Murder in the Rue Morgue.

KAY, EDWARD: Musical director, composer. b. New York City, Nov. 27, 1898; e. U. of Pennsylvania; later musical study. Became musical director, vaudeville, New York stage productions; similar work in radio. Musical director, Gaumont-British, 1932–33. To Hollywood, 1933, as musical director, Paramount, Allied Artists.
PICTURES INCLUDE: Highway Dragnet, Thunder Pass, Treasure of Ruby Hills, Big Tip Off, Las Vegas Shakedown, Betrayed Women, Night Freight, Toughest Man Alive.

KAY, GILBERT LEE: Director-Writer. b. Chicago, Ill., June 28. e. Los Angeles City Coll. Was asst. dir. at various studios from 1942–53; started directing on own in 1954. Formed Pearly Gate Productions, London.
PICTURES INCLUDE: Three Bad Sisters, The Tower, Ocean's 11 (s.p.), Comeback, (s.p.), Take Five (s.p.), Fame! (s.p.), Anything for Money (s.p.), The Wrong Mrs. Wright, Now It Can Be Told (s.p.), It Happened in Lisbon (s.p.), The Secret Door, A Harvest of Evil (s.p., dir.), Sometimes I Love You (s.p.), White Comanche, Ragan, Devil May Care, Maybe September (s.p.), Recent screenplays: The Oedipus Vendetta, The Lotus Affair, Candle in the Wind, Royal Flush.
TV: Directed: Treasury Men in Action, Man Behind the Badge, Reader's Digest, Passport To Danger, Hollywood Profile, Highway Patrol, Arabian Nights, Telephone Time, Silent Service, The Grey Ghost, Man with a Camera, Adventures in Paradise, Shotgun Slade, Perry Mason, Follow the Sun, Frontier Circus. Wrote: The Uncivil Engineer, 8:46 to Southampton
PLAYS: Directed: Two Faced Coin, Some Call It Love, French without Tears, Burlesque, London by Night, The Man from Madrid, Paris, With Love. Wrote and Directed: West End, Please Omit Flowers, The Girl From Soho.

KAY, GORDON: Producer. b. Montreal, Canada, Sept. 6, 1916; e. Williams Coll. (Mass.); married. Asst. prod. Republic 1946, assoc. prod., 1947. apptd. secy.-treas.; exec. asst. to head of prod. at Republic, Feb., 1951; prod., Univ. 1955; pres. Gordon Kay & Assoc., 1958.
PICTURES INCLUDE: Wild Frontier, Bandits of Dark Canyon, Oklahoma Badlands, Bold Frontiersman, He Rides Tall, Fluffy, Taggart, Gunpoint, Beardless Warriors.

KAYAMA, YUZO: Actor. b. April 11, 1937. e. law school Keio Univ. Debut Toho Studio 1959 in Man Against Man.
PICTURES INCLUDE: Westward desperado, Man from the East, Blood in the Sea, Three Dolls series, Bull of Ginza, Tsubaki Sanjuro.

KAYE, DANNY: Actor. b. Brooklyn N.Y., Jan. 18, 1913. On stage summer theatres; toured U.S. in act La Vie Paree; in night clubs; toured Orient, 1934; m.p. debut in Up in Arms, 1944; formed prod. co. Nov., 1952; Ambassador-at-Large UNICEF-U.N. Children's Fund. Recipient of Scopus Laureate for 1977; first award for International Distinguished Service from the United Nations Children's Fund; Honorary Degree from Colgate University, Doctor of Humane Letters.
TV: The Secret Life of Danny Kaye, numerous Danny Kaye Specials, Danny Kaye Show. Danny Kaye's "Look In" at Metropolitan Opera since 1973; televised, 1975. Won Emmy Award for Best Children's Special, 1975.
BROADWAY: Straw Hat Revue, Play's The Thing, Death Takes a Holiday, Lady in the Dark, Let's Face It, Two by Two, 1970–71.
PICTURES INCLUDE: Wonder Man, Kid from Brooklyn, Secret Life of Walter Mitty, Song Is Born, Inspector General, On the Riviera, Hans Christian Andersen, Knock on Wood, White Christmas, Court Jester, Merry Andrew, The Five Pennies, On the Double, Man for the Diners' Club, The Madwoman of Chaillot.

KAZAN, ELIA: Director. b. Constantinople, Turkey, Sept. 7, 1909; e. Williams Coll., Yale Dramatic School. With Group Theatre as apprentice & stage mgr.; on stage, 1934–41; plays include: Waiting for Lefty, Golden Boy, Gentle People, Five-Alarm Lilliom; m.p. acting debut in Blues in the Night, 1941; stage dir.; won Critics Award best direction of Skin of Our Teeth. All My Sons, Streetcar Named Desire, Death of a Salesman, Cat on a Hot Tin Roof; also dir.: One Touch of Venus, Harriet, Jacobowsky and the Colonel, Tea and Sympathy, Dark at the Top of the Stairs, J.B., Sweet Bird of Youth; co-dir., prod., Lincoln Center Repertory Theatre; dir., After The Fall, But For Whom Charlie; m.p. dir., 1944; Academy Award best direction, 1947, for Gentleman's Agreement; author novel, The Arrangement. 1967–68, novel, The Assassins; 1974 novel, The Understudy.
PICTURES INCLUDE: City of Conquest, Tree Grows in Brooklyn, Boomerang, Sea of Grass, Panic in the Streets, Pinky, Streetcar Named Desire, Viva Zapata, Man on a Tightrope, On the Waterfront (Acad. Award, best dir., 1954), East of Eden, Baby Doll, Face in the Crowd, Wild River, Splendor In the Grass, America, America, The Arrangement, The Visitors, The Last Tycoon.

KEACH, STACY: Actor. b. 1942. Began professional acting career as Marcellus, Player King in Joseph Papp's 1964 production of Hamlet in Central Park. Performances since have won three Obie Awards Vernon Rice Drama Desk Award for Macbird, Drama Desk Award, Tony Nomination for Indians.
CREDITS: The Heart Is A Lonely Hunter, End of the

Road, The Traveling Executioner, Brewster McCloud, Doc, Judge Roy Bean, The New Centurions, Fat City, Long Day's Journey Into Night (Stage), Macbird (Stage), Indians (Stage), Hamlet (Stage), Luther, Gravy Train, Caribe (TV Series), The Killer Inside Me, Conduct Unbecoming, Street People, The Squeeze, Gray Lady Down, The Adventures of Pedro and Man, The Ninth Configuration, Longriders, Directed Six Characters In Search Of An Author and Incident At Vichy; Deathtrap (stage), Hughie (stage), The Long Riders (star & exec. prod.), Road Games (film), Butterfly (film), Barnum (stage).

KEACH, SR., STACY: Executive. b. Chicago, May 29, 1914. Father of actors, Stacy and James. e. Northwestern Univ. (MA). Was instructor in theatre arts at Northwestern and Armstrong College and dir. at Pasadena Playhouse before entering industry. For 4½ yrs. was under contract at Universal Pictures; 3 yrs. at RKO; had own productions on NBC, CBS. In 1946 began producing and directing industrial stage presentations for Union Oil Co. and from then on became full-time prod. of m.p. and stage industrial shows. In 1946 formed Stacy Keach Productions, of which he is pres. In addition to directing, producing and writing he occasionally appears as actor in films.

KEATON, DIANE: Actress. b. So. Calif. e. Santa Ana College. Appeared in summer stock and studied at Neighborhood Playhouse in N.Y. Made prof. debut in Bdwy. prod. of Hair (1968); then co-starred with Woody Allen in Play It Again, Sam, repeating role for film version.
PICTURES INCLUDE: Lovers and Other Strangers, The Godfather, Play It Again, Sam, Sleeper, The Godfather, Part II, Love and Death, I Will . . . I Will . . . for Now, Harry and Walter Go to New York, Annie Hall, Looking for Mr. Goodbar, Interiors, Manhattan, Reds.

KEEL, HOWARD: Actor. r.n. Harold Keel. b. Gillespie, Ill., April 13, 1917. e. high school, Fallbrook, Calif.; m. Helen Anderson, dancer; d. Kaiya Liane. Stage and screen. Began career following George Walker scholarship award for singing, L.A.; appeared in plays, Pasadena Auditorium, concerts; won awards, Mississippi Valley and Chicago Musical Festivals. Stage debut: Carousel, 1945; principal role (Oklahoma). Screen debut, The Small Voice, London, 1948.
PICTURES INCLUDE: Annie Get Your Gun, Pagan Love Song, Three Guys Named Mike, Show Boat, Texas Carnival, Callaway Went Thataway, Lovely to Look At, Desperate Search, Ride Vaquero, Fast Company, Kiss Me Kate, Calamity Jane, Rose Marie, Seven Brides for Seven Brothers, Deep in My Heart, Jupiter's Darling, Kismet, Floods of Fear, Big Fisherman, Armored Command, Arizona Bushwackers.
PLAYS: Saratoga, No Strings.

KEESHAN, BOB: Performer. b. Lynbrook, N.Y., June 27, 1927. e. Fordham U. As network page boy became assistant to Howdy Doody's Bob Smith and originated role of Clarabelle the Clown; created children's programs Time for Fun (1953), Tinker's Workshop (1954), Captain Kangaroo (1955), Mister Mayor, 1965.

KEITEL, HARVEY: Actor. b. Brooklyn, N.Y. Served in U.S. Marine Corps. Over 10 yrs. experience in summer stock, repertory and little theatre after study at Actors Studio with Lee Strasberg and Frank Corsaro. Read for male lead in collegiate film prod. of Who's That Knocking at My Door?; Martin Scorsese, director, gave him the role. Has since become repertory member of Scorsese films.
PICTURES INCLUDE: Who's That Knocking at My Door?, Street Scenes, Mean Streets, Alice Doesn't Live Here Anymore, Taxi Driver, Mother, Jugs and Speed, Welcome to L.A., The Duellists, Fingers, Blue Collar, Bad Timing, The Border.

KEITH, BRIAN: Actor. b. Bayonne, N.J., Nov. 14, 1921; p. Robert Keith, actor. U.S. Marines, 1942–45; worked in stock co., radio shows, comm. films for TV; on B'way in Mr. Roberts, Darkness at Noon; dramatic TV shows including: Studio One, Suspense, Philco Playhouse, star of Crusader; m.p. debut in Arrowhead.
PICTURES INCLUDE: Jivaro, Alaska Seas, Bamboo Prison, Violent Men, Tight Spot, Five Against the House, Storm Center, Run of the Arrow, Nightfall, Sierra Baron, Those Calloways, The Raiders, The Young Philadelphians, Dino, A Tiger Walks, The Parent Trap, The Hallelujah Trail, Rare Breed, Nevada Smith, Reflections in a Golden Eye, Krakatoa, East of Java, Gaily, Gaily, Suppose They Gave a War and Nobody Came, McKenzie Break, Scandalous John, Something Big, The Yakuza, The Wind and the Lion, Nickelodeon, Hooper, Meteor, Charlie Chan and the Curse of the Dragon Queen.

KELLER, HARRY: Prod.-dir. b. Los Angeles, Calif., Feb. 22, 1913. Film ed., Nat'l Screen Service, Universal, Republic, 1936–50; dir., Republic 1950–53; TV dir., Loretta Young show, 1954–55; dir., Univ. Int., 1956–57; dir. Walt Disney, 1958–59; dir., Universal, 1960–62; for Univ. Prod-Dir. 1963–69; prod., WB, 1970–71.
PICTURES INCLUDE: The Unguarded Moment, Brass Bottle, Send Me No Flowers, That Funny Feeling, Voice in the Mirror, Tammy Tell Me True, Mirage, Texas Across the River, In Enemy Country, Skin Game.

KELLER, MARTHE: Actress. b. Switzerland. Started acting in France and attracted attention of U.S. directors after appearing in Claude Lelouch's And Now My Love.
PICTURES INCLUDE: The Devil by the Tail (debut), And Now my Love, Marathon Man, Black Sunday, Bobby Deerfield, Fedora, The Formula.

KELLERMAN, SALLY: Actress. b. Long Beach, Calif., June 2, 1936. e. Hollywood High. Studied acting in N.Y. at the Actors Studio and in Hollywood with Jeff Corey.
TV: Mannix, It Takes a Thief, Chrysler Theatre.
PICTURES INCLUDE: The Boston Strangler, The April Fools, M*A*S*H, Brester McCloud, Last of the Red Hot Lovers, Lost Horizon, Slither, Rafferty and the Gold Dust Twins, The Big Bus, Welcome to L.A., A Little Romance, Foxes, Loving Couples.

KELLERS, FREDERIC: Executive. b. Hackensack, N.J., Aug. 22, 1929. e. Wilmington College. Ohio. Wilmington Theatre. Ohio. 1947–49; Assoc. Film, Inc., N.Y., 1949–51; assistant box office treasurer, Radio City Music Hall, 1951–66; box office treasurer, 1966. Dir. Theas. Opera, 1967; v.p., operations, 1976; 1979, admissions mgr. Radio City Music Hall Productions; 1981, dir., house operations.

KELLEY, PATRICK: Executive. Joined MCA in 1950 as agent; with them for 20 years; named v.p. in chg. of talent for Universal's theatrical and TV project in 1964. Left MCA in 1970 to form First Artists with star partners Barbra Streisand, Sidney Poitier, Paul Newman (later joined by Steve McQueen and Dustin Hoffman). Resigned 1975 as F.A. bd. chmn. to head Pan Arts Corp., prod. co. of which George Roy Hill is bd. chmn.
PICTURES INCLUDE: A Little Romance (exec. prod.).

KELLOGG, PHILIP M.: executive. b. March 17, 1912, Provo, Wash. e. UCLA. Special feature writer for Hearst papers and magazines, 1933–34; MGM story dept., production dept., Irving Thalberg unit, 1934–35; Warner Bros. Film editor, 1935–41; Berg-Allenberg Agency, 1941–50; U.S. Naval Reserve officer, 1941–46; William Morris Agency, 1950-present, co-head of m.p. dept., dir. WMA, Ltd., London.

KELLY, DUKE: Actor-producer-director-writer. b. Dodge City, Kan., Aug. 26, 1936. e. Wichita State Univ., Fla. State Coll. (masters and doctorate); Masters in cinemagraphic arts. Phd, Archeology. Was in pro sports (football, baseball, boxing) and sgt. in U.S. Marines during Korean War. Also was deputy sheriff, high rigger, lumberjack and truck driver; professional parachutist for National Air Show, stock car driver, writer, lecturer and teacher. Began m.p. career as stunt man and horse wrangler; graduated to bit parts and eventually featured and supporting roles in both TV and m.p. Wrote scripts during first few years of career to supplement acting income. Did many legit theatre musicals, comedies and dramas. Worked radio and TV as personality; sang and danced in supper clubs, wrote music, etc Produced, wrote, directed and starred in seven features made by own co., Film Center Productions.
PICTURES INCLUDE: Producer-Director-Writer: Night of the Werewolf, Walk Home from Hell, Gunhawks, Gunner, Ride the Hot Wind, My Name is Legend, Conquest of Condor Valley. Director: Seven Men To Sundown, The Red Mask, Donovan. Actor: Ten Men from Now, Advise and Consent, The Alamo, Mustang, Walk Home from Hell, Gunhawks, Gunner, Ride the Hot Wind, Liberty Valance, Ride the High Country, My Name Is Legend, Conquest of Condor Valley, etc.
TV: Producer-Director-Writer: The Celebrity Gold Show, Movie Quiz, Box 777, The Costume Shop. Actor: 26 Men, Gunsmoke, F.B.I., Cannon, etc. Host of Celebrity Golf Show, Comedy Capers, Nightmare.

KELLY, GENE: Actor. b. Pittsburgh, Pa., Aug. 23, 1912. e. Penn State U., U. Pittsburgh. Bricklayer, concrete mixer, soda clerk, dance instructor before going on stage, in N.Y. prods. (Leave It to Me, One for the Money, The Time of Your Life, Pal Joey). On screen 1942 in For Me and My Gal; Summer Stock.
PICTURES INCLUDE: Pilot No. 5, Du Barry Was a Lady, Thousands Cheer, The Cross of Lorraine, Christmas Holiday, Anchors Aweigh, Cover Girl, Ziegfeld Follies, Pirate, Three Musketeers, Words and Music, Take Me Out to the Ball Game, Black Hand, On the Town, An American in Paris, Singin' in the Rain, It's A Big Country, Devil Makes Three, Brigadoon, Crest of the Wave, Deep in My Heart, Invitation to the Dance, It's Always Fair Weather, The Happy Road, Les Girls, Special Academy Award for advancing dance films, Gigot, A Guide for the Married Man, The Young Girls of Rochefort, Hello, Dolly!, The Cheyenne Social Club, 40 Carats, That's Entertainment, That's Entertainment, Part Two, Viva, Knievel!, Xanadu.

KELLY, GRACE: Actress. b. Philadelphia, Pa., Nov. 12; e. Amer. Acad. of Dramatic Arts, 1947–49. Amateur theatricals; several B'way plays; summer stock; TV plays; m.p. debut in 14 Hours.
PICTURES INCLUDE: High Noon, Mogambo, Dial M for Murder, Rear Window, Green Fire, Country Girl (Academy Award, best perf., 1954), Bridges at Toko-Ri, To Catch a Thief, The Swan, High Society.

KELLY, J. ARTHUR: Writer, producer, director. b. Brantford, Canada, 1922; served in Canadian Army Infantry (overseas), 1940–45; editor, publisher, Brantford Weekly Star, Brantford Merchants' Shopping Guide, 1947; sales mgr., C.K.P.C. Brantford, 1955; wrote, prod., This Most Gallant Affair, The Abbey On Monte Casino, The Samaritans, The Swimming Cases At Whistling, Ginsberg Isn't Guilty; wrote play, Wait For Me; own firm. Arthur J. Kelly Productions; dir., prod., Maytime Festival, Brantford, Ontario, Canada 1964. New play, He Hanged His Son. Dir. prod. Brantorama Historical Pageant Parade, 1967, Wrote, prod. dir. for Hotel Theatre, Onwonsyshon, 1968. 1970, wrote, prod., dir., Am I Really Joe Sholopka?; 1970, wrote, prod., dir., Standup and Shout About It; wrote, prod., dir., What In the Hell Ever Happened to Purgatory?; wrote, prod., dir., Is Mrs. Maloney' Our Only Hope?

KELLY, JIM: Actor. b. Paris, Ky. e. Univ. of Louisville. Studied karate at univ., winning trophies and int'l. middleweight championship. Opened school for karate in L.A. Did modelling and TV commercials. Was technical advisor for fight scenes on Melinda and played role in it.
PICTURES INCLUDE: Enter the Dragon (debut), Black Belt Jones, Three the Hard Way.

KELLY, NANCY: Actress. b. Lowell, Mass., March 25, 1921. e. Immaculate Conception Acad. N.Y.; St. Lawrence Acad., L.I.; Bentley Sch. for Girls. In number of pictures as child, and on stage in Susan and God (N.Y. prod. 1937). In 1938 on screen.
PICTURES INCLUDE: Torando, Women in Bondage, Gamblers Choice, Show Business, Double Exposure, Song of the Sarong, Woman Who Came Back, Murder in the Music Hall, Crowded Paradise, Bad Seed.
STAGE: Season in the Sun, 1950–51; Bad Seed, 1954–55.

KEMENY, JOHN: Producer. B. Budapest, Hungary. Producer for National Film Board of Canada, 1957–69. Formed International Cinemedia Center, Ltd. in 1969 in Montreal, as partner.
PICTURES INCLUDE: The Apprenticeship of Duddy Kravits, White Line Fever, Shadow of the Hawk, La Castles.

KEMP, JEREMY: Actor. b. Chesterfield, England, 1935. e. Abbottsholme School, Central School of Speech and Drama. Service with Gordon Highlanders. Early career on stage incl. Old Vic Theatre Company, 1959–61. Recent theatre: Celebration, Incident at Vichy, Spoiled, The Caretaker. National Theatre, 1979–80.
TV: Z Cars, The Lovers of Florence, The Last Reunion, Colditz, Brassneck, Rhinemann Exchange, Lisa, Goodbye, Henry VIII, St. Joan, The Winter's Tale.
PICTURES INCLUDE: Cast a Giant Shadow, Operation Crossbow, The Blue Max, Assignment K, Twist of Sand, Strange Affair, Darling Lilli, The Games, The Saltzburg Connection, The Blockhouse, The Bellstone Fox, 7% Solution, A Bridge Too Far, East of Elephant Rock, Caravans, The Prisoner of Zenda.

KEMP, MATTY: Producer, director, writer. b. Rockville Centre, N.Y. St. Paul's School, Hempstead, N.Y. Entered m.p. as actor-writer 1926 Universal Pictures; variously employed by Mack Sennett, Fox, Paramount, RKO, prod. Authors' Guild show for radio and musical shorts. Served in U.S. Army 1942–45 in signal corps pictorial, information education, and training film branch. Prod. dir. for series of musical shorts Universal 1945, formed Cameo Productions (Calif.) 1946; formed Masque Productions with Pene Raymond 1948.
PICTURES INCLUDE: (prod. dir.) Linda Be Good, Million Dollar Weekend; (Story) The French Line, (Writer, prod. dir.) Pan American Showtime, 13 half hour musical series, (prod. dir.) Meet the Family, series of 3 with Arthur Lake, (prod. dir.) Adventurous Hobby series 3 segments, (Story prod. dir.) The Birth of a Legend, documentary on Mary Pickford and Douglas Fairbanks, (Story, co-producer) America's Sweetheart feature documentary, Managing director production executive, Mary Pickford Co.

KEMP-WELCH, JOAN: Free lance, TV director-producer, actress. First appearance on stage 1927. Subsequently many stage parts and stage directorial assignments. First appeared in films 1938. Films included 60 Glorious Years, They Flew Alone, The Citadel, Busman's Honeymoon. West End theatre prods. include: Dead on Nine, Vicious Circle, Our Town, Desire Under the Elms. Since 1954 TV dir. Received TV Oscar for Light Entert., 1958. Desmond Davis Award for services to TV 1963. Silver Dove Monte Carlo Award, 1961. The Lover, awarded Prix Italia 1963 (drama). Many other productions, incl. musicals, ballet, drama, series, outside broadcasts. Dear Octopus, The Birthday Party, The Collection, View from the Bridge, Electra, 3 Sisters, A Midsummer Nights Dream, Dangerous Corner. Upstairs, Downstairs; prod. Armchair Theatre, 1973–74. 1974–75. French Without Tears, Wait Till Dark, The Price, Cranford Musical. The Other Side of the Swamp, Romeo & Juliet, The Circle, Hay Fever, It Happened in Harrods.
TV: The Piano. 1977–78 In South Africa, The Deep Blue Sea, The King Fisher, The Monkey Walk, Cause Celebre, Your Place or Mine, Home, The Unvarnished Truth, I Am Who I Am, Happy Birthday.

KENNEDY, ARTHUR: Actor. b. Worcester, Mass., February 17, 1914; p. Dr. and Mrs. J. T. Kennedy; e. Carnegie Institute of Technology; m. Mary Cheffey, prof. Has worked for George M. Cohan; Guthrie McClintic, Marc Connelly and others well known to theatregoers. Film debut: City for Conquest.
PICTURES INCLUDE: Champion, They Died With Their Boots On, High Sierra, Strange Alibi, Knockout, Highway West, Air Force, Devotion, Boomerang, The Window, Chicago Deadline, The Glass Menagerie, Red Mountain, Bright Victory, Bend of the River, Rancho Notorious, Girl in White, Lusty Men, Man from Laramie, Trial, Naked Dawn, Desperate Hours, Crashout, Rawhide Years, Peyton Place, Some Came Running, Claudelle Inglish, Adventures of a Young Man, Barabbas, Laurence of Arabia, Italiano Brava Gentle, Stay Away Joe, A Minute to Pray, a Second to Die, Hail Hero, Shark, My Old Man's Place, The Sentinel.

KENNEDY, BURT: Director. b. Muskegon, Mich. Began as writer of TV and film scripts, and most recently was writer, producer and director of Combat series.
PICTURES INCLUDE Mail Order Bride. The Rounders, The Money Trap, Return of the Seven, The War Wagon, Support Your Local Sheriff, Young Billy Young, The Devil's Backbone, Dirty Dingus Magee, Support Your Local Gunfighter, Hannie Caulder, The Train Robbers (also s.p.).

KENNEDY, DOUGLAS: Actor. b. Sept. 14, 1915. New York. e. Deerfield Academy, Mass., Amherst College. In armed services, World War II.
PICTURES INCLUDE: Opened By Mistake, Women Without Names, Those Were the Days, The Way of All Flesh, Northwest Mounted Police, The Round-Up, Nora Prentiss, Dark Passage, Possessed, The Unfaithful, Look for the Silver Lining, I Was An American Spy, Callaway Went Thataway, Fort Osage, Indiana Uprising, For Men Only, China Corsair, Last Train From Bombay, Ride the Man Down, Torpedo Alley, War Paint, Gun Belt, Sea of Lost Ships, Lone Gun, Massacre Canyon, Big Chase, Sitting Bull, Cry Vengeance, Wyoming Renegades, Eternal Sea, Strange Lady in Town, Wiretapper, Fastest Guitar Alive.
TV: Steve Donovan Western Marshal series.

KENNEDY, GEORGE: Actor. b. N.Y., N.Y., Feb. 18, 1927. f. orchestra leader at N.Y. Proctor Theatre, m. dancer with Le Ballet Classique in vaudeville. At 2 acted in touring co. of Bringing Up Father. At 7, disc jockey with his own radio show for children. Joined W.W. II Army at 17, earned two Bronze Stars and combat and service ribbons. In Army 16 years, became Capt. and Armed Forces Radio and TV officer. 1957, opened first Army Information Office, N.Y. Served as technical advisor to Phil Silver's Sergeant Bilko TV series. Began acting in 1959 when discharged from Army.
TV: Sugarfoot, Cheyenne.
PICTURES INCLUDE: Lonely Are the Brave, Strait Jacket, The Silent Witness, Island of the Blue Dolphins, The Man From the Diners Club, Little Shepherd of Kingdom Come, Mirage, See How They Run, McHale's Navy, Charade, In Harm's Way, The Sons of Katie Elder, Shenandoah, Hush . . . Hush Sweet Charlotte, The Dirty Dozen, Hurry Sundown, Cool Hand Luke (Acad. Award Best Supporting Actor), The Ballad of Josie, Jolly Pink Jungle, Bandolero!, The Boston Strangler, Guns of the Magnificent Seven, Gaily, Gaily, The Good Guys and the Bad Guys, Airport, . . . tick . . . tick . . . tick, Zigzag, Dirty Dingus Magee, Fool's Parade, Lost Horizon, Cahill, Thunderbolt and Lightfood, Airport 1975, Earthquake, The Human Factor, Airport '77, Death on the Nile, Brass Target, The Concorde—Airport '79, Death Ship.

KENNEDY, JOSEPH: Executive. b. New York City. Joined United Artists in 1968 as salesman out of Boston branch. Named Chicago sls. mgr. (1969–72); Jacksonville branch mgr. (1972–78); southern div. mgr. (1978–79). In 1980 named v.p. & asst. gen. sls. mgr. responsible for UA eastern sls. territories.

KENNEY, H. WESLEY: Producer-Director, stage, TV, film. b. Dayton, O., grad. Carnegie Tech. Four-time Emmy winner; 1974–75 dir., All in the Family; exec. prod. Days of Our Lives; 1979–81, dir., Ladies Man Filthy Rich.

KENT, JEAN: Actress. b. London, Eng., June 29, 1921. e. Marist Coll., Peekham, London; p. prof. Fields & Norrie. First stage appearance at 3 and at 10 played in parents' act; chorus girl at Windmill Theatre, London, 1935; 2 yrs. repertory; Screen debut: It's That Man Again, 1941.
PICTURES INCLUDE: Trottle True, Her Favorite Husband, The Reluctant Widow, The Woman in Question, The Browning Version, Big Frame, Before I Wake, Shadow of Fear, Prince and the Showgirl, Bon Jour Tristesse, Grip of the Strangler, Beyond This Place, Please Turn Over, Bluebeard's Ten Honeymoons, Shout At the Devil.
TV: A Call on the Widow, The Lovebird, The Morning Star, November Voyage, Love Her to Death, The Lion and the Mouse, The Web, Sir Francis Drake series, Yvette, Emergency Ward 10, County Policy, Coach 7, Smile On the Face of the Tiger, No Hiding Place, Kipling, This Man Craig, The Killers, Vanity Fair, A Night with Mrs. Da Tanka, United serial. The Family of Fred, After Dark, Thicker than Water series, The Young Doctors, Brother and Sister, Up Pompei, Steptoe and Son, Doctor at Large, Family at War, K is for Killing, Night School, Tycoon series.

KENT, JOHN B.: Theatre executive and attorney. b. Jacksonville, Fla., Sept. 5, 1939. e. Yale Univ., Univ. of Fla., Law School, N.Y. Univ. grad. school of law (LLM in taxation, 1964). Partner in Kent, Watts, Durden, Kent & Mickler (1967 to present); Pres. & dir., Kent Investments, Inc. (1977 to present); dir. and off. Kent Theatres, Inc. and affiliated corps (1961 to present); v.p. and gen. counsel, (1970 to present). Was pres. 1967–70 when resigned to devote full time to law practice. NATO dir. (1972) and Presidents' Advisory Cabinet, 1979 to present) v.p. NATO of Fla., 1968–72, dir; 1973 to present. Member of Rotary Club of Jacksonville, Fla. Bar Ass'n., American Bar Ass'n., American Judicature Society.

KENYON, CURTIS: Writer.
TV PLAYS: Cavalcade of America, Fireside Theatre, Schlitz Playhouse, U.S. Steel Hour, 20th Century-Fox Hour.
PICTURES INCLUDE: Woman Who Dared, Lloyds of London, Wake Up and Live, Love and Hisses, She Knew All the Answers, Twin Beds, Seven Days' Leave, Thanks for Everything, Princess and the Pirate, Bathing Beauty, Fabulous Dorseys, Tulsa, Two Flags West.

KERASOTES, GEORGE G.: Exhibitor. b. Springfield, Ill., Mar. 27, 1911; e. U. of Ill. 1929–33, Lincoln Coll. of Law 1934–36. Entire career in m.p. ind.; chmn. bd., United Theatres Owners of Illinois; pres., Kerasotes Theatres, Mo., Ill.: pres. Theatre Owners of America, 1959–60; chmn. of board of TOA 1960–62; chmn. ACE Toll TV com.; bd. mem. Code Appeal Board; mem. exec. comm. NATO, bd. of dir. NATO. chairman NATO non-theatrical committee, dir. Sangamon Home Savings Assoc. Pres. Kerasotes Theatres; pres., St. Anthony's Church, Springfield, Ill.

KERKORIAN, KIRK: Executive. b. Fresno, CA, June 6, 1917. e. Los Angeles public schools. Served as capt., transport command, RAF, 1942–44. Commercial air line pilot from 1940; founder Los Angeles Air Service (later Trans Intl. Airlines Corp.), 1948; Intl. Leisure Corp., 1968; controlling stockholder, Western Airlines, 1970; chief exec. officer, MGM, Inc., 1973–74; chm. exec. com., vice-chm. bd., 1974—

KERR, DEBORAH: Actress. b. Helensburgh, Scotland, Sept. 30, 1921; e. Phyllis Smale Ballet Sch.; m. Anthony Bartley. On stage 1939 in repertory. Began Brit. screen career 1940 in Major Barbara; voted "Star of Tomorrow" Motion Picture Herald-Fame Poll, 1942. Voted one of top ten British money-making stars in Motion Picture Herald-Fame Poll, 1947. B'way debut in Tea & Sympathy, 1953.
PICTURES INCLUDE: Major Barbara, Love on the Dole, Hatler's Castle, The Day Will Dawn, The Avengers, Perfect Strangers, Colonel Blimp, Black Narcissus, The Hucksters, If Winter Comes, Edward My Son, Please Believe Me, King Solomon's Mines, Quo Vadis, Thunder in the East, Prisoner of Zenda, Dream Wife, Julius Caesar, Young Bess, From Here to Eternity, End of the Affair, King and I, Proud and Profane, Tea and Sympathy, Heaven Knows Mr. Alison, Affair to Remember, Count Your Blessings, Beloved Infidel, Sundowners, The Grass Is Greener, The Innocents, The Naked Edge, The Chalk Garden, Night of the Iguana, Marriage On the Rocks, Casino Royale, Eye of the Devil, The Gypsy Moths, The Arrangement.

KERR, FRASER: Actor. b. Glasgow, Scotland, 1931. Early career in repertory. Tours of Canada and America. Ent. TV 1956. Series incl. Emergency Ward 10, Dixon of Deck Green, Murder Bag. Many Shakespeare plays. Radio: BBC Drama Rep. Co., 39 Steps, The Ringer, The Bible, What Every Woman Knows.
STAGE & TV: Night Must Fall, Never a Cross Word, The Inside Man, On the Buses, Dr. Finlay's Casebook, Wicked Woman, Madelaine July, Doctor in the House, Counterstrike, Waggoner's Walk, Juno and the Paycock, Aquarius, Ev, Upstairs and Downstairs, Cover to Cover, Janine, Robert the Bruce, Caliph of Bagdad, Watch it, Sailor!, The Fosters, Weekend World, Doctor at Sea, Dads Army, Algernon Blackwood, Waiting for Sheila, Weekend Show, Mind Your Language, Yes, Minister.
PICTURES INCLUDE: What a Whopper, Carry on Regardless, Way of McEagle, Thomasina, Theatre of Death, Tom, Dick and Harriet, Granny Gets the Point, Nothing But The Night, The Lord of the Rings, Kidnapped, The Derelict.
RECORD PRODUCER: Tales of Shakespeare Series, The Casket Letters of Mary Queen of Scots.

KERR, JOHN: Actor. b. New York, N.Y., Nov. 15, p. Geoffrey Kerr, actor, and June Walker, actress. e. Harvard U., B.A., Columbia U., M.A. Actor in summer stock, TV; on Broadway in Bernardine, Tea and Sympathy, All Summer Long.
PICTURES INCLUDE: The Cobweb, Gaby, Tea and Sympathy, The Vintage, South Pacific, Girl of the Night, Pit and the Pendulum, Seven Women from Hell.
TV: Peyton Place.

KERSHNER, IRVIN: Director. b. Phila., Pa., April 29, 1923. e. Tyler School Fine Arts of Temple Univ., 1946, Art Center School, Univ. of Southern Calif. Designer, photography, adv., documentary, architectural; doc. film maker, U.S.I.S., Middle East, 1950–52; dir., cameraman, TV doc., Confidential File, 1953–55; dir.-prod.-writer, Ophite Prod.
PICTURES INCLUDE: Stakeout on Dope Street, Young Captives, Hoodlum Priest, The Luck of Ginger Coffey, A Fine Madness, The Flim Flam Man, Loving, Up the Sandbox, S*P*Y*s, Return of a Man Called Horse, Raid on Entebbe. (TV in U.S.): Eyes of Laura Mars, The Empire Strikes Back.
TV: The Rebel, Naked City, numerous pilots and other nat'l shows.

KEYES, EVELYN: Actress. b. Port Arthur, Texas; e. high school; began career as a dancer in night clubs; m. Artie Shaw.
PICTURES INCLUDE: The Buccaneer, Union Pacific, Gone with the Wind, A Thousand and One Nights, The Jolson Story, Mating of Millie, Johnny OX'Clock, Enchantment, Mr. Soft Touch, The Prowler, The Killer That Stalked New York, Smuggler's Island, The Iron Man, One Big Affair, Shoot First, 99 River Street, Hell's Half Acre, Top of the World, Seven-Year Itch, Around the World in 80 Days.

KIBBEE, ROLAND: Writer, producer, director. b. Monongahela, Pa., Feb. 15, 1914; e. L.A. City Coll., 1932–33. Radio writer: The Grouch Club, Fred Allen Show, Groucho Marx Show, Fanny Brice Show; Play: Easy Does It; Magazine Humor: Atlantic, Saturday Review, Script, Holiday.
SCREENPLAYS: Night in Casablanca, Angel On My Shoulder, Ten Tall Men, The Crimson Pirate, Vera Cruz, The Devil's Disciple, Top Secret Affair, The Desert Song, Now You See It, Now You Don't, producer & writer; Valdez Is Coming, executive producer & writer; Brock's Last Case, producer; Moll Flanders, The Midnight Man, writer, producer, director.
TV: The Ford Show, Ford Startime, The Bob Cummings Show, The Virginian, The Hitchcok Hour; Madigan, The Bob Newhart Show, It Takes A Thief, The Deputy, producer, creator, writer; prod., Madigan, Columbo; prod., creator, writer, McCoy. Exec. prod., The Family Holvak; prod., Barney Miller.

KIDDER, MARGOT: Actress. b. Yellowknife, Canada, Oct. 17, 1948.
PICTURES: Gaily, Gaily, Quackser Fortune Has a Cousin in the Bronx, Sisters, The Great Waldo Pepper, Superman, Mr. Mike's Mondo Video, The Amityville Horror, Willy and Phil, Superman II.

KILEY, RICHARD: Actor. b. Chicago, Ill., Mar. 31, 1922; e. Loyola U. Started prof. career radio, Jack Armstrong, All American Boy.
STAGE: Streetcar Named Desire, Misalliance, Kismet, Time Limit, No Strings, Man of LaMancha, Her First Roman.
PICTURES INCLUDE: The Mob, The Sniper, Eight Iron Men, Pick-Up on South Street, Blackboard Jungle, Phoenix City Story, Spanish Affair, Pendulum, The Little Prince, Looking for Mr. Goodbar.

KILLIAM, PAUL: Producer-performer. b. Mass., Sept. 12, 1916. e. Harvard. News supervisor, WOR-Mutual; prod.-performer TV Hometown. Matinee in NY units for CBS-TV Morning Show, NBC-TV Home Show; ind. prod. cartoons, shorts, comedies: prod. film series, Paul Killiam Show, Movie Museum, Silents Please.

KIMBLEY, DENNIS: Sales Manager, Motion Picture Sales, Kodak Ltd., in U.K. Early career in Kodak Testing Dept. responsible for quality control motion picture films. Joined Marketing Division 1966. Chairman BKSTS Film 75 and film 79 Conference Committee. President BKSTS 1976–78.

KIMMINS, ANTHONY: Director, writer. b. Harrow Middlesex, England, Nov. 10, 1901; e. R.N. Coll., Dartmouth. Capt. in Royal Navy, World War II. Playwright (While Parents Sleep, Night Club Queen, Chase the Ace, Winter Sport, The Amorous Prawn.) Began screen career 1934 in How's

Chances?; since author variously orig., s.p., collab., dial., many Brit. features (Talk of the Devil, Come On George, Laburnum Grove, etc.); in 1936 dir. & s.p. All at Sea; pictures dir. since include I See Ice, It's in the Air, Trouble Brewing. In 1946 returned to m.p. to write, produce and direct.
PICTURES INCLUDE: Mine Own Executioner, Bonnie Prince Charlie, Flesh and Blood, Mr. Denning Drives North, Who Goes There (Passionate Sentry), Captain's Paradise, Smiley, Smiley Gets a Gun, The Amorous Prawn.

KING, ALAN: Actor-Producer. b. New York, N.Y. Author, Anybody Who Owns His Own Home Deserves It, Help I'm a Prisoner in a Chinese Bakery.
TV: The Tonight Show, Kraft Music Hall, Comedy is King. Prod-star NBC-TV specials. Hosts and guest star, Tonight Show. Stars semi-annually at Sands Hotel, Las Vegas, Seventh Avenue (mini-series).
STAGE: The Impossible Years, The Investigation, Dinner At Eight, A Lion in Winter, Something Different.
PICTURES INCLUDE: Bye Bye Braverman, Anderson Tapes, Just Tell Me What You Want. Producer: Happy Birthday, Gemini.

KING, ANDREA: Actress, r.n. Georgette Barry; b. Paris, France; e. Edgewood High School, Greenwich, Conn.; m. N.H. Willis, attorney. Started career on N.Y. stage, following high school; in Growing Pains & Fly Away Home, Boy Meets Girl, Angel Street (Boston); Life with Father (Chicago); signed by Warner, 1943. Screen debut: The Very Thought of You. TV actress, CBS.
PICTURES INCLUDE: My Wild Irish Rose, Ride the Pink Horse, Mr. Peabody and the Mermaid, Song of Surrender, Southside 1-1000l, Dial 1119, Lemon Drop Kid, Mark of the Renegade, World In his Arms, Red Planet Mars, Daddy's Gone A-Hunting.

KING, HENRY: Director "Champion of Champion Directors" in Fame ratings. b. Christiansburg, Va., Jan 24, 1896. Worked for Norfolk & Western R.R. in many depts.; toured in stock, circuses, vaudeville, burlesque; on N.Y. stage in "Top O' the Morning"; m.p. actor, Pathe studios; writer-dir.; producer; exec. head Inspiration Co.; with Fox Film Co., remaining after merger with 20th-Century Pictures Co.
PICTURES INCLUDE: 23½ Hours Leave, Tol'able David, Fury, White Sister, Stella Dallas, Romola, Winning of Barbara Worth, Woman Disputed, Over the Hill, Carolina, State Fair, Country Doctor, Ramona, Lloyds of London, In Old Chicago, Alexander's Ragtime Band, Jesse James, Little Old New York, Stanley and Lingstone, Maryland, Chad Hanna, Yank in the R.A.F., Black Swan, Song of Bernadette, Wilson, Bell for Adano, Margie, Captain from Castile, Deep Waters, Twelve O'Clock High, The Gunfighter, I'd Climb the Highest Mountain, David and Bathsheba, Wait Till the Sun Shines, Nelly, Snows of Kilimanjaro, O. Henry's Full House, King of the Khyber Rifles, Untamed, Love Is a Many Splendored Thing, Carousel, The Sun Also Rises, This Earth Is Mine, Tender Is the Night.

KING, HERMAN: Producer. b. Chicago, Ill. Was engaged in vending machine business; mfr. Hollywood Talkitone Soundie Projectors; org. prod. co., King Bros. Prod. (with bros. Maurice and Franklin), 1941.
PICTURES INCLUDE: When Strangers Marry, Dillinger, Suspense, The Gangster, The Dude Goes West, Badman of Tombstone, Gun Crazy, Southside 1-1000, Drums in the Deep South, Mutiny, The Ring, Carnival Story, The Brave One, Gorgo, Captain Sinbad, Maya, Return of The Gunfighter, Heaven with a Gun.
TV: Maya series, King International Corp.

KING, MURRAY J.: Prod.-Exec. b. Chelsea, Mass., Jan. 13, 1914. e. Columbia U., B.S., M.A., 1936; film cutter, 1936; film cutter, asst. ed., Monogram Pictures. Hollywood, 1937–41; Air Corps Captain, WW. II, 1942–46; own studio, Murray King Pictures, Inc., 1947–54; dubbing studio, Mexico, 1954–55; dubbing studio, Havana, Cuba, 1956–60; partner, Miller International Corp. 1961; pres., Miller International Corp., 1964. Pres. Vidata International Corp., 1968 to 1976; President of M-K Communications Corp. 1977 to date, produced feature pictures as follows: The Doll That Took the Town; Bad Girls Don't Cry; Television Series: The Adventures of Blinkey; Producer; Film Publishers Group, Inc.

KING, PEGGY: Singer. b. Greensburg, Pa., 1931. Worked as secretary, then singer with band; morning radio show, Cleveland; recording artist; featured singer, George Gobel Show.

KING, PERRY: Actor. b. Alliance, Ohio, e. Yale. Studied with John Houseman at Julliard.
PICTURES INCLUDE: The Possession of Joel Delaney (debut), Slaughterhouse-Five, Big Truck, Poor Clare, The Lords of Flatbush, Mandingo, The Wild Party, Lipstick, Andy Warhol's Bad, The Choirboys, A Different Story.
TV: Medical Center, Hawaii Five-O, Apple's Way, Cannon.

KING, PETER: Executive. b. London, Eng., 1928. e. Marlborough College, Oxford U.; bd., Shipman & King Cinemas Ltd., 1956; borough councillor, 1959–61; chmn., London & Home counties branch, CEA, 1962–63; pres., CEA, 1964; dir. film ind. Defense Organization dir., Grade Org. 1966–68; man. dir. Shipman & King Cinemas Ltd., 1959–68. Ch. man. dir. Paramount Pictures (U.K.) Ltd. Britain, 1968–70. Man., dir., EMI Cinemas and Leisure Ltd., 1970–74. Vice pres. National Youth Theatre, Governor, National Film School, Chairman: King Publications Ltd.

KINGMAN, DONG: Fine Artist. b. Oakland, Calif., Mar. 31, 1911. e. Hong Kong 1916–1920; 1928, mem. motion picture co., Hong Kong branch; 1935; began to exhibit as fine artist in San Francisco; promotional, advertising or main title artwork for following films: World of Suzie Wong, Flower Drum Song, 55 Days of Peking, Circus World, King Rat, The Desperados, The Sand Pebbles, Lost Horizon-1973. 1966–7, created 12 paintings for Universal Studio Tour for posters and promotion; 1968, cover painting for souvenir program for Ringling Bros.-Barnum and Bailey Circus; treasurer for Living Artist Production since 1954; 1968, became art director for Greater Amusements. Exec. V.P. 22nd-Century Films, Inc. since 1968, Prod. & dir. short, Hongkong Dong. Also short subject film Dong Kingman, filmed and directed by James Wong Howe.

KINGSLEY, DOROTHY: Writer. (Mrs. William W. Durney): b. New York City, Oct. 14. e. Detroit Arts and Crafts Acad. Radio writer for Bob Hope, 1938; Edgar Bergen, 1939–43.
PICTURES INCLUDE: Date With Judy, Neptune's Daughter, Two Weeks With Love, Angels in the Outfield, Texas Carnival, It's a Big Country, When In Rome, Small Town Girl, Dangerous When Wet, Kiss Me Kate, Seven Brides for Seven Brothers, Jupiter's Darling, Don't Go Near the Water, Pal Joey, Green Mansions, Can-Can, Pepe, Half a Sixpence, Valley of the Dolls.
TV: Created series, Bracken's World.

KINGSLEY, WALTER: Executive. b. New York, Oct. 20, 1923. e. Phillips Academy, Andover; Amherst Coll., B.A., 1947. Charter member Big Brothers of Los Angeles. WCOP, Boston, 1947–50; Ziv Television Programs, Inc., 1950–58; President, Independent Television Corp., 1958–61. Member bd. dir. Big Brothers of Amer.; pres. Kingsley Co., 1962–66; exec. v.p. Wolper Prods. Metromedia Prods. Corp., 1966–72; faculty, Inter-Racial Council of Business Opportunity, N.Y.C.; commercial real estate development, 1972-present.

KINOSHITA, KEISUKE: Director. b. Japan, 1912. Entered Shochiku studio as film processor and progressed to director.
PICTURES INCLUDE: Twenty-four Eyes, Sun and Rose, Wild Chrysanthemum, A Japanese Tragedy, Times of Joy and Sorrow, Snow Flurry, Candle in the Wind, Carmen's Pure Love.

KINOY, ERNEST: Writer. Started career in radio writing sci. fic. programs (X Minus One, Dimension X). Created and wrote religious radio show, The Eternal Light. Moved in to TV and wrote for nearly all dramatic shows, including Studio One, Philco Playhouse, Playhouse 90.
PICTURES INCLUDE: Brother John, Buck and the Preacher, Leadbelly.
TV: The Defenders, Naked City, Dr. Kildare, Jacob and Joseph (special), David, the King (special).

KIRK, PHYLLIS: Actress. b. Syracuse, N.Y., Sept. 18. Perfume repr. model, Conover Agcy.; B'way play debut in My Name Is Aquilon; actress, summer stock; screen debut in Our Very Own; B'way production of Point of No Return.
TV: The Thin Man.
PICTURES INCLUDE: A Life of Her Own, Two Weeks With Love, Mrs. O'Malley and Mr. Malone, Three Guys Named Mike, About Face, Iron Mistress, Thunder Over The Plains, House of Wax, Crime Wave, River Beat, Canyon Crossroads, City After Midnight.

KIRNER, MICHAEL: Actor. b. April 2, 1944. e. Drama schools Munich and Rome.
PICTURES INCLUDE: Dangerous Journey, The Hot Country, The Devil from Cape Town, Operation Yellow Viper, African Story, The Manipulator, Sfida Nella Citta del Oro, The Big Game, Snow From Hell.

KITT, EARTHA: Actress, singer. b. Columbia, S.C. Professional career started as dancer in Katherine Dunham group; toured U.S., Mexico & Europe with group, then opened night club in Paris; in Orson Welles stage prod. of Faust for European tour; N.Y. night clubs; stage in U.S., New Faces; at Macambo Hollywood, 1953; author, Thursday's Child; on screen, 1954, New Faces; 1958, The Mark of the Hawk, St. Louis Blues, Anna Lucasta.

KLEES, ROBERT E.: Executive. b. New York, N.Y., Feb. 21, 1927. e. Duke Univ. (1947–51); Univ. of Calif. Graduate School of Management (1973–75). U.S. Navy, 1944–46; Union Carbide Corp., 1951–57; director of communications, Beckman Instruments, Inc., 1957–69; co-founder and v.p., mktg., International Biophysics Corp., 1969–73; v.p., mktg., Deluxe General, Inc., 1975 to date. Member: Academy of m.p. Arts & Sciences. Academy of TV Arts & Sciences, American Film Institute, Hollywood Radio and TV Society, SMPTE. Associate Member: American Cinema Editors, American Society of Cinematographers.

KLEIN, ALLEN: Producer. b. Dec. 18, 1931. Pres. ABKCO industries, Inc.
PICTURES INCLUDE: Force of Impulse, Pity Me Not, Mrs. Brown, You've Got A Lovely Daughter, Stranger in Town, The Stranger Returns, Samuri on a Horse, The Silent Stranger, Pete, Pearl & The Pole, The Grand Bouffe, Come together, Let It Be, The Holy Mountain, El Topo, The Concert for Bangladesh, The Greek Tycoon.

KLEIN, EUGENE V.: Executive. b. New York, Jan. 29, 1921. e. New York University. U.S. Air Force 1941–46; Chmn. of bd. and chief executive officer. Nat'l Gen. Corp. various entertainment group subsidiaries, Nat'l Gen. Prods., Inc., Nat'l Gen. Pictures Corp., Nat'l Gen. TV Dist., Inc., Nat'l Gen. TV Prods., Inc., Edu. Film Prods., Inc., NGC Broadcasting Corp.

KLEIN, HAL: Producer. b. Chicago, Ill., Aug. 20, 1914. e. UCLA. Asst. Dir., 20th Fox; prod. mgr., Associate Producer features and TV, U.S., Europe, Africa. Presently, Aubrey Schenck Enterprises.
PICTURES INCLUDE: War Party, Forst Courageous, Convict Stage, Don't Worry, Ambush Bay, Kill a Dragon, Impasse, More Dead Than Alive, Barquero, Angel Unchained.

KLEIN, HAROLD J.: Executive. b. New York, N.Y., e. U. of West Virginia, N.Y. Law School; reviewer, sales staff. Showman's Trade Review; booker, Brandt Theatres; booker, later vice-pres., gen. mgr., J&J Theatres, 1941–59; account executive, Exec. vice-pres., dir. of world-wide sales, ABC Films, Inc., N.Y., Pres., Klein Film Assn.; Exec. Vice President, Plitt Theatres, Inc.

KLEIN, MALCOLM C.: Executive. b. L.A., Calif., Nov. 22, 1927. e. UCLA, grad., 1948, U. of Denver; Prod. dir. managment, KLAC-TV (KCOP), L.A., 1948–52; acct. exec., KABC-TV, 1952–56; asst. gen. sales mgr., KABC-TV, 1956–59; exec. vice-pres. gen. mgr., NTA Broadcasting, N.Y., 1959; v.p., gen. mgr., RKO-General-KHJ-TV, 1960; joined National General Corp. 1968, vice-pres. Creative Services and Marketing. Pres. National General Television Productions, Inc., Pres. NGC Broadcasting Corp.; 1971, pres. Filmways TV Presentations; 1972, pres. Malcolm C. Klein & Assoc. mgmt. & mktng. consultants; 1973 gen'l. exec. Sterling Recreation Organization & Gen'l Mgr. Broadcast Division; pres., American Song Festival. Now exec. v.p., Telease Inc. & American Subscription Television; and mgt. & mktng. consultant.

KLEIN, PAUL L.: Television Executive. e. Brooklyn Coll. Veteran of U.S. Army Air Corps. Research analyst with Biow Co., ad agency, 1953–54; research manager for Doyle Dane Bernbach ad agency, 1955–60. Started with NBC in 1961 as supervisor, ratings, and rose to position of v.p., audience measurement in October, 1965. In August, 1970, left NBC to found Computer Television Inc., first independent pay-per-view TV co. in world. Time, Inc. bought his interest in CTI. Returned to NBC in March, 1976, as v.p., network mktg. & planning; then named v.p., programs. Appointed exec. v.p., programs, NBC-TV, November, 1977.

KLEINER, HARRY: Writer, producer. b. Phila., Penna. e. Temple University (B.S.); Yale University (M.F.A.).
TV: Writer: The Virginian, Bus Stop, musical Robin Hood, Rosenberg Trial.
PICTURES INCLUDE: Screenplay: Carmen Jones, Street With No Name, Miss Sadie Thompson, Salome, Garment Jungle (also prod.), Fantastic Voyage, Bullitt (co-s.p.), Le Mans, Overlords.
TV: Robin Hoods, Rosenberg Trial.

KLINE, FRED W.: Publicist. b. Oakland, CA May 17, 1918. e. U. Calif. Berkeley; M.P. pub. rel. since 1934; pres. The Fred Kline Agency; pres. Kline Communications Corporation; Owner, Fred Kline Agency, Inc.; Kline Communications Corp.; Fred W. Kline Prod., Inc.; Capitol News Service, Sacramento; L.A. News Bureau; Capitol Radio News Service, Inc.; Commissioner, Motion Picture Council, State of California; Commissioner, Los Angeles Fire Services Commission.

KLINGER, HENRY: Eastern story editor, 20th Century-Fox Film Corp. b. New York, N.Y., Mar. 15, 1908. e. CCNY, BA; NYU, BBA. Assoc. ed., Chatterbox, nat'l mag., 1931–33; pres. Booklovers' Guild, 1930–33; freelance story dept. work, RKO, Cosmopolitan & Fox Films, 1931–33; story dept., pub. contract, Fox, 1934; asst. story ed., 20th-Fox, 1936; acting story ed., 1941; assoc. story ed., 1942; story ed., 1956; exec. story ed., 1964; exec. aide, 1971–73. Lecturer, Author.

KLINGER, MICHAEL: Producer. Managing Dir. Michael Klinger Ltd., Avton Film Productions.
PICTURES INCLUDE: Cul-de-Sac, Repulsion, Saturday Night Out, The Projected Man, Yellow Teddy Bears, Black Torment, Study in Terror, The Penthouse, Baby Love, Something to Hide, Get Carter, Pulp, Rachel's Man, Tomorrow Never Comes, Gold, Shout at the Devil. Exec. Prod. Confessions series, Blood Relatives.

KLUGMAN, JACK: Actor. b. Philadelphia, Pa., April 27. e. Carnegie Tech. m. Brett Somers. After several menial jobs appeared on Broadway in Saint John, Stevedore; later understudied in Mister Roberts, taking over the doctor role; recent stage work includes Gypsy.
PICTURES INCLUDE: Timetable, Twelve Angry Men, Cry Terror, The Scarface Mob, Days of Wine and Roses, I Could Go On Singing, The Yellow Canary, Act One, Hail Mafia, The Detective, The Split, Goodbye Columbus, Who Says I Can't Ride a Rainbow?, Two Minute Warning.
TV: The Defenders (Emmy Award for role in 'Blacklist' segment), The FBI, Ben Casey, 90 Bristol Court, etc.; The Odd Couple (Emmy, 1971 and 1973), Quincy.

KLYNN, HERBERT DAVID: Producer. b. Cleveland, O., Nov. 11, 1917; e. Ohio State U., 1937, Cleveland S. of Art Western Reserve U., 1939. Fine art painter, exhibitor, 1939; animation dir., U.S. Govt. m.p. unit, 1941–44; color dir., UPA, 1943–50; exec. prod. mgr., UPA, 1954–56; vice-pres., exec. prod., UPA, 1956–57; TV comm., industry films 1955–59; formed own co., Format Films, Oct. 1959; pres., Format Films, Inc., 1960–63; started Herb Klynn and Assoc., 1963; formed Format Productions Inc., 1964, pres., exec.

KNIGHT, ARTHUR: Critic, educator. b. Phila., Pa., Sept. 3, 1916. e. CCNY. BA 1940. Asst. curator Museum of Modern Art Film Library, 1939–49; film consultant CBS-color television, Omnibus, Odyssey, Seven Lively Arts, etc.; film courses at CCNY, New School for Social Research, Hunter College; contributor to Encyclopedia Britannica, Collier's Encyclopedia, etc.; author: "The Liveliest Art," The Hollywood Style; film critic, Westways; Director, Hollywood Reporter; formerly Saturday Review; prof. USC Cinema Division.

KNIGHT, DAVID: Actor. b. Niagara Falls, N.Y., Jan. 16, 1928. e. Whittier College, Calif.; 1944–52 teacher Putney School, Vermont; 1952–53 student Royal Academy of Dramatic Art.
LONDON STAGE: Trial of Mary Dugan, The Tenth Man, How to Succeed in Business Without Really Trying, Caine Mutiny Court Martial.
PICTURES INCLUDE: Young Lovers (Chance Meeting), Out of the Clouds, Lost, Eyewitness, Across the Bridge, V.I., The Story of David, Nightmare, Bedtime Story.
TV: Lincoln in Illinois, Berkeley Square, The Philadelphia Story, The Unquiet Spirit, A Question of Pride, Trial and Error.

KNIGHT, ESMOND: Actor, b. East Sheen, Surrey, Eng., May 4, 1906; m. Nora Swinburne, 1946. Stage debut 1925 in "Merchant of Venice," London; Birmingham Repertory; Paris, 1929 in "Maya"; N.Y. stage: The Emperor's Clothes.
LONDON STAGE: Age of Consent, Bell, Book and Candle, Caine Mutiny Court Martial, The Country Wife, The Russian, The Lady from the Sea, Becket, Two Star for Comfort, Last Old Vic Season, Getting Married, Martin Luther King, Spithead Mutiny, Mister, Three Sisters, Henry V, Agincourt—The Archer's Tale, Crime and Punishment, Family Reunion.
TV: Nicholas Nickleby, Our Mutual Friend, The Queen and the Welshman, The Silver Box, David and Broccoli, Barnaby Rudge, Knight Errant, Danger Man, The Age of Kings, A for Andromeda, Ghosts, Lysistrata, The Midnight Men, Murder in the Cathedral, To Bury Caesar, She's a Free Country, The Queen and the Rebel, Crime and Punishment, Jacobean Theatre, The Last Reunion, Gideon's Way, Dr. Finlay's Casebook, The Spies, Lord Ringo, The Rat Catchers, The Parachute, The Gamblers, More Best Sellers, The Troubleshooters, Z-Cars, The Casual Affair, Dr. Who, Faith, Queen Elizabeth I, Public Eye, Withered Arm, Shades of Green, Ballet Shoes, Quiller, I Claudius, Kilvert Diaries, Supernatural, Romeo & Juliet, Rebecca.
PICTURES INCLUDE: Dandy Dick, Waltzes from Vienna, Pagliacci, What Men Live By, Contraband, This England, Silver Fleet, Halfway House, Canterbury Tale, Henry V, Black Narcissus, Uncle Silas, Holiday Camp, Hamlet, Red Shoes, Wild Heart, The River, Helen of Troy, Richard III, Count of Monte Cristo, Prince and the Showgirl, Sailor of Fortune, O.S.S., Third Man on the Mountain, Sink the Bismarck, Peeping Tom, The Man Who Came In From The Cold, Anne of a Thousand Days, The Yellow Dog, Robin & Marian.

KNIGHT, SHIRLEY: Actress. b. Goessell, Kansas, 1937. e. University of Wichita, University of California. Joined the Pasadena Community Playhouse.
PICTURES INCLUDE: Five Gates to Hell (debut), Ice Palace, The Dark At the Top of the Stairs, (Acad. Award nom.) The Couch, Sweet Bird of Youth, House of Women, Flight from Ashiya, The Group, Dutchman, Petulia, The Counterfeit Killer, The Rain People, Juggernaut, Beyond the Poseidon Adventure, Endless Love.

KNIGHT, TED: Actor. b. Terryville, CT, Dec. 7, 1923. r.n., Tadeus Wladyslaw Konopka. e. Randall Sch. Dramatic Arts. U.S. Army, 1942–44. Started with various radio stations in New York City; emcee, newsman, host for late-night movies at TV stations. Hollywood film debut, 1957; Bdwy. stage debut, 1977. formed Kono Prods., 1976.
TELEVISION: Mary Tyler Moore Show, Ted Knight Show, Ted Knight Musical-Comedy Variety Special Special (CBS).

KNOECHEL, ROBERT F.: Executive. b. Cincinnati, Ohio. e. University of Cincinnati. ZIV Television—1953 budget, cost control-local auditor, 1958 controller of West Coast; Independent companies, 1962–64 location auditing; Ivan Tor Films—1965 controller, 1967 exec. and studio mgr. in Miami, Florida; Columbia Pictures—1969 assistant controller; ABC Pictures Corp.—1971 controller; Warner Bros.—1972 asst. studio controller; The Burbank Studios—1974 controller, 1979 treas. and member exec. comm. and EDP steering comm. Trustee SAG Pension & Welfare—Inception to 1964. Trustee & Chairman—Miami, Florida IA Pension Plan 1967–68.

KNOPF, EDWIN H.: Producer, writer. b. New York City, Nov. 11, 1899; e. MacKenzie School and Amherst. On the legitimate stage for eight years in New York and Germany. Entered picture business in 1928 with Paramount. Wrote Mr. Imperium, Free Love, etc.; collaborated on lyrics, Reckless; s.p. Picadilly Jim. Head of Scenario Department, MGM, 1936–39; prod. Samuel Goldwyn Pictures.
PICTURES INCLUDE: The Cross of Lorraine, Cry Havoc, Valley of Decision, Sailor Takes a Wife, Secret Heart, Cynthia, B.F.'s Daughter. Edward My Son, Malaya, Fearless Fagan, Scandal at Scourie, Lili, Great Diamond Robbery, Glass Slipper, King's Thief, Diane, Gaby, The Vintage, Tip on a Dead Jockey, Fast Company, Border Legion, Slightly Scarlet, Only Saps Work, Santa Fe Trail, Light of the Western Stars, Nice Women, Law and the Lady.

KNOTTS, DON: Actor. b. Morgantown, W.VA, July 21, 1924. e. West Virginia Univ., Univ. of Arizona. Drafted into US Army where became part of show called Stars and Grapes, teamed with comedian Mickey Shaughnessy. After schooling resumed, was offered teaching fellowship but went to New York to try acting instead. Appeared on radio and TV, leading to role in No Time for Sergeants on Bdwy.; appeared in film version.
PICTURES: It's a Mad, Mad, Mad, Mad World, The Incredible Mr. Limpet, The Shakiest Gun in the West, The Apple Dumpling Gang, Herbie Goes to Monte Carlo, The Apple Dumpling Gang Rides Again.
TELEVISION: Garry Moore Show, Steve Allen Show, Andy Griffith Show (played Barney Fife), The Don Knotts Show, Three's Company.

KNOWLES, PATRIC: Actor. r.n. Reginald Lawrence Knowles; b. Horsforth, Yorkshire, England, Nov. 11, 1911. Joined Abby Repertory Theatre, 1930; m.p. debut (Ireland) in Irish Hearts, 1934; on London stage in By Appointment; U.S. m.p. debut in Charge of the Light Brigade, 1936; served in Canadian RAF & as civilian instructor USAF, World War II.
PICTURES INCLUDE: Honours Easy, Mister Hobo, Two's Company, Give Me Your Heart, It's Love I'm After, Expensive Husbands, Adventures of Robin Hood, How Green Was My Valley, Forever and a Day, Of Human Bondage, Bride Wore Boots, Ivy, Kitty, Monsieur Beaucaire, Dream Girl, Big Steal, Quebec, Three Came Home, Mutiny, Tarzan's Savage Fury, Jamaica Run, Flame of Calcutta, World Ransom, Khyber Patrol, No Man's Woman, Band of Angels, Auntie Mame, The Way West, In Enemy Country, The Devil's Brigade, Chisum, The Man, Terror in the Wax Museum, Arnold.

KNOX, ALEXANDER: Actor. b. Strathroy, Ont., Jan. 16, 1907. e. Western Ontario U. Author, novels: Bride of Quietness, Night of the White Bear, The Enemy I Kill, Raider's Moon, The Kidnapped Surgeon; plays, Old Master, The Closing Door, Red On White. Many TV appearances.
STAGE: (NY) Romeo and Juliet, The Three Sisters, Jupiter Laughs, Jason, The Closing Door; (London) The Jealous God, Winter Journey, Henry VIII, Geneva, King of Nowhere, Return to Tyassi, Burnt Flower Bed, When Dead Awaken.
PICTURES INCLUDE: The Sea Wolf, This Above All, Commandos Strike at Dawn, None Shall Escape, Over 21, Wilson, Sign of the Ram, Judge Steps Out, Sister Kenny, I'd Climb the Highest Mountain, Saturday's Hero, Sleeping Tiger, Divided Heart, Crack in the Mirror, The Viking, The Night My Number Came Up, Reach for the Sky, High Tide at Noon, Davy, Operation Amsterdam, The Longest Day, Wreck of the Mary Deare, Man in the Middle, Woman of Straw, Mr. Moses, Accident, Villa Rides, Shalako, Skullduggery, Puppet on a Chain, Khartoum, Nicholas and Alexandra, Potsdam.

KNOX, GORDON: Executive. b. Greenville, Texas, May 10, 1909. e. Univ. of Missouri; married. In addition to making commercials for several years has been employed by Warner Bros., Caravel Films and Walter Wanger Prods. Joined Princeton Film Center in 1940; Pres. SKS Prod. Inc., Santa Fe, N.M. Pres., PAC Productions, Inc., Princeton, N.J.

KOBAYASHI, MASAKI: Director. b. Japan, Feb. 4, 1916. e. Waseda Univ. Joined Shochiku 194[illegible] as asst. Director. Army. Rejoined Shochiku 1946.
PICTURES INCLUDE: No Greater Love, Road to Eternity, A Soldier's Prayer, Black River, Room with Thick Walls, Somewhere Beneath Wide Sky, Fountainhead, I'll Buy You, The Human Condition, The Inheritance.

KOCH, HOWARD: Writer. b. New York City, Dec. 12, 1902. e. St. Stephen's Coll., '22. B.A.; Columbia Law School, '25. LL.B. Hon. degree Doctor of Human Letters, Bard Coll., 1972. Playwright (Give Us This Day, In Time To Come, Straitjacket). Began screen career collab. s.p. The Sea Hawk. Radio: Invasion from Mars radio play for (Orson Welles broadcast); book: The Panic Broadcast (Little, Brown & Co.); Academy Award best s.p. (Casablanca). As Time Goes By, Memoirs of a Writer in Hollywood, New York and Europe under contract to Harcourt, Brace and Jovanovich for publication in the summer of 1979.
PICTURES INCLUDE: The Letter, Shining Victory, In This Our Life, Casablanca Mission to Moscow, Letter From an Unknown Woman, The Thirteenth Letter, The War Lover, The Fox, Loss of Innocence, No Sad Songs for Me, Sergeant York, Three Strangers.

KOCH, HOWARD W.: Prod., dir. b. New York, N.Y., Apr. 11, 1916. Runner on Wal St., asst. cutter, 20th-Fox; asst. dir., 20th-Fox, Eagle Lion, MGM; 2nd unit dir., freelance; prod. Aubrey Schenck Prod.; 1961–64, exec. prod. Frank Sinatra Enterprises; v.p., chg. prod., Paramount Pictures Corp., 1965. Elected pres. of the Academy of Motion Picture Arts and Sciences, June 2, 1977. On June 11, 1977, elected to the National Board of Directors Guild of America for two year term.
TV: Director, Miami Undercover, The Untouchables, Maverick, Cheyenne, Hawaiian Eye.
PICTURES INCLUDE: War Paint, Beachhead, Yellow Tomahawk, Shield for Murder Big House USA, Crime Against Joe, Emergency Hospital, Ghost Town, Broken Star, Rebel in Town, Pharaoh's Curse, Three Bad Sisters, Fort Yuma, Desert Sands, Quincannon Frontier Scout, Black Sleep, Hot Cars, "X-1," Sergeants 3, The Manchurian Candidate, Come Blow Your Horn, Four For Texas, Robin and the Seven Hoods, None But the Brave, The Odd Couple, On A Clear Day You Can See Forever, Plaza Suite, Star Spangled Girl, (Prod.), Last of the Red Hot Lovers (Prod.), Badge 373 (Prod.-Dir.) 1973, (Prod.) Once Is Not Enough; MATI (prod.), 1977; Airplane! (prod.), 1980; Dragonslayer (exec. prod.), 1981; Some Kind of Hero (prod.), 1982.

KOHLER, FRED, JR.: Actor.
PICTURES INCLUDE: The Prisoner of Shark Island, Sins of Man, Pigskin Parade, Hold That Coed, Young Mr. Lincoln, Big Bonanza, Why Girls Leave Home, Feudin', Fussin' and A-Fightin', Spoilers of the Plains, Baron of Arizona, Hollywood Thrill-Makers, Racing Blood.

KOHN, HOWARD E., II: Executive. b. McKeesport, Pa. National dir. of adv., publicity, roadshow dept., United Artists; indep. prod., Hidden Fear, 1957; pres. Lioni-Warren-Kohn, Inc., 1958; national roadshow dir., Columbia Pictures, Porgy & Bess, 1959; World wide co-ordinator, national co-ordinator adv. & pub. for El Cid, June 1961; named world wide co-ordinator adv., pub. all Samuel Bronston Productions, April 1962; pres., Starpower Inc., 1968; exec. v.p., Avanti Films, 1970; v.p. Avariac Prods., 1971; pres., Blossom Films, 1973. Elected member of ASCAP, 1975. Pres., Avanti Associates, 1976.

KOHNER, FREDERICK: Writer. b. Teplitz-Schoenau, Czechoslovakia, Sept. 25, 1905; p. Julius and Helen Kohner; e. Ph.D., Sorbonne Univ., Paris, Univ. of Vienna, Austria; married. Was newspaper reporter and feature writer on Berliner Tageblatt and Prager-Czechoslovaki Tagblatt. wrote book, "Film is Art," history of film development.
NOVELS: Seven Rooms in Hollywood, Kiki of Montparnasse, The Skier, Gidget, Cher Papa, Gidget Goes Hawaiian, The Contintal Kick, Mister, Will You Marry Me, Affairs of Gidget, Gidget Goes to Rome, Gidget in Love, The Gremmie, Gidget Goes Parisienne, 1967, Gidget Goes New York. Biography: The Magician of Sunset Boulevard (story of Paul Kohner), Amanda, a love story.
TV: Revue, Four Star, Fireside, Schlitz Playhouse.
STAGE: (collab.) Bees and the Flowers, Stalin, Allee.
PICTURES INCLUDE: Burning Secret, La Crise Est Finie, Sins of Men, Mad About Music, It's a Date, Men in Her Life, Tahiti Honey, Patrick the Great, Lake Placid Serenade, Pan Americana, Three Daring Daughters, Bride for Sale, Nancy Goes to Rio, Hollywood Story, Never Wave at a Wac, Toy Tiger.

KOHNER, SUSAN: Actress, b. L.A., Calif., Nov. 11, 1936. e. Univ. of Calif., 1954–55.
TV: Alcoa Hour, Schlitz Playhouse, Four Star Theatre,

Matinee Theatre, Climax, Suspicion, Playhouse 90, Route 66, Dick Powell Theatre.
STAGE: Love Me Little, He Who Gets Slapped, A Quiet Place, Rose Tatoo, Bus Stop, St. Joan, Sunday in New York, Take Her She's Mine, Pullman Car, Hiawatha.
PICTURES INCLUDE: To Hell and Back, The Last Wagon, Dino, Imitation of Life, The Big Fisherman, The Gene Krupa Story, All the Fine Young Cannibals, By Love Possessed, Freud.

KONTOS, SPERO L.: Thea. Equip. Dealer. b. Chicago, Ill., Dec. 17, 1922. e. Illinois Inst. of Technology, B.S. BM., eng.; ECLA Ent. Ind. as Chicago exhib., 1937; thea. mgmt. until 1942. U.S. Army, 1942–46; sis. eng., Abbott Theatre Equip. Co., 1946–48; sis. mgr., 1948–51; gen. mgr. 1951–54; sis. mgr. John P. Filbert Co. Inc., 1954–59; hd. Century 70/35 mm. projector program, 1959–60; Filbert Co., v.p. chg. sis. & eng., 1960–64; pres., 1964; pres. Theatre Equipment Dealers Assoc., 1965–67; bd. chm., TEDA, 1967–69; chief barker, L.A. Variety, 1971–72; pres., Megaron Corp., 1973, Academy of Motion Picture Arts & Sciences, 1975.

KOPELMAN, JEAN R.: Producer. b. New York, N.Y., Apr. 5, 1927; e. Middlebury (Vt.) Coll. clerical work, CBS radio & TV, 1946–47; first girl prog. asst. in prog. dept., CBS-TV, 1948; prod. mgr., Goodson-Todman Prod.
TV: Winner Take All, It's News to Me, What's My Line, Name's the Same, I've Got a Secret, Two for the Money, Beat the Clock, Choose Up Sides, Number Please, The Match Game, (NBC-TV).

KORBAN, BERNARD: Executive. b. New York City, Nov. 28, 1923; e. RCA Inst. of Tech., NYU. U.S. Army 1942–46; 1951–58, Public Relations and Promotions for Davega Stores; 1959–62, exploitation fieldman for Universal Pictures; 1962–66, supvr. of fieldmen and exploitation activities; 1966–68, exec. in chrg. field activities; exec. assist. to v.p., adv. pub. and promotion; 1972 dir. of exploitation, National General Pictures; 1973, dir. of Marketing, promotion and worldwide distribution for Brut Prods; 1974, dir. of mkt., Avco Embassy Pictures, 1975, vice pres. of advertising/publicity/promotion, Cine Artists Pictures Corp.; advertising/publicity, United Artists Corp. West Coast ad. mgr.; now UA v.p., West Coast adv.

KORMAN, HARVEY: Actor-Director. b. Chicago, Feb. 15, 1927. e. Wright Junior College. Began dramatic studies at Chicago's Goodman School of Drama at the Arts Institute. Acted in small roles in Broadway plays and did TV commercials until break came as comedian for Danny Kaye Show on TV. Staged comedy sketches for Steve Allen variety series in 1967. Became Carol Burnett's leading man on her show in 1967 and continued as regular to present. Has directed two episodes of The New Dick Van Dyke Show.
PICTURES INCLUDE: Actor: Three Bites of an Apple, Lord Love a Duck, Last of the Secret Agents, The April Fools, Blazing Saddles, Huckleberry Finn, High Anxiety, Americathon, First Family, History of the World—Part I.
TV: The Danny Kaye Show, Carol Burnett Show.

KORTY, JOHN: Director, Producer, Writer, Cameraman, Animator. b. 1936; B.A. Antioch College 1959; President, Korty Films, Inc., Mill Valley, CA.
THEATRICAL FEATURES: Alex & the Gypsy (Fox 1976); Oliver's Story (Paramount 1978). TELEVISION FEATURES: The Autobiography of Miss Jane Pittman (CBS 1974), Emmy, DGA Award; Farewell to Manzanar (NBC 1976), Humanitas, Christopher Awards; Forever (CBS 1977); A Christmas Without Snow (CBS 1980). INDEPENDENT FEATURES: Crazy Quilt (1964); Funnyman (1967); Riverrun (1969.) DOCUMENTARIES: Who Are the DeBolts? (ABC 1978), Oscar, Emmy, DGA Award; Cant It Be Anyone Else? (ABC 1980); Stepping Out: the DeBolts Grow Up, (HBO 1980). SHORT FILMS: The Language of Faces (AFSC 1961); Imogen Cunningham, Photographer (AFI grant 1970); The Music School (PBS 1975). ANIMATION: Breaking the Habit (1964), Oscar Nominee; Various children's films; Segments for Sesame Street and The Electric Company.

KORVIN, CHARLES: Actor. r.n. Geza Korvin Karpathi: b. Czechoslovakia, Nov. 21; e. U. of Paris, Sorbonne, 1933–36. Cameraman, dir., documentary films in Europe; to U.S.A. in 1937; Barter Theatre, 1940–42; on B'way, Dark Eyes, 1943; m.p. debut in Enter Arsene Lupin, 1945.
PICTURES INCLUDE: This Love of Ours, Temptation, Berlin Express, Killer That Stalked N.Y., Lydia Bailey, Tarzan's Savage Fury, Sangaree, Ship of Fools, The Man Who Had Power Over Women.

KOSCINA, SYLVA: Actress. b. Yugoslavia, Aug. 22, 1933. Grew up in Italy; as model placed under contract for films by Carlo Ponti.
PICTURES INCLUDE: The Railroad Man (debut), Juliet of the Spirits, Deadlier Than the Male, The Hornet's Nest, Casanova & Co., Sunday Lovers.

KOSINER, HARRY: Executive. b. New York City, Jan. 29. e. City Coll. N.Y. Stenographer, Fox Films, 1923; asst. to pres., Deluxe Lab., N.Y. 1930–36; eastern rep. Walter Wanger Prod., 1937–40; assistant to western div. mgr. United Artists, 1941–42; on staff m.p. bur., OWI, 1943; apptd. same yr. east. rep. Edward Small Prods., hdqts. N.Y.; home office staff, Columbia Pictures, 1951; v.p. National Screen Service, 1965.

KOSLECK, MARTIN: Actor. b. Barkotse. Pommern Mar. 24, 1907. Six years with Max Reinhardt's dramatic school and theatre in Berlin.
PICTURES INCLUDE: Espionage Agent, Calling Philo Vance, Nurse Edith Cavell, Bomber's Moon, North Star, Secrets of Scotland Yard, Hitler Gang, Gangs of the Waterfront, The Spider, The Wife of Monte Cristo, Pursuit to Algiers, The Frozen Ghost, The Mummy's Curse, Crime of the Century, Assigned to Danger, Smuggler's Cove, Something Wild, 36 Hours, Morituri, The Flesh Eaters.

KOSTER, HENRY: Director. b. Berlin, May 1, 1905. Comm. artist, cartoonist, newspaperman, critic; scenarist, Ufa, Terra, Aafa, Berlin; dir. Berlin, Vienna, Budapest, to Hollywood, 1936.
PICTURES INCLUDE: 100 Men and a Girl, Rage of Paris, It Started With Eve, Spring Parade, Music for Millions, Two Sisters from Boston, The Bishop's Wife, Come to the Stable, Inspector General, My Blue Heaven, Harvey, No Highway in the Sky, Mr. Belvedere Rings the Bell, Elopement, seq. O. Henry's Full House, Stars & Stripes Forever, My Cousin Rachel, The Robe, Desiree, A Man Called Peter, Good Morning Miss Dove, Power and the Prize, My Man Godfrey, Flower Drum Song, Mr. Hobbs Takes a Vacation, Take Her, She's Mine, Dear Brigitte, The Singing Nun.

KOTCHEFF, WILLIAM THEODORE: Director. b. Toronto, Canada, 1931. Ent. TV ind. 1952. After five years with Canadian Broadcasting Crop. joined ABC-TV in London, 1957.
PLAYS INCLUDE: Of Mice and Men, Desperate Hours, The Human Voice, Edna the Inebriate Woman, Signalman's Apprentice, Lights Out, Rx for the Defence.
STAGE: Progress The Park, Play With A Tiger, Luv, Maggie May, The Au Pair Man, Have You Any Dirty Washing, Mother Dear?
PICTURES INCLUDE: Life At The Top, Two Gentlemen Sharing, Outback, Billy Two Hats, The Apprenticeship of Duddy Kravitz, Fun with Dick and Jane, Who Is Killing The Great Chefs of Europe?, North Dallas Forty (also co-s.p.), Captured.

KOTTO, YAPHET: Actor. Has many stage credits, including starring roles on Broadway in The Great White Hope and The Zulu and the Zayda. Off-Bdwy.: Blood Knot, Black Monday, In White America, A Good Place To Raise a Boy.
PICTURES INCLUDE: The Limit (star, prod.), Nothing But a Man, The Liberation of L.B. Jones, Live and Let Die, Across 110th Street, Truck Turner, Bone, Report to the Commissioner, Sharks' Treasure, Hey Good Lookin', Friday Foster, Drum, Blue Collar, Brubaker.

KOVAKS, LASZLO: Cinematographer.
PICTURES INCLUDE: Hell's Angels On Wheels, The Savage Seven, Targets, Easy Rider, That Cold Day In the Park, Getting Straight, Alex In Wonderland, The Last Movie, Marriage of a Young Stockbroker, The King of Marvin Gardens, Pocket Money, What's Up, Doc?, Paper Moon, Huckleberry Finn, For Pete's Sake, Freebie and the Bean, Shampoo, Baby Blue Marie, Harry and Walter Go to New York, New York, New York, F.I.S.T., The Last Waltz, Paradise Alley, Butch and Sundance: The Early Days, The Runner Stumbles, Heart Beat.

KRAMER, JEROME: Executive-Producer-Director. b. Los Angeles, 1945. Graduated USC Law School 1971. Joined Braverman Productions, Inc. 1971; named exec. v.p. Co-producer with Charles Braverman of network specials (21 Years of A.I.P., Horror Hall of Fame), show titles (Rhoda, Cher) corporate and promotional films for Xerox, Petersen Publishing, etc. Producer and director of various television commercials (for Chevrolet, United Artists Records, Gulf Oil), educational films for United States Information Agency, special films and montages.

KRAMER, LARRY: Writer, producer. b. Bridgeport, Conn. U.S.A., 1935. e. Yale Univ. B.A. 1957. Ent. m.p. ind. 1958. Story edit. Columbia Pictures, N.Y. London 1960–65. Asst. to David Picker and Herb Jaffe, UA, 1965. Assoc. prod. and additional dialogue Here We Go Round The Mulberry Bush, 1968. Writ. prod. Women in Love, 1969. Writer: Lost Horizon, 1971; A Sea Change, 1972.

KRAMER, SIDNEY: Sales executive. b. New York City. e. LL.B., New York Law School and C.C.N.Y., married. Gen. sales mgr., RKO Pathe, June 1953; dir. and v.p. Cellofilm Corp. 1941–56; foreign sales mgr., RKO Radio, 1954–59; v.p. Cinemiracle Intl. 1960–61; v.p. T.P.E.A., 1960–61; foreign sls. mgr., Cinerama, Inc., 1962–65; Exec. Commonwealth Theatres, Puerto Rico, Inc., 1965–68; Exec. v.p. Cobian Jr. Enterprises Inc. 1968. M.P. consultant-exhibition, dist., foreign and Caribbean area, Oct., 1968 to 1970. Pres. Coqui Internat'l. Inc.; 1970–1980; vice. pres. of UAPR, Inc., Puerto Rico, U.A. Eastern Theatres, Inc.

KRAMER, STANLEY E.: Executive producer, director. b. New York City, Sept. 29, 1913; e. N.Y.U., B.Sc. 1933. Entered m.p. ind. via back lot jobs; with MGM research dept.; film cutter 3 yrs.; film ed.; m.p. & radio writer; served in U.S. Signal Corps, 1st Lt.
PICTURES INCLUDE: Champion, Home of the Brave, The Men, Cyrano de Bergerac, Death of a Salesman, High Noon, My Six Convicts, The Sniper, The Four Poster, The Happy Time, Eight Iron Men, 5,000 Fingers of Dr. T, Wild One, The Juggler, Caine Mutiny, Not As a Stranger, Pride and the Passion; prod. dir., s.p., The Defiant Ones, On the Beach, Inherit the Wind, Judgment at Nuremberg; prod., dir., It's a Mad, Mad, Mad, Mad World, Invitation To A Gunfighter, prod., Ship of Fools, Prod. Dir. Guess Who's Coming to Dinner, The Secret of Santa Vittoria, R.P.N.*, Bless the Beasts and Children, Oklahoma Crude, The Domino Principle, The Runner Stumbles.

KRAMS, ARTHUR: Set decorator b. New York, N.Y., July 15, 1912. e. Parsons School of Design, New York and Paris, 1928–31. Interior designer, decorator, Chicago, 1931–34; New York, 1934–39; Los Angeles, 1939–45; set decorator, MGM, 1945–54; set decorator, Paramount, 1954.
PICTURES INCLUDE: Luxury Liner, Merry Widow, Lili, Story of Three Loves, Student Prince, To Catch A Thief, Rose Tattoo, Court Jester, Arists and Models, Man Who Knew Too Much, The Rainmaker, Wild Is the Wind, The Mating Game, Bye, Bye Birdie, The Carpetbaggers, Where Love Has Gone, Sylvia, Marriage On The Rocks, The Swinger, Good Times, Caper of The Golden Bulls, Barefoot in The Park, The President's Analyst, Hang 'Em High, How Sweet It Is.

KRANTZ, STEVE: Exec. b. New York, N.Y., May 20, 1923. e. Columbia U., B.A. Dir. progs., NBC, New York, 1953; dir. prog. dev., Screen Gems, N.Y., 1955; v.p., gen. mgr. Screen Gems, Canada, 1958; dir. int. sls., 1960; formed Krantz Films, Inc. 1964.
TV: Steve Allen Show, Kate Smith Show, Hazel, Dennis the Menace, Winston Churchill—The Valiant Years, Telefilms, Marvel Super Heroes, Rocket Robin Hood, Animated Films.
PICTURES INCLUDE: Fritz the Cat, Heavy Traffic (prod.).

KRASNA, NORMAN: Producer, writer. b. Corona, L.I. (N.Y.C.), Nov. 7, 1909; e. New York U. Columbia U., Brooklyn (N.Y.) Law School Asst. drama ed , N.Y. Morning World; drama ed., N.T. Graphic; on N.Y. staff Exhibitors Herald-World; asst. publicity dir., Warner Bros. Studios; from 1932 variously author & collab. orig. & s.p. many pictures.
STAGE PLAYS: Louder, Please, Small Miracle, Dear Ruth, John Loves Mary, Kind Sir, Who Was That Lady I Saw You With, Sunday in New York.
PICTURES INCLUDE: Fury, Richest Girl in the World, Hands Across the Table, Bachelor Mother, The Devil and Miss Jones, It Started With Eve, Academy Award, best orig. s.p. 1943 (Princess O'Rourke, Warner), Practically Yours, The Big Hangover, White Christmas, The Ambassador's Daughter, Indiscreet, Who Was That Lady?, My Geisha, Sunday in New York, I'd Rather Be Rich.
Screen Writers Guild Laurel Award, 1959, Joined U.S. Air Force; formed Wald-Krasna Prod., distrib. through RKO 1950; sold interest to Wald, May, 1952. Formed Monovale Prod. 1956.

KRAUSS, OSCAR: Executive. b Brooklyn, N.Y. e. City College of N.Y., Architecture; Pratt Institute, Advertising Design, Artist, Paramount Pictures, 1936; 20th Century-Fox, 1936–40; Buchanan & Co., 1940; art dir., Lord & Thomas, 1941; U.S. Coast Guard, 1942–46; art dir., Republic Pictures, 1946–48; Monroe-Greenthal, Co., 1948–58; Oscar Krauss Advertising, Inc., pres., owner.

KREIMAN, ROBERT T.: Executive. b. Kenosha, Wis., Sept. 16, 1924. Served W. W. II Capt Army Corps of Engineers-ETO. e. Stanford, 1943; Univ. of Wis., 1942 — 1946–49. Dir., sales training, mgr., audio visual sales, Bell & Howell Co., 1949–58; V.P., Argus Cameras, Inc., 1958–61; V.P., gen. mgr., Commercial & Educ. Div., Technicolor 1961–69, v.p. gen. mgr., The Suburban Companies; 1969–71: pres. and chief exec. officer, Deluxe General, Inc. Pres. and Director of Movietonews, Inc. Bd. chm. Keith Cole Photograph, Inc. 1972–1978. bd. chm., pres. and chief exec. officer, Pace International Corp., 1969 to present. past pres. of UCLA Executive Program Ass'n. Fellow SMPTE, Member M.P. Academy; TV Academy; assoc. mem., American Society of Cinematographers.

KRESS, HAROLD F.: Director. film editor. b. Pittsburgh, Pa., June 26, 1913. e. U. of Calif., L.A. Film ed., Command Decision, Madame Curie, Mrs. Miniver, The Yearling; crime shorts; 5-reel Army documentary short, Ward Care for Psychotic Patients. Member: Acad. of M.P. Arts and Sciences, Screen Directors Guild, Film Editors Guild.
PICTURES INCLUDE: Painted Hills, No Questions Asked, Apache War Smoke, Ride Vaquero, Saadia, Rose Marie, Valley of the Kings, The Cobweb, The Prodigal, I'll Cry Tomorrow, Teahouse of the August Moon, Silk Stockings, Until They Sail, Merry Andrew, Imitation General, The World, The Flesh and The Devil, Count Your Blessings, Home from the Hills, The Greatest Story Ever Told, Walk Don't Run, Alvarez Kelly, Academy Award for Film Editing on How the West Was Won.
FILM EDITOR: Poseidon Adventure, The Iceman Cometh, 99-44/100ths % Dead, The Towering Inferno.

KREUGER, KURT: Actor. b. St. Moritz Switzerland, July 23, 1917; e. U. Lausanne, U. London. Came to U.S. 1937, partner in travel bureau: acted in Wharf Theat. group. Cape Cod, 1939; Broadway debut in Candle in the Wind with Helen Hayes, 1941.
PICTURES INCLUDE: Mademoiselle Fifi, Hotel Berlin, Paris Underground, Dark Corner, Unfaithfully Yours, Fear, The St. Valentine's Day Massacre, What Did You Do in the War Daddy?

KRIER, JOHN N.: Executive. b. Rock Island, Ill. e. Augustana Coll. Joined A. H. Blank Theatres, 1924: Grad. Publix Theatres Manager Training School, 1930: managed theatres in Illinois, Iowa, Nebraska; joined Intermountain Theatres, Salt Lake City, 1937: appointed Purchasing Head, 1946: buyer-booker, 1952: v.p., gen. mgr., 1955: appt. v.p. gen. mgr. ABC Theas., Arizona, 1968: appt. v.p. director Film Buying ABC Theatres of California & ABC Intermountain Theatres, Feb. 1972. Became consultant ABC Southern Theatres, 1974. Joined Exhibitors Relations Inc. as partner, 1978.

KRIM, ARTHUR B.: Attorney. b. New York City, 1910. e. Columbia U., B.A., '32, 1932 became member law firm Phillips, Nizer, Benjamin & Krim, N.Y. elected pres. United Artists Feb. 20, 1951; chairman of bd., 1969 to January 1978; Ch. of Board, Orion Pictures Company, March, 1978.

KRISTOFFERSON, KRIS: Actor-singer. b. Brownsville, Texas, 1937. e. Pomona College, Oxford Univ. (Rhodes Scholar). Joined U.S. Army briefly and taught English Literature at West Point. Started writing songs (country music) and hits have included Me and Bobby McGee, Why Me, Lord, Sunday Mornin' Comin' Down, etc. Film debut in Cisco Pike, 1971. Continues to make records, do concert tours and appear in films.
PICTURES: Cisco Pike, Pat Garrett and Billy the Kid, Blume in Love, Bring Me the Head of Alfredo Garcia, Alice Doesn't Live Here Anymore, Vigilante Force, The Sailor Who Fell from Grace with the Sea, A Star Is Born, Semi-Tough, Convoy, Heaven's Gate. Roll-Over.

KRITZMAN, SERGE: production designer, architect. b. Yugoslavia, July 8, 1914. e. Royal U. of Zagreb, Royal Academy of Arts. Wrote, dir. and acted in featurette Dr. Faust, 1933; designed interior Royal Yugoslav Pavilion, N.Y. World's Fair, 1939. Public Relations officer, Yugoslav Embassy, 1940–42; tech. div. 20th Century Fox Studios, 1943; organized Century Theatres; art div., mgr. facilities KFJ-TV, 1953; member bd. of gov. Acad. TV Arts and Sciences, 1964–67; chairman P.R. committee Society of Motion Picture Art Directors; pres. Hollywood chapter TV Acad., 1967–68.
PICTURES INCLUDE: The Glass Wall, Ride the High Iron, Life of Christ, Crime in the Streets, Flight to Hong Kong, The Fuzzy Pink Nightgown, Time Limit, Porgy and Bess, The Cabinet of Dr. Caligari, Promise Her Anything, Miss Lonelyhearts, Gunfight at Dodge City, The Big Bounce.
TV: Schlitz Playhouse of Stars series, James Mason series, The Fugitive, Batman, Dick Tracy, City Beneath the Sea, Frank Sinatra, Roman and the Renegade.

KRONICK, WILLIAM: Writer-director. b. Amsterdam, N.Y. e. Columbia College, A.B.; U.S. Navy photography; wrote, dir. featurette, A Bowl of Cherries, 1961.
TV DOCS: Wrote, dir., prod. Mysteries of the Great Pyramid; George Plimpton Specials; National Geographic, et. al.
FEATURE FILMS: Horowitz in Dublin (dir., s.p.); King Kong (2nd unit dir.); The 500 Pound Jerk (dir., TV movie).

KRONSBERG, JEREMY: Writer/Actor/Associate Producer. b. Sept. 15, Denver. Former contract player with 20th Century Fox; former dialogue director for The Perry Mason Show.
SCREENPLAY: Every Which Way But Loose, Rule Golden.

KRUEGER, CARL: Producer, writer. b. Chicago, Ill., Oct. 8, 1908. U.S. Air Force, 1925–31 press agent for Olsen and Johnson, Max Reinhardt, others; exploit, dir., United Artists Corp.; in charge of spec. events. Paramount; writer. 20th Century-Fox, then freelance; served in U.S. Air Force, World War II.
PICTURES INCLUDE: Thunderbirds prod. Thunderbolt, Golden Gloves Story, Comanche, Sabre Jet.

KRUGER, HARDY: Actor. b. Berlin, Germany, April 12, 1928; ent. m.p. ind. 1943; on stage since 1945.
PICTURES INCLUDE: The One That Got Away, Bachelor of Hearts (Rank). German version of The Moon is Blue (U.A.). 1959: The Rest is Silence (German film of Hamlet), Blind Date (Britain). He has also starred in twenty-four

German films. 1960: Filming in France and Germany. 1961–62: Films include: Taxi Pour Tobrouk (France), Hatari (Paramount, Hollywood). 1962: Les Dimanches de Ville d'Avray (France), etc. 1963: Le Gros Coup (France). 1964: Les Pianos Mecaniques (France), Le Chant Du Monde (France). 1965: Flight of the Phoenix (Hollywood) 1966: The Defector (U.S.A.), La Grande Sauterelle (France). 1967: Le Franciscain de Bourge (France). 1968–9: The Battle of the Neretva, The Red Tent, The Nun of Monza, The Secret of Santa Vittoria. 1971: Night Hair Child. 1972: Death Of A Stranger (Israel). 1973–74: Le Solitaire (France), Barry Lyndon (England). 1974–75: Paper Tiger (England/Germany), Potato Fritz (Germany). 1976: A Bridge Too Far (England/U.S.A.), L'Autopsie d'un Monstre (France). 1977: The Wild Geese (England).

KRUGER, JEFFREY S.: Concert impresario., Prod. Film Dist., Music Publ. and gramophone records exec. b. London, Eng., Apr. 19, 1931. Produced feature films Rock You Sinners; Sweet Beat; The Amorous Sex. Distributor of over 34 feature films including I Want To Be A Woman, Dial Rat For Terror, Equinox, The Shooting, Smashing The Crime Syndicate, A Whale of A Tale, The Good Time Outlaws. Exec. Prod. and worldwide dist. musical specials (Prod. in assoc. BBC). TV productions include one-hour musical specials of Anne Murray, Glen Campbell, Jacksons, Marvin Gaye, David Soul, Jerry Lee Lewis and series The Other Broadway. Concert presentations incl. ownership Flamingo Club and The Temple Club and via the Kruger Organisation, Sarah Vaughan, Billy Eckstine, Chet Baker, Garland Wilson, Thelma Carpenter, Billie Holiday, Glen Campbell, Diana Trask, Charlie Rich, Susan Maughan, Desmond Dekker, Roy Clark, Anne Murray, Jack Benny, Charley Pride, P.J. Proby, Connie Smith, Faron Young, Gladys Knight and the Pips, Helen Reddy, Wayne Newton, Dionne Warwick, Frank Gorshin, Trini Lopez, Buddy Greco, Stylistics, Supremes, George Burns, Marty Allen, Tony Bennett, Lena Horne, Barry White, Love Unlimited, Marvin Gaye, Chi-Lites, Bill Anderson, Sonny Terry, Brownie McGhee, Ray Stevens, Loretta Lynn, Jack Jones, Edwin Starr, David Soul, Jacksons, War, Blood, Sweat and Tears, Frankie Valli and the Four Seasons. Music publ. through Songs For Today. Own record production and distribution via Bulldog Records and Visual and Audio Leisure Co. Ltd. Director The Kruger Organisation (Concerts) Ltd. Songs For Today Ltd. Ember Enterprises Inc. Visual and Audio Leisure Co. Ltd. Hillbrow Productions Ltd.

KUBRICK, STANLEY: Producer, director, writer. b. New York, N.Y., July 26, 1928; e. Taft H.S. Staff photog., Look magazine; writer, prod., dir., documentaries including Day of the Fight, Flying Padre; prod., dir., s.p., Fear and Desire, Killer's Kiss; dir., s.p., The Killing; writer-dir., Paths of Glory; dir. Spartacus; prod. dir., Lolita; prod. dir., writer, Dr. Strangelove; prod. dir., writer, 2001; A Space Odyssey; A Clockwork Orange; Barry Lyndon; The Shining.

KUERTZ, SR., CHARLES H.: Executive. b. Cincinnati, O., Nov. 1, 1923. e. Roger Bacon Catholic High School. Started as usher in theatre industry in 1939 with M. Marcus Enterprises. Ran theatres in Ft. Wayne and Cincinnati. From 1950 to 1957 co-owned and operated three drive-in theatres in Nashville. Built other theatres around that area for various cos. From 1957 to 1961 was with Holiday Amusement Co., Cincinnati, as gen. mgr. Holiday sold to Martin Theatres which he joined in 1961 starting as dist. mgr. in Nashville, running 44 theatres. Became exec. v.p. for Martin in July, 1972; elected pres., 1979.

KUHN, THOMAS G.: Executive. b. Chicago, Ill., Nov. 10, 1935. e. Northwestern Univ., 1957; Univ. of So. Calif. (MBA), 1966. Singer on Roulette Records, 1958–59. KNBC-TV sales 1960–62; NBC business affairs, 1962–64; NBC mgr. live night time progs., 1965–67; dir. live night time progs., 1968–69. Warner Bros. TV, v.p. program dev. 1970; v.p. TV prod., 1971; exec. prod., Alice (WB-CBS), 1976; exec. prod., The Awakening Land (WB-NBC), 1977; exec. v.p. for Alan Landsburg Prods., 1977; exec. prod., Torn Between Two Lovers, 1978; exec. prod., The Jayne Mansfield Story, 1980. Staff, v.p., west coast, for RCA Selectavision Video Discs, 1980.

KULIK, SEYMOUR (BUZZ): Producer-dir. b. N.Y.C., TV: Lux Video, Kraft; joined CBS-TV as prod.-dir., 1956; dir., You Are There, Climax, Playhouse 90, Defenders, Dr. Kildare, Twilight Zone, Dick Powell Playhouse; exec. prod., Kentucky Jones. 1964: v.p. chg. West Coast Prods., Bob Banner Associates Inc., 1965; 1967 Prod-Dir. with Paramount Studios, Brian's Song (TV); Shamus.
PICTURES: Warning Shot, The Riot, Villa Rides, The Hunter, Pursuit.

KUNO, MOTOJI: Sr. Managing Director. Tokyo Shibaure Electric Co., Ltd. (Toshiba), Tokyo, Japan. Graduated Law Dept., Tokyo Imperial Univ. Mar., 1923. Became Toshiba auditor June, 1941; dir. June, 1942; sr. managing dir. Nov., 1948, perm. auditor Apr. 1949; dir. Fem., 1950; exec. dir. May, 1952; sr. managing dir. May, 1958.

KUPPER, W. J.: Executive. b. New York, N.Y., Oct. 17, 1896. World Film Corp. 2 yrs.; h.o. 20th-Fox, Aug., 1919; asst. mgr. Dallas, 1920; br. mgr., Washington, Oct. 1921; later, br. mgr., Albany, Charlotte, Pittsburgh, Chicago; asst. gen. sales mgr., h.o., Jan., 1924; spcl. rep., 1928; west div. mgr., Aug., 1932; gen. sales mgr., U.S., Canada, 1942; mag. dir., 20th-Fox Films of Gt. Brit., 1947–54; v.p., gen. sales mgr. Chromart Colour Org., England; 1954. Member: M.P. Pioneers; Variety Club of Gt. Brit.

KURALT, CHARLES: TV News Correspondent. b. Wilmington, NC. e. Univ. of North Carolina. Reporter-columnist for Charlotte News until joining CBS News as writer in 1956. Promoted to news assignment desk in 1958. Became first host of CBS News series, Eyewitness, in 1960. Named CBS News chief Latin American correspondent (based in Rio de Janeiro) in 1961 Appt. CBS News chief west coast correspondent in 1963; transferred to New York, 1964. Has worked on CBS Reports, CBS News Specials, and On the Road series for CBS Evening News with Walter Cronkite (latter 11 years). Now anchor man for CBS News Sunday Morning.

KURI, EMILE: Set decorator. b. Mexico City, Mex., June 11, 1907; e. Chaminade Coll., 1924–27. Interior decorator Be Hennesey Art Studio, 1929–32; then set decorator property dept. dir.
PICTURES INCLUDE: I'll Be Seeing You, Silver Queen, Spellbound, Duel in the Sun, Paradine Case, The Heiress (Academy Award, 1949), A Place in the Sun, Carrie, Shane, The Actress, Executive Suite, 20,000 Leagues Under the Sea (Academy Award, 1954); in charge of interior exterior decorations Disneyland. Several Golden Chair Awards (L.A. Furniture Mart). Mem.: Nat'l Acad. of TV Arts and Sciences, American Institute of Interior Designers, Bd. of Gov. Acad. of M.P. Arts and Sciences (1959–69). Honorary Sir Knights of Royal Rosarians, State of Oregon (1970), Decorating Consultant, Disney World, Fla.

KURODA, TOYOJI: Executive. b. Tokyo. April 29, 1920. e. Waseda Univ. Joined Motion Picture Producers Association 1945; appointed inspectorate Board of Trade in export film division 1948. On occasion of founding of Association for Diffusion Japanese Films Abroad (UniJapan Film) became manager.

KUROSAWA, AKIRA: Director. b. Japan. March 23, 1910. Entered PCL Studio 1936 and made director 1943.
PICTURES INCLUDE: Sanshiro Sugata, The Most Beautiful, Those Who Tread on The Tiger's Tail, No Regrets For Our Youth, The Quiet Duel, Scandal, Stray Dog, Rashomon, The Seven Samurai, The Drunken Angel, The Lower Depths, The Idiot, Ikiru, I Live in Fear, The Hidden Fortress, Throne of Blood, Yojimbo, High and Low, Red Beard, The Bad Sleep Well, Sanjourno, Kagemusha.

KURTZ, GARY: Producer-Director. b. Los Angeles, 1941. e. Univ. of So. Calif., Cinema School. Began prof. career during college. Has worked as cameraman, soundman, editor for Roger Corman, other prods. Also worked on documentaries. Was asst. dir. for two Monte Hellman westerns, Ride in the Whirlwind, The Shooting. Was prod. supvr. for The Hostage before being drafted into Marines. Spent 2 yrs. in Photo Field as cameraman, editor and still photo.
PICTURES INCLUDE: Two-Lane Blacktop (asst. prod.), Chandler (asst. prod.), American Graffiti (co.-prod.); Star Wars (prod.), The Empire Strikes Back (prod.).

KUTNER, MARTIN: Executive. Joined Paramount Pictures in 1971 as eastern div. mgr.; has held various positions in sls. dept. Named v.p., gen. sls. mgr.; in 1980 appt. senior v.p., domestic distribution.

KWIT, NATHANIEL TROY, JR.: Executive. b. N.Y.C., 1941. B.A. Cornell Univ.; M.B.A. New York Univ. 1964–68, American Broadcasting Co., Inc., exec. asst. to pres. of ABC Films. 1968–71, National Screen Service Corp., New York branch mgr., asst. genl. sls. mgr. 1971, founder , CEO Audience Marketing, Inc., later acquired by Viacom International as operating subsidiary. 1974 named v.p. marketing services, Warner Bros., Inc. 1979, named v.p. in charge video and special markets division, United Artists Corp.; 1981, named sr. v.p. in chg. VA television, video, special market div.

KYO, MACHIKO: Actress. b. Osaka, Japan. e. Osaka, Japan. Dancer in Osaka and Tokyo music halls; entered films 1948 with Daiei Studio; has appeared in numerous Japanese films.
PICTURES INCLUDE: Rashomon, Gate of Hell, Golden Demon, Story of Shunkin, Tales of Genji, Street of Shame, Teahouse of the August Moon, Ugetsu.

L

LADD, JR., ALAN: Executive. b. Los Angeles, Oct. 22, 1937. Motion picture agent, Creative Management Associates, 1963–69. M.p. producer, 1969–73; produced 9 films in 4 yrs. Joined 20th Century-Fox in 1973 in chg. of creative affairs in feature div. Promoted to v.p., production, 1974. In 1975 named sr. v.p. for worldwide prod.; 1976, promoted to pres. of 20th Century-Fox Pictures. Resigned & formed The Ladd Co., 1979.

LADD, CHERYL: Actress. b. Huron, S.D., July 2, 1951. r.n. Cheryl Stoppelmoore. M. film actor David Ladd. Joined professional Music Shop Band while in elementary school; left with it on tour, ending up in Los Angeles. Cast as voice of Melody character in animated Josie and the Pussycats. Studied acting with Milton Katselas. Did TV commercials, small parts in TV, made film Jamaica Reef (unreleased).
TELEVISION: Series: Charlie's Angels. Guest star on Ben Vereen. . . . His Roots, General Electric's All-Star Anniversary, John Denver and the Ladies, Police Woman, Happy Days, Switch, etc. Had own special, Cheryl Ladd. TV movie: A New Start.

LAKE, ARTHUR: Actor. r.n. Arthur Silverlake; b. Corbin, Ky. Stage debut at 3 yrs. of age; m.p. debut in silent Western 1924; sound m.p. debut Air Circus 1928; in vaudeville; on radio in "Blondie and Dagwood" series; owner of plastics household products co.; org. Arthur Lake Productions.
PICTURES INCLUDE: Sweet Sixteen comedies, Cradle Snatchers, Dance Hall, Annapolis Salute, Everybody's Doing It, There Goes My Heart, Blondie series, Sixteen Fathoms Deep.

LAMARR, HEDY: Actress. b. Vienna. At 15 starred in Ecstasy (awarded top Italian film-prize).
PICTURES INCLUDE: Algiers, I Take This Woman, Lady of the Tropics, Boom Town, Comrade X, Come Live With Me, Ziegfeld Girl; H. M. Pulham, Esq.; Tortilla Flat, Crossroads, White Cargo, Heavenly Body, Conspirators, Strange Woman, Samson and Delilah, A Lady Without a Passport, Dishonored Lady, Experiment Perilous, Let's Live a Little, Copper Canyon, My Favorite Spy, Story of Mankind, The Female Animal.

LAMAS, FERNANDO. Actor-Director, b. Buenos Aires, Jan. 9, 1925. e. Argentina, Spain, Italy. Trained at Teatro Experimental. After over 24 European and Latin-American films came to Hollywood under contract to MGM in 1951. Starred in more than 20 Hollywood films.
PICTURES INCLUDE: Rich, Young and Pretty, Lady and The Law, The Merry Widow, Sangaree, The Girl Who Had Everything, Dangerous When Wet, Diamond Queen, Jivaro, Rose Marie, The Girl Rush, 100 Rifles, The Cheap Detective. Dir.: Magic Fountain, The Violent Ones.
Several Box Office Awards and The New York Drama Critics Award for his Broadway play, Happy Hunting, international award for Place In Glory (wrote s.p.). 1967; debut as director: several TV segments of Run for Your Life. Has directed segments of The Bold Ones, Mannix, Alias Smith & Jones, The Rookies, S.W.A.T., Carl Reiner Show, and Starsky and Hutch.

LAMB, GIL: Actor. b. Minneapolis, Minn., June 14, 1906; e. U. Minn. On stage before screen debut 1939 in The Fleet's In, Lambs Clubs (N.Y.) On screen 1943.
PICTURES INCLUDE: Riding High, Rainbow Island, Practically Yours, Hit Parade of 1947, Make Mine Laughs, Humphrey Takes a Chance, Bye Bye Birdie, The Gnome-Mobile, Blackbeard's Ghost, The Love Bug.

LAMOUR, DOROTHY: Actress. b. Dec. 10, 1914, New Orleans, La.; e. Spence's Business Sch. Miss New Orleans 1931; sang on radio programs; screen debut in Jungle Princess 1938.
PICTURES INCLUDE: Spawn of the North, St. Louis Blues, Man About Town, Disputed Passage, Johnny Apollo, Typhoon, Road to Singapore, Caught in the Draft, Star Spangled Rhythm, Road to Utopia, Practically Yours, Medal for Benny, Duffy's Tavern, My Favorite Brunette, Road to Rio, Wild Harvest, Miracle Can Happen, Lulu Belle, Girl from Manhattan, Lucky Stiff, Slightly French, Manhandled, The Greatest Show on Earth, Road to Bali, Road to Hong Kong, Donovan's Reef, The Phynx.

LANCASTER, BURT: Actor. b. New York City, Nov. 2, 1913; e. N.Y.U. Was circus acrobat, in vaudeville; served in U.S. Army Special Service in Italy & N. Africa, World War II; screen debut in The Killers, 1946. Formed Hecht-Lancaster Orgn. in partnership with Harold Hecht.
PICTURES INCLUDE: Desert Fury, I Walk Alone, Brute Force, Sorry Wrong Number, Kiss the Blood Off My Hands, Criss Cross, All My Sons, Rope of Sand, Mister 880, Flame and the Arrow, Vengeance Valley, Ten Tall Men, Jim Thorpe—All American, Crimson Pirate, Come Back Little Sheba, South Sea Woman, From Here to Eternity, His Majesty O'Keefe, Apache, Vera Cruz, The Kentuckian, Rose Tattoo, Trapeze, The Rainmaker, Gunfight at the OK Corral, Sweet Smell of Success, Separate Tables, The Devil's Disciple, Elmer Gantry, The Young Savages, Birdman of Alcatraz, The Leopard, Seven Days in May, Hallelujah Trail, The Swimmer, The Gypsy Moths, Castlekeep, Airport, Valdez Is Coming, Lawman, Ulzana's Raid, Scorpio, Executive Action, The Midnight Man (prod., dir. & star), Conversation Piece, Buffalo Bill and the Indians, The Cassandra Crossing, Twilight's Last Gleaming, The Island of Dr. Moreau, 1900, Go Tell the Spartans, Cattle Annie and Little Britches, Atlantic City.

LANCHESTER, ELSA: Actress and authoress. b. London, England; m. Charles Laughton. Acted in London, New York. At 16, started the Children's Theatre in London. Acted in London and New York with her husband. First British film was The Constant Nymph, silent version; The Private Life of Henry VIII, 1933, London Films. In 1935: David Copperfield, Naughty Marietta, MGM, and The Bride of Frankenstein, U. In 1936: The Ghost Goes West, Rembrandt, UA-London Films. In 1938: the Beachcomber, Mayflower-Par. In 1938: authoress of Charles Laughton and I. In 1939–40: Hollywood and on stage. In 1941: on screen, Ladies in Retirement, Col. In 1942: Son of Fury, Tales of Manhattan, 20th-Fox. In 1943: Forever and a Day, Lassie Come Home, Passport to Adventure, Witness for the Prosecution, Bell, Book and Candle.
PICTURES INCLUDE: The Spiral Staircase, The Razor's Edge, Northwest Outpost, Bishop's Wife, Come to the Stable, Secret Garden, Inspector General, Buccaneer's Girl, Mystery Street, Frenchie, Petty Girl, Androcles and the Lion, Dreamboat, Les Miserables, The Girls of Pleasure Island, Hell's Half Acre, Three Ring Circus, Glass Slipper, Mary Poppins, That Darn Cat, Blackbeard's Ghost, Easy Come Easy Go, Rascal, Me, Natalie, My Dog, The Thief, Willard, Arnold, Murder by Death.
TV: Studio One, Omnibus, Schlitz Playhouse, John Forsythe Show, Dick Cavett, Johnny Carson, Merv Griffin, David Frost Shows Guest appearances on Mannix, Bill Cosby, To Catch a Thief, Then Came Bronson, Lucy Show.

LANDAU, ELY A.: Executive. b. New York, N.Y., Jan. 20, 1920; formed National Telefilm Associates, Inc., 1954; org. NTA Film Network, 1956 pres., chmn. of bd., National Telefilm Associates, Inc., 1957; resigned, 1961; formed Ely Landau Company, Inc., 1963; dist. The Servant, King and Country, Umbrellas of Cherbourg; prod. Long Day's Journey Into Night, The Fool Killer, The Pawnbroker, A Face of War, The Madwoman of Chaillot. Prod. King—A Filmed Record—Montgomery to Memphis, 1968; organized, directed one-night simultaneous charity showing 633 theatres U.S., 1970; 1972 Formed American Film Theatre, and the Ely Landau Organization, Inc. 1972–74 produced Iceman Cometh, Rhinoceros, The Homecoming, A Delicate Balance, Luther, Lost In The Stars, Burley Galileo, In Celebration, The Man in the Glass Booth, 1978: The Greek Tycoon. 1980: Hopscotch.

LANDAU, MARTIN: Actor. b. N.Y.C., 1934. Cartoon and staff artist on N.Y. Daily News; studied 3 yrs. at Actors Studio.
PICTURES INCLUDE: Pork Chop Hill, North by Northwest, Gazebo, Stagecoach to Dancer's Rock, Cleopatra, The Hallelujah Trail, The Greatest Story Ever Told, Nevada Smith, A Town Called Hell, Black Gunn, Strange Shadows in an Empty Room.
TV: Mission Impossible (1966–68), etc.
STAGE: Middle of the Night, Uncle Vanya, Stalag 17, Wedding Breakfast, First Love, The Goat Song.

LANDAU, RICHARD H.: Writer. b. New York City, Feb. 21, 1914; e. U. Arizona, Yale U. With Small-Landau agency handling writers and stories; shorts dept. writer MGM 1939; writer for RKO Radio since 1942; wrote documentaries and raining films for U.S. Army.
PICTURES INCLUDE: Gun in His Hand, Strange Confession, Challenge in the Night, Back to Bataan, Little Iodine, Christmas Eve, Crooked Way, Johnny One Eye, Roadblock, Lost Continent, F.B.I. Girl, Stolen Face, Bad Blonde, Spaceways, Sins of Jezebel, Blackout, Deadly Game, A Race for Life, Pearl of the South Pacific, Creeping Unknown.

LANDERS, ALBERT F.: Executive. b. Goshen, Ind., May 22, 1920. e. Duke U., U. of Cincinnati Grad School. Crosley Broadcasting, dir. of TV film opers., 1946–52; dir. of TV film opers., ABC-TV, Hollywood; 1953–55; ass't. to pres., General Film Laboratories, 1955–64; v.p. West Coast opers., DeLuxe General, Inc., 1964 to 1969; asst. gen. mgr., MGM Laboratories, 1970–77

LANDON, MICHAEL: Actor-Writer-Director. b. Forest Hills, N.Y., Oct. 31. e. U.S.C. Was athlete before signed by Warner Bros. to attend acting school. Roles in films and TV followed; big break came with Bonanza.
TV: Actor: Restless Gun, Bonanza, Little House on the Prairie. Guest appearances on Variety Shows. Writer-

Director: Love Came Laughing (Love Story series). Director: Roy Campanella Story. Producer: Little House on the Prairie.

LANDRES, PAUL: Director. b. New York, N.Y., Aug. 21, 1912. e. U.C.L.A. Started as asst. Film Editor at Universal 1931. Editor 1937 to 1949 of many feature films. Director of feature films and TV since 1949. Under Directorial contract to Warner Bros. 1961–62. Twenty-two feature films for theatrical release. Last film "Son of a Gunfighter" made in Spain for MGM release.

PICTURES INCLUDE: Miracle of the Hills, Johnny Rocco, Vampire, Flame Barrier, Oregon Passage, Last of the Badmen.

TV: 91 hour episodes of various series including among many others: The Outcasts, Bonanza, Daktari, The Rifleman, 77 Sunset Strip, Maverick, Hawaiian Eye, etc.

LANDSBURG, ALAN: Executive, producer. b. New York City, May 10, 1933. e. New York University, Producer for NBC News Dept., 1951–59; producer-writer, CBS, 1959–60; exec. prod., Wolper Productions/Metromedia Producers Corp., 1961–70; pres., Alan Landsburg Productions, 1970–present.

LANG, CHARLES: Cinematographer. b. Bluff, Utah, March 27, 1902. e. Lincoln H.S., Los Angeles, U.S.C. Entered m.p. ind. with Paramount Film Laboratory, then asst. cameraman; dir. of photography, Paramount, 1929–52; then freelance.

PICTURES INCLUDE: A Farewell to Arms (Academy Award for best photography, 1933); Ghost and Mrs. Muir, A Foreign Affair, September Affair, Ace in the Hole, Sudden Fear, Sabrina, Queen Bee, Man from Laramie, The Rainmaker, Some Like It Hot, The Magnificent Seven, Facts of Life, One-Eyed Jacks, Summer and Smoke, Charade, Father Goose, Wait Until Dark, Inside Daisy Clover, Hotel, Flim Flam Man, The Stalking Moon, Cactus Flower, Bob & Carol & Ted & Alice, The Love Machine, Doctors' Wives, Butterflies Are Free.

LANG, DAVID: Writer. b. New York City, Nov. 30, 1913. Was in Merchant Marine three years. Joined Charles Mintz Studio (Columbia) as cartoonist. Moved to MGM cartoon dept. 1938–40. Radio writer, Calling All Cars, KNX, Los Angeles, 1941. Contract writer at MGM, 1941–43. Yank on the Burma Road, Gambler's Choice, Hired Gun, North West Mounted Police. For Par. Midnight Manhunt, People Are Funny, Caged Fury, One Exciting Night. For Warner Bros. Flaxy Martin, Smart Money. For Columbia, Chain of Circumstance, Ambush at Tomahawk, The Nebraskan, Black Horse Canyon for U.I. Also Screaming Eagles, Hellcats of the Navy, Buckskin Lady. For PRC, Queen of Burlesque.

TELEVISION: Cheyenne, 87th Precinct, Ford Theatre, Gallant Men, Westinghouse Theatre, Adventures in Paradise, Bonanza, Trackdown, Wanted Dead or Alive, Rifleman, Rawhide, Have Gun Will Travel. At least two hundred credits in above series. Novelist. Oedipus Burning. Publisher, Stein and Day.

LANG, JENNINGS: Executive. b. New York City, May 28, 1915. e. St. John's Univ. & Law School. m. actress-singer Monica Lewis. Went into law practice in 1937 with Seligsburg and Lewis, m.p. law specialists. In 1938 went to Hollywood and became 2nd asst. dir. at Grand National Studios. Opened own office as actor's agent; first client, comedian Hugh Herbert. In 1940 joined Jaffe Agency; made partner and v.p. in 1942. Was pres. from 1948 to May, 1950, when resigned to join MCA. Worked in all phases of MCA operations; in 1952 made v.p. of MCA TV Ltd., and bd. mem. Involved with prod. and sales of TV prods. from inception of Revue (now Universal City Studios) in 1950. Organized Revue's New Projects Dept., creator and exec. in chg. of prog. dev. Involved with creation and sales of such series as Wagon Train, The Robert Cummings Show, Bachelor Father, Wells Fargo, Mike Hammer, etc. Most recently has been supvr. of Universal's World Premiere films, airing on Tuesday Night at the Movies and Saturday Night at the Movies (NBC).

Made Executive Producer at MCA (Universal) for motion pictures. Executive Producer on following films: Winning, They Might Be Giants, Puzzle of a Downfall Child, Coogan's Bluff, The Beguiled, Act of the Heart, Tell Them Willie Boy is Here, Play Misty for Me, Pete 'N Tillie, High Plains Drifter, Slaughterhouse Five, Charley Varrick, Breezy, The Great Waldo Pepper, Airport '75, Earthquake, Joe Kidd, The Great Northfield Minnesota Raid, The Eiger Sanction, Airport 1977, The Front Page, The Hindenburg. Producer: Swashbuckler, Roller Coaster, House Calls, Nunzio, Airport '79—The Concorde, Little Miss Marker, The Nude Bomb.

LANG, OTTO: Producer-Director, Four Academy Award nominations for Cinemascope Specials, Twentieth Century-Fox Film Corp. Saga of Western Man; ABC-TV Specials—The Legend of Cortez; Beethoven: Ordeal and Triumph.

TV: Man from U.N.C.L.E.; Daktari; Iron Horse; Cheyenne; Dick Powell Show; Zane Gray Theatre; Ann Sothern Show; Rifleman; Bat Masterson; Seahunt; The Deputy; Surfside 6; Hawaiian Eye. Prod. Twentieth Century Fox Hour. Dir. Man and the Challenge; Aquanauts; World of Giants. Dir. feature for Cinerama: Search for Paradise, Lancer, Felony Squad.

PICTURES: Prod., Call Northside 777; Five Fingers; White Witch Doctor. Specialist for foreign locations. Many segments for This World of Ours; Wide, Wide World. 1969. Assoc. prod: Tora! Tora! Tora!, 20th Century-Fox.

LANGAN, GLENN: Actor. b. Denver, Colo., July 8, 1917; e. Wheatridge H.S. Started career as asst. mgr. Elitch Gardens, Denver; traveled to N.Y., worked at odd jobs, until walk-on part in play, Swing Your Lady; signed by Hollywood after appearance in A Kiss for Cinderella, oppos. Luise Rainer.

PICTURES INCLUDE: Riding High, Four Jills in a Jeep, Something For The Boys, Margie, Homestretch, Forever Amber, Iroquois Trail, Treasure of Monte Cristo, Rapture, Hangman's Knot, One Girl's Confession, 99 River Street, Big Chase, Mutiny in Outer Space.

LANGE, HOPE: Actress. r.n. Hope Elise Ross Lange; b. Redding Ridge, Conn., Nov. 28; e. Reed Coll., Portland, Ore. & Barmore Jr. Coll., N.Y. Parents: John Lange, musician (Arr. music for stage shows, including Show Boat); Minnette Buddecke Lange, actress. Prof. stage debut at age 12 in The Patriots on Broadway; then in The Hot Corner; seen on TV by 20th-Fox prod. Buddy Adler and signed for Bus Stop, her screen debut.

PICTURES INCLUDE: Jesse James, Peyton Place, The Young Lions, In Love and War, The Best of Everything, A Pocketful of Miracles, How the West Was Won, Love is a Ball, Jigsaw, Death Wish.

TV: The Ghost and Mrs. Muir, The New Dick Van Dyke Show, That Certain Summer, I Love You—Goodbye.

LANGELLA, FRANK: Actor. Studied acting at Syracuse Univ.; later in regional repertory, summer stock, and on and off Bdwy. Film debut in The Twelve Chairs (1970).

PICTURES: Diary of a Mad Housewife, Dracula, Those Lips, Those Eyes, Sphinx.

LANGFORD, FRANCES: Singer-actress. b. Lakeland, Fla.; e. Southern Coll. Stage experience in vaudeville, nightclubs, national radio programs. In 1935: collab. on lyrics and appeared in Every Night at Eight, Collegiate Broadway Melody of 1936, Palm Springs, Born to Dance, The Hit Parade, Hollywood Hotel, Dreaming Out Loud, Too Many Girls, The Hit Parade of 1941, All-American Coed, Swing It Soldier, Mississippi Gambler, Yankee Doodle Dandy, This Is the Army, Career Girl, The Girl Rush, Dixie Jamboree, Radio Stars on Parade, People Are Funny, Deputy Marshall, Purple Heart Diary, Glenn Miller Story; TV appearances with Don Ameche.

LANGNER, PHILIP: Producer, b. New York, N.Y., Aug. 24, 1926. e. Yale U. President of The Theatre Guild and Theatre Guild Films, Inc. Producer the Westport Country Playhouse 1947–53. Joined The Theatre Guild 1954. Produced 28 plays on Broadway for the Theatre Guild including the Matchmaker, Bells Are Ringing, The Tunnel of Love, Sunrise At Campobello, A Majority of One, The Unsinkable Molly Brown, A Passage To India, Seidman and Son, The Royal Hunt of the Sun, The Homecoming, Absurd Person Singular and Golda.

FILMS: Producer of The Pawnbroker, Slaves and Born To Win. Associate Prod., Judgment At Nuremberg, and A Child Is Waiting.

LANSBURY, ANGELA: Actress. b. London, England, Oct. 16, 1925; e. South Hampstead School for Girls, England; Academy of Music, London; Feagin Dramatic School, N.Y.; p. Moyna Macgill, actress; also rel. to Robert B. Mantell, actor, & Rt. Hon. George Lansbury, gov't. official. Screen debut in "Gaslight," 1943, which won Hollywood Foreign Correspondents' Assoc. award.

PICTURES INCLUDE: National Velvet, Picture of Dorian Gray, Harvey Girls, Hoodlum Saint, Till the Clouds Roll By, Tenth Avenue Angel, If Winter Comes, State of the Union, Three Musketeers, Red Danube, Samson and Delilah, Kind Lady, Mutiny, Remains to be Seen, Purple Mask, A Lawless Street, A Life at Stake, Court Jester, Please Murder Me, A Breath of Scandal, All Fall Down, In the Cool of the Day, The Manchurian Candidate, Mr. Buddwing, Dear Heart, Something for Everyone, Bedknobs and Broomsticks, Death on the Nile, The Mirror Crack'd.

BROADWAY: Mame, Dear World, Gypsy.

LANSBURY, BRUCE: Executive. b. London, England, Jan. 12, 1930. e. UCLA. m. actress Moyna Macgill. writer, prod. KABC-TV, Los Angeles, 1957–59; joined CBS-TV, 1959, was ass't. dir., program dev., Hollywood, director for daytime and nighttime programs, producer of Great Adventure series; and v.p., programs, New York; 1964–66, indep. prod., Broadway stage; 1966–69 producer, Wild Wild West, CBS series; 1969–72, prod. Mission: Impossible, Paramount Movies of Week; now v.p., creative affairs, Paramount TV.

LANSING, SHERRY: Executive. b. Chicago, July 31, 1944. e. Northwestern Univ. Taught math, English and drama in L.A. city high schools, 1966–69. Acted in films (Loving, Rio Lobo) and numerous TV shows. Story editor for Wagner Intl. Prod. Co., 1972–74. Talent Associates, in chg. West Coast development (all projects), 1974–75. Appt. MGM story editor, 1975. In 1977 named MGM v.p. of creative affairs, Nov., 1977, appointed vice pres., production, at Columbia Pictures. Sept., 1978, named sen. v.p., production, for Columbia Pictures. January, 1980, appointed pres., Twentieth Century-Fox Productions.

LANTZ, WALTER: Animated cartoon producer. b. New Rochelle N.Y., April 27, 1900. Producer and creater of Woody Woodpecker, Andy Panda, Chilly Willy. Started with Gregory La Cava, 1916 with Katzenjammer Kids, Happy Hooligan and Krazy Kat. Joined J. R. Bray in 1922, producing Col. Heeza Liar, Dinky Doodle. Started with Universal Pictures in 1928. Produced first Technicolor cartoon for Paul Whiteman's King of Jazz. Produced Oswald Rabbit. Created Woody Woodpecker in 1941. Produced the first Woody Woodpecker TV show in 1957. Toured the Pacific War Zone on a handshake tour for the USO, with wife Gracie, the voice of Woody Woodpecker. Has been awarded the Golden Globe Award, the ASIFA Award, and in 1979 the OSCAR for achievement in the field of animation. Now producing the Woody Woodpecker TV show.

LARDNER, RING W., JR.: Writer. b. Chicago, Aug. 19, 1915; married; p. Ring W. and Ellis A.; e. Phillips Acad. and Princeton U. Was reporter on New York Daily Mirror. Publ. writer, Selznick International. Shared Academy Award with Michael Kanin. Orig. Screenplay, Woman of the Year, 1942.
PICTURES INCLUDE: The Cross of Lorraine, Tomorrow the World, Forever Amber, Forbidden Street, Four Days Leave, Cloak and Dagger, The Cincinnati Kid, M*A*S*H, The Greatest.

LARKIN, JAMES J.: Executive. B. Brooklyn, N.Y., Nov. 2, 1925; e. Columbia U., 1947–52. U.S. Air Force, 1943–46; BOAC rep. to entertainment ind., 1948–60; pres., Transportation Counselors Inc., 1960–62; pres., Larkin Associates, Inc., 1962–65; exec. Radio N.Y. Worldwide, 1965–68, V.P. Grolier Educational Corp., 1968–69; V.P. Visual Informational Systems, 1969–73. Pres., Business Television Services, Inc., 1973; exec. prod., Madhouse Brigade, 1977–79; prod.-writer, All Those Beautiful Girls, 1979–80.

LARSEN, KEITH: Actor. e. Univ. of Utah. Little theatre work.
TV: Series: The Hunter, Brave Eagle.
PICTURES INCLUDE: Green Glove, Son of Belle Star, Chief Crazy Horse, Desert Sands, Night Freight, Arrow in the Dust, Dial Red O, Security Risk, Wichita, Women of the Prehistoric Planet.

LARSON, G. BENNETT: Executive. Joined KDYL, Salt Lake City as announcer, prog. mgr., 1926. Prod., dir., NBC, New York, 1929–34. Shows included Ed Wynn Texaco Show, Fleischmann House and Chase & Sanborn Hour. Radio exec. with Joseph Katz Co., Ruthrauff & Ryan, and Young & Rubicam, 1934–42. Gen. mgr. and part owner of WWDC, Washington, 1942–45. Mgr of WPEN, Phila., 1945–47. v.p. of TV, WCAU-TV 1947–1950, App'd v.p. and gen. mgr., WPIX, New York, 1950; pres., Intermountain Broadcasting Corporation, KDYL AM-FM & TV, Salt Lake City, Utah, Exec. v.p. and gen. mgr. of Flower City Television Corporation, WOKR-TV, Rochester, New York, 1979–80, pres. Larson/Walker & Co., media brokers, Los Angeles; pres., G. Bennett Larson & Associates (broadcast consultants), 1981.

LASKY, JESSE JR.: Writer. b. New York City, Sept. 19, 1910; p. Jessie and Bessie Gains Lasky; e. Blair Academy, Hun School of Princeton, Grand Central School of Art, U. Of Dijon; entered m.p. industry, foreign dept., Paramount, Spain, 1930. Was censorship editor for Paramount on story board. Assistant to Sol Wurtzel Fox.
NOVELS INCLUDE: Curtain of Life, No Angels in Heaven, Spendthrift, Naked in a Cactus Garden, The Offer (with Pat Silver); autobiography, Whatever Happened to Hollywood?
OTHER MATERIAL: Three books of verse. Short stories published in Cosmopolitan Magazine; The Gehenna of the Bone; (plays) Hope and Pray and Private Beach (in collab.); Ghost Town: (collab. s.p. or story) Secret Agent; The Red Head (s.p.); Back in the Saddle; The Singing Hill (collab. s.p.); Steel Against the Sky (film script); Union Pacific; Northwest Mounted Police; Reap the Wild Wind; (collab. narr.) Land of Liberty, dir. orig. story, s.p. Omaha Trail; U.S. Army, 1942–46; (collab. s.p.) Unconquered. Samson and Delilah; (orig. s.p.) Sickle or Cross Thief of Venice (Italy); (s.p.) Women Without Names, City of Violence (Italy); (s.p.) Lorna Doone, Mask of the Avenger, Never Trust a Gambler, The Brigand, The Silver Whip, (collab. s.p.) Mission Over Korea, (story) Salome, (collab. s.p.) Hell and High Water, Iron Glove, (s.p.) Pearl of the South Pacific, Hot Blood, (collab. s.p.) Ten Commandments, (collab. s.p.) The Buccaneer, John Paul Jones, (collab. with Pat Silver) Wizard of Bagdad, Pirates of Tortuga, River Gambler, 7 Women From Hell, The Land Grabbers (orig. story), One of Those Things (English version).
TV: Naked City, Rescue 8, Shannon, 7 of the New Breed (in collab. with Pat Silver), Avengers, Danger Man, The Saint, The World of Lowell Thomas, The Baron, The Protectors, and four musical specials for BBC; Ten Who Dared (serial with Pat Silver)

LASSER, LOUISE: Actress. b. New York City. e. Brandeis Univ., New Sch. for Social Research. Appeared on stage before theatrical film debut in 1965 with What's New, Pussycat?
PICTURES: What's Up, Tiger Lily?, Take the Money and Run, Bananas, Everything You Always Wanted to Know About Sex, Such Good Friends, Slither.
TELEVISION: Masquerade, Mary Hartman, Mary Hartman (series); The Lie, Isn't It Shocking? (movies).

LASTFOGEL, ABE: Talent agent, William Morris Agency, Inc., hdqts., Beverly Hills, Calif., bd. chmn., 1969.

LATIMER, JONATHAN: Writer.
PICTURES INCLUDE: Lady in the Morgue, Last Warning, Phantom Raiders, Topper Returns, Night Has a Thousand Eyes, Beyond Glory, Redhead and Cowboy, Plunger of the Sun, Botany Bay, Night in New Orleans, Glass Key, Nocturne, They Won't Believe Me, Big Clock, Sealed Victory, Alias Nick Beal, Copper Canyon.

LA TOREE, CHARLES: Actor. b. New York City, Apr. 15, 1900. e. Columbia U. Long career on stage in character parts. Screen debut in Louisiana Purchase.
PICTURES INCLUDE: My Sister Eileen, Life Begins at 8:30, Casablanca, Mission to Moscow, Song of Bernadette, Enger Arsene Lupin, Hairy Ape, Uncertain Glory, Panhandle, 711 Ocean Drive.

LATSIS, PETER C.: Publicist. b. Chicago, Ill., Mar. 9, 1919. e. Wright Junior College, Chicago. Newspaper reporter, Chicago Herald-American, 1942–45; Army, 1943; joined Fox West Coast Theatres, Los Angeles, in theatre operations 1945; adv.-pub. dept 1946; asst dir. adv.-pub. 1955; press rep. National Theatres, 1958; press relations dir., National General Corp., 1963 home office special field pub. repr., American International Pictures, 1973; Filmways Pictures, 1980.

LATTUADA, ALBERTO: Director. b. Milan, Italy, 1914. Son of Felice Lattuada, musician, opera composer, and writer of scores of many of son's films. Studied architecture; founded the periodical Comirare. Later founded Italian Film Library, of which he is still pres. Also, pres., Cinema D'Essay. First screen work as scriptwriter and asst. dir. of two films, 1940.
PICTURES INCLUDE: Mill on the Po, Anna, The Overcoat, La Lupa, Love in the City, White Sister, Flesh Will Surrender, Without Pity, The She Wolf, Tempest, The Unexpected, Mafioso, The Mandrake, Matchless, The Betrayal, The Steppe, Oh, Serafina, Stay As You Are.

LAUGHLIN, TOM: Actor-Producer-Director-Writer. b. Minneapolis e. Univ. of Indiana, Univ. of Minn. where had athletic scholarships. m. actress Delores Taylor. Travelled around world, studying in Italy with Dr. Maria Montessori. Established, ran a Montessori school in Santa Monica for several yrs. Worked his way to Hollywood, where acted in bit parts until stardom came in Born Losers in 1967. Produced and starred in Billy Jack and The Trial of Billy Jack, also writing s.p. with wife under pseudonym Frank Christina. Heads own prod. co., Billy Jack Enterprises.
PICTURES INCLUDE: South Pacific, Tea and Sympathy, Gidget, Born Losers, Billy Jack, The Trial of Billy Jack, The Master Gunfighter, Billy Jack Goes to Washington.

LAUNDER, FRANK: Writer, director, producer. b. Hitchen, Eng., 1907, e. Brighton, Brit. civil svce.; dramatist; entered m.p. ind. as writer, 1930; Scenario ed., Gainsborough; org. Individual Pictures, Ltd., 1944 now called Launder-Gilliat and Company.
PICTURES INCLUDE: Lady Vanished, Night Train, Young Mr. Pitt, Millions Like Us, 2,000 Women, Notorious Gentleman, Green For Danger, Dulcimer Street, Great Manhunt, The Story of Gilbert and Sullivan, The Constant Husband, the Adventurers, Captain Boycott, Blue Lagoon, Happiest Days of Your Life Lady Godiva Rides Again, The Belles of St. Trinians, Only Two Can Play, Joey Boy, The Great St. Trinian's Train Robbery.

LAURENCE, DOUGLAS: Producer. b. Totowa, N.J., Dec. 16, 1922. U.S. Air Force. 1947–50; wrote, prod., dir., shows for NBC Radio. Exec. prod. talent buyer for Teleways Radio, TV Prod. 1953–59; ent. dir., Flamingo Hotel. Advisor to Brazilian govt. on building film industry.
PICTURES INCLUDE: Quick Before It Melts, Mister Buddwing, Doctor You've Got to Be Kidding, Speedway, Stay Away Joe, Live a Little Love a Little, The Outside Man, Rio Man.
STAGE: General Motors Motorama, L.A. Home Show, Calif., Wisc., Texas State Fairs, Ford Motor Auto Shows, Stay Eastern Univ. Dance Troupe, Jazz America.

TV: Strange Wills, John Charles Thomas Show, Sons of he Pioneers, All Star Hit Parade, John Gunther's High Road.

LAURENTS, ARTHUR: Writer-Director. b. New York City, e., Cornell, B.A. 1937. First professional writing as radio script writer in 1939. In Army 1941–45.
STAGE PLAYS: Author: Home of the Brave, The Bird Cage, The Time of the Cuckoo, West Side Story, Gypsy, Hallelujah, Baby! Director: Invitation to a March, I Can Get It For You Wholesale. Author-Director: Anyone Can Whistle, Do I Hear a Waltz?
SCREENPLAYS: Rope, Caught, Anastasia, The Way We Were (from his own novel), The Turning Point.

LAURIE, JOHN: Actor. b. Dumfries, Scotland, Mar. 25, 1897. Architect; on stage.
PICTURES INCLUDE: Juno and the Paycock, Red Ensign, Thirty-Nine Steps, Convoy, Sailors Three, Dangerous Moonlight, Fanny, By Gaslight, Henry V., Way Ahead, Medal for the General, Caesar and Cleopatra, Agitator, I Know Where I'm Going, School for Secrets, Brothers, Jassy, Uncle Silas, Hamlet, Bonnie Prince Charlie, Mine Own Executioner, Flood Tide, Treasure Island, Trio, No Trace, Pandora & Flying Dutchman, Happy Go Lovely, Laughter in Paradise, Encore, Saturday Island, The Black Knight, Devil Girl from Mars, Richard III, Rockets Galore, Campbell's Kingdom, Kidnapped, Ladies Who Do, The Siege of the Saxons, The Reptile, Mr. 10½, Dad's Army, Doctor Phibes, One of Our Dinosaurs Is Missing.
TV INCLUDES: Dad's Army, Red Gauntlet, Tales of Mystery, Late Night Line Up, Call My Bluff, Quiz Ball, The Gnomes of Dulwich, Justice Is A Woman, Eyeless in Gaza, A Midsummer Night's Dream, Jackanory, Nation Wide.

LAURIE, PIPER: Actress. r.n. Rosetta Jacobs. b. Detroit, Mich. Jan. 22, 1932. e. Los Angeles H.S. Acted in school plays, signed by U.I. in 1949; m.p. debut in Louisa; Broadway, The Glass Menagerie (revival).
PICTURES INCLUDED: The Milkman, Francis Goes to the Races, Prince Who Was A Thief, Son of Ali Baba, Has Anybody Seen My Gal, No Room for the Groom, Mississippi Gambler, Golden Blade, Dangerous Mission, Johnny Dark, Dawn at Socorro, Smoke Signal, Ain't Misbehavin', Until They Sail, The Hustler, Carrie, Ruby.

LAVEN, ARNOLD: Director-Producer. b. Chicago, Ill., Feb. 23, 1922; with W.B. 1940–42; U.S. Army 1942–45; dialogue dir., 1946–50.
PICTURES INCLUDE: Without Warning, Vice Squad, Down Three Dark Streets, The Rack, Slaughter on 10th Ave., Anna Lucasta, The Glory Guys, Rough Night in Jericho, Sam Whiskey.
TELEVISION: Part creator and director TV pilots: The Rifleman, Robert Taylor's Detectives, The Plainsmen. Many TV films.

LAVERY, EMMET G., JR.: Producer. Poughkeepsie, N.Y., Aug. 10, 1927. e. UCLA: AB, 1950; LL.B., 1953. Private practice, law, dealing w. entertainment matters, 1953–64; resident counsel subscription TV, 1964–65; dir. of business affairs, 20th Century-Fox TV, 1965–67; v.p. in chg. business affairs, Paramount Television, 1967–71, exec. v.p., Paramount Television, 1972–1974, prod. Paramount Television and Paramount Pictures 1974–1975; prod. Serpico (series) for Paramount TV-NBC-TV, 1976, Prod.: The Ghost of Flight 401 (NBC) The Victim: An Anatomy of a Mugging (CBS).

LAVIN, LINDA: Actress. b. Portland, Me. Oct. 15. e. College of William & Mary. First professional job in chorus of Camden County (N.J.) Music Circus. Worked in plays both off and on Broadway before turning to TV, where guest-starred on such series as Family, Rhoda, Phyllis and Harry O; had recurring role in Barney Miller—all before advent of her own series, Alice.

LAW, JOHN PHILIP: Actor. b. Hollywood, 1937. e. University of Hawaii. Trained at Repertory Theatre of Lincoln Center under Elia Kazan. Did numerous Italian films before discovery in U.S.
PICTURES INCLUDE: The Russians Are Coming, The Hawaiians, Barbarella, Von Richtofen and Brown, The Last Movie, The Love Machine, The Golden Voyage of Sinbad, The Cassandra Crossing, Tarzan, the Ape Man.

LAWFORD, PETER: Actor. b. London, England, Sept. 7, 1923; e. priv. On screen at 7 in Brit. prod. "Old Bill"; U.S. m.p. debut in "Lord Jeff" 1938; parking lot attendant; usher Westwood Village Theatre, Hollywood.
PICTURES INCLUDE: Thunderbirds, Mrs. Miniver, Yank at Eton, Eagle Squadron, Someone to Remember, Man from Down Under, White Cliffs of Dover, Paris After Dark, Canterville Ghost, Picture of Dorian Gray, Two Sisters from Boston, Cluny Brown, My Brother Talks to Horses, It Happened in Brooklyn, Good News. On an Island With You, Easter Parade, Julia Misbehaves, Little Women, Red Danube, Please Believe Me, Royal Wedding, Kangaroo, Just This Once, Hour of Thirteen, You For Me, Rogue's March, It Should Happen to You, Never So Few, Exodus, Ocean's 11, Sergeants 3, Advise & Consent, Dead Ringer, Sylvia, Harlow, The Oscar, Salt and Pepper, Skidoo, Hook, Line and Sinker, Buona Sera, Mrs. Campbell, April Fools, One More Time, They Only Kill Their Masters, That's Entertainment, Rosebud.
TV: Dear Phoebe, The Thin Man; pres., Chrislaw Prods., Inc., A Step Out of Line, Ellery Queen, The Doris Day Show.

LAWRENCE, BARBARA: Actress. b. Carnegie, Okla., Feb. 24, 1930; e. UCLA; mother Berenice Lawrence. Child model; successful screen try-out, 1944; screen debut in Billy Rose Diamond Horse Shoe.
PICTURES INCLUDE: Margie, Captain from Castile, You Were Meant for Me, Give My Regards to Broadway, Street With No Name, Unfaithfully Yours, Letter to Three Wives, Mother is a Freshman, Thieves Highway, Two Tickets to Broadway, Here Come the Nelsons, The Star, Arena, Paris Model, Her 12 Men, Oklahoma, Man With a Gun, Pay The Devil, Joe Dakota.

LAWRENCE, JODY: Actress. r.n. Josephine Lawrence Goddard. b. Fort Worth, Tex., Oct. 19, 1930; e. Beverly Hills H.S.; Hollywood Professional Sch. Signed by Columbia; m.p. debut in Mask of Avenger.
PICTURES INCLUDE: Family Secret, Son of Dr. Jekyll, Ten Tall Men, The Brigand, All Ashore, Capt. John Smith & Pocahontas, Scarlet Hour, Leather Saint.

LAWRENCE, MARC: Actor. b. New York City; e. City Coll. N.Y. On stage in Sour Mountain, Waiting for Lefty.
PICTURES INCLUDE: White Woman, Little Big Shot, Dr. Socrates, Road Gang, San Quentin, I Am the Law, While New York Sleeps, Dillinger, Flame of Barbary Coast, Club Havana, Don't Fence Me In, The Virginian, Life with Blondie, Yankee Fakir, Captain from Castile, I Walk Alone, Calamity Jane and Sam Bass, The Asphalt Jungle, Hurricane Island, My Favorite Spy, Girls Marked Danger, Helen of Troy, Custer of the West, Nightmare in the Sun.

LAWRENCE, STEVE: Actor. b. New York City, July 8, 1935. Singer in nightclubs and on TV.
PICTURES INCLUDE: Stand Up and Be Counted, The Blues Brothers.
TELEVISION: Steve and Eydie Celebrate Irving Berlin.

LAWSON, JOHN HOWARD: Writer. b. 1894, New York City; e. Williams Coll., 1914; ed. American Red Cross Bulletin in Rome, 1918. Wrote Broadway plays Loud Speaker, Processional, and others. With MGM, then freelance with Wanger, Goldwyn, 20th-Fox.
PICTURES INCLUDE: Sahara, Counter-Attack, Smash-Up.

LAWSON, SARAH: Actress. b. London, Eng., Aug. 6, 1928. e. Heron's Ghyll School (Sussex). stage debut in Everyman (Edinburgh Festival) 1947; screen debut in The Browning Version, 1949.
PICTURES INCLUDE: Street Corner, You Know What Sailors Are, Blue Peter, It's Never Too Late, Links of Justice, Three Crooked Men, Man with a Dog, Night Without Pity, The World Ten Times Over.
TV: Face to Face, River Line, Whole Truth, Lady From the Sea, Mrs. Moonlight, Silver Card, An Ideal Husband, Love and Money, Rendezvous, Invisible Man, Saber Buccaneers, White Hunter, Flying Doctor, On the Night of the Murder, Haven in Sunset, The Odd Man, Zero 1 series.

LAWYER, M.H. (A.C.R.A.): Distributor. b. Dec. 3, 1909. Bank official, 1927–49; with United Film Services, Jal Hind Pictures & Indo-Pak Film Corp., Karachi, before 1947; former secy., chmn., pres., M.P. Society of Sind & Baluchistan; secy. & treas., Sind & Baluchistan Film distrib. Synd.; pres. Famous Films, Pakistan.

LAY, BEIRNE, JR.: Writer, producer. b. Berkeley Springs, W. Va., Sept. 1, 1909; e. Yale U., B.A. Graduate as 2nd Lt. Army Air Corps Res. (pilot) 1933; col., AFRes, 10 yrs. active duty, Commander 487 Bomb Gp., 8th Air Force; Contributor to Post, Harpers, Fortune, Reader's Digest.
BOOKS: I Wanted Wings, I've Had It, 12 O'Clock High, Someone Has To Make It Happen, Earthbound Astronauts, I'm an Endangered Species.
PICTURES INCLUDE: I Wanted Wings, 12 O'Clock High, Above and Beyond, Strategic Air Command, Toward the Unkown, The Gallant Hours.

LAYE, EVELYN: Actress, singer. b. London, England, July 10, 1900. London stage debut 1918 The Beauty Spot. Subseq. in numerous musicals, dramas, London, including The Merry Widow, Madame Pompadour, Lilac Time, Blue Eyes, The New Moon, Paganini, Wedding in Paris, The Amorous Prawn, Strike a Light!, Let's All Go Down the Strand, Phil the Fluter, No Sex Please, We're British (1971–72). B'way stage debut, 1929, Bitter Sweet; m.p. debut 1927 silent production The Luck of the Navy.
PICTURES INCLUDE: One Heavenly Night, Waltz Time, Princess Charming, Evensong, The Night is Young, Make Mine a Million, Theatre of Death, Within and Without, Say Hello to Yesterday.
TV: Stars and Garters, Not Just An Act; 1969: Late Night

Line Up, The Golden Shot, The Simon Dee Show, The David Niven Show, This Is Your Life, Ommnibus.
PUBLICATION: Boo To My Friends (Autobiography).

LAZARUS, ERNA: Assoc. Producer Screen Writer. e. Columbia University, M.P. Industry major studio credits include: Hollywood or Bust, Martin & Lewis, sole story sole s.p.; Flareup, starring Raquel Welch, assoc. producer; Dancing in Manhattan, sole story-sole screenplay; Little Miss Big, sole screenplay-story; Michael O'Halloran, sole screenplay; Blonde From Brooklyn, sole story-sole s.p. Nobody's Sweetheart Now, story-screenplay, Meet Me After the Show, Betty Grable, original story; and others.
TV: Petticoat Junction, Donna Reed Show, Bewitched; Hawaiian Eye; Surfside Six; Mr. & Mrs. North; Ford Theatre, Irene Dunne, Sister Veronica; associate producer-head writer, Thomas Mitchell series, Mayor of the Town; Robert Taylor Detectives, and others.
RADIO: New York and Hollywood top comedy and dramatic shows.

LAZARUS, PAUL N.: Executive. b. Brooklyn, N.Y., March 31, 1913. e. Cornell U., B.A., '33. In U.S. Army, World War II. Entered m.p. ind. 1933 as gen. asst., press book dept., Warner Bros.; pres., AMPA, 1939–40. Joined Buchanan & Co., 1942 as m.p. account exec. To United Artists 1943 as dir. adv. & pub. Named asst. to pres., July 1948; joined Columbia exec. staff, New York, Aug. 1950; elected v.p. Columbia, Jan., 1954–62; exec. vice-pres. Samuel Bronston Prods., 1962–64; v.p., chg. Motion Pictures, Subscription Television Inc., 1964; exec. officer and partner, Landau Releasing Organization, 1964–65, exec. v.p., member bd. of dir., Nat'l Screen Serv. Corp., 1965–75. Lecturer and consultant, Film Studies Dept., Univ. of California at Santa Barbara. 1975 to present.

LAZARUS III, PAUL N.: Motion Picture Executive. Third generation film exec. Began career with Palomar Pictures Int'l. as exec. v.p.; joined ABC Pictures Corp. as v.p. in chg. of creative affairs. Mng. dir., CRM Productions, maker of educational films; v.p. for motion pictures. Marble Arch Productions.
PICTURES INCLUDE: Extreme Close-Up, Westworld, Futureworld, Capricorn One, Hanover Street.

LAZARUS, THEODORE R.: Executive. b. Brooklyn, N.Y., Aug. 5, 1919. e. Yale U. B.A., 1940. Adv. mgr., Eagle Lion Classics, 1951; adv., sales prom. mgr., WMGM, New York, 1951; then with Donahue and Coe adv. agency; secy., treas., George Blake Enterprises, TV film prod. firm, 1955; v.p. Gommi-TV, 1956; Charles Schlaifer and Co., Inc., 1957; adv. mgr., Paramount Pictures Corp., 1964; pres., Cinema Lodge, B'nai B'rith, 1968–71.

LAZARUS, THOMAS L.: Executive. b. New York, N.Y., Oct. 5, 1942. e. Boston U., 1960–62; New School for Social Research, 1962; acct. exec. Donahue and Coe, Inc., 1962–63; asst. to adv. mgr., Twentieth Century-Fox, 1963–64; adv. mgr., Seven Arts Pictures, 1964–66; Acct. exec. Chas. Schaiffer & Co., 1966; exec. coordinator adv., MCA, 1966–70; dir. creative services, 1970; writer, prod., dir. for CRM Productions, 1971–4.
PICTURES: Just You and Me Kid (story credit)
TELEVISION: Screenplay: Revenge or Justice, Charlie's Angels (The Jade Trap), The President's Mistress, Columbo, The Survival of Melissa, The Boys Next Door, etc.

LEACHMAN, CLORIS: Actress. b. Des Moines, Iowa, June 30, 1930. e. Northwestern Univ. Broadway stage, television, motion pictures.
TV: Lassie, A Brand New Life, The Migrants, Phyllis (series).
PICTURES INCLUDE: Butch Cassidy and the Sundance Kid, The Steagle, The Last Picture Show, (Acad. award best supporting actress, 1971), Dillinger, Daisy Miller, Young Frankenstein, Crazy Mama, High Anxiety, The North Avenue Irregulars, Scavenger Hunt, History of the World—Part I.

LEADER, ANTON MORRIS: Director-producer. b. Boston, Mass., e. Hebrew Coll., New Eng. Conservatory of Music. Dir., announcer, actor, Boston radio stations; prod. mgr., WMCA, N.Y.; Army Air Corp.; prod. dir., NBC; winner, Variety Showmanship Award, Peabody Award for Words at War; winner Brotherhood Award, Peabody Award for Eternal Light; Western Heritage Award, dir. TV Film, The Horsefighter (The Virginian); inst., Fordham U., prod., dir., Suspense, CBS; prod., dir. writer Columbia Pictures; prod., U-1;
PICTURES INCLUDE: It Happens Every Thursday, Sally and St. Anne, Go Man Go , Children of the Damned, Cockeyed Cowbows of Calico County.
TV: The Virginian, Rawhide, Tarzan, Daniel Boone, Playhouse 90, Father of the Bride, It Takes a Thief, Ironside, Star Trek, Lost in Space, I Spy, Get Smart, Hawaii Five-O, Firehouse, Movin' On, This Is The Life.

LEAN, DAVID: Director. b. Croydon, England, Mar. 25, 1908; e. Leighton Park. Entered m.p. ind. 1928 as camera asst. Graumont Pictures; asst. dir. & ed.; co-dir. 1942; dir.
PICTURES INCLUDE: Escape Me Never, Pygmalion, Invaders, In Which We Serve, This Happy Breed, Blithe Spirit, Brief Encounter, Great Expectations, Oliver Twist, Madeleine, One Woman's Story. The Sound Barrier, Hobson's Choice, Summertime, The Bridge on the River Kwai (Academy Award 1957, best direction), Lawrence of Arabia, Dr. Zhivago, Ryan's Daughter.

LEAR, NORMAN: Producer-Director-Writer. Began in TV as co-writer of weekly one-hour variety show, The Ford Star Revue in 1950. Followed as writer for Dean Martin and Jerry Lewis on the Colgate Comedy Hour and for the Martha Raye and George Gobel TV shows. With partner, Bud Yorkin, created and produced such specials as Another Evening with Fred Astaire, Henry Fonda and the Family, An Evening with Carol Channing, and The Many Sides of Don Rickles. In 1965 their company also produced the original Andy Williams Show. Moved into motion pictures in 1963, writing and producing Come Blow Your Horn.
PICTURES: Never Too Late (prod.), Divorce-American Style (s.p.), The Night They Raided Minsky's (co. prod., co-s.p.), Start the Revolution without Me (co.-prod.), Cold Turkey, (s.p., prod., dir.).
RECENT TELEVISION: Exec. prod. of All in the Family, Maude, Good Times, The Jeffersons, Mary Hartman, Mary Hartman, One Day at a Time, The Baxters, Palmerstown.

LEARNED, MICHAEL: Actress. b. Washington, D.C. Studied ballet and dramatics in school. Many stage credits include Under Milkwood, The Three Sisters, A God Slept Here, etc.; resident performances with Shakespeare festivals in Canada, Stratford, CT, and San Diego, CA. Gained fame on hit TV series, The Waltons, as the mother, Olivia. Theatrical film debut in 1980, Touched by Love.
TELEVISION: Gunsmoke, Police Story, Widow, Little Mo, etc.

LE BORG, REGINALD: Director. b. Vienna, Austria Dec. 11; e. U. Vienna, B.A.; Sorbonne, Paris; Columbia U. Banker, playwright & stage dir., Europe; then staged musical and operatic sequences. Columbia. Paramount. To MGM as dir. short subjects. In 1941 joined Universal. Entered U.S. armed services, 1942; returned to Universal, 1944; Rec'd citation from OWL; free lance since 1946. Wrote Academy Award winning short Heavenly Music, 1945. directed two films in England, one in Mexico.
PICTURES INCLUDE: She's for Me, The Mummy's Ghost, Calling Dr. Death, Weird Woman, Dead Man's Eyes, Jungle Woman, Destiny, Honeymoon Ahead; San Diego, I Love You; (prod.), Philo Vance's Secret Mission; (dir.) Fighting Fools, Hold That Baby, The Squared Circle, Young Daniel Boone, Wyoming Mail, Triple Cross, G.I. Jane; Models, Inc., Bad Blonde, Great Jesse James Raid, Sins of Jezebel, Joe Palooka, Champ, Port Said, Fall Guy, Formed Cosmos Prod., prod., White Orchid (dir.), Black Sleep, Voodoo Island, War Drums, The Dalton Girls.
TV: for Sovereign, 4 Star, Schlitz Playhouse, Wire Service, Navy Log, Maverick, Desilu Revue; Court of Last Resort, Warner Bros., Screen Gems, MGM, Ziv; recent features: The Flight That Disappeared, Deadly Duo, The Diary of A Madman. The Eyes of Annie Jones, So Evil, My Sister.

LEDER, HERBERT JAY: Writer-director, producer, b. New York City, Aug. 15, 1922. e. B.A., PHD. Play Doctor on Broadway; Director TV dept., Benton and Bowles Adv. chg. all T.V. & Film production, 13 yrs.; Features: writer, Fiend Without a Face; writer-director co-prod., Pretty Boy Floyd; writer-director, co-producer, Nine Miles to Noon; writer, Love Keeps No Score of Wrongs; writer, prod-dir., The Frozen Dead, It; Mia, writer-dir., Candyman, writer-director; writer, The Winners, The Way It Is, The Cool Crazies. Sponsored Films: Child Molester, Bank Robber, Shoplifter, Untouchables.

LEDERER, RICHARD: Executive. b. New York, N.Y. Sept. 22, 1916; e. Univ. of Virginia. B.S., 1938; Free lance writer, 1939–41; U.S. Army. Cryptanalyst, Signal Intell. Serv 1941–45; Adv. copywriter, Columbia Pictures, 1946–50; Adv. copywriter, Warner Bros., 1950–53; copy chief, Warner Bros., 1950–53; copy chief, Warner Bros., 1953–57; Asst. Nat'l Adv. mgr., Warner Bros. studios, 1957–1959; Prod., theatrical, TV. Warner Bros. studios, 1959–60; Dir. of adv., publicity, Warner Bros. Pictures, 1960; v.p. Warner Bros. Pictures, 1963. V.P. production, Warner Bros. Studio, 1969–70; indep. prod. to May, 1971, when returned to WB as adv.-pub., v.p. Now independent producer. 1980: Hollywood Knights.

LEE, ANNA: Actress. b. England; seen in First a Girl, The Camels Are Coming, You're in the Army Now, King Solomon's Mines.
PICTURES INCLUDE: Flesh and Fantasy, Summer Storm, Abroad with Two Yanks, High Conquest, Ghost and Mrs. Muir, Best Man Wins, Fort Apache, Prison Warden, This Earth is Mine.
Many TV shows.

LEE, CHRISTOPHER: Actor. b. London, England, May 27, 1922. e. Wellington College. Served RAF 1940–46. Ent. m.p. ind. 1947. Over 40 TV films. Recent TV: How The West Was Won.

PICTURES INCLUDE: Corridor of Mirrors, Hamlet, Moulin Rouge, Moby Dick, River Plate, Truth About Women, Tale of Two Cities, Curse of Frankenstein, Dracula, Man Who Could Cheat Death, The Mummy, Too Hot to Handle, Beat Girl, City of the Dead, Two Faces of Dr. Jekyll, The Terror of the Tongs, The Hands of Orlac, Taste of Fear, The Devil's Daffodil, Pirates of Blood River, Devil's Agent, Red Orchid, Valley of Fear, Katharsis, Faust '63, The Virgin of Nuremberg, The Whip and the Body, Carmilla, The Devil Ship Pirates, The Gorgon, The Sign of Satan, The House of Blood, The Dunwich Horror, Dr. Terror's House of Horrors, She, The Skull, The Mask of Fu Manchu, Dracula, Prince of Darkness, Rasputin, Theatre of Death, Circus of Fear, The Brides of Fu Manchu, Five Golden Dragons, Diabolica, Vengeance of Fu Manchu, Night of the Big Heat, The Pendulum, The Face of Eve, The Devil Rides Out, The Blood of Fu Manchu, The Crimson Altar, Dracula Has Risen from the Grave, The Oblong Box, De Sade 70, Scream and Scream Again, The Magic Christian, Julius Caesar, One More Time, Count Dracula, Bloody Judge, Taste the Blood of Dracula, Private Lives of Sherlock Holmes, El Umbragolo, Scars of Dracula, House That Dripped Blood, I Monster, Hannie Caulder, Dracula 72, Horror Express, Creeping Flesh, Death Line, Nothing but the Night, The Wicker Man, Poor Devil, Dark Places, Dracula Is Dead?, Eulalie Quitte les champs, The Three Musketeers, Earthbound, Man With The Golden Gun, The Four Musketeers, Killer Force, Diagnosis-Murder, Whispering Death, The Keeper, To The Devil a Daughter, Pere Et Fils, Airport 77, Alien Encounter, The End of the World, Return from Witch Mountain, Caravans, The Silent Flute, The Passage, The Pirates, Jaguar Lives, Arabian Adventure.

LEE, JOHN: Actor. b. Launceston, Tasmania, March 31, 1928. Early career on Australian stage and radio. Ent. m.p. ind. in England 1957.

PICTURES INCLUDE: Dunkirk, Cat Girl, Silent Enemy, Flying Scot, Gypsy and The Gentleman, Under Ten Flags, The Liar, Seven Keys, The Secret Partner, Dr. Crippen, Stitch in Time, Go Kart Go, Space Flight.

TV: Adventures of the Sea Hawk, International Detective, Rendezvous, Flying Doctor, Danger Man, Shadow Squad, Man and Superman, A Man Involved, Aren't We All, Probation Officer, Ladies of the Corridor, Golden Girl, After the Crash, The Net, Emergency Ward 10, Zero One, Sapper Series, Dr. Fancy, The Shifting Heart, Sergeant Cork, Dr. Who, The Materialists, The July Plot.

LEE, PEGGY: Singer, actress. r.n. Norma Egstrom; b. Jamestown, N.D., May 26, 1920. Began career as night club vocalist in Fargo & radio singer, WDAY, then with Sev Olsen, bandleader, Minneapolis, Will Osborne; Benny Goodman; collab. (with Dave Barbour) popular songs, Manana, It's a Good Day, What More Can a Woman Do? Leading feminine vocalist. TV & records; screen debut in Mr. Music (Bing Crosby); singer on Bing Crosby program, CBS, TV appearances.

PICTURES INCLUDE: Jazz Singer, Pete Kelly's Blues, act., collab. songs, Lady and the Tramp.

LEE, PINKY: Performer, r.n. Pincus Leff. b. St. Paul, Minn. Started in musical comedy stock, vaudeville; with Gus Edwards troupe; then toured vaudeville circuits as solo act; appeared in burlesque, Broadway musicals; star, Earl Carrolls Theatre Restaurant, Hollywood, 1942–46; star of Pinky Lee Show, NBC network.

PICTURES INCLUDE: Lady of Burlesque, That's My Gal, In Old Amarillo, South of Caliente.

LEEDS, MARTIN N.: Film-TV Executive. b. New York City, Apr. 6, 1916. e. NYU, BS, 1936; J.D., 1938. Admitted N.Y. Bar, 1938, Calif. Bar, 1948; dir. indsl. relations Wabash Appliance Corp., 1943–44; indsl. bus. relations cons. Davis & Gilbert. 1944–45; dir. indsl. relations Flying Tiger Lines, 1947; dir. bus. affairs CBS. TV div., 1947–53; exec. v.p. Desilu Productions, Inc., 1953–60; v.p. Motion Picture Center Studios, Inc.: mem. Industry comm. War Manpower Comm., 1943; chmn. Com. to form Television Code of Ethics: U.S. Army 1941. Mem. Los Angeles Bar Assn.; exec. v.p. in chg. of West Coast oper. & member of bd. of dir. Talent Associates—Paramount Ltd., Hollywood, 1962; TV production consultant; exec. v.p., Electronovision Prods. Inc., 1964; TV prod. & MP prod. consultant, 1965. Pres., chief exec. officer member of bd., Beverly Hills Studios, Inc., 1969, senior v.p., American Film Theatre, 1973; 1975, motion picture and TV attorney & consultant.

LEENHOUTS, LEWIS GRANT: Executive. b. Los Angeles, California; m. Edith Hargrave. From 1934–1938 variously stunt man, unit manager, cutter & asst. dir., Collab. orig. story No More Women, wrote orig. Heroes at Leisure. In 1939 wrote training films, U.S. Officer of Education; 1940–41 assoc. prod., writer, dir., Caravel Films Inc.; 1942–45 head of planning and production U.S. Navy Training Film & Motion Picture branch, Navy Dept.; 1946–47 gen. mgr. in charge of prod., National Educational Films, Inc. Later coordinator-producer, Industry Film Project, Motion Picture Assoc.; prod. California's Golden Beginning, Paramount assoc. chief and head of prod., International M.P. Div., U.S. Dept. of State, 1950–51; exec. v.p., Cinerama, Inc., 1952–53; planning and prod. head, Ford Foundation Fund for Adult Education, 1954, v.p., prod. head Cinerama, Inc., 1955–60. Ind. producer; prod. consultant Macbeth, Compass Prods. and Grand Prize Films, London; pres., Leehouts Prods., Inc., 1961; prod., 20th Century Fox, Movietone Digest, Children's Digest, 1963; Prod. Consultant, Cascade Pics. of Calif., 1964–65; 1967, dir. prod., Breakfast At the Capitol; prod., New Sounds in Africa; prod. head, Eddie Albert, Inc., 1971.

LEE-THOMPSON, J.: Writer, Director, Producer. On Brit. stage; writer of stage plays including: Murder Without Crime, Cousin Simon, Curious Dr. Robson (collab.) Thousands of Summers, Human Touch. Writer and m.p. director.

PICTURES INCLUDE: The Yellow Balloon, Weak and the Wicked, Yield To The Night, Woman In The Dressing Gown, Ice Cold in Alex, No Trees in the Street, Tiger Bay, I Aim at the Stars, The Guns of Navarone, Taras Bulba, Cape Fear, Kings of the Sun, What A Way to Go, John Goldfarb, Please Come Home, Return From the Ashes, MacKenna's Gold, Battle for the Planet of the Apes, Before Winter Comes, Country Dance, Conquest of the Planet of the Apes, Huckleberry Finn, The Reincarnation of Peter Proud, St. Ives, The White Buffalo, The Greek Tycoon, The Passage, Caboblanco, Happy Birthday To Me.

LEEWOOD, JACK: Producer. b. New York City, N.Y., e. Upsala Coll., Newark U. and New York U. From 1926–31 with Gottesman-Stern circuit as usher, asst. and relief mgr.; 1931–43 Stanley-Warner mgr. of Ritz, Capitol and Hollywood theatres 1943–47. Joined Warner Bros. field forces in Denver-Salt Lake; Seattle-Portland, 1947–52. Dir. pub. & adv. Screen Guild Prod.; Lippert Productions; prod. exec. in chge. story dept.; 1953–56 Allied Artists; 1957–62 Prod. 20th Cent. Fox; 1965–68 prod., Universal; 1976–78. Affiliated Theatre Serrio; 1978–79. Aubrey/Hammer Prod.

PICTURES INCLUDE: Holiday Rhythm, Gunfire, Hi-Jacked; Roaring City, Danger Zone, Lost Continent, F.B.I. Girl, Pier 23, Train to Tombstone, I Shot Billy the Kid, Bandit Queen, Motor Patrol, Savage Drums, Three Desperate Men, Border Rangers, Western Pacific Agent, Thundering Jets, Lone Texan, Little Savage, Alligator, People, 13 Fighting Men, Young Jesse James, Swingin' Along, We'll Bury You, 20,000 Eyes, Thunder Island, the Plainsman, Longest 100 Miles, Escape to Mindanao, Dallas Cowboys Cheerleaders, When Hell Was in Session.

LEFFERTS, GEORGE: Producer, writer, director. b. Paterson, N.J., dir., numerous award-winning TV series, films, exec. prod.-Time-Life films 1977–78; prod./writer, Movie of the Week (NBC) 1977–78. Biog: Who's Who in America, Who's Who in the World. 1975 Emmy Award Benjamin Franklin Specials (CBS).

TV: 1969 Emmy Award, prod. Hallmark Hall of Fame "Teacher Teacher," 1963 Emmy Award, wrote, prod., directed Purex Specials for Women. 1965 Producer's Guild Award, exec. prod. "Breaking Point" series. Other shows: The Bill Cosby Show, Studio One, Kraft Theatre, Chrysler Theatre, Sinatra Show, Lights Out, Alcoa, The Harness (Movie of the Week), The Bold Ones, Exec. prod., Bing Crosby Productions, prod., NBC 10 yrs, Independent.

PICTURES INCLUDE: The Stake, Mean Dog Blues, 1977–78, The Living End, The Boat, The Teenager.

BROADWAY: The Boat, 1968, Hey, Everybody, 1970.

LEFKO, MORRIS E.: b. March 30, 1907. Entered m.p. ind. as poster clerk; booker, salesman, br. mgr., Indianapolis, RKO, June 1941; br. mgr., Pittsburgh, July 1944; East Central dist. mgr., July, 1948; appt. sales exec. of Ten Commandments Unit. Para. Film Dist. Corp., N.Y.C. 1956; v.p. in chg. sls., Michael Todd Co., 1958. Joined MGM, Inc., July 1960 sls. mgr., of road shows, Ben-Hur, King of Kings, Mutiny on the Bounty; vice pres., gen. sales mgr., MGM, 1963. Exec. consultant to Pres., Cinema 5 Ltd., Jan. 1970. v.p., Network Cinema Corp., 1972; American Film Theatre, Ely Lando Co., Nov., 1972; v.p., sls. mgr. Brut Prods., 1976.

LEFKOWITZ, NAT: Chairman Emeritus, William Morris Agency, Inc. b. July 24, 1905, Brooklyn. e. CCNY School of Business Adm., grad. 1926, CPA: Brooklyn Law School, grad. 1938-LLM: admitted to NYS bar, Nov., 1938. Community activities: United Jewish Appeal; Federation of Jewish Philanthropies; M.P. Pioneers; Variety Clubs International Cinema Lodge B'nai B'rith; member, National Academy of TV Arts & Sciences, board of trustees of Will Rogers Hospital and O'Donnell Memorial Research Laboratories; Boy Scouts of America; USO bd.; N.Y. City Athletic Club; Dellwood Country Club, Friars Club, Variety Club, Trustee, Natl. Conf. of Christians & Jews, 1980.

LE GALLIENNE, EVA: Actress—Producer. b. London, England, 1899. Stage career began in 1914. Founded Civic Repertory Theatre in 1926, starring in and directing such

classics as Hedda Gabler, Romeo and Juliet, Alice in Wonderland, Camille, The Cherry Orchard. Also co-founder of American Repertory Theatre; toured with National Repertory Theatr.
TELEVISION: Alice in Wonderland, The Corn Is Green, The Bridge of San Luis Rey, Mary Stuart, The Royal Family (Emmy award, 1977).
PICTURES: Prince of Players, The Devil's Disciple, Resurrection.

LEGG, STUART: Producer. b. London, England, August 31, 1910. m. e. Marlborough Coll., 1923–28; St. John's Coll., Cambridge, 1928–31. Began as dir. and ed., prod. of documentary and public information shorts. From 1931–32, with British Instructional Films; 1932–34, Empire Marketing Board film unit; 1934–37, G.P.O. Film Unit; 1937–38. Film Centre, London; 1938–39, Strand Films, London; 1939–42. National Film Board, Ottawa. In 1945, head N.Y. br., John Grierson's co.; editor, World in Action, shorts enterprises.

LEGRAND, MICHEL JEAN: Composer-Conductor. b. France, 1931. Son of well-known arranger, composer and pianist, Raymond Legrand. At 11 Michel, a child prodigy, entered Paris Cons. and graduated nine years later with top honors in composition and as solo pianist. In late Fifties turned to composing for films and has composed, orchestrated and conducted scores of more than 50 films.
PICTURES INCLUDE: Lola, Eva, Vivre Sa Vie, La Baie des Anges, The Umbrellas of Cherbourg, Banda a Part, Un Femme Mariee, Les Demoiselles de Rochefort, Ice Station Zebra, The Thomas Crown Affair, Pieces of Dreams, The Happy Ending, Wuthering Heights, The Go-Between, Summer of '42, Picasso Summer, The Nelson Affair, Breezy, The Three Musketeers, Sheila Levine, Gable and Lombard, The Hunter.
TV: Brian's Song.

LEHMAN, ERNEST: Writer-Producer, Director. b. New York City. e. City College of New York. Began career as financial editor and free-lance short story and novelette writer. Recently published first novel, The French Atlantic Affair. First screenplays were Executive Suite, Sabrina, Sweet Smell of Success, based on own novelette.
PICTURES INCLUDE: The King and I, Somebody Up There Likes Me, North by Northwest, From the Terrace, West Side Story, The Prize, The Sound of Music. Became a prod.-wr. for first time with Who's Afraid of Virginia Woolf? Prod.-wr. Hello Dolly; prod.-wr.-dir., Portnoy's Complaint; Family Plot (s.p.), Black Sunday (co.-s.p.).

LEHMAN, GLADYS: Writer, Scenarios identified with such Reginald Denny pictures as Out All Night, Clear the Deck, On Your Toes and Companionate Trouble; joined Columbia in 1929.
SCRIPTS INCLUDE: The Little Accident, A Lady Surrenders, The Cat Creeps, Many a Slip, Seed, Strictly Dishonorable, Back Street, Embarrassing Moments, Nice Women, Good Girls Go to Paris, Blondie, Two Girls and a Sailor, Thrill of a Romance, Her Highness and the Bellboy, This Time for Keeps, collab. s.p. Luxury Liner, Golden Girl.

LEHR, MILTON H.: Director, producer, writer. b. Sept. 10, 1918. m. Harriet Smerling, 1942. e. Ohio U., A.B., 1938; Ohio State U., B.S., 1939; Virginia Polytech., M.S., 1940; Tex. A.&M. cand., PhD., 1941. Pres., Int'l Prods., Puerto Rico, 1949–66; pres., Int'l Video Prods., Miami, London, Madrid, 1960–70.
PICTURES & TV: Legend of Juggler, 1970; The Ray Anthony Show, 26 one-hr. Films, 1969; Prof. Irwin Corey Special, 1969; Swinging Scene of Ray Anthony (RKO-General), 1968; 240 musical shorts for Scopitone, Color Sonics and Automatic Canteen, 1966–67; Robert Taylor Presents Cugat in Madrid, 1965; Command Performance with Edmundo Ros, 26 ½-hr. Progs. for BBC, 1964; The Latin Touch, 26 ½-hr. progs. for Rheingold and RKO-General, 1960–61; Showtime, 212 1-hr. progs. (General Motors in Spanish) 1957–60; Telemundo, Puerto Rico, 1953–57; Televisa, Caracas, 1952; CMQ, Havana, 1949–51.

LEIBMAN, RON: Actor. e. Ohio Wesleyan Univ. Joined Actor's Studio in N.Y.; first professional appearance in summer theatre production of A View from the Bridge.
STAGE: The Premise, We Bombed in New Haven, Cop Out, Room Service, I Oughta Be in Pictures.
PICTURES INCLUDE: Where's Poppa (debut), The Hot Rock, Slaughterhouse Five, Your Three Minutes Are Up, Super Cops, Won Ton Ton, Up the Academy, Zorro, the Gay Blade.
TV: A Question of Guilt (ff), Kaz (series).

LEIBOWITZ, SAM: Executive. b. New York City, Feb. 12, 1913. Service mgr., Reel Photos, 1931–4; svg. mgr., Consolidated Film Ind., 1935; Apco Photo Co., Inc., treas., 1936–62; Apco-Apeda Photo Co., Inc., 1962–present, pres. & chief operating officer.

LEIDER, GERALD J.: Producer. b. Camden, New Jersey, May 28, 1931. e. Syracuse Univ. 1953; Bristol Univ., Eng. 1954 Fulbright Fellowship in Drama. m. Susan Trustman, son Matthew Trustman, born April 14 1972, son Kenneth Harold, born June 18, 1975. 1955 MCA, Inc., N.Y.; 1956–59 theatre producer in N.Y., London; Shinbone Alley, Garden District, and Sir John Gielgud's Ages of Man. 1960–61; director of special programs, CBS/TV; 1961–62, dir. of program sales, CBS-TV; 1962–69, vice pres., television operation, Ashley Famous Agency, Inc. Sept. 1969–Dec. 1974, pres. Warner Bros. Television, Burbank. Jan. 1975–Dec. 1976, exec. vice pres. foreign production Warner Bros. Pictures, Rome. Jan. 1977–present, independent producer under Jerry Leider Productions.
PICTURES: Wild Horse Hank, The Jazz Singer (1980).
TELEVISION: And I Alone Survived, Willa, The Hostage Tower.

LEIGH, JANET: Actress, r.n. Jeanette Helen Morrison; b. Merced, Calif., July 6, 1927. e College of Pacific (music). Screen debut in Romance of Rosy Ridge, 1947.
PICTURES INCLUDE: If Winter Comes, Hills of Home, Words and Music, Act of Violence, Little Women, That Forsyte Woman, Red Danube, Doctor and the Girl, Holiday Affair, It's a Big Country, Two Tickets to Broadway, Strictly Dishonorable, Angels in the Outfield, Just This Once, Naked Spur, Confidentially Connie, Houdini, Walking My Baby Back Home, Prince Valiant, Living It Up, Black Shield of Falworth, Rogue Cop, My Sister Eileen, Pete Kelly's Blues, Safari, The Vikings, The Perfect Furlough, Psycho, The Manchurian Candidate, Bye, Bye, Birdie, Wives and Lovers, Kid Rodello, Grand Slam, Hello Down There, One Is A Lonely Number, Night of the Lepus, Boardwalk, The Fog.

LEIGH, SUZANNA: Actress. b. England, 1945. Studies at the Arts Educational School and Webber Douglas School. Film debut 1961 in Oscar Wilde, followed by Bomb in the High Street. 1964: TV series made in France, Three Stars. 1965–6: Under contract to Hal Wallis and Paramount, Boeing Boeing, Paradise Hawaiian Style, 1966: The Deadly Bees, Deadlier Than The Male. 1967: The Lost Continent. 1968: Subterfuge. TV film series in West Indies, One On An Island. 1969: TV play, The Plastic People. 1970.
PICTURES INCLUDE: To Love a Vampire, Beware My Brethren, Son of Dracula.
TV: The Persuaders.

LEISER, HENRI: Producer. b. Anbergue, France. Jan. 30, 1903; e. U. Strasburg, Berlin. Paris LL.D. Co-prod., prod. & pres. 1922–39 of Alpha Films, Inc., Monaco; Artem, Paris; Les Films, Paris; ARTIP, Cuba; assoc. Wm. Morris, Jr., personal repr. Villa Lobos; vice-pres. Villa Lobos Music Corp. & Teleconcert Corp ; independent prod. with Henry Souvaine, Inc.
PICTURES INCLUDE: Brothers Karamazoff, Living Corpse, One Night in London, Burning Heart, Incomplete Marriage, Variety, Three Valises.
TV: Tele Bingo, Sandy's Kiddy Club.

LELOUCH, CLAUDE: Director-Cinematographer. b. Paris, France, Oct. 30, 1937. Began m.p. career with short subjects, 1956; French military service, motion picture department, 1957–60; formed Films 13, 1960; publicity Films and Scopitones, 1961–62.
PICTURES INCLUDE: Le Propre de L'homme (Man's Own), L'amour avec des Si (Love With Ifs), Une fille et des Fusils (Guns and a Girl), To be A Crook, Un Homme et une Femme (A Man and A Woman), Vivre pour Vivre (Live for Life), Life, Love, and Death, Love Is a Funny Thing (dir., photog., s.p.) The Crook (dir., photog., s.p.) Adventure Is Adventure (prod., dir., s.p.). Smic, Smac, Smoc, La Bonne Anne, Another Man, Another Chance.

LEMLEIN, NEAL C.: Executive. e. Tulane Univ.; New York Univ. Began career at Young & Rubicam in N.Y.; later joined Doyle Dane Bernbach, Inc., where named snr. acct. exec. assigned to 20th-Fox feature films account and CBS-TV Entertaiment Division. In March, 1980, joined Universal Pictures as dir. of mktg.

LEMMON, JACK: Actor. b. Boston, Mass. Feb. 8; e. Harvard U. Stage debut as a child; radio actor on soap operas; stock companies; U.S. Navy, WWII; many TV shows; B'way debut in Room Service 1953 other Broadway shows include: Face of a Hero (1960) and Tribute (1978). Other legitimate theatre appearances include: Idiot's Delight (1970, Los Angeles), Juno and the Paycock (1975, Los Angeles), Tribute (1979, Denver and Los Angeles). m.p. debut in It Should Happen to You.
PICTURES INCLUDE: Three for the Show, Phfft, My Sister Eileen, Mr. Roberts (Academy Award, best supporting actor, 1955), You Can't Run Away From It, Fire Down Below, Operation Mad Ball, Bell Book and Candle, It Happened to Jane, Some Like It Hot, The Apartment, The Wackiest Ship in the Army, Notorious Landlady, Days of Wine and Roses, Irma La Douce, Under the Yum Yum Tree, Good Neighbor Sam, How To Murder Your Wife, The Great Race, The Fortune Cookie, Luv, The Odd Couple, The April Fools, The Out of Towners. Debut as director: Kotch, 1971, The War Between Men and Women, 's Wonderful, 's Marvelous, 's Gershwin (TV Emmy Award 1972),

Save the Tiger (Academy Award, Best Actor), Avanti, The Prisoner of Second Avenue, The Front Page, The Entertainer (TV in U.S.), Alex & The Gypsy, Airport '77, The China Syndrome (Cannes festival award for best actor, 1979), Tribute.

LEMONT, JOHN: Producer, director, screenwriter. e. Canada, U.S.A and England. Ent. m.p. ind. as actor and writer/technician. Early career working on features and documentaries. Service with Army Kinematograph Service as writ./dir. A.B.P. Director of Trans-Oceanic Telefilm Productions, Zodiac Productions Ltd., Venture Productions Ltd. V. Chmn. Writers Guild of Gt. Britain. Awards: WGGB Laurel, Edgar Allen Poe, NY/USA Festival Gold Award.
PICTURES INCLUDE: Vacation Rhythm, The Patrol, The Green Buddha, Cross Channel, Missing Person, Harvest of the Seas, People's Playground, Port of London, Mirror and Markheim, And Women Shall Weep, The Shakedown, Konga, The Frightened City, Deep Waters, Horse Called Jester, Woman On the Stair.
TV: Fabian of the Yard, British Heritage, Frances Drake, Strange Study, Errol Flynn Theatre, Ivanhoe; prod., Sixpenny Corner, Dir. The Glorification of Al Toolum; dir, Witness in The Dark; TV Playhouse series.

LEON, SOL: Executive. b. July 2, 1913, New York City. e. N.Y.U., CCNY, Brooklyn Law School, BBL, master of law. V.p., TV Dept., William Morris Agency, N.Y.C.

LEONARD, HERBERT B.: Producer. b. New York, Oct. 8, 1922. e. New York University. Columbia Pictures, 1940. Naval Aviator, 1941–46. Prod. mgr., Sam Katzman Prod., 1946–48; formed Herbert B. Leonard Prod., 1954. TV debut as prod., Rin Tin Tin Series. Formed Norber Prod., Inc., with Norman Blacburn, 1955, prod. Circus Boy, TV series. Formed Lancer Prod., prod. Tales of the Bengal Lancers, TV series, 1956. Formed Wilbert Prod., Inc. with Cinefilm, Inc., prod. Rescue 8 TV series, 1958. Formed Shelle Prod., Inc., prod. Naked City, TV series, 1958 and Route 66, TV series. Formed Leonard Films, Inc., prod. motion picture Popi for United Artists, 1968–69; joined Allied Artists Int'l., TV Prods.

LEONARD, SHELDON: Actor. r.n. Sheldon Leonard Bershad. b. New York, N.Y., Feb. 22, 1907; e. Syracuse U., B.A. Theatre mgr., Publix; N.Y. stage, 10 yrs.; sec., Directors Guild of America. 3 Emmy awards, Sylvania award, 4 TV Director of the Year nominations by D.G.A.
PICTURES INCLUDE: Another Thin Man, Tall, Dark and Handsome, Tortilla Flat, Rise and Shine, Lucky Jordan, Somewhere in the Night, Her Kind of Man, Its a Wonderful Life, The Gansters, If You Knew Susie, Sinbad the Sailor, My Dream Is Yours, Take One False Step, Iroquois Trail, Here Come the Nelsons, Young Man with Ideas, Stop You're Killing Me, Diamond Queen, Money from Home, Guys and Dolls; dir. Real McCoys, Pocketful of Miracles.
TV: 1953–56: dir. Make Room for Daddy, Damon Runyon, G.E. Theatre, Electric Theatre, Jewelers' Showcase, Jimmy Durante Show; prod.-dir. Danny Thomas Show; package & exec. prod., Andy Griffith Show, Dick Van Dyke Show; exec. prod., Gomer Pyle, U.S.M.C., I Spy, My World and Welcome To It. In 1975 starred in Big Eddie (CBS), 1977, exec. prod. and co-star in Top Secret.

LEONE, ALFRED: Producer, b. Nov. 30, 1926. In entertainment industry since 1958. Assistant producer for Braken Productions and Hallmark Productions. In 1968 formed Europa-America in Rome, Italy. 1974, formed Leone Intl. SPA in Rome. In 1978 opened offices in Beverly Hills. Produced and distributed over 35 films, among them: Vengeance, Four Times That Night, Baron Blood, Lisa And The Devil, House of Exorcism, Love by Appointment, Gold of the Amazon Women, The Snake.

L'EPINE-SMITH, ERIC: Producer, director, writer. b. Thornton Heath Eng. m. Brenda Cameron. Stage as actor, prod. playwright, formed own co. Aldwitch Films; joined Wieland's Agency; apptd. dir. of co.; casting dir. Warner Bros. 1st Natl.; assoc. prod. Edward Dryhurst Productions, then prod. Now artists' mgr., L'Epine-Smith, Ltd.
PICTURES INCLUDE: Down on the Farm, Pitchfork and Powder-Puff, Making a Man of Him, House of Silence, Five Pound Man, Coming of Age, While I Live, Noose, Romantic Age, Come Die My Love.

LERNER, ALAN JAY: Playwright, lyricist.
N.Y. STAGE PLAYS: What's Up, The Day Before Spring, Brigadoon (N.Y. Drama Critics Award 1947), Love Life With Weill, 1948, Paint Your Wagon, My Fair Lady, NY Drama Critic Award, Tony Award, Donaldson Award (1956), Camelot, with Frederick Loewe; Grammy Award, On A Clear Day You Can See Forever with Burton Lane; Coco with Andre Previn; Gigi, with Frederick Loewe (Tony Award best score 1973–74), 1600 Pennsylvania Avenue with Leonard Bernstein (1976); Carmelia (1979).
SCREEN PLAYS: Royal Wedding, An American In Paris (Academy Award, best s.p., best song, 1958), My Fair Lady, Camelot, Paint Your Wagon, On A Clear Day You Can See Forever, The Little Prince.

LERNER, JOSEPH: Producer, director, writer. m. Geraldine Lerner, film ed. radio stage mgr. & actor on Broadway; with RKO, Columbia and Republic as dir., dial. dir., writer, 2nd unit dir., test dir., dir.-writer & head of special productions U.S. Army Signal Corps Photographic Center; writer of commercial and educational films 1946–47; vice-pres. in chg. of prod. Visual Arts Productions 1947; vice-pres. in chg. prod. Laurel Films 1949; prod.-dir., TV Gangbusters series, Grand Tour series; Girl on the Run, com. ind. films, Three Musketeers series TV, dir.-prod, 1961; Director, producer, writer, MPO Videotronics, Inc. 1967–1973. President, The Place for Film Making, Inc.; pres., Astracor Associates Ltd. in production: The Ditch Digger's Daughter, The Little Hat, The Mapmakers. Also lecturer and instructor at New York University, Wm. Patterson College.
CREDITS INCLUDE: (writer-dir.) Fight Never Ends; (prod.-writer) Kings of the Olympics, Olympics Cavalcade; (prod.-dir.-writer) United Nations Case Book (for CBS-TV), C-Man; (prod.-dir.) Guilty Bystander, Mr. Universe, writer, director co-producer, The Dark of Day.

LEROY, MERVYN: Producer, director, b. San Francisco, Calif., Oct. 15, 1900. In vaudeville before entering m.p. ind. as contrib. comedy & other situations. From 1927 dir. many pictures, & from 1927 variously dir. & prod. In 1942 apptd. dir. supvr. Hollywood prod. for coordinator Inter-Amer. Affairs. In 1955 orig. own prod. co., Mervyn LeRoy Prods.; A "Champion Directors" in Fame ratings Special Acad. Award 1945 for dir. The House I Live In. In 1976 won the Irving Thalberg Award.
PICTURES INCLUDE: Wizard of Oz, King and the Chorus Girl, High Pressure, Thirty Seconds Over Tokyo, I Am a Fugitive from a Chain Gang, Hard to Handle, Gold Diggers of 1933, Five Star Final, Oil for the Lamps of China, Little Caesar, Tugboat Annie, Anthony Adverse, Three Men on a Horse, Escape, Johnny Eager, Random Harvest, Madame Curie, Without Reservations, Homecoming, Little Women, Any Number Can Play, East Side West Side, Quo Vadis, Lovely to Look At, Million Dollar Mermaid, Latin Lovers, Rose Marie, Mister Roberts, Strange Lady in Town, Bad Seed, Toward the Unknown, No Time for Sergeants, Home Before Dark, The FBI Story, Wake Me When It's Over, Majority of One, The Devil at 4 O'Clock.

LESLIE, ALEEN: Writer. b. Pittsburgh, Pa., Feb. 5, 1908; e. Ohio State U. Contributor to magazines; columnist Pittsburgh Press; orig. & wrote radio series A Date with Judy 1941–50. Bway play Slightly Married, 1943; wrote, prod. Date with Judy, TV series; author, The Scent of the Roses, The Windfall.
PICTURES INCLUDE: Doctor Takes a Wife, Affectionately Yours, Henry Aldrich Plays Cupid, Stork Pays Off, Henry Aldrich Gets Glamour, It Comes Up Love, Rosie the Riveter, Father Was a Fullback, Father Is a Bachelor.

LESLIE, JOAN: Actress. Star of Tomorrow, 1946. r.n. Joan Brodell; b. Detroit, Mich., January 26, 1925; p. Agnes and John Brodell; e. St. Benedicts, Detroit; Our Lady of Lourdes, Toronto; St. Mary's Montreal; Immaculate Heart. H.S., Los Angeles. Modeled for color photography advertising or Eastman Kodak Co. Bd. of dir. Damon Runyon Foundation. Now on bd. of dir., St. Anne's Hospital.
PICTURES INCLUDE: The Sky's the Limit, This Is the Army, Yankee Doodle Dandy, Thank Your Lucky Stars, Rhapsody in Blue, Cinderella Jones, Hollywood Canteen, Where Do We Go From Here?, Too Young to Know, Repeat Performance, Northwest Stampede, Born To Be Bad, Skipper Surprised His Wife, Man in the Saddle, Toughest Man in Arizona, Hellgate, Woman They Almost Lynched, Flight Nurse, Jubilee Trail, Hell's Outpost, Revolt of Mamie Stover, Police Story.
TV: Ford Theatre, G.E. Theatre, Queen for a Day, The Keegans (MOW), Various commercials.

LESSER, SEYMOUR H. Executive. b. New York City, Nov. 9, 1929. e. Pace U., C.C.N.Y. Public accounting, 1953–60; asst. corporate controller & mgr. of accounting, Metro-Goldwyn-Mayer, 1960–68; gen'l mgr., financial v.p., Robbins Music Corp., 1968–70; financial consultant for entertainment industry, 1970–73; treasurer, Allied Artists Pictures Corp.

LESTER, MARK: Actor. b. Oxford, England, 1958. Ent. m.p. ind. 1963.
PICTURES INCLUDE: Allez France, Our Mother's House, 1967; Oliver, 1968; Run Wild, Run Free, 1969; The Boy Who Stole the Elephant, Eye Witness, SWALK (Melody), Black Beauty, Whoever Slew Auntie Roo?, Redneck, Crossed Swords.
TV: Special for Disney in Hollywood. 1970: American series for Krofft Television Productions, Scalawag, 1972; Graduation Trip, 1972; Danza Alla Porto Gli Olmi (Italian Entry Berlin '75), Seen Dimly Before Dawn '75.
STAGE: The Murder Game '76; The Prince & The Pauper '76.

LESTER, RICHARD: Director, composer. b. Philadelphia, Pa., 1932. Early career: dir. and mus. dir. TV, CBS, Phila., CBC-TV, Toronto. Ent. m.p. ind. 1957. TV: dir. TV Goon Shows. Composed (with Reg. Owen) Sea War Series. Films: Composed and dir., Running, Jumping and Standing Still Film.
PICTURES INCLUDE: It's Trad, Dad, Mouse On the Moon, Hard Day's Night, The Knack, Help! A Funny Thing Happened on the Way to the Forum, How I Won The War, Petulia, The Bedsitting Room, The Three Musketeers, Juggernaut, The Four Musketeers, Royal Flash, Robin & Marian, The Ritz, Butch and Sundance: The Early Days, Cuba, Superman II.

LETTER, LOUIS N.: Executive, b. New York, N.Y., August, 1937. e. Brooklyn College Business Administration, v.p. and dir. of operations, Century Theatres, New Hyde Park, N.Y.

LEVEE, MICHAEL: Producer. e. Stanford Univ. Four years as U.S. Navy officer in World War II. After service went to work for father's company, M.C. Levee Agency. In 1947 formed Levee-Stark Agency with Ray Stark. Joined MCA in 1950; 1952, appt. v.p. in chg of TV division. When MCA phased out of agency business became exec. v.p. at CMA. Later joined Robert Radnitz as exec. v.p. of Radnitz/Mattel division of Mattel Toys, formed to finance family-oriented films. In 1975 rejoined Stark as exec. v.p. of Rastar Group. Left to produce independently.
PICTURES: The Black Bird (exec. prod.), Casey's Shadow (exec. prod.), Slow Dancing in the Big City (co.-prod.).

LEVEN, BORIS: Art director, production designer. b. Moscow, Russia. e. Certificate of Beaux Arts Inst. of Design, N.Y., U.S.C.—Bachelor of Architecture degree. Painter of water colors, many awards in architectural competitions; sketch artist, set designer, Paramount Studios, 1933–35; asst. art dir., Major Pictures Corp., 1936; art dir., 20th Century-Fox, 1937–38, 41–42, 45–46, Universal, 1947–48, freelance from 1948. Nominated for Acad. Award 8 times. Academy award in art direction for West Side Story. Member: Academy of M.P. Arts & Sciences, Soc. of M.P. & TV Art Dirs. Delta Phi Delta, Tau Sigma Delta, Skull & Dagger. U.S. Army Air Force 1942–45.
PICTURES INCLUDE: Alexander's Ragtime Band, Tales of Manhattan, Shanghai Gesture, Hello Frisco Hello, I Wonder Who's Kissing Her Now, Criss-Cross, Mr. Peabody and The Mermaid, The Senator Was Indiscreet, Sudden Fear, The Star, Silver Chalice, Giant, Anatomy of Murder, West Side Story, Two For the Seesaw, The Sound of Music, The Sand Pebbles, Star! A Dream of Kings, The Andromeda Strain, The New Centurions, Jonathan Livingston Seagull, Mandingo, New York, New York, The Last Waltz.

LEVEY, ARTHUR: Pres., Skiatron Electronics & Television Corp.; Skiatron Broadcasting Co. b. New York City, 1903. Film industrialist operating in London and New York. Developed unique inventions, including Skiatron dark-trace tube used in Berlin Airlift and atomic tests, etc. Ultrasonic system of large-screen TV. Also Subscriber-Vision system of subscription television, etc. Member: Masons (Anima Lodge, London).

LE VIEN, JACK: Producer/Director. b. N.Y., N.Y., 1918; film ed., reporter, Pathe News; military service, 1941–46; news ed., gen. mgr., v.p., Pathe News; chmn, American Newsreel Assoc., 1956–59; dir. of prod., Hearst Metrotone News; prod. in assoc., ABC-TV, The Valiant Years; exec. prod., Black Fox; prod. The Finest Hours, A King's Story; chmn., exec. prod., Le Vien Films Ltd.; prod.; Other World of Winston Churchill (NBC, BBC), The Gathering Storm, Walk With Destiny, The Amazing Voyage of Daffodil & Daisy, The Queens Drum Horse, Where The Lotus Fell, Churchill and the Generals.

LEVIN, IRVING H.: Executive. b. Chicago, Ill., Sept. 8, 1921. e. grad., U. of Ill. U.S. Air Force, 1943–45; entered m.p. ind. as partner of Kranz-Levin Pictures and Realart Pictures of Calif. Inc.; indep. exch. in 1948; formed Mutual Prod, 1952; pres., Filmakers Releasing Org., 1953; secy., Filmakers Prod., Inc., 1952; pres. AB-PT Pictures Corp., 1956; pres., exec. prod., Oakhurst Television Prod., Inc., Sindee Prod., Inc.; pres., exec. prod. Atlas Enterprises, Inc.; pres., prod. Atlantic Pictures, 1959; exec. vice-pres., mem. of bd. of dir., National General Corp. (formerly NT&T) 1961 and Pres. National General Prod. Inc. 1966 & Pictures Corp. 1967. Pres. and chief operating officer, NGC. In 1975 formed Levin-Schulman Prods.

LEVIN, IRVING M.: Executive. b. San Francisco, Oct. 1916. ent. Theatre work, managing showhouses 1932; div. dir., San Francisco Theatres, circuit, 1945–79; bd. of dir., S.F. Theatres, Inc., 1945–68; pres., bd. of dir. pres. Blind Babies Foundation, 1958–79; Chief Barker, Variety Club of Northern Calif., Tent 32, 1957–58; bd. of dir., S.F. Chamber of Commerce, 1961–63; Star of Solidarity Award—Italian Gov., 1960; founded S.F. Film Festival, 1957; exec., dir., of Festival, 1957–64. President, NATO of Northern California, 1971–72; Chairman, Northern California Theatre Assn.; Chm., Theatre Assn. California.

LEVIN, JACK H.: Executive. b. New York City, Aug. 19. e. Technical Inst. New York. Mutual Film Co. 1916; F.I.L.M. Club to 1923; Film Board Trade to 1927; founded Copyright Protection Bureau, 1927–45; orig. audit bureau for m.p. ind. 1941; founded Confidential Reports, Inc. 1944; founded Certified Reports 1951, founded Certified Investigations, 1952. Representing theatre owners-theatre inspections and ticket integrity tests. Also market research and industrial investigations.
AUTHOR: 1932: Bootlegging 1934: Arguments in favor of Federal Registration. 1939: Marriage, Morals and Mothballs. 1943: Marriage, Morals and War, for U.S. Army.
Pres. Assoc. of Licensed Detectives of N.Y. State (1962–63); Chief Barker, Inc. Variety Club Tent 35 (1964–66). Currently Nat'l. Comm. Vocational Service, B'nai B'rith.

LEVINE, HOWARD A.: Executive. Served in legal depts. of advertising agencies, BBDO and Ogilvy & Mather, before joined legal dept. of Paramount Pictures as counsel to mktg. div. in 1976. In 1979 appt. exec. dir.—prod. mktg. for Paramount's motion picture div.

LEVINE, JOSEPH E.: Executive. b. Boston, Massachusetts, Sept. 9, 1905. Entered motion picture industry as owner of theatre in New Haven, Connecticut. Entered national distribution in 1943. Named "Pioneer of the Year", Variety Club of N.Y.; "Showman of the Year", 1960. Named "Producer of the Year" by Allied States Assoc. of M.P. Exhibitions, 1962. Named "Master Showman of the World" by TOA, 1963; winner of Cecil B. DeMille Award for "Showman of the Year" from Hollywood Foreign Press, 1964; winner First Annual Award, Conference of Personal Mgrs., East, for outstanding contributions to World of Entertainment, 1965. President Avco Embassy Pictures. Resigned 1974 to form own co., Joseph E. Levine Presents.
PICTURES INCLUDE: Hercules, 1959; Two Women, 1962; Boccaccio '70, The Sky Above, The Mud Below, Long Day's Journey Into Night, Federico Fellini's 8½, The Easy Life, Yesterday, Today and Tomorrow, Marriage Italian Style, Divorce Italian Style, The Carpetbaggers, Zulu, Casanova '70, Darling, Woman Times Seven, The Graduate. The Producers, The Lion in Winter, Carnal Knowledge, Day of the Dolphin, A Bridge Too Far, Magic.

LEVINE, MARTIN: Executive. b. Brooklyn, N.Y., June 29, 1909. e. Syracuse U., Brooklyn Law. Prod. of The Roosevelt Story, doc.; pioneer in fgn. film dist. in U.S., pres. of Cinema Lodge, B'nai B'rith; former vice pres. & bd. member, Motion Picture Pioneers; Mem. MPP; former first asst. chief barker, Variety clubs of America, Tent No. 35. Active with J.D.A., A.D.L.; v.p. nat. fund cabinet of B'nai B'rith. Exec. v.p., William Brandt Theatres, 1957. Former chmn., A.D.L. m.p. industry, 1962. Mem. bd. dirs. I.T.O.A. Chmn. M.P. Boy Scout Drive, 1958–59; National Jewish Home, 1960–61. Chmn. N.Y. Regional Board A.D.L., 1962–63, Chmn., M.P. Federation of Jewish Philanthropies, 1962–63. Awards: 6 by foreign govts. for promotion of foreign films in U.S.; I.F.I.D.A., special citation; Torch of Liberty Award, A.D.L., 1968. Elected to bd. of dir., Motion Picture Pioneers, elected to Crew, Variety Tent 35; 1972, President, Brandt Theatres.

LEVINSON, BARRY: Producer. b. New York, 1932. Ent. m.p. ind. 1956. Chairman of Board of the Cinema Fund.
PICTURES INCLUDE: The Only Way, First Love, The Night Visitor, The Amazing Mr. Blunden, Catholics, Who, The Internecine Project (also wrote).

LEVINSON, NORM: Executive. b. New Haven, Conn., March 17, 1925. Started theatre business as usher for Loew's Theatres, 1940. U.S. Army, 1943–46. Returned Loew's Theatres managerial positions New Haven and Hartford, Connecticut. MGM press representative, Minneapolis, Jacksonville, Atlanta, Dallas. General Manager, Trans-Texas Theatres, Dallas. President, Academy Theatres, Inc., Dallas. Promoted Championship Boxing, Dallas and Johannesburg, South Africa. Executive Vice President, Cobb Theatres, Birmingham, Alabama. Member of the President's Advisory Committee of NATO. Chairman of the Alabama Advertising Committee for NATO.

LEVINSON, RUTH POLOGE: Executive. b. N.Y., N.Y., Oct. 5. e. CCNY, NYU; public relations, Arthur P. Jacobs, 1957; adv. publ. dept. J. Arthur Rank to 1959; assistant national dir. of advertising, publ., and exploitation, American International Pictures to 1979 when established Ruth Pologe Levinson Group Public Relations & Marketing & Personal Management.

LEVITT, RUBY REBECCA: Set decorator. b. Corinth, N.Y., Sept. 10, 1907; e. Pratt Inst., Brooklyn. Buyer, interior decorator, department stores; store mgr.; set decorator, Universal, since 1944.
PICTURES INCLUDE: Letter From an Unknown Woman, Magnificent Obsession, The Shrike, Six Bridges to Cross, Private War of Major Benson, This Earth Is Mine, Pillow Talk, 40 Pounds of Trouble, For the Love of Money, Sound of Music, Willie Boy, The Scavengers, Colossus, Change of Habit, Andromeda Strain, Vanished, Happy Birthday,

Wanda June, The Other, Freebie and The Bean, The Manchn Eagle, Chinatown, Once Is Not Enough, Let's Do It Again, Harry and Walter Go To New York, A Star Is Born, New York-New York, Looking for Mr. Goodbar, The One The Only, Matilda, Promises in the Dark, The Jazz Singer.

TV: Mr. Ed, Addams Family, Double Life of Henry Phyffe, Movies of the Week, ABC, Banyon TV Series, Genesis II, Name of the Game, Love Story, Having Babies, Return Engagement.

Staff Decorator for television commercials, Filmways of California, Inc. Staff of Filmways of Calif., 1966. Universal Studios, 1968.

LEVY, BERNARD: Executive. b. Jacksonville, Florida. e. Brooklyn Law School, L.L.B. Legal staff of Superintendent of Insurance of the State of New York in the rehabilitation and liquidation of guaranteed title and mortgage companies, 1934–36; private practice of law, 1936–46; legal staff, Paramount Pictures, Inc., 1946–50; legal staff, United Paramount Theatres, 1950–51; exec. asst. to Edward L. Hyman, vice-pres., American Broadcasting Companies, Inc., in chg. of theatre administration, north, 1951–62; apptd. exec. liaison officer for southern motion picture theatres, ABC, Inc., 1962–64; exec. liaison officer, m.p. theas., ABC, Inc., 1965–72; vice pres., ABC Theatre Division, 1973. Retired, 1976.

LEVY, BUD: Executive. b. Jackson Heights, N.Y. April 3, 1928. e. New York Univ. Exec. v.p. Trans-Lux Corp., N.Y. Member: Variety Clubs Int'l., M.P. Pioneers, bd. of Indep. Theatres Owners Ass'n.; dir., Trans-Lux Corp.; New York area chm., Will Rogers Memorial Fund; Member, President's Advisory Board-NATO; director: NATO, TOP, CATO. Elected pres., Trans-Lux Corp., 1980.

LEVY, DAVID: Executive. b. Philadelphia, Pa., January 2, 1913; e. Wharton, U. of Pa.; exec. v.p., Four Star International, Inc. As v.p. & assoc. dir., Young & Rubicam. Inc., 1938–59, acquisitions for clients include: Father Knows Best, Goodyear Playhouse, Life of Riley, Gunsmoke. Was v.p. in charge of network TV programs & talent, NBC, 1959–61.

TV: Bonanza, Dr. Kildare, Dick Powell Show, Saturday Night at the Movies, Bat Masterson, Klondike, Americans, Outlaws, The Addams Family, Double Life of Henry Phyffe, The Chameleons, The Gods of Foxcroft Carbor House, as well as numerous TV plays.

LEVY, HERMAN M.: Attorney, New Haven, Conn. b. New Haven, Conn., Sept. 27, 1904; e. Yale, BA, '27, Yale Law School, LL.B., '29; Phi Beta Kappa, was in legal dept. RCA Photophone; newspaper reporter; admitted to Connecticut bar, 1929. In 1939 elected exec. secy. of MPTO of Connecticut. In 1943: Elected general counsel MPTOA. Elected gen. counsel, Theatre Owners of America, 1947–63. Pres., New Haven County Bar Assn., 1964; legislative agent, Conn. Assn. of Theatre Owners.

AUTHOR: More Sinned Against . . . Natl. Bd. of Review Magazine, 1941. Proving the Death of a Non-Resident Alien, Conn. Bar Journal, 1950; Need for a System of Arbitration M.P. Ind., Arbitration Journal, 1950; reprint of Industry Case Digest, 20th Century-Fox vs. Boehm in the Journal (Screen Producers Guild); Book Review of Antitrust in the Motion Picture Industry, by Michael Conant (Univ. of Calif. Law Review).

LEVY, JULES V.: Producer. b. Los Angeles, Calif., Feb. 12, 1923; e. U.S.C. Property dept., W.B., 1941; Army Air Force, 1942–45.

PICTURES INCLUDE: The Vampire, Return of Dracula, Vice Squad, Without Warning, Down Three Dark Streets, Geronimo, Glory Guys, Clambake, Scalphunters, Sam Whiskey, Underground, McKenzie Break, The Hunting Party, The Honkers', McQ, Brannigan, White Lightning, Gator, Kansas City Bomber, Rally.

TV: Rifleman, Robert Taylor's Detectives, Law of the Plainsman, The Big Valley.

LEVY, NORMAN: Executive. b. Bronx, New York, Jan. 3, 1935. e. City College. In 1957 joined Universal Pictures, holding various sales positions; 1967, went to National General Pictures, ultimately being named v.p. and exec. asst. to pres.; 1974, Columbia Pictures, v.p., gen. sls. mgr. In 1975 named Columbia exec. v.p. in chg. of domestic sls.; 1977, exec. v.p., mktg; 1978. pres., Columbia Pictures Domestic Distribution. In 1980 joined 20th-Fox as pres. of Entertainment Group.

LEVY, ROBERT S.: Producer. b. New York, N.Y. Dec. 15, 1932; e. Columbia U., 1951–55, BA; mgt. consultant, Ernest L. Loen & Assoc., 1955–63; bd. of dir. American Marketing Assoc., 1960–61; v.p., 1963; prod. Lullaby, 1963; pres. Pebble Productions, prod., A Comedy Tale of Fanny Hill; assoc. prod., The Todd Killings, 1971; assoc. prod., Some Call It Loving, 1975.

LEWINE, ROBERT F.: Executive. b. New York, N.Y., Apr. 16, 1913; e. Swarthmore Coll. Worked for restaurant chain, in real estate; U.S. Navy, 1942; creative staff, Cine-Television Studios, Inc.; v.p. in charge of oper., 1946; formed own co., 1947, for prod. of TV comm., industrial m.p.; also eastern rep., Dudley Films; radio-TV dir., Hirshon-Garfield, Inc., 1953; eastern prog. dir., ABC, 1953; dir., ABC-TV network prog. dept., 1954; v.p. in charge of programming and talent ABC-TV network, 1956; v.p., NBC, prog. dept. chg. nighttime programming, 1957; v.p. network programs, 1958; v.p. Figaro, Inc., 1958; v.p. programs, CBS Films, 1959; v.p. programs, Hollywood CBS Television Network, Apr., 1962; officer Acad. TV Arts & Sciences from 1954, Exec. comm., dir., Acad. TV Arts & Sciences. Pres., N.Y. Chapter 1959; nat'l. pres., 1961; nat'l. trustee, 1961–63; National pres., Academy TV Arts & Sciences, 1961–63; first vice pres., dir., Academy TV Arts & Sciences Foundation; pres., 1964; exec. v.p., Creative Management Associates Ltd.; v.p., Warner Bros., TV Pres. Nat'l Acad. of Television Arts and Sciences; trustee, Columbia College, L.A., Calif.; trustee, American Women in Radio and Television Foundation. 1977-NBC Television Network, exec. prod. dir.; chairman, PAW Society (Preservation of Animal Wildlife); Member International Advisory Council, Population Institute; Chairman of the Board, Riverside Broadcasting Company.

LEWIS, ARTHUR: Producer—Director—Writer. b. New York City. e. Univ. of So. Calif. and Yale. Began career as writer and assoc. prod. on the Jones Family TV series. Five years in U.S. Army; returned to screenwriting before producing Three Wishes for Jamie on Broadway and producing and directing Guys and Dolls in London's West End. In mid-60s produced plays with Bernard Delfont at Shaftesbury Theatre in London.

PICTURES: Producer: Loot, Baxter, The Killer Elite, Brass Target.

TELEVISION: Brenner, The Asphalt Jungle (prod.writer).

LEWIS, BERNARD: Executive. b. Brooklyn, N.Y., May 18, 1912. e. Bklyn Coll., N.Y.U. Reporter, feature writer, Bklyn. Daily Eagle, 1935–37; A.P., 1937–40; pub. dept., Paramount, 1940–44; spec. events dir., then asst. to dir. of adv., 20th-Fox, 1944–46; first pub. mgr. then exploit., prom. mgr., I.F.E., 1952–55; adv. pub. dir., Kingsley International Pictures, 1956–60; own pub. rel. org., The Bernard Lewis Company, 1960 to present. Lecturer at N.Y.U. on motion pictures.

LEWIS, CLAUDE P. JR.: Executive. b. Hutchinson, Kans., e. Howard U.; NYU, B.A., 1949. adv. dept., 20th Century-Fox, 1950; staff copywriter, 1951; copy chief, 1962; asst. adv. mgr., 1966; adv. mgr.; div. of adv., East Coast, 1979.

LEWIS, DAVID: Executive, b. Mass. e. N.Y.U.; ent. m.p. ind., 1925; with MGM in Paris, Madrid, Lisbon; mgr., Central America MGM. 1930; mgr., Cuba: mgr. Japan; Argentina, 1937; Brazil, 1939; since 1945 in Europe, headquarters Paris; vice pres., MGM. International and Regional dir. for Europe. 1968, pres. Two World Enterprises.

LEWIS, EDWARD: Producer. Began entertainment career as script writer, then co-produced The Admiral Was a Lady and teamed with Marion Parsonnet to bring the Faye Emerson Show to TV. Subsequently prod. first Schlitz Playhouse and China Smith series. Was v.p. of Kirk Douglas' indep. prod. co., where was assoc. prod. and writer-prod. Collaborated with John Frankenheimer on 8 films.

PICTURES INCLUDE: Lizzie (assoc. prod.), The Careless Years (prod., s.p.), Spartacus, The Last Sunset, Lonely Are the Brave, The List of Adrian Messenger, Seconds, Grand Prix, The Fixer (exec. prod.), The Gypsy Moths (exec.), I Walk the Line (exec.), The Horsemen, The Iceman Cometh (exec.).

LEWIS, HAROLD G.: Executive. b. New York City, Sept. 18, 1938. e. Union College; graduated 1960, electrical engineer. Joined ATA Trading Corp. in 1960 and has been pres. since 1977. Producer of feature animation. Importer and exporter for theatrical and TV features, documentaries, series, classics, etc.

LEWIS, JERRY: Actor. b. Newark, N.J., Mar. 16, 1926. e. Irvington H.S. m. Patti Palmer, vocalist, Jimmy Dorsey & Ted Florita bands; has six sons. Began his career in school theatricals, at 14, awarded plaque by American Red Cross for amateur show, Mosque Theatre. Formed comedy-team with Dean Martin (singer, straight man) at 500 Club, Atlantic City, N.J., 1946; appeared Latin Casino, Phila., then other nightclubs, on NBC television; played many m.p. theatres; signed by Hal Wallis; team m.p. debut in My Friend Irma (see biog. Dean Martin). Voted Most Promising Male Star in Television in m.p. Daily's 2nd annual TV poll, 1950. Voted (as team) one of top ten money making stars in m.p. Herald-Fame poll, 1951, 1953–54–57. Number 1, 1952; best comedy team in m.p. Daily's 16th annual radio poll, 1951, 1952, 1953; 1956 formed Jerry Lewis Prods. Inc., functions as prod., dir., writer & star.

PICTURES INCLUDE: My Friend Irma Goes West, At War With the Army, That's My Boy, The Stooge, Sailor Beware, Jumping Jacks, The Caddy, Scared Stiff, Money From Home, Living It Up, Three Ring Circus, You're Never Too Young, Artists and Models, Pardners, Hollywood or Bust,

Delicate Delinquent, Sad Sack, Rock-A-Bye Baby, Geisha Boy, Don't Give Up the Ship, Visit to A Small Planet, Cinderfella, Bellboy, The Ladies Man, The Errand Boy, It's Only Money, The Nutty Professor, Who's Minding the Store?, The Patsy, The Disorderly Orderly, Boeing-Boeing, The Family Jewels, Three on a Couch, Way . . . Way Out, The Big Mouth, Don't Lower the Bridge Raise the River, Hook Line and Sinker, Which Way to the Front?, One More Time, Hardly Working (star, dir., co-s.p.).

LEWIS, JOSEPH H.: Director. b. New York, N.Y., Apr. 6. e. DeWitt Clinton H.S. Camera boy, MGM; then asst. film ed. in chge. film ed., Republic; dir. in chge. 2nd units; dir. Universal; U.S. Signal Corps., WW II; dir. RKO, Columbia, W.B.

PICTURES INCLUDE: My Name is Julia Ross, So Dark the Night, Jolson Story, The Swordsman, Return of October, Undercover Man, Gun Crazy, Lady Without Passport, Retreat Hell!, Desperate Search, Cry of the Hunted, Big Combo, A Lawless Street.

TV: Rifleman series, Barbara Stanwyck Show: The Big Valley.

LEWIS, JUDY: Actress. b. Los Angeles, Nov. 6, 1936. e. New York School of Interior Design, Duchesne Residence School, N.Y., Marymount School, L.A.; mother Loretta Young.

TV: The Outlaws, The Secret Storm, Policewoman, Apple's Way, Streets of San Francisco, FBI.

BROADWAY: Mary, Mary.

LEWIS, LESTER: talent rep. b. New York, N.Y. Sept. 12, 1912. e. H.S. of Commerce, N.Y. Pres. Lester Lewis Associates, Inc., talent agency.

LEWIS, MICHAEL J.: Composer. b. Wales, 1939. First film score 1968, The Mad Woman of Chaillot, won Ivor Novello Award for best film score. 1973: first Broadway musical, Cyrano, Grammy Nomination '74, Caesar & Cleopatra (T.V. '76).

PICTURES INCLUDE: The Man Who Haunted Himself, Julius Caesar. 1970: Upon This Rock,1971: Unman, Wittering and Zigo, Running Scared, Baxter, Theatre of Blood, 11 Harrowhouse, 92 In The Shade, Russian Roulette, The Stick-Up, The Medusa Touch, The Legacy, The Passage, The Unseen, Sphinx.

LEWIS, MONICA: Singer, actress. b. Chicago, Il., May 5. e. Hunter Coll., N.Y. p. Leon Lewis, concert pianist, former medical dir. CBS; Jessica Lewis, child star with Nazimova; member Ben Greet's Shakespearean players, sang leading roles. Chicago Opera Co.; now vocal teacher. Started career as radio singer, own show, WMCA, N.Y.; on Beat the Band, 1946; co-star Chesterfield program; sang leading role, Girl Crazy, Ford Sunday Evening Hour, Own program, Monica Makes Music co-star Revere Camera show. Among first ten feminine singers in country on recording polls. Vocalist: Stork Club, Astor Roof, Copacabana, Persian Room.

PICTURES INCLUDE: Inside Straight, Excuse My Dust, The Strip, Everything I Have is Yours, Affair With a Stranger.

LEWIS, ROGER H.: Executive. b. New York, N.Y., Mar. 14, 1918. e. Lafayette Coll., Easton, Pa. 1939; postgrad. U.C.L.A., Columbia. Apprentice in adv. pub. dept. Warner, 1939; U.S. Army, WW II; spec. asst. to dir. of adv., pub. & exploit., 20th-Fox; v.p. creative dir. Monroe Greenthal Co.; adv. mgr. United Artists, 1952–55; nat'l dir. of adv. pub. and exploit, 1956; v.p. adv. pub. expl. 1959. Resigned June 1961 to enter independent film production. Exec. v.p. Garrick Films; prod. The Pawnbroker; TV script, The Defenders. Exec. Horizon Pictures; prod. The Swimmer; prod. exec. National General; script consultant, Universal-TV; prod. exec. W.B.; exec. producer, Shaft, 1971; Co-Producer, Shaft's Big Score, 1972; producer, Shaft in Africa, 1973. Now exec. v.p., Max Youngstein Enterprises.

LEWIS, TOM: Producer, b. Troy, N.Y. e. Union College, Schenectady, N.Y. m. Loretta Young, Prod. dir., Schenectady WTAM; prod. head Cleveland, 1934; prog. mgr. NBC Cleveland Div. 1935; radio prod., Young and Rubicam, 1937; then radio supervisor, mgr. of radio dept. v.p. radio prod., Hollywood offices. Now v.p. stockholder, member of plan. bd., v.p. Audience Research, Inc., 1941; U.S. Army World War II; created and became commandant Armed Forces Radio Service; awarded Legion of Merit. Excellent Order of British Empire, degree of Honorary Office of His Majesty the King of England; Knight of Malta; introduced Aldrich Family on Kate Smith Hour, developed program's subsequent format; also assoc. with Screen Guild Theatre, Fred Allen Show, Jack Benny Show; prod. writer, Cause for Alarm, MGM. Pres. Lewislor Prod., prod. rewriter Loretta Young Show, 1953–55; v.p. and head of TV-Radio dept., C.J. LaRoche and Co., Inc., New York; dir. Nat. Soc. of TV prod.; dir. of Academy of TV Arts and Sciences, bd. of dir. Beverly Hills Hotel

LEWIS, WYNDHAM O.B.E. (L.R.A.M., A.T.C.L.): Exhibitor. b. Briton Ferry, S. Wales, 1910. e. Neath Grammar School. Studied music at Royal Academy of Music; entered m.p. industry at Regal Marble Arch. 1929. Joined Odeon Theatres 1937. Served as Flying Officer RAF, 1940–45; returned to take over as man. dir. West of England Cinemas and assoc. companies: now booking director of circuit of 9 cinemas. Justice of the Peace for the City of Cardiff. President Cinematograph Exhibitors Assoc., 1966–69.

LIBERACE: Performer, pianist, r.n. Wladziu Valentino Liberace. b. Milwaukee, Wis., May 16, 1920. e. Wisconsin College of Music. Guest soloist with Chicago Symphony at 16, then hotel, night club entertainer, concert and recording artist. Own TV shows since 1952; m.p. debut in Sincerely Yours.

LIBERMAN, FRANK P.: Publicist, b. N.Y., N.Y., May 29, 1917. e. Cheshire Acad., Conn. 1934; Lafayette College, Easton, Pa., B.A. 1938. m. Patricia Harris, casting dir. Worked as copy boy, NY Daily News, 1938–39. Began career as publicist at Warner Bros. home office as messenger, 1939, promoted to pressbooks dept., transferred to Warner's Chicago office as field exploitation man. U.S. Signal Corps, 1941, public relations officer, Army Pictorial Service, on temporary duty with War Dept., Bureau of Public Relations in Pentagon. Discharged as Capt., 1946. Rejoined Warner Bros. on coast 2 years, 1947, est. own public relations office. Owner, Frank Liberman and Associates, Inc.

LICCARDI, VINCENT G.: Executive. b. Brooklyn, N.Y., Started as messenger at Universal Pictures, adv. dept. handling co-operative adv. Universal, asst adv. mgr. on Around the World in 80 Days, Michael Todd Co., asst. to exec. co-ord. of States & Adv. on Spartacus; National Exploit. Dir., Continental; National Dir. of Adv. & Publ., Continental; Nat. Dir. Adv. & Publ., Braintree Prod., adv. pub. mgr. Allied Artists, ad. mgr. Paramount, National Dir. Adv.-Pub., UMC Pictures, Screenwriter, Playboy to Priest, The Rivals, The Rivals-Part II, The Greatest Disaster of All Time, The Lady on the 9:40.

LIDER, EDWARD W. Executive. b. New Bedford, Mass., March 13, 1922. e. Dartmouth, Harvard Law. Served as attorney-at-law, 1948–50. President & treasurer, Fall River Theatres Corp. & Nathan Yamins Enterprises, 1950 to present; member of bd., Theatre Owners of New England; past pres., Theatre Owners of New England; past member of bd. & past treas., Allied States Assoc. of M.P. Exhibitors; general manager of Sonny & Eddy's Theatres in Boston: Exeter St., Academy, Harvard Square, Central Square, Allston C1nema-C2nema and Galeria Theatres.

LIEBER, PERRY W.: Publicist. b. Pleasant Prairie, Wis., June 1, 1905; e. U. of Illinois. Was in adv. dept. of J. P. Seaburg Co., Chicago, Ill. Pub. dept. RKO Radio Studio 1930; publ. dir. 1939; nat'l dir. pub., exploit. supv. N.Y. office, 1953–55. 20th Century-Fox pub. dept. 1956; in chg. studio adv. pub. dept., 20th Century Fox Studio, Aug. 1962. 1966 pub. dept. 20th Century-Fox. Mgr./P.F. Summa Corp. since 1970, Consultant, P.R., Summa Corp since July, 1977.

LIEBERFARB, WARREN: Executive. e. Wharton School of Commerce and Finance, Univ. of Penna (BS, Economics); Univ. of Michigan (Master of Business Admin.). Started career in industry at Paramount Pictures as dir. of mktg. and exec. asst. to Stanley Jaffe, then pres. Later joined 20th-Fox as v.p.—special market dist. (cable, pay-TV, non-theatrical), also representing 20th in all MPAA matters on cable and pay/TV. Joined Warner Bros. as v.p., exec. asst. to Ted Ashley, bd. chm.; later named v.p., intl. adv.-pub. In 1979 joined Lorimar as v.p., of Lorimar Productions, Inc., the parent company, based in New York. Promoted to snr. v.p.

LIEBERSON, SANDY: Executive. b. 1936. 1979, named pres. of 20th-Fox Productions, which company he joined in 1977 as v.p.—European production. Previously an independent producer (Performance, Bugsy Malone, Jabberwocky, etc.). Prior to that exec. in chg. of European operations of Creative Management Associates. In 1980 named int'l. v.p. for Ladd Co., based in London.

LIEBLING, HOWARD: Executive. Began career in publicity with 20th-Fox in N.Y. as magazine contact, following which was N.Y. press repr. for Stanley Kramer, pub. mgr. for United Artists, and natl. pub. mgr. for MGM on West Coast. Rejoined Fox in July, 1975 as prod. pub. mgr.

LIEBMAN, MAX: Producer-director. m. Sonya; writer. Career in theatre; producer Tars and Spars for U.S. Coast Guard; discovered Danny Kaye, Sid Caesar, Betty Garrett, Jules Munshin, Imogene Coca, Mat and Hari, Jerome Robbins, etc.; prod.-dir. Your Show of Shows since 1949; prod. Bob Hope's TV debut and NBC series color spectaculars; also Max Liebman Presents. Look award for best prod., 1950–52; Emmy Awards 1951–52; U.S. Steel Specials 1960–61.

LIGHTMAN, M.A., JR.: Exhibitor. b. Nashville, Tenn., Apr. 21, 1915; e. Southwestern U., Vanderbilt U., 1936, B.A. bd. chmn. Malco Theatres, Inc., Memphis, Tenn.

LINDBLOM, GUNNEL: Actress-Director. b. Gothenburg, Sweden, 1931. Discovered by Ingmar Bergman while studying at drama school of Gothenburg Municipal Theatre (1950–

53); she moved to Malmo, where he was director of the local Municipal Theatre. Under Bergman's direction she played in Easter, Peer Gynt, Faust, etc. between 1954–59. Later appeared in many Bergman films. Since 1968 has been on staff of Stockholm's Royal Dramatic Theatre, assisting Bergman and then beginning to direct on her own. Made film debut as director with Summer Paradise in 1977.
PICTURES INCLUDE: Actress: The Seventh Seal, Wild Strawberries, The Virgin Spring, Winter Light, The Silence, Rapture, Loving Couples. Director: Summer Paradise.

LINDEN, HAL: Actor. b. Bronx, N.Y., March 20. e. City College New York. Began career as saxophone player and singer, playing with bands of Sammy Kaye, Bobby Sherwood, etc. Drafted and performed in revues for Special Services. After discharge enrolled at N.Y.'s American Theatre Wing; appeared on Bdwy. in Bells Are Ringing, replacing Sydney Chaplin. On many TV talk and variety shows.
THEATRE: On a Clear Day, Wildcat, Something More, Subways Are for Sleeping, Ilya Darling, Apple Tree, Three Men on a Horse, Pajama Game, The Rothschilds.
TELEVISION: Host on ABC series, Animals Animals Animals; Barney Miller.
PICTURE: When You Comin' Back Red Ryder?

LINDFORS, VIVECA: Actress. b. Uppsala, Sweden, Dec., 1920; e. Royal Dramatic School, Stockholm. Stage debut in Ann-Scofi Hedvig school prod. Screen debut in The Crazy Family, 1941; reached stardom in If I Should Marry the Minister. U.S. screen debut in Night Unto Night.
PICTURES INCLUDE: Adventures of Don Juan, Dark City, Flying Missile, Gypsy Fury, No Sad Songs For Me, Journey Into Light, Four In a Jeep, The Raiders, No Time for Flowers, Run for Cover, Captain Dreyfus, Coming Apart, Puzzle of a Downfall Child, The Way We Were, Welcome to L.A., Girl Friends, A Wedding, Voices, The Hand.

LINDNER, TERRELL M.: Executive. b. Dromana, Australia. Aug. 10, 1915. Entered m.p. ind. in pub. dep. Columbia Pictures, Melbourne, Australia, 1942. Columbia mgr. Western Australia, Dec. 1946. gen. mgr. Columbia, New Zealand, 1948. United Artists home office rep. for India, Burma, Pakistan, Ceylon, 1952. apptd. United Artists dist. supvr. for Southeast Asia. hdqts. Bombay. Dec. 1953. Rank overseas. 1957. Opened West Indies offices. man. dir. Rank Filmes do Brasil, 1958. man. dir. Filmcenter Internacional Ltda., Brazil, (successors Rank) 1970.

LINKLETTER, ART: Emcee, prod. b. Moose Jaw, Saskatchewan, Canada, July 17, 1912. e. San Diego State Coll. Radio prg. mgr., San Diego Exposition, 1935; radio pgm. mgr. S.F. World's Fair, 1937–39; freelance radio ann. and m.c. 1939–42; m.c. People are Funny since 1942. Starred Inside Beverly Hills, NBC-TV, 1955; exec. prod. host, NBC-TV spec. Salute to Baseball, 1956; host, Art Linkletter's Secret World of Kids, NBC-TV's Ford Startime, 1959; 1969 House Party series became the Linkletter Show.
Author of: The Secret World of Kids, 1959, Kids Say the Darndest Things, 1957, Linkletter Down Under, 1969, Yes, You Can, 1979.

LINSON, ART: Producer-Director. b. Chicago. e. UCLA. Was rock music manager before turning to film production. Debuted as director also with Where the Buffalo Roam.
PICTURES: Rafferty and the Gold Dust Twins, (co.-prod.), Wash, American Hot Wax (also co-s.p.), Melvin and Howard.

LIPPERT, ROBERT J., JR.: Producer. director, film editor. b. Alameda, Calif. Feb. 28, 1928. e. St Mary's College 1946; all conference Football 1947. Film Editor of 65 motion pictures. Produced and directed nine pictures for Lippert Pictures and 20th Century Fox. Present position is president, for Lippert Theatres headquartered in San Francisco.

LIPSTONE, HOWARD H.: Executive, producer. b. Chicago, Ill., Apr. 28. e. UCLA, Univ. of So. Calif. Ass't to gen. mgr. at KLTA, 1950–55; program dir. at KABC-TV, 1955–65; exec. ass't to pres. at Selmur Prods., ABC subsidiary, 1965–69. Ivan Tors Films & Studios as exec. v.p., 1969–70; pres., Alan Landsburg Prods., 1970 to present.

LIPTON, DAVID A.: Executive. b. Chicago, Nov. 6, 1906. U.S. Army, WW II. Entered m.p. ind. 1921 as office boy, Balaban & Katz, Chicago; in 1922 joined pub. dept.; in 1929 transf. Detroit; in 1930 joined Famous Players Canadian Corp., org. pub. dept.; ret'd to Chicago office 1931; res. 1933 to become publ. dir. for Sally Rand. In 1937 joined CBS, N.Y. as press relations counsel. Named publ. dir. Universal N.Y., 1938; later to West Coast as studio publ. dir. In 1941 to N.Y. as dir. adv. publ. & exploit., Columbia; returned to Universal as exec. coordinator of adv. & promotion, 1946; nat'l dir. adv. pub., Jan. 1949; elected v.p. in chge. adv. pub.; 1974, MCA Discovision, Inc., public relations director.

LISTER, MOIRA: Actress. b. Capetown, So. Africa. e. Holy Family Convent, Johannesburg. Stage debut at 6 yrs. of age in Vikings of Heligoland; screen debut in Shipbuilders, 1943. Numerous TV appearances.
PICTURES INCLUDE: Love Story, Wanted for Murder, Don Chicago, Uneasy Terms, So Evil My Love, Another Shore, Once a Jolly Swagman, Run for Your Money, Pool of London, White Corridors, Something Money Can't Buy, Cruel Sea, Grand National Night, Limping Man, Trouble in Store, John and Julie, Deep Blue Sea, Seven Waves Away, The Yellow Rolls Royce, Joey Boy, Double Man, Stranger in the House.

LITTO, GEORGE: Executive. b. Philadelphia, Dec. 9, 1930. e. Temple Univ. Joined William Morris Agency in New York and then became indep. literary agent. Opened own office in Hollywood, 1962. Packaged film and TV productions, including six films for Robert Altman. 1981, named chmn. bd. & CEO, Filmways.
PICTURES INCLUDE: That Cold Day in the Park, M*A*S*H, McCabe and Mrs. Miller, Images, Brewster McCloud, Thieves Like Us (debut as exec. prod.), Drive-In (exec. prod.), Dressed To Kill (prod.). Blow Out (prod.).
TV: Hawaii Five-O.

LITVINOFF, SI: Producer. b. New York City, April 5, 1929. e. Adelphi College (A.B.); NYU School of Law (LLB). Theatrical lawyer, personal and business manager in New York until 1967 when left firm of Barovick, Konecky & Litvinoff to produce plays and films.
PICTURES: The Queen, All the Right Noises, Walkabout, A Clockwork Orange (exec. prod.), Glastonbury Fayre (exec. in chg. prod.); The Man Who Fell to Earth (exec. prod.).
STAGE: Leonard Bernstein's Theatre Songs, Cry of the Raindrop, Girl of the Golden West, Little Malcolm and His Struggle Against the Eunuchs, I and Albert (London).

LIVINGSTON, BLANCHE: Advertising and publicity. b. New York City. e. Columbia University. m. Albert Levi. Began with trade paper, handled publicity for Fox Theatres in Manhattan and Brooklyn. Joined RKO Theatres publicity dept.; became nat'l. dir. adv.-pub., RKO Stanley-Warner theatres. Past secretary and v.p. AMPA. Retired July, 1977.

LIVINGSTON, JAY: Comp. & Lyric. b. McDonald, Pa., March 28, 1915; e. U. Pa., 1937, UCLA, 1964–65. Army, World War II. Accompanist and arranger for various NBC singers and singing groups 1940–42, N.Y.; author music and special material for Olsen & Johnson, including various editions of Hellzapoppin, and Sons O'Fun: began composing picture songs, 1944. Under contract to Paramount, 1945–55; then free lance. Composed songs for over 100 films.
SONG HITS INCLUDE: G'bye Now, Stuff Like That There, To Each His Own, Golden Earrings, Silver Bells, Buttons and Bows (Acad. Award, 1949), Mona Lisa (Acad. Award, 1951), Que Sera Sera (Acad. Award, 1957), Tammy (Acad. nom.), Almost In Your Arms (Acad. nom.), Bonanza (TV Theme), Mister Ed (TV Theme), Dear Heart, (Acad. nom.), Wish Me a Rainbow, In the Arms of Love, Never Let Me Go, As I Love You, All the Time, Maybe September, Collab. music and lyrics for B'way show Oh Captain! 1958; Let It Ride, 1961. Two songs for Sugar Babies, 1980.
PICTURES INCLUDE: The Paleface, Fancy Pants, The Lemon-Drop Kid, Houseboat, Tammy and the Bachelor, The Man Who Knew Too Much, Dear Heart, The Night of the Grizzly, This Property is Condemned, The Oscar, Never Too Late, Harlow, What Did You Do In The War Daddy?, Wait Until Dark, Red Garters, Sorrowful Jones.

LIVINGSTON, JEFFERSON: Advertising executive. With MGM as pub., 1937–42; U.S. Army Air Force, 1943–45. Joined J. Arthur Rank Org., 1946 and loaned to U.A. for roadshow release of Henry V. Joined Universal, 1947, charge road show and promotion, Hamlet, etc. In charge adv. & pub. J. Arthur Rank prod., released through U-1., 1950; apptd. Eastern adv. mgr. Universal, 1951; U-1. exec. coord. sales, adv. 1959; v.p., & dir. of adv. & publicity, Mirisch Company, 1962; exec. v.p., Harold Robbins International, 1972.

LIVINGSTONE, PERCY: Executive. b. Leeds, England, 1913. e. City of Leeds School, School of Accountancy. Entered m.p. industry Aug. 1931 in clerical capacity with old Fox Company. Joined sales force 3 years later; appointed Dublin Branch Manager, 1939; attached to Head Office, 1945 southern district mgr., 1948; assistant sales mgr., 1951; sales mgr., 1956; dir. of sales, 1957; bd. member; gen. sales mgr. Warner-Pathe Dist., 1959; man. dir. 20th Century-Fox, 1961; 20th Cent.-Fox Inter'l, N.Y., 1967; sr., v.p., 20th-Fox Intl. Corp., 1975; Chrm. 20th-Fox Film Co. Ltd. 1977; Chrm., 20th-Fox Productions Ltd.

LLOYD, EUAN: Producer. b. Rugby (Warwick), Dec. 6, 1923. e. Rugby. Entered m.p. ind. in 1939 as theatre manager, then pub. dir.; dir. of Publ. Rank, 1946; joined Associated British-Pathe, Ltd. in same capacity; 1952 asst. to prod., Warwick Film Prod. Ltd. v.p. Highroad Productions, 1962–64. Rep. Europe Goldwyn's Porgy & Bess 1959.
PICTURES INCLUDE: April in Portugal, Heart of Variety, Invitation to Monte Carlo, The Secret Ways, Genghis Khan, Poppy Is Also A Flower, Murderer's Row, Shalako, Catlow,

The Man Called Noon, Paper Tiger, The Wild Geese, The Sea Wolves.

LLOYD, NORMAN: Producer. b. Jersey City, N.J., Nov. 8, 1914. e. N.Y.U., 1932. Acted on B'wy in: Noah, Liberty Jones, Everywhere I Roam, 1935–44; in various stock companies.
PICTURES INCLUDE: (actor) Spellbound, The Southerner, Green Years, Limelight. Prod. asst. on Arch of Triumph and The Red Pony, 1946. Assoc. prod. on The Alfred Hitchcock Show, 1957 (TV) and exec. prod. 1963. Also prod. of Up Above the World, Universal feature film; prod. of TV package deal, 2 hr. film.
STAGE: The Cocktail Party, The Lady's Not for Burning, Madame Will You Walk, The Golden Apple.

LO BIANCO, TONY: Actor. Has performed on N.Y. stage as well as in films and TV. Won Obie Award for performance in Yanks 3, Detroit 0, Top of the Seventh. Also acted on stage in The Office, The Rose Tattoo, The Royal Hunt of the Sun, etc.
PICTURES INCLUDE: The French Connection, The Honeymoon Killers, F.I.S.T., Bloodbrothers.

LOCKE, SONDRA: Actress. Debut in The Heart Is a Lonely Hunter (1968).
PICTURES: The Second Coming of Suzanne, Willard, A Reflection of Fear, The Outlaw—Josey Wales, The Gauntlet, Every Which Way But Loose, Bronco Billy, Any Which Way You Can.

LOCKHART, JUNE: Actress. p. Gene and Kathleen Lockhart. On TV in Lassie series.
PICTURES INCLUDE: All This and Heaven Too (1940, Warner Bros.), Sergeant York, Miss Annie Rooney (1942, United Artists), Meet Me in St. Louis, Son of Lassie, White Cliffs of Dover, Keep Your Powder Dry, Bury Me Dead, T-Men, It's a Joke, Son, Time Limit.

LOCKWOOD, GARY: Actor. b. Van Nuys, Calif., 1937. Began in Hollywood as stuntman.
PICTURES INCLUDE: Tall Story, Splendor in the Grass, Wild in the Country, The Magic Sword, It Happened at the World's Fair, Firecreek, 2001: A Space Odyssey, They Came to Rob Las Vegas, Model Shop, The Body, R.P.M., Stand Up and be Counted.

LOCKWOOD, JULIA: Actress. b. Bournemouth, Enl., Aug. 23, 1941, daughter of Margaret Lockwood. M.P. debut in Hungry Hill, 1946. Stage and TV debut, 1953.
PICTURES INCLUDE: I Have a Teenage Daughter, Please Turn Over, No Kidding.
STAGE: Alice in Wonderland, Goldilocks, Peter Pan, Servant of Two Masters, Birds on the Wing, Jockey Club Stakes, The Mating Game, Out On A Limb, Sextet.
TV: Royalty and Life With Father series, Your Obedient Servant, The Exam, Six Proud Walkers, Compact, The Flying Swan (1971), Birds on the Wing.

LOCKWOOD, MARGARET: Actress. b. Karachi, Pakistan, Sept. 15, 1916. e. R.A.D.A. Stage experience; pictures, from 1935. Voted one of the top ten British money-making stars in Motion Picture Herald-Fame Poll, in 7 successive years, 1944–50; top actress, 1946; won National Film Award, 1946, 1947, 1949. London stage, 1954.
PICTURES INCLUDE: Wild Justice, Lorna Doone, Midshipman Easy, Amateur Gentleman, Beloved Vagabond, Street Singer, Bank Holiday, Lady Vanishes, Rulers of the Sea, Stars Look Down, Night Train, Two on a Weekend, Quiet Wedding, Girl in the News, Alibi, Man in Grey, Love Story, Place of One's Own, I'll Be Your Sweetheart, Wicked Lady, Bedelia, Hungry Hill, Jassy, The Bad Sister, Madness tf the Heart, Highly Dangerous, Trent's Last Case, Laughing Anne, Trouble in the Glen, Cast a Dark Shadow, Spiders Web, The Slipper & The Rose.
LONDON STAGE: Subway In the Sky, Suddenly It's Spring, Signpost to Murder, An Ideal Husband, The Others, On a Foggy Day, Lady Frederick, Relative Valves (1973), Double Edge.
TV: Last of Mrs. Cheyney, Murder Mistaken, Call It a Day, The Great Adventure, The Royalty, Your Obedient Servant, That Lady, Yorky, Palace of Strangers, The Human Jungle (series), The Flying Swan, Justice Is a Woman, Five Finger Exercise (1971), Justice series (1972–74).

LOCKWOOD, ROGER: Executive. b. Middletown, Conn., June 7, 1936. e. Ohio Wesleyan Univ. Sports writer for Akron Beacon Journal, 1960–62. On executive staff of Lockwood & Gordon Theatres; exec. v.p. SBC Theatres, 1969–73. In 1974 asst. to exec. v.p., General Cinema Corp. In 1975 formed Lockwood/Friedman Film Corp., buying-booking and distribution organization. Pres., Theatre Owners of New England, 1971–72; pres., Young NATO 1965–67; bd. of dir. NATO, 1962-present; pres., Jimmy Fund, present; 1979–80, Variety Club of New England, pres.

LOEFFLER, LOUIS: Film editor, with 20th-Fox many years.
PICTURES INCLUDE: 4 Men & a Prayer, Rose of Washington Square, Swanee River, Forever Amber, Iron Curtain, Call Me Mister, We're Not Married, My Cousin Rachel, Titanic, How to Marry a Millionaire, Women's World, Man with the Golden Arm, Tall Men.

LOEWE, FREDERICK: Composer. b. Vienna June 10, 1904. Studied piano at Stern Conservatory; at 13 was youngest solo pianist to appear with Berlin Symphony Orchestra. Came to U.S. in 1924; played organ at Keith Albee's, Brooklyn. Became partner with Alan Jan Lerner for series of stage musicals, many of them filmed.
PICTURES INCLUDE: Brigadoon, Paint Your Wagon, My Fair Lady, Gigi, Camelot, The Little Prince.

LOGAN, JIMMY: Com., exhib. b. Glasgow, Scotland, 1928. p. Jack Short, May Dalziel, vaudeville act. Started as asst. theatre manager. On stage in vaudeville with the Logan Family. Began in radio in It's All Yours, weekly series on Scottish BBC. Joined Howard & Wyndham Ltd. as principal comedian in pantomime and summer revue. Televised in network shows from London and Glasgow. Appeared in Royal Variety Performance at Palladium, London, Nov. 1957; Scottish Royal Variety Performance, 1958. Starred in film, Floodtide; promoted stage comedy, Wedding Fever, 1962, 1964; chmn. Logan Theatres Ltd., 1964.
TV: Own series BBC-TV; Jimmy Logan Entertains, The Jimmy Logan Show series.

LOGAN, JOSHUA: Director, writer, producer. b. Texarkana, Tex., Oct. 5, 1908; e. Princeton U., Moscow Art Theatre.
STAGE: Camille, On Borrowed Time, I Married an Angel, Knickerbocker Holiday, Star in Your Eyes, Morning's at Seven, Charley's Aunt, By Jupiter, Happy Birthday, John Loves Mary, South Pacific, Mister Roberts, Wish You Were Here, Picnic, Fanny.
PICTURES INCLUDE: Garden of Allah, History is Made at Night, I Met My Love Again, Mister Roberts (collab.), Picnic, Fanny, Camelot, Paint Your Wagon.

LOLLOBRIGIDA, GINA: Actress. b. Subiaco, Italy, 1928. e. Acad. of Fine Arts, Rome, m.p. debut (Italy) in Love of a Clown.
PICTURES INCLUDE: The City Defends Itself, The White Line, Fanfan the Tulip, Times Gone By, Beat the Devil, Crossed Swords, The Great Game, Beauties of the Night, Wayward Wife, Bread Love and Dreams, Bread Love and Jealousy, Young Caruso, World's Most Beautiful Woman, Trapeze, Hunchback of Notre Dame, Solomon and Sheba, Never So Few, Go Naked in the World, Come September, Imperial Venus, Woman of Straw, That Splendid November, Hotel Paradiso, Buona Sera, Mrs. Campbell, Bambole, Plucked.

LOM, HERBERT: Actor. b. Prague. e. Prague U. Stage training London Embassy, Old Vic—Sadlers Wells and Westminster Schools. First British Picture 1941 (Mein Kampf—My Crimes); on TV, The Human Jungle Series.
PICTURES INCLUDE: Tomorrow We Live, Secret Mission, Young Mr. Pitt, Dark Tower, Cage of Gold, Whispering Smith vs. Scotland Yard, Two on the Tiles, Mr. Denning Drives North, Hell is Sold Out, Gaunt Stranger, Rough Shoot, The Net, The Love Lottery, Star of India, Beautiful Stranger, The Ladykillers, War and Peace Action, Fire Down Below, Hell's Drives, Chase a Crooked Shadow, Passport to Shame, Roots of Heaven, The Big Fisherman, North-West Frontier, I Aim at the Stars, Spartacus, Mysterious Island, Mr. Topaz, The Frightened City, El Cid, Tiara Tahiti, The Phantom of the Opera, Horse Without a Head, A Shot in the Dark, Uncle Tom's Cabin, Return From the Ashes, Gambit, The Assignment, Three Faces of Eve, Villa Rides, Doppelganger, Mr. Jericho, Dorian, Mark of the Devil, Count Dracula, Murders in the Rue Morgue, Dark Places, Death in Persepolis, Return of the Pink Panther, The Pink Panther Strikes Again, Charleston, Revenge of the Pink Panther, The Man with Bogarts' Face, Hopscotch, The Acts of Peter and Paul.

LOMBARDO, GOFFREDO: Executive. b. Naples, Italy, May 13, 1920. President Titanus Films.
PICTURES INCLUDE: Rocco and His Brothers, Sodom and Gomorrah, The Leopard, Four Days of Naples.

LOMITA, SOLOMON: Executive. b. New York City, April 23, 1937. Entire industry career with United Artists Corp. as follows: adm., intl. dept., 1962 asst., intl. sales, same year. 1963, asst. intl. print mgr.; 1965, intl., print mgr. In 1973 appt. dir. of film services. 1981, v.p., film services.

LONDON, JULIE: Singer-Actress. b. Santa Rosa, Calif., Sept. 26, 1926. r.n. Julie Peck. Launched as actress by agent Sue Carol (wife of Alan Ladd who arranged screen test, followed by contract for 6 films. As singer has appeared in nightclubs and recorded.
PICTURES INCLUDE: The Red House, The Fat Man, The Great Man, Saddle the Wind, Man of the West, The Third Voice.
TELEVISION: Perry Como Show, Steve Allen Show, Ed Sullivan Show, Emergency (series).

LONDON, MILTON H. Exhibitor b. Detroit, Mich., Jan. 12, 1916. e. U. of Mich. BA, 1937. Wayne U. Law School, 1938. Theatre owner since 1937. U.S. Army 1943–46. Invented Ticograph system of positive admissions control for theatres, 1950; pres. Theatre Control Corp., 1950–62; secy-treas. Co-op. Theas. of Michigan Inc., 1956–63; exec. comm., Council of M.P. Organizations, 1957–66; dir. M.P. Investors, 1960–67; exec. dir. Allied States Assoc. of M.P. Exhib., 1961–1966; exec. dir. National Assoc. of Theatre Owners, 1966–1969 pres., NATO of Michigan, 1954–74; pres. Metropolitan Exhibitors, Inc., adv. & pub. rel. Mich. State Fire Safety Bd., Chief Barker, Variety Club of Detroit, Tent No. 5. 1975–76; Life Patron, Variety Clubs International; Mich. Chairman, Motion Picture Pioneers.

LONERGAN, ARTHUR: Art director. b. New York City, Jan. 23, 1906. e. Columbia U. Instructor in history of architecture, NYU; architect with N.Y. School of Decoration and Design; joined MGM as illustrator in art dept., 1938; free-lance art dir., 1945; pres. Soc. of M.P. Art Dir., 1952–54; Acad. Award nomination, art. dir., The Oscar. 1975-The Georges Meliès Award for outstanding cinematic achievement for art direction in Science Fiction category.
TV: Loretta Young show, The Falcon, Mr. and Mrs. North, Topper, Life of Riley, Adventures of Hiram Holiday, Hitchcock Presents, GE Theatre, M. Squad, Restless Gun. etc.
PICTURES INCLUDE: Song in My Heart, Intrigue, Tender Years, Maneaters of Kumaon, Pitfall, Outpost in Morocco, Ride Vaquero, The Actress, It's Always Fair Weather, Tender Trap, Ransom, Forbidden Planet, On the Double, My Geisha, Who's Got the Action, Papa's Delicate Condition, A New Kind of Love, Who's Sleeping in My Bed, Robinson Caruso On Mars, Tickle Me, Red Line 7000, The Caper of the Golden Bull, Yours Mine and Ours, How Sweet It Is, Che, M*A*S*H*, Plaza Suite.

LONGSTREET, STEPHEN: Writer. b. New York City, April 18, 1907; e. Rutgers U.; Parsons Coll.; Rand Sch., London, B.A. Humorist, cartoonist (New Yorker, Collier's, etc.) 1930–37; ed. Free World Theat., radio plays; ed. film critic, Saturday Review of Literature, 1940, U.S. at War, Time 1942–43; writer for screen from 1942. On staff U.C.L.A. Elected pres. Los Angeles Art Assoc. 1970. 1974: appointed Prof. English Dep. Univ. of So. Calif. Modern Writing Course.
WRITINGS: Decade, The Golden Touch, The Gay Sisters, Last Man Around the World, Chico Goes to the Wars, Pedlocks, Lion at Mornino, Promoters, Boy in the Model T, Sometimes I Wonder, Wind at My Back, The Young Men of Paris, The Wilder Shore, War Cries on Horseback, Yoshiwara, Geishas and Courtesans, Canvas Falcons, Men and Planes of World War I, We All Went to Paris. New publications in 1973: Chicago 1860–1919, (show business & society), Divorcing (a novel); The General (novel, 1974), All Star Cast, 1977; The Queen Bees (1979).
STAGE: High Button Shoes.
PICTURES INCLUDE: The Gay Sisters, Golden Touch, Stallion Road, Jolson Story, Silver River, Helen Morgan Story, First Traveling Saleslady, Untamed Youth, Duel in the Sun, Greatest Show on Earth, Streets of Montmarte, The Crime, Uncle Harry, Rider on a Dead Horse.
TV: Cassey Jones Clipper Ship, Agent of Scotland Yard, m.c. author of The Sea, NBC; m.c. Press & Clergy, 1960–63; Viewpoint, CBS, 1963–65; series Boy in the Model T, Young Man From Boston, 1967, Blue and the Grey, Donald O'Connor Show, John J. Anthony talk shows, L.A. Guest; Art Linkletter Show, CBS Boutique, Writers Guild, Comm. of Public Relations. Film and Book critic for Readers' Syndicate since 1970. Professor performing arts dept. Univ. So. Calif. since 1973, where he is in 1979, presenting 12 great silent films, The Art & Entertainment of Silent Films in the cinema section of the college.

LONSDALE, PAMELA: Producer/director. Born Pinner, England. Entered TV 1960. m. Reginald Collin. Series include Tempo, Lion, Witch And Wardrobe, Smith, Ace of Wands. Rainbow, Shadows. 1975 Won British Academy Award—best childrens programme. 1979: exec. prod. Children's Drama Thames TV.

LOORAM, MARY HARDEN: Executive of National Catholic Office for Motion Pictures; (now U.S. Catholic Conference Division of Film Broadcasting) Chairman, International Federation of Catholic Alumnae M.P. Dept. (official reviewing group NCOMP). b. Brooklyn, N.Y. Chmn., M.P. Bureau 1926–52, asst. nat'l chmn. 4 yrs.; chmn. since 1932. Hon. deg. Fordham U. 1948. Member: Attorney General's Juvenile Delinquency Panel, Washington; White House Conference, 1950; Nat'l Catholic Theatre Conference; Honor: Pro Ecclessia et Pontifice, 1942. Member bd. dir. St. Joseph's College, Emmitsburg; Lady of the Holy Sepulchre, 1957; Sienna Medal (outstanding Catholic woman of the year), 1958; Lady Grand Cross, Holy Sepulchre, 1961; Woman of Achievement—Women's International Exposition, 1961. Nat'l Laymen's Award, 1968. Retired Aug. 1970 after 35 yrs. as an exec. mem. of Legion, NCOMP.

LOOS, ANITA: Writer. b. Sissons, Calif. Author novels & plays.
PICTURES INCLUDE: Gentlemen Prefer Blondes, New York Hat, Ex-Bad Boy, The Struggle, Midnight Mary; Hold Your Man, Lady of the Night, Red Headed Woman, Blondes of the Follies, Biography of Bachelor Girl, Girl from Missouri, Social Register, Riff Raff, San Francisco, Mama Steps Out, Saratoga, The Women, Susan and God, They Met in Bombay, I Married an Angel.

LOOS, MARY: Writer. b. San Diego, Calif. May 6, 1914. e. Stanford U., 1933. Actress m.p.; in public relations field N.Y. 1938; jewelry designer for Paul Flato; author of novel Return in the Vineyard, 1945. secy. Voyager Films, Inc., literary exec. M. J. Frankovich Prod. Novel: The Beggars Are Coming, Bantam Books, 1974, Belinda, 1976; The Barstow Legend, 1978.
PICTURES INCLUDE: Rose Marie, Maytime, Crusades, Cleopatra, Mr. Belvedere Goes to College, Mother Was a Freshman, Ticket to Tomahawk, When Willie Comes Marching Home, Father Was a Fullback, I'll Get By, Meet Me After the Show, Let's Do It Again, The French Line, Gentlemen Marry Brunettes, Over-Exposed, Woman's World.

LORD, JACK: Actor-writer-artist-director. b. N.Y.C., Dec. 30, 1930. e. N.Y.U. (Chancellor Chase scholarship), B.S. Artist, represented in various museums including Metropolitan Museum of Art, Museum of Modern Art, Brooklyn Museum, in New York. Bibliotheque National, Paris; British Museum; Fogg Museum, Harvard University. St. Gandens Plaque for Fine Arts; creator of Tramp Ship, McAdoo, Yankee Trader, The Hunter TV series. On Bway in Traveling Lady (Theatre World Award), Cat on a Hot Tin Roof. Fame Award, new male star, 1963, mem. of Directors Guild of America.
PICTURES INCLUDE: Court Martial of Billy Mitchell, Williamsburg Story, Tip On a Dead Jockey, God's Little Acre, Man of the West, The Hangman, True Story of Lynn Stuart, Walk Like a Dragon, Doctor No, Doomsday Flight, Ride to Hangman's Tree, Counterfeit Killer.
TV: Leads on all networks, including Omnibus Constitution series, Playhouse 90, Goodyear Playhouse, Studio One, U.S. Steel, etc. (Film shows incl.) Have Gun Will Travel, Untouchables, Naked City, Rawhide, Bonanza, The Americans, Route 66, Gunsmoke, Stagecoach West, Dr. Kildare, Greatest Show on Earth, Combat, Chrysler Theatre, 12 O'Clock High, The Loner, Laredo, The FBI, The Invaders, The Fugitive, The Virginian, Man from UNCLE, High Chaparral, Ironside, Star of Stoney Burke and Hawaii Five-O series, also director of Death with Father, How to Steal a Masterpiece; Honor Is An Unmarked Grave, The Bells Toll At Noon, Top of the World, Why Won't Linda Die? Dir.: Who Says Cops Don't Cry? episodes of Hawaii Five-O. Creator, director, and exec. Producer of M Station: Hawaii (2-hr film for TV), 1979.

LORD, ROSEMARY: Actress—Writer. b. Born May 16; Taunton, Somerset England. Entered industry in 1965. After various theatre work did several television appearances: Softly, Softly; Spyder's Web; 30 Minute Theatre; Rivals of Sherlock Holmes; etc. Also TV & film work in France, Germany and Holland. Now living in Hollywood, writing for major publications in UK & USA, and developing television scripts. Member: British Equity, U.S. Equity, SAG, Publicists Guild.
PICTURES INCLUDE: Actress: The Whisperers, Dr. Jekyll & Sister Hyde, The Watchers, Touch of Class, Duchess and the Dirtwater Fox, Return from Witch Mountain.

LOREN, SOPHIA: Actress. b. Rome, Italy, Sept. 20, 1934. e. Naples. m.p. debut, 1952.
PICTURES INCLUDE: Africa Beneath the Seas, Village of the Bells, Good People's Sunday, Neapolitan Carousel, Day in the District Court, Pilgrim of Love, Aida, Two Nights with Cleopatra, Our Times, Attila, Scourge of God, Too Bad She's Bad, Pride and the Passion, Gold of Naples, Boy on a Dolphin, Scandal in Lorrinto, Miller's Beautiful Wife, Desire Under the Elms, Houseboat, The Black Ordeal, That Kind of Woman, Heller With a Gun, Anatomy of Love, Breath of Scandal, Heller in Pink Tights, Bay of Naples, Two Women, El Cid, Boccaccio 70, Il Coltello nello Piaga, French, The Fall of the Roman Empire, The Great Spy Mission, Lady L, Marriage Italian Style, Judith, Arabesque, The Countess from Hong Kong, Happily Ever After, More than a Miracle, Ghosts—Italian Style, Sunflower, The Priest's Wife, Lady Liberty, White Sister, Man of La Mancha, The Voyage, The Verdict, The Cassandra Crossing, A Special Day, Angela, Brass Target, Firepower.

LOSEY, JOSEPH: Director, writer, producer, teacher. Chevalier de l'Orde des Arts erdes Lettres (1957) Litterarum Humaniorum Doctorem (Darthmouth College 1973). b. La Crosse, Wis., Jan. 14, 1909, e. Dartmouth C., 1929, B.A.; Harvard Grad. Sch. of Arts & Sciences 1930. Play & book reviewer, N.Y. Times, N.Y. Herald-Tribune, theatre mags., Saturday Review of Literature, 1930–31. Stage mgr., actor, reader, N.Y. stage plays; for Theatre Guild, others, 1931–32; stage mgr. at opening of Radio City Music Hall, 1932; dir. & co-prod., Little Ol' Boy, Jayhawker, Galileo on Bdwy, among many others, also, Living Newspapers for Federal Theatre, N.Y.; became m.p. dir. in 1937. Documentary & comml. films (40) incl. A Child Went Forth (distrib. by State Dept.); Youth Gets a Break (Nat'l Youth Adminis.); Petroleum and His Cousins (New York World's

Fair, 1939); educational films (60) as dir. Rockefeller Foundation, Human Relations Commission Film Project; Gun in His Hand (Crime Doesn't Pay short subj.), The Boy With Green Hair, The Lawless, The Prowler, M, the Big Night, Encounter. To Britain in 1953; on London stage in: The Wooden Dish, A Night at the Ball.

PICTURES INCLUDE: Time Without Pity, Blind Date (Chance Meeting), The Criminal, The Damned, Eva, The Servant, King and Country, Modesty Blaise, Accident, Boom, Secret Ceremony, Figures in a Landscape, The Go-Between, (Palme d' Or Award, Cannes 1971), The Assassination of Trotsky, A Doll's House, Galileo, The Romantic Englishwoman. Mr. Klein, Les Routes du Lud, Mozart's Don Giovanni.

LOUIS, JEAN: Designer. b. Paris, France, Oct. 5, 1907. Head designer, Hattie Carnegie, 7 yrs., before accepting post as Chief Designer Columbia Pictures. Later Universal Studios. Free lance in m.p. & TV. Pres. Jean-Louis, Inc.

LOUISE, TINA: Actress. b. New York, N.Y. e. Miami U., N.Y. Neighborhood Playhouse, Actors Studio, B'way.

STAGE: Two's Company, The Fifth Season, Almanac, Li'l Abner.

PICTURES INCLUDE: God's Little Acre, For Those Who Think Young, The Wrecking Crew, How to Commit Marriage, The Happy Ending, The Stepford Wives.

TV: Gilligan's Island, Mannix, Ironside, Kung Fu, Police Story, Kojak.

LOVE, BESSIE: Actress. b. Midland, Texas, Sept. 10, 1898. Began screen career with Triangle, Vitagraph, Callahan films. Early pictures. On stage, repertory, London (Love in Idleness). On staff Ealing Studio as technician, 1943. Co-mgr. and entertainer, Red Cross Hospital Unit, Great Britain, 1944.

PICTURES INCLUDE: Penny of Top Hill, Lovey Mary, Human Wreckage, The Eternal Three, Gentle Julia, Torment, Those Who Dance, The Lost World, The Purple Dawn, Dress Parade, The King of Main Street, The Song and Dance Man, Has Anybody Here Seen Kelly, Sally of the Scandals, Broadway Melody, Hollywood Revue of 1929, The Idle Rich, The Road Show, Chasing Rainbows, Conspiracy, The Swellhead, Morals for Women, Live Again, Atlantic Ferry.

STAGE: Gone With the Wind (London, 1972).

LOVE, JAMES A.: Producer, b. May 1, 1918; began film career, Warner Bros., 1938–41; film ed., Motion Picture Div., U.S. Coast Guard, WW II. chief film ed., Princeton Film Center; prod-dir. Cineffects Productions; pres., Lalley & Love Inc., 1952; changed firm name to James Love Productions, 1955; founded Monitor Film Distributors, 1964.

TV: Professor Yes 'n' No, Mary Margaret McBride, commercials for Democratic and Republican Conventions (1956); prod.-dir., industrial films; assoc. prod., Ballad of Gavilan; assoc. prod., By Jupiter.

LOVY, ALEX: Producer, director. b. Passaic, N.J. dir. Univ. Studios; assoc. prod. dir. Flintstones, Yogi Bear, Huckleberry Hound; dir., What a Wonderful Feeling. pres., Alex Lovy Productions, Inc.

LOWE, PHILIP L.: Executive. b. Brookline, Mass. Apr. 17, 1917. e. Harvard. Army, 1943–46. Checker, Loew's 1937–39; treasurer, Theatre Candy Co., 1941–58; Pres., ITT Sheraton Corp., 1969–70; pres., Philip L. Lowe and Assoc.; dir., First Artists, Teleprompter

LOWE, PHILIP M.: Executive. b. 1944. e. Deerfield Academy, Harvard College (cum laude in psychology, 1966); Columbia Business School (1968). Work experience includes major marketing positions at General Foods, Gillette, Gray Advertising, and Estee Lauder Cosmetics before co-founding Cinema Centers Corp. and Theatre Management Services in Boston. s professor marketing at Bentley College; consultant to major multi-national corporations. In 1978 named new head of Natl. Assoc. of Concessionaires, previously serving NAC as treas., v.p.

LOY, MYRNA: Actress. r.n. Myrna Williams; b. Helena Mont., Aug. 2, 1905; e. Westlake Sch. for Girls. Appeared in stage presentations, Grauman's Chinese theatre, then Hollywood. On screen in What Price Beauty, 1925; thereafter in many pictures, variously starred, co-starred & featured. Voted one of the ten best Money Making Stars in Motion Picture Herald-Fame Poll, 1937, 38.

STAGE: Marriage-Go-Round, There Must Be A Pony, Good Housekeeping, Barefoot in the Park, Dear Love, The Women, Don Juan in Hell, Relatively Speaking.

PICTURES INCLUDE: Last of the Duanes, Transatlantic, Arrowsmith, Vanity Fair, Animal Kingdom, The Prizefighter and the Lady, The Thin Man, Evelyn Prentice, Wife Versus Secretary, After the Thin Man, Test Pilot, The Rains Came, Another Thin Man, Third Finger, Left Hand; Shadow of the Thin Man, The Thin Man Goes Home, Best Years of Our Lives, Mr. Blanding Builds His Dream House, Red Pony, If This Be Sin, Cheaper by the Dozen, Belles on Their Toes, Ambassador's Daughter, Lonely hearts, From the Terrace, Midnight Lace, The April Fools, Airport 1975, The End, Just Tell Me What You Want.

TV: Meet Me in St. Louis, Minerva, George Gobel, Perry Como, Happy Birthday—June Allyson Show, Family Affair, The Virginians, Do Not Spindle or Mutilate, Death Takes a Holiday, The Couple Takes a Wife, Ironside, The Elevator, It Happened at Lakewood Manor, Summer Solstice.

LUBCKE, HARRY R.: Cons. TV engineer, registered patent agent. b. Alameda, Calif., Aug. 25, 1905. Married to Jean MacRae. e. U. of Calif. 1929. B.S. Holds numerous U.S. and foreign patents on television. In 1931: station W6XAO went on air on what is now television Channel No. 2 to become first station of kind in nation. New Mt. Lee studios built at cost of $250,000 in 1941, housing then largest TV stage 100x60x30 ft. Pioneered present television standard of 525 line (Aug., 1940). In 1942, television programs to promote war bonds sale. 1942–46 dir. war research for which certificates of commendation were received from Army & Navy.

MEMBER: RTPB, Panel 6, on television standards and various sub-committees since inception, and of Nat'l Television Systems Committee, which preceded it. Pres., Acad. TV Arts & Sciences, 1949. Dir. TV Don Lee Broadcasting System to Dec. 31, 1950; Consulting eng., CBS, thereafter consulting practice restricted to TV eng. electronics, program prod., 1951; cons. TV engineer, 1951; registered patent agent, 1952. Life Fellow, 1951, IEEE, AAAS, SMPTE, 1967. Board of Governors, Patent Law Association of Los Angeles, 1974. Life Member National Academy of Television Arts & Sciences, member engineering Emmy Awards Committee. Member Blue Ribbon panel Emmy Awards Committee; member ANTA 1977, Life Member, Actor's Fund of America, 1978; Diamond Circle, of Pacific Pioneer Broadcasters, 1980.

LUBER, BERNARD: b. Philadelphia, PA., May 18, 1908. Dickinson College and Law School, Carlisle, PA., 1928. Law work in New York City until 1930. Admitted to New York Bar. National publicity and theater operations for Paramount Publix Theaters until 1937. Joined Paramount Hollywood Studios 1937 as studio attorney. 1942–45 assoc. head of Paramount talent and casting dept. Head of talent and casting dept. of Republic Studios until 1947. General Manager of Jack Benny's Amusement Enterprises to 1950. Organized and part-owner of Pine-Thomas Productions, producers of 75 feature films. Organized, part owner and co-producer (with R. Polk) World Artists, producers of series of concert films, featuring Heifetz, Rubenstein, Piatigorsky and Marian Anderson, etc., for NBC Network. Laurel Award for Of Men and Music, feature film featuring concert artists. 1951, produced 26 half-hour films of Superman featuring George Reeves. From 1952 to 1964 produced feature films and tv series (for ABC Exclusive) in Hollywood, New York, London, Paris, Rome, Munich and Athens, etc. From 1965 and continuing to present, pres. of Celebrity Concert Corporation, importing cultural music films and recordings, and USA representative of Melodiya and Mezhkniga, Soviet Recording Industry, exporting Soviet recordings from USSR and importing American Recordings to USSR.

LUBIN, ARTHUR: Director. b. Los Angeles, Calif. Since 1935 has directed numerous pictures.

PICTURES INCLUDE: Buck Privates in the Navy, Hold That Ghost, Keep Em Flying Ride 'Em Cowboy, Eagle Squadron, Phantom of the Opera, White Savage, Ali Baba and the Forty Thieves, Delightfully Dangerous, Francis, Queen for a Day, Francis Goes to the Races, Rhubarb, Francis Covers the Big Town, Francis Goes to West Point, It Grows on Trees, South Sea Woman, Star of India, Lady Godiva, Francis in the Navy, Footsteps in the Fog, First Traveling Saleslady, Escapade in Japan, The Thief of Baghdad, The Incredible Mr. Limpett, Rain for a Dusty Summer, Night in Paradise, The Spider Woman Strikes Back, New Orleans, Impact, Queen for a Day, Star of India, Hold On!

TV: Maverick (Henry Fonda episodes), 77 Sunset Strip, Bonanza, the entire Mister Ed series (prod.-dir.).

LUCAS, GEORGE: Producer-Director-Writer. b. 1945. e. USC (cinema). Made short film called THX and won National Student Film Festival Grand Prize, 1967. Signed contract with WB. Ass't. to Francis Ford Coppola on The Rain People, during which Lucas made 2-hr. documentary on filming of that feature. Debut as director with THX 1138 for WB.

PICTURES INCLUDE: THX 1138 (dir., co-s.p.), American Graffiti (dir., co-s.p.), Star Wars (dir., s.p.); More American Graffiti (exec. prod.); The Empire Strikes Back (exec. prod.); Raiders of the Lost Ark (exec. prod.).

LUDWIG, IRVING H.: Executive. b. Nov. 3, 1910. e. Savage School, Rivoli Theatre, N.Y., mgr., theatre oper., Rugoff and Becker, 1938–39; opened first modern art type theatre, Greenwich Village, 1940. With Walt Disney Prod. in charge of theatre oper. on Fantasia in various cities, 1940–

41; buyer-booker, Rugoff and Becker, 1942–45; film sales admin., Walt Disney Prod. home office, 1945–53; v.p. and domestic sales mgr., Buena Vista Dist. Co., 1953; pres. gen. sales mgr., 1959–80. Now exec. consultant.

LUFKIN, DAN W.: Executive. Chairman of Exec. Comm., Columbia Pictures Industries (appt. July, 1978). Co-founder of Donaldson, Lufkin & Jenrette Securities Corp., investment banking and brokerage firm. Served as first commissioner of Dept. of Environmental Protection for State of Connecticut. Joined Columbia board in November, 1977.

LUFT, HERBERT G.: Writer. b. Essen, Germany, Aug. 21. Started career as journalist; came to U.S. 1940, to Hollywood, 1943. Magazine writer, film ed., researcher, translator, prod. asst., 1944. Adapt. M: prod. mgr., collab. s.p. This Is Freedom, Secret of Giving, Springtime in Copenhagen, Hong Kong Affair, Why Must I Die, Night Raiders, Syndicated Hollywood Columnist, writer, prod. exec., chmn. Int'l Film Comm.; pres. Hollywood Foreign Press, 1966–1968. Production Exec. Getty Picture Corp. (1969) (treas.); Flare-Up, Zeppelin, 1970. Shellia, 1970; Rage, 1971/72, Associate producer, The Mutation, (England) 1972/73. Assoc. producer The Devil's Men (1975), Shoot (1976); vice pres. in charge of production, Getty Picture Corp. (1976).

LUKE, KEYE: Actor, artist. b. Canton, China, 1904. e. Franklin High Sch., Seattle; Formerly artist for Fox West Coast Theats. & RKO Studios; also technical advisor on Chinese films. Screen debut as actor in Painted Veil, 1935.
STAGE: Flower Drum Song (3 yr).
PICTURES INCLUDE: Charlie Chan series, Oil for the Lamps of China, King of Burlesque, The Good Earth, International Settlement, Sued for Libel, Disputed Passage, Dragon Seed, Three Men in White, Between Two Women, First Yank in Tokyo, Tokyo Rose, Sleep My Love, Hell's Half Acre, World for Ransom, Bamboo Prison, Love is a Many Splendored Thing, 80 Days Around the World, Their Greatest Glory, Battle Hell, Fair Winds to Java, Nobody's Perfect, Project X, The Chairman, The Hawaiians, Noon Sunday, Won Ton Ton, Amsterdam Kill, Just You and Me, Kid.
TV: Gunsmoke, Danger, December Bride, Crusader, Wireservice, Crossroads, Soldiers of Fortune, My Little Margie, Annie Oakley, Ray Milland Show, Medic, Citizen Chang, Climax, Jerry Jewis, Trackdown, Perry Mason Show, The Littlest Hobo, This Is the Life, Smothers Bros., I Spy, FBI, Wackiest Ship in Army, Mickey Rooney Show, Johnny Quest, Kentucky Jones, Never Too Young, Bob Hope Chrysler Show, Family Affair, Big Valley, Dragnet, It Takes a Thief, Star-Trek, Adventures of Huck Finn, The Outsider, Scooby Doo, Paris 7000, Johnny Carson Show, Marcus Welby, M.D., Adam 12, Hawaii 5-0, Dinah Shore Show, The Lucy Show, Anna and the King of Siam, Kung Fu, Amazing Chan and Chan Clan, Follow The Sun, Target, The Corruptors, Fair Exchange, Cannon, Cat Creature, Love American Style, Judgement—trial of Yamashita, Judge Dee, Khan, Harry O, MASH, Quincy, How the West Was Won, Meeting of Minds, Vegas, Battle of the Planets, The Yee Family, Might Man and Yukk, Rickety Racket, Tang Face, Charlie's Angels, Reach for the Sun, Fly Away Home, Brothers, Adventures of Goldie Gold, etc.

LUKE, PETER: Playwright director. Author of plays for TV: Small Fish Are Sweet, 1958; Pigs Ear with Flowers, 1960; Roll on Bloomin' Death, 1961; A Man on Her Back (with William Sansom), 1965; Devil a Monk Won't Be, 1966. Wrote and directed films for BBC-TV: Anach 'Cuan (about the late Sean O Riada) 1967; Black Sound—Deep Song (About Federico Garcia Lorca) 1968; Author of Stage play, Hadrian VII, first produced at Birmingham rep in 1967 and at time of going to press has been playing in one part of the world or another ever since. Stage Play, Bloomsbury. Author of autobiography, Sisyphus & Reilly, publ., 1972, Prod. Phoenix Theatre 1974.

LUMET, SIDNEY: Director. b. Philadelphia, Pa., June 15, 1924. e. Professional Children's Sch.; Columbia U. Child actor in plays: Dead End, George Washington Slept Here, My Heart's in the Highlands. U.S. Armed Forces, WW II, 1942–46; dir. summer stock, 1947–49; taught acting, H.S. of Prof. Arts. Assoc. dir. CBS, 1950, dir. 1951.
TV: Mama, Danger, You Are There, Omnibus, Best of Broadway, Alcoa, Goodyear Playhouse.
PICTURES INCLUDE: 12 Angry Men, Stage Struck, That Kind of Woman, Fugitive Kind, A View From the Bridge, Long Day's Journey into Night, Fail Safe, The Pawnbroker, The Hill, The Group, The Deadly Affair, Bye Bye Braverman, The Sea Gull, The Appointment, Last of the Mobile Hot-Shots, The Anderson Tapes, Child's Play, The Offence, Lovin' Molly, Serpico, Murder on the Orient Express, Dog Day Afternoon, Network, Equus, The Wiz, Just Tell Me What You Want, Prince of the City, Deathtrap.

LUPINO, IDA: Actress-director. b. London, England, Feb. 4, 1918. e. Royal Academy of Dramatic Art, London. Daughter of Stanley Lupino, English stage and screen comedian. Brit. m.p. debut in "Her First Affair," 1932; in U.S. m.p. 1934; ent. independent prod., variously dir., prod. collab. s.p. and acting.
PICTURES INCLUDE: Money for Speed, High Finance, High Sierra, Ladies in Retirement, Moontide, Hard Way, Devotion, Man I Love, Escape Me Never, Deep Valley, Road House, Lust for Gold, Not Wanted, Never Fear, Outrage, Hard Fast & Beautiful, On Dangerous Ground, Beware My Lovely, The Hitch-Hiker, Jennifer, The Bigamist, Private Hell 36, Women's Prison, Big Knife, While the City Sleeps, Trouble With Angels, Junior Bonner, The Devil's Rain, The Food of the Gods.
TV: Mr. Adams and Eve. No. 5 Checked Out, The Trial of Mary Surrat, Honey West, Virginian, I Love a Mystery, Sam Benedict, Untouchables, G. E. Theater, Have Gun Will Travel, Thriller, Mr. Novac, Hong Kong, The Rogues, Chrysler Theatre, Kraft Theatre, Gilligan's Island, The Ghost and Mrs. Muir, The Bill Cosby Show, To Catch a Thief, Mod Squad, Family Affair.

LURASCHI, LUIGI G.: Exec. b. London, Jan 7, 1906. e. U. of Zurich. Long Island Studio, Paramount, 1929; home officer mgr. For. dept. hd, For. & dom. Censorship; Hollywood to 1960. Asst. Prod., Dino De Laurentiis Prod. 1960–65; asst. to pres. for prod. activities, Paramount, 1965. 1967 continental prod. exec. Paramount-Rome. Now v.p., intl.

LYDON, JAMES: Actor. b. Harrington Park, N.J., May 30, 1923; e. St. Johns Mil. Sch. On NY stage in Prologue to Glory, Sing Out the News. On screen 1939, Back Door to Heaven.
PICTURES INCLUDE: Thoroughbreds, Naval Academy, Henry Aldrich series, Twice Blessed, Life With Father, Out of the Storm, Joan of Arc, Miss Mink of 1949, Tucson, Gasoline Alley, Island in the Sky, The Desperado, Battle Stations, My Blood Runs Cold (assoc. prod.), Brainstorm, An American Dream, A Covenant With Death, First to Fight, The Cool Ones, Chubasco, Countdown, Assignment to Kill, The Learning Tree, Scandalous John.
TV: Frontier Circus (assoc. prod.), Wagon Train, Alfred Hitchcock hour, McHale's Navy, 77 Sunset Strip, Mr. Roberts.

LYLES, A. C.: Producer. b. Jacksonville, Florida. May 17. e. Andrew Jackson High School; Paramount Publix's Florida Theatre, 1928; interviewed Hollywood celebrities, Jacksonville Journal, 1932; mail boy, Paramount Studios, Hollywood, 1937; publicity dept., 1938; hd. of adv., publ. dept., Pine-Thomas unit at Paramount, 1940; assoc. prod., The Mountain; prod., Short Cut to Hell; assoc. prod., Rawhide. President, A. C. Lyles Productions, Inc. (Paramount Pictures).
PICTURES INCLUDE: Raymie, The Young and the Brave, Law of the Lawless, Stage to Thunder Rock, Young Fury, Black Spurs, Hostile Guns, Arizona Bushwackers, Town Tamer, Apache Uprising, Johnny Reno, Waco, Red Tomahawk, Fort Utah, Buckskin, Rogue's Gallery, Night of the Lepus, The Last Day.

LYNCH, DAVID: Director-Writer. b. Montana, 1947. e. Pa. Academy of Fine Arts, where received an independent filmmaker grant from America Film Institute. Made 16mm film, The Grandmother. Accepted by Center for Advanced Film Studies in Los Angeles, 1970. Wrote and directed Eraserhead (with partial AFI financing) which became cult movie. Co-wrote and directed The Elephant Man (1980).

LYNCH, RICHARD: Actor. Made Bdwy. debut in The Devils, latter both on and off Bdwy in The Balcony, Lion in Winter, etc. Film debut in Scarecrow (1973).
PICTURES: The Premonition, Steel, The Formula.
TELEVISION: Vampire.

LYNCH, T. MURRAY: Executive. b. St. Thomas, Ontario, July 22, 1920. Capitol Theatre, St. Thomas, Ont.; R.C.A.F., Capitol Theatre, London, Ont.; Capitol Theatre, Halifax, N.S.: Paramount Theatre, Moncton, N.B.; Avalon Mall Theatre, St. John's Newfoundland: City supervisor, St. John's; Maritime and Newfoundland district mgr., Famous Players Canadian Corp., Sept. 1, 1968; office in Capitol Theatre, Halifax, N.S.

LYNDE, PAUL: Actor. b. Mt. Vernon, Ohio, June 13. e. Northwestern U. After club circuit appeared on Broadway in New Faces of 1952, then Bye Bye Birdie; films include The Glass Bottom Boat, Bye Bye Birdie, Under the Yum-Yum Tree, Send Me No Flowers, The Villain.
TV: Bewitched, The Paul Lynde Show, Hollywood Squares.

LYNDON, VICTOR: Producer. b. London. Ent. m.p. ind. as asst. dir., Gainsborough Pictures, 1946–56; prod. mgr., 26 feature pictures; since 1957 prod., assoc. prod., pictures for Columbia, United Artists, M-G-M, Paramount, British Lion.
PICTURES INCLUDE: as prod. mgr. The African Queen, Albert R.N., The Admirable Crichton, As assoc. prod., Dr. Strangelove, Darling, 2001 A Space Odyssey. As prod., Spare The Rod, Station Six—Sahara, The Optimists.

LYNN, ANN: Actress. b. London, England. Ent. films and TV, 1958.

PICTURES INCLUDE: Naked Fury, Piccadilly Third Stop, The Wind of Change, TV film series: The Cheaters, The Other Side of the Underneath.
TV: After The Show, All Summer Long, Trump Card, Man at the Top, The Expert, Hine, The Intruders, Too Far, King Lear, The Zoo Gang. Recent films incl: The Uncle, Morning Tide, Shot In The Dark, Black Torment, Four In The Morning, Baby Love, Hitler—The Last Ten Days, Estuary, Who Pays The Ferryman, The Professionals, Zeticula, Westway. The Perfect House, Minder, To the Sound of Guns, Crown Court.

LYNN, JEFFREY: Actor. b. Auburn, Mass., 1909: e. Bates Coll. m.p. debut in 1938.
PICTURES INCLUDE: Four Daughters, Yes My Darling Daughter, Daughters Courageous, Espionage Agent, Roaring Twenties, Four Wives, Child Is Born, Fighting 69th, It all Came True, All This and Heaven, Too; My Love Came Back, Four Mothers, Million Dollar Baby, Law of the Tropics, Body Disappears, For the Love of Mary, Black Bart, Letter to Three Wives, Strange Bargain, Home Town Story, Up Front, Captain China, Lost Lagoon, Tony Rome.
BROADWAY: (Revival) Dinner at Eight

LYNN, ROBERT: Director, producer. 100 TV films. Assoc. prod. for Mid-Atlantic Films. 1970 prod. The Railway Children. TV: Space 1999.
PICTURES INCLUDE: Dr. Crippen, Postman's Knock, Victim 5, Mozambique, Coast of Skeletons, Carnival of Killers, 2nd unit dir., Superman.

LYON, EARLE: Producer, Executive. b. Waterloo, Ia., April 9, 1923. e. UCLA. Entered m.p. industry 1947. Independent prod. 1954–55 and 1958–59; 20th Century-Fox prod., 1956–57; at Columbia, 1959–60; Universal, 1960–63; developed new TV projects with Bob Banner Assoc., 1963; v.p. and gen. mgr. United Pictures Corp.
PICTURES INCLUDE: Silent Raiders, The Lonesome Trail, The Silver Star, Stagecoach Fury, The Quiet Gun, The Rawhide Trail, The Rebel Set, The Destructors, Cyborg 2087, Dimension 5, Destination Inner Space, Haunting at Castle Montego, The Ten Billion Dollar Caper, Panic in the City.
TV: Tales of Wells Fargo.

LYON, FRANCIS D. "PETE": Director. b. Bowbells, N.D., July 29. e. Hollywood H.S., Univ. of Calif. at L.A.; writer, prod., dir., OWI, W.W. II; assoc. with training, exploitation and information films. Maj. U.S. Army Signal Corps.
PICTURES INCLUDE: Shape of Things to Come, Knight Without Armour, Rembrandt, Intermezzo, Adam Had Four Sons, The Great Profile, Four Sons, Daytime Wife, Body and Soul, He Ran All the Way, Crazylegs, The Bob Mathias Story, Walt Disney's The Great Locomotive Chase, The Oklahoman, Gunsight Ridge, Bailout at 43,000, Escort West, Cinerama South Seas Adventure, The Young and The Brave, Destination Inner Space, The Destructors, The Money Jungle, The Girl Who Knew Too Much, Tiger By the Tail.
TV: Laramie, Perry Mason, Zane Grey Theatre, Bus Stop, M. Squad, Wells Fargo, Draft Suspense Theatre, etc.

LYON, SUE: Actress. b. Davenport, Iowa, July 10, 1946. e. Hollywood Prof. School.
PICTURES INCLUDE: Lolita, Seven Women, Night of he Iguana, The Flim Flam Man, Evel Knievel.

LYONS, RICHARD E.: Producer. b. Boston, Mass., 1921; e. Titlon Jr. Coll. & Dartmouth College. Eastern sls. dept., MGM Pictures, 1533–41: Field Photographic Br., U.S. Navy, 1941–46; eastern artist, repertoire exec., MGM Records, 1946–55; story ed., Edward Small Prod., 1955–56; assoc. story ed., Univ., 1956–57; staff prod., MGM Studios, 1960.
PICTURES INCLUDE: Frontier Gun, The Sad Horse, The Miracle of the Hills, Ride The High Country, Mail Order Bride, The Rounders, The Plainsman, Winchester 73, Stranger on the Run, Something for a Lonely Man, Coogan's Bluff, Death of a Gunfighter, Dirty Dingus Magee.
TV: The Daughters of Joshua Cabe, The Jerico Mile (ABC-TV), Shootout in A One Dog Town (ABC-TV movie of the week, 1973). The Daughters of Joshua Cabe Return (ABC-TV M.O.W.); Welcome to Xanadu. (ABC-TV M.O.W.), Kate Bliss and the Ticker Tape Kid (ABC-TV M.O.W.); Roughnecks (4 hr. mini series), I Married Wyatt Earp (NBC-TV M.O.W.).

LYONS, STUART: Producer. b. Manchester, England, 1928. e. Manchester University. Ent. m.p. ind. 1955. Asst. dir. TV series 1955–56. Casting dir. Associated British, 1956/60. Freelance cast. dir., 1960/63. Joined 20th Century-Fox Productions as cast. dir., 1963. Appt. director 20th Century-Fox Productions Ltd., 1967, man. dir. 1968. 1971: left Fox on closure Europe prod. Joined Hemdale Group as head of production, May, 1972. Left Hemdale Aug. 1973 to resume indep. prod.
PICTURES INCLUDE: As casting director: Those Magnificent Men in Their Flying Machines, High Wind in Jamaica, Rapture, The Blue Max, Cleopatra, The Prime of Miss Jean Brodie, The Chairman. As indep. producer: The Slipper & The Rose, Meetings with Remarkable Men.

M

MAC ARTHUR, JAMES: Actor. b. Los Angeles, Calif., Dec. 8, 1937; e. Harvard. p. Helen Hayes, Charles MacArthur. Stage debut, summer stock; The Corn Is Green, 1945; Life With Father, 1953.
PICTURES INCLUDE: The Young Stranger, The Light in the Forest, The Third Man on the Mountain, Kidnapped, Swiss Family Robinson, The Interns, Spencer's Mountain, The Love-Ins, Cry of Battle, Angry Breed.
TV: Strike a Blow, Hawaii Five-O.

MACAULAY, FRED: Producer-Executive. b. Sollas, North Uist, in Western Isles of Scotland, 1925. e. Dunskellar and Paible public schools, Inverness Royal Academy, and University of Edinburgh. Was engaged 1950 to 1954 on the Linguistic Survey of Scotland for Edinburgh University. Actively supported over many years the School of Scottish Studies at Edinburgh, and is recognised for encouraging the use of the Gaelic language in Scotland. Associated with BBC Gaelic-language broadcasting since joining the BBC in 1954. Was head of the BBC's Gaelic dept. at Glasgow for 15 years. Manager of Radio Highland, BBC community station at Inverness, since January 1980.

MAC DONALD, PHILIP: Writer. b. Scotland; e. St. Paul's School, London. Novelist, playwright. Began screen career 1933.
PICTURES INCLUDE: Sahara, Action in Arabia, The Body Snatcher, Strangers in the Night, Dangerous Intruder, Man Who Cheated Himself, Circle of Danger, Mask of the Avenger, Ring of Fear Tobor the Great.

MAC GRAW, ALI: Actress. b. 1938. Was top fashion model.
PICTURES INCLUDE: Goodbye Columbus, Love Story, The Getaway, Convoy Players, Just Tell Me What You Want.

MACK, IRVING: Executive. b. Centerville, Iowa, Aug. 26, 1895. Started in show business as asst. pub. man. to Ralph Thomas Kettering, White City Amusement Park, Chicago; 3 yrs. ass't pub. dir., Jones, Linick & Schafer; in 1917 pub. exploit. mgr., Universal, Chicago In 1919 started Filmack Trailer Co.; at one time published local Chicago trade paper and was Chicago rep. for Film Daily. Chief barker Variety Club, 1951; member of Covenant Club, Chicago, Prudence Masonic Lodge, Cinema Lodge, B'nai B'rith. Now chm. bd. Filmack Corp. One of original founders of Variety Club of Ill. (Tent 26).

MACKERRAS, SIR CHARLES, CBE 1974. Knighted 1979: TV music conductor. b. Schenectady, N.Y., Nov. 17, 1925; e. Sydney Grammar School, (Australia); Sydney Conservatorium of Music, Prague Academy of Music. Principal oboist Sydney Symphony Orchestra, 1943–46; conductor, Sadlers Wells Opera, 1948–54; permanent conductor BBC Concert Orchestra, 1954–56; Ballet arrangements Pineapple Poll, Lady & The Fool. Now conducts regularly all principal British orchestras recordings, concerts, TV. etc.; also European Festivals & concert tours S. Africa, Australia, Russia, Scandinavia and USA. 1st Cond. Hamburg State Opera, 1966–70. Musical dir. English Nat. Opera Co. 1970–77. Chief guest conductor BBC Symphony Orch. 1977. Conductor many TV operas notably Otello, Billy Budd, La Boheme, Macbeth.

MAC LAINE, SHIRLEY: Actress. b. Richmond, Va., April 24, 1934; e. Washington and Lee H.S., Arlington, Va. Dancer, singer; signed by Hal Wallis; m p. debut in Trouble with Harry. Producer of film documentary on China, The Other Half of The Sky, Author of 2 best-selling books: Don't Fall off the Mountain and You Can Get There from Here.
PICTURES INCLUDE: Artists and Models, Around the World in 80 Days, The Matchmaker, Some Came Running, Ask Any Girl, Career, Can-Can, The Apartment, All in a Night's Work, Two Loves, My Geisha, The Children's Hour, Two for the Seesaw, Irma La Duce, What A Way To Go, John Goldfarb Please Come Home, The Yellow Rolls Royce, Gambit, Woman Times Seven, Sweet Charity, Two Mules for Sister Sara, Desperate Characters, The Possession of Joel Delaney, The Turning Point, Being There, Loving Couples, Change of Seasons.
TV: Shirley's World, Specials.

MAC LEAN, IAN: Actor. b. London England, 1894. On Brit. stage from 1918. Recent plays Fifty-Fifty, The Sleeping Clergyman. Screen debut in Brit. picture Brewsters Millions, 1930.
PICTURES INCLUDE: Marigold, Arsenal Stadium Mystery, A Week in Paris, Murder Will Out, Confidential Lady, Atlantic Ferry, Young Mr. Pitt, The Reporter, Shipbuilders, Dreamin', Twilight Hour, Carabana, Headline, Appoint-

ment With Crime, Calling Paul Temple, Story of Shirley Yorke, Floodtide.

MAC MAHON, ALINE: Actress. b. McKeesport, Pa., May 3, 1899; e. Erasmus Hall, Barnard College Stage experience; Once in a Lifetime, Candida, others. Screen debut 1931 in Five Star Final.
PICTURES INCLUDE: The Mouthpiece, One Way Passage, Gold Diggers of 1933, World Changes, Babbitt, Side Streets, Mighty McGurk, Search, Roseanna McCoy, Flame and the Arrow, Eddie Cantor Story, Man From Laramie, All the Way Home.

MAC MURRAY, FRED: Actor. b. Kankakee, Ill., Aug. 30, 1908; e. Carroll College, Wis. Sang and played in orchestra to earn tuition. To Hollywood with band; extra roles on screen; to N.Y. with comedy stage band, then joined "Three's a Crowd" revue; vaudeville circuits, night clubs. Began screen career 1935.
PICTURES INCLUDE: Grand Old Girl, Beyond Suspicion, Murder He Says, Double Indemnity, Oregon Trail, The Absent Minded Professor, The Happiest Millionaire, Follow Me Boys, Charlie and the Angel, Car 99, Gilded Lily, Hands Across the Table, Bride Comes Home, Alice Adams, Trail of the Lonesome Pine, Maid of Salem, Exclusive, True Confession, Men with Wings, Coconut Grove, Sing You Sinners, Cafe Society, Invitation to Happiness, Honeymoon in Bali, Smoky, Pardon My Past, Suddenly It's Spring, Egg and I, Singapore, A Miracle Can Happen, Miracle of the Bells, Don't Trust Your Husband, Family Honeymoon, Never a Dull Moment, Callaway Went Thataway, Millionaire for Christy, Fair Wind to Java, The Moonlighter, Caine Mutiny, Pushover, Woman's World, Far Horizons, There's Always Tomorrow, At Gunpoint, Rains of Rachipur, Gun For a Coward, Quantez, Good Day for a Hanging, The Shaggy Dog, Face of a Fugitive, The Oregon Trail, Bon Voyage, Son of Flubber, Kisses For My President, The Swarm.
TV: My Three Sons (12 years).

MACNAUGHTON, ROBERT: Actor. b. New York City, Dec. 19, 1966. Entered entertainment industry in 1979.
TELEVISION: Angel City.
STAGE: Critic's Choice, Thousand Clowns, Camelot.

MACQUITTY, WILLIAM: Prod. Author. b. Ireland.
PICTURES INCLUDE: Feature Prod., The Way We Live, Blue Scar, The Happy Family, Street Corner, The Beachcomber, Above Us The Waves, Black Tent, A Night to Remember, The Informers.
AUTHOR: Abu Simbel, Great Botanical Gardens of The World, Buddha, Persia the Immortal Kingdom, Tutankhamen the Last Journey, Princes of Jade, The World in Focus, Island of Isis, The Wisdom of the Ancient Egyptians, Rameses The Great, Master of the World, Our World in Colour, Inside China.

MAC RAE, GORDON: Singer, actor. b. East Orange, N.J., Mar. 12, 1921, p. Wee Willie Mac Rae, early radio star. Child actor on radio, in stock, on stage in Three to Make Ready, juvenile soloist in Ray Bolger revue, with Millpond Playhouse, Roslyn, L.I., radio, TV singer.
PICTURES INCLUDE: Big Punch, Look for the Silver Lining, Backfire, Return of the Frontiersman, Tea for Two, West Point Story, On Moonlight Bay, Standing, About Face, By the Light of the Silvery Moon, Desert Song, 3 Sailors and a Girl, Oklahoma, Carousel, Best Things in Life Are Free.

MAC RAE, JEAN: Actress. b. Montreal, Quebec, Mar. 24, 1930. M. Harry Lubcke. Worked for National Film Bd. of Canada, Canadian Bdcst. Corp., and repertory theatre in Canada; U.S. studios, including MGM, 20th-Fox, Columbia, and Paramount; as TV studios at Desilu and 20th-Fox.

MADDEN, BILL: Executive. b. New York City, March 1, 1915; e. Boston U. Joined Metro-Goldwyn-Mayer as office boy, 1930; student salesman, 1938; asst. Eastern div. mgr., 1939; U.S. Navy, 1942–46; Boston salesman, M.G.M., 1947–53; Philadelphia branch mgr., 1954–59; Midwest div. sales mgr., 1960–68; roadshow sales mgr., 1969; v.p., general sales mgr., M.G.M.; corp., v.p. & gen. sls. mgr., MGM, 1973; retired from MGM, 1975; 1976-present, exec. consultant to motion picture industry; lecturer at UCLA.

MADISON, GUY: Actor. r.n., Robert Moseley. b. Bakersfield, Calif., Jan. 19, 1922; e. Bakersfield Jr. Coll.; m. Sheila Connolly, U.S. Navy; m.p. debut in Since You Went Away, 1944; Wild Bill Hickok radio and TV shows, Star of Tomorrow, 1954.
PICTURES INCLUDE: Till the End of Time, Honeymoon, Texas, Brooklyn and Heaven, Massacre River, Drums in the Deep South, Red Snow, Charge at Feather River, The Command, Five Against the House, Beast of Hollow Mountain, Last Frontier, On the Threshold of Space, Hilda Crane, Bullwhip, Gunmen of The Rio Grande, Sandokan Fights Back, Sandokan Against the Leopard of Sarawak, Mystery of Thug Island, Shatterhand, Payment in Blood.

MAESTRI, CHARLES J.: Exhibitor. b. San Francisco, Calif., m.; e. Univ. of San Francisco, 1930. Paramount booker. San Francisco, 1930–33. In 1933 West Coast Theatres booker; pres., Oregon-California Theatres, Inc.

MAGEE, PATRICK: Actor. b. 1924, County Armagh, N. Ireland. Stage work before entering films.
PICTURES INCLUDE: The Birthday Party, Hard Contract, Barry Lyndon, Telefon, The Bronte Sisters, Rough Cut, Sir Henry at Rawlinson's End.

MAGILL, MORT: Sales executive. b. Philadelphia, Pa., Oct. 10, 1907; e. U. of Pennsylvania, B.S. 1927. Silk salesman in southern U.S. 9 yrs.; salesman United Artists, Harrisburg 1936; city salesman Philadelphia 1938; br. mgr. Pittsburgh 1941; br. mgr. Philadelphia; part-owner Principal Films 1949; district mgr. for M.P. Sales Corp., Philadelphia & Pittsburgh 1949; br. mgr., U.A. Philadelphia; dist. mgr. Buena Vista, (Phila., Pittsburgh, Washington) 1954; dist. mgr. Nat'l Gen. Pictures, 1971, independent distribution in partnership with John Turner-Firm called Turner-Magill, Inc., covering New York, Washington, Philadelphia and Pittsburgh. Firm now called Magill Films, Inc.

MAHIN, JOHN LEE: Writer. b. Evanston, Ill. From 1930 author, collaborator many screenplays MGM pictures. Received 3 Oscar nominations, 3 WGA nominations, Photoplay Gold Medal, Laurl Award (WGA).
PICTURES INCLUDE: Scarface, Beast of the City, The Wet Parade, Captains Courageous, Chained, Love on the Run, The Prizefighter and The Lady, Naughty Marietta, Small Town Girl, Wife vs Secretary, The Devil is a Sissy, The Last Gangster, Too Hot to Handle, Sabotage Agent, The Spiral Road, The Horse Soldiers, North To Alaska, Hell Below, Johnny Eager, Dr. Jekyll and Mr. Hyde, Tortilla Flat, Down to Sea in Ships, Love That Brute, Show Boat, Quo Vadis, My Son John, Mogambo, Elephant Walk, Lucy Gallant, Bad Seed, Eskimo, Treasure Island, Red Dust, Boom Town, No Time for Sergeants, Heaven Knows Mr. Allison, Moment to Moment.

MAHONEY, JOCK: Actor. r.n. Jacques O'Mahoney. b. Chicago, Ill., Feb. 7, 1919. e. U. of Iowa. Fighter pilot instructor, U.S.M.C.; enter m.p. ind. as stuntman; then started in westerns; TV show, Range Rider, many appearances on Loretta Young Show.
PICTURES INCLUDE: Away All Boats, Day of Fury, Showdown at Abilene, Battle Hymn, I've Lived Before, Land Unknown, Joe Dakota, Money, Women & Guns, Moro Witch Doctor, Walls of Hell, Tarzan Goes to India, Tarzan's Three Challenges.

MAIBAUM, RICHARD: Writer, Producer, b. New York City, May 26. educ. NYU, Univ. of Iowa, B.A., M.A., Ph. Beta Kappa. Plays on Broadway: The Tree, Birthright, Sweet Mystery of Life, See My Lawyer (1939), Member, Shakespearean Repertory Theatre in New York as player, 1933, From 1935 to 1942 worked on screenplays in Hollywood: MGM (They Gave Him a Gun, Stablemates, etc.); Columbia (Amazing Mr. Williams); Paramount (I Wanted Wings); 20th-Fox (Ten Gentlemen from West Point). Army, 1942–46, dir. Combat Film Div., final rank Lt. Col. To Para. prod.-writer, 1946, wrote-produced O.S.S., Song of Surrender; collab. & prod. The Great Gatsby. Prod. Sainted Sisters, The Big Clock, Bride of Vengeance, Dear Wife, No Man of Her Own, Capt. Carey, U.S.A.; free lance, 1953; collab. s.p., Paratrooper; adapt., Hell Below Zero; writer of Teleplays; exec. prod. MGM-TV, 1958–60.
PICTURES INCLUDE: Ransom, Cockleshell Heroes, Bigger Than Life, Zarak, Tank Force, Killers of Kilimanjaro, The Day They Robbed the Bank of England, Battle at Bloody Beach, Dr. No, From Russia With Love, Goldfinger, Thunderball, Chitty, Chitty, Bang Bang, On Her Majesty's Secret Service, Diamonds Are Forever, The Man With The Golden Gun, The Spy Who Loved Me, For Your Eyes Only.

MAIDMENT, GEORGE JOSEPH: Executive. Apptd. chief continental auditor, Fox Film Co., Ltd., 1924; was later comptroller of that co. and of British Movietonews, Ltd.; dir., Fox British Pictures, Ltd. 1933, treas. & secy., Columbia Pictures Corp., Ltd., Columbia Pictures Export Ltd., and Columbia British Productions, Ltd. In War Service as regional officer. Ministry Home Security, then on production planning for engineering and aircraft, finally on reorganization work. Ministry of Agriculture. 1944; gen. mgr., London Films Export, Ltd. 1947 gen. mgr., Omnia Films Ltd., Producers Representations Ltd., 1951 gen. mgr. Pendennis Pictures Corp. Ltd., Tower Films Ltd., Pax Films Ltd., Vicar Prod. Ltd., Orb. Prod. Ltd., Orb. Intl. Ltd. Fellow, Inst. of Chartered Accountants, Fellow Chartered Institute of Secretaries; Fellow, Institute of Linguists.

MAIN, DAVID: Writer, Producer. b. Essex, England, 1929. Extensive television experience in Britain producing and directing for A.T.V., Granada, A.B.C. and B.B.C. Emigrated to Canada in 1960. Directed Moment of Truth for N.B.C., and Quentin Durgens M.P. for C.B.C., Famous Jury Trials for 20th Century Fox. In 1977–78 directed King of

Kensington, Le Club, A Gift to Last for CBC. President of Velvet Screen Plays Ltd. a subsidiary of Quadrant Films Ltd.

PICTURES INCLUDE: Sunday In The Country, 1973 (story & co-writer), and It Seemed Like A Good Idea At The Time, 1975 (co-writer); Find The Lady, 1976 (story, co-writer, co-producer); Double Negative, 1978 (co-prod.); Nothing Personal, 1979 (co-prod.).

MAJORS, LEE: Actor. b. Wyandotte, Mich., April 23, 1940. m. actress-model Farrah Fawcett-Majors. Star athlete in high school; turned down offer from St. Louis Cardinals in final year at Eastern Kentucky State College to pursue acting career. In L.A. got job as playground supervisor for park dept while studying acting at MGM Studio.

PICTURES INCLUDE: Will Penny (debut), The Liberation of L. B. Jones.

TV: The Big Valley, The Man From Shiloh, The Six Million Dollar Man. Movies: The Ballad of Andy Crocker, Weekend of Terror, The Gary Francis Powers Story.

MALDEN, KARL: Actor. r.n. Malden Sekulovich. b. Gary, Ind., Mar. 22, 1914.

B'WAY PLAYS: Golden Boy, Key Largo, Flight to West, Missouri Legend, Uncle Harry, Counterattack, Truckline Cafe, All My Sons, Streetcar Named Desire, Desperate Hours, Desire Under the Elms.

PICTURES INCLUDE: Boomerang, The Gunfighter, Where the Sidewalk Ends, Hall of Montezuma, Streetcar Named Desire, (Acad. Award best supporting actor, 1951), The Sellout, Diplomatic Courier, Operation Secret, Ruby Gentry, I Confess, Take the High Ground, Phantom of the Rue Morgue, On the Waterfront, Bombers B-52, Desperate Hours, Fear Strikes Out, Egghead, The Hanging Tree, One Eyed Jacks, Parrish; dir., Time Limit, Billion Dollar Brain, Hot Millions, Blue Hotel, Patton, Cat O'Nine Tails, Wild Rovers, Summertime Killer, Beyond the Poseidon Adventure, Meteor.

TV: Streets of San Francisco, Captains Courageous.

MALICK, TERENCE: Producer-Writer-Director. b. Texas, 1945. e. Harvard. Attended Oxford on Rhodes scholarship. Worked for Newsweek, Life and The New Yorker; lectured for year in philosophy at M.I.T. Studied at American Film Institute in Bev. Hills and made short funded by AFI.

PICTURES INCLUDE: Badlands (prod., dir., s.p.), Days of Heaven (s.p., dir.).

MALLE, LOUIS: Director. b. Thumeries, France, 1932. Started in film industry as assistant to Robert Bresson. Began career somewhat ahead of most young French directors referred to as the Nouvelle Vague (New Wave). Became internationally known with Les Amants (The Lovers) in 1958. Has also acted in films.

PICTURES INCLUDE: The World of Silence (co-dir. with J. Y. Cousteau); A Man Escaped (asst. to Bresson); Elevator to the Gallows, The Lovers, Zazie in the Metro, Private Life, Vive Le Tour, The Fire Within, Bon Baisers de Bangkok, Viva Maria, The Thief, William Wilson, Phantom India, Murmur of the Heart, Humain, Trop Humain, Lacombe, Lucien, Black Moon, Pretty Baby, Atlantic City.

MALLERS, ANTHONY: Theatre Owner. b. Portland, Ind., Oct. 4, 1933. e. Indiana Univ. (B.S., business admin.). Entered industry in 1957. Now pres. of Mallers Theatres, headquartered in Muncie, Ind.

MALMUTH, BRUCE: b. Brooklyn. e. City Coll., Brooklyn Coll. Acted in college productions. Moved to California and obtained job as page at TV center. In Army assigned to special services as director; reassigned to New York. Upon release began 10-year production stint in TV commercials. Debut as director of feature with Nighthawks, 1981.

MALONE, DOROTHY: Actress. b. Chicago, Ill., Jan. 30, 1930, e. So. Methodist U. Screen debut in The Big Sleep, 1946.

PICTURES INCLUDE: Young at Heart, Battle Cry, Pillars of the Sky, Written on the Wind (Acad. Award, best supporting actress, 1956), Man of a Thousand Faces, The Last Voyage, The Last Sunset, Beach Party, Abduction, Golden Rendezvous, Winter Kills.

TV: Dick Powell Theatre, Dr. Kildare, Bob Hope Show, Jack Benny Show, The Untouchables, The Greatest Show On Earth, Peyton Place.

MALTZ, ALBERT: Writer. b. Brooklyn, N.Y., Oct. 28, 1908; e. Columbia U., B.A., 1930; Yale Sch. Drama. From 1932 author, playwright, novelist ("The Cross and the Arrow" 1944). From 1933–37, exec. bd., Theat. Union. Instructor in playwriting. Sch. Adult Educ., N.Y.U., 1937–41; also ed. Equality, 1937–38. Began screen writing 1941. Paramount. Acad. Award in documentaries (Moscow Strikes Back), 1943.

MEMBER: Author's League (council 1935–41); Amer. Arbitration Soc. (arbitor, from 1938).

PICTURES INCLUDE This Gun for Hire, Destination Tokyo, Pride of the Marines, Cloak and Dagger, Naked City, Two Mules for Sister Sara.

MAMOULIAN, ROUBEN: Director. b. Tiflis, Caucasus (Russia), Oct. 8, 1897; e. Lycee Montaigne, Paris, Moscow U., (law). Stage dir. since 1918; first English prod., Beating on the Door, London, 1922; came to U.S., 1923; prod. dir., Eastman Theatre, Rochester, N.Y., 1923–26; dir. operas, operetta, musicals; organizer, 1925, dir., 1925–26, Eastman Theatre School; dir. Oklahoma for Berlin Art Festival, 1951; dir. Oklahoma! for Paris, France (Salut a La France), 1955; Oklahoma! for Rome, Milan, Naples and Venice, Italy; author of short stories, verse, articles; pub. books: Abigayil, 1964; "Shakespeare's Hamlet, A New Version," 1966; 1966. prod. Hamlet; 1967. Tribute to R.M.'s 40th Anniv. on stage & screen at Mod. Art in N.Y.; 1968, Festival of R.M. Films at National Film Thea. in London, Washington, D.C., and Montreal. Tribute to Rouben Mamoulian: Retrospectives of all his films and personal appearances for discussion at: San Francisco International Film Festival, Science Center in Toronto, Canada, Museum of Science in Buffalo, New York, University of California at Los Angeles, University of South Florida, Yale University, New York (all in 1971). Seminar for USA Educators by Rouben Mamoulian at the American Film Institute, 1970. Guest of Honor of U.S.S.R. at the International Film Festival in Moscow, 1971. The Art of Survival—symposium at University of Southern California, 1972. Musical Mamoulian—Retrospective of R.M.'s musical films at the Academy of Motion Pictures Arts & Sciences, Hollywood, 1972. Publication: Foreword to the book Chevalier, 1973. Retrospective showing of Rouben Mamoulian's films at UCLA, 1973. Special showing of Blood and Sand and discussion at Filmex, Hollywood, 1973. President of the international jury at the International Film Festival of San Sebastian, Spain, 1973. Festival of Mamoulian's films in San Sebastian, 1973. Festival of Rouben Mamoulians films at the Cinematheque Francaise, Paris, France, 1973. Symposium on the Coming of Sound to the American Film 1925–1940—George Eastman House, Rochester, New York, 1973. Showing of Dr. Jekyll & Mr. Hyde and stage appearance Filmex 1974—Hollywood, 1974. Festival of Mamoulian's films in Sydney and Melbourne, Australia, 1974. Establishment of the "Rouben Mamoulian Award for the Best Australian Short Film" in perpetuity, June, 1974. Shtwing of Mamoulian muscals and appearance at the American Film Institute's "History of American Musicals" at Kennedy Center of Performing Arts, Washington, D.C., Sept., 1974. Guest of honor and member of international jury at International Film Festival, Tehran, Iran, 1974. Guest of Honor at festivities in Ispahan, Shiraz and Persipolis, December, 1974. Lecture on "The State of films and the state of the World" at Washington Club, Washington, D.C., December, 1974. Lectures, seminar and showing of Mamoulian films at University of California at Santa Barbara, February, 1975.

BROADWAY PLAYS: Porgy, Marco's Millions, Wings over Europe, Congai, A Month in the Country, Solid South, Farewell to Arms Porgy and Bess, Oklahoma, Sadie Thompson, Carousel, St. Louis Woman, Lost in the Stars, Arms and the Girl.

PICTURES INCLUDE: Applause, City Streets, Dr. Jekyll and Mr. Hyde, Love Me Tonight, Song of Songs, Queen Christine, We Live Again, Becky Sharp, Gay Desperado, High, Wide and Handsome, Golden Boy, Mark of Zorro, Blood and Sand, Rings On Her Fingers, Summer Holiday, Silk Stockings, Never Steal Anything Small (co-author) 1959.

MANASSE, GEORGE: Producer. b. Florence, Italy, Jan. 1, 1938; e. U. of N.C. Producer, GSF Productions, Inc.; for Cannon was assoc. prod., Joe, Jump; prod. Who Killed Mary What's Her Name; for GSF was exec. prod., Journey Through Rosebud, Change in the Wind.

MANBY, C. R.: Executive. b. Battle Creek, Mich., Feb. 24, 1920; e. Hillsdale Coll., Mich., A.B. 1942; Harvard Grad. School of Bus. Admin., 1953; v.p., RKO General Inc., 1955–58; pres. Show Corp of Amer., 1958–75. Now sr. v.p., RKO General, Inc. & pres., RKO Pictures, Inc.

MANCINI, HENRY: Composer. b. 1922. Arranged music for the Glen Miller Story, The Benny Goodman Story, then began composing scores.

PICTURES INCLUDE: Touch of Evil, High Time, Breakfast at Tiffany's (AA, 1961), Bachelor in Paradise, Hatari, The Pink Panther Charade, A Shot in the Dark, Dear Heart, What Did You Do in the War, Daddy?, Two for the Road, Visions of Eight, Oklahoma Crude, 99 44/100s½ Dead, The White Dawn, The Girl from Petrovka, The Great Waldo Pepper, W. C. Fields and Me, The Pink Panther Strikes Again, Silver Streak, Revenge of the Pink Panther, Who Is Killing the Great Chefs of Europe?, Prisoner of Zenda, Nightwing, "10," Little Miss Marker, Change of Seasons, Back Roads, "S.O.B.".

MANCUSO, FRANK G.: Executive. b. Buffalo, N.Y., July 25, 1933. e. State Univ. of New York. Film buyer and operations supvr. for Basil Enterprises, theatre circuit, from 1958 to 1962. Joined Paramount as booker in Buffalo branch, 1962. Named sls. repr. for branch in 1964 and branch mgr. in 1967. In 1970 appt. v.p./gen. sls. mgr., Paramount Pictures Canada, Ltd., becoming pres. in 1972. In 1976 relocated with Paramount in U.S. as western div. mgr. in L.A. In Jan., 1977, appt. gen. sls. mgr. of N.Y. office; two months later promoted to v.p.—domestic distribution; 1979, named exec. v.p., distribution & mktg.

MANDEL, HARRY: Theatre executive. b. New York City, dir., adv. & publicity, RKO Theatres, 1938; added duties asst. to v.p. theatre operations, 1952; v.p. theatre operations, 1957; pres., 1961–66.

MANDELL, ABE: Executive. e. U. of Cincinnati. Entered broadcasting as actor on Cincinnati radio station prior to WW II. Served U.S. Army in Southwest Pacific, 1942–45. Formed independent motion picture distribution company in the Far East. Company, which became the largest independent motion picture distributor in the Far East, also operated and owned motion picture theaters throughout the Phillipine Islands and Indonesia, 1946–56; network-regional sales exec., Ziv Television, 1956–58; dir. foreign operations, Independent Television Corporation, 1958; v.p.-foreign oper., 1960; v.p.-sales and adm., 1961; exec. v.p., 1962; pres. 1965. In 1976 ITC changed its corporate name from Independent Television Corp. to ITC Entertainment, Inc. President to date of ITC Entertainment.

MANGANO, SILVANA: Actress. b. Rome, Italy. e. Dance Academy of Jia Ruskaja. Model to 1949; m.p. debut in L'Elisir D'Amore.
PICTURES INCLUDE: Bitter Rice, Lure of Sila, Musolino, the Brigand; Anna, Ulysses, Mambo, Gold of Naples, This Angry Age, Tempest, Five Branded Women, Barabbas, Teorama, Death in Venice, Decameron, Ludwig, Conversation Piece.

MANHEIM, MANNIE: Producer, writer. b. Syracuse, N.Y., Nov. 13; e. Syracuse U. Newspaper ed.; m.p. & mag. writer; radio & TV writer & prod.

MANKIEWICZ, DON M.: Writer. b. Berlin, Germany, Jan. 20, 1922. p. Herman J. Mankiewicz e. Columbia, B.A., 1942; Columbia Law School. Served in U.S. Army, 1942–46; reporter, New Yorker magazine, 1946–48; author of novels, See How They Run, Trial, It Only Hurts a Minute; magazine articles, short stories.
TV: Studio One, On Trial, One Step Beyond, Playhouse 90, Profiles in Courage.
PICTURES INCLUDE: Trial, I Want to Live.
TV PILOTS: Ironside, Marcus Welby, M.D., Sarge; Lanigan's Rabbi (collaboration); Rosetti and Ryan (collab.)

MANKIEWICZ, JOSEPH L.: Writer, director. b. Wilkes-Barre, Pa., Feb. 11, 1909; e. Columbia U. Asst. corr. in Berlin, Chicago Tribune; Ufa studio, translating subtitles into Eng. for release in Eng. & U.S.; returned to U.S. 1929 to join brother, Herman, on Paramount writing staff; MGM, 1933; Fox, 1943–51; dir., La Boheme, Metropolitan Opera House, 1953.
PICTURES INCLUDE: Skippy, Million Dollar Legs, If I Had a Million, Alice in Wonderland, Fury, Gorgeous Hussy, Mannequin, Three Comrades, Shopworn Angel, Philadelphia Story, Woman of the Year, Keys of the Kingdom, Dragonwyck, Somewhere in the Night, No Way Out, Late George Apley, Ghost and Mrs. Muir, House of Strangers, Letter to Three Wives, All About Eve, 1950, People Will Talk, 5 Fingers, Julius Caesar, Barefoot Contessa, Guys and Dolls, Quiet American, Suddenly Last Summer, The Honey Pot, There was a Crooked Man, Sleuth.

MANKIEWITZ, TOM: Writer. Specializes in adventure/suspense films.
PICTURES INCLUDE: The Sweet Ride (debut), Diamonds Are Forever, Live and Let Die, The Man with the Golden Gun, Mother, Jugs and Speed, The Cassandra Crossing, The Eagle Has Landed.

MANKOWITZ, WOLF: Author, playwright, producer & impresario. B. London, 1924. Journalist. Ent. m.p. in 1952. Musical play based his story Expresso Bongo, 1958. Musical play, Make Me An Offer, 1959; Belle, 1961; Pickwick, 1963; Passion Flower Hotel.
PICTURES INCLUDE: Make Me An Offer, Kid For Two Farthings, The Bespoke Overcoat, Trapeze, Expresso Bongo, The Millionairess, The Long and Short and Tall, The Day The Earth Caught Fire, Where the Spies Are, Assassination Bureau 1969; Bloomfield, 1970: Black Beauty, Treasure Island. 1973: The Hireling.
TV: Make Me An Offer, The Baby, The Girl, It Should Happen To A Dog, Conflict, The Killing Stones, A Cure for Tin Ear, The Model Marriage, ABC of Show Business, The Battersea Miracle, Dickens of London, 1976–77.

MANN, ABBY: Writer. b. Philadelphia. e. N.Y.U. First gained fame on TV; adapted own teleplay Judgment at Nuremberg into theatrical film.
PICTURES INCLUDE: Judgment at Nuremberg, A Child Is Waiting, Ship of Fools, The Detective, Report to the Commissioner.

MANN, DANIEL: Director. b. New York, N.Y., Aug. 8, 1912. e. Erasmus Hall, Brooklyn; Professional Children's School. Started as musician in resort hotels; then in Canada, on road; U.S. Army, World War II; then received scholarship to Neighborhood Playhouse, N.Y.; dir. teaching, TV dir.
BROADWAY: Come Back Little Sheba, Rose Tattoo, Streetcar Named Desire (City Center prod.), Paint Your Wagon.
PICTURES INCLUDE: Come Back Little Sheba, About Mrs. Leslie, Rose Tattoo, I'll Cry Tomorrow, Teahouse of the August Moon, Hot Spell, Last Angry Man, Mountain Road, Butterfield 8, Ada, Judith; A Dream of Kings, Who's Got the Action?, Who's Been Sleeping in my Bed?, For Love of Ivy, Willard, The Revengers, Maurie, Interval, Lost in the Stars, Matilda.

MANN, DELBERT: Director, Producer. b. Lawrence, Kans., Jan. 30, 1920; e. Vanderbilt Univ., Yale Univ., U.S. Air Force, 1942–45; stage mgr., summer stock, dir. Columbia, S.C. Town Theatre, 1947–49; asst. dir., NBC-TV, 1949; dir., NBC-TV, 1949–55. Past pres. Directors Guild of America.
PROGRAMS: Philco-Goodyear TV Playhouse, Producer's Showcase, Omnibus, Playwrights '56, Playhouse 90, Ford Star Jubilee, Lights Out, Mary Kay and Johnny, The Little Show, Masterpiece Theatre, Ford Startime.
STAGE: A Quiet Place, Speaking of Murder, Zelda, opera: Wuthering Heights; New York City Center.
PICTURES INCLUDE: Marty (Academy Award, best director, 1955), Bachelor Party, Desire Under the Elms, Separate Tables, Middle of the Night, The Dark at the Top of the Stairs, The Outsider, Lover Come Back, That Touch of Mink, A Gathering of Eagles, Dear Heart, Quick Before It Melts, Mister Buddwing, Fitzwilly, The Pink Jungle, Kidnapped, Birch Interval, Night Crossing.
TV: Heidi, David Copperfield, Jane Eyre. The Man Without A Country, A Girl Named Sooner, Breaking Up, Tell Me My Name, Home To Stay, All Quiet on the Western Front, To Find My Sons.

MANN, MICHAEL: Producer-Director-Writer. b. Chicago. e. Univ. of Wisc., London Film School. From 1965–72 directed commercials and documentaries in England, including Insurrection for NBC in Paris during the 1968 riots. Directed short film, Jaunpuri, winner of Juris Prize at Cannes Film Festival. Returned to U.S. in 1972 to direct documentary, 18 Days Down the Line. Wrote for prime-time TV (episodes of Starsky and Hutch, Police Story, Vegas).
PICTURE: Thief (exec. prod., s.p., dir.).
TELEVISION: The Jericho Mile (s.p., dir.). (DGA, 1980 best director award).

MANN, TED: Chairman of the Board, Mann Theatres Corp. of California. b. Wishek, N.D. Acquired first theatre in St. Paul, MN; then acquired and built 25 theatres and drive-ins in Minnesota; acquired National General Theatres circuit of 266 theatres in 1973. Produced Buster and Billie, Lifeguard and Brubaker. Now in production with Dragons of Krull.

MANNE, S. ANTHONY: Executive. b. New York, N.Y., July 19, 1940. e. Wharton School Univ. of Pa. (B.S., economics). Joined Columbia Pictures 1963; international dept. 1964; asst. mgr., Brazil, 1968; mgr., Brazil, 1972–76. Joined JAD Films, 1976; appointed v.p. 1978, exec. v.p., 1979.

MANSON, ARTHUR: Executive. b. Brooklyn, N.Y., Feb. 21, 1928; e. CCNY, grad. (Inst. Film Technique) '45; editor, American Traveler, U.S. Army, 1946; Advance agent, co. mgr., Henry V, U.S., 1948–50; producer's publ. rep., Stanley Kramer Distributing Corp., Samuel Goldwyn Productions, Lopert Films, dir. of adv. and publ., MGM Pictures of Canada, Ltd., 1952–53; publ. and adv. rep., Cinerama Corp., 1953–58; worldwide ad-pub Cinerama (wide screen process) 1958–60; adv. mgr., Columbia Pictures, 1961–62; nat'l dir. of adv., publ., Dino De Laurentiis, 1962–64; exec. asst. to v.p. chg. adv. & pub., 20th Century-Fox, 1964–67; v.p. chg, adv. & pub. Cinerama. Inc., and Cinerama Releasing Corp.; 1967–74; exec. v.p., sales & marketing, BCP, service of Cox Broadcasting Corp., 1974–75; v.p. chg. worldwide adv.-pub. Warner Bros., 1976. In 1977 formed own company, Cinemax Mkt. & Dist. Corp. and is pres.

MANULIS, MARTIN: Prod.-dir. b. N.Y.C. May 30, 1915; e. Columbia Univ., B.A. 1935. Head of prod. John C. Wilson, 1941–49; mgr. dir., Westport Country Playhouse, 1945–50; dir. B'way plays; staff prod. & dir. CBS-TV, 1951–58; head prod. 20th-Fox Television. Now pres. Martin Manulis Prods. Ltd.
BROADWAY: Private Lives, Made in Heaven, The Philadelphia Story, Pride's Crossing, Laura, The Men We Marry, The Hasty Heart, The Show Off.
TV: Suspense, Studio One, Climax, Best of Broadway, Playhouse 90.
PICTURES INCLUDE: Days of Wine and Roses, The Out-Of-Towners, Luv, Duffy.

MANVELL, ROGER (PhD., Litt. D.): Author, Co. Director, broadcaster, screen, TV writer. b. Oct. 10, 1909; e. Kings School Peterborough, London Univ. During World War II with films div. of Min. of Information, mem. Screen Dir. Guild (U.S.); chmn. British Radio & Television Writer's Ass'n. 1953–56; dir. British Film Academy, 1947–59; Consultant Soc. Film and Television Arts (formerly British Film Academy), 1959–75. Visiting fellow Univ. of Sussex; gov. and hd. of Dept. Film Hist., London Film School, 1967–75. Author numerous books on films and T.V. New Cinema in U.S.A., 1968; New Cinema in Britain (1969). The German Cinema, Shakespeare and Film (1971); Films & The 2nd World War (1974), Charles Chaplin (1974). Editor, International Encyclopedia of Film (1972). Theater and film (1979). Ingmar Bergman (1980) Art & Animation (1980). Apptd. Bingham Prof. of Humanities, University of Louisville, 1973. Visiting Prof. Film, Boston Univ., Mass., U.S.A. 1975–81.

MARA, ADELLE: Actress. r.n. Adelaida Delgado; b. Dearborn, Michigan, April 28. m. Roy Huggins. Singer, dancer with Xavier Cugat orchestra; Col. contract.
PICTURES INCLUDE: Shut My Big Mouth, Blondie Goes to College, Alias Boston Blackie, You Were Never Lovelier, Riders of the Northwest Mounted, Magnificent Rogue, Passkey to Danger, Traffic in Crime, Exposed, The Trespasser, Blackmail, Campus Honeymoon, Sands of Iwo Jima, Sea Hornet, Count The Hours, Wake of the Red Witch, Back from Eternity.

MARAIS, JEAN: Actor. b. Cherbourg, France, Dec. 11, 1913; e. Coll. St. Germain, Lycee Janson de Sailly & Lycee Condorcet Painter; photog; stage actor; French Air Army; m.p. debut in Pavillon Brule.
PICTURES INCLUDE: Carmen, Eternal Return, Beauty and the Beast, Ruy Blas, Les Parents Terribles, Secret of Mayerling, Souvenir, Orpheus, Eagle with Two Heads, Inside a Girl's Dormitory, Royal Affairs in Versailles, Paris Does Strange Things, Le Capitan, Le Bossu La Princesse de Cleves, Le Capitaine Fracasse, Honorable Stanilleu, Agent Secret, Patute, Fantomas, Le Gentleman de Cocody.

MARANS, MARDI: Executive. b. 1949. ed. Univ. of Calif. (1971 graduate). Worked in L.A. office of Doyle Dane Bernbach ad agency. Joined Warner Bros. in March, 1975, as asst. to media director. In 1979 appt. v.p. and director of media for WB, responsible for planning and executing placement of all advertising in U.S. and Canada for films distributed by so. Also serves as consultant to intl. div. on use of media in other countries.

MARCH, ALEX: Producer. b. Brooklyn, N.Y. 1920. Actor in stock cos.; later turned to dir.; U.S. Air Force World War II; on Broadway in Look Ma I'm Dancing; then script ed., casting dir., Philco Playhouse; assoc. prod., Nash Airflyte Theatre; joined CBS-TV as script ed., You Are There; prod., Pentagon UCA, Studio One Summer Theatre.

MARCH, DONALD: Production Executive. Held senior programming positions with ABC-TV network and Robert Stigwood Organization before joining CBS in 1977 as director, special projects, motion pictures for TV and mini-series. Later promoted to v.p., motion pictures for TV. Left CBS to serve as pres. of Filmways' theatrical division in early 1979; later that year rejoined CBS as v.p., theatrical films, with responsibility for selection, development and production of pictures for theatrical release.

MARCUS, BEN D.: Chairman of the Board & Chief Executive Officer, The Marcus Corporation, parent company of Marcus Theatres Corporation, Milwaukee, Wis. b. Poland, Aug. 10, 1911. e. U. of Minn., 1934. Promotion, adv. dept., Minn. Journal, 1933–35; entered exhib. in Wis.; now owns & operates 83 theatres 13 drive-ins; pres., nat'l dir., Wis. Allied Theatre Owners, 1950–51–56–57; dir. Wis. Allied Theatre Owners, 1950–51; nat'l dir., 1952; state chmn., Movietime U.S.A., 1951; nat'l pres., Allied States Assn., 1954; chmn., Audience Poll award. State of Wis., 1955; mem., Prod. Code Review Board, 1962–65; chief barker Variety Club, Milwaukee, 1954; COMPO triumvirate rep. Allied States Assn., 1958–65; chmn. exec. comm., Allied States Assn., 1963–65. Treas. Nato; chmn. of bd. NATO of Wisc., treas, nat'l NATO, Code and Rating Appeals Bd., 1968–79; v.p., Variety International, 1977–79. Received NATO-Walt Disney Humanitarian Award, 1977.

MARCUS, LOUIS: Producer, Director, Writer. b. Cork, Ireland, 1936. e. Natl. Univ. of Ireland (B.A., 1959). Based in Dublin since 1959, where has made nearly 30 theatrical documentary films. Produces and directs for Louis Marcus Documentary Film Production of Dublin and Louis Marcus Films Ltd. of London. In 1964 appt. by Irish govt. as bd. mem. of Dublin's Abbey Theatre. In 1972 appt. by govt. as mem. of Cultural Relations Comm. of Dept. of Foreign Affairs. Elected mem. of Academy of M.P. Arts & Sciences (short subject branch) 1974.
PICTURES INCLUDE: Fleadh Cheoil, Horse Laughs, Woes of Golf, Children at Work, Conquest of Light.
AUTHOR: The Irish Film Industry (1968).

MARDEN, MICHAEL: Executive. e. Princeton Univ. Joined CBS in N.Y. in 1957. Later was TV prog. exec. for Benton & Bowles, Inc. Returned to CBS in April, 1966 as gen. prog. exec. for the network in N.Y. Named director, feature films, and was responsible for the CBS Late Movie (premiere: 1972). In June, 1976 named director, motion pictures for TV and mini-series, Hollywood.

MARENSTEIN, HAROLD: exec. b. New York, N.Y., e. C.C.N.Y., 1937. Shipping, picture checking service, Warner Bros., 1935–45; Booking, Loew's Inc., 1945–48; Booking, contracts, Selznick Rel. Org., 1948–51; contracts, Paramount, 1951–52; asst. sls. gr. International Rel. Org., 1952; asst. sls. mgr., Janus Films, 1961–64; sls. exec., Rizzoli Films, 1965; 1967, nat'l. sales dir., Continental Dist.; gen. sales mgr., Cinemation Industries, 1968. v.p.-sales, dir., Cinemation Industries, 1971; 1976, gen. sls. mgr., General National Films; 1980, gen. sls. mgr., Lima Productions.

MARGO: Actress. r.n. Maria Marguerita Guadalupe Boldao y Castilla. b. Mexico City, May 10, 1918. Starred as dancer by Xavier Cugat at Waldorf Astoria, elsewhere; on N.Y. stage (Winterset, Masque of Kings, The Outward Room).
PICTURES INCLUDE: Winterset, Lost Horizon, Crime Without Passion, The Leopard Man, Viva Zapata!, I'll Cry Tomorrow.

MARGOLIN, JANET: Actress. b. New York City, 1943. e. N.Y. High School of Performing Arts. Discovered by Frank Perry while playing in Bdwy. show, Daughter of Silence, and hired by him for lead in David and Lisa.
PICTURES: The Greatest Story Ever Told, Bus Riley's Back in Town, Morituri, Nevada Smith, Enter Laughing, Buena Sera, Mrs. Campbell, Take the Money and Run, Annie Hall, Last Embrace.

MARGRAF, GUSTAV B.: Executive. b. Cape Giradeau, Mo., May 14, 1915; e. Southeast Missouri State Coll. and Duke Univ. Law Sch.; attorney with Cahill, Gordon, Zachry & Reindel 1939–48; v.p., gen. atty., NBC, 1948–53; v.p. for talent NBC, 1953–55; gen. counsel, dir. and v.p. Reynolds Metal Co.; mgng. dir., British Aluminum Co., Ltd.; pres., chmn. bd., Canadian British Aluminum Co., Ltd.; v. chmn., Manicouagan Power Co., chmn., Aluminum Wire & Cable Co. Ltd.; chr. dir., Reynolds Jamaica Mines, Metal Development Corp., Caribbean Steamship Co., Haitian Mines Inc.; chmn. Aluminum Foils Ltd.; pres. Aluminum Fed. of Gt. Britain.

MARGULIES, IRWIN: Attorney, executive. b. Austria. June 24, 1907; e. N.Y.U., BA, 1929; N.Y.U. Law Sch., LL.B., 1931. Assoc. with father in m.p. theatres, 1918; treas., counsel or Jayem Management Corp.; MGM legal dept., N.Y., 1930–42; asst. to v.p.; gen. counsel J. Robert Rubin, 1943–47; asst. secy, counsel, MGM Intern'l Films. Inc., 1945–47; in private practive since 1947. Member Calif. bar since 1948; counsel for many m.p. prod. & dist., cos., literary gencies, radio & TV prod., dist., & personalities; 1954–67 v.p. Horizon Pict. Inc.; 1967, v.p. in charge of bus. affairs, Warner-7 Arts; 1969, with Harry Saltzman; exec. prod., "Digby—The Biggest Dog in The World."

MARGULIES, LYNNE (BARKETT): Actress-Film Maker. b. Feb. 28, 1957 Sylmar, CA. Began film career in 1977 working both in front of and behind the camera. Second ranking executive in Nautilus Films.
PICTURES: Teenage Cruisers, Highriders, The Aftermath.

MARGULIES, STAN: Producer. b. New York, N.Y. Dec. 14, 1920. e. De Witt Clinton High School, New York Univ., B.S., June, 1940; Army Air Force, May, 1942; pub. rels. Air Force and the Infantry, wrote service magazines, newspapers including Yank; spec. feature writer & asst. Sunday editor, Salt Lake City Tribune; publicist, RKO Studios, Hollywood, March, 1947; continued publicity work at CBS-Radio, 20th Century-Fox, Walt Disney Productions. Bryna Films, 1955; became vice-pres., 1958; also served exec. prod., TV series, Tales of the Vikings; prod. aide. Spartacus.
PICTURES INCLUDE: 40 Pounds of Trouble, Those Magnificent Men In Their Flying Machines, Don't Just Stand There, The Pink Jungle, If It's Tuesday, This Must Be Belgium, I Love My Wife, Willy Wonka and the Chocolate Factory, One Is A Lonely Number, Visions of Eight.
TV: The Morning After, I Will Fight No More Forever, Collision Course, Roots, Roots: The Next Generation, Moviola, Murder Is Easy.

MARIN, RICHARD: Actor-Writer. b. Los Angeles, July 13, 1946. Teamed with Tommy Chong in improvisational group. Has made comedy recordings with him.
PICTURES: Cheech and Chong's Up in Smoke, Cheech and Chong's Next Movie, Cheech and Chong's Nice Dreams.

MARKHAM, MONTE: Actor. b. Manatee, Fla., June 21. e. Univ. of Ga. Military service in Coast Guard after which joined resident theatre co. at Stephens College, Missouri, where also taught acting. Joined ACT Theatre, San Francisco, for three years. Made TV debut in Mission: Impossible episode then starred in two series.
PICTURES INCLUDE: One is a Lonely Number, Hour of the Gun, Guns of the Magnificent Seven, Midway, Airport '77.

TELEVISION: The Second Hundred Years (series), Mr. Deeds Goes to Town (series), The New Perry Mason (series), Visions (movie), The Astronaut (movie), Death Takes a Holiday (movie).
BROADWAY: Irene.

MARKLE, FLETCHER: Writer, director, producer. b. Canada, March 27, 1921. Writer, dir., prod. Canadian Broadcasting Co. & BBC, London, 1942–46; prod., dir. Studio One series, CBS, 1947–48; Ford Theatre, CBS, 1948–49; writer, ed. & narrator The Robot Bomb, prize-winning doc. short, 1941; first m.p. dir. Jigsaw, 1949; thereafter Night Into Morning, The Man With a Cloak; prod. Studio One series, CBS, 1952–53; prod., dir., Life with Father, CBS-TV, 1953–55; Front Row Center, 1955. Contributing dir. and/or prod. to various TV film series: Mystery Theatre, Panic, No Warning, Colgate Theatre, Lux Playhouse, M Squad, Buckskin, Rendezvous, Tales of the Vikings, Thriller, Hong Kong, 1956–61. Dir. Father of the Bride, 1961; dir. m.p., The Incredible Journey, 1962; Telescope series, CBC-TV, 1963–72. Head of TV Drama, Canadian Broadcasting Corp. 1970–73, Exec. prod., CBC-TV features and writer in residence. University of Toronto, 1974–75.

MARKOE, JERRY: Composer-Arranger-Conductor. b. New York, N.Y. March 22, 1941. e. Manhattan School of Music (BM, MM in composition, 1968); Julliard School of Music. Winner of ASCAP music for theatre awards.
MUSICALS: Macbird, Fair Play for Eve, Another Time.
MUSIC FOR FILM, THEATER and TV: Faust, The Chinese Wall, He Who Gets Slapped, Ethan Frome, The Lion in Winter, Indulgences in the Louisville Harem, The Cycle, The Ancient Art of Tai Chi.
CHILDREN'S MUSICALS FOR STAGE & TV: Alice in Wonderland, Punch & Judy, Alladin, Cinderella, The Black Princess, Forgotten Treasure, Androcles and the Lion.

MARKOWITZ, ROBERT: Director. Mostly on TV before theatrical debut with Voices, 1979.
TELEVISION: The Deadliest Season, Song of Myself, With All Deliberate Speed, The 34th Star, The Storyteller.

MARKS, ALFRED: Actor-Comedian. b. London, TV, own series, Alfred Marks Time with wife, comedienne Paddie O'Neil.
PICTURES INCLUDE: Desert Mice, There Was A Crooked Man, Weekend With Lulu, The Frightened City, She'll Have To Go.
TV: Blanding's Castle, Hobson's Choice, Paris 1900, The Memorandum.

MARKS, ARTHUR: Producer, Director, Writer, Film Executive. b. Los Angeles, Calif., Aug. 2, 1927. M. Has 4 children. At 19 began work at MGM Studios as Production Messenger. Became Asst. Dir. in 1951, youngest dir. member of Directors Guild of Amer., 1957. Prod. of Perry Mason TV Show from 1961–66. Dir. of over 100 Perry Mason episodes. Prod.-dir.-writer of feature film togetherness. Writer-director tf Class of '74, writer-director of Bonnie's Kids, writer-director of The Roommates, producer-director of Detroit 9000, producer-director of The Centerfold Girls, director of A Woman For All Men, executive producer of Wonder Woman, executive producer, The Candy Snatchers. President and board member of Arthur Prod., Inc. & Arm Service Co. Exec. head of prod., Henry Plitt Prods.
RECENT CREDITS: Bucktown (dir.), Friday Foster (prod., dir.) J.D.'s Revenge (prod.-dir.), Monkey Hustle (prod., dir.).

MARKSON, BEN: Writer. b. Creston, Ia., Aug. 6. Army Capt., W.W. II. Reporter, feature writer columnist, playwright (Is My Face Red?). Joined Warners, 1932 as writer; since author orig. s.p. dial., adapt., collab., many pictures. Writes many TV plays.
PICTURES INCLUDE: Here Comes the Navy, Woman-Wise, Danger, Love at Work, Is My Face Red?, Brides Are Like That, Half-Naked Truth, Lady Killer, That I May Live, White Cockatoo, Goodbye Again, Great Mr. Nobody, The Beautiful Cheat, A Close Call for Boston Blackie, Prison Ship, The Falcon in San Francisco, Mr. District Attorney, It Happened on Fifth Ave., Edge of Eternity, With My Face to The Enemy.

MARLEY, J. PEVERELL: Cameraman. b. San Jose, Calif., Aug. 14, 1901; Chief cameraman for C. B. DeMille for The Ten Commandments, The Volga Boatman, King of Kings.
PICTURES INCLUDE: Night and Day, Of Human Bondage, Two Mrs. Carrolls, Life with Father, Off Limits, House of Wax, Serenade.

MARLOW, LUCY: Actress. r.n. Lucy Ann McAleer; b. Los Angeles, Calif., Nov. 20, 1932; e. U. of Ariz.
PICTURES INCLUDE: A Star Is Born, Lucky Me, Tight Spot, My Sister Eileen, Queen Bee, Bring Your Smile Along.

MARLOWE, HUGH: Actor. r.n. Hugh Hipple. b. Philadelphia, Pa. Formerly radio announcer, Little Theatre actor; won recognition on London stage. In N.Y. plays. Began screen career 1936. TV series Ellery Queen.
N.Y. PLAYS: Kiss the Boys Goodbye, Margin for Error, Land Is Bright, Lady in the Dark.
PICTURES INCLUDE: Marriage Is a Private Affair, Meet Me in St. Louis, Mrs. Parkington, Come to the Stable, 12 O'Clock High, All About Eve, Rawhide, Mr. Belvedere Rings the Bell, Day the Earth Stood Still, Wait Till the Sun Shines Nellie, Bugles in the Afternoon, Way of a Gaucho, Monkey Business, Stand at Apache River, Casanova's Big Night, Garden of Evil, Illegal, Earth vs. the Flying Saucers, Castle of Evil, The Last Shot You Hear.

MARLOWE, LOUIS J.: Producer, director. r.n. L. J. Goetten. b. St. Cloud, Minn., Jan. 4, e. Hollywood H.S., USC. Prop man, asst. dir., unit mgr., dr. special effects, shorts, second units Warner Bros., freelance dir. comedy series; Director features both theatrical and TV programs.

MARLY, FLORENCE: Actress. b. Czechoslovakia, June 2; e. Sorbonne, Paris. Appeared in 9 films in France inc. L'Alibi, Cafe de Paris, Les Maudits; also films in Argentina, Czechoslovakia.
PICTURES INCLUDE: Sealed Verdict, Tokyo Joe, Tokyo File 212, Gobs and Gals, Queen of Blood, Games.

MARRIOTT, PETER: Executive. b. Isle of Wright, England, 1921. Ent. ind. 1949. Assoc. prod. TV series Douglas Fairbanks Presents. Assoc. prod. Chase A Crooked Shadow, Moment of Danger, etc. Man. dir. Desilu Sales Inc. Dir. U.K. and Continental TV Sales 20th Century-Fox. Now man. dir. and dir. European operations NBC International (G.B.) Ltd.

MARSHALL, E. G.: Actor. b. Minnesota. Acting debut with Oxford Players, 1933. Numerous TV appearances on all networks; m.p. debut in House on 92nd St.
BROADWAY: Jason, Jacobowsky and the Colonel, Skin of Our Teeth, Iceman Cometh, Woman Bites Dog, The Survivors, The Gambler, The Crucible, The Little Foxes.
PICTURES INCLUDE: 13 Rue Madeleine, Call Northside 77, Caine Mutiny, Pushover, Bamboo Prison, Broken Lance, Silver Chalice, Left Hand of God, Scarlet Hour, 12 Angry Men, Bachelor Party, Town Without Pity, The Chase, Tora, Tora, Tora, The Bridge at Remagen, The Pursuit of Happiness, Interiors, Superman II.
TV: The Defenders, The Bold Ones.

MARSHALL, LARRY: Comedian, actor, writer, songwriter. b. Clarkston, Scotland. Trained Rutherglen repertory theatre. Own daily lunchtime television show for eight years. Appeared in many major television drama series; own evening shows; Comedy and panel games.

MARSHALL, PENNY: Actress. b. New York City, Oct. 15. m. actor Bob Reiner. Dropped out of Univ. of New Mexico to teach dancing. Acted in summer stock before going to Hollywood to make TV debut in The Danny Thomas Hour.
TV: The Bob Newhart Show, The Mary Tyler Moore Show, Chico and the Man, The Odd Couple (regular), Laverne & Shirley (regular). Movie: Let's Switch.

MARSHALL, PETER: Actor-TV Show Host. b. Huntington, W. Va., March 30. r.n. Pierre La Cock. Began career as NBC page in N.Y.C. Did plays both on and off Bdwy., starring with Julie Harris in Skyscraper. Teamed with the late Tommy Noonan in comedy act for nightclubs, guesting on Ed Sullivan Show and other variety shows. In 1974 made Las Vegas stage debut and since has been headliner there and in Reno and Lake Tahoe.
PICTURES INCLUDE: Ensign Pulver, The Cavern, Americathon.
TV: Two of the Most (local N.Y. show), The Hollywood Squares (host).

MARSHALL, TRUDY: Actress. b. Brooklyn, N.Y. Commercial photographers' model before entering pictures, 1942.
PICTURES INCLUDE: Secret Agent of Japan, Girl Trouble, Sentimental Journey, Dragonwyck, Talk About a Lady, Boston Blackie and the Law, Alias Mr. Twilight, Too Many Winners, Joe Palooka in the Knockout, Key Witness, Fuller Brush Man, Disaster, Shamrock Hill, Mark of Gorilla, The President's Lady.

MARSHALL, WILLIAM: Actor. b. Chicago, Ill., Oct. 12, 1917. In 1936, vocalist with Fred Waring's Orchestra; from 1937 orch. leader; on stage 1943, Army Air show Winged Victory. Screen debut 1940 in Sante Fe Trail.
PICTURES INCLUDE: Flowing Gold, City for Conquest, Flying with Music, Murder in the Music Hall, Earl Carroll's Sketchbook, That Brennan Girl, Calendar Girl, Blackmail, Adventures of Captain Fabian, Seven Different Ways.

MARSHALL, ZENA: Actress. b. Kenya (East Africa). e. France and finishing school in Ascot (England). Made her stage debut in repertory. Many TV appearance U.S. and England including Bob Hope show, Harpers W.I., Ghost Squad.
PICTURES INCLUDE: Marry Me, Dark Interval, Blind Man's Bluff, Love's a Luxury, Deadly Nightshade, My Wife's Family, Bermuda Affair, Let's Be Happy, Dr. No, The Guilty Party, Those Magnificent Men In Their Flying Machines, The Terrornauts.
TV: International Detective, Invisible Man, Dial 999,

Danger Man, Sir Francis Drake, Man of The World, Human Jungle, Sentimental Agent, Court Martial.

MARTA, JACK: Cameraman, Republic Studios. b. Butte, Montana, 1905; e. Los Angeles public schools. Began as laboratory worker for Fox. Then to Columbia. Later with Republic.
PICTURES INCLUDE: In Old Sacramento, Bells of San Angelo, Bill and Cool, Fair Wind to Java, Jubilee Trail, Timberjack, Last Command, Come Next Spring, Maverick Queen.

MARTEL, GENE: Producer, director. b. New York, N.Y., June 19, 1916. e. C.C.N.Y., University of Alabama, Sorbonne, Paris; newspaperman, New York and Birmingham, Ala.; dancer, actor, choreographer, director Broadway; prod. dir., many documentaries; films for State Department, others; dir. for Paramount Pictures. Joined Princess Pictures 1952 to make films in Europe; formed own company, Martel Productions Inc., 1954.
PICTURES INCLUDE: Check-mate, Double-barrelled Miracle, The Lie, Double Profile, Sergeant and the Spy, Black Forest, Eight Witnesses, Fire One, Phantom Caravan, Doorway to Suspicion, Diplomatic Passport, Immediate Disaster.

MARTIN, AL: Writer since 1932, including: (s.p.) The Last Alarm; (orig., s.p.) Flying Wild; (co lab. orig., s.p.) Mississippi Gambler; (s.p.) Stagecoach Buckaroo, The Devil with Hitler; (joint orig. s.p.) Gang Busters, serial: s.p., Amazon Quest; orig. story, Standing Room Only, Adventures of Rusty; s.p., Nine Girls, Strange Mrs. Crane; collab. s.p. Racing Luck; s.p., Army Bound, In the Money.
TV: Trouble with Father, Mystery Theatre, Ramar of the Jungle, Joe Palooka, Hank McCune Show, Wild Bill Hickok, Craig Kennedy Series, Mayor of the Town, Ford Theatre, Frontier Doctor, Lone Ranger, Damon Runyon Theatre, Roy Rogers, Screen Gems, Dupont Theatre, Gray Ghost, Restless Gun, Whirlybird, Trouble Shooters, Huckleberry Finn (serial), Spark Tarzan, 77 Sunset Strip, Man Trap, Laramie, My Favorite Martian; co-creat., My Living Doll.

MARTIN, CHARLES E.: Writer, director, producer. e. NYU, N.J. Law Sch.; under contract to MGM, Universal and Twentieth Century Fox; pres., Forward Films, Inc.
PICTURES INCLUDE: My Dear Secretary, No Leave No Love, Death of a Scoundrel, If He Hollers Let Him Go, Remember Vivian Valentine, Seduction American Style, How to Seduce A Woman, Hotshot, The Cop Who Played God, One Man Jury, Dead on Arrival, Muffet Ann Marvel, And Now It Can Be Told.
TELEVISION: Tallulah Bankhead Show, Gertrude Lawrence Show, Philip Morris Playhouse. Wrote 500 scripts.

MARTIN, DEAN: Actor. b. Steubenville, Ohio, June 17, 1917. e. Steubenville High School. Was amateur prizefighter; worked at odd jobs, mill hand, gasoline attendant, prior to acting career. Joined Jerry Lewis, comedian at 500 Club, Atlantic City, N.J., as straight man-singer, 1946; played many theatres, night clubs (see biography: Jerry Lewis). Team m.p. debut: My Friend Irma. Voted (with Jerry Lewis) one of the top ten Money-Making Stars in Motion Picture Herald-Fame poll, 1951, 1953–55; Number One, 1952.
PICTURES INCLUDE: My Friend Irma Goes West, At War With the Army, That's My Boy, The Stooge, Sailor Beware, Jumping Jacks, The Caddy, Scared Stiff, Money from Home, Living It Up, Three Ring Circus, You're Never Too Young, Artists and Models, Pardners, Hollywood Or Bust, 10,000 Bedrooms, The Young Lions, Some Came Running, Rio Bravo, Career, Who Was That Lady?, Bells Are Ringing, Ocean's 11, Ada, Sergeants 3, Who's Got the Action?, Toys In the Attic, Who's That Sleeping in My Bed, Robin and the Seven Hoods, Kiss Me Stupid, Sons of Katie Elder, Rough Night in Jericho, How To Save a Marriage, Airport, The Ambushers, The Wrecking Crew, Something Big, Showdown, Mr. Ricco, The Cannonball Run.
TV: Club Oasis, Dean Martin Show. Golden Globe Award.

MARTIN, DEWEY: Actor. b. Katemcy, Texas, Dec. 8: e. U. of Ga., Louisiana State Teachers Coll. U.S. Navy, W.W. II; actor, little theatres & stock; m.p. debut in Knock on Any Door.
PICTURES INCLUDE: The Thing, Big Sky, Tennessee Champ, Prisoner of War, Men of the Fighting Lady, Land of the Pharaohs, Desperate Hours, Proud and Profane, 10,000 Bedrooms.

MARTIN, DON: Writer. b. Phila., Pa., April 3, 1911; e. Girard Coll., U. Delaware, B.A.; George Washington U. Law School; husband of Tamara, actress, Journalist, Washington Post, Star, 1932–33; gen. mgr., Diamond State Theat. Co., 1933–34; newspaper, magazine, radio writer, 1934–42. Successively with Columbia, United Artists, MGM, RKO-Radio publicity depts. Joined 20th-Fox writing dept., 1943.
PICTURES INCLUDE: Devil's Cargo, Appointment with Murder, Shakedown, Shed No Tears, The Creeper, Lost Tribe, Arrow in the Dust, Lone Gun, Shield for Murder, Stranger on Horseback, Double Jeopardy, No Man's Woman, Deadliest Sin, Quincannon Frontier Scout, Emergency Hospital, Hot Cars, Brass Legend, Storm Rider. The Pretender, Lighthouse, Hat Box Mystery, Triple Threat, Search for Danger, Destination Murder, The Man Is Armed, Violent Road, Operation Tobruk, The Four Deuces.
TV: Playhouse of Stars, Racket Squad, Fireside Theatre, Mr. District Attorney, Trouble With Father, Rebound, Celebrity Theatre, Cheyenne, Sheriff of Chochise, Whirlybirds, The Violent Road.

MARTIN, EDWIN DENNIS: Executive. b. Columbus, Ga., Jan. 30, 1920, e. grad., Univ. of Georgia, B.S., 1940; Anti Air Craft Artillery, past pres. TOA, International.

MARTIN, HENRY H.: Executive. b. Holcomb, Miss., March 22, 1912. Joined Universal poster clerk, Oklahoma City, 1935; then booker, then salesman. Military service 1943–46. Rejoined Universal as salesman, Oklahoma City, then branch manager; to Dallas as branch manager, 1948, to district manager, 1951, to New York as southern division manager 1956; general sales manager, 1957. Elected v.p. Universal Pictures Co., Inc., Nov. 2, 1959. Namedv.p. of MCA, Inc., Dec., 1972 and president of Universal, Jan. 1, 1973.

MARTIN, MARIAN: Actress. b. Philadelphia, June 7, 1916. e. Bayonne School, Switzerland. On stage in Follies of 1933, George White's Scandals, and others.
PICTURES INCLUDE: Sinners in Paradise, Deadline for Murder, Suspense, Cinderella Jones, Queen of Burlesque, Angel on My Shoulder, That's My Girl, That Brennan Girl, Lighthouse, Thunder in the Pines, Come to the Stable, Key to the City, Dakota Lil.

MARTIN, MARY: Actress. b. Wetherford, Texas, Dec. 1, 1914; e. Ward-Belmont Sch., Nashville, Tenn.; m.p. debut in 1939.
STAGE: Lute Song, Leave It to Me, One Touch of Venus, South Pacific, Annie Get Your Gun, Kind Sir, Peter Pan, Jennie, I Do I Do.
PICTURES INCLUDE: Great Victor Herbert, Rhythm on the River, Love Thy Neighbor, Kiss the Boys Goodbye, New York Town, Birth of the Blues, Happy Go Lucky, Night and Day.
TV: Ford anniversary show, Rodgers & Hammerstein Cavalcade show. Peter Pan, Together with Music (with Noel Coward).

MARTIN, MILLICENT: Actress, singer. Toured U.S. The Boy Friend, 1954–57.
STAGE: Expresso Bongo, The Crooked Mile, Our Man Crichton, Tonight at 8, The Beggars Opera, Puss 'n Boots, The Card, Absurd Person Singular, Aladdin, Side By Side By Sondheim, Move Over Mrs. Markham, Meet Mr. Stewart.
TV: International Detective Series, Millie, That Was the Week That Was, Harry Moorings; own series, Mainly Millicent, Kiss Me Kate; 1966 own series, Millicent TV; 1967: London Palladium Color Show USA, Danny Kaye, Piccadilly London; From a Bird's Eye View (own series); Tom Jones show, Englebert Humperdinck show.
PICTURES INCLUDE: The Horsemaster, The Girl on the Boat, Nothing But the Best, Alfie, Stop the World I Want To Get Off.

MARTIN, R. E.: Exhibitor b. Columbus, Ga., March 25, 1917. e. grad., Univ. of Georgia, B.S.C., 1939. Co-owner, Martin Theatre Circuit; co-owner, TV stations, WTVM, Columbus, Ga., WTVC, Chattanooga, Tenn. partner, The Marbro Co.; dir., Columbus Transportation Co.; exec. comm., TOA; pres., Junior Achievement; past pres., Columbus Chamber of Commerce; dir., Y.M.C.A; member, Rotary Club.

MARTIN, RICHARD: Producer-Director. b. New York City, Feb. 2, 1938. e. Harvard. Stage and TV actor in N.Y. from 1959 to 1969; writer from 1969–73.
PICTURES INCLUDE: The Bengal Tiger, The Chase Expedition, Jaws of Death (also s.p.).

MARTIN, ROY E., Jr.: Executive. b. Columbus, Ga., March 25, 1917; e. U. of Ga., (BSC); Troy State (MSC, 1979); Retired.

MARTIN, STEVE: Actor. b. Waco, TX, 1945. e. Long Beach Coll., UCLA. Writer for various TV comedy shows (Smothers Brothers, Sonny & Cher, etc.). Acted on TV in Steve Martin—A Wild and Crazy Guy, Steve Allen Comedy Hours, Comedy Is Not Pretty, etc.
PICTURES: The Kids Are Alright, The Muppet Movie, The Jerk.

MARTIN, TONY: Singer, musician, actor. b. Oakland, Calif., Dec. 25, 1913; e. Oakland H.S, St. Mary's Coll.; m. Cyd Charisse, actress-dancer. Sang played saxophone & clarinet in high school band, engaged by nearby theatres for vaudeville; with Five Red Peppers, jazz group at 14 yrs.; two yrs. later with band, Palace Hotel, San Francisco; radio debut Walter Winchell program, 1932; joined Tom Gerund's band, World's Fair Chicago, 1933; played night clubs. First starring radio show, Tune Up Time (singer & emcee); on Burns and Allen program; own show for Texaco, Carnation Contented Hour.
RECORDINGS INCLUDE: Begin the Beguine, Intermezzo, The Last Time I Saw Paris, I'll See You in My Dreams, Domino, September Song.

PICTURES INCLUDE: Sing Baby Sing, Follow The Fleet, You Can't Have Everything; Ali Baba Goes to Town, Music in My Heart, Ziegfield Girl, Till The Clouds Roll By, Cabash, Two Tickets to Broadway, Here Come the Girls, Easy to Love, Deep in My Heart, Hit the Deck, Quincannon Frontier Scout, Let's Be Happy, The Big Store.

MARTINI, ALLEN V.: Executive. b. San Francisco, Dec. 14, 1919. e. Stanford U. Since resigning from Air Force ent. m.p. ind., 1945. Executive capacity RKO Studios; partner Filmakers Prod., 1950; United Artists Television, 1953; associated with National General Corporation since 1963; man. dir., Carthay Center Productions Ltd., National General's subsidiary, London, 1966; vice-pres., National General Productions Inc.; formerly vice-president, National General Television Corporation.

MARTINS, ORLANDO: Actor, singer, b. Lagos, Nigeria; e. Eko High School, Lagos, Stage debut, Stevedore, 1937; creen debut; Men Of Two Worlds, 1946.
PICTURES INCLUDE: The Hasty Heart, Where No Vultures Fly, West Of Zanzibar, Simba, Safari, Seven Waves Away, Naked Earth, The Nun's Story, Sapphire, Frankie & Johnnie, Cry the Beloved Country, In Judea, This Book Is News, Killers of Kilimanjaro, Sammy Going South, Call Me Bwana, Mister Moses.

MARTON, ANDREW: Director. b. Budapest, Hungary, Jan. 26, 1904. Vita Film, Vienna, 1922; to Hollywood, with Ernst Lubitsch, 1923; started directing in Berlin, then in London, Wolf's Clothing, Secret of Stamboul, School for Husbands; in Hollywood, Little Bit of Heaven, Gentle Annie, Gallant Bess.
PICTURES INCLUDE: King Solomon's Mines, Wild North, Storm Over Tibet, Devil Makes Three, Gypsy Colt, Prisoner of War, Men of the Fighting Lady, Green Fire, Underwater Warrior, Cinerama Seven Wonders of the World, It Happened In Athens, The Longest Day, The Thin Red Line, Crack in the World, Clarence the Crosseyed Lion, Around the World Under The Sea, Birds Do It, Africa-Texas Style.
TV: Man and The Challenge, Daktari, Cowboy in Africa, The Sea Hunt.

MARVIN, LEE: Actor. b. New York, N.Y., Feb. 19, 1924. Played in summer stock cos.; Bway, debut in Billy Budd; m.p. debut in You're in the Navy Now.
PICTURES INCLUDE: Diplomatic Courier, We're Not Married, Down Among the Sheltering Palms, Eight Iron Men, Stranger Wore a Gun, Big Heat, Gun Fury, Wild One, Caine Mutiny, Gorilla at Large, The Raid, Bad Day at Black Rock, Life in the Balance, Violent Saturday, Not As a Stranger, Pete Kelly's Blues, I Died a Thousand Times, The Rack, Shack Out on 101, Attack!, The Killers, Ship of Fools, Cat Ballou, The Professionals, The Dirty Dozen, Point Blank, Sergeant Ryker, Paint Your Wagon, Monte Walsh, Pocket Money, Prime Cut, Emperor of the North, The Spikes Gang, The Iceman Cometh, The Klansman, Great Scout and Cathouse Thursday, Shout at the Devil, Death Hunt.

MARX, SAMUEL: Writer-producer.
PICTURES INCLUDE: Lassie Come Home, This Man's Navy, My Brother Talks to Horses, The Beginning or End, A Lady Without Passport, Grounds for Marriage, Kiss of Fire, Ain't Misbehavin', Waterloo, Rome, The Ravine.

MASEFIELD, JOSEPH R.: Executive, Producer, Director, Writer. b. N.Y., N.Y., June 20, 1933, 1950 American Academy of Dramatic Arts. Writer-performer, club work (as Steve Parker). Later, actor in stock; formed EEF Film Productions, 1956; prod. A Story Like Two (short), A City Eats (doc.). In m.p. as unit mgr., asst. dir. and prod. mgr. Later writer and film editor: Montage (Time-Life); Ages of Man (IBM); Festival of Two Worlds (Bell Telephone Hour) and special, The New Face of Israel. Asst. dir., prod. mgr., Mitgebracht Aus New York (German TV), The Devils Doubloon, (feature); Writer-director, Citizen Smith, feature documentary. Co-director and assoc. prod. Hear My Song, (Cavalier Films), 1969 producer The Spy. 1969, vice pres. in charge of prod. PCI, Inc. Public Relations director, The Max Steiner Music Society; pres., Joseph R. Masefield & Associates. In 1974, formed Majer Prods. with Steve Jerro.
PICTURES INCLUDE: A New Life, The Vanquished, Trio, Living Planet, The Burning Man.

MASLANSKY, PAUL: Producer.—Director—Writer. Studied filmmaking at Cinemateque Francais where made a documentary (Letter from Paris) which won Cannes Festival award. Asst. dir. on Counterfeit Traitor; followed by prod. mgr. on several films in Europe, including The Running Man and Jason and the Argonauts. Wrote and produced Gothic horror films, including Castle of the Living Dead. Covered Israeli war in 1967 with CBS film group; spent year in Russia on The Red Tent, followed by The Blue Bird. Filmed Big Truck and Poor Claire in Israel; Miracles Still Happen in Brazil.
PICTURES: Damnation Alley (co-prod.), When You Comin' Back, Red Ryder (co-prod.).

MASON, JAMES: Actor. b. Huddersfield, Yorkshire, England, May 15, 1909; e. Marlborough Coll.; Peterhouse Coll., Cambridge; m. Pamela Kellino, actress & writer. On stage from 1931, tour. repertory; London debut 1933 in "Gallows Glorious"; other appearances include "Bonnett Over the Windmill," "Road to Rome," "Divorce for Christabel," "Jupiter Laughs." Screen debut 1935 in "Late Extra"; since in numerous Brit. feature pictures. Voted one of top ten Brit. money-making stars in Motion Picture Herald-Fame Poll: 1944, 45, 46, 47, 48.
PICTURES INCLUDE: The Mill on the Floss, Fire Over England, The Return of the Scarlet Pimpernel, I Met a Murderer, This Man Is Dangerous, Hotel Reserve, They Were Sisters, Seventh Veil, Wicked Lady, Odd Man Out, Upturned Glass, Patient Vanishes, Man of Evil, Hatter's Castle, Place of One's Own, Caught, Pandora and The Flying Dutchman, Lady's Possessed, Desert Fox, 5 Fingers, Botany Bay, Story of Three Loves, Prisoner of Zenda, Face to Face (Secret Share), Julius Caesar, Desert Rats, Man Between, Prince Valiant, A Star Is Born, 20,000 Leagues Under the Sea, Forever Darling, Bigger than Life, Island in the Sun, North by Northwest, Journey to The Center of The Earth, A Touch of Larceny, The Decks Ran Red, Marriage-Go-Round, Lolita, The Pumpkin Eater, Lord Jim, Georgy Girl, The Deadly Affair, The Seagull, Mayerling, Age of Consent, Child's Play, The Last of Sheila, The Mackintosh Man, 11 Harrowhouse, Mandingo, Voyage of the Damned, Cross of Iron, Heaven Can Wait, The Boys from Brazil, Murder by Decree, The Passage, ffolkes.

MASON, KENNETH: Executive. b. Rochester, N.Y. Sept. 21, 1917. e. Washington and Jefferson College; Univ. of Rochester (graduate work). Began career with Eastman Kodak Co. in Kodak Park cine processing dept. in 1935; transferred following year to film developing dept. at Kodak Research Laboratories. Returned to college in 1938 to complete studies and returned to Kodak year later to same dept. Later joined film planning dept., remaining there until entering military service in 1943. Returned to Kodak in 1946 as staff engineer in Kodak Office motion picture film dept. In 1950 appt. mgr. of Midwest Division of m.p. film dept.; became gen. mgr., Midwest Division, m.p. products sales dept. in 1963; named sls. mgr. of New York City region in 1965; appt. regional sls. mgr., Pacific Southern Region, Hollywood, in 1970. On Jan. 1, 1974 appt. mgr., product programs and research, Motion Picture and Audiovisual Markets Division, Kodak Office. On March 18, 1974 named gen. mgr. of that division. Elected asst. v.p. of co. on March 28, 1974, v.p., Dec. 11, 1978.
MEMBER: Board of Trustees of Washington and Jefferson College; past pres. of Society of Motion Picture & Television Engineers, honorary fellow of British Kinematograph Sound & Television Society; mbr. of University Film Assn., Academy of TV Arts and Sciences, Motion Picture Academy, American Society of Cinematographers, Variety Club.

MASON, MARSHA: Actress. b. St. Louis. m. Neil Simon. Came to N.Y. to continue dramatic studies and embark on theatre career. Appeared off-Broadway (Happy Birthday, Wanda June, Deerpark) and on Broadway (Cactus Flower). Now member of American Conservatory Theatre, San Francisco.
PICTURES INCLUDE: Blume in Love (debut), Cinderella Liberty, Audrey Rose, The Goodbye Girl, The Cheap Detective, Promises in the Dark, Chapter Two, Only When I Laugh.
TV: Love of Life (series), Brewsie and Willie, Cyrano de Bergerac.

MASON, PAMELA: Actress, writer. b. Westgate. England. Mar. 10, 1918; Stage debut 1936, The Luck of the Devil, London; also playwright (in collab. James Mason, Flying Blind, Made in Heaven). Author novels This Little Hand, A Lady Possessed, The Blinds Are Down, Ignoramus, Began Brit. screen career 1938, I Met a Murderer (orig. story & s.p.; cast); also in They Were Sisters, 1944. In 1946 (s.p. & Cast) The Upturned Glass; (acted) Pandora and the Flying Dutchman; acted, collab. s.p. Lady Possessed, syndicated TV, Pamela Mason Show, author Marriage Is the First Step Toward Divorce. Syndicated TV The Weaker Sex?; author, The Female Pleasure Hunt; lectures at women's clubs countrywide.

MASSEN, OSA: Actress. b. Denmark, Copenhagen. Jan. 13, 1916.
PICTURES INCLUDE: Honeymoon in Bali, Honeymoon for Three, A Woman's Face, Accent on Love, You'll Never Get Rich, The Devil Pays Off, Ireland, Strange Journey, Night Unto Night, Deadline at Dawn, Gentleman Misbehaves, Rocketship XM.

MASSEY, DANIEL: Actor. b. London, Eng., Oct. 10, 1933. e. Eaton and King's Colleges f. Raymond Massey. Active on stage and TV.
PICTURES INCLUDE: Girls at Sea, Upstairs and Downstairs, The Queen's Guard, Go to Blazes, Moll Flanders,

The Jokers, Star!, Fragment of Fear, Mary, Queen of Scots, Victory.

MASSEY, RAYMOND: Actor. b. Toronto, Canada, Aug. 30, 1896; s. Chester D. and Anna Vincent Massey, e. Appleby School, Oakville, Ont.; Belliol Coll., Oxford; Litt. D. (hon) Lafayette U.; Hobart Coll., Smith Coll., LL.D. (hon.) Queens U., Kingston, Ont. m. Dorothy Ludengton. Canadian Field artillery officer, World War I; wounded in Belgium; then with Siberian Expeditionary Force; on stage in England, 1922; since then in many plays in England and U.S.
THEATRE: Ethan Frome, Abe Lincoln in Illinois, Pygmalion, John Brown's Body; dir. numerous stage plays.
PICTURES INCLUDE: Has appeared in more than 60 motion pictures. The Scarlet Pimpernel, Things to Come, The Prisoner of Zenda, Abe Lincoln in Illinois, Hurricane, Possessed, Mourning Becomes Electra, Fountainhead, Arsenic and Old Lace, Santa Fe Trail, Sugarfoot, Dallas, David and Bathsheba, Come Fill the Cup, Prince of Players, Battle Cry, East of Eden, Seven Angry Men, The Naked and the Dead, McKenna's Gold.
TV: Dr. Kildare.

MASSIE, PAUL: Actor. b. Ontario, Canada. July 7, 1932. Early career on Canadian stage. Attended Central School, London, 1952, later jnd. Scottish National Children's Theatre. Entered m.p. ind. 1954 in Orders to Kill.
PICTURES INCLUDE: High Tide at Noon, Sapphire, Libel, The Two Faces of Dr. Jekyll, The Rebel, The Pot Carriers, Raising the Wind.
STAGE: Cat On a Hot Tin Roof.
TV: The Mark of the Warrior, The Last of the Brave, Ring Around the Moon, Secret Mission and Her Romeo.

MASTORAKIS, NICO: Writer-Director. b. Athens, Greece, 1941. Writer of novels and screenplays. Books include Fire Below Zero and The Ephesus Scrolls (co-author).
PICTURES: The Next One (s.p., dir.).

MASTROIANNI, MARCELLO: Actor. b. Fontane Liri, Italy, Sept. 28, 1924. e. U. of Rome theatrical company. Draftsman in Rome, 1940–43. Theatrical debut in Rome in Angelica, 1948.
PLAYS INCLUDE: Death of a Salesman, Streetcar Named Desire, Ciao Rudy.
PICTURES INCLUDE: Three Girls from Rome, The Miller's Beautiful Wife, Fever to Live, The Ladykillers of Rome, Love a La Carte, Days of Love, La Dolce Vita, White Nights, Divorce Italian Style, La Notte, A Very Private Affair, Bell Antonio, Where the Hot Wind Blows, The Organizer, 8½, Yesterday, Today and Tomorrow, Marriage Italian Style, Casanova '70, The 10th Victim, The Poppy Is Also a Flower, Shoot Loud, Louder . . . I Don't Understand, The Stranger, A Place for Lovers, Leo the Last, Diamonds for Breakfast, Sunflower, Jealousy Italian Style, The Priest's Wife. Formed indep. prod. co., Master Films, 1966; What?, The Grande Bouffe, Massacre in Rome, Down the Ancient Stairs, The Sunday Woman, A Special Day, Stay As You Are, Blood Feud, City of Women.

MATHEWS, CAROLE: Actress. b. Montgomery, Ill., Sept. 13. e. Aurora, Ill. H.S. Started as night club, radio entertainer; to Hollywood, 1944.
PICTURES INCLUDE: Massacre River, Great Gatsby, Special Agent, Meet Me At the Fair, Swamp Woman, Port of Heil, Shark River, Treasure of Ruby Hills, Requirement for a Redhead, Look In Any Window, Thirteen Men, Female Fiend, Tender Is The Night, End of the Road.
TV: Steel Hour, Kraft Theatre, Lux Video, Hitchcock Presents, Studio One, Californians, Texan, 77 Sunset Strip, Perry Mason, Four Star Theatre, M-Squad, Death Valley Days, Guestward Ho, Two Faces West, Johnny Midnight, Pete & Gladys, 87th Precinct, Ben Casey.

MATLACK, JACK D.: Theatre executive, publicist. b. Manton, Calif., Feb. 22, 1914; e. Chico, Calif. State Teachers Coll. Entered m.p. ind. as doorman Criterion, Medford, Ore.; then mgr. dir.; joined J.J. Parker Theatres, Portland, Ore., 1943; as exec. asst. to pres. and adv. dir. eight theatres; now pres. Jack Matlack Promotions. Portland, Ore., N.W.P.R. for studios; well known showman, winner 20 exploit. awards including Grand (Silver) Quigley Award, 1943; Quigley War Showmanship Award, 1942–44; numerous Quigley Pub. citations; cited by U.S. Govt. for outstanding contrib. to World War II effort; active as civic worker; on bd. Portland, Ore. C. of C. Kiwanis, Portland City Planning Comm., Oregon Advertising Club; named First Citizen of Portland, 1944.

MATOFSKY, HARVEY: Executive. b. Bklyn., N.Y., Jan. 12, 1933. e. Washington Square Coll., N.Y.U., 1950–52; Brooklyn Coll., B.A., 1954; Sorbonne, 1956; New York Film Institute, 1959. New York Times newspaper. 1952–54; U.S. Army Intelligence, Austria. Italy. 1954–56. Ent. m.p. industry, reporter. Independent Film Journal; assoc. editor. 1957; publicity dept., United Artists, 1959; international publicity dir., Embassy Pictures, 1960; apptd., dir., adv. & pub., Morningside Worldwide Pictures, 1961; appt. European publicity coordinator Embassy Pictures. Rome, 1963; publicity coordinator: The Wonders of Aladdin, Jason and the Golden Fleece, The Old Dark House, Ballad of the Running Man, Gidget Goes to Rome, "8½", Ghost at Noon, prod. dir. motion picture shorts N.Y. Film Institute. Executive vice pres. Marstan Productions (1969), and co-producer of the Alberto Moravia film "Crime at the Tennis Club" and TV special Upon this Rock. Author, s.p. The Stranger (based on Lillian Ross novel). Formed and is pres. RHM Music and Records, Ltd. London. Pres. of RHM Productions. 1973: produced Zandy's Bride.

MATSOUKAS, NICHOLAS JOHN: Executive. b. Pylos-Navarino, Greece, June 13, 1903. e. U. of Chicago, Ph.D. 1929. Editorial staff of Chicago Daily News; ran own public relations office; adv., prom. dir., Cinema Art Theatre Chicago, 1932; pub. relations dir. and campaign mgr., Greek War Relief Assn. 1940–41; dir. adv. pub. and community Service dept., Skouras Theatres Corp. since 1941; nat'l. dir., adv., pub. & exploitation, Magna Theatres Corp.; nat'l. campaign dir., Will Rogers Memorial Hospital, Saranac Lake, N.Y., American Korean Foundation; producer, show, It's Up to You, National Food Admin., U.S. Dept. of Agriculture; adv., pub. dir., United Artists Theatre Circuit, Inc., Rivoli Theatre; pres., Movie Bonus Corp.

MATTHAU, WALTER: Actor. b. New York, Oct. 1, 1923. Served in Air Force WW II. Studied journalism at Columbia, then acted in summer stock. First Broadway role, 1948, in Anne of a Thousand Days.
THEATRE: Will Success Spoil Rock Hunter? and The Odd Couple.
PICTURES INCLUDE: A Face in the Crowd, The Kentuckian, Slaughter on Tenth Avenue, Indian Fighter, No Power on Earth, Middle of the Street, Onion Head, Voice in the Mirror, King Creole, Lonely Are the Brave, Strangers When We Meet, Who's Got the Action, The Gangster Story, Charade, Goodbye Charlie, Mirage, The Fortune Cookie, A Guide for the Married Man, The Odd Couple, Candy, Cactus Flower, Hello Dolly, A New Leaf, Plaza Suite, Pete 'n' Tillie, Charley Varrick, The Laughing Policeman, The Taking of Pelham 1,2,3, The Front Page, The Sunshine Boys, The Bad News Bears, Casey's Shadow, House Calls, California Suite, Little Miss Marker, Hopscotch, First Monday in October.

MATTHEWS, JESSIE: (C.B.E.) Actress, Singer. b. London, England, March 11, 1907. London stage debut 1919 in Bluebell in Fairyland. Subseq. in numerous musicals, dramas, London. Starred on U.S. stage in Andre Charlot's Revue of 1924, Wake Up and Dream (1929–30), The Lady Comes Across (1941), Jessie Matthews in Concert (1979).
STAGE (London) One Damn Thing After Another, This Year of Grace, Wake Up and Dream, Ever Green, Hold My Hand, Sally Who?, Come Out to Play, Wild Rose, Maid to Measure, Sweethearts and Wives, Sauce Tartare, Five Finger Exercise, A Share in the Sun, The Water Babies (1973), The Jessie Matthews Show (1976), Lady Windermere's Fan (1978–79).
PICTURES INCLUDE: The Beloved Vagabond, Straws in the Wind, This England, Out of the Blue, There Goes the Bride, The Man From Toronto, The Midshipmaid, The Good Companions, Friday the Thirteenth, Waltzes from Vienna, Evergreen, First A Girl, It's Love Again, Head Over Heels, Gangway, Sailing Along, Climbing High, Forever and a Day (in Hollywood), Candles At Nine, Victory Wedding (as director), Life is Nothing without Music, Making the Grade, Tom Thumb, The Hound of the Baskervilles (1978).
TELEVISION: This is Your Life, The Apple Cart, Stars and Garters, Dolly The Nicest Man in the World, Angels, Another Opening Another Show, Nanny's Boy (in the BBC series, Jubilee), Parkinson, Edward and Mrs. Simpson (1978), The Winter Ladies (1979), Tales of the Unexpected (1980).

MATURE, VICTOR: Actor. b. Louisville, Ky., Jan. 29, 1916. TV retail store; trainee, Pasadena Theatre, Playbox Theatre; on B'way in Lady in the Dark; U.S. Coast Guard, WW II.
PICTURES INCLUDE: Housekeeper's Daughter, One Million B.C., Captain Caution, No No Nanette, I Wake Up Screaming, Shanghai Gesture, Song of the Islands, My Gal Sal, Footlight Serenade, Seven Days Leave, My Darling Clementine, Moss Rose, Kiss of Death, Cry of the City, Red Hot and Blue, Fury at Furnace Creek, Easy Living, Samson and Delilah, Wabash Avenue, Stella, Gambling House, Las Vegas Story, Androcles and the Lion, Million Dollar Mermaid, Something for the Birds, Glory Brigade, Affair with a Stranger, The Robe, Veils of Bagdad, Dangerous Mission, Betrayed, Demetrius & The Gladiators, The Egyptian, Chief Crazy Horse, Violent Saturday, Last Frontier, Safari, Zarak Pickup Alley, Tank Roce, The Bandit of Zhobe, Escort Wst., Big Circus, Timbuktu, Hannibal, The Tartars, After the Fox, Every Little Crook and Nanny, Won Ton Ton, the Dog That Saved Hollywood, Firepower.

MAURER, MAURICE (Ziggy): Executive. b. New York City, Aug. 25, 1914; e. Alfred U. Started in m.p. theatre field as usher, Palace Theatre, N.Y.; mgr. Fox theatres there; owned &

oper. Stanley Theatre; dir. Lopert Films, Inc.; v.p. in charge of oper., Astor, Victoria, Bijou theatres, N.Y.; owner, oper., Central Theatre; dir. Mayand the Brave, House of The Seven Hawks, High Time.

MAUREY, NICOLE: Actress. b. France. Studied dancing; French films include Blondine, Pamela, Le Cavalier Noir; stage appearances in France; U.S. debut in Little Boy Lost.
PICTURES INCLUDE: Secret of the Incas, Bold and the Brave, House of The Seven Hawks.

MAY, ELAINE: Actress, Screenwriter. b. Philadelphia, Pa., April 21, 1932. Repertory theatre in Chicago, 1954; comedy team with Mike Nichols, 1955.
PICTURES INCLUDE: Luv, Enter Laughing. 1970: writer, dir., star, A New Leaf; dir., The Heartbreak Kid; dir., s.p., Mikey and Nicky. 1978; co-s.p. Heaven Can Wait; actress, California Suite.

MAYEHOFF, EDDIE: Comedian. b. Baltimore, Md., July 7; e. Yale Sch. of Music. Adv. salesman, 1932; dance band leader, 6 yrs.; on radio with Norman Corwin; own show, Eddie Mayehoff on the Town; night clubs; B'way shows.
THEATRE: Let Freedom Sing, Early to Bed, Rhapsody, Billy Rose's Concert Varieties, Season in the Sun, Visit to a Small Planet.
TV: Adventures of Fenimore J. Mayehoff.
PICTURES INCLUDE: That's My Boy, The Stooge, Off Limits, Artists and Models, How to Murder Your Wife, Luv.

MAYER, BEN: Executive. b. Nov. 22, 1925. e. Manchester All Saints School of Art (England); Royal College of Art. President, Ben Mayer Design, Inc.

MAYER, GERALD: Producer, director. b. Montreal, Canada; p. both deceased: Jerry G., mgr. MGM studio, and Rheba G. Mayer (later Mrs. Hal Elias); e. Stanford (Journalism); corr. for San Francisco Examiner; pres. Sigma Delta Chi, prof. journalism soc. Navy lieut. amphibious forces, World War II. Entered m.p. ind. in prod. dept. MGM studios; first dir. job Dial 1119.
PICTURES INCLUDE: Dial 1119, Inside Straight, Sellout, Holiday for Sinners, Bright Road (Christopher Award for direction), The Marauders, African Drumbeat, and The Man Inside (Canadian).
TV: Canadian Broadcasting Corp. (prod./dir., TV Drama), prod. The Swiss Family Robinson (British-Canadian TV series). Director for U.S. TV: One Last Ride (mini-series), Lou Grant, Eight Is Enough, Quincy, Logan's Run, Mannix, Mission Impossible, Police Surgeon, Cimarron Strip, Peyton Place, Judd for the Defense, Bonanza, The Fugitive, Chrysler Thea., Ben Casey, Slattery's People, Profiles in Courage, The Defenders, Gunsmoke, etc.

MAYER, MICHAEL F.: Attorney-Executive b. White Plains, N.Y., Sept. 8, 1917; Harvard College B.S., 1939; Yale Law School L.L.B., 1942. Armed Forces 1942–46, Air Medal (1945); vice-pres. Kingsley International Pictures Corp., 1954–62. Executive Director and General Counsel, Independent Film Importers and Distributors of America Inc. (IFIDA), 1959–67. Special Counsel, French Society of Authors, Composers and Publishers, 1961-to date; British Performing Rights Society, 1962–67. Author: Foreign Films on American Screens (1966); Divorce and Annulment (1967); What You Should Know About Libel and Slander (1968); Rights of Privacy (1972); The Film Industries (1973)—revised edition published in 1978). Lecturer on motion picture problems at N.Y.U., Stanford, Univ. of Pennsylvania, Dartmouth and New York State University (Albany). Teacher of courses on Business Problems in Film, New School (1971 to present). Is secty. of Film Society of Lincoln Center, Inc.

MAYER, ROGER LAURANCE: Executive. b. New York City, Apr. 21, 1926. e. Yale Univ. (1948 graduate); Yale Law School (1951 graduate). In 1952 was practicing attorney; joined Columbia Pictures that year as atty. and named corp. exec., 1957. Left in 1961 to join MGM Studio as asst. gen. mgr. Since with MGM as follows: v.p., operations, 1964; v.p., administration, 1975 to present. Also exec. v.p., MGM Laboratories, 1974 to present.
MEMBER: Los Angeles County Bar Assn., Los Angeles Copyright Society, Academy of Motion Picture Arts & Sciences. Chm. & trustee for both Directors Guild Pension Plan and Producers Guild Pension Plan.

MAYER, SEYMOUR R.: Executive. b. New York, N.Y., July 30, 1912; e. N.Y.C. schools. Mgr. Loew's Theatres; div. mgr., Loew's in-town theatres, 1933; Armed Forces, WW II, as Major in charge of overseas m.p. service; with Loew's International: 1959, 1st v.p.; MGM Int'l.: Jan., 1963: May, 1969, pres., MGM Int'l.; Worldwide sales Sagittarius Prods., N.Y., May, 1970; pres., MSD Int'l. worldwide sales.

MAYES, WENDELL: Writer. b. Hayti, Missouri e. Johns Hopkins, Columbia. Military service in Pacific, World War II. Began career as actor legit theatre, turned to writing for television. Gained television recognition before moving to Hollywood as motion picture writer.
PICTURES INCLUDE: Spirit of St. Louis, The Way to the Gold, Enemy Below, The Hunters, From Hell to Texas, The Hanging Tree, Anatomy of A Murder, Advise And Consent, In Harm's Way, Von Ryan's Express, Hotel, The Stalking Moon, The Poseidon Adventure, The Revengers, Bank Shot, Death Wish, Love & Bullets, Charlie, Go Tell The Spartans.

MAYO, VIRGINIA: Actress. r.n., Virginia Jones. b. St. Louis, Mo.; e. St. Louis dramatic school. With Billy Rose's Diamond Horseshoe; then N.Y. stage (Banjo Eyes).
PICTURES INCLUDE: Kid from Brooklyn, Best Years of Our Lives, Secret Life of Walter Mitty, Out of the Blue, Smart Girls Don't Talk, The Girl From Jones Beach, Flaxy Martin, Colorado Territory, Always Leave Them Laughing, Flame and the Arrow, West Point Story, Along the Great Divide, Captain Horatio Hornblower, Painting the Clouds with Sunshine, Starlift, She's Working Her Way Through College, Iron Mistress, She's Back on Broadway, South Sea Woman, Devil's Canyon, King Richard & the Crusaders, Silver Chalice, Pearl of the South Pacific, Great Day in the Morning, Proud Ones, Congo Crossing, Big Land, Young Fury, Fort Utah, Castle of Evil.

MAZURKI, MIKE: Actor. b. Ukrainian descent; Tarnopal, Austria, Dec. 25, 1909; e. Manhattan Coll., New York, B.A., 1930; married. Toured United States and Canada as heavyweight wrestler. Screen debut 1941 in Shanghai Gesture.
PICTURES INCLUDE: Killer Dill, I Walk Alone, Unconquered, Nightmare Alley, Relentless, Neptune's Daughter, Come to the Stable, Rope of Sand, Samson & Delilah, Light Touch, Criminal Lawyer, Ten Tall Men, My Favorite Spy, The Egyptian, New Orleans Uncensored, New York Confidential, Blood Alley, Kismet, Davy Crockett, King of the Wild Frontier, Comanche, Around the World in 80 Days, Mad, Mad, Mad, Mad World, Four for Texas, Cheyenne Autumn, 7 Women, Bullwhip Griffin, Challenge to be Free, The Magic of Lassie, The Man with Bogart's Face.

MAZURSKY, PAUL: Producer-Director-Writer-Actor. b. Brooklyn, N.Y. April 25, 1930. Started acting in 1951 in the theatre, TV and films. Was nightclub comic 1954–1960 and directed plays. Began association with Larry Tucker by producing, directing, writing and performing in Second City, semi-improvisational revue. For four years they wrote the Danny Kaye TV show and created and wrote the Monkees series. First theatrical film I Love You, Alice B. Toklas (1969) which he wrote with Tucker and both men acted as exec. prod.
PICTURES INCLUDE: Bob and Carol and Ted and Alice, (dir., co-s.p.), Alex in Wonderland (dir., co-s.p., actor), Blume in Love (dir., s.p., prod., actor), Harry and Tonto (prod., dir., co-s.p.); Next Stop, Greenwich Village (prod., dir., s.p.); "Unmarried Woman (prod., dir., s.p.); A Man, A Woman and A Bank (actor): Willie and Phil (prod., dir., s.p.).

McCALL, JOAN: Actress-Writer. b. Grahn, Kentucky. e. Berea College, Starred on Broadway in Barefoot In The Park, The Star Spangled Girl, A Race of Hairy Men, and road companies of Barefoot In The Park, Star Spangled Girl, and Don't Drink The Water, Los Angeles Company of Jimmy Shine.
PICTURES INCLUDE: Grizzly, Act of Vengeance, The Devil Times Five. Screenwriter: The Predator, The Concision, Fly Away Home. 1980: Staff writer for The Days of Our Lives, Currently staff writer for Another World (both daytime soaps) under the pen name Joan Pommer.

McCALLUM, DAVID: Actor. b. Scotland, Sept. 19, 1933. Early career in rep. theatres and dir. plays for Army. Ent. m.p. Ind. 1953.
PICTURES INCLUDE: The Secret Place, Hell Drivers, Robbery Under Arms, Violent Playground, A Night to Remember, The Long and the Short and the Tall, Billy Budd, Freud, The Great Escape, The Greatest Story Ever Told, To Trap A Spy, Three Bits of the Apple, Sol Madrid, Mosquito Squadron, Watcher in the Woods.
TV: The Man From Uncle, The Invisible Man.

McCALLUM, JOHN: Actor, Producer, Director. b. Brisbane, Australia, CBE. Mar. 14, 1918; e. Royal Academy Dramatic Arts; served in Australian Imperial Forces. World War II. Appeared in repertory with Old Vic & Stratford-on-Avon seasons. On stage, 1937 in Judgment Day, Australian tour, 1955–56; on screen first 1944, Australia is Like This. Asst. man. dir. J. C. Williamson Theatres, Ltd., Australia, 1958; man. dir., 1960–66. Resigned chmn. Williamson-Powell Int. Films, 1965; chmn. Fauna Prod.; dir. Relatively Speaking on stage in Australia, 1968. Dir.; Plaza Suite, My Fair Lady, for Australian stage, 1969–70. Prod. TV Series, Skippy, Barrier Reef, Boney, Shannons Mob; London Stage 1973–74 Constant Wife. 1974 Comedy Theatre, Melbourne. 1976 Chichester Fest. 1976–77 The Circle. Exec. prod. Bailey's Bird, TV series.
PICTURES INCLUDE: A Son Is Born, Joe Goes Back, Root of All Evil, The Loves of Joanna Godden, It Always Rains on Sunday, Miranda, The Calendar, A Boy, a Girl and a Bike, Traveler's Joy, The Woman in Question, Valley of the Eagles, Lady Godiva Rides Again, Derby Day (Four Against Fate), Trent's Last Case, The Long Memory, Melba,

Devil on Horseback, Trouble in the Glen, Smiley (in Australia), Safe Harbour, Nickel Queen (dir.); The Z Men (prod.).

McCAMBRIDGE, MERCEDES: Actress. b. Joliet, Ill., March 17; e. Mundelein College Chicago, Ill., B.A. Did some radio work while in college; opposite Orson Welles two seasons, on Ford Theatre, other air shows; New York stage in: Hope for the Best, 1945; others: Place of Our Own, Twilight Bar, Woman Bites Dog, The Young and Fair; left latter play for Hollywood. Screen debut: All the King's Men (Academy Award for best supporting actress, 1950); own radio show, 1952.
PICTURES INCLUDE: Lightning Strikes Twice, Inside Straight, The Scarf, Johnny Guitar, Giant, A Farewell to Arms, Suddenly Last Summer, Cimarron, Angel Baby, Last Generation, Jigsaw, 99 Women, Thieves, The Concord-Airport '79.
AUTHOR: The Two of Us.
TV: Numerous appearances.

McCARTHY, FRANK: Producer. b. Richmond, Va., June 8, 1912. e. Virginia Military Institute and U. of Virginia, serving at VMI as instructor and tactical officer, and as police reporter for Richmond News-Leader, 1933–37. Press agent for producer George Abbott, 1937–39. During WW II was Asst. Secy. and later Secy. of War Dept. General Staff and military secy. to General George C. Marshall. Decorated with U.S. Distinguished Service, U.S. Legion of Merit, Order of the British Empire. Brigadier General in Army Reserve, Retired. In 1945, named Asst. Secy. of State for Administration; joined MPAA 1946 as asst. to v.p., later European mgr. Signed by 20th Century-Fox two years later as exec. and prod. Co-prod. Decision Before Dawn; prod., Sailor of the King, A Guide for the Married Man, Patton, Fireball Forward; at Universal, MacArthur.

McCARTHY, KEVIN: Actor. b. Seattle, Wash., Feb. 15; e. Minnesota U. Acted in sch. plays, stock; B'way debut in Abe Lincoln in Illinois. In U.S. Army; m.p. debut: Death of A Salesman.
PLAYS INCLUDE: Flight to West, Winged Victory, Truckline Cafe, Joan of Lorraine, Biography, Death of a Salesman, Anna Christie.
PICTURES INCLUDE: Drive a Crooked Road, Gambler from Natchez, Stranger on Horseback, Annapolis Story, Nightmare, Invasion of the Body Snatchers, A Big Hand for the Little Lady, Hotel, Kansas City Bomber, Buffalo Bill and the Indians, Hero at Large, Those Lips, Those Eyes.

McCLORY, SEAN: Actor. b. Dublin, Ireland, March 8, 1924; e. Jesuit Coll., U. of Galway. With Gaelic Theatre, Galway; Abbey Theatre. Dublin. Brought to U.S., in 1946 under contract to RKO Pictures. Produced and directed numerous plays, member of the Directors Guild of America and is the author of the produced drama, Moment of Truth.
PLAYS INCLUDE: Shining Hour, Juno and the Paycock, Anna Christie, Escape to Autumn, King of Friday's Men, Lady's Not for Burning, Billy Budd, Dial M for Murder, The Winslow Boy, Shadow of A Gunman, Saint Joan.
PICTURES INCLUDE: Film debut: Dick Tracy vs. The Claw, Rough-Shod, Beyond Glory, Daughter of Rosie O'Grady, Storm Warning, Lorna Doone, What Price Glory, The Quiet Man, Diane, Island in the Sky, Ring of Fear, Them, Long Grey Line, Cheyenne Autumn, Plunder of the Sun, Anne of the Indies, I Cover the Underworld, Botany Bay, Man in the Attic, Guns of Fort Petticoat, Kings Thief, Moonfleet, Bandolero, Day of the Wolves, Valley of the Dragons, Follow Me Boys, Rogues March, The Gnomobile, Well of the Saints, In Search of the Historical Jesus, Roller Boogie.
TV: Matinee Theatre, Climax, Lost in Space, My Three Sons, Suspense, The Untouchables, Hitchcock, Thriller, Bonanza, Gunsmoke, Mannix, Little House On The Prairie, S.W.A.T., The New Daughters of Joshua Cabe, The Captains And The Kings, Once An Eagle, Fish, Columbo, How the West Was Won. Co-starred in two series: The Californians, also directing several episodes, and Kate McShane and General Hospital.

McCLURE, WENDY: Actress, Dancer. b. Glasgow, Scotland, March 9, 1934. m. David Knight. e. Lowther College, North Wales. On London stage from 1955 in The Punch Revue, Alice Through the Looking Glass, Zuleika. M.P. debut in 1956 in Helen of Troy.
PICTURES INCLUDE: Helen of Troy, It's A Wonderful World, The Girl on the Boat, The Comedy Man, A King in New York.
TV: Knock on Any Door, Dr. Finlay's Casebook, The Newcomers, Over to Una, Z-Cars, The Borderers, Young Man in Trouble.

McCLURE, WILLIAM K.: Producer-Director. b. Knoxville, Tenn., Sept. 8, 1922; e. U. Tennessee & Catholic U., Washington, D.C. Contact man Washington staff, Pathe Newsreel, 1942; Los Angeles staff, 1943; asst. dir., "This Is America" series, 1943. Combat photographer, U.S. Army, 1944–46; Pub. mgr., RKO Pathe, Inc., 1947; Cameraman, corr., Warner Pathe Newsreel, 1948–51 Cameraman-corr., in Europe, Middle East for See It Now, CBS-TV, 1951–52, London, 1953; dir., cameraman, See It Now, 1952–54, Report on Africa. 1955–56, Burma, Buddhism. Neutralism, 1957, Watch on the Ruhr, 1958; dir. oper. Europe Small World. 1958–59. Prod., CBS Reports, co-prod.: CBS Reports, Iran Brittle Ally 1959, Berlin: End of the line 1960. Prod., CBS Reports: Britain, Blood Sweat and Tears, plus 20 1961 prod., Mr. Europe and the Common Market; prod., Germany Since Hitler; Adenauer Sums-up; prod., DeGaulle: The Challenge, DeGaulle: Roots of Power, East Europe, Satellites Out of Orbit; prod., CBS News, Henry Moore: Man of Form; Eur. prod., Town Meeting of the World, 1965–66. Prod. 1967 If you're appalled at my Texas I'm bewildered with your England. Prod. 1968 Don't Count the Candles.

McCRARY, TEX: Performer. r.n. John Reagan. b. Calvert, Tex., 1910. e. Exeter Acad, Yale. m. Jinx Falkenburg. Began journalistic career on N.Y. World Telegram and Literary Digest; Asst. to Arthur Brisbane on N.Y. Mirror, ed. chief 1933–41; writer and dir. of newsreels for Pathe Films; U.S. Air Force, WW II; author of First of Many.

McCREA, JOEL: Actor. b. Los Angeles, Calif., Nov. 5, 1905; e. Pomona College; husband of Frances Dee, actress. Stage experience: amateur dramatics and community plays taking male lead in "The Patsy," "Laff That Off" and "The Little Journey." In many pictures since 1932.
PICTURES INCLUDE: In 1940 He Married His Wife, Primrose Path, Foreign Correspondent. In 1941 Reaching for the Sun, Sullivan's Travels. In 1942 The Great Man's Lady, The Palm Beach Story. In 1943 The More the Merrier, Virginian, Ramrod, Four Faces West, South of St. Louis, Outriders, Colorado, Territory, Stars in My Crown, Saddle Tramp, Cattle Drive, San Francisco Story, Lone Hand, Shoot First, Border River, Black Horse Canyon, Stranger on Horseback, Wichita, First Texan, The Oklahoman, Trooper Hook, Fort Massacre, The Gunfight at Dodge City, Ride the High Country, Mustang Country.

McCULLOUGH, JIMMIE B.: Producer-Director. b. Mansfield, La., May 12, 1929. e. UCLA (1955). U.S. Marine Corps Reserve 1948–50; Ziv Studio, Hollywood, 1949–56: Acted on TV: Highway Patrol, CBS Playhouse 90, The Shadow, Teenage Monster, Crime File. Officer-director, American Metal Window Co., 1959–63. Owned and operated Macco Prod. Co., Inc., 1966–72. Now heads Jim McCullough Prods., Inc., Ry-Mac Intl. Film Dist., Inc., and Macco Prods., Inc.
PICTURES: Shepherd of the Hills (prod.-dir.), Where the Red Fern Grows (co-prod.), Creature from Black Lake (prod.-dir.), Charge of the Model Ts (prod.-dir.).

McDOWALL, BETTY: Actress. b. Sydney, Australia; e. Mt. Bernard Convent, N. Sydney. Early career radio, stage in Australia; ent. BBC TV, 1952; since in West End plays, many TV and radio plays and films.
STAGE: Age of Consent, Ghost Train, The Kidders, The Dark Halo, Period of Adjustment, Rule of Three, Signpost to Murder, Hippolytus, The Winslow Boy, Woman in a Dressing Gown, As Long As It's Warm, Caprice—In A Pink Palazzo, Sweet Bird of Youth, There was an Old Woman, What the Butler Saw, Two Dozen Red Roses, A Boston Story, The Man Most Likely To . . . , Sleeping Partner.
TV: Mid-Level and Glorification of Al Toolum, The Black Judge, Phone Call for Matthew Quade, Thunder on the Snowy, Shadow of Guilt, Traveling Lady, Torment, Biography, Notes for a Love Song, Esther's Altar, The Corridor People, The Braden Beat, The Douglas Fairbanks, Ivanhoe, The Foreign Legion, Fab an of the Yard, Four Just Men, Flying Doctor, No Hiding Place, 'Z' Cars, Days of Vengeance, Flower of Evil, Outbreak of Murder, Call Me Sam, The Prisoner, Public Eye, The Forgotten Door, All Out for Kangaroo Valley, Barry Humphries Scandals, Castle Haven, Albert and Victoria, Follyfoot, The Man Who Came to Dinner, Anne of Avonlea, Little Lord Fauntleroy, The Bass Player and the Blond (4 plays), The Gingerbread Lady.
PICTURES INCLUDE: First lead in England, Timelock, She Didn't Say No, Jack the Ripper, The Shiralee, Jackpot, Dead Lucky, Spare the Rod, Golliwog, Echo of Diana, First Men in the Moon, Ballad in Blue, The Liquidators, Willy Wagtails by Moonlight.

McDOWALL, MALCOLM: Actor. b. Leeds, England, June 1943. Was spear-holder for the Royal Shakespeare Co. in season of 1965–66 when turned to TV and then to films. Made debut in small role in Poor Cow, 1967.
PICTURES INCLUDE: If. . . . , Figures in a Landscape, The Raging Moon, A Clockwork Orange, O Lucky Man!, Royal Flash, Aces High, Voyage of the Damned, Caligula, The Passage, Time after Time.

McDOWALL, RODDY: Actor (Star of Tomorrow, 1944.) b. London, England, Sept 17, 1928; e. St. Joseph's London. First appeared in Murder in the Family at age of 8. Later, You Will Remember, The Outsider, Just William, Hey, Hey, U.S.A., This England, all made in England. In 1940 signed by 20th Century-Fox.

PICTURES INCLUDE: Man Hunt, How Green Was My Valley, Confirm or Deny, Son of Fury, On the Sunny Side, The Pied Piper, My Friend Flicka, Lassie Come Home, White Cliffs of Dover, Macbeth, Act, assoc. prod., Rocky, Kidnapped, Big Timber, Tuna Clipper, Black Midnight, Killer Shark, Steel Fist, The Subterraneans, Midnight Lace, Cleopatra, The Longest Day, The Greatest Story Ever Told, Shock Treatment, That Darn Cat, The Loved Ones, The Third Day, Daisy Clover, Bullwhip Griffin, Lord Love A Duck, The Defector, It, The Cool Ones, Planet of the Apes, Hello, Down There, Midas Run, Escape from the Planet of the Apes, Conquest of the Planet of the Apes. Directorial Debut, The Devil's Widow, 1971. Actor: The Legend of Hellhouse, The Poseidon Adventure, Arnold, Funny Lady, The Cat from Outer Space, Scavenger Hunt, Charlie Chan and the Curse of the Dragon Queen.

THEATRE: On B'way in Misalliance, Escapade, Doctor's Dilemma, No Time for Sergeants, Good as Gold, Compulsion, Handful of Fire, Look After Lulu, The Fighting Cock, 1959–60, Camelot, 1960–61, The Astrakhan Coat, 1966.

RADIO & TV: Stratford Shakespearean festival, 1955.

McEVEETY, BERNARD: Director. Comes from film family; father was pioneer as unit mgr. at New York's Edison Studios; brothers Vincent, also a dir., and Joseph, writer are at Disney Studios. Bernard's career began in 1953 at Paramount where was asst. dir. for 6 yrs. Earned full dir. responsibility on The Rebel, TV series.

PICTURES INCLUDE: Napoleon and Samatha, One Little Indian, The Bears and I.

TV: Episodes on Bonanza, Gunsmoke, Combat and Cimarron Strip (also prod.)

McEVEETY, VINCENT: Director. Joined Hal Roach Studios in 1954 as second asst. dir. Then to Republic for The Last Command. First Disney assignments: Davy Crockett shows and Mickey Mouse Club. Moved to Desilu as first asst. dir. on The Untouchables; made assoc. prod. with option to direct. Did segments of many series, including 34 Gunsmoke episodes. First theatrical film: Firecreek, 1968.

PICTURES INCLUDE: $1,000,000 Duck, The Biscuit Eater, Charley and the Angel, Superdad, The Strongest Man in the World, Gus, Treasure of Matecumbe. Herbie Goes to Monte Carlo.

TELEVISION: High Flying Spy.

McFADDEN, FRANK: Publicist. b. San Diego, Calif., June 3, 1914. e. Southwestern Bus. Coll. Pub. dept., Warner Bros. Studio, 1933–35; asst. pub. dir., Universal Studio, 1936–53; adv., pub. dir., Leonard Goldstein Prod., 1954; president pub. rel. agency, McFadden and Eddy, Inc., 1955; partner, McFadden, Strauss & Irwin, 1963. TV, 1956. Became partner in ICPR in 1975.

McGANNON, DONALD H.: Executive. b. Sept. 9, 1920. New York, N.Y.; e. Fordham U., Fordham Sch. of Law. Network exec., gen. mgr., owned, operated stations, DuMont Television Network, 1952–55; pres. chmn. of bd., Westinghouse Broadcasting Co., Inc., Ind., Md., Del., 1955; pres. chmn. of bd. Television Advertising Representatives, Inc.; chmn. of bd., AM Radio Sales Co.; chmn., WBC Productions, Inc.; Chmn., WBC Program Sales, Inc.; Chmn. The Advertising Council; Pres., National Urban League; Chmn., Connecticut Commission for Higher Education; chmn., Permanent research Comm., 1964 dis. serv. award, National Association of Broadcasters; dir., The Advertising Council, Inc.; dir., Radio Advertising Bureau; dir., Academy of Television Arts and Sciences Foundation; dir., Radio Free Europe; Communications Subcommittee, Community Relation Service; advsr. Pontifical Commission for Communications Media. Adv. North American Vatican Radio.

McGAVIN, DARREN: Actor. b. San Joaquin Valley, Calif., May 7, 1922. e. C. of the Pacific.

PLAYS INCLUDE: Death of a Salesman, My Three Angels, The Rainmaker, The Lovers, Dinner at Eight (revived).

PICTURES INCLUDE: Summertime, The Man with the Golden Arm, Court Martial of Billy Mitchell, Beau James, Delicate Delinquent, The Great Sioux Massacre, Bulley For a Badman, Mission Mars. Directorial Debut: Happy Mother's Day . . . Love, George (1973). Actor: Airport '77.

TV: Mike Hammer, The Outsider, The Night Stalker.

McGOOHAN, PATRICK: Actor. b. New York. Early career in repertory in Britain. London stage 1954 in Serious Charge; 1955 Orson Wells' Moby Dick. Ent. films 1955.

PICTURES INCLUDE: Passage Home, High Tide at Noon, Hell Drivers, The Gypsy and the Gentleman, Nor the Moon by Night, Two Living, One Dead, All Night Long, The Quare Fellow, Thomasina, Dr. Syn, Ice Station Zebra, The Moonshine War, Brass Target, Scanners.

TV: Plays and star of ATV's Danger Man series; also directed some of these; Secret Agent, The Prisoner.

McGOWAN, BROTHERS: Dorrell & Stuart. Writers, Directors, Producers. (Dorrell) b. Winnetka, Ill., (Stuart) b. Wilmette, Ill. In m.p. ind. since 1930. Producers. Independent production since 1950. TV series: Death Valley Days, Sky King, The Littlest Hobo. Features: Tokyo File, Japan—Bashful Elephant, Austria; Hellfire, Showdown, The Littlest Hobo, Snowfire, Billion Dollar Hobo, U.S. Feature, Mister Too Little now in release.

McGOWAN, ROBERT A.: Writer. b. Denver, Colo., May 22, 1901. Cartoonist; from 1923–29, writer of Hal Roach "Our Gang" comedies for his uncle, Robert Francis McGowan. Directed under the name of Anthony Mack, s.p., Babes in Toyland, collab. s.p., Too Many Parents, Boy Friend shorts for Warner-Vitagraph, N.Y. Prod. manager, four technicolor shorts. Wrote new series "Our Gang" comedies for MGM.

McGRATH, THOMAS J.: Producer. b. New York City. e. Washington Square Col. of N.Y.U. (B.A. 1956), N.Y.U. School of Law (LLB 1960). Has practiced law in N.Y.C. from 1960 to date. Became indep. prod. with Deadly Hero in 1976; Author, Carryover Basis Under The 1976 Tax Reform Act, published in 1977.

McGREAL, E. B. "MIKE": Executive. b. Bellingham, Wash., Aug. 13, 1905. e. U. of So. Calif. Camera dept., MGM, 1929–34; exec. dir. of photog., Warner Bros., 1934–54; v.p., Houston Fearless Corp., pres., Houston Color Film Lab. of Calif., 1954–56; Exec. dir., spec. photographic effects, 20th Century Fox, 1957–59; pres. Producers Service Corp., 1960–70. Currently president and owner of B.M.C. Corporation, McGreal and Hunt. Member—Academy of Motion Picture Arts & Sciences, Society of Motion Picture & Television Engineers, British Kinematograph Sound & Television Society.

McGREGOR, CHARLES: Executive. b. Jersey City, N.J., April 1, 1927. e. N.Y.U. 1958–1969, Co-founder, pres. and chief exec. officer, Banner Films, Inc. (World Wide TV Distribution), 1955–1958, salesman and div. mgr., Flamingo Films (domestic TV Dist.). 1953–55; Professional mgr. ABC Music Publishing. 1951–53: Prod. and partner Telco Prods. and G&M Productions (prods. of network and local shows). 1969–77: exec. v.p. in chg. of w-w dist., WB-TV; 1977: pres. WB-TV Distribution.

McGUIRE, DON: Writer, director. b. Chicago, Ill., Feb. 28, 1919. U.S. Army, 4 yrs.; press agent, Chicago and Hollywood; newsman, Hearst papers, Chicago; then actor, writer.

PICTURES INCLUDE: Double Deal, Dial 1119, Meet Danny Wilson, Willie and Joe in Back at the Front, Walking My Baby Back Home, Three Ring Circus, Bad Day at Black Rock, Artists and Models, Johnny Concho, Delicate Delinquent, Suppose They Gave a War and Nobody Came.

TV: Writer, dir. & co-prod., Hennessey series, creator, series Not for Hire; creator, writer, prod. series, Don't Call Me Charlie.

AUTHOR: Novels, The Day Television Died, 16000 Floogle Street.

McGUIRE, DOROTHY: Actress. b. Omaha, Neb., June 14, 1919; e. Ladywood convent, Indianapolis; Pine Manor, Wellesley, Mass.

STAGE: Bachelor Born, Our Town, My Dear Children, Swinging the Dream, Claudia, Legend of Lovers, Night of the Iguana (1976).

PICTURES INCLUDE: Claudia, Tree Grows in Brooklyn, Enchanted Cottage, Spiral Staircase, Claudia and David, Gentleman's Agreement, Mister 880, Callaway Went Thataway, I Want You, Make Haste to Live, Invitation, 3 Coins in the Fountain, Trial, Friendly Persuasion, Old Yeller, This Earth Is Mine, Remarkable Mr. Pennypacker, Dark at The Top of the Stairs, Summer Place, Susan Slade, Flight of the Doves.

TELEVISION: Rich Man, Poor Man.

McHUGH, FRANK: Actor. b. Homestead, Pa., May 23, 1899. Child actor in For His Children's Sake; on stage N.Y. and London.

PLAYS INCLUDE: Fall Guy, Tenth Avenue, Excess Baggage, Conflict, Show Girl.

PICTURES INCLUDE: If Men Played Cards as Women Do 1928; Top Speed Bright Lights, Easy Come, Easy Go, Kiss Me Again, Dawn Patrol, Toast of the Legion, Back Street, Her Cardboard Lover, Going My Way, Gold Diggers of 1935, Footlight Parade, Marine Raiders, Bowery to Broadway, Medal for Benny, State Fair, Mighty Joe Young, Pace That Thrills, My Son John, It Happens Every Thursday, Lion is in the Streets, There's No Business Like Show Business.

McHUGH, JAMES: Agent/Manager. b. Boston, Mass., Oct. 21, 1915; e. Holy Cross Coll. Joined MCA 1939. U.S. Army 1944–46. MCA-British and European, 1945–50. Formed James McHugh, Talent Agency 1953. Pres., Artists Mgr. Corp., pres., Turquoise Prod., Inc., v.p. investment consultant, Miguel Prod. and Ebsen Ent., Inc. Present clients: B. Ebsen, Brian Keith, Leo Penn, Alan Fudge.

McINTIRE, JOHN: Actor. b. Spokane, Wash., June 27, 1907; e. Local schools; m. Jeanette Nolan, actress; Radio announcer, actor teamed with wife.
PICTURES INCLUDE: Asphalt Jungle, Francis, Saddle Tramp, Winchester '73, Ambush, Scene of the Crime, You're in the Navy Now, Under the Gun, Raging Tide, Westward the Women, World in His Arms, Glory Alley, Sally & St. Anne, Horizons West, Lawless Breed, Mississippi Gambler, The President's Lady, Lion Is In the Street, War Arrow, Apache, Four Guns to the Border, Yellow Mountain, Far Country, Stranger on Horseback, Scarlet Coat, Phoenix City Story, Backlash, The Spoilers, The Kentuckian, World in My Corner, Away All Boats, The Tin Star, Who Was That Lady, Flaming Star, Two Rode Together, Summer & Smoke, Herbie Rides Again.
TV: Wagon Train, Naked City, The Virginian.

McINTOSH, STANLEY: Educator, executive b. Stewardson, Ill., Mar. 7, 1908. e. East Ill. University, B.Ed.; Northwestern U., M.A. Tchr., prin. Evanston, Ill. pub. schls.; Asst. Supt. Cook Co., Ill. pub. schools; Instr. Northwestern U.; visitng lecturer, U. of Iowa; prof. lectr. American U.; Regional dir., war training films, U.S. Office of Educ.; assoc. research & prod., Encyclopaedia Britannica Films, Inc.; appt. to MPAA, 1947 as asst. dir. educational service; in 1958 made dir. educational & community service; since 1955 ex. dir. Teaching Film Custodians; org. pres., CINE, 1957–58; appt. Consultant Indiana U. on merger of TFC-Indiana U., May 1973. Retired 1976.

McKELLAR, KENNETH: Singer. b. Paisley, Scotland. singer, stage prods., radio & TV; appeared, concerts, Edinburgh International Festival; toured, Scots Concert groups; created role, Jamie in A Wish For Jamie, A Love for Jamie. Many BBC-TV and ITV appearances in V.K. Records regularly.

McKENNA, SIOBHAN: Actress. b. Belfast, Ireland, May 24, 1923; e. St. Louis Convent, Monaghan; National Univ. of Ireland, B.A. Stage debut at Abbey Theatre, 1943; then London stage in The White Steed, 1947. Screen debut: Hungry Hill, 1946. London film and stage parts, also stage in Dublin; interested in development of Gaelic Theatre in Galway for which she has translated several plays including G. B. Shaw's Saint Joan.
PICTURES INCLUDE: Daughter of Darkness, The Lost People, The Adventurers, King of Kings, Of Human Bondage, Dr. Zhivago.
STAGE: (London) Saint Joan; (N.Y.) Chalk Garden.
TV: appearances in U.S.

McKINNEY, BILL: Actor. b. Tennessee, Sept. 12.
PICTURES INCLUDE: Hollywood Secret File; She Freak; The Road Hustlers; Smack; Breakheart Pass; Josey Wales; Deliverance; Junior Bonner; Judge Roy Bean; Kansas City Bomber; The Outfit; Thunderbolt & Lightfoot; Parallax View; Every Which Way But Loose; The Gauntlet; For Pete's Sake.
THEATRE: No Time For Sergeants; A Cook For Mr. General; Time Limit; Bus Stop; Of Mice And Men; The Garbage Hustler; Roadside; As You Like It; Hassan Of Bagdad; No Place To Run; Career; Hat Full Of Rain; The Gamblers; The Devil's Disciple; All My Sons; Deadwood Dick; Chanticleer; Hoot And Holler; War And Peace; Grapes Of Wrath; El Camino Real; The Little Prince.
TV: Showcase for Success; The Monkeys; Holvak; Baretta; Bronk; Sarah; Bold Ones; Smith & Jones; McCloud; Columbo; Ironside; Pvt. Slovak; The Healers.

McLAGLEN, ANDREW V.: Dir. b. London, Eng., July 28, 1920. e. U. of Virginia, 1939–40. prod. dept., Lockheed Aircraft Corp., 1940–44; asst. m.p. dir., 1944–54; dir., 1955–65; dir., CBS-TV. 1956–63.
PICTURES INCLUDE: Man in the Vault, Gun the Man Down, The Abductors, Freckles, The Little Shepherd of Kingdom Come, McLintock!, Shenandoah, The Rare Bree, Seven Men From Now, The Way West, The Ballad of Josie, Monkeys, Go Home, Devil's Brigade, Bandolero, The Undefeated, Fool's Parade Something Big, One More Train To Rob, Cahill, U.S. Marshall, Mitchell, The Last Hard Men, The Wild Geese, ffolkes, The Sea Wolves.
TV: Gunsmoke, Have Gun—Will Travel, Perry Mason, Rawhide, The Lineup, The Lieutenant, etc.

McLERIE, ALLYN: Actress. b. Grand Mere, Quebec, Canada, Dec. 1, 1926; e. High School, N.Y. Dancer since 15 in many B'way shows.
SHOWS INCLUDE: One Touch of Venus, On the Town, Finian's Rainbow, Where's Charley, Miss Liberty, Time Limit.
PICTURES INCLUDE: Where's Charley, Desert Song, Calamity Jane, Phantom of the Rue Morgue, Battle Cry, They Shoot Horses, Don't They?, The Cowboys, Jeremiah Johnson, The Magnificent Seven Ride, Cinderella Liberty.

McMAHON, ED: Performer b. Detroit, Mich., March 6, 1923. e. Boston Coll.; Catholic U. of America, B.A., 1949. U.S. Marines, 1942–53.
TV: Tonight Show, Monitor, Fortune Phone, Who Do You Trust, The Missing Links, commercials.
THEATRE: Stock Bdwy., Impossible Years.
PICTURES INCLUDE: The Incident, Fun with Dick and Jane.

McMAHON, JOHN J.: Executive. b. Chicago, Ill., 1932. e. Northwestern U. Served with U S. Army in Korea, beginning career on WGN-TV, Chicago; associated with ZIV-United Artists TV Productions during 1950's; joined American Broadcasting Company in 1958; v.p. & gen. mgr., WXYTZ-TV, Detroit, then KABC-TV, Los Angeles, 1968; v.p., ABC, 1968–72; joined National Broadcasting Company in 1972 as v.p., programs, west coast, NBC-TV; president, Hollywood Radio & Television Society; board member, Permanent Charities Committee. June, 1980, named pres. of Carson Prods. (Johnny Carson's prod. co.).

McNALLY, STEPHEN: Actor r.n. Horace McNally; b. New York City, July 29; e. Fordham U., LL.B. In school dramatics; practiced law 2 yrs.; N.Y.; stage, films, 1942.
PICTURES INCLUDE: Winchester 73, Wyoming Mail, No Way out, Air Cadet Apache Drums, Raging Tide, Lady Pays Off, Devil's Canyon, Make Haste to Live, A Bullet Is Waiting, Man From Bitter Ridge, Tribute to a Bad Man, Once You Kiss a Stranger, Black Gunn.

McNAMARA, THOMAS J.: Executive. b. New York, N.Y. e. N.Y. schools, Pace University, C.P.A. Price Waterhouse, 1943–1947; chief accountant, Motion Picture Association of America, Inc., 1948; asst. treas., 1955; treas. AMPECA, 1961, and AFRAM Films, 1973; treas. MPAA and MPEA, 1977.

McNEIL, ROBERT A.: Exhibitor. b. San Francisco, Cal., Jan. 14, 1889; e. public schools of San Francisco; m. Started show career when 17. Sang on stage, also operated projection machine and played in three shows a week in San Francisco. Worked on Pantages circuit for two years. Expanded theatre enterprises after San Francisco fire, owning five houses. Joined E. H. Emmick in forming Golden State Circuit in 1927. Associated with Mike Naify of T. and D. Jr. Enterprises to form United California Theatres, Inc., sold interest to Naify, retired; maintains interest in small San Francisco chain.

McNELLIS, MAGGI: Performer; b. Chic. Ill.; e. Rosemont Coll., Mundelein Coll.; m. Clyde Newhouse; began as singer in Chic. and New York clubs; radio shows; Maggi's Private Wire, Maggi's Magazine; on various TV shows.
TV: Maggi's Private Wire, ABC, Leave It to the Girls, ABC, NBC, Say It with Acting, ABC, Crystal Room, NBC, Maggi McNellis Show, WABC, WABD, N. Y.

McNICHOL, KRISTY: Actress. Made debut at age of 7 performing in commercials. Given regular role in Apple's Way; began appearing on such series as Love, American Style and The Bionic Woman. Attracted attention of Spelling-Goldberg Productions, who cast her as Buddy Lawrence in Family series.
PICTURES: The End, Little Darlings, The Night the Lights Went Out in Georgia, Only When I Laugh.
TELEVISION: Movies: Like Mom, Like Me, Summer of My German Soldier, My Old Man.

McQUEENEY, ROBERT: Actor. b. Bridgeport, Conn., March 5. e. Taft School, Watertown, Bard College; lieutenant, U.S. Army; prof. debut, summer theatre group, Jennerstown, Pa.
SHOWS INCLUDE: Billy Budd, Dial M for Murder, Affairs of State, Fragile Fox, Macbeth, Rain, The Second Man, The Tender Trap, Brainstorm, The Glory Guys.
TV: Robert Montgomery Presents, Philco, Kraft Theatre, Studio One, Omnibus Playwrights '56, Love of Life, The Guiding Light, The Gallant Men.

McWILLIAMS, HARRY KENNETH: Publicity executive. b. Middlesboro, Ky., July 20, 1907 e. Colorado U.; Denver Sch. of Tech. 1930–32 sales mgr., Acme Film Co., Bowles; Ted Bates Adv., Inc. Entered m.p. ind. 1927 as theat. & pub. mgr., Harry E. Huffman Theats., Denver, Col.; in 1928 partner, Filmcraft Labs., Denver (trailers, commercial and scenic short subjects). In 1928 joined Publix Theats. Corp.; 1941 joined Columbia Picture exploit. dept.; in Jan. 1944 named exploit. mgr. dir. adv. pub. rel. Screen Gems, Inc., TV sub. Col.; Pict., Jan. 1953-May, 1954; pres., gen. sales mgr. Air Programs, Inc., May 1954; asst. dir. adv., pub., exploit. Magna Theatres Corp., 1955–56; ass't gen. man. Original Amateur Hour Program 1959–60; prod. Academy Award Exhibitor Kit, 1961–62, ass't to Oscar Doob "Ben-Hur" dept., MGM, 1959; prom. Samuel Bronston, 1959; Columbia, 1960; pub. rel. U. of Cincinnati, 1961; co-ordinator of Community Relations; co-ordinator, New York City, Movie News, MPA, ad-pub. committee, co-ordinator, Academy Award News, TOA., 1963; p.r. council. Currently p.r. consultant NATO. Publisher of theatre ad book, 1975.

MEAD, THOMAS: Newsreel editor, producer. b. Tecumseh, Mich., Dec. 7, 1904; e. Wabash Coll. Was reporter, Chicago Daily News, prior to entering m.p. ind.; prod. Uni-

versal-Internat'l Newsreel & Variety View short subjects since 1938.
PICTURES INCLUDE: Snow Capers, Cheating in Gambling, Four Years Before the Mast, Tiny Terrors Make Trouble, Thundering Rails, Fun at the Zoo (1950), Landscape of Silence & Perils of the Forest, Hottest 500, Blond Empire, Venice of the East; man. dir., Universal-International Newsreel, New York.

MEADES, KENNETH RICHARDSON: Communications Consultant. b. Montreal, Can., March 2, 1943; e. Cote des Neiges Comm. Coll., Geo. Williams U., Sta. Ana Coll., Western Tech. Inst. Obtained first class Radiotelephone lic. with FCC; eng. dept., CFCF Radio & TV, 1960–63; technician with Broadcast Engineering Services, Garden Grove, Calif.; presently priv. consultant.

MEANEY, DONALD V.: Executive. b. Newark, N.J. e. Rutgers Univ. School of Journ. Worked as reporter for Plainfield (N.J.) Courier-News, Newark Evening News. Became news director of radio station WCTC in New Brunswick N.J. and later for WNJR, Newark. Joined NBC in 1952 as news writer; two years later became nat'l. TV news editor. Promoted to mgr., national news, 1960 and mrg., special news programs, 1961. Appt. dir. of news programs 1962 and gen. mgr., NBC News, 1965; v.p., news programming, NBC, 1967; v.p. news, Washington, 1974; mrg. dir., affiliate & intl. liaison, 1979.

MEDAK, PETER: Director. b. Budapest, Hungary. Entered industry in 1956 in London with AB-Pathe as trainee. Worked sound, editing and camera depts. Later was asst. dir., second unit dir. on various action pictures. In 1963 under contract to Universal Pictures, where started directing TV films. In 1967 under contract to Paramount Pictures.
PICTURES INCLUDE: Kaleidoscope (assoc. prod., 2nd unit dir.), Funeral in Berlin (2nd unit dir.), Fathom (assoc. prod.; 2nd unit dir.) Negatives (dir.), Day in the Death of Joe Egg (dir.), Ruling Class (dir.), Ghosts in the Noonday Sun (dir.).
TV: Third Girl from the Left, The Babysitters, The Dark Secret of Black Bayou.

MEDAVOY, MIKE: Executive. b. Shanghai, China, Jan 21, 1941. Lived in China until 1947 when family moved to Chile. Came to U.S. in 1957. e. UCLA, graduating in 1963 with honors in history dept. Started working in mail room at Universal Studios and became a casting director, from which he went to work for Bill Robinson as an agent trainee. Two years later joined GAC and CMA where he was a v.p. in the motion picture department. In 1971 joined IFA as vice-president in charge of motion picture dept. Represented American and foreign creative talents, among whom were Jane Fonda, Donald Sutherland, Michelangelo Antonioni, Jean-Louis Trintignant, Karel Reisz, Steven Spielberg, Robert Aldrich, George Cukor, John Milius, Terry Malick, Raquel Welch, Gene Wilder and Jeanne Moreau. While at IFA was involved in packaging The Sting, Young Frankenstein, Jaws and others, before joining United Artists Corp. in May, 1974, as senior v.p. in chg. of West Coast prod. In 1978 named exec. v.p., Orion Pictures Co. Is also pres. of Filmex, Inc. Member: UCLA Chancellors and Associates, The American Film Institute, the Board of Crossroads School; Board of Governors, Academy of Motion Picture Arts & Science and member of the visiting committee to the Dept. of Public Relations of Boston Museum of Fine Arts.

MEDFORD, DON: Director. b. Detroit, Mich.; e. Purdue U., U. of N.C., B.A., Yale U., M.F.A., actor, stage mgr., summer theatres; producer, B'way stage, Christopher Award.
TV: Kraft Theatre, General Electric Theatre, Alfred Hitchcock Presents, Climax, Twilight Zone, U.S. Steel Hour, Dick Powell, Eleventh Hour, Dr. Kildare, Fugitive, 12 O'clock High, The FBI, Man From UNCLE, Cimarron Strip, Baretta, Police Story, Kaz, Streets of San Francisco, Dynasty, Odd Job, The Changeling, Zorro, The Gay Blade.
FILMS: The Hunting Party, The Organization, The November Plan, Incident in San Francisco, The Coach.

MEDINA, PATRICIA: Actress. b. London, Eng. In many British films.
PICTURES INCLUDE: Secret Journey, Hotel Reserve, Don't Take It to Heart, Waltz Time. U.S. screen debut: Secret Heart, 1946; others: Moss Rose, Foxes of Harrow, Sangaree, Drums of Tahiti, Phantom of the Rue Morgue, Black Knight, Pirates of Tripoli, Duel on the Mississippi, Uranium Boom, Stranger At My Door, The Killing of Sister George.

MEDLEY, WM. THOMAS: Actor-singer. b. Hollywood, Calif., Sept. 19, 1940. Mem., Righteous Bros. rec. team.
TV: Shindig, The Danny Kaye Show, Ed Sullivan Show.
PICTURES INCLUDE: Beach Ball, Swingin' Summer.

MEDMAN, EDWARD A: Executive. b. Philadelphia, Pa. Nov. 11, 1937. e. University of Pennsylvania (Wharton School) 1955–58; University Pennsylvania Law School 1958–61, JDS; Joined General Counsel's Staff National Labor Relations Board, Washington, D.C., 1962–66; Trial Attorney NLRB, New York 1966–69; Joined Law firm of Poletti, Freidin, Prashker, Feldman & Gartner, 1969; Senior Labor Attorney, National Broadcasting Company 1970–72; Joined The Burbank Studios, Dec. 1972, v.p., legal/business affairs and chm. of exec. committee.

MEDWIN, MICHAEL: Actor, writer, producer. b. Eng., 1925; e. Institut Fischer. Switzerland. Stage debut 1940; m.p. acting debut in Root of All Evil, 1946.
PICTURES INCLUDE: My Sister and I, Mrs. Christopher, Gay One, Children of Chance, Operation Diamond, Black Memory, Just William's Luck, Ideal Husband, Picadilly Incident, Night Beat, Courtney's of Curzon Street, Call of the Blood, Anna Karenina, William Comes to Town, Woman Hater, Look Before You Love, Forbidden, For Them That Trespass, Queen of Spades, Trottie True, Boys in Brown, Trio, Long Dark Hall, Curtain Up, Street Corner (Both Sides of the Law). I Only Asked, Carry on Nurse, Wind Cannot Read, Heart of a Man, Crooks Anonymous, It's All Happening, Night Must Fall, I've Gotta Horse, 24 Hours To Kill, Charlie Bubbles, If . . . , Spring and Port Wine, O Lucky Man! Prod., Gumshoe, Law and Disorder.
TV: Granada's Army Game, The Love of Mike, Three Live Wires, Memorial Films Ltd.
THEATRE: Spring and Port Wine, Joe Egg, Forget-me-not Lane, Chez Nous, Alpha Beta.

MEEKER, RALPH: Actor: b. Minneapolis, Minn., Nov. 21, 1920; e. Northwestern U. In school plays; served in U.S. Navy.
THEATRE: On B'way in Doughgirls, Strange Fruit, Cyrano de Bergerac, Mr. Roberts, Streetcar Named Desire (also road co.), Picnic, Mrs. Dally; off-B'way: House of Blue Leaves.
PICTURES INCLUDE: Teresa, Four in a Deep Shadow, Sky, Somebody Loves Me, Glory Alley, The Naked Spur, Jeopardy, Code Two, Big House U.S.A., Kiss Me Deadly, Desert Sands, Ada, Dirty Dozen, Gentle Giant, The DEVIL'S Eight, I Walk the Line, Anderson Tapes, The Happiness Cage, Food of the Gods, Winter Kills.
TV: Lost Flight (Special).

MELAMED, DAVID J.: Executive. b. N.Y., Dec. 11, 1911; e. NYU; accountant, Columbia Pictures Corp., 1933–35; acct., RKO Radio Pictures, 1935–40; comptroller, Pathe Labs., Inc., 1941–43; U.S. Army, 1943–46; treas., Eagle Lion Films, 1947–51; exec. v.p., dir., 1952–58; dir. of adm., asst. to chmn. of bd., Nat'l General Corp., 1959–61; v.p., finance, American International Pictures, 1961.

MELCHIOR, IB: Director, writer. b. Copenhagen, Denmark, Sept. 17, 1917. p. Lauritz Melchior; e. Coll., Stenhus, Denmark, 1936. Actor, stage mgr., English Players, 1937–38; co-dir., 1938; actor in 21 stage prod. in Europe and U.S. on radio; set designer; stage man. dept., Radio City Music Hall, 1941–42; U.S. Military Intelligence, 1942–45; writer, dir., m.p. shorts for TV, 1947–48; TV actor, 1949–50; assoc. dir., CBS-TV, July, 1950; assoc. prod., G-L Enterprises, 1952–53; dir., Perry Como Show, 1951–54; dir. March of Medicine, 1955–56; writer, dir. of M.P. & TV films. 1957. Documentary writ. & dir. awarded Top Award by Nat'l. Comm. for Films for Safety, 1960. Golden Scroll Award, Academy of Science Fiction, Best Writing, 1976.
PICTURES INCLUDE: When Hell Broke Loose, Live Fast Die Young, The Angry Red Planet, Reptilicus, Journey to the Seventh Planet, The Case of Patty Smith, Robinson Crusoe on Mars, The Time Travellers, Ambush Bay, Planet of the Vampires, Death Race 2000.
AUTHOR: Novel, Order of Battle, 1972, Sleeper Agent, 1975, The Haigerloch Project, 1977, The Watchdogs of Abaddon, 1979, The Marcus Device, 1980.

MELFORD, AUSTIN: Actor, author, director. b. Alverstoke. England, Aug. 24, 1884; Actor & prod., Theatre Royal, Drury Lane, London, Gaiety Theatre, London, and others.
STAGE PLAYS: (Author) It's a Boy, Battling Butler, Night of the Garter, Nippy, Bob's Your Uncle, Blue for a Boy.
PICTURES INCLUDE: Oh, Daddy, Car of Dreams, Getting Married, We'll Smile Again, Heaven Round the Corner, Champagne Charlie, Sailors Three.

MELNICK, DANIEL: Executive. b. N.Y.C., April 21, 1934. e. N.Y. Univ. In 1954 was staff prod. for CBS-TV; then exec. prod., East Side West Side and N.Y.P.D. Joined ABC-TV as v.p. in chg. of programming. Partner in Talent Associates. Joined MGM as v.p. in chg. of prod.; in 1974 named sr. v.p. & worldwide head of prod.; 1977 in charge of worldwide production, Columbia Pictures; named pres., 1978. Resigned to be indep. prod. (now at 20th-Fox).
PICTURES INCLUDE: Straw Dogs, That's Entertainment! (exec. prod.); That's Entertainment, Two! (co-prod.); All That Jazz (exec. prod.); Altered States (exec. prod.); First Family (prod.).

MELNIKER, BENJAMIN: Motion Picture Producer—Attorney; b. Bayonne, N.J., e. Brooklyn College; LL.B., Fordham Law School, Loew's Theatres usher, private law practice, employed Legal Department Metro-Goldwyn-Mayer, vice president and general counsel, 1954–1969, executive vice president, 1968–1970, resigned from MGM December 1971;

also member MGM bd. dir. and mem. MGM exec. com.; Adjunct associate professor, New York Law School; former motion picture chmn. Anti-Defamation League, B'nai B'rith; Mem. Am., N.Y. State bar assns., Bar Assn. City N.Y., Acad. Motion Picture Arts and Scis.

MELSON, JOHN: Writer-Lyricist. b. Anderson, Ind., Dec. 17, 1930. e. Univ. of Paris (La Sorbonne, 1952 B.A.) and Univ. of Madrid (P.G. studies, 1953–4). m. Maria Perschy, actress. Reporter, Detroit Labor News, 1951; taught French and Italian in various commercial schools in Paris, 1954–5; orchestra musician 1946–50. Script writer for films since 1962. Member: WGAW, BMI

PICTURES INCLUDE: Battle of the Bulge (co-story, co-s.p.); Savage Pampas (co-s.p.); Cauldron of Blood (co-story, co-s.p.); Simon Bolivar (co-story, co-s.p.); Special Delivery (co-story, co-s.p.); Where Time Began (co-story, co-s.p.); Love and Bullets, Charlie (co-s.p.); The Animal (co-s.p.).

TV: Hawaii Five-O.

SONG LYRICS: Like a Dream (from the Summertime Killer); Cotton Candy Kingdom and The Familiar Stranger (both from Where Time Began).

MENGERS, SUE: Talent Agent. b. Bronx, N.Y. Started as receptionist, secretary with Music Corp. of America in New York. In 1963 became agent in small partnership. Two years later joined Creative Management Associates and sent to Hollywood. Now with Intl. Creative Mgt. (co. formed by merger of CMA and Marvin Josephson Associates). Clients include Barbra Streisand, Gene Hackman, Ryan O'Neal, Tatum O'Neal, Ali MacGraw, Cybill Shepherd, Peter Bogdanovich, Sidney Lumet, Arthur Penn, Nick Nolte, Robin Williams, Roman Polanski, Michael Caine.

MERCHANT, LAWRENCE H., JR.: Producer. b. Cambridge, Mass. e. Boston Latin School, Hebron Academy, Maine, Columbia University, New York University. Pres., Educational Book Div., Prentice-Hall; pres., Pillsbury Productions; prod. exec., TV-Today, Home, Tonight, Milton Berle, Show of Shows, Hallmark Hall of Fame, Eye Witness, R. Montgomery Presents, Wide, Wide World, Walter Winchell; TV advisor, Gen. Dwight D. Eisenhower; prod. CBS films; pres., Kachina Productions.

PICTURES INCLUDE: Hands of Dr. Maniacal, Back Track, Present Tense of Love.

MERCOURI, MELINA: Actress, b. Athens; schooling and training in Athens, fluent in French, German and English; Stage debut on Athens stage in avant-garde work; early stage career in Paris. Also made vocal recordings.

PLAYS INCLUDE: Morning Becomes Electra, La Nuit de Samaracande, Les Compagnons de la Marjolaine, Il Etait une Gare, Le Moulin de la Galette; to Greece 1954, in Stella. Also: A Streetcar Named Desire, Helen or the Joy of Living, The Queen of Clubs, The Seven Year Itch, Sweet Bird of Youth, Ilya Darling.

PICTURES INCLUDE: Stella, The Gypsy and the Gentleman, The Law, Never on Sunday, Phaedra, The Victors, Topkapi, He Who Must Die, 10:30 P.M. Summer, A Man Could Get Killed, Gaily, Gaily, Promise at Dawn, Earthquake, Once is Not Enough, Nasty Habits, A Dream of Passion.

MEREDITH, BURGESS: Actor b. Cleveland, Ohio, Nov. 16, 1909; e. Amherst Coll., M.A. (hon.); m. Kaja Sundsten. Capt. U.S. Army Air Corps, World War II. On stage 1929 Civic Repertory Co., N.Y.

STAGE PLAYS: Little O ' Boy, She Loves Me Not, The Star Wagon, Winterset, High Tor, Remarkable Mr. Pennypacker, etc.

PICTURES INCLUDE: Began screen career 1936 in Winterset. Idiot's Delight, Of Mice and Men, Second Chorus, That Uncertain Feeling, Tom, Dick and Harry, Street of Chance, Miracles Can Happen, Story of G.I. Joe, Diary of a Chambermaid, Magnificent Doll, Mine Own Executioner, Man on the Eiffel Tower, Gay Adventure, Joe Butterfly, Advise and Consent, Hurry Sundown, Fortune Garden, Stay Away Joe, McKenna's Gold, Hard Contract, There Was a Crooked Man, The Clay Pigeon, Such Good Friends, Golden Needles, The Day of the Locust, 92 in the Shade, The Hindenburg, Burnt Offerings, Rocky, The Sentinel, The Manitou, Foul Play, Magic, The Great Georgia Bank Hoax, Golden Rendezvous, Rocky II, When Time Ran Out, Clash of the Titans, True Confessions, Rocky III.

MERKE, UNA: Actress. b. Covington, Ky., Dec. 10, 1903.

PLAYS INCLUDE: Two by Two. Poor Nut, Pigs, Gossipy Sex, Coquette, Salt Water, Ponder Heart, Take Me Along.

PICTURES INCLUDE: Eyes of the World, Abraham Lincoln, Command Performance, Daddy Long Legs, Private Lives, Reunion in Vienna. Bombshell, Day of Reckoning, Merry Widow, Born to Dance, Saratoga, Road to Zanzibar, Twin Beds, This Is the Army, It's a Joke Son, Bride Goes Wild, Emergency Wedding, My Blue Heaven, Rich, Young and Pretty, Golden Girl, Millionaire for Christy, With A Song in My Heart, I Love Melvin, The Kentuckian, Bundle of Joy, The Mating Game, Summer and Smoke, Summer Magic, A Tiger Walks.

TV: Studio 1, Kraft Theatre, Playhouse 90, Climax; I Spy.

MERMAN, ETHEL: Actress. r.n. Ethel Zimmerman; b. Jan. 16, 1909, Astoria, N.Y.; e. William Cullen Bryant H.S. Astoria. Started as secretary, then nightclubs, vaudeville.

THEATRE: Girl Crazy, George White's Scandals, Take a Chance, Red Hot and Blue, Stars in Your Eyes, Du Barry Was a Lady, Panama Hattie, Something for the Boys, Annie Get Your Gun, Call Me Madam, Happy Hunting, Hello Dolly!

PICTURES INCLUDE: We're Not Dressing, Big Broadcast of 1936, Anything Goes, Alexander's Ragtime Band, Call Me Madam, There's No Business Like Show Business, It's a Mad, Mad, Mad, Mad World.

MERRALL, MARY: Actress. b. Liverpool, England; e. convents in Belgium and London; m. Franklyn Dyall. First stage appearance in 1907. Managed Abbey Theatre, Dublin, in 1918 with Franklyn Dyall and produced several plays. Has toured Canada, U.S. and South Africa; many plays since. On screen in 1939–40; Dr. O'Dowd & since in many British films.

PICTURES INCLUDE: Squadron Leader X, Dead of Night, Pink String and Sealing Wax, This Man Is Mine, Nicholas Nickleby, They Made Me a Fugitive, Three Weird Sisters, Badgers Green, For Them That Trespass, Trio, The Late Edwina Black, Out of True, Tale of Five Cities, Encore, Meet Me Tonight, Pickwick Papers, Weak and the Wicked, Duel in the Jungle, Belles of St. Trinians, Pay-Roll Robbery, Green Buddha. Its Great To Be Young, Campbell's Kingdom, Family Doctor, Camp on Blood Island, Spare the Rod, Everything I Have, Moll Flanders, Futtocks End.

TV: The Guardians, Idle at Work, Cider with Rosie, I can See it All, His Lordship Entertains, Kate, Justice, Hunter's Walk, Willy.

MERRICK, DAVID: Producer. b. Hong Kong, 1911. Famed Broadway stage impressario with long record of hits, including Fanny, The Matchmaker, Hello, Dolly!, Look Back in Anger, The Entertainer, Jamaica, World of Suzie Wong, Gypsy, Take Me Along, Becket, Sunday in New York, Oliver!, I Do! I Do!, Private Lives, etc. Recent: 42nd Street.

PICTURES: Child's Play (debut), The Great Gatsby, Semi-Tough, Rough Cut.

MERRILL, DINA: Actress. r.n. Nedinia Hutton; b. N.Y.C. Dec. 9, 1925. e. George Washington Univ., 1940–41; American Academy of Dramatic Arts, 1942–44. Fashion model, 1941–42. Acting debut: Here Today Mrs. January and Mr. X, Newport 1942.

PLAYS INCLUDE: My Sister Eileen, Major Barbara, Misalliance.

PICTURES INCLUDE: Debut. Desk Set, Catch Me If You Can, Operation Petticoat, Butterfield 8, Young Savages, I'll Take Sweden, The Greatest, A Wedding, Just Tell Me What You Want.

TV: debut, Kate Smith Show 1956; Four Star Theatre, Playwrights '56, Climax!, Playhouse 90, Westinghouse Presents, The Investigators, Checkmate, The Rogues, Bob Hope Presents, To Tell the Truth.

MERRILL, GARY: Actor. b. Hartford, Conn.; e. Loomis Prep. School, Bowdoin Coll., Trinity Coll. Stage career started in 1937, minor role, stage play The Eternal Road; toured, Brother Rat co.; then, Morning Star, See My Lawyer; on air in Young Dr. Malone, Helen Hayes Theatre, Theatre Guild, Gangbusters, Superman; army service 1941–45; upon disch. to stage in Born Yesterday, At War With the Army. Screen debut: Slattery's Hurricane, supporting role, then Twelve O'Clock High.

PICTURES INCLUDE: Where the Sidewalk Ends, All About Eve, The Frogmen, Decision Before Dawn, Another Man's Poison, Phone Call From a Stranger, Girl in White, Night Without Sleep, Blueprint for Murder, Black Dakotas, Human Jungle, Pleasure of His Company, The Woman Who Wouldn't Die, Clambake, The Incident, The Last Challenge, The Power, Huckleberry Finn, Thieves.

TV: The Mask, Justice, Dr. Kildare.

MERSON, MARC: Producer. b. New York City, Sept. 9, 1931. e. Swarthmore. Entered Navy in 1953; assigned as publicist to Admiral's Staff of Sixth Fleet Command in the Mediterranean. Upon discharge joined trade paper Show Business as feature editor. Joined CBS-TV as asst. to casting director. Left after 3 yrs. to work for Ely Landau as casting dir., packager and sometime producer of The Play of the Week on TV. Returned to CBS for 3-yr. stint doing specials and live programs. Left to organize Brownstone Productions as indep. prod. Now partner with Alan Alda in Helix Productions to package and produce TV shows.

PICTURES INCLUDE: The Heart Is a Lonely Hunter, People Soup (short), Leadbelly.

TV: Stage 67, Androcles and the Lion, Dummler & Son (pilot), The David Frost Revue (synd. series), We'll Get By.

MESMER, MARIE: Drama, film reviewer. b. Newark, N.J. Feb. 14, 1920; e. Syracuse Univ., Brooklyn (N.Y.) Inst. of Arts & Sciences; Radio adv. Crowell Pub. Co., 2 yrs.; danced with Ted Lewis and orchestra on tour; exhibiting artist in art galleries throughout U.S. for number years; currently drama reviewer Los Angeles (Calif.) Daily News, specializing in covering jazz music & personalities. Member: Los Angeles Press Club.

MESSENGER, GARY L.: Executive, b. Fresno, Calif., Jan. 14, 1941. e. University of Southern California-Cinema. 1967, joined MCA, Inc., asst. to exec. v.p.; 1970, independent motion picture & television producer; 1976, v.p. director production services, The Burbank Studios. Member of USC Cinema Circulus, Producers Guild of America.

METZLER, ROBERT: Writer. b. Buffalo, N.Y., Dec. 26, 1914; e. U. of S. California. Joined MGM writing staff 1939. In 1941: collaborated on screenplay Riders of the Purple Sage, 20th-Fox. In 1942: Sundown Jim, Dr. Renault's Secret; (s.p.) Circumstantial Evidence; (adapt.) The Undercover Woman.

MEYER, RICHARD C.: Producer, exec. film editor, screen writer, b. Frankfurt, Ger., 1920. e. Herne Bay Coll., Eng.; UCLA; cameraman, George Stevens' Special Motion Picture Coverage Unit; film ed., re-recording spvr., Louis de Rochemont; film dir., Prockter Prods.; co-org., Robert Lawrence Prods., free lance film ed., TV & m.p.; v.p., International Operation, Ass. Prod. Dino de Laurentiis.
PICTURES INCLUDE: Lost Boundaries, Seven Angry Men, King Oedipus, Crime in the Streets, The Wild Party, Hot Rod Rumble, Pyro, The Castilian, Young Jesse James, Return of the Fly, Jazz Party, Forbidden Dreams, The Babysitter, Alaska Passage, Men In War, God's Little Acre, Anna Lucasta, Spartacus, Happy Anniversary, Three in the Attic, Winning, Waterloo, Capone, Butch Cassidy and the Sundance Kid.
TV: Jack Benny Show, The Big Story, Grand Old Opry, Dear Phoebe, Farmer's Daughter, Wackiest Ship in the Army, Danny Thomas Anthology, The Iron Horse, The Outcasts, The Bold Ones, Police Story, King.

MEYER, RUSS: Producer, director. b. Oakland, Calif.; in 1942 joined Army Signal Corps, learned m.p. photography and shot combat newsreels.
PICTURES INCLUDE: The Immoral Mr. Teas, Eve and the Handyman, Finders Keepers, Lovers Weepers, Goodmorning and Goodbye, Vixen, Beyond the Valley of The Dolls, The Seven Minutes, and Sweet Suzy.

MEYERS, ROBERT: Executive. b. Mount Vernon, N.Y., Oct. 3, 1934. e. N.Y.U. Entered m.p. industry as exec. trainee in domestic div. of Columbia Pictures, 1956. Sales and adv., 1956–60; transferred to sales dep't. Columbia Pictures International, N.Y., posts there included supervisor of international roadshows and exec. ass't. to continental mgr. Joined National General Pictures as v.p.-foreign sales, 1969. Created JAD Films International Inc. in Feb. 1974 for independent selling and packaging of Motion Pictures around the world. September, 1977, joined Lorimar Productions Inc. as senior vice pres. of Lorimar Distribution International. Became pres. in April, 1978. Joined Filmways Pictures in 1980, named pres. & co-o. Is pres. of American Film Mktg. Assn.

MICHAELI, JOHN EDWARD: Executive. b. Los Angeles, Calif., Nov. 20, 1938; e. Mt. Carmel H.S., Sta. Monica City Coll., S. California State University, B.A., 1965; M.A. 1977. Merch. buyer & dir. of Adv., Maymart div. of May Co.; police beat reporter, KGIL; pub. ed. Bunker-Ramo Corp.; Off. boy then publicist, MGM, 1965; v.p. communications, Hanna-Barbera Prods.

MICHAUD, HENRI: Executive. b. Ismailia, Egypt, Sept. 24, 1912. e. H.E.C. (Commercial Acdmy.), Paris, 1932. Warner Bros., Paris, 1934, 1944. Paramount Internat'l Films, 1944–68; pres. Para. Intern'l; co-chmn., Cinema Int'l Corp., July, 1970. Co-chairman A.M. Film Consultants Ltd. 1977.

MICHAUD, HENRY A.: Executive. b. Gastonia, N.C., Mar. 19, 1914. e. Loyola Coll., Tech. School (Science) 1932–37. Amateur photographer. Cameraman on animation and title stand, Associated Screen News, 1939; cameraman, Naval and Air Force training films, industrial and tourist films, 1940–47. Partner, founded, Phoenix Studios, 1947 which later was reorganized as Omega Prods., 1951; in chg. technical services, multiple camera shooting methods, rear-projection techniques, etc., using 16mm negative for TV shows, 1952; dir. on numerous industrial and government films. Pres., Assoc. of Motion Picture Producers and Labs. of Canada, 1961. v.p. Omega Prod. of Montreal.

MICHEL, WERNER: Producer. b. Detmold, Germany March 5, 1910. e. U. of Berlin, U. of Paris, Ph.D. Radio writer, dir., co-author two Broadway revues, 1938, 1940; dir. French feature films; dir. Broadcast Div., Voice of America, 1942–46; prod., dir., CBS, 1946–48; asst. prog. dir., CBS, 1948–50; dir. of Kenyon and Eckhart TV dept., 1950–52; prod., DuMont TV network, 1952–55; dir., Electronicam TV-Film Prod., 1955–56; prod., Benton and Bowles; Procter and Gamble, 1956–57; v.p. & dir. TV-radio dept., Reach, McClinton Advertising, Inc., 1957–62; consultant, TV Programming & Comm'l-Prod., N. W. Ayer & Son, Inc.; v.p., dir., TV dept., SSC&B Advertising, 1963, pgm. exec., ABC-TV, Hollywood, 1975; director, dramatic programs, 1976; sr. v.p., creative affairs, MGM-TV, 1977; exec. v.p., Wrather Entertainment Intl., 1979; sr. v.p., programs, MGM-TV, 1980.

MICHELET, MICHEL: Composer. b. Kieff, Russia. Prof., Kieff & Vienna Conserv.; composed concert compositions, ballets, stage music; ent. m.p. industry, composed scores 105 films, France; to U.S. 1941; author many concert compositions.
PICTURES INCLUDE: Voice in the Wind (AA nom.), Hairy Ape (AA nom.), Music for Millions, The Chase, Lured, Siren of Atlantis, Man on the Eiffel Tower, Once a Thief, Tarzan's Peril, Fort Algiers, Un Missionaire, Le Secret de Soeur Angele, Petersburger, Nachte, Challenge (Tribute to Modern Art). Also did scores for many U.S. Information Service documentaries; arr. of Russian music Anastasia; Afrodife (score); The January (orig. songs & arr. of Russian music). Member: French Soc. of Composers (SACEM).

MICHELL, KEITH: Actor. b. Adelaide, Australia. Early career as art teacher, radio actor; toured Australia with Stratford Shakespearean Co. 1952–53; Stratford Memorial Theatre 1954–55, Old Vic Theatre 1956–7. Irma la Douce, Chichester, Art of Seduction, The First 400 Years, Robert & Elizabeth, Kain, The King's Mare, 1969: Don Quijote in Man of La Mancha, London, N.Y. 1970: Abelard & Heloise, London; 1971, NY and LA, 1972, Hamlet, London, in October 1973 apptd. Artistic Director, Chichester Festival Theatre.
TV: Pygmalion, Act of Violence, Mayerling Affair, Tiger At The Gates, Traveller Without Luggage, Guardian Angel, Wuthering Heights, The Bergonzi Hands, Ring Round The Moon, Spread of the Eagle, The Shifting Heart, Loyalties, Soldier in Love, Hallmark Hall of Fame, series; Kain, The Ideal Husband, The Six Wives of Henry VIII (series). Keith Michell at various London theatres, Dear Love.
PICTURES INCLUDE: True As A Turtle, Dangerous Exile, Gypsy and the Gentleman, The Hellfire Club, All Night Long, Seven Seas to Calais, Prudence and the Pill, House of Cards, Henry VIII and his Six Wives, Moments.

MIDDLETON, RAY: Actor. b. Chicago, Ill.; e. Illinois (music); Julliard School of Music, N.Y.C. With Detroit Civic Opera House; started in play, Roberta; in Chicago Opera Co. 1936; to Hollywood.
PICTURES INCLUDE: Knickerbocker Holiday, George White's Scandal's, American Jubilee, Lady for a Night, Girl from Alaska, I Dream of Jeannie, Sweethearts on Parade, Jubilee Trail, I Cover the Underworld, Road to Denver, 1776.
STAGE: Annie Get Your Gun, South Pacific, Man of La Mancha.

MIDLER, BETTE: Actress-Singer. b. Honolulu, 1945. e. Univ. of Hawaii. Appeared on Bdwy. in Fiddler on the Roof; Salvation, 1970; Tommy, Seattle Opera Co., 1971. Gained fame as singer-comic in nightclubs and cabarets. Has toured extensively with own stage shows: Divine Miss M, Clams on the Half-Shell, etc.
PICTURES: The Rose, Divine Madness, Jinxed.

MIFUNE, TOSHIRO: Actor. b. Tsingtao, China, April 1, 1920. e. Japanese schools. Served five years Japanese army. Joined Toho Studio 1946.
PICTURES INCLUDE: Snow Trail, Drunken Angel, Eagle of Pacific, Seven Samurai, I Live in Fear, Legend of Musashi, Throne of Blood, Rikisha Man, Three Treasures, Last Gunfight, I Bombed Pearl Harbor, Rose in Mud, Rashomon, Yojimbo, Animus Trujano (Mexican), Kiska, Red Beard, High and Low, Judo Saga, The Lost World of Sinbad, Hell in the Pacific, Paper Tiger, Midway, Winter Kills, 1941, Shogun.

MIDGEN, CHESTER L.: Executive. b. New York, N.Y., May 21, 1921; e. CCNY, B.A., 1941, Columbia U. LLB 1947; Member New York Bar. Attorney for National Labor Relations Board 1947–51. Currently National Executive Secretary, Screen Actors Guild; v.p., California Labor Federation, AFL-CIO. Vice Pres., Associated Actors and Artists of America, AFL-CIO. Vice pres., Hollywood Film Council. Executive committee Motion Picture and Television Fund. Trustee, John L. Dales Scholarship Fund. Trustee, Producers-Screen Actors Guild Pension and Welfare Funds. Board Member, Partnership for the Arts. Vice-chairman, State of California Motion Picture Development Council.

MIKELL, GEORGE: Actor. b. Lithuania. In Australia 1950–6 acting with Old Vic Co. Ent. films 1955. TV 1957. To England 1957; since appeared in numerous film and TV prod.
TV: Counsel at Law, Six Eyes on a Stranger, The Mask of a Clown, Green Grows the Grass, Opportunity Taken, OSS Series, Espionage, The Danger Man, Strange Report,

The Survivors, The Adventurer, Colditz, The Hanged Man, Quiller, Martin Hartwell, Flambards, Sweeney, The Secret Army, Sherlock Holmes.
PICTURES INCLUDE: The Guns of Navarone, The Password Is Courage, The Great Escape, Deadline for Diamonds, Where The Spies Are, The Spy Who Came in From the Cold, I Predoni Del Sahara, Sabina, The Double Man, Attack on the Iron Coast, Zeppelin, Young Winston, Scorpio, The Tamarind Seed, Sweeney Two, The Sea Wolves, Escape to Victory.
STAGE: Five Finger Exercise, Altona, The Millionairess, Love from a Stranger, Portrait of A Queen.

MILES, LORD: C.B.E. Actor, dir., writer. b. Hillingdon, Middlesex, Sept. 27, 1907; e. Pembroke Coll., Oxford. Brit. stage debut 1930 in Richard III, London. Brit. screen debut in Channel Crossing. Awarded C.B.E., 1953; founded Mermaid Theatre, 1959.
PICTURES INCLUDE: Pastor Hall, The Big Blockade, This Was Paris; (collab. s.p.) Thunder Rock; The Day Will Dawn, One of Our Aircraft Is Missing, In Which We Serve; (author, co-dir., actor) Tawney Pipit; Carnival, Great Expectations, Fame is the Spur, Nicholas Nickleby, Guinea Pig, Never Let Me Go, Moby Dick, Man Who Knew Too Much, Zarak, Tiger In the Smoke, Fortune Is a Woman, Smallest Show on Earth, St. Joan, Tom Thumb, Sapphire, Heavens Above, Run Wild, Run Free, The Specialist.
TV: Treasure Island, etc.

MILES, SARAH: Actress. b. 1941; e. R.A.D.A.; film debut, Term of Trial.
PICTURES INCLUDE: The Servant, The Ceremony, Six-Sided Triangle, Those Magnificent Men In Their Flying Machines, Blowup, Ryan's Daughter, Lady Caroline Lamb, The Man Who Loved Cat Dancing, The Sailor Who Fell from Grace with the Sea, The Big Sleep.
THEATRE: Vivat! Vivat Regina!

MILES, VERA: Actress. b. Boise City, Okla., Aug. 23, 1930. e. public schools, Pratt and Wichita, Kans.
TV: Climax, Pepsi Cola Playhouse, Schlitz Playhouse, Ford Theatre.
PICTURES INCLUDE: For Men Only, Rose Bowl Story, Charge at Feather River, Pride of the Blue Grass, Wichita, The Searchers, 23 Paces to Baker Street, Autumn Leaves, Wrong Man, Beau James, Web of Evidence, FBI Story, Touch of Larceny, Five Branded Women, Psycho, The Spirit Is Willing, Gentle Giant, Sergeant Ryker, Kona Coast, It Takes All Kinds, Hellfighters, The Wild Country, One Little Indian.

MILFORD, JOHN: Actor.
TV: Delvecchio; Code R; Little House On The Prairie; ABC Circle Short Story Theatre; Hip; $6 Million Man; Bionic Woman; Police Story; Streets Of San Francisco; Gemini Man; The Hunter; Faith for Today; Barnaby Jones; Blue Knight; Policewoman; Cannon; Joe Forrester; Medical Story; Men In Green; The Name Of The Game; Gold Ones; Mannix; The Magician; Twelve O'Clock High; The Invaders; Firehouse; The Sound Of Anger; The Sixth Sense; The Plainsman; Tombstone Territory; The Lieutenants; Columbo; The FBI; Petrocelli; Longstreet; Mod Squad; Dan August; King Of Diamonds; Men In Space; High Chaparral; Family Affair; Torra; Ironside; Rifleman; Jesse James; Wyatt Earp; Gunsmoke; The Virginian; Big Valley; Bonanza; Hawkins; Guns Of Will Sonnett; Robert Taylor Detectives.
PICTURES INCLUDE: Brannigan's Rebels; The Ten Commandments; For Pete's Sake; Heart Of The Rebel; Marty; Gunfight At Commanche Creek.
THEATRE: What Every Woman Knows; Pygmalion; The Prime Of Miss Jean Brodie Happy Birthday, Wanda June; Silent Night, Lonely Night; Child's Play; Diary Of A Scoundrel; Miss Julie; Apollo Of Bellac; Do You Know The Milky Way; The Vigil; Portrait Of A Madonna; State Of The Union; Night Must Fall; The Long Voyage Home; Murder In The Cathedral; The Male Animal; The Admirable Crichton; Aria De Capo; Nothing But The Truth; The Happy Prince; The Purging of Simon Madden Peterson House; Where Is Thy Victory; Harry; Noon And Night; Room Service; The American Nightmare.

MILGRAM, HENRY: Theatre Executive. b. Philadelphia, April 20, 1926. e. Univ. of Pa., Wharton School. In industry 35 years; now exec. v.p. Milgram Theatres. Variety Club Board Member for past 15 years, past president and chairman of the Board of Variety Club of Phila.; presently Variety Club Intl. Ambassador. Board member: Moss Rehabilitation Hospital, Likoff Cardiovascular Institute, Hahnemann Medical College & Hospital of Phila.

MILIUS, JOHN: Writer-Director. b. 1945. e. L.A. City College, USC (cinema course). While at latter won National Student Film Festival Award. Started career as ass't. to Lawrence Gordon at AIP. Began writing screenplays, then became director with Dillinger.
PICTURES INCLUDE: Devil's 8, Evil Knievel, The Life and Times of Judge Roy Bean, Jeremiah Johnson (co-s.p.), Dillinger (s.p.-dir.), Magnum Force (o.-s.p.), The Wind and the Lion (s.pm., dir.), Big Wednesday (dir. co.-s.p.); Apocalypse Now (s.p.); 1941 (exec. prod.-co. s.p.)..

MILLAND, RAY: Actor. r.n. Reginald Truscott-Jones; b. Neath, Wales, Jan. 3, 1908; e. King's Col.
PICTURES INCLUDE: Ambassador Bill, Payment Deferred, Bolero, One Hour Late, Glass Key, Big Broadcast of 1937, Three Smart Girls, Easy Living, Ebb Tide, Wise Girl, Beau Geste, French Without Tears, Irene, Doctor Takes a Wife, Lady Has Plans, Reap the Wild Wind, Are Husbands Necessary, Major and the Minor, Crystal Ball, Forever and a Day, Lady in the Dark, Kitty, Acad. Award best perf. ("Lost Weekend") 1945; Well-Groomed Lady, California, Imperfect Lady, Trouble With Women, Golden Earrings, Big Clock, Wings Over Honolulu, So Evil My Love, Sealed Verdict, Alias Nick Beal, It Happens Every Spring, Life of Her Own, Copper Canyon, Woman of Distinction, Circle of Danger, Night Into Morning, Rhubarb, Close to My Heart, Something to Live For, Bugles in the Afternoon, The Thief, Jamaica Run, Let's Do It Again, Dial M for Murder, Girl on the Red Velvet Swing; act. A Man Alone, Lisbon: act., Three Brave Men, River's Edge, Premature Burial, Panic in Year Zero, Man with the X-Ray Eyes, Love Story, Frogs, The House in Nightmare Park, Gold, Escape to Witch Mountain, The Last Tycoon, Battlestar Galactica, Oliver's Story.
TV: Rich Man, Poor Man, Seventh Avenue, Testimony of Two Men.

MILLAR, STUART: Producer-Director. b. New York, N.Y., 1929. e. Stanford; Sorbonne, Paris. Ent. industry working for Motion Picture Branch, State Dept., Germany. documentaries, Army Signal Corps, Long Island, Germany; journalist, International News Service, San Francisco; assoc. prod.-dir., The Desperate Hours; assoc. prod.-dir., Friendly Persuasion.
PICTURES INCLUDE: The Young Stranger; Stage Struck; Birdman of Alcatraz. I Could Go On Singing, The Young Doctors, Stolen Hours, The Best Man, Paper Lion, Little Big Man, When The Legends Die, Rooster Cogburn.

MILLER, ANN: Actress. r.n. Lucille Ann Collier; b. Houston, Tex., Apr. 12, 1923; e. Albert Sidney Johson H.S., Houston; Lawler Prof., School, Hollywood. Studied dance as child; played West Coast vaudeville theatres, Screen debut: New Faces of 1937.
STAGE: George White's Scandals, 1940, Mame, 1969.
PICTURES INCLUDE: Life of the Party, Stage Door, New Faces of 1937, Radio City Revels, Having a Wonderful Time, Room Service, You Can't Take It With You, Too Many Girls, Time Out for Rhythm, Priorities on Parade, Reveille With Beverly, Jam Session, Eve Knew Her Apples, Thrill of Brazil, Easter Parade, The Kissing Bandit, On the Town, Watch the Birdie, Texas Carnival, Two Tickets to Broadway, Lovely To Look At, Small Town Girl, Kiss Me Kate, Deep in My Heart, Hit the Deck, Opposite Sex, Great American Pastime.

MILLER, BEVERLY: Distributor. b. Huntsdale, Mo., April 8, 1906. Entered m.p. ind. in Mo. as exhib., 1925; salesman, National Screen Service, in midwest 1926–36; opened own bus., 1936 Producers Releasing Co., 1941; sold to Pathe Industries, 1945; branch mgr., Producers Releasing Co.; dist. mgr., Eagle Lion to 1948. Operated drive-ins 1949 to 1968, sold out to Mid-America Theatres, St. Louis, Missouri. exec. pres. United M.P. Assoc.; Pres. Mercury Film Co., Inc. Also v.p. Four Associates Ltd. Cinema Project Number One, producing movies in the Philippines.

MILLER, CHERYL: Actress. b. Sherman Oaks, Calif., Feb. 4, 1943. e. ULCA, Los Angeles Conservatory of Music.
PICTURES INCLUDE: First film, Casanova Brown, age 19 days. Appeared in over 100 films as child, more recently in The Monkey's Uncle, Clarence the Cross-Eyed Lion, The Initiation, The Man From Clover Grove, Doctor Death.
TV: Perry Mason Bachelor Father, Flipper and as co-star in Daktari, Donna Reed, Leave it to Beaver, Farmers Daughter, Wonderful World of Color, Dobie Gillis, Bright Promise, Love American Style, Emergency, Cades County.

MILLER, DAVID: Director. b. Paterson, N.J., Nov. 28, 1909, U.S. Army, World War II. Film ed. 1930, Columbia; Walter Futter prods. In 1933 short subjects ed. MGM; then dir. short subjects. Dir. features 1941.
PICTURES INCLUDE: Billy the Kid, Sunday Punch, Flying Tigers, Love Happy, Top O The Morning, Our Very Own, Saturday's Hero, Sudden Fear, Twist of Fate, Diane, Opposite Sex, The Story of Esther Costello, Midnight Lace, Backstreet, Lonely are the Brave, Captain Newman, MD, Hammerhead, Hail Hero, Executive Action, Bittersweet Love, Best Place To Be, Goldie and The Boxer, Love for Rent.

MILLER, DICK (RICHARD): Actor/Writer. b. N.Y.C., Dec. 25, 1928. e. CCNY, Columbia Univ. Grd. NYU, Theater School of Dramatic Arts. Commercial artist, Psychologist; Bellevue Mental Hygiene Clinic, Queens General Hospital Psychiatric dept. U.S. Navy, WWII. Boxing champ, U.S. Navy.

Semi-pro football. Broadway stage, radio; Disc jockey, The Dick Miller Show, WMCA, WOR-TV Over 500 live shows. Did first live night talk show, Midnight Set, CBS Wrote, produced and directed many early radio and TV shows. Wrote Screenplays; T.N.T. Jackson, Which Way To The Front, Four Rode Out and others.

PICTURES INCLUDE: Has appeared in over 65 features, including: Not of This Earth, Thunder Over Hawaii, Rock All Night, Sorority Girl, The Terror, Bucket Of Blood, Little Shop Of Horrors, Targets, War Of The Satelittes, The Long Ride Home, St. Valentine Day Massacre, Capone, Executive Action, White Line Fever, Cannonball, Mr. Billion, N.Y., N.Y.

MILLER, JASON: Writer-Actor. Entered regional playwriting contest during high school in Scranton, Penna. and since has moved back and forth between acting and writing. Wrote That Championship Season, winner of N.Y. Drama Critics Best Play award, 1972, Tony Award, 1973, and Pulitzer Prize for Drama.

PICTURES INCLUDE: The Exorcist (actor), That Championship Season (s.p.), The Nickel Ride (actor), A Home of our Own (actor), Fitzgerald in Hollywood (actor), A Love Story (s.p.).

MILLER, J.P.: Writer. b. San Antonio, Tex., Dec. 18, 1919. e. Rice Univ., 1937–41, Yale Drama School, 1946–47. U.S. Navy, Lieut., 1941–46; pub. poetry, short stories.

ORIGINAL DRAMAS INCLUDE: Philco TV Playhouse: Hide and Seek, Old Tasslefoot, The Rabbit Trap, The Pardon-me Boy; Playhouse 90, Days of Wine and Roses, CBS Playhouse, The People Next Door, The Unwanted, The Lindbergh Kidnaping Case, Gauguin The Savage.

PICTURES INCLUDE: The Rabbit Trap, (story, s.p.) Days of Wine and Roses (story, s.p.) The Young Savages, (co-author, s.p.) Behold A Pale Horse, (s.p.) The People Next Door (story, s.p.), Helter Skelter (s.p.)

NOVELS: The Race for Home, Liv.

MILLER, JAMES P.: Executive. Began m.p./industry career in 1971 in legal dept. of United Artists (N.Y.). Left to go with Paramount Pictures in legal dept.; then moved to Columbia in 1977 as snr. counsel; later assoc. gen. counsel. In 1979 named Warner Bros. v.p.—studio business affairs.

MILLER, MARVIN: Actor, narrator, writer. b. St. Louis, Mo., July 18, 1913. e. Washington U., A.B., 1930–34. Radio stations KWK, WIL, KMOX, St. Louis 1931–39; since free lance; announcer & actor on numerous major network radio shows, Chicago & Los Angeles. Member: Mason, Scottish Rite; Legion of Honor Order of DeMolay; Shrine; poetry groups; Motion Picture, TV and Recording Academies, Pacific Pioneer Broadcasters.

PICTURES INCLUDE: Johnny Angel, Intrigue, Off Limits, Forbidden, Shanghai Story, Peking Express, The Naked Ape, Where Does it Hurt?, I Wonder Who's Killing Her Now, Prime Time, Kiss Daddy Goodbye.

Narrated many award-winning industrial, educational, military films. UPA cartoons, Disney, DePatie-Freleng Filmation, Hanna-Barbera. Actor or anncr. filmed and live TV.

TV: Star, Millionaire, CBS-TV; Narrator, FBI; Star, Aquaman, Fantastic Voyage; Narr., Electrawoman, Burrud Nature Specials, Bigfoot & Wildboy.

Recorded entire Talking Bible, other talking books, Audio Book; Two Grammy awards for Dr. Seuss Children's Records, RCA Victor. Synd. radio, Marvin Miller, Storyteller; Almanac.

MILLER, MAX B.: Executive. Father, Max Otto Miller, producer silent features and shorts. e. Los Angeles Valley Coll., UCLA, Sherwood Oaks Coll. Writer of articles on cinema for American Cinematographer and other publications. Owns and manages Fotos Intl., entertainment photo agency with offices in 46 countries. Recipient of Golden Globe Award in 1976 for Youthquake, documentary feature. Chm. of bd. of Hollywood Foreign Press Ass'n. since 1976.

MILLER, NAT: Executive. b. London, 1909. Joined Granada Theatres 1924; appointed chief booker. Own organization, ORB Productions Ltd., 1954; man. dir. Animated Motion Pics., Ltd.

PICTURES INCLUDE: Secret Tent and Grove Family Film series, Nudist Paradise, Death Over My Shoulder, That Kind of Girl, Yellow Teddybears.

TV: TV film series. UK Film Distribution.

MILLER, ROBERT ELLIS: Director. Worked on Broadway and TV before feature film debut with Any Wednesday (1966).

PICTURES INCLUDE: Sweet November, The Heart Is a Lonely Hunter, The Buttercup Chain, Big Truck and Poor Claire, The Girl from Petrovka, The Baltimore Bullet.

TV: The Voice of Charlie Pont, And James Was a Very Small Snail.

MILLER, RONALD W.: Producer. b. April 17, 1933, Los Angeles, e. U. of Southern Calif. Football player with Los Angeles Rams. Two years U.S. Army. 1957 joined Walt Disney Productions as 2nd asst. dir. Old Yeller. Assoc. prod. TV series Walt Disney Presents; assoc. or co-prod. additional 37 episodes Disney TV. Exec. prod. Walt Disney's Wonderful World of Color. 1980, named pres., Disney Productions.

PICTURES INCLUDE: Bon Voyage, Summer Magic, Son of Flubber, Moon Pilot, The Misadventures of Merlin Jones, A Tiger Walks, The Monkey's Uncle, That Darn Cat, Robin Crusoe, U.S.N., Monkey's Go Home, Prod. of: Never A Dull Moment, The Boatniks, The Wild Country, No Deposit, No Return, Treasure of Matecumbe, Freaky Friday, The Littlest Horse Thieves, Herbie Goes to Monte Carlo, Pete's Dragon, Candleshoe, Return from Witch Mountain, Cat from Outer Space, The North Avenue Irregulars, The Black Hole, Midnight Madness, Watcher in the Woods, The Last Flight of Noah's Ark.

Assisted Walt Disney, Pageant Direct 1960 U.S. Olympics. By 1968, v.p., exec. prod. mem. bd. of dir. Walt Disney Productions.

MILLER, WINSTON: Writer. b. St. Louis, Mo., June 22, 1910; e. Princeton U. In 1935 entered m.p. ind.

PICTURES INCLUDE: Good Morning, Judge, Song of Texas, Home in Indiana, One Body Too Many, They Made Me a Killer, Double Exposure, My Darling Clementine, Relentless, Station West, Last Outpost, Rocky Mountain, Hong Kong, Blazing Forest, Carson City, The Vanquished, Boy from Oklahoma, Jivaro, Bounty Hunter, Run for Cover, Far Horizons, Lucy Gallant, Tension at Table Rock, April Love, A Private's Affair, Hound Dog Man, Escapade In Japan, Mardi Gras.

MILLS, HAYLEY: Actress. b. London, Eng., April 18, 1946. e. Elmhurst Boarding School, Surrey. m.p. debut Tiger Bay. 1959; Pollyanna, 1960; signed Disney contract 1960.

PICTURES INCLUDE: The Parent Trap, Whistle Down the Wind, The Castaways, Summer Magic, The Chalk Garden, The Moonspinners, The Truth About Spring, Sky West and Crooked, Trouble With Angels, The Family Way, Pretty Polly, A Matter of Innocence, Twisted Nerve, Take a Girl Like You, Sillouettes, What Changed Charley Farthing, The Diamond Hunters.

MILLS, SIR JOHN: Actor, Producer. b. Suffolk, England, February 22, 1908, m. Mary Hayley Bell. Previously clerk. Film actor since 1933. In 1942: "The Young Mr. Pitt." "In Which We Serve," One of top ten money-making Brit. stars in Motion Picture Herald-Fame Poll, 1945, 1947, 1949–50, 1954, 1956–58, Oscar for Ryans Daughter.

PICTURES INCLUDE: We Dive at Dawn, This Happy Breed, Blue for Waterloo, Cottage To Let, Way to the Stars, Waterloo Road, Great Expectations, So Well Remembered, October Man, Scott of the Antarctic, Operation Disaster, Mr. Denning Drives North, Gentle Gunman, Long Memory, Hobson's Choice, End of the Affair, Colditz Story, Above Us the Waves, Escapade, It's Great to be Young, Around the World in 80 Days, War and Peace, Baby and the Battleship, Town on Trial, Monty's Double, Dunkirk, Summer of the 17th Doll, Tiger Bay, Swiss Family Robinson, Tunes of Glory, The Singer Not The Song, Flame In the Streets, Tiara Tahiti, The Valiant, The Chalk Garden, The Truth About Spring, The Great Spy Mission, King Rat, The Wrong Box, 1965; dir., Sky West and Crooked; acted, The Family Way, Cowboy in Africa, Chukka, A Black Veil For Lisa, Oh! What a Lovely War, Run Wild, Run Free, Ryan's Daughter, Adam's Woman, Dulcima, Oklahoma Crude, Young Winston, Lady Caroline, The Human Factor, Trial By Combat, Lamb, The Big Sleep.

STAGE: Good Companions, 1975; Great Expectations, 1976; Separate Tables, 1977.

MILNER, JACK: Exec. Prod. b. Jersey City, N.J., Nov. 2, 1910. e. Roosevelt H.S., L.A., Calif. M.P. industry 1927. Worked many phases from laboratory-camera-editorial dept. to financing and co-producing feature pictures; formed Milner Bros. Productions with brother Dan Milner, 1955; prod. Phantom from 10,000 Leagues, From Hell It Came, Jail Break, etc.

TV: prod. My Dog Sheppy, From Here to Now, Come As You Are.

MILNER, MARTIN: Actor. b. Detroit, Mich., Dec. 28. e. USC. m.p. debut in Life with Father, 1947; U.S. Army 1952–54, directed 20 training films.

PICTURES INCLUDE: Sands of Iwo Jima, The Halls of Montezuma, Operation Pacific, The Captive City, Battle Zone, Mr. Roberts, Pete Kelly's Blues, On the Threshold of Space, Gunfight at the O.K. Corral, Sweet Smell o' Success, Marjorie Morningstar, Too Much, Too Soon, Compulsion, 13 Ghosts, Valley of the Dolls.

TV: Route 66 (4 yrs.), Adam-12 (6 yrs), Swiss Family Robinson.

MILO, GEORGE: Set Decorator, r.n. George Milo Vescia. b. New York, N.Y., Dec. 19, 1909. e. High School, Art Schools; landscape, seascape, still life painter, portrait sculptor; set decorator since 1937.

TV: Republic Studios, Revue Studios; Dangerous Assignment, Stories of the Century, Thriller, Alfred Hitchcock, Checkmate, General Electric.

PICTURES INCLUDE: Wake of the Red Witch, Fair Wind

to Java, Borderline, Jubilee Trail, Make Haste to Live, Eternal Sea, Last Command, Come Next Spring, Psycho, Judgement at Nuremberg, The Last Cowboy, That Touch of Mink, The Birds.

MIMIEUX, YVETTE: Actress. b. Los Angeles, Calif., Jan. 8. e. Vine Street School, Le Conte Junior High, Los Angeles, Los Ninos Heroes de Chapultepec, Mexico City, Hollywood High School, Calif. Appeared with a theatrical group, Theatre Events; Sympn. Concert: Persphone, Oakland Orchestra, 1965, N.Y. Philharmonic, Linc. Ctr., L.A. Philharmonic, Hollywood Bowl.
PICTURES INCLUDE: debut, Time Machine, Where the Boys Are, The Four Horsemen of the Apocalypse, Light in the Piazza, The Wonderful World of the Brothers Grimm, Diamond Head, Toys In the Attic, Joy In The Morning, Reward, Monkeys Go Home Dark of the Sun, Caper of the Golden Bulls, Picasso Summer, Three in the Attic, Skyjacked, The Neptune Factor, Jackson County Jail, The Black Hole.
TV: Tyger Tyger, 1964. 1970: series, Most Deadly Game, Death Takes A Holiday, Black Noon.
STAGE: I Am A Camera, 1963; Owl and the Pussycat, 1966.
CONCERTS: Persephane—Houston Symphony, London Royal Philharmonic.

MINDLIN, MICHAEL, JR.: Executive. b. New York, N.Y. e. Columbia U., 1940–41; Duke U., 1941–43. U.S. Air Force; short subject pub., W.B., 1946–47; assoc., George & Dorothy Ross pub., 1947; pub., adv. dir., Ballet Russe de Monte Carlo, 1947–49; asst. to prod. Joseph Kaufman, 1950; with We the People, radio & TV show, 1950–51; Alfred Katz pub. & mgt., 1951–53; adv. & pub. dir., Lopert Films, 1953, pub., Hecht-Lancaster, 1955; adv., pub. dir., Figaro, Inc., 1956–58; assoc. prod. David Merrick Prod., 1958–59; formed prod. co. with Raoul Levy, 1959; v.p. chg. adv., publ., Filmways, Inc., 1963, v.p. prod., Filmways, Inc., 1968. Formed own prod. co. 1969, v.p., prod., Warren Bros., 1972–74; Paramount Pictures, Producer, 1974–75; Frank Yablans Enterprises, 1975.

MINER, WORTHINGTON C.: Theatrical producer. e. Buffalo, N.Y.; e. Kent School; Yale Univ.; Cambridge Univ., England. m. Frances Fuller, actress. Started theatrical career 1925; author and dir. RKO Radio Pictures, 1933–34; mem. of exec. bd. of Theatre Guild, 1937; entered TV, 1939; developed and produced TV shows.
TV: Studio One, Toast of the Town, the Goldbergs, Medic, Play of the Week; Ely Landau Associates. exec. prod., The Pawnbroker, The Fool Killer.
1964–68, pres. American Acad. of Dram. Arts.

MINNELLI, LIZA: Actress-singer. b. 1946; p. Judy Garland & Vincente Minnelli; After repertory work & off-broadway roles starred on Broadway in Flora, The Red Menace (Tony award).
PICTURES INCLUDE: Charlie Bubbles, The Sterile Cuckoo, Tell Me That You Love Me Junie Moon, Cabaret, That's Entertainment!, Lucky Lady, Silent Movie, A Matter of Time, New York, New York, Arthur.

MINNELLI, VINCENTE: Director. b. Chicago, Ill. Feb. 28. Toured with Minnelli Brothers Dramatic Tent Show as child; joined Balaban & Katz to assist in staging presentations; then to N.Y. Paramount in same work; stage Du Barry (N.Y.); art dir., Radio City Music Hall 3 yrs.; prod. At Home Abroad, Ziegfeld Follies, Very Warm for May (N.Y.). Screen debut, 1943.
PICTURES INCLUDE: Cabin in the Sky, I Dood It, Meet Me in St. Louis, The Clock, Ziegfeld Follies, Yolanda and the Thief; Undercurrent, The Pirate, Madame Bovary, Father of the Bride, Father's Little Dividend, An American in Paris, Bad & the Beautiful, Three Loves, The Bandwagon, Long Long Trailer, Brigadoon, The Cobweb, Kismet, Lust for Life, Tea and Sympathy, Designing Woman, Gigi, The Reluctant Debutante, Some Came Running, Home from the Hill, Bells Are Ringing, Four Horsemen of the Apocalypse, Two Weeks in Another Town, Courtship of Eddie's Father, Goodbye Charlie, The Sandpiper, On a Clear Day You Can See Forever, A Matter of Time.

MINSKY, HOWARD G.: Writer-Producer-Sales and Distribution Exec. Paramount-Twentieth Century-Fox. Agency Exec. Wm. Morris. Pres. Cinema Consultants. Produced Love Story. 1970.

MIRANDA, ISA: Actress. r.n. Ines Sampietro. b. Milan, Italy, July 5, 1917. m. Alfredo Guarini, producer. Typist; stage actress; made debut in 1934.
PICTURES INCLUDE: Darkness, Everybody's Lady, Red Passport, White Lady of Maharadscha, You Are My Happiness (German), Hotel Imperial & Adventure in Diamonds (U.S.), Malombra, Zaza, Mistake to Be Alive, Pact with the Devil, Walls of Malapaga, La Ronde, Seven Deadly Sins, Secret Conclave, Before the Deluge, Secret of Helen, Rasputin, We Women, Summertime, The Yellow Rolls Royce, The Great Train Robbery, The Shoes of the Fisherman.

MIRISCH, DAVID: Executive. b. Gettysburg, Pa., July 24, 1935. e. Ripon College. United Artists Corp., 1960–63; former exec. with Braverman-Mirisch adv. public rel. firm.

MIRISCH, MARVIN E.: Executive. b. New York, N.Y., March 19, 1918. e. College of the City of New York, B.A., 1940. Print dept., contract dept., asst. booker N.Y. exch.; head booker, Grand National Pictures, Inc., 1936–40; officer, gen. mgr., vending concession operation 800 theatres, Midwest, Theatres Candy Co., Inc., Milwaukee, Wisc., 1941–52; exec., corporate officer in chg., indep. producer negotiations, other management functions, Allied Artists Pictures, Inc., 1953–57; Chmn. of Bd., Chief Exec. Officer in chg. of all business affairs, admin. & financing, distr. liaison, The Mirisch Company, Inc., 1957 to present. Member of Board of Governors and currently vice-president of Academy of Motion Picture Arts & Sciences. Member Motion Picture Pioneers. Currently president of Academy of MPAA Foundation; vice chm., AMPTP.

MIRISCH, WALTER: Producer. b. New York City, Nov. 8, 1921; e. U. Wisconsin, B.A. 1942; Harvard Grad. Sch. Business Admin., 1943. In m.p. indust. with Skouras Theatres Corp., 1938–40; Oriental Theatre Corp. 1940–42. 1945 with Monogram, A.A.: apptd. exec. prod. Allied Artists, July, 1951; pres. and exec. head of prod. The Mirisch Corporation 1969; 1960–61; pres. Screen Prod. Guild; 1962, mem. bd. dir., MPAA; bd. Gvnrs., Academy of Motion Pictures Arts and Sciences, 1964; 1967, pres., Center Thea. Group of L.A.; bd. dir., Wisconsin Alum. Assn.; bd. of dir. Cedars-Sinai Medical Center, Bd. of Advisors, California State University—Northridge, Board of Governors, Academy of Motion Picture Arts & Sciences—1972. President, Permanent Charities Committee 1962–63; President, Academy of Motion Picture Arts & Sciences—1973–77.
PICTURES INCLUDE: By Love Possessed, Two for the Seesaw, Toys in the Attic, Hawaii, Fitzwilly, In the Heat of the Night, They Call Me Mister Tibbs, The Organization, Mr. Majestyk, Midway, Gray Lady Down, Same Time, Next Year, Prisoner of Zenda, Dracula.

MITCHELL, ANDREW: Producer. b. Giffnock, Scotland, 1925. e. Fettes College, Edinburgh; early career, banking; Associated British Picture Corporation, Elstree Studios; gen. mgr., Elstree Distributors, 1961 dir., Kenwood Films Ltd., 1963; assoc. prod., Hand in Hand, The Young Ones, Summer Holiday, French Dressing, Wonderful Life. 1965; Pro., Up Jumped a Swagman. Asst. to mgr. dir., Assoc. British Prods., Ltd. Prod. con. Associated British Productions Ltd. 1969–70: Film Finances, Mng. dir., Leslie Grade Film Ltd, 1973. Mng. Dir. E.M. . Elstree Studios. 1977 Prod. Are You Being Served?

MITCHELL, CAMERON: Actor. b. Dallastown, Pa., Nov. 4, 1918; e. Theatre School (New York City); New York Theatre Guild, 1938–40; on stage with Lunt & Fontanne, "Taming of the Shrew." Radio announcer, sportscaster before joining U.S. Army Air Forces 1942–44. Star of Tomorrow, 1954.
PICTURES INCLUDE: Mighty McGurk, High Barbaree, Cass Timberlane, Leather Gloves, The Sellout, Death of a Salesman, Japanese War Bride, Flight to Mars, Man in the Saddle, Outcasts of Poker Flat, Okinawa, Les Miserables, Pony Soldier, Powder River, Man on a Tightrope, How To Marry a Millionaire, Hell & High Water, Gorilla At Large, Garden of Evil, Desiree, Strange Lady in Town, Love Me Or Leave Me, House of Bamboo, Tall Men, View from Pompey's Head, Carousel, Monkey on My Back, Face of Fire, Inside The Mafia, The Unstoppable Man, The Last of the Vikings, Three Came to Kill, Blood and Black Lace, Ride in the Whirlwind, Hombre, Island of the Doomed, Nightmare in Wax, Buck and the Preacher, Slaughter, The Midnight Man, The Klansman, Viva, Knievel!, The Swarm.
TV: High Chapparel, Andersonville Trial, Swiss Family Robinson.

MITCHELL, GUY: Singer. actor. b. Detroit, Mich., Feb. 22, 1927 e. High Sch., San Francisco. Studied singing; U.S. Navy, 1945–46; singer, radio stations; vocalist with Carmen Cavallaro orch.; sings with Columbia records; m.p. debut in Those Redheads from Seattle, since in: Red Garters.
TV: Guy Mitchell TV Show, Whispering Smith Series NBC.

MITCHELL, JACK: Executive, b. Paris, Kentucky, Nov. 3, 1918; District mgr. for Schine Theatres; joined Wometco in 1963; now vice-pres. & General Manager for Wometco's Florida theatres and Director of Advertising & Promotion, Wometco Entertainment Division, Wometco Enterprises, Inc. Five-time winner, Box Office Showman of the Year Award. Member—Advertising Committee, NATO National; Board of Directors, NATO of Florida Member—A.D.S.; Serve on Action Committee—Greater Miami Chamber of Commerce; Member—Ad Club of Greater Miami; Chairman—NATO of Florida Advertising. National winner . . . 1st place . . . Film Day contest. Winner, Show-o-Rama Showman Award 1977. Elected pres. NATO of Fla., 1979; NATO nat'l. bd. of directors, 1980.

MITCHELL, JOHN H.: Executive. b. Rochester, N.Y., Apr. 27, 1918; e. University of Michigan. radio dept., Erwin Wasey & Co., 1939; mgr. sales service dept., Mutual Broadcasting System, 1939; U.S. Navy; WVET and Honel Corp., 1946; dir. of television, United Artists, 1948; v.p. & gen. mgr. UA television dept., to 1952; Screen Gems, Inc., 1952; on bd. dirs., 1961, and first v.p. Columbia Pictures Ind., Inc. 1969 Pres. Screen Gems Div. of Columbia Pictures Industries; 1973, pres. Columbia Pictures Television, Division of Columbia Pictures Industries, and executive vice president of Columbia Pictures Industries Inc. 1975, elected to board of directors of Columbia Pictures Industries; 1975, elected pres. of Hollywood Radio and Television Society. Resigned all offices-Columbia Pictures Industries 1977. Formed John H. Mitchell Co., Inc. 1977. Located in Los Angeles, television program and distribution consultants. First v.p.-Academy of Television Arts & Sciences, 1978–79; chmn., Academy of TV Arts & Sciences, forum luncheons, 1979–80; pres., Academy of TV Arts & Sciences, 1980–82.

MITCHELL, STEVE: Actor. b. New York, Dec. 15; e. Univ. of Wisconsin; grad. Feagin School of Drama, New York. U.S. Navy World War II and Korean War. Summer stock, Hilltop Md. Bway appearances Diamond Lil. Toured with Strike a Match. Screen debut in Walk East on Beacon St. 1952. M.P. and TV rep. State of Ark. (consultant).

PICTURES INCLUDE: Above and Beyond, Tender Hearts, Battle Circus, Fearless Feagin, Gypsy Colt, Gunsight Ridge, Big Combo, The Killing, Deep Six, Terror in a Texas Town, Seven Men From Now, China Doll, Submarine Sea Hawk, Girl in the Red Bikini, Most Dangerous Man Alive, Cause of Death, Once A Thief, Nevada Smith, Young Runaways.

MITCHUM, JIM: Actor. b. Los Angeles, May 8, 1941. m. actress Wendy Wagner. Son of Robert Mitchum. e. Univ. High School in L.A. Went directly from school to Hollywood Professional School. On-job prof. training at Barter Theatre in Virginia.

PICTURES INCLUDE: Thunder Road (debut), The Last Time I Saw Archie, The Victors, Ambush Bay, Tramplers, In Harm's Way, Invisible Six, Moonrunners, Beat Generation, Ride the Wild Surf, Trackdown.

MITCHUM, ROBERT: Actor. b. Bridgeport, Conn., Aug. 6, 1917. Odd jobs; to California; joined Long Beach Players Guild; appeared in Hopalong Cassidy series with William Boyd; in Westerns 8 yrs. RKO.

PICTURES INCLUDE: Story of G.I. Joe, Undercurrent, Pursued Locket, Til' the End of Time, Desire Me, Crossfire, Out of the Past, Rachel and the Stranger, Blood on the Moon, Red Pony, Big Steal, His Kind of Woman, Where Danger Lives, Her Forbidden Past, Macao, Racket, One Minute to Zero, Lost Men, Angel Face, White Witch Doctor, Second Chance, She Couldn't Say No, River of No Return, Track of the Cat, Night of the Hunter, Not as a Stranger, Man with the Gun, Foreign Intrigue, Bandido, Heaven Knows Mr. Allison, Fire Down Below, Wonderful Country, Home from the Hill, Sundowners, The Last Time I Saw Archie, Cape Fear, The Longest Day, List of Adrian Messenger, Two for the Seesaw, Rampage, Mr. Moses, El Dorado, Anzio, Villa Rides, Five Card Stud, Secret Ceremony, Young Billy Young, Good Guys and the Bad Guys, Ryan's Daughter, Going Home, Wrath of God, Friends of Eddie Coyle, The Yakuza, Farewell My Lovely, Midway, The Last Tycoon, Amsterdam Kill, The Big Sleep, Matilda.

MITOSKY, ALAN P.: Executive. b. Philadelphia, March 13, 1934. e. Univ. of Penna., School of Fine Arts (BFA, 1957). Was curator of museums in Phila. (1957–65) and later assoc. dir. of Philadelphia Civic Center, handling performing arts and film programs, promotion of trade shows, conventions, etc. In 1970–2 v.p., production, Madison Square Garden. In 1972 named pres. Leisure Presentations of America, packaging and producing film properties and rock and pop concerts. In 1974 named gen. mgr. Dragon Aire Ltd. and No Moss Films, handling distribution of theatrical properties. In 1975 pres. of Coliseum Films and Athena Films Ltd., natl. theatrical film dist., prod., foreign sales and prod. repr. in 1977—project administrator, National Endowment for the Arts Short Film Showcase—dist. of special films.

MIZIKER, RONALD D.: Producer. b. Cleveland, OH, Oct. 14, 1941. e. Univ. of New Mexico, Univ. of So. Calif. Began career as cameraman in 1961 at educational TV station in Albuquerque; soon after produced, wrote and directed variety of TV programs; two series nationally televised weekly on NET network. Following graduate work at Univ. of So. Calif. in 1965 became producer for 50-50 Club, syndicated daily talk show for AVCO Broadcasting Corp. Joined entertainment div. of Walt Disney Prods. in 1969; was dir. of show development, creating entertainment and parades for Disneyland, Walt Disney World, etc. Left in 1978 to be independent TV producer and director and to produce live events and spectaculars. Formed own company, R.D. Miziker & Associates. Recent credits: Perry Como's Christmas in New Mexico, 25 Years of Mouseketeers, Wonderful World of Disney segments.

MOBERLY, LUKE: Producer, director, writer. b. Richmond, Ky., Nov. 1, 1925. e. Univ. of Cincinnati, 1947, Univ. of Miami, 1951. Designed and built studio buildings; designer and set builder.

PICTURES INCLUDE: Country A-Go-Go, The Misfits, Gavilan, Devil's Sisters, Go-Go-A-Go-Go, etc. Wrote screenplay: Psychedelic, Rhinemans Seven, Ride Roxie Ride, In the Skin with Me, Easy Little Laura, Old Time Religion. Dir. and prod.: Moonlight Maniac, Assoc., prod.: That Nice Boy, writer, producer, director.: Sweet Talker, Skunk Ape, Moberly/Gordon Studios Inc. Production Company for feature films, 1975.

MOCIUK, YAR W.: Executive. b. Ukraine, Jan. 26, 1927. e. CCNY, World Univ., Peoples Univ. of Americas (Puerto Rico). Expert in field of m.p. care and repair; holds U.S. patent for method and apparatus for treating m.p. film. Has also been film producer and director. Founder and pres. of C&M Films Service, Inc. until 1973. Now chm. of bd. and pres. of Filmtreat International Corp. Member: M.P. & TV Engineers; Univ. Film Assn. Pres., Ukrainian Cinema Assn. of America.

MOGER, ART: Publicity executive-Author. b. Boston, Mass., April 4. e. Boston U. Coll. of Journalism (BS) Cartoonist & caricaturist, stage & screen attractions; feature writer. nat. mag.; script writer, guest appearances with Fred Allen, and many radio, TV shows (Mike Douglas Show, Merv Griffin); creator cartoon strip Seymour Shoze, dealing with m.p. and comic strip, Groucho, based on Groucho Marx: orig., About Faces, puzzle series; in 1937 syndicated cartoon series; pressbook cartoonist; handled personal appearances of celebrities. Pub. exec., Warner Bros. & other m.p. cos. Holds record for "My Favorite Jokes" in Parade Magazine. (8 consecutive times) First honorary trustee of Nat'l. Broadcasters Hall of Fame; member Sigma Delta Chi, professional journalism society.

AUTHOR: You'll Dial Laughing, Pros and Cons, Some of My Best Friends Are People, The Complete Pun Book, Everybody Loves an Underdog; pres., Advertising Enterprises, Boston; exec., Amuse-A-Menu Corp.; dorArt Enterprises; pres. and originator Tub Thumpers of Amer.

MOGER, STANLEY H.: Executive. Pres., SFM Entertainment, Exec. Vice Pres., SFM Media Corp. b. Boston, Nov. 13, 1936. e. Colby Coll., Waterville, ME (BA, 1958). m. Marcia Fleishman, May 29, 1960, children—Robin 19, Wendy 16; announcer/TV personality/WVDA and WORL (Boston) 1953–54; WGHM (Skowhegan) 1955–56; WTWO-TV (Bangor) 1955; WMHB (Waterville) 1956–57; WTVL (Waterville); US Army reserve, 1958–64, with calls to active duty in 1958–59 and 1961–62; account exec., NBC Films/California National Productions, Chicago; 1959–60, asst. sales manager, Midwest, Hollingbery Co., Chicago, 1960–63, and New York 1963–66; account executive, Storer Television Sales, 1966–69; co-founded SFM, Sept. 29, 1969. In 1978, named pres., SFM Entertainment which is responsible for the revival of Walt Disney's "Mickey Mouse Club," "The Adventures of Rin-Tin-Tin"; Mobil Showcase Network; SFM Holiday Network. Executive producer-"Television-Annual 1978–79," ABC-TV Network; Executive producer-"Your New Day" with Vidal Sassoon.

PROPERTIES INCLUDE: SFM Holiday Network, SFM Documentary Network, Car Care Central, Your New Day with Vidal Sassoon, March of Time series, Nightside, Sparrow, Co-Packager Superstars-ABC-TV and Battle of the Network Stars ABC-TV; Network Sales: Adventures of the Wilderness Family-ABC-TV; Across the Great Divide-CBS-TV; Mysteries from Beyond Earth-NBC-TV; To the Ends of the Earth-PBS.

MOLINARO, EDOUARD: Director. b. Bordeaux. Made amateur films at university and launched professional career via award-winning technical shorts. First feature film, Le Dos au Mur (1957).

PICTURES INCLUDE: Girls for the Summer, A Ravishing Idiot, Gentle Art of Seduction, La Cage Aux Folles (also Part II), Sunday Lovers.

MOLL, ELICK: Writer. b. New York City, Mar. 20, 1907; e. U. Illinois; U. Chicago. Short story writer, novelist, playwright; Seidman and Son, Memoir of Spring, Night Without Sleep, Image of Tallie; with Samuel Goldwyn, MGM, Fox; stories appear in magazines, texts on short story; won 2nd prize O'Henry Memorial Award Prize Stories 1937.

PICTURES INCLUDE: You Were Meant for Me, Wake Up and Dream, House on Telegraph Hill, Night Without Sleep, Storm Center, Spring Reunion.

TV: Playhouse 90, Chrysler Theatre, others.

MONACO, EITEL: Executive. b. Montazzoli, Chieti, Italy, May 16, 1903; e. LLD. Gen. mgr., film div. in State Dept. for Popular Culture to Italian armistice; vice-pres., ENIC; gen. secy., ANICA (Italian Association of Producers & Distributors); pres., 1950; pres., Unitalia, promoting Italian m.p. throughout the world.

MONASH, PAUL: Producer-Writer. b. New York City. e. Univ. of Wisconsin, Columbia Univ. Was in U.S. Army Signal Corps and Merchant Marine; newspaper reporter; high school teacher; and civilian employee of U.S. gov't. in Europe. Wrote two novels: How Brave We Live, The Ambassadors. Entered industry writing TV scripts on such shows as Playhouse 90, Studio One, Theatre Guild of the Air, Climax, etc. Authored two-part teleplay which launched The Untouchables. In 1958 won Emmy award for The Lonely Wizard, dramatization of life of German-born electrical inventor Charles Steinmetz. Made m.p. debut as exec. prod. of Butch Cassidy and the Sundance Kid, 1969.
PICTURES INCLUDE: Slaughterhouse-Five (prod.), The Friends of Eddie Coyle (prod., s.p.), The Front Page (prod.), Carrie (prod.).

MONICELLI, MARIO: Director. b. Rome, Italy, May 15, 1915. Ent. m.p. industry in production; later co-authored, collab., comedies. Dir.: Big Deal on Madonna Street, The Great War, Tears of Joy, Boccaccio '70, The Organizer.

MONKHOUSE, BOB: TV-radio-cabaret comedian & comedy writer. b. Beckenham, Kent, June 1, 1928; e. Dulwich College. Debut 1948 while serving in RAF, own radio comedy series 1949–75 (winters), own TV series, BBC 1952–56, ITV 1956–78. Several West End revues, Boys from Syracuse (Theatre Royal, Drury Lane, 2 yrs.); Come Blow Your Horn, (Prince of Wales); The Gulls; others. Films include: Carry On, Sergeant; Weekend with Lulu; Dentist in the Chair; She'll Have to Go; Bliss of Mrs. Blossom; others. Major cabaret attraction U.K., Australia, Hong Kong. British star of numerous TV series including: What's My Line?; Who Do You Trust?; Mad Movies; Quick On The Draw; Bob Monkhouse Comedy Hour; The Golden Shot (8 years); Celebrity Squares (3 years, ongoing); I'm Bob, he's Dickie! (1978 series). Regular dramatic guest spots ITV & BBC-TV.

MONKS, JOHN, JR.: Writer, actor, prod., dir. b. Brooklyn, N.Y. e. Virginia Military Institute, A.B.; actor, stock, b'way, radio, m.p. U.S. Marines, 1942; commissioned Major, 1945. Playwright Co-author Brother Rat.
PICTURES INCLUDE: Brother Rat, Co-author Brother Rat And A Baby, Strike Up the Band, The House on 92nd Street, 13 Rue Madeleine, Wild Harvest, Dial 1119., The West Point Story, People Against O'Hara. Where's Charley. So This Is Love, Knock On Any Door, No Man Is An Island.
TV: Climax, 20th Century-Fox Hour, Gen. Electric Theatre; CBS Special High Tor; Creator serial Paradise Bay.

MONTAGNE, EDWARD J.: Producer-director. b. Brooklyn; e. Loyola University, of Notre Dame. RKO Pathe, 1942; U.S. Army, 1942–46; prod. many cos. after army.
PICTURES INCLUDE: Tattooed Stranger, The Man With My Face, McHale's Navy, McHale's Navy Joins The Air Force, P.J.
TV: Man Against Crime, Cavalier Theatre, The Vaughn Monroe Show, The Hunter I Spy, McHale's Navy; exec. prod. of film-CBS-N.Y., Phil Silvers Show. Prod. & head of programming, Wm. Esty Adv. Co., 1950; Program consultant, William Esty Co.; v.p. Universal TV prod. & dir.: 5 Don Knotts features; prod.: Andy Griffith, Angel in My Pocket, Ellery Queen, A Very Missing Person, Short Walk to Daylight, Hurricane, Terror on the 40th Floor, Francis Gary Powers, Million Dollar Ripoff.

MONTALBAN, RICARDO: Actor b. Mexico City, Mex., Nov. 25, 1920. Appeared in Mexican pictures 1941–45; to U.S.
PICTURES INCLUDE: Fiesta, On an Island With You, Kissing Bandit, Neptune's Daughter, Battleground, Mystery Street, Right Cross, Two Weeks with Love, Across the Wide Missouri, Mark of Renegade, My Man and I, Sombrero, Border Incident, Latin Lovers, Saracen Blade, Life in The Balance, Sayonara, Hemingway's Adventures of a Young Man, The Reluctant Saint, Love Is a Ball, Sol Madrid, Blue, Sweet Charity, Conquest of the Planet of the Apes, The Train Robbers.
TELEVISION: Series: Fantasy Island.

MONTAND, YVES: Actor, singer. r.n. Yvo Livi. b Monsumano, Italy. Worked as a truck loader, waiter in dock-side bar, and barber. Performed in Marseilles as singer in clubs and music halls. m. Simone Signoret, actress, 1951.
PICTURES INCLUDE: Etoile Sans Lumiere, Les Portes de La Nuit, L'Idole, Souvenir Perdus, Le Salaire de la Peur (Wages of Fear), Nostri Tempi, Mar, Napoleon, Les Heros Sont Fatiques, Marguerite de la Nuit, Uomini e Lupi, Les Sorcieres de Salem (The Crucible), La Lunga Strada, Un Denomme Squarcio, Le Pere et L'enfant, La Loi (Where the Hotwind Blows), Let's Make Love, Sanctuary, My Geisha, Goodbye Again, The Sleeping Car Murders, La Guerre est Finie, Is Paris Burning?, Grand Prix, Live for Life, Z, On A Clear Day You Can See Forever, The Vow, State of Siege.

MONTGOMERY, DOREEN: Writer. e. Edinburgh University (M.A.).
PICTURES INCLUDE: Lassie from Lancashire, Room 13, Meet Mr. Penny, Dead Men Tell No Tales, Poison Pen, Just William, Bulldog Sees It Through, At Villa-Rose, House of the Arrow, Flying Squad, Man in Grey, Fanny by Gaslight, Dream of Olwen, Second Mrs. Elliott, Shadow of the Eagle, Scarlet Web, One Jump Ahead, Narrowing Circle, Dance Little Lady, You Can't Escape, Murder Reported, Island of Dreams.
TV: Fairbanks series Aggie series: Fabian of the Yard. Last of the Mohicans, Charlie Chan, William Tell, Invisible Man, Danger Man, Zero One, Dr. Finlay's Casebook, Taxi, The Avengers, No Hiding Place, Borderers, Who-dun-it.
BOOK: Voices in the Dark, Play, The Summer House.

MONTGOMERY, GEORGE: Actor. r.n. George Montgomery Letz; b. Brady, Mont., Aug. 29, 1916; e. U. of Montana. Armed Services, World War II.
PICTURES INCLUDE: Cisco Kid and the Lady (1939), Star Dust, Young People, Charter Pilot, Jennie, Cowboy and the Blonde, Accent on Love, Riders of the Purple Sage, Last of the Duanes, Cadet Girl, Roxie Hart, Ten Gentlemen from West Point, Orchestra Wives, China Girl, Brasher Doubloon, Three Little Girls in Blue, Lulu Belle, Belle Starr's Daughter, Girl From Manhattan, Sword of Monte Cristo, Texas Rangers, Indian Uprising, Cripple Creek, Pathfinder, Jack McCall Desperado, Fort Ti, Gun Belt, Battle of Rogue River, Lone Gun, Masterson of Kansas, Seminole Uprising, Robbers' Roost, Too Many Crooks, Stallion Trail, The Steel Claw, Watusi, Samar, Hallucination Generation, Hostile Guns.

MONTGOMERY, ROBERT: Actor, director, producer. r.n. Henry Montgomery, Jr.; b. Beacon N.Y., May 21, 1904. Captain, U.S. Naval Reserve, World War I. Entered m.p. for MGM 1929. Captain, U.S. Naval Reserve, World War II Exec. prod. Cagney-Montgomery Prod. Consultant to President of U.S. on all radio and TV, 1952–60.
PLAYS: Dawn Mack, Arlene O'Dare, One of the Family.
PICTURES INCLUDE: College Days, So This is College, Untamed, Their Own Desire, On the Set, Single Standard, Free and Easy, Sins of the Children, Big House, Private Lives, Letty Lynton, When Ladies Meet, Another Language, Night Flight, Riptide, Vanessa—Her Love Story, Petticoat Fever, Last of Mrs. Cheyney, Yellowjack, Cat and the Canary, Haunted Honeymoon, Rage in Heaven, Here Comes Mr. Jordan, They Were Expendable, Saxon Charm, Secret Land, June Bride, (dir. acted in) Lady in the Lake, Eye Witness, Ride the Pink Horse, Once More My Darling, The Gallant Hours.

MOONJEAN, HANK: Producer-Director. Began as asst. dir. at MGM. Later turned to producing.
PICTURES INCLUDE: Assoc. Prod.: The Great Gatsby, WUSA. Exec. Prod. The Fortune, The End. Producer; Hooper, Smokey and the Bandit II, The Incredible Shrinking Woman, Paternity.

MOORE, CONSTANCE: Actress. b. Sioux City, Iowa, Jan. 18, 1922. Sang on radio; Lockheed prog., 2 yrs; Jurgen's Show, 2 yrs. Screen debut 1938. TV shows, nightclubs. N.Y. Stage: The Boys From Syracuse, By Jupiter.
PICTURES INCLUDE: Prison Break, A Letter of Introduction, I Wanted Wings, Take A Letter Darling, Show Business, Delightfully Dangerous, Earl Carroll Sketchbook, In Old Sacramento, Hit Parade of 1947.

MOORE, DICKIE: Actor. b. Los Angeles, Cal., Sept. 12, 1925. Began picture career when only 11 months old, appearing with John Barrymore in The Beloved Rogue.
PICTURES INCLUDE: Oliver Twist, Peter Ibbetson, Dangerous Years, Out of the Past, Eight Iron Men, Member of the Wedding.

MOORE, DUDLEY: Actor-Writer-Musician. b. England. e. Oxford, graduating in 1958. Toured British Isles with jazz group before joining with three other Englishmen to put on revue, Beyond the Fringe, which was a hit in U.K. and N.Y. Appeared later with Peter Cook in N.Y. in Good Evening.
PICTURES: Monte Carlo or Bust, The Bed Sitting Room, Bedazzled, Foul Play, "10", Wholly Moses, Arthur.

MOORE, ELLIS: Executive. b. New York, N.Y., May 12, 1924. e. Washington and Lee U., 1941–43. Newspaperman in Arkansas, Tenn., 1946–52; Joined NBC 1952; mgr. of bus. pub., 1953; dir., press dept., 1954; dir., press & publicity, Dec. 22, 1959; vice-pres., 1961; pub. rel. dept., Standard Oil Co. (N.J.), 1963–66; v.p. press relations, ABC-TV Network, 1966–68; v.p. public relations ABC-TV Network, 1968–70; v.p. public relations, ABC 1970, v.p. public relations, ABC, Inc., 1972; v.p. corporate relations, ABC, Inc., 1979.

MOORE, GARRY: Performer. r.n. Thomas Garrison Morfit; b. Baltimore, Md., Jan. 31, 1915. Continuity writer, WBAL; announcer, sports commentator, KWK, St. Louis; comedian, writer, Club Matinee show, Chicago; Everything Goes, N.Y.; teamed with Jimmy Durante on radio to 1947; m.c., Take It or Leave It, Breakfast in Hollywood.
TV: Star of Garry Moore Show, I've Got A Secret. Best TV daytime show: Fame Poll, 1958; To Tell The Truth.

MOORE, KIERON: Actor. b. Skibereen, Co. Cork, Eire, 1925; e. St. Mary's Coll. (Dublin). Stage debut, 1945 in Desert Rats; appeared in Red Roses For Me. Film debut 1947 in A Man About The House.
PICTURES INCLUDE: Anna Karenina, Mine Own Executioner, Ten Tall Men, David and Bathsheba, Saints and Sinners, Naked Heart, Honeymoon Deferred, Man Trap (Man in Hiding), Conflict of Wings (Fuss Over Feathers), Green Scarf, Blue Peter, Satellite in the Sky, Three Sundays to Live, The Key, The Angry Hills, The Day They Robbed the Bank of England, League of Gentlemen, The Seige of Sidney Street, Faces of Evil, Lion of Sparta, Steel Bayonet, I Thank a Fool, Double Twist, The Day of the Triffids, The Thin Red Line, The Main Attraction, Crack in the World, Son of a Gunfighter, Never Love a Savage, Run Like a Thief, Custer of the West, Ryan International, The Dolmetch Story, Zoo Gang, The Progress of Peoples, The Parched Land.

MOORE, KINGMAN T.: Director. b. Macon, Ga., Sept. 13, 1919; e. Wharton Sch. of Finance, U. of Pa., BS; Member: Lambs Club, Phi Delta Theta, Actor, 1939–41; U.S. Army, 1941–46; asst. cutter, RKO Redio Pictures, 1946–47; production asst., 1947–48; CBS-TV, N.Y., 1948–49; NBC-TV, N.Y., Hollywood, 1949–52; dir. CBS shows include: Places Please, Backstage with Barry Wood, Kobbs Korner, Adventures in Jazz.
TV: Candid Camera, Lights Out, Your Show of Shows, Colgate Comedy Hour.

MOORE, MARY TYLER: Actress. b. Brooklyn. N.Y., Dec. 29, 1936; Began as professional dancer.
TV: Steve Canyon, Richard Diamond, 77 Sunset Strip, Hawaiian Eye, Bachelor Father; feature role on The Dick Van Dyke Show (2 emmy awards); star, The Mary Tyler Moore Show, Mary.
PICTURES INCLUDE: X-15, Thoroughly Modern Millie, Don't Just Stand There, What's So Bad About Feeling Good, Change of Habit, Ordinary People.

MOORE, ROBERT: Director. b. Washington, D.C. Studied and taught drama at Catholic University. Acted on Broadway as a teen-ager in Jenny Kissed Me, later appeared in Cactus Flower and Everything in the Garden. His staging of The Boys in the Band attracted the attention of Neil Simon, and he directed Simon's Promises, Promises, Last of the Red Hot Lovers and The Gingerbread Lady.
TV: Rhoda, Thursday's Game (movie).
PICTURES INCLUDE: Tell Me That You Love Me, Junie Moon (actor), Murder by Death (theatrical debut as director), The Cheap Detective (dir.); Chapter Two (dir.).

MOORE, ROGER: Actor. Dir. b. London, England, Oct. 14 e. Art School, London; Royal Acad. of Dramatic Art.
PLAYS: Mr. Roberts, I Capture the Castle, Little Hut, others.
BROADWAY: A Pin to See the Peepshow.
TV: Maverick, The Alaskans, Ivanhoe, The Saint, The Persuaders.
PICTURES INCLUDE: Last Time I Saw Paris, Interrupted Melody, King's Thief, Diane, The Miracle, Gold of the Seven Saints, Rachel Cade, Rape of the Sabines, No Mans Land, Crossplot, The Man Who Haunted Himself, Live and Let Die, Gold, The Man with the Golden Gun, That Lucky Touch, Street People, Shout at the Devil, Sherlock Holmes in New York, The Spy Who Loved Me, The Wild Geese, Escape To Athena, Moonraker, ffolkes, The Sea Wolves, Sunday Lovers, For Your Eyes Only, The Cannonball Run.

MOORE, TERRY: Actress. r.n. Helen Koford; b. Los Angeles, Cal., Jan. 1, 1932; mother Luella Bickmore, actress. Photographer's model as a child; on radio; with Pasadena Playhouse 1940; in m.p. 1933.
PICTURES INCLUDE: Gaslight, Son of Lassie, Sweet and Low Down, Shadowed, Devil on Wheels, Return of October, Mighty Joe Young, He's a Cockeyed Wonder, Gambling House, Two of a Kind, Sunny Side of the Street, Star of Tomorrow: 1958, Man on a Tightrope, Beneath the 12-mile Reef, King of the Khyber Rifles, Daddy Long Legs, Shack Out on 101, Postmark for Danger, Come Back Little Sheba, Bernardine, Why Must I Die?, Platinum High School, A Private's Affair, Cast a Long Shadow, City of Fear, Black Spurs, Town Tamer, Waco, A Man Called Dagger.

MOORE, THOMAS W.: Executive. e. U of Mo.; Naval aviator, USNR, 1940–45. Adv. dept., Star, Meridian, Miss.; v.p., adv. mgr., Forest Lawn Memorial Park; account exec., CBS-TV Film Sales, Los Angeles; gen. sales mgr., CBS-TV Film Sales, 1956; v.p. in chg. programming & talent, 1958; pres., ABC-TV Network, 1962; chmn. bd., Ticketron, 1968; pres., Tomorrow Entertainment, Inc. 1971.

MORAN, DOLORES: Actress. b. Stockton, Cal., 1926. Screen debut, 1941.
PICTURES INCLUDE: Yankee Doodle Dandy, The Hard Way, Old Acquaintances, Man I Love, Christmas Eve, Johnny One Eye, Count The Hours, Silver Lode.

MORE, KENNETH: Actor. b. Gerrards Cross (Buckinghamshire), England, Sept. 20, 1914; e. Victoria College, Jersey (Channel Islands). Stage debut at Windmill Vaudeville Theatre, 1937; screen debut in Scott of the Antarctic, 1948. Voted one of Top Ten British Moneymaking Stars in M.P. Herald-Fame poll, British and International poll, 1955–56. Many TV appearances, incl. On Approval.
STAGE: Our Man Crichton, The Forsythe Saga, The Secretary Bird, The Winslow Boy, Getting On, Signs of the Times.
PICTURES INCLUDE: Now Barabbas, The Clouded Yellow, No Highway, Appointment with Venus, Franchise Affair, Brandy for the Parson, Yellow Balloon, Never Let Me Go, Genevieve, Our Girl Friday (Adventures of Sadie), Doctor in the House, Raising a Riot, Deep Blue Sea, Reach for the Sky, Admirable Crichton, Sheriff of Fractured Jaw, Next to No Time, A Night To Remember, The 39 Steps, Northwest Frontier, Sink the Bismarck, The Man in the Moon, The Greengage Summer, Some People, We Joined the Navy, The Comedy Man, The Mercenaries, Fraulein Doktor, Oh! What a Lovely War, Battle of Britain, Scrooge, Slipper and the Rose, Leopold in the Snow.

MOREAU, JEANNE: Actress. b. Paris, France, 1928. e. Nat'l Conservatory of Dramatic Art. Stage debut with Comedie Francaise, acting there until 1952 when she joined the Theatre Nationale Populaire, Director: Lumiere (film), 1976.
PLAYS: A Month in the Country, La Machine Infernale, Pygmalion, Cat on a Hot Tin Roof.
PICTURES INCLUDE: The She-wolves, Elevator to the Scaffold, The Lovers, Le Dialogue Des Carmelites, Les Liaisons Dangereuses, Moderato Cantabile, La Notte, Jules and Jim, A Woman is a Woman, Eva, The Trial, Bay of Angels, The Victors, Le Feu Follet, Diary of a Chambermaid, The Yellow Rolls-Royce, The Train, Mata Hari, Viva Maria, Mademoiselle, Chimes at Midnight, Sailor From Gibraltar, The Bride Wore Black, The Immortal Story, Great Catherine, Monte Walsh, Alex in Wonderland, The Little Theatre of Jean Renoir, The Last Tycoon, Mr. Klein.

MORENO, RITA: Actress. b. Humacao, Puerto Rico, Dec. 11, 1931. Spanish dancer since childhood; night club entertainer; m.p. debut in 1950.
PICTURES INCLUDE: Pagan Love Song, Toast of New Orleans, Singin' in the Rain, The Ring, Cattle Town, Latin Lovers, Jivaro, Yellow Tomahawk, Garden of Evil, Untamed, Seven Cities of Gold, Lieutenant Wore Skirts, King and I, This Rebel Breed, Summer and Smoke, West Side Story, The Night of the Following Day, Marlowe, Popi, Carnal Knowledge, The Ritz, Happy Birthday, Gemini, The Four Seasons.

MORGAN, ANDRE: Producer. b. Morocco, 1953. e. Univ. of Kansas, majoring in Oriental languages-literature. Met Raymond Chow and joined his film production co. in Hong Kong. Produced Enter the Dragon, Bruce Lee kung-fu picture. Now acts as prod. on Golden Harvest's Chinese language films as well as new division established to produce English-language films.
PICTURES INCLUDE: Enter the Dragon, The Amsterdam Kill, The Boys in Company C.

MORGAN, DENNIS: Actor. r.n. Stanley Morner; b. Prentice, Wisc., Dec. 10, 1920; e. Carroll Coll. Started with State Lake Theat., Chicago. Toured midwest in Faust, sang in Empire Room of Palmer House, Chicago, appeared on NBC programs and sang lead in Xerxes. Screen debut, 1936.
PICTURES INCLUDE: Susy, The Fighting 69th, Three Cheers for the Irish, My Wild Irish Rose, Two Guys from Milwaukee, Two Guys from Texas, Cheyenne, Perfect Strangers, To the Victor, One Sunday Afternoon, Raton Pass, Pretty Baby, Painting the Clouds with Sunshine, Star of Tomorrow (1941), This Woman is Dangerous, Cattle Town, Gun That Won the West, Pearl of the South Pacific, Uranium Boom.

MORGAN, HARRY: Actor. r.n. Harry Bratsburg; b. Detroit, Mich., Apr. 10, 1915; e. U. Chicago. Screen debut 1942.
TV: December Bride, Pete and Gladys, MASH, Dragnet, etc.
PLAYS: Gentle People, My Heart's in the Highlands, Thunder Rock, Night Music, Night Before Christmas.
PICTURES INCLUDE: To the Shores of Tripoli, Loves of Edgar Allen Poe, Orchestra Wives, Dragonwyck, Appointment with Danger, The Highwayman, When I Grow Up, The Well, Blue Veil, Bend of the River, Scandal Sheet, My Six Convicts, Boots Malone, High Noon, What Price Glory, Stop You're Killing Me, Arena, Torch Song, Glenn Miller Story, About Mrs. Leslie, Forty-Niners, Far Country, Not As a Stranger, Backlash, Strategic Air Command, Support Your Local Sheriff, Charlie and the Angels, Snowball Express, The Apple Dumpling Gang, The Greatest, Cat from Outer Space.

MORGAN, MICHELE: Actress. r.n. Simone Roussel; b. Paris, France, Feb. 29, 1920; e. Dieppe, dramatic school, Paris. Decided on acting career at 15 yrs., won role at 17 opposite Charles Boyer in Gribouille (later filmed as The Lady in Question, Hollywood). Made several pictures abroad; to U.S. 1940. First Amer. film Joan of Paris, 1942.

PICTURES INCLUDE: The Chase, Symphonie Pastorale, Fallen Idol, Fabiola, 7 Deadly Sins, Moment of Truth, Daughters of Destiny, Naked Heart, Proud and the Beautiful, Grand Maneuver, Oasis, Lost Command, Benjamin.

MORGAN, TERENCE: Actor. b. London, Eng., Dec. 8, 1921; e. Ewell Castle, Surrey & R.A.D.A. Stage debut: There Shall Be No Night, London 1943 m.p. debut: Hamlet, 1948.
PICTURES INCLUDE: Capt. Horatio Hornblower, Encore, Mandy, It Started in Paradise, Steel Key, Street Corner, Turn the Key Softly, Always a Bride, Forbidden Cargo, Dance Little Lady, Femina, Svengali, They Can't Hang Me, March Hare, It's a Wonderful World, The Scamp, Tread Softly Stranger, The Shakedown, Picadilly Third Stop, The Curse of the Mummy's Tomb, The Penthouse, Hide and Seek, The Lifetaker.
TV: Sir Francis Drake, The Persuaders, etc.

MORIARTY, MICHAEL: Actor. b. Detroit, 1942. e. Dartmouth. Studied at London Academy of Music and Dramatic Arts. Appeared with New York Shakespeare Festival, Charles Street Playhouse (Boston), Alley Theatre (Houston) and Tyrone Guthrie Theatre (Minneapolis). Broadway debut in The Trial of the Catonsville Nine.
PICTURES: Glory Boy (debut), Hickey & Boggs, Shoot It, Bang the Drum Slowly, The Last Detail, Report to the Commissioner, Who'll Stop the Rain.
TV: The Glass Menagerie, Girls of Summer, The Deadliest Season, Holocaust.
STAGE: Find Your Way Home, Richard III

MORIN, ROBERT B.: Executive. b. New York, N.Y. Nov. 24. e. Harvard Coll., Merch, Mgr., ad agency. AAP sls., 1955; AAP mgr. program sls., 1956; eastern sls. mgr., MGM-TV, 1957; sls. mgr., Lopert-UA, 1959; v.p., Allied Artists TV; formed Scandia Films, pres., 1965; joined RKO, 1966, in chg. foreign operations; 1968, pres. Lin Medallion Pict. Corp.; 1970–73, Talent Associates; 1973–76, exec. v.p., Heritage Eng.; 1976–78, v.p., MGM TV; 1978 to present, pres. Lorimar Syndication; 1980, exec. v.p., 20th Century Fox.

MORITZ, MILTON I: Executive, b. Pittsburgh, Pa., Apr. 27, 1933; e. Woodbury College, grad. 1955. Owned, operated theatres in L.A., 1953–55; U.S. Navy 1955–57; American International Pictures asst. gen. sls. mgr., 1957; nat'l. dir. of adv. and publ., 1958; v.p. and bd. mem. of American International Pictures, 1967; 1975, named Senior v.p.; in 1980 formed own co., The Milton I. Moritz Co., Inc., Inc., mktg. & dist consultant. Pres. of Variety Club of Southern California Tent 25, 1975–76.

MORLEY, KAREN: Actress. r.n. Mabel Linton; b. Ottumwa, Iowa; e. Calif. U. Campus Baby Star, 1931; bit part in Scarface, 1932.
PICTURES INCLUDE: Arsene Lupin, Fast Life, Gabriel Over the White House, Dinner at Eight, Crime Doctor, Thunder in the Night, Devil's Squadron, Beloved Enemy, Outcast. Pride and Prejudice, Jealously, Unknown, 13th Hour, Framed, Six Gun Serendade, Code of the Saddle, M..

MORLEY, ROBERT: Actor, writer. b. Wiltshire, England, May 26, 1908; e. Royal Acad. Dram. Arts, London. U.S. Air Force, 1943; prod., assoc. prod., Captain Gallant, Arsen, Chicot, Forbidden Cargo; TV Series: Captain Gallant; Scene of the Crime; TV Advisor, Telepictures of Morroco, Telerama, Ltd., Eng., 1954–59; helped org., Alliance of TV Film Producers; prod., Telerama, Inc., Georgetown Films, Inc., 1960; prod., dir., TV film series closeup, Telefilm, Inc. Presently president, Creative Assoc., Inc. and co-prod. with Helen Ainsworth; exec. prod. with Leon Fromkess, The Long Corridor.
PLAYS: Brit. stage debut London 1929 in Treasure Island, also tour and repertory; N.Y. debut 1938 in Oscar Wilde; other appearances include Great Romancer, Pygmalion; 1948, Edward My Son, co-author, lead London and B'way stage; author several plays (Short Story, Goodness, How Sad, Staff Dance).
PICTURES INCLUDE: Marie Antoinette, You Will Remember, Major Barbara, Big Blockade, This Was Paris, Foreman Went to France, Young Mr. Pitt, I Live in Grosvenor Square, Outcast of the Islands, African Queen, Edward My Son, Curtain Up, The Final Test, Melba, Gilbert & Sullivan, Beat the Devil, Beau Brummell, Rainbow Jacket, Good Die Young, Quentin Durward, Loser Takes All, Six Months Grace, Full Treatment, Hippo Dancing, Around the World in 80 Days, Sheriff of Fractured Jaw, The Journey, The Doctor's Dilemma, Battle of the Sexes, Libel, Oscar Wilde, Go To Blazes, The Young Ones, Nine Hours to Rama, Murder at the Gallop, Topkapi, Amanda, Take Her She's Mine, Genghis Khan, Hot Enough for June, The Alphabet Murders, Sinful Davey, Hot Millions, Some Girls Do, The Trygon Factor, Song of Norway, The Blue Bird, Who Is Killing the Great Chefs of Europe?, Scavenger Hunt, The Human Factor, The Great Muppet Caper.
TV: Call My Bluff (series), Charge!

MORRICONE, ENNIO: Composer-Arranger. b. Rome. Studied with Goffredo Petrassi at the Academy of Santa Cecilia in Rome.
PICTURES INCLUDE: A Fistful of Dollars, El Greco, Fists in the Pocket, Matchless, Theorem, Once Upon a Time in the West, Investigation of a Citizen, The Bird with the Crystal Plumage, Cat O'Nine Tails, The Red Tent, Four Flies in Grey Velvet. The Decameron, The Black Belly of the Tarantula, Battle of Algiers, Burn, Partner, Orca, The Heretic, Exorcist II, 1900, Days of Heaven, The Island.

MORRIS, HOWARD: Actor. b. New York, N.Y., Sept. 4, 1919; e. N.Y.U. U.S. Army 4 yrs.; dir., Who's Minding the Mint, Don't Drink the Water.
BROADWAY: Hamlet, Call Me Mister, John Loves Mary, Gentlemen Prefer Blondes.
TV: Your Show of Shows, Caesar's Hour.

MORRIS, NELSON: Producer. b. Warsaw, Poland, July 25, 1920; e. Columbia Univ., 1939. Began in m.p. ind. in 1930 as child actor, Paramount Studio, Astoria, L.I.; photographer, Life Mag., 1939–43; WWII, training film div. Army Pictorial Service. After war continued career in advertising, industrial and mag. photography. Org., American Society of Magazine Photographers. Formed, Nelson Morris Prods., 1953; engaged in prod. of industrial and doc. film; prod. films for TV, theatrical feature films; producer, stage prod., Bicycle Ride to Nevada. Now pres., Astoria Pictures Corp.

MORRIS, OSWALD: Cinematographer. b. London. Left school at 16 to work for two years as camera dept. helper at studios. Was lensman for cameraman Ronald Neame who gave Morris first job as cameraman; in 1949 when Reame directed The Golden Salamander he made Morris dir. of photography.
PICTURES INCLUDE: The Golden Salamander, The Card, The Man Who Never Was, Mister Moses, Moulin Rouge, Beat the Devil, Moby Dick, Heaven Knows, Mr. Allison, The Roots of Heaven, Reflections in a Golden Eye, The Mackintosh Man, Oliver!, Scrooge, The Pumpkin Eater, The Hill, The Spy Who Came in from the Cold, A Farewell to Arms, The Key, The Guns of Navarone, Lolita, Term of Trial, Fiddler on the Roof, Sleuth, The Odessa File, The Man Who Would Be King, Seven Per Cent Solution, Just Tell Me What You Want, The Great Muppet Caper.

MORRIS, R.A.: Executive. b. London, England, 1925; e. Imperial Services College, Windsor. Served Royal Navy 1942–46. Ent. m.p. ind. Animation Dept. Polytechnic Studios 1947; AB-Pathe, 16MM salesman, 1948; assist. to overseas sales mgr., 1949; overseas sales mgr., 1952; joined ABC-TV develop world-wide dist. TV films, 1958; res. to form Programme Exchange Ltd., 1960; formed Anthony Morris (London) Ltd., 1968.

MORRIS, RICHARD: Director, writer. b. San Francisco, Cal., May 14, 1924; e. Burlingame High School, 1939–42, Chouinard Art Institute, 1946–47; Neighborhood Playhouse, 1947–48; U.S. Army special services, writing shows, Victory Bond; Universal-Int., talent dept., writing, directing skits, writing music for acting class and Korean War entertainment troups; s.p. Take Me to Town, Finders Keepers.
AUTHOR, STAGE: The Unsinkable Molly Brown, Golden Gate, Tomato Pie.
PICTURES INCLUDE: If a Man Answers, I Married a Psychiatrist, It Seems There Were These Two Irishmen, Thoroughly Modern Millie, The Man Who Was Magic, Change of Habit.
TV: Wrote & dir. teleplays, Loretta Young Show, wrote & dir. The Pearl (Loretta Young Show), wrote teleplays, Private Secretary, Ford Theatre, Kraft Television Theatre, dir., The Wild Swans (Shirley Temple Show-Christopher Award).

MORRIS, SEYMOUR: Executive. b. Aug. 26, 1906. Started in show business with B.S. Moss, juvenile theatricals; entered m.p. ind. with C.B.C. Pictures on adv. accessories; exp. dept., U.A., Columbia. Ran own theatres, N.Y. State; mgr., upstate N.Y. theatres, Skouras Bros., until 1933; then mgr., dist. mgr., Schine; dir. of adv., pub., pub. rel., Schine Ent. Inc., including theatres, hotels, radio station etc. In 1961 opened own adv. & p.r. agency, Seymour Morris Associates Inc.; also pres. of Northeast Film Enterp. (1967); pres. PMK Prod. Inc.

MORRIS, WILLIAM, JR.: Consultant, retired president, William Morris Agency, Inc. b. New York, Oct. 22, 1899.

MORRISSEY, PAUL: Writer-Director. b. New York City. e. Fordham Univ. Service in Army; worked for insurance co. and Dept. of Public Welfare before becoming film-maker. Was involved in indep. film prod. for four years prior to joining Andy Warhol with whom he worked on such films as Chelsea Girls, Four Stars, Bike Boy, Nude Restaurant, Lonesome Cowboys, Blue Movie, L'Amour, Women in Revolt.
PICTURES INCLUDE: Flesh, Trash, Heat, Andy Warhol's Frankenstein, Dracula.

MORRISON, BURT: Executive. Joined 20th Century-Fox in 1963; named snr. financial officer for 20th Century-Fox Pictures. In 1980 named v.p., finance & administration, for The Ladd Company.

MORROW, DON: Announcer, actor, narrator. b. Stamford, Conn., Jan. 29; e. Syracuse U., Southern Methodist U., Wesconn. Currently spokesman for Ford, and IBM.

MORROW, JEFF: Actor. b. New York, N.Y., Jan. 13; e. Pratt Institute. Worked as commercial illustrator. Many radio and TV appearances including TV, Iron Horse.
01 BROADWAY: Romeo and Juliet, St. Joan, January Thaw, Billy Budd, Three Wishes for Jamie, Lace on Her Petticoat, The Suspects.
PICTURES INCLUDE: The Robe, Flight to Tangier, Siege at Red River, Tanganyika, Sign of the Pagan, Captain Lightfoot, This Island Earth, World in My Corner, Creature Walks Among Us, Giant Claw, Story of Ruth, Harbor Lights.

MORROW, VIC: Actor, writer, director. b. Bronx, N.Y., Feb. 14, 1932; e. Florida Southern College, Paul Mann's N.Y. Actors' Workshop. Dir., off-Broadway, Deathwatch, The Maids; The Firstborn, Desilu Prof. Theatre Workshop, pres., The Firstborn; appeared in A Streetcar Named Desire, summer stock; dir., A Man Called Sledge, Columbia Pictures.
PICTURES INCLUDE: Blackboard Jungle, Survival, Tribute to a Badman, Men in War, God's Little Acre, Cimarron, Portrait of a Mobster, Last Year at Malibu, Sledge, Deathwatch, Step out of Line, Police Story, The Bad News Bears, Treasure of Matecumbe, Message from Space.
TV: Bonanza, The New Breed, The Untouchables, The Rifleman, G.E. Theatre, The Outlaws, Combat, world premiere, The Glass House (CBS Movie of the Week), Roots, Captain and the Kings.

MORSE, ROBERT: Actor. b. Newton, Mass., May 18. Served U.S. Navy. Studied with American Theatre Wing, New York, where he had small role in film The Proud and the Profane, 1956. Following radio work, appeared on Broadway stage in The Matchmaker, 1956.
BROADWAY: Say, Darling, Take Me Along, How to Succeed in Business Without Really Trying, Sugar.
PICTURES INCLUDE: The Matchmaker, Honeymoon Hotel, Quick Before it Melts, The Loved One, Oh Dad Poor Dad, How to Succeed in Business Without Really Trying, Guide for the Married Man, The Boatniks.
TV: That's Life series.

MORSE, TERRY: Director. b. St. Louis, Mo., Jan. 30, 1906. Started in cutting room dept., First National, 1922; film ed., dir., Warner Bros.
PICTURES INCLUDE: Lady in Ermine, Heart of New York, Misbehaving Ladies, Dragon Murder Case, Personality Kid, Massacre, I Love for Love, Front Page Woman, Case of the Curious Bride, Woman in Red, Stolen Holiday, Patient in Room 18, Crime School, Waterfront, On Trial, Smashing the Money Ring, Heart of the North, British Intelligence, No Place to Go, Dick Tracy, Fog Island, Tulsa, Rogue River, Journey Into Light, Ruby Gentry, Unknown World, The Moonlighter, Top Banana, Gang Busters, Godzilla King of the Monsters, Taffey & the Jungle Hunter, Young Dillinger.

MORTON, ARTHUR: Composer, arranger. b. Duluth, Minn., Aug. 8, 1908; e. U. of Minn., 1929. Composer for various film cos. including Universal, RKO, United Artists; with Columbia since 1948.
PICTURES INCLUDE: Night Life of the Gods, Princess O'Hara, Riding On Air, Fit for a King, Turnabout, Walking Hills, The Nevadan, Rogues of Sherwood Forest, Father is a Bachelor, Never Trust a Gambler, Harlem Globetrotters, Big Heat, Pushover, He Laughed Last.
ORCH. ARRANGEMENTS: Laura, Smokey, From Here to Eternity, Jolson Story, Salome, Phfft, No Sad Songs For Me, Born Yesterday, Long Gray Line, Man from Laramine, My Sister Eileen, Queen Bee, Picnic, Jubal, Autumn Leaves, Johnny Concho, Harder They Fall, 3:10 to Yuma, Full of Life, Garment Jungle, They Came to Cordura, Strangers When We Meet, Touch of Mink, Critics Choice, Diamond Head, Toys in the Attic, Man from the Diners' Club, Von Ryan's Express, The Saboteur, Code Name—Morituri, In Harm's Way, What a Way to Go, The New Interns, Rio Conchos, Dear Briggitte, Our Man Flint, Planet of the Apes, Flim Flam Man, Justine, Patton, Tora Tora Tora, Mephisto Waltz, Ballad of Cable Hogue, Traveling Executioner, Escape From the Planet of the Apes, Cold Turkey, Wild Rovers, The Other, Ace Eli, One Little Indian, The Don is Dead, Papillon, Chinatown, Breakout, The Wind and the Lion, Logan's Run, The Omen, Islands in the Stream, Passover Plot, Twilight's Last Gleaming, Damnation Alley, MacArthur, Capricorn One, Coma, The Swarm, Omen II (Damien), Boys from Brazil, Magic, Superman, Alien, Players, Meteor, Star Trek, Inchon, Masoda, The Final Conflict, Outland, Raggedy Man, Night Crossing.
TV: Black Saddle, Laramie, Bus Stop, Follow the Sun, My Three Sons, Peyton Place, Medical Center, Daniel Boone, Lancers, National Geographic, Say Goodbye, How to Stay Alive, Hooray For Hollywood, The Waltons, Apple's Way, Medical Story.

MOSES, CHARLES ALEXANDER: Executive, producer, publicist. b. Chicago, Ill., March 1, 1923; e. Northwestern U. School of Journ., Aernoautical U. (B.S.), Englewood Eve. Jr. Coll. Prod. radio-TV shows, pub. rel. dir., Goldblatt Bros., Chicago, 1944–50; Prod. Radio Free Europe, Munich, 1951–52; spec. rep., RKO Radio Pictures, 1952; European adv.-pub. dir., United Artists, Paris, 1953–56; dir., adv., pub. and exploit., Bel-Air Prod., 1956–58; pub. exec., Screen Gems, Warners, Mirisch; own p.r. firm, Hollywood, 1959; adv., pub. and p.r. dir., Sinatra Enterprises, 1962; dir., adv., pub. and exploit., Associates and Aldrich Co., 1964; exec.-in-charge, Universal Picts. N.Y. domestic and foreign dept., Universal Studios adv., pub., and exploit. rep., Europe, Universal Picts. associate studio pub. dir., 1966–68. Orig. Story, Frankenstein, 1970. Pres., Pub. Guild of America, 1962–67. Pres., Charles A. Moses Co., 1968.

MOSES, GILBERT: Director. New York Stage director; m.p. debut Willie Dynamite, 1973; The Fish That Saved Pittsburgh, 1979.
STAGE: Ain't Supposed to Die a Natural Death, Slave Ship.
TV: Roots.

MOSS, ARNOLD: Actor-dir. b. New York, Jan. 28; e. CCNY, BA; Columbia U., MA; New York U., Ph.D.; Teacher, B'klyn Coll. 1932–39; visiting prof., U. of Conn. 1973; Pace U., 1975; College of William and Mary, 1976; Purdue U., 1977. Neighborhood Playhouse School of the Theatre, 1974–76. Theatre, film & TV actor-director.
PICTURES INCLUDE: Temptation, Loves of Carmen, Reign of Terror, Border Incident, Kim, Mask of the Avenger, My Favorite Spy, Viva Zapata, Salome, Casanova's Big Night, Bengal Brigade, Hell's Island, Jump Into Hell, The 27th Day, The Fool Killer, Caper of the GOLDEN Bulls, Gambit.

MOSS, CHARLES B., JR.: Theatre Executive and Producer. b. New York City, Aug. 29, 1944. e. Univ. of Penn. (B.A., 1966), Boston Univ. School of Law (LIB, 1969). Was asst. prof. of law at Boston Univ. 1969–70. Entered ind. as v.p. of B.S. Moss Enterprises, 1970; now pres. Has produced three films: Let's Scare Jessica to Death, Stigma, Diary of the Dead.

MOSS, FRANK L.: Writer-producer. b. New York, N.Y.: e. Duke & Columbia U: Reporter, drama & film critic, N.Y.: U.S. Army Air Force, 1942–46.
PLAYS: author: Glamour Girl, Call To Arms, (colab), So Goes The Nation, (colab), Some People's Children, American Pastoral, City On A Hill.
PICTURES INCLUDE: The Unvanquished, Whiphand, Carribean, Sangaree, Papago Wells, The Half Breed, Sweetheart of Sigma Chi. MILITARY: 22 Air Force Training Films; 17 documentaries.
TV PILOTS: Outer Limits, Grand Jury, The Texan.
TV SCRIPTS: Telephone Hour, Four Star Playhouse, Winston Churchill's Valiant Years, Route 66, Wagon Train, Laramie, Wild, Wild West, The Texan, G.E. Theater, Wire Service, U.S. Marshall, M-Squad, Stoney Burke, Tales of The Texas Rangers, T.V. Reader's Digest, Sheriff of Cochise, Whirlybirds, Line-Up, Wyatt Earp, Rin Tin Tin, Walter Winchell File, Daniel Boone, Man Who Never Was, Felony Squad, Richard Diamond, Lassie, Like The Rich People, Hired Mother, Shenandoah, Counterspy, White Hunter.

PROD-STORY Ed: Screen Televideo, Sovereign Prod., Wire Service, T.V. Reader's Digest, Wyatt Earp.

MOUND, FRED: Executive. b. St. Louis, Mo., April 10, 1932. e. St. Louis Univ., Quincy Coll. 1946–52, assoc. with father, Charles Mound, at Park Theatre in Valley Park, Mo.; 1952–53, Universal Pictures (St. Louis); 1953, booker, UA, St. Louis; 1955 promoted to salesman in Kansas City; 1957, salesman, St. Louis; 1962, Indianapolis branch mgr. In 1967 named UA regional mgr., Dallas and in 1970 became S.W. Div. mgr; 1977–78, asst. gen. sls. mgr. for Southern, N.W. and S.W. division, operating out of Dallas. In June, 1978 appt. v.p., asst. gen. sls. mgr. of UA.

MOYLAN, WILLIAM J: (M.B.K.S.) Producer-director, script editor and researcher, journalist. Entered m.p. ind. 1927; credits include 59 features and over 200 shorts; in charge prod. in various studios 1933–42; prod. and ed. Indian News Parade for Indian Govt. and war corr. with India and South East Asia Commands, 1943–46; prod. information films of India 1947; Indep. prod. documentaries in Ireland 1948; dir. eastern language versions De Lane Lea Process Ltd., 1949–51; films advisor to Pakistan Govt. apptd. by Commonwealth Relations Office under Tech. Co-op. Scheme of the Colombo Plan, 1952–58; overseas adv. De Lane Lea Process, 1959–60. Founded Munster House Prods. in Ireland, 1961.

MUGGERIDGE, MALCOLM: Journalist, TV interviewer. b. Sanderstead Surrey, Mar. 24, 1903; e. Selhurst Grammar School & Selwyn Coll., Cambridge. Lecturer at Egyptian

Univ., Cairo, 1927–30; served on several newspapers including Manchester Guardian, Calcutta Statesman, Evening Standard; with Army (Legion of Honor, Croix de Guerre with Palm Medaille de la Reconnaissance Francaise), 1939–45; Daily Telegraph Washington Correspondent, 1946–47; deputy editor, Daily Telegraph, 1950–52; editor, Punch, 1953; ent. TV 1951 as interviewer on Panorama program.

MUHL, EDWARD E.: Executive and Producer. b. Richmond, Ind. Feb. 17, 1907; gen. mgr., Universal 1948–53; v.p., chg. production 1953–68.

MUIR, GRAEME: TV producer, actor. b. London, Feb., 1916; e. Oundle and St. John's, Oxford. Awarded Silver Medal Royal Academy of Dramatic Art 1938; first broadcast in Alcestis, 1940; many TV plays since; now freelance producer.
TV: Its Magic, The Dancing Years, Showtime, Perchance to Dream, Gilbert and Sullivan, Brothers In Law, Mr. Justice Duncannon, Christmas Night with the Stars, The Marriage Lines, Charley's Aunt, Comedy Playhouse, The Whitehall Worrier, Harry Worth, Not in Front of the Children, The Very Merry Widow, Bachelor Father, Birds on the Wing, My Wife Next Door, Oh Father!, Second Time Around, It Ain't Half Hot Mum, Oneupmanship, You're Only Young Twice, Life Begins at Forty, Hows' Your Mother?

MUIR, E. ROGER: Producer. b. Canada, Dec. 16, 1918. e. U. of Minn. Partner Minn Advertising Services Co., Photographer, Great Northern Railway; motion picture producer Army Signal corps; NBC TV producer, Howdy Doody, exec. producer, Concentration. Now pres. Nicholson-Muir Prods, TV program packager, U.S. Spin-Off, Pay Cards, Canada Pay Cards, Headline Hunters, Definition, Celebrity Dominoes; co-creator Newlywed Game, exec. prod. I Am Joe's Heart, I Am Joe's Lung, I Am Joe's Spine, I Am Joe's Stomach, The New Howdy Doody Show, Supermates, Second Honeymoon, Groaner, Generation Jury.

MULDAUR, DIANA: Actress. Began on New York stage then turned to films and TV, appearing on numerous major network shows.
PICTURES INCLUDE: The Swimmer, Number One, The Lawyer, One More Train To Rob, The Other, Chosen Survivors.

MULHOLLAND, ROBERT E.: Executive. e. Northwestern Univ. Joined NBC News as newswriter in Chicago in 1962. In 1964 made midwestern field producer for Huntley-Brinkley Report. In 1964 moved to London as European producer for NBC News; 1967, named Washington producer of Huntley-Brinkley Report. Transferred to L.A. in 1967 to be director of news, west coast. Named exec. prod. of NBC Nightly News. In 1973 appt. v.p., NBC news.; 1974 exec. v.p. of NBC News. In 1977 appt. pres. of NBC Television Network; also elected to board of directors.

MULLIGAN, ROBERT: Director. b. Bronx, N.Y.; e. Fordham U., A.B. With CBS-TV, asst. supvr. in radio oper.; prod. asst. on Suspense, TV; then asst. dir. & dir.: TV Playhouse, NBC, Alcoa-Goodyear, Studio One, Playhouse 90, Hallmark Hall of Fame.
PICTURES INCLUDE: Fear Strikes Out (1957), The Rat Race, The Great Imposter, Come September, To Kill a Mockingbird, The Spiral Road, Love with the Proper Stranger, Baby the Rain Must Fall, Inside Daisy Clover, Up the Down Staircase, Pursuit of Happiness, Summer of '42, The Other, The Nickel Ride, Bloodbrothers, Same Time Next Year.

MURPHY, BEN: Actor. b. Jonesboro, Ark., March 6. e. Univ. of Ill. (B.S., political science).
TV: Alias Smith and Jones, The Name of the Game, Grif, Gemini Man (all series).

MURPHY, GEORGE: U.S. Senator. b. New Haven, Conn., July 4, 1902; e. Yale U. Toolmaker for Ford Co., miner, real estate agent, nightclub dancer, actor. On stage from 1927 (Good News, Of Thee I Sing, Roberta, etc.). Screen debut 1934 in Kid Millions. Member: Screen Actors Guild (pres.); Nat'l. Com. WAC. Spcl. Academy Award "for interpreting m.p. ind. correctly to country at large," 1940; chmn., Hollywood Coordinating Com.; bd. mem., USO Inc.; v.p., Acad. of M.P. Arts & Sciences; div. of pub. rel., MGM, 1954–58; bd. mem. & past pres., Screen Actors Guild; bd. mem., M.P. Relief Fund. Joined Desilu Productions as v.p. in chg. public affairs, 1959; v.p., bd. of dir., Technicolor Corp. Elected U.S. Senator, California.
PICTURES INCLUDE: I'll Love You Always, You're a Sweetheart, London by Night, Broadway Melody of 1938, Risky Business, Two Girls on Broadway, A Girl, a Guy and a Gob, Tom, Dick and Harry, Ringside Maisie, Mayor of 44th Street, For Me and My Gal, Powers Girl, This Is the Army, Up Goes Maisie, Arnelo Affair, Cynthia, Tenth Avenue Angel, Big City, Battleground, Border Incident, No Questions Asked, Talk About a Stranger, It's a Big Country, Walk East on Beacon, Jamboree (Boy Scout film), This is the Army, Broadway Melody.

MURPHY, JOHN F.: Theatre executive. b. Brooklyn, N.Y., Mar. 25, 1905; e. CCNY. Entire career with Loew's Theatres; started over 45 yrs. ago as asst. mgr. Hillside & Valencia Theatres, Jamaica, N.Y.; apptd. gen. mgr. in chg. of out-of-town Theatres, 1942–54; v.p. Loew's Theatres, Aug., 1954; dir., Loew's Theatres, Oct. 1956; exec. v.p., Loew's Theatre, 1959; ret. July 1963; continues on bd. as advisor and director.

MURPHY, MARY: Actress. b. Washington, D.C., Jan. 26; e. University High Sch., Santa Monica, 1949; m.p. debut in Lemon Drop Kid.
PICTURES INCLUDE: Carrie, When Worlds Collide, Atomic City, Turning Point, Come Back Little Sheba, Off Limits, Houdini, Main Street to Broadway, Wild One, Beachhead, Mad Magician, Make Haste to Live, Sitting Bull, Hell's Island, Desperate Hour, A Man Alone, Maverick Queen, Zex, Harlow, Junior Bonner.

MURPHY, RICHARD: Writer, director. b. Boston, Mass.; e. Williams Coll. Entered m.p. ind. 1937. Capt. U.S. Army Sig. Corps, SWPA, 1942–45. Contract writer, 20th Century-Fox, 1945–54. Writer-prod., 20th-Fox 1964–72; pres. Cinecom World Ent. Ltd., 1974.
PICTURES INCLUDE: Boomerang, Deep Waters, Cry of the City, Panic in the Streets, You're in the Navy Now, Les Miserables, Desert Rats, Broken Lance, Three Strips in the Sun, Wackiest Ship in the Army, Compulsion, Last Angry Man.
TV: Our Man Higgins, Screen Gems, creator, TV series, The Felony Squad.

MURRAY, BARBARA: Actress. b. London, England, Sept. 27, 1929. Stage debut in Variety, 1946; screen debut in Badger's Green, 1948. Various TV appearances.
PICTURES INCLUDE: Passport to Pimlico, Don't Ever Leave Me, Boys in Brown, Poets Pub, Tony Draws a Horse, Dark Man, Frightened Man, Mystery Junction, Another Man's Poison, Hot Ice, Street Corner (Both Sides of the Law), Meet Mr. Lucifer, Doctor At Large, Campbell's Kingdom, A Cry from the Streets, Girls in Arms.

MURRAY, BILL: Actor. b. Chicago. Was pre-med student; left to join brother, Brian, in Second City, the Chicago improvisational troupe. Appeared with brother on radio, National Lampoon Show, and in off-Bdwy. revue, of same name. Hired by ABC for Saturday Night Live; then by NBC.
PICTURES: Meatballs (debut), Where the Buffalo Roam, Caddyshack, Stripes.

MURRAY, DON: Actor-director, writer. b. Hollywood, Calif., July 31, 1929.
PLAYS: Insect Comedy, Rose Tattoo, The Skin of Our Teeth, The Hot Corner, Smith (a musical), The Norman Conquests.
PICTURES INCLUDE: Bus Stop, Bachelor Party, Hatful of Rain, The Hoodlum Priest, Advise & Consent, Baby the Rain Must Fall, Sweet Love, Bitter, The Cross and the Switchblade, Conquest of the Planet of the Apes, One Man's Way, The Plainsman, Escape From East Berlin, Shake Hands with the Devil, From Hell to Texas, Confessions of Tom Harris, Call Me by My Rightful Name, The Borgia Stick, Deadly Hero, Damien, Endless Love.

MURRAY, JAN: Performer. b. New York, N.Y., Performed in nightclubs, vaudeville, B'way, radio, TV, films. Was m.c., Songs for Sale, and Sing It Again, CBS-TV; TV guest star many programs; on Dollar a Second, ABC; Jan Murray Time, NBC; Treasure Hunt, ABC.
PLAYS: A Funny Thing Happened on the Way to the Forum.
PICTURES INCLUDE: Who Killed Teddy Bear?

MURRAY, KEN: Actor. r.n. Don't Court. b. New York, N.Y. On N.Y. stage as master of ceremonies; also Hollywood stage in Blackouts of 1942–49. Screen debut 1929 in Half Marriage. Since in numerous features and short subjects.
PICTURES INCLUDE: Leathernecking, You're a Sweetheart, Swing Sister Swing, A Night at Earl Carroll's, Swing It Soldier, Juke Box Jenny, Special Academy award 1947 (Bill and Coo)., Follow Me Boys, The Power, The Man Who Shot Liberty Valance, Sound of Flubber.
TV: Ken Murray Show TV; p. d., act.: Marshall's Daughter; prod.: Where Were You, El Coyote, Hollywood, My Home Town, Hollywood Without Makeup.
BROADWAY: Ken Murray's Hollywood.
LITERARY: The Golden Days of San Simeon—Doubleday, 1971, Now in 8th Edition.

MUSANTE, TONY: Actor. b. Bridgeport, Conn., June 30, 1936. e. Oberlin College. Taught High School, directed local theatre, then appeared off-Broadway and on Dupont Show of the Month (Rice With Terror).
PICTURES INCLUDE: Once A Thief, The Incident, The Detective, A Professional Gun, The Love Circle, The Bird With The Crystal Plumage, The Last Run, Anonymous Venetian.
TV: Toma (series).

MUSCHAMP, THOMAS: Writer. b. London, Eng., May 27, 1917; e. Wm. Ellis School, Hampstead. Prof. prod. plays include: One Man's Meat, Bella Vista, Out of Season, Bridge of Sighs, Monday to Saturday, I Dreamt I Dwelt in Marble Halls, Stranger in Manhattan, Tie the Noose Tighter. TV scripts include: A Girl for George, Bridge of Sighs; in German, Gwendoline.

MUSTO, MICHAEL J.: Producer-writer B. N.Y.C. Jan. 16, 1917; e. S. Niagara U. Apprentice for J. J. Shubert. Comedy Workshop, N.Y.C. W-Co.P Hellzapoppin, Wrote, Prod. Dir. Burlesque shows and stage revues. Prod. Industrial Films, Commercials, Operas, Fund Raising Shows. Prod. Films: The Glass House, Phenomena 7-7, Down Tin Pan Alley, Single Room Furnished, Educated Heart, Agnes, Spring Fancy, The Atheist, Man Who Cried Wolf, Several Robert Montgomery Presents, Hallmark Hall of Fame, TV Specials, The Bold Ones, Strange Is The Wind. Wrote: Due-Bill Marriage, Spring Fancy, Granada, Seppi and His Brothers, Amato and The Yenta, The Cliffdwellers, Other. Owner. Empire Films and Cinema City Studios. Manager, Ruth Burch.

MUTSU, IAN YONOSUKE: Producer-Distributor. b. London, England, Jan. 14, 1907. e. Univ. of Birmingham, England. Journalist 1931–48. Daily Express, UPI, etc.; Japan repr. MGM-Hearst Metrotone News, U.S. Newsreel Pool, later Telenews, UPI-Movietone News, ITN London; 1952 pres. International Motion Picture Co. for news coverage, documentaries, custom films for worldwide clients; 1966 agent BBC TV sales, Modern Talking Picture Service NYC; 1973 founded Mutsu Inc. (Canada). Productions include numerous documentaries on Japan, Japanese background flms for overseas clients.

MYERS, JULIAN F.: Public relations. b. Detroit, Mich., Feb. 22, 1918; e. Wayne Univ., 1935–37, Univ. Southern Calif., 1937–39. Distribution, Loew's Detroit, 1941–42; asst. story editor, idea man, Columbia, 1942–46; publicist, 20th Century-Fox, 1948–62; public relations, Julian F. Myers, Inc., 1962; pres., Myers Studios, Inc., 1966; pres., New Horizons Broadcasting Corp., 1968–69; snr. publicist American Intl. Pictures, 1970–80. Executive Board Hollywood Press Club; member Variety Clubs; Academy of Motion Pictures Arts & Sciences; Board of Governors Film Industry Workshops, Inc. 1977, western vice-pres., The Publicists Guild; Recipient of Publicists Guild's Robert Yeager Award; 1979, re-elected western v.p., Publicists Guild. First male member Hollywood Women's Press Club. Co-founder HANDS (Hollywood Answering Needs of Disaster Survivors). Member, m.p. Pioneers. Winner, 1980 Publicists Guild Les Mason Award. Instructor in publicity, UCLA, 1979. Filmways Pictures, pub. dept., 1980–81.

MYERS, PETER S.: Executive. b. Toronto, Ont., Canada, May 13, 1920; e. U. of Toronto, R.C.A.F. 1941–46. Salesman, Warner, 1946; Toronto br. mgr., Eagle Lion, 1947; Toronto br. mgr., 20th Century-Fox, 1948; Canadian div. mgr., 1951; Canadian gen. mgr., 1955; man. dir. Canada, 1959; gen. sales mgr. in chg. of dom. distribution, 1968; v.p., 1969; named snr. v.p., domestic dist., 1979; snr. v.p., 20th-Fox Entertainment, 1980.

MYERSON, BERNARD: Executive. b. New York, N.Y., March 25, 1918. Entered m.p. ind. with Fabian Theatres, 1938–63, last position as exec. v.p.; joined Loew's Theatres as v.p., 1963; exec. v.p. and board member, Loew's Corp.; pres. Loew's Theatres, 1971. Mem. of Executive Committee Greater N.Y. Chapter, National Foundation of March of Dimes; Board of Directors and President of Will Rogers Memorial Fund; Mem. National Assn. Theatre Owners (dir., exec. com.); Motion Pictures Assn. Am. (dir.); Motion Picture Pioneers (pres.) Clubs: Variety (dir.).

MYERSON, BESS: Executive, entertainer. b. New York, N.Y.; e. Hunter Coll., 1945; hon. degree, Doctor of Humane Letters, L.I. Univ., Doctor of Public Service, Seton Hall, 1972. Miss America 1945; guest soloist with N.Y. Philharmonic, 1946; starred in several network TV programs; co-star, The Big Payoff; panelist, I've Got a Secret, Tournament of Roses, Miss America Pageant, Candid Camera. Commissioner, N.Y. City Dept. of Consumer Affairs, 1969–73. Consumer consultant, Citibank and Bristol-Myers. Syndicated columnist, contributing ed. Red Book Magazine, TV & Radio consumer documentaries.

N

NADEL, ARTHUR: Producer, director, writer. b. New York, N.Y. Film editor for Paramount, 20th Century-Fox, Walt Disney, U.S. Air Force m.p. div., United Artists; superv. editor, McCann-Erickson; prod., dir., writer, Universal; v.p., Levy-Gardner-Laven.
PICTURES INCLUDE: Clambake, Lola, Underground, No Trumpets, No Drum.
TV: The Rifleman, The Plainsman, Great Adventure, Arrest and Trial, Kraft Theatre, The Virginian, Big Valley, Daniel Boone, Cowboy in Africa, Bonanza, Delphi Bureau, Banyon, Streets of San Francisco: NBC Specials, Welcome Home (Emmy Winner): This Year in Jerusalem, Vortex in Oatmeal, Universal Studios, The Chase: NBC Special, Crime Without Punishment: executive consultant Filmation Studios, Shazam, The Secrets of Isis.

NADER, GEORGE: Actor. b. Pasadena, Calif.; e. Occidental College, B.A., Pasadena Playhouse. Served in U.S. Navy. Many TV appearances, m.p. debut in Monsoon. First novel, Chrome, recently published by Putnam's.
PICTURES INCLUDE: Carnival Story, Miss Robin Crusoe, Sins of Jezebel, Fours Guns to the Border, Six Bridges to Cross, Lady Godiva, Second Greatest Sex, Away All Boats, Congo Crossing, Unguarded Moment, Four Girls in Town, Man Afraid, Joe Butterfly, Nowhere to Go, The Secret Mark of D'artagnan, The Great Space Adventure, Zigzag, Walk by the Sea, The Human Duplicators, Sumuru, House of a Thousand Dolls, Alarm on 83rd Street, Murder at Midnight, Count-Down for Manhattan, Dynamite in Green Silk, The Check and Icy Smile, The Murder Club From Bklyn, Death in a Red Jaguar, End Station of the Damned, Bullets on Broadway, Beyond Atlantis.
TV: Letter to Loretta, Fireside Theatre, Chevron Theatre, Ellery Queen, Man and the Challenge, Shannon.

NAIFY, MARSHALL: Executive. b. Sacramento, Calif., March 23, 1920; e. U. So. Calif. U.S.A.F. Ch. exec. comm., United Artists Theatre Circuit Inc.; pres., Magna Pictures Corp.

NAIFY, ROBERT: Pres., United Artists Theatre Circuit Inc.; exec. v.p., Magna Pictures Corp.

NAKAMURA, MOTOHIKO: Executive. b. Tokyo, Japan, Aug. 10, 1929. e. U. of California at Berkeley, U. of Pennsylvania, Tokyo University. Joined Fuji Film Co., Ltd. in Tokyo, 1953; assigned to Japan Camera Center, New York City, 1956–67; North American rep. for Fuji Photo Film Co., Ltd., 1957–63; assistant export sales mgr., Fuji Photo Co., Ltd. in charge of international marketing of all products, 1964–71; exec. v.p., resident mgr., Fuji Photo Film U.S.A., Inc.

NALLE, BILLY: Theatre concert organist, popular field, ASCAP composer. b. Fort Myers, Fla.; graduate, The Julliard School. Over 5000 major TV shows from New York; now artist-in-residence, Wichita Theatre Organ, Inc. Reader's Digest, Telarc & WTO Records Artist. Public Relations: Billy Nalle Music, Wichita.

NAPIER, ALAN: Actor. b. In 1930, Caste, Nettleford; In a Monastery Garden, Hagen; Loyalties, ATP, 1933; Wings Over Africa, Premier Stafford, 1936; For Valour, Capitol, 1937; Wife of General Ling, Four Just Men, 1938.
PICTURES INCLUDE: The Uninvited, The Song of Bernadette, Action in Arabia, Ministry of Fear, The Hairy Ape, Mlle. Fifi, Dark Waters, Hanover Square, Isle of the Dead, Three Strangers, High Conquest, Driftwood, Forever Amber, Macbeth, Joan of Arc, Hills of Home, Tripoli, Highwayman, Strange Door, Blue Veil, Across the Wide Missouri, Big Jim McLain, Young Bess, Julius Caesar, Desiree, Moonfleet, Court Jester, Miami Expose, Mole People, Until They Sail, Journey to the Center of the Earth, Batman.
TV: Many TV dramatic shows and Batman series.

NARIZZANO, SILVIO: Producer-Director. b. Montreal, Canada, 1927. e. Univ. of Bishop's, Lennoxville, Quebec. (B.A.). Was active as actor-director in Canadian theatre before going to England for TV and theatrical film work.
PICTURES INCLUDE: Director: Under Ten Flags (co-dir.), Die! Die! My Darling!, Georgy Girl, Blue, The Man Who Had Power Over Women, Loot, Redneck, The Sky Is Falling. Producer: Negatives, Fadeout, Redneck.

NASH, N. RICHARD: Writer.
BROADWAY: author: Second Best Bed, The Young and Fair, See the Jaguar, The Rainmaker, Girls of Summer, Handful of Fire, Wildcat, 110 in the Shade, The Happy Time, Echoes.
PICTURES INCLUDE: Nora Prentiss, Vicious Years, The Rainmaker, Porgy and Bess, Sainted Sisters, Dear Wife, Welcome Stranger, Dragonfly.
TV: Many TV plays for Television Playhouse, U.S. Steel, General Electric.
NOVEL: Cry Macho, East Wind, Rain, The Last Magic, Aphrodite's Cave.
PHILOSOPHY: The Athenian Spirit, The Wounds of Sparta.

NATHAN, WYNN: Executive. e. U. of So. Calif. Casting dept., Universal, 1941; then U.S. Air Force 4 yrs.; asst. oper. mgr., Universal, 1946; talent agent, Wm. Morris Agency; talent agent, Century Artists; then United TV Programs; v.p., MCA, 1954; v.p. chg. syndication & bd. of dir., MCA; pres., TV Marketeers, 1961; v.p. & gen. mgr., Metromedia Program Sales, 1966; vice president, Worldwide Syndication Time-Life Television 1970-May, 1981. Now pres. & chief exec. officer, Lionheart TV Intl.

NATHANSON, MORTON: Publicist. b. Fayetteville, Tenn., July 27, 1918; e. N.Y.U. Legitimate stage press agent for Katharine Cornell, Group Theatre, Theatre Guild, Michael Todd, Playwrights' Co., others. Entered m.p. ind. as publicity staff writer, United Artists; then Eastern adv., pub. dir., Samuel Goldwyn Prod. and Liberty Films; pub. mgr., Paramount, Horizon; pub. dir., U.A., 1952–60; dir., internat'l adv., pub., U.A., 1960–62; headed own public relations co., 1962–67, then v.p., Ruder & Finn, Inc., 1967–71; public relations dir., Nat'l. Multiple Sclerosis Society, 1972–73.

NATHANSON, NAT: Executive. b. Detroit, Mich. Entered m.p. ind. 1935 with United Artists as salesman, Denver exch.; to Chicago, 1937; mgr., Milwaukee office, 1941–44; br. mgr., Chicago 1944; br. mgr., San Francisco 1945; br. mgr., Chicago 1947; apptd. Eastern & Canadian gen. sales mgr., Feb., 1950; sales exec., Allied Artists, 1952; midwest dist. mgr., Chicago, Sept. 1954; asst. gen. sales mgr., 1963; gen. sls. mgr., 1965. Member: Variety (chief Barker, Tent 26, Chic.; Charter mem. Tent 32, San Francisco), Variety Internat'l Property Master, 1954–55. Pres., Producers Releasing Organization, 1966; cent. div. mgr, 20th Century-Fox, 1967; Eastern & Canadian sales mgr., Four Star Excelsior, 1970; Central Divison mgr., Allied Artists Pictures Corporation headquartering in Chicago, 1970; International Ambassador Variety Club International, 1973–79.

NATWICK, MILDRED: Actress. b. Baltimore, Md., June 19, 1908; e. Bryn Mawr Sch., Baltimore, Bennett Sch., Millbrook. Prof. stage debut in Carry Nation, 1932; London debut in Day I Forget.

PLAYS: Wind and the Rain, Distaff Side, End of Summer, Love From a Stranger, Candida, Missouri Legend, Stars in Your Eyes, Grass Harp, Blithe Spirit (Barter Theatre award), Waltz of the Toreadors (nominated for Tony), The Firstborn, The Good Soup, Critic's Choice, Barefoot in the Park, Our Town, Landscape, 70, Girls 70 (nominated for Tony), Bedroom Farce.

PICTURES INCLUDE: Long Voyage Home, Enchanted Cottage, Yolanda and the Thief, Late George Apley, Woman's Vengeance, Three Godfathers, Kissing Bandit, She Wore a Yellow Ribbon, Cheaper by the Dozen, Quiet Man, Against All Flags, Trouble with Harry, Court Jester, Teenage Rebel, Tammy and the Bachelor, Barefoot in the Park (nominated for Oscar), If It's Tuesday This Must be Belgium, Trilogy, The Maltese Bippy, Daisy Miller, At Long Last Love.

TELEVISION: Blithe Spirit (nominated for Emmy), House Without an Xmas Tree, Thanksgiving Treasure, Money to Burn, The Snoop Sisters (Emmy award), The Easter Promise, Little Women, McMillan & Wife, Hawaii Five-O, Love Boat, You Can't Take It With You.

NAZARRO, CLIFF: Actor. b. New Haven, Conn., January 31, 1904. Stock, vaudeville, master of ceremonies in all leading theatres. Appeared with "Two Marjories" in Vitaphone No. 2116. In 1936: "Romance Rides the Range," Spectrum.

PICTURES INCLUDE: Outside of Paradise, a Desperate Adventure, King of the Turf, St. Louis Blues, Arise, My Love, In Old Colorado, Mr. Dynamite, Artists and Models Abroad, Dive Bomber, Night of January 16, World Premiere, Rookies on Parade, Sailors on Leave, Blondie Goes to College, Melody for Three, You'll Never Get Rich, Pardon My Stripes, Call of the Canyon, Hillbilly Blitzkrieg, Shanty-town, I'm from Arkansas, Swing Hostess, Gentleman Joe Palooka.

NEAGLE, ANNA (DAME): (D.B.E.) Actress, producer. r.n. Marjorie Robertson. b. Forest Gate, England, Oct. 20, 1904; m. Herbert Wilcox C.B.E. Teacher of dancing, world championship finalist. On London stage, m.p. debut 1929 in Mary Was Love (as Marjorie Robertson), then changed her name to Anna Neagle and appeared in Should a Doctor Tell? and The Chinese Bungalow. Then put under contract by Herbert Wilcox. Voted top British moneymaking actress Motion Picture Herald-Fame poll six years, 1947–52. Voted top box office star in the world 1949. International Gold Medal five times as Best Actress of the Year. National Film Award for Spring in Park Lane, 1948; Odette in 1951. Created C.B.E. in 1952. D. B. G. 1969.

PLAYS: London stage: In Andre Charlot and C. B. Cochran revues, including This Year of Grace and Wake Up and Dream (London and Broadway); starred in Stand Up and Sing; others: As You Like It, Twelfth Night, Peter Pan, Emma, The Glorious Days, The More The Merrier, 1965–71. Charlie Girl. 1973; No, No Nanette.

PICTURES INCLUDE: Goodnight Vienna, The Flag Lieutenant, The Little Damozel, Bitter Sweet, The Queen's Affaire, Nell Gwyn, Peg of Old Drury, Limelight, The Three Maxims, London Melody, Victoria the Great, Sixty Glorious Years, (to Hollywood 1939–42 for:) Nurse Edith Cavell, Irene, No No Nanette Sunny, and Forever and a Day. (Back to England to star in:) They Flew Alone (in U.S. Wings and the Woman), Queen Victoria, The Yellow Canary, I Live in Grosvenor Square (in U.S. A Yank in London), Piccadilly Incident, The Courtneys of Curzon Street, Spring in Park Lane, Elizabeth of Ladymead, Maytime in Mayfair, Odette, The Lady With a Lamp, Derby Day, Lilacs in the Spring (in U.S. Let's Make Up), King's Rhapsody, My Teenage Daughter, No Time for Tears, The Man Who Wouldn't Talk, The Lady Is a Square. Produced: These Dangerous Years, Wonderful Things, The Heart of a Man.

TV: The Spice of Life (BBC-TV), What's My Line (BBC-TV), Juke Box Jury (BBC TV), Day by Day (Southern Television), Sunday Night at the London Palladium (AtV), The Elstree Story (ABC), The Magic Box (ABC), The Golden Shot (ATV) and numerous interviews. Her two television drama appearances in A Letter From the General (Anglia), and Shadow on the Sun (Anglia) both had record viewing audiences of over 20 million.

NEAL, PATRICIA: Actress. b. Packard, Ky., Jan. 20, 1926; e. Northwestern U. Doctor's asst., cashier, hostess, model, jewelry store clerk. In summer stock; Broadway debut in "Another Part of the Forest" 1947, winning the Donaldson & Drama Critic Awards, also in Children's Hour. M.p. debut in John Loves Mary, 1948. Academy Award—Best Actress, 1963.

TV: The Bastard.

PICTURES INCLUDE: Fountainhead, Hasty Heart, Bright Leaf, Three Secrets, Breaking Point, Raton Pass, Operation Pacific, Day the Earth Stood Still, Weekend With Father, Diplomatic Courier, Washington Story, Something for the Birds, Face in the Crowd, Hud (Academy Award), Psych 59, In Harms Way, The Subject Was Roses, The Night Digger, Baxter!, Happy Mother's Day . . . Love, George, The Passage (1978), Ghost Story.

NEAME, RONALD: Cinematographer, producer, director. b. 1911; e. U. Coll. Sch., London; son, Elwin Neame, London photog. & Ivy Close, m.p. actress. Entered m.p. ind. 1928; asst. cameraman on first full-length Brit. sound "blackmail," dir. by Alfred Hitchcock 1929; became chief cameraman & lighting expert, 1934 in 1945 joint assoc. prod., Noel Coward Prods.

PICTURES INCLUDE: Elizabeth of England, Invitation to the Waltz, Brief Ecstasy, It's in the Air, Gaunt Stranger, Four Just Men, Major Barbara, A Yank in the R.A.F. (Brit. flying sequence), One of Our Aircraft is Missing, In Which We Serve, This Happy Breed, Blithe Spirit, Brief Encounter, Great Expectations, Oliver Twist, Passionate Friends, Take My Life, Magic Box, Golden Salamander, The Card (The Promoter), Million Pound Note (Man With a Million), Man Who Never Was, Seventh Sin, Windom's Way, The Horse's Mouth, Tunes of Glory, Escape From Zahrain, I Could Go on Singing, The Chalk Garden, Mister Moses, Gambit, The Prime of Miss Jean Brodie, Scrooge, The Poseidon Adventure, The Odessa File, Meteor, Hopscotch, First Monday in October.

NEBENZAL, SEYMOUR: Producer b. New York City. July 22, 1899. Began prod. 1925 in Berlin (M. Ariane, West Front 1918, Beggar's Opera): then prod. Mayerling, Betrayal in Paris. Prod. in U.S. from 1940; in 1945 pres. & exec. prod., Nero Films, Inc., Hollywood.

PICTURES INCLUDE: We Who Are Young, Summer Storm, Whistle Stop, The Chase, Heaven Only Knows, M.

NEEDHAM, HAL: Director. In industry over 25 years. Was popular stuntman and second unit action director before full directorial debut with Smokey and the Bandit.

PICTURES: Hooper, The Villain, Smokey and the Bandit II, The Cannonball Run.

TELEVISION: Hal Needham's Wild World of Stunts (syndicated series he wrote, directed and starred in): Directed Death Car on the Freeway (movie); Stunts Unlimited.

NEFF, HILDEGARDE: Actress, author. r.n. Hildegard Knef. b. Ulm, Germany, Dec. 28, 1925; e. Art Acad., Berlin. Film cartoonist for UFA, Berlin; on Berlin stage after war; appeared in German films: Murderers Are Among Us, Between Yesterday and Tomorrow, Film Without Title, The Sinner. On B'way in: Silk Stockings. U.S. m.p. debut in Decision Before Dawn. Author of best-selling autobiography, 1971.

PICTURES INCLUDE: Diplomatic Courier, Night Without Sleep, Snows of Kilimanjaro, Holiday for Henrietta, Man Between, Svengali, Subway in the Sky, Mozambique.

NEGULESCO, JEAN: Director. b. Rumania (Craiova), Feb. 29, 1900; e. Liceul Carol Univ., Rumania. Stage dir., artist, painter, Came to U.S. in 1927.

PICTURES INCLUDE: The Mask of Dimitrios, The Conspirators, Nobody Lives Forever, Three Strangers, Humoresque, Deep Valley, Johnny Belinda, Road House, Forbidden Street, Three Came Home, Under My Skin, Mudlark, Take Care of My Little Girl, The Full House, Phone Call From a Stranger, Lydia Bailey, Lure of the Wilderness, Titanic, Scandal at Scourie, How to Marry a Millionaire, The Rains of Ranchipur, Woman's World, Three Coins in The Fountain, Daddy Long Legs, Boy on a Dolphin, The Gift of Love, A Certain Smile, Count Your Blessings, The Best of Everything, Jessica, The Pleasure Seekers, Hello, Goodbye.

NELSON, BARRY: Actor. r.n. Robert Neilson. b. Oakland, Calif.; e. U. of Calif. London stage: No Time for Sergeants, 1957.
PICTURES INCLUDE: A Guy Named Joe, Winged Victory, Man with My Face, First Traveling Saleslady, Mary, Mary, The Borgia Stick, Airport, Pete 'n Tillie, The Shining.
BROADWAY: Light Up the Sky, Rat Race, Moon is Blue, Mary, Mary, Cactus Flower, Everything in the Garden, Seascape, The Norman Conquests, The Act.
TV: series: The Hunter, My Favorite Husband, Washington: Behind Closed Doors.

NELSON, DAVID: Actor. b. New York, N.Y., Oct. 24, 1936; e. Hollywood H.S., U. of Southern Calif.
PICTURES INCLUDE: Here Comes the Nelsons, Peyton Place, The Remarkable Mr. Pennypacker, Day of the Outlaw, The Big Circus, "30," The Big Show, No Drums, No Bugles, The Wheel, The Sinners.
TV: Adventures of Ozzie and Harriet; dir.: Easy To Be Free (special), OK Crackerby series.

NELSON, EDWARD, J.: Executive. b. July 15, 1911, St. Louis, Mo. e. Iowa State. Started as commercial m.p. & aerial photographer. With National Theatre Supply, 1934–42. AAFMPS, mgr., engineering E.TO., 1942–52. Joined Ballantyne of Omaha in 1952; made pres. 1973, which post now holds. Member: T.E.A. pres., May 1974–76. and bd. mbr., 1970–74. SMPTE. Variety Club. Nominated for Academy Award 1973 for designing VIP, automated projection system.

NELSON, GENE: Dancer, actor. r.n. Gene Berb. b. Seattle, Wash., March 24, 1920; e. Santa Monica, Calif. H.S. Began dancing and ice skating in school; joined Sonja Henie ice show; played in This is the Army, World War II; after discharge, featured in It Happens on Ice, Center Theatre, New York. To Hollywood for I Wonder Who's Kissing Her Now; joined Hollywood group prod. stage musical, Lend an Ear; to Warner for Daughter of Rosie O'Grady.
PICTURES INCLUDE: Tea for Two, West Point Story, Lullaby of Broadway, Painting the Clouds With Sunshine, Starlift, She's Working Her Way Through College, She's Back on Broadway, Three Sailors and a Girl, Crime Wave, Oklahoma, Way Out, Atomic Man, So This Is Paris. Directed, The Hand of Death, Hootenany Hoot, The Cool Ones, Your Cheatin' Heart, Kissin' Cousins, Wake Me When the War is Over, The Letters, Harum Scarum.
TV: Directed Mod Squad, I Dream of Jeannie, Hawaii Five-O, Farmer's Daughter, Donna Reed Show, Burke's Law, Felony Squad, Laredo, The Rifleman, The Wackiest Ship, Iron Horse, FBI, The Rookies.
Broadway: Follies, 1970; Music, Music, 1974.

NELSON, HARRIET: r.n. Harriet Hilliard. b. Des Moines, Iowa, July 18; e. High Sch., Kansas City, m. Ozzie Nelson. Appeared in dramatic & musical roles in shows; singer with Ozzie Nelson band; on radio shows: Believe It or Not, Seeing Stars, Red Skelton, Adventures of Ozzie & Harriet. Appeared in m.p.: Here Come the Nelsons.
TV: Adventures of Ozzie & Harriet, ABC-TV.
STAGE: Marriage-Go-Round; rec.: Ozzie & Harriet.
PLAYS: Impossible Years, State Fair.

NELSON, LORI: Actress. r.n. Dixie Kay Nelson; b. Santa Fe, N.M., Aug. 15, 1933; e. High Sch., L.A. Child actress; photographer's model; m.p. debut in Ma and Pa Kettle at the Fair.
PICTURES INCLUDE: Bend of the River, Francis Goes to West Point, All I Desire, All-American, Walking my Baby Back Home, Tumbleweed, Underwater, Destry, Revenge of the Creature, I Died a Thousand Times, Sincerely Yours, Mohawk, Day the World Ended, Pardners, Hot Rod Girl.
TV: How to Marry a Millionaire, Wagon Train, Laramie, Bachelor Father.

NELSON, RALPH: Producer, director, playwright, actor. b. New York City. Stage actor 1933–41; Wrote play, Angels Weep, for Cleveland Playhouse, 1943; also two Bdwy. plays: Mail Call (1944) and The Wind Is Ninety (1945). Directed plays: Here's Mama (1952); The Trouble-Makers (1953) and The Man in the Dog Suit (1957). From 1948 to 1960 directed over 1,000 dramatic shows on all TV networks. Has made film appearances in many pictures he produced and/or directed, now president Rainbow Productions. Honorary Doctor of Humane Letters, Columbia College, 1975.
PICTURES INCLUDE: Director: Requiem for a Heavyweight, Lilies of the Field, (also prod.) Soldier in the Rain, Fate Is the Hunter, Father Goose, Duel at Diablo, (also prod.) Charly, (also prod.) Counterpoint, Tic-Tic-Tic, Soldier Blue, Flight of the Doves, (also prod.) The Wrath of God, (also prod.) The Wilby Conspiracy, Embryo, A Hero Ain't Nothin' But a Sandwich, Because He's My Friend, You Can't Go Home Again. Screenplays: The Man in the Funny Suit, Flight of the Doves, The Wrath of God.
TV: Director: Playhouse 90, Studio One, Philco Playhouse, G.E. Theatre, Ford Startime, Omnibus, ABC Theatre, Elgin Hour, Dick Powell Theatre, Dupont Show of the Month, Front Row Center, Climax, Desilu Playhouse, and specials: Cinderella, Aladdin, This Happy Breed, Nutcracker Suite. Producer: Climax Series, Studio One, The Old Vic Hamlet, The Man in the Funny Suit, Doyle Against the House, Three Soldiers, John J. Digges, Farmer's Daughter (pilot). Christmas, Lilies of the Field.

NELSON, RAYMOND E.: Executive. b. Cleveland, O., Feb. 12, 1907; e. Baldwin-Wallace Univ. Eastern prod. mgr., NBC: supervisor of TV, Mutual; prod.-conductor, light operas; stock co. dir.; prod.-dir. TV shows on all networks: Television Follies, Fashions on Parade, Western Theatre, Boys from Boise, etc.; prod.-dir.-writer, Rod and Gun Club of the Air on Mutual, radio and TV; prod., dir., Story Time, ABC-TV. Pres., Keystone Ad Agency, Nelson Prod., Inc.; dir. of spot sales, TV Bureau of Adv.; v.p., gen. mgr., NTA Film Network.

NELSON, RICK: Actor, singer. r.n. Eric Hilliard Nelson. b. Teaneck, N.J., May 8, 1940; e. Hollywood H.S. Recording artist, Decca Records.
PICTURES INCLUDE: Here Comes the Nelsons, The Wackiest Ship in the Army, Rio Bravo, A Story of Three Loves, Love and Kisses.
TV: Adventures of Ozzie and Harriet.

NEPOMUNCENO, LUIS: Producer, director. b. Manila, Philippines, July 24, 1930; e. De La Salle College, Manila, Phil., New York Inst. of Photography, N.Y. Tech. Ass't. to pres., Nepomunceno Productions, 1946–54; gen. mgr., Mepro, Inc., 1954–59; pres., Film Advertising Media Exhibitions, Inc., 1959 to date; pres., Luis Nepomunceno & Sons, Inc., 1961 to date; pres., Filipino Theas. Ass'n., 1969. Awards include: Filipino Academy of Motion Pictures Arts & Sciences 3 Awards 1968, 8 Awards 1969. Grand Prix, 2nd & 3rd Manila Film Festivals; 6 Awards, 4th Manila Film Festival. Asian Film Festivals: 2 Awards 14th Festival, 4 Awards 15th Festival.
PICTURES INCLUDE: Tears in the Dark, Manila, Open City, Because of a Flower, Heaven on Earth, Mountain Maiden, The Beggar, A Time for Dying, The Hunted.

NETTER, DOUGLAS: Motion Picture and Television Producer. V.P., gen. mgr., Todd A.O., 1955–1957. Sam Goldwyn Prods., N.Y. gen. mgr., 1958–1960. Formed own company rep. producers, 1961–1967. V.P Jalem Prod., 1968–1969. Exec. v.p. MGM, 1969–1974. Produced Mr. Ricco in 1975. American Co-Producer, The Wild Geese, 1977. Producer—Louis L'Amour's The Sacketts, 4 hour mini-series for NBC 1978. Exec. Producer—The Buffalo Soldiers, NBC pilot 1979. Produced Brian Garfield's Wild Times, a 4 hour mini-series, Metromedia, 1980. Produced Roughnecks, 4 hour mini-series, Metromedia, 1980. L'Amour's Cherokee Trail for Walt Disney Productions and CBS, 1981.

NETTLETON, LOIS: Actress. e. Studied at Goodman Theatre, Chicago. Replaced Kim Hunter in Darkness At Noon on Broadway.
PLAYS INCLUDE: Cat on a Hot Tin Roof, Silent Night, Lonely Night, God & Kato Murphy, The Wayward Stork, The Rainmaker, A Streetcar Named Desire.
PICTURES INCLUDE: Period of Adjustment, Come Fly With Me, Mail Order Bride, Valley of Mystery, Bamboo Saucer, The Good Guys and the Bad Guys, Dirty Dingus Magee, The Sidelong Glances of a Pigeon Kicker, The Honkers, Echoes of a Summer.

NEUMAN, ALAN: Producer, director. b. New York City. July 25, 1924; e. N.Y. Univ., B.A. American Theatre Wing. Dir., NBC-TV, 1947; prod., dir. of over 4000 TV shows; winner of 48 Nat'l Awards for excellence in TV. Now pres., Alan Neuman Prods.
TV: Exploring the Unknown (NBC Big Event), Person to Person, America Pause-Coke Special, Wide Wide World, General Motors Special Show, Inner Sanctum film series, Colgate Comedy Hour, Lights Out, Blind Date, Chevrolet Teletheatre, We the People, Kate Smith Show, Believe It or Not, Little Show, Dagmar's Canteen, Meet the Press, Ellery Queen; exec. prod.: College of the Air, Beat the Odds, Let. T, Laugh-In, This Is Your Life, You Asked For It.

NEWBERY, CHARLES BRUCE: Executive. b. Melbourne, Australia; e. All Saints Grammar School, Melbourne and Melbourne U. Entered m.p. ind. with Hoyt's Theatres, Ltd., Melbourne, 1929; publ. mgr., Fox Studios, Eng., 1934; controller, Fox Newsreel Theatres, 1935; managing dir., Fox Films, India, 1937; Supvr., India, China, Malaya, Fox Films, 1940; Film advisor, Government of India, 1941; estab. film studios & Newsreel for Government of India. suprv. Far East & Australia, Republic, 1945; supvr. Middle, Near and Far East. Republic, 1947; supvr. Eng. & Europe, Republic, 1948; estab. Republic Productions Great Britain, 1952; v.p. bd. mem. dir. of sales, Republic, U.S.A., 1953; pres., Charles Newbery Assoc., Newbery-Warden Associates, London, 1960; mgr. dir., Santor Film Prods., Amerion Prods, Inc., 1964; sen. v.p., bd. mem., administration, In-Flight Motion Pictures Inc., v.p., Intransit Motion Pictures, Inc., 1965. Senior v.p. Sales and Marketing 1967, President and Chief Operating Office, May 1973; vice chm. bd., Inflight, June 1979; dir. of mktg., Life Services Co.,

of America June, 1979; vice chm. bd., Inflight and sls.—Europe, East and Africa, 1980.
PICTURES INCLUDE: Deadlier Than the Male, Catch Me if You Can, Kiss Her Goodbye.

NEWBROOK, PETER: Producer and director of photography. Entered m.p. ind. 1934, Warner Bros. Studios, Teddington. Chmn. Titan Film Distributors Ltd.
PICTURES INCLUDE: After working on such pictures as The Sound Barrier, The Captain's Paradise, Hobson's Choice, Summer Madness, The Deep Blue Sea, Anastasia, The Bridge on the River Kwai; became dir. photog.: Lawrence of Arabia (2nd unit), 1961; In The Cool of the Day, That Kind of Girl, 1962; The Yellow Teddybears, Saturday Night Out, 1963; The Black Torment, 1964. Prod. and photog; Gonks Go Beat, 1965; prod. and photog; The Sandwich Man, 1966; Press For Time, Corruption, The Smashing Bird I Used to Know, Bloodsuckers, She'll Follow You Anywhere, Crucible of Terror, The Asphyx, The Wonderful World of Greece.

NEWCOM, JAMES E.: Associate producer. b. Indianapolis, Ind., Aug. 29, 1907; e. U. California. Reader, MGM, 1926, then in stocks and bonds with E. F. Hutton; then actor and film editing dept., MGM, 1930; film ed. 1933; asst. prod., 1952.
PICTURES INCLUDE: Gone With the Wind, Rebecca, Since You Went Away, Annie Get Your Gun, Trial, Somebody Up There Likes Me, Wings of Eagles, Until They Sail, Farewell to Arms, The Inn of the Sixth Happiness; Revue Television; Paramount Studios. Nine Hours to Rama, The Impossible Years, editor: Tora Tora Tora.

NEWGARD, ROBERT M.: Executive. b. Des Moines, Iowa, Jan. 14, 1925; e. U. of Notre Dame, 1948; Mirisch Bros., Theatres Candy Co., built and operated theas. in midwest, 1948–50; operated record dist. business 1949–53, Columbia and MGM records; prod. shows at KTTV Hollywood, 1952; Interstate Television Subsidiary Allied Artists, dist. and prod'n of TV films, 1953–56; midwestern and western sales mgr., Screen Gems, 1956–66; v.p. world syndication, Paramount Television, 1967–73; now v.p., intl. mkt., Film Services Corp.

NEWHART, BOB: Actor-Comedian. b. Chicago, Sept. 5, 1929. e. Loyola Univ. In Army 2 yrs.; then law school; left to become copywriter and accountant. Acted with theatrical stock co. in Oak Park; hired for TV man-in-street show in Chicago. Recorded comedy album for Warner Bros. Record Co., The Button Down Mind of Bob Newhart, which was big hit. Followed by two more successful albums. Did series of nightclub engagements and then acquired own TV variety series in 1961. Frequently appears in Las Vegas and headlines college concerts. Has guested on most major TV variety and comedy series.
PICTURES INCLUDE: Cool Millions, Catch 22, Cold Turkey, First Family.
TV: The Bob Newhart Show, Thursday's Game (movie).

NEWLAND, JOHN: Director, actor. b. Cincinnati, Ohio. Musicals, many TV appearances. Actor, dir., Robert Montgomery Show, My Lover, My Son.

NEWLEY, ANTHONY: Actor, writer, composer, singer. b. Sept. 24, 1931.
PICTURES INCLUDE: Cockleshell Heroes, Battle of the River Plate, Port Afrique, Fire Down Below, Good Companions, X the Unknown, High Flight, No Time to Die, The Man Inside, The Bandit, The Lady is a Square, Idle on Parade, Killers of Kilimanjaro, Let's Get Married, Jazz Boat, In The Nick, The Small World of Sammy Lee, Dr. Dolittle, 1966; Can Hieronymus Merkin Ever Forget Mercy Humpee and Find True Happiness?, Sweet November, Quilp (star, music) Willie Wonka and the Chocolate Factory (score); Summer Tree (dir.).
TV: Sammy, Sunday Night Palladium, The Strange World of Gurney Slade (series), Saturday Spectaculars, The Johnny Darling Show, Hollywood Squares, Merv Griffin Show, The Tonight Show, Anthony Newley Special (London).
PLAYS: West End stage Stop The World—I Want to Get Off. N.Y. stage: Roar of the Greasepaint (wrote, composed with Leslie Bricusse), Good Old Bad Old Days.

NEWMAN, ALFRED S.: Executive. b. Brooklyn, N.Y., Nov. 16, 1940. e. New York University. Public relations work for Equitable Life Insurance, Trans World Airlines prior to joining Columbia Pictures in 1968 as writer in publicity dept.; named New York publicity mgr., 1970; national publicity mgr., 1972; joined MGM as East adv't-pub. dir., 1972; named director of adv't, pub. and promotion, 1974; named v.p., worldwide adv., pub., promo., 1978.

NEWMAN, DAVID: Writer. b. N.Y.C., Feb. 4, 1937. e. Univ. of Mich. (M.S., 1959). Was writer-editor at Esquire Magazine where he met Robert Benton, an art director, and formed writing partnership. All credits co-written with Benton.
PICTURES INCLUDE: Bonnie and Clyde, There Was a Crooked Man, Floreana, What's Up, Doc?, Money's Tight, Bad Company, Superman (co-s.p.), Superman II (co-s.p.).
STAGE: It's a Bird . . . It's a Plane . . . It's Superman (libretto), Oh! Calcutta (one sketch).

NEWMAN, EDWIN: News Correspondent. Joined NBC News in 1952, based in N.Y. since 1961. Reports news on NBC-TV and often assigned to anchor instant specials. Has been substitute host on Today, appeared on Meet the Press and has reported NBC News documentaries. Hosts interview series, Speaking Freely, on WNBC-TV, N.Y.C. station of NBC.

NEWMAN, JOSEPH M.: Producer, Director, writer. b. Logan, Utah. Aug. 7, 1909. Started as office boy MGM, 1925; jobs in production dept. to 1931; asst. to George Hill, Ernst Lubitsch, etc., 1931–37; asstd. in organization of MGM British studios 1937; dir. short subjects 1938; dir. "Crime Does Not Pay" series 1938–42; Major in U.S. Army Signal Corps 1942–46; dir. 32 Army Pictorial Service Pictures. Member: AMPAS, SDG Masons.
PICTURES INCLUDE: Northwest Rangers, Abandoned, Jungle Patrol, Great Dan Pitch, 711 Ocean Drive, Lucky Nick Cain, Guy Who Came Back, Love Nest, Red Skies of Montana, Outcasts of Poker Flat, Pony Soldier, Dangerous Crossing, Human Jungle, Kiss of Fire, This Island Earth, Flight to Hong Kong, Fort Massacre, Big Circus, Tarzan The Ape Man, King of the Roaring Twenties, Twenty Plus Two, The George Raft Story, Thunder of Drums.

NEWMAN, MARTIN H.: Executive. b. Brooklyn, N.Y. m. Nov. 16, 1913; e. N.Y.U., 1934, Certified Public Accountant (N.Y.); Century Theatres, 1936–1974, exec. v.p. 1966–1974; National Association of Theatre Owners, 1975–1977; exec. dir., Will Rogers Memorial Fund, 1977.

NEWMAN, NANETTE: Actress-writer. b. Northampton, England; m. to Brian Forbes. Ent. films in 1946 and TV in 1951.
TV: The Glorious Days, The Wedding Veil, Broken Honeymoon, At Home, Trial by Candlelight, Diary of Samuel Pepys, Faces in the Dark.
FILMS INCLUDE: The Personal Affair, The League of Gentlemen, The Rebel, Twice Around the Daffodils, The L-Shaped Room, Wrong Arm of the Law, Of Human Bondage, Seance on a Wet Afternoon, The Wrong Box, The Whisperers, Deadfall, The Madwoman of Chaillot, The Raging Moon, (U.S. title: Long Ago Tomorrow), The Stepford Wives, It's A 2'2" Above the Ground World (The Love Ban), Man at the Top, International Velvet.
RECENT TV: Balzac (BBC), Fun Food Factory, TV series, Stay with Me Till Morning.

NEWMAN, PAUL: Actor. Director. b. Cleveland, Ohio, Jan. 26, 1925. e. Kenyon Coll., Yale School of Drama. Summer stock; on Broadway in Picnic, The Desperate Hours, Sweet Bird of Youth, Baby Want A Kiss.
TV: Philco, U.S. Steel, Playhouse 90.
PICTURES INCLUDE: The Silver Chalice, The Rack, Somebody Up There Likes Me, The Long, Hot Summer, Cat on a Hot Tin Roof, Rally Around the Flag Boys, The Young Philadelphians, Exodus, The Hustler, Sweet Bird of Youth, Adventures of a Young Man, Hud, A New Kind of Love, The Prize, The Outrage, What a Way to Go, Lady L., Harper, Hombre, Cool Hand Luke, The Secret War of Harry Frigg, Winning, Butch Cassidy and the Sundance Kid. Dir. Rachel, Rachel. Formed First Artists Prod. Co. Ltd. 1969 with Sidney Poitier, Steve McQueen and Barbra Streisand, 1971; actor, director, Sometimes A Great Notion; Pocket Money (actor), 1971; Life & Times of Judge Roy Bean, 1972, The Effect of Gamma Rays (dir., 1972), The Mackintosh Man, The Sting, The Towering Inferno, The Drowning Pool, Buffalo Bill and the Indians, Slap Shot, Quintet, When Time Ran Out, Fort Apache, The Bronx, Absence of Malice.

NEWMAN, ROBERT V.: Executive. b. New Haven, Conn., Publicity director and mgr. for George Broadhurst, George M. Cohan, George White, etc.; dir. publicity "Abie's Irish Rose." Broadway stage prod., first 14 mos.; then prod. N.Y. plays ("Off Key," "The Sap from Syracuse," "The Poor Nut," "Twelve Miles Out," "Bad Girl," "Old Man Murphy," "Storm Song," "Trick for Trick," etc.). 1942–43 exec. asst., handling scripts & stories for Wm. Goetz, exec. prod., 20th-Fox; v.p. Republic Prod., Inc.; also asst. secy.-treas., Republic Pictures Corp. 1944–51; v.p. Samuel Goldwyn Prod., 1951–58; Pres. Batjac Prod. 1958 to 1963; founded San Remo Prods, 1963; Paramount Pictures 1964–67; exec. prod. & dir. of studio operations, 1967–70; in chg. of Quaker Oats m.p. div., 1970–72; exec. and gen. mgr. Hughes Productions, 1972–77; Executive Paramount Pictures, 1978.

NEWMAN, SAMUEL: Producer, writer, attorney. b. Jersey City, New Jersey, Dec. 14, 1919; e. N.Y.U., USFV College of Law, B.A., Juris Doctor; Captain, Armed Forces, WW II; writ. prod. Armed Forces Radio Service; freelance, contract writ. 30 m.p.; asst. exec. prod., Columbia Pictures; Pres., Samuel Newman Prods., Inc.; Head writer, story consultant, Perry Mason Show, CBS-TV 1960–64; head writer, story consultant, associate producer, Tarzan Series, NBC-TV, 1966–67; practicing licensed member, California and Federal Bar.

NEWMAN, SYDNEY.: (F.R.S.A.) Chief Creative Consultant for Canadian Film Development Corp. b. Toronto, Canada. Studied painting, drawing, commercial art at Central Techn. Schl. To Hollywood in 1938. Joined National Film Board of Canada under John Grierson. Prod. over 300 shorts. Later became exec. prod. all Canadian government cinema films, 1947–52; Canadian Broadcasting Corp., 1952, as dir. outside broadcasts, features and documentaries. Later became drama sup. and prod. Canadian Television Theatre. Joined ABC-TV in England, 1958. as sup. of drama and producer of Armchair Theatre: Head of Drama Group, TV, BBC, 1963. Commissioned and prod. first TV plays by Arthur Hailey, Harold Pinter, Alun Owen, Angus Wilson, Peter Lake. Fellow Society of Film & TV Arts, 1968; Prod. Associated British Pictures. SFTA award 1968; Zeta award, Writers Guild, Gt. Btn., 1970. 1970: Special advisor, ch. dir., Broadcast Programmes branch, Canadian Radio & TV Commission, Ottawa. Aug., 1970: Appt. Canadian Govt. Film Commissioner and chm., National Film Board of Canada; Trustee, National Arts Centre, Ottawa; board member, Canadian Broadcasting Corporation, Canadian Film Development Corp., Canadian Picture Pioneers Special Award. Special Advisor on Film to the Secretary of State for Canada, 1975–77; pres., Sydney Newman Enterprises.

NEWMAN, WALTER BROWN: Writer.
PICTURES INCLUDE: Ace in the Hole (co-s.p.), Underwater, The Man with the Golden Arm (co-s.p.), The True Story of Jesse James, Crime and Punishment, USA, The Interns (co-s.p.), Cat Ballou (co-s.p.).

NEWTON-JOHN, OLIVIA: Actress-Singer. b. Cambridge, England. Brought up in Melbourne, Australia, where won first talent contest at 15, with prize trip to England. Stayed there 2 yrs. performing as part of duo with Australian girl singer, Pat Carroll, in cabarets and on TV. Started recording; several hit records. Became a regular guest on TV series, It's Cliff Richard. Gained world-wide prominence as singer, winning several Grammys and other music awards.
PICTURES: Grease, Xanadu.

NEY, RICHARD: Actor, writer, prod., financier. b. N.Y. e. Columbia Univ., B.A. Acted in RCA TV demonstration, New York World's Fair; on stage in "Life with Father." On screen 1942 in Mrs. Miniver, War Against Mrs. Hadley, In armed services, World War II. Many TV shows. Financial advisor consultant, Richard Ney and Associates; financial advisor, lecturer; author, The Wall Street Jungle.
PICTURES INCLUDE: Late George Apley, Ivy, Joan of Arc, The Fan, Secret of St. Ives, Lovable Cheat, Babes in Bagdad, Miss Italia, Sergeant and The Spy.

NIBLEY, SLOAN: Writer. b. Ore.; e. U. of Utah, U.C.L.A., m. Linda Stirling, actress, Three yrs. U.S. Navy; employed as writer at major studios; contributor to mags.; wrote many western pictures incl.: Carson City, Springfield Rifle; prod., and writer many TV shows on film and live TV for Ralph Edwards.

NICHOLS, MIKE: Actor-Director. b. Berlin, Ger., Nov. 6, 1931. e. Chicago U. Compass Players, teamed with Elaine May; night clubs.
TV: Broadway, An Evening with Mike Nichols and Elaine May.
STAGE: Barefoot in the Park, The Knack, Luv, The Odd Couple, The Apple Tree, The Little Foxes, Plaza Suite, Uncle Vanya, The Prisoner of 2nd Avenue, Streamers, Comedians, The Gin Game. Produced Annie, 1976.
PICTURES INCLUDE: Who's Afraid of Virginia Woolf, The Graduate, Catch 22, Carnal Knowledge, The Day of the Dolphin, The Fortune.

NICHOLSON, ELWOOD J.: Cinematographer, research engineer, physicist. b. Dewitt Co., Clinton, Ill., Mar. 8, 1908; e. Chicago Eng. Coll., D.E.E. 1925; Amer. Eng. Coll., N.Y.C., Ph.D., 1929, 1927 cameraman & technician U.S. Army Air Corps, Panama; 1928 photog. & tech. Chas. A. Lindbergh; 1929 aerial photog. skywriting, Curtiss Field, L.I.; 1930 pilot & photog., newsreels & Curtiss Flying Service; 1931 newsreel cameraman, prod. & ed., Bermuda; 1932, Natl. Theat. Supply; 1934 designed equip. for Billy Rose's Jumbo, N.Y.C.; 1935 newsreels; 1936 designed equip. sound & lighting Max Reinhardt's Eternal Road, N.Y.C.; 1937–40 eng., George Teegue Process, Hollywood; 1940–44 research eng., U.S. Army Wright Field, Dayton, Ohio; 1944–45 photo research developing color tem. meter; color photog. & design eng., Wolff Studios, Hollywood. In 1947 designed & built new type 16mm sound on film recorder; invented "Cinematic" sound; mag. tape; electro devices.
PICTURES INCLUDE: Pansy the Horse, Clyde Beatty series, Operation Airstrip, Drink Up series.
MEMBER: SMPTE, IATSE. Sec'y, treas., Magna-Sound, Inc., 1954.

NICHOLSON, EMRICH: Art director. b. Shelburn, Ind., Sept. 4, 1913. e. Yale U., B.F.A.; Chouinard Art School. Pres., Nicholson-Maier Co. (architects); wrote Contemporary Shops in the U.S. 1943; joined art dept. Paramount 1941; apptd. art dir. 1946.
PICTURES INCLUDE: Black Bart, Highwayman, One Touch of Venus, Sleeping City, Walking My Baby Back Home, Magnificent Obsession, Sign of The Pagan. Retired, 1969.

NICHOLSON, JACK: Producer, director, actor, writer. b. Neptune, N.J.; April 22, 1936. Began career in cartoon department of MGM. Made acting debut in Hollywood stage production of Tea and Sympathy.
PICTURES INCLUDE: (Acting credits) The Shooting, Psych Out, Hell's Angels on Wheels, Little Shop of Horrors, The Raven, Ride the Whirlwind, Flight to Fury, Ensign Pulver, Too Young To Live, Studs Lonigan, Cry Baby Killer, Easy Rider, Five Easy Pieces, Carnal Knowledge, A Safe Place, (Writing credits) The Trip, Head, Flight to Fury, Ride the Whirlwind, Drive, He Said. (Producing credits) Ride the Whirlwind, The Shooting, Head, Drive, He Said. Directing debut with Drive, He Said, 1971. Acting: The Last Detail, 1973, Chinatown, Tommy, The Passenger, The Fortune, One Flew Over the Cuckoo's Nest, The Missouri Breaks, The Last Tycoon, Goin' South (also dir.), The Shining, The Postman Always Rings Twice, The Border, Reds.

NICHOLSON, ROBERT (NICK): Producer, actor, composer, arranger. b. Buffalo, N.Y. Musical dir., WBEN, WGR Buffalo, N.Y.; Musical dir., Howdy Doody, Ruff & Reddy, Top Dollar, Look To Win, How Do You Rate. Prod., Do, Re, Mi. VP., Nicholson-Muir Prod. Package Prod. creator Newlyweds Game, Matches 'n Mates, Big Spender, Pay Cards, Video Encyclopedia, Reader's Digest "Joe" series, The Joy of Sewing, Headline Hunters, Family Challenge, Ladies 'n Gentlemen.

NICKELL, PAUL: Director. e. Morehead, Ky., State Teachers College, U. of No. Carolina English instructor, North Carolina State College; then cameraman, asst. dir., dir. WPTZ, Philadelphia; dir. CBS-TV 1948.
TV: Studio One, Best of Broadway, Climax, Playhouse 90.

NICOL, ALEX: Actor, Director. b. Ossining, N.Y., Jan. 20, 1919; e. Fagin Sch. of Dramatic Arts, Actor's Studio; U.S. Cavalry.
THEATRE: Forward the Heart, Sundown Beach, Hamlet, Richard II, South Pacific, Mr. Roberts, Cat on a Hot Tin Roof.
PICTURES INCLUDE: Sleeping City, Target Unknown, Air Cadet, Raging Tide, Meet Danny Wilson, Red Ball Express, Because of You, Tomahawk, Redhead From Wyoming, Lone Hand, Law and Order, Champ for a Day, Black Glove, Heat Wave, About Mrs. Leslie, Dawn at Socorro, Strategic Air Command, Man from Laramie, Great Day in the Morning, Sincerely Yours, Five Branded Women, Via Margutta, Under 10 Flags, Gunfighters at Casa Grande. Dir., Sleeping Skull, Then There Were Three. Acted in: The Brutal Land, Bloody Mama, Homer, The Gilded Cage. Director: Point of Terror. Actor: Hells Black Night: Director: Screaming Skull, Then There Were Three.

NIELSEN, LESLIE: Actor. b. Regina, Sask., Canada, Feb. 11. e. Victoria H.S., Edmonton. Disc jockey, announcer for Canadian radio station; studied at Neighborhood Playhouse; N.Y. radio actor summer stock.
TV APPEARANCES: Studio One, Kraft, Philco Playhouse, Robert Montgomery Presents, Pulitzer Prize Playhouse, Suspense, Danger, Justice, Man Behind the Badge, Death of a Salesman, The New Breed, Swamp Fox, Peyton Place, Ben Casey, Wild Wild West, The Virginian, The Loner.
PICTURES INCLUDE: Vagabond King, Forbidden Planet, Ransom!, Opposite Sex, Hot Summer Night, Tammy and the Bachelor, Night Train To Paris, Harlow, Dark Intruder, Beau Geste, Gunfight in Abilene, The Reluctant Astronaut, Counterpoint, Rosie, Dayton's Devils, How to Commit Marriage, Change of Mind, The Resurrection of Zachary Wheeler, The Poseidon Adventure, Viva, Knievel!, City on Fire, Airplane!

NIEMEYER, HARRY: Publicist. b. New York, N.Y., May 31, 1909. Drama ed. St. Louis Times; pub. depts., Warner Bros., RKO, Paramount and Columbia. In January, 1946 appointed asst. pub. dir., Benedict Bogeaus Prods. Hollywood correspondent St. Louis Post-Dispatch. Apptd. pub. dir. Samuel Bischoff Prod., 1948; Universal-International studio pub. dept., 1949; asst. pub. dir., 1958; Free lance pub., major studios, 1965–71.

NIGH, JANE: Actress. r.n. Bonnie Lenora Nigh. b. Hollywood, Calif., Feb. 25, 1926; e. Polytechnic H.S., Long Beach, Calif. Defense plant worker.
PICTURES INCLUDE: Something for the Boys; State Fair, Dragonwyck, Unconquered, Give My Regards to Broadway, Red, Hot and Blue, Blue Grass of Kentucky, County

Fair, Captain Carey, U.S.A., Blue Blood, Disc Jockey, Fort Osage, Rodeo.

NILES, FRED A.: Executive. b. Milwaukee, Wisc., Sept. 12, 1918. e. Univ. of Wisc. Was radio news commentator and capt. of office of educ. and information of U.S. Army in World War II. Now heads own film studio, Fred A. Niles Communications Centers, Inc., Chicago.

NIMOY, LEONARD: Actor. b. Boston, Mass., 1931. Along with active career in films, TV and stage, has been writer and photographer. Author of three books of own photography and poetry, as well as autobiography, I Am Not Spock. Has also been speaker on college lecture circuit.
PICTURES: Queen for a Day, Rhubarb, The Balcony, Catlow, Invasion of the Body Snatchers, Star Trek—The Film.
TELEVISION: Star Trek, Mission: Impossible, etc.
STAGE: Equus, Sherlock Holmes, Vincent (one-man show).

NIMS, ERNEST: Executive. b. Des Moines, Iowa, Nov. 15, 1908; e. U. of Iowa; started in film cutting dept., Fox Films, 1930; film editor, with Walter Wanger, Republic and International Pictures, 1931–46; supervising film ed., head of ed. dept., Universal-International, 1947–52; ed. exec., U-I, 1952–62; assoc. prod., CBS, Rawhide, 1959–61; post-production exec. Universal Pictures, 1962–73, named v.p., 1966.

NIVEN, DAVID, JR.: Executive. b. London, Dec. 15, 1942. Joined William Morris Agency in Beverly Hills in 1963. Transferred same yr. to New York: in next five yrs. worked for agency's European offices in Rome, Madrid and London. In 1968 joined Columbia Pictures' U.K. office as a prod. exec.; 1972, named mg. dir. of Paramount Pictures in U.K. In 1976 became indep. prod., forming partnership with Jack Wiener.
PICTURES INCLUDE: The Eagle Has Landed, Escape to Athena.

NIVEN, DAVID: Actor. b. Kirriemuir, Scotland, Mar. 1; e. Sandhurst Coll. Served in Highland Light Infantry in Malta; advanced military training Eng.; in lumber camp Canada; newspaper writer briefly; laundry messenger; repr. of London wire firm in U.S., gunnery instructor to revolutionists in Cuba; serviced with Brit. Army 1939–45, rising from 2nd Lt. to Col.; m.p. debut 1935.
PICTURES INCLUDE: Without Regret, Rose Marie, Dodsworth, Beloved Enemy, Charge of the Light Brigade, Dinner at the Ritz, Four Men and a Prayer, Three Blind Mice, Dawn Patrol, Bluebeard's Eighth Wife, Wuthering Heights, Raffles, Eternally Yours, Kiss in the Dark, Enchantment, Kiss for Corliss, Happy Go Lovely, Toast of New Orleans, Soldiers Three, Lady Says No, Island Rescue, The Moon is Blue, Love Lottery, O'Leary Night (Tonight's the Night), King's Thief, Birds and the Bees, Around the World in 80 Days, My Man Godfrey, Separate Tables, Ask Any Girl, Happy Anniversary, Please Don't Eat the Daisies, Guns of Navarone, Guns of Darkness, Best of Enemies, Captive City, 55 Days At Peking, The Pink Panther, King of the Mountain, Conquered City, Island of the Blue Dolphins, Lady L, The Eye of The Devil, Casino Royale, The Extraordinary Seaman, Eye of the Devil, Before Winter Comes, The Brain, The Statue, Vampira, Paper Tiger, No Deposit, No Return, Murder by Death, Candleshoe, Death on the Nile, Escape to Athena, Rough Cut, The Sea Wolves.
AUTHOR: Round the Rugged Rocks, The Moon is a Balloon, 1972.

NIX, WILLIAM PATTERSON: Attorney. b. Philadelphia, April 10, 1948. e. Georgetown Univ. (A.B., 1970), Antioch Graduate School (M.A., 1971), Hofstra Univ. School of Law (J.D., 1976). New York University School of Law (LL.M., 1979). In 1976 joined the legal staff of the Motion Picture Association of America. Handles copyright and literary property matters, in addition to censorship legislation, piracy matters, and First Amendment litigation on behalf of producers, distributors and exhibitors. Serves as counsel to the MPAA Title Registration Bureau and counsel to the MPAA Classification and Rating Administration Code for Advertising.

NIZER, LOUIS: Attorney. b. London, Eng., Feb. 6, 1902; e. Columbia Coll., B.A., 1922; Columbia Univ. Law School, LL.B., 1924. At college won Curtis Oratorical Prize (highest award for oratory at Columbia) twice; won recognition from U.S. Govt. for oratory various Liberty Loan drives. Since 1920 member firm Phillips, Nizer, Benjamin, Krim & Ballon; attorney, exec. secy., New York Film Board of Trade 1928 and thereafter; attorney many important personalities in films, stage, opera; counsel to m.p. cos., theatrical prod., also film, stage, radio exec.; trial counsel many litigations involving contract, copyright disputes.
AUTHOR: Writer, legal and other subj.; author (books) My Life in Court, Between You and Me, What to Do With Germany, Thinking on Your Feet, New Courts of Industry. Books: The Jury Returns, Analysis and Commentary on the Official Warren Commission Report on the Assassination of President John F. Kennedy. Legal Essays; author (articles); My Most Unforgettable Character, How Nazi Courts Operate, Analysis of Standard Exhibition Contract, Analysis of Motion Picture Code, Freeing the Judge, New Concepts in the Law of Radio, An Analysis of the Wagner Act, Recent Developments in Law of Plagiarism, The Implosion Conspiracy, Reflections Without Mirrors.
MEMBER: American Bar Assn., Bar Assn. of New York; hon. member, AMPA; identified Red Cross drives, March of Dimes drives; was chmn., theatrical div. La Guardia & Dewey campaigns; gen. counsel to MPAA, 1966.

NOBLE, PETER: Writer, producer, actor, TV personality. b. London, Eng., June 18. e. Hugh Myddelton Sch.; Latymer Sch. Author several books on m.p. ind.; writer & conducts movie radio prog. for B.B.C. & Luxembourg (Film Time, Movie-Go-Round, Peter Noble's Picture Parade). Formed Peter Noble Productions, 1953; Acted in many pictures; Ed. Screen International 1975.
PICTURES INCLUDE: Production assoc., Runaway Bus; asst. prod., to Dorothy a Son; co-prod., s.p. Fun at St. Fanny's; s.p., Three Girls in Paris; assoc. prod., Lost; s.p., Captain Banner; prod. Strange Inheritance; London Columnist; Hollywood Reporter etc. 1967–75.
AUTHOR: Editor, British Film Year Book; author of biographies of Bette Davis, Erich Von Stroheim, Ivor Novello, Orson Welles. Author book, I Know That Face. Wrote screen plays, the King of Soho, Love in the Limelight, The Story of Ivor Novello, etc.
TV: Find the Link, Other Screen, Film Fanfare, Movie Memories, Yakity Yak, Startime, Thank Your Lucky Stars, Juke Box Jury, Simon Dee Show, Star Parade, Who's Whose, Movie Magazine, The Big Noise, The Name Game, Line Up, Tea Break, Today. Prod. consult. On The Braden Beat, The Frost Program, Dee Time. 1969–70 Prod. Cons. Simon Dee Show. Appeared on Anything You Can Do Looks Familiar, Password, Etc. 1971–1975. Wrote scripts and prod. consultant Movie Quiz (series). Appears frequently on Today TV series. 1976, Two's Company, Looks Familiar. Prod. con. Musical Time Machine BBC2 series, Talking about films on radio, including BBC Star Sound, Radio Luxembourg, Film Faces, etc.

NOIRET, PHILIPPE: Actor. b. France, 1931. Has played character roles in numerous international films.
PICTURES INCLUDE: Zazie dans le Metro, Les Copains, The Night of the Generals, Topaz, Murphy's War, A Time for Loving, The Serpent, Let Joy Reign Supreme.

NOLAN, LLOYD: Actor. b. San Francisco, Calif.; e. Santa Clara Prep. School; Stanford U. In 1927 joined Pasadena Community Theat.; later in road co. of play The Front Page; in such stage prods. as Cape Cod Follies, High Hat, Reunion in Vienna, Americans, One Sunday Afternoon, Caine Mutiny Court Martial. Screen debut in Stolen Harmony, 1934.
PICTURES INCLUDE: Guadalcanal Diary, Bataan, Circumstantial Evidence, Captain Eddie, The House on 92nd Street, Lady in the Lake, Somewhere in the Night, Wild Harvest, Two Smart People, Green Grass of Wyoming, Street With No Name, Sun Comes Up, Easy Living, Bad Boy, The Lemon Drop Kid, Island in the Sky, Crazylegs, Last Hunt, Santiago, Abandon Ship, Hatful of Rain, Peyton Place, Susan Slade, Circus World, Sergeant Ryker, Airport, Earthquake.
TV: Julia.

NOLTE, C. ELMER, JR.: Executive v.p., b. Baltimore, Md., Oct. 19, 1905. Managing dir., FOH. Durkee Enterprises, Baltimore. Pres., NATO of Md., 1955–56; treas., 1957–59; pres., 1952–66, 67–69.

NOLTE, NICK: Actor. b. Omaha, Nebr., 1942. Attended 5 colleges in 4 yrs. on football scholarships, including Pasadena City College and Phoenix City College. Joined Actors Inner Circle at Phoenix and appeared in several plays. Did stock in Colorado. In 1968 joined Old Log Theatre in Minneapolis and after 3 yrs. left for New York, appearing at Cafe La Mama. Went to L.A. and did several TV series before big break in mini-series Rich Man, Poor Man as Tom Jordache.
PICTURES INCLUDE: Return to Macon County Line, The Deep, Who'll Stop the Rain, North Dallas Forty, Heart Beat, Cannery Row.
TV: Guested on Medical Center, Gunsmoke, etc. Starred in Rich Man, Poor Man, Book I. Movies: Feather Farm.

NORDEN, CHRISTINE: Actress, singer. r.n. Mary Lydia Thornton. b. Sunderland, England, Dec. 28, 1924. m. Jack Clayton, director (q.v.). On British stage in Tell the World, 1942; Take It Easy, 1944; Cinderella, 1951–52; The Marriage-Go-Round, 1979–80. On Broadway in Tenderloin, 1960–61; The Butler and Egg Man, 1966; Marat de Sade, 1967; Scuba Duba, 1968 etc. Signed by Sir Alexander Korda 1945 to seven year contract with London Films and Metro Goldwyn Mayer.
PICTURES INCLUDE: Night Beat, Mine Own Executioner, An Ideal Husband, Ido of Paris, A Yank Comes Back (in U.S.: A G.I. Returns), Saints and Sinners, The

Interrupted Journey, Black Widow, A Case for PC 49, Reluctant Heroes.
TELEVISION: Angel's Ransom, The Virginia Graham Show, Playhouse 90, Give My Regards to Broadway, Time to Spare, Love and Marriage.

NORMAN, STANLEY: Producer-Director. b. Dec. 22, 1921. e. Filmacademy Berlin.
PICTURES INCLUDE: The Hot Country, Dangerous Journey, Land without Grace, The Devil from Cape Town, Nobodies Land, Operation Yellow Viper, The Man Who Came To Kill, El Cisco, Tests to Kill, The Spy Pit, Taste of Savage, The Manipulator, The Big Game, In a Colts Shadow.

NORRIS, CHARLES GLENN: Executive. b. Taylorsville, N.C., Nov. 24, 1906; e. Nat'l U. Law Sch., Wash., D.C. Asst. poster clerk, Fox Film Co., 1928; booker, Wash., D.C., 1934; ad-sales mgr., Aug., 1935; salesman, Phila.; Wash., July, 1937; Baltimore, July, 1944; br. mgr., Wash., Jan. 1946; dist. mgr., July, 1946; br. mgr., Wash., Apr., 1948; Atlantic div. mgr., Wash., Jan., 1952; eastern sales mgr., April, 1954; central Canadian sales mgr., April, 1956; asst. gen. sales mgr., April, 1959; gen. sls. mgr., 1960–69; exec. capacity in distribution, 20th Century-Fox, The Glenoris Corp.
MEMBER: Variety Club, English Speaking Union, M.P. Pioneers.

NORRIS, EDWARD: Actor. b. Philadelphia, Pa.; e. Culver Mil. Acad. Ran away to sea; later reporter on Philadelphia Ledger, Philadelphia Bulletin & Morning Record. Joined Philadelphia Professionals, then to Theatre Union Group, N.Y. On stage (Doomsday Circus). Screen debut Queen Christina, 1934.
PICTURES INCLUDE: Boys' Town, Dr. Ehrlich's Magic Bullet, Mystery of Marie Roget, Great Impersonation, Decoy, Heartaches, Trapped by Boston Blackie, Forgotten Women, Breakthrough, Killer Shark, Highway 301, Murder Without Tears, The Kentuckian.

NORRIS, KATHI: Performer. b. Newark, Ohio. e. Univ. of Chicago; banker; acct. exec., Grant Advertising, Inc., 1940–45; columnist and writer, 1944–46; star, TV, Kathi Norris Show, WABD, WNBC, 1949–54; Today Show, NBC, 1954; spokeswoman, General Electric television shows, NBC, CBS, ABC, 1954–57; hostess-narrator, True Story, NBC TV, 1957–61; panelist, Leave It to the Girls, 1962–63; radio-TV spokeswoman, Ralston Purina, 1962–64; syndication and commercials, 1964–66.

NORRIS, ROBERT: Executive. b. Santa Rosa, California, 1924. Ent. ind. 1955 as op. man, TV-Aud. Measurement, then in 1956 advt. controller and appt. to board ABC TV 1963 with responsibility company's TV film development. Then exec. director ABC Television Films Ltd., and appoint. man. director Associated British-Pathe Ltd., 1967; appt., man. dir., Anglo-EMI Film Dist., 1970; now man. dir., Film & TV Marketing Service Ltd. and dir. Walport Telmar Ltd., dir. Trantel Ltd.

NORTH, ALEX: Composer. b. Chester, Pa., Dec. 4, 1910; e. Curtis Inst., 1928–29; Julliard Sch. of Music, 1932–34. Composer for ballet, radio, TV, theatre; U.S. Army, 1942–46; Guggenheim Fellowship, 1947–48; comp., 40 documentary films, 1937–50; comp. Revue for Clarinet & Orch. for Benny Goodman, 1947. Member ASCAP; Dramatists Guild.
PICTURES INCLUDE: Streetcar Named Desire, Death of a Salesman, Viva Zapata, Les Miserables, Pony Soldier, Member of the Wedding, Go Man Go, Desiree, Unchained, The Racers, Man with the Gun, Rose Tatoo, I'll Cry Tomorrow, Cleopatra (Composers and Lyricists Award, best film score, 1964), Shoes of the Fisherman Golden Globe Award; best film score 1968), The Children's Hour, The Misfits, Spartacus, The Rainmaker, The Agony and the Ecstacy, Who's Afraid of Virginia Woolf?, A Dream of Kings, Willard, Pocketmoney, Rebel Jesus, Once Upon a Scoundrel, Lost in the Stars, (musical director) Journey Into Fear, Shanks, Bite the Bullet, Somebody Killed Her Husband, Wise Blood, Carny, Dragonslayer. 14 Acad. Awards nominations.

NORTH, EDMUND H.: Writer. b. New York City, e. Stanford Univ., U.S. Army Signal Corps, five yrs., World War II, sep. as major.
PICTURES INCLUDE: One Night of Love, I Dream Too Much, Dishonored Lady, Flamingo Road, Young Man with a Horn, In a Lonely Place. collab. Only the Valiant; s.p. Day the Earth Stood Still, Outcasts of Poker Flat; collab. s.p., Destry, Far Horizons, Proud Ones; s.p. Cowboy; screen story and s.p., Sink the Bismarck!; collab. s.p., H.M.S. Defiant; collab. story and s.p. Patton; story and collab. s.p. Meteor.

NORTH, SHEREE: Actress. r.n. Dawn Bethel. b. Los Angeles, Calif., Jan. 17. e. Hollywood H.S. Amateur dancer with USO at 11; prof. debut at 13; many TV appearances; on Broadway in Hazel Flagg.
PICTURES INCLUDE: Excuse My Dust, Living It Up, How To Be Very Very Popular, Lieutenant Wore Skirts, Best Things in Life Are Free, Way to the Gold, Destination Inner Space, The Gypsy Moths, Charley Varick, The Outfit, Breakout, The Shootist, Telefon.
TV: Eddie.

NOSSECK, MAX (Alexander M. Norris): Director. b. Nakel, Poland, Sept. 19, 1902. e. U. of Berlin; studied art in Vienna; stage actor four years Poland and Germany. In 1919: Actor, assist. dir., film ed., dir. (DLS Berlin, Biograph, U.F.A.). In 1931: dir. The Unhappy Man, DLS Berlin. In 1938: first picture in Portuguese; also dir. in Spain, France, Holland; dir. Gado Bravo, Overture to Glory, Girls Under 21. In U.S., 1939–40.
PICTURES INCLUDE: The Brighton Strangler, Dillinger, Black Beauty, Return of Rin-Tin-Tin, Kill or Be Killed, the Hoodlum.

NOVAK, KIM: Actress. r.n. Marilyn Novak. b. Feb. 13, 1933. e. Wright Junior Coll., Los Angeles City Coll. Started as model, named World's Favorite Actress, Brussels World Fair; m.p. debut in Pushover.
PICTURES INCLUDE: Phfft, Five Against the House, Picnic, Man with the Golden Arm, Eddy Duchin Story, Jeanne Eagels, Pal Joey, Middle of the Night, Bell, Book and Candle, Vertigo, Pepe, Strangers When We Meet, The Notorious Landlady, Boys' Night Out, Of Human Bondage, Kiss Me, Stupid, The Amorous Adventures of Moll Flanders, The Legend of Lylah Clare, The Great Bank Robbery, Tales That Witness Madness, The White Buffalo, The Mirror Crack'd.

NYKVIST, SVEN: Cinematographer. b. 1922, Sweden. Became internationally known through photographing most of Ingmar Bergman's pictures.
PICTURES INCLUDE: Sawdust and Tinsel, Karin Mansdotter, The Silence, Loving Couples, Persona, Hour of the Wolf, Cries and Whispers (AA), The Dove, Black Moon, Scenes from a Marriage, The Magic Flute, Face to Face, The Tenant, The Serpents' Egg, Pretty Baby, Autumn Sonata, King of the Gypsies, Hurricane, Starting Over, Willie and Phil, The Postman Always Rings Twice.

NYREN, DAVID O.: Prod. b. Des Moines, Iowa, Nov. 9, 1924. e. Columbia U., 1945–47. U.S. Marine Corps, 1943–45; radio & TV prod. supervision, W. Wallace Orr Adv. Agency, 1947–49; radio & TV dir., Ruthrauff & Ryan, Baltimore, 1949; TV supervisor & commercial prod. TV shows, Ruthrauff & Ryan, N.Y., 1950; also dir. of new program development & talent, 1951–52; radio-TV acct. exec., Dancer-Fitzgerald-Sample, 1952–55; vice-pres. in chge. of TV programming, 1955. Sr. V.P. in chg. TV Programs, 1966; David Nyren Prods., 1971.

O

OAKLAND, SIMON: Actor. b. N.Y.C., 1922. Was prof. violinist. Spent 3½ yrs. in military service, World War II. Studied Amer. Thea. Wing. First appeared on Bdwy. in Skipper Next to God. Has done summer stock and TV.
PICTURES INCLUDE: The Brothers Karamazov, I Want To Live, Who Was That Lady?, The Rise and Fall of Legs Diamond, Psycho, Murder, Inc., West Side Story, Follow That Dream, Third of a Man, Wall of Noise, The Raiders, The Satan Bug, The Plainsman, The Sand Pebbles, Tony Rome, Chubasco, Scandalous John, Chato's Land, Emperor of the North, Happy Mother's Day.

OATES, WARREN: Actor. b. Depoy, Kentucky.
PICTURES INCLUDE: The Rise and Fall of Legs Diamond, Private Property, Ride the High Country, The Land We Love, Mail Order Bride, The Rounders, Major Dundee, Return of the Seven, In the Heat of the Night, The Split, The Wild Bunch, The Shooting, There Was a Crooked Man, Two-Lane Blacktop, The Hired Hand, Tom Sawyer, Dillinger, The Thief Who Came to Dinner, Kid Blue, Badlands, The White Dawn, Bring Me The Head of Alfredo Garcia, Cockfight, Race with the Devil, 92 in the Shade, Drum, Brinks' Job, 1941, Stripes, The Border.

OBER, PHILIP: Actor. b. Fort Payne, Ala., March 23, 1902. e. Princeton Univ., 1919–21; Adv. bus., Brooklyn Edison Co., Liberty Magazine, Collier's and the Spur, 1922–32.
STAGE: The Animal Kingdom, She Loves Me Not, Personal Appearance, Kiss the Boys Goodbye, Mr. and Mrs. North, Junior Miss, Craig's Wife, Dear Ruth, Light Up the Sky.
PICTURES INCLUDE: The Secret Fury, Never A Dull Moment, The Magnificent Yankee, The Dull Knife, The Washington Story, Come Back, Little Sheba, My Mother and Mr. McChesney, From Here to Eternity, About Mrs. Leslie, Broken Lance, Tammy, Escapade in Japan, The High Cost of Loving, 10 North Frederick, Torpedo Run, North by Northwest, The Mating Game, Beloved Infidel, Let No Man Write My Epitaph, Elmer Gantry, The Facts of Life, Go Naked In The World. The Ugly American, The

Brass Bottle, The Ghost and Mr. Chicken, The Assignment.
TV: Lux Video Thea., Climax, Screen Gems, etc. 1952–67.
MEMBER: Council, Actor's Equity Assoc., 1947; Bd. of Governors Screen Actors Guild, 1954.

OBOLER, ARCH: Writer. b. Chicago, Ill.; e. U. Chicago. Began writing for radio 1936 as author Lights Out, "Arch Oboler's Plays," Plays for Americans; 5 vols. radio scripts published; collab. s.p. Escape.
PICTURES INCLUDE: Bewitched, Arnelo Affair, Five, The Twonky. Org. Arch Oboler Prod., August, 1952. Prod. Bwana Devil, 1952, first 3-D feature, winner Peabody Award. Writer, dir., exec. prod., 1 & -, 1961. Prod., writer, dir. The Bubble, (in space-vision).

O'BRIAN, HUGH: Actor. b. Rochester, N.Y., Apr. 19, 1930; r.n. Hugh J. Krampe; Univ. of Cincinnati; U.S. Marine Corps; actor, stock cos.; actor with many m.p. cos.
PICTURES INCLUDE: Young Lovers, Never Fear, Vengeance Valley, Little Big Horn, On the Loose, The Cimarron Kid, Red Ball Express, Sally and Saint Anne, The Raiders, The Lawless Breed, Meet Me at the Fair, Seminole, Man from the Alamo, Back to God's Country, Saskatchewan, Fireman Save My Child, Drums Across the River, Broken Lance, There's No Business Like Show Business, White Feather, The Fiend Who Won the West, Twinkle in God's Eye, Brass Legend, Rope Law, Come Fly With Me, Love Has Many Faces, In Harm's Way, Ten Little Indians, Ambush Bay, Cowboy in Africa, Harpy, Killer Force, The Shootist, Game of Death.
TV: Wyatt Earp, ABC; Special Dial M for Murder; Hallmark Hall of Fame, A Punt A Pass and A Prayer; It's a Man's World. TV movie: Wild Women, ABC, Probe, NBC series, Tomorrow is Now, Space In the Age of Aquarius, Search (NBC series) Murder on Flight 502 (ABC Movie), Fantasy Island (ABC Movie), Cruise Into Terror (ABC movie).
THEATRE: B'way Plays: Destry Rides Again, First Love, Guys and Dolls. National co. of Cactus Flower.
Pres. H.O.B. Inc. Bev. Hills, 1956. Pres., Flounder: Hugh O'Brian Youth Foundation; Nat'l Chmn., Cystic Fibrosis Research Foundation 1969–74, Pres. Thalians 1956–57; Founder Hugh O'Brian Annual Acting Awards at UCLA.

O'BRIAN, JACK: Journalist. b. Buffalo, N.Y., Aug. 16, 1914. Worked on various N.Y. and Buffalo papers in many capacities since 1932; N.Y. Herald Tribune, N.Y. World-Telegram drama critic & N.Y. columnist, AP 1943–49; Radio-TV Ed., N.Y. Journal American 1949–65 Voice of B'way columnist syndicated to more than 150 newspapers by King Features; author, Great Godfrey and many magazine articles. Since 1965, daily radio show via WOR, N.Y.C.

O'BRIAN, PETER: Producer. b. Toronto, 1947. e. University of Toronto, Emerson College, Boston. Assistant dir. in feature films, television series, and commercials. Prod. for Quadrant Films Ltd., 1975. Also pres. of Muddy York Motion Pictures Ltd.
PICTURES INCLUDE: Me (prod.) 1974; The Mourning Suit (Line Producer) 1974; Clown Murders (principal) 1975; Love At First Sight (prod.) 1975; Blood and Guts, 1977; Fast Company, 1978.

O'BRIEN, EDMOND: Actor. b. New York, Sept. 10, 1915; e. Fordham U. one year, won scholarship at Neighborhood Playhouse School of the Theatre; m. Olga San Juan, prof. On stage: Daughters of Atrios, Hamlet, Parnell Family Portrait, Henry IV. On Orson Welles Radio program. Signed by RKO.
PICTURES INCLUDE: The Hunchback of Notre Dame, A Girl, A Guy and A Gob, Parachute Battalion, Obliging Young Lady, Powder Town, Victory, The Killers, The Web, Double Life, Another Part of the Forest, Fighter Squadron, Act of Murder, For Love of Mary, White Heat, Redhead and the Cowboy, Admiral Was a Lady, DOA, 711 Ocean Drive, Between Midnight and Dawn, Two of a Kind, Silver City, Warpath, Denver and the Rio Grande, Turning Point, Hitch-Hiker, Julius Caesar, Man in the Dark, China Venture, The Bigamist; act., co-dir., Shield for Murder; act., Shanghai Story, Barefoot Contessa (Acad. Award, best supporting actor, 1954); Pete Kelly's Blues, D-Day the Sixth of June, The Rack, 1984 Girl Can't Help It, The Third Voice, The Last Voyage, The Great Impostor, The Man Who Shot Liberty Valance, Moon Pilot, Birdman of Alcatraz, Fantastic Voyage, The Viscount, The Love God, The Wild Bunch, To Commit a Murder, They Only Kill Their Masters, 99 and 44/100ths ½ Dead.
TV: Johnny Midnight, 1960; 333 Montgomery Street, MGM-TV, NBC, 1962. Flesh and Blood, Sam Benedict (series).

O'BRIEN, GEORGE: Actor. b. San Francisco, Calif., Apr. 19, 1900; e. Santa Clara Coll.; medical student, then U.S. Navy. In amateur theatricals. Entered m.p. ind. as asst. cameraman, stunt man, double finally actor.
PICTURES INCLUDE: The Iron Horse, The Man Who Came Back, The Roughneck, The Johnstown Flood, Is Zat So?, Sharpshooters, Honor Bound, Blindfold, Noah's Ark, My Wild Irish Rose; since 1939 in many Westerns (RKO); Fort Apache, She Wore a Yellow Ribbon, Gold Raiders, Cheyenne Autumn.

O'BRIEN, LIAM: Writer. b. New York, N.Y., March, 1913; e. Fordham U., Manhattan Coll., A.B., 1935. Author B'way play Remarkable Mr. Pennypacker, 1953.
PICTURES INCLUDE: Chain Lightning, Redhead and the Cowboy, Of Men and Music, Diplomatic Courier, Here Comes the Groom, The Stars Are Singing, Young at Heart.

O'BRIEN, MARGARET: Actress. b. Los Angeles, Calif., Jan. 15, 1937. Screen debut at 4 in Babes in Arms, Acad. Award best child actress, 1944. Voted one of ten best money-making stars in Motion Picture Herald-Fame Poll 1945–46.
PICTURES INCLUDE: Journey for Margaret, Dr. Gillespie's Criminal Case, Lost Angel, Thousands Cheer, Jane Eyre, The Canterville Ghost, Meet Me in St. Louis, Music for Millions, Our Vines Have Tender Grapes, Three Wise Fools, Unfinished Dance, Tenth Avenue Angel, Secret Garden, Big City, Little Women, Her First Romance, Anabelle Lee, Diabolic Wedding.

O'BRIEN, PAT: Actor. b. Milwaukee, Wis., Nov. 11, 1899; e. Marquette U.; Great Lakes Naval Training Station; Sargent School, N.Y. In stock; on Broadway in Man's Man, Up and Up; on the road in Coquette, Front Page; m.p. debut in "Front Page" 1932; org. with Phil Ryan, 1944, Terneen Productions.
PICTURES INCLUDE: Flying High, Final Edition, Virtue, Strange Case of Clara Deane, Air Mail, Destination Unknown, Bureau of Missing Persons, Flirtation Walk, Oil for the Lamps of China, Angels With Dirty Faces, Knute Rockne, Flowing Gold, Bombardier, The Iron Major, His Butler's Sister, Marine Raiders, Having Wonderful Crime, Secret Command, Man Alive, Perilous Holiday, Crack-Up, Riffraff, Fighting 69th, Fighting Father Dunne, The Boy With Green Hair, The Fireball, Johnny One Eye, Criminal Lawyer, People Against O'Hara, Okinawa, Jubilee Trail, Ring of Fear, Inside Detroit, The Last Hurrah, Town Tamer, The End, Ragtime.

O'BRIEN, VIRGINIA: Actress. b. Los Angeles, Calif. On stage in "Meet the people." Screen debut 1940 in Hullabaloo; now retired.
PICTURES INCLUDE: The Big Store, Lady Be Good, Ringside Maisie, Ship Ahoy, Panama Hattie, DuBarry Was a Lady, Thousands Cheer, Meet the People, Two Girls and a Sailor, The Harvey Girls, Ziegfeld Follies, Till Clouds Roll By, The Showoff, Merton of the Movies.

O'CONNELL, JACK: Producer-Director-Writer-Lyricist. b. Boston. After Germany in WWII got BA Princeton MBA Harvard. After being copy group head at D'Arcy and McCann advertising and doing 500 TV commercials entered feature films working with Fellini on La Dolce Vita, then asst. director to Antonioni on L'Avventura, then writer-producer-director Greenwich Village Story, Revolution, Christa (a.k.a. Swedish Fly Girls), City Women. Features have been invited by critics to represent U.S. at Cannes, Locarno, Berlin and Venice Festivals.

O'CONNOR, CARROLL: Actor. b. New York City, Aug. 2, 1925. e. University College, Dublin & U. of Montana. Three years with Dublin's Gate Theatre, then N.Y. where stage credits include Ullyses in Nighttown, Playboy of the Western World, The Big Knife; m.p. debut in Fever in the Blood, 1960.
PICTURES INCLUDE: By Love Possessed, Lonely Are The Brave, Cleopatra, In Harm's Way, What Did You Do In The War Daddy?, Hawaii, Not With My Wife You Don't, Waterhole No. 3, The Devil's Brigade, For Love Of Ivy, Kelly's Heroes, Doctors' Wives, Law and Disorder.
TV: US Steel Hour, Armstrong Circle Theatre, Kraft Theatre, All In The Family (Emmy Golden Globe awards), Of Thee I Sing.
AUTHOR: Ladies of Hanover Tower (play); Little Anjie Always, The Great Robinson (screenplays).

O'CONNOR, DONALD: Actor. Star of Tomorrow, 1943. b. Chicago, Ill., Aug. 28, 1925. In vaudeville with family and "Sons o' Fun" (Syracuse, N.Y.) before screen debut 1938 in Sing You Sinners; in number other pictures 1938–39 (Sons of the Legion; Tom Sawyer, Detective, Beau Geste, On Your Toes, etc.); in vaudeville 1940–41, then resumed screen career with What's Cookin'? 1942. Entered armed services, 1943.
PICTURES INCLUDE: Private Buckaroo, Give Out, Sisters, When Johnny Comes Marching Home, It Comes Up Love, Mr. Big, Top Man, Patrick the Great, Follow the Boys, The Merry Monahans, Bowery to Broadway, This Is the Life, Something in the Wind, Are You With It? Feudin', Fussin' and a-Fightin'. Yes Sir, That's My Baby, Francis series, Curtain Call at Cactus Creek, The Milkman, Double Crossbones, Singin' In the Rain, I Love Melvin, Call Me Madam, Walking My Baby Back Home, There's No Business Like Show Business, Anything Goes, Buster Keaton Story, Cry for Happy, That Funny Feeling, That's Entertainment, Ragtime.

TV: Colgate Comedy Hour, 1953–54. Voted best TV performer, M.P. Daily poll, 1953, The Donald O'Connor Show.

O'DONNELL, WILLIAM: Exhibitor. b. Chicago, Ill. Grad., 18 years on B'way; 19 yrs. Interstate Circuit, Dallas; Pres., Cinema Art Theatres, Inc., Dallas, 1956.

OFFENHAUSER, WILLIAM H., JR.: Executive, engineer; b. Brooklyn, N.Y., May 8, 1904; e. Columbia U., RCA, 1929–32; inventor MGM squeezetrack, 1929; contractor, Army Sig. Corps, 1933–34; sales eng., J. A. Maurer, Inc., 1936–39; mgr., Precisions Films Labs., N.Y., & v.p., mem. of bd., J. A. Maurer, 1939–43; project eng., Johns Hopkins U., 1942–43; consultant, Sig. Corps Photo Center, 1944–45; consultant, film in color TV, CBS, 1946–47, 1949–51; research project, Cornell U. Med. Coll., 1947–49; consultant, Telenews Prod., 1951–52; v.p., Andre Debrie of America, Inc., 1953; ind. consultant, films & TV, 1954. 1960–61, (Photo Staff) M.I.T. Lincoln Laboratory, Lexington, Mass. Author, 16mm Sound Motion Pictures—A Manual; co-author, Microrecording-Industrial and Library Microfilming; ind. consultant, Films, TV, biological Acoustics and Biophysics; 1968, pres. Radio Club of America; 1969–73 Bd. Dir. Radio Club of America.

O'HANLON, GEORGE: Actor. b. B'klyn., N.Y., Nov. 23, 1917; e. La Salle Acad., Providence, R.I. p. Sam Rice and Luly Beeson, prof., served in U.S. Air Corps, 1942–45. Appeared in stock, vaudeville and Broadway musicals. On screen first, 1941, New Wine.
PICTURES INCLUDE: The Hucksters, Headin' for Heaven, Triple Cross, Spirit of West Point, Are You With It? star in Joe McDoakes series: Park Row, Cattle Town, Tanks Are Coming, Lion & the Horse, Battle Stations, Million Dollar Duck.
TV: Real George, Life of Riley, Bop Girl, Desilu Playhouse, Pantomime Quiz.

O'HARA, GERRY: Director. b. Boston-Lincs., England 1924. e. St. Mary's Catholic School, Boston. Junior Reporter Boston Guardian. Entered industry in 1942 with documentaries and propaganda subjects. Dir. debut 1963 That Kind of Girl.
PICTURES INCLUDE: Game for Three Lovers; Pleasure Girls (wrote & dir.); Maroc 7; Love in Amsterdam; All the Right Noises (orig. screenplay & dir.).
TV: The Avengers; Man in a Suitcase, Journey into the Unknown.

O'HARA, MAUREEN: Actress. r.n., Maureen FitzSimons. b. Dublin. Abbey School of Acting. Won numerous prizes for elocution. Under contract to Erich Pommer-Charles Laughton. Co-starred, Abbey & Repertory Theatre: Jamaica Inn, Mayflower Pictures, 1939; Hunchback of Notre Dame, RKO. In 1940: A Bill of Divorcement, Dance, Girls, Dance, RKO. In 1941: They Met in Argentina, RKO; How Green Was My Valley, 20th-Fox. In 1942: To the Shores of Tripoli, Ten Gentlemen from West Point, The Black Swan.
PICTURES INCLUDE: The Fallen Sparrow, Buffalo Bill, The Spanish Main, Do You Love Me?, Miracle on 34th Street, Foxes of Harrow, The Homestretch, Sitting Pretty, Woman's Secret, Forbidden Street, Sentimental Journey, Sinbad the Sailor, Father Was a Fullback, Comanche Territory, Tripoli, Bagdad, Rio Grande, At Sword's Point, Kangaroo, Flame of Araby, Quiet Man, Against All Flags, Redhead from Wyoming, War Arrow, Fire Over Africa, Magnificent Matador, Lady Godiva, Long Gray Line, Everything But the Truth, Wings of Eagles, The Deadly Companions, Our Man in Havana, Mr. Hobbs Takes a Vacation, McLintock, Spencer's Mountain, The Parent Trap, The Rare Breed, The Battle of Villa Fiorita, How Do I Love Thee:, Big Jake, The Red Pony.

O'HERLIHY, DAN: Actor. b. Wexford Ireland; e. National U. of Ireland. Actor with Abbey Theatre, Dublin Gate, Langford Prod.; announcer on Radio; Eireann; on Broadway in The Ivy Green.
PICTURES INCLUDE: Odd Man Out, Macbeth, At Swords Point, Actors and Sin, Adventures of Robinson Crusoe, Black Shield of Falworth, Bengal Brigade, Purple Mask, Virgin Queen, That Woman Opposite, City After Midnight, 100 Rifles, The Tamarind Seed, MacArthur.

OHMART, CAROL: Actress. b. Salt Lake City, Utah, June 3. e. Lewis and Clark H.S., Spokane. Newspaper writer, model, radio actress.
TV: Versatile Varieties, Studio One, Philco Playhouse, Colgate Comedy Hour.
THEATRE: On Broadway in Kismet.
PICTURES INCLUDE: Scarlet Hour; Wild Party, House On Haunted Hill, Scavengers.

O'HORGAN, TOM: Director. e. DePaul Univ. At age 12 wrote opera, Doom on the Earth. Is also musician, singer, actor. Responsible for developing many revolutionary off-off Bdwy. artistic innovations in such productions as The Maids, Tom Paine, Futz.
PICTURES INCLUDE: Futz, Rhinoceros.
STAGE: Broadway: Hair, Lenny, Jesus Christ, Superstar, Inner City, Dude.

OHTANI, HIROSHI: Executive. b. Tokyo, Nov. 2, 1910. e. Kobe Commercial C. 1935; entered Shochiku Co., Ltd., 1936; exec. dir. 1937; man. dir. 1944; dir. Taisho-kan Theatre Co., Ltd. 1948; dir. Chuei Co., Ltd. 1953; man. dir. 1953; aud. Schochiku Co., 1954; pres. 1960–62; pres., Chugai motion picture chain, 1962.

O'KEEFE, ALLAN J.: Exhibitor. b. Minneapolis, Minn. e. St. Thomas Coll., St. Paul, Minn.; Univ. of Minnesota, Minneapolis; m. railroading with Soo Line; salesman, Belden-Evans Shirt Co., Minneapolis; Western Theatre Equipment Co., 1919–21; W. W. Hodkinson Corp., 1921; Robertson-Cole, 1922; Goldwyn Cosmopolitan and Metro Goldwyn 1923–24; Producers Distributing Corp. 1925; then br. mgr. Producers Distr. Corp; Pathe Exchange, Inc., Salt Lake City, Utah 1926–30; br. mgr. Pathe and RKO Pathe Dist. Corp., Los Angeles, 1930–31; br. mgr., Universal, Portland, Ore. 1932–33; Los Angeles 1933–May, 1938; west. dist. mgr. 1938; west div. sales mgr. 1943; asst. gen. sales mgr. 1946–50; west coast exhib., 1951–53; v.p. in chge. of sales, Polalite Co., July, 1953. In exhibition since 1953.

OKON, TED: Producer-director. b. New Kensington, Pa., Oct. 27, 1929. e. U. of Pittsburgh, BA 1949. Started career as radio announcer, disk jockey; John Harris Enterprises, theatres and Ice Capades, 1 year; prod. dir., WDTV-TV, Pittsburgh, formed Togo Productions prod. live and film TV shows: Reach, Yates and Matoon. TV-radio dir. 1956; TV-radio pro. dir. Reach McClinton & Co., 1957. sr. v.p. pro. TV comm. Benton and Bowles Adv. 1958–63; Exec. TV prog. dir. art., Ogilvy, Benson & Mather Adv. 1963–65; v.p. prog. dir. Van Praag Prod.; Exec. prod. Girl Game, TV show, 1966; Pres. Rough & Ready, 1966–69; Pres. prod-dir. Tape 16, Teletronics 1969–70; Pres. 1970's Productions; Exec. prod. Women's Clubhouse, TV show, Pres., exec. prod., Pennysaver Productions, Inc., TV programming syndication. Exec. prod. The Pennysaver Place, Dance Party '76, $50,000 Crossword, Sho-Biz-Quizz, Polka Party, Salsa Time, Northstage Theatre Restaurant.

OLEMERT, THEODORA: Producer/writer. Doctor of Law Criminology of Paris University. Entered films through documentaries on child delinquency. Worked during the war as assistant to Prof. René Cassin on Franco/British cultural relations including films. Afterwards joined Jean Benoit-Levy, chief of the United Nations Film Section. Formed Triangle Films Ltd. for international coproductions; associated with l'Editon Française Cinématographique (Paris).
PRODUCED: Leonardo da Vinci, G. B. Shaw, Chopin, Teiva, The Sixth Day of Creation, Salvador Dali, Is Venice Sinking?, Edith Piaf and Corsica. Co-produced: Midnight Episode, Van Gogh, Molière, Mont St. Michel and Chinese Theatre. Preparing further films for international releases.

OLIN, MILTON E.: Executive. b. Chicago, Ill., Jan. 25, 1913. e. U. of Chicago, Ph.B., 1934; American Theatre Wing, N.Y., 1946–47. Freelance nightclub MC. radio & TV writer, radio actor; charter member AFRA, Chicago; writer, World Broadcasting System, 1946; radio transcription sales, Ray Block Enterprises & Richard H. Ullman, Inc., 1947–51; div. sales mgr., United Artists TV Corp., 1951–54; TV sales rep., WABC-TV, N.Y., 1954; nat'l sales mgr., Telefilm Enterprises, TV exec., Walt Disney Prod., 1956–60.

OLIVER, ANTHONY: Actor, author. b. July 4, 1923. e. Monmouth, R.A.F., 1940–46; stage debut, 1946; numerous plays; pictures; TV debut, 1946; many TV plays for BBC & ITV. Writes & tells own stories for TV. Authority on antiques. Author of The Victorian Staffordshire Figure and a novel, The Pew Group.

OLIVIER, LORD. (Sir Laurence Olivier): Actor. b. Dorking, England, May 22, 1907. On stage, London, N.Y. since 1925. Recent plays include, Caesar & Cleopatra, Anthony & Cleopatra, London, 1951, N.Y. 1952; Venus Observed, London 1950; Sleeping Prince, London, 1953; Shakespeare Season, Stratford-upon-Avon, 1955; The Entertainer, London, N.Y., 1957–58; Titus Andronicus European Tour 1957; Coriolanus Stratford-upon-Avon, 1959. Rhinoceros, London, 1960; Becket, N.Y., 1960 & Tour 1961. dir., Chichester festival Theatre; 1962, dir., The Chances, The Broken Heart; Uncle Vanya, dir., acted, 1965; Othello, 1964. dir., National Theatre, 1962; Hamlet, Uncle Vanya. Recruiting Officer 1963–64. Othello Master Builder 1963–65; dir., The Crucible, 1965; Moscow, Berlin, Othello And Love For Love 1965; dir. Juno and the Paycock, 1965–66. Tour of Canada with National Thea. Co. in Love for Love, Dance of Death, Flea in Her Ear, 1967, Home and Beauty, 1969, Shylock in The Merchant of Venice, 1970; Long Day's Journey Into Night, 1972–74; Dir. Amphitryon 38, 1971; Saturday, Sunday, Monday, 1973; The Party, 1973; Eden End, 1974.
PICTURES INCLUDE: As You Like It, Fire Over England, Divorce of Lady X, Wuthering Heights, Rebecca, Pride and Prejudice, That Hamilton Woman, Demi Paradise,

Henry V (dir. & cast), Hamlet (prod., dir. & cast), Acad. Award, Best Actor, 1948, Carrie, Beggar's Opera, Richard III (prod., dir. & cast), Prince & the Showgirl, Devil's Disciple, Spartacus, The Entertainer, Term of Trial, Bunny Lake Is Missing, Othello, Khartoum, Shoes of the Fisherman, Oh! What a Lovely War, Battle of Britain, Dance of Death, David Copperfield. THREE Sisters, Sleuth, Nicholas & Alexandra, Lady Caroline Lamb, 7½ Solution, Marathon Man, A Bridge Too Far, The Betsy, The Boys from Brazil, A Little Romance, Dracula, Clash of the Titans, Inchon, The Jazz Singer.

TV: John Gabriel Borkman, London. The Moon and Sixpence, N.Y.; The Power & The Glory, N.Y., Uncle Vanya, Long Day's Journey Into Night, 1973, Merchant of Venice, 1973, World at War, 1973, Love Among the Ruins, 1975, Jesus of Nazareth 1976. Prod. and acted in The Collection, Cat on a Hot Tin Roof. Prod. and dir. Hindle Wakes. Pres. and acted in Come Back Little Sheba, Daphne Laureola. Prod. and acted in Saturday Sunday Monday, Brideshead Revisited.

O'LOUGHLIN, GERALD STUART: Actor. b. N.Y.C., Dec. 23, 1921. e. Blair Academy, Lafayette Coll., U. of Rochester, Neighborhood Playhouse; U.S. Marine, WW II.

THEATRE: Broadway: Streetcar, Shadow of a Gunman, Dark at the Top of the Stairs, A Touch of the Poet, Cook For Mr. General, One Flew Over The Cuckoo's Nest, Calculated Risk. Off Broadway: Who'll Save the Plowboy, Harry, Noon & Night.

PICTURES INCLUDE: Lovers and Lollypops, Cop Hater, Hatful of Rain, Ensign Pulver, A Fine Madness. In Cold Blood, The Valachi Papers, Desperate Characters, The Organization, Twilight's Last Gleaming.

TV: The Defenders, Ben Casey, Dr. Dildare, 12 O'clock High, For the People, Going My Way, Naked City, Alcoa Premiere.

OLSON, CARL: Executive. Joined United Artists in 1954 as branch mgr. for Des Moines. Named Minneapolis branch mgr., 1958; promoted to asst. gen. sls. mgr. in 1961. Named western div. mgr. in 1964. In 1960 appt. v.p. & asst. gen. sls. mgr. responsible for UA western regions.

OLSON, NANCY: Actress. b. Milwaukee, Wis.; e. U. of Wisconsin, Univ. of California, Los Angeles; m. Alan Jay Lerner composer, writer. No prof. experience prior to films.

PICTURES INCLUDE: Union Station, Canadian Pacific, Sunset Boulevard, Mr. Music, Submarine Command, Force of Arms, So Big, Boy From Oklahoma, Battle Cry, Pollyanna, The Absent-Minded Professor, Smith!, Airport 1975.

O'NEAL, FREDERICK: Actor-Director-Lecturer. b. Brooksville, Miss., Aug. 27, 1905. e. public schools of Brooksville and St. Louis, Mo., New Theatre School, American Theatre Wing. Acted primarily on stage. Is now pres. of Associated Actors and Artists of America.

PICTURES INCLUDE: Pinky, No Way Out, Something of Value, Anna Lucasta, Take a Giant Step.

TV: Car 54, Where Are You (series, 1961–62).

O'NEAL, PATRICK: Actor. b. Ocala, Fla., 1927. e. Univ. of Fla. Neighborhood Playhouse. In stock cos. before N.Y. TV, 1951.

PICTURES INCLUDE: The Mad Magician, The Black Shield of Falworth, From the Terrace, A Matter of Morals, The Cardinal, In Harm's Way, King Rat, Chamber of Horrors, A Fine Madness, Alvarez Kelly, Matchless, The Assignment, Where Were You When the Lights Went Out?, The Secret Life of an American Wife, Castle Keep, Stiletto, Corky, The Way We Were, The Stepford Wives.

O'NEAT, RON: Actor. b. Utica, N.Y. e. Ohio State Univ. Spent 8 yrs. at Karamu House in Cleveland (inter-racial theatre) from 1957 to 1966, taking part in 40 plays. 1967–68 spent in N.Y. teaching acting in Harlem. Appeared in all-black revue 1968, The Best of Broadway, then in summer stock. Acted off-Broadway in American Pastorale and The Mummer's Play. 1970 joined Joseph Papp's Public Theatre. Big break came with No Place To Be Somebody, which won him top acting honors: the Obie, the Clarence Derwent, the Drama Desk and the Theatre World Awards.

PICTURES INCLUDE: Move, The Organization, Super Fly, Super Fly TNT, The Master Gunfighter, When a Stranger Calls, The Final Countdown.

TV: The Interns (guest).

STAGE: Tiny Alive, The Dream of Monkey Mountain.

O'NEAL, RYAN: Actor, r.n. Patrick Ryan O'Neal. b. Los Angeles, Calif., April 20, 1941. Began as stand-in, then stuntman, then actor in "Tales of the Vikings," series, in Germany, 1959; freelanced in Hollywood; Screen Gems Pilots, Donny Dru, Our Man Higgins, and co-starred in series made from Empire; co-star, Peyton Place, ABC-TV.

PICTURES INCLUDE: The Big Bounce, The Games, Love Story, Wild Rovers, What's Up Doc, Paper Moon, The Thief Who Came To Dinner, Barry Lyndon, Nickelodeon, A Bridge Too Far, The Driver, Oliver's Story, Green Ice, Partners, So Fine.

O'NEAL, TATUM: Actress. b. Los Angeles, 1964. Daughter of Ryan O'Neal and Joanna Moore. Won A.A. for debut performance in Paper Moon.

PICTURES INCLUDE: Paper Moon, The Bad News Bears, Nickelodeon, International Velvet, Little Darlings.

O'NEIL, THOMAS F.: Executive. b. Kansas City, Mo., Apr. 18, 1915; e. Holy Cross Coll., 1933–37. Employed by General Tire and Rubber Co., 1937–41; U.S. Coast Guard, 1941–46; v.p., dir., Yankee Network, Boston, 1948–51; pres. chmn. of bd. RKO General, Inc., since 1952. Arranged purchase RKO Radio by General Teleradio, Inc. from Howard Hughes, July, 1955; chairman of the Board, RKO General, Inc.; chmn. & dir. General Tire & Rubber Co.

ONTKEAN, MICHAEL: Actor. b. Canada, 1950. e. Univ. of New Hampshire. son of Leonard and Muriel Cooper Ontkean, actors.

PICTURES: Slap Shot, Voices, Willie and Phil, Making Love.

TELEVISION: The Rookies (series).

OPATOSHU, DAVID: Actor b. N.Y.C., Jan. 30; e. Morris High School. U.S. Army, 1942–46; played character roles, The Group Theatre at 19; appeared on Broadway.

THEATRE: Me and Molly, Once More With Feeling, Silk Stockings, The Reclining Figure The Wall, Bravo Giovanni.

PICTURES INCLUDE: Cimmarron, Naked City, The Brothers Karamazov, Exodus, Act of Mercy, Best of Enemies, Enter Laughing, Romance of a Horse Thief.

OPOTOWSKY, STAN: Executive. e. Tulane Univ. Served in U.S. Marine Corps as combat corr. and later joined United Press, working in New Orleans, Denver, and New York. Published own weekly newspaper in Mississippi before returning to N.Y. to join New York Post as mgr. editor and traveling natl. corr. Is also cinematographer and film editor. Joined ABC News as TV assignment editor; named asst. assignment mgr. In 1974 named dir. of operations for ABC News TV Documentaries. In 1975 named dir. of TV News Coverage, ABC News.

AUTHOR: TV: The Big Picture, The Longs of Louisiana, The Kennedy Government, Men Behind Bars.

OPPENHEIMER, GEORGE: Writer. b. New York, N.Y., Feb. 7, 1900; e. Williams Coll. 1916–20; Harvard, 1921. Alfred A. Knopf, publishers, 1921–25; co-founder, Viking Press, 1925–33; WWII, 1942–45; playwright; short stories; radio.

PICTURES INCLUDE: Rendezvous, We Went to College, Libeled Lady, Day at the Races, Married Before Breakfast, Adventures of Don Juan; Anything Can Happen, Tonight We sing, Decameron Nights.

TV: 30 Topper plays.

OPPENHEIMER, JESS: Producer, director, writer. b. San Francisco, Calif., Nov 11, 1913. e. Stanford U.; radio, Packard Hour, Fred Astaire, 1937–39; head writer, Screen Guild Program, 1939–40; Writer: Rudy Vallee, John Barrymore Show, 1939–41; head writer, prod., dir., Baby Snooks, 1943–48; writer, prod., dir., Lucille Ball, My Favorite Husband, 1948–51.

TELEVISION: Creator, head writer, prod., I Love Lucy, The Glynis Johns Show. Prod., Get Smart, Gen. Motors 50th Anniversary Program, Danny Kaye Special, Emmy Show. Creator, prod., The Debbie Reynolds Show. Prod., writer: Ford Startime Special, Bob Hope Chrysler Hours (also dir.); creator, prod.-dir., Angel series. Writer, All in the Family.

OPPENHEIMER, PEER J.: Executive. e. Georgetown University, UCLA. Founder and pres. of O&O Productions, 1960–64, founder and pres. of HeiRaMatt Productions, 1964. Exec. prod. at CCC Films, Berlin 1965–66. Pres. of Elite Syncopation, Ltd. London, 1973. Producer of over 22 films including Operation CIA, Lady Hamilton, Kashmir Run, Nashville Girl, Emma Mae, Cassanova and Company. Cofounder and vice-pres. of PRO-International Films, 1975-

West coast editor of Family Weekly since 1954. Member, Writers' Guild of America. Guest lecturer, UCLA.

ORIOLO, JOSEPH: Executive. b. Union City, N.J., Feb. 21, 1913. e. Union Hill High, Cooper Union, Hudson Coll. of Art Animator, Max Fleisher Studios, 1932–42. Est. own studios, specializing in animated commercials, educationals, industrials and film strips, 1943–58. Vice-pres., exec. prod., Felix the Cat Prod., also Felix the Cat syndicated newspaper strip, 1958–61; prod., dir., Mighty Hercules series; exec. vice pres., Adventure Cartoons, Television, Inc.

ORKIN, AD: Executive. b. Jackson, Miss., Dec. 7, 1922. e. Univ. of Miss. With Trans World Airlines as flt. eng. 1945–50. Now co-owner of Orkin Amusements in Jackson.

ORLEBECK, LESTER: Director, film editor. b. Sheboygan, Wis., June 26, 1907, e. Carroll Coll., Waukesha, Wis. Entered m.p. ind. as lab. tech. with Bennett Laboratories 1925; with Mascot as film ed. 1930. Edited many outdoor & feature pictures including first Gene Autry, Roy Rogers & Don Barry series (Republic). Worked on many Westerns.

In 1942 dir. 60 MM Motor, Horsemanship Mounted for U.S. Army. Western films since.

O'ROURKE, JOHN J.: Executive. b. N.Y., N.Y., July 3, 1922. e. City College of New York, 1950. Entered the industry 1939 Music Hall/New York. 20th Century Fox Film Corp. 1941–59, asst. to dir. of exploitation, MGM, 1960–62; asst. exploitation mgr. Astor Pictures 1962–63; exploitation mgr. 1963–67; National dir. of exploitation Avco Embassy Pictures, 1967; national co-ordinator roadshows, United Artists, 1968; asst. roadshow mgr. Universal Pictures. 1969 joined Cinemation Industries as dir. advertising, publicity and exploitation. 1974, v.p., Harry K. McWilliams Assoc. Advertising, 1977, vice pres., Benjamin Philip Associates, Inc., Advertising.

ORR, WILLIAM T.: Executive. b. New York City, Sept. 27, 1917. e. Coburn School, Rumsey Hall, Philips Exeter Academy. Impersonator, Meet the People, revue. Contract, Warner Bros. Joined U.S. Air Force, 1942. Assigned, production duties. Air Force's first motion picture unit, 1945. Joined Warner Bros. staff, 1946. Entertained, various night clubs and acting on Broadway stage, New York. Returned to Warner Bros., 1947 as exec. talent dept. and shortly named asst. to Steve Trilling, exec. asst. to Mr. Warner. Chg., studio's TV opers., 1955 as exec. prod. vice-pres., Warner Bros. Pictures, Inc., Nov. 29, 1957–62; vice-pres. in chg. of prod. both features and television, March 1961 to March 1962; vice-pres. in chg. of television production, 1962–63; prod., Sex and the Single Girl; asst. to pres., exec. prod., TV div., J. L. Warner, 1963–65; formed Wm. T. Orr Co., 1966, for prods. of M.P.s and TV films.

PICTURES INCLUDE: My Love Came Back, Thieves Fall Out, Navy Blues and Three Sons O'Guns, The Mortal Storm, The Big Street, Unholy Parners, Wicked, Wicked (exec. prod.).

ORSATTI, VICTOR M.: Executive. b. Los Angeles, Cal., Nov. 25, 1905. e. U. Southern Calif., B.A., 1928. Played baseball with St. Louis Cardinals; test pilot at Lockheed during World War II; artists' rep., Hollywood, since 1933; pres., Sabre prod.; pres., Calhoun-Orsatti Enterprises.

ORTH, MARION: Writer. b. Chicago; e. Knickerbocker Hall, Indianapolis, Ind., and St. Helen's Hall, Portland, Ore. Sold her first story in 1918 to Lois Weber for a picture which took her to Los Angeles; began writing for the screen; has been connected with Universal, First National, Famous Players and DeMille; went to Fox to descript for Come to My House; later did scripts for Sharp Shooters, Street Angel, Hangman's House, and many others for major companies.

OSBORN, ANDREW: Actor-Prod., BBC-TV. b. London, Eng., April 9, 1912. e. Christ's Hospital. Stage debut with Channing Pollock's The Enemy, 1931; played leading parts many stage plays 1935–38; ran own theatre. TV debut as actor, 1936.

TV: Adventure Story, Ghosts, Justice, Mourning Becomes Electra, Richard of Bordeaux, Hamlet, Wandering Jew, High Fidelity. Starred in serials, incl. Stand By to Shoot, The Six Proud Walkers, The Naked Lady, Solo for Canary; as prod. and dir. over 70 stage prod. TV; prod. and dir., Gentle Maiden, People of the Night, Goldfish in the Sun, One Morning Near Troodos, The Devil Came from Dublin, Flight of the Dove, etc. Serials: Crime of the Century, Common Room, The Man Who Sold Death. TV series: Maigret, Dr. Finlay's Casebook. Head of series dept. BBC-TV, for ten years. Drama series incl: Troubleshooters, Colditz, Lotus Eaters, Softy Softly, Trial, Detective, Menace, Champion House, The Mask of Janus, Adam Adament, Doomwatch, Fall of Eagles, Onedin Line, The Borderers, Pike, The Expert, Sherlock Holmes, Vendetta, Cluff, The Spies, Shadow of the Tower, The regiment. 1974 returned to production. Prod. Walk With Destiny, When The Boat Comes In, series, The Aphrodite Inheritance, serial.

OSBORNE, JOHN: Dramatist. b. London, England, 1929.

AUTHOR: Plays, includ. 1956: Look Back in Anger. 1957: Epitaph for George Dillon. 1958: The Entertainer. 1959: The World of Paul Slickey. 1961: Luther. 1963: Plays for England. 1964: Inadmissible Evidence. Films of his plays include Look Back in Anger, The Entertainer. Film scripts: Tom Jones (Oscar, 1964). Plays: A Patriot for Me, A Bond Honoured, Time Present, The Hotel in Amsterdam, West of Suez, Hedda Gabler (adaptn.).

TV: The Right Prospectus, Very Like A Whale, A Subject of Scandal and Concern.

OSCO, WILLIAM; Producer. b. Akron, OH, Aug. 17, 1948. Pres., General National Enterprises (dist.), since 1967; pres., Bill Osco Productions (films) since 1969; pres., Cybelle Productions (films) since 1979.

PICTURES: Flesh Gordon, Alice in Wonderland, Easter Sunday.

OSGOOD, STANTON M.: Executive. b. Berlin, N.H., e. Dartmouth Coll., B.A. Mgr. Fox Metropolitan Playhouse, Paramount, and worked on short subjects with Fred Waller, 1936–37; market analyst, Young & Rubicam, 1939–41; U.S. Navy, 1942–46; supvr., shorts, RKO Pathe, 1946–48; dir. of TV prod., CBS-TV, 1948–50; NBC, 1950–56, asst. dir., TV oper.; mgr., film prod. & theatre TV; gen. mgr., Paramount Sunset Corp., 1956–58; Ted Bates & Co., adv. agency, 1958–62; bus. mgr., Will Rogers Memorial Hospital, Saranac Lake, N.Y., 1962; special asst. to the president, Pratt Institute, 1969; bond sales prom. rep., U.S. Treasury, Savings Bond Div., 1971.

OSMOND, DONNY: Singer/TV Host. b. Ogden, Utah, Dec. 9, 1958. Seventh of 9 children, he was fifth member of family to become professional singer. (Four brothers, Alan, Wayne, Merrill and Jay, are original and present members of Osmond Bros., who originally sang barbershop quartet.) Donny made debut at 4 on Andy Williams Show. Donny has had 12 gold albums. Is now co-host of Donny & Marie on TV with sister.

OSMOND, MARIE: Singer/TV Host. b. Ogden, Utah, Oct. 13, 1959. Began career at age of 7 while touring with her brothers. Her first album, Paper Roses, became a gold one. Appears as co-host with brother Donny on TV's Donny & Marie.

O'SULLVAN, KEVIN P.: Executive. b. N.Y.C., April 13, 1928. e. Queens College, Flushing, N.Y. Associated with television 33 yrs., initially as a talent; later as businessman. Entered entertainment world after winning first prize in Arthur Godfrey Talent Scouts competition in 1948. 1950–55—professional singer, actor on TV, in theatre, night clubs. 1955–57—on radio-TV promotion staff, Ronson Corp. 1958–61 salesman, Television programs of America. 1961–67—director of program services, Harrington, Righter and Parsons. In 1967 joined ABC Films, domestic sales div. as v.p. & gen. sales mgr. In Jan., 1969 named v.p., gen. mgr., ABC Films, Inc.; in April same yr. named pres. In July 1970 made pres., ABC Int'l. TV, while retaining position as pres., ABC Films. In April, 1973 become pres., chief operating officer, Worldvision Enterprises, Inc., co. formed to succeed ABC Films when FCC stopped networks from TV program dist.

O'SULLIVAN, MAUREEN: Actress. b. Boyle, Eire, May 17; e. convents in Dublin, London; finishing sch., Paris. On screen numerous pictures, various Hollywood cos. from 191. On many TV shows.

PICTURES INCLUDE: The Big Shot, MGM Tarzan series, Tugboat Annie, The Barretts of Wimpole Street, The Thin Man, David Copperfield, Anna Karenina, Cardinal Richelieu, The Voice of Bugle Ann, A Day at the Races, Big Clock, Bonzo Goes to College, All I Desire, Mission Over Korea, Duffy of San Quentin, Steel Cage, Never Too Late.

STAGE: The Front Page (1971), No Sex Please, We're British (1973).

OSWALD, GERD: Director. b. Berlin, Germany. Father Richard Oswald, Austrian prod. dir., founder of UFA. Child actor, Vienna 1938; asst. dir., Monogram, Republic, Goldwyn, Paramount; 1949–53, 20th-Fox, prod. mgr., 2nd unit dir., test dir. 1954, 20th-Fox, assoc., prod., Man on a Tightrope, Nightpeople. Prod. Oasis, 1955–58, dir., U.A., A Kiss Before Dying, San Sebastian Film Festival nominee; Brass Legend; Crime of Passion; Fury at Showdown; Valerie; Paris Holiday. 1959–67, dir., Columbia, Screaming Mimi. Prod.-dir., screenplay Germ., prod. The Day the Rains Came; J. Arthur Rank-Allied Artists: Brainwashed, Bombay, & Venice Film Festival nominee; dir., 20th-Fox, The Longest Day (St. Mere-Eglise Seq.). Prod. Dir. Italo-French, Co-prod., The Scarlet Eye; dir., Universal, Agent from H.A.R.M. 1970. Story prod.-dir. Warner Bros. 80 Steps to Jonah, outstanding merit award by motion picture council; 1971 prod.-dir., A.I.P., Bunny O'Hare; 1976, dir. European co-prod.; To The Bitter End.

TV: Ford Theatre, G.E. Hour, Playhouse 90, Perry Mason, Rawhide, Black Saddle, The Virginian, Outer Limits (Soldier episode won HUGO award at world Sci-Fi convention), Fugitive, Blue Light, Felony Squad, Michner's Adventure in Paradise, Star Trek, Shane, Daniel Boone, Gentle Ben, Bonanza, It Takes A Thief, James Garner's Nichols.

OSWALD, RICHARD: Producer, director. b. Vienna, Austria. e. Vienna Dramatic School. Actor, dir., Vienna stage; entered m.p. industry as prod. dir., 1914; formed own co. in 1916. Prod., dir. more than 100 films in Europe, came to Hollywood, 1933.

PICTURES INCLUDE: Lady Hamilton, Lilac Time, White Horse Inn, Lucrezia Borgia, Hound of the Baskervilles, Cagliostro, Dreyfus, My Song Goes Round the World, Storm of Asia, Captain of Kopenick, Isle of the Missing Men, Lovable Cheat, Devil on Broadway, Don Carlos, Countless Maritza, Mayerling (TV).

O'TOOLE, PETER: Actor. b. Ireland, Aug. 2, 1932. Studied at R.A.D.A. Early career with Bristol Old Vic. London Stage in The Long, the Short and the Tall. 1960, with the Strat-

ford-on-Avon Company. Ent. films 1959 in Kidnapped. Partner with Jules Buck, Keep Films, Ltd.
PICTURES INCLUDE: The Savage Innocents, The Day They Robbed the Bank of England, Lawrence of Arabia, Becket, Lord Jim, What's New Pussycat, How to Steal a Million, The Night of the Generals, The Bible, Great Catherine, The Lion in Winter, Goodbye Mr. Chips, Brotherly Love, Murphy's War, Under Milk Wood, The Ruling Class, Man of La Mancha, Rosebud, Man Friday, Foxtrot, Caligula, The Stunt Man.
TELEVISION: Masada (mini-series).

O'TOOLE, STANLEY: Producer. Earliest experience with production costs; worked on Cleopatra, Singer, Not the Song, No Love for Johnny, Victim, etc. In 1966 named chief cost acct. for Paramount in U.K.; 1967, promoted to prod. exec. Worked on Downhill Racer, Running Scared, etc. Produced The Last of Sheila in 1972 for W.B.; 1974–75 was in Prague working on Operation Daybreak. Produced The Seven-Per-Cent Solution for Universal. Formed own Martinat Co. and produced for WB The Squeeze. Recent credits: The Boys from Brazil, Nijinsky.

OTWELL, RONNIE RAY: Theatre Executive. b. Carrollton, Ga., Aug. 13, 1929. e. Ga. Inst. Tech. Entered industry as mgr., Bremen Theatre (Ga.), 1950; dir. pub., adv., Martin Theatres, Columbus (Ga.), 1950–63; v.p., dir. Martin Theatres of Ga., Inc., 1963, Martin Theatres of Ala., Inc., 1963; dir. Martin Theatres of Columbus, 1963; sr. v.p., Martin Theatres Companies, 1971. Member: NATO, Ga., NATO, Columbus C of C; Columbus Mus. Arts & Crafts; Assn. U.S. Army.

OWEN, ALUN: Writer. b. Liverpool, Eng., 1925.
STAGE: A Little Winter Love, Maggie May, Progress to the Park, The Rough and Ready Lot, There'll Be Some Changes Made, Norma (Mixed Doubles), Shelter, Fashion of Your Time.
TV: The Ruffian, Not Trams to Lime Street, After the Funeral, Lena O My Lena (for ITV's Armchair Theatre); The Rose Affair (two awards, 1961), Ways of Love, You Can't Win 'Em All, A Hard Knock, Dare to be a Daniel, The Stag, The Strain, A Local Boy, Ruth, Funny, Pal, Giants and Ogres, The Piano Player, The Web Flight, Buttons, Lucky, Left. Ronnie Barker and Forget-me-not series.
PICTURES INCLUDE: The Criminal, A Hard Day's Night, Minding the Shop, Park People, You'll Be the Death of Me, McNeil, Cornelius, Emlyn.

OWEN, BILL: Actor. b. Acton Eng., Mar. 14, 1915. Screen debut in Way to the Stars. Numerous TV appearances.
PICTURES INCLUDE: School for Secrets, Daybreak, Dancing With Crime, Easy Money, When the Bough Breaks, My Brother's Keeper, Martha, Parlor Trick, The Roundabout, Trottie True, Once a Jolly Swagman, A Day to Remember, You See What I Mean, Square Ring, Rainbow Jacket, Ship That Died of Shame, Not so Dusty, Davy, Carve her Name with Pride, Carry on Sergeant, Carry on Nurse, Night Apart, Shakedown, Hell Fire Club, Carry on Regardless, Carry on Cabby!, Secret of Blood Island, Georgy Girl, Headline Hunters, O Lucky Man!, Kadoyng, In Celebration, When The Screaming Stopped, Comeback.

OWENS, GARY: Performer, broadcaster. b. Mitchell, S.D., May 10, 1936; e. Wesleyan U. Disc Jockey, KMPC, Hollywood; winner of Gavin Poll, 1966–69; named TV Radio Mirror top D.J. of year, 1970; named top D.J. in U.S. by Billboard, 1971–72; among top three TV announcers in U.S. in 'FAME' poll for three years; regular performer, "Rowan & Martin's Laugh-In" since inception in 1968.

OXBERRY, JOHN: b. New Rochelle, N.Y., March 17, 1918. Executive. Asst. photog. to Margaret Bourke-White, 1931–32; assoc. A.P. Lane Film Studio, 1932–38; assoc., R. J. Irwin, prod. photog., 1938–41; U.S. Army Signal Corps., 1941–45; cameraman, Tech Film Studio, N.Y., 1945–46; head, camera dept. Mini Toons Inc., 1946–47; pres. Animation Equipment Co., 1947–61; pres., Oxberry Corp., 1962–69; pres. Ox Products, 1971.

P

PAAR, JACK: Actor. b. Canton, O., May 1, 1918; radio announcer in Cleveland & Buffalo; served in U.S. Armd Forces World War II; entertained in Pacific zone with 28th Special Service Div. On radio with own show; m.p. debut in Variety Time, 1948. CBS-TV; Jack Paar Show, NBC-TV.
PICTURES INCLUDE: Walk Softly Stranger, Footlight Varieties, Love Nest, Down Among the Sheltering Palms.

PACINO, AL: Actor gained attention as stage actor.
STAGE: The Indian Wants the Bronx, (Obie award), Does a Tiger Wear A Necktie? (Tony Award), The Connection, Hello Out There, Tiger at the Gates, The Basic Training of Pablo Hummel (Tony Award).
PICTURES INCLUDE: Panic in Needle Park (debut), The Godfather, Scarecrow, Serpico, The Godfather II, Dog Day Afternoon, Bobby Deerfield, And Justice for All, Cruising.

PACKARD, FRED M.: Producer. b. Los Angeles, Calif., March 14, 1919; e. Los Angeles City Coll.; m. Shelagh Mary Rank, d. of J. Arthur Rank, c. Susan & Fred Arthur Rank. Radio commentator, sports and news; personal mgr. Joe Reichman and orch.; adv. pub. rep. Bing Crosby; officer in chge. film ed. dept. & prod Army films, U.S. Army Pictorial Service in England, France, 1935–41; actor (bit, feat. roles) including East Side of Heaven; film ed. dept. RKO studio; Universal in various dept.; pres. Fred M. Packard Prod., thereafter and since, to Columbia as prod.

PAGANO, JO: Writer. b. Denver, Col , Feb. 5, 1906. From 1930, novelist, short story writer, screen writer and comm., illustrator, 1936 with Goldwyn. Over 100 filmed teleplays.
PICTURES INCLUDE: (collab orig. s.p. co-prod.) That's the Spirit, (novel basis, s.p.) Try and Get Me (adapted from own novel) The Condemned, (collab. s.p.) Hotel Berlin, Adventures in Silverado, Jungle Goddess, Murder Without Tears, Security Risk.

PAGE, GALE: Actress. r.n. Sally Rutter. b. Spokane, Wash., 1918.
PICTURES INCLUDE: Crime School, The Amazing Dr. Clitterhouse, Four Daughters, Heart of the North, You Can't Get Away With Murder, Daughters Courageous, Naughty But Nice, Indianapolis Speedway, Four Wives, A Child is Born, They Drive by Night, Knute Rockne—All American, Four Mothers, Time of Your Life, Anna Lucasta, About Mrs. Leslie.

PAGE, GERALDINE: Actress. b. Missouri. m. Rip Torn. After summer stock & off-broadway roles, to Broadway and m.p. Theatre awards include Drama Critics' Circle (2).
STAGE: Mid-Summer, The Immoralist, Summer and Smoke (revival), The Rainmaker, Separate Tables, Sweet Bird of Youth, Strange Interlude, The Three Sisters, Angela, Black Comedy, Sarah Siddons, Theatre World, Donaldson.
PICTURES INCLUDE: Hondo, Summer and Smoke, Sweet Bird of Youth (academy nominations), Toys in the Attic, Dear Heart, The Happiest Millionaire, You're a Big Boy Now, Trilogy, Whatever Happened to Aunt Alice, Beguiled, J.W. Coop, The Day of the Locust, Nasty Habits, Interiors, Honky Tonk Freeway.
TV: A Christmas Memory, The Thanksgiving Visitor (Emmy awards), Barefoot in Athens, The Name of the Game.

PAGE, PATTI: Performer. r.n. Clara Ann Fowler. b. Claremore, Okla. 1927. e. Univ. of Tulsa. Staff performer, radio stat. KTUL, Tulsa; m. Charles O'Curran. Appeared on CBS radio show; star Patti Page Show, TV film series, The Big Record; author, Once Upon a Dream.
PICTURES INCLUDE: Elmer Gantry, Dondi, Boys Night Out.

PAGET, DEBRA: Actress. r.n. Debralee Griffin. b. Denver, Colo., Aug. 19, 1933. e. drama & dancing privately. Stage debut in Marry Wives of Windsor, 1946; in Jeanne D'Arc little theatre prod.; m.p debut in Cry of the City, 1948.
PICTURES INCLUDE: House of Strangers, Broken Arrow, Fourteen Hours, Bird of Paradise, Anne of the Indies, Belles on Their Toes, Les Miserables, Stars & Stripes Forever, Prince Valiant, Demetrius & the Gladiators, Princess of the Nile, Gambler from Natchez, White Feather, Seven Angry Men, Last Hunt, Ten Commandments, Tales of Terror, The Haunted Palace.

PAIGE, BOB: Performer. r.n. John Arthur Page. b. Indianapolis, Ind., Dec. 2; e. UCLA. Singer, amateur theatricals, usher Fox West Coast theatres; singer, KGER, Long Beach, Cal., announcer; program dir., KMTR, Hollywood; played films Columbia, Paramount, Universal; 30 films.
TV: Fireside Theatre, Pepsi-Cola Playhouse, Lux Theatre, Studio One, Kraft; 1957, host on Bride and Groom; 1958 co-star Bess Myerson, Big Payoff.

PAIGE, JANIS: Actress (Star of Tomorrow, 1947). r.n. Donna Mae Jaden. b. Tacoma, Wash., Sept. 16, 1923. Sang with Tacoma Opera Co. m.p. debut, 1944, Hollywood Canteen, N.Y. stage in 195-, and TV in 1956.
STAGE: Two Gals and a Guy, Fugitive Lady, Pajama Game, It's Always Jan.
PICTURES INCLUDE: Of Human Bondage, The Time the Place and the Girl, Two Guys from Milwaukee, Her Kind of Man, Cheyenne, Love and Learn, Wallflower, Winter Meeting, One Sunday Afternoon, Romance on High Seas, House Across the Street, Younger Brothers, Mr. Universe, Remains to be Seen, Please Don't Eat the Daisies, The Caretakers, Welcome to Hard Times.

PAINE, CHARLES F. Executive. b. Cushing, Texas, Dec. 23, 1920. e. Stephen F. Austin University, pres. Tercar Theatre Company; pres., NATO of Texas, 1972–73. NATO board member, 1973 to present; Motion Picture Pioneers member; Variety Club of Texas member.

PAKULA, ALAN J.: Producer-director. b. New York, N.Y., April 7, 1928. e. Yale U. Prod. apprentice, MGM, 1950; prod. asst., Para. 1951; prod. Para., 1955. Own prod. co., Pakula-Mulligan Prod. Stage prod. and m.p. dir. prod.
STAGE: Comes a Day, Laurette, There Must Be a Pony.
PICTURES INCLUDE: Fear Strikes Out, To Kill a Mockingbird, Love with the Proper Stranger, Baby the Rain Must Fall, Inside Daisy Clover, Up the Down Staircase, The Stalking Moon, The Sterile Cuckoo, Klute, Love and Pain and the Whole Damned Thing, The Parallax View, All the President's Men, Comes a Horseman, Starting Over (co.-prod., dir.). Roll-over.

PALANCE, JACK: Actor. b. Lattimer, Pa., Feb. 18, 1920. e. U. of N. Car. Professional fighter; U.S. Air Corps. Broadway stage and m.p.
STAGE: The Big Two, Temporary Island, The Vigil, Streetcar Named Desire, Darkness at Head.
PICTURES INCLUDE: Star of Tomorrow, Flight to Tangier, Man in the Attic, Sign of the Pagan, Silver Chalice, Kiss of Fire, Big Knife, I Died a Thousand Times, Attack!, Lonely Man, House of Numbers, Ten Seconds to Hell, Warriors Five, Barabbas, Contempt, Torture Garden, Kill a Dragon, They Came to Rob Las Vegas, The Desperadoes, Che, The Mercenary, Justine, Legion of the Damned, A Bullet for Rommel, The McMasters, Monte Walsh, Companeros, The Horsemen, The Professionals, Oklahoma Crude, Craze, The Four Deuces, The Diamond Mercenaries.
TV: Dr. Jekyll and Mr. Hyde, Dracula, Bronk (series).

PALEY, WILLIAM S.: Corp. Officer. b. Chicago, Sept. 28, 1901. e. U. of Penn. Took over operation Columbia Broadcasting System (now CBS, Inc.) as pres. 1928; chairman of the board since January, 1946; built network to leading position and est. innovations in broadcasting. During war on leave to supervise OWI radio in Mediterranean area. Chief of radio of Psychological Warfare Division, SHAEF, 1944–45; Dep. Chief info. Control Div. of U.S.G.C.C. 1945; Colonel, A.U.S. Deputy Chief Psychological Warfare Division, SHAEF, 1945; pres. & dir., William S. Paley Foundation, Inc.; pres. & dir. Greenpark Foundation, Inc. Founder and Chairman of The Board of Trustees of The Museum of Broadcasting; trustee, Emeritus Columbia University; trustee and chairman, Museum of Modern Art. Trustee North Shore University Hospital 1949–73, Co-Chairman emeritus of the Board. Decorations include the Legion of Merit, Medal of Merit. Legion of Honor, Croix de Guerre with Palm. Chairman of President's Materials Policy Comm., 1951–52, which issued report, "Resources for Freedom."

PALLOS, STEVEN: Producer. b. Budapest, Hungary, Journalist before entering m.p. ind. Began prod. & dist. in France & Continental Europe. Entered partnership with Sir Alexander Korda in 1929; later formed with him London Film Prod. From 1943–46 Brit. Army, head Psychological Warfare Br., Film Div., Italy; dir. Pendennis Pictures Corp., Ltd., Gibraltar Prods., Ltd., Britannia Film Distributors, Ltd.
RECENT PICTURES INCLUDE: Where the Spies Are, Captain Nemo & The Underwater City, Catch Me a Spy, The Glorious Musketeers.

PALMER, BETSY: Actress. b. East Chicago, Ind., Nov. 1, 1929. e. De Paul U. Actress in stock; on Broadway in Grand Prize, Affair of Honor, Roar Like a Dove.
TV: Studio One, Philco Playhouse, Danger, U.S. Steel Hour, Martin Kane, Climax, Kraft Theatre, Today, The Garry Moore Show, I've Got a Secret.
PICTURES INCLUDE: Mister Roberts, Long Gray Line, Queen Bee, The Tin Star, The Last Angry Man.

PALMER, GREGG: Actor. r.n. Palmer Lee. b. San Francisco, Calif., Jan. 25, 1927; e. U. of Utah. U.S. Air Force, 1945–46; radio announcer, disc jockey; then to Hollywood; many TV appearances.
PICTURES INCLUDE: Cimarron Kid, Battle at Apache Pass, Son of Ali Baba, Red Ball Express, Francis Goes to West Point, Sally and St. Anne, The Raiders, Back at the Front, Redhead From Wyoming, Column South, Veils of Bagdad, Golden Blade, The All American, Taza Son of Cochise, Magnificent Obsession, Playgirl, To Hell and Back, Creature Walks Among Us, Hilda Crane, Zombies of Mora Tau, Revolt of Fort Laramie, Rebel Set, Thundering Jets, Forty Pounds of Trouble, Night Hunt, The Undefeated, Chisum, Rio Lobo, Big Jake, Providenza (Italy), Ci Risiamo Vero Providenza (Italy & Spain). The Shootist, The Man with Bogart's Face.
TV: Movie of the week: Go West Young Girl, Hostage Heart, True Grit.

PALMER, LILLI: Actress. b. Austria. m. Carlos Thompson, act. N.Y. stage in 1952, Bell Book & Candle; London stage (same play) in 1955. On TV in Lilli Palmer Presents.
STAGE: (London) Suite in 3 Keys. (N.Y.) Venus Observed, Love of 4 Colonels.
PICTURES INCLUDE: Secret Agent, Silent Barrier, Sunset in Vienna, Command Performance, Man With 10 Faces, Girl Must Live, Chamber of Horrors, Thunder Rock, Gentle Sex, English Without Tears, Cloak and Dagger, Notorious Gentleman, Body and Soul, Beware of Pity, My Girl Tisa, No Minor Vices, Her Man Gilbey, Wicked City, Long Dark Hall, The Four Poster, Main Street to Broadway, Is Anna Anderson Anastasia?, But Not For Me, Conspiracy of Hearts, Counterfeit Traitor, The Pleasure of His Company, Adorable Julia, The Flight of the White Stallions, The Great Spy Mission, Operation Crossbow, Le Tonnerre De Dieu, Le Voyage DuPere, Nobody Runs Forever, Hard Contract, Marquis De Sade, La Residencia, Oedipus the King, What the Peeper Saw, Murders in the Rue Morgue, The House That Screamed, The Boys from Brazil.

PALMER, MARIA: Actress. b. Vienna, Austria, Sept. 5, 1924. e. College de Bouffement, summa cum laude; studied interpretive dance, ballet and drama. Vienna Academy of Music & the Interpretive Arts. Award at International Dance Congress, 1936; youngest member, Reinhardt's Theatre in der Josefstadt, Vienna, 1938–39; m.p. debut in Mission to Moscow.
STAGE: The Moon is Down, The Vigil, The Girl on the Via Flaminia, The Heiress, A Streetcar Named Desire, Dial M for Murder, The Diary of Anne Frank, The Happy Time.
PICTURES INCLUDE: Mission to Moscow, Days of Glory, Lady on a Train, The Other Love, The Web, Surrender, Strictly Dishonorable, By the Light of the Silvery Moon, Nostradamus and the Queen, Three for Jamie Dawn, Flight Nurse, City of Women.
TV: The Diamond Necklace, Schlitz Playhouse, Lux Video, Montgomery Presents, Matinee, Playhouse 90, Desilu Playhouse, One Step Beyond, Rawhide, Sincerely Maria Palmer (wrote and starred).

PAM, JERRY: Publicist. b. London, England, Oct. 17, 1926. e. Cambridge, London Univ. Reporter, Paris & London; free lance writing, Australia; 1950–53. To U.S. in 1953, on Hollywood Reporter, drama ed. Beverly Hills Citizen, 1953–54; publicist, Moulin Rouge, MGM studios; drama ed., Valley Times 1959–61; partner, Pam and Joseph pub. rel. counsellors; est. Jerry Pam & Associates, pub. rel., April 1965; formed Guttman & Pam, Ltd., 1971. Exec. prod., Highpoint, 1979.

PAMPANINI, SILVANA: Actress. b. Rome, Italy, Sept. 25, 1925. e. Academy of St. Cecilia. Studied singing, several concert appearances. Elected Miss Italia of 1946–47; m.p. debut in Secret of Don Giovanni.
PICTURES INCLUDE: Second Ark, Twin Trouble, O.K. Nero, City Stands Trial, A Husband for Anna, Songs of Half a Century, Songs Songs Songs, Matrimony, Enchanting Enemy, A Day in District Court, Loves of Half a Century, Slave of Sin, Orient Express, Merry Squadron, Princess of the Canary Islands, Mademoiselle Gobette, Don Juan's Night of Love, Roman Tales.

PAN, HERMES: Dance director. In 1935, handled Roberta, RKO stage ensembles. In 1938 won Academy honors for dance direction Damsel in Distress.
PICTURES INCLUDE: Top Hat, Old Man Rhythm, In Person, I Dream Too Much, Follow the Fleet, Swing Time, Shall We Dance, Radio City Revels, Let's Dance, Three Little Words, Excuse My Dust, Texas Carnival, Lovely to Look At, Sombrero, Kiss Me Kate, Student Prince, Hit the Deck, Jupiter's Darling, Meet Me in Las Vegas, Porgy and Bess, Can-Can, Flower Drum Song, Cleopatra, My Fair Lady, Finian's Rainbow, Darling Lilli, The Lost Horizon.
TV: An Evening with Fred Astaire (Emmy Award, choreography), Astaire Time, Sounds of America, Star-times Academy Awards of Songs, Remember How Great, Frances Langford Show.

PANAMA, CHARLES A. (CHUCK): Publicist, b. Chicago, Ill., Feb. 2, 1925. e. Northwestern U., Beloit College, UCLA. Publicist, Los Angeles Jr. Chamber of Commerce; staff So. Calif. sports ed., Los Angeles bureau, INS; pub. 20th Century-Fox Studios; adv. pub. dir., Arcola Pics.; opened L.A. office, John Springer Associates. Assoc. with Jerry Pam & Assoc. Account exec. Rogers, Cowan & Brenner, Inc.; dir. m.p. div., Jim Mahoney & Assoc. v.p. Guttman & Pam, Ltd.; ast. pub. dir., 20th-Fox TV.

PANAMA, NORMAN: Writer, prod., dir. b. Chicago, Ill., April 21, 1918. e. U. Chicago, PhB., 1936. Writer, radio (Bob Hope, Phil Baker, Rudy Vallee); sketches N.Y. stage (Schubert revue Keep off the Grass, 1939). Began screen career as writer 1941; wrote, co-prod. & dir., with Melvin Frank, Li'l Abner, Facts of Life, The Road to Hong Kong, Mr. Blandings Builds His Dream House.
PICTURES INCLUDE: My Favorite Blonde, Happy Go Lucky, Star-Spangled Rhythm, Thank Your Lucky Stars, And the Angels Sing, Road to Utopia, Duffy's Tavern, Our Hearts Were Growing Up, Monsieur Beaucaire, It Had to Be You, Return of October, The Reformer and the Redhead, Strictly Dishonorable, Callaway Went Thataway, Above and Beyond, Knock on Wood, White Christmas, Court Jester, Not With My Wife You Don't, How to Commit

Marriage, Coffee, Tea, or Me. 1975 wrote and directed: I Will, I Will . . . For Now. 1977: directed, Barnaby and Me; 1978 wrote Fade In—Fade Out, The Stewardesses, Li'l Abner, for NBC-TV, 1979. Recent works: Co-authored The Glass Bed (novel), and two plays: A Talent for Murder & The Bats of Portobello.

PANTAGES, CLAYTON G.: Executive. b. Hartford, March 6, 1927. e. Trinity Coll. Served various executive posts for 11 years with 20th Century-Fox; gen. sales mgr. Magna Pictures; Pres., International Coproductions, Inc.

PAPAS, IRENE: Actress. b. near Corinth, Greece. Entered dramatic school at 12. At 16 sang and danced in variety shows. Film debut in 1951 Greek film, Lost Angels; 1958 Greek Popular theatre in Athens.

STAGE: The Idiot, Journey's End, The Merchant of Venice, Inherit the Wind, That Summer, That Fall, Iphigenia in Aulis.

PICTURES INCLUDE: Dead City, The Unfaithful, Atilla the Hun, Theodora, Whirlpool, Tribute to a Bad Man, The Guns of Navarone, Antigone (Best Actress Award, Salonika Film Festival), Electra (Best Actress Award, Salonika Film Festival), Zorba the Greek, The Brotherhood, Anne of a Thousand Days, Z, A Dream of Kings, A Ciascuno il Suo, The Odyssey, The Trojan Women, Moses, Mohammed, Messenger of God, Lion of the Desert.

PAPAZIAN, STEVEN J.: Executive. b. San Diego, CA., July 10, 1944. e. California State University at Northridge. 1968 joined Warner Bros. production cost and auditing dept. 1972, The Burbank Studios—exec. asst. to the president; 1975 director-management services; 1978 director-production services; 1979 v.p.-production services. Member of exec. comm.

PARK, ROBERT H.: Executive. b. Atlanta, Georgia, May 11, 1916. e. University of Texas; Attorney for Jefferson Amusement Co.; now bd. chm., Tercar Theatre Company.

PARKER, BENJAMIN R.: Producer-director. b. New Haven, Conn., Sept. 16, 1909. e. U.C.L.A., Los Angeles and Yale Drama Sch. In m.p. indus. 1939, indep. prod. George Washington Carver. During WW II, dir. 23 training films for Army Air Corps; transferred Signal Corps, Photographic Center, Astoria, L.I., made 39 one-reel musical subjects; many shorts as prod., dir., for General Film prod. Co.

PARKER, ELEANOR: Actress. B. Cedarville, Ohio, June 26, 1922. In Cleveland play group; in summer stock Martha's Vineyard; at Pasadena Community Playhouse.

PICTURES INCLUDE: They Died With Their Boots On, Buses Roar, Mission to Moscow, Between Two Worlds, Very Thought of You, Crime By Night, Last Ride, Never Say Goodbye, Pride of the Marines, Of Human Bondage, Escape Me Never, Woman in White, Voice of the Turtle, Chain Lightning, Caged, Three Secrets, Valentino, Millionaire for Christy, Detective Story, Scaramouche, Above and Beyond, Escape from Fort Bravo, Naked Jungle, Valley of the Kings, Many Rivers to Cross, Interrupted Melody, Man with the Golden Arm, King and Four Queens, Lizzie, Seventh Sin, Home from the Hill, Return to Peyton Place, Madison Avenue, The Oscar, An American Dream, Warning Shot, The Eye of the Cat, The Sound of Music, Sunburn.

TV: Bracken's World, Vanished.

PARKER, FESS: Actor. b. Fort Worth, Tex., Aug. 16, e. U. of Tex., U. of So. Calif. US Navy, 1943–46; national co., Mr. Roberts, 1951; m.p. debut, Untamed Frontier, 1952.

PICTURES INCLUDE: No Room for the Groom, Springfield Rifle, Thunder Over the Plains, Island in the Sky, Kid from Left Field, Take Me to Town, Them, Battle Cry, Davy Crockett, King of the Wild Frontier, Davy Crockett and the River Pirates, Great Locomotive Chase, Westward Ho the Wagons, Old Yeller, The Light in the Forest, The Hangman, The Jayhawkers, Hell Is For Heroes, Smoky.

TV: Mr. Smith Goes to Washington series, Daniel Boone series, Jonathan Winters, Walt Disney Presents, Ed Sullivan, Phyllis Diller, Joey Bishop, Dean Martin, Red Skelton, Glen Campbell.

PARKER, JAMESON: Actor. b. Baltimore, Md. e. Beloit College. Professional stage debut in Washington Theatre Club production, Caligula. Acted with Arena Stage in D.C.; worked in dinner theatres and summer stock. Moved to N.Y., working in TV commercials and touring in play, Equus. Feature film debut in The Bell Jar (1979).

PICTURES: A Small Circle of Friends.

TELEVISION: Series: Somerset, One Life to Live. Movies: Women at West Point, Anatomy of a Seduction.

PARKER, SUZY: Actress. r.n. Cecelia Parker. b. San Antonio, Tex. Oct. 28; e. Schools in N.Y. & Florida. Began career at 17 as fashion model; became known as highest paid fashion model and cover girl in U.S.: to Paris under contract to fashion magazine; m.p. debut as model in Funny Face; signed by 20th-Fox prod. chief Buddy Adler for part opposite Cary Grant in Kiss Them For Me.

PICTURES INCLUDE: Ten North Frederick, The Best of Everything, Circle of Deception, The Interns.

PARKER, WILLARD: Actor. r.n. Worster Van Eps. b. New York City, Feb. 5, 1912. Meter reader, tennis pro; on stage in Johnny Belinda, screen debut in Slight Case of Murder, 1938; in Westerns.

PICTURES INCLUDE: What a Woman, Fighting Guardsman, One Way to Love, Relentless, Wreck of the Hesperus, You Gotta Stay Happy, Calamity Jane and Sam Bass, Bandit Queen, Emergency Wedding, Hunt the Man Down, Caribbean, The Vanquished, Sangaree, Kiss Me Kate, Great Jesse James Raid, The Earth Dies Screaming, Waco, The Great Waltz.

PARKHURST, PEARCE: Owner, novelty adv., ent. b. Gloucester, Mass., Apr. 12, 1919; e. Gloucester H.S., 1936; entered m.p. ind. at Union Hill Theatre, Gloucester, 1931; mgr., New Star, Dover-Foxcroft, Me., 1937; mgr., 1st drive-in, Methuen, Mass. for E.M. Loew 1946; 1st Lt. Air Force; mgr. theatres in Karachi, India New England, Va., Indiana, Ohio, N.C. Mich., Fla., Md., Penn., Ky.; present post, 1964, 1st prize winner, nat'l pub m.p. contest, 1949 scroll winner, 2nd quarter, 1st place winner, 3rd, 4th qtrs. Quigley Awards for Showmanship, 1950. Member: Variety Club, 1950; Allied Theatres of Mich.; apptd. 2nd v.p. 1954; first v.p. 1955; Michigan Allied Chairman State Allied Convention 1953. Mgr. Round Table, M.P. Herald; Lansing C. of C.: Amer. Legion; Veterans of Fgn. Wars; Nat'l Geo. Soc.; Nat'l Travelers Club. Adv. Specialty Guild, Premium Adv. Assn. of America, Inc.; operator own spec. adv. business, Pearce Parkhurst Enterprises, 1968, catering to theatre trade; with COMPO regional committee, 1954–55; Idea Advertising Budco Quality Theatres, 1967; Chakeres Theatres, 1964; Florida State Theatres, 1968; adv. spec. distrib. to theatre trade, 1969–82. Also mail order business with complete line Elvis Presley souvenirs for trade and public, 1977–82.

PARKINSON, ALLEN: Executive. b. Rexburg, Idaho, Jan. 26, 1910. e. Utah Agricultural College, Logan, Utah; pres. Sleepeze Drug Co.; dir of sls., Mercury Record Company, pres. Movieland Wax Museum.

PARKS, BERT: Announcer, m.c. b. Atlanta, Ga. Announcer, then chief announcer in Atlanta radio station; announcer, network, N.Y., for Eddie Cantor; m.c. for Xavier Cugat's show; U.S. Army, WWII; radio shows include Break the Bank, Stop the Music, Double or Nothing.

TV: Break the Bank, Stop the Music.

PARKS, GORDON: Director, writer, photographer, composer. Photo-journalist.

AUTHOR: The Learning Tree; A Choice of Weapons; A Poet and His Camera; Whispers of Intimate Things; In Love; Born Black, Moments Without Proper Names, Flavio, To Smile in Autumn, Shannon.

PICTURES INCLUDE: The Learning Tree, Shaft, Shaft's Big Score, Super Cops, Leadbelly.

PARKS, MICHAEL: Actor. b. 1938. Made m.p. debut in Wild Seed, 1964.

PICTURES INCLUDE: Bus Riley's Back in Town, The Bible, The Idol, The Happening, The Last Hard Men, Sidewinder One. ffolkes, Hard Country.

TV: Along Came Bronson (series, numerous TV movies. 1973: Can Ellen Be Saved? 1976: Savage Bees.

PARKS, RICHARD: Executive. e. Duke University; University of North Carolina (M.A. in theatre arts). Taught theatre arts at Duke. Associated with Curtis Brown, Ltd., literary agency, for 8 years, specializing in m.p. rights for literary materials and representing authors. Joined United Artists in December, 1978 as exec. asst. to Steven Bach, then snr. v.p. for east coast production. In 1980 promoted to v.p.—production.

PARRISH, ROBERT R.: Director, producer. b. Columbus, Georgia, Jan. 4; actor before joining RKO in 1933, first as assistant director, then film editor. With various companies since, including 20th Century-Fox, Universal, Columbia, United Artists, J. Arthur Rank, etc. Won Academy Award, best film editing, Body and Soul, 1947. U.S. Navy 1941–45; won documentary Academy Award, 1942 and 1943 for Battle of Midway and December 7th. Formed own independent production company, Trimark Productions, Inc., 1955.

PICTURES INCLUDE: City Lights, All Quiet on the Western Front, The Divine Lady, A Double Life, Caught, No Minor Vices, All the King's Men, Cry Danger (dir.) The Mob, San Francisco Story, Assignment—Paris, My Pal Gus, Shoot First, The Purple Plain, Lucy Gallant, Fire Down Below, Saddle the Wind, The Wonderful Country, In the French Style, Up From the Beach, Casino Royale, The Bobo, Duffy, A Town Called Bastard, Flashman.

PARRY, GORDON: Director, producer. b. Aintree, Liverpool, Eng., July 24, 1908; p. Studied electrical eng. before entering m.p. ind. as actor in 1929. Joined Gaumont-British in 1931 as assist. dir. and remained for four years, then prod. mgr. Rock Studios. Returned to GB as assist. to Walter Forde on King of the Damned. Left GB to become prod. mgr. for Soskin Prod. Ltd. and Amalgamated Studios; prod. television in conjunction with E.M.I. Author radio plays and short stories. Unit mgr. Ealing Studios propaganda shorts department. In 1942 under contract to Two Cities Films, 2nd unit.

PICTURES INCLUDE: In Which We Serve, The Demi-Paradise, English Without Tears, The Way to the Stars, Beau Brummell, Bond Street, Third Time Lucky, Now Barabbas, Three Men and a Girl, Midnight Episode, Tom Brown's Schooldays, Night Was Our Friend, Women of Twilight, Innocents of Paris, Front Page Story, Fast and Loose, Yank in Ermine, Sailor Beware, Touch of the Sun, Surgeon's Knife, Tread Softly Stranger, Friends and Neighbors, The Navy Lark.

TV: Robin Hood.

PARSONS, ESTELLE: Actress. b. Marblehead, Mass. e. Connecticut College for Women, Bachelor's Degree in Political Science. Attended Boston U. Law School. Helped harvest crops in England with the Women's Land Army. Was active in politics; worked for the Committee for the Nation's Health in Wash. and the Republican Finance Committee in Boston. Was elected to public office in Marblehead, Mass. Joined NBC-TV's Today Show as prod. asst.; then writer, feature producer and commentator. Appeared in two Julius Monk revues, Jerry Herman's Nightcap and the Three-Penny Opera. Dramatic stage debut, Mrs. Dally Has A Lover (won Theatre World Award). Has appeared with the Lincoln Center Repertory Theatre, Mahagonny. Acad. Award, Best Supporting Actress for Bonnie and Clyde.

STAGE: Next Time I'll Sing to You (Obie Award), In the Summer House (Obie Award), Ready When You Are, C.B. Malcolm, The Seven Descents of Myrtle, and Miss Reardon Drinks a Little, The Norman Conquests, Ladies of the Alamo, Miss Margarida's Way, Pirates of Penzance.

PICTURES INCLUDE: Rachel, Rachel, Don't Drink the Water, Strangers, Watermelon Man, I Never Sang For My Father, I Walk the Line, Two People, For Pete's Sake.

PARSONS, HARRIET: Producer, writer. b. Burlington, Ia. e. Wellesley Coll.; daughter of the late Louella O. Parsons, Hearst syn. m.p. columnist. Joined MGM scenario staff 1928, assoc. ed. Photoplay Magazine, 1929–30; mag. contrib. 1930–33; columnist, Universal Service, International News Service, 1931–44; on m.p. staff Los Angeles Examiner, 1935–44; Hollywood commentator, NBC, 1938; author, prod., dir. Screen Snapshots, Columbia, 1933–40; in 1940 joined Republic as writer, prod., dir., commentator of Meet the Stars series; assoc. prod. Republic, 1941–43. prod., writer, RKO 1943–55; 1956 TCF-TV. Member: Wellesley Alumnae Assn., Zeta Alpha, Hollywood Women's Press Club, Producers Guild of America, Acad. of M.P. Arts & Sciences.

PICTURES INCLUDE: The Enchanted Cottage, Night Song, I Remember Mama, Never a Dull Moment, Clash by Night, Susan Slept Here.

PARSONS, LINDSLEY: Pres., Completion Service Co., Hollywood. Mng. dir., Linpar Cinema Completion Corp., Toronto. b. Tacoma, Wash., Sept. 12. e. U. Calif. (L.A.) On ed. staff City News Service, L.A.: Alhambra Post-Advocate; Calexico Chronicle; Santa Rosa Press Democrat; Humboldt Times; San Marino News (ed. & pub.). Joined Monogram 1931 as pub. dir. in 1933 author s.p. Sagebrush Trails; then wrote orig. s.p. Westerns for Monogram, Republic, Grand Nat'l. In 1939 assoc. prod. Tough Kid; from 1940 prod. numerous westerns; prod. Wayne Morris & James Oliver Curwood series for Allied Artists; prod. Motion Pictures Int'l, 1956–72.

PICTURES INCLUDE: Rocky Rhythm Inn, Casa Manana, Big Timber, Call of the Klondike, Sierra Passage, Yukon, Manhunt, Yellow Fin, Northwest Territory, Desert Pursuit, Torpedo Alley, Jack Slade, Loophole, Cry Vengeance, Finger Man, Return of Jack Slade, Come On, The Intruder, Cruel Tower, Dragon Wells Massacre, Portland Expose, Oregon Passage, Wolf Larsen, Crash Boat, The Purple Gang, Mara of the Wilderness, Good Times, The Big Cube, Bravo Hennessey, Coasts of War.

TV: Gray Ghost TV series.

PARTON, DOLLY: Singer-Composer-Actress. b. Sevierville, TN, Jan. 19, 1946. Gained fame as country music singer, composer and radio and TV personality. Many awards for recordings.

PICTURES: Nine to Five, The Best Little Whorehouse in Texas.

TELEVISION: Porter Wagoner Show, Cass Walker program, Bill Anderson Show, Wilbur Bros. Show.

PARTRIDGE, DEREK: Actor. Born London, England. 1935. Educated Charterhouse. Entered business in 1958 as documentary scriptwriter/assistant director with Film Producer's Guild. Films (as actor): Incident At Midnight, The Verdict, The Murder Game, The High Bright Sun, King and Country, Thunderball, Where The Spies Are. 1966–67: In Rome, films: 7 Donne Per I MacGregor, Madigan's Millions, Cjmango, Don Giovannino. 1968–70: In Hollywood, TV: Star Trek, Robin Hood, My Friend Tony. Films: The Killing of Sister George, Fiuggi (Rome). 1971–72: Acting in Britain (after 5 years abroad). Many TV plays and series. TV commercials, voice-overs, commentaries. 1971: Directed and filmed clay-pigeon shooting film in Europe for ICI.

PASSER, IVAN: Director-Writer. b. Czechoslovakia. Began career as writer.

PICTURES INCLUDE: Writer: Loves of Blonde, Fireman's Ball. Director: Intimate Lighting, Born To Win, Law and Disorder, Crime and Passion, The Silver Bears, Cutter and Bone.

PASSMORE, HENRY: Producer. b. London, Eng., Nov. 28, 1905; e. Univ. College School (London). Enter m.p. ind. in 1929; gen. mgr. A.C.T. Films, Ltd.; joined Grendon Films, 1954; org. Croydon Passmore Prod., Ltd.; prod., Delavine Affair, Children Galore; prod. supvr., Charter Film Prod., Ltd., Guild Holdings & Augusta Prod. Man. dir. Reporters International, 1969; consult. prod. to De Beers Ind. Diamond Div.

PICTURES INCLUDE: Brothers in Law, Lucky Jim.

PASTER, GARY M.: Executive. b. St. Louis, Missouri, July 4, 1943. e. University of Missouri—B.A., U.C.L.A., U.S.C.—Graduate School of Business. 1970 joined Columbia Pictures, assistant to v.p. of facilities and operations division; 1972 joined The Burbank Studios as assist. to the president and as treasurer. 1976 v.p.—administration and chairman of the exec. committee. September, 1977 president. Board of Directors/Trustees: Contract Services Administration Trust Fund, St. Joseph Medical Center Institute, Permanent Charities Committee of Entertainment Industry. Member: Academy of Motion Picture Arts and Sciences, Academy of Television Arts and Sciences, Los Angeles Film Development Council, Advisory Commission—California Motion Picture Council, Hollywood Radio and T.V. Society. Board member of the Permanent Charities Committee of the Entertainment Industry, Board of Trustees of the St. Joseph Medical Center Foundation Board of Trustees Oakwood School, and alternate Trustee of the Contract Services Administration Trust Fund.

PASTERNAK, JOE: Producer, b. Szilagysomlyo, Hungary, Sept. 19, 1901. 2nd asst. dir. Paramount 1923; asst. dir. Universal (Hollywood), 1926, then prod. mgr. Berlin; made pictures in Vienna and Budapest, returned to Hollywood 1937; assoc. prod. then prod. A "Champion of Champion producers" in Fame Ratings.

PICTURES INCLUDE: Zwei Menschen, Unter Falscher Flagge, Grosse Scehnsucht, Unsichtbare Front, Fraulein Paprika, Gruss Und Gruss, Veronika, Scandal in Budapest, Csibi, Spring Parade, Katherine, Three Smart Girls, 100 Men and a Girl, Mad About Music, That Certain Age, Three Smart Girls Grow up, Destry Rides Again, It's a Date, Nice Girl, It Started With Eve, Presenting Lily Mars, Thousands Cheer, Two Girls and a Sailor, Music for Millions, Thrill of a Romance, Anchors Aweigh, Her Highness and the Bellboy, Two Sisters from Boston, Holiday in Mexico, No Leave No Love, This Time for Keeps, Three Daring Daughters, On an Island With You, Date With Judy, In the Good Old Summertime, Big City, Unfinished Dance, Nancy Goes to Rio, Summer Stock, That Midnight Kiss, Toast of New Orleans, The Great Caruso, Rich, Young and Pretty, The Strip, Merry Widow, Skirts Ahoy!, Because You're Mine, Small Town Girl, Latin Lovers, Easy to Love, Flame and the Flesh, Student Prince, Athena, Hit the Deck, Love Me or Leave Me, Meet Me in Las Vegas, Opposite Sex, 10,000 Bedrooms, This Could Be the Night, Where the Boys Are, Jumbo, A Ticklish Affair, Girl Happy, Penelope.

PATERSON, NEIL: Novelist, screenwriter. b. Scotland, 1916. e. Edinburgh Univ. War Service Lt. R.N.V.R. Early career as novelist. Awarded Atlantic Award in Literature, 1946. Dir. Grampian TV Chmn. of Production; films of Scotland. Vice-chmn. Scottish Arts Council; Gov. Pitlochry Thea.; Gov. National Film School. Former Gov. British Film Institute. Member Arts Council Gt. Britain.

NOVELS: The China Run, Behold Thy Daughter and Delilah, Man on the Tight Rope.

PICTURES INCLUDE: Man on a Tight Rope, The Little Kidnappers, Woman for Joe, High Tide at Noon, The Shiralee, Innocent Sinners, Room at the Top (s.p. Acad. Award, 1960), The Spiral Road, The Golden Fool, The Forty Days of Musa Dagh, Keeper of My Heart.

PATRICK, C. L.: Theatre Executive. b. Honaker, Va., Dec. 6, 1918. Pres. of Fuqua Industries, which owns Martin Theatres and Gulf States Theatres. Prior to this was pres. and chm. of Martin Theatres. Member NATO exec. Comm.; International Ambassador of Variety; Motion Picture Pioneer of the Year in 1976.

PATRICK, GEORGE: Art director, b. Salida, Colo. May 21, 1905. e. U. of So. Calif., 1923–27. Entered m.p. ind. as junior draftsman, First National Pictures, 1928; asst. art dir., art dir. for many prod.; designed, supervised construction of Frontierland section of Disneyland, 1955.

PICTURES INCLUDE: Deadline U.S.A., Beneath the 12-Mile Reef, The Racers.

PATRICK, NIGEL: Actor, director. b. London, Eng., May 2, 1913; e. private. Stage debut in The Life Machine, 1932. Screen debut, Mrs. Pim of Scotland Yard 1939; served with H.M. Forces, 1939–46. Top ten British money making star, 1952, 1953.

PICTURES INCLUDE: Uneasy Terms, Spring in Park Lane, Noose, Silent Dust, Jack of Diamonds, The Perfect Woman, Operation Disaster, Trio, Pandora & the Flying Dutchman, The Browning Version, Young Wives Tale, Encore, The Sound Barrier, Who Goes There, Meet Me Tonight, Pickwick Papers, Grand National Night (Wicked Wife), Forbidden Cargo, Sea Shall Not Have Them, Prize of Gold, All for Mary, Raintree County, How to Murder a Rich Uncle, Count Five and Die, The Man Inside, Sapphire, League of Gentlemen, The Trials of Oscar Wilde, Johnny Nobody, The Informers, The Battle of Britain, The Virgin Soldiers, The Executioner, Tales from the Crypt, The Great Waltz, The Mackintosh Man, The Silver Bears.

TV: Zero One series.

PATTERSON, RICHARD L.: Executive. b. St. Louis, Mo., November 1924. B.A. U.C.L.A., 1950; graduate work and teaching Theatre Arts Dept., U.C.L.A. until 1952; film dept. and dir. TV until 1955; H.N. Swanson Literary Agency handling TV dept.; head literary dept., Wm. Morris Agency to 1959; chmn., man. dir. Paramount British Pictures, Ltd., 1959; chmn. Seven Arts Prods. (U.K.) Ltd.: dir. Bryanston Seven Arts Ltd., 1961; assoc. Of Human Bondage; prod. The Wild Affair. Joined Christopher Mann Ltd., 1964; int. man. dir., P.L. Representation, 1967; prod. exec. MGM, 1972.

PAUL, M. B.: Cameraman, director. r.n. Morrison Bloomfield Paul. b. Montreal, Canada. Sept. 30. e. De Paul U.; newsreel, publicity picture service; partner, Seymour Studios, 1930–33; film test biz. own studio. Hollywood, 1933–35; prod. adv. films, asst. in N.Y. E.W. Hammons, 1945–47; Acad. Award, one-piece color translucent background system, 1950. Dir. of photography, optical effects, Daystar, United Artists, Outer Limits, 1963; designed, patented, Scenoramic process, 1965. Camera, Paradise Road. Features, Film project supervision A/V consult. Sceno 360 surround system development. Mem. AMPAS, Friars, SMPTE.

PAUL, RALPH: Announcer, m.c. b. Denver, Oct. 11, 1920. e. U. of Denver, 1945. Prod., dir. news ed. announcer, m.c. & narrator for radio stations KVOD, Denver; KTSM, El Paso; WITH, Baltimore; WOR, N.Y.; now free-lance, radio & TV, N.Y.

TV: Strike It Rich, Sound Stage, Television Playhouse.

PAULEY, JANE: Program Hostess-Reporter. b. Indianapolis, Ind., Oct. 31, 1950. e. Indiana Univ. Involved in Indiana state politics before joining WISH-TV, Indianapolis, as reporter. Co-anchored midday news reports and anchored weekend news reports. Co-anchor of nightly news at WMAQ-TV, NBC station in Chicago. Joined Today show in October, 1976, as featured regular, prior to which had made guest appearances on that program.

PAVAN, MARISA: Actress, r.n. Marisa Pierangeli. b. Cagliari, Sardinia, June 19, 1932; e. Torquado Tasso coll.; s. Pier Angeli, actress. Came to U.S. 1950; m.p. debut in What Price Glory.

PICTURES INCLUDE: Down Three Dark Streets, Drum Beat, Rose Tattoo, Diane, Man in the Gray Flannel Suit, John Paul Jones, Solomon and Sheba, Midnight Story.

PAVLIK, JOHN M.: Executive. b. Melrose, Iowa, Dec. 3, 1939. e. Univ. of Minnesota, B.A. 1963. Reporter, Racine (Wis.) Journal-Times, San Bernardino (Calif.) Sun-Telegram, 1963–66; Writer, News Bureau, Pacific Telephone, Los Angeles, 1966–68; assistant director of public relations, Association of Motion Picture and Television Producers, 1968–72; director of public relations, 1972–78; vice-president, 1978–79; executive administrator, Academy of Motion Picture Arts and Sciences, 1979–present. Member, board of trustees and Executive Committee, and chairman, Awards and Honors Committee, Motion Picture and Television Fund; member, board of directors, Permanent Charities Committee of the Entertainment Industries; member, board of directors, Hollywood Chamber of Commerce; vice-president, Los Angeles Film Development Committee, 1977–78, member, executive council, 1974–; special consultant, California Motion Picture Council, 1974–79.

PAVLOW, MURIEL: Actress. b. June 27, 1924. e. England, France & Switzerland. Stage debut in Dear Octupus, 1938; screen debut in Quiet Wedding.

PICTURES INCLUDE: Night Boat to Dublin, Shop at Sly Corner, It Started in Paradise, The Net (Project M7), Malta Story, Conflict of Wings (Fuss Over Feathers), Doctor in the House, Simon and Laura, Reach for the Sky, Eye Witness, Tiger in the Smoke, Doctor at Large, Rooney, Whirlpool, Meet Miss Marple

PAXTON, JOHN: Writer. b. Kansas City Mo., Mar. 21, 1911. e. U. Missouri, B.J. 1934. Publicist, 1935–36; assoc. ed., Stage mag. 1937–39; free lance 1939–41; pub. New York Theatre Guild, 1941. To Hollywood as writer 1942.

PICTURES INCLUDE: Murder My Sweet, Cornered, So Well Remembered, Cross Fire, My Pal Wolf, Crack-Up, Rope of Sand, Of Men and Music. Fourteen Hours, Wild One, The Cobweb, Prize of Gold, Pick-up Alley, How to Murder a Rich Uncle, On the Beach, Kotch.

PAY, WILLIAM: Executive. b. London, England; joined London office Quigley Publications. Served in RAF, 1941–46; rejoined Quigley; dir. Burnup Service Ltd., 1951; London news ed., Quigley Pub., 1955. Dir., Quigley Pub. Ltd., 1961; appt. mgr. dir., 1963; mgr. dir., Burnup Company. Appt. Sec. British Kinematograph Sound & TV Society. Conference Co-ordinator biennial Intern. Technology Conferences in London, 1975–81.

PAYNE, JOHN: Actor. b. Roanoke, Va. e. Mercersburg Acad., Pa., Roanoke Coll. of Virginia and Columbia. On radio programs. Stage debut 1973, Good News.

PICTURES INCLUDE: Dodsworth, Wings of the Navy, Indianapolis Speedway Kid Nightingale, Star dust, Maryland, Great Profile, Tin Pan Alley, King of the Lumberjacks, Tear Gas Squad, Great American Broadcast, Sun Valley Serenade, Week-End in Havana, Remember the Day, To the Shores of Tripoli, Hello Frisco Hello, The Dolly Sisters, Sentimental Journey, Razor's Edge, Wake Up and Dream, Miracle on 34th Street, Larceny, Saxon Charm, El Paso, Crooked Way, Captain China, Eagle and Hawk, Passage West, Crosswinds, Blazing Forest, Caribbean, The Vanquished, Kansas City, Confidential, Raiders of the 7 Seas, 99 River Street, Rails into Laramie, Silver Lode, Hell's Island, Santa Fe Passage, Road to Denver, Tennessee's Partner, Slightly Scarlet Rebel in Town, Bail Out at 43,000, Hidden Fear.

PAYNE, NORMAN: Artists' and writers' manager. b. London, England. Ent. entertainment ind., 1939. Early career music, then formed talent agency, J.P. Productions, 1945. Later bought by MCA, 1951. Became dir. MCA and head of light ent. for theatres and TV throughout Europe. On MCA terminating reformed agency, Albemarle Scripts (London) Ltd. and Albemarle Films Int., including Norman Payne Agency, TV offices also in Germany & Australia.

PEACH, MARY: Actress. b Durban, South Africa, 1934. Ent. films 1957. TV appearances include Ghosts, The Rat Wife, The Master Builder, Love Story, Hadley, Somerset Maugham, Can You Keep a Secret?, Cat on a Hot Tin Roof, Blind Love, Disraeli, Children at the Gods, Fox.

PICTURES INCLUDE: Room at the Top, Follow That Horse, No Love for Johnnie, A Gathering of Eagles, Ballad in Blue, Scrooge.

RECENT THEATRE: Birthday Run, The Marriage, When the Music Stops, Hilda, Don't Gas the Blacks, Old Times, The Chairman, The Autumn Garden.

PEAKER, E. J.: Actress, singer, dancer. Edra Jeanne Peaker, b. Tulsa, Okla., Feb. 22. e. Univ. of New Mexico, Univ. of Vienna, Austria. Stage debut Bye, Bye Birdie; m.p. debut Hello, Dolly. Films include All American Boy.

TV: That's Life series, The Flying Nun, That Girl, Movie of Week, Love American Style, Odd Couple, Police Woman, Rockford Files, Get Christy Love.

PECK, GREGORY: Actor, producer. b. La Jolla, Calif. April 5, 1916; U. Calif.; Neighborhood Playhouse Sch. of Dramatics. On dramatic stage (The Doctor's Dilemma, The Male Animal, Once in a Lifetime, The Play's the Thing, You Can't Take It With You Sons and Soldiers, etc.); on screen 1944 in Days of Glory. Voted one of ten best Money-Making Stars Motion Picture Herald-Fame Poll, 1947, 1952. Co-prod. and starred in Big Country, for his company, Anthony Productions prod. the Trial of the Catonsville Nine, The Dove (St. George Productions).

PICTURES INCLUDE: Keys of the Kingdom, Valley of Decision, Spellbound, Yearling, Duel in the Sun, Macomber Affair, Gentleman's Agreement, Paradine Case, Yellow Sky, 12 O'Clock High, Great Sinner, The Gunfighter, Only the Valiant, David and Bathsheba, Captain Horatio Hornblower, World in His Arms, Snows of Kilimanjaro, Roman Holiday, Night People, Man With a Million, Purple Plains, Man in the Gray Flannel Suit, Moby Dick, Designing Woman, The Bravados, Pork Chop Hill, On the Beach, Beloved Infidel, Guns of Navarone, Cape Fear, To Kill a Mocking Bird, Captain Newman, M.D., Behold a Pale Horse, Mirage, Arabesque, MacKenna's Gold, Stalking Moon, The Chairman, Marooned, I Walk the Line, Shootout, Billy Two Hats, The Omen, MacArthur, The Boys from Brazil, The Sea Wolves.

PECKINPAH, SAM: Writer-director. b. Fresno, Calif., Feb. 21, 1925. e. B.A., Fresno State College; M.A. University of Southern Calif. TV writer, Gunsmoke, awarded Writers Guild nom. for Best Writing Achievement in TV for specific segment. 20th Century Fox Hour, nom. from Writers Guild Best Achievement in TV. Created the Rifleman (dir. first five shows). Wrote script for Dick Powell's Zane Grey Theatre which became the pilot for the Westerner, which he wrote, prod. and dir. The Westerner received Producers Guild Nom. as Best Filmed Series. Also wrote-dir., Klondike and Stage 1967 production of Noon Wine (Screen Directors Guild Award.) Member: Directors Guild of America, Writers Guild of America, Acad. of M.P. Arts & Sciences.

PICTURES INCLUDE: Deadly Companions, Ride the High Country (won Belgian Film Festival Grand Prize, Mexican Film Festival Award, Best Film); Major Dundee, The Glory, Guys (writer), Villa Rides (writer), The Wild Bunch, The Ballad of Cable Hogue, Straw Dogs, Junior Bonner, The Getaway, Pat Garrett and Billy the Kid, Bring Me the Head of Alfredo Garcia, The Killer Elite, The Cross of Iron, Convoy.

PEERCE, LARRY: Director. b. Bronx, N.Y.

PICTURES INCLUDE: One Potato, Two Potato, The Incident, Goodbye Columbus, The Sporting Club, The Big TNT Show, A Separate Peace, Ash Wednesday, The Other Side of the Mountain, Two Minute Warning, The Other Side of the Mountain-Part II, The Bell Jar, Why Would I Lie?

PELLATT, JOHN: Production Executive. Abandoned school for theatre at age of 14. Became stage manager before serving in H.M. Forces. Then became assistant director for Warner Bros., MGM, London Films, etc. Since 1955 worked as prod. manager/assoc. producer with British Lion, Columbia, Paramount, United Artists, Ivan Tors, American International. Also while assistant gen. prod. mgr. 20th Century Fox, associated with Inn of Sixth Happiness, Roots of Heaven, The Blue Angel, Sons and Lovers, Sink the Bismark.

TV: Series for Incorporated Television Espionage. Worked in Europe, Middle and Far East, India, East and North Africa, Congo, Bahamas, Australia, Hollywood, Southern Africa.

PICTURES INCLUDE: Noose, The Wooden Horse, Silent Dust, Lavender Hill Mob, Man in the White Suit, Ivanhoe, Innocents in Paris, Mogambo, Knights of the Round Table, A Kid for Two Farthings, Oh Rosalinda, The Twelve Days of Christmas, The Captain's Paradise, Zarak Khan, The Green Man, St. Trinians, Only Two Can Play, They're a Weird Mob, Age of Consent, Count Five and Die, Help, The Revolutionary, Wuthering Heights, Elephant Country, Who Slew Auntie Roo, Tower of Evil, The Zoo Gang TV series, Born Free TV series (Kenya), The Diamond Mercenaries (Killer Force in Us), One Away. (South & S.W. Africa).

PENN, ARTHUR: Director. b. Philadelphia. e. Black Mountain College, Asheville, N.C., and Universities of Perugia and Florence in Italy. Began as TV dir. in 1953, twice winner of Sylvania Award. Dir. stage plays Two for the Seesaw, Miracle Worker, Toys in the Attic, All the Way Home, Golden Boy. Entered m.p. as dir. of Left-Handed Gun in 1958.

PICTURES INCLUDE: Miracle Worker, Micky One, The Chase, Bonnie and Clyde, Alice's Restaurant, Little Big Man, Visions of Eight, Night Moves, The Missouri Breaks, Four Friends.

PEPPARD, GEORGE: Actor, b. Detroit, Mich., Oct. 1, e. Dearborn H.S., Carnegie Tech., BA, Fine Arts. U.S. Marine Corps. Legit. stage debut, Pittsburgh Playhouse, 1949. Worked as mason, construction laborer, fencing instructor, Braddock, Pa. Signed by Sam Spiegel to appear in The Strange One.

STAGE: Girls of Summer, The Pleasure of His Company.

PICTURES INCLUDE: Pork Chop Hill, Home from the Hill, The Subterraneans, Breakfast At Tiffany's, How the West Was Won, The Victors, The Carpetbaggers, The Third Day, Operation Crossbow, The Blue Max, Tobruk, Rough Night in Jericho, P.J., House of Cards, What's So Bad About Feeling Good, Pendulum, Cannon for Cordoba, The Executioner, One More Train to Rob, The Groundstar Conspiracy, Newman's Law, Damnation Alley, Five Days From Home, Your Ticket Is No Longer Valid, Battle Beyond the Stars, Race to the Yankee Zephyr.

TV: Little Moon of Alban, Suspicion, U.S. Steel Hour, Alfred Hitchcock Presents, Matinee Theatre, Alcoa-Goodyear Playhouse, Studio One, Hallmark Hall of Fame, Banacek, The Bravos, Doctors Hospital, Story of Dr. Sam Sheppard, Crisis in Mid-Air, Torn Between Two Lovers.

PEPPER, L. J.: Exhibitor. Entered industry with Wilby-Kincey org., Selma, Alabama, 1923, while still in Selma H.S. With Wilby-Kincey 23 years. Opened own Center Theatre, Kingsport, Tenn., 1949. Went to Birmingham to open Eastwood Mall Theatre, 1964. General mgr., Cobb Theatres, two years. General mgr. Jefferson Amusement Co. with theatres in Birmingham and Selma, Ala.

PEPPERCORN, CARL: Executive. b. New York. e. NYU. Ent. film ind. FBO (forerunner of RKO), held var. sls. positions, home office and branches. Lt. Comm. US Navy, WW II; pres. Military Bank of Naples, sales mgr. NY branch RKO; asst. to Eastern sales mgr., RKO; gen'l sales mgr., RKO Canada; pres. Dairy-maid Chocolate Co., Toronto; v.p. Fairweather Dept. Stores, Canada; chmn. bd. Andako Mining Co., Canada; v.p. & gen'l sales mgr. Continental Dist. Inc.; v.p. & gen'l sales mgr. Embassy Pictures, Inc.; exec. v.p. and gen'l sales mgr. Cinema V Inc, pres. & sls. mgr., U.M. Film Distributors and Peppercorn-Wormser Film Distributors; Diversified Film Representatives, Inc., Peppercorn Film Enterprises.

PERAKOS, SPERIE P.: Executive. b. N. Britain, Conn., Nov. 12, 1915. e. Cheshire Acad., Yale, Harvard Law School, Student mgr., Stanley-Warner thtrs., 1939–40; Perakos thtrs., 1940–41; Fellow, Pierson Coll., Yale, 1946–present; Yale Alumni Bd., 1949 to present; Yale Alumni Film Bd. 1952 to present; member Alumni Council for Yale Drama School; past pres. Yale Club of N. Britain, Conn.; dir. of Films & Filmings Seminars, Pierson Coll., Yale; prod. Antigone, 1962; pres. Norma Film Prod., Inc., 1962 to present. Past pres. and now chm. Yale's Peabody Museum Associates and member of the University Council of the Peabody Museum. Pres., Perakos Theatres.

PERCIVAL, LANCE: Actor, singer. b. Sevenoaks, Kent, England, 1933.

PICTURES INCLUDE: V.I.P.'s, Yellow Rolls Royce, Mrs. Brown You've Got A Lovely Daughter, Darling Lili.

TV: That Was The Week That Was, Lance Percival Show.

PEREIRA, HAL: A.I.A. Art Director. Unit art dir., Paramount, 1942–46, special assignments. Paramount Home Office exec. staff for domestic and international companies 1947–50; appointed super. art dir., all films Paramount Studios, 1950; art dir., Rose Tatoo (Academy Award, best black & white art dir., 1955). 29 Academy nominations; 1968 design consul. William L. Pereira & Assoc., consultant motion pictures, theatres, television, leisure parks. Staff: Loyola Marymount University. School of Communications & Fine Arts 1976–78.

PERGAMENT, HARVEY: Producer, distributor; b. Detroit, Mich. June 18. e. Ohio State U.; NYU. With Air Transport Command, Army Air Forces, World War II. From 1928–37 gen. mgr. & sec. Film Exchange, Inc., N.Y. in 1937 org. and was pres., Pictorial Film Library, Inc.; gen. foreign sales mgr., Natl. Pictures, Inc., N.Y. Resigned 1940 to form Cavalcade Pictures; also Crystal Pictures, Inc.; ed. foreign language versions After Mein Kampf, Adventure in Music; disposed of interest in Crystal Pictures on entering armed forces, 1944. In 1945 formed Cavalcade Pictures, Hollywood; pres. thereafter; Cavalcade Television Programs, Inc. & Cavalcade Pictures Int'l Corp., 1951; foreign sales rep.; Eve Prods. Inc.; Essence Prods.; Phaeton Int'l Pics.; Savoy Road Show Pics.; Bernard Glasser Prod., Balut Prod., Belish Prod. Inc., Unusual Films, Inc., Interfilm Nassau Ltd., Timely Prods., Inc., Fountain Films, Woo Woo Prods., Sudan Prods., Inc., Lyman Dayton Prods., Inc., IPC Prod. Corp., VIP, Ltd., Doc-Art Prods., Premier Investment Corp., Zenith Int'l Films, Iota Prod., Lau Film Co., (Hong Kong), Ausable Co., Inc., Samwa Co. (Korea), Cosa Nueva Prod., R.M. Film Intern'l, Ad-Art Agency, Bill Copeland Prod.

PERINAL, GEORGES: Cinematographer. b. Paris, France, 1897. Ent. m.p. ind. 1913 as cameraman, French pictures; silent pictures until 1929; Jean de la Lune, David Golder, etc.; from 1929 several pictures for Rene Clair, Souls les Toits de Paris, Le Million, A Nous La Liberte, 14 Juillet; in 1933 to England; since cameraman many feature pictures for various Brit. cos. (London Films, B&D, United Artists, 20th-Fox, MGM-Korda, etc.); Acad. Award, color photog., 1940, Thief of Bagdad.

PICTURES INCLUDE: The Private Life of Henry VIII, Catherine the Great, Escape Me Never, Rembrandt, I Claudius, Under the Red Robe, Drums, The Four Feathers, Dangerous Moonlight, First of the Few, Colonel Blimp, Perfect Strangers, Man About the House, Ideal Husband, Fallen Idol, Forbidden Street, The Mudlark, No Highway in the Sky, House on the Square, Three Cases of Murder, Man Who Loved Redheads, Woman for Joe, Satellite in the Sky, Loser Take All, Lady Chatterley's Lover, King in New York, St. Joan, Bonjour Tristesse, Tom Thumb, Honeymoon, Serious Charge, Once More With Feeling, The Day They Robbed the Bank of England, Oscar Wilde.

PERKINS, ANTHONY: Actor. b. New York City, Apr. 14, 1932. Columbia U., Rollins College. Broadway stage in Tea and Sympathy. Dr. Steambath, 1970. Wrote s.p. Last of Sheila, 1973.

TV: Kraft Theatre, Studio One, U.S. Steel Hour, Armstrong Theatre.

PICTURES INCLUDE: The Actress, Friendly Persuasion, The Lonely Man, Fear Strikes Out, The Tin Star, This Bitter Earth, Desire Under the Elms, The Matchmaker, Green Mansions, On the Beach, Tall Story, Psycho, Goodbye Again, Phaedra, The Trial, The Fool Killer, The Champagne Murders, Pretty Poison, Catch 22, Someone Behind the Door, Ten Days' Wonder, WUSA, Play It As It Lays, Lovin' Molly, Murder on the Orient Express, Mahogany, Remember My Name, Winter Kills, The Black Hole, ffolkes.

STAGE: Greenwillow, Look Homeward Angel, Steambath, Equus.

PERKINS, JOHN HENRY ROWLAND II: Executive. b. July 10, 1934, Los Angeles. e. Univ. So. Calif., UCLA. Joined William Morris Agency, 1959; exec. in TV dept. In January, 1975, co-founded Creative Artists Agency, of which is senior partner and first pres. Member Hollywood Radio & TV Society since 1964, and Academy of TV Arts & Sciences since 1962.

PERLMUTTER, DAVID M.: Producer. b. Toronto, Canada, 1934. e. University of Toronto. Pres., Quadrant Films Ltd.

PICTURES INCLUDE: The Neptune Factor, 1972; Sunday In The Country, 1973; It Seemed Like A Good Idea At The Time, 1974; Love at First Sight, Find the Lady, 1975; Blood and Guts, The Third Walker, Two Solitudes, 1977; Fast Company, 1978; Double Negative, Nothing Personal, 1979; Misdeal, Love, 1980.

PERREAU, GIGI: Actress. r.n. Ghislaine Perreau; b. Los Angeles, Calif. Feb. 6; sister of Janine Perreau & Richard Miles, prof. m.p. debut in Madam Curie 1943; many stage and TV guest appearances.

PICTURES INCLUDE: Dark Waters, Abigail, Dear Heart, Family Honeymoon, Roseanna McCoy, High Barbaree, Song of Love, Green Dolphin Street, Two Girls and a Sailor, Shadow on the Wall, My Foolish Heart, For Heaven's Sake, Never a Dull Moment, Reunion in Reno, Lady Pays Off, Weekend With Father, Has Anybody Seen My Gal, Bonzo Goes to College, There's Always Tomorrow, Man in the Gray Flannel Suit, Dance With Me Henry, Tammy Tell Me True, Journey to the Center of Time, Hell on Wheels.

PERRIN, NAT: Writer. With Arthur Sheekman wrote series of radio broadcasts for the Marx Brothers. With Sheekman in 1933 wrote additional dialogue on Duck Soup. Par. Collab. s.p. Roman Scandals, many others. CBS-TV West Coast, exec. prod., color programs.

PICTURES INCLUDE: Kid Millions, Rose of the Rancho, Dimples, Pigskin Parade, Stowaway, Don't Tell the Wife, New Faces of 1937, On Again—Off Again, Swing Fever, Whistling in Brooklyn, Abbott and Costello in Hollywood, Song of the Thin Man, Petty Girl, Tell It To The Judge, Miss Grant Takes Richmond, Emergency Wedding.

PERRINE, VALERIE: Actress. b. Phoenix, 1944. e. Univ. of Arizona. Was showgirl in Las Vegas before discovered by agent Robert Walker who got her contract with Universal Pictures. Film debut in Slaughterhouse Five.

PICTURES INCLUDE: The Last American Hero, Lenny (N.Y. Film Critics Award, best supporting actress), W. C. Fields and Me, Mr. Billion, Superman, The Electric Horseman, Can't Stop the Music, Superman II, The Border.

PERRY, ANTHONY: Producer. b. London, England, 1929. Ent. m.p. ind. 1948 with Two Cities story dept. asst. story ed., Rank Prods. Wrote orig. story and prod. asst. Simba. Prod., The Secret Place in 1957. Created and prod. TV series, Interpol Calling, 1959; founded Eyeline Films, prod. many Brit. prize-winning commercials. Sold Eyeline Films, 1963; wrote, dir. Emma, 1964–65, res. prod. Keep Films/Embassy Prods., London, 1966–67; prod. Dare I Weep, Dare I Mourn, for ABC-TV, and Fernancel TV series. Admin. Yellow Submarine, joined Trickfilm as man. dir. Chmn. Film & TV Copyrights, Ltd.

PICTURES INCLUDE: The Impersonator, Girl on Approval, The Party's Over

PERRY, EARL: Circuit Executive. b. Aug. 11, 1921. e. Tulane Univ. Spent 4 yrs. as Air Force officer in World War II. Worked for Twentieth Century Fox for four years; then entered exhibition in 1951 as vice pres. and gen. mgr. of Pittman Theatres; in 1965, formed Ogden-Perry Theatres, and as pres., operates theatres in Louisiana, Mississippi, Tennessee, and Florida. Past President of NATO of Louisiana; director-at-large for National Association of Theatre Owners; Past Chief Barker of Variety Club Tent 45.

PERRY, FRANK: Director, producer, writer. served as apprentice, Westport, Conn. Country Playhouse; spent nine years in stock as stage mgr., prod. mgr., and managing director. U.S. Army, 1952–54; director-observer, Actors Studio, 1955; Theatre Guild, 1956–60; s.p. (with Eleanor Perry): Somersault, David and Lisa, and others.

PICTURES INCLUDE: The Swimmer, Ladybug, Ladybug, David and Lisa, Last Summer, Trilogy, Diary of a Mad Housewife, Doc, Play It As It Lays, The Man on the Swing, Rancho Deluxe, Mommie Dearest.

PERSCHY, MARIA: Actress. b. Eisenstadt, Austria, Sept. 23, 1940. e. Max Rheinhardt Seminar (Vienna). m. John Melson, writer. Started in 1958 with German film (Nasser Asphalt) and has appeared in over 50 European and U.S. features. Has also appeared on European TV. Recipient of Laurel Award in 1963.

PICTURES INCLUDE: Man's Favorite Sport, Squadron 633, Ride the High Wind, Murders in the Rue Morgue, Last Day of the War, The Desperate Ones, The Tall Woman, Witch Without a Broom, etc.

TV: General Hospital, Hawaii Five-O.

PERSKY, LESTER: Executive. b. New York City, July 6, 1927. e. Brooklyn College. Officer in U.S. Merchant Marine, 1945–48. Founder and pres. of own adv. agency, 1951–1964. Theatrical stage producer, 1966–69. Produced Fortune and Men's Eyes for MGM in 1971. In 1973 creative director and co-owner Persky Bright Org. (owner-financier of numerous motion pictures for private investment group). Films include Last Detail, Golden Voyage of Sinbad, For Pete's Sake, California Split, The Man Who Would be King, The Front, Shampoo. Also financing/production services for Hard Times, Taxi Driver, Missouri Breaks, Bound for Glory, Sinbad and the Eye of the Tiger. Now pres. and chief exec. officer, Lester Persky Productions, Inc.

PICTURES: Produced Equus, Hair, Yanks, Lone Star.

PERSOFF, NEHEMIAH: Actor. b. Jerusalem, Israel, Aug. 14, 1920. e. Hebrew Technical Inst., 1934–37; Actors Studio electrician, 1937–39; signal maint., N.Y.C. Subway, 1939–41.

STAGE: Only in America, Galileo, Richard III, King Lear, Peter Pan, Peer Gynt, Tiger At the Gates, Colombe, Mont Serrat.

PICTURES INCLUDE: Fate is the Hunter, Al Capone, Some Like It Hot, The Harder They Fall, The Badlanders, Men In War, This Angry Age, The Big Show, The Greatest Story Ever Told, The Wild Party, The Power, Mrs. Pollifax—Spy, Red Sky at Morning, Psychic Killer, Voyage of the Damned.

TV: Philco-Goodyear Show, Kraft, Producers Showcase, Danger, You Are There, Untouchables, Route 66, Naked City, Wagon Train, Rawhide, Gunsmoke, Thriller, Hitchcock Thriller, Bus Stop, Five Fingers, Mr. Lucky.

PERTWEE, JON: TV performer. b. London, July 7, 1919. e. Sherborne, Royal Academy Dramatic Art. Early career, Arts League Traveling Theatre, 5 yrs. repertory; regularly on radio, TV, music hall, cabaret and circus.

PICTURES INCLUDE: Murder At the Windmill, Miss Pilgrim's Progress, Will Any Gentleman?, Gay Dog, It's A Wonderful Life, Mr. Drake's Duck, A Yank in Ermine, Ugly Duckling, Just Joe, Not a Hope in Hell, Nearly A Nasty Accident, Ladies Who Do, Carry on Cleo, I've Gotta Horse, Carry On Cowboy, Carry On Screaming, A Funny wthing Happened On the Way To The Forum, Up in the Air, The Hod, March of the Desert, One of our Dinosaurs is Missing, The House That Dripped blood, The Island of Young Tigers.

TV: Own series, Sunday Night, London Palladium Compere Variety Show, Doctor Who series, Who Dunnit, Worzel Gummidge.

RADIO: Navy Lark.

STAGE: See You Inside, A Funny Thing Happened On The Way To The Forum, There's A Girl In My Soup, Oh Clarence, My Dear Gilbert, The Bedwinner, Don't Just Lie There, Say Something, Irene, Touch It Light.

PERTWEE, MICHAEL: Writer. b. April 14, 1916. e. Sherborne School and France. Early career, journalist; ent. m.p. ind. 1937. Co-author, co-presentation many BBC plays and serials; many appearances panel games.

PICTURES INCLUDE: Laughter in Paradise, On Monday Next, Top Secret, Happy Ever After, Now and Forever, Against the Wind, Interrupted Journey, The Naked Truth, Too Many Crooks, Bottoms Up, Make Mine Mink, It Started in Naples, In the Doghouse, Mouse on the Moon, Ladies Who Do, Finders Keepers, Strange Bedfellows, A Funny Thing Happened on the Way to the Forum, Salt and Pepper, One More Time, Don't Just Lie There Say Something, Digby the Biggest Dog in the World.

TV: Rainy Day, Strictly Personal, Grove Family, Man in a Moon, The Frightened Man, Yakity Yak (ATV 1956), The Old Campaigner, Terry Thomas series, B and B series, Never a Cross Word, Six of Rix, Men of Affairs.

STAGE: The Four Musketeers, Drury Lane, Done It Again, Don't Just Lie There Say Something, Birds of Paradise, A Bit Between the Teeth, Six Of One, Ace In A Hole, Find the Lady.

PESCOW, DONNA: Actress. b. Brooklyn, NY, March 24. e. American Academy of Dramatic Arts. Started career on summer tour in Ah Wilderness in 1975. Did bit in ABC daytime series, One Life to Live. Film debut in Saturday Night Fever (1977). Regular on ABC-TV series, Angie.

PETERS, BERNADETTE: Actress. b. New York City, Feb. 28, 1944. Professional debut at age 5 on TV's Horn & Hardart Children's Hours. Stage debut with NY City Center production of The Most Happy Fella. Screen debut, Ace Eli and Rodger of the Skies.

PICTURES INCLUDE: W.C. Fields & Me, Vigilante Force, Silent Movie, The Jerk, Tulips, Pennies from Heaven, Heartbeeps.

PETERS, BROCK: Actor. b. Harlem, N.Y.C. e. Univ. of Chicago. Had numerous featured roles on and off Bdwy. in road and stock cos., nightclubs, TV. Toured with DePaur Infantry Chorus 3 seasons. Made m.p. debut, Carmen Jones, 1955.
PICTURES INCLUDE: Carmen Jones, Porgy and Bess, To Kill a Mockingbird, The L-Shaped Room, The Pawnbroker, Black Girl, Soylent Green, Lost in the Stars.
STAGE: King of the Dark Chamber, Othello, The Great White Hope, Lost in the Stars, Framed.

PETERS, JON: Producer. b. Van Nuys, CA, 1947. Started hair-styling business; built it into multi-million-dollar firm before turning film producer. Formed Jon Peters Organization. 1980, joined with Peter Guber and Neil Bogart to form The Boardwalk Co.
PICTURES INCLUDE: A Star Is Born, The Eyes of Laura Mars, The Main Event, Die Laughing, Caddyshack.

PETERS, ROBERT C.: Executive. With Paramount Pictures since 1967 holding variety of financial posts: controller, TV div.; v.p., finance-west coast; pres., non-theatrical & educational dist. div. in 1980 appt. snr. v.p. for Paramount, assuming responsibility for all west coast administration, studio operations, post-production, personnel, industrial relations and Paramount Sound Systems, Inc. Continues responsibility for Paramount merchandising and has taken over talent relations.

PETERSEN, CLIFF: TV producer-director. b. Oct. 20, 1906. Ashland, Wis. on radio, Duluth; mem. vocal groups; joined Don McNeill's Breakfast Club cast 1936 with vocal group; later joined ABC mid-west prod. staff; assigned to Breakfast Club, 1945; superviser, Don McNeill's TV Club.

PETERSON, S. DEAN: Executive. b. Toronto, Canada, December 18, 1923. e. Victoria College, University of Toronto. WWII service RCNVR; 1946 TV newsreel cameraman NBC; founded own production company in 1947; incorporated Dordean Realty Limited to acquire new studios 1959; formed Peterson Production Limited in 1957 to make TV commercials and sponsored theatrical shorts; has international awards as producer, director, director of photography; formed Studio City Limited in 1965 to produce television series and features acquiring an additional studio complex and backlot in Kleinberg, Ontario; 1972 formed SDP Communications Ltd. to package M.P. and TV; 1970 incorporated Intermedia Financial Services Limited to provide specialized financing and consultation to companies in M.P. and TV industries.
Past-President Canadian Film and Television Association, mbr. Variety Club, Tent 28; Canadian Society of Cinematographers; Directors Guild of America, Directors Guild of Canada, SMPTE.

PETERSON, EDGAR: Television producer. b. Mobile, Ala., Jan. 12, 1913. e. Harvard. Theat. & radio prod., N.Y., 1935–36; asst. to dir. div. info., Resettlement Admin., Wash.; specialist in Public Info.; writer, dir. documentary films, 1938–41; civ. assoc. to Col. Frank Capra. Prod. Why We Fight series, 1942–45; prod. asst. to Dore Schary, Vanguard Films; assoc. prod. Spiral Staircase; exec. asst. RKO 1947; spec. research & writing for Div. of Int'l M.P., U.S. State Dept.; prod. TV shows; secty./ treas., Cornwall Prods., prod. of Janet Dean Series; mgr. script & story dept. CBS-TV; assoc. prod., then prod., Climax, CBS-TV; CBS consultant, London; exec. prod. CBS-TV. Created format and prod. techniques for Winston Churchill series, the Valiant Years, N.Y., London prod. operations and supv. writers, Hollywood, N.Y., London, 1961; writer-dir., Jack Douglas Organizations, prod. in France for series, Keyhole. U.A. TV writer; NTA-GOETZ Series, The Giants, prod., creator, The Nobel Prize, 1962; wrote orig. story and s.p. The Black Veil; created, wrote TV series, A Letter for Me, Americans Abroad; script and book, musical version Tom Sawyer, orig. treatment, s.p., Islandia, 1963. Consult. to Subscription TV in 1964; 1965–67 s.p. Assault in Eden. TV Emmy for The Pharmacist's Mate, starring Gene Raymond and Brian Donlevy. Won first Emmy for hour length film made especially for TV for program, Pulitzer Prize Playhouse; writer-editor for U.S. History Society biographical encyclopedia, The People Book, Hollywood, 1973; writer/editor, Research Associates, 1973-present. Currently, TV/MP writer.
PICTURES INCLUDE: Spiral Staircase, Till the End of Time, The Farmer's Daughter, Bachelor & Bobby Soxer (all asst. prod.), Assault in Eden, Northward the Coast.
TV: Pulitzer Prize Playhouse, Amazing Mr. Malone, Wonderful Town. Prod., Climax (CBS), Front Row Center (CBS), exec. prod. Westinghouse Summer Theatre, (CBS).

PETERSON, PAUL: Actor. b. Glendale, Calif., Sept. 23, 1945. e. Valley Coll. Mouseketeer (TV).
PICTURES INCLUDE: Houseboat, This Could Be the Night.
TV: The Donna Reed Show, Playhouse 90, Lux Video Theatre, GE Theatre, The Virginian, Ford Theatre, Valentine's Day, Shindig.

PETERSON, RICHARD W.: Executive. b. Denver, Colo., June 15, 1949. e. Col. School of Broadcasting; Harper Coll. Joined Kennedy Theatres, Chicago, 1966. In 1968 went with Great States Theatres (now Plitt Theatres), Chicago. Was city mgr. of Crocker and Grove Theatres, Elgin, Ill. In 1973 joined American Automated Theatres, Oklahoma City, as dir. of adv., pub. Promoted to dir. of U.S. theatre operations.

PETRIE, DANIEL: Director.
PICTURES INCLUDE: Lifeguard, Buster and Billie; Spy With A Cold Nose; The Idol; Stolen Hours; Raisin in the Sun; The Bramble Bush, The Betsy, Resurrection, Fort Apache, The Bronx.
THEATRE: Shadow of My Enemy; Who'll Save The Plowboy?; Mornin' Sun; Monopoly, The Cherry Orchard, Volpone.
TV FILMS: Sybil, Eleanor and Franklin, Eleanor and Franklin: The White House Years, Silent Night, Lonely Night, Harry Truman, Plain Speaking.

PETROU, DAVID MICHAEL: Writer-Producer. b. Washington, D.C., Nov. 3, 1949. e. Univ. of MD (BA); Georgetown Univ. (MA). Publicity assoc., Psychiatric Institutes of America, Washington, D.C., 1971; assoc. dir. of publicity & film liaison, Random House, 1974; guest lecturer, screen writing & film production, The American University Consortium, Washington, D.C., spring, 1980; Woodrow Wilson Fellowship, 1971. Entered industry in 1975. Joined Salkind Organization in chg. of literary projects. Worked in numerous production capacities on Crossed Swords, Superman, Superman II. 1977, exec. in chg. of literary development, Salkind. Wrote Crossed Swords (1978) and The Making of Superman. Co-authored screenplay, Shoot to Kill. 1978–79, promotional development on Time after Time for Warner Bros.; working on novel, The Hostess; dir., special projects Joseph Kennedy Foundation. Organized U.S. premiere of Superman II.

PEVERALL, JOHN: Producer. b. Islington, England. Started in entertainment industry in 1945 in mail room of J. Arthur Rank Prods. Promoted to asst. dir. Time out for military service in Royal Air Force Air-Sea Rescue Unit. Resumed career as asst. dir. and unit prod. mgr. on several films produced in Britain and throughout Europe. In 1969 became associated with newly-formed Cinema Center Films as prod. exec. in London. When firm suspended activities became freelance as asst. dir.
PICTURES INCLUDE: Conduct Unbecoming and The Man Who Fell to Earth (both assoc. prod.), The Deer Hunter (prod.).

PEVNEY, JOSEPH: Director, actor. b. New York City; e. N.Y.U., 1933; m. Mitzi Green, former child star; actress, nightclub entertainer. Began career in vaudeville at 13 as jr. mem. song & dance team; later stage in Home of the Brave. US Army ETO WW II.
STAGE: Counsellor at Law, Key Largo, Native Son; (dir.) Swan Song, Let Freedom Sing.
PICTURES INCLUDE (acting) Nocturne, Outside The Wall, Body & Soul; (dir.) Counsellor at Law, Key Largo, Native Son, Stage Door, Lady from Texas, Meet Danny Wilson, Iron Man, Flesh and Fury, Just Across the Street, Because of You, Desert Legion, It Happens Every Thursday, Back to God's Country, Yankee Pasha, Playgirl, Three Ring Circus, Six Bridges to Cross, Foxfire, Female on the Beach, Away All Boats, Congo Crossing, Tammy and the Bachelor, Man of a Thousand Faces, Cash McCall, Crowded Sky, Night of the Grizzly.

PEYSER, JOHN: Producer, director. b. New York, N.Y., Aug. 10, 1916; e. Colgage U., 1938. In TV ind. since 1939, with Psychological Warfare Div., ETO., WW II; pres. Peypac-productions, Madrid.
TV: Director: Hawaii Five-0, Mannix, Movin On, Swiss Family Robinson, Bronk, Combat Untouchables, Rat Patrol, etc.
FEATURE FILMS: Spain, The Open Door: Kashmiri Run; Honeymoon with a Stranger; Four Rode Out; Massacre Harbor.

PHILBIN, JACK: Producer. b. Lynn, Mass., e. Boston U., U.C.L.A. Started in Hotel management. Personal mgr., Glenn Miller 1939; Johnny Long 1942; Perry Como, 1944; Jackie Gleason 1948; v.p. General Artists Corp. 3 years; exec. prod., Jackie Gleason Show, Stage Show, The Honeymooners, America's Greatest Bands, CBS prod., The Vic Damone Show, The Big Record, exec. prod. Jackie Gleason American Scene Magazine TV Show, CBS-TV, 1962–63, Jackie Gleason Show 1965–74; Jackie Gleason Specials 1975–1980.

PHILLIPS, BILL: Actor. b. Washington, D.C. e. George Washington U. On N.Y. stage, Alice in Wonderland, Lightnin', etc. On screen 1940 in City for Conquest.
PICTURES INCLUDE: Larceny, Inc., Sergeant York, Action in the North Atlantic, Johnny Come Lately, See Here, Private Hargrove, Thirty Seconds Over Tokyo, Harvey Girls, Holiday in Mexico, Till Clouds Roll By, Sea of Grass, Liv-

ing in Big Way, Easy Living, Big Jack, Johnny Allegro, Customs Agent, Chain Gang, Bugles in the Afternoon, New York Confidential.

PHILLIPS D. JOHN: Motion Picture Theatre Consultant. b. New York City. Advertising and publicity mgr. Borden Co. Produce Sales Div., 1933–36; adv. & pub. mgr., Paul R. Dillon Co., Inc., 1936–41. Became field exploitation rep., United Artists Corp., 1941–42; Short Subjects & Paramount News adv. & pub. mgr. Paramount Pictures, 1942–47; exec. dir. Metropolitan Motion Pictures Theatres Assn., 1947–79, New York.

PHILLIPS, JULIA: Producer. b. Brooklyn, N.Y. e. Mt. Holyoke College. Production asst. at McCall's Magazine; later became textbook copywriter for Macmillan. In 1970 she and her husband, Michael, formed a company with actor Tony Bill (Bill/ Phillips Productions) to develop film projects.
PICTURES INCLUDE: Steelyard Blues, The Sting, Taxi Driver, The Big Bus, Close Encounters of the Third Kind.

PHILLIPS, LESLIE: Actor, producer. b. London, April 20, 1924. Early career as child actor. Ent. m.p. ind. 1935.
PICTURES INCLUDE: High Flight, Les Girls, Smallest Show on Earth, Carry On Nurse, King Ferdinand of Naples, This Other Eden, The Navy Lark, Doctor in Love, Please Turn Over, Watch Your Stern, No Kidding, Week-End With Lulu, VIP, Carry on Constable, Raising the Wind, In the Doghouse, Crooks Anonymous, Fast Lady, Father Came Too, Doctor in Clover, You Must Be Joking, Maroc 7, Some Will Some Won't, Doctor in Trouble, The Magnificent 7 & Deadly Sins, Not Now Darling, Don't Just Lie There, Spanish Fly, Not Now Comrade.
TV: Our Man At St. Marks Impasse, The Gong Game, Time and Motion Man, Reluctant Debutante, A Very Fine Line, The Suit, The Culture Vultures (series), Edward Woodward Show, Casanova 74 (series).

PHILLIPS, MICHAEL: Producer b. Brooklyn, N.Y., Nov. 10, 1916. Mgr. & prof. pugilist (featherweight), 1930–36. Ent. m.p. ind. as secy to Ray Milland, 1937; apptd asst. dir., U.S. Army Special Service Div. 1942–45; script ed. & prod. activities, Eddie Bracken Radio Prod. until 1949 (orig. & s.p.) Double Cross; formed own prod. co. Demyrtha Prod., Inc. 1950. In 1970 formed prod. co. with wife, Julia, and Tony Bill.
PICTURES INCLUDE: Steelyard Blues, The Sting, Taxi Driver, The Big Bus, Close Encounters of the Third Kind, Heartbeeps, Cannery Row.

PICERNI, PAUL: Actor. b. New York, N.Y. Dec. 1, 1922. e. Lovola U., Los Angeles. U.S. Air Force 1943–46; head of drama dept. Mt. St. Mary's College, 1949–50; TV credits include Untouchables (co-star).
TV: Philco Playhouse, Climax, Lux, Loretta Young Show, Desilu, Kojak, Mannix, Police Story, Lucy Special.
PICTURES INCLUDE: Breakthrough, I Was a Communist for the FBI, Mara Maru, Desert Song, She's Back on Broadway, House of Wax, Shanghai Story, To Hell and Back, Miracle in the Rain, Bobby Ware is Missing, Omar Khayyam, Brothers Rico, Young Philadelphian, The Young Marrieds, The Scalphunters, Airport, Kotch.

PICKENS, SLIM: r.n. Louis Bert Lindley, Jr.; b. Kingsberg, Calif., June 29, 1919; e. High School, Hanford, Calif. With rodeos since age of 12; m.p. debut in Rocky Mountain; since in many Rex Allen Westerns.
PICTURES INCLUDE: Sun Shines Bright, Boy from Oklahoma, The Outcast, Santa Fe Passage, Last Command, When Gangland Strikes, Stranger At My Door, Great Locomotive Chase, One-Eyed Jacks, Dr. Strangelove, The Honkers, Pat Garrett and Billy the Kid, Blazing Saddles, Rancho Deluxe, White Line Fever, Mr. Billion, The Swarm, Beyond the Poseidon Adventure, 1941, Tom Horn, Honeysuckle Rose.

PICKER, ARNOLD M.: Executive. b. New York, Sept. 19, 1913; m.; p. Celia and David V. Picker; e. Coll. of City of New York, U. of London; joined Columbia's foreign dept. in June, 1935. Asst. to foreign mgr.; then vice-pres. 1945, Columbia International Corp.; v.p. charge foreign dist., Oct., 1951; exec. v.p. UA, and in charge all distribution, June 1961; chmn. exec. com., 1967. Retired.

PICKER, DAVID V.: Executive. b. New York City, May 14, 1931; e. Dartmouth Coll., B.A., 1953. Father Eugene Picker, exec. Loew's Theatres. Ent. industry in 1956 as adv. pub. & exploitation liaison with sls. dept., United Artists Corp.; exec. v.p. U.A. Records; asst. to Max Youngstein, v.p.; v.p. U.A.; first v.p. UA; pres. 1969. Resigned 1973 to form own production co. In 1976 joined Paramount Pictures as pres. of m.p. div. Now exec. v.p., Lorimar Productions.
PICTURES INCLUDE: Juggernaut, Lenny, Smile, Royal Flash, Won Ton Ton, The One and Only, Oliver's Story, The Jerk.

PICKER, EUGENE D.: Executive b. New York City, Nov. 17, 1903; married; p. David V. and Celia C. Picker; e. N.Y.U. and School of Business. Started with father in Bronx theatres; joined Loew's Inc., 1920; in charge circuit operations, New York City area, 1945; v.p. Loew's Theatres, Sept. 1954; member bd. of dir., 1956, exec. v.p. Sept. 1958; pres. Loew's Theatres, March 1959. Res. 1961 as pres. Loew's Theatres; joined U.A. as v.p., July 1961. Joined Trans-Lux Corp. as exec. v.p., Jan. 1967; pres. & chief oper. Officer of Entertainment Division of Trans-Lux Corp.; Exec. consultant motion picture industry Jan. 1, 1974 and Pres. E.D.P. Films Inc. as of June 1974; pres. NATO, 1969–71, ch. bd., 1971–72 Bd. dir., Will Rogers Hospital, bd. ch. At present member board of directors, Trans Lux Corp. and Foundation of Motion Picture Pioneers and Broadway Association.

PICKMAN, JEROME: Executive. b. New York, N.Y., Aug. 24, 1916; e. St. John's U.; B'klyn Law Sch. of St. Lawrence U., LL.B. Reporter N.Y. newspapers, 1930–40; U.S. Army, WW II, adv. dept. 20th-Fox, 1944–46; Eagle-Lion Films, 1947; Paramount, April 1949; v.p., dir., adv. & pub., Paramount Film Dist. Corp., 1951–60; v.p. domestic gen. sls. mgr., Paramount Film Dist. Co., N.Y. till 1962; sls. exec. Columbia Pictures 1963–67; pres. Continental Div. of Walter Reade Org., 1967–70; pres., Levitt-Pickman Film Corp., 1971; snr. v.p., domestic distribution, Lorimar Productions, 1979.

PICKMAN, MILTON EUGENE: Executive. b. N.Y.C. e. Columbia U., LL.B; St. Lawrence U. Reporter, Brooklyn Daily Eagle, 1925–29; asst. Florenz Ziegfeld Publicity, 1929–31; mgr. & agent, name bands, 1932–40; exec. M.C.A., 1933–36; U.S. Army, 1942–45; eastern studio rep. & asst. to Harry Cohn, Columbia, 1945; own agency, 1948–51; v.p. & gen. mgr., Wald-Krasna Prod., RKO, 1952; exec. asst. to Jerry Wald, v.p. in charge of prod., Columbia, 1952–54; v.p. of TV prog., Screen Gems Prod.; gen. mgr. Samuel Goldwyn Prod., 1958; prod. exec. and asst. to Buddy Adler, v.p., chg. of prod., 20th Century Fox Studios; exec. v.p., American Entertainment Enterprises.

PICKUS, ALBERT M.: Exhibitor. b New Haven, Conn., Mar. 20, 1903; e. Washington and Lee U., 1920–24, B.S. Accountant, auditor, New Haven; ent. m.p. ind., 1926; owner, Stratford Theatre, Stratford, Conn.; dir., MPTO of Conn., 1953–59; v.p., TOA, 1953–57; asst. pres. TOA rep. to COMPO, 1957–58; chmn. exec. com. TOA, 1959; exec. com., ACE, 1959; pres. TOA, 1959–61; chmn. of bd., 1962.

PIDGEON, WALTER: Actor. b. East St. John, New Brunswick, Canada, Sept. 23, 1898. Prof. career began with Elsie Janis in At Home, in which they toured U.S. & England. Made Victor phonograph records; then vaudeville with Elsie Janis. Pres., Screen Actors Guild, 1953.
STAGE: No More Ladies, Something Gay, The Night of January 16, There's Wisdom in Women, Happiest Millionaire.
PICTURES INCLUDE: Her Private Life, Bride of the Regiment, Mlle. Modiste, Going Wild, The Gorilla, Journal of a Crime, Big Brown Eyes, Fatal Lady, Happiest Millionaire, Youngest Profession, Madame Curie, Mrs. Parkington, Week-End at the Waldorf, Holiday in Mexico, Secret Heart, If Winter Comes, Julia Misbehaves, Command Decision, Red Danube, That Forsythe Woman, The Miniver Story, Soldiers Three, Calling Bulldog Drummond, Unknown Man, The Sellout, Million Dollar Mermaid, The Bad and the Beautiful, Scandal At Scourie, Dream Wife, Executive Suite, Men of the Fighting Lady, Last Time I Saw Paris, Deep in the Heart, Hit the Deck, Forbidden Planet, The Rack, Voyage to the Bottom of the Sea, Advise & Consent, Warning Shot, Funny Girl, Skyjacked, The Neptune Factor, Harry in Your Pocket, Two Minute Warning.

PIERCE, FREDERICK S: Executive. e. Bernard Baruch School of B.A., City College N.Y. Served with U.S. Combat Engineers in Korean War. Associated with Benj. Harrow & Son, CAP, before joining ABC in 1956. Served as analyst in TV research dep.; prom. to supvr. of audience measurements, 1957, named mgr. next year. In 1961 made dir. of research; 1962 dir. of research, sales devel. Named dir. of sales planning, sales devel April, 1962; elec. v.p., Feb. 1964 and made nat. dir. of sales for TV. In 1968 named v.p., planning; March 1970 named asst. to pres. In July 1972, named v.p. in chg. ABC TV planning and devel. and asst. to pres. ABC TV, March 1973. Named sr. v.p., ABC TV, Jan., 1974. Elected pres., ABC Television Division, October, 1974.

PIERSON, FRANK: Producer-Director-Writer. b. 1925. Was correspondent for Time Magazine before entering show business as story editor of TV series, Have Gun, Will Travel. Later served as both producer and director for show. Developed a number of properties for Screen Gems before writing theatrical screenplays.
PICTURES INCLUDE: Cat Ballou (co-s.p.), Cool Hand Luke, (s.p.), The Anderson Tapes, (s.p.), The Looking Glass War, (s.p., dir.) Dog Day Afternoon, (s.p.), A Star Is Born (dir., s.p.), King of the Gypsies (dir., s.p.).

PILCHER, TONY: Executive. b. Boston, England, 1936. e. Shrewsbury School. Ent. m.p. industry 1960 with Anglo-Scottish Pictures. Became prod. exec. 1961, German rep.

and exec. 1968. Joined AB-Pathe as German rep., TV and Advertising Films division 1964. TV prod. Heumann Ogilvy & Mather, Frankfurt, 1966; prod. Chambers and Partners; Guild TV Services and freelance, 1967–69. Prod. Wace Film Prods., 1970; Signal Films, 1971; Rayant TV, 1972; Filmshop, 1979.

PILE, SUSAN: Executive. Now director of West Coast adv. & pub. for m.p. division of Paramount Pictures. Had six-year stint with Diener, Hauser, Bates Advertising Agency (L.A.); worked at Lion Gates Films; headed own mktg. firm in association with Marykay Powell (clients including Cinema 5, 20th-Fox, Filmex 1976, etc.). Unit publicity on films for Warner Bros., Universal; also publicist at latter co.

PINCUS, IRVING: Producer. b. New York, N.Y. e. N.Y. Univ. Indep. TV prod., 1949–63; creator-prod.; The Real McCoys, Ellery Queen, Mr. Imagination. Theatre prod. The Good, co-author; The More The Merrier, Higher and Higher, ind. prod. Prod. feature, To Find a Man, Col., 1971.

PINE, HOWARD: Producer. b. New York, N.Y., April 4, 1917. e. Stanford U. Asst. dir., Pine-Thomas Prod., then prod. mgr.; prod. Universal.
PICTURES INCLUDE: Cult of the Cobra, Man from Bitter Ridge, Private War of Major Benson, Running Wild, Nightmare.

PINK, SIDNEY: Producer, director, writer. b. Pitts., Pa., Mar. 6, 1916. e. Univ. of Pitts., B.S., 1934–37; U.S.C., Law, 1940–41. Projectionist, mgr., booker, Warner Bros., Pa., Fox West Coast, United Artists Theatres, Calif. Prod. budget mgr., Something to Sing About and Lost Horizon.
PICTURES INCLUDE: Bwana Devil, Angry Red Planet, Green-Eyed Elephant, Reptilicus, Journey to The Seventh Planet, Operation Camel, Valley of the Swords, Madigan's Millions.

PINKHAM, RICHARD R.: Executive. b. New York, N.Y. April 11, 1914. e. Yale U., 1932–36. Time, Inc., 3 yrs.; Lord and Thomas, 1 yr.; Herald Tribune, 5 yrs.; mgr. of TV planning, NBC, 1951–52; exec. prod., Today, 1952–54; exec. prod. Home, Tonight, 1954; dir. of participating prog., 1954; v.p. participating prog., 1954–55; v.p. in charge of TV programs, 1955–56; v.p. of adv. 1957; senior v.p. in chg. media, programs, bd. Ted Bates & Company, Inc. senior v.p. in chg. media, programs, bd. dir. Vic Chairman of the Board, 1976, Ted Bates & Company, Inc. from which retired 1979. Now pres., Alden Rey Enterprises.

PINSKER, ALLEN: Executive. b. New York City, Jan. 23, 1930. e. N.Y. Univ. Mgr., Hempstead Theatre, 1950. In 1954 joined Island Theatre Circuit as booker-buyer; named head buyer 1958. In 1968 joined United Artists Eastern Theatres as film buyer; head buyer, 1969, v.p., 1970. Named v.p. United Artists Theatre Circuit, 1972. In 1973 named UAET exec. v.p., member bd., 1974.

PIPER, FREDERICK: Actor. b. London, England, Sept. 23, 1902. On British stage from 1922. m.p. debut 1932 in Brit. picture The Good Companions.
PICTURES INCLUDE: Four Just Men, Jamaica Inn, Spare a Copper, 49th Parallel, The Big Blockade, Pink String and Sealing Wax, Hue and Cry, Loves of Joanna Godden, The October Man, Master of Bankdam. It Always Rains on Sunday, Escape, My Brother's Keeper, Don't Ever Leave Me, Fly Away Peter, Passport to Pimlico, Look Before You Love, Blue Lamp, Brandy for the Parson, Hunted (Stranger in Between), Home at Seven (Murder on Monday), Escape Route, Conflict of Wings (Fuss Over Feathers), Rainbow Jacket, Lease of Life, Doctor at Sea, Man in the Road, Passionate Stranger, Suspended Alibi, Doctor at Large, Se sation, Second Fiddle, Birthday Present, Barnacle Bill, Dunkirk. A Touch of Larceny, The Day They Robbed the Bank of England, Dead Luckey, The Frightened City, VIP, What A Carve-Up, Only Two Can Play, Postman's Knock, Reach for Glory, Ricochet, Becket, One Way Pendulum, Catacombo, He Who Rides A Tiger, Burke and Hare.
TV: Gert, Maude, Hindle Wakes, Mornings at Seven, The Green Bay Tree, Time Out of Mind, The Breaking Point, The Citadel, Emergency Ward 10, Zero One, Harpers West One, The Human Jungle, The Teachers, Guilty Party, Sergeant, Corky, Rupert of Hentzav, The Bender, The Hidden Truth, Danger Man, The Brothers Karamazov, Tea Party, The Long House, 199 Park Lane, Dixon of Dock Green, the Idiot, Weavers Green, The Prisoner, The Happy Sacking, Great Expectations, The Man Behind You, Approach to Living.

PIROSH, ROBERT: Writer, director, producer. S.p. credits include Battleground (solo orig. s.p. Academy Award winner), Hell Is For Heroes, What's So Bad About Feeling Good, S.P.-dir. credits include Go For Broke, Washington Story, Valley of the Kings.
TV: (Wrote, prod. pilots) Laramie, Combat. (Writer) Hawaii Five-0, Ellery Queen, Mannix, Bonanza, Ironside, The Waltons, Barnaby Jones, The Bold Ones, etc.

PISIER, MARIE-FRANCE: Actress. b. Indochina. First discovered by Francois Truffaut who cast her in L'Amour a Vingt Ans. When film completed returned to school for degree in political science. Continued to work in films.
PICTURES: Trans-Europe Express, Stolen Kisses, Celine et Julie Vont en Bateau, Cousin Cousine, Souvenirs d'en France, Barocco, The Other Side of Midnight, French Postcards.

PLANCK, ROBERT: Cinematographer.
PICTURES INCLUDE: It's Great to be Alive, Life in the Raw, Jane Eyre, Last of the Mohicans, By Secret Command, Canterville Ghost, Secrets in the Dark, Cass Timberlane, It Happened in Brooklyn, Three Musketeers, Luxury Liner, Little Women, Madame Bovary, Summer Stock, Royal Wedding, Texas Carnival, Scandal at Scourie, Torch Song, Rhapsody.
TV: My Three Sons.

PLATT, MILT: Executive. b. New York City. e. CCNY, RCA Institute, Ohio State U.; US Army 1942–46; div. mgr., RKO Radio Pictures until 1957; gen. sls. mgr. Continental Dist., 1957–65; vice-pres. & gen. sls. mgr. Sherpix, 1965; v.p. & gen. sls. mgr., Comet Film Distributors, Inc., 1965; v.p. & sls. mgr., Times Film Corp., 1968. Pres. Eagle Amusement Co., 1970. Appointed member of the Appeals Board of the MPA rating system, 1971; pres. of Pisces Group, Ltd., 1972; v.p. & gen. sls. mgr., International Co-productions, Inc., 1974. pres., Milton Platt Co., 1975. Member governing committee, IFIDA.

PLEASENCE, DONALD: Actor. b. Worksop, England, Oct. 5, 1919. Repertory, first London appearance in Twelfth Night. RAF, WW II. Since London stage, NY stage, ent. m.p. ind. 1953.
STAGE: Vicious Circle, Saint's Day, Hobson's Choice, The Rules of the Game, The Lark, Ebb Tide, The Caretaker, Poor Bitos, The Man in the Glass Booth, Wise Child (N.Y.); voted actor of the year, 1958.
PICTURES INCLUDE: Manuela, The Man in the Sky, Heart of a Child, Tale of Two Cities, Battle of the Sexes, The Shakedown, The Horsemasters, Spare the Rod, No Love for Johnnie, The Caretaker, The Great Escape, The Greatest Story Ever Told, Hallelujah Trail, Fantastic Voyage, Cul-de-Sac, You Only Live Twice, Matchless, 13, Will Penny, Arthur! Arthur, THX 1138, Soldier Blue, Outback, Jerusalem File, Pied Piper, Innocent Bystanders, Death Line, Wedding in White, The Rainbow Boys, The Black Windmill, Journey Into Fear, Escape to Witch Mountain, Hearts of the West, The Devil Within Her, The Last Tycoon, Passover Plot, Trial by Combat, The Eagle Has Landed, Goldenrod, Oh God!, Fear, the Uncanny, Telefon, Escape from New York.
TV: Fate and Mr. Browne, Small Fish are Sweet, The Silk Purse, A House of His Own, The Traitor, The Millionairess, The Cupboard Machinal, The Hatchet Man, The Bandstand, Ambrose, Thou Good and Faithful servant, Call Me Daddy, Taste, The Fox Trot, Omnibus, Julius Caesar, Occupations, The Joke, The Cafeteria, Hindle Lakes, Sgt. Peppers Lonely Heart's Club Band, Halloween.

PLESHETTE, EUGENE: Executive. b. Jan. 7, Brooklyn, N.Y. e. CCNY & LaSalle U., Para. Pict. Acting School; stage actor; assoc. prod. and dir. three off-Broadway plays; treas. and house mgr. N.Y. Paramount; v.p. Reid-Singer Music; exec. mgr. Brooklyn Paramount thea., 1945; mgn. dir. 1953; v.p. in chg. of ABC Merchandising Inc., AB-PT, Inc. and American Broadcasting Co., 1962; exec. v.p., MSG-ABC Prods., Inc. 1965; exec. v.p. Don Reid TV Prod.; 1975, President, Pleshette Associates.

PLESHETTE, SUZANNE: Actress. b. New York, N.Y. Jan. 31. e. Performing Arts H.S., Finch College, Syracuse U. Broadway debut, Compulsion; m.p. debut Geisha Boy; major TV shows including The Bob Newhart Show, 1972.
PICTURES INCLUDE: Rome Adventure, The Birds, 40 Pounds of Trouble, Wall of Noise, A Rage to Live, Youngblood Hawke, A Distant Trumpet, The Ugly Dachshund, Bullwhip Griffin, Fate is the Hunter, Mr. Buddwing, Nevada Smith, Blackbeard's Ghost, The Power, If It's Tuesday This Must Be Belgium, Suppose They Gave a War and Nobody Came, Support Your Local Gunfighter, The Shaggy D.A., Oh, God! Book II.
STAGE: The Cold Wind and the Warm, The Golden Fleecing, The Miracle Worker, Compulsion, Two for the Seesaw.

PLESKOW, ERIC: Executive. b., Vienna, Austria. Served as film officer, U.S. War dept., 1946–48; entered industry in 1948 as asst. gen. mgr., Motion Picture Export Association, Germany; 1950–51, continental rep. for Sol Lesser Prods.; joined United Artists in 1951 as Far East Sales Mgr.; named mgr., S. Africa, 1952; mgr., Germany, 1953–58; exec. asst. to continental mgr., 1958–59; asst. continental mgr., 1959–60; continental mgr., 1960–62; v.p. in charge of foreign distribution, 1962; exec. v.p. & chief operating off., Jan. 1, 1973; pres. & chief exec. off., Oct. 1, 1973. Resigned in 1978 to becomes pres. and chief exec officer of Orion Pictures.

PLITT, HENRY G.: Executive. b. New York, Nov. 26, 1918. e. Syracuse U. St. Lawrence U. Law School. War service, 6 yrs.; Paramount Pictures International Corp.; United Detroit Theatres; asst. gen. mgr. North Ohio Theatres Corp.; div. mgr., then v.p., Paramount Gulf Theatres, New Orleans; pres. gen. mgr. Paramount Gulf Theatres; pres., ABC Films, 1959–65; Pres. ABC Great States Inc., Great States Theas. 1966; v.p. Prairie Farmer publications, 1971, v.p., ABC Theatre Holdings, Inc.; 1974, purchased Northern Theatre Circuit from American Broadcasting Company, consisting of 127 theatres, naming these theatres Plitt Theatres, Inc., of which he is pres. 1978, purchased rest of ABC theatres from American Broadcasting Company consisting of 272 screens, and will rename them Plitt Theatres.

PLUMMER, CHRISTOPHER: Actor. b. Toronto, Canada. Stage career started with English repertory group visiting Canada; toured U.S. in Nina, 1953; Bway debut in The Constant Wife, 1954; on road and in N.Y. with The Dark is Light Enough; American Shakespeare Festival at Stratford, Conn., in Julius Caesar and The Tempest; Shakespeare Festival at Stratford, Ontario, in Henry V, Twelfth Night, and Hamlet; on Bway in The Lark and Cyrano.
TV: Oedipus Rex, Omnibus, After the Fall, The Shadow Box.
PICTURES INCLUDE: Stage Struck, Inside Daisy Clover, The Sound of Music, Triple Cross, The Battle of Britain, The Royal Hunt of the Sun, Lock Up Your Daughters, The Phyx, The Return of the Pink Panther, Conduct Unbecoming, The Man Who Would Be King, International Velvet, The Silent Partner, Murder by Decree, Hanover Street, Somewhere in Time, Eyewitness.

PLUNKETT, PATRICIA: Actress. b. Streatham, Eng., Dec. 17, 1928; e. Royal Acad. Dram. Art. Stage debut in Pick Up Girl, 1946; m.p. debut in It Always Rains on Sundays.
STAGE: Girl Who Couldn't Quite, French Without Tears, The Brothers, Family Upstairs.
PICTURES INCLUDE: Bond Street, For Them That Trespass, Landfall, Murder Without Crime, Mandy (Crash of Silence), Crowded Day, Dunkirk, Three Survived.
TV: African Patrol, Jungle Boy series, The Big Thirst, Emergency Ward 10.

PLUNKETT, WALTER: Costume designer. b. Oakland, Calif., June 5, 1902. e. U. of Calif. Began career as actor, vaudeville & stock lo., San Francisco; designed first stage costumes for vaudeville chorus on tour, costume designer. N.Y. stage and Metropolitan Opera House.
PICTURES INCLUDE: Gone With the Wind, Lust for Life, Raintree County, Pollyanna, Bells are Ringing, How the West was Won.

PODELL, ALBERT N.: Executive. b. New York City, Feb. 25, 1937. e. Cornell U., U. of Chicago. Articles editor, Playboy Magazine, 1959–61; dir. of photog., Argosy Magazine, 1961–64; account exec. on 20th Century-Fox at Diener, Hauser, Greenthal, 1966–68; national advertising dir., Cinema Center Films, 1969; account supervisor on Columbia Pictures at Charles Schlaifer; creator & dir. of Annual Motion Picture Advertising Awards sponsored by Cinema Lodge, B'nai B'rith.

PODESTA, ROSSANA: Actress. b. Tripoli, 1934. Discovered in Italy by French director, Leonide Moguy, who starred her in Tomorrow Is Another Day. For 15 years appeared only in films by her husband, Marco Cicario; now divorced.
PICTURES INCLUDE: Cops and Robbers, La Red, Ulysses, Helen of Troy, Santiago, The Golden Arrow, Il Prete Sposato, Seven Golden Men, Pano, Burro e Marmelate, The Quiet Countrywoman, Sunday Lovers, Stasi.

PODHORZER, MUNIO: Executive. b. Berlin, Germany, Sept. 18. e. Jahn-Realgymnasium. U. of Berlin Medical School. U.S. Army, 1943–47; pres. United Film Enterprises, Inc.; formerly secy.-treas. 86th St. Casino Theatre, N.Y.; former v.p. Atlantic Pictures Corp.; former pres. Casino Films, Inc.; former pres. Film Development Corp.; former rep. Export-Union of the German Film Ind.; former U.S. rep. Franco-London Film, Paris; former pres., Venus Productions Corp.; former U.S. rep. Atlas Int'l Film GmbH, Munich; former U.S. rep. Bavaria Ateller Gesellschaft U.S. past rep. Israfilm Ltd., Tel-Aviv; past rep. Tigon British Film Prod., London; past rep. Elias Querejeta, P.C., Madrid; past rep. Equiluz Films, Madrid, past rep. Airport Cine, Hawaii. Member: Variety Club, Cinema Lodge, B'nai B'rith, Past Board of Governors IFIDA; past pres. CID Agents Assoc. Former gen. foreign sales mgr., theatrical division of National Telefilm Associates. Presently representing Atlas Film—AV, Germany; Barcino Films, S.A., Spain; Filme Scorpio, Mexico; Eagle Films Ltd., United Kingdom; Les Films Du Capricorne, S.R.L., Italy; Les Films Jacques Leitienne, France; Nero Films Classics U.S.A.; Schongerfilm, Germany; Europgroup Film Distributors of France; Profilmes, S.A., Spain; KFM Films, Inc. U.S.A.; Signo Producciones, Spain; Les Films D'Alma, France; Intra Films, Italy; Elektra Films, Greece.

PODHORZER, NATHAN: Executive. b. Brody, Poland, Nov. 27, 1919. e. City Coll. of N.Y., Rutgers Univ., U. of So. Calif. U.S. Army, 1942–46; documentary film prod., Israel, 1946–57; vice pres., secy., United Film Enterprises, Inc.

POITIER, SIDNEY: Actor. b. Feb. 24, 1924; e. Nassau. Formed First Artists Prod. Co. Ltd., 1969 with Paul Newman and Barbra Streisand.
STAGE: Lysistrata, Anna Lucasta, Freight.
PICTURES INCLUDE: No Way Out, Cry the Beloved Country, Red Ball Express, Go Man Go, Blackboard Jungle, Good-Bye My Lady, Edge of the City, Something of Value, Porgy and Bess, All the Young Men, Devil at Four O'Clock, A Raisin in the Sun, The Long Ships, Lilies of the Field, Slender Thread, A Patch of Blue, Duel at Diablo, To Sir With Love, In the Heat of the Night, Guess Who's Coming to Dinner, The Lost Man, They Call Me Mister Tibbs, Brother John, For Love of Ivy, Buck and the Preacher, A Sweet December (dir., star), Uptown Saturday Night (dir., star), The Wilby Conspiracy, Let's Do It Again (dir., star), A Piece of the Action (dir.-star), Stir Crazy (dir.).

POLANSKI, ROMAN: Director/Writer. b. Paris, France, 1933. Lived in Poland from age of three. Early career, Art School in Cracow; five years State Film College at Lodz. Wrote and dir. Two Men In a Wardrobe, Le Gros et Le Maigre, Mammals. From 1961 wrote and dir. Knife in the Water (Poland), Repulsion, Cul-De-Sac, The Vampire Killers, Rosemary's Baby, Macbeth, What?, Chinatown. Co-scripted & Dir. The Tenant (Paris).

POLIER, DAN A.: Executive. b. Atlanta, Ga. e. Ga. Mil. Acad. and U. of Ill. m. Leila Michalove. Sports columnist, Charlotte (N.C.) News; sports editor, army wkly, Yank; entered m.p. industry adv. dept. 20th Century-Fox; head of booking dept. Fox West Coast Theatres; v.p. and director of film buying NT&T Amusement Corp.; v.p. and co-director theatre operations, National General Corp. (formerly NT&T); named v.p. in chg. of production, National General Productions, Inc., 1967. Joined Radnitz/Mattel Productions in 1972 as v.p. in chg. of distribution for Sounder; v.p. Mann Theatres Corp. of Calif., 1973. Rejoined Radnitz/Mattel Productions, 1975, as vice pres. in charge of marketing. Now gen. sls. mgr. New World Pictures.
PICTURES INCLUDE: Cheyenne Social Club, Stalking Moon, Dream of Kings, Daddy's Gone Hunting, Grasshopper, El Condor, How Sweet It Is, Baby Maker.

POLL, MARTIN H.: Producer. b. New York. e. Wharton School, U. of Pa.; pres. Gold Medal Studios, 1956; prod. Love Is a Ball, Sylvia, The Lion in Winter, The Appointment, The Magic Garden of Stanley Sweetheart, The Man Who Loved Cat Dancing, Night Watch, Love and Death (exec. prod.); The Sailor who Fell From Grace with the Sea, Somebody Killed Her Husband, The Dain Curse, Nighthawks.

POLLACK, SYDNEY: Director, actor. b. South Bend, Ind., July 1. m. Claire Griswold. e. Neighborhood Playhouse, Assistant to Sanford Meisner at Neighborhood Playhouse. Appeared as actor on B'way in A Stone for Danny Fisher, The Dark is Light Enough. As TV actor: Playhouse 90 segments, Shotgun Slade, 15 Ben Caseys, A Cardinal Act of Mercy (won 5 Emmy nominations), The Game on Bob Hope-Chrysler Theatre (won Emmy for direction), Two is the Number. Dir. debut in 1960. Dir. play at UCLA, P.S. 193. Prepared the American version of The Leopard.
PICTURES INCLUDE: The Slender Thread (dir.), This Property is Condemned, The Scalphunters, Castle Keep, They Shoot Horses, Don't They?, Jeremiah Johnson, The Way We Were, The Yakuza, Three Days of the Condor, Bobby Deerfield, The Electric Horseman, Absence of Malice.

POLLEXFEN, JACK: Producer, director, writer. b. San Diego, Calif., June 10, 1918. e. Los Angeles City Coll. Newspaperman, magazine writer, playwright: prod. for RKO, United Artists, Columbia, Allied Artists.
PICTURES INCLUDE: Son of Sinbad, At Swords Point, Secret of Convict Lake, Desert Hawk, Lady in the Iron Mask, Dragon's Gold, Problem Girls, Captive Women, Captain Kidd and the Slave Girl, Neanderthal Man, Captain John Smith and Pocahontas, Return to Treasure Island, Sword of Venus, 1000 Years from Now, Daughter of Dr. Jekyll, Monstrosity.

POLONSKY, ABRAHAM: Director, writer. b. N.Y.; e. CCNY, BA., Columbia Law School. Taught at City College 1932 until war. Wrote s.p. Golden Earrings, I Can Get it For You Wholesale. Wrote novels: The Discoverers, The World Above, The Season of Fear. Wrote orig. story and s.p. Body and Soul; collab. s.p. and directed Force of Evil. Blacklisted from 1951–66. Coll. s.p. Madigan 1968; dir. and s.p. Tell Them Willie Boy is Here, 1970; dir. Romance of a Horse Thief, 1971.

PONTI, CARLO: Producer. b. Milan, Italy, Dec. 11, 1913. Univ. of Milan 1934. Prod. first picture in Milan, Little Old World; prod. Lux Film Rome; prod. first of a series of famous Toto pictures, Toto Househunting.
PICTURES INCLUDE: A Dog's Life, The Knight Has Ar-

rived. Musolino, The Outlaw, Romanticism, Sensuality, The White Slave, Europe 1951, Toto in Color, The Three Corsairs, Ulysses, The Woman of the River, An American of Rome, Attila, War and Peace, The Last Lover, The Black Orchid, That Kind of Woman, Marriage Italian Style, The Great Spy Mission, Happily Ever After, The Girl and the General, Sunflower, Best House in London, Lady Liberty, White Sister, What?, Andy Warhol's Frankenstein, The Passenger, The Cassandra Crossing, A Special Day.

POOLE, FRANK S.: Executive. b. London, England, 1913. e. Dulwich Coll, 1925–31. Ent. m.p. ind. 1931. Early career with Pathe Pictures, Twickenham Film Distributors, until joining 20th Century Fox as London Branch Office Supervisor 1939. War service 1940–46. Rejoined Fox 1946–53; appt. Leed Branch mgr. 1954–59; supv. 1959–61; asst. sls. mgr., 1961 until joined Frank Film Distrib. as asst. sls. mgr. 1962. Appt. sls. mgr. July 1965, and to board as dir. of sls. Aug. 1965. Appt. gen. mgr. 1968; jnt. man. dir. 1969; appt. man. dir. July 1970; appt. dir. Rank Overseas Film Dist. Ltd., 1972; apptd. co-chmn Fox-Rank Distributors Ltd., Dec. 1972; appt. vice-chairman Rank Film Distributors Ltd. 1977. 1975, elected to Committee of Cinema & TV Veterans. Oct., 1978, retired from Rank Organisation. Appt. chmn., Appeal Tribunal for the Film Industry. Appt. chm., Grebelands Mgt. Committee & to exec. council of CTBF, 1979, assoc. Ceoff Reeve & Associates. 1980, chm. & mng. dir., Omandry Intl. Ltd.

POPE, EDWARD J.: Producer, writer. b. Rye, N.Y., Aug. 3, 1919. e. Western Reserve Academy, 1938; Yale University, 1942. m. H.I.H. Princess Niloufer. World record, all-American swimmer, 1940–42; Commander, USNR, 1942–45; author, History of U.S. Navy, WW II; author, with Marechal Juin, La Defense Nationale, 1946; awarded Exceptional Civil Service Medal; Cross of Malta; Commandante, Grand Order of Merit Italian Republic; Commander, Order of Medjidiye; junior writer, MGM, 1946; prod., writer documentaries, 2 Academy Award nominations, 1947–48; prod. Seaboard Studios, ABC-TV studios, 1949; prod. film seq. Robert Montgomery Presents and 10 Million Bros. (Germany). Producer, Dumont TV, 1950; Auerbach Film Enterprises, 1950–51; ass. prod. 70 US TV films, Europe, 1951–53; v.p., exec. prod. Alpha Omega Films (France), 1954–55; chief int'l and pub. rel. U.S. Defense Dept. Europe, 1956–64; exec. prod., Compagnie Francaise De Coproduction Internationales (France) 1965–69; v.p. General Productions S.A., 1968–69; and Giroux Films, S.A. 1969; associate prod., Sweet Hunter; prod. 12 + 1; pres., H.I. Management SA (Holiday Inns) Europe.

POPKIN, HARRY M.: Exhibitor, producer. b. N.Y.C., e. Jarvis Collegiate Sch., Toronto, Canada; U. of Cal. Started career as cashier, ticket-seller, asst. to mgr., Gore Bros. Theatre Circuit, Los Angeles; became mgr. & 3 yrs. later theatre owner, L.A.; acq. other theatres, formed Eastland Circuit; purchases Gore Bros. interests; ent. sports world, oper. major auditorium, L.A., San Francisco, Oakland; first m.p. prod. all-Negro films, then, feature picture And Then There Were None; financed other ind. prod. Since 1949 org. ind. prod. unit, Hollywood; world-wide offices.
PICTURES INCLUDE: Impact, Champagne for Caesar, Second Woman, The Well, The Thief, Top Banana.

PORTER, DON: Actor. b. Miami, Okla. e. Oregon Inst. of Tech. Wide theatre work; then m.p. roles. U.S. Army, 3 yrs.
TV: Co-star, Private Secretary, Ann Sothern Show, Gidget, Bionic Woman, Hawaii Five-0, Switch, Love Boat, Three's Company, The President's Mistress, The Murder That Wouldn't Die, The Last Song, Dallas.
STAGE: The Best Man, Any Wednesday, Generation, Plaza Suite, The Price, How To Succeed in Business Without Really Trying, Harvey.
PICTURES INCLUDE: The Racket, The Savage, 711 Ocean Drive, Because You're Mine, Our Miss Brooks, Bachelor in Paradise, Youngblood Hawke, The Candidate, 40 Carats, Mame, White Line Fever, The Last Resort.

PORTER, RAND: Actor/writer/producer. b. Los Angeles, Nov. 9, 1933. e. Tulane Univ. Child actor on stage 1946–50; radio, 1950–51. 1955–64, radio, TV personality. Made films in France and U.S. 1949–56. Producer from 1964–present.

POST, TED: Producer, director. b. Brooklyn, N.Y., dir. many stage plays; dir. CBS-TV Repertoire Thea.; Producer-dir., NBC-TV Coney Island of the Mind.
TV: Studio One, Ford Theatre, Playhouse of Stars, Fred Astaire Show, Gunsmoke, Rawhide, Twilight Zone, Wagon Train, Combat, Peyton Place, Alcoa, Defenders, Route 66, Baretta, and Columbo. Movies: Dr. Cook's Garden, Girls in One Office, Cagney & Lacey, Night Stars, Five Desperate Women. etc.
MINI-SERIES: Rich Man, Poor Man.
PICTURES INCLUDE: Go Tell The Spartans, Magnum Force, Hang 'em High, Beneath The Planet Of The Apes, The Harrad Experiment, Good Guys Wear Black, Whiffs.

POTTER, BRIDGET: Executive. Married, author Robert Wool. Served as assoc. prod. of Dick Cavett Show; in chg. TV development for Palomar Pictures Intl. Joined ABC-TV, being named v.p. of prime time program development—east coast. In 1980 appointed v.p., Lorimar Productions, in chg. of m.p. and TV development for east coast.

POWELL, CHARLES, M.: Executive. b. New York, Feb. 17, 1934. B.S. NYU/Journalism. Columbia Pictures National Publicity Manager, National Exploitation Manager, 1959–69. Paramount Pictures, National Publicity Coordinator, 1963–64. WNBC-Radio/TV, Advertising/Promotion Manager 1965. Director adv., pub. for M. J. Frankovich 1969–71. Joined MGM as dir. adv. pub.-expl. in 1972; named div. v.p. & corp. v.p., 1974; Columbia Pictures, v.p. advertising/publicity/exploitation 1975. Universal Pictures, sr. v.p., 1976–80. Currently bd. chm., Powell & Young, m.p. consultants. Board of Governors, Acad. of Motion Picture Arts & Sciences, since 1973. Founder & President, Synagogue for the Performing Arts, L.A.

POWELL, DILYS C.B.E.: Journalist. b. Bridgenorth, England. e. Bournemouth H.S. & Somerville College, Oxford. Film critic, London Sunday Times 1939–1976; now critic films on TV; Film critic, Punch, frequent broadcaster, author of several books; member of Independent Television Authority, 1954–57. Mem., Cinematograph Films Council, 1965–69.

POWELL, JANE: Actress. r.n. Suzanne Burce. b. Portland, Ore., Apr. 1. Had own radio program over KOIN, Portland; singer on natl. networks; m.p. debut in Song of the Open Road, 1944, Star of Tomorrow, 1948.
PICTURES INCLUDE: Holiday in Mexico, Three Daring Daughters, Luxury Liner, Date With Judy, Nancy Goes to Rio, Two Weeks With Love, Royal Wedding, Rich Young and Pretty, Small Town Girl, Three Sailors and a Girl, Seven Brides for Seven Brothers, Athena, Deep in My Heart, Hit the Deck, Girl Most Likely, The Enchanted Island.
STAGE: Irene (Broadway, 1974).

POWELL, WILLIAM: Actor. b. Pittsburgh, Pa., July 29, 1892; e. American Acad. Dram. Arts, New York. First appearance in The Ne'er Do Well, N.Y., 1912; then in stock & touring in Within the Law, The King, and others; on N.Y. stage 1920 in Spanish Love. m.p. debut, 1920, Sherlock Holmes; numerous pictures since. Voted one of the best Money-Making Stars in M.P. Herald-Fame Poll 1937.
PICTURES INCLUDE: Four Feathers, One Way Passage, Evelyn Prentice, The Thin Man, Rendezvous, After the Thin Man, The Last of Mrs. Cheney, Shadow of the Thin Man, Crossroads, Youngest Profession, Heavenly Body, Thin Man Goes Home, The Senator Was Indiscreet, Life with Father, Mr. Peabody and the Mermaid, Take One False Step, Dancing in the Dark, It's a Big Country, Treasure of Lost Canyon, The Girl Who Had Everything, How to Marry a Millionaire, Mister Roberts.

POWER, MALA: Actress. r.n. Mary Ellen Powers. b. San Francisco, Calif., Dec. 29, 1921; p. George and Dell Powers, latter, dramatic coach. e. Max Reinhardt Dramatic School, Hollywood, Calif.; U. of Calif., Los Angeles. Pasadena Playhouse in For Keeps, 1946; Distant Isle; Actor's Lab, Hollywood; did considerable radio work.
PICTURES INCLUDE: Outrage, Edge of Doom, Cyrano de Bergerac, Rose of Cimarron, City Beneath the Sea, City That Never Sleeps, Geraldine, Yellow Mountain, Rage at Dawn, Bengazi, Tammy and the Bachelor, Storm Rider, Flight of the Lost Balloon, Daddy's Gone-A-Hunting.

POWERS, C. F. (MIKE) JR: Executive. b. San Francisco, March 6, 1923. e. at Park College, Mo., Columbia University, N.Y., graduated University of Oregon. Entered film business with P.R.C. in Portland, Oregon, 1947. Became Eagle Lion branch mgr. in Portland, 1950, and then United Artists. Moved to Seattle, Wash. as branch mgr. of 20th Century Fox, 1960. Was then western division mgr. for 20th Century Fox until 1967, then western division mgr. for Cinerama till 1973. Became exec. v.p., head film buyer for Robert L. Lippert Theatres, Transcontinental Theatres and Affiliated Theatres until 1978. Present position, western division mgr. for Filmways Pictures. President of Catholic Entertainment Guild of Northern Calif.; past Chief Barker of Variety Club Tent 32, San Francisco.

POWERS, STEFANIE: Actress. b. Hollywood, CA, Nov. 2, 1942. Theatrical m.p. debut in Among the Thorns. TV debut in The Girl from U.N.C.L.E. series.
PICTURES: Experiment in Terror, McClintock, Warning Shot, Escape to Athena.
TELEVISION: Feather and Father series, Washington: Behind Closed Doors (mini-series), Hart to Hart (series).

PRAGER, ALICE HEINECKE: Music executive. b. N.Y.C., Aug. 2, 1930; d. Paul and Ruth (Colin) Heinecke; grad. Russell Sage Coll., 1951; grad. study N.Y.U. With SESAC, Inc., N.Y.C., 1946– , now exec. v.p., mng. dir.; pres. The Personal Touch, Inc. Named hon. citizen, Tenn., Fla.; Zonta, International Radio and TV Execs. Soc., Int'l & Am. (mem. adv. bd) women in radio and TV, A.I.M., Broadcast Pioneers, Advertising Women of N.Y. (Editor, Ad Libber).

PRATLEY, GERALD: Commentator, b. and e. London, Eng. Joined Canadian Broadcasting Corp., 1946; writer, narrator and producer of The Movie Scene and Music from the Films; asst. member British Film Academy, dir. Canadian Film Institute 1953; chairman Toronto and District Film Council 1956; co-dir. & founder A.G.E. Film Society, Toronto; contributor to US and European film journals; film consult., Canadian Centennial Comm. Chmn., Canadian Film Awards; director, Stratford Film Festival; director, Ontario Film Institute, Toronto. Prof. of film, York Univ., Univ. of Toronto. Seneca College, McMaster Univ.

PRATT, CHARLES A.: Executive. b. Chicago, Oct. 17, 1923. e. Culver Military Academy, 1942; Amherst College, 1949. Worked for General Foods and Alberto-Culver before joining NBC. Now pres. Bing Crosby Productions.

PRELOCK, EDWARD P.: Executive. b. Cleveland, Ohio, Sep. 1, 1934. e. University of Kansas, B.A., 1958, J.D., 1963. Defensive end, Cleveland Brown and Saskatchewan Roughriders, 1958; personnel director, Transport Company of Texas, Honolulu, Hawaii, 1959–61; Justice of the Peace, Lawrence Township, Kansas, 1962–63; director of labor relations, Walt Disney Productions, 1964–75; was executive vice-president of Association of Motion Picture and Television Producers (AMPTP), 1975–77; first vice. pres., AMPTP, 1977– ; President, Central Casting Corporation, 1976. Chairman of the Board, Contract Services Administration Trust Fund, 1976–; president, Animated Film Producers Association, 1971–; member, board of directors, Permanent Charities Committee of the Entertainment Industries, 1976–; member, board of directors, AMPTP, 1973–.

PREMINGER, OTTO: Director, producer, actor. b. Vienna, Austria, Dec. 5, 1906. e. U. Vienna. At 17 became actor with Max Reinhardt troupe at Theatre-in-the-Josefstadt, Vienna; later dir. that theat.; 1935–40, stage prod. & dir. in U.S. and on faculty Drama Sch. Yale U.; NY stage dir. Autobiography, Preminger, published 1977 by Doubleday.
STAGE: Margin for Error, The Moon is Blue, Critics Choice, Full Circle.
PICTURES INCLUDE: Royal Scandal, Forever Amber, Laura, Centennial Summer, Daisy Kenyon, In the Meantime Darling, Fallen Angel, Whirlpool, Where the Sidewalk Ends, The 13th Letter, Angel Face, The Moon is Blue, River of No Return, Carmen Jones, Court Martial of Billy Mitchell, Man With the Golden Arm, St. Joan, Bonjour Tristesse, Porgy and Bess, Anatomy of a Murder, Exodus, Advise & Consent, The Cardinal, In Harm's Way, Bunny Lake is Missing, Hurry Sundown, Skidoo, Tell Me That You Love Me Junie Moon, Such Good Friends, Rosebud, The Human Factor.

PRENTISS, PAULA: Actress, r.n. Paula Ragusa. B. San Antonio, Texas, March 4; e. Northwestern U.; Bachelor degree in drama, 1959; on TV in He & She; on stage in As You Like It, Arf!
PICTURES INCLUDE: Where the Boys Are, The Honeymoon Machine, Bachelor in Paradise, Man's Favorite Sport, Catch 22, Move, The World of Henry Orient, In Harms Way, What's New Pussycat?, Scraping Bottom, Last of the Red Hot Lovers, The Parallax View, The Stepford Wives, The Black Marble.

PRESLE, MICHELINE: Actress. r.n. Micheline Chassagne. b. Paris, France, Aug. 22, 1922; e. Raymond Rouleau Dram. Sch. m.p. debut in Je Chante; on stage in Colinette. Am. Stram Gram, Spectacle des Allies; to U.S., 1945.
PICTURES INCLUDE: Jeunes Filles en Detresse, L'Histoire de Rire, La Nuit Fantastique, Felicie Nanteuil, Seul Amour, Faibalas, Boule de Suif, Jeux Sont Faix, Diable au Corps, Under My Skin, American Guerilla in the Philippines, Adventures of Captain Fabian, Sins of Pompeii, House of Ricordi.

PRESNELL, ROBERT R., JR.: Writer. b. Chicago, Ill., July 21, 1914. e. Weymouth Coll., England; U. of Geneva; Harvard U., Stanford U., B.A. m. Marsha Hunt. Police reporter, Milwaukee Journal, 1937–38; copywriter, Young and Rubicam, 1938–40; then freelance writer; author of novel, Edgell's Island.
TV: Dr. Kildare, McCloud, Twilight Zone, Alcoa Premiere, 11th Hour, The Virginians, World Premiere, Bracken's World, Universal 120, Ritual of Evil, Banacek, Columbo, Love Story, Secret Night Caller (NBC movie of week), Smashup on Interstate 5 (ABC 2-hr. film).
PICTURES INCLUDE: Man in the Attic, Legend of the Lost, Conspiracy of Hearts, Let No Man Write My Epitaph, The Third Day.

PRESSMAN, EDWARD R.: Producer. b. N.Y.C., ed. Fieldston School. grad., Stanford Univ.; studied at London School of Economics. Began career with film short, Girl, in collaboration with director Paul Williams in London. They formed Pressman-Williams Enterprises.
PICTURES INCLUDE: Out of It, The Revolutionary, Dealing: or the Berkeley to Boston Forty Brick, Lost Bag Blues, Sisters, Badlands (exec. prod.), Phantom of the Paradise; Paradise Alley (exec. prod.): Old Boyfriends (prod.); The Hand (prod.).

PRESSMAN, MICHAEL: Producer-Director. b. New York City, July 1, 1950. e. CA Inst. of Arts. Comes from show business family; was actor in college.
PICTURES INCLUDE: The Great Texas Dynamite Chase, The Bad News Bears Breaking Training, Boulevard Nights, Those Lips Those Eyes.
TELEVISION: Like Mom, Like Me.

PRESTON, ROBERT: Actor. b. Newton Highlands, Mass.; e. Pasadena Playhouse, m.p. debut, 1938 in King of Alcatraz.
PICTURES INCLUDE: Illegal Traffic, Disbarred, Union Pacific, Beau Geste, Typhoon, Moon Over Burma, Northwest Mounted Police, New York Town, Macomber Affair, Wild Harvest, Tulsa, Big City, Lady Gambles, Whispering Smith, Blood on the Moon, Sundowners, Best of the Bad Men, My Outlaw Brother, When I Grow Up, Cloudburst, Face to Face (Bride Comes to Yellow Sky), Last Frontier, Dark at the Top of the Stairs, The Music Man, Junior Bonner, Child's Play, Mame, Semi-Tough, "S.O.B," Victor/Victoria.
STAGE: The Music Man (Tony Award), A Lion in Winter, I Do, I Do (Tony Award), Mack and Mabel, Sly Fox.

PREVIN, ANDRE: Composer, Conductor. b. Berlin, Germany, Apr. 6, 1929; composed and conducted over 50 m.p. scores; currently music director, Pittsburgh Symphony Orchestra, & conductor emeritus of London Symphony Orchestra. Guest conductor of most major symphony orchestras in U.S. and Europe.
PICTURES INCLUDE: Three Little Words, Cause for Alarm, It's Always Fair Weather, Bad Day at Black Rock, Invitation to the Dance, Catered Affair, Designing Woman, Silk Stockings, Gigi (Academy Award), Porgy and Bess (Academy Award), Subterraneans, Bells are Ringing, Pepe, Elmer Gantry, Four Horsemen of the Apocalypse, One Two Three, Two for the Seesaw, Long Day's Journey Into Night, Irma LaDouce (Academy Award), My Fair Lady (Academy Award), Goodbye Charlie, Inside Daisy Clover, Fortune Cookie, Thoroughly Modern Millie, Valley of the Dolls, Paint Your Wagon, The Music Lover.

PREVIN, STEVE: Producer, executive. b. New York, 1925. Film editor Universal studios 1941; to Europe 1951 as dir. Foreign Intrigue, TV series. Also dir. TV Sherlock Holmes, Captain Gallant, The Vikings; 1960 joined Disney, dir. Almost Angels, Waltz King, Escapade in Florence; 1965 Eon Productions; London, as TV exec.; joined Paramount in London as European prod. exec.; 1967, Paramount, London, don, prod. exec.; 1969, Commonwealth United, London, Eur., prod. exec.; prod. exec. Don't Look Now; '74 from 1975 v.p. in charge of European production Filmways Pictures.

PRICE, FRANK: Executive. b. 1930. e. Michigan State Univ., following naval service. Joined CBS in N.Y. in 1951 as story editor and writer. Moved to Hollywood in 1954, serving as story editor first at Screen Gems and then NBC (Matinee Theatre). In 1959 joined Universal as an assoc. prod. and writer. In 1961 named exec. prod. of The Virginian TV series. Appt. exec. prod. of Ironside; later did It Takes a Thief and several World Premiere movies. In 1964 named v.p. of Universal TV; 1971, sr. v.p.; 1974, pres. Also v.p., MCA, Inc. In 1978 left to join Columbia as pres. of new company unit, Columbia Pictures Productions.

PRICE, ROGER: Performer. b. Charleston, W. Va., Mar. 6, 1920; e. U. of Michigan, American Acad. of Art, Max Reinhardt Dramatic Workshop. Appeared at many night clubs, many TV guest appearances; writing credits include the Don Knotts Show, Governor & J..., Bob Hope, The Partners, The Bluffers. V.P. of Price, Stern, Sloan Publishers.
TV: Toast of the Town, Arthur Godfrey's Friends, Gary Moore Show, This is Show Business, Jack Parr, Get Smart, Johnny Carson. creator of: The Kallikaks (NBC); The Waltons, McMillan and Wife, Mike Douglas.
PICTURES: Mame, Day of the Locust, The Strongest Man in the World, Mixed Company, At Long Last Love, Pete's Dragon, The Devil and Max Devlin, Love on the Run.

PRICE, VINCENT: Actor. b. St. Louis, Mo., May 27, 1911; e. Yale U., U. of London, Nuremberg U.
PICTURES INCLUDE: The Song of Bernadette, Buffalo Bill, The Eve of St. Mark, Wilson, The Keys of the Kingdom, Laura, A Royal Scandal, Leave Her to Heaven, Dragonwyck, Shock, Long Night, Moss Rose, Three Musketeers, Rogues Regiment, The Web, The Bribe, Baron of Arizona, Champagne for Caesar, Bagdad, His Kind of Woman, Adventures of Captain Fabian, Las Vegas Story, House of Wax, Dangerous Mission, Mad Magician, Son of Sinbad, Serenade, While the City Sleeps, Mysterious House of Usher, Return of the Fly, The Bat, The Tingler, House on Haunted Hill, House of Usher, Pit and Pendulum, Tales of Terror, The Mask of the Red Death, War Gods of the Deep, Dr. Gold Foot and the Sex Machine, The House of 1,000 Dolls, More Dead Than Alive, The Oblong Box,

Scream and Scream Again, Dr. Phibes, Dr. Phibes Rise Again!, Theatre of Blood, Madman, Scavenger Hunt.

PRIES, RALPH W.: Executive. b. Atlanta, Ga. August 31, 1919. Graduated Georgia Tech. Former President, Ogden Food Service Corporation. Board Member: First Federal Savings & Loan of Philadelphia; National Foundation—March of Dimes; Moss Rehabilitation Hospital; St. Christopher's Hospital for Children; William Likoff Cardiovascular Institute of Hahnemann Medical College and Hospital; Police Athletic League; Motion Picture Pioneers; Board Member and Past International President: Variety Clubs International; Board Member: Will Rogers Hospital, Hahnemann Medical College & Hospital.

PRIESTLEY, J.B.: Writer. b. Bradford, Yorkshire, England, Sept. 13, 1894. Many of his novels and plays made into films; also author several orig. s.p., add. dial., acted in advisory capacity on scripts both Hollywood and England. Pres. Screenwriters Assoc., London, 1944–45. Devised, appeared in Lost City on TV, and You Know What People Are, TV series 1955.
PICTURES INCLUDE: The Foreman Went to France, Britain at Bay, Priestley's Postcripts, Battle for Music, They Came to a City, Last Holiday, An Inspector Calls.

PRINCE, HAROLD: Director-Producer. b. New York, N.Y. Jan. 30, 1928. e. Pennsylvania U. Working as stage mgr. for George Abbott on three shows, later co-produced, produced and/or directed the following: The Pajama Game, Damn Yankees, New Girl In Town, West Side Story, Fiorello!, Tenderloin, Take Her, She's Mine, A Call on Kuprin, A Funny Thing Happened on the Way to the Forum, She Loves Me, The Matchmaker (revival), Fiddler On The Roof, Poor Bitos, Baker Street, Flora, The Red Menace, Superman, Cabaret, Zorba, Company, Follies, The Great God Brown, The Visit, Love for Love (the last three all revivals), A Little Night Music, Candide and Pacific Overtures, Side by Side by Sondheim, Some of My Best Friends, On The Twentieth Century and Evita (London, 1978, Broadway 1979, Los Angeles, Australia & Chicago, 1980; Vienna & Mexico City, 1981), Sweeney Todd (Broadway, 1978; London, 1980) and also directed the operas Ashmadei and Silverlake for N.Y. City Opera, Girl of Golden West for Chicago Lyric Opera Co. and San Francisco Opera and Willie Stark for Houston Grand Opera.
MOVIES DIRECTED: Something for Everyone and A Little Night Music. Authored: Contradictions, Notes on Twenty-Six Years in the Theatre (Dodd, Mead & Co., New York, 1974).

PRINCE, WILLIAM: Actor. b. Nichols, N.Y., Jan. 26, 1913; with Maurice Evans, actor, 2 yrs., radio announcer. On N.Y. stage, Ah Wilderness; m.p. debut in 1943.
STAGE: Guest in the House, Across the Board on Tomorrow Morning, The Eve of St. Mark, John Loves Mary, As You Like It, I Am a Camera, Forward the Heart, Affair of Honor, Third Best Sport, The Highest Tree, Venus at Large, Strange Interlude, The Ballad of the Sad Cafe, Mercy Street.
PICTURES INCLUDE: Destination Tokyo, Cinderella Jones, The Very Thought of You, Roughly Speaking, Objective Burma, Pillow to Post, Lust for Gold, Cyrano de Bergerac, Secret of Treasure Mountain, Macabre, Sacco and Vanzetti, The Heartbreak Kid, The Stepford Wives, Rollercoaster, The Cat from Outer Space, The Promise.

PRINZ, LE ROY: Producer, director. b. St. Joseph, Mo., July 14, 1895. Ran away from home at 15; shipped on freighter; joined French Foreign Legion, in French aviation corps & 94th Aerial Squad, World War I; remained after war; dir. dances Folies Bergere, Paris, then Spain & Switzerland; ret. to U.S., 1920; trained student aviators in Mexico; to US 1925 after So. America; back to dance direction; to Hollywood 1931, on choreography for C. B. DeMille, including Sign of the Cross, The Crusaders; many films for Paramount including several road series with Crosby and Hope; to Warner where conceived & dir. musical sequences in all Warner musicals 1940–53; dir. Boy and His Dog, Academy Award short subject, 1948; choreography, Ten Commandments; industrial films, 1959; New York Life—American Tel. & Tel. 1960–61; Prinz Prod., Inc., Japan, Hong-Kong films; exec. prod. Valley Music Hall Theatre Round, Salt Lake City; pres. 1966 prod. dir. for Arameta Enterprises; Films U.S. Navy, pres. Jai-Alai Arameta Enterprises; Films U.S. Navy, pres. Jai-Alai Corp. of Calif. 1969. Exposition 70, Tokyo, Japan, 1971, commercial films, Warner Bros., Seven Arts, Texaco; pres., Mini Fronton of Calif.
PICTURES INCLUDE: Yankee Doodle Dandy, Desert Song, This Is The Army, Night and Day, Helen Morgan Story, Sayonara, South Pacific.

PRIZEK, MARIO: TV Producer-director. b. Edmonton, Alberta, Canada, 1922. e. U. of Alberta, U. of British Columbia.
TV: Bamboula, Parade, Socrates, A Month in the Country, The Diary of a Scoundrel, Peer Gynt, Queen After Death, The Unburied Dead, Gianni Schicci (opera), Colombe, Ring Around the Moon, Three Sisters, The Zoo Story, The Dumb Waiter, Gallileo.

PROVINE, DOROTHY: Actress. b. Deadwood, S.D., Jan. 20, 1937; e. U. of Washington.
TV: The Alaskans, The Roaring 20's.
PICTURES INCLUDE: The Bonnie Parker Story, It's A Mad, Mad, Mad World, Good Neighbor Sam, The Great Race, That Darn Cat, Who's Minding the Mint?

PRYOR, RICHARD: Actor. At age 7 played drum with professionals. Appearances on TV (Johnny Carson, Merv Griffin, Ed Sullivan) established him as standup comic. Wrote TV scripts for Lily Tomlin and Flip Wilson; co-author of film, Blazing Saddles. Several albums are best-selling hits.
PICTURES INCLUDE: Actor: Lady Sings the Blues, Bingo Long and the Travelin' All Stars, Silver Streak, Greased Lightning, Which Way Is Up? Blue Collar, The Wiz, California Suite, Wholly Moses, In God We Trust, Stir Crazy, Bustin' Loose (star, co-prod.).

PRYOR, THOMAS M.: Journalist. b. New York, NY, May 22, 1912. Joined NY Times, 1929; m.p. dept. 1931 as reporter, editor, asst. film critic; Hollywood bureau chief, corres., NY Times, 1951–59; editor, Daily Variety, 1959.

PUCKHABER, RALPH L.: Executive. b. Flat Rock, N.C., June 28, 1921. e. Rollins Coll. Theatre. mgr., Orlando, West Palm Beach, Miami 1938–54; adv. pub. dir., Palm Beach, 1943–54; spec. prom & group sls rep for FST—Magna, Todd, MGM, Col., Fox, UA., 1954–63; dir. adv. pub., ABC Florida State Theatres, Inc., 1963. In 1974 named adv. p.r., publ dir. for ABC Southeastern Theatres (Fla.-Ga.-Ala.).

PULLING, M. J. L. (C.B.E.), M.A., C. ENG. FIEE. Engineer. b. Sandown, Isle of Wight, May 30, 1906; e. Marlborough Coll., King's Coll., Cambridge. With telephone and radio companies in various technical capacities 1929–34; joined BBC in eng. information dept., 1934; asst. to asst. chief eng. 1937; in charge of sound recording dept. 1941; senior supt. eng. TV service, 1950; con. TV service eng., 1956; asst. dir. eng., 1962, Deputy dir., eng. 1963; Chmn. Electronics and Comm. Sec. I.E.E. 1959–60; mem. council, 1963–66; retired from BBC, 1967; Consultant Telecommunications Engineering.

PURCELL, NOEL: Actor. b. Dublin, Eire. Dec. 23, 1900; e. Irish Christian Brothers, Dublin. Made stage debut as child actor; later played leading roles in most of Sean O'Casey & other Abbey Theatre plays at Gaietey, Dublin; also starred as comedian in vaudeville prod. including Something in the Air, Royal Flesh, others at Theatre Royal, Dublin. Screen debut, Odd Man Out, 1946.
PICTURES INCLUDE: Captain Boycott, The Blue Lagoon, Talk of a Million, No Resting Place, Encore, Appointment with Venus (Island Rescue), Crimson Pirate, Decameron Nights, Father's Doing Fine, Pickwick Papers, Grand National Night (Wicked Wife), The Seekers (Land of Fury), Doctor in the House, Svengali, Moby Dick, Doctor at Sea, Lust for Life, Jacqueline, Watch Your Stern, The Millionairess, Mutiny on the Bounty, Lord Jim, Drop Dead Darling, I Spy, Sinful Davy, Violent Enemy, Where's Jack?

PURDOM, EDMUND: Actor. b. Welwyn Garden City, England, Dec. 19. e. St. Ignatius Coll., London. Played leads, character roles for Northampton Rep. Co., Kettering Rep., two seasons at Stratford-On-Avon; London stage in Way Things Go, Malade Imaginaire, Romeo and Juliet, played in Caesar and Cleopatra, Anthony and Cleopatra, London and N.Y.; TV and radio appearances N.Y. & London.
PICTURES INCLUDE: Titanic, Julius Caesar, Student Prince, The Egyptian, Athena, The Prodigal, King's Thief, Moment of Danger, Rasputin, The Comedy Man, The Beauty Jungle.

PURDY, RAI: Producer. b. London, Eng., Nov. 10, 1910; e. Canada. Owned & operated Rai Purdy Prod., Toronto for 12 yrs. Canada's largest ind. radio prod. co.: dir. narr., Out of the Night; C.O. of Canadian Army Shows. Overseas, with rank of Lt. Col., 1950–56 prod. dir. CBS N.Y.: dir. prog. Scottish TV Ltd, Glasgow, 1957–60 TV prod. Toronto, Canada, 1960.

PYKE, REX: Film-Television Producer/Director. Recent productions include Akenfield, Landscape, Eric Clapton's Rolling Hotel, Van Morrison in Ireland and Woodstock in Europe 1979.

Q

QUAID, DENNIS: Actor. b. Houston, TX Apr. 9, 1954. e. Houston Univ. Appeared in Houston stage productions before leaving for Hollywood. Movie debut: September 30, 1955 (1978).
PICTURES INCLUDE: Crazy Mama, Our Winning Sea-

son, Seniors, Breaking Away, Gorp, The Long Riders, All Night Long, Caveman, The Night the Lights Went Out in Georgia.

QUAID, RANDY: Actor. Discovered by Peter Bogdanovich while still jr. at Drama Dept. at Univ. of Houston and cast in his The Last Picture Show (1971).
PICTURES INCLUDE: What's Up, Doc?, Paper Moon, Lolly-Madonna XXX, The Last Detail, The Apprenticeship of Duddy Kravitz, Breakout, The Missouri Breaks, Bound for Glory, The Choirboys, Midnight Express, The Long Riders.

QUALEN, JOHN: Actor. b. Vancouver, B.C.; m. Pearle Larson, 1924. Daughters: Meredith Kilpatrick, Kathleen Roberts, Elizabeth Bacon. e. studied declamation under Eliss Day; bachelor of oratory, also degree in dietetics. Played piano, flute & saxo U.S. phone in concert tours, Canada & On stage in Street Scene, Counsellor at Law, (later in film versions); m.p. debut in Arrowsmith.
TV: Mr. Ed, Andy Griffith Show, Hazel, Danny Thomas Show, Partridge Family.
PICTURES INCLUDE: Casablanca, Adventure, The Fugitive, Grapes of Wrath, My Girl Tisa, Big Steal, Jackpot, Woman on Run, Flying Missile, Belle Le Grand, Hans Christian Andersen, Ambush at Tomahawk Gap, I the Jury, Student Prince, High & the Mighty, Passion, Unchained, Sea Chase, At Gunpoint, The Searchers, Johnny Concho, Country Doctor, Two Rode Together, The Man Who Shot Liberty Valance, The Prize, The Seven Faces of Dr. Lao, Cheyenne Autumn, Sons of Katie Elder, Patch of Blue, Big Hand for the Little Lady, Firecreek, Criss Cross.

QUAYLE, ANTHONY: Actor, stage producer. b. Lancashire, England, 1913. Early career acting Old Vic Co., Elsinore and on tour Continent, North Africa. Army service, 1939–45; joined Stratford Memorial Theatre as director & actor 1948; many plays and acted Shakespearean roles; m.p. debut 1955, Oh Rosalinda; many TV credits in U.S. & England; Moses, David & Saul, Q.B. VII (Emmy). Recent TV: Ice Age, Henry IV.
PICTURES INCLUDE: Battle of the River Plate, The Wrong Man, No Time for Tears, Woman in a Dressing Gown, The Man Who Would Not Talk, Ice Cold in Alex, Serious Charge, Tarzan's Greatest Adventure, Guns of Navarone, H.M.S. Defiant, Lawrence of Arabia, The Fall of the Roman Empire, East of Sudan, Operation Crossbow, MacKenna's Gold, Before Winter Comes, Anne of A Thousand Days, Everything You Always Wanted to Know About Sex*, Bequest to the Nation, Tamarind Seed, The Eagle Has Landed, Murder by Decree.

QUEEN, ROBERT I.: Writer, author, publicist. b. Aug. 12, 1919; e. CCNY, 1942; reporter, Bronx Home News, 1936–41; information & editorial specialist, U.S. Civil Service (pub. rel.) 1942–55; admin. asst. to pres., Cinema Diorama Adv. Agency in chg., Radio-TV-Press Rel., 1955–56; night copy, headline ed., Journal of Commerce, 1956; dir., Radio-TV 1956–58, Greater New York Fund, writer of GNYF documentaries, GNYF variety shows and panel interview shows. Writer Nick Kenny WABD-TV shows; Operation Maverick, WPIX-TV Army Recruiting Service series; Suspense and The Web shows as free lance writer, CBS Radio Network Press Inf. Dept., copy ed. 1958, 1959 Press Consult. Cong. Alfred E. Santangello, 1960–63; Author: Emigres in Wartime, 1940; Creative PR in Planning Spec. Events Planning Special Events, 1958; co-author with wife, Elephant Comes to Play; own pub. rel. firm, 1955–73; p.r. counsel to NYC Transit Police PBA (1967–69); p.r. counsel-legislative aide, Hon. Thomas J. Manton (D. Queens) 1971; p.r. council, 1971–72, Councilman Morton Povman, 1971; p.r. counsel to State Sen. John R. Dunne (1969–70); teaching consult., Sch. of Contin. Prof. Studies, Pratt Institute; N.Y., N.J. Corres. Press Wire Services, Pub. rel. consultant. 1972, Thomas J. Manton Campaign for Congress; (1973–75) Public Affairs Officer, The Methodist Hospital in Brooklyn and concurrently, Community Relations-Public Relations Aide to State Assemblyman Leonard P. Stavisky, Chairman, Committee on Education; (1973–77) Public Relations Counsel to N.Y. State Senator Jeremiah B. Bloom; 1977 to 1979. Public relations counsel, Jeremiah B. Bloom for Governor Campaign, (1978). P.R. Counsel to N.Y. State Assemblyman Saul Weprin (1979 to date).

QUICK, ROBERT E.: Executive. b. Milwaukee, Wis., Aug. 8, 1917. Joined Loew's Inc., while attending Marquette U.; 1936–42, Loew's Inc., ad sales mgr. booker, apprentice salesman; 1943, adm. asst. Seattle and St. Louis Regional Offices, AAFMPS (formerly Army M.P. Service); 1945, mgr., New Orleans branch. AAFMPS: 1946, gen. man., MIDPAC M.P.S., Hawaii; 1948, ex. asst., HQ AAFMPS, Washington, D.C.; 1949, chief, European M.P.S., Nurenburg, Germany; 1952, dep. chief, AAFMPS Worldwide, Washington, D.C.; 1958, chief, AAFMPS Worldwide.

QUIGLEY, MARTIN, JR.: Editor, writer. b. Chicago, Ill., Nov. 24, 1917. e. A. B. Georgetown Un; M.A., Ed. D, Columbia U., M.P. Herald, Oct. 1939; spcl. ed. rep., M.P. Herald & M.P. Daily, May, 1941; wartime film work in U.S., England, Eire & Italy, Dec. 1941–Oct. 1945; assoc. ed., Quigley Pub., Oct. 1945; ed. M.P. Herald, July, 1949; also edit. dir. of all Quigley Pub., 1956; pres. Quigley Pub. Co., 1964; author, Great Gaels, 1944, Roman Notes, 1946, Magic Shadows—the Story of the Origin of Motion Pictures, 1948. Editor, New Screen Techniques, 1953; m.p. tech. section, Encyclopaedia Brit., 1956; co-author, Catholic Action in Practice, 1963. Co-author: Films in America, 1929–69, 1970. Pres., QWS, Inc., educational consultants, 1975. Village of Larchmont, N.Y., trustee, 1977–79; mayor, 1980–81.

QUIGLEY, WILLIAM J.: Executive. b. New York City, July 6, 1951. e. Wesleyan Univ. (BA). From 1973 to 1974 was advt. circulation mgr. for Quigley Publishing Co. Taught school in Kenya in 1974; returned to U.S. to join Grey Advt. as media planner. In 1975 joined Walter Reade Organization as asst. film buyer; promoted to head film buyer in 1977.

QUILLAN, EDDIE: Actor. b. Philadelphia, Pa., March 31, 1907; p. Sarah Owen and Joseph Quillan, professionals; stage training playing in the Quillan act with his family. On the screen in Up and At 'Em.
PICTURES INCLUDE: Night Work, Big Money, A Little Bit of Everything, The Big Shot, Dark Mountain, This Is the Life, Moonlight and Cactus, Song of the Sarong, A Guy Could Change, Sensation Hunters, Sideshow, Brigadoon, Did You Hear the One About the Traveling Saleslady?, Angel in My Pocket, How to Frame a Figg.

QUINE, RICHARD: Actor. b. Detroit, Mich., Nov. 12, 1920. In U.S. Coast Guard during WW II; in vaudeville 6 yrs.; also on radio in 1938. NY stage debut in Very Warm for May, then in My Sister Eileen. First appearance on screen in The World Changes, 1932; variously acted, prod. & dir. many films.
PICTURES INCLUDE: Babes on Broadway, Tish, My Sister Eileen, For Me and My Gal, Dr. Gillespie's New Assistant, Stand By for Action, We've Never Been Licked, Rookie Fireman, Sunny Side of the Street, Purple Heart Diary, Rainbow Round My Shoulder, Sound Off, All Ashore, Cruisin' Down the River, Pushover, So This is Paris, Solid Gold Cadillac, Full of Life, Mad Ball, Drive a Crooked Road, Bell Book and Candle, It Happened to Jane, Strangers When We Meet, The World of Suzie Wong, Notorious Landlady, Paris When It Sizzles, Sex and the Single Girl, How to Murder Your Wife, Synanon, Oh Dad Poor Dad Mama's Hung You In The Closet and I'm Feeling So Sad, Hotel (dir.), A Talent for Loving (dir.), "W" (dir.), Prisoner of Zenda (dir.).
TV: The Jean Arthur Show (exec. prod.).

QUINLAN, KATHLEEN: Actress. b. Pasadena, CA. Played small role in film, One Is a Lonely Number, while in high school. Also did bit in American Graffiti. Major role debut in Lifeguard (1976).
PICTURES: Airport '77, I Never Promised You a Rose Garden, The Promise, The Runner Stumbles, Sunday Lovers.

QUINN, ANTHONY: Actor. b. Mexico, Apr. 21, 1915. Began on screen, 1936 Acad. Award, best supp. actor, Viva Zapata, 1952, Lust for Life, 1956.
PICTURES INCLUDE: Guadalcanal Diary, Buffalo Bill, Irish Eyes Are Smiling, China Sky, Back to Bataan, Where Do We Go From Here?, Black Gold, Tycoon, The Brave Bulls, Mask of the Avenger, The Brigand, World in His Arms, Against All Flags, Ride Vaquero, City Beneath the ea, Seminole, Blowing Wild, East of Sumatra, Long Wait, Magnificent Matador, Ulysses, Naked Street, Seven Cities of Gold, La Strada, Attila the Hun, Lust for Life, Wild Party, Man from Del Rio, Ride Back, Hunchback of Notre Dame, The River's Edge, Hot Spell, Black Orchid, Last Train From Gun Hill, Warlock, Heller With a Gun, Heller in Pink Tights, Savage Innocents, The Guns of Navarone, Barabbas, Requiem for a Heavy-weight, Behold a Pale Horse, Zorba the Greek, High Wind in Jamaica, The Visit, Guns for San Sebastian, The Secret of Santa Vittoria, A Dream of Kings, Flap, A Walk in the Spring Rain, R.P.M.*, Across 110th Street, Deaf Smith and Johnny Ears, The Don Is Dead, Mohammad, Messenger of God, The Greek Tycoon, Caravans, The Passage, Lion of the Desert.

QUINN, STANLEY J., JR.: Producer, director. b. Brooklyn, N.Y., Mar. 18, 1915; e. Princeton U., 1932–36. Radio writer for Edgar Bergen show; mgr., J. Walter Thompson radio dept. in Australia, 1941–43; war corresp., 1943–45; radio prod. J. Walter Thompson, 1946; prod., dir., Kraft TV Theatre, NBC, 1947–53; ABC, Oct. 1953; vice pres., J. Walter Thompson, June 1954; exec. prod. Lux Video Theatre, 1954; pres. Quinn, McKenney Prod.; v.p. head radio, TV dept., D.C.S.&S.; prod., Grey Adv. for Revlon commercials. Dir. of Admin. MGM-TV Studios 1963–64; dir. Radio-TV Center U. of Conn. 1965–76. Commercial dir. Kraft Foods, 1964–72; Director, Rad o/TV Div., CIMT, Univ. Conn., 1976–80. Freelance TV dir., 1980.

R

RABINOVITZ, JASON: Executive. b. Boston, MA., Aug. 17, 1921. e. Harvard College where elected to Phi Beta Kappa. Following World War II service as military intelligence captain with paratroops, took MBA degree at Harvard Business School in 1948. Started in industry in 1949 as asst. to secty.-treas., United Paramount Theatres. Asst. controller, ABC, 1953; adm. v.p., ABC-TV, 1956; joined MGM as asst. treas., 1957; named MGM-TV gen. mgr., director of business & financial affairs, 1958; treas. & chief financial officer, MGM, Inc., 1963; advanced to financial v.p. & chief financial officer, 1967. In 1971 named exec. v.p., Encyclopedia Brittanica Education Corp.; sr. v.p., American Film Theatre, 1974–75. Rejoined MGM as v.p./financial asst. to the pres., 1976. Elected v.p. finance, 1979.

RACHMIL, LEWIS J.: New York City, July 3, 1908. e. NYU, Yale University School of Fine Arts, Art Director, producer at Paramount, RKO, Columbia, United Artists, director and studio exec. v.p. Ziv-United Artists; v.p. Mirisch-U.A.; v.p. ABC Pictures Corp.; presently v.p. and executive production manager, MGM.

RACKMIL, MILTON R.: Executive. b. New York City. e. NYU. Certified Public Accountant prior to assoc. with Brunswick Record Co. 1929; co-founder Decca Records, 1934; pres. Decca Records, 1949; pres. and member of board of dir. Universal Pictures, 1952, after Decca bought controlling stock interest in Universal; pres. emeritus, Universal, 1973.

RADIN, PAUL: Producer. b. N.Y.C. ed. N.Y.U. m. Writer Decla Dunning. After college went in adv. Became v.p. in chg of m.p. div. of Buchanan & Co. During the war posted in Middle East as film chief for Office of War Information for that area. On return to U.S. assigned by Buchanan to ad campaign for Howard Hughes' The Outlaw. Turned to talent mgr., joining the Sam Jaffe Agency. Then joined Ashley-Famous Agency. Became exec. prod. for Yul Brynner's indep. prod. co. based in Switzerland, with whom made such films as The Journey, Once More with Feeling, Surprise Package.
PICTURES INCLUDE: Born Free, Living Free, Phase IV, The Blue Bird.

RADNER, GILDA: Actress. Writer. Began career as part of Toronto's Second City improvisational group. Moved to New York when John Belushi invited her to work with him on National Lampoon Radio Hour for which she wrote and performed. Soon after joined TV's Saturday Night Live. Appeared in Bdwy. in one-woman show, Gilda Radner—Live from New York, which was filmed for theatrical movie.
PICTURES: First Family, Gilda Live.

RADNITZ, ROBERT B.: Producer. b. Great Neck, N.Y.; e. U. of Virginia. Taught 2 years at U. of Virginia, then became reader for Harold Clurman; wrote several RKO "This is America" scripts, then to Broadway where co-prod., The Frogs of Spring; prod. The Young and the Beautiful; to Hollywood working at United Artists, then as story consultant for 20th Century-Fox; prod. A Dog of Flanders (first feature) in 1960. Board of Directors, Producer Guild of America, last 3 years first producer with retrospective at Museum of Modern Art. First producer honored by joint resolution of both houses of Congress, 1973. Sounder received four Academy Award nominations: best picture, best actor, best actress, best screenplay. Pres. Radnitz Productions, Ltd.
PICTURES INCLUDE: Misty, Island of the Blue Dolphins, And Now Miguel, My Side of the Mountain, The Little Ark, Sounder, Where The Lilies Bloom, Birch, Interval, Sounder II, A Hero Ain't Nothin' But a Sandwich, Mary White (TV-Emmy for best screenplay, 1978).

RAFELSON, BOB: Producer-Director-Writer. b. N.Y.C., 1935. After Army Service did program promotion for a radio station; then hired by David Susskind to read scripts. Became script supervisor of Play of the Week for Susskind and Ely Landau; wrote 34 adaptations for the series. Joined Screen Gems in California, developing program idea for Jackie Cooper, then head of TV prod. arm of Columbia. Later formed BBS Productions with Burt Schneider and Steve Blauner; their first film, Head.
PICTURES INCLUDE: Head (co-s.p.); Five Easy Pieces (co-prod., dir.), The King of Marvin Gardens (prod.-dir.), Stay Hungry (co-prod., co-s.p., dir.), The Postman Always Rings Twice (co. prod.-dir.).

RAFFERTY, FRANCES: Actress. b. Sioux City, Iowa, June 26, 1922; e. U. of California, premedical student UCLA. TV show, December Bride, CBS.
PICTURES INCLUDE: Seven Sweethearts, Private Miss Jones, Girl Crazy, War Against Mrs. Hadley, Thousands Cheer, Dragon Seed, Honest Thief, Mrs. Parkington, Barbary Coast Gent, Hidden Eye, Abbott and Costello in Hollywood, Adventures of Don Coyote, Money Madness, Lady at Midnight, Old Fashioned Girl, Rodeo, Shanghai Story.

RAFFIN, DEBORAH: Actress. b. Los Angeles, Calif., March 13, 1953. e. Valley College. Was active fashion model before turning to acting when discovered by Ted Witzer. Made m.p. debut in 40 Carats (1973).
PICTURES INCLUDE: The Dove, Once Is Not Enough, The Sentinel, Touched by Love.
TELEVISION: Haywire.

RAGSDALE, CARL V.: Director, producer. b. Illmo, Missouri, May 16, 1925. e. Washington U., St. Louis and Denison U., Ohio. U.S. Navy M.P. Officer, WW II and Korea; CPT, US Naval Reserve. v.p. industrial m.p. prod., Depicto Films, 1954–61; started Carl Ragsdale Associates, 1962; pres. and exec. prod. Sun Dial Films, Inc. 1963–69. Winner of Academy Award for Best Documentary Short of 1966. Nominated for Academy Award for Best Documentary Short of 1967.

RAINES, ELLA: Actress. r.n. Ella Wallace Raines Olds. b. Snoquainde Falls, Wash., Aug. 6, 1921. e. U. Wash. Winner dramatic scholarships; actress little theatre groups; on screen 1943 in Corvette K-225.
PICTURES INCLUDE: Phantom Lady, Enter Arsene Lupin, Hail the Conquering Hero, The Suspect, Strange Affair of Uncle Harry, Time Out of Mind, The Web, Brute Force, Senator was Indiscreet, Impact, Dangerous Profession, Walking Hills, Fighting Coast Guard, Ride the Man Down.

RAINS, ROBERT H.: Executive. b. N.Y., N.Y., Aug. 12, 1921. e. USC, NYU: Air Force 3 yrs.; publicist, Int'l Pictures 1946; dir. of radio activities, Universal-Int'l Pictures, 1946; asst. casting dir., 1952; dir. Radio-TV activities, 1955; exec. in chg. of TV press dept., Universal City Studios, 1961; Universal v.p. 1966. Retired April 21, 1978.

RAKOFF, ALVIN: Producer, director. b. Toronto, Canada. e. Toronto U. Early career as journalist. Dir. in French & U.S. T.V. England, Canada Emmy Award Winner, 1968 for Call Me Daddy.
STAGE: Hamlet.
PICTURES INCLUDE: On Friday at 11, The Comedy Man, Crossplot, Hoffman, Say Hello to Yesterday, City of Fire, Death Ship, Dirty Tricks.
TV: The Caine Mutiny, Court Martial, Requiem for a Heavyweight, Our Town, The Velvet Alley, A Town Was Turned to Dust, Jokers Justice, Call Me Back, Day Before Atlanta, Heart to Heart, The Seekers, Sweet War Man, The Move after Checkmate, The Stars in My Eyes, Call Me Daddy, Summer & Smoke, Don Quixote, Shadow of a Gunman, The Impeachment of Andrew Johnson, Cheap in August, In Praise of Love, Nicest Man in the World, Dame of Sark, The Kitchen, Romeo and Juliet.

RAKSIN, DAVID: Composer. b. Philadelphia, Pa., Aug. 4, 1912. e. U. of Pennsylvania, studied music with Isadore Freed and Arnold Schoenberg. Composer of music for films, ballet, dramatic and musical comedy, stage, radio and TV, symphony orchestra and chamber ensembles. Arranger of music of Chaplin film, Modern Times; pres. Composers and Lyricists Guild of America, 1962–70; film cartoons include Madeline and The Unicorn in the Garden (UPA), professor of Music and Urban Semester, U.S.C., and Faculty, UCLA School of Music.
PICTURES INCLUDE: Laura, Secret Life of Walter Mitty, Smoky, Force of Evil, Across the Wide Missouri, Carrie, Bad and the Beautiful, Apache, Suddenly, Big Combo, Jubal, Hilda Crane, Separate Tables, Al Capone, Night Tide, Too-Late Blues, Best of the Bolshoi (music for visual interludes), Two Weeks in Another Town, The Redeemer, Invitation to a Gunfighter, Sylvia, A Big Hand for the Little Lady, Will Penny, Glass Houses, What's the Matter with Helen?
TV: Wagon Train, Five Fingers, Journey, Life With Father, Tender is the Night, Father of the Bride, Ben Casey, Breaking Point, Prayer of the Ages, Report from America, Medical Center, The Olympics (CBC).

RALSTON, GIL: Producer, writer. b. Los Angeles, 1912. e. American Acad. N.Y. stage, mgr. Broadway shows, then writer dir., NBC radio; radio dir., exec. prod. Proctor and Gamble, radio shows include Ransom Sherman Show, Junior Miss, Truth or Consequences, Lowell Thomas, Red Skelton Show independent TV prod, packaging firm exec. prod. 1948; exec. prod. CBS-TV dramatic shows, N.Y.; dir. network programs, N.Y. CBS-TV 1956. Exec. prod. High Adventure, 1957; writer, Naked City, Route 66, Ben Casey.
TV: Cavalcade of America, General Electric Theatre, Suspense.

RALSTON, RUDY Producer. b. Prague, Czechoslovakia, Jan. 30, 1918. e. grad. eng. Realka U. Came to U.S. & joined Consolidated Lab.; exec. Republic Prod.; prod. Republic, 1950.

PICTURES INCLUDE: No Man's Woman, Double Jeopardy, Terror at Midnight, Hell's Crossroads, The Lawless Eighties, Last Stagecoach West, Man Who Died Twice.

RAMPLING, CHARLOTTE: Actress. b. Sturmer, England, 1946. e. Jeanne D'Arc Academie pour Jeune Filles, Versailles; St. Hilda's, Bushey, England. Ent. m.p. ind. 1966. TV credits include: Six More for BBC, The Superlative Seven-Avenger series, Mystery of Cader Iscom, The Fantasists, What's in it for Henry, Zinotchka, Sherlock Holmes.
PICTURES INCLUDE: The Knack, Rotten to the Core, Georgy Girl, The Long Duel, Sequestro di Persona, The Damned, Three, Ski Bum, Corky, Tis Pity She's a Whore, The Six Wives of Henry VIII, Asylum, The Night Porter, Giordano Bruno, Zardoz, Caravan to Vaccares, Yuppi Dui La Chair De L'orchidee, Farewell My Lovely, Foxtrot, Orca, The Mauve Taxi, Stardust Memories.

RAMSAY, PATRICK: b. Bristol, Eng., 1926. e. Marlborough College and Jesus College, Cambridge (M.A. History), Eng. Served with Royal Naval Volunteer Reserve (Fleet Air Arm) 1944–46. Joined British Broadcasting Corporation 1949 as a report writer in monitoring service. Became assistant, appointments dept., 1953 and three years later re-joined BBC External Services as senior administrative assistant. In charge of news administration, radio and television 1958 to 1963, helping develop the U.K. regional television news network for BBC. 1963, appointed planning manager (projected arrangements), television. 1966 assistant controller, programme services, television, then assistant controller, program planning, television, from December 1969. Controller, programme services, television, April 1972. Appointed controller, BBC-Scotland, May 1979.

RAND, HAROLD: Executive. b. N.Y., Aug. 25, 1928. e. L.I.U., B.S. 1948–50; City Coll. of N.Y., 1945–46; U.S. Army 1946–48; ent. m.p. ind. 1950, pub. dept. 20th-Fox; variety of posts incl. writer, trade press, newspaper contacts; joined Walt Disney's Buena Vista pub. mgr., 1957; pub. mgr. Paramount Pictures, 1959; formed own pub. rel. firm, 1961; dir. of pub. Embassy Picture Corp. 1962; dir. of world pub. 20th Century Fox 1962; resigned 1963; dir. of adv. & pub., Landau Co, 1963; dir. world pub., Embassy Pictures, 1964; est. Harold Rand & Co., Inc., 1966, pres; Co-formed Talent Four Artists, Ltd. in 1975 to produce motion pictures. Harold Rand & Co. acting as feature film packager and producer's representative. Now pres. of Satellite Consultants, Ltd. (SACON), production & mktg. co. for TV and home video mkts.

RANDALL, TONY: Actor. b. Tulsa, Okla. e. Northwestern U. Prof. N.Y. debut as actor in Circle of Chalk; then in Candida and others; U.S. Army 1942–46; radio actor on many shows.
STAGE: Corn is Green, Anthony & Cleopatra, Caesar & Cleopatra, Inherit the Wind, Oh Men Oh Women, Oh Captain, The Sea Gull, The Master Builder.
PICTURES INCLUDE: Oh Men Oh Women, Will Success Spoil Rock Hunter, No Down Payment, The Mating Game, Pillow Talk, Adventures of Huckleberry Finn, Let's Make Love, Lover Come Back, Boys' Night Out, 7 Faces of Dr. Lao, Send Me No Flowers, Fluffy, The Alphabet Murders, Bang! You're Dead, Hello Down There, Everything You Always Wanted to Know About Sex*, Foolin' Around, Scavenger Hunt.
TV: One Man's Family, TV Playhouse, Mr. Peepers, Max Liebman Spectaculars, Sid Caesar, Dinah Shore, Playhouse 90, Odd Couple, The Tony Randall Show.

RANDOLPH, ELSIE: Actress, singer. b. London, England, Dec. 9, 1904. London stage debut 1919 in The Girl for the Boy. M.p. debut 1932 in Alfred Hitchcock's Rich and Strange.
STAGE (London): Battling Butler, Toni, Madame Pompadour, Boodle, Sunny, Peggy-Ann, That's A Good Girl, Follow Through, Wonder Bar, Stand Up and Sing, Mr. Whittington, This'll Make You Whistle, Room for Two, Top Hat and Tails, The Maid of the Mountains, It's Time to Dance, Great Day, Is Your Honeymoon Really Necessary?, No Other Verdict, The Lost Generation, What A Racket!, Lord Arthur Savile's Crime, The Cure for Love (1978).
PICTURES INCLUDE: Brother Alfred, Life Goes On, Rise and Shine, Night of the Garter, Yes, Mr. Brown, That's A Good Girl, This'll Make You Whistle, Smash and Grab, Cheer the Brave, Sky Riders, Frenzy (1972), Charleston (1977).
TV: Harper's West One, Stars and Garters, The Jimmy Logan Show, Fraud Squad, Tales of Piccadilly, Father Brown, Z-Cars, The Coffee Lace, Thriller, Within These Walls (1977), Edward and Mrs. Simpson (1978).

RANSOHOFF, MARTIN: Executive. b. New Orleans, La., 1927. e. grad., Colgate Univ. Adv., Young & Rubicam, 1948–49; slsmn, writer, dir., Gravel Films, 1951; formed own co., Filmways, 1952; industrial films, commercials; formed Filmways TV Prods., Filmways, Inc., Filmways of Calif. chmn. bd. Filmways, Inc., resigned from Filmways in 1972 and formed own independent motion picture and television production company.
TV: Mister Ed, The Beverly Hillbillies, Petticoat Junction, The Addams Family.
PICTURES INCLUDE: The Amereicanization of Emily, The Sandpiper, Boys Night Out, The Loved One, The Cincinnati Kid, See No Evil, Ten Rillington Place, King Lear, Topkapi, Fuzz, Castle Keep, Ice Station Zebra, Catch 22, Save The Tiger, The White Dawn, Silver Streak (exec. prod.), The Wanderers, Change of Seasons, American Pop.

RAPF, MATTHEW: Producer, writer. b. New York City, Oct. 22, 1920; e. Dartmouth Coll., B.A. 1942; p. Harry Rapf, producer. U.S. Navy, World War II as Lt. (j.g.).
PICTURES INCLUDE: Adventures of Gallant Bess s.p., co-prod., assoc. prod., story, The Sellout; prod., Desperate Search, Big Leaguer, Half a Hero.
TV: Loretta Young Show, Frontier, Great Gildersleeve, The Web, Jefferson Drum, Man From Blackhawk, Two Faces West, Ben Casey, Slattery's People, Iron Horse, Young Lawyers, Hardcase, Terror In the Sky, Shadow On the Land, Marcus-Nelson Murders, Kojak, Switch, Doctor's Hospital, Eischied, Oklahoma City Rolls, Gangster Chronicles.

RAPHAEL, FREDERIC: Writer. b. Chicago, 1931. e. Charterhouse, St. John's College, Cambridge. First novel pub., 1956, subsequently, The Earlsdon Way, The Limits of Love, A Wild Surmise, The Graduate Wife, The Trouble With England, Lindmann, Orchestra and Beginners. 1970: Like Men Betrayed. 1971: Who Were You With Last Night. 1972: April, June and November. 1973: Richard's Things. 1975: California Time. 1976: The Glittering Prizes. 1979: Sleeps Six & Other Stories. Ent. m.p. ind., 1956. Several plays for ATV, 1960–62. First feature film credits 1964. Nothing But The Best; 1965, Darling; Ac. Awd. for original screenplay. 1967: Two For The Road, Far From The Madding Crowd. 1971: A Severed Head. 1974: Daisy Miller. 1976: The Glittering Prices, Rogue Male. 1978: Roses, Roses . . . ; School Play, Something's Wrong, The Oresteia of Aeschylus. 1979: From the Greek and An Early Life (plays); Best of Friends (TV).

RAPHAELSON, SAMSON: Writer. b. New York City, March 30, 1896. e. U. Illinois. English instructor at U. Illinois; police reporter on N.Y. Times; has written short stories, plays, essays, screen plays.
WRITINGS INCLUDE: (plays) Jazz Singer, Young Love, Wooden Slipper (director), Accent on Youth, White Man, Skylark, dir., Jason, dir., Perfect Marriage, Hilda Crane; book: The Human Nature of Playwriting; (s.p.) Shop Around the Corner, Heaven Can Wait, Green Dolphin Street, That Lady in Ermine, Suspicion, Trouble in Paradise, The Merry Widow, Smiling Lieutenant, One Hour With You, Broken Lullaby (The Man I Killed) Angel, A Prince in Disguise. Visiting prof. of playwriting, Univ. Illinois, 1948. Adjunct Professor, Film Division, and Drama Department, Columbia University, since 1976. Received Laurel Award for lifetime screenwriting achievement from Writers Guild of America, 1977. Honorary degree, Doctor of Humane Letters, Col. Univ., 1981. Stories and articles in Esquire, New Yorker, Sat. Evening Post, etc.

RAPHEL, DAVID: Executive. b. Boulogne-s/Seine, France, Jan. 9, 1925; e. University in France. Entered m.p. ind. as asst. to sales mgr. in France, 20th-Fox, 1950–51; asst. mgr. in Italy, 1951–54; mgr. in Holland, 1954–57; asst. to European mgr. in Paris, 1957–59; European mgr. for TV activities in Paris, 1959–61; Continental mgr. in Paris, 1961–64, transferred to N.Y. as vice-pres. in chge. of international sales, 1964; named pres., 20th Century-Fox International, 1973. In Feb., 1975, also appointed senior vice-pres., worldwide marketing, feature film division, for 20th-Fox, headquartered in Los Angeles. In November, 1976, joined ICM, appointed director general of ICM (Europe) headquartered in Paris. In 1979 elected pres. ICM with headquarters in Los Angeles.

RAPPER, IRVING: Director. b. London, Eng.; e. New York U., B.A. Assoc. Gilbert Miller in Stage prod. dir. Animal Kingdom, Firebird, Five Star Final, etc.
PICTURES INCLUDE: Shining Victory, One Foot in Heaven, The Gay Sisters, The Adventures of Mark Twain, Rhapsody in Blue, The Corn is Green, Deception; Now, Voyager; Voice of the Turtle Anna Lucasta, The Glass Menagerie, Another Man's Poison, Forever Female, Bad For Each Other, The Brave One, Marjorie Morningstar, The Miracle, Joseph and His Brethren, Pontius Pilate, The Christine Jorgensen Story, Born Again, Justus.

RAPPOPORT, GERALD J.: Executive-Film Producer. b. New York, N.Y., 1925. e. New York University. U.S. Marine Corps. 1955–1958—pres., Major Artists Representatives Corp.; 1958–1960—director of Coast Sound Services, Hollywood; 1957 to present, pres., Sewan Music Publishers. 1961 to present, repre., major Eastern European film producers. 1970 to present, dir., International Film Exchange Ltd., New York and Paris, and pres. of Family Entertainment Corp.

RATHER, DAN: News Correspondent. Joined CBS News in 1962 as chief of southwest bureau in Dallas. White House Correspondent, 1964 to 1974. Covered top news events, from Democratic and Republican national conventions to President Nixon's trip to Europe (1970) and to Peking and Moscow (1972). Anchored CBS Reports, 1974–75. Presently co-editor of 60 minutes (since 1975) and anchors Dan Rather Reporting on CBS Radio Network (since 1977). Winner of numerous awards, including 5 Emmys.

RAVETCH, IRVING: Director, scenarist, producer. m. Harriet Frank.
PICTURES INCLUDE: The Long Hot Summer, The Sound and the Fury, Home from the Hill, The Dark at the Top of the Stairs, Hud, Hombre, The Reivers (also prod.), Conrack (co. s.p.), Norma Rae (co.-s.p.).

RAWLINSON, A. R.: Writer. b. London, August 9: e. Rugby School, Pembroke College. Cambridge. In Queen's Royal Regiment, World War I: ship repairer & coal exporter (Liverpool & London) 1919–26; professional writer since 1929; in m.p. since 1930; assoc. prod. 1936; scenario ed. Brit. Natl. Films, Ltd. 1938; Lt. Col. British Army, World War II; scenario ed. & assoc. prod. Nettlefold Films, Ltd. 1948.
WRITINGS INCLUDE: (plays) Five Farthings, Magnolia Street, This Desirable Residence, Chain Male, Private Life of Helen of Troy, Birth-mark, Scarlet Thread, Troubadour, (s.p. & orig. play) Leap Year.
PICTURES INCLUDE: Jew Suss, Man of the Moment, Man Who Knew Too Much, Strange Boarders, Spies of the Air, Gaslight, Chinese Bungalow, Face at the Window, This England, White Unicorn, My Sister & I, Calling Paul Temple, Story of Shirley Yorke, Dark Secret, Paul Temple's Triumph; assoc. prod. Wall of Death. Prod. There Was a Young Lady, Broken Horseshoe; prod., s.p. Black Riders, Stock Car, Cloak Without Dagger, Man, dir., Balblair Prod. Ltd. 1954; 1961 s.p. Gaolbreak.
TV: Scripts for Douglas Fairbanks Prod., CBS Assoc. Rediffusion: Grandada TV; TV serials B.B.C.; Diary of Samuel Pepys, Lorna Doone, Infamous John Friend, The Moonstone, Amelia, Kipling Series (BBC). Lord Glenaldy, serial BBC (sound). A Bottle of Rum, Captain Barnstead, Kipling Puck of Pook's Hill, Rewards and Fairies, Tales of The Supernatural, series. BBC: Light That Failed, Hursham Rew, Pilgrim's Way, The Gwaine Inheritance, Windows on the Beach, Three Kipling Love Stories, Traitor's Gate, M'sieur from Armenteers, A Theatre in the Family, Jacobs Ladder, Too Much Music, The Shoplifter, Four From The Days Work, A Quest for Witches.

RAY, ALDO: Actor. r.n. Aldo DaRe. b. Pen Argyl, Pa., Sept. 25, 1926: e. U. of Calif. U.S. Navy, June, 1944–May, 1946: constable. Crockett, Calif., Nov. 1950–Sept. 1951: m.p. debut in Saturday's Hero.
PICTURES INCLUDE: Star of Tomorrow. 1954: since in Marrying Kind, Pat and Mike, Let's Do It Again, Miss Sadie Thompson, Battle Cry, We're No Angels, Three Stripes in the Sun, Nightfall, Men in War, God's Little Acre, Four Desperate Men, Day They Robbed the Bank of England, Sylvia, What Did You Do in the War Daddy?, To Kill A Dragon, Dead Heat on a Merry-Go-Round, Welcome To Hard Times, The Power, Green Berets, The Violent Ones, Angel Unchained, And Hope To Die, Psychic Killer, Seven Alone.

RAY, MICHEL: Actor. b. Gerrard Cross. Buckinghamshire, England, 1945, e. Lycee Francais, London.
PICTURES INCLUDE: Divided Heart, Brave One.

RAYBURN, GENE: Performer, b. Christopher, Ill., Dec. 22. e. Knox Coll., Galesburg, Ill. NBC guide; with many radio stations as announcer in Baltimore, Philadelphia, N.Y.; U.S. Army Air Force, 1942–45, Rayburn and Finch, show, WNEW, N.Y., 1945–52; Gene Rayburn Show, NBC radio; TV shows: Many appearances as host-humorist on game shows, variety shows, drama shows. Summer stock: leads in comedies.
BROADWAY: Bye Bye Birdie, Come Blow Your Horn.
TV: Helluva Town, Amateur's Guide to Love, The Match Game.

RAYE, MARTHA: Actress. b. Butte, Mont.: p. Reed and Hooper, professionals. On stage: sang and did comedy with Paul Ash's orchestra; was in Earl Carroll's Sketch Book; Lew Brown's Calling All Stars. Appeared in night clubs.
PICTURES INCLUDE: Rhythm on the Range, The Big Broadcast of 1937, Hideway Girl, College Holiday, Par, Waikiki Wedding, Mountain Music, Artists and Models, Double or Nothing, Par. Pin Up Girl, Four Jills and a Jeep, Monsieur Verdoux, Pufnstuf, The Concorde—Airport '79.
TV: All Star Revue, Martha Raye Show, NBC.

RAYMOND, GENE: Actor. r.n. Raymond Guion, b. N.Y.C., Aug. 13, 1908. Air Force Reserve, WW II; formed ind. prod. co., Masque Prod., 1949; broadway starred Shadow of My Enemy, 1957, National Co., The Best Man 1960.
STAGE: Why Not?, The Potters, Cradle Snatchers, Take My Advice, Say When, Mirrors, Jones, Young Sinners.
PICTURES INCLUDE: Personal Maid, Ladies of the Big House, The Night of June 13th, Forgotten Commandments, If I Had A Million, Red Dust, Sadie McKee, Brief Moment, I am Suzanne, Flying Down to Rio, The House on 56th Street, Seven Keys to Baldpate, Hooray for Love, Zoo in Budapest, Behold My Wife, Mr. and Mrs. Smith, Walking On Air, Coming Out Party, Life of the Party, She's Got Everything, The Woman in Red, Ann Carver's Profession, Transient Lady, The Best Man.
TV: Star-host, TV Fireside Theatre and TV Reader's Digest; Lux Video Theatre, Robert Montgomery Presents, Climax, Playhouse 90, Kraft Theatre, Red Skelton, U.S. Steel Hour, Uncle, Girl from Uncle, Laredo, Ironsides, Julia, Judd, etc.

RAYMOND, PAULA: Actress. r.n. Paula Ramona Wright, b. San Francisco, Calif.: e. San Francisco Jr. Coll. 1942. Started career in little theatre groups, San Francisco; leading roles, Ah! Wilderness, Peter Pan, other plays; model, Meade-Maddick Agency; TV appearance 1949.
PICTURES INCLUDE: Devil's Doorway, Inside Straight, Duchess of Idaho, Crisis, Grounds For Marriage, Tall Target, Texas Carnival, The Sellout, Bandits of Corsica, City That Never Sleeps, Beast from 20,000 Fathoms, King Richard & the Crusaders, Human Jungle, Gun That Won the West, The Flight That Disappeared, 5 Bloody Graves, Blood of Dracula's Castle.

RAYNOR, LYNN S.: Prod. Executive. b. 1940. Child Actor 1944–54. Continued acting through 1963. Managed Civic Theatre, L.A. prod. The Balcony, The Crawling Arnold Review. Joined Television Enterprises, 1965; Commonwealth United Entertainment, 1968 as prod. supvr.; later headed business affair Europe. In 1973 opened London office for Vidtronics Co. 1974, formed Parazone Entertainment. In 1976–77 producer, Public Bdg. Network; 1976–78, producer, James Flocker Enterprises.
TELEVISON: Waiting for Godot (co.-prod., 1977); Camp Wilderness (series, 1978–79); Marilyn (ABC movie); Ghosts That Still Walk, Alien Encounters (both docu-dramas).

READER, RALPH (C.B.E.): Performer, producer. b. Crewekerne Somerset, May 25, 1903; e. Crewekerne School and St. John's, Cardiff. Early career graduating from New York chorus boy to dance dir., prod.; has produced 34 shows in London theatres; with R.A.F. throughout War as intelligence officer and show-producer; creator. Gang Shows; responsible for 151 productions at Royal Albert Hall, several films; author of three books; BBC disc jockey in Housewives Choice program.
TV: It's A Great Life series, BBC, Chance of A Lifetime, Radio Luxembourg, Ralph Reader Parade, Saturday spectacular, Window of the West, Flying High, U.S.A. TV Series, A.B.C. Calling At Random, Mid-day Special, Starlight Hour, Interview Time, Golden Shot, Festival of Remembrance, A Star Remembers, Scenes Familiar, Right with the Stars, This is your Life, Showman Ralph Reader, Pebble Mill at One.
RADIO: Startime, Radio Look Who's Here, Aut, & Comp. Summer Holiday, You Can't Go Wrong If You're Right, Happy Family, All For The Boys, David Coleman Series, Golden Shot, Be My Guest radio series, Pete Murray Show, Russ Conway Show, Portrait, Keith Fordyce program. Author of Plays: The Hill, Next Door, Great Oaks. 1972: variety show tour, Ralph Reader and the Stars of Tomorrow; 40th year, The Gang Show, Portrait (2nd edition), Nationwide. Series BBC W. "Showman."

READICK, ROBERT: Actor. b. 1926. New York; e. Prof. Children's Sch.
BROADWAY: George Washington Slept Here, All in Favor, Biggest Thief in Town.
PICTURES INCLUDE: Harrigan's Kid, Canterville Ghost.
TV: That Wonderful Guy, Plainclothesman, Two Girls Named Smith, Big Story.

REAGAN, CHARLES M.: b. Lawrenceburg, Ind., June 30, 1896; e. U. Notre Dame, B.S., '17. In Hotel business before entering m.p. ind. 1920 as salesman Cincinnati branch; apptd. mgr. Indianapolis exch., 1922; dist. mgr. Cincinnati, Indianapolis, Louisville, 1925; mgr. Chicago dist., 1932; west iv. mgr., 1934; asst. gen. sales mgr., 1941; vice-pres. & dir., Paramount Pictures, Inc., Paramount Film Distributing Corp.; pres. & dir., Paramount Films Service Ltd. (Canada). Domestic & Canadian distrib. & vice-pres., April, 1944. Resigned, Feb., 1949; joined MGM, sales executive; v.p. chge. dom. sales, 1952–57; vice pres. Denham Theatre Inc.; 540–15th St. Inc.; Empire Theatres, Inc.

REAGAN, RONALD: Actor. b. Tampico, Illinois, Feb. 6, 1911. e. high school and Eureka College; m. Nancy Davis; lifeguard. Wrote weekly sports column for a Des Moines, Iowa newspaper; broadcast sporting events. Signed as actor by Warner Bros. in 1937. In World War II. 1942–45, capt., USAAF. Actor until 1966 on TV as well. Program supvr., General Electric Theatre, Death Valley Days. Gov., California, 1967–74. Businessman and rancher. Elected Pres. of U.S., 1980.
PICTURES INCLUDE: Love Is On the Air, Submarine D-1, Sergeant Murphy, Swing Your Lady, Accidents Will

Happen, Cowboy from Brooklyn, Boy Meets Girl, Girls on Probation, Going Places, Dark Victory, Naughty but Nice, Hell's Kitchen, Kings Row, Juke Girl, Desperate Journey, This is the Army, The Killers, That Hagen Girl, Night Unto Night, Voice of the Turtle, John Loves Mary, Girl from Jones Beach, Hasty Heart, Louisa, Last Outpost, Bedtime for Bonzo, Storm Warning, Hong Kong, She's Working Her Way Through College, Winning Team, Tropic Zone, Law & Order, Prisoner of War, Cattle Queen of Montana, Tennessee's Partner, Hellcats of the Navy, The Killers (1964).

REARDON, BARRY: Executive. Began industry career with Paramount Pictures; named v.p.; left to join General Cinema Theatres Corp. as snr v.p. Now with Warner Bros. as exec. v.p. & gen. sls. mgr. of domestic distribution co.

REASON, REX: Actor. b. Berlin, Germany, Nov. 30, 1928. e. Hoover H.S., Glendale, Calif. Worked at various jobs; studied dramatics at Pasadena Playhouse.
PICTURES INCLUDE: Storm Over Tibet, Salome, Mission Over Korea, Taza Son of Cochise, This Island Earth, Smoke Signal, Lady Godiva, Kiss of Fire, Creature Walks Among Us, Raw Edge, Miracle of The Hills, The Rawhide Trail.

REASONER, HARRY: News correspondent. b. Dakota, Iowa. Apr. 17, 1923; e. Stanford Univ., Univ. of Minnesota. Beg. journalism career, reporter, Minneapolis Times 1941–43; U.S. Army, WW II. Ret. to Times, drama critic 1946–48. Author, book, Tell Me About Women, 1946; newswriter, radio station WCCO. CBS affiliate. Minn., 1950–51; writer U.S. Information Agency Manila, 1951–54; news-dir., KEYD-TV (now KMSP-TV). Minn., 1954; Joined CBS News, N.Y., 1956; ABC News 1970–78. Rejoined CBS 1978.

REDDY, HELEN: Singer. b. Australia, Oct. 25, 1942. Parents were producer-writer-actor Max Reddy and actress Stella Lamond. e. in Britain. Began career at age four as singer and had appeared in hundreds of stage and radio roles with parents by age of 15. Came to New York City in 1966, subsequently played nightclubs, appeared on TV. First single hit record: I Don't Know How To Love Him (Capitol). Grammy Award, 1973 as best female singer of year for I Am Woman. Other Gold singles: Delta Dawn, Leave Me Alone, Angle Baby. Gold Albums: Love Song for Jeffrey, Free & Easy, No Way to Treat a Lady, I Don't Know How To Love Him, Music, Music. Platinum albums: I Am Woman, Long Hard Climb, Helen Reddy's Greatest Hits. Most Played Artist by the music operators of America; American Music Award 1974; Los Angeles Times Woman of the Year (1975); No. 1 Female Vocalist in 1975 and 1976; Record World, Cash Box and Billboard; one of the Most Exciting Women in the World because of her strong convictions, beauty and intelligence—International Bachelor's Society (1976).
PICTURES: Airport 1975 (debut), Pete's Dragon.
TV: David Frost Show, Flip Wilson Show, Mike Douglas Show, etc. The Helen Reddy Show (Summer, 1973, NBC), Permanent host of Midnight Special. Appearances on Tonight Show, Mac Davis Show. Hosted Merv Griffin Show, Sesame St.—Children's Workshop Prod., Tonight Show, Muppet Show.

REDELINGS, LOWELL E.: Columnist & Editor. b. Park Ridge, Ill.; e. U.S.C., B.S. Sports writer Hollywood Citizen-News 1935; sports ed. 1939; film ed. 1943–62; motion picture editor, Evening Outlook, Santa Monica, Feb. 1962–64; Film Columnist, Evening Outlook, to 1965; film drama critic, Los Angeles Herald-Examiner, 1965–66; Holly. Correspondent, Quigley Pub. Co., 1967; Land Developer, 1968; 1970–75, amusements ed., Hicks-Deal Publications to May, 1975. Now retired.

REDFORD, ROBERT: Actor. b. Santa Monica, Calif., Aug. 18, 1937. University of Colorado, left to travel in Europe, 1957. Attended Pratt Institute and American Academy of Dramatic Arts. m. Lola Van Wangemen.
BROADWAY: Walk-on in Tall Story, also in The Highest Tree, Sunday in New York, Barefoot in the Park.
PICTURES INCLUDE: Warhunt, 1961; Situation Hopeless, But Not Serious, Inside Daisy Clover; The Chase, This Property Is Condemned, Barefoot in the Park, Tell Them Willie Boy is Here, Butch Cassidy and the Sundance Kid, Downhill Racer, The Crow Killer, The Hot Rock, The Candidate, Jeremiah Johnson, The Way We Were, The Sting, The Great Gatsby, The Great Waldo Pepper, Three Days of the Condor, All The President's Men, A Bridge Too Far, The Electric Horseman, Brubaker. Director: Ordinary People (AA).
TV: The Iceman Cometh, In the Presence of Mine Enemies, (Playhouse 90).

REDGRAVE, LYNN: Actress. b. London, England, 1943. Ent. m.p. and TV, 1962.
PICTURES INCLUDE: Tom Jones, Girl With Green Eyes, Georgy Girl, The Deadly Affair, Smashing Time, The Virgin Soldiers, The Last of the Mobile Hot-Shots, Viva la Muerta Tua, Every Little Crook and Nanny, Everything You Always Wanted to Know About Sex*, The National Health, The Happy Hooker, The Big Bus, Sunday Lovers.
TV: Pretty Polly, Ain't Afraid to Dance, The End of the Tunnel, I Am Osango What's Wrong with Humpty Dumpty, Egg On the Face of the Tiger, Blank Pages, A Midsummer Night's Dream, Pygmalion, Turn of the Screw, William, Vienna 1900, Daft as a Brush, Not For Women Only, Co-host U.S. talkshow (NBC).

REDGRAVE, SIR MICHAEL: Actor, author. b. Bristol, Eng., March 20, 1908; m. Rachel Kempson; p. Roy Redgrave & Margaret Scudamore, prof.; e. Clifton Coll., Magdalene Coll., Cambridge. Author of two plays, publications, 1959. The Mountebank's Tale (novel). The Aspern Papers (play). Public schoolmaster three years. Started stage career in 1934, Liverpool Repertory.
STAGE: In 1936; Old Vic Season. 1937; Thunder Rock, Beggars Opera, Tiger and the Horse 1960, Out of Bounds 1962, Uncle Vanya (Chichester Festival 1962–63); mem., Nat. Theatre Cox., 1963–64. In 1938 signed long-term contract with Gainsborough, G.B.
PICTURES INCLUDE: Browning Version, Importance of Being Earnest, Green Scarf, Sea Shall Not Have Them, Confidential Report, Dam Busters, Night My Number Came Up, Oh Rosalinda!, 1984, Town Without Pity, The Quiet American, Law and Disorder, Shake Hands with the Devil, The Wreck of The Mary Deare, No! My Darling Daughter, The Innocents, The Loneliness of the Long-Distance Runner, Young Cassidy, The Hill, The Heroes of Telemark, Oh What a Lovely War Goodbye Mr. Chips, The Lady Vanishes, Stolen Life, Stars Look Down, Way to the Stars, Dead of Night, Kipps Jeannie (Girl in Distress), Thunder Rock, The Go-Between.
Entered Royal Navy 1941. Voted among first ten Money Making Stars in British productions in Motion Picture Herald-Fame Poll, 1942, 1946, 1955.
STAGE: Tiger at the Gates, The Sleeping Prince, 1956; New York, 1961–62; The Complaisant Lover, Voyage Round My Father, 1972.

REDGRAVE, VANESSA: Actress. b. 1937. Early career with Royal Shakespeare Company. Ent. m.p. 1958 in Behind the Mask.
TV: A Farewell To Arms, Katherine Mansfield.
STAGE: Daniel Deronda, Cato Street, The Threepenny Opera, Twelfth Night, As You Like It, Taming of the Shrew, Cymbeline, The Sea Gull, The Prime of Miss Jean Brodie, Anthony & Cleopatra, Design for Living, Macbeth, Lady from the Sea.
PICTURES INCLUDE: Morgan, A Suitable Case for Treatment, A Man for All Seasons, Blow-up, Red and Blue, 1967 to Hollywood for Camelot, Charge of the Light Brigade, Isadora, Oh What a Lovely War, The Seagull, A Quiet Place in the Country, Drop Out, Trojan Women, The Devils, La Vacanza Mary Queen of Scots, Murder on the Orient-Express, Out of Season, Seven-per-cent Solution, Julia, Agatha, Yanks, Bear Island.

REDSTONE, EDWARD S.: Exhib. b Boston, Mass., May 8, 1928. e. Colgate U., B.A , 1949; Harvard Grad. School of Bus. Admin., M.B.A., 1952; v.p., treas., Northeast Drive-In Theatre Corp.; v.p., Theatre Owners of New England, 1962; chmn., advis. comm., mem. bd. dirs., TOA; gen. conven. chmn., joint convention TOA & NAC, 1962; pres. National Assn. of Concessionaires, 1963; chief barker. Variety Club of New England, 1963; pres., Theatre Owners of New England; gen. chmn. 35th annual reg. convention.

REDSTONE, MICHAEL: Exhibitor. b. Boston, Mass., Apr. 11, 1902. Built first drive-in in N.Y., 1938. Chief Barker of Variety Club of New England, 1957–58; treas. of Jimmy Fund; member of exec. comm. of the Children's Cancer Research Foundation. Chmn of the board and treas., Northeast Theatre Corp , Boston.

REDSTONE, SUMNER M.: Exhib. b. Boston, Mass., May 27, 1923. e. Harvard Coll., 1944; Harvard Law School, 1947. Law sec., U.S. Court of Appeals, 9th Circuit, 1947–48; instr., U. of San Francisco Law School, Labor Man. School, 1947; spec. asst. to U.S. Attorney Gen., 1948–51; partner, Ford, Bergson, Admans, Borkland and Redstone; President attny., Northeast Theatre Corp.; mem., asst. pres., TOA, 1960–63; pres., 1964–65. Chairman of The Board 1965–66.

REED, DONNA: Actress. r.n. Donna Mullenger. b. Denison, Ia., Jan. 27, 1921; e. Los Angeles City Coll. In school plays, then on screen.
PICTURES INCLUDE: The Getaway; Shadow of the Thin Man, Bugle Sounds, Courtship of Andy Hardy, The Human Comedy, The Man from Down Under, See Here, Private Hargrove, The Picture of Dorian Gray, Gentle Annie, They Were Expendable, Faithful in My Fashion, It's a Wonderful Life, Green Dolphin Street, Beyond Glory, Chicago Deadline, Saturday's Hero, Scandal Sheet, Hangman's Knot, Raiders of the 7 Seas, The Caddy, From Here to Eternity, (Best Supporting Actress Academy Award 1953), Gun Fury, They Rode West, Three Hours to Kill, The Last Time I Saw

Paris, Far Horizons, Benny Goodman Story, Ransom!, Backlash, Beyond Mombasa.
TV: Donna Reed Show.

REED, MAXWELL: Actor. b. Larn, Ireland, 1920. Has been sailor, Merchant Navy; during World War II served as pilot officer, R.A.F. Spent 18 months in repertory. TV appearances.
PICTURES INCLUDE: Screen debut in 1946 in The Years Between, Daybreak, The Brothers, Dear Murderer, Holiday Camp, Dark Man, Sea Devils, Square Ring, Brain Machine, Roadhouse Girl, Before I Wake (Shadow of Fear), Helen of Troy, A Time for Killing, Love is a Funny Thing, The Line Up.

REED, OLIVER: Actor. b. Wimbledon, England, Feb. 13, 1938.
PICTURES INCLUDE: The Rebel, His and Hers, Beat Girl, The Angry Silence, League of Gentlemen, Two Faces of Dr. Jekyll, Sword of Sherwood Forest, Bulldog Breed, Paranoic, No Love for Johnnie, Curse of the Werewolf, Pirates of Blood River, Curse of Captain Clegg, The Damned, The Party's Over, Scarlet Blade, Assassination Bureau, Shuttered Room, The System, Brigand of Kandahar, The Trap, I'll Never Forget What's His Name, Hannibal Brooks, The Jokers, Oliver, The Girl Getters, Women in Love, Take a Girl Like You, The Lady in the Car, Hunting Party, The Devils, Zero Population Growth, Sitting Target, Triple Echo, Fury Rides the Wind, Dirty Weekend, Revolver, Blue Blood, Three Musketeers, Death in Persepolis, Tommy, The Four Musketeers, Ten Little Indians, Royal Flash, Sell Out, Burnt Offerings, The Great Scout and Cathouse Thursday, The Prince and the Pauper, Assault on Paradise, Tomorrow Never Comes, The Big Sleep, The Class of Miss MacMichael, The Broad, Lion in the Desert.

REED, PETER: Executive. Ent. m.p. ind. 1937. Early career with Union Cinemas, General Film Distributors, Warner Bros. RAF Bomber Command 1939–45. UA Salesman, 1945–50. Paramount, 1950–70. Appt. sales man. Paramount 1963 and to brd. dir. 1966. Formed Carthay Center Distributors Ltd., now NGC Distrs. Ltd., subsidiary of National General Corporation, in May 1970. With NBC withdrawal from UK distribution in 1972, formed Eton Films as ind. venture. 1973: joined Anglo-EMI in exec. capacity. Man. Dir. Cathay Films (Uk) Ltd. 1974. UK Rep. Theatrical Dist. N.T.A.

REED, PHILIP: Actor. b. New York City; e. Cornell U. Left coll. in freshman yr. to join Hoboken, N.J. stock co.; later in vaudeville & on N.Y. stage (Grand Street Follies, Grand Hotel, Ziegfeld Follies of 1931, Serena Blandish, Melody, My Dear Children).
PICTURES INCLUDE: Female, The House on 56th Street, Lost Lady, Dr. Monica, Old Acquaintance, People Are Funny, Big Town After Dark, Song of Sheherizade, Pirates of Monterey, Bandit Queen, Tripoli, Davy Crockett, Indian Scout, Target, Take Me to Town, Girl in the Red Velvet Swing, Harem Scarem.

REEVE, CHRISTOPHER: Actor. b. New York City, Sept. 25, 1952. e. Cornell (BA); graduate work at Julliard. National tour with Irregular Verb to Love, co-starring with Celeste Holm. Went to England, visiting repertory theatres; hired for backstage work by Old Vic. Returned to U.S.; played role on daytime TV series, Love of Life. On Bdwy. with Katharine Hepburn in A Matter of Gravity.
PICTURES: Superman, Somewhere in Time, Superman II, Deathtrap.

REEVES, HAZARD, E.: Executive b. Baltimore, Md., July 6, 1906. e. Georgia Inst. of Technology, B.S. 1928; In research eng. dept., Columbia Phonograph Co., 1928; asst. chief eng., Stanley Recording Co., 1929; chief eng., 1930–31; consultant Harvard Film Foundation, 1930–32; chief eng., Standard Sound Co., 1932–33; founder, Reeves Sound Studios, Inc., 1933 (now Cinetel, Inc.); Audio Devices, Inc., 1937; Reeves Industries, Inc., 1946; co-founder, Reeves-Ely Laboratories, Inc., 1942 (now a division of Dynamics Corporation of America); co-founder, 1947, and first pres., Cinerama, Inc.; chmn. bd., Reeves Telecom Corp., Realtron Corp.

REEVES, STEVE: Actor. b. Glasgow, Mont., Jan. 21, 1926. Delivered newspapers. Mr. Pacific, Mr. America, Mr. World, Mr. Universe; ent. theatrical field, Kismet. Appeared The Vamp, Wish You Were Here.
PICTURES INCLUDE: Athena, Goliath and the Barbarians, Sword of Siracusa, Judos, David and Goliath, Hercules, Giant of Marathon, Last Days of Pompeii, Hercules Unchained, White Warrior, Morgan the Pirate, The Thief of Baghdad, The Trojan Horse, The Private Prince, A Long Ride From Hell.

REGAN, PHIL: Singer and actor. b. Brooklyn, N.Y., May 28, 1906. While a boy was a boatman at Charleston, S.C., navy yard. Was a clerk in an oil company; clerked in different parts of the country, then became a court clerk in New York, finally a detective. After voice was trained, sang for Columbia Broadcasting System. Was signed by WB.
PICTURES INCLUDE: The Key, Housewife, Dames in 1934, Sweet Rosie O'Grady, Swing Parade of 1946, Sunbonnet Sue, Three Little Words.
TV: Many appearances and several radio shows.

REHME, ROBERT G.: Executive. b. Cincinnati, Ohio, May 5, 1935. e. Univ. of Cincinnati. 1953, mgr., RKO Theatres, Inc., Cincinnati; 1961, adv. mgr., Cincinnati Theatre Co.; 1966, dir. of field adv., United Artists Pictures; 1969, named dir. of pub. and ass't gen. sls. mgr., United Artists Pictures; 1972, pres., B&R Theatres and v.p., April Fools Films; 1976, v.p. & gen. sls. mgr., New World Pictures; Feb. 1978, joined Avco Embassy Pictures as snr. v.p. & chief operating officer; Dec. 1978, named exec. v.p.; 1979, named pres., Avco Embassy Pictures, Inc. 1981, joined Universal Pictures as pres. of distribution & marketing.

REILLY, CHARLES E., JR.: Communications Executive. e. St. Joseph's College, Network Liaison, TV Guide Magazine; asst. to the v.p. & dir., corporate relations, Young & Rubicam, 1964–66; executive dir., National Catholic Office for Radio, 1966–71; Secretary, Catholic Communications Foundation (CCF) 1968–77; exec. v.p., Patrick Carr Associates 1971–73; corp. exec. Communispond Inc. of J. Walter Thompson Co., 1973–76; v.p., Speech Dynamics subsidiary, Ogilvy & Mather International, 1976–77; pres., In-Person Communications Inc. 1977–present.

REILLY, CHARLES NELSON: Actor-Director. b. New York City, Jan. 13, 1931. e. Univ. of Conn. On Broadway mostly in comedy roles before turning to TV and films. Recently directed stage plays.
PICTURES INCLUDE: A Face in the Crowd, Two Tickets to Paris, The Tiger Makes Out.
TELEVISION: Broadway of Lerner and Lowe, Ghost and Mrs. Muir, Dean Martin Show.

REINAUER, RICHARD: Executive. b. Chicago, Ill., April 28, 1926. e. U. of Ill., grad. 1952; prod., dir., free lance, 1952–59; bus. mgr., asst. prod., Showcase Theatre Evanston, 1952; prod., dir., NBC, Chicago, 1953–55; film dir., Kling Studios, 1956; asst. dir., Foote Cone & Belding, 1956–59; dir., radio, TV & m.p., American Medical Assoc., 1959–64; pres., Communications Counselors, 1963–64; exec. dir., TV Arts & Sciences Foundation, 1964; pres., Acad. of TV Arts & Sciences, Chicago Chapter, 1970–72. assoc. prod. & asst. dir. Wild Kingdom & asst. to pres., Don Meier Prods., 1965-present. Member-Illinois Nature Preserve Commission.

REINER, CARL: Performer; b. New York, N.Y., March 20, 1923. Comedian on B'way, Call Me Mr., Inside U.S.A., Alive and Kicking; m.p. The Russians are Coming; on TV in Your Show of Shows, first Bob Hope Show, Caesar's Hour; Sid Caesar Invites You, 1958; prod.-writer, The Dick Van Dyke Show, CBS (Emmy Award-writing Comedy) 1961–62. Prod., The New Dick Van Dyke Show 1973; Heaven Help Us, 1975–76.
PICTURES INCLUDE: The Gazebo, writ. orig. s.p. The Thrill of It All, dir. co-author & co-prod. The Comic; Generation. A Performer: It's a Mad, Mad, Mad, Mad World; Happy Anniversary, Gidget Goes Hawaiian, The End. Co-author & dir: Enter Laughing; Dir: Where's Poppa, Oh, God!, The One and Only, The Jerk.
RECORDINGS: Carl Reiner and Mel Brooks, The 2000 Year Old Man, The 2001 Year Old Man, 2013 Year Old Man.
AUTHOR: Enter Laughing, Something Different, B'way.

REINHARDT, GOTTFRIED: Producer, writer. b. Berlin, Germany, p. Max Reinhardt, noted theatrical prod.; Else Reinhardt, actress; brother, Wolfgang Reinhardt, prod.; e. Berlin. Began career at 19 as asst. to prod. Ernst Lubitsch, father's friend, with m.p. Design For Living; asst. to Walter Wanger; later to Bernard H. Hyman (Saratoga; San Francisco). Wrote orig. story, I Live My Life, The Great Waltz; collab. Bridal Suite; book for NY musicals, Rosalinda, Helen of Troy. U.S. Army service, Signal Corps. 1942–46.
PICTURES INCLUDE: Comrade X, Rage in Heaven, Two-Faced Woman, (co-prod.) Homecoming, Command Decision, The Great Sinner, The Red Badge of Courage, Invitation, Young Man With Ideas, (dir. 2 seq.) Story of Three Loves, Betrayed, Town Without Pity, Hitler: The Last Ten Days (prod., co. s.p.).

REINHART, ALICE: Actress. b. San Francisco, Calif., May 6. Legit. theatre work, 1928–49; appeared in stock, numerous Broadway plays; radio work since 1930; shows include Woman in My House, The Whistler, Life Can Be Beautiful, many other shows on all networks; on stage, American Dream in Hollywood.
TV: Stage 13, Man Behind the Badge, Big Town, O. Henry Series, Dragnet, Crusader, The Donna Reed Show.
PICTURES INCLUDE: Lieutenant Wore Skirts, The Iron Sheriff, Bachelor Flat.

REISENBACH, SANFORD E.: Executive. e. N.Y. University. Associated with Grey Advertising for 20 years; exec. v.p. and pres. of Grey's leisure/entertainment division in N.Y. In August, 1979, joined Warner Bros. as exec. v.p. in chg. of worldwide adv. & pub.

REISNER, ALLEN: Director. b. N.Y.C.
FEATURES: The Day They Gave Babies Away; St. Louis Blues.
FEATURES FOR TV: The Captain And the Kings; Mary Jane Harper Cried Last Night, Your Money Or Your Wife; To Die In Paris; The Cliff; Skag.
TV: Film; Kojak, Hawaii Five-0, Cannon, Mannix Etc. Live: Playhouse 90, Studio One, Climax, Etc.

REISS, JEFFREY C.: Executive. b. Brooklyn, N.Y., April 14, 1942. e. Washington Univ., St. Louis (B.A., 1963). Consultant at N.Y. Univ. and Manhattanville College and instructor at Brooklyn College before entering industry. Agt. in literary dept. for General Artists Corp., 1966. Supervised development in N.Y. of Tandem Prods. for Norman Lear, 1968. Produced off-Bdwy. plays 1968–70. Dir. of progm. devel. for Cartridge TV, Inc. (mfg. of video-players-recorders) 1970–73. Joined ABC Entertainment as director of network feature films, 1973–75. Joined Viacom International Inc. as vice pres. of Premium Television services in 1976, promoted to pres. of Showtime Entertainment, Inc., the pay television subsidiary, in 1977. Appt. exec. v.p., Viacom Entertainment Group, 1980.

REISS, STUART A.: Set decorator. b. Chicago, Ill., July 15, 1921. e. L.A. High Sch., 1939. Property man, 20th-Fox, 1939–42; U.S. Army Air Corps, 1942–45; set decorator, 20th-Fox since 1945; 6 Acad. nom.; 2 Acad. Awards, Diary of Anne Frank & Fantastic Voyage.
PICTURES INCLUDE: Titanic, How to Marry a Millionaire, Hell and High Water, There's No Business Like Show Business, Soldier of Fortune, Seven Year Itch, Man in the Grey Flannel Suit, Teen Age Rebel, What a Way to Go, Doctor Doolittle, Fantastic Voyage, Oh God! Swarm, Beyond the Poseidon Adventure, Carbon Copy, All the Marbles.

REISZ, KAREL: Director. b. Czechoslovakia, 1926. e. Britain. Wrt. Technique of Film Editing for British Film Academy. Worked with British Film Institute and National Film Library, 1954; co. dir. (with Tony Richardson) Momma Don't Allow, 1957; prod. Every Day Except Christmas, 1958; dir. We Are the Lambeth Boys, 1959–60; dir. Saturday Night & Sunday Morning, 1962; prod. This Sporting Life, 1963; dir. Night Must Fall, 1966 dir. Morgan, 1968; dir. Isadora; The Gambler. Dir. On the Road. (BBC-TV). Dir. Who'll Stop the Rain (1978), The French Lieutenant's Woman, 1981 (also co-prod.).

RELPH, MICHAEL: Producer, director, writer, designer. 1942 art dir. Ealing Studios then assoc. prod. to Michael Balcon on The Captive Heart, Frieda, Kind Hearts and Coronets, Saraband (also designed: nominated Oscar). 1948 appt. producer and formed prod/dir. partnership Basil Dearden (until 1972). 1971–76 Governor Brit. Film Institute. Chairman B.F.I. Prod. Board. Chairman Film Prod. Assoc. of G.B., member Films Council.
PICTURES INCLUDE: (For Ealing) The Blue Lamp (Brit. Film Academy: Best Brit. Film 1950), I Believe in You, The Gentle Gunman, The Square Ring, The Rainbow Jacket, Out of the Clouds, The Ship That Died of Shame, Davy (for Brit. Lion), The Smallest Show on Earth. (For Rank) Violent Playground, Rockets Galore (Island Fling U.S.) Sapphire (Brit. Film Academy: Best Brit. Film 1959) 1960 Founder Dir. Allied Film Makers: Prod. The League of Gentlemen, Man in The Moon (Co-author S.P.) Victim, Life For Ruth (Walk in the Shadow U.S.) Also produced: Secret Partner (For M.G.M.), All Night Long (For Rank), The Mind Benders (For Anglo). A Place To Go (Author S.P.) For United Artist Produced: Woman of Straw (Co-author S.P.) Masquerade (Co-author S.P.) For Paramount produced, wrote and designed The Assassination Bureau. For E.M.I. produced The Man Who Haunted Himself (Co-author S.P.). 1978 Exec prod., Scum, co-prod., An Unsuitable Job for a Woman. Exec. in chg. U.K. prod., Kendon Films Ltd.

RELYEA, ROBERT E.: Executive. b. Santa Monica, Calif., May 3, 1930. e. UCLA (B.S., 1952). In Army 1953–55. Started production of films in 1968 with Bullitt. Now exec. v.p. in chg. of worldwide prod. for Melvin Simon Productions.
PICTURES INCLUDE: Bullitt, The Reivers, Day of the Dolphin.

REMBUSCH, TRUEMAN T.: Exhibitor. b. Shelbyville, Ind., July 27, 1909; f. Frank J. Rembusch, pioneer exhibitor; e. Notre Dame U. Sch. of Commerce, 1928; m. Mary Agnes Finneran. Ent. m.p. ind., 1928, servicing sound equip., father's circuit; became mgr., 1932; elect. bd. of dir., Allied Theatre Owners of Ind., 1936–45, pres. 1945–51, 1952–53; dir. chmn. Allied TV Committee, 1945–50; pres. Allied States Assn., 1950–51; 1952, named by Allied as one of triumvirate heading COMPO; elected chmn. Joint Com. on Toll TV, 1954; Nov. 1953 named by Gov. of Indiana as dir. State Fair Board. Currently pres. Syndicate Thea., Inc., Franklin, Ind. member, In Notre Dame Club of Indianapolis (Man of Yr., 1950); BPOE, Antelope Club, 4th Degree K of C, Meridian Hills Country Club, Marco Island Country Club and Hillview Country Club of Franklin. American Radio Relay League (amateur & commerce, licenses); OX5 Aviation Pioneers; awarded patent, recording & counting device, 1951; dir. Theatre Owners of Indiana; dir. to NATO; dir. NATO member ad hoc comm 1972 chair., NATO Statistical Committee; presently chm., trade practice comm.; 1976-NITE Award service to Independent Exhibition.

REMICK, LEE: Actress. b. Boston, Dec. 14, 1937. e. Miss Hewitt's Sch., Barnard Coll. Started in summer stock; on tour in Jenny Kisses Me The Seven Year Itch, Paint Your Wagon; first N.Y. stage appearance, Be Your Age; major TV shows incl. Studio 1, Playhouse 90, Armstrong Circle Theatre; m.p. debut in A Face In the Crowd.
PICTURES INCLUDE: The Long Hot Summer, Anatomy of a Murder, Wild River, Sanctuary, Experiment in Terror, The Days of Wine and Roses, The Wheeler Dealers, Travelin' Lady, Baby the Rain Must Fall, Hallelujah Trail, Hard Contract, A Severed Head, Sometimes A Great Notion, A Delicate Balance, Hennessy, The Omen, Telefon, The Medusa Touch, The Europeans, The Competition, Tribute.

REPP, ED EARL: Writer; TV, White Savage, Incident at Yuma, many others.
PICTURES INCLUDE: West of Cheyenne, Rawhide Ranger, Silver City Raiders, Six Gun Gospel, Trigger Trail, Texas Panhandle, Fighting Frontiersman, Guns of Hate, many Westerns, currently featured in western history magazines.

RESNAIS, ALAIN: Director. b. Cannes, France, 1922. Began career as asst. dir. to Nicole Vedres on compilation of film for Paris 1900. During '50s worked as asst. editor and editor; experimented with making his own 16mm films. Did series of shorts on various painters, culminating with documentary on Van Gogh (1948), which he co-directed with Robert Hessens, with whom he later filmed Guernica. Co-directed The Statues Also Die with Chris Marker.
PICTURES INCLUDE: Night and Fog, Hiroshima, Mon Armour, Last Year at Marienbad, Muriel, La Guerre Est Finie, Je t'Aime, Je t'Aime, Stavisky, Providence, Mon Oncle d'Amerique.

RETCHIN, NORMAN: Writer, producer. b. Chicago. May 6, 1919. e. U. of Pa., 1940. U.S. Air Force, 1943–45; J. Walter Thompson adv. agency, 1946–51; story ed. & assoc. prod. Screen Televideo Co., 1951–55; story ed. Clover Prod. 1955; stage, orig. bk. and lyrics, The Genius Farm; TV, orig. prod., The Untouchables.
PICTURES INCLUDE: Uranium Boom, (collab.) Leather Saint, Ride Out for Revenge, Mission Mars, Into Thin Air, (orig. story).

RETTIG, TOMMY: Actor. b. Jackson Heights, N.Y., Dec. 10, 1941; on stage in Annie Get Your Gun; TV series, Lassie.
PICTURES INCLUDE: Gobs and Gals, Paula, Lady Wants Mink, 5000 Fingers of Dr. T., So Big, River of No Return, The Raid, The Egyptian, The Cobweb, At Gunpoint.

REVERE, ANNE: Actress. b. New York City, June 25, 1907; e. Wellesley Coll., B.A., m. Samuel Rosen, director. On stage, Stuart Walker Stock Co., 1928–29, Double Door, 1933–34. Children's Hour, 1934–37, org. & dir. Surry Theat., Maine, N.Y., 1936–39; Acad. Award best supporting role (National Velvet), 1945.
PICTURES INCLUDE: Double Door, Howards of Virginia, Men of Boys Town, Remember the Day, Star Spangled Rhythm, Song of Bernadette, National Velvet, Keys of the Kingdom, Sunday Dinner for a Soldier, Dragonwyck, Forever Amber, Body and Soul, Gentleman's Agreement, Place in the Sun, Great Missouri Raid, Tell Me That You Love Me Junie Moon, Birch Interval.

REVILL, CLIVE: Actor. b. Wellington, New Zealand, Ent. m.p. ind. 1965.
STAGE: Irma La Douce, The Mikado, Oliver, Marat/Sade, Jew of Malta, Sherry, Chichester Season, The Incomparable Max (N.Y.), Sherlock Holmes (NY).
PICTURES INCLUDE: Bunny Lake is Missing, Once Upon a Tractor, Modesty Blaise, A Fine Madness, Kaleidoscope, The Double Man, Fathom, Italian Secret Service, Nobody Runs Forever, Shoes of the Fisherman, Assassination Bureau, The Private Life of Sherlock Holmes, The Buttercup Chain, A Severed Head, Boulevard de Rhum, Avanti!, Flight to the Sun, The Legend of Hell House, The Little Prince, The Black Windmill, Ghost in the Noonday Sun, One of Our Dinosaurs Is Missing, Galileo, Matilda.
TV: Chicken Soup with Barley, Volpone, Bam, Pow, Zapp. Candida, Platonov, A Bit of Vision, Mill Hill, The Piano Player, Hopcroft in Europe, A Sprig of Broome, Ben Franklin in Paris, Pinocchio, The Great Houdini, Show Business Hall of Fame, Feather and Father, Winner Take All, The New Avenger, Licking Hitler, Columbo, Centennial, A Man Called Sloane, Nobody's Perfect, Marya, Moviola, Diary of Anne Frank.

REYNOLDS, BURT: Actor/Director b. Waycross, Ga., Feb. 11, 1936. Former Fla State U football star, won fame as actor on TV in series: Riverboat, Gunsmoke, Hawk and Dan August.
PICTURES INCLUDE: Armored Command, Angel Baby,

Operation CIA, Navajo Joe, Impasse, Shark, Sam Whiskey, 100 Rifles, Fade-In, Skullduggery, Everything You Wanted To Know About Sex, Fuzz, Deliverance, Shamus, White Lightning, The Man Who Loved Cat Dancing, The Longest Yard, W.W. & The Dixie Dancekings, At Long Last Love, Hustle, Lucky Lady, Gator (dir.-star), Silent Movie, Nickelodeon, Smokey & The Bandit, Semi-Tough, The End (dir-star), Hooper, Starting Over, Rough Cut, Smokey & The Bandit II, Cannonball Run, Paternity, Sharky's Machine (dir.-star).

REYNOLDS, CLARKE: Writer, producer. b. Bklyn, N.Y. e. U. of Missouri; pub. rel. NYC, pub. rel on WAAM-TV, Baltimore.
PICTURES INCLUDE: Disc-Jockey, Shotgun, Chisera, Gunmen of Laredo, Imperial Venus, Gunfighters of Casa Grande, Gringo, Genghis Khan, Son of A Gunfighter, The Viking Queen, The Viscount, Warhead, Shalako, The Marauders, The Red Sun, Private Navy, Operation Thunderbolt, Cruise Missile, Night Games.
TV: Celebrity Playhouse, Damon Runyon Theatre, Cheyenne, 20th Century-Fox Hour, Lux Video Hour, West Point, Whiplash, Playhouse 90, Walter Winchell File, Wells Fargo, Tombstone Territory, The Texan, Peter Gunn, Wagon Train, U.S. Marshall, Zane Grey Theatre, Broken Arrow (pilot), The Beachcomber, The Deputy, The Investigators, Cain's Hundred, Created Man Without a Gun.

REYNOLDS, DEBBIE: Actress. r.n. Mary Frances Reynolds. b. El Paso, Tex., April 1, 1932. e. Burbank & John Burroughs H.S., Burbank, Calif. With Burbank Youth Symphony during h.s.; beauty contest winner (Miss Burbank) 1948, signed by Warner Bros.; on stage in Personal Appearances, Blis-Hayden Theater, Star of Tomorrow, 1952.
PICTURES INCLUDE: The Daughter of Rosie O'Grady, Three Little Words, Two Weeks With Love, Mr. Imperium, Singing in the Rain, Skirts Ahoy, I Love Melvin, Give a Girl a Break, Affair of Dobie Gillis, Susan Slept Here, Athena, Hit the Deck, Tender Trap, Catered Affair, Bundle of Joy, Tammy and the Bachelor, The Mating Game, Say One for Me, It Started With A Kiss, The Gazebo, The Rat Race, Pleasure of His Company, Second Time Around, How the West Was Won, Goodbye Charlie, The Unsinkable Mollie Brown, The Singing Nun, Divorce, American Style, What's The Matter With Helen?, That's Entertainment!
TV: The Debbie Reynolds Show.
STAGE: Irene.

REYNOLDS, JOYCE: Actress. b. San Antonio, Texas, Oct. 7, 1924; e. UCLA. School plays including role of Alice in Alice in Wonderland at UCLA, contracted by Warner Bros. in 1942, appearing in Yankee Doodle Dandy, George Washington Slept Here, Mark Twain, Constant Nymph, Janie, Always Together, Wallflower, Girl School, Dangerous Inheritance.

REYNOLDS, MARJORIE: Actress. b. Buhl, Idaho, Aug. 12, 1921. On screen as child 1923 & later (Scaramouche, Svengali, Revelation, etc.).
PICTURES INCLUDE: Murder in Greenwich Village (1937), College Humor, Holiday Inn, Star-Spangled Rhythm, Dixie, Ministry of Fear, Up in Mabel's Room, Three Is a Family, Duffy's Tavern, Bring on the Girls, Meet Me on Broadway, Heaven Only Knows, Bad Men of Tombstone, Great Jewel Robber, Rookie Fireman, Home Town Story, No Holds Barred, Models, Inc., Silent Witness.
TV: The Life of Riley (series).

REYNOLDS, SHELDON: Writer, producer, director. b. Philadelphia, Pa., 1923. e. NYU. Radio-TV writer; programs include My Silent Partner, Robert Q. Lewis Show, We the People, Danger; writer, prod., dir. Foreign Intrigue film; prod. dir. s.p. collab. story, Foreign Intrigue.

REYNOLDS, STUART: Producer. b. Chicago, Ill., March 22, 1907. e. Chicago law schools. Adv. exec., Lord and Thomas, BBD & O. General Mills; sales exec. Don Lee-Mutual; formed Stuart Reynolds Prod., TV films.
TV: General Electric Theatre, Cavalcade of America, Your Jeweler's Showcase, Wild Bill Hickok, educational/training films; Eye of the Beholder.

RHEINER, SAMUEL: Executive. b. NYC. e. NYU, BCS; Fordham U. Joined Paramount as asst. to James R. Cowan; asst. org. Publix Stage Show dept.; then gen. mgr. Artists Booking Office (Paramount sub.); in chg. cost control Paramount theat., N.Y. & Bklyn, Rivoli, Rialto, Criterion theats., N.Y.; gen. mgr. scenic & costume depts., Publix Theats. In 1937 tr. to Paramount Studios, reorg. music dept. Resigned 1939 to enter independent prod. becoming v.p. Boris Morros Prods. (Flying Deuces); org. Nat'l Pictures Corp. of Calif. (Second Chorus, Tales of Manhattan). In 1945 assoc. prod. David L. Loew Prod. Artists, Inc.; The Southerner prod. Inner Sanctum; treas. Horizon Pictures, Inc.
PICTURES INCLUDE: The Southerner, The Stranger, Carnegie Hall, Inner Sanctum, The Prowler, When I Grow Up, African Queen, Melba, On the Waterfront, Bridge on the River Kwai, The Strange One, The Chase, Suddenly Last Summer, Lawrence of Arabia, Night of the Generals, The Swimmer, Nicholas and Alexandra, The Happening, The Last Tycoon.

RHODES, E. A.: Exhibitor. b. Cape Town, S.A. 1900. Father, one of first cinema exhibitors in Transvaal. Joined m.p. ind. 1916 with Schlesinger Org., later in Britain with P.D.C. & U.A., 1941 mgr. with The Classic Cinemas. Took control Booking Dept. 1945; apptd. chmn. and managing dir. 1962; retired from Classic 1973. Chmn. London CEA 1956–57, 1967–68, 1969–73. Pres., National Assoc., Bingo Clubs, 1967. Chmn, 1968. Pres. Cinema Theatre Assoc. 1972–77. Pres. Cinema Veterans Soc., 1975–76 Prop. Curzon Cinemas, Eastboume.

RICH, IRENE: Actress. b. Buffalo, N.Y. Oct. 13, 1897. e. St. Margaret's Episcopal School; painting. Real estate saleswoman; stock farmer; on radio, in night clubs, vaudeville.
PICTURES INCLUDE: Exalted Flapper, Lady Windemere's Fan, Lost Lady, Craig's Wife, They Had to See Paris, So This is London, Down to Earth, That Certain Age, Mortal Storm, Lady in Question, Queen of the Yukon, Keeping Company, This Time for Keeps, Angel and the Badman, Calendar Girl, New Orleans, Fort Apache, Joan of Arc.

RICH, JOHN: Producer-Director. b. Rockaway Beach, N.Y., July 6, 1925. e. U. of Michigan, B.A., Phi Beta Kappa, 1948; M.A. 1949 Sesquicentennial Award, 1967; bd. of dir., Screen Dir. Guild of America, 1954–1960; v.p. 1958–1960 Founder-Trustee, Producers-Directors Pension Plan, chmn. of bd. 1965, 1968, 1970; treasurer, Directors Guild of America, 1966–67; v.p. 1967–72.
TV: Academy Awards, The Dick Van Dyke Show, 1963; All in the Family, 1972 (director); All in the Family, 1973 (producer); Directors Guild Award, Most Outstanding Directorial Achievement, 1971.
PICTURES INCLUDE: Boeing-Boeing; The New Interns; Wives and Lovers; Roustabout; Easy Come, Easy Go.

RICHARD, CLIFF: Singer and actor. b. 1940. Ent. show business 1958 in TV series Oh Boy. Other TV includes Sunday Night at the London Palladium, several Cliff Richard Shows; film debut in Serious Charge, 1959; star, play Aladdin, London Palladium Theatre, 1964–65. Top British Singer, 1960–66.
PICTURES INCLUDE: Expresso Bongo, The Young Ones, Summer Holiday, Wonderful Life, Voted top box-office Star of Grt. Britain, 1962–63, 1963–64. Finder's Keepers, Two a Penny, Take Me High.

RICHARDS, JEFF: Actor. r.n. Richard Mansfield Taylor. b. Portland, Ore. Nov. 1. e. Tacoma, Wash. H.S., U.S.C. served in U.S. Navy, 1943–46; professional baseball.
PICTURES INCLUDE: Kill the Umpire, Tall Target, The Strip, Angels in the Outfield, The Sellout, Above and Beyond, Desperate Search, Code Two, Big Leaguer, Crest of the Wave, Seven Brides for Seven Brothers, Many Rivers to Cross, The Marauders, It's A Dog's Life, Waco, Johnny Belinda.

RICHARDS, THOMAS LLOYD: Producer-Writer. b. Cardiff, Wales. Entered industry in London at Stoll Studios in 1926, working in production and as 3rd asst. director. Named 1st asst. dir. for producer Herbert W. Wilcox, going to Hollywood to work on Wilcox films starring Anna Neagle: No, No, Nanette, Nurse Edith Cavell, Sunny, and Irene. Worked for RKO Studios as 1st asst. dir., acting in that capacity for Without Reservations, etc. Joined Four Star Productions (TV producers) as v.p. in chg. prod. Produced 24 in Untouchables series at Desilu. Prod. assoc. of many TV special and features for TV; also special consultant to producers of TV series and mini-series.

RICHARDSON, SIR RALPH: Actor. b. Cheltenham Gloucestershire, England, Dec. 19, 1902; m. Merial Forbes; made first appearance on stage in 1921 as Lorenzo in The Merchant of Venice; joined Birmingham Repertory Theatre 1926 and has appeared in many plays, particularly Shakespearean prod.; hobbies: Drawing, modeling, literature, aviation, squash, racquets. In 1934 appeared in Friday the 13th, GB; Retum of Bulldog Drummond, BIPUA. In 1942, The Day Will Dawn, Soskin-Gen. Film, The Avengers, Par. In 1943, The Silver Fleet, The Volunteer, Released from Fleet Air Arm; then repertory, Old Vic, London. Knighted in New Year's Honor's List, 1947, British money-making star, 1953.
PICTURES INCLUDE: School for Secrets, Anna Karenina, The Fallen Idol, Heiress, Outcast of the Islands, Home at Seven (Murder on Monday), Holly & the Ivy, The Sound Barrier, Richard III, Smiley, Passionate Stranger, Our Man in Havana, Exodus, Lion of Sparta, Long Day's Journey Into Night, Woman of Straw, Dr. Zhivago, Khartoum, The Wrong Box, 1968, Battle of Britain, Oh! What a Lovely War, The Bed Sitting Room, Looking Glass War, David Copperfield, 1972: Oh Lucky Man; Alice in Wonderland, A Doll's House, Rollerball, Dragonslayer.

RICHARDSON, TONY: Director. b. Shipley, Yorks; England, 1928. e. Wadham Coll., Oxford, where he dir. number prod. for O.U.D.S. Began career with BBC TV and directed such plays as Othello and The Gambler. In 1955 joined English

Stage Co. as Assoc. Artistic Dir. Started with Look Back in Anger, Member of the Wedding, 1958 at Shakespeare Memorial Theatre. 1960, N.Y co-dir., A Taste of Honey. Recent stage work, dir., Seagull and St. Joan of the Stockyards, Threepenny Opera, I Claudius. Ent. m.p. ind. 1958 to dir. prod. Look Back in Anger, The Entertainer, 1959–60; prod. Saturday Night and Sunday Morning.

PICTURES INCLUDE: A Taste of Honey, Loneliness of the Long Distance Runner, Tom Jones, The Loved One, Mademoiselle, The Sailor from Gibralter, Red and Blue, Charge of the Light Brigade, Laughter in the Dark, Hamlet, Ned Kelly, A Delicate Balance, Dead Cert, Joseph Andrews.

RICHMAN, MARK: Actor. b. Philadelphia, Pa. Stage credits incl. End as a Man, Masquerade, The Zoo Story, Detective Story, Rose Tattoo.

PICTURES INCLUDE: Friendly Persuasion, Dark Intruder, Agent for H.A.R.M., For Singles Only.

TV: Cain's Hundred.

RICHMOND, TED: Producer. b. Norfolk, Va., June 10, 1912; e. Mass. Inst. Tech. Ent. m.p. ind. as publicity dir., RKO Theats.; later mgr. Albany dist. Pub. dir. Fabian circuit, N.Y.: Paramount upper N.Y. state theats.; Grand Nat'l Pictures. Author Grand Nat'l series Trigger Pal, Six Gun Rhythm. Formed T. H. Richmond Prods., Inc., 1941.

PICTURES INCLUDE: Hit the Hay, The Milkman, Kansas Raiders, Shakedown, Smuggler's Island, Strange Door, Cimarron Kid, Bronco Buster, Has Anybody Seen My Gal, No Room for the Groom, Weekend with Father, The Mississippi Gambler, Desert Legion, Column South, Bonzo Goes to College, Forbidden, Walking My Baby Back Home, Francis Joins the Wacs, Bengal Brigade, (Formed Copa Prod. with Tyrone Power, 1954) Count Three and Pray, Nightfall, Abandon Ship, Solomon and Sheba, Charlemagne. Formed Ted Richmond Prod. Inc. for MGM release, 1959. Reactivated Copa Prod. Ltd., England, 1960. Produced for MGM: Bachelor in Paradise, Advance to the Rear; for Paramount, Pancho Villa; for Mirisch-U.S., Return of the 7; Independent, Red Sun; Producer, Papillon-Allied Artists, various distributors.

RICHTER, W. D.: Writer. b. Connecticut. e. USC Film School.

PICTURES: Slither, Peeper, Nickelodeon, Invasion of the Body Snatchers, Dracula, Brubaker, All Night Long.

RICKETSON, FRANK H., JR.: Executive. b. Leavenworth, Kan. e. U. Ky., LL.B. U. Denver. On ed. staff Kansas City Star, The Denver Post. Member exploitation Dept., Famous Players-Lasky Corp., 1921–24. Pres. Consolidated Theatres 1924 to 1929, Pres. Fox Inter-Mountain Theatres, 1934 to 1957, v.p. & gen. mgr. of Theatre Operations, National Theatres, Inc., 1956–58. Author of The Management of Motion Picture Theatres, Europe after the Marshall Plan, Opportunity Needs a Selling, Gems from a Thousand Sources. Retired director, Public Service Co. of Colorado, Continental Airlines, United Banks of Colorado, Cheyenne Newspapers Inc., Frontier Broadcasting Co., honorary bd. Central City Opera House Assn., Denver Museum of Natural History, first individual elected to Rocky Mountain Motion Picture Association Hall of Fame; member of the Colorado Motion Picture and Television Advisory Commission.

REISNER, DEAN: Writer. Began career as director: Bill and Coo, 1947. Has collaborated on following screenplays, among others: Coogan's Bluff, Dirty Harry, Play Misty for Me.

RIFKIN, JULIAN: Exhibitor. b. Boston, May 26, 1915. e. MIT. Member bd. of dir. Allied States Assoc. of M.P. Exhibitors, and Theatre Owners of America. Pres. Theatre Owners of New England 1961–63. Chairman bd. Theatre Owners of New England. 1964–65. Past pres. Allied Artists Corp. of New England. Pres., 1968–69, chmn. of bd. Nat'l Assoc. of Theatre Owners, 1970. Pres. Rifkin Theatres. Pres. Cinema Centers Corp. Chm. NATO Code and Rating Comm., 1968–79.

RIGG, DIANA: Actress. Formerly with the Royal Shakespeare Co. at Aldwych Theatre, 1962–64. Ent. TV in The Avengers series, 1965. Ent. films 1967.

PICTURES INCLUDE: The Assassination Bureau, On Her Majesty's Secret Service, The Great Muppet Caper. 1970–71, on London and NY stage.

RISI, DINO: Director. b. Italy, 1916. Studied medicine but left for film job as assistant on Mario Soldati's Giacomo L'Idealista. Interned in Switzerland in World War II. Returned home to make documentaries and short films before directing Vacanze col Gangster in 1952.

PICTURES INCLUDE: Sign of Venus, Poveri ma Bellis, Il Sorpasso, Scent of a Woman, Sunday Lovers.

RISSIEN, EDWARD L.: Executive. b. Des Moines, IA. e. Grinnell Coll., Stanford Univ. (B.A., 1949). Army Air Force, World War II. Bdwy. stage, mgr., 1950–53; v.p., Mark Stevens. Prods., 1954–56; prod., v.p., Four Star, 1958–60; prog. exec., ABC-TV, 1960–62; v.p., Bing Crosby Prods., 1963–66; v.p., Filmways TV Prods.; assoc. producer, Columbia, 1968–69; indep. producer, 1970; prod., WB, 1971; exec. v.p., Playboy Prods., 1972–80; consultant & indep. prod., 1981–82.

PICTURES INCLUDE: Snow Job (prod., WB); Castle Keep (exec. prod.); Julius Vrooder (proc., Fox); Saint Jack (exec. prod.).

TELEVISION: Minstrel Man (exec. prod., CBS); A Whale for the Killing (exec. prod., ABC); Ocean View Park (exec. prod., ABC); Big Bob Johnson (exec. prod., NBC).

RITCHIE, MICHAEL: Director. b. Waukesha, Wisc., 1938. e. Harvard Univ. where he directed first production of Arthur Kopit's play, Oh Dad, Poor Dad, Mama's Hung You in the Closet and I'm Feeling So Sad. Professional career began as ass't. to Robert Saudek on Ford Foundation's Omnibus TV series. Later became assoc. prod. and then dir. on Saudek's Profiles in Courage series. Then had dir. assignments on top series (Man from U.N.C.L.E., Dr. Kildare, The Big Valley, Felony Squad).

PICTURES INCLUDE: Downhill Racer, Prime Cut, The Candidate, Smile (also prod.), The Bad News Bears, Semi-Tough, An Almost Perfect Affair, The Island, Bette Midler's Divine Madness.

TELEVISION: Man from U.N.C.L.E., Dr. Kildare, The Big Valley, Felony Squad, Run for Your Life, The Outsider (pilot).

RITT, MARTIN: Dir. b. N.Y., N.Y. Mar. 2. e. Elon Coll., Burlington, Ky, Started as actor in Golden Boy. N.Y. stage; studied acting under Elia Kazan; dir. stage; Mr. Peebles and Mr. Hooker, The Man, Set My People Free, A View From the Bridge, N.Y.; acted in 150 and dir. 100 TV dramas.

TV: Danger.

PICTURES INCLUDE: Edge of the City (first screen dir.), No Down Payment, The Long Hot Summer, The Black Orchid, The Sound and the Fury, Adventures of a Young Man, Hud (also co-producer), The Outrage, Spy Who Came In From the Cold, Hombre, The Brotherhood, The Molly Maguires, The Great White Hope, Sounder, Pete 'n' Tillie, Conrack, The Front, Casey's Shadow, Norma Rae, Back Roads. Acted: End of the Game.

RITTER, JOHN: Actor. b. Hollywood. Father was late Tex Ritter, country-western star. Married, actress Nancy Morgan. Attended Hollywood High School. Interest in acting began at Univ. of So. Calif. in 1968. Appeared with college cast at Edinburgh Festival; later with Eva Marie Saint in Desire under the Elms. Gained fame as star of TV series, Three's Company.

PICTURES: Americathon, Hero at Large, Wholly Moses, They All Laughed.

TELEVISION: Leave Yesterday Behind.

RIVE, KENNETH: Executive. b. London, England, July 26, 1919. Early career as actor, radio compere, theatrical agent. Served in Intell. Corps. during WWII. After demob. theatre sup. and gen. man. cinema co. promoted dir. 1947. Started in continental exhibition forming Gala Film Distrib. Ltd., 1950. Now man. dir. Gala Film Dist., Carlton & Majestic Cinemas & Dance Hall Ltd., Cosmopolitan Film Dist. Ltd., Gala-Jacey Enterprises Ltd., Berkeley & Continentale (Holdings) Ltd., Gala TV Prod Ltd., Galaworldfilm Prod. Ltd., Gala Maintenance Ltd.; 1966, mgr. dir. Grade-Rive Ltd., Gala Co-Productions, Ltd., Gala Int'l Film & TV Enterprises, Ltd. Film Industry Defence Org., Malta United Film Corp. Ltd., Children's Film Foundation Ltd., International Films Theatre Ltd.

PICTURES INCLUDE: During One Night, The Boys, Devil Doll, Curse of Simba.

RIVERA, GERALDO: TV Reporter. b. N.Y.C., July 4, 1943. e. Univ. of Arizona, Columbia School of Journalism. Started legal career then switched to journalism, making several TV documentaries on such subjects as institutions for retarded, drug addiction, migrant workers, etc. Joined WABC-TV, New York, in 1970.

TV: Geraldo Rivera: Good Night America; Good Morning America (contributor). 20/20.

RIVKIN, ALLEN: Writer, producer b. Hayward, Wis., Nov. 20, 1903. Newspaperman, novelist, playwright. Authored 85 s.p. including Farmer's Daughter, Battle Circus, Prisoner of War, Joe Smith American, Eternal Sea, Big Operator.

TV: Prod. Troubleshooters series, 1960 Democratic Nat'l Convention; author (with Laura Kerr) Hello, Hollywood; past pres. scr. br. and currently dir. pub. rel., WGAW, liaison rep., Legal Advisory Committee on m.p. & TV of the American Bar Assn. Recipient of Morgan Cox and Valentine Davies awards.

RIVKIN, JOE: Production executive. b. Hartford, Conn. Exploit. dir., Pathe, also artist mgr. then to Hal Roach. In armed services Pres. Carthay Prod., v.p. gen. mgr. Wald & Krasna Prod., 1951; talent exec. Allied Artists Pictures. Now with Wm. Morris Agency.

RIX, BRIAN: C.B.E. Actor, manager. b. Yorkshire, England, 1924. e. Bootham Sch., York. Ent. m.p. ind. 1951.

TV: Farces, 1956–72, over 80. Men of Affairs (series); A Roof over My Head (series).
PICTURES INCLUDE: Reluctant Heroes, What Every Woman Wants, Up to His Neck, Dry Rot, Wanted on Voyage, And the Same to You, The Night We Dropped a Clanger, The Night We Got the Bird, Nothing Barred, Don't Just Lie There Say Something!

ROACH, HAL, JR.: President, owner Hal Roach Studios. Unit mgr. 20th Century-Fox, 1944–45; gen. mgr., Rainbow Prod., 1945–46. Member of Acad. of M.P. Arts & Sciences, M.P. Producers Assoc., Pres. Alliance of TV Film Producers; treas. & bd. member and former pres. of Acad. of TV Arts & Sciences; pres. Rabco Corp., Hal Roach Prod., chmn. exec. officer, dir., F. I. Jacobs, Detroit; past chmn. bd., Mutual Broadcasting System, bd. member, Vitapix Corp.; gen, exec., TV Prod., Seven Arts Associated.
PICTURES INCLUDE: Block Heads, A Chump at Oxford, One Million B.C., Road Show, All American Co-ed, Calaboose, Prairie Chickens, Army Training films, Physical Education & Military Training, Military Justice of Court Martials, Fighting Man series.
TV: Stu Erwin Show, Racket Squad, Public Defender, My Little Margie, Passport to Danger, Screen Directors Playhouse, Stories of John Nesbitt, Code 3, Charlie Farrell Show, Gale Storm Show, Blondie, Forest Ranger.

ROBARDS, JASON, JR.: Actor. b. Chicago, July 26, 1922. Served in Navy during WW II. Studied acting at Acad. of Dramatic Arts. Began with Children's World Theatre, radio parts, asst. stage mgr. on Stalag 17. First major break 1953 in play American Gothic, followed by The Iceman Cometh, Long Day's Journey into Night, Toys in the Attic. After the Fall, others; to Hollywood, 1958.
PICTURES INCLUDE: The Journey, By Love Possessed, Long Day's Journey Into Night, Tender is the Night, Act One, A Thousand Clowns, Any Wednesday, The St. Valentine's Day Massacre, The Night They Raided Minsky's, Hour of the Gun, The Loves of Isadora, Once Upon a Time in the West, Ballad of Cable Hogue, Fools, Johnny Got His Gun, Murders in the Rue Morgue, Pat Garrett and Billy the Kid, All the President's Men, Julia, Comes a Horseman, Hurricane, Raise the Titanic! Caboblanco.
TV: Washington: Behind Closed Doors.

ROBBIE, SEYMOUR MITCHELL: Producer, director.
TV: Studio One, Omnibus, Play of the Week, Jackie Gleason Show, The Man From U.N.C.L.E., Eleventh Hour, Dupont Show of the Week, F Troop, Mr. Roberts, Amos Burke, Honey West, Disney Wonderful World of Color, Felony Squad, Lost in Space, Name of the Game, Hawaii Five-0, It Takes a Thief, High Chapparal, Mannix, Mission: Impossible, Cannon (TV film), F.B.I., Mod Squad, Streets of San Francisco, Kojak, Moving On.
PICTURES INCLUDE: C.C. & Company, 1970, Marco.

ROBBINS, BURTON: Executive. Chm. & chief exec. officer, National Screen Service Corp. b. New York City, Feb. 2. e. New Rochelle, N.Y. public schools; Clark Prep, Hanover, N.H.; Vanderbilt University, Nashville, Tenn. Entered motion picture industry 1941 as trainee at National Screen Service Philadelphia branch. Subsequently served as salesman and asst. branch manager at NSS Albany, N.Y. and Detroit branches, respectively. Entered U.S. Army Air Corps in January, 1942, Anti-Aircraft Artillery O.C.S. Wilmington N.C. October 1943, U.S. Army Air Corps European theatre June 1944 to September 1945. Separated from Army of U.S. as First Lieutenant, December 1945. Rejoined NSS January 1946, subsequently appointed head of advertising accessories division. 1953, appointed asst. to the pres., and elected vice-pres. in charge of sales in September 1955. Elected president and chief executive officer, August 1963, and chairman of the board and chief executive officer February, 1979. Past Motion Picture Industry chairman Federation of Jewish Philanthropies; past trustee of Central Synagogue in New York City; past vice-pres. & current director, Will Rogers Memorial Fund; past treas. and current vice-pres. & director, Foundation of Motion Picture Pioneers; past pres., Cinema Unit B'nai B'rith in New York City; and currently int. chm., Variety Clubs International.

ROBBINS, MATTHEW: Writer-Director. e. Univ. of S. Calif. School of Cinema. Writes scripts in collaboration with Hal Barwood, Robbins branching out into directing also with Corvette Summer in 1978.
PICTURES INCLUDE: Scripts, all with Barwood: The wsugarland Express, The Bingo Long Traveling All-Stars and Motor Kings, Corvette Summer (also dir.); Dragonslayer (also dir.).

ROBERT, PATRICIA HARRISON. Executive. b. Atlanta, Ga., March 31, 1939. e. Manhattanville College of the Sacred Heart, Ecole Francaise (Paris), U. of Virginia Graduate School. Dir. of pub. & pub. relations, Gerald Rafshoon Advertising, 1965–69; drama critic, Atlanta Magazine, 1965–68; drama critic, feature writer, The Atlanta Constitution, 1968–69; asst. to publicity director, The Walter Reade Organization, 1969–70; director of publicity—public relations, Radio City Music Hall, 1970. Became Director of Advertising—Public Relations July, 1973.

ROBERTS, BEN: Writer. r.n. Ben Eisenberg. b. New York, N.Y. Mar. 23, 1916. e. N.Y.U., 1932–35. Free lance pub. rel. counsel, 1938; collab. on Broadway shows, Merry Widow, Jackpot, Dream with Music, Portrait in Black.
PICTURES INCLUDE: (collab. s.p.) White Heat, Goodbye My Fancy, Captain Horatio Hornblower, Come Fill the Cup, White Witch Doctor, O. Henry's Full House, King of the Khyber Rifles, Green Fire, Serenade.

ROBERTS, CURTIS: Producer. b. Dover, England. e. Cambridge U. Child actor. England, Germany; numerous pictures for Rank Org.; prod. England, on Broadway in Gertie, Island Visit; co-prod. on Broadway, Horses in Midstream, Golden Apple, Tonight or Never; tour and N.Y. 1965, The Journey, B'way, 1965.
TV: Rendezvous, Deadly Species, Top Secret, The Ilona Massey Show, When In Rome, Ethan Frome, Black Chiffon, Monaco.
PICTURES INCLUDE: An Actress in Love, La Die, Hypocrite, Jet Over the Atlantic, The Vixen, Farewell Party, Polly's Return, Rain Before Seven, Halloween, Malaga, My Dear Children, Bus Stop, Eve Arden Show, Rose-Marie Show.
BOOKS: History of Summer Theatre, The History of Vaudeville, Other Side of the Coin, 1969, History of Music (Popular) 1900–70. 1970, History of English Music Halls (1972), Latta (1972). Then There Were Some (1979).
TOURS: Blithe Spirit, Showboat, Kiss Me Kate, Generation, The Camel Bell, Farewell Party, Twentieth Century, Great Sebastians, Goodbye Charlie, Time of the Cuckoo, Under Papa's Picture, Everybody's Gal, Divorce Me Darling, Gingerbread Lady, September Son, 1975, Same Time Next Year.

ROBERTS, LYNNE: Actress (also known as Mary Hart); b. El Paso, Tex., Nov. 22, 1922. e. Lawler Prof. Sch. On stage prior to screen debut in Dangerous Holiday, 1937.
PICTURES INCLUDE: My Buddy, Big Bonanza, Girls of the Big House, The Chicago Kid, Winter Wonderland, Eyes of Texas, Sons of Adventure, Timber Trail, Trouble Preferred, Madonna of the Desert, Secret Service Investigator, Call of Klondike, Blazing Sun, Great Plane Robbery, The Blazing Forest, Because of You, Port Sinister.

ROBERTS, MARGUERITE: Writer. b. Clarks, Neb. m. John Sanford, novelist.
PICTURES INCLUDE: Honky Tonk, Dragon Seed, Ziegfeld Girl, Sea of Grass, If Winter Comes, Ambush, Escape, Soldiers Three, Diamond Head, Love Has Many Faces, Five Card Stud, Norwood, Shootout, Red Sky At Morning, True Grit.

ROBERTS, STANLEY: Writer. b. New York, N.Y. May 17, 1916. e. C.C.N.Y.
PICTURES INCLUDE: Sing While You're Able, Thanks for Listening, Anything for a Thrill, Under Western Skies, Penthouse Rhythm, Song of the Thin Man, Louisa, Curtain Call at Cactus Creek, Up Front, Death of a Salesman, Story of Will Rogers, Caine Mutiny.

ROBERTS, WILLIAM: Writer-Producer. b. Los Angeles. e. USC.
PICTURES INCLUDE: You For Me, Fast Company, Easy To Love, Her 12 Men, Private War of Major Benson, The Sheepman, The Mating Game, The Magnificent Seven, Wonderful World of the Brothers Grimm, Come Fly With Me, The Devil's Brigade, The Bridge At Remagen, One More Train to Rob, Red Sun, The Last American Hero, Posse.
TV: head writer, Life With Father; created Donna Reed Show.

ROBERTSON, CLIFF: Actor. b. La Jolla, Calif., 1925.
STAGE: Mr. Roberts, Late Love, The Lady and the Tiger, The Wisteria Tree, Orpheus Descending.
TV: Philco-Goodyear, Studio One, Robert Montgomery presents; Man Without a Country (movie); Washington: Behind Closed Doors.
PICTURES INCLUDE: Picnic, Autumn Leaves, Battle of the Coral Sea, As the Sea Ranges, Underworld, USA, The Big Show, Gidget, All in a Night's Work, The Interns, PT 109, The Best Man, 633 Squadron, Masquerade, The Honey Pot, The Devil's Brigade, Charly, (Academy Award, Best Actor, 1969), Too Late the Hero, The Great Northfield, Minnesota Raid, J. W. Coop (director, actor); Man On a Swing, Three Days of the Condor, Midway, Shoot.

ROBERTSON, DALE: Executive, actor, producer (Star of Tomorrow, 1951), r.n. Dayle; b. Oklahoma City, Okla., July 14, 1923; e. Oklahoma Military Coll. Prof. prizefighter; U.S. Army Sept. 1942–June 1945; m.p. debut in Fighting Man of the Plains.
PICTURES INCLUDE: Caribou Trail, Two Flags West, Call Me Mister, Take Care of My Little Girl, Golden Girl, Lydia Bailey, Return of the Texan, Outcasts of Poker Flat, O. Henry's Full House, Farmer Takes a Wife, Gambler from Natchez, Sitting Bull, Son of Sinbad, Day of Fury, Law of the Lawless, Blood on the Arrow, The Walking Major, The

Coast of Skeleton, The One-Eyed Soldier, The Last Ride of the Daltons, Dakota Incident, View from the Terrace, Fast and Sexy, Hell's Canyon.
TELEVISION: (series) Wells Fargo, The Iron Horse, Death Valley Days, Hollywood Palace. (films) Melvin Purvis, Kansas City Massacre.

ROBERTSON, F. J.: Producer. b. London, England, 1916. Early career as pianist, musical comedy, writer (composer) stage revues, pantomime. Theatre manager. Ent. TV 1954 as writer; created Goldhawk Studios of which man. dir. 1964, prod. The Little Ones, Columbia release.

ROBIN, DANY: Actress. b. Paris, France. Dancer since child. Played at the opera; acted on stage in Paris, then m.p.
PICTURES INCLUDE: Thirst of Men, Naughty Martine (L'Eventail); American Language debut in Act of Love, 1954. Holiday for Henrietta, Topaz.

ROBIN, LEO: Lyricist. b. Pittsburgh, Pa., April 6, 1900. e. U. of Pittsburgh. Became actor in New York after some time in Carnegie Tech. drama dept., then newspaperman publicist.
STAGE: Girl in Pink Tights, Hit the Deck, Greenwich Village Follies, Judy, Bubbling Over; to Hollywood 1929, formed musical partnerships with Richard Whitting, Ralph Rainger.
SONGS: Hallelujah, June in January, Louise, Prisoner of Love, Beyond the Blue Horizon, My Ideal, No Love No Nothin'.
PICTURES INCLUDE: Innocents of Paris, Monte Carlo, Little Miss Marker, Big Broadcast of 1935, Big Broadcast of 1937, Big Broadcast of 1938 (Award 1938 for best song, Thanks for the Memory), Waikiki Wedding, Paris Honeymoon, Gulliver's Travels, Coney Island, Gang's All Here, My Gal Sal, The Time The Place and the Girl, Just for You, Meet Me After the Show, Small Town Girl, Latin Lovers, Hit the Deck, My Sister Eileen.

ROBSON, FLORA: Actress. b. South Shields, England, March 28, 1902. e. Palmers Green High School, R.A.D.A. Before film career was factory welfare worker and on leg. stage.
PICTURES INCLUDE: Dance Pretty Lady, One Precious Year, Catherine the Great, Fire Over England, Farewell Again, Wuthering Heights, The Lion Has Wings, Poison Pen, We Are Not Alone, Invisible Stripes, The Sea Hawk, Bahama Passage, Saratoga Trunk, Great Day, Years Between, Black Narcissus, Frieda, Holiday Camp, Caesar and Cleopatra, Saraband for Dead Lovers, Good Time Girl, Tall Headlines, Malta Story, Romeo and Juliet, High Tide at Noon, No Time for Tears, The Gypsy and the Gentleman, Innocent Sinners, Murder at the Gallop, Guns at Ratasi, Young Cassidy, Those Magnificent Men in Their Flying Machines, Seven Women, The Kings Story, The Eye of the Devil, The Shuttered Room, Cry on the Wind, Fragment of Fear, Cry in the Wind, The Cellar, The Beloved, Dominque, Les Miserables, Clash of the Titans.
STAGE: Black Chiffon (1959), The Importance of Being Ernest, Ring Around the Moon.
TV: Corn is Green, Message for Margaret, The Return, Humanity, The Untouchable, Mother Courage, The Gentle Shade, The Human Jungle, The Misunderstanding, David Copperfield, The Old Ladies, Brighton Belle, Something for the Children, Heidi, Mr. Lollipop, The Shrimp and The Anemone, The Oresteia of Aeschylus, The Man Called Intrepid, Rebecca.

"ROCHESTER": See Eddie Anderson.

ROCKERT, JOHN F.: Executive. b. Kansas City, Mo., Oct. 29, 1924, e. Univ. So. Calif. In U.S. Navy, 1942–45. Joined Universal Pictures in 1951; left in 1957 to start independent productions. From 1960 to 1968 handled indep. roadshow distribution (4-walling). In 1969 formed CineWorld Corporation, natl. dist. co., of which he is pres. In 1975–76 did tax shelter financing for 13 films. Currently involved in distribution, production packaging and intl. co-production.

RODGERS, GABY: Actress. b. Frankfort, Germany, Mar. 29, 1928; e. Mt. Holyoke Coll., 1944–48. Barter Theatre Award, 1950; summer stock, many TV shows.
TV: Suspense, Philco Playhouse, Danger, Robert Montgomery, Pulitzer Prize Playhouse, Kraft TV Theatre, Omnibus.
PICTURES INCLUDE: Kiss Me Deadly.

ROEG, NICHOLAS: Director-Cameraman. b. London Aug. 15, 1928. Entered film industry through cutting rooms of MGM's British Studios, dubbing French films into English. Moved into prod. as clapper boy and part of photographer Freddie Young's crew. Next became camera operator (Trials of Oscar Wilde, The Sundowners). Had first experience as cameraman on TV series (Police Dog and Ghost Squad). Feature film debut as cameraman on Doctor Crippen. Debut as director on Performance; co-directed with Donald Cammell. First solo dir. film, Walkabout.
PICTURES INCLUDE: Cameraman: The System, Every Day's a Holiday, The Caretaker, Nothing But the Best, A Funny Thing Happened on the Way to the Forum, Fahrenheit 451, Far from the Madding Crowd, Petulia. Director-Cameraman: Performance (co.-dir.), Walkabout, Don't Look Now, The Man Who Fell To Earth, Bad Timing.

ROGELL, ALBERT S.: Producer and director. b. Oklahoma City, Okla., Aug. 21, 1901; e. public school and h.s., Spokane, Wash., Washington State Coll.; m. Irma Warner; camera man; cutter, titler, author, dir. Has been assoc. with First Nat'l, Universal, FBO and Tiffany. Ent. m.p. ind. 1916 with Washington M.P. Co., Spokane. Went to Hollywood, 1917, as assist. to George Loane Tucker, prod. of The Miracle Man; joined Andrew J. Callahan Prod. In 1921 made Hollywood's first coop. film, costing $1,200 which C.B.C. bought; prod. Hallroom Boys series for Columbia. Then joined Western Pictures Exploitation Co. In 1925 joined Universal for two years, dir. 8 pic. a year; to First Nat'l & dir. Shepherd of the Hills, Aloha, Mamba, Tiffany Tip Off.
PICTURES INCLUDE: Magnificent Rogue, Earl Carroll, Sketchbook, Heaven Only Knows, Song of India, Northwest Stampede, Admiral Was a Lady, Shadow of Fear.
TV: Ford Theatre, 20th Century Hour, Broken Arrow, etc.

ROGERS, CHARLES (BUDDY): Actor. b. Olathe, Kan., Aug. 13, 1904; m. Mary Pickford; p. Maude & Bert Henry Rogers; e. U. of Kansas, and was trained for screen in Paramount Picture School. Appeared in Fascinating Youth and others. In armed services WW II. In 1945 named v.p. & treas. Comet Prods., Inc. Assoc. prod. Sleep My Love, 1950, pres. PRB, Inc., prod. radio, video shows.
PICTURES INCLUDE: Wings, My Best Girl, Get Your Man, Abie's Irish Rose, The Lawyer's Secret, Road to Reno, Working Girls, This Reckless Age, Best of Enemies. Fox: Take a Chance, Dance Band, Old Man Rhythm, One In a Million, Let's Make a Night of It, This Way Please, Golden Hoofs, Mexican Spitfire's Baby, Sing for Your Supper, Mexican Spitfire at Sea, Mexican Spitfire Sees a Ghost, Don't Trust Your Husband.

ROGERS, FRED: Entertainer & producer. b. Latrobe, PA, 1928. e. Rollins College (BA, music composition). In 1951 served as asst. prod. of NBC-TV's The Voice of Firestone and NBC-TV Opera Theatre. Later promoted to network floor dir., supervising Your Lucky Strike Hit Parade, Kate Smith Hour, etc. In Nov., 1953, joined WQED-TV in Pittsburgh, educational TV station, to handle programming. In 1954 started Children's Corner series, writing, producing and performing; it ran 7 years. In 1963 was ordained minister of United Presbyterian Church dedicated to working with children and families through TV. Same year introduced character of Mister Rogers on Canadian Bdctg. Corp. of 15-min. daily program. Ran for one year—was similar in content to Neighborhood of Make-Believe segments on present half-hour program, Mister Rogers' Neighborhood. In 1964 programs were incorporated into larger, half-hour format on ABC affiliate in Pittsburgh. In 1966, 100 programs acquired by Easter Educational Network, broadcast in Pittsburgh, and seen for first time in other cities. Program now carried over 250 PBS stations.
AWARDS: George Foster Peabody, NET Special, Saturday Review TV, General Federation of Women's Club, Ralph Lowell, Gabriel, ten honorary doctoral degrees from universities. Emmy—(1980) Outstanding Individual Achievement in Children's Programming; Abe Lincoln Distinguished Communications Recognition Award.

ROGERS, GINGER: Actress. r.n. Virginia Katherine McMath; b. Independence, Mo., July 16, 1911. On stage in vaudeville, m.p. theat. presentations & musical comedy (Girl Crazy;). On screen from 1930. Academy Award, best performance by an actress. 1940 (Kitty Foyle, RKO Radio); voted among ten best Money-Making Stars in M.P. Herald-Fame Poll 1935, '37. In numerous pictures including Young Man of Manhattan, Gold Diggers of 1933, and others.
PICTURES INCLUDE: 42nd Street, Flying Down to Rio, The Gay Divorcee, Top Hat, Swing Time, The Story of Irene & Vernon Castle, Having Wonderful Time, Bachelor Mother, Tom Dick & Harry, Vivacious Lady, Stage Door, Primrose Path, Kitty Foyle, Roxie Hart, The Major & the Minor, Once Upon a Honeymoon, Lady in the Dark, Tender Comrade, I'll be Seeing You, Week-End at the Waldorf, Heartbeat, Magnificent Doll, It Had to Be You, Barkeleys of Broadway, Perfect Strangers, Groom Wore Spurs, Storm Warning, We're Not Married, Dream Boat, Monkey Business, Forever Female, Black Widow, Twist of Fate, Tight Spot, First Traveling Saleslady, Oh! Men, Oh! Women, Teenage Rebel, Harlow, The Confession.
TV: Como Show, Pontiac, Pat Boone, Dinah Shore, Bob Hope, Ed Sullivan, Hollywood Palace, Chrysler, Steve Allen, Jack Benny, Cinderella.

ROGERS, HENRY C.: Publicist. b. Irvington, N.J., April 19, 1914; e. U. of Pa., 1934, bd. chm., Rogers & Cowan, Inc.

ROGERS, JOHN W.: Producer. b. Buffalo, N.Y. 1916; p. Charles R. Rogers, prod.; e. U.S.C., 2nd asst. dir., 20th-Fox, then

asst. dir., 5 yrs.; assoc. prod., Pine-Thomas prod.; U.S. Army, 4 yrs.; indep. Prod.; assoc. prod., U-I, 1951; prod., 1952.
PICTURES INCLUDE: Powers Girl, Fabulous Dorseys, Buccaneer's Girl, Iron Man, Raging Tide, Here Come the Nelson's, Red Ball Express; prod., Spirit of West Point, Law and Order, War Arrow, Ride Clear of Diablo, Black Horse Canyon.

ROGERS, LAWRENCE H., II: Executive, b. Trenton, N.J. Sept. 6, 1921; e. Princeton U. 1942, U.S. Army, 1942–1946; WSAZ, Huntington, W. Va. Radio & TV, V.P. & Gen. Mgr., 1949–55; WSAZ, Inc., President, 1955–59; Taft Broadcasting Co., v.p., 1959–63; Taft Broadcasting Co., President, 1963–1976. Vice Chairman, Hanna-Barbera Productions, L.A., Calif. and Cinemobile Systems, Hollywood. Director: Cine Artists International, Hollywood; Cincinnati Financial Corp.; Inter-Ocean Insurance Co., Cinti.; Cardinal Fund, Ohio; Federal Reserve Bank of Cleveland, Cincinnati Branch; Theater Development Fund, New York; Greater Cincinnati Foundation; Rockford College, Rockford, Ill.

ROGERS, PETER: Executive. b. Rochester, Eng., Feb. 20, 1916; e. Kings School, Rochester, Journalist and in theatre and BBC; joined G. W. H. Productions 1941 as script writer; with Gainsborough Studios; asst. scenario ed. to Muriel Box; assoc. prod.; personal asst. to Sydney Box 1949.
PICTURES INCLUDE: Dear Murderer, Holiday Camp, When the Bough Breaks, Here Come the Huggetts, Huggetts Abroad, Vote for Huggett, It's Not Cricket, Marry Me, Don't Ever Leave Me, Appointment with Venus (Island Rescue), The Clouded Yellow, The Dog and the Diamonds (Children's Film Found.), Up to His Neck, You Know What Sailors Are, Cash on Delivery, To Dorothy A Son, Gay Dog, Circus Friends, Passionate Stranger, After the Ball, Time Lock, My Friend Charles, Chain of Events, Carry on Sergeant, Flying Scott, Cat Girl, Solitary Child, Carry On Teacher, Carry On Nurse, Carry On Constable, Please Turn Over, Watch Your Stern, The Tommy Steele Story, The Duke Wore Jeans, No Kidding, Carry On Regardless, Raising the Wind, Twice Around the Daffodils, Carry on Cruising, The Iron Maiden, Nurse on Wheels, Carry on Cabby, This is My Street, Carry On Jack, Carry On Spying, Carry On Cleo, The Big Job, Carry on Cowboy, Carry on Screaming, Don't Lose Your Head, Follow that Camel, Carry on Doctor, Carry on Up the Khyber, Carry on Camping, Carry on Assault, Carry on Henry, Quest, Revenge, Carry on At Your Convenience, All Coppers Are . . . Carry on Matron, Carry on Abroad, Bless This House, Carry on Girls, Carry on Dick, Carry on Behind, Carry on England, The Best of Carry On, Carry on Emmannuelle.
TV: Ivanhoe series, Carry on Laughing.

ROGERS, RICHARD H.: Executive. b. New York, Jan. 26, 1926. e. Colby Coll., Waterville, Me. Served in U.S. Army during WW II with Engineers. Joined Ziv TV in research and asst. to film ed. on Yesterday's Newsreel, 1948; sales dept., 1949; Flamingo Films, handling midwest sales, 1950; R. Monroe Prod., sales mgr. of TV properties, 1952; southeastern sales mgr. Sterling TV, 1954; PR dept., Shell Oil, 1955; acct. exec. Modern Talking Picture Service, 1957; elected v.p. and eastern sales mgr., 1962; v.p. for theatre division, 1965. Pres. of RHR Filmedia, Inc.

ROGERS, RODDY: Producer. b. Philadelphia; e. St. George's School. R.I.; U. of Pa. asst. acct. exec., Gray & Rogers Adv. Agency, Phila., 1946–48; stage mgr. and lighting coordinator at WFIL-TV, 1948; operations correlator, 1949; prod., dir., 1950; exec. prod. for WFIL-TV, 1951, prod. dir. Paul Whiteman TV Teen Club for ABC and Youth on the March, both on ABC-TV; dir. TV prod., WFIL-TV, 1953, mgr. radio, TV prod., Ward Wheelock Co., New York, 1954; prog. prod., radio-TV account supervisor N. W. Ayer & Son, 1955, N.Y.; chge. network sup. Ayer, N.Y. 1958; v.p., mgr. home office Ayer, 1959, pres. Agency Services Co.; pres. E. H. Rogers & Co., 1969.

ROGERS, ROY: Actor. Family name originally Leonard Slye. b. Cincinnati, O. Nov. 5, 1911. Radio singer; many m.p. from 1937. Voted No. 1 Money-Making Western Star in M.P. Herald-Fame, 1943–54 inclusive; also voted one of ten best money-making stars in 1945, '46. Acting & prod. TV films, 1952 with wife, Dale Evans; one-hour spectaculars, Chevy Show, 1959–60; contracted for several TV specials and for nationwide appearances with Roy Rogers touring show in Canada & U.S., 1962; State Fairs, Rodeo's since 1962; TV series.
PICTURES INCLUDE: Under Western Stars, The Old Barn Dance, Rep. in 1938; Billy the Kid Returns, Come On Rangers, Rough Riders, Round-Up, Frontier, Pony Express, Southward Ho!, in Old Caliente, Wall Street Cowboy, Heart of the Golden West, Sunset Serenade, Son of Paleface.

ROGERS, TED: Producer. b. Cleveland, O., Oct. 21, 1920. e. Cornell U.; dir. prog.; prod. I.T.C.

ROGERS, WILL, JR.: Actor. b. New York, N.Y. Oct. 20, 1912; p. late Will Rogers, actor; e. Stanford U., 1935. Publisher, ed., Beverly Hills Citizen; for. corresp.; elected congressman from Calif.; U.S. Army, WW II; m.p. debut in Story of Will Rogers, since in Eddie Cantor Story, Boy From Oklahoma.

ROHM, MARIA: Actress. b. Vienna, Austria, 1949. Early career with Burgtheater in Vienna. Ent. TV 1963, Films 1966. Films incl: Million Eyes of Sumuru, Five Golden Dragons, House of a Thousand Dolls, 99 Women, Blood of Fu Manchu, Justine, Venus in Furs, Eugenie, Black Beauty, Treasure Island, Call of the Wild, The Assassin Is Not Alone, Ten Little Indians, End of Innocence. Flight to Hell, TV Series, Quatermain.

ROLAND, GILBERT: Actor. r.n. Luis Alonso. b. Juarez, Mexico, Dec. 11, 1905; p. Mr. & Mrs. Francisco Alonso, father a bull-fighter in Spain. e. private schools in Mexico.
PICTURES INCLUDE: Captain Kidd, Pirates of Monterey, Dude Goes West, Malaya, We Were Strangers, Crisis, The Furies, The Torch, Bullfighter and Lady, Mark of Renegade, Ten Tall Men, My Six Convicts, Glory Alley, Miracle of Fatima, Apache War Smoke, Bad & the Beautiful, Thunder Bay, Diamond Queen, Beneath the 12-Mile Reef, French Line, Underwater, The Racers, That Lady, Treasure of Pancho Villa, Guns of the Timberland, The Wild Innocents, Eyes of Father Thomasino, The Big Circus, Samar, Cheyenne Autumn, The Reward, High Chaparral, Christian Licorice Store, Islands in the Stream, Caboblanco, Barbarosa.
TV: Bonanza, Alfred Hitchcock Presents, The FBI, Gunsmoke, The Fugitive, etc.

ROLEY, SUTTON: Director. b. Belle Vernon, Pa. Made dir. debut in live TV on Lights Out. Directed many live NBC-TV shows in "golden days" including Macbeth, with Judtih Anderson and Maurice Evans. Moved to L.A.: joined Ziv-UA and wrote, produced and directed Target. Was asst. prod. on Men of Annapolis.
PICTURES INCLUDE: Chosen Survivors.
TELEVISION: The Fugitive, Name of the Game, Gunsmoke, Rawhide, Mannix, Mission Impossible, Lost in Space, Bonanza, The Invaders, and many TV feature films.

ROLLE, ESTHER: Actress. b. Pompano Beach, Fla., Nov. 8. e. New School for Social Research. One of original members of Negro Ensemble Co. in N.Y. Has appeared both off and on Bdwy and in several TV series.
TELEVISION: Guest roles in N.Y.P.D., Like It Is, The Winners. Regular on series: Maude, Good Times. Film: I Know Why the Caged Bird Sings.

ROLLINS, JACK: Executive. Joined in partnership with Charles H. Joffe (see bio) to set up management-production org. Involved in production of many Woody Allen films with Joffe.
TV: Dick Cavett Show (exec. prod.).

ROMAN, LAWRENCE: Writer. b. Jersey City, N.J., May 30, 1921. e. UCLA, 1943.
PICTURES INCLUDE: (collab.) Drums Across the River, Vice Squad, (collab.) Naked Alibi, (collab.) One Desire, Man From Bitter Ridge, (s.p.) Kiss Before Dying, The Sharkfighters, (s.p.) Slaughter on Tenth Avenue, The Swinger, Author B'way play & collab. s.p. Under the Yum Yum Tree, Author B'way play P.S. I Love You. s.p. Paper Lion, collab. Red Sun, orig. s.p. A Warm December; wrote play, Buying Out, prod. in Buffalo, N.Y.; wrote play Crystal, Crystal Chandelier, prod. in Stockbridge, Mass. McQ, (orig. s.p.) The Mayflower Number; (orig. s.p.) Abracadabra (org. s.p), Skeletons (orig. s.p.); Lovers Three (orig. s.p.). Broadway play: If! If! If!, Omar Bradley (TV special movie).

ROMAN, RUTH: Actress. b. Boston, Mass., Dec. 23; p. professionals; e. Girls H.S., Boston; Bishop Lee Dramatic Sch. Started career with little theatre groups: New Eng. Repertory Co., Elizabeth Peabody Players. Screen debut in Universal serial, Queen of the Jungle, then minor roles; author stories, The Whip Son, The House of Seven Gables.
PICTURES INCLUDE: Good Sam, Belle Starr's Daughter, Whip Son, House of Seven Gables, The Window, Champion, Barricade, Beyond the Forest, Always Leave Them Laughing, Colt .45, Three Secrets, Dallas, Strangers on a Train, Tomorrow is Another Day, Invitation, Starlift, Mara Maru, Young Man With Ideas, Blowing Wild, Far Country, Shanghai Story, Tanganyika, Down Three Dark Streets, Joe Macbeth, Bottom of the Bottle, Great Day in the Morning, Rebel in Town, Bitter Victory, Look in Any Window, Miracle of the Cowards (Spanish prod.), Love Has Many Faces.
TV: Naked City, Route 66, The Defenders, Breaking Point, Eleventh Hour, Producers Showcase, Dr. Kildare, The Long Hot Summer, Go Ask Alice (movie).

ROMERO, CESAR: Actor. b. New York City, Feb. 15, 1907; e. Collegiate Sch., Riverdale Country Sch. In U.S. Coast Guard, WW II. In 1927 on N.Y. stage.
PICTURES INCLUDE: Wintertime, Coney Island, Captain from Castile, Show Them No Mercy, Beautiful Blonde

from Bashful Bend, Deep Waters, That Lady in Ermine, Once a Thief, The Jungle, Lost Continent, FBI Girl, Happy Go Lovely, Scotland Yard Inspector, Prisoners of the Casbah, Shadow Man, The Americano, Vera Cruz, The Racers, Leather Saint, Ocean's 11, The Computer Wore Tennis Shoes, Madigan's Millions, Now You See Him, Now You Don't, The Spectre of Edgar Allan Poe, The Strongest Man in the World, Carioca Tiger, and The Story of Father Kino.

ROONEY, MICKEY: Actor. r.n. Joe Yule, Jr.; b. Brooklyn, N.Y., Sept. 23, 1922; son of Joe Yule & Nell Carter, vaudeville performers. U.S. Army, WW II; in vaudeville during early infancy with parents and others before m.p. debut and after; from age of 5 to 12 created screen version of Fontaine Fox newspaper comic character Mickey McGuire in series of short subjects of that title, also appeared in number of features (Not to be Trusted, Orchids and Ermine, The King, etc.). Adopting name of Mickey Rooney, ret. to vaudeville; resumed screen career 1934. Special Academy Award 1940; for Andy Hardy characterization; voted among first ten Money-Making Stars in M.P. Herald-Fame Poll. 1938–43.

PICTURES INCLUDE: The Strip, Sound Off, Off Limits, All Ashore, Slight Case of Larcency, Drive a Crooked Road, Bridges at Toki-Ri, Atomic Kid, Twinkle in God's Eye, Bold and the Brave, Magnificent Roughnecks, The Last Mile, Big Operation, Private Lives of Adam and Eve, Platinum High School, King of the Roaring 20's, Breakfast at Tiffany's, It's a Mad, Mad, Mad, Mad World, Everything's Ducky. The Secret Invasion, The Extraordinary Seaman, The Comic, The Cockeyed Cowboys of Calico County, Skidoo, Pulp, Richard, B.J. Presents, That's Entertainment!, The Domino Principle, Pete's Dragon, The Black Stallion, Arabian Adventure.

TV: Playhouse 90, Pinocchio, Eddie, Somebody's Waiting, The Dick Powell Theater, Mickey.

ROONEY, PAT: Performer stage, nightclubs, pictures. Producer. e. Denver Univ.; Santa Monica College; Univ. Cal. Los Angeles; Marquette Univ. Entertainer vaudeville, theatres, stage, TV and pictures. Captain U.S. Army Air Corps, during Korean War. Entertained troops Far East Commands. Producer, 1960, C.B.S. Films Inc., producing TV pilots and series. 1962, formed Pat Rooney Prods. with Del E. Webb, hotel and construction exec. 1963–68, producer for Jerry Buss Prods. c/o Paramount Pictures, Hollywood, making feature films.

MAJOR PICTURES INCLUDE: Dime With A Halo, Danger Pass, Caged, Law of the Lawless, Requiem for a Gunfighter, Bounty Killer, Young Once, Hells Angels, Fools, Christmas Couple, Black Eye, Jan and Dean.

ROOT, WELLS: Writer. b. Buffalo, N.Y. e. Yale, Drama ed., NY World; dramatic and film critic, Time mag.; fiction, articles, various magazines; many TV dramas, various programs.

PICTURES INCLUDE: I Cover the Waterfront, Tiger Shark, Bird of Paradise, Prisoner of Zenda, Magnificent Obsession, Texas Across the River.

ROSE, ALEX: Producer. r.n. Alexandra Rose. Started in m.p. distribution with Medford Films, marketing low-budget pictures. Later became asst. sls. mgr. for New World Pictures, working in all phases of distribution. In 1972 met Tamara Asseyev and joined her in co-producing, starting with Drive-In, released by Columbia.

PICTURES INCLUDE: With Tamara Asseyev: Drive-In, I Wanna Hold Your Hand, Big Wednesday, Norma Rae.

ROSE, DAVID: Composer, conductor. b. London, England, June 15, 1910. To U.S. 1914; studied music, Chicago Coll. of Music; pianist with Chicago orchestras; staff arranger for several radio stations; mus. dir., West Coast network; while in U.S. Army, comp. dir. of Winged Victory; songs include Holiday for Strings, Our Waltz.

PICTURES INCLUDE: Texas Carnival, Rich Young & Pretty, The Clown, Bright Road, Jupiter's Darling, Port Afrique.

TV: Musical dir., Red Skelton Show; composer-conductor, Bonanza. Emmy, Fred Astaire special.

ROSE, DAVID E.: Producer. B. Kansas City, Mo. 1895. Ent. industry in 1930 in assoc. Douglas Fairbanks; v.p. gen. mgr. United Artists Studios; same posts with Goldwyn. Prod. and dir. UA Corp. until 1938 when he went to England to become chmn., man. dir. of all Paramount companies in Great Britain; currently prod. as David E. Rose Productions, Inc.

PICTURES INCLUDE: Island of Desire, Sea Devils, End of the Affair, Port Afrique, The Safecracker, The House of the Seven Hawks.

ROSE, JACK: Writer. b. Warsaw, Poland, Nov. 4, 1911; e. Ohio U. 1934, B.A. m. Audrey Mary Rose, writer, prod. Paramount Pictures, L.A.

PICTURES INCLUDE: Ladies Man, Sorrowful Jones, The Great Lover, It's A Great Feeling, Pale Face, My Favorite Brunette, Road to Rio, Daughter of Rosie O'Grady, Always Leave Them Laughing, On Moonlight Bay, Riding High, I'll See You in My Dreams, Room for One More, April in Paris, Trouble Along the Way, Living it Up, Seven Little Foys, Houseboat, Five Pennies, Beau James, It Started in Naples, Double Trouble, Papa's Delicate Condition, Who's Got the Action?, Who's Been Sleeping in My Bed?, A Touch of Class, the Duchess and the Dirtwater Fox.

ROSE, JACK: Exhibitor. b. Chicago, May 23, 1897. e. U. of Chicago, Ph.D. Clerk, stenographer, private secretary, in Navy during WW I, and accountant. Was assist. br. mgr. for Selznick Pictures, Film Classics of Illinois, Warner, then slsmn. for Warners and assoc. with James Coston as film buyer for Coston booking circuit; formed own booking and buying exchange; secretary and treas. Indiana-Illinois Theatres, Inc.; also film buyer, Chief Barker, Variety Club of Illinois, 1946–47 Nat'l Canvassman, 1947–48; Intern'l Rep. 1950–53. Member: Covenant Club of Ill., Exec. Club of Chicago, Masons, American Legion; trustee, Chicago Sinai Congreg., M.P. Pioneers.

ROSE, JOHN C.: Writer, producer. b. Chicago, Ill., Nov. 7, 1905. e. Dartmouth College Free lance commercial art, sls, prom. & adv. N.Y. 1929–33; jr. acct. exec. Young & Rubicam, Inc., N.Y. 1933; bus. mgr. radio dept. Compton Adv., Inc. N.Y. 1934–36; story ed. & story research dir., Walt Disney, 1936–42; headed Disney survey, South America, State Dept.; W.W. II worked on Frank Capra's "Why We Fight" films, orig. "G.I. Movies," officer-in-chg. motion pictures & graphics, U.S. Army Education Program, 1942–44; v.p. chg. prod., Pathescope Co. of America, 1945; Indie Prods., Hollywood, 1945–51; Chief of Health Education Films, Korean War, U.S. Navy, 1951–53; consultant, Indie TV Prod., 1954–61; formed & pres. Nugget Prods., Inc. 1962; Organized Film Industry Development Co. of Iran, 1975. Ret., 1976.

PICTURES INCLUDE: The Incredible Mr. Limpet.

ROSE, REGINALD: Writer, b. New York, 1921. e. C.C.N.Y. Worked as clerk, publicist, Warner Bros.; adv. acct. exec., copy chief; US Air Force, WW II; first TV play, Bus to Nowhere, 1951; since then numerous TV plays, Studio One, other programs.

PICTURES INCLUDE: Crime in the Streets, 12 Angry Men, The Man in the Net, Somebody Killed Her Husband, The Wild Geese, The Sea Wolves.

TELEVISION: Studs Lonigan.

ROSE, STEPHEN: Executive. Entered m.p. industry in 1964 with Columbia Pictures; named adv. dir. In 1970 joined Cinema V Distributing, Inc. as dir. of adv.; left in 1971 to take post at Cinemation Industries, where was named v.p. and bd. member. In 1975 joined Paramount Pictures as dir. of adv.; promoted to v.p./adv. In 1979 formed Barrick Prods. with Gordon Weaver, of which co. he is pres.

ROSE, SYDNEY: Producer. b. 1939. Ent. ind. 1956 as freelance writer/photographer, 1959–61: publicity director Films de France Ltd. 1961–3: independent film publicist. 1963–8: Artistes' management company in association with Frankie Vaughan. 1968: Formed independent artistes' agency. 1970–4: Executive agent M.A.M. Agency Ltd. 1975: Independent film producer. 1977: Exec. prod., The Kids are Alright. 1979: Co-prod., Five Star Five.

ROSE, WILLIAM: Screen writer. b Jefferson City, U.S.A. Ent. m.p. ind. in Britain, 1947.

PICTURES INCLUDE: Gift Horse (Glory at Sea), Genevieve, The Maggie (High and Dry), Touch and Go, The Ladykillers, Man in the Sky, Davy, The Smallest Show on Earth, It's A Mad, Mad, Mad, Mad, World, The Russians are Coming, The Russians are Coming, The Flim Flam Man, Guess Who's Coming to Dinner, The Secret of Santa Vittoria.

ROSENBERG, FRANK P.: Producer. b. New York City, Nov. 22, 1913. e. Columbia U., N.Y.U. Joined Columbia 1929; writer m.p. & radio; exploit, mgr., 1941; apptd. national dir. adv., publicity, exploitation, Columbia Pictures Feb. 1944. Pub. dir. M.P. Victory Loan, 1945; dir. pub. Columbia Pictures Studios, Hollywood, Jan. 1946. Resigned 1947 to enter production. Co-prod. Man Eater of Kumaon. Collab. adapt., assoc. prod. Where the Sidwalk Ends.

PICTURES INCLUDE: Secret of Convict Lake, Return of the Texan, The Farmer Takes a Wife, King of the Khyber Rifles, Illegal, Miracle in the Rain, Girl He Left Behind, One-Eyed Jacks, Critic's Choice, Madigan, exec. prod., The Steagle, prod. The Reincarnation of Peter Proud; sole adaptation, Gray Lady Down.

TV: Exec. prod. and prod. for Schlitz Playhouse programs during 1957–58; exec. prod., Arrest and Trial, 1963–64; exec. prod. Kraft Suspense Theatre, 1964–65; v.p. MCA Universal 1964.

ROSENBERG, MARK: Executive. Started career in magazine publishing field in New York. Adv. exec. with Seiniger & Associates; agent in m.p. dept. of IFA (later became ICM). With literary agency of Adams, Ray, Rosenberg. In 1978 joined Warner Bros., as v.p., prod.; in 1980 promoted to snr. v.p. of prod.

ROSENBERG, RICK: Producer. b. Los Angeles. e. L.A. City College, UCLA. Started career in mail room of Columbia Pictures, then asst. to prod. Jerry Bresler on Major Dundee and Love Has Many Faces. Asst. to Col. v.p., Arthur Kramer. Was assoc. prod. on The Reivers and in 1970 produced first feature, Adam at Six A.M., with Bob Christiansen, with whom co-produced all credits listed below.
PICTURES INCLUDE: Adam at Six A.M., Hide in Plain Sight.
TV: Features: Suddenly Single, The Glass House, A Brand New Life, The Man Who Could Talk to Kids, The Autobiography of Miss Jane Pittman, I love You . . . Goodbye, Queen of the Stardust Ballroom, Born Innocent, A Death in Canaan, Strangers.

ROSENBERG, STUART: Director-Producer. b. New York City, 1927. e. New York University.
PICTURES INCLUDE: Murder, Inc., Cool Hand Luke, The April Fools, WUSA, Pocket Money, The Laughing Policeman, The Drowning Pool, Voyage Of The Damned, The Amityville Horror, Brubaker. Has made over 300 TV shows for such series as The Untouchables, Naked City, The Defenders, Espionage. Chrysler Theatre, Emmy Award, 1962

ROSENCRANS, LEO S.: Writer. b. Linesville, Pa. e. Findlay Coll., B.A., O.B., LHD. NBC, 1931–34; radio ed. U.S. Office of Education, 1935–36; commercial m.p. since 1937; v.p., Creative Dept., Jerry Fairbanks Prods. of Calif., since 1951; more than 500 m.p., radio and TV credits.

ROSENFELT, FRANK E.; Executive. b. Peabody, Mass., Nov. 15, 1921. e. Cornell Univ. (B.S.) & Cornell Law School (L.L.B.). Served as atty. for RKO Radio Pictures, before joining MGM in 1955 as member of legal dept. Appt. secty. in 1966. Named v.p., gen. counsel in 1969 and pres. in 1973. In 1974 also named chief exec. officer. As of May, 1980, is bd. chm. & ch ef exec. officer. Member: Bd. of Governers, Academy of M.P. Arts & Sciences, bd. of trustees, American Film Institute.

ROSENFIELD, JONAS, JR.: Publicity executive. b. Dallas, Texas. e. U. of Miami, A.B. In U.S. Navy, World War II. Warner Bros. pressbook copy department, adv. mgr. Walt Disney, Adv. copywriter Donahue & Coe. Advertising copy chief 20th Century-Fox. Pres. N.Y. Screen Publicists Guild. In July 1942, ex-officio member industry's War Activities Committee. In 1945 apptd. asst. adv. mgr., 20th Cent.-Fox; adv. mgr., 1949–51; dir. of pub. rel. for Italian Films Export 1952; v.p. chg. adv. prom., pub. IFE Releasing Corp., 1953–55; exec. asst. to Paul Lazarus; v.p., of Columbia Pictures, 1955; exec. in chg., avd., pub. expl., Columbia, 1958; v.p. in chg. adv. pub. expl. Columbia, 1960; elected v.p., Columbia Int'l Pictures Corp., 1962; elected gen. exec. officer, Columbia Pictures, 1962; vice president, worldwide advertising, publicity and promotion, Twentieth Century-Fox, 1963–77; film mktg. consultant, 1977–78; lecturer in mktg., University of So. Calif., 1978–79; v.p. in chg. of worldwide mktg., Melvin Simon Productions, 1979. In 1980 named snr. v.p.; 1981, joined Filmways Pictures as exec. v.p., worldwide adv./pub., promo.

ROSENSTEIN, GERTRUDE: Director. b. New York, N.Y. e. Barnard Coll., B.A. Neighborhood Playhouse. exec. asst. to George Balanchine & Lincoln Kirstein, NYC Ballet. Assoc. with Gian Carlo Menotti, Festival of Two Worlds, Spoleto, Italy.
TV: Assoc. dir., NBC Opera, Emmy Awards, election coverages, Kennedy inauguration, Huntley-Brinkley Report, Kennedy Memorial Mass; dir., Concentration; TV staff dir., NBC. Now free-lance director.

ROSENTHAL, BUD: Executive. b. Brooklyn, N.Y., Mar. 21, 1934. e. Brooklyn College, B.A. 1954, New York U.; U.S. Army, 1954–56; college correspondent, N.Y. Times; ent. m.p. ind. as associate editor, Independent Film Journal, 1957. Joined Columbia Pictures publicity dept. as trade paper contact and news writer, 1959, newspaper and syndicate contact 1960; appointed national publicity mgr., Columbia Pictures Corp., 1962–67; asst. prod. Something For Everyone; pub. dir., Anderson Tapes, Such Good Friends; story ed. and casting dir., Sigma Prods., 1972–75; associate producer, Broadway play, Full Circle, 1973, assoc. prod., Rosebud, 1974; dir. intl. press relations, The Bluebird, 1975; Warner Bros. project coordinator, Superman, 1977–79.

ROSENTHAL, ROBERT M.: Producer. b. N.Y.C., Dec. 28, 1936. e. Lawrence Academy, Mass., 1952–56. U. of Penn., The Wharton School, 1956–60. Pictorial Officer, Prod.-dir. for U.S. Army Signal Corp., 1960–62; Chief, U.S. Army Production Facilities, France, 1961–62. Production mgr., Gurney Productions Inc., 1963; comptroller, Jalor Productions Inc., 1964; pres. Rosenthal Productions Inc., 1964; prod. Lieut. Wolf, I Wonder Why, Been Down So Long It Looks Like Up To Me.

ROSI, FRANCESCO: Director. b. Naples, Italy, 1922. Apprenticed as asst. to Visconti and Antonioni; directed first feature La Sfida (The Challenge) in 1958.
PICTURES INCLUDE: Salvatore Giuiliano, Hands Over the City, More Than a Miracle, Just Another War, Lucky Luciano, The Mattei Affair, Eboli.

ROSIAN, PETER F.: Executive. b. Persia, Sept. 11, 1902; e. NYU. Commonwealth Pictures, N.Y., 1923; later office mgr.; booker, MGM, 1925, then salesman, spcl. sales rep, br. mgr., Grand Nat'l, Cleveland, 1937; N.Y. 1938, spcl. sales rep., Universal, 1939; br. mgr., Cincinnati, 1940; dist. mgr. hdqts., Cincinnati, 1941; dist. mgr. 1951, Cleveland; regional sales mgr. 1958, Cleveland. Retired from MCA, 1970. Organized Personal Film Research Co.

ROSMARIN, CHARLES: Manager. b. London, England, Jan. 27, 1911. Insurance agent; district mgr. Columbia Pictures 1931; mgr. Venezuela office 1944; mgr. Monogram Pictures, Argentina 1946; mgr. RKO Radio Pictures, Argentina, 1949; gen. sales mgr. Europe—Near East, hdqtrs. Paris 1953; gen. mgr. Europe, Near East, 1958; gen. mgr. Metro-Goldwyn-Mayer, Italy, 1961; European rep., RKO General Inc., 1968.

ROSS, DIANA: Singer-Actress. B. Detroit, Mich. Formed musical group at age 14 with two friends, Mary Wilson and Florence Ballard. In 1960 they auditioned for Berry Gordy, head of Motown Record Corp. and were hired to sing backgrounds on records for Motown acts. After completing high school, the trio was named the Supremes and went on tour with Motor Town Revue. Over period of 10 yrs. Supremes had 15 consecutive hit records and once had five consecutive records in no. one spot on charts. In 1969 Diana Ross went on her own, appearing on TV and in nightclubs. Made film debut in Lady Sings the Blues, 1973.
PICTURES INCLUDE: Lady Sings the Blues, Mahogany, The Wiz.
TV: Diana! (spec.).

ROSS, FRANK: Producer-writer. b. Boston, Mass., Aug. 12, 1904; e. Exeter and Princeton. President, Frank Ross Inc. In 1939, asst. prod. Of Mice and Men, UA, Roach. In 1941, producer, The Devil and Miss Jones, RKO. Co-author of story and s.p. The More the Merrier. Spec. Acad. Award (1945) for prod. The House I Live In, short subject on tolerance.
PICTURES INCLUDE: The Lady Takes a Chance, Flame and the Arrow, The Lady Says No, The Robe, Demetrius and the Gladiators, Rain of Ranchipur, Kings Go Forth, Mr. Moses, Maurie (prod.).

ROSS, HERBERT: Director, b. New York City, May 13. Resident Choreographer A.B.T., Choreographer on Broadway for I Can Get It For You Wholesale, Tovarich, Anyone Can Whistle, Do I Hear a Waltz, On a Clear Day You Can See Forever, The Apple Tree. Broadway Director: Chapter Two. I Ought To Be in Pictures. Ent. m.p. ind. as choreographer/musical sequences dir.
PICTURES INCLUDE: As Director/Choreographer: Funny Girl. As director: Goodbye Mr. Chips, The Owl and the Pussycat, Play It Again, Sam, The Last Of Sheila (produced and directed), Funny Lady, The Sunshine Boys, The Seven-Per-Cent Solution (produced and directed), The Turning Point (produced and directed), The Goodbye Girl, California Suite, Nijinsky, Pennies from Heaven (prod.-dir.).
TV: Bell Telephone Hour, Fred Astaire Special.

ROSS, KATHARINE: Actress. b. Los Angeles, Calif., Jan. 29, 1943. e. Santa Rosa College. Joined the San Francisco Workshop, appeared in The Devil's Disciple, The Balcony, 1962, poetry festival. TV debut, 1962 in Sam Benedict segment.
TV: Doctors at Work, World Premiere, The Longest Hundred Miles, Ben Casey, The Bob Hope-Chrysler Theatre, The Virginian, Wagon Train, Kraft Mystery Theatre, the Lieutenant, The Road West. Acted in the UCLA theatre Arts prod. of King Lear.
PICTURES INCLUDE: Shenandoah, Mister Buddwing, The Singing Nun, Games, The Graduate (nom. for Acad. Award, Best Actress; voted Most Promising Female Newcomer, Golden Globe Award), Hellfighters, Butch Cassidy and the Sundance Kid, Tell Them Willie Boy is Here, They Only Kill Their Masters, The Stepford Wives, Voyage of the Damned, The Betsy, The Swarm, The Legacy, The Final Countdown.

ROSS, KENNETH: Writer. b. London, Sept. 16, 1941.
TV: The Roundelay, ATV Network, 1963. The Messenger, CBC Network, 1966.
THEATRE: The Raft, Hampstead Theatre, London, 1964. Under The Skin, Citizens' Theatre, Glasgow, 1968. Mr. Kilt & The Great I Am, Hampstead Theatre, London, 1970.
PICTURES INCLUDE: Entered industry 1970. Brother Sun, Sister Moon, 1970. Slag, 1970. Original story and screenplay, The Reckless Years, 1971. Abelard & Heloise, 1971. The Day of the Jackal (nominated for Writers' Guild, SFTA, and Golden Globe Awards), 1971–72. The Devil's Lieutenant, 1972. The Odessa File (nominated for Writers' Guild Award), 1973–74. Original story and screenplay,

Quest, 1974–75. Original story and screenplay, Lorenzo The Magnificent, 1975.

ROSS, LANNY: Singer, M.C., Actor. b. Seattle, Washington, Jan. 19, 1906; e. Yale 1928. Columbia Law, Juilliard, 1933. AAU title 1927–28; qualified 1928 Olympics, Star of Showboat NBC 1932–37; retired from track to sing. Town Hall debut, 1936; Showboat, 1932–36. Radio; Packard Mardi Gras, Hit Parade Franco American, Camel Caravan; U.S. Army, Major 1943–46; Ivory Show. First musical show on TV Swift Show 1947–49. Concert Tour of U.S. 1949–54. WCBS Radio, The Lanny Ross Show 1954–61, Telephone Hour, 1963; Today Show, 1964; Les Crane Show, 1965; Joe Franklin, 1973; Salvation Army Banquet, Palm Beach, 1973–75. Records include Jolly Doctor Dolliwell, Silver Dollar Country Album.
PICTURES INCLUDE: Melody in Spring, College Rhythm, The Lady Objects, Stage Door Canteen.

ROSSO, LEWIS, T.: Executive. b. Hoboken, N.J., Feb. 3, 1911. Ent. m.p. ind. 1930; prod. & mgt. for Consolidated Film Ind., 1930–44; Republic Prod., 1944–50; prod. mgr. Republic, 1950–55; asst. sec'y and asst. treas. Republic Pictures Corp., 1959; exec. asst to exec. prod. mgr., 20th Century-Fox Films, 1960; plant mgr., Samuel Goldwyn Studios, 1961–71; exec. admn. asst. to gen'l mgr., The Burbank Studios, 1972–81.

ROTH, CY: Writer, producer, director. b. Chicago, Ill., Mar. 16, 1912; e. U. of Calif. at L.A. Formed and headed Cy Roth Productions, later formed and was pres. Saturn Films, Inc.
PICTURES INCLUDE: Flight to Outer Space, Caballero Jones, Spanish Territory, Top Secret, Solar Rescue, Crawling Hand, Stone Dragon, Battle at Silver Pass, Passport to Freedom, Danger at Ghost Town; dir. Combat Squad; prod. dir. s.p. Air Strike, Fire Maidens of Outer Space.
TV: Prod., dir., writer. Adventures of Rainbow Riley; writer, Central Allied Intelligence, Tokyo Police, Attack Squadron, The Twisted Black Cross; dir., Rheingold Theatre, Outlaw Queen; prod., dir., s.p., Attack Squadron, Nuremberg, Growing Wild; prod., dir. Horrors of the Black Forest. Heads C.R. Ent.

ROTH, PAUL A.: Executive, born Asheville, N.C., March 28, 1930. U. of N.C., A.B. Political Science 1948–51. George Washington Univ. Law School 1951–52. U.S. Army 1952–55. Dist. Mgr. Valley Enterprises, Inc. 1955–56. Vice Pres. Roth Enterprises, Inc. 1956–65. Pres. Roth Enterprises, Inc. 1965-present. President NATO of Virginia 1971–73. Chairman of Board NATO of Virginia, 1973–1975. Member National NATO Board, 1971-present. Executive Committee NATO of Metro-D.C. 1970-present. Variety Club Tent 11 Board Member 1959–65. President National NATO, 1973–75; chairman National NATO board of directors 1975-77. Co-Chairman Motion Picture Energy Conservation Council, 1973-present. Member Foundation Motion Picture Pioneers, 1973-present. Director & advisory committee, Will Rogers Hospital, 1973-present. Trustee American Film Institute, 1973–75.

ROTH, RICHARD A.: Producer. b. Bev. Hills, 1943. e. Stanford U. Law School. Worked for L.A. law firm before beginning film career as lawyer and literary agent for Ziegler-Ross Agency. In 1970 left to develop s.p. Summer of '42 with Herman Raucher.
PICTURES INCLUDE: Summer of '42, Our Time, The Adventures of Sherlock Holmes Smarter Brother, Julia, Outland.

ROTHA, PAUL: Producer and director. b. London, 1907. e. London U., Slade School of Art. Ent. film ind. in art dept. of British Int'l in 1928. Visited America 1937–38 and advised General Education Board (Rockefeller Foundation) and Museum of Modern Art Film Library, N.Y., regarding documentary film technique. Managing dir. Paul Rotha Productions, Ltd., London; ex-chmn., 1953 vice-chmn. fellow, British Film Academy.
AUTHOR: (film books) The Film Till Now, Celluloid Documentary Film, Movie Parade, Television in the Making, Rotha on the Film, The Innocent Eye, The Biography of Robert J. Flaherty, Head of documentary BBC-TV, 1953–55.
PICTURES INCLUDE: Contact, Shipyard, Face of Britain, Today We Live, Future's in Air, World of Plenty, Night Shift, Children of the City, Land of Promise, A City Speaks, The World is Rich, World Without End (with Basil Wright), No Resting Place, Cat and Mouse, Cradle of Genius, The Life of Adolf Hitler, The Silent Raid.

ROTHKIRCH, Dr. EDWARD V.: Producer. b. July 30, 1919; e. Friedrich Wilheim U., Berlin; Rockhurst Coll., Midwestern College. Prod. asst., research, Pan American Prod., 1941; research, Pacific Films, 1942; U.S. Air Force, 1942–44; asst. prod., Pan American Productions, 1945; analyst, Cambridge Prod., 1947; assoc. prod., Pentagon Films, 1949; assoc. prod., Reelestic Pictures, 1950; assoc. prod. Cambridge-Meran Prod. Co., 1951; assoc. exec. prod., Cambridge Prod., 1954; also v.-p. Continental Prod. Services; assoc. exec. prod. Trinity Hill Productions, produced Pan-American Highway 1954, The Keepers TV series, 1953–58, Famous Women of the Bible, 1955–58; To the Stars TV series, 1954–58; also sec.-treas. Crusader Records and v.p. Orbit Records. Member of many professional societies, director International Association of Independent Producers, presently Executive Producer—Galaxie Productions, and Encore Records also Executive Editor, Intercontinental Media Service, Ltd.

ROTHMAN, MO: Executive. b. Montreal, Canada, Jan. 14, 1919. Royal Canadian A.F., WW II. Student trainee, Universal, India; mgr., Malaysia; mgr. prod. Worldwide Rep. for Edward Small, 1950; intern'l sales dept. U.A., 1952; sales mgr. for Continent, Near East, 1955; Cont'l mgr., 1957; v.p. int'l oper., 1960; exec. v.p., Columbia Pict. Int'l, 1960; v.p. world distrib. 1966; exclusive distributor all media, Chaplin Pictures, 1971

ROUNDTREE, RICHARD: Actor. b. New Rochelle, N.Y. e. Southern Illinois University. Former model, Ebony Magazine Fashion Fair; joined workshop of Negro Ensemble Company, appeared in Kongi's Harvest, Man, Better Man, Mau Mau Room; played lead role in Philadelphia road company of The Great White Hope before film debut.
PICTURES INCLUDE: Shaft, Shaft's Big Score, Shaft in Africa, Charley One-Eye, Earthquake, Man Friday, Diamonds, Escape to Athena.
TV: Shaft (series, 1973), Firehouse (movie), Roots.

ROUSE, RUSSELL: Prod., dir., writer. b. New York City. ed. U.C.L.A. Formed Greene-Rouse Prods. with Clarence Greene. Acad. Oscar co-orig. story Pillow Talk. Acad. award nom. co-orig. s.p. The Well. 2 Writers Guild nominations. Writers Guild award outstanding teleplay: One Day In The Life of Ivan Denisovitch.
PICTURES INCLUDE: Prod., collab. s.p. The Town Went Wild. Orig. story Yokel Boy. Collab. orig. s.p. The Home Front, D.O.A. Co-author Pillow Talk. Dir. and collab. s.p. The Well, The Thief, Wicked Woman, New York Confidential, Fastest Gun Alive, House of Numbers, Thunder in the Sun, A House Is Not A Home, The Oscar, Dir. Caper of the Golden Bulls. Dir. and collab. Prod.-writer T.V. series: Tightrope.

ROWE, ROY: Owner-operator, Rowe Amusement Co., Burgaw, N.C. b. Burgaw, May 29, 1905; e. U. of No. Car. Eng. instructor, private bus. coll., 1926–29; Publix Sch. for Mgrs., N.Y., 1930–31; mgr. theatres, Spartanburg, S.C.; Greensboro & Raleigh, N.C.; mgr., Warner Theatre, Pittsburgh, Pa., 1931–34; city mgr for Warner Theatres, Washington, Pa., 1934–35; opened own theatres in N.C. 1935; member N.C. Senate, 1937, 1941, 1945, 1949, 1957, 1965; House of Rep., 1943; Major, Civil Air Patrol, WWII; pres. Carolina Aero Club, 1943–44; chmn. N.C. Aeronautics Comm., 1941–49; dir. Theatre Owners No. & So. Car. 1943–45; pres., Theatre Owners of S.C. & N.C. 1944–45; pres. Assn. of Governing Boards of State Universities, 1964. Owned and operated motel, Carolina Beach, N.C., 1965–67., Rowe Insurance Agency, 1967–69. Mem. Exec. Bd., U. of N.C. Trustees, 1969. Principal Clerk N.C. Senate 1969–75. Retired. Now water color artist and world traveller.

ROWLAND, ROY: Director. b. New York City, Dec. 31; e. U. of So. California (law). Script clerk; asst. dir.; asst. to late W. S. Van Dyke on Tarzan pictures; dir. of shorts, "How to" Benchley series; Crime Does Not Pay series. Pete Smith Specialties.
PICTURES INCLUDE: Think First, Stranger in Town, Lost Angel, Our Vines Have Tender Grapes, Tenth Avenue Angel, Night patrol, Ski Soldier, Boys' Ranch, Romance of Rosy Ridge, Killer McCoy, Scene of the Crime, Outriders, Excuse My Dust, Two Weeks With Love, Bugles in Afternoon, 5000 Fingers of Dr. T. Affair with a Stranger, The Moonlighter, Witness to Murder, Rogue Cop, Many Rivers to Cross, Hit the Deck, Meet Me in Las Vegas, Slander, Somewhere I'll Find Him, Gun Glory, The Seven Hills of Rome, The Girl Hunters, Gunfighters of Casa Grande, They Called Him Gringo, Tiger of the Seven Seas, Thunder Over the Indian Ocean.

ROWLAND, WILLIAM: Producer, director. b. Philadelphia, Pa., Sept. 12, 1900; e. U. of Pa. Actor: vaude., prod.; play prod.; author; prod. shorts, Universal, 1931–32; prod. features Moonlight and Pretzels, Take a Chance, Sweet Surrender, I Loved an Actress; prod., dir. in chge. Spanish prod., RKO, 1937; Spanish Acad. Award for best prod., dir., Perfidis, 1939; won highest Spanish awards for prod. dir., author, Harvest of Hate, 1940; prod. dir., Int'l Forum series, Follies Girl.
PICTURES INCLUDE: Song for Miss Julie, Flight to Nowhere, Women in the Night.

ROWLANDS, GENA: Actress. b. Cambria, Wisc. June 19, 1936. e. Univ. of Wisc. Came to New York to attend American Academy of Dramatic Arts, where met and married John Cassavetes. Made Bdwy. debut as understudy and then succeeded to role of the Girl in The Seven Year Itch. Launched as star with part in The Middle of the Night,

which she played 18 mos. Then brought to Hollywood to make film debut in The High Cost of Living (1958).
PICTURES INCLUDE: Lonely Are the Brave, The Spiral Road, A Child Is Waiting, Tony Rome, Faces, Minnie and Moskowitz, A Woman Under the Influence, Two Minute Warning, The Brinks' Job, Gloria.
TV: The Philco TV Playhouse, Studio One, Alfred Hitchcock Presents, Dr. Kildare, Bonanza, The Kraft Mystery Theatre, Columbo.

ROWLEY, JOHN H.: Executive. b. San Angelo, Tex., Oct. 6, 1917. e. U. of Tex., 1935–39. v.p. United Artists Theat. Cir. Inc., Rowley United Div.: past president, NATO of Texas; part Int'l Chief barker, Variety Clubs Int'l; past pres., TOA; pres., Variety Foundation of Texas.

ROYLE, SELENA: Actress. b. New York City, Nov. 6, 1904; d. Edwin Milton Royle & Selena Fetter, pro. Made debut on N.Y. stage, 1921 in Launcelot & Elaine. Has appeared in numerous N.Y. prod.; on radio starred in Woman of Courage, Kate Hopkins, and many others. On screen 1941 in Stage Door Canteen.
PICTURES INCLUDE: The Sullivans, Mrs. Parkington, This Man's Navy, Main Street After Dark, Courage of Lassie, Green Years, Gallant Journey, No Leave, No Love, Night and Day, Summer Holiday, Wild Harvest, Cass Timberlane, Smart Woman, You Were Meant for Me, Joan of Arc, You're My Everything, My Dream Is Yours, Bad Boy, Big Hangover, Branded, Dammed Don't Cry, The Heiress, He Ran All the Way, Come Fill the Cup, Robot Monster, Murder Is My Beat.

ROZSA, MIKLOS: Composer. b. April 18, 1907, Budapest, Hungary. e. Leipzig Conservatory. Wrote great number of symphonic and chamber music works. Composed music for many m.p. In 1936, Knight Without Armor, Acad. Award best music scoring for drama or comedy (Spellbound) 1945; best music scoring for drama or comedy (Double Life) 1947; Ben-Hur, 1959. Pres., Screen Composers Assn. 1956. Cesar of French Academy for Providence, 1978.
PICTURES INCLUDE: Jungle Book, Song to Remember, Double Idemnity, The Killers, Madame Bovary, The Lost Weekend, Spellbound, Asphalt Jungle, Quo Vadis, Ivanhoe, Julius Caesar, Story of Three Loves, Plymouth Adventure, Young Bess, Knights of the Round Table, A Time to Love and a Time to Die, The World the Flesh and the Devil, Ben Hur, Lust for Life, Something of Value, King of Kings, El Cid, Sodom and Gomorra, The VIP's, The Power, The Green Berets, The Private Life of Sherlock Holmes, Sinbad's Golden Voyage, Providence, Secret Files of J. Edgar Hoover, Fedora, Last Embrace, Time After Time, Eye of the Needle.

RUBIN, STANLEY: Producer, writer. b. New York, N.Y. Oct. 8, 1917; ed. UCLA, 1933–37. Phi Beta Kappa. Writer—radio, magazines, pictures, 1937–41; U.S. Army Air Force, 1942–45; writer, prod., owner, Your Show Time, Story Theatre TV series; producer, RKO, 20th-Fox, U.I.
PICTURES INCLUDE: The Narrow Margin, My Pal Gus, Destination Gobi, River of No Return, Destry, Francis in the Navy, Behind the High Wall, Rawhide Years, Promise Her Anything, The President's Analyst, The Take.
TV: Ghost and Mrs. Muir, Bracken's World, The Man and the City.

RUDDY, ALBERT S.: Producer. b. Montreal, Canada, March 28, 1934. e. U. of So. Calif., B.S. in Design, School of Architecture, 1956.
PICTURES INCLUDE: The Wild Seed (Pennebaker); prod.; Little Fauss & Big Halsey (Alfran Prods.), Paramount Studios Prod.; Making It (Alfran Prods.—for 20th Century Fox); prod. The Godfather (Albert S. Ruddy Production, Inc. for Paramount). 1974; The Longest Yard, Coonskin, both for Para. 1978: Matilda. 1981: The Cannonball Run.

RUDIE, EVELYN: Actress, Singer, Songwriter. r.n. Evelyn Rudie Bernauer, b. Hollywood, Calif. March 28. e. Hollywood H.S., U.C.L.A. At 19, after childstar career in TV and films, stage debut at Gallery Theatre in Hollywood as songwriter, musical director, choreographer and star performer: Ostrogoths and King of the Schnorrers. Currently Producer, Artistic Director at Santa Monica Playhouse; founder of own repertoire Company; among major productions: Backstreet, Dreamplay, and musicals Anatol-Anatol, Attorney at Love, Author-Author, The Alchemist, etc. Screen debut as child performer Daddy Longlegs. Received Emmy Nomination for first TV leading role, Eloise, Playhouse 90, 1956.
PICTURES INCLUDE: The Wings of Eagles, Gift of Love. Filmdom's Famous Fives critics award, 1958.
TV: Hostess with the Mostess, Playhouse 90, Dinah Shore, Red Skelton Show, George Gobel Show, Omnibus, Matinee Theatre, Hitchcock presents, Gale Storm Show, Jack Paar, Wagon Train, G.E. Theatre, 77 Sunset Strip, etc.

RUGOFF, DONALD S.: Executive. b. Brooklyn, N.Y., Feb. 7, 1927. e. Woodmere Academy, Harvard; U.S. Navy, 1944–46; dir. Motion Picture Theatres Assoc.; pres. Cinema 5 Ltd.

RUGOLO, PETE: Composer, arranger. b. Sicily, Italy, Dec. 25, 1915. To U.S., 1919; e. San Francisco State Coll., Mills Coll., Oakland. Armed Forces, 1942–46; pianist, arr. for many orch. including Stan Kenton; m.p. and TV.
PICTURES INCLUDE: The Strip, Skirts Ahoy, Glory Alley, Latin Lovers, Easy to Love, Jack the Ripper.
TV: Richard Diamond, The Thin Man, Thriller.

RULE, ELTON H.: Executive. b. Stockton, CA, 1917. e. Sacramento Coll. With Amer. Bdg. Cos., Inc., since 1952; gen. sls. mgr., KABC-TV, 1953–60; gen. mgr., 1961–8; pres., ABC TV Network 1968–70; group v.p. Am. Bdg. Cos. Inc., 1969–72; pres., ABC div. 1970–72; pres., chief operating officer, mem. exec. comm., Amer. Bdg. Cos. since 1972.

RULE, JANICE: Actress. b. Cincinnati, O., Aug. 15, 1931; e. Wheaton & Glenhart H.S., Glen Ellyn, Ill. Dancer 4 yrs. in Chicago & New York nightclubs; stage experience in It's Great To Be Alive, as understudy of Bambi Lynn; in chorus of Miss Liberty; Picnic, 1953. Screen debut, Goodbye My Fancy.
PICTURES INCLUDE: Starlift, Holiday for Sinners, Rogue's March, Woman's Devotion, Gun for a Coward, Subterraneans, Invitation to a Gunfighter, The Chase, Welcome to Hard Times, The Ambushers, Kid Blue, 3 Women.

RUNCIMAN, ALEX: Director-producer. b. Santa Cruz, Calif., May 30, 1924. e. U. of So. Calif., B.A., 1949; M.A., 1951; Ph.D., 1959. Served in U.S. Army, 1943–46; radio work, 1946–49; in prod. department, CBS, 1953; staff dir., 1954; has dir. panel, news, educational, variety programs, Television Journal, 1959–60; prod. Eye on St. Louis, 1960–61; prod. award-winning public service programs; Host of Metroplex Series, 1963.

RUSH, BARBARA: Actress. b. Denver, Colo. Jan. 4. e. U. of Calif. First stage appearance at age of ten, Loberto Theatre, Santa Barbara, Calif., in fantasy, Golden Ball; won acting award in coll. for characterization of Birdie, (The Little Foxes); scholarship, Pasadena Playhouse Theatre Arts Coll.
PICTURES INCLUDE: The First Legion, Quebec, Molly, When Worlds Collide, Flaming Feather, Prince of Pirates, It Came From Outer Space, Taza Son of Cochise, Magnificent Obsession, Black Shield of Falworth, Captain Lightfoot, Kiss of Fire, World in My Corner, Bigger Than Life, Oh Men! Oh Women!, Harry Black and the Tiger, The Young Philadelphians, Bramble Bush, Strangers When We Meet, Come Blow Your Horn, Robin and the Seven Hoods, Hombre, Airport, The Man, Superdad, Can't Stop the Music.

RUSH, RICHARD: Director, producer, writer; wrote, dir. prod., Too Soon To Love (U.I.), 1960; Of Love and Desire (Fox), 1963; dir. A Man Called Dagger (MGM) 1965; dir. Fickle Finger of Fate, 1966; dir. Thunder Alley, 1966; dir. Hell's Angels on Wheels, 1967; dir. wrote Psych-Out, 1967; dir. Savage Seven, 1968; dir., prod. Getting Straight (Columbia), 1970, dir. prod. Freebie and the Bean (Warner Bros.) 1974; dir., prod. wrote, The Stunt Man (Fox), 1980.

RUSSELL, CHARLES W.: Producer, b. N.Y.C., Mar. 31, 1918. Actor, m.p. & radio.
PICTURES INCLUDE: The Purple Heart, The Late George Apley.
TV: CBS-TV prod., You Are There, Danger.

RUSSELL, EDWARD: Executive. b. Chicago, June 12, 1945. e. Columbia Coll. Started with industry in Chicago as asst. adv. mgr. for Plitt Theatres, 1968–70; asst. adv. mgr., midwest div., MGM, 1970–72; midwest adv. mgr., United Artists, 1972–74; midwest adv. mgr., Filmways (AIP), 1974–77; in 1977 made natl. dir. of cooperative adv. & field promo. for Filmways; promoted in 1980 to exec. asst. to the v.p. of worldwide adv., promo., pub. Member: Academy of M.P. Arts & Sciences.

RUSSELL, JACK: Performer. b. Saratoga Spr., N.Y., Sept. 22, 1919. e. Rollins College, 1945–47. U.S. Air Force, 1940–45; appeared in B'way shows including: Sleepy Hollow, Heaven on Earth, As the Girls Go, Alive & Kicking; many nightclubs as singer; on NBC-TV opera, Pagliacci, 1951; guest star N.Y.C. Opera, 1953; featured aqua show, A Night in Venice, 1952–53; singer, Your Show of Shows, NBC-TV, 1950–53.

RUSSELL, JANE: Actress. b. Bemidji, Minn., June 21, 1921; e. Max Reinhardt's Theatrical Workshop & Mme. Ouspenskaya. Photographer's model; m.p. debut in Outlaw, 1943.
PICTURES INCLUDE: Young Widow, Paleface, Montana Belle, His Kind of Woman, Double Dynamite, Macao, Son of Paleface, Las Vegas Story, Gentlemen Prefer Blondes, French Line, Underwater, Gentlemen Marry Brunettes, Foxfire, Tall Men, Hot Blood, Revolt of Mamie Stover, Fuzzy Pink Nightgown, Darker Than Amber, Born Losers, Fate Is The Hunter, Waco.

RUSSELL, JOHN: Actor. b. Los Angeles, Calif., Jan. 3, 1921; e. U. of Calif. Served in U.S. Marine Corps. 1942–44, as 2nd Lt. m.p. debut in Frame-Up.
PICTURES INCLUDE: Story of Molly X, Gal Who Took the West, Slattery's Hurricane, Yellow Sky, Sitting Pretty, Forever Amber, Somewhere in the Night, Within These Walls, Don Juan Quilligan, Bell for Adano, Barefoot Mailman, Man in the Saddle, Hoodlum Empire, Oklahoma Annie, Fair Wind to Java, Sun Shines Bright, Jubilee Trail, Hell's Outpost, Last Command, Rio Bravo, Yellowstone Kelly, Fort Utah.

RUSSELL, KEN: Director. b. Southampton, England, 1927. e. Walthamstow Art School. Early career as dancer, actor, stills photographer, TV documentary film-maker. Ent. TV ind. 1959. Made 33 documentaries for BBC-TV.
PICTURES INCLUDE: Prokofiev, Elgar, Bartok, The Debussy films, Isadora Duncan, Song of Summer—Delius, Dance of the Seven Veils, French Dressing, Billion Dollar Brain, Women in Love, The Music Lovers, The Devils, The Boy Friend, Savage Messiah, Mahler, Tommy, Lisztomania, Valentino, Altered States.

RUSSELL, ROBERT: Performer. b. Passaic, N.J., Baritone, Philadelphia Grand Opera Co.; then radio work as quizmaster, m.c., singer, gagwriter, composer, writer spec. material; m.c., writer, composer, dir. eleven Miss America pageants; staged, wrote, dir. five Miss Universe pageants.
TV: Birthday Party, Bonny Maid Show, It's in the Bag, Toast of the Town, Live Like a Millionaire; now m.c. Stand Up and Be Counted, CBS-TV; pres. Pageant Productions, Inc.

RUSSELL, THERESA: Actress. b. San Diego, CA, 1957. e. Burbank High School. Studied at Actors' Studio in Hollywood. Professional film debut in The Last Tycoon (1977).
PICTURES: Straight Time, Bad Timing/A Sensual Obsession.
TELEVISION: Blind Ambition (mini-series).

RUTHERFORD, ANN: Actress. b. Toronto, Canada, 1924. Trained by mother (cousin of Richard Mansfield); with parents in stock as child; later on Los Angeles radio programs. Screen debut, 1935.
PICTURES INCLUDE: Laramie Trail, Happy Land, Bermuda Mystery, Two O'Clock Courage, Bedside Manner, The Madonna's Secret, Murder in the Music Hall, Secret Life of Walter Mitty, Operation Haylift, Adventures of Don Juan, They Only Kill Their Masters.

RYAN, ARTHUR N.: Executive. Joined Paramount in N.Y. in 1967 as asst. treas; later made dir. of admin. and business affairs, exec. asst. to Robert Evans and asst. scty. In 1970 appt. v.p.-prod. adm. In 1975 named senior v.p. handling all prod. operations for Paramount's m.p. and t.v. divisions, including supvr. of physical prod., business affairs, studio operations and coordination with co.'s creative personnel and filmmakers. Named asst. to the Chairman and Chief Executive Officer 1976; Chairman and President Magicam, Inc.; Chairman Fortune General Corp.; Chairman Paramount Communications; co-chm. of scholarship comm. of Academy of Motion Picture Arts and Sciences; trustee of Univ. Film Study Center in Boston. Joined Technicolor in August 1976 as pres., chief operating officer and director. Chairman Technicolor Audio-Visual Systems International, Inc.; director Technicolor S.P.A.; director Technicolor Film International; and chairman of executive committee, Technicolor Graphic Services, Inc.; director, Technicolor, Inc.; chairman, Technicolor Fotografica, S.A.; Chairman Technicolor Film International Service Company, Inc.; director and Deputy Chairman Technicolor Limited; director, The Vidtronics Company, Inc.; Permanent charities committee of the Entertainment Industry; Hollywood Canteen Foundations. Dir., Calif. Institute of Arts.

RYAN, MITCHELL: Actor. b. Louisville, Ky. was New York stage actor working off-Bdwy. for Ted Mann and Joseph Papp; on Bdwy. in Wait Until Dark. Member of Arena Stage group in Washington.
PICTURES INCLUDE: Monte Walsh, The Hunting Party, My Old Man's Place, High Plains Drifter, The Friends of Eddie Coyle, Magnum Force.

RYAN, PEGGY: Actress, r.n. Margart Irene Ryan. b. Long Beach, Calif., August 28, 1924. Before entering pictures appeared in vaudeville; on TV in Hawaii Five-0.
PICTURES INCLUDE: Top of the Town, The Women Men Marry, The Flying Irishman, She Married a Cop, Girls Town, Miss Annie Rooney, Private Buckaroo, Give Out, Sisters, Top Man, Mr. Big, Patrick the Great, The Merry Monahans, Follow the Boys, Babes on Swing Street, Bowery to Broadway, Merrily We Sing, That's the Spirit, This Is the Life, Here Come the Co-eds.
STAGE: Everybody, Men in Her Diary, Shamrock Hill, There's a Girl in My Heart, All Ashore.

RYAN, SHEILA: Actres. r.n. Katherine Elizabeth McLaughlin; b. Topeka, Kansas, June 8, 1921; e. Le Conte Jr. H.S., Hollywood H.S.
PICTURES INCLUDE: Something for the Boys, The Caribbean Mystery, Lone Wolf in London, Cobra Strikes, Caged Fury, Western Pacific Agent, Mask of the Dragon, Jungle Manhunt, On Top of Old Smoky, Pack Train.

RYDELL, MARK: Producer-Director-Actor. b. 1934. Studied acting with Sanford Meisner of N.Y. Neighborhood Playhouse. Became member of Actors Studio. Was leading actor for six years on daytime CBS serial, As The World Turns. Made Broadway debut in Seagulls over Sorrento and film bow in Crime in the Streets. Went to Hollywood as TV director (Ben Casy, I Spy, Gunsmoke, etc.). Theatrical feature debut: The Fox. Partner with Sidney Pollack in Sanford Prods., film, TV prod. co.
PICTURES INCLUDE: Director: The Fox, The Receivers, The Cowboys, Cinderella Liberty (also prod.), Harry and Walter Go To New York. Actor: The Long Goodbye.

RYDER, LOREN L.: Technical executive. b. Pasadena, Calif., Mar. 9, 1900; e. U. Calif., B.A. research engineer, telephone, telegraph 1924–25; radio importer 1926; asst. mgr. ShermanClay radio-phono. depts. 1927–28; head of sound dept. Paramount Pictures Corp., 1936–57; pres. Ryder Sound Services Inc., Ryder Magnetic Sales Corp., v.p. Nagra Magnetic Recorders. War Research consultant, National Defense Research Committee and Office of Scientific Research & Development, Washington; past pres. Soc. of M.P. & Television Engineers; recipient War-Navy Research Award; member, Academy of M.P. Arts & Sciences; recipient of Academy Awards for sound quality, sound effects in Spawn of the North, first use of fine-grain film, introduction of magnetic recording, development of Vista-Vision; past national sec'y., Academy of Television Arts & Sciences.

RYLANDER, ALEXANDER S.: Executive. b. N.Y.C., Jan. 9. Columbia, 1941; organ, spec. events dept., 1946; expl. mgr., Jan. 1953; exploit. dir., NBC 1955; dir. prom. services, 1959; vice pres., promotion, 1961; staff v.p. spec. projects, RCA, 1966. v.p. news & information, 1967.

RYSKIND, MOORIE: Writer. e. Columbia U. Co-auth. Many plays including Of Thee I Sing, Animal Crackers, Louisiana Purchase, Currently columnist, syndicated by Human Events.
PICTURES INCLUDE: Palmy Days, Animal Crackers, Night at the Opera, Anything Goes, Ceiling Zero, Rhythm on the Range, My Man Godfrey, Luckiest Girl in the World, Room Service, Man About Town, Penny Serenade, Claudia, Where Do We Go From Here?, Heartbeat.

S

SACKS, SAMUEL: Attorney, agent. b. New York, N.Y., March 29, 1908. C.C.N.Y., St John's Law School, LL.B, 1930. Admitted Calif. Bar, 1943; priv. prac., law, N.Y. 1931–42; attorney, William Morris Agency, Inc., Sept. 1942; head of west coast TV business affairs, 1948–75. bd. of dir., Alliance of Television Film Producers, 1956–60. L.A. Copyright Society Treasurer, Beverly Hills Bar Assn., Los Angeles Bar Assn., American Bar Assn.; Academy of TV Arts & Sciences; Hollywood Radio & TV Society. Pres. Adat Shalom Synagogue, 1967–69 chmn. of bd., 1969–71; pres., American Field Service West L.A. Chapter 1970–72, United Synagogue of America (west region), v.p.; counsel, entertainment field, Simon & Sheridan, 1975– .

SAFER, MORLEY: News Correspondent. b. Toronto. e. Univ. of Western Ontario. Started as corresp. and prod. with Canadian Broadcasting Corp. Joined CBS News as head of Saigon Bureau, 1965. Chief of CBS London bureau 1967–70. Joined 60 minutes as co-editor in Dec., 1970.

SAFFLE, M.W. "BUD": Executive. b. Spokane, Wash., June 29, 1923. e. Univ. of Wash. In service 1943–46. Started in m.p. buiness as booker, 1948. Entire career with Saffle Theatre Service as buyer-booker; named pres. in 1970. Also pres. of Grays Harbor Theatres, Inc., operating theatres in Aberdeen, Wash. Also operates d.i. in Centralia, Wash. On bd. of NATO of Wash. for 15 yrs; pres. of same for 2 terms and secty.-treas. 6 yrs. Elected to National NATO bd. in 1972. Founder of Variety Tent 46, serving as chief barker three times.

SAFIR, SIDNEY: Executive. b. Vienna, Aust., Feb. 2, 1923. e. London U. Ent. m.p. ind. 1940, Shipman & King Cinemas; RKO Radio Picture, 1941; salesman, British Lion, 1943; European sls. mgr., Lion Int'l, 1958; gen. sls. mgr. Lion Int'l, 1960; president, Lion Int'l Inc., 1965; man. dir. Lion Int'l Ltd. 1969; dir., British Lione Film Ltd., 1972.

SAINT, EVA MARIE: Actress. b Newark, N.J., July 4, 1924. e. Bowling Green State U., Ohio. Radio, TV actress; on Broadway in Trip to Bountiful; m.p. debut in On the Waterfront (Acad. Award, best supporting actress, 1954).
PICTURES INCLUDE: That Certain Feeling, Raintree County, Hatful of Rain, North by Northwest, Exodus, All

Fall Down, Grand Prix, The Stalking Moon, Loving, Cancel My Reservation.
TV: How the West Was Won.

ST. JACQUES, RAYMOND: Actor-Director. Began career as actor, asst. dir. and fencing dir. for American Shakepeare Festival, Stratford, Conn. Made prof. acting debut in off-Bdwy. play, High Name Today. Made m.p. debut, Black Like Me, 1964.
PICTURES INCLUDE: The Pawnbroker, The Comedians, Mr. Moses, Madigan, Mister Buddwing, The Green Berets, Uptight, If He Hollers Let Him Go, Change of Mind, Cotton Comes to Harlem, Cool Breeze, Come Back Charleston Blue, Book of Numbers (also dir. debut), Lost in the Stars.
STAGE: The Blacks, Night Life, The Cool World, Seventh Heaven.

ST. JAMES, SUSAN: Actress. b. Los Angeles, Aug. 14, 1946. e. Conn. College for Women. Was model for six years; then signed to contract by Universal Pictures.
TELEVISION: The Name of the Game (series), McMillan & Wife (series), numerous shows.
PICTURES INCLUDE: What's So Bad About Feeling Good?, Jigsaw, P.J., Where Angels Go . . . Trouble Follows, Magic Carpet, Outlaw Blues, Love at First Bite, How to Beat the High Cost of Living, Carbon Copy.

ST. JOHN, BETTA: Actress. b. Hawthorne, Calif., Nov. 26. B'way debut in Carousel; then in South Pacific in N.Y. & London; m.p. debut in Dream Wife; since many more m.p., TV films.
PICTURES INCLUDE: All the Brothers were Valiant, The Robe, Dangerous Mission, Saracen Blade, Student Prince, Law vs. Billy the Kid, Alias John Preston, Naked Dawn, Tarzan and the Lost Safari, High Tide at Noon, City of the Dead.

ST. JOHNS, RICHARD R.: Executive Producer. b. Los Angeles, CA. e. Stanford University, B.A. 1953; Doctorate of Jurisprudence, Stanford Law School 1954. Joined law firm O'Melveny & Meyers 1954, specializing in entertainment law. 1963 became partner in law firm. 1968 became senior v.p., Filmways, Inc., becoming president and chief operating office in 1969. 1972, formed Richard R. St. Johns and Associates, independent management and packaging firm. Formed Guinness Film Group in 1975, branching out into full-scale motion picture production.
PICTURES: (exec. prod.): The Uncanny, Death Hunt, Matilda, The Silent Flute (Circle of Iron), Nightwing, The Wanderers, The Mountain Men, The Final Countdown, A Change of Seasons, Dead & Buried, American Pop and Venom.

SAKS, GENE: Director—Actor. b. N.Y.C., Nov. 8, 1921. e. Cornell Univ. m. actress Bea Arthur. Attended Dramatic Workshop of the New School for Social Research. Active in off-Broadway in 1948–49, forming cooperative theatre group at Cherry Lane Theatre. Joined Actor's Studio, followed by touring and stock. Also appeared in live TV dramas (Philco Playhouse, Producer's Showcase). Directed many Broadway plays before turning to film direction with Barefoot in the Park (1967).
PICTURES INCLUDE: Director: Barefoot in the Park, The Odd Couple, Last of the Red Hot Lovers, Mame, Cactus Flower. Actor: A Thousand Clowns, Prisoner of Second Avenue, The One and Only.
BROADWAY: Director; Enter laughing, Nobody Loves an Albatross, Generation, Half a Sixpence, Mame. Actor: Middle of the Night, Howie, The Tenth Man, A Shot in the Dark, A Thousand Clowns, Same Time, Next Year.

SALANT, RICHARD S.: Executive. b. New York, N.Y., Apr. 14, 1914; e. Harvard Coll. A.B., 1931–35; Harvard Law Sch., 1935–38; Atty. Gen.'s Com. on Admin. Procedure, 1939–41; Office of Solicitor Gen., U.S. Dept. of Justice, 1941–43; U.S. Naval Res., 1943–46; assoc., Roseman, Goidmark, Colin & Kave, 1946–48; then partner, 1948–51; pres. CBS news div., 1961–64; v.p. special asst. to pres. CBS, Inc., 1964; pres., CBS news div., 1966; mem. bd. of dir., CBS, Inc. 1964–69; vice chm., NBC bd., 1979–.

SALE, RICHARD: Writer, director. b. New York, N.Y., Dec. 17, 1911; e. Washington & Lee, A.B., 1934; m. Irma Foster, designer. Author novels, over 400 published stories; honor roll Best Short Stories, 1935; ent. m.p. ind. with Paramount, 1944; member: WGAW, DGA: Acad. of M.P. Arts & Sciences, Authors League of Amer., BMI: pres. Voyager Films, Inc.; v.p. Libra Productions, Inc.
PICTURES INCLUDE: Strange Cargo, Rendezvous with Annie, Spoilers of the North, Campus Honeymoon, Lady at Midnight, Calendar Girl, Inside Story, Dude Goes West, Mother is a Freshman, Father Was a Fullback, When Willie Comes Marching Home, Mr. Belvedere Goes to College, Ticket to Tomahawk, I'll Get By, Driftwoods Meet Me After the Show, Half Angel, Let's Make It Legal, Girl Next Door, My Wife's Best Friend, Let's Do It Again, Torpedo Run, French Line, Suddenly, Women's World, Gentlemen Marry Brunettes, Abandon Ship, The Oscar, The White Buffalo.

SALETRI, FRANK R.: Producer-Director-Writer. b. Chicago, Jan. 20, 1928. Is criminal trial lawyer, member of State of Calif. bar. Heads FRSCO Prods., Ltd., m.p. prod.
PICTURES INCLUDE: Black Frankenstein—Blackenstein (s.p., prod.); Black the Ripper (s.p., dir.); The Return of the Ghost of the Son of the Bride of the House of Frankenstein, 1984 (s.p., dir.), The Skid Row Slasher, s.p., dir; The Secret of the Maltese Falcon, s.p., dir; Annually prod., s.p. & dir. for The Academy of Science Fiction, Horror and Fantasy Films. Annually proe., s.p. & dir. for The Annual Count Dracula Society 'Ann Radcliffe' Awards. (Both are documentaries).

SALKIND, ALEXANDER: Producer. b. Danzig/Gdansk, of Russian extraction. Grew up in Berlin where father, Miguel, produced films. Went to Cuba with father to assist him in film production. First solo venture a Buster Keaton comedy, 1945. Returned to Europe where made many pictures in Spain, Italy, France and Hungary.
PICTURES INCLUDE: The Three Musketeers, The Four Musketeers, Superman.

SALKIND, ILYA: Producer. b. Mexico City. e. Univ. of London. Grew up in many countries as father, Alexander, produced films. First film job as production runner on The Life of Cervantes for father. Was assoc. prod. on Light at the Edge of the World.
PICTURES: The Three Musketeers, The Four Musketeers, Superman, Superman II (exec. prod.).

SALKOW, SIDNEY: Director, writer. b. New York City, June 16, 1911; e. City Coll. N.Y., B.A. Harvard Law Sch. Stage dir. & prod. asst. number N.Y. dram. prods. (Dir. Bloodstream, Black Tower, etc.) and mgr. summer theat. From 1933 variously dialogue dir., assoc. dir., writer & dir. numerous pictures Paramount, Universal, Republic, Columbia, etc.; dir. number of pictures in Lone Wolf series (for Columbia), Tillie the Toiler, Flight Lieutenant, etc. In armed service, WW II.
PICTURES INCLUDE: Millie's Daughter, Bulldog Drummond at Bay, Admiral Was a Lady, Fugitive Lady, Golden Hawk, Scarlet Angel, Pathfinder, Prince of Pirates, Jack McCall Desperado, Raiders of the 7 Seas, Sitting Bull, Robbers' Roost, Shadow of the Eagle, Las Vegas Shakedown, Toughest Man Alive, Chicago Confidential, Iron Sheriff, Great Sioux Massacre, Martin Eden.
TV: Created, prod. dir., This is Alice series for Desilu, Lassie, Fury, Wells Fargo series. Headed prod. for F&F Prod. in Rome, 1967–71.

SALMI, ALBERT: Actor. b. Coney Island, N.Y. After serving in World War II, studied—1948–54—with Dramatic Workshop, American Theatre Wing, Actors Studio. Appeared in many off-Bdwy. plays and live TV prods. Big hit on Bdwy. in Bus Stop. Other plays include The Rainmaker and The Brothers Karamazov. Has appeared in over 20 feature films and over 200 TV shows.
PICTURES INCLUDE: The Lawman, Something Big, The Deserter, The Crazy World of Jules Vrooder, Empire of the Ants, Dragonslayer.
TV: Gunsmoke, Barnaby Jones, etc.

SALT, WALDO: Writer. b. Chicago, Ill., Oct. 18, 1914. e. Stamford U., 1934, A.B. Taught dramatics & music, Menlo Jr. Coll., 1935; served as civilian consultant writer, Army films, OWI Overseas Film Burea, 1945.
PICTURES INCLUDE: Shopworn Angel, Tonight We Raid Calais, Rachel & the Stranger, M, Wild Man of Borneo, Mr. Winkle Goes to War, The Flame & the Arrow, Midnight Cowboy, Gang That Couldn't Shoot Straight, Serpico, The Day of the Locust, Coming Home.

SALTER, HANS J: Composer, conductor. e. U. Acad. of Music, Vienna-Austria, Mus. dir.: Volksopera, Vienna; State Oper. Berlin; Metropole Theatre, Berlin; comp., cond., UFA, Berlin, 1929–33; European br., Universal, 1934–36; to U.S., Univ. 1938–47, 1950–52, wrote over 150 scores.
PICTURES INCLUDE: It Started With Eve, His Butler's Sister, Scarlet Street, Magnificent Doll, The Spoilers, Frenchie, Flesh and Fury, Golden Horde, The Sign of the Ram, Frightened City, Ghost of Frankenstein, Black Friday, House of Frankenstein, The Wolfman, Hold That Ghost, The Invisible Man Returns, the Mummy's Hand, Man-Eater of Kumaon, This Island Earth, Tomahawk, The Battle of Apache Pass, Please Believe Me, Apache Drums, Untamed Frontier, Lover Come Back, Thunder on the Hill, Bend of the River, Against All Flags, Black Shield of Falworth, Sign of the Pagan, Far Horizons, Man Without a Star, Wichita, Autumn Leaves, Red Sundown, Hold Back the Night, Rawhide Years, The Oklahoman, Three Brave Men, Pay the Devil, Law of the Trigger, Female Animal, Raw Wind in Eden, The Wild and the Innocent, Bat Masterson Story, Man in the Net, Come September, Follow That Dream, If a Man Answers, Bedtime Story, The Warlord, Beau Geste, Return of the Gunfighter.
TV: Wichita Town, Laramie, The Law and Mr. Jones, The Virginian, Wagon Train, Lost in Space, Maya.

SALTZMAN, HARRY: Producer. b. October, 1915. St. John, N.B., Canada. Ent. film ind. 1945.

PICTURES INCLUDE: The Iron Petticoat, Look Back in Anger, The Entertainer, Saturday Night, Sunday Morning; formed Lowndes Prods. Ltd., 1964 and prod.: Dr. No, Call me Bwana, From Russia With Love, Goldfinger, The Ipcress File, Thunderball, Funeral in Berlin, You Only Live Twice, Billion Dollar Brain, Play Dirty, Battle of Britain, On Her Majesty's Secret Service, Tomorrow, The Man With the Golden Gun, Nijinsky.

TV: Prod. Robert Montgomery Show, Capt. Gallant of the Foreign Legion.

SALZBURG, JOSEPH S.: Producer, editor. b. New York, N.Y., July 27, 1917. Film librarian, then rose to v.p. in chge. of prod., Pictorial Films, 1935–42; civilian chief film ed. U.S. Army Signal Corps Photo Center, 1942–44; U.S. Army Air Forces, 1944–46; prod. mgr., Pictorial Films, 1946–50; prod. mgr. Associated Artists Prod., then M.P. for TV, 1950–51; org. m.p. prod. & edit. service for theatrical, non-theatrical & TV films; prod. mgr., dir. of films oper., official Films. Oct. 1956; prod. sup. tech. dir. Lynn Romero Prod. features and TV; assoc. prod. Lynn Romero Prod. TV series, Counterthrust; v.p., sec'y, B.L. Coleman Assoc., Inc. & Newspix, Inc. 1961; pres. National Production Assoc., Inc. 1962, chief of production, UPI Newsfilm, 1963. Prod. a/c Fred A. Niles Comm. Center, 1966. Appt. v.p., F.A. Niles Communications Centers, N.Y., 1969. In 1979 appointed in addition exec. producer & gen. mgr., F.A. Niles Communication Centers Inc., N.Y. studio.

SALZBURG, MILTON J.: Producer. b. New York, N.Y., Dec. 17, 1912. Book publishing field, N.Y.; film librarian; ed. instructional film, Dealers in Death; founder & pres. Pictorial Films, Inc., prod. instructional & topical shorts; pres. Academic Film Co. & Academic Prod. of Hollywood; sold Pictorial to Pathe Industries, remaining as pres., 1945; pres. 1946; sales mgr., Post Pictures, 1946; re-purchased Pictorial Films & pres. Nov. 1961; pres. Cornell Films; prod., Pattern for Survival, documentary film & Target: U.S.A., industrial film; pres. Novel Films, Dec. 1952; pres. Award TV Corp. (prod. Jimmy Demaret Golf series, Golden Era of Boxing), pres. Birad Corp., prod. dist. educat. films, 1960 pres. Productions Unlimited, Inc., producers of 16 mm educational films. Prod. 1963 Life Cycle of WASP, Cine Golden Eagle, Pre-Historic Animals, 1965 Story of Rice. Prod. Instincts of an Insect which has won Blue Ribbon, American Film Festival; Chris Award, Columbus Film Festival; Golden Eagle at Cine; Silver Hugo at Chicago Film Festival. 1967, v.p. & gen. sales mgr., Fleetwood Films, 1970, pres. Non-Theatrical Div., United Productions of America, DEI Industres, Inc. Left UPA in 1971 to reactivate Productions Unlimited, Inc., producing shorts and supplying films to govt. agencies.

SAMETH, JACK: Executive Producer, Producer, Director. Syracuse U., US Army. ABC TV 1950–63. NET 1963–67. ABC TV 1967–69 NET 1969–77. Executive Director, ABC TV News. Now exec. prod., special programs, WNET.

SAMPSON, LEONARD E.: Exhibitor. b. New York, N.Y. Oct., 1918. e. CCNY (B.B.A. '39). Entered m.p. industry as stagehand helper and usher, Skouras Park Plaza, Bronx 1932–36; asst. mgr. Gramercy Park, 1937–38; mgr., 5th Avenue Playhouse, 1939–41; mgr., Ascot Bronx, 1941–42. In Army 1942–46. On return entered into partnership with cousin Robert C. Spodick in Lincoln, a New Haven art house. Organized Nutmeg Theatres, operating 6 art and conventional theatres in Conn., associated with Norman Bialek in Westport and Norfolk. Sold Nutmeg in 1968 to Robert Smerling. Retains partnership with Spodick in New Haven's York Sq., Lincoln, and Crown. Built Groton, Conn., Cinemas I and II in 1970 and Norwich, Conn., Cinema I & II, 1976 and acquired Village Cinemas I & II, Mystic, in association with Spodick and William Rosen.

SAMUELS, ABRAM: Executive. b. Allentown, Pa., Sept. 15, 1920. e. Lehigh U. U.S. Army 1942–46; pres. Automatic Devices Co. 1946–76; named bd. chm. in 1976.

SAMUELSON, DAVID W.: F.B.K.S., A.R.P.S., B.S.C.: Executive. b. London, England, 1924. Son of early producer G.B. Samuelson. Joined ind. 1941 with British Movietone News. Later film cameraman, 1947. Left Movietone 1960 to join family company, Samuelson Film Service Ltd. Past president British Kinematograph Sound & TV Soc., Chmn, Inc. Ass. Kine. Mans. V.P. UNIATEC. Author of Motion Picture Camera and Lighting Equipment, Motion Picture Camera Techniques, Motion Picture Camera Data.

SAMUELSON, PETER GEORGE WYLIE: Producer. b. London, England, October 16, 1951. Masters Degree: English Literature, Cambridge University. Early career as interpreter and production assistant. Production Manager: 1973, Speed Merchants. 1974, High Velocity. 1975, One by One. 1976, Return of the Pink Panther. 1977, Santa Fe 1836. 1978, produced: A Very Big Withdrawal. 1979, partner: Kinesis Productions, is motion picture development and production company. Executive president and chief officer.

SAMUELSON, SYDNEY: C.B.E. Hon. F.B.K.S. B.S.C. Executive, b. London, England 1925. Early career as cinema projectionist 1939–42 Gaumont British News 1942–43. Royal Air Force 1943–47. Asst. cameraman, cameraman, director/cameraman. Founded Samuelson Film Service, 1955. Man. dir. Panavision Corp. of Calif. Ltd. 1965. Chmn. and Chief Executive dir. Samuelson Film Service Ltd., Trustee British Academy of Film and Television Arts (chm. of Council, 1973–76). Governor British Society of Cinematographers. Executive member of Council, Cinema and Television Benevolent Fund. Pres., Television Veterans, 1980–81.

SANDERS, DENIS: Writer-producer-director, executive SRS Productions. b. New York City, Jan. 21, 1929. e. UCLA (M.A. 1954). Awards include 2 Academy Awards, A Time Out of War 1954; Czechoslovakia, 1968 in 1970. Made over 70 documentaries including Trial: City and County of Denver v.s. Lauren R. Watson; Arbitration: The Truth of the Matter.

FEATURES INCLUDE: Elvis, That's The Way It Is, MGM: Soul To Soul, Cinerama; One Man's Way, UA: War Hunt, UA: Crime and Punishment, USA.

SANDERS, HONEY: Actress, Singer. b. N.Y.C. TV; Edge of Night, CBS/TV; film, radio commercials.

STAGE: Gentlemen Prefer Blondes, Plain and Fancy, High Button Shoes, Carousel, Pajama Game, Damn Yankees, Diary of Anne Frank, Matchmaker, View From the Bridge.

SANDERS, NAT: Executive. b. New York City, Mar 4, 1892; e. C.C.N.Y., 1912. Publicity dir. & pub. rel. counsel; independent business mgr. for m.p. stars and directors; with sales dept. Commonwealth Pictures, Inc.; gen. mgr. Sanders Films; gen. mgr. Film Alliance of U.S. Inc.; pres. The Baker's Wife, Inc.; pres. Leo Films, Inc.; vp. English Films, Inc.; pres. Film Renters, Inc.; gen. mgr., United Motion Pictures Org., gen. mgr., Times Film Corp.; southern dist. rep.: Lopert Films, Inc., Janus Films, Inc., Audubon Films, Continental Distribution Corp., Mishkin Motion Pictures, Inc.; member: M.P. Pioneers, Variety Clubs International Sect'y, National Assoc. M.P. Ind., Motion Picture Pioneers of Florida.

SANDERS, TERRY BARRETT: Producer, director, writer. b. New York, Dec. 20, 1931. e. UCLA, 1951; Co-prod., photographed, A Time Out of War, 1954. Academy award best two-reel subject, and won first prize Venice Film Festival, etc.; co-wrote The Day Lincoln Was Shot, CBS-TV; s.p. The Naked and the Dead; prod. Crime and Punishment—USA., prod. War Hunt; prod. and dir. Portrait of Zubin Mehta for U.S.I.A. Assoc. dean Film School, Calif. Instit. of the Arts.

TV: Prod. dir., Hollywood and the Stars, The Legend of Marilyn Monroe, National Geographic Society specials; exec. v.p., American Film Foundation.

SANDS, ERNEST: Executive. b. Newark, N.J., May 5, 1924. e. NYU, B.S., 1947. Sales exec., Warner Bros., 1947–59; film buyer, Fabian Theatres, 1960–61; gen. sales mgr., Astor Pic., 1962; gen. sales mgr., Allied Artists, 1963–64; asst. gen. sales mgr., Paramount, 1965; v.p. asst. gen. sales mgr., Paramount, resigned 1969. Ivan Tors Films Inc., sales exec., 1970; v.p. gen. sales mgr., Cannon Releasing Corp., 1970–1973; Sands Film Co. Inc. president, March, 1973, Distributor and Producer's Representative.

SANDS, JOHNNY: Actor r.n. John Harp; b. Lorenzo, Tex., Apr. 29, 1927. Started in school plays; screen debut in Stranger, 1946.

PICTURES INCLUDE: Till the End of Time, Affairs of Geraldine, Born to Speed, Blaze of Noon, Bachelor and the Bobby-Soxer, Fabulous Texan, Admiral Was a Lady, Lawless, Two Flags West, Aladdin and His Lamp, Basketball Fix.

SANDS, TOMMY: Singer. b. Chicago, Aug. 27, 1937; e. Schools there and Greenwood, La. Father, Benny Sands, pianist with stage bands. Started career as guitar player, singer when 10, at KWKH station, Shreveport, followed by other radio app.; TV with Ernie Ford, then Kraft TV show The Singin' Idol; recording contract won him million record sales of "Teen Age Crush."

PICTURES INCLUDE: Sing Boy Sing, Mardi Gras, Love in a Goldfish bowl, Babes in Toyland, The Longest Day.

SANFORD, CHARLES: Musical Director. b. June 17, 1905, N.Y.C. Has been conducting since age of 15; asst. Cond. NY Hippodrome; district musical supervisor, RKO; conducted various shows, assoc cond. to Alexander Smallens, Porgy and Bess.

TV: Musical dir., Admiral Broadway Review, 1948; Your Show of Shows, 1950–54; Bob Hope, 1950–51; Elgin American Show, 1950; Beatrice Lillie Show, 1950–51; Max Liebman Presents, 1954–56; Producer's Showcase, 1957–58; Jerry Lewis, Patrice Munsel Shows; 1959–60 Max Liebman Specials, Phil Silvers, Sid Caesar Specials.

SANGER, GERALD: (C.B.E., J.P.): Director, British Movietone News, Ltd. b. Su-biton, Eng., May 23, 1898; married. e. Shrewsbury School and Kebie Coll., Oxford. Lieut., Royal Marines, 1917–19. Secretary of Hon. Esmond Harmsworth, M.P. (Lord Rothermere), 1921–29. Ed. Brit. Movietone News, 1929–53; dir., Assoc. Newspapers Ltd., 1944–63; F. Hewitt & Son (1927) Ltd. 1954; News Holdings Ltd., 1954; Associated-Rediffusion Ltd., 1954–56; Daily Mail & Gen. Trust Ltd., 1963.

SANGSTER, JIMMY: Producer, director, screenwriter. Ent. m.p. ind. 1943. Prod. man. for Break in the Circle, Men of Sherwood Forest, X the Unknown.
PICTURES INCLUDE: Man on the Beach, The Curse of Frankenstein, The Trollenberg Terror, The Georkel, The Blood of the Vampire, Dracula, Intent to Kill, The Revenge of Frankenstein, Jack the Ripper, The Mummy, The Brides of Dracula, The Man Who Could Cheat Death, The Siege of Sydney Street, The Criminal, The Hell Fire Club, See No Exit, 1960 prod. and scripted Taste of Fear for Hammer Films 1961. Prod. The Savage Guns; 1962; prod. s.p. Maniac, Nightmare, Devil Ship Pirates; prod. Hysteria, s.p., The Giants, Paranoiac, 1963; s.p. Brainstorm, 1964; prod., s.p. The Nanny, s.p. Deadlier Than the Male, The Bridge of Newgate Jail, 1965; 1966 s.p. Java Weed, s.p. The Anniversary. 1967 prod. 1968 s.p. Doubled in Diamonds, Hide and Seek, Foreign Exchange, Private I (prod.) 1969; s.p. The Killing Game, The Claw, Touchfeather. 1970 wrote, prod. dir. The Horror of Frankenstein; dir. Lust for a Vampire, 1971; co-wrote s.p. Gingerbread House, Murder by Month Club, A Taste of Evil, The Goldfish Bowl, s.p. prod., dir. Fear in the Night, 1971. 1973: Screenplay: The Fairytale Man. 1974: s.p. The Monstrous Defect. 1977: s.p. The Legacy. 1980: s.p., Phobia.
TV: Writing credits include: Motive for Murder, The Assassins, I Can Destroy the Sun, 1972, exec. story consultant Screen Gems-NBC short story series, McCloud, Banacek, Cannon, Ironside, etc. 1977–78; prod. CBS series, Young Dan'l Boone; wrote & produced pilot for CBS: Ebony Ivory and Jade; wrote pilot for ABC: Adventure (Columbia TV); wrote pilot for NBC: Murder in Music City; The Concrete Cowboys (CBS-movie); Holiday (CBS pilot); Once Upon a Spy.

SANSOM, LESTER A.: Producer. b. Salt Lake City, Utah. e. University of Utah. Radio singer under name of Jack Allen, 1930; ent. m.p. ind. in editorial dept., Fox Film Corp., Dec. 1931; served in U.S. Navy as head of film library, Washington, D.C., 1942–45; head of edit. dept. & post-prod., Allied Artists, since 1953; assoc. prod. Skabenga; prod., co-writer, Battle Flame; assoc. prod. Hell to Eternity, exec. prod. The Thin Red Line, prod. Crack in the World; prod. Bikini Paradise, Battle of the Bulge, Custer of the West, Co-prod., Krakatoa—East of Java; exec. prod. 12 + 1.

SAPERSTEIN, HENRY G.: Executive. b. Chicago, Ill., June 2, 1918. e. U. of Chicago. Theatre owner, Chicago, 1943–45; pres. Television Personalities, Inc., 1955–67 Mister Magoo, Dick Tracy, TV shows, 1960–62; 1960–67 Glen Films, Inc.; prod., All-Star Golf, 1958–62; prod. Championship Bowling, 1958–60; prod. Ding Dong School, 1959–60; pres. owner, UPA Pictures, Inc., prod. Gay Purr-ee, animated features. Mr. Magoo, Dick Tracy cartoon series, Mr. Magoo's Christmas Carol. Prod. War of the Gargantuas, Invasion of the Astros, What's Up, Tiger Lily?, T-A-M-I, T.N.T. Show, Turnon, Tune In Drop Out, 1968; Hell in the Pacific. Pres. Screen Entertainment Co., Benedict Pictures Corp., United Prod. of America; pres. H.G. Saperstein & Associates. Producer: The Vaudeville Thing; Tchaikovsky Competition, Gerald McBoing Boing Show.

SARAFIAN, RICHARD C.: Director. b. New York City. Studied medicine and law before entering film industry with director Robert Altman doing industrial documentaries. Made documentaries before starting TV career.
TV: Gunsmoke, Bonanza, Guns of Will Sonnet, I Spy etc.
PICTURES: Andy (debut, 1965), Run Wild, Run Free, Ballad of a Badman, Fragment of Fear, Man in the Wilderness, Vanishing Point, Lolly Madonna (XXX), The Man Who Loved Cat Dancing, The Next Man (also co-prod.), Sunburn.

SARANDON, SUSAN: Actress. b. New York. e. Catholic University. Came to New York to pursue acting, first signing with Ford Model Agency. Made film debut in Joe. Also appeared on TV in A World Apart series. Co-produced film, The Last of the Cowboys.
PICTURES INCLUDE: The Great Waldo Pepper, The Rocky Horror Show, Lovin' Molly, The Front, assoc. cond. to Alexander Smallens, Porgy Page, Dragonfly, Walk Away Madden, The Other Side of Midnight, Pretty Baby, King of the Gypsies, Loving Couples, Atlantic City.

SARGENT, ALVIN: Writer. Began career as writer for TV, then turned to theatrical films.
PICTURES INCLUDE: The Stalking Moon, Gambit, The Sterile Cuckoo, The Effect of Gamma Rays on Man in the Moon Marigolds, Love and Pain (and the whole damn thing), Julia, Bobby Deerfield, Straight Time, Ordinary People.

SARGENT, DICK: Actor. b. Carmel, CA, 1933. Veteran of over 140 TV shows, 17 feature films and four TV series.
PICTURES: Captain Newman, M.D., Operation Petticoat, Mardi Gras, Bernadine, Hardcore, etc.
TELEVISION: Bewitched.

SARGENT, JOSEPH: Director.
PICTURES INCLUDE: The Taking of Pelham One, Two, Three, MacArthur, Goldengirl.
TV: The Marcus-Nelson Murders (Emmy award for tv movie that was pilot for Kojak series), The Night That Panicked America, Sunshine, Tribes, Hustling.

SARGENT, THORNTON: Executive. b. Wichita, Kans., June 22, 1902. e. U. of Mich., A.B. 1922, M.A., 1923. Instructor U. of Mo.; drama ed. St. Louis Times; adv.-pub. dir. Skouras Thea., 1932; Fox studio pub. 1933; pub. dir. Fox West Coast Thea., 1934; dist. mgr. L.A. first-runs 1935–43; training films div. Bur. Aeronautics and asst. officer in chg., Naval Photo Center, Hollywood; dir. adv. and pub. rel. Fox West Coast Thea., 1946; Lou Smith Associates, 1958; sales mgr. UPA Pictures, 1960; Sargent-Bourne Publicity, 1961–68. Public Relations consultant, 1968–81.

SARGOY, EDWARD A.: Attorney. b. New York City. e. C.C.N.Y., B.S., 1921, Columbia U. Law Sch., LL.B., 1924. Admitted to N.Y. bar. Specialist in copyright & m.p. law. Member Law firm of Sargoy & Stein, New York City, later known as Sargoy, Stein & Hanft, spec. counsel nationally for 11 major m.p. distrib. cos. in copyright infringement and distribution, from Jan. 1946 until his retirement, December, 1970. Now counsel. Compiler and annual reviser of U.S. Copyright Law Digest for Martindale—Hubbell Law Directory, 1948–76. Technical consultant to U.S. Delegation, for Inter-American Copyright Convention signed in Washington, D.C., June 1946, and Universal Copyright Conv., signed in Geneva, Sept. 1952. Chairman, Committee Copyrights Assn. Bar City of N.Y., 1957–60, formerly chairman of: Copyright Subsection of Amer. Bar Assn. and of its Committees on Copyright on Int'l Copyright, and on Program for Copyright Law Revision. Author of "U.C.C. Protection in the U.S.A." In June 1958 issue of N.Y. Univ. Law Review, and contributions to other legal periodicals. Lecturer on copyright and m.p. law in courses at Practicing Law Institute and in various law schools. Panelist in various bar association copyright symposia; pres. Copyright Society of the U.S.A., 1963–67.

SARLUI, ED: Executive. b. Amsterdam, Holland, Nov. 10, 1925. Owner, Peruvian Films, S.A.; pres., Radio Films of Peru, S.A.; pres. Bryant Films Educatoriana, S.A.; partner, United Producers de Colombia Ltd.; pres. Royal Film N.V.; pres., United Producers de Centroamerica, S.A.; pres. United Producers de Mexico, S.A.; pres., United Producers Int'l, Inc.

SARNOFF, ROBERT W.: Executive. b. New York, N.Y., July 2, 1918; Harvard U., B.A., 1939; Columbia Law Sch. 1940. In office of Coordinator of Info., Wash., D.C., Aug. 1941; the U.S. Navy, Mar. 1942; asst. to publisher, Gardner Cowles, Jr., 1945; mem. of staff Look Mag., 1946, with NBC, 1948–65; pres., Dec. 1955–58; chmn. bd., 1958; bd. of dir. RCA, 1957; chmn bd. chief exec. officer, NBC, 1958–65; pres. RCA, 1966; Chief Exec. Officer, 1968; bd. chmn., 1970–75. Mem., TV Pioneers, 1957; pres., 1952–53; International Radio & TV Society, Broadcasters Committee for Radio Free Europe. Am Home Products, Inc., dir., of Business Committee for the Arts.

SARNOFF, THOMAS: Executive. b. New York, N.Y., Feb. 23, 1927. e. Phillips Academy, Andover, Mass., 1939–43, Princeton U., 1943–45, Stanford U. grad. 1948, B.S. in E.E., Grad School of Bus. Admin. 1948–49. Sgt., U.S. Army Signal Corps, 1945–46; prod. & sales, ABC-TV, Hollywood, 1949–50; prod. dept. MGM, 1951–52; asst. to dir. of finance and oper., NBC, 1952–54; dir. of prod. and bus. affairs, 1954–57; vice pres., prod. and bus. affairs, 1957–60; v.p. adm. west coast, 1960–62; v.p. west coast, 1962; exec. v.p. 1965–77; bd. of dir., NBC prods 1961–77; bd of dir. Hope Enterprises 1960–75; dir. NABCAT, Inc. 1967–75; dir. Valley County Cable TV, Inc. 1969–75; Pres. NBC Entertainment Corp. 1972–77; Pres. Sarnoff International Enterprises, Inc. 1977– . Past pres. Research Foundation at St. Joseph Hospital of Burbank; past pres. Permanent Charities of the Entertainment Ind.; past ch. bd. of trustees, National Acad. of TV Arts and Sciences.

SARRAZIN, MICHAEL: Actor. r.n. Jacques Michel Andre Sarrazin. b. Quebec, Canada, May 22, 1940. Began acting at 17 on CBC TV; signed by Universal, 1965, appearing on several TV shows.
PICTURES INCLUDE: Gunfight in Abilene (debut), The Film-Flam Man, The Sweet Ride, Journey to Shiloh, A Man Called Gannon, Eye of the Cat, In Search of Gregory, They Shoot Horses, Don't They?, The Pursuit of Happiness, Sometimes a Great Notion, Believe in Me, Harry in Your

Pocket, For Pete's Sake, The Reincarnation of Peter Proud, Scaramouche, The Gumball Rally, Caravans.
TV: Chrysler Theatre, The Virginian, World Premiere.

SASSOWER, HARVEY L.: Advertising director. b. N.Y., July 28, 1945. e. CCNY (B.A. in adv.); 1968 asst. to adv. mg., United Artists. 1969, asst. to adv. dir., 20th Century-Fox, 1969, appointed adv. mgr. of ABC Pictures Corp., dir. of adv., ABC Pictures, 1970; pres., Universal Spectrum, Inc. (design & adv. studio). Art director; author.

SATENSTEIN, FRANK: Producer-director. b. N.Y.C., Oct. 7, 1924; e. Cornell Univ., Columbia Univ. U.S. Air Force; bd. chm. & dir., American Book—Stratford Press Inc.; dir., Standard Security Life Insurance Co. of New York; dir. Trans-Lux Corp.
THEATRE: Are You With It, Deep Are The Roots, The Greatest of These, Eagle-Lion.
PICTURES INCLUDE: Open Secret, Closeup.
TV: Ken Murray Show, Frank Sinatra Show, The Show Goes On, Strike It Rich, I've Got a Secret, Robert Q. Lewis, Godfrey's Talent Scouts, Jackie Gleason Show, What's My Line, Timex Show, Red Buttons.

SAUL, OSCAR: Writer. b. Brooklyn, N.Y., e. B'klyn Coll. Co-author play, Medicine Show; m.p. ed., U.S. Public Health Svce; numerous radio and tv plays.
PICTURES INCLUDE: collab. s.p., Once Upon a Time, Strange Affair; collab. story, Road House, Lady Gambles; s.p., Woman in Hiding, Secret of Convict Lake; adapt., Streetcar Named Desire; collab. sp. p., Thunder on the Hill, Affair in Trinidad; prod., Let's Do It Again; collab. s.p. Helen Morgan Story; s.p. Joker Is Wild; collab. story Naked Maja; collab. s.p. Second Time Around, Major Dundee; s.p. The Silencers; collab. s.p. Man and Boy. Novel: The Dark Side of Love (NBC movie).

SAVAGE, DAVID: Executive Producer & Advertising Executive. b. New York, N.Y., March 17, 1924. e. Columbia U., 1945. In research development & testing div., Eastman Kodak Co., 2 yrs.; adv. mgr. asst. nat'l sales mgr., Official Films; org., film dept. mgr. WCBS-TV; dir. of film procurement, CBS; mgr. of film procurement, NBC 1953; exec. v.p., Bernard L. Schulbert; pres., Theatrical Enterprises div.; prod., indiv. shows.; v.p., in chg. radio, TV, acct. sup., Lynn Baker Inc., 1959; v.p. in chg. m.p. prod., TV prog. Gerald Prods., 1962; pres., Gerald's Clef-10 Prods. Inc.; mgr. planning, merchandising, Recorded Tape Dept., RCA Records, 1966; Promo. Mgr. Special Products Marketing RCA Records Div; pres., Response Industries, Inc., (adv. agency), 1970 which became affiliate of McCann Erickson, and was v.p. of McCann Erickson in addition to continuing as pres. of R.I.

SAVAGE, JOHN: Actor. Studied at American Academy of Dramatic Arts. In Manhattan organized Children's Theatre Group which performed in Public Housing. Has appeared in many plays both on and off Bdwy. Won Drama Circle Award for performance in One Flew Over the Cuckoo's Nest in Chicago and Los Angeles.
PICTURES: Bad Company, Steelyard Blues, All the Kind Strangers, The Deer Hunter, Hair, The Onion Field, Cattle Annie and Little Britches, Inside Moves.
STAGE: Fiddler on the Roof, Ari, Siamese Connections, The Hostage, American Buffalo.
TELEVISION: Eric, Gibbsville (series).

SAVALAS, TELLY: Actor. b. Garden City, N.Y., Jan. 21, 1924. r.n. Aristotle Savalas, e. Col. Univ. (B.S.). Joined Information Services of State Dep't.; made exec. dir. Then named sr. dir. of news, special events for ABC, where created Your Voice of America series. Acting career began with debut in Bring Home a Baby on Armstrong Circle Theatre TV.
PICTURES INCLUDE: Birdman of Alcatraz, Young Savages, Cape Fear, Man from the Diner's Club, Battle of the Bulge, Greatest Story Ever Told, Beau Geste, Dirty Dozen, Buona Sera, Mrs. Campbell, Crooks and Coronets, Kelly's Heroes, On Her Majesty's Secret Service, Killer Force, Lisa and the Devil, Capricorn One, Escape to Athena, Beyond the Poseidon Adventure.
TELEVISION: Mongo's Back in Town (movie), Visions (movie), The Marcus-Nelson Murders (movie), Kojak (series).

SAWELSON, MEL: Executive b. Los Angeles, Calif., Sept. 5, 1929. e. USC 1947–48, UCLA 1948–49. Entered M.P. industry in 1947; mgr., Acme Film Laboratories, Inc., 1952; pres., Sports-TV, 1957–59; produced, Olympic Films, International Olympic Organization, 1956; produced, Big 10 Hilites, PCC Hilites, All American Game of the Week, 1957–59; 1st m.p. lab. exec. to install videotape, 1959; created Acme-chroma process of transferring color videotape to film; pres., Acme Film & Videotape Labs., 1967–71; v.p. Consolidated Film Industries, 1971, exec. v.p. 1972; pres., Glen Glenn Sound Co., 1972.

SAXON, JOHN: Actor. b. Brooklyn, N.Y., Aug. 5, 1935. Model; m.p. debut: Running Wild.
PICTURES INCLUDE: The Unguarded Moment, Rock Pretty Baby, Summer Love, The Restless Years, The Reluctant Debutante, Cry Tough, Portrait in Black, The Unforgiven, The Plunderers, Posse from Hell, War Hunt, Nightmare, Evil Eye, For Singles Only, Joe Kidd, Enter The Dragon, Black Christmas, Strange Shadows in an Empty Room, The Electric Horseman.

SCHACKER, MARSHALL: Producers representative. b. Los Angeles, Calif., Jan. 10, 1922. e. Univ. of Southern Calif., 1939–42. U.S. Army Signal Corps., 1942–45. Bus. rep., Indep. M.P. Prods. in Hollywood, 1945–48; talent, prog. dept. exec., National Broadcasting Co., 1948–52. Producers representative specializing in representation of European m.p. companies, 1952 to present.
CURRENTLY REPRESENTING: Italy: Titanus, Documento Films, Malenotti, Cirac, CCM; France: Terra Film, Les Films Agiman, Cormoran Film, Fides Film, Rialto Film, Arpa; England: Regal Films International, Ltd. Pres., Premiere Films, Inc., Premiere Video, Inc.

SCHAEFER, CARL: Publicist, b. Cleveland, Ohio, Sept. 2. e. U.C.L.A., contr. to mag., including Vanity Fair, Hollywood Citizen-News, 1931–35; Warner Bros., 1935.; Huesped de Honor, Mexico, 1943; OSS WWII, 1944–45; Int'l Comt. AMPS, chmn. 1966–67; Italian Order of Merit, 1957; Chevalier de l'ordre de la Couronne, Belgium, 1963. Pres., Foreign Trade Assn. of Southern Calif., 1954; chmn. of bd., 1955; British-American C. of C., Dir., 1962; Chevalier French Legion d'Honneur, 1955; Comm Hollywood Museum; dir., International Relations, Warner Bros. Seven Arts Int'l Corp., 1960; formed own firm, Carl Schaefer Enterprises, 1971. Dir. pub. rel., British-American Chamber of Commerce, 1971; dir. pub. rel. Iota International Pictures, 1971; dir. pub. rel. Lyric Films International, 1971; bureau chief (Hollywood) Movie/TV Marketing, 1971; managing dir., International Festival Advisory Council, 1971; dir. pub. rel. & adv. Francis Lederer Enterprises (Inc. American National Academy of Performing Arts, and Canoga Mission Gallery) 1974; West Coast rep. Angelika Films of N.Y. 1974, Hwd. rep Krowitz/Geiger Products. 1975– ; Hwd. corresp. Movie News, S'pore, & Femina, Hong Kong, 1974– ; member Westn. Publications Assn. 1975– ; field rep. Birch Records 1975; Hollywood rep. Antena Magazine, Buenos Aires; dir pub rel Style Magazine. Coordinator Hollywood Reporter Annual Key Art Awards; coordinator Hollywood Reporter Annual Marketing Concept Awards; executive committee & historian ShoWesT.

SCHAEFER, GEORGE: Producer, director. b. Wallingford, Conn., Dec. 16, 1920. e. Lafayette Coll., Yale Drama School. Dir. for Broadway stage; assoc. prod., Teahouse of the August Moon; prod. N.Y. City Center Theatre Co., 1949–52; dir., Dallas, Tex., State Fair musicals, 1952–58; prod., dir., B'way Play: Write Me a Murder, The Last of Mrs. Lincoln; dir. for TV.
TV: Including Macbeth, Little Moon of Alban; Victoria Reginia; The Magnificient Yankee, A War of Children; In This House of Brede, F. Scott Fitzgerald & The Last of the Belles, Truman At Potsdam, Sandburg's Lincoln, Amelia Earhart; Our Town; Blind Ambition; Here There Be Dragons, Voyage of the Mayflower, Barry Manilow: One Voice, The Bunker, People vs. Jean Harris; prod.-dir., Hallmark Hall of Fame. Pres. Compass Prod., Inc. Pres., Directors Guild of America.
PICTURES: Dir., Generation, Pendulum, Doctor's Wives, Once Upon a Scoundrel, An Enemy of the People.

SCHAEFER, JACK: Writer. b. Cleveland, O., Nov. 19, 1907. e. Oberlin Coll., Columbia U. edit. & pub., Theatre News, 1935–40, The Movies, 1939–41
AUTHOR: Short stories for TV anthologies; Shane, First Blood, Jeremy Rodock, Company of Cowards, Monte Walsh, The Great Endurance Horse Race.

SCHAFER, NATALIE: Actress. b. New York, N.Y. e. Merrill School and Hamilton Institute.
B'WAY: Lady in the Dark, Susan and God, The Doughgirls.
PICTURES INCLUDE: Marriage is a Private Affair, The Snake Pit, Molly and Me, Dishonored Lady, Susan Slade, 40 Carats, The Day of the Locust.
TV: I Love Lucy, Route 66, 77 Sunset Strip, Thriller, The Beverly Hillbillies, Gilligan's Island.

SCHAFFNER, FRANKLIN: Director b. Tokyo, Japan, May 30, 1920; e. Franklin & Marshall Coll. U.S. Navy, W.W. II; Dir. Person to Person, Studio One. Kaiser-Aluminum Hour, Playhouse 90; Emmy Best dir. 1945–55.
PICTURES INCLUDE: The Stripper, 1962, The Double Man, The Best Man, Planet of the Apes, Patton, Nicholas and Alexandra, Papillon, Islands in the Stream, The Boys from Brazil, The Sphinx.

SCHARY, DORE: Writer-producer-director. b. Newark, N.J., Aug. 31, 1905. Little theatre dir., actor in stock; playwright, Too Many Heroes, Sunrise at Campobello, The Highest Tree, The Devil's Advocate, Banderol; newspaper, mag. columnist author orig. & s.p. alone and in collab., numerous pictures from 1932.

Pres. & chief exec. off., Theatre Vision Inc., 1972; pres. & chief exec. officer, Schary Productions, Inc., 1973.

PICTURES INCLUDE: Boys Town, Young Tom Edison, Edison the Man; Acad. award, best orig., Boys Town, 1938; in 1942 appt. exec. prod. MGM; prod., Journey for Margaret, The War Against Mrs. Hadley, Lassie Come Home, Joe Smith, American, Lost Angel, Bataan, I'll Be Seeing You, Till the End of Time, The Spiral Staircase, The Farmer's Daughter, The Bachelor and the Bobby-soxer, Mr. Blandings Builds His Dream House, The Window, The Set-up, Crossfire, Bad Day at Black Rock, Go For Broke, The Swan, Designing Woman, Lonely Hearts, Sunrise at Campobello, writer, prod., dir., Act One.

Joined Vanguard Films, 1943, as prod. exec.; v.p. in chg. prod., RKO Radio Pictures, 1947; July 1948 to Dec. 1956, v.p. in chg. prod. MGM. A Champion of Champion Producer Fame ratings. Formed indep. co., 1958.

STAGE: Author, co-prod. with Theatre Guild, Sunrise at Campobello; author, co-prod. with Theatre Guild and dir., The Highest Tree; co-prod. with Theatre Guild, dir., The Unsinkable Molly Brown; dramatized, prod., dir., The Devil's Advocate; co-prod. with Theatre Guild and dir., Something About A Soldier; co-prod., dir., The Zulu & the Zayda. Dramatized and produced Herzl.

SCHATZBERG, JERRY: Director. b. N.Y.C. Early career in photography as asst. to Bill Helburn.

PICTURES INCLUDE: Puzzle of a Downfall Child, Panic in Needle Park, Scarecrow, Sweet Revenge (prod.-dir.), The Seduction of Joe Tynan, Huckleberry Rose.

SCHEIDER, ROY: Actor. b., Orange, N.J. e. Franklin and Marshall College where he won the Theresa Helburn Acting Award twice. First professional acting in N.Y. Shakespeare Festival 1961 prod. of Romeo and Juliet. Became member of Lincoln Center Repertory Co. and acted with Boston Arts Festival, American Shakespeare Festival, Arena Stage (Wash., D.C.) and American Repertory Co.

PICTURES INCLUDE: Loving, Paper Lion, Stiletto, Star!, Puzzle of a Downfall Child, Klute, The French Connection, The Outside Man, The Seven Ups, The Inheritor, Sheila Levine, Jaws, Marathon Man, Sorcerer, Jaws II, Last Embrace, All That Jazz.

TV: Hallmark Hall of Fame, Studio One, N.Y.P.D.

STAGE: Richard III, Stephen D, Sergeant Musgrave's Dance, The Alchemist.

SCHELL, MARIA: Actress. b. Vienna, 1926. Sister of Maximillian Schell. Made debut at 12 in Swiss film, The Gravel Pit. Subsequently appeared in many British and American films.

PICTURES: Angel with a Trumpet, The Heart of the Matter, So Little Time, The Magic Box, The Brothers Karamazov, The Hanging Tree, Cimarron, The Odessa File, Voyage of the Damned, The Twist, Superman.

SCHELL, MAXIMILIAN: Actor. b. Vienna, Dec. 8, 1930.

TV: Shows include: Playhouse 90, Judgment at Nuremberg, The Fifth Column, Turn The Key Deftly.

PICTURES INCLUDE: Children, Mother and the General, The Young Lions, Judgment at Nuremberg, Five Finger Exercise, The Condemned of Altona, Return from the Ashes, The Deadly Affair, Counterpoint, The Desperate Ones, The Castle, Krakatoa East of Java, The Odessa File. Dir., starred First Love; The Man in the Glass Booth (actor), End of the Game (dir., co-prod.), St. Ives, A Bridge Too Far, Cross of Iron, Julia, Players, The Black Hole.

SCHENCK, AUBREY: Producer. b. Brooklyn, N.Y., Aug. 26, 1908; e. Cornell U.; N.Y.U. With law firm of O'Brien, Driscoll & Raftery; buyer & attorney for Natl. Theaters, 1936; prod. for 20th Century-Fox 1945; exec. prod. Eagle Lion 1946; contract prod. Universal Internatl. 1948; Aubrey Schenck Productions, Inc.

PICTURES INCLUDE: Shock, Johnny Comes Flying Home, Strange Triangle, Repeat Performance, T-Men, Mickey, It's a Joke Son, Trapped, Port of New York, Wyoming Man, Undercover Girl, Fat Man, Target Unknown; formed own co. to prod. War Paint, Beachhead. Also: Yellow Tomahawk, Shield for Murder, Big House, U.S.A., Crime Against Joe, Emergency Hospital, Ghost Town, Broken Star, Rebels in Town, Pharaoh's Curse, Three Bad Sisters, Fort Yuma, Desert Sands, Quincannon, Frontier Scout, Black Sleep, Hot Cars, War Drums, Voodoo Island, Revolt at Fort Laramie, Tomahawk Trail, Untamed Youth, Girl in Black Stockings, Bop Girl Goes Calypso, Up Periscope, Violent Road, Reckless, Frankenstein 1970, Wild Harvest, Robinson Crusoe On Mars, Don't Worry, Ambush Bay, Kill a Dragon, Impasse, More Dead Than Alive, Barquero, Daughters of Satan.

TV: Miami Undercover, series.

SCHERICK, EDGAR J: Executive Producer. e. Harvard Univ.; elected to Phi Beta Kappa. Introduced Wide World of Sports on TV thru his co., Sports Programs, Inc. Was v.p. in chg. of network programming at ABC-TV. Now pres. of Palomar Pictures Int'l.

PICTURES INCLUDE: For Love of Ivy, The Birthday Party, Take the Money and Run, They Shoot Horses, Don't They?, The Killing of Sister George, Sleuth, The Heartbreak Kid, Law and Disorder, The Stepford Wives, I Never Promised You a Rose Garden.

SCHERMER, JULES: Associate producer. In 1942: True to the Army, Henry Aldrich, Editor, Par. In 1943: Henry Aldrich Gets Glamour, Par.; collab. s.p. The Sullivans.

PICTURES INCLUDE: Framed, Man From Colorado, Criss-Cross, Illegal Entry, Union Station, Pride of St. Louis, Lydia Bailey, Pickup on South Street, Pushover, Carthay Center Prods. Inc., 1965. Prod., A Dream of Kings.

SCHIAVONE, JAMES: Executive. b. Niagara Falls, N.Y., Nov. 14, 1917. e. U. of Michigan. Started career as newspaperman; general manager WWJ-AM-FM-TV, Detroit, 1952–68; v.p.-gen. mgr., KSAT-TV, San Antonio, Tex., 1969-present.

SCHICK, ELLIOT: Executive. b. Brooklyn, N.Y., Dec. 24, 1924. e. Brooklyn Coll. (B.A.), New School for Social Research (drama workshop, directing 1945–46). Author of book for Ballet Theatre, Manfred; book, The Administration of the Economic and Social Council. 1942–48, prod. & dir. radio shows for WNYC, N.Y.C., composed and arranged music for radio and stage production; 1946–48 dialogue director, Republic Studios; 1948–50, prod. & dir. TV shows and commercials, Nova Productions; 1950–51, editor, United Nations Film Div.; 1951–53, editor, Candid Camera; 1953–55, asst. studio mgr., American Natl. Studios; 1955–56, prod. mgr., Gross-Krasne; 1956–60, prod. & dir. industrials, commercials, documentaries, Fred Niles, Chicago; 1960–63, dir., commercials, Hollywood Film Commercials; 1964–65; prod. & dir. live and video tape shows for KCET; 1965–69, asst., dir. for TV series: 12 O'Clock High, Time Tunnel, Star Trek and features: Hombre, Hard Contract, Bloody Mama, Billy Jack, Tora, Tora, Tora; 1969–72, prod. mgr. for features: 3 in the Attic, Up in the Cellar, Bunny O'Hare, Honkers, Hickey and Boggs, Kansas City Bomber, White Lightning; 1973–77 spvr. prod. for AIP on Sugar Hill, Return to Macon County, Cooley High, Futureworld, Island of Dr. Moreau in 1977 joined EMI Films, Inc. as v.p. prod.; 1978, exec.-in-chg.-prod., Deer Hunter; 1980, produced The Earthling; 1981, prod., Private Benjamin (TV series), supvr. prod., Pippin (TV).

SCHIFRIN, LALO: Composer, b. Buenos Aires. Father was conductor of Teatro Colon in B.A. for 30 years. Schifrin studied with Juan Carlos Paz in Arg. and later Paris Cons. Returned to homeland and wrote for stage, modern dance, TV. Became interested in jazz and joined Dizzie Gillespie's band in 1962 as pianist and composer. Settled in L.A.

PICTURES INCLUDE: The Cincinnati Kid, The Liquidator, Cool Hand Luke, The President's Analyst, The Fox, Kelly's Heroes, Magnum Force, Man on a Swing, The Four Musketeers, Voyage of the Damned, The Eagle Has Landed, Rollercoaster, The Manitou, Boulevard Nights, Escape to Athena, The Amityville Horror, The Nude Bomb, The Competition, Caveman.

SCHILLER, FRED: Playwright, screen & TV writer, radio producer. Awarded: Literary Prize, for McCall magazine story Ten Men and a Prayer. National Member of Smithsonian Institution, Member of Societé Des Auteurs et Compositeurs Dramatiques. Was chief corres. European Newspaper Feature Services, exploitation director. RKO Radio Studios. Honored by the Univ. of Wyoming for literary achievements with a special Fred Schiller Collection for their library.

PICTURES FOR: MGM, Columbia, RKO, Republic and Henri Sokal Films, Paris.

TV: TV writer; NBC All Star Special adapt., George Bernard Shaw's play, Inca of Persualem. All in all 53 TV plays, Four Star Playhouse, The Millionaire, Loretta Young Show, Charlie Chan, Desilu, Colonel Flack, The Veil, Manhunt, The Islanders, The Beachcomber, Follow the Sun, The Third Man, The Americans You Don't Know; Two 90 min. TV specials, Finder Bitte Melden, Vienna, and Demandez Vickey!, Paris. TV Pilot, Comedy Series Diet Fever, ABC.

STAGE: Come On Up (U.S.), Anything Can Happen (London), Demandez Vicky (Paris), Finder Please Return (Athens, Madrid), Finder Bitte Melden (Berlin, Hamburg, Vienna).

SCHILLER, LAWRENCE J.: Producer-Director. b. New York City, Dec. 28, 1936. Photo-journalist with Life Magazine & Saturday Evening Post, 1958–70; collaborated on numerous books including three by Norman Mailer: The Executioner's Song, Marilyn, and The Faith of Graffiti; Muhammad Ali with Wilfrid Sheed; Minamata with Eugene Smith.

PICTURES: The Man Who Skied Down Everest (editorial concept & direction); Lady Sings the Blues, Butch Cassidy & the Sundance Kid (conceived and executed special still montages & titles); The American Dreamer (prod./dir.).

TELEVISION: Hey, I'm Alive (prod./dir.); The Trial of Lee Harvey Oswald (prod.); The Winds of Kitty Hawk (prod.).

SCHIMEL, ADOLPH: Lawyer. b. Vienna, Austria. U.S., 1904 naturalized. 1910 e. A.B. Coll, City of N.Y., 1920; LL.B., Harvard 1923; Admitted to N.Y. bar, 1924, since practiced in N.Y.C.; Sec'y., Universal Pictures since 1945; v.p., gen. counsel, Universal & subs ds., 1950 to 1972. Mem: law com. (chmn.), 1955 to 1972 MPA, Bar Assn. of the City of New York, American Bar Assn; consultant to Universal since 1972.

SCHIMMEL, HERBERT D.: Executive. b. New York, N.Y., October 15, 1927. e. University of Pennsylvania, Wharton School. C.P.A., partner with M. Schimmel and Co. prior to entering m.p. industry in 1966 as treasurer, Europix Consolidated Corp.; pres. 1968; pres., Europix International Ltd., 1971.

SCHINE, G. DAVID: Executive. Film exhibitor until 1966 in New York, Ohio, Kentucky, Maryland, Delaware, and West Virginia. Exec. prod. of French Connection, 1971. Writer, prod., dir. of That's Action!, 1977. Chief Exec. officer of Schine Productions (production) and Epic Productions, Inc. (distribution), and Studio Television Services, Inc.

SCHLAIFER, CHARLES: Executive. President, Charles Schlaifer and Company, Inc., advertising agency with offices in New York and Los Angeles. b. Omaha, Nebr. Reporter Daily News, World-Herald, (Omaha). In 1928 apptd. adv. mgr. Paramount theatres, Omaha; then of Publix theats., Omaha; then of Tri-State circuit, Neb., Iowa; 1936–42 managing dir. United Artists Theats. San Francisco; advisor, nat'l adv., United Artists prod. In 1942 apptd. adv. mgr. 20th Cent.-Fox; named asst. dir. adv., publicity & exploitation, 1944; appt'd. dir. of advertising, publicity, exploitation and radio, 1945. Resigned 20th-Fox post, Feb. 1, 1949, to establish own adv. agency. Pres., Charles Schlaifer & Co., Inc. Chmn. advertising advisory council, MPAA; instructor at New School of Social Research, N.Y., on m.p.; revised m.p. adv. code; permanent chmn. first MPAA pub rel. com. Member; Nat'l Advisory Mental Health Council to U.S. Surgeon General; vice-chmn. bd. of gov., Nat'l Assn. of Mental Health. Lecturer, writer on adv. & mental health bd. of gov., Menninger Foundation; founder, co-chmn., Nat'l Mental Health Comm., secy., treas., Joint Comm. on Mental Illness & Health; expert witness Congress, govt. hearings. Elected Hon. Fellow of the Amer. Psychiatric Assn., 1959; V. chmn., trustee in chg., Mental Health and Mental Retardation Facilities, N.Y. State, 1964; secy., treas., Joint Commission Mental Health for Children; vice chmn. bd. Foundation for Child Mental Welfare, 1963. Member bd. trustees Research Foundation 1966. Member White House Conference on the Handicapped, 1952–65; elected honorary fellow, Post Graduate Psychiatric Institute, 1968. Hon. Doctor of Letters, John F. Kennedy College, Wahoo, Neb., 1969; Chmn. N.Y. State Health and Mental Hygiene Facilities Improvement Corp., 1970; Hon. Fellow—American Ortho Psychiatric Assoc., 1970; Hon. Fellow British Royal Society of Health. Wisdom Award Hon. Wisdom Mag., 1969; social conscience award Karen Horney Clinic, 1972. Chairman, New York State Facilities Development Corp., 1973– ; Advisory Council to the National Institute of Mental Health & to the Surgeon General of the U.S. 1976–

SCHLANG, JOSEPH: Executive. b. N.Y.C. Feb. 24, 1911, e. N.Y.U. Owner and leader in N.Y.C. real estate, partner of N.Y. Stock Exchange firm and exec. dir. of many enterprises. Pres. of International Opera Co. & Opera Presentations, Inc. Has produced two weekly radio programs: Opera Stars of Tomorrow and 100 & More Ways to Improve N.Y.C. since April, 1973. Opera Presentations, Inc., a non-profit corp. distributes and exhibits opera, ballet and art films throughout America and the school system. Over 100 cultural films owned by Schlang are supplied to Opera Presentations free to use.

SCHLESINGER, JOHN: Director. b. London, England. 1960: Wrote and dir. Terminus for British Transport Films (Golden Lion, best doc., Venice); The Class. Some episodes The Valiant Years series.
PICTURES INCLUDE: 1961: A Kind of Loving (Golden Bear, Berlin); 1962–63: Billy Liar; 1964–65: Darling (New York Film Critics Award); 1966–67: Far From the Madding Crowd. 1968: Midnight Cowboy (Best dir. and film, S.F.T.A. and 'Oscars'), 1970: Sunday, Bloody Sunday (Best dir. and film, S.F.T.A.), Visions of Eight (sequence), The Day of the Locust, Marathon Man, Yanks, Honky Tonk Freeway.

SCHLONDORFF, VOLKER: Director. b. Wiesbaden, Germany, Mar. 31, 1939. Studied in France, acquiring degree in political science in Paris. Studied at French Intl. Film School, before becoming asst. to Jean-Pierre Melville, Alain Resnais, and Louis Malle Debut film, Der Junge Torless (Young Torless), 1965.
PICTURES: A Degree of Murder, 1966; Michael Kohlhass, 1968; The Sudden Fortune of the Poor People of Kombach, 1970; Die Moral der Rugh Halbfass, 1971; A Free Woman, 1972; The Lost Honor of Katharine Blum, 1975; Le Coup de Grace, 1976; Valeska Gert, 1977; The Tin Drum, 1979.

SCHLOSSBERG, JULIAN: Producer-Distributor-Radio Host. b. N.Y.C. Jan. 26, 1942. e. N.Y.U. Joined ABC-TV network 1964 as asst. acct. rep.; named act. rep. 1965; 1966, joined Walter Reade Organization as asst. v.p. chg. of TV; 1969, moved to WRO Theatre Div.; 1970, joined faculty of School of Visual Arts; 1971, named v.p. of WRO Theatres; 1972, Jan., 1976, joined Paramount Pictures as v.p. in charge of feature film acquisition. Since 1978 pres. & owner of Castle Hill Productions; 1974, prod. & moderated An Evening with Joseph E. Levine at Town Hall, N.Y.C.; 1974 to 1980, host of radio show Movie Talk on WMCA (N.Y.C.), WMEX (Boston), and WICE (Providence).
PICTURES INCLUDE: Ten from Your Show of Shows, No Nukes.
THEATRE: It Had To Be You.

SCHLOSSER, HERBERT S.: Executive. b. Atlantic City, N.J. e. Princeton U., Yale Law. Joined law firm of Phillips, Nizer, Benjamin, Krim & Ballon, 1954; attorney, California National Productions (subsidiary of National Broadcasting Company) 1957; v.p. & gen. mgr., 1960; director, talent & program administration, NBC television network, 1961; v.p., talent & program admin., 1962; v.p. programs, west coast, 1966–72; exec. v.p., NBC-TV, 1972; president, NBC Television Network, 1973, pres. & chief operating officer, NBC, April 1, 1974–76; Pres. & chief executive officer, 1977–78; exec. V.P. RCA 1978.

SCHLUSSELBERG, MARTIN: Film Executive. b. Sept. 1936. e. Yeshiva Univ. Booking clerk. UA. 1956; head booker, Citation Films, 1958; head booker, Desilu Dist., Co., 1961; head booker, and asst. to gen. sls. mgr., Medallion Pictures Corp., 1963; World Ent. Corp., 1966, Sales Mgr.; Crystal Pictures, 1978, sls. mgr., v.p.

SCHMIDT, WOLF: Producer-Distributor. b. Freiburg/Br., Germany, June 30, 1937. Came to U.S. 1962 as free-lance journalist. Started producing in 1969, distributing independently since 1972.
PICTURES INCLUDE: Ski Fever (prod.), Stamping Ground (co-prod.), Young Hannah (exec. prod.), Things Fall Apart (prod.), The Passover Plot (prod.), Run For the Roses (co-prod.).

SCHNEER, CHARLES H.: Producer, b. Norfolk, Va., May 5, e. Columbia College pres., Morningside Prods. Inc. & Pictures Corp.; 1956. Founded Andor Films 1974.
PICTURES INCLUDE: Proc. The 3 Worlds of Gulliver, The 7th Voyage of Sinbad, I Aim at the Stars, Face of a Fugitive, Good Day for a Hanging, Battle of the Coral Sea, Tarawa Beachhead, Mysterious Island, Jason and the Argonauts, First Men In The Moon, Half A Sixpence, Land Raiders, Valley of Gwangi, The Executioner, The Golden Voyage of Sinbad Sinbad & The Eye of the Tiger, Clash of the Titans.

SCHNEIDER, DICK: Prod.-dir. b. Cazadero, Calif., Mar. 7; e. Coll. of the Pacific; U.S. Navy, WWII.
TV: Dough Re Mi, NBC-TV Wide Wide World, NBC-TV, Colgate Comedy Hours, Beatrice Lilly & Jackie Gleason Comedy Hours, Henry Morgan Show, Kate Smith Show, Big Story, Treasury Men in Action, Doorway to Danger, Today Show, Home, Tonight Show, General Mills Circus; dir. coverage of political conventions; dir. NBC-TV coverage, Princess Margaret's Wedding and Paris Summit conference; dir. Eleanor Roosevelt Specials, 1959–60; Something Special 61, At This Very Moment, Inauguration, Gemini, Emmys, 1962, 1963, 1964; Papal Mass for all networks at Yankee Stadium, 1965–66; 1966–67 Tonight Show, Orange Bowl, Macy's Parade; Jr. Miss Pageant; 1967–70 College Queen, Emmy Award. Prod.; Macy's Parade, 1968–69 Orange Bowl Parade, 1968–69; Prod.-dir., NBC Expt. in TV, New Communication; prod., Big Sur; prod.-dir., Jr. Miss Pageant, 1968–69; dir. Dream House, ABC; dir. Who, What or Where, NBC; produced in 1970: Macy's Parade, Junior Miss, Orange Bowl Parade. 1971–79 Macy's Parade, Stars and Stripes; in 1972: Post Parade, Stars and Stripes 1973–75; Rose Parade 1974–80; Salute to Sir Lew; Jeopardy; NBC Star Salute, 1980; Emmy winner, 1980; Rose Parade, 1981; Star Salute, 1981.

SCHNEIDER, JOHN A.: Exec. b. Chicago, Ill., Dec. 4, 1926. e. Notre Dame, B.S U.S.N.R., 1943–47, Exec. assignments with CBS-TV, in Chicago and New York City 1950–58; VP, gen. mgr. WCAU-TV, Philadelphia 1958–64; WCBS-TV, New York City 1964–65; pres. CBS-TV Network 1965–66; pres. CBS/Broadcast Group 1966–69, 1971–77; exec. VP CBS Inc. 1969–71, TV and MP consultant 1977–79; consultant WCI, 1979; pres. CEO Warner Amex Satellite Entertainment Corp., 1980.

SCHNEIDER, ROMY: Actress. b. Vienna, Austria, Sept. 23, 1938. Parents Magda Schneider and Wolf Albach-Retty, of the stage. Debut on screen when 15. Appeared in 10 films produced on the Continent in four years. Introduced in U.S., 1958, as star of The Story of Vickie.
PICTURES INCLUDE: Forever, My Love; Boccaccio '70;

Good Neighbor Sam, Triple Cross, Otley, My Lover, My Son, The Hero, Ludwig, The Infernal Trio.

SCHNEIER, FREDERICK: Executive. b. New York, N.Y., May 31, 1927; e. N.Y. Univ., 1951, bus. admin., N.Y.U. Grad. Sch., MBA, 1953. Dir. Sls. planning, Mutual Broadcasting System, 1947–53; media research dir., RKO Teleradio, 1953–55; RKO Teleradio Advisory Comm., 1955–56; exec. staff RKO Teleradio & dir., marketing services, 1956–58; exec. vice-pres., Showcorporation, 1958–71; v.p. TV programming, RKO General, 1972–1973; v.p.; Hemdale Leisure Corporation, 1973–79. Joined Viacom Enterprises as v.p., feature films, 1979. Appt. sr. v.p., program acquisitions & motion pictures, 1980.

SCHNUR, JEROME: Producer-director. b. N.Y.C., July 30, 1923; e. Carnegie Tech. Film prod. & dir. many indep. cos., 1939–43; prod., dir., training films, U.S. Air Force 1943–46; indep. film prod., Hollywood, 1946–50: on bd. of dir., Alson Prod., Inc. & Burwood Pictures Corp.; prod., dir., CBS-TV, N.Y., 1950–51; prod., dir., Goodson-Todson Prod., 1951–56; pres., Holiday Prod., indep. TV prod., 1956; Metropole Prod., indep. film prod., 1956; exec. prod. chg. creative prog. Frank Cooper Assoc., 1958; pres., Jerome Schnur Prods., Inc. Packager and Producer TV and Film.
TV: It's News to Me, Two for the Money, The Name's the Same, What's My Line, Beat the Clock, Robert Q. Lewis Show, Judge for Yourself, Holiday film series, Fred Allen Show, Dotto, Make the Connection, Scene of the Crime, Who Pays, Shari Lewis Show, Bell Telephone Hour, U.S. Steel Spec., Private Eye—Private Eye, exec. prod., United Fund Simulcast This Is My Town; writer, dir., Mark of Cain, CBS-TV, Chronicle, Science Special, Tomorrow Was Yesterday; dir., L'Enfance du Christ, CBS-TV; exec. prod., Supermarket Sweep, ABC-TV; dir., Michelangelo's Pauline Chapel. CBS News spec.; dir., Jeptha's Daughter. St. Joan. TV baltet specials. Exec. Prod. Everybody's Talking; dir. Missions of San Antonio; Light in the Wilderness, and David Wept (all CBS). Exec. Prod. Politithon 1970—PBS. Exec. Prod. & Dir. "Threatened Paradise" (PBS) "A Time To Live" (PBS) 1972; Director "Questions of Abraham" Special Contata (CBS) "American Ballet Theatre Special" (PBS) "Luther" (CBS) 1973. Exec. prod., Musical Chairs, Strategy, others, 1974; Director, Joffrey ballet, PBS special; dir. Jerusalem Symphony, CBS spec.; 1977; dir. Song of Songs, CBS spec., 1978, Director, All Star-Jazz Show, CBS Special 1976. Recipient 1970–71 Peabody Award, 1970–71 Saturday Review Award, 1972 Emmy Citation; 1975 Ohio State Award.

SCHOENFELD, JOE: Executive. b. June 2, 1907. New York City. e. New York City Schools. Reporter, Hearst newspapers, 1924–32; reporter & editor, Variety, 1932–43; William Morris Agency, 1944–50; editor, Daily Variety, 1950–59; M.P. agency exec., William Morris, 1959–1974; film prod. present.

SCHOENFELD, LESTER: Executive. b. Brooklyn, N.Y., Dec. 6, 1916; e. C.C.N.Y., 1934–38, Asst. mgr., Randforce Amusement, 1936–38; mgr., Rugoff & Becker circuit, 1938–47; mgr., Golden & Ambassador Theatres, 1948; print & sales dept., Film Classics, 1948–50; chg. of Theatrical, non-theatrical & TV dist., Brit. Info. Serv.; est. Lester A. Schoenfeld Films, 1958.

SCHONFELD, NORMAN J.: Theater Executive, b. Newark, N.J., June 25, 1934, e. University of PA, BA 1955, Navigator, U.S.A.F., 1957, sls. rep. Bache & Co. 1963, v.p. Tiger Films, Inc., 1966 founder and president, Wood Theater Group.

SCHORR, JOSE: Advertising and Publicity Specialist; e. Cornell Univ. Contributor to McCall's, The New Yorker, Redbook, Fortune, This Week, Saturday Evening Post. Joined Columbia as adv. and pub. writer. Ed., Columbia Mirror. National mag. and syndicate pub. contact. Co-author Sensible Dieting (Knopf). Army, 1942–45; Sec'y. Soc. of Mag. Writers; adv. writer, 20th Century-Fox, Donahue & Coe. Universal, Solters & Sabinson.

SCHRADER, PAUL: Writer. b. Grand Rapids, Mich., 1946. e. Columbia Univ., U.C.L.A. (Master's in Cinema). Served as film critic for L.A. Free Press and Cinema 1970, 1971, 1972. First screenplay was The Pipeliner (unproduced).
PICTURES INCLUDE: The Yakuza, Taxi Driver, Close Encounters of the Third Kind, Rolling Thunder, Blue Collar (co-s.p., dir). Hardcore (s.p., dir.), Old Boyfriends (co-s.p.), American Gigolo (s.p., dir.), Raging Bull (co-s.p.).

SCHREIBER, EDWARD: Producer. b. Jersey City, N.J., Mar. 30, 1913. e. John Marshall Coll., Jersey City. Joined Warner Bros., 1936, press contact, then pub. dir., WB Vitaphone Studio, pub. and adv. dir., War Activities Comm., 1942. Joined Century Theatres, dir. adv., pub., 1946. Left ind., 1949 became builder 10 yrs. Ret., m.p. industry as independent prod., screenplay writer, Mad Dog Coll. Kathie's Lot (19 min.); Useless Mouths (20 min.); Cassals Conducts (28 min.); Acad. Award winner. Prod. off B'way: Unfair to Goliath.

SCHULBERG, BUDD WILSON: Writer. b. New York City, Mar. 27, 1914; son of B. P. Schulberg, prod.; e. Dartmouth Coll. Publicist, Paramount Pictures, 1931; writer for screen from 1932; addit. dial., A Star Is Born; collab. s.p., Little Orphan Annie; orig. & collab. s.p., Winter Carnival. Weekend for Three; collab. orig., City Without Men; adap. Government Girl. Armed services World War II. Author, 3 best-selling novels including The Disenchanted. Harder They Fall; story, s.p., On the Waterfront (Acad. Award & Writers Guild Award, 1954), Face in the Crowd, Wind Across the Everglades. Recent books: Everything That Moves (novel), Motion Pictures: Memories of a Hollywood Prince.

SCHULMAN, WILLIAM B.: Advertising executive. b. Dayton, O., July 2, 1916; e. Ohio State U., B.A., 1940; Columbia U. Lieut. Comm. U.S. Navy. WWII. In 1945 with Universal home office exploitation dept.; New England pub. and exploitation mgr., Universal. 1946. Became dir. adv. publ. & exploit., member bd. dir. Realart Pictures; gen. mgr. Screen Ad Adv. Agency, Inc.; adv. mgr. Sharmark Ent. Inc.

SCHUMAN, EDWARD L.: Executive vice president & partner, Quartet Films. b. Lisbon, Conn., Sept. 3, 1916. e. Wayne Univ., 1932–34, Univ. of Michigan, 1934–37. Mathematician, 1937–51, assistant actuary, Detroit City Employees Pension Fund. Pres., Studio Theatre Corp., Detroit, 1951 to 1975; pres., Studio 8 Theatre Corp., Detroit, 1953 to 1975; secy.-treas., Studio New Center Theatre Corp. of Detroit, 1975; v.p. and gen. mgr. Art Theatre Guild, 1954–60; v.p. Rugoff Theatres, N.Y., 1960–1963. V.p. and bd. member, Walter Reade Organization 1975: Associate producer Broadway play, Same Time Next Year, The Comedians, and Anna Christie. Assoc. prod. of First Monday in October.

SCHUSTER, HAROLD: Director. b. Cherokee, Ia., Aug. 1.
PICTURES INCLUDE: Marine Raiders, My Friend Flicka, So Dear to My Heart, Girl Trouble, Wings of the Morning, Zanzibar, The Tender Years, Postman Didn't Ring, A Very Young Lady, Exposed, Bombers Moon, Spring Tonic, Breakfast in Hollywood, Dinner at the Ritz, Strange Cargo, Framed, On the Sunnyside, Small Town Deb, Kid Monk Baroni, Ma He's Making Eyes at Me, Swing That Cheer, Jack Slade, Loophole, Security Risk, Port of Hell, Tarzan's Hidden Jungle, Finger Man, Return of Jack Slade, Black Beauty, Portland Expose, Dragon Wells, Massacre.
TV PLAYS: Playhouse of Stars, Ellery Queen, The Line-Up, Wire Service, Meet McGraw, The Gray Ghost, U.S. Marshall, Death Valley Days, Wichita Town, Man With a Camera, The Californians, The Outlaws, Laramie.
TV: Deacon St., Overland Trail, Death Valley Days, Tombstone Territory, Zane Grey Theatre, Twilight Zone, Surfside Six.

SCHUTE, MARTIN: Producer. Began as asst. dir. to Lewis Milestone; then graduated to prod. supervisor on Gentlemen Marry Brunettes for United Artists in 1955. Spent 3 years with Columbia Pictures in London. Became gen. mgr. of Otto Preminger's production co. during making of Exodus.
PICTURES INCLUDE: The Cardinal (assoc. prod.), Bunny Lake Is Missing (assoc. prod.), Maroc 7 (exec. prod.), The Lion in Winter (prod. exec.), Macho Callahan (co-prod.), Miss Julie (co-prod.), Not Now Darling (co-prod.), The Tempter (co-prod.), Not Now Comrade (prod.), Silver Bears (exec. prod.).

SCHWAB, LAURENCE, JR.: Producer, director, writer. b. Mar. 31, 1922, N.Y.C. Prod., dir., documentary films, TV films & shows.
PICTURES INCLUDE: American Inventory, Soldiers of Fortune, Stakeout.
TV: The Clock, Lights Out, Kate Smith Hour, Amazing Mr. Malone, NBC Matinee Theatre.

SCHWARTZ, ARTHUR: Producer, composer. b. Brooklyn, N.Y. e. New York U., J.D., A.B.; Columbia U., M.A.; Phi Beta Kappa. Member ASCAP, Authors League Amer., Dramatists Guild. Teacher N.Y. public schools; practiced law; then turned to music; composer songs & scores many N.Y. musicals.
MUSICALS INCLUDE: The Band Wagon, Flying Colors, Revenge With Music, Between the Devil, Stars in Your Eyes, Inside U.S.A., A Tree Grows in Brooklyn, By the Beautiful Sea.
PICTURES INCLUDE: Composing for screen includes scores for Girl From Paris, Under Your Spell, Navy Blues. score, collab. story Thank Your Lucky Stars; prod., Cover Girl; music, The Time, the Place and the Girl, Excuse My Dust, Dangerous When Wet, The Bandwagon, You're Never Too Young.

SCHWARTZ, ARTHUR H.: Lawyer, b. New York, N.Y., Nov. 18, 1903; e. Columbia Coll. & Law Sch. Asst. U.S. atty. for so. dist. of N.Y.; 1926–33; chief, Prohibition Div., 1931–33; assoc. with Nathan Burkan, 1933; member Schwartz & Frohlich since 1936 (became Schwartz, Burns, Lesser & Jacoby, 1970); now Special Counsel, Burns, Jackson, Miller, Summit & Jacoby; trial counsel in litigation affecting m.p.

ind. including copyright problems, contract, labor, work; counsel to N.Y. State Jt. Legislative Com. on Legislative Practices, Procedures & Expenditures, 1943–46; N.Y. State Republican campaign mgr., 1944; counsel to N.Y. State Temporary Comm. on Need for State U.; N.Y. State Comm. on Co-Ord., of State Affairs, 1947–60; counsel N.Y.S. Repub. Com.; member of N.Y.S. Law Revision Comm. 1960–1974 of N.Y.S. Comm. on Judicial & Legislation Salaries 1972-to-date Justice of N.Y. State Supreme Court 1952, former pres. N.Y. County Lawyers Assn. Chairman. NY 5 Board of Electors, 1974–75; Board of Editors, N.Y. Law Journal 1976–

SCHWARTZ, BERMUDA: Publicist. b. Deal, N.J., July 30, e. Seton Hall University; grad. Transylvania University, Newspaper editor, radio personality, Ice Capades; Mutual Radio Network; Broadway publicity; head, Schwartz, O'Heney, Thompson Inc., N.Y. pub./rel. firm. Motion picture publicity, Studio City, North Miami, Fla., Ivan Tors, Miami, Fla.

SCHWARTZ, BERNARD: Producer. Brought to Hollywood by the late Howard Hughes to watch his film interests; Schwartz teamed with atty. Greg Bautzer to package movie deals for clients. Re-cut number of Buster Keaton's silent movies into documentary anthologies (The Golden Age of Comedy, When Comedy Was King, etc.). Subsequently made TV series, One Step Beyond, followed by The Wackiest Ship in the Army, Miss Teen International specials, etc. Named pres., Joseph M. Schenck Enterprises, for which made Journey to the Center of the Earth, Eye of the Cat, A Cold Wind in August, I Passed for White, The Shattered Room, Trackdown. Presently partnered with Alan Silverman of Essaness Theatres. 1979, produced Coal Miner's Daughter.

SCHWARTZ, LESLIE R.: Exhibitor. b. New York City, June 7, 1915; e. Lehigh U., 1937. Entered m.p. ind. with Century Theatres Construction Co., 1937–39; film buyer, Century Theatres, 1940–42; personnel exec. 1942–43; U.S. Army service, 1943–45; with Andrews, Inc., Century Theat. concessions, 1945–48; apptd. gen. theatre mgr. 1948; pres., 1955. Member: Pi Lambda Phi; North Shore Country Club; pres., Metropolitan Motion Picture Assn., 1960–62; pres., 1955–present.

SCHWARTZ, SOL A.: Executive. b. New York City; e. New York University; m. Marion Phillips; s. Leonard. Started career with RKO circuit as assistant manager Alhambra Theatre, New York, 1922; mgr. Cameo, Momart, Orpheum & Prospect, Fordham, Albee theatre; western zone mgr. 1942; gen. mgr. RKO out-of-town theatres, 1944, then v.p. RKO theatres; pres., gen. mgr., RKO Theatres, Inc., Jan. 1, 1951; v.p. Columbia Pictures Corp., Mar. 3, 1961; v.p., theatres, 20th-Fox, 1968.

SCOFIELD, PAUL: Actor. b. Hurstpierpoint, England, Jan. 21, 1922. Gained greatest fame on London stage doing much Shakespeare and modern plays, including Staircase and Desire Under the Elms.

PICTURES INCLUDE: That Lady, Carve Her Name with Pride, The Train, A Man for All Seasons (Oscar, 1966), King Lear, Scorpio, A Delicate Balance.

SCOGNAMILLO, GABRIEL A.: Art director. b. New York, N.Y., Oct. 27, 1906; e. Royal Acad. of Fine Arts, Italy, 1922–25; N.Y. Sch. of Fine & Applied Arts, France, 1925. Architectural designer; stage scenery; Paramount, Long Island, 1928; Braunberger-Richebe Studios, Paris, 1930–33; MGM, 1934; in armed forces, 1942–46; returned to MGM, 1946.

PICTURES INCLUDE: Man Proof, Babes on Broadway, Thousands Cheer, Great Caruso, Lovely to Look At, Story of Three Loves, Latin Lovers, Strange Lady in Town, World of Tomorrow (Disneyland).

TV: Ozzie & Harriet, Westinghouse Playhouse, MGM-TV.

SCORSESE, MARTIN: Writer-Director-Editor. Began career while film arts student at N.Y. Univ., doing short, It's Not Just You, Murray.

PICTURES INCLUDE: Editor: Woodstock, Medicine Ball Caravan, Elvis. Director: Who's That Knocking at My Door? Boxcar Bertha, Mean Streets (also co-s.p.), Alice Doesn't Live Here Anymore, Taxi Driver, New York, New York, The Last Waltz (also actor), Raging Bull.

SCOTT, ADRIAN: Producer. b. Arlington, N.J., Feb. 6, 1912; e. Amherst Coll. Assoc. ed. stage mag. before entering m.p. prod. In 1940 collab. s.p. Keeping Company; later Parson of Panamint, We Go Fast, Mr. Lucky. In 1943 joined RKO-Radio as prod. (Farewell, My Lovely). In 1944 prod. My Pal, Wolf.

PICTURES INCLUDE: (collab. adapt.) Miss Susie Slagle's (prod.) Cornered, So Well Remembered, Crossfire.

SCOTT, ALLAN: Writer. In 1933 wrote play, Goodbye Again, Warner. In 1934: collaborated on screen play of Let's Try Again, By Your Leave, RKO-Radio.

PICTURES INCLUDE: s.p. So Proudly We Hail; orig. s.p. I Love a Soldier; collab. orig. s.p. Here Come the Waves; s.p. Let's Dance; collab. adapt., s.p. Wait Till the Sun Shines Nellie; assoc. prod., s.p. Four Poster; collab. s.p., 5000 Fingers of Dr. T.

SCOTT, BARBARA: Sr. Vice President & General Attorney, MPAA. e. Wellesley College, Yale U. School of Law. Began career as law clerk to Judge Charles Wyzanski of U.S. District Ct. in Mass. 1949, joined law firm of Poletti, Diamond, Roosevelt, Freiden, Mackay; 1950, associated w. Dwight, Royall, Harris, Koegel, Caskey. In 1959 joined staff Motion Picture Assoc. of America as Associate counsel specializing in censorship matters. Member bd. of dir. Wiltwyck, Federation of Protestant Welfare Agencies, New York Philharmonic. Member N.Y.C. Board of Ethics, N.Y. Convention Center Corp., N.Y. Community Trust, N.Y.C. Bar Assoc., Citizens Comm. for Children and American Civil Liberties Union. Successor Trustee Yale Corporation.

SCOTT, GEORGE C.: Actor, Director. b. Wise, Va., Oct. 18, 1927. Served 4 years Marine Corps. e. University of Missouri, appeared in varsity productions, summer stock, Shakespeare.

THEATRE: Off-Broadway in Richard III, As You Like It, Children of Darkness, Merchant of Venice, Desire Under the Elms, Antony and Cleopatra; Broadway in Comes a Day, The Andersonville Trial, The Wall, The Little Foxes, Plaza Suite, Uncle Vanya, director All God's Chillun Got Wings, director-actor Death of a Salesman, Sly Fox.

PICTURES INCLUDE: Film debut in Hanging Tree, 1959. Anatomy of a Murder, The Hustler, List of Adrian Messenger, Dr. Strangelove, The Bible, Film-Flam Man, Patton, They Might Be Giants, The Last Run, Hospital, The New Centurions, Oklahoma Crude, The Day of the Dolphin, The Hindenburg, Islands in the Stream, Crossed Swords, Movie Movie, Hardcore, The Changeling, The Formula, Taps.

Director-Actor, Rage; Director-Producer-AQCTOR, The Savage is Loose.

TV: Major TV playhouses including Dupont Show of the Month, Playhouse 90, Hallmark Hall of Fame, Kraft Theatre, Omnibus, Armstrong Theatre, Play of the Week, NBC Sunday Showcase, Dow Hour of Great Mysteries, Esso Theatre; East Side, West Side series; The Crucible, Jane Eyre, The Price, Fear on Trial, Beauty and the Beast; directed The Andersonville Trial.

SCOTT, GORDON: Actor. r.n. Gordon M. Werschkul. b. Portland, Ore., Aug. 3, 1927. e. Oregon U. U.S. Army, 1944–47; then worked as fireman, cowboy, life guard; signed by Sol Lesser Prod. for role of Tarzan; debut in Tarzan's Hidden Jungle; since in: Tarzan and the Lost Safari, Tarzan's Greatest Adventure, The Tramplers.

SCOTT, GORDON L. T.: Producer. b. Edinburgh, Scotland, January 3, 1920. e. George Watson's Boys College. Served H.M. Forces 1939–46. Ent. m.p. ind. 1946 as 3rd asst. dir. with Ealing Studios. 1948: 1st asst. dir. on Passport to Pimlico, Train of Events; joined Associated British 1949 and worked on The Dancing Years, The Franchise Affair, Laughter in Paradise, Angels One Five, Isn't Life Wonderful, Will Any Gentleman, Rob Roy; 1953 prod. man. The Weak and the Wicked, It's Great to be Young, The Dam Busters; 1956 appt. production exec. asst.; prod. Look Back in Anger, Sands of the Desert, Petticoat Pirates, The Pot Carriers, The Punch & Judy Man, Crooks in Cloisters, Forbush and the Penguins, Voices, The Maids, Snow Children, Out of Season, Hedda, The Abbess. Left Associated British 1969. Currently freelance. Recent Prods: Spectre Hanover Street.

TV: (Prod): International Detective Series, The Avengers, Pathfinders, The Maids, Snow Children, A Man Called Intrepid.

SCOTT, IAN: Mng. dir. Thames TV, Ltd. Early career as chartered accountant. Ent. m.p. ind. 1958.

SCOTT, JANETTE: Actress. b. Morecambe (Lancs), Eng., Dec. 14, 1938; e. Lycee Francaise De Londres. Stage debut in Third Person 1951; since in Peter Pan; screen debut in Went the Day Well at age of 2½.

PICTURES INCLUDE: No Place for Jennifer, The Galloping Major, No Highway, Background, Helen of Troy, As Long As They're Happy, Now and Forever, Good Companions, Happy Is the Bride Lady Is a Square, Devil's Disciple, School for Scoundrels, His and Hers, Double Bunk, The Old Dark House, Paranoiac, The Beauty Jungle, Crack in the World, The White Savage.

TV: Many appearances including Dashing White Sergeant, A Man's Woman, Cincerella, Six Wonderful Girls, Amos Burke.

SCOTT, LIZABETH: Actress. b. Scranton, Pa. e. Marywood College, Pennsylvania, and Alviene Sch. of Drama, New York. Toured with Hellzapoppin show; fashion model, understudied Tallulah Bankhead in N.Y. prod., Skin of Our Teeth. Screen debut in You Came Along, 1946.

PICTURES INCLUDE: Strange Love of Martha Ivers, Dead Reckoning, Desert Fury, I Walk Alone, Pitfall, Company

She Keeps, Red Mountain, Two of a Kind, Racket, Stolen Face, Scared Stiff, Bad for Each Other, Silver Lode, The Weapon, Loving You, Pulp.

Many TV shows.

SCOTT, MARGARETTA: Actress. b. London, England, Feb. 13, 1912; e. Conv. of Holy Child, London; Royal Acad. Dram. Art. In provincial rep. prior to London debut 1929 in, Her Shop; thereafter in numerous London productions. Screen debut 1934.

TV & PICTURES INCLUDE: Fanny by Gaslight, Man from Morocco, The First Gentleman, Mrs. Fitzherbert, Calling Paul Temple, Case of Shirley Yorke, Where's Charley, Landfall, Last Man to Hang, Town on Trial, Devil's General, Mayerling Story, You Never Can Tell, Long Sunset, The Heiress, Henry IV of Pimandello, Adventure Story, Second Mrs. Tanquery, Queen in Hamlet, Maigret, The Weather in the Streets, Member of the Family, Nicked at the Bottle, End of Term, File on Harry Jordan, The Importance of Being Earnest, Mr. Aitch, Upstairs and Downstairs, Elizabeth R., The Poisoning of Charles Bravo, The Duchess of Duke Street, What Every Woman Knows, All Creatures Great and Small, Tales of the Unexpected, Murder Trap, Racing Game, Together.

SCOTT, MARTHA: Actress. b. Jamesport, September 22, 1916; e. U. Michigan. In little theatres over U.S.; summer stock N.Y.; on radio with Orson Welles; film debut in "Our Town" 1940.

PICTURES INCLUDE: Cheers for Miss Bishop, They Dare Not Love, One Foot in Heaven, In Old Oklahoma; Hi Diddle Diddle; So Well Remembered, When I Grow Up, Desperate Hours, Ten Commandments, Airport 1975, Turning Point.

PLAYS INCLUDE: Soldier's Wife, Voice of the Turtle, The Number, Male Animal, Remarkable Mr. Pennypacker.

SCOTT, MORTON W.: Executive. b. San Francisco, Calif., Jan. 17; e. Stanford U. Pres., Studio City Television Productions, Inc.

SCOTT, RANDOLPH: Actor. b. Orange Co., Va., Jan. 23, 1903; e. U. of No. Car. On stage west coast; m.p. debut in Sky Bridge, 1931.

PICTURES INCLUDE: Roberto, Follow the Fleet, My Favorite Wife, Trail Street, The Gunfighters, Christmas Eve, Albuquerque, Return of the Bad Men, Coroner Creek, Canadian Pacific, Fighting Man of the Plains, Doolins of Oklahoma, Walking Hills, Cariboo Trail, The Nevadan, Colt .45, Sugarfoot, Sante Fe, Fort Worth, Man in the Saddle, Starlift, Carson City, Man Behind the Gun, Hangman's Knot, Thunder Over the Plains, Stranger Wore a Gun, Riding Shotgun, Bounty Hunter, Ten Wanted Men, Rage at Dawn, Tall Man Riding; act., assoc. prod., Lawless Street, Westbound, The Name's Buchanan. Ride Lonesome, Comanche Station, Ride the High Country.

One of Top Ten Money-Making Stars, M.P. Herald-Fame Poll, 1950–53. Retired.

SCOTT, WALTER M.: Set decorator. b. Cleveland, Ohio, Nov. 7. 1906. e. U.S.C., 1929; Chouinard School of Art. Interior decorator, Fred B. Martin Co., 1929–30, asst. mgr., United Studios, 1930–31; joined Fox Film Corp., 1931; asst. set decorator, 1932; set decorator, 1933; set decorator, 20th Century-Fox, 1935; supervising set decorator, 1952–72. Now consultant decorator for Fox.

PICTURES INCLUDE: Twenty-one Academy Nominations—6 oscars for Robe, King and I, Diary of Ann Frank, Cleopatra, Fantastic Voyage, Hello, Dolly, On Board of Governors of the Academy of Motion Picture Arts and Sciences 1968–76, 1980–81.

SCULLY, JOE: Talent Executive-Casting Director-Producer. b. Kearney, N.J., March 1, 1926. e. Goodman Memorial Theatre of the Art Institute of Chicago, 1946. m. Penelope Gillette. Acted until 1951. CBS-TV, N.Y. Casting Dir., Danger You Are There, Omnibus, The Web, 1951–56. 1956–60, CBS-TV, Associate Prod., Studio One, Dupont Show of the Month, Playhouse 90. 1962–64, CBS Stations Div. KNXT, Producer, Repertoire Workshop. 1965–70 Casting Dir. 20th Century-Fox Films. 1970–74 Independent Casting Director. 1974–75 Universal TV, Casting Dir. 1975, NBC-TV, Manager, Casting & Talent. 1978, Re-established Joe Scully-Casting independent Service to the industry.

PICTURES INCLUDE: Hello Dolly, In Like Flint, Valley of the Dolls, Planet of the Apes, The Flim-Flam Man, Sounder, Lady Sings the Blues, Play It As It Lays, The Stone Killer, Parallax View, Lifeguard, Man in the Glass Booth, Middle Age Crazy, Death Wish II.

TV: Peyton Place, Room 222; Pilots: Julia, The Ghost & Mrs. Muir. 1970, Joe Scully Casting, Indep., The Bill Cosby Show. 1971: TV Feature, Thief, Missiles of October, Earth II; Series: Search, Bonanza, Nichols, Snoop Sisters, Columbo, Switch, McMillan & Wife, Tales of the Unexpected.

SCULLY, PETER R.: Producer. b. Arlington, Mass., June 14, 1920; e. Mass. Inst. of Tech., 1939. With Ross Federal Research Corp. N.Y. (consumer studies) 1939–40; studied, Feagin School Dramatic Art. N.Y., to Universal as contract actor, then film ed., 1940–41; U.S. Marine Corps until 1945; film ed. Univ-Internat'l 1945–48; became m.p. prod. indep. cos., releasing through Monogram Prod., Inc. (Mayfair Pictures, Inc. & Motion Pictures, Inc.).

PICTURES INCLUDE: Father's Wild Game, Father Takes The Air, Son of Belle Star.

SCUSE, DENNIS GEORGE (M.B.E.): Executive. b. Ilford, England, May 19, 1921. e. Park School; Mercers' School, London. British Army, Royal Artillery, 1939; command entertainment officer, Ceylon Army, 1941–42; joined British Army Broadcasting Service, Italy, 1944–46; overseas div., BBC, 1947; loaned to Forces Broadcasting Service, Middle East, 1948–50; dir., British Forces network, Germany, 1950–57; ret., BBC-TV, senior planning asst., Jan. 1958; chief asst., light entertainment dept., BBC-TV,1960; apptd., chief asst. (TV) BBC N.Y. office, Sept. 1960; apptd., BBC rep. in U.S., July, 1962; gen. mgr., Television Enterprises, 1963; gen. mgr., Radio Enterprises 1968.

SEARS, HEATHER: Actress. b. London, 1935. Early career in repertory; m.p. debut, 1956.

PICTURES INCLUDE: Dry Rot, Story of Esther Costello (Golden Virgin), Room at the Top, The Siege of Pinchgut, Sons and Lovers, The Phantom of the Opera, Saturday Night Out, The Black Torment, Great Expectations.

TV: Death of the Heart, Our Town, Ring Around the Moon, The Paraguana Harp, Cymbals at Your Door, An Inspector Calls, A Reason for Staying, Trial Run, Giaconda Smile, First Love, The Informer, Footprints in the Jungle, Somerset Maugham series.

STAGE: Look Back in Anger, The Apollo de Bellac, The Seashell, South, The Door, The Caucasian Chalk Circle, Chichester Festival Theatre, How the Other Half Lives, Cause Celebre, Blithe Spirit, Antigone, War and Peace, Macbeth, etc.

SECOMBE, HARRY: Singer, comedian, actor. b. Swansea, Wales, Sept. 8, 1921; e. m.p. debut, Penny Points to Paradise, 1951; awarded, C.B.E., 1963. 1963–64: London stage starring in Pickwick. 1965: same role New York stage. 1967–68 The Four Musketeers, 1975 The Plumber's Progress.

PICTURES INCLUDE: Forces Sweetheart, Down Among the Z Men, Trilby, Davy, Jet Storm, Oliver, The Bed Sitting Room, Song of Norway, Rhubarb, Doctor in Trouble, The Magnificent Seven Deadly Sins, Sunstruck.

TV: Numerous appearances, incl. own series: Secombe and Friends, The Harry Secombe Show, etc. Also special version, Pickwick. Author of Twice Brightly, Goon for Lunch.

SEGAL, GEORGE: Actor. b. N.Y., N.Y., Feb. 13, 1934. m. Marion Sobol. e. Columbia U. 1955. B.A. Worked as janitor, ticket-taker, soft-drink salesman, usher and under-study at N.Y.'s Circle in the Square theatre. Acting debut: Downtown Theatre's revival of Don Juan.

THEATRICAL CREDITS INCLUDE: 1956 revival: The Iceman Cometh, N.Y. Shakespeare Festival's Antony and Cleopatra, Leave It To Jane, The Premise (satiric improvisational revue), Rattle of a Simple Man, The Knack.

Formed a nightclub singing act with Patricia Scott. Record album of ragtime songs and banjo music: The Yama, Yama Man. Dir. debut: Bucks County Playhouse prod. Scuba Duba.

TV: Death of A Salesman, Of Mice and Men, The Desperate Hours.

PICTURES INCLUDE: The Young Doctors (1961), The Longest Day, Act One, The New Interns, Invitation to A Gunfighter, Ship of Fools, King Rat, The Lost Command, Who's Afraid of Virginia Woolf? (Acad. Award nom.), The Quiller Memorandum, The St. Valentine's Day Massacre, Bye Bye Braverman, No Way to Treat a Lady, The Southern Star, The Bridge at Remagen, The Girl Who Couldn't Say No, Loving, The Owl and the Pussycat, Where's Poppa?, Born to Win, Hot Rock, A Touch of Class, Blume in Love, The Terminal Man, California Split, The Black Bird, Russian Roulette, The Duchess and the Dirtwater Fox, Fun with Dick and Jane, Rollercoaster, Who Is Killing The Great Chefs of Europe?, Lost and Found, The Last Married Couple in America, Carbon Copy.

SEGAL, MAURICE: Publicist. b. New York, N.Y., July 22, 1921; e. C.C.N.Y., 1937–41. Entered m.p. ind., adv. dept., 20th Fox, 1941–42; U.S. Army 1942–46; feature writer, pub. dept., 20th Fox, April, 1946; asst. to dir., adv., pub., Century Circuit, 1947; press book dept., Paramount, 1949; trade press rep. 1950; trade press rep. RKO Radio, Nov. 1952; res. to join Richard Condon-Kay Norton, publicists, May, 1953; adv., pub. dept., U-I. Sept. 1954; asst. pub. mgr., United Artists Apr. 1957. Hollywood pub.-exploit., coordinator, 1958; exec. in chg. of M.P. press dept., Universal City Studios, 1966; West Coast adv.-pub. dir., National Gen. Pictures, 1971; Pres., Maurice E. Segal Co., 1974; div., West Coast operations, Charles Schlaifer & Co., 1976; v.p., Max Youngstein Enterprises, 1979; exec. v.p., Taft Intl. Pictures, 1980.

SEIDMAN, LLOYD: Adv. exec. b. Brooklyn, N.Y., 1913. e. Columbia U., B.A., 1932; NYU School Business Administration. adv. dept., Montgomery Ward; M.P. dept., Lord & Thomas Adv. Agency, 1935; Donohue & Coe, 1936; v.p., 1942–61; pres., Educ. Science Div., U.S. Industries Inc., 1961; v.p., West, Weir & Bartel, 1963–65; exec. staff, v.p., Charles Schlaifer & Co., 1965; acct. superv. adv., Loew's Theatres; Jack Tinker & Partners, 1968; acct. superv. CBS Films (Cinema Center Films); vice-president, 1969, Blaine-Thompson Company.

SEIGENFELD, EDWARD P.: Executive. b. New York City, January 2, 1937. e. Fieldston School, Johns Hopkins, NYU School of Law. At National Broadcasting Company was asst. adv. mgr., 1961; mgr., W. Coast adv., 1962–65; national adv. mgr., 1965–67; adv. mgr., Paramount Pictures Corp., 1967–68; account supervisor, Charles Schlaifer & Co., 1968–71; At Allied Artists Pictures Corp.: dir. of adv. & pub., 1971; vice president, adv. & pub., 1971 to 1978 when joined United Artists as v.p. of adv./pub.

SELBY, DAVID: Actor. b. Morganstown, W. Va. e. W. Va. Univ. Acted in outdoor dramas in home state and did regional theatre elsewhere. Was ass't. instructor in lit. at So. Ill. Univ.

PICTURES INCLUDE: Up the Sandbox, Super Cops, Lady in Blue, Night of Dark Shadows, Rich Kids, Rich and Famous.

TV: Dark Shadows (series), Washington: Behind Closed Doors, Telethon, Family, etc.

SELF, WILLIAM: Producer. b. Dayton, Ohio, June 21, 1921. e. Univ. of Chicago, 1943; prod.-dir., Schlitz Playhouse of Stars, 1952–56; prod., The Frank Sinatra Show, 1957; exec. prod., CBS-TV, The Twilight Zone, Hotel De Paree. 1960-61 exec. prod., 20th Century-Fox TV. Hong Kong, Adventures in Paradise, Bus Stop, Follow The Sun, Margie; v.p. in chg. of prod., 20th Century-Fox TV, 1962; exec. v.p., 1964. Pres., FOX TV 1969; v.p. 20th Century Fox Film Corp., 1969; pres. of William Self Productions, Inc., partner, Frankovich/Self Productions; 1975; vice-pres., programs, Hollywood CBS Television Network, 1976; 1977, v.p. motion pictures for television and miniseries, CBS Television Network.

SELIG, ROBERT WILLIAM: Exhibitor. b. Cripple Creek, Colo., e. U. of Denver, 1932, B.A.; Doctorate, 1959. Vice pres., Pacific Theatres. Lifetime Trustee, U. of Denver. Member Kappa Sigma, Omicron Delta Kappa, Beta Gamma Sigma; Nat'l Methodist Church Foundation; Chairman, Theatre Association of California; board of directors and, Advisory Committee, Los Angeles Chamber of Commerce.

SELLARS, ELIZABETH: Actress. b. Glasgow, Scotland, May 6, 1923; e. Queenswood, Hatfield. Memb. Lincolns Inn. Stage debut with Wilson Barrett Co. in Laburnum Grove, Glasgow.

PLAYS INCLUDE: Remarkable Mr. Pennypacker, South Sea Bubble, Tea & Sympathy, The Sound of Murder; 1960–61: Stratford upon Avon

PICTURES INCLUDE: m.p. debut in Floodtide. Also: Madeleine, Prime of Miss Jean Brodie, Guilt Is My Shadow, Cloudburst, Hunted (Stranger in Between), Night Was Our Friend, Gentle Gunman, Long Memory, Barefoot Contessa, Desiree, Prince of Players, Three Cases of Murder, Last Man to Hang, Man in the Sky, The Shiralee, Jetstorm, The Day They Robbed the Bank of England, The Moment of Truth, Middle of Nowhere, 55 Days at Peking, The Chalk Garden, The Hireling.

TV: Many appearances include: Still Life, Too Late for the Mashed Potato, Harbour Island, The Happy Ones, The Sound of Murder, R 3 (series), Family Christmas, Person to Person, Person Unknown, Shades of Greene, Beasts.

SELLER, THOMAS: Writer. b. San Francisco, Calif. e. Stanford U., Yale Drama School. Armed Forces, World War II; contract writer. MGM, 8 years.

PICTURES INCLUDE: Andy Hardy Meets Debutante, Tish, The Man From Down Under, The Black Arrow.

TV WRITER FOR: Rawhide, Laramie, Combat, Private Secretary, The Ann Sothern Show, Playhouse of Stars, Riverboat, The Tall Man, Death Valley Days, Cimarron City, Cavalcade of America, The Lone Ranger, The Stu Erwin Show, Walt Disney's Wonderful World of Color, The FBI, Bonanza, The Virginian, Kentucky Jones.

SELLERS, ARLENE: Producer. In partnership as attorneys in L.A. for 20 years with Alex Winitsky before they turned to financing and later production of films.

PICTURES INCLUDE: (co-prod. with Winitsky) End of the Game, The Seven-Per-Cent Solution, Cross of Iron, Night Calls, Silver Bears, Cuba.

SELTZER, JULES: Executive. b. New York City, Sept. 4, 1908; e. U. of Penn. Adv. and pub. depart., Fox Theatres Corp., Roxy Theatre, N.Y. District pub. Publix Theatres, Florida. Publ. depart., Warner Brothers Theatres, Philadelphia, for seven years. In 1937; Quigley Silver Showmanship Awards winner. Adv. and pub. dir., Warner Brothers theatres, Philadelphia, for two years, then pub. representative for Hal Roach. In 1940; joined Western Advertising. Resigned March, 1943, to join U.S. Navy. Returned to Hal Roach Studios dir. pub., 1945: 20th-Fox Studios, 1949. Retired 1973.

SELTZER, WALTER: Executive. b. Philadelphia, Pa., Nov. 7, 1914; e. U. of Penn.; married. Publicity man for Warner Bros. Theatres, Philadelphia; Fox West Coast Theatres; to Hollywood with MGM 1936–39; Columbia, 1940–41. Enlisted U.S. Marine Corp., 1941–44. Pub. dir., Hal Wallis, 1945–54; v.p. in chg. adv & pub., Hecht-Lancaster Orgn., Feb., 1954–55; assoc. prod., The Boss; partner, Glass-Seltzer, pub. rel. firm; v.p. & exec. prod, Pennebaker Production. Vice pres., M.P. & TV Fund.

PICTURES INCLUDE: One-Eyed Jacks, Shake Hands With the Devil, Paris Blues, The Naked Edge, Man in the Middle, Wild Seed, War Lord, Beau Geste, Will Penny, Number One, Darker Than Amber, The Omega Man, Skyjacked, Soylent Green, The Cay, The Last Hard Men.

SEMEL, TERRY: Executive. b. New York City, Feb. 24, 1943. e. C.C.N.Y. (MBA, 1967). Was with C.P.A. accounting firm 1965–66. Entered ind. with Warner Bros. in 1966 as br. mgr., New York, Cleveland, Los Angeles. From 1971 to 1973 was domestic sls. mgr. for CBS—Cinema Center Films. In 1973 joined Buena Vista as v.p., gen. sls. mgr. In 1975 went to Warner Bros. as v.p., gen. sls. mgr. In 1978 named exec. v.p. and chief operating officer.

SEN, BACHOO: Producer/director/distributor. Born 1934, Calcutta, India. Entered industry 1950 in India and 1958 in U.K. Director of English Film Co. Ltd., English Film Co. (Exports) Ltd. and English Film Co. (Productions) Ltd. 1968: Produced: Her Private Hell, Loving Feeling. 1969: Produced: Love Is A Splendid Illusion. 1972: Sex Is Beautiful, Love-Hungry Girls. 1973: Tenderness. 1975: Erotic Inferno, The Intruders. 1970: Formed The English Film Co. 1974: Consultant, Superground, Ltd. and Active Video Ltd.

SENDREY, ALBERT: Music composer, arranger, conductor. b. Chicago, Ill., 1921; e. Trinity Coll. Music, London, U.S.C., Paris & Leipzig Conservatories. Winner, Chi. Symphony Orch. prize for First Symphony, 1941; 1948, Reichhold Award Detroit Symph. Orch. for 2nd Symphony. Ohio Sesquicentenn. Award for Overture, Johnny Appleseed, 1953; One Act Opera: The Telltale Stones, 1964; arr., Mary Martin at Radio City Music Hall, 1965. Compose, arr., orch. for many plays, films and TV.

MEMBER: M.P. Academy, ASCAP, CLGA, ASMA, TV Academy.

BROADWAY: orch., arr.: Peter Pan, Ziegfeld Follies, New Faces, At the Grand, Pink Jungle, The Great Waltz, Turn to the Right.

PICTURES INCLUDE: The Yearling, Three Musketeers, Father's Little Dividend, Duchess of Idaho, Royal Wedding, Easy to Love, Great Caruso, American in Paris, Brigadoon, Athena, Finian's Rainbow, Guys and Dolls, Meet Me in Las Vegas, Opposite Sex, High Society, Raintree County, Let's Be Happy, Ride the High Country, Hallelujah Trail, The Hook, The Comancheros, Nevada Smith, The Oscar, Thoroughly Modern Millie, Hello Down There, Private Navy of Sgt. O'Farrell, Hard Times, Soul of Nigger Charley.

TV: comp. music: Laramie, Wagon Train, Ben Casey, Wolper Documentaries, Americans Abroad, J. F. Kennedy Anthology; Young Man from Boston, collab. with Harry Sukman, High Chaparral, Bonanza (with D. Rose), The Monroes, Ken Murray's Hollywood, SWAT collab. B. de Vorzon, Hard Times, collab.

SERNAS, JACQUES: Actor-Producer. b. Lithuania, July 30, 1925. Became naturalized French citizen, studying medicine in Paris. Was amateur boxer when heard Jean Gabin needed an acting boxer for The Mirror, in which he made professional debut as actor. Has appeared in over 80 films, made in recent years primarily in Italy, now his home.

PICTURES INCLUDE: Lost Youth, The Golden Salamander, Helen of Troy, Jump into Hell, The First Night, La Dolce Vita, 55 Days in Peking, F.B.I.: Operation Baalbeck (also prod.), Operation Gold in the Balearic Islands (also prod.), Super Fly TNT.

TV: The School of the Painters of Paris (prod. only), The 18th Century Seen Through Its Famous Painters (prod. only), The Red Triangle (Ital. series).

SERPE, RALPH B.: Producer. b. Portland, Me., Dec. 23, 1914; e. Columbia U. Ind. thea. agent, 1936. U.S. Army, Spec. Services, 1942. v.p., Scalera Films, 1946; v.p., Italian Films Export, 1948; U.S. rep., assoc. prod., exec. asst., Dino De Laurentiis, 1952.

PICTURES INCLUDE: War and Peace, Ulysses, The Tempest, Under Ten Flags, Yovanka, Barabbas, The Bible, Drum, The Brinks' Job.

SETTON, MAXWELL: Producer. b. Cairo, Egypt, Oct. 24, 1909; e. Clifton Coll., London U., Sorbonne, Paris U. Called to bar, 1931; practiced before H.B.M. Supreme Court & International Mixed Courts of Egypt, 1931–35; to Mayflower Pictures Corp. as legal adviser, 1937; mem. bd. & joint

man. dir., 1939. Served with H. M. Forces, 1939–46. With Rank Org., 1949; with Aubrey Baring in Mayflower Pictures; formed Marksman Films Ltd. & prod. Twist of Fate. Man. dir. Bryanston Films Ltd. and Bryanston Seven Arts Ltd., Pax Films until June 1962; exec. in chg. prod., Columbia Pictures, Rome & Paris, 1962; mgr. dir., Columbia (British) Prods. Ltd., 1964; v.p. Columbia Pictures Intern'l., 1965–69; chmn, & man. dir., Paramount British Pictures Ltd. to 1972; now independent.
PICTURES INCLUDE: Prod: Spider and the Fly, Cairo Road, So Little Time, Golden Mask, Appointment in London, They Who Dare; (Marksman Films Ltd.) prod: Twist of Fate; co-prod: Footsteps in the Fog; prod: Wicked As They Come, Town on Trial, I Was Monty's Double, Beyond This Place.

SEVAREID, ERIC: News commentator. b. Velva, N.D., Nov. 26, 1912; e. U. of Minn., Paris. Started career as reporter: Minneapolis Journal, Paris Herald Tribune, United Press; joined CBS radio news staff in Paris at outbreak of World War II; with CBS radio, TV since; nat. corres; retired (1977).
BOOKS: author: Not So Wild a Dream, In One Ear, Small Sounds in the Night, This is Eric Sevareid.

SEWALL, BARBARA JEAN: Publicist. b. Holly, Colo., Jan. 6; e. UCLA; pub. dir. & adv. mgr., Saks Fifth Ave., Beverly Hills; Hollywood Athletic Club; U.S. rep., Chile films, Santiago; RKO, Goldwyn, MGM; Edward Small Prod., Warner Bros., 20th Century-Fox, Cinerama, owner public relations office Hollywood since 1971.

SEYMOUR, DAN: Actor. b. Chicago, Ill., Feb. 22, 1915. Performer burlesque, nightclubs. Stage, screen and TV actor.
PLAYS: Rain, Amphitryon 38, Room Service, and others.
PICTURES INCLUDE: Casablanca, To Have and Have Not, Confidential Agent, Intrigue, Key Largo, Johnny Belinda, Rancho Notorious, Maru Maru, Glory Alley, Face to Face, The System, Second Chance, Big Heat, Human Desire, Moonfleet, Buster Keaton Story, Sad Sack, Watusi, Return of the Fly, Leader of the Pack, The Mummy, The Way Things Were, Soft Touch, Escape from Witch Mountain, Rainbow Island.
TV: Casablanca, Restless Gun, Perry Mason, Untouchables, Holiday Inn, This Gun for Hire, 77 Sunset Strip, Hawaiian Eye, Get Smart, My Mother The Car, Batman, U.N.C.L.E., Beverly Hillbillies, Voyage to the Bottom of the Sea, My Favorite Martian, Don't Eat the Daisies, The Bob Hope Show, Kojak, The Chase, Barbary Coast, Fantasy Island.

SHADOWS, THE. Instrumental vocal group. Ent. films and TV, 1958 as accompanying group to Cliff Richard. Now stars in own right. Voted 3rd Most Popular International Artists in American Billboard 1963–64. Film debut in Serious Charge. Many TV shows in England and Ed Sullivan Show in America. EMI recording stars. Stars, score writers, play: Aladdin. London Palladium Theatre; 1966 Finders Keepers. 1967 Cinderella.
PICTURES INCLUDE: Expresso Bongo, The Young Ones, Summer Holiday, Wonderful Life; 1964, Christopher Miles dir. them as leads in Rhythms and Greens.

SHAFTEL, JOSEF: Producer, director, writer. b. Cleveland, Ohio, 1919. Early career violinst, musician, newspaper correspondent, photographer, screen writer, director. Ent. m.p. ind. 1949.
PICTURES INCLUDE: prod: Man Who Watched the Trains Go By, Naked Dawn, No Place to Hide, Naked Hills, The Biggest Bundle of Them All, The Bliss of Mrs. Blossom, The Last Grenade, Goodbye Gemini, Say Hello to Yesterday, The Statue, The Trojan Women, Soul to Soul, Assassination of Trotsky, Where Does it Hurt?, Redneck, Alice's Adventures in Wonderland.
TV: 1962–63 prod: The Untouchables.

SHAGAN, STEVE: Writer. b. New York City. Apprenticed in night school little theatres, film lab chores, stagehand jobs. Wrote, produced and directed film short, One Every Second; moved to Hollywood in 1959. Was IATSE man, working as grip, stagehand, electrician to support film writing. Was freelance advertising man and publicist; produced "Tarzan" TV show. In 1968 began writing and producing two-hour films for TV.
PICTURES INCLUDE: Save the Tiger (prod., s.p.), Hustler (s.p.), Voyage of the Damned (co.-s.p.), The Formula (s.p., prod.).
TV: Writer-producer: River of Mystery, Spanish Portrait, Sole Survivor, A Step Out of Line, etc.

SHALIT, GENE: b. N.Y.C., 1932. e. Univ. of Illinois. Started as free-lance writer; joined NBC Radio Network, working on Monitor. Has been book and film critic, sports and general columnist. In January, 1973 replaced Joe Garagiola as featured regular on Today Show.

SHAMROY, LEON: Cameraman. b. New York, N.Y., July 16, 1901; e. Cooper Union, N.Y., CCNY, Columbia U. (mech. eng.); m. Mary Anderson, actress. Asst. to Nicholas J. Shamrov, developer Lawrence motor. Turned to cinematography & since cameraman various Hollywood prods. Academy Awards: color cinemaphotography, Black Swan, 1943; Wilson, 1945; Leave Her to Heaven, 1946; Cleopatra, 1964. Look Awards: Wilson, 1945; 10 O'Clock High, 1949. Holds 20th-Fox contract.
PICTURES INCLUDE: Call Me Madam, The Robe, King of the Khyber Rifles, The Egyptian, There's No Business Like Show Business, Daddy Long Legs, Love Is a Many-Splendored Thing, King and I, South Pacific, Porgy and Bess, Cleopatra, The Cardinal, What a Way to Go!, John Goldfarb Please Come Home, The Agony and The Ecstasy, Planet of the Apes, Justine.

SHANE, MAXWELL: Writer, director, producer. b. Paterson, N.J.; e. U.C.L.A., Univ. of South. Calif. Law School. Hollywood film trade journal ed. and in exec. posts publicity & adv. depts., number of film cos. & theat. circuits, Los Angeles, Portland, Ore., N.Y.; later writer, Fanchon & Marco Revues; then pres., Hillman-Shane Adv. Agency Inc., Hollywood. Sole author or collab. orig s.p., dir., many feature films and TV shows.
PICTURES INCLUDE: Tornado, Gambler's Choice, The Navy Way, Minesweeper, Timber Queen, One Body Too Many, Cowboy in Manhattan, Dark Mountain, High Powered, Double Exposure, Scared Stiff, Tokyo Rose, People Are Funny, Seven Were Saved, Fear in the Night, Adventure Island, City Across the River, Wait Till The Sun Shines Nellie, The Glass Wall, Human Desire, Three Hours to Kill, Hell's Island, Naked Street, Nightmare.
TV: M Squad, Checkmate, Thriller, Virginian; wrote numerous other TV dramatic shows.

SHAPIRO, BYRON M.: Executive b. Hill City, Minn., Aug. 15, 1918. e. Univ. of Minn. (graduated 1940). Served in U.S. Army Air Corp., 1941–44. Entered m.p. industry as booker with Columbia in Minneapolis, 1944, later becoming branch mgr. in Des Moines (1951), St. Louis (1957), Minneapolis, (1958). Named western div. mgr. for Columbia in 1966. In 1974 appt. MGM v.p. domestic dist.

SHAPIRO, IRVIN: Executive. Formerly newspaperman; m.p. reviewer; director of national publicity campaigns. Publicity, RKO; active in ind. dist. & in export & import. Now pres., Films Around the World, Inc.; Filmworld Export Corp. which represents leading film producers and distributors throughout the world.

SHAPIRO, JACOB: Executive. b. Harbin, China, Aug. 26, 1928; e. Gakushuin Univ. (Peers' School), Tokyo, Japan, B.A. Pol. Sci. Joined Columbia Pictures Intl., N.Y., 1961; Japan sales manager, 1961–65; Puerto Rico gen. mgr., 1965–68; appointed Japan and South Korea gen. mgr., Columbia, 1968; from 1970 Columbia supv. for Philippines concurrently with Japan & South Korea gen. mgr.

SHAPIRO, KEN: Producer-Director-Writer-Actor. b. New Jersey, 1943. e. Bard College. Was child actor on tv and teacher in Brooklyn before opening "world's first video theatre" in East Village of Manhattan: Channel One, 90 mins. of TV lampoons and original material shown on TV monitors to live audience. Took 16 mm material on college dates with success, culminating in feature film: The Groove Tube (1974).

SHAPIRO, ROBERT K.: Executive. b. New York, N.Y., Feb. 12; e. CCNY, St. John's Coll. Usher, Rialto Theatre, N.Y., 1928; chief of service, 1929; Paramount Theatre, N.Y., 1930; treas. asst. mgr., Rivoli, Criterion, B'klyn Paramount Theatres, 1930–33; treas., Paramount Theatre, 1933; asst. mgr., 1936; house mgr., 1941; mgr., 1945; exec mgr., 1953; managing dir., 1954; coord., TV network progs., Western Div., American Broadcasting, 1964. Member: Variety Club of N.Y.; past pres., Cinema Lodge, B'nai B'rith, Trustee Agva Youth Fund; pres., Metropolitan Motion Picture Theatres Assn., 1962–64, Head of Int. Motion Pic. Div. 1974.

SHAPIRO, ROBERT W.: Executive. b. Brooklyn, N.Y., 1938. Joined William Morris Agency, Inc., 1958. dir. and head of motion picture dept., William Morris Agency (UK) Ltd., 1969; man. dir., 1970. 1974 vice president William Morris, Inc. In March, 1977 joined Warner Bros. as exec. v.p. in chg. of worldwide production. 1981, named WB pres., theatrical production div.

SHARIF, OMAR: Actor. b. Alexandria, Egypt, April 10, 1932. r.n. Michel Shahoub; e. Victoria College, Cairo. pres. of College Dramatic Society; m. Faten Hamama. Starred in 21 Egyptian and two French films prior to Lawrence of Arabia.
PICTURES INCLUDE: The Blazing Sun, 1955 (debut), Goha, 1959, Lawrence of Arabia, Ghengis Khan, The Fall of the Roman Empire, The Yellow Rolls-Royce, Behold a Pale Horse, Doctor Zhivago, Night of the Generals, More Than a Miracle, McKenna's Gold, Che!, Funny Girl, The Appointment, Mayerling, The Horsemen, The Burglars, The Tamarind Seed, The Mysterious Island of Captain Nemo, Juggernaut, Funny Lady, Crime and Passion, The Baltimore Bullet, Green Ice.

SHARP, ALAN: Writer. b. Glasgow, Scotland. Writes western screenplays.
PICTURES INCLUDE: The Hired Hand, Ulzana's Raid, Billy Two Hats, Night Moves.

SHARP, DON: Writer, director. b. Hobart, Tasmania, Australia, 1922. Early career as actor in Australia. Ent. m.p. ind. in England with Group Three as screenwriter, 1951. Began directing 1955 with doc. children's films, 2nd unit work and filmed TV series.
PICTURES INCLUDE: Kiss of the Vampire, It's All Happening, Devil Ship Pirates, Witchcraft, Those Magnificent Men in Their Flying Machines (2nd unit), Curse of the Fly, The Face of Fu Manchu, Rasputin—The Mad Monk, Our Man in Marrakesh, The Brides of Fu Manchu, Rocket to the Moon, Taste of Excitement, The Violent Enemy, Puppet on a Chain, Psychomania, Dark Places, Callan, Hennessy, The Four Feathers, The 39 Steps, Bear Island.

SHARPS, WALLACE S.: (B.Sc.): Executive. b. London, England, 1927. Manager, photographic subsidiary A.E.I. Group, visual aids consultant, exec. prod., T.V. Commercials, Ltd.; head of TV dept., Smee's Advertising Ltd.; founded Sharps Television Services Ltd., 1957. Consultant to TV Audience Measurement Ltd.; Univ. lect. on Int'l. Communications; Author of "Sharp's Colour Chart and Grey Scale," "Advertising and Production Techniques for Commercial Television," Dictionary of Cin. and Sound Recording, Tape Recorder Manual; gen. ed., Fountain Press Movie-books series, Cons. Ed. Broadcasting.
Member: NAEB of USA.

SHAVELSON, MELVILLE: Writer, director. b. B'klyn, N.Y., April 1, 1917; e. Cornell U. 1937, A.B. Radio writer: We The People, Bicycle Party, 1937, Bob Hope Show, 1938–43. Screen writer; apptd. prod., Warner Bros, 1951. Conceived for TV: Make Room for Daddy, My World and Welcome To It. Author: book, How To Make a Jewish Movie. Lualda, The Great Houdinis, The Eleventh Commandment, Ike. Pres., Writers Guild of America, West, 1969–71, 1979–81. Pres., Writers Guild Foundation, 1978–81.
PICTURES INCLUDE: Princess & The Pirate, Wonder Man, Kid From Brooklyn, Sorrowful Jones, It's a Great Feeling, Daughter of Rosie O'Grady, Always Leave Them Laughing, Where There's Life, On Moonlight Bay, I'll See You in My Dreams, Room For One More, April in Paris, Trouble Along the Way, Living It Up, Seven Little Foys, Beau James, Houseboat, It Started in Naples, The Five Pennies, On the Double, The Pigeon That Took Rome, A New Kind of Love, Cast a Giant Shadow, Yours Mine and Ours, The War Between Men and Women, Mixed Company.
TV FEATURES: The Legend of Valentino, The Great Houdinis, Ike.

SHAW, EDWARD S.: Producer-Actor-Executive. b. Buffalo, N.Y., July 13, 1939. e. El Camino College. Began as child actor on radio, joined CBS-TV in 1957, then owner/president of p.r./adv. firm under his own banner.
PICTURES INCLUDE: Actor; Sergeant Rutledge, Claudell English, Palm Springs Weekend, Rise & Fall of Legs Diamond, PT-109, Critics Choice. Production: Assoc. prod. Harlow, Give 'em Hell Harry; Exec. prod. Half A House, Aftermath; Co-prod., Hollywood Knight, Lady Taxi Driver; Prod., The Josephine Baker Story.

SHAW, SEBASTIAN: Actor. b. Holt, England, May 29, 1905; e. Gresham School, Slade School of Art before screen career; m. Margaret Delamere; p. Dr. Geoffrey Shaw (musician) and Mary Shaw. British m.p. debut 1930 in Caste. In R.A.F. World War II.
PICTURES INCLUDE: Get Your Man, Brewster's Millions, Ace of Spades, Jury's Evidence, Tomorrow We Live, Men Are Not Gods, Farewell Again, The Squeaker, Spy in Black, Too Dangerous to Live, The Flying Squad, Allas the Bulldog, East of Piccadilly, Glass Mountain, Landfall, Feature Story, Laxdale Hall, It Happened Here, 1968, Midsummer Night's Dream.

SHAWLEE, JOAN: Actress. b. Forest Hills, N.Y., March 5, 1929; e. Calif. & N.Y. high schools. Powers model, night club entertainer.
PLAYS: By Jupiter, Connecticut Yankee in King Arthur's Court.
TV: Abbott and Costello TV show, series: Aggie.
PICTURES INCLUDE: Marrying Kind, Something for the Birds, Conquest of Space A Star is Born, Francis Joins the Wacs, Some Like it Hot, The Apartment, Critic's Choice, Irma LaDouce, All the Fallen Angels, Tony Rome, Live a Little, Love a Little, Willard, One More Train To Rob.

SHAWN, DICK: Actor. r.n. Richard Schulefand. b. Buffalo, N.Y., December 1; e. U. of Miami. Actor, play: For Heaven's Sake, Mother, A Funny Thing Happened on the Way to the Forum. Worked night club circuit, 5 yrs.
PICTURES INCLUDE: Wake Me When it's Over, It's a Mad Mad Mad Mad World, The Producers, The Happy Ending.
TV: Max Liebman TV Spectaculars, Mary (series).

SHAYNE, ROBERT: Actor. b. Yonkers, N.Y. On N.Y. stage, on screen 1943 in Shine on Harvest Moon.
PLAYS: Claudia, Night of January 16th, Both Your Houses, Yellow Jack, Whiteoaks, Without Love, etc.
PICTURES INCLUDE: Mr. Skeffington, Christmas in Connecticut, Welcome Stranger, Neanderthal Man, North by Northwest, The Arrangement, Tora Tora Tora, Barefoot Executive, Million Dollar Duck
TV: Marcus Welby, M.D., Doris Day Show, S.W.A.T., Emergency, Superman series, etc.

SHEAFF, DONALD J.: Executive. b. Oct. 23, 1925; e. U.C.L.A., 1948, Pierce College, 1957. Served 4 yrs. during WW II in Navy Air Corps in South Pacific. 1946, joined Technicolor Motion Picture Div. in supervisory capacity; 1957, lab. supervisor, Lookout Mountain Air Force Station, handling Top Secret film for Air Force and Atomic Energy Commission. Est. and org. the installation of Vandenberg Air Force Base Lab. facilities, which Technicolor designed. 1961 joined Panaco or Corp., 1963 joined Pacific Title and Art Studio in charge of color control for special effects and titles. Returned to Technicolor Corp. 1966, app't. Plant Mgr. of Television Div., Oct. 1966, V.P. & Gen. Mgr. of the Television Div., July, 1973 appt v.p. & gen. mgr., Motion Picture Division; 1976, mgr., special visual effects, Universal City Studios. Member: SMPTE, Nat'l Academy of Television Arts & Sciences. Has conducted scientific seminars for SMPTE.

SHEDLO, RONALD: Producer. b. Los Angeles, 1940. Secretary to Errol Flynn, 1956. Asst. prod. Strangers in the City, 1961. Prod. The Whisperers, 1966. Prod. The Reckoning, Day of Locust, Letters from Frank, Back Roads.

SHEELER, MARK: Actor. r.n. Morris Sheeler. b. New York, N.Y., April 24, 1923; e. UCLA Disc jockey 1942–48. Air Corp. Photog. during war. Voices, animated cartoons, commercials. M.p. debut in Born Yesterday.
PLAYS: Hillbarn Thea., Calif : Harvey, The Happy Time, Time of Your Life, Amphitryon 38, etc.
PICTURES INCLUDE: The High and the Mighty, Blood Alley, It Came From Beneath the Sea, Apache Warrior, Tank Battalion, P.O.W., Book of Israel, Why Must I Die?, Irma La Douce, The Raven, Elmer Gentry, How the West Was Won, Mary Poppins, Unsinkable Molly Brown, Sound of Music, After the Fox, Elmer Gantry, See the Man Run, Capricorn I, Damien, Omen II, The Hand.
TV: Hitchcock Presents, Kraft Thea., Day in Court, Defenders, East Side, West Side Dr. Kildare, The Fugitive, Mr. Ed, Jack Benny Show, Life of Riley, Man from Uncle, Doctors & the Nurses, The Invaders, Batman, Here's Lucy, Dennis the Menace, Andy Griffith, Marcus Welby, Banacek, Moses the Lawgiver, All My Children, Kojak, Charlie's Angels, Police Woman, The Tonight Show, Mash, Fantasy Island, Photo Clinic (PBS), The C.C. Connection, Nero Wolfe, House Calls.

SHEEN, MARTIN: Actor. r.n. Ramon Estevez. b. Dayton, O., Aug. 3, 1940.
PICTURES INCLUDE: The Subject Was Roses, The Incident, Catch 22, No Drums, No Bugles, Rage, Pickup on 101, Badlands, The Cassandra Crossing, The Little Girl Who Lives Down the Lane, Apocalypse Now, The Final Countdown.
TV: That Certain Summer, Letters for Three Lovers, The Execution of Private Slovik, The California Kid, The Missiles of October, Sweet Hostage, etc.

SHEFFIELD, JOHN: Actor. b. Pasadena, Calif., April 11, 1931; e. UCLA. Stage debut at 7 in On Borrowed Time. Created screen role of Tarzan's son; in Tarzan pictures since.
PICTURES INCLUDE: Babes in Arms, Little Orvie, Bomba series, creating name part in Million Dollar Baby, Knute Rockne, Roughly Speaking, Cisco Kid.
TV: series: Bantu the Zebra Boy.

SHEFTER, BERT: Composer, conductor. b. Russia, May 15, 1904; e. Carnegie Inst. of Tech., Curtis Inst., Damrosch Inst. Member of piano team, Gould & Shefter, on radio & in theatres; org. own band; concert pianist; comp., cond. for many films and TV.
PICTURES INCLUDE: compose, conduct: Tall Texan, No Escape, Great Jesse James Raid, Sins of Jezebel, The Big Circus, The Fly, Lost World, Jack the Giant Killer, Monkey on My Back, Cattle King, Curse of the Fly, Last Man on Earth, Voyage to the Bottom of the Sea, The Bubble, Dog of Flanders.
TV: Written shows for Sunset Strip, Surfside, Hawaiian Eye, Maverick, Sugarfoot, Lawman, Bourbon St., Roaring 20's.

SHEINBERG, SIDNEY JAY: Executive. b. Corpus Christi, TX, Jan. 14, 1935. e. Columbia Coll. (Ab, 1955; LLB, 1958). Admitted to Calif. bar, 1958; assoc. in law U. Cal. Sch. Law, Los Angeles, 1958–59; Joined MCA, Inc, 1959. Pres., TV div., 1971–74; exec. v.p., parent co., 1969–73. Named MCA pres., 1973.

SHELDON, DAVID: Director-Writer-Producer. b. New York City. e. Yale Univ., School of Drama (M.F.A.), Principia College

(B.A.). Directed N.Y. & L.A. companies of The Star Spangled Girl, Alley Oop, Jimmy Shine, etc. Co-founded The Gateway Playhouse in N.Y. where produced & directed over 50 plays and musicals. From 1972–74 was exec. in chg. of development at American Intl. Pictures supervising various production & post-production aspects of such as Dillinger, The Frogs, Macon County Line, Slaughter, Dr. Phibes, Boxcar Bertha, Heavy Traffic, etc.

PICTURES INCLUDE: Producer-Writer, Sheba, Baby, Grizzly, The Evil, Project: Kill. Producer: Just Before Dawn. Director: Satellite, The Predator, Lovely But Deadly.

SHELDON, JAMES: Director. b. New York, N.Y.; e. U. of N.C. Page boy, NBC; announcer-writer-dir., NBC Internat'l Div.; staff dir., ABC radio; staff prod. dir., Young & Rubicam; free lance prod. dir. of many programs live and film, N.Y. and Hollywood.

TV: prod., dir.: Mr. Peepers, Armstrong Circle Theatre, Robert Montgomery Presents, Schlitz Playhouse, West Point, Zane Grey Theatre, The Millionaire, Desilu Playhouse, Perry Mason, Twilight Zone, Route 66, Naked City, The Virginian, Alfred Hitchcock Presents, Fugitive, Espionage, Defenders, Nurses, Bing Crosby Show, Family Affair, Wonderful World of Disney, Man From UNCLE, Felony Squad, That Girl, Ironside, My World and Welcome To It, To Rome With Love, Owen Marshall, Room 222, Gidget Grows Up (ABC Movie), Apple's Way, Love American Style, McMillan and Wife, Sanford and Son, Ellery Queen, Rich Man, Poor Man II, Family, MASH, Switch, Loveboat, With This Ring (ABC Movie), Sheriff Lobo, Gossip Columnist (movie), Knott's Landing, The Waltons, 240-Robert, Nurse.

SHELDON, SIDNEY: Writer-Producer-Novelist. b. Chicago, Feb. 11, 1917. e. Northwestern Univ. Novels made into films include Other Side of Midnight and Bloodline.

PICTURES INCLUDE: Bachelor and the Bobbysoxer, Easter Parade, Annie Get Your Gun, Dream Wife, Jumbo, Pardners, You're Never Too Young, Birds and the Bees, Three Guys Named Mike, Remains To Be Seen, Gambling Daughters, Dangerous Lady.

TELEVISION: Created Patty Duke Show; Created and produced I Dream of Jeannie and Nancy.

SHENSON, WALTER: Producer. b. San Francisco, Calif.; e. Stanford U., Calif.; Ent. m.p. ind. 1941; studio exec., writing, prod., prom. shorts, trailers, Columbia; sup. publ., expl., London, Columbia European production, 1955.

PICTURES INCLUDE: prod.: The Mouse That Roared, A Matter of WHO, The Mouse on the Moon, A Hard Day's Night, Help!, 30 is a Dangerous Age, Cynthia, Don't Raise the Bridge Lower the River, A Talent for Loving, Welcome to the Club (prod.-dir.), The Chicken Chronicles (co-prod.).

SHEPARD, SAM: Writer-Actor. b. Fort Sheridan, Ill. Grew up in California, Montana and South Dakota. Lives near San Francisco, where, in addition to writing, runs a drama workshop at the Univ. of California at Davis. Recipient of Brandeis Univ. Creative Arts Citation (1976) and American Academy of Arts and Letters Award (1975).

PLAYS INCLUDE: Cowboys and Rock Garden (double bill), Chicago, Icarus' Mother, and Red Cross (triple bill—1966 Obie Award), La Turista (1967 Obie), Forensic and the Navigators, Melodrama Play, Tooth of Crime (1973 Obie), Back Dog Beast Bait, Operation Sidewinder, 4-H Club, The Unseen Hand, Mad Dog Blues, Shaved Splits, Rock Garden, Curse of the Starving Class (1978 Obie), Buried Child, etc.

ACTOR: Days of Heaven, Resurrection.

SHEPHERD, CYBILL: Actress-singer. b. Memphis, Tenn., February 18. e. Hunter Col., NYU, USC. Was fashion model before acting debut in 1971. Debut record album, Cybill Does It . . . To Cole Porter (Paramount), 1974.

PICTURES INCLUDE: The Last Picture Show (debut), The Heartbreak Kid, Daisy Miller, At Long Last Love, Taxi Driver, Special Delivery, Silver Bears.

SHEPHERD, RICHARD: Producer. b. Kansas City, Mo., June 4, 1927. e. Stanford Univ. In U.S. Naval Reserve, 1944–45. Entered ent. field as exec. with MCA, 1948, functioning in radio, TV, and m.p. fields until 1956, with time out for U.S. Army, 1950–52. In 1956 became head of talent for Columbia Pictures. In 1962 joined CMA talent agency on its founding, becoming exec. v.p. in chg. of m.p. div. Left to join Warner Bros. in Aug., 1972, as exec. v.p. for prod. Resigned Oct. 1, 1974 to become indep. prod. In 1976 named MGM sr. vp. & worldwide head of theatrical prod.

PICTURES INCLUDE: Twelve Angry Men, The Hanging Tree, The Fugitive Kind, Breakfast at Tiffany's (prod.), Alex and the Gypsy, Robin and Marian.

SHER, ABBOTT J.: Executive. b. Duluth, Minn., Oct. 21, 1918; e. U. of Minnesota. Sls. mgr., Sher Packing Co., 1938–39; U.S. Army, 1942–45. Mgr., Jayhawk Thea., Kansas City, Kan., 1947–48; office mgr., Exhibitors Film Delivery & Service Co., 1949–54; sec., Exhibitors Film Delivery, 1955; theatre operator; vice-pres., sec., Exhibitors Film Delivery & Service Co., Inc., 1960; pres., M.P. Ass'n. of Greater Kansas City, 1960; bd. of dir., United Theatre Owners Heart of America, 1960–61; bd. of dir., Northtown Theatre Corp., Kansas City, Mo., 1961–62; secy. United Theatre Owners Heart of America, 1963; treas., 1964; v.p., 1965; pres. United Motion Pict. Assoc., 1967–68; Dough Guy, Variety Clubs International, Kansas City Tent No. 8, 1974; asst. chief barker, 1975–76; chief barker, Variety Clubs Int. Kansas City Tent 8, 1978–79.

SHER, JACK: Writer, producer, director. b. Minneapolis, Minn. e. U. of Minnesota. Syndicated newspaper columnist; contrib. mag. articles (fiction & non-fiction). U.S. Army, WW II.

BROADWAY: The Perfect Setup, 1963.

BOOKS: author: The Cold Companion, 12 Sports Immortals, 12 More Sports Immortals.

PICTURES INCLUDE: collab. s.p.: My Favorite Spy, Off Limits; s.p.: Kid from Left Field; addit. dial.: Shane; s.p., collab. story: World in My Corner; collab. s.p.: Walk the Proud Land, Joe Butterfly; s.p., dir.: Four Girls in Town; s.p., dir.: Kathy O, The Wild and the Innocent; collab. s.p., dir.: 3 Worlds of Gulliver; collab. s.p.: Paris Blues; s.p., dir.: Love in a Goldfish Bowl; s.p.: Critic's Choice; collab. s.p.: Move Over Darling, produced and wrote "Goodby Raggedy Ann," produced "Slither."

SHER, LOUIS K.: Executive. b. Columbus, Ohio, Feb. 25, 1914; e. Ohio State Univ., 1933. Exec., Stone's Grills Co., 1934–37; owned & operated, Sher Vending Co., 1937–43. U.S. Army, 1943–46. V.p., Sons Bars & Grills, 1947–54; org. & pres., Art Theatre Guild, 1954; opened art theatres for first time in many cities, org. opera film series, film classic series and similar motion picture activities in many cities. Org., Film Festival at Antioch Coll., 1960; pioneer in fighting obscenity laws in Ohio; operates 43 theatres in midwest and western states. Co-producer of the musical broadway production Shenandoah and American Dance Machine.

SHERDEMAN, TED: Writer, producer, director. b. Lincoln, Neb.; e. U. of Nebraska, Creighton U.

PICTURES INCLUDE: Lust For Gold, Scandal Sheet, Breakthrough, Retreat Hell, Winning Team, Eddie Cantor Story, Riding Shotgun, Them, McConnell Story, Toy Tiger, Away All Boats, Maracaibo, St. Louis Blues, Hell to Eternity, A Dog of Flanders, The Big Show, Misty, Island of the Blue Dolphins, And Now Miguel, My Side of the Mountain, Latitude Zero.

TV: Wagon Train, Zane Grey, Californians, Astronaut, Hazel, Men Into Space, My Favorite Martian, Bewitched, Family Affair, The Monroes, Flying Nun.

SHERMAN, AL: Producers Representative, Film and public relations consultant. b. New York. N.Y., April 27; e. Columbia U., CCNY. Film & pub. consultant, Royal Norwegian Information Service, Wash., D.C.; Wash.; pres., Sherman Films, Inc.; American rep., Films of Scotland, Edinburgh, Scotland; Patria Film Prods. Ltd., London; British Films Ltd., London. Member: Silurians; Educational Film Library Association; American Federation of Film Societies; M.P. Pioneers, American Film Institute.

SHERMAN, EDWARD: Talent agent. b. Russia, Jan. 12, 1903. Entered vaudeville agency field Philadelphia and New York 1920; estab. Edward Sherman Agency, New York; in 1941 estab. branch Hollywood, Calif.; org. & pres., MPM Corp. to distribute films for television, 1947; also org. & pres., Glove TV Corp., 1952; formed Edward Sherman (personal mgt.), 1956.

SHERMAN, GEORGE: Director. b. New York, N.Y.

PICTURES INCLUDE: The Bandit of Sherwood Forest, Red Canyon, Yes Sir That's My Baby, Sword in the Desert, Comanche Territory, Sleeping City, Spy Hunt, Tomahawk, Target Unknown, Golden Horde, Steel Town, Raging Tide, Against All Flags, Battle at Apache Pass, The Lone Hand, Veils of Bagdad, War Arrow, Border River, Johnny Dark, Dawn at Socorro, Chief Crazy Horse, Count Three and Pray, Treasure of Pancho Villa, Comanche, Reprisal, Flying Fontaines, Enemy General. Formed Shergari Corp. with F.H. Ricketson, Jr., and Ted R. Gamble made For the Love of Mike; dir.: Panic Button, Wounds of Hunger, Jaquin Murieta, Smokey, Big Jake. In 1976 wrote, produced, and directed Artie-Charley and Friend, indep. feature.

TV: Prod. and/or dir. for 20th Century-Fox, and NBC, CBS. 1978, prod., Little Mo (TV movie NBC), Man O'War and Trudy (TV movie ABC).

SHERMAN, RICHARD M.: Composer, lyricist. b. New York, N.Y., June 12, 1928; e. Bard Coll. B.A. 1949. Info. & Educ. Br., U.S. Army, 1953–55. Songwriter, composer, Walt Disney Prods.

SONGS: Things I Might Have Been, Tall Paul, Christmas in New Orleans, Mad Passionate Love, Midnight Oil, You're Sixteen, Pineapple Princess, Lets Get Together, Maggie's Theme, Chim Chim Cheree, Comedy Album: Smash Flops.

PICTURES INCLUDE: Nightmare, The Cruel Tower, Absent Minded Professor, The Parent Trap, Big Red, The Castaways, Moon Pilot, Bon Voyage, Legend of Lobo, Summer Magic, Miracle of the White Stallions, The Sword

in the Stone, Merlin Jones, Mary Poppins, Those Calloways, The Monkey's Uncle, That Darn Cat, Symposium of Popular Songs, Winnie the Pooh, Bedknobs & Broomsticks, Tom Sawyer (s.p. and score.), Huckleberry Finn.
TV: Wonderful World of Color. Bell Telephone Hour.

SHERMAN, ROBERT B.: Composer, lyricist. b. New York, N.Y., Dec. 19, 1925; e. Bard Coll., B.A., 1949. U.S. Army, WWII, 1943–45. Songwriter, 1952–60; pres., Music World Corp., 1958; songwriter, composer, Walt Disney, 1960–65.
SONGS: Things I Might Have Been, Tall Paul, Young & In Love, Midnight Oil, Pineapple Princess. You're 16, Let's Get Together, Maggie's Theme, Chim Chim Cheree.
PICTURES INCLUDE: Absent Minded Professor, The Parent Trap, Big Red, The Castaways, Moon Pilot, Bon Voyage, Legend of Lobo, Miracle of the White Stallions, Summer Magic, The Sword in the Stone, Merlin Jones, Mary Poppins, Those Calloways, The Monkey's Uncle, That Darn Cat, Winnie the Pooh, Symposium of Popular Songs, Bedknobs & Broomsticks, Tom Sawyer (s.p. and score.), Huckleberry Finn.
TV: Wonderful World of Color, Bell Telephone

SHERMAN, ROBERT M.: Executive. Entered ind. as agent for MCA; later joined Arthur P. Jacobs pub. rel. firm. Became acc't. exec. when Jacobs merged with Rogers & Cowan. In 1964 joined CMA; 1967 made v.p. in m.p. div., serving both in London and Hollywood. In 1972 formed own prod. co., Layton Prods., with first film, Scarecrow for WB. In 1973 prod. Night Moves, also for WB. In 1974 named v.p., prod., for 20th-Fox. Returned to independent prod.: The Missouri Breaks, Convoy.

SHERMAN, SAMUEL M.: Producer, director, writer. b. New York, N.Y.; e. CCNY, BA. Entered m.p. ind. as writer, cameraman, film ed., neg. & sound cutter; nat'l mag, ed., Westerns Magazine, 1959; pres., Signature Films; prod., dir., TV pilot, The Three Mesquiteers, 1960; prod., Pulse Pounding Perils, 1961; helped create, ed.,dir., Screen Thrills Illustrated; exec. prod., Screen Thrills; v.p., Golden Age Films, 1962; prod., Joe Franklin's Silent Screen, 1963; N.Y. rep., Victor Adamson Productions; owns world rights; The Scarlet Letter; 1965; N.Y. rep., Tal Prods., Hlywd.; adv. & pub., Hemisphere Pictures; ed., autobiog., Joe Bonomo; prod., writer, Chaplin's Art of Comedy, The Strongman; prod., Hollywood's Greatest Stuntman; story adapt., Fiend With the Electronic Brain. 1967, prod. Spanish version Chaplin Su Arte Y Su Comedia; tech. consul., Hal Roach Studios, NBC, Music from the Land; 1968, N.Y. rep. East West Pict. of Hollywood. 1963, N.Y. rep., Al Adamson Prods. of Hollywood; post-prod'n consultant: The Fakers, Creatures of the Prehistoric Planet; sales U.K. of U.S. product; promo. consultant, Love is a Woman, Blood Fiend, Brides of Blood, The Ghastly Ones. 1969, post prod'n and promo. consultant, Blood Demon, Mad Doctor of Blood Island. Ed.-in-chief, bk., The Strongman. Prod'n consultant, Satan's Sadists, The Blood Seekers. Pres., Independent-International Pictures Corp, assoc. prod. Horror of the Blood Monsters, Blood of Ghastly Horror; Producer-Writer Brain of Blood; prod. supervisor Dracula vs. Frankenstein; Exec. prod. Angels, Wild Women; pres., Producers Commercial Productions Inc.; producer-writer, The Naughty Stewardesses; producer, Girls For Rent; producer TV special, Wild, Wild World of Comedy; executive producer, The Dynamite Brothers; producer-writer, Blazing Stewardesses; production consultant, In Search of Dracula; executive producer, Cinderella 2000; producer & orig. story, Team-Mates; Chairman of Creditors' Committee, Allied Artists Television Corp.; president, Independent-International Entertainment, TV div. Independent-International Pictures Corp.

SHERMAN, VINCENT: Director. b. Vienna. Ga., July 16, 1906; e. Oglethorpe U. B.A. Writer, actor, dialogue dir., then prod. dir.
PICTURES INCLUDE: Dir.: Underground, Saturday's Children, Flight from Destiny, Old Acquaintances, In Our Time, Mr. Skeffington, Pillow to Post, Nora Prentiss, The Unfaithful, Adventures of Don Juan, Somewhere in the City, Hasty Heart, Damned Don't Cry, Harriet Craig, Goodbye, My Fancy, Lone Star, Assignment—Paris; prod. dir.: Affair in Trinidad, The Young Philadelphians, The Naked Earth, Second Time Around.
TV: 25 episodes of Medical Center; 3 Westside Medical; 6 Baretta; 4 Waltons; 2 Doctors Hospital; The Last Hurrah (movie); Women at West Point (movie); The Yeagers (pilot), Bogey, The Dream Merchants.

SHERRIN, NED: Producer, director, writer. Early career writing plays and musical plays. Prod., dir., ATV Birmingham, 1955–57; prod., Midlands Affairs, Paper Talk, etc. Joined BBC-TV 1957 and produced many TV talk programmes. Novels: (with Caryl Brahms) Cindy-Ella or I Gotta Shoe (also prod. as stage play), Rappell 1910, Benbow Was His Name.
TV: England: prod.: Ask Me Another, Henry Hall Show, Laugh Line, Parasol. Assoc. prod.: Tonight series. Little Beggars, 1962; prod., creator: That Was The Week That Was, 1962–63; prod., dir.: Benbow Was His Name (co-author), 1964; Take a Sapphire (co-author), The Long Garden Party, The Long Cocktail Party. ABC of Britain revue. Prod., dir.: three-weekly series Not So Much a Programme, More a Way of Life, 1964–65. Appearances inc. Your Witness, Quiz of The Week, Terra Firma, Who Said That, The Rather Reassuring Programme, Song by Song.
PICTURES INCLUDE: prod.: The Virgin Soldiers (with Leslie Gilliat), Every Home Should Have One, Up Pompeii, Girl Stroke Boy (co-author with Caryl Brahms), Up the Chastity Belt, Rentadick, The Garnet Saga, Up the Front, The National Health, The Cobblers of Umbridge (dir. with Ian Wilson).

SHIELDS, BROOKE: Actress. b. New York City, May 31, 1965. Discovered at age 11 months by photographer Francesco Scavullo to pose in Ivory Snow ads. Later became Breck girl in commericals; appeared in Richard Avenon's Colgate ads for 3 yrs.
PICTURES INCLUDE: Alice Sweet Alice (Communion), Pretty Baby, Tilt, King of the Gypsies, Wanda Nevada, Just You and Me, Kid, The Blue Lagoon, Endless Love.
TV: The Prince of Central Park, After the Fall.

SHIELDS, WILLIAM A.: Executive. b. New York City, 1946. e. El Camino College and California State College. Entered the motion picture industry in 1966 when he went to work for Pacific Theatres. Other industry affiliations included: MGM sales department, Los Angeles and Denver, 1970; New World Pictures, Western Division manager, 1972; branch manager, 20th Century-Fox, Washington, 1973; New York district manager, 20th Century-Fox, 1973–75. Joined Mann Theatres Corporation of California as head booker in 1975. Gen. sls. mgr.'Far West Films, 1977–79; joined Avco Embassy as Western div. mgr., promoted to asst. gen. sls. mgr., 1980; promoted to v.p.-gen. sls. mgr., January, 1981.

SHIMA, KOJI: Director. b. Japan. Entered m.p. ind. as actor, 1930; then asst. dir. director. Dir., Daiei M. P. Co. (now defunct). Presently indep. director.
PICTURES INCLUDE: Golden Demon, Phantom Horse.

SHINBACH, BRUCE D.: Executive. b. South Bend, Ind., June, 1939. e. Univ. of Colorado, (B.A. 1963), N.Y. Institute of Finance, Northwestern Univ. (M.A., 1965). Stockbroker for Harris, Upham & Co., 1964, shopping center developer, Dixie Associates, 1966 to present. Pres., Monarch Theatres.

SHIPMAN, BARRY: Writer. b. S. Pasadena, Cal., Feb. 24, 1912. Parents professional. M.p. writer since 1936. U.S.M.C.R. photog. officer 1942–46. 43 sole s.p. credits including Montana Territory for Col., 1946–52.

SHIPMAN, KENNETH: Producer. b. London, England, 1930. Electro-mechanical engineer, entered m.p. ind. as assistant dir., Riverside Studios, 1949; mgr., Southhall Studios; dept. chmn., Bryanston Films; formed Kenneth Shipman Productions, 1964; prod., The System, exec. prod., Burke and Hare, Au Pair Girls.

SHIRE, DAVID: Composer.
PICTURES INCLUDE: Farewell My Lovely, The Conversation, All the President's Men, Saturday Night Fever, Old Boyfriends, Norma Rae, The Night the Lights Went Out in Georgia.

SHIRE, TALIA: Actress. Raised on road by her father, arranger-conductor Carmine Coppola, who toured with Broadway musicals. After 2 yrs. at Yale School of Drama she moved to L.A. where appeared in many theatrical productions. Is sister of Francis Ford Coppola.
PICTURES INCLUDE: The Dunwich Horror, Gas-s-s, The Christian Licorice Store, The Outside Man, The Godfather, The Godfather, Part II, Rocky, Old Boyfriends, Rocky II, Windows.
TV: Rich Man, Poor Man, Kill Me If You Can.

SHIVAS, MARK: Producer/Director. TV credits incl: Margins Of The Mind. Presenter of Cinema. The Six Wives of Henry VIII, Casanova, The Edwardians, The Evacuees, The Glittering Prizes, Abide With Me Rogue Male, 84 Charing Cross Road, The Three Hostages, She Fell Among Thieves, Professional Foul, Telford's Change, On a Giant's Shoulders. Film: Exec. prod. Henry VIII and His Six Wives, Richards' Things.

SHONFELD, PHIL: Executive. Began career in sls. dist. at Universal Pictures. Later asst. to western sls. mgr. for National General and western dist. mgr. for Buena Vista. Joined Warners in November, 1976 as asst. to v.p. & gen. sls. mgr. In 1979 promoted to v.p.—sls. admin., continuing to head playdate and contract depts. while becoming involved in all aspects of sls. dept.

SHORE, DINAH: Singer. r.n. Frances Rose Shore. b. Winchester, Tenn., Mar. 1, 1920; e. Vanderbilt U., B.A. 1939; m. George Montgomery. Became singer WNEW, N.Y., 1938; joined NBC as sustaining singer, 1938; started contract RCA-Victor, 1940; star Chamber Music Soc. of Lower Basin

St. program, 1940; joined Eddie Cantor radio pgm., 1941; star own radio program, General Foods, 1943; entertained troops European Theatre of operations, 1944; radio program, Procter & Gamble. Star TV show, Chevrolet, 1951–61; Dinah Shore Specials, 1964–65. 1969: Dinah Shore Special, Like Hep. 1970–71: Dinah's Place, Dinah!
AWARDS: Awarded New Star of Radio Motion Picture Daily Poll and World Telegram-Scripps-Howard Poll, 1940; Best Popular Female Vocalist M.P. Daily Fame's Annual Poll Radio and TV 1941–61; Michael Award Best Female Vocalist, Radio and TV, 1950, 51, 52; Billboard Award; Favorite Female Vocalist in radio, 1949; Billboard Award Favorite Female Vocalist in records, 1947; Gallup Poll One of Most Admired Women in the World, 1958–61; 6 Emmy Awards 1954 to 59. Recording of songs by Decca; Los Angeles Times Woman of the Year Award, 1957; TV-Radio Mirror mag. award, best female singer, radio, 1952, 53, 56, 57, 58; TV-Radio Mirror mag. award, TV's Best Musical Variety Show, 1956, 58, 59; Peabody TV Award, 1957; Fame's Critics' Poll, Best Female Vocalist, 1958, 63; Hollywood Foreign Press Assn's Golden Globe Award, 1959; Radio-TV Daily, Female Vocalist of the Year, 1949, 56.

SHORE, SIG: Producer. b. New York City. Served as navigator in Air Force, World War II. First job in films in pub. dept. at Warner Bros. Formed own ad agency on West Coast; then turned to TV production with The Errol Flynn Theatre. Engaged by David O. Selznick Films to dist. its films to TV outlets. Entered theatrical distribution, importing Hiroshima, Mon Amour, The 400 Blows, etc. In mid-50s became involved in cultural exchange program of US State Dept., importing and distributing Soviet films. Returned to TV production, turning out over 250 shows, including The Outdoor World for Shell Oil. Headed co. for Ivan Tors which made Flipper, Daktari, Gentle Ben series. In 1970 formed Plaza Pictures for theatrical dist.
PICTURES INCLUDE: Super Fly, Super Fly TNT, That's The Way of the World.

SHOWALTER, MAX: Actor, composer. s.n. Casey Adams. b. Caldwell, Kans., June 2, 1917; e. Caldwell H.S.; Pasadena Playhouse. Composed background music for films: Vicki, Return of Jack Slade. Arrangements for records.
BROADWAY: Knights of Song, Very Warm for May, My Sister Eileen, Showboat, John Loves Mary, Make Mine Manhattan, Lend an Ear.
PICTURES INCLUDE: Always Leave Them Laughing, With a Song in My Heart, What Price Glory, My Wife's Best Friend, Niagara, Destination Gobi, Dangerous Crossing, Vicki, Night People, Naked Alibi, Never Say Goodbye, Bus Stop, Down Three Dark Streets, Designing Woman, Female Animal, Voice In the Mirror, The Naked and the Dead, It Happened in June, Elmer Gantry, Return to Peyton Place, Summer and Smoke, Music Man, Smog, Bon Voyage, My Six Loves, Lord Love a Duck, The Anderson Tapes.

SHULMAN, MAX: Writer. b. St. Paul, Minn., March 14, 1919; e. U. of Minn., B.A., 1942.
NOVELS: Barefoot Boy with Cheek, Feather Merchants, Zebra Derby, Sleep Till Noon, Rally Round the Flag Boys.
BROADWAY: Barefoot Boy with Cheek, Tender Trap.
PICTURES INCLUDE: collab. story: Always Leave Them Laughing; Collab. story, s.p.: Confidentially Connie; story, s.p.: Affairs of Dobie Gillis, Half a Hero.
TV: Creator: The Many Loves of Dobie Gillis.

SHUMATE, HAROLD: Writer. b. Austin, Texas; e. Texas and Missouri. Wrote short stories, worked on newspaper; sold first story to motion pictures 1916; came to Los Angeles 1923; written over 100 motion pictures for major studios, several books, innumerable television stories and scripts.

SHUPERT, GEORGE T.: b. Alpena, Mich., July 24, 1904; e. U. Mich., 1926. Investment banking, Detroit, Mich., 1926–38; found George T. Shupert & Co., 1936; dir. of sales, commercial film div., Paramount Pictures, 1940–42; TV dept., 1942–51; v.p., dir., commercial TV oper., Paramount TV Prod., Inc.; v.p. Peerless TV Prod., Inc.; v.p., gen. mgr., United Artists Television Corp.; v.p., ABC Film Syndication, Aug., 1953; pres., March, 1954; v.p. chg., TV, MGM, 1959–61; named director, TV sales, v.p., 20th Century-Fox TV, 1961–64; pres., Sunrise Broadcasting Co., 1965–77. Director, American Video Corp., (Cable TV) Pompano Beach, Fla.

SHURPIN, SOL: Executive. b. New York, N.Y., Feb. 22, 1914; e. Pace Institute, 1936. Law stenog., 1932–33; Joe Hornstein, Inc., 1933–41; National Theatre Supply, 1941–48; purchased interest in Raytone Screen Corp., became v.p., 1948; pres., Raytone, 1952; pres., Technikote Corp., which succeeded Raytone Screen, 1956-present; sole owner, Technikote Corp., 1962.

SHYLEN, BEN: Editor emeritus of Boxoffice.

SIDARIS, ANDY: Producer, director. b. Chicago, Ill., Feb. 20, 1932. e. Southern Methodist University, B.A. in Radio-TV. Began television career in 1950 in Dallas, Texas as a director at station WFAA-TV; now pres., The Sidaris Company.
PICTURES INCLUDE: Dir., Stacey, The Racing Scene, M*A*S*H football sequences, Seven (prod.-dir.).
TV: Dir., The Racers/Mario Andretti/Joe Leonard/Al Unser, ABC's Championship Auto Racing, ABC's NCAA Game of the Week, 1972 Summer Olympics, 1968 Summer Olympics (won Emmy Award), Wide World of Sports, The Racers/Craig and Lee Breedlove, dir.: The Burt Reynolds Late Show, dir., Kojak episode, Nancy Drew, Dukes of Hazzard.

SIDNEY, GEORGE: Director, producer. b. New York, N.Y. 1916; son of L. K. Sidney, veteran showman and v.p. MGM and Hazel Mooney, actress. From 1932 at MGM as test, second unit and short subjects dir. Several Academy Awards for shorts, Our Gang Comedies, Pete Smith, etc. In 1941 made feature dir., MGM. Pres., Directors' Guild of America, 16 yrs; spec. presidential assignment to Atomic Energy Commission and U.S. Air Force; 1961–66, Pres., Hanna-Barbera Productions; Doctorate of Science Hanneman Medical University and Hospital. Mem. ASCAP
PICTURES INCLUDE: dir., prod.: Free and Easy, Pacific Rendezvous, Pilot No. 5, Thousands Cheer, Bathing Beauty, Anchors Aweigh, Harvey Girls, Cass Timberlane, Three Musketeers, Red Danube, Key to the City, Annie Get Your Gun, Holiday in Mexico, Show Boat, Scaramouche, Young Bess, Kiss Me Kate, Jupiter's Darling, Eddie Duchin Story, Jeanne Eagels, Pal Joey, Who Was That Lady, Pepe, Bye Bye Birdie, A Ticklish Affair, Viva Las Vegas, Who Has Seen the Wind?, U.N. special; The Swinger, Half A Sixpence.

SIDNEY, SYLVIA: Actress. b. New York, N.Y., Aug. 8, 1910; e. Theatre Guild School. On stage, then screen debut in Through Different Eyes.
PLAYS: Nice Women, Crossroads, Bad Girl, The Gentle People, etc.
PICTURES INCLUDE: City Streets, Five Minutes from the Station, Ladies of the Big House, Confessions of a Co-Ed, An American Tragedy, The Miracle Man, Merrily We Go to Hell, Madame Butterfly, Blood on the Sun, Mr. Ace, Searching Wind, Love from a Stranger, Les Miserables, Violent Saturday, Behind the High Wall, Summer Wishes, Winter Dreams, I Never Promised You a Rose Garden, Damien-Omen II.
TV: Raid on Entebbe.

SIEGEL, DON: Director. b. Chicago, Ill., Oct. 26, 1912; e. Jesus Coll., Cambridge U. (England). Has appeared with the Royal Academy of Dramatic Arts, London, and Contemporary Theatre Group, Hollywood. Joined Warner Bros. as asst. film librarian, 1934; became asst. cutter and head of insert dept. Organized montage dept.; wrote and dir. all montages.
PICTURES INCLUDE: dir.: Star in the Night and Hitler Lives (two Academy Awards for distinctive achievement for shorts, 1945). dir.: The Verdict, Night Unto Night, Big Steal, Duel at Silver Creek, No Time For Flowers, Count the Hours, China Venture, Riot in Cell Block 11, Private Hell 36, Annapolis Story, Invasion of the Body Snatchers, Madigan, Coogan's Bluff, Two Mules for Sister Sara, The Beguiled, Dirty Harry; appeared in Play Misty for Me, Charley Varrick, Black Windmill, The Shootist, Telefon, Escape from Alcatraz, Rough Cut.

SIEGEL, SIMON B.: Financial vice-president, treas., American Broadcasting Paramount Theatres & mem. bd. of dir. Started career in ind. 1929, comptroller's staff Paramount Pictures, Inc.; asst. to comptroller of the company's theatre div., 1941; comptroller, 1949; joined United Paramount Theatres, 1950; in 1953 named treasurer when ABC & UPT were merged; named financial v.p. and treas., 1957; exec. vice-pres., ABC-Paramount Theatres.

SIEGEL, SOL C.: Producer. b. New York City, Mar. 30, 1903; e. Columbia U. Joined Republic Pictures 1934 as exec. prod. To Paramount 1940 as exec. prod. of unit. In 1944 entered indep. prod. with Kiss and Tell. Returned to Paramount 1945. Prod. Blue Skies, Welcome Stranger, Perils of Pauline; prod. 20th Century-Fox, 1946; indep. prod. releasing through MGM, 1955; became vice-president in charge of production MGM Studios, April, 1958. A Champion of Champion Producers, in Fame ratings. Independent Prod. Unit 1964 Columbia Pict.; and Paramount Pict. 1966.
PICTURES INCLUDE: Letter to Three Wives, House of Strangers, I Was a Male War Bride, Panic in the Streets, Prince of Foxes, Stella, My Blue Heaven, Fourteen Hours, Deadline U.S.A., What Price Glory, Dream Boat, Monkey Business, Call Me Madam, The President's Lady, Gentlemen Prefer Blondes, Three Coins in the Fountain, Broken Lance, There's No Business Like Show Business, High Society, Man on Fire, Les Girls, Merry Andrew, Some Came Running, The World, The Flesh and the Devil, Home from the Hill, Alvarez Kelly, Walk Don't Run, No Way to Treat a Lady.

SIGNORET, SIMONE: Actress. r.n. Simone Kaminker. b. Wiesbaden, 1921. Appeared in roles as extra. First break in Les

Visiteurs Du Soir. Attended classes at Solange Sicard's school.

PICTURES INCLUDE: Le Couple Ideal, 1945, Les Demons de L'Aube, 1945, Macadam, 1946, Fantomas, 1947, Dedee D'Anvers, 1948, Guerriers Dans L'Ombre, 1948, L'Impasse des Deux Anges, 1948, Maneges, 1949, Swiss Tour B, XV, 1949, La Ronde, 1950. Ombre & Lumiere, 1950, Casque D'or (Golden Marie—Acad. Award, London, 1953), Therese Raquin, 1953, Les Diaboliques, 1954, The Crucible, Room at the Top, 1958, The Day and The Hour, Ship of Fools, The Sleeping Car Murders, The Deadly Affair, Games, The Seagull, The Confession, Le Chat, Madame Rosa.

STAGE: Les Sorcieres de Salem.

SILBER, ARTHUR: Executive. b. Phila., Pa., Feb. 13, 1919. e. Univ. of Penna. Engineer, U.S. Navy. Owner, Silabert Corp. operator of Lawrence Park theatre, also Able-Silber Theatres & Concessions, Phila., Pa.; pres., Silber-Stoltz Prods.; 1966, Westcoast film buyer and western division manager for General Cinema; 1972-v.p., Avco Triplex Cinema (General Cinema of California) June, 1972, purchased Exhibitor Service Inx.; January, 1975, exec. v.p., Mann Theatres; June, 1975, pres. of Four Stars Cinema Corp. & Four Stars of Ohio; v.p., film, Pacific Theatres, 1977.

SILBERKLEIT, WILLIAM B.: Executive. e. New York University, Full Colonel, USAF. Pres., Tommy J Productions, independent producer: Bonnie's Kids, Roommates, Centerfold Girls, Linda Lovelace for President, Alice Cooper's Welcome to My Nightmare, A Woman for All Men. In 1975 became chm. and co-founder of Pro-International Films. Executive producer for PRO-International: Young Lady Chatterly, Once Upon a Girl, Detroit 9000, Emma Mae.

SILLIPHANT, STIRLING: Executive-Writer. b. Detroit, Mich., Jan. 16, 1918. e. U. Southern California, B.A., 1938; married. On pub. staff, Walt Disney Productions, Burbank, 1938–41; 1941–42, exploit. & pub., Hal Horne Org. for 20th Century-Fox in New York & other key cities, 1942–43, asst. to Spyros P. Skouras. U.S. Navy, World War II. Since 1946, 20th-Fox; in chge. special events and promotions, June 1949; apptd. Eastern pub. mgr. Aug, 13, 1951; prod., Joe Louis Story; co-prod., collab. s.p., 5 Against the House; screenwriter, producer, Naked City, Route 66, The Slender Thread, In the Heat of the Night; s.p., Marlowe; collab., s.p., The Liberation of L. B. Jones; s.p., A Walk in the Spring Rain; p., Shaft, Shaft in Africa, (s.p.). The New Centurions (s.p.), The Poseidon Adventure, (s.p.), The Towering Inferno (s.p.), The Killer Elite (co.-s.p.), The Enforcer (co.-s.p.), Telefon (co-s.p.), The Swarm, When Time Ran Out.

SILVER, JOAN MICKLIN: Writer-Director. b. Omaha, Nebr., 1935. e. Sarah Lawrence College. Began career as writer for educational films. Original s.p., Limbo, purchased by Universal Pictures. In 1972 Learning Corp. Of Am. commissioned her to write and direct a 30-min. documentary, The Immigrant Experience. Also wrote and directed two children's films for same co. First feature was Hester Street, which she wrote and directed.

PICTURES INCLUDE: Hester Street (s.p., dir.); Bernice Bobs Her Hair (short, s.p.-dir., later shown on TV), Between the Lines (dir.); Head Over Heels (s.p., dir.).

SILVER, LEON J.: Executive. b. Boston, Mass., March 25, 1918. e. U. of So. Calif., 1935–38. Independent prod. of short subjects, 1939; story analyst, Paramount, 1940, film writer, U.S. Army Pictorial Service, 1941–45; free lance writer, 1946; film writer. prod., U.S. Public Health Service, 1946–51, asst. chief, foreign film prod., U.S. Dept. of State, 1951–54; acting chief, domestic film prod., U.S. Information Agency, 1955. Division Chief, Worldwide Documentary Film & Television Product, U.S. Information Agency, Apr. 6, 1968, 1978 to present, sr. advisor iv, film production. Resigned, 1980. Now TV network writer-producer-novelist.

SILVER, MILTON: Advertising executive. b. New York City. U.S. Army World War I. Co-ed. Who's Who on Screen and Little Movie Mirror books; co-author Broadway stage production, The Mystery Ship; dir. adv., exploit., Universal Pictures, trailer ed, adv. manager, National Screen Service; exec. asst. to dir. adv. pub. Republic Pictures; to Souvaine Selective Pictures as adv., pub. dir., 1951; adv. pub. dept. United Artists 1953. Free lance writer since 1960.

SILVER, PAT: Writer. co-s.p., Wizard of Baghdad, Pirates of Tortuga, orig. story, s.p., 7 Woman From Hell, The Golden Touch, Dare To Be Free; prod., Ghost Town.

TV: Mables Fables Series, Rescue 8, New Breed, Shannon, Paladin.

SILVER, RAPHAEL D.: Producer. b. Cleveland, Ohio, 1930. e. Harvard College and Harvard Graduate School of Business Adm. Is pres. of Midwestern Land Devel. Corp. and Hodgson Houses, Inc. In 1973 formed Midwest Film Productions to produce Hester Street, written and directed by his wife, Joan Micklin Silver. Also distributed film independently. Also produced Between the Lines, directed by wife. Directed On the Yard.

SILVERMAN, FRED: b. N.Y.C. Sept., 1937. e. Syracuse Univ., Ohio State (Master's in TV and theatre arts). Joined WGN-TV, indep. sta. in Chicago. Came to N.Y. for exec. post at WPIX-TV, where stayed only six weeks when CBS-TV hired him as dir. of daytime programs. Named v.p.—programs June, 1970. In May, 1975 left CBS to become pres., ABC Entertainment. In June, 1978, named pres. and chief exec. officer of NBC.

SILVERMAN, RON: Producer-writer. b. Los Angeles, Calif., June 13, 1933. e. UCLA, 1951–53, U. of Arizona, 1953–55; reporter reviewer, Daily Variety, 1957–61; asst. to prod.-dir. Mark Robson, Red Lion Films, 20th Century-Fox, 1961–62; assoc. prod., Daystar Productions, 1961; v.p., 1964; assoc. prod. Crackerby TV series, 1965. Prod. exec., Warner Bros. TV, 1966; prod. & exec. Ted Mann Prods., 1967.

PICTURES INCLUDE: Buster and Billie (prod.), 1974, Lifeguard, Brubaker.

TV: Wild Wild West (writer), 1967.

SILVERMAN, SYD: Executive. b. N.Y., Jan. 23, 1932. e. The Manlius School, 1946–50; Princeton Univ., 1950–54. Lt., U.S. Army, 1954–56. Publisher, Variety and Daily Variety, executive editor, Weekly Variety.

SILVERS, PHIL: Actor. b. Brooklyn, N.Y., May 11, 1912; in vaudeville as boy tenor, later comedian in burlesque, then on dramatic stage, See My Lawyer, Yokel Boy; Tom, Dick and Harry. On screen 1940 in The Hit Parade.

PICTURES INCLUDE: You're in the Army Now, Roxie Hart, My Gal Sal, Footlight Serenade, Just Off Broadway, Coney Island, Cover Girl, Four Jills in a Jeep, A Lady Takes a Chance, Something for the Boys, Take It or Leave It, Diamond Horseshoe, Don Juan Quilligan, A Thousand and One Nights, If I'm Lucky, Summer Stock, Top Banana, Lucky Me, You'll Never Get Rich, 1955; 40 Pounds of Trouble, Mad, Mad, Mad World. A Guide for the Married Man, Buona Sera, Mrs. Campbell, The Boatniks, The Strongest Man in the World, Won Ton Ton, The Chicken Chronicles, The Cheap Detective.

TV: Noted Best Comedian and Best Television Performer in M.P. Daily's 2nd annual television poll, 1950.

THEATRE: Broadway Shows: High Button Shoes, Top Banana (Tony Award, best male star, 1951), Do Re Mi, A Funny Thing Happened On the Way to the Forum (Tony Award, best male actor-musical, 1972).

SILVERS, SID: Actor, writer, song writer. b. Brooklyn, N.Y., Jan. 1, 1908. On stage with Phil Baker for 7 yrs., as stooge & collab., many songs; began screen writing, 1930; Oh, Sailor, Behave; Follow The Leader, Par.

PICTURES INCLUDE: (collab. story, s.p. & appeared) Bottoms Up, Born To Dance; (collab. s.p. & appeared) Broadway Melody of 1936, (collab. story) Broadway Melody of 1938; (collab. music) Walking On Air; (add. dial. & appeared) 52nd Street; (collab. s.p.) Kentucky Kernels, The Gorilla, The Fleet's In, For Me and My Gal. Collab. s.p., Two Tickets to Broadway; collab. story, The Stooge.

SILVERSTEIN, MAURICE: Executive. b. Syracuse, N.Y., March 1, 1912; booker, salesman, MGM domestic dep't; International Dep't, MGM; supervisor Southeast Asia Hdqts. Singapore, MGM, 1938–42; OWI chief, film distribution for Europe, hdqts. London, during World War II; asst. sales supervisor, Far East MGM; regional director, Latin America, 1947; liaison exec. to handle independent productions MGM, 1956; vice-pres., MGM International, 1957; first vice-pres., 1958; pres., MGM International, 1963; vice-pres., parent company, Metro-Goldwyn-Mayer Inc. 1970; Silverstein Int'l Corp., pres.

SILVESTER, VICTOR: (O.B.E.) Musical director. b. Wembley, England. e. St. John's Leatherhead, John Lyons, Harrow. Exclusive BBC contract; with his orchestra, Records exclusively for Pye Records.

SIMMONS, JEAN: Actress. b. London, Jan. 31, 1929. e. Aida Foster Sch., London. m. Richard Brooks, director, writer. Screen debut 1944, in Give Us the Moon. Voted one of top ten British money-making stars in M.P. Herald-Fame Poll, 1950–51.

PICTURES INCLUDE: Mr. Emmanuel, Kiss the Boys Goodbye, Sports Day, Caesar and Cleopatra, Way to the Stars, Great Expectations, Hungry Hill, Black Narcissus, The Women In the Hall, Blue Lagoon, Hamlet, Adam and Evalyn, Trio, So Long as the Fair, Cage of Gold, The Clouded Yellow, Androcles and the Lion (U.S. film debut), Angel Face, Young Bess, Affair with a Stranger, The Actress, The Robe, She Couldn't Say No, A Bullet Is Waiting, The Egyptian, Desiree, Footsteps in the Fog, Guys and Dolls, Hilda Crane, This Could Be the Night, Until They Sail, Home Before Dark, Spartacus, The Grass Is Greener, All the Way Home, Elmer Gantry, Divorce American Style, Rough Night in Jericho, The Happy Ending (Acad. Award nom.).

SIMMONS, JOHN: Producer, director, writer, creative consultant. e. St. Clement Danes, the Univ. of London, the Temple. Assoc. with many adv. doc. feature & TV films. Numerous International Festival awards, incl. Oscar nomin.; creative dir., Cinevista Ltd.; wrote, The Blue Bird, Loganberry Fair (lyrics, etc.), devised adv. campaigns, The Guns of Navarone, Summer Holiday; creative dir., John Simmons Creative Consultants Ltd. Join The Tea Set, Shell, Ovaltine, Waddington, Tide, Terylene, Marks & Spencer, Schweppes, Crown, Cleveland, Gold Camera Award (1st Place) for The Bosch Equation, U.S. Int. Ind. Film Festival, 1974. 1st Prize San Francisco Fest. 1975. Dev. & Dir. Cinema Ad. Awards, 1976. Corr. to Fin. Times, Variety, Campaign etc. Gold Camera Award 1977 for Stop Her, Silver Award, New York Fest. Consultant, Rank Advertising Awards (cinema) and commercial radio. Best radio commercial Award 1977. Award, Cannes 1978, 'A Clear Edge.'

SIMMONS, STANLEY: Producer, director. b. New York City, May 8, 1915. e. McBurney Preparatory School. Entered m.p. ind. 1930; pres. own distributing co.; advance man on roadshow m.p. 5 yrs.; ed. Hollywood shorts & features; apptd. managing dir. Amer. Museum of Natural History's exhibit at N.Y. World's Fair; film ed.; prod.-dir. Carnival in Brazil, 1941; served in U.S. Army 1942–45, as prod.-dir.-ed. over 100 documentaries & head of special trailer dept. ETO; supvr. color Western feature series; prod. head of Realart Productions, Inc.; lab. & exchange rep. for Soundies Distributing Corp.; prod. mgr. North Amer. Video Productions, Inc.

PICTURES INCLUDE: (prod.-dir. for video) It's a Woman's World series, Album of Songs (musical series), spots for Life Cigarettes and others. 9 Acapello musicals.

SIMMS, FRANK: Announcer. b. Tulsa, Okla., June 8, 1921. e. U. of Tulsa, 1946. Radio announcer, U.S. Air Force, WWII; announcer, KVOO, Tulsa, to N.Y., 1953.

TV: I Love Lucy, My Little Margie, Public Defender, Garry Moore Show.

SIMON, MELVIN: Executive. b. N.Y.C., Oct. 21, 1926. e. City College, New York (BBA, 1949); graduate work at Indiana Univ. Law School. Owns and operates, in partnership with two brothers, over 90 shopping centers in U.S. In 1978 formed Melvin Simon Productions, privately owned corp., to finance films.

MEMBER: Friars Club; N.Y. div.; 1978, v.p., In-tl. Council of Shopping Centers; 1978, commerce and industry chm. of muscular dystrophy; mem. bd., Indiana Repertory Theatre 1978, corporate sponsor: Indianapolis 500 Festival, Indianapolis Museum of Arts, Indianapolis Children's Museum; Indianapolis Zoological Society.

SIMON, NEIL: Playwright. b. Bronx, N.Y. Wrote comedy for radio with brother, Danny, also TV scripts for Sid Caesar, Red Buttons, Jackie Gleason.

PLAYS INCLUDE: Barefoot in the Park, The Odd Couple, Sweet Charity, The Star Spangled Girl, Plaza Suite, Promises, Promises, The Last of the Red Hot Lovers, The Gingerbread Lady. Adapted several of own plays to screen and wrote original s.p., The Out-of-Towners.

PICTURES: Sunshine Boys, Murder by Death, The Goodbye Girl, The Cheap Detective, California Suite, Seems Like Old Times, Only When I Laugh.

SIMONE, SIMONE: Actress. b. April 23, 1914, Marseilles, France. Played in many films in Europe, among them Les Beaux Jours and Lac aux Dames. On stage in Toi C'est Moi, and others.

PICTURES INCLUDE: Girl's Dormitory, Ladies in Love, Seventh Heaven, Love and Kisses, Josette, Johnny Doesn't Live Here Any More, Silent Bell, Temptation, Harbor, Lost Women, La Ronde, Pit of Loneliness, Le Plaisir.

SIMPSON, GARRY: Producer, director, writer. e. Stanford U. Major shows with NBC-TV: Jimmy Durante Show, Armstrong Circle Theatre, Ed Wynn Show, Philco TV Playhouse, Ballet Theatre. Awards: Academy of TV Arts & Sc., Sylvania. Documentary film writer-producer awards: International Film & TV Festival, Chicago Film Festival, Broadcast Media Awards. Currently, pgm mgr., Vermont Ed. TV Network.

SIMPSON, DONALD C.: Executive. b. Alaska, 1946. e. Univ. of Oregon (Phi Beta Kappa). Began m.p. career as Warner Bros.' acct. exec. with Jack Woodell Agency in San Francisco in 1969. In 1971 joined WB as member of adv./mktg. group, handling special projects. In 1975 joined Paramount Pictures as prod. exec. for Motion Picture Division. Promoted in 1977 to v.p. in chg. of prod. for division. 1980, appt. snr. v.p., in chg. of world-wide prod.

SIMS, JOAN: Actress. b. London, England, 1930.

PICTURES INCLUDE: Dry Rot, Off the Record, No Time for Tears, Just My Luck, The Naked Truth, The Captain's Table, Passport to Shame, Emergency Ward 10, Most of the 'Carry On' films, Doctor in Love, Watch Your Stern, Twice Round The Daffodils, The Iron Maiden, Nurse on Wheels, Doctor in Clover, Doctor in Trouble, The Garnett Saga, Not Now Darling, Don't Just Lie There Say Something, Love Among the Ruins, One of Our Dinosaurs Is Missing.

TELEVISON: Born and Bred, Worzel Gummidge.

SINATRA, FRANK: Actor, singer. b. Hoboken, N.J., 1915. Sportswriter; then singer on radio various N.Y. stations; joined Harry James orchestra, later Tommy Dorsey. On screen 1943 in Reveille with Beverly. Spec. Academy Award 1945 for acting in The House I Live In, short subject on tolerance; Academy Award 1953.

PICTURES INCLUDE: Higher and Higher, Step Lively, Anchors Aweigh, It Happened in Brooklyn, Till the Clouds Roll By, Miracle of the Bells, Kissing Bandit, Take Me Out to the Ball Game, On the Town, Double Dynamite, Meet Danny Wilson, From Here to Eternity (Acad. Award. best supporting actor, 1953), Suddenly, Young at Heart, Not as a Stranger, Guys and Dolls, Tender Trap, Man With the Golden Arm, Johnny Concho, High Society, Around The World in 80 Days, Pride and the Passion, The Joker is Wild, Pal Joey, Kings Go Forth, Some Came Running, A Hole in the Head, Never So Few, Ocean's 11, Devil at Four O'Clock, Sergeants 3, The Manchurian Candidate, Come Blow Your Horn, The List of Adrian Messenger, 4 for Texas, Robin and the Seven Hoods, None But the Brave (dir.), Von Ryan's Express, Marriage on the Rocks, Cast a Giant Shadow, Assault on a Queen, The Naked Runner, Tony Rome, The Detective, Lady in Cement, Dirty Dingus Magee, That's Entertainment!, The First Deadly Sin.

TV: The Frank Sinatra Show, numerous specials, etc. Won both an Emmy and a Peabody Award. Hersholt Humanitarian Award, 1971. 1977: Contract on Cherry Street (movie).

SINCLAIR, ANDREW: Director, writer. Early career as novelist and historian, playwright. Published 18 books in U.K., U.S.A. Entered m.p. ind. 1968.

PICTURES INCLUDE: s.p.: Before Winter Comes, Adventures in the Skin Trade, The Voyage of the Beagle, You?; dir., writer: The Breaking of Bumbo, 1970; dir., writer: Under Milk Wood, 1971; prod. Malachi's Cove, 1973. Writ. The Representative, 1976; The Scarlet Letter; Martin Eden.

SINCLAIR, MARY: Actress. r.n. Mary Cook. b. San Diego, Nov. 15. Model; summer stock; TV debut, CBS, 1949. Screen debut in Arrowhead, 1953.

TV: Fireside Theatre, Studio One, Suspense, Pulitzer Prize Playhouse. Under contract CBS TV, 1951.

SINDELL, BERNARD I.: Talent agent b. Nob. 27, 1912. e. Western Reserve Univ.; vice-pres., Jaffe Agency, 1944–57; talent agent, the Sindell Agency, 1957.

SINDEN, DONALD: Actor. b. Plymouth, England, Oct. 9, 1923. Stage debut 1941 in fit-up shows; TV debut 1948; screen debut in 1953, Cruel Sea.

PICTURES INCLUDE: The Cruel Sea, Mogambo, A Day to Remember, You Know What Sailors Are, Doctor in the House, The Beachcomber, Mad About Men, Simba, Above Us the Waves, Josephine and Men, An Alligator Named Daisy, Black Tent, Eyewitness, Tiger in the Smoke, Doctor at Large, Rockets Galore, The Captain's Table, Operation Bullshine, Your Money or Your Wife, The Siege of Sydney Street, Twice Around the Daffodils, Mix Me a Person, Decline and Fall, Villain, Rentadick, The National Health, The Day of the Jackal, The Island at the Top of the World.

TV: Bullet in the Ballet, Road to Rome, Dinner With the Family, Odd Man In, Love from Italy, The Frog, The Glove, The Mystery of Edwin Drood, The Happy Ones, The Comedy of Errors, The Wars of the Roses, The Red House, Blackmail, A Bachelor Gray, Our Man at St. Marks (3 series), The Wind in the Tall Paper Chimney, A Woman Above Reproach, Call My Bluff, Relatively Speaking, Father Dear Father, The 19th Hole, Seven Days in the Life of Andrew Pelham (serial), The Assyrian Rejuvenator, The Organization (serial), The Confederacy of Wives, Tell It to The Chancellor, Two's Company (4 series), All's Well That Ends Well.

SINGER, A: Managing Director BBC Radio 1978. b. Bradford, York, England, Jan. 21, 1927; e. Giggleswick High School, Bradford Grammer Sch. Entered m.p. ind. 1943; dir. 8 films; worked as unit mgr. in Austria; TV as BBC prod., 1949; prod. 80 programs of events and features. Joined Scottish Region 2-year attachment. N.Y. office, BBC as TV officer working with American TV networks and local stations, Oct., 1953; ret. to London, 1956; prod. BBC-TV, Eye on Research and Science Intl. series; Asst. Head Outside Broadcasts 1959; exec. producer for E.B.U.; on first programme exchange between Europe and America 1962; Head of Outside Broadcasts Feature and Scientific Programmes, 1963. Head of features group, BBC Television, 1967. Controller BBC 2, 1974.

SINGLETON, PENNY: Actress. r.n. Dorothy McNulty. b. Philadelphia, Pa., September 15; e. Columbia U. First Broadway success came as top comedienne in Good News., exec. pres. AGVA.

BROADWAY: Follow Through, Hey Nonny Nonny.

PICTURES INCLUDE: Innumerable films in Blondie series; Young Widow.

SINN, ROBERT S.: Executive. b. Philadelphia, Pa., March 9, 1930; e. U. of Pa., A.B., M.S. Pres., Ultronic Systems Corp., 1960–70; chmn., The Wall Street Venture Capital Corp.; chmn., Microwave Semiconductor Corp.; chmn., Medical Electroscience & Pharmaceuticals Inc.; backer, Joe. Founder GSF Productions, Inc.

SIODMAK, CURT: Director. e. U. of Zurich. Engineer, newspaper reporter, writer in Berlin; novelist, including "F.P.1 Does Not Answer," adapt. 1932 for Ufa. Originals and screenplays in France and England including France (Le Bal), Transatlantic Tunnel GB.
PICTURES INCLUDE: In U.S. originals and screenplays, many pictures including Her Jungle Love, Aloma of the South Sea, Par.; Invisible Woman; basis: Son of Dracula; s.p.: The Mantrap; (orig.) House of Frankenstein; collab. orig. s.p.: Shady Lady; s.p.: Beast with Five Fingers; collab. s.p., story: Berlin Express; collab. s.p.: Tarzan's Magic Fountain, Four Days Leave; dir.: Bridge of the Gorilla; collab. s.p., dir.: The Magnetic Monster; s.p.: Riders to the Stars; story, s.p.: Creature with the Atom Brain; story: Earth vs. the Flying Saucers.

SIRK, DOUGLAS: Director, writer. b. Denmark, April 26, 1900; e. U. Hamberg, Munich. Prod. for UFA, stage actor & producer, newspaperman. From 1944, dir., Hollywood.
PICTURES INCLUDE: Sleep My Love, Lured, Slightly French, Shockproof, Mystery Submarine; prod. dir.: The First Legion; dir.: Thunder on the Hill, Lady Pays Off, Weekend With Father, Has Anybody Seen My Gal?, No Room for the Groom, Meet Me at the Fair, Take Me to Town, All I Desire, Taza Son of Cochise, Magnificient Obsession, Sign of the Pagan, Captain Lightfoot, There's Always Tomorrow, All That Heaven Allows, Battle Hymn, Written on the Wind, A Time to Love and a Time to Die, Imitation of Life.

SKELTON, RED: Actor, comedian. r.n. Richard Skelton. b. Vincennes, Ind. Joined medicine show at 10; later in showboat stock, minstrel shows, vaudeville, burlesque, circus. Screen debut 1939 in Having Wonderful Time. On radio from 1936. Red Skelton Show, TV, since 1950. Composer of music, writer of short stories and painter.
PICTURES INCLUDE: Flight Command, Lady Be Good, The People vs. Dr. Kildare, Whistling in the Dark, Whistling in Dixie, Ship Ahoy, Maisie Gets Her Man, Panama Hattie, Du Barry Was a Lady, Thousands Cheer, I Dood It, Whistling in Brooklyn, Bathing Beauty, Ziegfeld Follies, Fuller Brush Man, Southern Yankee, Neptune's Daughter, Yellow Cab Man, Three Little Words, Watch the Birdie, Excuse My Dust, Texas Carnival, Lovely to Look At, The Clown, Half a Hero, Great Diamond Robbery, Public Pigeon No. 1, Those Magnificent Men and Their Flying Machines.

SKILES, MARLIN: Composer. b. Harrisburg, Pa., Dec. 17, 1906; e. Froehlich Sch. of Music, Harrisburg. Pianist for several dance orch.; mus. dir., pianist, arr. for Paul Whiteman, Irving Aaronson, others; to Hollywood, 1932; comp., arr. for radio & many m.p.
PICTURES INCLUDE: com., arr.: Over 21, Jolson Story, Gilda, Tonight and Every Night, Dead Reckoning, Callaway Went Thataway, Rose Bowl Story, Battle Zone, The Maze, Fighter Attack, Pride of the Blue Grass, Arrow in the Dust, Dial Red O, Annapolis Story, many Bowery Boys features, Sudden Danger.

SKIRBALL, JACK H.: Producer. b. Homestead, Pa., June 23, 1896. (LLD. & D.H.L.) Was with Alco Films, Cincinnati, and Metro. Later on advisory bd., div. mgr., Educational; assoc. with Wm. N. Skirball in Skirball Brothers Ohio Theatres; sales mgr. Educational, 1932. In 1932–38: in chg. prod. & distrib. for Educational. Resigned July, 1939, to take up indep. prod. In 1944–45 formed Skirball-Manning Prod. with B. Manning. Pres., Films for Television, Inc. Produced stage play, N.Y.: Jacobowsky and the Colonel, 1944. General Intl. Film Co., Inc. Recipient M.P. Pioneers Award 1925.
PICTURES INCLUDE: producer, independently: Half a Sinner (U.); Miracle on Main Street (Col.); El Milagro de la Calle Major (20th-Fox). Franchise holder with William N. Skirball for Ohio on WB, Columbia, Educational. In 1940, assoc. prod., Frank Lloyd's The Howards of Virginia (Col.); In 1941: The Lady From Cheyenne, This Woman is Mine (U.). In 1942 formed Jack H. Skirball Prod., Inc.; prod.: Saboteur (U.) In 1943, Shadow of a Doubt, It's in the Bag, then; Guest Wife, So Goes My Love, The Secret Fury, Payment on Demand, A Matter of Time.

SKIRBALL, WILLIAM N.: Exhibitor. b. Homestead, Pa. Began career with Metro in Des Moines & Chicago; states rights distrib., Cleveland; assoc. with brother, Jack H. Skirball (of Skirball-Manning prod. team), educational films br. mgr.; also brother of Joseph Skirball, first nat'l franchise owner, Pittsburgh. Currently partner Skirball Bros. circuit (10 theatres, Ohio), hdqts., Cleveland, Ohio.

SKLAR, MICHAEL: Producer. b. Philadelphia, e. Temple U. Beg. career in broadcasting field, free lancing, script writer, radio: Grand Central, First Nighter, Superman, 1939; producer, ed., writer, WMCA, N.Y., Five Star Final; joined CBS-TV, 1948 You Are There; New World A'Coming, 1947 radio prod.; wrote-prod., The Greatest Drama, series of film biog. of famous people, Fox Movietone; Eye on the World, Talkaround, The Good Morning Show, CBS-TV; prod., New Frontiers, 1956; prod., Conquest, 1957 presented in cooperation with the American Assoc. for the Advancement of Science, 1960; prod., writer, Intertel, U.S.A. 1961–62; Britain Postscript to Empire, Mexico—The Unfinished Revolution; prod., writer, Exploring The Universe, 1962; prod., Mayer-Sklar Prods., 1962–63; prod., writer, Intertel, U.S.A., 1964; This Question of Color; prod. exec., Confidential—For Women, 1966, prod., How it Happened.

SKOLSKY, SIDNEY: Columnist, Producer. b. New York, N.Y., May 2, 1905; e. N.Y. Univ.; p. Louis and Mildred Skolsky. Press agent on Broadway for Sam H. Harris, George White, Earl Carroll and others; weekly feature Tintypes for the N.Y. Sun and then to New York Daily News as columnist. In 1935, collab. on story, The Daring Young Man, 20th-Fox. In 1937 resigned as Hollywood columnist for Daily News; to New York Daily Mirror, then New York Post. Column syndicated by United Features.
PICTURES INCLUDE: prod.: The Jolson Story, 1946; Eddie Cantor Story, 1953.
TV: special: prod.: Hollywood, the Golden Era, 1961.

SLATER, DAPHNE: Actress. b. Bayswater, London, England, March 3, 1928; e. Haberdashers' Askes Sch.; Royal Acad. Dramatic Arts. Stage debut: The Rising Generation, 1945; several plays including King Lear; m.p. debut in Courtneys of Curzon Street, 1947; TV debut for BBC in I Want To Be An Actor, 1946.
TV: Emma, Shout Aloud Salvation, All the Year Round, They Fly by Twilight, Pride and Prejudice, The Affair at Assino, Beau Brummell, Jane Eyre, Precious Bane, The Dark Is Light Enough, Mary Rose, Julius Caesar, Berkeley Square, Less Than Kind, The Burning Glass, Persuasion, The Winslow Boy She Stoops to Conquer, Nothing to Pay, The Father, The Bald Prima Donna, The Big Breaker, The Cocktail Party Photo Finish, The Seagull, Love Story, Emergency Ward 10, Jackanory, The First Freedom, Man of Our Times, The Jazz Age, Callan, The Piano Tuner, Happy Ever After, The Pretenders, Virtue, Elizabeth R, The Staff Room, Footprints in the Sand.

SLATZER, ROBERT FRANKLIN: Writer, Director, Producer, Author; b. Marion, Ohio; April 4, 1927; e. Ohio State Univ., UCLA, 1947; Radio news commentator sportscaster, wrote radio serials; adv. dir., Brush-Moore Newspapers; feature writer, Scripps-Howard Newspapers; adv. exec., The Columbus Dispatch; syndicated columnist, New York Journal-American; wrote several guest columns for Walter Winchell and Dorothy Kilgallen; author of several western short stories and novels; wrote, directed, produced many industrial films, documentaries, sports specials and commercials; 1949–51, writer for Harry "Pop" Sherman Productions, Monogram Pictures, Republic Studios, Eagle-Lion Films; 1951, publicist, Hope Enterprises; Pub. Dir., Paramount Pictures; 1952, personal manager to Marilyn Monroe, Ken Maynard, James Craig, Gail Russell and other stars; 1953, story editor and assoc. prod., Joe Palooka Productions; 1953–57, staff writer Universal Studios, RKO Radio Pictures, MGM, Columbia and Paramount Studios; 1958, formed Robert F. Slatzer Productions; 1960, exec. in chg. of prod., Jaguar Pictures Corp.; 1963–65, president, Slatzer Oil & Gas Company; 1966–67, board director, United Mining & Milling Corp.; 1967–70, wrote and directed several feature films; 1970–74, exec., Columbia Pictures Corp.; 1974, resumed producing and financing features and television films 1976, honored as "Fellow", Mark Twain Institute.
PICTURES INCLUDE: White Gold, The Obsessed, Mike and the Heiress, Under Texas Skies, They Came To Kill, Trail of the Mounties, Jungle Goddess, Montana Desperado, Pride of the Blue, Green Grass of Wyoming, The Naked Jungle, Warpaint, Broken Lance, Elephant Walk, South of Death Valley, The Big Gusher, Arctic Flight, The Hellcats, Bigfoot, John Wayne's 'No Substiture for Victory', Joniko—Eskimo Boy, Operation North Slope, Don't Go West, Mulefeathers, The Unfinished, Single Room Furnished, Viva Zapata.
TELEVISION: The Great Outdoors, Adventures of White Arrow, Let's Go Boating, The Joe Palooka Story, Amos & Andy, I am the Law, Files of Jeffrey Jones, Fireside Theatre, the Unser Story, Year of Opportunity (Award winning spec.), The Big Ones, Ken Maynard's West, Where are They Now?, The Groovy Seven, The Untouchables, The Detectives, Wild Wild West, Wagon Train, Playhouse 90, Highway Patrol, David Frost Special, Today Show.
AUTHOR: (novels) Desert Empire, Rose of the Range, Rio, Rawhide Range, The Cowboy and the Heiress, Daphne, Campaign Girl, Scarlet, The Dance Studio Hucksters, Born to be Wild, Single Room Furnished, The West is Still Wild,

Gusher, The Young Wildcats; (Biographies) The Life and Curious Death of Marilyn Monroe, The Life and Legend of Ken Maynard, Who Killed Thelma Todd?, The Duke of Thieves, Bing Crosby—The Hollow Man.

SLAVIN, GEORGE: Writer. b. Newark, N.J., 1916; e. Bucknell U., Drama Dept., Yale U.
PICTURES INCLUDE: story, collab. s.p., Intrigue; collab. story, Woman on Pier 18; collab. s.p., The Nevadan, Mystery Submarine; collab. story & s.p. Peggy, Red Mountain, City of Bad Men; collab. story, Weekend with Father, Thunder Bay, Rocket Man; collab. story, collab s.p., Smoke Signal, Uranium Boom; collab. s.p., Desert Sands.

SLEZAK, WALTER: Actor. b. Vienna, Austria, May 3, 1902; son of Leo Slezak, opera singer. In banking business before becoming actor Ufa; also on Berlin stage. In 1930 to U.S. for N.Y. stage production Meet My Sister; other roles N.Y. stage followed. First U.S. screen role in Once Upon a Honeymoon, RKO; on B'way, in My 3 Angels, Fanny.
PICTURES INCLUDE: Fallen Sparrow, Lifeboat, And Now Tomorrow, The Spanish Main, Cornered, Sinbad the Sailor, Born to Kill, Riffraff, Pirate, Inspector General, Yellow Cab Man, Bedtime for Bonzo, People Will Talk, Confidentially Connie, Call Me Madam, White Witch Doctor, Steel Cage, The Gazebo, Come September, Swinger's Paradise, Emil and the Detectives, 24 Hours To Kill, Dr. Coppelius, Black Beauty.

SLOAN, JOHN R.: Producer. e. Merchiston Castle, Edinburgh, 1932–39; asst. dir. and prod. man. Warners, London, Hollywood; 1939–46, Army Service.
PICTURES INCLUDE: Sea Devils, The End of the Affair, Port Afrique, Abandon Ship, The Safecracker, Beyond this Place, The Killers of Kilimanjaro, Johnny Nobody, The Reluctant Saint, The Running Man, The Last Command, To Sir With Love, Fragment of Fear, Dad's Army, Lord Jim, No Sex Please We're British, The Odessa File, Force From Navarone.

SLOCOMBE, DOUGLAS: Cinematographer. b. England, 1913. Former journalist.
PICTURES: Dead of Night, The Captive Heart, Hue and Cry, The Loves of Joanna Godden, It Always Rains on Sunday, Saraband for Dead Lovers, Kind Hearts and Coronets, Cage of Gold, The Lavender Hill Mob, The Man in the White Suit, The Titfield Thunderbolt, Man in the Sky, The Smallest Show on Earth, Tread Softly, Stranger, Circus of Horrors, The Young Ones, The L-Shaped Room, Freud, The Servant, Guns at Batashi, A High Wind in Jamaica, The Blue Max, Promise Her Anything, The Vampire Killers, Fathom, Robbery, Boom, The Lion in Winter, The Italian Job, The Music Lovers, Murphy's War, Jesus Christ Superstar, The Great Gatsby, Rollerball, Hedda, Nasty Habits, Julia, Lost and Found, Nijinsky, Raiders of the Lost Ark.
TELEVISION: Love Among the Ruins, The Corn Is Green, etc.

SMAKWITZ, CHARLES A.: e. Syracuse University, named by Warner Bros. as dist. mgr. Albany, Troy and Utica, and named as zone mgr. Warner Bros., New York State Theatres. Made zone mgr. of Stanley Warner Theatres for the State of New Jersey and New York State Theatres added to N.J. Zone and made headquarters in Newark, N.J. in 1955, transferred to New York City home office as national director of public relations, publicity, advertising for Stanley Warner Corp. nation-wide. In the various areas was Mayor's Rehabilitation and Urban Development Com. of Albany, State Albany Chairman War Activities Committee, National Public Dir. Red Cross Fund Organized Albany V.C. Tent #9 and served as Chief Barker 3 years, 1949–1952, elected National Representative V.C. of America. Member N.Y. State Program and Planning Committee, Pres. Heart Assn. of Albany County, N.J. State Chairman Nat. Conf. of Christian and Jews 1955–62. Organized and Pres. of Syr. Un. of Greater, N.Y. Served as Office of V.C. of N.Y. Assistant to Spyros R. Skoveas 1969–71, served as International-Ambassador at large of V.C. International Consultant exhibition, production, promotion, real estate and public relations.

SMART, RALPH: Writer, producer, director. b. London, Eng., 1908. Dir. documentaries and wrote s.p. for features 1927–40; prod. films for Royal Australian A.F. 1940–44; assoc. prod. The Overlanders, prod. dir. and s.p. Bush Christmas in Australia; ret. to England to dir. Boy, Girl and a Bike, Bitter Springs, Never Take No for An Answer, Always a Bride; dir. TV series Robin Hood, 1955–56; exec. prod. Incorporated Television programme Co., 1957; series William Tell, The Invisible Man, prod. author. 2 series, Danger Man; now freelancing.

SMART, ROY L.: Exhibitor. b. Clayton, Ala. e. U. of Ala., drama critic, city ed. Birmingham Age-Herald; U.S. Naval Air Corps, mgr. theatres, Ala., Fla., Carolinas; dist. mgr. Wilby-Kincey theatres, N.C. & S.C. assistant gen. mgr., hdqts., Charlotte, N.C.; during WW II Naval Aviation Cadet Selection Bd., Naval Officer Procurement Comm., co-ordinator War Activities Comm. for N.C. & S.C. Variety Club, past chief barker; American Legion, past vice commander; Kiwanis, past pres.; City School Commissioner, Charlotte, N.C.; past pres. Charlotte City Club; v.p. Charlotte Country Club; m.p. pioneer; bd. member, Charlotte Symphony Orch.

SMIGHT, JACK: Director. b. Minneapolis, Minn., 1926
PICTURES INCLUDE: I'd Rather Be Rich, The Third Day, Harper, Kaleidoscope, The Secret War of Harry Frigg, No Way to Treat a Lady, Strategy of Terror, The Illustrated Man, The Travelling Executioner, Rabbit Run, Dr. Frankenstein, Frankenstein: The True Story, Airport 1975, Midway, Damnation Alley, Fast Break, Loving Couples.
TV: Banacek, Columbo, etc.

SMITH, ALEXIS: Actress. b. Penticton, Can., June 8, 1921; e. Los Angeles City Coll. In summer stock British Columbia; star in coll. prod., Night of January 16th; in m.p. for Warner Bros.
PICTURES INCLUDE: Smiling Ghost, Dive Bomber, Steel Against the Sky, Gentleman Jim, Thank Your Lucky Stars, Constant Nymph, Conflict, Adventures of Mark Twain, Rhapsody in Blue, Horn Blows at Midnight (Star of Tomorrow, 1943), One More Tomorrow, Night and Day, Of Human Bondage, Stallion Road, Two Mrs. Carrolls, Woman in White, Decision of Christopher Blake, Whiplash, South of St. Louis, Any Number Can Play, One Last Fling, Undercover Girl, Wyomying Hall, Montana, Here Comes the Groom, Cave of the Outlaws, Turning Point, Split Second, Sleeping Tiger, Eternal Sea, The Young Philadelphians, Once Is Not Enough, The Little Girl Who Lives Down the Lane, Casey's Shadow.
BROADWAY: Follies, 1971, Platinum, 1978.

SMITH, A. BERKELEY: TV commentator, producer. b. Bournemouth, Eng., Dec. 8, 1918. e. Blundell's Sch., St. Andrews U. Royal Artillery, 1939–46; lecture tour, U.S., 1945; prod.-comment., BBC North American service, 1946; to U.N. Assembly, New York, for BBC home service, 1949; TV as BBC outside broadcasts prod., 1949; devised, prod. sports & farming series with other documentaries; Comm. State Occasions. Theatre Shows, etc., 1951; prod. Show Places series; prod. exchange program from R.T.F. Paris; commentator Coronation & all State visits, 1953; prod. Arenascope, programs radio show, 1954; asst. head TV outside broadcasts. Feb. 1955; head outside broadcasts Southern Television Ltd., 1958; asst. Con. Programs, 1961; Con. Programs (Southern TV Ltd), 1964; dir. Southern TV, 1965, Left Dec. 75, Dir. ITV Prog. Planning, 1976.

SMITH, CHARLIE MARTIN: Actor. b. 1955. e. Calif. State Univ. f. is animation artist Frank Smith.
PICTURES INCLUDE: The Culpepper Cattle Company, Fuzz, The Spikes Gang, American Graffiti, More American Graffiti.
TELEVISION: The Brady Bunch, Monte Nash.

SMITH, CONSTANCE: Actress. r.n. Constance Smyth. b. Limerick, Ire.; e. St. Louis Convent. Began career as model for Noel Mayne; winner of beauty contest Screen Magazine. Free-lance actress, studios in Eng., prior to Hollywood m.p. debut in The Mudlark.
PICTURES INCLUDE: Brighton Rock, Now Barrabas, Trottie True, Murder at the Window, Blackmail, The Perfect Woman, Room to Let, The 13th Letter, I'll Never Forget You, Red Skies of Montana, Treasure of the Golden Condor, Lure of the Wilderness, Taxi, Man in the Attic, Big Tipoff.

SMITH, HOWARD K.: News commentator. b. Ferriday, La., May 12, 1914. e. Tulane Univ., 1936. Heidelberg U., Germany; Oxford U., Rhodes Scholarship. United Press, London, 1939; United Press Bureau, Copenhagen; United Press, Berlin, 1940; joined CBS News, Berlin corr., 1941. Reported on occupied Europe from Switzerland to 1944; covered Nuremberg trials, 1946; ret. to U.S., moderator, commentator or reporter, CBS Reports, Face the Nation, Eyewitness to History, The Great Challenge, numerous news specials. Sunday night news analysis. CBS News Washington corr., 1957; chief corr. & mgr., Washington Bureau, 1961; joined, ABC News, Jan. 1962. News and Comment, ABC news. Anchorman and Commentator, ABC Evening News. Author: Last Train from Berlin, 1942, The State of Europe, 1949. Washington, D.C.—The Story of Our Nation's Capital, 1967.

SMITH, HY: Executive. b. New York City, June 3, 1934. e. Baruch School, CCNY, BBA. Joined Paramount Pictures 1967, foreign ad.-pub coordinator; 1969-joined United Artists as foreign ad-pub mgr., named int'l. ad-pub dir., 1970; named v.p.; intl. adv-pub. 1976; Appointed vice pres. worldwide Advertising, publicity and promotion, July 1978; 1981, named first v.p., adv./pub./promo.

SMITH, JACLYN: Actress. b. Houston, Texas, Oct. 26. Started acting while in high school and studied drama and psychology at Trinity Univ. in San Antonio. Appeared in many commercials as model.
PICTURES INCLUDE: Bootleggers, The Adventures.

TV: McCloud, Get Christy Love, The Rookies, World of Disney, Switch, Charlie's Angels (regular).

SMITH, JACQUELINE: Executive. b. Phila., Pa., May 24, 1933. e. Antioch College, 1954. m. William Dale Smith, novelist (pseudonym David Anthony). Actress, Bermudiana Theatre in Bermuda, The Antioch Theatre in Ohio and several little theatres in Washington, D.C.; with the Stanford U. Players. Did research and writing for RCA, worked as a nursery school teacher and directed Little Theatre groups in the West Coast area. 1957: wrote and produced over 100 weekly children's programs at KPIX, San Francisco. Promotion director at WPIX-TV, N.Y. Worked with CBS TV Network for over 5 yrs. as general prog. exec. and exec. prod. of CBS West Coast Daytime Programs. Currently director of special projects at Warner Bros. Television Dept. and executive producer of Warner Bros. Animation Division.

SMITH, JOSEPH P.: Executive. b Bklyn, N.Y. e. Columbia Univ. Started career Wall Street; joined RKO Radio Pictures, served in sales and managerial posts; exec. vice-pres., Lippert Productions, Hollywood; formed, president, Telepictures, Inc; pres., Pathe Pictures, Inc., Pathe News, Inc.

SMITH, KATE: Performer. r.n. Kathryn Elizabeth S. b. Greenville, Va., May 1, 1909; on Broadway in musicals, Honeymoon Lane, Hit the Deck, Flying High; joined Ted Collins for radio prog., 1931; various variety programs followed; started daytime prog. as commentator, 1938, now has daytime Kate Smith Hour and weekly Kate Smith Evening Hour on NBC-TV in N.Y.

SMITH, KENT: Actor. On screen from 1943 in Forever and a Day, Hitler's Children, This Land is Mine; N.Y. stage, The Autumn Garden and The Wild Duck.
PICTURES INCLUDE: Youth Runs Wild, The Cat People, Three Russian Girls, The Spiral Staircase, Nora Prentiss, Magic Town, Voice of the Turtle, Fountainhead, Damned Don't Cry, My Foolish Heart, Paula, Commanche, The Balcony, Covenant with Death, Games, Kona Coast, Assignment to Kill, Pete an' Tillie, Cops and Robbers.
TV: Richard II, Profiles in Courage, The Outer Limits, Peyton Place.

SMITH, MAGGIE: Actress. b. Ilford, England, 1934. Early career Oxford Playhouse. Numerous TV appearances Britain, America.
PICTURES INCLUDE: Nowhere to Go, The V.I.P.s, Young Cassidy, Othello, The Honey Pot, Hot Millions, The Prime of Miss Jean Brodie (Acad. Award, Best Actress), Oh What a Lovely War, Love and Pain, Travels with my Aunt, Murder by Death, Death on the Nile, California Suite (AA), Clash of the Titans.

SMITH, MAURICE: Producer, director, writer. b. London, England, May 12, 1939. e. St. Ignatius College. Prior to entering m.p. industry, worked in bank in England, on newspaper in Canada, pool hustler, general contractor in Los Angeles, Calif.
PICTURES INCLUDE: The Glory Stompers, Scream Free, Cycle Savages, Hard Trail, Diamond Stud, Love Swedish Style, November Children, How Come Nobody's On Our Side, Joys of Jezebel.

SMITH, RAY E.: Executive. v.p., asst. sect. & asst. treas., buyer & booker, union negotiator, Jamestown Amusement Co., Inc. & Shea Enterprises, New York, N.Y., b. Oct. 10, 1908. e. Northhampton Commercial College, 1930.

SMITH, ROGER: Actor. b. South Gate, Calif., Dec. 18, 1932. e. U. of Arizona, started career at age 7, one of the Meglin Kiddies, appearing at the Mayan Theater, Wilshire, Ebell.
PICTURES INCLUDE: No Time to be Young, Crash Landing, Operation Madball, Man of a Thousand Faces, Never Steal Anything Small, Aunti Mame, Rogues Gallery.
TV: The Horace Heidt Show, Ted Mack Original Amateur Hour, 77 Sunset Strip, writer, ABC-TV, sings, composes, American folk songs.

SMITH, WILLIAM: Actor.
PICTURES: Twilight's Last Gleaming, The Frisco Kid, Any Which Way You Can.
TELEVISION: Rich Man, Poor Man, Hawaii 5-0

SMOLEN, DONALD E.: Executive. b. New York City, Aug. 10, 1923; e. NYU 1943, Pratt Institute 1947, Ecole Des Beaux Arts 1949. Art dept. Fox, 1940–41; art dept. Kayton Spiero Advtg., 1942; illustrator designer, Gilbert Miller Studios 1946–49; free lance illustrator, designer, 1951–65 servicing such accounts as UA, Fox, Warner Bros., TWA, Ford Motors; dir. of adv., UA. Resigned 1974 to form own co., Donald E. Smolen and Associates, consultant for m.p. adv.; 1975 merged to form Smolen, Smith and Connolly, advertising and marketing consultants to m.p. industry. Created ad campaigns for All the Presidents' Men, Fiddler on the Roof, Rocky, Superman, Star Wars.

SNAPER, WILBUR: Exhibitor. b. Perth Amboy, N.J., Oct. 30, 1911. e. Yale U. 1933, N.J. Law, 1935. Pres. Snaper Theatres, New York, nat'l pres. Allied States Assn.; partner 1957. Member: Yale Club, Variety Club, N.Y. Bridge Club; pres. Triangle Theatres; pres., AIT Theatres, N.J.; pres. Triangle Theatres.

SNELL, PETER R.E.: Producer. b. 1941. Prod.: 1967: Winters Tale, Some May Live, A Month in the Country, Carnaby 68. 1968: Subterfuge. 1969–70 Juliet Ceasar, Goodbye Gemini. 1971: Anthony and Cleopatra. 1972: The Wicker Man. Appt. head of prod. and man. dir. British Lion 1973. Joined Robert Stigwood group 1975. Hennessy. 1976: Returned to indep. prod. 1978: Bear Island. 1980: Goodbye California.

SNODGRESS, CARRIE: Actress. b. Chicago, IXLL., Oct 27; e. Northern Il The nois U. Acted with Goodman Theatre where plays include All Way Home, Oh What a Lovely War, Caesar and Cleopatra and Tartuffe (Sarah Siddons Award, 1966).
PICTURES INCLUDE: Rabbitt, Run, Diary of a Mad Housewife, The Fury.
TV: World Premier (Silent Night, Lonely Night, The Whole World is Watching), The Outsider, The Virginian, Judd for the Defense, Medical Center, Marcus Welby, M.D.

SNOWDEN, ALEC CRAWFORD: Independent producer b. Leeds. Ent. m.p. ind as asst. dir. with Stoll Film Co. Later with Fuller Hayward Theatre Corp., New Zealand; subsequently with British Films Ltd., G-B Screen Services, Spectator Films; Merton Park Studios Ltd. Instigated and prod. Edgar Lustgarten Scotland Yard series.
PICTURES INCLUDE: Brain Machine, Case of the Red Monkey, Confession (Deadliest Sin), Timeslip, Dial 999 (Way Out), Intimate Stranger (The Guilty Secret), The Hypnotist, Counterfeit Plan, Man in the Shadow, The Key Man, Escapement, The Strange Awakening, Broth of a Boy, This Other Eden, Over the Odds.

SNOWDON, LEIGH: Actress. b. Memphis, Tenn., June 28, 1932. e. Lambeth College, Jackson, Tenn. Model, TV appearances.
PICTURES INCLUDE: Kiss Me Deadly, All That Heaven Allows, Francis in the Navy, Rawhide Years, Square Jungle, Creature Walks Among Us, Outside the Law.

SNYDER, TOM: Newscaster, show host. b. Milwaukee, May 12, 1936. e. Marquette Univ. First job in news dept. of WRIT, Milwaukee. Subsequently with WSAV-TV, Savannah; WAII-TV, Atlanta; KTLA-TV, Los Angeles; and KYW-TV, Philadelphia, before moving to KNBC in Los Angeles in 1970 as anchorman for weeknight newscast. Named host of NBC-TV's Tomorrow program in Oct., 1973, and moved to N.Y.C. in Dec., 1974, at which time began as anchorman of one-hour segment of NewsCenter 4. In Aug., 1975, inaugurated the Monday-thru-Friday editions of NBC News Update, one-minute prime time news spot. Now host for Tommorow talk show.

SNYDER, WILLIAM L.: Producer, executive. b. Baltimore, Md., Feb. 14, 1920; e. B.A. The John Hopkins U., 1940. Lt. Comnder., U.S. Navy WWII, 1941–45. Established Rembrandt Films, N.Y.C., 1948 as importer and dist. of foreign films. Co-prod. White Mane (UA, dom.), winner Cannes Fest. Grand Prize, seven other int'l awards, 1945. Prod. 13 Tom & Jerry Cartoons (MGM) and more than 100 Miss Nightingale, She, Little Lord Fauntleroy, The Secret Army (2 series). Popeye and Krazy Kat cartoons for U.S. TV plus number cartoons (Para.), 1956. Won Academy Award for cartoon Munro (Para.), 1961. Rec'd Academy nominations for Self Defense for Cowards (1962), The Game (1963), How to Avoid Friendship (1964) and Nudnik (1965). Prod. cartoon feature Alice in Paris (Childhood Prodns.), 1966. Prod. I a Woman II, 1968 (Chevron Picts.), and The Daughter (Chevron Picts.), 1970.

SOAMES, RICHARD: Executive. b. London, England, 1936. Joined Film Finances Ltd. 1972; Appt. director Film Finances Ltd., 1977; Appt. man. dir. 1979. Appt. pres. Film Finances Canada Ltd.

SOBLE, RON: Actor. b. Chicago, March 28, 1932. e. Univ. of Michigan. Served U.S. Army, 11th Airborne, in Japan. Studied acting in New York and was member of Jose Quintero's Circle in the Square Players. Acted in such plays as Romeo and Juliet, Murder in the Cathedral, The Petrified Forest; prod. assoc. on TV series Suspense and Danger, and appeared in 56 series. Co-star in The Monroes TV show.
PICTURES INCLUDE: Navajo Run, Al Capone, The Cincinnati Kid, Joe Kidd.

SOBOL, LOUIS: New York columnist, N.Y. World-Journal-Tribune and other papers. b. New Haven, Conn., Aug. 10, 1896. Universal short series, Down Memory Lane, 1933. Contributor of articles and short stories to nat'l mags. Broadcast Lucky Strike Cigarettes for 10 weeks; Ludwig Baumann 39 weeks, Borden's 26 weeks; ABC network 45 weeks; first novel published, Six Lost Women, also Some Days Were Happy & Along the B'way Beat, 1951. Appeared several times at Loew's State Theatre, New York, with own unit; also one week at Palace Theatre. Broadway columnist, N.Y. Evening Graphic, 1929–31; joined Journal-American and Hearst newspapers 1931. Author s.p., Bill's Gay Nineties, collab. s.p., Need for Each Other; Broadway Beat & The Longest Street; ABC network.

SOCHIN, IRVING: Executive. b. Boston, Mass., Mar. 15, 1910; e. English H.S., 1927. In vaude., night clubs with own band & act; salesman, U.A., May, 1938; city salesman, U.A., Cincinnati, 1940; 20th Fox, Cincinnati, 1941; 20th Fox sales mgr., Cincinnati, 1942–46; acting br. mgr., Indianapolis, 1946. Headed booking, buying co., Theatre Owners Corp., to Aug. 1947; then Cincinnati br. mgr., Universal to Jan. 1, 1950; sales mgr., U.I. spcl. films div. asst. to gen. sales mgr. 1952–53. Short subjects sales mgr., gen. sales man. for Rank Film Dist. of America, Inc., 1957–58. Member: Masonic Lodge; Variety Club (past chief barker), Cincinnati Tent 3, 1947–48; spec. rep., Continental Pictures, 1960; v.p., gen. sls. mgr.; Times Film Corp., 1961–64; v.p., gen. sls. mgr., Rizzoli Film Distribs. Inc. 1964; West div. mgr. 20th-Fox, 1969; Fox western div. mgr., 1969–75; Fox special films head, 1975; West div. mgr. Ganma III, 1976–77; 1978, vice pres. and gen'l sales mgr., Vadib International Films, San Francisco based distribution co. Now retired.

SOFAER, ABRAHAM: Actor. b. Rangoon, Burma, Oct. 1, 1896. Stage debut 1921 in Chas. Doran Shakespeare Co.'s first London appearance, Glorianna, 1925; starred in several London productions & subsequently Old Vic & Stratford-on-Avon; broadcaster & TV star; played in many British films; Broadway stage and films in Hollywood.
PICTURES INCLUDE: His Majesty O'Keefe, Elephant Walk, The Naked Jungle, Out of the Clouds, Bhowani Junction, Sinbad, King of Kings, Head.

SOLDATI, MARIO: Director. b. Turin, Italy, 1906. e. Superior Institute of the History of Art, Rome. Lecturer, Columbia U.; author of America First Love, other books; m.p. dir. since 1939.
PICTURES INCLUDE: O.K. Nero, Wayward Wife, Stranger's Hand, Don Juan's Night of Love.

SOLO, ROBERT H.: Executive. Early career as agent with Ashley-Famous; later production as exec. asst. to Jack Warner and Walter MacEwen at Warner Bros. In London prod. Scrooge for Cinema Center Films, 1970; co-prod. The Devils, 1971; 1971, named WB v.p., foreign production 1974, named exec. v.p., prod. at Burbank Studio. Now indep. prod. Invasion of the Body Snatchers, 1978, for United Artists; The Awakening (EMI-Orion), 1980; I, The Jury (American Cinema), 1981.

SOLOMON, FLORENCE: Executive. b. New York City. e. Hunter College. Research mgr., Ligon Johnson's copyright title search firm, 1937–65; dir. for research, Entertainment Copyright Research Co., Inc., 1965–1968; ASCAP (Music Index-Library) 1968-present. Pres. Cinema History Correspondent Society assoc. ed. TV Movies, research associate, The Great Movie Series, The Slapstick Queens, The Paramount Pretties, The MGM Stock Co.; The RKO Gals, The Glamour Girls, The Great Spy Pictures, Actors' Television Credits, 1950–1972; Good Dames, Hollywood's Great Love Teams, Liza! Elvis!, The Debonairs, The Swashbucklers, Hollywood Players: The Forties, The Great Gangster Pictures, Hollywood Players: The Thirties, The Great Western Pictures, the All-Americans, Leading Ladies, The Great Science Fiction Pictures, Hollywood Character Actors, Film Actors Guide: Western Europe, Hollywood on Hollywood, Hollywood Beauties, The Funsters, The Forties Gals, Actors TV Credits: Supplement I, The Hollywood Regulars, Best of MGM: The Golden Years.

SOLOMON, MICHAEL JAY: TV Executive. b. Jan. 20, 1938. e. Emerson in Boston; N.Y.U. Evening School of Commercial while working at first job with United Artists in 1956. Became student booker. Hired by Seymour Florin as booker for one year; returned to UA in intl. dept. January, 1960 went to Panama as asst. to mgr. of Central America. Transferred to Bogota after one yr.; made mgr. of UA in Peru and Bolivia. In 1964 joined MCA Latin American div., reorganizing office in Mexico and later opening office in Brazil. Made v.p., 1968, of MCA-TV. Supervised Latin American and Caribbean div.; 1973 made head of feature film sales to tv stations internationally while still supervising MCA business in Latin America. Resigned 1977 to form Michael Jay Solomon Films International, Inc. & Solomon International TV Newsletter. In 1978 formed Telepictures Corporation to distribute films worldwide and enter production.

SOLOMON, THEODORE GEORGE: Executive. b. Jan. 5, 1920, e. LSU, 1941. Advisor/Consultant to Gulf States Theatres, New Orleans, LA, operating theatres in Alabama, Louisiana, Mississippi, and Texarkana; Past chm. of the National Association of Theatre Owners, past president of the National Association of Theatre Owners.

SOLOW, SIDNEY PAUL: Executive. b. Jersey City, N.J., Sept., 15, 1910. e. New York Univ., 1926–30; chief chemist, Consolidated Film Industries, 1932–36; plant supt., 1937–42; gen. mgr., 1942; Lecturer, University of Southern California (USC), Department of Cinema, 1947–66; Pacific Coast Section, SMPTE, 1948–49; assoc. member, American Society of Cinematographers, 1948-present; chmn. docum. comm., Acad. of M.P. Arts & Sciences, 1951–56; bd. of dir., Acad. of TV Arts & Sciences, 1953; pres., Consolidated Film Industries, 1954 to 1977 when named chrn. exec. comm.; v.p. Republic Pictures Corp., 1954; chmn. scientific & technical awards comm., Acad. of M.P. Arts & Sciences; treas., SMPTE, 1959–60; Permanent Charities Comm., 1958; campaign chmn, Permanent Charities, 1961; mot. pic. section chmn, United Jewish Welfare Fund, 1966; professor, USC, 1966 to present; Honarary Life Member, American Cinema Editors, 1965; Honorary Member, Delta Kappa Alpha, Nat'l Cinema Fraternity, 1972; Winner of Progress Medal, SMPTE, 1974. Elected fellow British Kinematograph & TV Society, 1978.

SOLT, ANDREW: Writer. b. Budapest, Hungary, June 7, 1916; e. St. Stephen's Coll.
PICTURES INCLUDE: They All Kissed the Bride, Without Reservations, Joan of Arc, Little Women, Jolson Story, In a Lonely Place, Thunder on the Hill, Lovely to Look At; orig. s.p. For the First Time.
TV: Hitchcock Presents, Ford Theatre, Douglas Fairbanks Presents, Wire Service, General Electric Theatre, BBC and Stuttgart TV.
STAGE: 1973, Geld In Der Tasche, produced by Theatre am Kurfurstendamm, West Berlin.

SOLTZ, CHARLENE E.: Executive. b. New London, Conn. e. Mitchell College and University of Bridgeport. Entered motion picture industry in 1970 as assistant to director of press information for Motion Picture Association of America, Inc. In 1970 named director of press information.

SOMERS, SUZANNE: Actress. b. San Bruno, CA, Oct. 16, 1946. e. Lone Mountain, San Francisco College for Women. Pursued modeling career; worked as regular on Mantrap, syndicated talk show. Did summer stock and theatrical films. Wrote books: Touch Me Again, Some People Live More Than Others. Biggest TV success in Three's Company series.
PICTURES: Bullitt, Daddy's Gone A-Hunting, Fools, Magnum Force, American Graffiti, Yesterday's Hero, Nothing Personal.
TELEVISION: One Day at a Time, Lotsa Luck, The Rockford Files, Starsky & Hutch, The Rich Little Show, Battle of the Network Stars, Us Against the World, Love Boat, etc. TV movies: Happily Ever After, The Princess and the Lumberjack, Zuma Beach, Ants.

SOMMER, ELKE: Actress. b. Germany, 1941. To Britain 1956 English. Ent. films in Germany, 1958, and since made films in Germany and Italy incl. Friend of the Jaguar, Traveling Luxury, Heaven and Cupid, Ship of the Dead. 1960: made debut in British films.
PICTURES INCLUDE: Don't Bother to Knock, The Victors, The Money Trap, Love the Italian Way, A Shot in the Dark, The Venetian Affair, Deadlier than the Male, Frontier Hellcat. Under contract to ABPC; The Corrupt Ones, The Wicked Dreams of Paula Schultz, They Came to Rob Las Vegas, The Wrecking Crew, Baron Blood, Percy, Zeppelin, Ten Little Indians, Lisa and the Devil, The Prisoner of Zenda.

SONDERGAARD, GALE: Actress. b. Litchfield, Minn.; m. Herbert Biberman dir.; e. Univ. of Minnesota, A.B. On stage in Strange Interlude, Red Rust, and others. In 1936: Anthony Adverse, W.B. In 1937: Maid of Salem, Par.; Seventh Heaven, 20th-Fox. Ranked first in Academy 1936 awards as outstanding actress in supporting role for Anthony Adverse, The Life of Emile Zola, W.B.
PICTURES INCLUDE: Isle of Forgotten Sins, The Strange Death of Adolph Hitler, The Climax, The Invisible Man's Revenge, Gypsy Wildcat, Christmas Holiday, Enter Arsene Lupin, Anna and the King of Siam, Road to Rio, Pirates of Monterey, East Side, West Side, Return of a Man Called Horse.

SONENFELD, OTTO: Producer, distributor. Former br. mgr., RKO, Czechoslovakia, ind. prod. German & Czech films, latter including Erotikon (silent), Ecstasy, first with Hedy Lamarr, Revolution of Blood and Mud, first Czech feature-length doc.; first Israeli prod. Adama (Tomorrow is a Wonderful Day); oper. as Forum Films, hdqts., Tel-Aviv. Also rep. Korda-London Film, London & Lux Films, Rome and Israel; member Int'l Fed. of Film Prod., Paris.

SORIANO, DALE: Publicist, b. Brooklyn, N.Y. Ent. m.p. Ind. as a reelboy in projection room, later usher, sign painter, manager, film booker stagehand (NBC-CBS-ABC). Orchestra leader, director of info. for various N.Y.C. depts., licenses, real estate, law, public events and firearms control board. Publicist for Americana festivals, various veteran organizations. V.F.W., A.L., D.A.V., C.W.V. and Army & Navy Union and many philanthropic organizations. World War II served with 1st Marine Div. U.S.M.C. in S.W. Pacific thea. of oper. 1960–68 v.p. Publicists Association, I.A.T.S.E.

conducted many TV and vet columns for the Brooklyn Eagle and independent newspapers. Show*Biz Column 1st for the Brooklyn Daily Bulletin. Also stage, records, movies, radio and TV Editor. Pres. of Lighthouse Productions and pres. of Flatlands Chamber of Commerce; pres. National Council of Civic Assn. Member of Motion Picture Pioneers, National Publicists Association, and American Newspaper Guild.

SORRELL, HELENA: Dramatic coach. b. Edinburgh, Scotland, Dec. 30. e. Royal Academy of Dramatic Arts, London, Eng., and Edinburgh, Scotland. Toured England, Ireland, Scotland and Wales as dramatic actress and director. With Jessie Bonstelle, Detroit, six seasons as actress and director in Bonstelle Prod.; dir of Bonstelle Dramatic School, Detroit. Studied singing, piano, music and dancing abroad and in America. Played and toured all over America and Canada in such stage plays as The Fool, Autumn Crocus, Dinner at Eight, Death Takes a Holiday, Peter Ibbesten, Design for Living, Cynara, Philip Goes Forth. Picture career; dramatic coach, RKO Radio Studios, Selznick, 20th-Fox.

SORVINO, PAUL: Actor. b. New York City. Acted on Bdwy.; broke into films with Where's Poppa in 1970.
PICTURES: Day of the Dolphin, Made for Each Other, The Gambler, A Touch of Class, Oh, God, Bloodbrothers, Slow Dancing in the Big City, The Brink's Job, Lost and Found, Cruising.
TELEVISION: Seventh Avenue, Tell Me Where It Hurts, etc.

SOTHERN, ANN: Actress. b. Valley City, N.D., Jan. 23, 1923; e. Washington U.; p. Annette Yde-Lake, opera singer. On stage, in m.p. since 1934.
PICTURES INCLUDE: Let's Fall in Love, Melody in Spring, Three Hearts for Julia, Swing Shift Maisie, Cry Havoc, Thousands Cheer, Maisie series, April Showers, Letter to Three Wives, Judge Steps Out, Words and Music, Nancy Goes to Rio, Blue Gardenia, Lady in a Cage, Sylvia, Chubasco, The Killing Kind, Golden Needles, Crazy Mama, The Manitou.
TV: Private Secretary, Ann Sothern Show, Captain and the Kings.

SOUL, DAVID: Actor. b. Chicago, Aug. 28. Attended several colleges but gave up studies to pursue a career in music. Made 25 singing appearances on The Merv Griffin Show where was spotted and given screen test. Signed contract with Screen Gems and given starring role in ABC-TV series, Here Come the Brides.
PICTURES INCLUDE: Johnny Got His Gun, Magnum Force, Dog Pound Shuffle.
TV: Guest Star: The Streets of San Francisco, Cannon, Medical Center, The Rookies, Ironside, Star Trek, McMillan and Wife, Dan August, Circle of Fear, Owen Marshall, Counselor at Law. Movies: The Disappearance of Flight 412, Intertect, Movin On, Little Ladies of the Night. Series: Starsky and Hutch.

SPACEK, SISSY: Actress. b. Quitman, Texas. Christened Mary Elizabeth. Cousin of actor Rip Torn. Was photographic model; attended acting classes in New York under Lee Strasberg.
PICTURES INCLUDE: Prime Cut (debut), Ginger in the Morning, Badlands, Carrie, Welcome to L.A., 3 Women, Heart Beat, Coal Miners' Daughter, Raggedy Man.
TV: The Girls of Huntington House, The Migrants, Katherine (all TV movies), two episodes of The Waltons.

SPALLA, JOSEPH SALVATORE "RICK": Executive, producer. b. Sandusky, Ohio, May 22, 1923. e. Ohio State U. 1943 and 1946; U. of So. Calif., B.A., 1953. m. Shirley Egland. Studied voice while teenager, soloist with Sandusky High a cappella choir, Ohio State U. Chorus and Men's Glee Club & OSU Symphonic Choir. To Hollywood, 1946, as City Messenger. Member Screen Extras Guild, Screen Actors Guild. Co-owner of cab co. 1953–54, cameraman with Hollywood Prod. Co., travel series: The Open Road and document series, Before Your Eyes. 1954–56, one of the founders, prod. head and v.p., Bill Burrud Prod., Inc. Started half-hour weekly series, Wanderlust (later syndicated as Vagabond). Prod. supervisor and co-creator of Assignment America, 1956–57, v.p. and gen. mgr. of Jet TV and Film Prods., Inc. Exec. prod. of TV: The Open Road. Creator and exec. prod. of High Road to Danger, 1957–62, formed RSVP, Rick Spalla Video Productions. Prod. Those Faraway Places, Enchanted Lands. Created series, Guest Shot, 1962–67, formed Hollywood Newsreel Syndicate, Inc., sub. of RSVP; created, prod. Hollywood Backstage TV series, prod., dir. Hollywood's Fantastic Artists and Models Ball (TV special); producer-director of TV series, Portrait of a Star, (12 hour-long specials) also TV series Holiday on Wheels. Owns millions of feet of film on stars, newsreel, travelogue footage etc. Producer of TV commercials and industrial films.

SPARKS, ROBERT: Producer. b. West Union, Ohio; e. U. of Cincinnati Law School; assoc. prod. Para., 1932; story ed., RKO, 1935; prod Columbia, 1938; U.S.M.C., 1942; Univ., 1946; exec. prod. RKO, 1947; exec. prod. CBS TV network, 1955–59; dir syndication, Screen Gems, Hollywood, 1959–62; exec. prod., CBS 1962; exec. prod., MGM TV, 1963.

SPECKTOR, FREDERICK: Executive. b. April 24, 1933, Los Angeles. e. Univ. of So. Calif., UCLA. M.P. agent, Ashley Famous Agency, 1962–64; Artists Agency Corp., 1964–68; exec. M.P. dept., William Morris Agency, 1968-78; exec. Creative Artists Agency, 1978–present.

SPELLING, AARON: Executive. Was actor/writer before becoming producer at Four Star in 1957. In 1967, formed Thomas/Spelling Productions to produce several series and movies, including Mod Squad. In 1969, formed his own company, Aaron Spelling Productions, and then in 1972, partnered with Leonard Goldberg to produce The Rookies, Charlie's Angels, Fantasy Island, Starsky and Hutch, Hart to Hart, and under his own company banner, Love Boat, Vegas, Dynasty, and a total of 92 movies for television.

SPENCER, LESLIE B.: Exhibitor. b. Boston, Nov. 17, 1938. Father was late exhibitor Leslie Beudsler. Pres., Community Playhouse, Inc., Wellesley Hills, MA.

SPENGLER, PIERRE: Producer. b. Paris, 1947. Went on stage at 15; returned to language studies at Alliance Francaise. Entered film industry as production runner and office boy. Teamed for first time with friend Ilya Salkind on The Light at the Edge of the World, produced by Alexander Salkind.
PICTURES INCLUDE: Bluebeard, The Three Musketeers, The Four Musketeers, Crossed Swords, Superman, Superman II.

SPENSER, JEREMY: Actor. b. Ceylon, 1937; e. Downshill School, Farnham, England. Ent. films 1947 in Anna Karenina.
PICTURES INCLUDE: It's Great To Be Young, The Prince and the Showgirl, Wonderful Things, Ferry to Hong Kong, Roman Spring of Mrs. Stone, Vengeance, King and Country, He Who Rides a Tiger.

SPERLING, MILTON: Producer. b. New York City. e. C.C.N.Y., Shipping clerk Paramount Newsreel; messenger boy, Paramount, Astoria studio; script clerk, United Artists studio, Hollywood, secy. to Darryl Zanuck, Warner Bros.; secy. to Hal B. Wallis, assoc. prod., Edward Small Prod.; asst. to Winfield Sheehan, Fox Films; screen writer 20th-Fox, Four Sons, Sing, Baby, Sing, Thin Ice, Cisco Kid; apptd. prod.; served in U.S. Marines as Capt. 3½ yrs; prod. Sun Valley Serenade, Hello, Frisco, Hello, Crash Dive, Rings on Her Fingers; To the Shores of Tripoli; Iwo Jima; documentary m.p.; org. & pres. of U.S. Pictures, 1945.
PICTURES INCLUDE: Cloak and Dagger, Pursued, Three Secrets, The Enforcer, Distant Drums, Retreat Hell, Blowing Wild, prod. collab. story & s.p. Court Martial of Billy Mitchell; supv. prod. Top Secret Affair, Marjorie Morningstar; prod. collab. s.p. The Bramble Bush; prod. collab. s.p., Merrill's Marauders; collab. prod. s.p. Battle of the Bulge; prod., collab. Captain Apache.
TV: Black Market Baby (MOW, ABC), Brave New World (exec. prod., NBC. mini-series); The Dream Merchants (exec. prod., OPT series).

SPEWACK, BELLA: Writer. b. Hungary, 1899. e. Washington Irving H.S., 1917. m. Samuel Spewack, 1922; writer. Reporter N.Y. Call, N.Y. Mail, N.Y. World, N.Y. Evening World; feature writer, N.Y. Herald Tribune, N.Y. Times; nat'l pub. dir. Camp Fire Girls, then Girl Scouts; asst. to husband for corr. for N.Y. World, Russia, 1922–26; author plays with husband include: Solitaire Man, Poppa, War Song, Clear All Wires, Spring Song, Boy Meets Girl, Leave It to Me, Kiss Me Kate, My Three Angels, Festival. Member: Dramatists Guild; SWG; pres., N.Y. Girls Scholarship Fund.
PICTURES INCLUDE: Clear All Wires, Boy Meets Girl, Cat and the Fiddle, Rendezvous, The Nuisance, Three Loves of Nancy, My Favorite Wife, When Ladies Meet, Weekend at the Waldorf, Move Over Darling, We're No Angels (based on play My Three Angels).
TV: Mr. Broadway, Kiss Me Kate, My Three Angels, The Enchanted Nutcracker, Kiss Me Kate, BBC.

SPIEGEL, LARRY: Producer-Writer-Director. b. Brooklyn, New York. e. Ohio Univ. With CBS-TV 1960–66; Benton & Bowles, 1966–67; Wells, Rich, Green, 1967–71; BBD&O, 1972–73. Now co-heads Spiegel-Bergman Productions.
PICTURES INCLUDE: Hail (s.p.); Book of Numbers (s.p.); Death Game (prod.); Stunts (prod.); Spree (direc./s.p.); Phobia (prod.); Pure Escape (prod.).
TV: Alexander (s.p., prod.); Incredible Indelible Magical Physical Mystery Trip (s.p.); Bear That Slept Through Christmas (s.p.); Never Fool With A Gypsy Ikon (s.p.); Mystery Trip Through Little Red's Head (s.p.); Planet of Apes (animated) (s.p.).

SPIEGEL, SAM: Producer. b. Jaroslau, Poland, Nov. 11, 1903. e. University of Vienna; with Universal 1930–33, producing German and French films in Berlin for European dis-

tribution. Independent producer 1933–37 in Europe. In 1941, co-producer with Boris Morns of Tales of Manhattan.
PICTURES INCLUDE: The Stranger, 1945; We Were Strangers with John Huston (Horizon Prod.), The Prowler, When I Grow Up, African Queen, Melba, On the Waterfront (Acad. Award, best picture, 1954); The Strange One, Bridge on the River Kwai (Acad. Award, best picture, 1957); Suddenly Last Summer, 1959; Lawrence of Arabia, Dangerous Silence, The Chase, The Night of the Generals, The Happening, The Swimmer, Nicholas and Alexandra, The Last Tycoon.

SPIEGEL, TED: Publicist. b. New York City. e. N.Y.U., B.S. 1948. Joined Columbia Pictures 1948 in adv. copy, exploitation, pub. depts.; foreign public rel., Columbia International 1956. Spec. adv., pub. rep., Kingsley Int'l Pictures, 1960; spec. asst. to pres. Kingsley Int'l Pictures, 1962; mgr. dir., theatre operations, The Landau Co., 1963; pub. exec., Embassy Pics. Corp., 1964; dir. pub., Avco Embassy Pictures, 1968. In 1976 joined A. Stirling Gold as dir. adv./pub. relations. In 1978 named acct. exec. for Solters & Roskin.

SPIELBERG, STEVEN: b. Cincinnati, 1947. e. Calif. State Coll. Made home movies as child; completed first film with story and actors at 12 yrs. old in Phoenix. At 13 won film contest for 40-min. war movie, Escape to Nowhere. At 16 made 140-min. film, Firelight. At CSC made five films. First professional work, Amblin', 20 min. short which led to signing contract with Universal Pictures.
TV: Duel, Something Evil.
PICTURES INCLUDE: The Sugarland Express, Jaws, Close Encounters of the Third Kind, I Wanna Hold Your Hand (exec. prod.), 1941, Used Cars (exec. prod.), Raiders of the Lost Ark; Poltergeist (prod.-s.p.), E.T. and Me (dir.).

SPIER, WILLIAM: Producer, director, novelist, radio-TV writer; m. June Havoc. Network exec., producer; prod. radio shows including March of Time, Columbia Work Shop, Adventures of Sam Spade, Philip Morris Playhouse, Suspense; co-dir. film, Lady Possessed.
TV: The Clock, 1951; Fred Waring, Omnibus, 1952–53; Medallion Theatre, Willy, 1954–55; Perry Mason, 1956; scripts for Omnibus, The Clock, 20th Century-Fox Hour, The Lineup, Reader's Digest, Passing Parade, Alfred Hitchcock, Steve Canyon, Peter Gunn, Desilu, Playhouse 90, The Untouchables, 77 Sunset Strip, Bourbon Street, Dragnet, Richard Diamond, CIC, The Islanders, Thin Man, etc. Winner of Peabody Award, Look Magazine, Mystery Writers of America, Writers' Guild Award, best hour script of 1961.

SPIGELGASS, LEONARD: Writer. Early pictures include: collb. s.p. Letter of Introduction, Service DeLuxe, Unexpected Father, Boys from Syracuse; collab. story, s.p. All Through the Night, Tight Shoes; s.p. Butch Minds the Baby, Big Street, Youngest Profession; U.S. armed forces, WW II.
PICTURES INCLUDE: Perfect Marriage, So Evil My Love, I Was a Male War Bride, Mystery Street, Night Into Morning, Law and the Lady, Because You're Mine, Scandal at Scourie, Athena, Deep in My Heart, Ten Thousand Bedrooms, Silk Stockings, A Majority of One, Gypsy.
TV: Climax, Playhouse 90.
STAGE: A Majority of One, Dear Me, The Sky is Falling, The Wrong Way Light Bulb, Look to the Lillies.

SPIKINGS, BARRY: Executive. b. 1939. Ent. m.p. ind. 1963. Jnt. Mang. Dir. British Lion Films Ltd., 1975. Co-prod. Conduct Unbecoming, Man Who Fell To Earth, 1976. Appt. jnt. man. dir. EMI Films Ltd., 1977. Prod. The Deer Hunter, Convoy. 1979, appt. chm. & chief exec., EMT Film & Theatre Corp.; chm. & chief exec, EMI Films, Ltd., chm. EMI Cinemas, Ltd.; chm., Elstree Studios, Ltd. Named dir. of Associated Film Dist. (USA). Chm. EMI-TV Programs, Inc., 1980, appt. chm. chief exec., EMI Films Group.

SPIRES, JOHN B.: Executive. b. New York, N.Y. e. N.Y.U. Assistant manager Tribune Theatre, N.Y., 1934–36; Paramount Pictures lab., 1936–42; Capt. U.S. Army, 1942–46; Major U.S. Reserves, 1946–48; foreign rep., RKO, Europe, 1946; asst. foreign mgr., United World Films, 1947–48; in charge 16mm foreign oper., Universal 1949; asst. to European gen. mgr., Universal International, 1950; European gen. mgr., Continental Europe, Near East, Universal International, 1955; gen. mgr., foreign film sales, MCA-TV, 1958; dir., TV sis., UK, Europe, MGM-TV, 1961; dir., Int'l TV sis., MGM-TV, 1964; named v.p. OF MGM-TV, 1973; snr. v.p., MGM-TV, 1978. Founded Phoenix Intl. TV Associates, 1980; pres., Phoenix Intl.

SPITZ, HENRY: Producer. b. Paterson, N.J., Mar. 4, 1905. e. N.Y.M.A. Gen. mgr., Quackenbush Dept. Stores; on bd. Preakness Hill Country Club & Barnet Memorial Hosp.; v.p. Marshall-Matherson dept. store, 1930–34; ent. m.p. ind. as prod. & bus. mgr., Cameo, 1935; assoc. prod., Cameo-Imperial Pictures, 1936; asst. prod. mgr. & asst. dir. Conn Prod.; asst., prod. mgr. Universal.
PICTURES INCLUDE: I Demand Payment, High Hat, Here's Flash Casey, Headleys at Home, Hawaiian Nights, Private Affairs; assoc. prod. Two Dollar Bettor, Sabre Jet, Commanche.

SPITZER, MARIAN: Writer. b. New York City. e. NYU. m. Harlan Thompson, writer. Was reporter on Brooklyn Times, then N.Y. Globe, in dramatic dept. Did publicity for B.F. Keith Circuit, Shubert & Edgar Selwyn. First short story pub. in 1924 by Smart Set. Novels include Who Would Be Freed? and A Hungry Young Lady. Was in editorial dept. Fox, then contract writer. Became asst. to chmn. of ed. board. Paramount.
PICTURES INCLUDE: Dolly Sisters, Look for the Silver Lining.

SPIVAK, LAWRENCE E.: TV-radio producer. b. N.Y. e. Harvard U. LL.D. (hon.) Wilberforce U.; Litt.D. (hon.) Suffolk U. L.H.D. (hon.) Tampa U. Began as bus. mgr., Antiques Mag., 1921–30; asst. to the pub., Hunting and Fishing, Nat. Sportsman mags., 1930–33; bus. mgr., American Mercury, 1934–39; pub., 1939–44; editor, pub., 1944–50; founder, pub. until 1954, Ellery Queen's Mystery Mag., The Mag. of Fantasy and Science Fiction, Mercury Mystery Books, Bookseller Mysteries, Jonathan Press Books; producer, panel member, TV program, Meet the Press. Recipient two Peabody Awards. Winner of Emmy Award for outstanding achievement in Coverage of Special Events and Honor Award from University of Missouri for Distinguished Service in Journalism.

SPODICK, ROBERT C.: Exhibitor. b. New York City, Dec. 3, 1919; e. CCNY, 1940; ent. m.p. ind. as errand boy Skouras Park Plaza, Bronx 1932–33; reel boy, asst. mgr., Loew's Theatres; mgr., Little Carnegie and other art theatres; exploitation man, United Artists. Acquired Lincoln, New Haven art house in 1945 in partnership with cousin Leonard E. Sampson; developed Nutmeg Theatre circuit, which was sold in 1968 to Robert Smerling. Now operates in Groton, Ct., Cinemas I and II; Norwich Cinemas I and II and Village Cinemas I & II in Mystic in Partnership with Sampson and William Rosen; also Whalley Lincoln, Crown, York Sq., Theatres in New Haven. Pres., Allied of Conn., 1962–64; Pres. NATO of Conn. 1968–73. Past chm. exec. comm., Conn. Ass'n of Theatre Owners.

SPRINGSTEEN, R. G.: Director. b. Tacoma, Wash., Sept. 8, 1904; e. U. Washington. With Universal Studios 1930; asst. dir. Fox; asst. dir. & dir. Republic.
PICTURES INCLUDE: Out of the Storm, Hellfire, Red Menace, Singing Guns, Honeychile, Toughest Man in Arizona, Gobs & Gals, Perilous Journey, Geraldine, I Cover the Underworld, Cross Channel, When Gangland Strikes, Come Next Spring, Track the Man Down, Secret Venture, Double Jeopardy, Johnny Reno, Red Tomahawk.

STACK, ROBERT: Actor. b. Los Angeles, Calif., Jan. 13, 1919; e. U. So. Calif.; dramatic school. In U.S. Armed Forces, WW II.
PICTURES INCLUDE: First Love, When the Daltons Rode, Mortal Storm, Little Bit of Heaven, Nice Girls, Badlands of Dakota, To Be or Not To Be, Eagle Squadron, Men of Texas, Fighter Squadron, Date With Judy, Miss Tatlock's Millions, Mr. Music, Bullfighter and the Lady, My Outlaw Brother, Bwana Devil, War Paint, Conquest of Cochise, Sabre Jet, Iron Glove, High & the Mighty, House of Bamboo, Good Morning Miss Dove, Great Day in the Morning, Written on the Wind, John Paul Jones, Last Voyage, Killers of Killimanjaro, The Caretakers, The Corrupt Ones, Story of a Woman, 1941, Airplane!.
TV: The Untouchables, Name of the Game.

STAFFORD, BRENDAN J. (B.S.C.): Director lighting cameraman. Has served as photographer various documentary companies, including Student Nurse, Sports Day, Journey for Jeremy, Riders of the Forest, etc. Directed Mediaeval Monastery, A Nation Once Again, Proud Canvas, A Car is Born, etc., 1947–48. Photography, Fortune Lane, Via London, Cape Cargoes, etc. Directed, Our Country: 1948–49, directed Stranger at My Door; directed and photographed approx. 40 TV films; 1949–50. Photographed Paul Temple Triumphs, E.C.A. film on Britain; dir. Marshall Plan for Ireland; dir. and photo. 8 TV films, 1950–51; formed Irish Film Prods. Made Portrait of Dublin; photographed The Promise, 1951–55; formed Meridan Films, Ltd.; dir. The Armchair Detective, Men Against the Sun, etc.; photo. The Wallet, Circumstantial Evidence, The Night Won't Talk, etc.; dir. of photo. to Douglas Fairbanks series and Fabian of the Yard series; photo. Eight O'Clock Walk, Bang You're Dead, Checkmate, Profile, One Jump Ahead, Find the Lady, Hide-Up, Murder Reported, There's Always a Thursday, The Hostage, Curse of Nostradamus, The Shakedown, Woman Shall Weep, Witness in the Dark, Your Money or Your Wife. 1969: Crossplot, The Promise, Dangerous Hands.
TV: Gay Cavalier, Files of the O.S.S., Errol Flynn, White Hunter, William Tell, Invisible Man, Danger Man series, 1960–61. The Pursuers, One Step Beyond, Sir Francis Drake series; 1962, part Third Man series, Man of the World series; Sentimental Agent series; 1963, Danger Man; 1964,

Danger Man (Secret Agent), 1965–66, The Man Who Never Was, The Prisoner, 1966–67; 1968, The Saint, 1969 UFO Series; 1970 2nd UFO series; 1971, The Protectors series; 1972, The Pathfinders series and 2nd The Protectors series; 1975 The Protectors, Space 1999; 1976/7 Commercials and Documentaries.

STAGGS, JACK E.: Executive Director, Motion Picture and Television Fund. b. Los Angeles, August 10, 1920. A registered pharmacist, he is a Fellow of the American College of Apothecaries and Member of American College of Hospital Administrators. Educated Idaho Southern and Columbia Universities. Served in World War II as Chief Pharmacist Mate in the Navy. Former chief administrator of So. Nevada Memorial Hospital of Las Vegas. Joined Motion Picture and Television Fund Sept. 1, 1968.

STAHL, AL: Executive. Syndicated newspaper cartoonist; asst. animator, Max Fleischer, gag ed. Terrytoons; U.S. Signal Corps; opened own studios, 1946; prod. first animated TV cartoon show; pres., Animated Prod., prod. live and animated commercials; member of bd. NTFC.

STALLONE, SYLVESTER: Actor-Writer. b. New York City, July 6, 1946. After high school taught at American College of Switzerland instructing children of career diplomats, young royalty, etc. Returned to U.S. in 1967 and graduated from Univ. of Miami in 1969. Came to New York to seek acting career, taking part-time jobs, including usher for Walter Reade Theatres. Then turned to writing, selling several TV scripts. Back to acting with lead role in The Lords of Flatbush.

PICTURES INCLUDE: Capone, Death Race 2000, Rocky (also s.p.), F.I.S.T. (actor, co. s.p.), Paradise Alley (actor, s.p., dir). Rocky II (actor, s.p., dir.), Nighthawks, Victory, Rocky III.

STAMP, TERENCE: Actor. b. London, England, 1938. After stage experience. Alfie On Broadway. Recent stage: Dracula, The Lady from the Sea.

PICTURES INCLUDE: Billy Budd, Term of Trial, The Collector, Modesty Blaise, Far from the Madding Crowd, Poor Cow, Blue, Tales of Mystery, Theorem, The Mind of Mr. Soames, A Season in Hell, Hu-Man, The Divine Creature, Strip-Tease, Superman, Meetings With Remarkable Men, The Thief of Baghdad, Together, Superman II.

STANDER, LIONEL: Actor. b. New York City. e. coll.; N.Y. stage, 1952.

PICTURES INCLUDE: Scoundrel, Page Miss Glory, Gay Deception, Music Goes 'Round, Mr. Deeds Goes to Town, Meet Nero Wolfe, Guadalcanal Diary, Big Show-Off, Specter of the Rose, In Old Sacramento, Kid From Brooklyn, Gentleman Joe Palooka, Pal Joey, Mad Wednesday, Call Northside 777, Unfaithfully Yours, Trouble Makers, Two Gals and a Guy, St. Benny the Dip, Cul de Sac, A Dandy in Aspic, The Gang That Couldn't Shoot Straight, Pulp, The Black Bird, The Cassandra Crossing, New York, New York, Matilda, 1941.

STANFILL, DENNIS C.: Executive. b. Centerville, Tenn., April 1, 1927; e. Lawrenceburg H.S., Annapolis, Oxford U., M.A. After active duty in the Navy served in foreign affairs section, office of Chief of Naval Operations; resigned commission & joined Lehman Brothers, as corporate finance specialist, 1959–65; v.p., finance, Times Mirror Company, 1965–69; ent. m.p. ind. as exec. v.p., finance, 20th Century-Fox & member of Fox board & exec. committee, 1969; named pres. 20th Century-Fox, March, 1971; named chmn., chief exec. officer, Sept. 1971. Resigned, 1981.

STANG, ARNOLD: Performer, b. Chelsea, Mass., Sept. 28, 1927. Radio, 1935–50; on B'way, in five plays and in m.p. and short subjects; guest appearances on TV shows.

TV: Milton Berle, Danny Thomas, Perry Como, Ed Sullivan, Red Skelton, Frank Sinatra, Wagon Train, Top Cat, Batman, Bonanza, Bob Hope, Danny Kaye, Broadside, Jackie Gleason, Emergency, Feeling Good, Chico & The Man, Super Jaws & Catfish.

PICTURES INCLUDE: Man with the Golden Arm, The Wonderful World of the Brothers Grimm, It's a Mad, Mad, Mad, Mad World, Pinocchio in Outer Space, Dondi, Alakazam the Great, Hello Down There, Skidoo, Walt Disney's Aristocats, Raggedy Ann & Andy, Gang That Couldn't Shoot Straight, We Go Polo, That's Life.

STANLEY, PHYLLIS: Actress. b. London, England, Oct. 30. e. Beauclair Sch., London. On London stage in 1932–35.

STAGE: The Miracle, Words & Music, The Town Talks, Oh! You Letty, Happy Returns, Lights Up.

PICTURES INCLUDE: Side Street Angel (m.p. debut), St. Martin's Lane, Jeannie, One Exciting Night, Next of Kin, The Case of Lady Brooke, Lovely to Look At, Take Me To Town, Her 12 Men, Strange Lady in Town, Black Sleep, Seventh Veil.

TV: Fireside Theatre, Ford Theatre, Four Star Playhouse (Schlitz Playhouse, Lux Video Theatre, Matinee Theatre).

STANWYCK, BARBARA: Actress. r.n. Ruby Stevens; b. Brooklyn, N.Y., July 16, film debut, Locked Door, 1929.

PICTURES INCLUDE: Ladies of Leisure, Night Nurse, So Big, Bitter Tea of General Yen, Brief Moment, Woman in Red, Annie Oakley, Plough and the Stars, Stella Dallas, Mad Miss Manton, Union Pacific, Golden Boy, Meet John Doe, Two Mrs. Carrolls, B.F.'s Daughter, Sorry, Wrong Number, Lady Gambles, East Side, West Side, Thelma Jordan, The Furies, No Man of Her Own, To Please a Lady, Man With a Cloak, Clash by Night, Jeopardy, All I Desire, Titanic, Blowing Wild, The Moonlighter, Executive Suite, Witness to Murder, Violent Men, Double Indemnity, Ball of Fire, Cattle Queen of Montana, Escape to Burma, There's Always Tomorrow, Maverick Queen, These Wilder Years, Crime of Passion, Trooper Hook, Forty Guns, Walk on the Wild Side, Roustabout, The Night Walker.

TV: Guest, Jack Benny, Ford, Zane Grey, Alcoa-Goodyear; 4 yrs., 1965–69, The Big Valley.

STAPLETON, JEAN: Actress. b. New York City, Jan. 19, 1923. e. Wadleigh High School. Summer stock in New Hampshire, Maine, Mass. and Penna. First Broadway appearance in In the Summer House.

STAGE: Damn Yankees, Bells Are Ringing, Something Wild, Up the Down Staircase, Cold Turkey, Klute.

TV: All in the Family.

STAPLETON, MAUREEN: Actress. b. Troy, N.Y., June 21, 1925. Studied acting with Herbert Berghof in N.Y. in 1944 and became member of Actors Studio. Broadway debut, 1946, in The Playboy of the Western World. Other early roles in: Antony and Cleopatra, Detective Story, The Bird Cage. Became a star in 1951 in The Rose Tattoo and later acted in 27 Wagons Full of Cotton, Orpheus Descending, The Glass Menagerie. m.p. debut 1949 in Lonelyhearts. Recent play: Gingerbread Play (Tony Award), 1970–71.

PICTURES INCLUDE: Fugitive Kind, View from the Bridge, Bye Bye Birdie, Airport, Plaza Suite, Interiors, The Runner Stumbles, On the Right Track, Reds.

TV: Tell Me Where It Hurts (movie), Queen of the Stardust Ballroom, Cat on a Hot Tin Roof.

STARGER, MARTIN: Executive. e. N.Y. City Coll. Serviced in U.S. Army Signal Corp., where prod. training films. Joined BBDO, starting in TV prod. dept.; later made v.p. & assoc. dir. of TV. Joined ABC in April 1966, as v.p. of programs, ABC-TV, East Coast. In March, 1968, prom. to v.p. and natl prog. dir; in 1969 named v.p. in chg. progr. Named pres., ABC Entertainment, July 17, 1972. In June, 1975 formed Marstar Productions Inc., M.P. & TV production company of which he is pres. In March 1978 formed Marble Arch Productions, of which he is pres.

STARK, RAY: Producer. e. Rutgers U. Began career after WWII as agent handling radio scripts, and later literary works for such writers as Costain, Marquand and Hecht. Joined Famous Artists Agency, where he represented such personalities as Marilyn Monroe, Kirk Douglas and Richard Burton; in 1957, resigned exec. position to form Seven Arts Prods. with Eliot Hyman, serving as exec. v.p. and head of production until July, 1966, when he left to take on personal production projects, including Reflections in a Golden Eye and Funny Girl.

PICTURES: The Sunshine Boys, Robin and Marian, Murder by Death, Casey's Shadow, The Cheap Detective, California Suite, The Electric Horseman, Chapter Two, Seems Like Old Times.

STARR, EVE: Columnist. b. Chicago, Ill., May 1. e. Columbia Univ., Miami Univ., Julliard Sch. of Music. Radio-TV playwright, feature writer, actress. Hollywood foreign correspondent; m.p. writer, newspapers, mags.; m.p. columnist, Hollywood Citizen-News; Daily syndicated TV column, Inside TV—General Features Synd.; feature column, Tell It To Eve, International Newspapers; Traveling With Eve, Nat'l Syndicated Travel Column.

STEEL, ANTHONY: Actor. b. London, Eng., May 21, 1920; e. private schools, South Ireland & Cambridge. Film debut in Saraband for Dead Lovers.

PICTURES INCLUDE: Portrait from Life, Christopher Columbus, Helter Skelter, Poet's Pub, The Blue Lamp, The Wooden Horse, Laughter in Paradise, The Mudlark, Where No Vultures Fly (Ivory Hunter), Another Man's Poison, Emergency Call, Something Money Can't Buy, The Planter's Wife (Outpost in Malaya), Malta Story, Master of Ballantrae, Albert, R.N. (Break to Freedom), West of Zanzibar, Sea Shall Not Have Them, Passage Home, Storm Over the Nile, Black Tent, Checkpoint, Valerie, A Question of Adultery, Harry Black, Honeymoon, The Switch.

STEELE, TOMMY: Performer. b. London, 1936. Early career Merchant Navy. First TV and film appearances, 1956. Composed and sang title song for The Shiralee.

PICTURES INCLUDE: Kill Me Tomorrow, The Tommy Steele Story, The Duke Wore Jeans, Tommy the Toreador, Light Up the Sky, It's All Happening, The Happiest Millionaire, Half A Sixpence, Finian's Rainbow, Where's Jack?

TV: Tommy Steele Spectaculars, Richard Whittington Esquire (Rediffusion), Ed Sullivan Show, Gene Kelly Show, Perry Como Show, Twelfth Night, The Tommy Steele Hour,

Tommy Steele in Search of Charlie Chaplin, Tommy Steele And A Show, Quincy's Quest.

STEIGER, ROD: Actor. b. Westhampton, N.Y., Apr. 14, 1925; e. Westside H.S., Newark, N.J. Served in U.S. Navy, then employed in Civil Service; studied acting at N.Y. Theatre Wing Dramatic Workshop Actors' Studio; numerous TV plays; on Broadway in ANTA prod. of Night Music; m.p. debut in On the Waterfront.

PICTURES INCLUDE: Big Knife, Oklahoma! Court Martial of Billy Mitchell, Jubal, Harder They Fall, Back from Eternity, Run of the Arrow, Unholy Wife, Al Capone, Seven Thieves, The Mark, Reprieve, 13 West Street, Hands Upon the City, The Time of Indifference, The Pawnbroker, In the Heat of the Night, The Girl and the General, No Way to Treat a Lady, And There Came a Man, The Illustrated Man, Three Into Two Won't Go, Duck, You Sucker, Happy Birthday, Wanda June, Waterloo, The Lolly-Madonna War, Lucky Luciano, Hennessy, W. C. Fields and Me, F.I.S.T., The Amityville Horror, Lion of the Desert, Cattle Annie and Little Britches.

STEIN, JOSEPH L.: Attorney. b. Portland, Me.; e. NYU, NYU Law Sch., LL.B. Admitted to N.Y. Bar. Specialist in copyright and m.p. law. Member of Sargoy & Stein, New York until Dec. 1970 when he retired, becoming counsel to firm. Compilers and annual revisers of U.S. Copyright Law Digest for Martindale-Hubbell Law Directory. Special counsel nationally for ten major m.p. dist. cos. in copyright & dist. matters. Former member: American Bar Assn. and of its Copyright Subsection's Committees on Anti-Trust Matters, 1951–54, on Copyright Office Affairs. 1953–55, 1959–63, 1966–67, on Copyright Law Revision, 1955–56, 1963–65, on Int. Copyrights, 1956–57, 1966–67, on Cooperation with Other Sections, 1956–57, on authors, 1959–61, and on Related Rights, 1957–59; American Patent Law Assn.: American Judicature Society; National Panel of Arbitrators of American Arbitration Assn.; American Society of International Law; American Foreign Law Soc.; Assn. of the Bar of City of New York; N.Y. County Lawyers' Assn.; N.Y. State Bar Assn.; Fed. Bar Assn. of N.Y., N.J. & Conn.; dir., N.Y. Tuberculosis & Health Assn., Inc.

STEIN, RONALD: Composer, conductor, pianist, teacher, filmmaker. b. St. Louis, Mo., April 12, 1930. e. Washington Univ., St. Louis, 1947–51, Yale Univ., New Haven, 1951–52, U.S.C., Los Angeles, 1960–63 U.S. Army, 1952–54. Music director's asst. Municipal Opera, St. Louis, 1950–51, 1954; asst. music dir., Greek Theatre, 1955; asst. prof. of music, Calif. State Univ. at Northridge, 1960–61; assoc. mus. dir., Phoenix Star Theatre, 1964. 78 m.p. scores since 1955; Documentary film director, writer, editor, photographer. U.S. Educational Films (Administrator). Professor of Music, Univ. Colorado, 1980.

PICTURES INCLUDE: The Littlest Hobo, Raymie, Legend of Tom Dooley, Premature Burial, Of Love and Desire, Dime With A Halo, Psych-Out, Dementia 13, The Rain People, Getting Straight, The Prisoners, Dateline Yesterday TV series, (Editorial Supervisor). National Parks—A Road For The Future (Photographer, Editor), The Eye of The Mind (Writer, Director).

STEINBERG, HERB: b. New York, N.Y. July 3, 1921; e. CCNY, 1937–41. Capt. U.S. Army, 1942–46; pub. PRC, 1946, Eagle Lion, 1946–49, Paramount 1949; pub. mgr. 1951; expl. mgr., 1954; studio adv. & pub. dir., 1958; exec. chg. of spec. proj., press dept., Universal City Studio, 1963; dir., adv. & mkt. Universal Studio Tours, 1971; 1974 v.p., MCA Recreation Services advertising & marketing; 1978, v.p., Universal Pictures.

STELLINGS, ERNEST G.: Exhibitor. b. Petersburg, Ind.; e. Washington, Ind., H.S. Ent. m.p. ind. 1912; theatre mgr., Washington, Ind., 1918; mgr. Wilmington, N.C., 1920; mgr., Wilmington Theatres, 1923–1939; joined information of Everett Enterprises, Charlotte, N.C., 1939; with Everett Enterprises. Stewart and Everett Theatres, 1939–62; leased two Charlotte theatres, 1951; v.p. Stellings-Gossett Theatres. Dir. Theatre Owners of N.C. and S.C.; pres. TOA, 1956–58; lifetime mem., TOA exec. comm.

STELLOFF, SKIP: Producer. b. N.Y.C., Aug. 10, 1925, e. U.S. Naval Academy; U.S. Navy, 1943–51; WWDC Radio, Wash., D.C., 1951–52; Ziv Television, 1952–56; sls. mgr., C&C Television, 1956–58; pres., Heritage Productions, 1958–60.

STEMBLER, JOHN H.: Executive. b. Miami, Fla., Feb. 18, 1913; e. U. of Florida Law School, 1937. Asst. U.S. att., South. dist. of Fla., 1941; U.S. Air Force, 1941–45; pres. Georgia Theatre Co., 1957; NATO finance chm. member exec. comm. and past pres.; Major Gen. USAF (Ret); bd. chm., National Bank of Georgia.

STEPHEN, SUSAN: Actress. b. London, Eng., July 16, 1931; e. Eastbourne and Royal Acad. of Dramatic Art. Screen debut in His Excellency, 1951.

PICTURES INCLUDE: Stolen Face, Treasurer Hunt, Father's Doing Fine, Luxury Girls, The Red Beret (Paratrooper), House Across the Lake, Private Man, Golden Ivory, For Better For Worse, As Long As They're Happy, Value for Money, His Excellency, Pacific Destiny, Barretts of Wimpole Street, Carry on Nurse.

TV: Little Women, Birdcage Room, No Hero, Pillars of Midnight.

STERLING, JAN: Actress. r.n. Jane Sterling Adriance. b. April 3, 1923. e. Private tutors; Fay Compton Sch. of Dramatic Art, London. N.Y. stage debut: Bachelor Born.

STAGE: Panama Hattie, Present Laughter, John Loves Mary, Two Blind Mice, Front Page, Over 21, Born Yesterday.

PICTURES INCLUDE: Johnny Belinda (debut), Appointment with Danger, Mating Season, Union Station, Skipper Surprised His Wife, Big Carnival, Caged, Rhubarb, Flesh and Fury, Sky Full of Moon, Pony Express, The Vanquished, Split Second, Alaska Seas, High & the Mighty, Return From the Sea, Human Jungle, Women's Prison, Female on the Beach, Man with the Gun, 1984, The Harder They Fall, Love in a Goldfish Bowl, The Incident, The Minx.

STERLING, ROBERT: Actor. r.n. William Sterling Hart; b. Newcastle, Pa., Nov. 13, 1917; e. U. of Pittsburgh; m. Anne Jeffreys, actress. Fountain pen salesman, day laborer, clerk, industrial branch credit mgr., clothing salesman on West Coast; served as pilot-instructor U.S. Army Corps. 3 yrs.

PICTURES INCLUDE: Blondie Meets the Boss, Only Angels Have Wings, Manhattan Heartbeat, Yesterday's Heroes, Gay Caballero, Penalty, I'll Wait for You, Get-Away, Ringside Maisie, Two-Faced Woman, Dr. Kildare's Victory, Johnny Eager, This Time for Keeps, Somewhere I'll Find You, Secret Heart, Roughshod, Bunco Squad, Sundowners, Show Boat, Column South.

TV: Topper.

STERN, ALFRED E. F.: Public relations executive. b. Boston, Mass., Aug. 4. e. Boston U. Reporter, editor, Lowell Sun, Quincy Patriot-Ledger, Dartmouth News; publicist, RKO Radio Pictures, 1946–54; publicity dir., 1955; West Coast publicity dir., NTA, 1958; own public relations org., Alfred E. F. Stern Co., 1960.

STERN, EDDIE: Film buyer. b. New York City, Jan. 13, 1917. e. Columbia Sch. of Journalism; head film buyer and booker, specializing in art theatres, for Rugoff and Becker, N.Y.; Captain, USAF; joined Wometco Ent. in 1952 as asst. to film buyer; now v.p. motion picture theatre film buying and booking, Wometco Enterprises, Inc.

STERN, EZRA E.: Attorney. b. New York, N.Y., Mar. 22, 1908. e. Southwestern U. 1930, LL.B.; pres., Wilshire Bar Assn. Former legal counsel for So. Calif. Theatre Owners Association. Member: Calif. State Bar; member, Int'l Variety Clubs; former chief barker, Variety Club So. Calif. Tent 25; pres., Variety Int'l Boys' Club; board of dir., Los Angeles Metropolitan Recreation & Youth Services Council; bd. of trustees, Welfare Planning Council, Los Angeles Region; former mem. Los Angeles Area Council, Boys' Club of America; pres., Variety International Boys' Club 1976–77 and 1979–80. Member bd., Will Rogers Institute, M.P. Pioneers.

STERN, STANLEY L.: Executive. b. Philadelphia, Dec. 31, 1917. Major, USAF, World War II & Korea. Entire working career with Wometco Enterprises, starting as usher in 1933. Served as theatre mgr., m.p. booker, in chg. of real estate and insurance, before taking over Wometco Theatre Division in 1973. Elected v.p., 1959, sr. v.p., in 1967, dir. in 1972. Placed in chg. Wometco Entertainment Division, 1976.

MEMBER: VCI, Tent 33, Motion Picture Pioneers. Past v.p. and dir., American Society of Insurance Management; v.p. and dir. NATO of Florida; mem. bd., NATO.

STERN, STEWART: Writer. b. New York, N.Y., Mar. 22, 1922. e. Ethical Culture Schools, 1927–40; U. of Iowa, 1940–43; Rifle Squad Leader, S/Sgt. 106th Inf. Div., 1943–45; actor, asst. stage mgr., The French Touch, B'way, 1945–46; dialogue dir. Eagle-Lion Studios, 1946–48. 1948 to date: screenwriter.

TV: (Plays) Crip, And Crown Thy Good, Thunder of Silence, Heart of Darkness, Sybil.

PICTURES INCLUDE: Teresa, Benjy (orig. s.p.) Rebel Without a Cause, The Rack, The James Dean Story, The Outsider, The Ugly American, Rachel, Rachel, The Last Movie, Summer Wishes—Winter Dreams (orig. s.p.).

STERNAD, RUDOLPH: Producer, designer, art director. b. N.Y., Oct. 6, 1905. e. Cooper Union Inst., 1922–23; Beaux Arts, 1924; Chouinand, 1925; arch. designer, Society M.P. Art Directors; designer, MGM, Fox, United Artists, Univ. Studios, 1926–34; art dir., 20th-Fox, Old Chicago, Suez, etc. 1934–40; art dir. Columbia Studios, More the Merrier, Talk of the Town, Craig's Wife, 1001 Nights, You Were Never Lovelier, etc.; prod. designer, Stanley Kramer Prod., 1948–62.

PICTURES INCLUDE: Champion, Home of the Brave, The Men, Cyrano de Bergerac, High Noon, Death of a Salesman, Four Poster, Caine Murder, Not as a Stranger, The Defiant Ones, On the Beach, Judgment at Nuremberg, Pressure Point, A Child Is Waiting, It's a Mad, Mad, Mad, Mad World.

STEVENS, ANDREW: Actor. b. Memphis, Tenn., 1956. Son of actress Stella Stevens. e. Memphis Messick H.S. Studied acting with Strasberg, David Craig, Vincent Chase, and Robert Easton. Began on TV, later playing in feature films.
PICTURES INCLUDE: Shampoo, Day of the Animals, Vigilante Force, The Boys in Company C, The Fury, Death Hunt.
TV: Adam-12, Apple's Way, The Quest, Police Story, Shazan, Once an Eagle (mini-series), Oregon Trail, Pumper One. Movies: S Secrets, The Last Survivors, The Werewolf of Woodstock, The Bastard (mini-series), The Rebels (mini-series), Beggarman Thief, Topper.

STEVENS, CONNIE: Actress. r.n. Concerta Ann Ingolie. b. Brooklyn, N.Y., August 8, 1938. e. Sacred Heart Academy, Hollywood Professional School; began career as winner of several talent contests in Hollywood; prof. debut, Hollywood Repertory Theatre's prod. Finian's Rainbow; recordings include: Kookie, Kookie, Lend Me Your Comb, 16 Reasons, What Did You Make Me Cry For, From Me to You, They're Jealous of Me, A Girl Never Knows.
TV: Hawaiian Eye, Wendy and Me, Sex Symbol (movie).
PICTURES INCLUDE: Eighteen and Anxious, Young and Dangerous, Drag Strip Riot, Rock-a-Bye Baby, Parish, Susan Slade, Palm Springs Weekend, Cruise A-Go-Go, The Grissom Gang, Last Generation.

STEVENS, CRAIG: Actor. b. Liberty, Mo.; e. U. Kansas. Played in coll. dramatics. On screen 1941 in Affectionately Yours.
PICTURES INCLUDE: Since You Went Away, The Doughgirls, Roughly Speaking, Too Young to Know, Humoresque, The Man I Love, That Way With Women, Night Unto Night, Love and Learn, Lady Takes a Sailor, Phone Call from a Stranger, French Line, Where the Sidewalk Ends, Duel on the Mississippi, The Name's Buchanan, Gunn, Limbo Line, The Snoop Sisters, "S.O.B.".
TV: Lux Video Theatre, Four Star Playhouse, Loretta Young Show, Schlitz Playhouse, Peter Gunn, Dinah Shore, Ernie Ford Shows, Chevvy Show, Summer on Ice, Person to Person, Man of the World series (ATV England); Mr. Broadway, The Millionaire, Name of the Game, The Bold Ones.
STAGE: Here's Love, King of Hearts, Plain and Fancy, Critics Choice, Mary Mary; Nat. Co. Cactus Flower.

STEVENS, GEORGE, JR.: Director, producer. b. Los Angeles, Calif., Apr. 3, 1932. e. Occidental College, 1949–53, B.A., 1st Lieut. U.S. Air Force; TV dir., Alfred Hitchcock Presents, Peter Gunn, 1957–61; prod. asst. Giant Productions, 1953–54; prod. asst. Mark VII, Ltd., 1956–57; assoc. prod., dir., The Diary of Anne Frank, 1957–59; assoc. prod., dir., The Greatest Story Ever Told, 1959–62; dir. M.P. Service, U.S. Information Agency 1962–67; chmn., U.S. deleg. to Film Festivals at Cannes (1962, 1964), Venice (1962, 1963), Moscow (1963); prod. John F. Kennedy; Years of Lightning, Day of Drums; co-chm., American Film Institute. Producer, America at the Movies (1976). Executive producer of The American Film Institute's Salute to James Cagney for which he was awarded an Emmy by National Academy of Television Arts and Sciences on May 19, 1975; producer/writer, American Film Institute's Salutes, 1973–81; executive producer, The Stars Salute America's Greatest Movies, 1977; producer/writer, The Kennedy Center Honors, 1978–80; producer/writer, America Entertains Vice Premier Deng, 1978.

STEVENS, K. T.: Actress. r.n. Gloria Wood. b. Hollywood, Calif.; e. U. So. Calif.; d. late Sam Wood, director.
STAGE: You Can't Take It With You, The Man Who Came to Dinner, My Sister Eileen, Nine Girls, St. Joan.
PICTURES INCLUDE: (debut) Peck's Bad Boy, The Great Man's Lady, 1942; Nine Girls, Address Unknown, Kitty Foyle, Harriet Craig, Vice Squad, Tumbleweed, Missile to the Moon.

STEVENS, LEE: Executive Administrator. b. March 10, 1930, New York City. e. NYU School of Commerce, grad. 1951 BS; NYU Law School, grad. 1957, Jurist Doctor, member of NY bar. With William Morris Agency 28 years. Now exec. v.p. & chief operating officer.

STEVENS, LESLIE: Executive, writer, producer, director. b. Wash., D.C., Feb. 3, 1924. e. Westminister School, London, Yale Drama School, American Theatre Wing, N.Y.C. Sold first play, The Mechanical Rat at 15; wrote six plays for summer stock groups, 1941–42; U.S. Air Force, 1943; pres., exec. prod. Daystar Prods.; writer, producer, director, film TV and features.
STAGE: Bullfight (off B'way) 1953–54; wrote, Broadway: Champagne Complex, 1954; The Lovers, 1956; The Marriage-Go-Round, 1958; The Pink Jungle, 1959.
TV: For Playhouse 90 Invitation to a Gunfighter, Charley's Aunt, Rumors of Evening, The Violent Heart, Portrait of a Murderer, The Second Man; Kraft TV Theatre, Duel; Four Star Playhouse, Award, Producers Showcase, Bloomer Girl; created, prod., dir. Stoney Burke, ABC series, 1962, created, prod., dir. Outer Limits, pilot It Takes a Thief; exec. prod. series, ABC; prod.-writer pilot McCloud, NBC; exec. prod. series, exec. prod. Men From Shiloh series; exec. prod., writer-dir., Name of the Game, Warner Bros. (Leslie Stevens Productions), 1971, prod.-writer, pilot of Search, exec. prod. of NBC: series creator, CATV series Movie of Today, Paperback Playhouse, Earthside Missile Base. Exec. prod., Universal Studios, Invisible Man, 1975. Supervising prod. Gemini Man, 1976 pilot and series.
PICTURES INCLUDE: author, The Left-Handed Gun 1958; co-prod.-wrote-dir. Private Property 1959; wrote, The Marriage-Go-Round 1960; wrote-prod.-dir. Hero's Island 1961.

STEVENS, MARK: Actor. r.n. Richard Stevens. b. Cleveland, Ohio, Dec. 13; e. privately; Beaux Arts and Sir George Williams Sch. of Fine Arts, Montreal. Had varied career before appearing on stage and radio in Canada; later joined station WAKB in Akron; then prod. mgr., WJW, Akron. Screen debut in Objective Burma.
PICTURES INCLUDE: God Is My Co-Pilot, Pride of the Marines, From This Day Forward, The Dark Corner, I Wonder Who's Kissing Her Now, Between Midnight and Dawn, Katie Did It, Little Egypt, Reunion in Reno, Mutiny, Big Frame, Torpedo Alley, Jack Slade, Cry Vengeance, Timetable, September Storm, Fate Is the Hunter, Frozen Alive, Sunscorched.
TV: Martin Kane, NBC; Big Town, NBC; prod., starred, script. Formed Mark Stevens Prod., Mark Stevens Television Prod., 1955.

STEVENS, STELLA: Actress. b. Hot Coffee, Mississippi, Oct. 1, 1938. e. Attended Memphis State University. Modeled in Memphis when she was discovered by talent scouts. Was briefly a term contract actress at 20th Century-Fox, later under exclusive contract to Paramount, then Columbia.
PICTURES INCLUDE: Say One For Me, The Blue Angel, Li'l Abner, Too Late Blues, Man Trap, The Courtship of Eddie's Father, The Nutty Professor, Advance to the Rear, Synanon, The Secret of My Success, The Silencers, The Ballad of Cable Hogue, Stand Up & Be Counted, Slaughter, The Poseidon Adventure, Arnold, Las Vegas Lady, Cleo Jones and the Casino of Gold, Nickelodeon, The Manitou.
TV: Five episodes of Ben Casey.

STEVENSON, PARKER: Actor. b Philadelphia, June 4. e. Princeton Univ. Began professional acting career by starring in film, A Separate Peace, while high school senior, having attracted attention through work on TV commercials.
PICTURES INCLUDE: A Separate Peace (debut), Our Time, Lifeguard.
TV: The Streets of San Francisco, Gunsmoke, Hardy Boys Mysteries. Guest and host on several talk shows and specials.

STEVENSON, ROBERT: Director, writer. b. London, 1905. Educated Shrewsbury, Cambridge U. Under contract British companies 1929–39; Selznick 1939–43. American citizenship 1940. U.S. Army, War Department, Italian campaign, 1943–46. Novel, Darkness in the Land; scripted Tudor Rose. Selznick, 1946–49 RKO 1949–52. TV scripting 15 shows; directed over 100 TV shows.
TV: Cavalcade of America, Ford Theatre, Playhouse of Stars, G.E. Theatre, Hitchcock Presents, 20th Century Fox Hour, Gunsmoke, 1952–56. Walt Disney, 1956 to present.
PICTURES INCLUDE: Nine Days a Queen, To the Victor, Back Street, Jane Eyre, To the Ends of the Earth, Walk Softly Stranger, My Forbidden Past, Dishonored Lady, Las Vegas Story, Johnny Tremaine, Old Yeller, In Search of the Castaways, The Mis-Adventures of Merlin Jones, The Absent-Minded Professor, Son of Flubber, That Darn Cat, Kidnapped, Darby O'Gill and the Little People, Blackbeard's Ghost, The Gnome-Mobile, The Monkey's Uncle, Mary Poppins, The Love Bug, Bedknobs and Broomsticks, Herbie Rides Again, Island at the Top of the World, One of Our Dinosaurs is Missing, The Shaggy D.A.

STEWART, ELAINE: Actress. b. Montclair, N.J., May 31, 1929. Usherette & cashier, m.p. theatre, Montclair; model, Conover Agcy., 1948; many TV shows; screen debut in Sailor Beware; Star of Tomorrow, 1954.
PICTURES INCLUDE: Sky Full of Moon, The Bad and the Beautiful, Desperate Search, Code Two, Slight Case of Larceny, Young Bess, Take the High Ground, Brigadoon, Adventures of Hajji Baba, Tattered Dress, Rise and Fall of Legs Diamond, Most Dangerous Man Alive.

STEWART, HUGH: Producer. b. Falmouth, Cornwall, Dec. 14, 1910. e. Clayesmore School, St. John's College, Cambridge; ent. m.p. ind. 1932; editor, Gaumont British, 1932.
PICTURES INCLUDE: Man Who Knew Too Much, London Films: Dark Journey, Storm in a Teacup, Action for Slander, South Riding, Q Planes, Spy in Black. Army Service, 1939–45. Co-producer with Frank Capra of Anglo-US War Film, Tunisian Victory; True Glory. An Ideal Husband, Trottie True, Night Without Stars, Long Memory,

Up To His Neck, Man of the Moment, Up in the World, Just My Luck, Innocent Sinners, The Square Peg, Follow a Star, Make Mine Mink, The Bulldog Breed, In the Doghouse, On the Beat, A Stitch In Time, The Intelligence Men, The Early Bird, That Riviera Touch, The Magnificent Two, All at Sea, Mr. Horatio Knibbles, Anoop and The Elephant, The Flying Sorcerer, The Chiffy Kids (two series), High Rise Donkey.

STEWART, JAMES: Actor b. Indiana, Pa., May 20, e. Mercersburg Acad.; Princeton U. With Falmouth Stock Co., Cape Cod; on N.Y. stage in Goodbye Again; stage mgr. for Camille with Jane Cowl (Boston). Won Acad. Award for best performance 1940; joined U.S. Air Forces 1942, commissioned Col. 1944. Voted one of top ten money-making stars; M.P. Herald-Fame poll, 1950, 52, 54, 57; No. 1 Money-Making Star, 1955. 1968, Screen Actors Guild Award. Mem.: Bd. of Trustees, Princeton U. Trustee, Claremont College; exec. bd. of Los Angeles Council of Boy Scouts of America, bd. of dirs., Project Hope.

STAGE: Spring in Autumn, All Good Americans, Yellow Jack, Journey at Night, Harvey.

PICTURES INCLUDE: Murder Man, Rose Marie, Wife vs. Secretary, Next Time We Love, Speed, Gorgeous Hussy, Born to Dance, Seventh Heaven, You Can't Take It With You, Vivacious Lady, Mr. Smith Goes to Washington, Philadelphia Story, It's A Wonderful World, Destry Rides Again, Shop Around the Corner, Mortal Storm, It's a Wonderful Life, Magic Town, Call Northside 777, Rope, You Gotta Stay Happy, Stratton Story, Malaya, Winchester '73, Broken Arrow, Harvey, Jackpot, No Highway in the Sky, Greatest Show on Earth, Carbine Williams, Bend of the River, Naked Spur, Thunder Bay, Glenn Miller Story, Far Country, Rear Window, Strategic Air Command, Man From Laramie, Man Who Knew Too Much, Spirit of St. Louis, Night Passage, Vertigo, Bell Book and Candle, FBI Story, The Mountain Road, Two Rode Together, Man Who Shot Liberty Valance, Mr. Hobbs Takes a Vacation, How the West Was Won, Take Her, She's Mine, Cheyenne Autumn, Dear Brigitte, Shenandoah, The Rare Breed, Flight of the Phoenix, Firecreek, Bandolero, Cheyenne Social Club, Fool's Parade, That's Entertainment, The Shootist, Airport '77, The Big Sleep.

TV: Hawkins, 1973.

STEWART, JAY: Broadcast Announcer. b. Sept. 6, Summitville, Indiana. Was Monty Hall's sidekick on "Let's Make A Deal" for 14 years, currently is doing The Joker's Wild and Tic Tac Dough.

STEWART, MARILYN: Public Relations Executive. b. New York City. e. Hunter College. Singer, composer, artists mgr.—latter two hobbies; first was profession that began prior to entry into industry. Entered Ind. as scty. in 1960 to Si Seadler, MGM dir. of adv. Promoted in 1963 to asst. to Seadler, who was then dir. adv.-pub.-prom. for special projects at MGM. Left in 1964 to become prom.-pub. dir. for Verve/Folkways Records; duties also included a&r and talent-scouting. In 1966 joined 20th-Fox as radio/tv pub. coordinator. In 1969 went to Para. Pictures as mag. pub. coordinator. In Aug. '70 named worldwide director of pub. for Para., including creation of overall mkt. concepts, becoming 1st woman to be appointed to that position at major co. Campaigns included Love Story and The Godfather. In 1972 opened own p.r. firm, The Mediary, specializing in m.p. pub. Headquarters in N.Y.; repr. in L.A. Has represented American Film Theatre, The Lords of Flatbush, Bang the Drum Slowly, A Very Big Withdrawal, The Kids Are Alright, Autumn Sonata, The Tin Drum, L.A. Film Exposition, Michael Moriarty, etc.

STEWART, MARTHA: Actress. r.n. Martha Haworth, b. Bardwell, Ky., Oct. 7, 1922; m. Joe E. Lewis, prof. vocalist, Glenn Miller and Claude Thornhill bands; appeared in Copacabana, N.Y.; star Lucky Strike Hit Parade, 1943. On screen first, 1945 Doll Face.

PICTURES INCLUDE: Johnny Comes Flying Home, I Wonder Who's Kissing Her Now, Daisy Kenyon, Are You With It?, Convicted, In a Lonely Place, Aaron Slick from Punkin Crick.

STEWART, PEGGY: Actress. b. West Palm Beach, Fla., June 5, 1923; in 1933 was champion swimmer, Georgia; acq. honors in horseback riding. Taught riding in Calif. at age of 14. On San Francisco stage: Great American Family, Goodnight Ladies.

PICTURES INCLUDE: Wells Fargo, That Certain Age, Back Street, All This and Heaven Too, Pride of Maryland, The Longhorn, Vengeance Trail, Kansas Territory, Montana Incident, The Way West, The Animals, Pickup on 101.

STIGWOOD, ROBERT: Executive. b. Adelaide, Australia, 1934. e. Sacred Heart College. Began career as copywriter for Aust. ad agency; at 21 left home for England. Series of first jobs led to his opening a London theatrical agency. Began casting commercials for TV; produced records for clients. Became first independent record producer in Great Britain. In mid '60s joined forces with Brian Epstein, manager of Beatles, to become co-mgr. of NEMS Enterprises. At Epstein's death formed own company, launching careers of such artists as Bee Gees, Cream, etc. Moved into theatre production in London: Hair, Jesus Christ Superstar, Pippin, Oh! Calcutta! Evita. Entered film production with Jesus Christ Superstar. Formed RSO Records in 1973.

PICTURES INCLUDE: Jesus Christ Superstar, Tommy, Saturday Night Fever, Grease, Sgt. Pepper's Lonely Hearts Club Band, Moment by Moment, Times Square (co.-prod.), The Fan.

STOCKWELL, DEAN: Actor. b. Hollywood, Calif., Mar. 5; p. Harry and Betty Veronica Stockwell; e. L.I. public schools and Martin Milmore, Boston. On stage in Theatre Guild prod. Innocent Voyage. Has appeared on radio in Death Valley Days and Dr. Christian.

PICTURES INCLUDE: (debut) Anchors Aweigh, The Valley of Decision, Abbott and Costello in Hollywood, The Green Years; named in 1949 M.P. Herald-Fame Stars of Tomorrow poll; Home Sweet Homicide, The Mighty McGurk, The Arnelo Affair, The Romance of Rosy Ridge, Song of the Thin Man, Gentleman's Agreement, Deep Waters, Down to Sea in Ships, Boy With Green Hair, Secret Garden, Happy Years, Kim, Stars in My Crown, Cattle Drive, Compulsion, Sons and Lovers, Long Day's Journey Into Night, Psych-out, The Dunwich Horror, The Last Movie, The Loners.

STODDARD, BRANDON: Executive. e. Yale Univ., Columbia Law School. Was program ass't. at Batton, Barton, Durstine and Osborn before joining Grey Advertising, where was successively, program operations supvr., dir. daytime programming, v.p. in chg. of TV, radio programming. Joined ABC in 1970; named v.p. daytime programs for ABC Entertainment, 1972; v.p. children's programs, 1973. Named v.p., motion pictures for TV, 1974. In 1976 named v.p. dramatic programs and motion pictures for television. In June, 1979, named pres., ABC Motion Pictures.

STOLBER, DEAN: Executive. b. Philadelphia, Penna., Sept. 2, 1944. e. Harvard (A.B., 1966); NYU School of Law (J.D., 1969). Acted on Broadway and in TV before starting career in law and business affairs in films. 1979, sr. v.p., business affairs, for United Artists Corp.; 1981 named exec. v.p.

STOLL, GEORGE: Musical director, MGM. b. Minneapolis, Minn., May 7, 1905. Was musical director Bing Crosby Woodbury program, CBS for two years; Camel Cigarette program, CBS for two years; Shell Oil program, NBC, 2 years. Violin soloist for a number of years touring U.S. and Canada. In 1936 appeared in MGM short product; musical dir. numerous films for MGM. Acad. Award best scoring musical picture (Anchors Aweigh) 1945.

PICTURES INCLUDE: Neptune's Daughter, Two Weeks with Love, I Love Melvin, Dangerous When Wet, Rose Marie, Flame & the Flesh, Student Prince, Athena, Hit the Deck, Love Me or Leave Me, Meet Me in Las Vegas.

STOLLER, MORRIS: Executive. b. New York, N.Y., Nov. 22, 1915. e. C.C.N.Y., BBA 1933; N.Y.U., CPA 1935, Brooklyn Law School, LLB, 1937; joined William Morris agency in 1936; president, Permanent Charities Comm., 1962; director and officer, Artist's Mgr. Guild, 1950–67; dir. St. Jude Hospital, 1955–67; presently bd. chm., chief operating officer, William Morris Agency.

STOLNITZ, ART: Executive. b. Rochester, N.Y. March 13, 1928. e. University of Tennessee. U.S. Navy Air Force. Legal dept., William Morris Agency, 1953, dir. business affairs, ZIV, 1959; dir. new program development, ZIV-United Artists, 1960; literary agent, MCA, 1961; dir. business affairs, Selmur Productions, Selmur Pictures, 1963; v.p. ABC Pictures, 1969; v.p. Metromedia Producers Corporation, 1974, executive v.p. Metromedia Producers Corporation; 1975 exec. v.p. and prod. Charles Fries Prods. 1976, prod. Edgar J. Scherick Productions; 1976–77 prod., Grizzly Adams (TV); 1977, v.p. business affairs, Warner Bros.-TV; 1980, snr. v.p., business affairs.

STOLOFF, VICTOR: Producer, writer, director, editor; e. French Law U. Ac. Fines Arts. Prod. dir. writer of award winning documentaries (Warner Bros. release); Prod. dir. writer first U.S. film made in Italy, When in Rome; First U.S. film made in Egypt; Collaborator William Dieterle films; Contract; writer, dir. to Sidney Buchman, Columbia, S.p. Volcano, The Sinner, Shark Reef.

TV: Ford Theater, Lloyd Bridges series, NBC, National Velvet, CBS High Adventure, with Lowell Thomas, Journey Around the World, Prod. Writ. Of Love and Desire; producer-director, Intimacy, Prod. on location 22, Hawaii Five-0, The 300 Year Weekend, Why? Director (orig. s.p.). Created Woman of Iran, first of TV series. 1977: producer-director: The Washington Affair.

STOLTZ, ARNOLD T.: Publicity executive. b. New York; e. Washington, U., St. Louis. Worked as reporter, publicity adv. man, 1933–37. Was pub. dir. and mgr., Warner and Loew theatres. Was radio commentator under name of Solly Wood. Pen name, Franc Arnold. Winner Quigley Silver Grand Award for exploitation, 1941. In 1942 named exploitation mgr., United Artists. In July, 1945 named natl. dir. and pub. PRC Pictures; then sales rep. Resigned 1947 to enter prod.; exec. prod. Movie Star, American Style. Joined Famous Players Corp.; adv., pub. dir., 1965; v.p. & gen. sales mgr., 1966. pres. Movie-Rama Color Corp., 1967; exec. prod. Tell Me in the Sunlight, The Golden Poodle.

STONE, ANDREW L.: Producer, director. b. Calif., July 16, 1902; e. U. Calif. Ent. ind. 1918 at Universal San Francisco exch.; later author, prod., dir. series of pictures for Paramount; prod., dir. for Sono-Art; 1932–36. org. and oper. Race Night company; prod., dir., The Girl Said No, 1936; Stolen Heaven, Say It in French, The Great Victor Herbert, Magician Music, 1940. Dir. Stormy Weather; formed Andrew Stone Prods., 1943.

PICTURES INCLUDE: Hi Diddle Diddle, Sensations of 1945, Bedside Manner, Bachelor's Daughter, Fun on a Weekend, Highway 301, Confidence Girl, Steel Trap, Blueprint for Murder, Night Holds Terror, Julie, Cry Terror, The Decks Ran Red, The Last Voyage, Ring of Fire, Password is Courage, Never Put It in Writing, Secret of My Success, Song of Norway, The Great Waltz.

STONE, BURTON S.: Executive. b. Feb. 16, 1928; e. Florida So. Coll; Pres., Deluxe General, Inc. and Deluxe Laboratories. Was film ed., Hollywood Film Co. 1951–53; serv. mgr., sales mgr. and gen. mgr., Consolidated Film Inds., 1953–61; nat'l sales mgr., Movielab, 1961–63; pres., All Service Film Laboratories, 1963–68; v.p. Technicolor, Inc., 1968–70. Pres., Precision Film Labs., 1970–78.

MEMBER: Board of directors, Will Rogers Hospital; member, Academy of M.P. Arts & Sciences; assoc. member, American Society of Cinematographers; member, SMPTE; trustree, Motion Picture Laboratory Technicians, Local 702 IATSE, Pension and Welfare Fund; member, Cinema Lodge, B'nai B'rith, Variety Club 1035, N.Y., Motion Picture Pioneers. University Film Producers, Association of Cinema Laboratories.

STONE, EZRA C.: Actor, director, producer. b. New Bedford, Mass., Dec. 2, 1917; e. American Acad. of Dramatic Arts, N.Y., 1934–35; actor: National Junior Theatre, 1931; WCAU, Phila., 1930–34; vaudeville, Steel Pier, Atlantic City, N.J., 1931–32; WMHA & Plays & Players Productions, Phila., 1928–30; Broadway: Parade, Ah Wilderness, Oh Evening Star, Three Men on a Horse, Room Service, Brother Rat; created Henry Aldrich, What a Life, The Alchemist, She Stoops to Conquer; prod. asst. to George Abbott, 1935–40; created Henry Aldrich on radio's Aldrich Family, 1938; starred, Those Were the Days, Paramount, 1939, This is the Army, Warners 1943, USAAF, 1941–45; directed on Broadway: See My Lawyer, 1939, Reunion in New York, 1940.

STAGE: This is the Army, 1942, January Thaw, 1946, At War With the Army, 1949, To Tell You the Truth, 1947, Me and Molly, 1948, Wake Up Darling, 1956, Make a Million, 1958, The Man That Corrupted Hadleyburg, 1951, The Pink Elephant, 1953, Dear Ruth, 1961, Come Blow Your Horn, 1962, God Bless Our Bank, Fallen Angels, 1963; Finishing Touches (1978); Dracula—The Vampire King (1978).

PICTURES INCLUDE: (docum.) I.B.M., American Heart Assn. Chapman Coll., University of Judaism, Jewish Theological Seminary; dir. live action sequences for The Daydreamer; The Forty Million (producer, director, co-narrator).

TV: Aldrich Family, Danny Thomas, Ed Wynn, Ezio Pinza, Martha Raye, Fred Allen, Herb Shriner, Life With Father, Sid Caesar, Joan Davis, dir. Joe and Mabel, prod., dir. Bachelor Father, Angel, The Hathaways, spec. Affairs of Antol, Shari Lewis, Bob Hope, My Living Doll, Munsters, Karen, Tammy, O.K. Crackerby, Please Don't Eat the Daisies; dir., writer Woody Allen pilot, Loredo, Pistols & Petticoats, Petticoat Junction, Phyllis Diller Show, Lost in Space, Tammy Grimes show, Julia, Flying Nun, Debbie Reynolds, The Jimmy Stewart Show, Lassie, Sandy Duncan Show, Tribute to the Lunts, Love American Style, Bob Newhart, Space Academy, ABC Circle Playhouse Project UFO, Actor, (PBS Paul Muni, biography film). Did 200 documentary films for IBM. Has over 25 intl. film awards; Grand Prize, Barcelona Int'l Film Festival.

AUTHOR: Coming Major, 1945, co-author; Deems Taylor; Puccini Opera, 1951; contributor to: Variety, Magazine Digest, N.Y. Post; Equity Magazine, etc. Teacher: American Acad. Dramatic Arts, assoc. dir., American Theatre Wing; American College Theatre Festival, Princeton, Yale, U.C.L.A.; dir., David Library of Amer. Revolution.

STONE, MARIANNE: Actress. b. London, England. Studied Royal Acad. of Dramatic Art, making her first appearance West End stage in The Kingmaker, 1946. Since then has appeared in numerous films.

TV: Maigret, Bootsie and Snudge, Jimmy Edwards Show, Wayne and Schuster Show, Roy Hudd Show, Harry Worth Show, Steptoe and Son, Informer, Love Story, Father Dear Father, Bless This House, The Man Outside, Crown Court, Public Eye, Miss Nightingale, She, Little Lord Fauntleroy, The Secret Army (2 series).

PICTURES INCLUDE: Brighton Rock, Seven Days to Noon, The Clouded Yellow, Wrong Arm of the Law, Heavens Above, Stolen Hours, Nothing But the Best, Curse of the Mummy's Tomb, Hysteria, The Beauty Jungle, A Hard Day's Night, Battle of a Simple Man, Echo of Diana, Act of Murder, Catch Us If You Can, You Must Be Joking, The Countess from Hong Kong, The Wrong Box, To Sir With Love, 1967–68; The Bliss of Mrs. Blossom, Here We Go Round the Mulberry Bush, Carry on Doctor, The Twisted Nerve, The Best House in London, Oh! What a Lovely War, 1970; The Raging Moon, There's a Girl in My Soup, All the Right Noises, Assault, Carry On at Your Convenience, All Coppers are . . . , Carry on Girls, Penny Gold, The Vault of Horror, Percy's Progress, Confessions of a Window Cleaner, Carry on Dick, That Lucky Touch, Sarah, Carry on Behind, Confessions From a Holiday Camp, The Chiffy Kids, What's Up Superdoc?; The Class of Miss McMichael, The Human Factor, Dangerous Davies.

STONE, PETER: Writer. Won Oscar in 1964 for Father Goose; Emmy for segment of The Defenders.

PICTURES INCLUDE: Charade, Arabesque, Mirage, Sweet Charity, The Taking of Pelham One Two Three, 1776 (adapted own stage musical book to screen), Silver Bears, Who Is Killing the Great Chefs of Europe?, Why Would I Lie?

STONEMAN, JAMES M.: Executive. b. Jan. 16, 1927. Pres., Interstate Theatres Corp., Boston.

STOREY, FREDERICK: Executive. b. Columbus, Georgia, Nov. 12, 1909. e. Georgia Tech. Adv. staff Atlanta Journal, 1933–38; adv. staff C. P. Clark Adv. Agcy., 1938; partner 1940; U.S. Navy, 1941–46; staff Georgia Theatre Co., 1946; v.p. 1947–52. Founded Storey Theatres Inc., Atlanta, Ga.; 1952, now chm. exec. comm./(formerly pres.) of Georgia State Theatres; dir. numerous theatre cos.; officer, dir. radio, TV firms; v.p. treas., dir., Motion Picture Theatre Owners of Georgia, member NATO, member, pres., advisory council.

STORKE, WILLIAM F.: Executive. b. Rochester, N.Y. e. Univ. of Calif. In Navy in World War II First position with NBC Hollywood guest relations staff, 1948. Moved to continuity acceptance dept. as comm. editor. Prom. to asst. mgr, comm. supvr. before joining NBC West Coast sales dept., 1953. Transferred to N.Y. as progr acct. exec., 1955; named administrator, participating prog. sales, Nov., 1957. Named dir., participating program sales, 1959. Named dir., program adm., NBC-TV, Jan., 1964; in Feb. elected v.p., program adm. In 1967 named v.p., programs, East Coast; Nov. 1, 1968, appt. v.p., special programs, NBC TV Network; 1979, pres., Claridge Group, Ltd.

STORM, GALE: Actress. r.n. Josephine Cottle. b. Bloomington, Texas, April 5, 1922. In high school became interested in dramatics. To films, 1940.

PICTURES INCLUDE: Between Midnight and Dawn, Underworld Story, Curtain Call at Cactus Creek, Al Jennings of Oklahoma, Texas Rangers, Woman of the North Country.

TV: My Little Margie, Oh Susanna, radio show.

STRAIGHT, BEATRICE: Actress. b. Old Westbury, N.Y. 1918. Trained in classics; won Tony award early in career for best actress in Arthur Miller's The Crucible. Many films and TV programs.

PICTURES: Patterns, The Nun's Story, Phone Call from a Stranger, Network (AA), The Promise, The Formula, Endless Love.

STAGE: A Streetcar Named Desire, Phedre, The Heiress, Ghosts, The Lion in Winter etc.

TELEVISION: Beacon Hill, many specials.

STRAMER, HUGO: Executive. b. Budapest, Hungary. Up to 1957 man. dir. of RKO in various Latin American territories. Joined Paramount Pictures 1957 as man. dir. for Argentine, Uruguay and Paraguay. 1960: man. dir. Paramount Pictures Germany. 1963–69 Continental man. for Paramount in Paris. Then man. dir. Cinema Int'l Corp. (UK). 1972, joined MPEAA in London. 1973, apptd CIC man. dir. for Argentina, Uruguay and Paraguay.

STRASBERG, LEE: Acting Instructor-Director-Actor. b. Nov. 17, 1901, Bodanov (then Austria-Hungary, now in U.S.S.R.). At 8 moved to New York with family. Began acting in teens at settlement house; soon began directing. Studied at American Laboratory Theatre under Richard Boleslavski and Maria Ouspenskaya, who had studied with Stanislavski, founder of Moscow Art Theatre. Founded Group Theatre in early '30s; since 1949 artistic director of Actors Studio, known for acting technique called The Method.

STAGE CREDITS: Supervised productions of Marathon '33, Baby Wants a Kiss, Blues for Mr. Charlie, revivals of Strange Interlude, The Three Sisters, etc.

PICTURES: Actor: The Godfather, Part II (AA nominee), And Justice for All, Going in Style, Boardwalk.

STRASBERG, SUSAN: Actress. b. New York, N.Y.; e. there. Parents: Lee Strasberg, stage dir. & dir. of Actors Studio, and Paula Miller, actress. Made stage debut in off-Bway.

production of Maya; on TV in series, The Marriage, The Duchess and the Smugs, Catch a Falling Star; starred on Bway. in The Diary of Anne Frank.
PICTURES INCLUDE: Picnic, Stage Struck, Scream of Fear, Adventures of a Young Man, The Trip, Psych-Out, The Name of the Game is Kill, Rollercoaster, The Manitou, In Praise of Older Women.

STRASSBERG, STEPHEN: Publicist. B. New York, N.Y.; e. C.C.N.Y., B.S.S., joined Loew's Inc., 1940; served in U.S. Army, WW II; publicist with Republic Pictures, 1946–49; asst. nat'l adv. dir., Film Classics, 1949–50; pub. dir., Eagle Lion Classics, 1950; publicist Lopert Films, Inc., 1951; dir. of adv., pub., Imperial Films, 1953; pub. dir., WABC, WABC-TV, N.Y., 1955; asst. dir., Press Information; ABC, 1957; dir. press info. ABC-TV Network, 1958; dir., News Information, ABC-TV Network, 1975.

STRATTON, JOHN: Actor. b. Clitheroe, England, 1925; e. Royal Grammar School, Clitheroe. Early career in repertory; m.p. and TV debut 1948.
TV: The First Mrs. Fraser, The Confidential Clerk, Adams Apple, You Know What People Are (series). Death of a Salesman, Quatermass in The Pit, The Wind and the Rain, The Secret Kingdom, Kipps, The Dobson Fund, The Problem of Mary Winshaw, A Perfect Woman, Climate of Fear, Thank You and Goodnight, What's In It for Walter, A Free Weekend, Workshop Limits, 24 Hour Call, 2 Cars, The Odd Man, It's Dark Outside, For the West, Man in Room 17, Julie's Gone, The Trouble Shooters, Turn out the Lights, Mr. Rose, The Black Doctor, Sir Arthur Conan Doyle, The Newcomers, Tickle Time, Letters from the Dead, Z Car, Wanted Single Gentleman, Softly Softly, Fall of the Goat, City '68, Artists in Crime, The First Lady, Print and Be Damned, The Expert, Resurrection (serial), Sherlock Holmes, Measure of Malice, The Elusive Pimpernel (serial), The Pallisers, Fall of the Eagles, Clayhanger, Trinity Tales, Witch of Pendle, When We Are Married, Just William, Backs to the Land.
PICTURES INCLUDE: Small Back Room, Seven Days to Noon, Appointment With Venus (Island Rescue), Happy Family, Cruel Sea, Long Arm, Man In the Sky, Seven Waves Away (Abandon Ship), The Challenge, Strangler's Web.

STRAUSBERG, SOLOMON M.: Exhibitor, attorney. b. N.Y.C. Feb. 16, 1907. e. U. of P., 1927, B.S. in Econ.; Columbia U. Law Sch., 1930, J.D.; St. John's U. Law Sch. 1931, L.L.M. U.S. Navy, Lt. Comdr. Supply Corps, 1942–45; chmn. of bd. & pres. Interboro Circuit Inc., New York City. Member: Variety Club, Cinema Lodge, M.P. Pioneers; pres. Metropolitan M.P. Theatres Assn., Inc., 1956; chmn. of bd., Metro. M.P. Theatres Assn., 1958–61.

STRAUSS, JOHN: Publicist. b. N.Y. Apr. 2, 1913; e. Yale U., B.A., 1935. Securities trader & analyst, Mabon & Co., N.Y. 1935–41; New Eng. factory rep., Star Rubber Co., 1941; city sales rep., W.B., Buffalo, 1941–45; unit pub., W.B., Burbank, 1945; asst. pub. dir., PRC., 1946; unit pub., Enterprise Prod., 1946–47; pub., Columbia Studios, 1947–48; Pres., McFadden, Strauss & Irwin, Inc., pub. rel. counsellors for personalities, m.p. TV & indust. accts., Los Angeles, New York, Palm Springs, London, Paris.

STRAUSS, PETER: Actor. b. Croton-on-Hudson, N.Y., 1947. e. Northwestern Univ. Spotted at N.U. by talent agent and sent to Hollywood. Debut in Hail Hero (1969). Has acted at Mark Taper Theatre in Dance Next Door, The Dirty Man.
PICTURES: Hail, Hero!, Soldier Blue, The Trail of the Catonsville Nine, The Last Tycoon.
TELEVISION: Man Without a Country, Attack on Terror, Rich Man, Poor Man, The Forgotten Kennedy, Masada.

STRAUSS, PETER E. Executive. b. Oct. 7, 1940. e. Oberlin College, London School of Economics, Columbia University School of Law (L.L.B., 1965). Vice pres., University Dormitory Development Co., 1965–68; v.p., Allart Cinema 16, 1968–69; v.p. Production, Allied Artists, 1969–70; v.p. operations and member of board, Allied Artists Pictures Corp., 1970; June 14, 1974 elected exec. v.p. Joined Rastor Films; left to become independent.

STRAUSS, RICHARD: b. Chicago, Ill.; e. Northwest U., Fordham U., Univ. of Illinois (also graduate work U. of Ill., U.S.C.). President of Bevelite-Adler Co. Past pres. TESMA; bd. secty. TEA, Calif. Sign Ass'n., Exhibit rules of Nat'l. Electric Sign Ass'n.; Variety Club, life member; bd. of dir., National Electric Sign Assn.

STREEP, MERYL: Actress. b. Bernardsville, N.J. April 22, 1949. Gained fame on N.Y. stage in such plays as Trelawny of the Wells, The Taming of the Shrew, The Cherry Orchard, Happy End, etc.
PICTURES: Julia, Manhattan, The Deer Hunter, The Seduction of Joe Tynan, Kramer vs. Kramer (AA, best supporting actress, 1979), The French Lieutenant's Woman, Stab.
TELEVISION: Holocaust.

STREISAND, BARBRA: Singer, actress. b. New York, April 24, 1942. e. Erasmus H.S., Brooklyn. Appeared in New York night clubs. On Broadway in I Can Get It For You Wholesale, debut 1962; Funny Girl.
PICTURES INCLUDE: (debut) Funny Girl, Hello Dolly, On A Clear Day You Can See Forever, The Owl and the Pussycat, What's Up Doc?, Up the Sandbox, The Way We Were, For Pete's Sake, Funny Lady, A Star is Born, (star, prod.), The Main Event (star, co-prod.), All Night Long.
TV: My Name is Barbra, Color Me Barbra, Bell of 14th Street, A Happening in Central Park.

STRICKLYN, RAY: Actor. b. Houston, Texas, October 8; e. University of Houston.
PLAYS INCLUDE: Broadway debut in Moss Hart's The Climate of Eden. Tour: Stalag 17. Off-B'way: The Grass Harp. Alley Theatre, Houston: Mourning Becomes Electra, Long Day's Journey Into Night, etc. Los Angeles: Starred in Compulsion and The Caretaker.
FILMS INCLUDE: The Proud and the Profane, Crime In the Streets, Somebody Up There Likes Me, Catered Affair, The Last Wagon, Return of Dracula, 10 North Frederick, The Remarkable Mr. Pennypacker, The Big Fisherman, The Plunderers, The Lost World, Young Jesse James, Track of Thunder, Arizona Raiders, Shuffle.
AWARDS: Broadway Theatre World Award; two Hollywood Foreign Press Golden Globe nominations (10 North Frederick and The Plunderers).

STROCK, HERBERT L.: Prod.-writer, dir., film editor. b. Boston, Mass., Jan. 13, 1918. e. Univ. of So. Calif., A.B., M.A. in Cinema; prof., of Cinema, Univ. of So. Calif., 1941. Started career, publicity leg man, Jimmy Fidler, Hollywood columnist; editorial dept., MGM, 1941–47; pres., IMPPRO, Inc., 1955–59; assoc. prod.-supv. film ed., U.A.; dir., AIP; dir., Phoenix Films. Pres., Herbert L. Strock Prods.
PICTURES INCLUDE: Storm Over Tibet, Magnetic Monster, Riders to the Stars, The Glass Wall, Gog, Battle Taxi, Donovan's Brain, Rider on a Dead Horse, Devil's Messenger, Brother on the Run, One Hour of Hell, Witches Brew, Blood of Dracula, I Was a Teenage Frankenstein, The Crawling Hand. Director; Brother On the Row, Monster. Writer-film editor, Hurray for Betty Boop (cartoon).
TV: Highway Patrol, Harbor Command, Men of Annapolis, I Led Three Lives, The Veil, Dragnet, 77 Sunset Strip, Maverick, Cheyenne, Bronco, Sugarfoot, Colt 45, Science Fiction Thea., Seahunt, Corliss Archer, Bonanza, Hallmark Hall of Fame, The Small Miracle, Hans Brinker, The Inventing of America (specials) for World Vision Intl.; What Will We Say to a Hungry World (5 hr. Telethon), They Search for Survival (special). NBC specials, Flipper (series). Documentaries: Atlantis, Legends, UFO Journals.

STROSS, RAYMOND: Producer, exhibitor. b. Leeds, England, May 22, 1916. e. Abington, Oxford. Ent. m.p. ind. 1933, worked at Sound City, other studios as exec. until 1938; then prod., exhibitor; formed Raymond Stross Theatres; re-entered production, 1950.
PICTURES INCLUDE: Hell is Sold Out, Tall Headlines, Man Who Watched the Trains Go By, Rough Shoot (Shoot First), Star of India, Jumping for Joy, As Long as They're Happy, An Alligator Named Daisy, A Touch of the Sun, The Flesh is Weak, A Question of Adultery, The Angry Hills, A Terrible Beauty, The Mark, Very Edge, The Leather Boys, Ninety Degrees in the Shade, The Fox, I Want What I Want, The Woman Who Rode Away.

STRUDWICK, SHEPPERD: Actor. b. Hillsboro, N.C., Sept. 22, 1907; e. U. of N.C. 1928, B.A. Stage experience since 1933; member Phi Beta Kappa Naval officer 1942–45.
STAGE: Both Your Houses, Let Freedom Ring, End of Summer, As You Like It, Christopher Blake, Affairs of State, Ladies of the Corridor.
PICTURES INCLUDE: (debut) Congo Maizie, 1940; Flight Command, Remember the Day, Edgar Allen Poe, Joan of Arc, Enchantment, The Red Pony, All the King's Men, A Place in the Sun, Eddie Duchin Story, Autumn Leaves, That Night, Sad Sack, Psychomania, The Daring Game, Cops and Robbers.

STRUTHERS, SALLY: Actress. b. Portland, OR, 1948.
PICTURES: The Phynx, Charlotte, Five Easy Pieces, The Getaway.
TELEVISION: Summer Bros. Smothers Show, Tim Conway Comedy Hour, All in the Family (series). Films: The Great Houdinis, Aloha Means Goodbye, Hey, I'm Alive.

STULBERG, GORDON: Executive. b. Toronto, Canada, Dec. 17, 1923; e. U. of Toronto, B.A., Cornell Law, LLB. Was assoc. & member, Pacht, Ross, Warne & Bernhard; ent. m.p. ind. as exec. asst. to v.p., Columbia Pictures Corp., 1956–60; v.p. & chief studio admin. off., 1960–67; pres. of Cinema Center Films (div. of CBS) 1967–71; pres. 20th Century-Fox, Sept. 1971–75; 1980, named pres. & chief operating officer, PolyGram Pictures. Member of N.Y. & Calif. bars.

STURDIVANT, B. VICTOR: Circuit executive. b. Gurdon, Ark., April 5, 1901; e. Crossett, Ark., Baylor U., Waco, Tex. Specialized journalism followed by newspaper work as reporter several Texas papers, news ed., Memphis (Tenn.)

News-Scimitar, city ee., Jacksonville, Fla. Journal, staff corresp. and then mgr. southern bureau (Atlanta, Ga.) International News Service. Three yrs. adv. & pub. counsel automobile ind. Became affiliated with Skouras-Publix Theat. in 1930. Joined Fox West Coast theatres August 1932, oper. various districts throughout So. Calif. until June, 1942, when apptd. div. mgr. No. Calif. with headquarters in San Francisco. Originator-organizer Los Angeles Theatre Defense Bureau Dec. 7, 1941; national campaign dir., M.P. Ind. Fourth War Loan; v.p. Calif. War Chest; pres., California Theatres Assn., Inc. Latin Amer. mgr., Nat'l Theats. Amusement Co.; org. Espectaculos Victor Sturdivant, Rio de Janerio, Brazil, dist. m.p. & TV films, manage talent Latin America Owner-mgr. Silver Crest Ent.; pres. Sturdivant, Inc., Yuma, Ariz.; pres. Yuma City Cham. Comm.; Yuma City Man of the Year, 1967; chmn. exec. comm., Citizens' Conference of Arizona Courts; mem. Gov. of Arizona Advis. Bd.; pres. NATO of Arizona.

STURGES, JOHN ELIOT: Director. b. Jan. 3, Oak Park, Ill. e. Marin Jr. Coll. asst. in blueprint dept., RKO-Radio Pictures, 1932; art dept; asst. film ed.; prod. asst., David O. Selznick; film ed.; Captain Air Corps, WW II; directed, edited, 45 documentaries, training films.
PICTURES INCLUDE: The Man Who Dared, Shadowed, Alias Mr. Twilight, For the Love of Rusty, Keeper of the Bees, Best Man Wins, Sign of the Ram, The Walking Hills, The Capture, Mystery Street, Right Cross, The Magnificent Yankee, Kind Lady, The People vs. O'Hara, The Girl in White, Jeopardy, Fast Company, Escape from Fort Bravo, Bad Day at Black Rock, Gunfight at the O.K. Corral, The Old Man and the Sea, Last Train from Gun Hill, Never So Few, The Magnificent Seven, By Love Possessed, Sergeants Three, A Girl Named Tamiko, The Great Escape, The Satan Bug, Hallelujah Trail, Hour of the Gun, Ice Station Zebra, Marooned, Joe Kidd, McQ, The Eagle Has Landed.

STURGIS, NORMAN: Director, actor, writer. b. Dallas, Tex. 1922; prod., Your Navy Sings, Evanston; TV dir. Space X, Theaterama, Plays Anthology, The Viewers, TV commercials, educational films and newsfilms.
TV: The Web, Gunsmoke, Twilight Zone, Alcoa Goodyear Theatre, Bat Masterson, The Untouchables.
PICTURES INCLUDE: Solid Gold Cadillac, Bernardine, Mardi Gras, Compulsion.

STYNE, JULE: Composer, producer, r.n. Jules Stein. b. London, Eng. To U.S. as a child; guest piano soloist with Chicago Symph. Orch. at 8; played with many dance bands; gen. mus. dir. Granada & Marbro Theat., Chicago; vocal coach, arranger, conductor & comp. for several m.p. studios; entertainment consult.
SONGS: I've Heard That Song Before, It's Magic, I'll Walk Alone, It's Been a Long, Long Time; Let It Snow, 3 Coins in the Fountain (Acad. Award in collab. Sammy Cahn, 1954), People.
STAGE: High Button Shoes, Gentlemen Prefer Blondes, Two on the Aisle, Hazel Flagg, Gypsy, Sugar, Peter Pan, Bells Are Ringing, Funny Girl.
PICTURES INCLUDE: Kid from Brooklyn, It Happened in Brooklyn, Romance on the High Seas, It's a Great Feeling, West Point Story, Meet Me After the Show, Living It Up, My Sister Eileen.

SUBOTSKY, MILTON: Writer, producer. b. New York, 1921. Early career studying engineering. Wrote, dir. and edited doc. & educational films. Ent. m.p. ind. 1938. Wrote and prod. live TV programmes 1941. Wrote & prod. TV film series Junior Science 1954. Wrote & co-prod. Rock, Rock, Rock; Jamboree, The Last Mile, Timegate, in USA. In England: City of the Dead, It's Trad Dad, Just for Fun, Dr. Terror's House of Horrors, Dr. Who and The Daleks, Daleks' Invasion Earth 2150 A.D., The Skull, The Psychopath, The Deadly Bees, The Terrornauts, They Came From Beyond Space, Torture Garden, Danger Route, The Birthday Party, Thank You All Very Much (Brit. title: A Touch of Love), Scream and Scream Again, The Mind of Mr. Soames, The House That Dripped Blood, I Monster, What Became of Jack and Jill, Tales From the Crypt, Asylum, Vault of Horror, And Now the Screaming Starts, Madhouse, From Beyond the Grave, The Beast Must Die, The Land That Time Forgot, At The Earth's Core, The Uncanny, Dominique, The Martian Chronicles, The Monster Club.

SUCHER, HENRY: Writer. b. New York City, Sept. 3, 1900; e. N.Y.U. With A.E.F., WW I. M. Pearl Krex, Reporter, N.Y. Globe, then mag. ed. (Nugents, Garment Weekly, The Brooklynite); publicity adv. copywriter; contrib. light satirical verses and humorous bits to the Sat. Eve. Post, Colliers, The New Yorker and other nat'l mags. Began screen career 1941 with collab. orig. s.p. The Miracle Kid. In 1942 wrote Soundies for Roosevelt-Coslow-Mills Corp. Wrote serials, Jack Holt, G. Man, and others for Larry Darmoor at Columbia Studios. Contract writer, Eagle-Lion Studios.
PICTURES INCLUDE: Captive Wild Woman, The Mummy's Tomb, Mug Town, The Mummy's Ghost (adapt. & collab. orig.), The Frozen Ghost, Jungle Woman, Your Fate Is In Your Hands, Lines of Destiny; short subjects released by RKO; now v.p. Edward Finney Prod., v.p. Boots and Saddles Pictures, Inc.

SUGAR, JOSEPH M.: Executive. b. New York City. e. N.Y.U. Ent. m.p. ind. 1938 in contract dept., Republic Pictures; ent. U.S. Army Air Forces, 1942 asst. contract mgr. PRC, then mgr.; when PRC merged with Eagle Lion continued as mgr. contract dept.; then exec. asst. to dist. v.p.; to U.A.N.Y. exchange, Jan. 1, 1953; v.p. sales, Magna Theatre Corp., 1959; member: Cinema Lodge, B'nai B'rith; road show mgr., 20th Century-Fox, 1962; v.p. chg. dom. distrib., 1963; exec. v.p. Warner-Seven Arts, 1967; pres. Cinerama Rel. Corp., 1968; pres. Gamma III Dist. Co., 1975; May, 1976: Joe Sugar, Inc., pres.; 1978, joined AIP as exec. v.p., worldwide sls. and pres. of AIP Distributing Co. which is now Filmways Pictures

SUGARMAN, HAROLD: Executive. b. New York City, Oct. 17, 1905; m.; e. New York Univ. Joined Paramount foreign sales dept., 1927. In charge foreign versions and adapt. at Paramount Astoria studio, Long Island City, 1928–31. Dist. mgr., UA Latin American distribution 1932–36. Prod. ind. pictures in Hollywood, 1936–40. Org. and head Universal Studio foreign dept., 1941–46. Transf. to Universal New York office, 1946 to org. and head 16 mm. overseas oper. Apptd. v.p. in chge. of export, United World Films, Inc., 1947; apptd. superv. So. Europe, Universal-International, hdqts., Rome, Italy, 1950; prod. & sales exec. Archer Television Prod., 1952. dist. & prod. exec. dir., of Chancellor Films.

SUHOSKY, BOB: Publicist. b. Philadelphia, Pa., Nov. 23, 1928. Marine Corps 1946–57; 20th Century Fox Television, publicist, 1959, publicity director 1964–70; pres. Bob Suhosky Assoc., public relations, 1971–77. Chairman, Suhosky & Hardiman Public Relations, 1977 to date. Wrote s.p. for Lone Star Girls (theatrical film) and Code R (TV).

SULLIVAN, BARRY: Actor. r.n. Patrick Barry. b. New York, N.Y., Aug. 29, 1912; e. N.Y.U., Temple U. Usher in theat.; buyer for dept. stores; N.Y. stage: The Man Who Came to Dinner, Brother Rat, Idiot's Delight, The Land is Bright, Caine Mutiny Court Martial, etc.
PICTURES INCLUDE: Woman of the Town, Lady in the Dark, Rainbow Island, Two Years Before the Mast, And Now Tomorrow, Duffy's Tavern, Three Guys Named Mike, Cause for Alarm, Grounds for Marriage, Life of Her Own, Nancy Goes to Rio, Inside Straight, Payment on Demand, Mr. Imperium, No Questions Asked, Unknown Man, Skirts Ahoy, Bad & the Beautiful, Jeopardy, Cry of the Hunted, China Venture, Loophole, Her 12 Men, Miami Story, Playgirl, Queen Bee, Texas Lady, Maverick Queen, Strategic Air Command, Purple Gang, Seven Ways from Sundown, Light in the Piazza, War Lords of Outer Space, Stage to Thunder Rock, Buckskin, Tell Them Willie Boy Is Here, Earthquake, The Human Factor, Oh, God!, Caravans.

SULLIVAN, J. CHRISTOPHER: Actor. b. Greenville, TX, Sept. 15, 1932. e. Prairie View A & M Univ. Stage debut in The Sign in Sidney Brustein's Window. Also appeared in Hatful of Rain, Anna Lucasta, Ceremonies in Dark Old Men. First black to teach at any predominantly white university in the south; is instructor of speech and communications at Univ. of Texas (Austin) since 1964.
PICTURES INCLUDE: Night Call Nurses, The Venetian Affair, The Lost Man, Body Heat.
TELEVISION: The White Shadow, Jeffersons, Good Times, One Day at a Time, General Hospital, Serpico, etc.

SULLIVAN, REV. PATRICK J., S.J. S.T.D.: Director, Office for Film & Broadcasting, U.S. Catholic Conference. b. New York, N.Y., March 25, 1920; e. Regis H.S.: Georgetown U., A.B., 1943; Woodstock Coll. M.A., 1944; Fordham U., 1945–47; S.T.L. Weston Coll., 1947–51; S.T.D. Gregorian U. (Rome), 1952–54; Prof. of Theology, Woodstock Coll., 1954–57; editor-in-chief, Film & Broadcasting Review; prod. TV and radio network religious programming. Consultor, Pontifical Commission for Social Communications; member executive committee, International Catholic Film Organization (O.C.I.C.); member, Board of Management, International Catholic Radio and Television Association (UNDA).

SULTAN, ROGER H.: Executive. b. Algiers, C.P.A.; e. U. of Algiers, LL.B. pub. adv. slsmn., 1934–35; thea. exploit., Algiers, 1936–39; French Army, 1939–40, 1943–45; bkr., slsmn., film distrib. co., N. Africa, 1941–42, 1945–46; auditor, office mgr., asst. supvsr. Atlantic Div., Warner Bros., Brazil, 1947–51; br. mgr., spvsr., No. Div., Allied Artists, Brazil, 1953–57; mgr., Allied Artists Chile, 1957–62; asst. gen. sls. mgr. (for.), Allied Artists Int., New York, 1962–64; v.p., gen. sls. mgr., 1965; pres., gen. sls. mgr., 1968–70; producer's rep.

SUMMERFIELD, ELEANOR: Actress. b. London, Eng., Mar. 7, 1921. Stage debut in Cornelius, 1939; screen debut in London Belongs to Me (in U.S.A., Dulcimer Street), 1947.
PICTURES INCLUDE: Take My Life, All Over the Town,

No Way Back, The Third Visitor, Scrooge, Laughter in Paradise, Mandy, Top Secret, Isn't Life Wonderful, Street Corner (Both Sides of the Law), Face the Music, Murder by Proxy, Lost, It's Great to be Young, Odongo, No Road Back, A Cry from the Streets, The Millionairess, Spare the Rod, Don't Bother to Knock, Petticoat Pirates, On the Fiddle, Act of Mercy, On the Beat, The Running Man, The Yellow Hat, Private Eye, Foreign Exchange, Some Will, Some Won't, Watcher in the Woods.

TV: (BBC, indep.) My Wife's Sister, The Two Charles (series), The Invasion, Dickie Henderson Shows, Night Train to Surbiton, Many A Slip, World of Wooster, David Copperfield, The Suede Jacket, Haunted, You're Lovely in Black, Misleading Cases, East of Eden, Tennis, Elbow, Foot Game, Murder at the Panto, Now Take My Wife, Madly in Love, Kate, Simon Fenton's Story, Kept on a String, Password, Husband of the Year, The Overnight Bag, The Rather Reassuring Show, Murder at the Wedding, Pig in the Middle.

SUNDBERG, CLINTON: Actor, b. Appleton, Minn., Dec. 7; e. Hamline U. Teacher of English literature; in New England stock cos.; on London and N.Y. stage.

STAGE: Silent Witness, She Loves Me Not, Boy Meets Girl, Room Service, Arsenic and Old Lace, Over 21, Diary of Anne Frank, Mary, Mary.

PICTURES INCLUDE: Undercurrent, Undercover Maisie, Living in a Big Way, Song of Love, Hucksters, Kissing Bandit, Easter Parade, Good Sam, Date With Judy, Words and Music, Command Decision, In the Good Old Summertime, Key to City, Annie Get Your Gun, Toast of New Orleans, Fat Man, As Young As You Feel, Belle of New York, Girl Next Door, Main Street to Broadway, The Caddy, Birds and the Bees, Wonderful World of Brothers Grimm, How the West Was Won.

SUNSHINE, MORTON: International Executive Director, Variety Clubs International, 1975 to present. b. Brooklyn, N.Y., Sept. 20, 1915. B.S. in S.S. (1935), LL.B. (1938), J.S.D. (1939); admitted to N.Y. Bar 1939. Federal Bar 1941. Practicing attorney, 1941–44; Federal investigator with U.S.C.S.C. 1944–45; business mgr., Independent Theatre Owners Assn. 1945–46; exec. dir. since 1946; member COMPO tax & legis. com., 1950; sp. rep. Org. of M.P. Ind. of City of N.Y., 1952–55; exec. coord. Tony Awards dinner Amer. Theatre Wing, 1958–60, Sophie Tucker Golden Jubilee; ind. trib. to Jimmy Durante; Will Rogers; A. Montague; Herman Robbins; S. H. Fabian; Rodgers-Hammerstein; Eric Johnston; Diamond Jubilee, Stagehands, 1962. Danny Thomas; indus. pub. rel. consultant; exec. coord. IFIDA Int'l Film Awards Dinners; pres. cabinet, Nat'l Assn. of Theatre Owners, Exec. dir., Variety Club of N.Y., 1970; Consultant to American Film Institute, 1973–75; executive director, Motion Picture Pioneers, 1970–75; Editor-Publisher, Independent Film Journal, 1947–75.

SURTEES, BRUCE: Cinematographer. Son of Robert L. Surtees.

PICTURES INCLUDE: Play Misty for Me, Dirty Harry, The Great Northfield Minnesota Raid, The Beguiled, High Plains Drifter, Blume in Love, Joe Kidd, Lennie (A.A. nomination), Leadbelly, Sparkle, Big Wednesday, Escape from Alcatraz.

SURTEES, ROBERT L.: Cinematographer. b. Covington, Ky., Aug. 9, 1906; e. Withrow H.S., Cincinnati, Ohio, 1925. Ent. m.p. ind. as asst. cameraman, Universal Pictures Corp., 1927–29; Universal, Berlin, 1929–30; ret'd to U.S.; since with First Nat'l, Warner Bros., Pathe, MGM; won Look Mag. Award for Film Achievement, 1950; member: ASC. 13 A.A. nominations; winner 3 times.

PICTURES INCLUDE: Our Vines Have Tender Grapes, Thirty Seconds Over Tokyo, Intruder in the Dust. Acad. Award, best color photog., King Solomon's Mines, 1950; best photog. The Bad and the Beautiful, 1952; Merry Widow, Mogambo, Trial, Oklahoma, The Swan, Raintree County, Merry Andrew, Ben Hur (AA), Mutiny on the Bounty, The Hallelujah Trail, The Collector, The Satan Bug, Lost Command, Doctor Dolittle, Sweet Charity, Summer of '42, The Cowboys, The Other, The Sting, The Great Waldo Pepper, The Hindenburg, A Star is Born, The Turning Point, Same Time Next Year, Bloodbrothers.

SUSSAN, HERBERT: Producer. b. New York, N.Y., Feb. 24, 1921. e. N.Y.U. U. of So. Calif. Radio writer, prod. 1937–42; series included American Rhapsody, Theme and Variations; m.p. prod., Army Air Forces, 1942; supervised prod., dir. films in Far East on effects of atomic bombs in Japan, life and culture, 1945–46; dir. film coverage, Louis de Rochemont, 1947; author of 3 photo books on Japan; assoc. dir., CBS TV 1948, worked on Suspense, Mama, Studio One, others. prod. dir., 1949: dir. 54th St. Revue, Ken Murray Show, Eddie Fisher Show, Christmas with the Stars; prod., dir. Guy Lombardo Show; sen, prod. Wide Wide World, 1955–58; natl. dir. sp. prog. NBC-TV, 1958–60; pres. Herbert Sussman Ent.

SUSSKIND, DAVID: Producer—TV Talk Show Host. b. New York City, 1920. Producer of TV shows, founder of production co., Talent Associates; host of own syndicated interview program (The David Susskind Show; formerly called Open End.) Notable for producing TV adaptations of famous plays: Glass Menagerie, The Crucible, Death of a Salesman, etc. In 1970 Norton Simon, Inc. acquired his co., renaming it Talent Associates—Norton Simon for theatrical as well as TV film production. Sold in 1977 to Time-Life Films.

SUTHERLAND, DONALD: Actor b. St. John, New Brunswick, Canada, July 17. m. Shirley Douglas, actress. e. U. of Toronto. At 14 became a radio announcer and disc jockey. Worked in a mine in Finland. Theatre includes: The Male Animal (debut), The Tempest (Hart House Theatre, U. of Toronto), Two years at London Acad. of Music and Dramatic Art. Spent a year and a half with the Perth Repertory Theatre in Scotland, the repertory at Nottingham, Chesterfield, Bromley and Sheffield. London stage debut: August for the People.

STAGE: On a Clear Day You Can See Canterbury, The Shewing Up of Blanco Posnet, The Spoon River Anthology.

PICTURES INCLUDE: The Castle of the Living Dead, Dr. Terror's House of Horrors, The Dirty Dozen, Oedipus the King, Interlude, Joanna. The Split, Start the Revolution Without Me, The Act of the Heart, M⅛A⅛S⅛H⅛, Kelly's Heroes, Little Murders, Alex in Wonderland, Klute, F.T.A., Steelyard Blues, Lady Ice, Don't Look Now, S⅛P⅛Y⅛S⅛, The Day of the Locust, Casanova, The Eagle Has Landed, 1900, National Lampoon's Animal House, Invasion of the Body Snatchers, Murder by Decree, The Great Train Robbery, Nothing Personal, Ordinary People, Gas, Eye of the Needle.

TV: (British) Gore Vidal's Marching to the Sea, Albee's The Death of Bessie Smith, Hamlet at Elsinore, The Saint, The Avengers, Gideon's Way, The Champions.

SUTTON, JAMES T.: Executive. b. Calif., Sept. 13 e. Columbia U.; film inspector, U.S. Government; overseas m.p. service, WW II; co-owner, gen. mgr., Hal Davis Studios; hd. TV commercial div., Allan Sandler Films; Academy Art Pictures; pres., chmn. of bd., exec. prod., Royal Russian Studios, Inc., western hemisphere div.; pres. exec. prod. Gold Lion Prods., Inc.; pres. exec. prod. James T. Sutton-John L. Carpenter Prods.; pres., exec. dir., Airax Corp.; pres. of Skyax (div. of Airax).

SUTTON, JOHN: Actor. b. Rawalpindi, India, Oct. 22, 1908; e. Wellington, Coll.; Sandhurst (England). Spent 8 yrs. hunting in Africa; became veldt rancher; managed tea plantation; lived in China, Malaya, Philippines; studied voice in Los Angeles; became tech. adv. on films with British background; then actor. On screen 1937.

PICTURES INCLUDE: Jane Eyre, The Hour Before the Dawn, Claudia and David, Captain from Castile, Three Musketeers, Bagdad, Second Woman, Second Face, Payment on Demand, David and Bathsheba, Thief of Damascus, Captain Pirate, Golden Hawk, My Cousin Rachel, Sangaree, East of Sumatra, The Bat.

SUTTON, REGINALD: Sound technician. b. London, Oct. 24, 1916. e. Acton County School, Kent College, Canterbury, Regent Street Polytechnic. Early career with BBC engineering div.; ent. m.p. ind. with British Movietonews; following news-reel work recorded whole Royal Tour of Australasia 1953–54 for Flight of the White Heron in CinemaScope; other CinemaScope recordings include Supersonic Age, Birthday Parade, Pastimes and Pleasures; TV features, commercials. Chief of sound, Samuelson Film Service Ltd., 1960. Location facilities and Westrex Magnetic Transfer suite, Formed own company, Soundways Ltd., recording sound for m.p., 1960.

SWALLOW, NORMAN: Producer. b. Manchester, Eng., Feb. 17, 1921. e. Manchester Grammar Sch., Keble Coll. (Oxford U.). British Army 1941–46; BBC as writer-prod. of docum., 1946; wrote 3 docum. films, 1948; TV as docum. prod., 1950; prods. include American Looks at Britain, with Howard K. Smith for CBS, Wilfred Pickles at Home series; orig. Speaking Personally series with appearance of people like Bertrand Russell; co-prod. TV coverage of Britain's general election, 1951; ed. prod. BBC monthly prog. Special Inquiry, 1952–56; World is Ours, 1954–56; study tour Middle East, India, Pakistan, Ceylon, 1956–57; writer, prod. Line of Defense, I Was a Stranger; asst. head films for BBC, 1957; writer-prod., On Target, 1959; apptd. chief asst. (doc. & gen.), BBC-TV, 1960; asst. editor, Panorama, BBC-TV, 1961. Joined Denis Mitchell films, May 1963; writer, prod., Pomp and Pageantry, The Right to Health, A Wedding on Saturday, 1964; The End of a Street, 1964; exec. prod., Report from Britain, 1964; writer, prod., Youth, British, Football; co-prod. This England. prod. A Railwayman for Me, 1966; co-prod. Ten Days That Shook the World, 1967. Co-prod. The Long Bridge, 1968. prod., dir., The Three Happiest Years; exec. prod. Omnibus series, 1968–70. Writer, prod., dir. To Leningrad With Love; exec. prod. Omnibus Series, 1968–72. Writer, co-prod. Eisenstein, 1970. BBC-TV Head of Arts Features 1972–4.

Prod-dir. series A Lasting Joy 1972. Exec. Prod. Granada TV since 1974 of The Christians, This England, Clouds of Glory. Winner Desmond Davis Award (UK) 1977.

SWANN, FRANCIS: Writer. b. Annapolis, Md., July 17, 1913; e. Princeton U.; Johns Hopkins U. 1935. Stage experience, 8 yrs., actor, musician, dir., wrote plays: Out of the Frying Pan, Family Man, Bad Ange , Follies of 1910, Paradise Island, Dear God, Send Us a Producer, Country Bumpkins; fiction writer. Novels pub.: The Brass Key, Greenwood, Royal Street, You'll Hang, My Love, Hermit's Island, Hellgate Plantation, Hacienda Triste, House of Terror; nat'l pub.; U.S. Navy, 1942–46. Member: Dramatists Guild, Writers Guild of America, Authors League of America.
PICTURES INCLUDE: Shine on Harvest Moon, The Gay Intruders, Time Place & the Girl, Love & Learn, Jungle Patrol, Tarzan's Peril, 711 Ocean Drive, Barefoot Mailman, One Big Affair, Force of Impulse, Instant Love, Candy Man, A Very Rich Man, Cover Up, Make Your Own Bed.

SWANSON, GLORIA: Actress. b. Chicago, Ill., Mar. 27, 1899; e. schools in Key West, Puerto Rico, Chicago Art Inst. Screen debut at old Essanay Studios (Elvira Farina, The Meal Ticket); then to Keystone (The Nick of Time Baby, Teddy at the Throttle, Haystacks and Steeples, etc.); next with Sennett Comedies (Pullman Bride) 1917; Triangle (Her Decision, You Can't Believe Everything, Wife or Country, Secret Code, etc.).
PICTURES INCLUDE: (under Cecil B. DeMille) Don't Change Your Husband, For Better For Worse, Male and Female, Why Change Your Wife, The Affairs of Anatol; later in Paramount films The Great Moment, The Gilded Cage, My American Wife, Prodigal Daughter, Bluebeard's Eighth Wife, others; among Alan Dawn prods. The Humming Bird, Society Scandal, Manhandled, Wages of Virtue, Madame Sans Gene, Stage Struck. In 1926 formed own corp. to prod. pictures for release through United Artists; first film The Lovers of Sonya; later Sadie Thompson, The Trespasser, What a Widow, Indiscreet, Tonight or Never. In 1935 Music in the Air (Fox); in 1941 Father Takes a Wife (RKO), 1949, Sunset Boulevard; Three for Bedroom C, When Comedy Was King, Airport-1975.
THEATRE: New York stage: Twentieth Century, 1951; Nina, 1952; Butterflies Are Free, 1970–73.
TV: Killer Bees (movie of the week), 1974.

SWANSON, ROBERT E.: Musician. b. New York, Dec. 1, 1920. e. N.Y.U. Arranging, performing in bands 10 yrs.; singing commercial for Piel's Beer, 1948; since creating, prod. musical com. radio, TV. pres. Robert Swanson Prod. Inc.

SWAYZE, JOHN CAMERON: Reporter, commercial spokesman. b. Wichita, Kansas, Apr. 4, 1906; e. Univ. of Kansas; Anderson-Milton Dramatic Sch. Reporter and ed., Kansas City, Mo. Journal-Post; news dept. KMBC, Kansas City; head of news, NBC western network, Hollywood; NBC radio news reports, N.Y.C.; covered political conv. TV 1948–52; began News Caravan on NBC in 1949–56; panel mem. Who Said That, 1949–51; Watch the World on NBC-TV, 1948–50; Sightseeing With The Swayzes (with family) NBC-TV, 1953; news program, ABC, 1957. Host: Circle Theatre, NBC, and panel member, To Tell the Truth, CBS. Voted best news commentator, M.P. Daily TV Poll, 1951–55.
MEMBER: Lambs and Players clubs; National Press Club, Greenwich, Conn., CC.

SWEDROE, JEROME D.: Executive. b. N.Y.C., September 7, 1925. e. NYU, B.S. Entered industry 1949 bookkeeping dept. Skouras Theatres Corp. Advanced through dept. to Comptroller in March 1962 and v.p. June 1964. Appointed exec. v.p. in charge operations of the chain, name had been changed to United Artists Eastern Theatre Circuit, Inc., in April, 1967; promoted v.p. parent company, UATC, January 1968. Resigned to join Cinecom March 3, 1969. Elected director Aug. 1969; joined Redstone Management Co., June, 1971; named v.p., operations, U.S. Cinema Corp., December, 1971, consultant October, 1973.

SWENSON, S.A.G.: Producer, distributor. b. Stockholm, Sweden. e. Stockholm, London & N.Y. Entered m.p. industry in N.Y., 1921; Swedish mgr. Warner, 1923–35, pres. Wirefilm, Swedish prod. & distrib. Co., since 1935; prod. 50 feature films; Swedish rp. Korda London Films 1947–60; Disney 1954–56. Member: Various trade ass'ns Sweden, France, U.S.A. Now operates TV & theatrical import & export in Sweden & New York: Scandinavia—Wire—Films.

SWERLING, JO: Writer. b. Russia, Apr. 8, 1897. Newspaper & mag. writer; author vaude. sketches; co-author plays, The Kibitzer, Guys and Dolls.
PICTURES INCLUDE: s.p., The Kibitzer, Guys and Dolls; s.p., Platinum Blonde, Washington Merry-Go-Round, Dirigible, Man's Castle; collab. s.p., Whole Town's Talking; s.p., No Greater Glory, Pennies from Heaven, Double Wedding, Made for Each Other; collab. s.p., The Westerner; s.p., Confirm or Deny, Blood and Sand; collab. s.p., Pride of the Yankees; story, Lady Takes a Chance; s.p., Crash Dive, Lifeboat, Leave Her to Heaven, Thunder in the East.
TV: collab. The Lord Don't Play Favorites, NBC.

SWIFT, ALLEN: Performer. b. New York City, Jan. 16, 1924; Actor vaudeville, legitimate stage, radio and TV shows, including Bob Hope, Eddie Cantor, Jackie Gleason shows; voices for Mighty Mouse and UPA features and commercials; now announcer-master of ceremonies.

SWIFT, DAVID: Producer, director, writer. b. Minneapolis, 1919. Served with 8th Air Force in England, WW II. Entered m.p. ind. in Walt Disney animation dept. After service, comedy writer for radio. Later starting in 1949, TV drama writer for Philco Playhouse, Studio One, Kraft Theatre, Omnibus. Created Mr. Peepers, Jamie. Writer-dir. Playhouse 90, Rifleman, Wagon Train, Climax, others. First feature film Pollyana (writer-dir.).
PICTURES INCLUDE: The Parent Trap, The Interns, Love Is a Ball, Under the Yum Yum Tree; pr.-dir.-writer, Good Neighbor Sam, How to Succeed in Business Without Really Trying.

SWIFT, LELA: Director.
TV: Studio One, Suspense, The Web, Justice, Dupont Show of the Week, Purex Specials For Women, (Emmy Award) Dark Shadows, Norman Corwin Presents, ABC Late Night 90 min. Specials, ABC Daytime 90 min. Play Break. Won three Emmy awards for best director of day-time serial: Ryan's Hope (1977, 1979, 1980).

SWIMMER, SAUL: Producer-director. b. Uniontown, Pa. e. Carnegie Tech Drama Dept. Producer-director of Melephant. Winner, Gold Leaf Venice Award, 1959. Director: Force of Impulse. Exec. prod. The Wounds of Hunger; prod. dir. Around the World of Mike Todd (Embassy Pictures TV Special, 1967), 1965–67; TV series development for Screen Gems. Advisor: CBS-TV Children's Film Festival (1967). In Charge of Production: Mrs. Brown, You've Got a Lovely Daughter, Only Lovers Left Alive; Stranger in the Night.
PICTURES INCLUDE: Cometogether (co-prod., dir.), The Concert for Bangladesh (dir.), Blindman (prod.).

SWINK, ROBERT E.: Film editor, Director. b. Rocky Ford, Colo., June 3, 1918; married. Joined editorial dept., RKO Radio, 1936; apptd. film ed., 1941. In U.S. Army Signal Corps, 1944–45; supv. editor, Fox studio. Edited numerous productions.
PICTURES INCLUDE: Detective Story, Carrie, Roman Holiday, Desperate House, Friendly Persuasion, The Big Country, The Diary of Anne Frank, The Young Doctors, The Children's Hour, The Best Man, The Collector, How To Steal A Million, Flim Flam Man, Funny Girl, The Liberation of L. B. Jones, The Cowboys, Skyjacked, Lady Ice, Papillion, Three the Hard Way, Rooster Cogburn, Midway, Islands in the Stream, Gray Lady Down, The Boys From Brazil, The In-Laws, Going in Style, The Sphinx.

SWIT, LORETTA: Actress. b. Passaic, NJ, Nov. 4, 1937. Stage debut in Any Wednesday. Toured in Mame for year. Arrived in Hollywood in 1969 and began TV career. Theatrical film debut in Stand Up and Be Counted, 1972.
PICTURES: Freebie and the Bean, Race with the Devil, "S.O.B."
TELEVISION: MASH, Gunsmoke, Mannix, Hawaii Five-0, Mission: Impossible, The Doctors, Cade's County. Films: Shirts/Skin, Coffeeville, Valentine, Mirrors, Mirror, The Walls Came Tumbling Down.

SWOPE, HERBERT BAYARD, JR.: Director, producer, commentator. b. N.Y. e. Horace Mann Sch., Princeton Univ. U.S. Navy, 1941–46; rejoined CBS-TV as remote unit dir., 1946 directing many "firsts" in sportscasting; winner, Variety Show Management Award for sports coverage & citation by Amer. TV Society, 1948; joined NBC as dir., 1949; prod. dir., 1951; Lights Out, The Clock, The Black Robe, dir., Robt. Montgomery Presents; winner, Sylvania, 1952. TV Award Outstanding Achievement in Directorial Technique; became executive producer, NBC-TV in charge of Wide, Wide World; directed Arsenic & Old Lace on Best of Broadway, CBS-TV, 1955; dir. Climax, CBS-TV, Film prod., 20th Century-Fox, Hilda Crane, Three Brave Men, True Story of Jesse James, The Bravados, The Fiend who Walked the West; 1960–62; exec. prod. 20th-Fox TV; Many Loves of Dobie Gillis, Five Fingers; dir. co-prod. on Broadway, Step On A Crack, Fragile Fox, Fair Game for Lovers. 1970–72 exec. at NYC Off-Track Betting Corp. 1973–74; vice-president, Walter Reade Organization, Inc.; 1975–76 producer-host, This Was TV, Growth of a Giant; 1976–77 commentator, Swope's Scope (radio); Critic (TV); column Swope's Scope, Palm Beach (FLA) pictorial.

SYKES, ERIC: Scriptwriter, comedian. b. Oldham, England, 1924. Early career actor; ent. TV industry, 1948; wrote first three series, BBC's Educating Archie and radio, TV comedy series for Frankie Howard, Max Bygraves, Harry Secombe. BBC panel show member. Longterm contract ATV, 1956; own BBC series, 1958–65, Sykes Versus TV, The Frankie Howard Show; Sykes and a Big, Big Show, 1971,

1978 19th Year BBC-TV series. Toured extensively in Big Bad House. 1977 Summer Show Sykes. 1978 Tour of Sykes to Rhodesia, Australia, Canada, The Plank, Rhubarb.
PICTURES INCLUDE: Watch Your Stern, Invasion Quartet, Village of Daughters, Kill or Cure, Heavens Above, The Bargee, One Way Pendulum, Those Magnificent Men in Their Flying Machines, Rotten to the Core, The Liquidator, Spy With The Cold Nose. 1967, Dir. s.p. The Plank, Shalako, The Monte Carlo Rally, Theatre of Blood.

SYLBERT, ANTHEA: Executive. b. New York. e. Barnard (B.A.); Parsons School of Design (M.A.). Early career in costume design with range of Bdwy., off-Bdwy. and m.p. credits (Rosemary's Baby, F.I.S.T., Shampoo, The Fortune, A New Leaf, The Heartbreak Kid. Two A.A. nominations for creative costume designs for Julia and Chinatown. Joined Warner Bros. in October, 1977, as v.p., special projects, acting as liaison between creative executives, production dept., and creative talent producing films for company. In October, 1978, named v.p., production (projects included One Trick Pony, Personal Best, etc.). In March, 1980 appointed v.p.—production, for United Artists, working on Stab, Jinx, Swing Shift.

SYMS, SYLVIA: Actress. b. London, 1934. e. Convent and Grammar Schools. Film debut, 1955, My Teenage Daughter.
PICTURES INCLUDE: No Time For Tears, Birthday Present, Woman In A Dressing Gown, Ice Cold in Alex, The Devil's Disciple, Moonraker, Bachelor of Hearts, No Trees in the Street, Ferry to Hong Kong, Expresso Bongo, Conspiracy of Hearts, The World of Suzie Wong, Flame in the Streets, Victim, Quare Fellow, Punch & Judy Man, The World Ten Times Over, East of Sudan, The Eliminator, Operation Crossbow, The Big Job, Hostile Witness, The Marauders, The White Cold, Danger Route, Run Wild, Run Free, The Desperados, The Tamarind Seed, Give Us This Day, There Goes the Bride.
TV: The Human Jungle (series), Something to Declare, The Saint (series), The Baron (series), Bat Out of Hell, Department in Terror, Friends and Romans, Strange Report, Half-hour Story, The Root of All Evil, The Bridesmaid, Clutterbuck, Movie Quiz, My Good Woman, Looks Familiar, Love and Marriage, The Truth About Verity, I'm Bob, He's Dickie, Blankety Blank.

T

TAFFNER, DONALD L.: Executive. b. New York City. e. St. Johns University. William Morris Agency, 1950–59; Paramount Pictures. 1959–63; D. L. Taffner Ltd., 1963-present.

TALBOT, JOSEPH T. (Brud): Producer-Director-Actor. b. NYC, Sept. 16, 1938. e. Deerfield Academy, University of Virginia, Welsh Drama Workshop. Co-producer, actor: Force of Impulse, 1961; Without Each Other, 1962; Wounds of Hunger, 1963; co-star, Finger on the Trigger, 1964; Production supervisor TV special "Around the World of Mike Todd." 1965. Producer, Soundblast '66, musical spectacular Yankee Stadium, 1966. Producer-director TV commercials & documentaries, Filmex, Inc., 1966–69. Producer-director TV Documentary, The Singlehanders. President Dana Films Ltd., 1969–72; prod. Run Before the Wind, (Official m.p. rep. to Bicentennial Council of the Thirteen Original States), 1972–73 Executive in charge of Production Cinemation Industries; 1976–78 pres., Truesdale Pictures Ltd.
OTHER PICTURES: The Case of the Smiling Stiff (prod., dir.); The Black Pearl (co.-prod.); Willie Sutton, Bank Robber (Prod.)

TALBOT, LYLE: Actor. r.n. Lysle Hollywood. b. Pittsburgh, Pa., Feb. 8, 1904. In Army Air Corps, World War II. First screen appearance in Vitaphone short; then in Love Is a Racket, 1932.
PICTURES INCLUDE: Up in Arms, Sensations of 1945, One Body Too Many, Dixie Jamboree, Gambler's Choice, Strange Impersonation, Vicious Circle, Mutineers, Sky Dragon, The Jackpot, Sea Tiger, Down Among the Sheltering Palms, Star of Texas, Capt. Kidd & the Slave Girl, Tobor the Great, Steel Cage, There's No Business Like Show Business, Jail Busters, Sudden Danger.
MEMBER: Masonic Lodge (Shriner), Lambs, Masquers, American Legion.

TAMBLYN, RUSS: Actor. b. Los Angeles, Calif., Dec. 30. e. No. Hollywood H.S. West Coast radio shows; on stage with little theatre group; song-and-dance act in Los Angeles clubs, veterans hospitals.
PICTURES INCLUDE: Boy with Green Hair, Reign of Terror, Samson and Delilah, Deadly Is the Female, Kid from Cleveland, Captain Carey, U.S.A., Father of the Bride, As Young As You Feel, Father's Little Dividend, Winning Team, Retreat Hell, Take the High Ground, Seven Brides for Seven Brothers, Many Rivers to Cross, Hit the Deck, Last Hunt, Fastest Gun Alive, Young Guns, Don't Go Near the Water, Peyton Place, High School Confidential, Tom Thumb, Cimarron, West Side Story, Wonderful World of the Brothers Grimm, The Haunting, Long Ships, Son of Gunfighter, The Last Movie.

TANDY, JESSICA: Actress. b. London, Eng., June 7, 1909; wife of Hume Cronyn. On London, N.Y. stage, 1928–42.
THEATRE: N.Y. stage. Streetcar Named Desire, Hilda Crane, The Four Poster, Coward In Two Keys; The Way of the World, Eve and A Midsummer Night's Dream at Stratford Festival 1976, Canada; limited tours of Many Faces of Love 1974, 75, 76 and for CBC, Canada 1977; performed in The Gin Game Long Wharf Thea., New Haven, Conn. June 1977; Long Day's Journey Into Night, Theatre London, London, Canada 1977. Appeared in The Gin Game, Pulitzer Prize Winning Play at The Golden Thea. N.Y.C. 1977–78, and toured with it in U.S., Toronto, London and U.S.S.R., 1979.
PICTURES INCLUDE: The Green Years, Forever Amber, A Woman's Vengeance, September Affair, The Desert Fox, Adventures of a Young Man, Butley, A Light In The Forest, The Birds, Honky Tonk Freeway, Stab, Garp.

TANEN, NED: Executive. b. Los Angeles. e. UCLA (law degree). Joined MCA, Inc. 1954; Appd. v.p. in 1968. Brought Uni Records, since absorbed by MCA Records, to best-seller status with such artists as Neil Diamond, Elton John, Olivia Newton-John. First became active in theatrical film prod. in 1972. In 1975 began overseeing feature prod. for Universal. In 1976 named pres. of Universal Theatrical Motion Pictures, established as div. of Universal City Studios to encompass all theatrical film prod. activities.

TANKERSLEY, ROBERT K.: Executive. b. Decatur, Ill., July 31, 1927. In U.S. Navy, 1945–46; Marine Corps, 1949–55. With Natl. Theatre Supply as salesman in Denver 13 yrs. 1959 to present, pres. Western Service & Supply, Denver, theatre equip. form. Also is v.p. theatre chain, Theatre Operators, Inc., Bozeman, Mont. Member: Theatre Equipment Assn. (past pres.), National NATO (presently dir.), Rocky Mt. Motion Picture Assn. (past pres.), SMPTE, Motion Picture Pioneers.

TANNER, WINSTON R.: Exhibitor. b. Appomattox, Va., Feb. 10, 1905. e. U. of Richmond. Entered m.p. ind., 1939 as asst. to owner, mgr. Free State Victoria Theatres, Kenbridge, Va.; dir. partner Kendig-Tanner Theatres, 1942; bought partner's interest in 1957; operator Tanner Theatres 7 houses, in Va.; operates Winston R. Tanner booking-buying service.

TANTO, GYULA PAL: Director, journalist, publicist. b. Budapest, Hungary, 1927. Ent. m.p. ind. 1947 as press officer MGM, Budapest; film critic, Nepsport (Hungarian daily newspaper), 1949–56; worked for Cinevision Film Makers Inc. London as pub. dir. several films, 1960; sports tech. adviser Olympic films, Rome, 1960; Tokyo, 1964; Mexico City, 1968; editor of Balkan Press, film section since 1969.
DIRECTOR: Football World Cup, Olympia Games (1972) Munich (1976) Montreal Miss World, 1971–74 1975–76 (TV), World Swimming Championship, Belgrade; 1st Gymnastic World Cup, Sport in Scotland.

TAPS, JONIE: Producer. Executive. Columbia Studio.
PICTURES INCLUDE: Jolson Story, Down to Earth, Thrill of Brazil. Produced: When You're Smiling, Sunny Side of Street, Sound Off, Rainbow Round My Shoulder, All Ashore, Cruisin' Down the River, Drive a Crooked Road, Three for the Show, Bring Your Smile Along, He Laughed Last, Shadow on the Window.
MEMBER: Friars Club, Hillcrest Country Club.

TARADASH, DANIEL: Writer. Director. b. Louisville, Ky., Jan. 29, 1913. e. Harvard Coll., BA, 1933; Harvard Law Sch., LL.B., 1936. Passed N.Y. Bar, 1937; won nationwide playwriting contest, 1938; U.S. Army W.W. II.
PICTURES INCLUDE: collab. s.p. Golden Boy, A Little Bit of Heaven, Knock on Any Door; s.p., Rancho Notorious, Don't Bother to Knock, From Here to Eternity (Acad. Award 1953); Desiree; collab. story, s.p. and dir., Storm Center; s.p., Picnic, Bell Book and Candle. The Saboteur Code Name—Morituri; col. s.p. Hawaii; collab. s.p. Castle Keep, s.p. Doctors' Wives; Collab. s.p. The Other Side of Midnight. Pres. Screen Writers Branch, WGA, 1955–56; v.p., Writers Guild of America, West 1956–59; mem. Writers Guild Council, 1954–65; mem., bd. of govnrs. M.P. Acad. of Arts & Sciences, 1964–74, and v.p. 1968–70; chmn., trustee, Producers-Writers Guild Pension plan 1960–73. Mem. Bd. of Trustees of American Film Institute 1967–69. WGA's Valentine Davies Award, 1971. Pres., Academy M.P. Arts & Sciences, 1970–73, member of Board of Trustees, Entertainment Hall of Fame Foundation. Member, Public Media General Programs panel for the National Foundation for the Arts, Pres. Writers Guild of America, West, 1977–1979. Nat'l. chm., Writers Guild of America, 1979–81.

TARNOFF, JOHN B.: Executive. b. New York City, Mar. 3, 1952. e. UCLA (motion pictures & TV, 1973–74); Amherst College (BA, 1969–73). Named field exec. with Taylor-Laughlin Distribution (company arm of Billy Jack Enterprises) 1974; left in 1975 to be literary agent with Bart/Levy, Inc.; later with Michael Levy & Associates, Paul Kohner/Michael Levy Agency; Headed TV dept., Kohner/Levy, 1979. Now v.p., motion picture development, for MGM.

TASCO, RAI: Actor, announcer. r n. Ridgeway Tasco. b. Boston, Mass., Aug. 12, 1917. e. Boston English High. 1935. U.S. Army, 1935–45. Grad., Cambridge School of Radio & TV, New York, 1950. Appeared in most TV and radio dramatic shows, New York and Hollywood; Broadway & films; dramatic instructor.

TATUM, DONN B.: Executive. b. Los Angeles, January 9, 1913. e. Stanford, Oxford. Chairman of Board, Walt Disney Prods. 1943, lawyer, RCA and ABC. 1949, v.p., counsel, Dir. of Don Lee Companies. Gen. Mgr., KABC, dir. of Television, Western Division ABC. 1956, prod. business mgr. for Walt Disney Productions; exec. v.p. Disneyland, v.p. TV sales; v.p. and adm. asst. to pres. and exec. comm.; exec. v.p.; president; now Chrmn. of Ed. Member of Bd. of Dir. of Bank of America, NT&SA; a director of Greyhound Corp, Union Oil Co. of Calif.; director of Kings County Development Co.; director and pres. John Tracy Clinic; director of Community Television of So. Calif. (Ch. 28); dir. & v.p. Community Building Funds of So. Calif.; trustee of Calif. Institute of the Arts, the Salk Institute & the Disney Foundation.

TAVERNIER, BERTRAND: Director-Writer. b. Lyon, April 25, 1941. After 2 yrs. of law study, quit to become film critic. Partner for 6 yrs. with Pierre Rissient in film promotion company, during which time he studied all aspects of filmmaking. Debut picture: The Watchmaker of St. Paul in 1974.
PICTURES INCLUDE: Let Joy Reign Supreme (dir., co-writer), The Judge and the Assassin.

TAYLOR, ANTHONY: Producer. b. Los Angeles, Calif., Feb. 5, 1931. e. U. of So. Cal.; USAF 1954–56; mem. Chicago Board of Trade, 1962–65; columnist, L.A. Herald Examiner, 1964–65; also mem. N.Y Mercantile Exchange, Chicago Mercantile Exchange; partner commodity Futures Co., Westwood, Calif.; 1966, produced feature Incubus; 1966, award for motion picture excellence, San Francisco Int'l Film Festival; 1967, award Incubus, Cork Int'l Film Festival, Ireland; 1968: prod. feature, Possession.

TAYLOR, DELORES: Actress-Writer. b. Winner, So. Dak. e. Univ. of So. Dak, studying commercial art. m. Tom Laughlin. First TV experience was heading art dept. at RCA wholesale center in Milwaukee. Made feature film debut as actress in Billy Jack in 1971. Wrote s.p. with husband for that and the sequel, The Trial of Billy Jack, under pseudonym Teresa Christina.
PICTURES INCLUDE: Billy Jack, The Trial of Billy Jack.

TAYLOR, DON: Actor-Director. b. Freeport, Pa., Dec. 13. e. Pennsylvania State Univ. Appeared in Army Air Corps' Winged Victory; on stage & screen; author short stories, radio dramas, one-act plays, TV shows, both ½ hour and hour.
PICTURES INCLUDE: Actor: Girl Crazy, Naked City, For the Love of Mary, Battleground, Father of the Bride, Father's Little Dividend, Submarine Command, Flying Leathernecks, Blue Veil, Japanese War Bride, Stalag 17, The Girls of Pleasure Island, Destination Gobi, Johnny Dark, I'll Cry Tomorrow, Bold and the Brave. Director: Jack of Diamonds, Five Man Army, Escape from The Planet of the Apes, Tom Sawyer, Echoes Of A Summer, The Great Scout and Cathouse Thursday, The Island of Dr. Moreau, Damien-Omen II, The Final Countdown.

TAYLOR, ELIZABETH: Actress. b. London, Eng., Feb. 27, 1932; e. Bryon House, London, When 3 years old danced before Princess Elizabeth, Margaret Rose. Came to U.S. on outbreak World War II. In 1943–44 screen debut Lassie Comes Home.
PICTURES INCLUDE: National Velvet, Life with Father, Cynthia, Courage of Lassie, Little Women, White Cliffs of Dover, Jane Eyre, Date With Judy, Conspirator, Big Hangover, Father of the Bride, Father's Little Dividend, Love Is Better Than Ever, Place in the Sun, Ivanhoe, The Girl Who Had Everything, Rhapsody, Elephant Walk, Beau Brummell, Last Time I Saw Paris, Giant, Raintree Country, Suddenly, Last Summer, Butterfield 8, Cleopatra, The V.I.P.'s, The Night of the Iguana, Who's Afraid of Virginia Woolf, The Taming of the Shrew, The Sandpiper, Doctor Faustus, The Comedians, Reflections In A Golden Eye, Boom!, Secret Ceremony, The Only Game in Town, X, Y, and Zee, Hammersmith Is Out, Night Watch, Ash Wednesday, That's Entertainment!, The Driver's Seat, The Blue Bird, A Little Night Music, The Mirror Crack'd.

TAYLOR, KENT: Actor. r.n. Louis Weiss. b. Nashua, Iowa, May 11, 1907. In many feature productions from 1931.
PICTURES INCLUDE: Merrily We Go to Hell, Blonde Venus, Death Takes a Holiday, Mrs. Wiggs of the Cabbage Patch, Alaska, Bomber's Moon, Roger Touhy—Gangster, Notorious Gentleman, Dangerous Millions, Deadline for Murder, Second Chance, Crimson Key, Half Past Midnight, Western Pacific Agent, Payment on Demand, Playgirl, Slightly Scarlet, Track the Man Down, Secret Venture, Ghost Town, Iron Sheriff.
TV: Boston Blackie.

TAYLOR, LAWRENCE: Writer. b. Atlanta, Ga.; e. Northwestern U., '33. Reporter, dramatic critic, foreign correspondent 1934–40; writer radio, m.p., 1940–42; U.S.A.F., 1942–45; writer, m.p. story consultant. TV since 1945, U.S. and England.
PICTURES INCLUDE: Dixie Jamboree, Philo Vance's Secret Mission, Devil Ship, Bulldog Drummond Strikes Back; (story basis) Lady Without a Passport; (collab. s.p.) Jackie Robinson Story, Candelario (coll. with son, Heath).

TAYLOR, MICHAEL: Executive. b. New York City. e. Univ. of Miami (English, Drama). Worked at producing for stage before joining United Artists in 1973 as trainee. Named asst. to v.p. in chg. of production. Appointed exec. in chg. of production in London, where worked in acquisition and production of properties, working on such pictures as The Spy Who Loved Me, Valentino, The Pink Panther Strikes Again, etc. Left U.A. to form Taylor/Wigutow Productions with former U.A. executive Dan Wigutow. First film of new company: Last Embrace (1979).

TAYLOR, ROD: Actor. b. Sydney, Australia, Jan. 11, 1929. e. Fine Arts College. Started out as artist then turned to acting on stage. After co-starring in film Long John Silver, to Hollywood in 1954. First film The Virgin Queen, followed by Giant, Separate Tables and Step Down to Terror. Also made a number of TV appearances. Formed own company, Rodler, Inc., for TV-film production.
PICTURES INCLUDE: Ask Any Girl, The V.I.P.s, The Birds, Sunday in New York, Young Cassidy, The Time Machine, Seven Seas to Calais, A Gathering of Eagles, 36 Hours, Do Not Disturb, The Glass Bottom Boat, Hotel, Chuka, Dark of the Sun, The Hell with Heroes, Zabriskie Point, The Train Robbers, Trader Horn, The Deadly Trackers.

TEITELBAUM, PEDRO: Executive. b. Porto Alegre, Rio Grande Do Sul, Brasil, Nov 21, 1922. e. Colegio Uniao, Brasil, 1942 (CPA); Univ. of Porto Alegre, Brasil, 1945 (economics & business admin.) 1939; Columbia Pictures; 1943, Warner Brothers; 1958, Latin-American supervisor for Republic Pictures; 1957, producer, distributor, exhibitor in Brasil; 1968, area supervisor for United Artists; 1973, v.p. intl. sales; 1975, v.p. international sales & distribution; 1976, senior v.p. and foreign manager. In Jan., 1977, joined CIC as senior exec. v.p. Named pres., July, 1977.

TELLER, IRA: Executive, b. New York, N.Y., July 3, 1940. e. CCNY 1957–61; NYU Graduate Sch. of Arts, 1961–62; publicist, Pressbook Dept., 20th Century Fox., 1961–62; asst. to adv. mgr., Embassy Pictures Corp., 1962–63; asst. adv. mgr., Columbia Pictures Corp. 1963; adv. mgr., Columbia Pictures Corp., 1964, 1964–65; asst. to chmn. of bd., Diener, Hauser, Greenthal Agy., 1966; adv. mgr., 20th Century-Fox, 1966–67; 1967, adv. dir. 20th Cent.-Fox; Dir. of Adv., Nat'l General Pictures Corp., 1969; eastern dir., adv.-pub., 1972; National Director, Adv-Pub., 1973; Bryanston Distributors, Inc. v.p. adv.-pub., 1974; Cine Artists Pictures Corp. v.p. adv-pub., 1975; Lorimar Productions, v.p., adv.-marketing, 1975–77. 1977-present, pres. Ira Teller and Company, Inc.

TEMPLE, SHIRLEY JANE: Actress. b. Santa Monica, Calif., April 23, 1929. In 1932 screen debut, Red Haired Alibi. In 1933 To the Last Man; then leading figure in Baby Burlesque series Educational shorts until Stand Up and Cheer (1934), which resulted in career as child star.
PICTURES INCLUDE: Baby Takes a Bow, Bright Eyes, Now I'll Tell, Charge of Heart, Little Miss Marker, Now and Forever, The Little Colonel, Our Little Girl, Curly Top, The Littlest Rebel, Captain January, Poor Little Rich Girl, Dimples, Stowaway, Wee Willie Winkle, Heidi, Rebecca of Sunnybrook Farm, Little Miss Broadway, Just Around the Corner, Little Princess, Susannah of the Mounties, The Blue Bird, Young People. In 1941 Kathleen. In 1942 Miss Annie Rooney. Voted one of ten best Money-Making Stars in Motion Picture Herald-Fame Poll, 1934–39. Since You Went Away, I'll Be Seeing You, Kiss and Tell, That Hagen Girl, Honeymoon, Fort Apache, Bachelor and the Bobby-Soxer, Mr. Belvedere Goes to College, Adventure in Baltimore, Story of Seabiscuit, Kiss for Corliss.
TV: Hostess, fairy tale series.

TENNANT, WILLIAM: Executive. Partner in literary agency of Ziegler, Ross, and Tennant. Turned to m.p. production with Cleopatra Jones for Warner Bros., following with writing and producing of sequel, Cleopatra and the Casino of Gold. In 1975 joined Columbia Pictures as v.p.-prod., headquartering at Burbank Studios. Named pres., Casablanca Filmworks; now pres., PolyGram Pictures m.p. division.

TERRY, SIR JOHN: Film Finance Consultant. b. London, England, 1913; e. Mill Hill School. Early career as solicitor.

Entered m.p. ind. Film Producers Guild 1946–47; then legal dept. Rank Organization until 1949; joined National Film Finance Corporation; its chief solicitor 1949–57; sec., 1956–57; appt. man. dir., 1958–78.

TERRY-THOMAS: Actor, comedian. r.n. Thomas Terry Hoar Stevens. b. London, England, July 14, 1911; e. Ardingly College, Sussex.
PICTURES INCLUDE: Private's Progress, The Green Man, Lucky Jim, Tom Thumb, I'm All Right, Jack, Too Many Crooks, Make Mine Mink, School for Scoundrels, A Matter of WHO, Bachelor Flat, Operation Snatch, Kill or Cure, The Mouse on the Moon, It's a Mad, Mad, Mad, Mad World, Wild Affair, Strange Bedfellows, How to Murder Your Wife, Those Magnificent Men in Their Flying Machines, Rocket to the Moon, Guide for the Married Man, Kiss The Girls and Make Them Die, The Perils of Pauline, Blast Off, 2000 Years Later, Where Were You When the Lights Went Out?, Don't Raise the Bridge, Lower the River, How Sweet It Is!, Arabella, Dr. Phibes, and many films made in Europe, Side By Side, Spanish Fly, The Amorous Adventures of Tom Jones, The Last Remake of Beau Geste, Hound of the Baskervilles.
TV: (U.S.) Danny Kaye Show, Garry Moore Show, Andy Williams Show, Jack Paar Show, Judy Garland Show, Burke's Law (series) Red Skelton Show.

TESICH, STEVE: Writer. b. Yugoslavia. e. Indiana Univ., Columbia. Came to U.S. at age 14. While doing graduate work in Russian literature at Columbia left to begin writing. Taken up by American Place Theatre which did his play, The Carpenters, in 1970 and then six others. Wrote screenplays; first produced was Breaking Away, 1979.

TETZLAFF, TED: Director. b. Los Angeles, Calif., June 3, 1903. Joined camera dept. Fox Studios, became first cameraman; dir., 1940; served in U.S. Air Corps as a Major, World War II.
PICTURES INCLUDE: cameraman: Enchanted Cottage, Notorious; dir.: World Premiere, Riffraff, Fighting Father Dunne, Window, Johnny Allegro, Dangerous Profession, Gambling House, White Tower, Under the Gun, Treasure of Lost Canyon, Terror on a Train, Son of Sinbad.

TEWKESBURY, JOAN: Writer. e. Univ. of So. Calif. Directed and choreographed Little Theatre prods. in L.A. area; taught in theatre arts depts. of three universities: Univ. of So. Calif., Immaculate Heart and Mount St. Mary's. Became script supvr. for Robert Altman on McCabe & Mrs. Miller.
PICTURES INCLUDE: Thieves Like Us (co. s.p.), Nashville.

THALHIMER, JR., MORTON G.: Theatre Executive. b. Richmond, Va., June 27, 1924. e. Dartmouth Coll. (1948, B.A.); Univ. of Va. (1959). In Navy in World War II. Joined Century Theatres as trainee 1948; Jamestown Amusement, 1949–50. Now pres. Neighborhood Theatre, Inc. Charter member of Theatre Owners of America; continuing member of NATO, served on finance comm. and Trade Practice comm. bd. member and past president of NATO of Va. (1973–75.) Mem. Variety Club Int'l., Tent 11; patron life member, Variety Club of Israel, Tent 51.

THATCHER, HEATHER: Actress. b. London, England. London stage debut in Girl From Ciro's, 1916. Entered m.p. ind. as extra in 1916.
PLAYS: Naughty Princess, Sally, Constant Wife, School for Husbands, Lucky Break, Full House.
PICTURES INCLUDE: (Hollywood pictures since 1936): Hour of Thirteen, Duel in the Jungle.

THACHER, RUSSELL: Producer-Writer. b. Hackensack, N.J. e. Bucknell Univ., N.Y.U. Author of novels: The Captain, The Tender Age, A Break in the Clouds. Editor Omnibook Magazine, 1946–58; Book of the Month Club, 1958–63. Exec. story editor, MGM, 1963–69. Exec. prod., MGM, 1969–72.
PICTURES INCLUDE: Travels with My Aunt (assoc. prod.), Soylent Green, The Cay, Last Hard Men.

THAXTER, PHYLLIS: Actress. b. Portland, Me., Nov. 20, 1921; e. St. Genevieve School, Montreal. Screen debut in Thirty Seconds Over Tokyo.
PLAYS: What A Life, There Shall Be No Night, Claudia. Take Her She's Mine.
PICTURES INCLUDE: Week-End at the Waldorf, Bewitched, Tenth Avenue Angel, Sign of the Ram, Blood on the Moon, The Breaking Point, Fort Worth, Jim Thorpe—All American, Come Fill the Cup, She's Working Her Way Through College, Operation Secret, Springfield Rifle, Women's Prison, World of Henry Orient, Superman.
TV: Wagon Train, Alfred Hitchcock, Twilight Zone, Purex Specials For Women, Playhouse 90, The Fugitive, Defenders, etc.

THOMAS, BILL: Fashion designer. b. Chicago, Ill., Oct. 13, 1921; e. U. of S.C., A.B. 1941. Art Center, Los Angeles, 1941. Started career as designer on assignment to USO shows; Theatre Royle, Brisbane, Australia; from duty with U.S. A.F., 1941–46. Asst. to Irene, Irene Inc., 1948; asst. costume designer MGM Studios 1947–48; U-I Studios 1949–59; currently free lance.
PICTURES INCLUDE: High Time, Beloved Infidel, By Love Possessed, Babes in Toyland.

THOMAS, DANNY: Actor. r.n. Amos Jacobs. b. Deerfield, Mich., Jan. 6. Night club entertainer; on radio and TV, films.
TV: Own TV show, The Danny Thomas Show, CBS-TV, 1953–64; NBC-TV specials 1964–66, Danny Thomas Show, 1967–68; CBS-TV Specials 1967 & 1970. Make Room for Grandaddy series 1970–71, ABC-TV; The Practice, 1976–77; NBC-TV.
PICTURES INCLUDE: Unfinished Dance, Big City, Call Me Mister, I'll See You in My Dreams, Jazz Singer.

THOMAS, GERALD: Producer, director. b. Hull, England, 1920. Entered m.p. industry 1946.
PICTURES INCLUDE: Tony Draws a Horse, Appointment With Venus, Venetian Bird, Sword and the Rose, A Day to Remember, Mad About Men, Doctor in the House, Above Us the Waves, A Novel Affair, After the Ball, Timelock, Vicious Circle, Chain of Events, Solitary Child, The Duke Wore Jeans, Carry On Sergeant, Carry on Nurse, Carry on Teacher, Please Turn Over, Carry on Constable, Watch Your Stern, No Kidding, Carry On Regardless, Raising The Wind, Twice Around the Daffodils, Carry on Cruising, The Iron Maiden, Nurse On Wheels, Call Me a Cab, Carry on Jack, Carry on Spying, Carry On Cleo, The Big Job, Carry On Cowboy, Carry on Screaming, Don't Lose Your Head, Follow That Camel, Carry on Doctor, Carry On Up The Khyber, Carry on Up the Jungle, Carry on Loving, Carry on Camping, Carry On Again, Doctor, Carry On, Henry, Carry On At Your Convenience, Carry On Matron, Carry on Abroad, Bless This House, Carry On Girls, Carry On Dick, Carry On Behind, Carry On England, That's Carry On, Carry On Emmannuelle.
TV: Prod. and dir. Rob Roy, serial. Prod. Carry on Christmas for Thames TV. Dir. Carry On Dick, Prod., Carry On Laughing. Dir. Best of Carry On. Prod. Odd Man Out, series.

THOMAS HARRY E.: Exhibitor. b. Monroe, La., May 22, 1920; e. L.S.U. 1938–41. Psychological Branch of Army Air Force, 1942–46. Past pres., secy., and treas. of NATO of Mississippi. Dir. of Design & Const. & Sec. Gulf State Theatres Inc. Retired 1978.

THOMAS, HOWARD: Executive. b. CWM. Monmouthshire. Entered m.p. ind. in 1944. Newspaper columnist & feature writer; adv. copyrwriter; writer of revue sketches, books, radio plays; mgr., sponsored radio dept., London Press Exchange, Ltd.; BBC prod. & orig. The Brains Trust. Prod.-in-chief, Associated British-Pathe, Ltd.; managing dir., ABC Television Ltd., 1956. Appt. dir., Associated British Picture Corp., 1964; div. dir., EMI, Ltd., 1970; man. dir., Thames TV, 1968–74; Thames TV Int. Ltd., pres., Cinema TV Veterans, 1979–80. Dir., Argus Press Ltd.

THOMAS, JAMES: Critic, producer. b. Huddersfield, York, England, June 9, 1922; e. Huddersfield Coll. Reporter, Huddersfield Daily Examiner, 1946–47; Yorkshire Evening Post, 1947–48; London, News-Chronicle, 1948–50; radio & TV corres. and spec. writer on entertainment, News Chronicle, 1950; TV editor, Daily Express, 1957. Assoc. ed., The Time, The Place and The Camera, ATV series, 1961–62. Ed., The Braden Beat, 1963. Prod., On the Braden Beat, 1964–65. 1965–67 TV columnist, Daily Express.

THOMAS, LOWELL: Commentator. b. Woodington, Darke County, Ohio, Apr. 6, 1892; e. Northern Indiana Univ., Denver Univ., Princeton Univ., Kent College of Law. Instructor, Eng. Lit., Princeton, 1914; lecturer, biographer, historian and film producer until 1930 when broadcaster for CBS; commentator on first TV news program for NBC, 1939; & on first daily TV prog., 1940; foreign commentator during WWII; now, commentator on own daily prog. for CBS. Chmn. bd., then vice chmn., Cinerama Prod. Corp., 1952; pres., 1953–54; exec. prod., High Adventure, TV series; TV series in conjunct'n with BBC, The World of Lowell Thomas; co-founder, Capital Cities Broadcasting Corp.

THOMAS, MARLO: Actress. b. Detroit, Nov. 21, 1938. f. is Danny Thomas. e. U. of Southern Calif. Started career with small TV roles, summer stock. Appeared in London stage prod. of Barefoot in the Park. Debut in own TV series, That Girl, 1966. Most Promising Newcomer Awards from both "Fame" and "Photoplay."
PICTURES INCLUDE: Jenny, Thieves.

THOMAS, RALPH: Director. b. Hull, Yorkshire, England, Aug. 10, 1915; e. Tellisford College, Clifton and University College, London. Journalist in early career, entered m.p. ind. 1932 as film ed.; service with 9th Lancers, 1939–45; then film director.
PICTURES INCLUDE: prod.: The Clouded Yellow, Appointment with Venus (Island Rescue), Day to Remember, Travellers' Joy, Venetian Bird, Once Upon a Dream, Doctor in the House, Mad about Men, Above Us the Waves, Doctor At Sea, Iron Petticoat, Checkpoint, Doctor at Large,

Campbell's Kingdom, A Tale of Two Cities, The Wind Cannot Read, The 39 Steps, Upstairs and Downstairs, Conspiracy of Hearts, Doctors in Love, No Love for Johnnie, No, My Darling Daughter, A Pair of Briefs, The Wild & The Willing, Doctor in Distress, Hot Enough for June, The High Bright Sun, Agent 008¾, Doctor in Clover, Deadlier Than the Male, Nobody Runs Forever, Some Girls Do, Doctor in Trouble, Percy, Quest, The Love Ban, Percy's Progress, A Nightingale Sang in Berkeley Square.

THOMAS, RICHARD: Actor. b. N.Y.C., June 13, 1951. e. Col. Univ. Made TV Debut at age 7 and featured in several series.
PICTURES INCLUDE: Last Summer, Red Sky at Morning, The Todd Killings, Cactus in the Snow, You'll Like My Mother, September 30th, 1955.
TELEVISION: Medical Center, Marcus Welby, M.D., The F.B.I., The Waltons (regular).

THOMAS, ROBERT G. ("BOB"): Producer-Director. b. Glen Ridge, N.J., July 21, 1943. e. Univ. of Bridgeport, Fairleigh Dickinson Univ. Produced educational radio programs, 1962, WPKN-FM. Asst. stage manager at Meadowbrook Dinner Theatre, 1963. In March, 1964 began career as TV cameraman for New York stations. Worked both full-time and free-lance for major TV and video tape studios. In January, 1968, started Bob Thomas Productions and began producing business/sales films and TV commercials. Has 8 awards from natl. film festivals; nominated for 5 Emmys for TV series called The Jersey Side he produced for WOR-TV.
PICTURES: Shorts: Valley Forge with Bob Hope, New Jersey—200 Years. Road-Eo '77.
TV: The Jersey Side (talk/entertainment), Jersey People (weekly talk/entertainment program); $10,000 Touchdown (game show). Sir Reginald and the 3rd Dimension (Children's Show) 3-D Television—for local TV station promotion "The First Annual Comedy Awards" (TV Special).

THOMAS, ROBERT J. ("BOB"): Columnist, Associated Press, Hollywood. b. San Diego, Cal., Jan. 26, 1922; p. George H. Thomas, publicist; e. U. of Calif., L.A. Joined Associated Press staff, Los Angeles, 1943; corr. Fresno, 1944; Hollywood since 1944. Writer mag. articles; appearances, radio; orig. story Big Mike.
BOOKS: author: The Art of Animation, King Cohn, Thalberg, Selznick, Winchell, Secret Boss of California, The Heart of Hollywood, Howard, The Amazing Mr. Hughes, Weekend '33, Marlon, Portrait of the Rebel as an Artist, Walt Disney, An American Original; Bud and Lou, The Abbott and Costello Story; The Road to Hollywood (with Bob Hope), The One and Only Bing, Joan Crawford.

THOMAS, WILLIAM C.: Producer, writer, director. b. Los Angeles, Calif., Aug. 11, 1903; e. U. of So. Calif. From 1925–30 publicity Pantages Theats.; Fox West Coast Theats.; pub. dept., Columbia; writer, Paramount, RKO; prod. from 1936; partner, Pine Thomas Prod.
PICTURES INCLUDE: Scared Stiff, co-prod.: Dangerous Passage, Double Exposure High Powered; co-prod. & dir.: Midnight Manhunt; dir. & co-prod.: They Made Me a Killer; co-prod. & dir.: Big Town, Tokyo Rose, Hot Cargo, Swamp Fire, People Are Funny, I Cover Big Town, Fear in the Night, Seven Were Saved, Danger Street, Jungle Flight, Albuquerque, El Paso, Capt. China, Eagle and the Hawk, The Lawless, Tripoli, Last Outpost, Crosswinds, Passage West, Hong Kong, Caribbean, Blazing Forest, Tropic Zone, The Vanquished, Jamaica Run, Sangaree, Those Redheads from Seattle, Jivaro, Hell's Island, Run for Cover, Far Horizons, Lucy Gallant, Nightmare, The Big Caper, Bailout at 43,000, High Seas Hijack (co-producer with Toho Films Ltd.); Crusin' High (prod. writer).

THOMOPOULOS, ANOTHONY D.: Executive. b. 1938. e. Georgetown Univ. Began career in broadcasting at NBC, 1959, starting as mailroom clerk and moving to radio division in prod. & admin. Shortly named to post in International Division Sales, involved with programming for stations and in developing TV systems for other nations. Joined Four Star Entertainment Corp. as dir. of foreign sales, 1964; named v.p., 1965; exec. v.p., 1969. In 1970 joined RCA SelectaVision Div. as dir. of programming. In 1971 joined Tomorrow Entertainment as v.p. In 1973 joined ABC as v.p., prime-time programs in N.Y.; 1974, named v.p., prime-time TV creative operations, ABC Entertainment. In 1975 named v.p. of special programs, ABC Entertainment; 1976 made v.p., ABC-TV, assisting pres. Frederick S. Pierce in supervising all activities of the division. In Feb., 1978 named pres. of ABC Entertainment.

THOMPSON, CARLOS: Actor. b. Buenos Aires, Argentina, June 7; e. National U., Buenos Aires. Stage & screen actor in Argentina; Hollywood m.p. debut in Fort Algiers, 1953.
PICTURES INCLUDE: Flame and the Flesh, Valley of the Kings, Magic Fire.

THOMPSON, JACK: Actor. b. Sydney, Australia, Aug. 31, 1940. e. Queensland Univ. Joined drama workshop at school; first part was in TV soap opera as continuing character.
PICTURES INCLUDE: A Sunday Too Far, The Chant of Jimmy Blacksmith, Breaker Morant (Australian award), The Earthling.

THOMPSON, MARSHALL: Actor. r.n. James Marshall Thompson; b. Peoria, Ill., Nov. 27; e. Occidental Coll., L.A. In school dramatics; studied for clergy; wrote play Faith, prod. by Westwood Players; in Westwood Players as actor; m.p. debut in Reckless Age, 1944.
PICTURES INCLUDE: They Were Expendable, Gallant Bess, Valley of Decision, Homecoming, B.F.'s Daughter, Words and Music, Command Decision, Roseanna McCoy, Battleground, Dial 1119, Devil's Doorway, Mystery Street, Tall Target, Basketball Fix, My Six Convicts, Rose Bowl Story, The Caddy, Battle Taxi Port of Hell, Cult of the Cobra, Crashout, To Hell and Back, Clarence, The Cross-eyed Lion, Around the World Under the Sea.

THOMPSON, SADA: Actress. b. Des Moines, Sept. 27, 1929. e. Carnegie Inst. of Technology, Pittsburgh. First N.Y. stage appearance in Under Milkwood with Dylan Thomas. Bdwy. career has produced many awards topped by The Effects of Gamma Rays, for which she won Obie, Drama Desk, Variety Poll.
TV: Sandburg's Lincoln, The Entertainer, Our Town. Series: Family (regular).

THOMSON, ALAN: Studio Publicity manager, EMI Elstree Studios Ltd. e. in Scotland. Early career in Glasgow with his father, well-known cinema manager. Ent. journalism on staff film magazine, other Scottish publications. Ent. m.p. with Associated British Productions 1953. Appt. studio pub. mgr. 1956.

THOR, JEROME: Performer. b. Brooklyn, N.Y., Jan. 5, 1920; e. Neighborhood Playhouse. On B'way in He Who Gets Slapped, My Sister Eileen, Doughgirls, Golden Boy, etc.; on TV in Studio One, Suspense, Danger, Foreign Intrigue, Rogue for Hire, Riot in Juvenile Prison.

THORPE, RICHARD: Director. b. Hutchinson, Kan., Feb. 24, 1896; m. Belva Kay, prof. In vaudeville, stock & musical comedy, 1915–18. Now retired.
PICTURES INCLUDE: cast, Torchy Comedies, Three O'Clock in the Morning, Burn 'Em Up Barnes, Flame of Desire; dir. since 1933. dir., Night Must Fall, Ivanhoe, Double Wedding, Crowd Roars, Earl of Chicago, Huckleberry Finn, White Cargo, Two Girls and a Sailor, Sun Comes Up, Big Jack, Challenge to Lassie, Malaya, Black Hand, Three Little Words, Vengeance Valley, The Great Caruso, Unknown Man, It's a Big Country, Carbine Williams, Prisoner of Zenda, The Girl Who Had Everything, All the Brothers Were Valiant, Knights of the Round Table, Student Prince, Athena, Quentin Durward, The Prodigal, The Tartars, Honeymoon Machine, Horizontal Lieutenant, Follow The Boys, The Truth About Spring, That Funny Feeling, Scorpio Letters, Last Challenge, Retired.

THULIN, INGRID: Actress, director. b. Solleftea, Sweden, Jan. 27, 1929. m. Harry Schein, founder and head of Sweden's Film Institute. Made acting debut at 15 at the Municipal Theatre in Norrkoping. Studied at Stockholm's Royal Dramatic Theatre. Worked with Malmo repertory. Appeared on Swedish stage in Gigi, Peer Gynt, Two for the Seesaw, Twelfth Night, Miss Julie. Has directed plays in Stockholm. N.Y. stage debut, 1967: Of Love Remembered.
PICTURES INCLUDE: For Ingmar Bergman: Wild Strawberries, Brink of Life (Best Actress Award, Cannes Film Festival), The Magician, Winter Light, The Silence, The Hour of the Wolf, The Ritual, Night Games, The Bathers, Adelaide, La Guerre Est Finie The Four Horsemen of the Apocalypse, Return From Ashes, The Damned, Cries and Whispers, Moses, The Cassandra Crossing. Dir. short film: Devotion.

THUNDER, CLOUD, CHIEF: Actor r.n. Victor Daniels. b. Muskogen, Okla., April 12, 1899. Worked on cattle ranches, in mining, as boxer; became rodeo performer, stunt man. Entered pictures 1929. In many films since.
PICTURES INCLUDE: Fighting Seabees, Buffalo Bill, Sonora, Stagecoach, Romance of the West, Senator Was Indiscreet, Ambush, Ticket to Tomahawk, Colt 45.

THURSTON, CAROL: Actress. b. Forsyth, Mont.; e. Montana U.; Bliss-Hayden School of Acting. With father in amateur theatricals prior to screen debut.
PICTURES INCLUDE: Swamp Fire, Jewels of Brandenburg, Last Round-Up, Rogue's Regiment, Arctic Manhunt, Apache Chief, Flaming Feather, Arctic Flight, Conquest of Cochise, Killer Ape, Yukon Vengeance.

TIDYMAN, ERNEST: Writer. b. Cleveland. Began crime reporting at 14. After long experience as newspaperman started writing fiction at 42. Wrote the novels on which the Shaft films were based.
PICTURES INCLUDE: The French Connection (AA), Shaft, Shaft's Big Score, High Plains Drifter, Report to the Commissioner (Co-s.p.), Street People (Co-s.p.).

TIERNEY, GENE: Actress. b. Brooklyn, N.Y., Nov. 20, 1920; e. St. Margaret's School, Brilmont, Switzerland; Miss Farmer's School, Farmington, Conn.
PICTURES INCLUDE: Return of Frank James, Hudson's Bay, Tobacco Road, Belle Starr, Sundown, Shanghai Gesture, Son of Fury, Heaven Can Wait, Laura, Bell for Adano, Leave Her to Heaven, Dragonwyck, Razor's Edge, Ghost and Mrs. Muir, Iron Curtain, That Wonderful Urge, Whirlpool, Where the Sidewalk Ends, Night and the City, Mating Season, On the Riviera, Secret of Convict Lake, Way of a Gaucho, Close To My Heart, Plymouth Adventure, Never Let Me Go, Personal Affair, The Egyptian, Black Widow, Left Hand of God, Advise & Consent.

TIERNEY, LAWRENCE: Actor. b. Brooklyn, N.Y., Mar. 15, 1919; e. Manhattan Coll. Track athlete (natl. championship Cross Country Team, N.Y. Athletic Club). On stage as actor. Screen debut 1943 in The Ghost Ship.
PICTURES INCLUDE: Youth Runs Wild, Dillinger, Mama Loves Papa, Badman's Territory, Step By Step, San Quentin, Devil Thumbs a Ride, Born to Kill, Bodyguard, Kill or Be Killed, Best of the Bad Men, Shakedown, Greatest Show on Earth, Hoodlum, Bushwackers, Steel Cage, Custer of the West, Such Good Friends.

TIFFIN, PAMELA: Actress. b. Oklahoma City, Okla., Oct. 13. e. Hunter College. Started modeling as a teenager. Film debut in Summer and Smoke.
PICTURES INCLUDE: One Two Three, State Fair, Come Fly With Me, For Those Who Think Young, The Pleasure Seekers, The Hallelujah Trail, Harper, Paranoia, Kiss The Other Sheik, Viva Max.
PLAY: Dinner at Eight.

TILLSTROM, BURR: Performer. b. Chicago, Oct. 13, 1917. e. Univ. of Chicago. With WPA-Chicago Park Dist. Theatre, 1936; traveled with puppet, marionette, stock shows; mgr. puppet exhibits and marionette Theatre, Marshall Field & Co., Chicago (performed on experimental television for RCA), 1939; sent to Bermuda by RCA to do ship to shore television broadcasts 1940; creator and impresario puppet troupe headed by Kukla and Ollie on show "Kukla, Fran and Ollie" with Fran Allison since 1947 (pioneered in television when he presented puppets on Balaban and Katz 1st telecast. Stat. WBKB. Chicago, 1941). Winner two Peabody awards and three national Academy of TV Arts & Science Emmy awards; hon. Doctorate of Letters, Hope College, Holland, Mich., 1972.

TINKER, GRANT A.: Executive. b. Stamford, Conn., Jan. 11, 1926. e. Dartmouth College, 1947. m. Mary Tyler Moore. Joined NBC radio prog. dept. 1949. In 1954 with McCann-Erickson ad agency, TV dept. In 1958, Benton & Bowles Ad Agency, TV dept. From 1961–66 with NBC, v.p., programs, West Coast; v.p. in chg. of programming, N.Y., 1966–67. Joined Universal Televison as v.p., 1968–69; 20th-Fox, v.p., 1969–70. Became pres. MTM Enterprises, Inc. 1970. Named NBC pres., 1981.

TINLING, JAMES: Director. b. Seattle, Wash., May 8, 1889; e. U. Washington. Started in m.p. as prop boy for Century Studios; wrote comedies 2 yrs.; stunt man; dir.
PICTURES INCLUDE: Silk Legs, Don't Marry, Very Confidential, Exalted Flapper, True Heaven, Words and Music, Fox, Flood, Mr. Moto's Gamble, Boy Friend, Riders of the Purple Sage, Crime Buster, Rendezvous, Deadline for Murder, Strange Journey, Dangerous Millions, Roses Are Red, Second Chance, Night Wind, Trouble Preferred, Tales of Robin Hood.
TV films.

TISCH, LAURENCE A.: Executive. b. Bklyn, N.Y., March 5, 1923. e. N.Y. Univ., 1941; Univ. of Pa. Wharton School, 1942; Harvard Law School, 1946. Pres. Tisch Hotels, Inc., 1950–59; Pres. Americana Hotel, Inc., Miami Beach, 1956–59; Chmn. of bd. and chief executive officer of Loews Corp since 1960. Also chmn. of bd. of CNA Financial Corp since 1974.

TISCH, PRESTON ROBERT: Executive. b. Bklyn., N.Y., April 29, 1926. e. Bucknell U. Lewisberg, Pa., 1943–44; B.A, U. Mich., 1948. Pres. Loew's Corporation.

TODD, ANN: Actress. e. Central School of Speech Training & Dramatic Art.
BRITISH STAGE PLAYS INCLUDE: Service, When Ladies Meet, Man in Half-Moon Street, Peter Pan, Brit., Lottie Dundass.
PICTURES INCLUDE: Squeaker, Seventh Veil, Perfect Strangers, Gaiety George, Hollywood debut in Paradine Case. Ret. to theatre Feb., 1951 in stage version, Seventh Veil, So Evil My Love; Old Vic. Theatre, 1954–55: Doctor's Dilemma, Four Winds, Duel of Angels, One Woman's Story, Daybreak, Madeleine, The Sound Barrier, Green Scarf, Time Without Pity, Taste of Fear, Son of Captain Blood, Ninety Degrees in The Shade, The Human Factor.
TV: Many appearances and TV films, New York, Hollywood. London incl.: Camille, The Vortex, The Door, Snows of Kilimanjaro, TV film, Hollywood. 1964: Prod., travelogue in Nepal, Love Story, Reading for Glory. Makes own Diary Documentaries and appears in them. Films for cinema and TV incl. Thunder in Heaven, Thunder of Gods, Thunder of Kings. Persian Fairy Tale. Force in the Sun, Thunder of Silence. Recent TV appearance, The Last Target.

TODD, RICHARD: Actor. b. Dublin, Eire, June 11, 1919. e. Shrewsbury. In repertory, 1937; founder-member, Dundee Repertory Theatre, 1939; distinguished war service, 1939–46; Dundee Repertory, 1946–48; screen debut, 1948; For Them That Trespass, 1948. 1970 Founder-Director Triumph Theatre Productions.
PICTURES INCLUDE: The Hasty Heart, Lightning Strikes Twice (USA), Robin Hood, The Venetian Bird, Sword and the Rose, Rob Roy, A Man Called Peter (USA), Virgin Queen (USA), The Bed, Dam Busters, D-Day the Sixth of June (USA), Marie Antoinette, Yangtse Incident, Chase a Crooked Shadow, The Naked Earth, Danger Within, The Long the Short and the Tall, The Hellions, Never Let Go, The Longest Day, The Boys, The Very Edge, exec. prod., star own prod. Don't Bother to Knock, Operation Crossbow, Coast of Skeletons, Asylum, The Big Sleep.
STAGE: 1966–67, An Ideal Husband; 1967–68, Dear Octopus. Co-founder, Triumph Theatre Prods., Ltd. Plays since 1970: Roar Like a Dove, Grass Is Greener, The Marquise, 1972. (USA). Sleuth, 1972–73 (Australia). Thunder by Numbers 1973, The Hollow Crown 1974–75, Equus. 1975, Sleuth, 1976, On Approval, Quadrille, 1977; This Happy Breed, 1980.

TODDY, TED: President, Toddy Picture Co., Toddy Prod., Inc., Dixie National Pictures, Inc., Consolidated National Film Exchanges; v.p., Acceptance Film Corp. b. New York City, June 15, 1912. Worked three years at Universal. For 12 years Columbia, Atlanta. Then formed Dixie Film Exchange in Atlanta. In 1940 purchased Million Dollar Pictures of Supreme Pictures and formed Consolidated National Film Exchanges. In 1941: consolidated as Toddy Pictures Co., prod. under Dixie National Pictures banner, Mr. Washington Goes to Town, Lucky Ghost, Up Jumped the Devil, Professor Creeps. Under Sepia Art Co., prod. One Round Jones, etc.; own ex. in Atlanta, Dallas, Chicago, New York and Los Angeles. Member: War Prod. Bd.

TODMAN, HOWARD: Executive. b. N.Y.C., Nov. 24, 1920; e. Hamilton College 1941; dir. business affairs, Goodson-Todman Productions; treas., Goodson-Todman Associates, Inc.; v.p. & treas., Goodson-Todman Enterprises, Ltd.; Treasurer, Peak Prods., Inc.; Treas. Goodson-Todman Bcstg. Inc.; v.p. Price Productions, Inc.; v.p. Celebrity Productions, Inc.; chm., N.Y. Cancer Crusade, radio & TV.

TOGNAZZI, UGO: Actor. b. Cremona, Italy, March 23, 1922. Graduate of law. Started entertainment career in 1945 as comic in music hall revues. Film career began in 1950 with Les Cadets de Gascogne. Has produced four films and a detective series for TV. Also acted on stage.
PICTURES INCLUDE: His Women, The Fascist, Queen Bee, The Magnificent Cuckold, Question of Honor, Barbarella, Property Is No Longer a Theft, La Cage Aux Folles, La Cage II, Sunday Lovers.

TOM, C.Y.: Cinematographer and distributor. b. Toy Shan, Kwangtung, China, Nov. 6, 1907; graduated N.Y. Institute of Photography 1926; photographed newsreels for The Great Wall Film Co. of Shanghai; in New York, 1926–29; in charge of production, Shanghai, 1929–32. Studied production techniques in Hollywood. Toured Europe, managing Chinese vaudeville, 1934–35. Studio mgr. and dir. photography for Chi Ming Motion Picture Co., 1935–41. President, Chinamerica Film Exchange and Chinamerica Film Studio, Hong Kong and Shanghai. Distributor, Monogram, Film Classics and Telenews, Hong Kong, Macao and China; asst. man. dir., Capitol Theatre, Hong Kong, 1948–59.

TOMBES, ANDREW: Actor. b. Ashtabula, Ohio; p. nonprofessionals; e. public schools and Phillips Exeter Academy; married. Appeared in vaudeville and in musical comedies including The O'Brien Girl, Ziegfeld Follies of 1922, and other stage musicals. In 1933 appeared in Moulin Rouge, UA-20th-Cent. From 1934 in many productions, principal Hollywood companies.
PICTURES INCLUDE: Can't Help Singing, Singing Sheriff, Night Club Girl, Goin' to Town, Don't Fence Me In, Frontier Gal, G.I. Honeymoon, Lake Placid Serenade, Oh You Beautiful Doll, Oklahoma Annie, I Dream of Jeanie, How To Be Very Very Popular.

TOMBRAGEL, MAURICE: Writer.
PICTURES INCLUDE: Legion of Lost Flyers, Horror Island, Mutiny in the Arctic, Two Senoritas from Chicago, Lone Wolf in Mexico, Return of the Whistler, Prince of Thieves, The Creeper, Highway 13, Thunder in the Pines, Sky Liner, Arson Inc., Motor Patrol, Fort Bowie, Moon Pilot, s.p. Monkeys Go Home; v.p. Running Wild, Golden Circle Prods., 1973.
TV: Wild Bill Hickcok, Stories of the Century, Annie Oakley, Soldiers of Fortune, Western Marshal, Wyatt Earp,

Frontier Doctor, Texas Rangers, Sergeant Preston, Adventures of Jim Bowie, Bat Masterson, Walt Disney's Elfego Baca, John Slaughter TV series, Life of Johann Strauss, Escapade in Florence, Bristle Face. Gallegher series, The Tenderfoot series. For Disney, The Treasure of San Marco (2 parts), The Gentle Ben Series.

TOMLIN, LILY: Actress. b. Detroit e. Wayne State Coll. Started inventing characters for comedy sketches in college, used them in cafe and night club dates. In December, 1969, first appeared on Laugh-In, TV series, gaining national attention.
TV: Laugh-In, The Lily Tomlin Show, Lily.
PICTURES INCLUDE: Nashville, The Late Show, Moment by Moment, Nine to Five, The Incredible Shrinking Woman.
RECORDS: This Is a Recording, And That's The Truth.

TOMLINSON, DAVID: Actor. b. May 17, 1917; e. Tonbridge Sch. On stage first, 1934 in "Outward Bound." Screen debut, 1938, "Name, Rank and Number."
PICTURES INCLUDE: Way to the Stars, School for Secrets, Quiet Wedding, Miranda, Easy Money, Sleeping Car to Trieste, Chiltern Hundreds, So Long at the Fair, Wooden Horse, Broken Journey, Hotel Sahara, Castle in the Air, Made in Heaven, Landfall, Three Men in a Boat, Carry on Admiral, Up the Creek, Further Up the Creek, Follow That Horse, Tom Jones, Mary Poppins, The Truth About Spring, City In the Sea, The Liquidator, The Love Bug, Bed Knobs & Broomsticks, From Hong Kong with Love, Wambling Free, Dominique, Fiendish Plot of Dr. Fu Manchu.

TOOMEY, REGIS: Actor. b. Pittsburgh, Pa., Aug. 13, 1902; e. U. Pittsburgh; Carnegie Inst. of Technology (drama); on N.Y. & London stage 5 yrs.; film debut in Alibi, 1929.
PICTURES INCLUDE: Spellbound, Big Sleep, Her Sister's Secret, Guilty, High Tide, Magic Town, Bishop's Wife, Boy With Green Hair, Mighty Joe Young, Come to the Stable, Cry Danger, Tall Target, People Against O'Hara, Show Boat, My Six Convicts, Battle at Apache Pass, Just For You, My Pal Gus, Never Wave at a Wac, It Happens Every Thursday, High and the Mighty, Top Gun, Guys and Dolls, Great Day in the Morning, 3 for Jamie Down, Dakota Incident, Warlock, Guns of the Timberland, The Day of the Gun, The Last Sundown, Journey to the Bottom of the Sea, Man's Favorite Sport, Peter Gunn, Change of Habit, Run Shadow Run, The Carey Treatment.
TV: Four Star Theatre, December Bride, Hey Mulligan, Dodsworth, Richard Diamond (series), Shannon (series), Burke's Law (series), Petticoat Junction (series).

TORME, MEL: Singer, actor. b. Chicago, Ill., Sept. 13. Singing debut at age of 4; won radio audition 1933; on radio; composed song "Lament to Love"; with Chico Marx's orchestra as drummer, arranger & vocalist 1942; served in U.S. Army, World War II; m.p. debut in "Higher and Higher" 1943; org. vocal group "Meltones"; many recordings; in night clubs.
PICTURES INCLUDE: Pardon my Rhythm, Good News, Let's Go Steady, Janie Gets Married, Night and Day, Good News, Words and Music, Duchess of Idaho, Girls Town, Walk Like A Dragon.

TORN, RIP: Actor. b. Temple, Tex., Feb. 6, 1931. e. Texas A&M, Univ. of Texas. Served in army. Signed as understudy for lead in Cat on a Hot Tin Roof on Broadway.
PICTURES INCLUDE: Baby Doll, A Face in the Crowd, King of Kings, The Cincinnati Kid, You're a Big Boy Now, Beach Red, Sol Madrid, Beyond the Law, Coming Apart, Maidstone, Tropic of Cancer, Payday, Crazy Joe, Birch Interval, The Man Who Fell to Earth, Nasty Habits, Coma, The Seduction of Joe Tynam, One Trick Pony, First Family.

TORS, IVAN: Writer, producer, director. b. Budapest, Hungary, June 12, 1916; e. U. of Budapest and Fordham U., N.Y. While in Europe wrote numerous plays, incl. Mimi, Keep Your Distance, Wind Without Rain, During World War II with Army Air Forces and O.S.S. Wrote orig. story, Below the Deadline; (collab. s.p.) Song of Love; w)s.p.) That Forsyte Woman; (s.p.) In Good Old Summertime; (collab. s.p.) Watch the Birdie; story, 49 Men; co-prod. collab. s.p. Storm Over Tibet; prod. collab. s p. The Glass Wall, The Magnetic Monster; prod., Riders to the Stars; prod., story G.O.G.; co-prod., Battle Taxi; prod. Underwater Warriors. For Ziv-TV, prod. Science Fiction Theatre, Ripcord, Aquanauts, Man and the Challenge, Sea Hunt. Member of Faculty, Department of Psychiatry and Behavioral Sciences, University of California, Irvine.
PICTURES INCLUDE: Flipper, Rhino, Flipper's New Adventure, Zebra in the Kitchen, Clarence the Cross-eyed Lion, Around the World Under the Sea, Birds Do It, Namu the Killer Whale; Africa, Texas Style, Gentle Giant, Daring Game, Hello Down There, Island of the Lost, The Aquarians. directed, produced March of the Desert; Where the Lions Rule.
RECENT PICTURES: Escape from Angola (exec. prod.), Galyon File (exec. prod., dir.).
TV: Flipper, Daktari, Cowboy in Africa, Gentle Ben, Jambo, Primus, Elephant Country, NBC special. 26 segments Encyclopedia of Wildlife; 3 one-hour Specials, Orientation and Navigation; Labyrinth of Odor; Pain in Nature, Danny and The Mermaid, CBS.

TOTTER, AUDREY: Actress. b. Joliet, Ill., Dec. 20. In many stage plays. On radio 1939–44; film debut in "Main Street" 1944.
PLAYS INCLUDE: Copperhead, Stage Door, Late Christopher Bean, My Sister Eileen.
PICTURES INCLUDE: Her Highness and the Bellboy, Dangerous Partners, Sailor Takes a Wife, Cockeyed Miracle, Lady in the Lake, High Wall, Beginning or the End, Unsuspected, Alias Nick Beal, Saxon Charm, Any Number Can Play, Tension, Set-Up, Under the Gun, Blue Veil, Sellout, F.B.I. Girl, Assignment-Paris, My Pal Gus, Woman They Almost Lynched, Cruisin' Down the River, Man in the Dark, Mission Over Korea, Champ for a Day, Massacre Canyon, Women's Prison, A Bullet for Joey, Vanishing American, Chubasco.
TV: Medical Center.

TOWERS, HARRY ALAN: Executive and producer. b. London, England, 1920. Prod. and wrote: 1963: Sanders of the River; 1964: Code Seven Victim Five.
PICTURES INCLUDE: City of Fear, Mozambique, Coast of Skeletons, Sandy the Seal, 24 Hours to Kill, The Face of Fu Manchu, Ten Little Indians, Marrakesh, Circus of Fear, The Brides of Fu Manchu, Sumuru, Five Golden Dragons, The Vengeance of Fu Manchu, Jules Verne's Rocket to the Moon House of a Thousand Dollas, The Face of Eve, Blood of Fu Manchu, 99 Women, Girl From Rio, Marquis de Sade's Justine, Castle of Fu Manchu, Venus in Furs. 1969: Philosophy in the Boudoir, Eugenie, Dorian Gray, Count Dracula, The Bloody Judge. Black Beauty, Night Hair Child, The Call of the Wild, Treasure Island, White Fang, Death in Persepolis, Ten Little Indians, End of Innocence, Black Cobra, Black Velvet-White Silk, Night of The High Tide, King Solomon's Treasure.

TOWNE, ROBERT: Writer. Was member of Warren Beatty's production staff on Bonnie and Clyde and made contributions to that screenplay.
PICTURES INCLUDE: Villa Rides, The Tomb of Ligeia, The Last Detail, Chinatown, Shampoo (co-s.p.), The Yazuka (co-s.p.).

TOWNSEND, CLAIRE: b. 1953. e. Princeton. Joined 20th Century-Fox in 1976; named west coast story editor & v.p., creative affairs. Left in 1978 to go to United Artists, where named v.p. of production, responsible for managing the acquisition, development and production of feature films.

TOWNSEND, LEO: Writer. b. Faribault, Minn., May 11, 1908; e. Minnesota U. Formerly radio and feature story magazine writer. In 1942 collab. s.p., It Started With Eve, collab. orig. s.p. Seven Sweethearts.
PICTURES INCLUDE: The Amazing Mrs. Halliday, Can't Help Singing, Chip Off the Old Block, Night and Day, Southside E-1000, One Big Affair, Dangerous Crossing, Siege at Red River, A Life in the Balance, White Feather, Running Wild, Fraulein, The Black Hand, Shadow On the Window, Flight to Hong Kong That Way With Women, Seven Sweethearts, I'd Rather Be Rich, Bikini Beach, Beach Blanket Bingo, How to Stuff a Wild Bikini, Fireball 500.
TV: Jane Wyman Theatre, Wagon Train, Maverick, 77 Sunset Strip, Hawaiian Eye, Dinah Shore Chevy Show, Shirley Temple Show, Perry Mason, Bourbon Street Beat, Beulah, Bachelor Father, Surfside Six, Destry, Patty Duke Show, Gidget, Man From U.N.C.L.E., My Three Sons, The Munsters, Batman, Andy Griffith Show, Bewitched.

TOWNSEND, PAULINE SWANSON: Writer. b. Athen, Ohio, e. Ohio University. Former Hollywood head of Tom Fizdale and Robert S. Taplinger p.r. agencies. Fiction and non-fiction writer for most national magazines.
TV: Death Valley Days, Patty Duke Show, Gidget, The Munsters, Batman, Andy Griffith Show, Bewitched.

TOYE, WENDY: Director. e. privately. Early career, ballet actress, dir., plays and musicals; entered m.p. industry 1952; dir., Stranger Left No Card (Award winning Short Cannes Festival 1953).
PICTURES INCLUDE: Three Cases Of Murder, Teckman Mystery, Raising A Riot, On The 12th Day (Nom. Best Short America 1955), All For Mary, True as a Turtle, We Joined The Navy, The Kings Breakfast.
STAGE: As You Like It, Majority Of One, Fledermaus, Orpheus in the Underworld, La Vie Parisienne, Virtue in Danger, Midsummer Night's Dream, Robert & Elizabeth, On the Level. Dir., Tribute to Noel Coward, Phoenix Th., Great Waltz, Soldiers Tale, Show Boat, Cowardly Custard, Stand and Deliver, She Stoops to Conquer, R Loves J, The Confederacy, Italian Girl in Algiers, Made in Heaven, Follow the Star, Make Me a World, Oh Mr. Porter, Once More With Music, Colette.
TV: Esme Divided, Chelsea at Nine, Orpheus in the Underworld, The Soldiers Tale, Cliff in Scotland, Girls Wanted,

Istanbul, A Goodley Manor for a Song, Girls, Girls, Girls, All Star Golden Gala ATV, Follow the Star.
OPERAS: Seraglio, Don Pasquale, The Impresario, Italian Girl in Algiers, The Merry Widow.

TRAMBUKIS, WILLIAM J.: Executive. b. July 26, 1926. Began career as usher with Loews in Providence, R.I. 1941. Served 1943–46 with Navy Seabees. Recipient of Quigley Awards. Managed/supervised Loews Theatres in several New England cities, Harrisburg, Pa., Syracuse, Rochester, Buffalo, N.Y., Washington, D.C., Richmond, Norfolk, Va., Toronto, Canada, Atlanta, Ga. Appt. Loews NorthEastern Division mgr. 1964, Loews gen. mgr. 1975: v.p. in 1976.

TRAVERS, BILL: Actor, Producer, Director. b. Newcastle-on-Tyne, England. Actor in repertory co.; London stage in Cage Me a Peacock, Damask Cheek, Square Ring, I Captured the Castle, 1961; A Cook for Mr. General (Broadway); Royal Shakespeare Theatre Co., 1962. Abraham Cochrane, 1964. Peter Pan, 1970.
PICTURES INCLUDE: Square Ring, Romeo and Juliet, Geordie, Footsteps in the Fog, Bhowani Junction, Barretts of Wimpole Street, Smallest Show on Earth, Seventh Sin, Passionate Summer, Bridal Path, Gorgo, The Green Helmet, Two Living—One Dead, Born Free, Duel at Diablo, A Midsummer Night's Dream, Ring of Bright Water, Boulevard du Rhum, The Belstone Fox, River of Sand, Bloody Ivy.
TV: A Cook for the General (Kraft), Episode, A Giant Is Born (U.S.A.), Espionage, Rawhide, CBS Voice of America (Rome), Lorna Doone, The Admirable Crichton, The Lion at World's End, Christian the Lion. Prod. dir. The Wild Doges of Africa, Baboons of Gombe, A Prospect of Whales, The Hyena Story, Death Trap, The Lions of the Serengeti, The Queen's Garden.

TRAVIS, RICHARD: Actor. r.n. William Justice. b. Carlsbad, N.M., Apr. 17, 1913. Started career in radio as announcer, sportscaster; writer and publisher of Cinemag. (theatrical tabloid); 3 yrs. U.S. Army.
PICTURES INCLUDE: The Man Who Came To Dinner, Postman Didn't Ring, Big Shot, Escape from Crime, Busses Roar, Truck Busters, Mission to Moscow, Last Ride, Jewels of Brandenburg, Alaska Patrol, Sky Liner, Operation Haylift, Passage West, Roaring City, Mask of Dragon, Lost Women, Girl in the Red Velvet Swing, Speed to Spare, Pier 21, Fingerprints Don't Lie, Annapolis Story, City of Shadows.
TV: Lucky Strike Theatre, Racket Squad, Lone Ranger, Schlitz Playhouse, Loretta Young Show, Treasury Men in Action, Pepsi Cola Playhouse, Ray Milland Show, Pride of the Family, The Falcon, Fury, Crossroads; host-narrator, Code 3, Missile to the Moon.

TRAVOLTA, JOHN: Actor. b. Englewood, N.J., Feb. 18, 1955. Quit school at 16 to pursue theatre career; first stage role in Who Will Save the Plowboy? Did off-Bdwy prod. of Rain; next to Broadway in Grease. Toured with latter for 10 months. Also in Over Here on Bdwy. with Andrew Sisters for 10 months.
PICTURES INCLUDE: Carrie, Saturday Night Fever, Grease, Moment by Moment, Urban Cowboy, Blow Out.
TV: Emergency, Owen Marshall, The Rookies, Medical Center, Welcome Back, Kotter (regular). Movie: The Boy in the Plastic Bubble.

TREE, DOROTHY: Actress. b. Brooklyn, N.Y. May 21, 1909; e. Cornell U.; m. Michael Uris, m.p. writer. On stage in: Street Scene, Clear All Wires, Torch Song, m.p. debut in "Life Begins" 1932; retired from screen 1943; wrote for m.p. in collab. with Michael Uris.
PICTURES INCLUDE: (cast) East of Fifth Avenue, Woman in Red, Night at the Ritz, Great Garrick, Having Wonderful Time, Confessions of a Nazi Spy, City in Darkness, Abe Lincoln in Illinois, Hitler Dead or Alive, Edge of Darness, Men, Asphalt Jungle, No Sad Songs for Me.
Under contract to Stanley Kramer Productions 1950.

TREEN, MARY LOU: Actress. b. St. Louis, Mo.; p. Helene Sullivan Treen, prof.; e. Ramona Convent; Denishawn School; Marion Morgan Dancers. With Fachon & Marco revues; on circuits; in early Bryan Foy short subjects; on stage in Little Boy Blew In; screen debut in Hot Tires.
PICTURES INCLUDE: Happiness Ahead, Strange Impersonation, From This Day Forward, She Wouldn't Say Yes, One Exciting Week, Let's Live a Little, The Stooge, Sailor Beware, Let's Do It Again, The Caddy, When Gangland Strikes, Birds and the Bees, Rockabye Baby, Paradise Hawaiian Style.

TREMAYNE, LES: Actor. b. London, England, Apr. 16, 1913; Northwestern U., Chicago Art Institute; Columbia, UCLA. First professional appearance in British m.p., 1916, with mother; stock, little theatres, vaudeville, 1925–40; entered radio field, 1931; numerous shows on all networks.
SHOWS INCLUDE: Woman in My House (with wife, Alice Reinheart), Errand of Mercy, You Are There, One Man's Family, Heartbeat Theatre, The First Nighter (lead 7 yrs.); on Broadway in Heads or Tails, Detective Story.
TV: Lux Video Theatre, 20th Century-Fox Hour, Navy Log, One Man's Family, Meet Millie, The Millionaire, The Whistler, Truth or Consequences, NBC Matinee, The Girl, O'Henry series, Rin Tin Tin, Bachelor Father, The Texan, Adventures of Ellery Queen, Court of Last Resort, Rifleman, State Trooper, Rescue 8, June Allyson-Dupont Show, Wagon Train, M Squad, Hitchcock Presents, Mr. Ed., Perry Mason.
PICTURES INCLUDE: The Racket, Blue Veil, Francis Goes to West Point, It Grows on Trees, I Love Melvin, Under the Red Sea, Dream Wife, War of the Worlds, Susan Slept Here, Lieutenant Wore Skirts, Unguarded Moment, Everything But the Truth, Monolith Monsters, Perfect Furlough, North by Northwest, Say One for Me, The Gallant Hours, The Angry Red Planet, The Story of Ruth.
Blue ribbon award for best perf. of the month for A Man Called Peter; dir. Hollywood Repertory Theatre, 1957; pres. Hollywood Actors' Council, 1951–58; various stage appearances, narrator of Air Force film, 1958; chmn. Actors Div. workshop com. Acad. TV Arts & Sciences; Member of The Workshop Comm. of the Hollywood M.P. & TV Museum Commission.

TRENT, JOHN: Director, Producer, Writer. b. London, England, 1935. e. London. Emigrated to Toronto, Canada in 1961. Pres. of Five Continents Prods. Ltd. Extensive television production including: Moment of Truth for NBC, Wojeck, Quentin Durgens M.P., and Whiteoaks of Jalna for CBC.
PICTURES INCLUDE: The Bushbaby, 1968, for MGM (prod.-dir.); Homer, 1969 for Palomar/Cinema Centre (dir.); The Man Who Wanted To Live Forever, 1970 for ABC/National General (dir.), Dead Of Night, 1972 for Quadrant (exec. prod.), Sunday In The Country, 1973 for Quadrant (dir.-writer), It Seemed Like A Good Idea At The Time, 1975 for Quadrant (dir.-co-writer), Middle Age Crazy, Misdeal.
TV: The Fighting Men and Riel for CBC.

TREVOR, CLAIRE: Actress. b. New York City; e. Amer. Acad. of Dram. Arts; Columbia U.; m. Milton H. Bren. On Broadway in Party's Over, Whistling in the Dark, Big Two.
PICTURES INCLUDE: Life in the Raw, Last Trail, Mad Game, Jimmy and Sally, Stagecoach, Allegheny Uprising, Dark Command, Murder, My Sweet, Johnny Angel, Crack-Up, Bachelor's Daughters, Born to Kill, Raw Deal, Valley of the Giants, Babe Ruth, Velvet Touch, Key Largo (Acad. Award), Lucky Stiff, Best of the Bad Men, Border Line, Hoodlum, Empire, Hard, Fast and Beautiful, My Man and I, Stop, You're Killing Me, Stranger Wore a Gun, High and the Mighty, Man Without a Star, Lucy Gallant, The Mountain, Majorie Morningstar, Two Weeks in Another Town, The Stripper, How to Murder Your Wife, Capetown Affair.
TV: Dodsworth (Academy Emmy).

TREXLER, CHARLES B.: Exhibitor. b. Wadesboro, N.C., Feb. 8, 1916. From 1937 to Nov. 1948 was practicing CPA except for 2 yrs. in U.S. Army in World War II. Joined Stewart & Everett Theatres in 1948 as controller. In March, 1953 named gen. mgr.; Jan. 1, 1954, exec. v.p., treas.; May, 1962 named pres. Retains title as treas. bd. mbr., NATO of North and South Carolina; v.p. & bd. mbr., National NATO.

TRIKONIS, GUY: Director. b. New York City. Started career in chorus of West Side Story on Bdwy. Turned to direction, making low-budget "weekenders" (films shot in 12 days only on weekends).
PICTURES: Moonshine County Express, The Evil, Touched by Love.
TELEVISION: Dark Side of Terror, Dressed To Kill, The Last Convertible (final three hours).

TRINDER, TOMMY: TV performer. b. London, March 24, 1909. Early career, music halls. Starred London Palladium 1939, Top Of The World 1940, Best Bib and Tucker 1942, Happy And Glorious 1944, Here, There And Everywhere 1947, Fancy Free 1951, Variety and several Command Performance shows. Entered m.p. industry 1936.
PICTURES INCLUDE: Almost A Honeymoon, Laugh It Off, Sailors Three, The Foreman Went To France, Champagne Charlie, Fiddlers Three, Bitter Springs, You Lucky People.
TV: Master of Ceremonies, Sunday Night At The London Palladium (ATV), My Wildest Dream, BBC, The Trinder Box.

TRINTIGNANT, JEAN-LOUIS: Actor. b. Aix-en-Provence, France, Dec. 11, 1930. m. Nadine Marquand. Theatre debut: 1951, To Each According to His Hunger. Then Mary Stuart, Macbeth (at the Comedie de Saint-Etienne). 1955 screen debut.
PICTURES INCLUDE: Si Tous Les Gars Du Monde, La Loi Des Rues, And God Created Woman, Club De Femmes, Les Liaisons Dangereuses, L'Ete Violent, Austerlitz, La Millieme Fenetre, Plein Feux Sur L'Assasin, Coeur Battant, L'Atlantide, The Game of Truth, Horace 62, Les Sept Peches Capitaux (7 Capital Sins), Le Combat Dans L'Ile, The Easy Life, Il Successo, Nutty, Naughty Chateau, Les

Pas Perdus, La Bonne Occase, Mata-Hari, Meurtre A L'Italienne, La Longue Marche, Le 17eme Ciel, Un Jour A Paris, Is Paris Burning?, The Sleeping Car Murder, A Man and a Woman, Enigma, Safari Diamants, Trans-Europ-Express, Mon Amour, Mon Amour, Un Homme A Abattre, La Morte Ha Fatto L'Uovo, Les Biches, Grand Silence, Z, Ma Nuit Chez Maud (My Night at Maud's), The Conformist, The Crook, Without Apparent Motive, The Outside Man, The French Conspiracy, Les Violons du Bal, The Sunday Woman.

TROELL, JAN: Writer-Director-Cinematographer. b. Sweden, 1931. Was teacher before entering industry. In early 60s photographed Bo Widerberg's first film, The Pram. Became apprentice in TV; made m.p. debut as director in 1965 with Stay in the Marshland.
PICTURES INCLUDE: Here Is Your Life, Eeny, Meeny, Miny, Mo., The Emigrants, The New Land, Zandy's Bride.

TROSPER, GUY: Writer. b. Lander, Wyo. Started as reader, Samuel Goldwyn; then story ed.; screen writer since 1941.
PICTURES INCLUDE: collab. s.p., Stratton Story; s.p., Devil's Doorway, Inside Straight; story; Pride of St. Louis; collab. s.p., Many Rivers to Cross; s.p., The Americano, Girl He Left Behind.

TRUFFAUT, FRANCOIS: Director, writer, critic. b. Paris, France, Feb. 6, 1932. Started working at 14 as office boy, messenger, and welder in a factory. Org. and ran small cine-clubs in Paris. Journalist in the m.p. section of "Work and Culture." Served in French Army 1951–53. Joined the m.p. section of the Ministry of Agriculture. Critic with Les Cahiers Du Cinema. 1955, 16mm short: Une Viste. 1956: became asst. to Roberto Rossellini, prepared three films (never produced). 1966: wrote The Cinema of Alfred Hitchcock. Acted in Close Encounters of the Third Kind.
FILMS DIRECTED INCLUDE: 1958, Les Mistons (won Dir. Award, Brussels film festival; Prix des Jeunes Spectactuers (Belgium); Germany's Gold Medal; U.S.A. Blue Ribbon Award); 1959 Histoire D'eau (short); 1959: The Four Hundred Blows (Grand Prix Cannes; Grand Prix Fr. Catholic Office Cannes, N.Y. Critics Award for Best Foreign Film; David O. Selznick Award, Oscar for best story written directly for screen). Shoot the Piano Player (Prize for photography in Germany; New Critics Prize France); Jules and Jim (Prix del la Mise en Scene, Mar del Plata and Acapulco Festivals; Danish Oscar Bodil for best European film of the year); 1962, Love at Twenty; 1963, Soft Skin; The Bride Wore Black; Stolen Kisses; Mississippi Mermaid; The Wild Child; Bed and Board; Two English Girls, Such a Gorgeous Kid Like Me, La Nuit Americain, The Story of Adele H, The Man Who Loved Women.

TRYON, TOM: Actor. b. Hartford, Conn., Jan. 14. 1926. e. Yale U. Served in U.S. Navy, World War II, studied at Art Students League; with Cape Playhouse, Dennis, Mass., as set painter, asst. stage mgr., actor; prod. asst., CBS; then TV actor.
PLAYS INCLUDE: Wish You Were Here, Cyrano de Bergerac, Richard III.
PICTURES INCLUDE: m.p. debut in Scarlet Hour; since in Screaming Eagles, Three Violent People, Moon Pilot, Marines Let's Go, The Cardinal, The Glory Guys.
AUTHOR: The Other, Crowned Heads.

TSUKASA, YOKO: Actress. b. Tottori, Japan, Aug. 20, 1934. e. Kyoritsu College; joined Toho Studio 1954 after period as magazine cover girl.
PICTURES INCLUDE: Don't Die My Darling, Blue Beast, Eternity of Love, End of Summer, Three Treasures, Yojimbo (The Bodyguard), Women of Design.

TUCHOCK, WANDA: Writer, producer. b. Pueblo, Colo., Mar. 20; e. U. of Calif., B.A. Wrote many pictures, directed Finishing School.
PICTURES INCLUDE: Hallelujah!, The Champ, Sporting Blood, Susan Lennox, Nob Hill, Sunday Dinner for a Soldier, The Homestretch, Foxes of Harrow.
Member: SWG; Acad. of M.P. Arts & Sciences, NSTP.

TUCKER, FORREST: Actor. b. Plainfield, Ind., Feb. 12, 1919; e. George Wash. U., Wash., D.C. Screen debut, 1940, in The Western, foll. by 42 feature roles, 1940–51, first starring role, Rock Island Trail, Rep., 1950. Star of Tomorrow, 1952. Member: Masquers Club of Hollywood, Lakeside Golf.
PICTURES INCLUDE: The Yearling, The Howards of Virginia, Never Say Goodbye, Warpath, Sands of Iwo Jima, Oh! Susanna; Crosswinds, Fighting Coast Guard, Wild Blue Yonder, Flaming Feather, Bugles in the Afternoon, Hoodlum Empire, Montana Belle, Hurricane Smith, Ride the Man Down, San Antone, Flight Nurse, Pony Express, Jubilee Trail, Laughing Anne, Trouble in the Glen, Rage at Dawn, Vanishing American, Finger Man, Night Freight, Paris Follies of 1956, Break in the Circle, Counterplot, The Night They Raided Minsky's, Auntie Mame, Barquero, Chisum, The Wild McCullocks.
TV: Crunch and Des, F Troop; Rad. Show: WCFL, Chicago.

Chicago Co., Plaza Suite, Oct. 1968 to June, 1969. Now assoc. prod. Drury Lane Theatre, Ill. Mem., Lambs Club, Friars Club, N.Y. Opened Nat'l Co., Music Man, L.A., on tour.

TUCKER, MELVILLE: Executive. b. New York City, Mar. 4, 1916; e. Princeton U. Asst. purchasing agent Consolidated Laboratories, N.Y., 1934–36; sound effects & picture ed., Republic Productions, Inc. 1936–8; then asst. production mgr. & first asst. dir., 1938–42; served in U.S. Army 1942–46; asst. prod. Republic 1946; assoc. producer, 1947–52; prod., Universal 1952–54; prod. exec. v.p., Universal, 1955–70; prod.-Verdon Prods. 1971–present.
PICTURES INCLUDE: The Missourians, Thunder in God's Country, Rodeo King and the Senorita, Utah Wagon Train. U-I prod., 1953: Drums Across the River, Black Shield of Falworth; production exec. U-, 1954–71. prod., Verdon Prod., prod. A Warm December 1972; prod., Uptown Saturday Night, Producer, Let's Do It Again, 1975; prod., A Piece of the Action, 1977; exec. prod., Stir Crazy, 1980.

TULIPAN, IRA H.: Publicist. b. New York City. e. N.Y.U. 1934, B.S. Entered m.p. ind. as theatre mgr., Boston, 1934; pub. dept. Warner Bros. home office, 1935–40; joined 20th Century-Fox 1942; U.S. Army service 1943–46; returned to Fox upon disch., feature writer, press book ed.; trade paper contact; newspaper contact; pub. mgr. 1955; ass't. dir. adv. pub. expl. Columbia Pictures. 1960; exec. adm. asst. to adv., publ., v.p., Columbia, 1963; dir. overseas prod. pub., 1966. Returned to U.S. in 1978 as eastern pub.-coordinator.

TULLY, MONTGOMERY: Director-writer. b. Ireland. e. Kings College, U. London, B. Sc. Before entering m.p. indus. in 1929 as writer, was in advertising bus.; wrote short stories. Apptd. dir.-writer, 1937; documentaries, 1929–44.
PICTURES INCLUDE: (dir. & s.p.) Murder in Reverse, Spring Song, Mrs. Fitzherbert; (s.p.) Waltz Time, For You Alone, and Lisbon Story, Punchdrunk (s.p.); Boys in Brown (dir. s.p.); Tale of Two Cities, Five Women (dir.); orig. s.p. Berlin Story; dir. Girdle of Gold, Small Town Story, 36 Hours, The Diamond, Five Days, Devil's Point, The Hypnotist, Counterfeit Plan; dir., s.p., Dial 999 (Way Out), No Road Back, One Man's Secret; dir. Scotland Yard and Man From Interpol series. Dir. Jackpot, Dead Lucky, The House on Marsh Road, The Man Who Was Nobody, 1959–61; Third Alibi, She Knows Y'Know, Out of the Fog, Checkmate, 1961–62; Clash By Night, Boy With A Flute, 1964–65. Dir. Who Killed The Cat? The Terrornauts, Battle Beneath the Earth, 1966. s.p. The Hawks; s.p. The Ballbreakers, 1967–68. The Fighting Contessa, 1969.

TULLY, TOM: Actor. b. Durango, Colo., Aug. 21. Appeared in many New York plays.
PLAYS INCLUDE: Time of Your Life, White Steed, Jason and Ah, Wilderness.
PICTURES INCLUDE: Came to Hollywood 1943. Since: Till the End of Time, Kiss and Tell, I'll Be Seeing You, The Unseen, Destination Tokyo, Intrigue, Scudda Hoo Scudda Hay, Killer McCoy, Illegal Entry, Branded, Tomahawk, Caine Mutiny, Love Me Or Leave Me, Soldier of Fortune, Behind the High Wall, 10 North Frederick, The Wackiest Ship in the Army, Blood on the Moon, Coogan's Bluff.
TV: The Lineup (network series), Philco Playhouse, Ford Theatre, Celebrity Playhouse, Front Row Center, San Francisco Beat, Alfred Hitchcock. 1959–61: dir., s.p., Strange Awakening, I Only Asked, Price of Silence, Man From Interpol (series), Jackpot, Dead Lucky, The House In March Road, The Man Who Was Nobody. 1962: Third Alibi, She Knows Y'Know, Out of The Fog. 1963–64: Checkmate, Clash By Night, Boy With A Flute.

TUNBERG, KARL: Writer. b. Spokane, Wash. From 1937 collab. many Hollywood s.p.
PICTURES INCLUDE: You Can't Have Everything, My Lucky Star, Hold That Co-Ed. Down Argentine Way, Yank in the RAF. In 1942 (collab. s.p.) My Gal Sal, Orchestra Wives, Lucky Jordan, (collab. s.p.) Dixie, Standing Room Only, Bring on the Girls; (prod. & collab. s.p.) Kitty; (prod.) You Gotta Stay Happy; (collab. s.p.) Love That Brute, Night Into Morning, Law and the Lady, Because You're Mine, Scandal at Scourie, Valley of the Kings; s.p. Beau Brummell, Scarlet Coat, Seventh Sin, Ben Hur, s.p. Count Your Blessings; s.p. Libel, Taras Bulba; s.p., I Thank A Fool; s.p. The Seventh Dawn, Harlow (story); col. s.p., Where Were You When the Lights Went Out?

TUNICK, EUGENE: Executive. b. Cincinnati, Ohio, Oct. 21, 1920. e. Hughes High School, 1934–38; Univ. of Cincinnati, 1938–39. 1941, shipping clerk, RKO, Cincinnati. 1942, Armed Forces. 1944–45, Third Booker, Second Booker, Office Manager-RKO Cincinnati, 1946, Salesman, West Virginia and Ohio for RKO, Cincinnati. 1948, Sales-manager, RKO. 1949, Salesmanager, Eagle Lion, N.Y.C. Oct. 1949–51, Branch Manager Eagle Lion, Indianapolis. 1951–52, Lippert Branch Mgr. Indianapolis. 1952–53, Lippert franchise holder, Cincinnati. Dec. 1954, United Artists Branch Mgr. Philadelphia. 1957, District Mgr. U.A., Phila. Sept. 1961, Eastern and Canadian Div. Mgr. U.A., N.Y. June 1968, Assistant General Sales Mgr., National General Pictures. June 1969, V.P. and General Sales Mgr. National General Pictures. Jan. 1, 1970, Executive V.P. General Sales Mgr. NGP; exec. v.p., Mid States Theatres, 1971; exec. v.p., National Amusements (Redstone), 1975; 1977, v.p. General Cinema Theatres (Western Div.); 1978, v.p. in charge of Special Projects, Amer. Int'l Pictures; later in year named v.p. & gen. sls. mgr.; 1979, v.p. & gen. sls. mgr., Filmways; promoted to exec. v.p. & gen. sls. mgr.; 1981.

TUNICK, IRVE: Writer. b. New York, N.Y.; e. N.Y.U. and Georgetown U. Commenced writing career with radio scripts; later senior writer (TV) for Cavalcade of America and You Are There, stories produced by Studio One and Armstrong Circle Theatre; five radio and TV scripts won Peabody Awards; script editor for Princess Pictures 1953–55; orig. s.p. for Lady of Vengeance; s.p. for High Hell.

TURELL, SAUL J.: Writer, Producer, Executive. b. N.Y.C., Jan. 20, 1921. e. City Coll., U. of Chicago. B.A.—Univ. of The State of N.Y. Started Sterling Films, 1946; merged Walter Reade-Sterling, 1961; pres., 1961–65. Presently: pres., Janus Films Inc. Adjunct Assoc. Prof. N.Y.U.
PICTURES INCLUDE: The Great Chase, The Love Goddesses, My Home Is Copacabana, The Art of Film Series, Robeson: A Film Tribute. (Oscar, 1980).
TV: King's Crossroads, Adventure Theatre, Killiam Show, Movie Museum, Silents Please, The Big Moment, Hollywood The Golden Years, The Legend of Valentino.
PRIZE WINNING SHORT SUBJECTS INCLUDE: Sandy Steps Out, Hot Ice, The Princess and the Dragon, Death in the Arena, The Fun Factory.

TURFKRUYER, MARC: Journalist. Pres. Antwerp Sec. of the Belgian Film Press Assn.; editor of Weekblad Cinema, the oldest Belgian trade paper; contributor to the daily paper Volksgazet (Antwerp) and to the weeklies ABC and Het Toneel, Motion Picture Herald (New York); Pres., International Film Press Assn.

TURMAN, LAWRENCE: Producer. b. Los Angeles, 1926. e. UCLA. In textile business 5 years, then joined Kurt Frings Agency; left in 1960 to form Millar-Turman Prods.
PICTURES INCLUDE: (prod.) The Young Doctors, I Could Go on Singing, The Best Man. Formed own prod. co., Lawrence Turman, Inc., to make The Flim-Flam Man, The Graduate, Pretty Poison, The Drowning Pool, First Love, Heroes, Walk Proud, Caveman (co-prod.).

TURNER, CLIFFORD: Producer. b. Leeds, England, 1913. ent. m.p. industry as cutting room asst. Gaumont British. Edited number early British pictures, before going to Hollywood in 1935. Subsequently edited for Warners, Columbia, Universal, Fox. Returned to England 1948 to edit The Small Back Room. Dir. and exec. producer since 1950 in Hollywood and New York. Formed Boulevard Film Productions Ltd., Screen Biographies International Inc., Television Enterprises Inc. Four Against the Bank of England. 1972: Utrillo. 1975, Rose of Cimarron, 1976 Mystery of the General Grant; La Cicatrice, The Valadon Story.

TURNER, LANA: Actress. b. Wallace, Idaho, Feb. 8, 1921; p. Virgil Turner, prof.
PICTURES INCLUDE: They Won't Forget, Great Garrick, Adventures of Marco Polo, Love Finds Andy Hardy; Rich Man, Poor Girl; Dramatic School, Calling Dr. Kildare, These Glamour Girls, Dancing Coed. Two Girls on Broadway, We Who Are Young, Ziegfeld Girl, Dr. Jekyll and Mr. Hyde, Johnny Eager, Slightly Dangerous, Youngest Profession, Marriage Is a Private Affair, Keep Your Powder Dry, Week-End at the Waldorf, Postman Always Rings Twice, Green Dolphin Street, Cass Timberlane, Homecoming, Three Musketeers, Life of Her Own, Mr. Imperium, The Merry Widow, Bad & the Beautiful, Latin Lovers, Flame & the Flesh, Betrayed, The Prodigal, Sea Chase, Rains of Ranchipur, Diane, Lady and the Flyer, Imitation of Life, Portrait in Black, By Love Possessed, Bachelor in Paradise, Who's Got the Action?, Love Has Many Faces, Madame X, The Big Cube, Persecution, Bittersweet Love.
TV: The Survivors.

TUSHINGHAM, RITA: Actress. b. Liverpool, England, 1942. Student at Liverpool Playhouse, Motion pic. debut 1961 in A Taste of Honey.
PICTURES INCLUDE: The Leather Boys, A Place to Go, Girl With Green Eyes, The Knack, Dr. Zhivago, The Trap, Smashing Time, Diamonds for Breakfast, The Guru, The Bedsitting Room, Straight on 'til Morning, Situation, Instant Coffee, The Human Factor, Rachel's Man, The Slum Boy, Green Eyes, The Black Journal, Bread, Butter and Jam, Mysteries.

TUTIN, DOROTHY: Actress. b. London, Eng., Apr. 8, 1930; e. St. Catherine's School Bramley, Guildford (Surrey). Stage debut in The Thistle & the Rose, 1949.
PLAYS INCLUDE: Much Ado About Nothing, The Living Room, I Am a Camera, The Lark, Wild Duck, Juliet, Ophelia, Viola, Portia, Cressida, Rosalind, The Devils, Once More With Feeling, The Cherry Orchard, Victoria Regina-Portrait of a Queen, Old Times, Peter Pan, What Every Woman Knows, Month in the Country, Macbeth, Antony and Cleopatra, Undiscovered Country, Reflections.
PICTURES INCLUDE: Screen debut in The Importance of Being Earnest. Also: The Beggar's Opera, A Tale of Two Cities, Cromwell, Savage Messiah.
TV: Living Room, Victoria Regina, Invitation to a Voyage, Antigone, Colombe, Carrington V.C., The Hollow Crown, Scent of Fear, From Chekhov With Love, Anne Boleyn in The Six Wives of Henry VIII, Flotsam and Jetsam, Mother & Son, South Riding, Willow Cabins, Ghosts, Sister Dora.

TWIGGY: Recording Artiste. b. London, Sept. 19, 1949. r.n. Leslie Hornby. m. actor Michael Witney. At 17 regarded as world's leading high fashion model. Made m.p. debut in The Boy Friend (1971).
TV: Hosted and starred in major American & British music shows including Twiggy (UK), Twiggy and Friends (UK), and Juke Box (US).
PICTURES INCLUDE: W, There Goes the Bride, The Blues Brothers.

TWYMAN, ALAN P.: Executive. b. Dayton, Ohio, May 30, 1934. e. Univ. of Cincinnati. Twyman Films, Inc. sales 1958, vice pres.-pres., 1975. NAVA board of directors, 1964–1969; pres. 1970, chm. of bd., 1972.

TYLER, BEVERLY: Actress. r.n. Beverly Jean Saul. b. Scranton, Pa., July 5, 1928; e. Scranton H.S. Church choir singer; on radio, TV.
PICTURES INCLUDE: Green Years, My Brother Talks to Horses, Beginning or End, The Palomino, The Fireball, Cimarron Kid, Battle at Apache Pass.

TYRRELL, SUSAN: Actress. b. San Francisco, 1946. Made first prof. appearance with Art Carney in summer theatre tour prod. of Time Out for Ginger. Worked in off-Bdwy. prods. and as waitress in coffee house before attracting attention in Lincoln Center Repertory Co. prods. of A Cry of Players, The Time of Your Life, Camino Real.
PICTURES INCLUDE: The Steagle, Shoot Out, Fat City, Catch My Soul, Islands in the Stream, Andy Warhol's Bad, I Never Promised You A Rose Garden, Another Man, Another Chance, September 30, 1955.

TYSON, CICELY: Actress. b. New York City, Dec. 19, 1933.
PICTURES INCLUDE: A Man Called Adam, The Comedians, The Heart Is a Lonely Hunter, Sounder, A Hero Ain't Nothin, But a Sandwich, The Concorde—Airport '79.
TELEVISION: The Autobiography of Miss Jane Pittman, Roots, East Side, West Side, A Woman Called Moses, King.

U

UGGAMS, LESLIE: Singer. b. New York, May 25, 1943. e. Professional Children's Sch., grad., 1960. Juilliard. Beg. singing career age 5. TV debut: age 7, Johnny Olsen's TV kids. Records for Columbia Records.
STAGE: Hallelujah Baby.
RADIO: Peter Lind Hayes-Mary Healy Show, Milton Berle, Arthur Godfrey, Star Time.
TV: Milton Berle Show, Name That Tune, Jack Paar Show, Garry Moore, Sing Along With Mitch.

ULLMAN, ELWOOD: Writer. b. Memphis, Tenn.; e. Washington U., St. Louis, BA. Newspaper reporter, St. Louis Post-Dispatch; m.p., TV & mag. writer.
PICTURES INCLUDE: Martin and Lewis, Abbott and Costello, Bowery Boys films, Kettle films, Battle Flame, Three Stooges, Tickle Me, Dr. Goldfoot and The Bikini Machine.
FICTION: Country Gentleman, Sat. Eve. Post, Esquire, etc.

ULLMANN, LIV: Actress. b. Japan of Norwegian parents. Accompanied parents to Canada when World War II began and later returned to Norway. Was catapulted to fame in a succession of Swedish films directed by Ingmar Bergman.
PICTURES INCLUDE: Swedish: Persona (debut), Hour of the Wolf, Shame, The Passion of Anna, The Emigrants, Cries and Whispers; American: The Devil's Imposter (formerly Pope Joan), Lost Horizon, 40 Carats, Zandy's Bride, The Abdication, Face to Face, A Bridge Too Far, The Serpent's Egg, Autumn Sonata.

UNGER, ANTHONY B.: Executive Producer. b. New York, Oct. 19, 1940. e. Duke Univ. and Univ. of Southern Calif. Prod. ass't Third Man, TV series, 1961. v.p. Unger Productions, Inc., 1964; v.p. Landau-Unger Co., Inc., 1965; v.p. Commonwealth United Entertainment in London, 1968.
PICTURES INCLUDE: 1966: assoc. prod., The Desperate Ones. 1968: assoc. prod., The Madwoman of Chaillot. 1969: exec. prod., The Battle of Neretva, The Magic Christian, Julius Caesar. 1970: exec. prod., The Devil's Widow. 1973: exec. prod., Don't Look Now.

UNGER, KURT: Producer. b. Berlin, Jan. 10, 1922. Entered ind. in 1939 in chg. m.p. entertainment British troops in Middle East. Subsequently distributor for United Artists Corp. in Israel and Italy.
PICTURES INCLUDE: Judith, Best House in London, Puppet on a Chain, Pope Joan (The Devil's Imposter).

UNGER, STEVEN A.: Executive. Started as independent prod. and dist. of theatrical and TV films. In June, 1978, joined Universal Pictures Intl. Sales as foreign sls. mgr. Named v.p. Universal Theatrical Mot on Pictures in 1979, responsible for licensing theatrical or TV features not handled by C.I.C. in territories outside U.S. & Canada. In 1981 joined CBS Theatrical Films as int'l. v.p., sls.

URICH, ROBERT: Actor. b. Toronto, Ohio, Dec. 19. e. Florida State Univ. (BA in radio and TV communications). Appeared in university plays. Was sales account executive at WGN Radio, Chicago, before turning to stage acting.
TELEVISION: The FBI, Gunsmoke, Kung Fu, Marcus Welby, MD, S.W.A.T., Boy & Carol & Ted & Alice, Soap, Tabitha, The Love Boat, Vega$.
PICTURE: Magnum Force

URQUHART, ROBERT: Actor, writer. b. Scotland, October 16, 1922; e. George Heriots, Edinburg. Served in Merchant Navy 1938–45; stage debut, Park Theatre, Glasgow; screen debut: You're Only Young Twice, 1951.
PICTURES INCLUDE: Isn't Life Wonderful, The House Of The Arrow, Knights Of The Round Table, Happy Ever After (Tonight's The Night), Golden Ivory, The Dark Avenger, You Can't Escape, Yangtse Incident, Curse of Frankenstein, Dunkirk, The Trouble with Eve, Danger Tomorrow, Foxhole in Cairo, Murder in Mind, The Bulldog Greed, 55 Days At Peking, The Break, Murder at the Gallup, The Syndicate, The Limbo Line, The Looking Glass War, Brotherly Love (Country Dance).
TV: Tamer Tamed, Infinite Shoeblack, Morning Departure, The Human Touch, The Iron Harp, Sleeping Clergyman, The Naked Lady, For Services Rendered, The Bright One, Jango, Murder Swamp, She Died Young, Plane Makers series, Reporter, Inheritors series; 1976 series: Mr. Goodall, The Nearly Man, The Button Man, Happy Returns.
AUTHOR: (Wrote TV) House of Lies, End of the Tether, Landfall, The Touch of a Dead Hand.

USTINOV, PETER: Actor, writer, dir. b. London, Eng., 1921; e. Westminster Sch. In Brit. Army, World War II. On Brit. stage from 1937. Screen debut 1941 in Brit. picture Mein Kampf, My Crimes.
PICTURES INCLUDE: The Goose Steps Out, One of Our Aircraft Is Missing, Let the People Sing, The Way Ahead, The True Glory, (collab. s.p.) The Way Ahead; (wrote, dir. & co-prod.) School for Secrets, Vice Versa; (adapt., dir., co-prod.) Private Angelo, Odette, Quo Vadis, Hotel Sahara, The Egyptian, Beau Brummell, We're No Angels, Lola Montez, The Spies, An Angel Flew Over Brooklyn, The Sundowners, Spartacus; prod., scripted, acted, Romanoff and Juliet; dir., prod./scpt., acted, Billy Budd, Topkapi, John Goldfarb Please Come Home; prod. dir., Lady L., Blackbeard's Ghost, The Comedians, Hot Millions, Viva Max. Dir. and acted, Hammersmith Is Out, acted Big Truck, Poor Clare, Logan's Run, Treasure of Matecumbe, One of Our Dinosaurs Is Missing, Le Taxi Mauve, The Last Remake of Beau Geste, Doppio Delitto, Death on the Nile, Charlie Chan and the Curse of the Dragon Queen, The Great Muppet Caper.
STAGE: Romanoff and Juliet, N.Y., London; and 17 other plays. Dir., acted, Photo Finish; wrote, Life In My Hands, The Unknown Soldier and His Wife, Half Way Up The Tree.

V

VACCARO, BRENDA: Actress. b. Brooklyn, N.Y., Nov. 18, 1939. e. Thomas Jefferson High School, Dallas; studied two yrs. at Neighborhood Playhouse in N.Y. Was waitress and model before landing first Bdwy. role in Everybody Loves Opal. Toured in Tunnel of Love and returned to N.Y. for role in The Affair. Won Tony for best supporting actress in Cactus Flower.
PICTURES INCLUDE: Midnight Cowboy (debut), I Love My Wife, Summertree, Going Home, Once Is Not Enough, Airport '77, House by the Lake, Capricorn One, The First Deadly Sin, Zorro, the Gay Blade.
TV: The F.B.I., The Name of the Game, The Helen Reddy Show, The Shape of Things (special), Sunshine (tv movie), etc.

VALE, EUGENE: Writer. b. April 11, 1916. e. Zurich, Switzerland. m. Evelyn Wahl. Story and s.p., The Second Face, The Shattered Dream, 1954 SWG award nom., best written telefilm; The Dark Wave. 1957, m.p. academy award nominations.
TV: Four Star Playhouse, Fireside Theatre, 20th Century Fox Hour, Schlitz Playhouse, Hollywood Opening Night, NBC, Crusader, Lux Video Theatre, Danger, CBS, Chevron Theatre, Douglas Fairbanks, Pepsicola Playhouse, Waterfront, Christophers, Cavacade of America, Hallmark Hall of Fame.
AUTHOR: Technique of Screenplay Writing.

VALENTE, CATERINA: Singer. b. Paris, France.
RECORDINGS INCLUDE: Malaguena, in German; I Happen to Like New York.
TV: The Bing Crosby Show, The Ed Sullivan Show, The Entertainers.

VALENTI, JACK J.: Executive. b. Sept. 5, 1921. e. Univ. of Houston, B.A., 1946; Harvard, MBA, bus. admin., 1948. Air force pilot in European theatre, WWII; adv. and pub. rel. exec. in Houston; special asst. and advisor to Pres. Lyndon B. Johnson, 1963–66; pres., MPAA; elected pres., MPAA, MPEA and AMPTP, June, 1966.

VALENTINO, THOMAS J.: Chairman, Thomas J. Valentino, Inc. b. New York, N.Y., Apr. 27, 1907; e. high sch. Cartoonist; piano & organ tuner R. Wurlitzer Co., then Starr Piano Co., 1924–29; Gennett Recording Co., N.Y.; synchronized sound shorts, feature films; sound consultant, Major Records, N.Y.; owns large library of mood music & sound effects. Elected v.p., Recording Ass'n of America, 1969–71.

VALLEE, RUDY: Actor, r.n. Hubert Prior Vallee; b. Island Pond, Vt., July 28, 1901; e. U. of Maine, Yale U. After college began career in entertainment as singer with dance orchestra; org. own band, Connecticut Yankees, popular dance orchestra, later also feature various New York musicals and on radio. First screen dramatic work in The Vagabond Lover, RKO; later various shorts and feature appearances with orchestra. In 1942 role in The Palm Beach Story, Par. In 1943, Happy Go Lucky, Par. In World War II chief petty officer (bandmaster) U.S. Coast Guard. In 1945 formed (with Leslie Charteris) Saint Enterprises, Inc.
PICTURES INCLUDE: Man Alive, People Are Funny, It's in the Bag, The Fabulous Suzanne, Bachelor and Bobby Soxer, Beautiful Blonde from Bashful Bend, Father Was a Fullback, Mother Is a Freshman, Admiral Was a Lady, Mad Wednesday, Ricochet Romance, Gentlemen Marry Brunettes, How to Succeed in Business Without Really Trying, Live a Little, Love a Little.

VALLI, ALIDA: Actress. b. Pola, Italy, May 31, 1921. e. M.P Academy, Rome (dramatics); m. Oscar de Mejo, pianist-composer. In Italian m.p.; won Venice Film Festival Award in "Piccolo Monde Antico" (Little Old World).
PICTURES INCLUDE: Vita Ricomincia, Giovanna; to U.S. 1947; U.S. m.p. debut in Paradine Case, 1947; later pictures: Miracle of the Bells, The Third Man, Walk Softly Stranger, White Tower, Lovers of Toledo, Stranger's Hand, The Castilian, Ophelia, Spider's Stratagem, The Cassandra Crossing, Susperia, 1900.

VALLIN, RICK: Actor. b. Russia; son of Nedja Yatsenko, ballerina. In stage stock, radio; joined Pasadena Playhouse, 1942. Screen debut 1942.
PICTURES INCLUDE: King of the Stallions, Nearly Eighteen, Corregidor, Lady from Chungking, A Night for Crime, Panther's Claw, Northwest Outpost, Two Blondes and a Redhead, Jungle Jim, Shamrock Hill, Comanche Territory, Aladdin and His Lamp, Magic Carpet, Woman in the Dark, Jungle Manhunt, Strange Fascination, Voodoo Tiger, Star of Texas, many Westerns, Bowery to Bagdad, Treasure of Ruby Hills, Dial Red O.

VALLONE, RAF: Actor. b. Turin, Italy; e. U. of Turin. Newspaper writer; m.p. debut in Bitter Rice.
PICTURES INCLUDE: Under the Olive Tree, Anna, Path of Hope, White Line, Rome 11 O'Clock, Strange Deception, Anita Garibaldi, Daughters of Destiny, Teresa Raquin, Riviera, The Secret Invasion. 1961: Two Women, El Cid, A View From the Bridge, Phaedra, Kiss The Girls and Make Them Die, The Desperate Ones, The Cardinal, The Italian Job, The Kremlin Letter, Summertime Killer, Rosebud, The Human Factor, The Other Side of Midnight, The Greek Tycoon Lion of the Desert.

VAN CLEEF, LEE: Actor. b. Somerville, N.J., Jan. 9, 1925. e. Somerville H.S., 1942. Joined U.S. Navy, 1942; asst. mgr. in summer camp, public accountant; then joined little theatre group.
PICTURES INCLUDE: High Noon, Beast from 20,000 Fathoms, Vice Squad, The Nebraskan, Gypsy Colt, Arrow in the Dust, Yellow Tomahawk, Dawn at Socorro, Princess of the Nile, Ten Wanted Men, The Conqueror, Big Combo, Treasure of Ruby Hills, I Cover the Underworld, Road To Denver, Posse from Hell, The Man Who Shot Liberty Valance, A Man Alone, Vanishing American, Tribute to a Bad Man, For a Few Dollars More, The Good, The Bad and The Ugly, The Big Gundown, Death Rides a Horse, Day of Anger, Barquero, El Condor, The Magnificent Seven Ride, Return of Sabata, Escape from New York.

VAN DOREN, MAMIE: Actress. r.n. Joan Lucille Olander. b. Rowena, S.D., Feb. 6, 1933. e. Los Angeles High Sch. Secy. law firm, L.A.; prof. debut as singer with Ted Fio Rita orch.
THEATRE: Appeared in many stock plays incl.: Once in a Lifetime, Boy Meets Girl, Come Back Little Sheba.
PICTURES INCLUDE: (m.p. debut) Forbidden; All American, Yankee Pasha, Francis Joins the Wacs, Ain't Misbehavin, Second Greatest Sex, Running Wild, Star in the Dust, Untamed Youth, Girl in Black Stockings, Teachers Pet, The Navy Vs. The Night Monsters.

VAN DYKE, DICK: Actor. b. West Plains, Mo., Dec. 18, 1925. U.S.A.F., WWII. partner, ad. agency, nightclub acts: Merry Mutes, Eric and Van.
THEATRE: The Girls Against the Boys, Bye Bye Birdie.
PICTURES INCLUDE: Bye Bye Birdie, Mary Poppins, What a Way to Go. The Art of Love, Lieutenant Robinson Crusoe, Divorce American Style, What's New Fitzwilly?, Chitty Chitty Bang Bang, Some Kind of Nut, The Comic, Cold Turkey, The Runner Stumbles.
TV: The Merry Mute Show, The Music Shop, The Dick Van Dyke Show (variety prog.), Cartoon Theater, The New Dick Van Dyke Show, The Carol Burnett Show.

VAN FLEET, JO: Actress. b. Oakland, Calif. e. College of the Pacific. Neighborhood Playhouse.
THEATRE: On Broadway in Winter's Tale, Whole World Over, Closing Door, King Lear, Flight into Egypt, Camino Real, Trip to Bountiful (Tony Award); Look Homeward Angel (Critics Award); The Glass Menagerie; Off Broadway: The Alligators, Oh Dad, Poor Dad, Mama's Hung You In The Closet and I'm Feeling So Sad.
PICTURES INCLUDE: East of Eden (Academy Award, best supporting actress, 1955); I'll Cry Tomorrow (Look Award), Rose Tattoo, This Angry Age, King and Four Queens, Gunfight at the OK Corral; Wild River, Cool Hand Luke, I Love You, Alice B. Toklas, Gang Who Couldn't Shoot Straight, The Tenant.
TELEVISION: Cinderella, Bonanza, Mod Squad, Power, etc.

VAN HEUSEN, JIMMY: Composer. b. Syracuse, N.Y., Jan. 26, 1919. e. Syracuse U. Pianist with publishing houses; songs for many m.p. including Sunday, Monday or Always, Swinging on a Star (Acad. Award 1944), Sunshine Cake, etc.
PICTURES INCLUDE: Road to Rio, Emperor Waltz, Connecticut Yankee, Mr. Music, Riding High, Road to Bali, Little Boy Lost, All the Way (Academy Award), Call Me Irresponsible (Academy Award), won Emmy for Love and Marriage; Songwriters Hall of Fame.

VAN PALLANDT, NINA: Actress. b. Copenhagen, Denmark. e. Univ. of So. Calif. Returned to Denmark where married Baron Frederik Van Pallandt with whom she had appeared as folk singer throughout Europe. Made 3 films with him; went on world tour together. Now divorced. Has appeared in New York as singer.
PICTURES: The Long Goodbye, A Wedding, Quintet, American Gigolo, Cloud Dancer, Cutter and Bone.
TELEVISION: The Sam Shepherd Murder Case.

VAN PATTEN, DICK: b. N.Y.C., Dec. 9. Began career as child actor with Bdwy. debut at 7 yrs., playing son of Melvyn Douglas in Tapestry in Gray. Has worked since in stage, radio, TV, films.
PICTURES INCLUDE: Making It, Joe Kidd, Soylent Green, Dirty Little Billy, Westworld, Gus, Treasure of Matecumbe, High Anxiety.
TV: Series: Mama, The Partners, The New Dick Van Dyke Show, Eight Is Enough. Guest Roles: Arnie, The Rookies, Cannon, Banyon, The Little People, The Streets of San Francisco, When Things Were Rotten.
STAGE: The Lady Who Came to Stay, O Mistress Mine, On Borrowed Time, Ah, Wilderess, Watch on the Rhine, The Skin of Our Teeth, Kiss and Tell, Mister Roberts, I Was Dancing.

VAN PEEBLES, MELVIN: Director, writer, composer. Was navigator-bombardier in Air Force, drove cable car in San Francisco.
AUTHOR: A Bear for the F.B.I., Don't Play Me Cheap (play) won 1st prize, Belgian Festival; and photo-essay book on cable cars. Albums of his own music: Brer Soul, Ain't Supposed to Die a Natural Death. Composed music for Watermelon Man.
PICTURES INCLUDE: The Story of a Three-Day Pass, Watermelon Man, Sweet Sweetback.
MEMBER: French Director's Guild; one of first Black directors to be member of Directors Guild of America.
STAGE: Ain't Supposed to Die a Natural Death, Don't Play Us Cheap.

VAN PRAAG, WILLIAM: Executive, Producer, Director, Writer, Editor. Advertising Consultant. b. N.Y.C., Sept. 13, 1924. e. CREI, Columbia Univ.; U.S. Army, 1942. Paramount, 1945; Brandt Bros. Prods., 1946; NBC, 1947; v.p. Television Features, 1948. Devlpd. vidicon system in m.p. prod., 1949. Started, pres., Van Praag Prod. Inc. 1951. Formed Ernst-Van Praag, Inc. 1971, a communications and marketing counseling firm (N.Y., Brussels, Tokyo). Pres., International Film, TV and A-V Producers Ass'n, 1969, Creator of Van-O-Vision. Winner of commercial, short subject and feature theatrical awards. Author of Color Your Picture, Primer of Creative Editing, and Van Praag's Magic Eye. Past pres., Film Producer's Ass'n, mem. DGA, SAG, 71 IATSE, National Academy of TV Arts and Sciences, International Radio and TV Executive Society and Soc. of MP and TV Engineers.

VAN RIKFOORD, HAROLD C.: Producer. b. New York, Jan. 1, 1935. e. Bryant H.S., grad., 1952; N.Y. Univ. Errand bty, stage mgr., N.Y., 1954–57. Co-prod. two Broadway plays. Prod., The Long Ride Home, Rome, 1957; prod., short subject films; prod., Moment of Crisis, 1959; prod., television pilot films; exec. prod., Jonathan Shields Prod.

VAN TAALINGEN, J.TH.: b. Amsterdam, Holland, Oct. 16, 1921. Joined Nederlandse Bioscoopbond 1946. Apptd. general director Nederlandse Bioscoopbond 1970.

VAN THAL, DENNIS: Executive. b. London, June 4, 1909. e. University College. Early career musical dir.; war service, Royal Navy; entered m.p. industry 1946; casting dir. Pinewood Studios; dir., Myron Selznick Ltd.; joined dir. Alexander Korda as prod. exec., 1953; apptd. dir., Big Ben Films, 1955; joined Ealing Films, 1956; prod. assoc. Anastasia; assoc. prod. The Admirable Crichton, Barnacle Bill, The Scapegoat; mgr. dir. London Management, Ltd.

VAN VOLKENBURG, J. L.: Executive. b. Sioux City, Ia., Dec. 6, 1903. e. U. of Minn. Batten, Barton, Durstine, Osborne adv. agency, 1928; CBS, Inc., since 1932; v.p. CBS and pres. CBS-TV Div.; ret. Jan., 1957; pres. McCann, Erickson Prod., Inc., 1960.

VANCE, DENNIS: TV Producer, director. b. Birkenhead, England, 1924. Service in Merchant Navy & Fleet Air Arm. Early career with BBC and Rank Organization making film debut in Hungry Hill. Others: Hamlet, Saraband for Dead Lovers, Christoper Columbus, Landfall, Shadow of the Eagle; 1952, prod. for BBC Inc. Public Prosecutor, Mr. Betts Runs Away; 1956 drama supv., and later prod. supv. ABC-TV, 1959–60 dir., U.S. Steel Hour and play of The Week, N.Y.; 1961: dir. Henry IV Part 1, Old Vic. 1961–64 ATV prod. dir. 1964: Gen. man. Caribbean Broadcasting Corp. set up TV in West Indies. 1966: Returned to ATV and Thames TV. Contract prod. dir.
TV: The Misfit, Man at the Top, Spyders Web.

VANCE, LEIGH: Scriptwriter and producer. b. Harrogate, England 1922. Early career; reporter, critic. Ent. TV 1951, then films, many TV scripts; 1961, won Edgar Allan Poe Award, 1969, brought to Hollywood by Paramount. In 1973 exec. story consultant and producer, Cannon; 1974, exec. story consultant, Caribe; 1975, prod., Bronk. 1976: Exec. prod., Baretta & Switch, 1977/8: Prod., Switch.
PICTURES INCLUDE: The Flesh Is Weak, Heart of a Child, Picadilly Third Stop, Women Shall Weep, The Shakedown, Eyes of Youth, The Frightened City, It's All Happening Dr. Crippen, Outcast, Walk Like A Man, Cross Plot, Tall Cool Girl, The Black Windmill.
TV: Mannix, Mission Impossible, many pilots, The Avengers, The Saint, Cannon, Caribe, Bronk, Baretta, Switch.

VARDA, AGNES: Photographer-Director-Writer. b. France, 1928. Started as still photographer for The Theatre National Populaire de Jean Vilar. Became a photo-journalist. In 1954 wrote and directed first film, La Pointe Courte. Afterwards made both documentaries and features.
PICTURES INCLUDE: Cleo from 5 to 7, Le Bonheur, Les Creatures, Lion's Love, One Signs, The Other Doesn't.

VAUGHN, ROBERT: Actor. b. New York City, 1932. Played Napoleon Solo in The Man From U.N.C.L.E. TV series.
CREDITS: Hell's Crossroads, No Time to Be Young, Unwed Mother, Good Day for a Hanging, The Young Philadelphians, The City Jungle (Acad. Award nom.), The Magnificent Seven, The Big Show, The Caretakers, To Trap a Spy, The Spy With My Face, One Spy too Many, The Venetian Affair, How to Steal the World, Bullitt, The Bridge at Remagen, The Mind of Mr. Soames, If It's Tuesday, This Must Be Belgium, Julius Caesar, The Statue, The Clay Pigeon, One of Our Spies Is Missing (TV, Great Britain), The Spy In the Green Hat (TV, Great Britain), The Towering Inferno, Starship Invasions, "S.O.B.".
AUTHOR: Only Victims, 1972.
TV: Washington: Behind Closed Doors.

VEITCH, JOHN: Executive. b. New York City, June 22, 1920. Started production career as asst. director and moved through ranks as prod. mgr., assoc. prod., prod., second unit dir., and exec. prod. mgr. Appointed Columbia Pictures exec. asst. prod. mgr., 1961; became exec. prod. mgr., 1963. Named v.p. & exec. prod. mgr., 1966; promoted in 1977 to exec. v.p. & exec. prod. mgr. of worldwide physical productions. In 1979 named pres. of Columbia Pictures Productions.

VELAISE, ROBERT: Executive. e. Zurich U.; ent. production 1957; formed Vauban Productions of Paris, Robert Velaise Productions, Ltd., of London, 1961; prod., Les Oeufs de l'Autruche, Nina, Geliebte Hochstaplerin, Give Me Ten Desperate Men, The Wrong Arm of the Law, The Go-Between.

VELAZCO, ROBERT E.: r.n. Emil Velazco, Jr. b. Dallas, Texas, Jan 1, 1924. e. Columbia Univ., Business Administration 1942–43. Started in film business with father, Emil Velazco, 1942. Following yrs. worked for Emil Velazco, Inc., NTA, Ross Gaffney, Inc. During this time worked on over 8,500 productions from 5 sec. to feature films. In charge of Velazco, Inc., Kansas City, Mo., 1947–49. Owner and pres., Musifex Co., Inc., 1958. Opened Musifex Inc. Arlington, Va., 1972. Two films scored by Velazco nominated for Academy Award, 1978 and 1980.

VELDE, DONALD L.: Executive. b Tremont, Ill., Sept. 10, 1902. e. Bradley Prep, Swarthmore Coll. Paramount Pictures, 1922–40; National Screen Service, 1940–48; chairman, Donald L. Velde, Inc.

VELDE, JAMES R.: Executive. b. Bloomington, Ill., e. Illinois Wesleyan Univ. Entered m.p ind. as night shipper Paramount ex. Detroit, 1934; then city salesman, office mgr. until joining Army, 1943, rejoining same ex. upon dischge., 1946; to Paramount, Washington as Baltimore city salesman, same yr.; br. mgr. Selznick Rel. Org. Pittsburgh, 1948; salesman Eagle-Lion Classics, Pittsburgh, 1949; br. mgr. ELC, Des Moines, 1949; br. mgr., ELC, Detroit, 1950; west coast dist. mgr., United Artists, April, 1951; Western div. mgr. UA, 1952; gen. sales mgr., 1956; v.p., 1958; dir., UA, 1968; sr. v.p., 1972. Retired, 1977.

VENTURA, LINO: Actor. b. Italy, 1918. French resident since age of 8. Was professional wrestler until injury ended career in ring. First acting appearance in Touchez pas au Grisbi (1953) opposite Jean Gabin. Veteran of over 50 films.
PICTURES: Marie Octobre, Crooks in Clover, Les Aventuriers, The Valachi Papers, Wild Horses, Sunday Lovers, etc.

VERNON, ANNE: Actress. r.n. Edith Antoinette Alexandrine Vignaud. b. Paris, Jan. 24. e. Ecole des Beaux Arts, Paris. Worked for French designer; screen debut in French films; toured with French theatre group; first starring role, Le Manneuin Assassine 1948.
PICTURES INCLUDE: Ainsi Finit La Nuit, A Warning to Wantons, Patto Col Diavolo, A Tale of Five Cities, Shakedown, Song of Paris, Edward and Caroline, Terror on a Train.

VERONA, STEPHEN: Director. e. School of Visual Arts. Directed some 300 commercials before turning to feature films in 1972, which he wrote as well.
PICTURES: Pipe Dreams, Boardwalk.

VERSOIS, ODILE: Actress. b. Paris, June 15, 1930. e. Paris. Early career in ballet at Paris Opera; worked on stage in France, Belgium, Switzerland and North Africa; screen debut in 1947 with films in France and Italy.
PICTURES INCLUDE: (In England) Into the Blue, A Day to Remember, Young Lovers (Chance Meeting), To Paris With Love, Checkpoint, Ruler Without Crown, Passport to Shame, Docks, Nude in a White Car.

VETTER, RICHARD: Exec. b. San Diego, Calif., Feb. 24, 1928. e. Pepperdine Coll., B.A., 1950; San Diego State Coll., M.A., 1953, U.C.L.A., Ph.D., 1959. U.S. Navy: aerial phot., 1946–48, reserve instr., San Diego County Schools, 1951–54; asst. prof., audio-vis. commun., U.C.L.A., 1960–63. Inventor, co-dev., Dimension 150 Widescreen Process. 1957–63: formed D-150 Inc., 1963; exec. v.p. mem.: SMPTE, Technical & Scientific Awards Committee, AMPAS.

VICTOR, HARVEY L.: Executive. b. New York. e. Carnegie Inst. of Technology, N.Y. University. Acct. exec., Television Programs of America; acct. supv., Official Films Inc.; acct. exec., various ad. agencies; N.Y. mgr., Special Devices, Inc.; vice-pres., gen. sls. mgr., Jayark Films Corporation 1958; pres., Harvey Victor Films, Ltd., 1964.

VIDOR, KING WALLIS: Director. b. Galveston, Tex., Feb. 8, 1895. e. Peacock Military Academy, San Antonio, Tex.; Tome College, Port Deposit, Md.
PICTURES INCLUDE: Turn in the Road, La Boheme, The Crowd, Big Parade, Hallelujah, Billy the Kid, Street Scene, The Champ, Cynara, Bird of Paradise, Stranger's Return, Our Daily Bread. In 1936 produced, directed and collaborated on screen story Texas Rangers. Stella Dallas, The Citadel, Northwest Passage, Comrade X, H.M. Pulham, Esq.; (prod. dir.) An American Romance. In Jan., 1949, signed term contract with Warners TV show; Lights Diamond Jubilee, 1954. Duel in the Sun, Fountainhead, Ruby Gentry, Man Without a Star, War and Peace, Solomon and Sheba.

VIETHEER, GEORGE C.: Executive. b. New York, N.Y., Feb. 27, 1910. e. Norwich U., B.S , 1932; Syracuse U, M.S.. 1933. U.S. Army officer, 1933–37; chief service bureau, later Wash. rep. Panama Canal & R.R. Co., 1937–41; deputy dir., personnel, Office of Emergency Mgt., 1941–43; U.S. Army, 1943–46; mem. Nat'l Comm. on Manpower Shortages, 1946; deputy dir., personnel, U.S. Dept. of Commerce, 1946–50; U.S. rep. to U.N. Public Admin. Conf., 1950; deputy asst. admin. Econ. Stab. Agency, 1951; mgr., Washington office, MPAA, 1952; subsequently v.p., MPEAA and also v.p. American Motion Picture Export Co. (Africa) and v.p., Afram Films, Inc. Author: various pubs., articles on bus. mgt. Retired since 1977.

VILAS, WILLIAM H.: Executive. e. Harvard U.: U. Minnesota, B.A. Dunwoody Institute. In adv. & sales management; wrote, prod. or supervised over 100 productions for TV, industrial or theatre release; organized & supervised film training projects for U.S. Armed Forces, World War II: v.p., Wilding Pictures Prod.; special consultant for OSS; prod. of theatrical, TV & non-theatrical m.p. For pub. rel.; recent TV: Terry and the Pirates, series on film, dir., motion picture & television dept. & account exec., J. M. Mathes, Inc., N.Y.; chmn. bd. gov. Yankee Highway Assn.; dir. Citizens Assn. of Conn.; pub. Brotherhood Press. Clubs; Pequet Yacht, Southport, Conn., Harvard, New Canaan, Conn.

VILLECHAIZE, HERVE: Actor. b. Paris, France, April 23, 1943. Sought career as artist, studying in Paris and then coming to New York to the Art Students League. Studied acting with Julie Bovasso. First film, The Guitar, shot in Spain. On Broadway in Elizabeth the First and Gloria and Esperenze. Also performed mime in N.Y. City Opera productions.
PICTURES: Hollywood Blvd. No. 2, Hot Tomorrow, The Man with the Golden Gun, Crazy Joe, The Gang That Couldn't Shoot Straight, Seizure, The One and Only.
TELEVISION: Fantasy Island (regular).

VINCENT, JAN-MICHAEL Actor. b. Denver, Colo., July 15. e. Ventura City (Calif.) College as art major. Joined National Guard. Discovered by agent Dick Clayton. Hired by Robert Conrad to appear in his film, Los Bandidos, made in Mexico. Signed to 6-mo. contract by Universal, for which made Journey to Shiloh. Then did pilot TV movie for 20th-Fox based on Hardy Boys series of book. Originally called self Michael Vincent; changed after The Undefeated.
PICTURES INCLUDE: Los Bandidos, Journey to Shiloh, The Undefeated, Going Home, The Mechanic, The World's Greatest Athlete, Buster and Billie, Bite the Bullet, White Line Fever, Baby Blue Marine, Vigilante Force, Shadow of the Hawk, Damnation Alley, Big Wednesday, Hooper, Defiance, Hard Country.
TV: Lassie, Bonanza, The Banana Splits Adventure Hour, The Survivors (series), Tribes (movie released theatrically abroad), The Catcher (movie), Sandcastle (movie).

VINCENT, JR., FRANCIS T: Executive. President & chief exec. officer of Columbia Pictures Industries (appt. July, 1978). e. William College, Yale Law School. 1968–78, partner in law firm of Caplin & Drysdale, specializing in corporate banking and securities matters. Since March 1, 1978, assoc. dir. of, Division of Corporation Finance of Securities & Exchange Commission. Trustee of William College & Hotchkiss School.

VINCENT, JUNE: Actress r.n. Dorothy June Smith. b. Harrods, Ohio. Concert pianist; photographer's model. On dramatic stage, The Family. Screen debut Second Honeymoon, 1937.
PICTURES INCLUDE: The Climax, Ladies Courageous, Can't Help Singing, Shed No Tears, Creeper, Counterspy Meets Scotland Yard, Secrets of Monte Carlo, Colorado Sundown, Night Without Sleep, Wac From Walla Walla, Clipped Wings, City of Shadows.

VINCENT, KATHARINE: Actress. r.n. Ella Vincenti. b. St. Louis, Missouri, May 28, 1919. e. Two years of high school, left in 1937 to go on the stage. m Pandeno Descanto, producer.
THEATRE: Broadway shows include: Love or Bust, 1938; Could She Tell?, 1939; Banners of 1939; Czarina Smith, 1940. Numerous road show tours.
PICTURES INCLUDE: Peptipas Waltz, 1942 (debut), Error in Her Ways, 1943, Skin Deep, 1944, The Hungry, Voodoo Village, Welcome to Genoa, 1950, Unknown Betrayal, 1956, The Hooker, 1962 (Descanto films).
TV: The Untouchables, Moses, The Lawgiver.

VINER, MICHAEL: Producer-Writer. b. 1945. e. Harvard, Georgetown. Served as aide to Robert Kennedy; was legman for political columnist Jack Anderson. Settled in Hollywood, where worked for prod. Aaron Rosenberg, first as prod. asst. on three Frank Sinatra films; then asst. prod. on Joaquin Murietta. In music industry was record producer, manager, executive, eventually heading own division, at MGM. Debut as writer-producer in 1976 with TV special, Special of the Stars. Theatrical film debut as prod.-co-writer of Touched by Love, 1980.

VITALE, JOSEPH A.: Actor. b. New York City, Sept. 6, 1901. In 1924, on dramatic stage (Hold on to Your Hats, Page Miss

Glory, All Editions, I'd Rather Be Right, Common Ground). Screen debut, 1943.

PICTURES INCLUDE: None But the Lonely Heart, Lady Luck, Road to Rio, Where There's Life, Connecticut Yankee, Illegal Entry, Red Hot and Blue, Paleface, Fancy Pants, My Friend Irma Goes West, Stop You're Killing Me, Stranger Wore a Gun, Square Jungle, Rumble On the Docks, Apache Rifles.

TV: Climax, Lineup, Bengal Lancers, Wagon Train, Schlitz Playhouse, Cimmaron City, Telephone Time, Wyatt Earp, Rawhide, Red Skelton, The Thin Man, M Squad, Dawson, Ben Casey, Empire, Hazel, Mr. Ed, To Rome With Love, Fisherman's Wharf (pilot).

VITALE, MILLY: Actress. b. Rome, Italy. July 16, 1938. e. Lycee Chateaubriand, Rome. Has appeared in numerous Italian films.

PICTURES INCLUDE: The Juggler, Seven Little Foys, Breath of Scandal.

VLECK, JOSEPH V.: Executive, b. Fort Morgan, Colorado. e. U. of Denver, School of Theatre and Graduate School Business Adm. Fox-Intermountain Theatres Denver, Ogden, Pocatello, La Junta; foreign film program director. Promotion manager, 1962 National General Theatres home office advertising department, appointed director adv. and publicity, 1965; now with Mann Theatres as dir. of adv.

VOGEL, JESSE: Writer, Composer. b. New York, N.Y., Oct. 24, 1925. e. CCNY, B.S.Sc., 1947; Paris Conservatory of Music, 1949; dir., Vogel Films, Paris, post-production services, English and Foreign versions. Screenplays: Carmen, Baby; Therese and Isabelle. Produced: Who's Harriet? My Pleasure Is My Business.

VOHS, JOAN: Actress. r.n. Elinor Joan Vohs. b. St. Albans, L.I., N.Y., July 30, 1931; e. Andrew Jackson H.S., Professional Children's School. Radio City Music Hall Rockette at 16; Bdwy play; Parlor Story; many TV shows.

TV: Readers Digest, Stage Seven, Dear Phoebe, Ray Bolger Show, Robert Cummings Show, Stage 7, Frontier, Schlitz Playhouse, Lux Video Theatre.

PICTURES INCLUDE: My Dream Is Yours, Girl from Jones Beach, Inspector General, Yes Sir That's My Baby, Vice Squad, Fort Ti, Crazylegs, Sabrina, Cry Vengeance, Fort Yuma, Terror at Midnight.

VOIGHT, JON: Actor. b. Yonkers, N.Y. Dec. 29, 1938. e. Archbishop Stepinac High School, White Plains, N.Y.: Catholic University, Bachelor of Fine Arts Degree, 1960; studied acting at the Neighborhood Playhouse and in private classes with Sanford Meisner, four yrs. Off-Broadway in: A View From the Bridge (revival), 1964. Won the Theatre World Award for Broadway prod. That Summer, That Fall. Played Romeo, San Diego Shakespeare Festival.

TV: Public Broadcast Lab.'s The Dwarf, also Gunsmoke, Cimarron Strip.

PICTURES INCLUDE: Fearless Frank, Hour of the Gun, Midnight Cowboy (Acad. Award nom., Best Actor), Out of It, Catch 22, The Revolutionary, All American Boy, Deliverance, Conrack, The Odessa File, End of the Game, Coming Home, The Champ, Lookin' To Get Out (also co-s.p.).

VOLONTE, GIAN MARIA: Actor. b. Milan, April 9, 1933. e. Rome's National Academy of Dramatic Art (1957 graduate). Entered on professional theatrical career, playing Shakespeare and Racine, along with modern works, Sacco and Vanzetti and The Deputy. On TV in Chekov's Uncle Vanya and Dostoyevsky's The Idiot. First major film roles in Un Uomo da Bruciare (1961) and Il Terrorista (1963). Called self John Welles in credits for "spaghetti westerns."

PICTURES INCLUDE: For a Fistful of Dollars, For a Few Dollars More, Investigation of a Citizen Above Suspicion, The Working Class Goes to Heaven, Wind from the East, Sacco and Vanzetti, L'Attenat, Slap the Monster on Page One, Just Another War, The Mattei Affair, Lucky Luciano, Eboli.

VON SYDOW, MAX: Actor. b. Lund, Sweden, April 10. m. Keratin Olin, actress, 1951. Theatrical debut in a Cathedral School of Lund prod. of The Nobel Prize. Served in the Swedish Quartermaster Corps two yrs. Studied at Royal Dramatic Theatre School in Stockholm. Tour in municipal theatres. Has appeared on stage in Stockholm, London, Paris and Helsinki in Faust, The Legend and The Misanthrope. 1954 won Sweden's Royal Foundation Cultural Award.

PICTURES INCLUDE: Wild Strawberries, Brink of Life, The Magician, The Seventh Seal, The Virgin Spring, Through a Glass Darkly, Winter Light, Hawaii, The Greatest Story Ever Told, The Reward, Hour of the Wolf, Shame, The Kremlin Letter, The Passion of Anna, The Immigrants, Night Visitor, The Emigrants, The New Land, Three Days of The Condor, Voyage of the Damned, Exorcist II: The Heretic, March or Die, Flash Gordon, Victory.

VONDERHAAR, RAYMOND T.: Executive. b. Rugby, No. Dak., Nov. 1, 1919. e. St. Cloud State. Spent entire career in exhibition. Pres., Allied Theatre Assn., 1963–64; pre., NATO North Central States, 1965–76; National NATO bd., 1965–80; also memb. exec. comm. 1965–80; NATO No. Central, chm., 1976–80.

VORHAUS, BERNARD: Director. In 1933, wrote and directed Money for Speed (UA-British).

PICTURES INCLUDE: 1934, directed The Ghost Camera (Radio-British), Crime on the Hill (BIP) and Night Club Queen (Universal-British); 1935: Broken Melody (APD), Blind Justice (Universal-British), Ten Minutes Alibi (BL), Street Song (Radio-British). In 1936; directed Last Journey, Twickenham; Dark World, Fox British. Associate producer Broken Blossoms; director, Dusty Ermine, Twickenham; Cotton Queen, Rock Studios, Bury Me Dead, Winter Wonderland, The Spiritualist, So Young, So Bad, Pardon My French.

VYE, MURVYN: Actor, singer. b. Quincy, Mass., July 15, 1913. e. Yale U., 1937.

THEATRE: Hamlet, The Live Wire, The Number, All In Fun, Something for the Boys, Oklahoma, Carousel.

PICTURES INCLUDE: Golden Earrings, Connecticut Yankee, Road to Bali, Destination Gobi, Pickup on South Street, Green Fire, Escape to Burma, Pearl of the South Pacific, Best Things in Life are Free, Al Capone, Pay or Die.

TV: Studio One, U.S. Steel, Ford Theater, Lux Video, Celebrity Playhouse, Schlitz Playhouse, etc.

W

WADLEIGH, MICHAEL: Director. b. Akron, Ohio, Sept. 24, 1941. e. Ohio State U., B.S., B.A., M.A., Columbia Medical School. Directed Woodstock.

WAGENHEIM, CHARLES: Actor. From 1941 in Meet Boston Blackie, Fingers at the Window, Halfway to Shanghai, Sin Town, I Escaped from the Gestapo.

PICTURES INCLUDE: Colonel Effingham's Raid, The House on 92nd Street, Summer Storm, Dark Corner, Pirates of Monterey, Scudda-Hoe! Scudda-Hay, Cry of City, Lady Without Passport, House on Telegraph Hill, Vicki, Beneath the 12-Mile Reef, Canyon Crossroads, Blackjack Ketchum, Desperado.

WAGGONER, LYLE: Actor. b. Kansas City, Kan., April 13, 1935. e. Washington Univ., St. Louis. Was salesman before becoming actor with road co. prod. of Li'l Abner. Formed own sales-promo co. to finance trip to Calif. for acting career in 1965. Did commercials, then signed by 20th-Fox for new-talent school.

TELEVISION: The Carol Burnett Show, It's Your Bet (host), The New Adventures of Wonder Woman.

WAGNER, LINDSAY: Actress. b. Los Angeles, June 22. Appeared in school plays in Portland, Oregon; studied singing and worked professionally with rock group. In 1968 went to L.A. and decided to make acting her career. Signed to Universal contract in 1971.

PICTURES INCLUDE: Two People, Paper Chase, Second Wind, Nighthawks.

TV: The F.B.I., Owen Marshall, Counselor at Law, Night Gallery, The Bold Ones, Marcus Welby, M.D., The Rockford Files, The Six Million Dollar Man, The Bionic Woman.

WAGNER, RAYMOND JAMES: Executive. b. College Point, L.I., N.Y., Nov. 3, 1925. e. Middlebury College & Williams College. Joined Young & Rubicam, Inc., as radio-TV commercial head in Hollywood, 1950–59. Head of pilot development, Universal Studios, 1960–65. Produced Petulia for Warner Bros. (1967) and was exec. prod. of Loving for Columbia (1969). Presently v.p. of production (features) or MGM.

WAGNER, ROBERT: Actor. b. Detroit, Feb. 10, 1930; e. Saint Monica's H.S. Film debut, Halls of Montezuma.

PICTURES INCLUDE: The Frogmen, Let's Make It Legal, With A Song in My Heart, What Price Glory, Stars and Stripes Forever, The Silver Whip, Titanic, (Star of Tomorrow, 1953). Beneath the 12-Mile Reef, Prince Valiant, Broken Lance, White Feather, Kiss Before Dying, The Mountain, True Story of Jesse James, Stopover Tokyo, The Hunters, In Love and War, Say One for Me, Between Heaven and Hell, All the Fine Young Cannibals, Sail a Crooked Ship, The Longest Day, The War Lover, The Condemned of Altona, Harper, Banning, The Biggest Bundle of Them All, Don't Just Sit There, Winning, The Towering Inferno, Midway, The Concord—Airport '79.

TV: It Takes A Thief series; Madame Sin; Switch (series), Cat on a Hot Tin Roof, Pearl, Hart to Hart.

WAKELY, JIMMY: Actor, singer, songwriter. b. Mineola, Ark., Feb. 16, 1914. Began as farmer, then singer, radio, Oklahoma City, Decca recordings. Author lyrics & music numerous hillbilly songs (Too Late, I'll Never Let You Go). Screen debut 1939 in Saga of Death Valley; collab. on music Six Lessons from Madame La Zonga, 1941. From 1939 in many Westerns various cos. (Paramount, Universal, Republic, etc.); producing Westerns independently; act., Marshal's Daughter, Arrow in the Dust.

WALD, MALVIN: Writer, producer. b. New York City, N.Y., Aug. 8. e. Brooklyn College, B.A., J.D. Woodland University College of Law, graduate work Columbia U., New York University U.S.C. Newspaper reporter and editor, publicist, social worker, radio actor. Screenplays and original stories for Columbia, 20th-Fox, UA, MGM, WB; U.S. Air Force; tech. sgt., wrote 30 documentary films for film unit.
PICTURES INCLUDE: The Naked City (Acad. Award nomination, best story); Behind Locked Doors, The Dark Past, Ten Gentlemen from West Point, The Powers Girl, Two in a Taxi, Undercover Man, Outrage, On the Loose; (assoc. producer and sec.-treas., Filmakers Pictures, Inc.); Battle Taxi, Man on Fire, Al Capone, Venus in Furs. In Search of Historic Jesus. Legend of Sleepy Hollow. Shorts: Employees Only (Academy Award nomin., best sht. doc.), Boy Who Owned a Melephant (Venice Children's Film Festival gold medal), Unarmed in Africa, The Policeman, James Weldon Johnson, Me An Alcoholic?
TV: Many credits including Playhouse 90, Hollywood: The Golden Years, The Rafer Johnson Story, D-Day, Project: Man in Space, Biography of A Rookie, Alcoa-Goodyear Hour, Climax, Shirley Temple Storybook, Peter Gunn, Perry Mason, Dobie Gillis, Combat, Daktari, (Associate Prod.) Primus, California Tomorrow, prod. Mod Squad, Untamed World, Around the World of Mike Todd, The Billie Jean King Show, Life and Times of Grizzly Adams, Mark Twain's America, Greatest Heroes of the Bible, Littlest Hobo. Exec. prod., 20th Century Fox TV Documentary Unit, 1963–64 writer-prod. U.S.I.A., 1964–65; writer-prod., Ivan Tors Films, 1965–69; prof., U.S.C., Division of Cinema; Writer's Guild of America, Acad. of Motion Picture Arts and Sciences, co-author of book, Three Major Screenplays. Contributor to book, Close-Ups. Published s.p., Naked City.

WALD, RICHARD C.: Executive. b. New York City. e. Columbia College, Clare College (Cambridge). Joined the New York Herald Tribune in 1951 as Columbia College correspondent; religion editor, political reporter; foreign correspondent (London, Bonn), 1959–63; associate editor, 1963–65; managing editor from 1965 until paper ceased publication in 1966; Sunday editor, World Journal Tribune, 1966; managing editor, Washington Post, 1967; vice president, Whitney Communications Corp., 1967–68; joined National Broadcasting Company as vice president, NBC News, 1968; exec. v.p., 1972; president, NBC News, 1973.

WALDMAN, WALTER: Publicist. b. New York City. e. CCNY, Columbia U. Newspaper reporter; magazine writer; radio-television-film critic; radio news writer, Current Events, Triangle Publications; The Bronx Home News; Grolier Society; Netherlands Information Bureau, station WLIB; free-lance New York Times and New York Herald Tribune; writer, press book dept., 20th Century-Fox; copy writer, advertising dept., Republic Pictures; writer, Variety, Boxoffice Magazine; publicity writer, copy chief, publicity dept., Paramount Pictures; publicity copy chief, contact, United Artists.

WALKEN, CHRISTOPHER: Actor. b. Astoria, N.Y., Mar. 31, 1943. Began career in off-Bdwy. musical revue, Best Foot Forward, starring Liza Minnelli. Continued in musicals until cast in original Bdwy. procuction of The Lion in Winter, winning Clarence Derwent Award for performance as King Philip. Switched to dramatic roles, winning Obie Award for title role in Kid Champion and Theater World Award for performance in N.Y. City Center revival of Rose Tattoo.
PICTURES INCLUDE: The Anderson Tapes (debut), Next Stop, Greenwich Village, Roseland, Santa Fe—1936, The Sentinel, Annie Hall, The Deer Hunter, Heaven's Gate, The Dogs of War, Pennies from Heaven.

WALKER, CARDON E.: Executive. b. Rexburg, Idaho, Jan. 9, 1916. e. UCLA, Los Angeles; B.A. 1938. Four years Officer, U.S. Navy. Started with Walt Disney Productions 1938; camera, story, unit director short subjects, budget control. Headed, 1950, adv. & pub. 1956, v.p. in chg. of adv. & sales. 1960 member bd. of dir. & exec. comm. 1965 v.p., mkt. 1967 exec. v.p. operations. 1968, exec. v.p. and chief operating officer; pres., 1971; Nov. 1976 pres. and chief executive officer; June, 1980, named bd. chm. & chief exec. officer.

WALKER, CLINT: Actor. b. Hartford, Ill., May 30, 1927. e. schools there. Joined Merchant Marine 1944, worked as sheet metal worker, carpenter, other jobs in Alton, Ill.; set out with wife and infant daughter for oil fields in Texas; decided to try acting. Got screen test at Paramount Studios for Cecil B. De Mille; later landed contract to star in Cheyenne TV films at Warner.
PICTURES INCLUDE: Fort Dobbs, Yellowstone Kelly, None But the Brave, Pancho Viva, Maya, Night of the Grizzly, Gold of the Seven Saints, Send Me No Flowers, The Dirty Dozen, The Great Bank Robbery, Sam Whisky, More Dead Than Alive, etc.

WALKER, DAVID PAT: b. Glasgow Scotland, 1928. Joined British Broadcasting Corporation as studio manager in Glasgow, became radio producer, then a production assistant in television. Joined BBC in Northern Ireland as that station's only television producer in 1960. Became assistant head of programmes, BBC, Northern Ireland, then rejoined BBC Scotland in similar post 1970. Head of programmes, Scotland, 1973, and Head of network productions at Scotland 1978. Comes from theatrical family, sister Amanda Walker being an actress and his mother, Madeleine Christie, actress in TV, films, theatre and radio. Assistant controller, BBC Scotland, June 1979.

WALKER, JOSEPH: Cameraman. From 1928 cinematographer many Columbia productions. Member Amer. Soc. Cinematographers (second vice-pres. 1944–45). Inventor of Electra-Zoom Lens used in TV stations; chief eng., Radio Optical Research Co., Hollywood, retired.
PICTURES INCLUDE: It Happened One Night, Mr. Deeds Goes to Town, Lost Horizon, Here Comes Mr. Jordan, Jolson Story, Affair in Trinidad, Born Yesterday.

WALKER, MARTIN: Singer, actor. b. Dundee, Scotland, Jan. 6, 1938. e. N.J., Yonkers, N.Y. high schools. B'way stage play 3 Wishes for Jane; winner Godfrey Talent Scouts, CBS-TV, Paul Whiteman Show, ABC-TV; guest on Morey Amsterdam Show, Johnny Olsen Show, Johnny Andrew Show etc.; recording artist, Pinky Records.

WALKER, NANCY: Actress-Director. b. Philadelphia, May 10, 1922. As child toured Europe with her parents, The Barto and Mann vaudeville team. Pursued career as serious vocalist until George Abbott steered her toward comedy when she auditioned for Broadway prod., Best Foot Forward. Directed film, Can't Stop the Music (1980).
STAGE: On the Town, Barefoot in the Park, Phoenix '55, Fallen Angels, Copper and Brass, Pal Joey, Wonderful Town, Girls Against the Boys, Look Ma, I'm Dancing, Along Fifth Avenue, Lucky Me.
PICTURES INCLUDE: Best Foot Forward, Girl Crazy, Broadway Rhythm, Meet the People, Stand Up and Be Counted, Forty Carats, Won Ton Ton, Murder by Death.
TV: Nearly every major TV show, plus series: McMillan & Wife, Rhoda, The Nancy Walker Show, Blansky's Beauties.

WALLACE, CHARLES: Producer, writer, director. b. Temple, Texas, May 5, 1930. e. New Mexico Military Institute; Arizona State Univ. Ent. industry, announcer engineer, KTYL, Phoenix, 1948; prod. dir., KVAR, Phoenix, 1950; prod. dir., Wide Wide World, NBC-TV, 1955–58; prod. dir., writer, NBC News overseas, U.S. Air Force Power for Peace, 1956; assoc. prod. writer, Twenty-Six Men, ABC-TV, 1958; prod. dir., writer, American Pauses spectaculars, CBS-TV Zane Grey Theatre, Johnny Ringo, Wyatt Earp, Robert Taylor Show, 1959; prod. writer, Four Star pilots, Hardcase-Yuma, 1959; writer, Wells Fargo, NBC-TV, Ichabod and Me, 1960; writer, Death Valley Days, U.S. Air Force Classified Documentaries; writer, Screen Gems, Whirlpool pilot, 1961; prod., Lum 'n Abner, Four Star TV, 1962.

WALLACE, IRVING: Writer. b. Chicago, Ill., March 19, 1916. e. Williams Inst., Berkeley, Calif., 1935–36. For. correspondent in Japan & China, Liberty Mag., 1940; U.S. Army doing m.p. & photog.; covered France, Germany, Spain for Sat. Evening Post, Collier's, Reader's Digest, 1946, 1947, 1949, 1953.
AUTHOR: The Fabulous Originals, 1955; The Square Pegs, 1957; The Sins of Philip Fleming, 1959; The Fabulous Showman, 1959; The Chapman Report, 1960; The Twenty-Seventh Wife, 1961; The Prize, 1962; The Three Sirens, 1963; The Man, 1964; The Sunday Gentleman, 1965; The Plot, 1967. The Writing of One Novel, 1968; The Seven Minutes, 1969; The Nympho and Other Maniacs, 1971; The Word, 1972, The Fan Club, 1974; The People's Almanac, 1975; The R Document, 1976.
PICTURES INCLUDE: story, collab. s.p., West Point Story; s.p., Meet Me at the Fair; collab. s.p., Desert Legion, Gun Fury; collab. story and s.p., Split Second; s.p., The Burning Hills, Bombers B-52; novel (basis for film), The Chapman Report, The Prize, The 7 Minutes, The Man.

WALLACE, JEAN: Actress. b. Chicago, Oct. 12, 1930. As teenager signed by Paramount as contract player and studied at Actors Laboratory (Method School) in Hollywood. Later, switched to 20th Century-Fox. In 1951 went to Argentina to star in first film, Native Son. Board of Directors Permanent Charities Committee and (Womens Auxiliary), Antans (The Womens Auxiliary of the American National Theater and Academy); hospitality chm., Beverly Hills P.T.A.
PICTURES INCLUDE: Storm Fear, The Big Combo, The Man on the Eiffel Tower, The Devil's Hairpin, Maracaibo, The Sword of Lancelot, Beach Red, No Blade of Grass. bd. of dir. of Screen Smart Set.

WALLACE, MIKE: TV commentator, interviewer. b. Brookline, Mass., May 9, 1918. e. Univ. of Michigan; 1939. Night Beat,

WABD, N.Y., 1956; The Mike Wallace Interview, ABC, 1956–58; Newspaper col., Mike Wallace Asks, N.Y. Post, 1957–58; News Beat, WNTA-TV, 1959–61; The Mike Wallace Interview, WNTA-TV, 1959–61; Biography, 1962; correspondent, CBS News, 1963, CBS Radio: Personal Closeup, Mike Wallace at Large; Co-editor, 60 Minutes, CBS News.

WALLACH, ELI: Actor. b. Brooklyn, N.Y., Dec. 7, 1915. e. U. of Texas. Capt. in Medical Admin. Corps during WW II. After college acting, appeared in Summer stock. Made Broadway debut in Skydrift, 1945, followed by Antony & Cleopatra, The Rose Tattoo, Major Barbara, Rhinoceros, Luv. Charter member, Actors Studio in 1947.

PICTURES INCLUDE: first film Baby Doll. Also: The Magnificent Seven, Seven Thieves, The Misfits, The Victors, Hemingway's Adventures of A Young Man, How the West Was Won, Act One, Genghis Khan, The Moonspinners, Lord Jim, How to Steal a Million, The Good, the Bad and the Ugly, The Tiger Makes Out, Band of Gold, How to Save a Marriage and Ruin Your Life, A Lovely Way to Die, Ace High, The Brain, Zigzag, The People Next Door, Romance of a Horse Thief, Cinderella Liberty, Crazy Joe, Movie Movie, The Hunter.

TV: Studio 1, Philco Playhouse, Playhouse 90; Won Emmy for A Poppy Is Also a Flower.

WALLACH, GEORGE: Producer, writer, dir. b. New York, N.Y., Sept. 25, 1918; e. N.Y.U. Actor in theatre & radio 1938–45; U.S. Navy 1942–45; supvr. radio-TV Div. of Amer. Thea. Wing 1946–48; prod.-dir. WNEW, 1947–48; program mgr. WNYC 1948–50; prod.-writer-dir. for WNBC-WNBT, 1950; Dir., news, spec. events WNBT-WNBC, 1951–52; prod. mgr., NBC Film Div. since 1953; CBS-TV; formed George Wallach Prod., 1956; prod. dir. It Happened in Havana, appt. TV officer, U.S.I.A., 1957. Film-TV-film officer American Embassy, Bonn, Germany, 1961. Film-TV officer American Embassy; Tehran, Iran, 1965–66; m.p. Prod. Officer, JUSPAO, American Embassy, Saigon, 1966; prod.-dir.-wr., Greece Today, 1967–68. Exec. prod.-dir., George Wallach Productions, spec. doc., travel, and industrial films, chairman, Film-TV Dept., N.Y. Institute of Photography, 1968–75; Prof. film-TV-radio, Brooklyn Coll., 1975–80; presently special projects officer, DGA, N.Y.

PICTURES INCLUDE: NBC-producer. Inner Sanctum, The Falcon, His Honor Homer Bell, Watch the World; assoc. prod., prod. mgr., Bwana Devil; dir., Wanted, CBS-TV series.

WALLIS, HAL B.: Producer. b. Chicago, Ill.; m. Martha Hyer, actress. Entered m.p. ind. 1922, learning exhib. & distrib. as mgr. Los Angeles theat. Later in pub. dept., Warner Bros. Then variously with Principal Pictures Corp.; First Nat'l (studio mgr., prod. mgr.); Warner Bros. in chge. prod. From 1933 prod. many features; established Hal Wallis Prods., 1944. Academy-Irving Thalberg Award, 1938, 1943. Partner with Joseph Hazen in Wallis-Hazen Prod. rel. via Paramount. Wallis-Hazen dissolved June 1953.

PICTURES INCLUDE: Little Caesar, Five Star Final, Dawn Patrol, Midsummer Night's Dream, The Story of Louis Pasteur, Anthony Adverse, Zola, Jezebel, King's Row, Sergeant York, The Male Animal, In This Our Life, Yankee Doodle Dandy, Desperate Journey, Now Voyager, Casablanca, Saratoga Trunk, Love Letters, The Affairs of Susan, You Came Along, Strange Love of Martha Ivers, Searching Wind, Sorry Wrong Number, The Accused, Bitter Victory, The Furies, Desert Fury, I Walk Alone, Rope of Sand, My Friend Irma, My Friend Irma Goes West, Dark City, September Affair, Red Mountain, That's My Boy, The Stooge, Peking Express, Sailor Beware, Jumping Jacks, Come Back Little Sheba, Scared Stiff, Cease Fire, Money from Home, About Mrs. Leslie, Three Ring Circus, Rose Tattoo, Artists and Models, Hollywood or Bust, The Rainmaker, Gunfight at the OK Corral, Loving You, Wild Is the Wind, Sad Sack, Last Train from Gun Hill, Don't Give Up the Ship, Hot Spell, King Creole, Career, Visit to A Small Planet, All In A Night's Work, G.I. Blues, Summer and Smoke, Blue Hawaii, A Girl Named Tamiko, Girls! Girls! Girls!, Wives & Lovers, Fun in Acapulco, Becket, Roustabout, The Sons of Katie Elder, Boeing-Boeing, Easy Come, Easy Go, Barefoot in the Park, 5 Card Stud, True Grit, Norwood, Anne of the Thousand Days, Red Sky at Morning, Shoot Out, Mary, Queen of Scots, The Partner with Joseph Hazen in Wallis-Hazen Prod. rel. via Public Eye, The Nelson Affair, The Don Is Dead, Rooster Cogburn.

WALSH, DERMOT: Actor, playwright. b. Dublin, Eire, Sept. 10, 1924; e. St. Mary's Coll., University Coll., Four Courts (Dublin), Abbey School of Acting. With Longford Productions. Dublin Gate Theatre, screen debut in Hungry Hill 1947.

PICTURES INCLUDE: Jassy, Mark of Cain, To the Public Danger, My Sister and I, Third Time Lucky, Torment, Frightened Many, Ghost Ship, Counterspy, Floating Dutchman, Straw Man, Blue Parrot, On the Night of the Full Moon, Bond of Fear, The Hide-Out, On the Stroke of Nine, Sea Fury, Sea of Sand, The Bandit, The Flesh and the Fiends, Shoot to Kill, The Challenge, The Clock Struck Three, Tarnished Heroes, Tell Tale Heart, The Breaking Point, The Trunk, Bedelia, Woman of Mystery, Chain of Events, The Crowning Touch, Make Mine a Million, Crash Dive, The Witness, Emergency, The Cool Mikado, The Switch, Echo of Diana, Infamous Conduct.

TV: Richard The Lionheart (series), Too Late for the Mashed Potato, Court Martial, Love Me Like A Brother, Softly Softly.

WALSH, KAY: Actress. b. London, Eng., 1915. Dancer in Andre Charlot revues; screen debut in How's Chances, 1934.

PICTURES INCLUDE: In Which We Serve, This Happy Breed, The October Man, Vice Versa, Oliver Twist, Stage Fright, Last Holiday, The Magnet, Encore, Meet Me Tonight (Tonight at 8:30), Young Bess, Stranger in Between, Lease Of Life, Rainbow Jacket, Cast a Dark Shadow, Now and Forever, The Horse's Mouth, Tunes of Glory, Reach for Glory, 80,000 Suspects, A Study in Terror, Beauty Jungle, Gideon's Way, The Baron, The Witches, He Who Rides A Tiger, The Ruling Class.

WALSH, RICHARD F.: President Emeritus International Alliance Theatrical Stage Employees and M.P. Machine Operators. b. Brooklyn, N.Y., 1900. In 1917 became apprentice stage electrician; 1920 full member Local 4, Brooklyn stagehand union IATSE & MPMO. Stage electrician various theatres. In 1924 elected pres. Local 4. 1926 business agent: 1934 int'l vice-pres. IATSE & MPMO; Nov., 1941 named pres. by exec. board; elected pres. by convention June, 1942; re-elected International pres. by convention thereafter; v.p., AFL-CIO, 1956. ch. of bd., Will Rogers Hospital, pres., Union Label Dept, retired as IATSE president in 1974.

WALSTON, RAY: Actor-Director. b. New Orleans, Louisiana, Nov. 22, 1918. Dir. of Bdwy-musical, Damn Yankees, 1974.

TV: You Are There, Producers Showcase, There Shall Be No Night, Studio One, Playhouse 90, My Favorite Martian.

PICTURES INCLUDE: South Pacific, Damn Yankees, Kiss Them For Me, Say One for Me, The Tall Story, The Apartment, Portrait In Black, Convicts Four, Wives and Lovers, Who's Minding the Store, Kiss Me Stupid, Caprice, Paint Your Wagon, The Sting, Popeye.

WALTER, JESSICA: Actress. b. Brooklyn, N.Y., Jan. 31. e. High School of Performing Arts. Studied at Bucks County Playhouse and Neighborhood Playhouse. Many TV performances plus lead in series, For the People. Broadway debut in Advise and Consent, 1961. Also, Photo Finish (Clarence Derwent Award), Night Life, A Severed Head.

PICTURES INCLUDE: Lilith, The Group, Grand Prix, Bye Bye Braverman, Number One, Play Misty For Me, Going Ape.

TV: Love of Life, Ironside, Amy Prentiss.

WALTERS, BARBARA: TV Personality. b. Boston, 1931. e. Sarah Lawrence Coll. Began working in TV right after graduation. Joined Today in 1961 as writer, making occasional on-camera appearances. In 1963 became full-time on camera. In April, 1974, named permanent co-host. Also hosts own synd. prog., Not for Women Only. In 1976 joined ABC-TV.

WALTERS, CHARLES: Director. b. Pasadena, Calif., Nov. 17; e. Univ. Southern California. On stage in Fanchon & Marco shows, 1934; in dance team Versailles Club, 1935.

STAGE: (cast) Musical Parade, Jubilee, Transatlantic Rhythm, Show Is On, Between the Devil, I Married An Angel, DuBarry Was a Lady; stage dir.: Let's Face It, Banjo Eyes.

PICTURES INCLUDE: (dir. dance sequences in m.p.) Seven Days' Leave, Presenting Lily Mars, Girl Crazy, Best Foot Forward, Meet Me in St. Louis, Abbott & Costello in Hollywood, Ziegfeld Folies, Summer Holiday; dir. Good News, Easter Parade, Barkleys of Broadway, Summer stock, Three Guys Named Mike, Texas Carnival, Belle of New York, Lili, Dangerous When Wet, Torch Song, Easy to Love, Glass Slipper, Tender Trap, High Society, Don't Go Near the Water, Ask Any Girl, Please Don't Eat the Daisies, Two Loves, Jumbo, The Unsinkable Molly Brown, Walk, Don't Run.

TV: Here's Lucy and The Governor and J.J.; 2 Lucille Ball Specials.

WANAMAKER, SAM: Actor, stage producer, film director. b. Chicago, 1919. e. Drake University. On Broadway stage as actor, producer, presenter. Hollywood films incl.: My Girl Tisa. British films incl: Give Us This Day, Mr. Denning Drives North. London West End stage: Winter Journey, The Big Knife, The Threepenny Opera, The Rainmaker. 1955: starred in film The Secret. 1956–58: Acting, producing London and Liverpool stage. 1959–60: Acting at Stratford-on-Avon. Numerous TV plays incl.: The Big Wheel, The White Death and dir. A Young Lady of Property, 1960: Film, The Criminal. 1961–62: Acting on New York stage in The Far Country. TV productions: The Defenders, Oedipus Rex, etc. Film: Taras Bulba. 1963: prod. dir. Children From Their Games, N.Y. stage.

TV: CBS special, Russian Self Impressions. Man of World, TV series. 1962: Stage, Washington D.C., Rhinoceros, Arms and The Man. Opera productions, King Priam, Forza Del Destino, etc. 1963: Dir., A Case of Libel (N.Y.). TV: Espionage, Outer Limits. Film: Man in the Middle (London). 1964–65: Acted in Macbeth (Chicago) and films, Those Magnificent Men In Their Flying Machines, The Spy Who Came In From The Cold, The Warning Shot, The Day The Fish Came Out, Voyage of the Damned. Dir. The Executioner. Dir. several The Defenders TV series (N.Y.), 1968; The Eliminator, The Chinese Visitor. 1971; Catlow. 1972; Sinbad and the Eye of the Tiger, 1977; Founder and exec. dir. Globe Playhouse Trust; starred BBC film Arturo Ui. 1973. Dir. Sydney Opera House Opening. War & Peace. 1974. Acted in The Law. Prod. Bankside Festival. U.S. Tour Shakespeare's globe. 1980: Private Benjamin, The Competition.

WARBURTON, JOHN: Actor. b. Liverpool, Eng., e. Oxford U. With British Army, World War I; then on stage Liverpool Repertory Theatre. To U.S. 1920.

STAGE: Journey's End, Bird In the Hand, Dishonorable Lady.

PICTURES INCLUDE: on screen 1933 in Cavalcade. Also: Chi Ci, Becky Sharp, Sisters, Saratoga Trunk, The White Cliffs, Marriage Is a Private Affair, Confidential Agent, Living in a Big Way, City Beneath the Sea, East of Sumatra, Royal African Rifles.

WARD, BILL: O.B.E., f.r.s.a. b. Plymouth, Devon, Jan. 19, 1916. e. Hoe Grammar School. Early career eng. asst., BBC; entered TV industry 1936; technical asst., 1936–37; War service; studio manager BBC 1946; presentation asst., 1946; prod. 1947; senior prod. light ent., 1950–55; joined Associated Television Ltd., 1955; prod. controller 1956. Judge at Golden Rose Festival, Montreux, Covered Mexico Olympics, ITV 1966, Exec. prod. ITV World Cup, delegate Prix Italia Festival. 1967, Judge Cannes O.B. Festival. 1968 Apptd. exec. dir. & dir. of programs, ATV Network. 1971 chmn. ITV Sports committee, Dep. Man. Dir. & Dir. Progs. ATV Network.

TV: Sunday Night at the Palladium, Golden Hour, Royal Variety Show, 1964.

WARD, BURT: Actor/Executive. b. July 6, California. Is pres. of Entertainment Management Corp. and Entertainment Licensing Corp., fan club and merchandising companies, respectively.

TV: Batman.

WARD, SARAH E.: Executive. b. Warrenton, Va., April 5, 1920. 1949–1962, office clerical positions with RKO, 20th Century-Fox, Todd AO Corp; v.p. sales mgr. for Europix Consolidated Corp., 1968; v.p.—sec'y for Europix International Ltd., 1971.

WARD, SIMON: Actor. b. London, England, 1941. Ent. ind. 1964.

TV: Spoiled, Chips With Everything, 1969. Films incl. Frankenstein Must Be Destroyed, I Start Counting, Young Winston, Hitler—The Last Ten Days, The Three Musketeers, The Four Musketeers, Children of Race, Deadly Strangers, All Creatures Great & Small, Aces High, The Battle Flag, Holocaust 2000.

WARDEN, JACK: Actor. b. Newark, N.J., 1925. With Margo Jones theatre in Dallas.

TV: Philco Goodyear Producer's Showcase, Kraft, Raid on Entebbe.

THEATRE: View from the Bridge, Very Special Baby.

PICTURES INCLUDE: From Here to Eternity, 12 Angry Men, Edge of the City, Bachelor Party, Escape from Zahrain, The Thin Red Line, Summertree, Who Is Harry Kellerman?, The Sporting Club, Welcome to the Club, Billy Two Hats, The Apprenticeship of Duddy Kravitz, Shampoo, All the President's Men, Heaven Can Wait, Death on the Nile, And Justice for All, Being There, The Great Muppet Caper, Carbon Copy.

WARING, FRED: Musician. b. Tyrone, Pa., June 9, 1900; e. Penn State. Choral & orchestral conductor, star & producer many seasons concert, radio, stage & television; winner many national awards; strongly identified with music education through Waring Music Workshop & Shawnee Press at Delaware Gap in Pennsylvania. Member Lambs, N.Y.

WARNER, DAVID: Actor. b. Manchester, England, 1941. Made London stage debut in Tony Richardson's version of A Midsummer Night's Dream. Four seasons with Royal Shakespeare Co.

PICTURES INCLUDE: Tom Jones, Morgan, Work Is a Four Letter Word, The Bofors Gun, The Fixer, The Seagull, Michael Kohlhaas, The Ballard of Cable Hogue, Perfect Friday, Straw Dogs, A Doll's House, Tales from the Crypt, Providence, Cross of Iron, Silver Bears. The Concorde—Airport '79, Time After Time, The Island.

WARNER, JACK JR.: Producer. b. San Francisco, Calif., Mar. 27, 1916; p. Jack L. Warner, and Mrs. Albert S. Rogell; e. Beverly Hills High Sch., B.A., U. of Southern Calif. Entered Warner New York office studying distrib. and exhib. for 1½ years. Transferred to prod. dept. at West Coast studios, then to short subject dept. as assoc. prod. As Reserve Officer called to active duty in 1942 and served as combat photo unit officer in 164th Signal Photo Co. for one year. Transf. to Signal Corps Photographic Center, Astoria, N.Y., where participated in prod. Army Signal Corps Training Films. Was asst. to chief of training films prod. In 1944 assigned to Hq. First U.S. Army Group to assist in planning combat photography for invasion of Europe. Until cessation of hostilities was asst. chief Photo Branch Office of Chief Signal Officer in 12th Army Group and on fall of Germany was on staff of General Eisenhower in Frankfurt as asst. and acting photo officer, Office of the Chief Signal Officer, Theatre Service Forces European Threatre (TSFET). Released from active duty April 20, 1945. Commissioned Lt. Col. Signal Corps Reserve. In 1947 with Warner Bros. Pictures Distrib. Corp., making survey of exhib. and distrib. as related to prod.; liaison between Warner and Assoc. Brit. Pictures on The Hasty Heart, 1948–49; org. Jack M. Warner Prod., Inc., 1949; first film, The Man Who Cheated Himself, distrib. by 20th Cent.-Fox; prod. dept., Warner Bros., 1951; prod. exec. Warners 1953. In charge of TV film prod. for Warners 1955; exec. in charge of television comm. and ind. film dept., Warner Bros. 1956; v.p., Warner Bros. Pictures, Inc., Jan. 1958; Warner association terminated Jan. 1959. Reactivated independent motion picture company Jack M. Warner Prod. Future activities production of feature television and industrial films; pres., Jack Warner Prods., Inc., prod. theatrical films, 1961; prod., dir., Brushfire; Commissioned Colonel, Signal Corps. U.S. Army Reserve 1962; 1977–78: Producer, TV series & films for theatrical and TV, Jack Warner Pdns; writer, 1979–81.

WARREN, CHARLES MARQUIS: Director, producer, writer b. Baltimore, Md.; e. Baltimore City College. Commander U.S. Navy, World War II; then to Hollywood. Writer for screen, magazines author of books.

BOOKS INCLUDE: Only the Valiant, Valley of the Shadow, Wilderness, Deadhead.

PICTURES INCLUDE: Only the Valiant, Beyond Glory, Redhead and the Cowboy, Streets of Laredo, Springfield Rifle, Day of the Evil Gun, Little Big Horn, Hellgate, Pony Express, Arrowhead, Flight to Tangier, 7 Angry Men, Trooper Hook, Tension at Table Rock, Cattle Empire, Charro!, Down to the Sea, Time of the Furies.

TV: Creator of Gunsmoke, Rawhide, Gunslinger, and the Virginian, series as well as prod.-dir.-writer. Exec. prod. on Iron Horse series.

WARREN, GENE: Executive. Pres. of Excelsior Prods., production co. specializing in special effects and animation. Has headed 2 other cos. of similar nature over past 20 years, functioning at various times as prod., dir., studio prod. head and writer. Producer-Director of following shorts: The Tool Box, Suzy Snowflake, Santa and the Three Dwarfs, Land of the Midnight Sun and these documentaries/training films: Mariner I, Mariner III, Apollo, U.S. Navy titles.

Special effects on theatrical features including, Black Sunday, McNamara's Band, Satan's School for Girls, My Name Is John, The Power, 7 Faces of Dr. Lao, Wonderful World of the Brothers Grimm, The Time Machine, Tom Thumb, etc. TV series include The Man from Atlantis, Land of the Lost, Star Trek, Outer Limits, Twilight Zone, Mission Impossible, etc.

WARREN, HARRY: Song writer, director. b. Brooklyn, N.Y., Dec. 24, 1893. Asst. dir. Vitagraph Studio, Brooklyn; song writer various N.Y. shows: Sweet and Low, Crazy Quilt; writer songs in many pictures since sound, including several WB Gold Diggers series, Wonder Bar, Down Argentine Way, Tin Pan Alley, also dramatic pictures. Academy Award for best motion picture song, Lullaby of Broadway, from Gold Diggers, 1935; (with Mack Gordon) for best motion picture song, 1943 (You'll Never Know from Hello Frisco Hello); 1946 (On the Atchison Topeka & Santa Fe (from Harvey Girls).

SONGS INCLUDE: (m.p.) Serenade in Blue, At Last, That's Amore, Chattanooga Choo-Choo.

PICTURES INCLUDE: Texas Carnival, Just for You, Skirts Ahoy, The Caddy, Artists and Models, Birds and the Bees.

WARRICK, RUTH: Actress. b. St. Louis, June 29. In 1941: "Citizen Kane," "Obliging Young Lady," RKO; "The Corsican Brothers," UA-Small. In 1942: "Journey Into Fear," RKO. In 1943: "Forever and a Day," RKO. On TV in Peyton Place.

PICTURES INCLUDE: The Iron Major, Secret Command, Mr. Winkle Goes to War, Guest in the House, China Sky, Driftwood, Daisy Kenyon, Arch of Triumph, The Great Dan Patch, Make Believe, Ballroom, Three Husbands, Let's Dance, One Too Many, Roogie's Bump.

WASHBURN, DERIC: Writer. b. Buffalo, N.Y. e. Harvard Univ. (English lit.). Has written number of plays, including The Love Nest and Ginger Anne.
PICTURES: Silent Running (co-s.p.), The Deer Hunter, The Border.

WASILEWSKI, VINCENT, T.: Executive. b. Athens, Ill. e. Univ. of Ill., Coll. of Engineering, 1940; U.S. Air Force, Sept. 1942–Oct. 1945; ret. to Univ. of Ill., bachelor degree in political science, 1948, degree of Doctor of Jurisprudence, 1949; admitted to the Illinois Bar, 1950; joined staff of the National Association of Broadcasters, Oct. 12, 1949; chief attorney, Feb. 1953; mgr., government relations, Aug. 1955; v.p., governmental affairs, NAB, June 1960; exec. v.p., NAB, Aug. 1961; pres., Jan. 1965.

WASSERMAN, DALE: Writer, producer. b. Rhinelander, Wis., Nov. 2, 1917. Stage: lighting designer, dir., prod.; dir. for. attractions, S. Hurok; began writing, 1954. Founding member & Trustee of O'Neill Theatre Centre; Artistic Director Midwest Playwrights Laboratory; member Motion Picture Academy; awards include Emmy, Tony, Critics Circle (Broadway), Outer Circle; Writers Guild, etc.
TV: The Fog, The Citadel, The Power and the Glory, Engineer of Death, The Lincoln Murder Case, I Don Quixote, Elisha and the Long Knives, and others.
PLAYS: Livin' the Life, 998, One Flew Over the Cuckoo's Nest, The Pencil of God, Man of La Mancha, Play With Fire.
PICTURES INCLUDE: Cleopatra, The Vikings, The Sea and the Shadow, Quick, Before It Melts, Mister Buddwing, A Walk With Love and Death, Man of La Mancha.

WASSERMAN, LEW: Chairman of the bd., Chief Executive Officer, MCA, Inc., Universal City Studios, Universal City, Calif.

WATERBURY, RUTH: Executive. b. Rensselaer, N.Y.; e. private and public schools, Albany, N.Y.; m. late Harold Cary. Feature writer, New York World, New York Daily News ed. Photoplay Mag. In 1938 superv. ed. for films, all Macfadden publications; Louella Parsons Hollywood staff, 1945–64; Books: Elizabeth Taylor, 1964, Richard Burton, 1965.

WATERHOUSE, KEITH: writer. b. Leeds, England, 1929. Early career as journalist, novelist. Author of There is a Happy Land, Billy Liar, Jubb, The Bucket Shop. Ent. m.p. ind. 1960, s.p. (with Willis Hall).
PICTURES INCLUDE: Whistle Down The Wind, A Kind of Loving, Billy Liar, Man in the Middle, Pretty Polly, Lock Up Your Daughters, The Valiant, West Eleven.
TV: (series): Inside George Webley, Queenie's Castle, Budgie, Billy Liar, There is a Happy Land.

WATERS, GEORGE, W.: Writer, producer, executive. b. Washington, D.C., Dec. 24, 1908; e. Catholic Univ. of America, 1923. Reporter, Times-Herald, Wash., D.C., 1923–25; Reporter, state ed., Washington Evening Star, 1925–30; City ed., Wash. Times-Herald, 1936–43; city ed., Miami Daily News, 1943–47; dir. of publicity Int'l Polo, Miami Seahawks pro football team, night clubs, etc., 1950; senatorial accts., press relations, Wash., D.C., 1950–55; co-author, assoc. prod., Tank Battalion, 1958; co-author, pictures: Speed Crazy, Why Must I Die?
TV: Scene of the Crime; Segments; Violence Comes Easy, Collector's Item. Georgetown Films.

WATERSTON, SAM: Actor. b. Cambridge, Mass., Nov. 15, 1940. e. Yale Univ. Began acting on stage in New York (Indians, Halfway Up a Tree, The Trial of the Catonsville Nine) and in Los Angeles (Meeting by the River, Volpone, etc.). Won Drama Desk Award as best actor for role in Joseph Papp production of Much Ado About Nothing (1972–73 Broadway season).
PICTURES INCLUDE: Three, Savages, The Great Gatsby, Rancho de Luxe, Journey into Fear, Capricorn One, Interiors, Heaven's Gate.
TV: Much Ado About Nothing, The Glass Menagerie.

WATKINS, GARTH: Actor. b. London, Aug. 8, 1922. e. University College School. Served in Royal Air Force 1944–45. Entered films in 1945 with role in The Captive Heart.
PICTURES INCLUDE: Bedelia, Gaiety George, A Matter of Life and Death, The Hanging Judge, Goodbye Mr. Chips, Cromwell, The Rise and Fall of Michael Rimmer, Virgin Witch, Fright, Naughty, Twins of Evil, Mary, Queen of Scots, Steptoe and Son, Henry VIII, Cinderella, The Omen.
TV: People in Conflict (Canada).

WATLING, JACK: Actor. b. Chingford, London, Eng., Jan. 13, 1923. e. Italia Conti Stage Sch. Stage debut in Where the Rainbow Ends, 1936; m.p. debut in 60 Glorious Years, 1937.
PICTURES INCLUDE: Goodbye, Mr. Chips, We Dive at Dawn, Day Will Dawn, Demi Paradise, Way Ahead, Journey Together, Easy Money, Winslow Bay, Quartet, Courtneys of Curzon Street, Once a Sinner, Naked Heart, White Corridors, Under Capricorn, Private Information, Father's Doing Fine, Meet Mr. Lucifer, Trouble in The Glen, Sea Shall Not Have Them, Confidential Report, Admirable Crichton, A Night to Remember, Sink the Bismarck, The Night We Sprang A Leek, Nearly A Nasty Accident, Mary Had A Little, Three On A Spree, Edgar Wallace, The Nanny, 11 Harrow House.
TV: Crime on My Hands, No Medals, Disturbance, Manor at Northstead, Devil Makes Three, Aren't We All, World of Tim Fraser, No Hiding Place, Zero I, Hancock Half Hour, Anna Karenina, Both Ends Meet, Loyalties, Ghost Squad, Tales of Mystery, Boyd Q.C., The Planemakers, Love Story, Six of the Best, The Power Game, Dixon of Dock Green, Mickey Dunne, Emergency Ward 10, Dr. Who, Softly Softly, No That's Me Over Here, The Power Game, The Pathfinders, Spike Milligan, The Mackinnons, Many Wives of Patrick, The Cedar Tree, The Doctors' Daughters.

WATT, HARRY: Director. b. Edinburgh, Scotland, Oct. 18, 1906. e. Edinburgh Acad.; Edinburgh U. Joined Empire Marketing Board Film Unit, 1931; asst. Robert Flaherty on Man of Aran. Wrote and dir. number of documentaries including Six-thirty Collection, Night Mail, North Sea. Since outbreak of war worked with Crown Film Unit on propaganda films, including Squadron 992, Britain at Bay, Dover Front Line, Target for Tonight, London Can Take It, and Christmas Under Fire. In 1942;: associated Ealing Studios (s.p., dir.) Nine Men; (dir.) Fiddlers Three, 1946; wrote and dir. The Overlanders in Australia, 1947; Eureka Stockade, Australia, 1949; Where No Vultures Fly (Ivory Hunter), West of Zanzibar; in Africa, 1950–54; exec. prod., Granada TV, 1955; rejoined Ealing Films; wrote, dir., People Like Maria; s.p. dir., Siege of Pinchgut; The Lion.

WAX, MO.: Publisher and editor of Film Bulletin. b. Philadelphia. e. Villanova U. Also pres., Audienscope, Inc., Entertainment research organization.

WAYLAND, LEN: Actor. b. California, Dec. 28. e. Junior College, Modesto, California. Wrote, prod. weekly radio series 1939–41, KPAS, KTRB, Calif. Service, radar navigator, 1941–45; en. theatre, Tobacco Road, 1946; 1973, formed Len Wayland Prods. for production of theatrical pictures and TV series. In 1976–77: produced/directed, Don't Let It Bother You. 1978, pro./dir., You're not there yet, for own company.
THEATRE: Played summer and winter stock 1947–49. B'way, Streetcar Named Desire, 1949, and tour; toured as lead, Heaven Can Wait, 1951; My Name Is Legion, 1952–53; toured, Love of Four Colonels, 1953, Stalag 17, 1954. Lead, NBC-TV serial, A Time to Live, 1954. NBC-TV, First Love, 1955; Armstrong Circle Theatre, Justice, Sgt. Bilko, Kraft Theatre, etc., 1956–58; NBC-TV, Dr. Weaver, From These Roots, 1958–61. Off Broadway prod., USA, 1960; on Broadway; A Man For All Seasons, 1962.
TV: Profiles In Courage, Dr. Kildare, Gunsmoke, Slattery's People, Ben Casey, A Noise in The World, 1967, Love Is A Many Splendored Thing, 1969; Dragnet, Outsider, 1970; Ironside, Name of the Game, The Bold Ones, Daniel Boone, The Virginian, Project U.F.O., Sam.

WAYNE, DAVID: Actor, r.n. Wayne McKeekan; b. Travers City, Mich., Jan. 30, 1916; e. Western Michigan land, 1936; in marionette shows, 1937.
THEATRE: Finian's Rainbow, Mister Roberts, Teahouse of the August Moon.
PICTURES INCLUDE: Portrait of Jennie, Adam's Rib, Reformer and the Redhead, My Blue Heaven, Stella, M, Up Front, As Young As You Feel, With a Song in My Heart, Wait 'Til the Sun Shines, Nellie; Down Among the Sheltering Palms, The I Don't Care Girl, We're Not Married, O. Henry's Full House, Tonight We Sing, How to Marry a Millionaire, Hell and High Water, Tender Trap, Last Angry Man, The Big Gamble, The Andromeda Strain, The African Elephant (narrator), Huckleberry Finn, The Front Page, The Apple Dumpling Gang.

WAYNE, JOEL: Executive. Had career as art director before joining Grey Advertising in 1962, where headed staff of 250 and was responsible for all TV, radio and print campaigns in behalf of General Foods, Procter and Gamble, B.F. Goodrich, Revlon, etc., as exec. v.p. & creative director. Left in 1980 to become Warner Bros., v.p. in chg. of creative advertising, headquartered in Burbank.

WAYNE, MICHAEL A.: Executive. r.n. Michael A. Morrison. b. Los Angeles, Calif., Nov. 23, 1934. f. John Wayne. e. Loyola H.S., Loyola Univ., B.B.A. Asst. dir., various companies, 1955–56; asst. dir., Revue Prods., 1956–57; pres. Batjac Prods. and Romina Prods., 1961; asst. to producer: China Doll, 1957; Escort West, 1958; The Alamo, 1959; prod., McClintock, 1962; co-prod., Cast Giant Shadow, 1965; prod. The Green Berets, 1967; exec. prod. Chisum, 1968; prod. Big Jake, 1970; prod. The Train Robbers, 1972; prod. Cahill, U.S. Marshall, 1972; exec. prod. McQ, 1973, Brannigan, 1974.

WAYNE, PATRICK: Actor. b. 1939. Son of John Wayne. Made film debut at age 11 in Rio Grande.
PICTURES INCLUDE: The Searchers, The Alamo, The Comancheros, McClintock, The Bears and I, Mustang Country, Sinbad and the Eye of the Tiger.

WEAVER, DENNIS: Actor, director. b. Joplin, Mo., June 4, 1925. e. Univ. of Oklahoma, B.A. fine arts, 1948.
TV: Chester, in Gunsmoke, 1955–64; title role, Kentucky Jones, NBC series, 1964–65; Gentle Ben, 1967–69; McCloud, 1970–76, The Forgotten Man, Duel, The Rolling Man, The Great Man's Whiskers, McCloud, (1970–77), Terror on the Beach, Intimate Strangers (ABC) 1977, Pearl (1978), Centennial (1978); Stone (series, 1979–80). Movies: Amber Waves, Dr. Mudd (1980).
PICTURES INCLUDE: Duel at Diablo, Way Way Out, Gentle Giant, A Man Called Sledge, What's the Matter with Helen?, The Gallant Hours Batangas.

WEAVER, MARJORIE: Actress. b. Grossville, Tenn., Mar. 2, 1913; e. U. of Kentucky, Indiana U. Played with McCauley Stock Co.; Billy Rose's Shrine Minstrels. In 1936 appeared in China Clipper, WB; from 1937 in many 20th Century-Fox pictures.
PICTURES INCLUDE: Let's Face It, Pardon My Rhythm, You Can't Ration Love, The Great Alaskan Mystery, Shadows of Suspicion, Fashion Model, Leave It to Blondie, We're Not Married.

WEAVER, SYLVESTER L., JR.: Executive. b. Los Angeles, Calif., Dec. 21, 1908; e. Dartmouth Coll. CBS, Don Lee Network, 1932–35; Young & Rubicam adv. agency, 1935–38; adv. mgr., American Tobacco Co., 1938–47; v.p. Young & Rubicam, 1947–49; joined NBC as v.p., chg. TV, 1949; appt'd v.p. chg. NBC Radio & TV networks, 1952; vice-chmn. bd., NBC, Jan., 1953; pres., NBC, Dec., 1953; bd. chmn., Dec. 1955; own firm, 430 Park Avenue., N.Y.C., 1956; chmn. of bd. McCann-Erickson Corp. (Intl.), 1959; pres., Subscription TV, Inc. Comm. Consultant in Los Angeles, Cal. and President, Weaver Productions, Inc.

WEBB, JACK: Actor, Producer, Director. b. Santa Monica, California, Apr. 2, 1920; e. Belmont H.S., Los Angeles. Began radio career, San Francisco, 1945, announcer, writer, actor. Army Air Forces WWII. 1942–45; freelance radio actor, title role, Pat Novak, 1946 in San Francisco; Johnny Modero, Pier 23, 1947, Los Angeles; created "Dragnet" 1949; created "Pete Kelly's Blues" 1950; "Dragnet" winner Best Mystery Show and Most Popular Star of Tomorrow categories, M.P Daily's 15th annual radio poll 1950–51; Best Mystery Show, TV & radio, 1953–54; winner Best Mystery Show, Academy of Television Arts and Sciences 1952–54; winner Look magazine award, Best Director, 1954. Director, star, "Dragnet" TV film series; prod. TV series, Pete Kelly's Blues, D.A..'s Man, True, Dragnet, 1967–70; Adam-12, 1968–74; The D.A., 1970–71; O'Hara, U.S. Treasury, 1970–71; Emergency, 1971–75; Chase, 1973, The Rangers, 1974–75 and Mobile Two, 1975; Project UFO (1978).
PICTURES INCLUDE: Sunset Boulevard, Appointment With Danger, The Men, Halls of Montezuma, You're in the Navy Now; prod. act., dir., Dragnet, Pete Kelly's Blues, The D.I.-30-; The Last Time I Saw Archie.

WEBB, JERRY: Editor. b. Indianapolis, Ind. Feature editor, supervision, Fox Studio, 1929–33; 20th Century-Fox, 1934–57. supv. film editor, Bonjour Tristese, London, 1957; film editor, Yovanka, Under Ten Flags, Roma, 1959–60; Esther and The King, Roma, 1960; Lion of Sparta, Athens, London, 1961; supv. editor, Damon & Pythias, Roma, 1962. 1964 entered commercial field. 1968: invented new anamorphic method of photography and printing of film: Vertiscope.

WEBB, ROBERT A.: Film editor, writer. b. Springfield, Ill., May 27, 1911. Film editor, Goldwyn, 20th-Fox; served 3½ years with John Ford's Naval Photog. Unit & headed film editorial unit at Nurnberg Internat'l Military Trials; ed. many short film sketches, Omnibus (TV), 1953; ed., Patti Page Shows. Member: M.P. Film Editors, Soc. of M.P. Film Editors, Acad. of TV Arts & Sciences.

WEBB, ROBERT D.: Director
PICTURES INCLUDE: assoc. prod., Lure of the Wilderness; dir., Glory Brigade, Proud Ones. Prod., dir.: Threshold of Space, Seven Cities of Gold, Beneath the 12 Mile Reef, White Feather, The Jackals, Capetown Affair, Love Me Tender.

WEBBER, ROBERT: Actor. b. Santa Ana, Calif., Oct. 14. e. Belmont High, Van Nuys High, Compton Jr. Coll. USMC, 1943–45. Summer stock. Co-starred, Wonderful Town, American Pavilion, Brussels World Fair, 1958.
THEATRE: Two Blind Mice, Goodbye My Fancy, Royal Family, No Time for Sergeants, Orpheus Descending, Fair Game, A Loss of Roses, Period of Adjustment.
PICTURES INCLUDE: Highway 301, Twelve Angry Men, The Stripper, The Sandpiper, The Third Day, Harper, The Silencers, The Hired Killer, Dead Heat on a Merry-Go-Round, Don't Make Waves, The Dirty Dozen, The Big Bounce, The Great White Hope, The French Mistress, $(Dollars), Every Man Is My Enemy, Bring Me the Head of Alfredo Garcia, Midway, Casey's Shadow, The Choirboys; The Revenge of the Pink Panther, Ten, Private Benjamin, S.O.B., Sunday Lovers, Wrong Is Right.
TV: Over 400 shows from 1947 through 1979.

WEBSTER, DAVID: Dir. Public Affairs, BBC. b. Taunton, England, 1931. Joined BBC News in 1953. Moved to BBC Television Current Affairs 1959 and worked on Face to Face and Lifeline before going to Panorama in 1959. Panorama producer until start of BBC-2 in 1964, when he left to run Enquiry, Encounter on BBC-2. Returned to Panorama as Deputy Editor, becoming Editor in 1967. Appt. Assistant Head of Current Affairs Group, Television 1970. For five years (197-–76) BBC rep. in the USA. Chairman of the International Council of the National Academy of Television Arts and Sciences 1974–75. In 1976 returned to England to become the BBC's Controller of Information Services. Appt. Director, Public Affairs 1977.

WEBSTER, M. COATES: Writer. b. East Orange, N.J. December 15, 1906. e. U. Pa., Wesleyan U. On stage 8 yrs. Wrote 3 novels (published) Showboy, Derelict Alley, Strange Fraternity.
PICTURES INCLUDE: Klondike Kate; collab., story, s.p., Adventures of a Rookie; (s.p.) Blonde Ransom; (collab. s.p.) Jungle Captive; (s.p.) Strange Confession, I'll Remember April, Song of Arizona, Frisco Tornado, Rough Riders of Durango Desert of Lost Men, Captive of Billy the Kid, Leadville Gunslinger, Montana Belle.

WECHSLER, LAZAR: Producer. b. Poland, 1896; e. Polytechnical Inst. of Zurich. Set up first film co. in Switzerland making documentary & pub. shorts.
PICTURES INCLUDE: Wings Over Ethiopia, This Is China, Marie Louise (Acad. Award, best s.p.), The Last Chance, The Search, Four in a Jeep, The Village, Heidi, Heidi and Peter, It Happened in Broad Daylight, The Marriage of Mr. Mississippi, Shadows Growing Longer, The Right to Be Born (The Doctor Speaks Out).

WEILER, GERALD E.: Producer. b. Mannheim, Germany, May 8, 1928. e. Harvard, 1946–48; Columbia, B.S., 1949–51; NYU Grad. Sch., 1951–53; writer, WHN, N.Y. writer, sports ed., news ed., Telenews Prod., Inc., 1948–52; asst. to prod., Richard de Rochemont, Vavin, Inc., 1952; U.S. Army, 1953–55; v.p., 1955–73; President, Weiler Communications Inc. 1973.

WEINBERG, HERMAN G.: Writer, film editor. b. New York City, Aug. 6, 1908. e. Inst. Musical Art (N.Y.). Entered m.p. ind. 1927 as musical consultant in scoring of silent foreign films for various little theats., N.Y. Then U.S. correspondent various European film publications; Close-Up (Switzerland), 1928–33; Film Art (London) 1929–30; Cinema Quarterly (Edinburgh) 1930–31; Cinea Pour Tous (Paris) 1931; Intercine (official publication of film div., League of Nations, Rome) 1932; Sight & Sound, British Film Institute, London; ed. and titled many foreign prod., dir., Autumn Fire, 1930. In 1950, U.S. correspondent Filmkunst, Vienna; Bianco e Nero, Rome; Cinema, Rome; also ed. series Indexes on Directors for British Film Institute; U.S. corr. Cahiers du Cinema; Filmkunst, Vienna; Film, Venice; Film, Munich; former film ed. Liberty Mag.; prod. supvr. The Knife Thrower; ed., U.S. edition, 50 Years of Italian Cinema, 50 Years of Opera and Ballet in Italy; syndicated column, Coffee, Brandy and Cigars; contrib. to newspapers, mags.; author: Sin & Cinema; cont. to anthologies, 20 Years of Cinema in Venice. Introd. to Art of the Movies, The Cinema Between Two World Wars; film critic; Film Culture, Film Quarterly, Playboy Magazine, 1960–76, Instructor, film history. The City College, N.Y.: Juror at San Francisco film festival, 1960; Vancouver film festival, 1961; prep. exhibit, Homage to Erich Von Stroheim, Montreal Film Festival, 1964. Work in progress: an autobiography.
AUTHOR: Josef Von Sternberg. Editions Seghers, Paris, 1966 and American edition by E. P. Dutton, 1967, The Lubitsch Touch, E. P. Dutton, 1968, Saint Cinema, D.B.S. Publications, 1970. The Complete Greed, Arno Press, 1972. Revised edition, E. P. Dutton, 1973. Revised edition "Saint Cinema" (Dover) 1973, Stroheim: A Pictorial Record (Dover 1974), The Complete 'Wedding March' (Little, Brown) 1975. Revised edition, "The Lubitsch Touch" (Dover) 1977.

WEINBLATT, MIKE: Television Executive. e. Syracuse Univ. Served in Army as counter-intelligence agent, mostly in Japan (1952–53). Joined NBC in 1957; has headed two major TV network functions—talent/program admin. & sls. Joined network business affairs dept. in 1958 as mgr., business affairs, facilities operations; rose to post of director, pricing & financial services before moving to sales in November, 1962, as mgr., participating program sales. Named v.p., eastern sales, NBC-TV, 1968. Named v.p., talent & program admin., October, 1968; promoted to v.p. sales, February, 1973. January, 1975 named snr. v.p., sales; later became exec. v.p. Appointed exec. v.p. & gen. mgr. of NBC TV network in August, 1977.

WEINSTEIN, HENRY T.: Executive prod. b. New York, N.Y. July 12, 1924. e. C.C.N.Y., Carnegie Institute of Technology. Entered theatre as dir., 1950. Stage mgr. for Herman Shumlin. Dir. of: the Brattle Theatre, Cambridge, Mass., Margo Jones Theatres, Dallas, Theatre, Theatre in the Round, Houston, Texas. Proc. for Theatre Guild, N.Y. 1957–59. 1960–65 producer 20th Century-Fox, M.G.M. 1970–75 executive in charge of production, American Film Theatre. Presently executive in charge of production for L.C. Entertainment.

PICTURES INCLUDE: Tender is the Night, Joy in the Morning, Cervantes, Madwoman of Chailliot, The Battle of Neretva, Magic Christian, A Delicate Balance, The Homecoming, The Iceman Cometh, Lost In the Stars, Butley, Luther, Rhinoceros, Galileo, The Man In the Glass Booth, In Celebration.

TV: Play of the Week series, prod. 1959–60.

WEINSTEIN, PAULA: Executive. Started as theatrical agent with William Morris and International Creative Management. With Warner Bros. 1976–78 as production v.p.; left to go to 20th-Fox in same capacity. Named Fox snr. v.p., worldwide production. In 1980 appointed v.p.-production of The Ladd Company, headquartering at Burbank Studios offices of firm.

WEINTRAUB, FRED: Executive, producer. b. April 27, 1928, Bronx, N.Y. e. U. of Pennsylvania Wharton School of Business, Owner of The Bitter End Coffeehouse to 1971. Personal management, Campus Coffee House Entertainment Circuit; TV Production Hootenanny, Popendipity; syndicated TV show host: From The Bitter End; motion picture production; v.p., creative services, Warner Bros. 1969, executive in charge of Woodstock; producer motion pictures, Weintraub-Heller Productions, 1974, Enter The Dragon, Rage, Black Belt Jones, Truck Turner, Golden Needles, Animal Stars, Hot Potato, The Ultimate Warrior, Dirty Knights Work, Those Cuckoo Crazy Animals, Crash, Outlaw Blues, The Pack, The Promise, Tom Horn, Battle Creek Brawl, Force: Five.

WEINTRAUB, SY: Executive. b. New York City, 1923. e. Univ. of Missouri (B.A., Bach, of Journalism, 1947); graduate of American Theater Wing. Started career in 1949 forming with associates a TV syndication co., Flamingo Films, Inc., which merged with Associated Artists to form Motion Pictures for Television, Inc., largest syndicator at that time. He originated Superman and Grand Ol'Opry series for TV. In 1958 bought Sol Lesser Prods., owners of film rights for Tarzan, and began producing and distributing Tarzan films through Banner Productions, Inc. Also formerly chmn. of bd. of Panavision, Inc.; bd. mem. and pres. of National General Television Corp., and pres. of KMGM-TV in Minneapolis. In 1978 named chm. of Columbia Pictures Industries' new Film Entertainment Group, also joining Office of the Chief Executive of CPI.

WEIS, DON: Writer, director, producer. b. Milwaukee, Wis., May 13, 1922; e. U. of Southern Calif. No prior bus. experience.

PICTURES INCLUDE: (dial. dir.) Body & Soul, Red Pony, Champion, Home of the Brave, The Men; (dir.) Letter From a Soldier, sequence in It's a Big Country, Bannerline, Just This Once, You for Me, I Love Melvin, Remains To Be Seen, A Slight Case of Larceny, Half a Hero, Affairs of Dobie Gillis, Adventures of Haiji Baba, Ride the High Iron, Catch Me If You Can, Gene Krupa Story, Critic's Choice, Looking for Love, The King's Pirate, Repo.

TV: Dear Phoebe. Best TV director, 1956, 1958. Screen Dir. Guild, The Longest Hundred Miles, It Takes a Thief, Ironside, M.A.S.H., Happy Days, Planet of the Apes, Bronk, Petrocelli, The Magician, Mannix, Night Stalker, Barbary Coast, Courtship of Eddie's Father, Starsky & Hutch, Hawaii Five-O, Chips, Charlie's Angels, Love Boat, Fantasy Island

WEIS JACK: Producer-Director-Writer-Cinematographer-Film Editor. b. Tampa, Fla., October 1, 1932. e. Notre Dame (BS); Master Univ. of Chicago. Was in U.S. Air Force six yrs. Founded Associated Productions/Associated Advertising Productions, Inc. in New Orleans in Aug., 1967, and has been involved in over 120 films and approx. 1,500 commercials. Produced, directed and wrote original s.p.s for several HEW youth rehabilitation pictures. Member: Cinematography Local 666, Chicago IATSE; Film Editor Local 780, IATSE, Chicago; Directors Guild of America; Writers Guild of America.

PICTURES INCLUDE: Quadroon (dir.), Storyville (prod., dir., s.p.), Damballa (prod., dir., s.p.); creature from Hony Island Swamp (prod.-dir., s.p.); Lehia (prod., dir., camera), You Never Gave Me Roses, (s.p., prod., dir.); The Perfect Circle (TV-s.p., prod., dir.); Crypt of Dark Secrets (prod., dir., s.p.); Mardi Gras Massacre, prod., dir., s.p., editor).

WEISBERG, BRENDA: Writer. b. Rowne, Poland. Magazine writer, social service, public health, drama instructor. Married to the late Morris Meckler.

PICTURES INCLUDE: s.p., China Sky; s.p., When a Girl's Beautiful, Burning Cross, Port Said, Rusty series, Girl's School; collab. s.p., Isle of Samoa; collab. orig., Reunion in Reno, Alias Mr. Twilight, King of the Wild Horses, Little Tough Guy, s.p. Shadowed. Collaborator, Scarlet Claw, Ding Dong Williams, Babes On Swing Street, Weird Woman, The Mummy's Ghost, The Mad Ghoul, Keep 'Em Slugging, Mug Town, Tough As They Come, Mob Town, There's One Born Every Minute, Hit the Road, Sing Another Chorus, You're Not So Tough.

TV: Fireside Theatre, Philco Theatre, Matinee Theatre.

AUTHOR: U(short stores) Woman's Home Companion, Collier's. American Mercury (Mencken's) Plain Talk, Forum

WEISBORD, SAM: Executive, b. New York, N.Y., Sept. 21, 1911. William Morris Agency, Inc., 1929; Bd. of Directors, Cedars-Sinai, Medical Center bd. of dirs. Menninger Foundation; Board of Directors Motion Picture and Television Fund. Exec. member Board of Directors, President & Chief Exec. Officer, William Morris Agency.

WEISER, MARTIN: Publicist. b. New York City, Oct. 27, 1916. e. U. of Rochester, 1936. Pub. dept. Warner Bros., N.Y., 1936–40; exploitation dir., western states, 1940–45; studio exploitation head, 1945–50; adv. pub. head, Lippert Pictures, 1950. Partner, Weiser and Thomas, Inc., 1954; pres., Marty Weiser Co., prom. Columbia Studio, 1959; Pub. dept., Warner Bros. Studios, 1961; Paramount Studios, 1962; asst. ad. pub. dir., Allied Artists Pictures, 1963; studio exploitation director Warner Bros. 1964; west coast advertising and exploitation co-ordinator for Warner Bros. Inc., 1973.

WEISS, FRITZ: Producer, director. Dir. of first anti-Hitler film, Vagabond, 1930. Dir. I am a Gypsy, 1934 film was confiscated in Vienna with its negative by Goebbels. Emigrated to England in 30's. Worked during WWII as prod. in the English theatre and as director-editor for the British Ministry of Information on anti Nazi propaganda and documentary films. Prod. mgr. of The Journey, Our Man in Vienna, and Mayerling. Director of Austrian TV Screen play for The Sex Commandment and Naked Odyssey.

WEISSMAN, MURRAY: Executive. b. New York City, Dec. 23, 1925. e. U. of Southern California. As an editor and news reporter for Fairchild Publications worked on Women's Wear Daily, Footwear News, Supermarket News, 1949–52; promotion mgr., TV Guide, 1952; asst. publicity director, KABC-TV, 1953–60; asst. dir. of press info., CBS, 1960–66; mgr., TV press dept., Universal Studio, 1966–68; executive in charge of m.p. press dept., Universal Studios & asst. secty., Universal Pictures, 1968–76; marketing exec., Columbia Pictures, 1976–77; vice pres. of advertising & publicity, Lorimar Productions, 1977; vice pres., ICPR Public Relations Company, 1978.

WEISSMAN, SEYMOUR J.: Executive, producer, director. Weissman Franz Productions. b. Brooklyn, N.Y., May 28, 1931. e. Kenyon Coll., Eng. Lit., A.B., 1953; Univ. of Southern Calif., Cinema, 1955. Unity Films, 1954; Henry Strauss & Co., 1954; Dir. of motion pictures, White Sands Proving Grounds, N.M., 1954–55; M.P.O., 1955; Coleman Prod., 1956; prod. dir., Dynamic Films, Inc., 1958–59; prod., dir., Viston Assoc., 1959–64; dir., VPI Prods., 1966.

WEISSMULLER, JOHNNY: Actor. b. Chicago, Ill., June 2, 1904. e. Chicago U. Champion swimmer. Screen debut 1932 in Tarzan, the Ape Man.

PICTURES INCLUDE: Tarzan's Desert Mystery, Swampfire, Tarzan and the Amazons, Tarzan and the Mermaids, Jungle Jim, The Lost Tribe, Fury of the Congo, Jungle Manhunt, Forbidden Land, Others in Jungle Jim film & TV series.

WEITMAN, ROBERT M.: Executive. b. New York City, August 18, 1905. e. Cornell University. Attended Paramount Managers' Training School. Assigned to Rialto Theatre, assistant manager, in 1926. Promoted to manager, Brooklyn Paramount. City manager of New York Paramount Public Theatres. Appointed managing director of New York Paramount Theatre in 1935. Managing dir. of N.Y. and Brooklyn Paramount Theatres. Vice-pres. United Paramount Theatres, supervising southern and Philadelphia houses; v.p. American Broadcasting-Paramount Theatres, Inc. in chge. of programming and talent; v.p. in charge of program development, CBS-TV, 1956; v.p. chge. prod. MGM-TV, 1960; v.p. in chg. prod., MGM, Inc., 1962; member MGM bd. directors; resigned MGM, July, 1967; v.p. in charge of studio production, Columbia, July, 1967. Dec., 1969 formed own prod. company.

WEITZNER, DAVID: Executive. b. New York, Nov. 13, 1938. e. Michigan State U. Entered industry in 1960 as mem. Columbia Pictures adv. dep't; later with Donahue and Coe as ass't exec. and Loew's Theatres adv. dep't; later with Embassy Pictures, adv. mgr.; dir. of adv. and exploitation for Palomar Pictures Corp.; v.p. in charge of adv., pub., and exploitation for ABC Pictures Corp.; v.p., entertainment/leisure div., Grey Advertising; v.p., worldwide adv., 20th Century Fox; now exec. v.p. adv./pub./promo., Universal Pictures.

WELCH, PETER: Actor. b. London, England, 1922. Early career as draughtsman. Formed own rep. company, later 5 years in various reps. Ent. m.p. ind. 1955.

TV: Three Tough Guys, Robin Hood, Doctor Morelle, The Chigwell Chicken, A Candle to the Madonna, Twelfth Night, Othello, Shadow Squad, William Tell, The Queen's

Corporal, No Hiding Place, Life of Falstaff, The Enemy, The Odd Man, The Lark, Dixon on Dock Green, Z Cars, The Plane Makers, Hancock Cause of Death, The Discharge of Trooper Lusby, North and South, The Expert, The Trouble-Shooters, Spytrap, Smith, The Strange Report. The Carnforth Practice, Softly, Softly, The Healing, Dickens of London, Snacker Emmerdale Farm, Law and Order, series. Limbo Connection, serial; Fox series.

THEATRE: Rights of Man, Broken Sword, All in a Good Time, Cromwell at Drogheda, A Man and his Wife, Candida, Little Foxes, Whose Life Is It Anyway?

PICTURES INCLUDE: Doctor Syn, The Secret of Blood Island, The Legend of Young Dick Turpin, Astronaut at the Court of King Arthur.

WELCH, RAQUEL: Actress. r.n. Raquel Tejada. b. Chicago, Ill. e. La Jolla H.S. Fashion and photographic modeling. Co-hostess. Hollywood Palace.

PICTURES INCLUDE: Roustabout, A House is Not a Home, Swinging Summer, Fantastic Voyage, Shoot Louder . . . I Don't Understand, One Million Years B.C., Fathom, Bedazzled, The Biggest Bundle of Them All, The Queens, Bandolero, Lady in Cement, 100 Rifles, Flare Up, The Magic Christian, Myra Breckinridge, Hannie Caulder, Kansas City Bomber, Fuzz, The Last of Sheila, The Three Musketeers, The Four Musketeers, The Wild Party, Mother, Jugs and Speed, Crossed Swords.

WELCH, ROBERT L.: Producer, writer. b. Chicago, Ill., Nov. 23, 1910. e. Northwestern U., U. of Illinois. In stock as actor-prod. Hedgerow Theatre, Phila.; with Pasadena Community Playhouse & Padua Hills Theatre; on radio wrote & prod. Kate Smith, Fred Allen & Jack Benny shows, originated Henry Aldrich shows with Young & Rubicam, N.Y.; prod. all U.S. Armed Forces radio shows, World War II; writer-prod. Paramount Pictures Corp.

PICTURES INCLUDE: (co lab. s.p.) Variety Girl; (prod.) Paleface, Sorrowful Jones, Top O' the Morning, Fancy Pants, Mr. Music, The Lemon Drop Kid, Son of Paleface. Writer, dir., prod., NBC-TV, 1953.

MEMBER: Lambs, AMPAS, Screen Writers Guild, Producers Guild, American Legion.

WELD, TUESDAY: Actress. b. New York, Aug. 27, 1943. e. Hollywood Professional School. Began modeling at 4 yrs. Film debut, Rock, Rock, Rock.

PICTURES INCLUDE: Rally Round The Flag, Boys! The Five Pennies, The Private Lives of Adam and Eve, Return to Peyton Place, Wild in the Country, Bachelor Flat, Lord Love a Duck, Pretty Poison I Walk the Line, A Safe Place, Play It As It Lays, Looking for Mr. Goodbar, Who'll Stop the Rain, Thief.

TV: The Many Loves of Dobie Gillis.

WELDEN, BEN: Actor. b. Toledo, Ohio, 1901. First appeared on the stage in 1926 at the Central Park Theatre, New York; then on London stage. Has appeared in many British and Hollywood film productions.

PICTURES INCLUDE: It's in the Bag, Appointment With Murder, Search for Danger, Fighting Fools, Sorrowful Jones, Tales of Robin Hood, All Ashore, Killers from Space, Steel Cage, Ma and Pa Kettle at Waikiki, Hidden Guns.

WELDON, JOAN: Actress. b. San Francisco, Calif., Aug. 5, 1933. e. San Francisco Conservatory of Music, U. of Calif. at Berkeley, City Coll. in San Francisco. Sang in chorus of San Francisco Grand Opera Co.; then L.A. Civic Light Opera Co.; m.p. debut in The System, 1953.

PICTURES INCLUDE: So This Is Love, Stranger Wore a Gun, The Command, Riding Shotgun, Them, Deep in My Heart.

WELK, LAWRENCE: Orchestra leader. b. Strasburg, N.D., March 11. Played accordion community dances, church socials, etc. Started own group. Biggest Little Band in America. Played hotels, ballrooms, music became known as Champagne Music. Signed Aragon Ballroom, Pacific Ocean Park, Calif., 1951, with weekly television show. Champagne Music Makers, ABC-TV, July 2, 1955; The Lawrence Welk Show, ABC; signed lifetime contract, Hollywood Palladium, July 1961. Recording: Calcutta, 1961; syndicated network show started 1971.

WELLES, ORSON: Producer, writer, actor. b. Kenosha, Wis.; p. Richard Head Welles, inventor, Beatrice Ives Welles, concert pianist; e. Todd School at Woodstock, Ill.

RADIO: March of Time, The Shadow, Campbell Playhouse. co-producer of Julius Caesar, Doctor Faustus, Horse Eats Hat, Macbeth, Shoemaker's Holiday, Heartbreak House.

PICTURES INCLUDE: in 1941: producer, director, writer and star of Citizen Kane, RKO. In 1942: produced, wrote and directed, The Magnificent Ambersons, RKO. Produced and wrote Journey into Fear, RKO, in 1943: Jane Eyre, 20th-Fox. In 1945 author column Orson Welles Almanac in N.Y. Post. Tomorrow Is Forever, Lady from Shanghai, (s.p., dir., prod., acted in) Macbeth; Prince of Foxes, Black Magic, Third Man, Black Rose, Trent's Last Case, Trouble in the Glen, Three Cases of Murder; act, prod., dir., Othello; act. Moby Dick, Compulsion, The Tartars; dir. star, Mr. Arkadin; prod., dir., wrt., The Trial; prod., dir., Chimes at Midnight (Falstaff); dir., starred, The Immortal Story. Acted in: Casino Royale, The Sailor From Gibraltar, I'll Never Forget What's 'is Name, Oedipus the King, The Southern Star, House of Cards, The Kremlin Letter, Start the Revolution Without Me, Catch 22, A Safe Place, Necromancy, F for Fake, Voyage of the Damned.

TV: The Man Who Came to Dinner.

WELLS, FRANK G.: Executive. b. March 4, 1932, Calif. e. Pomona College, 1949–53; Oxford U., 1953–55; Rhodes Scholarship Jurisprudence; U.S. Army, Infantry first lieutenant, 1955–57; Stanford Law School, 1957–59; joined Gang, Tyre & Brown (entertainment industry law firm) 1959; partner, 1962–69; mem., State Bar of Calif., American Bar Assoc., Los Angeles County Bar Assoc., pres., Warner Bros. Inc.

WELSH, CHARLES L.: b. New York, N.Y., June 7, 1906. e. Morris H.S., 1923. Press dept., N.Y. Hippodrome, 1919; B.F. Keith's Theats., 1921–29; United Trailer Svce., 1930; Meyer-Rigger Lab., 1931; partner Welgot Trailer Svce., 1932; mgr. special film div. Nat'l Screen Service, 1959. Pres., Summit Screen Service, 1967.

WELTMAN, PHILIP: Executive. b. Brooklyn, N.Y., July 16, 1908. e. Brooklyn College, 1923–27, N.Y.U., 1928–30, N.Y. Stock Exch. Inst., 1934–36. New York stock exchange firm, Chas. E. Quincey & Co., 1930–38; joined William Morris Agency, N.Y. office, 1938; trans. to coast, 1947; U.S. Armed Forces, 1943–45, battlefield commission, rec'd Croix de Guerre; transferred to Coast, 1947; Sr. exec., TV div.

WELTNER, GEORGE: Executive. b. Chicago, Ill., Aug. 16, 1901; e. B.S., Columbia, 1922. Entered foreign dept. of Paramount, 1922; asst. to head, Paramount Internat'l, 1934; asst. mgr., v.p., 1944; pres., Paramount Internat'l Films 1945; dir. of domestic and foreign sales, distrib., Paramount Pictures; pres., Paramount Film Dist. Corp., 1955; v.p. Paramount Pictures Corp., exec. vice-pres., Paramount Pictures Corp., 1962; pres., 1964; resigned, 1967, rank of cmmdr. in Order of Merit, Italian Republic, 1966. La Medaille de Vermeil, Consei Municipal de Paris, 1966; Medal of Honor Veterans of Foreign Wars, 1966.

WENDERS, WIM: Director. b. Dusseldorf, Germany, 1945. Studied film from 1967 to 1970 at Filmhochschule in Munich. Worked as film critic. In 1967 made first short films (Schauplatze) and three others before first feature, Summer in the City, in 1970.

PICTURES INCLUDE: Die Angst DesTormanns Beim Elfmeter, The Scarlet Letter, Aus Der Familie Der Panzerechsen, Falsche Bewegung, Alice in the Cities, The Goalie's Anxiety at the Penalty Kick, Kings of the Road, The American Friend.

WERLE: Costume designer. Real name Dan Werle. b. Chicago, Ill. e. Chicago Art Institute. Started as designer for New York wholesale house; later, Hattie Carnegie; Chez Paree in Chicago; opened own business, California, 1945. Designer for seven seasons Loretta Young hostess-gowns on The Loretta Young Show, NBC-TV. Also Rosemary Clooney Show; Playhouse 90; free lance, consul. designer; pres., man. designer, Werle Originals Inc. mem., Costume Designers Guild; has designed for television appearances Merle Oberon, Claudette Colbert, Ava Gardner, etc. Contracted to design for Barbara Stanwyck Theatre, Rosemary Clooney Show, Playhouse 90; now making clothes for special shows, also selling to top specialty stores throughout U.S. and Canada. Opened new salon feat. Coutoure and Custom Gowns, Dresses, Suits, boutique div., 1971. Designed for That Girl (ABC); Lana Turner's Town Hall & TV appearances, 1975.

WERNER, OSKAR: Actor, director, producer. b. Vienna, Austria, Nov. 13, 1922. Actor with repertory gp., Burg Theatre, Vienna.

PICTURES INCLUDE: Angel With a Trumpet, Eroica, Wonder of Our Days, Wonder Boy, Decision Before Dawn, Last Ten Days, Jules Et Jim, Ship of Fools, Interlude, The Shoes of the Fisherman, Voyage of the Damned.

TV: prod., dir., cast: A Certain Judas.

WERTMULLER, LINA: Writer-Director. b. Rome. After graduating high school enrolled in drama school. Began working in theatre in 1951; spent decade doing everything from puppetry to stage managing to writing and directing for radio and TV. Began film career as asst. to Fellini on 8½ in 1962. Following year wrote and directed first film, The Lizards. Had big TV success with series called Gian Burasca and then returned to theatre for a time.

PICTURES: This Time, Let's Talk About Men, The Seduction of Mimi, Love and Anarchy, All Screwed Up, Swept Away, Seven Beauties, A Night Full of Rain.

WESTCOTT, HELEN: Actress. r.n. Myrthas Helen Hickman. b. Hollywood, Calif; e. L. A. Jr. Coll., 1946. In play The Drunkard, at 7, for 9 yrs.; many radio shows.

PICTURES INCLUDE: Midsummer Night's Dream (as

child), Adventures of Don Juan, Girl from Jones Beach, Mr. Belvedere Goes to College, Whirlpool, Dancing in the Dark, The Gunfighter, Three Came Home, Secret of Convict Lake, Phone Call from a Stranger, Return of Texan, With a Song in My Heart, Loan Shark, A&C Meet Dr. J. & Mr. H., Charge at Feather River, Hot Blood, The Last Hurrah.

WESTON, JACK: Actor. b. 1915. Began career in 1934 in children's division of Cleveland Playhouse. In Army in World War II. Success came in Broadway hit, Season in the Sun. Was frequent performer in top TV shows during 1950s. Film debut in Stage Struck in 1958.

PICTURES INCLUDE: Stage Struck, Please Don't Eat the Daisies, All in a Night's work, The Honeymoon Machine, It's Only Money, Palm Springs Weekend, The Incredible Mr. Limpet, Mirage, The Cincinnati Kid, Wait Until Dark, The Thomas Crown Affair, The April Fools, Cactus Flower, A New Leaf, Fuzz, Marco, Gator, The Ritz, Cuba, The Four Seasons.

TELEVISION: Studio One, Philco Theatre, Kraft Playhouse, Rod Browning of the Rocket Rangers.

WESTON, JAY: Producer. b. New York City, March 9, 1929. e. NYU. Operated own pub. agency before moving into film prod. In 1965 launched Weston Production; sold orig. s.p., The War Horses, to Embassy Pictures; acquired and marketed other properties. Became prod. story exec. for Palomar-ABC Pictures in 1967.

PICTURES INCLUDE: For Love of Ivy (co.-prod.), Lady Sings the Blues (co.-prod.), W. C. Fields and Me, Chu Chu and the Philly Flash.

STAGE: Does a Tiger Wear a Necktie (co.-prod.).

WESTON, ROBERT R.: Executive. b. N.Y., e. Peekskill Military Acad., Fordham Univ., Publicity dir., WFUV-FM, 3 yrs.; copy writer, Columbia Pictures, asst. accnt. exec., Donahue & Coe, asst. adv. mgr., United Artists adv. dir., Embassy Pictures, v.p., asst. to exec. v.p.; v.p. asst. to pres., resigned to become independent film producer, 1971. Presently pres., Harold Robbins Int'l., film prod. co. Prod.: The Betsy.

WEVER, WARREN: Talent agent. b. Dennison, Texas, Feb. 20; e. Col. Univ., 1938–39; UCLA, 1939–41, B.A. casting dir., Warner Bros., 1942–53; talent agent, Wynn Rocamora Agency, 1953–60; the Mitchell J. Hamilburg Agency, 1960. Formed own agency, Warren Wever Artists' Mgt., 1970.

WEXLER, HASKELL: Cinematographer, director. b. 1926. Photographed educational and industrial films before features.

PICTURES INCLUDE: The Savage Eye, The Bus, The Hoodlum Priest, Angel Baby, A Face In the Rain, America, America, The Best Man, The Loved One, Who's Afraid of Virginia Woolf, In the Heat of the Night, The Thomas Crown Affair. Co-produced, directed, wrote, Medium Cool. 1972: Trial of Catonsville Nine, Interview Allende, American Graffiti, One Flew Over the Cuckoo's Nest, Introduction to the Enemy, Bound for Glory, Days of Heaven.

WHEELER, LYLE: Art director. b. Woburn, Mass., Feb. 12, 1905; e. U. So. Calif. Mag. illustrator, industrial designer before entering m.p. ind. as art dir. of Garden Of Allah. In 1944 apptd. supervising art dir. 20th Century-Fox. Academy Award, in collab. art-direction black & white for Anna and the King of Siam; Gone With The Wind, color art dir., collab., The Robe, 1953. Love Is a Many-Splendored Thing, Daddy Longlegs, The Diary of Anne Frank (AA, 1959), Journey to the Center of the Earth, The Cardinal.

WHELDON, SIR HUW. P. (O.B.E., M.C.): Producer, executive, author. Consultant, NBC, personnel and electronics, etc. b. Prestatyn (Wales), May 7, 1916; e. Friars Bangor; Soester Gymnasium (Germany); London Univ. With Airborne Division during War; dir for Wales Arts Council, 1946; entered TV 1952, produced TV programmes. Head of Documentary and Music Programmes, BBC TV, 1964; controller of programs BBC TV, 1965–68; managing director of television, BBC, 1969. 1975. Chrm. Court of Governors, LSE., pres., Royal TV Society (Wales); Governor Royal Coll. of Art, National Film School, London Festival Ballet. Trustee: National Portrait Gallery, Knighted 1976. Hon. D. Litt (Ulster), Hon. Fellow (LSE).

TV: Television Talks, 1954; also: Press Conference, Is This Your Problem, Orson Welles Sketchbook, All Your Own, Men in Battle, Men of Action, Monitor, Portraits of Power, Royal Heritage, Library of Congress.

WHITE, JESSE: Actor. b. Buffalo, N.Y., Jan. 13, 1919; e. Akron, O., H.S. Did odd jobs, then salesman; radio, vaudeville, burlesque, nightclubs and little theatre work; Broadway stage debut in Moon is Down, 1942; other shows include Harvey, Born Yesterday, etc. Has appeared on numerous radio and TV shows, regular on Danny Thomas, Ann Southern Shows.

PICTURES INCLUDE: Harvey, Death of a Salesman, Callaway Went Thataway, Million Dollar Mermaid, Witness to Murder, Forever Female, Not As a Stranger, Bad Seed, Back from Eternity, Designing Woman, Marjorie Morningstar, Legs Diamond, Fever in the Blood, Sail a Crooked Ship, It's Only Money, The Yellow Canary, It's a Mad, Mad, Mad, Mad World, Looking For Love, A House Is Not a Home, Bless the Beasts and Children, The Cat from Outer Space.

WHITE, JULES J.: Producer, director. Started as juvenile actor, film editor, assist. cameraman, gag writer, dir., shorts and features; exec. prod., short subjects, Columbia, 1933–58; prod. and/or dir., over six hundred shorts, including the Three Stooges several features; co-prod., CBS-TV; prod., free lance.

WHITE, LAWRENCE R.: Executive. b. 1926. e. Syracuse U. Began career as producer-director for the Dumont Television Network in 1948. Dir. of programming, Benton & Bowles, Inc., 1951; joined CBS TV network as v.p., daytime programming, 1959; dir. of program development, CBS, 1963; joined NBC television network in 1965 as v.p., daytime programs; v.p. programs, east coast, 1969; v.p. programs, NBC-TV, 1972; ind. prod. affiliated with CPT, 1975.

WHITE, LEONARD: Producer, director, actor. b. Sussex, Eng. TV dir., prod., CBS-TV (Canada), T.W.W. Ltd. T.T. TV & ABC-TV; Jupiter Thea., Inc.; Crest Theatre; Royal Alexandra Thea.; Toronto Thea., 1953–57. England, Playhouse, Oxford, Perth Repertory Thea., Hornchurch, Guilford Repertory Thea. Belgrade Thea., Coventry. Actor: U.S.A. debut in A Sleep of Prisoners, 1951–52; London West End. In the White Devil, He Who Gets Slapped, Macbeth, Still She Wished for Company, Point of Departure.

PICTURES INCLUDE: The Dark Man, The Large Rope, River Beat, Hunted, Martin Luther, Passage Home, Breakout, Circumstantial Evidence, At the Stroke of Nine, etc.

TV: All networks, G. Britain and CBC (Canada). Prod., ABC-TV, 1960–68. Inside Story series, Armchair Mystery Thea., series. Police Surgeon series, The Avengers, series, Out of This World, series, Armchair Theatre series. Prod., 1968–69; prod., Thames Television, 1969–70. Drama consultant CBC-TV Toronto.

WHITE, PAUL: Executive, producer, director. b. New York, N.Y.; e. Columbia U. Worked for N.Y. Times, N.Y. & Wash., D.C., managing ed., Nation-Wide News Service; publisher of first consumer weekly radio magazine "Radio Review"; ed.-publisher of "Key Magazine"; home office exec. Paramount Pictures. Author, alone & in collab., books including I Find Treason. Served in U.S. Marine Corps as officer-in-chg. combat photography in the Pacific WWII, prod. & dir. wire recording of first sound ever made of actual war conditions under battle fire in the Marshalls. Joined Selznick Releasing Org. as gen. mgr. of European operations 1946; formed own co., Paul White Productions, Inc. 1948; v.p. & dir., Procter Prod., Inc.; pres., PSI-TV 1953. v.p. Subscription TV in chg. pgmg; pioneer cablevision and Pay-TV; pres. MCI; created 11 audio-visual inventions; exec. consultant posts; RCA; Impresario S. Hurok and his org; chm. exec. comm, The Education Guild; 1979–80, bd. chm. Palolite 4D Systems.

PICTURES INCLUDE: Battle of the Marshalls, Saipan, Tinian, To the Shores of Iwo Jima, Song of Siam, Pearl of the Orient, Land of Fair Dinkum, Unusual Sports, Flying Doctor.

TV: film series started & supervised incl.: Playhouse of Stars, China Smith, Play of The Week, Orient Express, prod. & dir. "The Keys to Peace" (Pope Paul's U.S. Visit); exec. prod. two 90-min TV Specials, "The Bolshoi Ballet" Bicentennial project; OP Sail 1976, The Tall Ships; Int'l. Productions (features & TV).

WHITE, ROY B.: Executive, exhibitor. b. Cincinnati, Ohio. e. University of Cincinnati. Flight engineer, U.S. Air Force during World War II; worked in sales department of 20th Century-Fox, 1949–52; began in exhibition, 1952; pres., Mid-States Theatres; past president, National Association of Theatre Owners, Chairman of the Board, National Association of Theatre Owners: Board of Trustees—American Film Institute, NATO of Ohio, v.p., Motion Picture Pioneers, Will Rogers Hospital, Nat'l. Endowment for Arts.

WHITE, WILFRID HYDE: Actor. b. Gloucestershire, Eng.; e. Marlborough Coll. Stage debut, 1925; screen debut in Rembrandt, 1936. TV appearances.

PICTURES INCLUDE: The Winslow Boy, Britannia Mews, Adam and Evelyn, The Third Man, Golden Salamander, Trio, Mr. Drake's Duck, The Browning Version, Outcast of the Islands, Mr. Denning Drives North, Gilbert & Sullivan, Million Pound Note (Man With a Million), Betrayed, Duel in the Jungle, Rainbow Jacket, To Dorothy, A Son, Quentin Durward, March Hare, The Silken Affair, Libel, North West Frontier, Two-Way Stretch, Let's Make Love, On the Double, Ada, The Castaways, Crooks Anonymous, My Fair Lady (Hollywood), 1964; You Must Be Joking, Ten Little Indians, The Sandwich Man, Marrakesh.

WHITELAW, BILLIE: Actress. b. Coventry, England. Acted on radio and television since childhood. Winner of the TV Actress of the Year; Variety Club Actress of the Year 1960.

British Academy Award 1969. U.S. National Society of Film Critics Award for Best Supporting Actress, Variety Club, 1977. Evening New, Best Film Actress, 1977.
STAGE: 3 years with National Theatre of Great Britain. Revue: England, My England. Stage plays include: Progress to the Park, A Touch of the Poet, Othello, Tretawney of the Wells, After Haggerty, Not I, Alphabetical Order, Footfalls, Molly, The Greeks, Happy Days.
PICTURES INCLUDE: No Love for Johnnie, Mr. Topaze, Hell Is a City, Payroll, Charlie Bubbles, The Adding Machine, Twisted Nerve, Start the Revolution Without Me, Leo the Last, Eagle in a Cage, Gumshoe, Frenzy, Nightwatch, The Omen, Leopard in the Snow, The Water Babies.
TV: Over 100 leading roles incl. No Trains to Lime Street, Lady of the Camelias, Resurrection, The Pity of it All, You and Me, A World of Time, Dr. Jekyll and Mr. Hyde, Poet Game, Sextet (8 plays for BBC), Wessex Tales, The Fifty Pound Note, Supernatural (2 plays), Three plays by Samuel Beckett, Eustace and Hilda, The Oresteia of Aeschylus, The Haunted Man, The Bad Soldier Schultz.

WHITFIELD, RICHARD ALLEN: Producer-Executive. b. Goldsboro, N.C., 1946. e. Univ. of North Carolina. Adv. writer-producer for American Brands, Pepsicola, 1970–73; TV-film producer, industrial-educational, 1973–77; v.p., adv., Independents International Films, Inc., distributor, 1978; v.p.-producer, Rick Friedberg & Associates, 1979; pres. Golden Image Motion Picture Corp., feature production, 1980.
PICTURES INCLUDE: K-GOD (exec. prod.), Used Cars (video segments prod.), Bones of Peking (prod.).

WHITMAN, STUART: Actor. b. San Francisco, Calif., Feb. 1; e. Los Angeles City Coll. Army Engineers June 1945 to Sept. 1948, Engineers at Fort Lewis, Wash.; while in army, became light heavyweight boxer. Studied drama, L.A. City Coll., 2 yrs.; attended Ben Bard's training sch. for actors, became member of Chekhov group, and Arthur Kennedy famed local Stage Society. Joined tent show, Here Comes Mr. Jordan, then TV and pictures.
PICTURES INCLUDE: When Worlds Collide, The Day The Earth Stood Still, China Doll, Darby's Rangers, Johnny Trouble, Hell Bound, Seven Men from Now, No Sleep Till Dawn, Ten North Frederick, These Thousand Hills, The Deck Ran Red, The Sound and The Fury, Hound Dog Man, The Story of Ruth, Murder, Inc., Francis of Assisi, The Fiercest Heart, The Mark, The Comancheros, Reprieve, The Day and the Hour, Signpost to Murder, Rio Conchos, Sands of Kalahari, An American Dream, Last Escape, Last Generation, Night of the Lepus, Crazy Mama, Call Him Shatter, Las Vegas Lady, Strange Shadows in an Empty Room.
TV: The Crowd Pleaser (Alcoa-Goodyear), Highway Patrol, Dr. Christian, Hangman's Noose (Zane Grey).

WHITMORE, JAMES: Actor. r.n. James Allen Whitmore, Jr. b. White Plains, N.Y., Oct. 1; e. Yale U. In Yale Drama School players; co-founder Yale radio station, 1942; U.S. Marine Corps, World War II; in USO, in American Wing Theatre school, in stock. Broadway debut in Command Decision, 1947; m.p. debut in Undercover Man. Star of Tomorrow.
PICTURES INCLUDE: Battleground, Asphalt Jungle, Next Voice You Hear, Mrs. O'Malley and Mr. Malone, Outriders, Please Believe Me, Across the Wide Missouri, It's a Big Country, Because You're Mine, Above and Beyond, Girl Who Had Everything, All the Brothers Were Valiant, Kiss Me Kate, The Command, Them, Battle Cry, McConell Story, Last Frontier, Oklahoma, Face of Fire, Eddie Duchin Story, Who Was That Lady?, Black Like Me, Chuka, Water Hole No. 3, Nobody's Perfect, Planet of the Apes, Madigan, The Split, Guns of the Magnificent Seven, Chato's Land, Where the Red Fern Grows, Give 'em Hell, Harry, The Serpent's Egg, The First Deadly Sin.
TV: The Law and Mr. Jones.

WHITTAKER, WILLIAM: Producer. Entered m.p. ind. as asst. dir. prod. man. with Gaumont British 1932. War service 1939–45. Joined Associated British Picture Corporation as prod. man 1946 Producer 1954. From 1960 Con. of AB Productions. 1970: independent producer.
PICTURES INCLUDE: (from 1946) prod.: Dam Buster, No Time for Tears, Ice Cold in Alex, The Rebel, The Cracksman, The Bargee, Mister Ten Per Cent, Judith, Universal Soldier.

WHITTINGHAM, JACK: Writer producer, b. London, England, 1910; e. Charterhouse and Lincoln College, Oxford. Formerly journalist Daily Express, Morning Post.
PLAYS: Return Ticket, Ill Wind.
PICTURES INCLUDE: Q. Planes (Clouds Over Europe); orig. s.p.: Escape to Danger, Welcome Mr. Washington, Twilight Hour, Waltz Time, Green Fingers, Hunted; orig. story and s.p.: Kiss the Bride Goodbye, The Lisbon Story, Counterblast, The Dancing Years, Cage of Gold, Pool of London, I Believe in You, The Divided Heart, The Prince and the Pauper, While the Storm Lasts; prod., story, s.p.: The Birthday Present orig. story collab. s.p.: Mandy, West of Zanzibar; collab. story, orig. s.p.: Thunderball.

WICKES, MARY: Actress. e. Washington U., Doctor of Arts (hon.), 1969.
THEATRE: (B'way) The Man Who Came to Dinner, Town House, Stage Door, Danton's Death. Dramatic & musical comedy stock includes: St. Louis Municipal Opera, Starlight Theatre, Houston Music Theatre.
PICTURES INCLUDE: The Man Who Came to Dinner, Now, Voyager, White Christmas, The Music Man, The Trouble With Angels, Where Angels Go Trouble Follows, Napoleon and Samantha, Snowball Express.
TV: The Halls of Ivy, The Danny Thomas Show, Dennis the Menace, Bonino, Hitchcock Presents, Studio One, Playhouse 90.

WIDEM, ALLEN M.: Author, editor. b. Hartford, Conn., Feb. 14, 1925; e. Univ. of Connecticut, B.S. In U.S. Army, 1943–46, as Army corr., European Theatre of Oper. Author, m.p. scripts, radio and TV properties and stage plays. Amusements-TV editor and columnist, Hartford, Connecticut, Times.

WIDMARK, RICHARD: Actor. b. Sunrise, Minn., Dec. 26; e. Lake Forest U. Instructor, 1938. On radio, then stage, films.
PICTURES INCLUDE: Kiss of Death, Cry of the City, Road House, Street With No Name, Yellow Sky, Down to the Sea in Ships, Slattery's Hurricane, Night and the City, Panic in the Streets, No Way Out, Halls of Montezuma, The Frogmen, Red Skies of Montana, Don't Bother to Knock, O. Henry's Full House, My Pal Gus, Destination Gobi, Pickup on South Street, Take the High Ground, Hell & High Water, Broken Lance, Prize of Gold, The Cobweb, Backlash, Last Wagon, Saint Joan, Warlock, Kingdom of Man, The Long Ships, Run for the Sun, The Alamo, Judgment at Nuremberg, How the West Was Won, The Way West, Madigan, Death of a Gunfighter, When The Legends Die, Murder on the Orient Express, Twilight's Last Gleaming, The Domino Principle, Rollercoaster, Coma, The Swarm.

WIENER, JACK: Executive. b. Paris, France, June 8, 1926. Pub. rep., MGM, New Orleans, Jacksonville, Fla., 1952–56; joined Columbia Pictures Int. Corp., 1956, continental publicity mgr., Columbia, Paris; v.p., Columbia Pics. Int'l., 1966; continental prod. exec., 1968; v p. Columbia Pictures Corp., 1970. In 1972 became indep. prod., making Vampira and The Eagle Has Landed.

WIESEN, BERNARD: Producer, Director, Executive. b. New York City, Oct. 6. e. College of the City of New York (B.B.A.) Pasadena Playhouse College of Theatre (Master of Theatre Arts), Dramatic Workshop of New School. Is pres. of M.A.N. International Productions.
PICTURES INCLUDE: Producer-Director: Fear No More. Asst. Director on films, including The King and I, The Left Hand of God, The Rains of Ranchipur, To Catch a Thief, The Trouble with Harry.
TV: Director: How to Marry A Millionaire, Valentine's Day. Assoc. Producer: Valentine's Day, Three on an Island, Cap'n Ahab, Sally and Sam. Assoc. Prod: Daniel Boone. Producer/Director: Julia, Co-Producer-Director: The Jimmy Stewart Show. Prod. Exec.: Executive Suite (pilot). Exec. Paramount TV, director of current programming.
STAGE: First Monday in October (Bdwy.)

WIESENTHAL, SAM: Executive. b. New York, N.Y., Jan. 26, 1909; e. N.Y.U., 1929. In story dept., Universal Pictures 1929–31; asst. to pres., 1931–39; v.p., gen mgr. & dir. of General Service Studios & producing companies since 1942, exec. v.p., assoc. producer many films. pres., Olympic Prod., 1964.
PICTURES INCLUDE: Bridge at San Luis Rey, Dark Waters, The Macomber Affair, Cry Danger, with W. R. Frank; Second Chance (RKO) The Americano, Bengazi, Tension at Table Rock, The Day They Gave Babies Away, The Jack Dempsey Story.

WILCOX, CLAIRE: Actress. b. Toronto, Canada, 1955. Photographer's model, 3 yrs. of age; appeared on numerous national magazine covers; TV commercials; signed contract, Curtis Enterprises.
PICTURES INCLUDE: 40 Pounds of Trouble, Wives and Lovers.
TV: Harris Versus the World.

WILCOXON, HENRY: Actor. b. British West Indies, Sept. 8, 1905. r.n. Harry Wilcoxon. On London stage from 1925; made his first sound film appearance in 1931 in The Perfect Lady (BL). Named assoc. prod. by Cecil B. DeMille on Greatest Show on Earth. Produced The Buccaneer (1959), in which also acted.
PICTURES INCLUDE: The Flying Squad, 1931, The Lovelorn Lady, 1931, (BIP); A Taxi to Paradise, and many others. Since 1942: The Man Who Wouldn't Die, (20th-Fox), Mrs. Miniver, Johnny Doughboy, Unconquered, Crusades, Samson and Delilah, Miniver Story, Scaramouche, The War Lord, The Private Navy of Sgt. O'Farrell, Man in the Wilderness.

WILD, JACK: Actor. b. Lancashire, England, 1952. m.p. debut in Oliver, 1968. To Hollywood 1969 to star in H. R. Pufnstuf TV series for Krofft Television Productions. Under contract to Hemdale Group.

PICTURES INCLUDE: Pufnstuf, Oliver!, Swalk, Flight of the Doves, Pied Piper, The 14.

TV: Our Mutual Friend, The Government Inspector.

WILDE, ARTHUR L.: Executive. b. San Francisco, Calif., May 27, 1918. S.F. Daily News; Matson Lines; pub. dept., Warner Bros., 1936; dir. exploitation, CBS; pub. dir., Hal Wallis Prod.; pub. dept., Paramount; pub., Hecht-Hill-Lancaster; v.p., Arthur Jacobs, public rel.; Blowitz-Maskell Publicity Agency; pub. dir., C. V. Whitney Pictures; gen. v.p., 1958; owner, pub.-ad. agency, The Arthur L. Wilde Co., 1961–65; freelance publicist, 1965–66; pub. rel. consultant, Marineland of Florida, 1965; unit publicity dir., United Artists, National General, Paramount, 1966–69; free lance publicity, 1971; unit publicist, MGM, Paramount, United Artists, 1972–74; staff position; Features Publicity at Paramount Pictures, 1973. Free lance unit publicist again in 1976 at Universal, Paramount and Lorimar Productions. 1978–79, Columbia Pictures & Universal Studios; 1980, Marble Arch. Prods. & Northstar Intl. Pictures; 1981, studio pub. mgr., 20th Century-Fox.

WILDE, CORNEL: Actor, producer, director, author. b. New York, Oct. 13, 1918; e. Columbia & City Coll., N.Y. Art Sch., Budapest; Columbia Med. School; studied drama with Strasberg, Bulgakov, Chekov. First screen role, 1941, High Sierra.

STAGE: Moon Over Mulberry Street, Love is Not So Simple, Daughters of Altreus, Having a Wonderfuyl Time, Romeo & Juliet.

PICTURES INCLUDE: Wintertime, A Song to Remember, The Perfect Snob, A Thousand and One Nights, Leave Her to Heaven, Bandit of Sherwood Forest, Centennial Summer, Forever Amber, The Homestretch, It Had to be You, Walls of Jericho, Roadhouse, Four Days Leave, Two Flags West, Greatest Show on Earth, At Sword's Point, Saadia, Passion, Woman's World, Big Combo, Scarlet Coat, Hot Blood, Star of India, Beyond Mombasa, Omar Khayyam, Edge of Eternity, Constantine the Great; act., prod., dir.: Storm Fear, Maracaibo; act., prod., dir., cowriter: The Devil's Hairpin, The Sword of Lancelot; act., prod., dir.: The Naked Prey; act., prod., dir., co-author s.p.: Beach Red; prod., dir., co-author s.p.: No Blade of Grass; prod., dir., act., screenplay: Sharks' Treasure; actor: The Fifth Musketeer.

WILDER, BILLY: Producer, director, writer. b. Austria, June 22, 1906. Newspaperman active Vienna, Berlin; then author screen story People on Sunday, s.p.: Emil and the Detectives (in Museum of Modern Art), UFA. Wrote, dir., Mauvaise Graine, Paris. To Hollywood 1934, (collab. s.p) Bluebeard's Eighth Wife, Ninotchka, What a Life, Midnight, Arise My Love, Ball of Fire, Hold Back the Dawn. Head Film Section, Psych. Warfare Div., U.S. Army, 1945, Am. Zone, Germany.

PICTURES INCLUDE: collab. s.p., dir.) The Major and the Minor, Five Graves to Cairo, Double Indemnity, The Emperor Waltz, A Foreign Affair. Academy Award for dir., best s.p. (collab.) for Lost Weekend, 1945. Academy Award for best story & s.p. (collab.) for Sunset Boulevard, 1950; (collab. s.p., dir., prod) The Big Carnival, Stalag 17, Sabrina, Love in the Afternoon, Some Like it Hot, The Apartment (Academy Award for best direction, best story, and s.p., best picture, 1960); One, Two, Three, Irma La Douce; (collab. s.p., dir. co-prod.) Seven Year Itch; (collab., dir.) Spirit of St. Louis; (collab. s.p. & dir.) Witness for the Prosecution; (prod., dir., collab.) Kiss Me, Stupid, The Fortune Cookie, The Private Life of Sherlock Holmes, Avanti, The Front Page, Fedora.

WILDER, GENE: Actor. b. Milwaukee, Wis., 1935. e. University of Iowa. Joined Bristol Old Vic company in England, became champion fencer; in New York, worked as chauffer, fencing instructor, etc. Before N.Y. off-Broadway debut in Roots.

BROADWAY: The Complaisant Lover, Mother Courage, Luv.

PICTURES INCLUDE: Bonnie and Clyde, The Producers, Start the Revolution Without Me, Quacker Fortune Has an Uncle in the Bronx, Willie Wonka and the Chocolate Factory, Everything You Always Wanted to Know About Sex*, Blazing Saddles, Rhinoceros, Young Frankenstein, The Little Prince, Adventures of Sherlock Holmes Smarter Brother (s.p., dir., star), Silver Streak, The World's Greatest Lover (s.p., dir., star), The Frisco Kid, Stir Crazy, Sunday Lovers, (dir., s.p., act.).

WILDER, W. LEE: Producer, director. b. Austria; e. U. Vienna. Awards: First prize 1950 Venice Film Festival musical documentary category.

PICTURES INCLUDE: The Vicious Circle, Shadows of Fire, The Pretender, Once a Thief, Three Steps North, Phantom from Space, Killers from Space, Snow Creature, Big Bluff, Manfish, Man Without a Body, Fright, Spy in the Sky, Bluebeard's Ten Honeymoons.

WILK, TED: Theatrical agent. b. Minneapolis, Minn., Jan. 5, 1908. e. U. of Mich., 1926–30. Publix Theatres, Duluth, Minn., 1930; Warner Bros., Minneapolis, 1932–33; Film Daily, Hollywood, 1934–40; U.S. Army, 1941–46; Lou Irwin agency 1946–61; Ted Wilk Agency, since 1961.

WILKINSON, ROBERT: Executive. Came to Universal in 1941 as student booker in Memphis branch. Spent 4 years in Marines; then returned to Memphis as salesman, branch mgr.; later mgr. of Dallas branch, then district and regional mgr. Named Universal asst. gen. sls mgr. in 1964. Appt. exec. v.p. in chg of domestic dist. in 1978.

WILLIAMS, BERT: Executive, actor. b. Newark, N.J., April 12, 1922. e. USC Navy, 1942–45. Summer Stock, 1940–41; world's prof. diving champion, 1945–48; star diver, Larry Crosby, Buster Crabb, Johnny Weismuller, Dutch Smith Shows, 1945–48; writer, asst. prod., Martin Mooney Prods., PRC, Goldwyn Studios; pres., Bert Prods., Bert Williams Motion Picture Producers and Distributors, Inc. Member, M.P. Academy of Fine Arts & TV Academy of Arts & Science.

THEATRE: roadshow plays include: Cat on a Hot Tin Roof, Hamlet, Run From The Hunter, Sugar and Spice.

PICTURES INCLUDE: Actor, Angel Baby; prod., dir., actor, The Nest of the Cuckoo Birds; actor, Around the World Under the Sea; s.p. auth. Deathwatch 28; dir., Twenty Eight Watched, prod. dir. Adventure To Treasure Reef. s.p. Knife Fighters. orig. story & s.p.; Black Freedom; A Crime of Sex; The Masters (actor). Crazy Joe, Serpico, Lady Ice, The Klansman, Report to the Commissioner, Tracks, All the President's Men, From Noon Till Three. (all actor), While Buffalo (actor), Helter Skelter (actor), Shark Bait (writer), The Big Bus (actor). Actor: Wanda Nevada, Horn, Sunnyside, Cuba, The Last Resort, The All Night Treasure Hunt. Recent acting: Tom Horn, Kill Castro, Midnight Madness.

TV: Flipper, Sea Hunt, prod., Speargun, Gentle Ben, The Law (pilot) and Police Story (actor). Recent actor: Get Christy Love, General Hospital, Columbo, Brenner for the People (actor), Mayday 40,000 Feet (actor) Jigsaw John (Blue Knight episode). Actor: Police Woman, Chips, Mobil One, Street Killing, East of Eden, etc. Produced, directed & wrote pilot, Fifth St. Gym. Appeared on Mike Douglas Show & Johnny Carson Show.

WILLIAMS, BILL: Actor. r.n. William Katt. b. Brooklyn, N.Y. e. Pratt Inst., Brooklyn. In U.S. Army, World War II. Began as professional swimmer; later with Municipal Opera House, St. Louis; then on vaudeville tour, U.S. & England. Screen debut 1944 in "Murder in the Blue Room."

PICTURES INCLUDE: Thirty Seconds Over Tokyo, Those Endearing Young Charms, Blue Blood, Great Missouri Raid, Operation Haylift, Cariboo Trail, Havana Rose, Rose of Cimarron, Son of Paleface, Bronco Buster, Pace That Thrills, Torpedo Alley, Racing Blood, Outlaw's Daughter, Apache Ambush, Hell's Horizon, Wiretapper, Broken Star, Dogs Best Friend, Buckskin, Tickle Me, Scandalous John.

WILLIAMS, BILLY DEE: Actor. b. New York City, April 6, 1937. Studied acting with Paul Mann and Sidney Poitier at actor's workshop in Harlem. Was child actor; Broadway adult debut in The Cool World in 1961.

STAGE: A Taste of Honey, Hallelujah, Baby.

PICTURES INCLUDE: The Last Angry Man (debut), The Out-of-Towners, The Final Comedown, Lady Sings the Blues, Hit!, Mahogany, The Bingo Long Travelling All-Stars, The Empire Strikes Back, Nighthawks.

TV: Brian's Song, The Glass House and appearances on series: The F.B.I., The Interns, Mission Impossible, Mod Squad.

WILLIAMS, BOB: Writer. b. Sioux Falls, S.D., Oct. 19, 1913.

PICTURES INCLUDE: collab. s.p., Treat 'Em Rough, Overland Mail Robbery, Pride of the Plains, Youth on Trial, Sergeant Mike; s.p., Lone Texas Ranger, Trail to Vengeance, many Westerns; colab. s.p., Stage to Tucson.

WILLIAMS, CARA: Comedienne. r.n. Bernice Kamiat. b. Brooklyn, N.Y. e. Holywood Professional Sch.; ent. ind., 20th Century Fox, child actress.

PICTURES INCLUDE: Boomerang, Something For the Boys, Meet Me In Las Vegas, Never Steal Anything Small, The Defiant Ones, The Man From the Diners' Club, Doctors' Wives.

TV: Pete and Gladys, Alfred Hitchcock Presents, Desilu Playhouse, The Jackie Gleason Show, Henry Fonda Special, The Cara Williams Show.

WILLIAMS, CARL W.: Executive. b. Decatur, Ill., March 9, 1927. e. Illinois State Normal U., B.S., 1949; UCLA, M.A., 1950. dir. adv. photo., Clark Equipment Co., 1951–54; film dir., WKAR-TV, E. Lansing, Mich., 1954–56; Prod., dir., Capital Films, E. Lansing, Mich., 1957; dir., A-V Laboratory, UCLA, 1957–63; co-dev. Dimension 150 Widescreen process, 1957; formed D-150 Inc., 1963; Filbert Co., 1970, v.p., 1977.

MAMPAS, SMPTE, AFI, Variety Club, Tent 25.

WILLIAMS, CHARLES: Actor, writer. b. Albany, N.Y., Sept. 27, 1898. Appeared on stage in "Rio Rita," dialogue for vaudeville and stage. To Fox Studios, Hollywood, 1931. In 1933 appeared in Gambling Ship. In 1935 collab. music and lyrics, Gigolette; collab. orig. s.p. Arm of the Law, Midnight Limited.
PICTURES INCLUDE: The Girl from Monterey, Where Are Your Children? (collab orig. & s.p.) Hollywood and Vine; (cast) End of the Road Identity Unknown, Doll Face, Heldorado, Due Goes West, Grand Canyon.

WILLIAMS, CINDY: Actress. b. Van Nuys, Calif. Aug. 22. e. L.A. City Coll. Appeared in high school and college plays; first prof. role in Roger Corman s film Gas. Made TV debut in Room 222 and had continuing role.
PICTURES INCLUDE: Beware the Blob, Drive, He Said, The Christian Licorice Store, Travels with My Aunt, American Graffiti, The Conversation, Mr. Ricco, The First Nudie Musical, More American Graffiti.
TV: Episodes of The Funny Side, The Neighbors, Barefoot in the Park, My Word and Welcome to It, Love, American Style, Nanny and the Professor, The Bobby Sherman Show—Getting Together; The Migrants (movie), Laverne and Shirley.

WILLIAMS, DERICK: Cinematographer. b. Nottingham, Eng., 1906. e. Brighton Coll.; Nottingham U. Cinematog., 1929; first cameraman & experimental work on Evans color process, 1930; head camera dept., Gainsborough Pictures, 1932; org. camera dept., Nat'l Studios, Sydney, Australia, 1935.
PICTURES INCLUDE: In Which We Serve, Way Ahead, Way to the Stars, Beware of Pity, For Them That Trespass, Lady Possessed; mgr. dir., Derick Williams Prod., Ltd.; Voice of London, Ltd.

WILLIAMS, DIAHN: Actress. b Gainesville, Fla., June 30. e. Univ. of Miami (BA, psychology, speech), also attended Univ. of Florida, UCLA. m. Thomas J. McGrath, prod. & atty. Started as fashion model in New York, France, Germany. Top woman exec. as assoc. dir. of pub. rel. at Chesebrough-Ponds, 1971–72.
PICTURES INCLUDE: Chair de Poule, Another Nice Mess, Deadly Hero.
TV: Harry's Girls (series, 1963–64), Somerset (series, 1973–74) and guest-starred in many TV shows, including I Spy, Get Smart, Tarzan, Here Comes the Brides, Andy Griffith Show, G.E. Theatre, etc.

WILLIAMS, ELMO: Film editor. b. Oklahoma City, Okla., Apr. 30, 1913. Film editor 1933–39, with British & Dominion Studio, England. Since then with RKO-Radio as film editor for numerous major productions; Academy Award, best film ed. (collab.) High Noon, 1952; dir. film ed., Tall Texan; prod. dir., ed., The Cowboy; ed., 20,000 Leagues Under the Sea; dir. Apache Kid; second unit dir., film ed. The Vikings; dir. 2nd Unit DFZ prod., The Big Gamble; assoc. prod., The Longest Day; mgr., dir., 20th Century Fox Prod. Ltd. Producer: Tora! Tora! Tora! v.p., worldwide production, 20th Century-Fox Film, 1971. President Ibex Films. Exec. v.p., Gaylord Prods., 1979.
TV: co-prod. dir., Tales of the Vikings.

WILLIAMS, ESTHER: Actress. b. Los Angeles, Calif., Aug. 8, 1923. e. U. of So. Calif. Swimmer San Francisco World's Fair Aquacade; professional model. On screen 1942 in Andy Hardy Steps Out.
PICTURES INCLUDE: A Guy Named Joe, Bathing Beauty, Thrill of a Romance, This Time for Keeps, Ziegfeld Follies, Hoodlum Saint, Easy to Wed, Fiesta, On an Island With You, Take Me Out to the Ball Game, Neptune's Daughter, Pagan Love Song, Duchess of Idaho, Texas Carnival, Skirts Ahoy!, Million Dollars Mermaid, Dangerous When Wet, Easy to Love, Jupiter's Darling, Unguarded Moment, The Big Show. Voted one of Top Ten Money-Making Stars in M.P. Herald-Fame poll, 1950.

WILLIAMS, GEORGE EMLYN: (C.B.E.) Writer, actor, producer. b. Mostyn, Flintshire, Wales, Nov. 26, 1905. e. Christ Church, Oxford, M.A., LL.D. (Hon.). On London stage from 1927 (And So to Bed, The Case of the Frightened Lady, Wild Decembers, Night Must Fall, The Corn is Green, etc.); also author many plays (Glamour, The Late Christopher Bean, Night Must Fall, He Was Born Gay, The Corn is Green, The Light of Heart, etc.). Began Brit. screen career 1932 as actor in The Case of the Frightened Lady; author variously adapt. or dial. Friday the Thirteenth. Evergreen, This England, In 1945 on Brit. stage; author & cast the Wind of Heaven, Trespass, Accolade, Charles Dickens Mixed Bill & Bleak House, Someone Waiting.
PICTURES INCLUDE Broken Blossoms, The Citadel, Jamaica Inn, The Stars Look Down, The Girl in the News, Major Barbara, You Will Remember, Hatter's Castle, Last Days of Dolwyn, Three Husbands, Scarf, Another Man's Poison, Ivanhoe, Deep Blue Sea, I Accuse, Beyond This Place, The Wreck of the Mary Deare, The L-Shaped Room, The Eye of the Devil.
TV: The Winslow Boy, The Defenders, Yob and Nabob.
BOOKS: George, Beyond Belief, Emlyn, Doctor Crippen's Diary, Headlong.

WILLIAMS, GRANT: Actor. b. New York, N.Y., Aug. 18, 1930. e. Queens Coll., U. of Ill., N.Y.C.; U.S. Air Force, 1943–52. Pub. work. MCA acted with Barter Theatre, Abingdon, Va., summer stock, road cos.; many TV dramatic shows; m.p. debut in Red Sundown.
PICTURES INCLUDE: Showdown at Abilene, Written on the Wind, Outside the Law, Four Girls in Town, Incredible Shrinking Man, Monolith Monsters, The Leech.

WILLIAMS, JOHN: Composer. b. New York City, Feb. 8, 1932. e. UCLA, Juilliard School. Worked as session musician in '50s; began career as film composer in late '50s. Considerable experience as musical director and conductor as well as composer.
PICTURES INCLUDE: I Passed for White, Because They're Young, The Secret Ways, Bachelor Flat, Diamond Head, Gidget Goes to Rome, The Killers, None But the Brave, John Goldfarb Please Come Home, The Rare Breed, How To Steal A Million, The Plainsman, Not with My Wife You Don't, Penelope, A Guide for the Married Man, Fitzwilly, Valley of the Dolls, Daddy's Gone A-Hunting, Goodbye Mr. Chips (mus. supvr. & dir.), The Reivers, Jane Eyre, Fiddler on the Roof (musc. dir.). The Cowboys, Images, Pete 'n' Tillie, The Poseidon Adventure, Tom Sawyer (musc. supvr.), The Long Goodbye, The Man Who Loved Cat Dancing, The Paper Chase, Cinderella Liberty, Conrack, The Sugarland Express, Earthquake, The Towering Inferno, The Eiger Sanction, Jaws, Family Plot, The Missouri Breaks, Midway, Black Sunday, Star Wars, Encounters of the Third Kind, The Fury, Jaws II, Meteor, Quintet, Dracula, Raiders of the Lost Ark.
TV: Once Upon a Savage Night, Sergeant Ryker, Heidi.

WILLIAMS, MURIEL: Actress. e. Leland Powers Dramatic School, Boston. Played in stock. Cape Playhouse, Dennis, Mass.; with nat'l co. Autumn Garden, Heaven Can Wait; with late husband formed model, prod. agency, Hart Model Agency; commentator for fashion shows.
TV: frequent appearances; series: Brighter Day, CBS-TV.

WILLIAMS, OSCAR: Writer-Producer-Director. e. San Francisco State Univ., getting degree in film, TV. Was director's intern on The Great White Hope (directed by Martin Ritt) through the American Film Institute.
PICTURES INCLUDE: The Final Comedown (s.p., prod. dir.); Black Belt Jones (s.p., assoc. prod.) Five on the Black Hand Side (dir.), Truck Turner (s.p.), Hot Potato (s.p., dir.).

WILLIAMS, PAUL: Composer. b. Omaha. Began career at studios as set painter and stunt parachutist. Bit and character parts in commercials followed. Seen briefly in The Chase and The Loved One. Became song writer, collaborating briefly with Biff Rose and later with Roger Nichols, with whom wrote several best-sellers, including We've Only Just Begun, Rainy Days and Mondays, Just an Old-Fashioned Love Song, etc.
PICTURES INCLUDE: Cinderella Liberty (score), Phantom of the Paradise (actor, score), Bugsy Malone (score), A Star Is Born (songs).

WILLIAMS, PAUL: Director. First gained attention as director of film short, Girl, which won Golden Eagle award. Made in collaboration with producer Edward R. Pressman, with whom he formed Pressman-Williams Enterprises.
PICTURES INCLUDE: Out of It, The Revolutionary, Dealing: or the Berkeley to Boston Forty Brick, Lost Bag Blues, Nunzio.

WILLIAMS, RICHARD: Producer-painter-film animator. b. March, 1933, Toronto, Canada. Entered industry in 1955. Founded Richard Williams Animation Ltd. in 1962, having entered films by producing The Little Island (1st Prize, Venice Film Festival) in 1955. His company produces TV commercials, entertainment shorts and animated films. Also designed animated feature titles for such films as What's New Pussycat?, A Funny Thing Happened On The Way To The Forum, Casino Royale, etc. (20 feature titles in 6 years). 1969: Animated sequences: Charge of the Light Brigade (Woodfall). 1971: A Christmas Carol, animated TV special for ABC-TV. His company makes TV commercials for England, America, France and Germany. His films have won awards at Festivals at Venice, Edinburgh, Mannheim, Montreal, Trieste, Melbourne, West Germany, New York, Locarno, Vancouver, Philadelphia, Zagreb, Hollywood, Cork, Los Angeles. 1973: Won Hollywood Oscar, Best Cartoon.

WILLIAMS, ROBERT: Actor. b. Glencoe, Ill.; Dartmouth U., B.A. Member: Psi Upsilon, Dartmouth Club, University Club (Chicago).
THEATRE: Burlesque, Strictly Dishonorable, Let Freedom Ring, Sailor Beware, Room Service, Eve of St. Mark.
PICTURES INCLUDE: The Girl in the Case, Secret Command, That Wonderful Urge, Apartment for Peggy, Dark Past, It Happens Every Spring, Oh, You Beautiful Doll, The Lawless, Mister 880, Groom Wore Spurs.

WILLIAMS, ROBERT X., JR.: Exhibitor, attorney. b. Taylor, Miss., Nov. 16, 1900. e. U. of Miss. Law Sch.; m. owner & mgr. Lyric Theatre, Oxford, Miss. since 1917; active in oper. to prevent adverse legislation (theatres); del., Nat'l Democratic Conventions, Phila. & Chicago; chmn. campaign funds for Democratic Party, 2nd Congressional Dist., Miss.; chmn. Roosevelt Electors of Oxford; treas. Pat Harrison campaign, Lafayette Co.; pres. De Soto Memorial Highway Assn. of Miss.; secy.-treas. Jackson to Jackson Highway Ass'n.

MEMBER: MPTO of Ark., Miss. & Tenn. (pres.); MPTO of Miss. (pres.); mayor, City of Oxford; Rotary (dir. Oxford); Shriner; Soldiers of Soil Div. of Lafayette Co. (sgt.); U. of Miss. Alumni Assn., 2nd Congressional Dist. (chmn.).

WILLIAMS, ROBIN: Actor—Comedian. b. Edinburgh, Scotland. Family moved to U.S. when he was one year old. e. Claremont Men's College (Calif.), College of Marin (Calif.), studying acting at latter. Continued studies at Julliard in New York. As San Francisco club performer appeared at Holy City Zoo, Intersection, the Great American Music Hall and The Boardinghouse. In Los Angeles performed as stand-up comedian at Comedy Store, Improvisation, and The Ice House. First TV appearance on Laugh In, followed by The Great American Laugh Off.

TELEVISION: Mork and Mindy, America Tonight, Ninety Minutes Live, The Alan Hamel Show.

PICTURES: The Last Laugh, Popeye, The World According to Garth.

WILLIAMS, ROGER: Pianist, concert-film-TV personality. b. Omaha, Oct. 1. e. Drake U., Idaho State Coll. Hon. Ph.D. Midland and Wagner Colls. Served U.S. Navy WWII. Appeared as guest artist in number of films. Public debut on TV's Arthur Godfrey Talent Scouts and Chance of a Lifetime. Other TV appearances include Ed Sullivan, Hollywood Palace, Kraft Summer Series, Celanese Special. Tours in addition to U.S. and Australia. Concert Halls—Japan, Mexico, Union of South Africa. Recorded 50 Albums, Kapp (now MCA) Records, with sales over 15 million albums.

WILLIAMS, TENNESSEE: Playwright. r.n. Thomas Lanter Williams. b. Columbus, Miss., Mar. 26, 1914. e. U. of Mo., 1931–33; Washington U., St. Louis, 1936–37; U. of Iowa, A.B., 1938.

AUTHOR: (plays) Battle of Angels, Glass Menagerie, You Touched Me; Pulitzer Prize for Streetcar Named Desire, 1947 and Cat On a Hot Tin Roof, Camino Real 1954; Summer and Smoke, Rose Tattoo, Orpheus Descending; Sweet Bird of Youth, Period of Adjustment, Slapstick Tragedy, The 7 Descents of Myrtle, In a Bar of a Tokyo Hotel, Small Craft Warnings, Novel: The Roman Spring of Mrs. Stone.

PICTURES INCLUDE: collab. s.p., Glass Menagerie; s.p. Streetcar Named Desire, Rose Tattoo, Night of the Iguana.

WILLIAMS, TREAT: Actor. b. Rowayton, Conn. e. Franklin and Marshall College. Landed role on Bdwy. in musical, Over There! also played leading role in Grease on Bdwy. Film debut in The Ritz.

PICTURES: Deadly Hero, The Eagle Has Landed, Hair, 1941, Why Would I Lie?, Pursuit, Prince of the City.

WILLIAMSON, FRED: Actor. b. Gary, Ind., March 5. e. Northwestern Univ. Spent 10 yrs. playing pro football before turning to acting.

PICTURES INCLUDE: M*A*S*H*, Tell Me That You Love Me, Junie Moon, The Legend of Nigger Charley, Hammer, Black Caesar, The Soul of Nigger Charley, Hell Up in Harlem, That Man Bolt, Crazy Joe, Three Tough Guys, BLACK Eye, Three the Hard Way, Bucktown.

TV: Julia (series), Police Story, Monday Night Football.

WILLIAMSON, NICOL: Actor. b. England, 1939. Has played many classical roles with Royal Shakespeare Co., including Macbeth, Molvolio, and Coriolanus. Starred on Broadway in Inadmissible Evidence; musical debut in Rex.

PICTURES INCLUDE: Six Sided Triangle, Inadmissible Evidence, The Bofors Gun, Laughter in the Dark, The Reckoning, Hamlet, The Jerusalem File, The Seven-Per-Cent Solution. The Goodbye Girl (cameo), The Cheap Detective, The Human Factor, Excalibur.

WILLIAMSON, PATRICK: President-Columbia Pictures International Corp., New York Apptd. June 1978. b. England, Oct. 1929. Joined Columbia Pictures London office 1944-career spanned advertising & publicity responsibilities until 1967 when apptd. managing director Columbia Great Britain in 1971. Also man. dir. on formation of Columbia-Warner. Promoted to executive position in Columbia's home office, New York, April, 1973 and headed international operations since Feb. 1974.

WILLIS, GORDON: Cinematographer. Acted two summers in stock at Gloucester, MA, where also did stage settings and scenery. Photographer in Air Force; then cameraman, making documentaries. In TV did commercials and documentaries.

PICTURES INCLUDE: End of the Road, Loving, The Landlord, Little Murders, Bad Company, Klute, Up the Sandbox, The Paper Chase, The Godfather, The Parallax View, The Godfather, Part II, All the President's Men, September 30, 1955, Annie Hall, Comes a Horseman, Interiors. Director: Windows (1980; debut).

WILLIS, TED: Screenwriter and dramatist.

PICTURES INCLUDE: The Blue Lamp (orig. treatment), Good-Time, Trouble in Store, One Good Turn, Top of the Form, Up to His Neck, Woman in a Dressing Gown, The Young and the Guilty, Great to Be Young, No Trees in the Street, The Horsemasters, Flame in the Streets, Bitter Harvest.

TV: Dixon of Dock Green, Big City, Look in Any Window, Strictly for the Sparrows, Scent of Fear, Inside Story, Hot Summer Night, Sergeant Cork, The Four Seasons of Rosie Carr, The Sullivan Brothers, Knock On Any Door, Virgin of the Secret Service, Crime of Passion, Black Beauty, Hunter's Walk.

WILLOUGHBY, GEORGE W.: (GFPE). Producer. b. Bergen, Norway. e. Oslo University. First, writer and journalist; ent. m.p. industry as assistant dir. and film editor; dir. pictures in Sweden, Norway; formed distribution cos. in Oslo and Stockholm. Prod. in England since 1950. Consultant for Guaranty of Completion Company, City Share Trust Ltd.

PICTURES INCLUDE: prod. Bar Barson, I Killed; dir. Panique, assoc. prod. Valley of Eagles, Hell Below Zero, Action of the Tiger; prod. Passionate Summer, Beat Girl; assoc. prod., Nothing but the Best; prod., Masque of The Red Death, The City Under the Sea, Love in Amsterdam, Who'd Want to Kill a Nice Girl Like You?, Squeeze a Flower, The Man Outside, Outback, Wake in Fright, Summer Rain, Age of Innocence, Boardwalk.

WILLS, SIR, JOHN SPENCER: Executive. b. London, Aug. 10, 1904. e. Merchant Tailors' School. Became connected with automobile industry 1922 and appointed dir. (later man. dir. and deputy chmn.), chmn. of British Electric Traction Co., Ltd., 1939; formerly chairman, Rediffusion, Ltd. chmn., Rediffusion Television Ltd.; apptd. program contractor to Independent Television Authority; chmn., dir. many other public companies; governor, Royal Shakespeare Theatre, 1946–74. Chartered Inst. Transport (pres., 1950–51); mem. council, Soc. of the Royal Opera House, 1962–74; v. patron, Theatre Royal Windsor Trust, 1965.

WILSON, FLIP: Performer. b. Newark, N.J. r.n. Clerow Wilson. Left school at 16 to join Air Force; played clubs in Fla. & Bahamas until 1965 when guest appearance on NBC began; The Flip Wilson Show debuted 1970–71 season, NBC.

PICTURE: The Fish That Saved Pittsburgh.

WILSON, NORMAN: (OBE). Editor. b. Edinburgh, Scotland. ed. The Living Cinema, formerly ed. Cinema Quarterly; assoc. ed. World Film News; author scripts for British documentaries and book Presenting Scotland: A Film Survey; attached to British Ministry of Information, World War II; ed. Film Forum, Documentary. Chmn. Edinburgh Film Guild, Edinburgh International Film Festival, 1947–60; Films of Scotland Com.; Governor, Brit. Film Inst.

WILSON, RICHARD: Producer, director. b. McKeesport, Pa., Dec. 25, 1915. e. Denver U. actor, announcer, Denver radio stations; radio actor, N.Y.; actor, asst. stage mgr., stage mgr., prod. asst. with Mercury Theatre, 1937–38; mgr., prod., summer theatres, 1939–40; assoc. with all Orson Welles films & radio shows to 1951; U.S. Air Force, 1942–45; assoc., prod., then prod., U-I.

PICTURES INCLUDE: assoc. prod., Lady from Shanghai, Macbeth, Ma and Pa Kettle on Vacation, Ma and Pa Kettle Go to Waikiki, Redhead from Wyoming; prod., Ma and Pa Kettle at Home, Golden Blade, Man with a Gun, Kettles in the Ozarks; dir. The Big Boodle; co-writer, dir. Man with a Gun; co-writer, dir., Raw Wind in Eden; dir., Al Capone; collab. s.p., dir. Invitation to a Gunfighter, Three in the Attic (prod.-dir.).

WILSON, SCOTT: Actor. b. Atlanta, Ga., 1942. Was college athlete on basketball scholarship when injured and had to leave school. Moved to L.A. and enrolled in local acting class. Made film debut, In Cold Blood, 1968.

PICTURES INCLUDE: The Grissom Gang, The Gypsy Moths, Castle Keep, In the Heat of the Night, The New Centurions, Lolly-Madonna XXX, The Great Gatsby.

WILSON, WARREN: Writer, producer. b. Boston, Mass., May 11, 1909. On stage under name Warren Burke. Began writing for screen 1941; screenplay Blondie Goes to College, Col. in 1942; Tanks a Million, UA; Sing Your Worries Away, RKO; Helzapoppin, Strictly in the Groove.

PICTURES INCLUDE: (collab., s.p.) She Gets Her Man; (story, prod.) Her Lucky Night; (collab. s.p.) If You Knew Susie; (org. s.p.) Square Dance Katy; (s.p.) Big Timber.

WINCHELL, PAUL: Performer. b. New York, N.Y., 1924. e. School of Industrial Arts. At 13 won first prize Major Bowes Radio Amateur Hour; signed by Ted Weems; created Jerry Mahoney when 17; ventriloquist & star own Paul Winchell-Jerry Mahoney show, NBC-TV.

WINDSOR, MARIE: Actress. b. Maryvale, Utah, Dec. 11, 1924. r.n. Emily Marie Bertelson. Winner of beauty contests, including Miss Utah. Worked as telephone girl, dancing teacher. Trained for acting by Maria Ouspenskaya.
PICTURES INCLUDE: Song of the Thin Man, Force of Evil, Dakota Lil, Little Big Horn, The Narrow Margin, The Eddie Cantor Story, The Bounty Hunter, Swamp Woman, The Killing, The Story of Mankind, Critics Choice, Mail Order Bride, Chamber of Horrors, Support Your Local Gunfighter, One More Train To Rob, Cahill, U.S. Marshall, The Outfit, Freaky Friday.

WINELAND, LLOYD G.: Theatre Executive. b. Washington, D.C., 1917. e. George Washington Univ. (A.B.), 1937. Pres., Wineland Theatres. Bd. mbr., TOA, 1962066; on bd. of NATO of Metropolitan D.C., 1939–81. Circuit owns and operates four Maryland drive-ins and triple indoor cinema.

WINITSKY, ALEX: Producer. In partnership as attorneys in L.A. for 20 years with Arlene Sellers before they turned to financing and later production of films.
PICTURES INCLUDE: (co-prod. with Winitsky) End of the Game, The Seven-Per-Cent Solution, Cross of Iron, Night Calls, Silver Bears, Cuba.

WINKLER, HENRY: Actor. b. New York City, Oct. 30, 1946. e. Emerson College, Yale School of Drama. Appeared with Yale Reportory Co.; returned to N.Y. to work in radio. Did 30 TV commercials before starring in The Great American Dream Machine and Masquerade on TV.
PICTURES INCLUDE: The Lords of Flatbush (debut), Crazy Joe, Heroes, The One and Only.
TV: The Mary Tyler Moore Show, The Bob Newhart Show, The Paul Sand Show, Rhoda, Happy Days (regular), Laverne & Shirley. Movie: Katherine.

WINKLER, IRWIN: Producer. b. N.Y.C. e. N.Y. Univ. With I. Robert Chartoff formed production co., Chartoff-Winkler Prods. All films co-produced with Chartoff.
PICTURES INCLUDE: Double Trouble, Point Blank, The Split, They Shoot Horses, Don't They?, The Strawberry Statement, Leo The Last, Believe in Me, The Gang That Couldn't Shoot Straight, The New Centurions, Up the Sandbox, The Mechanic, Busting, SPY's, The Gambler, Breakout, Rocky, Nickelodeon, New York, New York, Valentino, Comes a Horseman, Uncle Joe Shannon, Raging Bull.

WINNER, MICHAEL: Producer, director, writer. b. London, Eng. 1935. e. Cambridge U.; ent. m.p. ind. as columnist, dir., Drummer Films.
TV: White Hunter series, Dick and the Duchess series.
PICTURES INCLUDE: orig. s.p., Man With A Gun; prod. dir., writ., Shoot to Kill, Swiss Holiday, Climb Up the Wall, Out of the Shadow, Some Like it Cool, Girls, Girls, Girls, It's Magic, Behave Yourself; formed, Scimitar Films; prod. dir., Haunted England, Play It Cool, The Mikado; dir., West 11; co-prod., dir., The System; co-author, dir. You Must Be Joking; co-prod. author. dir. The Jokers; prod. dir. I'll Never Forget What's 'Is Name; prod. dir., Hannibal Brooks, co-prod. dir., The Games prod, dir. Lawman, prod. dir., The Nightcomers, Chato's Land; dir. The Mechanic, Scorpio, The Stone Killer 1973, Death Wish, (prod. dir.), 1974 Won Ton Ton, The Dog That Saved Hollywood (prod., dir.), 1975. Prod. dir. The Sentinel, 1976, The Big Sleep, 1977; Firepower, 1978.

WINSTEN, ARCHER: Motion picture critic, New York Post. b. Seattle, Wash., September 18, 1904; p. Harry J. Winsten and Nell Archer. e. Princeton University, 1926. Varied newspaper experience before joining New York Post as columnist 1933; present post, 1936; ski ed., N.Y. Post since 1942.

WINTERS, DEBORAH: Actress. b. Los Angeles. e. Professional Children's School, New York City, began studying acting with Stella Adler at age 13. Acting debut at age six in TV Commercials.
CREDITS: Me, Natalie, Hail Hero!, The People Next Door, Kotch, The People Next Door (TV production) Class of '44.

WINTERS, DAVID: Choreographer, actor, director. b. London, April 5. Acted in both Broadway and m.p. version of West Side Story. Directed and acted in number of TV shows. Choreography credits include films Viva Las Vegas, Billie, Send Me No Flowers, Tickle Me, Pajama Party, Girl Happy, The Swinger, Made in Paris, Easy Come, Easy Go, Was choreographer for TV series Hullabaloo and Steve Allen Show, and several TV specials.

WINTERS, JERRY: Producer, director. b. Waterbury, Conn., Aug. 18, 1917. e. Antioch College, B.A., 1940. Photog., 1940–42; U.S. Air Force, 1942–46; photog., Hollywood, 1946–47; prod. assoc. Tonight on Broadway, CBS-TV, 1949; assoc. prod., College Bowl, ABC-TV, 1950–51; in charge N.Y. film prod., Television Varieties, Inc., 1951–54; Production head Eldorado Int'l Pictures Corp., 1964–67; vice president, Edutornics Corp., 1968; pres. Giralda Pros., 1971.
PICTURES INCLUDE: prod., Renoir; prod.-dir., Herman Melville's Moby Dick, Speak to Me Child; prod; English version, The Loves of Liszt.

WINTERS, JONATHAN: Performer. b. Dayton, Ohio, Nov. 11, 1925. e. Kenyon College. Disc jockey, Dayton and Columbus stations; night club comedian.
TV: And Here's the Show, Columbus—TV, NBC Comedy Hour, Jonathan Winters Show, NBC-TV.
PICTURES INCLUDE: It's A Mad, Mad, Mad, Mad World, The Loved One, Viva Max, The Fish That Saved Pittsburgh.

WINTERS, RICHARD J.: Executive. b. New York City, Oct. 3, 1928. e. Columbia Univ. Started in film industry in 1952 in pub. dep't., RKO Pictures; joined 20th-Fox in 1954 in pub. dept, first as nat'l. mag. contact then dir. of special events; went to MGM 1957 as nat'l. mag. contact, named N.Y. pub. mgr., 1959; nat'l. pub. dir., 1961.; in 1969 named nat'l. pub. dir. for Columbia Pictures. Worked on production pub. assignments for MGM and Paramount in 1972. Named nat'l. dir. adv.-pub.-prom., Paramount TV, 1973; promoted to exec. dir. adv./pub./promo., 1977.

WINTERS, ROLAND: Actor. b. Boston, Mass., Nov. 22, 1905. Appeared on stage and in stock, 1923–33; numerous radio programs, 1933–47. On screen in 13 Rue Madaleine, Return of October. Starred as Charlie Chan in pictures for Monogram.
PICTURES INCLUDE: West Point Story, Follow the Sun, Inside Straight, She's Working Her Way Through College, Jet Pilot, So Big.

WINTERS, SHELLEY: Actress. r.n. Shirley Schrift. b. St. Louis, Mo., Aug. 18, 1922. e. Wayne U. Clerked in 5 & 10 cent store; in vaudeville; on dramatic stage (Conquest, Night Before Christmas, Meet the People, Rosalinda, etc.). On screen 1944 in Nine Girls, Sailor's Holiday. N.Y. stage; A Hatful of Rain, Girls of Summer, Minnie's Boys.
PICTURES INCLUDE: Double Life, Cry of the City, Larceny, Take one False Step, Johnny Stool Pigeon, Great Gatsby, South Sea Sinner, Winchester '73, Place in the Sun, Untamed Frontier, My Man and I, Tennessee Champ, Executive Suite, Saskatchewan, Playgirl, Mambo, Night of the Hunter, I Am a Camera Big Knife, Treasure of Pancho Villa, I Died a Thousand Times, Cash on Delivery, Diary of Anne Frank, Young Savages, Lolita, Chapman Report, A House is Not a Home, A Patch of Blue, Alfie, Enter Laughing, The Scalphunters, Wild in the Streets, Buena Sera Mrs. Campbell, The Mad Room, Bloody Mama, What's the Matter with Helen?, The Poseidon Adventure, Cleopatra Jones, Something To Hide, Blume in Love, Diamonds, Next Stop Greenwich Village, The Tenant, Pete's Dragon, City on Fire, "S.O.B"

WINTMAN, MELVIN R.: Theatre Executive, b. Chelsea, Mass., June 28, 1918. e. Univ. of Mass., Northeastern Univ. (degree of juris doctor). Major, infantry, AUS, World War II. Attorney. Now exec. v.p. & dir. General Cinema Corp., Boston, pres. & dir. GCC Theatres, Inc., Boston, dir. Will Rogers Memorial Fund. Former pres. Theatre Owners of New England (1969–70); past dir. NATO (1969–70); treas., Nat'l Assoc. of Concessionaires (1960).

WISBERG, AUBREY: Writer, producer, director. b. London, Eng., Oct. 20; e. Columbia U. Newspaper writer, radio, TV dramatist U.S. Eng., Australia, radio diffusionist, Paris, France. Author novels Bushman at Large, Patrol Boat 999, This Is the Life; plays Virtue Inc., Whiphand.
PICTURES INCLUDE: s.p. prod. The Man From Planet X, Captive Women, Sword of Venus, The Neanderthal Man, Capt. John Smith & Pocahontas, Problem Girls, Dragon's Gold, Capt. Kidd & the Slave Girl, Return to Treasure Island, Murder is My Beat, The Women of Pitcairn Island, Port Sinister, Submarine Raider, Counter Espionage, Escape in the Fog, U-Boat Prisoner, Power of the Whistler, Adventures of Rusty, After Midnight, The Wreck of the Hesperus, The Big Fix, Betrayal from the East. The Falcon's Adventure, Son of Sinbad, At Swords Point, Hit Parade, They Came to Blow Up America, Bombers Moon, Rendezvous 24, The Lady in the Iron Mask, The Steel Lady, Treasure of Monte Cristo, Road to the Big House, The Burning Cross, St. Montana Mike, Casanova's Big Night, So Dark the Night, The Desert Hawk, s.p. Just Before Dawn, Out of the Depths, Target Minus Forty, Mission Mars, Ride the Wild Wind, Hercules in N.Y., Evil in the Blood.

WISDOM, NORMAN: Actor, singer, comedian. Musical and legit. b. London, Eng. Many London West End Stage shows including Royal Command Performances. New York Broadway shows include Walking Happy and Not Now Darling. Two Broadway awards. Films include Trouble in Store (Academy Award) and 19 others mostly for the Rank Organization and United Artists. Latest American film, The Night They Raided Minsky's.

WISE, ROBERT: Producer, director. b. Winchester, Ind., Sept. 10, 1914; e. Franklin Coll., Franklin, Ind. No prior bus. experience. Ent. m.p. ind. in cutting dept. RKO, 1933; sound cutter, asst. ed.,; film ed., 1939; dir., 1943; to 20th Century-

Fox, 1949; ass'n. Mirisch Co. independent prod. 1959; assn. MGM independent prod., 1962; assn. 20th Century Fox Independent Prod. 1963. Partner, Filmakers Group, The Tripar Group.
PICTURES INCLUDE: The Body Snatchers, Blood on the Moon, The Set Up, Day the Earth Stood Still, Captive City, So Big, Executive Suite, Helen of Troy, Tribute to a Bad Man, Somebody Up There Likes Me, Until They Sail, Run Silent Run Deep, I Want to Live, Odds Against Tomorrow, West Side Story, Two For the Seesaw, The Haunting, The Sound of Music, The Sand Pebbles, Star!, The Andromeda Strain, Two People, The Hindenburg, Audrey Rose, Star Trek.

WISEMAN, JOSEPH: Actor. b. Montreal, Canada. Began acting in the thirties, including Bdwy. stage, radio, m.p. and later TV.
PICTURES INCLUDE: Viva Zapata, Les Miserables, The Silver Chalice, The Garment Jungle, Dr. No, Bye Bye Braverman, The Night They Raided Minsky's The Valachi Papers, The Apprenticeship of Duddy Kravitz, The Betsy.
STAGE: King Lear, Golden Boy, The Diary of Anne Frank, Uncle Vanya, The Last Analysis, Enemies, etc.

WITHERS, GOOGIE: Actress. b. Karachi, India, Mar. 12, 1917. Trained as a dancer under Italia Conti, Helena Lehmiski & Buddy Bradley; stage debut Victoria Palace in Windmill Man, 1929. Best Actress Award, Deep Blue Sea, 1954. Began screen career at 18. TV also. Theatrical tours Australia, Sun Award, Best Actress, 1974.
PICTURES INCLUDE: Traveler's Joy, Night and the City, White Corridors, Derby Day, Devil on Horseback, Safe Harbor, Nickel Queen.
STAGE: (Britain) Winter Journey, Deep Blue Sea, Hamlet, Much Ado About Nothing. (Australian) Plaza Suite, Relatively Speaking, Beckman Place, Woman in a Dressing Gown, The Constant Wife, First Four Hundred Years, Roar Like a Dove, The Cherry Orchard, An Ideal Husband. (London) Getting Married, Exit the King. (New York) The Complaisant Lover. Chichester Festival Theatre and Haymarket, London, in The Circle, The Kingfisher, Importance of Being Earnest.
TV: Series, Within These Walls.

WITHERS, JANE: Actress. b. Atlanta, Ga. By 1934 attracted attention as child player on screen, after radio appearance in Los Angeles and experimental pictures parts, in 1934 in Fox production Bright Eyes, Ginger; thereafter to 1942 featured or starred in numerous 20th-Fox prod. Voted Money-Making Star M.P. Herald-Fame Poll 1937, 1938.
PICTURES INCLUDE: North Star, Johnny Doughboy, My Best Gal, Faces in the Fog, Dangerous Partners, Affairs of Geraldine, Danger Street, Giant, The Right Approach, Captain Newman.

WIZAN, JOE: Producer. Started in industry as agent for William Morris Agency. Left to form London Intl. Artists, Ltd. in association with Richard Gregson, Alan Ladd, Jr. and Mike Gruskoff. When firm dissolved joined Creative Management Associates as v.p. in chg. of creative services. In 1969 formed own indep. prod. co. 1981, named pres., CBS Theatrical Film Div.
PICTURES: Jeremiah Johnson, Junion Bonner, Prime Cut, The Last American Hero, Audrey Rose, Voices, And Justice for All.

WIZEMAN, JR. DONALD G.: Executive. b. Fort Smith, Ark., Nov. 17, 1944. e. Old Dominion Univ. Formed Wizeman & Associates, Ltd. Advertising Agency and Artus Specialty Company in 1967; is pres. of both. In 1971 formed Filmakers, Ltd. which also heads.
PICTURES INCLUDE: Moonchild (exec. prod.), Come Out of the Bathroom Hannibal Fry (exec. prod.).

WOLCOTT, JAMES L.: Executive. b. Wilmington, Del. e. Williams Coll., Harvard Bus. Sch.; Wilmington Trust Co., banking dept.; Fox Film Corp. New York legal dept.; Fox Movietonews cutter, asst. ed. short subjects dept.; Fox Film Corp. publicity dept.; deLuxe Lab.; 10th Century-Fox Studios asst. dir., head of test dept.; March of Time prod. mgr., ed. board; Audience Research, Inc. v.p. and gen. mgr.; Pathe Laboratories, Inc. sales v.p., exec. v.p. and director; Pathecolor, Inc. pres. and director; Transamerica Releasing Corp. pres. and director; Dorado Int'l Corp. pres. and director; James Wolcott Productions, Inc. pres. and director.

WOLF, EMANUEL, L.: Executive. b. Brooklyn, N.Y., Mar. 27, 1927. e. Syracuse U.B.A., 1950; Maxwell School, Syracuse U. M.A. 1952; Maxwell Scholar in Public Admin.-Economics; Chi Eta Sigma (Econ. Hon.). 1952-55. Management consultant, exec. office of Secretary of Navy & Dept. of Interior, Wash., D.C. 1956. National dir. of Program & Admin. of a Veterans Org. 1957–61. Pres. E. L. Wolf Associates, Washington, D.C. 1961–Jan. 1965. Treasurer, Kalvex, Inc. Dec. 1962. Director Kalvex, Inc. March 1963. Dir. Allied Artists Pictures Corp., Jan. 1965. Pres. Kalvex, Inc. April 1966–present, pres. & chmn. of the Bd. Kalvex, Inc.; Chmn. of the Bd. Vitabath, Inc.; Chmn. of the Bd. Lexington Instruments; pres. & chairman of the Bd. Pharmaceutical Savings Plan, Inc. Syracuse U. Corporate Advisory Board, American Committee for the Weizmann Institute of Science (Bd. of Directors). Pres. and Chmn. of Bd., Allied Artists Pictures Corp: January, 1976: pres., bd. chm. & chief executive officer of Allied Artists Industries Inc., created by Merger of Allied Artists Pictures Corp., Kalvex Inc. and PSP, Inc.

WOLF, HERBERT: Producer. b. New York, N.Y., July 11, 1917. e. N.Y.U., 1937. Radio & TV prod. Wolf Presentations, Inc.
TV: (N.Y.) Masquerade Party, Break the Bank, Hold That Note, Keep Talking, Window Shopping.

WOLF, THOMAS HOWARD: TV news exec. b. New York, N.Y., April 22, 1916; e. Princeton, B.A., magna cum laude, 1937. Time & Life Mag. 1937–39; 1937–39 NEA (Scripps-Howard) 1940–46; European mgr., NEA, 1942–46. War correspondent, ETO, MTO) NBC radio correspondent, Paris, 1944–45; co-owner, pres., Information Prod., Inc. founded 1951; co-owner, chairman. Butterfield & Wolf, Inc. founded 1955; prod. CBS series. Tomorrow, 1960; exec. prod., CBS daily live Calendar Show, 1961–62; senior prod., ABC News Report, 1963; exec. prod., ABC Scope, 1964–66. v.p. dir. of TV Documentaries, 1966; v.p., dir. of TV Public Affairs, 1974; dir. TV Cultural Affairs, 1976.

WOLFF, LOTHAR: Producer, director, president: Lothar Wolff Productions Inc. b. Bromberg, Ger. Asst. prod., March of Time. WWII; Officer in Charge, U.S. Coast Guard M.P. section; head, Marshall Plan film section, Paris, 1949–50; M.P. advisor to Indonesian Govt. 1953–55; v.p. & prod., Louis de Rochemont Assoc.; prod. & co-author, Martin Luther, produced: Question Seven, 13 half hour Planet Earth Films (Natl Academy of Sciences), Fortress of Peace (Swiss Govt. Cinerama); assoc. prod. Lost Boundaries, Walk East on Beacon, Windjammer, Roman Spring of Mrs. Stone. Series executive producer for Time-Life Films (The World We Live In; Other people, Other Places; Wild, Wild World of Animals). 1976–77: Consultant Eductl. Film Division, Natl. Geographic; Editorial Director, I Sought my Brother, Harvard U./WQED; Producer, The Joy of Bach, Luthern Film Associates, a 90-minute special.

WOLFSON, MITCHELL: Executive. b. Key West, Fla., Sept. 13, 1900. Ent. m.p. ind. as exhibitor, Miami, 1924; later co-owner Wometco Circuit; bd. chm. & pres. Wometco Ent. Inc.; TV stations. WTVJ Miami: WLOS-TV, Asheville, N.C.; WLOS-FM, Asheville, KVOS-TV, Bellingham, Wash.; KVOS-TV (BC) Ltc., Vancouver, B.C.; Seaquarium, Miami; bottling: Coca Cola and WZZM-TV, Grand Rapids, Mich., WWHT, Newark, N.J.; STV and CATV. Automatic vending firms. Founder, Audubon House, Key West, 1939 elected City Councilman, Miami Beach; 1943 Mayor; bd. chm.; chm. City of Miami Off-Street Parking Board; in 1949 chmn. TOA TV comm. Clubs; Rotary, Elks, Shriners, Lt. Col. U.S. Army, WW II; trustee Mt. Sinai Hospital; LL.D., U. of Miami; bd. chm., Financial Federal Savings & Loan Assn.; director and member, executive board, National Association of Theatre Owners; mem. Variety Club.

WOLFSON, RICHARD: Executive. b. New York City, Jan. 7, 1923. e. Harvard Coll. Yale Law School, 1945–47, law sect'y to Justice Wiley Rutledge, U.S. Supreme Court; law instructor at NYU Law School; later received Guggenheim Fellowship; 1952, joined Wometco Ent. as counsel and asst. to pres.; named v.p. and dir. in 1959 and senior v.p. in 1962; named exec. v.p. and general counsel in 1973; named chm., exec. comm., 1976; co-author of Jurisdiction of the Supreme Court of the United States and author of articles in various legal publications.

WOLPER, DAVID L.: Executive. b. New York, N.Y., Jan. 11, 1928. e. Drake U., U. of So. Calif., Treas., Flamingo Films, 1948; merged with Associated Artist to form M.P. for TV, Inc., acting as v.p. in chge of West Coast oper., 1950; v.p. reactivated Flamingo Films, 1954; also pres. Harris-Wolper Pictures, Inc.; pres. Wolper Prod. 1958; pres. Dawn Prod.; v.p. Bd. Dir. Metromedia, 1965; pres. Wolper Pictures Ltd. 1967; ch. of bd. Wolper Prod., Inc., 1967; pres. Wolper Pictures, 1968; pres. Wolper Productions, 1970; pres. & ch. of bd. of dir. The Wolper Organization, Inc., 1971; married Gloria Diane Hill, July 11, 1974.
TV & M.P.: The Race For Space, 1958; Story of . . . series, 1962–63; Biography series, 1962–63; Hollywood and the Stars series, 1963–64; The Making of the President, 1960, 1964, 1968; Men in Crisis series, 1964; Four Days in November feature, 1965; National Geographic Society Specials, 1965–68, 1971–75; The March of Time series, 1965–66; The Rise and Fall of the Third Reich, 1967–68; The Undersea World of Jacques Cousteau, 1967–68; If It's Tuesday, This Must Be Belgium feature, 1968; Plimpton specials, 1970–72; Say Goodbye, 1971; Willy Wonka and the Chocolate Factory feature, 1971; Appointment With Destiny series, 1971–73; Visions of Eight feature, 1973; American Heritage specials, 1973–74; Primal Man specials, 1973–75; Get Christie Love! series, 1974; Judgment specials, 1974; Birds Do It, Bees Do It feature, 1974; Chico and the Man series, 1974–; Smithsonian Specials, 1974–;

Sandburg's Lincoln, 1974–76; Welcome Back, Kotter series, 1975–.

WOOD, NATALIE: Actress. b. San Francisco, Calif., 1938. p. father, noted designer, set decorator, Hollywood studios; mother, ballet dancer. Screen debut: Tomorrow is Forever.

PICTURES INCLUDE: Ghost and Mrs. Muir, Miracle on 34th Street, Driftwood, Bride Wore Boots, Green Promise, One Desire, Scudda-Hoo ScuddaHay; Chicken Every Sunday, Father Was a Fullback, No Sad Songs for Me, Jackpot, Never a Dull Moment, Our Very Own, Blue Veil, Just for Your, The Star, Rebel Without a Cause, Cry in the Night, The Searchers, Burning Hills, Girl He Left Behind, Bombers B-52, Marjorie Morningstar, Kings Go Forth, Cash McCall, Splendor in the Grass, West Side Story, Gypsy, Love with a Proper Stranger, Sex and the Single Girl, Inside Daisy Clover, This Property is Condemned, Penelope, Bob and Carol and Ted and Alice, Peeper, Meteor, The Last Married Couple in America.

TV: Cat on a Hot Tin Roof.

WOODHAM-SMITH, GEORGE IVON: Solicitor. b. London, England, Dec. 20, 1895. 1st Lt., Machine Gun Corps., Brit. Army, World War I, studied law; also journalist. Since 1928 partner in Richards Butler & Co., London law firm. In charge of legal work for the Rank Org.

WOODLAND, NORMAN: Actor. b. Dusseldorf, Germany, March 16, 1910; e. Edward VI School. Stratford-on-Avon. Spent several years in repertory cos.; joined BBC 1939 and spent 6 yrs. as commentator; stage career largely assoc. with Shakespearean and classical plays; TV appearances; screen debut in Hamlet.

PICTURES INCLUDE: Escape, I Know You, All Over the Town, Background, Romeo and Juliet, Master Plan, Richard III, Teenage Daughter, Guilty, No Road Back, The Bandit, Rough and the Smooth.

WOODS, DONALD: Actor.

PICTURES INCLUDE: Watch on the Rhine, Roughly Speaking, 13 Ghosts, Kissin' Cousins, Moment to Moment, A Time to Sing.

STAGE: Two for the Seesaw, L.A., 1961; Rosmersholm, NYC, 1962; One by One, NYC, 1964; Soldier, You Can't Take It With You, Chicago, 1969; Twelfth Night, Assassination 1865, Chicago, 1969–71.

TV: G.E. Theatre, Wagon Train, Thrillers, Sunset Strip, Ben Casey, Laramie, The Rebel, The Law and Mr. Jones, The Roaring 20's, Wild Wild West, Bonanza.

WOODWARD, EDWARD: O.B.E. Actor, Singer. b. Croydon, England. LPs. 1 Gold Disc.

TV: Sword of Honour, Callan series. Over 300 TV plays latest: Bassplayer and Blonde, Saturday, Sunday, Monday, and 1990 series. Television Actor of the Year, 1969–70; also Sun Award Best Actor 1970, 71, 72.

PICTURES INCLUDE: Becket, File on the Golden Goose, Murders in the Rue Morgue, Julius Caesar, The Listener, Young Winston, Sitting Target, Hunted, Wicker Man, Callan, Stand Up Virgin Soldiers, Breaker Morant.

THEATRE: 16 West End Plays and Musicals. Latest: On Approval.

WOODWARD, JOANNE: Actress. b. Thomasville, Ga., Feb. 27; e. La. State U. Studied at Neighborhood Playhouse Dramatic Sch. m. Paul Newman, N.Y.; appeared in many TV dramatic shows; on B'way in Picnic; m.p. debut in Count Three and Pray.

PICTURES INCLUDE: Kiss Before Dying, Three Faces of Eve (Academy Award 1957, best actress); Long Hot Summer, No Down Payment, Rally Round the Flag Boys, The Sound and the Fury, From the Terrace, Fugitive Kind, Paris Blues, The Stripper, A New Kind of Love, Signpost to Murder, A Big Hand for the Little Lady, Rachel, Rachel (Acad. Award nom.); Winning, USA, The Effect of Gamma Rays on Man-in-the-Moon Marigolds, Summer Wishes, Winter Dreams, The Drowning Pool, The End.

TV: Sybil, See How She Runs.

WOODY, CHESTER F.: Actor. b. Zoar Valley, Colo., Aug. 13, 1934; e. U. of Colo. Writer, stories, songs, lyrics used on radio, TV.

PICTURES INCLUDE: The Glenn Miller Story, The Mob, Monsters of the Deep, Hit the Deck, The Rangers, The Big Circus, Gangland, Adventure Is our Destiny, Freed, Distant Field Are Greener, And Now My Love, Murder at Midnight, The Queen's Guard, Look Back and Cry.

WOOLF, JOHN: Chairman Romulus Films, Ltd. Man. dir. British and American Film Holdings. Exec. dir., Anglia Television Ltd. Mem. Cinematograph Films Council. Bd. of man. Services Kinema Corp.

PICTURES INCLUDE: Pandora and the Flying Dutchman, African Queen, Moulin Rouge, Beat the Devil, Carrington V.C., I am a Camera, Sailor Beware, Three Men in a Boat, The Story of Esther Costello, Room at the Top, The L-Shaped Room, Life At The Top, Oliver! (Oscar for best film 1969), The Day of the Jackal, The Odessa File.

WOOLNER, LAWRENCE H.: Executive. b. St. Louis, Mo. e. Fairfax High School, L.A. After discharged from U.S. Army at end of World War II started drive-in theatre in New Orleans. Formed Woolner Bros. Pictures there in 1955. Later moved to Hollywood, entering distribution and production as Woolner Bros. Pictures. In 1970 started New World Pictures as pres. in chg. sales, dist. In 1971 started own company, Dimension Pictures.

WORKER, ADRIAN: Producer. b. Kempston, England, 1916. e. public school. Bedford. Early career as accountant; joined Rank Organization, then proc. sup. Highbury Studios; studio man., Islington studios; freelance prod. man. 1949.

PICTURES INCLUDE: Another Man's Poison, Mr. Drake's Duck. Joined Warwick on formation 1952. Prod. sup. Red Beret (Paratrooper), Hell Below Zero, The Black Knight, Prize of Gold; prod., Safari, Beyond Mombasa. Formed Zonic Prod. 1956; prod. Naked Earth, Intent to Kill; assoc. prod. Danger Within, I'm All Right Jack; man. dir. Shepperton Studios, 1959; 1976, prod. Fern, the Red Deer for Children's Film Foundation; 1978, Kidnapped (TV series).

WORMSER, IRVING: Executive. b. Feb. 1, 1900. Booking, sales, Pathe Films, 1916–24; state right operation, 1924–30; Columbia Pict. sales, 1930–44; eastern sales mgr., Film Classics. 1944–46; exec. sales, gen. sales mgr., Columbia, 1946–54; co-owner, St. Cloud circuit, 1949–55; v.p. gen. sales mgr., Dist. Corp. of Amer., 1954–58; pres. Continetal Dist. Inc., 1958–66; prod. rep. for Palomar Pictures; now v.p.-tres. of both U.M. Film Distributors and Peppercorn-Wormser.

WORMSER, RICHARD: Writer. b. New York, N.Y., Jan. 30, 1908. e. Princeton U. Publicist. 1927–32; fiction writer, Street & Smith, 1932–34; freelance fiction writer, 1934–41; forest officer. U.S. Forest Service. 1941–45; then free lance m.p., magazine writer; novels include Pass through Manhattan, Lonesome Quarter, Road to Carmichael's.

PICTURES INCLUDE: s.p., In Self Defense; collab. orig., The Showdown; orig. s.p. Fort Dodge Stampede.

WORTH, IRENE: Actress. b. 1916 Bdwy. debut in The Two Mrs. Carrolls, after which went to London where made her home. Appeared with Old Vic and Royal Shakespeare Co.; returned to U.S. to appear on Bdwy. in the Cocktail Party, Toys in the Attic, Sweet Bird of Youth, Cherry Orchard; etc.

PICTURES: Orders to Kill (British AA as best actress), King Lear, Nicholas and Alexander, Rich Kids, Eyewitness.

TELEVISION: The Lady from the Sea, The Duchess of Malfi, The Way of the World, Prince Orestes, etc.

WOWCHUK, HARRY N.: Actor, Writer, Photographer, Producer, Executive. b. Phila. Pa. Oct. 16, 1948. e. Santa Monica City College, U.C.L.A. Theater Arts, 1970. Started film career as actor, stunt-driver-photographer. T.V. and commercial credits include: Warner Bros.; Columbia Records; R.C.A.; Playboy Magazine; T.V. Guide; Seal Test; Camel Cigarettes; Miller High Life; American Motors; Camera V; A&W Rootbeer; Harold Robbins Productions. Former exec. v.p. International Cinema, in charge of prod. and distribution; V.P. J. Newport Film Productions; pres., United West Productions.

PICTURES INCLUDE: The Lost Dutchman, Las Vegas Lady, This Is A Hijack, Tidal Wave, Tunnel Vision, Incredible 2-Headed Transplant, Jud, Bad Charleston Charlie, Some Call It Loving, Summer School Teachers, Five Minutes of Freedom, Pushing Up Daisies, Money-Marbles-Chalk, The Models, Love Swedish Style, Up-Down-Up, Sunday's Child, Soul Brothers, Freedom Riders, Perilous Journey, Claws of Death, Georgia Peaches.

WOWCHUK, NICHOLAS: Executive, Producer, Writer, Editor, Financier. b. Philadelphia, Pa. e. St. Basil's College, U.C.L.A. Founder-publisher: All-American Athlete Magazine; Sports and Health Digest. The Spectator. Former sports writer: Phila. Evening Public Ledger; Phila. Daily Record; Phila. Inquirer. Founder & board chairman: Mutual Realty Investment Co.; Mutual Mortgage Co., Beverly Hills, Calif. President: Mutual General Films, Bev. Hills, Calif.; Abbey Theatrical Films, N.Y.C.; Mutual Film Distribution Co.; Mutual Recording & Broadcasting Enterprises.

PICTURES INCLUDE: Exec Prod.: Perilous Journey; Incredible 2-Headed Transplant; Pushing Up Daisies; Money-Marbles-Chalk; Five Minutes Of Freedom; The Campaign; Claws of Death. Prod.: Scorpion's Web; Pursuit; Brave Men; Sea of Despair; The Hetman; Cossack's In Battle; The Straight White Line; Tilt, Rooster, To Live . . . You Gotta Win.

WRATHER, JACK: Producer. b. Amarillo, Texas, May 24, 1918; e. U. of Texas; m. Bonita Granville, actress. Oil producer since 1938. During WW II, U.S. Marine Corps. Producer Jack Wrather Productions, Inc. & Freedom Prod.

PICTURES INCLUDE: The Guilty, High Tide, Perilous Waters, Strike It Rich, Guilty of Treason, Lone Ranger, The Magic of Lassie, The Legend of the Lone Ranger. co-owner: KFMB-TV, San Diego; KERO-TV, Bakersfield, Calif.;

Muzak Corp.; pres. Lone Ranger, Television, Inc., Lassie Television, Inc., Sergeant Preston of the Yukon; owner Disneyland Hotel. pres., Wrather Corp. Director, Tele-PromTer. Chm. of board, Wrather Entertainment Intl. Director, Continental Airlines.

WRAY, FAY: Actress. b. Alberta, Canada, Sept. 10, 1907. m. Robert Riskin, writer. On stage in Pilgrimage Play, Hollywood, 1923; m.p. debut in Gasoline Love; thereafter in many m.p. for Paramount to 1930; then in films for various Hollywood and Brit. prod.
PICTURES INCLUDE: Small Town Girl, Treasure of the Golden Condor, Rock Pretty Baby, Tammy, Out of Time.
TV: Pride of the Family, ABC-TV.

WRIGHT, BASIL: Producer and direcor. b. London, June 12, 1907. Educated at Sherborne school, Corpus Christi College, Cambridge. Assistant, 1929, to John Grierson and was associated with him in creation of "documentary" film in England.
PICTURES INCLUDE: The Country Comes to Town, O'er Hill and Dale, Windmill in Barbadoes, Cargo from Jamaica, and Liner Cruising South for Empire Marketing Bd. in 1931–33. In 1934 made The Song of Ceylon and awarded First Prize and Prix de Gouvernement at Brussels Film Festival in 1935. With Henry Watt directed Night Mall for GPO Film Unit, under prod. of John Grierson. In 1937 founded Realist Film Unit. Exec. prod., Film Centre, Ltd., 1939–44. In 1945, prod. in chge, Crown Film Unit. In 1946 adv. to Dir. Gen. of M.O.I.; prod. dir. Waters & Time for Festival of Britain, 1950–51; made World Without End for Unesco. 1953: with Gladys Wright prod. The Immortal Land and Greek Sculpture; prod. dir. A Place for Gold, 1960, 1962–68 Lecturer at U. of C., Los Angeles, 1975 Guest of Honour Melbourne Film Festival; 1975–76 Lecture Tour USA, Canada. 1977–78 Visiting Lecturer Temple Univ., Philadelphia.

WRIGHT, TERESA: Actress. b. New York City, Oct. 27, 1918. m. Divorced from Niven Busch and Robert Anderson. Ent. m.p. 1941, The Little Foxes, Goldwyn-RKO. In 1942: Pride of the Yankees, Goldwyn-RKO, Mrs. Miniver, MGM Acad. Award 1942 supporting role (Mrs. Miniver).
PICTURES INCLUDE: Shadow of a Doubt, Casanova Brown, Best Years of Our Lives, Trouble with Women, Pursued, Imperfect Lady, Enchantment, The Capture, The Men, Something to Live For, California Conquest, Steel Trap, Count the Hours, The Actress, Track of the Cat, Hail Hero, The Happy Ending, Roseland, Somewhere in Time.
STAGE: Tours: Mary, Mary, Tchin-Tchin, The Effect of Gamma Rays on Man-in-the-Moon Marigolds, Noel Coward in Two Keys, The Master Builder. Regional Theatre: Long Days, Journey into Night, You Can't Take It With You, All The Way Home. New York: Life with Father, Dark at the Top of the Stairs, I Never Sang for My Father, Death of a Salesman, Ah, Wilderness!, Morning's at Seven

WRIGHT, TONY: Actor. b. London, Dec. 10, 1925. Stage debut in repertory, South Africa; screen debut; Flanagan Boy, 1951.
PICTURES INCLUDE: A Toi De Jouer Callaghan, Plus De Whiskey, Pour Callaghan (France), Jumping for Joy, Jacqueline, Tiger in the Smoke, Seven Thunders, Broth of a Boy, Faces in the Dark, In the Wake of a Stranger, Journey to Nowhere, The Liquidator.
TV: Compact, Marriage Lines, No Hiding Place, The Saint, Mystery Theatre, Curtains for Sheila, Crossroads, Make Me A Widow, Wednesday's Train, Onedin Line, 6 Saints, Persuaders, The Jensen Code, Follow Me, Kidnapped.

WRIGHT, WILLIAM H.: Producer. b. Lawrenceburg, Ind., Apr. 29, 1902; e. Indiana U. Reporter, Cincinnati Post, Indianapolis Star; exploit. dept., Paramount, Columbus, Ohio, 1925; Long Island studio pub. dept., 1926; Hollywood studio, 1927; asst. to Selznick, Schulberg, others, 1931; asst. to Selznick, MGM, 1935; to Harry Cohn, Columbia, 1936; prod. asst. to Selznick, 1937; free lance writer, 1940; staff writer, MGM, 1941; prod., 1944–53; prod. Columbia, 1955; Warner Bros., 1963–64; Paramount, 1964. Trustee, M.P. & TV Fund.
PICTURES INCLUDE: Letter for Evie, Three Wise Fools, Bride Goes Wild, Act of Violence, Skipper Surprised His Wife, Black Hand, Stars in My Crown, Mrs. O'Malley and Mr. Malone, People Against O'Hara, Shadow in the Sky, Love Is Better Than Ever; co-prod. Young Man with Ideas; prod. The Clown, The Naked Spur, Dead Ringer, A Distant Trumpet; documentary, Williamsbury, The Story of a Patriot; collab. s.p., The Sons of Katie Elder.
TV: prod. Adventures of Jim Bowie, Barbara Stanwyck Theatre; co-creator, Kentucky Jones, NCB.

WUNSCH, ROBERT J.: Executive. e. Yale. Literary agent—first with Adams, Ray & Rosenberg; later with Intl. Famous Agency. Entered indep. motion picture production; coproducer of Slap Shot, and exec. producer of Defiance. Named dir. of motion pictures for TV at CBS. Left to join United Artists as v.p., production, West Coast, in June, 1979.

WYATT, JANE: Actress. b. Aug. 10, 1913, New York: e. Miss Chapin's School, Barnard College; m. Edgar B. Ward. Joined Apprentice School, Berkshire Playhouse, Stockbridge, Mass. Understudied in Tradewinds and The Vinegar Tree. Appeared in Give Me Yesterday and the Tadpole. In 1933 succeeded Margaret Sullivan in Dinner at Eight. m.p.: New York stage, The Autumn Garden, 1951; other plays, The Bishop Misbehaves, Conquest, Eveninsong, The Mad Hopes.
PICTURES INCLUDE: Great Expectations, The Navy Comes Through, The Kansan, None But the Lonely Heart, Boomerang, Gentlemen's Agreement, No Minor Vices, Bad Boy, Canadian Pacific, Pitfall, Task Force, Our Very Own, My Blue Heaven, Man Who Cheated Himself, Criminal Lawyer, Lost Horizon, Never Too Late, Treasure of Matecumbe.
TV: Father Knows Best, winner 3 Emmy Awards (6 yrs.), Bob Hope Chrysler Theater, The Virginian, Wagon Train, U.S. Steel Hour, Bell Telephone Hour, Hostess moderator, Confidential For Women, My Father My Mother. Guest star—Star Trek, Barefoot in the Park Pilot CBS-TV, The Ghost and Mrs. Muir, Here Come the Brides, Love American Style, ABC Movie of Week, Guest star, the Men of Shiloh, Alias Smith and Jones, Kraft Music Hall; co-star, Hollywood TV Theater, Katherine (guest star), Tom Sawyer; Father Knows Best Reunion, A Love Affair, Amelia Earhart, Superdome, Fantasy Island, Love Boat, The Nativity, The Millionaire.

WYMAN, JANE: Actress. r.n. Sarah Jane Fulks. b. St. Joseph, Mo., Jan. 4, 1914. In 1936: My Man Godfrey, U; Cain and Mabel, Smart Blonde, Warner. From 1937 in many Hollywood pictures. In 1942: Larceny, Inc., My Favorite Spy, Footlight Serenade. Voted one of top ten money-making stars in M.P. Herald-Fame poll 1954. TV show: Fireside Theatre.
PICTURES INCLUDE: Princess O'Rourke, Doughgirls, Make Your Own Bed, Crime by Night, Lost Weekend, One More Tomorrow, Night and Day, The Yearling, Cheyenne, Magic Town; academy award, best actress, Johnny Belinda, 1948, Three Guys Named Mike, Here Comes the Groom, Blue Veil, Just for You, Story of Will Rogers, Let's Do It Again, So Big, Magnificent Obsession, Lucy Gallant, All That Heaven Allows, Miracle in the Rain, Pollyanna, Holiday for Lovers, Bon Voyage, How to Commit Marriage.

WYMORE, PATRICE: Actress. b. Miltonvale, Kans., Dec. 17; p. James A. Wymore, oper. exhib. film delivery service throughout Kans.; ret. Began career as child performer, tent shows, county fairs, vaudeville; later, toured night clubs in middle west, own song & dance act; modelled, Chicago understudy Betty Bruce, U Up in Central Park, played role Hollywood Bowl; then, NY stage, Hold It! All For Love; radio & TV roles.
PICTURES INCLUDE: Screen debut: Tea for Two, 1950, then Rocky Mountain, I'll See You In My Dreams, Star-Lift, Big Trees, Man Behind the Gun, She's Working Her Way Through College, She's Back on Broadway, Chamber of Horrors.

WYNN, KEENAN: Actor. b. New York City, July 27, 1916. s. of Ed Wynn (died June 19, 1966). e. St. John's Military Acad. New York stage, radio stock in Skowhegan, Me.
PICTURES INCLUDE: See Here, Private Hargrove, Since You Went Away, Marriage Is a Private Affair, Between Two Women, The Clock, Without Love, Week-End at the Waldorf, What Next Corporal Hargrove?, The Hucksters, B.F.'s Daughter, Three Musketeers, Neptune's Daughter, Annie Get Your Gun, Three Little Words, Royal Wedding, Kind Lady, Angela in the Outfield, Texas Carnival, Phone Call from a Stranger, Fearless Fagan, Sky Full of Moon, Desperate Search, Battle Circus, Code Two, All the Brothers Were Valiant, Kiss Me Kate, Long Long Trailer, Tennessee Champ, Men of the Fighting Lady, Glass Slipper, The Marauders, Running Wild, Man in the Gray Flannel Suit, Shack Out on 101, Johnny Concho, Great Man, Some Came Running, A Hole in the Head, King of the Roaring Twenties, Absent-Minded Professor, Warning Shot, Welcome to Hard Times, The War Wagon, Point Blank, Finian's Rainbow, McKenna's Gold, Smith, Once Upon A Time in the West, 80 Steps to Jonah, Cancel My Reservation, Pretty Maids All in a Row, The Animals, B.J. Presents, The Mechanic, Snowball Express, Wild in the Sky, Herbie Rides Again, Nashville, The Devil's Rain, The Shaggy D.A., Orca, Just Tell Me What You Want.

WYNN, MAY: Actress. r.n. Donna Lee Hickey. b. New York, Jan. 8, 1930; e. Newtown H.S., Elmhurst, L.I. Model; chorus girl, then singer, dancer; played bits in Hollywood; then The Caine Mutiny, They Rode West, Violent Men.

WYNN, TRACY KEENAN: Writer. Fourth generation of show business; son of actor Keenan Wynn; grandson of Ed Wynn; great-grandson of Frank Keenan Wynn, Irish Shakespearean actor who made Bdwy. debut in 1880.
PICTURES INCLUDE: The Longest Yard, The Deep (co. s.p.).

TV: The Glass House, Tribes, The Autobiography of Miss Jane Pittman.

WYNNE, DONALD. Production executive and writer. b. Surrey, England 1918. Early career: actor, camera prod. assistant. Ent. m.p. industry 1935; served in Army 1938–46; since prod. man., assoc, prod. Concanen, Nettlefold, Pan Productions.
PICTURES INCLUDE: Fabian of the Yard series, On Such a Night, prod. sup. Rank Screen Services. Wrote, dir. prod. for African Film Prod. in S. and E. Africa; Donald Wynne Prods. Ltd.

WYNTER, DANA: Actress. b. London, England. June 8. e. Rhodes U. On stage in London; TV appearances include Robert Montgomery Show, Suspense, Studio One, U.S. Steel Hour.
PICTURES INCLUDE: Invasion of the Body Snatchers, View from Pompey's Head, D-Day, The Sixth of June, Something of Value, Fraulein, Shake Hands with the Devil, In Love and War, Sink the Bismarck, The List of Adrian Messenger, Airport.
TV: Playhouse 90, Dick Powell Show, Wagon Train, Virginian, Burkes Law, Bob Hope Presents, Alfred Hitchcock, Twelve O'clock High, The Rogues, Ben Casey, FBI Story, My Three Sons, Wild Wild West, The Man Who Never Was.

Y

YABLANS, FRANK: Executive. B. Brooklyn, N.Y. Aug. 27, 1935. Ent. m.p. ind. as Warner Bros. booker, 1957. Warner Bros. salesman in N.Y., Boston, Milwaukee, Chicago, 1957–59. Milwaukee br. mgr. Buena Vista, 1959–66. Midwest sales mgr., Sigma III, 1966. Eastern sales mgr., 1967, sales v.p. 1968. V.P. general sales mgr., Paramount Pic. Corp., 1969; v.p.-dist., April 1970; sr. v.p.-mkt., Oct., 1970; exec. v.p., April 1971; named pres. May, 1971. In Jan., 1975, became an independent producer, his company called, Frank Yablans Presentations Inc.
PICTURES: Silver Streak (exec. prod.), The Other Side of Midnight (prod.), The Fury (prod.), North Dallas Forty (prod. & co-s.p.s.).

YABLANS, IRWIN: Executive. b. Brooklyn, N.Y. Began career in industry at WB in 1956 after two-yr. stint in Germany with U.S. Army. Held m.p. sales posts in Washington, D.C., Albany, Detroit, Milwaukee and Portland. In 1962 joined Paramount as L.A. mgr.; in 1964 made western sales mgr. In 1972 entered production as assoc. prod. on Howard W. Koch's Badge 373. First prod. on own: The Education of Sonny Carson (1974). Now pres. of Compass Int'l. Pictures.
PICTURES: Halloween, Roller Boogie (both exec. prod.), Roller Boogie (exec. prod., story), Fade To Black (story).

YATES, PETER: Producer, director. b. 1929. Ent. m.p. ind. as studio man and dubbing asst. with De Lane Lea. Asst. dir. The Entertainer, The Guns of Navarone, A Taste of Honey. Stage: dir. The American Dream, The Death of Bessie Smith.
TV: Danger Man, The Saint (series).
PICTURES INCLUDE: Summer Holiday, One Way Pendulum, Robbery, Bullitt, John and Mary, Murphy's War, The Hot Rock, The Friends of Eddie Coyle, For Pete's Sake, Mother, Jugs and Speed, The Deep, Breaking Away, Eyewitness.

YAWITZ, PAUL A.: Writer. b. St. Louis, Mo. Feb. 5, 1905; e. U. Missouri; U. Washington. Member: Zeta Beta Tau. Publicist, Ned Wayburn Enterprises, Fox Film Corp.; Broadway columnist, N.Y. Mirror; writer, prod., RKO. Started screen career 1937.
PICTURES INCLUDE: (orig. s.p.) She Has What It Takes, They Knew What They Wanted (orig. s.p.); Affairs of Annabelle (orig. s.p.); Go Chase Yourself (orig. s.p.); The Chance of a Lifetime; (collab. s.p.) The Racket Man, Beauty for Sale; Breakfast for Two; (s.p.) I Love a Bandleader, Boston Blackie Booked on Suspicion; (orig.) A Close Call for Boston Blackie; (collab. story) Walk Softly, Stranger; collab. s.p. Models, Inc. (orig. s.p.); The Black Scorpion.
Editor Beverly Hills Courier.

YORDAN, PHILIP: Writer. b. Chicago, Ill. e. University of Ill., B.A., Kent Coll. Law, LL.D. Author, producer, playwright (Anna Lucasta). Began screen writing 1942 with collab. s.p. Syncopation.
PICTURES INCLUDE: Unknown Guest, Johnny Doesn't Live Here, When Strangers Marry, Dillinger, Whistle Stop, The Chase, Suspense; play & s.p. Anna Lucasta, House of strangers; s.p. Edge of Doom; collab. s.p. Detective Story, (Acad. Award nomination) Mary Maru, s.p., Houdini, Blowing Wild; collab. s.p., Man Crazy, Naked Jungle; s.p., Johnny Guitar; story, Broken Lance (Acad Award 1954); adapt. Conquest of Space; collab. s.p. Man from Laramie, Last Frontier; prod. s.p. Harder They Fall, Men In War, No Down Payment, God's Little Acre; s.p. Bravados, Time Machine, The Day of the Outlaw, Studs Lonigan, King of Kings; collab., El Cid, 55 Days at Peking, Fall of the Roman Empire; prod., Crack in the World, Battle of the Bulge, Royal Hunt of the Sun, Brigham, Cataclysm.

YORK, DICK: Actor. r.n. Richard Allen York. b. Fort Wayne, Ind., Sept. 4, 1928. e. De Paul U. Drama School. Appeared on radio shows; commercial m.p. actor; on Broadway in Tea and Sympathy, Bus Stop; TV appearances.
TV: Omnibus, Robert Montgomery Show, Mr. D.A., The Web, Wagon Train, Alfred Hitchcock, Twilight Zone, Route 66, Going My Way, Bewitched.
PICTURES INCLUDE: Inherit the Wind, Cowboy, They Came to Cordura, My Sister Eileen, Three Stripes in the Sun.

YORK, MICHAEL: Actor. b. Fulmer, England, 1942. Early career with Oxford University Dramatic Society and National Youth Theatre; later Dundee Repertory, National Theatre.
TV: The Forsyte Saga, Rebel in the Grave, Great Expectations, Jesus of Nazareth, True Patriot, Much Ado About Nothing, A Man Called Intrepid.
STAGE: Any Just Cause, Hamlet, Outcry (Broadway), Ring Round The Moon (Los Angeles).
PICTURES INCLUDE: The Taming of the Shrew, Accident, Red and Blue, Smashing Time, Romeo and Juliet, The Strange Affair, The Guru, Alfred The Great, Justine, Something for Everyone, Zeppelin, La Poudre D'Escampette, Cabaret, England Made Me, Lost Horizon, The Three Musketeers, Murder on the Orient Express, The Four Musketeers, Conduct Unbecoming, Logan's Run, Seven Nights in Japan, The Last Remake of Beau Geste, The Island of Dr. Moreau, Fedora, The Riddle of the Sands.

YORK, SUSANNAH: Actress. b. London, 1942. Ent. TV 1959. Appearances include: The Crucible The Rebel and the Soldier, The First Gentleman, The Richest Man in the World. Ent. films in 1960.
PICTURES INCLUDE: Tunes of Glory, There Was a Crooked Man, Greengage Summer, Freud, Tom Jones, While the Tiger Sleeps, The Seventh Dawn, Scene Nun—Take One, Sands of Kalahari, Scruggs, Kaleidoscope, A Man for All Seasons, Sebastian, The Killing of Sister George, Oh What a Lovely War, The Battle of Britain, Lock Up Your Daughters, They Shoot Horses Don't They? (Acad. Award nom.), Brotherly Love, Zee & Co., Happy Birthday Wanda June, Images, The Maids, Gold, Conduct Unbecoming, Heaven Save Us From Our Friends, Sky Riders, The Silent Partner, Superman, The Shout, Superman II, Falling in Love Again, Alice Loophole, The Awakening.
TV: Slaughter of St. Teresa's Day, Kiss On A Grass Green Pillow, Fallen Angles, Prince Regent, Second Chance.
STAGE: A Cheap Bunch of Flowers, Wings of the Dove, A Singular Man, Man and Superman, Mrs. Warren's Profession, Peter Pan, Appearances.

YORKIN, BUD: Producer-director. r.n. Alan "Bud" Yorkin. b. Washington, Pa., Feb. 22, 1926. e. Carnegie Tech., Columbia U. U.S. Navy, 1942–45; prod., dir., NBC-TV.
TELEVISION: Song at Twilight, Martin & Lewis Show, Abbott and Costello Show, Ritz Bros. Show, Spike Jones Show; writer, prod. dir., Tony Martin Show, 1954–55; dir., George Gobel Show, 1954–55; prod. dir., The Ernie Ford Show 1956–57; owner. co-prod All In The Family; Sanford and Son; Maude; Good Times; What's Happening!!; Carter Country.
PICTURES: Dir. Never Too Late; Divorce American Style; Inspector Clouseau prod.-dir. Start the Revolution Without Me; Thief Who Came to Dinner; co-prod.-dir. Come Blow Your Horn; exec.-prod. The Night They Raided Minsky's; Cold Turkey.

YOSHISAKA, KIYOJI: Executive. b. Shanghai, China 1908. e. Thomas Hambury College, Shanghai. Ent. ind. 1930 RCA Victor of China, 1932 Victor Talking Machine, 1938 Manchuria Talking Machine, 1940 Victor of Japan. Left Photophone Div. of Victor in 1950 for Tokyo Theatre Supply 1950–58; 1959–63 Rhythm Friend Corp.; exec. dir.s Nichior 1964–66; became consultant Nihon Eiga Shizai 1966 and after name changed to Toshiba Photo Phone became mgr. trade. div. 1969 then dir. and gen. mgr. of trade div. in 1970. Retired.

YOUNG, ALAN: Actor. r.n. Angus Young; b. North Shield, Northumberland, England, Nov. 19, 1919. Sign painter; cartoonist, acted first as monologuist at 13 years in Canada; radio comedian 10 yrs. in Canada and U.S.; served in Canadian Navy as sub-lt. 1939–42; wrote, dir. and acted in comedy broadcasts; TV: Alan Young Show, Mister Ed; m.p. debut in Margie 1946.
PICTURES INCLUDE: Chicken Every Sunday, Mr.Belvedere Goes to College, Aaron Slick from Punkin Crick, Androcles and the Lion, Gentlemen Marry Brunettes, Tom Thumb, Time Machine.

YOUNG, BUDDY: Executive. b. New York City, June 15, 1935; e. CCNY, UA publicity dept. 1952; asst. pub. mgr., UA 1963;

pub. dir. Fox, Oct. 1965; west coast coordinator of adv. and publicity UA. In 1975 joined Columbia Pictures as worldwide dir. of adv.-pub.-exp. In 1976 named MGM adv.-pub. co-ordinator. Joined Universal Pictures in 1976 as dir. of pub. & promo. promoted to v.p., of advertising, publicity and promotion; 1980, partner in m.p. consulting firm, Powell & Young.

YOUNG, BURT: Actor-Writer. Worked at variety of jobs (boxer, trucker, etc.) before turning to acting and joining Actor's Studio. Appeared in off-Bdwy. plays which led to Hollywood career.
PICTURES INCLUDE: Cinderella Liberty, The Gambler, The Killer Elite, Chinatown, Rocky, The Choirboys, Convoy, Uncle Joe Shannon (actor, s.p.), All the Marbles.
TV: MASH, Baretta, etc.

YOUNG, CARROLL: Writer. b. Cincinnati, Ohio, Oct. 9, 1908; e. St. Xavier Coll. Publicist for Pathe Studios, Fox West Coast Theatres, RKO Studios, MGM, 1930–35; story ed. Sol Lesser Prod., Ernst Lubitsch Prod., 1936–40; asst. to exec. prod., RKO, 1941; U.S. Army Air Force, 1942–44.
PICTURES INCLUDE: story, collab. s.p. Tarzan Triumphs; story, Tarzan's Desert Mystery; story, s.p. Tarzan and Leopard Woman, Tarzan and Mermaids; s.p. many in Jungle Jim series, Hidden City, The Jungle; story, Lost Continent; collab. s.p. Tarzan and the She-Devil; story s.p. Cannibal Attack; collab. story & s.p. Apache Warrior; collab. s.p. She-Devil, The Deerslayer; collab. s.p. Machete. Asst. to prod. Lewis Milestone Prods.

YOUNG, FREDDIE: Cinematographer. b. 1902. Ent. m.p. ind. 1917. Early career as cameraman and directing for Gaumont Film Co., British and Dominion Film Corp., MGM British, Army Training Films, now freelance.
PICTURES INCLUDE: Victoria The Great, Sixty Glorious Years, Nell Gywn, Goodbye Mr. Chips, Edward My Son, Ivanhoe (Oscar Nom.), Bhowani Junction, Greengage Summer, Lawrence of Arabia (Oscar award), Macbeth, Doctor Zhivago (Oscar award), The Deadly Affair, You Only Live Twice, Battle of Britain, Ryan's Daughter (Oscar), Nicholas and Alexandra (Oscar Nom.), Luther, The Tamarind Seed, Great Expectations, The Blue Bird, Permission to Kill, The Man In the Iron Mask, Rough Cut.

YOUNG, HAROLD: President, Harold Young Prod., Inc. e. Columbia U. Film ed., Warner Bros., MGM, First Natl. Studios; co-writer, Brit. & Dominions; superv. ed., Paramount Studios, Paris; also Alexander Korda's Brit. and French prod. units. Dir. The Scarlet Pimpernel, for Korda; then to Hollywood.
PICTURES INCLUDE: Time Bomb, Hi Ya Chum, Mexican Fiesta, The Frozen Ghost, The Three Caballeros (live action parts), I'll Remember April, Jungle Captive, Song of the Sarong, Citizen Saint, Roogie's Bump, Carib Gold.

YOUNG, IRWIN: Executive. b. N.Y.C. e. Perkiomen School, Lehigh Univ., B.S., 1950. Pres., Du Art Film Laboratories, Inc.

YOUNG, LORETTA: Actress. r.n. Gretchen Young; b. Salt Lake City, Utah, Jan. 6, 1913. e. Ramona Convent, Alhambra, Calif. Immaculate Heart College, Hollywood. After small part in Naughty But Nice, lead in Laugh Clown, Laugh. Played in almost 100 films. Won Academy Award best perf. (Farmer's Daughter) 1947; NBC-TV Loretta Young Show, 1953–61; CBS-TV New Loretta Young Show 1962; won TV Acad. Awards, 1954–56–59. Has won innumerable other awards.

YOUNG, MORRIS: Executive. b. South Shields, England, 1919. Ent. m.p. industry 1935. Managing dir. United Artists until 1977. Appt. man. dir. Rank Film Distributors Ltd.

YOUNG ROBERT: Actor. b. Chicago, Ill., Feb. 22, 1907. Star numerous pictures before and after sound; from 1932 in many productions, various Hollywood producers; star. weekly TV show, Father Knows Best.
PICTURES INCLUDE: Claudia, Sweet Rosie O'Grady, The Canterville Ghost, Secrets in the Dark, The Enchanted Cottage, Those Endearing Young Charms, Claudia and David, Lady Luck, They Won't Believe Me, Crossfire, Sitting Pretty, Adventures in Baltimore, And Baby Makes Three, Bride for Sale, Second Woman, That Forsyte Woman, Goodbye My Fancy, Half-Breed, Secret of the Incas.
TV: Marcus Welby, M.D.

YOUNG, TERENCE: Director, writer. b. Shanghai, China, June 20, 1915; e. Cambridge U. Served with Guards Armoured Div. WWII; ent. m.p. ind. 1936 at BIP Studios.
PICTURES INCLUDE: On the Night of the Fire, On Approval, Dangerous Moonlight; Theirs is the Glory, dir., Corridor of Mirrors, One Night With You, Woman Hater, They Were Not Divided, The Valley of the Eagles, Red Beret (Paratrooper), That Lady, Safari, Storm Over the Nile, Action of the Tiger, Serious Charge, Black Tights, Dr. No, From Russia With Love, Moll Flanders, Thunderball, Triple Cross, The Rover, Wait Until Dark, Mayerling, The Christmas Tree, The Red Sun, Grand Slam, The Valachi Papers, War Goddess, The Klansman, Jackpot, Bloodline.

YOUNGSTEIN, MAX E.: Executive, b. New York City, March 21, 1913; e. Fordham, U., LL.B.; Brooklyn Law School, LL.M.; Business mgr., Hal Horne Org., 1941–42; asst. dir. adv., pub. & exploit. 20th Fox, 1941–3; asst. dir. studio special services dept. 20th Fox, 1943. Pub. dir., Motion Picture & Special events section. War Finance Div., U.S. Treas., 1944–5. Dir., ad. pub. and exploitat., 1945–46, exec. v.p., Stanley Kramer Story Prods.; Eagle Lion Films, 1946. Named v.p. in charge adv., pub. and exploitat., 1949. Res. May, 1949 to become dir. adv., pub. exp. & member exec. com. Paramount Pictures; v.p. & dir. Paramount Film Dist. Corp., June, 1950; v.p. & nat'l dir. adv. pub. & partner United Artists, Feb. 26, 1951; appointed gen. v.p., member of board, 1955. Pres. United Artists Records & United Artists Music, 1957–62; exec. vice pres. Cinerama, Inc., pres. Cinemiracle, Jan. 1962, indep. producer, 1972; President, Max E. Youngstein Enterprises Inc.; exec. prod., Best of Cinerama, Man In The Middle; prod. Fail Safe, The Money Trap, Co-Prep. The Dangerous Days of Kiowa Jones, Welcome to Hard Times, Prod. Young Billy Young. 1972, v.p. Todd-AO Corp.; 1973, pres., Taylor-Laughlin Dist. Co.; senior v.p., Billy Jack Prods. National Student Film Corp.; v.p. Billy Jack Records & Publishing. 1975, Distrib., prod., merchandising, and marketing consultant to Filmation, Optimus Productions, Moonrunners Ltd., AFC Dist. Co., Grey Advertising., and Geo. Litto Productions. Consultant to New World Pictures on Tidal Wave; to High Key Limited Partnership on Big Mo; to Sequoia Pictures on Stunts; to Shirley MacLaine Enterprises. 1976, appointed general film consultant to Reader's Digest. Consultant to the Barry Film Company on Suzanne. Consultant to George Litto Productions on Obsession and Drive-In. 1977 consultant to Bart-Palevsky productions on Fun with Dick and Jane and Island in the Stream. Advisor on The Amsterdam Kill. Consultant to Rico-Lion for Olly, Olly Oxen Free with Katherine Hepburn. Consultant to Golden Harvest International Ltd., Hong Kong (Paragon Films, Golden Films). Co-executive producer of the Diahann Carroll Television Specials. 1978, consultant George Peppard's Five Days From Home; Robert Mitchum's Amsterdam Kill and Power Play with Peter O'Toole. 1979, consultant Roy Disney's Shamrock Productions Pacific High, Rank Film Distributors, Taft Broadcasting (Hanna-Barbera), Encore Productions' Celebration, Ralph Waite's On The Nickel, Bobrun Productions' Running and Selkirk Films' "Bear Island." 1980–81, bd. chm. & chief exec. officer, Taft International Pictures, Inc. Member of the Board of Governors of the Producers Guild. Lecturer on motion pictures at the University of Southern California, Stanford University and Dartmouth University. Member of the Motion Picture Pioneers Foundation, and one of the founders of the Children's Asthmatic and Research Home in Denver, Colorado. Member of the Board of Governors of Albert Einstein College of Medicine in New York City.

Z

ZAMPA, LUIGI: Director. b. Rome, Italy, 1905. Playwright studied at Experimental Film Center, Rome, 1935–38; script writer, director.
PICTURES INCLUDE: American on Vacation, To Live in Peace, Difficult Years, Angelina, The White Line, City on Trial, Two Gentlemen in a Carriage, His Last 12 Hours, We Women, Woman of Rome, Art of Getting Along.

ZAMPI, GIULIO: Associate producer. b. London, Eng., Sept. 29, 1923; p. Mario Zampi; e. Rome U. Ed. to Marcel Varnel, then to Mario Zampi; dir. of Anglofilm Ltd. & Transocean Films Ltd.
PICTURES INCLUDE: Phantom Shot, Fatal Night, Shadow of the Past, Come Dance with Me, Third Time Lucky, Laughter in Paradise, Top Secret, I Have Chosen Love, Happy Ever After (Tonight's the Night), Now and Forever, The Naked Truth, Too Many Crooks, Bottoms Up, Five Golden Hours.

ZANUCK, RICHARD DARRYL: Executive. b. L.A., Calif., Dec. 13, 1934; e. Stanford U. 1952–56; f. Darryl Zanuck. Story dept., 20th Century Fox, 1954; N.Y pub. dept., 1955; asst. to prod., Island in the Sun, 1957; The Sun Also Rises, 1956; v.p., Darryl F. Zanuck Prod. 1958; prod., Compulsion, 1959, DFZ Prod.; 20th Century Fox; prod., Sanctuary, 1961; prod., The Chapman Report, 1962; asst. to prod., The Longest Day, 1962; president's prod. rep., 20th Century Fox Studio 1963; v.p. charge prod., 20th Fox; pres., 20th Fox TV exec. v.p. chge. prod., 20th Fox, 1967. 1968: Chmn. of Bd., Television div., 20th Century Fox, 1969: Pres., 20th Century Fox Film Corp. Joined Warner Bros., March 1971, as sr. exec. v.p.; Resigned July, 1972 to form Zanuck-Brown Production Company, Universal Pictures.

Films under Z/B Company released: "Sssssss," 1973; "The Sugarland Express," 1974; "Willie Dynamite," 1973; "The Sting," 1973; "The Black Windmill," 1974, The Girl from Petrovka, 1974; Jaws, 1975; Jaws 2, 1978; The Island, 1980; Neighbors, 1981.

ZEFFIRELLI, FRANCO: Director. b. Italy, 1922. Was stage director before entering film industry.
PICTURES INCLUDE: The Taming of the Shrew, Romeo and Juliet, Brother Sun, Sister Moon, The Champ, Endless Love.

ZEMECKIS, ROBERT: Director-Writer. b. Chicago 1952. e. USC's School of Cinema. At USC wrote, produced and directed a 14-minute film, A Field of Honor, which won special jury award at Second Annual Student Film Awards sponsored by M.P. Academy of Arts & Sciences, plus 15 international honors. Has film editing background, having worked as cutter on TV commercials in Illinois. Also cut films at NBC News, Chicago, as summer job. After schooling went to Universal to observe on set of TV series, McCloud. Wrote script for that series in collaboration with Bob Gale. Turned to feature films, directing I Wanna Hold Your Hand and co-writing s.p. with Gale and co-writing 1941 with him.

ZENS, WILL: Producer-director. b. Milwaukee, Wis., June 26, 1920. e. Marquette Univ. and Univ. of Southern Calif. BA, MA. Wrote, produced and directed many TV shows. Formed Riviera Productions in 1960 to produce theatrical motion pictures.
TV: Punch & Trudy, Your Police, Aqua Lung Adventures.
PICTURES INCLUDE: Capture That Capsule, The Starfighters, To the Shores of Hell, Road to Nashville, Hell on Wheels, From Nashville with Music, Yankee Station, Help Me . . . I'm Possessed!

ZETTERLING, MAI: Actress, director. b. Sweden, May 24, 1925. e. Ordtuery Sch. & Theater Sch. First m.p.: Sweden, Frenzy. Has made numerous stage and screen appearances since in Sweden. British screen debut, Frieda. Since 1969 directing plays & films in Sweden. Won Golden Lion at Venice in 1964 for The War Game.
PICTURES INCLUDE: Bad Lord Byron, Quartet, Portrait from Life, Romantic Age, Blackmailed, Hell is Sold Out, Desperate Moment, Knock on Wood, Dance Little Lady, Prize of Gold, Seven Waves Away (Abandon Ship), The Truth About Women, Jetstorm, Faces in the Dark, Piccadilly Third Stop, Offbeat, Only Two Can Play, The Main Attraction. 1965–66 wrote, dir. Loving Couples, Night Games. 1968: Dr. Glas, The Girls, The Rain Hat, We Have Many Names.
TV: Idiot's Delight, Mayerling, Doll's House, Dance of Death, etc. dir. doc. for BBC and in Sweden.

ZIEFF, HOWARD: Director. Started as artist and photographer, working as newsreel photographer for L.A. TV station. Went to N.Y. to do still photography; became top photo artist in advertising. Turned to film direction with Slither in 1972.
PICTURES INCLUDE: Slither, Hearts of the West, House Calls, The Main Event.

ZIFFER, FELIX C.: Lawyer. b. Staniscawow, Poland. Aug. 8, 1930. e. Yale U., 1947–51; Columbia U. Law School, 1952–55; law practice, since 1955; rep. many firms in music & record fields inc. United Artists (spec. counsel), J. Arthur Rank Organization (gen. counsel); pres., International Entertainment Corp.

ZIFKIN, WALTER: Executive. b. July 16, 1936. New York City. e. UCLA A.B. 1958; USC LL.B. 1961. CBS legal dept., 1961–63; William Morris Agency 1963-present; exec. vice-pres. WMA 1974.

ZIGMOND, JERRY: Exhibitor. b. Denver, Colo.; e. U. of Denver, and Chicago; m. Joined Paramount 1926; managed theatres in Los Angeles and Denver; later with J. H. Cooper Circuit, Lincoln, Neb.; managing dir., Newman Theatre, Kansas City; then exec. theatre dept., Paramount Theatres, home office; in charge of Paramount San Francisco Theatres; div. mgr. West Coast, United Paramount, L.A. ABC-TV dir. of Promotional Activities for M.P. Acad. Awards Telecast (1961–64) and ABC-TV dir. of exploitation, Western div.

ZIMBALIST, EFREM, JR.: Actor. b. New York, Nov. 30, 1923. Fay School, Southboro, Mass., St. Paul's, Concord, N.H., Yale. Studied drama, Neighborhood Playhouse, N.Y. Stage debut, The Rugged Path. Shows with American Repertory Theatre; Henry VIII, Androcles and the Lion, What Every Woman Knows, Yellow Jack, Appeared, Hedda Gabler. Co-prod., The Medium, The Telephone, The Consul, (Critics Award, Pulitzer Prize) Barrymore Theatre. m.p. debut, House of Strangers. Gave up acting after death of his wife and served as asst. to father, Curtis Institute of Music for 4 yrs. Returned to acting, stock co., Hammonton, N.J., 1954. Appeared in TV series.
TV: Philco, Goodyear Playhouse, U.S. Steel Hour, Maverick, 77 Sunset Strip, The FBI.
PICTURES INCLUDE: Bomber B-52, Band of Angels, The Deep Six, Violent Road, Girl on the Run, Too Much Too Soon, Home Before Dark, The Crowded Sky, A Fever in the Blood, By Love Possessed, Chapman Report, The Reward, Harlow (electronovision), Airport 1975.

ZINNEMAN, FRED: Director. b. Vienna, Australia, Apr. 29, 1907; e. Vienna U. (law). Studied violin as a boy; after law, studied photographic technique, lighting & mechanics (Paris); asst. cameraman 1 yr. Paris; came to U.S. 1929; extra in m.p. All Quiet on the Western Front, 1930; asst. to Berthold Viertel, script clerk & asst. to Robert Flaherty, 1931; dir. Mexican documentary The Wave; short subjects dir., MGM, winning Academy Award for That Mothers Might Live, 1938; feature dir. 1941; winner of first Screen Directors' Award 1948 with The Search. 4 N.Y. Film Critics Awards; 2 Director's Guild Annual Awards; 3 Acad. Awards.
PICTURES INCLUDE: The Seventh Cross, The Search, The Men, Teresa, High Noon (N.Y. Film Critics Award), Benjy, short for L.A. Orthopedic Hosp. for which Academy Award, best doc. short, 1951; Member of the Wedding, From Here to Eternity (Acad. Award, best dir., 1953, N.Y. Film Critics Award, Directors Guild Award), Oklahoma, Hatful of Rain, Nun's Story (N.Y. Film Critics Award), Sundowners, Behold a Pale Horse, A Man for All Seasons, Day of the Jackal, Julia.

ZINNEMANN, TIM: Producer. Began career in industry as film editor; then asst. dir. on several films. Production mgr. for several projects; assoc. prod. on The Cowboys and Smile. Produced Straight Time for Warners with Stanley Beck.
PICTURES: A Small Circle of Friends, The Long Riders.
TELEVISON: The Jericho Mile (ABC).

ZOUARY, MAURICE H.: Executive. b. Brooklyn, N.Y., July 17, 1921; e. Sch. of Industrial Design & Art, 1937; Trans-Lux Theatres, 1941; U.S. Armed Forces, 1943; production, acct. exec. ad agencies; Bud Gamble Prod., 1948; TV dept., Edward S. Kellogg Agency, L.A.; formed Zouary TV-Film Prod., 1950; prod. supvr. Films for Industry, 1950–52; TV prod., Grey Adv. Agency, 1952–54; prod. new commercials div., Guild Films, 1955–56; prod. Dore Prod., 1956–57; reactivated Zouary TV Film Prod., 1957; formed Film-video Releasing Corp. film stock shot library, 1957. Prod.: Freedom (feature film); Prod. Kiddie Camera (TV series); Prod.: The Vaudevillains (TV special); Prod.: Dr. DeForest. In 1975 formed TV National Releasing Corp., supplier of TV programming.

ZSIGMOND, VILMOS: Cinematographer. b. Hungary. e. National Film School. Began career photographing Hungarian Revolution of 1956. Later escaped from Hungary with friend Laszlo Kovacs, also a cinematographer.
PICTURES: McCabe and Mrs. Miller, Images, Deliverance, Scarecrow, The Long Goodbye, Cinderella Liberty, Sugarland Express, Obsession, Close Encounters of the Third Kind, The Deer Hunter, Heaven's Gate.

ZUGSMITH, ALBERT: Producer, director, writer. b. Atlantic City, N.J., Apr. 24, 1910. e. U. of Va. Pres. Intercontinental Broadcasting Corp.; ed. publ. Atlantic City Daily World; v.p. Smith Davis Corp.; Chrmn. of bd., Continental Telecasting Corp., Television Corp. of America; assoc. ed. American Press; pres. World Printing Co.; exec. CBS; pres., American Pictures Corp.
PICTURES INCLUDE: Written on the Wind, Man in the Shadow, Red Sundown, Star in the Dust, Tarnished Angels, The Incredible Shrinking Man, The Girl in the Kremlin, The Square Jungle, Female on the Beach, Touch of Evil, Captive Women, Sword of Venus, Port Sinister; Invasion U.S.A., Top Banana, Paris Model, Slaughter on Tenth Avenue, The Female Animal, High School Confidential, Night of the Quarter Moon, Beat Generation, The Big Operator, Girls Town, Platinum High School, Private Lives of Adam and Eve, Dondi, College Confidential, Confessions of an Opium Eater, The Great Space Adventure, On Her Bed of Roses, Fanny Hill, The Rapist! author, Private Lives of Adam and Eve, The Beat Generation, How to Break Into the Movies, The Chinese Room, Street Girl, The President's Girl Friend, The Phantom Gunslinger, Sappho, Darling, Menage A Trois, Two Roses and a Goldenrod, The Friend Neighbors, Why Me, God?, Tom Jones Rides Again, etc.

Pictures

*CREDITS FOR FEATURE FILMS, SEPTEMBER, 1980–AUGUST, 1981

*FEATURE RELEASES, 1955–1980

CREDITS FOR FEATURE FILMS

September, 1980–August, 1981

Major credits for feature films screened for the trade press in New York and reviewed in Quigley Publications' newsletter, Motion Picture Product Digest.

All Night Long .... Universal

Producers: Leonard Goldberg and Jerry Weintraub. Director: Jean-Claude Tramont. Screenplay: W.D. Richter. In color. Running time: 99 minutes. MPAA Rating: R. Release date: March 6, 1981.
Players: Gene Hackman, Barbra Streisand, Diane Ladd, Dennis Quaid, Kevin Dobson, William Daniels, Ann Doran, Jim Nolan.

Altered States ....Warner Bros.

Executive Producer: Daniel Melnick. Producer: Howard Gottfried. Director: Ken Russell. Screenplay: Sidney Aaron, from the novel of the same name by Paddy Chayefsky. In color. Running time: 105 minutes. MPAA Rating: R. Release date: December, 1980.
Players: William Hurt, Blair Brown, Bob Balaban, Charles Haid, Thaao Penghlis, Miguel Godreau, Dori Brenner, Peter Brandon, Charles White Eagle.

An American Werewolf in London .... Universal

A Lycanthrope Films Limited production for PolyGram Pictures. Producer: George Folsey, Jr. Executive Producers: Peter Guber and Jon Peters. Director-Writer: John Landis. Music: Elmer Bernstein. Cinematography: Robert Paynter. Editor: Malcolm Campbell. In color. Running time: 97 minutes. MPAA Rating: R. Release date: August 21, 1981.
Players: David Naughton, Jenny Agutter, Griffin Dunne, John Woodvine, Brian Glover, David Schofield, Lila Kane, Paul Kember, Don McKillop, Frank Oz.

Any Which Way You Can ....

Executive Producer: Robert Daley. Producer: Fritz Manes. Director: Buddy Van Horn. Screenplay: Stanford Sherman, based on characters created by Jeremy Joe Kronsberg. In color. Running time: 115 minutes. MPAA Rating: PG. Release date: December 17, 1980.
Players: Clint Eastwood, Sondra Locke, Geoffrey Lewis, William Smith, Harry Guardino, Ruth Gordon, Michael Cavanaugh, Barry Cobin, Roy Jenson, Bill McKinney, William O'Connell, John Quade, Al Ruscio.

American Pop .... Columbia

Bakshi Productions/Aspen Productions. Producers: Martin Ransohoff and Ralph Bakshi. Director: Mr. Bakshi. Screenplay: Ronnie Kern. In color. Running time: 98 minutes. MPAA Rating: R. Release date: February, 1981.
Players: Featuring the Voices of Ron Thompson, Marya Small, Jerry Holland, Lisa Jane Persky, Jeffrey Lippa, Roz Kelly.

Arthur ....Warner Bros.

An Orion Pictures release. Producer: Robert Greenhut. Executive Producer: Charles H. Joffe. Writer-Director: Steve Gordon. Music: Burt Bacharach. Cinematography: Fred Schuler. Editor: Susan E. Morse. In color. Running time: 96 minutes. MPAA Rating: PG. Release date: July 17, 1981.
Players: Dudley Moore, Liza Minnelli, John Gielgud, Geraldine Fitzgerald, Jill Eikenberry, Stephen Elliott, Red Ross, Barney Martin, Thomas Barbour, Anne de Salvo, Marjorie Barnes.

Atlantic City ....Paramount

International Cinema Corp. A John Kemeny—Denis Heroux production. Executive Producers: Joseph Beaubien and Gabriel Boustant. Producer: Mr. Heroux. Director: Louis Malle. Screenplay: John Guare. In color. Running time: 105 minutes. MPAA Rating: Release date: April, 1981.
Players: Burt Lancaster, Susan Sarandon, Kate Reid, Michel Piccoli, Hollis McLaren, Robert Joy, Al Waxman, Robert Goulet, Moses Znaimer, Harvey Atkin, Eleanor Beecroft, Norma Dell'Agnese, Louis Del Grande.

The Awakening ....Warner Bros.

Orion Pictures Company. Producer: Robert Solo. Director: Mike Newell. Screenplay: Allan Scott, Chris Bryant and Clive Exton, based on the novel, "The Jewel of the Seven Stars," by Bram Stoker. In color. Running time: 100 minutes. MPAA Rating: R. Release date: October, 1980.
Players: Charlton Heston, Susannah York, Jill Townsend, Patrick Drury, Bruce Myers, Nadim Sawalha, Miriam Maygolyes, Ian.

Back Roads ....Warner Bros.

A CBS Theatrical Films presentation. Meta-Films Associates, Inc. Produced in association with Marion Rosenberg. Producer: Ronald Shedlo. Director: Martin Ritt. Screenplay: Gary DeVore. In Panavision and color. Running time: 94 minutes. MPAA Rating: R. Release date: March, 1981.
Players: Sally Field, Tommy Lee Jones, David Keith, Miriam Colon, Michael Gazzo, Dan Shor, M. Emmet Walsh, Barbara Babcock, Nell Carter, Alex Colon, Lee de Broux, Ralph Seymour, Royce Applegate.

Bad Timing/A Sensual Obsession ....World Northal

A Rank Organization Production. A Sondra Gilman/Louise Westergaard presentation. Producer: Jeremy Thomas. Director: Nicolas Roeg. Screenplay: Yale Udoff. In color. Running time: 122 minutes. MPAA Rating: X. Release date: October, 1980.
Players: Art Garfunkel, Theresa Russell, Harvey Keitel, Denholm Elliott, Daniel Massey, Dana Gillespie.

Beyond the Reef .... Universal

A Dino DeLaurentiis Co. production. Producer: Raffaella DeLaurentiis. Director: Frank C. Clark. Screenplay: Louis LaRusso II and Jim Carabatsos, from the novel by Clement Richer. In color. Running time: 91 minutes. MPAA Rating: PG. Release date: May, 1981.
Players: Dayton Ka'ne, Maren Jensen, Kathleeen Swan, Keahi Farden, Oliverio Maciel Diaz, Joseph Ka'ne.

Blow Out .... Filmways

Producer: George Litto. Writer-Director: Brian De Palma. Music: Pino Donaggio. Cinematography: Vilmos Zsigmond. Editor: Paul Hirsch. In color. Running time: 108 minutes. MPAA Rating: R. Release date: July 24, 1981.
Players: John Travolta, Nancy Allen, John Lithgow, Dennis Franz, Peter Boyden, Curt May, Ernest McClure, Dave Roberts, Maurice Copenald, Claire Carter, John Aquino, John Hoffmeister, Patrick McNamara, Terrence Currier, Tom McCarthy, Dean Bennett.

Body Heat .... Warner Bros.

A Ladd Company release. Producer: Fred T. Gallo. Director-Writer: Lawrence Kasdan. Music: John Barry. Cinematography: Richard H. Kline. Editor: Carol Littleton. In color. Running time: 113 minutes. MPAA Rating: R. Release date: August 28, 1981.

Players: William Hurt, Kathleen Turner, Richard Crenna, Ted Danson, J.A. Preston, Mickey Rourke, Kim Zimmer, Jane Hallaren, Lanna Saunders, Carola McGuinness, Michael Ryan, Larry Marko.

Borderline .... Associated Film Dist.

A Lord Grade presentation. Producer: James Nelson. Director: Jerrold Freedman. Screenplay: Steve Kline and Mr. Freedman. In Panavision and color. Running time: 106 minutes. MPAA Rating: PG. Release date: October, 1980.

Players: Charles Bronson, Bruno Kirby, Bert Remsen, Michael Lerner, Kenneth McMillan, Ed Harris, Karmin Murcelo, Enrique Castillo, A. Wilford Brimley, Norman Alden, James Victor, Panchito Gomez, John Ashton, Lawrence Casey.

Breaker Morant
.... New World Pictures/Quartet Films

A South Australian Film Corp. presentation. Producer: Matt Carroll. Director: Bruce Beresford. Screenplay: Mr. Beresford, Jonathan Hardy, and David Stevens, from a play by Kenneth Ross. In color. Running time: 108 minutes. MPAA Rating: PG. Release date: February, 1981.

Players: Edward Woodward, Jack Thompson, John Waters, Bryan Brown, Charles Tingwell, Frank Wilson, Terence Donovan, Vincent Ball, Lewis Fitzgerald, Rod Mullinar.

Bustin' Loose .... Universal

A Richard Pryor Production. Producers: Richard Pryor and Michael S. Glick. Director: Oz Scott. Screenplay: Roger L. Simon. Adaptation: Lonne Elder, III. Story: Richard Pryor. In color. Running time: 94 minutes. MPAA Rating: R. Release date: May, 1981.

Players: Richard Pryor, Cicely Tyson, Robert Christian, George Coe, Bill Quinn, Roy Jenson, Fred Carney, Peggy McCay, Alphonso Alexander, Kia Cooper, Edwin DeLeon, Jimmy Hughes, Edwin Kinter, Tami Luchow, Angel Ramirez, Janet Wong.

Caddie .... Atlantic Releasing Corp.

Australian Film Commission. Producer: Anthony Buckley. Director: Donald Crombie. Screenplay: Joan Long, based on the book, "Caddie, A Sydney Barmaid—an autobiography written by herself." In color. Running time: 107 minutes. No MPAA Rating. Release date: February, 1981.

Players: Helen Morse, Takis Emmanuel, Jack Thompson, Jacki Weaver, Melissa Jaffer, Ron Blanchard, Drew Forsythe, Kirrili Nolan, Lynette Currin.

The Cannonball Run ...... 20th Century-Fox

A Golden Harvest presentation. Producer: Albert S. Ruddy. Executive Producer: Raymond Chow. Director: Hal Needham. Screenplay: Brock Yates. Music: Al Capps. Cinematography: Michael Butler. Editors: Donn Cambern and William D. Gordean. In color. Running time: 95 minutes. MPAA Rating: PG. Release date: June, 1981.

Players: Burt Reynolds, Roger Moore, Farrah Fawcett, Dom DeLuise, Dean Martin, Sammy Davis, Jr., Jack Elam, Adrienne Barbeau, Terry Bradshaw.

Cattle Annie and Little Britches .. Universal

A King-Hitzig Production. Producers: Rupert Hitzig and Alan King. Executive Producers: John Daly and Derek X. Dawson. Director: Lamont Johnson. Screenplay: David Eyre and Robert Ward, based on a screen story by Ward from his novel. In color. Running time: 97 minutes. MPAA Rating: PG. Release date: May, 1981.

Players: Burt Lancaster, John Savage, Rod Steiger, Diane Lane, Amanda Plummer, Scott Glenn, Redmond Gleeson, William Russ, Ken Call, Buck Taylor, John Quade.

Caveman .... United Artists

A Turman-Foster Company production. Producers: Lawrence Turman and David Foster. Director: Carl Gottlieb. Screenplay: Rudy De Luca and Mr. Gottlieb. In color. Running time: 91 minutes. MPAA Rating: PG. Release date: April, 1981.

Players: Ringo Starr, Barbara Bach, Dennis Quaid, Shelley Long, John Matuszak, Avery Schreiber, Jack Gilford.

A Change of Seasons ..... 20th Century-Fox

A Film Finance Group presentation. Executive Producer: Richard R. St. Johns. Producer: Martin Ransohoff. Director: Richard Lang. Screenplay: Erich Segal, Ronni Kern, and Fred Segal, based on a story by Erich Segal. In color. Running time: 102 minutes. MPAA Rating: R. Release date: December, 1980.

Players: Shirley MacLaine, Anthony Hopkins, Bo Derek, Michael Brandon, Mary Beth Hurt, Ed Winter, Paul Regina.

Cheech and Chong's
Nice Dreams
.... Columbia

A C. & C. Brown Production. Producer: Howard Brown. Director: Thomas Chong. Screenplay: Mr. Chong and Richard "Cheech" Marin. Music: Harry Betts. Cinematography: Charles Correll. Editors: Tom Avildsen and Tony Lombardo. In color. Running time: 97 minutes. MPAA Rating: R. Release date: June 5, 1981. Players: Thomas Chong, Richard "Cheech" Marin, Stacy Keach, George Craig Brown, Jimmy Fame, Evelyn Guerrero, Peter Jason, Suzanne Kent, Michael Masters, Paul Reubens, Tim Rossovich, Dr. Tomothy Leary, James Faracci, Jeff Pomerantz.

Chu Chu and the Philly Flash
.... 20th Century-Fox

A Jay Weston production from Melvin Simon Productions. Producer: Mr. Weston. Director: David Lowell Rich. Screenplay: Barbara Dana from a story by Henry Barrow. Muic: Pete Rugolo. Cinematography: Victor J. Kemper. Editor: Argyle Nelson. In color. Running time: 100 minutes. MPAA Rating: PG. Release date: August, 1981.

Players: Alan Arkin, Carol Burnett, Jack Warden, Danny Aiello, Adam Arkin, Danny Glover, Sid Haig, Vincent Schiavelli, Ruth Buzzi, Vito Scotti, Lou Jacobi.

City of Women .... New Yorker Films

Gaumont/Opera production. Director: Federico Fellini. Screenplay: Fellini, Bernardino Zappoini and Brunello Rondi. In color. Running time: 138 minutes No MPAA Rating. Release date: April, 1981.

Players: Marcello Mastroianni, Ettore Manni, Anna Brucnal, Bernice Stegers, Donatella Maniani, Sara Tafuri, Jole Silvani, Carla Terizzi, Katren Gabelein.

Clash of the Titans .... United Artists

A Metro-Goldwyn-Mayer presentation. Producers: Charles H. Schneer and Ray Harryhausen. Director: Desmond Davis, Screenplay: Beverley Cross. Music: Laurence Rosenthal. Cinematography: Ted Moore. Editor: Timothy Gee. In color. Running time: 118 minutes. MPAA Rating: PG. Release date: June, 1981.

Players: Harry Hamlin, Judi Bowker, Burgess Meredith, Maggie Smith, Ursula Andress, Claire Bloom, Sian Phillips, Flora Robson, and Laurence Olivier as Zeus.

Coast to Coast .... Paramount

A Steve Tisch—Jon Avnet production. Director: Joseph Sargent. Screenplay: Stanley Weiser. In color. Running time: 95 minutes. MPAA Rating: PG. Release date: October 3, 1980.

Players: Dyan Cannon Robert Blake, Quinn Redeker, Michael Lerner, Maxine Stuart, Bill Lucking, Ellen Gerstein, Patricia Conklin, David Moody.

The Competition .... Columbia

A Rastar/William Sackheim production. Producer: Mr. Sackheim. Director: Joel Oliansky. Screenplay: Mr. Oliansky, from a story by himself and Mr. Sackheim. In color. Running time: 129 minutes. MPAA Rating: PG. Release date: December, 1980.

Players: Richard Dreyfuss, Amy Irving, Lee Remick, Sam Wanamaker, Joseph Cali, Ty Henderson, Vickie Kriegler, Adam Stern, Bea Silvern, Philip Sterling, Gloria Stroock, Delia Salvi.

Condorman .... Buena Vista

Walt Disney Productions. Executive Producer: Ron Miller. Producer: Jan Williams. Director: Charles Jarrott. Screenplay: Marc Stirdivant, suggested by "The Game of X" by Robert Sheckley. Music: Henry Mancini. Cinematography: Charles F. Wheeler. Editor: Gordon D. Brenner. In Panavision and color. Running time: 90 minutes. MPAA Rating: PG. Release date: August, 1981.

Players: Michael Crawford, Oliver Reed, Barbara Carrera, James Hampton, Jean-Pierre Kalfon, Dana Elcar, Vernon Dobtcheff, Robert Arden.

Cutter and Bone .... United Artists

A Gurian Entertainment Production. Producer: Paul R. Gurian. Director: Ivan Passer. Screenplay: Jeffrey Alan Fiskin, based on the novel by Newton Thornburg. In color. Running time: 109 minutes. MPAA Rating: R. Release date: March, 1981.

Players: Jeff Bridges, John Heard, Lisa Eichhorn, Ann Dusenberry, Stephen Elliott, Arthur Rosenberg, Nina Van Pallandt, Patricia Donahue, Geraldine Baron.

Deadly Blessing .... United Artists

An InterPlanetary Production from PolyGram Pictures. Producers: Micheline and Max Keller and Pat Herskovic. Director: Wes Craven. Screenplay: Glenn M. Benest, Matthew Barr and Mr. Craven, based on a story by Benest and Barr. Music: James Horner. Cinematography: Robert Jessup. Editor: Richard Bracken. In color. Running time: 102 minutes. MPAA Rating: R. Release date: August 14, 1981.

Players: Maren Jensen, Susan Buckner, Sharon Stone, Jeff East, Lisa Hartman, Lois Nettleton, Ernest Borgnine, Coleen Riley.

Death Hunt ....20th Century-Fox

An Albert S. Ruddy Production. Producer: Murray Shostak. Director: Peter Hunt. Screenplay: Michael Grais and Mark Victor. In color. Running time: 96 minutes. MPAA Rating: R. Release date: May, 1981.

Players: Charles Bronson, Lee Marvin, Andrew Stevens, Carl Weathers, Ed Lauter, Scott Hylands, Angie Dickinson, Henry Beckman, William Sanderson, John Cedar, James O'Connell, Len Lesser, Dick Davalos.

The Devil and Max Devlin ......Buena Vista

A Walt Disney production. Producer: Jerome Courtland. Director: Steven Hilliard Stern. Screenplay: Mary Rodgers. In color. Running time: 96 minutes. MPAA Rating: PG. Release date: February, 1981.

Players: Elliott Gould, Bill Cosby, Susan Anspach, Adam Rich, Julie Budd, David Knell.

Dirty Tricks .... Avco Embassy

A Filmplan International production. Executive Producers: Victor Solnicki, Pierre David, and Arnold Kopelson. Producer: Claude Heroux. Director: Alvin Rakoff. Screenplay: William Norton, Sr., Eleanor Elias Norton, Thomas Gifford and Camille Gifford, from a novel by Thomas Gifford. In color. Running time: 91 minutes. MPAA Rating: PG. Release date: May, 1981.

Players: Elliott Gould, Kate Jackson, Arthur Hill, Rich Little, Nick Campbell, Angus McInnes, Michael Kirby, Michael McNamara, Martin McNamara.

Divine Madness ....Warner Bros.

A Ladd Company release. Producer-Director: Michael Ritchie. Executive Producer: Howard Jeffrey. Writers: Jerry Blatt, Bette Midler, and Bruce Vilanch. In Panavision and color. Running time: 94 minutes. MPAA Rating: R. Release date: September, 1980.

Players: Bette Midler and the Harlettes: Jocelyn Brown, Ula Jedwig, Diva Gray.

The Dogs of War .... United Artists

A Norman Jewison/Patrick Palmer production. Producer: Larry DeWaay. Director: John Irvin. Screenplay: Gary DeVore and George Malko, based on a novel by Frederick Forsyth. In Panavision and color. Running time: 101 minutes. MPAA Rating: R. Release date: February, 1981.

Players: Christopher Walken, Tom Berenger, Colin Blakely, Hugh Millais, Paul Freeman, Jean-Francois Stevenin, Jobeth Williams, Robert Urquhart, Winston Ntshona.

Dragonslayer ....Paramount

A Barwood/Robbins Production. Producer: Hal Barwood. Director: Matthew Robbins. Screenplay: Mr. Barwood and Mr. Robbins. Music: Alex North. Cinematography: Derek Vanlint. Editor: Tony Lawson. In Panavision and color. Running time: 110 minutes. MPAA Rating: PG. Release date: June 26, 1981.

Players: Peter MacNicol, Caitlin Clarke, Ralph Richardson, John Hallam, Peter Eyre, Albert Salmi, Sydney Bromley, Chiloe Salaman, Emrys James, Roger Kemp, Ian McDiarmid, Ken Shorter, Jason White.

The Earthling .... Filmways

A Samuel Z. Arkoff presentation. Executive Producer: Stephen W. Sharmat. Producers: Elliot Schick and John Strong. Director: Peter Collinson. Screenplay: Lanny Cotler. In color. Running time: 98 minutes. MPAA Rating: PG. Release date: February, 1981.

Players: William Holden, Ricky Schroder, Jack Thompson, Olivia Hammett, Jane Harders, Alwyn Kurts, Pat Evison, Redmond Phillips, Willie Fennell, Allan Penney, Walter Pym, Harry Neilson, Tony Barry.

The Elephant Man ....Paramount

A Brooksfilms production. Executive Producer: Stuart Cornfeld. Producer: Jonathan Sanger. Director: David Lynch. Screenplay: Christopher DeVore, Eric Bergren, and Mr. Lynch, based on "The Elephant Man and Other Reminiscences" by Sir Frederick Treves and in part on "The Elephant Man: A Study in Human Dignity," by Ashley Montagu. In Panavision and black-and-white. Running time: 125 minutes. MPAA Rating: PG. Release date: October, 1980.

Players: Anthony Hopkins, John Hurt, Anne Bancroft, John Gielgud, Wendy Hiller, Freddie Jones, Michael Elphick, Hannah Gordon, Helen Ryan, John Standing, Dexter Fletcher, Lesley Dunlop, Phoebe Nicholls, Pat Gorman, Claire Davenport.

Endless Love .... Universal

A Polygram Pictures presentation. Producer: Dyson Lovell. Executive Producer: Keith Barish. Director: Franco Zeffirelli. Screenplay: Judith Rascoe, based on the book by Scott Spencer. Music: Jonathan Tunick. Cinematography: David Watkin. Editor: Michael J. Sheridan. In color. Running time: 110 minutes. MPAA Rating: R. Release date: July 17, 1981.

Players: Brooke Shields, Martin Hewitt, Shirley Knight, Don Murray, Richard Kiley, Beatrice Straight, Jimmy Spader, Ian Ziering, Robert Moore, Penelope Milford, Jan Miner.

Escape from New York...... Avco Embassy

A Debra Hill Production. Producers: Larry Franco and Ms. Hill. Director: John Carpenter. Screenplay: Mr. Carpenter and Nick Castle. Music: Mr. Carpenter in association with Alan Howarth. Cinematography: Dean Cundey. Editor: Todd Ramsay. In color. Running time: 102 minutes. MPAA Rating: R. Release date: July, 1981.

Players: Kurt Russell, Lee Van Cleef, Ernest Borgnine, Donald Pleasence, Isaac Hayes, Season Hubley, Harry Dean Stanton, Adrienne Barbeau.

Excalibur .... Warner Bros.

An Orion Pictures presentation. Producer-Director: John Boorman. Screenplay: Rospo Pallenberg and Mr. Boorman, adapted from Malory's "Le Morte D'Arthur" by Mr. Pallenberg. In color. Running time: 140 minutes. MPAA Rating: R. Release date: April 10, 1981.

Players: Nigel Terry, Helen Mirren, Nicholas Clay, Cherie Lunghi, Paul Geoffrey, Nicol Williamson, Robert Addie, Gabriel Byrne, Keith Buckley, Katrine Boorman, Liam Neeson, Corin Redgrave, Niall O'Brien, Patrick Stewart, Clive Swift, Ciarin Hinds, Liam O'Callaghan, Michael Muldoon, Charley Boorman.

Eye of the Needle .... United Artists

A Kings Road production. Producer: Stephen Friedman. Director: Richard Marquand. Screenplay: Stanley Mann, based on the novel by Ken Follett. Music: Miklos Rozsa. Cinematography: Alan Hume. Editor: Sean Barton. In color. Running time: 111 minutes. MPAA Rating: R. Release date: July 24, 1981.

Players: Donald Sutherland, Kate Nelligan, Ian Bannen, Christopher Cazenove, Faith Brook, Barbara Ewing, David Hayman, Alex McCrindle, Philip Martin Brown, Rupert Frazer, Jonathan Nicholas Haley.

Eyes of a Stranger .... Warner Bros.

A Georgetown Production. Producer: Ronald Zerra. Director: Ken Wiederhorn. Screenplay: Mark Jackson and Eric L. Bloom. In Panavision and color. Running time: 85 minutes. MPAA Rating: R. Release date: March 27, 1981.

Players: Lauren Tewes, Jennifer Jason Leight, John DiSanti, Peter DuPre, Gwen Lewis, Kitty Lunn, Timothy Hawkings, Ted Richert.

Eyewitness .... 20th Century-Fox

Producer-Director: Peter Yates. Screenplay: Steve Tesich. In color. Running time: 102 minutes. MPAA Rating: R. Release date: March, 1981.

Players: William Hurt, Sigourney Weaver, Christopher Plummer, James Woods, Irene Worth, Kenneth McMillan, Pamela Reed, Albert Paulsen, Steven Hill, Morgan Freeman. Alice Drummond, Sharon Goldman, Chao-Li Chi.

Fade to Black .... American Cinema

A Leisure Investment Company & Movie Ventures Ltd. production. Executive producers: Irwin Yablans and Sylvio Tabet. Producers: George G. Braunstein and Ron Hamady. Writer-Director: Vernon Zimmerman. In color. Running time: 100 minutes. MPAA Rating: R. Release date: October, 1980.

Players: Dennis Christopher, Tim Thomerson, Gwynne Gilford, Linda Kerridge, Morgan Paull, James Luisi, Eve BrentAshe, John Steadman, Marcie Barkin, Mickey Rourke, Peter Horton, Hennen Chambers, Melinda Fee, Norman Burton.

The Fan .... Paramount

RSO Films, Ltd. Producer: Robert Stigwood. Executive Producer: Kevin McCormick. Director: Edward Bianchi. Screenplay: Priscilla Chapman and John Hartwell, based on the novel by Bob Randall. In color. Running time: 95 minutes. MPAA Rating: R. Release date: May 15, 1981.

Players: Lauren Bacall, James Garner, Maureen Stapleton, Hector Elizondo, Michael Biehn, Anna Maria Horsford, Kurt Johnson, Feiga Martinez, Reed Jones, kaiulani Lee, Charles Blackwell.

Fear No Evil .... Avco Embassy

A LaLoggia Productions picture. Producers: Frank LaLoggia and Charles M. LaLoggia. Director-Writer: Frank LaLoggia. In color. Running time: 99 minutes. MPAA Rating: R. Release date: January, 1981.

Players: Stefan Arngrim, Elizabeth Hoffman, Kathleen Rowe McAllen, Frank Birney, Daniel Eden, Jack Holland, Barry Cooper, Alice Sachs, Paul Haber, Roslyn Gugino.

The Final Conflict .... 20th Century-Fox

A Harvey Bernhard production in association with Mace Neufeld. Producer: Mr. Bernhard. Director: Graham Baker. Screenplay: Andrew Birkin, based on characters created by David Seltzer. In Panavision and color. Running time: 109 minutes. MPAA Rating: R. Release date: March 20, 1981.

Players: Sam Neill, Rossano Brazzi, Don Gordon, Lisa Harrow, Barnaby Holm, Mason Adams, Robert Arden, Tommy Dugan, Leueen Willoughby, Louis Mahoney, Marc Boyle, Richard Oldfield, Milos Kirek, Tony Vogel.

The First Deadly Sin ..... Filmways Pictures

An Elliott Kastner presentation of an Artanis-Cinema VII production. Executive Producers: Frank Sinatra and Mr. Kastner. Producers: George Pappas and Mark Shanker. Director: Brian Hutton, Screenplay: Mann Rubin, based on the novel by Lawrence Sanders. In color. Running time: 112 minutes. MPAA Rating: R. Release date: October, 1980

Players: Frank Sinatra, Faye Dunaway, David Dukes, George Coe, Brenda Vaccaro, James Whitmore, Martin Gabel, Joe Spinell, Jeffrey de Munn, Anthony Zerbe, Fred Fuster.

First Family .... Warner Bros.

An Indie-Prod production. Producer: Daniel Melnick. Director-Writer: Buck Henry. In color. Running time: 97 minutes. MPAA Rating: R. Release date: December 25, 1980.

Players: Gilda Radner, Bob Newhart, Madeline Kahn, Richard Benjamin, Bob Dishy, Harvey Korman, Austin Pendleton, Rip Torn, Fred Willard, Julius Harris, John Hancock, Maurice Sherbanee, Buck Henry, Shatsmi Sarume Dance Troupe.

First Monday in October .... Paramount

A Paul Heller production. Producers: Paul Heller and Martha Scott. Director: Ronald Neame Screenplay: Jerome Lawrence and Robert E. Lee, based on their play. Music: Ian Fraser. Cinematography: Fred J. Koenekamp. Editor: Peter E. Berger. In color. Running time: 95 minutes. MPAA Rating: R. Release date: August 21, 1981.

Players: Walter Matthau, Jill Clayburgh, Barnard Hughes, Jan Sterling, James Stephens, Joshua Bryant, Wiley Harker, F.J. O'Neill, Charles Lampkin, Lew Palter, Richard McMurray, Herb Vigran.

Flash Gordon .... Universal

A Dino De Laurentiis presentation. Director: Mike Hodges. Screenplay: Lorenze Semple, Jr., from an adaptation by Michael Allin, based on characters created by the late Alex Raymond. In color. Running time: 113 minutes. MPAA Rating: PG. Release date: December 5, 1980.

Players: Sam Jones, Melody Anderson, Topol, Max Von Sydow, Ornelia Muti, Timothy Dalton, Brian Blessed, Peter Wyngrade, Mariangela Melato, Richard O'Brien.

For Your Eyes Only .... United Artists

Eon Productions, Ltd. Executive Producer: Michael G. Wilson. Producer: Albert R. Broccoli. Director: John Glen. Screenplay: Richard Maibaum and Mr. Wilson Music: Bill Conti. Cinematography: Alan Hume. Editor: John Grover. In Panavision and color. Running time: 128 minutes. MPAA Rating: PG. Release date: June 26, 1981.

Players: Roger Moore, Carole Bouquet, Topol, Lynn-Holly Johnson, Julian Glover, Cassandra Harris, Jill Bennett, Michael Gothard, John Wyman, Jack Hedley, Louis Maxwell, Desmond Llewelyn, Geoffrey Keen, Walter Gotell, and John Wells and Janet Brown.

The Formula .... United Artists

An MGM presentation. Producer-Writer: Steve Shagan, script based on his own book. Director: John Avildsen. In color. Running time: 117 minutes. MPAA Rating: R. Release date: December, 1980.

Players: George C. Scott, Marlon Brando, Marthe Keller, John Gielgud, G.D. Spradlin, Beatrice Straight, Richard Lynch, John Van Dreelen, Robin Clarke, Ike Eisenmann.

Fort Apache, The Bronx . . . 20th Century-Fox

Time-Life Films. Produced in association with the Producer Circle Co. Executive Producer: David Susskind. Producers: Martin Richards and Gill Champion. Director: Daniel Petrie. Screenplay: Heywood Gould. In color. Running time: 120 minutes. MPAA Rating: R. Release date: February, 1981.

Players: Paul Newman, Edward Asner, Ken Wahl, Danny Aiello, Rachel Ticotin, Pam Grier, Kathleen Beller, Tito Goya, Miguel Pinero, Jaime Tirelli, Lance William Guecia, Ronnie Clanton, Clifford David.

The Four Seasons . . . . Universal

Producer: Martin Bregman. Director-Writer: Alan Alda. In color. Running time: 108 minutes. MPAA Rating: PG. Release date: May 22, 1981.

Players: Alan Alda, Carol Burnett, Len Cariou, Sandy Dennis, Rita Moreno, Jack Weston, Bess Armstrong, Elizabeth Alda, Beatrice Alda, Robert Hitt, Kristi McCarthy, David Stackpole.

The Fox and the Hound . . . . Buena Vista

A Walt Disney production. Producer: Wolfgant Reitherman. Executive Producer: Ron Miller. Directors: Art Stevens, Ted Berman, and Richard Rich. Story: Larry Clemmons, Ted Berman, Peter Young, Steve Hulett, David Michener, Burny Mattinson, Earl Kress, and Vance Gerry, from the book by Daniel P. Mannix. In color. Running time: 83 minutes. MPAA Rating: G. Release date: July, 1981.

Players: Voices of Mickey Rooney, Kurt Russell, Pearl Bailey, Jack Albertson, Sandy Duncan, Jeannette Nolan, Pat Buttram, John Fiedler, John McIntire, Dick Bakalyan, Paul Winchell, Keith Mitchell, Corey Feldman.

Friday the 13th Part II . . . . Paramount

Producers: Steve Miner and Dennis Murphy. Director: Mr. Miner. Screenplay: Ron Kurz, based on characters created by Victor Miller. In color. Running time: 87 minutes. MPAA Rating: R. Release date: May, 1981.

Players: Amy Steel, John Furey, Adrienne King, Kirsten Baker, Stu Charnc, Warrington Gillette, Walt Gorney, Marta Kober, Tom McBride, Bill Randolph, Lauren-Marie Taylor, Russell Todd.

Gallipoli . . . . Paramount

A Robert Stigwood and Rupert Murdoch presentation for Associated R&R Films Pty. Ltd. Producers: Mr. Stigwood and Patricia Lovell. Director: Peter Weir. Screenplay: David Williamson, based on a story by Mr. Weir. Music: Brian May. Cinematography: Russell Boyd. Editor: William Anderson. In Panavision and color. Running time: 110 minutes. MPAA Rating: PG. Release date: August 28, 1981.

Players: Mark Lee, Bill Kerr, Ron Graham, Harold Hopkins, Charles Yunupingu, Heath Harris, Gerda Nicolson, Brian Anderson, Mel Gibson, Robert Grubb, Tim McKenzie.

Gas . . . . Paramount

Producer: Claude Heroux. Executive Producers: Victor Solnicki and Pierre David. Director: Les Rose. Screenplay: Richard Wolf. In color. Running time: 94 minutes. MPAA Rating: R. Release date: July, 1981.

Players: Donald Sutherland, Susan Anspach, Howie Mandel, Sterling Hayden, Sandee Currie, Peter Aykroyd, Keith Knight, Helen Shaver, Alf Humphries.

Gloria . . . . Columbia

Producer: Sam Shaw. Director-Writer: John Cassavetes. In color. Running time: 121 minutes. MPAA Rating: PG. Release date: October, 1980.

Players: Gena Rowlands, Buck Henry, Julie Carmen, John Adames, Tony Knesich, Gregory Cleghorne, Lupe Garnica, Jessica Castillo, Tom Noonan, Ronald Maccone, John Finnegan, Basilio Franchina.

Going Ape . . . . Paramount

A Hemdale Production. Producer: Robert L. Rosen. Director-Writer: Jeremy Joe Kronsberg. In color. Running time: 89 minutes. MPAA Rating: PG. Release date: April 10, 1981.

Players: Tony Danza, Jessica Walter, Stacey Nelkin, Danny De Vito, Art Metrano, Frank Sivero, Rick Hurst, Howard Mann.

The Great Muppet Caper . . . . Universal

A Lord Grade presentation. An Associated Film Distribution Corp. release. Executive Producer: Martin Starger. Producers: David Lazer and Frank Oz. Director: Jim Henson. Screenplay: Tom Patchett, Jay Tarses, Jerry Juhl, and Jack Rose. Music and Lyrics: Joe Raposo. Cinematography: Oswald Morris. Editor: Ralph Kemplen. In color. Running time: 95 minutes. MPAA Rating: G. Release date: June 26, 1981.

Players: Charles Grodin, Diana Rigg, John Cleese, Robert Morley, Peter Ustinov, Jack Warden, and the Muppet Performers by Jim Henson, Frank Oz, Dave Goelz, Jerry Nelson and Richard Hunt.

The Hand . . . . Warner Bros.

An Orion Pictures release. A Jon Peters presentation of an Edward R. Pressman production. Producer: Mr. Pressman. Director-Writer: Oliver Stone based upon the novel, "The Lizard's Tale," by Marc Brandel. In color. Running time: 108 minutes. MPAA Rating: R. Release date: April, 1981.

Players: Michael Caine, Andrea Marcovicci, Annie McEnroe, Viveca Lindfors, Nicholas Hormann, Rosemary Murphy, Dave Maddow, Bruce McGill.

Happy Birthday To Me . . . . Columbia

A John Dunning—Andre Link Production. Director: J. Lee Thompson. Screenplay: John Saxton, Peter Jobin and Timothy Bond, from a story by Mr. Saxton. In color. Running time: 108 minutes. MPAA Rating: R. Release date: May, 1981.

Players: Melissa Sue Anderson, Glenn Ford, Lawrence Dan, Sharon Acker, Frances Hyland, Tracy Bregman, Lisa Langlois.

Hard Country . . . . Universal

An Associated Film Distribution release. Producers: David Greene and Mack Bing. Director: Mr. Greene. Screenplay: Michael Kane. In color. Running time: 104 minutes. MPAA Rating: PG. Release date: March, 1981.

Players: Jan-Michael Vincent, Kim Basinger, Michael Parks, Gailard Sartain, Tanya Tucker, Sierra Pecheur, John Chappell, Daryl Hannah.

Hardly Working . . . . 20th Century-Fox

Producers: James M. McNamara and Igo Kantor. Director: Jerry Lewis. Screenplay: Michael Janover and Mr. Lewis, from a story by the former. In color. Running time: 91 minutes. MPAA Rating: PG. Release date: April 3, 1981.

Players: Jerry Lewis, Susan Oliver, Roger C. Carmel, Deanna Lund, Harold J. Stone, Steve Franken, Buddy Lester, Leonard Stone, Jerry Lester, Billy Barty.

Heaven's Gate . . . . United Artists

Joann Carelli, producer. Michael Cimino, writer-director. In 70mm Panavision and color. Running time: 205 minutes, plus intermission. MPAA Rating: R. Release date: November, 1980, (limited); February, 1981, (general).

Players: Kris Kristofferson, Christopher Walken, John Hurt, Sam Waterston, Brad Dourif, Isabelle Huppert, Joseph Cotten, Jeff Bridges, Roseanne Vela, Ronnie Hawkins, Geoffrey Lewis, Nicholas Woodeson, Stefan Shcherby, Waldemar Kalinowski, Terry Quinn, John Conley, Margaret Benczak, James Knoveloch, Erika Petersen, Paul Koslo, Robin Barlett, Tom Noona.

He Knows You're Alone . . . . . United Artists

A Metro-Goldwyn-Mayer presentation. A Lansbury/Beruh production. Executive Producers: Edgar Lansbury and Joseph Beruh. Producer: George Manasse. Director: Armand Mastrianni.

Screenplay: Scott Parker. In color. Running time: 94 minutes. MPAA Rating: R. Release date: September, 1980.
Players: Don Scardino, Caitlin O'Hearney, Elizabeth Kemp, Tom Rolfing, Lewis Arlt, Patsy Pease, James Rebhorn, Tom Hanks, Dana Barron, Joseph Leon.

History of the World—Part I .... 20th Century Fox

Brooksfilms Limited. Written, Produced and directed by Mel Brooks. In Panavision and color. Running time: 86 minutes. MPAA Rating: R. Release date: June 12, 1981.
Players: Mel Brooks, Dom DeLuise, Madeline Kahn, Harvey Korman, Cloris Leachman, Ron Carey, Sid Caesar, Gregory Hines, Pamela Stephenson, Shecky Greene Mary-Margaret Humes.

Honky Tonk Freeway .... Universal

An Associated Film Distribution Corp. release. An EMI Films production. Producers: Don Boyd and Howard W. Koch, Jr. Director: John Schlesinger. Screenplay: Ed Clinbon. In color. Running time: 107 minutes. MPAA Rating: PG. Release date: August 21, 1981.
Players: Beau Bridges, Beverly D'Angelo, William Devane, Teri Garr, Paul Jabra, Hume Cronyn, Jessica Tandy, Geraldine Page, Daniel Stern, Howard Hesseman, Deborah Rush, Joe Grifasi, George Dzunda, Ron Frazier, David Rasche.

The Howling .... Avco Embassy

International Film Investors and Wescom Productions. Executive Producers: Daniel H. Blatt and Steven A. Lane. Producers: Michael Finnell and Jack Conrad. Director: Joe Dante. Screenplay: John Sayles and Terence H. Winkless, based on the novel by Gary Brandner. In color. Running time: 91 minutes. MPAA Rating: R. Release date: March, 1981.
Players: Dee Wallace, Patrick Macnee, Dennis Dugan, Christopher Stone, Belinda Balaski, Kevin McCarthy, John Carradine, Slim Pickens, Elizabeth Brooks, Robert Picerdo, Margie Impert.

The Idolmaker .... United Artists

A Koch/Kirkwood Production. Producers: Gene Kirkwood and Howard W. Koch, Jr. Director: Taylor Hackford. Screenplay: Edward Di Lorenzo. In color. Running time: 119 munutes. MPAA Rating: PG. Release date: November, 1980.
Players. Ray Sharkey, Tovah Feldshuh, Paul Land, Peter Gallagher, Joe Pantoliano, Maureen McCormick, John Aprea, Richard Bright, Olympia Dukakis, Steven Apostlee Peck.

The Incredible .... Shrinking Woman

Universal
A Lija Production. Executive Producer—Writer: Jane Wagner. Producer: Hank Moonjean. Director: Joe Schumacher. In color. Running time: 88 minutes. MPAA Rating: PG. Release date: January 30, 1981.
Players: Lily Tomlin, Charles Grodin, Ned Beatty, Henry Gibson, Elizabeth Wilson, Mark Blankfield, Maria Smith, Pamela Bellwood, John Glover, Nicholas Hormann, James McMullan, Shelby Balik, Justin Dana, Richard A. Baker and Mike Douglas as himself.

In God We Trust .... Universal

A Howard West/George Shapiro production. Director: Marty Feldman. Screenplay: Mr. Feldman and Chris Allen. In color. Running time: 97 minutes. MPAA Rating: PG. Release date: September 26, 1980.
Players: Marty Feldman, Peter Boyle, Louis Lasser, Wilfred Hyde-White, Richard Pryor, Andy Kaufman, Severn Darden, Eddie Parkes, Stephanie Ross, Richard A. Roth.

Inside Moves ....Associated Film Dist.

A Goodmark Production. Producers: Mark M. Tanz and R. W. Goodwin. Director: Richard Donner. Screenplay: Valerie Curtin and Barry Levinson, based on the novel by Todd Walton. In color. Running time: 113 minutes. MPAA Rating: PG. Release date: December, 1980.
Players: John Savage, David Morse, Diana Scarwid, Amy Wright, Tony Burton, Bill Henderson, Steve Kahan, Jack O'Leary, Bert Remsen, Harold Russell, Pepe Serna, Herold Sylvester, Arnold Williams.

It's My Turn .... Columbia

A Rastar—Martin Elfand production. Producer: Mr. Elfnad. Director: Claudia Weill. Screenplay: Eleanor Bergstein. In color. Running time: 91 minutes. MPAA Rating: R. Release date: October, 1980.
Players: Jill Clayburgh, Michael Douglas, Charles Grodin, Beverly Garland, Steven Hill, Teresa Baxter, Joan Copeland, John Gabriel, Charles Kimbrough, Roger Robinson, Jennifer Salt, Daniel Stern, Diane Wiest.

The Jazz Singer ......Associated Film Dist.

EMI Films. Producer: Jerry Leider. Director: Richard Fleischer. Screenplay: Herbert Baker, as adapted by Stephen H. Foreman from the play by Samson Raphaelson. In color. Running time: 115 minutes. MPAA Rating: PG. Release date: December 19, 1980.
Players: Neil Diamond, Laurence Olivier, Lucie Arnaz, Catlin Adams, Franklyn Ajaye, Paul Nicholas Sully Boyar, Mike Kellin, James Booth, Luther Waters, Oren Waters, Rod Gist, Walter Janowitz, Janet Brandt.

Kagemusha ....20th Century-Fox

Toho Co., Ltd. and Kurosawa Productions. Executive producers: Akira Kurosawa and Tomoyuki Tanaka. Executive Producers, international version: Francis Ford Coppola and George Lucas. Director: Mr. Kurosawa. Screenplay: Mr Kurosawa and Masato Ide. Running time: 161 minutes. MPAA Rating: PG. Release date: October, 1980. In Japanese with English titles.
Players: Tatsuya Nakadai, Tsutomu Yamazaki, Kenichi Hagiwara, Jinpachi Nezu, Shuji Otaki.

King of the Mountain .... Universal

A PolyGram Pictures production. Producer: Jack Frost Sanders. Director: Noel Nosseck. Screenplay: H.R. Christian. In color. Running time: 90 minutes. MPAA Rating: PG. Release date: April, 1981.
Players: Harry Hamlin, Joseph Bottoms, Deborah Van Valkenburgh, Richard Cox, Dennis Hopper, Dan Hagerty, Seymour Cassel, Jon Sloan.

Knightriders ....United Film Dist. Co.

A Laurel Production. Producer: Richard P. Rubinstein. Director-Writer: George A. Romero. In color. Running time: 145 minutes. MPAA Rating: R. Release date: April 10, 1981.
Players: Ed Harris, Gary Lahti, Tom Savini, Amy Ingersoll, Patricia Tallman, Christine Forrest, Werner Shook, Brother Blue, Cynthia Adler, John Ampeas, Ken Hixon, John Hostetter, Albert Amerson.

La Cage Aux Folles II .... United Artists

A French—Italian co-production. Les Productions Artistes Associes (Paris); DM. Produzione s.r.l. (Rome). Executive Producer: Marcello Danon. Director: Edouard Molinaro. Screenplay: Francis Veber, based on a story by himself, Jean Poiret, and M. Danon. In color. Running time: 100 minutes. MPAA Rating: R. Release date: February, 1981. In French with English subtitles.
Players: Ugo Tognazzi, Michel Serrault, Marcel Bozzuffi, Paola Borboni, Giovanni Vettorazzo, Glauco Onorato, Roberto Bisacco, Benny Luke, Michael Galabru.

The Legend of The Lone Ranger .... Universal

A Lord Garde/Jack Wrather Presentation. A Martin Starger Production. Producer: Walter Coblenz. Director: William A. Fraker. Screenplay: Ivan Goff & Ben Roberts and Michael Kane and Wil-

liams Roberts. Adaptation: Jerry Derloshon. In color. Running time: 98 minutes. MPAA Rating: PG. Release date: May, 1981.
Players: Klinton Spilsbury, Michael Horse, Christopher Lloyd, Matt Clark, Juanin Clay, Jason Robards, John Bennett Perry, David Hayward, John Hart, Richard Fransworth, Lincoln Tate, Ted Flicker.

Lion of the Desert .....United Film Dist. Co.

A Falcon International production. Producer-Director: Moustapha Akkad. Screenplay: H.A.L. Craig. In Panavision and color. Running time: 162 minutes. MPAA Rating: PG. Release date: April, 1981.

Players: Anthony Quinn, Oliver Reed, Rod Steiger, John Gielgud, Irene Papas, Raf Vallone, Gastone Moschin, Takis Emmanuel, Stefano Patrizi, Sky Dumont, Robert Brown, Eleonora Stathopoulous, Andrew Keir, Rodolfo Bigotti.

Loving Couples ....20th Century-Fox

A Time—Life Films presentation. Executive Producer: David Susskind. Producer: Renee Valent. Director: Jack Smight. Screenplay: Martin Donovan. In color. Running time: 96 minutes. MPAA Rating: PG. Release date: October, 1980.

Players: Shirley MacLaine, James Coburn, Susan Sarandon, Stephen Collins, Sally Kellerman, Nan Martin, Shelly Batt, Bernard Behrens, Anne Bloom, Fred Carney, Helena.

Melvin and Howard .... Universal

A Linson—Phillips—Demme production. Producers: Art Linson and Don Phillips. Director: Jonathan Demme. Screenplay: Bo Goldman. In color. Running time: 93 minutes. MPAA Rating: R. Release date: November, 1980.

Players: Paul Le Mat, Jason Robards, Mary Steenburgen, Michael J. Pollard, Jack Kehoe, Pamela Reed, Dabney Coleman, John Glover, Charles Napier, Charlete Holt, Gloria Grahame.

Middle Age Crazy ....20th Century-Fox

A Sid and Marty Krofft presentation. Producers: Robert Cooper and Ronald Cohen. Produced with participation of the Canadian Film Development Corp. Director: John Trent. Screenplay: Carl Kleinschmitt. In color. Running time: 95 minutes. MPAA Rating: R. Release date: August, 1980.

Players: Bruce Dern, Ann-Margret, Graham Jarvis, Eric Christmas, Helen Hughes, Geoffrey Bowes, Michael Kane, Diane Dewey, Vivian Reis, Patricia Hamilton, Anni Lantuch, Deborah Wakeham, Gina Dick.

The Mirror Crack'd ....Associated Film Dist.

EMI Films. Producers: John Brabourne and Richard Goodwin. Director: Guy Hamilton. Screenplay: Jonathan Hales and Barry Sandler, based on the book by Agatha Cristie. In color. Running time: 105 minutes. MPAA Rating: PG. Release date: December 19, 1980.

Players: Angela Lansbury, Geraldine Chaplin, Tony Curtis, Edward Fox, Rock Hudson, Kim Novak, Elizabeth Taylor, Marella Oppenheim, Wendy Morgan, Margaret Courtenay, Charles Gray, Maureen Bennett.

Modern Romance ....

Producers: Andrew Scheinman and Martin Shafer. Director: Albert Brooks. Screenplay: Mr. Brooks and Monica Johnson. In color. Running time: 93 minutes. MPAA Rating: R. Release date: March, 1981.

Players: Albert Brooks, Kathryn Harrold, Tyann Means, Bruno Kirby, Jane Hallaren, Karen Chandler, Dennis Kort, Bob Einstein, Virginia Feingold, Thelma Leeds, Candy Castillo.

Mon Oncle d'Amerique .... New World Pictures

Producer: Philippe Dussart in co-production with Andrea Films and T.F.I. Director: Alain Resnais. Screenplay: Jean Gruault, based on the works of Henri Laborit. In color. Running time: 123 minutes. No MPAA Rating. Release date: December, 1980.

Players: Gerard Depardieu, Nicole Garcia, Roger-Pierre, Marie Dubois, Nelly Borgeaud, Pierre Arditi, Gerard Darrieu, Philippe Laudenbach, Professor Henri Laborit.

My Bloody Valentine ....Paramount

Produced with participation of the Canadian Film Development Corp. and Famous Players Ltd. Producers: John Dunning, Andre Link and Stephen Miller. Director: George Mihalka. Screenplay: John Beaird. In color. Running time: 91 minutes. MPAA Rating: R. Release date: February, 1981.

Players: Paul Kelman, Lori Hallier, Neil Affleck, Keith Knight, Alf Humphreys, Cynthia Dale, Helene Udy, Rob Stein, Tom Kovacks, Terry Waterland, Carl Marotte, Jim Murchison, Gina Dick, Peter Cowper, Don Francks, Patricia Hamilton, Larry Reynolds, Jack Van Evera.

Nighthawks .... Universal

A Martin Poll/Production Company production. Producer: Mr. Poll. Director: Bruce Malmuth. Screenplay: David Shaber, from a story by himself and Paul Sylbert. In color. Running time: 98 minutes. MPAA Rating: R. Release date: April 10, 1981.

Players: Sylvester Stallone, Billy Dee Williams, Lindsay Wagner, Persis Khambatta, Nigel Davenport, Rutger Hauer, Hilarie Thompson, Joe Spinell, Walter Mathews, E. Brian Dean, Caesar Cordova, Charles Duval, Tony Munafo.

The Night the Lights Went Out in Georgia ....Avco Embassy

G.L. Productions, Ltd. Producers: Elliot Geisinger, Howard Kuperman, Ronald Saland and Howard Smith. Executive Producers: William and Carole Blake. Director: Ronald F. Maxwell. Screenplay: Bob Bonney. Music: David Shire. Cinematography: Bill Butler. Editor: Anne Goursaud. In color. Running time: 120 minutes. MPAA Rating: PG. Release date: June, 1981.

Players: Kristy McNichol, Dennis Quaid, Mark Hamill, Sunny Johnson, Don Stroud, Arlen Dean Snyder, Barry Corbin, Lulu McNichol, Royce Clark.

Nine to Five ....20th Century-Fox

An IPC Films production. Producer: Bruce Gilbert. Director: Colin Higgins. Screenplay: Mr. Higgins and Patricia Resnick, from a story by the latter. In color. Running time: 110 minutes. MPAA Rating: PG. Release date: December 19, 1980.

Players: Jane Fonda, Lily Tomlin, Dolly Parton, Dabney Coleman, Sterling Hayden, Elizabeth Wilson, Henry Jones, Lawrence Pressman, Marian Mercer.

Nobody's Perfect .... Columbia

Producer: Mort Engelberg. Executive Producer: Ted Swanson. Director: Peter Bonerz. Screenplay: Tony Kenrick, based on his novel. In color. Running time: 96 minutes. MPAA Rating: PG. Release date: July, 1981.

Players: Gabe Kaplan, Alex Karras, Robert Klein, Susan Clark, Paul Stewart, Alex Rocco, Arthur Rosenberg, James Cromwell, Bobby Ramsen, John Di Santi, Ric Applewhite, Will Knickerbocker, Peter Bonerz.

Oh God! Book II ....Warner Bros.

Producer-Director: Gilbert Cates. Screenplay: Josh Greenfeld, Hal Goldman, Fred S. Fox, Seaman Jacobs, and Melissa Miller, based on a story by Mr. Greenfeld. In color. Running time: 94 minutes. MPAA Rating: PG. Release date: October, 1980.

Players: George Burns, Suzanne Pleshette, David Birney, Louanne, John Louie, Conrad Janis, Anthony Holland, Hugh Downs.

One-Trick Pony ....Warner Bros.

Producers: Michael Tannen and Michael Hausman. Director: Robert M. Young. Screenplay: Paul Simon. In color. Running time: 98 minutes. MPAA Rating: R. Release date: October 3, 1980.

Players: Paul Simon, Blair Brown, Rip Torn, Joan Hackett, Allen Goorwitz, Mare Winningham, Michael Pearlman, Lou Reed, Steve Gadd, Eric Gale, Tony Levin, Richard Tee, Harry Shearer, The B-52s, Sam and Dave, The Lovin' Spoonful, and Tiny Tim.

Ordinary People .... Paramount

A Wildwood Enterprises Production. Producer: Ronald L. Schwary. Director: Robert Redford. Screenplay: Alvin Sargent, based on the novel by Judith Guest. In color. Running time: 124 minutes. MPAA Rating: R. Release date: September, 1980.

Players: Donald Sutherland, Mary Tyler Moore, Judd Hirsch, Timothy Hutton, M. Emmet Walsh, Elizabeth McGovern, Dinah Manoff, Fredric Lehne, James B. Sikking, Basil Hoffman, Quinn Redeker, Mariclare Costello, Meg Mundy, Elizabeth Hubbard, Adam Baldwin, Scott Doebler.

Outland .... Warner Bros.

A Ladd Company Release. Producer: Richard A. Roth. Writer-Director: Peter Hyams. In Panavision and color. Recorded in Dolby Stereo. Running time: 109 minutes. MPAA Rating: R. Release date: May, 1981.

Players: Sean Connery, Peter Boyle, Frances Sternhagen, James B. Sikking, Kika Markham, Clarke Peters, Steven Berkoff.

Popeye .... Paramount

A presentation of Paramount Pictures and Walt Disney Productions. Executive producer: C.O. Erickson. Producer: Robert Evans. Director: Robert Altman. Screenplay: Jules Feiffer. In Technovision and color. Running time: 114 minutes. MPAA Rating: PG. Release date: December 12, 1980.

Players: Robin Williams, Shelley Duvall, Ray Walston, Paul Dooley, Paul L. Smith, Richard Libertini, Donald Moffat, MacIntyre Dixon, Roberta Maxwell, Donovan Scott, Allan Nicholls, Wesley Ivan Hurt.

The Postman Always Rings Twice .... Paramount

Lorimar Productions, Inc. A Northstar International picture made in association with CIP-Europaische Treuhand AG, Germany. Producers: Charles Mulvehill and Bob Rafelson. Director: Mr. Rafelson. Screenplay: David Mamet, based on the novel by James M. Cain. In color. Running time: 125 minutes. MPAA Rating: R. Release date: March (special); April, 1981 general.

Players: Jack Nicholson, Jessica Lange, John Colicos, Michael Lerner, John P. Ryan, Anjelica Huston, William Traylor, Tom Hill, Jon Van Ness, Brian Farrell, Raleigh Bond, William Newman, Albert Henderson.

Prince of the City .... Warner Bros.

An Orion Pictures Company release. Producer: Burtt Harris. Executive Producer: Jay Presson Allen. Director: Sidney Lumet. Screenplay: Ms. Allen and Mr. Lumet, based on the book by Robert Daley. Music: Paul Chihara. Cinematography: Andrzej Bartkowiak. Editor: John J. Fitzstephens. In color. Running time: 167 minutes. MPAA Rating: R. Release date: August, 1981 (limited); October, (general).

Players: Treat Williams, Jerry Orbach, Richard Foronjy, Don Billett, Kenny Marino, Carmine Caridi, Tony Page, Norman Parker, Paul Roebling, Bob Balaban, James Tolkan, Steve Inwood, Lindsay Crouse, Matthew Laurance, Tony Turco, Ron Maccone, Ron Karabatsos, Tony DiBenedetto, Tony Munafo, Robert Christian.

Private Benjamin .... Warner Bros.

A Goldie Hawn/Nancy Meyers Production. Producers-Writers: Ms. Meyers, Charles Shyer, and Harvey Miller. Director: Howard Zieff. In color. Running time: 110 minutes. MPAA Rating: R. Release date: October 10, 1980.

Players: Goldie Hawn, Eileen Brennan, Armand Assante, Robert Webber, Sam Wanamaker, Barbara Barrie, Albert Brooks, Mary Kay Place, Harry Dean Stanton.

Raging Bull .... United Artists

A Robert Chartoff—Irwin Winkler Production. Produced in association with Peter Savage. Director: Martin Scorsese. Screenplay: Paul Schrader and Mardik Martin, based on the book of the same name by Jake La Motta, with Joseph Carter and Peter Savage. In black-and-white with sequences in color. Running time: 128 minutes. MPAA Rating: R. Release date: November, (limited); December, 1981 (general).

Cast: Robert De Niro, Cathy Moriarty, Joe Pesci, Frank Vincent, Nicholas Colasanto, Theresa Saldana, Frank Adonis, Mario Gallo, Frank Topham, Lori Anne Flax, Joseph Bono, James V. Christy, Bernie Allen, Bill Mazer, Bill Hanrahan, Rita Benett, Mike Miles.

Raiders of the Lost Ark .... Paramount

Lucasfilm Ltd. Producer: Frank Marshall. Director: Steven Spielberg. Screenplay: Lawrence Kasdan, from a story by George Lucas and Philip Kaufman. Music: John Williams. Cinematography: Douglas Slocombe. Editor: Michael Kahn. In Panavision and color. Running time: 115 minutes. MPAA Rating: PG. Release date: June 12, 1981.

Players: Harrison Ford, Karen Allen, Paul Freeman, Ronald Lacey, John Rhys-Davies, Denholm Elliott, Wolf Kahler, Anthony Higgins, Alfred Molina, Vic Tablian, Don Fellows.

Resurrection .... Universal

Producers: Renee Missel and Howard Rosenman. Director: Daniel Petrie. Screenplay: Lewis John Carlino. In color. Running time: 103 minutes. MPAA Rating: PG. Release date: October, 1980.

Players: Ellen Burstyn, Sam Shepard, Richard Farnsworth, Roberts Blossom, Clifford David, Pamela Payton-Wright, Jeffrey DeMunn, Eva Le Gallienne, Lois Smith, Madeleine Thornton-Sherwood, Richard Hamilton, Carlin Glynn, Lane Smith, Penelope Allen.

Sam Marlow, Private Eye . 20th Century-Fox

Melvin Simon Productions. Producer-Writer: Andrew J. Fenady, based on his novel. Director: Robert Day. In color. Running time: 111 minutes. MPAA Rating: PG. Release date: October, 1980.

Players: Robert Sacchi, Franco Nero, Michelle Phillips, Olivia Hussey, Misty Rowe, Victor Buono, Herbert Lom, Sybil Danning, Dick Bakalyan, Gregg Palmer, Jay Robinson, George Raft, Yvonne de Carlo.

Savage Harvest .... 20th Century-Fox

Producers: Ralph Helfer and Sandy Howard. Director: Robert Collins. Screenplay: Robert Blees and Robert Collins, based on a story by Ralph Helfer and Ken Noyle. In color. Running time: 86 minutes. MPAA Rating: PG. Release date: May, 1981.

Players: Tom Skerritt, Michelle Phillips, Shawn Stevens, Ann-Marie Martin, Derek Partridge, Arthur Malet, Tana Helfer.

Scanners .... Avco Embassy

A Filmplan International Production. Executive Producers: Victor Solnicki and Michael Ironside. Producer: Claude Heroux. Director-Writer: David Cronenberg. In color. Running time: 103 minutes. MPAA Rating: R. Release date: January, 1981.

Players: Jennifer O'Neill, Stephen Lack, Patrick McGoohan, Lawrence Dane, Charles Shamata, Adam Ludwig, Michael Ironside, Victor Desy, Robert Silverman.

The Sea Wolves .... Paramount

A Lorimar presentation. Made by Richmond Light Horse Productions Ltd. for Varius A. G. An Anglo-American-Swiss Enterprise. Producer: Euan Lloyd. Director: Andrew V. McLaglen. Screenplay: Reginald Rose, based on "Boarding Party" by James Leasor. In color. Running time: 120 minutes. MPAA Rating: PG. Release date: June, 1981.

Players: Gregory Peck, Roger Moore, David Niven, Trevor Howard, Barbara Kellerman, Patrick Macnee, Kenneth Griffith, Patrick Allen, Wolf Kahler, Robert Hoffmann, Dan Van Husen.

Second Hand Hearts .... Paramount

A Lorimar Pictures production. Producer: James William Guercio. Director: Hal Ashby. Screenplay: Charles Eastman. In color. Running time: 102 minutes. MPAA Rating: PG. Release date: May, 1981.

Players: Robert Blake, Barbara Harris, Collin Boone, Ambert Rose Gold, Jessica Standbury, Bert Remsen, Sondra Blake, Shirley Stoler, Woodrow Chambliss.

Seems Like Old Times .... Columbia

A Rastar Production. Producer: Ray Stark. Director: Jay Sandrich. Screenplay: Neil Simon. In color. Running time: 102 minutes. MPAA Rating: PG. Release date: December 19, 1980.

Players: Goldie Hawn, Chevy Chase, Charles Grodin, Robert Guillaume, Harold Gould, George Grizzard, Yvonne Wilder, T.K. Carter, Judd Omen, Marc Alaimo.

Somewhere in Time .... Universal

A Rastar/Stephen Deutsch production. Producer: Stephen Deutsch. Director: Jeannot Szwarc. Screenplay: Richard Matheson, based on his novel, "Bid Time Return." In color. Running time: 103 minutes. MPAA Rating: PG. Release date: October 3, 1980.

Players: Christopher Reeve, Jane Seymour, Christopher Plummer, Teresa Wright, Bill Erwin, George Foskovec, Susan French, John Alvin, Eddra Gale.

"S.O.B." .... Paramount

A Lorimar Pictures presentation. Producers: Blake Edwards and Tony Adams. Writer-Director: Blake Edwards. Music: Henry Mancini. Cinematography: Harry Stradling. Editor: Ralph E. Winters. In color. Running time: 121 minutes. MPAA Rating: R. Release date: July 1, 1981.

Players: Julie Andrews, William Holden, Marisa Berenson, Larry Hagman, Robert Loggia, Stuart Margolin, Richard Mulligan, Robert Preston, Craig Stevens, Loretta Swit, Robert Vaughn, Robert Webber, Shelly Winters.

Sphinx .... Warner Bros.

An Orion Pictures production. Executive Producer: Franklin J. Schaffner. Producer: Stanley O'Toole. Director: Mr. Schaffner. Screenplay: John Byrun, adapting the novel by Robin Cook. In Panavision and color. Running time: 119 minutes. MPAA Rating: PG. Release date: February 11, 1981.

Players: Lesley-Anne Down, Frank Langella, Maurice Ronet, Sir John Gielgud, Vic Tablian, Martin Benson, John Rhys-Davies, Nadim Sawalha, Tutte Lemkow, Safed Jaffrey, Eileen Way, William Hootkins, Mark Kingston.

Stardust Memories .... United Artists

A Jack Rollins-Charles H. Joffe production. Producer: Robert Greenhut. Director-Writer: Woody Allen. In black-and-white. Running time: 88 minutes. MPAA Rating: PG. Release date: October, 1980.

Players: Woody Allen, Charlotte Rampling, Jessica Harper, Marie-Christine Barrault, Tony Roberts, Daniel Stern, Amy Wright, Helen Hanft, John Rothman, Anne DeSalvo, Joan Neuman, Ken Chapin, Leonardo Cimino.

Stir Crazy .... Columbia

Producer: Hannah Weinstein. Director: Sidney Poitier. Screenplay: Bruce Jay Friedman. In color. Running time: 114 minutes. MPAA Rating: R. Release date: December 12, 1980.

Players: Gene Wilder, Richard Pryor, Georg Stanford Brown, Jobe th Williams, Miguelangel Suarez, Craig T. Nelson, Barry Corbin, Charles Weldon, Nicholas Coster, Joel Brooks, Jonathan Banks, Erland Van Lidth De Jeude.

Stripes .... Columbia

An Ivan Reitman Film. Producers: Mr. Reitman and Dan Goldberg. Director: Mr. Reitman. Screenplay: Len Blum, Mr. Goldberg, and Harold Ramis. Music: Elmer Bernstein. Cinematography: Bill Butler. Editors: Eva Ruggiero, Michael Luciano, and Harry Keller. In color. Running time: 105 minutes. MPA Rating: R. Release date: June 26, 1981.

Players: Bill Murray, Harold Ramis, Warren Oates, P.J. Soles, Sean Young, John Candy, John Larroquette, John Voldstad, John Diehl, Lance LeGault, Roberta Leighton.

Student Bodies .... Paramount

A Universal Cinema Inc. production. Producer: Allen Smithee. Executive Producers: Jerry Belson and Harvey Miller. Director-Writer: Mickey Rose. Music: Gene Hobson. Cinematography: Robert Ebinger. Editor: Kathryn Ruth Hope. In color. Running time: 86 minutes. MPAA Rating: R. Release date: August, 1981.

Players: Kristen Riter, Matthew Goldsby, Richard Brando, Joe Flood, Joe Talarowski, Mimi Weddell, Carl Jacobs, Peggy Cooper, Janice E. O'Malley.

The Stunt Man .... 20th Century-Fox

A Melvin Simon production. Producer-Director: Richard Rush. Screenplay: Lawrence B. Marcus, from an adaptation by Mr. Rush of the novel by Paul Brodeur. In color. Running time: 129 minutes. MPAA Rating: R. Release date: October, 1980.

Players: Peter O'Toole, Steve Railsback, Barbara Hershey, Allen Goorwitz, Alex Rocco, Sharon Farrell, Adam Roarke, Phillip Bruns, Chuck Bail.

Sunday Lovers .... United Artists

A Metro-Goldwyn-Mayer presentation. Viaduc Productions, S.A. Producer: Leo Fuchs. Directors: Edouard Molinaro, Bryan Forbes, Dino Risi and Gene Wilder. Screenplay: Francis Veber, Leslie Bricusse, Age & Scarpelli, and Mr. Wilder. In color. Running time: 125 minutes. MPAA Rating: R. Release date: February, 1981.

Players: England: Roger Moore, Denholm Elliott, Lynn Redgrave, Priscilla Barnes. America: Gene Wilder, Kathleen Quinlan, Dianne Crittenden. France: Lino Ventura, Robert Webber, Catherine Salviat. Italy: Ugo Tognazzi, Rossana Podesta, Sylva Koscina, Beba Loncar.

Superman II .... Warner Bros.

An International Film Production Inc. picture. An Alexander Salkind presentation. An Alexander and Ilya Salkind production. Producer: Pierre Spengler. Director: Richard Lester. Screenplay: Mario Puzo, David Newman, and Leslie Newman, from a story by Mr. Puzo. In Panavision and color. Running time: 127 minutes. MPAA Rating: PG. Release date: June 19, 1981.

Players: Gene Hackman, Christopher Reeve, Ned Beatty, Jackie Cooper, Sarah Douglas, Margot Kidder, Jack O'Halloran, Valerie Perrine, Susannah York, Clifton James, E.G. Marshall, Marc McClure, Terence Stamp.

Take This Job and Shove It .... Avco Embassy

A Cinema Group presentation. Producer: Greg Blackwell. Executive Producers: William J. Immerman and J. David Marks. Director: Gus Trikonis. Screenplay: Barry Schneider, from a story by Jeffrey Bernini and Mr. Schneider. Music: Billy Sherrill. Cinematography: James Devis. Editor: Richard Belding. In color. Running time: 100 minutes. MPAA Rating: PG. Release date: May, 1981.

Players: Robert Hays, Art Carney, Barbara Hershey, David Keith, Tim Thomerson, Martin Mull, Eddie Mull, Eddie Albert, Penelope Milford.

Tarzan, The Ape Man .... United Artists

A Metro-Goldwyn-Mayer presentation. A Svengali Production. Producer: Bo Derek, Director: John Derek. Screenplay: Tom Rowe and Gary Goodard, based on characters created by Edgar Rice Burroughs. Music: Perry Botkin. Cinematography: John Derek. Editor: James B. Ling. In color. Running time: 112 minutes. MPAA Rating: R. Release date: July 24, 1981.

Players: Bo Derek, Richard Harris, John Phillip Law, Miles O'Keefe, Akushula Selayah, Steven Strong, Maxime Philoe, Leonard Bailey, Wilfrid Hyde-White, Laurie Mains, Harold Ayer.

The Tempest .... World Northal

A Boyd's Co. production. Producers: Sarah Radclyffe, Guy Ford and Mordecai Schreiber. Executive Producer: Don Boyd. Director-Adapter: Derek Jarman. In color. Running time: 95 minutes. Release date: October, 1980.

Players: Heathcote Williams, Karl Johnson, Toyah Willcox, Peter Bull, Richard Warwick, Elisabeth Welch, Jack Birkett, Ken Campbell, David Meyer, Christopher Biggins, Neil Cunningham, Peter Turner.

Terror Train .... 20th Century-Fox

An Astral Bellevue-Pathe production. Producer: Harold Greenberg. Director: Roger Spottiswoode. Screenplay: T. Y. Drake. In color. Running time: 97 minutes. MPAA Rating: R. Release date: October, 1980.

Players: Ben Johnson, Jamie Lee Curtis, Hart Bochner, David Cooperfield, Derek MacKinnon, Sandee Currie, Timothy Webber, Anthony Sherwood, Howard Busgang, Steve Michaels, Greg Swanson, D. D. Winters, Joy Boushel, Victor Knight.

Tess .... Columbia

A Franco-British co-production: Renn Productions (France) and Burrill Productions (England) with the participation of the Societe Francaise De Production (S.F.P.). Producers: Claude Berri and Timothy Burrill. Executive Producer: Pierre Grunstein. Director: Roman Polanski. Screenplay: Gerard Brach, Mr. Polanski and John Brownjohn, based on the novel, "Tess of the D'Ubervilles" by Thomas Hardy. In color. Running time: 170 minutes. MPAA Rating: PG. Release date: February, 1981.

Players: Nastassia Kinski, Peter Firth, Albert Simono, Brigid Erin Bates, Jeanne Biras, John Bett, Leigh Lawson.

Thief .... United Artists

Michael Mann/Cann Productions. Producers: Jerry Bruckheimer and Ronnie Caan. Director: Michael Mann. Screenplay and Screen Story: Mr. Mann, based on "The Home Invaders" by Frank Hohimer. In color. Running time: 118 minutes. MPAA Rating: R. Release date: March, 1981.

Players: James Caan, Tuesday Weld, Willie Nelson, James Belushi, Robert Prosky, Tom Signorelli, Dennis Farina, Nick Nickeas, W. R. (Bill) Brown, Norm Tobin, John Santucci, Gavin MacFadyen.

This Is Elvis .... Warner Bros.

A David L. Wolper Production. Produced, directed, and written by Malcom Leo and Andrew Solt. In color and black-and-white. Running time: 109 minutes. MPAA Rating: PG. Release date: May, 1981.

Players: Elvis Presley, David Scott, Paul Boensch 3rd, Johnny Harra, Lawrence Kobler, Rhonda Lyn, Debbie Edge, Larry Raspberry. Narration: Ral Donner, Joe Esposito, Linda Thompson, Lisha Sweetnam, Virginia Kiser, Michael Tomack.

Times Square .... Associated Film Dist.

An EMI Films release. Producers: Robert Stigwood and Jacob Brackman. Director: Alan Moyle. Screenplay: Mr. Brackman, from a story by Mr. Moyle and Leanne Unger. In Panavision and color. Running time: 111 minutes. MPAA Rating: R. Release date: October 17, 1980.

Players: Tim Curry, Trini Alvarado, Robin Johnson, Peter Coffield, Herbert Berghof, David Marguilies, Anna Maria Horsford, Michael Margotta, J. C. Quinn, Miguel Pinero.

Tribute .... 20th century-Fox

A Turman-Foster Company production. Producers: Joel B. Michaels and Garth H. Drabinsky. Director: Bob Clark. Screenplay: Bernard Slade, based on his play. In color. Running time: 125 minutes. MPAA Rating: PG. Release date: December, 1980 (limited); February, 1981 (general).

Players: Jack Lemmon, Robby Benson, Lee Remick, Colleen Dewhurst, John Marley, Kim Cattrall, Gale Garnett, Teri Keane.

Under the Rainbow .... Warner Bros.

An Orion Pictures Company release. An Innovisions/ECA presentation. Producer: Fred Bauer. Director: Steve Rash. Screenplay: Pat McCormick, Harry Hurwitz, Martin Smith, Pat Bradley and Fred Bauer, based on a story by Bauer and Bradley. Music: Joe Renzetti. Cinematography: Frank Stanley. Editor: David Blewitt. In color. Running time: 98 minutes. MPAA Rating: PG. Release date: July 31, 1981.

Players: Chevy Chase, Carrie Fisher, Eve Arden, Joseph Maher, Robert Donner, Billy Barty, Mako, Cork Hubbert, Pat McCormick, Adam Arkin, Richard Stahl, Freeman King.

Victory .... Paramount

A Lorimar Pictures presentation. Producer: Freddie Fields. Director: John Huston. Screenplay: Evan Jones and Yabo Yablonsky, from a story by Mr. Yablonsky, Djordje Millcevic, and Jeff Maguire. Music: Bill Conti. Cinematography: Gerry Fisher. Editor: Roberto Silvie. In color. Running time: 117 minutes. MPAA Rating: PG. Release date: July, 1981.

Players: Sylvester Stallone, Michael Caine, Pele, Max Von Sydow, Daniel Massey, Carole Laure.

Wolfen .... Warner Bros.

An Orion Pictures release. Producer: Rupert Hitzig. Executive Producer: Alan King. Director: Michael Wadleigh. Screenplay and Story: David Eyre and Mr. Wadleigh, based on the novel by Whitley Strieber. Music: James Horner. Cinematography: Gerry Fisher. Editors: Chris Lebenzon, Dennis Dolan, Martin Bram, and Marshall M. Borden. In Panavision and color. Running time: 115 minutes. MPAA Rating: R. Release date: July 24, 1981.

Players: Albert Finney, Diane Venora, Edward James Olmos, Gregory Hines, Tom Noonan, Dick O'Neill, Dehl Beri, Peter Michael Goetz, Sam Gray, Ralph Bell, Max M. Brown, Annie Marie Photamo, Sarah Felder.

Zorro, The Gay Blade .. ..20th Century-Fox

A Melvin Simon Production. Producers: George Hamilton and C. O. Erickson. Executive Producer: Mr. Simon. Director: Peter Medak. Screenplay: Hal Dresner, based on a story by himself, Greg Alt, Don Moriarty, and Bob Randall. Music: Ian Fraser. Cinematography: John A. Alonzo. Editor: Hillary Jane Kranze. In color. Running time: 93 minutes. MPAA Rating: PG. Release date: July 17, 1981.

Players: George Hamilton, Lauren Hutton, Brenda Vaccaro, Ron Leibman, Donovan Scott, James Booth, Helen Burns, Clive Revill, Eduardo Noriega.

FEATURE PICTURES

January 1, 1955—August 1, 1980

[In the following listings (R) indicates release dates; (CS) indicates CinemaScope; (VV) indicates VistaVision; (SS) indicates Super-Scope; (T) indicates Technirama; (P) Panavision; (D-150) Dimension-150; (D) Dyalscope; (F) Franscope; (U) Ultrascope; (VS) VistaScope; (M) Megascope; (R) Regalscope; (H) Hammerscope; (TE) Techniscope. All films released starting in 1974 in color unless otherwise indicated.]

Length In Mins.

a.k.a. CASSIUS CLAY 85
UNITED ARTISTS. (R) November, 1970. Documentary on Mohammad Ali.

AARON LOVES ANGELA 98
COLUMBIA. (R) December, 1975. Kevin Hooks, Irene Cara.

ABANDON SHIP 100
COLUMBIA. (R) May, 1957. Tyrone Power, Mai Zetterling.

ABBOTT AND COSTELLO MEET THE MUMMY 79
UNIVERSAL. (R) June, 1955. Abbott and Costello.

ABBY 91
AMERICAN INTERNATIONAL. (R) December, 1974. William Marshall, Carole Speed.

ABDICATION, THE 103
WARNER BROS. (R) October, 1974. Peter Finch, Liv Ullmann.

ABDUCTORS, THE 80
TWENTIETH CENTURY-FOX. (R) July, 1957. Victor McLaglen, George Macready.

ABDULLAH'S HAREM 88
TWENTIETH CENTURY-FOX. (R) June, 1956. Gregory Ratoff, Kay Kendall. Color.

ABOMINABLE DR. PHIBES, THE 93
AMERICAN INTERNATIONAL. (R) May, 1971. Vincent Price, Joseph Cotten. Color.

ABOMINABLE SNOWMAN, THE 85
TWENTIETH CENTURY-FOX. (R) October, 1957. Forrest Tucker, Peter Cushing.

ABOVE US THE WAVES 92
REPUBLIC. (R) October 26, 1956. John Mills, John Gregson.

ABSENT MINDED PROFESSOR, THE 97
BUENA VISTA. (R) April, 1961. Fred MacMurray, Nancy Olson.

ABYSSES, LES (Fr., Eng. titles) 90
KANAWHAX. (R) November, 1964. Francine Berge, Colette Berge.

ACCIDENT 105
CINEMA V. (R) April, 1967. Dirk Bogarde, Stanley Baker, Color.

ACCURSED, THE 78
ALLIED ARTISTS. (R) July, 1958. Robert Bray, Donald Wolfit.

ACCUSED OF MURDER 74
REPUBLIC. (R) December, 1956. David Brian, Vera Ralston. Color.

ACE ELI AND RODGER OF THE SKIES 92
TWENTIETH CENTURY-FOX. (R) April, 1973. Cliff Robertson, Pamela Franklin. Color.

ACE HIGH 123
PARAMOUNT. (R) October, 1969. Eli Wallach, Terrence Hill, Bud Spencer. Color.

ACROSS 110TH STREET 102
UNITED ARTISTS. (R) December, 1972. Anthony Quinn, Anthony Franciosa. Color.

ACROSS THE BADLANDS 55
COLUMBIA. (R) September 14, 1950. Charles Starrett, Smiley Burnette.

ACROSS THE BRIDGE (Brit.) 103
RANK. (R) January, 1958. Rod Steiger, Maria Landi.

ACROSS THE RIVER 85
DEBEMA. (R) April, 1965. Lou Gilber, Kay Doubleday.

ACT OF THE HEART 103
UNIVERSAL. (R) October 1970. Genevieve Bujold, Donald Sutherland. Color.

ACT ONE 110
WARNER BROS. (R) December, 1963. George Hamilton, Jason Robards, Jr.

ACTION OF THE TIGER (Cs) 93
METRO-GOLDWYN-MAYER. (R) August, 1957. Van Johnson, Martine Carol.

ACTIVIST, THE 86
REGIONAL. (R) December, 1969. Michael Smith, Leslie Gilbrum. Color.

Length In Mins.

ADA (Cs) 108
METRO-GOLDWYN-MAYER. (R) August, 1961. Susan Hayward, Dean Martin. Color.

ADALEN 31 (Swed., Eng. titles) 115
PARAMOUNT. (R) October, 1969. Peter Schildt, Kerstin Tidelius. Color.

ADAM AT 6 A.M. 100
NATIONAL GENERAL. (R) September, 1970. Michael Douglas, Lee Purcell. Color.

ADDING MACHINE, THE 100
REGIONAL. (R) September, 1969. Phyllis Diller, Milo O'Shea, Billie Whitelaw. Color.

ADELAIDE (Fr., Eng. titles) 86
SIGMA III. (R) February, 1969. Ingrid Thulin, Jean Sorel, Sylvia Fennec. Color.

ADORABLE CREATURES 106
CONTINENTAL. (R) January, 1956. Martine Carol, Danielle Darrieux.

ADORABLE JULIA 94
SEE-ART. (R) April, 1964. Lilli Palmer, Charles Boyer.

ADRIFT 102
MPO. (R) July, 1971. Rade Markovic, Paula Pritchett. Color.

ADVANCE TO THE REAR (P) 97
METRO-GOLDWYN-MAYER. (R) April, 1964. Glenn Ford, Stella Stevens.

ADVENTURE OF SHERLOCK HOLMES' SMARTER BROTHER 91
TWENTIETH CENTURY-FOX. (R) December, 1975. Gene Wilder, Madeline Kahn.

ADVENTURERS, THE 171
PARAMOUNT. (R) March, 1970. Bekim Fehmiu, Ernest Borgnine, Candice Bergen. Color.

ADVENTURES OF A YOUNG MAN (Cs) 145
TWENTIETH CENTURY-FOX. (R) June, 1962. Richard Beymer, Diane Baker. Color.

ADVENTURES OF BULLWHIP GRIFFIN, THE 110
BUENA VISTA. (R) March, 1967. Roddy McDowall, Suzanne Pleshette. Color.

ADVENTURES OF HUCKLEBERRY FINN, THE (Cs) 107
METRO-GOLDWYN-MAYER. (R) May, 1960. Tony Randall, Patty McCormack. Color.

ADVENTURES OF SADIE (Brit.) 88
TWENTIETH CENTURY-FOX. (R) May, 1955. Joan Collins, Kenneth More. Color.

ADVISE AND CONSENT (P) 139
COLUMBIA. (R) June, 1962. Henry Fonda, Charles Laughton.

AFFAIR IN HAVANA 70
ALLIED ARTISTS. (R) October, 1957. John Cassavetes, Raymond Burr.

AFFAIR IN RENO (N) 75
REPUBLIC. (R) February 15, 1957. John Lund, Doris Singleton.

AFFAIR OF THE SKIN, AN 102
ZENITH. (R) January, 1964. Kevin McCarthy, Herbert Berghof.

AFFAIR TO REMEMBER, AN (Cs) 114
TWENTIETH CENTURY-FOX. 1957. Cary Grant, Deborah Kerr. Color

AFRICA ADDIO (TS) 125
RIZZOLI. (R) January, 1967. Documentary. Color.

AFRICA-TEXAS STYLE! 106
PARAMOUNT. (R) May, 1967. Hugh O'Brian, John Mills. Color

AFRICAN ELEPHANT, THE 92
NATIONAL GENERAL. (R) October, 1971. Documentary.

AFRICAN LION, THE 75
BUENA VISTA. (R). October 10, 1955. Documentary. Color.

AFRICAN MANHUNT 65
REPUBLIC. (R) January 5, 1955. Myron Healey, Karin Booth.

AFRICAN SAFARI 98

Length In Mins.

CROWN INTERNATIONAL. (R) February, 1969. Documentary. Color.

AFTER THE FOX (P) **103**
UNITED ARTISTS. (R) December, 1966. Peter Sellers, Victor Mature. Color.

AFTER YOU COMRADE **84**
CONTINENTAL. (R) April, 1967. Jamie Uys, Bob Courtney. Color.

AGATHA **98**
WARNER BROS. (R) February, 1979. Dustin Hoffman, Vanessa Redgrave.

AGE OF CONSENT **98**
COLUMBIA. (R) January, 1970. James Mason, Helen Mirren. Color.

AGENT FOR H.A.R.M. **84**
UNIVERSAL. (R) February, 1966. Mark Richman, Wendell Corey. Color.

AGENT 8¾ **98**
CONTINENTAL. (R) June, 1965. Dirk Bogarde, Sylvia Koscina. Color.

AGONY AND THE ECSTASY, THE (Todd-AO) **140**
TWENTIETH CENTURY-FOX. (R) October, 1965. Charlton Heston, Rex Harrison Color.

AIN'T MISBEHAVIN' **82**
UNIVERSAL. (R) July, 1955. Rory Calhoun, Piper Laurie, Jack Carson, Mamie Van Doren. Color.

AIR PATROL **70**
TWENTIETH CENTURY-FOX. (R) July, 1962. Willard Parker, Merry Anders.

AIR STRIKE **67**
LIPPERT. (R) May 6, 1955. Richard Denning, Gloria Hean,

AIRPLANE **88**
PARAMOUNT. (R) July, 1980. Robert Hayes, Julie Hagerty.

AIRPORT (Todd-AO) **137**
UNIVERSAL. (R) March, 1970. Burt Lancaster, Dean Martin, Jean Seberg.

AIRPORT 1975 .... **106**
UNIVERSAL. (R) October, 1974. Charlton Heston, Karen Black.

AIRPORT '77 (P) **113**
UNIVERSAL. (R) May, 1977. . Jack Lemmon, Lee Grant.

AL CAPONE **105**
ALLIED ARTISTS. (R) March 26, 1959. Rod Steiger, Fay Spain.

ALAKAZAM THE GREAT **85**
AMERICAN INTERNATIONAL. (R) July, 1961. Animated Cartoon. Color.

ALAMO, THE **199**
UNITED ARTISTS. (R) October, 1960. John Wayne, Richard Widmark. Color.

ALASKA PASSAGE (RS) **71**
TWENTIETH CENTURY-FOX. (R) February, 1959. Bill Williams, Lyn Thomas.

ALBERT SCHWEITZER **80**
DE ROCHEMONT. (R) January, 1957. Documentary. Color.

ALEX IN WONDERLAND **109**
MGM. (R) December, 1970. Donald Sutherland, Jeanne Moreau. Color.

ALEXANDER THE GREAT (CS) **141**
UNITED ARTISTS. (R) April, 1956. Richard Burton, Frederic March, Claire Bloom. Color.

ALF AND FAMILY **100**
SHERPIX. (R) August, 1972. Warren Mitchell, Randy Nichols. Color.

ALFIE (TS) **114**
PARAMOUNT. (R) October, 1966. Michael Caine, Shelley Winters. Color.

ALFRED THE GREAT (P) **125**
MGM. (R) October, 1969. David Hemmings, Michael York. Color.

ALREDO ALFREDO **100**
PARAMOUNT. (R) December, 1973. Dustin Hoffman, Stefania Sandrelli.

ALI BABA AND THE SEVEN MIRACLES OF THE WORLD (Cs) **79**
AMERICAN INTERNATIONAL. (R) November, 1961. Cartoon Feature. Color.

ALIAS JESSE JAMES **92**
UNITED ARTISTS. (R) March, 1959. Bob Hope, Rhonda Fleming. Color.

ALIAS JOHN PRESON (Brit.) **71**
ASSOCIATED ARTISTS. (R) November, 1955. Alexander Knox, Betta St. John.

ALICE DOESN'T LIVE HERE ANYMORE **113**
WARNER BROS. (R) January, 1975. Ellyn Burstyn, Kris Kristofferson.

ALICE IN WONDERLAND IN PARIS AND WHITE MANE **90**
CHILDHOOD PRODUCTIONS. (R) February, 1966. Animated.

ALICE'S ADVENTURES IN WONDERLAND **100**
AMERICAN NATIONAL ENTERPRISES. (R) November, 1972. Ralph Richardson, Fiona Fullerton, Peter Sellers. Color.

ALICE'S RESTAURANT **111**
UNITED ARTISTS. (R) August, 1969. Arlo Guthrie, Pat Quinn, James Broderick. Color.

ALIEN **124**
TWENTIETH CENTURY-FOX. (R) May, 1979. Tom Skeritt, Sigourney Weaver.

ALL-AMERICAN BOY, THE (P) **118**
WARNER BROS. (R) November, 1973. Jon Voight, Carol Androsky.

ALL AT SEA (Brit.) **87**
METRO-GOLDWYN-MAYER. (R) January, 1958. Alec Guiness, Irene Brownee.

ALL FALL DOWN **111**
METRO-GOLDWYN-MAYER. (R) April, 1962. Eva Marie Saint, Warren Beatty.

ALL HANDS ON DECK (Cs) **98**
TWENTIETH CENTURY-FOX. (R) April, 1961. Pat Boone, Buddy Hackett. Color.

ALL IN A NIGHT'S WORK (VV) **94**
PARAMOUNT. (R) March, 1961. Dean Martin, Shirley MacLaine. Color.

ALL MINE TO GIVE **87**
UNIVERSAL. (R) January, 1958. Glynis Johns, Cameron Mitchell. Color.

ALL NEAT IN BLACK STOCKINGS **96**
NATIONAL GENERAL. (R) October, 1969. Victor Henry, Susan George. Color.

ALL SCREWED UP (Ital., English titles) **105**
NEW LINE CINEMA. (R) January, 1976. Luigi Diberti, Lina Polito.

ALL THAT HEAVEN ALLOWS **89**
UNIVERSAL. (R) January, 1956. Jane Wyman, Rock Hudson. Color.

ALL THAT JAZZ **123**
TWENTIETH CENTURY-FOX. (R) December, 1979. Roy Scheider, Jessica Lange.

ALL THE FINE YOUNG CANNIBALS (Cs) **122**
METRO-GOLDWYN-MAYER. (R) September, 1960. Natalie Wood, Robert Wagner. Color.

ALL THE LOVING COUPLES **85**
U-M. (R) October, 1969. Norman Aldren, Gloria Manon, Scott Graham. Color.

ALL THE PRESIDENT'S MEN **135**
WARNER BROS. (R) April, 1976. Robert Redford, Dustin Hoffman.

ALL THE WAY HOME **103**
PARAMOUNT. (R) October, 1963. Jean Simmons, Robert Preston.

ALL THE YOUNG MEN **86**
COLUMBIA. (R) August, 1960. Alan Ladd, Sidney Poitier.

ALL THESE WOMEN **80**
JANUS. (CR) October, 1964. Jark Kulle, Georg Funkquist. Color.

ALLIGATOR NAMED DAISY, AN (VV) **84**
RANK. (R) October, 1957. Donald Sinden, Diana Dors. Color

ALLIGATOR PEOPLE, THE (Cs) **74**
ALLIED ARTISTS. (R) July, 1959. Beverly Garland, George Macready.

ALMOST ANGELS **93**
BUENA VISTA. (R) October, 1962. Peter Weck, Sean Scully. Color.

ALMOST PERFECT AFFAIR, AN **93**
PARAMOUNT. (R) April, 1979. Keith Carradine, Monica Vitti.

ALOHA, BOBBY AND ROSE **89**
COLUMBIA. (R) April, 1975. Paul Le Mat, Dainne Hull.

ALONE AGAINST ROME (Totalscope) **100**
MEDALLION. (R) 1960. Jeffrey Lang, Rosanna Podesta. Color.

ALPHABET MURDERS, THE **90**
METRO-GOLDWYN-MAYER. (R) May, 1966. Tony Randall, Anita Ekberg.

ALPHAVILLE (Fr., Eng. titles) **100**
PATHE CONTEMPORARY. (R) October, 1965. Eddie Constantine, Anna Karina.

ALVEREZ KELLY (P) **116**
COLUMBIA. (R) October, 1966 William Holden, Richard Widmark, Color

ALVIN PURPLE **92**
SANDS FILM CO. (R) October, 1975. Graeme Blundell, George Whaley.

AMAZING COLOSSAL MAN, THE **81**
AMER. INT. PICS. (R) September, 1957. Glenn Langan, Cathy Downs.

AMAZING GRACE **99**
UNITED ARTISTS (R) August, 1974. Moms Mabley, Slappy White.

AMAZING TRANSPARENT MAN, THE **60**

Length In Mins.

AMERICAN INTERNATIONAL. (R) July, 1960. Marguerite Chapman, Douglas Kennedy.

AMAZON TRADER, THE 41
WARNER BROTHERS. (R) September 8, 1956. John Sutton, Maria Fernandez. Color.

AMBASSADOR's DAUGHTER, THE (Cs) 102
UNITED ARTISTS. (R) September, 1956. Olivia de Havilland, John Forsythe, Myrna Loy. Color.

AMBUSH AT CIMARRON PASS (Rs) 87
TWENTIETH CENTURY-FOX. (R) March, 1958. Scott Brady, Margia Dean.

AMBUSH BAY 109
UNITED ARTISTS. (R) September, 1966. Hugh O'Brian, Mickey Rooney. Color.

AMBUSHERS, THE 102
COLUMBIA. (R) December, 1967. Dean Martin, Senta Berger. Color.

AMERICA, AMERICA 174
WARNER BROS. (R) February, 1964. Stathis Giallelis, Frank Wolff.

AMERICAN DREAM, AN 107
WARNER BROS. (R) October, 1966. Stuart Whitman, Janet Leigh. Color.

AMERICAN FRIEND, THE 127
NEW YORKER FILMS. (R) October, 1977. Bruno Ganz, Dennis Hopper.

AMERICAN GIGOLO 115
PARAMOUNT. (R) February, 1980. Richard Gere, Lauren Hutton.

AMERICANIZATION OF EMILY, THE 117
METRO-GOLDWYN-MAYER. (R) December, 1964. James Garner, Julie Andrews.

AMERICANO, THE 85
RKO RADIO. (R) January 29, 1955. Glenn Ford, Ursula Theiss. Color.

AMERICATHON 85
UNITED ARTISTS. (R) August, 1979. Peter Riegert, Harvey Korman.

AMITYVILLE HORROR, THE 117
AMERICAN INTERNATIONAL. (R) July, 1979. James Brolin, Margot Kidder.

AMOROUS ADVENTURES OF MOLL FLANDERS, THE (P) 126
PARAMOUNT. (R) May, 1965. Kim Novak, Richard Johnson. Color.

AMSTERDAM KILL, THE (P) 92
COLUMBIA. (R) March, 1978. Robert Mitchum, Bradford Dillman.

ANASTASIA (Cs) 105
TWENTIETH CENTURY-FOX. (R) December, 1956. Ingrid Bergman, Yul Brynner, Helen Hayes. Color.

ANATOMY OF A MARRIAGE (Fr., Eng. titles) .... 96 & 97
JANUS. (R) October, 1964. Marie-Jose Nat, Jacques Charrier. (Two-Parts).

ANATOMY OF A MURDER 160
COLUMBIA. (R) July, 1959. James Stewart, Lee Remick.

AND HOPE TO DIE 99
TWENTIETH CENTURY-FOX. (R) November, 1972. Robert Ryan, Jean-Louis Trintingant, Lea Massari. Color.

AND JUSTICE FOR ALL 120
COLUMBIA. (R) October, 1979. Al Pacino, Jack Warden.

AND NOW FOR SOMETHING COMPLETELY DIFFERENT 88
COLUMBIA. (R) August, 1972. Graham Chapman, John Cleese. Color.

AND NOW MIGUEL 95
UNIVERSAL. (R) June, 1966. Pat Cardi, Michael Ansara.

AND NOW MY LOVE (Fr., Eng. titles) 121
AVCO-EMBASSY. (R) March, 1975. Marthe Keller, Andre Dussolier.

AND NOW THE SCREAMING STARTS 87
CINERAMA. (R) April, 1973. Peter Cushing. Color.

AND SO TO BED (Germ., Eng. titles) 112
MEDALLION. (R) May, 1965. Hildegard Knef, Lilli Palmer.

AND SOON THE DARKNESS 98
LEVITT-PICKMAN. (R) April, 1971. Pamela Franklin, Michele Dotrice. Color.

AND SUDDENLY IT'S MURDER (Cs) 90
ROYAL. (R) January, 1964. Alberto Sordi, Dorian Gray.

AND THERE CAME A MAN 90
BRANDON. (R) April, 1968. Rod Steiger, Adolfo Celi. Color.

ANDERSON TAPES, THE 98
COLUMBIA. (R) June, 1971. Sean Connery, Dyan Cannon. Color.

ANDROMEDA STRAIN, THE (P) 130
UNIVERSAL. (R) March, 1971. Arthur Hill, David Wayne.

ANDY 86
UNIV. (R) February, 1965. Norman Alden, Tamara Day Karhanova.

ANDY HARDY COMES HOME 80
METRO-GOLDWYN-MAYER. (R) August, 1958. Mickey Rooney, Patricia Breslin.

ANDY WARHOL'S BAD 105
NEW WORLD. (R) May, 1977. Carroll Baker, Perry King.

ANDY WARHOL'S DRACULA 106
BRYANSTON. (R) January, 1975. Joe Dallesandro, Udo Kier.

ANDY WARHOL'S FRANKENSTEIN 95
BRYANSTON. (R) May, 1974. Joe Dallesandro, Monique Van Vooren.

ANGEL BABY 97
ALLIED ARTISTS. (R) May, 1961. George Hamilton, Mercedes McCambridge.

ANGEL IN MY POCKET 105
UNIVERSAL. (R) April, 1969. Andy Griffith, Jerry Van Dyke, Kay Medford. Color.

ANGEL LEVINE, THE 104
UNITED ARTISTS. (R) August, 1970. Harry Belafonte, Zero Mostel, Ida Kaminska. Color.

ANGEL WHO PAWNED HER HARP, THE (Brit.) 73
ASSOCIATED ARTISTS. (R) February, 1956. Felix Aylmer, Diane Cilento.

ANGEL WORE RED, THE 99
METRO-GOLDWYN-MAYER. (R) September, 1960. Ava Gardner, Dirk Bogarde.

ANGELA 81
TWENTIETH CENTURY-FOX. (R) April, 1955. Dennis O'Keefe, Mara Lane.

ANGELS FROM HELL 86
AMERICAN INTERNATIONAL. (R) June, 1968. Tom Stern, Arlene Martel. Color.

ANGELS OF DARKNESS (Ital., Eng. dial.) 84
EXCELSIOR. (R) February, 1957. Linda Darnell, Anthony Quinn.

ANGRY HILLS, THE (Cs) 105
METRO-GOLDWYN-MAYER. (R) June, 1959. Robert Mitchum, Stanley Baker.

ANGRY RED PLANET, THE 83
AMER. INT'L. PICS. (R) 1960. Gerald Mohr, Les Tremayne. Color.

ANIMAL FARM (Brit.) 75
DISTRIBUTORS CORP. OF AMERICA. (R) January, 1955. Cartoon. Color.

ANIMAL WORLD, THE 82
WARNER BROTHERS. (R) June 23, 1956. Documentary. Color.

ANIMALS ARE BEAUTIFUL PEOPLE 92
WARNER BROS. (R) March, 1975. Documentary.

ANNA LUCASTA 97
UNITED ARTISTS. (R) March, 1959. Eartha Kitt, Sammy Davis, Jr.

ANNAPOLIS STORY 81
ALLIED ARTISTS. (R) April 10, 1955. John Derek, Diana Lynn. Color.

ANNE AND EVE 89
CHEVRON. (R) July, 1970. Gio Petre, Marie Liljedhal. Color.

ANNE OF THE THOUSAND DAYS (P) 145
(R) January, 1970. Richard Burton, Genevieve Bujold. Color.

ANNIE HALL 94
UNITED ARTISTS. (R) April, 1977. Woody Allen, Diane Keaton.

ANNIVERSARY, THE 95
TWENTIETH CENTURY-FOX. (R) Feb., 1968. Bette Davis, Christian Roberts. Color.

ANOTHER MAN, ANOTHER CHANCE 136
UNITED ARTISTS. (R) November, 1977. James Caan, Genevieve Bujold

ANOTHER TIME, ANOTHER PLACE (VV) 98
PARAMOUNT. (R) June, 1958. Lana Turner, Barry Sullivan.

ANY GUN CAN PLAY 105
GOLDEN EAGLE. (R) September, 1969. Edd Byrnes, Gilbert Roland. Color.

ANY NUMBER CAN WIN (D) 112
METRO-GOLDWYN-MAYER. (R) October, 1963. Jean Gabin, Alain Delon.

ANY WEDNESDAY 109
WARNER BROS. (R) December, 1966. Jane Fonda, Jason Robards. Color.

ANYONE CAN PLAY 88
PARAMOUNT. (R) September, 1968. Ursula Andress, Virna Lisi, Claudine Auger. Color.

ANYTHING GOES (VV) 106
PARAMOUNT. (R) April, 1956. Bing Crosby, Donald O'Connor, Jeanmarie, Mitzi Gaynor. Color.

ANZIO 117
COLUMBIA. (R) June, 1968. Robert Mitchum, Peter Falk, Earl Holliman. Color.

APACHE AMBUSH 68

Length In Mins.

COLUMBIA. (R) September, 1955. Bill Williams, Richard Jaeckel.

APACHE GOLD (Cs) **91**
COLUMBIA. (R) May, 1965. Lex Barker, Mario Adorf. Color.

APACHE RIFLES **92**
TWENTIETH CENTURY-FOX. (R) November, 1964. Audie Murphy, Michael Dante, Color.

APACHE TERRITORY **75**
COLUMBIA. (R) September, 1958. Rory Calhoun, Barbara Bates.

APACHE UPRISING (Techniscope) **90**
PARAMOUNT. (R) January, 1966. Rory Calhoun, Corine Calvet. Color.

APACHE WARRIOR (Cs) **73**
TWENTIETH CENTURY-FOX. (R) August, 1957. Keith Larsen, Jim Davis.

APACHE WOMAN **83**
AMERICAN INTERNATIONAL. (R) October, 1955. Lloyd Bridges, Joan Taylor. Color.

APARTMENT, THE (P) **125**
UNITED ARTISTS. (R) July, 1960. Jack Lemmon, Shirley MacLaine.

APE WOMAN, THE (Ital., Eng. titles) **97**
EMBASSY. (R) September, 1964. Ugo Tognazzi, Annie Girardot.

APOCALYPSE NOW (70mm) **146**
UNITED ARTISTS. (R) August, 1979. Marlon Brando, Robert Duvall, Martin Sheen.

APPALOOSA, THE (TS) **98**
UNIVERSAL. (R) October, 1966. Marlon Brando, Anjanette Comer. Color.

APPLE DUMPLING GANG, THE **100**
BUENA VISTA. (R) July, 1975. Bill Bixby, Susan Clark.

APPOINTMENT IN LONDON (Brit.) **96**
ASSOCIATED ARTISTS. (R) November, 1955. Dirk Bogarde, Dinah Sheridan.

APPOINTMENT WITH A SHADOW (Cs) **73**
UNIVERSAL. (R) November, 1958. George Nader, Joanna Moore.

APPRENTICESHIP OF DUDDY KRAVITZ, THE (P) **120**
PARAMOUNT. (R) July, 1974. Richard Dreyfuss, Micheline Lanctot.

APRIL FOOLS, THE **95**
NATIONAL GENERAL. (R) June, 1969. Jack Lemmon, Catherine Deneuve.

APRIL LOVE (Cs) **91**
TWENTIETH CENTURY-FOX. (R) November, 1957. Pat Boone, Shirley Jones. Color.

ARABELLA **91**
UNIVERSAL. (R) September, 1969. Virna Lisi, James Fox, Margaret Rutherford, Terry-Thomas. Color.

ARABESQUE **118**
UNIVERSAL. (R) July, 1966. Gregory Peck, Sophia Loren. Color.

ARABIAN ADVENTURE **98**
ASSOCIATED FILM DIST. (R) November, 1979. Christopher Lee, Milo O'Shea.

ARISTOCATS, THE **78**
BUENA VISTA. (R) December, 1970. Feature length cartoon. Color.

ARIZONA BUSHWHACKERS **86**
PARAMOUNT. (R) March, 1968. Howard Keel, Scott Brady. Color.

ARIZONA RAIDERS (TE) **88**
COLUMBIA. (R) July, 1965. Audie Murphy, Michael Dante. Color.

ARMORED COMMAND **99**
ALLIED ARTISTS. (R) July, 1961. Tina Louise, Howard Keel.

ARNOLD **93**
CINERAMA. (R) November, 1973. Stella Stevens, Roddy McDowall.

AROUND THE WORLD IN 80 DAYS (Todd-AO) **178**
UNITED ARTISTS. (R) October, 1956. David Niven, Shirley MacLaine. Color.

AROUND THE WORLD UNDER THE SEA (P) **117**
METRO-GOLDWYN-MAYER. (R) June, 1966. Lloyd Bridges, Shirley Eaton. Color.

ARRANGEMENT, THE (P) **129**
WARNER BROS. (R) December, 1969. Kirk Douglas, Faye Dunaway, Deborah Kerr. Color.

ARRIVEDERCI, BABY! (P) **105**
PARAMOUNT. (R) December, 1966. Tony Curtis, Rosanna Schiaffino. Color.

ARSON FOR HIRE **67**
ALLIED ARTISTS. (R) March, 1959. Steve Brodie, Lyn Thomas.

ARTISTS AND MODELS (VV) **109**
PARAMOUNT. (R) January, 1956. Dean Martin, Jerry Lewis. Color.

ARTURO'S ISLAND **90**
METRO-GOLDWYN-MAYER. (R) December, 1962. Reginald Kernan, Key Meersman

AS LONG AS THEY'RE HAPPY **76**
RANK. (R) November, 1957. Jack Buchanan, Jeanette Scott.

AS LONG AS YOU'RE NEAR ME (Ger.-Eng. titles) **101**
WARNER BROTHERS. (R) June 9, 1956. Maria Schell, O.W. Fisher.

AS THE SEA RAGES **74**
COLUMBIA. (R) September, 1960. Maria Schell, Cliff Robertson.

AS YOUNG AS WE ARE **76**
PARAMOUNT. (R) September, 1958. Robert Harland, Pippa Scott.

ASH WEDNESDAY **99**
PARAMOUNT. (R) November, 1973. Elizabeth Taylor, Henry Fonda.

ASK ANY GIRL (Cs) **101**
METRO-GOLDWYN-MAYER. (R) June, 1959. David Niven, Shirley MacLaine. Color.

ASSASSINATION BUREAU, THE **110**
PARAMOUNT. (R) March, 1969. Oliver Reed, Diana Rigg, Telly Savalas, Curt Jurgens. Color.

ASSASSINATION OF TROTSKY, THE **103**
CINERAMA. (R) October, 1972. Richard Burton, Alain Delon, Romy Schneider. Color.

ASSAULT ON A QUEEN (P) **106**
PARAMOUNT. (R) Summer, 1966. Frank Sinatra, Virna Lisi. Color.

ASSIGNMENT K **97**
COLUMBIA. (R) June, 1968. Stephen Boyd, Camilla Sparv. Color.

ASSIGNMENT: OUTERSPACE **79**
AMERICAN INT'L. PICS. (R) 1962. Archie Savage, Rik Von Hutter. Color.

ASSIGNMENT TO KILL **102**
WARNER BROS.-7 ARTS. (R) January, 1969. Patrick O'Neal, Joan Hackett, Herbert Lom. Color.

ASTONISHED HEART, THE (Brit.) **92**
UNIVERSAL. (R) March, 1950. Noel Coward, Celia Johnson.

ASYLUM **88**
CINERAMA. (R) October, 1972. Peter Cushing, Britt Eckland. Color.

AT GUNPOINT (Cs) **81**
ALLIED ARTISTS. (R) December 25, 1955. Fred MacMurray, Dorothy Malone. Color.

AT LONG LAST LOVE **115**
20th CENTURY-FOX. (R) March, 1975. Burt Reynolds, Cybill Shepherd.

AT THE EARTH'S CORE **90**
AMERICAN INTERNATIONAL. (R) June, 1976. Doug McClure, Peter Cushing.

ATLANTIS, THE LOST CONTINENT **90**
METRO-GOLDWYN-MAYER. (R) May, 1961. Anthony Hall, Joyce Taylor. Color.

ATOMIC MAN, THE **78**
ALLIED ARTISTS. (R) March 4, 1956. Gene Nelson, Faith Domergue.

ATOMIC SUBMARINE, THE **73**
ALLIED ARTISTS. (R) January, 1960. Arthur Franz, Dick Foran.

ATRAGON (Colorscope) **88**
AMER. INT'L. PICS. (R) 1965. Kenji Sawara, Akemakita. Color.

ATTACK **107**
UNITED ARTISTS. (R) October, 1958. Jack Palance, Eddie Albert.

ATTACK OF THE CRAB MONSTERS **64**
ALLIED ARTISTS. (R) March 3, 1957. Richard Garland, Pamela Duncan.

ATTACK OF THE 50 FOOT WOMAN **65**
ALLIED ARTISTS. (R) May, 1958. Allison Hayes, William Hudson.

ATTACK OF THE GIANT LEECHES **62**
AMER. INT'L. PICS. (R) 1959. Ken Clark, Yvette Vickers. b&w.

ATTACK OF THE JUNGLE WOMEN **72**
BARJUL INT. (R) September, 1959. William Phillips. Color.

ATTACK OF THE PUPPET PEOPLE **78**
AMERICAN INTERNATIONAL. (R) June, 1958. John Hoyt, John Agar.

ATTACK ON THE IRON COAST **89**
UNITED ARTISTS. (R) March, 1968. Lloyd Bridges, Andrew Keir. Color.

AUDREY ROSE (P) **113**
UNITED ARTISTS. (R) April, 1977. Marsha Mason, Anthony Hopkins.

AUNTIE MAME (T) **143**
WARNER BROS. (R) December, 1958. Rosalind Russell, Forrest Tucker. Color.

AUTUMN LEAVES **108**

	Length In Mins.
COLUMBIA. (R) August, 1956. Joan Crawford, Cliff Robertson.	
AUTUMN SONATA NEW WORLD PICTURES. (R) October, 1979. Ingrid Bergman, Liv Ullmann.	97
AVANTI UNITED ARTISTS. (R) December, 1972. Jack Lemmon, Juliet Mills. Color.	144
AVENGER, THE MEDALLION. (R) January, 1965. Steve Reeves, Cupia Marlier. Color.	108
AWAY ALL BOATS (VV) UNIVERSAL. (R) August, 1956. Jeff Chandler, Julie Adams, George Nader, Lex Barker. Color.	114

B

	Length In Mins.
B.S., I LOVE YOU 20TH CENTURY-FOX. (R) March, 1971. Peter Kastner, Joanna Barnes. Color.	99
BABES IN TOYLAND BUENA VISTA. (R) December, 1961. Ray Bolger, Rommy Sands. Color.	105
BABETTE GOES TO WAR (Cs) COLUMBIA. (R) March, 1960. Brigette Bardot, Jacques Charrier. Color.	103
BABY BLUE MARINE COLUMBIA. (R) April, 1976. Jan Michael-Vincent, Glynnis O'Connor.	90
BABY DOLL WARNER BROTHERS. (R) December 29, 1956. Karl Malden, Carroll Baker, Eli Wallach.	114
BABY FACE NELSON UNITED ARTISTS. (R) November, 1957. Mickey Rooney, Carolyn Jones.	85
BABY LOVE AVCO-EMBASSY. (R) March, 1969. Ann Lynn, Keith Barron, Linda Hayden. Color.	102
BABY MAKER, THE NATIONAL GENERAL. (R) October, 1970. Barbara Hershey, Collin Wilcox-Horne. Color.	109
BABY THE RAIN MUST FALL COLUMBIA. (R) January, 1965. Lee Remick, Steve McQueen.	100
BABYSITTER, THE CROWN INTERNATIONAL. (R) October, 1969. Patricia Wymer, George E. Carey.	90
BACHELOR FLAT (Cs) TWENTIETH CENTURY-FOX. (R) January, 1961. Tuesday Weld, Richard Beymer. Color.	92
BACHELOR IN PARADISE (Cs) METRO-GOLDWYN-MAYER. (R) November, 1961. Bob Hope, Lana Turner. Color.	109
BACHELOR PARTY, THE UNITED ARTISTS. (R) April, 1957. Don Murray, E.G. Marshall.	93
BACK DOOR TO HELL TWENTIETH CENTURY-FOX. (R) January, 1965. Jimmie Rodgers, Conrad Maga.	68
BACK FROM ETERNITY UNIVERSAL. (R) September, 1956. Robert Ryan, Anita Ekberg, Rod Steiger.	97
BACK FROM THE DEAD (Cs) TWENTIETH CENTURY-FOX. (R) September, 1957. Arthur Franz, Peggie Castle.	75
BACK STREET UNIVERSAL. (R) November, 1961. Susan Hayward, John Gavin.	107
BACKFIRE (Fr., Eng. titles) ROYAL. (R) April, 1965. Jean-Paul Belmondo, Jean Seberg.	97
BACKLASH UNIVERSAL. (R) April, 1956. Richard Widmark, Donna Reed. Color.	84
BAD COMPANY PARAMOUNT. (R) October, 1972. Jeff Bridges, Barry Brown. Color.	91
BAD DAY AT BLACK ROCK (Cs) METRO-GOLDWYN-MAYER. (R) January 7, 1955. Spencer Tracy, Anne Francis. Color.	81
BAD MAN'S COUNTRY WARNER BROS. (R) August, 1958. George Montgomery, Buster Crabbe.	68
BAD NEWS BEARS, THE PARAMOUNT. (R) April, 1976. Walter Matthau, Tatum O'Neal.	105
BAD NEWS BEARS IN BREAKING TRAINING, THE PARAMOUNT. (R) August, 1977. William Devane, Clifton James.	97
BAD SEED, THE WARNER BROTHERS. (R) September 29, 1956. Nancy Kelly, Patty McCormick.	129
BADGE #73 PARAMOUNT. (R) August, 1973. Robert Duvall, Verna Bloom.	115
BADGE OF MARSHAL BRENNAN, THE ALLIED ARTISTS. (R) May 26, 1957. Jim Davis, Carl Smith.	78
BADLANDERS, THE (Cs) METRO-GOLDWYN-MAYER. (R) August, 1958. Alan Ladd, Ernest Borgnine. Color.	83
BADLANDS WARNER BROS. (R) May, 1974. Martin Sheen, Sissy Spacek.	95
BADLANDS OF MONTANA TWENTIETH CENTURY-FOX. (R) May, 1957. Rex Reason, Margia Dean.	75
BAIL OUT AT 43,000 UNITED ARTISTS. (R) May, 1957. John Payne, Karen Steel.	78
BALCONY, THE CONTINENTAL. (R) March, 1963. Shelley Winters, Peter Falk.	84
BALLAD OF A SOLDIER (Russian: Eng. titles) ARTKINO. (R) December, 1960. b&w.	89
BALLAD OF CABLE HOGUE, THE WARNER BROS. (R) March, 1970. Jason Robards, Stella Stevens. Color.	120
BALLAD OF JOSIE, THE (T) UNIVERSAL. (R) February, 1968. Doris Day, Peter Graves. Color.	102
BALLAD OF LOVE, A (Russian, Eng. Titles) ARTKINO. (R) February 19, 1966. Victoria Fyodorova, Valentin Smirnitsky.	45
BALTIMORE BULLET, THE AVCO EMBASSY. (R) April, 1980. James Coburn, Omar Sharif.	103
BAMBOLE (Ital., with English Titles) ROYAL FILMS INT'L. (R) June, 1965. Virna Lisi, Nino Manfredi.	111
BAMBOO PRISON, THE COLUMBIA. (R) January, 1955. Robert Francis, Dianne Foster.	80
BANANA PEEL PATHE-CONT. (R) January, 1965. Jeanne Morreau, Jean-Paul Belmondo.	97
BANANAS UNITED ARTISTS. (R) May, 1971. Woody Allen, Louise Lasser. Color.	82
BAND OF ANGELS WARNER BROS. (R) August, 1957. Clark Gable, Yvonne DeCarlo. Color.	127
BAND OF OUTSIDERS (Fr., Eng. Titles) ROYAL FILMS INTERNATIONAL. (R) March, 1966. Anna Karina, Sami Frey.	95
BANDIDO (Cs) UNITED ARTISTS. (R) September, 1956. Robert Mitchum, Ursula Thiess. Color.	92
BANDIT OF ZHOBE, THE (Cs) COLUMBIA. (R) April, 1959. Victor Mature, Anthony Newly. Color.	80
BANDITS OF ORGOSOLO PATHE-CONT. (R) May, 1964. Michele Lossu, Peppaddu Cuccu.	98
BANDOLERO TWENTIETH CENTURY-FOX. (R) July, 1968. James Stewart, Dean Martin, Raquel Welch, George Kennedy. Color.	106
BANG, BANG, YOU'RE DEAD AMERICAN INTERNATIONAL. (R) August, 1966. Tony Randall, Senta Berger. Color.	92
BANG THE DRUM SLOWLY PARAMOUNT. (R) August, 1973. Michael Moriarty, Robert de Niro.	97
BANK SHOT UNITED ARTISTS. (R) July, 1974. George C. Scott, Joanna Cassidy.	85
BANNING (TS) UNIVERSAL. (R) August, 1967. Robert Wagner, Anjanette Comer. Color.	102
BARABAS (T-70) COLUMBIA. (R) December, 1962. Anthony Quinn, Silvana Mangano. Color.	134
BARBARELLA PARAMOUNT. (R) October, 1968. Jane Fonda, John Phillip Law, Anita Pallenberg. Color.	98
BARBARIAN AND THE GEISHA, THE (Cs) TWENTIETH CENTURY-FOX. (R) October, 1958. John Wayne, Elko Ando. Color.	104
BARBARIANS, THE ALLIED ARTISTS. (R) December, 1957. Pierre Cresoy, Helene Remy.	80
BAREFOOT BATTALION (Germ., Eng., titles)	63

Length In Mins.

TWENTIETH CENTURY-FOX. (R) July, 1956. Maria Costi, Nicos Fermas.

BAREFOOT EXECUTIVE, THE 95
BUENA VISTA. (R) April, 1971. Kurt Russel, Joe Flynn. Color.

BAREFOOT IN THE PARK. 106
PARAMOUNT. (R) June, 1967. Robert Redford, Jane Fonda. Color.

BARON BLOOD 90
AMERICAN INTERNATIONAL. (R) October, 1972. Elke Sommer, Joseph Cotten. Color.

BARQUERO 115
UNITED ARTISTS. (R) May, 1970. Lee Van Cleef, Warren Oates. Color.

BARRETTS OF WIMPOLE STREET, THE (Cs) 105
METRO-GOLDWYN-MAYER. (R) February 1, 1957. Jennifer Jones, Bill Travers. Color.

BARRY LYNDON 185
WARNER BROS. (R) December, 1975. Ryan O'Neal, Marisa Berenson.

BARTLEBY 78
MARON. (R) February, 1972. Paul Scofield, John McEnery. Color.

BASHFUL ELEPHANT, THE 82
ALLIED ARTISTS. (R) February, 1962. Molly Mack, Helmut Schmid.

BAT, THE 80
ALLIED ARTISTS. (R) August, 1959. Vincent Price, Agnes Moorehead.

BATMAN 105
20TH CENTURY-FOX. (R) August, 1966. Adam West, Burt Ward. Color.

BATTLE AT BLOODY BEACH 60
TWENTIETH CENTURY-FOX. (R) June, 1961. Audie Murphy, Sal Mineo. Color.

BATTLE BENEATH THE SEA 92
MGM. (R) May, 1968. Kerwin Mathews, Vivian Ventura. Color.

BATTLE BEYOND THE SUN 75
AMERICAN INTERNATIONAL. (R) April, 1963. Ed Perry, Arla Powell.

BATTLE CRY (Cs) 148
WARNER BROTHERS. (R) March 12, 1955. Van Heflin, Aldo Ray. Color.

BATTLE FLAME 75
ALLIED ARTISTS. (R) April, 1959. Ronny Pennell, Ronnie Knox.

BATTLE FOR THE PLANET OF THE APES, THE 90
TWENTIETH CENTURY-FOX. (R) June, 1973. Roddy McDowall, Claude Akins. Color.

BATTLE HELL 112
DCA DIST. (R) 1956. Richard Todd, Akim Tamiroff. b&w.

BATTLE HYMN (Cs) 108
UNIVERSAL. (R) March, 1957. Rock Hudson, Martha Hyer. Color.

BATTLE IN OUTER SPACE 90
COLUMBIA. (R) June, 1960. Ryo Ikebe, Kyoko Angai. Color.

BATTLE OF ALGIERS (Algerian, Eng. titles) 123
ALLIED ARTISTS. (R) 1967. Jean Martin, Yacef Saadi. b&w.

BATTLE OF BLOOD ISLAND 64
THE FILM GROUP. (R) May, 1960. Richard Devon, Ron Kennedy.

BATTLE OF BRITAIN, THE (P) 133
UNITED ARTISTS. (R) October. Laurence Olivier, Ralph Richardson, Michael Caine. Color.

BATTLE OF NERETVA, THE (P) 102
AMERICAN INTERNATIONAL (R) February, 1971. Yul Brynner, Orson Welles. Color.

BATTLE OF THE BULGE (Ultra Panavision) 163
WARNER BROTHERS. (R) January, 1966. Henry Fonda, Robert Shaw. Color.

BATTLE OF THE CORAL SEA, THE 80
COLUMBIA. (R) November, 1959. Cliff Robertson, Gia Scala.

BATTLE OF THE VILLA FIORITA, THE (P) 111
WARNER BROS. (R) June, 1965. Maureen O'Hara, Rossano Brazzi. Color.

BATTLESTAR GALACTICA 120
UNIVERSAL. (R) May, 1979. Lorne Green, Richard L. Hatch.

BATTLE STATIONS 81
COLUMBIA. (R) February, 1956. John Lund, William Bendix, Keefe Brasselle.

BATTLE TAXI 82
UNITED ARTISTS. (R) January, 1955. Sterling Hayden. Arthur Franz.

BAWDY ADVENTURES OF TOM JONES, THE 94
UNIVERSAL. (R) 1976. Nicky Henson, Trevor Howard.

BAXTER! 100
NATIONAL GENERAL. (R) February, 1973. Scott Jacoby, Patricia Neal, Britt Eckland Color.

BAY OF THE ANGELS (Fr., Eng. titles) (Cs) 85
PATHE-CONT. (R) November, 1964. Jeanne Moreau, Claude Mann.

BAYOU 84
UNITED ARTISTS. (R) June, 1957. Peter Graves, Lita Milan.

BEACH BALL 83
PARAMOUNT. (R) October, 1965. Edd Byrnes, Chris Noel.

BEACH BLANKET BINGO (P) 98
AMERICAN INTERNATIONAL. (R) April, 1965. Frankie Avalon, Annette Funicello. Color.

BEACH PARTY (P) 100
AMERICAN INTERNATIONAL. (R) August, 1963, Bob Cummings, Dorothy Malone. Color.

BEACH RED 105
UNITED ARTISTS. (R) September, 1967. Cornel Wilde, Rip Torn. Color.

BEACHCOMBER, THE (Brit.) 82
UNITED ARTISTS. (R) January, 1955. Glynis Johns, Robert Newton. Color.

BEAR AND THE DOLL, THE 90
PARAMOUNT. (R) October, 1971. Brigitte Bardot, Jean-Pierre Cassel. Color.

BEARS AND I, THE 89
BUENA VISTA. (R) October, 1974. Patrick Wayne, Chief Dan George.

BEAST FROM A HAUNTED CAVE 75
THE FILM GROUP. (R) October, 1959. Michael Forest, Sheila Carol.

BEAST IN THE CELLAR 104
CANNON. (R) April, 1971. Beryl Reid, Flora Robson. Color.

BEAST OF BUDAPEST 72
ALLIED ARTISTS. (R) February, 1958. Gerald Milton, John Hoyt.

BEAST OF HOLLOW MOUNTAIN, THE (Cs) 79
UNITED ARTISTS. (R) August, 1956. Guy Madison, Patricia Medina. Color.

BEAST WITH 1,000,000 EYES 71
AMERICAN RELEASING CORP. (R) January, 1956. Paul Birch, Lorna Thayer.

BEASTS OF MARSEILLES, THE 70
LOPERT. (R) July, 1959. Stephen Boyd, James Robertson Justice.

BEAT GENERATION, THE (Cs) 95
METRO-GOLDWYN-MAYER. (R) July, 1959. Steve Cochran, Mamie Van Doren.

BEATSVILLE 72
ALLIED ARTISTS. (R) June, 1959. Gregg Palmer, Kathleen Crowley.

BEAU GESTE (T) 105
UNIVERSAL. (R) July 20, 1966. Guy Stockwell, Doug McClure. Color.

BEAU JAMES (VV) 105
PARAMOUNT. (R) July, 1957. Bob Hope, Vera Miles. Color.

BEAUTIFUL BUT DANGEROUS 103
TWENTIETH CENTURY-FOX. (R) February, 1958. Gina Lollobrigida, Victorio Gassman. Color.

BEBO'S GIRL 106
READE-STERLING. (R) November, 1964. Claudia Cardinale, George Chakiris.

BECAUSE THEY'RE YOUNG 102
COLUMBIA. (R) March, 1960. Dick Clark, Michael Callan.

BECKET (P-70) 148
PARAMOUNT. (R) Special March, 1964. Richard Burton, Peter O'Toole. Color.

BED AND BOARD (Fr., with English Titles) 95
COLUMBIA. (R) January, 1971. Jean-Pierre Leaud, Claude Jade. Color.

BED SITTING ROOM, THE 90
UNITED ARTISTS. (R) October, 1969. Rita Tushingham, Ralph Richardson, Peter Cook. Color.

BEDAZZLED (P) 107
TWENTIETH CENTURY-FOX. (R) December, 1967. Peter Cook, Raquel Welch. Color.

BEDEVILLED (Cs) 85
METRO-GOLDWYN-MAYER (R) April 22, 1955. Anne Baxter, Steve Forrest. Color.

BEDFORD INCIDENT, THE 102
COLUMBIA. (R) October, 1965. Richard Widmark, Sidney Poitier.

BEDKNOBS AND BROOMSTICKS 117
BUENA VISTA. (R) November, 1971. Angela Lansbury, David Tomlinson. Color.

BEDTIME STORY 99
UNIVERSAL. (R) July, 1964 Marlon Brando, David Niven. Color.

BEEN DOWN SO LONG IT LOOKS LIKE UP TO ME 90

Length In Mins.

PARAMOUNT. (R) October, 1971. Barry Primus, Linda De Coff. Color.

BEFORE THE REVOLUTION (Ital., Eng. titles) 112
NEW YORKER. (R) July, 1965. Adriana Asti, Francesco Barilli.

BEFORE WINTER COMES 102
COLUMBIA. (R) March, 1969. David Niven, Topol, Anna Karina, John Hurt. Color.

BEGINNING OF THE END 73
REPUBLIC. (R) June 1957. Peter Graves, Peggi Castle.

BEGUILED, THE 105
UNIVERSAL. (R) April, 1971. Clint Eastwood, Geraldine Page. Color.

BEHIND THE GREAT WALL (Totalscope) 96
CONTINENTAL. (R) 1959. Documentary. Narrated by Chet Huntley. Color.

BEHIND THE HIGH WALL 85
UNIVERSAL. (R) July, 1956. Tom Tully, Sylvia Sidney.

BEHOLD A PALE HORSE 113
COLUMBIA. (R) August, 1964. Gregory Peck, Anthony Quinn.

BEING THERE 125
UNITED ARTISTS. (R) December, 1979. Peter Sellers, Shirley MacLaine.

BELIEVE IN ME 88
METRO-GOLDWYN-MAYER. (R) September, 1971. Michael Sarrazin, Jacqueline Bisset. Color.

BELL, BOOK AND CANDLE 103
COLUMBIA. (R) January, 1959. James Stewart, Kim Novak. Color.

BELLBOY, THE 72
PARAMOUNT. (R) July, 1960. Jerry Lewis, Alex Gerry.

BELLE DE-JOUR (Fr.) 100
ALLIED ARTISTS. (R) April, 1968. Catherine Deneuve, Genevieve Page. Color.

BELLE SOMMERS 63
COLUMBIA. (R) 1962. Polly Bergen, David Janssen.

BELLES OF ST. TRINIAN'S, THE (Brit.) 90
ASSOCIATED ARTISTS. (R) January, 1955. Alastair Sim, Joyce Grenfell.

BELL JAR, THE 112
AVCO EMBASSY. (R) March, 1979. Marilyn Hassett, Julie Harris.

BELLS ARE RINGING (Cs) 127
METRO-GOLDWYN-MAYER. (R) July, 1960. Judy Holliday, Dean Martin. Color.

BELOVED INFIDEL (Cs) 123
TWENTIETH CENTURY-FOX. (R) November, 1959. Gregory Peck, Deborah Kerr. Color.

BEN 95
CINERAMA. (R) June, 1972. Lee Harcourt Montgomery, Joseph Campanella. Color.

BEN-HUR (MGM Camera 65) 212
METRO-GOLDWYN-MAYER. (R) Special. November, 1959. Charlton Heston, Jack Hawkins. Color.

BENEATH THE PLANET OF THE APES (P) 95
20TH CENTURY-FOX. (R) July, 1970. James Franciscus, Kim Hunter. Color.

BENGAZI (SS) 79
RKO RADIO. (R) September 14, 1955. Richard Conte, Richard Carlson, Mala Powers.

BENJAMIN (Fr.) 100
PARAMOUNT. (R) March, 1968. Pierre Clementi, Michele Morgan. Color.

BENNY GOODMAN STORY, THE 110
UNIVERSAL. (R) February, 1956. Steve Allen, Donna Reed. Color.

BERNADINE (Cs) 95
TWENTIETH CENTURY-FOX. (R) July, 1957. Pat Boone, Terry Moore. Color.

BERSERK! 96
COLUMBIA. (R) Jan., 1968. Joan Crawford, Ty Hardin. Color.

BEST HOUSE IN LONDON, THE 94
MGM. (R) September, 1969. David Hemmings, Joanna Pettet, George Sanders. Color.

BEST MAN, THE 102
UNITED ARTISTS. (R) May, 1964. Henry Fonda, Cliff Robertson.

BEST OF ENEMIES, THE (T) 104
COLUMBIA. (R) September, 1962. David Niven, Alberto Sordi. Color.

BEST OF EVERYTHING, THE (Cs) 121
TWENTIETH CENTURY-FOX. (R) October, 1959. Hope Lange, Stephen Boyd. Color.

BEST OF WALT DISNEY'S TRUE-LIFE ADVENTURES .. 89
BUENA VISTA. (R) October, 1975. Documentary.

BEST THINGS IN LIFE ARE FREE, THE (Cs) 104
TWENTIETH CENTURY-FOX. (R) September, 1956. Gordon McRae, Ernest Borgnine, Dan Dailey, Sheree North. Color.

BETRAYED WOMEN 70
ALLIED ARTISTS. (R) July 17, 1955. Carole Mathews, Tom Drake.

BETSY, THE 120
ALLIED ARTISTS. (R) February, 1978. Laurence Olivier, Robert Duvall.

BETTER A WIDOW 105
UNIVERSAL. (R) March, 1969. Virna Lisi, Peter McEmery, Gabrielle Ferzetti. Color.

BETWEEN HEAVEN AND HELL (Cs) 94
TWENTIETH CENTURY-FOX. (R) October, 1956. Robert Wagner, Terry Moore. Color.

BETWEEN THE LINES 101
MIDWEST FILM PRODS. (R) April, 1977. John Heard, Lindsay Crouse

BETWEEN TIME AND ETERNITY 96
UNIVERSAL. (R) September, 1950. Lilli Palmer, Willy Birgel. Color.

BEWARE OF CHILDREN 80
AMERICAN INTERNATIONAL. (R) April, 1961. Leslie Phillips, Geraldine McEwan.

BEYOND A REASONABLE DOUBT 80
UNIVERSAL. (R) September, 1956. Dana Andrews, Joan Fontaine.

BEYOND ALL LIMITS 115
OMAT. (R) May, 1961. Jack Palance, Maria Felix. Color.

BEYOND CONTROL 89
MISHKIN. (R) June, 1971. Anthony Baker.

BEYOND MOMBASA 90
COLUMBIA. (R) June, 1957. Donna Reed, Cornel Wilde. Color.

BEYOND THE DOOR 100
FILM VENTURES INTERNATIONAL. (R) May, 1975. Juliet Mills, Richard Johnson.

BEYOND THE GREAT WALL (Mandarin) 100
FRANK LEE. (R) August, 1967. Lin Dai, Chao Lei. Color.

BEYOND THE LAW 96
GROVE. (R) October, 1968. Rip Torn, George Plimpton, Norman Mailer, Micky Knox. b&w.

BEYOND THE POSEIDON ADVENTURE 122
WARNER BROS. (R) May, 1979. Michael Caine, Sally Field

BEYOND THE TIME BARRIER 80
AMERICAN INTERNATIONAL. (R) July, 1960. Robert Clarke, Darlene Tompkins.

BEYOND THE VALLEY OF THE DOLLS (P) 109
20TH CENTURY-FOX. (R) June, 1970. Dolly Read, Cynthia Myers. Color.

BHOWANI JUNCTION (Cs) 110
METRO-GOLDWYN-MAYER. (R) June 8, 1956. Ava Gardner, Stewart Granger. Color.

BIBLE . . . IN THE BEGINNING, THE (D-150) 175
20th CENTURY-FOX. (R) September, 1966. George C. Scott, Ava Gardner. Color.

BIG BANKROLL, THE 106
ALLIED ARTISTS. (R) September, 1961. Stephen Boyd, Juliette Greco.

BIG BEAT, THE 82
UNIVERSAL. (R) March, 1958. William Reynolds, Andra Martin. Color.

BIG BLUFF 70
UNITED ARTISTS. (R) June, 1955. John Bromfield, Martha Vickers.

BIG BOODLE, THE 83
UNITED ARTISTS. (R) January, 1957. Errol Flynn, Rosanna Rory.

BIG BOUNCE, THE 102
WARNER BROS.-7 ARTS. (R) March, 1969. Ryan O'Neal, Leigh Taylor Young, James Daly. Color.

BIG BUS, THE 88
PARAMOUNT. (R) July, 1976. Joseph Bologna, Stockard Channing.

BIG BUST OUT, THE 80
NEW WORLD PICTURES. (R) March, 1973. Vonetta McGee, Gordon Mitchell. Color.

BIG CAPER, THE 84
UNITED ARTISTS. (R) April, 1957. Rory Calhoun, Mary Costa.

BIG CIRCUS, THE (Cs) 108/12
ALLIED ARTISTS. (R) July, 1959. Victor Mature, Red Buttons. Color.

BIG COMBO, THE 89
ALLIED ARTISTS. (R) February 13, 1955. Cornel Wilde, Richard Conte.

BIG COUNTRY, THE (T) 168
UNITED ARTISTS. (R) August, 1958. Gregory Peck, Jean Simmons. Color.

BIG CUBE, THE 91
WARNER BROS.-7 ARTS. (R) March, 1969. Lana Turner, George Chakiris, Richard Egan. Color.

BIG FISHERMAN, THE 180
BUENA VISTA. (R) August, 1959. Howard Keel, Susan Kohner.

	Length In Mins.
BIG FIX, THE	108
UNIVERSAL. (R) October, 1979. Richard Dreyfuss, Susan Anspach.	
BIG GUNDOWN, THE	90
COLUMBIA. (R) August, 1968 Lee Van Cleef, Thomas Milan, Luisa Rivelli. Color.	
BIG HAND FOR THE LITTLE LADY, A	95
WARNER BROTHERS. (R) June 1, 1966. Henry Fonda, Joanne Woodward. Color.	
BIG HOUSE, U.S.A.	82
UNITED ARTISTS. (R) March, 1955. Broderick Crawford, Ralph Meeker.	
BIG JAKE (P)	110
NATIONAL GENERAL. (R) June, 1971. John Wayne, Richard Boone. Color.	
BIG KNIFE, THE	111
UNITED ARTISTS. (R) November, 1955. Jack Palance, Ida Lupino, Wendell Cory, Shelley Winters.	
BIG LAND, THE	93
WARNER BROTHERS. (R) February 23, 1957. Alan Ladd, Virginia Mayo, Edmond O'Brien. Color.	
BIG MOUTH, THE	107
COLUMBIA. (R) July, 1967. Jerry Lewis, Harold J. Stone. Color.	
BIG NIGHT, THE	74
PARAMOUNT. (R) February, 1960. Randy Sparks, Venetia Stevenson.	
BIG OPERATOR, THE (Cs)	91
METRO-GOLDWYN-MAYER. (R) September, 1959. Mickey Rooney, Steve Cochran. b&w.	
BIG RED	93
BUENA VISTA. (R) June, 1962. Walter Pidgeon, Gilles Payant. Color.	
BIG RED ONE, THE	113
UNITED ARTISTS. (R) July, 1980. Lee Marvin, Mark Hamill.	
BIG SHOW, THE (Cs)	118
TWENTIETH CENTURY FOX. (R) May, 1961. Esther Williams, Cliff Robertson. Color.	
BIG SLEEP, THE (P)	100
UNITED ARTISTS. (R) March, 1978. Robert Mitchum, Sarah Miles.	
BIG T.N.T. SHOW, THE	98
AMERICAN INTERNATIONAL. (R) January, 1966. Folk n' Rock Musical.	
BIG TIP OFF	78
ALLIED ARTISTS. (R) March 20, 1955. Richard Conte, Constance Smith.	
BIG WEDNESDAY (P)	125
WARNER BROS. (R) Summer, 1978. Jan-Michael Vincent, William Katt.	
BIGGER THAN LIFE (Cs)	95
TWENTIETH CENTURY-FOX. (R) August, 1956. James Mason, Barbara Rush. Color.	
BIGGEST BUNDLE OF THEM ALL, THE (P)	100
MGM. (R) Jan., 1968. Edward G. Robinson, Vittorio De Sica. Color.	
BIKINI BEACH (P)	100
AMERICAN INTERNATIONAL. (R) July, 1964. Frankie Avalon, Annette Funicello.	
BILLIE	87
UNITED ARTISTS. (R) September, 1965. Patty Duke, Jim Bakus. Color.	
BILLION DOLLAR BRAIN (P)	108
UNITED ARTISTS. (R) December, 1967. Michael Caine, Francoise Dorleac. Color.	
BILLY BUDD (Cs)	112
ALLIED ARTISTS. (R) June, 1962. Peter Ustinov, Robert Ryan.	
BILLY JACK	112
WARNER BROS. (R) May, 1971. Tom Laughlin, Delores Taylor. Color.	
BILLY LIAR	96
CONTINENTAL. (R) November, 1963. Tom Courtenay, Julie Christie.	
BILLY ROSE'S JUMBO (P)	125
METRO-GOLDWYN-MAYER. (R) December, 1962. Doris Day, Stephen Boyd. Color.	
BILLY TWO HATS	99
UNITED ARTISTS. (R) March, 1974. Gregory Peck, Desi Arnaz, Jr.	
BIMBO THE GREAT	96
WARNER BROS. (R) June, 1961. Charles Holm, Eleanor Marlowe. Color.	
BINGO LONG TRAVELING ALL-STARS AND MOTOR KINGS, THE	111
UNIVERSAL. (R) July, 1976. Billy Dee Williams, James Earl Jones.	
BIRCH INTERVAL	105
GAMMA III. (R) May, 1976. Eddie Albert, Rip Torn.	
BIRD WITH THE CRYSTAL PLUMMAGE, THE (Te)	98
UMC PICTURES. (R) September, 1970. Tony Musante, Suzy Kendall. Color.	
BIRDMAN OF ALCATRAZ	143
UNITED ARTISTS. (R) August, 1962. Burt Lancaster, Karl Malden.	
BIRDS, THE	120
UNIVERSAL. (R) April. 1963. Rod Taylor, Jessica Tandy. Color.	
BIRDS AND THE BEES, THE (VV)	94
PARAMOUNT. (R) May, 1956. George Gobel, Mitzi Gaynor, David Niven. Color.	
BIRDS DO IT	95
COLUMBIA. (R) Aug. 1966. Soupy Sales, Tab Hunter. Color.	
BIRDS IN PERU	95
REGIONAL. (R) November, 1968. Jean Seberg, Maurice Ronet, Piette Brasseur Danielle Darrieux. Color.	
BIRDS, THE BEES, AND THE ITALIANS, THE (Ital., Eng. Titles)	115
SEVEN ARTS. (R) April, 1967. Virna Lisi, Gastone Moschin.	
BIRTHDAY PARTY, THE	127
CONTINENTAL. (R) December, 1968. Robert Shaw, Patrick McGee, Dandy Nichols. Color.	
BITE THE BULLET (P)	131
COLUMBIA. (R) June, 1975. Gene Hackman, Candice Bergen.	
BITTER VICTORY (Cs)	83
COLUMBIA. (R) March, 1958. Richard Burton, Curt Jurgens.	
BITTERSWEET LOVE	92
AVCO-EMBASSY. (R) October, 1976. Lana Turner, Robert Lansing.	
BLACK AND WHITE IN COLOR (Fr., Eng. titles)	90
ALLIED ARTISTS. (R) April, 1977. Jean Carmet, Jacques Dufilho.	
BLACK BEAUTY	106
PARAMOUNT. (R) November, 1971. Mark Lester, Walter Slezak. Color.	
BLACK BELT JONES	87
WARNER BROS. (R) January, 1974. Jim Kelly, Gloria Henry.	
BLACK BIRD, THE	98
COLUMBIA. (R) December, 1975. George Segal, Stephane Audran.	
BLACK CAESAR	93
AMERICAN INTERNATIONAL. (R) February, 1973. Fred Williamson, D'Urville Martin. Color.	
BLACK DUKE, THE	90
PRODUCTION RELEASING. (R) July, 1964. Cameron Mitchell, Gloria Milland. Color.	
BLACK EYE	98
WARNER BROS. (R) May, 1974. Fred Williamson, Rosemary Forsyth.	
BLACK GIRL	97
CINERAMA. (R) January, 1973. Brock Peters, Leslie Uggams, Claudia McNeil. Color.	
BLACK GOLD	98
WARNER BROS. (R) June, 1963. Philip Carey, Diane McBain.	
BLACK GUNN	94
COLUMBIA. (R) December, 1972. Jim Brown, Martin Landau. Color.	
BLACK HOLE, THE	97
BUENA VISTA. (R) December, 1979. Maxmillian Schell, Anthony Perkins.	
BLACK KLANSMAN, THE	88
U.S. FILMS. (R) June 26, 1966. Richard Glidden, Rima Kutner.	
BLACK LIKE ME	107
READE-STERLING. (R) May, 1974. James Whitmore, Dan Priest.	
BLACK MAMA, WHITE MAMA	87
AMERICAN INTERNATIONAL. (R) January, 1973. Margaret Markov, Pam Grier. Color.	
BLACK MARBLE, THE	113
AVCO EMBASSY. (R) March, 1980. Robert Foxworth, Paula Prentiss.	
BLACK MOON	100
TWENTIETH CENTURY-FOX. (R) November, 1975. Cathryn Harrison, Therese Giehse.	
BLACK ON WHITE	80
AUDUBON. (R) October, 1969. Anita Sanders, Nino Segurino, Terry Carter. Color.	
BLACK ORCHID, THE (VV)	93
PARAMOUNT. (R) March, 1959. Sophia Loren, Anthony Quinn.	
BLACK ORPHEUS (Brazilian, Eng. titles)	98
LOPERT. (R) 1960. Bruno Mello, Marpessa Dawn. Color.	
BLACK SABBATH	99
AMERICAN INTERNATIONAL. (R) May, 1964. Boris Karloff, Mark Damon. Color.	
BLACK SCORPION, THE	83

Length In Mins.

WARNER BROS. (R) October, 1957. Richard Denning, Mara Corday.
BLACK SLEEP, THE ... 83
UNITED ARTISTS. (R) June, 1956. Basil Rathbone, Akim Tamiroff.
BLACK SPURS ... 81
PARAMOUNT. (R) May, 1965. Rory Calhoun, Terry Moore. Color.
BLACK STALLION, THE ... 125
UNITED ARTISTS. (R) October, 1979. Kelly Reno, Mickey Rooney.
BLACK SUNDAY (P) ... 145
PARAMOUNT. (R) April, 1977. Robert Shaw, Bruce Dern.
BLACK SUNDAY ... 84
AMERICAN INTERNATIONAL. (R) February, 1961. Barbara Steele, John Richardson.
BLACK TENT, THE (VV) ... 85
RANK. (R) July, 1957. Anthony Steel, Donald Sinden.
BLACK THURSDAY (Fr., Eng. titles) ... 92
LEVITT-PICKMAN. (R) December, 1974. Christine Pascal, Christian Rist.
BLACK TUESDAY ... 80
UNITED ARTISTS. (R) January, 1955. Edward G. Robinson, Peter Graves.
BLACK VEIL FOR LISA ... 88
COMMONWEALTH UNITED. (R) October, 1969. John Mills, Luciana Paluzzi. Color.
BLACK WHIP, THE ... 77
TWENTIETH CENTURY-FOX. (R) December, 1956. Hugh Marlowe, Colleen Gray.
BLACK WINDMILL, THE (P) ... 106
UNIVERSAL. (R) June, 1974. Michael Caine, Donald Pleasence.
BLACK ZOO, THE ... 88
ALLIED ARTISTS. (R) May, 1963. Michael Gough, Jeanne Cooper. Color.
BLACKBEARD'S GHOST ... 107
BUENA VISTA. (R) February, 1968. Peter Ustinov, Suzanne Pleshette. Color.
BLACKBOARD JUNGLE, THE ... 101
METRO-GOLDWYN-MAYER. (R) March 25, 1955. Glenn Ford, Anne Francis.
BLACKJACK KETCHUM, DESPERADO ... 76
COLUMBIA. (R) April, 1956. Howard Duff, Maggie Mahoney. Color.
BLACULA ... 92
AMERICAN INTERNATIONAL. (R) July, 1972. William Marshall, Denis Nichols. Color.
BLADE ... 90
JOSEPH GREEN PICTURES. (R) December, 1973. John Marley, Jon Cypher.
BLAST OF SILENCE ... 77
UNIVERSAL. (R) July, 1961. Allen Baron, Molly McCarthy.
BLAZING SADDLES (P) ... 94
WARNER BROS. (R) February, 1974. Cleavon Little, Gene Wilder.
BLESS THE BEASTS AND CHILDREN ... 109
COLUMBIA. (R) July, 1971. Bill Mumy, Barry Robins. Color.
BLINDFOLD (P) ... 102
UNIVERSAL. (R) June, 1966. Rock Hudson, Claudia Cardinale. Color.
BLINDMAN ... 105
20TH CENTURY-FOX. (R) March, 1972. Tony Anthony, Ringo Starr. Color.
BLISS OF MRS. BLOSSOM, THE ... 93
PARAMOUNT. (R) October, 1968. Shirley MacLaine, Richard Attenborough, James Booth. Color.
BLOB!, THE ... 82
PARAMOUNT. (R) October, 1958. Steve McQueen, Aneta Corseaut. Color.
BLONDE BAIT ... 71
ASSOCIATED FILM. (R) April 2, 1956. Beverly Michaels, Jim Davis.
BLONDE BLACKMAILER ... 58
ALLIED ARTISTS. (R) January, 1958. Richard Arlen, Susan Shaw.
BLONDE SINNER, THE (Brit.) ... 73
ALLIED ARTISTS. (R) Nov. 18, 1956. Diana Dors, Michael Craig.
BLOOD ALLEY (Cs) ... 118
WARNER BROTHERS. (R) October 1, 1955. John Wayne, Lauren Bacall. Color.
BLOOD AND BLACK LACE ... 90
ALLIED ARTISTS. (R) 1965. Cameron Mitchell, Eva Bartok. Color.
BLOOD AND LACE ... 87
AMERICAN INTERNATIONAL. (R) March, 1971. Gloria Grahame, Milton Selzer. Color.
BLOOD AND ROSES (T) ... 74
PARAMOUNT. (R) September, 1961. Mel Ferrar, Elsa Martinelli. Color.
BLOOD AND STEEL ... 63
TWENTIETH CENTURY-FOX. (R) November, 1959. John Lupton, James Edwards.
BLOOD BATH ... 69
AMERICAN INTERNATIONAL. (R) March, 1966. William Campbell, Marissa Mathes.
BLOOD BROTHERS ... 116
WARNER BROS. (R) October, 1979. Paul Sorvino, Tony Lo Bianco, Richard Gere.
BLOOD FIEND ... 70
HEMISPHERE. (R) January, 1969. Christopher Lee, Julian Glover.
BLOOD OF DRACULA ... 68
AMERICAN INTERNATIONAL. (R) November, 1957. Sandra Harrison, Louise Lewis.
BLOOD OF DRACULA'S CASTLE ... 84
CROWN INTERNATIONAL. (R) December, 1969. John Carradine, Paula Raymond.
BLOOD OF THE VAMPIRE ... 87
UNIVERSAL. (R) November, 1958. Donald Wolfit, Barbara Shelly. Color.
BLOOD ON SATAN'S CLAW ... 100
CANNON. (R) April, 1971. Patrick Wymark, Michele Dotrice. Color.
BLOOD ON THE ARROW ... 91
ALLIED ARTISTS. (R) November, 1964. Dale Robertson, Martha Hyer. Color.
BLOOD ON THE BALCONY (Ital., Eng. titles) ... 92
JILLO. (R) August, 1964. Documentary.
BLOODTHIRSTY BUTCHERS ... 79
MISHKIN. (R) January, 1970. John Miranda, Anabella Wood. Color.
BLOODY BROOD ... 111
ALLIED ARTISTS. (R) October, 1960. Barbara Lord, Jack Bett.
BLOODY MAMA ... 92
AMERICAN INTERNATIONAL PICTURES. (R) March, 1970. Shelley Winters, Don Stroud, Pat Hingle. Color.
BLOODY PIT OF HORROR ... 74
PACEMAKER PICTURES. (R) March, 1969. Mickey Hagitay.
BLOW-UP ... 110
METRO-GOLDWYN-MAYER. (R) April, 1967. Vanessa Redgrave, David Hemmings. Color.
BLUE ... 103
PARAMOUNT. (R) April, 1968. Terence Stamp, Joanna Pettit. Color.
BLUE ANGEL, THE (Cs) ... 107
TWENTIETH CENTURY-FOX. (R) September, 1959. Curt Jurgens, May Britt. Color.
BLUE BIRD, THE ... 95
TWENTIETH CENTURY-FOX. (R) May, 1976. Elizabeth Taylor, Jane Fonda.
BLUE COLLAR ... 110
UNIVERSAL (R) February, 1978. Richard Pryor, Harvey Keitel.
BLUE DENIM (Cs) ... 89
TWENTIETH CENTURY-FOX. (R) August, 1959. Carol Lynley, Brandon de Wilde.
BLUE HAWAII (P) ... 101
PARAMOUNT. (R) November, 1961. Elvis Presley, Joan Blackman. Color.
BLUE LAGOON, THE ... 104
COLUMBIA. (R) June, 1980. Brooke Shields, Christopher Atkins.
BLUE MAX, THE (Cs) ... 156
20TH-CENTURY FOX. (R) June, 1966. George Peppard, James Mason. Color.
BLUE PETER, THE (Alt. title: NAVY HEROES) ... 93
DCA DIST. (R) 1957. Greta Gynt, Kieron Moore. Color.
BLUE WATER, WHITE DEATH ... 100
NATIONAL GENERAL. (R) June, 1971. Documentary.
BLUEBEARD ... 125
CINERAMA. (R) August, 1972. Richard Burton, Raquel Welch, Joey Heatherton. Color.
BLUEBEARD (Alt. title: LANDRU) ... 118
Michele Morgan, Hildegarde Neff. Color.
BLUEBEARD'S TEN HONEYMOONS ... 92
ALLIED ARTISTS. (R) May, 1960. George Sanders, Corinne Calvert.
BLUEPRINT FOR ROBBERY ... 87
PARAMOUNT. (R) January, 1961. Jay Barney, J. Pat O'Malley.
BLUES BROTHERS, THE ... 133
UNIVERSAL. (R) June, 1980. John Belushi, Dan Aykroyd.
BLUES FOR LOVERS ... 89
20TH CENTURY-FOX. (R) September, 1966. Ray Charles, Tom Bell.
BLUME IN LOVE ... 117

Length In Mins.

WARNER BROS. (R) July, 1973. George Segal, Susan Anspach. Color.

BOARDWALK 98
ATLANTIC RELEASING. (R) November, 1979. Ruth Gordon, Lee Strasberg.

BOATNIKS, THE 100
BUENA VISTA. (R) July, 1970. Robert Morse, Stephanie Powers. Color.

BOB & CAROL & TED & ALICE 104
COLUMBIA. (R) November, 1969. Robert Culp, Natalie Wood, Elliott Gould, Dyan Cannon. Color.

BOBBIE JO AND THE OUTLAW 89
AMERICAN INTERNATIONAL. (R) April, 1976. Marjoe Gortner, Lynda Carter.

BOBBIKINS (Cs) 89
TWENTIETH CENTURY-FOX. (R) June, 1960. Max Bygraves, Shirley Jones.

BOBBY DEERFIELD (P) 123
COLUMBIA, (R) October, 1977. Al Pacino, Marthe Keller.

BOBBY WARE IS MISSING 66
ALLIED ARTISTS. (R) October 23, 1955. Neveille Brand, Arthur Franz.

BOBO, THE 105
WARNER BROS.-SEVEN ARTS. (R) September, 1967. Peter Sellers, Britt Ekland. Color.

BOCCACCIO 70 145
AVCO EMBASSY, (R) 1962. Sophia Loren, Anita Ekberg. Color.

BODY, THE 111
MGM. (R) February, 1971. Documentary.

BOEING, BOEING 102
PARAMOUNT. (R) Christmas, 1965. Tony Curtis, Jerry Lewis. Color.

BOFORS GUN, THE 106
UNIVERSAL. (R) September, 1968. Nicol Williamson, Jan Holm, David Warner, Richard O'Callaghan. Color.

BOLD AND THE BRAVE, THE (SS) 90
RKO RADIO. (R) April 18, 1956. Wendell Corey, Mickey Rooney, Don Taylor, Nicole Maurey.

BOLSHOI BALLET, THE 99
RANK FILM DISTRIBUTORS OF AMERICA. (R) January, 1958. Galina Ulanova. Color.

BOLSHOI BALLET 67 76
PARAMOUNT. (R) September, 1966. Ballet documentary. Color.

BOMB FOR A DICTATOR, A. 73
MEDALLION. (R) 1960. Pierre Fresnay, Michel Auclair. b&w.

BOMBAY TALKIE 110
MERCHANT/IVORY. (R) November, 1970. Shashi Kapoor.

BOMBERS B-52 (Cs) 106
WARNER BROS. (R) November, 1957. Natalie Wood, Karl Malden. Colo.

BON VOYAGE 130
BUENA VISTA. (R) June, 1962. Fred MacMurray, Jane Wyman. Color.

BONJOUR TRISTESSE (Cs) 94
UNITED ARTISTS (R) February, 1958. Deborah Kerr, David Niven. Color.

BONNE SOUPE, LA (Cs) 97
INTERNATIONAL CLASSICS. (R) March, 1964. Annie Girardot, Marie Bell.

BONNIE AND CLYDE 111
WARNER BROS.-SEVEN ARTS. (R) September, 1967. Warren Beatty, Faye Dunaway. Color.

BONNIE PARKER STORY, THE (Superama) 79
AMER. INT. PICS. (R) May, 1958. Dorothy Provine, Jack Hogan.

LE BONHEUR 87
CLOVER FILMS. (R) May, 1966 Jean-Claude Drouot, Claire Drouot. Color.

BOOK OF NUMBERS 80
AVCO-EMBASSY. (R) April, 1973. Raymond St. Jacques, Freda Payne, Color.

BOOM 113
UNIVERSAL. (R) July, 1968. Elizabeth Taylor, Richard Burton, Noel Coward. Color.

BOP GIRL 79
UNITED ARTISTS. (R) July, 1957. Mary Kaye Trio, Mudy Tyler.

BORA BORA 90
AMER. INT. PICS. (R) February, 1970. Haydee Politoff, Corrado Pani, Color.

BORN AGAIN 110
AVCO EMBASSY. (R) October, 1978. Dean Jones, Anne Francis.

BORN FREE (P) 95
COLUMBIA. (R) April, 1966. Virginia McKenna, Bill Travers. Color.

BORN LOSERS 112
AMERICAN INTERNATIONAL. (R) September, 1967. Jeremy Slate, Tom Laughlin. Color.

BORN RECKLESS 79
WARNER BROS. (R) May, 1959. Mamie Van Doren, Jeff Richards.

BORN TO WIN 90
UNITED ARTISTS. (R) October, 1971. George Segal, Paula Prentiss, Karen Black. Color.

BORSALINO (Fr. titles) 123
PARAMOUNT. (R) October, 1970. Jean-Paul Belmondo, Alain Delon. Color.

BOSS, THE 89
UNITED ARTISTS. (R) October, 1956. John Payne, Doe Avedon.

BOSTON STRANGLER, THE 116
TWENTIETH CENTURY-FOX. (R) October, 1968. Tony Curtis, Henry Fonda, George Kennedy. Color.

BOTTOM OF THE BOTTLE, THE (Cs) 88
TWENTIETH CENTURY-FOX. (R) January, 1956. Van Johnson, Joseph Cotten, Ruth Roman, Jack Carson. Color.

BOUCHER, LE 90
CINERAMA. (R) December, 1971. Jean Yanne, Stephane Audran. Color.

BOULEVARD NIGHTS 102
WARNER BROS. (R) March, 1979. Richard Yniguez, Danny De La Paz.

BOUND FOR GLORY 147
UNITED ARTISTS. (R) February, 1977. David Carradine, Ronny Cox.

BOUNTY KILLER, THE (TE) 92
EMBASSY. (R) July, 1965. Dan Duryea, Rod Cameron. Color.

BOWERY TO BAGDAD 84
ALLIED ARTISTS. (R) January 2, 1955. Bowery Boys.

BOY . . . A GIRL, A 85
CINEMA J. (R) April, 1969. Dino Martin Jr., Karen Steele.

BOY AND THE PIRATES, THE 82
UNITED ARTISTS. (R) April, 1960. Charles Herbert, Susan Gordon. Color.

BOY CRIED MURDER, THE 86
UNIVERSAL. (R) March, 1966. Veronica Hurst, Phil Brown. Color.

BOY, DID I GET A WRONG NUMBER 99
UNITED ARTISTS. (R) June, 1966. Bob Hope, Elke Sommer, Phyllis Diller. Color.

BOY FRIEND, THE 110
METRO-GOLDWYN-MAYER. (R) December, 1971. Twiggy, Christopher Gable. Color.

BOY NAMED CHARLIE BROWN, A. 85
NATIONAL GENERAL CORP. (R) March, 1970. Cartoon. Color.

BOY ON A DOLPHIN (Cs) 111
TWENTIETH CENTURY-FOX. (R) April, 1957. Sophia Loren, Alan Ladd. Color.

BOY TEN FEET TALL, A (Cs) 88
PARAMOUNT. (R) January, 1965 Edward G. Robinson, Fergus McClelland. Color.

BOY WHO CAUGHT A CROOK 72
UNITED ARTISTS. (R) October, 1961. Wanda Hendrix, Don Beddoe.

BOY WHO CRIED WEREWOLF, THE 93
UNIVERSAL. (R) August, 1973. Kerwin Mathews, Elaine Devry.

BOY WHO STOLE A MILLION 84
PARAMOUNT. (R) September, 1960. Virgillo Texera, Maurice Reyna.

BOYS FROM BRAZIL, THE 124
TWENTIETH CENTURY-FOX. (R) October, 1979. Gregory Peck, Laurence Olivier.

BOYS IN COMPANY C, THE (R) 125
COLUMBIA. (R) February, 1978. Stan Shaw, Andrew Stevens.

BOYS IN THE BAND, THE 120
NATIONAL GEN. CORP. (R) April, 1970. Kenneth Nelson, Leonard Frey, Cliff Gorman, Laurence Luckinbill. Color.

BOYS' NIGHT OUT (Cs) 115
METRO-GOLDWYN-MAYER. (R) June, 1962. Kim Novak, James Garner. Color.

BOYS OF PAUL STREET, THE 108
TWENTIETH CENTURY-FOX. (R) March, 1969. Anthony Kemp, William Burleigh, John Moulder-Brown. Color.

BRAIN, THE 100
PARAMOUNT. (R) October, 1969. David Niven, Jean-Paul Belmondo, Bourvil. Color.

BRAIN EATERS, THE 60
AMERICAN INTERNATIONAL. (R) October, 1958. Edwin Nelson, Alan Frost.

BRAIN MACHINE, THE (Brit.) 72

Length In Mins.

RKO RADIO. (R) February 15, 1956. Patrick Barr, Elizabeth Allan.
BRAINSTORM (P) **114**
WARNER BROS. (R) May, 1965. Jeff Hunter, Ann Francis.
BRAINWASHED **102**
ALLIED ARTISTS. (R) June, 1961. Curt Jurgens, Claire Bloom.
BRAMBLE BUSH, THE **105**
WARNER BROS. (R) February, 1960. Richard Burton, Barbara Rush. Color.
BRAND X. **87**
C.M.B. FILMS. (R) May, 1970. Taylor Mead, Sally Kirkland.
BRANNIGAN (P) **111**
UNITED ARTISTS. (R) March, 1975. John Wayne, Richard Attenborough.
BRASS BOTTLE, THE **89**
UNIVERSAL. (R) May, 1964. Tony Randall, Burl Ives. Color.
BRASS LEGEND, THE **79**
UNITED ARTISTS. (R) December, 1956. Hugh O'Brien, Nancy Gates.
BRASS TARGET (P) **112**
UNITED ARTISTS. (R) December, 1979. Sophia Loren, John Cassavetes.
BRAVADOS, THE **98**
TWENTIETH CENTURY-FOX. (R) June, 1958. Gregory Peck, Joan Collins. Color.
BRAVE ONE, THE (Cs) **100**
UNIVERSAL. (R) October, 1956. Michel Ray. Color.
BREAD AND CHOCOLATE (Ital., Eng. titles) **106**
WORLD NORTHAL (R) July, 1978. Nino Manfredi, Anna Karina.
BREAK, THE **75**
MANCUNIAN. (R) September, 1963. Tony Britton, William Lucas.
BREAK IN THE CIRCLE (Brit.) **69**
TWENTIETH CENTURY-FOX. (R) April, 1967. Forrest Tucker, Eva Bartok.
BREAK TO FREEDOM (Brit.) **88**
UNITED ARTISTS. (R) June, 1955. Anthony Steel, Jack Warner.
BREAKFAST AT TIFFANY'S **115**
PARAMOUNT. (R) October, 1961. Audrey Hepburn, George Peppard. Color.
BREAKING AWAY **99**
TWENTIETH CENTURY-FOX. (R) August, 1979. Dennis Christopher, Dennis Quaid.
BREAKOUT **96**
COLUMBIA. (R) May, 1975. Charles Bronson, Robert Duvall.
BREATH OF SCANDAL, A. **98**
PARAMOUNT. (R) December, 1960. Sophia Loren, John Gavin. Color.
BREEZY **108**
UNIVERSAL. (R) November, 1973. William Holden, Kay Lenz.
BREWSTER McCLOUD (P) **104**
MGM. (R) December, 1970. Bud Cort, Sally Kellerman. Color.
BRIDE OF THE BEAST, THE **78**
ALLIED ARTISTS. (R) March, 1958. Charlotte Austin, Lance Fuller.
BRIDE OF THE MONSTER **67**
BANNER. (R) February, 1956. Bela Lugosi.
BRIDE WORE BLACK, THE **107**
LOPERT. (R) June, 1968. Jeanne Moreau, Jean-Claude Brialy, Michel Bouquet. Color.
BRIDES OF BLOOD **77**
ALLIED ARTISTS. (R) January, 1969. John Ashley, Kent Taylor.
BRIDES OF DRACULA **85**
UNIVERSAL. (R) June, 1960. Peter Cushing, Freda Jackson. Color.
BRIDES OF FU MANCHU, THE **94**
SEVEN ARTS. (R) December, 1966 Christopher Lee, Douglas Wilner. Color.
BRIDGE, THE **102**
ALLIED ARTISTS. (R) May, 1961. Fritz Wepper, Michael Hinz.
BRIDGE AT REMAGEN, THE **115**
UNITED ARTISTS. (R) June, 1969. George Segal, Robert Vaughn, Ben Gazzara. Color.
BRIDGE ON THE RIVER KWAI, THE (Cs) **161**
COLUMBIA. (R) November, 1957. William Holden, Alec Guinness. Color.
BRIDGE TO GLORY **93**
JILLO. (R) July, 1964. Reunato Salvatori, Nino Castelnuovo.
BRIDGE TO THE SUN **112**
METRO-GOLDWYN-MAYER. (R) October, 1961. Carroll Baker, James Shigeta.

Length In Mins.

BRIDGE TOO FAR, A (P) **175**
UNITED ARTISTS. (R) June, 1977. Dirk Bogarde, James Caan.
BRIDGES AT TOKO-RI, THE **103**
PARAMOUNT. (R) February, 1955. Fredric March, William Holden, Grace Kelly. Color.
BRIEF VACATION, A (Italian; Eng. titles) **106**
ALLIED ARTIST. (R) February, 1975. Florinda Bolkan, Renato Salvatori.
BRIGAND OF KANDAHAR (Cinemascope) **81**
COLUMBIA. (R) May, 1966. Ronald Lewis, Oliver Reed. Color.
BRIGHTY OF THE GRAND CANYON **89**
FEATURE FILM CORP. OF AMERICA. (R) June, 1967. Jiggs, Joseph Cotten. Color.
BRING ME THE HEAD OF ALFREDO GARCIA **112**
UNITED ARTISTS. (R) August, 1974. Warren Oates, Isela Vega.
BRING YOUR SMILE ALONG **83**
COLUMBIA. (R) August, 1955. Frankie Laine, Keefe Brasselle. Color.
BRINK'S JOB, THE **118**
UNIVERSAL. (R) December, 1978. Peter Falk, Peter Boyle.
BROKEN LAND (Cs) **59**
TWENTIETH CENTURY-FOX. (R) April, 1962. Kent Taylor, Dianna Darrin. Color.
BROKEN STAR, THE **82**
UNITED ARTISTS. (R) April, 1956. Howard Duff, Lita Baron.
BRONCO BILLY **117**
WARNER BROS. (R) June, 1980. Clint Eastwood, Sondra Locke.
BROTHER JOHN **94**
COLUMBIA. (R) April, 1971. Sidney Poitier, Will Geer, Color.
BROTHER OF THE WIND **87**
SUN INTERNATIONAL. (R) October, 1972. Dick Robinson. Color.
BROTHER SUN, SISTER MOON **121**
PARAMOUNT. (R) April, 1973. Graham Faulkner, Judi Bowker. Color.
BROTHERHOOD, THE **98**
PARAMOUNT. (R) December, 1968. Kirk Douglas, Alex Cord, Irene Papas, Susan Strasberg. Color.
BROTHERLY LOVE **112**
MGM. (R) April, 1970. Peter O'Toole, Susannah York, Michael Craig. Color.
BROTHERS **104**
WARNER BROS. (R) April, 1977. Bernie Casey, Vonetta McGee.
BROTHERS IN LAW (Brit.) **90**
CONTINENTAL. (R) October, 1957. Ian Carmichael, Richard Attenborough.
BROTHERS KARAMAZOV, THE **146**
METRO-GOLDWYN-MAYER. (R) February, 1958. Yul Brynner, Maria Schell. Color.
BROTHERS RICO, THE **81**
COLUMBIA. (R) September, 1957. Richard Conte, Dianne Foster.
BRUBAKER **130**
TWENTIETH CENTURY-FOX. (R) June, 1980. Robert Redford, Yaphet Kotto.
BRUTE AND THE BEAST, THE **97**
AIP. (R) December, 1968. Franco Nero, George Hilton, Nino Castelnuovo. Color.
BUCCANEER, THE (VV) **119**
PARAMOUNT. (R) January, 1959. Yul Brynner, Charlton Heston. Color.
BUCHANAN RIDES ALONE **78**
COLUMBIA. (R) August, 1958. Randolph Scott, Craig Stevens. Color.
BUCK AND THE PREACHER **102**
COLUMBIA. (R) April, 1972. Sidney Poitier, Harry Belafonte. Color.
BUCKET OF BLOOD, A **65**
AMERICAN INTERNATIONAL. (R) October, 1959. Dick Miller, Barbara Morris.
BUCK ROGERS IN THE 25TH CENTURY **86**
UNIVERSAL. (R) March, 1979. Gil Gerard, Pamela Hensley.
BUCKSKIN **97**
PARAMOUNT. (R) May, 1968. Barry Sullivan, Joan Caulfield, Wendell Corey. Color.
BUCKSKIN LADY, THE **68**
UNITED ARTISTS. (R) June, 1957. Patricia Medina, Richard Denning.
BUDDY HOLLY STORY, THE **113**
COLUMBIA. (R) June, 1978. Gary Busey, Don Stroud
BUFFALO BILL AND THE INDIANS, OR SITTING BULL'S HISTORY LESSON **125**
UNITED ARTISTS. (R) July, 1976. Paul Newman, Joel Grey.

	Length In Mins.
BUG	**100**
PARAMOUNT. (R) June, 1975. Bradford Dillman, Joanna Miles.	
BUGSY MALONE	**94**
PARAMOUNT. (R) September, 1976. Scott Baio, Florrie Dugger.	
BULLET FOR A BADMAN	**80**
UNIVERSAL. (R) October, 1964. Audie Murphy, Darren McGavin, Color.	
BULLET FOR JOEY, A.	**85**
UNITED ARTISTS. (R) April, 1955. Edward G. Robinson, George Raft.	
BULLET FOR PRETTY BOY, A	**91**
AMERICAN INTERNATIONAL. (R) September, 1970. Fabian Forte, Jocelyn Lane. Color.	
BULLET FOR SANDOVAL, A.	**91**
UMC PICTURES. (R) June, 1970. Ernest Borgnine, George Hilton. Color.	
BULLET FOR THE GENERAL, A	**114**
AVCO-EMBASSY. (R) April, 1969. Gian Maria Volonte, Klaus Kinski, Martine Beswick Color.	
BULLFIGHT (Span.-Eng. narr.)	**76**
JANUS. (R) July, 1956. Documentary.	
BULLITT	**113**
WARNER BROS.-7 ARTS. (R) December, 1968. Steve McQueen, Robert Vaughn, Jacqueline Bisset, Don Gordon. Color.	
BULLWHIP (Cs)	**80**
ALLIED ARTISTS. (R) May 25, 1958. Guy Madison, Rhonda Fleming. Color.	
BUNDLE OF JOY	**98**
UNIVERSAL. (R) December, 1956. Debbie Reynolds, Eddie Fisher. Color.	
BUNNY LAKE IS MISSING (P)	**107**
COLUMBIA. (R) October, 1965. Laurence Olivier, Carol Lynley. b&w.	
BUNNY O'HARE	**92**
AMERICAN INTERNATIONAL. (R) July, 1971. Bette Davis, Ernest Borgnine. Color.	
BUONA SERA, MRS. CAMPBELL	**113**
UNITED ARTISTS. (R) February, 1969. Gina Lollobrigida, Shelley Winters, Phil Silvers, Peter Lawford. Color.	
BURGLAR, THE	**90**
COLUMBIA. (R) June, 1957. Dan Duryea, Jayne Mansfield.	
BURGLARS, THE	**117**
COLUMBIA. (R) June, 1972. Omar Sharif, Jean-Paul Belmondo. Color.	
BURMESE HARP, THE (Jap.; Eng. Titles)	**116**
BRANDON. (R) April, 1967. Bentaro Mikuni, Shoji Yasui.	
BURN!	**112**
UNITED ARTISTS. (R) December, 1970. Marlon Brando, Evaristo Marquez. Color.	
BURN, WITCH, BURN	**90**
AMERICAN INTERNATIONAL. (R) April, 1962. Janet Blair, Peter Wyngarde.	
BURNING HILLS, THE (Cs)	**92**
WARNER BROTHERS. (R) September 1, 1956. Natalie Wood, Tab Hunter. Color.	
BURNT OFFERINGS	**116**
UNITED ARTISTS. (R) October, 1976. Karen Black, Oliver Reed, Bette Davis.	
BUS RILEY'S BACK IN TOWN	**93**
UNIVERSAL. (R) April, 1965. Ann-Margaret, Michael Parks. Color.	
BUS STOP (Cs)	**94**
TWENTIETH CENTURY-FOX. (R) August, 1956. Marilyn Monroe, Don Murray. Color.	
BUSHBABY, THE	**100**
MGM. (R) November, 1970. Margaret Brooks, Louis Gossett. Color.	
BUSTER AND BILLIE	**100**
COLUMBIA. (R) June, 1974. Jan-Michael Vincent, Joan Goodfellow.	
BUSTER KEATON STORY, THE (VV)	**97**
PARAMOUNT. (R) May, 1957. Donald O'Connor, Ann Blyth.	
BUSTING	**92**
UNITED ARTISTS. (R) February, 1974. Elliott Gould, Robert Blake.	
BUSY BODY, THE (TS)	**101**
PARAMOUNT. (R) February, 1967. Sid Caesar, Robert Ryan. Color.	
BUT NOT FOR ME (VV)	**105**
PARAMOUNT. (R) October, 1959. Clark Gable, Carroll Baker.	
BUTCH AND SUNDANCE: THE EARLY DAYS (P)	**107**
TWENTIETH CENTURY-FOX. (R) June, 1979. William Katt, Tom Berenger.	
BUTCH CASSIDY AND THE SUNDANCE KID (P)	**110**
20TH CENTURY-FOX. (R) October, 1969. Paul Newman, Robert Redford, Katherine Ross. Color.	
BUTLEY	**127**
AFT DISTRIBUTING. (R) 1974, special. Alan Bates, Jessica Tandy.	
BUTTERCUP CHAIN, THE (P)	**95**
COLUMBIA. (R) April, 1971. Hywell Bennett, Leigh Taylor-Young. Color.	
BUTTERFIELD 8 (Cs)	**109**
METRO-GOLDWYN-MAYER. (R) November, 1960. Elizabeth Taylor, Laurence Harvey. Color.	
BUTTERFLIES ARE FREE	**109**
COLUMBIA. (R) July, 1972. Goldie Hawn, Edward Albert. Color.	
BY LOVE POSSESSED	**115**
UNITED ARTISTS. (R) July, 1961 Lana Turner, Efrem Zimbalist, Jr. Color.	
BYE BYE BIRDIE (P)	**112**
COLUMBIA. (R) June, 1963. Janet Leigh, Dick Van Dyke. Color.	
BYE, BYE BRAVERMAN	**94**
WARNER BROS.-SEVEN ARTS. (R) April, 1968. George Segal, Jessica Walter. Color.	

C

	Length In Mins.
C.C. AND COMPANY	**94**
AVCO-EMBASSY. (R) October, 1970. Joe Namath, Ann-Margaret. Color.	
CABARET	**118**
ALLIED ARTISTS. (R) February, 1972. Liza Minelli, Michael York, Joel Gray. Color.	
CABINET OF CALIGARI, THE (Cs)	**95**
TWENTIETH CENTURY-FOX. (R) June, 1962. Glynis Johns, Dan O'Herlihy	
CACTUS FLOWER	**103**
COLUMBIA. (R) December, 1969. Walter Matthau, Ingrid Bergman, Goldie Hawn. Color.	
CADDYSHACK	**93**
WARNER BROS. (R) July, 1980. Chevy Chase, Ted Knight.	
CAGE OF EVIL	**70**
UNITED ARTISTS. (R) July, 1960. Ronald Foster, Pat Blair.	
CAHILL, UNITED STATES MARSHALL (P)	**103**
WARNER BROS. (R) June, 1973. John Wayne, Gary Grimes. Color.	
CAIRO	**91**
METRO-GOLDWYN-MAYER. (R) January, 1963. George Sanders, Richard Johnson.	
CALIFORNIA DREAMING	**92**
AMERICAN INTERNATIONAL. (R) April, 1979. Glynnis O'Connor, Dennis Christopher.	
CALIFORNIA SPLIT (P)	**111**
COLUMBIA. (R) August, 1974. George Segal, Elliott Gould.	
CALIFORNIA SUITE	**103**
COLUMBIA. (R) December, 1978. Alan Alda, Michael Caine, Jane Fonda.	
CALL ME BWANA	**103**
UNITED ARTISTS. (R) June, 1963. Bob Hope, Anita Ekberg. Color.	
CALLING HOMICIDE	**61**
ALLIED ARTISTS. (R) September 30, 1956. Bill Elliott, Kathleen Case.	
CALTIKI, THE IMMORTAL MONSTER	**76**
ALLIED ARTISTS. (R) September, 1960. John Merivale, Didi Sullivan.	
CALYPSO HEAT WAVE	**86**
COLUMBIA. (R) June, 1957. Johnny Desmond, The Triniers.	
CALYPSO JOE	**76**
ALLIED ARTISTS. (R) May 12, 1957. Herb Jeffries, Angie Dickinson.	
CAMELOT (P)	**178**
WARNER BROS.-SEVEN ARTS. (R) October, 1967. Special. Richard Harris, Vanessa Redgrave. Color.	
CAMILLE 2000 (P)	**115**
AUDUBON. (R) July 1969. Danielle Gaubert, Nino Castelnuovo. Color.	
CAMP ON BLOOD ISLAND, THE (M)	**81**
COLUMBIA. (R) July, 1958. Andre Morell, Carl Mohner.	
CAN HIERONYMUS MERKIN EVER FORGET MERCY HUMPPE AND FIND TRUE HAPPINESS	**106**
REGIONAL. (R) March, 1969. Anthony Newley, Joan Collins, Milton Berle, George Jessel. Color.	
CANADIANS (Cs)	**85**
TWENTIETH CENTURY-FOX. (R) March, 1961. Robert Ryan, Torin Thatcher. Color.	
CAN-CAN (Todd-AO)	**130**
TWENTIETH CENTURY-FOX. (R) March, 1960. Special. Frank Sinatra, Shirley MacLaine. Color.	
CANCEL MY RESERVATION	**99**

Length In Mins.

WARNER BROS. (R) October, 1972. Bob Hope, Eva Marie Saint. Color.

CANDIDATE, THE 110
WARNER BROS. (R) July, 1972. Robert Redford, Don Porter. Color.

CANDLESHOE 101
BUENA VISTA. (R) February, 1978. David Niven, Helen Hayes.

CANDY 126
CINERAMA. (R) December, 1968. Ewa Aulin, Charles Aznavour, Marlon Brando, Richard Burton, James Coburn. Color.

CANNIBAL GIRLS 80
AMERICAN INTERNATIONAL. (R) October, 1972. Eugene Levy, Andrea Martin. Color.

CANNON FOR CORDOBA 104
UNITED ARTISTS. (R) October, 1970. George Peppard, Giovanna Ralli. Color.

CAN'T STOP THE MUSIC 118
ASSOCIATED FILM DIST. (R) June, 1980. Village People, Valerie Perrine.

CANYON CROSSROADS 83
UNITED ARTISTS. (R) February, 1955. Richard Basehart, Phyllis Kirk.

CANYON RIVER (Cs) 80
ALLIED ARTISTS. (R) August 5, 1965. George Montgomery, Marcia Henderson. Color.

CAPE FEAR 105
UNIVERSAL. (R) May, 1962. Gregory Peck, Robert Mitchum.

CAPER OF THE GOLDEN BULLS, THE 104
EMBASSY. (R) May, 1967. Stephen Boyd, Yvette Mimieux. Color.

CAPONE 101
20TH CENTURY-FOX. (R) April, 1975. Ben Gazzara, Susan Blakely.

CAPRICE (CS) 98
TWENTIETH CENTURY-FOX. (R) June, 1967. Doris Day, Richard Harris. Color.

CAPRICORN ONE 127
WARNER BROS. (R) June, 1978. Elliott Gould, James Brolin.

CAPTAIN FROM KOEPENICK, THE 93
DCA DIST. (R) 1956. Heinz Ruhmann. Color.

CAPTAIN KRONOS: VAMPIRE HUNTER 91
PARAMOUNT. (R) June, 1974. Horst Janson, John Carson.

CAPTAIN LIGHTFOOT (Cs) 91
UNIVERSAL. (R) March, 1955. Rock Hudson, Barbara Rush. Color.

CAPTAIN NEMO AND THE UNDERWATER CITY (P) 105
MGM. (R) March, 1970. Robert Ryan, Chuck Conners, Nanette Newman. Color.

CAPTAIN NEWMAN, M.D. 126
UNIVERSAL. (R) April, 1964. Gregory Peck, Tony Curtis. Color.

CAPTAIN SINBAD 85
METRO-GOLDWYN-MAYER. (R) June, 1963. Guy Williams, Heidi Bruhl. Color.

CAPTAIN'S TABLE, THE 90
TWENTIETH CENTURY-FOX. (R) September, 1960. John Gregson, Peggy Cummins. Color.

CAR, THE (R) 95
UNIVERSAL. (R) May, 1977. James Brolin, Kathleen Lloyd.

CAR WASH 97
UNIVERSAL. (R) September, 1976. Franklyn Ajaye, Sully Boyar.

CARAVANS 127
UNIVERSAL. (R) November, 1978. Anthony Quinn, Michael Sarrazin.

CARAVAN TO VACCARES 84
BRYANSTON. (R) February, 1976. Charlotte Rampling, David Birney.

CARDINAL, THE (P-70) 175
COLUMBIA. (R) Special in December, 1963. Tom Tryon, Carol Lynley. Color.

CAREER 105
PARAMOUNT. (R) November, 1959. Dean Martin, Anthony Franciosa.

CAREER GIRL 61
ASTOR-WOODBURN FILMS. (R) May, 1960. June Wilkinson, Charles Robert Keane. Color.

CAREER GIRL 69
PBC PICTURES. (R) January 11, 1944. Frances Langford, Craig Wood.

CARELESS YEARS, THE 70
UNITED ARTISTS. (R) September, 1957. Dean Stockwell, Natalie Trundy.

CARESSED 81
JOSEPH BRENNER ASSOCIATES. (R) October, 1965. Robert Howay, Angela Gann.

CARETAKER, THE (formerly THE GUEST) 105
JANUS. (R) January, 1964. Alan Bates, Donald Pleasence, Robert Shaw.

CARETAKERS, THE 97
UNITED ARTISTS. (R) August, 1963. Robert Stack, Polly Bergen.

CAREY TREATMENT, THE 100
METRO-GOLDWYN-MAYER. (R) March, 1972. James Coburn, Jennifer O'Neil. Color.

CARMEN JONES (Cs) 107
TWENTIETH CENTURY-FOX. (R) January, 1955. Dorothy Dandridge, Harry Belafonte. Color.

CARNAL KNOWLEDGE 100
AVCO EMBASSY. (R) July, 1971. Jack Nicholson, Art Garfunkle. Color.

CARNIVAL OF CRIME 83
CROWN. (R) June, 1964. Jean-Pierre Aumont.

CARNY 108
UNITED ARTISTS. (R) May, 1980. Gary Busey, Jodie Foster.

CAROLINA CANNONBALL 74
REPUBLIC. (R) January 28, 1955. Judy Canova.

CAROUSEL (Cs) 128
TWENTIETH CENTURY-FOX. (R) February, 1956. Gordon MacRae, Shirley Jones, Cameron Mitchell. Color.

CARPETBAGGERS, THE 150
PARAMOUNT. (R) April, 1964. Carroll Baker, George Peppard. Color.

CARRIE 98
UNITED ARTISTS. (R) November, 1976. Sissy Spacek, Piper Laurie.

CARRY IT ON 80
MARON. (R) September, 1970. Documentary. Joan Baez, David Harris.

CARRY ON CLEO 92
GOVERNOR FILMS. (R) October, 1965. Sidney James, Amanda Barrie. Color.

CARRY ON CONSTABLE 86
GOVERNOR. (R) November, 1961. Sidney James, Eric Ranker.

CARRY ON DOCTOR 95
AMERICAN INTERNATIONAL. (R) November, 1972. Francis Bigger, Kenneth Williams. Color.

CARRY ON NURSE 87
GOVERNOR. (R) June, 1960. Terence Longdon, Kenneth Connor.

CARRY ON SPYING 87
GOVERNOR. (R) February, 1965. Barbara Windsor, Kenneth Williams.

CARRY ON TEACHER 86
GOVERNOR. (R) October, 1962. Kenneth Connor, Charles Hawtry.

CARTHAGE IN FLAMES 95
COLUMBIA. (R) March, 1961. June Suarez, Anne Heywood. Color.

CARTOUCHE 73
RKO. (R) August, 1957. Richard Basehart, Patricia Roc.

CARTOUCHE (D) 124
EMBASSY. (R) July, 1964. Jean-Paul Belmondo, Claudia Cardinale. Color.

CASANOVA 70 113
EMBASSY. (R) July, 1965. Marcello Mastroianni, Virna Lisi. Color.

CASE AGAINST BROOKLYN, THE 82
COLUMBIA. (R) June, 1958. Darren McGavin, Maggie Hayes.

CASE OF THE NAVES BROTHERS, THE 93
EUROPIX INTERNATIONAL. (R) September, 1972. Raoul Cortez, Juca De Olivere. Color.

CASE OF THE RED MONKEY (Brit.) 73
ALLIED ARTISTS. (R) June 26, 1955. Richard Conte, Rona Anderson.

CASEY'S SHADOW (P) 117
COLUMBIA (R) March, 1978. Walter Matthau, Alexis Smith.

CASH McCALL 102
WARNER BROS. (R) January, 1960. James Garner, Natalie Wood. Color.

CASH ON DELIVERY 82
RKO RADIO. (R) January 15, 1956. Shelley Winters, Peggy Cummins.

CASH ON DEMAND 84
COLUMBIA. (R) February, 1962. Peter Cushing, Andre Morrell.

CASINO ROYALE (P) 130
COLUMBIA. (R) April, 1967. Peter Sellers, Ursula Andress and all-star cast. Color.

CASSANDRA CROSSING, THE (P) 125
AVCO-EMBASSY. (R) February, 1977. Sophia Loren, Richard Harris.

CAST A DARK SHADOW (Brit.) 84
DIST. CORP. OF AMER. (R) November, 1957. Dirk Bogarde, Margaret Lockwood.

Length In Mins.

CAST A GIANT SHADOW (P) **142**
UNITED ARTISTS. (R) April, 1966. Kirk Douglas, Senta Berger. Color.
CAST A LONG SHADOW **82**
UNITED ARTISTS. (R) July, 1959. Audie Murphy, Terry Moore. Color.
CASTILIAN, THE **129**
WARNER BROS. (R) October, 1963. Spartaco Santony, Teresa Velasquez.
CASTLE, THE **90**
CONTINENTAL. (R) March, 1969. Maximillian Schell, Cordula Tranto W, Trudik Daniel. Color.
CASTLE KEEP (P) **105**
COLUMBIA. (R) July, 1969. Burt Lancaster, Patrick O'Neal, Jean-Pierre Aumont. Color.
CASTLE OF EVIL **80**
UNITED PICTURES. (R) October, 1969. Scott Brady, Virginia Mayo.
CAT, THE **87**
EMBASSY. (R) Summer, 1966. Roger Perry, Peggy Ann Gardner. Color.
CAT AND MOUSE (German; Eng. titles) **92**
GROVE. (R) February, 1970. Lars Brandt, Peter Brandt, Wolfgang Neuss. b&w.
CAT BALLOU **97**
COLUMBIA. (R) August, 1965. Jane Fonda, Lee Marvin. Color.
CAT GIRL, THE **67**
AMER. INT. PICS. (R) September, 1957. Barbara Shelley, Robert Ayers.
CAT ON A HOT TIN ROOF **108**
METRO-GOLDWYN-MAYER. (R) September, 1958. Elizabeth Taylor, Paul Newman. Color.
CAT O'NINE TAILS, THE **112**
NATIONAL GENERAL. (R) May, 1971. Karl Malden, James Franciscus. Color.
CATCH MY SOUL **95**
CINERAMA. (R) March, 1974. Richie Havens, Lance Le Gault.
CATCH 22 (P) **121**
PARAMOUNT. (R) June, 1970 Alan Arkin, Richard Benjamin, Jon Voight, Orsen Welles. Color.
CATERED AFFAIR, THE **93**
METRO-GOLDWYN-MAYER. (R) June 22, 1956. Bette Davis, Ernest Borgnine, Debbie Reynolds, Barry Fitzgerald.
CATHERINE & CO. (French; English titles) **99**
WARNER BROS. (R) March, 1976. Jane Birkin, Patrick Dewaere.
CATLOW **101**
UNITED ARTISTS. (R) October, 1971. Yul Brynner, Richard Crenna. Color.
CATS, THE (Swedish; Eng. titles) **93**
NSF. (R) February, 1969. Eva Dahlbeck, Isa Quensel, Gio Petre. b&w.
CATTLE EMPIRE (Cs) **82**
TWENTIETH CENTURY-FOX. (R) February, 1958. Joel MacCrea, Gloria Talbott. Color.
CATTLE KING **88**
METRO-GOLDWYN-MAYER. (R) June, 1963. Robert Taylor, Joan Caufield.
CAVERN, THE **83**
TWENTIETH CENTURY-FOX. (R) November, 1965. Rosanna Schiaffino, Joan Saxon.
CELEBRATION AT BIG SUR **82**
20TH CENTURY-FOX. (R) April, 1971. Musical documentary. Joan Baez, John Sebastian. Color.
CELL 2455, DEATH ROW **77**
COLUMBIA. (R) May, 1955. William Campbell, Robert Campbell.
CEREMONY, THE **105**
UNITED ARTISTS. (R) December, 1963. Laurence Harvey, Sarah Miles.
CERTAIN SMILE (Cs) **106**
TWENTIETH CENTURY-FOX. (R) August, 1958. Rossano Brazzi, Joan Fontaine. Color.
CESAR AND ROSALIE **110**
CINEMA V. (R) December, 1972. Yves Montand, Romy Schneider. Color.
CHA-CHA-CHA-BOOM! **72**
COLUMBIA. (R) October, 1956. Stephen Dunne, Alix Talton.
CHAIN OF EVIDENCE **64**
ALLIED ARTISTS. (R) January 6, 1957. Bill Elliott, James Lydon.
CHAIRMAN, THE (P) **102**
20th CENTURY-FOX. (R) July, 1969. Gregory Peck, Anne Heywood, Arthur Hill. Color.
CHALK GARDEN, THE **105**
UNIVERSAL. (R) June, 1964. Deborah Kerr, Hayley Mills. Color.
CHALLENGE FOR ROBIN-HOOD, A **85**
TWENTIETH CENTURY-FOX. (R August, 1969. Barrie Ingham, James Hayter, Leon Greene, Peter Blythe. Color.
CHAMBER OF HORRORS **99**
WARNER BROS. (R) September, 1966. Patrick O'Neal, Cesare Danova. Color.
CHAMP, THE **121**
UNITED ARTISTS. (R) April, 1979. Jon Voight, Faye Dunaway.
CHAMPAGNE MURDERS, THE **98**
UNIVERSAL. (R) March, 1968. Anthony Perkins, Maurice Ronet. Color.
CHANCE MEETING **96**
PARAMOUNT. (R) March, 1960. Hardy Kruger, Micheline Presle.
CHANDLER **88**
METRO-GOLDWYN-MAYER. (R) December, 1971. Warren Oates, Leslie Caron. Color
CHANGE IN THE WIND, A **87**
CINERAMA. (R) November, 1972. William Devane, Anne Meara. Color.
CHANGE OF HABIT **93**
UNIVERSAL. (R) November, 1969. Elvis Presley, Mary Tyler Moore, Barbara McNair. Color.
CHANGE OF MIND **96**
CINERAMA. (R) October, 1969. Raymond St. Jacques, Susan Oliver, Leslie Nielson. Color.
CHANGES **93**
CINERAMA. (R) February, 1969. Kent Lane, Michele Carey, Manuela Thiess.
CHAPMAN REPORT, THE **125**
WARNER BROS. (R) October, 1962. Efram Zimbalist, Jr., Shelley Winters. Color.
CHAPPAQUA **82**
REGIONAL. (R) November, 1967. Jean-Louis Barrault, Conrad Rooks.
CHAPTER TWO **124**
COLUMBIA. (R) December, 1979. James Caan, Marsha Mason.
CHARADE **114**
UNIVERSAL. (R) January, 1964. Cary Grant, Audrey Hepburn. Color.
CHARGE OF THE LIGHT BRIGADE, THE **128**
UNITED ARTISTS. (R) October, 1968. Trevor Howard, Vanessa Redgrave, John Gielgud. Color.
CHARLEY AND THE ANGEL **93**
BUENA VISTA. (R) March, 1972. Fred MacMurray, Cloris Leachman. Color.
CHARLEY ONE-EYE **107**
PARAMOUNT. (R) May, 1973. Richard Roundtree, Roy Thinnes. Color.
CHARLEY VARRICK (P) **111**
UNIVERSAL. (R) October, 1973. Walter Matthau, Joe Don Baker.
CHARLIE BUBBLES **91**
REGIONAL. (R) February, 1968. Albert Finney, Liza Minnelli. Color.
CHARLIE THE LONESOME COUGAR **75**
BUENA VISTA. (R) October, 1967. Ron Brown, Brian Russell. Color.
CHARLOTTE (French; Eng. titles) **100**
GAMMA III. (R) June, 1975. Sirpa Lane, Roger Vadim.
CHARLOTTE'S WEB **94**
PARAMOUNT. (R) March, 1973. Cartoon. Color.
CHARLY **106**
CINERAMA. (R) October, 1968. Cliff Robertson, Claire Bloom, Leon Janney. Color.
CHARRO! **98**
NGC. (R) April, 1969. Elvis Presley, Ina Balin, Lynn Kellogg. Color.
CHARTROOSE CABOOSE **75**
UNIVERSAL. (R) August, 1969. Molly Bee, Ben Cooper. Color.
CHASE, THE (P) **135**
COLUMBIA. (R) February, 1966. Marlon Brando, Jane Fonda. Color.
CHASE A CROOKED SHADOW **87**
WARNER BROS. (R) March, 1958. Richard Todd, Anne Baxter.
CHASTITY **85**
AMER. INT. PICS (R) June, 1969. Cher, Barbara London. Color.
CHATOS'S LAND **100**
UNITED ARTISTS. (R) May, 1972. Charles Bronson, Jack Palance. Color.
CHE! (P) **96**
20th CENTURY-FOX. (R) June, 1969. Omar Sharif, Jack Palance. Color.
CHEAP DETECTIVE, THE **92**
COLUMBIA. (R) June, 1978. Peter Falk, Ann-Margaret.
CHECKPOINT **84**
RANK. (R) June, 1957. Anthony Steel, Odile Versois. Color.

Length In Mins.

CHEECH AND CHONG'S NEXT MOVIE 99
UNIVERSAL. (R) July, 1980. Richard Marin, Thomas Chong.

CHERRY, HARRY & RAQUEL 71
EVE PRODUCTIONS. (R) December, 1969. Linda Ashton, Charles Napier. Color.

CHEYENNE AUTUMN (Super P-70) 160
WARNER BROS. (R) December, 1964. Richard Widmark, Caroll Baker. Color.

CHEYENNE SOCIAL CLUB 103
NATIONAL GENERAL. (R) June, 1970. James Stewart, Henry Fonda, Shirley Jones. Color.

CHICAGO CONFIDENTIAL 74
UNITED ARTISTS. (R) September, 1957. Brian Keith, Beverly Garland.

CHICKEN CHRONICLES, THE 94
AVCO EMBASSY. (R) October, 1977. Phil Silvers, Ed Lauter.

CHICAGO SYNDICATE 84
COLUMBIA. (R) July, 1955. Dennis O'Keefe, Abe Lane.

CHIEF CRAZY HORSE (Cs) 86
UNIVERSAL. (R) April, 1955. Victor Mature, Suzan Ball. Color.

CHILD IS WAITING, A 102
UNITED ARTISTS. (R) January, 1963. Burt Lancaster, Judy Garland.

CHILDISH THINGS 85
FILMWORLD. (R) June, 1969. Don Murray, Linda Evans.

CHILDREN OF THE DAMNED 90
METRO-GOLDWYN-MAYER. (R) January, 1964. Ian Henry, Alan Badel.

CHILDREN'S HOUR, THE 109
UNITED ARTISTS. (R) March, 1962. Audrey Hepburn, Shirley MacLaine.

CHILD'S PLAY 100
PARAMOUNT. (R) October, 1972. Robert Preston, Beau Bridges, James Mason. Color.

CHINA! 65
JANUS. (R) May, 1965. Documentary. Color.

CHINA DOLL 85
UNITED ARTISTS. (R) August, 1958. Victor Mature, Lili Hus.

CHINA GATE (Cs) 97
TWENTIETH CENTURY-FOX. (R) May, 1957. Nat King Cole, Gene Barry.

CHINA IS NEAR (Ital.) 108
ROYAL. (R) January, 1968. Glauco Mauri, Elda Tattoli.

CHINA SYNDROME, THE 122
COLUMBIA. (R) March, 1979. Jane Fonda, Jack Lemmon, Michael Douglas.

CHINATOWN (P) 130
PARAMOUNT. (R) July, 1974. Jack Nicholson, Faye Dunaway.

CHINESE CONNECTION, THE 105
NATIONAL GENERAL. (R) April, 1973. Bruce Lee, James Tien. Color.

CHISUM 111
WARNER BROS. (R) July, 1970. John Wayne, Christopher George, Ben Johnson. Color.

CHITTY CHITTY BANG BANG 142
UNITED ARTISTS. (R) December, 1968. Dick Van Dyke, Sally Ann Howes, Lionel Jeffries. Color.

CHLOE IN THE AFTERNOON 97
COLUMBIA. (R) September, 1972. Bernard Verly, Zouzou. Color.

CHOIRBOYS, THE 119
UNIVERSAL. (R) December, 1977. Charles Durning, Lewis Gossett, Jr.

C.H.O.M.P.S. 89
AMERICAN INTERNATIONAL. (R) December, 1979. Wesley Eure, Valerie Bertinelli.

CHOSEN SURVIVORS 99
COLUMBIA. (R) May, 1974. Jackie Cooper, Alex Cord.

CHRISTIAN LICORIC STORE, THE 90
NATIONAL GENERAL. (R) November, 1971. Beau Bridges, Maud Adams. Color.

CHRISTINE JORGENSON STORY, THE 98
UNITED ARTISTS. (R) June, 1970. John Hansen, Joan Tompkins. Color.

CHRISTMAS THAT ALMOST WASN'T, THE 95
CHILDHOOD. (R) November, 1966. Rossano Brazzi, Paul Tripp. Color.

CHRISTMAS TREE, THE 110
CONTINENTAL. (R) October, 1969. William Holden, Virna Lisi. Color.

CHROME AND HOT LEATHER 91
AMERICAN INTERNATIONAL. (R) August, 1971. William Smith, Tony Young. Color.

CHRONICLE OF A SUMMER (Fr., Eng. titles) 88
PATHE-cont. (R) May, 1965. Documentary.

CHUBASCO (P) 100
WARNER BROS.-7 ARTS. (R) June, 1968. Richard Egan, Susan Strasberg. Color.

Length In Mins.

CHUKA 105
PARAMOUNT. (R) May, 1967. Rod Taylor, Ernest Borgnine. Color.

CID, EL (T-70) 160
ALLIED ARTISTS. (R) December, 1961. Charlton Heston, Sophia Loren. Color.

CIMARRON (Cs) 147
METRO-GOLDWYN-MAYER. (R) March, 1961. Glenn Ford, Maria Schell. Color.

CINCINNATI KID, THE 113
METRO-GOLDWYN-MAYER. (R) November, 1965. Steve McQueen, Edward G. Robinson. Color.

CINDERELLA LIBERTY (P) 117
TWENTIETH CENTURY-FOX. (R) February, 1974. James Caan, Marsha Mason.

CINDERFELLA 88
PARAMOUNT. (R) December, 1960. Jerry Lewis, Ed Wynn. Color.

CIRCLE OF DECEPTION (Cs) 100
TWENTIETH CENTURY-FOX. (R) February, 1961. Bradford Dillman, Suzy Parker.

CIRCLE OF LOVE (Fr., Eng. titles) 105
READE-STERLING. (R) March, 1965. Jane Fonda, Maurice Ronet. Color.

CIRCUS GIRL 88
REPUBLIC. (R) April 20, 1956. European Cast. Color.

CIRCUS OF HORRORS 88½
AMERICAN INTERNATIONAL. (R) May, 1960. Anton Diffring, Erika Remberg. Color.

CIRCUS WORLD. (Cinerama, Super T-FO) 135
PARAMOUNT. (R) June, 1964. John Wayne, Claudia Cardinale. Color.

CISCO PIKE 100
COLUMBIA. (R) October, 1971. Karen Black, Gene Hackman. Color.

CITY AFTER MIDNIGHT 84
STATER RIGHTS-RKO. (R) September, 1959. Phyllis Kirk, Dan O'Herlihy.

CITY OF FEAR 81
COLUMBIA. (R) February, 1959. Vince Edwards, Lyle Talbot.

CITY OF FEAR 90
ALLIED ARTISTS. (R) May, 1965. Terry Moore.

CITY OF SHADOWS 70
REPUBLIC. (R) June 1, 1955. Victor McLaglen, John Baer.

CITY ON FIRE 101
AVCO EMBASSY. (R) September, 1979. Barry Newman, Susan Clark.

CLAIRE'S KNEE 110
COLUMBIA. (R) February, 1971. Jean-Claude Brialy, Aurora Cornu, Beatrice Romand. Color.

CLAMBAKE (P) 99
UNITED ARTISTS. (R) November, 1967. Elvis Presley, Shelley Fabares. Color.

CLARENCE THE CROSS-EYED LION 98
METRO-GOLDWYN-MAYER. (R) April, 1965. Marshall Thompson, Betsy Drake. Color.

CLASS OF '44 (P) 105
WARNER BROS. (R) April, 1973. Gary Grimes, Gerald Houser. Color.

CLAUDELLE INGLISH 99
WARNER BROS. (R) September, 1961. Diane McBain, Arthur Kennedy.

CLAUDINE 92
TWENTIETH CENTURY-FOX. (R) April, 1974. Diahann Carroll, James Earl Jones.

CLAY PIGEON 97
METRO-GOLDWYN-MAYER. (R) August, 1971. Telly Savales, Robert Vaughn. Color.

CLEOPATRA (TODD-AO) 243
TWENTIETH CENTURY-FOX. (R) Special. June, 1963. Elizabeth Taylor, Richard Burton. Color.

CLEOPATRA JONES (P) 89
WARNER BROS. (R) July, 1973. Tamara Dobson, Bernie Casey. Color.

CLEOPATRA JONES AND THE CASINO OF GOLD (P) .. 96
WARNER BROS. (R) July, 1975. Tamara Dobson, Stella Stevens.

CLEOPATRA'S DAUGHTER (U) 93
McDALLION (R) February, 1963. Debra Paget, Robert Alda. Color.

CLIMAX, THE (Ital.) 97
LOPERT. (R) September, 1967. Ugo Tognazzi, Stefania Sandrelli.

CLOCKWORK ORANGE 137
WARNER BROS. (R) December, 1971. Malcolm McDowell, Patrick Magee. Color.

CLOPORTES (Fr., Eng. titles) (Cs) 102
INTERNATIONAL CLASSICS. (R) April, 1966. Lino Ventura, Charles Aznavour. Color.

CLOSE ENCOUNTERS OF THE THIRD KIND (P) 135

Length In Mins.

COLUMBIA. (R) November, 1977. Richard Dreyfuss, Francois Trauffaut.

CLOSELY WATCHED TRAINS (Czech.) **89**
SIGMA III. (R) October, 1967. Vaclav Neckar, Jitka Nendova.

CLOUDS OVER ISRAEL **102**
HAROLD CORNSWEET-ISRAEL. (R) May, 1966. Co-Prod. Yiftach Spector, Ehud Banal.

CLOWNS, THE **91**
LEVITT-PICKMAN. (R) June, 1971. Documentary. Color.

C'MON, LET'S LIVE A LITTLE (TS) **85**
PARAMOUNT. (R) April, 1967. Bobby Lee, Jackie DeShannon. Color.

COALMINER'S DAUGHTER **125**
UNIVERSAL. (R) March, 1980. Sissy Spacek, Tommy Lee Jones.

COAST OF SKELETONS (T) **90**
SEVEN ARTS. (R) November, 1965. Richard Todd, Derek Nimmo. Color.

COBRA, THE (TS) **93**
AMERICAN INTERNATIONAL. (R) January, 1968. Anita Ekberg, Dana Andrews. Color.

COBWEB, THE (Cs) **124**
METRO-GOLDWYN-MAYER. (R) July 15, 1955. Richard Widmark, Lauren Bacall, Charles Boyer, Gloria Grahame. Color.

COCKEYED COWBOYS OF CALICO COUNTY, THE **99**
UNIVERSAL. (R) April, 1970. Dan Blocker, Nanette Fabray, Jim Backus. Color.

COCKLESHELL HEROES, THE (Cs) **97**
COLUMBIA. (R) May, 1956. Jose Ferrer, Trevor Howard. Color.

CODE 7, VICTIM 5! (TS) **83**
COLUMBIA. (R) February, 1965. Lex Barker, Ann Smyrner. Color.

COFFY **90**
AMERICAN INTERNATIONAL. (R) May, 1973. Pam Grier. Color.

COLD TURKEY **106**
UNITED ARTISTS. (R) February, 1971. Dick Van Dyke, Bob Newhart. Color.

COLE YOUNGER GUNFIGHTER (Cs) **79**
ALLIED ARTISTS. (R) February, 1958. Frank Lovejoy, Abby Dalton. Color.

COLLECTOR, THE **119**
COLUMBIA. (R) June, 1965. Terence Stamp, Samantha Eggar. Color.

COLLEGE CONFIDENTIAL **90**
UNIVERSAL. (R) August 1960. Steve Allen, Jayne Meadows.

COLOR ME DEAD **97**
COMMONWEALTH UNITED. (R) October, 1969. Tom Tryon, Carolyn Jones. Color.

COLOSSUS OF NEW YORK, THE (VV) **70**
PARAMOUNT. (R) June, 1958. Ross Martin, Mala Powers.

COLOSSUS OF RHODES, THE (Supertotalscope) (Ital.: dubbed) **127**
MGM. (R) 1960. Rory Calhoun, Lea Massi.

COLOSSUS: THE FORBIN PROJECT **100**
UNIVERSAL. (R) May, 1970. Eric Braeden, Susan Clark. Color.

COMA (P) **112**
UNITED ARTISTS. (R) February, 1978. Genevieve Bujold, Michael Douglas.

COMANCHE (Cs) **87**
UNITED ARTISTS. (R) March 1956. Dana Andrews, Linda Cristal. Color.

COMANCHE STATION (Cs) **74**
COLUMBIA. (R) March, 1960.

COMANCHEROS, THE (Cs) **107**
TWENTIETH CENTURY-FOX. (R) November, 1961. John Wayne, Stewart Whitman. Color.

COME BACK CHARLESTON BLUE **100**
WARNER BROS. (R) June, 1972. Godfrey Cambridge, Raymond St. Jacques. Color.

COME BLOW YOUR HORN (P) **112**
PARAMOUNT. (R) June, 1963 Frank Sinatra, Lee J. Cobb. Color.

COME FLY WITH ME (P) **109**
METRO-GOLDWYN-MAYER. (R) April, 1963. Dolores Hart, Hugh O'Brian. Color.

COME NEXT SPRING **92**
REPUBLIC. (R) March 9, 1956. Ann Sheridan, Steve Cochran. Color.

COME ON, THE (SS) **83**
ALLIED ARTISTS. (R) April 15 1955. Ann Baxter, Sterling Hayden.

COME SEPTEMBER (Cs) **112**
UNIVERSAL. (R) June, 1961. Rock Hudson, Gina Lollobrigida. Color.

COME SPY WITH ME **85**

20th CENTURY-FOX. (R) January, 1967. Troy Donahue, Andrea Dromm. Color.

COMEDIANS, THE (P) **160**
MGM. (R) December, 1967. Richard Burton, Elizabeth Taylor. Color.

COMEDY OF TERRORS, THE (P) **88**
AMERICAN INTERNATIONAL. (R) December, 1963. Vincent Price, Peter Lorre. Color.

COMES A HORSEMAN **119**
UNITED ARTISTS. (R) October, 1978. James Caan, Jane Fonda.

COMIC **95**
COLUMBIA. (R) November, 1969. Dick Van Dyke, Michelle Lee, Mickey Rooney. Color.

COMING APART **110**
(R) Ooctober, 1969. Rip Torn, Viveca Linfors, Sally Kirkland. b&w.

COMING HOME **127**
UNITED ARTISTS. (R) February, 1978. Jane Fonda, Jon Voight.

COMMITTEE, THE **90**
COMMONWEALTH UNITED ENTERTAINMENT. (R) March, 1969. Don Sturdy, Carl Gottlieb, Christopher Ross. Color.

COMPANEROS **107**
CINERAMA. (R) April, 1972. Franco Nero, Jack Palance. Color.

COMPANY OF KILLERS **83**
UNIVERSAL. (R) August, 1970. Van Johnson, Ray Milland. Color.

COMPULSION (Cs) **103**
TWENTIETH CENTURY-FOX. (R) April, 1959. Orson Welles, Diana Varsi.

COMPUTER WORE TENNIS SHOES, THE **90**
BUENA VISTA. (R) February, 1970. Kurt Russell, Caesar Romero. Color.

CONCERT FOR BANGLADESH, THE **100**
20th CENTURY-FOX. (R) April, 1972. Rock Concert. Documentary. Color.

CONCORDE—AIRPORT '79, THE **109**
UNIVERSAL. (R) August, 1979. Alain Delon, Susan Blakely.

CONDEMNED OF ALTONA, THE (Cs) **114**
TWENTIETH CENTUFY-FOX. (R) October, 1963. Sophia Loren, Maximillian Schell.

CONDUCT UNBECOMING **107**
ALLIED ARTISTS. (R) October, 1975. Michael York, Richard Attenborough.

CONFESSION, THE (French; Eng. titles) **138**
PARAMOUNT. (R) December, 1970. Yves Montand, Simone Signoret. Color.

CONFESSIONS OF A WINDOW CLEANER **90**
COLUMBIA. (R) November, 1975. Robin Askwith, Anthony Booth.

CONFESSIONS OF AN OPIUM EATER **85**
ALLIED ARTISTS. (R) May, 1962. Vincent Price, Linda Ho.

CONFORMIST, THE (Ital.; Eng. titles) **110**
PARAMOUNT. (R) December, 1970 Jean-Louis Trintignant, Stefania Sandrelli. Color.

CONGO CROSSING **83**
UNIVERSAL. (R) July, 1956. Virginia Mayo, George Nader. Color.

CONGRESS DANCES (Ger.) (Cs) **90**
REPUBLIC. (R) January, 1957. Hannelore Ballman. Color.

CONJUGAL BED, THE **90**
AVCO EMBASSY. (R) 1963. Ugo Tognazzi, Marina Vlady. b&w.

CONQUERED CITY **91**
AMERICAN INTERNATIONAL. (R) January, 1965. David Niven, Ben Gazzara.

CONQUEROR, THE (Cs) **111**
RKO RADIO. (R) March 28, 1956. John Wayne, Susan Hayward, Pedro Armendariz. Color.

CONQUEROR WORM, THE **87**
AMERICAN INTERNATIONAL. (R) May, 1968. Vincent Price, Ian Ogilvy. Color.

CONQUEST OF SPACE **81**
PARAMOUNT. (R) March, 1955. Walter Brooks, Eric Fleming. Color.

CONQUEST OF THE PLANET OF THE APES **88**
20th CENTURY-FOX. (R) June, 1972. Roddy McDowall, Don Murray. Color.

CONRACK (P) **106**
TWENTIETH CENTURY-FOX. (R) March, 1974. Jon Voight, Paul Winfield.

CONSTABLE AND THE CROSS (Tota scope) **114**
AVCO EMBASSY. (R) 1962. Cornel Wilde, Belinda Lee. Color.

CONTEMPT (Cs) **103**
EMBASSY. (R) October, 1964. Brigitte Bardot, Jack Palance. Color.

Length In Mins.

CONTEST GIRL (Cs) ... 102
CONTINENTAL. (R) May, 1966. Ian Hendry, Janette Scott. Color.

CONVERSATION, THE ... 112
PARAMOUNT. (R) April, 1974. Gene Hackman, John Cazale.

CONVICT STAGE ... 71
TWENTIETH-CENTURY-FOX. (R) June, 1965. Harry Lauter, Donald Berry.

CONVOY (P) ... 111
UNITED ARTISTS. (R) June, 1978. Kris Kristofferson, Ali MacGraw.

COOGAN'S BLUFF ... 94
UNIVERSAL. (R) November, 1968. Clint Eastwood, Lee J. Cobb, Susan Clark. Color.

COOL AND THE CRAZY, THE ... 78
AMERICAN INTERNATIONAL PICTURES. (R) March, 1958. Gigi Pereau.

COOL BREEZE ... 101
METRO-GOLDWYN-MAYER. (R) April, 1972. Thalmus Rasulala, Judy Pace. Color.

COOL HAND LUKE ... 129
WARNER BROS.-SEVEN ARTS. (R) October, 1967. Paul Newman, George Kennedy. Color.

COOL ONES, THE (P) ... 95
WARNER BROS. (R) April, 1967. Roddy McDowall, Debbie Watson. Color.

COOL WORLD, THE ... 105
CINEMA V. (R) April, 1964. Hampton Clanton, Carl Lee.

COOLEY HIGH ... 107
AMERICAN INTERNATIONAL. (R) June, 1975. Glynn Turman, Lawrence-Hilton Jacobs.

COONSKIN ... 83
BRYANSTON. (R) August, 1975. Barry White, Charles Gordone.

COP, THE ... 100
AUDUBON. (R) May, 1971. Michel Bouquet, John Garko. Color.

COP HATER ... 75
UNITED ARTISTS. (R) September, 1958. Robert Loggia, Gerald O'Laughlin.

COP OUT ... 95
CINERAMA. (R) January, 1968. James Mason, Geraldine Chaplin. Color.

COPPER SKY (Cs) ... 77
TWENTIETH CENTURY-FOX. (R) September, 1957. Jeff Morrow, Coleen Gray.

COPS AND ROBBERS ... 89
UNITED ARTISTS. (R) August, 1973. Cliff Gorman, Joe Bologna.

CORNBREAD EARL AND ME ... 95
AMERICAN INTERNATIONAL. (R) June, 1975. Moses Gunn, Keith Wilkes.

CORRIDORS OF BLOOD ... 89
METRO-GOLDWYN-MAYER. (R) April, 1963. Boris Karloff, Betta St. John.

CORRUPT ONES, THE (TE) ... 92
WARNER BROS. (R) February, 1967. Robert Stack, Elke Sommer. Color.

CORRUPTION ... 91
COLUMBIA. (R) December, 1968. Peter Cushing, Sue Lloyd, Noel Trevarthon. Color.

CORVETTE SUMMER (P) ... 110
UNITED ARTISTS. (R) June, 1978. Mark Hamill, Annie Potts.

COSMIC MAN, THE ... 72
ALLIED ARTISTS. (R) January, 1959. Bruce Bennett, John Carradine.

COSSACKS, THE (Totalscope) ... 113
UNIVERSAL. (R) May, 1960. Edmund Purdom, John Drew Barrymore. Color.

COTTON COMES TO HARLEM ... 97
UNITED ARTISTS. (R) June, 1970. Godfrey Cambridge, Raymond St. Jacques, Calvin Lockhart. Color.

COUCH, THE ... 100
WARNER BROS. (R) March, 1962. Grant Williams, Shirley Knight.

COUNT FIVE AND DIE ... 92
TWENTIETH CENTURY-FOX. (R) March, 1958. Jeffrey Hunter, Anne Marie Duringer.

COUNT THREE AND PRAY (Cs) ... 102
COLUMBIA. (R) October, 1955. Van Heflin, Joanne Woodward. Color.

COUNT YORGA, VAMPIRE ... 91
AMERICAN INTERNATIONAL. (R) June, 1970. Robert Quarry, Roger Perry. Color.

COUNT YOUR BLESSINGS (Cs) ... 102
METRO-GOLDWYN-MAYER. (R) April, 1959. Deborah Kerr, Rossano Brazzi. Color.

COUNTDOWN (P) ... 101
WARNER BROS.-SEVEN ARTS. (R) February, 1968. James Caan, Joanna Moore. Color.

COUNTDOWN AT KUSINI ... 101
COLUMBIA. (R) April, 1976. Ruby Dee, Greg Morris.

COUNTERFEIT CONSTABLE, THE ... 86
SEVEN ARTS. (R) November, 1966. Robert Dhery. Color.

COUNTERFEIT KILLER, THE ... 95
UNIVERSAL. (R) January, 1970. Jack Lord, Shirley Knight.

COUNTERFEIT PLAN, THE ... 86
WARNER BROTHERS. (R) May 11, 1957. Zachary Scott, Peggie Castle.

COUNTERFEIT TRAITOR, THE ... 140
PARAMOUNT. (R) July, 1962. William Holden, Lilli Palmer. Color.

COUNTERFEITERS OF PARIS, THE ... 90
METRO-GOLDWYN-MAYER. (R) July, 1962. Jean Gabin, Bernard Blier.

COUNTERPLOT ... 76
UNITED ARTISTS. (R) October, 1959. Forrest Tucker, Allison Hayes.

COUNTESS DRACULA ... 94
TWENTIETH CENTURY-FOX. (R) October, 1972. Ingrid Pitt, Nigel Green. Color.

COUNTESS FROM HONG KONG, A ... 108
UNIVERSAL. (R) April, 1967. Marlon Brando, Sophia Loren. Color.

COUNTRY GIRL, THE ... 104
PARAMOUNT. (R) March, 1955. Bing Crosby, Grace Kelly, William Holden.

COUNTRY MUSIC ... 93
UNIVERSAL. (R) November, 1972. Marty Robbins, Sammy Jackson. Color.

COUNTRY MUSIC HOLIDAY ... 81
PARAMOUNT. (R) May, 1958. Ferlin Husky, Zsa Zsa Gabor.

COUPLE, THE (Russ.) ... 103
ARTKINO. (R) August, 1967. Ivan Marin, Vera Kuznetsoya.

COURT JESTER, THE (VV) ... 101
PARAMOUNT. (R) March, 1958. Danny Kaye, Glynis Johns. Color.

COURT MARTIAL (Brit.) ... 105
KINGSLEY. (R) August, 1955. David Niven, Margaret Leighton.

COURT MARTIAL OF BILLY MITCHELL, THE (Cs) ... 100
WARNER BROTHERS. (R) December 31, 1955. Gary Cooper, Charles Bickford, Ralph Bellamy, Rod Steiger. Color.

COURTSHIP OF EDDIE'S FATHER, THE (P) ... 117
METRO-GOLDWYN-MAYER. (R) March, 1963. Glenn Ford, Shirley Jones. Color.

COUSIN, COUSINE (French with English titles) ... 95
LIBRA FILMS. (R) September, 1976. Marie-Christine Barrault, Victor Lamoux.

COUSINS, THE (French; Eng. titles) ... 112
CONTINENTAL. (R) 1959. Claude Brialy, Gerard Blain.

COVENANT WITH DEATH, A ... 97
WARNER BROS. (R) February, 1967. George Maharis, Laura DeVon. Color.

COVER GIRL KILLER! ... 61
FANFARE. (R) May, 1960. Harry H. Corbett, Felicity Young.

COVER ME, BABE (P) ... 89
20th CENTURY-FOX. (R) November, 1970. Robert Forster, Sondra Locke. Color.

COWBOY ... 92
COLUMBIA. (R) March, 1958. Glenn Ford, Jack Lemmon. Color.

COWBOYS, THE ... 128
WARNER BROS. (R) February, 1972. John Wayne. Color.

CRACK IN THE MIRROR (CS) ... 97
TWENTIETH CENTURY-FOX. (R) May, 1960. Orson Welles, Juliette Greco.

CRACK IN THE WORLD ... 96
PARAMOUNT. (P) February, 1965. Dana Andrews, Janette Scott. Color.

CRASH LANDING ... 76
COLUMBIA. (R) 1958. Gary Merrill, Nancy Davis.

CRASHING LAS VEGAS ... 62
ALLIED ARTISTS. (R) April 22, 1956. Leo Gorcey, Hunts Hall.

CRASHOUT ... 82
FILMAKERS. (R) April, 1955. William Bendix, Arthur Kennedy.

CRAZE ... 96
WARNER BROS. (R) July, 1974. Jack Palance, Diana Dors.

CRAZY JOE ... 100
COLUMBIA. (R) February, 1974. Peter Boyle, Paula Prentiss.

CRAZY MAMA ... 82

	Length In Mins.
NEW WORLD. (R) July, 1975. Cloris Leachman, Stuart Whitman.	
CRAZY PARADISE (Danish, Eng. titles)	**95**
SHERPIX. (R) August, 1965. Dirch Passer, Hans W. Petersen. Color.	
CRAZY QUILT	**75**
CONTINENTAL. (R) October, 1966. Tom Rosqui, Ina Mela.	
CRAZY WORLD OF JULIUS VROODER, THE (P)	**100**
20TH CENTURY-FOX. (R) October, 1974. Timothy Bottoms, Barbara Seagull.	
CREATURE WALKS AMONG US, THE	**78**
UNIVERSAL. (R) April, 1956. Jeff Morrow, Leigh Snowden, Rex Reason.	
CREATURE WITH THE ATOM BRAIN	**78**
COLUMBIA. (R) July, 1955. Richard Denning, Angela Stevens.	
CREATURES THE WORLD FORGOT	**94**
COLUMBIA. (R) July, 1971. Julie Ege, Tony Bonner. Color.	
CREEPING FLESH, THE	**87**
COLUMBIA. (R) October, 1972. Christopher Lee, Peter Cushing. Color.	
CREEPING UNKNOWN, THE (Brit.)	**78**
UNITED ARTISTS. (R) June, 1956. Brian Donlevy, Margia Dean.	
CRESCENDO	**83**
WARNER BROS. (R) November, 1972. Stefanie Powers, James Olson. Color.	
CRIES AND WHISPERS	**94**
NEW WORLD PICTURES. (R) February, 1973. Liv Ullmann, Harriet Andersson. Color.	
CRIME AGAINST JOE	**69**
UNITED ARTISTS. (R) May, 1956. John Bromfield, Julie London.	
CRIME AND PUNISHMENT, U.S.A.	**96**
ALLIED ARTISTS. (R) May, 1959. George Hamilton, Mary Murphy.	
CRIME BENEATH THE SEA	**66**
ALLIED ARTISTS. (R) September, 1957. Mara Corday, Pat Conway.	
CRIME IN THE STREETS	**91**
ALLIED ARTISTS. (R) June 19, 1956. James Whitmore, John Cassavetes, Sal Mineo.	
CRIME OF PASSION	**84**
UNITED ARTISTS. (R) February, 1957. Barbara Stanwyck, Sterling Hayden.	
CRIMSON BLADE, THE (M)	**81**
COLUMBIA. (R) March, 1964. Lionel Jeffries, Oliver Reed. Color.	
CRIMSON CULT, THE	**87**
AMER. INT. PICS. (R) April, 1970. Boris Karloff, Christopher Lee. Color.	
CRIMSON KIMONO, THE	**82**
COLUMBIA. (R) October, 1959. Victoria Shaw, Gleen Corbett.	
CRITIC'S CHOICE (P)	**100**
WARNER BROS. (R) April, 1963. Bob Hope, Lucille Ball. Color.	
CROMWELL (P)	**141**
COLUMBIA. (R) October, 1970. Richard Harris, Alec Guinness. Color.	
CROOK, THE (French; Eng. titles)	**120**
UNITED ARTISTS. (R) May, 1971. Jean-Louis Trintingant, Christine Lelouch. Color.	
CROOKED CIRCLE, THE	**72**
REPUBLIC. (R) November, 1957. John Smith, Fay Spain.	
CROOKED WEB, THE	**77**
COLUMBIA. (R) December, 1955. Frank Lovejoy, Mari Blanchard.	
CROSS AND THE SWITCHBLADE, THE	**106**
ROSS. (R) June, 1970. Pat Boone, Erik Estrada. Color.	
CROSS CHANNEL (Brit.)	**90**
REPUBLIC. (R) September 29, 1955. Wayne Morris, Yvonne Furneaux.	
CROSS OF IRON	**120**
AVCO-EMBASSY. (R) May, 1977. James Coburn, Maximilian Schell.	
CROSSED SWORDS (P)	**113**
WARNER BROS. (R) March, 1978. Oliver Reed, Raquel Welch.	
CROSSUP	**83**
UNITED ARTISTS. (R) January, 1958. Larry Parks, Constance Smith.	
CROWDED PARADISE	**93**
TUDOR. (R) June, 1956. Hume Cronyn, Nancy Kelly.	
CROWDED SKY, THE	**104**
WARNER BROS. (R) September, 1960. Dana Andrews, Rhonda Fleming. Color.	
CRUEL TOWER, THE	**80**
ALLIED ARTISTS. (R) October 28, 1956. John Ericson, Mari Blanchard.	

	Length In Mins.
CRUISING	**106**
UNITED ARTISTS. (R) February, 1980. Al Pacino, Paul Sorvino.	
CRUNCH	**90**
AMERICAN INTERNATIONAL. (R) May, 1970. Harold Leipnitz, Sybille Maar. Color.	
CRY BABY KILLER, THE	**62**
ALLIED ARTISTS. (R) August, 1958. Carolyn Mitchell, Jack Nicolson.	
CRY FOR HAPPY (Cs)	**110**
COLUMBIA. (R) March, 1961. Glenn Ford, Donald O'Connor. Color.	
CRY IN THE NIGHT, A	**75**
WARNER BROTHERS. (R) September 15, 1956. Edmond O'Brien, Natalie Wood, Brian Donlevy.	
CRY OF BATTLE	**99**
ALLIED ARTISTS. (R) October, 1963. Van Heflin, Rita Moreno.	
CRY OF THE BANSHEE	**87**
AMERICAN INTERNATIONAL. (E) July, 1970. Vincent Price, Essy Persson. Color.	
CRY TERROR	**96**
MGM. (R) April, 1958. James Mason, Rod Steiger.	
CRY TOUGH	**83**
UNITED ARTISTS. (R) August, 1959. John Saxon, Linda Cristal.	
CUBA	**122**
UNITED ARTISTS. (R) December, 1979. Sean Connery, Brooke Adams.	
CUL-DE-SAC	**110**
SIGMA III. (R) November, 1966. Donald Pleasance, Francoise Dorleac.	
CULPEPPER CATTLE CO., THE	**92**
20th CENTURY-FOX. (R) April, 1972. Gary Grimes, Billy "Green" Bush. Color.	
CULT OF THE COBRA	**82**
UNIVERSAL. (R) May 1955. Faith Domergue, Richard Long.	
CULT OF THE DAMNED, THE (formerly Angel, Angel Down We Go)	**93**
AMER. INT. PICS. (R) August, 1969. Jennifer Jones, Jordon Christopher. Color.	
CURE FOR LOVE, THE (Brit.)	**97**
ASSOCIATED ARTISTS. (R) November, 1955. Robert Donat.	
CURSE OF THE DEMON	**83**
COLUMBIA. (R) July, 1958. Dana Andrews, Peggy Cummins	
CURSE OF THE FACELESS MAN, THE	**66**
UNITED ARTISTS. (R) August, 1958. Richard Anderson, Elaine Edwards.	
CURSE OF THE FLY (Brit.)	**86**
AVCO EMBASSY. (R) 1965. Brian Donlevy, Carole Grey, b&w.	
CURSE OF THE LIVING CORPSE, THE	**83**
LIPPERT. (R) 1964. Helen Waren, Roy R. Shelder.	
CURSE OF THE MUMMY'S TOMB, THE (TS)	**81**
COLUMBIA. (R) February, 1965. Terence Morgan, Ronald Howard. Color.	
CURSE OF THE UNDEAD	**79**
UNIVERSAL. (R) July, 1959. Eric Fleming, Michael Pate.	
CURSE OF THE WEREWOLF, THE	**91**
UNIVERSAL. (R) June, 1961. Clifford Evans, Yvonne Romaine. Color.	
CURUCU, BEAST OF THE AMAZON	**76**
UNIVERSAL. (R) December, 1956. John Bromfield, Beverly Garland. Color.	
CUSTER OF THE WEST (Cinerama)	**140**
CINERAMA. (R) February, 1968. Robert Shaw, Mary Ure. Color.	
CYBORG 2087	**86**
FEATURE FILM CORP. OF AMER. (R) 1966. Michael Rennie, Wendell Corey, Karen Steele. Color.	
CYCLE SAVAGES, THE	**82**
AMERICAN INTERNATIONAL. (R) April, 1970. Bruce Dern, Melody Patterson. Color.	
CYCLOPS	**75**
ALLIED ARTISTS. (R) July, 1957. James Craig, Gloria Talbott.	

D

D-DAY THE SIXTH OF JUNE (Cs)	**108**
TWENTIETH CENTURY-FOX. (R) June, 1958. Robert Taylor, Richard Todd, Dana Wynter, Edmond O'Brien. Color.	
DADDY-LONG LEGS (Cs)	**126**
TWENTIETH CENTURY-FOX. (R) May, 1955. Fred Astaire, Leslie Caron. Color.	
DADDY O	**74**

Length In Mins.

AMERICAN INTERNATIONAL. (R) March, 1959. Dick Contino, Sandra Giles.

DADDY'S GONE A-HUNTING **108**
NATIONAL GENERAL. (R) July, 1969. Carol White, Paul Burke. Color.

DAGMAR'S HOT PANTS, INC. **94**
AMERICAN INTERNATIONAL. (R) November, 1971. Diana Kjaer, Anne Grete. Color.

DAISY MILLER **91**
PARAMOUNT. (R) May, 1974. Cybill Shepherd, Barry Brown.

DAKOTA INCIDENT **88**
REPUBLIC. (R) July 23, 1956. Linda Darnell, Dale Robertson. Color.

DALTON GIRLS, THE **71**
UNITED ARTISTS. (R) December, 1957. Merry Anders, Lisa Davis.

DAMIEN—OMEN II (P) **110**
20th CENTURY-FOX. (R) June, 1978. William Holden, Lee Grant.

DAM BUSTERS, THE (Brit.) **102**
WARNER BROTHERS. (R) July 16, 1955. Richard Todd, Ursula Jeans, Michael Redgrave.

DAMN CITIZEN **88**
UNIVERSAL. (R) March, 1958. Keith Anders, Lisa Davis.

DAMN THE DEFIANT (Cs) **101**
COLUMBIA. (R) September, 1962. Alec Guinness, Dirk Bogarde. Color.

DAMN YANKEES **110**
WARNER BROS. (R) September, 1958. Tab Hunter, Gwen Verdon. Color.

DAMNATION ALLEY (P) **92**
20th CENTURY-FOX. (R) October, 1977. Jan-Michael Vincent, George Peppard.

DAMNED, THE **160**
WARNER BROS. (R) January, 1970. Dirk Bogarde, Ingrid Thulin, Helmet Berger. Color.

DAMON AND PYTHIAS **99**
METRO-GOLDWYN-MAYER. (R) September, 1962. Guy Williams, Don Burnett. Color.

DANCE LITTLE LADY (Brit.) **87**
TRANS-LUX. (R) January, 1956. Mai Zetterling, Terence Morgan. Color.

DANCE WITH ME HENRY **80**
UNITED ARTISTS. (R) December, 1956. Abbott and Costello.

DANDY IN ASPIC. A (P) **107**
COLUMBIA. (R) April, 1968. Laurence Harvey, Tom Courtenay. Color.

DANGER: DIABOLIK **102**
PARAMOUNT. (R) May, 1968. John Phillip Law, Marisa Mell. Color.

DANGER ROUTE **90**
UNITED ARTISTS. (R) February, 1968. Richard Johnson, Diana Dors. Color.

DANGEROUS EXILE (VV) **90**
RANK FILM DISTRIBUTORS OF AMERICA. (R) October, 1958. Louis Jourdan, Belinda Lee. Color.

DANGEROUS YOUTH **98**
WARNER BROTHERS (R) June, 1958. George Baker, Frankie Vaughn.

DANIEL BOONE, TRAIL BLAZER **76**
REPUBLIC. (R) September 14, 1956. Bruce Bennett, Lon Chaney. Color.

DARBY O'GILL AND THE LITTLE PEOPLE **93**
BUENA VISTA. (R) July, 1959. Janet Munro, Albert Sharpe. Color.

DARBY'S RANGERS **121**
WARNER BROS. (R) January, 1958. James Garner, Etchika Choureau.

DARING GAME **100**
PARAMOUNT. (R) April, 1968. Lloyd Bridges, Nico Minardos. Color.

DARK AT THE TOP OF THE STAIRS, THE **124**
WARNER BROS. (R) October, 1960. Robert Preston, Dorothy McGuire. Color.

DARK INTRUDER **59**
UNIVERSAL. (R) October, 1965. Leslie Nielsen, Gilbert Green.

DARK OF THE SUN **105**
MGM. (R) July, 1968. Rod Taylor, Yvette Mimieux, Jim Brown. Color.

DARK PURPOSE **90**
UNIVERSAL. (R) February, 1964. Shirley Jones, Rossano Brazzi. Color.

DARK RIVER (ARG.) **90**
TIMES. (R) February, 1956. Adriana Bennett.

DARKER THAN AMBER **97**
NATIONAL GENERAL. (R) August, 1970. Rod Taylor, Suzy Kendall. Color.

DARLING (Brit.) **122**
EMBASSY. (R) August, 1965. Laurence Harvey, Dirk Bogarde.

DARLING LILI (P) **136**
PARAMOUNT. (R) January, 1970. Julie Andrews, Rock Hudson. Color.

DARWIN ADVENTURE, THE **91**
TWENTIETH CENTURY-FOX. (R) October, 1972. Nicholas Clay, Ian Richardson. Color.

DAUGHTER OF DR. JEKYLL **71**
ALLIED ARTISTS. (R) July, 1957. John Agar, Gloria Talbott.

DAUGHTERS OF DARKNESS **87**
MARON. (R) July, 1971. Delphine Seyrig, John Karlen. Color.

DAUGHTERS OF SATAN **92**
UNITED ARTISTS. (R) October, 1972. Tom Selleck, Barra Grant. Color.

DAVID AND GOLIATH (Cs) **95**
ALLIED ARTISTS. (R) April, 1961. Orson Welles, Ivo Payer. Color.

DAVID AND LISA **94**
CONTINENTAL. (R) February, 1963. Keir Dullea, Janet Margolin.

DAVY CROCKETT AND THE RIVER PIRATES **85**
BUENA VISTA. (R) July 17, 1956. Fess Parker, Buddy Ebsen. Color.

DAVY CROCKETT, KING OF THE WILD FRONTIER **95**
BUENA VISTA. (R) June, 1955. Fess Parker, Buddy Ebsen. Color.

DAWN OF THE DEAD **125**
UNITED FILM DIST. (R) April, 1979. David Enge, Ken Foree.

DAY AND THE HOUR, THE (Franscope) **104**
METRO-GOLDWYN-MAYER. (R) February, 1964. Simone Signoret, Stuart Whitman.

DAY FOR NIGHT **116**
WARNER BROS. (R) October, 1973. Jacqueline Bisset, Valentina Cortese.

DAY IN THE DEATH OF JOE EGG, A **106**
COLUMBIA. (R) June, 1972. Alan Bates, Janet Suzman. Color.

DAY MARS INVADED EARTH, THE (Cs) **70**
TWENTIETH CENTURY-FOX. (R) February, 1963. Kent Taylor, Marie Windsor.

DAY OF ANGER (P) **109**
NATIONAL GENERAL. (R) November, 1969. Lee Van Cleef. Color.

DAY OF FURY, A **78**
UNIVERSAL. (R) May, 1956. Dale Robertson, Mara Corday. Color.

DAY OF THE BAD MAN, THE (Cs) **81**
UNIVERSAL INTERNATIONAL (R) April, 1953. Fred MacMurray, Joan Waldon. Color.

DAY OF THE DOLPHIN, THE **104**
AVCO-EMBASSY. (R) December, 1973. George C. Scott, Trish Van Devere.

DAY OF THE EVIL GUN **95**
MGM. (R) March, 1968. Glenn Ford, Arthur Kennedy. Color.

DAY OF THE JACKAL, THE **142**
UNIVERSAL. (R) May, 1973. Edward Fox, Delphine Seyrig. Color.

DAY OF THE LOCUST, THE **140**
PARAMOUNT. (R) May, 1975. Donald Sutherland, Karen Black.

DAY OF THE OUTLAW **90**
UNITED ARTISTS. (R) July, 1959. Robert Ryan, Burl Ives.

DAY OF THE TRIFFIDS, THE (Cs) **119**
ALLIED ARTISTS. (R) July, 1962. Howard Keel, Nicole Maurey. Color.

DAY THE EARTH CAUGHT FIRE, THE **90**
UNIVERSAL. (R) May, 1962. Edward Judd, Janet Murno, Leo McKern.

DAY THE FISH CAME OUT, THE **109**
INTERNATIONAL CLASSICS. (R) October, 1967. Tom Courtenay, Candice Bergen. Color.

DAY THE HOT LINE GOT HOT, THE **100**
COMMONWEALTH UNITED. (R) December, 1969. George Chakiris, Robert Taylor, Color.

DAY THE WORLD ENDED, THE **90**
AMERICAN RELEASING CORP. (R) January, 1956. Richard Denning, Lori Nelson.

DAY THEY ROBBED THE BANK OF ENGLAND, THE **85**
METRO-GOLDWYN-MAYER. (R) August, 1960. Aldo Ray, Elizabeth Sellars.

DAY TO REMEMBER, A (Brit.) **72**
REPUBLIC. (R) March 29, 1955. Stanley Holloway, Joan Rice.

DAYDREAMER, THE **101**
EMBASSY. (R) Summer, 1966. Animagic-Live action/animation. Color.

DAYS OF HEAVEN **95**

Length In Mins.

PARAMOUNT. (R) September, 1978. Richard Gere, Brooke Adams, Sam Shepard.

DAYS OF THRILLS AND LAUGHTER 93
TWENTIETH CENTURY-FOX. (R) March, 1961. Douglas Fairbanks, Charlie Chaplin.

DAYS OF WINE AND ROSES 117
WARNER BROS. (R) February, 1963. Jack Lemmon, Lee Remick.

DAYTON'S DEVILS 100
COMMONWEALTH UNITED. (R) February, 1969. Rory Calhoun, Leslie Neilson. Color.

DE SADE 113
AMER. INT. PICS. (R) September, 1969. Keir Dullea, Senta Berger, Lilli Palmer. Color

DEAD ARE ALIVE, THE 103
NATIONAL GENERAL. (R) May, 1972. Samantha Eggar, Alex Cord. Color.

DEAD HEAT ON A MERRY-GO-ROUND 104
COLUMBIA. (R) October, 1966. James Coburn, Camilla Sparv. Color.

DEAD RINGER 115
WARNER BROS. (R) February, 1964. Bette Davis, Karl Malden.

DEADFALL 120
TWENTIETH CENTURY-FOX. (R) September, 1968. Michael Caine, Giovanna Ralli, Eric Portman. Color.

DEADLIER THAN THE MALE (Fr.) 104
CONTINENTAL (R) November, 1957. Jean Gabin, Danniele Delorme.

DEADLIER THAN THE MALE (TS) 97
UNIVERSAL. (R) March, 1967. Richard Johnson, Elke Sommer. Color.

DEADLIEST SIN, THE 75
ALLIED ARTISTS. (R) January 29, 1956. Sydney Chaplin, Audrey Dalton.

DEADLY AFFAIR, THE 106
COLUMBIA. (R) February, 1967. James Mason, Simone Signoret. Color.

DEADLY BEES, THE 84
PARAMOUNT. (R) May, 1967 Suzanna Leigh, Guy Doleman. Color.

DEADLY COMPANIONS, THE 90
PATHE-AMERICAN. (R) June, 1961. Maureen O'Hara, Brian Keith. Color.

DEADLY HERO 96
AVCO-EMBASSY. (R) March, 1976. Don Murray, Diahn Williams.

DEADLY MANTIS, THE 73
UNIVERSAL. (R) May, 1967. Craig Stevens, Aux Talton.

DEADLY TRACKERS, THE 105
WARNER BROS. (R) December, 1973. Richard Harris, Rod Taylor.

DEADLY TRAP, THE 91
NATIONAL GENERAL. (R) October, 1972. Faye Dunaway, Frank Langella. Color.

DEAF SMITH AND JOHNNY EARS 90
METRO-GOLDWYN-MAYER. (R) May, 1973. Anthony Quinn, Franco Nero. Color.

DEALING 88
WARNER BROS. (R) February, 1972. Barbara Hershey, Robert F. Lyons. Color.

DEAR BRIGITTE (Cs) 100
TWENTIETH CENTURY-FOX. (R) January, 1965. James Stewart, Billy Mumy. Color.

DEAR HEART 114
WARNER BROS. (R) January, 1965. Glenn Ford, Geraldine Page.

DEAR JOHN (Swedish, Eng. Titles) 115
SIGMA III. (R) March, 1966. Jarl Kulle, Christina Schollin.

DEATH IN SMALL DOSES 79
ALLIED ARTISTS. (R) August, 1957. Peter Graves, Mala Powers.

DEATH IN VENICE 130
WARNER BROS. (R) June, 1971. Dirk Bogarde, Silvana Mangano. Color.

DEATH OF A GUNFIGHTER 94
UNIVERSAL. June, 1969. Richard Widmark, Lena Horne, John Saxon, Michael McGreevey. Color.

DEATH OF A SCOUNDREL 119
UNIVERSAL. (R) October, 1956. George Sanders, Yvonne de Carlo, Zsa Zsa Gabor.

DEATH OF TARZAN, THE (Czech; Eng. titles) 72
BRANDON. (R) July, 1968. Rudolf Hrusinsky, Jana Stepkanova. b&w.

DEATH ON THE NILE 140
PARAMOUNT. (R) September, 1978. Peter Ustinov, Bette Davis, Mia Farrow.

DEATH RIDES A HORSE 114
UNITED ARTISTS. (R) September, 1969. Lee Van Cleef, John Phillip Law. Color.

DEATH SHIP 91
AVCO EMBASSY. (R) April, 1980. George Kennedy, Richard Crenna.

DEATH WISH 91
PARAMOUNT. (R) July, 1974. Charles Bronson, Hope Lange.

DECAMERON, THE 101
UNITED ARTISTS. (R) December, 1971. Franco Citti, Ninetto Davoli. Color.

DECISION AGAINST TIME 87
METRO-GOLDWYN-MAYER. (R) August, 1957. Jack Hawkins, Elizabeth Sellers.

DECISION AT SUNDOWN 77
COLUMBIA. (R) November, 1957. Randolph Scott, John Carroll. Color.

DECKS RAN RED, THE 84
MGM. (R) October, 1958. James Mason, Dorothy Dandridge. Color.

DECLINE AND FALL OF A BIRD WATCHER, THE 113
TWENTIETH CENTURY-FOX. (R) January, 1969. Rodney Bewes, Colin Blakely, Patience Collier. Color.

DEEP, THE (P) 123
COLUMBIA. (R) June, 1977. Robert Shaw, Jacqueline Bisset, Nick Nolte.

DEEP BLUE SEA, THE (Cs) 99
TWENTIETH CENTURY-FOX. (R) November, 1955. Vivian Leigh, Kenneth More. Color.

DEEP END 88
PARAMOUNT. (R) September, 1971. Jane Asher, John Moulder Brown. Color.

DEEP SIX, THE 110
WARNER BROS. (R) January, 1958 Alan Ladd, Diane Foster. Color.

DEER HUNTER, THE (P) 183
UNIVERSAL. (R) February, 1979. Robert De Niro, John Savage, Christopher Walken.

DEERSLAYER, THE (Cs) 76
TWENTIETH CENTURY-FOX. (R) September, 1957. Lex Barker, Rita Moreno. Color.

DEFECTOR, THE 106
SEVEN ARTS. (R) November, 1966. Montgomery Clift, Hardy Kruger. Color.

DEFIANCE 103
AMERICAN INTERNATIONAL. (R) March, 1980. Jan Michael Vincent, Art Carney

DEFIANT ONES, THE 97
UNITED ARTISTS. (R) August, 1958. Tony Curtis, Sidney Poitier.

DELICATE BALANCE, A. 132
AFT DISTRIBUTING. (R) 1973, special. Katherine Hepburn, Paul Scofield.

DELICATE DELINQUENT, THE (VV) 101
PARAMOUNT. (R) July, 1957. Jerry Lewis, Darren McGavin. Color.

DELINQUENTS, THE 71
UNITED ARTISTS. (R) March, 1957. Tommy Laughlin, Peter Miller.

DELIVERANCE 109
WARNER BROS. (R) August, 1972. Jon Voight, Burt Reynolds. Color.

DELTA FACTOR, THE 91
CONTINENTAL. (R) July, 1970. Yvette Mimieux, Christopher George. Color.

DEMENTIA 13 81
AMER. IN'L. PICS. (R) 1964. William Campbell, Luana Anders. b&w.

DEMON SEED (P) 94
UNITED ARTISTS. (R) April, 1977. Julie Christie, Fritz Weaver.

DERBY 91
CINERAMA. (R) April, 1971. Mike Snell, Charlie O'Connell. Color.

DESERT ATTACK 80
TWENTIETH CENTURY-FOX. (R) December, 1960. John Mills, Sylvia Syms.

DESERT HELL (Rs) 82
TWENTIETH CENTURY-FOX. (R) June, 1958. Brian Keith, Barbara Hale.

DESERT PATROL 78
UNIVERSAL. (R) March, 1962. Richard Attenborough, John Gregson.

DESERT SANDS (SS) 87
UNITED ARTISTS. (R) September, 1955. Ralph Meeker, Maria English. Color.

DESERT WARRIOR 87
MEDALLION. (R) 1960. Ricardo Montalban, Anna Maria Ferrero. Color.

DESERTER, THE (P) 99
PARAMOUNT. (R) April, 1971. Bekim Fehmiu, Richard Crenna. Color.

DESERTER AND THE NOMADS (Czech.-Ital. with Eng. titles) 103
ROYAL. (R) October, 1969. Ferenc Geijza, Helena Gorova. Color.

Length In Mins.

DESIGNING WOMAN (Cs) **118**
METRO-GOLDWYN-MAYER. (R) April, 1957. Gregory Peck, Lauren Bacall. Color.
DESIRE IN THE DUST (Cs) **105**
TWENTIETH CENTURY-FOX. (R) October, 1960. Raymond Burr, Martha Hyer.
DESIRE UNDER THE ELMS (VV) **114**
PARAMOUNT. (R) March, 1958. Sophia Loren, Anthony Perkins.
DESK SET (Cs) **103**
TWENTIETH CENTURY-FOX. (R) May, 1957. Spencer Tracy, Katharine Hepburn. Color.
DESPERADOES, THE **90**
COLOMBIA. (R) April, 1969. Vince Edwards, Jack Palance, George Maharis. Color.
DESPERADOES ARE IN TOWN, THE **73**
TWENTIETH CENTURY-FOX. (R) November, 1956. Robert Arthur, Rex Reason.
DESPERATE CHARACTERS **88**
PARAMOUNT. (R) September, 1971. Shirley Maclaine, Kenneth Mars, Sada Thompson. Color.
DESPERATE HOURS, THE (VV) **112**
PARAMOUNT. (R) November, 1955. Humphrey Bogart, Fredric March.
DESPERATE ONES, THE **104**
COMMONWEALTH UNITED. (R) October, 1969. Maximillian Schell, Raf Vallone, Irene Papas.
DESTINATION INNER SPACE **83**
MAGNA. (R) June, 1966. Scott Brady, Sheree North, Gary Merrill. Color.
DESTINATION 60,000 .... **65**
ALLIED ARTISTS. (R) June, 1957. Preston Foster, Coleen Gray.
DESTRY **95**
UNIVERSAL. (R) January, 1955. Audie Murphy, Mari Blanchard. Color.
DETECTIVE, THE (Brit.) **86**
COLUMBIA. (R) March, 1955. Alec Guinness, Joan Greenwood.
DETECTIVE, THE **114**
TWENTIETH CENTURY-FOX. (R) July, 1968. Frank Sinatra, Lee Remick, Jacqueline Bisset. Color.
DETECTIVE BELLI **103**
PLAZA PICTURES. (R) December, 1970. Franco Nero, Florinda Bolkan. Color.
DETOUR **78**
BRANDON. (R) August, 1969. Nevena Kipokanova, Ivan Andonov. b&w.
DEVIL AT FOUR O'CLOCK **126**
COLUMBIA. (R) October, 1961. Spencer Tracy, Frank Sinatra. Color.
DEVIL BY THE TAIL, THE (French; Eng. titles) **93**
LOPERT. (R) June, 1969. Yves Montand, Maria Schell. Color.
DEVIL DOLL **80**
ASSOCIATED. (R) September, 1964. Bryant Haliday, William Sylvester.
DEVIL GODDESS **70**
COLUMBIA. (R) October, 1955. Johnny Weissmuller, Angela Stevens.
DEVIL IN LOVE **97**
WARNER BROS.-7 ARTS. (R) May, 1968. Vittorio Gassman, Mickey Rooney. Color.
DEVIL IS A WOMAN, THE **105**
TWENTIETH CENTURY-FOX. (R) October, 1975. Glenda Jackson, Lisa Harrow.
DEVILS, THE **109**
WARNER BROS. (R) July, 1971. Vanessa Redgrave, Oliver Reed. Color.
DEVIL'S ANGELS (P) **83**
AMERICAN INTERNATIONAL. (R) April, 1967. John Cassavettes, Beverly Adams. Color.
DEVIL'S BRIDE, THE **95**
TWENTIETH CENTURY-FOX. (R) July, 1968. Chris Lee, Charles Grey, Nike Arrighi. Color.
DEVIL'S BRIGADE, THE (P) **130**
UNITED ARTISTS. (R) May, 1968. William Holden, Vince Edwards. Color.
DEVIL'S DISCIPLE, THE **82**
UNITED ARTISTS. (R) August, 1959. Burt Lancaster, Kirk Douglas.
DEVIL'S EIGHT, THE **97**
AMERICAN INTERNATIONAL PICTURES. (R) March, 1969. Christopher George, Tom Nardini, Leslie Parrish. Color.
DEVIL'S GENERAL, THE (Ger.) **120**
STEBBINS. (R) March, 1957. Curt Jurgens, Marianne Cook.
DEVIL'S HAIRPIN, THE (VV) **83**
PARAMOUNT. (R) October, 1957. Cornel Wilde, Jean Wallace. Color.
DEVILS OF DARKNESS **88**
TWENTIETH CENTURY-FOX. (R) May, 1965. William Sylvester, Hubert Noel. Color.
DEVIL'S OWN, THE **90**
20th CENTURY-FOX. (R) February, 1967. Joan Fontaine, Kay Walsh. Color.
DEVIL'S RAIN, THE **85**
BRYANSTON. (R) July, 1975. Ernest Borgnine, Eddie Albert.
DEVIL'S TRAP, THE **85**
SALISBURY. (R) August, 1964. Vitezslan Vejrazka, Vit Olmer.
DEVIL'S WIDOW, THE **92**
AMERICAN INTERNATIONAL. (R) November, 1972. Ava Gardner, Ian McShane. Color.
DEVIL-SHIP PIRATES, THE (M) **89**
COLUMBIA. (R) May, 1964. Christopher Lee, Andrew Keir. Color.
DIABOLIQUE (Fr.) **107**
UMPO. (R) November, 1955. Simone Signoret, Vera Clouzot.
DIAL RED O **63**
ALLIED ARTISTS. (R) March 13, 1955. Bill Elliott, Keith Larsen.
DIAMOND HEAD (P) **107**
COLUMBIA. (R) January, 1963. Charlton Heston, Yvette Mimieux. Color.
DIAMOND SAFARI **67**
TWENTIETH CENTURY-FOX. (R) January, 1958. Kevin McCarthy, Audie Morell.
DIAMONDS **108**
AVCO-EMBASSY. (R) October, 1975. Robert Shaw, Richard Roundtree.
DIAMONDS ARE FOREVER **119**
UNITED ARTISTS. (R) December, 1971. Sean Connery, Jill St. John, Charles Gray. Color.
DIANE (Cs) **110**
METRO-GOLDWYN-MAYER. (R) January 6, 1956. Lana Turner, Pedro Armendariz, Roger Moore. Color.
DIARY OF A BACHELOR **88**
AMERICAN INTERNATIONAL. (R) October, 1964. William Traylor, Dagne Crane.
DIARY OF A CHAMBERMAID (F) **97**
INTERNATIONAL CLASSICS. (R) March, 1965. Jeanne Moreau, Michel Piccoli.
DIARY OF A HIGH SCHOOL BRIDE **80**
AMERICAN INTERNATIONAL. (R) July, 1959. Anita Sands, Ronald Foster.
DIARY OF A MAD HOUSEWIFE **94**
UNIVERSAL. (R) August, 1970. Carrie Snodgress, Richard Benjamin. Color.
DIARY OF A MADMAN **96**
UNITED ARTISTS. (R) March, 1963. Vincent Price, Nancy Kovack. Color.
DIARY OF A SCHIZOPHRENIC GIRL **108**
ALLIED ARTISTS. (R) April, 1960. Margarita Lozano, Ghislaine D'Orsay. Color.
DIARY OF ANNE FRANK, THE (Cs) **180**
TWENTIETH CENTURY-FOX. (R) March, 1959. Millie Perkins, Joseph Schildkraut.
DID YOU HEAR THE ONE ABOUT THE TRAVELING SALESLADY? **97**
UNIVERSAL. (R) March, 1968. Phyllis Diller, Bob Denver. Color.
DIE! DIE! MY DARLING! **97**
COLUMBIA. (R) May, 1965. Tallulah Bankhead, Stephanie Powers. Color.
DIE LAUGHING **108**
WARNER BROS. (R) March, 1980. Robby Benson.
DIE MONSTER DIE! (Colorscope) **80**
AMERICAN INTERNATIONAL. (R) October, 1965. Boris Karloff, Nick Adams. Color.
DIFFERENT STORY, A **107**
AVCO EMBASSY. (R) June, 1978. Perry King, Meg Foster.
DIG THAT URANIUM **61**
ALLIED ARTISTS. (R) January 8, 1956. Leo Gorcey, Huntz Hall.
DILLINGER **106**
AMERICAN INTERNATIONAL. (R) June, 1973. Warren Oates, Ben Jonson. Color.
DIME WITH A HALO **94**
METRO-GOLDWYN-MAYER. (R) May, 1963. Barbara Luna, Paul Langton.
DIMENSION 5 .... **88**
UNITED PICTURE CORP. (R) 1956. Jeffrey Hunter, France Nuyen. Color.
DINGAKA **98**
EMBASSY. (R) May, 1965. Stanley Baker, Julliet Prowse. Color.
DINO **94**
ALLIED ARTISTS. (R) July, 1957. Sal Mineo, Susan Kohner.
DINOSAURUS! (Cs) **85**

Length In Mins.

UNIVERSAL. (R) July, 1960. Ward Ramsey, Kristina Hanson. Color.

DIONYSUS IN '69 ... 90
SIGMA III. (R) March, 1970. William Finley, William Shephard. b&w.

DIRTY DINGUS MAGEE (P) ... 91
MGM. (R) November, 1970. Frank Sinatra, George Kennedy. Color.

DIRTY DOZEN, THE ... 150
MGM. (R) July, 1967. Lee Marvin, Ernest Borgnine. Color.

DIRTY GAME, THE ... 91
AMERICAN INTERNATIONAL (R) April, 1966. Henry Fonda, Robert Ryan, Vittorio Gassman.

DIRTY HARRY ... 102
WARNER BROS. (R) March, 1972. Clint Eastwood, Harry Guardino. Color.

DIRTY LITTLE BILLY ... 93
COLUMBIA. (R) January, 1973. Michael J. Pollard, Lee Purcell. Color.

DIRTY MARY CRAZY LARRY ... 93
TWENTIETH CENTURY-FOX. (R) May, 1974. Peter Fonda, Susan George.

DISCREET CHARM OF THE BOURGEOISIE, THE ... 100
TWENTIETH CENTURY-FOX (R) November, 1972. Fernando Rey, Delphine Seyrig. Color.

DISEMBODIES, THE ... 65
ALLIED ARTISTS. (R) July, 1957. Paul Burke, Allison Hayes.

DISORDER (Ital., Eng. titles) ... 105
PATHE-CONT. (R) June, 1964. Louis Jordan, Susan Strasberg.

DISORDERLY ORDERLY, THE ... 89
PARAMOUNT. (R) December, 1964. Jerry Lewis, Glenda Farrell. Color.

DISTANT TRUMPET, A (P) ... 117
WARNER BROS. (R) May, 1964. Troy Donahue, Suzanne Pleshette. Color.

DIVIDED HEART, THE (Brit.) ... 80
REPUBLIC. (R) August 11, 1955. Cornel Borchers, Yvonne Mitchell, Alexander Knox.

DIVORCE AMERICAN STYLE ... 109
COLUMBIA. (R) July, 1967. Dick Van Dyke, Debbie Reynolds. Color.

DIVORCE ITALIAN STYLE ... 104
AVCO EMBASSY. (R) 1962. Marcello Mastroianni, Daniela Rocca. b&w.

DO NOT DISTURB (Cs) ... 102
TWENTIETH CENTURY-FOX. (R) Christmas, 1965. Doris Day, Rod Taylor. Color.

DO YOU KEEP A LION AT HOME? (Czech, Eng. titles) ... 81
BRANDON. (R) December, 1966. Ladislaw Ocanasek, Josef Filip. Color.

DOC ... 95
UNITED ARTISTS. (R) August, 1971. Stacy Keach, Faye Dunaway, Harris Yulin. Color.

DOC SAVAGE . . . THE MAN OF BRONZE ... 100
WARNER BROS. (R) June, 1975. Ron Ely.

DOCTOR AT LARGE (VV) ... 98
UNIVERSAL. (R) June, 1957. Dirk Bogarde, Muriel Pavlow. Color.

DOCTOR AT SEA (Brit.) (VV) ... 92
REPUBLIC. (R) February 23, 1956. Dirk Bogarde, Brenda De Banzi, Brigitte Bardot. Color.

DOCTOR BLOOD'S COFFIN ... 92
UNITED ARTISTS. (R) May, 1961. Kieron Moore, Hazel Court.

DOCTOR COPPELIUS ... 97
CHILDHOOD. (R) December, 1968. Walter Slezak, Claudia Corday, Eileen Elliott. Color.

DOCTOR CRIPPEN ... 97
WARNER BROS. (R) February, 1964. Donald Pleasence, Coral Browne.

DOCTOR DEATH: SEEKER OF SOULS ... 93
CINERAMA. (R) November, 1973. John Considine, Barry Coe.

DOCTOR FAUSTUS ... 95
COLUMBIA. (R) February, 1968. Richard Burton, Elizabeth Taylor. Color.

DOCTOR GLAS (Danish; Eng. titles) ... 83
TWENTIETH CENTURY-FOX. (R) April, 1969. Peter Oscarrson, Ulf Palms, Lone Hertz. b&w.

DOCTOR GOLDFOOT AND THE BIKINI MACHINE (P) ... 90
AMERICAN INTERNATIONAL. (R) November, 1965. Vincent Price, Frankie Avalon. Color.

DOCTOR GOLDFOOT AND THE GIRL BOMBS ... 85
AMERICAN INTERNATIONAL. (R) November, 1966. Vincent Price, Fabian. Color.

DOCTOR IN DISTRESS ... 103
GOVERNOR. (R) June, 1964. Dirk Bogarde, James Robertson Justice.

DOCTOR IN LOVE ... 93
GOVERNOR. (R) April, 1962. Michael Graig, Virginia Maskell. Color.

DOCTOR IN THE HOUSE (Brit.) ... 92
REPUBLIC. (R) February 2, 1955. Dirk Bogarde, Muriel Pavlow. Color.

DOCTOR JEKYLL AND SISTER HYDE ... 95
AMERICAN INTERNATIONAL. (R April, 1972. Ralph Bates, Martine Beswick.

DOCTOR NO ... 111
UNITED ARTISTS. (R) May, 1963. Sean Connery, Ursula Andress. Color.

DOCTOR PHIBES RISES AGAIN! ... 89
AMERICAN INTERNATIONAL. (R July, 1972. Vincent Price, Terry-Thomas. Color.

DOCTOR STRANGELOVE OR: HOW I LEARNED TO STOP WORRYING AND LOVE THE BOMB (P) ... 93
COLUMBIA. (R) January, 1964. Peter Sellers, George C. Scott. b&w.

DOCTOR TERROR'S HOUSE OF HORRORS (T) ... 98
PARAMOUNT. (R) April, 1965. Peter Cushing, Christopher Lee. Color.

DOCTOR WHO AND THE DALEKS (TE) ... 85
CONTINENTAL. (R) Summer, 1966. Peter Cushing, Roy Castle, Jennie Linden. Color.

DOCTOR, YOU'VE GOT TO BE KIDDING (P) ... 94
METRO-GOLDWYN-MAYER. (R) April, 1967. Sandra Dee, George Hamilton. Color.

DOCTOR ZHIVAGO (P) ... 197
METRO-GOLDWYN-MAYER. (R) Special. December, 1965. Omar Sharif, Julie Christie. Color.

DOCTOR'S DILEMMA, THE ... 98
METRO-GOLDWYN-MAYER. (R) January, 1959. Leslie Caron, Dick Bogarde. Color.

DOCTORS' WIVES ... 100
COLUMBIA. (R) February, 1971. Dyan Cannon, Richard Crenna. Color.

DOG DAY AFTERNOON ... 130
WARNER BROS. (R) October, 1975. Al Pacino, John Cazale.

DOG OF FLANDERS, A (Cs) ... 97
TWENTIETH CENTURY-FOX. (F) March, 1960. David Ladd, Donald Crisp. Color.

(DOLLARS) ... 119
COLUMBIA. (R) December, 1971. Warren Beatty, Goldie Frobe. Color.

DOLL'S HOUSE, A ... 95
PARAMOUNT. (R) May, 1973. Claire Bloom, Anthony Hopkins. Color.

DOMINO KID, THE ... 78
COLUMBIA. (R) October, 1957. Rory Calhoun, Kristine Miller.

DOMINO PRINCIPLE, THE (P) ... 100
AVCO-EMBASSY. (R March, 1977. Gene Hackman, Candice Bergen.

DON IS DEAD, THE ... 115
UNIVERSAL. (R) November, 1973. Anthony Quinn, Frederic Forrest.

DON JUAN (Ger.) ... 85
TIMES. (R) March, 1956. Cesare Danova, Josef Neinrad. Color.

DON JUAN'S NIGHT OF LOVE (Ital.-Eng. dubbed) ... 71
REPUBLIC. (R) May 26, 1955. Raf Vallone, Silvana Pampanini.

DON QUIXOTE ... 106
METRO-GOLDWYN-MAYER. (R) January, 1961. Color.

DONDI ... 100
ALLIED ARTISTS. (R) April, 1961. David Janssen, Patti Page.

DONOVAN'S REEF ... 109
PARAMOUNT. (R) July, 1963. John Wayne, Lee Marvin. Color.

DON'T DRINK THE WATER ... 98
AVCO EMBASSY. (F) November, 1969. Jackie Gleason, Estelle Parsons. Color.

DON'T GIVE UP THE SHIP ... 89
PARAMOUNT. (R) July, 1959. Jerry Lewis, Dina Merrill.

DON'T GO NEAR THE WATER (CS) ... 107
METRO-GOLDWYN-MAYER. (R) November, 1956. Glenn Ford, Gia Scala. Color.

DON'T JUST STAND THERE ... 99
UNIVERSAL. (R) May, 1968. Robert Wagner, Mary Tyler Moore. Color.

DON'T KNOCK THE ROCK ... 84
COLUMBIA. (R) January, 1957. Bill Haley, Alan Dale.

DON'T KNOCK THE TWIST ... 87
COLUMBIA. (R) April, 1962. Chubby Checker, Gene Chandler.

DON'T LOOK BACK ... 90
LEACOCK-PENNEBAKER. (R) September, 1967. Bob Dylan, Joan Baez.

DON'T LOOK NOW ... 116

Length In Mins.

BUENA VISTA. (R) February, 1969. Terry-Thomas, Bourvil Louis de Funes. Color.

DON'T LOOK NOW (P) ... 110
PARAMOUNT. (R) January, 1974. Julie Christie, Donald Sutherland.

DON'T MAKE WAVES (P) ... 97
MGM. (R) June, 1967. Tony Curtis, Claudia Cardinale. Color.

DON'T RAISE THE BRIDGE, LOWER THE RIVER ... 99
COLUMBIA. (R) June, 1968. Jerry Lewis, Jacqueline Pearce. Color.

DON'T TEMPT THE DEVIL (Fr., Eng. titles) ... 106
UMPO. (R) May, 1964. Marina Vlady, Bouril.

DON'T WORRY, WE'LL THINK OF A TITLE ... 88
UNITED ARTISTS. (R) May, 1966. Morey Amsterdam, Rose Marie.

DORIAN GRAY ... 93
AMERICAN INTERNATIONAL. (R) November, 1970. Helmut Berger, Richard Todd. Color.

DOUBLE JEOPARDY ... 70
REPUBLIC. (R) June 23, 1955. Rod Cameron, Gale Robbins.

DOUBLE MAN, THE ... 105
WARNER BROS.-7 ARTS. (R) April, 1968. Yul Brynner, Britt Ekland. Color.

DOUBLE TROUBLE (P) ... 90
METRO-GOLDWYN-MAYER. (R) June, 1967. Elvis Presley, Annette Day. Color.

DOULOS—THE FINGER MAN ... 108
PATHE CONTEMP. (R) April, 1969. Jean Paul Belmondo, Michael Piccoli. b&w.

DOVE, THE ... 105
PARAMOUNT. (R) September, 1974. Joseph Bottoms, Deborah Raffin.

DOWN THE ANCIENT STAIRS (Italian; English titles) ... 110
TWENTIETH CENTURY-FOX. (R) October, 1975. Marcello Mastroianni.

DOWNHILL RACER ... 102
PARAMOUNT. (R) November, 1969. Robert Redford, Gene Hackman, Camilla Sparv. Color.

DRACULA (P) ... 109
UNIVERSAL. (R) July, 1979. Frank Langella, Laurence Olivier.

DRACULA A.D. 1972 ... 100
WARNER BROS. (R) November, 1972. Christopher Lee, Peter Cushing. Color.

DRACULA HAS RISEN FROM THE GRAVE ... 92
WARNER BROS.-7 ARTS. (R) Feb., 1969. Christopher Lee, Rupert Davies, Veronica Carlson. Color.

DRACULA—PRINCE OF DARKNESS ... 90
TWENTIETH CENTURY-FOX. (R) January, 1966. Christopher Lee, Barbara Shelley. Color.

DRAGON FLY ... 95
AMERICAN INTERNATIONAL. (R) April, 1976. Beau Bridges, Susan Sarandon.

DRAGON SKY ... 95
LOPERT. (R) September, 1964. Naria Hem, Sam El. Color.

DRAGOON WELLS MASSACRE (Cs) ... 88
ALLIED ARTISTS. (R) April 28, 1957. Barry Sullivan, Mona Freeman, Dennis O'Keefe, Katy Jurado. Color.

DRAGSTRIP GIRL ... 69
AMERICAN INTERNATIONAL. (R) April, 1957. Fay Spain, Steven Terrell.

DRAGSTRIP RIOT ... 68
AMERICAN INTERNATIONAL PICTURES. (R) March, 1958. Yvonne Lime, Gary Clark.

DRANGO ... 92
UNITED ARTISTS. (R) January, 1957. Jeff Chandler, Joanne Dru.

DREAM MAKER, THE ... 87
UNIVERSAL. (R) March, 1964. Tommy Steele, Michael Medwin. Color.

DREAM OF KINGS, A ... 107
NATIONAL GENERAL. (R) December, 1969. Anthony Quinn, Irene Papas, Inger Stevens. Color.

DREAMER ... 90
TWENTIETH-CENTURY FOX. (R) May, 1979. Tim Matheson, Susan Blakely.

DREAMER, THE (Hebrew; Eng. titles) ... 86
CANNON. (R) April, 1970. Tuvia Tavi, Berta Litvina.

DREAMS OF GLASS ... 83
UNIVERSAL. (R) August, 1970. John Denos, Caroline Barrett.

DRESSED TO KILL ... 105
FILMWAYS. (R) July, 1980. Michael Caine, Angie Dickinson.

DRIVE, HE SAID ... 90
COLUMBIA. (R) June, 1971. William Tepper, Karen Black. Color.

DRIVE-IN ... 96
COLUMBIA. (R) May, 1976. Lisa Lemole, Glenn Morshower.

Length In Mins.

DRIVER, THE ... 91
20th CENTURY-FOX. (R) July, 1978. Ryan O'Neal, Bruce Dern.

DRIVER'S SEAT, THE ... 101
AVCO-EMBASSY. (R) October, 1975. Elizabeth Taylor, Ian Bannen.

DROWNING POOL, THE (P) ... 108
WARNER BROS. (R) July, 1975. Paul Newman, Joanne Woodward.

DRUM ... 101
UNITED ARTISTS. (R) August, 1976. Warren Oates, Isela Vega.

DRUMS OF AFRICA. ... 92
METRO-GOLDWYN-MAYER. (R) May, 1963. Frankie Avalon, Mariette Hartley. Color.

DUCHESS AND THE DIRTWATER FOX, THE ... 105
TWENTIETH CENTURY-FOX. (R) April, 1976. George Segal, Goldie Hawn.

DUCK! YOU SUCKER ... 138
UNITED ARTISTS. (R) June, 1972. Rod Steiger, James Coburn. Color.

DUEL, THE ... 87
ARTKINO. (R) September, 1964. Oleg Stristhenov, Vladimir Druzhnikov. Color.

DUEL AT APACHE WELLS (N) ... 70
REPUBLIC. (R) January, 1957. Anna Maria Alberghetti, Ben Cooper.

DUEL AT DIABLO ... 103
UNITED ARTISTS. (R) May, 1966. James Garner, Sidney Poitier, Dennis Weaver. Color.

DUEL OF CHAMPIONS ... 93
MEDALLION. (R) June, 1964. Alan Ladd, Franca Bettoja. Color.

DUEL OF THE TITANS (Cs) ... 90
PARAMOUNT. (R) June, 1963. Steve Reeves, Gordon Scott. Color.

DUEL ON THE MISSISSIPPI ... 72
COLUMBIA. (R) October, 1955. Lex Barker, Patricia Medina. Color.

DUELLISTS, THE ... 100
PARAMOUNT. (R) January, 1978. Keith Carradine, Harvey Keitel.

DUFFY ... 101
COLUMBIA. (R) September, 1968. James Coburn, James Mason, James Fox, Susannah York. Color.

DULCIMA ... 93
CINEVISION. (R) September, 1972. Carol White, John Mills. Color.

DUNKIRK ... 113
METRO-GOLDWYN-MAYER. (R) September, 1958. John Mills, Robert Urquhart.

DUNWICH HORROR, THE ... 90
AMERICAN INTERNATIONAL. (R) January, 1971. Sandra Dee, Dean Stockwell. Color.

DUSTY AND SWEETS McGEE ... 94
WARNER BROS. (R) June, 1971.

DUTCHMAN ... 85
CONTINENTAL. (R) February, 1967. Shirley Knight, Al Freeman, Jr.

DYNAMITERS, THE (Brit.) ... 74
ASTOR. (R) May, 1956. Wayne Morris, James Kenney, Patrick Holt.

E

EAGLE HAS LANDED, THE (P) ... 123
COLUMBIA. (R) March, 1977. Michael Caine, Donald Sutherland.

EAGLE IN A CAGE ... 98
NATIONAL GENERAL. (R) January, 1972. Kenneth Haigh, Billie Whitelaw. Color.

EARTH VS. THE FLYING SAUCERS ... 83
COLUMBIA. (R) July, 1956. Joan Taylor, Hugh Marlowe.

EARTHQUAKE (P) ... 122
UNIVERSAL. (R) November, 1974. Charlton Heston, Ava Gardner.

EAST OF EDEN (Cs) ... 115
WARNER BROTHERS. (R) April 9, 1955. James Dean, Julie Harris, Raymond Massey. Color.

EASY COME, EASY GO ... 95
PARAMOUNT. (R) April, 1967. Elvis Presley, Dodie Marshall. Color.

EASY LIFE, THE (It., Eng. titles) ... 105
AVCO EMBASSY. (R) Vittorio Gassman, Catherine Spaak. b&w.

EASY RIDER ... 94
COLUMBIA. (R) July, 1969. Peter Fonda, Dennis Hopper, Jack Nicholson.

EBOLI (Ital., Eng. titles) ... 120

Length In Mins.

FRANKLIN MEDIA. (R) March, 1980. Gian Maria Volonte.

ECHOES OF A SUMMER 95
CINE ARTISTS. (R) May, 1976. Richard Harris, Jody Foster.

EDDY DUCHIN STORY, THE (Cs) 123
COLUMBIA. (R) July, 1956. Tyrone Power, Kim Novak. Color.

EDGE, THE 100
FILMMAKER'S DIST. CENTER. (R) May, 1969. Jack Rader, Tom Griffin.

EDGE OF ETERNITY (Cs) 80
COLUMBIA. (R) November, 1959. Cornel Wilde, Victoria Shaw. Color.

EDGE OF FURY 77
UNITED ARTISTS. (R) May, 1958. Michael Higgins, Louis Holmes.

EDGE OF HELL 76
UNIVERSAL. (R) September, 1956. Hugo Haas, Francesca De Scaffa.

EDGE OF THE CITY 65
METRO-GOLDWYN-MAYER. (R) January 4, 1957. John Cassavettes, Sidney Poitier.

EDUCATION OF SONNY CARSON, THE (P) 105
PARAMOUNT. (R) July, 1974. Rony Clanton, Don Gordon.

EFFECT OF GAMMA RAYS ON MAN-IN-THE-MOON MARIGOLDS, THE 101
TWENTIETH CENTURY-FOX. (R) January, 1973. Joanne Woodward, Nell Potts, Roberta Wallach. Color.

EIGER SANCTION, THE (P) 128
UNIVERSAL. (R) May, 1975. Clint Eastwood, George Kennedy.

8½ (It.: Eng. titles) 135
AVCO EMBASSY, (R) 1963. Marcello Mastrianni, Claudia Cardinale. b&w.

EIGHT O'CLOCK WALK (Brit.) 87
ASSOCIATED ARTISTS. (R) May, 1955. Richard Attenborough, Cathy O'Donnell.

EIGHT ON THE LAM 106
UNITED ARTISTS. (R) May, 1967. Bob Hope, Phyllis Diller. Color.

EIGHTEEN AND ANXIOUS 93
REPUBLIC. (R) November, 1957. M. Webster, William Campbell.

80 STEPS TO JONAH 105
WARNER BROS. (R) November, 1969. Wayne Newton, Jo Van Fleet, Keenan Wynn. Color.

EL CONDOR 102
NATIONAL GENERAL. (R) June, 1970. Jim Brown, Lee Van Cleef, Patrick O'Neal. Color.

EL DORADO 126
PARAMOUNT. (R) June, 1967. John Wayne, Charlene Holt. Color.

EL GRECO (Cs) 95
20th CENTURY-FOX. (R) October, 1966. Mel Ferrer, Rosanna Schiaffino. Color.

EL TOPO 123
ABCKO. (R) November, 1971. Alexandro Jodorowsky, Mara Lorenzio. Color.

ELEANOR ROOSEVELT STORY, THE 90
ALLIED ARTISTS. (R) November, 1965. Documentary.

ELECTRA GLIDE IN BLUE 113
UNITED ARTISTS. (R) September, 1973. Robert Blake, Billy Green Bush.

ELECTRIC HORSEMAN, THE 120
COLUMBIA. (R) December, 1979. Robert Redford, Jane Fonda.

ELECTRONIC MONSTER, THE 72
COLUMBIA. (R) May, 1960. Rod Cameron, Mary Murphy.

11 Harrowhouse (P) 98
TWENTIETH CENTURY-FOX. (R) August, 1974. Charles Grodin, Candice Bergen.

ELMER GANTRY 162
UNITED ARTISTS. (R) July, 1960. Burt Lancaster, Jean Simmons. Color.

ELVIRA MADIGAN (Swedish) 89
CINEMA V. (R) November, 1967. Pia Degermark, Thommy Berggren. Color.

ELVIS ON TOUR 93
METRO-GOLDWYN-MAYER. (R) November, 1972. Documentary starring Elvis Presley. Color.

ELVIS—THAT'S THE WAY IT IS 97
MGM. (R) November, 1970. Documentary. Elvis Presley. Color.

EMBRYO 108
CINE ARTISTS. (R) May, 1976. Rock Hudson, Diane Ladd.

EMERGENCY HOSPITAL 82
UNITED ARTISTS. (R) August, 1956. Margarot Lindsay, Bryon Palmer.

EMIGRANTS, THE 148
WARNER BROS. (R) October, 1972. Max von Sydow, Liv Ullmann. Color.

EMIL AND THE DETECTIVES 99
BUENA VISTA. (R) December, 1964. Walter Slezak, Heins Schubert. Color.

EMMANUELLE, THE JOYS OF A WOMAN (French: English subtitles) 100
PARAMOUNT. (R) February, 1976. Sylvia Kristel, Umberto Orsini.

EMPEROR OF THE NORTH 120
TWENTIETH CENTURY-FOX. (R) June, 1973. Lee Marvin, Ernest Borgnine. Color.

EMPIRE OF THE ANTS 89
AMERICAN INTERNATIONAL. (R) July, 1977. Joan Collins, Robert Lansing.

EMPIRE STRIKES BACK, THE (P) 118
TWENTIETH CENTURY-FOX. (R) May, 1980. Mark Hamill, Harrison Ford Carrie Fisher.

EMPTY CANVAS, THE 118
EMBASSY. (R) March, 1964. Bette Davis, Horst Buchholz.

ENCHANTED ISLAND (C) 87
WARNER BROS. (R) November, 1958. Dana Andrews, Jane Powell.

END, THE 100
UNITED ARTISTS. (R) May, 1978. Burt Reynolds, Dom DeLuise.

END OF THE AFFAIR, THE 106
COLUMBIA. (R) May, 1955. Deborah Kerr, Van Johnson.

END OF THE GAME 106
TWENTIETH CENTURY-FOX. (R) May, 1976. Jon Voight, Jacqueline Bisset.

END OF THE LINE 83
JEROME BALSAM FILMS. (R) August, 1959. Alan Baxter, Barbara Shelley.

END OF THE ROAD 100
ALLIED ARTISTS. (R) February, 1970. Stacy Keach, Harris Yulin, Dorothy Tristan, James Earl Jones. Color.

END OF THE WORLD IN OUR USUAL BED IN A NIGHT FULL OF RAIN, THE 104
WARNER BROS. (R) February, 1978. Giancarlo Giannini, Candice Bergen.

ENDLESS SUMMER, THE 85
BRUCE BROWN. (R) July, 1966. Documentary. Narrated by Brown. Color.

ENEMY BELOW, THE (Cs) 98
TWENTIETH CENTURY-FOX (R) December, 1957. Robert Mitchum, Curt Jurgens. Color.

ENEMY FROM SPACE 84
UNITED ARTISTS. (R) September, 1957. Brian Donlevy, Sydney James.

ENEMY GENERAL, THE 78
COLUMBIA. (R) September, 1960. Van Johnson, Jean Pierre Aumont.

ENFORCER, THE 97
WARNER BROS. (R) December, 1976. Clint Eastwood.

ENGAGEMENT ITALIANO (Ital.: eng. titles) 85
SEDGEWAY. (R) July, 1966. Rossano Brazzi, Annie Girardot.

ENGLAND MADE ME (P) 100
ATLANTIC PRODUCTIONS. (R) Fall, 1973. Peter Finch, Michael York.

ENOUGH ROPE (Fr., Eng. titles) 104
ARTIXO. (R) May, 1966. Gert Frobe, Marina Vlady.

ENSIGN PULVER 104
WARNER BROS. (R) June, 1964. Robert Walker, Burl Ives. Color.

ENTER LAUGHING 112
COLUMBIA. (R) July, 1967. Jose Ferrer, Shelley Winters. Color.

ENTER THE DRAGON (P) 98
WARNER BROS. (R) August, 1973. Bruce Lee, John Saxon.

ENTERTAINER, THE 97
CONTINENTAL. (R) October, 1960. Laurence Olivier, Brenda de Banzie.

ENTERTAINING MR. SLOANE 94
CONTINENTAL. (R) June, 1970. Beryl Reid, Harry Andrews, Peter McEnery. Color.

EQUUS 138
WARNER BROS. (R) October, 1977. Richard Burton, Peter Firth.

ERIC SOYA'S "17" (Danish, Eng. titles) 87
PEPPERCORN-WORMSTER/FILM ENTERPRISES. (R) January, 1967. Ole Soltoft, Haas Petersen. Color.

ERIC THE CONQUEROR (Colorscope) 81
AMERICAN INTERNATIONAL. (R) September, 1963. Cameron Mitchell, Alice Kessler. Color.

EROICA (Polish, Eng. titles) 82
AMERPOL ENTERPRISE FILMS. (R) February, 1966. Edward Dziewonski, Barbara Polomska.

ERRAND BOY, THE 92

	Length In Mins.
PARAMOUNT. (R) December, 1961. Jerry Lewis, Brian Donlevy.	
ESCAPADE IN JAPAN	**92**
UNIVERSAL. (R) November, 1957. Teresa Wright, Cameron Mitchell.	
ESCAPE BY NIGHT	**75**
ALLIED ARTISTS. (R) September, 1964. Terence Longdon, Jennifer Jayne.	
ESCAPE FROM ALCATRAZ	**112**
PARAMOUNT. (R) June, 1979. Clint Eastwood.	
ESCAPE FROM EAST BERLIN	**94**
METRO-GOLDWYN-MAYER. (R) November, 1962. Don Murray, Christine Kaufmann.	
ESCAPE FROM HELL	**80**
CROWN. (R) June, 1964. Mark Stevens.	
ESCAPE FROM REDROCK (RS)	**75**
TWENTIETH CENTURY-FOX. (R) January, 1958. Brian Donlevy, Eilene Janssen.	
ESCAPE FROM SAN QUENTIN	**81**
COLUMBIA. (R) November, 1957. Johnny Desmond, Merry Anders.	
ESCAPE FROM TERROR	**70**
COOGAN-ROGERS. (R) March, 1960. Jackie Coogan, Mona Knox. Color.	
ESCAPE FROM THE PLANET OF THE APES	**98**
TWENTIETH CENTURY-FOX. (R) June, 1971. Roddy McDowell, Kim Hunter. Color.	
ESCAPE FROM ZAHRAIN	**93**
PARAMOUNT. (R) June, 1962. Yul Brynner, Sal Mineo. Color.	
ESCAPE TO ATHENA (P)	**100**
ASSOCIATED FILM DIST. (R) June, 1979. Roger Moore, Telly Savalas.	
ESCAPE TO BURMA (SS)	**87**
RKO RADIO. (R) April 9, 1955. Barbara Stanwyck, Robert Ryan. Color.	
ESCAPE TO NOWHERE (French; Eng. titles)	**118**
PEPPERCORN-WORMSER. (R) August, 1974. Lina Ventura, Lea Massari.	
ESCAPE TO THE SUN	**105**
CINEVISION. (R) October, 1972. Laurence Harvey, Josephine Chaplin. Color.	
ESCAPE TO WITCH MOUNTAIN	**98**
BUENA VISTA. (R) March, 1975. Eddie Albert, Ray Milland.	
ESCORT WEST (Cs)	**75**
UNITED ARTISTS. (R) January, 1959. Victor Mature, Elaine Stewart.	
ESTHER AND THE KING (Cs)	**109**
TWENTIETH CENTURY-FOX. (R) December, 1960. Joan Collins, Richard Egan. Color.	
ETERNAL SEA, THE	**103**
REPUBLIC. (R) May 5, 1955. Sterling Hayden, Alexis Smith.	
EVA	**115**
TIMES FILM (R) June, 1965. Jeanne Moreau, Stanley Baker.	
EVE	**97**
COMMONWEALTH UNITED. (R) April, 1969. Celeste Yarnall, Robert Walker, Herbert Lom.	
EVE AND THE DRAGON (Cs)	**80**
AMERICAN INTERNATIONAL. (R) February, 1960. Color.	
EVEL KNIEVEL	**90**
FANFARE. (R) July, 1971. George Hamilton, Sue Lyon. Color.	
EVENING WITH THE ROYAL BALLET, AN	**90**
SIGMA III. (R) September, 1965. Rudolph Nureyev, Margot Fonteyn. Color.	
EVERY BASTARD A KING	**93**
CONTINENTAL. (R) April, 1970. Pier Angeli, William Berger. Color.	
EVERY DAY IS A HOLIDAY	**76**
COLUMBIA. (R) July, 1966. Marisol, Angel Peralta. Color.	
EVERY LITTLE CROOK AND NANNY	**92**
METRO-GOLDWYN-MAYER. (R) June, 1972. Victor Mature, Lynn Redgrave. Color.	
EVERYTHING BUT THE TRUTH	**83**
UNIVERSAL. (R) December, 1956. Maureen O'Hara, John Forsythe, Tim Hovey. Color.	
EVERYTHING YOU ALWAYS WANTED TO KNOW ABOUT SEX AND WERE AFRAID TO ASK	**87**
UNITED ARTISTS. (R) August, 1972. Woody Allen, Gene Wilder, Lynn Redgrave. Color.	
EVERYTHING'S DUCKY	**81**
COLUMBIA. (R) November, 1961. Mickey Rooney, Buddy Hackett.	
EVERY WHICH WAY BUT LOOSE	**114**
WARNER BROS. (R) December, 1978. Clint Eastwood, Sondra Locke.	
EVIL OF FRANKENSTEIN	**86**
UNIVERSAL. June, 1964. Peter Culshing, Kiwi Kingston. Color.	
EXECUTIONER, THE	**107**
COLUMBIA. (R) June, 1970. George Peppard, Joan Collins. Color.	
EXECUTIVE ACTION	**91**
NATIONAL GENERAL. (R) November, 1973. Burt Lancaster, Robert Ryan.	
EXODUS (P)	**212**
UNITED ARTISTS. (R) December, 1960. Paul Newman, Eva Marie Saint. Color.	
EXORCIST, THE	**121**
WARNER BROS. (R) December, 1973. Ellen Burstyn, Max von Sydow.	
EXORCIST II: THE HERETIC (P)	**110**
WARNER BROS. (R) June, 1977. Linda Blair, Richard Burton.	
EXPERIMENT IN TERROR	**123**
COLUMBIA. (R) April, 1973. Glenn Ford, Lee Remick.	
EXPLOSION	**96**
AMER. INT. PICS. (R) March, 1970, Don Stroud, Gordon Thomson. Color.	
EXPLOSIVE GENERATION, THE	**89**
UNITED ARTISTS. (R) October, 1961. William Shatner, Patty McCormack.	
EXPRESSO BONGO	**108**
CONTINENTAL. (R) April, 1960. Laurence Harvey, Sylvia Syms.	
EXTERMINATING ANGEL, THE (Span.)	**91**
ALTURA. (R) August, 1967. Silvia Pinal, Jose Baviera.	
EXTRAORDINARY SEAMAN, THE	**80**
METRO-GOLDWYN-MAYER. (R) January, 1969. David Niven, Faye Dunaway, Alan Alda, Mickey Rooney. Color.	
EXTREME CLOSE-UP	**90**
NATIONAL GENERAL. (R) May, 1973. James McMullan, Kate Woodville, Color.	
EYE FOR AN EYE, AN	**92**
EMBASSY. (R) June, 1966. Robert Lansing, Pat Wayne, Slim Pickens. Color.	
EYE OF THE CAT	**102**
UNIVERSAL. (R) June, 1969. Michael Sarrazin, Gayle Hunnicutt, Eleanor Parker. Color.	
EYE OF THE DEVIL	**92**
MGM. (R) September, 1967. David Niven, Deborah Kerr.	
EYE OF THE NEEDLE, THE (Italian with English titles)	**97**
ELDORADO. (R) July, 1965. Vittorio Gassman, Annette Stroyberg.	
EYES OF ANNIE JONES, THE	**85**
20th CENTURY-FOX. (R) 1964. Richard Conte, Francesca Annis.	
EYES OF LAURA MARS (P)	**104**
COLUMBIA. (R) August, 1978. Faye Dunaway, Tommy Lee Jones	

F

F.B.I. CODE 98	**94**
WARNER BROS. (R) June, 1964. Robert P. Cannon, Fred Vitale.	
F.B.I. STORY, THE	**149**
WARNER BROS. (R) October, 1959. James Stewart, Vera Miles. Color.	
F.I.S.T	**143**
UNITED ARTISTS. (R) April, 1978. Sylvester Stallone, Rod Steiger.	
FM	**110**
UNIVERSAL. (R) April, 1978. Michael Brandon, Eileen Brennan	
F.T.A.	**94**
AMERICAN INTERNATIONAL. (R) July, 1972. Jane Fonda, Donald Sutherland, Peter Boyle. Color.	
FABULOUS WORLD OF JULES VERNE	**95**
WARNER BROS. (R) June, 1961. Lou Tuck, Ernie Navara.	
FACE IN THE CROWD, A	**126**
WARNER BROS. (R) June, 1957. Andy Griffith, Patricia Neal.	
FACE OF A FUGITIVE	**81**
COLUMBIA. (R) May, 1959. Fred MacMurray. Color.	
FACE OF FIRE	**80**
ALLIED ARTISTS. (R) August, 1959. Cameron Mitchell, James Whitmore.	
FACE OF FU MANCHU, THE (TE)	**96**
SEVEN ARTS. (R) October, 1965. Christopher Lee, Nigel Green. Color.	
FACE OF WAR, A	**77**
FEATURE FILM. (R) March, 1968. Documentary.	
FACE TO FACE (Swedish; English sub-titles.)	**136**

Length In Mins.

PARAMOUNT. (R) April, 1976. Liv Ullmann, Erland Josephson.

FACES **127**
CONTINENTAL. (R) November, 1968. John Marley, Gena Rowlands, Fred Draper, Lynn Carter. b&w.

FACES IN THE DARK **84**
STATES RIGHTS. (R) September, 1964. John Gregson, Mai Zetterling.

FACTS OF LIFE, THE **103**
UNITED ARTISTS. (R) December, 1960. Bob Hope, Lucille Ball.

FAHRENHEIT 451 .... **111**
UNIVERSAL. (R) January, 1967. Oskar Werner, Julie Christie. Color.

FAIL SAFE **111**
COLUMBIA. (R) October, 1964. Dan O'Herlihy, Walter Matthau.

FAIR IS THE KEY **105**
PARAMOUNT. (R) March, 1973. Suzy Kendall, Barry Newman. Color.

FALL GIRL **83**
MEDALLION. (R) 1961. John Agar, Greta Chi. b&w.

FALL OF THE HOUSE OF USHER, THE (Cs) **85**
AMERICAN INTERNATIONAL. (R) July, 1960. Vincent Price, Mark Damon. Color.

FALL OF THE ROMAN EMPIRE, THE (P) **180**
PARAMOUNT. (R) Special in March, 1964. Sophia Loren, Stephen Boyd. Color.

FALSTAFF **115**
PEPPERCORN-WORMSER-SALTZMAN. (R) February, 1967. Orson Welles, Jeanne Moreau, Keith Baxter, John Gielgud.

FAME **127**
UNITED ARTISTS. (R) May, 1980. Eddie Barth, Irene Cara.

FAMILY HONOR **104**
CINERAMA. (R) February, 1973. Anthony Page. Color.

FAMILY JEWELS, THE **100**
PARAMOUNT. (R) June, 1965. Jerry Lewis, Donna Butterworth. Color.

FAMILY PLOT **120**
UNIVERSAL. (R) April, 1976. Karen Black, Bruce Dern.

FAMILY WAY, THE **115**
WARNER BROS-SEVEN ARTS. (R) June, 1967. Hayley Mills, John Mills. Color.

FANDO & LIS (Spanish; Eng. titles) **82**
CANNON. (R) February, 1970. Sergio Klainer, Diana Mariscal.

FANNY **133**
WARNER BROS. (R) July, 1961. Maurice Chevalier, Leslie Caron. Color.

FANNY HILL **91**
CINEMATION. (R) Diana Kjaer, Hans Ernbank.

FANTASTIC PLASTIC MACHINE, THE **93**
CROWN. (R) May, 1969. Nat Young, Bob McTavish, George Greenough. Color.

FANTASTIC VOYAGE (Cs) **105**
20th CENTURY-FOX. (R) October, 1966. Stephen Boyd, Raquel Welch. Color.

FANTOMAS (Cs) **105**
LOPERT PICTURES. (R) April, 1966. Marais, Lonis De Funes. Color.

FANTOME DE LA LIBERTE, LE (French; Eng. titles) **104**
20th CENTURY-FOX. (R) November, 1974. Adrianna Asti, Julien Bertheau.

FAR COUNTRY, THE **97**
UNIVERSAL. (R) February, 1955. James Stewart, Ruth Roman. Color.

FAR FROM THE MADDING CROWD (70mmP) **175**
MGM. (R) Special. September 1967. Julie Christie, Terence Stamp. Color.

FAR HORIZONS, THE (VV) **106**
PARAMOUNT. (R) June, 1955. Charlton Heston, Fred MacMurray. Color.

FAREWELL, MY LOVELY **97**
AVCO-EMBASSY. (R) August, 1975. Robert Mitchum, Charlotte Rampling.

FAREWELL TO ARMS, A (Cs) **150**
TWENTIETH CENTURY-FOX. (R) December, 1957. Rock Hudson, Jennifer Jones. Color.

FARMER, THE (P) **98**
COLUMBIA. (R) March, 1977. Gary Conway, Angel Tompkins.

FASCIST, THE (Ital. with English Titles) **102**
EMBASSY. (R) June, 1965. Ugo Tognazzi, Georges Wilson.

FAST AND SEXY (T) **96**
COLUMBIA. (R) September 1960. Gina Lollobrigida, Dale Robertson. Color.

FAST BREAK **107**
COLUMBIA. (R) March, 1979. Gabriel Kaplan, Harold Sylvester.

FASTEST GUITAR ALIVE, THE **85**
METRO-GOLDWYN-MAYER. (R) May, 1967. Roy Orbison, Sammy Jackson. Color.

FASTEST GUN ALIVE, THE **89**
METRO-GOLDWYN-MAYER. (R) July 6, 1956. Glenn Ford, Jeanne Crain

FAT CITY **96**
COLUMBIA. (R) August, 1972. Stacy Keach, Jeff Bridges, Susan Tyrell. Color.

FATE IS THE HUNTER (CS) **106**
TWENTIETH CENTURY-FOX. (R) October, 1964. Glenn Ford, Nancy Kwan.

FATHER GOOSE **115**
UNIVERSAL. (R) January, 1965. Cary Grant, Leslie Caron. Color.

FATHER OF A SOLDIER (Russian, Eng. titles) **83**
ARTKINO. (R) February 19, 1966. Georgi Makharashvilli, Sergo Zakhariadze.

FATHOM (CS) **99**
TWENTIETH CENTURY-FOX. (R) August, 1967. Raquel Welch, Tony Franciosa. Color.

FATSO **94**
TWENTIETH CENTURY-FOX. (R) February, 1980. Dom DeLuise, Anne Bancroft.

FEAR, THE (Greek) **102**
TRANS-LUX. (R) October, 1967. Alexis Damianos, Mary Chronopoulou.

FEAR STRIKES OUT (VV) **100**
PARAMOUNT. (R) March, 1957. Anthony Perkins, Karl Malden.

FEARLESS FRANK **79**
AMERICAN INTERNATIONAL. (R) December, 1969. Jon Voight, Monique Van Vooren. Color.

FEARLESS VAMPIRE KILLERS, OR PARDON ME, BUT YOUR TEETH ARE IN MY NECK (P) **94**
MGM. (R) November, 1967. Sharon Tate, Jack MacGowran. Color.

FEARMAKERS, THE **83**
UNITED ARTISTS. (R) October, 1958. Dana Andrews, Dick Foran.

FEDORA **116**
UNITED ARTISTS. (R) April, 1979. William Holden, Marthe Keller.

FELLINI SATYRICON (P) **127**
UNITED ARTISTS. (R) March, 1970. Martin Potter, Hiram Keller. Color.

FELLINI'S AMARCORD (Italian; Eng. titles) **127**
NEW WORLD. (R) October, 1974. Magali Noel, Bruno Zanin.

FELLINI'S CASANOVA **158**
UNIVERSAL. (R) February, 1977 Donald Sutherland, Tina Aumont.

FELLINI'S ROMA **120**
UNITED ARTISTS. (R) October, 1972. Federico Fellini, Stefano Majore. Color.

FEMALE ANIMAL, THE (CS) **84**
UNIVERSAL INTERNATIONAL. (R) April, 1958. Hedy Lamarr, Jane Powell.

FEMALE FIENDS **71**
CINEMA ASSOCIATES. (R) April, 1960. Lex Barker, Carole Mathews.

FEMALE JUNGLE **71**
AMERICAN INTERNATIONAL. (R) June 15, 1966. Jayne Mansfield, Lawrence Tierney.

FEMALE ON THE BEACH **97**
UNIVERSAL. (R) September, 1955, Joan Crawford, Jeff Chandler.

FERNANDELL THE DRESSMAKER (FR.) **84**
UNION. (R) August, 1957. Fernandel, Suzy Delair.

FERRY CROSS THE MERSEY **86**
UNITED ARTISTS. (R) February, 1965. Gerry and the Pacemakers.

FERRY TO HONG KONG (CS) **118**
TWENTIETH CENTURY-FOX. (R) April, 1961. Curt Jergens, Orson Welles. Color.

FESTIVAL **90**
PEPPERCORN-WORMSER. (R) November, 1967. Documentary.

FEVER HEAT **100**
PARAMOUNT. (R) May, 1968. Nick Adams. Color.

FEVER IN THE BLOOD, A **117**
WARNER BROS. (R) January, 1961. Efrem Zimbalist, Jr., Angie Dickinson.

FFOLKES **99**
UNIVERSAL. (R) April, 1980. Roger Moore, James Mason.

FIDDLER ON THE ROOF **181**
UNITED ARTISTS. (R) November, 1971. Topol, Norma Crane, Leonard Frey Molly Picon. Color.

FIEND WHO WALKED THE WEST, THE (CS) **101**
TWENTIETH CENTURY-FOX. (R) August, 1958. Hugh O'Brian, Robert Evans.

FIEND WITHOUT A FACE **94**

Length In Mins.

METRO-GOLDWYN-MAYER. (R) June, 1958. Marshall Thompson, Terrence Kilburn.

FIENDISH PLOT OF DR. FU MANCHU, THE **108**
WARNER BROS. (R) August, 1980. Peter Sellers, Helen Mirren.

FIERCEST HEART, THE (CS) **91**
TWENTIETH CENTURY-FOX. (R) April, 1961. Stuart Whitman, Juliet Prowse. Color.

FIFTH HORSEMAN IS FEAR, THE (Czech) **100**
GIGMA III. (R) May, 1968. Miraslav Machacek, Olga Schenplugova.

55 DAYS AT PEKING (T-70) **150**
ALLIED ARTISTS. (R) Special. May, 1963. Charlton Heston, Ava Gardner. Color.

FIGHTING CHANCE, THE **70**
REPUBLIC. (R) December 15, 1955. Rod Cameron, Julie London, Ben Cooper.

FIGHTING PRINCE OF DONEGAL, THE **89**
BUENA VISTA. (R) October, 1966. Peter McEnery, Susan Hampshire. Color.

FIGHTING TROUBLE **61**
ALLIED ARTISTS. (R) September 16, 1956. Huntz Hall, Stanley Clements.

FIGURES IN A LANDSCAPE (P) **109**
NATIONAL GENERAL. (R) July, 1971. Robert Shaw, Malcolm McDowell. Color.

FILE OF THE GOLDEN GOOSE **105**
UNITED ARTISTS. (R) November, 1969. Yul Brynner, Charles Gray. Color.

FILLMORE **105**
20th CENTURY-FOX. (R) June, 1972. Rock Concert Documentary. Color.

FINAL COUNTDOWN, THE **103**
UNITED ARTISTS. (R) July, 1980. Kirk Douglas, Martin Sheen.

FINDERS KEEPERS **89**
UNITED ARTISTS. (R) April, 1967. Cliff Richard. Color.

FINE MADNESS, A **112**
WARNER BROTHERS. (R) June 25, 1966. Sean Connery, Joanne Woodward, Patrick O'Neill. Color.

FINE PAIR, A **89**
NATIONAL GENERAL. (R) May, 1969. Rock Hudson, Claudia Cardinale, Thomas Milan. Color.

FINEST HOURS, THE **114**
COLUMBIA. (R) October, 1964. Documentary on Sir Winston Churchill. Color.

FINGER MAN **82**
ALLIED ARTISTS. (R) June 19, 1955. Frank Lovejoy, Peggie Castle.

FINGER OF GUILT (Brit.) **71**
RKO. (R) November, 1956. Richard Basehart, Mary Murphy.

FINGER ON THE TRIGGER (T) **87**
ALLIED ARTISTS. (R) November, 1965. Rory Calhoun, James Philbrook. Color.

FINIAN'S RAINBOW **145**
WARNER BROS.-7 ARTS. (R) October, 1968. Fred Astaire, Petula Clark, Don Francks. Color.

FIRE DOWN BELOW (CS) **116**
COLUMBIA. (R) July, 1957. Rita Hayworth, Robert Mitchum. Color.

FIREPOWER **104**
ASSOCIATED FILM DIST. (R) April, 1979. Sophia Loren, James Coburn.

FIRE WITHIN, THE **110**
NEW YORKER FILMS. (R) May, 1968. Maurice Ronet, Lena Skerla, Jeanne Moreau.

FIREBALL 500 (P) **92**
AIP. (R) June 29, 1966. Frankie Avalon, Chill Wills, Annett Funicello. Color.

FIRECREEK (P) **104**
WARNER BROS.-SEVEN ARTS. (R) February, 1968. James Stewart, Henry Fonda. Color.

FIREMAN'S BALL, THE **73**
CINEMA V. (R) October, 1968. Jan Vostroil, Josef Sebanek.

FIREMAN'S BALL, THE (Czech; Eng. titles) **73**
CINEMA V. (R) October, 1968. Vaclay Stockel, Josef Svet, John Vostrcil. Color.

FIRST CIRCLE, THE **98**
PARAMOUNT. (R) January, 1973. Gunther Malzacher, Elzbieta Czyzewska. Color.

FIRST LOVE **88**
PARAMOUNT. (R) November, 1977. William Katt, Susan Dey.

FIRST LOVE **90**
UMC PICTURES. (R) October, 1970. Maxmilian Schell, Dominique Sanda. Color.

FIRST MAN INTO SPACE **76**
METRO-GOLDWYN-MAYER. (R) February, 1959. Marshall Thompson, Maria Landi.

FIRST MEN IN THE MOON (P) **107**
COLUMBIA. (R) October, 1964. Edward Judd, Lionel Jeffries. Color.

FIRST NUDIE MUSICAL, THE **100**
PARAMOUNT. (R) March, 1976. Stephen Nathan, Cindy Williams.

FIRST TEXAN, THE (CS) **82**
ALLIED ARTISTS. (R) June 29, 1956. Joel McCrea, Felicia Farr. Color.

FIRST TIME, THE **90**
UNITED ARTISTS. (R) April, 1969. Jacqueline Bisset, Wes Stern, Rick Kelman, Wink Roberts. Color.

FIRST TO FIGHT (P) **97**
WARNER BROS. February, 1967. Chad Everett, Marilyn Dein. Color.

FIRST TRAVELING SALESLADY, THE **92**
UNIVERSAL. (R) August, 1956. Ginger Rogers, Barry Nelson. Color.

FISH THAT SAVED PITTSBURGH, THE **104**
UNITED ARTISTS. (R) November, 1979. Julius Erving, Jonathan Winters.

FISTFUL OF DOLLARS, A (T) **98**
UNITED ARTISTS. (R) January, 1967. Clint Eastwood, Marianne Koch. Color.

FISTS IN THE POCKET (Italian) **105**
PEPPERCORN-WORMSER. (R) December, 1967. Lou Castel, Paola Pitagora.

FISTS OF FURY **103**
NATIONAL GENERAL. (R) May, 1973. Bruce Lee, Tien Foong. Color.

FITZWILLY (P) **102**
UNITED ARTISTS. (R) December, 1967. Dick Van Dyke, Barbara Feldon. Color.

FIVE AGAINST THE HOUSE **84**
COLUMBIA. (R) June, 1955. Guy Madison, Kim Novak.

FIVE BOLD WOMEN **82**
CITATION FILMS. Jeff Morrow, Merry Anders.

FIVE BRANDED WOMEN **108**
PARAMOUNT. (R) May, 1960.

FIVE CARD STUD **103**
PARAMOUNT. (R) July, 1968. Dean Martin, Robert Mitchum, Inger Stevens. Color.

FIVE EASY PIECES **96**
COLUMBIA. (R) September, 1970. Jack Nicholson, Karen Black. Color.

FIVE FINGER EXERCISE **108**
COLUMBIA. (R) June, 1962. Rosalind Russell, Jack Hawkins.

FIVE FINGERS OF DEATH **104**
WARNER BROS. (R) March, 1973. Lo Lieh, Wang Ping. Color.

FIVE GATES TO HELL (CS) **107**
TWENTIETH CENTURY-FOX. (R) October, 1959. Neville Brand, Benson Fong.

FIVE GOLDEN HOURS **90**
COLUMBIA. (R) June, 1961. Cyd Charisse, Ernie Kovaks.

FIVE GUNS TO TOMBSTONE **71**
UNITED ARTISTS. (R) February, 1961. James Brown, John Wilder.

FIVE GUNS WEST **78**
AMERICAN RELEASING CORP. (R) April, 1955. John Lund, Dorothy Malone. Color.

FIVE MAN ARMY, THE **107**
MGM. (R) February, 1970. Peter Graves, James Daly. Color.

FIVE MILES TO MIDNIGHT. **110**
UNITED ARTISTS. (R) March, 1963. Sophia Loren, Anthony Perkins.

FIVE MILLION YEARS TO EARTH **98**
TWENTIETH CENTURY-FOX. (R) February, 1968. James Donald, Barbara Shelley. Color.

FIVE ON THE BLACK HAND SIDE **96**
UNITED ARTISTS. (R) October, 1973. Clarice Taylor, Leonard Jackson.

FIVE PENNIES, THE (VV) **157**
PARAMOUNT. (R) August, 1959. Barbara Bel Geddes, Danny Kaye. Color.

FIVE STEPS TO DANGER **80**
UNITED ARTISTS. (R) January, 1957. Sterling Hayden, Ruth Roman.

FIVE WEEKS IN A BALLOON (CS) **101**
TWENTIETH CENTURY-FOX. (R) August, 1962. Red Buttons, Fabian. Color.

FIXER, THE **132**
METRO-GOLDWYN-MAYER. (R) December, 1968. Alan Bates, Dirk Bogarde, Georgia Brown. Color.

FLAME AND THE FIRE **80**
CONTINENTAL. (R) March, 1966. Documentary. Color.

FLAME OF THE ISLANDS **90**
REPUBLIC. (R) January 6, 1958. Yvonne De Carlo, Zachary Scott, Howard Duff. Color.

FLAME OVER INDIA (CS) **130**

Length In Mins.

TWENTIETH CENTURY-FOX. (R) May, 1960. Kenneth More, Lauren Bacall. Color.

FLAMING FRONTIER 70
WARNER BROS.-7 ARTS. (R) Feb., 1968. Stewart Granger. Color.

FLAMING STAR (CS) 92
TWENTIETH CENTURY-FOX. (R) December, 1960. Elvis Presley, Barbara Eden. Color.

FLAP (P) 106
WARNER BROS. (R) November, 1970. Anthony Quinn, Shelley Winters. Color.

FLAREUP 100
MGM. (R) November, 1969. Raquel Welch, James Stacy, Luke Askew. Color.

FLEA IN HER EAR, A 94
TWENTIETH CENTURY-FOX. (R) October, 1968. Rex Harrison, Rosemary Harris, Louis Jordan. Color.

FLESH 95
FACTORY FILM. (R) June, 1968. Joe Dallesandro, Geraldine Smith.

FLESH AND THE SPUR 60
AMERICAN INTERNATIONAL. (R) September 25, 1956. John Agar, Maria English. Color.

FLIGHT FROM ASHIYA (P) 100
UNITED ARTISTS. (R) April, 1964. Yul Brynner, Richard Widmark. Color.

FLIGHT OF THE DOVES 101
COLUMBIA. (R) April, 1971. Ron Moody, Jack Wild. Color.

FLIGHT OF THE PHOENIX, THE 148
TWENTIETH CENTURY-FOX. (R) February, 1966. James Steward, Richard Attenborough. Color.

FLIGHT TO HONG KONG 88
UNITED ARTISTS. (R) October, 1956. Rory Calhoun, Dolores Donlon.

FLIM-FLAM MAN, THE (P) 104
TWENTIETH CENTURY-FOX. (R) September, 1967. George C. Scott, Sue Lyon. Color.

FLIPPER 90
METRO-GOLDWYN-MAYER. (R) May, 1963. Chuck Connors, Luke Halpin. Color

FLIPPER'S NEW ADVENTURE 103
METRO-GOLDWYN-MAYER. (R) June, 1964. Luke Halpin, Pamela Franklin. Color.

FLOAT LIKE A BUTTERFLY, STING LIKE A BEE 94
GROVE. (R) December, 1969. Documentary on Muhammad Ali.

FLOOD TIDE (CS) 81
UNIVERSAL. (R) February, 1955. George Nader, Cornell Borchers.

FLOODS OF FEAR 82
UNIVERSAL. (R) May, 1959. Howard Keel, Anne Heywood.

FLOWER DRUM SONG (P) 133
UNIVERSAL. (R) January, 1962. Nancy Kwan, James Shigeta. Color.

FLYING FONTAINES, THE 84
COLUMBIA. (R) December, 1959. Michael Callan, Evy Noslund. Color.

FOG, THE 91
AVCO EMBASSY. (R) February, 1980. Adrienne Barbeau, Hal Holbrook.

FOLLOW ME 86
CINERAMA. (R) June, 1969. Bob Purrey, Claude Codgen, Mary Lou McGinnis. Color.

FOLLOW ME BOYS 131
BUENA VISTA. (R) Christmas, 1966. Fred MacMurray, Vera Miles. Color.

FOLLOW THAT DREAM (P) 110
UNITED ARTISTS. (R) April, 1962. Elvis Presley, Arthur O'Connel. Color.

FOLLOW THE BOYS (P) 95
METRO-GOLDWYN-MAYER. (R) March, 1963. Connie Francis, Paula Prentiss. Color.

FOOD OF THE GODS, THE 88
AMERICAN INTERNATIONAL. (R) June, 1976. Marjoe Gortner, Pamela Franklin.

FOOLIN' AROUND 101
COLUMBIA. (R) April, 1980. Gary Busey, Annette O'Toole.

FOOLS 93
CINERAMA. (R) December, 1970. Katharine Ross, Jason Robards. Color.

FOOLS' PARADE 98
COLUMBIA. (R) July, 1971. James Stewart, George Kennedy. Color.

FOOTSTEPS IN THE FOG 90
COLUMBIA. (R) September, 1955. Jean Simmons, Stewart Granger. Color.

FOOTSTEPS IN THE NIGHT 62
ALLIED ARTISTS. (R) March 24, 1957. Bill Elliott, Don Haggerty.

FOR A FEW DOLLARS MORE (T) 130
UNITED ARTISTS. (R) May, 1967. Clint Eastwood, Lee Van Cleef. Color.

FOR BETTER FOR WORSE (Brit.) 83
STRATFORD. (R) June 22, 1955. Dirk Bogarde, Cecil Parker. Color.

FOR LOVE OF IVY 112
CINERAMA. (R) July, 1968. Sidney Poitier, Abby Lincoln, Beau Bridges. Color.

FOR LOVE OR MONEY 108
UNIVERSAL. (R) October, 1963. Kirk Douglas, Mitzi Gaynor. Color.

FOR PETE'S SAKE 90
COLUMBIA. (R) June, 1974. Barbra Streisand, Michael Sarrazin.

FOR SINGLES ONLY 91
COLUMBIA. (R) May, 1968. John Saxon, Mary Ann Mobley. Color.

FOR THE FIRST TIME (T) 97
METRO-GOLDWYN-MAYER. (R) September, 1959. Mario Lanza, Zsa Zsa Gabor. Color.

FOR THE LOVE OF MIKE (CS) 84
TWENTIETH CENTURY-FOX. (R) August, 1960. Richard Basehart, Stu Erwin. Color.

FOR THEM THAT TRESPASS (Brit.) 95
STRATFORD. (R) July 6, 1959. Richard Todd, Patricia Plunkett.

FOR THOSE WHO THINK YOUNG (CS) 96
UNITED ARTISTS. (R) June, 1964. James Darren, Pamela Tiffin. Color.

FORBIDDEN CARGO (Brit.) 83
JACON. (R) June, 1956. Nigel Patrick, Elizabeth Sellars.

FORBIDDEN ISLAND 66
COLUMBIA. (R) March, 1959. Jon Hall, Nan Adams. Color.

FORBIDDEN PLANET (SS) 98
METRO-GOLDWYN-MAYER. (R) March 30, 1956. Walter Pidgeon, Anne Francis. Color.

FORCE 10 FROM NAVARONE 118
AMERICAN INTERNATIONAL. (R) December, 1978. Robert Shaw, Harrison Ford.

FOREIGN INTRIGUE (CS) 100
UNITED ARTISTS. (R) May, 1956. Robert Mitchum, Genevieve Page. Color.

FOREVER DARLING 91
METRO-GOLDWYN-MAYER. (R) February 10, 1956. Lucille Ball, Desi Arnaz. Color.

FOREVER MY LOVE 147
PARAMOUNT. (R) April, 1962. Romy Schneider, Karl Boehm.

FORT BOWIE 80
UNITED ARTISTS. (R) February, 1958. Ben Johnson, Jan Harrison.

FORT COURAGEOUS 72
TWENTIETH CENTURY-FOX. (R) May, 1965. Fred Bier, Harry Lauter.

FORT DOBBS 90
WARNER BROS. (R) January, 1958. Clint Walker, Virginia Mayo.

FORT MASSACRE (CS) 80
UNITED ARTISTS. (R) May, 1958. Joel McCrea, Forrest Tucker. Color.

FORT UTAH (TE) 83
PARAMOUNT. (R) September, 1967. John Ireland, Virginia Mayo. Color.

FORT YUMA 79
UNITED ARTISTS. (R) October, 1955. Peter Graves, Joan Vohs. Color.

FORTUNE, THE (P) 86
COLUMBIA. (R) June, 1975. Stockard Channing, Jack Nicholson, Warren Beatty.

FORTUNE AND MEN'S EYES 102
MGM. (R) June, 1971. Wendell Burton, Michael Greer. Color.

FORTUNE COOKIE, THE (P) 125
UNITED ARTISTS. (R) October, 1966. Jack Lemmon, Walter Matthau. b&w.

40 CARATS 110
COLUMBIA. (R) July, 1973. Liv Ullmann, Edward Albert, Gene Kelly. Color.

FORTY GUNS (CS) 79
TWENTIETH CENTURY-FOX. (R) September, 1957. Barbara Stanwyck, Barry Sullivan.

40 GUNS TO APACHE PASS 95
COLUMBIA. (R) 1966. Audie Murphy, Michael Burns. Color.

40 POUNDS OF TROUBLE (P) 106
UNIVERSAL. (R) February, 1963. Tony Curtis, Phil Silvers. Color.

FOUL PLAY 116
PARAMOUNT. (R) July, 1978. Goldie Hawn, Chevy Chase.

FOUR AGAINST FATE (Brit.) 84

	Length In Mins.
ASSOCIATED ARTISTS. (R) June, 1955. Anna Neagle, Michael Wilding.	
FOUR BAGS FULL (Fr.)	84
TRANS-LUX. (R) September, 1957. Jean Gabin, Bourvil.	
FOUR BOYS AND A GUN	78
UNITED ARTISTS. (R) January, 1957. Frank Sutton, Tarry Treen.	
FOUR CLOWNS	97
20th CENTURY-FOX. (R) September, 1970. b&w.	
"4-D MAN"	85
UNIVERSAL. (R) October, 1959. Robert Lansing, Lee Meriwether. Color.	
FOUR DAYS IN NOVEMBER	120
UNITED ARTISTS. (R) November, 1964. Documentary.	
FOUR DAYS OF NAPLES	124
METRO-GOLDWYN-MAYER. (R) March, 1963. Regina Blanchi, Aldo Giuffre.	
FOUR DESPERATE MEN	96
CONTINENTAL. (R) 1960. Aldo Ray, Heather Sears. b&w.	
FOUR FAST GUNS	72
UNIVERSAL. (R) February, 1960. James Craig, Martha Vickers.	
FOUR FLIES ON GREY VELVET (TE)	110
PARAMOUNT. (R) August, 1972. Michael Brandon, Mimsy Farmer. Color.	
4 FOR TEXAS	124
WARNER BROS. (R) December, 1963. Frank Sinatra, Dean Martin. Color.	
FOUR GIRLS IN TOWN (CS)	85
UNIVERSAL. (R) January, 1957. George Nader, Julie Adams. Color.	
FOUR HORSEMEN OF THE APOCALYPSE (CS)	158
METRO-GOLDWYN-MAYER. (R) February, 1962. Glenn Ford, Ingrid Thulin. Color.	
FOUR MUSKETEERS, THE (P)	103
20th CENTURY-FOX. (R) March, 1975. Oliver Reed, Raquel Welch, Richard Chamberlain.	
FOUR SKULLS OF JONATHAN DRAKE, THE	70
UNITED ARTISTS. (R) June, 1959. Eduard Franz, Grant Richards.	
FOUR WAYS OUT (Ital.-Eng. dubbed)	77
CARROLL. (R) September, 1956. Gina Lollobrigida.	
FOX, THE	110
CLARIDGE. (R) January, 1968. Anne Heywood, Sandy Dennis, Keir Dulles. Color.	
FOXES	106
UNITED ARTISTS. (R) March, 1980. Jodie Foster, Scott Baio.	
FOXFIRE	91
UNIVERSAL. (R) July, 1955. Jeff Chandler, Jane Russell. Color.	
FOXHOLE IN CAIRO	79
PARAMOUNT. (R) January, 1961. James Robertson Justice, Albert Lieven.	
FOXIEST GIRL IN PARIS, THE (Fr.)	100
TIMES. (R) 1957. Martine Carol, Miscla Auer. b&w.	
FRAGMENT OF FEAR	96
COLUMBIA. (R) September, 1971. David Hemmings, Gayle Hunnicutt. Color.	
FRAMED	106
PARAMOUNT. (R) August, 1975. Joe Don Baker, Conny Van Dyke.	
FRANCIS IN THE HAUNTED HOUSE	80
UNIVERSAL. (R) August, 1956. Mickey Rooney, Virginia Welles.	
FRANCIS IN THE NAVY	80
UNIVERSAL. (R) August, 1955. Donald O'Connor, Martha Hyer.	
FRANCIS OF ASSISI (CS)	111
TWENTIETH CENTURY-FOX. (R) August, 1961. Bradford Dillman, Dolores Hart. Color.	
FRANKENSTEIN-1970 (CS)	83
ALLIED ARTISTS. (R) July 6, 1958. Boris Karloff, Jana Lund.	
FRANKENSTEIN AND THE MONSTER FROM HELL	93
PARAMOUNT. (R) June, 1974. Peter Cushing, Shane Briant.	
FRANKENSTEIN CONQUERS THE WORLD (Tohoscope) Japan	87
AMER. INT'L. PICS. (R) Nick Adams.	
FRANKENSTEIN CREATED WOMAN	92
20th CENTURY-FOX. (R) March, 1967. Peter Cushing, Susan Denberg. Color.	
FRANKENSTEIN MEETS THE SPACE MONSTER	78
ALLIED ARTISTS. (R) 1965. James Karen, Nancy Marshall. b&w.	
FRANKENSTEIN MUST BE DESTROYED	101
WARNER BROS. (R) February, 1970. Peter Cushing, Simon Ward. Color.	
FRANKIE AND JOHNNY	87
UNITED ARTISTS. (R) April, 1966. Elvis Presley, Donna Douglas. Color.	
FRANTIC (French)	90
TIMES. (R) June, 1961. Jeanne Moreau, Maurice Ronet.	
FRATERNITY ROW	101
PARAMOUNT. (R) April, 1977. Peter Fox, Gregory Harrison.	
FRAULEIN (CS)	90
TWENTIETH CENTURY-FOX. (R) May, 1958. Mel Ferrer, Dana Wynter. Color.	
FRAULEIN DOKTOR	102
PARAMOUNT. (R) May, 1969. Suzy Kendall, Kenneth More, Capucine, James Booth. Color.	
FREAKY FRIDAY	95
BUENA VISTA. (R) February, 1977. Barbara Harris, Jodie Foster.	
FRECKLES (CS)	84
TWENTIETH CENTURY-FOX. (R) September, 1960. Martin West, Carol Christensen. Color.	
FREEBIE AND THE BEAN (P)	113
WARNER BROS. (R) December, 1974. Alan Arkin, James Caan.	
FRENCH CANCAN (Fr.)	98
UMPO. (R) April, 1956. Jean Gabia, Francis Armoul. Color.	
FRENCH CONNECTION, THE	104
20th CENTURY-FOX. (R) October, 1971. Gene Hackman, Fernando Rey, Roy Scheider. Color.	
FRENCH CONNECTION II	118
20th CENTURY-FOX. (R) June, 1975. Gene Hackman, Fernando Rey.	
FRENCH CONSPIRACY, THE	125
CINE GLOBE. (R) November, 1973. Jean Louis Trintignant, Michel Piccoli.	
FRENCH POSTCARDS	95
PARAMOUNT. (R) October, 1979. Miles Chapin, Blanche Baker.	
FRENZY	116
UNIVERSAL. (R) June, 1972. Jon Finch, Barry Foster, Alec McCowen. Color.	
FREUD	139
UNIVERSAL. (R) December, 1962. Montgomery Clift, Susannah York.	
FRIDAY FOSTER	89
AMERICAN INTERNATIONAL. (R) December, 1975. Pam Grier, Yaphet Kotto.	
FRIEND OF THE FAMILY (Fr., English titles)	95
INTERNATIONAL CLASSICS. (R) November, 1965. Jean Marais, Danielle Darrieux.	
FRIENDLY PERSUASION	137
ALLIED ARTISTS. (R) November 25, 1956. Gary Cooper, Dorothy McGuire. Color.	
FRIENDS	102
PARAMOUNT. (R) April, 1971. Sean Bury, Anicee Alvina. Color.	
FRIENDS OF EDDIE COYLE, THE (P)	103
PARAMOUNT. (R) July, 1973. Robert Mitchum, Peter Boyle. Color.	
FRIGHTENED CITY, THE	97
ALLIED ARTISTS. (R) October, 1962. Herbert Lom, John Gregson.	
FRISCO KID, THE	122
WARNER BROS. (R) July, 1979. Gene Wilder, Harrison Ford.	
FRISKY (Ital.)	96
DISTRIBUTORS CORP. OF AMERICA. (R) November, 1955. Gina Lollobrigida, Vittorio De Sica.	
FRITZ THE CAT	78
CINEMATION. (R) April, 1972. Animated Adult Feature. Color.	
FROGS	91
AMERICAN INTERNATIONAL. (R) March, 1972. Ray Milland, Adam Roarke. Color.	
FROM EAR TO EAR (WS)	81
CINEMATION. (R) January, 1971. Nicole De Bonne, Daniele Argence. Color.	
FROM HELL IT CAME	71
ALLIED ARTISTS. (R) August, 1957. Tod Andrews, Tina Carver.	
FROM HELL TO TEXAS (CS)	100
TWENTIETH CENTURY-FOX. (R) June, 1958. Don Murray, Diane Varsi. Color.	
FROM HERE TO ETERNITY	118
COLUMBIA. (R) September, 1953. Burt Lancaster, Montgomery Clift, Deborah Kerr, Frank Sinatra, Donna Reed.	
FROM NOON TILL THREE	100
UNITED ARTISTS. (R) August, 1976. Charles Bronson, Jill Ireland.	
FROM RUSSIA WITH LOVE	118
UNITED ARTISTS. (R) April, 1964. Sean Connery, Daniela Bianchi. Color.	
FROM THE MIXED-UP FILES OF MRS. BASIL E. FRANKWEILER	105

Length In Mins.

CINEMA V. (R) October, 1973. Ingrid Bergman, Sally Prager.

FROM THE TERRACE (CS) 144
TWENTIETH CENTURY-FOX. (R) July, 1960. Paul Newman, Joanne Woodward. Color.

FRONT, THE (P) 94
COLUMBIA. (R) October, 1976. Woody Allen, Zero Mostel.

FRONT PAGE, THE (P) 105
UNIVERSAL. (R) December, 1974. Jack Lemmon, Walther Matthau, Carol Burnett.

FRONT PAGE STORY (Brit.) 95
ASSOCIATED ARTISTS. (R) April, 1955. Jack Hawkins, Elizabeth Allan.

FRONTIER GAMBLER 76
ASSOCIATED FILM. (R) September, 1956. John Bromfield, Coleen Gray.

FRONTIER GUN (RS) 70
TWENTIETH CENTURY-FOX. (R) December, 1958. John Agar, Joyce Meadows.

FRONTIER HELLCAT (CS) (dubbed) 98
COLUMBIA. (R) 1966. Stewart Granger, Elke Sommer. Color.

FRONTIER UPRISING 68
UNITED ARTISTS. (R) March, 1961. James Davis, Nancy Hadley.

FRONTIER WOMAN 80
TOP. (R) July, 1956. Cindy Carson, Lance Fuller. Color.

FROZEN DEAD, THE 95
WARNER BROS.-SEVEN ARTS. (R) September, 1967. Dana Andrews, Anna Palik.

FUEGO 81
HAVEN INTERNATIONAL. (R) November, 1969. Isabel Sarli, Armanda Bo. Color.

FUGITIVE KIND, THE 119
UNITED ARTISTS. (R) May, 1960. Marlon Brando, Anna Magnani.

FULL IN ACAPULCO 97
PARAMOUNT. (R) November, 1963. Elvis Presley, Ursula Andress. Color.

FULL OF LIFE 91
COLUMBIA. (R) March, 1957. Judy Holliday, Richard Conte.

FUN WITH DICK AND JANE 102
COLUMBIA. (R) February, 1977. George Segal, Jane Fonda.

FUNERAL IN BERLIN (P) 102
PARAMOUNT. (R) February, 1967. Michael Caine, Eva Renzi. Color.

FUNNIEST MAN IN THE WORLD, THE 95
GROVE. (R) December, 1969. Narrated by Douglas Fairbanks Jr. b&w.

FUNNY FACE (VV) 105
PARAMOUNT. (R) April, 1957. Audrey Hepburn, Fred Astaire. Color.

FUNNY GIRL 155
COLUMBIA ROADSHOW. (R) special. September, 1968. Barbra Streisand, Omar Sharif, Kay Medford, Anne Francis. Color.

FUNNY LADY (P) 136
COLUMBIA. (R) March, 1975. Barbra Streisand, James Caan.

FUNNY THING HAPPENED ON THE WAY TO THE FORUM, A 99
UNITED ARTISTS. (R) October, 1966. Zero Mostel, Phil Silvers. Color.

FURTHER PERILS OF LAUREL AND HARDY, THE 99
TWENTIETH CENTURY-FOX. (R) October, 1967. Stan Laurel, Oliver Hardy.

FURY, THE 118
20th CENTURY-FOX. (R) March, 1978. Kirk Douglas, John Cassavetes.

FURY AT GUNSIGHT PASS 68
COLUMBIA. (R) February, 1956. David Brian, Richard Long.

FURY AT SHOWDOWN 75
UNITED ARTISTS. (R) April, 1957. John Derek, John Smith.

FURY AT SMUGGLER'S BAY (Panascope) 92
EMBASSY. (R) December, 1963. Peter Cushing, John Fraser.

FURY IN PARADISE 77
FILMAKERS. (R) October, 1955. Peter Thompson, Rea Iturbi. Color.

FURY RIVER 74
MGM. (R) 1961. Keith Larsen, Buddy Ebsen.

FUTUREWORLD 107
AMERICAN INTERNATIONAL. (R) July, 1976. Peter Fonda, Blythe Danner.

FUTZ 92
COMMONWEALTH UNITED. (R) November, 1969. Seth Allen, John Bakos, La Mama Troupe. Color.

FUZZ 92
UNITED ARTISTS. (R) May, 1972. Yul Brynner, Burt Reynolds, Raquel Welch. Color.

FUZZY PINK NIGHTGOWN, THE 87
UNITED ARTISTS. (R) August, 1957. Jane Russell, Keenan Wynn.

G

G.I. BLUES 104
PARAMOUNT. (R) November, 1960 Elvis Presley, Juliet Prowse. Color.

GABLE AND LOMBARD 129
UNIVERSAL. (R) February, 1976. James Brolin, Jill Clayburgh.

GABY (Cs) 97
METRO-GOLDWYN-MAYER. (R) May 11, 1956. Leslie Caron, John Kerr. Color.

GAILY, GAILY (P) 110
UNITED ARTISTS. (R) December, 1969. Beau Bridges, Melina Mercouri, Brian Keith. Color.

GALIA (Fr.; Eng. titles) 105
ZENITH INTERNATIONAL. (R) December, 1966. Mireille Darc, Venantino Venantini.

GALILEO 145
AFT. (R) 1975. Topol, Edward Fox.

GALLANT HOUR, THE 115
UNITED ARTISTS. (R) May, 1960. James Cagney, Dennis Weaver.

GAMBIT (T) 109
UNIVERSAL. (R) December, 1966 Shirley MacLaine, Michael Caine, Color.

GAMBLER, THE 111
PARAMOUNT. (R) October, 1974 James Caan, Paul Sorvino.

GAMBLER WORE A GUN, THE 66
UNITED ARTISTS. (R) May, 1961. James Davis, Merry Anders.

GAMBLERS, THE 93
U-M. (R) January, 1970. Don Gordon, Suzy Kendall. Color.

GAME IS OVER, THE (P) (Fr.; Eng. Titles) 96
ROYAL INTERNATIONAL. (R) January, 1967. Jane Fonda, Peter McEnery Color.

GAMES 98
UNIVERSAL. (R) October, 1967. Simone Signoret, James Caan, Katherine Ross. Color.

GAMES, THE (P) 95
20th CENTURY-FOX. (R) May, 1970. Michael Crawford, Stanley Baker, Ryan O'Neal. Color

GAMMA PEOPLE, THE 79
COLUMBIA. (R) December, 1956. Paul Douglas, Eva Bartok.

GANG BUSTERS 78
VISUAL DRAMA. (R) March, 1955. Myron Healey, Don C. Harvey.

GANG THAT COULDN'T SHOOT STRAIGHT, THE 96
METRO-GOLDWYN-MAYER. (R) December, 1971. Jerry Orbach, Leigh Taylor-Young. Color.

GANG WAR (Rs) 74
TWENTIETH CENTURY-FOX. (R) July, 1958. Charles Bronson, Kent Taylor.

GARDEN OF THE FINZI-CONTINIS 95
CINEMA 5. (R) December, 1971. Lino Capolicchio, Dominique Sanda, Helmut Berger. Color.

GARMENT JUNGLE, THE 88
COLUMBIA. (R) June, 1957. Lee J. Cobb, Richard Boone.

GARNET BRACELET, THE (Russian; Eng. Titles) 90
ARTKINO. (R) June, 1966. Ariadna Shengelaya, Igor Oserov. Color.

GATHERING OF EAGLES, A 115
UNIVERSAL. (R) July, 1963. Rock Hudson, Rod Taylor, Color.

GATOR 116
UNITED ARTISTS. (R) July, 1976. Burt Reynolds, Jack Weston.

GAUNTLET, THE (P) 110
WARNER BROS. (R) December, 1977. Clint Eastwood, Sondra Locke

GAY DECEIVERS, THE 90
FANFARE. (R) June, 1969. Kevin Coughlin, Brook Bundy, JoAnn Harris. Color.

GAY PURR-EE 86
WARNER BROS. (R) November, 1962. Judy Garland, Robert Goulet. Color

GAZEBO, THE (Cs) 102
METRO-GOLDWYN-MAYER. (R) January, 1960. Glenn Ford, Debbie Reynolds.

GEISHA BOY, THE (VV) 98
PARAMOUNT. (R) December, 1958. Jerry Lewis, Marie MacDonald. Color.

Length In Mins.

GENE KRUPA STORY, THE **101**
COLUMBIA. (R) December, 1959. Sal Mineo, Suzan Kohner.
GENERALE DELLA ROUVERE **129**
CONTINENTAL. (R) 1960. Vittorio DeSica, Hannes Messmer. b&w.
GENERATION **104**
AVCO EMBASSY. (R) December, 1969. David Janssen, Kim Darby, Carl Reiner. Color.
GENTLE GIANT **93**
PARAMOUNT. (R) October, 1967. Dennis Weaver, Vera Miles. Color.
GENTLE RAIN, THE **110**
COMET. (R) May, 1966. Christopher George, Lynda Day, Fay Spain.
GENTLE TOUCH, THE **86**
RANK. (R) August, 1957. George Baker, Belinda Lee.
GENTLEMEN MARRY BRUNETTES (Cs) **97**
UNITED ARTISTS. (R) October, 1955. Jane Russell, Jeanne Crain. Color.
GEORGE RAFT STORY **106**
ALLIED ARTISTS. (R) December, 1961. Ray Danton, Jayne Mansfield.
GEORGIA, GEORGIA **91**
CINERAMA. (R) March, 1972. Diana Sands, Dirk Benedict. Color.
GEORGY GIRL **100**
COLUMBIA. (R) October, 1966. James Mason, Lynn Redgrave.
GERONIMO! **101**
UNITED ARTISTS. (R) April, 1962. Chuck Connors, Kamala Devi. Color.
GERVAISE (Fr.) **89**
CONTINENTAL. (R) November, 1957. Maria Schell, Francoise Perier.
GET CARTER **110**
MGM. (R) February, 1971. Michael Caine, John Osborne. Color.
GET ON WITH IT (Brit.) **88**
GOVERNOR. (R) August, 1965. Bob Monkhouse, Kenneth Connor.
GET TO KNOW YOUR RABBIT **93**
WARNER BROS. (R) June, 1972. Tom Smothers, Orson Welles. Color.
GET YOURSELF A COLLEGE GIRL **86**
METRO-GOLDWYN-MAYER. (R) November, 1964. Mary Ann Mobley, Joan O'Brien. Color.
GETAWAY, THE (TAO) **122**
NATIONAL GENERAL. (R) December, 1972. Steve McQueen, Ali MacGraw. Color.
GETTING STRAIGHT **124**
COLUMBIA. (R) June, 1970. Elliott Gould, Candice Bergen. Color.
GHOST AND MR. CHICKEN, THE (T) **90**
UNIVERSAL. (R) May, 1966. Don Knotts, Joan Staley. Color.
GHOST DIVER **76**
TWENTIETH CENTURY-FOX. (R) October, 1957. James Craig, Audrey Totter.
GHOST IN THE INVISIBLE BIKINI, THE (P) **82**
AMERICAN INTERNATIONAL. (R) April, 1966. Tommy Kirk, Deborah Walley. Color.
GHOST OF DRAGSTRIP HOLLOW **65**
AMERICAN INTERNATIONAL. (R) July, 1959. Jody Fair, Martin Braddock.
GHOST OF THE CHINA SEA **79**
COLUMBIA. (R) September, 1958. David Brian, Lynn Bernay.
GHOST TOWN **75**
UNITED ARTISTS. (R) March, 1956. Kent Taylor, Marian Carr.
GHOSTS, ITALIAN STYLE **92**
METRO-GOLDWYN-MAYER. (R) January, 1969. Sophia Loren, Vittorio Gassman, Marlo Adorf. Color.
GIANT **201**
WARNER BROTHERS. (R) November 24, 1956. Rock Hudson, Elizabeth Taylor, James Dean. Color.
GIANT BEHEMOTH, THE **83**
ALLIED ARTISTS. (R) March, 1959. Gene Evans, Andre Morell.
GIANT CLAW, THE **76**
COLUMBIA. (R) June, 1957. Jeff Morrow, Mara Corday.
GIANT FROM THE UNKNOWN **77**
ASTOR. (R) 1958. Buddy Baer, Bob Steel. b&w.
GIANT GILA MONSTER, THE **73**
McLENDON. (R) November, 1959. Don Sullivan. Lisa Simone.
GIANT OF MARATHON, THE (D) **92**
METRO-GOLDWYN-MAYER. (R) May, 1960. Steve Reeves, Mylene Demongeot. Color.
GIDEON OF SCOTLAND YARD **91**
COLUMBIA. (R) February, 1959. Jack Hawkins, Anna Massey. Color.

Length In Mins.

GIDGET (Cs) **95**
COLUMBIA. (R) April, 1959. James Darren, Sandra Dee. Color.
GIDGET GOES HAWAIIAN **102**
COLUMBIA. (R) June, 1961. James Darren, Michael Callan. Color.
GIDGET GOES TO ROME **101**
COLUMBIA. (R) August, 1963. Cindy Carol, James Darren. Color.
GIDRAH, THE THREE-HEADED MONSTER (TO) **85**
CONTINENTAL. (R) October, 1965. Yosuke Natsuki, Yuriko Hoshi. Color.
GIFT OF LOVE, THE (Cs) **105**
TWENTIETH CENTURY-FOX. (R) February, 1958. Lauren Bacall, Robert Stack. Color.
GIGANTIS **78**
WARNER BROS. (R) June, 1959. Hiroshi Koizumi, Sotsuko Wakayama.
GIGI (Cs) **116**
MGM. (R) December, 1958. Maurice Chevalier, Leslie Caron. Color.
GIGOT **104**
TWENTIETH CENTURY-FOX. (R) November, 1962. Jackie Gleason, Katherine Kath. Color.
GILDA LIVE **91**
WARNER BROS. (R) March 28, 1980. Gilda Radner, Father Guido Sarducci.
GIMME SHELTER **91**
CINEMA V. (R) December, 1970. Documentary. The Rolling Stones. Color.
GIRL, AND THE GENERAL, THE **105**
MGM. (R) October, 1967. Rod Steiger, Virna Lisi. Color.
GIRL CAN'T HELP IT, THE (Cs) **97**
TWENTIETH CENTURY-FOX. (R) December, 1956. Tom Ewell, Jayne Mansfield, Edmond O'Brien. Color.
GIRL FRIENDS **86**
WARNER BROS. (R) August, 1978. Melanie Mayron, Eli Wallach
GIRL FROM PETROVKA, THE (P) **104**
UNIVERSAL. (R) August, 1974. Goldie Hawn, Hal Holbrook.
GIRL GETTERS, THE **93**
AIP. (R) June, 1966. Oliver Reed, Jane Merrow.
GIRL HAPPY (P) **96**
METRO-GOLDWYN-MAYER. (R) April, 1965. Elvis Presley, Shelley Fabares. Color.
GIRL HE LEFT BEHIND, THE **103**
WARNER BROTHERS. November 10, 1956. Tab Hunter, Natalie Wood.
GIRL HUNTERS, THE (P) **103**
COLORAMA. (R) June, 1963. Mickey Spillane, Shirley Eaton.
GIRL IN BLACK STOCKINGS, THE **73**
UNITED ARTISTS. (R) October, 1957. Lex Barker, Anne Bancroft.
GIRL IN THE KREMLIN, THE **81**
UNIVERSAL. (R) May, 1957. Lex Barker, Zsa Zsa Gabor.
GIRL IN THE RED VELVET SWING, THE (Cs) **109**
TWENTIETH CENTURY-FOX. (R) October, 1955. Ray Milland, Joan Collins, Farley Granger. Color.
GIRL MOST LIKELY, THE **98**
UNIVERSAL. (R) February, 1958. Jane Powell, Cliff Robertson. Color
GIRL NAMED TAMIKO, A (P) **110**
PARAMOUNT. (R) February, 1963. Laurence Harvey, France Nuyen. Color.
GIRL OF THE NIGHT **93**
WARNER BROS. (R) October, 1960. Anne Francis, Lloyd Nolan.
GIRL ON A MOTORCYCLE, THE **91**
CLARIDGE. (R) November, 1968. Alain Delon, Marianne Faithfull, Roger Mutton. Color.
GIRL RUSH, THE (VV) **65**
PARAMOUNT. (R) September, 1955. Rosalind Russell, Fernando Lamas. Color.
GIRL WHO COULDN'T SAY NO, THE **83**
20th CENTURY-FOX. (R) November, 1969. Virna Lisi, George Segal, Lila Kedrova. Color.
GIRL WHO KNEW TOO MUCH, THE **96**
COMMONWEALTH UNITED. (R) September, 1969. Adam West, Nancy Kwan. Color.
GIRL WITH GREEN EYES **91**
LOPERT. (R) August, 1964. Rita Tushingham, Peter Finch.
GIRLS! GIRLS! GIRLS! **106**
PARAMOUNT. (R) November, 1962. Elvis Presley, Stella Stevens. Color.
GIRLS IN PRISON **87**
AMERICAN INTERNATIONAL. (R) July 15, 1956. Richard Denning, Joan Taylor.
GIRLS ON THE BEACH **80**
PARAMOUNT. (R) May, 1965. Martin West, Noreen Corcoran. Color.

	Length In Mins.
GIRLS TOWN METRO-GOLDWYN-MAYER. (R) October, 1959. Mel Torme, Mamie Van Doren.	92
GIT! AVCO EMBASSY. (R) June, 1965. Jack Chaplain, Heather North, Leslie Bradley.	92
GIVE 'EM HELL, HARRY! THEATRE TELEVISION. (R) 1976. James Whitmore.	102
GLADIATORS SEVEN (T) METRO-GOLDWYN-MAYER. (R) May, 1964. Richard Harrison, Lordana Nusciak. Color.	92
GLASS BOTTOM BOAT (P) METRO-GOLDWYN-MAYER. (R) July, 1966. Doris Day, Rod Taylor. Color.	110
GLASS HOUSES COLUMBIA. (R) January, 1972. Bernard Barrow, Jennifer O'Neill. Color.	90
GLASS SLIPPER, THE METRO-GOLDWYN-MAYER. (R) April 8, 1955. Leslie Caron, Michael Wilding. Color.	94
GLASS TOMB, THE LIPPERT. (R) April 15, 1955. John Ireland.	59
GLEN AND RANDA UMC. (R) August, 1971. Steven Curry, Shelley Plimpton. Color.	94
GLOBAL AFFAIR, A METRO-GOLDWYN-MAYER. (R) February, 1964. Bob Hope, Lilo Pulver.	84
GLORY (SS) RKO RADIO. (R) January 11, 1956. Margaret O'Brien, Walter Brennan, Charlotte Greenwood. Color.	100
GLORY GUYS, THE (P) UNITED ARTISTS. (R) July, 1965. Tom Tryon, Harve Presnell. Color.	112
GLORY STOMPERS, THE (Colorscope) AMERICAN INTERNATIONAL. (R) November, 1967. Dennis Hopper, Chris Noel. Color.	85
GNOME-MOBILE, THE BUENA VISTA. (R) July, 1967. Walter Brennan, Karen Dotrice. Color.	90
GO-BETWEEN, THE COLUMBIA. (R) Fall, 1971. Julie Christie, Alan Bates, Margaret Leighton, Michael Redgrave. Color.	116
GO GO MANIA (T) AMERICAN INTERNATIONAL. (R) May, 1965. The Beatles, The Animals. Color.	70
GO NAKED IN THE WORLD (Cs) METRO-GOLDWYN-MAYER. (R) February, 1961. Gina Lollobrigida, Anthony Franciosa. Color.	103
GO TELL THE SPARTANS AVCO EMBASSY. (R) June, 1978. Burt Lancaster, Marc Singer.	114
GOAL! (TS) COLUMBIA. (R) February, 1967. Soccer Documentary. Color.	106
GOD IS MY PARTNER (R) TWENTIETH CENTURY-FOX. (R) July, 1957. Walter Brennan, John Hoyt.	82
GODDESS, THE COLUMBIA. (R) May, 1958. Kim Stanley, Lloyd Bridges.	105
GODDESS OF LOVE, THE (Cs) TWENTIETH CENTURY-FOX. (R) December, 1960. Belinda Lee, Jack Sernas. Color.	65
GODFATHER, THE PARAMOUNT. (R) March, 1972. Marlon Brando, Al Pacino, James Caan. Color.	78
GODFATHER, PART II, THE PARAMOUNT. (R) December, 1974. Al Pacino, Robert Duvall.	200
GOD'S LITTLE ACRE UNITED ARTISTS. (R) May, 1958. Aldo Ray, Robert Ryan.	110
GODSPELL COLUMBIA. (R) April, 1973. Victor Garber, David Haskell. Color.	103
GODZILLA, KING OF THE MONSTERS (Jap.; Eng. dubbed) EMBASSY. (R) April, 1956. Raymond Burr.	80
GODZILLA vs. THE THING (Tohoscope) AMERICAN INTERNATIONAL. (R) October, 1964. Akira Takarada, Yuriko Hoshi. Color.	90
GOIN' DOWN THE ROAD CHEVRON. (R) October, 1970. Doug McGrath, Paul Bradley. Color.	90
GOIN' SOUTH PARAMOUNT. (R) October, 1978. Jack Nicolson, Mary Steenburgen.	105
GOING HOME METRO-GOLDWYN-MAYER. (R) November, 1971. Robert Mitchum, Brenda Vaccaro. Color.	100
GOING IN STYLE WARNER BROS. (R) December 25, 1979. George Burns, Art Carney, Lee Strasberg.	97
GOING STEADY COLUMBIA. (R) February, 1958. Molly Bee, Alan Reed, Jr	79
GOLD (P) ALLIED ARTISTS. (R) October, 1974. Roger Moore, Susannah York.	120
GOLD FOR THE CAESARS (CinemaScope) METRO-GOLDWYN-MAYER. (R) June, 1964. Jeffrey Hunter, Mylene Domongeot. Color.	86
GOLD OF NAPLES, THE (Ital.) DISTRIBUTORS CORPORATION OF AMERICA. (R) February, 1957. Silvana Mangano, Vittorio DeSica, Sophia Loren, Toto.	107
GOLD OF THE SEVEN SAINTS (Warnerscope) WARNER BROS. (R) April, 1961. Clint Walker, Roger Moore.	83
GOLDEN AGE OF COMEDY, THE DCA DIST. (R) 1958. Laurel & Hardy, Charlie Chaplin. b&w.	78
GOLDEN ARROW, THE (T) METRO-GOLDWYN-MAYER. (R) April, 1964. Tab Hunter, Rossana Podesta. Color.	91
GOLDENGIRL AVCO EMBASSY. (R) June, 1979. Susan Anton, James Coburn.	104
GOLDEN DEMON, THE (Jap.) HARRISON. (R) January, 1956. Jun Negami. Color.	95
GOLDEN NEEDLES (P) AMERICAN INTERNATIONAL. (R) July, 1974. Joe Don Baker, Elizabeth Ashley.	93
GOLDEN VOYAGE OF SINBAD, THE COLUMBIA. (R) March, 1974. John Philip Law, Caroline Munro.	105
GOLDFINGER UNITED ARTISTS. (R) December, 1964. Sean Connery, Gert Ferbe. Color.	108
GOLDSTEIN ALTURA. (R) May, 1965. Lou Gilbert, Ellen Madison.	82
GOLIATH AND THE BARBARIANS (Totalscope) AMERICAN INTERNATIONAL. (R) December, 1959. Steve Reeves, Chelo Alonso. Color.	90
GOLIATH AND THE DRAGON (Colorscope) AMERICAN INTERNATIONAL. (R) November, 1960. Mark Forrest, Broderick Crawford. Color.	90
GOLIATH AND THE VAMPIRES (Totalscope) AMERICAN INTERNATIONAL. (R) May, 1964. Gordon Scott, Gianna Maria Canale.	91
GONE ARE THE DAYS! HAMMER. (R) September, 1963. Ruby Dee, Ossie Davis.	97
GONE WITH THE WIND (70mm) MGM. (R) October, 1967. Clark Gable, Vivian Leigh. Color.	220
GONG SHOW MOVIE, THE UNIVERSAL. (R) May, 1980. Chuck Barris, Robin Altman.	89
GOOD DAY FOR A HANGING COLUMBIA. (R) January, 1959. Fred MacMurray, Maggie Hayes. Color.	85
GOOD DIE YOUNG, THE UNITED ARTISTS. (R) February, 1955. Richard Basehart, Gloria Grahame.	100
GOOD GUYS AND THE BAD GUYS, THE (P) WARNER BROS. (R) November, 1969. Robert Mitchum, George Kennedy, Tina Louise. Color.	91
GOOD MORNING, MISS DOVE (Cs) TWENTIETH CENTURY-FOX. (R) November, 1955. Jennifer Jones, Robert Stack. Color.	107
GOOD NEIGHBOR SAM COLUMBIA. (R) June, 1964. Jack Lemmon, Romy Schneider. Color.	130
GOOD, THE BAD, AND THE UGLY, THE (T) UNITED ARTISTS. (R) December, 1967. Clint Eastwood, Eli Wallach, Lee Van Cleef. Color.	161
GOOD TIMES COLUMBIA. (R) May, 1967. Sonny and Cher. Color.	91
GOOD TIMES, WONDERFUL TIMES ROGOSIN. (R) August, 1966. Documentary.	70
GOODBYE AGAIN UNITED ARTISTS. (R) August, 1961. Ingrid Bergman, Yves Montand.	120
GOODBYE CHARLIE (Cs) TWENTIETH CENTURY-FOX. (R) December, 1964. Tony Curtis, Debbie Reynolds. Color.	117
GOODBYE, COLUMBUS PARAMOUNT. (R) June, 1969. Richard Benjamin, Ali MacGraw, Jack Klugman. Color.	105
GOODBYE GEMINI CINERAMA. (R) September, 1970. Judy Geeson, Martin Potter. Color.	90
GOODBYE GIRL, THE	110

Length In Mins.

WARNER BROS. (R) December, 1977. Richard Dreyfuss, Marsha Mason.

GOODBYE MR. CHIPS (P) **151**
MGM. (R) November, 1969. Peter O'Toole, Petula Clark, Sian Phillips, Michael Redgrave. Color.

GOODBYE MY LADY **95**
WARNER BROS. (R) May 12, 1956. Brandon de Wilde, Walter Brennan.

GORDON'S WAR (P) **90**
TWENTIETH CENTURY-FOX. (R) August, 1973. Paul Winfield, Carl Lee.

GORGO **76**
METRO-GOLDWYN-MAYER. (R) February, 1961. Bill Travers, Will Sylvester. Color.

GORGON, THE **83**
COLUMBIA. (R) February, 1965. Peter Cushing, Christopher Lee. Color.

GOSPEL ACCORDING TO ST. MATTHEW, THE (Ital., Eng. titles) **135**
CONTINENTAL DISTRIBUTING. (R) February 17, 1966. Enrique Irazoque, Margherita Caruso.

GOSPEL ROAD, THE **83**
TWENTIETH CENTURY-FOX. (R) March, 1973. Johnny Cash, Robert Elfstrom. Color.

GRADUATE, THE (P) **105**
EMBASSY. (R) December, 1967. Anne Bancroft, Dustin Hoffman. Color.

GRAND BOUFFE, THE **122**
ABKCO FILMS. (R) October, 1973. Marcello Mastroianni, Michel Piccoli.

GRAND MANEUVER, THE (Fr.) **107**
UMPO. (R) October, 1956. Michele Morgan. Gerard Philipe. Color.

GRAND OLYMPICS **120**
TIMES FILM. (R) April, 1964. 1960 Rome Olympics. Color.

GRAND PRIX (Super Panavision) **175**
METRO-GOLDWYN-MAYER. (R) December, 1966. James Garner, Eva Marie Saint. Color.

GRAND SLAM (TE) **121**
PARAMOUNT. (R) January, 1968. Edward G. Robinson, Janet Leigh. Color.

GRAND SUBSTITUTION, THE (Shawscope) (Chinese; Eng. titles) **116**
FRANK LEE INTERNATIONAL. (R) November, 1965. Li Li-Hus, Ivy Ling Po. Color.

GRASS IS GREENER, THE (T) **102**
UNIVERSAL. (R) January, 1961. Cary Grant, Jean Simmons. Color.

GRASSHOPPER, THE **95**
NATIONAL GENERAL. (R) May, 1970. Jacqueline Bisset, Jim Brown, Joseph Cotten. Color.

GRAVY TRAIN, THE **95**
COLUMBIA. (R) July, 1974. Stacy Keach, Frederic Forest.

GRAYEAGLER **104**
AMERICAN INTERNATIONAL. (R) December, 1977. Ben Johnson, Iron Eyes Cody.

GRAY LADY DOWN (P) **110**
UNIVERSAL. (R) March, 1978. Charlton Heston, David Carradine.

GRAZIE, ZIA **93**
AVCO-EMBASSY. (R) January, 1969. Lisa Gastoril, Lou Castel, Gabriel Ferzetti. b&w.

GREASE (P) **110**
PARAMOUNT. (R) June, 1978. John Travolta, Olivia Newton–John.

GREASED LIGHTNING **96**
WARNER BROS. (R) July, 1977. Richard Pryor, Beau Bridges.

GREAT ADVENTURE, THE (Swed.) **75**
DE ROCHEMONT. (R) September, 1955. Non-Professional Cast.

GREAT AMERICAN PASTIME, THE **89**
METRO-GOLDWYN-MAYER. (R) December, 1956. Tom Ewell, Anne Francis.

GREAT ESCAPE, THE (P) **168**
UNITED ARTISTS. (R) July, 1963. Steve McQueen, James Garner. Color.

GREAT GATSBY, THE **148**
PARAMOUNT. (R) April, 1974. Robert Redford, Mia Farrow.

GREAT GEORGIA BANK HOAX, THE **88**
WARNER BROS. (R) November, 1978. Richard Basehart, Ned Beatty.

GREAT IMPOSTOR **112**
UNIVERSAL. (R) February, 1961. Tony Curtis, Edmond O'Brien.

GREAT LOCOMOTIVE CHASE, THE (Cs) **85**
BUENA VISTA. (R) June 20, 1956. Fess Parker, Jeffrey Hunter. Color.

GREAT MAN, THE **92**
UNIVERSAL. (R) February, 1957. Jose Ferrer, Dean Jagger, Keenan Wynn.

Length In Mins.

GREAT NORTHFIELD MINNESOTA RAID, THE **91**
UNIVERSAL. (R) April, 1972. Cliff Robertson, Robert Duvall. Color.

GREAT RACE, THE (P) **170**
WARNER BROS. (R) Special. June, 1966. Jack Lemmon, Tony Curtis. Color.

GREAT SANTINI, THE **118**
WARNER BROS. (R) October, 1979. Robert Duvall, Blythe Danner, Michael O'Keefe.

GREAT SCOUT AND CATHOUSE THURSDAY, THE **102**
AMERICAN INTERNATIONAL. (R) June, 1976. Lee Marvin, Oliver Reed.

GREAT SIOUX MASSACRE, THE (Cs) **91**
COLUMBIA. (R) September, 1965. Joseph Cotten, Darren McGavin. Color.

GREAT SIOUX UPRISING, THE **80**
U-I. (R) July, 1958. Jeff Chandler, Faith Domergue. Color.

GREAT SPY CHASE, THE **87**
AMERICAN INTERNATIONAL. (R) May, 1966. Lino Ventura, Bernard Blair.

GREAT ST. LOUIS BANK ROBBERY, THE **86**
UNITED STATES. (R) February, 1959. Steve McQueen, David Clarke.

GREAT TRAIN ROBBERY, THE **111**
UNITED ARTISTS. (R) February, 1979. Sean Connery, Donald Sutherland.

GREAT WALDO PEPPER, THE (T-AO 35) **106**
UNIVERSAL. (R) March, 1975. Robert Redford, Bo Svenson.

GREAT WALTZ, THE **135**
METRO-GOLDWYN-MAYER. (R) October, 1972. Mary Costa, Horst Buchholz, Rossano Brazzi. Color.

GREAT WAR, THE (Cs) **118**
UNITED ARTISTS. (R) October, 1961. b&w.

GREAT WHITE HOPE, THE **103**
TWENTIETH CENTURY-FOX. (R) December, 1970. James Earl Jones, Jane Alexander. Color.

GREATEST, THE **100**
COLUMBIA. (R) May, 1977. Muhammad Ali, Ernest Borgnine.

GREATEST STORY EVER TOLD, THE (Cinerama, P-70) **195**
UNITED ARTISTS. (R) Specialty. Max Von Sydow, Charlton Heston. Color.

GREED IN THE SUN (Cs) **122**
METRO-GOLDWYN-MAYER. (R) September, 1965. b&w.

GREEK TYCOON, THE **106**
UNIVERSAL. (R) May, 1978. Anthony Quinn, Jacqueline Bisset.

GREEN BERETS, THE **141**
WARNER BROS.-7 ARTS. (R) July, 1968. John Wayne, David Janssen, Jim Hutton. Color.

GREEN BUDDHA, THE (Brit.) **61**
REPUBLIC. (R) July 9, 1955. Wayne Morris, Mary Germaine.

GREEN FIRE (Cs) **100**
METRO-GOLDWYN-MAYER. (R) January 21, 1955. Grace Kelly, Stewart Granger. Color.

GREEN HELMET, THE **88**
METRO-GOLDWYN-MAYER. (R) April, 1961. Bill Travers, Nancy Walters.

GREEN MAGIC (Ital.-Eng. narration) **83**
I.F.E. (R) May, 1955. Documentary. Color.

GREEN MANSIONS **104**
METRO-GOLDWYN-MAYER. (R) April, 1959. Audrey Hepburn, Anthony Perkins.

GREEN SCARF, THE (Brit.) **96**
ASSOCIATED ARTISTS. (R) January, 1955. Michael Redgrave, Ann Todd, Leo Glenn.

GREEN SLIME, THE **85**
METRO-GOLDWYN-MAYER. (R) May, 1969. Robert Horton, Luciana Paluzzi. Color.

GREEN-EYED BLONDE, THE **73**
WARNER BROS. (R) December, 1957. Susan Oliver, Linda Plowman.

GREENGAGE SUMMER **93**
COLUMBIA. (R) April, 1961. Kenneth Moore, Danielle Darrieux. Color.

GREENWICH VILLAGE STORY **95**
SHAW INT. (R) July, 1963. Robert Hogan, Melinda Plank.

GREETINGS **88**
SIGMA III. (R) December, 1968. Robert De Niro, Jonathan Warden.

GREYFRIARS BOBBY **90**
BUENA VISTA. (R) October, 1961. Donald Crisp, Laurence Nals.

GRIMM'S FAIRY TALES FOR ADULTS ONLY **76**
CINEMATION. (R) February, 1971. Marie Liljedhal, Ingrid Von Bergen. Color.

GRISSOM GANG, THE **125**

Length In Mins.

NATIONAL GENERAL. (R) 1971. Kim Darby, Scott Wilson. Color.

GROOVE TUBE, THE 75
LEVITT-PICKMAN. (R) June, 1974. Ken Shapiro, Richard Belzer.

GROUNDSTAR CONSPIRACY, THE 96
UNIVERSAL. (R) May, 1972. George Peppard, Michael Sarrazin. Color.

GROUP, THE 150
UNITED ARTISTS. (R) March, 1966. Candice Bergen, Joan Hackett. Color.

GUESS WHAT WE LEARNED IN SCHOOL TODAY 96
CANNON. (R) May, 1971. Richard Carballo, Zachary Haines. Color.

GUESS WHO'S COMING TO DINNER 108
COLUMBIA. (R) February, 1968. Spencer Tracy, Katharine Hepburn, Sidney Poitier. Color.

GUEST, THE 105
JANUS. (R) January, 1964. Alan Bates, Donald Pleasence.

GUESTS ARE COMING (Pol.; Eng. titles) 110
MITCHELL KOWAL. (R) January, 1967. Mitchell Kowal, Kazimeriz Opalinski.

GUIDE, THE 120
STRATTON. (R) February, 1965. Dev Anand, Waheeda Renmann. Color.

GUIDE FOR THE MARRIED MAN, A (P) 89
TWENTIETH CENTURY-FOX (R) June, 1967. Walter Matthau, Robert Morse. Color.

GUILT (Swed.) 90
CROWN INTERNATIONAL. (R) November, 1967. Sven Bertil Taube, Helena Brodin.

GULLIVER'S TRAVELS BEYOND THE MOON (Animated) 85
CONTINENTAL. (R) June 1, 1966. Color.

GUMBALL RALLY, THE 106
WARNER BROS. (R) August, 1976. Michael Sarrazin, Normann Burton.

GUMSHOE 88
COLUMBIA. (R) March, 1972. Albert Finney, Billie Whitelaw, Frank Finlay. Color.

GUN BROTHERS 79
UNITED ARTISTS. (R) September, 1956. Buster Crabbe, Neville Brand.

GUN DUEL IN DURANGO 73
UNITED ARTISTS. (R) May, 1957. George Montgomery, Ann Robinson.

GUN FEVER 81
UNITED ARTISTS. (R) January, 1958. Mark Stevens, John Lupton.

GUN FOR A COWARD (Cs) 88
UNIVERSAL. (R) March, 1957. Fred MacMurray, Jeffrey Hunter. Color.

GUN GLORY (Cs) 88
METRO-GOLDWYN-MAYER. (R) August, 1957. Stewart Granger, Rhonda Fleming. Color.

GUN HAWK, THE 92
ALLIED ARTISTS. (R) October, 1963. Rory Calhoun, Rod Cameron. Color.

GUN RUNNERS, THE 83
UNITED ARTISTS. (R) September, 1958. Audie Murphy, Eddie Albert.

GUN STREET 62
UNITED ARTISTS. (R) December, 1961. James Brown, John Clark.

GUN THAT WON THE WEST, THE 71
COLUMBIA. (R) September, 1955. Dennis Morgan, Paula Raymond, Richard Denning. Color.

GUN THE MAN DOWN 78
UNITED ARTISTS. (R) November, 1956. James Arness, Angie Dickinson.

GUNBATTLE AT MONTEREY 76
ALLIED ARTISTS. (R) September, 1957. Sterling Hayden, Ted de Corsia.

GUNFIGHT 68
UNITED ARTISTS. (R) May, 1961. James Brown, Joan Staley. Color.

GUNFIGHT, A. 94
PARAMOUNT. (R) June, 1971. Kirk Douglas, Johnny Cash. Color.

GUNFIGHT AT COMANCHE CREEK (P) 90
ALLIED ARTISTS. (R) December, 1963. Audie Murphy, Ben Cooper. Color.

GUNFIGHT AT DODGE CITY, THE 81
UNITED ARTISTS. (R) May, 1959. Joel McCrea.

GUNFIGHT AT THE OK CORRAL (VV) 122
PARAMOUNT. (R) May, 1957. Burt Lancaster, Kirk Douglas, Rhonda Fleming.

GUNFIGHT IN ABILENE 86
UNIVERSAL. (R) May, 1967. Bobby Darin, Emily Banks, Color.

GUNFIGHTERS OF ABILENE 122
UNITED ARTISTS. (R) 1960. Buster Crabbe, Barton MacLane.

GUNFIGHTERS OF CASA GRANDE (Cs) 92
METRO-GOLDWYN-MAYER. (R) May, 1965. Alex Nicol, Jorge Mistral. Color.

GUNMAN'S WALK (Cs) 96
COLUMBIA. (R) July, 1958. Van Heflin, Tab Hunter. Color.

GUNMEN FROM LAREDO 67
COLUMBIA. (R) March, 1959. Robert Knapp, Jana Davi. Color.

GUNMEN FROM THE RIO GRANDE 86
ALLIED ARTISTS. (R) 1965. Guy Madison, Madeline Lebeau. Color.

GUNN 95
PARAMOUNT. (R) June, 1967. Craig Stevens, Laura Devon. Color.

GUNPOINT 86
UNIVERSAL. (R) May, 1966. Audie Murphy, Joan Staley. Color.

GUNS AT BATASI (Cs) 108
TWENTIETH CENTURY-FOX. (R) October, 1964. Richard Attenborough, Flora Robson.

GUNS DON'T ARGUE 90
VISUAL DRAMA. (R) October, 1957. Myron Healey, Jean Harvey.

GUNS FOR SAN SEBASTIAN (Franscope) 111
MGM. (R) April, 1968. Anthony Quinn, Anjanette Comer. Color.

GUNS, GIRLS AND GANGSTERS 70
UNITED ARTISTS. (R) January, 1959. Mamie Van Doren, Gerald Mohr.

GUNS OF AUGUST 99
UNIVERSAL. (R) January, 1965. Documentary.

GUNS OF DARKNESS 95
WARNER BROS. (R) August, 1962. David Niven, Leslie Caron.

GUNS OF FORT PETTICOAT, THE 82
COLUMBIA. (R) April, 1957. Audie Murphy, Kathryn Grant. Color.

GUNS OF HATE 62
RKO RADIO. (R) June 18, 1948. Tim Holt, Nan Leslie.

GUNS OF NAVARONE, THE (Cs) 157
COLUMBIA. (R) July, 1961. Gregory Peck, David Niven. Color.

GUNS OF THE BLACK WITCH (Colorscope) 81
AMERICAN INTERNATIONAL. (R) February, 1962. Don Megowan, Silvana Pampanini. Color.

GUNS OF THE MAGNIFICENT SEVEN 106
UNITED ARTISTS. (R) May, 1969. George Kennedy, James Whitmore, Monte Markham. Color.

GUNS OF THE TIMBERLAND 91
WARNER BROS. (R) March, 1960. Alan Ladd, Jeanne Crain.

GUNSIGHT RIDGE 85
UNITED ARTISTS. (R) September, 1957. Joel MacCrea, Mark Stevens.

GUNSLINGER, THE 83
AMERICAN INTERNATIONAL. (R) June 15, 1956. John Ireland, Beverly Garland.

GUNSMOKE 79
UNIVERSAL. (R) March, 1953. Audie Murphy, Susan Cabot. Color.

GUNSMOKE IN TUCSON (Cs) 79
ALLIED ARTISTS. (R) December 28, 1958. Mark Stevens, Gale Robbins. Color.

GURU, THE 112
TWENTIETH CENTURY-FOX. (R) February, 1969. Michael York, Saeed Jaffrey, Utpal Dutt. Color.

GUS 96
BUENA VISTA. (R) July, 1976. Edward Asner, Don Knotts.

GUYS AND DOLLS (Cs) 150
METRO-GOLDWYN-MAYER. (R) Special. 1955. Marlon Brando, Jean Simmons, Frank Sinatra, Vivian Blaine. Color.

GYPSY (T) 149
WARNER BROS. (R) January, 1963. Rosalind Russell, Natalie Wood. Color.

GYPSY AND THE GENTLEMAN, THE 89
RANK FILM DISTRIBUTORS OF AMERICA. (R) October, 1955. Melina Mercouri, Keith Mitchell. Color.

GYPSY GIRL 102
CONTINENTAL. (R) Summer, 1966. Hayley Mills, Ian McShane. Color.

GYPSY MOTHS, THE 110
MGM. (R) November 1969. Burt Lancaster, Deborah Kerr, Gene Hackman. Color.

GYPSY WILDCAT 77
UNIVERSAL. (R) September 1, 1944. Maria Montez, Jon Hall. Color.

	Length In Mins.

H

H-MAN (Cs) (Japan, dubbed) 79
COLUMBIA. (R) 1958. Kenji Sahara, Yumi Shirakawa.

HAIL HERO 100
NATIONAL GENERAL. (R) October, 1969. Arthur Kennedy, Teresa Wright, Michael Douglas. Color.

HAIR (P) 118
UNITED ARTISTS. (R) March, 1979. John Savage, Treat Williams.

HALF A SIXPENCE (P) 148
PARAMOUNT. (R) February, 1968. Tommy Steele, Julia Foster. Color.

HALLELUJAH, THE HILLS 82
VERMONT. (R) December, 1963. Peter H. Beard, Marty Greenbaum.

HALLELUJAH TRAIL, THE (Utra Panavision) 156
UNITED ARTISTS. (R) June, 1965. Special. Burt Lancaster, Lee Remick. Color.

HALLIDAY BRAND, THE 77
UNITED ARTISTS. (R) January, 1957. Joseph Cotten, Viveca Lindfors.

HALLS OF ANGER 96
UNITED ARTISTS. (R) April, 1970. Calvin Lockhart, Janet MacLachlan. Color.

HAMLET (Russian, Eng. titles) 148
UNITED ARTISTS. (R) March, 1966. Innokenti Smoktunovsky, Mikhall Nazvanov.

HAMLET 114
COLUMBIA. (R) January, 1970. Nicol Williamson, Gordon Jackson, Judy Parfitt, Marianne Faithful. Color.

HAMMER 91
UNITED ARTISTS. (R) September, 1972. Fred Williamson, Bernie Hamilton. Color.

HAMMERHEAD 99
COLUMBIA. (R) July, 1968. Vince Edwards, Judy Geeson, Peter Vaughan. Color.

HAMMERSMITH IS OUT 108
CINERAMA. (R) October, 1972. Elizabeth Taylor, Richard Burton, Peter Ustinov. Color.

HAND, THE 61
AMERICAN INTERNATIONAL. (R) April, 1961. Derek Bond, Ronald Lee Hunt.

HAND IN HAND 75
COLUMBIA. (R) January, 1961. John Gregson, Dame Sybil Thorndike.

HANDLE WITH CARE 82
MGM. (R) April, 1958. Dean Jones, Joan O'Brien.

HANDS OF ORLAC, THE 82
CONTINENTAL. (R) 1961. Mel Ferrer, Dany Carrell. b&w.

HANG 'EM HIGH 114
UNITED ARTISTS. (R) August, 1968. Clint Eastwood, Inger Stevens, Ed Begley. Color.

HANG YOUR HAT ON THE WIND 86
BUENA VISTA. (R) December, 1969. Ric Natoli, Judson Pratt.

HANGING TREE, THE 106
WARNER BROS. (R) February, 1959. Gary Cooper, Maria Schell. Color.

HANGMAN, THE 86
PARAMOUNT. (R) May, 1959. Robert Taylor, Tina.

HANNIBAL (Supercinescope) 108
WARNER BROS. (R) June, 1960. Victor Mature, Rita Gam. Color.

HANNIBAL BROOKS 101
UNITED ARTISTS. (R) April, 1969. Oliver Reed, Michael J. Pollard, Karin Baal. Color.

HANNIE CAULDER 85
PARAMOUNT. (R) June, 1972. Raquel Welch, Robert Culp, Ernest Borgnine. Color.

HANOVER STREET 105
COLUMBIA. (R) May, 1979. Harrison Ford, Lesley-Anne Down.

HAPPENING, THE 101
COLUMBIA. (R) March, 1967. Anthony Quinn, George Maharis. Color.

HAPPIEST MILLIONAIRE, THE 154
BUENA VISTA. (R) October, 1967. Fred MacMurray, Greer Garson. Color.

HAPPINESS CAGE, THE 90
CINERAMA. (R) June, 1972. Christopher Walken, Ronny Cox, Ralph Meeker. Color.

HAPPY ANNIVERSARY 81
UNITED ARTISTS. (R) December, 1959. David Niven, Mitzi Gaynor.

HAPPY BIRTHDAY, GEMINI 107
UNITED ARTISTS. (R) May, 1980. Madeline Kahn, Rita Moreno.

HAPPY BIRTHDAY, WANDA JUNE 105
COLUMBIA. (R) December, 1971. Rod Steiger, Susannah York. Color.

HAPPY END (Czech, Eng. titles) 73
CONTINENTAL. (R) June, 1968. Vladimir Mensik, Jaroslavia Obermaierova, Joseph Abraham. b&w.

HAPPY ENDING, THE (P) 116
UNITED ARTISTS. (R) December, 1969. Jean Simmons, John Forsythe, Lloyd Bridges, Shirley Jones. Color.

HAPPY MOTHER'S DAY . . . LOVE, GEORGE 90
CINEMA V. (R) August, 1973. Patricia Neal, Cloris Leachman.

HAPPY NEW YEAR 112
AVCO-EMBASSY. (R) November, 1973. Lino Ventura, Francoise Fabian.

HAPPY ROAD, THE 100
METRO-GOLDWYN-MAYER. (R) February, 1957. Gene Kelly, Barbara Laage.

HAPPY THIEVES, THE 88
UNITED ARTISTS. (R) January, 1962. Rex Harrison, Rita Hayworth.

HARAKIRI 135
TOHO. (R) August, 1964. Tatsuya Nakadai.

HARBOR LIGHTS (Cs) 68
TWENTIETH CENTURY-FOX. (R) July, 1963. Kent Taylor, Miriam Colon.

HARD CONTRACT 106
TWENTIETH CENTURY-FOX. (R) May, 1969. James Coburn, Lee Remick, Lilli Palmer. Color.

HARD DAY'S NIGHT, A 85
UNITED ARTISTS. (R) August, 1964. The Beatles.

HARD MAN, THE 79
COLUMBIA. (R) December, 1957. b&w.

HARD RIDE, THE 93
AMERICAN INTERNATIONAL. (R) April, 1971. Robert Fuller, Sherry Bain. Color.

HARD TIMES 97
COLUMBIA. (R) October, 1975. Charles Bronson, James Coburn.

HARD TO HANDLE (formerly Citizens Band) 98
PARAMOUNT. (R) May, 1977. Paul Le Mat, Candy Clark

HARDCORE 106
COLUMBIA. (R) February, 1979. George C. Scott, Peter Boyle.

HARDER THEY COME, THE 100
NEW WORLD PICTURES. (R) February, 1973. Jimmy Cliff, Janet Barrity. Color.

HARDER THEY FALL, THE 109
COLUMBIA. (R) April, 1956. Humphrey Bogart, Rod Steiger, Jan Sterling.

HARLOW (ELECTRONOVISION) 109
MAGNA. (R) May, 1965. Carol Lynley, Ginger Rogers.

HARLOW (P) 121
PARAMOUNT. (R) June, 1965. Carroll Baker, Martin Balsam. Color.

HAROLD AND MAUDE 91
PARAMOUNT. (R) December, 1971. Ruth Gordon, Bud Cort, Vivian Pickles. Color.

HAROLD LLOYD'S WORLD OF COMEDY 94
CONTINENTAL. (R) June, 1962.

HARPER (P) 121
WARNER BROTHERS. (R) March, 1966. Paul Newman, Lauren Bacall. Color.

HARRAD SUMMER 103
CINERAMA. (R) August, 1974. Robert Reiser, Laurie Walters.

HARRY AND TONTO 115
20th CENTURY-FOX. (R) September, 1974. Art Carney, Ellen Burstyn.

HARRY AND WALTER GO TO NEW YORK 123
COLUMBIA. (R) June, 1976. James Caan, Elliott Gould.

HARRY BLACK AND THE TIGER (Cs) 117
TWENTIETH CENTURY-FOX. (R) September, 1958. Stewart Granger, Barbara Rush. Color.

HARRY IN YOUR POCKET 103
UNITED ARTISTS. (R) August, 1973. James Coburn, Michael Sarrazin.

HARUM SCARUM 95
METRO-GOLDWYN-MAYER. (R) November, 1965. Elvis Presley, Mary Ann Mobley. Color.

HARVEY MIDDLEMAN, FIREMAN 75
COLUMBIA. (R) July, 1965. Gene Troob Nick, Hermione Gingold. Color.

HATARI! 159
PARAMOUNT. (R) June, 1962. John Wayne, Hardy Kruger. Color.

HATFUL OF RAIN, A (Cs) 109
TWENTIETH CENTURY-FOX. (R) July, 1957. Eva Marie Saint, Don Murray.

HAUNTED STRANGLER, THE 81
METRO-GOLDWYN-MAYER. (R) June, 1958. Boris Karloff, Anthony Dawson.

HAUNTING, THE (P) 112

Length In Mins.

METRO-GOLDWYN-MAYER. (R) September, 1963. Julie Harris, Claire Bloom.

HAVE ROCKET WILL TRAVEL 76
COLUMBIA. (R) August, 1959. Moe Howard, Larry Fine.

HAVING A WILD WEEKEND 91
WARNER BROTHERS. (R) June, 1965. Dave Clark Five.

HAWAII (P) 189
UNITED ARTISTS. (R) October, 1966. Julie Andrews, Max Von Sydow. Color.

HAWAIIANS, THE 134
UNITED ARTISTS. (R) June, 1970. Charlton Heston, Geraldine Chaplin, John Phillip Law. Color.

HAWKS AND THE SPARROW, THE (Ital.) 91
BRANDON. (R) July, 1967. Toto, Ninetto Davoli.

HE LAUGHED LAST 77
COLUMBIA. (R) August, 1956. Frankie Laine, Lucy Marlowe. Color.

HE RIDES TALL 84
UNIVERSAL. (R) April, 1964. Tony Young, Dan Duryea.

HE WHO MUST DIE 129
LOPERT. (R) May, 1965. Melina Mercouri.

HEAD 86
COLUMBIA. (R) November, 1968. Peter Tork, David Jones, Michael Nesmith, Mickey Dolenz. Color.

HEAD OF A TYRANT (Totalscope) 94
UNIVERSAL. (R) June, 1960. Massimo Girotti, Isabelle Corey. Color.

HEADLESS GHOST, THE (D) 63
AMERICAN INTERNATIONAL. (R) April, 1959. Richard Lyon, Lilian Sottane.

HEADLINE HUNTERS 90
REPUBLIC. (R) September 15, 1955. Rod Cameron, Julie Bishop.

HEAD OVER HEELS 99
UNITED ARTISTS. (R) October, 1979. John Heard, Mary Beth Hurt.

HEAR ME GOOD (VV) 80
PARAMOUNT. (R) October, 1957. Hal March, Joe E. Ross. Color.

HEART BEAT 109
WARNER BROS. (R) January, 1980. Nick Nolte, Sissy Spacek.

HEART IS A LONELY HUNTER, THE 125
WARNER BROS.-7 ARTS. (R) October, 1968. Alan Arkin, Leurinda Barrett, Stacy Keach, Jr. Color.

HEART OF SHOW BUSINESS, THE 40
COLUMBIA. (R) June, 1957. All Star Cast.

HEART OF THE MATTER, THE (Brit.) 100
ASSOCIATED ARTISTS. (R) November, 1954. Trevor Howard, Maria Schell.

HEARTBREAK KID, THE 104
TWENTIETH CENTURY-FOX. (R) December, 1972. Cybill Shepherd, Charles Grodin, Eddie Albert. Color.

HEARTS AND MINDS 112
WARNER BROS. (R) March, 1975. Documentary.

HEARTS OF THE WEST 102
UNITED ARTISTS. (R) October, 1975. Jeff Bridges, Andy Griffith.

HEAT 100
WARHOL. (R) October, 1972. Joe Dallesandro, Sylvia Miles. Color.

HEAVEN CAN WAIT 100
PARAMOUNT. (R) June, 1978. Warren Beatty, Julie Christie.

HEAVEN KNOWS, MR. ALLISON (Cs) 107
TWENTIETH CENTURY-FOX. (R) March, 1957. Deborah Kerr, Robert Mitchum. Color.

HEAVEN ON EARTH 84
JB FILMS. (R) January, 1961. Barbara Florian, Charles Fawcett. Color.

HEAVEN WITH A GUN 101
METRO-GOLDWYN-MAYER. (R) May, 1969. Glenn Ford, Carolyn Jones, Barbara Hershey. Color.

HEAVY TRAFFIC 82
AMERICAN INTERNATIONAL. (R) August, 1973. Joseph Kaufman, Beverly Hope Atkinson.

HEDDA 100
BRUT PRODS. (R) February, 1976. Glenda Jackson, Timothy West.

HEIDI 95
WARNER BROS.-7 ARTS. (R) March, 1968. Eva Maria Singhammer, Gertraud Mittermayr. Color.

HEIDI AND PETER (Swiss-Eng. dubbed) 89
UNITED ARTISTS. (R) December, 1955. Elisbeth Sigmund, Heinrich Gretler. Color.

HELEN MORGAN STORY, THE (Cs) 118
WARNER BROS. (R) September, 1957. b&w.

HELEN OF TROY (Cs) 118
WARNER BROTHERS. (R) February 11, 1956. Rossana Podesta, Jack Sernas. Color.

HELGA 87
AMERICAN INTERNATIONAL. (R) August, 1968. Ruth Gassman. Color.

HELL BENT FOR LEATHER (Cs) 82
UNIVERSAL. (R) February, 1960. Audie Murphy, Felicia Farr. Color.

HELL BOUND 79
UNITED ARTISTS. (R) October, 1957. John Russell, June Blair.

HELL CANYON OUTLAWS 72
REPUBLIC. (R) October, 1957. Dale Robertson, Brian Keith.

HELL DRIVERS (VV) 91
RANK FILM DISTRIBUTORS OF AMERICA. (R) May, 1958. Stanley Baker, Herbert Lom.

HELL, HEAVEN OR HOBOKEN 100
NTA. (R) September, 1959. M.E. Clifton Jones, Joan Mills.

HELL IN KOREA 82
HAL ROACH. (R) July 1959. George Baker, Harry Andrews.

HELL IN THE PACIFIC 103
CINERAMA. (R) January, 1969. Lee Marvin, Toshiro Mifune. Color.

HELL IS A CITY 96
COLUMBIA. (R) November, 1960. Stanley Baker, Maxine Audley.

HELL IS FOR HEROES 90
PARAMOUNT. (R) June, 1962. Steve McQueen, Bobby Darin.

HELL ON DEVIL'S ISLAND (Cs) 74
TWENTIETH CENTURY-FOX. (R) September, 1957. Helmut Dantine, William Talman.

HELL ON FRISCO BAY (Cs) 98
WARNER BROTHERS. (R) January 28, 1956. Alan Ladd, Edward G. Robinson, Joanne Dru. Color.

HELL SHIP MUTINY 66
REPUBLIC. (R) December, 1957. Jon Hall, John Carradine.

HELL SQUAD 64
AMERICAN INTERNATIONAL. (R) July, 1958. Wally Campo, Frederic Gavlin.

HELL TO ETERNITY 132
ALLIED ARTISTS. (R) August, 1960. Jeffrey Hunter, David Janssen.

HELL WITH HEROES, THE 102
UNIVERSAL. (R) September, 1968. Rod Taylor, Claudia Cardinale. Keven McCarthy. Color.

HELLCATS OF THE NAVY 82
COLUMBIA. (R) May, 1957. Ronald Reagan, Nancy Davis.

HELLFIGHTERS, THE 121
UNIVERSAL. (R) January, 1969. John Wayne, Katherine Ross, Jim Hutton. Color.

HELLO, DOLLY! (Todd-AO) 148
20th CENTURY-FOX. (R) December, 1969. Barbara Streisand, Walter Matthau, Michael Crawford. Color.

HELLO DOWN THERE 98
PARAMOUNT. (R) March, 1969. Jim Backus, Tony Randall, Janet Leigh. Color.

HELLO-GOODBYE 107
20th CENTURY-FOX. (R) July, 1970. Michael Crawford, Curt Jurgens, Genevieve Gilles. Color.

HELL'S ANGELS 69
Tom Stern, Jeremy Slate, Conny Van Dyke.

HELL'S BELLES 95
AMERICAN INTERNATIONAL. (R) April, 1969. Jeremy Slate, Adam Roarke, Jocelyn Lane. Color.

HELL'S CROSSROADS (N) 73
REPUBLIC. (R) March 8, 1957. Stephen MacNally, Peggie Castle.

HELL'S FIVE HOURS 76
ALLIED ARTISTS. (R) February, 1958. Steven McNally, Coleen Gray.

HELL'S HORIZON 90
COLUMBIA. (R) December, 1955. John Ireland, Maria English.

HELL'S ISLAND (VV) 84
PARAMOUNT. (R) May, 1955. John Payne, Mary Murphy.

HELLSTROM CHRONICLE, THE 90
CINEMA 5. (R) July, 1971. Insect documentary. Color.

HELP! 90
UNITED ARTISTS. (R) August, 1965. The Beatles. Color.

HEMINGWAY'S ADVENTURES OF A YOUNG MAN (Cs) 145
TWENTIETH CENTURY-FOX. (R) June, 1962. Richard Beymer, Diane Baker. Color.

HENNESSY 103
AMERICAN INTERNATIONAL. (R) August, 1975. Rod Steiger, Lee Remick.

HENRY VIII AND HIS SIX WIVES 125
LEVITT-PICKMAN. (R) September, 1973. Keith Michell, Donald Pleasence.

	Length In Mins.
HERBIE GOES TO MONTE CARLO BUENA VISTA. (R) July, 1977. Dean Jones, Don Knotts.	105
HERBIE RIDES AGAIN BUENA VISTA. (R) Summer, 1974. Helen Hayes, Ken Berry.	89
HERCULES WARNER BROS. (R) July, 1959. Steve Reeves.	103
HERCULES IN THE HAUNTED WORLD (Totalscope 100) WOOLNER. (R) October, 1964. Reg Park, Christopher Lee. Color.	91
HERCULES, SAMSON AND ULYSSES (Ital. dubbed) MGM. (R) 1965. Kirk Morris, Richard Lloyd. Color.	85
HERCULES UNCHAINED (Cs) EMBASSY-WARNER BROS. (R) July, 1960. Steve Reeves, Sylva Koscina. Color.	97
HERE COMES THE JETS (R) TWENTIETH CENTURY-FOX. (R) June, 1959. Steve Brodie, John Doucette. b&w.	71
HERE WE GO ROUND THE MULBERRY BUSH LOPERT. (R) March, 1968. Barry Evans, Judy Geeson. Color.	96
HERO, THE AVCO-EMBASSY. (R) August, 1972. Richard Harris, Romy Schneider. Color.	97
HERO AT LARGE UNITED ARTISTS. (R) February, 1980. John Ritter, Anne Archer	99
HEROD THE GREAT (Totalscope) ALLIED ARTISTS. (R) December, 1960. Edmund Purdom, Sylvia Lopez. Color.	89
HEROES UNIVERSAL. (R) November, 1977. Henry Winkler, Sally Field.	113
HEROES DIE YOUNG ALLIED ARTISTS. (R) November, 1960. Erika Peters, Scott Borland.	76
HEROES OF TELEMARK, THE (P) COLUMBIA. (R) January, 1966. Kirk Douglas, Richard Harris. Color.	131
HERO'S ISLAND (P) UNITED ARTISTS. (R) September, 1962. James Mason, Neville Brand. Color.	91
HESTER STREET MIDWEST FILMS. (R) September, 1975. Steven Keats, Carol Kane.	90
HEY BOY! HEY GIRL! COLUMBIA. (R) May, 1959. Louis Prima, Keely Smith.	81
HEY, LET'S TWIST PARAMOUNT. (R) December, 1961. Joey Dee, Teddy Randazzo.	80
HEY THERE, IT'S YOGI BEAR COLUMBIA. (R) June, 1964. Animated Feature. Color.	90
HI, MOM! SIGMA 3. (R) April, 1970. Robert DeNira, Jennifer Salt. Color.	87
HICKEY AND BOGGS UNITED ARTISTS. (R) September, 1972. Bill Cosby, Robert Culp. Color.	111
HIDDEN FEAR UNITED ARTISTS. (R) July, 1957. John Payne, Conrad Nagel.	83
HIDDEN GUNS REPUBLIC. (R) January 29, 1956. Bruce Bennett, Richard Arlen.	66
HIDDEN HOMICIDE REPUBLIC. (R) February, 1959. Griffith Jones, Patricia Lafran.	70
HIDE AND SEEK UNIVERSAL. (R) February, 1964. Ian Carmichael, Janet Munro.	90
HIDE IN PLAIN SIGHT (P) UNITED ARTISTS. (R) March, 1980. James Caan, Jill Eikenberry.	93
HIDING PLACE, THE WORLD WIDE PICTURES. (R) Fall, 1975. Julie Harris, Eileen Heckart.	145
HIGH BRENNER. (R) April, 1969. Astri Thorvik, Lanny Beckman. b&w.	82
HIGH ANXIETY 20th CENTURY-FOX. (R) February, 1978. Mel Brooks, Madeline Kahn.	91
HIGH BALLIN' AMERICAN INTERNATIONAL. (R) May, 1978. Peter Fonda, Jerry Reed.	100
HIGH COMMISSIONER CINERAMA. (R) September, 1968. Rod Taylor, Christopher Plummer, Lili Palmer. Color.	93
HIGH COST OF LIVING, THE (Cs) METRO-GOLDWYN-MAYER. (R) March, 1958. Jose Ferrer, Anthony Newley.	87
HIGH FLIGHT (Cs) COLUMBIA. (R) April, 1958. Ray Milland, Anthony Newley.	85
HIGH HELL PARAMOUNT. (R) January, 1958. John Derek, Elaine Stewart.	87
HIGH INFIDELITY (Ital. with English Titles) MAGNA. (R) June, 1965. Nino Manfredi, Fulvia Franco.	120
HIGH PLAINS DRIFTER UNIVERSAL. (R) April, 1973. Clint Eastwood, Verna Bloom. Color.	105
HIGH POWERED RIFLE, THE TWENTIETH CENTURY-FOX. (R) September, 1960. Willard Parker, Allison Hayes.	60
HIGH SCHOOL BIG SHOT THE FILM GROUP. (R) July, 1959. Tom Pittman, Virginia Aldridge.	70
HIGH SCHOOL CONFIDENTIAL! (Cs) MGM. (R) June, 1958. Russ Tamblyn, Jan Sterling.	85
HIGH SCHOOL HELL CATS AMERICAN INTERNATIONAL. (R) July, 1959. Yvonne Line, Brett Halsey.	68
HIGH SOCIETY (VV) METRO-GOLDWYN-MAYER. (R) Special, 1957. Bing Crosby, Grace Kelly, Frank Sinatra. Color.	107
HIGH SOCIETY ALLIED ARTISTS. (R) April 17, 1955. Leo Gorcey, Huntz Hall.	61
HIGH TERRACE, THE (Brit.) ALLIED ARTISTS. (R) December 9, 1956. Dale Robertson, Lois Maxwell.	70
HIGH TIME (Cs) TWENTIETH CENTURY-FOX. (R) September, 1960. Bing Crosby, Fabian. Color.	103
HIGH, WILD AND FREE AMERICAN INTERNATIONAL. (R) April, 1968. Documentary. Color.	105
HIGH WIND IN JAMAICA, A (Cs) TWENTIETH CENTURY-FOX. (R) June, 1965. Anthony Quinn, James Coburn. Color.	135
HILDA CRANE (Cs) TWENTIETH CENTURY-FOX. (R) April, 1956. Jean Simmons, Guy Madison, Jean Pierre Aumont. Color.	87
HILL, THE METRO-GOLDWYN-MAYER. (R) October, 1965. Sean Connery, Harry Andrews.	123
HILL 24 DOESN'T ANSWER (Israel-Eng. dial.) CONTINENTAL. (R) November, 1955. Michael Wager, Haya Hararit.	102
HILLS RUN RED, THE (T) UNITED ARTISTS. (R) September, 1967. Thomas Hunter, Henry Silva. Color.	89
HINDENBURG, THE UNIVERSAL. (R) December, 1975. George C. Scott, Anne Bancroft.	125
HIPPODROME Gerhard Reidmann, Willy Birgel. Color.	90
HIRED GUN, THE (Cs) METRO-GOLDWYN-MAYER. (R) September, 1957. Rory Calhoun, Anne Francis.	85
HIRED HAND, THE UNIVERSAL. (R) August, 1971. Peter Fonda, Warren Oates. Color.	93
HIRED KILLER, THE (TS) PARAMOUNT. (R) March, 1967. Robert Webber, Franco Nero. Color.	94
HIRELING, THE COLUMBIA. (R) June, 1973. Sarah Miles, Robert Shaw. Color.	108
HIS EXCELLENCY (Brit.) JOSEPH BRENNER. (R) February, 1956. Eric Portman, Cecil Parker.	84
HIT! (P) PARAMOUNT. (R) October, 1973. Billy Dee Williams, Richard Pryor.	134
HIT AND RUN UNITED ARTISTS. (R) March, 1957. Cleo Moore, Hugo Hass.	84
HIT THE DECK (Cs) METRO-GOLDWYN-MAYER. (R) March 4, 1955. Jane Powell, Tony Martin, Debbie Reynolds. Color.	112
HITLER ALLIED ARTISTS. (R) July, 1961. Richard Basehart.	78
HITLER: THE LAST TEN DAYS PARAMOUNT. (R) May, 1972. Alec Guiness, Simon Ward. Color.	106
HOA BINH TRANSVUE. (R) September, 1971. Phi Lan, Huynh Cazenas. Color.	93
HOLD BACK THE NIGHT ALLIED ARTISTS. (R) July 29, 1956. John Payne, Mona Freeman.	75
HOLD BACK TOMORROW	75

Length In Mins.

UNIVERSAL. (R) November, 1955. Cleo Moore, John Agar.

HOLD ON (P) 85
METRO-GOLDWYN-MAYER. (R) April, 1966. Peter Blaire Noone, Karl Green. Color.

HOLD THAT HYPNOTIST 61
ALLIED ARTISTS (R) February 24, 1957. Huntz Hall, Stanley Clements.

HOLE IN THE HEAD, A (Cs) 120
UNITED ARTISTS. (R) July, 1959. Frank Sinatra. Edward G. Robinson. Color.

HOLIDAY FOR HENRIETTA (Fr.) 103
ARDEE. (R) January, 1955. Dany Robin, Hildegarde Neff.

HOLIDAY FOR LOVERS (Cs) 102
TWENTIETH CENTURY-FOX. (R) July, 1959. Clifton Webb, Jane Wyman. Color.

HOLLYWOOD KNIGHTS, THE 91
COLUMBIA. (R) May, 1980. Robert Wuhl, Fran Drescher.

HOLLYWOOD OR BUST (VV) 95
PARAMOUNT. (R) December, 1956. Dean Martin, Jerry Lewis, Anita Ekberg. Color.

HOMBRE (P) 111
20th CENTURY-FOX. (R) April, 1967. Paul Newman, Frederic March. Color.

HOME BEFORE DARK 137
WARNER BROS. (R) November, 1958. Jean Simmons, Dan O'Herlihy.

HOME FROM THE HILL (Cs) 150
METRO-GOLDWYN-MAYER (R) March, 1960. Robert Mitchum, Eleanor Parker. Color.

HOME IS THE HERO 83
SCOA. (R) February, 1961. Arthur Kennedy, Walter Macken.

HOME OF YOUR OWN, A (Brit.) 43
CINEMA V. (R) July, 1965. Comedy.

HOMEBODIES 96
AVCO-EMBASSY. (R) September, 1974. Peter Brocco, Frances Fuller.

HOMECOMING, THE 116
AFT DISTRIBUTING. (R) Special, 1973. Cyril Cusack, Ian Holm.

HOMER 95
NATIONAL GENERAL. (R) September, 1970. Don Scardino, Alex Nicol, Tisa Farrow. Color.

HOMICIDAL 87
COLUMBIA. (R) June, 1961. Glenn Corbett, Patricia Breslin.

HONEY POT, THE 131
UNITED ARTISTS. (R) May, 1967. Rex Harrison, Susan Hayward. Color.

HONEYMOON HOTEL (P) 89
METRO-GOLDWYN-MAYER. (R) August, 1964. Robert Goulet, Nancy Kwan. Color.

HONEYMOON KILLERS, THE 108
CINERAMA. (R) February, 1970. Shirley Stoler, Tony LoBianco, Mary Jane Higby. b&w.

HONEYMOON MACHINE, THE (Cs) 100
METRO-GOLDWYN-MAYER. (R) August, 1961. Steve McQueen, Brigid Bazlen. Color.

HONEYSUCKLE ROSE (P) 119
WARNER BROS. (R) July, 1980. Willie Nelson, Dyan Cannon.

HONG KONG AFFAIR 79
ALLIED ARTISTS. (R) May 11 1958. Jack Kelly, May Wynn.

HONG KONG CONFIDENTIAL 64
UNITED ARTISTS. (R) November, 1958. Gene Barry, Beverly Tyler.

HONKERS, THE 110
UNITED ARTISTS. (R) February, 1972. James Coburn, Louis Nettleton, Slim Pickens. Color.

HOODLUM PRIEST, THE 101
UNITED ARTISTS. (R) April, 1961. Don Murray, Larry Gates.

HOOK, THE (P) 98
METRO-GOLDWYN-MAYER. (R) February, 1963. Kirk Douglas, Robert Walker.

HOOK LINE AND SINKER 92
COLUMBIA. (R) April, 1969. Jerry Lewis, Peter Lawford, Anne Francis. Color.

HOOKED GENERATION, THE 92
ALLIED ARTISTS. (R) April, 1969. Jeremy Slate, Steve Alaimo, John Davis Chandler. Color.

HOOPER 97
WARNER BROS. (R) July, 1978. Burt Reynolds, Jan-Michael Vincent.

HOOTENANNY HOOT 91
METRO-GOLDWYN-MAYER. (R) August, 1963. Peter Breck, Ruta Lee.

HORIZONTAL LIEUTENANT, THE (Cs) 90
METRO-GOLDWYN-MAYER. (R) March, 1962. Jim Hutton, Paula Prentiss. Color.

HORNET'S NEST 100
UNITED ARTISTS. (R) September, 1970. Rock Hudson, Sylva Koscina. Color.

HORROR HOUSE 79
AMER. INT. PICS. (R) May, 1970. Frankie Avalon, Jill Haworth, Dennis Price. Color.

HORROR OF DRACULA 82
UNIVERSAL INTERNATIONAL. (R) June, 1958. Peter Cushing, Melisa Stribling. Color.

HORROR OF PARTY BEACH, THE 78
20th CENTURY. (R) 1964. John Scott, Alice Lyon.

HORROR OF THE BLACK MUSEUM (Cs) 95
AMERICAN INTERNATIONAL. (R) April, 1959. Michael Gough, Graham Curnow. Color.

HORSE IN THE GREY FLANNEL SUIT 113
BUENA VISTA. (R) November, 1968. Dean Jones, Diane Baker, Lloyd Bochner. Color.

HORSE SOLDIERS, THE 119
UNITED ARTISTS. (R) July, 1959. John Wayne, William Holden. Color.

HORSEMEN, THE (P) 109
COLUMBIA. (R) June, 1971. Omar Sharif, Leigh Taylor-Young. Color.

HORSE'S MOUTH, THE 93
UNITED ARTISTS. (R) March, 1959. Alec Guinness. Kay Walsh. Color.

HOSPITAL, THE 104
UNITED ARTISTS. (R) December, 1971. George C. Scott, Diana Rigg. Color.

HOSTILE GUNS (TE) 91
PARAMOUNT. (R) July, 1967. George Montgomery, Yvonne deCarlo. Color.

HOT ANGEL, THE 73
PARAMOUNT. (R) December, 1958. Jackie Loughery, Edward Kemmer.

HOT BLOOD (Cs) 85
COLUMBIA. (R) March, 1956. Jane Russell, Cornel Wilde. Color.

HOT CAR GIRL 71
ALLIED ARTISTS. (R) August, 1958. Richard Bakalyan, June Kenney.

HOT CARS 60
UNITED ARTISTS. (R) August, 1956. John Bromfield, Joi Lansing.

HOT MILLIONS 120
METRO-GOLDWYN-MAYER. (R) October, 1968. Peter Ustinov, Maggie Smith, Karl Malden. Color.

HOT ROCK, THE 101
20th CENTURY-FOX. (R) January 1972. Robert Redford, George Segal, Zero Mostel. Color.

HOT ROD GANG 72
AMER. INT. PICS. (R) June, 1958. John Ashley, Jody Fair.

HOT ROD GIRL 75
AMERICAN INTERNATIONAL. (R) July 15, 1956. Lori Nelson, John Smith.

HOT ROD RUMBLE 79
ALLIED ARTISTS. (R) May 12, 1957. Leigh Snowden, Richard Hartunian.

HOT RODS TO HELL 92
METRO-GOLDWYN-MAYER. (R) March, 1967. Dana Andrews, Jeanne Crain. Color.

HOT SHOTS 61
ALLIED ARTISTS. (R) December 23, 1956. Huntz Hall, Stanley Clements.

HOT SPELL (VV) 86
PARAMOUNT. (R) June, 1958. Shirley Booth, Anthony Quinn.

HOT SUMMER NIGHT 86
METRO-GOLDWYN-MAYER. (R) February 15, 1957. Leslie Nielsen, Colleen Miller.

HOTEL 124
WARNER BROS. (R) March, 1967. Rod Taylor, Catherine Spaak. Color.

HOTEL PARADISO (P) 100
METRO-GOLDWYN-MAYER. (R) November, 1966. Alec Guinness, Gina Lollobrigida. Color.

HOUND DOG MAN (Cs) 87
TWENTIETH CENTURY-FOX. (R) November, 1959. Fabian, Carol Lynley. Color.

HOUND OF THE BASKERVILLES 84
UNITED ARTISTS. (R) June, 1959. Peter Cushing.

HOUR OF DECISION (Brit.) 74
ASTOR. (R) January, 1957. Jeff Morrow, Hazel Court.

HOUR OF THE GUN (R) 100
UNITED ARTISTS. (R) October, 1967. James Garner, Jason Robards. Color.

HOUR OF THE WOLF (Swed.) 88
LOPERT. (R) April, 1968. Liv Ullmann, Max von Sydow.

HOURS OF LOVE, THE (Ital. English Titles) 89

Length In Mins.

CINEMA V. (R) September, 1965. Vgo Tognazzi, Emmanuela Riva.

HOUSE BY THE LAKE 89
AMERICAN INTERNATIONAL. (R) March, 1977. Brenda Vaccaro, Don Stroud.

HOUSE CALLS 104
UNIVERSAL. (R) March, 1978. Walter Matthau, Glenda Jackson.

HOUSE IS NOT A HOME, A 90
EMBASSY. (R) August, 1964. Shelley Winters, Robert Taylor.

HOUSE OF BAMBOO (Cs) 102
TWENTIETH CENTURY-FOX. (R) July, 1955. Robert Ryan, Robert Stack. Color.

HOUSE OF CARDS 105
UNIVERSAL. (R) March, 1969. George Peppard, Inger Stevens, Orson Welles. Color.

HOUSE OF DARK SHADOWS 98
MGM. (R) September, 1970. Jonathan Frid, Joan Bennett. Color.

HOUSE OF FRIGHT (M) 80
AMERICAN INTERNATIONAL. (R) May, 1961. Paul Massie, Dawn Addams. Color.

HOUSE OF NUMBERS (Cs) 92
METRO-GOLDWYN-MAYER. (R) September, 1957. Jack Palance, Barbara Lang.

HOUSE OF 1,000 DOLLS 78
AMERICAN INTERNATIONAL. (R) October, 1967. Vincent Price, Martha Hyer. Color.

HOUSE OF RICORDI (Ital.) 117
BAKER-BRILL. (R) March, 1956. Marta Toren, Micheline Prelle, Danielle Delorme. Color.

HOUSE OF THE DAMNED (Cs) 62
TWENTIETH CENTURY-FOX. (R) March, 1963. Ronald Foster, Merry Anders.

HOUSE OF THE SEVEN HAWKS 92
METRO-GOLDWYN-MAYER. (R) November, 1959. Robert Taylor, Nicole Maurey.

HOUSE OF USHER 85
AMERICAN INTERNATIONAL. (R) July, 1960. Vincent Price, Mark Damon. Color.

HOUSE OF WOMEN 85
WARNER BROS. (R) May, 1962. Shirley Knight, Andrew Duggan.

HOUSE ON HAUNTED HILL 75
ALLIED ARTISTS. (R) January, 1959. Vincent Price, Carol Ohmart.

HOUSE THAT DRIPPED BLOOD, THE 101
CINERAMA. (R) MARCH, 1971. Peter Cushing, Christopher Lee. Color.

HOUSE THAT SCREAMED, THE (P) 89
AMERICAN INTERNATIONAL. (R) March, 1971. Lilli Palmer, Cristina Galbo. Color.

HOUSEBOAT (VV) 110
PARAMOUNT. (R) November, 1958. Cary Grant, Sophia Loren.

HOUSTON STORY, THE 79
COLUMBIA. (R) February, 1958. Gene Barry, Barbara Hale, Edward Arnold.

HOW DO I LOVE THEE? (WS) 110
CINERAMA. (R) October, 1970. Jackie Gleason, Maureen O'Hara. Color.

HOW I WON THE WAR 109
UNITED ARTISTS. (R) November, 1967. John Lennon, Michael Crawford. Color.

HOW NOT TO ROB A DEPARTMENT STORE (Fr., Eng. Titles) 95
ARTIXO PRODUCTIONS, LTD. (R) December, 1965. Jean-Claude Brialy, Marie Laforet.

HOW SWEET IT IS! 90
NATIONAL GENERAL. (R) July, 1968. James Garner, Debbie Reynolds, Maurice Ronet. Color.

HOW THE WEST WAS WON (Cinerama) 162
METRO-GOLDWYN-MAYER. (R) November, 1962. Special. All Star Cast. Color.

HOW TO BEAT THE HIGH COST OF LIVING 102
FILMWAYS. (R) June, 1980. Susan Saint James, Jane Curtin.

HOW TO BE VERY, VERY POPULAR (Cs) 89
TWENTIETH CENTURY-FOX. (R) July, 1955. Betty Grable, Robert Cummings, Sheree North. Color.

HOW TO FRAME A FIGG 103
UNIVERSAL. (R) February, 1971. Don Knotts, Joe Flynn. Color.

HOW TO MAKE A MONSTER 75
AMERICAN INTERNATIONAL. (R) July, 1958. Robert Harris, Paul Brinegar.

HOW TO MURDER A RICH UNCLE 80
COLUMBIA. (R) January, 1958. Charles Coburn, Migel Patrick.

HOW TO MURDER YOUR WIFE 118
UNITED ARTISTS. (R) February, 1965. Jack Lemmon, Virna Lisi. Color.

Length In Mins.

HOW TO SAVE A MARRIAGE—AND RUIN YOUR LIFE (P) 108
COLUMBIA. (R) March, 1968. Dean Martin, Stella Stevens. Color.

HOW TO SEDUCE A WOMAN 106
CINERAMA. (R) January, 1974. Angus Duncan, Angel Tompkins.

HOW TO STEAL A MILLION (P) 127
20th CENTURY-FOX. (R) August, 1966. Audrey Hepburn, Peter O'Toole. Color.

HOW TO STUFF A WILD BIKINI 90
AMERICAN INTERNATIONAL. (R) July, 1965. Annette Funicello, Dwayne Hickman. Color.

HOW TO SUCCEED IN BUSINESS WITHOUT REALLY TRYING (P) 119
UNITED ARTISTS. (R) March, 1967. Robert Morse, Michele Lee. Color.

HUCKLEBERRY FINN (P) 118
UNITED ARTISTS. (R) April, 1974. Jeff East, Paul Winfield.

HUD (P) 112
PARAMOUNT. (R) May, 1963. Paul Newman, Melvyn Douglas.

HUGS AND KISSES (Swedish; Eng. titles) 93
EMBASSY. (R) August, 1963. Agneta Ekmanner, Sven Bertil-Taube, Hakan Serner. b&w.

HUK 84
UNITED ARTISTS. (R) August, 1956. George Montgomery, Mona Freeman. Color.

HUMAN DUPLICATORS, THE 82
ALLIED ARTISTS. (R) March, 1965. George Nader, Barbara Nichols. Color.

HUMAN FACTOR, THE 95
BRYANSTON. (R) October, 1975. George Kennedy, John Mills.

HUMAN FACTOR, THE 115
UNITED ARTISTS. (R) February, 1980. Richard Attenborough, John Gielgud.

HUNCHBACK OF NOTRE DAME, THE (Cs) 103
ALLIED ARTISTS. (R) November, 1957. Gina Lollobrigida, Anthony Quinn. Color.

HUNCHBACK OF ROME 84
ROYAL. 1963 Gerard Blain, Anna Maria Ferrero.

HUNT, THE (Ital.) 93
TRANS-LUX. (R) June, 1967. Ismael Merlo, Alfredo Mayo.

HUNTER, THE 97
PARAMOUNT. (R) August, 1980. Steve McQueen, Eli Wallach.

HUNTERS, THE (Cs) 108
TWENTIETH CENTURY-FOX. (R) September, 1958. Robert Mitchum, Robert Wagner. Color.

HUNTERS OF THE DEEP 64
DISTRIBUTORS CORP. OF AMERICA. (R) January, 1955. Documentary. Color.

HUNTING PARTY, THE 108
UNITED ARTISTS. (R) July, 1971. Candice Bergen, Oliver Reed. Color.

HURRICANE (T-AO) 119
PARAMOUNT. (R) April, 1979. Jason Robards, Mia Farrow.

HURRY SUNDOWN (P) 146
PARAMOUNT. (R) March, 1967. Michael Caine, Jane Fonda. Color.

HURRY UP, OR I'LL BE 30 88
AVCO-EMBASSY. (R) November, 1973. John Lefkowitz, Linda de Coff.

HUSBANDS 138
COLUMBIA. (R) December, 1970. Ben Gazzara, Peter Falk, John Cassavetes. Color.

HUSH . . . HUSH SWEET CHARLOTTE 133
TWENTIETH CENTURY-FOX. (R) March, 1965. Bette Davis, Olivia de Havilland.

HUSTLE 120
PARAMOUNT. (R) December, 1975. Burt Reynolds, Catherine Deneuve.

HUSTLER, THE (Cs) 135
TWENTIETH CENTURY-FOX. (R) October, 1961. Paul Newman, Piper Laurie.

HYPNOTIC EYE, THE 77
ALLIED ARTISTS. (R) February, 1960. Jacques Bergerac, Merry Anders.

HYSTERIA 6
METRO-GOLDWYN-MAYER. (R) September, 1965. Robert Webber, Anthony Newlands.

I

I ACCUSE (Cs) 99
METRO-GOLDWYN-MAYER. (R) March, 1958. Jose Ferrer, Anton Walbrook.

Length In Mins.

I AIM AT THE STARS 107
COLUMBIA. (R) October, 1960. Curt Jurgens, Victoria Shaw.

I AM A CAMERA (Brit.) 98
DISTRIBUTORS CORP. OF AMERICA. (R) August, 1955. Julie Harris, Laurence Harvey, Shelley Winters. Color.

I AM CURIOUS (BLUE) 103
GROVE. (R) May, 1970. Lena Nyman, Borje Ahlstedt, Peter Lindgren. b&w.

I AM CURIOUS (YELLOW) 120
EVERGREEN. (R) March, 1969. Lena Nyman, Borje Ahlstedt, Peter Lindgren. b&w.

I BURY THE LIVING 75
UNITED ARTISTS. (R) July, 1958. Richard Boone, Theodore Bikel.

I COULD GO ON SINGING (P) 99
UNITED ARTISTS. (R) April, 1963. Judy Garland, Dirk Bogarde. Color.

I COVER THE UNDERWORLD 70
REPUBLIC. (R) May 19, 1955 Sean McClory, Joanne Jordan.

I DEAL IN DANGER 89
20TH CENTURY-FOX. (R) October, 1966. Robert Goulet, Christine Carere. Color.

I DIED A THOUSAND TIMES (Cs) 106
WARNER BROTHERS. (R) November 12, 1955. Jack Palance, Shelley Winters. Color.

I ESCAPED FROM DEVIL'S ISLAND 89
UNITED ARTISTS. (R) September, 1973. Jim Brown, Christopher George.

I LIKE MONEY (MR. TOPAZE) (Cs) 97
TWENTIETH CENTURY-FOX. (R) September, 1962. Peter Sellers, Nadia Gray. Color.

I LIVE IN FEAR (Jap; Eng. titles) 105
BRANDON. (R) January, 1967. Toshiro Mifune, Eiko Miyoshi.

I LOVE MY WIFE 95
UNIVERSAL. (R) December, 1970. Elliott Gould, Brenda Vaccaro. Color.

I LOVE YOU, ALICE B. TOKLAS 94
WARNER BROS.-7 ARTS. (R) October, 1969. Peter Sellers, Jo Van Fleet, Leigh Taylor-Young. Color.

I MARRIED A MONSTER FROM OUTER SPACE 78
PARAMOUNT. (R) October, 1958. Tom Tryon, Gloria Talbott.

I MARRIED A WOMAN 84
UNIVERSAL. (R) May, 1958. George Gobel, Diana Dors.

I MOBSTER (Cs) 80
TWENTIETH CENTURY-FOX. (R) February, 1959. Steve Cochran, Lita Milan.

I NEVER PROMISED YOU A ROSE GARDEN 96
NEW WORLD. July, 1977. Bibi Andersson, Kathleen Quinlan, Sylvia Sidney.

I NEVER SANG FOR MY FATHER 90
COLUMBIA. (R) November, 1970. Melvyn Douglas, Gene Hackman. Color.

I SAW WHAT YOU DID 82
UNIVERSAL. (R) September, 1965. Joan Crawford, John Ireland.

I THANK A FOOL (Cs) 100
METRO-GOLDWYN-MAYER. (R) October, 1962. Susan Hayward, Peter Finch. Color.

I WALK THE LINE 95
COLUMBIA. (R) October, 1970. Gregory Peck, Tuesday Weld. Color.

I WANNA HOLD YOUR HAND 104
UNIVERSAL. (R) April, 1978. Nancy Allen, Bobby Di Cicco.

I WANT TO LIVE 120
UNITED ARTISTS. (R) December, 1958. Susan Hayward, Simon Oakland.

I WANT WHAT I WANT 91
CINERAMA. (R) February, 1972. Anne Heywood, Harry Andrews. Color.

I WAS A TEENAGE FRANKENSTEIN 72
AMERICAN INTERNATIONAL. (R) November, 1957. Whit Bissell, Phyllis Coates.

I WAS A TEENAGE WEREWOLF 75
AMER. INT. PICS. (R) June, 1957. Yvonne Lime, Michael Landon.

I WILL, I WILL. . . FOR NOW 107
TWENTIETH CENTURY-FOX. (R) February, 1976. Elliott Gould, Diane Keaton.

ICE CASTLES 113
COLUMBIA. (R) February 1979. Robby Benson, Coleen Dewhurst.

ICE PALACE 143
WARNER BROS. (R) July, 1960. Richard Burton, Robert Ryan. Color.

ICE STATION ZEBRA 148
METRO-GOLDWYN-MAYER. (R) Special. 1969. Rock Hudson, Ernest Borgnine, Patrick McGoohan, Jim Brown. Color.

ICEMAN COMETH, THE 239
AFT DISTRIBUTING. (R) Special, 1973. Lee Marvin, Fredric March, Robert Ryan.

I'D RATHER BE RICH 96
UNIVERSAL. (R) October, 1964. Sandra Dee, Robert Goulet. Color.

IDOL, THE 107
EMBASSY. (R) July, 1966. Jennifer Jones, Michael Parks.

IF 111
PARAMOUNT. (R) February, 1969. Malcolm McDowell, David Wood, Richard Warwick. Color.

IF A MAN ANSWERS 102
UNIVERSAL. (R) November, 1962. Sandra Dee, Bobby Darin. Color.

IF ALL THE GUYS IN THE WORLD (Fr) 95
BUENA VISTA. (R) April, 1957. Jean Gaven, Helen Perdriere.

IF EVER I SEE YOU AGAIN 105
COLUMBIA. (R) May, 1978. Joe Brooks, Shelley Hack.

IF HE HOLLERS LET HIM GO 106
CINERAMA. (R) October, 1968. Dana Wynter, Raymond St. Jacques, Kevin McCarthy. Color.

IF IT'S TUESDAY, THIS MUST BE BELGIUM 99
UNITED ARTISTS. (R) May, 1969. Susanne Pleshette, Ian McShane, Mildred Natwick. Color.

I'LL CRY TOMORROW 117
METRO-GOLDWYN-MAYER. (R Special. December, 1955. Susan Hayward, Eddie Albert, Richard Conte.

I'LL NEVER FORGET WHAT'S 'IS NAME 99
REGIONAL. (R) April, 1968. Oliver Reed, Carol White. Color.

I'LL TAKE SWEDEN 96
UNITED ARTISTS. (R) June, 1965. Bob Hope, Tuesday Weld. Color.

ILLEGAL 99
WARNER BROTHERS. (R) October, 1955. Edward G. Robinson, Nina Foch.

ILLUSTRATED MAN, THE 103
WARNER BROS.-7 ARTS. (R) April, 1969. Rod Steiger, Claire Bloom, Robert Drivas. Color.

I'M ALL RIGHT, JACK 104
COLUMBIA. (R) October, 1960. Peter Sellers, Ian Carmichael.

IMAGE OF LOVE 88
GREEN RELEASING ORGANIZATION. (R) June, 1965. Documentary. Color.

IMAGES 100
COLUMBIA. (R) December, 1972. Susannah York, Rene Aubertjonois. color.

IMITATION GENERAL (Cs) 88
METRO-GOLDWYN-MAYER. (R) July, 1958. Glenn Ford, Red Buttons.

IMITATION OF LIFE (Cs) 124
UNIVERSAL. (R) April, 1959. Lana Turner, John Gavin. Color.

IMMORAL TALES (French, English sub-titles) 90
NEW LINE CINEMA. (R) March, 1976. Paloma Picasso, Lise Danvers.

IMMORTAL BACHELOR (Ital., Eng. titles) 95
S. J. INTL. (R) February, 1980. Giancario Giannini, Monica Vitti.

IMMORTAL STORY, THE 93
ALTURA FILMS. (R) August, 1969. Orson Welles, Jeanne Moreau, Roger Coggio, Norman Ashley. Color.

IMPASSE 100
UNITED ARTISTS. (R) May, 1969. Burt Reynolds, Anne Francis, Lyle Bettger. Color.

IMPOSSIBLE ON SATURDAY (Fr., Eng. titles) 120
MAGNA. (R) February, 1966. Robert Hirsch, Dahlir Friedland.

IMPOSSIBLE YEARS, THE 92
METRO-GOLDWYN-MAYER. (R) December, 1968. David Niven, Lola Albright, Chad Everett. Color.

IN-BETWEEN AGE, THE 76
ALLIED ARTISTS. (R) August 3, 1958. Mary Steele, Lee Patterson.

IN-LAWS, THE 103
WARNER BROS. (R) June, 1979. Alan Arkin, Peter Falk.

IN CELEBRATION 131
AFT. (R) 1975. Alan Bates, Brian Cox.

IN COLD BLOOD (P) 134
COLUMBIA. (R) February, 1968. Robert Blake, Scott Wilson.

IN ENEMY COUNTRY 107
UNIVERSAL. (R) July, 1968. Tony Franciosa, Anjanette Comer, Guy Stockwell. Color.

IN HARM'S WAY (P) 165
PARAMOUNT. (R) April, 1965. John Wayne, Kirk Douglas.

IN LIKE FLINT (CS) 114

	Length In Mins.
20TH CENTURY-FOX. (R) March, 1967. James Coburn, Lee J. Cobb. Color.	
IN PRAISE OF OLDER WOMEN AVCO EMBASSY. (R) February, 1979. Tom Berenger, Karen Black.	108
IN SEARCH OF GREGORY UNIVERSAL. (R) Julie Christie, Michael Sarrazin, John Hunt. Color.	90
IN SEARCH OF THE CASTAWAYS BUENA VISTA. (R) December, 1962. Hayley Mills, Maurice Chevalier. Color.	100
IN THE COOL OF THE DAY (P) METRO-GOLDWYN-MAYER. (R) June, 1963. Jane Fonda, Peter Finch. Color.	89
IN THE FRENCH STYLE COLUMBIA. (R) September, 1963. Jean Seberg, Stanley Baker.	105
IN THE HEAT OF THE NIGHT ALLIED ARTISTS. (R) August, 1967. Rod Steiger, Sidney Poitier. Color.	109
IN THE MONEY ALLIED ARTISTS. (R) February, 1957. Huntz Hall, Stanley Clements.	81
IN THE YEAR OF THE PIG PATHE. (R) November, 1969. Documentary. b&w.	101
INADMISSIBLE EVIDENCE PARAMOUNT. (R) June, 1968. Nicol Williamson, Eleanor Fazan, Jill Bennet. b&w.	96
INCIDENT AT PHANTON HILL UNIVERSAL. (R) July, 1966. Robert Fuller, Dan Duryea. Color.	88
INCIDENT IN AN ALLEY UNITED ARTISTS. (R) 1961. Charles Warfield, Erin O'Donnell.	84
INCREDIBLE JOURNEY, THE BUENA VISTA. (R) October, 1963. Emile Genest, John Drainie. Color.	80
INCREDIBLE MELTING MAN, THE AMERICAN INTERNATIONAL. (R) February, 1978. Alex Rebar, Burr DeBenning.	86
INCREDIBLE MR. LIMPET, THE WARNER BROS. (R) March, 1964. Donn Knotts, Carole Cook. Color.	99
INCREDIBLE PETRIFIED WORLD, THE GOVERNOR FILMS. (R) April, 1960. John Carradine, Robert Clarke.	70
INCREDIBLE SHRINKING MAN, THE UNIVERSAL. (R) April, 1957. Grant Williams, Randy Stuart.	81
INCREDIBLE 2-HEADED TRANSPLANT, THE AMERICAN INTERNATIONAL. (R) April, 1971. Bruce Dern, Pat Priest. Color.	87
INCREDIBLY STRANGE CREATURES, THE FAIRWAY. (R) 1964. Cash Flagg, Carolyn Brandt. Color.	82
INDESTRUCTIBLE MAN, THE ALLIED ARTISTS. (R) March 18, 1956. Lon Chaney, Casey Adams.	78
INDIAN FIGHTER (Cs) UNITED ARTISTS. (R) December, 1955, Kirk Douglas, Elsa Martinelli. Color.	83
INDIAN PAINT EAGLE-INTERNATIONAL. (R) February, 1966. Johnny Crawford, Jay Silverheels. Color.	91
INDISCREET WARNER BROS. (R) July, 1958. Cary Grant, Ingrid Bergman.	100
INFORMATION RECEIVED UNIVERSAL. (R) July, 1962. Sabina Sesselman, Wiliam Sylvester.	77
INHERIT THE WIND UNITED ARTISTS. (R) November, 1960. Spencer Tracy, Fredric March.	127
INHERITOR, THE HERA FILMS. (R) October, 1973. Jean-Paul Belmondo, Carla Gravina.	111
INN OF THE SIXTH HAPPINESS, THE (Cs) TWENTIETH CENTURY-FOX. (R) December, 1958. Ingrid Bergman, Curt Jurgens. Color.	156
INNOCENT BYSTANDERS PARAMOUNT. (R) January, 1973. Stanley Baker, Geraldine Chaplin. Color.	111
INNOCENT, THE (Italian, Eng. titles) ANALYSIS FILM RELEASING. (R) February, 1979. Giancarlo Giannini, Laura Antonelli.	112
INNOCENTS, THE (Cs) TWENTIETH CENTURY-FOX. (R) January, 1962. Deborah Kerr, Michael Redgrave.	100
INNOCENTS IN PARIS (Brit.) TUDOR. (R) March, 1955. Alastair Sim, Claire Bloom, Margaret Rutherford.	90
INSECT WOMAN, THE (Jap., Eng. titles) NIKKATSU. (R) June, 1964. Sachiko Hidari, Jitsuko Yoshimura.	123
INSERTS UNITED ARTISTS. (R) February, 1976. Richard Dreyfuss, Jessica Harper.	117
INSIDE DAISY CLOVER (P) WARNER BROTHERS. (R) February, 1966. Natalie Wood, Christopher Plummer. Color.	128
INSIDE DETROIT COLUMBIA. (R) January, 1956. Dennis O'Keefe, Pat O'Brien.	82
INSIDE OUT WARNER BROS. (R) November, 1975. Telly Savalas, Robert Culp, James Mason.	97
INSPECTOR CLOUSEAU UNITED ARTISTS. (R) June, 1968. Alan Arkin, Delia Boccardo, Frank Finlay. Color.	94
INTENT TO KILL (CS) TWENTIETH CENTURY-FOX. (R) February, 1959. Richard Todd, Betsy Drake.	89
INTERLUDE COLUMBIA. (R) July, 1968. Oscar Werner, Barbara Ferris, Virginia Meskell. Color.	113
INTERLUDE (Cs) UNIVERSAL. (R) September, 1957. June Allyson, Rossano Brazzi. Color.	90
INTERNATIONAL VELVET UNITED ARTISTS. (R) July, 1978. Tatum O'Neal, Christopher Plummer.	126
INTERNS, THE COLUMBIA. (R) August, 1962. Michael Callan, Cliff Robertson.	121
INTERRUPTED MELODY (Cs) METRO-GOLDWYN-MAYER. (R) July 1, 1955. Glenn Ford, Eleanor Parker. Color.	106
INTERVAL AVCO-EMBASSY. (R) June, 1973. Merle Oberon, Robert Wolders. Color.	84
INTIMACY GOLDSTONE FILM ENTERPRISES, INC. (R) May, 1966. Jack Ging, Nancy Malone.	87
INTRUDER, THE (Brit.) ASSOCIATED ARTISTS. (R) January, 1955. Jack Hawkins, George Cole.	84
INVASION OF THE BODY SNATCHERS UNITED ARTISTS. (R) December, 1978. Donald Sutherland, Brooke Adams.	114
INVASION OF THE BODY SNATCHERS (SS) ALLIED ARTISTS. (R) February 5, 1956. Kevin McCarthy, Dana Wynter.	80
INVASION OF THE SAUCERMEN AMERICAN INTERNATIONAL. (R) June, 1957. Steve Terrell, Gloria Castillo.	75
INVASION OF THE STAR CREATURES AMER. INT'L. PICS. (R) 1963. Bob Ball, Frankie Ray. b&w.	70
INVASION 1700 MEDALLION. (R) January, 1965. Jeanne Crain, John Drew Barrymore. Color.	112
INVASION QUARTET METRO-GOLDWYN-MAYER. (R) October, 1961. Bill Travers, Spike Milligan.	87
INVESTIGATION OF A CITIZEN ABOVE SUSPICION (Italian, Eng. titles) COLUMBIA. (R) December, 1970. Gian Maria Volonte, Florinda Bolkan. Color.	115
INVINCIBLE GLADIATOR, THE (TS) 7 ARTS. (R) 1963. Richard Harrison, Isabel Corey. b&w.	96
INVINCIBLE SIX, THE CONTINENTAL. (R) July, 1970. Stewart Whitman, Elke Sommer. Color.	96
INVISIBLE AVENGER REPUBLIC. (R) December, 1958. Richard Derr, Helen Westcott.	90
INVISIBLE BOY, THE METRO-GOLDWYN-MAYER. (R) October, 1957. Richard Eyer, Phillip Abbot.	85
INVISIBLE INFORMER, THE REPUBLIC. (R) August, 1956. Linda Stirling, William Henry.	87
INVISIBLE INVADERS UNITED ARTISTS. (R) June, 1959. John Agar, Jean Byran.	67
INVITATION TO A GUNFIGHTER UNITED ARTISTS. (R) November, 1964. Yul Bryner, Janice Rule. Color.	91
INVITATION TO THE DANCE METRO-GOLDWYN-MAYER. (R) March 1, 1957. Gene Kelly, Nora Kaye. Color.	92
IPCRESS FILE, THE (TE) UNIVERSAL. (R) October, 1965. Michael Caine, Nigel Green. Color.	106
IRMA LA DOUCE (P) UNITED ARTISTS. (R) July, 1963. Jack Lemmon, Shirley MacLaine. Color.	149

Length In Mins.

IRON PETTICOAT, THE (VV) 87
METRO-GOLDWYN-MAYER. (R) December 21, 1956. Bob Hope, Katharine Hepburn. Color.

IRON SHERIFF, THE 75
UNITED ARTISTS. (R) April, 1957. Sterling Hayden, Constance Ford.

IS PARIS BURNING? (P) 173
PARAMOUNT. (R) November, 1966. All Star Cast.

ISABEL 108
PARAMOUNT. (R) July, 1968. Genevieve Bujold, Mark Strange. Color.

ISLAND, THE (P) 114
UNIVERSAL. (R) June, 1980. Michael Caine, David Warner.

ISLAND AT THE TOP OF THE WORLD, THE 93
BUENA VISTA. (R) December, 1974. David Hartman, Donald Sinden.

ISLAND IN THE SUN (Cs) 119
TWENTIETH CENTURY-FOX. (R) June, 1957. James Mason, Joan Fontaine. Color.

ISLAND OF DR. MOREAU, THE 98
AMERICAN INTERNATIONAL. (R) July, 1977. Burt Lancaster, Michael York.

ISLAND OF LOST WOMEN 67
WARNER BROS. (R) May, 1959. June Blair, Venetia Stevenson.

ISLAND OF LOVE (P) 101
WARNER BROS. (R) May, 1963. Robert Preston, Tony Randall. Color.

ISLAND OF TERROR 90
UNIVERSAL. (R) February, 1967. Peter Cushing, Edward Judd. Color.

ISLAND OF THE BLUE DOLPHINS 99
UNIVERSAL. (R) August, 1964. Celia Kaye, Larry Domasin. Color.

ISLAND WOMEN 72
UNITED ARTISTS. (R) May, 1958. Marie Windsor, Vince Edwards.

ISLANDS IN THE STREAM (P) 105
PARAMOUNT. (R) March, 1977. George C. Scott, David Hemmings, Claire Bloom.

ISTANBUL (Cs) 84
UNIVERSAL. (R) February, 1957. Errol Flynn, Cornel Borchers. Color.

IT 95
WARNER BROS.-SEVEN ARTS. (R) January, 1967. Roddy McDowell, Jill Haworth. Color.

IT CAME FROM BENEATH THE SEA 80
COLUMBIA. (R) July, 1955. Kenneth Tobey, Faith Domergue.

IT CONQUERED THE WORLD 75
AMERICAN INTERNATIONAL. (R) July 25, 1956. Peter Graves, Beverly Garland.

IT HAPPENED AT THE WORLD'S FAIR (P) 105
METRO-GOLDWYN-MAYER. (R) May, 1963. Elvis Presley, Joan O'Brien. Color.

IT HAPPENED HERE 95
LOPERT. (R) August, 1966. Pauline Murray, Sebastian Shaw.

IT HAPPENED IN ATHENS (Cs) 105
TWENTIETH CENTURY-FOX. (R) June, 1962. Jayne Mansfield, Trax Colton. Color.

IT HAPPENED IN ROME (T) 105
RANK FILM DISTRIBUTORS OF AMERICA. (R) February, 1959. June Laverick, Vittorio De Sica. Color.

IT HAPPENED IN THE PARK (Fr.) 81
ELLIS. (R) September, 1957. Gerard Philipe, Micheline Presle.

IT HAPPENED TO JANE 98
COLUMBIA. (R) June, 1959. Doris Day, Jack Lemmon.

IT LIVES AGAIN 96
WARNER BROS. (R) May, 1978. Frederic Forrest, Kathleen Lloyd.

IT ONLY HAPPENS TO OTHERS 68
CINERAMA. (R) October, 1971. Marcello Mastroianni, Catherine Deneuve. Color.

IT STARTED IN NAPLES (VV) 100
PARAMOUNT. (R) August 1960. Clark Gable, Sophia Loren. Color.

IT STARTED WITH A KISS (Cs) 104
METRO-GOLDWYN-MAYER. (R) September, 1959. Glenn Ford, Debbie Reynolds. Color.

IT, THE TERROR FROM BEYOND SPACE 67
UNITED ARTISTS. (R) August, 1958. Marshall Thompson, Shawn Smith.

IT WON'T RUB OFF, BABY! 86
U-M FILM DIST. (R) November, 1969. Don Murray, Dick Gregory, Diane Varsi Robert Hocks.

ITALIAN CONNECTION, THE 87
AMERICAN INTERNATIONAL. (R) October, 1973. Mario Adorf, Henry Silva.

ITALIAN JOB, THE (P) 101
PARAMOUNT. (R) September, 1969. Michael Caine, Noel Coward, Raf Vallone. Color.

ITALIANO BRAVA GENTE (Eng. titles) 156
EMBASSY. (R) October, 1965. Arthur Kennedy, Peter Falls.

IT'S A DOG'S LIFE (Cs) 83
METRO-GOLDWYN-MAYER. (R) December 13, 1955. Jeff Richards, Jarma Lewis. Color.

IT'S A MAD, MAD, MAD WORLD (Ultra Panavision) 162
UNITED ARTISTS. (R) November, 1963. Spencer Tracy, Milton Berle. All Star Cast. Color.

IT'S A WONDERFUL WORLD 89
BRENNER. (R) October, 1959. Mylene Demongeot, Terrence Morgan. Color.

IT'S ALWAYS FAIR WEATHER (Cs) 102
METRO-GOLDWYN-MAYER. (R) September 2, 1955. Gene Kelly, Dan Dailey, Cyd Charisse. Color.

IT'S NEVER TOO LATE 96
SEVEN ARTS. (R) October, 1963 Phyllis Calvert, Patrick Barr. Color.

IT'S ONLY MONEY 84
PARAMOUNT. (R) December, 1962. Jerry Lewis, Zachary Scott.

IT'S SHOWTIME 87
UNITED ARTISTS. (F) April, 1976. Documentary.

I'VE LIVED BEFORE 82
UNIVERSAL. (R) September, 1956. Jock Mahoney, Leigh Snowden.

J

J.D.'S REVENGE 96
AMERICAN INTERNATIONAL. (R) August, 1976. Glynn Turman, Joan Pringle.

J.W. COOP 112
COLUMBIA. (R) January, 1972. Cliff Robertson, Geraldine Page. Color.

JACK FROST 79
EMBASSY. (R) Fall, 1966. Natasha Sedykh, Alexander Khvylya.

JACK OF DIAMONDS 105
MGM. (R) October, 1967. George Hamilton, Joseph Cotten. Color.

JACK, THE GIANT KILLER 94
UNITED ARTISTS. (R) July, 1962. Kerwin Mathews, Judi Meredith. Color.

JACK THE RIPPER 85
PARAMOUNT. (R) February, 1960. Lee Patterson, Eddie Byrne.

JACQUELINE 92
RANK. (R) September, 1957. John Gregson, Kathleen Ryan.

JACQUES BREL IS ALIVE AND WELL AND LIVING IN PARIS 98
AFT. (R) 1975. Elly Stone, Mort Shuman, Joe Masiell.

JAGUAR 80
REPUBLIC. (R) January 20, 1956. Sabu, Chiquita.

JAIL BUSTERS 61
ALLIED ARTISTS. (R) September 18, 1955. Leo Gorcey, Huntz Hall.

JAILBREAKERS, THE 64
AMERICAN INTERNATIONAL. (R) June, 1960. Robert Hutton, Mary Castle.

JAILHOUSE ROCK (Cs) 96
METRO-GOLDWYN-MAYER. (R) November, 1957. Elvis Presley, Judy Tyler.

JAMBOREE 86
WARNER BROS. (R) December, 1957. Fats Domino, Jerry Lee.

JAMES DEAN STORY, THE 82
WARNER BROS. (R) August, 1957. Documentary.

JANIS 96
UNIVERSAL. (R) February, 1975. Janis Joplin, Big Brother & The Holding Company.

JASON AND THE ARGONAUTS 104
COLUMBIA. (R) June, 1963. Todd Armstrong, Nancy Kovack. Color.

JAWS (P) 124
UNIVERSAL. (R) June, 1975. Roy Scheider, Robert Shaw, Richard Dreyfuss.

JAWS 2 (P) 117
UNIVERSAL. (R) June, 1978. Roy Scheider, Lorraine Gary.

JAYHAWKERS, THE (VV) 100
PARAMOUNT. (R) October, 1959. Jeff Chandler, Fess Parker. Color.

JAZZ BOAT (Cs) 90
COLUMBIA. (R) November, 1960. Anthony Newley, Anne Aubrey.

JEANNE EAGELS 109

Length In Mins.

COLUMBIA. (R) August, 1957. Kim Novak, Jeff Chandler.

JENNIFER ON MY MIND 90
UNITED ARTISTS. (R) November, 1971. Michael Brandon, Tippy Walker. Color.

JENNY 88
CINERAMA. (R) January, 1970. Marlo Thomas, Alan Alda. Color.

JEREMIAH JOHNSON (P) 107
WARNER BROS. (R) December, 1972. Robert Redford. Color.

JERK, THE 93
UNIVERSAL. (R) December, 1979. Steve Martin, Bernadette Peters.

JERUSALEM FILE, THE 94
METRO-GOLDWYN-MAYER. (R) March, 1972. Bruce Davison, Nicol Williamson. Color.

JESSE JAMES MEETS FRANKENSTEIN'S DAUGHTER 88
EMBASSY. (R) 1966. John Lupton, Estelita.

JESSICA 112
UNITED ARTISTS. (R) April, 1962. Angie Dickinson, Maurice Chevalier. Color.

JESSIE'S GIRLS 86
MANSON DISTRIBUTING. (R) April, 1975. Ben Frank, Sandra Currie.

JESUS 117
WARNER BROS. (R) October, 1979. Brian Deacon.

JESUS CHRIST SUPERSTAR (TAO) 106
UNIVERSAL. (R) July, 1973. Ted Neeley, Carl Anderson. Color.

JET ATTACK 68
AMERICAN INT. PICS. (R) February, 1958. John Agar, Audrey Totter.

JET PILOT 112
UNIVERSAL. (R) October, 1957. John Wayne, Janet Leigh. Color.

JIGSAW 97
BEVERLY PICTURES. (R) June, 1965. Jack Warner, Ronald Lewis.

JIM—THE WORLD'S GREATEST 90
UNIVERSAL. (R) February, 1976. Gregory Harrison, Robbie Wolcott.

JIMI HENDRIX 102
WARNER BROS. (R) October, 1973. Documentary.

JOANNA 107
TWENTIETH CENTURY-FOX. (R) November, 1968. Genevieve Waite, Calvin Lockhart, Christian Doermer. Color.

JOCK PETERSEN 97
AVCO-EMBASSY. (R) November, 1975. Jack Thompson, Jacki Weaver.

JOE 107
CANNON. (R) July, 1970. Peter Boyle, K. Callan, Dennis Patrick. Color.

JOE BUTTERFLY (Cs) 90
UNIVERSAL. (R) July, 1957. Audie Murphy, George Nader. Color.

JOE DAKOTA 79
UNIVERSAL. (R) September, 1957. Jock Mahoney, Luana Patten. Color.

JOE HILL 114
PARAMOUNT. (R) October, 1971. Thommy Berggren, Anja Schmidt. Color.

JOE KIDD 88
UNIVERSAL. (R) July, 1972. Clint Eastwood, Robert Duvall, John Saxon. Color.

JOE MACBETH 90
COLUMBIA. (R) February, 1956. Paul Douglas, Ruth Roman.

JOHN AND MARY (P) 92
20th CENTURY-FOX. (R) December, 1969. Dustin Hoffman, Mia Farrow, Michael Tolan. Color.

JOHN F. KENNEDY: YEARS OF LIGHTNING, DAY OF DRUMS 87
EMBASSY. (R) April, 1966. Documentary.

JOHN GOLDFARB, PLEASE COME HOME (Cs) 96
TWENTIETH CENTURY-FOX. (R) April, 1965. Shirley MacLaine, Peter Ustinov. Color.

JOHN PAUL JONES (T) 126
WARNER BROS. (R) July, 1959. Robert Stack, Macdonald Carey. Color.

JOHNNY CASH—THE MAN, HIS WORLD, HIS MUSIC .. 94
CONTINENTAL. (R) September, 1969. Documentary. Color.

JOHNNY CONCHO 84
UNITED ARTISTS. (R) July, 1956. Frank Sinatra, Phyllis Kirk.

JOHNNY COOL 101
UNITED ARTISTS. (R) October, 1963. Henry Silva, Elizabeth Montgomery.

JOHNNY GOT HIS GUN 112
MARKETING & DISTRIBUTION CO. (R) August, 1971. Timothy Bottoms, Donald Sutherland. Color.

Length In Mins.

JOHNNY NOBODY 88
MEDALLION. (R) November, 1965. Nigel Patrick, Yvonne Mitchell.

JOHNNY RENO (TE) 83
PARAMOUNT. (R) March, 1966. Dana Andrews, Jane Russell. Color.

JOHNNY ROCCO 84
ALLIED ARTISTS. (R) December, 1958. Stephen McNally, Richard Eyer.

JOHNNY TIGER 102
UNIVERSAL. (R) April, 1966. Robert Taylor, Geraldine Brookes. Color.

JOHNNY TREMAIN 80
BUENA VISTA. (R) July, 1957. Hal Stalmaster, Luana Patten. Color.

JOHNNY TROUBLE 80
WARNER BROS. (R) September, 1957. Ethel Barrymore, Cecil Kellaway.

JOKER, THE 86
UNITED ARTISTS. (R) October, 1961. Francois Maistre, Anne Tonietti.

JOKER IS WILD, THE (VV) 123
PARAMOUNT. (R) October, 1957. Frank Sinatra, Mitzi Gaynor.

JOKERS, THE (Brit.) 94
UNIVERSAL. (R) June, 1967. Michael Crawford, Oliver Reed. Color.

JONATHAN LIVINGSTON SEAGULL (P) 115
PARAMOUNT. (R) December, 1973. Documentary.

JORY 95
AVCO-EMBASSY. (R) November, 1972. John Marley, Robby Benson. Color.

JOURNEY, THE 125
METRO-GOLDWYN-MAYER. (R) February, 1959. Yul Brynner, Deborah Kerr. Color.

JOURNEY THROUGH ROSEBUD 93
CINERAMA. (R) March, 1972. Robert Forster, Kristoffer Tabori. Color.

JOURNEY TO FREEDOM 60
REPUBLIC. (R) June, 1957. Jacques Scott, Genevieve Aumont.

JOURNEY TO JERUSALEM, A 95
SIGMA III. Leonard Bernstein, Isaac Stern.

JOURNEY TO SHILOH 101
UNIVERSAL. (R) June, 1968. James Caan, Michael Sarrazin, Brenda Scott. Color.

JOURNEY TO THE CENTER OF THE EARTH (Cs) 132
TWENTIETH CENTURY-FOX. (R) December, 1959. Pat Boone, James Mason. Color.

JOURNEY TO THE FAR SIDE OF THE SUN 99
UNIVERSAL. (R) September, 1969. Roy Thinnes, Ian Hendry, Patrick Wymark. Color.

JOURNEY TO THE LOST CITY 95
AMERICAN INTERNATIONAL. (R) October, 1960. Debra Paget, Paul Christian. Color.

JOURNEY TO THE SEVENTH PLANET 80
AMER. INT'L. PICS. (R) 1962. John Agar, Greta Thyssen. Color.

JOY HOUSE (Franscope) 98
METRO-GOLDWYN-MAYER. (R) November, 1964. Alain Delon, Jane Fonda.

JOY IN THE MORNING (P) 133
METRO-GOLDWYN-MAYER. (R) June, 1965. Richard Chamberlain, Yvette Mimieux. Color.

JOY RIDE 85
ALLIED ARTISTS. (R) October, 1958. Rad Fulton, Ann Doran.

JOYRIDE 92
AMERICAN INTERNATIONAL. (R) June, 1977. Desi Arnaz, Robert Carradine.

JUBAL (Cs) 101
COLUMBIA. (R) May, 1956. Glenn Ford, Ernest Borgnine, Rod Steiger. Color.

JUDEX 91
CONTINENTAL. (R) April, 1966. Michel Vitold, Channing Pollock.

JUDGE AND THE SINNER, THE 94
CASINO. (R) June, 1964. Heinz Ruehmann, Karin Baal.

JUDGEMENT AT NUREMBERG 176
UNITED ARTISTS. (R) Special. October, 1961. Spencer Tracy, Burt Lancaster.

JUDITH (P) 109
PARAMOUNT. (R) February, 1966. Sophia Loren, Peter Finch. Color.

JUGGERNAUT 109
UNITED ARTISTS. (R) September, 1974. Richard Harris, Omar Sharif.

JUKE BOX RACKET 62
BRENNER. (R) May, 1960. Steve Karmen, Arlene Corwin.

JUKE BOX RHYTHM 81
COLUMBIA. (R) April, 1959. Jack Jones, Jo Morrow.

JULIA 117

Length In Mins.

20th CENTURY-FOX. (R) October, 1977. Jane Fonda, Vanessa Redgrave.

JULIE **99**
METRO-GOLDWYN-MAYER. (R) November, 1956. Doris Day, Louis Jourdan, Barry Sullivan, Frank Lovejoy.

JULIET OF THE SPIRITS **137**
RIZZOLI. (R) November, 1965. Guilietta Masina, Sandra Milo. Color.

JULIUS CAESAR **117**
AMER. INT. PICS. (R) Charlton Heston, Jason Robards, John Gielgud. Color.

JUMP **97**
CANNON. (R) June, 1971. Tom Ligon, Sally Kirkland. Color.

JUMP INTO HELL **93**
WARNER BROTHERS. (R) May 14, 1955. Jack Sernas, Kurt Kasznar.

JUNGLE BOOK, THE **78**
BUENA VISTA. (R) October, 1967. Cartoon. Color.

JUNGLE CAT **70**
BUENA VISTA. (R) October 1960. Wildlife Feature. Color.

JUNGLE HEAT **75**
UNITED ARTISTS. (R) August, 1957. Lex Barker, Mari Blanchard.

JUNGLE MOON MEN **70**
COLUMBIA. (R) April, 1955. Johnny Weissmuller, Jean Byron.

JUNIOR BONNER **100**
CINERAMA. (R) June, 1972. Steve McQueen, Ida Lupino, Robert Preston. Color.

JUPITER'S DARLING (Cs) **95**
METRO-GOLDWYN-MAYER. (R) February 18, 1955. Esther Williams, Howard Keel. Color.

JURY OF ONE **93**
AVCO-EMBASSY. (R) October, 1975. Sophia Loren, Jean Gabin.

JUST BEFORE NIGHTFALL (French; English sub-titles) **110**
LIBRA FILMS CORP. (R) September, 1975. Stephane Audran, Michel Bouquet.

JUST FOR FUN **72**
COLUMBIA. (R) June, 1963 Mark Wynter, Cherry Roland.

JUST TELL ME WHAT YOU WANT **112**
WARNER BROS. (R) February, 1980. Ali MacGraw, Alan King.

JUST YOU AND ME, KID **93**
COLUMBIA. (R) July, 1979. George Burns, Brooke Shields.

JUSTINE (P) **116**
20th CENTURY-FOX. (R) August, 1969. Anouk Aimee, Dirk Bogarde, Robert Foster, Ann Karina. Color.

K

KALEIDOSCOPE **103**
WARNER BROS. (R) October, 1968. Warren Beatty, Susannah York. Color.

KANCHENJUNGHA **102**
EDWARD HARRISON. (R) August, 1966. Chhabi Biswas, Karuna Banerji. Color.

KANSAS CITY BOMBER **99**
METRO-GOLDWYN-MAYER. (R) August, 1972. Raquel Welch, Kevin McCarthy. Color.

KAPO **116**
LIONEX. (R) June, 1964. Susan Strasberg, Laurent Terzieff.

KARATE **80**
ALLIED ARTISTS. (R) 1963. Joel Holt, Reiko Okada. b&w.

KATHY O' **99**
UNIVERSAL. (R) June, 1958. Dan Duryea, Jan Sterling.

KELLY AND ME (Cs) **86**
UNIVERSAL. (R) April, 1957. Van Johnson, Piper Laurie. Color.

KELLY'S HEROES **145**
MGM. (R) July, 1970. Clint Eastwood, Telly Savalas, Don Rickles, Donald Sutherland. Color.

KENNER **100**
MGM. (R) August, 1971. Jim Brown, Madlyn Rhue, Robert Coote.

KENTUCKIAN, THE (Cs) **104**
UNITED ARTISTS. (R) August, 1955. Burt Lancaster, Diana Lynn, Diane Foster. Color.

KENTUCKY FRIED MOVIE, THE **90**
UNITED FILM. (R) August, 1977. Bill Bixby, George Lazenby.

KENTUCKY RIFLE **82**
HOWCO. (R) July, 1956. Chill Wills, Lance Fuller. Color.

Length In Mins.

KES **109**
UNITED ARTISTS. (R) September, 1970. David Bradley, Colin Welland. Color.

KETTLES IN THE OZARKS, THE **81**
UNIVERSAL. (R) April, 1956. Marjorie Main, Arthur Hunnicutt.

KETTLES ON OLD MAC DONALD'S FARM **81**
UNIVERSAL. (R) June, 1957. Marjorie Main, Parker Fennelly.

KEY, THE **125**
COLUMBIA. (R) July. 1958. William Holden, Sophia Loren, Trevor Howard

KEY WITNESS (Cs) **81**
METRO-GOLDWYN-MAYER. (R) October, 1960. Jeffrey Hunter, Pat Crowley.

KHARTOUM (Ultra Panavision) **134**
UNITED ARTISTS. (R) Summer 1966. Charlton Heston, Lawrence Olivier. Color.

KID BLUE **100**
TWENTIETH CENTURY-FOX. (R) May, 1973. Dennis Hopper, Warren Oates. Color.

KID FOR TWO FARTHINGS, A (Brit.) **91**
LOPERT. (R) April, 1956. Celia Johnson, Diana Dors. Color.

KID GALAHAD **95**
UNITED ARTISTS. (R) August, 1962. Elvis Presley, Gig Young. Color.

KID RODELO **91**
PARAMOUNT. (R) January, 1966. Don Murray, Janet Leigh.

KIDNAPPED **97**
BUENA VISTA. (R) April, 1960. Peter Finch, James MacArthur. Color.

KIDNAPPED **100**
AMERICAN INTERNATIONAL. (R) December, 1971. Michael Caine, Trevor Howard, Lawrence Douglas. Color.

KILL A DRAGON **91**
UNITED ARTISTS. (R) November, 1967. Jack Palance, Fernando Lamas. Color.

KILL, KILL, KILL **90**
CINERAMA. (R) May, 1972. Stephen Boyd, Jean Seberg. Color.

KILL OR CURE **88**
METRO-GOLDWYN-MAYER. (R) November, 1962. Terry Thomas, Eric Sykes.

KILLER ELITE, THE **123**
UNITED ARTISTS. (R) December, 1975. James Caan, Robert Duvall.

KILLER FORCE **101**
AMERICAN INTERNATIONAL. (R) December, 1975. Telly Savalas, Peter Fonda.

KILLER IS LOOSE, THE **73**
UNITED ARTISTS. (R) February, 1956. Joseph Cotten, Rhonda Fleming, Wendell Corey.

KILLER SHREWS, THE **72**
McLENDON. (R) November, 1959. James Best, Ingrid Goude.

KILLERS, THE **95**
UNIVERSAL. (R) October, 1964. Lee Marvin, Angie Dickinson. Color.

KILLER'S KISS **67**
UNITED ARTISTS. (R) November, 1955. Frank Silvera, Jamie Smith.

KILLERS OF KILIMANJARO (Cs) **91**
COLUMBIA. (R) April, 1960. Robert Taylor, Anthony Newley. Color.

KILLERS THREE **88**
AMERICAN INTERNATIONAL. (R) November, 1968. Robert Walker, Diane Versi, Dick Clark. Color.

KILLING, THE **83**
UNITED ARTISTS. (R) July 1956. Sterling Hayden, Coleen Gray.

KILLING GAMES, THE **94**
REGIONAL. (R) September, 1968. Jean-Pierre Cassel, Claudine Auger, Michael Duchaussoy. Color.

KILLING OF A CHINESE BOOKIE, THE **135**
FACES DIST. CORP. (R) February, 1976. Ben Gazzara, Timothy Carey.

THE KILLING OF SISTER GEORGE **138**
CINERAMA. (R) January, 1969. Beryl Reid, Susannah York, Coral Browne. Color.

KIMBERLY JIM **82**
EMBASSY. (R) April, 1965. Jim Reeves, Madeline Usher. Color.

KIND OF LOVING, A **112**
GOVERNOR. (R) September, 1962. Alan Bates, June Ritchie.

KING: A FILMED RECORD . . . MONTGOMERY TO MEMPHIS **182**
MARON. (R) September, 1969. Documentary.

KING AND COUNTRY **86**

Length In Mins.

ALLIED ARTISTS. (R) September, 1965. Dirk Bogarde, Tom Courtney.

KING AND FOUR QUEENS, THE (Cs) **86**
UNITED ARTISTS. (R) December, 1956. Clark Gable, Eleanor Parker. Color.

KING AND I, THE (Cs) **133**
TWENTIETH CENTURY-FOX. (R) July, 1956. Deborah Kerr, Yul Brynner. Color.

KING CREOLE (VV) **116**
PARAMOUNT. (R) July, 1958. Elvis Presley, Carolyn Jones.

KING DINOSAUR **50**
LIPPERT. (R) June 17, 1955. Bill Bryant, Wanda Curtiss.

KING IN NEW YORK, A **105**
CLASSIC ENTERTAINMENT. (R) Special, 1973. Charles Chaplin, Dawn Addams.

KING KONG (P) **134**
PARAMOUNT. (R) December, 1976. Jeff Bridges, Charles Grodin, Jessica Lange.

KING KONG ESCAPES **96**
UNIVERSAL. (R) August, 1968. Rhodes Reason, Mie Harna, Linda Miller. Color.

KING KONG VS. GODZILLA **90**
UNIVERSAL. (R) July, 1963. Michael Keith, James Yagl.

KING OF HEARTS (TS) (Fr.) **102**
LOPERT. (R) June, 1967. Alan Bates, Pierre Brasseur. Color.

KING OF KINGS (T-70) **163**
METRO-GOLDWYN-MAYER. (R) September, 1961. Jeffrey Hunter, Siobhen McKenna. Color.

KING OF MARVIN GARDENS, THE **103**
COLUMBIA. (R) October, 1972. Jack Nicholson, Bruce Dern, Ellen Burstyn. Color.

KING OF THE CORAL SEA (Aust.) **74**
ALLIED ARTISTS. (R) June 24, 1956. Chips Rafferty, Charles Tingwell.

KING OF THE GRIZZLIES **93**
BUENA VISTA. (R) February, 1970. John Yesno, Chris Wiggins. Technicolor.

KING OF THE GYPSIES **110**
PARAMOUNT. (R) December, 1978. Sterling Hayden, Shelley Winters, Eric Roberts.

KING OF THE WILD STALLIONS (Cs) **76**
ALLIED ARTISTS. (R) May, 1959. George Montgomery, Diane Brewster.

KING RAT **133**
COLUMBIA. (R) November, 1965. George Segal, Tom Courtenay.

KINGS GO FORTH **109**
UNITED ARTISTS. (R) July, 1958. Frank Sinatra, Tony Curtis.

KINGS OF THE SUN (P) **108**
UNITED ARTISTS. (R) December, 1963. Yul Brynner, George Chakaris. Color.

KING'S PIRATE, THE **100**
UNIVERSAL. (R) August, 1967. Doug McClure, Jill St. John. Color.

KING'S STORY A **100**
CONTINENTAL. (R) May, 1967. Documentary. Color.

KING'S THIEF, THE (Cs) **79**
METRO-GOLDWYN-MAYER. (R) August 5, 1955. Edmund Purdom, Ann Blyth, David Niven, George Sanders. Color.

KISMET (Cs) **113**
METRO-GOLDWYN-MAYER. (R) December 23, 1955. Howard Keel, Ann Blyth. Color.

KISS AND KILL **101**
COMMONWEALTH UNITED. (R) December, 1969. Christopher Lee, Richard Greene, Shirley Eaton.

KISS BEFORE DYING, A (Cs) **94**
UNITED ARTISTS. (R) June, 1956. Robert Wagner, Jeffrey Hunter, Virginia Leith. Color.

KISS ME DEADLY **103**
UNITED ARTISTS. (R) May, 1955. Ralph Meeker, Cloris Leachman.

KISS ME STUPID (P) **126**
LOPERT. (R) December, 1964. Dean Martin, Kim Novak.

KISS OF FIRE **87**
UNIVERSAL. (R) October, 1955. Jack Palance, Barbara Rush. Color.

KISS OF THE VAMPIRE **88**
UNIVERSAL. (R) September, 1963. Clifford Evans, Noel Willman. Color.

KISS THE GIRLS AND MAKE THEM DIE **101**
COLUMBIA. (R) January, 1967. Michael Connors, Dorothy Provine. Color.

KISS THE OTHER SHEIK (Italian, English titles) **86**
METRO-GOLDWYN-MAYER. (R) August, 1968. Marcello Mastroianni, Pamela Tiffin, Virna Lisi. Color.

KISS THEM FOR ME (Cs) **105**
TWENTIETH CENTURY-FOX. (R) December, 1957. Cary Grant, Jayne Mansfield. Color.

KISSES FOR MY PRESIDENT **113**
WARNER BROS. (R) August, 1964. Fred MacMurray, Polly Bergen.

KISSIN' COUSINS (P) **96**
METRO-GOLDWYN-MAYER. (R) March, 1964. Elvis Presley, Arthur O'Connell. Color.

KITTEN WITH A WHIP **83**
UNIVERSAL. (R) December, 1964. Ann-Margret, John Forsythe.

KLANSMAN, THE **112**
PARAMOUNT. (R) November, 1974. Lee Marvin, Richard Burton.

KLUTE **114**
WARNER BROS. (R) July, 1971. Jane Fonda, Donald Sutherland. Color.

KNACK . . . AND HOW TO GET IT, THE **84**
UNITED ARTISTS. (R) July, 1965. Rita Tushingham, Ray Brooks.

KONA COAST **93**
WARNER BROS.-7 ARTS. (R) May, 1968. Richard Boone, Vera Miles. Color.

KONGA **90**
AMERICAN INTERNATIONAL. (R) March, 1961. Michael Gough, Margo Johns. Color.

KONOS **78**
TWENTIETH CENTURY-FOX. (R) April, 1957. Barbara Lawrence, John Emery.

KOTCH **114**
CINERAMA. (R) October, 1971. Walter Matthau, Deborah Winters. Color.

KRAKATOA, EAST OF JAVA **101**
CINERAMA. (R) May, 1969. Maximillian Schell, Diane Baker, Brian Keith. Color.

KRAMER VS. KRAMER **105**
COLUMBIA. (R) December, 1979. Dustin Hoffman, Meryl Streep.

KREMLIN LETTER, THE (P) **113**
20th CENTURY-FOX. February, 1970. Bibi Anderson, Richard Boone, Nigel Green. Color.

KWAIDAN (Japanese, Eng. titles) **125**
CONTINENTAL. (R) November, 1965. Oriental Tales. Color.

L

L'IMMORTELLE (Fr.; Eng. titles) **95**
GROVE. (R) November, 1969. Francoise Brion, Jacquese Daniel-Viacroze. b&w.

L-SHAPED ROOM, THE **125**
DAVIS-ROYAL. (R) June, 1963. Leslie Caron, Tom Bell, Brock Peters.

LA BOHEME **107**
WARNER BROTHERS. (R) October-November, 1965. La Scala Opera Co. Color.

LA BONNE SOUPE (French; Eng. titles) (Cs) **197**
TWENTIETH CENTURY-FOX. (R) March, 1964. b&w.

LA CAGE AUX FOLLES (French; Eng. titles) **99**
UNITED ARTISTS. (R) May, 1979. Ugo Tognazzi, Michel Serrault.

LA CHAMADE (Fr.; Eng. titles) **103**
LOPERT. (R) July, 1969. Catherine Deneuve, Michel Piccoli, Roger Van Hool. Color.

LA DOLCE VITA (Totalscope) **175**
AMERICAN INTERNATIONAL. (R) October, 1966. Re-released English-dubbed version. Marcello Mastroianni, Anita Ekberg.

LA FEMME INFIDELE (Fr.; Eng. titles) **100**
ALLIED ARTISTS. (R) November, 1969. Stephane Audran, Michael Bouquet, Maurice Ronet. Color.

LA FUGA (Ital., Eng. titles) **92**
INTERNATIONAL CLASSICS. (R) March, 1966. Giovanna Ralli, Anouk Aimee.

LA GUERRE EST FINIE (Fr.; Eng. titles) **121**
BRANDON. (R) January, 1967. Yves Montand, Ingrid Thulin.

LA PRISONNIERE (French; Eng. titles) **105**
AVCO-EMBASSY. (R) March, 1969. Laurent Terzieff, Bernard Fresson, Elizabeth Wiener. Color.

LA SORCIERE (Fr.) **97**
ELLIS. (R) December, 1956. Marina Vlady.

LA STRADA (Ital.) **115**
TRANS-LUX (R) July, 1956. Anthony Quinn, Richard Basehart, Giuletta Masins.

LA TRAVIATA **110**
ROYAL. (R) Feb., 1968. Anna Moffo, Franco Bonisolli. Color.

LA VIE DE CHATEAU (Fr.; Eng. titles) **92**
ROYAL FILMS. (R) March, 1967. Catherine Deneuve, Philippe Noiret.

LA VISTA **115**

Length In Mins.

PROMENADE. (R) August, 1966. Sandra Milo, Francois Perier.

LACOMBE, LUCIEN (French; Eng. titles) **141**
20TH CENTURY-FOX. (R) October, 1974. Pierre Blaise, Aurore Clement.

LAD: A DOG **98**
WARNER BROS. (R) June, 1962. Peggy McKay, Peter Breck.

LADIES MAN, THE **106**
PARAMOUNT. (R) June, 1961. Jerry Lewis, Helen Traubel. Color.

LADIES WHO DO **85**
CONTINENTAL. (R) November, 1963. Robert Morley, Peggy Mount.

LADY AND THE TRAMP (Cs) **75**
BUENA VISTA. (R) July, 1955. Cartoon. Color.

LADY CAROLYN LAMB **127**
UNITED ARTISTS. (R) February. 1972. Sarah Miles, Jon Finch, Richard Chamberlain. Color.

LADY GODIVA **80**
UNIVERSAL (R) November, 1955. Maureen O'Hara, George Nader. Color.

LADY ICE **93**
NATIONAL GENERAL. (R) July. 1973. Donald Sutherland, Jennifer O'Neill. Color.

LADY IN A CAGE **93**
PARAMOUNT. (R) June, 1964. Olivia de Havilland, Ann Southern.

LADY IN CEMENT **94**
TWENTIETH CENTURY-FOX. (R) November, 1968. Frank Sinatra, Raquel Welch, Richard Conte. Color.

LADY IN THE CAR WITH GLASSES AND A GUN, THE (P) **105**
COLUMBIA. (R) January, 1971. Samantha Eggar, Oliver Reed. Color.

LADY KILLERS, THE (Brit.) **90**
CONTINENTAL. (R) 1956. Alec Guinness, Peter Sellers. Color.

LADY L. (P) **107**
MGM. (R) July, 1966. Sophia Loren, Paul Newman. Color.

LADY LIBERTY **95**
UNITED ARTISTS. (R) May, 1972. Sophia Loren.

LADY OF MONZA, THE **98**
TOWER. (R) May, 1970. Anne Heywood, Antonio Sabato, Hardy Kruger. Color.

LADY OF VENGEANCE **83**
UNITED ARTISTS. (R) August, 1957. Dennis O'Keefe, Ann Sears.

LADY SINGS THE BLUES (P) **144**
PARAMOUNT. (R) November, 1972. Diana Ross, Billy Dee Williams. Color.

LADY TAKES A FLYER, THE (Cs) **94**
UNIVERSAL. (R) February, 1958. Lana Turner, Jeff Chandler.

LADYBUG, LADYBUG **81**
UNITED ARTISTS. (R) December, 1963. Jane Connell, William Daniels.

LADYKILLERS, THE (Brit.) **93**
CONTINENTAL. (R) March, 1956. Alec Guinness, Cecil Parker. Color.

LAFAYETTE (T) **110**
MACO FILMS. (R) April, 1964. Jack Hawkins, Orson Wells. Color.

LAFAYETTE ESCADRILLE **93**
WARNER BROS. (R) March, 1958. Tab Hunter, Etchika Choureau.

LANCELOT AND GUINEVERE (P) **116**
UNIVERSAL. (R) June, 1963. Cornel Wilde, Jean Wallace. Color.

LAND OF FURY (BRIT.) **82**
UNIVERSAL. (R) March, 1955. Jack Hawkins, Glynis Johns. Color.

LAND OF THE PHARAOHS (Cs) **105**
WARNER BROTHERS. (R) July 2, 1955. Jack Hawkins, Joan Collins. Color.

LAND RAIDERS **100**
COLUMBIA. (R) September, 1969. Telly Savalas, George Maharis. Color.

LAND THAT TIME FORGOT, THE **91**
AMERICAN INTERNATIONAL. (R) April, 1975. Doug McClure, John McEnery.

LAND UNKNOWN, THE (Cs) **78**
UNIVERSAL. (R) August, 1957. Jock Mahoney, Shawn Smith.

LANDLORD, THE **113**
UNITED ARTISTS. (R) May, 1970. Beau Bridges, Lee Grant, Diana Sands, Pearl Bailey. Color.

LAS VEGAS SHAKEDOWN **79**
ALLIED ARTISTS. (R) May 15, 1955. Dennis O'Keefe, Thomas Gomez.

LAST AMERICAN HERO, THE **100**
TWENTIETH CENTURY-FOX. (R) July, 1973. Jeff Bridges, Valerie Perrine. Color.

LAST ANGRY MAN, THE **100**
COLUMBIA. (R) November, 1959. Paul Muni, David Wayne.

LAST BLITZKRIEG, THE **84**
COLUMBIA. (R) January, 1959. Van Johnson, Kerwin Mathews.

LAST BRIDGE, THE **90**
UNION. (R) August, 1957. Maria Schell, Bernard Wicki.

LAST CHALLENGE, THE (P) **105**
MGM. (R) October, 1967. Glenn Ford, Angie Dickinson. Color.

LAST CHAPTER, THE **90**
BEN-LAR. (R) February, 1966. Documentary.

LAST COMMAND, THE **110**
REPUBLIC (R) August 3, 1955. Sterling Hayden, Anna Marie Alberghetti. Color.

LAST DAYS OF POMPEII, THE (Cs) **103**
UNITED ARTISTS. (R) July, 1960. Steve Reeves, Cristina Kauffman. Color

LAST DETAIL, THE **105**
COLUMBIA. (R) February, 1974. Jack Nicholson, Otis Young, Randy Quaid

LAST EMBRACE (P) **102**
UNITED ARTISTS. (R) May, 1979. Roy Scheider, Janet Margolin.

LAST ESCAPE, THE **90**
UNITED ARTISTS. (R) July, 1970. Stuart Whitman, John Collin, Pinkas Braun. Color.

LAST FLIGHT OF NOAH'S ARK **97**
BUENA VISTA. (R) July, 1980. Elliott Gould, Ricky Schroder.

LAST FRONTIER, THE (Cs) **98**
COLUMBIA. (R) January, 1956. Victor Mature, Guy Madison, Robert Preston. Color.

LAST GRENADE, THE (P) **94**
CINERAMA. (R) March, 1970. Stanley Baker, Alex Cord. Color.

LAST HARD MEN, THE **96**
TWENTIETH CENTURY-FOX. (R) April, 1976. Charlton Heston, James Coburn.

LAST HUNT, THE (Cs) **108**
METRO-GOLDWYN-MAYER. (R) February 24, 1956. Robert Taylor, Stewart Granger. Color.

LAST HURRAH, THE **121**
COLUMBIA. (R) November, 1958. Spencer Tracy, Jeffrey Hunter.

LAST MAN TO HANG, THE (Brit.) **75**
COLUMBIA. (R) December, 1956. Tom Conway, Eunice Gayson.

LAST MARRIED COUPLE IN AMERICA, THE **100**
UNIVERSAL. (R) February, 1980. George Segal, Natalie Wood.

LAST MILE, THE **81**
UNITED ARTISTS. (R) January, 1959. Mickey Rooney, Clifford David.

LAST MOVIE, THE **108**
UNIVERSAL. (R) October, 1971. Dennis Hopper, Peter Fonda. Color.

LAST OF SHEILA, THE **120**
WARNER BROS. (R) June, 1973. Richard Benjamin, James Coburn, Dyan Cannon. Color.

LAST OF THE BADMEN (Cs) **80**
ALLIED ARTISTS. (R) February 17, 1957. George Montgomery, Keith Larsen. Color.

LAST OF THE DESPERADOS **71**
ASSOCIATED FILM. (R) December 1, 1955. James Craig, Jim Davis.

LAST OF THE FAST GUNS (Cs) **82**
UNIVERSAL. (R) July, 1958. Jock Mahoney, Gilbert Roland. Color.

LAST OF THE MOBILE HOT-SHOTS **108**
WARNER BROS. (R) January, 1970. James Coburn, Lynn Redgrave, Robert Hooks. Color.

LAST OF THE PONY RIDERS **80**
COLUMBIA. (R) November, 1963. Gene Autry, Smiley Burnette.

LAST OF THE RED HOT LOVERS **98**
PARAMOUNT. (R) June, 1972. Alan Arkin, Paula Prentiss, Sally Kellerman. Color.

LAST OF THE SECRET AGENTS?, THE **90**
PARAMOUNT. (R) May, 1966. Marty Allen, Steve Rossi. Color.

LAST OF THE SKI BUMS, THE **86**
U-M. (R) November, 1969. Ron Funk, Mike Zuettel, Ed Ricks. Color.

LAST OF THE VIKINGS (Dyaliscope) (Ital.) **102**
MEDALLION. (R) 1960. Cameron Mitchell, Edmund Purdom.

LAST PICTURE SHOW, THE **118**
COLUMBIA. (R) October, 1971. Timothy Bottoms, Jeff Bridges, Cybil Shepherd, Ben Johnson.

Length In Mins.

LAST REMAKE OF BEAU GESTE, THE **83**
UNIVERSAL. (R) July, 1977. Ann-Margret, Marty Feldman, Michael York.

LAST RUN, THE (P) **92**
MGM. (R) July, 1971. George C. Scott, Tony Musante. Color.

LAST SAFARI, THE **115**
PARAMOUNT. (R) November, 1967. Kaz Garas, Stewart Granger. Color.

LAST SHOT YOU HEAR, THE **87**
LIPPERT. (R) 1969. Hugh Marlowe, Zena Walker.

LAST STAGECOACH WEST **67**
REPUBLIC. (R) July, 1957. Jim Davis, Mary Castle.

LAST SUMMER **97**
ALLIED ARTISTS. (R) 1969. Barbara Hershey, Richard Thomas, Bruce Davison, Cathy Burns. Color.

LAST SUNSET, THE **115**
UNIVERSAL. (R) July, 1961. Rock Hudson, Kirk Douglas. Color.

LAST TANGO IN PARIS **132**
UNITED ARTISTS. (R) February, 1972. Marlon Brando, Maria Schneider. Color.

LAST TEN DAYS, THE **113**
COLUMBIA. (R) April, 1956. Documentary.

LAST TIME I SAW ARCHIE, THE **98**
UNITED ARTISTS. (R) June, 1961. Robert Mitchum, Jack Webb.

LAST TRAIN FROM GUN HILL (VV) **94**
PARAMOUNT. July, 1959. Kirk Douglas, Anthony Quinn. Color.

LAST TYCOON, THE **122**
PARAMOUNT. (R) February, 1977. Robert DeNiro, Tony Curtis.

LAST VALLEY, THE (Todd-AO) **126**
CINERAMA. (R) February, 1971. Michael Caine, Omar Sharif. Color.

LAST VOYAGE, THE **91**
METRO-GOLDWYN-MAYER. (R) February, 1960. Robert Stack, Dorothy Malone. Color.

LAST WAGON, THE (Cs) **99**
TWENTIETH CENTURY-FOX. (R) September, 1956. Richard Widmark, Felicia Farr. Color.

LAST WALTZ, THE **115**
UNITED ARTISTS. (R) April, 1978. Bob Dylan, Joni Mitchell.

LAST WAVE, THE **106**
WORLD NORTHAL. (R) Feburary, 1979. Richard Chamberlain.

LAST WOMAN OF SHANG **107**
SHAW. (R) December, 1964. Lin Dai, Nam Koong-Woon. Color.

LAST WORLD, THE (Cs) **90**
TWENTIETH CENTURY-FOX. (R) July, 1960. Michael Rennie, Jill St. John. Color.

LATE SHOW, THE **94**
WARNER BROS. (R) February, 1977. Art Carney, Lily Tomlin.

LATITUDE ZERO **99**
NATIONAL GENERAL. (R) May, 1970. Joseph Cotten, Cesar Romero. Color.

LAUGHING POLICEMAN, THE **112**
TWENTIETH CENTURY-FOX. (R) December, 1973. Walter Matthau, Bruce Dern.

LAUGHTER IN THE DARK **101**
LOPERT. (R) May, 1969. Nicol Williamson, Anna Karina, Jean-Claude Durot. Color.

LAUREL AND HARDY'S LAUGHING 20'S **91**
METRO-GOLDWYN-MAYER. (R) September, 1965. Stan Laurel, Oliver Hardy.

LAW AND DISORDER (P) **103**
COLUMBIA. (R) October, 1974. Carroll O'Connor, Ernest Borgnine.

LAW AND JAKE WADE, THE (Cs) **86**
MGM. (R) June, 1958. Robert Taylor, Richard Widmark. Color.

LAW OF THE LAWLESS (T) **87**
PARAMOUNT. (R) March, 1964. Dale Robertson, Yvonne de Carlo. Color.

LAWLESS EIGHTIES **70**
REPUBLIC. (R) May, 1958. Buster Crabbe, John Smith.

LAWLESS STREET, A **78**
COLUMBIA. (R) December, 1955. Randolph Scott, Angela Lansbury. Color.

LAWMAN **99**
UNITED ARTISTS. (R) August, 1971. Burt Lancaster, Robert Ryan, Lee J. Cobb. Color.

LAWRENCE OF ARABIA (P) **221**
COLUMBIA. (R) Special. December, 1962. Peter O'Toole, Sir Alex Guinness. Color.

LAWYER, THE **120**
PARAMOUNT. (R) January, 1970. Barry Newman, Harry Gould, Diana Muldaur. Color.

LAY THAT RIFLE DOWN **71**
REPUBLIC. (R) July 7, 1955. Judy Canova, Robert Lowery.

LE MANS (P) **106**
NATIONAL GENERAL. (R) July, 1970. Steve McQueen. Color.

LE VIOL **81**
FREENA. (R) September, 1968. Bibi Anderson, Bruno Cremer, Frederic de Pasquale. Color.

LEADBELLY **126**
PARAMOUNT. (R) March, 1976. Roger E. Mosley, Paul Benjamin.

LEARNING TREE, THE (P) **107**
WARNER BROS. (R) September, 1969. Kyle Johnson, Alex Clarke, Estelle Evans. Color.

LEASE OF LIFE (Brit.) **93**
I.F.E. (R) March, 1956. Robert Donat, Kay Walsh. Color.

LEATHER BOYS, THE (Br. T.) **105**
RLP PICTURES. (R) November, 1965. Rita Tushingham, Colin Campbell.

LEATHER SAINT, THE (VV) **86**
PARAMOUNT. (R) June, 1956. John Derek, Paul Douglas, Jody Lawrence, Cesar Romero.

LEECH WOMAN, THE **77**
UNIVERSAL. (R) June, 1960. Coleen Gray, Grant Williams.

LEFT HAND OF GOD, THE (Cs) **87**
TWENTIETH CENTURY-FOX. (R) September, 1955. Humphrey Bogart, Gene Tierney. Color.

LEFT HANDED GUN, THE **102**
WARNER BROTHERS. (R) May, 1958. Paul Newman, Lita Milan.

LEGACY, THE **102**
UNIVERSAL. (R) September, 1979. Katharine Ross, Sam Elliott.

LEGEND OF HELL HOUSE, THE **90**
TWENTIETH CENTURY-FOX. (R) June, 1973. Roddy McDowall, Pamela Franklin. Color.

LEGEND OF LOBO, THE **67**
BUENA VISTA. (R) November, 1962. Documentary. Color.

LEGEND OF LYLAH CLARE, THE **130**
METRO-GOLDWYN-MAYER. (R) November, 1968. Kim Novak, Peter Finch, Ernest Borgnine. Color.

LEGEND OF NIGGER CHARLEY, THE **99**
PARAMOUNT. (R) May, 1972. Fred Williamson, Durville Martin. Color.

LEGEND OF THE LOST (T) **109**
UNITED ARTISTS. (R) December, 1957. John Wayne, Sophia Loren.

LEGEND OF TOM DOOLEY **79**
COLUMBIA. (R) June, 1959. Michael Landon, Jack Hogan.

LEGION OF THE DOOMED **75**
ALLIED ARTISTS. (R) September 21, 1958. Bill Williams, Dawn Richard.

LEGIONS OF THE NILE (Cs) **91**
TWENTIETH CENTURY-FOX. (R) December, 1960. Linda Cristal, Georges Marchal. Color.

LEMONADE JOE (Czech.) **84**
ALLIED ARTISTS. (R) November, 1967. Karel Fiala, Olga Schoberova.

LENNY **112**
UNITED ARTISTS. (R) November, 1974. Dustin Hoffman, Valerie Perrine. b&w.

LEO THE LAST **103**
UNITED ARTISTS. (R) May, 1970. Marcello Mastroianni, Billie Whitelaw, Calvin Lockhart. Color.

LEOPARD, THE (Cs) **165**
TWENTIETH CENTURY-FOX. (R) October, 1963. Burt Lancaster, Alain Delon. Color.

LEPKE (P) **110**
WARNER BROS. (R) June, 1975. Tony Curtis, Anjanette Comer.

LES BICHES (Fr.) **99**
VIP. (R) 1968. Stephane Audran, Jacqueline Sassard, Jean Louis Trintignant. Color.

LES CREATURES **93**
NEW YORKER FILMS. (R) April, 1969. Catherine Deneuve, Michael Piccoli, Eva Dahlbeck.

LES GAULOISES BLEUES (French, Eng. titles) **93**
LOPERT. (R) May, 1969. Annie Giardot, Jean Pierre Kalfon, Nella Bielski. Color.

LES GIRLS (Cs) **114**
METRO-GOLDWYN-MAYER. (R) November, 1957. Gene Kelly, Mitzi Gaynor. Color.

LES LIASONS DANGEREUSES **105**
ASTOR. (R) December, 1961. Gerard Philipe, Jeanne Moreau.

LET IT BE **80**
UNITED ARTISTS. (R) May, 1970. The Beatles. Color.

LET JOY REIGN SUPREME (Fr., Eng. titles) **120**

	Length In Mins.
SPECIALTY FILMS. (R) September, 1977. Philippe Noiret, Jean Rochefort.	
LET NO MAN WRITE MY EPITAPH	106
COLUMBIA. (R) October, 1960. Burl Ives, Shelley Winters.	
LET THE GOOD TIMES ROLL	99
COLUMBIA. (R) June, 1973. Chuck Berry, Little Richard. Color.	
LET'S BE HAPPY (Cs)	103
ALLIED ARTISTS. (R) June, 1957. Tony Martin, Vera-Ellen. Color.	
LET'S DO IT AGAIN	112
WARNER BROS. (R) October, 1975. Sidney Poitier, Bill Cosby.	
LET'S KILL UNCLE	92
UNIVERSAL. (R) October, 1966. Nigel Green, Mary Badham. Color.	
LET'S MAKE LOVE (Cs)	118
TWENTIETH CENTURY-FOX. (R) September, 1960. Marilyn Monroe, Yves Montand. Color.	
LET'S MAKE UP (Brit.)	94
UNITED ARTISTS. (R) April, 1955. Errol Flynn, Anna Neagle. Color.	
LET'S ROCK	79
COLUMBIA. (R) June, 1958. Julius La Rosa, Phyllis Newman.	
LET'S SCARE JESSICA TO DEATH	89
PARAMOUNT. (R) September, 1971. Zohra Lampert, Barton Heyman. Color.	
LET'S TALK ABOUT WOMEN	108
EMBASSY. (R) September, 1964. Vittorio Gassman, Maria Fiore.	
LETTERS FROM MY WINDMILL (Fr.)	116
TOHAN. (R) December, 1955. Henri Velbert, Rene Sarvil.	
LIBEL	100
METRO-GOLDWYN-MAYER. (R) October, 1959. Olivia de Havilland, Dirk Bogarde.	
LIBERATION OF L.B. JONES, THE	101
COLUMBIA. (R) April, 1970. Lee J. Cobb, Roscoe Lee Brown, Lola Falana, Anthony Zerbe. Color.	
LIBERTINE, THE	90
AUDUBON. (R) June, 1969. Catherine Spaak, Jean Louis Trintignant. Color.	
LICKERISH QUARTET, THE	90
AUDUBON. (R) October, 1970. Silvana Venturelli, Frank Wolff. Color.	
LIES MY FATHER TOLD ME	102
COLUMBIA. (R) October, 1975. Yossi Yadin, Len Birman.	
LIEUTENANT ROBIN CRUSOE, U.S.N.	110
BUENA VISTA. (R) July, 1966. Dick Van Dyke, Nancy Kwan. Color.	
LIEUTENANT WORE SKIRTS, THE (Cs)	99
TWENTIETH CENTURY-FOX. (R) January, 1956. Tom Ewell, Sheree North. Color.	
LIFE AND TIMES OF JUDGE ROY BEAN, THE (P)	120
NATIONAL GENERAL. (R) December, 1972. Paul Newman, Ava Gardner. Color.	
LIFE AT STAKE, A	78
FILMAKERS. (R) December, 1955. Angela Lansbury, Keith Andes.	
LIFE AT THE TOP	117
ROYAL INTERNATIONAL. (R) February, 1966. Laurence Harvey, Jean Simmons	
LIFE BEGINS AT 17	75
COLUMBIA. (R) July, 1958. Mark Damon, Dorothy Johnson.	
LIFE IN DANGER	63
ALLIED ARTISTS. (R) July, 1964. Darren Nesbitt, Julie Hopkins.	
LIFE IN THE BALANCE, A	74
TWENTIETH CENTURY-FOX. (R) July, 1955. Ricardo Montalban, Anne Bancroft.	
LIFE OF BRIAN	91
WARNER BROS. (R) September, 1979. Graham Chapman, Terry Gilliam.	
LIFE OF OHARU	146
TOHO. (R) April, 1964. Kinuyo Tanaka, Toshiro MiFune.	
LIFEGUARD	96
PARAMOUNT. (R) June, 1976. Sam Elliott, Anne Archer.	
LIGHT AT THE EDGE OF THE WORLD, THE (P)	120
NATIONAL GENERAL. (R) July, 1971. Kirk Douglas, Yul Brynner. Color.	
LIGHT IN THE FOREST, THE	93
BUENA VISTA. (R) July, 1958. Carol Lynley, James MacArthur. Color.	
LIGHT IN THE PIAZZA (Cs)	101
METRO-GOLDWYN-MAYER. (R) January, 1962. Olivia de Havilland, Rossano Brazzi. Color.	
LIGHT TOUCH, THE	85

	Length In Mins.
UNIVERSAL. (R) November, 1958. Jack Hawkins, Margaret Johnston. Color.	
LI'L ABNER (VV)	144
PARAMOUNT. (R) December, 1959. Peter Palmer, Leslie Parrish. Color.	
LILI	81
MGM. (R) July 10, 1953. Leslie Caron, Mel Ferrer, Jean Pierre Aumont. Color.	
LILIES OF THE FIELD	94
UNITED ARTISTS. (R) September, 1963. Sidney Poitier, Lilia Skala.	
LILITH	114
COLUMBIA. (R) October, 1964. Warren Beatty, Jean Seeberg.	
LIMBO	112
UNIVERSAL. (R) November, 1972. Kate Jackson, Katherine Justice, Kathleen Nolan. Color.	
LINEUP, THE,	86
COLUMBIA. (R) June, 1958. Eli Wallach, Robert Keith.	
LION, THE (Cs)	96
TWENTIETH CENTURY-FOX. (R) November, 1962. William Holden, Capucine. Color	
LION IN WINTER, THE	134
AVCO-EMBASSY. (R) October, 1968. Peter O'Toole, Katherine Hepburn, Jane Merrow. Color.	
LION'S LOVE	110
MAX L. RABB. (R) 1969. Viva, James Rado, Gerome Ragni, Shirley Clarke. Color.	
LIPSTICK	90
PARAMOUNT. (R) April, 1976. Margaux Hemingway, Chris Sarandon.	
LIQUIDATOR, THE (P)	105
METRO-GOLDWYN-MAYER. R) November, 1966. Rod Taylor, Trevor Howard. Color.	
LISA (Cs)	112
TWENTIETH CENTURY-FOX. (R) June, 1962. Stephen Boyd, Dolores Hart.	
LISBON (N)	90
REPUBLIC. (R) August 17, 1956. Ray Milland, Maureen O'Hara. Color.	
LIST OF ADRIAN MESSENGER, THE	98
UNIVERSAL. (R) June, 1963. George C. Scott, Dana Wynter.	
LISTEN, LET'S MAKE LOVE	91
LOPERT. June, 1969. Pierre Clementi, Beba Lancar. Color.	
LISZTOMANIA	106
WARNER BROS. (R) October, 1975. Roger Daltrey, Sara Kestelman.	
LITTLE ARK	100
NATIONAL GENERAL. (R) March, 1972. Theodore Bikel, Philip Frame. Color.	
LITTLE BIG MAN (P)	150
NATIONAL GENERAL. (R) December, 1970. Dustin Hoffman, Faye Dunaway, Chief Dan George. Color.	
LITTLE CIGARS	92
AMERICAN INTERNATIONAL. (R) May, 1973. Billy Curtis, Angel Tompkins.	
LITTLE DARLINGS (P)	92
PARAMOUNT. (R) March, 1980. Tatum O'Neal, Kristy McNichol.	
LITTLE FAUSS AND BIG HALSEY (P)	97
PARAMOUNT. (R) October, 1970. Robert Redford, Michael J. Pollard. Color.	
LITTLE GIRL WHO LIVES DOWN THE LANE, THE	94
AMERICAN INTERNATIONAL. (R) May, 1977. Jodie Foster, Martin Sheen.	
LITTLE HUT, THE	90
METRO-GOLDWYN-MAYER. (R) May 24, 1957. Ava Gardner, Stewart Granger, David Niven. Color.	
LITTLE LAURA AND BIG JOHN	84
CROWN INTERNATIONAL. R) April, 1973. Fabian Forte, Karen Black. Color.	
LITTLE MISS MARKER	100
UNIVERSAL. (R) March, 1980. Walter Matthau, Julie Andrews.	
LITTLE MOTHER	90
AUDUBON. (R) October, 1972. Christiane Kruger, Ivan Desny. Color.	
LITTLE MURDERS	110
TWENTIETH CENTURY-FOX. (R) February, 1971. Elliott Gould, Lou Jacobi, Alan Arkin. Color.	
LITTLE NUNS, THE (Ital. Eng. dubbed)	101
EMBASSY. (R) September, 1965. Catherine Spaak, Sylva Koscina.	
LITTLE PRINCE, THE	88
PARAMOUNT. (R) December, 1974. Richard Kiley, Steven Warner.	
LITTLE ROMANCE, A	108
ORION PICTURES. (R) May, 1979. Laurence Olivier, Arthur Hill.	
LITTLE SAVAGE, THE (R)	73

Length In Mins.

TWENTIETH CENTURY-FOX. (R) March, 1959. Pedro Armendariz, Rudolfo Hoyos.

LITTLE SHEPHERD OF KINGDOM COME, THE (Cs) 108
TWENTIETH CENTURY-FOX. (R) January, 1961. Jimmie Rodgers, Luana Patten. Color.

LITTLEST HOBO, THE 77
ALLIED ARTISTS. (R) July, 1958. Ruddy Hart, Wendy Stuart.

LITTLEST HORSE THIEVES, THE 105
BUENA VISTA. (R) April, 1977. Alastair Sim, Peter Barkworth.

LITTLEST OUTLAW, THE 75
BUENA VISTA. (R) February, 1956. Joseph Calleia, Pedro Armendariz. Color.

LIVE A LITTLE, LOVE A LITTLE 90
METRO-GOLDWYN-MAYER. (R) November, 1968. Elvis Presley, Michele Carey, Don Porter. Color.

LIVE AND LET DIE 121
UNITED ARTISTS. (R) June, 1973. Roger Moore, Jane Seymour.

LIVE FAST, DIE YOUNG 82
UNIVERSAL. (R) May, 1958. Mary Murphy, Norma Eberhardt.

LIVE FOR LIFE (Fr.) 130
UNITED ARTISTS. (R) December, 1967. Yves Montand, Candice Bergen. Color.

LIVELY SET, THE 95
UNIVERSAL. (R) November, 1964. James Darren, Pamela Tiffin. Color.

LIVING FREE 91
COLUMBIA. (R) April, 1972. Nigel Davenport, Susan Hampshire. Color.

LIVING IDOL (Cs) 101
METRO-GOLDWYN-MAYER. (R) July, 1957. Steve Forrest, Lillian Montevecchi. Color.

LIZZIE 81
METRO-GOLDWYN-MAYER. (R) March 15, 1957. Eleanor Parker, Richard Boone.

LOCK UP YOUR DAUGHTERS 102
COLUMBIA. (R) October, 1969. Christopher Plummer, Susannah York, Glynis Johns. Color.

LOGAN'S RUN 118
UNITED ARTISTS. (R) June, 1976. Michael York, Richard Jordan.

LOLITA 152
METRO-GOLDWYN-MAYER. (R) June, 1962. James Mason, Shelley Winters.

LOLLY MADONNA (XXX) 103
METRO-GOLDWYN-MAYER. (R) February, 1973. Rod Steiger, Robert Ryan, Jeff Bridges. Color.

LONE RANGER, THE 86
WARNER BROTHERS. (R) February 25, 1956. Clayton Moore, Jay Silverheels, Lyle Berger, Bonita Granville. Color.

LONE RANGER AND THE LOST CITY OF GOLD, THE 80
UNITED ARTISTS. (R) June, 1958. Clayton Moore, Jay Silverheels. Color.

LONE TEXAN, THE (Rs) 70
TWENTIETH CENTURY-FOX. (R) March, 1959. Willard Parker, Grant Williams.

LONELINESS OF THE LONG DISTANCE RUNNER 103
CONTINENTAL. (R) January, 1963. Tom Courtenay, Michael Redgrave.

LONELY ARE THE BRAVE 107
UNIVERSAL. (R) June, 1962. Kirk Douglas, Gena Rowlands.

LONELY HEARTS 101
UNITED ARTISTS. (R) March, 1959. Montgomery Clift, Robert Ryan.

LONELY MAN, THE (VV) 87
PARAMOUNT. (R) June, 1957. Jack Palance, Anthony Perkins.

LONESOME TRAIL, THE 73
LIPPERT. (R) July 1, 1955. Wayne Morris, John Agar.

LONG AGO TOMORROW 100
CINEMA 5. (R) October, 1971. Malcolm McDowell, Nanette Newman. Color.

LONG AND THE SHORT AND THE TALL, THE 102
CONTINENTAL. (R) December, 1961. Richard Todd, Laurence Harvey. b&w.

LONG DAY'S DYING, THE 93
PARAMOUNT. (R) June, 1968. David Hemmings, Tom Bell, Tony Beckley. Color.

LONG DUEL, THE (P) 115
PARAMOUNT. (R) October, 1967. Yul Brynner, Trevor Howard. Color.

LONG GRAY LINE, THE (Cs) 138
COLUMBIA. (R) Special, 1955. Tyrone Power, Maureen O'Hara. Color.

LONG HAUL, THE 88
COLUMBIA. (R) December, 1957. Victor Mature, Diana Dors.

LONG HOT SUMMER, THE (Cs) 115
TWENTIETH CENTURY-FOX. (R) March, 1958. Paul Newman, Joanne Woodward. Color.

LONG JOHN SILVER (Cs) 109
DISTRIBUTORS CORP. OF AMERICA. (R) February, 1955. Robert Newton. Color.

LONG RIDE FROM HELL, A 94
CINERAMA (R) March, 1970. Steve Reeves, Wayde Preston. Color.

LONG RIDERS, THE 98
UNITED ARTISTS. (R) May, 1980. David Carradine, Keith Carradine.

LONG ROPE, THE (Cs) 61
TWENTIETH CENTURY-FOX. (R) February, 1961. Hugh Marlowe, Alan Hale. Color.

LONG SHIPS, THE (T-70) 125
COLUMBIA. (R) June, 1964. Richard Widmark, Sidney Poitier. Color.

LONGEST DAY, THE (Cs) 180
TWENTIETH CENTURY-FOX. (R) October, 1962. Eddie Albert, Arletty.

LONGEST YARD, THE 123
PARAMOUNT. (R) October, 1974. Burt Reynolds, Eddie Albert.

LOOK BACK IN ANGER 99
WARNER BROS. (R) September, 1959. Richard Burton, Claire Bloom.

LOOK IN ANY WINDOW 85
ALLIED ARTISTS. (R) March, 1961. Paul Anka, Ruth Roman.

LOOKING FOR DANGER 61
ALLIED ARTISTS. (R) October, 1957. Huntz Hall, Stanley Clements.

LOOKING FOR LOVE (P) 83
METRO-GOLDWYN-MAYER. (R) August, 1964. Connie Francis, Jim Hutton. Color.

LOOKING FOR MR. GOODBAR 136
PARAMOUNT. (R) October, 1977. Diane Keaton, Tuesday Weld.

LOOKING GLASS WAR, THE (P) 108
COLUMBIA. February, 1970. Christopher Jones, Pia Degenmark, Ralph Richardson. Color.

LOOKING UP 94
LEVITT-PICKMAN. (R) April, 1977. Marilyn Chris, Dick Shawn.

LOOT 90
CINEVISION. (R) May, 1972. Richard Attenborough, Lee Remick. Color.

LOOTERS, THE 87
UNIVERSAL. (R) May, 1955. Rory Calhoun, Julie Adams.

LORD JIM (Super P-70) 154
COLUMBIA. (R) Special. February, 1965. Peter O'Toole, James Mason. Color.

LORD LOVE A DUCK 95
UNITED ARTISTS. (R) February, 1966. Roddy McDowall, Tuesday Weld.

LORD OF THE FLIES 90
CONTINENTAL. (R) August, 1963. James Aubrey, Tom Chapin.

LORD OF THE JUNGLE 69
ALLIED ARTISTS. (R) June, 1955. Johnny Sheffield, Wayne Morris.

LORD OF THE RINGS, THE 131
UNITED ARTISTS. (R) November, 1978. Cartoon.

LORDS OF FLATBUSH, THE 85
COLUMBIA. (R) May, 1974. Perry King, Sylvester Stallone.

LOS TARANTOS 81
SIGMA III. (R) July, 1964. Sara Lezana, Daniel Martin. Color.

LOSERS, THE 95
FANFARE. (R) May, 1970. William Smith, Bernie Hamilton, Adam Roarke. Color.

LOSS OF INNOCENCE 99
COLUMBIA. (R) October, 1961. Kenneth More, Danielle Darrieux. Color.

LOST AND FOUND 106
COLUMBIA. (R) June, 1979. George Segal, Glenda Jackson.

LOST BATTALION, THE 83
AMERICAN INTERNATIONAL. (R) September, 1961. Leopold Salcedo, Diane Jergens.

LOST COMMAND 129
COLUMBIA. (R) May, 1966. Anthony Quinn, Alain Delon. Color.

LOST CONTINENT, THE 89
TWENTIETH CENTURY-FOX. (R) June, 1968. Eric Porter, Hildegard Knef, Suzanna Leigh. Color.

LOST CONTINENT (Ital.-Eng. narr.) (Cs) 64
LOPERT. (R) March, 1957. Documentary. Color.

LOST HORIZON 151
COLUMBIA. (R) March, 1973. Peter Finch, Liv Ullmann, Michael York. Color.

Length in Mins.

LOST IN THE STARS **114**
AFT DISTRIBUTING. (R) Special, 1974. Brock Peters, Melba Moore.

LOST LAGOON **79**
UNITED ARTISTS. (R) February, 1958. Jeffrey Lynn, Peter Donat.

LOST MAN, THE **122**
UNIVERSAL. (R) July, 1969. Sidney Poitier, Joanna Shimkus, Al Freeman. Color.

LOST MISSILE, THE **70**
UNITED ARTISTS. (R) December, 1958. Robert Loggia, Ellen Parker.

LOST WORLD, THE (Cs) **98**
TWENTIETH CENTURY-FOX. (R) July, 1960. Michael Rennie, Jill St. John. Color.

LOST WORLD OF SINBAD, THE **95**
AMERICAN INTERNATIONAL. (R) March, 1965. Toshiro Mifume, Makoto Satoh. Color.

LOVE A LA CARTE (Ital., Eng. titles) **98**
PROMENADE. (R) January, 1965. Simone Signoret, Marcello Mastroianni.

LOVE AFFAIR (Yugoslav.) **90**
BRANDON. (R) Feb., 1968. Eva Ras, Ruzica Sokic.

LOVE AND DEATH **82**
UNITED ARTISTS. (R) June, 1975. Woody Allen, Diane Keaton.

LOVE AND KISSES **87**
UNIVERSAL. (R) November, 1965. Rick Nelson, Jack Kelly. Color.

LOVE AND MARRIAGE (Ital.; Eng. titles) **106**
EMBASSY. (R) September, 1966. Lando Buzzanca, Sylva Koscina.

LOVE AND PAIN (AND THE WHOLE DAMN THING) **110**
COLUMBIA. (R) May, 1973. Maggie Smith, Timothy Bottoms. Color.

LOVE AT FIRST BITE **96**
AMERICAN INTERNATIONAL. (R) April, 1979. George Hamilton, Susan Saint James.

LOVE BUG, THE **107**
BUENA VISTA. (R) April, 1969. Dean Jones, Michele Lee, David Tomlinson. Color.

LOVE GOD, THE **101**
UNIVERSAL. (R) August, 1969. Don Knotts, Anne Francis, Edmund O'Brien. Color.

LOVE GODDESSES, THE **87**
CONTINENTAL. (R) March, 1965. Marlene Dietrich, Lillian Gish.

LOVE HAS MANY FACES **105**
COLUMBIA. (R) February, 1965. Lana Turner, Cliff Robertson. Color.

LOVE IN A GOLDFISH BOWL **88**
PARAMOUNT. (R) June, 1961. Tommy Sands, Fabian. Color.

LOVE IN 4 DIMENSIONS (Ital.; Eng. titles) **108**
ELDORADO PICTURES INTERNATIONAL. (R) September, 1965. Michele Mercier, Sylva Koscina.

LOVE IN THE AFTERNOON **125**
ALLIED ARTISTS. (R) July, 1957. Gary Cooper, Audrey Hepburn.

LOVE-INS, THE **91**
COLUMBIA. (R) July, 1967. James McArthur, Susan Oliver. Color.

LOVE IS A BALL (P) **111**
UNITED ARTISTS. (R) March, 1963. Glenn Ford, Hope Lange. Color.

LOVE IS A FUNNY THING (Fr.: Eng. titles) **110**
UNITED ARTISTS. (R) March, 1970. Jean-Paul Belmondo, Annie Girardot. Color.

LOVE IS A MANY SPENDORED THING (Cs) **102**
TWENTIETH CENTURY-FOX. (R) August, 1955. William Holden, Jennifer Jones. Color.

LOVE ISLAND **68**
ASTOR PICTURES. (R) May, 1960. Paul Valentine, Eva Gabor. Color.

LOVE MACHINE, THE **108**
COLUMBIA. (R) August, 1971. John Phillip Law, Dyan Cannon, Robert Ryan. Color.

LOVE ME OR LEAVE ME (Cs) **108**
METRO-GOLDWYN-MAYER. (R) June 10, 1955. Doris Day, James Cagney. Color.

LOVE ME TENDER (Cs) **89**
TWENTIETH CENTURY-FOX. (R) November, 1956. Elvis Presley, Richard Egan, Debra Paget.

LOVE ON A PILLOW (Cs) **102**
ROYAL FILMS. (R) December, 1963. Brigitte Bardot, Robert Hossen. Color.

LOVE SLAVES OF THE AMAZONS **81**
UNIVERSAL. (R) December, 1957. Don Taylor, Grianna Segale. Color.

LOVE STORY **90**
PARAMOUNT. (R) December 1970. Ali MacGraw, Ryan O'Neil, John Marley, Ray Milland. Color.

LOVE WITH THE PROPER STRANGER **100**
PARAMOUNT. (R) February, 1964. Natalie Wood, Steve McQueen.

LOVED ONE, THE **116**
METRO-GOLDWYN-MAYER. (R) October, 1965. Robert Morse, Jonathan Winters.

LOVELY WAY TO DIE, A **104**
UNIVERSAL. (R) October, 1968. Kirk Douglas, Sylva Koscina, Eli Wallach. Color.

LOVEMAKER, THE (Spain) **99**
TRANS-LUX. (R) March 1958. Betsy Blair, Jose Suarez.

LOVER BOY (Brit.) **85**
TWENTIETH CENTURY-FOX. (R) October, 1955. Gerard Philipe, Valerie Hobson, Joan Greenwood.

LOVER COME BACK **107**
UNIVERSAL. (R) March, 1962. Doris Day, Rock Hudson. Color.

LOVERS AND LOLLIPOPS **80**
TRANS-LUX. (R) April, 1956. Lori March, Gerald O'Loughlin.

LOVERS AND OTHER STRANGERS **106**
CINERAMA. (R) August, 1970. Gig Young, Anne Jackson. Color.

LOVES AND TIMES OF SCARAMOUCHE, THE **91**
AVCO-EMBASSY. (R) March, 1976. Michael Sarrazin, Ursula Andress.

LOVES OF A BLONDE, THE (Czech; Eng. titles) **88**
PROMINENT. (R) December, 1966. Hana Brejchova, Vladimir Pucholt.

LOVES OF ISADORA, THE **131**
UNIVERSAL. (R) June, 1969. Vanessa Redgrave, James Fox, Ivan Tchenko, Jason Robards. Color.

LOVES OF ONDINE, THE **106**
FACTORY FILMS. (R) December, 1969. Ondine, Viva, Joe Dallesandro.

LOVES OF SALAMMBO (Cs) **72**
TWENTIETH CENTURY-FOX. (R) October, 1962. Jeanne Valerie, Jacques Sernas. Color.

LOVIN' MOLLY **98**
COLUMBIA. (R) March, 1974. Anthony Perkins, Beau Bridges, Blythe Danner.

LOVING **90**
COLUMBIA. (R) March, 1970. George Segal, Eva Marie Saint. Color.

LOVING COUPLES (Swedish; Eng. titles) **113**
PROMINENT. (R) September, 1966. Harriet Anderson, Gunnel Lindblom.

LOVING YOU (VV) **101**
PARAMOUNT. (R) July, 1957. Elvis Presley, Lizabeth Scott. Color.

LUCK OF GINGER COFFEY, THE **100**
READE-STERLING. (R) September, 1964. Robert Shaw, Mary Ure.

LUCKY LADY **117**
TWENTIETH CENTURY-FOX. (R) December, 1975. Liza Minnelli, Gene Hackman, Burt Reynolds.

LUCKY LUCIANO **110**
AVCO-EMBASSY. (R) 1975. Gian-Maria Volonte, Rod Steiger.

LUCY GALLANT (VV) **104**
PARAMOUNT. (R) December, 1955. Jane Wyman, Charlton Heston, Claire Trevor, Thelma Ritter. Color.

LUDWIG **173**
METRO-GOLDWYN-MAYER. (R) March, 1973. Helmut Berger, Trevor Howard, Romy Schneider. Color.

LUNA **141**
TWENTIETH CENTURY-FOX. (R) October, 1979. Jill Clayburgh, Matthew Barry.

LURE OF THE SWAMP (Cs) **75**
TWENTIETH CENTURY-FOX. (R) June, 1957. Marshall Thompson, Joan Vohs.

LUST FOR LIFE (Cs) **122**
METRO-GOLDWYN-MAYER. (R) September 15, 1956. Kirk Douglas, Anthony Quinn. Color.

LUTHER **112**
AFT DISTRIBUTING. (R) Special, 1974. Stacy Keach, Patrick Magee.

LUV (P) **95**
COLUMBIA. (R) August, 1967. Peter Falk, Jack Lemmon, Elaine May. Color.

M

M*A*S*H (P) **116**
20TH CENTURY-FOX. April, 1970. Elliott Gould, Donald Sutherland, Sally Kellerman. Color.

MA AND PA KETTLE AT WAIKIKI **79**
UNIVERSAL. (R) April, 1955. Marjorie Main, Percy Kilbride.

MA BARKER'S KILLER BROOD **89**
FILM SERVICE. (R) June, 1960. Lurene Tuttle, Tris Coffin.

Length In Mins.

MACABRE 73
ALLIED ARTISTS. (R) March, 1958. William Prince, Jacqueline Scott.
MACABRO 87
TRANS AMERICAN. (R) July, 1966. Documentary. Color.
MACARTHUR 130
UNIVERSAL. (R) July, 1977. Gregory Peck.
MACBETH 139
COLUMBIA. (R) December, 1971. John Finch, Francesca Annis. Color.
MACHETE 75
UNITED ARTISTS. (R) December, 1958. Mari Blanchard, Albert Dekker.
MACHINE GUN KELLY 84
AMERICAN INTERNATIONAL. (R) May, 1958. Charles Bronson, Susan Cabot.
MACHINE GUN McCANN 94
COLUMBIA. (R) September, 1970. John Cassavetes, Peter Falk. Color.
MACHO CALLAHAN (P) 99
AVCO-EMBASSY. (R) August, 1970. David Janssen, Jean Seberg. Color.
MACK, THE 100
CINERAMA. (R) March, 1973. Max Julien, Carol Speed. Color.
MACKENNA'S GOLD 128
COLUMBIA. (R) May, 1969. Gregory Peck, Omar Sharif, Telly Savalas. Color.
MACKINTOSH MAN, THE 105
WARNER BROS. (R) August, 1973. Paul Newman, James Mason.
MACON COUNTY LINE 89
AMERICAN INTERNATIONAL. (R) June, 1974. Alan Vint, Cheryl Waters.
MACUMBA LOVE 86
UNITED ARTISTS. (R) June, 1960. Ziva Rodann, William Wellman, Jr.
MAD ADVENTURES OF "RABBI" JACOBS, THE 96
TWENTIETH CENTURY-FOX. (R) September, 1973. Louis de Funes, Suzy Delair.
MAD AT THE WORLD 71
FILMAKERS. (R) May, 1955. Frank Lovejoy, Keefe Brasselle.
MAD DOG COLL 86
COLUMBIA. (R) May, 1961. John Chandler, Neil Nephew.
MAD-DOGS AND ENGLISHMEN (Todd-AO) 114
MGM. (R) April, 1971. Rock Music. Documentary. Joe Cocker. Color.
MAD LITTLE ISLAND 94
RANK FILM DISTRIBUTORS OF AMERICA. (R) February, 1959. Jeannie Carson, Donald Sinden. Color.
MAD ROOM, THE 92
COLUMBIA. (R) March, 1969. Stella Stevens, Shelley Winters, Skip Ward. Color.
MADAM WHITE SNAKE (Shawscope) (Chinese, Eng. titles) 105
FRANK LEE. (R) October, 1965. Lin Dai, Chao Lei. Color.
MADAM X 100
UNIVERSAL. (R) March, 1966. Lana Turner, John Forsythe. Color.
MADAME BUTTERFLY (Ital.) 114
I.F.E. (R) May, 1956. Kaoru Yachigusa, Michiko Tanaka. Color.
MADAME ROSA (Fr., Eng. titles) 105
ATLANTIC RELEASING. (R) March, 1978. Simone Signoret, Claude Dauphin.
MADDALENA (Ital.-Eng. dubbed) 90
I.F.E. (R) December, 1955. Marta Toren, Gino Cervi. Color.
MADDALENA 105
INT'L CO-RODS. (R) November, 1971. Lisa Gastoni, Eric Wolfe. Color.
MADE FOR EACH OTHER 107
20TH CENTURY-FOX. (R) December, 1971. Rene Taylor, Joseph Bologna. Color.
MADE IN ITALY (Ital.; Eng. titles) (TE) 101
ROYAL INTERNATIONAL. (R) May, 1967. Anna Magnani, Virna Lisi. Color.
MADE IN PARIS (P) 103
METRO-GOLDWYN-MAYER. (R) February, 1966. Ann-Margret, Louis Jourdan. Color.
MADEMOISELLE (P) 103
LOPERT. (R) August, 1966. Jeanne Moreau, Ettore Manni.
MADIGAN (T) 101
UNIVERSAL. (R) May, 1968. Richard Widmark, Inger Stevens. Color.
MADIGAN'S MILLIONS 79
AMERICAN INTERNATIONAL. (R) December, 1969. Dustin Hoffman, Elsa Martinelli. Color.

Length In Mins.

MADISON AVENUE (Cs) 94
TWENTIETH CENTURY-FOX. (R) January, 1962. Dana Andrews, Barbara Eden.
MADWOMAN OF CHAILLOT, THE 145
WARNER BROS. (R) October, 1969. Katharine Hepburn, Charles Boyer, Edith Evans. Color.
MAEDCHEN IN UNIFORM (German with English titles) 91
SEVEN ARTS. (R) June, 1965. Lilli Palmer, Romy Schneider.
MAFISO 100
ZENITH. (R) July, 1964. Alberto Sordi, Norma Bengell.
MAGIC 106
TWENTIETH CENTURY-FOX. (R) November, 1978. Anthony Hopkins, Ann-Margret.
MAGIC BOY 83
METRO-GOLDWYN-MAYER. (R) April, 1962. Cartoon. Color.
MAGIC CHRISTIAN, THE 93
COMMONWEALTH UNITED. (R) February, 1970. Peter Sellers, Ringo Starr, Raquel Welch. Color.
MAGIC FIRE 85
REPUBLIC. (R) March 29, 1956. Yvonne DeCarlo, Carlos Thompson, Rita Gam. Color.
MAGIC FLUTE, THE (Swedish; Eng. titles) 136
SURROGATE RELEASING. (R) October, 1975. Ulrik Cold, Josef Kostlinger.
MAGIC GARDEN OF STANLEY SWEETHEART, THE 113
MGM. (R) June, 1970. Don Johnson, Diane Hull, Holly Near. Color.
MAGIC SWORD, THE 80
UNITED ARTISTS. (R) April, 1962. Basil Rathbone, Estelle Winwood. Color.
MAGIC WEAVER, THE 87
ALLIED ARTISTS. (R) 1965. Color.
MAGIC WORLD OF TOPO GIGIO, THE 75
COLUMBIA. (R) December, 1965. Fantasy. Color.
MAGNIFICENT CONCUBINE, THE (Chinese, Eng. titles) 97
FRANK LEE. (R) June, 1966. Li Li-Hus, Yen Chuan. Color.
MAGNIFICENT CUCKOLD, THE (Ital., Eng. titles) 113
CONTINENTAL. (R) April, 1965. Claudia Cardinale, Ugo Tognazzi.
MAGNIFICENT MATADOR, THE (CS) 94
TWENTIETH CENTURY-FOX. (R) June, 1955. Maureen O'Hara, Anthony Quinn. Color.
MAGNIFICENT ROUGHNECKS 73
ALLIED ARTISTS. (R) July 22, 1956. Jack Carson, Mickey Rooney.
MAGNIFICENT SEVEN, THE (P) 126
UNITED ARTISTS. (R) November, 1960. Yul Brynner, Eli Wallach. Color.
MAGNIFICENT SEVEN RIDE, THE 100
UNITED ARTISTS. (R) August, 1972. Lee Van Cleef, Michael Callan. Color.
MAGNUM FORCE (P) 124
WARNER BROS. (R) December, 1973. Clint Eastwood, Hal Holbrook.
MAGUS, THE 116
TWENTIETH CENTURY-FOX. (R) Jan., 1969. Anthony Quinn, Michael Caine, Candice Bergen. Color.
MAHLER 115
MAYFAIR FILM GROUP. (R) February, 1975. Robert Powell, Georgina Hale.
MAIDEN FOR A PRINCE, A (Ital.) 92
ROYAL INTERNATIONAL. (R) September, 1967. Vittorio Gassman, Virna Lisi. Color.
MAIDS, THE 95
AFT. (R) 1975. Glenda Jackson, Susannah York.
MAIL ORDER BRIDE (P) 83
METRO-GOLDWYN-MAYER. (R) February, 1964. Buddy Ebsen, Keir Dullea. Color.
MAIN ATTRACTION, THE 96
METRO-GOLDWYN-MAYER. (R) November, 1962. Pat Boone, Nancy Kwan.
MAIN CHANCE, THE 60
EMBASSY. (R) July, 1966. Gregoire Aslan, Tracy Reed.
MAIN EVENT, THE 112
WARNER BROS. (R) June, 1979. Barbara Streisand, Ryan O'Neal.
MAJOR DUNDEE (P) 124
COLUMBIA. (R) April, 1965. Charlton Heston, Richard Harris. Color.
MAJORITY OF ONE, A 153
WARNER BROS. (R) February, 1962. Rosalind Russell, Alec Guinness. Color.
MAKE ME AN OFFER (Brit.) 88
ASSOCIATED ARTISTS. (R) February, 1956. Peter Finch, Adrienne Corri. Color.
MAKE MINE MINK 101
CONTINENTAL. (R) December, 1960. Terry-Thomas, Athene Seyler.
MAKING IT 97

Length In Mins.

TWENTIETH CENTURY-FOX. (R) April, 1971. Kristoffer Tabori, Marlyn Mason. Color.

MALAGA (Brit.) 97
WARNER BROS. (R) 1962. Trevor Howard. Dorothy Dandridge.

MALAMONDO 79
MAGNA. (R) November, 1964. Documentary. Color.

MALCOLM X 92
WARNER BROS. (R) May, 1972. Documentary Biography. Color.

MALE COMPANION (Fr., Eng. titles) 92
INTERNATIONAL CLASSICS. (R) January, 1966. Jean-Pierre Cassel, Catherine Deneuve. Color.

MALIZIA 98
PARAMOUNT. (R) February, 1974. Laura Antonelli, Turi Ferra.

MALTESE BIPPY, THE (P) 92
MGM. (R) June, 1969. Dan Rowan, Dick Martin, Carol Lynley. Color.

MAMBO 94
PARAMOUNT. (R) April, 1955. Silvana Mangano, Shelley Winters, Vittorio Gassman.

MAME (P) 132
WARNER BROS. (R) April, 1974. Lucille Ball, Beatrice Arthur.

MAN, A WOMAN, AND A BANK, A 100
AVCO EMBASSY. (R) October, 1979. Donald Sutherland, Brooke Adams.

MAN, THE 94
PARAMOUNT. (R) August, 1972. James Earl Jones, Martin Balsam. Color.

MAN AFRAID (CS) 64
UNIVERSAL. (R) June, 1957. George Nader, Phyllis Thaxter.

MAN ALONE, A 96
REPUBLIC. (R) October 17, 1955. Ray Milland, Mary Murphy. Color.

MAN AND A WOMAN, A 102
ALLIED ARTISTS. (R) July, 1966. Anouk Aimee, Jean-Louis Trintignant. Color.

MAN AND BOY 98
LEVITT-PICKMAN. (R) March, 1972. Bill Cosby, Leif Erickson. Color.

MAN CALLED ADAM, A 102
EMBASSY. (R) July, 1966. Sammy Davis, Louis Armstrong.

MAN CALLED DAGGER, A 82
MGM. (R) January, 1968. Terry Moore, Jan Murray. Color.

MAN CALLED FLINTSTONE, THE 87
COLUMBIA. (R) August, 1966. Animation. Color.

MAN CALLED GANNON, A 105
UNIVERSAL. (R) May, 1969. Tony Franciosa, Michael Sarrazin, Judi West. Color.

MAN CALLED HORSE, A (P) 114
NATIONAL GENERAL. (R) April, 1970. Richard Harris, Judith Anderson. Color.

MAN CALLED NOON, THE 97
NATIONAL GENERAL. (R) September, 1973. Richard Crenna, Stephen Boyd.

MAN CALLED PETER, A (CS) 119
TWENTIETH CENTURY-FOX. (R) April, 1955. Richard Todd, Jean Peters. Color.

MAN CALLED SLEDGE, A 90
COLUMBIA. (R) March, 1971. James Garner, Dennis Weaver. Color.

MAN COULD GET KILLED, A (P) 99
UNIVERSAL. (R) April, 1966. James Garner, Melina Mercouri. Color.

MAN ESCAPED, A (Fr.) 94
CONTINENTAL. (R) November, 1957. Robert Bresson.

MAN FOR ALL SEASONS, A 120
COLUMBIA. (R) December, 1966. Paul Scofield, Wendy Hiller. Color.

MAN FROM BITTER RIDGE, THE 80
UNIVERSAL. (R) June, 1955. Lex Barker, Mara Corday. Color.

MAN FROM BUTTON WILLOW, THE 84
UNITED SCREEN ARTS. (R) January, 1965. Animated Western. Color.

MAN FROM DEL RIO 93
UNITED ARTISTS. (R) October, 1956. Anthony Quinn, Katy Jurado.

MAN FROM GALVESTON, THE 57
WARNER BROS. (R) January, 1964. Jeff Hunter, Preston Foster.

MAN FROM GOD'S COUNTRY (CS) 70
ALLIED ARTISTS. (R) January, 1958. George Montgomery, Randy Stuart. Color.

MAN FROM LARAMIE, THE (CS) 104
COLUMBIA. (R) August, 1955. James Stewart, Arthur Kennedy. Color.

MAN FROM O.R.G.Y., THE 92
CINEMATION. (R) April, 1970. Robert Walker, Steve Rossi. Color.

MAN FROM THE DINERS' CLUB, THE 96
COLUMBIA. (R) April, 1963. Danny Kaye, Cara Williams.

MAN IN A COCKED HAT 88
LION INTERNATIONAL. (R) May, 1960. Terry-Thomas, Peter Sellers.

MAN IN THE DARK 80
UNIVERSAL. (R) March, 1965. William Sylvester, Barbara Shelley.

MAN IN THE GRAY FLANNEL SUIT, THE (CS) 152
TWENTIETH CENTURY-FOX. (R) March, 1956, Gregory Peck, Jennifer Jones, Fredric March. Color.

MAN IN THE MIDDLE (CS) 94
TWENTIETH CENTURY-FOX. (R) January, 1964. Robert Mitchum, France Nuyen.

MAN IN THE NET, THE 97
UNITED ARTISTS. (R) April, 1959. Alan Ladd, Carolyn Jones.

MAN IN THE ROAD 83
REPUBLIC. (R) April 12, 1957. Derek Farr, Ella Raines.

MAN IN THE SHADOW (CS) 80
UNIVERSAL. (R) January, 1958. Jeff Chandler, Orson Welles.

MAN IN THE STORM 112
TOHO CO. (R) August, 1969. Toshiro Mifune.

MAN IN THE VAULT 73
UNIVERSAL. (R) December, 1956. Anita Ekberg, William Campbell.

MAN IN THE WILDERNESS 105
WARNER BROS. (R) November, 1971. Richard Harris, John Bindon. Color.

MAN INSIDE, THE (CS) 89
COLUMBIA. (R) December, 1958. Jack Palance, Anita Ekberg.

MAN IS ARMED, THE 76
REPUBLIC. (R) October 19, 1956. Dane Clark, William Talman, May Wynn.

MAN OF A THOUSAND FACES (CS) 122
UNIVERSAL. (R) October, 1957. James Cagney, Dorothy Malone.

MAN OF LA MANCHA 130
UNITED ARTISTS. (R) December, 1972. Sophia Loren, Peter O'Toole, Color.

MAN OF THE WEST (CS) 100
UNITED ARTISTS. (R) October, 1958. Gary Cooper, Julie London. Color.

MAN ON A STRING 92
COLUMBIA. (R) May, 1960. Ernest Borgnine, Kerwin Mathews.

MAN ON A SWING (P) 110
PARAMOUNT. (R) March, 1974. Cliff Robertson, Joel Grey.

MAN ON FIRE 96
METRO-GOLDWYN-MAYER. (R) July, 1957. Bing Crosby, Inger Stevens.

MAN ON THE PROWL 86
UNITED ARTISTS. (R) December, 1957. Mala Powers, James Best.

MAN ON THE ROOF (Swedish with English titles) 110
CINEMA 5. (R) April, 1977. Carl-Gustaf Lindstedt, Hakan Serner.

MAN OR GUN (N) 79
REPUBLIC. (R) May, 1958. Macdonald Carey, Audrey Totter.

MAN OUTSIDE, THE 97
ALLIED ARTISTS. (R) March, 1969. Van Heflin, Heidelinde Weis, Pinkas Braun. Color.

MAN-TRAP (P) 90
PARAMOUNT. (R) September, 1961. Jeffrey Hunter, David Janssen.

MAN UPSTAIRS, THE 89
LION INT. (R) August, 1959. Richard Attenborough, Bernard Lee.

MAN WHO COULD CHEAT DEATH, THE 83
PARAMOUNT. (R) September, 1959. Anton Diffring, Hazel Court. Color.

MAN WHO COULDN'T WALK 63
FALCON. (R) March, 1964. Eric Pohlman, Peter Reynolds.

MAN WHO DIED TWICE, THE (N) 70
REPUBLIC. (R) June, 1958. Rod Cameron, Vera Ralston.

MAN WHO HAD POWER OVER WOMEN, THE 89
AVCO EMBASSY. (R) November, 1970. Rod Taylor, Carol White. Color

MAN WHO KNEW TOO MUCH, THE (VV) 120
PARAMOUNT. (R) June, 1956. James Stewart, Doris Day. Color.

MAN WHO LOVED CAT DANCING, THE (P) 114
METRO-GOLDWYN-MAYER. (R) July, 1973. Burt Reynolds, Sarah Miles. Color.

MAN WHO LOVED REDHEADS, THE (Brit.) 89

Length In Mins.

UNITED ARTISTS. (R) July, 1955. Moira Shearer, John Justin. Color.

MAN WHO NEVER WAS, THE (CS) **103**
TWENTIETH CENTURY-FOX. (R) February, 1956. Clifton Webb, Gloria Grahame. Color.

MAN WHO SHOT LIBERTY VALANCE **122**
PARAMOUNT. (R) April, 1962. James Stewart, John Wayne.

MAN WHO TURNED TO STONE, THE **71**
COLUMBIA. (R) March, 1957. Victor Jory, Ann Doran.

MAN WHO UNDERSTOOD WOMEN (CS) **105**
TWENTIETH CENTURY-FOX. (R) October, 1959. Leslie Caron, Henry Fonda. Color.

MAN WITH CONNECTIONS, THE (French; Eng. titles) .. **93**
COLUMBIA. (R) March, 1971. Guy Bedos, George Geret. Color.

MAN WITH THE BALLOONS, THE (Italian, Eng. titles) .. **85**
SIGMA III. (R) June, 1968. Marcello Mastroianni, Catherine Spaak. Color.

MAN WITH THE GOLDEN ARM, THE **119**
UNITED ARTISTS. (R) January, 1956. Frank Sinatra, Eleanor Parker, Kim Novak.

MAN WITH THE GOLDEN GUN, THE **125**
UNITED ARTISTS. (R) December, 1974. Roger Moore, Christopher Lee.

MAN WITH THE GUN, THE **83**
UNITED ARTISTS. (R) November, 1955. Robert Mitchum, Jan Sterling.

MAN WITHOUT A STAR **89**
UNIVERSAL. (R) April, 1955. Kirk Douglas, Jeanne Crain. Color.

MANCHURIAN CANDIDATE, THE **126**
UNITED ARTISTS. (R) November, 1962. Frank Sinatra, Laurence Harvey.

MANDINGO **127**
PARAMOUNT. (R) May, 1975. James Mason, Susan George, Perry King.

MANDRAGOLA (Ital., Eng. titles) **98**
EUROPIX-CONSOLIDATED. (R) June, 1966. Rosanna Schiaffino, Philippe Leroy.

MANFISH **76**
UNITED ARTISTS. (R) February, 1956. John Bromfield, Victor Jory. Color.

MANHATTAN (P) (black-and-white) **105**
UNITED ARTISTS. (R) May, 1979. Woody Allen, Diane Keaton.

MANHUNT IN THE JUNGLE (C) **79**
WARNER BROTHERS. (R) May, 1958. Robin Hughes, Luis Alvarez.

MANIAC (M) **87**
COLUMBIA. (R) October, 1963. Kerwin Mathews, Nadia Gray.

MANITOU, THE (P) **104**
AVCO EMBASSY. (R) March, 1978. Tony Curtis, Michael Ansara.

MAN'S FAVORITE SPORT? **120**
UNIVERSAL. (R) March, 1964. Rock Hudson, Paula Prentiss. Color.

MANY RIVERS TO CROSS (CS) **92**
METRO-GOLDWYN-MAYER. (R) February 4, 1955. Robert Taylor, Eleanor Parker. Color.

MARA OF THE WILDERNESS **90**
ALLIED ARTISTS. (R) January, 1965. Adam West, Linda Saunders. Color.

MARACAIBO (VV) **88**
PARAMOUNT. (R) May, 1958. Cornel Wilde, Jean Wallace. Color.

MARAT/SADE **115**
UNITED ARTISTS. (R) February, 1967. Royal Shakespeare Company. Color.

MARAUDERS, THE **80**
METRO-GOLDWYN-MAYER. (R) May 20, 1955. Dan Duryea, Keenan Wynn. Color.

MARCELINO (Span.) **90**
UMPO. (R) October, 1956. Pablito Calvo, Raefael Rivelles.

MARCH OR DIE **106**
COLUMBIA. (R) August, 1977. Gene Hackman, Terence Hill.

MARCO **109**
CINERAMA. (R) December, 1973. Desi Arnaz, Jr., Jack Weston, Zero Mostel.

MARCO POLO (CS) **100**
AMERICAN INTERNATIONAL. (R) August, 1962. Rory Calhoun, Yoka Tani. Color.

MARCO THE MAGNIFICENT! (Franscope) **110**
METRO-GOLDWYN-MAYER. (R) Sept., 1966. Horst Buchholz, Anthony Quinn. Color.

MARDI GRAS (CS) **107**
TWENTIETH CENTURY-FOX. (R) November, 1958. Pat Boone, Christine Cerere. Color.

MARILYN (CS) **83**
TWENTIETH CENTURY-FOX. (R) June, 1963. Marilyn Monroe. Color.

MARINES, LET'S GO (CS) **104**
TWENTIETH CENTURY-FOX. (R) August, 1961. Tom Tryon, David Hedison. Color.

MARJOE **88**
CINEMA 5. (R) July, 1972. Cinema-Verite Documentary of Pop Evangelist. Color.

MARJORIE MORNINGSTAR (C) **125**
WARNER BROTHERS. (R) April, 1958. Gene Kelly, Natalie Wood.

MARK, THE (CS) **127**
CONTINENTAL. (R) October, 1961. Maria Schell, Stuart Whitman.

MARK OF THE HAWK, THE (SS) **85**
UNIVERSAL. (R) December, 1958. Sidney Poitier, Juano Hernandez. Color.

MARLOWE **100**
MGM. (R) October, 1969. James Garner, Gayle Hunnicutt, Carroll O'Connor. Color.

MARNIE **129**
UNIVERSAL. (R) August, 1964. Tippi Hedren, Sean Connery. Color.

MAROC 7 (P) **92**
PARAMOUNT. (R) January, 1968. Gene Barry, Cyd Charisse. Color.

MAROONED (P) **134**
COLUMBIA. (R) December, 1969. Gregory Peck, Richard Crenna, David Janssen. Color.

MARRIAGE CAME TUMBLING DOWN, THE (French, Eng. titles) **88**
ROYAL. (R) November, 1968. Michel Simon, Marie Dubois, Yves Lefebre. Color.

MARRIAGE-GO-ROUND, THE (CS) **98**
TWENTIETH CENTURY-FOX. (R) January, 1961. Susan Hayward, James Mason. Color.

MARRIAGE ITALIAN STYLE **102**
EMBASSY. (R) December, 1964. Sophia Loren, Marcello Mastroianni. Color.

MARRIAGE OF A YOUNG STOCKBROKER **95**
20TH CENTURY-FOX. (R) September, 1971. Richard Benjamin, Joanna Shimkus. Color.

MARRIAGE ON THE ROCKS (P) **109**
WARNER BROTHERS. (R) October, 1965. Frank Sinatra, Deborah Kerr. Color.

MARRIED COUPLE, A **97**
AQUARIUS. (R) January, 1970. Documentary. Color.

MARRIED WOMAN, THE (Fr., English titles) **94**
ROYAL FILMS INTERNATIONAL. (R) August, 1965. Macha Meril, Bernard Noel.

MARRY ME! MARRY ME! **87**
ALLIED ARTISTS. (R) July, 1969. Elisabeth Wiener, Regine. Color.

MARTY **94**
UNITED ARTISTS. (R) March, 1955. Ernest Borgnine, Betsy Blair.

MARTYRS OF LOVE (Czech, Eng. titles) **73**
NEW LINE. (R) February, 1969. Peter Kopriva, Hans Kuberova, Josef Konicek. b&w.

MARY HAD A LITTLE **79**
UNITED ARTISTS. (R) 1961. Agnes Laurent.

MARY, MARY **126**
WARNER BROS. (R) November, 1963. Debbie Reynolds, Barry Nelson. Color.

MARY POPPINS **140**
BUENA VISTA. (R) January, 1965. Julie Andrews, Dick Van Dyke. Color.

MARY, QUEEN OF SCOTS **128**
UNIVERSAL. (R) December, 1971. Vanessa Redgrave, Glenda Jackson. Color.

MARYJANE **104**
AMERICAN INTERNATIONAL. (R) February, 1968. Patty McCormick, Kevin Coughlin. Color.

MASCULINE FEMININE (Fr.; Eng. titles) **103**
ROYAL. (R) October, 1966. Jean-Pierre Leaud, Chantal Goya.

MASK, THE **83**
WARNER BROS. (R) October, 1961. Paul Stevins, Claudette Nevins.

MASQUE OF THE RED DEATH, THE (P) **90**
AMERICAN INTERNATIONAL. (R) June, 1964. Vincent Price, Hazel Court. Color.

MASQUERADE **101**
UNITED ARTISTS. (R) May, 1965. Cliff Robertson, Jack Hawkins. Color.

MASSACRE **70**
TWENTIETH CENTURY-FOX. (R) June, 1956. Dane Clark, James Craig. Color.

MASSACRE IN ROME **105**
NATIONAL GENERAL. (R) October, 1973. Richard Burton, Marcello Mastroianni.

MASTER GUNFIGHTER, THE **121**

Length In Mins.

TAYLOR-LAUGHLIN. (R) October, 1975. Tom Laughlin, Ron O'Neal.

MASTER OF THE WORLD **104**
AMERICAN INTERNATIONAL. (R) June, 1961. Vincent Price, Charles Bronson. Color

MASTER PLAN, THE (Brit.) **77**
ASTOR. (R) February, 1955. Wayne Morris.

MASTER SPY, THE **71**
ALLIED ARTISTS. (R) July, 1964. Stephen Murray, June Thornburn.

MASTER TOUCH, THE **101**
WARNER BROS. (R) May, 1974. Kirk Douglas, Florinda Bolkan.

MASTERS OF THE CONGO JUNGLE (CS) **80**
TWENTIETH CENTURY-FOX. (R) December, 1959. Documentary. Color.

MASTERSON OF KANSAS **73**
COLUMBIA. (R) January, 1955. George Montgomery, Nancy Gates. Color.

MATCHLESS **104**
UNITED ARTISTS. (R) October, 1967. Ira Furstenberg, Patrick O'Neal. Color.

MATCHMAKER, THE (VV) **101**
PARAMOUNT. (R) August, 1958. Shirley Booth, Anthony Perkins.

MATILDA **105**
AMERICAN INTERNATIONAL. (R) July, 1978. Elliott Gould, Robert Mitchum.

MATING GAME, THE (CS) **96**
METRO-GOLDWYN-MAYER. (R) March, 1959. Debbie Reynolds, Tony Randall. Color.

MATTEI AFFAIR, THE **118**
PARAMOUNT. (R) May, 1973. Gian Maria Volonte, Luigi Squarzina. Color.

MATTER OF INNOCENCE, A (TS) **102**
UNIVERSAL. (R) February, 1968. Hayley Mills, Trevor Howard. Color.

MAURIE **113**
NATIONAL GENERAL. (R) August, 1973. Bernie Casey, Bo Swenson.

MAURIETA **108**
WARNER BROTHERS. (R) September, 1965. Jeffrey Hunter, Arthur Kennedy. Color.

MAVERICK QUEEN, THE (Naturama) **90**
REPUBLIC. (R) May 3, 1956. Barbara Stanwyck, Barry Sullivan, Scott Brady, Mary Murphy. Color.

MAYA (P) **91**
METRO-GOLDWYN-MAYER. (R) June, 1966. Clint Walker, Jay North. Color.

MAYERLING **140**
METRO-GOLDWYN-MAYER. (R) February, 1969. Omar Sharif, Catherine Deneuve, James Mason. Color.

McCABE AND MRS. MILLER **120**
WARNER BROS. (R) July, 1971. Warren Beatty, Julie Christie. Color.

McCLINTOCK (P) **127**
UNITED ARTISTS. (R) November, 1963. John Wayne, Maureen O'Hara. Color.

McCONNELL STORY, THE (CS) **107**
WARNER BROTHERS. (R) September 3, 1955. Alan Ladd, June Allyson. Color.

McGUIRE, GO HOME! **101**
CONTINENTAL. (R) April, 1966. Dirk Bogarde, George Chakiris. Color.

McHALE'S NAVY **90**
UNIVERSAL. (R) September, 1964. Ernest Borgnine, Joe Flynn. Color.

McHALE'S NAVY JOINS THE AIR FORCE **92**
UNIVERSAL. (R) July, 1965. Joe Flynn, Tim Conway. Color.

McKENZIE BREAK, THE **103**
UNITED ARTISTS. (R) October, 1970. Brian Keith, Helmut Griem. Color.

McMASTERS **89**
CHEVRON. (R) June, 1970. Burl Ives, Brock Peters, David Carradine. Color.

McQ (P) **116**
WARNER BROS. (R) February, 1974. John Wayne, Eddie Albert.

ME AND THE COLONEL **109**
COLUMBIA. (R) October, 1958. Danny Kaye, Curt Jurgens.

ME, NATALIE **111**
NATIONAL GENERAL. (R) August, 1969. Patty Duke, James Farentino, Salome Jens. Color.

MEAN STREETS **110**
WARNER BROS. (R) October, 1973. Robert de Niro, Harvey Keitel.

MEATBALLS **92**
PARAMOUNT. (R) July, 1979. Bill Murray, Harvey Atkin.

MECHANIC, THE **95**
UNITED ARTISTS. (R) October, 1972. Charles Bronson, Jan-Michael Vincent. Color

MEDICINE BALL CARAVAN **90**
WARNER BROS. (R) August, 1971. Rock Concert Tour Documentary. Color.

MEDITERRANEAN HOLIDAY (Wonderama) **130**
CONTINENTAL. (R) March, 1964. Travelogue. Color.

MEDIUM COOL **110**
PARAMOUNT. (R) September, 1969. Robert Forster, Vera Bloom, Peter Bonerz. Color.

MEDUSA TOUCH, THE **110**
WARNER BROS. (R) April, 1978. Richard Burton, Lee Remick.

MEET ME IN LAS VEGAS (CS) **112**
METRO-GOLDWYN-MAYER. (R) March 9, 1956. Dan Dailey, Cyd Charisse. Color.

MEIN KAMPF **110**
COLUMBIA. (R) May, 1961. Documentary.

MELINDA **109**
METRO-GOLDWYN-MAYER. (R) August, 1972. Calvin Lockhart, Rosalind Cash. Color.

MELODY **103**
LEVITT-PICKMAN. (R) April, 1971. Jack Wild, Mark Lester. Color.

MEN IN WAR **104**
UNITED ARTISTS. (R) February, 1957. Robert Ryan, Aldo Ray.

MEN OF SHERWOOD FOREST (Brit.) **77**
ASTOR. (R) September, 1956. Don Taylor, Eileen Moore. Color.

MENACE IN THE NIGHT **78**
UNITED ARTISTS. (R) October, 1958. Griffith Jones, Lisa Gastoni.

MEPHISTO WALTZ, THE **115**
TWENTIETH CENTURY-FOX. (R) March, 1971. Alan Alda, Jacqueline Bisset. Color.

MERCENARY, THE **105**
UNITED ARTISTS. (R) March, 1970. Franco Nero, Tony Musante, Jack Palance. Color.

MERRILL'S MARUADERS **98**
WARNER BROS. (R) June, 1962. Jeff Chandler, Ty Hardin. Color.

MERRY ANDREW (CS) **103**
MGM. (R) April, 1958. Danny Kaye, Pier Angeli. Color.

MERRY-GO-ROUND (Hungary) **90**
TRANS-LUX. (R) April, 1958. Bela Barsi, Manyi Kiss.

MERRY WIVES OF WINDSOR, THE **97**
SIGMA III. (R) March, 1966. Norman Foster, Colette Boxy. Color.

MESSAGE FROM SPACE **105**
UNITED ARTISTS. (R) November, 1978. Vic Morrow, Sonny Chiba.

METEOR (P) **103**
AMERICAN INTERNATIONAL. (R) October, 1979. Sean Connery, Natalie Wood.

MGM'S BIG PARADE OF COMEDY **109**
METRO-GOLDWYN-MAYER. (R) September, 1964. Stars of Past and Present.

MIAMI EXPOSE **73**
COLUMBIA. (R) September, 1956. Lee J. Cobb, Patricia Medina.

MICHAEL AND HELGA **87**
AMERICAN INTERNATIONAL. (R) April, 1969. Ruth Gassmann, Felix Franchy. Color.

MICKEY ONE **93**
COLUMBIA. (R) October, 1965. Warren Beatty, Hurd Hatfield, Alexandra Stewart.

MIDAS RUN **103**
CINERAMA. (R) May, 1969. Richard Crenna, Anne Heywood, Fred Astaire. Color.

MIDDLE OF THE NIGHT **118**
COLUMBIA. (R) July, 1959. Kim Novak, Fredric March.

MIDNIGHT COWBOY **119**
UNITED ARTISTS. (R) May, 1969. Dustin Hoffman, Jon Voight, Sylvia Miles. Color.

MIDNIGHT EXPRESS **120**
COLUMBIA. (R) October, 1978. Brad Davis, Bo Hopkins.

MIDNIGHT LACE **108**
UNIVERSAL. (R) November, 1960. Doris Day, Rex Harrison. Color.

MIDNIGHT MADNESS **110**
BUENA VISTA. (R) February, 1980. David Naughton, Debra Clinter.

MIDNIGHT MAN, THE **117**
UNIVERSAL. (R) May, 1974. Burt Lancaster, Susan Clark.

MIDNIGHT STORY, THE (CS) **89**
UNIVERSAL. (R) August, 1957. Tony Curtis, Marisa Pavan.

MIDSUMMER NIGHT'S DREAM, A (F) **93**
COLUMBIA. (R) Special. January, 1962. Suzanne Farrell, Edward Villella. Color.

Length In Mins.

MIGHTY URSUS, THE (Ital.-dubbed) **90**
UNITED ARTISTS. (R) 1962. Ed Fury, Christina Gajoni. Color.

MIKADO, THE **125**
WARNER BROS. (R) March, 1967. D'Oyly Carte Opera Company. Color.

MIKEY & NICKY **118**
PARAMOUNT. (R) January, 1977. Peter Falk, John Cassavetes.

MILKY WAY, THE **86**
U-M. December, 1969. Paul Frankeur, Laurent Terzieff.

$1,000,000 DUCK **92**
BUENA VISTA. (R) July, 1971. Dean Jones, Sandy Duncan. Color.

MILLION EYES ON SUMARU, THE (TE) **90**
AMERICAN INTERNATIONAL. (R) May, 1967. Frankie Avalon, Shirley Eaton. Color.

MILLIONAIRES, THE (CS) **90**
TWENTIETH CENTURY-FOX. (R) March, 1961. Sophia Loren, Peter Sellers. Color.

MIND BENDERS, THE **98**
AMERICAN INTERNATIONAL. (R) May, 1963. Dirk Bogarde, Mary Ure.

MIND OF MR. SOAMES, THE **95**
COLUMBIA. (R) October, 1970. Terence Stamp, Robert Vaughn. Color.

MINI-SKIRT MOB, THE **82**
AMERICAN INTERNATIONAL. (R) May, 1968. Jeremy Slate, Sherry Jackson. Color.

MINNIE AND MOSKOWITZ **114**
UNIVERSAL. (R) December, 1971. Gena Rowlands, Seymour Cassel. Color.

MINOTAUR, THE **92**
UNITED ARTISTS. (R) April, 1961. Bob Mathias, Rosanna Schiaffino. Color.

MINUTE TO PRAY, SECOND TO DIE, A **103**
CINERAMA. (R) April, 1968. Alex Cord, Robert Ryan. Color.

MINX, THE **84**
CAMBIST FILMS. (R) September, 1969. Jan Sterling, Robert Rodan. Color.

MIRACLE, THE (T) **121**
WARNER BROS. (R) December, 1959. Carroll Baker, Roger Moore. Color.

MIRACLE IN THE RAIN **107**
WARNER BROTHERS. (R) April 7, 1956. Jane Wyman, Van Johnson.

MIRACLE OF LOVE, THE **83**
TIMES. (R) January, 1969. Biggi Freyer, Katarina Haertel. b&w.

MIRACLE OF THE HILLS (CS) **73**
TWENTIETH CENTURY-FOX. (R) July, 1959. Rex Reason, Nan Leslie.

MIRACLE OF THE WHITE STALLIONS **118**
BUENA VISTA. (R) April, 1963. Robert Taylor, Lilli Palmer.

MIRACLE WORKER, THE **108**
UNITED ARTISTS. (R) May, 1962. Anne Bancroft, Patty Duke.

MIRAGE **109**
UNIVERSAL. (R) June, 1965. Gregory Peck, Diane Baker. Color.

MISADVENTURES OF MERLIN JONES **88**
BUENA VISTA. (R) January, 1964. Tommy Kirk, Annette Funnicello. Color

MISFITS, THE **124**
UNITED ARTISTS. (R) February, 1961. Clark Gable, Marilyn Monroe.

MISSILE MONSTERS **75**
REPUBLIC. (R) May, 1958. Walter Reed, Lois Collier.

MISSILE TO THE MOON **78**
ASTOR. (R) October, 1959. Richard Travis, Cathy Downs.

MISSION STARDUST **95**
TIMES. (R) November, 1968. Essy Pearson, Lang Jeffries, John Karelsen. Color.

MISSISSIPPI MERMAID, THE (Fr.; Eng. titles) **110**
UNITED ARTISTS. (R) April, 1970. Jean-Paul Belmondo, Catherine Deneuve, Michel Bouquet. Color.

MISSOURI TRAVELER, THE **103**
BUENA VISTA. (R) January, 1958. Brandon de Wilde, Lee Marvin. Color.

MR. BILLION **93**
TWENTIETH CENTURY-FOX. (R) March, 1977. Terence Hill, Valerie Perrine.

MR. HOBBS TAKES A VACATION (CS) **116**
TWENTIETH CENTURY-FOX. (R) July, 1962. James Stewart, Maureen O'Hara. Color.

MR. KLEIN **122**
QUARTET FILMS. (R) November, 1977. Alain Delon, Jeanne Moreau.

MR. MAJESTYK **103**
UNITED ARTISTS. (R) July, 1974. Charles Bronson, Al Lettieri.

MR. RICCO (P) **98**
UNITED ARTISTS—MGM. (R) February, 1975. Dean Martin.

MR. SARDONICUS **89**
COLUMBIA. (R) October, 1961. Oscar Homolka, Ronald Lewis.

MISTER BUDDWING **100**
METRO-GOLDWYN-MAYER. (R) October, 1966. James Garner, Jean Simmons.

MISTER COREY (CS) **92**
UNIVERSAL. (R) March, 1957. Tony Curtis, Martha Hyer. Color.

MISTER MOSES (P) **113**
UNITED ARTISTS. (R) May, 1965. Robert Mitchum, Carroll Baker. Color.

MISTER ROBERTS (CS) **123**
WARNER BROTHERS. (R) July 30, 1955. Henry Fonda, James Cagney, William Powell, Jack Lemmon. Color.

MISTER ROCK AND ROLL **86**
PARAMOUNT. (R) September, 1957. Alan Freed, Lionel Hampton.

MISTRESS FOR THE SUMMER, A (Fr., Eng. titles) **80**
AMERICAN. (R) May, 1964. Pascale Petit, Micheline Presle. Color.

MISTY (CS) **92**
TWENTIETH CENTURY-FOX. (R) July, 1961. David Ladd, Arthur O'Connell. Color.

MITCHELL **95**
ALLIED ARTISTS. (R) June, 1975. Joe Don Baker, Martin Balsam.

MIXED COMPANY **109**
UNITED ARTISTS. (R) October, 1974. Barbara Harris, Joseph Bologna.

MOBY DICK **116**
WARNER BROTHERS. (R) June 30, 1956. Gregory Peck, Richard Basehart, Leo Genn, Orson Welles. Color.

MODEL MURDER CASE, THE **90**
CINEMA V. (R) November, 1964. Ian Hendry, Ronald Fraser.

MODEL SHOP **95**
COLUMBIA. (R) February, 1969. Anouk Aimee, Gary Lockwood, Alexandra Hay. Color.

MODESTY BLAISE **119**
TWENTIETH CENTURY-FOX. (R) June, 1966. Monicavitti, Terence Stamp, Dirk Bogarde. Color.

MOHAMMAD—MESSENGER OF GOD (P) **180**
TARIK FILM DISTRIBUTORS. (R) March, 1977. Anthony Quinn, Irene Papas.

MOHAWK **79**
TWENTIETH CENTURY-FOX. (R) April, 1956. Scott Brady, Rita Gam. Color.

MOLE PEOPLE, THE **78**
UNIVERSAL. (R) December, 1956. John Agar, Cynthia Patrick.

MOLLY MAGUIRES, THE (P) **123**
PARAMOUNT. (R) February, 1970. Richard Harris, Sean Connery, Samantha Eggar. Color.

MOMENT BY MOMENT **102**
UNIVERAL. (R) December, 1978. Lily Tomlin, John Travolta.

MOMENT OF TRUTH, THE (Span., Eng. titles) **110**
RIZZOLI. (R) August, 1965. Miguel Mateo Miguelin. Color.

MOMENT TO MOMENT **108**
UNIVERSAL. (R) February, 1966. Jean Seberg, Honor Blackman. Color.

MONDO CANE **105**
TIMES. (R) April, 1963. Documentary. Color.

MONDO PAZZO **96**
RIZZOLI. (R) January, 1965. Documentary. Color.

MONEY JUNGLE, THE **95**
COMMONWEALTH UNITED. (R) November, 1968. John Ericson, Lola Albright, Leslie Parrish. Color.

MONEY, MONEY, MONEY **115**
CINERAMA. (R) March, 1972. Lino Ventura, Jacques Brel. Color.

MONEY TALKS **87**
UNITED ARTISTS. (R) August, 1972. Allen Funt Documentary. Color.

MONEY TRAP, THE (P) **92**
METRO-GOLDWYN-MAYER. (R) February, 1966. Glenn Ford, Elke Sommer.

MONEY, WOMEN AND GUNS (CS) **80**
UNIVERSAL. (R) January, 1959. Jock Mahoney, Kim Hunter. Color.

MONIQUE **86**
AVCO EMBASSY. (R) April, 1970. Sibylla Kay, Joan Alcom, David Sumner. Color.

MONITORS, THE **92**
COMMONWEALTH UNITED. (R) November, 1969. Guy Stockwell, Susan Oliver. Color.

Length In Mins.

MONKEY IN WINTER ... 104
METRO-GOLDWYN-MAYER. (R) February, 1963. Jean Gabin, Jean-Paul Belmondo.

MONKEYS, GO HOME! ... 101
BUENA VISTA. (R) February, 1967. Maurice Chevalier, Dean Jones. Color.

MONKEY ON MY BACK ... 93
UNITED ARTISTS. (R) May, 1957. Cameron Mitchell, Dianne Foster.

MONKEY'S UNCLE, THE ... 87
BUENA VISTA. (R) July, 1965. Tommy Kirk, Annette. Color.

MONOLITH MONSTERS, THE ... 77
UNIVERSAL. (R) December, 1957. Grant Williams, Lola Albright.

MONSTER ON THE CAMPUS ... 76
UNIVERSAL. (R) November, 1958. Arthur Franz, Joanna Moore.

MONSTER THAT CHALLENGED THE WORLD, THE ... 83
UNITED ARTISTS. (R) June, 1957. Tina Holt, Audrey Dalton.

MONTE CARLO STORY, THE (Technirama) ... 99
UNITED ARTISTS. (R) August, 1957. Marlene Dietrich, Vittorio de Sica. Color.

MONTE WALSH (P) ... 106
NATIONAL GENERAL. (R) October, 1970. Lee Marvin, Jeanne Moreau. Color.

MONTEREY POP ... 80
LEACOCK PENNEBAKER. (R) Otis Redding, Scott Mackenzie, Jimi Hendrix. Color.

MONTY PYTHON AND THE HOLY GRAIL ... 90
CINEMA V. (R) May, 1975. Graham Chapman, John Cleese.

MOON PILOT ... 98
BUENA VISTA. (R) April, 1962. Tom Tryon, Brian Keith. Color.

MOON-SPINNERS, THE ... 118
BUENA VISTA. (R) July, 1964. Hayley Mills, Peter McEnery. Color.

MOON WOLF ... 85
ALLIED ARTISTS. (R) 1966. Carl Moehner, Ann Savo. b&w.

MOON ZERO TWO ... 100
WARNER BROS. (R) March, 1970. James Olson, Catherina Von Schell. Color.

MOONFLEET (CS) ... 89
METRO-GOLDWYN-MAYER. (R) June 24, 1955. Stewart Granger, Viveca Lindfors. Color.

MOONRAKER (P) ... 126
UNITED ARTISTS. (R) June, 1979. Roger Moore, Lois Chiles.

MOONRUNNERS ... 102
UNITED ARTISTS. (R) July, 1975. Jim Mitchum, Kiel Martin.

MOONSHINE WAR, THE (P) ... 100
MGM. (R) August, 1970. Patrick McGoohan, Richard Widmark. Color.

MORALS SQUAD ... 57
BRENNER. (R) May, 1960. Bob O'Connell, Beverly Bennett.

MORE ... 120
CINEMA V. (R) December, 1969. Mimsy Farmer, Heinz Engelmann, Klaus Grunberg. Color.

MORE AMERICAN GRAFFITI (P) ... 111
UNIVERSAL. (R) August, 1979. Candy Clark, Bo Hopkins, Ron Howard.

MORE DEAD THAN ALIVE ... 104
UNITED ARTISTS. (R) May, 1969. Clint Walker, Vincent Price, Anne Francis. Color.

MORE THAN A MIRACLE (Franscope) ... 105
MGM. (R) November, 1967. Sophia Loren, Omar Sharif. Color.

MORGAN ... 97
CINEMA V. (R) April, 1966. Vanessa Redgrave, David Warner.

MORGAN THE PIRATE (CS) ... 95
METRO-GOLDWYN-MAYER. (R) July, 1961. Steve Reeves, Valerie Larange. Color.

MORITURI (CS) ... 123
TWENTIETH CENTURY-FOX. (R) August, 1965. Marlon Brando, Yul Brynner.

MOSQUITO SQUADRON ... 90
UNITED ARTISTS. (R) November, 1970. David McCallum, Suzanne Neve. Color.

MOST DANGEROUS MAN ALIVE ... 82
COLUMBIA. (R) June, 1961. Ron Randell, Debra Paget.

MOTHER KUSTERS GOES TO HEAVEN (German with English titles) ... 108
NEW YORKER FILMS. (R) March, 1977. Brigette Mira, Ingrid Caven.

MOTHRA ... 100
COLUMBIA. (R) May, 1962. Franky Sakari, Hiroshi Koizymix. Color.

MOTORCYCLE GANG ... 78
AMERICAN INTERNATIONAL. (R) October, 1957. Steve Terrell, Ann Neyland.

MOUNTAIN, THE (VV) ... 105
PARAMOUNT. (R) November, 1956. Spencer Tracy, Robert Wagner, Claire Trevor. Color.

MOUNTAIN MEN, THE ... 102
COLUMBIA. (R) July, 1980. Charlton Heston, Brian Keith.

MOUNTAIN ROAD, THE ... 102
COLUMBIA. (R) June, 1960. James Stewart, Lisa Lu.

MOUSE THAT ROARED, THE ... 83
COLUMBIA. (R) November, 1959. Peter Sellers, Jean Seberg. Color.

MOVE (P) ... 90
20TH CENTURY-FOX. (R) August, 1970. Elliott Gould, Genevieve Waite, Paula Prentiss. Color.

MOVE OVER DARLING (CS) ... 103
TWENTIETH CENTURY-FOX. (R) December, 1963. Doris Day, James Garner. Color.

MOVIE MOVIE ... 105
WARNER BROS. (R) January, 1979. George C. Scott, Trish Van Devere.

MOZAMBIQUE (TE) ... 98
SEVEN ARTS. (R) February and March, 1966. Steve Cochran, Hildegarde Neff. Color.

MRS. BROWN YOU'VE GOT A LOVELY DAUGHTER ... 95
METRO-GOLDWYN-MAYER. (R June, 1968. Peter Noone, Barry Whitwam, Keith Hopwood. Color.

MRS. POLLIFAX, SPY ... 110
UNITED ARTISTS. (R) March, 1971. Rosalind Russell, Darren McGavin. Color.

MUGGER, THE ... 74
UNITED ARTISTS. (R) November, 1958. Kent Smith, Nan Martin.

MUMMY'S SHROUD, THE ... 90
20TH-FOX-SEVEN ARTS. (R) March, 1967. Andre Morell, John Phillips. Color.

MUMSY, NANNY, SONNY, GIRLY ... 101
CINERAMA. (R) May, 1970. Michael Bryant, Ursula Howells, Vanessa Howard. Color.

MUNSTER GO HOME! ... 96
UNIVERSAL. (R) July, 1966. Fred Gwynne, Yvonne DeCarlo. Color.

MUPPET MOVIE, THE ... 94
ASSOCIATED FILM DIST. (R) June, 1979. Guest Stars.

MURDER AHOY! ... 93
METRO-GOLDWYN-MAYER. (R) September, 1964. Margaret Rutherford, Lionel Jeffries.

MURDER AT 45 R.P.M. (Fr.) ... 105
MGM. (R) 1960. Danielle Darrieux, Michael Auclair.

MURDER AT THE GALLOP (P) ... 81
METRO-GOLDWYN-MAYER. (R) July, 1963. Margaret Rutherford, Robert Morley.

MURDER BY CONTRACT ... 81
COLUMBIA. (R) December, 1958. Vince Edwards, Michael Granges.

MURDER BY DECREE ... 120
AVCO EMBASSY. (R) February, 1979. Christopher Plummer, James Mason.

MURDER CZECH STYLE (Czech.; Eng. titles) ... 90
ROYAL. (R) August, 1968. Rudolf Hrusinsky, Kveta Fialova, Vaclav Voska. b&w.

MURDER GAME, THE ... 75
20TH CENTURY-FOX. (R) April, 1966. Ken Scott, Maralandi.

MURDER, INC. ... 103
TWENTIETH CENTURY-FOX. (R) June, 1960. Stuart Whitman, May Britt.

MURDER IS MY BEAT ... 77
ALLIED ARTISTS. (R) February 27, 1955. Barbara Payton, Paul Langton.

MURDER MUST FOUL (P) ... 90
METRO-GOLDWYN-MAYER. (R) September, 1964. Margaret Rutherford, Stringer Davis.

MURDER ON APPROVAL (Brit.) ... 70
RKO RADIO. (R) May 30, 1956. Tom Conway.

MURDER ON THE CAMPUS ... 61
COLORAMA. (R) September, 1963. Terence Longdon, Donald Gray.

MURDER ON THE ORIENT EXPRESS (P) ... 127
PARAMOUNT. (R) December, 1974. Albert Finney, Lauren Bacall, Ingrid Bergman.

MURDER SHE SAID ... 87
METRO-GOLDWYN-MAYER. (R) January, 1962. Margaret Rutherford, Arthur Kennedy.

MURDERERS' ROW ... 108
COLUMBIA. (R) December, 1966. Dean Martin, Ann-Margaret. Color.

MURDERS IN THE RUE MORGUE ... 86
AMERICAN INTERNATIONAL. (R) July, 1971. Jason Robards, Christine Kaufman, Herbert Lom. Color.

MURIEL ... 115

Length In Mins.

LOPERT. (R) November, 1963. Delphine Seyrig, Jean-Pierre Kerien. Color.

MURMUR OF THE HEART 118
READE-PALOMAR. (R) October, 1971. Lea Massari, Benoit Ferreux. Color.

MURPH, THE SURF 101
AMERICAN INTERNATIONAL.(R) March, 1975. Robert Conrad, Don Stroud.

MURPHY'S WAR (P) 108
PARAMOUNT. (R) July, 1971. Peter O'Toole, Sian Phillips, Philippe Noiret. Color.

MUSCLE BEACH PARTY (P) 94
AMERICAN INTERNATIONAL. (R) March, 1964. Frankie Avalon, Annette Funicello. Color.

MUSIC BOX KID, THE 74
UNITED ARTISTS. (R) June, 1960. Ronald Foster, Luane Pattern.

MUSIC LOVERS, THE 122
UNITED ARTISTS. (R) 1971. Richard Chamberlain, Glenda Jackson. Color.

MUSIC MAN (T) 151
WARNER BROS. (R) April, 1962. Robert Preston, Shirley Jones. Color.

MUSTANG 73
UNITED ARTISTS. (R) February, 1959. Jack Beutel.

MUTATIONS, THE 92
COLUMBIA. (R) June, 1974. Donald Pleasance, Tom Baker.

MUTINY IN OUTER SPACE 81
ALLIED ARTISTS. (R) 1964. Glenn Langan, Susannah York. b&w.

MUTINY ON THE BOUNTY (P) 179
METRO-GOLDWYN-MAYER. (R) Special. November, 1962. Marlon Brando, Trevor Howard. Color.

MY BLOOD RUNS COLD (P) 104
WARNER BROS. (R) March, 1965. Troy Donahue, Joey Heatherton.

MY BODYGUARD 96
TWENTIETH CENTURY-FOX. (R) July, 1980. Chris Makepeace, Adam Baldwin.

MY DOG BUDDY 77
COLUMBIA. (R) August, 1960. Ken Curtis.

MY FAIR LADY (Super P-70) 170
WARNER BROS. (R) Special. October, 1964. Audrey Hepburn, Rex Harrison. Color.

MY GEISHA (T) 120
PARAMOUNT. (R) July, 1962. Shirley MacLaine, Yves Montand. Color.

MY GUN IS QUICK 88
UNITED ARTISTS. (R) August, 1957. Robert Bray, Pamela Duncan.

MY LOVER, MY SON 98
MGM. (R) March, 1970. Romy Schneider, Donald Houston. Color.

MY MAN GODFREY (CS) 92
UNIVERSAL. (R) December, 1957. June Allyson, David Niven. Color.

MY NAME IS NOBODY (P) 115
UNIVERSAL. (R) July, 1974. Terence Hills, Henry Fonda.

MY NIGHT AT MAUD'S (French.; Eng. titles) 105
PATHE CONTEMPORARY. (R) February, 1971. Jean-Louis Trintignant, Francoise Fabian. b&w.

MY SEVEN LITTLE SINS (Fr.) 98
KINGSLEY. (R) October, 1956. Maurice Chevalier, Della Scala.

MY SIDE OF THE MOUNTAIN 100
PARAMOUNT. (R) March, 1969. Teddy Eccles, Theodore Bikel, Tudi Wiggins. Color.

MY SISTER EILEEN (CS) 108
COLUMBIA. (R) October, 1955. Janet Leigh, Jack Lemmon, Betty Garrett. Color.

MY SISTER, MY LOVE (Swed.; Eng. titles) 96
SIGMA III. (R) February, 1967. Per Oscarsson, Bibi Andersson.

MY SIX LOVES (VV) 101
PARAMOUNT. (R) April, 1963. Debbie Reynolds, Cliff Robertson. Color.

MY SON, THE HERO 111
UNITED ARTISTS. (R) October, 1963. Pedro Armend Ariz, Jacqueline Sassard. Color.

MY SWEET CHARLIE 97
UNIVERSAL. (R) February, 1970. Patty Duke, Al Freeman, Jr. Color.

MY WAY 95
JOSEPH BRENNER ASSOCIATES. (R) February, 1974. Joe Stewardson, Richard Loring.

MY WIFE'S HUSBAND (Franscope) 85
LOPERT. (R) February, 1965. Fernandel, Bourvil. b&w.

MYRA BRECKINRIDGE 94
20TH CENTURY-FOX. (R) June, 1970. Mae West, Raquel Welch, John Houston, Rex Reed. Color.

MYSTERIANS, THE (CS) 85
METRO-GOLDWYN-MAYER. (R) May, 1959. Kenji Sahara, Yumi Shirakawa.

MYSTERIOUS ISLAND 101
COLUMBIA. (R) August, 1961. Michael Craig, Joan Greenwood. Color.

MYSTERIOUS ISLAND OF CAPTAIN NEMO, THE 96
CINERAMA. (R) June, 1974. Omar Sharif, Philippe Nicaud.

MYSTERY OF THE BLACK JUNGLE 72
REPUBLIC. (R) October 20, 1955. Lex Barker.

MYSTERY OF THUG ISLAND (Euroscope) 88
COLUMBIA. (R) May, 1966. Guy Madison, Peter Van Eyck. Color.

MYSTERY SUBMARINE 90
UNIVERSAL. (R) February, 1963. Edward Judd, James Robertson Justice.

N

NADA GANG, THE (French; Eng. titles) 106
NEW LINE CINEMA. (R) November, 1974. Fabio Testi, Maurice Garrel.

NAKED AFRICA 69
AMER. INT. PICS. (R) August, 1957. Quentin Reynolds. Color.

NAKED AMAZON (Braz.-Eng. narration) 67
TIMES. (R) July, 1955. Documentary.

NAKED AMONG THE WOLVES (Ger.; Eng. titles) 100
LOPERT. (R) January, 1967. Erwin Geschonneck, Fred Delmare.

NAKED AND THE DEAD, THE 131
WARNER BROS. (R) August, 1958. Aldo Ray, Cliff Robertson.

NAKED ANGELS 81
FAVORITE FILMS. (R) July, 1969. Michael Greene, Jennifer Gan.

NAKED APE, THE 85
UNIVERSAL. (R) August, 1973. Johnny Crawford, Victoria Principal.

NAKED DAWN, THE 82
UNIVERSAL. (R) November, 1955. Arthur Kennedy, Betta St. John. Color.

NAKED EARTH (Cs) 96
TWENTIETH CENTURY-FOX. (R) June, 1959. Richard Todd, Juliette Green.

NAKED EDGE 99
UNITED ARTISTS. (R) May, 1961. Gary Cooper, Deborah Kerr.

NAKED GUN, THE 69
ASSOCIATED FILM. (R) November, 1956. Willard Parker, Mara Corday.

NAKED HEART, THE (Brit.) 96
ASSOCIATED ARTISTS. (R) March, 1955. Michele Morgan, Kieron Moore.

NAKED HILLS, THE 78
ALLIED ARTISTS. (R) June 17, 1956. David Wayne, Keenan Wynn. Color.

NAKED IN THE SUN 72
ALLIED ARTISTS. (R) September, 1957. James Craig, Barton MacLaine.

NAKED KISS, THE 93
ALLIED ARTISTS. (R) 1964. Constance Towers, Anthony Eisley.

NAKED MAJA, THE (T) 111
UNITED ARTISTS. (R) April, 1959. Ava Gardner, Anthony Franciosa. Color.

NAKED PARADISE 71
AMERICAN INTERNATIONAL. (R) January, 1957. Richard Denning, Beverly Garland. Color.

NAKED PREY, THE (P) 94
PARAMOUNT. (R) April, 1966. Cornel Wilde, Gert Van Der Berg. Color.

NAKED RUNNER, THE (TS) 103
WARNER BROS.-SEVEN ARTS. (R) July, 1967. Frank Sinatra, Peter Vaughan. Color.

NAKED SEA 70
RKO RADIO. (R) December 13, 1955. Documentary. Color.

NAKED STREET, THE 64
UNITED ARTISTS. (R) August, 1955. Farley Granger, Anne Bancroft, Anthony Quinn.

NAKED WORLD 92
TIMES. (R) November, 1968. Documentary. Color.

NAME OF THE GAME IS KILL, THE 88
FANFARE. (R) May, 1968. Jack Lord, Susan Strasberg. Color.

NAMU THE KILLER WHALE 89
UNITED ARTISTS. (R) July, 1966. Robert Lansing, John Anderson. Color.

NANAMI (Japanese; Eng. titles) 87

Length In Mins.

GOLDEN EAGLE. (R) April, 1969. Akio Takahashi, Kuniko Ishii, Koji Mitsui. b&w.
NANNY, THE 93
TWENTIETH CENTURY-FOX. (R) November, 1965. Bette Davis, Wendy Craig.
NAPOLEON AND SAMANTHA 91
BUENA VISTA. (R) July, 1972. Johnny Whitaker, Jodie Foster, Michael Douglas. Color.
NASHVILLE (P) 159
PARAMOUNT. (R) June, 1975. Ronee Blakley, Keith Carradine, Lily Tomlin.
NASTY HABITS 91
BRUT PRODS. (R) March, 1977 Glenda Jackson, Melina Mercouri, Geraldine Page.
NASTY RABBIT, THE (T) 85
FAIRWAY INT. (R) December, 1964. Mischa Terr, Arch Hall, Jr. Color.
NATIONAL LAMPOON'S ANIMAL HOUSE 109
UNIVERSAL. (R) July, 1978. John Belushi, Tim Matheson.
NATURAL ENEMIES 100
CINEMA 5. (R) November, 1979. Hal Holbrook, Louise Fletcher.
NAVAJO RUN 79
AMERICAN INTERNATIONAL. (R) December, 1964. Johnny Seven, Virginia Vincent.
NAVY VS. THE NIGHT MONSTER, THE 90
FEAT. FILM CORP. OF AMER. (R) 1965. Anthony Eisley, Mamie Van Doren. Color.
NAVY WIFE 83
ALLIED ARTISTS. (R) May 20, 1956. Joan Bennett, Gary Merrill, Shirley Yamaguchi.
NAZARIN (Spanish; Eng. titles) 92
ALTURA. (R) June, 1968. Francisco Rabal, Rita Macedo, Marga Lopez. b&w.
NEA (French; Eng. titles) 101
LIBRA FILMS. (R) August, 1978 Samy Frey, Ann Zacharias.
NEARLY A NASTY ACCIDENT 86
UNIVERSAL. (R) April, 1962. Jimmy Edwards, Kenneth Connor.
NECROMANCY 82
CINERAMA. (R) November, 1972. Orson Welles, Pamela Franklin. Color.
NED KELLY 100
UNITED ARTISTS. (R) June, 1970. Mick Jagger, Allen Bickford, Geoff Gilmour. Color.
NEGATIVES 90
CONT. (R) November, 1968. Peter McEnery, Diane Cilento, Glenda Jackson. Color.
NELSON AFFAIR, THE 118
UNIVERSAL. (R) May, 1972. Glenda Jackson, Peter Finch. Color.
NEPTUNE FACTOR, THE (P) 97
TWENTIETH CENTURY-FOX. (R) May, 1973. Ben Gazzara, Yvette Mimieux. Color.
NETWORK 121
UNITED ARTISTS. (R) February, 1977. Faye Dunaway, William Holden.
NEVADA SMITH (P) 130
PARAMOUNT. (R) July, 1966. Steve McQueen, Karl Malden. Color.
NEVER A DULL MOMENT 100
BUENA VISTA. (R) June, 1968. Dick Van Dyke, Edward G. Robinson, Dorothy Provine. Color.
NEVER LET GO 90
CONTINENTAL. (R) June, 1963. Richard Todd, Peter Sellers.
NEVER LOVE A STRANGER 91
ALLIED ARTISTS. (R) May 4, 1958. John Drew Barrymore, Lita Milan.
NEVER ON SUNDAY 97
LOPERT PICTURES. (R) October, 1960.
NEVER PUT IT IN WRITING 93
ALLIED ARTISTS. (R) April, 1964. Pat Boone, Milo O'Shea.
NEVER SAY GOODBYE 97
UNIVERSAL. (R) March, 1956. Rock Hudson, Cornel Borchers, George Sanders. Color.
NEVER SO FEW (Cs) 124
METRO-GOLDWYN-MAYER. (R) December, 1959. Frank Sinatra, Gina Lollobrigida. Color.
NEVER STEAL ANYTHING SMALL (Cs) 92
UNIVERSAL. (R) March, 1959. James Cagney, Shirley Jones. Color.
NEVER TOO LATE (P) 105
WARNER BROTHERS. (R) December, 1965. Paul Ford, Maureen O'Sullivan. Color.
NEW ANGELES, THE (Ital., Eng. titles) 94
PROMENADE. (R) May, 1965. Episodes on Italian Life.
NEW CENTURIONS, THE 103
COLUMBIA. (R) August, 1972. George C. Scott, Stacy Keach, Jane Alexander. Color.

Length In Mins.

NEW INTERNS, THE 123
COLUMBIA. (R) June 1964. Michael Callan, Barbara Eden.
NEW KIND OF LOVE, A 110
PARAMOUNT. (R) October, 1963. Paul Newman, Joanne Woodward. Color.
NEW LAND, THE 161
WARNER BROS. (R) October, 1973. Max von Sydow, Liv Ullmann.
NEW LEAF, A 102
PARAMOUNT. (R) April, 1971. Walter Matthau, Elaine May. Color.
NEW ORLEANS UNCENSORED 70
COLUMBIA. (R) March, 1955. Arthur Franz, Beverly Garland.
NEW YORK CONFIDENTIAL 87
WARNER BROTHERS (R) March 12, 1955. Broderick Crawford, Marilyn Maxwell.
NEW YORK, NEW YORK 155
UNITED ARTISTS. (R) June, 1977. Liza Minnelli, Robert De Niro.
NEWMAN'S LAW 98
UNIVERSAL. (R) May, 1974. George Peppard, Roger Robinson.
NEXT MAN, THE 108
ALLIED ARTISTS. (R) November, 1976. Sean Connery, Cornelia Sharpe.
NICE GIRL LIKE ME, A 90
AVCO EMBASSY. (R) December, 1969. Barbara Ferris, Harry Andrews, Gladys Cooper. Color.
NICE LITTLE BANK THAT SHOULD BE ROBBED, A (Cs) 87
TWENTIETH CENTURY-FOX. (R) December, 1958. Tom Ewell, Mickey Rooney.
NICHOLAS AND ALEXANDRA 187
COLUMBIA. (R) December, 1971. Michael Jayston, Janet Suzman. Color.
NICKEL RIDE, THE 99
20TH CENTURY-FOX. (R) January, 1975. Jason Miller.
NICKELODEON (P) 122
COLUHBIA. (R) December, 1976. Ryan O'Neal, Burt Reynolds.
NIGHT AMBUSH (VV) 93
RANK FILM DISTRIBUTORS OF AMERICA. (R) July, 1958. Dick Bogarde, Marius Goring.
NIGHT CREATURES 81
UNIVERSAL. (R) June, 1962. Peter Cushing, Yvonne Romain. Color.
NIGHT DIGGER, THE 110
MGM. (R) May, 1971. Patricia Neal, Pamela Brown. Color.
NIGHT FIGHTERS, THE 74
ALLIED ARTISTS. September, 1960. Robert Mitchum, Dan O'Herlihy.
NIGHT FREIGHT 79
ALLIED ARTISTS. (R) August 28, 1955. Forrest Tucker, Barbara Britton.
NIGHT GAMES (Swed.; Eng. titles) 104
MONDIAL. (R) December, 1966. Ingrid Thulin, Keve Hjelm.
NIGHT HOLDS TERROR, THE 86
COLUMBIA. (R) September, 1955. Jack Kelly, Hildy Parks.
NIGHT MOVES (P) 100
WARNER BROS. (R) June, 1975. Gene Hackman.
NIGHT MUST FALL 105
METRO-GOLDWYN-MAYER. (R) March, 1964. Albert Finney, Susan Hampshire
NIGHT OF DARK SHADOWS 97
METRO-GOLDWYN-MAYER. (R) September, 1971. David Shelby, Grayson Hall. Color.
NIGHT OF THE BLOOD FEAST 65
AMERICAN INTERNATIONAL. (R) August, 1958. Michael Emmet, Angela Greeney.
NIGHT OF THE FOLLOWING DAY, THE 93
UNIVERSAL. (R) March, 1969. Marlon Brando, Rita Moreno, Richard Boone. Color.
NIGHT OF THE GENERALS, THE (P) 148
COLUMBIA. (R) February, 1967. Peter O'Toole, Omar Sharif. Color.
NIGHT OF THE GRIZZLY (TE) 102
PARAMOUNT. (R) May, 1966. Clint Walker, Martha Hyer. Color.
NIGHT OF THE HUNTER, THE 93
UNITED ARTISTS. (R) September, 1955. Robert Mitchum, Shelley Winters.
NIGHT OF THE IGUANA, THE 125
METRO-GOLDWYN-MAYER. (R) August, 1964. Richard Burton, Ava Gardner.
NIGHT OF THE LEPUS 88
METRO-GOLDWYN-MAYER. (R) October, 1973. Stuart Whitman, Janet Leigh. Color.
NIGHT OF THE LIVING DEAD, THE 90

Length In Mins.

Length In Mins.

CONT. (R) October, 1968. Judith O'Dea, Russell Streiner. b&w.

NIGHT OF THE QUARTER MOON (Cs) 96
METRO-GOLDWYN-MAYER. (R) February, 1959. Julie London, John Drew Barrymore.

NIGHT PASSAGE (T) 90
UNIVERSAL. (R) August, 1957. James Stewart, Audie Murphy. Color.

NIGHT PORTER, THE 117
AVCO-EMBASSY. (R) October, 1975. Dirk Bogarde, Charlotte Rampling.

NIGHT RUNNER, THE 79
UNIVERSAL. (R) February, 1957. Ray Danton, Colleen Miller.

NIGHT THE WORLD EXPLODED, THE 64
COLUMBIA. (R) June, 1957. Kathryn Grant, William Leslie.

NIGHT THEY KILLED RASPUTIN, THE 87
BRIGADIER FILMS. (R) 1962. Edmund Purdom. John Drew Barrymore.

NIGHT THEY RAIDED MINSKY'S, THE 100
UNITED ARTISTS. (R) December, 1968. Jason Robards, Britt Eklund. Color.

NIGHT TIDE 84
AMERICAN INTERNATIONAL. (R) April, 1963. Dennis Hopper, Linda Lawson.

NIGHT TO REMEMBER, A 123
RANK FILM DISTRIBUTION OF AMERICA. (R) March, 1959. Kenneth More, Ronald Allen.

NIGHT TRAIN TO PARIS 65
TWENTIETH CENTURY-FOX. (R) November, 1964. Leslie Nelson, Alizia Gur.

NIGHT VISITOR, THE 102
UMC PICTURES. (R) February, 1971. Max von Sydow, Trevor Howard. Color.

NIGHT WALKER, THE 86
UNIVERSAL. (R) January, 1965. Robert Taylor, Barbara Stanwyck.

NIGHT WATCH, THE (Fr. Eng. titles) 118
CONSORT/ORION. (R) April, 1964. Michael Constantin, Jean Keraudy.

NIGHT WATCH 99
AVCO-EMBASSY. (R) August, 1973. Elizabeth Taylor, Laurence Harvey.

NIGHTCOMERS, THE 95
AVCO EMBASSY. (R) February, 1972. Marlon Brando, Stephanie Beacham. Color.

NIGHTFALL 78
COLUMBIA. (R) February, 1957. Aldo Ray, Brian Keith.

NIGHTMARE (H) 83
UNIVERSAL. (R) June, 1964. David Knight, Moira Redmond.

NIGHTMARE 89
UNITED ARTISTS. (R) June, 1956. Edward G. Robinson, Kevin McCarthy.

NIGHTMARE CASTLE 93
ALLIED ARTISTS. (R) July, 1966. Barbara Steele, Paul Miller.

NIGHTMARE IN THE SUN 81
ZODIAC. (R) January, 1965. Ursula Andress, John Derek. Color.

NIGHTMARE IN WAX 87
CROWN INTL. (R) December, 1969. Scott Brady, Cameron Mitchell.

NIGHTWING 105
COLUMBIA. (R) June, 1979. Nick Mancuso, David Warner.

NIGHTS OF CABIRIA (Ital.) 110
LOPERT. (R) November, 1957. Guiletta Massina, Francois Ferrier.

NIGHTS OF LUCRETIA BORGIA, THE (Totalscope) 106
COLUMBIA. (R) August, 1960. Belinda Lee, Jacques Sernas. Color.

NIJINSKY 125
PARAMOUNT. (R) March, 1980. Alan Bates, George De La Pena.

NIKKI, WILD DOG OF THE NORTH 74
BUENA VISTA. (R) July, 1961. Jean Coutu, Emile Genest. Color.

NINE DAYS OF ONE YEAR 102
ARTKINO. (R) December, 1964. Alexi Batalov, Innokenty Smoktunovsky.

NINE HOURS TO RAMA (Cs) 125
TWENTIETH CENTURY-FOX. (R) April, 1963. Horst Buchholz, Jose Ferrer. Color.

NINE LIVES OF FRITZ THE CAT, THE 76
AMERICAN INTERNATIONAL. (R) July, 1974. Cartoon.

1900 245
PARAMOUNT. (R) December, 1977. Robert De Niro, Gerard Depardieu.

1984 91
COLUMBIA. (R) September, 1956. Edmond O'Brien, Jan Sterling.

1941 (P) 118
UNIVERSAL. (R) December, 1979. Dan Aykroyd, Ned Beatty.

90 DEGREES IN THE SHADE 90
LANDAU/UNGER. (R) October, 1966. Anne Heywood, James Booth.

99 AND 44/100% DEAD (P) 97
TWENTIETH CENTURY-FOX. (R) July, 1974. Richard Harris, Edmond O'Brien.

99 WOMEN 90
COMMONWEALTH UNITED. (R) April, 1969. Maria Schell, Mercedes McCambridge. Color.

92 IN THE SHADE 91
UNITED ARTISTS. (R) August, 1975. Peter Fonda, Warren Oates.

NO BLADE OF GRASS 97
MGM. (R) October, 1970. Nigel Davenport, Jean Wallace. Color.

NO DOWN PAYMENT (Cs) 105
TWENTIETH CENTURY-FOX. (R) October, 1957. Joanne Woodward, Sherry North.

NO DRUMS, NO BUGLES 85
CINERAMA. (R) August, 1971. Martin Sheen. Color.

NO MAN IS AN ISLAND 114
UNIVERSAL. (R) October, 1962. Jeffrey Hunter, Marshall Thompson. Color.

NO MAN'S WOMAN 70
REPUBLIC. (R) October 27, 1955. Marie Windsor, Patrick Knowles.

NO MORE EXCUSES 82
IMPACT FILMS. (R) January, 1968. Robert Downey, Allen Noel.

NO, MY DARLING DAUGHTER! 96
ZENITH. (R) February, 1964. Michael Redgrave, Michael Craig.

NO NAME ON THE BULLET (Cs) 77
UNIVERSAL. (R) February, 1959. Audie Murphy, Charles Drake. Color.

NO NUKES 103
WARNER BROS. (R) July, 1980. Jackson Browne, David Crosby.

NO PLACE TO HIDE 72
ALLIED ARTISTS. (R) August 26, 1956. David Brian, Keenan Wynn. Color.

NO PLACE TO LAND (N) 78
REPUBLIC. (R) October, 1958. John Ireland, Mari Blanchard.

NO TIME FOR BREAKFAST (French; Eng. titles) 100
BOURIA FILM ENTERPRISES. (R) September, 1978. Annie Girardot, Jean-Pierre Cassel.

NO TIME FOR SERGEANTS 111
WARNER BROS. (R) July, 1958. Andy Griffith, Myron McCormick.

NO TIME TO BE YOUNG 82
COLUMBIA. (R) August, 1957. Robert Vaughn, Roger Smith.

NO TREE IN THE STREET 96
SEVEN ARTS. (R) January, 1964. Sylvia Syms, Herbert Lom.

NO WAY TO TREAT A LADY 108
PARAMOUNT. (R) March, 1968. Rod Steiger, Lee Remick. Color.

NOBODY WAVED GOODBYE 80
CINEMA V. (R) April, 1965. Peter Kastner, Julie Biggs.

NOBODY'S PERFECT (TE) 103
UNIVERSAL. (R) January, 1968. James Whitmore, Doug McClure. Color.

NONE BUT THE BRAVE (P) 110
WARNER BROS. (R) February, 1965. Frank Sinatra, Clint Walker. Color.

NOOSE FOR A GUNMAN 69
UNITED ARTISTS. (R) May, 1960. Jim Davis, Lyn Thomas.

NORMA RAE 110
TWENTIETH CENTURY-FOX. (R) March, 1979. Sally Field, Beau Bridges, Ron Leibman.

NORMAN . . . IS THAT YOU? 91
UNITED ARTISTS. (R) September, 1976. Redd Foxx, Pearl Bailey.

NORTH AVENUE IRREGULARS, THE 100
BUENA VISTA. (R) February, 1979. Edward Herrmann, Barbara Harris.

NORTH BY NORTHWEST (VV) 136
METRO-GOLDWYN-MAYER. (R) July, 1959. Cary Grant, Eva Marie Saint. Color.

NORTH DALLAS FORTY 120
PARAMOUNT. (R) August, 1979. Nick Nolte, Mac Davis.

NORTH TO ALASKA (Cs) 122
TWENTIETH CENTURY-FOX. (R) November, 1960. John Wayne, Stewart Granger. Color.

Length In Mins.

NORWOOD ... 96
PARAMOUNT. (R) May, 1970. Glen Campbell, Kim Darby, Joe Nameth. Color.

NOSFERATU, THE VAMPYRE (Ger., Eng. titles) ... 105
TWENTIETH CENTURY-FOX. (R) October, 1979. Klaus Kinski, Isabelle Adjani.

NOT AS A STRANGER ... 138
UNITED ARTISTS. (R) July, 1955. Robert Mitchum, Olivia de Havilland, Frank Sinatra, Gloria Grahame.

NOT MINE TO LOVE (Hebrew: Eng. titles) ... 90
MEADOW. (R) March, 1969. Oded Kotler, Shuy Osherov, Judith Soleh. b&w.

NOT OF THIS EARTH ... 87
ALLIED ARTISTS. (R) March 3, 1957. Paul Birch, Beverly Garland.

NOT ON YOUR LIFE (Sp., Eng. titles) ... 90
PATHE-CONT. (R) March, 1965. Nino Manfredi, Emma Penelia.

NOT WITH MY WIFE, YOU DON'T ... 118
WARNER BROS. (R) November, 1966. Tony Curtis, Virna Lisi. Color.

NOTHING BUT A MAN ... 92
CINEMA V. February, 1965. Ivan Dixon, Abbey Lincoln.

NOTHING BUT THE BEST ... 99
ROYAL. July, 1964. Alan Bates, Millicent Martin. Color.

NOTORIOUS LANDLADY, THE ... 128
COLUMBIA. (R) July, 1962. Kim Novak, Jack Lemmon.

NOW YOU SEE HIM, NOW YOU DON'T ... 88
BUENA VISTA. (R) July, 1972. Kurt Russell, Cesar Romero, Joe Flynn. Color.

NOWHERE TO GO ... 87
METRO-GOLDWYN-MAYER. (R) March, 1959. George Nader, Bernard Lee.

NUDE BOMB, THE ... 94
UNIVERSAL. (R) May, 1980. Don Adams, Sylvia Kirstel.

NUMBER ONE ... 105
UNITED ARTISTS. (R) August, 1969. Charlton Heston, Jessica Walter, Bruce Dern. Color.

NUN AT THE CROSSROADS, A ... 100
UNIVERSAL. (R) March, 1970. Rosanna Schiaffino, John Richardson. Color.

NUN'S STORY, THE ... 154
WARNER BROS. (R) July, 1959. Audrey Hepburn, Peter Finch. Color.

NUNZIO ... 86
UNIVERSAL. (R) May, 1978. David Proval, James Andronica.

NURSE ON WHEELS ... 96
JANUS. (R) November, 1963. Juliet Mills, Ronald Lewis.

NUTTY, NAUGHTY CHATEAU ... 102
LOPERT. (R) October, 1964. Monica Vitti, Curt Jurgens. Color.

NUTTY PROFESSOR, THE ... 107
PARAMOUNT. (R) July, 1963. Jerry Lewis, Stella Stevens. Color.

O

O LUCKY MAN ... 177
WARNER BROS. (R) June, 1973. Malcolm McDowell, Ralph Richardson, Rachel Roberts. Color.

O.S.S. 117 IS NOT DEAD (D) ... 81
REPUBLIC. (R) February, 1959. Magali Noel, Ivan Desney.

OASIS (Cs) ... 84
TWENTIETH CENTURY-FOX. (R) December, 1956. Michele Morgan, Piere Brasseuer, Cornell Borchers. Color.

OBLONG BOX, THE ... 95
AMER. INT. PICS. (R) June. 1969. Vincent Price, Christopher Lee, Alaistair Williamson. Color.

OBSESSION ... 98
COLUMBIA. (R) August, 1976. Cliff Robertson, Genevieve Bujold.

OBSESSION (Swedish; Eng. titles) ... 104
O.R.P. (R) August, 1968. Matthias Hendrickson, Maude Adelson, Lars Lind. Color.

OCEAN DRIVE ... 103
COLUMBIA. (R) July, 1959. Edmund O'Brien, Joanne Dru.

OCEAN'S ELEVEN (P) ... 128
WARNER BROS. (R) August, 1960. Frank Sinatra, Dean Martin. Color.

ODD COUPLE, THE ... 105
PARAMOUNT. (R) July, 1968. Jack Lemmon, Walter Matthau. Color.

ODE TO BILLY JOE ... 106
WARNER BROS. (R) June, 1976. Robby Benson, Glynnis O'Connor.

ODESSA FILE, THE (P) ... 128
COLUMBIA. (R) October, 1975. Jon Voight, Maximillian Schell.

ODONGO (Cs) ... 83
COLUMBIA. (R) November, 1956. MacDonald Carey, Rhonda Fleming. Color.

OEDIPUS THE KING ... 97
UNIVERSAL. (R) December, 1968. Christopher Plummer, Orson Welles, Lilli Palmer. Color.

OF HUMAN BONDAGE ... 98
METRO-GOLDWYN-MAYER. (R) September, 1964. Kim Novak, Laurence Harvey.

OF LIFE AND LOVE (Ital.; Eng. titles) ... 103
DCA DIST. (R) 1958. Anna Magnani, Walter Chiara. b&w.

OF LOVE AND DESIRE (Cs) ... 97
TWENTIETH CENTURY-FOX. (R) September, 1963. Merle Oberon, Steve Cochran. Color.

OFFENCE, THE ... 112
UNITED ARTISTS. (R) May, 1973. Sean Connery, Trevor Howard. Color.

OH DAD, POOR DAD, MAMMA'S HUNG YOU IN THE CLOSET AND I'M FEELIN SO SAD ... 86
PARAMOUNT. (R) February, 1967. Rosalind Russell, Robert Morse. Color.

OH, GOD! ... 104
WARNER BROS. (R) October, 1977. George Burns, John Denver.

OH HEAVENLY DOG ... 103
TWENTIETH CENTURY-FOX. (R) August, 1980. Chevy Chase, Benji.

OH, MEN! OH, WOMEN! (Cs) ... 90
TWENTIETH CENTURY-FOX. (R) February, 1957. Ginger Rogers, Dan Dailey, David Niven. Color.

OH! WHAT A LOVELY WAR (P) ... 139
PARAMOUNT. (R) October, 1969. Laurence Olivier, Michael Redgrave, Maggie Smith. Color.

OHAYO (Japanese; Eng. titles) ... 93
SCHOCHIKV FILMS OF AMERICA. (R) February, 1966. Kuniko Miyake, Ehisv Ryu.

OKLAHOMA! (Cs) ... 140
TWENTIETH CENTURY-FOX. (R) December, 1956. Gordon MacRae, Shirley Jones, Gloria Grahame, Gene Nelson. Color.

OKLAHOMA! (Todd-AO) ... 148
MAGNA. (R) Special, 1955. Gordon MacRae, Gloria Grahame, Gene Nelson, Charlotte Greenwood, Eddie Albert, Shirley Jones. Color.

OKLAHOMA CRUDE (P) ... 112
COLUMBIA. (R) June, 1973. George C. Scott, Faye Dunaway. Color.

OKLAHOMA WOMAN ... 80
AMERICAN INTERNATIONAL. (R) June 15, 1956. Richard Denning, Peggie Castle.

OKLAHOMAN, THE (Cs) ... 81
ALLIED ARTISTS. (R) May 19, 1957 Joel McCrea, Barbara Hale. Color.

OLD BOYFRIENDS ... 109
AVCO-EMBASSY. (R) April, 1979. Talia Shire, Richard Jordan.

OLD DARK HOUSE, THE ... 86
COLUMBIA. (R) October, 1963. Tom Poston, Robert Morley.

OLD DRACULA ... 88
AMERICAN INTERNATIONAL. (R) September, 1975. David Niven, Teresa Graves.

OLD MAN AND THE SEA, THE (C) ... 86
WARNER BROS. (R) October, 1958. Spencer Tracy, Felipe Pazos.

OLD YELLER ... 82
BUENA VISTA. (R) January, 1958. Dorothy McGuire, Fess Parker. Color.

OLIVER! ... 153
COLUMBIA. (R) December, 1968. Ron Moody, Shani Wallis, Oliver Reed. Color.

OLIVER'S STORY ... 90
PARAMOUNT. (R) December, 1978. Ryan O'Neal, Candice Bergen.

OLYMPICS IN MEXICO, THE ... 120
COLUMBIA. (R) Not set. Documentary. Alan Jeffreys narrator. Color.

OMAR KHAYYAM (VV) ... 101
PARAMOUNT. (R) August, 1957. Cornel Wilde, Michael Rennie. Color.

OMEGA MAN, THE ... 98
WARNER BROS. (R) August, 1971. Charlton Heston, Rosalind Cash, Anthony Zerbe. Color

OMEN, THE ... 110
TWENTIETH CENTURY-FOX. (R) July, 1976. Gregory Peck, Lee Remick.

ON A CLEAR DAY YOU CAN SEE FOREVER (P) ... 130
PARAMOUNT. (R) June, 1970. Barbra Streisand, Yves Montand, Bob Newhart. Color.

ON HER MAJESTY'S SECRET SERVICE (P) ... 140

Length In Mins.

UNITED ARTISTS. (R) December, 1969. George Lazenby, Diana Rigg, Telly Savalas. Color.

ON MY WAY TO THE CRUSADES I MET A GIRL WHO ... 93
WARNER BROS. (R) September, 1969. Tony Curtis, Monica Vitti. Color.

ON THE BEACH 133
UNITED ARTISTS. (R) December, 1959. Gregory Peck, Ava Gardner.

ON THE DOUBLE 92
PARAMOUNT. (R) July, 1961. Danny Kaye, Dana Wynter, Color.

ON THE THRESHOLD OF SPACE (Cs) 96
TWENTIETH CENTURY-FOX. (R) March, 1956. Guy Madison, Virginia Leith. Color.

ON THE YARD 102
MIDWEST FILMS. (R) January, 1979. John Heard, Thomas Waites.

ONCE A THIEF (P) 107
METRO-GOLDWYN-MAYER. (R) September, 1965. Alain Delon, Ann-Margret.

ONCE IN PARIS 100
LEIGH & McLAUGHLIN. (R) November, 1978. Wayne Rogers, Gayle Hunnicutt.

ONCE IS NOT ENOUGH (P) 120
PARAMOUNT. (R) June, 1975. Kirk Douglas, Alexis Smith.

ONCE MORE WITH FEELING (Technirama) 92
COLUMBIA. (R) February, 1960. Yul Brynner, Kay Kendall. Color.

ONCE UPON A HORSE (Cs) 85
UNIVERSAL. (R) September, 1958. Dan Rowan, Dick Martin.

ONCE UPON A TIME IN THE WEST (P) 165
PARAMOUNT. (R) July, 1969. Henry Fonda, Jason Robards, Charles Bronson. Color.

ONCE YOU KISS A STRANGER 106
WARNER BROS. (R) December, 1969. Paul wburke, Carol Lynley, Martha Hyer. Color.

ONE AND ONLY, THE 95
PARAMOUNT. (R) February, 1978. Henry Winkler, Kim Darby.

ONE AND ONLY, GENUINE, ORIGINAL FAMILY BAND, THE 110
BUENA VISTA. (R) April, 1968. Walter Brennan, Buddy Ebsen. Color.

ONE DAY IN THE LIFE OF IVAN DENISOVICH 100
CINERAMA. (R) June, 1971. Tom Courtenay, Alfred Burke. Color.

ONE DESIRE 94
UNIVERSAL. (R) August, 1955. Anne Baxter, Rock Hudson. Color.

ONE-EYED JACKS (VV) 141
PARAMOUNT. (R) Special. March, 1961. Marlon Brando, Karl Malden. Color.

ONE FLEW OVER THE CUCKOO'S NEST 129
UNITED ARTISTS. (R) November, 1975. Jack Nicholson, Louise Fletcher.

ONE FOOT IN HELL (Cs) 90
TWENTIETH CENTURY-FOX. (R) July, 1960. Alan Ladd, Don Murray. Color.

101 DALMATIANS 80
BUENA VISTA. (R) March, 1961. Feature Cartoon. Color.

100 RIFLES 110
TWENTIETH CENTURY-FOX. (R) April, 1969. Jim Brown, Racquel Welch, Burt Reynolds. Color.

ONE LITTLE INDIAN 90
BUENA VISTA. (R) June, 1973. James Garner. Color.

ONE MAN'S WAY 100
UNITED ARTISTS. (R) February, 1964. Don Murray, Diana Hyland.

ONE MILLION YEARS B.C. 91
20TH CENTURY-FOX. (R) April, 1967. Raquel Welch, John Richardson. Color.

ONE MORE TIME 93
UNITED ARTISTS. (R) May, 1970. Sammy Davis, Jr., Peter Lawford, Ester Anderson. Color.

ONE MORE TRAIN TO ROB 108
UNIVERSAL. (R) May, 1971. George Peppard, Diana Muldaur.

ONE OF OUR DINOSAURS IS MISSING 97
BUENA VISTA. (R) July, 1975. Helen Hayes, Peter Ustinov.

ONE ON ONE 98
WARNER BROS. (R) August, 1977. Robby Benson, Annette O'Toole.

ONE POTATO, TWO POTATO 92
CINEMA V. July, 1964. Barbara Barrie, Bernie Hamilton.

ONE SINGS, THE OTHER DOESN'T (Fr., Eng. titles) 105
CINEMA 5. (R) October, 1977. Valerie Mairesse, Therese Liotard.

ONE SPY TOO MANY 102
METRO-GOLDWYN-MAYER. (R) September, 1966. Robert Vaughn, David McCallum. Color.

ONE THAT GOT AWAY, THE (Brit.) 106
RANK. (R) March, 1958. Hardy Kruger, Colin Gordon.

1,001 ARABIAN NIGHTS 70
COLUMBIA. (R) December, 1959. Mister Magoo. Color.

1000 CONVICTS AND A WOMAN 92
AMERICAN INTERNATIONAL. (R) October, 1971. Alexandra Hay, Sandor Eles. Color.

ONE, TWO, THREE (P) 108
UNITED ARTISTS. (R) December, 1961. James Cagney, Arlene Francis.

ONE WAY OUT 60
RANK. (R) December, 1957. Jill Adams, Lyndon Brook.

ONE WAY PENDULUM 85
LOPERT. (R) January, 1965. Eric Sykes, George Cole.

ONIBABA (TOHOSCOPE) 99
TOHOX. (R) February, 1965. Nobuko Otowa, Jitsuko Yoshimura.

ONION FIELD, THE 126
AVCO-EMBASSY. (R) September, 1979. John Savage, James Woods.

ONIONHEAD 110
WARNER BROS. (R) October, 1958. Andy Griffith, Felicia Farr.

ONLY GAME IN TOWN, THE (P) 113
20TH CENTURY-FOX. (R) February, 1970. Elizabeth Taylor, Warren Beatty. Color.

ONLY ONE NEW YORK 75
EMBASSY. (R) October, 1964. Documentary.

ONLY TWO CAN PLAY 106
KINGSLEY INT. (R) March, 1962.

ONLY WHEN I LARF 104
PARAMOUNT. (R) February, 1969. Richard Attenborough, David Hemmings, Alexandra Stewart. Color.

OPERATION AMSTERDAM 105
TWENTIETH CENTURY-FOX. (R) June, 1960. Peter Finch, Eva Bartok.

OPERATION BIKINI 83
AMERICAN INTERNATIONAL. (R) April, 1963. Tab Hunter, Frankie Avalon.

OPERATION BOTTLENECK 78
UNITED ARTISTS. (R) February, 1961. Ron Foster, Milko Takax.

OPERATION CAMEL 70
AMERICAN INTERNATIONAL. (R) August, 1961. Nora Hayden, Louis Reynard. Color.

OPERATION C.I.A. 90
ALLIED ARTISTS. (R) 1965. Burt Reynolds, John Hoyt. b&w.

OPERATION CONSPIRACY 69
REPUBLIC. (R) September, 1957. Phillip Friend, Mary Mackenzie.

OPERATION CROSSBOW (P) 116
METRO-GOLDWYN-MAYER. (R) July, 1965. Color.

OPERATION DAMES 74
AMERICAN INTERNATIONAL. (R) March, 1959. Eve Meyer, Check Henderson.

OPERATION EICHMANN 92
ALLIED ARTISTS. (R) March, 1961. Werner Klemperer, Ruta Lee.

OPERATION KID BROTHER (TS) 104
UNITED ARTISTS. (R) October, 1967. Neil Connery, Daniela Bianchi. Color.

OPERATION MAD BALL 105
COLUMBIA. (R) November, 1957. Jack Lemmon, Ernie Kovacs.

OPERATION MALAYA (Brit.) 80
AMERICAN RELEASING CORP. (R) October, 1955. Documentary.

OPERATION PETTICOAT 124
UNIVERSAL. (R) December, 1959. Cary Grant, Tony Curtis. Color.

OPERATION SNAFU 89
AMERICAN INTERNATIONAL. (R) February, 1965. Sean Connery, Stanley Holloway.

OPERATION SNATCH 83
CONTINENTAL. (R) September, 1962. Terry-Thomas, George Sanders.

OPPOSITE SEX, THE (Cs) 117
METRO-GOLDWYN-MAYER. (R) October 26, 1956. June Allyson, Ann Sheridan, Joan Collins, Dolores Gray. Color.

OPTIMISTS, THE 110
PARAMOUNT. (R) October, 1973. Peter Sellers, Donna Mullane.

ORCA (P) 92
PARAMOUNT. (R) July, 1977. Richard Harris, Charlotte Rampling.

OREGON PASSAGE (Cs) 81
ALLIED ARTISTS. (R) January, 1958. John Erickson, Lola Albright. Color.

OREGON TRAIL, THE (Cs) 88

Length In Mins.

TWENTIETH CENTURY-FOX. (R) September, 1959. Fred MacMurray, William Bishop. Color.

ORGANIZATION, THE ... 108
UNITED ARTISTS. (R) October, 1971. Sidney Poitier, Barbara McNair. Color.

ORGANIZER, THE (Ital., Eng. titles) ... 126
READE-STERLING. (R) May, 1964. Marcello Mastroianni, Renato Salvatori.

OSCAR, THE ... 119
EMBASSY. (R) March, 1966. Stephen Boyd, Elke Sommer. Color.

OSS 117-MISSION FOR A KILLER (F) ... 84
EMBASSY. (R) September, 1966. Frederick Stafford, Mylene Demongeot. Color.

OTHELLO (P) ... 170
WARNER BROTHERS. (R) February, 1966. Laurence Olivier, Frank Finlay. Color.

OTHELLO ... 92
UNITED ARTISTS. (R) June, 1955. Orson Welles, Suzanne Cloutier.

OTHER, THE ... 100
20TH CENTURY-FOX. (R) May, 1972. Uta Hagen, Christopher and Martin Udvarnoky. Color.

OTHER SIDE OF MIDNIGHT, THE ... 163
TWENTIETH CENTURY-FOX. (R) July, 1977. Marie-France Pisier, John Beck, Susan Sarandon.

OTHER SIDE OF THE MOUNTAIN, THE ... 103
UNIVERSAL. (R) March, 1975. Marilyn Hassett, Beau Bridges.

OTHER SIDE OF THE MOUNTAIN—PART 2, THE ... 105
UNIVERSAL. (R) February, 1978. Marilyn Hassett, Timothy Bottoms.

OTHER VOICES ... 100
DHS FILMS. (R) November, 1969. Documentary. b&w.

OTLEY ... 90
COLUMBIA. (R) March, 1969. Tom Courtenay, Romy Schneider, Alan Badel. Color.

OUR MAN FLINT (Cs) ... 107
TWENTIETH CENTURY-FOX. (R) January, 1966. James Coburn, Lee J. Cobb. Color.

OUR MISS BROOKS ... 85
WARNER BROTHERS. (R) March 3, 1956. Eve Arden, Gale Gordon.

OUR MOTHER'S HOUSE ... 105
MGM. (R) October, 1967. Dirk Bogarde, Margaret Brooks. Color.

OUR TIME ... 88
WARNER BROS. (R) June, 1974. Pamela Sue Martin, Betsy Slade.

OUT OF IT ... 95
UNITED ARTISTS. (R) November, 1969. Barry Gordon, Jon Voight, Lada Edmund, Jr. b&w.

OUT OF SIGHT ... 87
UNIVERSAL. (R) June, 1966. Jonathan Daly, Karen Jensen. Color.

OUT OF THE CLOUDS ... 79
RANK. (R) June, 1957. Anthony Steel, Robert Beatty. Color.

OUT OF TOWNERS, THE ... 98
PARAMOUNT. (R) June, 1970. Sandy Dennis, Jack Lemmon. Color.

OUTBACK ... 105
UNITED ARTISTS. (R) October, 1971. Donald Pleasence, Gary Bond. Color.

OUTFIT, THE ... 103
UNITED ARTISTS. (R) April, 1974. Robert Duvall, Karen Black, Joe Don Baker.

OUTLAW BLUES ... 100
WARNER BROS. (R) July, 1977. Peter Fonda, Susan Saint James.

OUTLAW JOSEY WALES, THE ... 135
WARNER BROS. (R) July, 1976. Clint Eastwood, Chief Dan George.

OUTLAW TREASURE ... 67
AMERICAN RELEASING CORP. (R) July, 1955. John Forbes, Frank "Red" Carpenter.

OUTLAWS IS COMING, THE ... 89
COLUMBIA. (R) January, 1965. The Three Stooges.

OUTLAWS OF TEXAS ... 80
MONOGRAM. (R) December 10, 1958. Whip Wilson, Andy Clyde.

OUTLAW'S SON ... 87
UNITED ARTISTS. (R) July, 1957. Dane Clark, Lori Nelson.

OUTRAGE, THE ... 97
METRO-GOLDWYN-MAYER. (R) November, 1964. Paul Newman, Laurence Harvey.

OUTRAGEOUS ... 96
CINEMA 5. (R) August, 1977. Craig Russell, Hollis McLaren.

OUTSIDE IN ... 90
ROBBINS INTERNATIONAL. (R) September, 1972. Darrel Larson, Heather Manzies. Color.

OUTSIDE MAN, THE ... 104
UNITED ARTISTS. (R) November, 1973. Jean-Louis Trintignant, Ann-Margret.

OUTSIDE THE LAW ... 80
UNIVERSAL. (R) June, 1956. Ray Danton, Leigh Snowden.

OUTSIDER, THE ... 108
UNIVERSAL. (R) April, 1962. Tony Curtis, James Franciscus.

OVER-EXPOSED ... 80
COLUMBIA. (R) April, 1956. Cleo Moore, Richard Crenna.

OVER THERE 1914—1918 ... 90
PATHE-CONT. (R) March, 1965. French Documentary.

OVERCOAT, THE (Rus., Eng. titles) ... 78
CINEMASTER'S INT. (R) March, 1965. Roland Bykov, Y. Tolubeyev.

OWL AND THE PUSSYCAT, THE ... 95
COLUMBIA. (R) December, 1970. Barbra Streisand, George Segal. Color.

P

P.J. (TS) ... 109
UNIVERSAL. (R) March, 1968. George Peppard, Gayle Hunnicut. Color.

P.O. BOX 303 ... 72
ALLIED ARTISTS. (R) June, 1959. Van Johnson, Vera Miles.

P.T. RAIDERS ... 79
CONTINENTAL. (F.) September, 1959. Richard Attenborough, George Baker.

PACIFIC VIBRATIONS ... 92
AMERICAN INTERNATIONAL. (R) June, 1971. Surfing documentary. Color.

PAD (AND HOW TO USE IT), THE ... 86
UNIVERSAL. (R) September, 1966. Brian Bedford, James Farentino. Color.

PADDY ... 97
ALLIED ARTISTS. (R) April, 1970. Milo O'Shea, Des Cave, Deborah Molloy. Color.

PAGANS, THE ... 80
ALLIED ARTISTS (R) June 1, 1958. Pierre Cressoy, Helene Remy.

PAINT YOUR WAGON (P) ... 166
PARAMOUNT. (R) October, 1969. Lee Marvin, Clint Eastwood, Jean Seberg. Color.

PAIR OF BRIEFS, A ... 90
TWENTIETH CENTURY-FOX. (R) Tony Stevens, Francis Pibright.

PAJAMA GAME, THE ... 101
WARNER BROS. (R) August 1957. Doris Day, John Raitt. Color.

PAL JOEY ... 111
COLUMBIA. (R) December 1957. Rita Hayworth, Frank Sinatra. Color.

PALACES OF A QUEEN ... 80
UNIVERSAL. (R) November, 1967. Documentary. Color.

PALM SPRINGS WEEKEND ... 100
WARNER BROS. (R) November, 1963. Troy Donahue, Connie Stevens. Color.

PANAMA SAL ... 70
REPUBLIC. (R) October, 1957. Elena Verdugo, Edward Kemmer.

PANIC BUTTON ... 90
GORTON. (R) May, 1964. Maurice Chevalier, Jayne Mansfield.

PANIC IN NEEDLE PARK, THE ... 110
TWENTIETH CENTURY FOX. (R) July, 1971. Al Pacino, Kitty Winn. Color.

PANIC IN THE PARLOR ... 81
DIST. CORP. OF AMER. (R) December, 1957. Peggy Mount, Cyril Smith.

PANIC IN THE STREETS ... 96
TWENTIETH CENTURY-FOX. (R) September, 1950. Richard Widmark, Paul Douglas.

PANIC IN YEAR ZERO ... 95
AIP. (CR) June, 1962. Ray Milland, Jean Hagen.

PANTALOONS (Fr.) ... 93
UMPO. (R) December, 1956. Fernandel, Carmen Sevilla. Color.

PAPA, MAMA, THE MAID AND I (FR.) ... 94
KINGSLEY. (R) October, 1956. Fernand Ledoux.

PAPA'S DELICATE CONDITION ... 98
PARAMOUNT. (R) March, 1963. Jackie Gleason, Glynis Johns. Color.

PAPER CHASE, THE (P) ... 112
TWENTIETH CENTURY-FOX. (R) October, 1973. Timothy Bottoms, Lindsay Wagner.

PAPER LION ... 107

	Length In Mins.
UNITED ARTISTS. (R) October, 1968. Alan Alda, Lauren Hutton, Sugar Ray Robinson. Color.	
PAPER MOON	**102**
PARAMOUNT. (R) May, 1973. Ryan O'Neal, Tatum O'Neal.	
PAPILLON (P)	**150**
ALLIED ARTISTS. (R) December, 1973. Steve McQueen, Dustin Hoffman.	
PARADISE ALLEY	**108**
UNIVERSAL. (R) November, 1978. Sylvester Stallone.	
PARADISE, HAWAIIAN STYLE	**91**
PARAMOUNT. (R) Summer, 1966. Elvis Presley, Suzanna Leigh.	
PARADISE LAGOON (The Admirable Crichton)	**94**
COLUMBIA. (R) April, 1958. Kenneth Moore, Diane Cilento. Color.	
PARALLAX VIEW, THE (P)	**100**
PARAMOUNT. (R) June, 1974. Warren Beatty, Paula Prentiss.	
PARANOIA (Cromoscope)	**91**
COMMONWEALTH UNITED. (R) October, 1969. Carroll Baker, Lou Castel. Color.	
PARANOIAC	**80**
UNIVERSAL. (R) May, 1963. Janette Scott, Oliver Reed.	
PARATROOP COMMAND	**83**
AMERICAN INTERNATIONAL. (R) February, 1959. Richard Bakalyan, Jack Hogan.	
PARDNERS (VV)	**85**
PARAMOUNT. (R) August, 1956. Dean Martin, Jerry Lewis. Color.	
PARENT TRAP, THE	**124**
BUENA VISTA. (R) June, 1961. Hayley Mills, Maureen O'Hara. Color.	
PARIS BLUES	**98**
UNITED ARTISTS. (R) October, 1961. Paul Newman, Joanne Woodward.	
PARIS DOES STRANGE THINGS	**86**
WARNER BROTHERS. (R) March 2, 1957. Ingrid Bergman, Mel Ferrer. Color.	
PARIS FOLLIES OF 1958	**78**
ALLIED ARTISTS. (R) November 27, 1955. Forrest Tucker, Margaret Whiting. Color.	
PARIS HOLIDAY (T)	**100**
UNITED ARTISTS. (R) March, 1958. Bob Hope, Fernadel.	
PARIS IN THE MONTH OF AUGUST	**94**
TRANS-LUX. (R) March, 1968. Charles Aznavour, Susan Hampshire.	
PARIS SECRET	**84**
CINEMA V. (R) September, 1965. Documentary. Color.	
PARIS WHEN IT SIZZLES	**110**
PARAMOUNT. (R) April, 1964. William Holden, Audrey Hepburn. Color.	
PARISIENNE, LA	**85**
UNITED ARTISTS. (R) August, 1958. Brigitte Bardot, Charles Boyer. Color.	
PARRISH	**137**
WARNER BROS. (R) July, 1961. Troy Donahue, Claudette Colbert. Color.	
PARSON AND THE OUTLAW, THE	**71**
COLUMBIA. (R) September, 1957. Anthony Dexter, Sonny Tufts.	
PART 2 WALKING TALL	**109**
AMERICAN INTERNATIONAL. (R) July, 1975. Bo Svenson, Luke Askew.	
PARTY, THE (P)	**99**
UNITED ARTISTS. (R) April, 1968. Peter Sellers, Claudine Longet. Color.	
PARTY CRASHERS, THE	**76**
PARAMOUNT. (R) September, 1958. Connie Stevens, Robert Driscoll.	
PARTY GIRL (Cs)	**99**
METRO-GOLDWYN-MAYER. (R) September, 1958. Robert Taylor, Cyd Charisse. Color.	
PARTY'S OVER, THE (Brit.)	**94**
ALLIED ARTISTS. (R) 1966. Oliver Reed, Catherine Woodville.	
PASSAGE, THE (T)	**99**
UNITED ARTISTS. (R) March, 1979. Anthony Quinn, James Mason, Malcolm McDowell.	
PASSAGES FROM FINNEGAN'S WAKE	**94**
GROVE PRESS. (R) November, 1967. Martin J. Kelly, Jane Reilly.	
PASSENGER, THE	**122**
UNITED ARTISTS-MGM. (R) April, 1975. Jack Nicholson, Maria Schneider.	
PASSION OF ANNA, THE (Swedish; Eng. titles)	**99**
UNITED ARTISTS. (R) May, 1970. Liv Ullman, Bibi Anderson, Max Von Sydow. Color.	
PASSPORT TO CHINA	**75**
COLUMBIA. (R) March, 1961. Richard Basehart, Lisa Gastoni.	
PASSPORT TO TREASON (Brit.)	**70**
ASTOR. (R) June, 1956. Rod Cameron, Lois Maxwell.	
PASSWORD IS COURAGE, THE	**116**
METRO-GOLDWYN-MAYER. (R) January, 1963. Dirk Bogarde, Maria Perschy.	
PAT GARRETT AND BILLY THE KID (P)	**106**
METRO-GOLDWYN-MAYER. (R) May, 1973. James Coburn, Kris Kristofferson. Color.	
PATCH OF BLUE, A (P)	**105**
METRO-GOLDWYN-MAYER. (R) March, 1966. Sidney Poitier, Shelley Winters.	
PATHS OF GLORY	**88**
UNITED ARTISTS. (R) January, 1958. Kirk Douglas, Ralph Meeker.	
PATSY, THE	**103**
PARAMOUNT. (R) July, 1964. Jerry Lewis, Ina Balin. Color.	
PATTERNS	**83**
UNITED ARTISTS. (R) March, 1956. Van Heflin, Ed Begley, Everett Sloane.	
PATTON (D-150)	**170**
20th CENTURY-FOX. (R) February, 1970. George C. Scott, Karl Malden. Color.	
PAUL AND MICHELLE (P)	**103**
PARAMOUNT. (R) May, 1974. Anicee Alvina, Sean Bury, Keir Dullea.	
PAWNBROKER, THE	**110**
ALLIED ARTISTS-LANDAU. (R) April, 1965. Rod Stieger, Geraldine Fitzgerald.	
PAWNEE	**80**
REPUBLIC. (R) September, 1957. George Montgomery, Bill Williams. Color.	
PAY OR DIE	**111**
ALLIED ARTISTS. (R) July, 1960. Ernest Borgnine, Zohra Lampert.	
PAYDAY	**103**
CINERAMA. (R) February, 1973. Rip Torn, Anna Capri. Color.	
PAYMENT IN BLOOD	**90**
COLUMBIA. (R) December, 1968. Guy Madison, Edd Byrnes, Louise Barrett. Color.	
PAYROLL	**94**
ALLIED ARTISTS. (R) July, 1962. Michael Craig, Francoise Prevost.	
PEACEMAKER, THE	**82**
UNITED ARTISTS. (R) November, 1956. James Mitchell, Risemari Bowe.	
PEACH THIEF	**84**
BRANDON. (R) September, 1969. Nevena Kokanova, Mikhail Mikhailov. b&w.	
PEARL OF THE SOUTH PACIFIC (SS)	**85**
RKO RADIO. (R) July 4, 1955. Virginia Mayo, Dennis Morgan, David Farrar. Color.	
PEEPER	**87**
TWENTIETH CENTURY-FOX. (R) September, 1975. Michael Caine, Natalie Wood.	
PENDULUM	**106**
COLUMBIA. (R) January, 1969. George Peppard, Jean Seberg, Richard Riley. Color.	
PENELOPE (P)	**97**
METRO-GOLDWYN-MAYER. (R) December, 1966. Natalie Wood, Ian Bannen. Color.	
PENTHOUSE, THE	**100**
PARAMOUNT. (R) October, 1967. Suzy Kendall, Terence Morgan. Color.	
PEOPLE NEXT DOOR, THE	**93**
AVCO EMBASSY. (R) September, 1970. Julie Harris, Eli Wallach, Deborah Winters. Color.	
PEOPLE THAT TIME FORGOT, THE	**90**
AMERICAN INTERNATIONAL. (R) July, 1977. Patrick Wayne, Doug McClure.	
PEPE (Cs)	**195**
COLUMBIA. (R) Special. December, 1960. Cantinflas, Dan Dailey. Color.	
PERCY	**103**
MGM. (R) April, 1971. Hywel Bennett, Elke Sommer. Color.	
PERFECT COUPLE, A	**110**
TWENTIETH CENTURY-FOX. (R) April, 1979. Paul Dooley, Marta Heflin.	
PERFECT FRIDAY	**94**
CHEVRON. (R) October, 1970. Stanley Baker, Ursula Andress. Color.	
PERFECT FURLOUGH, THE (Cs)	**93**
UNIVERSAL. (R) January, 1959. Tony Curtis, Janet Leigh. Color.	
PERFORMANCE	**106**
WARNER BROS. (R) August, 1970. James Fox, Mick Jagger. Color.	
PERILS OF PAULINE	**90**
UNIVERSAL. (R) July, 1967. Pat Boone, Pamela Austin. Color.	
PERIOD OF ADJUSTMENT	**112**

Length In Mins.

METRO-GOLDWYN-MAYER. (R) November, 1962. Tony Franciosa, Jane Fonda.

PERMISSION TO KILL ... 96
AVCO-EMBASSY. (R) January, 1976. Dirk Bogarde, Ava Gardner.

PERRI ... 74
BUENA VISTA. (R) November, 1957. Live Action & Animation. Color.

PERSONA (Swed.; Eng. titles) ... 81
LOPERT. (R) March, 1967. Bibi Anderson, Liv Ullmann.

PERSUADER, THE ... 72
ALLIED ARTISTS. (R) June, 1957. James Craig, William Talman.

PETE KELLY'S BLUES (Cs) ... 95
WARNER BROTHERS. (R) August, 1955. Jack Webb, Janet Leigh, Edmond O'Brien, Peggy Lee. Color.

PETE 'N' TILLIE (P) ... 100
UNIVERSAL. (R) December, 1972. Carol Burnett, Walter Matthau. Color.

PETE SEEGER—A SONG AND A STONE ... 85
THEATRE EXCHANGE. (R) February, 1972. Documentary Biography. Color.

PETER RABBIT AND TALES OF BEATRIX PORTER ... 90
MGM. (R) June, 1971. The Royal Ballet. Color.

PETE'S DRAGON ... 134
BUENA VISTA. (R) December, 1977. Helen Reddy, Jim Dale.

PETULIA ... 105
WARNER BROS.-7 ARTS. (R) August, 1968. Julie Christie, George C. Scott, Richard Chamberlain. Color.

PEYTON PLACE (Cs) ... 157
TWENTIETH CENTURY-FOX. (R) December, 1957. Lana Turner, Lee Philips. Color.

PHAEDRA ... 115
UNITED ARTISTS. (R) October, 1962. Melina Mercouri, Anthony Perkins.

PHANTASM ... 90
AVCO EMBASSY. (R) May, 1979. Michael Baldwin, Bill Thornbury.

PHANTOM FROM 10,000 LEAGUES, THE ... 80
AMERICAN RELEASING CORP. (R) January, 1956. Kent Taylor, Cathy Downs.

PHANTOM OF THE JUNGLE ... 75
LIPPERT. (R) May 20, 1955. Jon Hall, Anne Gwynne.

PHANTOM OF THE OPERA ... 84
UNIVERSAL. (R) April, 1962. Herbert Lom, Heather Sears. Color.

PHANTOM OF THE PARADISE ... 92
TWENTIETH CENTURY-FOX. (R) November, 1974. Paul Williams, William Finely.

PHANTOM STAGECOACH, THE ... 69
COLUMBIA. (R) April, 1957. William Bishop, Kathleen Crowley.

PHARAOH'S CURSE ... 66
UNITED ARTISTS. (R) February, 1957. Mark Dana, Ziva Shapir.

PHARAOH'S WOMAN, THE (T) ... 88
UNIVERSAL. (R) May, 1961. John Drew Barrymore, Jr., Linda Cristal. Color.

PHASE IV ... 86
PARAMOUNT. (R) September, 1974. Nigel Davenport, Lynne Frederick.

PHEONIX CITY STORY, THE ... 100
ALLIED ARTISTS. (R) August 14, 1955. Richard Kiley, Kathryn Grant.

PHYNX, THE ... 91
WARNER BROS. (R) May, 1970. Michael Miller, Ray Chippeway. Color.

PICKPOCKET ... 88
NEW YORKER. (R) April, 1969. Marin Lassalle, Pierre Leymarie.

PICKUP ALLEY ... 92
COLUMBIA. (R) August, 1957. Victor Mature, Anita Ekberg.

PICNIC (Cs) ... 115
COLUMBIA. (R) February, 1956. William Holden, Rosalind Russell, Kim Novak, Arthur O'Connell. Color.

PICTURE MOMMY DEAD ... 88
EMBASSY. (R) September, 1966. Don Ameche, Martha Hyer. Color.

PIECE OF THE ACTION, A ... 135
WARNER BROS. (R) October, 1977. Sidney Poitier, Bill Cosby.

PIECES OF DREAMS ... 108
UNITED ARTISTS. (R) September, 1970. Robert Forster, Lauren Hutton.

PIED PIPER, THE ... 90
PARAMOUNT. (R) June, 1972. Donovan, Jack Wild, Donald Pleasence. Color.

PIER 5, Havana ... 67
UNITED ARTISTS. (R) July, 1959. Cameron Mitchell, Allison Hayes.

PIERROT LE FOU (French; Eng. titles) ... 110
PATHE. (R) January, 1969. Jean Paul Belmondo, Anna Karina, Dirk Sanders. Color.

PIGEON THAT TOOK ROME, THE (P) ... 101
PARAMOUNT. (R) October, 1962. Charlton Heston, Elsa Martinelli.

PIGEONS ... 87
MGM. (R) November, 1970. Jordan Christopher, Jill O'Hara. Color.

PILGRIMAGE FOR PEACE: POPE PAUL VI VISITS AMERICA, A ... 56
JOSEPH L. ROBERTS. (R) July, 1966. Documentary. Color.

PILLARS OF THE SKY (Cs) ... 95
UNIVERSAL. (R) October, 1956. Jeff Chandler, Dorothy Malone. Color.

PILLOW TALK (Cs) ... 105
UNIVERSAL. (R) October, 1959. Rock Hudson, Doris Day. Color.

PINK JUNGLE, THE ... 104
UNIVERSAL. (R) August, 1968. James Garner, Eva Renzi, George Kennedy. Color.

PINK PANTHER, THE (T) ... 113
UNITED ARTISTS. (R) March, 1964. David Niven, Peter Sellers. Color.

PINK PANTHER STRIKES AGAIN, THE (P) ... 103
UNITED ARTISTS. (R) December, 1976. Peter Sellers, Herbert Lom.

PINOCCHIO IN OUTER SPACE ... 71
UNIVERSAL. (R) December, 1965. Animated. Color.

PIPE DREAMS (P) ... 89
AVCO-EMBASSY. (R) November, 1976. Gladys Knight, Barry Hankerson.

PIRATES OF BLOOD RIVER, THE ... 87
COLUMBIA. (R) August, 1962 Kerwin Matthews, Glenn Corbett. Color.

PIRATES OF TORTUGA (Cs) ... 97
TWENTIETH CENTURY-FOX. (R) October, 1961, Ken Scott, Letita Roman. Color.

PIRATES OF TRIPOLI ... 72
COLUMBIA. (R) February, 1955. Paul Henreid, Patricia Medina. Color.

PISTOL FOR RINGO, A (TE) ... 97
EMBASSY. (R) November, 1966 Montgomery Wood, Fernando Sancho. Color.

PIT AND THE PENDULUM, THE (P) ... 95
AMERICAN INTERNATIONAL. (R) August, 1961. Vincent Price, John Kerr. Color.

PIZZA TRIANGLE, THE (Ital.; Eng. titles) ... 99
WARNER BROS. (R) November 1970. Marcello Mastroianni, Monica Vitti. Color.

PLACE CALLED GLORY, A (TE) ... 92
EMBASSY. (R) August, 1966. Lex Barker, Pierre Brice. Color.

PLACE FOR LOVERS, A ... 90
MGM. (R) August, 1969. Faye Dunaway, Marcello Mastroianni. Color.

PLAGUE OF THE ZOMBIES, THE ... 90
TWENTIETH CENTURY-FOX (R) January 1966. Andre Morelli, Diane Clare. Color.

PLAINSMAN, THE ... 92
UNIVERSAL. (R) September, 1966. Don Murray, Guy Stockwell. Color.

PLAN 9 FROM OUTER SPACE ... 76
D.C.A. (R) July, 1959. Gregory Walcott, Mona McKinnon.

PLANET OF THE APES (P) ... 112
TWENTIETH CENTURY-FOX. (R) April, 1968. Charlton Heston, Kim Hunter. Color.

PLANET OF THE VAMPIRES (Colorscope) ... 86
AMERICAN INTERNATIONAL. (R) November, 1965. Barry Sullivan, Norma Bengel. Color.

PLATINUM HIGH SCHOOL ... 93
METRO-GOLDWYN-MAYER. (R) May, 1960. Mickey Rooney, Terry Moore.

PLAY DIRTY ... 117
UNITED ARTISTS. (R) January, 1969. Michael Caine, Nigel Davenport, Nigel Green. Color.

PLAY IT AGAIN, SAM ... 86
PARAMOUNT. (R) April, 1972. Woody Allen, Tony Roberts, Diane Keaton. Color.

PLAY IT AS IT LAYS ... 98
UNIVERSAL. (R) October, 1972. Tuesday Weld, Anthony Perkins. Color.

PLAY IT COOL ... 74
ALLIED ARTISTS. (R) May 1963. Billy Fury, Helen Shapiro.

PLAY MISTY FOR ME ... 102
UNIVERSAL. (R) October, 1971 Clint Eastwood, Jessica Walter. Color.

PLAYBOY OF THE WESTERN WORLD, THE ... 100
JANUS. (R) January, 1963. Siobhan McKenna, Gary Raymond. Color.

PLAYERS ... 115

Length In Mins.

PARAMOUNT. (R) June, 1979. Ali MacGraw, Dean-Paul Martin.

PLAYGIRL AND THE WAR MINISTER, THE 90
UNION. (R) November, 1963. Joan Greenwood, Cecil Parker.

PLAZA SUITE 115
PARAMOUNT. (R) June, 1971. Walter Matthau, Maureen Stapleton. Color.

PLEASE DON'T EAT THE DAISIES (Cs) 111
METRO-GOLDWYN-MAYER. (R) April, 1960. Doris Day, David Niven. Color.

PLEASE! MR. BALZAC (Fr.) 99
DIST. CORP. OF AMER. (R) October, 1957. Daniel Gelin, Brigitte Bardot.

PLEASE MURDER ME 78
DISTRIBUTORS CORP. OF AMERICA. (R) March, 1956. Angela Lansbury, Raymond Burr.

PLEASE TURN OVER 86
COLUMBIA. (R) December, 1960. Ted Ray, Jean Kent.

PLEASURE OF HIS COMPANY, THE 114
PARAMOUNT. (R) August, 1961. Fred Astaire, Debbie Reynolds. Color.

PLEASURE SEEKERS, THE (Cs) 107
TWENTIETH CENTURY-FOX. (R) January, 1965. Ann-Margaret, Carol Lynley. Color.

PLUNDER ROAD (R) 72
TWENTIETH CENTURY-FOX. (R) December, 1957. Gene Raymond, Carolyn Craig.

PLUNDERERS, THE 94
ALLIED ARTISTS. (R) November, 1960. Jeff Chandler, John Saxon.

PLUNDERERS OF PAINTED FLATS (N) 72
REPUBLIC. (R) January, 1959. Corine Calvet, John Carroll.

POACHER'S DAUGHTER, THE 74
SHOW CORP. OF AMERICA. (R) June, 1960. Julie Harris, Harry Brogan.

POCKET MONEY 102
NATIONAL GENERAL. (R) February. 1972. Paul Newman, Lee Marvin. Color.

POCKETFUL OF MIRACLES (P) 136
UNITED ARTISTS. (R) December, 1961. Glenn Ford, Bette Davis. Color.

POE'S TALES OF TERROR 90
AMERICAN INTERNATIONAL. (R) August, 1962. Vincent Price, Peter Lorre. Color.

POINT BLANK (P) 92
MGM. (R) October, 1967. Lee Marvin, Angie Dickinson. Color.

POINT OF ORDER 97
POINT FILMS. (R) January, 1964. McCarthy hearings.

POLICE DOG STORY, THE 62
UNITED ARTISTS. (R) February, 1961. Jim Brown, Merry Anders.

POLICE NURSE 64
TWENTIETH CENTURY-FOX. (R) May, 1963. Ken Scott, Merry Anders.

POLLYANA 134
BUENA VISTA. (R) July, 1960. Hayley Mills, Jane Wyman. Color.

POOR COW 104
NATIONAL GENERAL CORPORATION. (R) February, 1968. Terence Stamp, Carol White. Color.

POP CORN 85
SHERPIX. (R) December, 1970. Mick Jagger & The Rolling Stones, Jim Hendrix. Color.

POP! 115
UNITED ARTISTS. (R) June, 1969. Alan Arkin, Rito Moreno, Miguel Alehandro. Color.

POPPY IS ALSO A FLOWER, THE 100
COMET. (R) October, 1966. Senta Berger, Stephen Boyd. Color.

PORGY AND BESS (Todd-AO) 146
COLUMBIA. (R) Special. June, 1959. Sidney Poitier, Dorothy Dandridge. Color.

PORK CHOP HILL 97
UNITED ARTISTS. (R) May, 1959. Gregory Peck.

PORT AFRIQUE 92
COLUMBIA. (R) October, 1956. Phil Carey, Pier Angeli. Color.

PORTLAND EXPOSE 72
ALLIED ARTISTS. (R) August, 1957. Edward Binns, Carolyn Craig.

PORTNOY'S COMPLAINT 101
WARNER BROS. (R) July, 1972. Richard Benjamin, Karen Black, Lee Grant. Color.

PORTRAIT IN BLACK 112
UNIVERSAL. (R) July, 1960. Lana Turner, Anthony Quinn. Color.

PORTRAIT OF A MOBSTER 108
WARNER BROS. (R) April, 1961. Vic Morrow, Leslie Parrish.

PORTRAIT OF A SINNER 96
AMERICAN INTERNATIONAL. (R) February, 1961. Nadja Tiller, Tony Britton.

PORTRAIT OF AN UNKNOWN WOMAN 86
UNIVERSAL. (R) May, 1958. Ruth Leuwerik, O.W. Fisher.

POSEIDON ADVENTURE, THE (P) 117
TWENTIETH CENTURY-FOX. (R) December, 1972. Gene Hackman, Shelley Winters, Ernest Borgnine. Color.

POSSE FROM HELL 89
UNIVERSAL. (R) May, 1969. Audie Murphy, John Saxon. Color.

POSSESSION OF JOEL DELANEY, THE 105
PARAMOUNT. (R) May, 1972. Shirley MacLaine, Perry King. Color.

POSTMARK FOR DANGER 84
RKO RADIO. (R) January 1956. Terry Moore. Robert Beatty.

POUND 92
UNITED ARTISTS. (R) August, 1970. Joe Madden, James Green. Color.

POWER, THE (P) 109
MGM. (R) February, 1968. George Hamilton, Suzanne Pleshette. Color.

POWER AND THE PRIZE, THE (Cs) 89
METRO-GOLDWYN-MAYER. (R) October, 1956. Robert Taylor, Elizabeth Mueller.

PREMATURE BURIAL 8
AMERICAN INTERNATIONAL. (R) March, 1962. Ray Milland, Hazel Court. Color.

PREMONITION, THE 95
AVCO-EMBASSY. (R) February, 1976. Sharon Farrell, Richard Lynch.

PRESIDENT'S ANALYST, THE (P) 104
PARAMOUNT. (R) December, 1967. James Coburn, Godfrey Cambridge. Color.

PRESSURE POINT 91
UNITED ARTISTS. (R) September, 1962. Sidney Poitier, Bobby Darin.

PRETTY BABY 109
PARAMOUNT. (R) April, 1978. Brooke Shields, Keith Carradine.

PRETTY MAIDS ALL IN A ROW 92
MGM. (R) May, 1971. Rock Hudson, Angie Dickinson. Color.

PRETTY POISON 89
TWENTIETH CENTURY-FOX. (R) September, 1968. Anthony Perkins, Tuesday Weld, Beverly Garland. Color.

PRICE OF FEAR, THE 79
UNIVERSAL. (R) May, 1956. Merle Oberon, Lex Barker.

PRIDE AND THE PASSION, THE (VV) 182
UNITED ARTISTS. (R) July, 1957. Cary Grant, Frank Sinatra. Color.

PRIEST'S WIFE, THE 106
WARNER BROS. (R) February, 1971. Marcello Mastroianni, Sophia Loren. Color.

PRIME CUT 91
NATIONAL GENERAL. (R) June, 1972. Gene Hackman, Lee Marvin. Color.

PRIME OF MISS JEAN BRODIE, THE 116
TWENTIETH CENTURY-FOX. (R) March, 1969. Maggie Smith, Robert Stephens, Pamela Franklin. Color.

PRIME TIME, THE 70
ESSANJAY. (R) May, 1960. JoAnn Le Compte, Frank Roche.

PRINCE AND THE SHOWGIRL, THE 117
WARNER BROS. (R) July, 1957. Marilyn Monroe, Laurence Olivier. Color.

PRISONER, THE (Brit.) 91
COLUMBIA. (R) March, 1956. Alec Guinness, Jack Hawkins.

PRISONER OF SECOND AVENUE, THE (P) 99
WARNER BROS. (R) March, 1975. Jack Lemmon, Anne Bancroft.

PRISONER OF THE IRON MARK 80
AMER. INT'L PICS. (R) 1969. Michael Lemoine, Wandisa Guida. Color.

PRISONER OF THE VOLGA (Totalscope) 102
PARAMOUNT. (R) May, 1960. John Derek, Elsa Martinelli. Color.

PRISONER OF ZENDA, THE 108
UNIVERSAL. (R) May, 1979. Peter Sellers, Lionel Jeffries.

PRIVATE LIFE OF SHERLOCK HOLMES, THE (P) 125
UNITED ARTISTS. (R) November, 1970. Robert Stephens, Colin Blakely. Color.

PRIVATE LIVES OF ADAM AND EVE, THE 87
UNIVERSAL. (R) December, 1960. Mickey Rooney, Mamie Van Doren.

PRIVATE NAVY OF SGT. O'FARRELL, THE 92

Length In Mins.

UNITED ARTISTS. (R) June, 1968. Bob Hope, Phyllis Diller. Color.

PRIVATE PROPERTY. ... 79
CITATION FILMS. (R) April, 1960. Kate Manx, Corey Allen.

PRIVATE WAR OF MAJOR BENSON, THE ... 105
UNIVERSAL. (R) August, 1955. Charlton Heston, Julie Adams. Color.

PRIVATE'S AFFAIRS, A (Cs) ... 105
TWENTIETH CENTURY-FOX. (R) August, 1959. Sal Mineo, Christine Carere. Color.

PRIVATE'S PROGRESS (Brit.) ... 96
DISTRIBUTORS CORPORATION OF AMERICA. (R) September, 1956. Richard Attenborough, Dennis Price, Ian Carmichael.

PRIVILEGE ... 103
UNIVERSAL. (R) September, 1967. Paul Jones, Jean Shrimpton. Color.

PRIZE, THE ... 136
METRO-GOLDWYN-MAYER. (R) December, 1963. Paul Newman, Edward G. Robinson. Color.

PRIZE OF GOLD, A ... 98
COLUMBIA. (R) June, 1955. Richard Widmark, Mai Zetterling. Color.

PRODIGAL, THE (Cs) ... 110
METRO-GOLDWYN-MAYER. (R) May 6, 1955. Lana Turner, Edmund Purdom. Color.

PRODUCERS, THE ... 88
EMBASSY. (R) June, 1968. Zero Mostel, Gene Wilder. Color.

PROFESSIONALS, THE (P) ... 117
COLUMBIA. (R) November, 1966. Burt Lancaster, Lee Marvin. Color.

PROFESSIONALS, THE ... 68
AMERICAN INTERNATIONAL (R) July, 1961.

PROJECT X ... 97
PARAMOUNT. (R) June, 1968. Christopher George, Greta Baldwin, Monte Markham. Color.

PROJECTED MAN, THE (TE) ... 77
UNIVERSAL. (R) February, 1967. Bryant Haliday, Mary Peach. Color.

PROJECTIONIST, THE ... 85
TED MARON. (R) January, 1971. Chuck McCann, Ina Balin. Color.

PROMISE, THE (P) ... 97
UNIVERSAL. (R) March, 1979. Kathleen Quinlan, Stephen Collins.

PROMISE AT DAWN ... 103
AVCO EMBASSY. (R) January, 1971. Melina Mercouri, Assaf Dagan. Color.

PROMISE HER ANYTHING ... 98
PARAMOUNT. (R) March, 1966. Warren Beatty, Leslie Caron. Color.

PROMISES IN THE DARK ... 115
WARNER BROS. (R) November, 1979. Marsha Mason, Ned Beatty.

PROUD AND THE BEAUTIFUL, THE (Fr.) ... 94
KINGSLYE. (R) June, 1956. Michele Morgan, Gerard Philipe.

PROUD AND THE PROFANE, THE (VV) ... 111
PARAMOUNT. (R) August, 1956. William Holden, Deborah Kerr.

PROUD ONES, THE (Cs) ... 94
TWENTIETH CENTURY-FOX. (R) May, 1956. Robert Ryan, Virginia Mayo, Jeffrey Hunter. Color.

PROUD REBEL, THE ... 105
BUENA VISTA (DISNEY.) (R) June, 1958. Alan Ladd, Olivia De Haviland. Color.

PROVIDENCE ... 104
CINEMA 5. (R) February, 1977. Ellen Burstyn, Dirk Bogarde, John Gielgud.

PRUDENCE AND THE PILL ... 92
TWENTIETH CENTURY-FOX. (R) July, 1968. Deborah Kerr, David Niven, Robert Coote. Color.

PSYCHE 59 ... 94
ROYAL. (R) April, 1964. Curt Jergens, Patricia Neal.

PSYCHE OUT ... 98
AMERICAN INTERNATIONAL. (R) April, 1968. Susan Strasberg, Jack Nicholson. Color.

PSYCHIC KILLER ... 90
AVCO EMBASSY. (R) December, 1975. Jim Hutton, Julie Adams.

PSYCHO ... 109
PARAMOUNT. (R) August, 1960. Anthony Perkins, Vera Miles.

PSYCHO CIRCUS ... 65
AMERICAN INTERNATIONAL. (R) May, 1967. Christopher Lee, Leo Genn.

PSYCHOUT FOR MURDER (WS) ... 88
TIMES FILM (R) January, 1971. Rosanno Brazzi, Adrienne La Russa. Color.

PSYCOPATH, THE (TE) ... 83
PARAMOUNT. (R) May 1966. Patrick Wymark, Margaret Johnson. Color.

PT 109 (P) ... 140
WARNER BROS. (R) July, 1963. Cliff Robertson, Ty Hardin. Color.

PUBLIC EYE, THE ... 90
UNIVERSAL. (R) August, 1972. Mia Farrow, Topol, Michael Jayston. Color.

PUBLIC PIGEON NO. ONE ... 79
UNIVERSAL. (R) June, 1957. Red Skelton, Vivian Blaine. Color.

PUFNSTUF ... 94
UNIVERSAL. (R) June. 1970. Jack Wild, Billie Hayes, Martha Raye. Color.

PULP ... 95
UNITED ARTISTS. (R) October, 1972. Michael Caine, Mickey Rooney. Color.

PUMPKIN EATER, THE ... 110
ROYAL INT. (R) November, 1964. Ann Bancroft, Peter Finch.

PUPPET ON A CHAIN ... 97
CINERAMA. (R) April, 1972. Sven-Bertil Taube, Barbara Parkins. Color.

PURPLE GANG, THE ... 83
ALLIED ARTISTS. (R) January, 1960. Barry Sullivan, Robert Blake.

PURPLE HILLS, THE (CS) ... 60
TWENTIETH CENTURY-FOX. (R) October, 1961. Gene Nelson, Joanna Barnes. Color.

PURPLE MASK, THE (Cs) ... 82
UNIVERSAL. (R) July, 1955. Tony Curtis, Colleen Miller. Color.

PURPLE PLAIN, THE (Brit.) ... 100
UNITED ARTISTS. (R) April, 1955. Gregory Peck, Win Min Than. Color.

PURSUIT OF HAPPINESS, THE ... 98
COLUMBIA. (R) February, 1971. Michael Sarrazin, Barbara Hershey. Color.

PURSUIT OF THE GRAF SPEE (VV) ... 106
RANK. (R) November, 1957. John Gregson, Anthony Quayle. Color.

PUSHER, THE ... 81
UNITED ARTISTS. (R) January, 1960. Kathy Carlyle, Felice Orlandi.

PUSSYCAT, PUSSYCAT, I LOVE YOU ... 99
UNITED ARTISTS. (R) March, 1970. Ian McShane, John Gavin, Severn Darden. Color.

PUTNEY SWOPE ... 84
CINEMA 5. (R) July, 1969. Stanley Gottlieb, Allen Garfield. b&w.

PUZZLE OF A DOWNFALL CHILD ... 104
UNIVERSAL. (R) November, 1970. Faye Dunaway, Barry Primus. Color.

PYRO ... 99
AMERICAN INTERNATIONAL. (R) April, 1964. Barry Sullivan, Martha Hyer.

PYX, THE (P) ... 111
CINERAMA. (R) October, 1973. Karen Black, Christopher Plummer.

Q

QUACKSER FORTUNE HAS A COUSIN IN THE BRONX 90
UMC. (R) July, 1970. Gene Wilder, Margot Kidder. Color.

QUANTEZ (Cs) ... 80
UNIVERSAL. (R) October, 1957. Fred MacMurray, Dorothy Malone. Color.

QUANTRILL'S RAIDERS (Cs) ... 75
ALLIED ARTISTS. (R) April 27, 1958. Steve Cochran, Gale Robbins. Color.

QUARE FELLOW, THE ... 85
AJAY PICTURES. (R) October, 1966. Patrick McGoohan, Sylvia Syms.

QUE SEAS FELIZ (Span.) ... 102
COLUMBIA. (R) March, 1960. Miguel Acoves Mevia, Rosita Quintana.

QUEEN, THE ... 68
GROVE PRESS. (R) June, 1968. Documentary. Color.

QUEEN BEE ... 95
COLUMBIA. (R) November, 1955. Joan Crawford, Barry Sullivan.

QUEEN OF BABYLON (Ital.-Eng. dial.) ... 109
TWENTIETH CENTURY-FOX (R) July, 1956. Rhonda Fleming, Ricardo Montalban. Color.

QUEEN OF BLOOD ... 81
AMERICAN INTERNATIONAL. (R) March, 1966. John Saxon, Basil Rathbone. Color.

QUEEN OF OUTER SPACE (Cs) ... 80

Length In Mins.

ALLIED ARTISTS. (R) September, 1958. Zsa Zsa Gabor, Eric Fleming. Color.

QUEEN OF THE PIRATES 80
COLUMBIA. (R) August, 1961. Gianna Maria Canale, Massimo Serato.

QUEENS, THE 110
ROYAL. (R) March, 1968. Raquel Welch, Claudia Cardinale. Color.

QUENTIN DURWARD (Cs) 101
METRO-GOLDWYN-MAYER. (R) October 21, 1955. Robert Taylor, Kay Kendall, Robert Morley. Color.

QUEST FOR A LOST CITY 60
RKO RADIO. (R) May 4, 1955. Dana and Ginger Lamb. Color.

QUICK BEFORE IT MELTS (CP) 98
METRO-GOLDWYN-MAYER. (R) February, 1965. George Maharis, Robert Morse. Color.

QUICK GUN, THE (T) 88
COLUMBIA. (R) May, 1964. Audie Murphy, Merry Anders. Color.

QUIET AMERICAN, THE 120
UNITED ARTISTS. (R) January, 1958. Audie Murphy, Michael Redgrave. Color.

QUIET GUN, THE 77
TWENTIETH CENTURY-FOX. (R) January, 1957. Forrest Tucker, Cleo Moore.

QUIET PLACE IN THE COUNTRY, A 106
LOPERT. (R) August, 1970. Franco Nero, Vanessa Redgrave. Color.

QUILLER MEMORANDUM (P) 105
TWENTIETH CENTURY-FOX. (R) January, 1967. George Segal, Alec Guinness. Color.

QUINCANNON, FRONTIER SCOUT 83
UNITED ARTISTS. (R) May, 1956. Tony Martin, Peggie Castle. Color.

QUINTET 122
TWENTIETH CENTURY-FOX. (R) February, 1979. Paul Newman, Vittorio Gassman.

R

R.P.M.* 92
COLUMBIA. (R) September, 1970. Anthony Quinn, Ann-Margret. Color.

RA EXPEDITIONS, THE 93
UNIVERSAL. (R) August, 1974. Documentary.

RABBIT, RUN 94
WARNER BROS. (R) October, 1970. James Caan, Anjanette Comer. Color.

RABBIT TRAP, THE 75
UNITED ARTISTS. (R) June, 1959. Ernest Borgnine, David Brian.

RACE FEVER 90
ALLIED ARTISTS. (R) March, 1965. Joe Morrison, David Blanchard. Color.

RACE WITH THE DEVIL 88
TWENTIETH CENTURY-FOX. (R) June, 1975. Peter Fonda, Warren Oates.

RACERS, THE (Cs) 112
TWENTIETH CENTURY-FOX. (R) February, 1955. Kirk Douglas, Bella Darvi, Gilbert Roland. Color.

RACHEL, RACHEL 101
WARNER BROS.-7 ARTS. September, 1968. Joanne Woodward, James Olsen, Kate Harrington. Color.

RACK, THE 100
METRO-GOLDWYN-MAYER. (R) November 2, 1956. Paul Newman, Walter Pidgeon, Wendell Corey, Edmond O'Brien, Anne Francis.

RAFFERTY AND THE GOLD DUST TWINS (P) 92
WARNER BROS. (R) February, 1975. Alan Arkin, Sally Kellerman.

RAGE (P) 100
WARNER BROS. (R) November, 1972. George C. Scott, Richard Basehart. Color.

RAGE 103
COLUMBIA. (R) December, 1966. Glenn Ford, Stella Stevens. Color.

RAGE AT DAWN 87
RKO RADIO. (R) March 26, 1955. Randolph Scott, Mala Powers. Color.

RAGE TO LIVE, A (P) 101
UNITED ARTISTS. (R) May, 1965. Suzanne Pleshette, Ben Gazzara. b&w.

RAGGEDY ANN & ANDY (P) 84
TWENTIETH CENTURY-FOX. (R) April, 1977. Animated cartoon.

RAID ON ROMMEL 99
UNIVERSAL. (R) February, 1971. Richard Burton, Wolfgang Preiss.

RAIDERS, THE 80
UNIVERSAL. (R) November 1952. Richard Conte, Viveca Lindfors. Color.

RAIDERS, THE 75
UNIVERSAL. (R) May, 1964. Robert Culp, Brian Keith. Color.

RAIDERS FROM BENEATH THE SEA, THE 78
TWENTIETH CENTURY-FOX. (R) January, 1965. Ken Scott, Merry Anders.

RAIDERS OF LEYTE GULF, THE 80
HEMISPHERE. (R) July, 1963. Jennings Sturgeon, Efren Reyes.

RAIDERS OF OLD CALIFORNIA 72
REPUBLIC. (R) November, 1957. Jim Davis, Arleen Whelan.

RAILROAD MAN, THE (Ital., Eng. Titles) 105
CONTINENTAL. (R) October, 1965. Pietro Germi, Luisa Della Noce.

RAILWAY CHILDREN, THE 108
UNIVERSAL. (R) October, 1971. Dinah Sheridan, Bernard Cribbens, Jenny Agutter. Color.

RAIN PEOPLE, THE 101
(R) September, 1969. James Caan, Shirley Knight, Robert Duvall. Color.

RAINMAKER, THE (VV) 121
PARAMOUNT. (R) February, 1957. Katharine Hepburn, Burt Lancaster. Color.

RAINS OF RANCHIPUR, THE (Cs) 104
TWENTIETH CENTURY-FOX. (R) December, 1955. Lana Turner, Richard Burton, Fred MacMurray, Joan Caulfield, Michael Rennie. Color.

RAINTREE COUNTY (Camera 65) 185
METRO-GOLDWYN-MAYER. (R) Special. October, 1957. Elizabeth Taylor, Montgomery Clift. Color.

RAISE THE TITANIC (T) 122
ASSOCIATED FILM DIST. (R) August, 1980. Jason Robards, Richard Jordan.

RAISIN IN THE SUN, A 128
COLUMBIA. (R) April, 1961. Sidney Poitier, Claudia McNeil.

RAISING A RIOT (Brit.) 90
CONTINENTAL. (R) May, 1957. Kenneth More, Ronald Squire. Color.

RALLY 'ROUND THE FLAG BOYS! (Cs) 106
TWENTIETH CENTURY-FOX. (R) February, 1959. Paul Newman, Joanne Woodward. Color.

RAMPAGE 98
WARNER BROS. (R) October, 1963. Robert Mitchum, Elsa Martinelli. Color.

RAMPARTS OF CLAY 85
CINEMA 5. (R) February, 1971. Leila Schenna. Color.

RANCHO DELUXE 93
UNITED ARTISTS. (R) March, 1975. Jeff Bridges, Sam Waterston.

RANSOM! 104
METRO-GOLDWYN-MAYER. (R) January 20, 1956. Glenn Ford, Donna Reed.

RAPE OF MALAYA, THE (Brit.) 107
LOPERT. (R) 1956. Virginia McKenna, Peter Finch.

RAPTURE (Cs) 105
INTERNATIONAL CLASSICS. (R) September, 1965. Melvyn Douglas, Dean Stockwell.

RARE BREED, THE (P) 97
UNIVERSAL. (R) February, 1966. James Stewart, Maureen O'Hara. Color.

RASCAL 85
BUENA VISTA. June, 1969. Steve Forrest, Bill Mumy, Elsa Lanchester. Color.

RASPUTIN—THE MAD MONK (Cs) 92
TWENTIETH CENTURY-FOX. (R) April, 1966. Christopher Lee, Barbara Shelley. Color.

RAT FINK 90
CINEMA DISTRIBUTORS OF AMERICA. (R) February, 1966. Schuyler Hayden.

RAT RACE, THE 105
PARAMOUNT. (R) July, 1960. Tony Curtis, Debbie Reynolds. Color.

RATTLE OF A SIMPLE MAN 96
READE-STERLING. (R) December, 1964. Harry H. Corbett, Diane Cilento.

RAVAGERS, THE 79
HEMISPHERE. (R) December, 1965. John Saxon, Fernando Poe, Jr.

RAVEN'S END (Danish; Eng. Titles) 100
NEW YORKER. (R) August, 1970. Thommy Berggren, Keve Hjelm.

RAW EDGE 76
UNIVERSAL. (R) September, 1956. Rory Calhoun, Yvonne de Carlo. Color.

RAW WIND IN EDEN (Cs) 89
UNIVERSAL. (R) October, 1958. Esther Williams, Jeff Chandler. Color.

RAWHIDE BREED 68

Length In Mins.

ALLIED ARTISTS. (R) January, 1958. Rex Reason, Nancy Gates.

RAWHIDE YEARS, THE 85
UNIVERSAL. (R) July, 1956. Tony Curtis, Colleen Miller. Color.

RAWMIE 72
ALLIED ARTISTS. (R) July, 1960. David Ladd, Julie Adams.

RAZZIA (Fr.) 105
KASSLER. (R) December, 1958. Jean Gabin, Magli Noel.

REACH FOR GLORY 89
ROYAL. (R) September, 1963. Harry Andrews, Kay Walsh.

REACH FOR THE SKY (Brit.) 104
RANK. (R) June, 1957. Kenneth More, Muriel Pavlow.

REAL LIFE 99
PARAMOUNT. (R) March, 1979. Charles Grodin, Frances Lee McCain.

REASON TO LIVE . . . A REASON TO DIE, A 92
K-TEL MOTION PICTURES. (R) Special, 1973. James Coburn, Telly Savalas.

REBEL IN TOWN 78
UNITED ARTISTS. (R) July, 1956. John Payne, Ruth Roman.

REBEL SET, THE 72
ALLIED ARTISTS. (R) August, 1959. Gregg Palmer, Kathleen Crowley.

REBEL WITHOUT A CAUSE (Cs) 111
WARNER BROTHERS. (R) October 29, 1955. James Dean, Natalie Wood. Color.

REBELLION (Japanese; Eng. Titles) 120
TOHO. (R) October, 1968. Toshiro Mifune, Takeshi Kato, Yoko Tsukasa. Color.

RECKONING, THE 109
COLUMBIA. (R) October, 1969. Nichol Williamson, Rachel Roberts, Paul Rogers. Color.

RED AND THE WHITE, THE (Hungarian; Eng. Titles) ... 92
BRANDON. (R) March, 1969. Tatyana Konyukova, Krystyna Mikolaiewska, Mikhail Kozakov. b&w.

RED BEARD 185
FRANK LEE. (R) December, 1968. Toshiro Mifune, Yuzo, Kayama, Yoshio Tsukasa. b&w.

RED DESERT 116
RIZZOLI. (R) February, 1965. Monica Vitti, Richard Harris. Color.

RED LANTERNS (Germ., Eng. Titles) 85
TIMES. (R) March, 1965. Jenny Karezi, George Foondas.

RED LINE 7000 110
PARAMOUNT. (R) Thanksgiving, 1965. James Caan, Laura Devon. Color.

RED SKY AT MORNING 113
UNIVERSAL. (R) Richard Thomas, Catherine Burns, Desi Arnaz, Jr. Color.

RED SUN 112
NATIONAL GENERAL. (R) June, 1972. Charles Bronson, Ursula Andress. Color.

RED SUNDOWN 82
UNIVERSAL. (R) March, 1956. Rory Calhoun, Martha Hyer. Color.

RED TENT, THE 121
PARAMOUNT. (R) July, 1971. Sean Connery, Peter Finch, Claudia Cardinale. Color.

RED TOMAHAWK 82
PARAMOUNT. (R) January, 1967. Howard Keel, Joan Caulfield. Color.

REDEEMER, THE (Span., Eng. Titles) 93
EMPIRE FILMS. (R) March, 1966. Luis Alvarez, Maruchi Fresno. Color.

REFLECTIONS IN A GOLDEN EYE (P) 108
WARNER BROS.-7 ARTS. (R) October, 1967. Elizabeth Taylor, Marlon Brando. Color.

REFORM SCHOOL GIRL 71
AMER. INT. PICS. (R) June, 1959. Gloria Castillo, Ross Ford.

REINCARNATION OF PETER PROUD, THE (P) 104
AMERICAN INTERNATIONAL. (R) April, 1975. Michael Sarrazin, Jennifer O'Neill.

REIVERS, THE (P) 107
NATIONAL GENERAL. (R) December, 1969. Steve McQueen, Sharrell Farrell, Rupert Crosse. Color.

RELUCTANT ASTRONAUT, THE 102
UNIVERSAL. (R) May, 1967. Don Knotts, Arthur O'Connell. Color.

RELUCTANT DEBUTANTE, THE (Cs) 96
METRO-GOLDWYN-MAYER. (R) August, 1958. Rex Harrison, Kay Kendall. Color.

RELUCTANT SAINT, THE 105
COLUMBIA. (R) December, 1962. Maximilian Schell, Ricardo Montalban.

REMARKABLE MR. PENNYPACKER, THE (Cs) 87
TWENTIETH CENTURY-FOX. (R) March, 1950. Clifton Webb, Dorothy McGuire. Color.

REPORT TO THE COMMISSIONER 112
UNITED ARTISTS. (R) February, 1975. Michael Moriarty, Yaphet Kotto.

REPRIEVE 135
ALLIED ARTISTS. (R) May, 1962. Ben Gazzara, Stuart Whitman.

REPRISAL 74
COLUMBIA. (R) November, 1956. Guy Madison, Felicia Farr. Color.

REPTILE, THE 90
TWENTIETH CENTURY-FOX. (R) April, 1966. Noel Williams, Jennifer Daniels. Color.

REPTILICUS (C) 90
AMERICAN INTERNATIONAL. (R) October, 1961. Bodil Miller, Carl Ottosen. Color.

REPULSION 105
ROYAL FILMS INTERNATIONAL. (R) October, 1965. Catherine Deneuve, Ian Hendry.

REQUIEM FOR A GUNFIGHTER (T) 91
EMBASSY. (R) June, 1965. Rod Cameron, Stephen McNally. Color.

REQUIEM FOR A HEAVYWEIGHT 87
COLUMBIA. (R) October, 1962. Anthony Quinn, Jackie Gleason.

RESCUERS, THE 76
BUENA VISTA. (R) June, 1977. Animated Cartoon.

RESTLESS BREED, THE 81
TWENTIETH CENTURY-FOX. (R) May, 1957. Scott Brady, Anne Bancroft.

RESTLESS NIGHT, THE 102
CASINO. (R) May, 1965. Bernhard Wicki, Ulla Jacobson.

RESTLESS YEARS, THE (Cs) 86
UNIVERSAL. (R) December, 1958. John Saxon, Sandra Dee.

RETURN FROM THE ASHES (P) 105
UNITED ARTISTS. (R) November, 1965. Maximilian Schell, Samantha Eggar.

RETURN FROM WITCH MOUNTAIN 94
BUENA VISTA. (R) March, 1978. Bette Davis, Christopher Lee.

RETURN OF COUNT YORGA, THE 97
AMERICAN INTERNATIONAL (R) August, 1971. Robert Quarry, Mariette Hartley. Color.

RETURN OF DON CAMILLO, THE (Ital.) 155
I.F.E. (R) December, 1955. Fernandel, Gino Cervi.

RETURN OF DRACULA 77
UNITED ARTISTS. (R) June, 1958. Francis Lederer, Norma Eberhardt.

RETURN OF JACK SLADE, THE (SS) 79
ALLIED ARTISTS. (R) October 9, 1955. John Ericson Mari Blanchard.

RETURN OF MR. MOTO, THE 71
TWENTIETH CENTURY-FOX (R) December, 1965. Henry Silva, Terence Longdon.

RETURN OF THE FLY (Cs) 80
TWENTIETH CENTURY-FOX (R) August, 1950. Vincent Price, Brett Halsey.

RETURN OF THE MAN CALLED HORSE, THE 129
UNITED ARTISTS. (R) August, 1976. Richard Harris, Gale Sondergaard.

RETURN OF THE PINK PANTHER, THE 113
UNITED ARTISTS. (R) May, 1975. Peter Sellers, Christopher Plummer.

RETURN OF THE SEVEN (P) 95
UNITED ARTISTS (R) October, 1966. Yul Brynner, Robert Fuller. Color.

RETURN OF THE STREETFIGHTER, THE 75
NEW LINE. (R) July, 1975. Sonny Chiba.

RETURN TO MACON COUNTY 89
AMERICAN INTERNATIONAL. (R) July, 1975. Nick Nolte, Don Johnson.

RETURN TO PEYTON PLACE (Cs) 122
TWENTIETH CENTURY-FOX. (R) May. 1961. Carol Lynley, Jeff Chandler. Color.

RETURN TO WARBOW 67
COLUMBIA. (R) January, 1958. Phil Carey, Catherine Macleod. Color.

REVENGE OF FRANKENSTEIN, THE 94
COLUMBIA. (R) July, 1958. Peter Cushing, Francis Matthews.

REVENGE OF THE CONQUERED 84
AMERICAN INTERNATIONAL. (R) February, 1964. Burt Nelson, Wandisa Guida.

REVENGE OF THE CREATURE (3D) 82
UNIVERSAL. (R) May, 1955. John Agar, Lori Nelson.

REVENGE OF THE GLADIATORS 100
PARAMOUNT. (R) September, 1965. Roger Browne, Scilla Gabel. Color.

REVENGE OF THE PINK PANTHER 98
UNITED ARTISTS. (R) July, 1978. Peter Sellers, Herbert Lom.

REVENGERS, THE 112

Length In Mins.

NATIONAL GENERAL. (R) June, 1972. William Holden, Ernest Borgnine. Color.

REVOLT AT FORT LARAMIE **78**
UNITED ARTISTS. (R) March, 1957. John Dhener, Gregg Palmer. Color.

REVOLT IN THE BIG HOUSE **79**
ALLIED ARTISTS. (R) November, 1958. Gene Evans, Robert Blake.

REVOLT OF MAMIE STOVER, THE (Cs) **98**
TWENTIETH CENTURY-FOX. (R) April, 1956. Jane Russell, Richard Egan. Color.

REVOLT OF THE SLAVES, THE **99**
UNITED ARTISTS. (R) May, 1961. Rhonda Fleming, Lang Jeffries. Color.

REVOLUTION **86**
LOPERT. (R) August, 1968. Documentary. Color.

REVOLUTIONARY, THE **100**
UNITED ARTISTS. (R) July, 1970. Jon Voight, Jennifer Salt, Seymour Cassel. Color.

REWARD, THE (Cs) **92**
TWENTIETH CENTURY-FOX. (R) October, 1965. Max Von Sydow, Yvette Mimieux. Color.

RHINO! **91**
METRO-GOLDWYN-MAYER. (R) May, 1964. Harry Guardino, Robert Culp. Color.

RHINOCEROS **101**
AFT DISTRIBUTING. (R) Special, 1974. Zero Mostel, Gene Wilder, Karen Black.

RICH KIDS **96**
UNITED ARTISTS. (R) August, 1979. Trini Alvarado, Jeremy Levy.

RICHARD III (Brit.) (VV) **155**
LOPERT. (R) March, 1956. Laurence Olivier, Claire Bloom, Ralph Richardson, John Gielgud. Color.

RIDE A CROOKED TRAIL (Cs) **87**
UNIVERSAL. (R) September, 1958. Audie Murphy, Gia Scala. Color.

RIDE A VIOLENT MILE **80**
TWENTIETH CENTURY-FOX. (R) December, 1957. John Agar, Penny Edwards.

RIDE A WILD PONY **91**
BUENA VISTA. (R) April, 1976. Michael Craig, John Meillon.

RIDE BACK, THE **79**
UNITED ARTISTS. (R) May, 1957. Anthony Quinn, William Conrad.

RIDE BEYOND VENGEANCE **108**
COLUMBIA. (R) April, 1966. Chuck Connors, Michael Rennie. Color.

RIDE LONESOME (Cs) **73**
COLUMBIA. (R) February, 1959. Randolph Scott, Karen Steele. Color.

RIDE OUT FOR REVENGE **79**
UNITED ARTISTS. (R) November, 1957. Rory Calhoun, Gloria Grahame.

RIDE THE HIGH COUNTRY (Cs) **94**
METRO-GOLDWYN-MAYER. (R) May, 1962. Randolph Scott, Joel McCrea. Color.

RIDE THE HIGH IRON **74**
COLUMBIA. (R) January, 1957. Don Taylor, Sally Forrest.

RIDE THE WILD SURF **101**
COLUMBIA. (R) July, 1964. Fabian, Tab Hunter. Color.

RIDE TO HANGMAN'S TREE, THE **90**
UNIVERSAL. (R) July, 1967. Jack Lord, James Farentino. Color.

RIDE, VAQUERO **90**
MGM. (R) July 17, 1958. Robert Taylor, Ava Gardner, Howard Keel. Color.

RIDER ON A DEAD HORSE **72**
ALLIED ARTISTS. (R) May, 1962. John Vivyan. Lisalu.

RIDER ON THE RAIN (French; Eng. Titles) **115**
AVCO EMBASSY. (R) May, 1970. Marlene Jobert, Charles Bronson, Annie Cordy. Color.

RIFIFI (Fr.) **118**
UMPO. (R) June, 1956. Jean Servais, Carl Mohner.

RIFIFI IN TOKYO **89**
METRO-GOLDWYN-MAYER. (R) April, 1963. Karl Boehm, Michael Vitold.

RIGHT APPROACH, THE (Cs) **92**
TWENTIETH CENTURY-FOX. (R) May, 1961. Frankie Vaughn, Juliet Prowse.

RING-A-DING RHYTHM! **78**
COLUMBIA. (R) September, 1962. Helen Shapiro, Craig Douglas.

RING OF BRIGHT WATER **107**
CINERAMA. (R) June, 1969. Bill Travers, Virginia McKenna. Color.

RING OF FIRE **91**
METRO-GOLDWYN-MAYER. (R) May, 1961. David Janssen, Joyce Taylor. Color.

RING OF TREASON **89**
PARAMOUNT. (R) June, 1964. Bernard Lee, Margaret Tyzack.

RINGS AROUND THE WORLD **98**
COLUMBIA. (R) September, 1966. Circus Documentary narrated by Don Ameche. Color.

RIO BRAVO (C) **141**
WARNER BROS. (R) April, 1959. John Wayne, Dean Martin.

RIO CONCHOS (Cs) **107**
TWENTIETH CENTURY-FOX. (R) October, 1964. Richard Boone, Stuart Whitman. Color.

RIO LOBO (P) **114**
NATIONAL GENERAL. (R) December, 1970. John Wayne, Jorge Rivero. Color.

RIOT **97**
PARAMOUNT. (R) January, 1969. Jim Brown, Gene Hackman, Ben Carruthers. Color.

RIOT IN JUVENILE PRISON **71**
UNITED ARTISTS. (R) April, 1959. John Hoyt, Marcia Henderson.

RIOT ON SUNSET STRIP **85**
AMERICAN INTERNATIONAL. (R) March, 1967. Aldo Ray, Mimsy Farmer. Color.

RIP OFF **90**
J. CINEMAX. (R) September, 1972. Don Scardino, Ralph Endersby. Color.

RISE AND FALL OF LEGS DIAMOND, THE **101**
WARNER BROS. (R) February, 1960. Ray Danton, Karen Stecie.

RISING OF THE MOON, THE **81**
WARNER BROS. (R) August, 1957. Frank Lawton, Dennis O'Day.

RITUAL, THE (Swedish; Eng. Titles) **75**
JANUS. (R) September, 1969. Ingrid Thulin, Anders Ek.

RITZ, THE **91**
WARNER BROS. (R) September, 1976. Jack Weston, Rita Moreno.

RIVALS **101**
AVCO-EMBASSY. (R) August, 1972. Joan Hackett, Scott Jacoby. Color.

RIVER CHANGES, THE **91**
WARNER BROTHERS (R) March 24, 1956. Rossana Rory, Harold Maresch.

RIVERRUN **87**
COLUMBIA. (R) April, 1970. John McLiam, Louise Ober, Mark Jenkins. Color.

RIVERS EDGE, THE (Cs) **87**
TWENTIETH CENTURY-FOX. April, 1957. Ray Milland, Debra Paget, Anthony Quinn. Color.

RIVIERA (Ital.-Eng. dubbed) **68**
I.F.E. (R) September. 1956. Martine Carol. Color.

ROAD HOUSE GIRL (Brit.) **68**
ASTOR. (R) June, 1955. Sandra Dorn, Marwell Reed.

ROAD TO DENVER **90**
REPUBLIC. (R) June, 16, 1955. John Payne, Mona Freeman. Color.

ROAD TO HONG KONG, THE **91**
UNITED ARTISTS. (R) June, 1962. Bing Crosby, Bob Hope.

ROAD TO SALINA (P) **96**
AVCO-EMBASSY. (R) February, 1971. Rita Hayworth, Robert Walker. Color.

ROADIE (P) **105**
UNITED ARTISTS. (R) June, 1980. Meat Loaf, Kaki Hunter.

ROADRACERS, THE **78**
AMERICAN INTERNATIONAL. (R) March, 1959. Joel Lawrence, Marian Collier.

ROBBER'S ROOST **82**
UNITED ARTISTS. (R) May, 1955. Robert Montgomery, Richard Boone.

ROBBERY **113**
EMBASSY. (R) October, 1967. Stanley Baker, Joanna Pettet. Color.

ROBBERY UNDER ARMS **83**
RANK FILM DISTRIBUTORS OF AMERICA. (R) May, 1958. Peter Finch, Ronald Lewis. Color.

ROBIN AND MARIAN **107**
COLUMBIA. (R) April, 1976. Sean Connery, Audrey Hepburn.

ROBIN AND THE SEVEN HOODS (P) **123**
WARNER BROS. (R) June, 1964. Frank Sinatra, Dean Martin. Color.

ROBIN HOOD **83**
BUENA VISTA. (R) December, 1973. Cartoon.

ROBINSON CRUSOE ON MARS (Ts) **110**
PARAMOUNT. (R) June, 1964. Paul Mantee, Vic Lundin. Color.

ROCK-A-BYE (VV) **108**
PARAMOUNT. (R) July, 1958. Jerry Lewis, Marilyn Maxwell. Color.

ROCK ALL NIGHT **65**

Length In Mins.

AMERICAN INTERNATIONAL. (R) April, 1957. Dick Miller, Abby Dalton.

ROCK AROUND THE CLOCK 77
COLUMBIA. (R) April, 1956. Bill Haley, Johnny Johnston.

ROCK AROUND THE WORLD 71
AMER. INT. PICS. (R) August, 1957. Tommy Steele, Abby Dalton.

ROCK, PRETTY BABY 89
UNIVERSAL. (R) January, 1957. Sal Mineo, John Saxon.

ROCK, ROCK, ROCK 83
DISTRIBUTORS CORPORATION OF AMERICA. (R) December, 1956. Alan Freed.

ROCKABILLY BABY (R) 82
TWENTIETH CENTURY-FOX. (R) October, 1957. Virginia Field, Douglas Kennedy.

ROCKY 121
UNITED ARTISTS. (R) January, 1977. Sylvester Stallone, Talia Shire.

ROCKY II 119
UNITED ARTISTS. (R) June, 1979. Sylvester Stallone, Talia Shire.

RODAN! (Jap.) 70
DIST. CORP. OF AMER. (R) August, 1957. Kenji Sawara, Yumi Shirakawa. Color.

ROLLERBALL 128
UNITED ARTISTS. (R) June, 1975. James Caan.

ROLLERCOASTER (P) 119
UNIVERSAL. (R) June, 1977. George Segal, Richard Widmark, Timothy Bottoms.

ROLLING THUNDER 99
AMERICAN INTERNATIONAL. (R) October, 1977. William Devane, Tommy Lee Jones.

ROMAN SPRING OF MRS. STONE, THE 104
WARNER BROS. (R) December, 1961. Vivien Leigh, Warren Beatty. Color.

ROMANCE OF A HORSETHIEF 101
ALLIED ARTISTS. (R) July, 1971. Yul Brynner, Eli Wallach. Color.

ROMANOFF AND JULIET 103
UNIVERSAL. (R) March, 1961. John Gavin, Sandra Dee. Color.

ROMANTIC ENGLISHWOMAN, THE 117
NEW WORLD PICTURES. (R) November, 1975. Glenda Jackson, Michael Caine.

ROME ADVENTURE 119
WARNER BROS. (R) March, 1962. Troy Donahue, Angie Dickinson. Color.

ROMEO AND JULIET 90
W.E. (R) September, 1968. Gerald Meynier, Rosemarie Dexter, Carlos Estrada. Color

ROMEO AND JULIET 138
PARAMOUNT. (R) October 1968. Leonard Whiting, Olivia Hussey, Michael York. Color.

ROMEO AND JULIET (Brit.) 140
UNITED ARTISTS. (R) December, 1954. Laurence Harvey, Susan Shentall. Color.

ROMEO AND JULIET 126
EMBASSY. (R) October, 1966. Royal Ballet Performance. Color.

ROOKIE, THE (Cs) 84
TWENTIETH CENTURY-FOX. (R) January, 1960. Tommy Noonan, Pete Marshall.

ROOM AT THE TOP 113
CONTINENTAL. (R) April, 1959. Laurence Harvey, Simone Signoret.

ROONEY 88
RANK FILM DISTRIBUTORS OF AMERICA. (R) June, 1958. Barry Fitzgerald, John Gregson.

ROOSTER COGBURN 107
UNIVERSAL. (R) October, 1975. John Wayne, Katharine Hepburn.

ROOTS OF HEAVEN, THE (Cs) 125
TWENTIETH CENTURY-FOX. (R) November, 1958. Errol Flynn, Juliett Greco. Color.

ROSANNA (Mex.-Eng. dubbed) 72
JACON. (R) June, 1956. Rosanna Podesta.

ROSE, THE 134
TWENTIETH CENTURY-FOX. (R) November, 1979. Bette Middler, Alan Bates.

ROSE FOR EVERYONE, A (Ital.) 107
ROYAL FILMS. (R) July, 1967. Claudia Cardinale, Nino Manfredo. Color.

ROSE TATTOO, THE (VV) 117
PARAMOUNT. (R) February, 1956. Burt Lancaster, Anna Magnani.

ROSEBUD (P) 126
UNITED ARTISTS. (R) April, 1975. Peter O'Toole, Richard Attenborough.

ROSELAND 103
CINEMA SHARES. (R) October, 1977. Teresa Wright, Lou Jacobi.

ROSEMARY'S BABY 136
PARAMOUNT. (R) June, 1968. Mia Farrow, John Cassavettes, Ruth Gordon. Color.

ROSIE! (TE) 98
UNIVERSAL. (R) November, 1967. Rosalind Russell, Brian Aherne. Color.

ROTTEN TO THE CORE (Brit.) 90
CINEMA V. (R) July, 1965. Anton Rodgers, Eric Sykes.

ROUGE ET NOIR (Fr.) 145
DIST. CORP. OF AMER. (R) April, 1958. Gerard Philipe, Danielle Darrieux. Color.

ROUGH CUT 111
PARAMOUNT. (R) June, 1980. Burt Reynolds, Lesley-Anne Down.

ROUGH NIGHT IN JERICHO (TS) 104
UNIVERSAL. (R) October, 1967. Dean Martin, Jean Simmons. Color.

ROUND TRIP 86
CONTINENTAL DISTRIBUTING. (R) July, 1967. Ventantino Venantini, Ellen Faison. Color.

ROUNDERS, THE (P) 85
METRO-GOLDWYN-MAYER. (R) February, 1965. Glenn Ford, Henry Fonda. Color.

ROUNDUP, THE (Hungarian; Eng. Titles) no time given
ALTURA. (R) May, 1969. Janos Gorbe, Tibor Motnar, Andras Kazak. b&w.

ROUSTABOUT (TS) 101
PARAMOUNT. (R) November, 1964. Elvis Presley, Barbara Stanwyck. Color.

ROYAL AFFAIRS IN VERSAILLES (Fr.) 152
TIMES. (R) March, 1957. Claudette Colbert, Orson Welles. Color.

ROYAL FLASH 99
TWENTIETH CENTURY-FOX. (R) October, 1975. Malcolm McDowell, Alan Bates.

ROYAL HUNT OF THE SUN (P) 113
NATIONAL GENERAL. (R) October, 1969. Christopher Plummer, Robert Shaw, Nigel Davenport. Color.

RULING CLASS, THE 150
AVCO-EMBASSY. (R) October, 1972. Peter O'Toole, Alistair Sim. Color.

RUMBLE ON THE DOCKS 82
COLUMBIA. (R) December, 1956. James Darron, Laurie Carroll.

RUN, ANGEL, RUN! 94
FANFARE. (R) April, 1969. William Smith, Valerie Starett, Gene Shane. Color.

RUN FOR COVER (VV) 92
PARAMOUNT. (R) April, 1955. James Cagney, Viveca Lindfors. Color.

RUN FOR THE SUN (SS) 99
UNITED ARTISTS. (R) August, 1956. Richard Widmark, Jane Greer. Color.

RUN FOR YOUR WIFE (TS) 97
ALLIED ARTISTS. (R) November, 1966. Ugo Tognazzi, Marina Vlady. Color.

RUN OF THE ARROW 85
UNIVERSAL. (R) September, 1957. Rod Steiger, Sarita Montiel. Color.

RUN SILENT, RUN DEEP 93
UNITED ARTISTS. (R) April, 1958. Clark Gable, Burt Lancaster.

RUN WILD, RUN FREE 100
COLUMBIA. (R) June, 1969. John Mills, Mark Lester, Sylvia Sims. Color.

RUNAWAY DAUGHTERS 90
AMERICAN INTERNATIONAL. (R) October 31, 1956. Maria English, Anna Sten.

RUNNER STUMBLES, THE 109
TWENTIETH CENTURY-FOX. (R) November, 1979. Dick Van Dyke, Kathleen Quinlan.

RUNNING 100
UNIVERSAL. (R) November, 1979. Michael Douglas, Susan Anspach.

RUNNING MAN, THE (P) 103
COLUMBIA. (R) October, 1963. Laurence Harvey, Lee Remick. Color.

RUNNING TARGET 83
UNITED ARTISTS. (R) November, 1956. Doris Dowling, Arthur Franz.

RUNNING WILD 80
UNIVERSAL. (R) December, 1955. Mamie Van Doren, William Campbell, Keenan Wynn.

RUSH TO JUDGMENT 122
IMPACT. (R) May, 1967. Documentary on Kennedy Assassination.

RUSS MEYER'S VIXEN 105
EE PRODUCTIONS. (R) April, 1969. Ericka Gavin, Harrison Page.

RUSSIAN ROULETTE 93
AVCO-EMBASSY. (R) July, 1975. George Segal, Cristina Raines.

RUSSIANS ARE COMING THE RUSSIANS ARE COMING, THE 120

Length In Mins.

UNITED ARTISTS. (R) May, 1966. Carl Reiner, Eva Marie Saint, Alan Arkin.

RX MURDER (Cs) **65**
TWENTIETH CENTURY-FOX. (R) August, 1958. Rick Jason, Marius Goring.

S

S.O.S. PACIFIC **92**
UNIVERSAL. (R) July, 1960. Eddie Constantine, Pier Angeli.

S*P*Y*S **87**
TWENTIETH CENTURY-FOX. (R) June, 1974. Elliott Gould, Donald Sutherland.

SABAKA **81**
UNITED ARTISTS. (R) February, 1955. Boris Karloff, Reginald Denny. Color.

SABATA (TE) **107**
UNITED ARTISTS. (R) October, 1970. Lee Van Cleef, William Berger. Color.

SABU AND THE MAGIC RING **61**
ALLIED ARTISTS. (R) March, 1958. Sabu, William Marshall.

SACCO AND VANZETTI **120**
UMC. (R) November, 1971. Gian Maria Volonte, Riccardo Cucciolla. Color.

SAD HORSE, THE (Cs) **81**
TWENTIETH CENTURY-FOX. (R) March, 1959. David Ladd, Chill Wills. Color.

SAD SACK, THE (VV) **98**
PARAMOUNT. (R) December, 1957. Jerry Lewis, David Wayne.

SADDLE THE WIND (Cs) **84**
METRO-GOLDWYN-MAYER. (R) March, 1958. Robert Taylor, Julie London. Color.

SAFARI (Cs) **91**
COLUMBIA. (R) June, 1956. Victor Mature, Janet Leigh. Color.

SAFE AT HOME **83**
COLUMBIA. (R) April, 1962. Mickey Mantle, Roger Maris.

SAFE CRACKER, THE **96**
METRO-GOLDWYN-MAYER. (R) January, 1958. Ray Milland, Barry Jones.

SAFE PLACE, A **94**
COLUMBIA. (R) October, 1971. Tuesday Weld, Orson Welles. Color.

SAGA OF HEMP BROWN, THE (Cs) **79**
UNIVERSAL. (R) October, 1958. Rory Calhoun, Beverly Garland. Color.

SAIL A CROOKED SHIP **88**
COLUMBIA. (R) January, 1962. Robert Wagner, Dolores Hart.

SAILOR FROM GIBRALTAR, THE **89**
LOPERT. (R) April, 1967. Jeanne Moreau, Ian Bannen.

SAILOR WHO FELL FROM GRACE WITH THE SEA, THE **105**
AVCO-EMBASSY. (R) April, 1976. Sarah Miles, Kris Kristofferson.

ST. IVES **93**
WARNER BROS. (R) August, 1976. Charles Bronson, Jacqueline Bisset.

SAINT JOAN **110**
UNITED ARTISTS. (R) June, 1957. Richard Widmark, Richard Todd.

SAINTLY SINNERS **78**
UNITED ARTISTS. (R) February, 1962. Don Beddoe, Ellen Corby.

SALESMAN **90**
MAYSLES. (R) April, 1969. Paul Brennan, Charles McDevitt, James Baker. b&w.

SALT AND PEPPER **101**
UNITED ARTISTS. (R) July, 1968. Sammy Davis, Jr., Peter Lawford, Michael Bates. Color.

SALO, 120 DAYS OF SODOM (It., Eng. titles) **117**
ZEBRA RELEASING. (R) October, 1977. Paolo Bonacelli, Giorgio Cataldi.

SALT OF THE EARTH **104**
INDEPENDENT PRODUCTIONS. (R) October, 1965. Rosvara Revueltas, Juan Chacon.

SALTO (Polish; Eng. titles) **104**
KANAWHA. (R) August, 1966. Zbigniew Cybulski, Gustaw Holoubek.

SALUT L'ARTISTE (French; English sub-titles) **96**
ARCHER KING. (R) February, 1976. Marcello Mastroianni, Francoise Fabian.

SAM WHISKEY **96**
UNITED ARTISTS. (R) March, 1969. Burt Reynolds, Clint Walker, Ossie Davis. Color.

SAMAR **89**
WARNER BROS. (R) May, 1962. George Montgomery, Gilbert Roland. Color.

SAME TIME, NEXT YEAR **119**
UNIVERSAL. (R) March, 1979. Ellen Burstyn, Alan Alda.

SAMSON AND THE SEVEN MIRACLES OF THE WORLD (Colorscope) **78**
AMER. INT'L. PICS. (R) 1963. Gordon Scott, Yoko Tani. Color.

SAMSON AND THE SLAVE QUEEN (Colorscope) **86**
AMER. INT'L. PICS. (R) 1964. Alan Steel, Pierre Brice. Color.

SAMURAI (Jap.) **92**
FINE ARTS. (R) November, 1955. Toshiro Mifune. Color.

SAMURAI ASSASSIN **122**
TOHO-MIFUNE. (R) March, 1965. Toshiro Mifune, Keiju Kobayashi.

SANCTUARY (Cs) **90**
TWENTIETH CENTURY-FOX. (R) March, 1961. Lee Remick, Yves Montand.

SAND PEBBLES, THE (P) **195**
20TH CENTURY-FOX. (R) December, 1966. Steve McQueen, Richard Attenborough. Color.

SANDOKAN THE GREAT (TS) (Ital., dubbed) **105**
MGM. (R) 1965. Steve Reeves, Genevieve Grad.

SANDPIPER, THE (P) **116**
METRO-GOLDWYN-MAYER. (R) June, 1965. Elizabeth Taylor, Richard Burton. Color.

SANDRA (Ital., Eng. titles) **100**
ROYAL FILM INTERNATIONAL. (R) January, 1966. Claudia Cardinale, Michael Craig.

SANDS OF BEERSHEBA **90**
LANDAUUNGER. (R) May, 1966. Diane Baker, David Opatoshu.

SANDS OF KALAHARI (P) **119**
PARAMOUNT. (R) November, 1965. Stuart Whitman, Stanley Baker. Color.

SANTA CLAUS CONQUERS THE MARTIANS **80**
EMBASSY. (R) November, 1964. John Call, Leonard Hicks. Color.

SANTA FE PASSAGE **90**
REPUBLIC. (R) May 12, 1955. John Payne, Faith Domergue. Color.

SANTIAGO **93**
WARNER BROTHERS. (R) July, 1956. Alan Ladd, Rossana Podesta. Color.

SAPPHIRE **92**
UNIVERSAL. (R) November, 1959. Nigel Patrick, Yvonne Mitchell. Color.

SASQUATCH **96**
NORTH AMERICAN PRODUCTIONS. (R) January, 1976. George Lauris.

SATAN BUG, THE (P) **114**
UNITED ARTISTS. (R) March, 1965. George Maharis, Richard Basehart. Color.

SATAN NEVER SLEEPS (Cs) **126**
TWENTIETH CENTURY-FOX. (R) March, 1962. William Holden, Clifton Webb. Color.

SATAN'S SATELLITES **70**
REPUBLIC. (R) May, 1958. Judd Holdren, Aline Towne.

SATCHMO THE GREAT **63**
UNITED ARTISTS. (R) September, 1957. Louis Armstrong, Leonard Bernstein.

SATELLITE IN THE SKY (Brit.) (Cs) **85**
WARNER BROTHERS. (R) July 21, 1956. Kieron Moore, Louis Maxwell. Color.

SATURDAY MORNING **82**
COLUMBIA. (R) May, 1971. Group Therapy Documentary. Color.

SATURDAY NIGHT FEVER **120**
PARAMOUNT. (R) December, 1977. John Travolta, Karen Lynn Gorney.

SATURDAY NIGHT AND SUNDAY MORNING **90**
CONTINENTAL. (R) March, 1961. Albert Finney, Shirley Anne Field.

SAVAGE! **81**
NEW WORLD PICTURES. (R) February, 1973. James Inglehart, Carol Speed. Color.

SAVAGE GUNS, THE (Metroscope) **90**
MGM. (R) 1962. Richard Basehart, Don Taylor. Color.

SAVAGE INNOCENTS (T-170) **110**
PARAMOUNT. (R) February, 1961. Anthony Quinn, Yoko Tani. Color.

SAVAGE IS LOOSE, THE **114**
CAMPBELL DEVON. (R) November, 1974. George C. Scott, Trish van Devere.

SAVAGE MESSIAH **100**
METRO-GOLDWYN-MAYER. (R) October, 1972. Dorothy Tutin, Scott Anthony. Color.

Length In Mins.

SAVAGE PRINCESS (Indian) 101
UNITED ARTISTS. (R) October 1, 1955. Dilip Kumar, Mimmi. Color.

SAVAGE SAM 103
BUENA VISTA. (R) June, 1963. Brian Keith, Tommy Kirk. Color.

SAVAGE SEVEN, THE 97
AMERICAN INTERNATIONAL. (R) May, 1968. Robert Walker Jr., Adam Roarke. Color.

SAVAGE WI, THE 103
AMERICAN INTERNATIONAL. (R) January, 1970. Documentary. Color.

SAVE THE CHILDREN 123
PARAMOUNT. (R) September, 1973. Documentary.

SAVE THE TIGER 100
PARAMOUNT. (R) February, 1972. Jack Lemmon, Jack Gilford. Color.

SAY HELLO TO YESTERDAY 91
CINERAMA. (R) February, 1971. Jean Simmons, Leonard Whiting. Color.

SAY ONE FOR ME (Cs) 119
TWENTIETH CENTURY-FOX. (R) June, 1959. Bing Crosby, Debbie Reynolds. Color.

SAYONARA (Technirama) 147
WARNER BROS. (R) December, 1957. Marlon Brando, Patricia Owen. Color.

SCALAWAY 93
PARAMOUNT. (R) October, 1973. Kirk Douglas, Mark Lester.

SCALP HUNTERS, THE 102
UNITED ARTISTS. (R) March, 1968. Burt Lancaster, Ossie Davis. Color.

SCANDAL, INC. 79
REPUBLIC. (R) October 12, 1956. Robert Hutton, Paul Richards.

SCANDALOUS JOHN (P) 113
BUENA VISTA. (R) July, 1971. Brian Keith, Alfonso Arau. Color.

SCAPEGOAT, THE 92
METRO-GOLDWYN-MAYER. (R) August, 1959. Alec Guinness, Bette Davis.

SCARECROW (P) 115
WARNER BROS. (R) April, 1973. Gene Hackman, Al Pacino. Color.

SCARLET COAT, THE (Cs) 101
METRO-GOLDWYN-MAYER. (R) August 19, 1955. Cornel Wilde, Michael Wilding, Anne Francis, George Sanders. Color.

SCARLET HOUR, THE (VV) 95
PARAMOUNT. (R) April, 1956. Carol Ohmart, Tom Tryon, Jody Lawrence.

SCAVENGER HUNT 112
TWENTIETH CENTURY-FOX. (R) December, 1979. Richard Benjamin, James Coco.

SCAVANGERS, THE 79
ROACH. (R) October, 1959. Vince Edwards, Carol Ohmart.

SCENES FROM A MARRIAGE (Swedish; Eng. titles) 168
CINEMA V. (R) October, 1974. Liv Ullmann, Erland Josephson.

SCENT OF A WOMAN (Italian; English sub-titles) 103
TWENTIETH CENTURY-FOX. (R) February, 1976. Vittorio Gassman.

SCHIZOID 93
AMERICAN INTERNATIONAL. (R) February, 1972. Florinda Bolkan, Stanley Baker. Color.

SCHOOL FOR SCOUNDRELS 94
CONTINENTAL. (R) July, 1960. Ian Carmichael, Terry Thomas.

SCORPIO 114
UNITED ARTISTS. (R) April, 1973. Burt Lancaster, Alain Delon. Color.

SCRATCH HARRY 94
CANNON. (R) August, 1969. Harry Walker Staff, Victoria Wilde, Christine Kelly. Color.

SCREAM AND SCREAM AGAIN 95
AMER. INT. PICS. (R) February, 1970. Vincent Price, Christopher Lee, Peter Cushing. Color.

SCREAM, BLACULA, SCREAM 96
AMERICAN INTERNATIONAL. (R) June, 1973. William Marshall, Don Mitchell. Color.

SCREAM OF FEAR 81
COLUMBIA. (R) September, 1961. Susan Strasberg, Ronald Lewis.

SCREAMING EAGLES 81
ALLIED ARTISTS. (R) May 27, 1956. Tom Tryon, Jan Merlin.

SCREAMING MIMI 79
COLUMBIA. (R) April, 1958. Anita Ekberg, Phil Carey.

SCREAMING SKULL, THE 70
AMERICAN INTERNATIONAL. (R) August, 1958. Alex Nicol, John Hudson.

SCROOGE (P) 111
NATIONAL GENERAL (R) November, 1970. Albert Finney, Alec Guinness. Color.

SEA CHASE, THE (Cs) 117
WARNER BROTHERS. (R) June 4, 1955. John Wayne, Lana Turner. Color.

SEA GULL, THE 141
WARNER BROS.-7 ARTS. (R) December, 1968. James Mason, Simone Signoret Vanessa Redgrave. Color.

SEA GYPSIES, THE 101
WARNER BROS. (R) April, 1978. Robert Logan, Heather Rattray.

SEA PIRATE, THE 83
PARAMOUNT. (R) September, 1967. Antonella Lualdi, Terence Morgan. Color.

SEA SHALL NOT HAVE THEM, THE (Brit.) 91
UNITED ARTISTS. (R) June, 1955. Michael Redgrave, Dirk Bogarde.

SEA WIFE, THE (Cs) 82
TWENTIETH CENTURY-FOX. (R) August, 1957. Richard Burton, Joan Collins. Color.

SEANCE ON A WET AFTERNOON 115
ARTIXO. (R) November, 1964. Kim Stanley, Richard Attenborough.

SEARCH FOR BRIDEY MURPHY, THE (VV) 84
PARAMOUNT. (R) October, 1956. Louis Hayward, Teresa Wright.

SEARCHERS, THE (VV) 119
WARNER BROTHERS. (R) May 26, 1956. John Wayne, Jeffrey Hunter, Vera Miles. Color.

SEASON OF PASSION 92
UNITED ARTISTS. (R) October, 1961. Ernest Borgnine, Anne Baxter.

SEBASTIAN 100
PARAAANT. (R) January, 1968. Dirk Bogarde, Susannah York. Color.

SECOND BEST SECRET AGENT IN THE WHOLE WORLD, THE (Brit.) 96
EMBASSY. (R) November, 1965. Tom Adams, Israel Stepanek. Color.

SECOND GREATEST SEX, THE (Cs) 87
UNIVERSAL. (R) December, 1955. Jeanne Crain, George Nader. Color.

SECOND TIME AROUND (Cs) 99
20TH CENTURY-FOX. (R) November, 1961. Debbie Reynolds, Steve Forrest. Color.

SECONDS 106
PARAMOUNT. (R) October, 1966 Rock Hudson, Salome Jens.

SECRET AGENT FIREBALL 89
AMERICAN INTERNATIONAL. (R) December, 1965. Richard Harrison, Dominique Boschero. Color.

SECRET AGENT SUPER DRAGON 95
UNITED SCREEN ARTS. (R) May 30, 1966. Ray Danton, Marisa Mell. Color.

SECRET CEREMONY 109
UNIVERSAL. (R) January, 1969. Elizabeth Taylor, Mia Farrow, Robert Mitchum. Color.

SECRET DOOR, THE 89
ALLIED ARTISTS. (R) 1963. Robert Hutton, Martin Benson. Color.

SECRET INVASION, THE (P) 95
UNITED ARTISTS. (R) October, 1964. Stewart Granger, Raf Vallone. Color.

SECRET LIFE OF AN AMERICAN WIFE, THE 92
TWENTIETH CENTURY-FOX. (R) August, 1968. Walter Matthau, Anne Jackson, Patrick O'Neal. Color.

SECRET OF DEEP HARBOR 70
UNITED ARTISTS. (R) October, 1961. Ron Foster, Barry Kelly.

SECRET OF LIFE 73
BUENA VISTA. (R) November 20, 1956. True Life Adventure. Color.

SECRET OF MAGIC ISLAND 63
EMBASSY. (R) April, 1965. Children's Fantasy. Color.

SECRET OF MY SUCCESS, THE (P) 112
METRO-GOLDWYN-MAYER. (R) October, 1965. Shirley Jones, Stella Stevens. Color

SECRET OF SANTA VITTORIA (P) 139
UNITED ARTISTS. (R) October, 1969. Anthony Quinn, Anna Magnani, Virra Lisi. Color.

SECRET OF THE PURPLE REEF, THE 80
TWENTIETH CENTURY-FOX. (R) October, 1960. Jeff Richards, Margia Dean. Color.

SECRET OF TREASURE MOUNTAIN 68
COLUMBIA. (R) June, 1956. Raymond Burr, William Prince.

SECRET PARTNER, THE 91
METRO-GOLDWYN-MAYER (R) March, 1961. Stewart Granger, Haya Harareet.

SECRET PLACE, THE 81
RANK. (R) February, 1958. Belinda Lee, Ronald Lewis.

SECRET SEVEN, THE 94

Length In Mins.

METRO-GOLDWYN-MAYER. (R) May, 1966. Tony Russell, Helga Line. Color.
SECRET VENTURE (Brit.) 68
REPUBLIC. (R) November 10, 1955. Kent Taylor, Jane Hylton.
SECRET WAR OF HARRY FRIGG, THE (TE) 119
UNIVERSAL. (R) April, 1968. Paul Newman, Sylvia Koscina. Color.
SECRET WAYS, THE 112
UNIVERSAL. (R) April, 1961. Richard Widmark, Sonja Ziemann.
SECRET WORLD 94
20th CENTURY-FOX. (R) July, 1969. Jacqueline Bisset, Giselle Pascal, Pierre Zimmer. Color.
SECRETS OF MONTE CRISTO, THE (Dyaliscope) 70
MGM. (R) 1961. Rory Calhoun. Color.
SECRETS OF THE REEF 72
CONTINENTAL. (R) July, 1956. Life Underwater. Color.
SEDUCED AND ABANDONED (Ital., Eng. Titles) 125
CONTINENTAL. (R) July, 1964. Stefania Sandrelli, Saro Urzi.
SEDUCTION OF JOE TYNAN, THE 107
UNIVERSAL. (R) August, 1979. Alan Alda, Barbara Harris.
SEE NO EVIL 89
COLUMBIA. (R) September, 1971. Mia SQARROW, Dorothy Alison. Color.
SEMI-TOUGH 108
UNITED ARTISTS. (R) November, 1977. Burt Reynolds, Kris Kristofferson.
SEMINOLE UPRISING 74
COLUMBIA. (R) May, 1955. George Montgomery, Karin Booth. Color.
SEND ME NO FLOWERS 100
UNIVERSAL. (R) November, 1964. Rock Hudson, Doris Day. Color.
SENIOR PROM 82
COLUMBIA. (R) January, 1959. Jill Corey, Paul Hampton.
SENSE OF LOSS, A 135
CINEMA V. (R) October, 1972. Documentary. Color.
SENTINEL, THE 89
UNIVERSAL. (R) February, 1977. Chris Sarandon, Cristina Raines.
SEPARATE PEACE, A 105
PARAMOUNT. (R) October, 1972. John Heyl, Parker Stevenson. Color.
SEPARATE TABLES 98
UNITED ARTISTS. (R) February, 1959. Deborah Kerr, Rita Hayworth.
SEPTEMBER STORM (Cs) 99
TWENTIETH CENTURY-FOX. (R) October, 1960. Joanne Dru, Mark Stevens. Color.
SEPTEMBER 30, 1955 101
UNIVERSAL. (R) January, 1978. Richard Thomas, Susan Tyrell.
SERAFINO (Ital., Eng. Titles) 96
ROYAL. (R) May, 1970. Adriano Celentano, Francesca Romana Coluzzi. Color.
SEREGENTI SHALL NOT DIE 84
ALLIED ARTISTS. (R) October, 1961. Documentary. Color.
SERENADE 121
WARNER BROTHERS. (R) April 21, 1956. Mario Lanza, Joan Fontaine, Sarita Montiel. Color.
SERGEANT, THE 107
WARNER BROS.-7 ARTS. (R) January, 1969. Rod Steiger, John Phillip Law, Ludmila Mikael. Color.
SERGEANT DEADHEAD (P) 89
AMERICAN INTERNATIONAL. (R) August, 1965. Frankie Avalon, Deborah Walley. Color.
SERGEANT RUTLEDGE 111
WARNER BROS. (R) May, 1960.
SERGEANT RYKER 86
UNIVERSAL. February, 1968. Lee Marvin, Vera Miles. Color.
SERGEANT WAS A LADY, THE 72
UNIVERSAL. (R) November, 1961. Martin West, Venetia Stevenson.
SGT. PEPPER'S LONELY HEARTS CLUB BAND. (P) 111
UNIVERSAL. (R) July, 1978. Peter Frampton, Barry Gibb.
SERGEANT'S 3 (P) 112
UNITED ARTISTS. (R) February, 1962. Frank Sinatra, Dean Martin.
SERIAL 91
PARAMOUNT. (R) March, 1980. Martin Mull, Tuesday Weld.
SERPENT'S EGG, THE 120
PARAMOUNT. (R) February, 1978. Liv Ullmann, David Carradine.
SERPICO 130
PARAMOUNT. (R) February, 1974. Al Pacino, John Randolph.
SERVANT, THE 115
LANDAU. (R) 1964. Dirk Bogarde, James Fox.
SEVEN ANGRY MEN 90
ALLIED ARTISTS. (R) March 27, 1955. Raymond Massey, Debra Paget.
SEVEN BEAUTIES (Italian; English sub-titles) 115
CINEMA 5. (R) January, 1976. Giancarlo Giannini, Fernando Rey.
SEVEN CITIES OF GOLD (Cs) 103
TWENTIETH CENTURY-FOX. (R) September, 1955. Richard Egan, Michael Rennie, Jeffrey Hunter, Anthony Quinn, Rita Moreno. Color.
SEVEN DAYS FROM SUNDOWN 86
UNIVERSAL. (R) September, 1960. Audie Murphy, Barry Sullivan. Color.
SEVEN DAYS IN MAY 120
PARAMOUNT. (R) February, 1964. Burt Lancaster, Kirk Douglas.
SEVEN DWARFS TO THE RESCUE, THE 84
CHILDHOOD. (R) January, 1965. Rosanna Podesta, Roberto Risso.
SEVEN FACES OF DR. LAO 100
METRO-GOLDWYN-MAYER. (R) March, 1964. Tony Randall, Arthur O'Connell. Color.
SEVEN GOLDEN MEN 87
WARNER BROS.-7 ARTS. (R) May, 1969. Rosanna Podesta, Philippe Leroy, Gastone Moschin. Color.
SEVEN GUNS TO MESA 69
ALLIED ARTISTS. (R) March, 1958. Charles Quinlivan, Lola Albright.
SEVEN HILLS OF ROME (T) 107
METRO-GOLDWYN-MAYER. (R) January, 1958. Mario Lanza, Renato Rascel. Color.
SEVEN LITTLE FOYS, THE (VV) 95
PARAMOUNT. (R) July, 1955. Bob Hope, Milly Vitale. Color.
SEVEN MEN FROM NOW 78
WARNER BROTHERS. (R) August 4, 1956. Randolph Scott, Gail Russell. Color.
SEVEN MINUTES, THE 115
20TH CENTURY-FOX. (R) July, 1971. Wayne Maunder, Marianne McAndrew. Color.
SEVEN-PER-CENT SOLUTION, THE 113
UNIVERSAL. (R) October, 1976. Alan Arkin, Vanessa Redgrave, Robert Duvall.
SEVEN SEAS TO CALAIS (Cs) 99
METRO-GOLDWYN-MAYER. (R) March, 1963. Rod Taylor, Keith Michell.
SEVEN SURPRISES 77
QUARTET INT. (R) September, 1964. Claude Jutra, Wallace Jensen. Color.
SEVEN THIEVES (Cs) 102
TWENTIETH CENTURY-FOX. (R) January, 1960. Edward G. Robinson, Rod Steiger.
SEVEN-UPS, THE 103
TWENTIETH CENTURY-FOX. (R) December, 1973. Roy Scheider, Tony Lo Bianco.
SEVEN WOMEN (P) 93
METRO-GOLDWYN-MAYER. (R) January, 1966. Anne Bancroft, Sue Lyon. Color.
SEVEN WOMEN FROM HELL (Cs) 88
TWENTIETH CENTURY-FOX. (R) October, 1961. Patricia Owens, Denise Darcel.
SEVEN YEAR ITCH, THE (Cs) 105
TWENTIETH CENTURY-FOX. (R) June, 1955. Marilyn Monroe, Tom Ewell. Color.
1776 (P) 148
COLUMBIA. (R) November, 1972. William Daniels, Howard DaSilva, Ken Howard. Color.
7TH CAVALRY 75
COLUMBIA. (R) December, 1956. Randolph Scott, Barbara Hale. Color.
7TH DAWN, THE 128
UNITED ARTISTS. (R) July, 1964. William Holden, Susannah York. Color.
SEVENTH JUROR, THE (Fr., Eng. titles) 90
TRANS-LUX. (R) January, 1964. Daniele DeLorme, Bernard Blier.
SEVENTH SIN, THE (Cs) 94
METRO-GOLDWYN-MAYER. (R) June, 1957. Eleanor Parker, Bill Travers.
7TH VOYAGE OF SINBAD, THE (Dynamation) 89
COLUMBIA. (R) December, 1958. Kerwin Mathews, Kathryn Grant. Color.
SEVERED HEAD A 96
COLUMBIA. (R) April, 1971. Lee Remick, Richard Attenborough. Color.
SEX AND THE SINGLE GIRL 114
WARNER BROS. (R) January, 1965. Tony Curtis, Natalie Wood. Color.
SEX KITTENS GO TO COLLEGE 94

Length In Mins.

ALLIED ARTISTS. (R) October, 1960. Mamie Van Doren, Tuesday Weld.

SEX OF ANGELS, THE (TE) 104
LOPERT. (R) December, 1969. Rosemarie Dexter, Laura Troschel. Color.

SHACK OUT ON 101 80
ALLIED ARTISTS. (R) December 4, 1955. Terry Moore, Frank Lovejoy, Keenan Wynn.

SHADOW MAN 77
LIPPERT. (R) October 16, 1963. Cesar Romero.

SHADOW OF FEAR (Brit.) 76
UNITED ARTISTS. (R) June, 1956. Mona Freeman, Jean Kent.

SHADOW OF THE CAT, THE 79
UNIVERSAL. (R) June, 1961. Andre Morell, Barbara Shelley.

SHADOW OF THE EAGLE (Brit.) 93
UNITED ARTISTS. (R) July, 1955. Richard Greene, Valentina Cortesa.

SHADOW ON THE WINDOW, THE 73
COLUMBIA. (R) March, 1957. Phil Carey, Betty Garrett.

SHAFT 100
MGM. (R) July, 1971. Richard Roundtree, Gwen Mitchell. Color.

SHAFT IN AFRICA (P) 113
METRO-GOLDWYN-MAYER. (R) June, 1973. Richard Roundtree, Frank Finlay. Color.

SHAFT'S BIG SCORE 105
METRO-GOLDWYN-MAYER. (R) June, 1972. Richard Roundtree, Moses Gunn. Color.

SHAGGY D.A., THE 92
BUENA VISTA. (R) December, 1976. Dean Jones, Tim Conway, Suzanne Pleshette.

SHAGGY DOG, THE 104
BUENA VISTA. (R) April, 1959. Fred MacMurray, Jean Hagen. Color.

SHAKE HANDS WITH THE DEVIL 110
UNITED ARTISTS. (R) June, 1959. James Cagney, Don Murray.

SHAKE, RATTLE AND ROCK 77
AMERICAN INTERNATIONAL. (R) October 31, 1956. Fats Domino, Lisa Gaye.

SHAKEDOWN, THE 91
UNIVERSAL. (R) February, 1961. Terence Morgan, Hazel Court.

SHAKESPEARE WALLAH 115
MERCHANT IVORY PRODUCTIONS. (R) March 22, 1966. Shashi Kapoor, Felicity Kendal.

SHAKIEST GUN IN THE WEST, THE (T) 101
UNIVERSAL. (R) May, 1968. Don Knotts, Barbara Rhoades. Color.

SHALAKO 113
CINERAMA. (R) November, 1968. Sean Connery, Brigitte Bardot, Stephen Boyd. Color.

SHAME (Swedish, Eng. titles) 100
LOPERT. (R) January, 1969. Liv Ullman, Max Von Sydow, Gunnar Bjornstrand. Color.

SHAMELESS OLD LADY, THE (Fr., Eng. titles) 95
CONTINENTAL. (R) September, 1966. Sylvie, Malka Ribovska.

SHAMPOO 112
COLUMBIA. (R) February, 1975. Warren Beatty, Julie Christie, Goldie Hawn.

SHAMUS 98
COLUMBIA. (R) January, 1973. Burt Reynolds, Dyan Cannon. Color.

SHANKS 93
PARAMOUNT. (R) October, 1975. Marcel Marceau.

SHARK FIGHTERS, THE (Cs) 73
UNITED ARTISTS. (R) November, 1956. Victor Mature, Karen Steele. Color.

SHARK'S TREASURE 95
UNITED ARTISTS. (R) May, 1975. Cornel Wilde, Yaphet Kotto.

SHE (Cs) 105
METRO-GOLDWYN-MAYER. (R) June, 1965.

$ta04SHE-CREATURE 77
AMERICAN INTERNATIONAL. (R) July 25, 1956. Maria English, Chester Morris.

SHE DEVIL, THE 77
TWENTIETH CENTURY-FOX. (R) April, 1957. Mari Blanchard, Albert Dekker.

SHE GODS OF SHARK REEF 65
AMERICAN INTERNATIONAL. (R) August, 1958. Don Durant, Lisa Montell.

SHE PLAYED WITH FIRE 95
COLUMBIA. (R) September, 1958. Jack Hawkins, Arlene Dahl.

SHEBA, BABY 90
AMERICAN INTERNATIONAL. (R) March, 1975. Pam Grier, Austin Stoker.

Length In Mins.

SHEEP HAS FIVE LEGS, THE (Fr.) 93
UMPO. (R) August, 1955. Fernandel.

SHEEPMAN, THE (Cs) 85
MGM. (R) May, 1958. Glenn Ford, Shirley MacLaine. Color.

SHEILA LEVINE IS DEAD AND LIVING IN NEW YORK (P) 112
PARAMOUNT. (R) February, 1975. Jeannie Berlin, Roy Scheider.

SHELL SHOCK 75
PARADE. (R) February, 1964. Beach Dickerson, Carl Crow.

SHENANDOAH 105
UNIVERSAL. (R) April 1965. James Stewart, Doug McClure. Color.

SHEPHERD GIRL, THE (Shawscope) (Chinese, Eng. titles) 105
FRANK LEE. (R) September, 1965. Julie Yeh Feng, Kwan Shan. Color.

SHERIFF OF FRACTURED JAW, THE (Cs) 103
TWENTIETH CENTURY-FOX. (R) January, 1959. Kenneth More, Jayne Mansfield. Color.

SHINBONE ALLEY 84
ALLIED ARTISTS. (R) April, 1971 Animated Feature. Voices of Eddie Bracken, Carol Channing. Color.

SHINING, THE 146
WARNER BROS. (R) May, 1980. Jack Nicholson, Shelley Duvall.

SHIP OF FOOLS 150
COLUMBIA (R) September, 1965 Oskar Werner, Simone Signoret.

SHIP THAT DIED OF SHAME, THE (Brit.) 79
CONTINENTAL. (R) August, 1956. Richard Attenborough, George Baker.

SHIP WAS LOADED, THE 81
GEORGE K. ARTHUR. (R) September, 1959. David Tomlinson, Peggy Cummins.

SHOCK CORRIDOR 101
ALLIED ARTISTS. (R) August, 1963. Peter Breck, Constance Towers.

SHOCK TREATMENT (Cs) 94
TWENTIETH CENTURY-FOX. (R) February, 1964. Stuart Whitman, Lauren Bacall.

SHOCK TROOPS (Fr., Eng. titles) 106
UNITED ARTISTS. (R) September, 1968. Jean Claude Brialy, Bruno Cremer, Jacques Perrin. Color.

SHOES OF THE FISHERMAN 157
METRO-GOLDWYN-MAYER. (R) Special. Anthony Quinn, David Janssen, Oskar Werner. Color.

SHOOT 94
AVCO-EMBASSY. (R) June, 1976. Cliff Robertson, Ernest Borgnine.

SHOOTIST, THE 99
PARAMOUNT. (R) July, 1976. John Wayne, Lauren Bacall.

SHOOTOUT 94
UNIVERSAL. (R) June, 1971. Gregory Peck, Robert F. Lyons. Color.

SHOOT-OUT AT MEDICINE BEND 87
WARNER BROS. (R) May, 1957. Randolph Scott, James Craig.

SHOP ON MAIN STREET, THE (Czech., Eng. titles) 128
PROMINENT. (R) January, 1966. Josef Kroner, Ida Kaminska.

SHORT CUT TO HELL (VV) 89
PARAMOUNT. (R) September 1957. Robert Ivers, Georgann Johnson.

SHOT GUN 80
ALLIED ARTISTS. (R) April 24, 1955. Sterling Hayden, Yvonne de Carlo. Color.

SHOT IN THE DARK, A (P) 101
UNITED ARTISTS. (R) July, 1964. Peter Sellers, Elke Sommer. Color.

SHOUT AT THE DEVIL (P) 128
AMERICAN INTERNATIONAL. (R) November, 1976. Lee Marvin, Roger Moore, Barbara Parkins.

SHOWDOWN 99
UNIVERSAL. (R) June, 1973. Dean Martin, Rock Hudson. Color.

SHOWDOWN 79
UNIVERSAL. (R) May, 1963. Audie Murphy, Kathleen Crowley.

SHOWDOWN AT ABILENE 80
UNIVERSAL. (R) October, 1956. Jock Mahoney, Martha Hyer. Color.

SHOWDOWN AT BOOT HILL (Rs) 71
TWENTIETH CENTURY-FOX. (R) June, 1958. Charles Bronson, Robert Hutton.

SHRIKE, THE 88
UNIVERSAL. (R) September, 1955. Jose Ferrer, June Allyson.

SHUTTERED ROOM, THE 100

Length In Mins.

WARNER BROS.-SEVEN ARTS. (R) January, 1968. Gig Young, Carol Lynley. Color.

SICILIAN CLAN, THE (P) (Fr., Eng. titles) 12
20TH CENTURY-FOX. (R) June, 1970. Jean Gabin, Alain Delon. Color.

SIDDHARTHA 86
COLUMBIA. (R) July, 1973. Shashi Kapoor, Simi Garewal. Color.

SIDECAR RACERS 100
UNIVERSAL. (R) August, 1975. Ben Murphy, Wendy Hughes.

SIDEWINDER ONE 97
AVCO-EMBASSY. (R) August, 1977. Marjoe Gortner, Michael Parks.

SIDNEY SHELDON'S BLOODLINE 116
PARAMOUNT. (R) June, 1979. Audrey Hepburn, Ben Gazzara.

SIEGE OF SYRACUSE (D) 97
PARAMOUNT. (R) February, 1962. Rossano Brazzi, Tina Louise. Color.

SIEGE OF THE SAXONS 85
COLUMBIA. (R) August, 1963. Janette Scott, Ronald Lewis. Color.

SIERRA BARON (Cs) 80
TWENTIETH CENTURY-FOX. (R) July, 1958. Brian Keith, Rick Jason. Color.

SIERRA STRANGER 74
COLUMBIA. (R) May, 1957. Howard Duff, Gloria McGhee.

SIGN OF THE GLADIATOR (Colorscope) 92
AMERICAN INTERNATIONAL. (R) October, 1959. Anita Ekberg, George Marshall. Color.

SIGN OF ZORRO, THE 91
BUENA VISTA. (R) June, 1960. Guy Williams.

SIGNPOST TO MURDER (P) 74
METRO-GOLDWYN-MAYER. (R) February, 1965. Joanne Woodward, Stuart Whitman.

SILENCE, THE (Sw., Eng. titles) 95
JANUS. (R) February, 1964. Ingrid Thulin, Gunnel Lindblom.

SILENCERS, THE 105
COLUMBIA. (R) March, 1966. Dean Martin, Stella Stevens. Color.

SILENT ENEMY, THE 91
UNIVERSAL. (R) November, 1958. Laurence Harvey, Dawn Addams.

SILENT MOVIE 86
TWENTIETH CENTURY-FOX. (R) July, 1976. Mel Brooks, Marty Feldman.

SILENT NIGHT, EVIL NIGHT 100
WARNER BROS. (R) May, 1975. Olivia Hussey, Keir Dullea.

SILENT RUNNING 90
UNIVERSAL. (R) March, 1972. Bruce Dern, Cliff Potts. Color.

SILENT WORLD, THE 86
COLUMBIA. (R) January, 1957. Deep Sea Documentary. Color.

SILK STOCKINGS (Cs) 117
METRO-GOLDWYN-MAYER. (R) July, 1957. Fred Astaire, Cyd Charisse. Color.

SILKEN AFFAIR, THE (Brit.) 96
DIST. CORP. OF AMER. (R) November, 1957. David Niven, Genevieve Page.

SILVER BEARS 113
COLUMBIA. (R) April, 1978. Michael Caine, Cybill Shepherd.

SILVER CHALICE, THE (Cs) 137
WARNER BROTHERS. (R) February 5, 1955. Virginia Mayo, Jack Palance. Color.

SILVER STAR, THE 73
LIPPERT. (R) March 25, 1955. Edgar Buchanan, Marie Windsor.

SILVER STREAK 110
TWENTIETH CENTURY-FOX. (R) December, 1976. Gene Wilder, Jill Clayburgh.

SIMBA (Brit.) 99
LIPPET. (R) September 9, 1955. Dirk Bogarde. Color.

SIMON 97
WARNER BROS. (R) March, 1980. Alan Arkin, Madeline Kahn.

SIMON AND LAURA (Brit.) 91
UNITED ARTISTS. (R) July, 1956. Peter Finch, Kay Kendall. Color.

SIMON KING OF THE WITCHES 89
FANFARE. (R) May, 1971. Andrew Pine, Brenda Scott. Color.

SIMON OF THE DESERT (Mex. dubbed) 45
ALTURA FILMS. (R) August, 1968. Silvia Pinal, Claudio Brook. b&w.

SINBAD AND THE EYE OF THE TIGER (Dynarama) 114
COLUMBIA. (R) May, 1977. Patrick Wayne, Taryn Power.

Length In Mins.

SINCERELY YOURS 115
WARNER BROTHERS. (R) November 26, 1955. Liberace, Joanne Dru, Dorothy Malone. Color.

SINFUL DAVEY 95
UNITED ARTISTS. (R) March, 1969. John Hurt, Pamela Franklin, Nigel Davenport. Color.

SING AND SWING 75
UNIVERSAL. (R) December, 1964. David Hemmings, John Pike.

SING BOY SING (Cs) 91
TWENTIETH CENTURY-FOX. (R) January, 1958. Tommy Sands, Lili Gentle.

SINGER, NOT THE SONG, THE (Cs) 129
PARAMOUNT. (R) January, 1962. John Mills, Dirk Bogarde. Color.

SINGING NUN, THE(P) 98
METRO-GOLDWYN-MAYER. (R) April, 1966. Debbie Reynolds, Ricardo Montalban. Color.

SINK THE BISMARCK! (Cs) 97
TWENTIETH CENTURY-FOX. (R) February, 1960. Kenneth More, Dana Wynter.

SINS OF POMPEII (Ital., Eng. dubbed) 73
VISUAL DRAMA. (R) December, 1955. Micheline Prelle, George Marchal.

SINS OF RACHEL CADE, THE 123
WARNER BROS. (R) April, 1961. Angie Dickinson, Peter Finch, Color.

SISTERS 94
AMERICAN INTERNATIONAL. (R) March, 1973. Margot Kidder, Jennifer Salt. Color.

SISTERS, THE (Greek, Eng. titles) 95
BRENNER. March, 1969. Petros Fissoun, Elli Fotiou, Nicos Rizos. b&w.

SITTING DUCKS 90
UNITED FILM DIST. (R) April, 1980. Michael Emil, Zack Norman.

SITTING TARGET 93
METRO-GOLDWYN-MAYER. (R) February, 1972. Oliver Reed, Jill St. John. Color.

SITUATION HOPELESS-BUT NOT SERIOUS 97
PARAMOUNT. (R) October, 1965. Alec Guinness, Michael Connors.

SIX BLACK HORSES 80
UNIVERSAL. (R) June, 1962. Audie Murphy, Dan Duryea. Color.

SIX BRIDGES TO CROSS 96
UNIVERSAL. (R) February, 1955. Tony Curtis, Julie Adams.

SIX DAYS TO ETERNITY 88
P.W. Documentary. (R) October, 1968. Color.

633 SQUADRON (P) 101
UNITED ARTISTS. (R) June, 1964. Cliff Robertson, George Chakiris. Color.

SIX PACK ANNIE 88
AMERICAN INTERNATIONAL. (R) December, 1975. Lindsay Bloom, Jana Bellan.

SKABENGA 61
ALLIED ARTISTS. (R) May 29, 1955. Documentary.

SKI BUM, THE 94
AVCO EMBASSY. (R) January, 1971. Zalman King, Charlotte Rampling. Color.

SKI FEVER 98
ALLIED ARTISTS. (R) March, 1969. Martin Miller, Claudia Martin, Frankie Avalon. Color.

SKI ON THE WILD SIDE 104
SIGMA III. (R) August, 1967. Documentary. Color.

SKI PARTY (P) 90
AMERICAN INTERNATIONAL. (R) June, 1965. Frankie Avalon, Dwayne Hickman, Color.

SKI TROOP ATTACK 61
THE FILM GROUP. (R) May, 1960. Frank Wolff, Michael Forrest.

SKIDOO 98
PARAMOUNT. (R) December, 1968. Jackie Gleason, Carol Channing, Frankie Avalon. Color.

SKIN GAME 102
WARNER BROS. (R) October, 1971. James Garner, Lou Gossett. Color.

SKULL, THE (TE) 83
PARAMOUNT. (R) September, 1965. Peter Cushing, Patrick Wymark. Color.

SKULLDUGGERY (P) 105
UNIVERSAL. (R) March, 1970. Burt Reynolds, Susan Clark. Color.

SKY ABOVE, THE MUD BELOW, THE 90
AVCO EMBASSY. (R) 1962. Narrated by William Peacock. Color.

SKY RIDERS 91
TWENTIETH CENTURY-FOX. (R) March, 1976 James Coburn, Susannah York.

SKYJACKED 100
METRO-GOLDWYN-MAYER. (R) May, 1972. Charlton Heston, Yvette Mimieux, Walter Pidgeon. Color.

	Length In Mins.
SLAMS, THE METRO-GOLDWYN-MAYER. (R) August, 1973. Jim Brown, Judy Pace.	91
SLANDER METRO-GOLDWYN-MAYER. (R) January 18, 1957. Van Johnson, Ann Blyth.	81
SLAP, THE (French; English sub-titles) SILVER SCREEN. (R) January, 1976. Lina Ventura, Annie Giradot.	98
SLAP SHOT UNIVERSAL. (R) March, 1977. Paul Newman, Strother Martin.	122
SLAUGHTER AMERICAN INTERNATIONAL. (R) August, 1972. Jim Brown, Stella Stevens. Rip Torn. Color.	92
SLAUGHTER ON TENTH AVENUE UNIVERSAL. (R) November, 1957. Richard Egan, Jan Sterling.	82
SLAUGHTERHOUSE FIVE UNIVERSAL. (R) April, 1972. Michael Sacks, Ron Leibman. Color.	104
SLAUGHTER'S BIG RIP OFF AMERICAN INTERNATIONAL. (R) July, 1973. Jim Brown, Brock Peters. Color.	92
SLAVE, THE (Cs) METRO-GOLDWYN-MAYER. (R) May, 1963. Steve Reeves, Jacques Sernas. Color.	110
SLAVE TRADE IN THE WORLD TODAY READE-STERLING. (R) November, 1964. Documentary. Color.	86
SLAVES CONTINENTAL. (R) September, 1969. Stephen Boyd, Dionne Warwick, Ossie Davis. Color.	110
SLEEPER UNITED ARTISTS. (R) December, 1973. Woody Allen, Diane Keaton.	88
SLEEPING BEAUTY (Ger., Eng. dubbed) CHILDHOOD PRODUCTIONS. (R) October, 1965. Fantasy. Color.	70
SLEEPING BEAUTY (Ballet) ROYAL FILMS INTL. (R) April, 1966. Alla Sizova, Yuri Solovien. Color.	90
SLEEPING BEAUTY (T) BUENA VISTA. (R) March, 1959. Animated. Color.	75
SLEEPING CAR MURDER, THE (Fr., Eng. titles) SEVEN ARTS. (R) February, 1966. Simone Signoret, Yves Montand.	90
SLENDER THREAD, THE PARAMOUNT. (R) January, 1966. Sidney Poitier, Anne Bancroft.	98
SLEUTH TWENTIETH CENTURY-FOX. (R) December, 1972. Laurence Olivier, Michael Caine. Color.	138
SLIGHTLY SCARLET (SS) RKO RADIO. (R) February 8, 1956. John Payne, Arlene Dahl, Rhonda Fleming. Color.	90
SLIM CARTER UNIVERSAL. (R) November, 1957. Jock Mahoney, Julie Adams.	82
SLIPPER AND THE ROSE, THE (P) UNIVERSAL. (R) February, 1977. Richard Chamberlain, Gemma Craven.	128
SLITHER METRO-GOLDWYN-MAYER. (R) March, 1973. James Caan, Peter Boyle, Sally Kellerman. Color.	97
SLOGAN (Fr., Eng. titles) ROYAL. (R) April, 1970. Serge Gainsbourg, Jane Birkin. Color.	94
SLOW DANCING IN THE BIG CITY UNITED ARTISTS. (R) November, 1978. Paul Sorvino, Anne Ditchburn.	101
SMALL CHANGE (French with English titles) NEW WORLD. (R) October, 1976. Geory Desmoudeaux, Philippe Goldman.	104
SMALL CIRCLE OF FRIENDS, A UNITED ARTISTS. (R) March, 1980. Brad Davis, Karen Allen.	112
SMALL TOWN IN TEXAS, A AMERICAN INTERNATIONAL. (R) June, 1976. Timothy Bottoms, Susan George.	95
SMALL WORLD OF SAMMY LEE, THE SEVEN ARTS (R) August, 1963. Anthony Newley, Julia Foster.	105
SMALLEST SHOW ON EARTH, THE (Brit.) TIMES. (R) November, 1957. Bill Travers, Virginia McKenna.	80
SMIC, SMAC, SMOC CINERAMA. (R) January, 1972. Catherine Allegret, Amidou. Color.	90
SMILE UNITED ARTISTS. (R) July, 1975. Bruce Dern, Barbara Feldon.	113

	Length In Mins.
SMILES OF A SUMMER NIGHT (Swed.) RANK. (R) February, 1958. Vlla Jacobson, Eva Dahlbeck.	108
SMILEY (Aust.) (Cs) TWENTIETH CENTURY-FOX. (R) January, 1957. Ralph Richardson, Colin Peterson. Color.	91
SMILEY GETS A GUN (Cs) TWENTIETH CENTURY-FOX. (R) January, 1959. Keith Calvert, Dame Sybil Thorndike. Color.	89
SMITH! BUENA VISTA. (R) April, 1969. Glenn Ford, Nancy Olson, Dean Jagger. Color.	112
SMOKE SIGNAL UNIVERSAL. (R) March, 1955. Dana Andrews, Piper Laurie. Color.	88
SMOKEY AND THE BANDIT UNIVERSAL. (R) May, 1977. Burt Reynolds, Sally Field.	97
SMOKEY AND THE BANDIT II UNIVERSAL. (R) August, 1980. Burt Reynolds, Sally Fields.	102
SMOKY 20TH CENTURY-FOX (R) September, 1966. Fess Parker, Diana Hyland, Color.	103
SNAKE WOMAN, THE UNITED ARTISTS. (R) 1961. John McCarthy, Susan Travers.	68
SNIPER'S RIDGE (Cs) TWENTIETH CENTURY-FOX. (R) March, 1961. Jack Ging, Stanley Clements.	61
SNOOPY COME HOME NATIONAL GENERAL. (R) July, 1972. Animated Feature. Color.	90
SNORKEL, THE COLUMBIA. (R) July 1958. Peter Van Eyck, Betta St. John.	74
SNOW JOB WARNER BROS. (R) February, 1972. Jean-Claude Killy, Daniele Gaubert. Color.	90
SNOW QUEEN, THE UNIVERSAL. (R) May, 1960. Cartoon Feature. Color.	70
SNOW TREASURE ALLIED ARTISTS. (R) October, 1968. James Franciscus, Bente Nielsen, Paul Austad. Color.	85
SNOW WAS BLACK, THE (Fr.) CONTINENTAL. (R) November, 1956. Daniel Gelin.	104
SNOW WHITE (Ger., Eng. dubbed) CHILDHOOD PRODUCTIONS. (R) October, 1965. Fantasy. Color.	74
SNOW WHITE AND THE THREE STOOGES (Cs) TWENTIETH CENTURY-FOX. (R) June, 1961. Carol Heiss, The Three Stooges. Color	107
SNOWBALL EXPRESS BUENA VISTA. (R) December, 1972. Dean Jones, Nancy Olson. Color.	99
SNOWFIRE ALLIED ARTISTS. (R) May 18, 1958. Molly McGowan, Don McGownan. Color.	73
SO THIS IS PARIS UNIVERSAL. (R) January, 1955 Tony Curtis, Gloria de Haven. Color.	96
SODOM AND GOMORRAH (Cs) TWENTIETH CENTURY-FOX. (R) January, 1963. Stewart Granger, Pier Angeli. Color.	154
SOFT SKIN, THE (Fr. Eng. titles) CINEMA V. (R) October, 1964. Jean Desailly, Francoise Dorleac.	120
SOL MADRID (P) MGM. (R) February, 1968. David McCallum, Stella Stevens. Color.	90
SOLDIER BLUE (P) AVCO EMBASSY. (R) August, 1970. Candice Bergen, Peter Strauss. Color.	112
SOLDIER IN THE RAIN ALLIED ARTISTS. (R) November, 1963. Jackie Gleason, Steve McQueen.	88
SOLDIER OF FORTUNE (Cs) TWENTIETH CENTURY-FOX. (R) June, 1955. Clark Gable, Susan Hayward. Color.	96
SOLID GOLD CADILLAC, THE COLUMBIA. (R) October, 1956. Judy Holliday, Paul Douglas.	99
SOLOMON AND SHEBA (T) UNITED ARTISTS. (R) December, 1959. Yul Brynner, Gina Lollobrigida. Color.	139
SOME CALL IT LOVING CINE GLOBE. (R) November, 1973. Zalman King, Carol White.	103
SOME CAME RUNNING (Cs) METRO-GOLDWYN-MAYER. (R) January, 1959. Frank Sinatra, Shirley MacLaine. Color.	127
SOME KIND OF NUT UNITED ARTISTS. (R) November, 1969. Dick Van Dyke, Angie Dickinson, Rosemary Forsyth. Color.	89
SOME LIKE IT HOT	120

Length In Mins.

UNITED ARTISTS. (R) March, 1959. Marilyn Monroe, Tony Curtis.

SOMEBODY KILLED HER HUSBAND (P) **97**
COLUMBIA. (R) September, 1978. Farrah Fawcett, Jeff Bridges.

SOMEBODY UP THERE LIKES ME **113**
METRO-GOLDWYN-MAYER. (R) Special. 1957. Paul Newman, Pier Angeli.

SOMEONE BEHIND THE DOOR **97**
CINERAMA. (R) December, 1971. Charles Bronson, Anthony Perkins. Color.

SOMETHING BIG **108**
NATIONAL GENERAL. (R) November, 1971. Dean Martin, Brian Keith. Color.

SOMETHING FOR EVERYONE **112**
NATIONAL GENERAL. (R) September, 1970. Angela Lansbury, Michael York. Color.

SOMETHING OF VALUE **113**
METRO-GOLDWYN-MAYER. (R) June, 1957. Rock Hudson, Dana Wynter.

SOMETHING WILD **112**
UNITED ARTISTS. (R) May, 1961. Carroll Baker, Ralph Meeker.

SOMETIMES A GREAT NOTION **114**
UNIVERSAL. (R) November, 1971. Paul Newman, Henry Fonda, Lee Remick. Color.

SON OF A GUNFIGHTER (Cs) **92**
METRO-GOLDWYN-MAYER. (R) May, 1966. Russ Tamblyn, Kieron Moore. Color.

SON OF CAPTAIN BLOOD, THE **88**
PARAMOUNT. (R) April, 1964. Sean Flynn, Allessandra Panaro. Color.

SON OF FLUBBER **100**
BUENA VISTA. (R) February, 1963. Fred MacMurray, Nancy Olson.

SON OF ROBIN HOOD, THE (Cs) **81**
TWENTIETH CENTURY-FOX. (R) July, 1959. Al Hedison, June Laverick. Color.

SON OF SAMSON (Total Scope) **89**
MEDALLION. (R) 1961. Mark Forest, Chelo Alonso. Color.

SON OF SINBAD (SS) **88**
RKO RADKO. (R) June 1, 1955. Dale Robertson, Sally Forrest. Color.

SON OF SPARTACUS, THE (Ital. dubbed) **102**
MGM. (R) 1962. Steve Reeves, Jacques Sernas.

SONG AND THE SILENCE, THE **80**
CLOVERHOUSE. (R) February, 1969. Annita Koutsouveli, Harry Rubin, Jim Murphy. b&w.

SONG OF NORWAY (Super Panavision) **138**
CINERAMA. (R) November, 1970. Florence Henderson, Toraly Maurstad. Color.

SONG OVER MOSCOW **94**
ARTKINO. (R) November, 1964. Herbert Rappaport, Olga Zabotinka. Color.

SONG REMAINS THE SAME, THE **136**
WARNER BROS. (R) October, 1976. Led Zeppelin.

SONG WITHOUT END (Cs) **130**
COLUMBIA. (R) August, 1960. Dirk Bogarde, Capucine. Color.

SONNY & JED **87**
K-TEL MOTION PICTURES. (R) Special, 1973. Thomas Milian, Telly Savalas.

SONS AND LOVERS (Cs) **103**
TWENTIETH CENTURY-FOX. (R) August, 1960. Trevor Howard, Dean Stockwell.

SONS OF KATIE ELDER, THE (P) **122**
PARAMOUNT. (R) June, 1965. John Wayne, Dean Martin. Color.

SORCERER **121**
PARAMOUNT and UNIVERSAL. (R) June, 1977. Roy Scheider.

SORCERERS, THE **87**
ALLIED ARTISTS. (R) January, 1968. Boris Karloff, Catherine Lacey. Color.

SORORITY GIRL **60**
AMER. INT. PICS. (R) October, 1957. Susan Cabot, Dick Miller.

SORROW AND THE PITY, THE **260**
CINEMA V. (R) March, 1972. Documentary.

SOUL OF NIGGER CHARLIE, THE **98**
PARAMOUNT. (R) June, 1973. Fred Williamson. Color.

SOUL TO SOUL **96**
CINERAMA. (R) August, 1971. Black Concert Documentary. Color.

SOUND AND THE FURY, THE (Cs) **115**
TWENTIETH-CENTURY-FOX. (R) March, 1959. Yul Brynner, Joanne Woodward. Color.

SOUND OF LAUGHTER, THE **75**
UNION. (R) December, 1963. Ed Wynn, Bing Crosby and All Star Cast.

SOUND OF MUSIC, THE (Todd-AO) **171**
TWENTIETH CENTURY-FOX. (R) Special. March, 1965. Julie Andrews, Christopher Plummer. Color.

SOUNDER **105**
TWENTIETH CENTURY-FOX. (R) October, 1972. Cicely Tyson, Paul Winfield. Color.

SOUTH PACIFIC (Todd-AO) **171**
TWENTIETH CENTURY-FOX. (R) March, 1958. Special. Rossano Brazzi, Mitzi Gaynor. Color.

SOUTHERN STAR, THE **104**
COLUMBIA. (R) July, 1969. George Segal, Ursula Andress, Orson Wells. Color.

SOYLENT GREEN **97**
METRO-GOLDWYN-MAYER. (R) April, 1973. Charlton Heston, Edward G. Robinson. Color.

SPACE CHILDREN, THE **69**
PARAMOUNT. (R) June, 1958. Adam Williams, Michel Ray.

SPACE FLIGHT I-CI (Brit.) **65**
20TH CENTURY-FOX. (R) 1965. Bill Williams, Norma West.

SPACE MASTER X-7 (Rs) **71**
TWENTIETH CENTURY-FOX. (R) June, 1958. Bill Williams, Lyn Thomas.

SPACEWAYS **76**
LIPPERT. (R) August 7, 1953. Howard Duff, Eva Bartok.

SPANISH AFFAIR, THE (VV) **95**
PARAMOUNT. (R) October, 1957. Richard Kiley, Carmen Savilla. Color.

SPANISH GARDENER, THE (VV) **95**
RANK. (R) October, 1957. Dirk Bogarde, Jon Witeley.

SPARKLE **99**
WARNER BROS. (R) April, 1976. Philip M. Thomas, Irene Cara.

SPARROWS CAN'T SING **93**
JANUS. (R) May, 1963. James Booth, Barbara Windsor.

SPARTACUS (T-70) **198**
UNIVERSAL. (R) Special. October, 1960. Kirk Douglas, Laurence Olivier. Color.

SPECIAL DAY, A (It., Eng. titles) **110**
CINEMA 5. (R) October, 1977. Sophia Loren, Marcello Mastroianni.

SPECIAL DELIVERY **99**
AMERICAN INTERNATIONAL. (R) July, 1976. Bo Svenson, Cybill Shepherd.

SPECIAL DELIVERY **86**
COLUMBIA. (R) September, 1955. Joseph Cotten, Eva Bartok.

SPECIAL SECTION (French; English sub-titles) **110**
UNIVERSAL. (R) December, 1975. Louis Seigner, Michel Lonsdale.

SPECTRE OF EDGAR ALLAN POE, THE **85**
CINERAMA. (R) April, 1974. Robert Walker, Jr., Cesar Romero.

SPEED CRAZY **75**
ALLIED ARTISTS. (R) May, 1959. Brett Halsey, Yvonne Lime.

SPEEDWAY **100**
MGM. (R) June, 1968. Elvis Presley, Nancy Sinatra. Color.

SPIDER, THE **72**
AMERICAN INTERNATIONAL. (R) October, 1958. Ed Kemmer, June Kenny.

SPIKES GANG, THE **96**
UNITED ARTISTS. (R) April, 1974. Lee Marvin, Gary Grimes.

SPINOUT (P) **95**
METRO-GOLDWYN-MAYER. (R) November, 1966. Elvis Presley, Shelley Fabres. Color.

SPIRAL ROAD, THE **145**
UNIVERSAL. (R) August, 1962. Rock Hudson, Burl Ives. Color.

SPIRIT IS WILLING, THE **100**
PARAMOUNT. (R) July, 1967. Sid Caesar, Vera Miles. Color.

SPIRIT OF ST. LOUIS, THE (Cs) **135**
WARNERS. (R) April 20, 1957. James Stewart. Color.

SPIRITS OF THE DEAD **117**
AMER. INT. PICS. (R) July, 1969. Brigitte Bardot, Alain Delon, Jane Fonda. Color.

SPLENDOR IN THE GRASS **124**
WARNER BROS. (R) October, 1961. Natalie Wood, Pat Hingle. Color.

SPLIT, THE **91**
METRO-GOLDWYN-MAYER. (R) October, 1968. Jim Brown, Diahnn Carroll, Julie Harris. Color.

SPOILERS, THE **84**
UNIVERSAL. (R) January, 1956. Jeff Chandler, Anne Baxter, Rory Calhoun. Color.

SPOILERS OF THE FOREST **68**
REPUBLIC. (R) April 5, 1957. Rod Cameron, Vera Ralston. Color.

SPOOK CHASERS **62**

	Length In Mins.
ALLIED ARTISTS. (R) June, 1957. Huntz Hall, Stanley Clements.	
SPOOK WHO SAT BY THE DOOR UNITED ARTISTS. (R) September, 1973. Lawrence Cook, Paula Kelly.	102
SPORTING CLUB, THE AVCO-EMBASSY. (R) February, 1971. Robert Fields, Maggie Blye. Color.	105
SPRING REUNION UNITED ARTISTS. (R) March, 1957. Betty Hutton, Dana Andrews.	79
SPY CHASERS ALLIED ARTISTS. (R) July 31, 1955. Leo Gorcey, Huntz Hall.	61
SPY IN THE SKY ALLIED ARTISTS. (R) July, 1958. Steve Brodie, Sandra Francis.	74
SPY IN YOUR EYE AMERICAN INTERNATIONAL. (R) December, 1965. Brett Halsey, Pier Angeli. Color.	88
SPY WHO CAME IN FROM THE COLD, THE PARAMOUNT. (R) February, 1966. Richard Burton, Claire Bloom.	112
SPY WHO LOVED ME, THE (P) UNITED ARTISTS. (R) July, 1977. Roger Moore.	125
SPY WITH A COLD NOSE, THE EMBASSY. (R) December, 1966. Laurence Harvey, Lionel Jeffries. Color.	93
SPY WITH MY FACE, THE METRO-GOLDWYN-MAYER. (R) February, 1966. Robert Vaughn, Senta Berger. Color.	88
SQUAD CAR TWENTIETH CENTURY-FOX. (R) September, 1960. Paul Bryar, Vici Raaf.	62
SQUARE JUNGLE, THE UNIVERSAL. (R) January, 1956. Tony Curtis, Ernest Borgnine, Pat Crowley.	86
SQUARE RING, THE REPUBLIC. (R) January 28, 1955. Jack Warner, Robert Beatty.	73
SQUIRM AMERICAN INTERNATIONAL. (R) August, 1976. John Scardino, Patricia Pearcy.	92
ST. VALENTINE'S DAY MASSACRE, THE (P) TWENTIETH CENTURY-FOX. (R) August, 1967. Jason Robards, George Segal. Color.	100
Sssssss UNIVERSAL. (R) July, 1973. Strother Martin, Dirk Benedict. Color.	90
STAGE STRUCK BUENA VISTA. (R) March, 1958. Henry Fonda, Susan Fonda, Susan Strasberg. Color.	95
STAGE TO THUNDER ROCK (TS) PARAMOUNT. (R) June, 1964. Barry Sullivan, Marilyn Maxwell. Color.	82
STAGECOACH (Cs) TWENTIETH CENTURY-FOX. (R) May-June, 1966. Ann-Margret, Red Buttons, Bing Crosby.	114
STAGECOACH TO DANCER'S ROCK UNIVERSAL. (R) October, 1962. Warren Stevens, Martin Landau.	72
STAGECOACH TO FURY TWENTIETH CENTURY-FOX. (R) October, 1956. Forrest Tucker, Mari Blanchard.	76
STAIRCASE (P) 20TH CENTURY-FOX. (R) August, 1969. Richard Burton, Rex Harrison. Color.	96
STAKEOUT ON DOPE STREET WARNER BROTHERS. (R) April, 1958. Yale Wexler, Jonathan Haze.	83
STALKING MOON NATIONAL GENERAL. (R) January, 1969. Eva Marie Saint, Gregory Peck, Robert Forster. Color.	109
STAND UP AND BE COUNTED COLUMBIA. (R) May, 1972. Jacqueline Bisset, Stella Stevens. Color.	99
STAR! TWENTIETH CENTURY-FOX. (R) October, 1968. Julie Andrews, Richard Crenna, Michael Craig. Color.	175
STAR IN THE DUST UNIVERSAL. (R) June, 1956. John Agar, Mamie Van Doren. Color.	80
STAR IS BORN (Cs) WARNER BROTHERS. (R) January 22, 1955. Judy Garland, James Mason. Color.	154
STAR IS BORN, A (P) WARNER BROS. (R) December, 1976. Barbra Streisand, Kris Kristofferson.	140
STAR OF INDIA UNITED ARTISTS. (R) June, 1956. Cornel Wilde, Jean Wallace. Color.	84
STAR TREK—THE MOTION PICTURE PARAMOUNT. (R) December, 1979. William Shatner, Leonard Nimoy.	132
STAR WARS (P) TWENTIETH CENTURY-FOX. (R) June, 1977. Mark Hamill, Harrison Ford, Carrie Fisher.	123
STARDUST COLUMBIA. (R) March, 1975. David Essex, Adam Faith.	107
STARSHIP INVASIONS WARNER BROS. (R) January, 1978. Robert Vaughn, Christopher Lee.	89
START THE REVOLUTION WITHOUT ME WARNER BROS. (R) February, 1970. Gene Wilder, Donald Sutherland. Color.	98
STARTING OVER PARAMOUNT. (R) October, 1979. Burt Reynolds, Jill Clayburgh.	101
STATE FAIR (Cs) TWENTIETH CENTURY-FOX. (R) April, 1962. Pat Boone, Bobby Darin. Color.	118
STATE OF SIEGE CINEMA V. (R) April, 1973. Yves Montand, Renato Salvatori. Color.	115
STATION SIX SAHARA ALLIED ARTISTS. (R) August, 1964. Carroll Baker, Ian Bannen.	99
STATUE, THE CINERAMA. (R) February, 1971. David Niven, Vierna Lisi. Color.	72
STAY AS YOU ARE (Ital., Eng. titles) NEW LINE CINEMA. (R) December, 1979. Marcello Mastroianni, Nastassia Kinski.	95
STAY AWAY JOE MGM. (R) April, 1968. Elvis Presley, Burgess Meredith. Color.	102
STAY HUNGRY UNITED ARTISTS. (R) May, 1975. Jeff Bridges, Sally Field.	103
STEAGLE, THE AVCO EMBASSY. (R) September, 1971. Richard Benjamin, Chill Wills, Cloris Leachman. Color.	91
STEEL BAYONET, THE (H) UNITED ARTISTS. (R) March, 1958. Leo Genn. Kieron Moore.	84
STEEL CLAW, THE WARNER BROS. (R) May, 1961. George Montgomery, Charlto Luna. Color.	96
STEEL JUNGLE, THE WARNER BROTHERS. (R) March 31, 1956. Perry Lopez, Beverly Garland.	86
STEELYARD BLUES WARNER BROS. (R) February, 1973. Jane Fonda, Peter Boyle. Color.	92
STEP DOWN TO TERROR UNIVERSAL. (R) March, 1959. Colleen Miller, Charles Drake.	75
STEPFORD WIVES, THE (P) COLUMBIA. (R) February, 1975. Katharine Ross, Paula Prentiss.	115
STERILE CUCKOO, THE PARAMOUNT. (R) October, 1969. Liza Minelli, Wendell Burton. Color.	107
STEWARDESSES, THE (3-D) SHERPIX. (R) November, 1970. Christina Hart, Michael Garrett. Color.	93
STIGMA CINERAMA. (R) August, 1972. Philip M. Thomas, Harlan Poe. Color.	93
STILETTO AVCO EMBASSY. (R) August, 1969. Alex Cord, Britt Eckland, Patrick O'Neal. Color.	99
STING, THE UNIVERSAL. (R) December, 1973. Paul Newman, Robert Redford.	129
STOLEN HOURS UNITED ARTISTS (R) October, 1963. Susan Hayward, Michael Craig. Color.	100
STOLEN KISSES (French; Eng. titles) LOPERT. (R) February, 1969. Jean Pierre Leaud, Delphine Seyrig, Michael Lonsdale. Color.	90
STONE KILLER, THE COLUMBIA. (R) August, 1973 Charles Bronson, Martin Balsam.	95
STOOGE, THE PARAMOUNT. (R) February, 1953. Dean Martin, Jerry Lewis.	100
STOP! LOOK! AND LAUGH! COLUMBIA. (R) July, 1960. The Three Stooges.	78
STOP ME BEFORE I KILL! (M) COLUMBIA. (R) June, 1961 Claude Dauphin, Diane Cilento.	105
STOP THE WORLD—I WANT TO GET OFF	98

Length In Mins.

WARNER BROTHERS. (R) May, 1966. Tony Tanner, Millicent Martin. Color.

STOP TRAIN 349 .... **95**
ALLIED ARTISTS. (R) July, 1964. Jose Ferrer, Sean Flynn.

STOPOVER TOKYO (Cs) **100**
TWENTIETH CENTURY-FOX. (R) October, 1957. Robert Wagner, Joan Collins. Color.

STORM CENTER **87**
COLUMBIA. (R) September, 1956. Bette Davis, Kim Hunter.

STORM FEAR **87**
UNITED ARTISTS. (R) January, 1956. Cornel Wilde, Jean Wallace, Dan Duryea.

STORM OVER THE NILE (Cs) **107**
COLUMBIA. (R) June, 1956. Anthony Steel, Laurence Harvey. Color.

STORM RIDER, THE **70**
TWENTIETH CENTURY-FOX. (R) March, 1957. Scott Brady, Mala Powers.

STORY OF A WOMAN **90**
UNIVERSAL. (R) February, 1970. Bibi Andersson, Robert Stack, James Farentino. Color.

STORY OF ADELE H., THE (French; English sub-titles) **95**
NEW WORLD PICTURES. (R) December, 1975. Isabelle Adjani, Bruce Robinson.

STORY OF ESTHER COSTELLO, THE **127**
COLUMBIA. (R) October, 1957. Joan Crawford, Rossano Brazzi.

STORY OF MANKIND, THE **100**
WARNER BROS. (R) November, 1957. Ronald Colman, Hedy Lamarr. Color.

STORY OF O, THE (French; Eng. titles) **97**
ALLIED ARTISTS. (R) November, 1975. Corinne Clery, Udo Keir.

STORY OF RUTH, THE (Cs) **132**
TWENTIETH CENTURY-FOX. (R) June, 1960. Stuart Whitman, Tom Tryon. Color.

STORY OF THE COUNT OF MONTE CRISCO, THE (D) .. **132**
WARNER BROS. (R) June, 1962. Louis Jourdan, Yvonne Forneau.

STORY OF VICKIE (Ger.) **109**
BUENA VISTA. (R) February, 1958. Romy Schneider, Adrian Haven. Color.

STORY ON PAGE ONE, THE (Cs) **123**
TWENTIETH CENTURY-FOX. (R) January, 1960. Rita Hayworth, Anthony Franciosa.

STOWAWAY GIRL **87**
PARAMOUNT. (R) September, 1957. Trevor Howard, Pedro Armendariz.

STOWAWAY IN THE SKY **82**
UNITED ARTISTS. (R) June, 1962. Jack Lemmon, Andre Gille. Color.

STRAIGHT TIME **114**
WARNER BROS. (R) March, 1978. Dustin Hoffman, Gary Busey.

STRAIT-JACKET **93**
COLUMBIA. (R) January, 1964. Joan Crawford, Diane Baker.

STRANGE ADVENTURE, A **70**
REPUBLIC. (R) August 24, 1956. Joan Evans, Ben Cooper.

STRANGE AFFAIR, THE **106**
PARAMOUNT. (R) September, 1968. Michael York, Jeremy Kamp, Susan George. Color.

STRANGE AFFECTION **87**
BRENNER. (R) November, 1959. Richard Attenborough, Dorothy Leigh.

STRANGE BEDFELLOWS **99**
UNIVERSAL. (R) February, 1965. Rock Hudson, Gina Lollobrigida. Color.

STRANGE CASE OF DR. MANNING, THE **75**
STATES-RIGHTS-REPUBLIC. (R) September, 1959. Ron Randall, Greta Gynt.

STRANGE INTRUDER **82**
ALLIED ARTISTS. (R) September 2, 1956. Ida Lupino, Edmund Purdom.

STRANGE LADY IN TOWN (Cs) **112**
WARNER BROTHERS. (R) April 30, 1955. Greer Garson, Dana Andrews. Color.

STRANGE ONE, THE **97**
COLUMBIA. (R) May, 1957. Ben Gazzara, Julie Wilson.

STRANGE SHADOWS IN AN EMPTY ROOM (P) **99**
AMERICAN INTERNATIONAL. (R) February, 1977. Stuart Whitman, John Saxon.

STRANGER, THE **105**
PARAMOUNT. (R) January, 1968. Marcello Mastroianni, Anna Karina. Color.

STRANGER AT MY DOOR **85**
REPUBLIC. (R) April 6, 1956. Macdonald Carey, Patricia Medina.

STRANGER IN MY ARMS, A (Cs) **83**
UNIVERSAL. (R) February, 1959. June Allyson, Jeff Chandler.

STRANGER IN TOWN (Brit.) **74**
ASTOR. (R) May, 1957. Alex Nicol, Colin Tapley.

STRANGER IN TOWN, A **90**
MGM. (R) April, 1968. Tony Anthony, Frank Wolf. Color.

STRANGER KNOCKS, A (Dan., Eng. titles) **81**
TRANS LUX. (R) March, 1965. Preben Lerdorff Rye, Brigitte Federspiel.

STRANGER ON HORSEBACK **66**
UNITED ARTISTS. (R) March, 1955. Joel McCrea, Miroslava. Color.

STRANGER RETURNS, THE **90**
METRO-GOLDWYN-MAYER. (R) August, 1968. Tony Anthony, Dean Vadis, Daniele Vargas. Color.

STRANGER'S HANDS, THE (Brit.) **88**
DISTRIBUTORS CORP. OF AMERICA. (R) March, 1955. Trevor Howard, Alida Valli.

STRANGERS WHEN WE MEET (Cs) **117**
COLUMBIA. (R) July, 1960. Kirk Douglas, Kim Novak. Color.

STRANGLER, THE **89**
ALLIED ARTISTS. (R) April, 1964. Victor Buono, David McLean.

STRATEGIC AIR COMMAND (VV) **114**
PARAMOUNT. (R) July, 1955. James Stewart, June Allyson. Color.

STRAW DOGS **113**
CINERAMA. (R) December, 1971. Dustin Hoffman, Susan George, David Warner. Color.

STRAWBERRY STATEMENT, THE **103**
MGM. (R) June, 1970. Bruce Davison. Kim Darby. Color.

STRAY DOG (Jap., Eng. titles) **122**
TOHO. (R) March, 1964. Toshiro Mifune, Takashi Shimura.

STREET FIGHTER **71**
SAVAGE-BRENNER. (R) October, 1959. Vic Savage, Ann Atmar.

STREET OF DARKNESS. **60**
REPUBLIC. (R) June, 1958. Robert Keys, John Close.

STREET OF SINNERS **76**
UNITED ARTISTS. (R) September, 1957. George Montgomery, Geraldine Brooks.

STREET PEOPLE **92**
AMERICAN INTERNATIONAL. (R) October, 1976. Roger Moore, Stacy Keach.

STREETCAR NAMED DESIRE **122**
WARNER BROS. (R) March 22, 1952. Vivien Leigh, Marlon Brando.

STRIPPER, THE (Cs) **95**
TWENTIETH CENTURY-FOX. (R) June, 1963. Joanne Woodward, Richard Beymer.

STRONGEST MAN IN THE WORLD, THE **92**
BUENA VISTA. (R) February, 1975. Kurt Russell, Joe Flynn.

STUDS LONIGAN **95**
UNITED ARTISTS. (R) September, 1960. Christopher Knight, Venetia Stevenson.

STUDY IN TERROR, A **94**
COLUMBIA. (R) April, 1966. John Neville, Donald Houston. Color.

SUBJECT WAS ROSES, THE **107**
METRO-GOLDWYN-MAYER. (R) November, 1968. Patricia Neal, Jack Albertson, Martin Sheen. Color.

SUBMARINE X-I **89**
UNITED ARTISTS. (R) July, 1969. James Caan, David Sumner. Color.

SUBMARINE SEAHAWK **87**
AMERICAN INTERNATIONAL. (R) February, 1959. John Bentley, Jan Brooks.

SUBTERRANEANS, THE (Cs) **86**
METRO-GOLDWYN-MAYER. (R) August, 1960. Leslie Caron, George Peppard. Color.

SUBWAY IN THE SKY **85**
UNITED ARTISTS. (R) September, 1959. Van Johnson, Hildegarde Neff.

SUCCESS, THE (Ital., Eng. titles) **103**
AVCO EMBASSY. (R) 1965. Vittorio Gassmann, Anouk Aimee. b&w.

SUCCUBUS (Ger., Eng. titles) **93**
AMERICAN. (R) April, 1969. Janine Reynaud, Jack Taylor, Howard Vernon. Color.

SUCH A GORGEOUS KID LIKE ME **95**
COLUMBIA. (R) March, 1973. Bernadette Lafont, Claude Brasseur. Color.

SUCH GOOD FRIENDS **101**
PARAMOUNT. (R) December, 1971. Dyan Cannon, James Coco, Lawrence Luckenbill. Color.

SUCKER, THE **101**
ROYAL FILMS INTERNATIONAL. (R) June, 1966. Bourvil, Louis De Funes. Color.

	Length In Mins.
SUDDEN DANGER ALLIED ARTISTS. (R) December 18, 1955. Bill Elliott, Tom Drake.	65
SUDDEN TERROR NATIONAL GENERAL. (R) March, 1971. Mark Lester, Lionel Jeffries. Color.	95
SUDDENLY, LAST SUMMER COLUMBIA. (R) January, 1960. Elizabeth Taylor, Katharine Hepburn.	114
SUGARLAND EXPRESS, THE (P) UNIVERSAL. (R) April, 1974. Goldie Hawn, Ben Johnson.	108
SUICIDE BATTALION AMER. INT. PICS. (R) February, 1958. Michael Connors, Jewell Lain.	75
SUICIDE MISSION COLUMBIA. (R) November, 1956. Leif Larsen, Michael Aldridge.	70
SULLIVAN'S EMPIRE UNIVERSAL. (R) August, 1967. Martin Milner, Linden Chiles. Color.	85
SUMMER AND SMOKE PARAMOUNT. (R) February, 1962. Geraldine Page, Laurence Harvey. Color.	116
SUMMER LOVE UNIVERSAL INTERNATIONAL. (R) April, 1958. John Saxon, Molly Bee.	85
SUMMER MAGIC BUENA VISTA. July, 1963. Hayley Mills, Burl Ives. Color.	109
SUMMER OF '42 WARNER BROS. (R) April, 1971 Jennifer O'Neill, Gary Grimes. Color.	102
SUMMER PARADISE (Swed., Eng. titles) CINEMA 5. (R) March, 1978. Birgktta, Valberg, Sif Ruud.	113
SUMMER PLACE, A WARNER BROS. (R) November, 1959. Richard Egan, Dorothy McGuire. Color.	130
SUMMER WISHES, WINTER DREAMS COLUMBIA. (R) October, 1973. Joanne Woodward, Martin Balsam.	90
SUMMERTIME UNITED ARTISTS. (R) June, 1955. Katharine Hepburn, Rossano Brazzi. Color.	99
SUMMERTIME KILLER AVCO-EMBASSY. (R) September, 1973. Christopher Mitchum, Karl Malden.	100
SUMMERTREE COLUMBIA. (R) June, 1971. Michael Douglas, Jack Warden, Brenda Vaccaro. Color	88
SUN ALSO RISES, THE (Cs) TWENTIETH CENTURY-FOX. (R) September, 1957. Tyrone Power, Ava Gardner. Color.	129
SUNBURN PARAMOUNT. (R) August, 1979. Farrah Fawcett, Charles Grodin.	101
SUNDAY BLOODY SUNDAY UNITED ARTISTS. (R) October, 1971. Peter Finch, Glenda Jackson, Murray Head. Color.	106
SUNDAY IN NEW YORK. METRO-GOLDWYN-MAYER. (R) February, 1964. Cliff Robertson, Jane Fonda. Color.	105
SUNDAY WOMAN, THE TWENTIETH CENTURY-FOX. (R) September, 1976. Marcello Mastroianni, Jacqueline Bisset.	110
SUNDOWNERS, THE WARNER BROS. (R) December, 1960. Deborah Kerr, Robert Mitchum. Color.	133
SUNFLOWER AVCO-EMBASSY. (R) September, 1970. Sophia Loren, Marcello Mastroianni. Color.	101
SUNRISE AT CAMPOBELLO WARNER BROS. (R) November, 1960. Ralph Bellamy, Greer Garson. Color.	143
SUNSHINE BOYS, THE UNITED ARTISTS. (R) December, 1975. Walter Matthau, George Burns.	111
SUPER COPS, THE UNITED ARTISTS. (R) March, 1974. Ron Liebman, David Selby.	94
SUPER FLY WARNER BROS. (R) August, 1972. Ron O'Neal, Carl Lee. Color.	98
SUPER FLY T.N.T. PARAMOUNT. (R) June, 1973 Ron O'Neal, Roscoe Lee Brown. Color.	87
SUPERBEAST UNITED ARTISTS. (R) October, 1972. Craig Littler, Antoinette Bower. Color.	91
SUPERCHICK CROWN INTERNATIONAL. (R) May, 1973. Joyce Jelson, Louis Quinn. Color.	92

	Length In Mins.
SUPERDAD BUENA VISTA. (R) February, 1974. Bob Crane, Barbara Rush.	95
SUPERMAN (P) WARNER BROS. (R) December, 1978. Marlon Brando, Gene Hackman, Christopher Reeve.	143
SUPPORT YOUR LOCAL GUNFIGHTER UNITED ARTISTS. (R) May, 1971. James Garner, Suzanne Pleshette. Color.	92
SUPPORT YOUR LOCAL SHERIFF UNITED ARTISTS. (R) April, 1969. James Garner, Joan Hackett, Walter Brennan. Color.	92
SUPPOSE THEY GAVE A WAR AND NOBODY CAME CINERAMA. (R) June, 1970. Tony Curtis, Brian Keith, Ernest Borgnine. Color.	113
SURF PARTY TWENTIETH CENTURY-FOX. (R February, 1964. Bobby Vinton, Patricia Morrow.	68
SURPRISE PACKAGE COLUMBIA. (R) November, 1960. Yul Brynner, Mitzi Gaynor.	100
SURRENDER HELL! ALLIED ARTISTS. (R) August, 1959. Keith Andes, Susan Cabot.	85
SURVIVE PARAMOUNT. (R) July, 1976. Hugo Stiglitz, Norma Lazareno.	85
SUSAN SLADE WARNER BROS. (R) November, 1961. Troy Donahue, Connie Stevens. Color.	110
SUSPIRA TWENTIETH CENTURY-FOX. (R) August, 1977. Jessica Harper, Joan Bennett.	92
SVENGALI (Brit.) METRO/GOLDWYN-MAYER. (R) September 9, 1955. Hildegarde Neff, Derek Bond. Color.	82
SWAMP WOMAN WOOLNER BROS. (R) April, 1956. Beverly Garland, Marie Windsor, Carole Mathews. Color.	72
SWAN, THE (Cs) METRO-GOLDWYN-MAYER. (R) April 27, 1956. Grace Kelly, Alec Guinness, Louis Jourdan. Color.	112
SWAN LAKE COLUMBIA. (R) January, 1960. M. Plisetskaya, N. Fadeychev. Color.	81
SWARM, THE WARNER BROS. (R) July, 1978. Michael Caine, Katharine Ross.	116
SWASHBUCKLER UNIVERSAL. (R) July, 1976. Robert Shaw, James Earl Jones.	100
SWEDEN—HEAVEN AND HELL AVCO EMBASSY. (R) August, 1969 Semi-documentary. Color.	90
SWEDISH WEDDING NIGHT (Swed., Eng. titles) ROYAL FILMS INTERNATIONAL. (R) November, 1965. Jarl Kulle, Lena Hansson.	96
SWEET AND SOUR (Fr., Eng. titles) PATHE-CONTEMPORARY. (R) December, 1964.	93
SWEET BIRD OF YOUTH (Cs) METRO-GOLDWYN-MAYER. (R) March, 1962. Paul Newman, Geraldine Page. Color.	120
SWEET BODY OF DEBORAH, THE WARNER BROS.-7 ARTS. (R) March, 1969. Carroll Baker, Jean Sorel, Evelyn Stewart. Color.	105
SWEET CHARITY UNIVERSAL. (R) February, 1969. Shirley MacLaine, Sammy Davis, Jr., Ricardo Montalban. Color.	151
SWEET CREEK COUNTY WAR, THE IMAGERY FILMS. (R) March, 1979. Richard Egan, Albert Salmi.	99
SWEET JESUS, PREACHER MAN METRO-GOLDWYN-MAYER. (R) May, 1973. Roger E. Morley, William Smith Color.	106
SWEET LIGHT IN A DARK ROOM (Czech., Eng. titles) PROMENADE. (R) June 1966. Dana Smutna, Ivan Mistrik.	93
SWEET LOVE, BITTER FILM 2 ASSOCIATES (R) January, 1967. Dick Gregory, Don Murray.	92
SWEET NOVEMBER WARNER BROS-SEVEN ARTS. (R) April, 1968. Sandy Dennis, Anthony Newley. Color.	114
SWEET REVENGE (P) UNITED ARTISTS. (R) May, 1977. Stockard Channing, Sam Waterston.	85
SWEET RIDE, THE TWENTIETH CENTURY-FOX. (R) June, 1968. Tony Franciosa, Michael Sarrazin, Jacqueline Bisset. Color.	110
SWEET SMELL OF SUCCESS UNITED ARTISTS. (R) June, 1957. Burt Lancaster, Tony Curtis.	96
SWEET SWEETBACK'S BAADASSSSS SONG	97

Length In Mins.

CINEMATION. (R) April, 1971. Melvin Van Peebles. Color.

SWEETHEARTS ON PARADE 90
REPUBLIC. (R) July 15, 1953. Ray Middleton, Eileen Christy. Color.

SWEPT AWAY . . . BY AN UNUSUAL DESTINY IN THE BLUE SEA OF AUGUST 116
CINEMA V. (R) September, 1975. Giancarlo Giannini, Mariangela Melato.

SWIMMER, THE 94
COLUMBIA. (R) May, 1968. Burt Lancaster, Janice Rule. Color.

SWIMMING POOL, THE 87
AVCO-EMBASSY. (R) August, 1970. Alain Delon, Romy Schneider. Color.

SWINGER, THE 81
PARAMOUNT. (R) November, 1966. Ann-Margret, Tony Franciosa. Color.

SWINGERS PARADISE (T) 82
AMERICAN INTERNATIONAL. (R) May, 1965. Cliff Richard, Walter Slezak. Color.

SWINGIN' MAIDEN, THE 81
COLUMBIA. (R) January, 1964. Michael Craig, Anne Helm. Color.

SWINGIN' SUMMER (TS) 80
UNITED SCREEN ARTS. (R) April, 1965. James Stacey, William A. Wellman, Jr. Color.

SWINGING ALONG (Cs) 74
LIPPERT. (R) 1962. Tommy Noonan, Pete Marshall, Barbara Eden. Color.

SWISS FAMILY ROBINSON (P) 128
BUENA VISTA. (R) December, 1960. John Mills, Dorothy McGuire. Color.

SWORD IN THE STONE, THE (T) 75
BUENA VISTA. (R) November, 1963. Animation. Color.

SWORD OF ALI BABA, THE 81
UNIVERSAL. (R) May, 1965. Peter Mann, Jocelyn Lane. Color.

SWORD OF DOOM (T) (Jap.) 122
TOHO. (R) May, 1967.

SWORD OF EL CID 85
PRODUCTION RELEASING. (R) July, 1964. Roland Carey, Chantel Deberg. Color.

SWORD OF SHERWOOD FOREST (M) 80
COLUMBIA. (R) January, 1961. Richard Greene, Peter Cushing. Color.

SWORD OF THE CONQUEROR (Cs) 95
UNITED ARTISTS. (R) September, 1962. Jack Palance, Eleanora Rossi Drago. Color.

SWORDSMAN OF SIENNA (Cs) 92
METRO-GOLDWYN-MAYER. (R) November, 1962. Stewart Granger, Sylva Koscina. Color.

SYLVIA 115
PARAMOUNT. (R) February, 1965. Carroll Baker, George Maharis.

SYMPHONY FOR A MASSACRE (Fr., Eng. Titles) 110
SEVEN ARTS. (R) May, 1965. Michel Auclair, Claude Dauphin.

SYNANON 107
COLUMBIA. (R) May, 1965. Edmond O'Brien, Chuck Connors.

T

T-BIRD GANG 75
THE FILM GROUP. (R) July, 1959. Ed Nelson, John Brinkley.

THX 1138 88
WARNER BROS. (R) March, 1971. Robert Duvall, Donald Pleasance. Color.

TABOOS OF THE WORLD 97
AMERICAN INTERNATIONAL. (R) May, 1965. Documentary. Color.

TAFFY AND THE JUNGLE HUNTER 90
ALLIED ARTISTS. (R) February, 1965. Jacques Bergerac, Manuel Padilla. Color.

TAGGART 85
UNIVERSAL. (R) February, 1965. Tony Young, Dan Duryea. Color.

TAKE, THE 93
COLUMBIA. (R) May, 1974. Billy Dee Williams, Eddie Albert.

TAKE A GIANT STEP 100
UNITED ARTISTS. (R) February, 1960. Johnny Nash, Estelle Hemsley.

TAKE A GIRL LIKE YOU 96
COLUMBIA. (R) October, 1970. Hayley Mills, Oliver Reed. Color.

TAKE A HARD RIDE 103
TWENTIETH CENTURY-FOX. (R) July, 1975. Jim Brown, Lee Van Cleef.

Length In Mins.

TAKE DOWN 107
BUENA VISTA. (R) March, 1979. Edward Herrmann, Lorenzo Lamas.

TAKE HER SHE'S MINE 98
TWENTIETH CENTURY-FOX. (R) November, 1963. James Stewart, Sandra Dee. Color.

TAKE IT ALL 90
LOPERT PICTURE. (R) April, 1966. Johanne, Claude Jutra.

TAKE THE MONEY AND RUN 85
CINERAMA. (R) August, 1969. Woody Allen, Janet Margolin, Marcel Hillaire. Color.

TAKING OF PELHAM ONE TWO THREE, THE (P) 102
UNITED ARTISTS. (R) October, 1974. Walter Matthau, Robert Shaw.

TAKING OFF 102
UNIVERSAL. (R) April, 1971. Buck Henry, Lynn Carlin. Color.

TALE OF TWO CITIES, A 117
RANK FILM DISTRIBUTORS OF AMERICA. (R) November, 1958. Dirk Bogarde, Dorothy Tutin.

TALES FROM THE CRYPT 92
CINERAMA. (R) March, 1972. Ralph Richardson, Joan Collins, Peter Cushing. Color.

TALES THAT WITNESS MADNESS 90
PARAMOUNT. (R) October, 1973. Kim Novak, Georgia Brown.

TALL BLOND MAN WITH ONE BLACK SHOE, THE 90
CINEMA V. (R) September, 1973. Pierre Richard, Bernard Blier.

TALL MAN RIDING 68
WARNER BROTHERS. (R) June 18, 1955. Randolph Scott, Dorothy Malone. Color.

TALL MEN, THE 121
TWENTIETH CENTURY-FOX. (R) October, 1955. Clark Gable, Jane Russell, Robert Ryan. Color.

TALL STORY 91
WARNER BROS. (R) April, 1960. Anthony Perkins, Jane Fonda.

TALL STRANGER, THE 83
ALLIED ARTISTS. (R) November, 1957. Joel McCrea, Virginia Mayo. Color.

TALL T, THE 78
COLUMBIA. (R) April, 1957.

TALL WOMEN, THE 92
ALLIED ARTISTS. (R) March, 1967. Anne Baxter, Maria Perschy. Color.

TAMAHINE (Cs) 85
METRO-GOLDWYN-MAYER. (R) March, 1964. Nancy Kwan, John Fraser. Color.

TAMANGO (Cs) 98
HAL ROACH. (R) August, 1959. Curt Jurgens, Dorothy Dandridge. Color.

TAMARIND SEED, THE (P) 123
AVCO-EMBASSY. (R) July, 1974. Julie Andrews, Omar Sharif.

TAMING OF THE SHREW, THE (P) 126
COLUMBIA/ROYAL FILMS. (R) March, 1967. Elizabeth Taylor, Richard Burton. Color.

TAMING SUTTON'S GAL (Naturma) 71
REPUBLIC. (R) September, 1957. John Lupton, Gloria Talbott.

TAMMY AND THE BACHELOR (Cs) 89
UNIVERSAL. (R) August, 1957. Debbie Reynolds, Leslie Neilsen. Color.

TAMMY AND THE DOCTOR 88
UNIVERSAL. (R) June, 1963. Sandra Dee, Peter Fonda. Color.

TAMMY AND THE MILLIONAIRE 80
UNIVERSAL. (R) July, 1967. Debbie Watson, Frank McGrath. Color.

TAMMY TELL ME TRUE 97
UNIVERSAL. (R) July, 1961. Sandra Dee, John Gavin. Color.

TANK BATTALION 89
AMERICAN INTERNATIONAL. (R) July, 1958. Don Kelly, Marjorie Hellen.

TANK COMMANDOS 79
AMERICAN INTERNATIONAL. (R) March, 1959. Robert Barron, Maggie Lawrence.

TANK FORCE (Cs) 81
COLUMBIA. (R) August, 1958. Victor Mature, Leo Genn. Color.

TARANTULA 80
UNIVERSAL. (R) December, 1955. John Agar, Mara Corday.

TARAS BULBA (P) 122
UNITED ARTISTS. (R) December, 1962. Tony Curtis, Yul Brynner. Color.

TARAWA BEACHHEAD 77
COLUMBIA. (R) November, 1958. Kerwin Mathews, Julie Adams.

TARGET ZERO 92

	Length In Mins.
WARNER BROTHERS. (R) December 10, 1955. Richard Conte, Peggie Castle.	
TARGETS	92
PARAMOUNT. (R) August, 1968. Boris Karloff, Tim O'Kelley, Nancy Houeh. Color.	
TARNISHED ANGELS, THE (Cs)	91
UNIVERSAL. (R) January, 1958. Rock Hudson, Robert Stack.	
TARTARS, THE	83
METRO-GOLDWYN-MAYER. (R) June, 1962. Victor Mature, Orson Welles. Color.	
TARZAN AND THE GREAT RIVER (P)	88
PARAMOUNT. (R) September, 1967. Mike Henry, Jan Murray. Color.	
TARZAN AND THE JUNGLE BOY	90
PARAMOUNT. (R) May, 1968. Mike Henry, Rafer Johnson. Color.	
TARZAN AND THE LOST SAFARI	80
METRO-GOLDWYN-MAYER. (R) May 3, 1957. Gordon Scott, Yolande Donland. Color.	
TARZAN GOES TO INDIA (Cs)	86
METRO-GOLDWYN-MAYER. (R) July, 1962. Jock Mahoney, Mark Dana. Color.	
TARZAN, THE APE MAN	82
METRO-GOLDWYN-MAYER. (R) October, 1959. Denny Miller, Joanna Barnes. Color.	
TARZAN THE MAGNIFICENT	88
PARAMOUNT. (R) July, 1960. Gordon Scott, Betta St. John.	
TARZAN'S DEADLY SILENCE	82
NATIONAL GENERAL. (R) April, 1970. Ron Ely, Jock Mahoney. Color.	
TARZAN'S FIGHT FOR LIFE (Cs)	86
METRO-GOLDWYN-MAYER. (R) July, 1958. Gordon Scott, Eve Brent. Color.	
TARZAN'S GREATEST ADVENTURE	90
PARAMOUNT. (R) July, 1959. Gordon Scott, Anthony Quayle. Color.	
TARZAN'S HIDDEN JUNGLE	73
RKO RADIO. (R) February 16, 1955. Gordon Scott, Vera Miles.	
TARZAN'S JUNGLE REBELLION	92
NATIONAL GENERAL. (R) May, 1970. Ron Ely, Manuel Padilla, Jr. Color.	
TARZAN'S THREE CHALLENGES (D)	92
METRO-GOLDWYN-MAYER (R) June, 1963. Jock Mahoney, Woody Strode. Color.	
TASTE OF HONEY, A	100
CONTINENTAL. (R) May, 1962. Dora Bryan, Rita Tushingham.	
TASTE THE BLOOD OF DRACULA	95
WARNER BROS. (R) September, 1970. Christopher Lee, Linda Hayden. Color.	
TATTERED DRESS, THE (Cs)	93
UNIVERSAL. (R) April, 1957. Jeff Chandler, Jeanne Crain.	
TATTOOED POLICE HORSE, THE	48
BUENA VISTA. (R) December, 1964. Sandy Sanders, Charles Seel. Color.	
TAXI DRIVER	112
COLUMBIA. (R) February, 1976. Robert De Niro, Jodie Foster.	
TAXI FOR TOBRUK (D)	90
SEVEN ARTS. (R) March, 1965. Charles Aznavour, Lino Ventura.	
TCHAIKOVSKY	105
TIOMKIN. (R) December, 1971. Innokenti Smoktunovsky, Antonina Shuranova. Color.	
TEA AND SYMPATHY (Cs)	122
METRO-GOLDWYN-MAYER. (R) Sept. 28, 1956. Deborah Kerr, John Kerr. Color.	
TEACHER'S PET (VV)	120
PARAMOUNT. (R) April, 1958. Clark Gable, Doris Day.	
TEAHOUSE OF THE AUGUST MOON, THE (Cs)	123
METRO-GOLDWYN-MAYER. (R) Dec., 1956. Marlon Brando, Glenn Ford. Color.	
TEARS FROM SIMON (Brit.)	91
REPUBLIC. (R) January, 1957. David Farrar, David Knight. Color.	
TECKMAN MYSTERY, THE	90
ASSOCIATED ARTISTS. (R) August, 1955. Margaret Leighton, John Justin.	
TEENAGE BAD GIRL (Brit.)	100
DIST. CORP. OF AMER. (R) December, 1957. Anna Neagle, Sylvia Sims.	
TEENAGE CAVE MAN	65
AMERICAN INTERNATIONAL. (R) July, 1958. Robert Vaughn, Leslie Bradley.	
TEENAGE CRIME WAVE	77
COLUMBIA. (R) November 1955. Tommy Cook, Mollie McCart.	
TEENAGE DOLL	71
ALLIED ARTISTS. (R) September, 1957. June Kenney, Fay Spain.	
TEENAGE MILLIONAIRE	84
UNITED ARTISTS. (R) August, 1961. Jimmy Clayton, Rocky Graziano.	
TEENAGE REBEL (Cs)	94
TWENTIETH CENTURY-FOX. (R) November, 1956. Ginger Rogers, Michael Rennie.	
TEENAGE REBELLION	81
TRANS AMERICAN. (R) May, 1967 Documentary.	
TEENAGE WOLF PACK (Ger.)	89
DIST. CORP. OF AMER. (R) December, 1958. Henry Bookholt, Karen Beal.	
TEENAGE ZOMBIES	73
GOVERNOR FILMS. (R) April, 1960. Don Sullivan, Katherine Victor.	
TEENAGERS FROM OUTER SPACE	85
WARNER BROS. (R) June, 1959. David Love, Dawn Anderson.	
TELEFON (P)	103
UNITED ARTISTS. (R) December, 1977. Charles Bronson, Lee Remick.	
TELL ME IN THE SUNLIGHT	86
MOVIE-RAMA. (R) May, 1967. Steve Cochran, Shary Marshall.	
TELL ME THAT YOU LOVE ME JUNIE MOON	113
PARAMOUNT. (R) June, 1970. Liza Minelli, Robert Moore, Ken Howard. Color.	
TELL THEM WILLIE BOY IS HERE (P)	96
UNIVERSAL. (R) February, 1970. Robert Redford, Robert Blake, Katharine Ross. Color.	
"10"	122
WARNER BROS. (R) October, 1979. Dudley Moore, Julie Andrews.	
TEN COMMANDMENTS, THE (VV)	219
PARAMOUNT. (R) Special, 1957. Charlton Heston, Yul Brynner. Color.	
TEN DAYS TO TULARE	77
UNITED ARTISTS. (R) November, 1958. Sterling Hayden, Grace Raynor.	
TEN DAYS WONDER	100
LEVITT-PICKMAN. (R) April, 1972. Orson Welles, Anthony Perkins. Color.	
TEN FROM YOUR SHOW OF SHOWS	92
WALTER READE. (R) February, 1972. Sid Caesar, Imogene Coca.	
TEN LITTLE INDIANS	92
SEVEN ARTS. (R) January, 1966. Hugh O'Brian, Shirley Eaton.	
TEN LITTLE INDIANS	98
AVCO-EMBASSY. (R) March, 1975. Oliver Reed, Elke Sommer.	
TEN NORTH FREDERICK (Cs)	102
TWENTIETH CENTURY-FOX. (R) May, 1958. Gary Cooper, Diane Varsi.	
10 RILLINGTON PLACE	109
COLUMBIA. (R) May, 1971. Richard Attenborough, Judy Geeson. Color.	
TEN SECONDS TO HELL	93
UNITED ARTISTS. (R) July, 1959. Jeff Chandler, Jack Palance.	
10:30 P.M. SUMMER	85
LOPERT. (R) November, 1966. Melina Mercouri, Romy Schneider. Color.	
TEN THOUSAND BEDROOMS (Cs)	114
METRO-GOLDWYN-MAYER. (R) March 29, 1957. Dean Martin, Anna Maria Alberghetti. Color.	
TEN WANTED MEN	80
COLUMBIA. (R) February, 1955. Randolph Scott, Jocelyn Brando. Color.	
TEN WHO DARED	92
BUENA VISTA. (R) November, 1960. Brian Keith, John Beal. Color.	
TENANT, THE	125
PARAMOUNT. (R) June, 1976. Roman Polanski, Isabelle Adjani.	
TENDER IS THE NIGHT (Cs)	146
TWENTIETH CENTURY-FOX (R) February, 1962. Jennifer Jones, Jason Robards, Jr. Color.	
TENDER SCOUNDREL (Fr.)	99
EMBASSY. (R) August, 1967. Jean-Paul Belmondo, Nadja Tiller. Color.	
TENDER TRAP, THE (Cs)	111
METRO-GOLDWYN-MAYER. (R) November 4, 1955. Frank Sinatra, Debbie Reynolds David Wayne, Celeste Holm. Color.	
TENNESSEE'S PARTNER (SS)	87
RKO RADIO. (R) September, 1955. John Payne, Ronald Reagan, Rhonda Fleming. Color.	
TENSION AT TABLE ROCK	93
UNIVERSAL. (R) October, 1956. Richard Egan, Dorothy Malone. Color.	
TENTH VICTIM, THE	92

Length In Mins.

EMBASSY. (R) December, 1965. Marcello Mastroianni, Ursula Andress. Color.

TEOREMA (Ital. dubbed) **98**
CONTINENTAL. (R) May, 1969. Terence Stamp, Silvana Mangano. Color.

TERM OF TRIAL **113**
WARNER BROS. (R) February, 1963. Laurence Olivier, Simone Signoret.

TERMINAL MAN, THE **100**
WARNER BROS. (R) June, 1974. George Segal, Joan Hackett.

TERRACE, THE (Span., Eng. titles) **90**
ROYAL. (R) November, 1964. Graciela Borges, Leonardo Favio.

TERROR, THE (VS) **81**
AMERICAN INTERNATIONAL. (R) September, 1963. Boris Karloff, Jack Nicholson. Color.

TERROR AT MIDNIGHT **70**
REPUBLIC. (R) April 27, 1956. Scott Brady, Joan Vohs.

TERROR FROM THE YEAR 5,000 .... **68**
AMERICAN INTERNATIONAL. (R) August, 1958. Joyce Holden, Ward Costello.

TERROR IN A TEXAS TOWN **80**
UNITED ARTISTS. (R) September, 1958. Sterling Hayden, Sebastian Cabot.

TERROR IN THE JUNGLE **82**
CROWN INTERNATIONAL. (R) March, 1969. Robert Burns, Fawn Silver.

TERROR IN THE WAX MUSEUM **95**
CINERAMA. (R) May, 1973. Ray Milland, Maurice Evans. Color.

TERROR IS A MAN **89**
VALIANT. (R) December, 1959. Francis Lederer, Greta Thyssen.

TERROR OF THE TONGS **80**
COLUMBIA. (R) May, 1961. Christopher Lee, Geoffrey Toone. Color.

TERROR STREET **83**
LIPPERT. (R) December 4, 1958. Dan Duryea, Alsy Albin.

TERRORNAUTS, THE **75**
AVCO EMBASSY. (R) June, 1969. Simon Oates, Zena Marshall. Color.

TERRY WHITMORE, FOR EXAMPLE **98**
GROVE. (R) November, 1969. Documentary. b&w.

TESS OF THE STORM COUNTRY (Cs) **85**
TWENTIETH CENTURY FOX. (R) December, 1960. Diane Baker, Lee Philips. Color.

TEXAS ACROSS THE RIVER (T) **101**
UNIVERSAL. (R) November, 1966. Dean Martin, Alain Delon. Color.

TEXAS CHAIN SAW MASSACRE, THE **87**
BRYANSTON. (R) November, 1974. Marilyn Burns, Allen Danziger.

TEXAS LADY (SS) **80**
RKO RADIO. (R) November 30, 1955. Claudette Colbert, Barry Sullivan. Color.

TEXICAN, THE (TS) **86**
COLUMBIA. (R) 1966. Audie Murphy, Broderick Crawford. Color.

THANK GOD IT'S FRIDAY **90**
COLUMBIA. (R) May, 1978. Hilary Beane, John Friedrich.

THANK HEAVEN FOR SMALL FAVORS (Fr., Eng. titles) **84**
INT. CLASSICS. (R) January, 1965. Bourvil, Francis Blancho.

THANK YOU VERY MUCH **106**
COLUMBIA. (R) August, 1969. Sandy Dennis, Ian McKellen, Michael Coles. Color.

THAT CERTAIN FEELING (VV) **103**
PARAMOUNT. (R) July, 1956. Bob Hope, Eva Marie Saint. Color.

THAT COLD DAY IN THE PARK **113**
COMMONWEALTH UNITED. (R) June, 1969. Sandy Dennis, Michael Burns, John Garfield, Jr.

THAT DARN CAT **116**
BUENA VISTA. (R) December, 1965. Hayley Mills, Dean Jones. Color.

THAT FUNNY FEELING **93**
UNIVERSAL. (R) September, 1965. Sandra Dee, Bobby Darin. Color.

THAT KIND OF WOMAN **92**
PARAMOUNT. (R) September, 1959. Sophia Loren, Tab Hunter.

THAT LADY (Cs) **100**
TWENTIETH CENTURY-FOX. (R) May, 1955. Olivia de Havilland, Gilbert Roland. Color.

THAT MAN BOLD **105**
UNIVERSAL. (R) December, 1973. Fred Williamson, Byron Webster.

THAT MAN FROM RIO **114**
LOPERT. (R) June, 1964. Jean-Paul Belmondo, Francoise Dorleac. Color.

THAT MAN GEORGE **92**
ALLIED ARTISTS. (R) October, 1967. George Hamilton, Claudine Auger. Color.

THAT MAN IN ISTANBUL (T) **117**
COLUMBIA. (R) February, 1966. Horst Bucholz, Sylvia Koscina. Color.

THAT NIGHT **88**
UNIVERSAL. (R) September, 1957. John Beal, Augusta Dabney.

THAT SPLENDID NOVEMBER **93**
UNITED ARTISTS. (R) June, 1971. Gina Lollobrigida, Paolo Turco. Color.

THAT TENNESSEE BEAT **84**
20TH CENTURY-FOX. (R) October, 1966. Sharon DeBord, Minnie Pearl. Color.

THAT TOUCH OF MINK (P) **99**
UNIVERSAL. (R) July, 1962. Cary Grant, Doris Day. Color.

THAT'S ENTERTAINMENT! **132**
UNITED ARTISTS. (R) June, 1974. Fred Astaire, Bing Crosby, Gene Kelly.

THAT'S ENTERTAINMENT, PART 2 .... **133**
UNITED ARTISTS. (R) May, 1976. Fred Astaire, Gene Kelly.

THEATRE OF BLOOD **104**
UNITED ARTISTS. (R) April, 1973. Vincent Price, Diana Rigg. Color.

THERE WAS A CROOKED MAN **125**
WARNER BROS. (R) December, 1970. Kirk Douglas, Henry Fonda. Color.

THERE'S A GIRL IN MY SOUP **95**
COLUMBIA. (R) December, 1970. Peter Sellers, Goldie Hawn. Color.

THERE'S ALWAYS TOMORROW **84**
UNIVERSAL. (R) February, 1956. Barbara Stanwyck, Fred MacMurray, Joan Bennett.

THERE'S ALWAYS VANILLA **91**
CAMBIST. (R) February, 1972. Roy Laine, Judith Streiner. Color.

THERESE (Fr., Eng. titles) **107**
PATHE-CONTEMPORARY. (R) November, 1963. Jean-Paul Belmondo, Francoise Dorleac.

THERESE AND ISABELLE (Fr.) **118**
AUDUBON. (R) May, 1968. Essy Persson, Anna Gael.

THESE ARE THE DAMNED **77**
COLUMBIA. (R) July, 1965. MacDonald Carey, Shirley Anne Field.

THESE THOUSAND HILLS (Cs) **96**
TWENTIETH CENTURY-FOX. (R) February, 1959. Don Murray, Richard Egan. Color.

THESE WILDER YEARS **91**
METRO-GOLDWYN-MAYER. (R) Aug. 3, 1956. James Cagney, Barbara Stanwyck, Walter Pidgeon.

THEY ALL DIED LAUGHING **90**
CONTINENTAL. (R) March, 1964. Leo McKern, Janet Murro.

THEY CALL ME MISTER TIBBS! **108**
UNITED ARTISTS. (R) August, 1970. Sidney Poitier, Martin Landau, Barbara McNair. Color.

THEY CAME TO CORDURA (Cs) **123**
COLUMBIA. (R) October, 1959. Gary Cooper, Rita Hayworth. Color.

THEY CAME TO ROB LAS VEGAS **130**
WARNER BROS.-7 ARTS. (R) February, 1969. Gary Lockwood, Elke Sommer, Lee J. Cobb, Jack Palance.

THEY MIGHT BE GIANTS **88**
UNIVERSAL. (R) March, 1971. George C. Scott, Joanne Woodward. Color.

THEY ONLY KILL THEIR MASTERS **97**
METRO-GOLDWYN-MAYER. (R) November, 1972. James Garner, Katharine Ross. Color.

THEY SHOOT HORSES, DON'T THEY? **129**
CINERAMA. (R) December, 1969. Jane Fonda, Michael Sarrazin, Susannah York, Gig Young. Color.

THEY WERE SO YOUNG **80**
LIPPERT. (R) January 7, 1955. Scott Brady, Raymond Burr.

THEY WHO DARE (Brit.) **101**
ASSOCIATED ARTISTS. (R) November, 1955. Dirk Bogarde, Akim Tamiroff. Color.

THIEF OF BAGHDAD, THE (Cs) **90**
METRO-GOLDWYN-MAYER. (R) August, 1961. Steve Reeves, Georgia Moll. Color.

THIEF OF PARIS (Fr.) **119**
LOPERT. (R) August, 1967. Jean-Paul Belmondo, Genevieve Bujold. Color.

THIEF WHO CAME TO DINNER, THE **105**
WARNER BROS. (R) March, 1972. Ryan O'Neal, Jacqueline Bisset. Color.

THIEVES **103**
PARAMOUNT. (R) February, 1977. Marlo Thomas, Charles Grodin.

THIEVES LIKE US **122**

Length In Mins.

UNITED ARTISTS. (R) February, 1974. Keith Carradine, Shelley Duvall.

THIN RED LINE, THE (Cs) ... 99
ALLIED ARTISTS. (R) June, 1964. Keir Dullea, Jack Warden.

THING THAT COULDN'T DIE, THE (Cs) ... 99
UNIVERSAL INTERNATIONAL. (R) June, 1958. William Reynolds, Andra Martin. Color.

THING WITH TWO HEADS, THE ... 93
AMERICAN INTERNATIONAL. (R) July, 1972. Ray Milland, Rosy Grier. Color.

THINGS OF LIFE, THE (French; Eng. titles) ... 88
COLUMBIA. (R) August, 1970. Romy Schneider, Michel Piccoli. Color.

THIRD DAY, THE (P) ... 119
WARNER BROTHERS. (R) August, 1965. George Peppard, Elizabeth Ashley. Color.

THIRD KEY, THE ... 84
RANK. (R) July, 1957. Jack Hawkins, John Stratton.

THIRD MAN ON THE MOUNTAIN, THE ... 105
BUENA VISTA. (R) November, 1959. Michael Rennie, James MacArthur. Color.

THIRD OF A MAN ... 81
UNITED ARTISTS. (R) October, 1962. James Drury, Jan Shepard.

THIRD SECRET, THE (Cs) ... 103
TWENTIETH CENTURY-FOX. (R) April, 1964. Stephen Boyd, Jack Hawkins.

THIRD VOICE, THE (Cs) ... 80
TWENTIETH CENTURY-FOX. (R) January, 1960. Edmund O'Brien, Julie London.

THIRTEEN FIGHTING MEN (Cs) ... 71
TWENTIETH CENTURY-FOX. (R) April, 1960. Grant Williams, Brad Dexter.

13 FRIGHTENED GIRLS ... 89
COLUMBIA. (R) July, 1963. Kathy Dunn, Murray Hamilton. Color.

13 GHOSTS ... 88
COLUMBIA. (R) July, 1960. Charles Herbert, Jo Morrow. Color.

13 WEST STREET ... 80
COLUMBIA. (R) May, 1962. Alan Ladd, Rod Steiger.

—30— ... 96
WARNER BROS. (R) November, 1959. Jack Webb, William Conrad.

30 FOOT BRIDE OF CANDY ROCK, THE ... 73
COLUMBIA. (R) August, 1959. Lou Costello, Dorothy Provine.

30 IS A DANGEROUS AGE, CYNTHIA ... 85
COLUMBIA. (R) March, 1968. Dudley Moore, Suzy Kendall. Color.

30 YEARS OF FUN ... 85
TWENTIETH CENTURY-FOX. (R) February, 1963. Film Clips.

39 STEPS, THE (Cs) ... 95
TWENTIETH CENTURY-FOX. (R) July, 1960. Kenneth More, Taina Elg. Color.

36 HOURS (P) ... 115
METRO-GOLDWYN-MAYER. (R) January, 1965. James Garner, Eva Marie Saint.

THIS ANGRY AGE (R) ... 100
COLUMBIA. (R) May, 1958. Anthony Perkins, Silvano Mangano. Color.

THIS BEAST MUST DIE! ... 93
CINERAMA. (R) April, 1974. Calvin Lockhart, Peter Cushing.

THIS COULD BE THE NIGHT (Cs) ... 105
METRO-GOLDWYN-MAYER (R) May 17, 1957. Jean Simmons, Paul Douglas, Anthony Franciosa.

THIS EARTH IS MINE (Cs) ... 125
UNIVERSAL. (R) July, 1959. Rock Hudson, Jean Simmons. Color.

THIS HAPPY FEELING (Cs) ... 92
UNIVERSAL. (R) June, 1958 Debbie Reynolds, Curt Jurgens. Color.

THIS IS RUSSIA ... 67
UNIVERSAL. (R) January, 1958. Documentary. Color.

THIS ISLAND EARTH ... 87
UNIVERSAL. (R) June, 1955. Bart Roberts, Faith Domergue. Color.

THIS MAN MUST DIE (French; Eng. titles) ... 115
ALLIED ARTISTS. (R) August, 1970. Michel Duchaussoy, Caroline Cellier. Color.

THIS PROPERTY IS CONDEMNED ... 110
PARAMOUNT. (R) Summer, 1966. Natalie Wood, Robert Redford. Color.

THIS REBEL BREED ... 90
WARNER BROS. (R) March, 1960. Rita Moreno, Mark Damon.

THIS SPECIAL FRIENDSHIP (Fr.) ... 99
PATHE CONTEMPORARY. (R) November, 1967. Lacombrade, Didier Haudepin.

THIS SPORTING LIFE ... 129
CONTINENTAL. (R) July, 1963. Richard Harris, Rachel Roberts.

THIS STRANGE PASSION (Mex.) ... 80
OMNIFILMS. (R) December, 1955. Arturo de Cordova.

THOMAS CROWN AFFAIR, THE ... 102
UNITED ARTISTS. (R) August, 1968. Steve McQueen, Faye Dunaway, Paul Burke. Color.

THOMASINE AND BUSHROD ... 95
COLUMBIA. (R) April, 1974. Max Julien, Venetta McGee.

THOROUGHLY MODERN MILLIE ... 138
UNIVERSAL. (R) April, 1967. Julie Andrews, Mary Tyler Moore. Color.

THOSE CALLOWAYS ... 131
BUENA VISTA. (R) November, 1964. Brian Keith, Vera Miles. Color.

THOSE DARING YOUNG MEN IN THEIR JAUNTY JALOPIES (P) ... 125
PARAMOUNT. (R) August, 1969. Peter Cook, Tony Curtis, Bourvil. Color.

THOSE FANTASTIC FLYING FOOLS (P) ... 92
AMERICAN INTERNATIONAL. (R) June, 1967. Burl Ives, Troy Donahue. Color.

THOSE LIPS, THOSE EYES (P) ... 106
UNITED ARTISTS. (R) August, 1980. Frank Langella.

THOSE MAGNIFICENT MEN IN THEIR FLYING MACHINES (Todd-AO) ... 152
TWENTIETH CENTURY-FOX. (R) June, 1956. Special. Stuart Whitman, Sarah Miles. Color.

THOUSAND CLOWNS, A ... 118
UNITED ARTISTS. (R) April, 1966. Jason Robards, Barbara Harris.

1,000 PLANE RAID, THE ... 94
UNITED ARTISTS. October, 1969. Christopher George, J.D. Cannon. Color.

THREAT, THE ... 66
WARNER BROS. (R) March, 1960. Robert Knapp, Linda Lawson.

THREE ... 104
UNITED ARTISTS. (R) December, 1969. Charlotte Rampling, Robie Porter, Sam Waterston. Color.

THREE BAD SISTERS ... 75
UNITED ARTISTS. (R) January, 1956. Maria English, Kathleen Hughes, Sara Shane, John Bromfield.

THREE BITES OF THE APPLE (P) ... 105
METRO-GOLDWYN-MAYER. (R) February, 1967. David McCallum, Sylva Koscina. Color.

THREE BRAVE MEN (Cs) ... 88
TWENTIETH CENTURY-FOX. (R) January, 1957. Ernest Borgnine, Ray Milland.

THREE CAME TO KILL ... 70
UNITED ARTISTS. (R) April, 1960. Cameron Mitchell, John Lupton.

THREE CASES OF MURDER (Brit.) ... 90
ASSOCIATE ARTISTS. (R) March, 1955. Orson Welles, Alan Badel.

THREE DAY PASS ... 103
SIGMA III. (R) May, 1968. Harry Barid, Nicole Berger.

THREE DAYS OF THE CONDOR ... 118
PARAMOUNT. (R) September, 1975. Robert Redford, Faye Dunaway.

THREE FACES OF EVE, THE (Cs) ... 91
TWENTIETH CENTURY-FOX. (R) October, 1957. Joanne Woodward, David Wayne.

THREE FOR JAMIE DAWN ... 81
ALLIED ARTISTS. (R) July 8, 1956. Laraine Day, Ricardo Montalban.

THREE FOR THE SHOW (Cs) ... 93
COLUMBIA. (R) April, 1955. Betty Grable, Marge and Gower Champion. Color.

THREE GIRLS DOWN ... 87
NEW WORLD PICTURES. (R) April, 1973. Robert Collins, Isabel Jewel. Color.

THREE GUNS FOR TEXAS ... 99
UNIVERSAL. (R) June, 1968. Neville Brand, Peter Brown. Color.

300 SPARTANS, THE (Cs) ... 114
TWENTIETH CENTURY-FOX. (R) September, 1962. Richard Egan, Sir Ralph Richardson. Color.

THREE INTO TWO WON'T GO ... 93
UNIVERSAL. (R) July, 1969. Claire Bloom, Rod Steiger, Judy Geeson. Color.

THREE LIVES OF THOMASINA, THE ... 97
BUENA VISTA. (R) December, 1963. Patrick McGoohan, Susan Hampshire. Color.

THREE MURDERESSES ... 90
TWENTIETH CENTURY-FOX. (R) February, 1960. Alain Delon, Mylene Demongeot

THREE MUSKETEERS, THE (P) ... 105
TWENTIETH CENTURY-FOX. (R) June, 1974. Oliver Reed, Raquel Welch, Richard Chamberlain.

THREE ON A COUCH ... 109

Length In Mins.

COLUMBIA. (R) Summer, 1966. Jerry Lewis, Janet Leigh. Color.

THREE ON A SPREE ... 83
UNITED ARTISTS. (R) September, 1961. Jack Watling, Carole Lesley.

THREE OUTLAWS, THE ... 74
ASSOCIATED FILM. (R) August, 1956. Neville Brand, Bruce Bennett.

THREE RING CIRCUS (VV) ... 104
PARAMOUNT. (R) January, 1955. Dean Martin, Jerry Lewis. Color.

THREE SISTERS, THE (Russian; Eng. titles) ... 112
BRANDON. (R) January, 1969. Lyubov Sokolva, Margarita Volodina, Tatyana Malchenko. Color.

THREE SISTERS, THE ... 165
AFT DISTRIBUTING. (R) Special, 1974. Laurence Olivier, Joan Plowright.

THREE STOOGES GO AROUND THE WORLD IN A DAZE ... 94
COLUMBIA. (R) September, 1963. The Three Stooges.

THREE STOOGES IN ORBIT, THE ... 87
COLUMBIA. (R) July, 1962. The Three Stooges.

THREE STOOGES MEET HERCULES, THE ... 89
COLUMBIA. (R) January, 1962. The Three Stooges.

THREE STRIPES IN THE SUN ... 93
COLUMBIA. (R) November, 1955. Aldo Ray, Phil Carey.

3:10 TO YUMA ... 92
COLUMBIA. (R) September, 1957. Glenn Ford, Van Heflin.

THREE THE HARD WAY ... 93
ALLIED ARTISTS. (R) June, 1974. Jim Brown, Fred Williamson, Jim Kelly.

THREE TOUGH GUYS ... 92
PARAMOUNT. (R) March, 1974. Lino Ventua, Isaac Hayes.

THREE VIOLENT PEOPLE (VV) ... 100
PARAMOUNT. (R) January, 1957. Charlton Heston, Anne Baxter. Color.

THREE WARRIORS ... 110
UNITED ARTISTS. (R) February, 1978. Charles White Eagle, McKee "Kiko" Redwing.

3 WOMEN (P) ... 126
TWENTIETH CENTURY-FOX. (R) April, 1977. Shelly Duvall, Sissy Spacek, Janice Rule.

THREE WORLDS OF GULLIVER, THE (Superdynamation) ... 100
COLUMBIA. (R) December, 1960. Kerwin Mathews, Jo Morrow. Color.

THRILL OF IT ALL, THE ... 108
UNIVERSAL. (R) August, 1963. Doris Day, James Garner. Color.

THUMB TRIPPING ... 84
AVCO-EMBASSY. (R) October, 1972. Michael Burns, Meg Foster. Color.

THUNDER ALLEY (P) ... 90
AMERICAN INTERNATIONAL. (R) March, 1967. Fabian, Annette Funicello. Color.

THUNDER IN CAROLINA ... 92
HOWCO INTERNATIONAL. (R) July, 1960. Rory Calhoun, Alan Hale. Color.

THUNDER IN THE SUN ... 81
PARAMOUNT. (R) May, 1959. Susan Hayward, Jeff Chandler. Color.

THUNDER ISLAND (Cs) ... 65
TWENTIETH CENTURY-FOX. (R) November, 1963. Gene Nelson, Fay Spain.

THUNDER OF DRUMS, A (CS) ... 97
METRO-GOLDWYN-MAYER. (R) August, 1961. Color.

THUNDER OVER ARIZONA ... 75
REPUBLIC. (R) August 4, 1956. Skip Homeier, Kristine Miller, George Macready. Color.

THUNDER OVER SANGOLAND ... 73
LIPPERT. (R) April 8, 1955. Jon Hall, Marjorie Lord.

THUNDER ROAD ... 92
UNITED ARTISTS. (R) May, 1958. Robert Mitchum, Gene Barry.

THUNDERBALL (P) ... 132
UNITED ARTISTS. (R) December, 1965. Sean Connery, Claudine Auger. Color.

THUNDERBIRDS ARE GO ... 92
UNITED ARTISTS. (R) June, 1968. Puppet Fantasy. Color.

THUNDERBOLT AND LIGHTFOOT (P) ... 115
UNITED ARTISTS. (R) May, 1974. Clint Eastwood, Jeff Bridges.

THUNDERING JETS (Rs) ... 73
TWENTIETH CENTURY-FOX. (R) May, 1958. Audrey Dalton, Rex Reason.

THUNDERSTORM ... 61
ALLIED ARTISTS. (R) May 6, 1956. Linda Christian, Carlos Thompson.

TIAJUANA STORY, THE ... 72
COLUMBIA. (R) October, 1957. Rodolfo Acosta, James Darren.

TIARA TAHITI ... 100
ZENITH. (R) November, 1963. James Mason, John Mills.

. . . TICK . . . TICK . . . TICK (P) ... 98
MGM. (R) January, 1970. Jim Brown, George Kennedy, Frederic March. Color.

TICKLE ME (P) ... 90
ALLIED ARTISTS. (R) June, 1965. Elvis Presley, Jocelyn Lane. Color.

TICKLISH AFFAIR, A (P) ... 88
METRO-GOLDWYN-MAYER. (R) July, 1963. Shirley Jones, Gig Young. Color.

TIGER AND THE FLAME, THE ... 97
UNITED ARTISTS. (R) May, 1955. Sohrab Modi, Mehtab. Color.

TIGER AND THE PUSSYCAT, THE ... 100
EMBASSY. (R) October, 1967. Vittorio Gassman, Ann-Margret. Color.

TIGER BY THE TAIL ... 99
COMMONWEALTH UNITED. (R) January, 1970. Christopher George, Tippi Hedren. Color.

TIGER MAKES OUT ... 94
COLUMBIA. (R) October, 1967. Eli Wallach, Anne Jackson. Color.

TIGER WALKS, A ... 91
BUENA VISTA. (R) March, 1964. Brian Keith, Vera Miles. Color.

TIGHT SPOT ... 97
COLUMBIA. (R) May, 1965. Ginger Rogers, Edward G. Robinson.

TIKO AND THE SHARK ... 88
METRO-GOLDWYN-MAYER. (R) Summer, 1966. Marlene-Among, Al Kauwe. Color.

TIMBERJACK ... 94
REPUBLIC. (R) February 28, 1955. Sterling Hayden, Vera Ralston. Color.

TIMBUKTU ... 81
UNITED ARTISTS. (R) October, 1959. Victor Mature, Yvonne De Carlo.

TIME AFTER TIME (P) ... 112
WARNER BROS. (R) October, 1979. Malcolm McDowell, David Warner.

TIME BOMB ... 92
ALLIED ARTISTS. (R) April, 1961. Curt Jurgens, Mylene Demongeot.

TIME FOR KILLING, A (P) ... 88
COLUMBIA. (R) October, 1967. Glenn Ford, Inger Stevens. Color.

TIME IS MY ENEMY (Brit.) ... 64
REPUBLIC. (R) May 24, 1957. Dennis Price, Renee Asherson.

TIME LIMIT ... 95
UNITED ARTISTS. (R) October, 1957. Richard Widmark, Richard Basehart.

TIME LOST AND TIME REMEMBERED ... 91
CONTINENTAL. (R) August, 1966. Sarah Miles, Cyril Cusack.

TIME MACHINE, THE ... 103
METRO-GOLDWYN-MAYER. (R) August, 1960. Rod Taylor, Alan Young. Color.

TIME OF INDIFFERENCE ... 84
CONTINENTAL. (R) November, 1965. Rod Steiger, Claudia Cardinale.

TIME TO LOVE, A (Cs) ... 133
UNIVERSAL. (R) July, 1958. John Gavin, Lilo Pulver. Color.

TIME TO SING, A ... 92
METRO-GOLDWYN-MAYER. (R) September, 1968. Hank Williams, Jr., Shelley Fabares, Ed Begley. Color.

TIME TRAVELERS, THE ... 82
AMERICAN INTERNATIONAL. (R) October, 1964. Preston Foster, Phillip Carey. Color.

TIMETABLE ... 79
UNITED ARTISTS. (R) April, 1956. Mark Stevens, King Calder.

TIN DRUM, THE (Ger., Eng. titles) ... 142
NEW WORLD PICTURES. (R) April, 1980. David Bennent, Mario Adorf.

TIN STAR, THE (VV) ... 93
PARAMOUNT (R) November, 1957. Henry Fonda, Anthony Perkins.

TINGLER, THE ... 80
COLUMBIA. (R) October, 1959. Vincent Price, Judith Evelyn.

TIP ON A DEAD JOCKEY (Cs) ... 99
METRO-GOLDWYN-MAYER. (R) August, 1957. Robert Taylor, Dorothy Malone.

TIS PITY SHE'S A WHORE ... 91
INTERNATIONAL CO-PRODUCTIONS. (R) April, 1974. Charlotte Rampling, Oliver Tobias.

TO BE A CROOK (Fr.; Eng. titles) ... 93

Length In Mins.

COMET. (R) February, 1967. Jean-Pierre Kalfon, Amidou, Janine Magnan.

TO BED . . . OR NOT TO BED (Ital., Eng. titles) 100
READE-STERLING. (R) December, 1963. Alberta Sordi, Barbara Wastenson.

TO CATCH A THIEF (VV) 97
PARAMOUNT. (R) September, 1955. Cary Grant, Grace Kelly. Color.

TO COMMIT A MURDER 91
CINERAMA. (R) January, 1970. Louis Jourdan, Senta Berger, Edmund O'Brien. Color.

TO DIE IN MADRID 85
ALTURA FILMS INTERNATIONAL. (R) September, 1965. Documentary.

TO DIE OF LOVE 110
METRO-GOLDWYN-MAYER. (R) February, 1972. Annie Girardot, Bruno Pradal. Color.

TO FIND A MAN 90
COLUMBIA. (R) January, 1972. Pamela Sue Martin, Darren O'Connor. Color.

TO HELL AND BACK (Cs) 100
UNIVERSAL. (R) October, 1955. Audie Murphy, Marshall Thompson. Color.

TO KILL A CLOWN 72
TWENTIETH CENTURY-FOX. (R) August, 1972. Alan Alda, Blythe Danner. Color.

TO KILL A MOCKINGBIRD 129
UNIVERSAL. (R) March, 1963. Gregory Peck, Mary Badham.

TO LOVE (Swed., Eng. titles) 88
PROMINENT. (R) November, 1964. Harriet Anderson, Zbigniew Cybulski.

TO PARIS WITH LOVE (Brit.) 78
CONTINENTAL. (R) April, 1955. Alec Guinness, Odile Versolfs. Color.

TO SIR WITH LOVE 105
COLUMBIA. (R) June, 1967. Sidney Poitier, Judy Geeson. Color.

TO TRAP A SPY 92
METRO-GOLDWYN-MAYER (R) February, 1966. Robert Vaughn, David McCallum. Color.

TOBRUK (T) 110
UNIVERSAL. (R) February, 1967. Rock Hudson, George Peppard. Color.

TOBY TYLER 96
BUENA VISTA. (R) February, 1960. Kevin Corcoran, Henry Calvin. Color.

TODD KILLINGS, THE 93
NATIONAL GENERAL. (R) August, 1971. Robert F. Lyons, Richard Thomas. Color.

TOKYO AFTER DARK 80
PARAMOUNT. (R) January, 1959. Richard Long, Michi Kobi.

TOKYO OLYMPIAD (Cs) 93
JACK DOUGLAS ASSOCIATES. (R) January, 1966. Documentary. Color.

TOM HORN (P) 98
WARNER BROS. (R) March, 1980. Steve McQueen, Linda Evans.

TOM JONES 131
LOPERT. (R) October, 1963. Albert Finney, Susannah York. Color.

TOM SAWYER 100
UNITED ARTISTS. (R) June, 1973. Johnny Whitaker, Celeste Holm. Color.

TOM THUMB 98
METRO-GOLDWYN-MAYER. (R) December, 1958. Russ Tamblyn, Alan Young. Color.

TOMAHAWK TRAIL 60
UNITED ARTISTS. (R) February, 1957. Chuck Connors, Susan Cummings.

TOMB OF LIGEIA, THE 81
AMERICAN INTERNATIONAL. (R) January, 1965. Vincent Price, Elizabeth Shepherd. Color.

TOMBOY AND THE CHAMP 92
UNIVERSAL. (R) Apr., 1961. Candy Moore, Ben Johnson. Color.

TOMMY 111
COLUMBIA. (R) March, 1975. Ann-Margret, Oliver Reed.

TOMORROW AT TEN 80
GOVERNOR. (R) April, 1964. John Gregson, Robert Shaw.

TONKA 97
BUENA VISTA. (R) December, 1958. Sal Mineo, Jerome Courtland. Color.

TONY ROME (P) 109
TWENTIETH CENTURY-FOX. (R) November, 1967. Frank Sinatra, Jill St. John. Color.

TOO BAD SHE'S BAD (Ital.) 95
KINGSLEY. (R) December 1955. Sophia Loren, Vittorio De Sica.

TOO LATE BLUES 100
PARAMOUNT. (R) January, 1962. Bobby Darin, Stella Stevens.

TOO LATE THE HERO (P) 133
CINERAMA. (R) July, 1970. Michael Caine, Cliff Robertson. Color.

TOO MUCH TOO SOON 121
WARNER BROTHERS. (R) May, 1958. Dorothy Malone, Errol Flynn.

TOO SOON TO LOVE 85
UNIVERSAL. (R) March, 1960. Jennifer West, Richard Evans.

TOO YOUNG FOR LOVE (Ital.-Eng. dubbed) 88
I.F.E. (R) April, 1955. Marina Vlady, Pierre Michel Beck.

TOP GUN 73
UNITED ARTISTS. (R) December, 1955. Sterling Hayden, William Bishop.

TOP OF THE WORLD 90
UNITED ARTISTS. (R May, 1955. Dale Robertson, Frank Lovejoy.

TOP SECRET AFFAIR 100
WARNER BROTHERS. (R) February 9, 1957. Kirk Douglas, Susan Hayward.

TOPAZ 125
UNIVERSAL. (R) December, 1969. John Forsythe, Michel Piccoli, Frederick Stafford. Color.

TOPKAPI 120
UNITED ARTISTS. (R) October, 1964. Melina Mercouri, Maximillian Schell. Color.

TORA! TORA! TORA! (P) 143
20TH CENTURY-FOX. (R) October, 1970. Jason Robards, Martin Balsam, E.G. Marshall. Color.

TORERO! 75
COLUMBIA. (R) November, 1957. Louis Precuna.

TORMENTED 75
ALLIED ARTISTS. (R) November 1960. Richard Carlson, Susan Gordon.

TORN CURTAIN 128
UNIVERSAL. (R) August, 1966. Paul Newman, Julie Andrews. Color.

TORPEDO RUN (Cs) 98
METRO-GOLDWYN-MAYER. (R) October, 1958. Glenn Ford, Ernest Borgnine.

TORTURE GARDEN 93
COLUMBIA. (R) March, 1968. Jack Palance, Burgess Meredith. Color.

TOUCH, THE 113
CINERAMA. (R) July, 1971. Elliott Gould, Bibi Anderson, Max von Sydow. Color.

TOUCH AND GO (Brit.) 85
UNIVERSAL. (R) April, 1956. Jack Hawkins, Margaret Johnston. Color.

TOUCH AND GO (French; Eng. titles) 110
LIBRA FILMS. (R) July, 1975. Marlene Jobert, Michael York.

TOUCH OF CLASS, A (P) 105
AVCO-EMBASSY. (R) June, 1973. Glenda Jackson, George Segal. Color.

TOUCH OF EVIL 95
UNIVERSAL. (R) February, 1960. Orson Welles, Charlton Heston. b&w.

TOUCH OF HELL, A 87
GOVERNOR. (R) February, 1964. Anthony Quayle, Sarah Churchill.

TOUCH OF LARCENY, A 93
PARAMOUNT. (R) January, 1960. James Mason, George Sanders.

TOUCHABLES, THE 97
TWENTIETH CENTURY-FOX (R) November, 1968. Judy Huxtable, Esther Anderson. Color.

TOUCHED BY LOVE 95
COLUMBIA. (R) April, 1980. Deborah Raffin, Diane Lane.

TOUGHEST GUN IN TOMSTONE 72
UNITED ARTISTS. (R) May, 1958. George Montgomery, Beverly Tyler.

TOUGHEST MAN ALIVE 72
ALLIED ARTISTS. (R) November 6, 1955. Dane Clark, Lita Milan.

TOWARD THE UNKNOWN 115
WARNER BROTHERS. (R) October 20, 1956. William Holden, Lloyd Nolan. Color.

TOWER OF LONDON 79
UNITED ARTISTS (R) 1962. Vincent Price, Joan Freeman.

TOWERING INFERNO, THE (P) 165
20TH CENTURY-FOX. (R) December, 1974. Paul Newman, Steve McQueen.

TOWN OF TRIAL 96
COLUMBIA. (R) August, 1957. Charles Coburn, Barbara Bates.

TOWN TAMER (T) 89

	Length In Mins.
PARAMOUNT. (R) November, 1965. Dana Andrews, Terry Moore. Color.	
TOWN THAT DREADED SUNDOWN, THE AMERICAN INTERNATIONAL. (R) February, 1977. Ben Johnson, Andrew Prine.	90
TOWN WITHOUT PITY UNITED ARTISTS. (R) October, 1961. Kirk Douglas, E. G. Marshall.	105
TOY TIGER UNIVERSAL. (R) July, 1956. Jeff Chandler, Laraine Day, Tim Hovey. Color.	88
TOYS IN THE ATTIC (P) UNITED ARTISTS. (R) August, 1963. Dean Martin, Geraldine Page.	90
TRACK OF THUNDER (TS) UNITED ARTISTS. (R) February, 1969. Tom Kirk, Ray Stricklyn, Faith Domergue. Color.	83
TRACK THE MAN DOWN (Brit.) REPUBLIC. (R) January 27, 1956. Kent Taylor, Petula Clark.	75
TRACKDOWN UNITED ARTISTS. (R) April, 1976. Jim Mitchum, Karen Lamm.	98
TRADER HORN METRO-GOLDWYN-MAYER. (R) June, 1973. Rod Taylor, Anne Heywood. Color.	106
TRAFFIC COLUMBIA. (R) January, 1973. Jacques Tati. Color.	89
TRAIN, THE UNITED ARTISTS. (R) April, 1965. Burt Lancaster, Paul Scofield.	133
TRAIN ROBBERS, THE (P) WARNER BROS. (R) February, 1972. John Wayne, Ann-Margret. Color.	92
TRAITORS, THE UNIVERSAL. (R) August, 1963. Patrick Allen, James Maxwell.	71
TRAITORS' GATE (Brit.) COLUMBIA. (R) 1966. Gary Raymond, Albert Lieven.	80
TRAMPLERS, THE EMBASSY. (R) July, 1966. Joseph Cotten, Gordon Scott. Color.	105
TRAP, THE (VV) PARAMOUNT (R) February, 1959. Richard Widmark, Lee J. Cobb. Color.	84
TRAPEZE (Cs) UNITED ARTISTS. (R) July, 1956. Burt Lancaster, Gina Lollobrigida, Tony Curtis. Color.	105
TRAPP FAMILY, THE TWENTIETH CENTURY-FOX. (R) March, 1961. Ruth Leuwerick, Hans Holt. Color.	106
TRAPPED IN TANGIERS (Cs) TWENTIETH CENTURY-FOX. July, 1960. Edmund Purdom, Genevieve Page.	74
TRAVELING EXECUTIONER, THE MGM. (R) November, 1970. Stacy Keach, Mariana Hill, Bud Cort. Color.	94
TRAVELS WITH MY AUNT METRO-GOLDWYN-MAYER. (R) December, 1972. Maggie Smith, Alec McGowen. Color.	109
TREAD SOFTLY STRANGER BENTLEY FILMS. (R) September, 1959. Diana Dors, George Baker.	85
TREASURE ISLAND NATIONAL GENERAL. (R) October, 1972. Orson Welles, Kim Burfield. Color.	93
TREASURE OF MATECUMBE BUENA VISTA. (R) July, 1976. Robert Foxworth, Joan Hackett, Peter Ustinov.	117
TREASURE OF PANCHO VILLA, THE (SS) RKO RADIO. (R) October 19, 1955. Shelley Winters, Rory Calhoun. Color.	96
TREASURE OF RUBY HILLS, THE ALLIED ARTISTS. (R) January 16, 1955. Zachary Scott, Carole Matthews.	71
TREASURE OF SAN GENNARO (Ital.) PARAMOUNT. (R) March, 1968. Nino Manfredi, Senta Berger. Color.	102
TREASURE OF SILVER LAKE, THE (Cs) COLUMBIA. (R) August, 1965. Lex Barker, Gotz George. Color.	82
TRIAL METRO-GOLDWYN-MAYER. (R) October 7, 1955. Glenn Ford, Dorothy McGuire, Arthur Kennedy.	105
TRIAL AND ERROR METRO-GOLDWYN-MAYER. (R) November, 1962. Peter Sellers, Richard Attenborough.	110
TRIAL OF BILLY JACK, THE (P) TAYLOR-LAUGHLIN. (R) November, 1974. Tom Laughlin, Delores Taylor.	170
TRIBUTE TO A BAD MAN (Cs) METRO-GOLDWYN-MAYER. (R) April 13, 1956. James Cagney, Stephen McNally. Color.	95
TRILOGY ALLIED ARTISTS. (R) November, 1969. Mildred Natwick, Maureen Stapleton, Martin Balsam. Color.	100
TRIO INFERNAL, LE (French; Eng. titles) LEVITT-PICKMAN. (R) November, 1974. Romy Schneider, Michel Piccoli.	106
TRIP, THE AMERICAN INTERNATIONAL. (R) August, 1967. Peter Fonda, Susan Strasberg. Color.	85
TRIPLE CROSS WARNER BROS.-SEVEN ARTS. (R) May, 1967. Christopher Plummer, Romy Schneider. Color.	126
TRIPLE DECEPTION (VV) RANK. (R) July, 1957. Michael Craig, Julia Arnal. Color.	86
TRIPLE ECHO ALTURA FILMS. (R) November, 1973. Glenda Jackson, Oliver Reed.	90
TROG WARNER BROS. (R) September, 1970. Joan Crawford, Michael Gough. Color.	91
TROJAN WOMEN, THE CINERAMA. (R) November, 1971. Katharine Hepburn, Vanessa Redgrave. Color.	105
TROOPER HOOK UNITED ARTISTS. (R) June, 1957. Joel McCrea, Barbara Stanwyck.	81
TROPIC OF CANCER PARAMOUNT. (R) March, 1970. Rip Torn, James Callahan. Color.	87
TROUBLE IN STORE (Brit.) REPUBLIC. (R) January 12, 1955. Norman Wisdom, Margaret Rutherford.	85
TROUBLE IN THE SKY UNIVERSAL. (R) July, 1961. Michael Craig, Peter Cushing.	76
TROUBLE WITH ANGELS, THE COLUMBIA. (R) April 6, 1966. Rosaline Russell, Binnie Barnes. Color.	112
TROUBLE WITH GIRLS, THE METRO-GOLDWYN-MAYER. (R) May, 1969. Elvis Presley, Marilyn Mason, Dabney Coleman. Color.	104
TROUBLE WITH HARRY, THE (VV) PARAMOUNT. (R) January, 1956. John Forsythe, Edmund Gwenn. Color.	99
TROUBLEMAKER, THE JANUS. (R) August, 1964. Thomas Aldredge, Joan Darling.	80
TROUBLEMAN TWENTIETH CENTURY-FOX. (R) November, 1972. Robert Hooks, Paul Winfield. Color.	99
TRUCK TURNER AMERICAN INTERNATIONAL. (R) June, 1974. Isaac Hayes, Yaphet Kotto.	91
TRUE GRIT PARAMOUNT. (R) July, 1969. John Wayne, Glen Campbell, Kim Darby. Color.	128
TRUE STORY OF JESSE JAMES, THE (Cs) TWENTIETH CENTURY-FOX. (R) February, 1957. Robert Wagner, Jeffrey Hunter. Color.	93
TRUE STORY OF LYNN STUART COLUMBIA. (R) March, 1958. Betsy Palmer, Jack Lord.	78
TRUNK, THE COLUMBIA. (R) October, 1961. Phil Carey, Julia Arnall.	72
TRUNK TO CAIRO AMERICAN INTERNATIONAL. (R) January, 1967. Audie Murphy, George Sanders. Color.	80
TRUTH ABOUT SPRING, THE UNIVERSAL. (R) April, 1965. Hayley Mills, John Mills. Color.	102
TRYGON FACTOR, THE WARNER BROS.-7 ARTS. (R) January, 1969. Stewart Granger, Susan Hampshire, Robert Morley. Color.	90
TSAR TO LENIN HERMAN AXELBANK. (R) January, 1966. Documentary.	86
TSAR'S BRIDE, THE ARTKINO. (R) April, 1966. Raissa Medashkovskaya, Natalya Rudava.	95
TUNES OF GLORY UNITED ARTISTS. (R) December, 1960. Alec Guinness, John Mills. Color.	106
TURNING POINT, THE 20th CENTURY-FOX. (R) November, 1977. Anne Bancroft, Shirley MacLaine.	118
12 ANGRY MEN UNITED ARTISTS. (R) April, 1957. Henry Fonda, Lee J. Cobb.	95
TWELVE CHAIRS, THE UMC PICTURES. (R) December, 1970. Ron Moody, Frank Langella, Dom DeLuise. Color.	94
TWELVE HOURS TO KILL (Cs) TWENTIETH CENTURY-FOX. (R) June, 1960. Niko Minardos, Barbara Eden.	88

Length In Mins.

12 TO THE MOON **74**
COLUMBIA. (R) June, 1960. Ken Clark, Michi Kobi.
28 PACES TO BAKER STREET (Cs) **103**
TWENTIETH CENTURY-FOX. (R) May, 1956. Van Johnson, Vera Miles. Color.
25th HOUR, THE (F) **134**
METRO-GOLDWYN-MAYER. (R) February, 1967. Anthony Quinn, Virna Lisi. Color.
TWENTY MILLION TO EARTH **82**
COLUMBIA. (R) July, 1957. William Hopper, Joan Taylor.
TWENTY PLUS TWO **102**
ALLIED ARTISTS. (R) October, 1961. David Janssen, Jeanne Crain.
27th DAY, THE **75**
COLUMBIA. (R) July, 1957. Valerie French, Gene Barry.
20,000 EYES (Cs) **60**
TWENTIETH CENTURY-FOX. (R) July, 1962. b&w.
TWICE TOLD TALES **119**
UNITED ARTISTS. (R) October, 1963. Vincent Price, Sebastian Cabot. Color.
TWILIGHT FOR THE GODS **120**
UNIVERSAL. (R) August, 1958. Rock Hudson, Cyd Charisse. Color.
TWILIGHT OF HONOR (P) **115**
METRO-GOLDWYN-MAYER. (R) October, 1963. Richard Chamberlain, Nick Adams.
TWILIGHT'S LAST GLEAMING **146**
ALLIED ARTISTS. (R) February, 1977. Burt Lancaster, Richard Widmark.
TWINKLE IN GOD'S EYE, THE **73**
REPUBLIC. (R) October 13, 1955. Mickey Rooney, Coleen Gray.
TWIST ALL NIGHT **85**
AMERICAN INTERNATIONAL. (R) March, 1962. Louis Prima, June Wilkinson
TWIST AROUND THE CLOCK **82**
COLUMBIA. (R) December, 1961. Chubby Checker, Dion.
TWIST OF SAND, A **90**
UNITED ARTISTS. (R) September, 1968. Richard Johnson, Honor Blackman, Jeremy Kemp. Color.
TWISTED NERVE **116**
NATIONAL GENERAL. (R) February, 1969. Hayley Mills, Hywell Bennet, Billie Whitelaw. Color.
TWO (Danish; Eng. titles) **85**
CHEVRON. (R) February, 1969. Gio Petrie, Lars Lunde, Juordis Peterson. Color.
TWO ARE GUILTY (Franscope) **117**
METRO-GOLDWYN-MAYER. (R) March, 1964. Anthony Perkins, Jean-Claude Brialy.
TWO BEFORE ZERO **78**
ELLIS. (R) April, 1963. Documentary.
TWO ENGLISH GIRLS **108**
JANUS FILMS. (R) October, 1972. Jean-Pierre Leaud, Kika Markham. Color.
TWO FOR THE ROAD (P) **112**
TWENTIETH CENTURY-FOX. (R) July, 1967. Albert Finney, Audrey Hepburn. Color.
TWO FOR THE SEESAW (P) **120**
UNITED ARTISTS. (R) December, 1962. Robert Mitchum, Shirley MacLaine.
TWO GENTLEMEN SHARING **105**
AMER. INT. PICS. (R) October, 1969. Robin Phillips, Hal Fredericks, Judy Geeson. Color.
TWO GROOMS FOR A BRIDE **73**
TWENTIETH CENTURY-FOX. (R) June, 1957. Virginia Bruce, John Carroll. Color.
TWO GUN LADY **70**
ASSOCIATED FILM. (R) October 15, 1956. Peggie Castle, William Talman, Marie Windsor.
TWO HEADED SPY, THE **93**
COLUMBIA. (R) March, 1959. Jack Hawkins, Gia Scala.
200 MOTELS **93**
UNITED ARTISTS. (R) November, 1971. Frank Zappa's Mothers, Theodore Bikel, Ringo Starr. Color.
TWO-LANE BLACKTOP **102**
UNIVERSAL. (R) July, 1971. James Taylor, Warren Oates. Color.
TWO LITTLE BEARS, THE (Cs) **81**
TWENTIETH CENTURY-FOX. (R) December, 1961. Eddie Albert, Jane Wyatt.
TWO LOVES (Cs) **100**
METRO-GOLDWYN-MAYER. (R) May, 1961. Shirley MacLaine, Laurence Harvey. Color.
TWO MEN OF KARAMJOA **102**
TOMORROW ENTERTAINMENT. (R) May, 1974. Documentary.
TWO-MINUTE WARNING (P) **115**
UNIVERSAL. (R) November, 1976. Charlton Heston, John Cassavetes.
TWO MULES FOR SISTER SARA (P) **105**
UNIVERSAL. (R) June, 1970. Shirley MacLaine, Clint Eastwood. Color.
TWO OF US, THE (FR.) **86**
CINEMA V. (R) February, 1968. Michel Simon, Alain Cohen.
TWO ON A GUILLOTINE (P) **107**
WARNER BROS. (R) February, 1965. Connie Stevens, Dean Jones.
TWO PEOPLE **100**
UNIVERSAL. (R) March, 1973. Peter Fonda, Lindsay Wagner. Color.
TWO RODE TOGETHER **109**
COLUMBIA. (R) July, 1961. James Stewart, Richard Widmark. Color.
2001: A SPACE ODYSSEY (Super Panavision) **160**
MGM. (R) Special. April, 1968. Keir Dullea, Gary Lockwood. Color.
2000 YEARS LATER **80**
WARNER BROS.-7 ARTS. (R) April, 1969. Terry-Thomas, Edward Everrett Horton, Pat Harrington. Color.
TWO TICKETS TO PARIS **78**
COLUMBIA. (R) October, 1962. Joey Dee, Gary Crosby.
TWO WEEKS IN ANOTHER TOWN (Cs) **107**
METRO-GOLDWYN-MAYER. (R) August, 1962. Kirk Douglas, Edward G. Robinson. Color.
TWO WEEKS IN SEPTEMBER (F) **96**
PARAMOUNT. (R) September, 1967. Brigitte Bardot, Laurent Terzieff. Color.
TWO WOMEN (Ital.; Eng. titles) **99**
AVCO EMBASSY. (R) 1961. Sophia Loren, Jean Paul Belmondo. b&w.

U

UGLY AMERICAN, THE **120**
UNIVERSAL. (R) April, 1963. Marlon Brando, Eiji Okada. Color.
UGLY DACHSHUND, THE **93**
BUENA VISTA. (R) February, 1966. Suzanne Pleshette, Dean Jones. Color.
UGLY ONES, THE **96**
UNITED ARTISTS. (R) September, 1968. Richard Wyler, Thomas Milian, Ella Karin. Color.
ULYSSES **104**
PARAMOUNT. (R) October, 1955. Kirk Douglas, Silvana Mangano. Color.
ULYSSES (P) **123**
CONTINENTAL. (R) March, 1967. Barbara Jefford, Milo O'Shea.
ULZANA'S RAID **103**
UNIVERSAL. (R) November, 1972. Burt Lancaster. Color.
UMBERTO D (Ital.) **89**
HARRISON. (R) November, 1955. Carlo Battisi.
UMBRELLAS OF CHERBOURG **90**
ALLIED ARTISTS-LANDAU. (R) December, 1964. Catherine Deneuve Nino Castelnuovo. Color.
UNCHAINED **75**
WARNER BROTHERS. (R) February 26, 1955. Elroy Hirsch, Barbara Hale.
UNCLE, THE **87**
LENART. (R) July, 1966. Rupert Davies, Brenda Bruce.
UNCLE JOE SHANNON **108**
UNITED ARTISTS. (R) December, 1979. Burt Young, Doug McKeon.
UNDEAD, THE **75**
AMERICAN INTERNATIONAL. (R) March, 1957. Alison Hayes, Richard Garland.
UNDEFEATED, THE (P) **119**
20TH CENTURY-FOX. (R) November, 1969. John Wayne, Rock Hudson. Color.
UNDER THE YUM-YUM TREE **110**
COLUMBIA. (R) November, 1963. Jack Lemmon, Carol Lynley. Color.
UNDER TWO FLAGS **92**
PARAMOUNT. (R) September, 1960. Van Heflin, Charles Laughton.
UNDERCOVERS HERO **95**
UNITED ARTISTS. (R) August, 1975. Peter Sellers, Lila Kedrova.
UNDERGROUND **100**
UNITED ARTISTS. (R) October, 1970. Robert Goulet, Daniele Gaubert. Color.
UNDERWATER! (SS) **93**
RKO RADIO. (R) February 9, 1955. Jane Russell, Gilbert Roland. Color.
UNDERWATER CITY, THE **78**
COLUMBIA. (R) February, 1962. William Lundigan, Julie Adams.

	Length In Mins.
UNDERWATER WARRIOR (Cs) MGM. (R) 1958. Dan Daily, Claire Kelly. b&w.	90
UNDERWORLD INFORMERS (Brit.) CONTINENTAL. (R) October, 1965. Nigel Patrick, Catherine Woodville.	105
UNDERWORLD, U.S.A. COLUMBIA. (R) March, 1961. Cliff Robertson, Dolores Dorn.	99
UNEARTHLY, THE REPUBLIC. (R) June, 1957. John Carradine, Allison Hayes.	73
UNFAITHFULS, THE ALLIED ARTISTS. (R) November, 1960. Gina Lollobrigida, May Britt.	89
UNFORGIVEN, THE (P) UNITED ARTISTS. (R) April, 1960. Burt Lancaster, Audrey Hepburn. Color.	125
UNGUARDED MOMENT, THE UNIVERSAL. (R) November, 1956. Esther Williams, George Nader. Color.	95
UNHOLY ROLLERS, THE AMERICAN INTERNATIONAL. (R) October, 1972. Claudia Jennings. Color.	88
UNHOLY WIFE, THE UNIVERSAL. (R) October, 1957. Diana Dors, Rod Steiger. Color.	94
UNIDENTIFIED FLYING OBJECTS UNITED ARTISTS. (R) May, 1956. Documentary. Color.	91
UNKNOWN TERROR, THE (Cs) TWENTIETH CENTURY-FOX. (R) September, 1957. John Howard, Mala Powers.	76
UNMAN, WITTERING AND ZIGO PARAMOUNT. (R) June, 1971. David Hemmings, Carolyn Seymour. Color.	100
UNMARRIED WOMAN, AN 20th CENTURY-FOX. (R) March, 1978. Jill Clayburgh, Alan Bates.	124
UNSINKABLE MOLLY BROWN, THE (P) METRO-GOLDWYN-MAYER. (R) June, 1964. Debbie Reynolds, Harve Presnell. Color.	128
UNTAMED (Cs) TWENTIETH CENTURY-FOX. (R) March, 1955. Tyrone Power, Susan Hayward. Color.	111
UNTAMED YOUTH WARNER BROTHERS. (R) May 18, 1957. Mamie Van Doren, Lori Nelson.	80
UNTIL THEY SAIL (Cs) METRO-GOLDWYN-MAYER. (R) October, 1957. Jean Simmons, Paul Newman.	94
UNWED MOTHER ALLIED ARTISTS. (R) July 20, 1958. Norma Moore, Robert Vaughn.	74
UP FROM THE BEACH (Cs) TWENTIETH CENTURY-FOX. (R) June, 1965. Cliff Robertson, Red Buttons.	99
UP IN SMOKE ALLIED ARTISTS. (R) Dec. 1957. Huntz Hall, Stanley Clements.	64
UP IN SMOKE PARAMOUNT. (R) September, 1978. Cheech Marin, Tommy Chong.	86
UP PERISCOPE (Cs) WARNER BROS. (R) February, 1959. James Garner, Edmund O'Brien.	111
UP THE ACADEMY WARNER BROS. (R) June, 1980. Ron Liebman, Wendell Brown.	96
UP THE CREEK DOMINANT PICTURES. (R) August, 1959. David Tomlinson, Peter Sellers.	83
UP THE DOWN STAIRCASE WARNER BROS.-SEVEN ARTS. (R) July, 1967. Sandy Dennis, Patrick Bedford. Color.	124
UP THE JUNCTION PARAMOUNT. (R) March, 1968. Suzy Kendall, Dennis Waterman. Color.	119
UP THE MACGREGORS COLUMBIA. (R) November, 1967. Agatha Flory, Leo Anchoriz. Color.	93
UP THE SANDBOX NATIONAL GENERAL. (R) December, 1972. Barbara Streisand, David Selby. Color.	97
UP TO HIS EARS LOPERT. (R) May, 1966. Jean-Paul Belmondo, Ursula Andress. Color.	109
UPPER HAND, THE (F) PARAMOUNT. (R) July, 1967. Jean Gabin, George Raft. Color.	86
UPSTAIRS AND DOWNSTAIRS TWENTIETH CENTURY-FOX. (R) March, 1961. Mylene Demongeot, Michael Craig. Color.	100
UPTIGHT PARAMOUNT. (R) September, 1968. Raymond St. Jacques, Ruby Lee, Frank Sinatra. Color.	104
UPTOWN SATURDAY NIGHT WARNER BROS. (R) Summer, 1974. Sidney Poitier, Bill Cosby.	104
URANIUM BOOM COLUMBIA. (R) March, 1956. Dennis Morgan, Patricia Medina.	67
UTAH BLAINE COLUMBIA. (R) February, 1957. Rory Calhoun, Susan Cummings.	75

V

	Length In Mins.
V.I.P.'S, THE (P) METRO-GOLDWYN-MAYER. (R) September, 1963. Elizabeth Taylor, Richard Burton. Color.	119
VAGABOND KING, THE (VV) PARAMOUNT. (R) September, 1956. Kathryn Grayson, Oreste. Color.	88
VALDEZ IS COMING UNITED ARTISTS. (R) March, 1971. Burt Lancaster, Susan Clark. Color.	90
VALERIE UNITED ARTISTS. (R) August, 1957. Sterling Hayden, Anita Ekberg.	84
VALIANT, THE (Cs) UNITED ARTISTS. (R) June, 1962. John Mills, Ettore Manni.	89
VALLEY OF GWANGI WARNER BROS.-7 ARTS. (R) June, 1969. James Franciscus, Gila Golan, Richard Carlson. Color.	95
VALLEY OF MYSTERY UNIVERSAL. (R) June, 1967. Richard Egan, Peter Graves. Color.	94
VALLEY OF THE DRAGONS COLUMBIA. (R) November, 1961. Cesare Dandus, Sean McCuory.	79
VALLEY OF THE REDWOODS TWENTIETH CENTURY-FOX. (R) May, 1960. John Hudson, Lynn Bernay.	63
VALUE FOR MONEY (VV) RANK. (R) July, 1957. John Gregson, Diana Dors. Color.	84
VAMPIRE, THE UNITED ARTISTS. (R) July, 1957. b&w.	86
VAMPIRE CIRCUS TWENTIETH CENTURY-FOX. (R) October, 1972. John Moulder-Brown, Lynne Frederick. Color.	87
VANISHING AMERICAN, THE REPUBLIC. (R) November 17, 1955. Scott Brady, Audrey Totter.	90
VANISHING POINT 20TH CENTURY-FOX. (R) March, 1971. Barry Newman, Cleavon Little. Color.	107
VANISHING WILDERNESS PACIFIC INTERNATIONAL. (R) Special, 1974. Documentary.	90
VARIETY LIGHTS (Ital.; Eng. titles) PATHE-CONT. (R) May, 1965. Peppino de Filippo, Carla del Poggio.	93
VAULT OF HORROR CINERAMA. (R) March, 1973. Terry-Thomas, Glynis Johns, Daniel Massey. Color.	93
VENETIAN AFFAIR, THE (P) METRO-GOLDWYN-MAYER. (R) January, 1967. Robert Vaughn, Elke Sommer. Color.	92
VENGEANCE OF FU MANCHU WARNER BROS.-SEVEN ARTS. (R) January, 1968. Christopher Lee, Tony Ferrer.	91
VENGEANCE OF SHE, THE TWENTIETH CENTURY-FOX. (R) May, 1968. John Richardson, Olinka Berova. Color.	101
VENUS IN FURS AMER. INT. PICS. (R) May, 1970. James Darren, Barbara McNair. Color.	86
VERBOTEN RANK FILM DISTRIBUTORS OF AMERICA. (R) April, 1959. James Best, Susan Cummings.	87
VERTIGO (VV) PARAMOUNT. (R) June, 1958. James Stewart, Kim Novak. Color.	127
VERY CURIOUS GIRL, A REGIONAL. (R) December, 1971. Bernadette La Font, Georges Geret. Color.	107
VERY HANDY MAN (Ital.; Eng. titles) RIZZOLI. (R) December, 1966. Ugo Tognazzi, Giovanna Ralli.	95
VERY HAPPY ALEXANDER (French; Eng. titles) CINEMA V. (R) February, 1969. Philippe Noiret, Francoise Brion, Marlene Jobert. Color.	94
VERY NATURAL THING, A	80

Length In Mins.

NEW LINE CINEMA. (R) July, 1974. Robert Joel, Curt Gareth.

VERY PRIVATE AFFAIR, A 94
METRO-GOLDWYN-MAYER. (R) September, 1962. Brigitte Bardot, Marcello Mastroianni. Color.

VERY SPECIAL FAVOR, A 104
UNIVERSAL. (R) August, 1965. Rock Hudson, Leslie Caron.

VICE AND VIRTUE (F) 108
METRO-GOLDWYN-MAYER. (R) March, 1965. Annie Girardot, Catherine Deneuve. b&w.

VICE RAID 71
UNITED ARTISTS. Feb., 1960. Edward G. Robinson.

VICTORS, THE (P) 175
COLUMBIA. (R) December, 1963. Vincent Edwards, George Hamilton.

VIEW FROM POMPEY'S HEAD, THE (Cs) 97
TWENTIETH CENTURY-FOX. (R) November, 1955. Richard Egan, Dana Wynter, Cameron Mitchell. Color.

VIGILANTE FORCE 89
UNITED ARTISTS. (R) April, 1976. Kris Kristofferson, Jan-Michael Vincent.

VIKING QUEEN, THE 91
TWENTIETH CENTURY-FOX. (R) September, 1967. Don Murray, Carita.

VIKING WOMEN, THE 70
AMER. INT. PICS. (R) December, 1957. Abby Dalton, Susan Cabot.

VIKINGS, THE (T) 114
UNITED ARTISTS. (R) July, 1958. Kirk Douglas, Tony Curtis. Color.

VILLA! (Cs) 72
TWENTIETH CENTURY-FOX. (R) October, 1958. Brian Keith, Cesar Romero. Color.

VILLA RIDES (P) 125
PARAMOUNT. (R) June, 1968. Yul Brynner, Robert Mitchum, Grazia Buccella. Color.

VILLAGE OF THE DAMNED, THE 78
METRO-GOLDWYN-MAYER. (R) January, 1961. George Sanders, Barbara Shelley.

VILLAGE OF THE GIANTS 81
EMBASSY. (R) October, 1965. Tommy Kirk, Johnny Crawford. Color.

VILLAIN (P) 90
MGM. (R) May, 1971. Richard Burton, Nigel Davenport. Color.

VILLAIN, THE 93
COLUMBIA. (R) July, 1979. Kirk Douglas, Ann-Margret.

VINTAGE, THE (Cs) 92
METRO-GOLDWYN-MAYER. (R) April 19, 1957. Mel Ferrer, Pier Angeli, John Kerr. Color.

VIOLATORS, THE 76
UNIVERSAL. (R) December, 1957. Arthur O'Connell.

VIOLENT FOUR, THE (Italian; Eng. titles) 98
PARAMOUNT. (R) September, 1968. Gian Marie Volonte, Thomas Milan, Margaret Lee. Color.

VIOLENT MEN, THE (Cs) 96
COLUMBIA. (R) January, 1955. Glenn Ford, Barbara Stanwyck. Color.

VIOLENT ROAD 86
WARNER BROTHERS. (R) May, 1958. Brian Keith, Dick Foran.

VIOLENT SATURDAY (Cs) 90
TWENTIETH CENTURY-FOX. (R) April, 1955. Victor Mature, Sylvia Sidney. Color.

VIOLONS DU BAL, LES (French; Eng. titles) 110
LEVITT-PICKMAN. (R) December, 1974. Marie-Josee Nat, Jean-Louis Trintignant.

VIRGIN AND THE GYPSY, THE 92
CHEVRON. (R) June, 1970. Joanna Shimkus, Franco Nero, Honor Blackman. Color.

VIRGIN QUEEN, THE (Cs) 92
TWENTIETH CENTURY-FOX. (R) August, 1955. Bette Davis, Richard Todd, Joan Collins. Color.

VIRGIN SOLDIERS, THE 96
COLUMBIA. (R) January, 1970. Lynn Redgrave, Hywel Bennett, Nigel Davenport. Color.

VISCOUNT, THE (Ts) 98
WARNER BROS. (R) May, 1967. Kerwin Matthews, Edmund O'Brien. Color.

VISIONS OF EIGHT 105
CINEMA V. (R) August, 1973. Documentary.

VISIT, THE (Cs) 90
TWENTIETH CENTURY-FOX. September, 1964. Ingrid Bergman, Anthony Quinn.

VISIT TO A SMALL PLANET 85
PARAMOUNT. (R) April, 1960. Jerry Lewis, Joan Blackman.

VISITORS, THE 87
UNITED ARTISTS. (R) February, 1972. Patrick McVey, Patricia Joyce. Color.

Length In Mins.

VITELLONI (THE YOUNG AND THE PASSIONATE) (Ital.) 103
JANUS. (R) October, 1956. Franco Fabrizi, Alberto Sordi.

VIVA KNIEVEL! 106
WARNER BROS. (R) June, 1977. Evel Knievel, Gene Kelly.

VIVA LAS VEGAS (P) 86
METRO-GOLDWYN-MAYER. (R) June, 1964. Elvis Presley, Ann-Margaret. Color.

VIVA MARIA (P) 119
UNITED ARTISTS. (R) Spring, 1966. Brigitte Bardot, Jeanne Moreau. Color.

VIVA MAX! 92
COMMONWEALTH UNITED. (R) December, 1969. Peter Ustinov, Pamela Tiffen, Jonathan Winters. Color.

VOICE IN THE MIRROR (Cs) 102
UNIVERSAL. (R) August, 1958. Richard Egan, Julie London.

VOICE OF THE HURRICANE 80
SELECTED PICTURES. (R) April, 1964. Muriel Smith, David Cole. Color.

VOICES 106
UNITED ARTISTS. (R) March, 1979. Michael Ontkean, Amy Irving.

VOLCANO 100
CINEMA 5. (R) October, 1977. Documentary.

VON RICHTOFEN AND BROWN 97
UNITED ARTISTS. (R) June, 197_. John Phillip Law, Don Stroud. Color.

VON RYAN'S EXPRESS (Cs) 117
TWENTIETH CENTURY-FOX. June, 1965. Color.

VOODOO ISLAND 76
UNITED ARTISTS. (R) February, 1957. Boris Karloff, Beverly Tyler.

VOODOO WOMAN 75
AMERICAN INTERNATIONAL. (R) March, 1957. Marla English, Tom Conway.

VOYAGE OF SILENCE 88
LOPERT. (R) September, 1968. Marco Pico, Ludmila Mikael. b&w.

VOYAGE OF THE DAMNED (P) 158
AVCO-EMBASSY. (R) February, 1977. Faye Dunaway, Max Von Sydow.

VOYAGE TO THE BOTTOM OF THE SEA (Cs) 105
June, 1961. Walter Pidgeon, Joan Fontaine. Color.

VOYAGE TO THE END OF THE UNIVERSE 75
AMERICAN INTERNATIONAL. (R) September, 1964. Dennis Stephans, Francis Smolen.

VULTURE, THE 91
PARAMOUNT. (R) May, 1967. Robert Hutton, Akim Tamiroff.

W

W.C. FIELDS AND ME 111
UNIVERSAL. (R) April, 1976. Rod Steiger, Valerie Perrine.

WUSA (P) 115
PARAMOUNT. (R) October, 1970. Paul Newman, Joanne Woodward. Color.

W.W. AND THE DIXIE DANCEKINGS 91
TWENTIETH CENTURY-FOX. (R) March, 1975. Burt Reynolds.

WACKIEST SHIP IN THE ARMY, THE (Cs) 99
COLUMBIA. (R) December, 1960. Jack Lemmon, Ricky Nelson. Color.

WACO (TS) 85
PARAMOUNT (R) September, 1966. Howard Keel, Jane Russell. Color.

WAGES OF FEAR, THE (Fr.) 100
DISTRIBUTORS CORP. OF AMERICA. (R) January, 1956. Yves Montand, Vera Clouzot.

WAIT UNTIL DARK 107
WARNER BROS.-SEVEN ARTS (R) December, 1967. Audrey Hepburn, Alan Arkin. Color.

WAITING FOR CAROLINE 83
LOPERT. (R) July, 1969. Alexandria Stewart, Francoise Tasse. Color.

WAKAMBA 65
RKO RADIO. (R) June 29, 1955. African adventure. Color.

WAKE ME WHEN IT'S OVER (Cs) 126
TWENTIETH CENTURY-FOX. (R) April, 1960. Ernie Kovacs, Margo Moore. Color.

WALK A TIGHTROPE 69
PARAMOUNT. (R) June, 1964. Dan Duryea, Patricia Owens.

WALK, DON'T RUN (P) 117
COLUMBIA. (R) July, 1966. Cary Grant, Samantha Eggar. Color.

Length In Mins.

WALK IN THE SHADOW (Brit.) 93
CONTINENTAL. (R) February, 1966. Michael Craig, Patrick McGoohan.
WALK IN THE SPRING RAIN, A (P) 100
COLUMBIA. (R) May, 1970. Ingrid Bergman, Anthony Quinn. Color.
WALK LIKE A DRAGON 95
PARAMOUNT. (R) June, 1960. Jack Lord, Nobu McCarthy.
WALK ON THE WILD SIDE 113
COLUMBIA. (R) February, 1962. Laurence Harvey, Capucine.
WALK PROUD 102
UNIVERSAL. (R) May, 1979. Robby Benson, Sarah Holcomb.
WALK TALL (Cs) 60
TWENTIETH CENTURY-FOX. (R) September, 1960. Willard Parker, Joyce Meadows. Color.
WALK THE DARK STREET 74
ASSOCIATED ARTISTS. (R) April, 1956. Chuck Connors, Don Ross.
WALK THE PROUD LAND (Cs) 88
UNIVERSAL. (R) September, 1956. Audie Murphy, Pat Crowley. Color.
WALK WITH LOVE AND DEATH, A 90
20TH CENTURY-FOX. (R) October, 1969. Anjelica Huston, Assaf Dayan. Color.
WALKABOUT 95
20TH CENTURY-FOX. (R) June, 1971. Jenny Agutter, David Gumpili. Color.
WALKING STICK, THE (P) 101
MGM. (R) April, 1970. David Hemmings, Samantha Eggar, Emlyn Williams. Color.
WALKING TALL 125
CINERAMA. (R) February, 1973. Joe Don Baker, Elizabeth Hartman. Color.
WALKING TARGET 74
UNITED ARTISTS. (R) August, 1960. Joan Evans, Ronald Foster.
WALL OF NOISE 112
WARNER BROS. (R) September, 1963. Suzanne Pleshette, Ty Hardin.
WALLS OF HELL, THE 88
HEMISPHERE. (R) October, 1964. Jock Mahony, Fernando Poe, Jr.
WALTZ OF THE TOREADORS 105
CONTINENTAL. (R) August, 1962. Peter Sellers, Dany Robin. Color.
WANDA 105
BARDENE INTERNATIONAL. (R) March, 1971. Barbara Loden, Michael Higgins.
WANDA NEVADA 102
UNITED ARTISTS. (R) May, 1979. Peter Fonda, Brooke Shields.
WANDERER, THE (French, Eng. titles) 103
(R) February, 1969. Brigitte Epssey, Jean Blaise, Alain Libolt. Color.
WANDERERS, THE 113
ORION PICTURES. (R) July, 1979. Ken Wahl, John Friedrich.
WAR AND PEACE (VV) 208
PARAMOUNT. (R) November, 1956. Audrey Hepburn, Henry Fonda, Mel Ferrer. Color.
WAR AND PEACE 132
CONTINENTAL. (R) Special. April, 1968. Ludmila Savelyeva, Sergei Bondarchuk. Color.
WAR BETWEEN MEN AND WOMEN, THE 108
NATIONAL GENERAL. (R) May, 1972. Jack Lemmon, Barbara Harris. Color.
WAR DRUMS 75
UNITED ARTISTS. (R) April, 1957. Lex Barker, Joan Taylor. Color.
WAR GAME, THE 47
PATHE CONTEMPORARY. (R) March, 1967. BBC Documentary by Peter Watkins.
WAR-GODS OF THE DEEP (Colorscope) 85
AMERICAN INTERNATIONAL. (R) May, 1965. Vincent Price, Tab Hunter. Color.
WAR HUNT 81
UNITED ARTISTS. (R) May, 1962. John Saxon, Robert Redford.
WAR IS HELL! 81
ALLIED ARTISTS. (R) February, 1964. Tony Russell, Baynes Barron.
WAR—ITALIAN STYLE (TS) 84
AMERICAN INTERNATIONAL. (R) January, 1967. Buster Keaton, Franco and Ciccio. Color.
WAR LORD, THE (P) 123
UNIVERSAL. (R) November, 1965. Charlton Heston, Richard Boone. Color.
WAR LOVER, THE 105
COLUMBIA. (R) November, 1962. Steve McQueen, Robert Wagner.

Length In Mins.

WAR OF THE COLOSSAL BEAST 68
AMERICAN INTERNATIONAL. (R) July, 1958. Dean Parkin, Sally Fraser.
WAR OF THE SATELLITES 72
ALLIED ARTISTS. (R) April 6, 1958. Susan Cabot, Dick Miller.
WAR OF THE ZOMBIES 85
AMERICAN INTERNATIONAL. (R) March, 1965. John Barrymore, Jr. Color.
WARKILL 100
UNIVERSAL. (R) May, 1968. George Montgomery, Tom Drake. Color.
WAR PARTY 91
TWENTIETH CENTURY-FOX. (R) March, 1965. Michael T. Miker, Davey Davidson
WAR WAGON, THE (P) 101
UNIVERSAL. (R) June, 1967. John Wayne, Kirk Douglas. Color.
WARLOCK (Cs) 121
TWENTIETH CENTURY-FOX. (R) April, 1959. Richard Widmark, Henry Fonda. Color.
WARM DECEMBER, A 100
NATIONAL GENERAL. (R) May, 1973. Sidney Poitier. Color.
WARNING SHOT 100
PARAMOUNT. (R) January, 1967. David Janssen, Ed Begley. Color.
WARRENDALE 100
GROVE PRESS. (R) September, 1968. b&w.
WARRIOR AND THE SLAVE GIRL, THE (Supercinescope) .. 89
COLUMBIA. (R) November, 1959. Gianna Maria Canale, Georges Marchal. Color.
WARRIOR EMPRESS, THE (Cs) 97
COLUMBIA. (R) May, 1961. Kerwin Mathews, Tina Louise. Color.
WARRIORS, THE 92
PARAMOUNT. (R) February, 1979. Michael Beck, James Remar.
WARRIORS, THE (Cs) 85
ALLIED ARTISTS. (R) September 11, 1955. Errol Flynn, Joanne Dru. Color.
WASP WOMAN, THE 73
THE FILMGROUP. (R) October, 1959. Susan Cabot, Fred Eisley.
WATERHOLE #3 (TS) 100
PARAMOUNT. (R) October, 1967. James Coburn, Margaret Blye. Color.
WATERLOO (P) 123
PARAMOUNT. (R) April, 1971. Rod Steiger, Christopher Plummer. Color.
WATERMELON MAN 97
COLUMBIA. (R) August, 1970. Godfrey Cambridge, Estelle Parsons. Color.
WATERSHIP DOWN 92
AVCO EMBASSY. (R) November, 1978. Cartoon.
WATTSTAX 102
COLUMBIA. (R) February, 1973. Isaac Hayes, Carla Thomas. Color.
WATUSI (Cs) 85
METRO-GOLDWYN-MAYER. (R) May, 1959. George Montgomery, Taina Elg. Color.
WAY OUT, THE 78
RKO RADIO. (R) April 11, 1956. Gene Nelsen, Mona Freeman.
WAY TO THE GOLD, THE (Cs) 94
TWENTIETH CENTURY-FOX. (R) May, 1957. Sheree North, Jeffrey Hunter. Color.
WAY . . . WAY OUT (Cs) 106
20TH CENTURY-FOX. (R) October, 1966. Jerry Lewis, Connie Stevens. Color.
WAY WE LIVE NOW, THE 110
UNITED ARTISTS. (R) May, 1970. Nicholas Pryor, Linda Simon, Joanne Miles. Color.
WAY WE WERE, THE (P) 118
COLUMBIA. (R) October, 1973. Barbara Streisand, Robert Redford.
WAY WEST, THE (P) 122
UNITED ARTISTS. (R) June, 1967. Kirk Douglas, Lola Albright. Color.
WAYWARD BUS, THE (Cs) 90
TWENTIETH CENTURY-FOX. (R) June, 1957. Jayne Mansfield, Dan Dailey.
WAYWARD GIRL, THE 71
REPUBLIC. (R) September, 1957. Marchia Henderson, Peter Walker.
WAYWARD WIFE, THE (Ital.—Eng. dubbed) 95
I.F.E. (R) January, 1955. Gina Lollobrigida, Alda Mangini.
WE ARE ALL MURDERERS (Fr.) 113
KINGSLEY. (R) January, 1957. Marcel Mouloudji, Raymond Pellegrin.
WE STILL KILL THE OLD WAY (Ital.) 90

Length In Mins.

LOPERT. (R) February, 1968. Gian Maria Volonte, Irene Papas. Color.

WEAPON, THE 80
REPUBLIC. (R) May 27, 1957. Steve Cochran, Lizabeth Scott.

WEB OF EVIDENCE 88
ALLIED ARTISTS. (R) September, 1959. Van Johnson, Vera Miles.

WEDDING, A 125
TWENTIETH CENTURY-FOX. (R) October, 1978. Carol Burnett, Mia Farrow, Lillian Gish.

WEDDING IN BLOOD 98
NEW LINE CINEMA. (R) May, 1974. Stephane Audran, Michael Piccoli.

WEDDING IN WHITE 101
AVCO-EMBASSY. (R) March, 1973. Donald Pleasence, Carol Kane. Color.

WEDDING NIGHT 99
AMERICAN INTERNATIONAL. (R) May, 1970. Dennis Waterman, Tessa Wyatt. Color.

WEDDING PARTY, THE 90
ONDINE. (R) April, 1969. Jill Clayburgh, Charles Pfluger, Valda Settonfield. b&w.

WEDNESDAY'S CHILD 108
CINEMA V. (R) October, 1972. Sandy Ratcliff, Bill Dean. Color.

WEE GEORGIE (Brit.) 94
TIMES. (R) October, 1956. Alastair Sim, Bill Travers. Color.

WEEKEND (Dan., Eng. Titles) 84
CINEMA-VIDEO. (R) May, 1964. Jens Osterhom, Birgit Bruel.

WEEKEND AT DUNKIRK 101
TWENTIETH CENTURY-FOX. (R) May, 1966. Jean-Paul Belmondo, Catharine Spaak. Color.

WEEKEND WITH LULU 75
COLUMBIA. (R) December, 1961. Bob Monkhouse, Leslie Phillips.

WELCOME HARD TIMES 105
METRO-GOLDWYN-MAYER. (R) May, 1967. Henry Fonda, Janice Rule. Color.

WELCOME HOME, SOLDIER BOYS 91
20TH CENTURY-FOX. (R) February, 1972. Joe Don Baker, Alan Vint. Color.

WELCOME TO L.A. 106
LION'S GATE. (R) April, 1977. Keith Carradine, Sally Kellerman.

WELCOME TO THE CLUB 90
COLUMBIA. (R) September 1971. Brian Foley, Jack Warden. Color.

WE'LL BURY YOU 75
COLUMBIA. (R) November, 1962. Documentary.

WE'RE NO ANGELS (VV) 103
PARAMOUNT. (R) July, 1955. Humphrey Bogart, Aldo Ray, Peter Ustinov, Joan Bennett. Color.

WEREWOLF, THE 80
COLUMBIA. (R) July, 1956. Steve Rich, Joyce Holden.

WEST OF ZANZIBAR (Brit.) 94
UNIVERSAL. (R) January, 1955. Anthony Steel, Sheila Sim. Color.

WEST SIDE STORY (P-70) 155
UNITED ARTISTS. (R) October, 1961. Natalie Wood, Richard Beymer. Color.

WESTBOUND (C) 96
WARNER BROS. (R) April, 1959. Randolph Scott, Virginia Mayo.

WESTWARD HO THE WAGONS (Cs) 96
BUENA VISTA. (R) December 25, 1957. Fess Parker, Kathleen Crowley. Color.

WESTWORLD (P) 88
METRO-GOLDWYN-MAYER. (R) 1973. Yul Brynner, Richard Benjamin.

WETBACKS 89
BANNER. (R) May, 1956. Lloyd Bridges, Nancy Gates. Color.

WHAT? 112
AVCO-EMBASSY. (R) October, 1973. Marcello Mastroianni, Sydne Rome.

WHAT A WAY TO GO! (Cs) 111
TWENTIETH CENTURY-FOX. (R) June, 1964. Shirley MacLaine, Paul Newman. Color.

WHAT BECAME OF JACK AND JILL? 93
20TH CENTURY-FOX. (R) June, 1972. Vanessa Howard, Mona Washbourne. Color.

WHAT DID YOU DO IN THE WAR, DADDY? (P) 119
UNITED ARTISTS. (R) July, 1966. James Coburn, Dick Shawn. Color.

WHAT DO YOU SAY TO A NAKED LADY? 90
UNITED ARTISTS. (R) February, 1970. Color.

WHAT EVER HAPPENED TO AUNT ALICE? 101
CINERAMA. (R) July, 1969. Geraldine Page, Ruth Gordon. Color.

WHAT EVER HAPPENED TO BABY JANE? 132
WARNER BROS. (R) November, 1962. Bette Davis, Joan Crawford.

WHAT'S GOOD FOR THE GOOSE 105
NATIONAL SHOWMANSHIP. (R) December, 1969. Norman Wisdom, Sally Geeson. Color.

WHAT'S NEW, PUSSYCAT 108
UNITED ARTISTS. (R) June, 1965. Peter Sellers, Peter O'Toole. Color.

WHAT'S SO BAD ABOUT FEELING GOOD? 94
UNIVERSAL. (R) June, 1968. George Peppard, Mary Tyler Moore, Dom de Luise. Color.

WHAT'S THE MATTER WITH HELEN? 101
UNITED ARTISTS. (R) June, 1971. Debbie Reynolds, Shelley Winters. Color.

WHAT'S UP DOC? 94
WARNER BROS. (R) March, 1972. Barbra Streisand, Ryan O'Neal, Madeleine Kahn. Color.

WHAT'S UP TIGER LILY? 80
AMERICAN INTERNATIONAL. (R) September, 1966. Woody Allen-dubbed farce of straight Japanese film. Color.

WHEELER DEALERS, THE (P) 106
METRO-GOLDWYN-MAYER. (R) November, 1963. James Garner, Lee Remick. Color.

WHEN A STRANGER CALLS 97
COLUMBIA. (R) October, 1979. Charles Durning, Carol Kane.

WHEN ANGELS GO—TROUBLE FOLLOWS 95
COLUMBIA. (R) April, 1968. Stella Stevens, Rosalind Russell. Color.

WHEN COMEDY WAS KING 81
20TH CENTURY-FOX. (R) 1960. Charlie Chaplin, Buster Keaton, etc. b&w.

WHEN DINOSAURS RULED THE EARTH 96
WARNER BROS. (R) February, 1971. Victoria Vetri, Robin Hawdon. Color.

WHEN EIGHT BELLS TOLL (P) 94
CINERAMA. (R) May 1971. Anthony Hopkins, Robert Morley. Color.

WHEN GANGLAND STRIKES 70
REPUBLIC. (R) March 15, 1956. Raymond Greenleaf, Marjie Miller.

WHEN HELL BROKE LOOSE 78
PARAMOUNT. (R) November, 1958. Charles Bronson, Richard Jaeckel.

WHEN THE BOYS MEET THE GIRLS (P) 97
METRO-GOLDWYN-MAYER. (R) December, 1965. Connie Francis, Harve Presnell. Color.

WHEN THE LEGENDS DIE 105
TWENTIETH CENTURY-FOX. (R) August, 1972. Richard Widmark, Frederic Forest. Color.

WHEN TIME RAN OUT (P) 115
WARNER BROS. (R) March, 1980. Paul Newman, William Holden.

WHEN YOU COMIN' BACK, RED RYDER 118
COLUMBIA. (R) February, 1979. Marjoe Gortner, Hal Linden.

WHERE DOES IT HURT? 90
CINERAMA. (R) August, 1972. Peter Sellers, Jo Ann Pflug. Color.

WHERE EAGLES DARE 158
METRO-GOLDWYN-MAYER. (R) April, 1969. Richard Burton, Clint Eastwood, Mary Ure. Color.

WHERE IT'S AT 104
UNITED ARTISTS. (R) May, 1969. David Janssen, Robert Drivas, Rosemary Forsyth. Color.

WHERE LOVE HAS GONE (TS) 114
PARAMOUNT. (R) October, 1964. Susan Hayward, Bette Davis. Color.

WHERE THE BOYS ARE (Cs) 90
METRO-GOLDWYN-MAYER. (R) December, 1960. George Hamilton, Dolores Hart. Color.

WHERE THE BULLETS FLY 88
EMBASSY. (R) November, 1966. Tom Adams. Dawn Addams. Color.

WHERE THE HOT WIND BLOWS 120
METRO-GOLDWYN-MAYER. (R) November, 1960. Gina Lollobrigida, Yves Montand.

WHERE THE LILIES BLOOM (P) 97
UNITED ARTISTS. (R) February 1974. Julie Gholson, Jan Smithers.

WHERE THE SPIES ARE (P) 110
METRO-GOLDWYN-MAYER. (R) January, 1966. David Niven, Francoise Dorleac. Color.

WHERE THE TRUTH LIES (D) 83
PARAMOUNT. (R) October, 1962. Juliette Greco, Jean Marc Bory.

WHERE WERE YOU WHEN THE LIGHTS WENT OUT? .. 94
METRO-GOLDWYN-MAYER. (R) July, 1968. Doris Day, Robert Morse, Terry-Thomas. Color.

WHERE'S JACK? 120
PARAMOUNT. (R) May, 1969. Tommy Steele, Stanley Baker, Flora Lewis. Color.

	Length In Mins.
WHERE'S POPPA? UNITED ARTISTS. (R) December, 1970. George Segal, Ruth Gordon. Color.	87
WHICH WAY IS UP? UNIVERSAL. (R) November, 1977. Richard Pryor, Lonette McKee.	94
WHICH WAY TO THE FRONT? WARNER BROS. (R) July, 1970. Jerry Lewis, Jan Murray. Color.	90
WHIFFS TWENTIETH CENTURY-FOX. (R) September, 1975. Elliott Gould, Jennifer O'Neill.	90
WHILE THE CITY SLEEPS RKO RADIO. (R) May 30, 1958. Dana Andrews, Ida Lupino, George Sanders, Rhonda Fleming.	100
WHIRLPOOL CINEMATION. (R) August, 1970. Karl Launchburg, Vivian Neves. Color.	75
WHISPERERS, THE LOPERT. (R) July, 1967. Dame Edith Evans, Eric Portman.	105
WHITE DAWN, THE PARAMOUNT. (R) July, 1974. Warren Oates, Timothy Bottoms.	110
WHITE FEATHER (Cs) TWENTIETH CENTURY-FOX. (R) February, 1955. Robert Wagner, Debra Paget. Color.	102
WHITE HUNTRESS AMER. INT. PICS. (R) August, 1957. Susan Stephan, John Bentley.	60
WHITE LIGHTNING UNITED ARTISTS. (R) July, 1973. Burt Reynolds, Jennifer Billingsley. Color.	101
WHITE LINE FEVER COLUMBIA. (R) July, 1975. Jan-Michael Vincent, Kay Lenz.	89
WHITE SHEIK, THE (Ital.) JANUS. (R) May, 1956. Alberto Sordi, Brunello Bovo.	86
WHITE SISTER COLUMBIA. (R) March, 1973. Sophia Loren. Color.	96
WHITE SQUAW, THE COLUMBIA. (R) November, 1956. David Brian, May Wynn.	73
WHITE VOICES (Te) FRANLA RIZZOLI. (R) April, 1965. Paolo Ferrari, Sandra Mild. Color.	93
WHITE WARRIOR WARNER BROS. (R) March, 1961. Steve Reeves, Georgia Moll. Color.	80
WHITE WILDERNESS BUENA VISTA. (R) October, 1958. True Life Adventure. Color.	72
WHO IS HARRY KELLERMAN? NATIONAL GENERAL. (R) August, 1971. Dustin Hoffman, Barbara Harris. Color.	108
WHO IS KILLING THE GREAT CHEFS OF EUROPE? WARNER BROS. (R) October, 1978. George Segal, Jacqueline Bisset.	115
WHO KILLED MARY WHAT'S ERNAME? CANNON. (R) November, 1971. Red Buttons, Sylvia Miles. Color.	90
WHO KILLED TEDDY BEAR? MAGNA. (R) October, 1965. Juliet Prowse, Sal Mineo.	91
WHO SAYS I CAN'T RIDE A RAINBOW? TRANSVUE. (R) November, 1971. Jack Klugman, Norma French. Color.	85
WHO SLEW AUNTIE ROO? AMERICAN INTERNATIONAL. (R) December, 1971. Shelley Winters, Mark Lester. Color.	89
WHO WAS THAT LADY? COLUMBIA. (R) January, 1960.	116
WHOLE TRUTH, THE COLUMBIA. (R) September, 1958. Stewart Granger, Donna Reed.	84
WHOLLY MOSES COLUMBIA. (R) June, 1980. Dudley Moore, James Coco.	105
WHO'LL STOP THE RAIN UNITED ARTISTS. (R) August, 1978. Nick Nolte, Tuesday Weld.	126
WHO'S AFRAID OF VIRGINIA WOOLF? WARNER BROS. (R) July, 1966. Elizabeth Taylor, Richard Burton.	129
WHO'S BEEN SLEEPING IN MY BED? (P) PARAMOUNT. (R) January, 1964. Dean Martin, Elizabeth Montgomery.	103
WHO'S GOT THE ACTION? (P) PARAMOUNT. (R) January, 1963. Dean Martin, Lana Turner. Color.	93
WHO'S MINDING THE MINT? COLUMBIA. (R) October, 1967. Jim Hutton, Dorothy Provine. Color.	97
WHO'S THAT KNOCKING AT MY DOOR? BRENNER. (R) September, 1968. Zina Bethune, Harvey Keitel. Color.	90
WHY HALLMARK. (R) November, 1972. Alberto Sordi, Elga Andersen. Color.	102
WHY BOTHER TO KNOCK (Cs) SEVEN ARTS. (R) November, 1964. Richard Todd, Nicole Maurey. Color.	88
WHY MUST I DIE? AMERICAN INTERNATIONAL. (R) July, 1960. Terry Moore, Debra Paget.	86
WHY WOULD I LIE? UNITED ARTISTS. (R) August, 1980. Treat Williams, Lisa Eichhorn.	106
WICHITA (Cs) ALLIED ARTISTS. (R) July 3, 1955. Joel McCrea, Vera Miles. Color.	81
WICKED AS THEY COME COLUMBIA. (R) February, 1957. Arlene Dahl, Phil Carey.	94
WICKED DREAMS OF PAULA SCHULTZ UNITED ARTISTS. (R) January, 1968. Elke Sommer, Bob Crane. Color.	113
WICKED, WICKED METRO-GOLDWYN-MAYER. (R) April, 1973. Tiffany Bolling, Scott Brady. Color.	95
WICKED WIFE (Brit.) ALLIED ARTISTS. (R) April 8, 1956. Nigel Patrick, Moira Lister.	75
WILBY CONSPIRACY, THE UNITED ARTISTS. (R) August, 1975. Sidney Poitier, Michael Caine.	101
WILD AND THE INNOCENT, THE (Cs) UNIVERSAL. (R) May, 1959. Audie Murphy, Joanne Dru. Color.	84
WILD ANGLES, THE (P) AMERICAN INTERNATIONAL. (R) July, 1966. Peter Fonda, Nancy Sinatra. Color.	93
WILD BUNCH, THE (P) WARNER. (R) William Holden, Robert Ryan, Ernest Borgnine. Color.	145
WILD CHILD, THE (French; Eng. titles) UNITED ARTISTS. (R) September, 1970. Francois Truffaut, Jean-Pierre Cargol.	85
WILD COUNTRY, THE BUENA VISTA. (R) February, 1971. Steve Forrest, Vera Miles. Color.	100
WILD DAKOTAS, THE ASSOCIATED FILM. (R) February, 1956. Bill Williams, Coleen Gray.	73
WILD GEESE, THE ALLIED ARTISTS. (R) November, 1978. Richard Burton, Roger Moore, Richard Harris.	132
WILD HERITAGE (Cs) UNIVERSAL. (R) August, 1958. Will Rogers, Jr., Maureen O'Sullivan. Color.	78
WILD IN THE COUNTRY (Cs) TWENTIETH CENTURY-FOX. (R) June, 1961. Elvis Presley, Jean Simmons. Color.	114
WILD IN THE STREETS AMERICAN INTERNATIONAL. (R) May, 1968. Shelley Winters, Christopher Jones. Color.	97
WILD IN THE WIND (VV) PARAMOUNT. (R) February, 1958. Anna Magnani, Anthony Quinn.	114
WILD McCULLOCHS, THE AMERICAN INTERNATIONAL. (R) May, 1975. Forrest Tucker, Max Baer.	93
WILD OAT, THE (Fr.) CARROLL. (R) June, 1956. Fernandel.	107
WILD ON THE BEACH TWENTIETH CENTURY-FOX. (R) August, 1965. Teenage Comedy.	77
WILD PARTY, THE AMERICAN INTERNATIONAL. (R) June, 1975. James Coco, Raquel Welch.	100
WILD PARTY, THE UNITED ARTISTS. (R) December, 1956. Anthony Quinn, Caol Ohmart.	81
WILD RACERS, THE AMERICAN INTERNATIONAL. (R) March, 1968. Fabian, David Landers. Color.	91
WILD RIVER (Cs) TWENTIETH CENTURY-FOX. (Cs) (R) June, 1960. Lee Remick, Montgomery Clift. Color.	105
WILD ROVERS (P) MGM. (R) June, 1971. William Holden, Ryan O'Neal. Color.	109
WILD WESTERNERS, THE COLUMBIA. (R) June, 1962. James Philbrook, Nancy Kovack. Color.	70
WILD, WILD PLANET, THE	93

Length In Mins.

METRO-GOLDWYN-MAYER. (R) June, 1967. Tony Russell, Connie Gastoni. Color.
WILD, WILD WINTER (Te) 80
UNIVERSAL. (R) January, 1966. Gary Clarke, Chris Noel. Color.
WILL ANY GENTLEMAN? (Brit.) 84
STRATFORD. (R) January 3, 1955. George Cole, Veronica Hunt.
WILL PENNY 109
PARAMOUNT. (R) March, 1968. Charlton Heston, Joan Hackett. Color.
WILL SUCCESS SPOIL ROCK HUNTER? (Cs) 94
TWENTIETH CENTURY-FOX. (R) August, 1957. Jayne Mansfield, Tony Randall. Color.
WILLARD 95
CINERAMA. (R) June, 1971. Bruce Davison, Ernest Borgnine. Color.
WILLIE AND PHIL 116
TWENTIETH CENTURY-FOX. (R) August, 1980. Michael Ontkean, Ray Sharkey.
WILLIE DYNAMITE 102
UNIVERSAL. (R) January, 1974. Roscoe Orman, Diana Sands.
WILLY McBEAN AND HIS MAGIC MACHINE 94
MAGNA. (R) August, 1965. Puppet Fantasy.
WILLY WONKA AND THE CHOCOLATE FACTORY 110
PARAMOUNT. (R) June, 1971. Gene Wilder, Jack Albertson, Peter Ostrum. Color.
WIND ACROSS THE EVERGLADES 98
WARNER BROS. (R) September, 1958. Burl Ives, Christopher Plummer.
WIND AND THE LION, THE (P) 119
UNITED ARTISTS—MGM. (R) June, 1975. Sean Connery, Candice Bergen.
WIND CANNOT READ, THE 107
TWENTIETH CENTURY-FOX. (R) February, 1960. Dick Bogarde, Yoko Tani.
WINDJAMMER, (Cinemiracle & Cinerama) 127
LOUIS DE ROCHEMONT. (R) April, 1958. Travellogue Color.
WINDOM'S WAY 108
RANK FILM DISTRIBUTORS OF AMERICA. (R) November, 1958. Peter Finch, Mary Ure. Color.
WINDOWS 94
UNITED ARTISTS. (R) January, 1980. Talia Shire, Joseph Cortese.
WINGS OF CHANCE 76
UNIVERSAL. (R) April, 1961. Jim Brown, Frances Rafferty. Color.
WINGS OF EAGLES, THE 110
METRO-GOLDWYN-MAYER. (R) Feb. 22, 1957. John Wayne, Maureen O'Hara. Color
WINK OF AN EYE 72
UNITED ARTISTS. (R) June, 1958. Doris Dowling, Jonathan Kidd.
WINNER, THE 82
NOELLE GILMORE AND ROBERT KINGSLEY. (R) June, 1965. Abdoulaye Faye, Marcel Bruchard.
WINNING 123
UNIVERSAL. (R) wjune, 1969. Paul Newman, Joanne Woodward, Robert Wagner. Color.
WINTER A GO-GO 88
COLUMBIA. (R) October, 1965. James Stacy, William Wellman, Jr. Color.
WINTER KILLS 97
AVCO EMBASSY. (R) May, 1979. Jeff Bridges, John Huston.
WINTERHAWK 98
HOWCO INTERNATIONAL. (R) 1975. Leif Ericson, Woody Strode.
WIRETAPPER 80
EMBASSY. (R) February, 1956. Bill Williams.
WISE GUYS, THE 100
UNIVERSAL. (R) June, 1969. Bourvil, Lino Ventura. Color.
WITCH WITHOUT A BROOM, A 86
PRODUCERS RELEASING ORG. (R) May, 1967. Jeff Hunter, Maria Perschy. Color.
WITCHCRAFT 79
20th CENTURY-FOX. (R) 1964. Lon Chaney, Jack w qedley. Color.
WITCHMAKER, THE 86
EXCELSIOR. (R) May, 1969. Anthony Eisley, Thordis Bandi.
WITH SIX YOU GET EGGROLL 95
NATIONAL GENERAL. (R) July, 1968. Doris Day, Brian Keith, Pat Caroll, Barbara Hershey. Color.
WITHOUT A STITCH (Danish; Eng. titles) 96
VIP. (R) January, 1970. Anne Grete, Ib Mossin. Color.
WITHOUT APPARENT MOTIVE 102
20th CENTURY-FOX. (R) May, 1972. Jean-Louis Trintignant, Dominique Sanda. Color.
WITNESS FOR THE PROSECUTION 114
UNITED ARTISTS. (R) February, 1958. Tyrone Power, Marlene Dietrich.
WITNESSES, THE 82
ALTURE FILMS INTERNATIONAL. (R) August, 1967. Documentary.
WIVES AND LOVERS 103
PARAMOUNT. (R) October, 1963. Janet Leigh, Van Johnson.
WIZ, THE 133
UNIVERSAL. (R) October, 1978. Diana Ross, Michael Jackson.
WIZARD OF BAGHDAD, THE (Cs) 92
TWENTIETH CENTURY-FOX. (R) December, 1960. Dick Shawn, Diane Baker Color.
WIZARDS 81
TWENTIETH CENTURY-FOX. (R) March, 1977. Animated cartoon.
WOLF DOG 69
TWENTIETH CENTURY-FOX. (R) July, 1958. Jim Davis, Allison Hayes.
WOLF LARSEN 83
ALLIED ARTISTS. (R) October, 1958. Barry Sullivan, Gita Hall.
WOMAN IN A DRESSING GOWN 93
WARNER BROS. (R) November, 1957. Yvonne Mitchell, Sylvia Syms.
WOMAN IN THE DUNES (Jap., Eng. titles) 128
PATHECONT. (R) October, 1964. Eiji Okada, Kyoko Kistlida.
WOMAN IS A WOMAN, A (Fr., Eng. titles) 80
PATHE-CONT. (R) November, 1964. Jean-Claude, Brialy, Ann Karina. Color.
WOMAN OBSESSED (Cs) 102
TWENTIETH CENTURY-FOX. (R) May, 1959. Susan Hayward, Stephen Boyd.
WOMAN OF STRAW 117
UNITED ARTISTS. (R) September, 1964. Gina Lollobrigida, Sean Connery. Color.
WOMAN OF THE RIVER 92
COLUMBIA. (R) September, 1957. Sophia Loren, Gerald Oury. Color.
WOMAN TIMES SEVEN 99
EMBASSY. (R) June, 1967. Shirley MacLaine, Alan Arkin. Color.
WOMAN WHO WOULDN'T DIE, THE 84
WARNER BROS. (R) May, 1965. Raymond Garth, Alice Taylor.
WOMAN'S DEVOTION, A 88
REPUBLIC. (R) November 9, 1956 Ralph Meeker, Janice Rule. Color.
WOMEN IN CAGES 78
NEW WORLD PICTURES. (R) January, 1973. Jennifer Gan, Judy Brown. Color.
WOMEN IN CELL BLOCK 7 100
AQUARIUS RELEASING. (R) August, 1974. Anita Strinberg, Eva Czemeys.
WOMEN IN LOVE 129
UNITED ARTISTS. (R) March, 1970. Alan Bates, Jennie Linden, Oliver Reed, Glenda Jackson. Color.
WOMEN OF PITCAIRN ISLAND 72
TWENTIETH CENTURY-FOX. (R) December, 1956. James Craig, Lynn Bari
WOMEN OF THE PREHISTORIC PLANET 91
FEATURE FILM CORP. OF AMER. (R) 1965. Wendell Corey, Keith Larsen. Color.
WOMEN OF THE WORLD 107
EMBASSY. (R) July, 1963. Documentary. Color.
WOMEN'S PRISON 80
COLUMBIA. (R) February, 1955. Ida Lupino, Howard Duff.
WON TON TON, THE DOG WHO SAVED HOLLYWOOD . 92
PARAMOUNT. (R) May, 1976. Madeline Kahn, Bruce Dern.
WONDERFUL COUNTRY, THE 96
UNITED ARTISTS. (R) October, 1959. Robert Mitchum, Julie London.
WONDERFUL TO BE YOUNG (Cs) 92
PARAMOUNT. (R) October, 1962. Cliff Richard, Robert Morley. Color.
WONDERFUL WORLD OF THE BROTHERS GRIMM, THE (Cinerama) 129
METRO-GOLDWYN-MAYER. (R) Special. July, 1962. Laurence Harvey, Claire Bloom. Color.
WONDERS OF ALADDIN, THE (Cs) 93
METRO-GOLDWYN-MAYER. (R) December, 1961. Donald O'Connor, Noelle Adam. Color.
WOODSTOCK (P) 184
WARNER BROS. (R) March, 1970. Color.
WORLD BY NIGHT 103
WARNER BROS. (R) September, 1961. International Acts. Color.
WORLD BY NIGHT NO. 2 (T) 118
WARNER BROS. (R) July, 1962. Documentary. Color.

Length In Mins.

WORLD IN MY CORNER **82**
UNIVERSAL. (R) March, 1956. Audie Murphy, Barbara Rush.

WORLD IN MY POCKET **93**
METRO-GOLDWYN-MAYER. (R) March, 1962. Rod Steiger, Nadja Tiller.

WORLD OF ABBOTT & COSTELLO, THE **75**
UNIVERSAL. (R) May, 1965. Bud Abbott, Lou Costello.

WORLD OF HANS CHRISTIAN ANDERSEN, THE (Cs) **75**
UNITED ARTISTS. (R) May, 1971. Animated. Color.

WORLD OF HENRY ORIENT, THE (P) **106**
UNITED ARTISTS. (R) June, 1964. Peter Sellers, Paula Prentiss. Color.

WORLD OF SUZIE WONG, THE **129**
PARAMOUNT. (R) February, 1961. William Holden, Nancy Kwan. Color.

WORLD, THE FLESH, AND THE DEVIL, THE (Cs) **95**
METRO-GOLDWYN-MAYER. (R) May, 1959. Harry Belafonte, Inger Stevens.

WORLD WAS HIS JURY, THE **82**
COLUMBIA. (R) January, 1958. Edmond O'Brien, Mona Freeman.

WORLD WITHOUT END (Cs) **86**
ALLIED ARTISTS. (R) March 25, 1956. Hugh Marlowe, Nancy Gates. Color.

WORLD WITHOUT SUN **93**
COLUMBIA. (R) January, 1965. Documentary. Color.

WORLD'S GREATEST ATHLETE, THE **93**
BUENA VISTA. (R) February, 1973. Jan-Michael Vincent, Tim Conway. Color.

WORLD'S GREATEST LOVER, THE **89**
20th CENTURY-FOX. (R) December, 1977. Gene Wilder, Carol Kane.

WORST AGENTS!, THE (Ital.) **83**
ALLIED ARTISTS. (R) 1966. Franco Franchi.

WRATH OF GOD, THE **112**
METRO-GOLDWYN-MAYER. (R) June, 1972. Robert Mitchum, Rita Hayworth, Frank Langella. Color.

WRECK OF THE MARY DEARE, THE (Cs) **105**
METRO-GOLDWYN-MAYER. (R) December, 1959. Gary Cooper, Charlton Heston. Color.

WRECKING CREW, THE **105**
COLUMBIA. (R) February, 1969. Dean Martin, Elke Sommer, Sharon Tate, Nancy Kwan. Color.

WRITTEN ON THE WIND **98**
UNIVERSAL. (R) January, 1957. Rock Hudson, Lauren Bacall, Robert Stack, Dorothy Malone. Color.

WRONG ARM OF THE LAW **91**
CONTINENTAL. (R) April, 1963. Peter Sellers, Lionel Jeffries.

WRONG BOX, THE **105**
COLUMBIA. (R) October, 1966. John Mills, Michael Caine. Color.

WRONG MAN, THE **105**
WARNER BROTHERS. (R) January 26, 1957. Henry Fonda, Vera Miles.

WUTHERING HEIGHTS **105**
AMERICAN INTERNATIONAL. (R) Janury, 1971. Timothy Dalton, Anna Calder-Marshall. Color.

WYOMING RENEGADES **78**
COLUMBIA. (R) March, 1955. Phil Carey, Martha Hyer. Color.

X

"X" **80**
AMERICAN INTERNATIONAL. (R) September, 1963. Ray Milland, Diana Van Der Vlis. Color.

X-15 (P) **107**
UNITED ARTISTS. (R) November, 1961. David McLean, Charles Bronson. Color.

XANADU **93**
UNIVERSAL. (R) August, 1980. Olivia Newton-John, Gene Kelly.

X THE UNKNOWN (Brit.) **81**
WARNER BROS. (R) July, 1957. Dean Jagger, Edward Chapman.

X Y AND ZEE **110**
COLUMBIA. (R) January, 1972. Elizabeth Taylor, Michael Caine. Color.

Y

YAKUZA, THE (P) **111**
WARNER BROS. (R) March, 1975. Robert Mitchum, Brian Keith.

YANCO **85**
HOFFMAN. (R) June, 1964. Richardo Ancons, Maria Bustamente.

Length In Mins.

YANG KWEI FEI (Jap.) **90**
BUENA VISTA. (R) September, 1956. Machiko Kyo.

YANK IN VIET-NAM, A **80**
ALLIED ARTISTS. (R) February, 1964. Marshall Thompson, Enrique Magalona.

YANKS **140**
UNIVERSAL. (R) September, 1979. Richard Gere, Vanessa Redgrave.

YAQUI DRUMS **71**
ALLIED ARTISTS. (R) October 14, 1956. Rod Cameron, Mary Castle.

YELLOW CANARY, THE (Cs) **93**
TWENTIETH CENTURY-FOX. (R) May, 1963. Pat Boone, Barbara Eden.

YELLOW ROLLS ROYCE, THE (P) **122**
METRO-GOLDWYN-MAYER. (R) July, 1965. Color.

YELLOW SUBMARINE **85**
UNITED ARTISTS. (R) November, 1968. Animation. Color.

YELLOWNECK **83**
REPUBLIC. (R) March 22, 1955. Lin McCarthy, Stephen Cortleigh. Color.

YELLOWSTONE KELLY **91**
WARNER BROS. (R) September, 1959. Clint Walker, Edward Byrnes.

YESTERDAY, TODAY AND TOMORROW **119**
EMBASSY. (R) March, 1964. Sophia Loren, Marcello Mastroianni. Color.

YESTERDAY'S ENEMY (M) **95**
COLUMBIA. (R) November, 1959. Stanley Baker, Guy Rolfe.

YOLANTA **82**
ARTKINO. (R) December, 1964. Natalya Rudnaya, Yuri Petrov. Color.

YOU ARE WHAT YOU EAT **75**
P.W. (R) June, 1968. Tiny Tim, Paul Butterfield, Peter Yarrow. Color.

YOU CAN'T RUN AWAY FROM IT (Cs) **95**
COLUMBIA. (R) November, 1956. June Allyson, Jack Lemmon. Color.

YOU CAN'T WIN 'EM ALL **97**
COLUMBIA. (R) July, 1970. Tony Curtis, Charles Bronson. Color.

YOU LIGHT UP MY LIFE **90**
COLUMBIA. (R) August, 1977. Didi Conn, Joe Silver.

YOU MUST BE JOKING (Brit.) **100**
COLUMBIA. (R) October, 1965. Michael Callan, Lionel Jeffries.

YOU ONLY LIVE TWICE (P) **117**
UNITED ARTISTS. (R) June, 1967. Sean Connery.

YOU ONLY LOVE ONCE (Fr., Eng. titles) **95**
SIGMA III. (R) June, 1969. Karen Blanguernon, Leslie Bedos. Color.

YOU'LL LIKE MY MOTHER **92**
UNIVERSAL. (R) October, 1972. Patty Duke, Richard Thomas. Color.

YOUNG AMERICANS **104**
COLUMBIA. (R) October, 1967. The Young Americans. Color.

YOUNG AND DANGEROUS (R) **78**
TWENTIETH CENTURY-FOX. (R) October, 1957. Leli Gentle, Mark Damon.

YOUNG AND THE BRAVE, THE **84**
METRO-GOLDWYN-MAYER. (R) May, 1963. Rory Calhoun, William Bendix.

YOUNG AND WILD (N) **69**
REPUBLIC. (R) April, 1958. Gene Evans, Scott Marlowe.

YOUNG AND WILLING **110**
UNIVERSAL. (R) February, 1964. Ian McShane, Virginia Maskell.

YOUNG ANIMALS, THE **100**
AMERICAN INTERNATIONAL. (R) October, 1968. Tom Nardini, Patty McCormack, David Macklin. Color.

YOUNG AT HEART **117**
WARNER BROTHERS. (R) January 1, 1955. Doris Day, Frank Sinatra. Color.

YOUNG BESS **112**
MGM. (R) May 29, 1958. Jean Simmons, Stewart Granger, Deborah Kerr. Color.

YOUNG BILLY YOUNG **88**
UNITED ARTISTS. (R) September, 1969. Robert Mitchum, Angie Dickinson, Robert Walker. Color.

YOUNG CAPTIVES, THE **66**
PARAMOUNT. (R) February, 1959. Steven Marlo, Luana Patten.

YOUNG CASSIDY **110**
METRO-GOLDWYN-MAYER. (R) March, 1965. Rod Taylor, Flora Robson. Color.

YOUNG DILLINGER **102**
ALLIED ARTISTS. (R) May, 1965. Nick Adams, Robert Conrad.

YOUNG DOCTORS, THE **100**

Length In Mins.

UNITED ARTISTS. (R) September, 1961. Fredric March, Ben Gazarra.

YOUNG DON'T CRY, THE **69**
COLUMBIA. (R) August, 1957. Sal Mineo, James Whitmore.

YOUNG FRANKENSTEIN **98**
20TH CENTURY-FOX. (R) December, 1974. Gene Wilder, Peter Boyle. b&w.

YOUNG FURY (TS) **80**
PARAMOUNT. (R) February, 1965. Rory Calhoun, Virginia Mayo. Color.

YOUNG GIRLS OF ROCHEFORT, THE (Fr.) **124**
WARNER BROS.-SEVEN ARTS (R) January, 1968. George Chakiris, Catherine Deneuve. Color.

YOUNG GUNS, THE **84**
ALLIED ARTISTS. (R) August 12, 1956. Russ Tamblyn, Gloria Talbott.

YOUNG GUNS OF TEXAS (Cs) **78**
TWENTIETH CENTURY-FOX. (R) February, 1963. Jim Mitchum, Jody McCrea. Color.

YOUNG JESSE JAMES (Cs) **73**
TWENTIETH CENTURY-FOX. (R) August, 1960. Ray Stricklyn, Merry Anders.

YOUNG LAND, THE **89**
COLUMBIA. (R) May, 1959. Pat Wayne, Yvonne Craig. Color.

YOUNG LIONS, THE (Cs) **167**
TWENTIETH CENTURY-FOX. (R) April, 1958. Marlon Brando, Montgomery Clift.

YOUNG LOVERS, THE **110**
METRO-GOLDWYN-MAYER. (R) November, 1964. Peter Fonda, Sharon Hugueny.

YOUNG PHILADELPHIANS, THE **136**
WARNER BROS. (R) May, 1959. Paul Newman, Barbara Rush.

YOUNG RACERS, THE **85**
AMER. INT'L. PICS. (R) 1963. Mark Damon, William Campbell. Color.

YOUNG RUNAWAYS, THE **91**
METRO-GOLDWYN-MAYER. (R) October, 1968. Brooke Bundy, Kevin Coughlin, Lloyd Bochner. Color.

YOUNG SAVAGES, THE **103**
UNITED ARTISTS. (R) May, 1961. Burt Lancaster, Shelley Winters.

YOUNG SINNER, THE **81**
UNITED SCREEN ARTS. (R) August, 1965. Tom Laughlin, Stephanie Powers.

YOUNG STRANGER, THE **84**
UNIVERSAL. (R) May, 1957. Kim Hunter, James MacArthur.

YOUNG SWINGERS, THE **71**
TWENTIETH CENTURY-FOX. (R) January, 1964. Rod Lauren, Molly Bee.

YOUNG, THE EVIL, AND THE SAVAGE, THE **82**
AMERICAN INTERNATIONAL. (R) July, 1968. Mike Damon, Eleanor Brown, Michael Rennie. Color.

YOUNG TORLESS **90**
KANAWHA. (R) July, 1968. Matthieu Carriere, Bernard Tischer, Marian Seidowsky. b&w.

YOUNG WARRIORS, THE (P) **93**
UNIVERSAL. (R) May, 1967. James Drury, Steve Carlson. Color.

YOUNG WINSTON (P) **160**
COLUMBIA. (R) October, 1972. Simon Ward, Robert Shaw, Anne Bancroft. Color.

YOUNGBLOOD HAWKE **137**
WARNER BROS. (R) November, 1964. James Franciscus, Suzanne Pleshette.

YOUR CHEATING HEART (P) **99**
METRO-GOLDWYN-MAYER. (R) January, 1965. George Hamilton, Susan Oliver.

YOUR PAST IS SHOWING! **87**
RANK FILM DISTRIBUTORS OF AMERICA. (R) July, 1958. Terry Thomas, Peter Sellers.

YOUR SHADOW IS MINE **90**
CONTINENTAL. (R) 1963. Jill Haworth, Michael Ruhl. Color.

YOUR THREE MINUTES ARE UP (P) **93**
CINERAMA. (R) August, 1973. Beau Bridges, Ron Liebman.

YOU'RE A BIG BOY NOW **96**
SEVEN ARTS. (R) February, 1967. Elizabeth Hartman, Geraldine Page. Color.

YOU'RE NEVER TOO YOUNG (VV) **102**
PARAMOUNT. (R) August, 1955. Dean Martin, Jerry Lewis, Diana Lynn. Color.

YOURS, MINE AND OURS **111**
UNITED ARTISTS. (R) May, 1968. Lucille Ball, Henry Fonda. Color.

Z

Z (French: Eng. titles) **127**
CINEMA V. (R) December, 1969. Yves Montand, Irene Papas, Jean Louis Trintignant. Color.

Z.P.G. **95**
PARAMOUNT. (R) May, 1972. Oliver Reed, Geraldine Chaplin, Diane Cilento. Color.

ZABRISKIE POINT (P) **112**
MGM. (R) February, 1970. Mark Frechette, Daria Halprin. Color.

ZACHARIAH **94**
CINERAMA. (R) February, 1971. John Rubinstein, Patricia Quinn. Color.

ZANDY'S BRIDE (P) **116**
WARNER BROS. (R) July, 1974. Gene Hackman, Liv Ullmann.

ZANZABUKU **64**
REPUBLIC. (R) April 13, 1956. Documentary. Color.

ZARAK (Cs) **99**
COLUMBIA. (R) January, 1957. Victor Mature, Anita Ekberg, Michael Wilding. Color.

ZARDOZ (P) **105**
TWENTIETH CENTURY-FOX. (R) March, 1974. Sean Connery, Charlotte Rampling.

ZEBRA IN THE KITCHEN **93**
METRO-GOLDWYN-MAYER. April, 1965. Jay North, Martin Milner. Color.

ZEPPELIN (P) **101**
WARNER BROS. (R) May, 1971. Elke Sommer, Michael York. Color.

ZERO HOUR **81**
PARAMOUNT. (R) November, 1957.

ZIGZAG (P) **104**
MGM. (R) April, 1970. George Kennedy, Anne Jackson, Eli Wallach. Color.

ZITA (French; Eng. titles) **92**
REGIONAL. (R) July, 1968. Joanna Simkus, Katina Paxinou, Suzanne Flon. Color.

ZOMBIES OF MORA-TAU **70**
COLUMBIA. (R) March, 1957. Allison Hayes, Gregg Palmer.

ZORBA THE GREEK **142**
INTERNATIONAL CLASSICS. (R) February, 1965. Anthony Quinn, Alan Bates.

ZORRO RIDES AGAIN **68**
REPUBLIC. (R) January, 1959. John Carroll, Helen Christian.

ZOTZ! **87**
COLUMBIA. (R) July, 1962. Tom Poston, Julia Meade.

ZULU (T) **138**
EMBASSY. (R) June, 1964. Stanley Baker, Michael Caine. Color.

Services

Film Distributors in Key Cities

ALBANY

Riverside Film Distributing, P.O. Box 255, Yorktown Hts., NY (914) 245-2910; Craig Clark.
United Artists Corp., 991 Broadway, 72204. (518) 462-5385; Franklin R. Meadow, br. mgr.
Wesco Film Service, Inc., 24 North 3 St., 12204. (518) 434-1289.

ATLANTA

Alliance International, 161 Spring St., NY 30303. (404) 524-7579.
American National, 3070 Presidential Drive, 30340; (404) 451-7861. Harvey Edwards.
Avco Embassy Pictures, 2812 New Spring Road. 30339. (404) 434-1391. Robert Benefield, div. mgr.
Benton Film Forwarding Co., 168 Baker St., N.W., (404) 577-2821. W. B. Langston, mgr.
Bradley Films, 1876 De Foor Ave., N.W., 30318. (404) 352-3316. Gordon Bradley.
Buena Vista Distribution Co., 3445 Peachtree Rd., N.E., 30326; (404) 266-8454.
Chappell Releasing Co., 2814 New Spring Rd. 30339; (404) 432-3361. Wayne Chappell, pres.
Clark Film, Inc., 2200 Northlake Pkwy., Tucker, GA 30084. (404) 491-7766. Lewis Owens.
Columbia Pictures, 2600 Century Parkway, N.E., 30345. (404) 325-9525. Henry Harrell, mgr.
Craddock Films, Inc., Atlanta Film Bldg., 161 Spring St., N.W.; Suite 417. (404) 523-5653.
Films Inc., 277 Pharr Rd., N.E.; Tel: (404) 237-0341.
General Film Distributing, 3950 Peachtree Rd., N.E. 30319; (404) 261-5363. C. L. Autry.
Harnell Independent Productions, 6065 Roswell Rd., 30328; (404) 256-3464. Walter Powell, gen. sls. mgr.
Independent Film Dist., Inc., 161 Spring St. N.E., Phone: 524-7579. Sam Davis, pres., Steve Davis, mgr./booker.
JACO Productions Inc., 207 Luckie St., N.W.; 524-4218.
Kay Exchange, 201 Luckie Street; 524-0261. Ike Katz, pres. Harry Katz, booker.
K-Tel Motion Pictures, 1645 Tully Circle, N.E., 30329; (404) 321-5973. Jim Bello.
Major Film Distributors Inc., 161 Spring St., N.W.; Tel: (404) 688-1339.
Modern TV, 4705 F. Bakers Ferry Rd.; (404) 696-2026. Bob Cole, TV manager.
New World Pictures, 1587 Northeast Expressway, 30313. (404) 321-2910, Jack Rigg.
Paramount Pictures Corp., 17 Executive Park Dr., N.E., 30329; (404) 325-7674. M. V. McAfee.
Toddy Pictures Co., Box 150, 30301; (404) 355-9654. Ted Toddy, pres.
Transvue Pictures Corp., 161 Spring St., N.W.; Tel: (404) 523-6566.
20th Century-Fox, 2200 Century Parkway, 30345. Jerry Smith, mgr.
United Artists, 401 W. Peachtree NE, Suite 1760, 30308; (404) 552-6386. Larry Terrell, mgr.
Universal Film Exchange, 205 Walton St., N.W., 30303; (404) 523-5081. Weber Howell, mgr.
Warner Bros. 2600 Century Pkwy., 30345. (404) 325-0301. Roy Donaldson.
Clem Williams Films Inc., 1277 Spring St., N.W.; (404) 872-5353.

BOSTON

Association Films, 410 Great Road, Littleton, MA 01460. (617) 486-3518.
Atlantic Releasing Corp., 585 Boylston St., 02116; (617) 266-5400.
Avco Embassy Pictures, 509 Statler Office Bldg., 02116; (617) 482-3325. Marc Halpern.
Buena Vista Distribution Co., 31 St. James Ave., 02116. (617) 426-9360.
Columbia Pictures Corp., 20 Providence St., 02116. (617) 426-8980. Gasper Urban, mgr.
Filmways Pictures, 31 St. James Ave., 02116. (617) 542-0677.
Ellis Gordon Films, 46 Church St., 02116. Ellis Gordon, mgr.
Lockwood & Friedman Film Corp., 430 Park Sq. Bldg., 02116; (617) 482-9717.
National Film Service Operating Corp., 621 E. 1st St., 02127. (617) 268-6510. Jimmie Choukas, mgr.
New England Film Distributing, 31 St. James Ave., 02116; (617) 482-9025.
Paramount Film Distributing Corp., 31 St. James Avenue, 02116. (617) 426-1070. Joe Rathgeb.
Jud Parker Films, 46 Church St. 02116; (617) 542-0744.
Topar Films, Inc., 771 Truman Parkway, Hyde Park, MA 02136, (617) 361-1640.
20th Century-Fox, 31 St. James Ave., 02116; (617) 426-2180. Cary Brokaw.
United Artists Corp., Park Square Bldg., 31 St. James Ave., 02116. (617) 426-6540. Joseph Griffin, br. mgr.
Universal Film Exchanges, Inc., 44 Winchester St., 02116. (617) 426-8760. Joan Corrado, mgr.
Warner Bros. Pictures Dist. Corp., 31 St. James Ave., 02116; (617) 482-3290. Jeff Goldstein, mgr.

BUFFALO

Frontier Amusements Corp., 505 Pearl St., 14202; (716) 854-6752. Mannie Brown, pres.
Paramount Pictures, Inc., 300 Delaware Ave., 14202. (716) 856-3758. Anthony J. Mercurio, mgr.
United Artists Corp., 300 Delaware Ave., 14202. (716) 854-1500. William Abrams, br. mgr.
Wesco Film Service, 108 Gruner Rd., 14225. (716) 897-0467.

CHARLOTTE

Avco-Ambassy Pictures, 230 S. Tryon St., 28202; (704) 375-9827. Donald Osley.
Buena Vista Dist. Co. Inc., 230 S. Tryon St., 28202. (704) 373-0724.
Carolina Booking Service, 291 So. Church St., 28202; (704) 375-7787. Frank Lowry.
Carolina Film Service Inc., 5012 Hovis Rd.; (704) 394-3129. E. J. Poole (ship); Marg. Miller (off.)
Charlotte Booking & Film Dist. Services, 221 So. Church St., 28201; (704) 376-5569. J. R. McClure.
Christian Film Service Corp., 2308 E. 7 St.; 333-5249.
Cinema Film Distributing, 222 S. Church, 28201; (704) 332-8539. Ron Witherspoon.
Clark Film Inc., 230 S. Tryon St., 28202; (704) 376-5569. Bob McClure, branch mgr.
Columbia Pictures Corp., 230 S. Tryon St., 28202. (704) 332-2502. Ed McLaughlin.
Dominant Pictures Co., 230 S. Tryon, 28202; (704) 334-1391. Harry Kerr, mgr.
Galaxy Films, 222 S. Church St.; (704) 372-6747. W. R. James, mgr.
Howco Exchange, 301 So. McDowell, 28204; (704) 834-8510. J. Francis White.
International Amusement Corp., 222 So. Church St., 28202; (704) 332-4163.
JACO Productions, 222 So. Church St., 28202; (704) 375-2519. K. Rogers, Charlie Mincey, branch mgrs.
Paramount Film Dist. Corp., 230 S. Tryon St., 28202. (704) 374-0193. Joe L. Cutrell, mgr.
Premier Pictures, 221 So. Church St., 28202; (704) 332-5101. Jerry Helms.
Pyramid Films, 221 So. Church St., 28202. (704) 333-2894.
Queens City Advertising & Amusement, Inc., 221 So. Church St., 28202; (704) 375-4419. Hugh M. Sykes, Jr.
20th Century-Fox Film Corp., 230 So. Tryon St., 28202. Larry Jameson, mgr.
United Artists Corp. 230 So. Tryon St., 28202. (704) 332-5070. Philip Cutrell, mgr.
Universal Film Exchanges, Inc., 705 E. Morehead St.; (704) 333-5564. Bill McClure, mgr.
Variety Films Inc., 1170 Northwestern Bank Bldg, 28202. (704) 374-1611. Bob McClure, mgr.
Warner Bros. Pictures, 330 S. Tryon St., 28202. (704) 376-5611. Harold Duckett. br. mgr.

CHICAGO

Apache Films, 32 W. Randolph, 60601; (312) 782-5620. Harry Goodman.
Avco-Embassy Pictures Corp. 190 N. State St., 60601. (312) 346-6122. Haywood Mitchusson, br. mgr.
Azteca Films, Inc., 1233 S. Wabash Ave., 60605; WAbash 2-6186. Edward G. Edwards, br. mgr.
Buena Vista Distribution Co. Inc., 8550 W. Bryn Mawr Ave., 60631. (312) 693-8580.

Don Buhrmester & Assoc., Inc., 203 N. Wabash Ave., 60601; (312) 782-0988. Don Buhrmester.
Clark Service, Inc., 2265 W. St. Paul Ave. (312) 342-3140. Jack Bailey.
Columbia Pictures Corp., 8550 W. Bryn Mawr Ave., 60631. (312) 693-3500. Evan Williams.
Dudelson Film Distributing, 1325 So. Wabash Ave., 60605; (312) 922-3546. Moe Dudelson.
Gilbreth Film Co., 32 West Randolph St., 60601; (312) 726-6220. Jack Gilbreth.
J.M.G. Film Co., 32 West Randolph St., 60601; (312) 346-6916. Virgil Jones.
Kaplan-Continental Pictures, 203 No. Wabash Ave., 60601. (312) 782-8413. Sam Kaplan.
William Lange & Associates, 32 West Randolph St., 60601; (312) 332-1734. Bill Lange.
Modern Talking Picture Service, 2020 Prudential Plaza; (312) 337-3252. Edwin Swanson, vice pres.
Modern TV, 1587 Elmhurst Rd., Elk Grove Village, IL 60007; (312) 593-3256.
Paramount Pictures Corp., 111 E. Wacker Dr., 60601; (312) 565-1990. Jeff Blake, br. mgr.
Select Film Co., 32 West Randolph St., 60601; (312) 236-1233. Sam Seplowin.
Teitel Film Corp., 600 S. Michigan, 60605; 427-4551. Chas. Teitel, bd. chmn.
Topar Films, Inc., 345 Fullerton Parkway, 60614; (312) 348-0210. John McLaurin.
Twentieth Century-Fox Film Corp., 35 E. Wacker Dr., 60601. Ray Russo.
United Artists Corp., 203 N. Wabash Ave., 60601; (312) 236-7390. Lou Aurelio, br. mgr.
Universal Film Exchanges, 425 N. Michigan Ave., 60611. (312) 822-0513. W. Gehring, mgr.
Warner Bros. Dist. Corp., 550 W. Jackson Blvd., 60606; (312) 726-1658. Seymour Hite, div. mgr.

CINCINNATI

Avco Embassy Pictures, 35 E. Seventh St. 45202. (513) 381-8480. Robert Rosen, mgr.
Buena Vista Distribution Co., Inc., 636 Northland Rd., 45240. (513) 742-0900. Bob Bruce, mgr.
J.M.G. Film Co., 636 Northland Blvd., 45240. (513) 851-9933. Jay Goldberg, mgr.
MYCO Films, 617 Vine St., 45202; (513) 579-8090. Jo Harrison.
Paramount Pictures Corp., 414 Walnut St., 45202. Mike Share.
Regency Film Distributing, 617 Vine St., 45202; (513) 621-2955. William Blum.
C.J. Ruff Film Distributing Co., 1620 Harrison Ave., 45214; (513) 921-8200.
Edward Salzberg, Inc., 35 East Seventh St., 45202; (513) 241-3671.
States Film Service, 421 Bauer St. (513) 621-4240. Harry Jansen.
20th Century-Fox Film Corp., 617 Vine Street; 241-6460. Tony Knollman, br. mgr.
United Artists Corp., 35 E. 7th St., 45202. (513) 241-1546. Tom Morris, br. mgr.
Universal Film Exchanges, Inc., 1628 Central Parkway, 45210. (513) 421-3820. D. Coons, br. mgr.
Warner Bros. Pictures Dist. Corp., 414 Walnut St., 45202. (513) 241-6824. Bill Waynberg.
Zipp Film Distributors, 617 Vine. (513) 241-5548.

CLEVELAND

Academy Film Service, Inc., 2108 Payne Ave.; 696-0661. Blair Mooney, pres. M. M. Blaettner, mgr.
Avco Embassy Pictures, Three Common Park Sq., 44122; (216) 292-3973. Gordon Bugie, br. mgr.
Buena Vista Distribution Co., Inc., 29001 Cedar Rd., 44124. (216) 442-3270. Jerry Pokorski, mgr.
Cine-Pix, 2108 Payne Ave., 44114; (216) 781-0622. Jack Kaufman.
Columbia Pictures Corp., 23200 Chagrin Blvd., 44122. (216) 292-3610. Richard Sands, br. mgr.
Imperial Pictures, 2108 Payne Ave., 44114. (216) 621-9376.
SelecPictures Corp., 29001 Cedar Rd., Lyndhurst, OH 44124; (216) 461-9770. Jay Schultz, pres.
States Film Service, Inc., Warner Bldg., 2336 Payne Ave.; (216) 771-3723. Harry Lyman, mgr.
20th Century-Fox Film Corp., 29001 Cedar Rd., Lyndhurst, OH 44124. William Anderholt.
United Artists, Corp., 29001 Cedar Rd., Lyndhurst, OH 44124. (216) 461-1266. D. Buckley, mgr.
Universal Film Exchanges Inc., 1721 Superior Ave., 44114; (216) 771-0413. J. Ryan, br. mgr.
Warner Bros. Pictures Distrib. Corp., 29001 Cedar Road, Lyndhurst, 44124. (216) 473-0560. Andrew Silverman.

DALLAS

American National Enterprises, 11422 Harry Hines Blvd., 75229; (214) 243-5171. A. H. Watts.
Avco Embassy Pictures, 10300 N. Central Expressway, 75231; (214) 739-0327. Gene Haufler, br. mgr
Buena Vista Distribution, 6060 N. Central Expressway, 75206; (214) 363-9494. Sebe Miller, dist. mgr.
Central Shipping & Inspection, 2500 S. Harwood St. (214) 421-5411. Wallace Jack, mgr.
Columbia Pictures, 10830 N Central Expressway, 75231. (214) 750-0100. Jack Foley.
Crump Distributors, Inc., 1712 Commerce St., 75201. (214) 741-3370. Jim Crump.
Dal Art Film Exchange, 2017 Young St., 75201; (214) 748-8342. Fred Beiersdorf, pres.
Dimension General Eric Distributing, 10830 N. Central Expressway, 72531; (214) 692-7744. Eric Deneve, Don Scruggs.
Film Booking Service, 500 So. Ervay, Suite 603, 75201; (214) 744-3165. Bennie Lunch.
Films Incorporated, 1414 Dragon St., Dallas 75207; RI 1-4071. Estelle Redd, br. mgr.
G&M Film Distributors, 500 So. Ervay St., 75201; (214) 748-7093. Frank Meyers, Jake Guiles.
Jaco Productions, 500 So. Ervay St., 75201; (214) 748-6145. Bill Hill.
K-Tel Motion Pictures, 1010 W. Mockingbird, 75247; (214) 634-1726. Jerry Malone.
Major Film Distributors, 1907 Elm St. 75201; (214) 744-4069. Jack Durrell.
Orbit Films, 4411 N. Central Exp. 75205; (214) 522-7780. Al Weiner.
Paramount Pictures, 8350 N. Central Expressway, 75206. (214) 369-4600. Terry Kierzek.
Sack Amusement Enterprises, 1710 Jackson St., 75201. (214) RI 2-9445.
Southern Enterprises, 2344 Farrington St., 75207; (214) 634-2690. Carl Sims, gen. mgr.
Starline Pictures, 3220 Lemmon West, 75204; (214) 522-8300. James Prichard.
Topar Films, Inc., 1 No. Park East. (214) 691-7660. M. Parker, J. Kauffman.
Twentieth Century-Fox, 11351 Forest Central Dr., Richard King, br. mgr.
United Artists Corp., Suite 438, 6060 N. Central Expressway, 75206. (214) 692-0777. Enoch Stevens, mgr.
Universal Film Exchanges, Inc., 11551 Forest Central Dr., 75243. Bob Harris, br. mgr.
Variety Film Distributors, 4308 N. Central Expressway, 75206; (214) 827-7800. Vern Fletcher.
Warner Bros. Distributing Corp., 7424 Greenville Ave., 75231. (214) 691-6101. Jackie Stanley.

DENVER

Buena Vista Dist. Co., Inc., 88 Steele St., 80206. (303) 321-1200. Al Hemingway, mgr.
Columbia Pictures Industries, 1860 Lincoln, Suite 250. (303) 534-6341. William Kreigenhofer, br. mgr.
Crest Films, 1860 Lincoln St., 80203; (303) MA 3-1221. Jack Felix, branch mgr.
Denver Shipping & Inspection, 2118 Stout St. (303) 222-5616. Frank Norris, mgr.
Favorite Film of California, Inc., 1860 Lincoln, Suite 416; 623-1221. Dick Notti, br. mgr.
L. E. "Bill" Hobson, 130 Pearl St., 80203; (303) 722-9443. Bill Hobson.
Roy Hunt, 921 21st St, 80205. (303) 893-3998.
J&B Independent Films, 1860 Lincoln, 80203. (303) 255-0495.
McGee Film Distributing, 655 S. Alton Way. (303) 343-3413.
Mountain State Film Dist., 2140 Champa St., 80205; (303) 623-1377, Bates Farley
Paramount Pictures Corp., 158 Fillmore St., 80206; Robert Fox.
20th Century-Fox Film Corp., 55 Steele St., Gary Erickson, br. mgr.
United Artists Corp., 88 Steele St., 80206. (303) 320-4907. Edna Gallagher, br. mgr.
Universal Film Exchanges, Inc., 801 21st St.; (303) 623-3281. Jack Box, br. mgr.
Warner Bros. Dist. Corp., 88 Steele St., 80206. (303) 355-4431. Jay Pekos.

DES MOINES

Columbia Pictures Corp., 1501 Ingersoll Ave., 50309. (515) 243-0105. Glenn Abrams, br. mgr.
D&D Enterprises, Inc, 311 Eleventh St., 50309. (515) 288-6006.

D&H Distributors, P.O. Box 4907, 50306. (515) 276-4263. Dick Haes.
Iowa Film Depot, 3123 Delaware Ave. (515) 265-1469. Art Trombley, mgr.
McCulloch Film Distributing, 500-A Produce Bank Bldg., 55403; (612) 333-2281. Stan McCulloch, owner.
National Screen Service Corp., 1005 High St.; 244-3911. Carl Sokolof, representative.
Paramount Film Distributing Corp., 1535 Linden St., 50309. (515) 288-3638. Charles A. Caliguiri, br. mgr.
Producers Distributing Co., 1219 Paramount Bldg., 50309; (515) 282-5157. Ken Weldon, Mike Weldon.
20th Century-Fox Film Corp., 3737 Woodland Ave., Mary Ann Shaughnessy.
United Artists Corp., 1213 Grand Ave., 50309. (515) 283-0481. George Bloxham.

DETROIT

American National Enterprises, 19400 W. 10 Mile Rd., Southfield, MI 48075; (313) 352-4282. Larry Bishop.
Avco Embassy Pictures, 23300 Greenfield Rd., Oak Park, 48237; (313) 968-3350. Pat Woolcott, bk. mgr.
Columbia Pictures Corp., 24100 Southfield Rd., Southfield 48075; (313) 557-2150.
D&R Distributors, 13131 Fenkell Ave., 48277; (313) 491-2180. Al Dezel, George Rossman.
Gail Film Distributors, 16300 W. 9 Mile Road, Southfield, MI 48075; (313) 557-5024. Arthur Weisberg, pres.
J.M.G. Films, 2330 Greenfield Road, Oak Park, 48237. (313) 968-0500.
Levin Film Distributors, 29501 Greenfield Rd., Southfield, MI, 48072; (313) 559-1101. Nate Levin.
National Film Service, 6111 Concord, 48211; (313) 923-2150. Ed McCauley, mgr.
Paramount Pictures, 23300 Greenfield Rd., Oak Park, 48237; (313) 968-8137. James Goldschlager.
R. Distributors, 13131 Fenkell, 48227. (313) 491-2180.
Regency Film Distributing, 24655 Southfield Road, Southfield, 48075; (313) 354-3245. Herb Gillis.
C.J. Ruff Films, 23300 Greenfield Road, Oak Park, 48237; (313) 968-7770. Dennis Glenn.
United Artists Corp., 24100 Southfield Rd., Southfield, MI 48075; (313) 557-5770. Robert Kapolintz, mgr.
Universal Film Exchanges, Inc., 55 Park St., Troy, MI 48099. (313) 583-1720. D. Gonda.
Warner Bros. Dist. Corp., 20820 Greenfield Road, Oak Park, MI 48237; (313) 564-5826. Don E. Martin, br. mgr.

HONOLULU, HI

Hawaii Nichibel Film Co. Inc., 728 9th Ave., 96816; (808) 737-3455.
Pacific Motion Picture Co. Ltd., 1190 Nuuanu Ave., 96817; (808) 538-1035.
Shochiku Films, 1387 N. Beretania St.; 964-444; 733-8181.

HOUSTON, TX

Conner, Charles M. Productions, 4713 Braeburn Drive, Bellaire, TX (713) 668-9900.

INDIANAPOLIS

Howco Films, 6385 N. Park Ave., 46220; (317) 251-5070. R. V. Jones.
Paramount Pictures, Inc., 428 Illinois Bldg.; 634-7563. John Kane, mgr.
States Film Service, 429 No. Senate Ave. (317) 638-3531. Robert Ahart, mgr.
United Artists Corp., 7002 No. Graham Rd., 46220. (317) 849-2714. Tom Morris.
Zipp Film Distributors, 718 N. Senate Ave. (317) 632-6873.

JACKSONVILLE, FL

Avco-Embassy Pictures, 101 Century 21 Dr., 32216. (904) 721-8365. James Dixon, br. mgr.
Buena Vista Distribution Co., Inc., Seaboard Coastline Bldg., 32202. (904) 358-1221. Bob Pollard, mgr.
Clark Film Inc., 905 North St., 32211; (904) 721-2122. Harry Clark, pres. Belton Clark, gen. mgr.
Columbia Pictures Corp., 9550 Regency Sq. Blvd., 32211. (904) 725-8891. Terry H. Tharpe.
Robert Farber Films, Guaranty Life Bldg.; 356-4232. Robert Farber, mgr.
General Film Productions, 128 E. Forsyth St.; 358-3641. Walter Powell, mgr.
Independent Film Exchange, 128 East Forsyth St.; 356-1475. Mrs. O. Glenn Gryder, mgr., 1011 Houston St. (shipping station.)
Jacksonville Film Service, 2208 West 21 St., (904) 355-5447. B. D. Benton, Bert Benton, Jr.
Paramount Film Dist. Corp., 9550 Regency Square Blvd., 32211. (904) 725-6470. Al Stout, mgr.
South Eastern Entertainment, 10348 Atlantic Circle, 32216. (904) 356-7216. Robert Capps.
20th Century-Fox Film Corp., One Regency Place, 32211. Woodrow Townsend, mgr.
United Artists, Inc., 9550 Regency Square Blvd., Suite 203, 32211. (904) 724-4290. Joseph Kennedy, mgr.
Universal Film Exchanges, Inc., 331 E. Bay St., 32202; (904) 721-1250. G. F. Byro, br. mgr.
Warner Bros. Pictures Dist. Corp., 9550 Regency Square Blvd., 32211. (904) 721-0480. R. E. Heffner.

KANSAS CITY, MO

Avco-Embassy Pictures, 4638 J. C. Nichols Pkwy., 64112. (816) 931-4526. Jerry Brethour, mgr.
Buena Vista Distribution Co., 4210 Johnson Dr., 66205. (913) 362-9500. Jim Witcher, mgr.
Central Shipping & Inspection, 101 West 10th Ave.; (816) 471-0884. Earl E. Jameson, Jr.
Columbia Pictures Corp., 3130 Broadway; 64111. (816) 561-3021. James Domlon, mgr.
Marcus Film Distributing, 3773 W. 95th St., Overland Park, KS, 66206; (913) 381-6222. Ben Marcus, pres.
Mercury Film Co., 3865 W. 95th St., KS, Overland Park, 66206. (913) 383-3880. Bev. Miller.
Midwest Films, 3879 W. 95th St., Shawnee Mission, KS 66206; (913) 381-2058. Gene Irwin.
Motion Picture Booking Agency, 9427 Bluejacket; 888-6590.
Paramount Pictures Corp., 4200 Johnson Dr., Shawnee Mission, KS, 66205. (913) 831-9212. Mike Klein, br. mgr.
Sun International Productions, 25 East 12th St. (816) 471-0415.
Thomas/Shipp Film Distributing, 110 W. 18th St., 64108; (816) 421-1692. Howard Thomas, John Shipp.
20th Century-Fox Film Corp., 106 W. 34th St. Neill Blatt, br. mgr.
United Artists Corp., 1703 Wyandotte, 64108. (816) 471-1123. Morton Truog, br. mgr.
United National Distributors, 1703 Wyandotte, 64108. (816) 474-5330. Gene Irwin.
Universal Film Exchanges, Inc., 1700 Wyandotte St., 64108. (816) 421-5624. S. Miller, br. mgr.
Warner Bros. Distributing Corp., 2440 Pershing Rd., 64108 (816) 474-9909. Frank Rhodes, br. mgr.

LOS ANGELES

American National Enterprises, 6115 Selma Ave., (213) 467-8624.
Amerikana Film Co., 6255 West Sunset Blvd., L.A., 90046. (213) 464-3131.
Avco-Embassy Pictures, 956 Seward St., 90038. (213) 460-7200. Ross Merrin, br. mgr.
Azteca Films, Inc., 555 N. La Brea Ave., 90036; (213) 938-2413.
BFA Educational Media, 2211 Michigan Ave., Santa Monica, 90212. (213) 829-2901.
Seymour Borde Associates, 292 S. LaCienega, Beverly Hills, 90035; (213) 652-6785.
Beuna Vista Dist. Co. Inc., 350 S. Buena Vista St., Burbank, 91521. (213) 841-1000.
Carlyle Films Ltd., 6430 Sunset Blvd., 90028; (213) 466-0864.
Cavalcade Pictures, Inc., 959 N. Fairfax Ave., 90046; (213) 654-4144.
Columbia Pictures Corp., 8671 Wilshire Blvd., 90211. (213) 657-6410. Robert Humak, br. mgr.
Counselor Films, 8816 Sunset Blvd., L.A.; 90069; (213) 659-5720.
Crest Film Distributors, 116 No. Robertson Blvd., 90052; (213) 652-8844. J. Persell.
Crown International Pictures, 292 S. La Cienega Blvd., B.H. 90211; (213) 657-6700.
Dart Enterprises, 5421 Santa Monica Blvd., L.A.; (213) 464-9283.
Far West Films, 116 No. Robertson Blvd., 90048; (213) 659-5161.
Favorite Films of California, Inc., 292 La Cienega Blvd., B.H.; (213) 657-6700.
Film Investment Corp., 333 S. Beverly Drive, B.H.; 90211; (213) 553-5806.
Hollywood International Film Corp. of America, 1044 S. Hill St., L.A. 90015; (213) 749-2067.
Manson International, 9145 Sunset Blvd., 90069; (213) 273-8640. Michael F. Goldman, pres.
Modern Talking Picture Service, 1145 N. McCadden Place, 90038; (213) 462-2202. Jack Whalen, manager.

National Screen Service Corp., 2001 S. Cienega Blvd., L.A. 90034; (213) 871-0598.
Paramount Film Dist. Corp., 9440 Santa Monica Blvd., Beverly Hills, 90210. (213) 550-8600. Walter Lange.
Services Marketing & Distribution Co, P.O. Box 85, Ross, CA 94957.
Summit Film Distributors, 116 No. Robertson Blvd., 90048. (213) 652-7702.
Toho Co., Ltd., 834 S. La Brea Ave., 90036; (213) 933-5877.
Topar Films, Inc., 116 N. Robertson Blvd., 90048. (213) 657-5901. Rick Parker.
Tower Film Co., 8400 W. Sunset Blvd., L.A. 90069; (213) 654-4414.
Twentieth Century-Fox Film Corp., 9440 Santa Monica Blvd., 90210 (213) 550-1044. Donna Littman.
United Artists, 116 No. Robertson Blvd., 90048. (213) 657-7000. Robert Wood, mgr.
United Film Distributing Co., 291 S. La Cienega Blvd., B.H., 90211; (213) 657-6210. J. Crotly.
Universal Film Exchanges, Inc., 8901 Beverly Blvd., 90048. (213) 550-7461. J. Finn, mgr.
Warner Bros. Pictures Distributing Corp., 8484 Wilshire Blvd., 90211. (213) 653-9600. Shirley Becker.

MEMPHIS

Blue Ribbon Pictures, 942 Normandy Ave., 38137. (901) 683-3949.
Clark/Pabst Films, Inc., 1188 Perkins Rd. S. 38117; (901) 683-8182.
Dal-Art Films, Exchange, 4942 Normandy St., 39317. (901) 683-3949.
Dimension General, 138 Huling St., 38102; (901) 278-4442. Jeff Williams.
Don Kay Enterprises, 138 Huling St., 38102; (901) 527-4023. Fordyce Kaiser, branch mgr.
Film Transit, Inc., 291 Hernando St.; 525-6894.
Howco Exchange, 399 South 2nd St.; 526-8328. Charles Arendall, br. mgr.
Memphis Film Service Inc., Homewood Rd., (901) 365-7550. A. S. Crews, G. L. Brandon.
Paramount Pictures, 100 N. Main St., 38103. (901) 527-9575. Tom Donahue.
Starline Pictures, 100 N. Main Bldg., 38103; (901) 527-9424. Bailey Prichard, owner.
Tab Films, 1364 No. Watkins, 38108; (901) 274-6491. Charles Arendall.
United Artists, 1437 Central Ave., 38104. K. Keifer.

MILWAUKEE

Buena Vista Dist. Co., 212 W. Wisconsin Ave., 53203. (414) 273-5111. Carole Sutter, br. mgr.
Independent Film Distributor, 6421 Milwaukee Ave.; 771-9470. Fern Anderson, mgr.
Mescop Distributing, 9235 W. Capitol Dr., 53222; (414) 466-1700. Fred Florence, pres.
Milwaukee Film Center, Inc., 333 No. 25th St.; (414) 344-0300. Oliver Trampe, John Prostinak.
Theatres Service Co., 9235 W. Capitol Dr.; 462-7970. Don Perlewitz, mgr.
United Artists Corp., 212 W. Wisconsin Ave., 53203. (414) 271-6529. John McKenna, br. mgr.

MINNEAPOLIS

Action Films Inc., 7400 Metro Blvd., 941-7769.
Avco Embassy Pictures, 15 S. 9th St. (612) 339-2719. Dean Lutz, mgr.
Buena Vista Distribution Co., Inc., 6950 Wayzata Blvd., 55426. (612) 546-8533. Irving Marks, br. mgr.
Columbia Pictures Corp., 711 Hennepin Ave., 55403. (612) 333-6227. Jack Ignatowicz.
Independent Film Distributors, 2249 Pennsylvania Ave., So.; 545-7015. Abbott Swartz, owner.
Independent Film Service, 245 Second Ave. No.; (612) 335-2203. Jack Bradley, mgr.
William H. Lange & Associates, 2901 Pleasant Ave. So., 55408. (612) 827-5371. Leroy Smith.
McCulloch Film Distributing, 500-A Produce Bank Bldg., 55403; (612) 333-2281. Stan McCulloch, owner.
Mid West Entertainment, 704 Hennepin Ave., 55043; (612) 332-4523. James Payne.
Northwest Cinema, 1602 Midwest Plaza Bldg., 55402; (612) 332-3456. Irving Braverman.
Paramount Pictures Corp., 6950 Wayzata Blvd., 55426. (612) 544-0112. F. C. Myers, br. mgr.
Leroy Smith, 2901 Pleasant Ave., S., 55408. (612) 827-5371.
20th Century-Fox, 600 So. County Rd. #18. Aaron Rosen, mgr.
United Artists Corp., 711 Hennepin Ave., 55403. (612) 333-7276. Walter Badger, mgr.
Universal Film Exchanges, Inc., 6800 Shingle Creek Parkway, 55430. (612) 566-9620. Frank Zanotti, br. mgr.
Viking Films, 1228 Wagon Wheel Rd., Hopkins, Minn., 55343; (612) 338-3841. Jack Kelvie.

NEW HAVEN

New Haven Film Service Inc., 90 Woodmont Rd., Milford, CT, 06460; (203) 878-1465. William Rosen, mgr.

NEW ORLEANS

Avco-Embassy Pictures, 1030 W. Central Expressway, 70130; (504) 739-0327. John Trickett, mgr.
Blue Ribbon Pictures, Inc., International Trade Mart, Suite 1400, 2 Canal St., 70130; 522-8788-89. George Pabst, owner.
Buena Vista Dist. Co., Inc., International Trade Mart, Suite 1438. 2 Canal St., 70130; (504) 525-2258. Larry Fine, mgr.
Clark/Pabst Film, Inc., P O. Box 7865, Metairie, LA, 70010; (504) 733-3555.
Columbia Pictures Corp., 1001 Howard Ave., 70113 (504) 529-2461. Gene Gibbons.
Delta Visual Service, Inc., 715 Girod St.; 525-9061. F. J. Didier, pres.
Film Inspection Service, Inc., 2411 Edenborn Ave., Metairie, LA; (504) 883-5552. Weldon C. Wade mgr. Dan M. Brandon, pres. & gen. mgr.
Goodrow, F. F. Distributor, 214 S. Liberty, 523-1474. F. F. Goodrow, owner.
I.F.I. Films, P.O. Box 8507, Metairie, LA 70011; (504) 837-6107. Charles Varnado.
Independent Films, Box 8007, Metairie, LA 70011. (504) 837-6106.
Jaco Productions, 822 Perdido, 70112; (504) 524-4218. Ken Rodgers.
Don Kay Enterprises, Inc., 1034 Carondelet, 524-2796.
Masterpiece Pictures, Inc., 215 S. Liberty St.; 522-8703. Mrs. Mamie Dureau, pres.
Paramount Pictures Corp., 3525 N. Causeway Blvd., Metairie, 70002. (504) 837-7751. David Garfinkle.
Southern Film Distributing, 143 N. Rampart St., 70112. (504) 581-6700.
Twentieth Century-Fox Distrib. Corp., 3829 Veterans Memorial Blvd., 70003. Jeff Lee, br. mgr.
United Artists Corp., 4900 Veterans Memorial Blvd., Metairie, LA 70002; (504) 885-7333. Greg Perash.
Universal Film Exchanges, Inc., 734A Martin Behrman Ave., Metairie, LA 70005. (504) 837-2631. N. Gazaway, mgr.
Warner Bros. Dist. Corp., 4539 Ino Road, Metairie, LA 70002; (504) 456-1751. David Chinich mgr.

NEW YORK

Audio-Brandon Films, 34 MacQuesten P'way S., Mt. Vernon, NY 10550; (914) 644-5051.
Avco Embassy Pictures Corp., 300 E. 42nd St., 10017; (212) 949-8900. Jerome Horowitz, br. mgr.
Bonded Film Distributors, (Division of Novo Communications, Inc.) 55 Main Street, Fort Lee, NJ 07024; (212) 557-6738. Joseph Marcy, mgr.
Bonded Services, (Division of Novo Communications, Inc.) 733 Third Avenue, New York, NY 10017; (212) 557-6700. Emanuel Kandel, exec. v.p.
Buena Vista Distribution Co., 477 Madison Ave., 10022. (212) 593-8900.
Cinema 5 Ltd., 595 Madison Ave., 421-5555.
Columbia Pictures, 711 Fifth Ave. (212) 751-4400. Joseph Curtin.
ITM Releasing Corp., 321 W. 44 St., 10036; (212) 582-6946.
Marvin Films, 1501 Broadway, 10036; (212) 354-5700. Marvin Friedlander, pres.
Modern Talking Picture Service, 45 Rockefeller Plaza, 10111; (212) 765-3100.
Modern TV, 2323 New Hyde Park Road, New Hyde Park, 11040; (516) 437-6300. Peter Glick, manager.
Paramount Film Distributing Corp, 1 Gulf & Western Plaza; 333-4600. Nat Stern.
Times Film Corp., 144 West 57 St., 10019; PL 7-6980.
Trans-Lux Distributing Corp., 625 Madison Ave., 10022; PLaza 1-3110.
20th Century-Fox Film Corp.; 40 W. 57th St., 10019 (212) 977-5500. Leo Fisch, mgr.
United Artists Corporation, 729 Seventh Ave., 10019. (212) 575-4897. Robert Shein, mgr.
United Film Dist. Co., 11 Middleneck Rd., Great Neck, NY 11021; (212) 895-7100. R. Hassanein.
Universal Film Exchanges, Inc., 445 Park Ave., 10022. (212) 759-7500. A. Quaedvlieg, mgr.

Warner Bros. Distributing Corporation, 75 Rockefeller Plaza, 10019. (212) 484-8960. Herb Gaines, NY sales mgr.

OKLAHOMA CITY

Conner, Charles M., Productions, 4713 Braeburn Dr., Bellaire, TX; (713) 668-9945.
O&A Film Lines, 708 W. Sheridan, 232-9900.
Okla. City Shipping & Inspection Bureau, 809 S.W. 7th; (405) 235-2553. Alfred Baird, mgr.
Screen Guild Productions of Oklahoma, Inc., 708 W. Sheridan; 232-4623. Harry McKenna, mgr.
United Artists Corp., 2000 Classen Blvd., 73106. (405) 528-2888. T. Dyksterhuis.

OMAHA

Buena Vista Film Dist. Co., Inc., 203 572; 393-2321.
Modern Sound Pictures Inc., 1402 Howard St., 68102; (402) 341-8476
Omaha Film Depot, Inc., office 1441 No. 11th St.; (402) 342-6576.

PHILADELPHIA

Alan Pictures, 900 Kings Highway North, Cherry Hill, NJ 08034; (215) 561-0800. Alan Strulson, pres.
Avco Embassy Pictures Corp., 425 15th St., 19102. (215) 563-8345. Tom Mihok, br. mgr.
Buena Vista Distribution, One Cherry Hill, Cherry Hill, NJ 08002. (609) 779-8804.
Capital Film Exchange, 309 N. 13th St. 19107. (215) 567-2698.
Columbia Pictures Corp. 1612 Market St., 19103. (215) 568-3889. Abe Dortheiner.
Filmways Pictures, 1530 Chestnut St., 19102; (215) 568-6684.
Magill Films, 1612 Market St., 19103; (215) 563-7428. Mort Magill.
M.Y. Film Co., Fox Theatre Bldg., 1612 Market St., 19103; (215) 665-9052. Emanuel Youngerman, pres.
Paramount, Fox Bldg., 1612 Market St., 19103. (215) 567-3672. Ralph Garman, br. mgr.
20th Century-Fox Film Corp., 1429 Walnut St., Louis Korte, br. mgr.
United Artists Corp., 117 S. 17th St., 19103. (215) 563-9500. Lew O'Neil.
Universal Film Exchanges, 1165 Marlkress Rd., Cherry Hill, NJ 08003. (609) 424-5045. P. Ciccotta, mgr.
Warner Bros. Pictures Dist. Corp., 400 Market St., 99106. (215) 928-9000. Frank Carroll.

PITTSBURGH

Cinema Consultants, Box 191, Grove City, PA 16127. (412) 458-8503.
New World Pictures of Pittsburgh, 107 6th St., (412) 391-0370. Morris Zyd, div. mgr.
John O'Glaus Agency, Box 18072, 15236. (412) 653-5493.
Paramount Pictures, Inc., Fulton Bldg., 107 6th St.; 281-9270; Don Hicks, mgr.
S. Perilman Films, Fulton Bldg., 107 6th St.; (412) 471-5535.
Pittsburgh Film Service, Inc., Bldg. 16, Nichol Ave., McKees Rocks, PA; (412) 771-2665. Peter Dana, Harold Tinker.
Screenguild Pictures, 415 Van Braam, 15219; (412) 281-1630. William Jenkins.
United Artists Corp., Fulton Bldg., 107 Sixth St.; 471-8960. David Litto, mgr.
Wheeler Films, 625 Fulton Bldg., 107 Sixth St., 15222. (412) 471-8960. David Litto.
Wheeler Film Co., 107 Sixth St., 15222; (412) 471-8225. Ross Wheeler, Jr.

PORTLAND, OR

Portland National Film Service, 2369 N.W. Quimby St. (503) 244-6205. Herb Kirsch, mgr.

ST. LOUIS

Avco-Embassy Pictures, 1750 S. Brentwood Blvd., 63144; (314) 962-7373. Jeannine Wieczorek, broker.
Kahan Film Distributors, 3974 Page Blvd.; (314) 371-6572. Meyer Kahan, mgr.
Kemp Film Distributors Corp., 539 No. Grand Blvd., 63103; (314) 535-1159. Stan Smith.
National Screen Service, 1001 Hanley Industrial Ct., 968-1730.
Paramount Pictures Corp., 539 N. Grand, 63103. (314) 533-4231. Barry Florin.
Thomas/Shipp Films, 539 N. Grand. (314) 535-1117.
United Artists Corp., 1750 So. Brentwood Blvd., 63144. (314) 533-0346. Edward Stevens.

SALT LAKE CITY

American Cinema, 555 East Fourth South. (801) 521-8161. Phillip G. Catherall, gen. sls. mgr.
American National Enterprises, 550 E. 200 South, 84102; (801) 521-9400. Ron Rodgers.
Avco Embassy Pictures, 2095 Douglas St. 84015. (801) 487-4700. Lyle Livsey, br. mgr.
Ed Brinn, Box 1714, 84410. (801) 355-4611.
JD Theatre Service, 1223 S. 20th St., East, 84108; (801) 466-1554. John Dahl.
20th Century-Fox Film Corp., 515 S. 700 East, 84102. Gordon Larsen.
United Artists, P.O. Box 11525, 84147; (801) 532-1224. Robert Loftis.

SAN FRANCISCO

Avco Embassy Pictures, 988 Market St.; 94102. (415) 928-7400. Margaret Rykowski, br. mgr.
Buena Vista Dist. Co., Inc., 680 Beach St., 94109. (415) 441-7114. L. Pilmaier, mgr.
Cardinal Films, 1255 Post St., 94102; (415) 776-2626. R. Stafford.
Columbia Pictures, 595 Market St., 94105. (415) 546-7211. Robert Capps, Jr., mgr.
Favorite Films of Calif. Inc., 230 Hyde St.; (415) 776-4408. N. P. Jacobs, pres. Andy Anderson, br. mgr.
Modern TV, 149 New Montgomery St.; (415) 543-1666. Henry Henn, TV manager.
Pacific Film Enterprises, 1 Hallidie Plaza, 94102; (415) 479-8223. Paul Williams.
Paramount Pictures, Inc., 1700 Montgomery St., 94111; (415) 433-8660. R. Litvin.
Ted Reisch Enterprises, 1325 Laurel St., San Carlos, 94070. (415) 593-7060.
Thomas Brothers, 30 Berry St., 94107. (415) 664-1901.
20th Century-Fox Film Corp., 605 Market St., J. Edward Shugrue, mgr.
United Artists Corp., 681 Market St., 94105. (415) 441-5500. Mike Musarra.
United Film Dist. Co., 127 Golden Gate Ave., 94102; (415) 928-3200. James Peirson.
Universal Film Exchange, 693 Sutter St., 94102. (415) 776-3660. James B. Mooney, mgr.
Warner Bros. Pictures Distributing Corp., 544 Golden Gate Ave., 94102. (415) 474-5938. Mike Timko.

SEATTLE

Buena Vista Dist. Co. Inc., 2419 Second Ave., 98121. (206) 624-0186. Homer Schmitt, br. mgr.
Favorite Films, 2318 Second Ave., 98121; (206) 624-6234. Pete Tolins, br. mgr.
Northwest Diversified Entertainment, 2318 2nd Ave., MA 3-5380.
Paramount Pictures Corp., 975 John St., 98109; (206) 747-1176. Joe vogel.
Parnell Film Distributors Inc., 2318 2nd Ave., 98121; (206) 624-2882.
Saffle's Theatre and TV Service, 975 John St., 98109; (206) 623-5177. M. W. "Bud" Saffle, pres.
Seattle National Film Services, 900 Maynard Ave., So; (206) 682-6685. Norman Jones, mgr.
20th Century-Fox Film Corp., 100 W. Harrison St., 98119. William Robinson.
United Artists Corp., 225-108th Ave., N.E. Bellevue, WA 98004. S. Amato, mgr.
U.S. Navy Dist. Booking Office, 13th Naval Dist.; LA 7-3863. L. D. Palermini, dr. spec. services.
Universal Film Exchange Inc., Lake Villa Apts., 25010 106th St., Apt 8-201E, Kent, WA. (206) 859-1429.

WASHINGTON, DC

Associated Pictures Co., 8601 Liberty Rd., Randallstown, MD; (301) 922-6000, Philip Glazer, mgr.
Avco Embassy Picture Corp., 9470 Annapolis Rd., Lanham, MD (301) 577-2700. Fritz Goldschmidt, br. mgr.
Association-Sterling Films, 1701 N. Fort Meyer Drive, Arlington, VA; 525-4475. C. Edgar Bryant, sales mgr.
Buena Vista Dist. Co., 5205 Leesbury Pike, Baileys Crossroads, VA 22041; (703) 931-6876.
Clark Service, Inc., 3194 Bladensburg Rd. N.E.; (202) 526-0733. Dan McClafferty, mgr.

Columbia Pictures Corp., 2011 I St., N.W., 20006; (202) 466-6570. Martin Zeidman, br. mgr.
DOS Films, Box 87, Nokesville, VA 22123; (703) 368-4429. Doris Steffey.
Key Theatre Ent., 1325½ Wisconsin Ave., N.W., 20007; (202) 965-4401, David Levy.
Modern Talking Picture Service, Inc., 2000 "L" St. N.W. 20036; (202) 293-1222. Robert Kelley, pub. aff. dir.
Paramount Film Dist. Co., 5249 Duke St., Suite 306, Alexandria, VA 22304; (703) 751-1733. George Kelly.
Peerless Dist. Co., 4620 Wisconsin Ave., N.W.; EM 2-8712. Teddy Shull, owner.
20th Century-Fox Film Corp., 1156 15th Street, Suite 701, N.W.; 223-6320. Francis X. Gornley, br. mgr.
United Artists Corp., 7315 Wisconsin Ave., N.W., 20014. (301) 652-6658. Doug Potash, br. mgr.
Universal Film Exchanges, Inc. 6801 Kenilworth Ave. Riverdale, MD 20840. (301) 699-3200. Steve Turner, br. mgr.
Warner Bros. Dist. Corp., 7315 Wisconsin Ave., 20814; (301) 657-9221. Charles Jordan.
Wheeler Film Co., 4701 42nd St. N.W., 20016. (202) 244-1500. Ross Wheeler, Sr.

CANADA

CALGARY, ALTA.

Astral Films, Ltd., 3904—1st St. NE, T2R2E3. (403) 230-3345.
Bellevue Film Distributors, Ltd., 522—11th Ave., SW, T2R068 (403) 264-4660.
Columbia Pictures of Canada, Ltd., 3904—1st St. NE (403) 230-3345
Empire Films Ltd., 3811 Edmonton Trail. Maws Kitigawa, mgr.
Paramount Film Service Ltd., 1410 11th Ave., S.W. (403) 245-4306. Donald Popou, mgr.
United Artists Corp. Ltd., 1003 11th Ave., S.W. T2R0N4. (403) 264-0384. Barry Newstead, mgr.
Universal Films Ltd., 3811 Edmonton Trail; (403) 276-4466. Al Genaske, mgr.
Warner Bros. Pictures Dist. Co. Ltd., 1147 17th Ave., S.W., T2X-0B7. (403) 245-3933. Blain Covert, mgr.

MONTREAL, QUE.

Astral Films, Ltd., 175 Montpellier Blvd., (514) 748-6541.
Bellevue Film Distributors, Ltd., 708 Walnut St.
Cine Pix, Inc., 8275 Maynard St., H4P 2C8. (514) 342-2340.
Columbia Pictures of Canada, Ltd., 175 Montpellier Blvd.; (514) 748-6976.
France Film Company, 1405 Maisonneuve St. J. A. DeSeve pres. and gen. mgr.
International Film Distributors Ltd., 5801 Monkland Ave.; (514) 486-7355.
Paramount Film Service Ltd., 5887 Monkland, (514) 488-9186.
United Artists Corp. Ltd., 3290 Cavendish Blvd., H4B 2M7. (514) 489-8256. Alvin Himelfarb.
Universal Films, Ltd., 8444 St. Laurent Blvd.; (514) 382-8475. J. Hurtubise, br. mgr.
Warner Bros. Distributing Co. Ltd., 5890 Monkland Ave., Rm. 107. (514) 481-2763. Archie Cohen, mgr.

SAINT JOHN, N.B.

Astral Films, Ltd., 55 Bentley St.; (506) 693-3877. Darly Madill, mgr.
Bellevue Film Distributors, Ltd., 77 Germain St.; (506) 693-8601. Vince Winchester, mgr.
Columbia Pictures of Canada, Ltd., 55 Bentley St., L. Simon, mgr.
Paramount Film Service Ltd., 77 Germaine St. (506) 657-5660.
United Artists Corp., Ltd., 77 Germain St., E2L-4S4. Isadore Davis, br. mgr.
Universal Films Ltd., 77 Germain St.; (506) 652-1600. Don McKelvie, mgr.

TORONTO, ONT.

Ambassador Film Distributors, Ltd., 175 Bloor St. E, M4W1C8. (416) 960-3180. Leonard Herberman.
Astral Films Ltd., 720 King St , West, M5V 2T3. (416) 364-3894.
Bellevue Film Distributors, Ltd., 40 Lesmill Rd., Don Mills, Ont. M3B2T5; (416) 449-9322. Herbert S. Mathers, pres.
Columbia Pictures of Canada, Ltd., 720 King St. West. M5V 2T3. Sandra Genesove.
Frontier Amusements, Ltd., 3636 Victoria Park Ave; (416) 499-1106. Morey Hamat.
International Film Distributors, Ltd., 41A Avenue Rd.; (416) 968-2180.
Marden Film Distributors, 4750 Yonge St.; (416) 225-1113.
Modern Talking Picture Service, 143 Sparks Ave., Ontario M2H 255; (416) 498-7290.
Paramount Film Service Ltd. 146 Bloor St.; (416) 922-3600. Robert Lightstone, gen. sales mgr.
Saguenay Films, Ltd., 44 Eglinton Ave., W.; (416) 487-2421.
Twentieth Century-Fox Corp. Ltd., 720 King St., West. Cullen Hulse.
United Artists Corp. Ltd., 2180 Yonge St., M4S2B9. (416) 487-5371.
Universal Films, 2450 Victoria Park Ave.; (416) 491-3000.
Warner Bros. Distributing Co. Ltd., 70 Carlton St.; (416) 922-5145.

VANCOUVER, B.C.

Astral Films Ltd., 1644 W 75 St., Larry Strick.
Bellevue Film Distributors, 788 Beatty St.; (604) 669-4738.
Columbia Pictures of Canada Ltd., 1644 W. 75 Ave.
Paramount Film Service, Ltd., 1646 West 75 Ave. (604) 263-2574. Chris Sullivan, mgr.
Prima Film, Inc., 1734 West Broadway, Suite 7; (604) 732-7421. Miss V. Hosford.
Sovereign Film Distributors, Ltd., 2416 Granville St.; MArine 3843. Fred Stone, mgr.
United Artists Corp., Ltd., 1682 West 7th Ave., V6K155. Thomas Lightburn, mgr.
Universal Films Ltd., 1200 W. 73 St.; (604) 263-1908. Bryan-Rudston Brown, mgr.
Warner Bros. Pictures Dist. Co. Ltd., 8041 Granville St. Roland Rickard, mgr.

WINNIPEG, MAN.

Astral Films, 20 Stevenson Rd.; (204) 632-1270.
Bellevue Film Distributors, 583 Ellice Ave.; (204) 774-4451.
Columbia Pictures of Canada Ltd., 20 Stevenson Rd.; (204) 632-1270
Paramount Film Service Ltd., 315 Donald St. (204) 943-1435.
United Artists Corp., Ltd., 365 Hargrave St., R3B2K3. Meyer Nackimson.
Universal Films Ltd., 583 Ellice Ave.; (204) 786-3397.
Warner Bros. Pictures Distributing Co., Ltd., 583 Ellice Ave. (204) 786-5513. Florent Boulet, mgr.

Distributors of Trailers

ATLANTA

NATIONAL SCREEN SERVICE, 1325 Logan Circle, N.W.; 351-1416. Robert Sedlak, br. mgr.
CINEMA CONCEPTS THEATRE SERVICE, INC., P.O. Box 720576, Atlanta 30328. (404) 396-1055.

BALTIMORE

ALPHA FILM LABORATORIES, Falls Rd.; CL 2-4150. Ernest Wood.
COLORAMA PRODT., 2805 Greenmount Ave.; 889-6777.
FOLKEMER PHOTO SERVICE, 9041 Chevrolet Drive, Ellicott City, MD; 465-7788.
HALLMARK FILMS, 1511 E. North Ave.; Ve 7-3516.
IDEAL PICTURES, 102 W. 25th St.; TU 9-9962. Nelson White.
KUNZ INC., 207 E. Patapsco Ave.; 355-7220.
LEWY FILM & SOUND STUDIOS, 853 N. Eutaw St.; 728-7000.
MONUMENTAL FILMS INC., Rockrose and Maiden Road; 462-1550.
QUALITY FILM LABS., 3010 Greenmount Ave.; 467-1744.
STANSBURY PHOTO FILMS, 2031 Merritt Ave.; 284-4625.
STARK FILMS, Patterson Village Shopping Center; 358-5044.

BOSTON

MASTER MOTION PICTURE COMPANY, 50 Piedmont St.; HAncock 6-3592. Abner Racoo, mgr.
NATIONAL SCREEN SERVICE, 95-97 Broadway, 02116. (617) 542-4476. Joe Rossi, sls. mgr.

BUFFALO

CLARK SERVICE, INC., 108 Gruner Road, 14225; (716) 897-0467. Robert Neffke, mgr.

CHARLOTTE

NATIONAL SCREEN SERVICE, 624 Anderson St., 28205; (704) 333-6628. Milton Lindner, mgr.

CHICAGO

FILMACK CORP., 1327 S. Wabash Ave., 60605; (312) 427-3395. Joseph R. Mack, pres.
NATIONAL SCREEN SERVICE CORP., 1322 Wabash Ave., 60605; (312) 427-8211. Paul Ayotte, sls. mgr.
UNIVERSAL IMAGES, LTD., 9504 E. 63rd Street, Kansas City, MO 64133; (816) 358-6166. Garrett Tuck, president, resident manager—R. M. Upham, P. O. Box 289, Clarendon Hills, IL 60514; (312) 323-6896.

CINCINNATI

NATIONAL SCREEN SERVICE, 1403 Central Parkway; (513) 621-8900. David Fletcher, sls. mgr.

DALLAS

JAMIESON FILM CO., 3825 Bryan St.; TA 3-8158. Bruce Jamieson, pres.; Hugh V. Jamieson, Jr., exec. vice-pres.
NATIONAL SCREEN SERVICE, 7138 Envoy Court. (214) 634-0101. Seymour Kaplan, sls. mgr.

DES MOINES

PARROT FILM SERVICE, 2125 Forest Ave.; 282-4211. Haydon Peterson, mgr.

DETROIT

NATIONAL SCREEN SERVICE, 2943 Cass Ave., 48201; (313) 831-0790. Ivan Clavet, mgr.

KANSAS CITY, MO

NATIONAL SCREEN SERVICE CORP., 1800 Baltimore Ave., 64108; 842-5893. William Lustig, sls. mgr.
UNIVERSAL IMAGES, LTD., 9504 E. 63rd Street, Kansas City, MO 64133; (816) 358-6166. Garrett Tuck, president.

LOS ANGELES

MOONDIAL MANUFACTURING CORP. (dba) Barnett Film Service, 8969 Sunset Blvd., Hollywood, CA 90069. (213) 271-4448. Gabriel Barnett, pres.; Matilda Barnett, v.p.; Morton Bregman, sec.-treas. (Sponsored intermission music trailers).
NATIONAL SCREEN SERVICE, 2001 S. La Cienega Blvd., 90034; 836-1505. Richard Barnes, sls. mgr.
UTA, INC., 1658 Cordova St., 9007; (213) 734-0510. Joan Atamian Bodin, pres.

MEMPHIS

TRI-STATE THEATRE SERVICE, 151 Vance Ave.; 525-8249. Bob Blank, owner.

MINNEAPOLIS

UNIVERSAL IMAGES, LTD., 9504 E. 63rd Street, Kansas City, MO 64133; (816) 358-6166. Garrett Tuck, president, resident manager—J. H. McDonald, 913 Horn Drive, Minnetonka, MN 55343; (612) 545-8617.
SLY-FOX FILMS, INC., 1025 Currie Ave.: 336-6777; Earle C. Sly, vice-pres.

NEW ORLEANS

NATIONAL SCREEN SERVICE, 5624 Jefferson Highway, Harahan, LA 70123; (504) 837-3570. Joseph Moll, sls. mgr.

NEW YORK CITY

MOTION PICTURE ADVERTISING CORP., 641 Lexington Ave.; 935-3434.
NATIONAL SCREEN SERVICE, 1600 Broadway, 10019; (212) 246-5700

PITTSBURGH

NATIONAL SCREEN SERVICE, 107 Sixth St., (depot only). 281-1808.

ST. LOUIS

CINE-GRAPHIC FILMS LABS INC., 101 N. 17th St.; GA 1-5827.
E. H. GOLDBERGER, 1210 Tamm; MI 7-7112.
HARDCASTLE FILM ASSOC., 7319 Wise; MI 7-4200.
PREMIER FILM AND RECORDING CORP., 3033 Locust; JE 1-3555.
SHELBY STORCK AND CO. INC., 4746 McPherson Ave.; FO 1-4200.
TECHNISONIC STUDIOS INC., 1201 S. Brentwood Blvd.; PA 7-1055.

SAN FRANCISCO

MOTION PICTURE SERVICE CO., 125 Hyde St.; ORdway 3-9162; Gerald L. Karski, pres.
NATIONAL FILM SERVICE, 35 Guy Place; 362-2182.

SEATTLE

NATIONAL SCREEN SERVICE CORP., 2413 Second Ave.; MAin 4-2882; Scott Dingler, br. mgr.

Film Carriers

ALBANY

CLARK SERVICE, INC., 24 North Third St., 12204; Tel.: (518) 434-1289; John Pemberton, mgr.
FIRLIK EXPRESS SERVICE INC., 141 South Hawk St.; Tel.: (518) 463-3712; John Firlik, pres.
WESCO FILM SERVICE, INC., 24 N. Third St., 12204. (518) 434-1289. John Pemberton.

ATLANTA

BENTON BROTHERS, 168 Baker St. N.W., 30313; (404) 577-2821.
THEATRES SERVICE CO., 830 Willoughby Way, N.E., (404) 577-2821. W. B. Langston.

BOSTON

NFS OPERATING CORP., 621 East 1st St., 02127. (617) AN 8-6510. Jimmie Choukas.

BUFFALO

OLIN FILM DELIVERY, 141 Nassau Ave., Kenmore; 875-4398.
SMITH & HOWELL FILM SERVICE, INC., 108 Gruner Road, Cheektowaga.
SOUTHWESTERN FILM DELIVERY, 141 Nassau Ave., Kenmore; Tel.: 875-4398.
WESCO FILM SERVICE, INC., 108 Gruner Rd., 14225. (716) 897-0467. Robert Neffke.

CHARLOTTE

CAROLINA FILM SERVICE, INC., 5012 Hovis Rd., 28203. (704) 394-3129. John H. Vickers.
CAROLINA DELIVERY SERVICE CO., INC., 1336 S. Graham St.; 333-5196.
M&M SHIPPING SERVICE, 1336 So. Graham St., 28201.
OBSERVER TRANSPORTATION CO., 1600 W. Independence Blvd.; Tel.: 377-5431.

CHICAGO

ALLIN EXPRESS SERVICE, 2101 S. Peoria, 60608; Tel.: 666-5407; J. B. Allin, owner.
CLARK FILM SERVICE, 222 No. Laflin St., 60607. (312) 829-3700. William Saley.
CONSOLIDATED FILM DELIVERY, 2270 S. Archer Ave.; 326-4120. Ralph McLaughlin, pres.
EMERY AIRFREIGHT, 686-7300; Chicago O'Hare Airport; Zip 60666, O'Hare Field; George Brown, mgr.
LAVIN BROS. FILM DELIVERY SERVICE, 6328 N. Richmond Ave.: BR 4-6450; William Lavin and Lewis Lavin, owners.
NOVO AIR FREIGHT, 2707 Coyle Ave., 593-7300; Donald C. Galloy, v.p. central region; Sid Weintraub, city mgr.
STANDARD TRUCKING CO., 2270 Archer Ave.; 225-1458; Richard McLaughlin, pres.

CINCINNATI

CLARK FILM SERVICE, 421 Bauer St., 45214. (513) 621-4240. Harry Jansen.
FILM SERVICE CO., 1717 Logan St.; 241-5986; Joseph Larkin, mgr.: 471-5848.
LAHMANN FILM SERVICE CO., 214 W. Elder St.; 421-2823; Lawrence Lahmann, mgr.
STATES FILM SERVICE, INC., 421 Bauer St., Tel.: 621-4240; Harry Jansen, mgr.

CLEVELAND

CLEVELAND FILM SERVICE, INC., 2336 Payne Ave., 44114. (216) 771-3723. Harry Lyman.
FILM TRANSIT CO., 2514 Bridge; 771-7636; P. L. Tanner, 3327 Montecello Blvd., Cleveland Heights, Yellowstone 2-4595; Art Marchland, 6545 Aylesworth Dr., TUxedo 5-0635.
STATES FILM SERVICE, INC., 2336 Payne Ave.; Tel.: 771-3723; Harry Lyman, mgr.; K. Wessel, owner.

COLUMBUS, OH

COLUMBUS-CINCINNATI TRUCKING CO., 939 N. 20th St.; 258-2913.

DALLAS

CENTRAL SHIPPING & INSPECTION, 2500 S. Harwood, 75215. (214) 421-5411; Wallace Jack, mgr.
FILM TRANSFER, 1066 W. Mockingbird La.; 637-6690.

DENVER

DENVER SHIPPING AND INSPECTION, 2118 Stout St., 80205. (303) 222-5616; Frank Norris, mgr.
SOUTHWESTERN FILM SERVICE, 2118 Stout St.; 244-2287; Sidney Johnson, mgr.

DES MOINES

D&H DISTRIBUTING, 421 Bower St., 45202.
IOWA FILM DEPOT, 3123 Delaware Ave., 50313. (515) 265-1469; Arthur Trombley, mgr.
NEWS FILM AGENCY, 2500 S. Harwood; HA 8-5181; Bill Durrett, mgr.
IOWA FILM DEPOT, 3123 Delaware, 50313. (515) 265-1469.
KING, H. W. DELIVERY SERVICE, 1320 Grand Ave.; 243-5269; Floyd L. King, mgr.

DETROIT

JAY G. TRUCKING SERVICE, 9237 Warwick; VE 8-5462; Jack Stahley, mgr.
NFS OPERATING CORPORATION, 6111 Concord Ave. 48211. (313) 923-2150. Ed McCauley.
PEP LINES TRUCKING CO., 15120 Third, Highland Park, TUlsa 3-3200.
ROE FILM SERVICE, Bad Axe, MI COngress 9-7927.
SULLIVAN'S FILM SERVICE. 2598 Chalmers, Detroit 48215; VA 1-4713; Ed Sullivan.

GRAND ISLAND, NB

RAPID FILM SERVICE, E. Lincoln Highway; Tel.: 382-4058.

HOBOKEN, NJ

HIGHWAY EXPRESS LINES, INC., Telephone in New York City, WHitehall 3-9188.

HOUSTON, TX

BLUE BONNET EXPRESS AGENCY INC., 5009 Rusk; S. C. Kirby, mgr. WA 3-9101.
FILM EXPRESS AGENCY, 1402 Palmer; CA 5-4170; A. M. Westbrook, mgr.

INDIANAPOLIS

BRADFORD FILM TRANSIT CO., 718 N Senate Ave., 46202. (317) 636-2800. Don A. Pierce.
CLARK TRANSFER, INC., 429 N. Senate Ave., 46204. (317) 638-3531.
INDIANA TRANSIT SERVICE, 4300 W. Morris St., 46241. (317) 241-9321; Wm. R. Smith, mgr.

JACKSONVILLE

JACKSONVILLE FILM SERVICE, 2208 West 21st St. 32209; (904) 355-5447. B. D. Benton, mgr.

KANSAS CITY, MO

CENTRAL SHIPPING & INSPECTION, 101 West Tenth Ave. No., 64116. (816) 471-0884; Earl Jameson, Jr.
INDEPENDENT FILM SHIPPERS, 1800 Central, 64108. (816) 471-3348.

LINCOLN, NB

MILLS FILM TRANSFER, 1234 S. 9th St.; Tel.; 432-1197.

LOS ANGELES

AIR SEA FORWARDERS, 9009 La Cienega Blvd., L.A. 90045; (213) 776-1611.
B&W FILM DELIVERY, 1972 W. Washington Blvd., Los Angeles, CA; 735-8383.
BARNETT INTERNATIONAL FORWARDERS, INC. OF CALIF., 8635 Aviation Blvd., Inglewood, 90301; (213) 776-1178.
BARNETT/NOVO INTERNATIONAL CORP., P.O. Box 91100, Worldway Postal Center, Los Angeles, CA 90009.
BEEKAY FILM DELIVERY, 1972 W. Washington Blvd.; 733-0233.
BLISS FILM DELIVERY, 1120 S. Second St., Alhambra, CA; 282-9532.
FILM TRANSPORT COMPANY OF CALIFORNIA, 1525 West 23rd Street; 734-4141.
GARDNER'S DELIVERY SERVICE, INC., 11229 S. Prairie, Inglewood, CA; 978-9018.
GILBOY, INC., 8536 National Blvd., 90230; (213) 559-2722. James Hunter, mgr.
RAY HACKIE FILM SERVICE, 1613½ W. 20th, 90007.
NOVO INTERNATIONAL CORP., 8635 Aviation Blvd., Inglewood, 90301; (213) 776-1178.
SANTA BARBARA SPECIAL DELIVERY, 1972 W. Washington Blvd., Los Angeles; REpublic 4-5590.

MEMPHIS

FILM TRANSIT, INC., 291 Hernanco St.; Tel.: 525-6894; Guilbert Brandon, president.
MEMPHIS FILM SERVICE, INC., 3931 Homewood Rd., 38118. (901) 365-7550. A. S. Crews.

MILWAUKEE

FILM CENTER, INC. 333 N. 25th St.; 342-3717; Oliver Trampe, mgr.
MILWAUKEE FILM CENTER, INC., 333 No. 25th St., 53233. (414) 344-0300. Oliver Trampe.
NOVO INT'L AIR FREIGHT CORP., 4930 S. 2, 482-3200.

MINNEAPOLIS

CENTURY MOTOR FREIGHT, INC., 3245-4th St. S.E. 55414; 645-9484; Steve Bonello, pres.
HYMAN FREIGHTWAYS, INC., 2690 Prior Ave. N., St. Paul; 633-7310; Eugene Pikovsky, pres.
INDEPENDENT FILM SERVICE, INC., 245 Second Ave. No., 55401. (612) 335-2203. Jim Perrin.

MIDDLEWEST MOTOR EXPRESS, Willmar. Minn.; Barney Johnson, owner.
MIDWEST MOTOR EXPRESS, 2778 Cleveland Ave. No., St. Paul; 633-5653; W. J. Greenstein, vice-president.
MIX TRANSFER, 701 N. 4th St.; 339-9646; Roy W. Mix, owner.
NOVO AIR FREIGHT, INC., 701 N. 4th St.; 336-5341.
PITTS FILM SERVICE, INC., 149 Jackson Ave. N., Hopkins, Minn.; 938-7290; Lawrence Johnson, pres.
RINGSBY UNITED SYSTEM, 2508 Kennedy St., N.E.; Tel.: 331-4330; Jerry Flanagan, terminal mgr.
TWIN CITIES NEWSPAPER SERVICE, INC., 1161 Selby Ave., St. Paul, 644-8810; Leo Loomis, Ed Riley, owners.
TWIN CITY FREIGHT, 2280 Ellis, St. Paul; 645-3661; W. E. Elshotz, Sr., pres.

NEW HAVEN

NEW HAVEN FILM SERVICE, 90 Woodmont Rd., 06460, Milford. (203) 878-1465. William Rosen.
ROSEN FILM DELIVERY SYSTEM, 1890 Dixwell Ave., Hamden: ATwater 8-9161.

NEW ORLEANS

FILM INSPECTION SERVICE, INC., 2411 Edenborn Ave., Metairie, LA 70001. (504) 833-5552. Ian Brandon.
NOVO AIR FREIGHT, 2411 Edenboro Ave., Metairie, LA; 837-0922.
SCHAFFER FILM SERVICE, INC., P.O. Box 9344, Metairie, LA; Tel.: 833-9670; Buddy Schaffer, owner/mgr.
TRANSWAY, INC., (A subsidiary of Novo).

NEW YORK CITY

BONDED SERVICES DIVISION, 733 Third Ave., 10017; (212) 557-6700.
HIGHWAY EXPRESS LINES, INC., 360-16 St., Hoboken, NJ; NY 267-5990.
HUDSON FILM DELIVERY CORP., 560 Main, Fort Lee, NJ; 201-947-5200.
STATE FILM DELIVERY, 560 Main. Fort Lee, NJ; 201-947-5200.
PRUDENTIAL FILM DISTRIBUTORS CORP., 630 Ninth Avenue; COlumbus 5-6884.
RAPID FILM DISTRIBUTORS, INC., 37-02 27th St., Long Island City, 11101; (212) 786-4600.
TRANS-WORLD INTERNATIONAL INC., 767 Fifth Ave., 755-9415.

OKLAHOMA CITY

MAGIC EMPIRE EXPRESS, 920 S.W. 2 Street; Tel.: CE 5-8543; Nolan Nuckols.
MISTLETOE EXPRESS SERVICE, INC., 111 N. Harrison; CE 6-1482; Jack LaMonte, mgr.
O&A FILM LINES, 706 W. Sheridan; CE 2-9900; Roy Avery, mgr.
OKLAHOMA CITY SHIPPING & INSPECTION, 8009 So. West 7th St., 73125. (405) 235-2553. Charles Baird.

OMAHA

OMAHA FILM DEPOT, 1441 N. 11th St., 68102. (404) 342-6576. Charles Janousek.

PHILADELPHIA

CLARK SERVICE INC., 829 No. 29th Street; (215) 232-3500.
HIGHWAY EXPRESS LINES, INC., 3200 S. 20th St., DE 6-6060.
STATES FILM SERVICE, INC., 130 Ferry Ave., Camden, NJ 08104. (215) LO 30303. Meyer Adelman.

PIERCE, NB

PIERCE FILM SERVICE, 512 S. Brown St., Tel.: 329-6365; Dale Hitz.

PITTSBURGH

EXHIBITORS SERVICE CO., 85 Helen St. McKees Rocks; 771-5010; George Callahan, Jr., pres.
KALLY EXPRESS EXHIBITORS SERVICE, 239 Calmar Dr., Verona, PA 15147; 731-9018; Ed M. Duva, owner.
PITTSBURGH FILM SERVICE, Building 16, Nichol Ave., McKees Rocks, 15136. (412) 771-2665; Pete Dana, mgr.

PORTLAND, OR

ADAMSON THEATRES, 909 N.W. 19th; CA 227-5684; Art Adamson, owner; Mark McDougal, mgr.
BELL, HOWARD FILM SERVICE, 907 N.W. 19th St.; 227-2932.
LOVETT INTERSTATE THEATRE SERVICE, 935 N.W. 19th Ave.; CA 4-6108; J. G. (Jack) Lovett, mgr.
PORTLAND NATIONAL FILM SERVICE, 2369 N.W. Quimby St., 97210. (503) 224-6205. Herb Krisch.

PORTSMOUTH, OH

HUNTINGTON/CINCINNATI TRUCKING LINES, LTD., 1006 Gay Street; Tel.: EL 3-7359.

ST. LOUIS

KAHAN FILM DISTRIBUTORS, 3974 Page Blvd., 63113. (314) 371-6572; Meyer Kahan, mgr.
LEWTON FILM SERVICE INC., 12721 San Clemente, 739-0717.

SALT LAKE CITY

NFS OPERATING CORP., 350 West Sixth St., 84110. (801) 364-7729. Rulon Hammer.
WYCOFF CO., INC., 560 S2W; 322-1361; Robert Andrus, mgr.

SAN ANTONIO, TEX.

TEXAS FILM SERVICE, 518 S. Main Ave.; Tel.: CA 7-9295.
VALLEY FILM SERVICE, 518 S. Main Ave.; Phone CA 7-9295.

SAN FRANCISCO

FILM MESSENGER SERVICE, 215 Golden Gate Ave.; 431-4074; Stan Draper, owner.
NFS OPERATING CORP., 35 Guy Place, 94105. (415) 362-2182. H. Przyborowski.

SEATTLE

INLAND NORTHWEST FILM SERVICE, INC., 2201 6th Ave., S.; Tel.: MU 2-4766; R. C. Kercheval, pres.
LOCAL FILM DELIVERY, INC., 900 Maynard South; MU 2-6663; Chas. D. Lawson, pres.
NATIONAL SCREEN SERVICE CORP., 2413 Second Ave.; MAin 4-2882; Arch McGlunchey, br. mgr.
NORTHWEST FILM SERVICE, INC., 79 S. Dearborn St.; MA 2-0241; Paul R. Grunewald, pres.
SEATTLE NATIONAL FILM SERVICE, 900 Maynard Ave., So., 98134. (206)

SYRACUSE, NY

SMITH & HOWELL FILM SERVICE, INC., 505 Burnet Ave.; 2-6836.

TOLEDO, OH

THEATRE TRANSPORT CO., 205 First St.; Tel.: OX 1-0711.

WASHINGTON, DC

MOLITCH FILM SERVICE, 3194 Bladensburg Rd., N.E. 20018. (202) 526-0733. Dan McClafferty.

Canada

CALGARY, ALTA.

FREDDIE'S TRANSPORT, Strathmore, Alta.; 382-3033; J. F. Teare, mgr.

MONTREAL, QUE.

MONTREAL MESSENGER SERVICE REG'D, 444 Notre Dame St. East; 288-1286; Harry A. Cohen, pres.

SAINT JOHN, N.B.

FITZPATRICK TRANSFER, 178 King St. E.; 693-5649; J. P. Fitzpatrick, mgr.

TORONTO, ONT.

MAVETY FILM DELIVERY LTD., 277 Victoria St.; EMpire 3-2413; William Pfaff.

Screening Rooms

ATLANTA

ATLANTA PREVIEW THEATRE, 161 Spring St., N.W.; (404) 523-0663.
20TH-CENTURY FOX EXCHANGE, 2200 Century Parkway, 30345.

BOSTON

E.M. LOEW THEATRES, 164 Tremont St., Boston, 02111. (617) 482-9200.
UNIVERSAL, 44 Winchester St., 02116. (617) 426-8760.

CHICAGO

H. & E. BALABAN THEATRES, 190 North State St., Chicago 60601; (312) 372-2262.
CHICAGO INTERNATIONAL FILM FESTIVAL, 235 Eugenie, Chicago 60614; (312) 644-3400.
DOUGLAS FILM INDUSTRIES, 10 W. Kinzie, 60610. (312) 664-7455.
ESSANESS THEATRES CORPORATION, 54 West Randolph, Chicago 60601. (312) 332-7465.
UNIVERSAL FILM EXCHANGES, INC., 425 North Michigan Avenue, Chicago 60601; (312) 822-0513.

CINCINNATI

20TH CENTURY-FOX FILM CORP., 617 Vine St.; (513) 241-6460.

DENVER

CENTURY SCREENING ROOM, 2100 Stout St.; (303) 534-7611.
THE FLICK THEATRE, Larimer Sq., 80202; (303) 244-9155.
WESTERN CINE SERVICE INC., 312 S. Pearl St., 80209; (303) 744-1017.

KANSAS CITY

COMMONWEALTH THEATRES, 215 W. 18th St., 64108. (816) 471-2390.
DICKINSON OPERATING CO., 5913 Woodson Rd., Mission, KS 66202. (316) 432-2334.

LOS ANGELES

ACADEMY AWARD THEATRE, 8949 Wilshire Blvd.; 278-8990.
AIDIKOFF, CHAS., SCREENING ROOM, 9255 W. Sunset Blvd., L.A.; 274-0866.
BEVERLY HILLS HOTEL, 9641 Sunset Blvd., Beverly Hills, CA 90210; 276-2251.
CENTRE FILMS INC., 1103 N. El Centro Ave., Hollywood 90038; (213) 466-5123.
CONSOLIDATED FILM INDUSTRIES, 959 Seward St., L.A. 90038; 462-3161.
DE LUXE GENERAL, 1546 N. Argyle Ave., L.A. 90028; 462-6171.
DIRECTORS GUILD OF AMERICA, 7950 W. Sunset Blvd., Los Angeles 90046; Tel.: 656-1220.
SAM GOLDWYN STUDIOS, 1041 N. Formosa Ave., L.A. 90046; 851-1234.
HOLLYWOOD SCREENING ROOM, 1800 N. Highland Ave., Suite 511, Hollywood 90028; (213) 466-1888.
DONALD H. NOSSECK THEATRE, 9229 Sunset Blvd., L.A. 90069; 274-4888.
MARTIN NOSSECK PROJECTION THEATRE, 9118½ Sunset Blvd., Hollywood, 90069. (213) 275-3037.
PREVIEW HOUSE, 7655 Sunset Blvd., L.A. 90046; (213) 876-6600.
JOE SHORE'S SCREENING ROOM, 9229 W. Sunet Blvd., L.A. 90046; (213) 274-4888.

MILWAUKEE

CENTRE THEATRE BLDG., 212 W. Wisconsin Ave., 53203, (608) 272-6020.
RKO-STANLEY WARNER THEATRES, INC., 5341 W. Fond du Lac Avenue (Capitol Court Theater); (608) 871-2213.

NEW YORK

(All major producers-distributors have screening rooms at their home offices in New York for their own use. Most also have screening room facilities at their New York exchanges).

CINE-METRIC THEATRE INC., 25 W. 45th St., 869-8670.
CINEMA 405 INC., 600 Madison Ave.; 759-5700.
FELDMAN, ARTHUR, 1585 Broadway, NY; 245-9738.
FIFTH AVENUE SCREENING ROOM, 4 West 56th St.; 541-5454.
MOVIELAB THEATRE SERVICE, INC., 619 W. 54th St.; 586-0360.
MUSEUM OF MODERN ART FILM LIBRARY, 11 W. 53rd St.; 956-7236.
PARK AVENUE SCREENING, INC., 445 Park Ave.; 688-3277.
PRECISION FILM LABS' SCREENING THEATRE, 630 Ninth Avenue. 489-8800.
PREVIEW THEATRE, 1600 Broadway; 246-0865.
RIZZOLI SCREENING ROOM, 712 5th Ave.; 397-3738.
ROSS-GAFFNEY INC., 21 W. 46 St.; 582-3744.
WESTSIDE SCREENINGS, INC., 311 West 43rd St.; 246-3838.

SAN FRANCISCO

MOTION PICTURE SERVICE CO., 125 Hyde St., (415) 673-9162.
THOMAS BROTHERS, 30 Berry St., 94107; (415) 664-1901.

TAMPA

POST PRODUCTION SERVICES, INC., 3808 San Nicholas St.; (813) 253-0400.

WASHINGTON

BYRON, INC., 65 K St., N.E.; (202) 783-2700.
CAPITAL FILM LABORATORIES, INC., 470 "E" St., S.W ; (202) 347-1717.
KOSTER FILM FACILITIES, INC., 1017 "E" St., S.W.; 544-4410.
NATIONAL CINE LABORATORIES, 4319 Rhode Island Ave., Brentwood, MD; 779-6800.

Canada

MONTREAL

ASSOCIATED SCREEN INDUSTRIES LTD., 2000 Northcliffe Ave.; (514) 484-1186.
CONSOLIDATED THEATRES LTD., 5887 Monkland Ave.; (514) 489-8461.
EMPIRE FILMS LTD., 8440 St. Lawrence; (514) 381-5961.
FRANCE FILM COMPANY, 1405 Maisonneuve St.; (514) 526-5971.
UNITED AMUSEMENT CORP. LTD., United Bldg., 5887 Monkland Ave.; (514) 489-8461.

TORONTO

FAMOUS PLAYERS CANADIAN CORP. LTD., 130 Bloor St., W.; (416) 964-5800.
FILM HOUSE, 22 Front St. W.; (416) 363-4321.
ODEON THEATRES (Canada) LTD., 225 Consumers Rd., Willowdale, Ont. (416) 491-1660.
PARAMOUNT FILM SERVICE, 111 Bond St.; (416) 366-8811.
TWENTIETH-CENTURY FOX CORP. LTD., 277 Victoria St., (416) 364-3471.

Film Processing, Raw Stock, Preservation, Storage

Film Labs, Processing

ATLANTA

CINEMA PROCESSORS, INC., 2156 Faulkner Rd.; 30324. (404) 633-1448.
COLOR-GRAPHIC, 3184 Roswl Rd., N.W.; 30305. (404) 233-2174.
SOUTHERN FILM LAB INC., 2381 John Glenn Dr., Chambler, GA 30341. (404) 458-0026.
WOMETCO & CO. FILM LABS, INC., 3630 Peachtree Rd., N.W., 30305. (404)233-5387.

BOSTON

CINE SERVICE LABORATORIES, INC., 51 Kondazian St.; Watertown 02172; (617) 926-0210.
D-4 FILM STUDIOS, INC., 109 Highland Ave., Needham Hts., MA 02194. (617) 444-0226.
FILM SERVICE LAB, 62 Berkeley St.; 02116. (617) 542-1238.
MASTER MOTION PICTURE CO., 50 Piedmont St., 02116; (617) 426-3592.
SPORTS FILM LAB, 361 W. Broadway, 02127. (617) 268-8388.

CHICAGO

ASTRO COLOR LAB, 61 W. Erie, 60610. (312) 280-5500.
CINEMA VIDEO PROCESSORS, 211 East Grand Ave.; 60611. (312) 527-4050.
COLBURN LABORATORY, GEORGE W., INC., 164 North Wacker Drive, 60606. (312) 332-6286.
DOUGLAS FILM INDUSTRIES, INC., 10 W. Kinzie, 60630. (312) 664-7455.
EAGLE FILM LABORATORY, INC., 4971 N. Elston, 60630. (312) 282-7161.
EDITEL, INC., 301 E. Erie (312) 440-2360.
ESKAY FILM SERVICE, INC., 540 N. Lakeshore, 60611; (312) 467-6250.
SUPERIOR BULK FILM CO., 442 N. Wells; SUperior 7-4448.
TITLE-CRAFT, 2722 W. Lawrence, 60625. (312) 271-5731.

CINCINNATI

CINCINNATI FM LABS, 3705 Lonsdale St., 45227; (513) 271-5540.
FILM ART, INC., 2436 Vine St., 45220: (513) 621-4930.
MARATHON MOVIE LABORATORY, 2436 Vine St., 45219; (513) 621-5313.

CLEVELAND

FILM LAB SERVICE, 4117 Prospect Ave., 44103; (216) 881-4510.

COLUMBIA, SC

SOUTHEASTERN FILM COMPANY, 3604 Main St.; 29203; (803) 252-3753.

COLUMBUS, OH

JOHN R. BENNETT, 1617 Aberdeen Ave., 43211. (614) 263-7007.

DALLAS

PSI FILM LAB., 3011 Diamond Park Dr., 75247. (214) 631-5670.
SOUTHWEST FILM LABORATORY, INC., 3024 Fort Worth Ave., 75211. (214) 331-8347.

DAYTON

FILM ASSOCIATES, INC., 4600 S. Dixie Ave., 45439; (513) 293-2164.

DETROIT

DYNAMIC FILM SERVICE, 3028 E. Grand Blvd.; (313) 873-4720.
FILM CRAFT LAB., INC., 66 Sibley, 48201; (313) 962-2611.
THE JAM HANDY ORGANIZATION, 2792 E. Grand Boulevard 48211; (313) 845-2450.
PRODUCERS COLOR SERVICE, 2921 E. Grand Blvd.; (313) 874-1112.

HOLLYWOOD-LOS ANGELES

AMERICAN FILM IND., 1138 N. La Brea Ave., Hollywood 90038; (213) 467-1118.
AUDIO VISUAL HEADQUARTERS CORP., 515 S. Olive St., L.A. 90013; (213) 629-3661.
CINEMA RESEARCH CORP., 6860 Lexington Ave., L.A. 90038; (213) 461-3235.
CINE-CRAFT FILM LAB., 8764 W. Beverly, Hwd., 90048; 213-652-7357.
CINESERVICE, INC., 1459 N. Seward St., 90028. (213) 463-3178.
COLOR REPRODUCTION COMPANY, 7936 Santa Monica Blvd., Los Angeles 90046; 654-8010.
CONSOLIDATED FILM INDUSTRIES, 959 No. Seward St., Los Angeles 90038; (213) 462-3161.
CREST FILM LABORATTRIES, 1141 N. Seward St., L.A. (90038); (213) 462-6696.
DELUXE GENERAL INC., 1546 N. Argyle Ave., Hollywood, CA 90028; 462-6171; 1377 N. Serrano Ave., Hollywood, CA 90027; 462-6171
EASTMAN KODAK LABORATORY, 1017 N. Las Palmas Ave., Los Angeles 90038; (213) 465-7151.
FILMSERVICE LABORATORIES, INC., 6327 Santa Monica Blvd., Hollywood 90038; (213) 464-5141.
FLORA COLOR, 1715 N. Mariposa St., Hollywood 90027; (213) 663-2291.
FOTO-KEM INDUSTRIES, INC., 2800 W. Olive Ave., Burbank, 91505. (213) 846-3101.
HOLLYWOOD FILM ENTERPRISES, INC., 6060 Sunset Blvd., Los Angeles 90028; (213) 464-2181.
MANN, HAL, 7070 Santa Monica Blvd., Hollywood, CA 90038; (213) 466-3231.
METRO-GOLDWYN-MAYER LABS., 10202 W. Washington Bl. C.C. 90230; (213) 836-3000.
METRO-KALVAR, INC., 8927 W. Exposition Bl. L.A. 90034; (213) 870-1602.
MOVIELAB, 6823 Santa Monica Blvd., Los Angeles 90038; (213) 469-5981.
NEWSFILM LABORATORY, INC., 516 N. Larchmont Bl., LA 90004; (213) 462-8292.
PACIFIC TITLE AND ART STUDIO, 6350 Sánta Monica Blvd., Los Angeles 90038; (213) 464-0121.
PEERLESS FILM PROCESSING CORP., 730 Salem, Glendale; (213) 242-2181.
REVERSALS UNLIMITED, 6903 Melrose Ave., L.A. 90038; (213) 938-3641.
TECHNICOLOR INC. (Cinema Systems Division), 4050 Lankershim Bl., North Hollywood, 91608; (213) 769-8500.
UNITED COLOR LAB, INC., 835 N. Seward, 90038. (213) 461-9921.
YALE LABS, 1509 N. Gordon St., L.A. 90028; (213) 464-6181.

HOUSTON

A-V CORP., 2518 No. Blvd., 77006. (713) 523-6701.
THE PHOTOGRAPHIC LABORATORIES, 1926 W. Gray, 77019; (713) 529-5846.

INDIANAPOLIS, IND.

FILMCRAFT LABORATORIES, 5323 W. 86th St., 46268. (317) 299-7070.

MEMPHIS, TENN.

MOTION PICTURE LABORATORIES, INC., 781 S. Main St.; 38106. (901) 774-4944.

MIAMI, FLA.

CAPITAL FILM LABORATORIES, INC., 1998 N.E. 150 St., 33161. (305) 949-4252.
WOMETCO FILM LABS, INC., 65 N.W. Third St., 33128. (305) 377-2611.

MILWAUKEE, WISC.

CENTRAL FILM LABORATORY & PHOTO SUPPLY, 1003 North Third St.; 53203. (414) 272-0606.

NEW ORLEANS

PAN AMERICAN FILMS, 822 N. Rampart St., 70116; (504) 522-5364.

NEW YORK CITY

ACCURATE FILM LABS, 45 W. 45th St., 10036. (212) 730-0555.
BEBELL INC., 416 W. 45th St., New York 10036. (212) 245-8900.
BERKEY FILM PROCESSING, 77 E. 13 St., 10003. (212) 475-8700.
CINEFFECTS VISUALS INC., 115 West 45th St., 10036. (212) 575-5151.
CINELAB CORP., 475 Tenth Ave., 10019. (212) 244-7400.
CONTROL FILM SERVICE, 421 W. 54th St. 10019. (212) 245-1574.
DELUXE GENERAL INC., 630 Ninth Ave., 10036. (212) 489-8800.
DU-ART FILM LABORATORIES, 245 W. 55th St., 10019. (212) 757-4580.
FILMTRONICS LABS, 231 West 54th Street, 10019. (212) 586-3150.
GUFFANTI FILM LABORATORIES, INC., 630 Ninth Ave., 10036; (212) 265-5530.
HUEMARK FILMS INC., 277 East 44th Street, 10017. (212) 986-5066.
J & D LABS INC., 421 West 54th Street, 10019. (212) 581-4725.
KIN-O-LUX, INC., 17 West 45th Street, 10036. (212) 586-1880.
MEDIA FILM SERVICES, INC., 351 West 52nd Street, 10019. (212) 581-4995.
MOVIELAB, INC., 619 W. 54th St., 10019. (212) 586-0360.
PRECISION FILM LABORATORIES, 630 Ninth Ave., 10036. (212) 489-8800.
QUALITY FILM LABORATORIES CO., 619 W. 54 St., 10019. (212) 586-4912.
RAPID FILM TECHNIQUE, INC., 37-02 27th Street, L.I.City, 11101. (212) 786-4600.
TVC LABORATORIES, INC., NEW YORK, 311 W. 43rd St., New York, 10036. (212) 397-8600.
TECHNICOLOR INC., 321 W. 44th St., 10036. (212) 582-7310.
U.S. PHOTO EQUIPMENT, 40-13 104th St., Corona, 11368. (212) 672-3140.
WOMETCO FILM LABS, INC., 310 Cedar Lane, Teaneck, NJ 07666. (201) 836-6706.

NIAGARA FALLS, N.Y.

NIAGARA PHOTO CENTER, 726 Division Ave. 14305; (716) 285-2914.

OMAHA

CORNHUSKER FILM PROCESSING LAB., 1817 Vinton St., 341-4290.

PHILADELPHIA

NEWS REEL LABORATORY, 1733 Sansom St., Philadelphia 3; RIttenhouse 6-3892; Louis W. Kellman, pres.

PITTSBURGH

WRS MOTION PICTURE LAB., 210 Semple St., Pittsburgh 15213; (412) 683-6300.

PORTLAND, ORE.

FORDE MOTION PICTURE LABS., 2153 NE Sandy Blvd., 97232; (503) 234-0553.
KING FILM LAB., 1501 SW Jefferson St., 97201; (503) 224-8620.

SALT LAKE CITY, UTAH

STOCKDALE & CO., INC.—PHOTO-TECH. LAB., 200 E. 1st South St.; 521-3505.

SAN FRANCISCO

DINER, LEO FILMS INC., 332-350 Golden Gate Ave. 94102; (415) 775-3664.
HIGHLAND LABS., 90 Tehama St., 94105; (415) 986-5480.
D.D. MONACO FILM LABORATORY, 234 Ninth St., 94103. (415) 864-5350.
MULTICHROME LABORATORIES, 760 Gough St., 94102. (415) 431-6567.
PALMER, W.A., FILMS, INC., 611 Howard St., 94105. (415) 986-5961.

SEATTLE, WASH.

FORDE MOTION PICTURE LABORATORY, 306 Fairview Ave. N., 98109. (206) 682-2510.

SHREVEPORT, LA.

DELTA FILMS, 327 Market Street, 71101; (318) 423-2679.

SPRINGFIELD, MASS.

BAY STATE FILM PRODUCTIONS, INC., 35 Springfield St., Agawarm 01101. (413) 734-3164.

TAMPA, FLA.

BEACON FILM LABORATORY, 3705 N. Nebraska Ave., Tel.: (813) 248-4935.

WASHINGTON, D.C.

BYRON, INC., 65 K Street, N.E., 20002. (202) 783-2700.
CAPITAL FILM LABORATORIES, INC., 470 E St., S.W., 20024. (202) 347-1717.
WASHINGTON FILM LABORATORIES, INC., 1042 Wisconsin Ave., N.W., 20007. (202) 333-1162.

CANADA

MONTREAL, QUE.

ASSOCIATED SCREEN INDUSTRIES, LTD., 2000 Northcliffe Avenue. Montreal. Subsidiary of Du-Art Film Labs, Inc.
BELLEVUE-PATHE LTD., 2000 Northcliffe Ave., (514) 484-1186.
BIER, DAVID STUDIOS & FILM LABS., 265 Vitre St., West, Montreal.
QUEBEC FILM LABS, 1085 Rue St. Alexandre. (514) 861-5483.

OTTAWA, ONT.

GRAPHIC FILMS LTD., 19 Fairmont Ave.; (613) 728-3513.

TORONTO, ONT.

BELLEVUE-PATHE LTD., 9 Brookhouse Rd., (416) 259-7811.
BEST FILM LAB., 344 Jarvis St. (416) 921-2137.
NORTHERN MOTION PICTURE LABORATORIES, LTD., 65 Granby St.; (416) 362-7631.

Film Preservation

THE CINECURE COMPANY, 1112 Lancaster Ave., Rosemont Pennsylvania, PA 19010 (215) 527-1855 Walter E. Shank, pres.
THE DURAFILM CO., 137 No. La Brea Ave., Hollywood, CA 90036. (213) 936-1156. Madelon Cohen, pres.
FILMLIFE INCORPORATED, Filmlife Bldg., 141 Moonachie Road, Moonachie, NJ 07074. (201) 440-8500. Marvin A. Bernard, bd. chm. & pres.
FILMTREAT INTERNATIONAL CORP., 250 W. 64 St., New York, NY 10023. (212) 799-2500. Y. W. Mociuk, pres. (See isplay ad on page 413.)
PEERLESS FILM PROCESSING CORPORATION, 250 W. 64th St., New York, NY 10023. (212) 242-2181. Y. M. Mociuk, pres.
RESEARCH TECHNOLOGY, INC., 4700 Chase Ave., Lincolnwood, IL 60646. 677-3000.

Photo Reproduction Labs

APCO-APEDA PHOTO CO., INC., 250 West 54th St., NY 10019, NY, (212) 586-5755.
BEBELL INC., 416 West 45th St., New York 10036; (212) 245-8900. 16mm. 35mm. reversal processing, black & white and color release prints. Slides and slide films, camera masters, dupes. Giant size color prints, plus transparencies.
FRANKLIN PHOTOS, INC., 353 West 48th St., N.Y., NY 10036. (212) CI 6-4255.

Film Storage Vaults

ATLANTA

ATLANTA FILM SERVICE, 161 Spring St., N.W.; Tel.: 524-5380. Geoffrey Tyers, pres. R. D. (Curly) Burns, mgr.
BENTON FILM FORWARDING CO., 168 Baker St., N.W.; 577-2821. William B. Langston, mgr.

HOLLYWOOD-LOS ANGELES

(Hollywood studios have own storage vaults)

BEKINS VAN AND STORAGE COMPANY, 1025 North Highland Avenue, (213) 466-9271.

BONDED SERVICES, INC., 8290 Santa Monica Blvd., 90046; (213) 654-7575.

CONSOLIDATED FILM INDUSTRIES, 959 No. Seward Street, Hollywood, CA 90038. (213) 462-3161. (Stores only film which Consolidated Laboratory is handling.)

EVCO REFRIGERATED FILM VAULTS, 838 N. Seward, Hollywood, 90038. (213) 464-9252.

HOLLYWOOD FILM CO., 956 N. Seward St. 90038; (213) 462-3284.

PRODUCERS FILM CENTER, 948 N. Sycamore Ave., 90038; (213) 851-1122.

NEW YORK CITY

BEKINS ARCHIVAL SERVICES, INC., 609 W. 51 St., 10019. (212) 489-7890.

BONDED SERVICES, 733 Third Avenue, New York, NY 10017; (212) 557-6700.

BONDED FILM STORAGE, 550 Main Street, NJ 07024; (212) 557-6732.

RAPID FILM TECHNIQUE, INC., 37-02 27 St., L.I. City 11101. (212) 786-4600.

TORONTO

BONDED FILMTREAT, 205 Richmond St. W.; (416) 363-6414.

CENTRAL FILM SHIPPING & STORAGE COMPANY, 277 Victoria St., Toronto; EM. 3-2909.

PEACOCK FILM STORAGE CO., Sheppard Ave., Agincourt, Ont.; HIckory 4-6135.

SPONSOR FILM SERVICES CORP. LTD., 443 Jarvis St., Toronto; WA 2-3159.

SYNDICATED FILM SERVICES LTD., 103 Church St., Toronto; EM. 4-1475.

Raw Stock Manufacturers

AGFA-GEVAERT, INC., (manufacturer, distributor 35mm 16mm color, black, white raw stock. Executive offices: 275 North St., Teterboro, NJ; 212 LO 5-7850 or 201 288-4100. Branches: 1025 Grand Central Ave., Glendale, CA, 91201, 213 246-8141; 6901 N. Hamlin Ave., Chicago, IL, 60645, 312 675-6000; 1019 Collier Rd., N.W., Atlanta, GA, 30318, 404-355-7450; 460 Tottem Pond Rd., Waltham, MA, 02154, 617-890-5430; 1355 Conant St., Dallas, TX, 75207, 214-ME1-7290; 31 East 28th St., New York City, 10016. (212) 685-6641; 1 West Hill Drive, Crocker Industrial Park, Brisbane, CA 94005. (415) 467-2330.

EASTMAN KODAK CO., 343 State St., Rochester, NY Tel.: (716) 325-2000. 1901 W. 22nd St., Oakbrook, IL 60521. Tel.: (312) 654-5300. 6706 Santa Monica Blvd., Hollywood, CA. Tel.: (213) 464-6131. 1133 Ave. of Americas, New York City 10036. Tel.: (212) 262-6000.

FUJI PHOTO FILM U.S.A., INC., (distributor of Fuji Professional Motion Picture Film) 350 Fifth Ave., N.Y., NY, 736-3335.

GAF CORP., Binghamton, NY 13902; (607) 729-6555.

GEVAERT COMPANY OF AMERICA, INC., (see Agfa-Gevaert).

ILFORD, INC., 70 W. Century Rd., Paramus, NJ 07562; (201) 265-6000.

3M COMPANY (MINNESOTA MINING & MANUFACTURING CO.), Photographic Products Division, manufacturer and distributor of Ferrania 3M motion picture films; 2501 Hudson Rd., St. Paul, Minn. 55101, (612) 733-0020; 845 Third Ave., New York, NY, (212) LT-1-1416; 6023 So. Garfield Ave., Los Angeles, CA 90022, (213) 685-9600.

RESEARCH TECHNOLOGY, INC., 4700 Chase Ave., Lincolnwood, IL 60646. (312) 677-3000.

Stock-Shot Film Libraries

HOLLYWOOD-LOS ANGELES

BUDGET FILMS, 4590 Santa Monica Blvd., L.A. 90029; 660-0187.

COLOR STOCK LIBRARY, INC., 910 N. Citrus Ave., (213) 466-1344.

EVCO LIBRARY, 838 N. Seward, Hollywood 90038; (213) 464-9252.

SHERMAN GRINBERG FILM LIBRARIES, INC., 1040 N. McCadden Pl., Hollywood, CA 90038; (213) 464-7491.

METRO-GOLDWYN-MAYER, 10202 W. Washington Blvd., Culver City, CA 90230; (213) 836-3000.

PRODUCERS LIBRARY SERVICE, 7325 Santa Monica Blvd., Hollywood 90046 (213) 851-2201.

ALLAN SANDLER FILM LIBRARIES, INC., 1001 No. Poinsetta Place, Hollywood; (213) 876-2021.

TRANS AMERICA FILM CORP., 1680 N. Vine St., L.A. 90028; 466-7575.

UNITED AIRLINES, (stock footage), 626 Wilshire Blvd., L.A. 90017; (213) 482-2000.

UNIVERSAL CITY STUDIOS, 100 Universal City, 91608; (213) 985-4321.

NEW YORK CITY

FOTOSONIC INC., 15 West 46th St., 10036. (212) JU 6-0355.

HEARST METROTONE NEWS, 235 E. 45 St. 10017. (212) 682-7690.

MOVIETONEWS, INC. FILM LIBRARY, (Subsidiary of 20th Century Fox), 1345 Ave. of Americas, 10019 (212) 397-8548.

MUSEUM OF MODERN ART FILM LIBRARY, 11 West 53rd Street, 10019. (212) CIrcle 5-8900.

SHERMAN-GRINBERG FILM LIBRARIES, INC., 630 Ninth Avenue, NY 10036; (212) 765-5170.

STOCK SHOTS TO ORDER, 521 Fifth Ave., New York, NY 10017. (212) 682-5844.

STRATFORD INTERNATIONAL FILM SEARCHERS, INC., 250 West 57th St., New York, JU 6-3828.

TELENEWS FILM CORP., 235 E. 45 St. 10019. (212) 682-5600.

Services for Producers

Advertising, In-Theatre

MOONDIAL MANUFACTURING CORP. (dba) Barnett Film Service, 8969 Sunset Blv., Hollywood, CA 90069. (213) 271-4448. Gabriel Barnett, pres.; Matilda Barnett, v.p.; Morton Bregman, sec.-treas. (Sponsored theatre clocks especially designed and perfected for advertising use in motion picture theatres.)

MOVIE MEDIA NETWORK, INC., 2600 Douglas Road, Coral Gables, FL 33134. (305) 448-0008. David I. Weiss, pres. (National program of in-theatre poster advertising; company installs posters of products and services for the general consumer public on the walls of theatre lobbies)

Animals and Trainers

EAST COAST

ANIMAL ACTORS INTERNATIONAL, INC., R.D.3 Box 221, Washington, NJ 07882 (201) 689-7539. Professionally trained animals for all media, natural filming locations.

ANIMAL TALENT SCOUTS, INC., 331 W. 18th St., New York, NY 10011. (212) 243-2700.

DAWN ANIMAL AGENCY, 160 W. 46th St., New York, NY 10036. (212) 757-9396.

WEST COAST

ANIMAL ACTORS OF HOLLYWOOD, 864 W. Carlisle Rd., Thousand Oaks, CA 91361 (805) 495-2122.

ANIMAL TRAINERS UNLIMITED, 12617 Foothill Blvd., Sylmar, CA 91342 (213) 896-0068.

LION ANIMAL RENTAL, Box 259, Ventura, CA 93001 (805) 648-6550.

PUMA PAW RANCH, P.O. Box 606, Rifle, CO 81650 (303) 625-2786.

WORKING WILDLIFE, Box 65, Acton, CA 93510 (805) 947-4041.

Animation

CALIFORNIA

BOSUSTOW, STEPHEN, PRODUCTIONS, 1649 Eleventh St., Santa Monica, CA 90404. (213) 394-0218.

HANNA BARBERA, 3400 West Cahuenga Blvd., Hollywood, CA 90028. (213) 466-1371.

IMAGINATION INC., 443 Jackson St., San Francisco, CA (415) 433-5480.

MONTROY SUPPLY CO., 4165 Beverly Blvd., Hollywood, CA (213) 662-2111.

QUARTET FILMS, INC., 5631 Hollywood Blvd., Los Angeles CA 90046. (213) 464-9225. Arthur Babbitt.

RUBY-SPEARS PRODUCTIONS, 3400 Cahuenga Blvd., W. Los Angeles, 90068. (213) 764-7700.

THOMAS BROTHERS, 30 Berry St., San Francisco, CA 94107. (415) 664-1901.

NEW YORK

ANIMATED CAMERA WORKSHOP, 51 East 42 St., N.Y., NY 10017 (212) 687-5009.

ANIMATED PRODUCTIONS, INC., 1600 Broadway, N.Y., NY 10019 (212) 265-2942.

BEBELL & BEBELL COLOR LABORATORIES, INC., 416 West 45th St., N.Y., NY 10036 (212) 245-8900.

ELINOR BUNIN PRODUCTIONS, INC., 30 East 60 St., N.Y., NY 10022 (212) 688-0759.

CLARK, IAN, 229 E. 96 St., New York, NY 10028 (212) 289-0998.

CREATIVISION, INC., 295 West 4th St., N.Y., NY 10012 (212) 924-3935.

GRANATO ANIMATION PHOTOGRAPHY INC., 15 West 46th St., N.Y., NY 10036 (212) 869-3231

HARVEY FAMOUS CARTOONS, 15 Columbus Circle, N.Y., NY 10024 (212) 582-2244.

HUBLEY STUDIO, 815 Park Ave., N.Y., NY 10021 (212) 744-8050.

LEO ANIMATION CAMERA SERVICE, 15 West 46th Street, N.Y., NY 10036 (212) 582-2515.

NATIONAL SCREEN SERVICE CORP., TV & INDUSTRIAL FILMS DIVISION, 1600 Broadway, N.Y., NY 10019 (212) CI 6-5700.

THE OPTICAL HOUSE, 25 West 45th St., N.Y., NY 10036 (212) 757-7840.

OVATION FILMS, 33 W 46 St., N.Y., NY 10036 (212) 581-4406.

RAINBOW FILM EFFECTS, INC., 45 West 45th Street, N.Y., NY 10036 (212) 247-1676.

STAHL, AL, PRODUCTIONS, 1600 Broadway, New York, NY 10019 (212) CO 5-2942.

TELEMATED M.P., 51 East 42, N.Y., NY 10017 (212) 682-3434.

ZEPLIN PRODUCTIONS, 850 Seventh Ave., N.Y., NY 10019. (212) 582-6633.

Camera Sales, Rental, Repairs

NEW YORK

ARRIFLEX CO. OF AMERICA, P.O. Box 1050, Woodside, NY 11377 (212) 932-3403.

CAMERA MART, INC., 465 West 55th St., 10019 (212) 757-6977.

CAMERA SERVICE CENTER, 625 W. 54th St., 10019 (212) 757-0906.

CINE 60 INC., 630 Ninth Ave.; 10019 (212) JU 6-8782.

F&B/CECO/SOS, 315 W. 43rd St., New York 10036 (212) 586-1420.

FILM EQUIPMENT RENTAL CO., 419 W. 54 St., 10019 (212) 581-5474.

GENERAL CAMERA, 471 11th Ave.; 10018 (212) 594-8700.

NATIONAL CINE EQUIPMENT, INC., 4140 Austin Blvd., Island Park; (212) 799-4602.

OLDEN CAMERA & LENS CO., INC., 1265 Broadway; (212) 725-1234.

WILLOUGHBY CAMERA STORES, 110 W. 32 St., 10001 (212) LO 4-1600.

HOLLYWOOD

AKKAD INTERNATIONAL PRODS. INC., 8730 Sunset Blvd., CA 90069; (213) 657-7670.

AREMAC CAMERA CO., 5609 Sunset Blvd., L.A. 90028; (213) 469-2988.

ARMISTEAD, MARK INC., 1041 N. Formosa, L.A. 90046; (213) 467-6181.

BACH AURICON, INC., 6950 Romaine St., Hollywood 90038 (213) 462-0931.

BELL & HOWELL, 623 Rodier Drive, Glendale, CA; (213) 245-6631.

BIRNS & SAWYER, 1026 N. Highland Ave., Hollywood 90038; (213) 466-8211.

F&B/CECO/SOS, 7051 Santa Monica Blvd., L.A. 90038; (213) 466-9361.

GALLI, HERMAN, CAMERA SERVICE, 6804 Melrose Ave., L.A. 90038; (213) 931-4111.

GORDON ENTERPRISES, 5362 N. Cahuenga Blvd., N. Hollywood 91601; (213) 985-5500.

HOLLYWOOD CAMERA, 1255 No. La Brea Ave., 91603; (213) 783-1134.

HOLLYWOOD FILM CO., 956 Seward St., Hollywood, 90038; (213) 462-3824.

KENWORTHY SNORKEL CAMERA SYSTEMS, INC., P.O. Box 49851, L.A. 90049; (213) 476-4100.

MOLE-RICHARDSON CO., 937 N. Sycamore Ave., Hollywood, 90038; (213) 851-0111.

PILL, JACK & ASSOCIATES INC., 6370 Santa Monica Blvd., Hollywood, CA 90038; (213) 466-5391.

TECH-CAMERA RENTALS, INC., 6370 S. Monica Blvd., 90038; (213) 466-3238.

UPS STUDIO RENTALS, 6561 Santa Monica Blvd., Hollywood, CA 90038; (213) 461-1442.

Checking Theatre Attendance

CERTIFIED MARKETING SERVICES, INC., 120 N. Main St., New City, NY 10956; (914) 638-1021. National field interviews; in-theatre and telephone. William P. Smith, pres.
CERTIFIED REPORTS, INC., (CRI) Kinderhook, NY 12106; (518) 758-6403. Theatre checking; industry research. Harold Roth, pres.
VERIFIED REPORTS, INC., Hatfield Road, R. D. 1, Mahopac, NY 10541; (914) 628-6412. Blind (covert) theatre checking; national coverage. Diane Rosenblith, pres.

Commercial Jingles

AEOLUS PRODUCTIONS, INC., 119 West 57th St., N.Y., NY 10019 (212) 541-9388.
FAILLACE PRODUCTIONS, 207 East 45th Street, New York, NY; 687-9080. Tony Faillace.
MAMORSKY, ZIMMERMANN & HAMM, INC., 56 West 45th St., N.Y., NY 10036 (212) 986-1084.
MUSIC MAKERS, INC., 200 Central Park So., N.Y., NY 10019 (212) 245-3737.

Consultants

AMERICAN FILM PRODUCTIONS INC., 1540 Broadway, N.Y., NY 10019 (212) 582-1900.
BROADCAST INFORMATION BUREAU, 30 East 42 St., N.Y., NY 10017 (212) 687-1668.
FILM COUNSELORS, INC., 500 Fifth Ave., N.Y., NY 10022 (212) 736-7220.
FILM PLANNING ASSOCIATES, INC., 305 East 46th St., N.Y., NY 10017 (212) 725-9316.
NATIONAL SCREEN SERVICE CORP., TV & INDUSTRIAL FILMS DIVISION, 1600 Broadway, N.Y., NY 10019 (212) CI 6-5700.
RADIO FEATURES INC., 527 Madison Ave., N.Y., NY 10022 (212) EL 5-4616.
ROSS-GAFFNEY, 21 West 46th St., N.Y., NY 10036 (212) JU2-3744.

Costumes & Uniforms

NEW YORK

ANIMAL COSTUME CO., 236 West 55th Street, 10019 (212) 245-5199.
BROOKS-VAN HORN COSTUME CO. INC., 117 West 17th St., 10011 (212) 989-8000.
CHENKO STUDIO, 108 W. 45th st., 10036. (212) 582-1646.
EAVES COSTUME COMPANY, INC., 151 W. 46th Street, 10019 (212) 757-3730.
KARINSKA, MME., 6 West 61 St., 10023 (212) 247-3341.

WEST COAST

A-CINEMA HOLLYWOOD COSTUMERS, 5514 Hollywood Blvd., 90028; (213) 464-9894.
ACUNA, FRANK, 8212 Sunset Blvd., L.A. 90046; (213) 656-1413.
ADELE'S OF HOLLYWOOD, 5059 Hollywood Blvd., L.A. 90038; (213) 662-2231.
IC COSTUME RENTALS, 6121 Santa Monica Blvd., L.A. 90038; (213) 496-2056.
TUXEDO CENTER, 7360 Sunset Blvd., L.A., 90046 (213) 874-4200.
WESTERN COSTUME CO., 5335 Melrose Ave., Hollywood, CA 90038 (213) 469-1451.

Cutting Rooms

LOS ANGELES

F&B/CECO EDITING CENTER, 7051 Santa Monica, L.A., CA 90038 (213) 466-9361.
HORIZONTAL EDITING STUDIOS, 6253 Hollywood Blvd., Hollywood, CA 90028; (213) 461-4643.

NEW YORK

ANIMATED PRODS., INC., 1600 Broadway, New York, NY 10019. (212) 265-2942.
CAMERA MART, THE., 456 W. 55 St., New York, NY 10019 (212) 757-6977.
CRITERION FILM LABORATORIES, INC., 415 West 55th St., 10019 (212) 265-2180.
DE LUXE GENERAL INC., 630 Ninth Ave., New York, NY 10036 (212) 489-8800; 1546 N. Argyle Ave., L.A. 90038; (213) 462-6171.
MOVIELAB THEATRE SERVICE, 619 West 54th Street, New York, NY 10019 (212) 586-0360; 6823 Santa Monica Blvd., L.A.; (213) 469-5981.
PREVIEW HOUSE, 7655 W. Sunset Blvd., L.A. 90046; (213) 876-6600.
PREVIEW THEATRE, 1600 Broadway, New York, NY 10019 (212) 246-0865.
ROSS-GAFFNEY, INC., 21 West 46th St., New York, NY 10036 (212) 582-3744.

Editing Equipment

CIRO EQUIPMENT CORPORATION, 6820 Romaine St., Hollywood, CA (213) 467-1296. (8, 16, 35 & 70mm self-perforating film splicers and splicing tape) Barry Green, pres.
KLM ASSOCIATES, INC., 11810 Charen Lane, Potomac, MD. (301) 299-7259 (sells and rents film editing equipment; horizontal editing tables and related hardware) Mar McElroy, pres.
TWENTY-FOURTH FRAME, 303 Depot., N.W., P.O. Box 2167, Christiansburg, VA 24073. (703) 382-4135 (manufactures 35mm, 16mm, two-picture 3-sound, 8-plate flatbed editing machines) Ike Jeanes, owner.

Lighting Equipment

BARDWELL & McALLISTER, 7269 Santa Monica Blvd., Hollywood, CA 90038; (213) 876-4133.
BATES LIGHTING CO., 1215 Bates Ave., L.A. 90029; (213) 661-1262.
BIRNS & SAWYER, 1026 N. Highland Ave., L.A. 90038; (213) 466-8211.
BERKEY-COLORTRAN, INC., (A division of Berkey Photo, Inc.), 1015 Chestnut St., Burbank, CA 91502; (213) 843-1200; 842 Broadway, N.Y., 475-8700. (Berkey Photo).
BLAKE FILMS, 104 West Concord, St., Boston, MA 02118. (617) 267-7713.
BOKEN, INC., 349 West 48th St., N.Y., NY 10036 (212) LT 1-5626.
THE CAMERA MART, 456 West 55th St., N.Y., NY 10019; (212) 757-6977.
F&B/CECO/SOS, 315 W. 43rd St., N.Y. 10036; (212) 586-1420; 7051 Santa Monica Blvd.; 90038 (213) 466-9361.
ALAN GORDON ENTS., CO., 5342 Cahuenga, No. Hollywood, CA (213) 985-5500.
LEONETTI CINE RENTALS, 5609 Sunset Blvd., L.A., CA 90028; (213) 469-2987.
LIGHTING & PRODUCTION EQUIPMENT, INC., 1081 Memorial Drive, S.E. (rear) Atlanta, GA 30316. (404) 681-0130. W. Bruce Harlan, pres.
LOS ANGELES STAGE LIGHTING CO. LTD., 1451 Venice Blvd., L.A., CA 90006; (213) 384-1241.
MOLE-RICHARDSON CO., 937 N. Sycamore Ave., L.A. 90038; (213) 851-0111.
CHARLES ROSS, INC., 333 W. 52nd St., N.Y., NY 10019 (212) CI 6-5470.
TIMES SQUARE THEATRICAL & STUDIO SUPPLY CORP., 318 West 47th St., New York, NY 10019 (212) 245-4155.

Market Research

MARKET RESEARCH CORP. OF AMERICA, 122 E. 42nd St., New York, NY 10017. (212) 953-7240; 624 South Michigan Ave., 60605 (312) 939-1500.

NIELSEN, A.C., CO., Nielsen Plaza, Northbrook, IL 60062. (312) 498-6300; 1290 Ave. of Americas, N.Y., NY 10019. (212) 956-2500.
OPINION RESEARCH CORP., Research Park, Princeton, NJ 08540. (212) 489-1955.
POLITZ, ALFRED, RESEARCH, INC., 300 Park Ave. So., N.Y., NY 10010. (212) 982-7600.
THE PULSE, INC., 730 5th Ave., N.Y., NY 10019. (212) 586-3316.
ROPER ORGANIZATION, THE, 1 Park Ave., N.Y., NY 10017. (212) 679-3523.
SAN FRANCISCO RESEARCH GROUP, 2200 Sacramento St., Suite 1206, San Francisco, CA 94115.
SINDLINGER & CO., INC., Harvard & Yale Aves., Swarthmore, PA 19081. (215) 544-9000.
STARCH INRA HOOPER, INC., East Boston Post Rd., Mamaroneck, NY 10543 (914) 698-0800.
TRENDEX, INC., 477 Madison Ave., New York, NY 10017 (212) 752-7160.
VIDEODEX, INC., 342 Madison Ave., New York, NY 10017 (212) 687-8837.

Merchandisers

DISNEY, WALT, PRODUCTIONS, (Character Merchandising Division), 477 Madison Ave., N.Y., NY 10022 (212) 593-8900.
HARVEY FAMOUS CARTOONS, 15 Columbus Circle, N.Y., NY 10024 (212) 582-2244.
HODGE CO., MARGARET, 488 Madison Ave., N.Y., NY 10022; (212) 758-9230.

Music, Music Libraries and Music Cutting

ARIES SOUND INTERNATIONAL, 245 E. 63rd St., N.Y., NY 10021 (212) 838-4940.
EMIL ASCHER, INC., 666 Fifth Ave., N.Y., NY 10019 (212) 581-4504.
AUDIO DIRECTORS, INC., 325 West 19th St., N.Y., NY 10011 (212) 924-5850.
CHAPELL MUSIC LIBRARY, 810 Seventh Ave., New York, NY 10019 (212) 977-7213.
CORELLI-JACOBS FILM MUSIC INC., 25 West 45th St., N.Y., NY 10036 (212) 586-6673.
DE WOLFE MUSIC LIBRARY, 25 W. 45th St., N.Y., NY 10036 (212) 586-6673.
EDWARDS, CLEVE (composer-producer), 311 Pall Mall St., Columbia, SC 29201. (803) 252-3504.
FILM SCORES BY JERRY MARKOE, 11 Fort George Hill 13c, New York, NY 10040. (212) 567-6223. Jerry Markoe, composer-conductor.
FILMSOUNDS, INC., 128 East 41st Street, N.Y., NY 10017 (212) 867-0330.
MacGREGOR, C.P., CO., 729 South Western Ave., Los Angeles 90005, (213) 384-4191.
MUSIC FOR FILMS INC., 49 West 45th St., N.Y., NY 10036 (212) 247-3577.
MUSIC HOUSE INC., 16 East 48th St., N.Y., NY 10017 (212) 758-7773.
M.S.T.S. MUSIC, 1600 Broadway, N.Y., NY 10019 (212) 246-4687.
PIC COLOR CORP., 25 W. 45th St., New York, NY 10036 (212) 575-5600.
PICTURE SCORES, INC., 115 West 45th St., N.Y., NY 10036 (212) JU 6-1845.
ROSS-GAFFNEY, INC., 21 W. 46th St., N.Y. 10036 (212) 582-3744.
SCORE PRODUCTIONS, INC., 249 East 49th St., N.Y., NY 10021 (212) 751-2510.
THOMAS BROTHERS, 30 Berry St., Los Angeles, CA 94107 (415) 664-1901.
VALENTINO, THOMAS J., INC, 150 W. 46th St., N.Y., NY 10036 (212) 246-46675.

Properties and Scenery

BASCH, JOS. GALLERIES CO., 5755 Santa Monica Blvd., Hollywood 90038 (213) 463-5116.
CINEMA MERCANTILE CO., 5857 Santa Monica Blvd., L.A. 90038; (213) 467-1151.
CINEMA PROPS CO., 6161 Santa Monica Blvd., Hollywood, 90038 (213) 464-3191.
FANTASY FAIR, 4310 San Fernando Rd., Glendale, CA (213) 245-7367.
GROSH, R. L. (Scenic Studios), 4114 Sunset Blvd.; L.A., CA, 90028 (213) 662-1134.
HOUSE OF PROPS., 1117 Gower St., Hollywood, CA; 463-3166.
METRO-GOLDWYN-MAYER, INC., 10202 West Washington Blvd., Culver City, CA 90230. (213) 836-3000.
OMEGA STUDIO RENTALS, 5757 Santa Monica, Blvd., L.A., 90038 (213) 466-8201.
ROSCHU, 6514 Santa Monica Blvd., L.A., 90038 (213) 469-2749.
SILVESTRI STUDIOS, 1733 Cordova, Hollywood; (213) 735-1481.
WOOLF BILLY HOUSE OF MOTION PICTURE ACCESSORIES, INC., 1112 N. Beachwood Dr., Hollywood; (213) 469-5335.

Rental Studios and Production Facilities

ATLANTA

LIGHTING & PRODUCTION EQUIPMENT, INC., 1081 Memorial Drive, S.E. (rear), 30316. (404) 681-0130.

LOS ANGELES

R.L. BEVINGTON, INC., 550 N. Bronson, Hollywood, CA 90004; (213) 466-7778.
BURBANK STUDIOS, 4000 Warner Blvd., Burbank, CA 91522; 213-843-6000.
CARTHAY STUDIOS, INC., 5903-07 W. Pico Blvd., L.A. 90035. (213) 938-2106.
CBS STUDIO CENTER, 4024 N. Radford Ave., N. Hollywood, CA 91604; (213) 763-8411.
CINEMOBILE SYSTEMS, INC., 11165 Gault St., No. Hollywood (213) 764-9900.
COLUMBIA PICTURES CORP., Colgems Sq., Burbank, 91505; 213-843-6000.
CULVER CITY STUDIOS, INC., 9336 W. Washington 90230; (213) 871-0360.
FALCON STUDIOS, 5526 Hollywood Blvd., L.A. 90028; (213) 462-9356.
F&B/CECO OF CALIFORNIA, INC. 7051 Santa Monica Blvd., Hollywood 90038; (213) 466-9361.
GOLDWYN, SAMUEL STUDIOS, 1041 N. Formosa Ave., L.A. 90046. (213) 851-1234
HOLLYWOOD GENERAL STUDIOS, 1040 N. Las Palmas Ave., L.A., 90038. (213) 469-9011.
METRO-GOLDWYN-MAYER INC., 10202 W. Washington Blvd., Culver City, CA 90230; (213) 836-3000.
MOLE-RICHARDSON CO., 937 N. Sycamore Ave., L.A. 90038; (213) 654-3060.
PARAMOUNT PICTURES CORP., 5451 Marathon St., 90038. (213) 463-0100.
PRODUCERS STUDIO, INC., 650 North Bronson Ave., Los Angeles 90004; (213) 466-8111.
U.P.A. PICTURES, INC., 4440 Lakeside Drive, Burbank, CA, 91505. (213) 849-6666.
UNIVERSAL CITY STUDIOS, 100 Universal City Plaza, Universal City Plaza, Universal City, 91608. (213) 985-4321.

MIAMI

LUKE MOBERLY STUDIOS AND PRODUCTIONS, 4810 S.W. 54, Ft. Lauderdale, FL; Tel.: 581-7508.

NEW YORK

ASTORIA MOTION PICTURE AND TELEVISION PRODUCTION CENTER, 35-11 35th Ave., Astoria. (516) 936-6827.
BOKEN, INC., 349 West 48th St., NY 10036; (212) 581-5507.
CINEMOBILE SYSTEMS, INC., 601 West 29th St., (212) 695-7585.
FILMWAYS, INC., 246 E. 127 (212) 876-7788.
NATIONAL SCREEN SERVICE CORPORATION, TV AND INDUSTRIAL FILMS DIVISION, 1600 Broadway, New York, NY 10017; (212) 246-5700.
PRODUCTION CENTER, 221 West 26th St., 10001; (212) 675-2211.
TITRA SOUND CORPORATION, 1600 Broadway, NY 10019; (212) PL 7-6681.

UNITED FILM ENTERPRISES, INC., 1546 Broadway, New York, NY 10036; (212) 586-1442.

TAMPA, FLORIDA

POST PRODUCTION SERVICES, 3808 San Nicholas St. (813) 253-0402.

Slides-Telops

BEBELL INC., 416 W. 45th St., N.Y., NY 10036 (212) 245-8900.
TWO BY TWO SLIDES, INC., 2 West 45 St., 10036. (212) 972-0744.

Sound and Recording Services

HOLLYWOOD

ALTEC SERVICE CORP., 1515 S. Manchester Ave., Anaheim (714) 774-2900.
ARTISAN SOUND RECORDERS, 1600 Wilcox Ave. Hollywood 90028. (213) 461-2751.
AUDIO EFFECTS COMPANY, 1600 N. Western Ave., Hollywood 90027 (213) 469-3692. (Complete recording services).
BURBANK STUDIOS, THE, 4000 Warner Blvd., Burbank, CA 91522; (213) 843-6000.
CINESOUND, 915 North Highland Ave., Hollywood, CA 90038; (213) 464-1155.
COLUMBIA PICTURES TELEVISION, 300 S. Colgens Square, Burbank, 91522; (213) 954-6000.
CONSOLIDATED FILM INDUSTRIES, 959 N. Seward St., L.A. 90038; (213) 462-3161.
DOLBY LABORATORIES, 5248 Buffalo Ave., Van Nuys, 91401. (213) 981-4123.
GLEN GLENN SOUND CO., 6624 Romaine St., Hollywood 90038. (213) 469-7221.
INDEPENDENT PRODUCERS SERVICE, 7370 Melrose Ave., Hollywood 90046; (213) 655-3599.
METRO-GOLDWYN-MAYER, 10202 W. Washington Blvd., Culver City 90230; (213) 836-3000.
NEWJACK SOUND RECORDERS, 1717 N. Highland Ave., L.A. 90038; (213) 466-6141.
PARAMOUNT STUDIOS, 5451 Marathon St., Hollywood 90038; (213) 468-5000.
PRODUCERS SOUND SERVICE, INC., 1223 N. Highland; (213) 462-6535.
QUALITY SOUND, 5625 Melrose Ave. 90038; (213) 467-7154.
RYDER SOUND SERVICES, INC., 1161 Vine St., Hollywood (213) 469-3511.
SCOTTSOUND, INC., 6110 Santa Monica Blvd. 90038; (213) 462-6981.
TODD-AO SOUND RECORDING SCORING, 1021 N. Seward St., Hollywood, 90038. (213) 463-1136.
TV RECORDERS, 6054 Sunset Blvd., Hollywood; (213) 469-8201.
UNIVERSAL CITY STUDIOS, 100 Universal City, Plaza, UC 91608; (213) 985-4321.
WARNER HOLLYWOOD, 1041 N. Formosa, Los Angeles, 90046; (213) 650-2500.
ZOETROPE STUDIOS, 1040 N. La Palmas, Hollywood 90038; (213) 469-9011.

NEW YORK

ADD SOUND, INC., 16 W. 46 St., 10036 (212) 575-0719.
AQUARIUS TRANSFER, 12 E. 46 St., 10017 (212) 581-0123.
AURA RECORDING, INC., 136 W. 52nd St., 10019 (212) 582-8105.
HIT FACTORY, THE, INC., 353 W. 48 St., 10036 (212) 581-9590.
IMAGE SOUND STUDIO, INC., 1619 Broadway, 10036 (212) 581-6717.
MAGNO SOUND, INC., 212 W. 48; 10036 (212) 757-8855.
MUSIC SOUND TRACK SERVICE, 1600 Broadway, 10019 (212) CIrcle 6-4687-8.
NATIONAL RECORDING STUDIOS, 730 Fifth Ave. 10019 (212) 757-6440.
RKO SOUND STUDIOS, 1440 Broadway, 10018 (212) 564-8000.
ROSS-GAFFNEY, INC., 21 W. 46th St., 10018 (212) 582-3745.
SOUND SHOP, THE, 304 E. 44 St., 10017 (212) 757-9800.
TITRA SOUND CORP., 1600 Broadway, 10019 (212) 757-6681.
THOMAS J. VALENTINO, 151 West 46th St., 10036 (212) 246-4675.

SAN FRANCISCO

DOLBY LABORATORIES, 731 Sansome St., 94111. (415) 392-0300.
THOMAS BROTHERS, 30 Berry St., 94107. (415) 664-1901.

WASHINGTON, DC

BYRON MOTION PICTURES INC., 65 K St., N.E.; 20002 (202) 783-2700.
CAPITAL FILM LABORATORIES, INC., 470 E. St., S.W.; 20024 (202) 347-1717.

Special Effects

NEW YORK

ACTION PRODUCTIONS, INC., 16 West 46 St., New York, NY 10036 (212) 391-2747.
ANIMATED CAMERA WORKSHOP, 51 East 42 St., N.Y., NY 10017. (212) 687-5009.
ANIMATED PRODUCTIONS, INC., 1600 Broadway, New York NY 10019 (212) 265-2942.
EUE/SCREEN GEM PRINTS, 222 East 44 St., New York, NY 10017 (212) 867-43030.
EXCEPTIONAL OPTICALS, INC., 17 East 45 St., New York, NY 10017 (212) 972-1760.
FILM OPTICALS, INC., 144 East 44 St., New York, NY 10017 (212) 697-4744.
I.F. STUDIOS, INC., 328 East 44 St., New York, NY 10017 (212) 683-4747.
MINI-EFFECTS, INC., 15 West 44 St., New York, NY 10036 (212) 869-0370.
NATIONAL SCREEN SERVICE CORP., TV & INDUSTRIAL FILMS DIVISION, 1600 Broadway, New York, NY 10019 (212) 246-5700.
OPTICAL HOUSE, INC., 25 West 45 St., New York, NY 10036 (212) 757-7840.
PROFESSIONAL OPTICALS, INC., 20 East 46 St., New York, NY 10017 (212) 682-1757.
RAINBOW FILMS EFFECTS, INC., 45 West 45 St., New York, NY 10036 (212) 247-1676.
TRI-PIX FILM SERVICE, INC., 49 West 45 St., New York, NY 10036 (212) JU 2-0650.
WORLD EFFECTS, INC., 20 East 46 St., New York, NY 10017 (212) 687-7070.

HOLLYWOOD

ANDERSON, HOWARD A., CO., Main Office: Paramount Studios, 5451 Marathon St., Hollywood, CA 90038 (213) 463-2336.
BRAVERMAN PRODS., INC., 8961 Sunset Blvd., L.A. 90069; (213) 278-5444.
CINEMA SERVICE, INC., 5725 Santa Monica Blvd., (213) 462-1376.
COAST SPECIAL EFFECTS, 4907 N. Lankershim Blvd., N. Hollywood, 91602. (213) 876-2021.
CONSOLIDATED FILM INDUSTRIES, 959 N. Seward St., 90038. (213) 462-3161.
FILM EFFECTS OF HOLLYWOOD, 1140 N. Citrus, Hollywood; (213) 469-5808.
MERCER RAY & CO., 4241 Normal Ave., (213) 663-9331.
MODERN FILM EFFECTS, 6824 Melrose Ave., 90030; (213) 938-2155.
OPTICAL PRINT SERVICES, INC., 1132 N. Vine St., Los Angeles 90028 (213) 462-7446.
PACIFIC TITLE & ART STUDIO, 6350 Santa Monica Blvd., Los Angeles 90038 (213) 464-0121.
VIDEOTAPE ENTERPRISES, 8610 Sunset Blvd., Los Angeles, 90069 (213) 659-4801.

Stop Watches

DUCOMMUN, M., CO., 48 Main St., Warwick, NY 10990.(914) 986-5757.
MARCEL WATCH CORP., 1180 Sixth Ave., N.Y., NY 10036. (212) 245-3355.
RACINE, JULES & CO., 85 Executive Rd., Elmsford, NY 10523. (914) 592-4760.

Sub-titles

WHITE ROSE CAPTION, INC., Sundvall Bldg., 83-90 Broadway, Elmhurst, Queens, NY 11373. (212) 271-3388. Negative subtitle cards—super-imposed negative and translations in most languages for motion pictures and TV. Ms. Rose Robles, pres.

Talent and Literary Agencies

Hollywood

ABRAMS-RUBALOFF & ASSOC., INC.,
9012 Beverly Blvd. (48), 273-5711
AGENCY FOR PERFORMING ARTS,
9000 Sunset Blvd. #.315 (69), 273-0744
AIMEE ENTERTAINMENT ASSOCIATES,
13749 Victory Blvd., Van Nuys, CA 91401; 994-9354.
ALVARADO, CARLOS.
8820 Sunset Blvd. (69), 652-0272
AMARAL AGENCY,
1000 Riverside Dr., Toluca Lake, 980-1013
ASSOCIATED BOOKING CORP.,
292 S. La Cienga Blvd., BH; 855-8051
AUER, MILES BOHM,
8344 Melrose Ave., Suite 29 (69). 462-6416
BARSKIN AGENCY,
11240 Magnolia Blvd., No. Hollywood; 985-2992.
BLAKE AGENCY, LTD.,
409 N. Camden Dr., BH, 278-6885
BLANCHARD, NINA,
1717 N. Highland (28), 462-7274
BLUMENTHAL ARTISTS AGENCY,
8833 Sunset Blvd., (69) 855-1403.
BREWIS AGENCY, ALEX,
8721 Sunset Blvd. (69), 274-9874
CALDER AGENCY,
4063 Radford, 91604; 508-7200.
CAREY-PHELPS-COLVIN,
1407 N. La Brea Ave. (28) 874-7780.
CENTURY ARTISTS, LTD.
9744 Wilshire Blvd. BH (90212) 273-4366
CHARTER MGMT.,
9000 Sunset Blvd. (69), 278-1690
CHARTWELL ARTISTS, LTD.,
1901 Ave. of Stars (67) 553-3600.
CHASIN-PARK-CITRON,
9255 Sunset Blvd., (69), 273-7190
COLTON & ASSOC., KINGSLEY,
321 S. Beverly Drive, BH, 277-5491
COMMERCIAL TALENT AGENCY,
8500 Wilshire, BH 90211; 855-0422
COMMERCIALS UNLTD.,
7461 Beverly Blvd. (36), 937-2220
CONNOR-CORFINO ASSOC., INC.,
14241 Ventura Blvd., Sherman Oaks, 981-1133
CONTEMPORARY-KORMAN,
132 Lasky Dr., BH, 274-7888
COUGLIN AGENCY, KERWIN,
10850 Riverside Dr., N. Hwd. 980-6100
CRONIN AGENCY, BERNYCE,
439 S. La Cienega Blvd. (48), 273-8144
CUMBER, LIL,
6515 Sunset Blvd. (28), 469-1919
CUNNINGHAM & ASSOC.,
261 S. Robertson, BH 90211; 855-1700
DIAMOND ARTISTS, LTD.,
8400 Sunset Blvd. (69), OL 4-5960
EISENBACH-GREENE-DUCHOW, INC.
760 N. LaCienega Blvd. L.A. (69), 659-3420
ENTERTAINMENT ENT.,
1680 Vine St. (28), 462-6001
F.C.A. AGENCY, INC.
1800 Century Park East (67) 277-8422
FELBER, WM.,
2126 Cahuenga Blvd. (68) 466-7629
FIELDS ASSOC., JACK,
9255 Sunset Blvd. (69), 278-1333
FLAIRE AGENCY,
8693 Wilshire, BH, 90211; 659-6721
FRINGS, KURT,
9440 Santa Monica Blvd., BH, 274-8881
GARRICK, DALE,
8831 Sunset Blvd. (69), 65 7-2661
GERSH AGENCY, PHIL.
222 N. Canon Dr., BH 274-6611
GIBSON, J. CARTER,
9000 Sunset Blvd. (69), 274-8813
GORDEAN-FRIEDMAN AGENCY, INC.,
9570 Wilshire, BH 90212; 273-4195
GRANITE AGENCY,
1920 S. La Cienega Blvd. (34), 934-8383
GREENVINE AGENCY,
9021 Melrose Av., (69), 278-5800
GREEN AGENCY, IVAN,
1888 Century Park East (67), 277-1541
GROSSMAN-STALMASTER AGENCY,
8730 Sunset Blvd. (69), 65 7-3040
HAMILBURG, MITCHELL,
292 S. La Cienga, BH, 657-1501
HECHT, BEVERLY,
8949 Sunset Blvd. (69), 273-3544
HUNT, GEORGE B.,
8350 Santa Monica Blvd. (46), 65 4-6600
HUSSONG AGENCY,
8271 Melrose Ave. (46), 655-2534
INGERSOLL, GEORGE,
6513 Hollywood Blvd. (28), 874-6434
INT'L CREATIVE MANAGEMENT,
8899 Beverly Blvd., (48), 550-4000
JUNIOR ARTIST UNLTD.,
4914 Lankershim Blvd., No. Hollywood. 762-9000
KELMAN, TONI AGENCY,
7813 Sunset Blvd. (46) 851-8822
KING AGENCY, HOWARD,
118 S. Beverly Dr., BH, 271-7294
KOHNER, PAUL,
9169 Sunset Blvd. (69), 271-5165
LAZAR, IRVING PAUL,
211 S. Beverly Dr., BH, 27 5-6153
LEONETTI, CAROLINE,
6526 Sunset Blvd. (28), 462-2345
LONGNECKER, ROBERT,
11704 Wilshire (25), 477-0039
LOO, BESSIE,
8730 Sunset Blvd. (69), 657-5888
LYONS, GRACE,
8730 Sunset Blvd. (69), 652-5290
McHUGH AGENCY, JAMES
8150 Beverly Blvd. (48), 651-2770
McMILLAN, HAZEL,
9426 Santa Monica Bl., BH, 276-9823
MEIKLEJOHN, WM.,
9250 Wilshire Blvd., Suite 412, BH 273-2566
MESSENGER, FRED,
8235 Sunset Blvd. (46), 654-3800
MISHKIN AGENCY,
9255 Sunset Blvd. (69), 274-5261
MOORE, LOLA D.,
13848 Vanowen St., Van Nuys 91405; 276-6097
MORRIS, WILLIAM,
151 El Camino Dr., BH, 274-7451
M.T.A. ARTISTS' MGR
4615 Melrose Ave. (29), 661-9888
MURPHY, MARY,
10701 Riverside Dr., NH, 985-4241
OLIVER & ASSOCS., MARINE,
8746 Sunset Blvd. (69), 657-1250
OTIS AGENCY, DOROTHY DAY,
6430 Sunset Blvd. (28), 461-4911
PEARSON, BEN,
606 Wilshire Blvd. SM (90401) 451-8414
PREMIERE ARTISTS & PROD.,
6399 Wilshire Blvd. (48), 651-338.
PRESCOTT, GUY,
8920 Wonderland Ave., (46), 656-1963
RAPER ENTERPRISES AGENCY
9441 Wilshire Blvd., (12) 273-7704
ROBINSON & ASSOC.,
132 S. Rodeo Dr. BH, 275-6114
ROBINSON-WEINTRAUB & ASSOCIATES,
554 S. San Vicente (48), 653-5802
ROSE, HAROLD,
8530 Wilshire Blvd., BH, 652-3961
RUSH, ART,
10221 Riverside Dr., No. Hwd., 985-3033
SACKHEIM AGENCY,
9301 Wilshire Blvd BH (90210) 276-3151
SANDERS, NORAH,
9255 Sunset Blvd. (69) 550-0797

SCHWARTZ & ASSOC., DON,
8721 Sunset Blvd. (69), 657-8910
SHAW, GLENN,
330 Barkam Blvd. (68), 851-6262
SHERRELL AGENCY LEW,
7060 Hollywood Blvd. (28), 461-9955
SHIPLEY-ISHIMOTO,
9163 Sunset Blvd. (69), CR 6-6251
SHREVE, DOROTHY,
1525 Superior Ave., Newport Beach, 92663; 642-3050
TANNEN & ASSOC. HERB,
6640 Sunset Blvd. (28), 461-3055
TOBIAS & ASSOCIATES,
1901 Ave. of Stars (67), 277-6211
TWENTIETH CENTURY ARTISTS,
13273 Ventura Bl., Studio City (91604) 990-8580
WEINER, JACK,
8721 Sunset Blvd. (69), 652-1140
WEVER, WARREN,
1104 S. Robertson Blvd. (35), 276-7065
WILK, TED,
9172 Sunset Blvd. (69), 273-0801
WOSK AGENCY, SYLVIA,
439 S. La Cienega (48), 274-8063
WRIGHT ASSOC., ANN,
8422 Melrose Pl., (69), 655-5040

New York

AGENCY FOR PERFORMING ARTS,
888 Seventh Ave., (10106) 582-1500.
ALEXANDER INC., WILLARD,
660 Madison Ave., (21), 751-7070
AMERICAN-INT'L TALENT,
166 W. 125th St., (27), 663-4626
ANDERSON, BEVERLY,
1472 Broadway, Ste. 806, (36), 279-5553
ASSOCIATED BOOKING CORP.,
1995 Broadway (23), 874-2400
ASTOR, RICHARD,
119 W. 57th St., (19), 581-1970
BISHOP, LOLA M.,
160 W. 46 (36) 997-1848
CASE, BERTHA,
345 W. 58 (19) 541-9451
COLEMAN-ROSENBERG AGCY.,
667 Madison Ave., (21), 838-0734
COOPER ASSOC., BILL,
16 E. 52nd St., (22), 758-6491
DEACY, JANE,
300 E. 75 (21) 652-4865
DIAMOND ARTISTS, LTD.,
119 W. 57th St., (19), 247-3025
DRAPER, STEPHEN AGENCY,
27 W. 57th St., (19), 421-5780
FORD—TALENT GROUP
344 E. 59th St., (22), 688-8628
GAGE GROUP, THE,
1650 Broadway, (19), 541-5250
HARTIG, MICHAEL,
527 Madison (22), 759-9163
HENDERSON/HOGAN AGENCY,
200 W. 57, (19), 765-5190
HUNT, DIANA,
44 W. 44th St., (36), 391-4971
HUNTER, JEFF,
119 W. 57th St., (19), 757-4995
INT'L CREATIVE MANAGEMENT,
40 W. 57th St., (19), 556-5600
JACOBSON/WILDER, INC.,
400 Madison Ave., (17), 759-0860
JAN J. AGENCY,
224 E. 46th St., (17), 490-1875
JORDAN, JOE TALENT AGCY.,
400 Madison Ave., (17), 838-4910
KAHN INC.,
853-7th Ave., (19), 582-1280
KENNEDY ARTISTS REP.,
881 7th Ave., (19), 675-3944
KING, ARCHER,
777 Seventh Ave. (19), 581-8513
KROLL, LUCY,
390 West End Ave., (24), 877-0556
LARNER, LIONEL LTD.,
850 7th Ave., (19), 246-3105
LEIGH, SANFORD,
527 Madison Ave., (22), 752-4450
LEWIS ASSOC., LESTER,
156 E. 52nd St., (22), 753-5082
McDERMOTT, MARGE,
214 E 39 (16) 889-1583
MARTINELLI ATTRACTIONS,
888 8th Ave., (36), 586-0963
M.E.W. COMPANY,
370 Lexington Ave., (17), 889-7272
MORRIS AGENCY, WM.,
1350 Ave. of Americas, (19), 586-5100
OPPENHEIM-CHRISTIE ASSOC.,
565 5th Ave., (17), 661-4330
OSCARD/FIFI AGENCY,
19 W. 44th St., (36), 764-1100
OSTERTAG, BARNA,
501 Fifth Ave., (17), 697-6339
PITT AGENCY, JOEL,
144 W. 57th St., (19), 765-6373
PREMIER TALENT ASSOC.,
3 E. 54 (22) 758-4900
RUBENSTEIN, BERNARD,
342 Madison Ave., (17), 986-1317
RYAN, CHARLES VERNON,
200 W. 57 (19) 245-2225
SCHULLER AGENCY, WM.,
667 Madison Ave., (22), 758-1919
SILVER, MONTY,
200 W. 57 (19) 765-4040
STEWART ARTISTS CORP.,
140 E. 63 (21) 752-0944
STROUD MGMT.,
18 E. 48th St. (17), 688-0226
TALENT REPS., INC.,
20 E. 53rd St., (22), 752-1835
THOMAS, JEAN,
342 Madison (17), 490-3954
THOMAS, MICHAEL,
22 E. 60th St., (22), 755-2616
TROY, GLORIA TALENT AGENCY,
1790 Broadway, (19), 582-0260
WATERS AGENCY, BOB,
510 Madison Ave., (22), 593-0543
WINANT III, WILLIAM A.,
30 Sutton Place; (22), 68 8-0768
WOLTERS, HANNS,
342 Madison Ave., (17), 867-9177
WRIGHT REPS., ANN,
137 E. 57th St., (22), 832-0110

Advertising & Publicity Representatives

Hollywood-Los Angeles-Beverly Hills

ABRAMS, BOB, AND ASSOCIATES, 2030 Prosser Ave., Los Angeles, 90025 475-7739
BARLOR ASSOCIATES, 428 North Palm Dr., Beverly Hills; 278-1998.
BLOWITZ & CANTON, INC., 9350 Wilshire Blvd., Beverly Hills, 90210; 275-9443.
BRAVERMAN-MIRISCH, INC., 9255 Sunset Blvd., Los Angeles 90069; 272-8608.
TOM BROCATO & ASSOCIATES, INC., 6380 Wilshire Blvd., Suite 1411, Los Angeles 90048 (213) 653-9570.
DICK BROOKS UNLIMITED, 9465 Wilshire Blvd., Beverly Hills, CA 90212; (213) 273-8477.
BYOIR, CARL, AND ASSOCIATES, INC., 900 Wilshire Blvd., Los Angeles, 90019; 627-6421.
DRUXMAN, MICHAEL B. & ASSOCIATES, 8831 Sunset Blvd., Los Angeles, 90069; 652-5592.
EDELMAN, DANIEL, INC., 1901 Ave. of the Stars, Los Angeles 90067; 555-1560.
FRANKEN PUBLIC RELATIONS, 1101 S. Robertson; 272-7577.
GARRETT, MICKEY, ASSOCIATES, 3349 Cahuenga Boulevard West; Los Angeles 90068; 851-5075.
GELFOND, FORDON AND ASSOCIATES, 9808 Wilshire Blvd., Beverly Hills 90212; 273-1440.
GOODMAN/NEMOY, 6464 Sunset Blvd., Los Angeles, 90028, 464-1135.
HANSON & SCHWAM, 9229 Sunset Blvd., Los Angeles 90069; 278-1255.

HARSHE-ROTMAN & DRUCK, INC., 3345 Wilshire Blvd., Los Angeles; 90036; 385-5271.
JENNEY, NOREEN PUBLIC RELATIONS, 1901 Ave. of Stars, Los Angeles 90067. (213) 553-1983.
LEE & ASSOCIATES, 8727 West 3rd, Los Angeles 90048; 278-5300.
LEWIS & ASSOCIATES, 3440 Wilshire Blvd., Los Angeles; 381-7641.
LEWIS HERM & ASSOCIATES, 316 W. Rossmore, Los Angeles; 464-8189.
LIBERMAN, FRANK, AND ASSOCIATES, 9021 Melrose Ave., Los Angeles 90069; 278-1993.
MAHONEY, JIM & ASSOCIATES, 120 El Camino, Beverly Hills; 272-7861.
McKENZIE, KING & GORDON, 1800 N. Argyle, Hollywood 90028 (213) 466-3421.
RUDER & FINN OF CALIFORNIA, INC., 9300 Wilshire Blvd., Beverly Hills; 274-8303.
SAMUELSON, CHIEF, & ASSOCIATES, 7260 Hillside Ave., Los Angeles 90046; 966-6106.
SOLTERS, ROSKIN, AND FRIEDMAN, INC. 9255 W. Sunset Blvd., Los Angeles 90069; 275-5303.
SUHOSKY, BOB ASSOCIATES, 9110 W. Sunset Blvd., Los Angeles 90069; 272-2612.
TAMKIN ASSOCIATES, INC., 1122 Robertson Blvd., 90035. (213) 275-6122.
THOMPSON, J. WALTER CO., 10100 Santa Monica Blvd., Los Angeles, 90067; 553-8383.

New York

BECK, MYER P., 729 7th Ave., NY 10019; 24 5-5552.
BILLINGS, MARION, 200 West 57th St., NY; 581-4493.
BYOIR, CARL & ASSOCIATES, INC., 380 Madison Ave., 986-6100.
CANAAN, LEE ASSOCIATES LTD., 205 East 42nd St., 10017. 682-4155.
DAVIS, AL, PUBLICITY, 711 Third Ave., 557-1045
EISEN, MAX, 234 West 44th St., 391-1072.
GIFFORD/WALLACE, INC., 1211 Park Ave. 427-7600.
EDWARD GLASS/BARRY SCHOOR, 1600 Broadway, Suite 1003M, New York, NY 10019; (212) 246-6650; (212) 586-1846. Creators and producers of advertising campaigns for motion pictures.
GOODMAN, FRANK, 1776 Broadway; 246-4180.
GREY ADVERTISING AGENCY, 777 Third Ave., 751-3500.
HARSHE-ROTMAN & DRUCK, INC., 300 East 44th St., 661-3400.
MARCH FIVE, INC., 30 East 42nd St., 68 7-3484.
MASLANSKY-KOENIGSBERG, INC., **250 West 57th St., NY 10019 489-7390.**
POST, MYRNA ASSOCIATES, 9 E. 53 St., 935-7122.
ROFFMAN, RICHARD H., ASSOCIATES, 697 West End Ave., NY; 74 9-3647.
ROGERS & COWAN, INC, 3 East 54, 486-7100.
RUDER & FINN, INC., 110 East 59th St.; 593-6400.
SOLTERS, ROSKIN, AND FREIDMAN, INC., 62 West 45th St., NY 10036; 840-3500.
WAX, MORTON D. PUBLIC RELATIONS, 1650 Broadway; NY; 247-2159.
WOLHANDER, JOE, ASSOCIATES, INC., 211 East 51st St., NY 10022; 759-2050.
YOUNG & RUBICAN INTERNATIONAL, 285 Madison; 59 53-2000.

Equipment and Supplies

* MANUFACTURERS & SERVICES
* STUDIO EQUIPMENT
* REFRESHMENT PRODUCTS, EQUIPMENT
* MANUFACTURERS & SERVICES (By Product Classification)
* SUPPLY DEALERS IN U.S. AND CANADA

Manufacturers & Services

A-Wagner Zip Change, Inc.
(formerly WAGNER Sign Service)
3100 Hirsch Street, Melrose Park, IL 60160. Tel.: (312) 681-4100.
WAGNER brand slotted dimensional marquee letters in 4", 6", 8", 10" and 17" letters in red, green, blue or black plastic. Wagner brand marquee background with stainless steel frames and unbreakable Enduronamel Panel Assembly, low cost metal marquee for 8" or larger letters, ready to install. Zip-Change plastic or metal flat marquee letters in 6", 8", 10", 12", 15", 17", and 24". Letter changing Mechanical Hands. Letter storage cabinets. Rating Units (PG, R, G). Alumabar letter track. Sell thru theatre suppliers only.

Altec Lansing
1515 S. Manchester Avenue, Anaheim, CA 92803. Tel.: (714) 774-2900.
Manufacturers of amplifiers, loudspeakers, electronic devices, microphones and transformers' "Voice of the theatre" speaker systems.

American Desk Manufacturing Co.
P.O. Box 429, Temple, TX; Tel.: (817) 773-1776.
Manufacturers of theatre seating.

American Playground Device Co.
Main Offices: 1801-1831 So. Jackson, P.O. Drawer 2599, Anderson, IN 46011. Tel.: (317) 642-0288.

Branch Offices: Nahma Blvd., Nahma, MI 49864. Tel.: (317) 642-0288.
Manufacturers of playground equipment, park benches, picnic tables, charcoal grills, flood lights and swimming pool equipment for drive-in theatres, and bicycle racks for theatres. Complete line of American approved extra-heavy-duty Park, Picnic, Playground, Swimming Pool, Beach and Dressing Room Equipment.

American Seating Company
Grand Rapids, MI 49504. Tel.: (616) 456-0433.

(District offices in principal cities)
Manufacturers of auditorium and theatre chairs and portable stacking chairs.

BRANCH OFFICES

CALIFORNIA: Manhattan Beach, 90266, 802 First St., P.O. Box 221; (213) 374-5330; E. D. Thompson. Mountain View, 94043, 2440 Charleston Rd.; (415) 964-6600; J. D. Laird.
GEORGIA: Marietta, 30067, 473 Indian Hills Trail; (404) 973-0385; J. DeFrancesco.
MASSACHUSETTS: Hingham, 02043, 12 Richard St.; (617) 749-7197; J. T. Corrigan.
MICHIGAN: Grand Rapids, 49504, 901 Broadway Ave., NW, (Main Office); (616) 456-0433, J. E. Sherry, (616) 456-0522, Q. R. Van Dore, (616) 456-0432; G. L. Courtney.
MISSOURI: Parkville, 64152, 10307 Mirror Lake Dr.; (816) 741-8195; J. W. Lamkin.
NEW JERSEY: Leonia, 07605, 135 Fort Lee Rd.; (201) 947-9506; R. K. Kenaga, D. A. Spadola
TEXAS: Richardson, 75080; 312 N Central Expressway, Suites 215-217; (214) 231-1419; J. K. Godwin.

ASC Technical Services Corp. (formerly Altec Service Corporation)
Executive offices: P.O. Box 706, Plano, TX 75074 (214) 422-2160.
Nationwide sound service and full booth maintenance.

DIVISION OFFICES

Nick Markanich, regional supervisor, 7 Van Riper Ave., Pompton Plains, NJ 07444 (201) 483-5305.
Indianapolis, IN 46226: LaMar Michael, regional supervisor, 5857 N. Winthrop; (317) 257-1807.
Houston, TX 77092: B. R. Emerson, regional supervisor, 7119 Pine Grove Drive, (713) 523-9077

Atlantic Audio Visual Corp.
630 Ninth Avenue, New York, NY 10036.
Projection system for mini or maxi theatres. Exlcusive U.S. Distributors for Zeiss Ikon Projection Equipment.

Automatic Devices Co.
2121 South 12th St., Allentown, Pa. 18103. Tel.: (215) 797-6000.
Manufacturers of stage, television, drapery and cubicle hardware.

Ballantyne of Omaha, Inc.
(A Subsidiary of Canrad-Hancvia, Inc.)

1712 Jackson Street, Omaha, NB 68102. Tel.: (402) 342-4444.
Complete sound and projection systems, from engineering and planning to construction and financing. Projectors, sound-heads, pedestals, rewind and automation. Transistorized or tube amplification. The complete package for mini, maxi, and multi-screened theatres, including platters. Sold only through selected theatre supply dealers.

Bauer Division, AIC Photo, Inc.
168 Glen Cove Road, Carle Place, NY 11514.
Super 8 movie cameras and projectors, 16 mm projectors.

Bausch & Lomb Incorporated
Scientific Optical Products Division

1400 North Goodman Street, Rochester, NY 14692. Tel.: (716) 338-6000.

Branch offices in principal cities.
Manufacturers of optical instruments of all types. BAL-cold glass reflectors.

Bevelite-Adler
14824 South Main Street, Gardena, CA 90248. Tel.: (800) 421-1256, (213) 321-5641.
Changeable marquee sign letters with Snap-Lok plastic 4", 6", 8", 10", 12", 15", 17", 24", red, green, blue, black; Stainless Steel frames with Plexiglas, Adlerite or glass; stainless steel, plastic, or aluminum track for mounting on any backgrounds, all of the above; for drive-ins, theatre marquees, and pylons. Also Pronto changeable letters and numbers from 3" to 42" plus Regency Injection Molded Changeable and permanent letters and numbers.

Brandt, Inc.
705 12th St., Watertown, WI 53094. Tel.: (414) 261-1780.
Manufacturers of coin sorting and counting machines, coin counting and packaging machines, automatic coin wrappers, currency and document counters.

Canrad-Hanovia, Inc.
100 Chestnut Street, Newark, NJ 07105.
Manufacturers of projection bulbs, Ballantyne projectors and Strong lamps.

CEMCORP (Consolidated Engineering & Manufacturing Corp.)
1515 Melrose Lane, Forest Hill, MD Tel.: 383-0036 and 879-3022.
Manufacturers and designers of ticket issuing equipment: "Autoticket" line, including Ticketaker control boxes, drive-in equipment, and ticket control systems.

Century Projector Corporation
1690 Oak, Lakewood NJ (201) 367-4434.
Manufacturers and distributors of motion picture equipment, sound reproducing equipment, automatic controls, and accessories.

Christie Electric Corp.
-8 3410W.67thSt., Los Angeles, CA 90043. Tel.: (213) 750-1151.
Manufacturers of Xenolite consoles (lamphouse, rectifier power supply and projector based housed in one unit), dual sound system, and automatic system.

Cinematograph International, Inc.
341 W. 44th St., New York, NY 10036. Tel.: (212) 246-6285.
Distributors of automated theatre equipment and supplies; Cinemeccanica Victoria 70/35mm projectors in U.S.; complete Drive-In theatre equipment; projection-sound equipment and supplies; speaker systems; projection arclamps-rectifiers-generators; Xenon light systems; stage lighting, spotlighting, stage draperies; auditorium seats, carpets; display frames, marquee letters, boxoffice equipment; equipment for Studios, Schools, Auditoriums.

Dolby Laboratories Inc.
731 Sansome St., San Francisco, CA 94111. Tel.: (415) 392-0300.
Manufacturer of noise reduction and equalization equipment for theatres and studios.

Econo-Pleat—Phil Sperling
(A division of Eastwest Carpet Mills)

2664 So. La Cienga Blvd., Los Angeles, CA 90034. Tel: (213) 871-1690.
Manufacturers of acoustical fabric wall coverings.

Electronics Div. of Drive-In Theatre Mfg. Co.
709 North 6th St., Kansas City, KS 66101. Tel.: (913) 321-3978.
Manufacturers of drive-in speakers, junction boxes, directional lighting, box office electronic security systems, custom electronic design. Platter systems—automation.

EPRAD, Inc.
P.O. Box 4712, 123 W. Woodruff, Toledo, OH 43620. Tel.: (419) 243-8106.
Manufacturers of stereo-optical (SVA) sound processors, digital dimmers, drive-in speakers, car counters, electronic ordering equipment, all transistor optical sound amplifier systems, transistorized head amplifiers, burglar alarms, speaker baskets, boxoffice cash controls, in car heaters, junction boxes, car-check boxoffice cash control systems, all transistor drive-in theatre power amplifiers, programatic automation systems, sword-large reel film handling devices, sound heads, xenon lamphouses.

Forest Bay Construction Corp.
2 Lawson Ave., East Rockaway, NY 11518. Tel.: (516) 599-0070.
Experts in theatre design.

Goldberg Brothers, Inc.
P.O. Box 5345, Denver, CO 80217.
Manufacturers of automatic rewind equipment, including reels, tables, film cabinets and splicers. Also box office and lobby accessories.

Griggs International Inc.
Box 630, Belton, TX 76513. Tel.: (817) 939-3761.
Manufacturers of conventional and Push-Back theatre chairs, lobby and office furniture.

Gulistan Carpet, Division of J. P. Stevens & Co., Inc.
1185 Avenue of the Americas, New York, NY 10036. Tel.: (212) 575-2282.
Manufacturers of Gulistan carpets.

Hertner Divison, General Battery Corp.
12690 Elmwood Avenue, Cleveland, OH 44111. Tel.: (216) 252-4242.
Manufacturers of Hertner TransVerteRs, ballast rheostats, and control panels.

Heywood-Wakefield Company
National Sales Office, Public Seating Division: 3010 Tenth St., Menominee, MI 49858.
Manufacturers of theatre seating and lounge furniture.

Hurley Screen Company, Inc.
26 Sarah Drive, Farmingdale, NY 11735. Tel.: (516) 293-7140.
Manufacturers of Super-Optica, Super-Glo, Lenticlite 20, Silver-Glo, and MW-16 projection screens and screen frames.

Ideal Seating Corporation
North Berwick, ME (207) 676-2271.
Manufacturers of Mark IV & System 4 theatre seating.

The Imperial Electric Company
64 Ira Avenue, Akron, OH 44309. Tel.: (216) 253-9126.

Indiana Cash Drawer Company

P.O. Box 236, Shelbyville, IN 46176. Tel.: (317) 398-6643.
Manufacturers of cash drawers.

Irwin Seating Company
3251 Fruitridge Road, P.O. Box 2429, Grand Rapids, MI 49501. Tel.: (616) 784-2621.
Manufacturers of theatre seating.

Kalart Victor Corporation
Hultenius Street, Plainville, CT 06062. Tel.: (203) 747-1663.
Manufacturers of audio-visual equipment and educational television systems including: 16mm sound and silent motion picture projectors: Tele-Beam large screen TV projectors; 16mm TV projectors; 16mm TV uniplex film chain; Multiplexer; Craig film editing and viewing equipment for super 8mm and 16mm film.

K-Hill® Signal Company, Inc.
326 West Third St., Uhrichsville, OH 44683. Tel.: (614) 922-0421.
Manufacturers of drive-in traffic counters. "One way" for boxoffice and flashing light for "exit" sneak-in protection.

Kinotone, Inc.
257 Tenth Avenue, Paterson, NJ 07524. Tel.: (201) 279-9700.
Distributors of projection and sound equipment.

Kneisley Electric Company
2501-9 Lagrange Street, P.O. Box 4692, Toledo, OH 43620. Tel.: (419) 241-1219.
Manufacturers of silicon rectifiers for carbon arcs; also power supplies and lamphouses for Xenon lamps. Silicon replacement stacks for selenium rectifiers, "sil-tubes" (silicon tubes) to replace gas-filled

tubes in rectifiers and in exciter lamp supplies. Silicon exciter lamp suppliers, Xenon conversions for 16mm projectors, slide projectors, and for strong trouper. Xenon conversions for carbon arc lamphouses. Xenon "super-spot" lamp and Xenon conversion for strong follow spots.

Kollmorgen Corporation

347 King St., Northampton, MA 01060. Tel.: (413) 586-2330.

Manufacturers of special optics, and optical instruments.

Koneta Matting Division

700 Lunar Drive, Wapakoneta, OH 45895. Tel.: (419) 738-2155.

Manufacturers of rubber and vinyl entrance mats and matting, nylon top-vinyl back wipe off mats.

LaVezzi Machine Works, Inc.

900 North Larch Avenue, Elmhurst, IL 60126. Tel.: (312) 832-8990.

Manufacturers of parts for projector mechanisms and soundheads, including complete intermittent movement assemblies. Also special high precision machine parts for all types of motion picture equipment.

Manko Fabrics Company, Inc.

50 West 36th Street, New York, NY 10018. Tel.: (212) OX 5-7470.

Specialists in manufacturing seat covers, backs, etc. Large inventory nylons, vinyls, mohair fabrics.

The Marble Company, Inc.

P.O. Box 8218, Nashville, TN 37207 (615) 227-7772.

National manufacturer of Marble "Double Eagle" carbons, X-Cel Xenon bulbs and products, projections lens, projection reflectors, and Polecat (in-car radio sound).

Massey Seating Company, Inc.

100 Taylor Street, Nashville, TN 37208. (615) 242-2561.

Manufacturers of theatre seating, polyurethene foam cushions, tailored upholstery units and general seating supplies. Products distributed by leading theatre equipment dealers.

Miracle Equipment Co.

Box 275, Grinnell, IA 50112. Tel.: (515) 236-7536.

Manufacturers of drive-in playground equipment, outdoor fiberglass picnic tables and benches. Also coin-operated carousels, trackless trains, and liter-liminaters (trash receptacles).

National Theatre Supply Company

(Division of National Screen Service Corp.)

1600 Broadway, New York, NY 10019. Tel.: (212) 246-5700.

Vice Pres., Managing Director, Dan Miller.

Products: Theatre equipment and supplies.

National Ticket Company

1650 Broadway, New York, NY 10019 (212) 757-1426.

Ticket printers to theatres; roll-machine-reserved.

Neumade Products Corporation

720 White Plains Rd., Scarsdale, NY 10583. Tel.: (914) 725-4900. PLANT: 2485 Walden Ave., Buffalo, NY 14225. Tel.: (716) 684-0051.

Manufacturers of motion picture, audio visual, and television equipment and accessories, 8mm, 70mm, 35mm, and 16mm fields.

Nick Mulone & Son

Pittsburgh St., Cheswick, PA 15024. Tel.: (412) 274-6646 or 274-5994.

Manufacturers of wide screen frames, all types.

Novelty Scenic Studios

40 Sea Cliff Ave., Glen Cove, NY 11542. Tel.: (212) 895-8668, (516) 671-5940.

Stage curtains and drapes, wall coverings, tracks and stage equipment.

Optical Radiation Corporation

1300 Optical Drive, Azusa, CA 91702. Tel.: (213) 969-3344.

Manufacturer of ORCON lamphouses, Xenon bulbs, automation sound systems, and platters.

Osram Sales Corporation

P.O. Box 7062, R.D. No. 3, Jeanne Drive, Newburgh NY 12550. Tel.: (914) 564-6300

U.S. distributor of OSRAM Xenon bulbs.

Panavision, Inc.

18618 Oxnard St., Tarzana, CA 91356.

Manufacturers and lessors of motion picture processes—Ultra Panavision 70, Super Panavision 70, Panavision 70 and Panavision 35.

Perey Turnstiles

535 Fifth Avenue, New York, NY 10017. (212) 599-0077.

Manufacturers of theatre turnstiles. Application engineering available

Poblocki and Sons

620 South First St., Milwaukee, Wis. 53204. Tel.: (414) 273-3333.

Manufacturers of marquees and canopies, signs, poster cases, etc.

Prestoseal Manufacturing Corp.

545 Meacham Ave., Elmont, NY 11003. Tel.: (516) 488-1155.

Manufacturers of motion picture and microfilm splicers.

Projected Sound, Inc.

469 Avon Ave., P.O. Box 112, Plainfield, Ind. 46168. Tel.: (317) 839-4111.

Manufacturers and suppliers of In-Car Speakers, Junction Boxes, cone units and component parts.

RCA

30 Rockefeller Plaza, New York, NY 10020. Tel.: (212) 689-7200.

RCA Service Company, Cherry Hill Offices, Camden, NJ 08101. Tel. (609) 963-8000.

Theatre Service Field Offices

JERSEY CITY, 43 Edward J. Hart Road, Liberty Industrial Park; W. P. Browski, mgr.
CLEVELAND, OH, 20338 Progress Drive, Strongsville, OH, M. J. Duffield, mgr.
ATLANTA, 4508 Bibb Blvd., Tucker, GA, L. N. Howard, mgr.
SKOKIE, IL, 7620 Gross Pointe Rd.; R. H. Bunting, mgr.
HOLLYWOOD, 1501 Beach St., Montebello, CA, M. B. Solomon Jr., mgr.

Rangertone Research Inc.

509 Madison Ave., New York, NY 10022.

Manufacturers and distributors of 16 mm and 35 mm motion picture sound recorders, projectors, interlock systems, telecine and film chain projectors.

The Rank Organisation (Rank Film Equipment)

P.O. Box 70, Great West Rd., Brentford, Middlesex, England. Tel.: 01-568-92222; Telex: 24408; Cables: RANKAUDIO BRENTFORD.

(See under companies, in Great Britain Section.)

Schneider/ISCO

185 Willis Avenue, Mineola, NY 11501 (516) 747-5100.

Manufacturers of the CINELUX line of 70mm, 35mm, 16mm theatre projection lenses. Optical accessories include: outdoor lenses, Zoom lenses (16mm only), Magna-Com 65, ½x reverse and 2X anamorphics, 2″ and 1″ spacers, and thread changers. Special applications and designs also available upon request.

Selby Industries, Inc.

P.O. Box 277, Richfield, OH 44286. Tel.: (216) 659-6631.

Manufacturers and erectors of drive-in screen towers and prefabricated box offices; drive-in design consultants; speaker posts; canopies; and fencing.

Alexander Smith Carpets, Inc.

Sales Division: 919 Third Avenue, New York, NY 10016. Tel.: (212) MU 9-1300.

Manufacturers of carpeting.

Soundfold, Inc., A. Sickels

Box 2125, Dayton, OH 45429. Tel.: (513) 228-3773.

Manufacturers of low cost, acoustical and insulative fabric wall coverings.

Special Optics Heyer Shultz Div./Metal Optics

101 East Main St., Little Falls, NJ 07424.

Manufacturers of metal projection arc reflectors.

Stewart Filmscreen Corp.

1161 W. Sepulveda Blvd., Torrance, CA 90502. Tel.: (213) 326-1422.

Manufacturers of seamless, professional projection screens for motion pictures and TV.

Strand Century, Inc.

20 Bushes Lane, Elmwood Park, NJ 07407; Tel.: (201) 791-7000.

5432 W. 102nd St., Los Angeles, CA 90045; Tel.: (213) 776-4600.

Manufacturers of studio, stage and auditorium lighting fixtures and all types of dimmer control equipment.

Strong Electric Corporation

(A subsidiary of Canrad-Hanovia, Inc.)

P.O. Box 1003, 87 City Park Avenue, Toledo, OH 43697. Tel.: (419) 248-3741.

Manufacturers of carbon and xenon arc motion picture projection lamphouse for 16 and 35mm film, 2 × 2 and 3¼ × 4 xenon slide projectors. Associate DC power supplies. Tufcold and silver reflectors. Carbon and xenon arc, quartz halogen and incandescent follow spotlights.

Technikote Corporation

63 Seabring Street, Brooklyn, NY 11231. Tel.: (212) MAin 4-6429/6430.

Manufacturers of seamless projection Screens—Anti-static XR-171 Pearlescent Dust-resistant screen, Jetwhite screen, Lenticolar Screens, Hilux metallic screen, Matte white screen; drive-in screen paints, wood & metal screen frames; rope and pulley screens.

Theatrical Electronics

Box 706, Matthews, NC 28105.

Reel brackets, rewinds plastic reels, sankor lens, rebuilt used equipment. Everything for the booth.

Union Carbide Corp.—Carbon Products Division

270 Park Avenue, New York, NY 10017. Tel.: (212) 551-2345.

Sales Office: 120 So. Riverside Plaza, Chicago, IL 60606. Tel.: (312) 454-2000.

In Europe: Union Carbide Europe S.A., Rue Pedro-Meylan 5, CH-1211, Geneva 17, Switzerland. Tel.: (022) 47.44.11.

Manufacturers of carbons for motion picture projectors, spotlights and effect lights; brushes, carbon, graphite and metal-graphite for electric motors, generators and converters.

Williams Screen Company

1674 Summit Lake Blvd., Akron, OH 44314. Tel.: 745-0521.

Manufacturers of projection screens.

Edw. H. Wolk, Inc.

1241 S. Wabash Ave., Chicago, IL 60605. Tel.: (312) WE 9-2720-21-22(-)45.

Manufacturers of replacement parts for all American made projectors, arc lamps and sound reproducers, 26½″ Reel Arms for regular and automated projection. Deluxe & "Moto-Matic" Hand Rewinders. "Solfonic" Solar Cells & Power Pacs. Distributor of Naren Spot Lights, Ex-Cell Sand/Water Urns, G.E. & Westinghouse Projection and Exciter Lamps, Cetron Photo-Electric Cells, Bausch & Lomb Glass Silvered Reflectors & "Wolk" Rectifier Bulbs; Sav-A-Reel kits.

Studio Equipment

Ampex Corporation

401 Broadway, Redwood City, CA 94063 Tel.: 367-2011 TWX: 415-364-9034.

Glendale, CA: 500 Rodier Dr. 91201 213-245-9373.

Dallas, TX: 1615 Prudential Dr. 75235 214-637-5200.

Elk Grove Village, IL: 2201 Estes Ave. 60007 312-593-6000.

Hackensack, NJ: 75 Commerce Way 07601 201-489-7400.

Atlanta, GA: 1680 Tully Circle, N.E. 30329 404-633-4131.

Audio/Video recording equipment for industry/studios/radio & TV stations.

Amusement Supply Co.

208 West Montcalm, Detroit 48207. Tel.: WO 1-3440.

R. Ruben, owner.

Automatic Devices Co.
2121 South 12th St., Allentown, PA 18103. Tel.: (215) 797-6000.
Manufacturers of stage and drapery hardware.

Bach Auricon, Inc.
6950 Romaine St., Hollywood, CA 90038. Tel.: Hollywood 2-0931.
Manufacturers of "Filmagnetic" sound-on-film recording equipment and "Auricon" motion picture cameras.

Bell & Howell Company (Professional Equipment Division
7100 McCormick Road, Chicago, IL 60645. (312) 262-1600.
Professional film printers, duplicators and related accessories for motion picture film and video laboratories.

The Camera Mart, Inc.
456 W. 55th St., New York, NY 10019. Tel.: (212) PLaza 7-6977.
Sale & Rental of motion picture and TV equipment, both film & videotape.

F&B/CECO, Inc.
315 West 43rd St., New York, NY 10036. Tel.: (212) 974-4600. Cable: 'CINEQUIP'
Rental, sale and service of photographic equipment for motion picture and TV studios. Manufacturer of professional tripods, tripod heads, editing viewers, camera motors, editing tables and other professional motion picture production equipment. Distributor for Hajnal Snorkel lens systems.

Fiberbilt Photo Products Division of Ikelheimer-Ernst, Inc.
601 West 26th St., New York, NY 10001. Tel.: (212) OR 5-5820.
Manufacturers of cases and trunks for all types of motion picture equipment; also film cases.

Jack A. Frost
234 Piquette Avenue, Detroit, MI 48202. Tel.: 873-8030.
Studio and motion picture lighting equipment.

Galaxy Productions
5233 N.E. 15th St., Des Moines, IA 50313. Tel.: (515) 265-4208/09.
Lease mobile sleeping, and eating facilities, a power plant and heavy duty rigs for location work.

Grosh Scenic Studios
4114-4122 Sunset Blvd., Hollywood, CA 90029. Tel.: (213) 662-1134.
Design, manufacture and installation of theatrical stage draperies; custom and standard theatrical hardware.

Kliegl Bros. Universal Electric Stage Lighting Company, Inc.
32-32 48th Ave., Long Island City, NY 11101. Tel.: (212) 786-7474.
Manufacturers of Klieglights and lighting control equipment for TV studios, stage, auditorium and architural lighting specialties.

Magnasync-Moviola Corp., Subsidiary of Craig Corp.
5539 Riverton Avenue, North Hollywood, CA 91601. Tel.: (213) 763-8441.
Manufacturers of sound equipment, magnetic film recorders, reproducers and accessories, film to videotape transfer system, communication tape recorders and film editing equipment and accessories.

Oxberry Berkey Technical
25-15 50th St., Woodside, L.I., NY Tel.: (212) 721-5555.

Branch Offices: 1015 Chestnut St., Burbank, CA; Burrell Way, Thetford, Norfolk, England.
Manufacturers of animation stands, optical printers, cameras, registration devices.

RCA
30 Rockefeller Plaza, New York, NY 10020. Tel.: (212) 598-5900.

Commercial Communications Systems Division, Broadcast Systems, Front and Cooper St., Camden, NJ 08102. Tel.: (609) 963-8000.

Commercial Communications Systems Division, Photophone Systems, 2700 West Olive Ave., Burbank, CA 91505. Tel.: (213) 849-6741.

Complete theatre service.

Ringold Theatre Equipment Co.
952 Ottawa N.W., Grand Rapids, Mich. 49503. Tel.: (616) 454-8852; Nights: (616) 949-0124. Paul Voudouris, president.

Branch Offices: 29525 Ford Rd., Garden City, MI 48135. Tel.: (313) 522-4650; Nights: (313) 722-8247. 7976 Broadview Rd., Broadview Heights, OH 44147. Tel.: (216) 526-2783.

Ritter Co. Studio Equipment

3325 Cahuenga Blvd., Hollywood, CA 90068. Tel.: (213) 851-3325.

Gas and Electric Wind Machines.

Roctronics Entertainment Lighting and Special Effects, Inc.

100-MPA Roctronics Park, Pembroke, MA 02359. (617) 826-8888.

Designers, manufacturers and sellers of entertainment lighting electronics (dimmers and color synthesizers) and special effects (foggers, bubblers, stroboscopes, laser beam scanners, animated displays).

Ross, Inc.

333 West 52nd St., New York, NY 10019. Tel.: (212) 246-5470.

Rental, sale and service of motion picture and TV lighting, grip equipment and generators. Sole distributor of Mole-Richardson Co. products in the State of New York.

S.O.S. Photo-Cine-Optics, Inc.

Main Office: 315 West 43 St., New York, NY 10036. (212) 974-4600.

Distributors, manufacturers and exporters of television production equipment. Includes 16 and 35mm studio and newsreel cameras; lenses and viewfinders; motion picture and TV accessories; animation equipment; lighting equipment for studio and video; sound recording equipment and accessories; theatre, preview and TV background projection equipment; editing and cutting room supplies; film handling equipment; printing and processing and laboratory testing equipment. All types of stage scenic materials.

Stewart Filmscreen Corp.

1161 W. Sepulveda Blvd., Torrance, CA 90502. Tel.: (213) 326-1422.

Manufacturers of complete line of Stewart Seamless, professional projection screens for rear projected backgrounds of all sizes in motion picture and TV studios, including screen frames and accessories; also seamless front projection screens for theatres and viewing rooms of all types.

Tru-Roll Corporation

622 Sonora Ave., Glendale, CA 91201. Tel.: (213) 243-9567 and 245-5741.

Designers and manufacturers of custom and standard theatrical hardware. Curved and straight drapery tracks, contour drapery rigging, motion picture screen frames, top and side screen masking rigging, motors and controls, motorized revolving stages, motorized bandwagons, orchestra enclosures, sliding panels, disappearing motion picture screens.

Westrex-Litton Industries, Inc.

2629 W. Olive Ave., Burbank, CA 91505. (213) 846-3394.

Westrex Recorders, re-recorders, mixers.

Branch Offices

Westrex Company Limited, Unit 1, Fairway Drive,, Bilton Fairway Industrial Estate, Greenford, Middlesex, UB6 8PW, England.
Westrex Company, Asia, 1302 Luk Hoi T-ng Bldg., 31, Queen's Road Central, Hong Kong.
Westrex Company, Asia, International House, Sec. 3, Hsin Yee Road, Taipei, Taiwan.
Westrex Company Italy, 65 via Constantine Maes, 00162 Rome, Italy, Cable: Westrex Rome, Tel.: 83 92 990.
Westrex Company Orient, C.P.O. Box 760, Tokyo, Japan, cable: Westrex Tokyo, Telex: J22612, Tel.: 211-6791.

XeTRON Corp.

10 Saddle Road, Cedar Knolls, NJ 07927 Tel.: (201) 267-8200.

Lorraine Carbon Division: Carbon electrodes for all motion picture applications. XeTRON Products Division: Professional 16, 35, and 70 mm projection equipment and accessories. Cinemaccanica 35mm and 70mm projectors, film, transports, sound systems, XeTRON xenon lamphouses, bulbs, platter transport, continuous loop transport, dimmer systems, projection lenses. Hortson and Fumeo 16mm projection systems.

Refreshment Products and Equipment

American Can Company—Dixie and Marathon Products

24th and Dixie Avenue, Easton, PA 18042. Tel.: 250-1400, and American Lane, Greenwich, CT 96830.

Manufacturers of paper products: plates, cups, bowls, towels, tissues, etc.

Paul F. Beich Co.

Bloomington, IL 61701. Tel.: (309)828-1311.

Manufacturers of 10¢ Bike Whiz. Complete line of other 10¢ bars and 15¢ Milk Chocolate caramel filled shells.

Blevins Popcorn Co.

Popcorn Village, Memphis, TN 38117. 813 Ridge Lake Blvd., P.O. Box 171233. Tel.: (901) 761-2062.

Processing Plants in Ridgway, IL and North Bend, NE

Concession supply branches in Atlanta, GA 30325, P.O. Box 20176 Sta. N., 2023 Hills Ave., N.W.: Cincinnati, OH 45246, 11705 Chesterdale Rd.: Dallas, TX 75226, P.O. Box 26264, 1901 S. Good-Latimer: Houston, TX 77206, P.O. Box 10243, 202 W. 38th St.: Jacksonville, FL 32203, P.O. Box 2752, 633 N. Myrtle Ave.: Louisville, KY 40214, 6224 Strawberry Lane: Memphis, TN 38116, P.O. Box 16834, 3143 Fleetbrook Dr.: Miami, FL 33152, P.O. Box 552901, 7851 N.W. 32nd Street: Nashville, TN 37209, P.O. Box 90766, 3098 Charlotte Ave.: New Orleans, LA 70181, P.O. Box 10585, 1017 Jefferson Hwy.: Omaha, NE 68101, P.O. Box 1282, 1608 Cuming St.: Tampa, FL 33601, P.O. Box 5461, 1212 N. 39th St.: West Terre Haute, IN 47885, Highway 40 West

Warehouses in Little Rock, AR, Knoxville, TN & North Loup, NE

Processors of raw popcorn and distributors of all high profit concession supplies and equipment.

Butler Fixture & Mfg. Co.

3939 So. Williams St., Denver, CO 80205. Tel.: (303) 572-3939.

Builders of theatre concession stands.

Canada Dry Corp. (Syrup Division)

100 Park Ave., New York, NY 10017. Tel.: (212) LE 2-4300.

Manufacturers of beverage syrups: Ginger Ale, Cola, Tahitian Treat, Hi-Spot, Orange, Cactus Cooler, Root Beer and others.

Castleberry's Food Co.

Box 1010, August, GA 30903.
Fast foods, sauces, packed foods.

Coca-Cola USA

P.O. Drawer 1734, Atlanta, GA 30301. Tel.: (404) 897-2121.
Manufacturers of soft drink syrups, Coca-Cola, Tab, Sprite, Fanta flavors, Fresca.

Continental Can Co. Bondware Division

633 Third Ave., New York, NY 10017. Tel.: 551-7000.
Paper cups for hot and cold drinks.

Crush International, Inc.

2201 Main St., Evanston, IL Tel.: (312) DAvis 8-8850 and MAin 5-0400.
Manufacturers of Hires Root Beer, Crush fruit syrups and Old Colony syrups.

Dr Pepper Company

5523 East Mockingbird Lane, P.O. Box 225086, Dallas, TX 75222. Tel.: (214) 824-0331.
Manufacturers of Dr Pepper and Sugar-Free Dr Pepper concentrates and fountain syrups.

Goetze's Candy Co., Inc.

3900 E. Monument St., Baltimore, MD 21205.
Candy designed exclusively for theatres; 3½ oz. 12 pack.

Greer Enterprises, Inc.

281 N. Grant Ave., Columbus, OH 43215. Tel.: CA 1-3245. Factory and Sales 31 Chicago St., Quincy, MI 49082. Tel.: (517) 639-9825.
Manufacturers of combination frankfurter barbecue unit and bun warmer, Glenray manufacturers of Moist Heat Portable Steam Table (steamette).

Henry Heide, Inc.

P.O. Box 271, New Brunswick, NJ 08903. Tel.: (201) 846-2400.
Manufacturers of candy—Jujyfruits, Jujubes, Chocolate Flavor Babies, Candy Corn, Red Hot Dollars and drops.

Hershey Foods Corp. Hershey Chocolate and Confectionery Division

19 East Chocolate Avenue, Hershey, PA 17033. Tel.: (717) 534-4200
Manufacturers of chocolate and cocoa.

M.J. Holloway & Company

308 West Ontario, Chicago, IL 60610. Tel.: (312) 642-2700.
Manufacturers of candy—Milk Duds, Slo-Poke and Black Crow Suckers, Hi-noon bars, Zooper Dooper bars.

Hollywood Brands, Inc.

A subsidiary of Consolidated Foods Corp.

836 South Chestnut Street, Centralia, IL 62801. Tel.: (618) 532-4767.
Manufacturers of Pay Day, Butternut, Milk Shake, and Zero.

Jet Spray Corp.

195 Bear Hill Road, Waltham, MA 02154. Tel.: (617) 890-7700.
Manufacturers of visual display, electrically refrigerated carbonated and non-carbonated beverage dispensers, hot chocolate dispensers, freeze-dried coffee dispensers, coin operated coffee and chocolate dispensers, instant mashed potato dispensers, hot and cold tea dispenser.

Manley, Inc.

1920 Wyandotte, Kansas City 8, MO (Mail: P.O. Box 1006, Kansas City, MO 64141). Tel.: (816) 421-6155.
Manufacturers of popcorn machines, popcorn, popcorn seasoning and beverage dispensers. Theatre and Drive-In theatre concession stands.

M & M/Mars

High Street, Hackettstown, NJ 07840.
Manufacturers of M & M's Plain Chocolate Candies, Milky Way Milk Chocolate Bar, M & M's Peanut Chocolate Candies, 3 Musketeers bar, Snickers, Mars Almond, Mars Double Crunch, Sprint Chocolate Wafer, Snickers Peanut Munch, Milky Way Dark Chocolate Bar.

National Rejectors, Inc.

P.O. Box 1550, Hot Springs, AR 71901.
Coin and currency handling equipment. Branches in major cities.

Ogden Food Service

254 Franklin St., Buffalo, NY. Tel.: 852-3339.

Ogden Foods, Inc.

3660 S. Lawrence St. Philadelphia, PA 19148
Operators of Theatre Food Concessions—both Indoor Houses and Drive-Ins.

Pepsi-Cola Company

Purchase, NY
Manufacturers of Pepsi-Cola syrup and bottlers of Pepsi-Cola, Diet-Pepsi, Teem.

Peter Paul Codbury, Inc.

New Haven Road, Naugatuck, CT 06770.
Manufacturers of candy; Mounds, Almond Joy, York Peppermint Patties, Codbury chocolate.

Quaker City Chocolate & Confectionery Company, Inc.

2901 Grant Ave., Philadelphia, PA 19114. Tel.: (215) 677-6500.
Manufacturers of 5¢ and 10¢ candies: Good and Plenty, Spice Drops, Spearmint Leaves, Orange Slices and Good 'N Fruity. Exclusive distributors for Bassett's Liquorice All Sorts.

H. B. Reese Candy Company, Inc., Subsidiary of Hershey Chocolate Co.

Hershey, PA 17033. Tel.: (717) 534-4100.
Manufacturers of Peanut Butter Cups.

Rex Packaging Company—A Division of Standard International Corp.

10 Minue Street, Carteret, NJ 07008. Tel.: (201) 969-1234.
Manufacturers of noiseless paper popcorn bags and a complete line of outdoor concession bags.

Roll-A-Grill Corp. of America
645 First Ave., New York, NY 10016. Tel.: (212) CH 4-5000.
Manufacturers of Roll-A-Grill automatic frankfurter machines, Roll Warmers, Sandwich Grills, Toas-Treat Sandwich Makers, Roll-A-Grill Speed Fryer, Roll-A-Spreader, compactor, miracle freezer, Roll-A-Slicer, Pressure Fryer.

Rowe International, Inc., A Subsidiary of Triangle Industries, Inc.
75 Troy Hills Road, Whippany, NJ 07981. Tel.: (201) 887-0400.
Manufacturers of Automatic Venders, music systems and bill and coin changers.

Royal Crown Cola Co.
1000 Tenth Avenue, P.O. Box 1440, Columbus, GA 31902. Tel.: (404) 322-4431.

Branches: 108-08 Queens Blvd., Forest Hills, NY 11375. Tel.: (212) 268-3100; 4849 Scott Street, Suite LL322, Schiller Park, IL 60176. Tel.: (312) 671-2565; Suite 127, 1145 Empire Central Plaza, Dallas, Texas 75247. Tel.: (214) 631-0755; Suite 1005, 3450 Wilshire Blvd., Los Angeles, CA 90005. Tel.: (213) 386-2711.
Manufacturers of soft drinks.

Scotsman Ice Systems
505 Front St., Albert Lea, MN 56007. Tel.: 373-3961.
Manufacturers of Scotsman ice cube and flake machines, combination ice machines and drink dispensers.

The Seven-Up Company
121 South Meramec, St. Louis, MO 63105. (314) 889-7777.
7UP and Diet 7UP are available in bottles, cans, pre-mix, and post-mix from the local 7UP Bottling company. The 7UP Bottler is a source for pre-mix and post-mix dispensing equpment and bottle and can venders.

Simonin's, C. F. Sons, Inc.
Tioga & Belgrade Streets, Philadelphia, PA 19134. Tel.: (215) GA 6-2300.
Manufacturers of popcorn oils and colored and flavored seasonings.

Smithfield Ham & Products Company
Smithfield, VA 23430. Tel.: (804) 357-2121.
James River brand barbecued meats and other genuine Smithfield Foods.

Stein Industries, Inc.
22 Sprague Avenue, Amityville, NY 11701. (516) 691-2222.
Manufacturers and distributors of popcorn warmers and concession stands. Doorman stub recepticles, box offices, display frames.

Sunkist Soft Drinks, Inc.
2600 Century Parkway, Atlanta, GA 30345. (404) 321-1776; A division of General Cinema Corp.
Suppliers of Sunkist Orange Soda Fountain syrup, bottles, cans.

Vendo Company
10500 Barkley, Overland Park, KS 66212. Tel.: (913) 341-1300.
Manufacturers of automatic venders for food, beverages (hot and cold), candy and cigarettes; distributors of stereo music systems, butter dispensers, etc.

Virga's Pizza Crust Company
2236 Conner Street, New York, NY 10466, Tel.: XX 4-9100.
Frozen pizza and ingredients.

Manufacturers and Services By Product Classification

Animation Equipment
Oxberry Berkey Technical
S.O.S. Photo-Cine-Optics, Inc.

Attraction Advertising Equipment
A-Wagner Zip Change
Bevilite-Adler

Beverages and Beverage Dispensers
Canada Dry Corp.
Coca-Cola USA
Crush International
Dr Pepper Co.
Jet Spray Corp.
Manley, Inc.
Pepsi-Cola Co.
Royal Crown Cola Corp.
Seven-Up Company
Sunkist Soft Drinks
Vendo Co.

Camera Equipment
Bell & Howell Company
Camera Mart, Inc.
F&B/CECO, Inc.
Mauer, J. A., Inc.
Mitchell Camera Corp.
Oxberry Berkey Technical
S.O.S. Photo-Cine-Optics, Inc.

Candy and Candy Machines
Beich, Paul F., Co.
Heide, Henry, Inc.
Hershey Chocolate Corp.
Holloway, M. J. Co.
Hollywood Brands, Inc.
M&M/Mars
Quaker City Chocolate & Confectionery Co., Inc.
H. B. Reese Candy
Vendo Co.

Carbons, Carbon Savers & Carbon Coolers
The Marbe Co., Inc.
Union Carbide Corp.
Xetron

Coin Changing Machines
National Rejectors, Inc.

Concession Stands
Butler Fixture & Mfg. Co.

Cups & Trays
American Can Company
Continental Can Co.
Lily-Tulip Cup Corp.

Curtains and Curtain Controls and Acoustical Drapes
Automatic Devices Co.
Econo-Pleat
Novelty Scenic Studios
Soundfold, Inc.

Drive-In Admissions Control
Eprad
K-Hill Signal Corp.

Drive-In Screens and Towers
Ballantyne of Omaha
Selby Industries, Inc.
Technikote

Equipment Cases
Fiberbilt, Division of Ikelheimer-Ernst, Inc.

Film Editing
A.V.E. Corporation
S.O.S. Photo-Cine-Optics, Inc.
Westrex-Litton Industries, Inc.

Film Splicers
Prestoseal Mfg. Corp.

Floor Coverings
Gulistan Carpet
Koneta Matting
Smith, Alexander, Inc.

Food and Food Concessions Suppliers
Castleberry's Food Co.
Rowe Mfg. Co.
Smithfield Ham & Products Co.
Tri-State Refreshments, Inc.
Vendo Co.
Virga's Pizza Crust Company

Food Equipment
Greer Enterprises, Inc.

In-Car Speakers and Heaters
Ballantyne of Omaha
Eprad, Inc.
The Marble Co., Inc.
Projected Sound, Inc.

Interior Decoration
Novelty Scenic Studios

Lenses, Projection
Atlantic Audio Visual Corp.
Bausch & Lomb, Inc.
Kollmorgen Corp.
The Marble Co., Inc.
Panavision, Inc.
Schneider Corp. of America

Lighting Equipment, Studio
Kliegl Bros. Universal Electric Stage Lighting Co., Inc.
Charles Ross, Inc.

Playground Equipment
American Playground Device Co.
Herschell, Allan, Co., Inc.
Miracle Space Equipment Co.

Popcorn Machines and Supplies
Blevins Popcorn Co.
Manley, Inc.
Rex Packaging
Simonin's, C. F. Sons, Inc.
Stein Industries, Inc.

Projection Light Equipment
Ballantyne of Omaha
Canrad Hanovia Inc.
Imperial Electric Co.
Hertner Division General Battery Corp.
The Marble Co., Inc.
Kneisley Electric Co.
Optical Radiation Corp.
Osram Sales Corporation
RCA
Rank Organisation
Strong Electric Co.
Xetron

Projection Room Accessories
Neumade Products Corp.
Theatre Equipment Co.

Projectors and Parts
A.V.E. Corporation
Ballantyne of Omaha
Century Projector Corp
Kinotone, Inc.
La Vezzi Machine Works
National Theatre Supply
RCA
Simplex Equipment Corp.
Edward H. Wolk

Recording Equipment
Ampex Corporation
Bach Auricon, Inc.
Magnasync Corp.
Maurer, J. A., Inc.
RCA
Reeves Equipment Corp.
S.O.S. Photo-Cine-Optics, Inc.
Westrex-Litton Industries, Inc.

Screen Frames
Stewart Filmscreen Corp.

Screens and Screen Paint

Hurley Screen Co.
Stewart Filmscreen Corp.
Technikote Corp.
Williams Screen Co.

Seating for Auditoriums

American Desk Co.
American Seating Co.
Griggs Int'l. Inc.
Heywood-Wakefield Co.
Ideal Seating Corp.
Irwin Seating Co.
Mannk Fabrics Co., Inc.
Massey Seating Co.

Sound Equipment

Altec-Lansing
Ballantyne of Omaha
Century Projector Corp.
Cinematograph International, Inc.
Kinotone, Inc.
RCA
Simplex Equipment Corp.
Westrex-Litton Industries, Inc.

Spotlights

Kneisley Electric Co.
Strong Electric Co.

Stage Lighting

Century Strand, Inc.
Kliegl Bros.

Studio Lighting

Century Strand, Inc.
Frost, Inc., Jack A.
S.O.S. Photo-Cine-Optics, Inc.
Charles Ross, Inc.

Theatre Design

Forest Bay Construction

Ticket Machines and Boxes

Eprad

Theatre Supply Dealers

IN THE UNITED STATES AND CANADA

ALABAMA

Birmingham

QUEEN FEATURE SERVICE, 2409 First Ave., North. Tel.: (205) 251-8665

ARIZONA

Phoenix

ARIZONA THEATRE EQUIPMENT, 1410 E. Washington. Tel.: (602) 437-0215

CALIFORNIA

Culver City

BUDD THEATRE SUPPLY, INC., 8537 West Washington Blvd. Tel.: (213) 870-9301

Inglewood

CARTER EQUIPMENT CO., INC., 1050 W. Florence Ave. Tel.: (213) 677-6117

Los Angeles

AUTOMATIC DEVICES CO., 553 Oak Knoll Ave., Pasadena. Tel.: 681-4338
BARRETT-ROBINSON CO., 9165 Las Tunas Dr., Temple City. Tel.: 285-1229
CONSOLIDATED SEATING, 1314 Scott Road, Bradbury, L.A. Tel.: 849-7637
CURRAN PRODUCTIONS, 1215 Bates Ave., L.A. Tel.: 662-8129
FILBERT, JOHN P., 1100 Flower, Gelendale: Tel.: 247-6550
GROSH, R. L., & Sons, 4114 Sunset Blvd. Tel.: 662-1134
NATIONAL THEATRE SUPPLY, 2001 S. La Cienega Blvd. 90034. Tel.: (213) 838-1821
OLESEN CO., 1535 Ivar Ave. L.A. Tel. 461-4631
PEMBREX THEATRE SUPPLY, 1100 Flower, Glendale. Tel.: 247-6550
THEATRE SERVICE & SUPPLY, INC., 1250 E. Walnut St., Pasadena, 91106. (213) 792-7158
THEATRE UPHOLSTERING CO., 1338 W. 24th, L.A. Tel.: 734-0740

San Francisco

PACIFIC THEATRE EQUIPMENT CO., 142 Leavenworth St. Tel.: (415) 771-2950
WESTERN THEATRICAL EQUIPMENT, 187 Golden Gate Ave. Tel.: (415) 861-7571

COLORADO

Denver

WESTERN SERVICE & SUPPLY, 2100 Stout St. Tel.: (303) 534-7611. R. K. Tankersley, mgr

DISTRICT OF COLUMBIA

Washington

BEN LUST, 623 Sligo Ave., Silver Spring, MD 20910; 589-6606. Irwin Lust, pres
R&S THEATRE SUPPLY CO., 4701 42nd St. N.W., Tel.: (202) 244-1500
WILMO CORP., 3322 M St., N.W. Tel.: (202) 337-6680

FLORIDA

Jacksonville

JACKSONVILLE FILM SERVICE, INC., 2208 West 21st. Tel.: (904) 355-5447

Miami

JOE HORNSTEIN, INC., 759 W. Flagler St. Tel.: (305) 545-5842

GEORGIA

Albany

DIXIE THEATRE SERVICE & SUPPLY, 1010 N. Slappey Dr

Atlanta

AMERICAN THEATRE SUPPLY, P.O. Box 54553. (30303) Tel.: (404) 875-3167
CAPITAL CITY SUPPLY, 2124 Jackson Parkway, N.W. Tel.: (404) 792-9424

NATIONAL THEATRE SUPPLY, 1325 Logan Circle, N.W. 30318. Tel.: (404) 351-1419
WIL-KIN THEATRE INC., 301 No Avenue, N.E. Tel.: (404) 876-0342

Savannah

RHODES SOUND & PROJECTION SERVICE, 218 E. 56th St

ILLINOIS

Chicago

ABBOTT THEATRE EQUIPMENT, 1311 S. Wabash Ave. (312) 427-7573
NATIONAL THEATRE SUPPLY (National Screen Service), 1322 S. Wabash Ave. 60605. Tel.: (312) 427-8211

INDIANA

Indianapolis

GER-BAR, INC., 339 N. Capitol Ave. Tel.: (317) 634-1727

IOWA

Des Moines

DES MOINES THEATRE SUPPLY, 1121 High St. Tel.: (515) 243-6520

KANSAS

Kansas City

MID-CONTINENT THEATRE SUPPLY CORP., 1800 Wyandotte. Tel.: (816) 221-0480

Wichita

SOUTHWEST THEATRE EQUIPMENT, 118½ W. Douglas Ave., P.O. Box 2138

KENTUCKY

Louisville

FALLS CITY THEATRE EQUIPMENT, 427 S. Third St
HADDEN THEATRE SUPPLY, 909 Emerson Ave., P.O. Box 4151

LOUISIANA

New Orleans

NATIONAL THEATRE SUPPLY, 5624 Jefferson Hwy., Harahan, LA 70123
SOUTHERN THEATRE SUPPLY, INC., 3822 Airline Highway, Metairie, Tel.: (504) 831-1001

MARYLAND

Baltimore

ALLIED THEATRE EQUIPMENT CO., INC., 12 E. 25th St. Tel.: (301) 235-2747
DUSMAN MOTION PICTURE SUPPLIES, 12 E. 25th St. Tel.: 235-4747
NATIONAL THEATRE SUPPLY, 6707 Whitestone Rd. Tel.: (301) 944-8844

MASSACHUSETTS

Boston

MAJOR THEATRE EQUIPMENT CORP., 44 Winchester St. Tel.: (617) 542-0445
MASSACHUSETTS THEATRE EQUIPMENT, 20 Piedmont St. Tel.: (617) 542-9814
NATIONAL THEATRE SUPPLY, 95 Broadway 02116. Tel.: (617) 542-2663

MICHIGAN

Detroit

ALTEC SERVICE CO., 401 Commerce Bldg. Tel.: 963-3180
AMUSEMENT SUPPLY CO., 208 W. Montcalm. Tel.: WO 1-3440
NORTHWEST STUDIOS, INC., 16209 W. McNicholis. Tel.: BR 3-2650

Garden City

RINGOLD THEATRE EQUIPMENT, 32647 Ford Road, Tel.: (313) 522-4650

Grand Rapids

RINGOLD THEATRE EQUIPMENT, 952 Ottawa, N.W. Tel.: (616) 454-8852

MINNESOTA

Minneapolis

MINNEAPOLIS THEATRE SUPPLY CO., 51 Glenwood Ave. Tel.: (612) 335-1166

MISSOURI

Kansas City

DRIVE-IN THEATRE MANUFACTUFING CO., 709 N. Sixth St.: 321-3978
MID CONTINENT THEATRE SUPPLY, 1800 Wyandotte. Tel.: 221-0480
NATIONAL THEATRE SUPPLY, 1800 Baltimore 64108. (816) 221-9858
SHREVE THEATRE SUPPLY, 709 North Sixth. Tel.: 321-3978

St. Louis

RINGOLD CINEMA EQUIPMENT CORP., 8421 Gravois Road, 63123 Tel.: (314) 352-2020

NEBRASKA

Omaha

QUALITY THEATRE SUPPLY, 1515 Davenport. Tel.: (402) 341-7253
SLIPPER THEATRE SUPPLY, INC. 1502 Davenport St. Tel.: (402) 341-3921

NEVADA

Las Vegas

PEMBREX THEATRE SUPPLY CO , 3519 Algonquin Dr. Tel.: (702) 735-5278

NEW JERSEY

Camden

MID-ATLANTIC THEATRE EQUIPMENT CO., 2600 Mt. Ephriam Ave., Tel.: (609) 962-6632
NATIONAL THEATRE SUPPLY CO , 130 Ferry Ave. 08104. Tel.: (609) 962-9200

NEW YORK

Albany

ALBANY THEATRE SUPPLY, 443 N. Pearl St. Tel.: (518) 465-8894. Jack McGrath

Kingston

SUMMIT ENTERPRISES, Gov. Clinton Hotel. Tel.: (914) 338-5095

Tarrytown

MOTION PICTURE ENTERPRISES, P.O. Box 276, 10591. (212) 245-0969. Herbert R. Pilzer, pres.

New York City

CAPITOL MOTION PICTURE SUPPLY, CORP., 630 Ninth Ave. Tel.: (212) 757-4510
JOE HORNSTEIN, INC., 341 W. 44th St. Tel.: (212) 246-6285
NATIONAL THEATRE SUPPLY, 1600 Broadway. Tel.: (212) 245-6900
STAR CINEMA SUPPLY, 217 W. 21st Street, Tel.: 675-3515
TIMES SQUARE THEATRICAL & STUDIO SUPPLY CORP., 318 W. 47th Street, Tel.: 245-4155

NORTH CAROLINA

Charlotte

AMERICAN THEATRE SUPPLY CO., 529 N. Tryar St. Tel.: 333-5076
CHARLOTTE THEATRE SUPPLY CO., 227 S. Church St. Tel.: (704) 333-9651
HARRIS THEATRE SALES, INC., 315 S. Church St.
STANDARD THEATRE SUPPLY, 1624 West Independence Blvd. Tel.: (704) 375-6008
WIL-KIN, INC., 800 S. Graham St., Tel.: (704) 333-6101

Greensboro

STANDARD THEATRE SUPPLY CO., 125 Higgins Street. Box 20660, Tel.: (919) 272-6165

OHIO

Cincinnati

NATIONAL THEATRE SUPPLY, 1105 W. 8th St., 45203. Tel.: (513) 621-8903

Cleveland

OHIO THEATRE SUPPLY CO., 7976 Broadview Rd. (216) 526-2783

Toledo

TOLEDO THEATRE SUPPLY, 3916 Secor Rd

OKLAHOMA

Oklahoma City

CAPITOL THEATRE SUPPLY, 5900 Mosteller Drive, Tel.: (405) 842-0860
OKLAHOMA THEATRE SUPPLY, 628 W. Sheridan Ave. Tel.: (405) 236-8691

OREGON

Portland

WEST COAST THEATRE SERVICE, 909 N.W. 19th St., Tel.: (503) 227-2932

PENNSYLVANIA

Philadelphia

ALLIED THEATRE EQUIPMENT CO., INC., 157 N. 12th St. Tel.: (215) 567-2047
BLUMBERG BROS., 1305 Vine St. Tel.: (215) WA 5-7240

Pittsburgh

ATLAS THEATRE SUPPLY, 1519 Forbes Ave
NATIONAL THEATRE SUPPLY, 107 6th St. (Fulton Bldg.) Tel.: (412) 471-4630
THEATRE EQUIP. & SERVICE CO., 100 Lighthill St. Tel.: (412) 322-4600

RHODE ISLAND

Providence

RHODE ISLAND SUPPLY, 357 Westminster St

TENNESSEE

Memphis

NATIONAL THEATRE SUPPLY, 412 S. Second St. Tel.: (901) 525-6616
TRI-STATE THEATRE SUPPLY, 151 Vance Ave. Tel.: 525-8249

Nashville

CAPITAL CITY SUPPLY CO, 713 Seduken Bldg. Tel.: (615) 256-0347
MID-SOUTH THEATRE SERVICE, 439 Brewer Drive. Tel.: (615) 832-5660

TEXAS

Dallas

HARDIN THEATRE SUPPLY, 714 S. Hampton Rd
HERBER BROS., 408 S. Harwood St.
MODERN SALES & SERVICE, 2200 Young St. Tel.: (214) 747-3191
NATIONAL THEATRE SUPPLY, 7138 Envoy Court, 75247. Tel.: (214) 634-7150
PINKSTON'S SALES AND SERVICE, 4207 Lawnview Ave. Tel.: (214) 388-1550
LOU WALTERS SALES & SERVICE, 4207 Lawnview Ave. Tel.: (214) 388-1550

Houston

SOUTHWESTERN THEATRE EQUIPMENT, 1702 Rusk Ave. Tel.: (713) 222-9461

San Antonio

INDEPENDENT THEATRE SUPPLY, 2750 E. Houston
TEXAS THEATRE SUPPLY, 915 S. Alamo St. 78205. Tel.: (214) 222-1002

UTAH

Salt Lake City

L&S SUPPLY CO., 214 E. First South St. Tel.: (801) 328-1641. Herb Schoenhardt, mgr
PETERSON THEATRE SUPPLY, 19 E. 2nd South, Tel.: (801) 822-3685
UNIVERSAL THEATRE SUPPLY, 264 East First South St., Tel.: (801) 328-1641

VIRGINIA

Roanoke

PERDUE MOTION PICTURES EQUIPMENT, INC., 2315 Williamson Rd., N.E. Tel.: (703) 366-0295

WASHINGTON

Seattle

ALTEC SERVICE CORP., White Henry Stuart Bldg. MA 3-8221
AMERICAN GENERAL SUPPLY CO., INC., 2300 1st Avenue. MA 4-4572
BERESFORD, CHAS. H. INC., 308 W. Republican. AT 4-6658
BROCKLIND'S INC., 901 Olive Way. MU 2-5898
BURNS, S.F. & CO., INC., 2319 2nd. (206) 624-2515
CHAMPION CRAFT & COSTUME SUPPLY, 216 Stewart. MA 3-1925
DISPLAY & STAGE LIGHTING INC., 2410 1st
FACTORY DIRECT DRAPERIES INC., 8300 Aurora N. LA 5-7932
FAIRBANKS HARLAN CO., 1405 Elliott W. 284-7420
GIBSON SOUND & LIGHT CO., 501 N. 36th
J. SCHLEIS THEATRICAL & DISPLAY SERVICES CO., 123 Belmont E. EA 2-1478
NATIONAL SCREEN SERVICE CORP., 2413 2nd Ave. (201) 265-2700
NATIONAL THEATRE SUPPLY CO., 2413 2nd Ave. 98121. (206) 624-7710
SLATER, W.A. CO., 2323 2nd. MU 2-0154
STAGECRAFT INDUSTRIES, 10237 Main, Bellevue. GL 4-3089

WEST VIRGINIA

Charleston

CHARLESTON THEATRE SUPPLY, 506 Lee St
MOORE THEATRE EQUIPMENT, 213 Delaware Ave., P.O. Box 782, Tel.: (304) 344-4413

WISCONSIN

Milwaukee

HARRY MELCHER ENTERPRISES, 3607 W. Fond Du Lac Ave. Tel.: (414) 442-5020

CANADA

Calgary

GENERAL SOUND & THEATRE EQUIPMENT, LTD., 6812-C 6th St. S.W. Calgary T2H 2K4. (403) 252-3773.

Montreal

GENERAL SOUND & THEATRE EQUIPMENT, LTD., 5887 Monkland Ave., H4A 1G6 (514) 589-5331

St. John, N.B.

GENERAL SOUND & THEATRE EQUIPMENT, LTD., 55 Bentley St., E2K 1B2 (506) 693-2819

Vancouver

GENERAL SOUND & THEATRE EQUIPMENT, LTD., 2182 W. 12th St., Vancouver B.C. V6K-2N4 (604) 736-5506

Winnipeg

GENERAL SOUND & THEATRE EQUIPMENT, LTD., 430 Kensington St., Winnipeg, Man. R3J 1J7 (204) 888-7543

British-Based Equipment/Services

ACMADE INTERNATIONAL
Oakside Oxford Road, Uxbridge, Mddx. UB9 4DX (89-36313) Telex: 8954606.

AEG-TELEFUNKEN (UK) LTD.
217 Bath Road, Slough, Berks. SL1 4AW (2-872101)

AGFA-GEVAERT LTD.
Great West Road, Brentford, Middx: (OL 560 2131)

AKG ACOUSTICS LTD.
191 The Vale, London, W3 7QS (01 749 2042)

ALLOTROPE LTD.
36/38 Lexington Street, London, W1R 3HR (01 437 1892) Telex: 21624 ALO FFD

AMPEX INTERNATIONAL
Acre Road, Reading, Berks, (0734 85200)

ANIMATION EQUIPMENT ENGINEERING LTD.
Wembley Commercial Centre, Unit 17, East Lane, Wembley, Middx. (01-904 0611)

ASCOTTS VIDEO
3 Soho Square, London, W.1. (01-734 2116)

ASTON ELECTRONIC DEVELOPMENTS LIMITED
125 Deepcut Bridge Road, Deepcut, Camberley, Surrey (02516 6221)

AUDIX LTD.
Station Road, Wenden, Saffron Walden, Essex.

AV DISTRIBUTORS (LONDON) LTD.
26 Park Road, London NW1 4SH (01-935 8161)

BALLANCROFT FILM & TV EQUIPMENT CO. LTD.
292 Worton Road, Isleworth, Middlesex, (01-560 494)

F.W.O. BAUCH LTD.
49 Theobald Street, Boreham Wood, Herts. (01-953 0091)

BELL & HOWELL LTD.
Alperton House, Bridgewater Road, Wembley, Middlesex HAO 1EG (01-902 8812)

BERKEY COLORTRAN (UK) LTD.
P.O. Box 5, Burrell Way, Thetford, Norfolk (0842-2484)

ROBERT BOSCH LTD.
PO Box 166, Rhodes Way, Watford.

BRABURY ELECTRONICS LTD.
119a Loverock Road, Reading, Berks RG3 1MS.

BOSTON INSULATED WIRE (UK) LTD.
1 Canbury Park Road, Kingston-upon-Thames, Surrey.

BRITISH FILMS LTD.
Carlyle House, 235 Vauxhall Bridge Road, London, SW1V 1EJ. (01 828 7965)

CANDA TELEVISION EQUIPMENT LTD.
5 Hanover Trading Estate, 1 North Road, London, N7 9HD. (01-607 9961) Telex: 28604 Mono Ref. 477 LDN

CINEFOCUS RENTALS
1 Pavilion Parade, Wood Lane, London, W12 0HQ. (01-743 2552)

CINTRON GROUP
Grove House, 551 London Road, Isleworth, Middlesex TW7 4DS (01-568 0131) Telex: 935054

DOEL ENGINEERING LTD.
Station Road, Hatfield Peverel, Chelmsford, Essex CM3 2DX (0245 380151) Telex: 987562 COCHAC.

DOLBY LABORATORIES INC.
346 Clapham Road, London SW9 9AP (01-720 1111).

EDRIC AUDIO VISUAL LTD.
34-36 Oak End Way, Gerrards Cross, Bucks. SL9 8BR (02813 84646 and 86521)

ELECTROSONIC LTD.
815 Woolwich Road, London, SE7 8LT (01 855 1101)

ELF AUDIO VISUAL LTD.
836 Yeovil Road, Trading Estate, Slough, Berks. (75-36123)

ENGLISH ELECTRIC VALVE COMPANY LIMITED
Chelmsford, Essex CM1 2QU, (0245 61777)

EUMIG (UK) LTD.
14 Priestley Way, Eldon Wall Trading Estate, London NW2 7TN (01-450 8070)

EVERSHED POWER-OPTICS LTD.
Bridge Wharf, Bridge Road, Chertsey, Surrey KT16 8LJ (09328-61181)

FILM FACILITIES (MAGNETIC).
3 Springbridge Mews, Springbridge Road, Ealing Broadway, London, W5 2AB (01-567 3613/4)

FUJI PHOTO FILM (UK) LTD.
99 Baker Street, London, W1R 3HR (01-487 5711)

FUTURE FILM DEVELOPMENTS
36-38 Lexington Street, London W1R 3HR. (01-437 1892) Telex: 21624 ALO FED

GEC (LAMPS AND LIGHTING) LTD.
PO Box 17, East Lane, Wembley, Middlesex HA9 7PG (01-904 4321).

GORDON AUDIO VISUAL LTD.
28 Market Place, Oxford Circus, London W1N 8PH (01-580 9191)

HARKNESS SCREENS LTD.
The Gate Studios, Station Road, Boreham Wood, Herts. (01-953 3611)

HAYDEN LABORATORIES LTD.
Hayden House, Chiltern Hill, Chalfont St. Peter Gernards Cross, Bucks. SL9 9EW. (02813-88447)

HEYDEN AND SONS LTD.
Spectrum House, Hillview Gardens, London NW4 (01 2035171)

HEYDEN-KEM ELECTRONIC MECHANISMS LTD.
24 Vivian Avenue, London NW4 3XP (01-202 0244)

INDEPENDENT CAMERA AND EQUIPMENT CO. LTD.
60 Farringdon Road, London EC1R 3BP. (01-251 3885)

INTERNATIONAL VIDEO CORPORATION (UK) LTD.
10 Portman Road, Reading, Berks., RG3 IJR, (0734 585421)

KADEK VISION
P.O. No. 21, Shepperton Studio Centre, Studios Road, Shepperton, Middx. TW17 0QD. (093 2866941)

JOHN KING FILMS, LTD.
FILM HOUSE, 71 East Street, Brighton BN1 1NZ. Brighton 202671)

KODAK LTD.
P.O. Box 66, Kodak House, Station Road, Hemel Hempstead, Herts. (0442-62331)

LEE ENTERPRISES LTD.
128 Wembley Park Drive, Wembley, Mddx. (01-969 9521)

LEE FILTERS LTD.
Walworth Industrial Estae, Andover, Hants. SP10 5AN (0264-66245)

LIPSNER-SMITH CO. LTD.
Unit 7, Cowley Mill Trading Estate, Longbridge Way, Uxbridge, Middx: (89 52191)

3M UNITED KINGDOM LTD.
3M HOUSE, P.O. Box 1, Bracknell, Berks. RG12 1JU (0344 26726)

MARCONI COMMUNICATION SYSTEMS LTD.
Marconi House, New Street, Chelmsford, Essex CM1 1PL. (0245 53221)

NEILSON-HORDELL LTD.
Central Trading Estate, Staines, Middlesex (81-56456)

NEVE ELECTRONICS INTERNATIONAL
Cambridge House Melbourn, Nr. Royston, Herts: SG8 6AU (0763 60776)

OPTICAL & TEXTILE LTD.
Barnet Trading Estate, Park Road, High Barnet, Herts. (01-442 2199) Telex: 8955869.

OSRAM (GEC) LTD.
P.O. Box 17, East Lane, Wembley, Middx. HA9 7PG. (01-904 4321)

JOHN PAGE LTD.
Wesley House, 75 Wesley Avenue, London, N.W. 10. (01 961 4181)

P.A.G. FILMS LTD., & P.A.G. POWER LTD.
565 Kingston Road, London SW20 85A (01-543 3131)

PHOTOGRAPHIC ELECTRICAL CO. LTD.
71 Dean Street, London W1V 6DE. (01-437 4633)

PHOTOMEC (LONDON) LTD.
(Humphries Group) Valley Road Industrial Estate, St. Albans, AL3 6NU, Herts. (56 50711).

PYE TVT LTD.
(The Broadcasting Company of Phillips)
PO Box 41, Coldhams Lane, Cambridge, CB1 3JU, (0223 45115)

RANK AUDIO VISUAL
P.O. Box 70, Great West Road, Brentford, Middlesex TW8 9HR (01-568 9222)

RANK CINTEL
Watton Road, Ware, Hertfordshire, (0920 3939)

RANK FILM LABORATORIES LTD.
North Orbital Road, Denham, Uxbridge, Middlesex UB9 5HQ (01-332 2323)

RANK STRAND STUDIO DIVISION
P.O. Box 54, Great WEst Road, Brentford, Middx. TW8 9HR. (01-568 9222)

RANK TAYLOR HOBSON
P.O. Box 36, Guthlaxton Street, Leicester LE2 OSP (Leicester 23801)

RANK PHICOM VIDEO GROUP
Film House, 142 Wardour Street, London W1V 4BU (01-734 2235-9) Telex: 261237

RCA INTERNATIONAL LTD.
Lincoln Way, Windmill Road, Sunbury-on-Thames, Middx: (76 85511)

ROBERT RIGBY LTD.
Premier Works, Northington Street, London WC1N 2JN (01-405 2944)

SAMCINE SALES LTD.
303 Cricklewood Broadway, London NW2 6PQ (01-450-4557)

SAMUELSON FILM SERVICE LTD.
303-315 Cricklewood Broadway, London NW2 6PQ. (01-452 8090)

SAMUELSON SIGHT & SOUND LTD.
303-315 Cricklewood Broadway, London NW2 6PQ (01-452 8090)

SCREEN ELECTRONICS LTD.
19 Anson Road, Martlesham Heath, Nr. Ipswich, Suffolk. (0473 623748) Telex 983732.

SHURE ELECTRONICS LTD.
Eccleston Road, Maidstone, Kent, ME15 6AU (0622 59881)

SONY BROADCAST LTD.
City Wall House, Basing View, Basingstoke, Hants. RG21 2LA (0256 55011)

TELEVISION INTERNATIONAL OPERATIONS LTD.
9-11 Windmill Street, London W1P 1HF (01-637 2477)

THOMSON-CSF ELECTRONIC TUBES LTD.
Ringway House, Bell Road, Daneshill, Basingstoke.

THORN EMI LTD.
Thorn House, Upper St Martin's Lane, London WC2H 9ED (01-836 2444)

VALIANT ELECTRICAL WHOLESALE CO
20 Lettice Street, Fulham, London SW6 (01-736 8115)

W. VINTEN LTD.
Western Way, Bury St. Edmunds, Suffolk IP33 3TB. (0284-2121)

WESTREX COMPANY LTD.
Bilton Fairway Estate, Long Drive, Greenford, Middx. (01-578 0950)

Government Film Bureaus

DEPARTMENT OF AGRICULTURE:

Motion Picture Division, Room 1614 South Bldg., 14th & Independence Ave., S.W., Washington, DC 20250; 447-6072; V. Buddy Renfro, Chief, Motion Picture Division.

This office produces motion pictures designed to aid in dissemination of information on the needs for agricultural products, their production, distribution, utilization and research concerning them. The films demonstrate improved methods in agriculture, stress the need for conservation of soil and other resources, explain the farm credit system, forestry, marketing, and similar subjects.

DEPARTMENT OF COMMERCE:

Bureau of Domestic Commerce, Consumer Goods and Services Division, 14th and E. Streets, N.W., Washington, DC 20230; 967-3793. SOL PADWO, Director.

This division studies and reports statistics of the industry at home and abroad. The material gathered is published in department reports and is used with information from other industries to help fulfill the function of "representing business to the government and the Government to business."

DEPARTMENT OF DEFENSE

The Pentagon Building, Washington, DC 20301.

Assistant Secretary for Public Affairs, Thomas B. Ross.

Audiovisual Division. Chief, Norman T. Hatch, 2E773, OX 7-4162. This Division is responsible for dissemination to the public through radio, television, still photos, motion pictures, and magazines and books information regarding the activities, aims and accomplished purposes of the Department of Defense (includes Army, Navy, Marine Corps, and Air Force). It also is the primary point of contact in the Department of Defense for approval of assistance to the aforementioned media on Department of Defense activities.

Audio-Visual Documentary Branch. Chief, Russell Wagner, 2E773, OX 5-0168. Acts on all requests for Department of Defense assistance from any segment of the motion picture and television industry concerned with documentary productions. Also includes television commercials and participation of armed forces personnel, bands and musical groups on TV variety programs.

Audio-Visual Production Branch. Chief, Donald E. Baruch, 2E773, OX 7-4596. Acts on all requests for Department of Defense assistance from any segment of the motion picture and television industry concerned with dramatic productions.

Audio-Visual Acquisitions Branch. Chief, Maj. Peter Friend, 2E773, OX 7-6161. Obtains and releases Department of Defense still and motion picture film to news and other media outlets. Arranges technical aspects of news conferences and media coverage at Pentagon.

ARMY AND AIR FORCE EXCHANGE SERVICE MOTION PICTURE DIVISION

(*Under Board of Directors, Army & Air Force Exchange Service*) Main Office: HQ Army & Air Force Exchange Service, 3911 S. Walton Walker Blvd. (mail address: Dallas, TX 75222 (214) 330-2622, R. L. Herdeman, director; W. R. Wood, Film Procurement Branch.

DEPARTMENT OF HEALTH EDUCATION AND WELFARE:

330 Independence Avenue, S.W., Washington, DC 20201. (202) 245-6296.

Information Office: 963-4568. Sanford H. Winston, director; York Onnen, asst. director; Harry Bell, dir., (Audio Visual Services), 963-5319.

Health Services & Mental Health Administration: Dr. Vernon E. Wilson, administrator: Edward J. McVeigh, Director of Information; 443-2086.

HS and MHA produces and distributes film, audio and video tape, and multi-media materials on problems of mental and physical health by a combination of contract and in-house facilities.

National Institute of Health: Dr. Robert S. Stone, Director; Storm Whaley, Associate Director for Communications, 496-4461.

NIH's Audio Visual facilities are responsible for the production and distribution of motion pictures, slides and other materials depicting biomedical research and its implications.

Environmental Protection Agency: William D. Ruckelaus, administrator, 755-2700; Thomas Hart, dir. of public affairs, 755-2700; Luke C. Hestor, films, 443-4804.

EPA produces films on the environment and information lists for the public.

SOCIAL SECURITY ADMINISTRATION:

James B. Cardwell, Commissioner; Russell R. Jalbert, asst. commr., Public Affairs (301) 594-1990; Robert Fenwick, Audio Visual Staff Chief, 944-5587.

SSA produces and distributes films for use throughout the country describing social security and its benefits.

SOCIAL AND REHABILITATION SERVICE:

James S. Dwight, Administrator; Robert A. Wilson, Director of Public Affairs, 962-4655.

SRS produces and distributes films on a wide variety of subjects, such as social welfare services, Medicaid, rehabilitation, juvenile delinquency and related topics.

OFFICE OF EDUCATION:

Terrel H. Bell, commissioner; Melinda Vpp, chief, office of public affairs, 254-5800.

Audio-Visual division: The Office of Public affairs is responsible for clearance and monitoring of all motion picture and other audio-visual materials generated by OE contracts with, or grants to companies or non-profit organizations and Institutions. Distribution of these materials is made through the National Audio-Visual Center (located at the National Archives and Records Services) or by special arrangements with producers. Public use of OE audiovisual materials is approved by the HEW Office of Public Affairs.

Caption films division: Malcolm Norwood, asst. chief, 963-5230.

This division purchases already produced films and captions them for use by the deaf.

DEPARTMENT OF THE INTERIOR

Bureau of Mines, Motion Picture Booking Section, 4800 Forbes Ave., Pittsburgh, Pa. 15213; Tel.: 412-621-4500. This is the main distributing center; however copies of some films are deposited at cooperating educational film distribution centers throughout the country.

The Bureau maintains a library of motion pictures depicting mining and metallurgical operations and related manufacturing processes. They are loaned free of charge, except for return postage (book rate) to educational institutions, industries, training workers, engineering, and scientific societies, civic and business associations, and other responsible organizations. The cost of production is paid by co-operating industrial concerns but the pictures do not carry trademarks, trade names or other direct advertising materials. Dr. A. Nicholas Vardac (202-634-1003), officer in charge Bureau of Mines Audio-Visual Programs, Washington, is in charge of production.

DEPARTMENT OF JUSTICE:

Constitution Avenue and 10th Street, N.W., Washington, DC 20530.

William Smith, Attorney General; Terence B. Adamson, Director of Public Information; 739-2014.

SMITHSONIAN INSTITUTION:

Washington, DC 20560; 628-4422.

S. Dillon Ripley, secretary; Nazaret Cherkezian, Telecommunications coordinator, 381-2414.

Coordinates production on all motion picture and TV filming.

DEPARTMENT OF STATE:

21st Stret and Virginia Avenue, Washington, D.C. 20530. Alexander Haig, Secretary of State.

OFFICE OF MEDIA SERVICES:

Paul E. Auerswald, director, 632-2083. Provides State Department assistance on film and TV production. Produces an average of one film a year on foreign policy subjects and distributes them for public service TV and classroom use.

OFFICE OF INTERNATIONAL TRADE:

Dept. of State; 632-2534.

(Concerned with commercial aspects of film industry, trade treaties, quotas, restrictions etc.)

EXECUTIVE AGENCIES

FEDERAL COMMUNICATIONS COMMISSION:

1919 M St., N.W., Washington, D.C. 20554—Charles R. Ferris,

Chairman; Samuel M. Sharkey Jr., Public Information Officer. (202) 632-7260.

Charged with regulating interstate and foreign communications including radio, television, satellite, cable and wire services. Licenses all standard, FM, and television broadcasting stations and services.

FEDERAL TRADE COMMISSION:

Sixth Street and Pennsylvania Avenue, N.W., Washington, D.C. 20580—Michael Pertschuk, chairman; Theodore O. Cron, director of information. 523-3625.

Administers statutes designed to promote fair competition; institutes proceeding to prevent unfair or deceptive practices, combinations in restraint of trade, false advertising, and illegal price discrimination. Supervises associations of exporters under the Export Trade Act.

SECURITIES AND EXCHANGE COMMISSION:

500 N. Capital St., Washington, D.C. 20549; Harold M. Williams, chairman; George A. Fitzsimmons. Secretary (information), 755-1160.

Registers security offerings and supervises exchange trading in securities.

U.S. INFORMATION AGENCY:

1750 Pennsylvania Avenue, N.W., Washington, D.C. 20547—John E. Reinhardt, director. (202) 724-9042.

Motion Picture and Television Service: 376-7806. McKinney Russell, assistant director; Stanley Moss, deputy assistant director. Certifies exemption of import duties if film is educational in nature.

USIA produces and acquires about 200 film and television documentaries annually for information and cultural programs in 117 countries. In addition, close to 350 targeted, and 150 worldwide new clips are made for use on foreign television. These are seen abroad in commercial theatres, in television, in schools and community centers, and by clubs, universities, and other audiences.

LIBRARY OF CONGRESS

10 First Street, S.E., Washingtin, D.C. 20540. 202-426-5000. Daniel J. Boorstin, Librarian of Congress; Mrs. Mary C. Lethbridge, Information Officer, 202-426-5108.

This agency serves as the national library and includes the U.S. Copyright Office as one of its departments. In addition to registering films for copyright and publishing a record of copyright registrations, the Library catalogs motion pictures and filmstrips, publishes printed catalog cards, book catalogs and machine-readable tapes for such materials, collects current film selectively, maintains an extensive archive of motion pictures from 1894 to date, and provides related bibliographic services.

Copyright Office—Barbara Ringer, Register of Copyrights.

Robert D. Stevens, Chief of the Copyright Cataloging Division, 703-557-8715, supervises the preparation of the semi-annual *Catalog of copyright Entries: Motion Pictures and Filmstrips*, which is distributed by the Superintendent of Documents, Government Printing Office, Washington, D.C. 20402. This publication contains descriptive data for all theatrical and nontheatrical pictures registered for copyright during each 6-month period and includes renewal registrations.

Processing Department—Joseph H. Howard, Director.

Vivian L. Schrader, Head of the Audiovisual Section of the Descriptive Cataloging Division, 202-426-5238, supervises the cataloging of approximately 5,000 motion pictures, filmstrips, and sets of slides and transparencies a year, largely on the basis of information supplied by producers and media libraries.

Research Department—Alan Fern, Director.

Motion Picture Section in the Prints and Photographs Division, 202-426-5840, which has custody of the Library's collection of over 60 000 motion pictures. The collection is primarily an archive. It contains educational, scientific, religious, industrial, and documentary films, as well as newsreels and features. Chiefly American, it includes German, Italian, and Japanese films, and in it are historically important films preserved by the Library from the early days of the motion-picture industry. Since 1942, about 1,500 representative titles a year have been retained.

Distributors of 16mm Feature Films

Following is a listing of distributors having substantial selections of 16mm films for lease or rental. Additionally, some of the companies may have prints available for outright purchase. Inquiries for catalogs listing complete product should be made to the addresses given below.

AUDIO BRANDON FILMS, INC.
34 MacQuesten Parkway South, Mt. Vernon, NY 10550.

AVCO EMBASSY PICTURES CORP.
300 East 42nd St., New York, NY 10017.

BENCHMARK FILMS
145 Scarborough Road, Briarcliff Manor, NY 10510.

BLACKHAWK FILMS
The Eastin-Phelan Corp., Davenport, IA 52808.

BUDGET FILMS
4590 Santa Monica Boulevard, Los Angeles, CA 90029.

CAROUSEL FILMS
1501 Broadway, New York, NY 10036.

CINEMA 5
1500 Broadway, New York, NY 10036.

COLUMBIA PICTURES
711 Fifth Avenue, New York, NY 10022.

CONTINENTAL 16
241 East 34th Street, New York, NY 10016.

CORINTH FILMS
410 East 62nd St., New York, NY 10021.

DISNEY PRODUCTIONS
500 South Buena Vista St., Burbank, CA 91521.

FILM CLASSIC EXCHANGE
1914 S. Vermont Avenue, Los Angeles, CA 90007.

FILM-MAKERS COOPERATIVE
175 Lexington Avenue, New York, NY 10016.

FILMS INCORPORATED
1144 Wilmette Ave., Wilmette, IL 60091.

GROVE PRESS, INC. FILM DIVISION
196 West Houston, New York, NY 10014.

HISTORICAL FILMS
Box 46505, Hollywood, CA 90046.

HURLOCK CINE-WORLD, INC.
13 Arcadia Rd., Old Greenwich, CT 06870.

INSTITUTIONAL CINEMA, INC.
10 First, Saugerties, NY

JANUS FILMS
745 Fifth Avenue, New York, NY 10022.

LEACOCK-PENNEBAKER
56 West 45th Street, New York, NY 10036.

MOVIE WONDERLAND
6116 Glen Tower, Hollywood, CA 90028.

MUSEUM OF MODERN ART FILM LIBRARY
11 West 53rd STreet, New York, NY 10019.

NATIONAL CINEMA SERVICE
333 West 57th Street, New York, NY 10019.

NEW LINE CINEMA
853 Broadway, New York, NY 10003.

NEW YORKER FILMS
16 West 61st St., New York, NY 10023.

OPERA PRESENTATIONS, INC.
75 Maiden Lane, New York, NY 10038.

PARAMOUNT NON-THEATRICAL
5451 Marathon St., Hollywood, CA 90038.

PAULIST PRODUCTIONS
P.O. Box 1057, Pacific Palisades, CA 90272.

PYRAMID FILMS
2801 Colorado Ave., Santa Monica, CA 90404.

RADIM FILMS, INC.
1034 Lake St., Oak Park, IL 60301.

ROA'S FILMS
1696 No. Astor Street, Milwaukee, Wisc. 53202.

SELECT FILM LIBRARY
115 West 31st St., New York, NY 10001.

STANDARD FILM SERVICE
14710 W. Warren Avenue, Dearborn, MI 48120.

SWANK MOTION PICTURES
201 S. Jefferson Ave., St. Louis, MO 63166.

"THE" FILM CENTER
915 12th Street N.W., Washington, D.C. 20005.

TRANS-WORLD FILMS
332 S. Michigan Ave., Chicago, IL 60604.

TWYMAN FILMS, INC.
Box 605, Dayton, OH 45401.

UNIFILM
419 Park Avenue South, New York, NY 10016.

UNITED ARTISTS FILM RENTALS
729 Seventh Avenue, New York, NY 10019.

UNITED FILMS, INC.
1122 South Cheyenne, Tulsa, OK 74119.

WARNER BROS., INC.
Non-Theatrical Division, 75 Rockefeller Plaza, New York, NY 10019.

Companies

* NETWORKS
* SET MANUFACTURERS
* MAJOR PRODUCERS
* MAJOR GROUP STATIONS OWNERS

Companies

Admiral Group

1701 E. Woodfield Rd., Schaumburg, IL 60172; (312) 884-2600. (Set manufacturer.)

PRESIDENT
John J. Henry
FINANCIAL VICE PRESIDENT & CONTROLLER
Robert D. Brown
VICE PRESIDENT PERSONNEL
Arthur J. Bazeley
INTERNATIONAL VICE PRESIDENT & GENERAL MANAGER
Vincent Barreca
VICE PRESIDENT—MARKETING & SALES
Frank L. DiLeo
PLANTS:
ELECTRONIC PRODUCTS: Harvard, Illinois
APPLIANCES: Midwest Mfg. Corp., Galesburg, IL
CABINETS: Shelbyville, IN
DISTRIBUTING BRANCHES: Boston, Buffalo, Chicago, Denver, Detroit, Los Angeles, Miami, Newark, Pittsburgh, San Antonio, San Francisco.

American Broadcasting Companies, Inc.

1330 Ave. of Americas, New York, NY 10019. (212) 887-7777; 2040 Avenue of the Stars, Century City, CA 90067; (213) 557-7777.

CHM. OF BOARD & CHIEF EXEC. OFFICER
Leonard H. Goldenson
PRESIDENT & CHIEF OPERATING OFFICER
Elton H. Rule
EXECUTIVE VICE PRESIDENT AND PRESIDENT, ABC TELEVISION
Frederick S. Pierce
SENIOR VICE PRESIDENT & GENL COUNSEL
Everett H. Erlick
VICE PRESIDENT AND CHIEF FINANCIAL OFFICER
Michael P. Mallardi
VICE PRESIDENT BUSINESS PLANNING
Ann Maynard Gray
VICE-PRESIDENT & SECRETARY
Jerome Golden
ASSISTANT SECRETARY
Franklin Feinstein, Philip R. Farnsworth
OFFICERS
Leonard H. Goldenson, Elton H. Rule, Frederick S. Pierce, Everett H. Erlick, James Abernathy, Peter Cusack, Richard L. Freund, Jerome B. Golden, Robert T. Goldman, Michael P. Mallardi, Ellis O. Moore, Warren D. Schaub, Alfred R. Schneider, Ann Maynard Gray, William E. Pierson, Edward D. Williams, Franklin Feinstein, John T. Dealy, Arnold H. Agree, Herbert A. Granath, David J. Vondrak, Vincent Romano, Philip R. Farnsworth, Alan J. Rowan, John J. Wolters
COMPTROLLER
William E. Pierson
BOARD OF DIRECTORS
Ray C. Adam, Frank T. Cary, John T. Connor, Donald C. Cook, Everett H. Erlick, Leonard H. Goldenson, Jack Hausman, Ulric Haynes, Jr., Leon Hess, George P. Jenkins, Thomas Macioce, Michael P. Mallardi, Frederick S. Pierce, Elton H. Rule, Martin J. Schwab, Simon B. Siegel

FINANCIAL (ALL ABC, INC.)

VICE PRESIDENT AND CHIEF FINANCIAL OFFICER
Michael P. Mallardi
VICE PRESIDENT, BUSINESS PLANNING
Ann Maynard Gray
VICE-PRESIDENT, CORPORATE AFFAIRS
James L. Abernathy
DIRECTOR, CORPORATE PROJECTS
James MacGregor
VICE-PRESIDENT OF MANAGEMENT INFORMATION SYSTEMS
Edward D. Williams
TREASURER
David Vondrak
VICE PRESIDENT, FINANCE
Warren D. Schaub
COMPTROLLER
William E. Pierson
ASSISTANT COMPTROLLER
Vincent Romano, Alan J. Rowan, John J. Wolters
DIRECTOR OF INTERNAL AUDIT
Joseph H. Lock
VICE PRESIDENT, TAXES
Arnold Agree

ADMINISTRATION

VICE-PRESIDENT, ADMINISTRATION, ABC, INC.
Robert T. Goldman
VICE PRESIDENT, LABOR RELATIONS, ABC, INC.
Richard L. Freund
VICE PRESIDENT, LABOR RELATIONS, EAST COAST, ABC
William S. Gennerich
VICE PRESIDENT, LABOR RELATIONS, WEST COAST, ABC
Irving Novick
VICE PRESIDENT, HUMAN RESOURCES, ABC, INC.
Peter Cusack
VICE PRESIDENT, CORPORATE PERSONNEL RELATIONS, ABC, INC.
Stephen J. Solomon
DIRECTOR, CORPORATE SERVICES
Robert Sammon
VICE PRESIDENT OF REAL ESTATE AND CONSTRUCTION, ABC
Fred Neuwirth
VICE PRESIDENT, EMPLOYEE BENEFITS AND HUMAN RESOURCES INFORMATION SYSTEMS
Thomas J. Gorey
VICE PRESIDENT, ADMINISTRATION, WEST COAST, ABC
Paul Masterson
VICE PRESIDENT, PERSONNEL, WEST COAST, ABC
Carole Foster

PUBLIC RELATIONS

VICE PRESIDENT, PUBLIC RELATIONS, ABC TELEVISION
Richard J. Connelly
VICE PRESIDENT, PUBLIC RELATIONS, PLANNING AND DEVELOPMENT
Patricia Matson
DIRECTOR, PROGRAM INFORMATION
Tom Mackin
DIRECTOR, NEWS PUBLIC RELATIONS, ABC
Alan Raymond
DIRECTOR, NEWS INFORMATION, WASHINGTON
Katherine F. Halpin
DIRECTOR, BUSINESS INFORMATION
Jeffrey R. Tolvin
DIRECTOR, EDITORIAL SERVICES
Kaye Pullen
DIRECTOR, PUBLIC RELATIONS, VIDEO ENTERPRISES,
Roann Levinsohn
MANAGER, NEWSPAPER, PUBLICITY
Charles L. Franke
MANAGER, ENTERTAINMENT PUBLICITY
David Horowitz
MANAGER, SPORTS INFORMATION
Irv Brodsky
MANAGER, BROADCAST PUBLICITY
Maxine Levinson
MANAGER, AUDIENCE INFORMATION
Gary Gerard
MANAGER, BUSINESS INFORMATION
Veronica Pollard
MANAGER, MAGAZINE DIVISION
Vic Ghidalia
MANAGER, NEWS PUBLIC RELATIONS, ABC
Constance Kastelnik
VICE PRESIDENT, PUBLIC RELATIONS, WEST COAST, ABC
Phil Kriegler
DIRECTOR, PUBLIC RELATIONS, BROADCASTING, WEST COAST
Bob Wright
MANAGER, PRESS RELATIONS, WEST COAST
Joe Maggio
MANAGER, BUSINESS INFORMATION, WEST COAST
Jeff Duclos
DIRECTOR, SPECIAL PROJECTS, WEST COAST
Jerry Hellard
MANAGER, SPECIAL PROJECTS, WEST COAST
John Chavez

LEGAL

SENIOR VICE PRESIDENT AND GENERAL COUNSEL, ABC, INC.
Everett H. Erlick

VICE PRESIDENT, CORPORATE LEGAL AFFAIRS
John T. Dealy
VICE PRESIDENT AND GENERAL ATTORNEY, ABC
Robert J. Kaufman
VICE PRESIDENT, AND GENERAL ATTORNEY, ABC-OWNED TV STATIONS AND ABC RADIO NETWORK
Mark Roth
VICE PRESIDENT, GENERAL ATTORNEY AND DIRECTOR OF BUSINESS AFFAIRS, ABC NEWS
Samuel Antar

CORPORATE RELATIONS

VICE PRESIDENT
Ellis O. Moore
VICE PRESIDENT, PLANNING AND VISUAL COMMUNICATION
Rick Giacalone

American Broadcasting Company

(The Broadcast Division of American Broadcasting Companies, Inc.)

ABC TELEVISION

EXECUTIVE VICE PRESIDENT, ABC, INC.
Frederick S. Pierce
SENIOR VICE PRESIDENT AND VICE PRESIDENT, ABC, INC.
Mark Cohen
PRESIDENT, ABC TELEVISION
John Severino
VICE PRESIDENT, PRODUCTION, ABC ENTERTAINMENT AND ABC MOTION PICTURES
Herb Jellinek
VICE PRESIDENT
Donal (cq) L. Flynn
VICE PRESIDENT
Charles M. Smith
VICE PRESIDENT, RESEARCH SERVICES
Marvin Mord
VICE PRESIDENT, AUDIENCE RESEARCH
George Keramidas
VICE PRESIDENT, AFFILIATE RESEARCH
Paul Sonkin
VICE PRESIDENT, PRIMARY AND SOCIAL RESEARCH
Melvin A. Goldberg
VICE PRESIDENT, FINANCIAL PLANNING AND CONTROLS, ABC TELEVISION NETWORK AND ABC ENTERTAINMENT
Albert Rubin
VICE PRESIDENT, BUSINESS ANALYSIS AND FINANCIAL PLANNING
P. Thomas Van Schaick
VICE PRESIDENT, ADMINISTRATION
Leonard E. Maskin
DIRECTOR, AFFILIATE CONTRACT ADMINISTRATION
Arnold Marfoglia
VICE PRESIDENT, FINANCIAL CONTROLS, ABC TELEVISION NETWORK & ABC ENTERTAINMENT
Murray Resnick
DIRECTOR, FINANCIAL CONTROLS, EAST COAST, ABC TV NETWORK & ABC ENTERTAINMENT
Patrick Lofaro
DIRECTOR, TELEVISION FINANCIAL CONTROLS, WEST COAST, ABC TV NETWORK & ABC ENTERTAINMENT
James E. O'Grady
VICE PRESIDENT, BUSINESS AFFAIRS AND CONTRACTS, ABC-TV
Ronald B. Sunderland
VICE PRESIDENT AND ASSISTANT TO THE SENIOR VICE PRESIDENT, THEATRICAL MOTION PICTURES AND TELEVISION AFFAIRS
Charles Smiley
VICE PRESIDENT, BUSINESS AFFAIRS, EAST COAST
Joe Morris
VICE PRESIDENT, CONTRACTS
Richard N. Burns
DIRECTOR, CONTRACTS, WEST COAST
Jerome Berger
VICE PRESIDENT, BUSINESS AFFAIRS, DAYTIME AND SPECIALS, EAST COAST
Anthony Farinacci
VICE PRESIDENT, TAPE PRODUCTION, WEST COAST
Albert J. Simon

ABC TELEVISION NETWORK

PRESIDENT
James E. Duffy
VICE PRESIDENT AND GENERAL MANAGER
George Newi
VICE PRESIDENT, COMMUNICATIONS AND ADMINISTRATION
Mary Jane Raphael
DIRECTOR, BUSINESS AFFAIRS
Bertrand Fainberg
VICE PRESIDENT
Richard N. Savage
DIRECTOR OF AFFILIATE DEVELOPMENT PLANNING
Susan Kusiel

AFFILIATE RELATIONS

VICE PRESIDENT IN CHARGE OF AFFILIATE RELATIONS
Robert F. Fountain
VICE PRESIDENT AND DIRECTOR OF AFFILIATE OPERATIONS
Mario Cucinotta
VICE PRESIDENT AND DIRECTOR OF STATION RELATIONS
Richard Kozak
VICE PRESIDENT, AFFILIATE PLANNING AND SPECIAL PROJECTS
Joe Giaquinto
MANAGER, TELEVISION NEWS ADVISORY SERVICE
Don Dunphy
DIRECTOR, TELEVISION NEWS ADVISORY SERVICE
Bryce Rathbone

NETWORK SALES

VICE PRESIDENT IN CHARGE OF SALES
H. Weller (Jake) Keever
VICE PRESIDENT AND GENERAL MANAGER, SALES
Warren J. Boorom
VICE PRESIDENT AND NATIONAL SALES MANAGER
Robert I. Silberberg
VICE PRESIDENT, DAYTIME SALES
John Tiedemann, Jr.
VICE PRESIDENT, SPORTS SALES
John Lazarus
VICE PRESIDENT, NEWS SALES
William Breen
VICE PRESIDENT, SALES ADMINISTRATION
Charles C. Allen
VICE PRESIDENT EARLY MORNING SALES
Robert Cagliero
VICE PRESIDENT, PRIME TIME SALES PROPOSALS
Marvin Goldsmith
VICE PRESIDENT, MARKETING
Jerry Solomon
VICE PRESIDENT, BUSINESS DEVELOPMENT, SALES
Anthony Rocco
VICE PRESIDENT, DETROIT SALES MANAGER
Robert Wozniak
VICE PRESIDENT, SALES, WESTERN DIVISION
Doug Streff
VICE PRESIDENT, SALES MANAGER, CENTRAL DIVISION
John Grace
VICE PRESIDENT, NEWS SALES, CENTRAL DIVISION, CHICAGO
William Gillogly
EASTERN SALES MANAGER
Ed Ryan
DIRECTOR, NEWS SALES
James E. Burnette
DIRECTOR, DAYTIME SALES
Alfred A. Smith
DIRECTOR, REGIONAL SALES
Josephine Maggio
DIRECTOR, SALES SERVICE, NIGHTTIME
Harold Geary
DIRECTOR, SPORTS SALES ADMINISTRATION AND SALES PROPOSALS
John McCarthy
DIRECTOR, PRIME TIME SALES PROPOSALS
Robert T. Wallen
DIRECTOR, DAYTIME AND SPORTS SALES SERVICE
Michael Kay
DIRECTOR, SPORTS MERCHANDISING AND PROMOTION
Al Cohen
DIRECTOR, SALES PROPOSALS AND DEVELOPMENT, CENTRAL DIVISION
Robert Sedlachek

ADVERTISING & PROMOTION

VICE PRESIDENT, CREATIVE SERVICES, ABC TV NETWORK
Symon B. Cowles
VICE PRESIDENT, ADVERTISING
John T. Curry, Jr.
VICE PRESIDENT AND NATIONAL ADVERTISING DIRECTOR
Paul W. (Pete) Barrett
ART DIRECTOR
Henry Epstein
VICE PRESIDENT, SPECIAL PROJECTS, CREATIVE SERVICES
Joel Cohen
DIRECTOR OF CONFERENCE PLANNING
Milton Carney

DIRECTOR OF ON-AIR ADVERTISING, EAST COAST
Frank Cuciti
DIRECTOR OF ADVERTISING ADMINISTRATION
Andre Gebstaedt
DIRECTOR, SALES DEVELOPMENT
Allen Hirschman
MANAGER, SALES DEVELOPMENT
Ralph Broitman
DIRECTOR, ON-AIR ADVERTISING, WEST COAST
Don Wollman

ABC SPORTS, INC.

PRESIDENT, ABC NEWS & SPORTS
Roone Arledge
SENIOR VICE PRESIDENT
James Spence
VICE PRESIDENT, PROGRAMMING AND ASSISTANT TO THE PRESIDENT
John Martin
VICE-PRESIDENT, PROGRAM PRODUCTION
Chuck Howard
VICE PRESIDENT, FINANCIAL AND TALENT AFFAIRS, ABC NEWS AND SPORTS
Irwin Weiner
VICE PRESIDENT, SYNDICATION AND MARKETING
Peter Dimmock
DIRECTOR OF SALES SERVICES
John Fitzgerald
VICE PRESIDENT, PROGRAM PLANNING
Charles Lavery, Jr.
DIRECTOR, PRODUCTION, CREATIVE SERVICES
Chet Forte
DIRECTOR, PRODUCTION COORDINATION
Dennis Lewin
DIRECTOR OF FILM PRODUCTION & DEVELOPMENT
John Wilcox
DIRECTOR OF ADMINISTRATION AND FINANCIAL CONTROL
Bob Apter
VICE PRESIDENT, OLYMPICS OPERATIONS
Marvin Bader
DIRECTOR, LEGAL AND BUSINESS AFFAIRS, 1984 OLYMPICS
Don Gadsden

ABC TELEVISION SPOT SALES, INC.

VICE PRESIDENT AND GENERAL MANAGER
Walter Flynn
GENERAL SALES MANAGER
Elio Betty
EASTERN SALES MANAGER
Michael Rosenberg
NEW YORK SALES MANAGER
Tom Kane
SALES MANAGER, DETROIT
Mike Kincaid
SALES MANAGER, SAN FRANCISCO
James Sherbert
EASTERN REGIONAL SALES MANAGER
Gary Schineller
MIDWEST REGIONAL SALES MANAGER
Gary Brooks
DIRECTOR, MARKET PLANNING
Keith Ritter
DIRECTOR OF PLANNING & ADMINISTRATION
Georgia Breza
BUSINESS MANAGER
Larry Perl
DIRECTOR, RESEARCH AND SALES PROMOTION
Murray Rosenblum

ABC OWNED TELEVISION STATIONS DIVISION

PRESIDENT
Richard A. O'Leary
VICE PRESIDENT, PROGRAMMING
Phil Boyer
VICE PRESIDENT AND GENERAL MANAGER, PLANNING AND DEVELOPMENT
David Johnson
VICE-PRESIDENT, PLANNING & RESEARCH
Walter Stein
VICE PRESIDENT, ADMINISTRATION & PLANNING
James Allegro
VICE-PRESIDENT, SALES
John S. Bonanni
VICE PRESIDENT, CREATIVE SERVICES
Carol Bauer
DIRECTOR, FINANCIAL PLANNING
Robert "Kit" Scripps
DIRECTOR, PROGRAM BUDGETING AND ADMINISTRATION
Nancy Royce
VICE-PRESIDENT AND GENERAL MANAGER, WABC-TV
Bill Fyffe
VICE-PRESIDENT AND GENERAL MANAGER, WXYZ-TV
Jeanne Findlater
VICE-PRESIDENT AND GENERAL MANAGER, WLS-TV
Peter Desnoes
VICE-PRESIDENT AND GENERAL MANAGER, KABC-TV
Tom Van Amberg
VICE-PRESIDENT AND GENERAL MANAGER, KGO-TV
James Osborn

ABC VIDEO ENTERPRISES

VICE PRESIDENT, ABC VIDEO ENTERPRISES
Herbert Granath
VICE PRESIDENT, PRODUCT DEVELOPMENT AND ACQUISITIONS
John Healy
VICE PRESIDENT, VIDEO SALES
Archie Purvis
VICE PRESIDENT IN CHARGE OF MARKETING
James G. Shaw
VICE PRESIDENT AND GENERAL MANAGER
S. Arthur Schimmel
DIRECTOR OF FINANCE AND PLANNING
Robert Wong
DIRECTOR, ABC WIDE WORLD OF LEARNING, INC.
Peter Schillaci
DIRECTOR, THEATRICAL SALES
Doug Brunger
VICE PRESIDENT, TELEVISION AND THEATRICAL SALES
Jack Singer
DIRECTOR, PROGRAM AND SALES SERVICES
Edith Hills

ABC ENTERTAINMENT

PRESIDENT
Anthony D. Thomopoulos
PRESIDENT OF ABC MOTION PICTURES AND SENIOR VICE PRESIDENT, ABC ENTERTAINMENT
Brandon Stoddard
VICE PRESIDENT, WORLDWIDE PRODUCTIONS ABC THEATRICAL MOTION PICTURES
Robert Bookman
VICE PRESIDENT, CREATIVE AFFAIRS, ABC THEATRICAL MOTION PICTURES
Philip Blumberg
SENIOR VICE PRESIDENT, PRIME TIME SERIES
Tom Werner
VICE PRESIDENT
Seymour Amlen
SENIOR VICE PRESIDENT AND ASSISTANT TO THE PRESIDENT
Lewis Erlicht
VICE PRESIDENT, MOTION PICTURES FOR TELEVISION
Stu Samuals
VICE PRESIDENT, PROGRAM PLANNING AND SCHEDULING
Gus Lucas
DIRECTOR, FEATURE FILM CREATIVE POST PRODUCTION
Andre de Szekely
VICE PRESIDENT, LIMITED SERIES AND NOVELS FOR TELEVISION, EAST COAST
Jacqueline Bubbin
DIRECTOR, PRIME TIME PROGRAM DEVELOPMENT, EAST COAST
Susan Nerenberg
EXECUTIVE PRODUCER, MOTION PICTURES FOR TELEVISION
Ronald W. Taylor
DIRECTOR, CURRENT COMEDY PROGRAMS
Lance Taylor
DIRECTOR, CREATIVE SERVICES, MOTION PICTURES AND NOVELS FOR TELEVISION
Dottie Gagliano
DIRECTOR, STORY DEPARTMENT
Maggie Duffy
VICE PRESIDENT, SPECIAL PROJECTS AND SENIOR EXECUTIVE IN CHARGE OF TALENT
Gary L. Pudney
VICE PRESIDENT, CHILDREN'S AND EARLY MORNING PROGRAMMING
Squire D. Rushnell
VICE PRESIDENT
Chuck King
DIRECTOR
Leslie Benson
DIRECTOR OF PROGRAMMING
Ruth Meyer
DIRECTOR, CHILDREN'S PROGRAMS
Sandy Russell Gartin
DIRECTOR, CURRENT DRAMATIC SERIES
Peter Roth
VICE PRESIDENT, DAYTIME PROGRAMS
Jackie Smith
VICE PRESIDENT, DAYTIME PROGRAMS, EAST COAST
Jo Ann Emmerich

DIRECTOR, DAYTIME PROGRAMS, EAST COAST
Randi LeWinter
DIRECTOR, DAYTIME PROGRAMS, WEST COAST
Wally Weltman
DIRECTOR, EARLY MORNING PROGRAMMING
Amy Dorn Kopelan
DIRECTOR, FEATURE FILMS AND LATE NIGHT SERIES
Paul Coss
DIRECTOR, PROGRAM ADMINISTRATION
Stephen K. Nenno
DIRECTOR, PROGRAM SERVICES
Judy Ahlborg
DIRECTOR, PROGRAM ADMINISTRATION, WEST COAST
Hank Miller
DIRECTOR, PROGRAM ADMINISTRATION, EAST COAST
Jack Sharkey
DIRECTOR, POST PRODUCTION, EAST COAST
Al Ludington

ABC RADIO

PRESIDENT
Ben Hoberman
VICE PRESIDENT, FINANCE AND ADMINISTRATION
Philip Giordano
VICE-PRESIDENT, PROGRAMMING
Rick Sklar
DIRECTOR, FINANCE AND ADMINISTRATION
Joseph Sherikjian
DIRECTOR, MARKETING
Robert Cambridge
MANAGER, INFORMATION AND PUBLIC RELATIONS
Henry Kavett

ABC RADIO NETWORKS

PRESIDENT
Edward F. McLaughlin
VICE PRESIDENT AND GENERAL MANAGER
Robert M. Chambers
VICE PRESIDENT OF PLANNING, FINANCE, AND ADMINISTRATION
William J. Battison
DIRECTOR OF FINANCE
Larry Friedman
DIRECTOR OF SPORTS, AND EXECUTIVE PRODUCER OF SPORTS PROGRAMMING
Shelby Whitfield
MANAGER, PUBLIC SERVICE & PROGRAM PROJECTS
Nancy Pfeffer

SALES

VICE PRESIDENT, DIRECTOR OF SALES
Louis A. Severine
VICE PRESIDENT, NATIONAL SALES MANAGER
Arthur Sulzburgh
VICE PRESIDENT, EASTERN SALES MANAGER
Robert Watson
VICE PRESIDENT, WESTERN SALES MANAGER
Frank Boehm
VICE PRESIDENT, CENTRAL SALES MANAGER
Dennis Glynn
VICE PRESIDENT, DETROIT SALES MANAGER
Joseph F. Kelly
DIRECTOR OF ADMINISTRATION MANAGER, SOUTHWEST SALES
Rhonda Hack
DIRECTOR, SALES ADMINISTRATION
Linda Meilan
MANAGER, SPECIAL PROGRAMMING SALES
Tom Rocco
SALES MANAGER, ENTERTAINMENT & INFORMATION NETWORKS
David Logan
SALES MANAGER, CONTEMPORARY & FM NETWORKS
Blaise Leonardi

CREATIVE SERVICES

DIRECTOR OF CREATIVE SERVICES
Michael A. Winter
DIRECTOR, MARKETING SERVICES
Leo Collins

MANAGER, ADVERTISING AND PROMOTION
Sam Patterson
MANAGER, PROMOTION AND PUBLICITY
Suzanne Banks

AFFILIATE AFFAIRS

V.P. & DIRECTOR
Richard McCauley
DIRECTOR, STATION ADMINISTRATION
Frank Atkinson
DIRECTOR, FM NETWORK
Virginia Westphal
DIRECTOR, ABC CONTEMPORARY NETWORK
Darryl Brown
DIRECTOR, ENTERTAINMENT RADIO NETWORK
Leslie Benson
DIRECTOR, INFORMATION RADIO NETWORK
Stuart G. Krane

ENGINEERING

DIRECTOR, ENGINEERING AND PROGRAM OPERATIONS
Ken Coughlin
MANAGER, TECHNICAL OPERATIONS
Harry Curtis
MANAGER, TECHNICAL OPERATIONS, Washington, DC
Richard Martinez
MANAGER, BROADCAST SERVICES
Bob Donnelly

RESEARCH

DIRECTOR OF RESEARCH
Bill McClenaghan
MANAGER RESEARCH AND MEDIA PLANNING
Sandra Kennedy

ABC OWNED AM RADIO STATIONS

PRESIDENT
Charles DeBare
DIRECTOR, BUSINESS ANALYSIS & PLANNING
Robert Holmgren
DIRECTOR, RESEARCH AND SALES SYSTEMS
Bruce Hoban
VICE PRESIDENT AND GENERAL MANAGER, WABC, New York
Alfred Racco
VICE PRESIDENT AND GENERAL MANAGER, KABC, Los Angeles
George Green
VICE PRESIDENT AND GENERAL MANAGER, WXYZ, Detroit
Charles Fritz
VICE PRESIDENT AND GENERAL MANAGER, KGO, San Francisco
Michael Luckoff
EXECUTIVE VICE PRESIDENT WMAL-AM, Washington, DC
Andrew Ockershausen

ABC OWNED FM RADIO STATIONS

PRESIDENT
Martin Greenberg
DIRECTOR, FINANCE AND ADMINISTRATION
Andrew Fisher
VICE PRESIDENT AND GENERAL MANAGER, WPLJ, New York
Joseph Parrish
VICE PRESIDENT & GENERAL MANAGER, KLOS, Los Angeles
Bill Sommers
V.P. & GENERAL MANAGER, WLS-AM, Chicago
Don Bouloukos
VICE PRESIDENT & STATION MANAGER, WLS-FM
John Gehron
V.P. & GENERAL MANAGER, WRIF, Detroit
John Hare
VICE PRESIDENT AND GENERAL MANAGER KSFX, San Francisco
Ron Denman
VICE PRESIDENT AND GENERAL MANAGER, KSRR, Houston
Jay Hoker
VICE PRESIDENT AND GENERAL MANAGER, WRQX-FM, Washington, DC
Ernest Fears
VICE PRESIDENT
A. Thomas Plant
DIRECTOR
Virginia Westphal
MANAGER OF STATION RELATIONS
Louis Callahan

ABC NEWS

PRESIDENT, ABC NEWS & SPORTS
Roone Arledge
SR. VICE PRESIDENT
Richard Wald
VICE PRESIDENT & ASSISTANT TO THE PRESIDENT
David Burke
VICE PRESIDENT, TELEVISION NEWS SERVICES
Nicholas Archer
V.P. & EXECUTIVE PRODUCER, "WORLD NEWS TONIGHT" AND SPECIAL EVENTS
Jeff Gralnick
VICE PRESIDENT, OF ABC NEWS FOR PROGRAM DEVELOPMENT
Av Westin
BUREAU CHIEF, WASHINGTON BUREAU
Bill Knowles
VICE PRESIDENT AND EXECUTIVE PRODUCER OF "ABC NEWS NIGHTIME"
William Lord

VICE PRESIDENT, EXECUTIVE PRODUCER, DOCUMENTARIES
Pam Hill

BROADCAST OPERATIONS AND ENGINEERING

PRESIDENT
Julius Barnathan
VICE PRESIDENT AND GENERAL MANAGER, TV FACILITIES & SERVICES, EAST COAST
Philip J. Levens
VICE PRESIDENT, PLANNING AND SPECIAL PROJECTS, WEST COAST
Norman H. Grant
V.P., BROADCAST ENGINEERING
R. LaVerne Pointer
DIRECTOR, TELEVISION NETWORK OPERATIONS
Richard W. Stubbe
V.P., PLANNING, ADMINISTRATION, & FINANCE
Timothy Millhiser
VICE PRESIDENT AND GENERAL MANAGER, TV FACILITIES AND SERVICES, WEST COAST
Robert Trachinger
VICE PRESIDENT, RADIO OPERATIONS
John Hidle
VICE PRESIDENT & DIRECTOR, EUROPEAN TECHNICAL PRODUCTION AND BROADCAST OPERATIONS
Jacques Lesgards
DIRECTOR NETWORKING SERVICES
John Gilmore
MANAGER, ALLOCATIONS AND LICENSING
John Serafin
DIRECTOR OF PRODUCTION SERVICES, New York
David Eschelbacher
DIRECTOR OF FILM SERVICES, New York
Edward Messina
DIRECTOR OF PRODUCTION PLANNING & CONTROLS, New York
Charles Baldour
DIRECTOR OF TRAFFIC
William Hynes
CHIEF OF ENGINEERING, WABC-TV, New York
James Baker
CHIEF ENGINEER, WABC-RADIO, New York
Winston Loyd
DIRECTOR OF TV OPERATIONS, WEST COAST
Randy Hooper
DIRECTOR OF FILM SERVICES, Los Angeles
Robert Hudson
DIRECTOR OF PRODUCTION SERVICES, Los Angeles
David Loring
DIRECTOR OF ENGINEERING, KABC-TV, Los Angeles
William Dryer
DIRECTOR OF ENGINEERING, WLS-TV, Chicago
Joe Kresnicka
DIRECTOR OF ENGINEERING, KGO-TV, San Francisco
Jim Von Striver
DIRECTOR OF ENGINEERING, WXYZ-TV, Detroit
Wallace Rodammer
DIRECTOR, EQUIPMENT PLANNING
Michael Fisher
DIRECTOR, AUDIO/VIDEO SYSTEMS
Max Berry
DIRECTOR, ALLOCATIONS AND R.F. SYSTEMS
Tony Uyttendaele
GENERAL MANAGER, TECHNICAL OPERATIONS CENTER
Herb Kraft

BROADCAST STANDARDS, MERCHANDISING MUSIC, QUALITY CONTROL

VICE PRESIDENT, ABC, INC.
Alfred R. Schneider

BROADCAST STANDARDS & PRACTICES

VICE PRESIDENT, EAST COAST
Julie Tarachow Hoover
DIRECTOR, EAST COAST
Jeffrey Edelstein
VICE PRESIDENT, WEST COAST
Thomas Kersey
DIRECTOR, WEST COAST
Don Ross Bay

QUALITY CONTROL

DIRECTOR
Vivian Weller
(See ad & promo on pg. #4.)

ABC INTERNATIONAL TELEVISION, INC.

PRESIDENT
Richard A. O'Leary

Avco Embassy Pictures Corp. Television

956 Seward Street, Los Angeles, CA 90038 (213) 460-7200.
PRESIDENT
Frank Capra, Jr.
SENIOR VICE PRESIDENT IN CHARGE OF LEGAL AND BUSINESS AFFAIRS
Peter Bierstedt
VICE PRESIDENT/TELEVISION
Lois Luger
NATIONAL TV SALES MANAGER
Charles Britt
EASTERN DIVISION TV SALES MANAGER
Merritt Sticker

British Broadcasting Corporation

630 Fifth Avenue, New York NY 10020 (212) 581-7100
DIRECTOR, UNITED STATES
David Webster
HEAD OF COMMERCIAL OPERATIONS
John Stringer
TELEVISION PROGRAM PRODUCTION
Peter Foges
Anne Watson
Barbara Benedek
CHIEF ENGINEER
Phil Laven
NEW YORK CORRESPONDENT
Paul Reynolds
UNITED NATIONS OFFICE
United Nations Building, New York; (212) 688-6266
CORRESPONDENT
Keith Hindell
WASHINGTON OFFICE
CBS Bldg.; 2030 M St. N.W.; Washington, DC 20036; (202) 223-2050
CORRESPONDENTS
Martin Bell-TV
Clive Small-Radio

CBS Inc.

51 W. 52nd St., New York, NY 10019 (212) 975-4321.
BOARD OF DIRECTORS
William S. Paley, Chairman; William A. M. Burden, Courtney C. Brown, Roswell L. Gilpatric, Franklin A. Thomas, Henry B. Schacht, Marietta Tree, John Phillips, Donald R. Osborn, Walter R. Yetnikoff, John R. Purcell, James R. Houghton, Gene F. Jankowski, Benno C. Schmidt, Thomas H. Wyman, Michel Bergerac.
CHAIRMAN
William S. Paley
PRESIDENT
Thomas H. Wyman
VICE PRESIDENT & PRESIDENT, CBS/BROADCAST GROUP
Gene F. Janowski
VICE PRESIDENT & PRESIDENT, CBS/RECORDS GROUP
Walter R. Yetnikoff
VICE PRESIDENT & PRESIDENT, CBS/COLUMBIA GROUP
Cornelius Keating
VICE PRESIDENT & TREASURER
Carl Anton Muller
VICE PRESIDENT, CHIEF ECONOMIST
David M. Blank
VICE PRESIDENT, FACILITIES
Drew Brinckerhoff
VICE PRESIDENT & GENERAL AUDITOR
Alvin H. Fenichel
VICE PRESIDENT, PERSONNEL
Joan Showatter
VICE PRESIDENT, FINANCE
Thomas M. Kirwan
VICE PRESIDENT, ADMINISTRATION
Haskell P. Macowatt
VICE PRESIDENT, CORPORATE AFFAIRS
E.K. Meade, Jr.
VICE PRESIDENT & GENERAL COUNSEL
James K. Parker
VICE PRESIDENT AND CONTROLLER
Louis J. Rauchenberger, Jr.
VICE PRESIDENT, INDUSTRIAL RELATIONS
James F. Simmons
VICE PRESIDENT, WASHINGTON
William Lillay, III
VICE PRESIDENT, PLANNING
Harry E. Smith
VICE PRESIDENT, ACQUISITIONS
Nathan Snyder

VICE PRESIDENT
Leonard Spinrad
VICE PRESIDENT, MANAGEMENT INFORMATION SYSTEMS
James A. Walsh
VICE PRESIDENT AND DIRECTOR OF TAXES
Marilyn Walsh
SECRETARY
Charles T. Bates
ASSISTANT SECRETARY
Nicoline G. Mix
ASSISTANT SECRETARY
Jack P. Purcell
ASSISTANT CONTROLLER
William F. Crean
ASSISTANT TREASURER
Joseph Rathman

DEPARTMENTAL OFFICERS

VICE PRESIDENT, NEGOTIATION SERVICES DEPARTMENT, INDUSTRIAL RELATIONS
Noel B. Berman
VICE PRESIDENT, INVESTOR RELATIONS, CORPORATE AFFAIRS
J. Garrett Blowers
VICE PRESIDENT, FACILITIES ENGINEERING AND BUILDING OPERATIONS
Norman C. Brodesser
VICE PRESIDENT, PERSONNEL PLANNING & DEVELOPMENT
Guy E. Fleming
VICE PRESIDENT, CORPORATE STAFF SERVICES
Gilroye A. Griffin, Jr.
VICE PRESIDENT, CORPORATE CONTRIBUTIONS & CIVIC AFFAIRS, CORPORATE AFFAIRS
John W. Kiermaier
VICE PRESIDENT, CORPORATE INFORMATION DEPARTMENT, CORPORATE AFFAIRS
Paula Gottschalk
VICE PRESIDENT, INDUSTRIAL RELATIONS, WEST COAST
Orison S. Marden
CONTROLLER, STAFF ORGANIZATION
Keith Sachs
VICE PRESIDENT, REAL ESTATE
Simon Volinsky
VICE PRESIDENT, PLACEMENT & SYSTEMS RESEARCH
Thomas Giordano

CBS/BROADCAST GROUP (CBG)

PRESIDENT
Gene F. Jankowski
VICE PRESIDENT AND ASSISTANT TO THE PRESIDENT
Gene P. Mater
VICE PRESIDENT RESEARCH
Jay Eliasberg
VICE PRESIDENT, PROGRAM PRACTICES
Donn O'Brien
VICE PRESIDENT, MANAGEMENT DEVELOPMENT
Harvey J. Struthers
SENIOR VICE PRESIDENT, CREATIVE DIRECTOR
Louis Dorfsman
DIRECTOR, COMMUNICATIONS
George Schweitzer

CBS TELEVISION NETWORK DIVISION

PRESIDENT
James H. Rosenfield
VICE PRESIDENT & ASST. TO PRESIDENT
David Fuchs
VICE PRESIDENT, SALES
Paul Isacsson
VICE PRESIDENT, AFFILIATE RELATIONS
Robert Hosking
VICE PRESIDENT, PRODUCTION FACILITIES & ENGINEERING
George Shannon
VICE PRESIDENT & GENERAL MANAGER, AFFILIATE RELATIONS
Bruce Bryant
VICE PRESIDENT, ADMINISTRATION & LEGAL SERVICES, AFFILIATE RELATIONS
Donald Clancy

CBS ENTERTAINMENT

PRESIDENT
Robert A. Daly
VICE PRESIDENT, PROGRAMS
B. Donald Grant
VICE PRESIDENT, PROGRAM ADMINISTRATION
Harvey Shephard
VICE PRESIDENT, CHILDREN'S PROGRAMS & DAYTIME PROGRAMS
Michael Ogiens
VICE PRESIDENT, PROGRAMS, MOTION PICTURES FOR TV AND MINI-SERIES
William Self
VICE PRESIDENT, CHILDREN'S PROGRAMS
Faith Frenz Heckman
VICE PRESIDENT, SPECIAL PROGRAMS
Bernie Sofronski
VICE PRESIDENT, PROGRAMS, NEW YORK
Alan Wagner
VICE PRESIDENT, BUSINESS AFFAIRS
Alan Levin
VICE PRESIDENT, PROGRAM PLANNING
Herman Keld
VICE PRESIDENT, MOTION PICTURES FOR TV
Steve Mills
VICE PRESIDENT, ADVERTISING & PROMOTION
Steve Sohmer
VICE PRESIDENT, PRESS INFORMATION
Barrie Richardson

CBS TELEVISION STATIONS DIVISION

PRESIDENT
Thomas F. Leahy
VICE PRESIDENT, ASS'T. TO THE PRESIDENT
George Dessart
VICE PRESIDENT, FINANCE & PLANNING
William G. Barker, Jr.
VICE PRESIDENT, PROGRAMMING
E. Hal Hough
VICE PRESIDENT, SALES
Gail Trell
EXECUTIVE NEWS DIRECTOR
Sam Zelman
VICE PRESIDENT, GENERAL MANAGER CBS/NATIONAL SALES
Thoren Schroeck
DIRECTOR, COMMUNICATIONS
Alan R. Morris
DIRECTOR, PROGRAM DEVELOPMENT, HOLLYWOOD
Gary Hoffman
GENERAL MANAGER, ENGINEERING & OPERATIONS TECHNOLOGY
Jerry Stahler
VICE PRESIDENT, GENERAL MANAGER WCBS-TV
Neil Derrough
VICE PRESIDENT, GENERAL MANAGER WCAU-TV
Daniel E. Gold
VICE PRESIDENT, GENERAL MANAGER WBBM-TV
Ed Joyce
VICE PRESIDENT, GENERAL MANAGER KMOX-TV
John McKay
VICE PRESIDENT, GENERAL MANAGER KNXT
Van Gordon Sauter
STATION MANAGER WCAU-TV
Richard Goldstein
STATION MANAGER KNXT
Robert Shay

CBS RADIO DIVISION

PRESIDENT
Sam Cook Digges
VICE PRESIDENT & ASST. TO THE PRESIDENT
Andrew Subbiondo
VICE PRESIDENT, GENERAL MANAGER, NETWORK
Richard M. Brescia
VICE PRESIDENT, NETWORK SALES
James A. Joyella
VICE PRESIDENT, PROGRAMS
Frank Miller
VICE PRESIDENT AND DIRECTOR, AFFILIATE RELATIONS
Cornelius Knox
VICE PRESIDENT, DIVISION SERVICES
W. Thomas Dawson
VICE PRESIDENT, CBS OWNED AM STATIONS
Peter Lund
VICE PRESIDENT, CBS OWNED FM STATIONS
J. Robert Coles
VICE PRESIDENT & CONTROLLER
Peter Keegan
VICE PRESIDENT, GENERAL MANAGER, CBS RADIO SPOT SALES
Nancy Widman
REGIONAL VICE PRESIDENT
Robert F. Hyland
VICE PRESIDENT, GENERAL MANAGER, WCBS-AM
John Lack
VICE PRESIDENT, GENERAL MANAGER, WCBS-FM
James McQuade
VICE PRESIDENT, GENERAL MANAGER, WBBM-AM
William C. O'Donnell
VICE PRESIDENT, GENERAL MANAGER, WBBM-FM
Louis M. Gredell
VICE PRESIDENT, GENERAL MANAGER, KNX-AM
George Nicholaw
VICE PRESIDENT, GENERAL MANAGER, KNX-FM
Robert L. Nelson
VICE PRESIDENT, GENERAL MANAGER, WCAU-AM
Robert B. Sherman

VICE PRESIDENT, GENERAL MANAGER, WCAU-FM
James M. Keating
VICE PRESIDENT, GENERAL MANAGER, KCBS-AM
Ray Barnett
VICE PRESIDENT, GENERAL MANAGER, KCBS-FM
Thomas H. Pierce
VICE PRESIDENT, GENERAL MANAGER, WEEI-AM
Eugene Lothery
VICE PRESIDENT, GENERAL MANAGER, WEEI-FM
John E. Baker

CBS NEWS DIVISION

PRESIDENT
William A. Leonard
SENIOR VICE PRESIDENT, DIRECTOR OF NEWS
Burton Benjamin
VICE PRESIDENT DIRECTOR, PUBLIC AFFAIRS BROADCASTS
Robert Chandler
VICE PRESIDENT, ADMINISTRATION ASST. TO PRESIDENT
Kay Wight
VICE PRESIDENT, PUBLIC AFFAIRS BROADCASTS
Margery Baker
VICE PRESIDENT, SPECIAL EVENTS & POLITICAL COVERAGE
Sylvia Westerman
VICE PRESIDENT, DEPUTY DIRECTOR, NEWS
John Lane
VICE PRESIDENT, OPERATIONS
Marshall B. Davidson
VICE PRESIDENT, WASHINGTON BUREAU CHIEF
Ed Touhy
VICE PRESIDENT, RADIO NEWS
Emerson L. Stone
VICE PRESIDENT & CONTROLLER
Joel Hymson

Capital Cities Communications, Inc.

24 East 51st St., New York, NY 10022. (212) 421-9595. (Group stations owner—Lowell Thomas Group.)

CHAIRMAN
Thomas S. Murphy
PRESIDENT
Daniel B. Burke
EXECUTIVE VICE PRESIDENT & PRESIDENT, BROADCAST DIVISION
Joseph P. Dougherty
EXECUTIVE VICE PRESIDENT
John B. Fairchild
EXECUTIVE VICE PRESIDENT
John B. Sias
SENIOR VICE PRESIDENT
Robert K. King
VICE PRESIDENT, ENGINEERING
Robert O. Niles
VICE PRESIDENT, CHIEF FINANCIAL OFFICER
Ronald J. Doerfler
SECRETARY
Gerald Dickler
TREASURER
Robert W. Gelles

Children's Television Workshop

1 Lincoln Plaza, New York, NY (212) 595-3456. (Educational Broadasting.)

PRESIDENT
Joan Ganz Cooney
EXECUTIVE VICE PRESIDENT
Paul B. Firstenberg
SENIOR VICE PRESIDENT, CORPORATE AFFAIRS
David V.B. Britt
VICE PRESIDENT, PRODUCTION, WEST COAST
David D. Connell
VICE PRESIDENT, PRODUCTION, EAST COAST
Alfred Hyslop
VICE PRESIDENT, COMMUNITY EDUCATION SERVICES
Evelyn P. Davis
VICE PRESIDENT, PUBLIC AFFAIRS
Robert A. Hatch
VICE PRESIDENT, RESEARCH
Dr. Edward L. Palmer
ASSISTANT SECRETARY
Iris Morales
PRESIDENT, CTW PRODUCTS AND INTERNATIONAL TELEVISION GROUPS
William F. Whaley

Columbia Pictures Television

(A Division of Columbia Pictures Industries, Inc.), Columbia Plaza, Burbank, CA 91505 (213) 954-6000.

PRESIDENT
Herman Rush
SENIOR VICE PRESIDENT, ADMINISTRATION
Paul B. Stager
SENIOR VICE PRESIDENT & EXECUTIVE PRODUCTION MANAGER
Seymour Friedman
VICE PRESIDENT, MUSIC
Richard Berres
VICE PRESIDENT, FINANCIAL ANALYSIS
Jim Claiborne
VICE PRESIDENT, DAYTIME
Noreen Conlin
VICE PRESIDENT, PUBLICITY & PROMOTION
Doug Duitsman
VICE PRESIDENT, SERIES PROGRAMS
Christine Foster
VICE PRESIDENT, LIVE AND TAPE PRODUCTION
Bill Fischer
VICE PRESIDENT, POST PRODUCTION
Christina Friedgen
VICE PRESIDENT, CREATIVE AFFAIRS
Stephen Girard
VICE PRESIDENT IN CHARGE OF BUSINESS AFFAIRS
Sid Kalcheim
VICE PRESIDENT, TALENT & CASTING
Meryl O'Laughlin
VICE PRESIDENT, LABOR RELATIONS
Orison Marden
VICE PRESIDENT, MOVIES & MINISERIES
Fernando Roca
DIRECTOR, TALENT & CASTING
Fran Bascom
DIRECTOR, LITERARY AFFAIRS
Deborah Cutler
DIRECTOR, BUSINESS AFFAIRS
Michael H. Lauer
DIRECTOR, EMPLOYEE RELATIONS
Harvey Lehman
DIRECTOR, CURRENT PROGRAMMING
Kathleen St. Johns
DIRECTOR, DRAMATIC DEVELOPMENT
Rachel Tabori

COLUMBIA PICTURES TELEVISION DOMESTIC DISTRIBUTION

15250 Ventura Blvd., Sherman Oaks, CA 91403 (213) 995-1300.

PRESIDENT
Henry A. Gillespie
VICE PRESIDENT, DOMESTIC SALES
Joseph Indelli
VICE PRESIDENT, ADVERTISING & PROMOTION
Steve Astor
VICE PRESIDENT, FINANCIAL ADMINISTRATION
Phil Cuppett
VICE PRESIDENT, FEATURE MARKETING
Dalton Danon
VICE PRESIDENT, DOMESTIC ADMINISTRATION
Irma Flamenco
VICE PRESIDENT, SALES AND RESEARCH DEPARTMENT
John Walden
VICE PRESIDENT, PROGRAMMING
Edward A. Warren
DIRECTOR, INFORMATION SERVICES
Richard Gire
DIRECTOR, SYNDICATED PROGRAM DEVELOPMENT
Brandy French
CREATIVE DIRECTOR, ADVERTISING & PROMOTION
Janet Radeck

NEW YORK OFFICE

711 Fifth Avenue, New York, NY 10022 (212) 751-4400

VICE PRESIDENT, SALES, EAST
Herbert Weiss
EASTERN SALES ACCOUNT EXECUTIVE
Ken Doyle

SYNDICATION SALES FIELD REPRESENTATIVES

Midwest (Chicago, 312-329-1650)

DIRECTOR/SALES ACCOUNT EXECUTIVE
Stuart Stringfellow

Southwest (Dallas, 214-967-2184/742-5060)

VICE PRESIDENT
Jack Ellison

Southeast (Atlanta, 404-256-3007)
VICE PRESIDENT
Don Bryan
PROGRAMMING EXECUTIVE
Tom Holland

West (Los Angeles, 213-995-1300)
VICE PRESIDENTS
Dick Campbell
Joe Abruscato

COLUMBIA PICTURES INTERNATIONAL

711 Fifth Avenue, New York, N.Y. 10022 (212) 751-4400
SENIOR VICE PRESIDENT, INTERNATIONAL TELEVISION
J. Brian McGrath

INTERNATIONAL OFFICES AND AGENTS

Argentina—Columbia Pictures of Argentina, Inc., LaValle 1878, Buenos Aires, Argentina, Armando Cortez, General Manager

Australia—Columbia Pictures Television Pty., Ltd., 45 MacQuarie St., Sydney, 2000 Australia, Hugh Broun, Managing Director

Canada—Columbia Pictures Television Canada, 365 Bloor Street East, Suite 1602, Toronto, Ontario M4W 3L4, David G. McLaughlin, General Manager

Television—Columbia Pictures Television Canada Limited, 1245 Rue Sherbrooke Quest, Montreal H3G 1G2, John Verge, Director of French Operations

Egypt—14 Gawad Hosni Street, 4th Floor, Apt. 47, Cairo, A.R.E., Galal Rizk, Agent

France—Columbia Pictures Television, 20 Rue Troyon, 75017 Paris, France, Jacques Porteret, Director of French Sales

Germany—Columbia Television Vertriebsgesellschaft, m.b.H., Ickstattstrasse 1, 8000 Munich 5, West Germany, Max Kimental, Managing Director

Greece—Hellas Television International, 96 Akadimias Street, Athens 141, Greece, George Michaelides, Agent, Bill Protopsaltis, Agent

Italy—TV Overseas S.R. L., Via Sicilia N. 50, 00187, Rome, Italy, Jimmy Manca, Agent

Japan—Japan International Enterprises, Inc., 4th Floor, Yoneda Building, 17-20 Shimbashi 6-chome, Minato-Ku, Tokyo Pref., 105 Japan, Toru Ohnuki, General Manager

Mexico—Columbia Pictures Television International, Darwin 68, Office 301, Mexico City, Mexico 5, D.F., Alvaro Mutis, General Manager

Philippines—Columbia Pictures Television International, 3rd Floor, Room 307, Philippine, President Lines Bldg., 1000 United Nations Avenue, P.O. Box 3373, Manila, Philippines, Conrado Javier, Sales Representative

Spain—San Lucas 21.-1, Madrid 4, Spain, Antonio Sanchez-Gijon, Agent

U.K. and Eire—Columbia Pictures Television, St. Margaret's House, 19/23 Wells Street, London WIP, 3 FP, England, Tim Vignoles, Managing Director

COLUMBIA PICTURES COMMUNICATIONS

P.O. Box 10612, 2402 E. Arizona Biltmore Circle, Phoenix, AZ 85016 (602) 956-8940
PRESIDENT
Karl Eller
SENIOR VICE PRESIDENT, OPERATIONS/FINANCE
Murray H. Topham
PRESIDENT, RADIO GROUP
Edward R. Boyd, Columbia Plaza, Burbank, CA 91505 (213) 954-6000

KCPX-AM & FM

1760 Freemont Drive, Salt Lake City, Utah 84104 (801) 972-3030.
VICE PRESIDENT & GENERAL MANAGER
Bruce Cummings
BUSINESS MANAGER & CONTROLLER
Randy Kimball

WWVA-AM/WCPI-FM

Capitol Music Hall, 1015 Main Street, Wheeling, WVA 26003 (304) 232-1170.
VICE PRESIDENT & GENERAL MANAGER
Ross Felton
BUSINESS MANAGER & CONTROLLER
Robert Haggerty

WYDE

2112 Eleventh Avenue South, Birmingham, AL 35205 (205) 322-4511.
VICE PRESIDENT & GENERAL MANAGER
Berkley Fraser
BUSINESS MANAGER & CONTROLLER
Robert Haggerty

New York Subways Advertising Co., 750 Third Avenue, N.Y., NY 10017 (212) 867-4700
PRESIDENT
Marvin Schwartz
EXECUTIVE VICE PRESIDENT
William A. Apfelbaum

EUE/SCREEN GEMS

222 East 44th Street, N.Y., NY 10017 (212) 867-4030; 3701 Oak Street, Burbank, CA 91505 (213) 954-3000.
PRESIDENT
Steve Elliot
EXECUTIVE VICE PRESIDENT & GENERAL MANAGER
George Cooney
VICE PRESIDENT, DIRECTOR OF OPERATIONS
Nicholas Bavaro
VICE PRESIDENT, FINANCE
John Clarke
VICE PRESIDENT AND SALES MANAGER
Eli Feldman
VICE PRESIDENT
Errol Linderman
VICE PRESIDENT
Thomas Whitesell
SENIOR VICE PRESIDENT & GENERAL MANAGER (WEST COAST)
Jerry Bernstein
VICE PRESIDENT AND EXECUTIVE PRODUCER (WEST COAST)
Larry DeLeon
VICE PRESIDENT, MIDWEST SALES REPRESENTATIVE (Chicago, 312-641-0808)
Daniel O'Brien
DIRECTORS (EAST COAST)
Bruce Dowad
Michael Elliot
Stephen Elliot
Bob Forgione
John McShane
Mickey Trenner
DIRECTORS (WEST COAST)
Richard Black
Dick Cunha
Stan Dragoti
David Dryer
Paul Herriot
Remi Kramer
Michael Zingale

EUE VIDEO

222 East 44th Street, N.Y., NY 10017 (212) 867-4030
EXECUTIVE VICE PRESIDENT & GENERAL MANAGER
Dan Rosen
EXECUTIVE VICE PRESIDENT, OPERATIONS
Shirley Danko
VICE PRESIDENT, ENGINEERING
Steve Walsh
VICE PRESIDENT, VIDEO SYNDICATION
Richard H. Smith
VICE PRESIDENT, VIDEO ACCOUNT EXECUTIVE
Tim Timpanaro
VIDEO ACCOUNT EXECUTIVE, KINESCOPE
Jerry Cantwell
DIRECTOR OF POST PRODUCTION SERVICES
Mary Gibney
ACCOUNT EXECUTIVE, POST PRODUCTION SERVICES
Jill Debin

ALTON FILMS

400 Madison Avenue, New York, N.Y. 10017 (212) 753-7522.
EXECUTIVE PRODUCER
Jordan Caldwell

INDEPENDENT ARTISTS

16 W. 56 Street, New York, N.Y. 10019 (212) 765-4640.
EXECUTIVE PRODUCER
Herb Sidel
SALES PRODUCER
Paul Rosen
MIDWEST SALES REPRESENTATIVE (Chicago, 312-943-0606)
Al Gay

FRED LEVINSON & CO.

12 East 82 Street, New York, N.Y. 10028 (212) 472-8888.
PRODUCERS
Stephen Brind
Michael Sitzer

TESTPATTERNS

222 E. 44 Street, New York, N.Y. 10017 (212) 551-8480.
VICE PRESIDENT & EXECUTIVE PRODUCER
Robert Elliott

COLUMBIA PICTURES VIDEOCASSETTE SERVICES

1501 Landmeier Road, Elk Grove Village, IL 60007 (302) 640-2350
EXECUTIVE VICE PRESIDENT
George Ricci
DIRECTOR—TECHNICAL SERVICES
Michael Sterling
GENERAL MANAGER
T. Martin Rennels
MANAGER—INDUSTRAL SALES
Pamela M. Arnest
MANAGER—PAY TV SALES
Richard Oliver
CONTROLLER
Irene Janczkowski
OPERATIONS MANAGER
Alex Rafferty

EDITEL

301 East Erie, Chicago, Illinois 60611 (312) 440-2360.
EXECUTIVE VICE PRESIDENT
Doyle Kaniff
GENERAL MANAGER
Lenard Pearlman
OPERATIONS MANAGER
Sarah Swiskow

EDITEL

729 North Highland Ave., Hollywood, Calif. 90038 (213) 931-1821
EXECUTIVE VICE PRESIDENT
Vince Lyons
GENERAL MANAGER
William Johnson
VICE PRESIDENT—ENGINEERING
Sam Holtz
CONTROLLER
Michael Vallens

COLUMBIA PICTURES PAY TELEVISION

Columbia Plaza South, Burbank, CA 91505 (213) 954-6000
VICE PRESIDENT AND GENERAL MANAGER
Anthony J. Lynn

COLUMBIA PICTURES HOME ENTERTAINMENT

Columbia Plaza, Burbank, CA 91505 (213) 954-600
VICE PRESIDENT AND GENERAL MANAGER
Robert Blattner II

Corinthian Broadcasting Corporation

280 Park Avenue, New York, NY 10017 (212) 949-3900
PRESIDENT
Philip J. Lombardo
EXECUTIVE VICE PRESIDENT, TELEVISION STATIONS DIVISION
James M. King
VICE PRESIDENT FINANCE & ADMINISTRATION
Jerry L. Gabert
VICE PRESIDENT, GENERAL COUNSEL
Lee C. Kitchin
VICE PRESIDENT PROGRAMMING
Lee S. Eden
VICE PRESIDENT MARKETING SERVICES
Elissa Lebeck
VICE PRESIDENT ADVERTISING & PUBLIC RELATIONS
Gene A. Davis
VICE PRESIDENT SALES
Burke E. Liburt

Cox Broadcasting Corp.

White Columns, 1601 W. Peachtree St., N.E. Atlanta, GA 30309. (404) 897-7000. (Group stations owner.) Includes Cox Cable Communications.
CHAIRMAN OF THE BOARD
Clifford M. Kirtland, Jr.
PRESIDENT AND CHIEF OPERATING OFFICER; PRESIDENT—BROADCASTING DIVISION
William A. Schwartz
SENIOR VICE PRESIDENT
Stanley G. Mouse
EXECUTIVE VICE PRESIDENT; PRESIDENT—COX CABLE COMMUNICATIONS
Robert C. Wright
VICE PRESIDENT—PERSONNEL & PUBLIC RELATIONS
E. William Bohn
VICE PRESIDENT
Thomas J. Frawley
VICE PRESIDENT—BROADCAST OPERATIONS
Michael S. Kievman
VICE PRESIDENT, SECRETARY AND TREASURER,
Raymond J. Tucker
VICE PRESIDENT; EXECUTIVE VICE PRESIDENT—RADIO, BROADCASTING DIVISION
James W. Wesley, Jr.

Walt Disney Productions

500 S. Buena Vista Street, Burbank, CA 91521. (Programming producer.)
CHAIRMAN OF THE BOARD AND CHIEF EXECUTIVE OFFICER
E. Cardon Walker
CHAIRMAN OF THE EXECUTIVE COMMITTEE
Donn B. Tatum
PRESIDENT AND CHIEF OPERATING OFFICER
Ronald W. Miller
SENIOR VICE PRESIDENT—FINANCE
MMitchal L. Bagnall
SENIOR VICE PRESIDENT—BUSINESS AND LEGAL AFFAIRS
Ronald J. Cayo
SENIOR VICE PRESIDENT—WALT DISNEY MARKETING DIVISION
Vincent H. Jefferds
SENIOR VICE PRESIDENT—ADVERTISING, PUBLICITY, PROMOTION AND PUBLIC RELATIONS
Jack B. Lindquist
VICE PRESIDENT—EPCOT
Carl G. Bongirno
VICE PRESIDENT—CONSUMER PRODUCTS AND MERCHANDISING
Barton K. Boyd
VICE PRESIDENT—TAX ADMINISTRATION AND COUNSEL
Jose M. Deetjen
VICE PRESIDENT—STUDIO OPERATIONS
Robert W. Gibeaut
VICE PRESIDENT—CORPORATE AND STOCKHOLDER AFFAIRS
Luther R. Marr
VICE PRESIDENT—GENERAL COUNSEL
Richard T. Morrow
VICE PRESIDENT—WALT DISNEY OUTDOOR RECREATION DIVISION
Richard A. Nunis
VICE PRESIDENT—TELEVISION
Lennart Ringquist
VICE PRESIDENT—CONSTRUCTION AND CONTRACT ADMINISTRATION AND PURCHASING
Howard Roland
VICE PRESIDENT AND EASTERN COUNSEL
Franklin Waldheim
SECRETARY
Doris A. Smith
TREASURER
Donald A. Escen
CONTROLLER
Bruce F. Johnson
ASSISTANT SECRETARY-TREASURER
Leland L. Kirk
ASSISTANT SECRETARY
Neal E. McClure
ASSISTANT TREASURER
Alvin L. Shelbourn
ASSISTANT TREASURER
Donald E. Tucker
ASSISTANT CONTROLLER
Douglas E. Houck

ASSISTANT CONTROLLER
Joe E. Stevens

Eastman Kodak Company

343 State Street, Rochester, NY 14650. (716) 724-4000; 1901 W. 22nd St., Oakbrook, IL 60521. (312) 654-5300; 6706 Santa Monica Boulevard, Hollywood, CA 90038. (213) 464-6131; and 1133 Ave. of the Americas, New York, NY 10036. (212) 262-6000. (Motion Picture Film offices & laboratories.)

Also located at 3250 Van Ness Avenue, San Francisco, CA 94109. (415) 776-6055; 6300 Cedar Springs Road, Dallas, TX 75235. (214) 351-3221.

CHAIRMAN OF THE BOARD & CHIEF EXECUTIVE OFFICER
Walter A. Fallon
PRESIDENT
Dolby H. Chandler
VICE PRESIDENT, DIRECTOR, CORPORATE COMMERCIAL AFFAIRS
David S. Greenlaw
VICE PRESIDENT, DIRECTOR, PHOTOGRAPHIC STRATEGIC PLANNING
R. Frederick Porter
ASST. VICE PRESIDENT, DIRECTOR GOVERNMENT RELATIONS
T. Hiatt
VICE PRESIDENT, DIRECTOR CORPORATE COMMUNICATIONS
D.G. Metz
VICE PRESIDENT & DIRECTOR CORPORATE PLANNING
R.B. Murray

U.S. AND CANADIAN PHOTOGRAPHIC DIVISION

EXEC. VICE PRESIDENT, GENERAL MANAGER, U.S. & CANADIAN PHOTOGRAPHIC DIV.
Douglass C. Harvey
EXECUTIVE VICE PRESIDENT & GENERAL MANAGER PHOTOGRAPHIC DIV.
K.R. Whitmore
VICE PRESIDENT, INTERNATIONAL PHOTOGRAPHIC DIVISION
Anthony Frothingham
ASST. VICE PRESIDENT, GENERAL MANAGER, MARKETING DIV.
J. Phillip Samper
ASST. VICE PRESIDENT, GENERAL MANAGER, CUSTOMER EQUIPMENT SERVICES DIV.
John H. Barnes
VICE PRESIDENT, GENERAL MANAGER, HEALTH SCIENCES DIV.
John C. Fink
VICE PRESIDENT, GENERAL MANAGER, MOTION PICTURE AND AUDIOVISUAL MARKETS DIV.
Kenneth M. Mason
ASST. VICE PRESIDENT, DIRECTOR, ADVERTISING AND PROMOTION
William K. Pedersen
VICE PRESIDENT, GENERAL MANAGER, CONSUMER/PROFESSIONAL & FINISHING MARKETS DIV.
John R. Robertson
VICE PRESIDENT, GENERAL MANAGER, GRAPHICS MARKETS DIV.
William E. Sherman
VICE PRESIDENT, GENERAL MANAGER, BUSINESS SYSTEMS MARKETS DIV.
J. Raymond Sutcliffe
VICE PRESIDENT, GENERAL MANAGER, KODAK PARK DIVISION
Richard T. Kramer
VICE PRESIDENT, ASST. GENERAL MANAGER, KODAK PARK DIV.
Wendel W. Cook
VICE PRESIDENT, GENERAL MANAGER, DISTRIBUTION DIV.
George A. Snyder
VICE PRESIDENT, GENERAL MANAGER, KODAK APPARATUS DIV.
J.M. Sewell
VICE PRESIDENT, ASST. GENERAL MANAGER, KODAK APPARATUS DIV.
Richard C. Kleinhans
PRESIDENT AND GENERAL MANAGER, KODAK CANADA LTD.
Bramwell B. Coles

INTERNATIONAL PHOTOGRAPHIC DIVISION

GROUP VICE PRESIDENT, GENERAL MANAGER, INTL. PHOTOGRAPHIC DIV.
C.J. Murphy
VICE PRESIDENT, GENERAL MANAGER, EUROPEAN REGION
David W. Hunt
CHAIRMAN & MANAGING DIRECTOR, KODAK LTD. (United Kingdom)
F. James Moorfoot
GENERAL MANAGER, KODAK A.G. (West Germany)
Helmut Nagel
PRESIDENT & GENERAL MANAGER, KODAK-PATHE S.A. (FRANCE)
Paul L. Vuillaume
VICE PRESIDENT, GENERAL MANAGER, ASIAN, AFRICAN, & AUSTRALIAN REGION
John E. Riggs
VICE PRESIDENT, GENERAL MANAGER, LATIN AMERICAN REGION
R.L. Smith

EASTERN CHEMICALS DIVISION

EXEC. VICE PRESIDENT, GENERAL MANAGER, EASTMAN CHEMICALS DIV.
T.F. Reid
VICE PRESIDENT, ASST. GENERAL MANAGER, EASTMAN CHEMICALS DIV.
Robert C. Petrey
VICE PRESIDENT, ASST. GENERAL MANAGER, EASTMAN CHEMICALS DIV.
J.H. Sanders
VICE PRESIDENT, GENERAL MANAGER'S STAFF, EASTMAN CHEMICALS DIV.
James A. Mitchell
PRESIDENT, ARKANSAS EASTMAN CO.
Marvin L. Gerner
PRESIDENT, TENNESSEE EASTMAN CO.
R.C. Hart
PRESIDENT, CAROLINA EASTMAN CO.
E.M. Olson
PRESIDENT, TEXAS EASTMAN CO.
Jack C. Vander Woude

EASTMAN TECHNOLOGY, INC.

PRESIDENT, EASTMAN TECHNOLOGY, INC.
David S. Greenlaw
PRESIDENT, SPIN PHYSICS, INC.
James U. Lemke

KODAK RESEARCH LABORATORIES

SENIOR VICE PRESIDENT, DIRECTOR, KODAK RESEARCH LABORATORIES
Leon J. Thomas

Legal

SENIOR VICE PRESIDENT, GENERAL COUNSEL
Kendall M. Cole
SECRETARY
Leonard S. Zartman
ASST. SECRETARY
James K. Robinson

Finance and Administration

SENIOR VICE PRESIDENT, DIRECTOR, FINANCE & ADMINISTRATION
Robert A. Sherman
ASST. VICE PRESIDENT, DIRECTOR, ADMINISTRATIVE SERVICES
Roy N. Holmes
ASST. VICE PRESIDENT, DIRECTOR, CORPORATE PHOTOGRAPHIC PRICING
Collins M. McKelvey
GENERAL COMPTROLLER
L.G. Pierce
ASST. COMPTROLLERS
Wayne K. Gilman, Lowrie G. Piercy Donald R. Whitney
TREASURER
Donald E. Snyder
ASST. TREASURER
Harold C. Passer

Corporate Relations

SENIOR VICE PRESIDENT, DIRECTOR, CORPORATE RELATIONS
James S. Bruce
VICE PRESIDENT, ASST. DIRECTOR, CORPORATE RELATIONS
William L. Sutton
DIRECTORS
Walter A. Fallon, Roger E. Anderson, James S. Bruce, Colby H. Chandler, Kendall M. Cole, Charles T. Duncan, Douglass C. Harvey, Robert S. Hatfield, J. Paul Lyet, T.F. Reid, Robert A. Sherman, W. Allen Wallis, Gerald B. Zornow, Juanita M. Kreps
EXECUTIVE COMMITTEE
Walter A. Fallon, James S. Bruce, Colby H. Chandler, Kendall M. Cole, Douglass C. Harvey, T.F. Reid, Robert A. Sherman
AUDIT COMMITTEE
R.E. Anderson, chairman; C.T. Duncan, R.S. Hotfield, J.P. Lyet, W.A. Wallis.

Filmways, Inc.

2049 Century Park East, Suite 3500, Los Angeles, CA 90067 (213) 557-8700. 540 Madison Ave., New York, NY 10022 (212) 758-5100.

CHAIRMAN OF THE BOARD, PRESIDENT & CHIEF EXECUTIVE OFFICER
Richard L. Bloch
VICE CHAIRMAN OF THE BOARD
L. Douglas Nolan
EXECUTIVE VICE PRESIDENT
Robert A. Grunburg
EXECUTIVE VICE PRESIDENT & CHIEF OPERATING OFFICER
Gerald S. Armstrong
SENIOR VICE PRESIDENT, CHIEF FINANCIAL OFFICER
Michael Garstin
SENIOR VICE PRESIDENT
Salvatore J. Iannucci
SENIOR VICE PRESIDENT
Stanley S. Sills
VICE PRESIDENT, TREASURER, CONTROLLER
Eugene B. Boylan
GENERAL COUNSEL, SECRETARY
Alan R. Markizon
ASSISTANT SECRETARY
David Z. Rosensweig

General Electric Broadcasting Company, Inc.

1400 Balltown Rd., Schenectady, NY 12309. (Group station owner.)

PRESIDENT
A. W.P. Baker, Jr.
VICE PRESIDENT AND GENERAL COUNSEL
J.J. Smith
TREASURER AND MANAGER—FINANCE
C.O. Yeates
VICE PRESIDENTS AND GENERAL MANAGERS
J.J. Delmonico (Schenectady)
E.T. Shadburne (Nashville)
Brian E. Cobb (Denver)
MANAGERS—TV PROGRAMMING
Arthur R. Garland (Schenectady)
Richard R. Cartier (Nashville)
Marcia A. Oeste (Denver)
MANAGERS—OPERATIONS
Charles B. King (Schnectady)
Daniele Fiorello (Denver)
Robert J. Hafner (Nashville)

International Creative Management

40 West 57th St., New York, NY 10019. Tel.: (212) 556-5600. 8899 Beverly Blvd., Los Angeles, CA 90048. Tel.: (213) 550-4000. (Representatives of performing and creative talent in the entertainment industry.) A division of Marvin Josephson Associates, Inc., the agency is the successor by merger of Creative Management Associates and International Famous Agency.

CHAIRMAN
Ralph Mann
PRESIDENT
Jeff Berg

Lorimar Productions

c/o Metro-Goldwyn-Mayer, 10202 W. Washington Blvd., Culver City, CA 90230 (213) 836-3000.

PRESIDENT
Lee Rich
CHAIRMAN OF THE BOARD
Merv Adelson

MCA Television

(Distributor a subsidiary of MCA, Inc. as is Universal Television, the production arm. See listing.) 445 Park Ave., New York, NY 10022 (212) 759-7500. 100 Universal Plaza, Universal City, CA 91608 (213) 985-4321.

FILM SYNDICATION DIVISION

CHAIRMAN
Louis N. Friedland
PRESIDENT, MCA TV
Don Menchel
PRESIDENT, MCA TELEVISION GROUP
Al Rush
SENIOR VICE-PRESIDENT, MARKETING
Shelly Schwab
SENIOR VICE-PRESIDENT, SALES MANAGER
Carl A. Russell
DIRECTOR OF MEDIA SALES
Jack B. Allen
VICE PRESIDENT, MCA TV ENTERPRISES
Chuck Gerber
VICE-PRESIDENT, CREATIVE SERVICES
Morton Slakoff
DIRECTOR, ADVERTISING, PROMOTION & PUBLICITY
Joe Ondrick
ASSISTANT DIRECTOR, PROMOTION & PUBLICITY
Gerri Paré
VICE PRESIDENT, RESEARCH & SALES DEVELOPMENT
Bob Davis
VICE PRESIDENT, MANAGER OF RESEARCH
Bert Herbert
VICE PRESIDENT, LOWER MIDWEST AREA
Jack Robertson
VICE PRESIDENT, MID-CENTRAL AREA
Phil Conway
VICE PRESIDENT, SOUTHEAST AREA
Carl Runge
VICE PRESIDENT, WESTERN AREA
Marc Grayson
VICE-PRESIDENT, UPPER MIDWEST AREA
Paul Hoffman
VICE-PRESIDENT, SOUTHWEST AREA
Tom Maples
GENERAL CONSULTANT
DeArv Barton
PRESIDENT, MCA TV INTERNATIONAL
Robert Bramson
EXECUTIVE VICE PRESIDENT AND DIRECTOR OF INTERNATIONAL SALES
Colin Davis
VICE-PRESIDENT, GERMANY
Karl de Vogt
SALES EXECUTIVE, GERMANY
Hendrick Van Daalen
VICE-PRESIDENT, JAPAN
Keinosuke Kuragaki
VICE-PRESIDENT, UNITED KINGDOM
Don Gale
VICE-PRESIDENT, AUSTRALIA AND FAR EAST
Neville Thorsen
VICE-PRESIDENT, FRANCE
Claude Perrier, Roger Cordjohn
VICE-PRESIDENT, BRAZIL
Wanderley Fucciolo
SALES EXECUTIVE, MIDDLE EAST
Kamal Sayegh
VICE-PRESIDENT, LATIN AMERICA
Francisco Padilla
VICE-PRESIDENT, CANADA
Peter Hughes
SALES EXECUTIVE, CANADA
Jeff Schlesinger
VICE-PRESIDENT, FRENCH CANADA
Gilles Morin
BRANCH OFFICES:
Atlanta, Chicago, Cleveland, Dallas, New York, St. Louis, Universal City; Montreal, Canada; Toronto, Canada; London, England; Paris, France; Amsterdam, Netherlands; Tokyo, Japan; Sydney, Australia; Munich, W. Germany; Sao Paulo, Brazil; Mexico City, Beirut, Lebanon.

MGM Television

(A Division of the MGM Film Co.) 10202 West Washington Blvd., Culver City, CA 90230 (213) 558-5000.

TELEVISION PRODUCTION DIVISION

PRESIDENT
Thomas D. Tannenbaum
SENIOR VICE PRESIDENT—BUSINESS AFFAIRS & ADMINISTRATION
Arthur Joel Katz
SENIOR VICE PRESIDENT—PROGRAMS
Werner Michel
VICE PRESIDENT—PROGRAM DEVELOPMENT
Dori Weiss
VICE PRESIDENT—CURRENT PROGRAMS
Jerome H. Stanley

VICE PRESIDENT—EXECUTIVE PRODUCTION MANAGER
Richard Birnie
VICE PRESIDENT—ADMINISTRATION
Leslie H. Frends
VICE PRESIDENT—TALENT
Joseph D'Agosta
DIRECTOR OF MOVIES AND MINI-SERIES
Virginia K. Hegge
DIRECTOR OF BUSINESS AFFAIRS
Steven P. Maier
DIRECTOR OF BUSINESS AFFAIRS
David G. Stanley
DIRECTOR OF COMEDY DEVELOPMENT
Phyllis A. Golden-Gottlieb
STORY EDITOR
Frederique Becker
ASSISTANT TO THE PRESIDENT
William S. Todman, Jr.

Television Sales Division

EXECUTIVE VICE PRESIDENT—WORLDWIDE SYNDICATION—N.Y.
Lawrence E. Gershman
SENIOR VICE PRESIDENT—DOMESTIC SALES—N.Y.
Joseph C. Tirinato
VICE PRESIDENT—DOMESTIC SYNDICATION—N.Y.
Neil Russell
SALES MANAGER—MID-WESTERN DIVISION—CHICAGO
Robert J. Horen
SALES MANAGER—WESTERN DIVISION—L.A.
William A. Kunkel
SALES MANAGER—SOUTHERN DIVISION—ATLANTA
Virgil B. Wolff
SALES MANAGER—EASTERN DIVISION—N.Y.
Philip L. Smith
DIRECTOR OF CREATIVE MARKETING—N.Y.
Susan Swimer
DIRECTOR OF MARKETING RESEARCH—N.Y.
Jean Goldberg
DIRECTOR OF INTERNATIONAL ADMINISTRATION/CLIENT RELATIONS
Sheryl S. Hardy
MARKETING RESEARCH ANALYST—N.Y.
Anna Greenberg
CONTRACT ADMINISTRATOR
Stacey Valenza
MANAGER OF WORLDWIDE FILM SERVICES
Dennis Murphy

Ancillary Rights Division

VICE PRESIDENT—ANCILLARY RIGHTS DIVISION
Peter W. Kuyper
PAY TV SERVICES COORDINATOR
Meyer Shwarzstein

MTM Enterprises, Inc.

4024 Radford Avenue, Studio City, CA Tel.: (213) 760-5000 (TV program production).

CHAIRMAN OF THE BOARD
Mary Tyler Moore
PRESIDENT
Arthur Price
EXECUTIVE VICE PRESIDENTS
Mel D. Blumenthal, Stuart Erwin, Jr.
VICE PRESIDENTS
Tom Palmieri, Andrew M. Siegel, Abner E. Singer
DATE OF ORGANIZATION
1970

The Magnavox Company

Executive Offices: 100 East 42nd St., New York, NY 10016. Corporate Offices: 1700 Magnavox Way, Fort Wayne, IN 46804. (Set manufacturer.)

PRESIDENT
Robert H. Platt
SR. VICE PRES.
Alfred di Scipio
SR. VICE PRES.
Jack W. Schrey
SR. VICE PRES.
Daniel F. Minahan
VICE PRESIDENT
Mark B. Bollman
VICE PRES.
Myrl W. Deitch
VICE PRES. & CONTROLLER
Brack W. Duker
VICE PRES.
Jack F. Feddersen
VICE PRES.
John M. Fauth
VICE PRESIDENT
Donald L. Hamilton
VICE PRESIDENT
Samuel H. Newnam
VICE PRESIDENT, SECRETARY & GENERAL COUNSEL
Samuel J. Rozel
VICE PRESIDENT
James T. Smith
VICE PRESIDENT
Arthur P. Stern
VICE PRESIDENT
Frank Van Steenberg, Sr.
VICE PRESIDET & TREASURER
James M. Whelan
ASSISTANT SECRETARY
Jacob Kanter

Metromedia, Inc.

Executive offices: 1 Harmon Plaza, Secaucus, NJ 07094 (201) 348-3244. (Group station owner.)

BOARD OF DIRECTORS
John W. Kluge, Chairman of the Board and President, Metromedia, Inc.; Lloyd M. Bauman; John F. Crisconi, President, Philadelphia Motor Car Co.; John C. Griswold, Dr. T.T. Goldsmith, Jr., consulting physicist; Mrs. Jane Pickens Hoving, executive of various civic organizations; Warren H. Lasher, pres., W.H. Lasher Co., Inc.; John B. Lomenzo, Attorney James H. Manges, Kuhn, Loeb & Co.
CHAIRMAN OF THE BOARD AND PRESIDENT
John W. Kluge
SENIOR VICE PRESIDENT
Ross Barrett
SENIOR VICE-PRESIDENT AND GENERAL COUNSEL
Alfred L. Schwartz
SECRETARY
Len Pack
VICE-PRESIDENT—FINANCE
Stuart Subotnick
VICE PRESIDENT—PUBLIC AFFAIRS
Mark Evans
SENIOR VICE PRESIDENT
Stanley Landon
VICE PRESIDENT—OPERATIONS
George H. Duncan
VICE PRESIDENT—CONTROLLER
Robert Maresch
VICE PRESIDENT, REGULATORY AFFAIRS & ASSOCIATE GENERAL COUNSEL
Thomas J. Dougherty
ASSISTANT TO THE PRESIDENT
Alfred R. Johnson
ASSISTANT SECRETARIES
Preston, Padden, David H. Frederickson, Gerald Friedman, Arnold Wadler, Muriel Reis Monica Link
DIRECTOR OF PERSONNEL ADMINISTRATION
Peggy Stallings
VICE PRESIDENT & DIRECTOR OF LABOR RELATIONS
Kenneth Cohen
ASSISTANT TREASURERS
Fred Cige, Mario Catuogno, Diana Brainerd, Kenneth Green

BROADCAST GROUP

METROMEDIA TELEVISION

PRESIDENT
Thomas Tilson
VICE PRESIDENT AND CONTROLLER
Harold Christiansen
VICE PRESIDENT, PROGRAMMING
Richard Ballinger

METROMEDIA RADIO

PRESIDENT
George Duncan

METRO TV SALES

PRESIDENT
William Tynan

FOSTER & KLEISER

PRESIDENT
Frank Lott

ENTERTAINMENT GROUP

METROMEDIA PRODUCERS CORPORATION

PRESIDENT
Robert D. Wood

ICE CAPADES,

PRESIDENT
Richard Palmer
HARLEM GLOBETROTTERS PRESIDENT
Arthur Harvey

ENTERTAINMENT GROUP

METROMEDIA PRODUCERS CORPORATION

PRESIDENT
Robert D. Wood

ICE CAPADES

PRESIDENT
Richard Palmer
HARLEM GLOBETROTTERS PRESIDENT
Arthur Harvey

Modern Talking Picture Service

General offices: 5000 Park St., North, St. Petersburg FL 33709 (813) 541-7571. (Distributor of free-loan sponsored films and video-cassettes.)

PRESIDENT
Carl H. Lenz
EXECUTIVE VICE PRESIDENT
William M. Oard
VICE PRESIDENT & GENERAL MANAGER
Dan Kater
VICE PRESIDENT, OPERATIONS
James D. McPoland
DATE OF ORGANIZATION
1937

Divisions

Modern TV—Distributor of free-loan shorts for television stations

Modern Cable Network—Distributor of programs and series for Cable Television Stations, Modern Satellite Network—Distributor of programs and series.

Regional Sales Offices:

New York: 45 Rockefeller Plaza, New York, NY 10111 (212) 765-3100

Chicago: 2020 Prudential Plaza, Chicago, IL 60601 (312) 337-3252

Washington, DC: 1901 L. Street, N.W. Washington, DC. 20036 (202) 293-1222

Los Angeles: 1145 N. McCadden Place, Los Angeles, CA 90038 (213) 462-2202

Canada: 143 Sparks Ave., Willowdale, Ontario, Canada M2H 2S5 (416) 498-7290

Regional Booking Offices:

St. Petersburg, FL 33709: 5000 Park St., North (800) 237-8913, Peter Schmidt, TV Manager.

National Broadcasting Co., Inc.

30 Rockefeller Plaza, New York, NY 10020; (212) 664-4444, Registered Telegraphic Address: NATBROCAST, NY

CHAIRMAN AND CHIEF EXECUTIVE OFFICER
Grant Tinker
VICE CHAIRMAN OF BOARD
Richard Salant
BOARD OF DIRECTORS
Richard Salant, Vice Chairman; Grant Tinker, John R. Petty, Dr. Cecily C. Selby, Donald B. Smiley, Thornton Bradshaw

OFFICERS

CHAIRMAN AND CHIEF EXECUTIVE OFFICER
Grant Tinker
PRESIDENT AND CHIEF OPERATING OFFICER
Robert Mulholland
VICE CHAIRMAN OF THE BOARD
Richard Salant
PRESIDENT, NBC-TV NETWORK
Ray Timothy
PRESIDENT, NBC TELEVISION
Irwin Segelstein
PRESIDENT, NBC ENTERTAINMENT
Brandon Tartikoff
PRESIDENT, NBC NEWS
William Small
PRESIDENT, NBC SPORTS
Arthur Watson
PRESIDENT, NBC TELEVISION STATIONS
Robert Walsh
EXECUTIVE VICE PRESIDENT FINANCE
Robert Butler
EXECUTIVE VICE PRESIDENT, OPERATIONS & TECHNICAL SERVICES
Richard Sonnenfeldt
EXECUTIVE VICE PRESIDENT, PUBLIC INFORMATION
Bud Rukeyser
VICE PRESIDENT, TV SALES
Robert Blackmore
VICE PRESIDENT, RESEARCH
William S. Rubens
VICE PRESIDENT, CORPORATE PLANNING
Kathryn Pelgrift
VICE PRESIDENT, MARKETING
Richard Golden
VICE PRESIDENT, BROADCAST RESEARCH
Al Ordover
EXEC. VICE PRESIDENT AND GENERAL COUNSEL
Cory Dunham
EXEC. VICE PRESIDENT, EMPLOYEE RELATIONS
George Black
VICE PRESIDENT, BUSINESS AFFAIRS & PRODUCTION OPERATIONS
Donald Carswell
VICE PRESIDENT, LABOR RELATIONS
Eugene McQuire
VICE PRESIDENT & GENERAL MANAGER, WRC-TV, WASHINGTON, D.C.
John Rohrbeck
VICE PRESIDENT, EMPLOYEE RELATIONS & LAW WEST COAST
Bernard Gold
VICE PRESIDENT & GENERAL MANAGER, KNBC, LOS ANGELES
Tom Straszewski
VICE PRESIDENT, BROADCAST STANDARDS EAST COAST
Rick Gitter
VICE PRESIDENT, PRESS INFORMATION
Alan Baker
VICE PRESIDENT, WASHINGTON
Peter B. Kenney
VICE PRESIDENT, FINANCE
John Baziotis
VICE PRESIDENT TAX ADMINISTRATION
Arthur Angstreich
VICE PRESIDENT, GENERAL MANAGER, WMAQ-TV, CHICAGO
Monte Newman
VICE PRESIDENT & CONTROLLER
Richard Raburn
VICE PRESIDENT, AFFILIATE RELATIONS
Pier Mapes
VICE PRESIDENT & TREASURER
James White
VICE PRESIDENT, CENTRAL SALES, TELEVISION NETWORK
Richard Schade
VICE PRESIDENT, DAYTIME SERIAL PROGRAMS
Jeff Ryder
VICE PRESIDENT, AVDTG & CREATIVE SERVICES
W. Watts Biggers
VICE PRESIDENT, SPECIAL PROGRAMS
Peter Calabrese
EXECUTIVE VICE PRESIDENT, OPERATIONS AND TECHNICAL SERVICES
Richard Sonnenfeldt
VICE PRESIDENT, EUROPEAN PRODUCTION SPORTS
Geoff Mason
VICE PRESIDENT, OPERATIONS AND ENGINEERING
Jack Kennedy
VICE PRESIDENT, PROGRAMS, EAST COAST
Josh Kane
SENIOR VICE PRESIDENT, PROGRAMS, WEST COAST
Perry Lafferty
VICE PRESIDENT BROADCAST PLANNING
Lee Currlin
SENIOR VICE PRESIDENT PROGRAMS, EAST COAST
Irv Wilson
VICE PRESIDENT, PRESS, EAST COAST
Curt Block
VICE PRESIDENT, PRESS, WEST COAST
Gene Walsh

TREASURY DIVISION

EXECUTIVE VICE PRESIDENT
Robert Butler
VICE PRESIDENT & CONTROLLER
George Blank
VICE PRESIDENT, FINANCE
John Baziotis
VICE PRESIDENT, TAX ADMINISTRATION
Arthur Angstreich
VICE PRESIDENT, MANAGEMENT INFORMATION SYSTEMS
Eric Koopman
VICE PRESIDENT, FINANCE, PLANNING & ANALYSIS
Barbara Landes

VICE PRESIDENT, STAFF FINANCE & ADMINISTRATION
Robert Small
VICE PRESIDENT, AUDITING
Arthur Dignam
VICE PRESIDENT, FINANCIAL STUDIES
Gerald Coleman
VICE PRESIDENT & TREASURER
James White
CONTROLLER, WEST COAST
Joseph L. Kubin

EXECUTIVE DIVISION

CHAIRMAN AND CHIEF EXECUTIVE OFFICER
Grant Tinker
PRESIDENT AND CHIEF OPERATING OFFICER
Robert Mulholland
VICE CHAIRMAN
Richard Salant

Station Relations

VICE PRESIDENT, AFFILIATE RELATIONS
Pier Mapes
VICE PRESIDENT, STATIONS RELATIONS ADMIN.
Raymond T. O'Connell
VICE PRESIDENT, AFFILIATE OPERATIONS
Anthony Cervini
EXECUTIVE VICE PRESIDENT, PUBLIC INFORMATION
Bud Rukeyser

Corporate Information

VICE PRESIDENT, PUBLIC INFORMATION
Alan Baker
VICE PRESIDENT, PRESS, EAST COAST
Curt Block
VICE PRESIDENT, PUBLIC INFORMATION, WEST COAST
Gene Walsh
VICE PRESIDENT, CORPORATE RELATIONS
Betty Hudson
DIRECTOR, CORPORATE COMMUNICATIONS
Albert Kroeger
VICE PRESIDENT, INFORMATION SERVICES
Bettye K. Hoffmann

Research

VICE PRESIDENT, RESEARCH
William S. Rubens
VICE PRESIDENT, AUDIENCE RESEARCH
George Hooper
VICE PRESIDENT, PROGRAM RESEARCH
Lenore Cantor
VICE PRESIDENT, BROADCAST RESEARCH
Alfred Ordover
VICE PRESIDENT NEWS & SOCIAL RESEARCH
Ron Milausky

Public Information, Press & Publicity

EXECUTIVE VICE PRESIDENT, PUBLIC INFORMATION
Bud Rukeyser
VICE PRESIDENT, PRESS INFORMATION
Alan Baker
VICE PRESIDENT, PRESS, EAST COAST
Curt Block
DIRECTOR, NEWS PUBLICITY
Ron Najman
MANAGER, NEWS PUBLICITY
Howard Vandermuelen
DIRECTOR, NATIONAL PRESS
Owen Comora
MANAGER, CORPORATE PRESS
Helen Manasian
MANAGER, NATIONAL PRESS
Hal Bender
MANAGER, MAGAZINE PRESS
Natalie Tiranno
DIRECTOR, SPORTS PUBLICITY
Michael Cohen
MANAGER, SPORTS PUBLICITY
Tom Merritt

Advertising

VICE PRESIDENT, ADVERTISING & CREATIVE SERVICES
W. Watts Biggers
VICE PRESIDENT, ADVERTISING OPERATIONS
Francis Pintauro
VICE PRESIDENT, PRINT & DESIGN
Charles Blake
VICE PRESIDENT, ON-AIR PROMOTION
Paula Mermelstein

Promotional Services

VICE PRESIDENT, TALENT RELATIONS
Jay Michelis
MANAGER, GUEST RELATIONS
Peter M. Tintle

Standards and Practices

VICE PRESIDENT, BROADCAST STANDARDS
Ralph Daniels
VICE PRESIDENT, BROADCAST STANDARDS, WEST COAST
Jerome Stanley
VICE PRESIDENT, BROADCAST STANDARDS, EAST COAST
Rick Gitter

Legal

EXECUTIVE VICE PRESIDENT AND GENERAL COUNSEL
Corydon Dunham
VICE PRESIDENT, LAW
Jay Gerber
ASSISTANT GENERAL ATTORNEYS
George Hiltzik, Marshall Wellburn, Stanley Kaufman, John Shute, Steve Stander, Eleanor O'Hara, Carl Trunk
SENIOR COUNSEL
Ellen Agress, John Rose, Barbara Hering, William Abbott, Jeff Rosen, John Shute
SENIOR ATTORNEY, WASHINGTON
John Sturm
VICE PRESIDENT, LAW, WEST COAST
Donald Zachary
VICE PRESIDENT, LAW, WASHINGTON
Howard Monderer
VICE PRESIDENT, COMPLIANCES AND PRACTICES
Alan Gerson
VICE PRESIDENT, EMPLOYEE RELATIONS, LAW, WEST COAST
Bernard Gold
DIRECTOR, GOVERNMENT RELATIONS, WASHINGTON, DC
Sallie Forman
VICE PRESIDENT, WASHINGTON
Peter Kenney
DIRECTOR, GOVERNMENT RELATIONS
Robert Hynes

News Division

PRESIDENT, NBC NEWS
William Small
EXECUTIVE VICE PRESIDENT
Richard Fischer
VICE PRESIDENT & EXECUTIVE ASSISTANT TO THE PRESIDENT
Sylvia Westerman
VICE PRESIDENT, SPECIAL PROGRAMS
Leslie Midgley
VICE PRESIDENT, NEWS, WASHINGTON
Donald Meaney
VICE PRESIDENT, NEWS OPERATIONS
Don Dunkel
GENERAL MANAGER, NEWS
William T. Corrigan
VICE PRESIDENT, RADIO NEWS
Jo Moring
GENERAL MANAGER, NEWS SERVICES
James Kitchell
VICE PRESIDENT NEWS PLANNING
Gordon Manning

Personnel Division

EXEC. VICE PRESIDENT
George Black
VICE PRESIDENT, HUMAN RESOURCE PLANNING & DEVELOPMENT
John Noon
VICE PRESIDENT, STAFF LABOR RELATIONS
Robert Hurford
VICE PRESIDENT, LABOR RELATIONS
Richard N. Goldstein
VICE PRESIDENT, PERSONNEL ADMINISTRATION
Roberta Romberg
VICE PRESIDENT, PERSONNEL & LABOR RELATIONS, WEST COAST
Stanley Klubech
COMPANY PHYSICIAN
Dr. Manuel Smith
DIRECTOR, ORGANIZATION & DEVELOPMENT
Al Jackson
DIRECTOR, PERSONNEL
Sandra DuPilka
DIRECTOR, LABOR RELATIONS, WEST COAST
Bernard Behan

NBC Entertainment Division

PRESIDENT, NBC ENTERTAINMENT
Brandon Tartikoff
VICE PRESIDENT, ADVERTISING & CREATIVE SERVICES
W. Watts Biggers
VICE PRESIDENT, ENTERTAINMENT ACQUISITION
Irwin Moss
VICE PRESIDENT, PROGRAM & TALENT NEGOTIATION
John Agoglia

Television Network Programs

VICE PRESIDENT, THEATRICAL FEATURES
Josh Kane
VICE PRESIDENT, SPECIAL PROGRAMS
Peter Calabrese
VICE PRESIDENT, MINI-SERIES AND NOVELS-FOR-TV
Ethel Winant
VICE PRESIDENT, DAYTIME SERIAL PROGRAMS
Jeff Ryder
VICE PRESIDENT, PROGRAM PLANNING
Lee Currlin
VICE PRESIDENT, COMEDY & VARIETY SHOWS
Saul Islin
VICE PRESIDENT, PROGRAM ADMINISTRATION & OPERATIONS, WEST COAST
Perry Massey, Jr.
VICE PRESIDENT, SPECIAL PROJECTS, WEST COAST
Peter Andrews
VICE PRESIDENT CASTING
Ed Foy III
VICE PRESIDENT, PRIME TIME SERIES
Paul King
VICE PRESIDENT, TALENT ADMINISTRATION
Al Trescony
SENIOR VICE PRESIDENT, PROGRAMS AND TALENT, UWEST COAST
Perry Lafferty
VICE PRESIDENT, STORY DEPARTMENT
Patrick Betz
VICE PRESIDENT, DRAMA DEVELOPMENT
Stuart Sheslow

Television Network Sales

VICE PRESIDENT, SALES
Robert Blackmore
VICE PRESIDENT, SALES ADMINISTRATION
Joe Iaricci
VICE PRESIDENT, NATIONAL SALES
Aaron Cohen
VICE PRESIDENT, EASTERN SALES
Larry Hoffner
VICE PRESIDENT, CENTRAL SALES
Richard Schade
VICE PRESIDENT, WESTERN SALES
Kristin Fitzgerald
VICE PRESIDENT, DETROIT SALES
Joseph R. Sutton
VICE PRESIDENT, PARTICIPATING PROGRAM SALES
Robert F. Conrad
VICE PRESIDENT, DAYTIME AND PARTICIPATING PROGRAM SALES
Joan Leahy
VICE PRESIDENT, SPORTS SALES
Bert Zeldin
VICE PRESIDENT, MARKETING
Richard Golden
VICE PRESIDENT, SALES SERVICE
Stephen Flynn
VICE PRESIDENT, SALES ADMINISTRATION
Joseph Iaricci
ART DIRECTOR
Robert Greenwell
MANAGER, OFFICE ADMINISTRATION
Hamilton Robinson

Television Network-General Manager Business Affairs

VICE PRESIDENT, CORPORATE BUSINESS AFFAIRS
Donald Carswell
VICE PRESIDENT, TELEVISION PRODUCTION AND BUSINESS AFFAIRS WEST COAST
Richard C. Welsch
DIRECTOR, BUSINESS AFFAIRS & ADMINISTRATION
James Dullaghan

Television Network-General Manager Operations & Engineering

EXECUTIVE VICE PRESIDENT, OPERATIONS & TECHNICAL SERVICES
Richard Sonnenfeldt
VICE PRESIDENT, OPERATIONS AND ENGINEERING
Jack Kennedy
VICE PRESIDENT, MANAGEMENT INFORMATION SYSTEMS
Eric Koopman
VICE PRESIDENT, OPERATIONS & SYSTEMS SOFTWARE
Al Grover
DIRECTOR, BUSINESS AFFAIRS OPERATIONS AND ENGINEERING
Victor F. Schwebius
DIRECTOR, PRODUCTION SERVICES
Bernard I. Paulson
VICE PRESIDENT, TELEVISION PROGRAM PRODUCTION OPERATIONS
Robert Galvin
VICE PRESIDENT, TV PROD & BUSINESS AFFAIRS, WEST COAST
Richard C. Welsch
MANAGER, RADIO TECHNICAL OPERATIONS, WASHINGTON
Martin Enghauser
MANAGER, FACILITIES ADMINISTRATIONS
Albert Humbert
VICE PRESIDENT, ENGINEERING
Daniel Walters

NBC SPORTS DIVISION

PRESIDENT
Arthur Watson
VICE PRESIDENT, ADMINISTRATION
EXECUTIVE PRODUCER
Don Ohlmeyer
VICE PRESIDENT, EUROPEAN PRODUCTION
Geoff Mason
VICE PRESIDENT & ASST. GENERAL ATTORNEY
Ben Raub
VICE PRESIDENT, SPORTS OPERATIONS
Alan Conal

NBC RADIO

PRESIDENT, NBC RADIO
Richard Verne
EXECUTIVE VICE PRESIDENT, AM RADIO
Robert Mounty
EXECUTIVE VICE PRESIDENT, FM RADIO
Walter Sabo
EXECUTIVE VICE PRESIDENT, NBC RADIO NETWORK
Chuck Penwick
VICE PRESIDENT, RADIO & INDUSTRY RELATIONS
Marion Stephenson
VICE PRESIDENT, SALES
Kevin Cox
VICE PRESIDENT, ADMINISTRATION
Frank Osborne
VICE PRESIDENT, AFFILIATE RELATIONS, THE SOURCE
Dan Forth

NBC Radio-Owned Stations

VICE PRESIDENT AND GENERAL MANAGER, NEW YORK, WNBC-AM
Bob Sherman
GENERAL MANAGER, WYNY-FM
Dan Griffin
VICE PRESIDENT, CHICAGO, WMAQ
Bert Sherwood
GENERAL MGR, WKQX
Richard Penn
VICE PRESIDENT & GENERAL MANAGER, SAN FRANCISCO, KNBR
William Dwyer
GENERAL MANAGER KYUU
John Hayes
VICE PRESIDENT, WASHINGTON, WRC
Frank Scott
GENERAL MANAGER, WKYS-FM
Raymond Yorke

Radio Sales

VICE PRESIDENT, SALES
Kevin Cox
VICE PRESIDENT, MARKET DEVELOPMENT
Robert Rush

Radio Network Programs

VICE PRESIDENT, RADIO NEWS
Morrie Trumble, dir. sports prog.
Buddy Young, dir. prog. operations

NBC OWNED STATIONS & NBC SPOT SALES DIVISION

NBC Owned Stations

PRESIDENT, NBC OWNED TELEVISION STATIONS
Robert Walsh
VICE PRESIDENT, PROGRAMS
Weston J. Harris
VICE PRESIDENT & GENERAL MANAGER, WRC-TV, Washington
John Rohrbeck
VICE PRESIDENT & GENERAL MANAGER, KNBC, Los Angeles
Tom Straszowski
VICE PRESIDENT & GENERAL MANAGER, WMAQ-TV, Chicago
Monte Newman
VICE PRESIDENT & GENERAL MANAGER, WKYC-TV, Cleveland
Neal A. Van Ells

NBC Spot Sales-Television

MANAGER, EASTERN SPOT SALES
Larry Tompkins
MANAGER, TV SPOT SALES, San Francisco office
Robert H. Anderson
MANAGER, TV SPOT SALES, Los Angeles office
Carl R. Schutz
DIRECTOR, MIDWEST SPOT SALES
James Martin
MANAGER, SPOT SALES, Chicago office
Don Douglas
MANAGER, Detroit
Richard E. Taylor

National Telefilm Associates, Inc.

12636 Beatrice Street, Los Angeles, CA 90066 (213) 306-4040.

CHAIRMAN OF THE BOARD
George C. Hatch
PRESIDENT
Aubrey "Bud" Groskopf
VICE PRESIDENT & WORLDWIDE SALES
Arthur S. Gross
DIRECTOR INTERNATIONAL SALES
Larry Cervantes
ACCT. EXEC. SPECIAL MARKETING
Neil Evans
ACCT. EXEC. MID-WEST DIVISION
John Herrin
ACCT. EXEC. WESTERN DIVISION
Barry Bernard
ACCT. EXEC. SOUTHWEST DIVISION
"Chips" Barrabee
ACCT. EXEC. EASTERN DIVISION
Burton Rosenburgh
CANADA: Peter Mandell (416) 868-1538.
FRANCE: 53, RUE CHARLOT, 75003, Paris March Hare, Ltd.
UK: 21 Cherry Garden Lane, Folkestone Kent, England Bernard Shaw, Sales Rep.

Osmond Communications

1420 East 800 No., Orem, UT 84059 (801) 224-4444. (Parent company for the operation divisions listed below)

PRESIDENT & CHIEF EXECUTIVE OFFICER
Reed R. Callister
EXECUTIVE VICE PRESIDENT
Alan Osmond
SENIOR VICE PRESIDENTS
Wayne Osmond, Merrill Osmond
VICE PRESIDENT FINANCE
John Taylor
VICE PRESIDENT ACCOUNTING
Karen Gordon
CORPORATE SECRETARY
Gordon Bench

OSMOND INTERNATIONAL

PRESIDENT
Michael Wuergler
VICE PRESIDENT
Phillip Catherall

OSMOND TELEVISION SALES

PRESIDENT
David Sifford

OSMOND TELEVISION

PRESIDENT
Toby Martin
VICE PRESIDENT
Dennis Johnson
VICE PRESIDENT—COMPTROLLER
Karen Gordon
VICE PRESIDENT—FILM PRODUCTION
Jack Reddish

OSMOND COMMERCIAL PRODUCTIONS

PRESIDENT
David Anderson
SENIOR VICE PRESIDENT
Dean Proctor

Paramount Television & Video Distribution Division

(A Division of Paramount Pictures Corporation) 5555 Melrose, Hollywood, CA 90038 (213) 468-5000. One Gulf & Western Plaza, New York, NY 10023 (212) 333-7600.

PRESIDENT, PARAMOUNT TELEVISION & VIDEO DISTRIBUTION
Richard Frank (CA)

DOMESTIC DISTRIBUTION/ADVERTISING AND PROMOTIONAL SERVICES

PRESIDENT, TELEVISION DOMESTIC DISTRIBUTION
M. Randolph Reiss
SENIOR VICE PRESIDENT SALES
Robert Jacquemin
VICE PRESIDENT, OPERATIONS (NY)
Leonard J. Grossi
MANAGER, ADMINISTRATION
Helen Ricketts
DIRECTOR, ADVERTISING & PROMOTIONAL SERVICES, PARAMOUNT TELEVISION & VIDEO DISTRIBUTION
Rick Weidner
SENIOR VICE PRESIDENT, PROGRAMMING
John Goldhammer
DIRECTOR, RESEARCH
Garrett Hart
RESEARCH ANALYST
Bruce Rosenblum
SALES OFFICES
New York—1 Gulf & Western Plaza, New York, NY 10023 (212) 333-4600.
Chicago—625 N. Michigan Ave., Suite 711, Chicago, IL 60611 (312) 372-0638.
Dallas—3131 Turtle Creek, Suite 222, Dallas, TX 75219 (214) 526-2222.
Los Angeles—5555 Melrose, Los Angeles, CA 90038 (213) 468-5000

INTERNATIONAL DISTRIBUTION (NY)

PRESIDENT, INTERNATIONAL SALES AND WORLDWIDE ACQUISITIONS
Bruce Gordon
DIRECTOR, OPERATIONS, INTERNATIONAL SALES
Joseph Lucas
SALES OFFICES
United Kingdom—Paramount Television Limited, 23 Berkeley House, Hay Hill, London, W1X 8JB, England. 01-499-5837
Australia—Paramount Pictures (Australia) Pty. Limited, Suite 3209, Australia Square, Box 4272, G.P.O., Sydney, 2001, N.S.W., Australia, 27-9367.
Canada—Paramount Pictures Corporation (Canada) Ltd., 146 Bloor Street West, Toronto, Ontario M5E 1M4, (416) 922-2002.
Latin America—Paramount Pictures Corp., 5555 Melrose, Los Angeles, CA., 90038 (213) 468-5000

PARAMOUNT VIDEO (CA)

PRESIDENT
Mel Harris
SENIOR VICE PRESIDENT, VIDEO
Robert Klingensmith

ADMINISTRATION (CA)

SENIOR VICE PRESIDENT
Ron Nelson

PRODUCTION DIVISION

PRESIDENT
Gary Nardino
SENIOR VICE PRESIDENT, BUSINESS AFFAIRS
Richard Weston
SENIOR VICE PRESIDENT, PRODUCTION
Robert Rosenbaum
VICE PRESIDENT, CURRENT PROGRAMS
Mark Ovitz
VICE PRESIDENT, PROGRAM DEVELOPMENT
Jeffrey Benson
VICE PRESIDENT, BUSINESS AFFAIRS
Cecelia Andrews
VICE PRESIDENT, LEGAL AFFAIRS
Howard Barton
VICE PRESIDENT, FINANCE
Warren Larson
VICE PRESIDENT, COMEDY DEVELOPMENT
Lawrence Lyttle
EXECUTIVE CASTING DIRECTOR
Hoyt Bowers
EXECUTIVE DIRECTOR, PUBLICITY/ADVERTISING
Richard J. Winters

Public Broadcasting Service

475 L'Enfant Plaza, West, SW Washington, DC 20024 (202) 488-5000; New York Office: 609 Fifth Ave., New York, NY 10017 (212) 753-7373 (A non-profit

corporation providing national programming for public TV stations)
CHAIRMAN OF THE BOARD
Dallin H. Oaks

PRESIDENT
Lawrence K. Grossman
SENIOR VICE PRESIDENT/CORPORATE SECRETARY
Michael Hobbs
SENIOR VICE PRESIDENT/PTV-1
Suzanne Weil
SENIOR VICE PRESIDENT/PTV-2
Wayne Godwin
SENIOR VICE PRESIDENT/PTV-3
William T. Reed
SENIOR VICE PRESIDENT/PROGRAM SUPPORT GROUP
Peter Downey
VICE PRESIDENT/PROGRAM ADMINISTRATION
Eric Sass
SENIOR VICE PRESIDENT/GENERAL COUNSEL
Eric Smith
VICE PRESIDENT/TREASURER, FINANCE
James Klutz
SENIOR VICE PRESIDENT/MARKETING AND ADMINISTRATION
Neil Mahrer
SENIOR VICE PRESIDENT/ENGINEERING AND OPERATIONS
Daniel Wells

QM Productions

1041 No. Formosa, Hollywood, CA 90046 (213) 650-2640 (Producer of programming).
PRESIDENT
Merrill H. Karpf

RCA Corporation

30 Rockefeller Plaza, New York, NY 10020. (Set manufacturer.)
CHAIRMAN AND CHIEF EXECUTIVE OFFICER
T.F. Bradshaw
EXECUTIVE VICE PRESIDENT, RESEARCH AND ENGINEERING
William C. Hittinger
EXECUTIVE VICE PRESIDENT, VIDEO DISCS
Herbert S. Schlosser
EXECUTIVE VICE PRESIDENT, INDUSTRIAL RELATIONS
George H. Fuchs
EXECUTIVE VICE PRESIDENT, CORPORATE AFFAIRS
Kenneth W. Bilby
STAFF VICE PRESIDENT, NEWS AND INFORMATION
Henry J. Bechtold
DIRECTOR, NEWS AND INFORMATION
Howard C. Enders
DIRECTOR, NEWS AND INFORMATION
Charles J. Smith

RKO General, Inc.

1440 Broadway, New York, NY 10018; (212) 764-7000. (Group station owner.)
VICE CHAIRMAN OF THE BOARD
John B. Poor
CHAIRMAN OF THE BOARD
Thomas F. O'Neil
PRESIDENT
Frank Shakespeare
EXECUTIVE VICE PRESIDENT-FINANCE & LEGAL
Hubert J. DeLynn
SENIOR VICE PRESIDENTS
Henry V. Greene, Jr., C. Robert Manby
VICE PRESIDENTS
J. Howard Carter, Jr., Kenneth R. Frankl, Robert L. Glaser, Arnold Kaufman, Patrick J. Winkler, Robert J. Williamson, Eugene J. Caser, Lloyd D. Ruth, Henry P. Sabatell, Charles A. Carrature, John B. Fitzgerald, John J. Spencer, Shane O'Neil, Jeffrey Ruthizer
CONTROLLER
John B. Fitzgerald
ASSISTANT CONTROLLER
Gerald O'Connell
TREASURER
Hubert J. DeLynn
ASSISTANT TREASURER
Charles A. Carrature
SECRETARY & GENERAL COUNSEL
Kenneth R. Frankl
ASSISTANT SECRETARIES
Patrick J. Winkler, Joseph Shortall, Oliver J. Janney

RADIO DIVISION-PRESIDENT
Robert J. Williamson
TELEVISION DIVISION—PRESIDENT
Robert L. Glaser
VICE PRESIDENT & GENERAL MANAGER—WNAC-TV
Pat A. Servodidio
VICE PRESIDENT & GENERAL MANAGER—WRKO (AM)
Charles Goldmark
VICE PRESIDENT & GENERAL MANAGER—KHJ (AM)
Neal Rocoff
VICE PRESIDENT & GENERAL MANAGER—KRTH (FM)
Allan Chlowitz
VICE PRESIDENT & GENERAL MANAGER—KHJ-TV
Charles Velona
VICE PRESIDENT & GENERAL MANAGER—WOR-TV
Robert Fennimore
VICE PRESIDENT & GENERAL MANAGER—WOR (AM)
Robert Devlin
VICE PRESIDENT & GENERAL MANAGER—WXLO (FM)
Lee Simonson
VICE PRESIDENT & GENERAL MANAGER—WHBQ-TV
Alex Bonner
VICE PRESIDENT & GENERAL MANAGER—WHBQ (AM)
Ron Thompson
PRESIDENT OF THE FM RADIO DIVISION & GENERAL MANAGER—WGMS/WGMS-FM
Jerry R. Lyman
VICE PRESIDENT & GENERAL MANAGER—KFRC (AM)
Patrick Norman
VICE PRESIDENT & GENERAL MANAGER—WAXY (FM)
Douglas Donoho
VICE PRESIDENT & GENERAL MANAGER—WFYR (FM)
James Barker
VICE PRESIDENT & GENERAL MANAGER—WROR (FM)
Thomas Baker
PRESIDENT—RKO RADIO SALES DIVISION
William J. Hogan

RKO RADIO NETWORKS

PRESIDENT
Thomas F. Burchill

Sony Corp. of America

9 West 57th St., New York, NY 10019. (212) 371-5800. (Set manufacturer)
CHAIRMAN
Kazuo Iwama
PRESIDENT
H. Schein
SENIOR VICE PRESIDENTS
Robert E. Dillon, Shiro Yamada
EXECUTIVE VICE PRESIDENT
Kenji Tamiya
SR. VICE PRESIDENT, MARKETING
R. Steiner
VICE PRESIDENT, FINANCE & ADMINISTRATION
R. Dillon, Jr.
VICE PRESIDENT, VIDEO PRODUCTS
R. O'Brion
VICE PRESIDENT, TECHNICAL OPERATIONS
R. Traeger
VICE PRESIDENT, BUSINESS PLANNING & DEVELOPMENT
M. Fink
VICE PRESIDENT, San Diego Plant
S. Yamada

Spelling-Goldberg Productions

c/o 20th Century-Fox Film Corp. P.O. Box 900, Beverly Hills, CA 90213 (213) 203-2021. (Program producers.)
EXECUTIVE PARTNERS
Aaron Spelling
Leonard Goldberg

Storer Broadcasting Company

1177 Kane Concourse, Miami Beach, FL; 33154; (305) 866-0211. (Group station owner.)
CHAIRMAN OF THE BOARD AND CHIEF FINANCIAL POLICY OFFICER
Bill Michaels
VICE CHAIRMAN OF THE BOARD AND CHIEF EXECUTIVE OFFICER
Peter Storer
PRESIDENT AND CHIEF OPERATING OFFICER
Terry Lee

PRESIDENT, TELEVISION STATIONS DIVISION
Ken Bagwell
VICE PRESIDENT, TREASURER AND CHIEF FINANCIAL ADMINISTRATIVE OFFICER
Kenneth R. Mosher
VICE PRESIDENT, COUNSEL AND SECRETARY
Abiah A. Church
VICE PRESIDENT, ADVERTISING AND PUBLIC RELATIONS
Bob Edell
PRESIDENT, CABLE DIVISION
Arno W. Mueller
VICE PRESIDENTS
Bill Flynn
Robert J. McBride
Henry Davis
Paul Raymon
Joseph Dimino
James Haviland
H. W. Ray
Warren Zwicky
Leonard Spragg
James Long
Andy Potos

T.A.T. Communications Company

1901 Avenue of the Stars, Los Angeles, CA 90067 (213) 553-3600. Motion picture and TV production; TV distribution)
CHAIRMAN & CHIEF EXECUTIVE OFFICER
Jerry Perenchio
PRESIDENT & CHIEF OPERATING OFFICER
Alan Horn
PRESIDENT, SYNDICATION
Robin French
EXECUTIVE VICE PRESIDENT
Charles Weber

Taft Broadcasting Co.

1906 Highland Ave., Cincinnati, OH 45219. (513) 721-1414. (Group stations owner)
CHAIRMAN OF THE BOARD
Charles S. Mechem, Jr.
VICE CHAIRMAN
David S. Ingalls
PRESIDENT
Dudley S. Taft
EXECUTIVE VICE PRESIDENT, TV GROUP
Ro Grignon
VICE PRESIDENT, SALES
Donald L. Chapin
VICE PRESIDENT, TELEVISION NEWS
Terrence J. Connelly
VICE PRESIDENT, TELEVISION OPERATIONS
Edward E. Herbert
VICE PRESIDENT, BUSINESS AFFAIRS
Jerome L. Kersting
VICE PRESIDENT, TELEVISION ENGINEERING
John R. Owen
VICE PRESIDENT, TELEVISION AND CABLE PROGRAMMING
Lucille S. Salhany

Taft Entertainment Company, The

10100 Santa Monica Blvd., Suite 2440 Los Angeles, CA 90067. (213) 557-0388.
PRESIDENT & CHIEF EXECUTIVE OFFICER
Sy Fischer
EXECUTIVE VICE PRESIDENT
Karl Honeystein
SENIOR VICE PRESIDENT, CREATIVE
Joel Cohen

DIVISIONS INCLUDE:

Cine Guarantors, 6311 Romaine St., Suite 7204, Hollywood, CA 90038. (213) 567-3006.

Hanna-Barbera Productions, Inc., 3400 Cahuenga Blvd., Hollywood, CA 90068. (213) 851-5000.

QM Productions, 1041 N. Formosa, Hollywood, CA 90046. (213) 650-2674.

Ruby-Spears Enterprises, 3459 Cahuenga Blvd., Hollywood, CA 90068. (213) 874-5100.

The Sy Fischer Company, Inc., 10100 Santa Monica Blvd., Suite 2440, Los Angeles, CA 90067. (213) 557-0388.

One East 57th Street, New York, NY 10022. (212) 486-0426.

Taft Merchandising Group, 660 Madison Avenue, 3rd Floor, New York, NY 10021 and 1506 Highland Ave, Cincinnati, OH 45219. (513) 421-9611.

Taft Pictures International, 10100 Santa Monica Blvd., Suite 2440, Los Angeles, CA 90067. (213) 557-0388 and P.O. Box 3039, Park City, VT 84060. (801) 649-7300.

Titus Productions, 211 East 51st St., New York, NY 10022. (212) 752-6460.

Tandem Productions

1901 Ave. of the Stars, Los Angeles, CA 90067. Tel.: (213) 553-3600. (Producers and distributors of TV programs)
OWNERS
Alan Yorkin and Norman Lear
CHAIRMAN AND CHIEF EXECUTIVE OFFICER
Jerry Perenchio
PRESIDENT & CHIEF OPERATING OFFICER
Alan Horn
EXECUTIVE VICE PRESIDENT
Arthur Warshaw
SENIOR VICE PRESIDENT, FINANCE
Daryl Egerstrom

Twentieth Century-Fox Television, Inc.

P.O. Box 900, Los Angeles, CA 90213; (213) 277-2211. A division of Twentieth Century-Fox Film Corp.
CHAIRMAN OF THE BOARD
Harris Katleman
GROUP EXECUTIVE VICE PRESIDENT
Edward B. Gradinger
EXECUTIVE VICE PRESIDENT, PRODUCTION
Andrea Bayne
VICE PRESIDENT, DEVELOPMENT
Peter Grad
VICE PRESIDENT AND EXECUTIVE PRODUCER SPECIAL PROJECTS
Barry Lowen
VICE PRESIDENT, CREATIVE AFFAIRS
Tony Spinner
VICE PRESIDENT, CURRENT PROGRAMS
Lea Stalmaster
VICE PRESIDENT, DAYTIME, SYNDICATED PROGRAMS & SPECIALS
George Paris
VICE PRESIDENT, TV MOVIES & MINI-SERIES
Richard Rosetti
VICE PRESIDENT, BUSINESS AFFAIRS
Dayna A. Kalins
VICE PRESIDENT, STUDIO BUSINESS AFFAIRS
Lee Zuckerman
VICE PRESIDENT, NETWORK BUSINESS AFFAIRS
Allan L. Rice
VICE PRESIDENT, FINANCE
George Sefeotis
VICE PRESIDENT, PRODUCTION MANAGEMENT
Mark Evans
VICE PRESIDENT, TV POST PRODUCTION
Joseph Silver
SR. VICE PRESIDENT SYNDICATION
Robert Morin
SR. VICE PRESIDENT, INTERNATIONAL SYNDICATION
Richard Harper
VICE PRESIDENT, DOMESTIC SYNDICATION
Steven R. Orr
VICE PRESIDENT AND GENERAL SALES MANAGER (NEW YORK)
Joe Greene
VICE PRESIDENT, ADVERTISING, PUBLIC RELATIONS, PROMOTION
Jerry Greenberg
DIRECTOR OF PROGRAM DEVELOPMENT
Arvin Kaufman
DIRECTOR OF CURRENT PROGRAMS
Ruth Slawson
DIRECTOR, DAYTIME & SYNDICATED PROGRAM DEVELOPMENT
Craig Kellem
DIRECTOR, MARKETING, ADVERTISING AND PROMOTION
Jeff Shadlow
DIRECTOR OF BUSINESS AND SALES ADMINISTRATION
Stanely DeCovnick
ASSOC. DIRECTOR, MOVIES & MINI-SERIES
Harriet Brown
ASSOCIATE DIRECTOR, BUSINESS AFFAIRS, SYNDICATION
Michael Shapiro

WESTERN DIVISION SALES MANAGER
David Skillman
DIRECTOR OF PRODUCTION, DAYTIME & SYNDICATED TELEVISION
Robert Braitwaite
TV BUDGET AND COST ADMINISTRATOR
Ed Lahti
DIRECTOR, SALES ADMINISTRATION, INTERNATIONAL SYNDICATION
Matt Barbera
DIRECTOR OF OPERATIONS, INTERNATIONAL SYNDICATION
Edwin Greenberg
DIRECTOR OF SERVICING, INTERNATIONAL SYNDICATION
Lourdes Chaves
ASSISTANT TO VICE PRESIDENT, PRODUCTION MANAGEMENT
Ted Butcher
DIRECTOR OF PLANNING AND ADMINISTRATION
Laurie Lindner

BRANCH OFFICES:

- NEW YORK OFFICE: 40 West 57th Street, New York, New York 10019 (212) 977-5500; Joseph Greene, Vice President and General Sales Manager; James Puffer, Northeastern Division Sales Manager; Harry Mulford, Director, Advertising Sales.
- CHICAGO OFFICE: 35 East Wacker Drive, Suite 2160, Chicago, Illinois 60601 (312) 372-1589; J. Stuart Stringfellow, Central Division Sales Manager; Dennis Juravic, Central Division Sales Representative.
- DALLAS OFFICE: 11551 Forest Central Drive, Suite 300, Dallas, Texas 75243 (214) 343-9252; Tom Maples, Southwestern Division Sales Manager.
- ATLANTA OFFICE: 2200 Century Parkway, Suite 560, Atlanta, Georgia 30345 (404) 634-0011; Tony Bauer, Southeastern Division Sales Manager.
- LOS ANGELES OFFICE: Box 900, Beverly Hills, California 90213 (213) 277-2211; David Skillman, Western Division Sales Manager.

FOREIGN SALES OFFICES

Twentieth Century-Fox Corp. Ltd., 45 Charles Street East, Sixth Floor, Toronto M4Y, 1S2, Ontario, Canada. Lawrence Keller, Sales Manager, English Markets.

Twentieth Century-Fox Television, 1440 St. Catherine Street West, Suite 616, Montreal, Quebec, Canada. Gerald Ross, Sales Manager, French Canada.

Twentieth Century-Fox France, Inc., 114 Rue La Boetie, 75008 Paris, France. William Saunders, Vice President—United Kingdom, Europe, Middle/Near East and Africa.

Twentieth Century-Fox Film Company, Ltd., 31-32 Soho Square, London W. 1, England. Michael Doury, Sales Representative—United Kingdom, Scandinavia.

Twentieth Century-Fox Film Corp. (Australia), P.O. Box Q301, Queen Victoria Building, Sydney 2000, Australia. Peter Broome, Vice President—Australia, Far East.

Twentieth Century-Fox Television Japan, Fukide Building, 1-13 Toranomon 4—Chome Minato-Ku, Tokyo 105, Japan. Goro Uzaki, Sales Representative, Japan.

Fox Interamericana, S.A., Praga 56-202, Mexico, 6 D.F. Jose Luis Cano, Sales Manager—Mexico and Specific Spanish Language Markets.

Fox Film do Brazil—TV Division, Rua Dr. Costa Jr 230 C.E.P., 05002 Sao Paulo, Brazil. Elie Wahba, Vice President, Latin America.

United Artists Television, Inc.

729 Seventh Avenue, New York, NY 10019; (212) 575-3000. (Distributor of feature films: producer and dist. of filmed series & special programs.)

PRESIDENT & CHIEF EXECUTIVE OFFICER
Martin J. Robinson
VICE PRESIDENT-MARKETING
Andy Coscia
VICE PRESIDENT & TREASURER
Ben Broyde
CONTROLLER
James Waltz
GENERAL MANAGER, STATION SERVICES
Don Meyer
DIRECTOR, CONTRACT & TV SYSTEMS
Daniel J. Lobel
DIRECTOR, RESEARCH & PRINT ADMINISTRATION
Lester Furman
DIRECTOR, TECHNICAL SERVICES
Joseph Trentin
MANAGER, SALES ADMINISTRATION
Ellen Dana
MANAGER, SALES TRAFFIC
Noami Tomesanyi
DIRECTOR, ADVERTISING & PROMOTION
Helen Killeen
SALES DIRECTORS & OFFICES

Central Division: Selwyn Ginsler, 520 North Michigan Avenue, Chicago, IL 60611. (312) 527-1030.

Eastern Division: Paul Kalvin, 729 7th Avenue, New York, NY 10019. (212) 575-4739.

Northeastern Division: 729 7th Avenue, New York, NY 10019. (212) 575-4741.

Western Division: 10202 West Washington Blvd., Culver City, CA 90230. (213) 559-3450.

Southwestern Division: Fred A. Watkins, P.O. Box 6095, Cedar Grove Station, Shreveport, LA 71106. (318) 865-3144.

Canada: Bill Wineberg, gen. mgr., UATV of Canada, 2 Carlton Street, Toronto 2, Ontario, Canada.

Print & Station Service Center: Don Meyer, general manager in chg of station services, 2904 Woodburn Avenue, Cincinnati, OH 45206. (513) 751-1323.

UNITED ARTISTS TELEVISION INT'L INC.

729 Seventh Ave., New York, NY 10019. (212) 575-4785.

VICE PRESIDENT & DIRECTOR OF INTERNATIONAL OPERATIONS
Raul Lefcovich
FOREIGN OFFICES:

Brazil—Rua Senador Dantas, 73-3 Andar ZC-06, 20.000 Rio de Janeiro, Brazil, Francisco Serrador, Director of Brazil. (224-5322).

Germany—Shesemannallee 13, Postfach 16 740, 6000 Frankfurt/Main 70, Germany. (6102/53548).

France—25-27 Rue d'Astorg, 75008 Paris, France, Mary Jane Fourniel, Dir. of France & French Speaking Terr. (265-45-90).

Mexico—Presidente Mazarik 61 3er. Piso, Mexico 5, D.F., Gustavo Montaudon, Dir. Latin America. (250-16 99).

United Kingdom—37-41 Mortimer Street, London W1A 2JL, England, Peter Pawsey, Dir. of Europe, Africa & M.E. (01-636 1655).

Australia—Town Hall House, 456 Kent Street, Sydney, N.S.W. 2000, Australia. Peter Press, Director of Far East. (328-1179).

Japan—No. 15 Mori Building, 8-10, Toranomon 2-chome, Minato-ku,Tokyo, 105 Japan. (503-3628; 503-3629).

Colombia—Apartado Aereo 91 188, Bogota 8, Colombia, Juan Jorge Jaeckel. (257-85-18).

Italy—Via Banchi Vecchi 41, Rome, Italy, Lionello Torossi. (565-766).

Spain—VIDEOPROGRAMA S.A., Martires de Alcala 4, Madrid 8, Spain, Jose Augustin. (247-6253).

Kuwait—RAWFID COMMERCIAL ESTABLISHMENT, Lanabour-Cr. 16121, P.O. Box Safat 3309, Kuwait, Sami Khattab.

Greece—United Artists of Greece, E.P.E., Ganvetta Street, No. 4, Athens, 142, Greece. Yanni Tacaziadis

Universal Television

(A division of Universal City Studios Inc, a subsidiary of MCA Inc.)

Universal City Studios, Universal City, CA 91608 (213) 985-4321. (Producer of TV programs)

PRESIDENT
Robert A. Harris
VICE PRESIDENTS
Burt Astor, Earl Bellamy, Gerry Gottlieb, Milt Hamerman, Stefanie Kowal, Richard Lindheim, Kerry McCluggage, Paul Miller, Joni Robbins.
VICE PRESIDENT, PUBLICITY, PROMOTION AND ADVERTISING
Ben Halpern

Warner Bros. Television

4000 Warner Blvd., Burbank CA 91522 (213) 843-6000. 75 Rockefeller Plaza, New York, NY 10019 (212) 484-8000.

TELEVISION PROGRAMMING

PRESIDENT
Alan Shayne
EXECUTIVE VICE PRESIDENT
Barry M. Meyer
SENIOR VICE PRESIDENT, CREATIVE AFFAIRS
Franklin Barton
SENIOR VICE PRESIDENT, BUSINESS AFFAIRS
Art Stolnitz
SENIOR VICE PRESIDENT IN CHARGE OF PRODUCTION
Richard Kobritz
VICE PRESIDENT, PUBLICITY, ADVERTISING & PROMOTION
David Horowitz

VICE PRESIDENT CURRENT PROGRAMMING
David Sacks
VICE PRESIDENT, MOVIES AND MINI-SERIES
Chuck McLain
VICE PRESIDENT, BUSINESS AFFAIRS
Johnna Levine
DIRECTOR, PROGRAM DEVELOPMENT
Jane Auerbach
DIRECTOR OF COMEDY DEVELOPMENT
Diana Dreiman
DIRECTOR, LEGAL AFFAIRS
Daniel J. Schiffer
DIRECTOR, STORY DEPT.
Gus Blackmon
VICE PRESIDENT POST-PRODUCTION
Fred Talmage
VICE PRESIDENT, PRODUCTION
Robert Doudell
DIRECTOR, CASTING
Vivian McRae
DIRECTOR, ESTIMATING
Rosalee Berman
DIRECTOR, POST-PRODUCTION
Karen Pingatore
VICE PRESIDENT, LEGAL AFFAIRS
Milton Segal
DIRECTOR, EAST COAST TELEVISION
Elaine Cohen

NETWORK FEATURES, PAY TV, & ANIMATION

EXECUTIVE VICE PRESIDENT
Edward Bleier
VICE PRESIDENT PAY TV MARKETING
Stanley Solson
VICE PRESIDENT NETWORK RELATIONS
Lorna Bitensky
EXECUTIVE PRODUCER, VICE PRESIDENT WARNER BROS. CARTOONS
Hal Geer
SENIOR EXECUTIVE PRODUCER, ANIMATION
Friz Freleng
DIRECTOR OF ANIMATION PROGRAMMING
Jean MacCurdy
DIRECTOR PAY TV
Adriana Perovich
MANAGER PLANNING AND DEVELOPMENT
Charles Jeffrey Calman
MANAGER ADVERTISING/PUBLICITY
Eric D. Frankel
MANAGER CLIENT SERVICES
Margaret Jelcich
MANAGER RESEARCH AND ADMINISTRATION
Charlotte Mulford

WORLDWIDE DISTRIBUTION

PRESIDENT
Charles D. McGregor
VICE PRESIDENT, BUSINESS AFFAIRS AND ADMINISTRATION
Arthur Kananack
VICE PRESIDENT, DOMESTIC SALES MANAGER
George Mitchell
VICE PRESIDENT, INTERNATIONAL SALES & ADMINISTRATION
John E. Whitesell
VICE PRESIDENT, ADVERTISING, PROMOTION & PUBLICITY
Gordon A. Hellmann
DIRECTOR, RESEARCH
Paul Simon
DIRECTOR, ADVERTISING, PROMOTION AND PUBLICITY
Joel Kaplan
DIRECTOR, FINANCIAL ADMINISTRATION
John Chickering
DIRECTOR OF PRODUCT COORDINATION
Dee Eulberg

630 Ninth Avenue, New York, NY 10036; (212) 484-8000.
MANAGER, TECHNICAL SERVICES
Erwin Markisch
MANAGER, BOOKING SERVICES
Clara Falkoff
MANAGER, PROGRAM INFORMATION
Louis B. Marino

DOMESTIC BRANCHES

New York: 57 Spring Court, Syosset, New York 11791; (516) 364-9198.
VICE PRESIDENT & EASTERN SALES DIRECTOR
Peter Affe

Covington, La.: Lowe Davis Rd. P.O. Box 999, (504) 892-2703.
SOUTHEASTERN SALES DIRECTOR
Bill Seiler

Los Angeles: 5161 W 58th Place, Los Angeles, CA 90056; (213) 843-6000.
WESTERN SALES DIRECTOR
Ed Donaldson

Sun Prairie: 360 Broadway Ave., Sun Prairie, WI 53590. (608) 249-9100.
NORTH CENTRAL SALES DIRECTOR
Robert Mitchell

Ft. Mitchell: 3058 Brookwood Circle, Ft. Mitchell, KY 41017; (606) 341-2585.
MIDWESTERN SALES DIRECTOR
John H. Louis

Westinghouse Broadcasting Co. (Group W)

(Subsidiary of Westinghouse Electric Corp.) 90 Park Ave., New York, NY; 983-6500.
PRESIDENT AND CHAIRMAN OF THE BOARD
Donald H. McGannon
PRESIDENT, TV STATION GROUP
Winthrop Baker
PRESIDENT, RADIO STATION GROUP
Richard Harris
CHAIRMAN, GROUP W PRODUCTIONS
David Salzman
SENIOR VICE-PRESIDENT
Joseph K. Mikita, Financing
AREA VICE-PRESIDENTS
Lamont L. Thompson
Ed Wallis

William Morris Agency, Inc.

1350 Avenue of the Americas, New York, NY 10019; 586-5100. Beverly Hills Office 151 El Camino; 274-7451. Nashville Office: 2325 Crestmoor Rd., 385-0310.
CHAIRMEN OF THE BOARD, EMERITI
Abe Lastfogel (Beverly Hills)
Nat Lefkowitz (New York)
CHAIRMAN OF THE BOARD
Morris Stoller (Beverly Hills)
PRESIDENT
Sam Weisbord (Beverly Hills)
EXECUTIVE VICE PRESIDENTS
(Beverly Hills)
Norman Brokaw
Roger Davis
Tony Fantozzi
Stan Kamen
Walter Zifkin
(New York)
Lee Stevens
Lou Weiss
VICE PRESIDENT, TREASURER
Larry Lewis (New York)

MOTION PICTURE DEPARTMENT
Beverly Hills—Flo Allen, Allan Badiner, Ed Bondy, Dennis Brody, Martin Caan, Roger Davis, Debra Greenfield, Annette Handley, Leonard Hirshan, Joan Hyler, Stan Kamen, Ed Limato, Gary Lucchesi, Ron Mardigan, Peter Meyer, Ed Mitchell, Mike Peretzian, John Ptak, Arnold Stiefel, Peter Turner, Irene Webb, Mike Zimring.
New York—Martin Bauer, George Lane, Carmen La Via, Biff Liff, Fred Milstein, Gilbert Parker, Johnnie Planco, Ed Robbins, Janet Roberts, Katie Rothacker, Esther Sherman, Elle Shushan, Lee Stevens.

TELEVISION DEPARTMENT
Beverly Hills—Larry Auerbach, Arthur Axelman, Jeralyn Badgley, Alan Berger, Norman Brokaw, Bruce Brown, Lee Cohen, Robert Crestani, Ruth Englehardt, Tony Fantozzi, Bob Goodman, Sam Haskell, Martin Hurwitz, Alan Iezman, Jerry Katzman, Steve Konow, Sol Leon, Debbie Miller, Marc Pariser, Pam Prince, Joe Rivkin, John Rosenmayer, Hal Ross, David Schiff, Michael Shaw, Steve Weiss, Fred Westheimer, Jeff Witjas, Paul Yamamoto.
New York—Lucy Aceto, Leo Bookman, Art Fuhrer, Jim Griffin, Leon Memoli, Jonathan Russo, James Sarnoff, Lee Stevens, Lou Weiss, Ron Yatter.

Zenith Radio Corporation

1000 Milwaukee Ave., Glenview, IL., 391-7000. (Set manufacturer.)

CHAIRMAN OF THE BOARD
Joseph S. Wright
PRESIDENT AND CHIEF EXECUTIVE OFFICER
Revone W. Kluckman
EXECUTIVE VICE PRESIDENT, SALES AND MARKETING
Walter C. Fisher
EXECUTIVE VICE PRESIDENT & GENERAL MANAGER, PICTURE TUBE AND COMPONENTS OPERATIONS
Karl H. Horn
SENIOR VICE PRESIDENT AND GENERAL MANAGER, BLACK-AND-WHITE TV, AUDIO AND VIDEO DEVICES OPERATIONS
Joseph P. Fiore
SENIOR VICE PRESIDENT AND GENERAL MANAGER, COLOR TELEVISION OPERATIONS
Robert B. Hansen
SENIOR VICE PRESIDENT, ADMINISTRATION AND PUBLIC AFFAIRS
Eugene M. Kinney
SENIOR VICE PRESIDENT, FINANCE
Jerry K. Pearlman
VICE PRESIDENT, GENERAL COUNSEL
Philip J. Curtis
VICE PRESIDENT, HUMAN RESOURCES
David W. Denton
VICE PRESIDENT, RESEARCH AND DEVELOPMENT, PICTURE TUBE & COMPONENTS OPERATIONS
Leonard Dietch
VICE PRESIDENT, SALES
J. D. Dougherty
VICE PRESIDENT, MARKETING
John L. McCallister
VICE PRESIDENT, COLOR TELEVISION MANUFACTURING
Melvin L. Moore
VICE PRESIDENT, AND GENERAL MANAGER, COLOR TELEVISION ENGINEERING
Carl A. Michelotti
VICE PRESIDENT, COLOR TELEVISION MANUFACTURING
Melvin L. Moore
VICE PRESIDENT AND GENERAL MANAGERS, VIDEO OPERATIONS
George A. Schupp
VICE PRESIDENT, PURCHASING
James A. Stark
VICE PRESIDENT AND GENERAL MANAGER, PICTURE TUBE OPERATIONS
A. John Torre
CONTROLLER
William A. Krein
SECRETARY
Dennis J. Waldeck
TREASURER
David S. Levin
DIRECTOR, PUBLIC RELATIONS
William A. Nail

Cable Television

* STATISTICS AND HISTORY

* MAJOR CABLE SYSTEM OPERATORS

* MAJOR PROGRAM SUPPLIERS

Cable Television

STATISTICS

HISTORICAL DATES

FIRST OVER-THE-AIR TOLL-TV EXPERIMENTS: 1950, Skiatron system tested over WOR-TV in New York City. 1951, Telemeter system tested over KTLA-TV, Los Angeles; Zenith tested its system over its own experimental station in Chicago.

FEBRUARY 10, 1955: Federal Communications Commission started proceedings to determine whether it should authorize TV stations to transmit programs paid for on a subscription basis. Study involved 92 docket volumes reflecting the views of over 25,000 persons on legal, technical and policy questions.

MAY 23, 1957: FCC concluded it had authority to authorize use of TV broadcast frequencies for subscription operations if it found it were in the public interest to do so. Did not define whether this service should be classified as "broadcasting" or some other.

OCTOBER 17, 1957: FCC adopted its first report on the proceedings and declared it would consider applications to conduct trial pay-TV operations over TV stations provided certain conditions were met.

MARCH 24, 1959: FCC issued third report saying it was ready to consider any pay-TV application by a commercial TV station conforming with requirements, among which were limitation of three-year trial authorizations to markets where there were at least four commercial TV stations (including the pay-TV one), and one trial system allowed per market.

JUNE 22, 1960: Hartford Phonevision Co. (later RKO General Phonevision Co.), a licensee of WHCT (Channel 18), Hartford, CT, applied for authority to conduct a three-year trial subscription-TV operation. Application granted February 23, 1961. Operations began June 29, 1962, and concluded in January, 1969.

JULY 3, 1967: FCC's subscription TV committee submitted a fourth report proposing an over-the-air subscription TV service. Rules were designed to integrate subscription TV into the total TV system so good programming would continue to be available over regular TV stations.

DECEMBER 12, 1968: FCC formally established over-the-air subscription TV as a regular broadcast service.

Statistics for 1979 (supplied by FCC)

Number of cable companies: 2,992.
Number of communities served: 8,500
Subscribers (homes served): 15,760,000 (up almost 12 per cent over the 14.11 million in 1978)
Financial: Revenues totalled $1,800,000,000 for 1979, up 20 per cent from previous year. Industry assets totalled $3,200,000,000, up 12 per cent. Expenses in 1979 climbed 20 per cent, from $918,000,000 to $1,100,000,000.
Average subscription rate for pay-cable: $8.73 a month, compared with average $7.37 a month for basic service. In 1978 these figures were $8.60 and $7.03, respectively.
Average cable company is small: 5,250 subscribers and revenues of $607,000.

State	Number of CATV Systems	Number of Communities Served	Total CATV Subscribers	Average Monthly Rate	Average Installation Rate
Alabama	55	117	272,344	$ 7.52	$18.17
Alaska	10	13	13,779	18.35	23.50
Arizona	38	67	75,701	8.49	20.76
Arkansas	62	105	173,548	6.49	14.97
California	168	479	1,700,164	7.34	15.96
Colorado	33	77	98,316	7.61	17.85
Connecticut	16	54	180,602	8.41	20.46
Delaware	8	42	98,516	7.61	23.11
Florida	84	307	606,383	7.38	19.98
Georgia	61	149	319,645	7.19	15.03
Hawaii	8	16	57,165	6.83	32.46
Idaho	22	47	46,328	7.40	22.41
Illinois	72	181	386,549	7.03	16.43
Indiana	52	113	221,705	6.65	16.26
Iowa	35	77	154,065	7.34	14.81
Kansas	83	122	205,687	7.49	12.50
Kentucky	104	266	199,753	6.65	17.38
Louisiana	36	80	259,034	7.21	13.81
Maine	19	61	83,553	7.02	14.78
Maryland	25	93	79,041	7.12	16.12
Massachusetts	34	98	272,651	7.45	17.66
Michigan	56	245	355,332	6.35	16.36
Minnesota	87	135	172,571	7.13	16.18
Mississippi	49	90	190,754	7.03	15.67
Missouri	64	110	171,312	6.35	13.58
Montana	27	53	95,496	8.47	20.27
Nebraska	32	52	83,241	7.22	12.38
Nevada	9	14	36,051	6.99	13.20
New Hampshire	16	55	75,430	6.74	18.47
New Jersey	33	215	452,022	6.50	20.42
New Mexico	29	61	115,800	7.91	15.79
New York	119	544	971,729	6.36	16.46
North Carolina	45	92	255,031	6.93	13.17
North Dakota	28	33	41,203	7.61	15.75
Ohio	116	491	725,684	6.81	14.45
Oklahoma	78	92	196,522	7.03	11.77
Oregon	59	128	159,074	7.40	18.95
Pennsylvania	179	1,172	1,107,301	6.06	15.58
South Carolina	31	71	141,690	7.74	15.44
South Dakota	14	27	50,849	7.25	13.50
Tennessee	44	101	153,838	7.01	13.63
Texas	179	299	735,465	6.84	14.45
Utah	3	7	16,338	7.55	15.62
Vermont	18	55	49,403	6.13	18.16
Virginia	55	128	191,653	6.83	14.12
Washington	62	207	346,281	7.71	18.91
West Virginia	87	271	197,749	6.56	16.06
Wisconsin	38	99	143,823	7.18	15.13
Wyoming	23	50	86,233	7.48	19.99
Territories	5	9	35,400	11.90	22.25
Totals:	2610	7470	12,857,804	7.03	16.35
Totals (projected to 100%)	2865	8200	14,113,945	7.03	16.35

(Figures for 1978 from FCC)

Cable System Operators

American TV & Communications Corp.

20 Inverness Place East, Englewood, CO 80112. (303) 773-3411.
PRESIDENT & CHIEF EXECUTIVE OFFICER
Monroe M. Rifkin
VICE PRESIDENT/EASTERN OPERATIONS
Joseph J. Collins
VICE PRESIDENT/WESTERN OPERATIONS
Michael J. McCrudden
VICE PRESIDENT/PLANNING & DEVELOPMENT
David D. Kinley
VICE PRESIDENT/MARKETING
Trygve E. Myhren

Cablecom-General, Inc.

8880 E. Arapahoe Re., Box 1818, Englewood, CO 80150. (303) 770-7500.
CHAIRMAN, PRESIDENT & CHIEF EXECUTIVE OFFICER
Richard A. Forsling
VICE PRESIDENT/OPERATIONS
Charles F. Caldara

Cablevision Systems Development Co.

366 N. Broadway, Jericho, NY 11753. (516) 822-4774.
PRESIDENT
Charles P. Dolan
VICE PRESIDENT
John Tatta

Comcast Corp.

1 Belmont Ave., Bala Cynwyd, PA 19440. (215) 667-4200.
PRESIDENT
Ralph J. Roberts
VICE PRESIDENTS
Daniel Aaron, Joseph G. Kennelly, D. Gerald Scott

Continental Cablevision, Inc.

54 Lewis Wharf, Boston, MA 02110. (617) 742-8265.
PRESIDENT
H. I. Grousbeck
EXECUTIVE VICE PRESIDENT & TREASURER
Amos B. Hostetter, Jr.
VICE PRESIDENT/CORPORATE DEVELOPMENT
William E. Clancy
VICE PRESIDENT/MARKETING
Thomas P. Willett

Cox Cable Communications

53 Perimeter Center East, Suite 300, Atlanta, GA 30346. (404) 393-0480
CHAIRMAN
J. Leonard Reinsch
PRESIDENT
Henry W. Harris
VICE PRESIDENTS
Thurber M. Foreman III, G. Lewis Davenport, William A. Pitney, Donald O. Williams
CORPORATE COMMUNICATIONS DIRECTOR
Thomas F. McGuire

General Electric Cablevision Corp.

1400 Balltown Rd., Schenectady, NY 12309 (518) 385-1321.
PRESIDENT
Reid L. Shaw
VICE PRESIDENT/GENERAL MANAGER
A. C. "Sam" Belanger
ENGINEERING MANAGER
Frank C. Baxter
MARKETING MANAGER
Paul F. Schonewolf

Heritage Communications, Inc.

2195 Ingersoll Ave., Des Moines, IA 50312. (515) 245-7585.
PRESIDENT
James M. Hoak, Jr.
EXECUTIVE VICE PRESIDENT/DEVELOPMENT
James S. Cownie
VICE PRESIDENT/ENGINEERING
Arthur C. Hutzler
VICE PRESIDENT/PUBLIC RELATIONS
William J. Riley
VICE PRESIDENT/MARKETING
J. Randall Steward

Liberty Communications, Inc.

2225 Coburg Rd., Eugene, OR 97401. (503) 485-5611.
PRESIDENT
1 Donald E. Tykeson
VICE PRESIDENTS
Carolyn S. Chambers, R. R. Rickers
DIRECTOR OF ENGINEERING
Jerome E. Partch
DIRECTOR OF CORPORATE DEVELOPMENT
Rex Bradley

Midwest Video Corp.

860 Tower Bldg., Little Rock, AR 72201. (501) 375-8885.
PRESIDENT
G. R. Morrell
VICE PRESIDENT
Lloyd B. Sands

New Channels Corp.

3 Northern Concourse, North Syracuse, NY 13212 (315) 455-5826.
PRESIDENT
E. R. Vadeboncoeur
FIRST VICE PRESIDENT/GENERAL MANAGER
Joel Fleming
EXECUTIVE VICE PRESIDENT
Robert Miron
SYSTEMS DIRECTOR
Leo A. Calistri

Sammons Communications, Inc.

Box 225728, Dallas, TX 75265. (214) 670-9599.
PRESIDENT
James Whitson
VICE PRESIDENT/CORPORATE DEVELOPMENT
William B. Strange
VICE PRESIDENT/WESTERN OPERATIONS
C. Erwin Sharp
VICE PRESIDENT/EASTERN OPERATIONS
Mark S. Weber

Service Electric Cable TV, Inc.

Main & Pine Sts., Mahanoy City, PA 17948. (717) 773-2585.
PRESIDENT/TREASURER
John Walson
SECRETARY
Margaret Walson

Storer Cable Communications

One Storer Ave., Thousand Oaks, CA 91360. (805) 497-3931. Eastern Headquarters: 5205 Fruitville Rd., Sarasota, FL 33578. (813) 3716700. Broadcasting Headquarters: 1177 Kane Concourse, Miami, FL 33154. (305) 866-0211.

CHAIRMAN
Bill Michaels
PRESIDENT
Peter Storer
EXECUTIVE VICE PRESIDENT
Terry H. Lee
VICE PRESIDENT/TV DIVISION
Kenneth L. Bagwell
TREASURER & VICE PRESIDENT/CATV DIVISION
Arno W. Mueller

TeleCable Corp.

Box 720, 740 Duke St., Norfolk, VA 235100. (804) 446-2000.

CHAIRMAN
Frank Batten
PRESIDENT
Richard D. Roberts
VICE PRESIDENT
E. C. Oldfield

Tele-Communications, Inc.

54 Denver Technological Center, Box 22595, Denver, CO 80222. (303) 771-8200.

CHAIRMAN
Bob Magness
VICE CHAIRMAN
George C. Hatch
PRESIDENT & CHIEF EXECUTIVE OFFICER
John C. Malone

Teleprompter Corporation

888 Seventh Avenue, New York, NY 10019; (212) 247-8700.

CHAIRMAN OF THE BOARD & CHIEF EXECUTIVE OFFICER
Jack Kent Cooke
PRESIDENT & CHIEF OPERATING OFFICER
Russell Karp
SENIOR VICE PRESIDENT & PRESIDENT OF THE CABLE TELEVISION DIVISION
William J. Bresnan
VICE PRESIDENT & GENERAL COUNSEL
Barry P. Simon
VICE PRESIDENT-FINANCE AND TREASURER
Richard M. Sykes
CONTROLLER
M. Jay Walkingshaw
SECRETARY
Bernard D. Fischman
ASSISTANT CONTROLLER
Vincent J. Bova
ASSISTANT SECRETARY
Claire Feldman
DIRECTORS
Jack Kent Cooke, Russell Karp, William J. Bresnan, Donald Ralph Cooke, Edward Patterson, James Lacher, Monte E Livingston, Philip L. Lowe, Charles Luckman, Dr Maurice B Mitchell, Dr Allen E Puckett, John H Richardson, William A Shea, Jack Wrather.

Times Mirror Cablevision

Times Mirror Square, Los Angeles, CA 90053. (213) 625-2345. (A subsidiary of the Times Mirror Co.)

PRESIDENT
Ralph J. Swett
VICE PRESIDENT
David Lewine, Richard E. Young
VICE PRESIDENT/ENGINEERING
Gilbert L. Tash
VICE PRESIDENT/OPERATIONS
William D. Guerena

Texas Community Antennas Group

3027 S.E. Loop 323, Tyler, TX 75701. (214) 595-3701.

PRESIDENT
Robert Rogers
VICE PRESIDENTS
A W. Riter, Wayne McKinney

UA—Columbia Cablevision, Inc.

49 Riverside Ave., Westport, CT 06880. (203) 227-9581.

CHAIRMAN
David H. Strassler
PRESIDENT
Robert M. Rosencrans
VICE PRESIDENTS
Kenneth S. Gunter, Steven L. Sinn
VICE PRESIDENT/OPERATIONS
Marvin Jones

United Cable TV Corp.

40 DTC East, 7995 E. Prentice Ave., Englewood, CO 80111. (303) 779-5999.

PRESIDENT & CHIEF EXECUTIVE OFFICER
Gene W. Schneider
CHAIRMAN
Van Irvine
VICE PRESIDENT/ENGINEERING
Richard C. Schneider
VICE PRESIDENT/OPERATIONS
Robert J. Ball

Viacom Communications

1211 Ave. of the Americans, New York, NY 10036. (212) 575-5175. (A division of Viacom)

PRESIDENT
Douglas H. Dittrich
VICE PRESIDENT/MARKETING
Samuel Street
VICE PRESIDENT/DEVELOPMENT
James Kingsdale

Vision Cable Communications, Inc.

501 Madison Ave. New York, NY 10022. (212) 759-1500.

CHAIRMAN
Sidney R. Knafel
PRESIDENT
George Lindemann
VICE PRESIDENTS
Michael S. Willner, Richard M. White

Warner Amex Cable Communications, Inc.

75 Rockefeller Plaza, New York, NY 10019. (212) 484-6711. Cable Address: Warncable.

CHAIRMAN & CHIEF EXECUTIVE OFFICER
Gustave M. Hauser
PRESIDENT & CHIEF OPERATING OFFICER
John D. Lockton

Western Communications, Inc.

Suite 225, 2855 Mitchell Dr., Box 4610, Walnut Creek CA. (415) 935-3055.

CHAIRMAN
Richard T. Thieriot
PRESIDENT
Edward M. Allen

Cable and Pay TV Program Suppliers

Cinemax

1271 Avenue of the Americas, New York, NY 10020 (212) 484-1000. A second pay-TV service created by Home Box Office, Inc. and designed to complement the foundation HBO service. (See HBO listing)

Columbia Pictures Pay Television

Columbia Plaza South, Burbank, CA 91505. (213) 954-6000.
VICE PRESIDENT & GENERAL MANAGER
Anthony J. Lynn

Entertainment & Sports Programming Network, Inc. (ESPN)

ESPN Plaza, Briston, CO 06010. (203) 584-8477. (203) 584-8477.
PRESIDENT & CHIEF OPERATING OFFICER
Chet Simmons
EXECUTIVE VICE PRESIDENT/PROGRAMMING & OPERATIONS
Scotty Connal
EXECUTIVE VICE PRESIDENT/ADMINISTRATION & MARKETING
Evan Baker
VICE PRESIDENT/AFFILIATE MARKETING
Jim Cavazzini
VICE PRESIDENT/ADVERTISING & SALES
Mike Presbrey

Hollywood Home Theatre

49 West 57th St., New York, NY 10019 (212) 750-1208.
PROGRAMMING DIRECTOR
Janet Rogers
DIRECTOR CORPORATE DEVELOPMENT
Jim Epstein

Home Box Office, Inc.

1271 Avenue of the Americas, New York, NY 10020 (212) 484-1000. (Films, sports & entertainment specials for pay TV) A subsidiary of Time, Inc. See also *Cinemax.*
CHAIRMAN OF THE BOARD
Gerald M. Levin
PRESIDENT
James O. Heyworth
SENIOR VICE PRESIDENTS
Winston H. Cox, Michael Fuchs
VICE PRESIDENT & TREASURER
Kevin Senie
VICE PRESIDENT & GENERAL COUNSEL
John Redpath

Home Theater Network, Inc.

465 Congress St., Portland, ME 04101. (207) 774-0300.
PRESIDENT
Peter M. Kendrick
VICE PRESIDENT/ENGINEERING
Raymond E. Murdough
VICE PRESIDENT/SALES
Stephen C. Broydrick
VICE PRESIDENT/ADMINISTRATION
Kathryn R. Petersonn

Optical Systems Corp.

970 A Detroit Ave., Concord, CA 94518. (415) 676-0225. (TV systems utilizing proprietary technology and specialized programming produced and developed by Optical. Operates leased channels on various cable systems)
PRESIDENT
Alan H. Greenstadt
VICE PRESIDENT/OPERATIONS
Thomas J. Zimmerman

Prism

Seven Penn Center Plaza, 17th & Market Sts., Philadelphia, PA 19103. (215) 665-9400.
PRESIDENT
Jack L. Williams
VICE PRESIDENT/MARKETING
Bruce Sellers
VICE PRESIDENT/PROGRAMMING
Sam Schroeder
DIRECTOR CORPORATE DEVELOPMENT
Jim Epstein

Pay Television Corp.

390 Plandome Rd., Manhasset, NY 11030. Tel.: (516) 627-7440. (Holds exclusive worldwide rights to operate and franchise others to use subscription TV systems developed by Zenith Radio Corp., Pay Television Corp. and System Development Corp.
PRESIDENT
George F. Wiemann

Satori Productions

250 W. 57th St., New York, NY 10019. (212) 581-8450.
PRESIDENT
Ernest G. Sauer
VICE PRESIDENT
Gary P. Connor

Showtime Entertainment, Inc.

1633 Broadway, New York, NY 10019 (A joint venture of Viacom International Inc. and Teleprompter Corporation which supplies films, entertainment specials and marketing services to the pay-cable industry)
PRESIDENT
Mike Weinblatt
SENIOR VICE PRESIDENT, PROGRAMMING
Greg Nathanson
SENIOR VICE PRESIDENT, MARKETING & AFFILIATES
John J. Sie
VICE PRESIDENT, FINANCE & ADMINISTRATION
Ron Bernard
VICE PRESIDENT, SALES & AFFILIATE MANAGEMENT
Jack Heim
VICE PRESIDENT, PROGRAM DEVELOPMENT
Michael J. Clark
VICE PRESIDENT, PROGRAM DEVELOPMENT, WEST COAST
Loreen Arbus
VICE PRESIDENT, PROGRAM PLANNING & ADMINISTRATION
Jim Miller
VICE PRESIDENT, OPERATIONS & PRODUCTION SERVICES
Stephan Schulte
VICE PRESIDENT, MARKETING & PUBLIC RELATIONS
Sue Denison
VICE PRESIDENT, PROMOTION
Thomas Furr
VICE PRESIDENT, RESEARCH
Earl Marsh
VICE PRESIDENT, NATIONAL ACCOUNTS
Richard Sullivan
VICE PRESIDENT, BUSINESS DEVELOPMENT
Pat Grotto

REGIONAL OFFICES:

Western—8 Crow Canyon Court, Suite 200, San Ramon, CA 94583 (415) 820-1046, Reva Melniker; South Central—5757 Alpha Road, Suite 221, Dallas, TX 75240 (214) 239-7212, Bishop "Butch" Henley; Eastern—(516) 671-6104, 70 Forest Ave., Glen Cove, NY 11542, Art Gusow; Southeast—(404) 633-4326, 1800 Century Blvd., Atlanta, GA 30345, Bob Mason; North Central—360 N. Michigan Ave., Chicago, IL 60601, Dick Ingebrand.

Telemation Program Services, Inc.

Time & Life Bldg., Rockefeller Center, New York, NY 10020. Tel.: (212) 484-1631. (Film buying organization representing cable television and pay TV systems)

Twentieth Century-Fox Telecommunications, Inc.

Box 900, Beverly Hills, CA 90213. 10201 W. Pico Blvd., Los Angeles, CA 90035. (213) 227-2211.

PRESIDENT
Stephen Roberts
VICE PRESIDENT/SALES
Ed Michalove

Wometco Home Theatre, Inc.

286 Eldridge Rd., Fairfield, NJ 07006. (201) 227-8700.

EXECUTIVE VICE PRESIDENT & CHIEF OPERATING OFFICER
John S. Auld
VICE PRESIDENT/OPERATIONS
Stafford Hopwood
VICE PRESIDENT/PROGRAMMING
Barry Teiman

Producers-Distributors

* PROGRAMS
* COMMERCIALS
* FEATURE FILMS
* SHORTS

PRODUCERS AND DISTRIBUTORS

A

AFP Distributors, Inc.

1540 Broadway, New York, NY 10036; 582-1900. Sheldon Abromowitz, pres.; Barbara Brazong, v.p.; John Wengen, ag. liaison.

A.P.A. Studios, Inc.

230 West 10th St., New York, NY 10014; 929-9436. Organized 1963. (Special effects motion picture firm working in commercials, features, industrials; stop-motion, 3-D Animation hi-speed, time-lapse, locals and graphics, total production). Lee Howard, pres.; David Rogers, art director; Robert Self, production mgr.

Abbey Theatrical Films

9399 Wilshire Blvd., Beverly Hills, CA 90210. (213) 274-8646. Nicholas Wowchuk, pres.; Harry Wowchuk, exec. v.p.; Wallace Davidson, financial v.p.; Miles Holman, gen. mgr.; Dorothy Ringer, secty./treas.; Harry Stern, sls. mgr. (TV distribution)

Academy Film Productions, Inc.

3918 West Estes Ave., Lincolnwood, IL 60645. (312) 674-2132. Bernard Howard, pres. (Produce live and filmed or videotaped shows and commercials, slides, animation, slide films and industrial films).

Adelphi Company

3 Key St., Eastport, ME, Eastport 308. Tamar Lane, production mgr. Mal Merrit, business manager. (Produce filmed programs, commercials and musicals.)

Afrasian Films

145 West 96th St., Suite 'A' New York, NY 10025; (201) 783-5934. Abida Chaudhri, gen. mgr.; (Producers and distributors of feature, documentary, TV and educational films.) Also complete shooting facilities & rental of Arriflex Cameras, Nagras, zooms & lighting equipment. Shooting and production services in India-Pakistan).

Alan Enterprises, Inc.

26170 Pacific Coast Highway, Malibu, CA; 90265. (213) 456-3676. Alan L. Gleitsman, pres.; Chris Bachanan, operations mgr.; Cheri Rosche, v.p.

- Contemporary Cinema Features (17 features); Cinema Classics (40 features); Fine Arts Cinema (29 features); The Edward Small Package (30 features); Special Cinema (35 features); War and Peace (four 90-min. programs); The Abbott and Costello Show (52 half-hours); Felix the Cat (260 color cartoons); The Mighty Hercules (130 color cartoons); Speed Racer (52 half-hours).

Alden Films

7820-20th Ave., Brooklyn, NY 11214; (212) 331-1045. Jerome Schpiro, mgr. (Distributors of Av-Ed and Israel and Judaic films; color and b/w.)

Allend'or Productions, Inc.

15036 Valley Vista Blvd., Sherman Oaks, CA 91403; (213) 986-4622. Algernon G. Walker, pres.; Irene J. Brooks, v.p.; Marcia J. Walker, secy., treas.; (Motion-picture and TV production.)

American Film Productions, Inc.

1540 Broadway, NY 10036; 582-1900. Robert Gross, pres.; Sheldon Abromowitz, pres.; Barbara Brazong, v.p.: (Produce and distribute film and videotape TV commercials, industrial, public relations, sales, educational training, medical, merchandising and documentary program films.)

Animated Productions, Inc.

1600 Broadway, New York, NY 10019; Columbus 5-2942. Al Stahl, pres.; Richard Stahl, v.p.; Shirlee Debrier, production dir.; Peter Puzzo, exec. dir. (Producers of programs, television commercials, cartoons and shorts, industrial films, specialists in animation, stop motion, and fotomation.) Fully-animated—with punched tape—Animation stand, precision Super 8 blow-up optical bench, Super 8 with opticas.)

Arlington Movie Studios

149-33 Hawthorne Ave., Flushing NY 11355 (212) 358-9224. Organized 1962. (Produce all types sound and silent motion picture films.) Producer: Frank Fuzz.

Artscope, Ltd.

145 West 96th St., Suite 'A', New York, NY 10025; (201) 783-5138. Amin Qamar Chaudhri, pres. (Producers of features, TV films, commercial and educational films. Also sound transfer facilities editing rooms for rental, and screening facilities.)

Arzto Pictures, Inc.

15 East 61st Street, New York, NY 10021; (212) 753-1050. Organized 1967. (Production of motion pictures, commercials, industrials, also editing services including negative matching.) Anthony Arzt, pres., prod. dir.

Associated Press, The

50 Rockefeller Plaza, New York, NY 10020; 621-1500. Keith Fuller, pres. & gen. mgr.; Roy Steinfort, v.p. for Broadcast Services. (Daily news and picture services for television.)

Association Films, Inc.

866 Third Avenue, New York, NY 10022; (212) 935-4210. Exchanges in Ridgefield, NJ; La Grange, IL; Dublin, CA. William N. Farlie, Jr., pres.; Annette Minkalis, TV mgr. (Distributors public service film programs.) Canadian affiliate; Association Films, 333 Adelaide St., Toronto, Ontario M5V 1R6.

- Over 1,000 public service films:

Audio Productions

(A division of Reeves Teletape, Inc.) 227 East 45 St., N.Y., NY 10017. (212) 573-8656. Peter J. Mooney, pres.; (Producers of Educational-Documentary and Industrial Films).

Austin Productions, Inc.

P.O. Box 713, Lima, OH 45802; (419) 224-9311. B. Otto Austin, Jr., pres.; N. F. Austin, vice-president; L. A. Pugh, sec. (Producers of film commericals and industrial films.)

Avco Embassy Pictures Corp.

956 Seward Street, Los Angeles, CA 90038; (213) 460-7200. Branch Offices: 300 East 42nd St., New York, NY 10017; (212) 949-8900; 506 Statler Building, Boston, MA 02116; (617) 482-3325. Lois Luger, v.p.—TV; Charles Breitt, ntl. sls. mgr.; Merritt Sticker, east. div. mgr.; Martin Weisman, sls. rep.

- *ENTERTAINER OF THE YEAR*—1981 release—15 titles: The Baltimore Bullet, Born Again, The Chicken Chronicles, A Different Story, The Fog, Jock Petersen, The Lion In Winter, A Man, A Woman and A Bank, Murder By Decree, Nightkill, The Onion Field, Phantasm, Prom Night, Promise At Dawn and Winter Kills. *THE NEW ONES II*—1979 release—16 titles—The Manitou, Stingray, Sidewinder One, Diamonds, Bittersweet Love, Scalpel, etc.; *THE NEW ONES*—1977 release—18 titles. Farewell My Lovely, Russian Roulette, Man Friday, Tamarind Seed, Day of the Dolphin, etc.; *V.I.P. GROUP*—1974 release—18 titles. The Graduate, Generation, People Next Door, They Call Me Trinity, C.C. and Company, etc.; *PLUS TWELVE*—The Producers, Robbery, Tiger and the Pussycat, Catch as Catch Can, etc.; *28 FOR '68*—Nevada Smith, The Oscar, Hellbenders, Sands of Kalahari, etc. *TOP TIME*—Marriage Italian Style, Darling, Yesterday, Today and Tomorrow, Boccaccio '70, etc.; *NIGHTMARE THEATRE*—13 horror films—Fury of the Wolfman, Murder Mansion, Dear Dead Delilah, Doomwatch, Marta, etc.; *SERIES*—Pete 'n Gladys—72½ hour episodes. (b/w—tape)

Award Films

20 N. Wacker Drive, Chicago, IL 60606; (312) 236-8955. Al D. LeVine, gen. mgr.; Robert Eirinberg, secty. (Producers and distributors of TV commercials; Industrials; also King Studios Syndicated Film Series & Packaged Commercials.)

- Bank, Savings & Loan, Beer, Soft Drink, Milk, Bread, Meat, Moving & Storage Series TV Spots (Animated and Live); Paradox (26) (O'Henry Stories); Hormel Girls (44); Jack Drees Boxing from Rainbo (26); All American Wrestling (52); Adventure Out Doors (26); Travelbug (13); TV Kitchen (26); Let's Go Golfing (13); Indpls 500-(23)-16mm color/sound; Sports Year (64-73); All American Football (26); Golf Tips From Top Pros (13); Courtesy on the Golf Course plus 5 golf instruction color films and swing irons (Pitch, Putt, Swing, Grip & Stance); U.S. Open '71-Trevino at Merion; U.S. Open at Pebble Beach 7; U.S. Open-"Miracle at Oakmont"; Ski Alberta; Scuba-Dive Nova Scotia; Gray Ghost of Saskatchewan; '76 U.S. Open (Jerry Pate); Tennis Everyone (John Newcombe); Heisman Trophy—The Possible Dream, (Staubach & Plunkett); WCT Tennis (Borg, Connors), Serve (Gary Thorne); Helix-Ski Canadian Rockies; Kodak Series Travel Canada (6); Gone Fishing; Minawanamut (Sailing); Ski East . . . for the Fun of It; Calgary Stampede (Rodeo); Waterways to Explore Yukon Country.

B

BBC (Television Enterprises)

Head Office BBC Television Centre, London W. 12, England, (01) 743-8000, Peter Dimmock cvo obe, General Manager Television Enterprises. Tel.: 743-8000 Ext. 500. Canada, John Ridge, BBC P.O. Box 500, Terminal A, Toronto, Ont., (416) 925-3311. Australia, Basil Sands, BBC, 177 Elizabeth St., Sydney, 61-9059. (Distributes all types of BBC TV Programming.), licenses BBC titles for merchandising and produces BBC records, cassettes and cartridges from the best of BBC TV and radio.

BBI Productions (Boston Broadcasters, Inc.)

5 TV Place, Boston, MA 02192; (617) 449-0400. Robert M. Bennett, exec. v.p.; Paul Rich, special asst. to exec. v.p.; Jack Duffield, dir. of syndication; Bruce Marson, prog. mgr.

Baldwin Organization, Inc., The

10 West 65th St., New York, NY 10023; (212) 877-6607. Ted Baldwin, pres. (Public relations for motion pictures and television shows.)

Banner, Bob, Associates, Inc.

Pacific Design Center, 8687 Melrose Ave., Suite M-20, Los Angeles CA 90069; (213) 657-6800. Bob Banner, pres.; J. William Hayes, first v.p.; Stephen Pouliot, v.p. (Production of television programs, motion pictures and theatrical productions.)

Barbre Productions, Inc.

Div. of Combined Communications Corporation, P.O. Box 5667, 1089 Bannock Street, Denver, CO 80217, (303) 266-3601. Alvin G. Flanagan, pres.; Jon D. Ackelson, producer/director. (Producer, distributor of filmed shows, producer of film commercials and business films.)

- Prairie World of the Kit Fox (1); Valley of the Standing Rocks (1); Harry Jackson—A Man and His Art; The Troopers Are Coming; Rocky Mountain Conquest.

Barry & Enright Productions

1888 Century Park, E., Los Angeles, CA, 90067 (213) 277-3414. Jack Barry, Dan Enright, principals.

Ben Barry & Associates, Inc.

159 West 53rd St., New York, NY 10019; (212) 753-3666. Ben Barry, pres.

- Feature films, including The Family, Honey Comb, Sabra, Action Man, Assassination, I Killed Rasputin, Peking Blonde, Singapore, Singapore, Johnny Banco, Black Sun, Restless Breed, Magnificent Matador, Sword of Monte Cristo, One Russian Summer, etc.
- Feature films, under House of Horrors heading, including The She Beast, Death Dream, Fangs of the Living Dead, The Night Evelyn Came Out of the Grave, Murder Clinic, Children Shouldn't Play with Dead Things, Blood Spattered Bride, Don't Look in the Basement, Nightmare Hotel, Kiss of the Tarantula, Invisible Terror, Zombies. Science fiction features: Star Pilot, Invaders from Mars, Fantastic Invasion from Planet Earth, No Survivors Please, Electronic Monster, Fabulous World of Jules Verne.

Beck, Alexander, Films, Inc.

1540 Broadway, New York, NY 10036; (212) 575-9494. Alexander Beck, president. (Distributes feature films, shorts and Westerns.)

The Behrens Company, Inc.

170 S.E. 14 St., Miami, FL 33131; (305) 371-6077. Robert A. Behrens, pres.; Elizabeth H. Behrens, secty.

Best Film & Video Corporation

98 Cutter Mill Road, Great Neck, NY 11021; (516) 487-4515. Roy Winnick, pres. (Cable, pay TV, free TV, cassette and disc distribution and production).

Milton Blackstone/Associates

P.O. Box 1892, La Jolla, CA 92037; (714) 459-8255. Milton H. Blackstone, exec. prod. (Live and film and tape (VTR) packages. Also, Talent, promotion and special events direction, true-life commercials, pr. films.)

- Comedians' Golf Classic; Comedy Hall of Fame (special); Wonderful World of Water (special); Beach Ball!; (Pro) Am Team Golf Championship; Mother Goose Parade (special); Jr. World Golf Championship; Mother Goose Parade (special); Jr. World Golf Championship (special); Defending Champion (series); Let There Be Balls (special); Golf Derby; Show Biz Salutes (series and specials); Sports City (in preparation) '76 Andy Williams Open (90 min.); Feature Film: Cockeyed Charley; New Series in Development; Hobo; Pilot; Something to Celebrate.

Borden Productions

Great Meadows Road, Concord, MA 01742 (617) 369-5030. Dick Borden, pres.; William Sweney, Gordon M. Beck, v.p.

Brandon Films, Inc.

34 MacQuesten Parkway, So., Mount Vernon, NY 10550; (914) 664-5051; Cable Brand-films. Leonard P. Weg, gen. mgr. (Distributes foreign feature films.)

Braverman Productions, Inc.

1237 Seventh St., Santa Monica, CA 90401. (213) 451-9762. Charles Braverman, pres.; Michael Roach, exec. prod.; Marshall Harvey, editor. (Producers of TV, educational and corporate films, TV commercials, titles and special montages, TV documentaries and specials, specialists in kinestasis animation.)

Bray Studios, Inc.

630 Ninth Ave., New York, NY 10036; (212) CL 5-4582. Paul Bray, Jr., pres.; Monroe Oakley, Jr., prod. mgr.; R. McDonald, film dist. mgr. (Producers and distributors of films, film strips, slides and multi-media presentations on wide variety of subject matter; services for producers include animation art and photography as well as location crews and equipment. Self-sponsored films for TV in general distribution.

Brickyard, Ltd., The, Motion Picture Arts Div.

P.O. Box 1752; 414 West 300 South, Salt Lake City, UT 84101; (801) 363-3757. Kent F. Johnson, pres. & gen. mgr.; Susan D. Sandberg, v.p.; L. D. Johnson, v.p. in chg. sls. (Motion picture production, producers services, equipment rentals, sales and service.)

Broadcasting Foundation of America

52 Vanderbilt Ave., New York, NY 10017; 986-6448. George E. Probst, chm. of board; Howard L. Kany, exec. dir.; Catherine D. Probst, prog. dir. (Produces and distributes news, public affairs, cultural and educational talk programs & international music festivals from overseas to radio stations, schools, colleges & universities, and libraries, abroad to stations, colleges and universities.)

Brown, Barry, Brillig Prod., Inc.

300 Central Park West, New York, NY 10024; (212) 595-5454. Organized 1959; Barry Brown, pres. (Feature motion pictures, industrial films & TV commercials.)

Byron Motion Pictures, Inc.

65 K Street, N.E., Washington, DC 20002; (202) 789-1100; Robb Roudabush, pres.; Michael T. D'Ambra, exec. v.p.

C

CPC Associates, Inc.

6309 Eleanor Ave., Hollywood, CA 90038 (213) 467-5900. Ron Seawright, pres.-gen. mgr. William Sterling, exec. v.p.; David Russell, v.p.; Harry Walton, special effects director; Howard Magwood, director; Francis Grumman, director/cameraman; Frank Furlong, animation director; Sam Pal, laser lights; Ted Eckelberry, editor. (Special Effects Stage, Optical & Editorial Departments. Film, Tape & Animated Commercials. Industrial films.)

Campaign Coaches of Long Island

(A division of F.L.A., Inc.) 80 Winifred Drive, North Merrick, NY 11566. (516) 378-4627. Frank C. Beckert, creative director. Richard E. Stember, Miami Beach office manager. (305) 893-1411. (Advertising/public relations/marketing & sales programs.)

Steve Campus Productions, Inc.

14 Madison Ave., P.O. Box 206, Valhalla, NY 10595 (914) 946-4343. Steve Campus, pres. (Producer of educational, industrial, medical, documentary films, and multi-media productions.)

George Carlson & Associates

113 Battery St., Seattle, WA 98121; (206) 623-8045. George Carlson, producer. (Program producers.)

- The Traveler/Northwest Traveler series.

Carthay Sound Stage

5907 West Pico Blvd., Los Angeles, CA (213) 938-2101. (Complete film videotape services & rental center.) Contact: Helen Miles.

Cavalcade Pictures, Inc., and Cavalcade Television Programs

959 No. Fairfax Ave., Hollywood, CA 90046 (213) 654-4144; Harvey Pergament, pres. (Producer, distributor and importer of filmed programs.)

- Famous Guests (13); Unk & Andy ABC Adventures (26); Operation Tomorrow (4); Adventures of Patches (26); Marvin Miller Story Teller (13); Junior Aces (13); Popular Science (78); Adventure Safari Album (26).

Cinecraft, Inc.

2515 Franklin Blvd., Cleveland, OH 44113; (216) 781-2300. Paul Culley, pres. (Producer of TV commercials, industrial motion pictures, slide films and recordings.)

Cinema Shares International Television, Ltd.

450 Park Ave., New York, NY 10022 (212) 421-3161. Kenneth M. Israel, pres.; Eric Ablog, v.p., finance; Beverly J. Partridge, v.p. program services. (TV distribution.)

Cinetudes Film Productions, Ltd.

295 West 4th St., New York, NY 10019 (212) 966-4600. Christine Jurzykowski, pres.; Neal Marshad, v.p.; Gale Goldberg, prod. mgr. (Production of TV programs, motion pictures and theatrical productions)

City Film Productions, City Film Center Inc.

64-12 65th Place, Middle Village, Queens, N.Y.C. 11379; (212) 456-5050; John R. Gregory, exec. prod. (Producers of 8mm/16mm/35mm motion pictures, for the fields of business, industry, advertising, sales, science, religion, education, health, entertainment and TV. Creative, consultation and production facilities.)

The Claridge Group, Ltd.

540 Madison Ave., New York, NY 10022 (212) 355-7070. William F. Storke, pres.; Janet H. Gardner, dir. of development. U.S. production arm of Trident TV. Ltd. (MOW's and dramatic specials.)

Coast Productions

1001 N. Pointsettia Place, Hollywood, CA 90046 (213) 876-2021. Jack Yopp, chm.; Jay Grandy, pres.; Patrick Collins, v.p./gen. mgr.; Jim Jordan, Niki Minter, Bob Gips, George Koblaska, Ray Rivas, Gary Hall, Howard Morris, directors; (Producers of TV commercials).

Coast Special Effects

4883 North Lankershim Blvd, North Hollywood, CA 91602 (213) 762-1182. Phil Kellison, special effects creative dir.; William Barnett, v.p. of Special Effects. Special Effects Directors: Rick Clemente, Gordon Rayner, Laine Liska. (Special Effects for tv commercials and feature films.)

Colbert Television Sales

1888 Century Park East, Los Angeles, CA 90067 (213) 277-7751.

- 15 Lassie movies (12 in color); 132 Lassie half-hours (164 in color); The Joker's Wild (260 half-hours); Tic Tac Dough (260 half-hours); Bullseye (game strip). Rifleman (168); Specials (1 hr. each): Frank Sinatra, Jr. in concert; Bert Convy, Juliet Prowse in Las Vegas.

Columbia Pictures Television

A Division of Columbia Pictures Industries, Inc.—15250 Ventura Blvd., Sherman Oaks, CA (213) 995-1300. (For officers and divisions see listing in Companies section.)

- Features: Features of the 80's, Volume I (29); Features of the 70's Volume I (30 features); Volume 2 (27); Volume 3 (30); Post 60 Volume I (55 features); Volume 2 (50); Volume 3 (51); Volume 4 (33). Volume 5 (39); Volume 6 (26); Volume 7 (30); Post 50 Volume 1 (73); Volume 2 (58); Post 48 (210). SERIES AVAILABLE: Charlies Angels (110); Starsky & Hutch (90); What's Happening!! (65); Carter Country (44); L.A.T.E.R. (65); Barney Miller (124); Fish (35); Special Edition (24); Partridge Family (96); Bewitched (252); I Dream Of Jeannie (139); Dennis The Menace (146); Father Knows Best (191); Farmer's Daughter (101); Here Come The Brides (52); Wackiest Ship In The Army (29); Gidget (32); Donna Reed Show (175); George Burns and Gracie Allen (239); Hazel (154); Monkees (58); 2 Reel Comedies (162); Serial Package (51); Three Stooges (190); Quick Draw McGraw (135); Batfink; Hanna-Barbera Cartoons; Ruff'n Reddy; Magilla Gorilla; Peter Potamus; Huckleberry Hound and Yogi Bear (276); Flintstones (166); Celebrity Charades (45); Dealer's Choice; Diamond Head Game Show; New Quiz Kids; Wild Bill Hickok; Jungle Jim; Circus Boy; Adventures of Seaspray; Casey Jones; Jet Jackson; Ivanhoe; Tales of the Texas Rangers; Manhunt; Rescue 8; Tightrope; Route 66 (116); Talahassee 7000; Two Faces West; Shannon; Behind Closed Doors; All Star Theatre; Award Theatre; Celebrity Playhouse; Showcase 39; Big Playback; Decision (26); Conflicts of Harry S. Truman; Winston Churchill, The Valiant Years.

Communication Commission of National Council of the Churches of Christ in the USA.

475 Riverside Drive, New York, N.Y. 10027. (212) 870-2567. William F. Fore, assistant general secretary for communication; D. Williams McClurken, executive director for broadcasting and film; David W. Pomerby, director, media resources; Lois J. Anderson, director placement and distribution; L. Franklin Devine, director ecumedia news service and broadcast news.

- Davey and Goliath (64 films for children); Davey and Goliath specials: Christmas, Lost and Found, Happy Easter, Halloween Who-Dun-It, New Year Promise, To the Rescue, School—Who Needs It!, 13 for 30 (13 half-hour films), Learning to Live (8), Campaign for Human Development and Jesus spots; others (including information about Network Religious TV programs)

Company of Artists Productions, Inc.

3518 W. Cahuenga Blvd., Suite 301, Hollywood, CA 90068, (213) 851-5811; Tom J. Corradine, exec. prod.; Kyle C. Thomas, v.p. & gen'l. mgr, prod. supervisor; Jay O. Lawrence & John P. Travasso, v p. video cassette div.; Edward Reagen, v.p., sls. mktg. (Producers of motion pictures & commercials.)

Condor Pictures, Inc.

1536 Viewsite Terrace, Hollywood, CA 90069, (213) OL 2-7447. Milton Simon, gen. mgr. (TV pilot films. Films made to order.)

Contempo Communications Inc.

1841 Broadway, Suite 1111, New York, NY 10023. (212) 247-4444. Joan F. Marshall, v.p. (Theatrical, Business, Documentary, award-winning Live Shows, Mixed-Media, Motion Pictures, Sales Mtgs.)

Continental International Pictures

448 Harrison Ave., Greenville, OH 45331; 548-2791. Walter D. McFarland, executive producer. (Producers of 16mm. black

& white and color sound documentary films for business, industry, government agencies, bureaus, and institutions.)

Corradine, Tom J., & Associates

3518 W. Cahuenga Blvd., Suite 301, Hollywood, CA., 90068; (213) 851-5811. Branch office in New York. Tom Corradine, president; Kyle V. Thomas, v.p.; Roger Danforth, head booker; Edward Reagan, v.p., marketing; John P. Travasso, v.p., video cassette div. (Distribute filmed series, feature films. Western representative for Weiss Global Enterprises, TV Cinema Sales Corp., and Bloom Film Group. National reps for TV National Releasing Corp., Lippert Pictures, Inc., and Film Video Releasing Corp.)

- 700 feature films, 192 Westerns, 500 cartoons, 6 series.

Cox, William G., Film Enterprises

9726 Edward Drive, Sun City, AZ 85351; William G. Cox, owner-producer. (Produce commercials, news clips informational & public relations films.)

Craven, Thomas, Film Corporation

316 East 53rd St., New York, NY 10022; (212) 688-1585. Thomas Craven, pres.; Willis Briley, exec. producer; Frynne Hamden, secty. (Producers of filmed and videotape series. Producers of TV specials, documentary films, Public Relations and Information Films. Producers of films for several Govt. Agencies.)

- Representative Productions: Search for Achievement-Management film; Vision USA 26 films in USIA series for TV; Legal Aid Society film; films for Peace Corps; Wood's Hole Oceanographic Institute; VW of America (New Car Introductory Films); film series for Oceanographer of Navy; Specials for TV: Coleridge; Dylan Thomas; Films for United Nations Fund for Population Activities.

Creativision Incorporated

112 Bentwood Dr., Stamford, CT 06903 (203) 329-7795. Victor Kayfotz, distribution mgr. (Sale of 16mm films, film loops, & videocassettes on the use of Hand Tools & Power Tools for Woodworking; Precision measuring tools for Metal-Working; & the Metric System.)

Crosby, Bing Productions, Inc.

6311 Romaine St., Hollywood, CA. 90038. (213) 465-3114. J. R. Rodgers, v.p., finance & business affairs; Ray Tucker, secty-treas.; Al Chunka, asst. secty. & asst. treas. (Motion pictures, television (film & tape) syndication.)

Crossover Programming Company

1237 7th St., Santa Monica, CA 90401 (213) 451-9762. Charles Braverman, pres.; Michael Roach, v.p. (Producers of pay television series, specials, variety shows. Specialists in kinestasis animation.)

Crystal Pictures, Inc.

1560 Broadway, New York, NY 10036 (212) 757-5130. F. Feinstein, president; S. Tager, dir. of sls. (Distribute features, Westerns, packages & reissue TV series.) Telex: 620852CRYSPC

- 60 features; 200 short subjects; comedies (Charlie Chaplin & other old time comedies) (112); TV series (one hour and ½ hour) (410).

Cummings Productions

% C. N. Weber, 5 Whitelaw Place, Baltimore, MD 21236. Ralph Waldo Cummings. (Creation and production of musical commercials, backgrounds, scores for promotional & entertainment projects.)

D

Day, Gordon M., Productions

15 W. 44 St., New York, NY 10036. (212) 840-2866. Gordon M. Day, pres.; Carol Bulkley, vice-pres.; (Write-produce musical commercials, jingles, copy, and dramatized spots.)

Demby Productions, Inc.

770 Lexington Ave., New York, NY 10021; (212) 421-7040; Emanuel Demby, president; Betty Jeffries, vice-pres. (Produce television and motion picture features.) Consultants, TV news documentaries.

- Cavalcade of Music; Carnival; Witch On Wheels; The Shadow; World in Space; The Creative Consumer; Smash, Crash, Pow!

Depicto Films Corp.

1 East Ridgewood Ave., Paramus, NJ 07652 (201) 652-9101. J. R. von Maur, president. (Producers of filmed commercials and industrial motion pictures. Sales meeting and conventions, road shows.)

Disney, Walt, Productions

500 So. Buena Vista, Burbank, CA 91421; (213) 845-3141; 849-3411; 477 Madison Avenue, New York, N.Y. 10022; (212) 759-3800. For officers and personnel see listing under Companies.

Documentary Films, Inc.

159 West 53 Street, New York, NY 10019 (212) 582-4318 Cable: Geraldrap, NY Telex: 420748. Gerald J. Rappoport, Christopher Wood, exec. officers. (Distributor of feature and short documentary films to theatres, television, and non-theatrical markets).

Dubie-Do Productions, Inc.

New York City (212) 765-4240 or 1 Laurie Drive, Englewood Cliffs, NJ 07632. (201) 568-4214. Richard S. Dubelman, pres. (TV production.)

Dunn, Cal. Studios, Inc.

141 W. Ohio St., Chicago 10, IL (312) 644-7600. Michael H. Dunn, pres. (Producers of motion pictures, multi-screen presentations, TV commercials, slidefilms for sales, training, promotion, indoctrination.)

E

The E. O. Corp.

P.O. Box 184, Shelby, N.C. 28150. (704) 482-0611. Earl Owensby, pres.; Gene Blackburn, v.p.; Linda Comer, secty.; Michael Putnam, sls. mgr.

Edwards, Ralph, Productions

1717 North Highland Ave., Hollywood, CA 90028; Ralph Edwards, owner. (Creator, packager, producer television shows.)

- The Cross-Wits, Metro-Media & syndication.
- Name That Tune: syndication.

F

Fairbanks, Jerry, Productions

826 N. Cole Avenue, Hollywood, CA. 90038; (213) 462-1101. Jerry Fairbanks, owner; Leo S. Rosencrans, creative dept.

Family Films

14622 Lanark Street, Panorama City, CA, 91402. Norman E. Walter, dir. of development; Robert Garmatz, marketing; Iria Austin, finance and credit.

Doris Faye Productions, Inc.

325 W. 45th St., New York, NY 10036; (212) CI 6-0430. Organized 1959; Doris Faye, pres. (TV and film packagers; specialists in "educating through comedy" programs.)

Fedderson, Don, Productions

12735 Ventura Blvd., Suite 28, Studio City, CA. 91604; (213) 985-4736. Don Fedderson, chairman of the board; Charles Spira, pres.

- My Three Sons (network); Family Affair (network); To Rome, With Love (network), The Smith Family (network). Also TV Consultants for Lawrence Welk Syndicate, The Lawrence Welk Show.

The Film Company

1600 Broadway, New York, NY 10019 Ronald Saland, Marcel Broekman, Elliot Geisinger, Howard Kuperman. (Motion picture production.) Mark Schneider (featurette industrial, commercial division).

Film Investment Corp.

333 South Beverly Drive, Beverly Hills, CA 90212. (213) 553-5806. Adrian Weiss, pres.; Steven A. Weiss, secty-treas.;

Mary Virginia Head, operations manager (Distributor of filmed series, features, shorts.)

- Full-length Westerns (60); full-length features (24); Episode serials (20 min. per episode) (16); Half-hour: Craig Kennedy, Criminologist (26); Thrill of Your Life (13); Twenty-minute; Wally Wales Westerns (8); Modern Bible Stories (11 min.), (3); approximately 6½ to 7 minute cartoons (15).

Filmack Studios

1327 S. Wabash Ave., Chicago, IL 60605; (312) 427-3395. Joseph R. Mack, pres.; Robert Mack, v.p. (Motion picture film producer for TV and industry, as well as filmstrips and slides for education.)

Films Five, Inc.

42 Overlook Rd., Great Neck, NY 11020; (516) 487-5865. Walter Bergman, head of studio oper. (Producers of industrials, documentaries, commercials live film, videotape & animation. Editorial department & massprint distribution on premises.)

Films of the Nations

7820 20th Ave., Brooklyn, NY 11214 (212) 331-1045. Jerome Schpiro, mgr. (Distributors of TV, industrial and educational free sponsored films.) Color and b/w.

Filmways, Inc.

2049 Century Park East—Suite 3500, Los Angeles, CA 90067 (213) 557-8700. 540 Madison Ave., New York, NY 10022. Tel.: (212) 758-5100. Richard L. Bloch, pres. & chief executive officer; L. Douglas Nolan, vice chairman of board; Robert A. Grunburg, executive v.p., chief financial officer; Salvatore J. Iannucci, senior v.p.; Eugene B. Boylan, v.p., treasurer, controller; Alan R. Markizon, general counsel, secretary, Lee Moselle, v.p.

Filmways Pictures, Inc.

Product distributed by Filmways Enterprises, Inc., 2049 Century Park East, Los Angeles, CA 90067 (213) 557-8700. For officers and personnel, see listing under Companies.

- Features: Films for 80's (45 color features); A Winning Hand (33 color features); World of Macabre (8 in color); Films for 70's (26 color). Ghoul-A-Rama (16, color); Ghoul-a-Rama II (11 color features); Special Actions Features (12 color); Startime Theatre (17; 5 color); Fantastic Sci-Fi Theatre (14, color); Children's Showtime (4 color animated, 2 live-action features); Young Adult Thea. (26, color); New Science-Fiction (13 color, 2 b&w); Amazing Sci-Fi 1 (6 color, 10 b&w); Amazing Sci-Fi II (7 color, 13 b&w); Amazing Sci-Fi III (24 color, 2 b&w); Top Secret Adventure (5 b&w); Dominant 10 (2 color, 8 b&w); Holiday Storybook of Fables (7 color, 1 b&w); Real Life Adventures (4 color documentaries); Strongmen of the World (23 color features); New Color Adventures (32); Adventures (8 color, 12 b&w); A.I.P. (26 b&w features); Selma (42 b&w features); A.I.P./Selma (3 color, 22 b&w features). HOUR SERIES. The Avengers (57 color, 26 b&w); Touch of Music (6 musical specials). HALF HOUR SERIES: Adventure of Ozzie and Harriet (200 b&w); Johnny Sokko & His Flying Robot (26 color); Prince Planet (52 b&w, animated); Comeback (documentary, 26-half hours); Jukebox (26 color); The Racers (22 color); Sports Challenge (18 color). CARTOONS: Sinbad, Jr. (130 five-min. each), Touch of Music (6-one hour musical specials.)

Fima Noveck Productions

1560 Broadway, New York, NY 10036. (212) 221-2695. Fima Noveck, pres.

Firestone Program Syndication Co.

1200 West Broadway, Hewlett, NY 11557. (516) 569-6900. Len Firestone, pres.

- The Gong Show (access); The Gong Show (strip); The New Candid Camera; The Dating Game (strip); Three's a Crowd (access; one a week and/or strip); Prisoner: Cell Block H (260 half-hours, strip); Card Sharks (strip); Young Medics (strip-serial).

Fordel Films, Inc.

2454 Fifth Ave., San Diego, CA 92101. Jack Tobin, pres., John K. Culley, v.p. & dir. of photog. (Producers of oceanographic films, documentaries, commercials.)

Format Productions, Inc.

4253 Reyes Drive, Tarzana, CA 91356; (213) 987-3900. Herbert Klynn, pres.; Marvin L. Klynn, exec. v.p. Selma Klynn, v.p.; (Animation/action films for TV commercials, TV programming, theatrical shorts, feature and industrial films.)

Four Star International, Inc.

19770 Bahama St., Northridge, CA 91324. (213) 709-1122. David B. Charnay, pres. and chmn. of the board; Joseph J. Doyle, v.p.; Hervey S. Shaw, v.p., bus. affairs & finance, treasurer; Martha Arnold, v.p. & secretary.

- Series: The Zane Grey Theatre Law of the Plainsman; Stagecoach, West; Richard Diamond; Wanted-Dead or Alive; Black Saddle; The Law & Mr. Jones; Johnny Ringo; Peter Loves Mary; The DuPont Show With June Allyson; Dante's Inferno; Robert Taylor's Detectives; The Tom Ewell Show; Target; The Corruptors; The David Niven Show; The Westerner; Trackdown; The Gertrude Berg Show; The Dick Powell Theatre; Jeanne Carson Show; Lloyd Bridges Show; Ensign O'Toole; McKeever and the Colonel; Saints & Sinners; Celebrity Game; The Rogues; Big Valley; Honey West; Smothers Brothers Show; Amos Burke; Secret Agent; Portrait; Here Come the Stars; Thrillseekers, Year 2000, Wonderful World of Magic; Lohman & Barkley; Monty Nash; Target: The Impossible Flying Fisherman; Boomerang.
- Specials: Twenty Years of Rock 'n Roll; Seals & Crofts; San Francisco; Blackstone Magic; David Niven Magic; America Screams; Voyage of Yankee Clipper.
- Features: 4 Pippi Features: Once Upon A Time; Man Against the Organization; Funeral for an Assassin; Mary, Mary, Bloody Mary; Calaglostro; Madron; Last of Mohicans; Violent Patriot; Della; Boy of Two Worlds, Secret Agent-Super Dragon; Bully; The Head of the Family; The Man From Nowhere; Weekend-Italian Style; The Battle of the Mods; Diary of a Telephone Operator; Summerdog; Legend of the Northwest; Little Mermaid; Dunderklumpen; Adventures of Pinocchio; Magic Pony Mr. Too Little.

Sandy Frank Film Syndication, Inc.

645 Madison Ave., New York City 10022 (212) 628-2770. Sandy Frank, pres.

G

Galaxie Productions, Inc.

P.O. Box 1933, Washington, D.C. 20013; (202) 638-5595. B. Blazes, President; Dr. Edward v. Rothkirch, exec. v.p. & exec. Producer; N. A. Caron, secy.-treas.; Paul Malec, chief engineer (Produce & distribute films for non-theatrical motion pictures & television: commercials, travelogues, documentaries, language dubbing on film, complete sound stage for audio recording, custom recording on tape or disc. Fully equipped audio visual, motion picture, & VTR remote truck for sound, motion picture, VTR, & still photography. Branch offices: c/o A.L. Weintraub, 5th Floor, 2250 S.W. 3rd Ave., Miami, FL 33129 & C.P.O. Box 1711, Tokyo 100-91 Japan; Suite 1108, 110 W. 40th St., New York, NY 10018.

Glen Glenn Sound Co. (A Division of Republic Corp.)

6624 Romaine Street, Hollywood, CA. 90038 (213) HO9-7221. Joseph D. Kelly, pres.; Tom Kobayashi, v.p. business affairs; Brad Blake, v.p./mktg; Richard Larson, v.p., ops.; Jack Phillips, dir. sales. (Complete film and video-tape sound recording services for feature motion pictures, television programs, commercials, and industrial or educational programs.)

Glenar Studios

211 S. Rose, Burbank, CA., 91505 (213) 848-0408. Sid Glenar, owner. (Produces films, commercials & animation.)

Gold Key Entertainment

(A Division of The Vidtronics Co., Inc.) Tower 53, 159 West 53rd St., New York, NY, 10019 (212) 486-9116; 6922 Hollywood Blvd., Suite 321 Hollywood, CA, 90028 (213) 463-7151; Offices in Toronto, London, Paris, Stockholm, and Tokyo.

Domestic Sales: Robert B. Muller, pres.; Leonard Soglio, v.p. and natl. sls. mgr.; Ben Barry, pres. intl. sls.

- Series: Krofft Super Stars (20 live-action half-hour children's shows); Popeye Cartoons (220); Beatles Cartoons (116); Cool McCool (60); Krazy Kat (50); Barney Google (50); Beetle Bailey (50); Perspective on Greatness.
- Movies: The Main Event (15 new films, 90 mins. each); The Gamma Chronicles (15 mysteries, monsters, etc., 90 mins. each); Blondie (28 features, 75 mins. each); Galaxy One (15 sci-fi features); Galaxy II (15 sci-fi features); The Neptune

Journals (12 features on mysteries and monsters of deep); The Beta Chronicles (12 features); The Alpha Chronicles (10 features); Bill Burrud Features (6 action/adventure features); Rainbow IV Outdoor Adventures (10 features); Rainbow III Outdoor Adventures (15 features); Rainbow II Outdoor Adventures (10 features); Rainbow I Outdoor Adventures (10 features); Thirteen TV Premiers (13 features); Scream Theater (32 features); Awards Theatre (40 features); Action Theatre I (20 features); Action Theatre II (20 features) Astor Classic Features (10 features) Astor Sci-Fi Features (6 features); The Late Show (30 features); Cisco Kid Features (13 features).

- Specials: The Palace (13 one-hour specials hosted by Jack Jones); Bobby Vinton's Rock 'n' Rollers (one-hour musical variety); Superman (2-hour musical); Alice's Adventures in Wonderland (101-min. feature); The Two Kennedys—A View from Europe (2-hour special); In Search of Ancient Astronauts (one-hour special); In Search of Ancient Mysteries (one-hour special); Loggins and Messina "In the Attic" (one-hour rock music special); Walt Wagner Show with Peggy Fleming (one-hour special); Dick Tracy Serials (4 serials, 15 episodes each; 4 features); Flash Gordon Serials (3 serials in 40 half-hour episodes).
- News Programming: TV Time Capsule.

Gold, Melvin L. Enterprises

301 East 48th St., New York, NY (212) MU 8-0897. Mel Gold, pres. (Consultant, producer and packager of live and film TV programs, industrial films, commercials, features.)

Goodson-Todman Productions

375 Park Ave., New York NY, 10020 (212) PLaza 1-0600. 6430 Sunset Blvd., Hollywood CA 90028; (213) 464-4300. William S. Todman, Mark Goodson, partners; Giraud Chester, exec. v.p.; Howard F. Todman, v.p., dir. of business affairs; Gil Fates, exec. prod.; Howard Felsher, Frank Wayne, Ira Skutch, Chester Feldman, producers; Paul Alter, Lloyd Gross, Marc Breslow, George Choderker, directors.

- The Rebel (film); The Richard Boone Show, Branded (film); Beat the Clock, To Tell The Truth, What's My Line, Password, The Price Is Right, Match Game PM, Tattletales, I've Got A Secret (all tape), Family Feud, Card Sharps, Match Game, Mind Readers, Password Plus.

Goulding-Elliott-Greybar Productions, Inc.

420 Lexington Avenue, New York, NY 10017 (212) LE 2-9014. Organized 1955. Ray Goulding, pres./sec.; Bob Elliott, vice-pres./treas. (Production of radio and TV commercials and radio & TV programs).

Gray-Schwartz Enterprises, Inc.

Teleflix Division, 21777 Ventura Blvd., Woodland Hills, CA 91364 (213) 702-9888. Marv Gray, pres.

Greatest Fights of the Century, Inc.

9 East 40th St., New York, NY 10016; William D. Cayton, pres.; Jim Jacobs, dir. (Produce and distribute fight films.)

- Big Fights of the Decades Series (formerly titled Greatest Fights of the Century) 400 quarter-hour programs from 1900–1980, featuring all the world champions, each in his greatest, and most memorable fight, including George Foreman, Joe Frazier, Mohammad Ali, Floyd Patterson, Rocky Marciano, Joe Louis, Jack Dempsey, Sugar Ray Robinson, Rocky Graziano, Archie Moore, etc.

Group W Productions (Westinghouse Broadcasting Company)

70 Universal City Plaza, Universal City, CA (213) 760-8800. Edwin T. Vane, pres. and chief exec. officer; George E. Resing Jr., senior v.p.; Peter F. Yaman, v.p. sls. (NY); Owen S. Simon, v.p., creative services

- The John Davidson Show (5 shows weekly; 90 or 60 min.); Hour Magazine (5 shows weekly, 60 mins.); PM Magazine Program Service; PM Special Editions (6 one-hour); Fight Back with David Horowitz (weekly, 30 mins.).

Gus Productions, Inc.

10889 Wilshire Blvd., Suite 606; Los Angeles, CA (213) 477-3046. (Motion picture and TV production) Alan J. Pakula, pres.; Hannah C. Pakula, v.p.; Eric Weissmann, secty.

Gutman, Leo A., Inc.

230 Park Ave., New York, NY (212) 682-5652.

- 14 Sherlock Holmes movies.
- Movies: 14 Sherlock Holmes; 11 Charlie Chan; 9 Mr. Moto; 13 East Side Kids; 7 Hollywood; 3 International; 2 Ingrid Bergman; 5 Westerns; 5 Cliffhangers; 23 Red Ryder; 26 Great Western Theatres (Chuck Connors), 90 mins. each or 39/60 mins.

Guymark Studios, Inc.

3019 Dixwell Ave., Hamden, CT 06518. Anthony Guarino, Jr., pres.; Guy Guarino, v.p., chg. of photo; Mark Guarino, v.p., chg. of audio & video. (Producers facilities and technical services for sound film, video, animation, product photography).

H

H-R Productions, Inc.

159 West 53 St., New York, NY 10019; (212) 541-8015. Herbert Rosen, president. (Producer of films, package shows.)

Handel Film Corporation

8730 Sunset Blvd., W. Hollywood, CA 90669 (213) 657-8990. Leo A. Handel, pres.; Peter Mertens, producer. (Produces filmed series.)

- Magic of the Atom (10); The Age of the Atom (color special); Sweden-Vikings Today Style (color special); (color special); Police Dog; 1-hr TV special (1972); Art in America (series 1981–10).

Jam Handy Organization, The

2900 East Grand Blvd., Detroit, MI 48202 (202) TRinity 5-2450. Jamison Handy, pres. (Producers of motion pictures, videotapes, filmstrips, slides, training programs, business meetings and consultation services.)

Hankinson Studio, Inc.

1156 Avenue of the Americas, New York, NY 10036. (212) 730-0434. Walter Klas, pres. (Producers of animation and graphic test commercials)

Hanna-Barbera Productions, Inc.

(A Division of The Taft Entertainment Company) 3400 Cahuenga Blvd., Hollywood, CA 90068. (213) 851-5000. Joseph Barbera, pres.; William Hanna, sr. v.p.; Patricia A. Healy, v.p. bus. affairs; Margaret Loesch, exec. v.p., children's programs & specials; John Michaeli, v.p. communications; William Scott, v.p. exec. in charge of special projects; Ross Sutherland, v.p. dir., personnel & labor relations; Iwao Takamoto, v.p. creative design; Martyn Weinberg, v.p. & gen. mgr.

- Series: (ABC) Superfriends; The Fonz and Laverne and Shirley Hour, Scooby Doo and Richie Rich Show; (CBS) Kwicky Koala Show (Kwicky Koala, Dirty Dawg, Crazy Claws, Bungle Brothers), The Trollkins, The Popeye/Olive Oyl Show (Popeye, Prehistoric Popeye, Private Olive Oyl); (NBC) The Flintstone Comedy Show (Flintstone Family Adventures; Dino and the Cavemouse; Pebbles, Dino and Bamm Bamm; Bedrock Cops, The Frankenstones), The Smurfs, Space Stars (Space Ghost, Herculoids, Teen Force, Astro and the Space Mutts, Space Stars Finale), HB Jokebook.
- Animated Specials: The Harlem Globetrotters Meet Snow White (NBC), The Smurfs (NBC), Rock Odyssey (ABC).

Larry Harmon Pictures Corp.

649 North Bronson Ave., Hollywood, CA 90004 (213) 463-2331. Larry Harmon, pres.; H.M. Alpert, secty.-treas. New York office: 60 W. 57th St. 10019; (212) 247-6295.

Harris-Tuchman Productions, Inc.

1226 West Olive Ave., Burbank, CA 91506 (213) 841-4100. Ralph G. Tuchman, pres. and gen. mgr.; Fran Harris, vice pres. and creative dir. (Producers of sales, training and Industrial Films.)

Harriscope Broadcasting Corporation

10889 Wilshire Blvd., Suite 1240, Los Angeles, CA 90024. Burt I. Harris, pres.; Jack Rosenthal, exec. v.p. & pres., broadcast div.; Harvey Simpson, finance dir.; Ruth Bender, exec. asst. (Production distribution of filmed programs and radio & TV broadcasting and CATV.)

- Jalopy Races From Hollywood (26); Main Event Wrestling (65); Beyond The Yukon (13); Christie Comedies (104); All Girl Wrestling (26); Blackstone, The Magician (39); Vaudeville-Musicals (160); So This Is Hollywood (24).

Hartley Film Foundation Inc.

59 Cat Rock Road, Cos Cob, CT 06807; (203)-869-1818. Elda Hartley, pres. (Producer of educational films on philosophy, psychology, religion, and health.)

Henley, Arthur, Productions

234 Fifth Avenue, New York, NY 10001 (212) 263-0136. Arthur Henley, president. (Production, writing consultation, live and film.)

- Make Up Your Mind.

Hollywood Center for the Audio-Visual Arts

(A facility of the Los Angeles Cultural Affairs Department—412 South Parkview Street—Los Angeles, CA, 90057; (213) 383-7342.); Clarence Inman, dir.; Walter J. Daugherty, Senior museum preparator.

Raymond Horn Productions

Route 6, Box 255, New Bern, NC 28560; (919) 633-3333. Raymond Horn, pres.; Gloria Horn, v.p.

Howco International

1013 Cameron Brown Building, 301 S. McDowell Street, Charlotte, NC 28204. (704) 375-6051; J. Francis White, gen. sls. mgr. (Dist. of features for television.)

The Hubley Studio, Inc.

815 Park Ave., New York, NY 10021 (212) 744-8050. John Hubley, pres.; Faith Hubley, v.p. (Producers of animated theatrical, industrial and TV films.)

I

Independent-International Entertainment

Executive offices: 223 Route 18, East Brunswick, NY 08816. (201) 249-8982. International Division: 165 West 47th St., New York, NY 10036. (212) 869-0333. Dan Q. Kennis, dir. network sls. & syndication; Samuel M. Sherman, in chg. of program development; Al Adamson, in chg. of production (West Coast); Steven Jacobson, dir. creative services. (A division of Independent-International Pictures Corp. Television distributor and packager-producer of feature film packages, TV specials, TV series)

Independent-International Pictures Corp.
Independent-International Communications

1440 Broadway, New York, NY 10018 (202) 869-9333. Lawrence P. Casey, exec. v.p.; Robert A. Kwartler dir., project development. (Producer/distributor of feature films and general programming for non-theatrical, theatrical, cable TV, pay TV, network TV, video cassettes and all subsidiary markets.)

Interlingual Television, K.K.

Mori Building No. 7, No. 2 Nishikubo Tomoecho, Shiba, Minato-Ku Tokyo, Japan. Tel.: 434-2506; Cable Interlingual Tokyo; Telex-J22862, Interlin, Tokyo. Largest independent distributor of Television films in the Far East, with affiliate offices in Australia, USA Europe. Sub-licenses in Japan and other countries over 4,000 half-hours of telecasting film of U.S., European and Australian origin.

- Riptide, Men in Danger Adventures of Long John Silver, Space Angel Cartoons, It's A Small World, Four Winds to Adventure, Roger Ramjet Cartoons, Four Seasons of Japan, Speed and Action, Big Challenge, Golf Around the World, Wings to Adventure, Let's Travel, World's Great Adventures, Animal Kingdom, Camera Eye, Pilopat and Patafil, Popeye the Sailor, Funny Company, Joe & The Bees, New Avengers, Spunky & Tadpole, Zoorama.

International Film Bureau Inc.

332 S. Michigan Ave., Chicago, IL 60604 (312) 427-4545. Wesley Greene, pres.; E.N. Nelsen, v.p., director of distribution. (Distributes live action, animated shorts, documentaries. Curriculum and general audience films for non-profit and commercial television.)

International Film Exchange, Ltd.

159 West 53rd Street, New York, NY 10019; (212) 582-4318; Cable: IFEXREP—New York; Telex: 420748. Paris Office: 45 Rue de Ponthieu, Paris 8e; 359-9600; Cable; FILMEXCHANGE, PARIS. Gerald J. Rappoport, Christopher Wood, Beulah Rappoport, Micheline Charest, Ron Weinberg, Christine Kieffer (Paris), exec. officers. (International distributors and producers of theatrical, non-theatrical, educational and television films.) Exclusive representative for: Ceskoslovensky Filmexport, Romania Films, Filmbulgaria, Defa Films

ITC Entertainment, Inc.

115 East 57th St., New York, NY 10022 (212) 371-6660. Abe Mandell, president; Leonard Kornblum, sr. exec. v.p., corporate administration, finance and business affairs.

Domestic Distribution: Pierre Weis exec. v.p., syndication; Joseph Ceslik, v.p., domestic syndication sales administration and operations; James C. Stern v.p., eastern division sales manager; S. Allen Ash, midwestern division sales manager; Al Lanken, southern division sales manager; Charles Keys, western division sales manager; Cy Kaplan, v.p., special projects.

International Distribution: Armando Nunez, exec. v.p., foreign sales; Valerie Rowley, v.p. and general manager ITC of Canada; Philip Jones, director of television sales, eastern hemisphere.

Creative & Production Services: Joseph Fusco, Jr., exec. v.p., advertising, publicity and promotion; Murray Horowitz, v.p., advertising-public relations, television division; Robert Mandell, director of creative services; Nat Leipziger, exec. v.p., programming and production.

- Series: The Muppet Show (24 half-hours, color); Let's Rock (24 half-hours, color); Return of The Saint (22, color); Edward The King (13 one-hours color); Space: 1999 (44 science-fiction hours, color); When Havoc Struck (12 half-hours, color).
- Features: Entertainment, Volume Two (16 features in color, including "The Boys from Brazil," "Capricorn One," and "Movie, Movie"); Entertainment, Volume One (15 features in color, including "The Return of the Pink Panther," "Cassandra Crossing," "The Eagle Has Landed," and "Voyage of the Damned"); The Thrillers (43 made-for-television movies, color).
- Specials: Superlative Seven (musical-variety hour specials, color, starring Julie Andrews, Steve Lawrence, Eydie Gorme, Ethel Merman, etc.); Super Six (musical-variety hour specials, color, starring Elton John, Julie Andrews, Muppets, Robert Goulet, etc.); Extraordinary 8 (musical variety hour specials, color, starring Julie Andrews, Dick Van Dyke, Jackie Gleason, etc.).
- Series: The Saint (114 hours); Crimes of Passion (30 hours); The Protectors (52 half-hours); My Partner the Ghost (26 hours); Department S (28 hours); UFO (26 hours); The Persuaders (24 hours); The Adventurer (26 half-hours); The Champions (30 hours); The Baron (26 hours); Man in a Suitcase (28 hours); The Prisoner (17 hours); Secret Agent (45 hours); Dangerman (39 half-hours); The Gale Storm Show (125 half-hours); Fury (114 half-hours).

Film/Janel Productions, Inc.

119 West 57th St., New York, NY 10019 (212) 765-1300. Gil Cates, pres. (Producer theatre, films, TV.)

J

JODAV Productions

161 West 54th St., Suite 403, New York, NY 10019 (212) 245-8298. Herman Krawitz, pres.; Robert Weiner, v.p. (Producer of films and TV programs.)

JWT Syndication

420 Lexington Ave., New York, NY 10017 (212) 867-1000. (Producers and distributors of specials and series for cash, barter and time banking) Marie Luisi, senior v.p.; Michael Tremper, v.p.; John Davis, v.p.; Norm Varney, v.p.

- World of Survival (½ hr.-21 orig.; 21 repeats); Comedy Shop (25 orig., 25 repeats); Newsweek News Service (inserts-52 weeks; 13-2½ min. inserts weekly); Today's Woman (inserts-52 weeks; 10 bimonthly 2½-minute inserts); Update on Health with Dr. Timothy Johnson (inserts-3 weekly; 1½ min.); House Call-½ hr.; 26 orig.; 26 repeats.; 4 1-hr. medical specials; Cartoon-a-torial (inserts-52 wks., 5 weekly, 15 second inserts); Take Five with Stiller & Meara (inserts-17 weeks of orig.; 5 weekly, 5 min. inserts); Magic of Mark Wilson (13 weeks of orig.; 5 weekly, 5 min. inserts or 13 orig. half hours.)
- Among the roster of specials are: Billboards Disco Party (top Disco stars; two 1-hour specials); All You Need is Love (5 1-hours on history of popular music); Words & Music-Sammy Cahn (2 1-hour specials from Broadway Show); Country & Western (2 1-hour specials with Wayne Newton & Lynn Anderson); For A Better World (Wildlife/Environmental, with star hosts).

K

K&H Productions, Inc. (Formerly Keitz & KHerndon, Inc.)

3601 Oak Grove, Dallas, TX 75204 (214) 526-5268. R.K. Keitz, pres.; Tom Young, exec. v.p.; John Bronaugh, v.p. (Producers of TV commercials, industrial, education and business films; A/V presentations; producer's services including animation, titles and optical printing.)

Stacy Keach Productions

5216 Laurel Canyon Blvd., North Hollywood, CA 91607 (213) 877-0472. Stacy Keach, Sr., pres.; Mary Keach, v.p.; Andrew J. Davis, secty. (Producer of films, commercials, public information films.)

- Verdict (film/tape series.)

M.A. Kempner, Inc.

4699 North Federal Highway, Pompano Beach, FL 33064 (305) 946-7660. Marvin A. Kempner, pres.

Killiam Shows, Inc.

6 E. 39th St., New York, NY 10016 (212) OR 9-8230. Paul Killiam, pres.; Patrick Montgomery, v.p. (Produces and distributes programs about silent films.)

King Features/Hearst Metrotone Television and Motion Pictures

235 East 45th St., New York City 10017 (212) 682-5600. J.F. D'Angelo, pres.; Benson M. Srere, v.p. & gen. mgr.; Charles Shutt, op., dir. TV & motion pictures; Samuel Gang, intl. sis. mgr.

King International Corporation

124 Lasky Drive, Beverly Hills, CA 90212 (213) 274-0333. Frank King, pres.; Herman King, v.p. mot. pict. div. (Produce TV series.)

- Maya.

King World Productions, Inc.

480 Morris Ave., Summit, NJ 07901 (201) 522-0100. Robert A. King, pres.; Roger M. King, v.p.; Michael G. King, v.p., sls.; Howard Henson, senior sls. rep. (Distributor of syndicated series and producer of TV features.)

- Properties: The Little Rascals (series); Hopalong Cassidy features & series (east of Mississippi); Abbott & Costello (cartoons); The Butcher Show with Merle Ellis (insert program); House Call Show with Dr. Lendon Smith (insert program); Good Growing Show with Allan Swenson (insert program); Carport Show with Bill Baker (insert program).

Klein, Walter J. Company, Ltd.

6301 Carmel Road, Box 220766, Charlotte, NC 28222 (704) 542-1403. Walter J. Klein, pres.; Richard A. Klein, v.p.; Robert P. Klein, v.p.; Vincent Hogan, lab manager; David Jordan, sales dir.; Jeff Smith, script dir.; Betsy Klein, distribution dir.; David Halliburton, Charles Shedd, Kelly McGowen, Casey Jones, Sandy Newton, Leigh Ann Gray, Tinecy Burnette, Jim Cook, Laura Dabney, James Dawkins, John Edwards, David Hackney, Denise Joseph, Steve Smith, Betty Stephens, Allyson Stephens Nancy Beam. (Production and free distribution of sponsored films for industry, TV, associations, government. 16,000 square feet, three locations, two mobile units. Member IQ, SMPTE, CINE, IFPA, PPofA, PRSA, IABC, AFI, WFC.)

L

Alan Landsburg Productions

1554 South Sepulveda Blvd., Los Angeles, CA 90025. (213) 473-9641. Alan Landsburg, chm. of bd.; Howard Lipstone, pres. (Producer for TV and theatrical films).

Lane, Don, Pictures, Inc.

35 West 45th St., N.Y., NY 100369 (212) 840-6355. Donald J. Lane, pres.; John Armstrong, producer/director; Carol Laufer, prod. mgr. (Produces commercial motion pictures, slide films and video cassettes; news films.)

Laufman, Herbert S., & Company

8140 Ridgeway, Skokie, IL 60076; (312) 675-4578. Herbert S. Laufman, pres. (Producers and distributors of live & film programs.)

- It's Baby Time (52).

Lawrence, Robert, Productions, Inc.

342 Madison Avenue, New York, NY 10173 (212) 697-7268. Robert L. Lawrence, pres.

Herbert Leonard Enterprises, Inc.

5300 Fulton Ave., Van Nuys, CA 91401 (213) 783-0457. Herbert B. Leonard, pres.; Walter Bernstein, v.p.; James P. Tierney, Esq., secty. (Producers of filmed series and feature films.)

- Herbert B. Leonard produced Rin Tin Tin (164); Naked City (138); Route 66 (116); Popi (U.A.); produced and directed Going Home (MGM).

Lexington Broadcast Services Co.

800 Third Ave., New York, NY 10022 (212) 838-1185. Henry Siegel, pres.; Roger Lefkon, sr. v.p. A subsidiary of Grey Advertising, Inc. (Program syndication).

Liberty Studios, Inc.

238 East 26th Street, New York, NY 10010 (212) 532-1865. Organized 1961. Anthony Lover, pres., producer-director. (Complete live action and special effects production. Owns and operates fully equipped sound stages, editing/optical facilities, camera, lighting, grip and sound equipment, equipment transfer trucks for sound stage as well as location shooting.

Lieb, Jack, Productions, Inc.

200 East Ontario St., Chicago, IL 60611. Warren H. Lieb, pres.; Charles Kite, editor; Susan Schrier, prod. mgr. (Audio-visual producers).

Lott Video Productions

The Lott Building, P.O. Box 1107, Santa Monica, CA (213) 397-4217. D.N. Lott, owner. (Producers of film commercials.)

M

MCA TV International

445 Park Ave., New York, NY 10022 (212) 759-7500. Ralph Franklin, pres. Foreign Offices: London, England, 170 Piccadilly, London W.1., Mayfair 7211; Paris, France, 8 Rue La Boetie, 75, Paris, Tel: 2659780; Canada: 1851 Est. Rue Serrbrooke, Suite 607, Montreal H2K 4L5 P.Q., Canada; (514) 524-1169; Toronto: MCA Bldg., 2450 Victoria Park Ave., Willowdale, 425, Ontario. Tel.: (416) 491-3000; Munich, Germany: Ismaningerstrasse 98, 8 Munich 80, 98-93-72. Tokyo, Japan: Maison Hirakawa Bldg., 2-Chome Chiyoda-Ku; Tokyo 102, Japan; Tel.: Japan 265-5726. Sydney, Australia: Universal House, Poplar & Pelican Sts., N.S.W. 61-9844. 305 Rue Said Aiach, Sao Paulo, Brazil: Ph. 549-8311. Amsterdam, Netherlands, Rijswijkstraat 175 Amsterdam W.3, Netherlands; (020) 17-52-22. Bogota, Colombia: Apartado Aereo 92471, Bogota, Colombia, 35-58-28. San Juan, Puerto Rico: P.O. Box 5787, Puerta de Tierra Station, San Juan, Puerto Rico 00906; (809) 781-2250.

- Jack Benny Show (104); Run For Your Life (85); Suspense Theatre (53); Checkmate (70); Thriller (67); Frontier Circus (26); Riverboat (44); Cimarron City (26); Overland Trail (17); Suspicion (21); M. Squad (117); Love That Bob (173); Restless Gun (77); D.A.'s Man (26); Staccato (27); S.A.-7 (26); Shotgun Slade (78); Johnny Midnight (39); Coronado 9 (39); Mickey Spillane's Mike Hammer (78); State Trooper (104); Follow That Man (82); Crusader (52); Anthologies (Famous Playhouse) (314); Biff Baker, USA (26); City Detective (65); Pride of the Family (40); Rosemary Clooney (30); Adventures of Kit Carson (104); Ray Milland Show (76); Soldiers of Fortune (52); Leave It to Beaver (234); Bachelor Father (157); Tall Man (75); Calvin & Colonel (26); Wells Fargo (167); Alfred Hitchcock Presents (267 hrs.; 93 ½-hrs.); Arrest & Trial (30); McHale's Navy (138); The Munsters (70); Wide Country (28); Wagon Train (252); Paramount pre-'48 feature films (500); Universal feature films (563); The Name of the Game (76; 90 min.); Ironside (198); Outsider (26); Family Affair (Int'l only) (138); It Takes A Thief (65); Adam 12 (174); The Survivors (26); To Rome, With Love, (26); Marcus Welby, M.D. (172); The Bold Ones (98); World Premieres (114); Mr. Terrific (17); Owen Marshall (69); O'Hara (22); The Partners (20);

Mystery Movie (155); Sarge (15); Alias Smith & Jones (43); Toma (22); Emergency (136); Dragnet (98); Griff (12); Run For Your Life (85); Six Million Dollar Man (108); Kojak (118); Baretta (58); Rockford Files (113); Night Gallery/Sixth Sense (97); Virginian (225); Chase (22); Emergency Plus 4 (26); Sierra (11); Sunshine (12); Sons & Daughters (9); Night Stalker (20); Get Christie Love (22); The Law (3); Lucas Tanner (22); Don Adams Screen Test (24); Rich Man, Poor Man (12); Doctors Hospital (13); Holvak (10); Invisible Man (12); Fay (9); Mobile 1 (13); Sara (13); City of Angeles (13); Bionic Woman (58); Novels II (33); Novels I (30 hrs); Universal Grand (50); Operation Prime Time '78 (The Bastard, Evening in Byzantium, The Immigrants); Operation Prime Time '77 (Testimony of Two Men); Woody Woodpecker and Friends (185).

MGM Television (A Division of Metro-Goldwyn-Mayer Inc.)

10202 W. Washington Blvd., Culver City, CA 90230 (213) 836-3000. (For officers and personnel see listing under Companies).

- Available programming: Medical Center (170); How the West Was Won (24); Tom & Jerry (263); Thirteen Tailor Mades (13); Eight More Tailor Mades (8); Extra Extra Movies (25); Conquest (6); The Rise & Fall of the Third Reich (3); An Evening With Gene Kelly (1); Courtship of Eddie's Father (73); Please Don't Eat the Daisies (58); Daktari (89); Then Came Bronson (26); Man From U.N.C.L.E. (128).

MGS SERVICES

a subsidiary of Viacom International Inc., New York, NY 619 W. 54th St. 10019 (212) 765-4500; Chicago, IL 201 E. Erie St. 60611 (312) 337-3761; Los Angeles, CA 6769 Lexington Ave. (213) 463-1124.

MPO Videotronics, Inc.

820 Second Ave., New York, NY 10017 (212) 557-9260. Arnold Kaiser, pres.; 640 N. La Salle St., Chicago, IL 60610 (312) 527-3680; Bill Bailey, mgr.; 941 No. Highland, Hollywood, CA (212) HO 6-3341.

MRC Films

71 West 23rd Street, New York, NY 10010 (212) 989-1754. Lawrence Mollot, exec. producer. (Producer of industrial, documentary and television films and videocassettes. Also sound slide and filmstrip presentations and multi media productions.)

Madison Square Garden Television

2 Pennsylvania Plaza, New York, NY 10121 (212) 563-8950. Joseph M. Cohen—pres. & chief operating officer, Madison Square Garden Communications Network; Allan Rubinstein, exec. v.p., Madison Square Garden Communications Network; Marlene H. Meyers, dir., operations and traffic, Madison Square Garden Television; John Tagliaferro, gen. mgr., Hughes Television Network; John Koushouris, exec. prod., Hughes Television Network.

- Live Events: National Invitation Tournament (Basketball); Regional College Basketball; Soap Opera Revue; Sportspage.

Manhattan Cable Television Inc.

120 East 23rd St., New York, NY 10010 (212) 598-7200. John F. Gault, pres. CATV service to area south of East 86th St. and south of West 79th St. in Manhattan.

Manley Productions, Inc.

111 West 57th St., N.Y., NY 10019 (212) 541-7733. Walter H. Manley, mgr. Cable: Waltmanly Telex: 421832. (Producers and distributors.)

Marathon International Productions, Inc.

10 E. 49th St., New York, NY 10017 (212) 688-1130. Konstantin Kalser, pres. and exec. producer. (Produce & distribute motion pictures and public information films.)

Mark Three Graphics, Inc.

150 Chesnut St., Providence, R.I. 02903 (401) 421-9242. Tony Petri, pres.; Paul Savacemo, dir. photog.; (Produces film commercials and industrial films, color & B&W.)

Medallion TV Enterprises, Inc.

8831 Sunset Blvd., W. Hollywood, CA 90069; OLdfield 2-8100. Canada: Mel Lefko, 712 Conacher Drive, Willowdale, Ontario M2M 3N5. Miami, Florida—Frank Stone, 7922 East Dr., PL 4-3686. John A. Ettlinger, production. (Producers and distributors of film programs for TV, commercials, packages of live programs, local and network.)

- Celebrity Billiards (30); View the Clue (52); Wrestling Stars of the 60's; Sew East (26); High Road To Danger (39); Medallion Cartoon Package (50); Kingdom of the Sea (41); Wonders of the World (117); Treasurers in Faith (45); Star Route (26); Creeping Terror Package—Volumes, I, II, III & IV, 20/20 Feature Package; Las Vegas Fight of the Week (26).

- Metromedia Producers Corporation

- 485 Lexington Avenue, New York, NY 10017 (212) 682-9100. 5746 Sunset Blvd., Los Angeles CA 90028 (213) 462-7111. CABLE: METPROEAST; TELEX: 424532; Lennart Ringquist, pres.

- Charlie's Angels, Family, Starsky and Hutch, S.W.A.T. (35), Movin' On (44), On The Rocks (22), Chopper One (13), Dusty's Trail (26), Firehouse (13), Here We Go Again (13), My Favorite Martian (107), My Favorite Martian (animated, 16), Primus (26); Made for TV movies: Premium Packages 1-2-3, Spelling-Goldberg Features, Aaron Spelling Features; Anthology Features, Jane Goodall and The World of Animal Behavior (4), The Undersea World of Jacques Cousteau (36), National Geographic Series, Untamed World (156), Beach Bums, Jackie Gleason Specials, Living Together, The Eva Gabor Show, The Cross-Wits, The Merv Griffin Show, Vaudeville, The Ice Palace, Future Shock, The Natural History of Our World: The Time of Man, Lorne Greene Specials (3), That Girl (136), Truth or Consequences, Mayberry R.F.D., For A Better Life, The Donna Fargo Show, David Frost Presents The Guiness Book of World Records (6), Fantasy Island, Groovie Goolies & Friends, Vebas, Friends.

Milner-Fenwick, Inc.

2125 Greenspring Drive, Timonium, MD 21093 (301) 252-1700. David Milner, pres.; Richard Milner, v.p.; Michael Quitt, sales mgr., v.p. (Producers of Educational, training nd documentary films; 16 mm and Super 8mm. Full motion picture services including sound recording, animation, live photography and editing.)

Mode-Art Pictures

3075 West Liberty Avenue, Pittsburgh, PA 15216 (412) 343-8700. James L. Baker, chm. of bd. and chief exec. officer; Robert L. Stone, pres.; (Producers of documentary and industrial films and television commercials.)

Modern Sound Pictures, Inc.

1402 Howard St., Omaha, Nebraska, 68102 (610) 341-8476. Keith T. Smith, pres. (Distributor of over 100 feature films and shorts.)

Modern Talking Picture Service, Inc.

General Offices: 5000 Park St., North St. Petersburg, FL 33709 (813) 541-7571. (Distributor of free-loan sponsored films and video-cassettes, and collateral materials.)

Divisions

Modern TV: Distributor of free-loan 16mm-sound and color motion pictures and videocassettes, and TV spot newsclips and public service announcements for television stations.
Modern Cable Network: Distributor of programs and series for Cable Television Stations.
Modern Satellite Network: Distributor of programs and services.

Affiliations

Modern Telecommunications, Inc.
One Dag Hammarskjold Plaza, New York, NY 10017 (212) 355-0510. (A post-production company specializing in videotape services.) Robert Weisgerber, president.

Morris, Nelson, Productions

11 East 48th Street, New York, NY 10017 (212) PL 3-1484. Nelson Morris. (Producer of feature motion pictures.)

Motion Picture Service Co., Inc.

125 Hyde St., San Francisco, CA 94102 (415) 673-9162. Lino Kwong, operations mgr. (Theatre advertising, special trailers, animated film strips, optical sound services, color film lab.

Movietonews, Inc. (20th Century-Fox Film Corp.)

460 W. 54 St., New York, NY 10019 (212) 977-5500. (Stock film library, international newsreels.)

Mutual General Film Co.

9399 Wilshire Blvd., Beverly Hills, CA 90210 (213) 274-8646. Nicholas Wowchuk, pres.; Harry Wowchuk, exec. v.p.; Miles Holman, gen. mgr.; Harry Stern, sls. mgr. (TV distribution.)

N

NTN Entertainment, Inc.

295 Madison Ave., Rm. 621, New York, NY; (212) 532-6167. Edward P. Noyes, pres.; Robert A. Sulfhorst, exec. v.p.

National Telefilm Associates, Inc.

12636 Beatrice Street, Los Angeles, CA 90066 (213) 390-3663. New York Office: 141 E. 56th St. #3-G 10022. (212) 752-4982. George C. Hatch, chm. of the bd.; Aubrey W. Groskopf, pres.; Arthur Gross, v.p. in charge of worldwide sales; Dee Hopkins, director of intl. sales; Ralph T. Smith, v.p. finance & treas.; Peter Nicholson, controller; Ken Harris, v.p. Film Services; Charles Chaplin, exec. v.p., Canada. Foreign Offices and Reps: Toronto M4W 1C8, Ontario, Canada, 175 Bloor Street, East, Charles S. Chaplin, exec. v.p.; Athens 516, Greece, 46, King Constantine Ave., Panos Spyropoulos, sales rep.; London (Kent), England, 21 Cherry Garden Lane, Bernard Shaw, sales rep.; France, 53, Rue Charlot, 75003, March Hare, Ltd.

- First run half hour syndicated series: Arthur of the Britons (26 action-adventures); Theatre Macabre (26 half hours); First Run: Sports Programming: Billie Jean King (5 half hours).
- New Product 1979: The Best of Dean Martin (26 hours); Fllip Wilson Show (26); Music Country, USA (20); Stranger Than Fiction (26 half-hours); Future Flipper (26 half hours).
- Half-hour series (no. episodes in parenthesis): Blue Angels (39); Boots and Saddles (39); Californians (69); Captured (26); Car 54 (60); Continental Classroom (160); David Cash Show (15); Falcon (39); Get Smart (138); Golddiggers (22); Great Gildersleeve (39); It's Your Bet (390); Jim Backus (39); Loretta Young (250); Not for Hire (39); Panic (31); Pony Express (39); Silent Service I (39); Silent Service II (39); Steve Donovan-Western Marshal (39); T.H.E. Cat (26); Theatre Macabre (26); Victory at Sea (26); Uncommon Valor (26); Animal Secrets (24).
- Hour Shows: Bonanza (310); Cameo Theatre (26); High Chaparral (96); I Spy (82); Laramie (60); Laredo (56); Outlaws (50); Richard Boone (25); Greatest (7); Theatre of Stars (17); News Documentaries (20); Great Movie Cowboys (26 hours) Roy Rogers hosts series of famous cowboy stars including Hoot Gibson, John Wayne, Johnny Mack Brown, Buster Crabbe, etc.
- Third Man (39); Grand Jury (39); Man Without a Gun (52); How To Marry A Millionaire (52); This Is Alice (39); African Patrol (39); Byline Steve Wilson (39); Christie Comedies (107 13-min. shows); Commando Cody (12); Fireside Theatre (100); Frontier Doctor (39); George Jessel's Show Business (26); Glen Cannon (39); Palmer Theater (35); Man from Cochise (156); Mantovani (38); Schlitz Playhouse (26); Stories of the Century (39); Stryker of Scotland Yard (13); William Tell (39); International Playhouse (26); about 2,200 feature films; 1,400 cartoons and short subjects.

Neal, Stanley, Productions, Inc.

(See Don Lane Pictures, Inc.)

Nemeth, Ted, Studios

381 Park Avenue South, New York, NY 10010; 255-4589. Ted Nemeth, producer, director, cinematographer; M.E. Bute, art dir. (Producers of theatrical films and TV commercials.)

New Merritt Enterprises, Inc.

(See Nicholson-Muir Prods., Inc.)

Newsfilm, Inc.

10 East 49th Street, New York, NY 10017 (212) 688-1130. Konstantin Kalser, pres. & exec. prod. (Produces and distributes newsfilms and public information films.)

Nicholson-Muir Productions Inc.

1890 Palmer Avenue, Larchmont, NY 10538. (212) 824-1614. (914) 834-3005. E. Roger Muir, pres.; Nick Nicholson, exec. v.p.

- Generation Jury, Groaners, Second Honeymoon and Supermates, The New Howdy Doody Show, Readers Digest Specials, Headline Hunters (Canada), Definition (Canada, W. Germany and England), Clue Me In, Bluffin Game, Strategy.

Niles, Fred A. Communications Centers, Inc.

1058 W. Washington Blvd., Chicago, IL 60607 (312) 738-4181. Fred A. Niles, pres.; James Miller, in charge of produc.; Norman C. Lindquist, Rolf Brandis, William Harder, 04 v.ps. Allen, James Morgan, vice-pres. Suzanne James, sls. mgr.; Bert Lindberg, chief charge man. Branch offices: Hollywood: 8530 Wilshire Blvd., Suite 500, Beverly Hills, CA. 90211; (213) 659-7392; New York: 240 W. 60th St., N.Y., NY 10023 (212) 586-2333. Joe Kane, gen. mgr.; Jos. Salzburg, v.p.; (Marketing including sales meetings, sales promotion, pictures for TV, theatre distribution, broadcast creative services and production; complete videotape facilities; multi-media presentations; slide films, close circuit TV; business theater presentations; merchandising; editing; media assistance.)

O

Odyssey Productions, Inc.

24 E. 51st St., New York, NY 10022 (212) 421-9595. Milton Fruchtman, pres.

- Lowell Thomas Television series, High Adventure (color-11) World of Lowell Thomas (half-hour—41).
- Official Films, Inc.
- 6 E. 45th Street, New York, NY 10017 (212) 687-8470. Cyndy Wynne, v.p., films. (Producer-distributor of film television series.)
- Disaster! Battle Line, Biography, Survival, The Adventures of Robin Hood, Biography of John F. Kennedy, The Big Story, The Buccaneers, Cartoons, Colonel March of Scotland Yard, Cross Current, Dateline Europe, Decoy, The Hunter, International Detective, The Invisible Man, Mr. Lucky, My Hero, Marilyn Monroe, Overseas Adventure, Peter Gunn, Star Performance, The Star & The Story, Sword of Freedom, Wire Service, Yancy Derringer.
- Okun, Lillian, Productions
- 307 East 44th St., New York, NY, 10017 (212) 661-3958, Ext. 311N. (Live shows—children and teen-age, women's.)
- Owen Murphy Productions, Inc.
- One Turkey Hill Road, South Wesport, CT 06880 (203) 226-4241. Paul Cohen, pres. (Producers of film shows, audio-visual presentations.)
- Owens, Harry, Television Productions
- 6710 Melrose Ave., Hollywood, CA 90038 (213) Webster 9-1491. Harry Owens, pres.; Homer O'Donnell, v.p. (Film producer.)

P

P.A.T. Film Services Inc.

630 Ninth Ave., N.Y., NY 10036 (212) 247-0900. Peter M. Piech, pres.; Ervin Rosenfeld, exec. v.p.; Charles Haydon, v.p. secretary; George A. Fausel, v.p. treasurer. Produces and distributes TV shows and commercials.

Pacific Motion

7243 Santa Monica Blvd., Hollywood, CA 90046. (213) 876-3045. Animation Producer/Directors: Sam Cornell and Gary Katona. (Animation producer for TV commercials.)

Paramount Television

5555 Melrose St., Hollywood, CA 90038: (213) 468-5000. For officers and personnel see listing under Companies.

- Features: Preview 1 (14); Portfolio IX (30); Portfolio VIII (37); Portfolio VII (30); Portfolios VI (30); Portfolio V (35); Portfolio IV (41); Portfolio III (60); Portfolio II (41); Portfolio I (46); Marquee III (18); Marquee III (18); Marquee II (17); Paramount Action Theatre (33); The Untouchables (3).
- Operation Prime-Time (OPT) Projects: Solid Gold '79 (2 hrs.); The Girl, The Gold Watch & Everything (2 hrs.); The Top of the Hill (4 hrs.); The Way They Were (2 hrs.); Golda (4 hrs.); Smiley's People (6 hrs.); Solid Gold (48 hrs.); The Girl, The Gold Watch & Dynamite (2 hrs.).
- Series: Entertainment Tonight (first-run, live, 52 wks.); Taxi (66 half-hrs.); Solid Gold (48 hrs.); Laverne & Shirley (134 half hrs.); Happy Days Again (189 half-hrs.); Make Me Laugh (195 half-hrs.); The Odd Couple (114 half-hrs.); The Brady Bunch (117 half-hrs.); Star Trek (79 hrs.); Love, American Style (224 half-hrs.); Star Trek Animated (22 half-hrs.); The Brady Kids (22 half-hrs.); Mission: Impossible (171 hrs.); The Untouchables (114 hrs.); The Lucy Show (156 half-hrs.); The Best of the West (13 half-hrs.); Bosom Buddies (19 half-hrs.);

The Brady Brides (10 half-hrs.); Fonz & The Happy Days Gang (13 half-hrs.); Foul Play (10 hrs.); The Greatest American Hero (8 hrs.); Mean Jeans (13 half-hrs.); Nero Wolfe (14 hrs.); Happy Days (189 half-hrs.); Here's Boomer (24 half-hrs.); Mork & Mindy (73 half-hrs.); Number 96 (6 hrs.); Taxi (66 half-hrs.).

- Mini-Series: (hours in parenthesis): A Town Like Alice (60); Backstairs at the White House (90); Evita Peron (4); Flesh and Blood (4); Golda (4); The Last Outlaw (8); The Moneychangers (6); Shogun (12); Smiley's People (6); The Timeless Land (8); Tinker, Tailor, Soldier, Spy (6); The Top of the Hill (4); Washington: Behind Closed Doors (12); Winds of War (16).
- Movies of the Week (hrs. in parenthesis): An Act of Love (2); The Greatest American Hero (2); Midnight Offerings (2); Tenspeed and Brown Shoe (2); A Time for Love (2); Act of Violence (2); The Alternative (1½); Assault on the Wayne (1½); Big Bob Johnson (2); Call to Danger (1½); Barbary Coast (2); The Caper (1½); Case for the Defense (1½); Chopper Squad (1½); Cindy (2); Clone Master (2); The Cops and Robin (2); The Deadly Game (2); Death in the Family (1½); The Defection of Simas Kudirka (2); Delancey Street (1½); and others.

Pathe News, Inc.

250 West 57th St., New York, NY 10019 (212) 582-6155. Joseph P. Smith, pres.; Samuel A. Costello, v.p., secty. & treas.; James J. Harrington, v.p. & gen. counsel. (Producer and distributor of documentaries, educational films and children's programs.

- Milestones of the Century (365); Men of Destiny (130); Pathe Educational Shorts (103); Harlem Variety Revue (13); Musical Parade of Stars (1100); Captain David Grief (39); Showtime (39).

John Pearson International

6330 San Vicente Blvd., Suite 301, Los Angeles, CA 90048 (213) 933-5879. John E. Pearson, chm. of bd., Arnie Frank, pres.; Stephen F. Bratcher, dir. of mktg. & sls. (Distribution of TV series, features, specials, etc.)

- Series: (all half-hour and in color) Insight (85), The Andy Williams Show (26), That Girl (136), So the Story Goes (15), The Mike Douglas Variety Show (26).
- Cartoons: Animation Wonderland (18—5–14 minute episodes) Lone Ranger Cartoons (Excl. Australia & Far East, but inclu. Japan, 26 half hour episodes) Mini-Musical Cartoons (14), Laura, Little Girl on the Prairie (26 Half hour Eps.)
- Features: Dan Curtis Dramatic Movies, including Dr. Jekyll and Mr. Hyde, Frankenstein, Portrait of Dorian Gray, Turn of the Screw, Invasion of Carol Enders, Shadow of Fear, Come Die With Me, Nightmare at 43 Hillcrest. Also Flight to Holocaust, Hells Angels 69, Minstrel Man, Shinbone Alley, Dead of Night, Dracula, Panic in Tokyo.
- Sports Programs: (all half-hour and in color) Andy at Indy (13 episodes), Chase the Wind, King Richard the Racer, Iron Men Steel Machines, Rodeo Cowboy, Twilight Cheat, Wings of the Wind. Hour programs in color: Roller Super Stars (26 eps.), World Championship Kick-Boxing (26 eps.), Salute to Baseball, Soccer U.S.A., To Win the Gold.
- Family Programs: (all half-hour and in color) Lassie (140 episodes and 15 features), The Lone Ranger (39 episodes and 13 features), Sergeant Preston of the Yukon (67 episodes), To Lassie With Love (Hour Special), Trick and Treat (130 episodes).
- Children's Programs: The Fabulous Storybook Lady (13 half hours), Young at Heart (15 half hours), Bloomin' Human (15 episodes), Marlo and the Magic Movie Machine (39 hour episodes), Domingo (30 half-hour episodes).
- Entertainment Specials: Steve Allen's Laughback, Lucille Ball Specials (7), Bell Telephone Jubilee Spectacular (90 min.), Johnny Cash-Ridin' the Rails (60 min.), Inside Hollywood (30 min.), Music of the Movies (60 min.), Something Special (22 eps.—60 mins.), Soul to Soul (60 min.), Rock and Roll Revival (67 min.), Count Basie at the Hollywood Palladium (60 min.), Your Choice For the Oscars (120 lmins.), First 100 Years of Recorded Music (120 min.), Happy Birthday Las Vegas (120 min.), Televisa Musical Specials (4—60 min.), Superstars—Las Vegas! including The Ben Vereen Special (86 min.), The John Davidson Special (70 min.), The Strange Case of Alice Cooper (72 min.), Beachboys Twentieth Anniversary Special (60 min.).
- Dramatic Specials: The Going Up of David Lev (90 min.), Macbeth (107 min.), The Shadow (30 min.), Teacher, Teacher (90 min.), American Film Institute Anthology (10 films—various).
- Documentary Specials: Karate, Kung-Fu (60 min.), Sharks (60 min.), To America (97 min.), Giants of the Deep (60 min.), Other Roads, Other Ways (30 min.), Dolphin (60 min.), Shanghai Shadows (60 min.).
- Special Interest Programs: Claymation (8 films—various), Barr Film Catalogue (50 films—various), Country Rhapsody (7 films—various), B.F.A Film/Video Catalogue (700 films—various), Encyclopaedia Britannica Film Catalogue (1000 films—various).

Peckham Productions, Inc.

9 East 48th St., New York, NY 10017 (212) PL 8-0490. Peter H. Peckham, pres.; William R. Littlefield, creative dir.; Hoyt Griffith, editorial dir.; Michael Ferrante, production mgr.

Petersen, Robert E., Productions

Sunset Boulevard, Los Angeles, CA 90069 (213) 657-5100. Robert E. Petersen, chm. of the bd.

Pohlman Film Productions, Inc.

527 No. 27th St., Milwaukee, WI 53208 (302) DIvision 2-6363. (Produces product commercials; industrial motion pictures, multi-media slide shows; slide product.)

Post-Newsweek Programs

(A Division of Post-Newsweek Stations) 1232 31st St. N.W., Washington, D.C. 20007 (202) 342-3800. Joel Chaseman, pres.; Leonard Giarraputo, exec. v.p.

- Shorts: Two one-hour children's entertainment and information specials each year with award-winning short subjects on film and tape and famous personalities in guest appearances.
- Go Tell It . . . Ben Hooks Reports (Six half-hours each year, taped on location with Benjamin Hooks, Executive Director of the NAACP, covering subjects of national importance.
- The Presidents: 80 Years on Camera (Distributed by Gould Entertainment Corporation.) Four one-hour specials on the 20th century Presidents from McKinley to Carter in glimpses of personal and public moments—all on motion picture film. Host is James Garner.
- The American Documents (Distributed by Gould Entertainment Corporation.) Thirteen one-hour specials each focusing on a chapter from America's history with Jean Stapleton, Theodore Bikel, Gloria Swanson, William Shatner, James Whitmore, Lowell Thomas. Narrated by Alexander Scourby.

Premier Film & Recording Corp.

3033 Locust St., St. Louis, MO 63103 (314) 531-3555. Wilson Dalzell, pres. (Producers of motion pictures, TV spots, strip films & carts, recording and record pressings.)

Prime T.V. Films, Inc.

527 Madison Avenue, New York, NY 10022 (212) 421-2170. Alec Campbell, Jr., pres.; Constance Campbell, v.p.

- Features: March of the Wooden Soldiers, When Comedy Was King, Golden Age of Comedy, Lady Takes a Chance, D'Jango (color, western), Day the Sky Exploded, Incredible Petrified World, Teenage Zombies, Ape Man, Corpse Vanishes, Limping Man, White Fire, Memory of Love, Orient Express, Stars Look Down, Spotlight Scandals, Nine East Side Kids: East Side Kids, Boys of the City, That Gang of Mine, Flying Wild, Bowery Blitzkrieg, Mr. Wise Guy, Let's Get Tough, Smart Alecks, Neath Brooklyn Bridge.
- Series: Charlie Chaplin Comedy Theatre—26 half hours, The Goldbergs—39 half hours, Sherlock Holmes—39 half hours, Courageous Cat (130 episodes, color, 5½ mins.)

Program Syndication Services, Inc.

405 Lexington Ave., New York City 10017 (212) 532-1560.

Q

Q.E.D. Productions, Inc.

21 West 46th St., New York, NY 10036 (212) JU2-4291. Robert Baron, pres., prod. dir.; Robert Naud and John Whited, directors; West Coast studio facilities, Helen Miles, mgr. & prod.; Jules Brenner & Gayne Rascher, cameramen; Angie Ross, supervising editor. (Producers of TV commercials, industrials, sales promotion films & multi-media presentations; complete video tape facilities).

Quality Program Sales, Inc.

824 Rome Drive, Los Angeles, CA 90065 (213) 222-8803. Kyle C. Thomas, pres. (Distributes films for TV.)

R

RFB Enterprises, Inc.

% Blumofe, 1100 Alta Loma Rd., Los Angeles, CA. 90069. (213) 657-5000. (Motion picture and TV production) Robert F. Blumofe, pres.-tres.; Kenneth Ziffren, secty.

Radim Films, Inc.

1034 Lake St., Oak Park, IL 60301 (312) 386-4826. Arthur B. Brown, pres.

Radio and Television Packagers, Inc.

9 East 40th St., New York, NY 10021 (212) LE 2-1717. William Cayton, pres. (Prod., dist. filmed series.)

- Jungle (59); Cartoon Classics; Series I (104); Series II (112); Series III (104); The Wild Swans (20); Mr. E. From Tau Ceti (20); The Woodcutter's Wish; The Amazing Gift; The Brave Duckling; The Enchanted Princess; Gunnar the Sailor; The Ice Witch; The Strange Circus; The Valiant Knight; Tale of the Northern Lights; Journey to the Beginning of Time; The Space Explorers; New Adventures of the Space Explorers; The Underseas Explorers; The Frog Princess; Beauty and the Beast; The Fisherman and the Fish; The Magic Antelope; The Tiny Oxen; The Fire Bird; Wanda and the Wicked Princess; Omar and the Ogres. New Features: Treasure Island Revisited, The Adventures of Mutt & Jeff and Bugoff; When Funnies Were Funny (new group of 42 cartoons; Animatoons (32 subjects).

Ragsdale, Carl Associates, Ltd.

4725 Stillbrooke, Houston, TX 77035 (713) 729-6530. Carl V. Ragsdale, pres. (Producer of documentaries and industrial motion pictures, television commercials, and theater short subjects.)

- Branches: Suite 200, 2120 "L" St. NW, Wash. D.C. 20037, Tel: (202) 347-7095, Arthur Neuman, exec. prod. in charge; 16036 Tupper Street, Sepulveda, CA 91343, Tel: (213) 894-6291, Frank Coghlan, exec. prod. in charge; Piazza Cairoli, 113, Roma, Italy 00186, Tel: 654-5182, Douglas R. Fleming, exec. prod. in charge; Suite 414, 630 Ninth Ave. New York City, NY 10036, Tel: (212) PL 7-8699, (Production facilities—Tom Carroll, exec. prod. in charge.)

Rasky, Harry, Productions, Inc.

(Production of feature films and TV specials.) CBC, P.O. Box 500, Terminal A, Toronto, Canada (416) 925-3311. (Worldwide co-productions.)

Redlin Productions, Inc.

19116 Lanark, Reseda, CA (213) 885-8462. William A. Redlin, pres.; Emil F. Redlin, secty., treas. (Producer live and film programs, features.)

Regency Productions, Inc.

10 Columbus Circle, New York, NY 10019 (212) 489-9310; Robert Seidelman, pres. (Producers and distributors.)

Rhodes Productions

6535 Wilshire Blvd., L.A., CA 90048 (213) 655-2900. A division of Filmways, Inc. Jack E. Rhodes, pres.; Ralph V. Cunningham, exec. v.p.; William C. Rhodes, senior v.p.; J. Richard Deitsch, v.p.; Bruce A. Genter, v.p.; Nicholas A. Fasola, acct. exec.

- Hollywood Squares, current nighttime series, 78/26, color, tape. (30 min.); Second City Television-Series I, 26/26, color, tape. (30 min.); Second City Television-Second Season, 26/26, color, tape. (30 min.); Cisco Kid, 156 episodes, color, film, multiple runs. (30 min.); Circle of Stars, 12/12 one-hour country-western musical programs, color, tape; Celebrity Concert Specials, 24/24 hour musical programs, color, tape; Green Acres, 170 episodes, color, film, multiple runs. (30 min.); Addams Family, 64 episodes, black-and-white, film, mutliple runs. (30 min.); Mister Ed, 143 episodes, black-and-white, film, multiple runs. (30 min.); Divorce Court, 260 episodes, color, tape. (30 min.); Gambit, 130/130, color, tape. (30 min.); Keystone Komedies, 79 shorts, various lengths, black-and-white, film; Plants Are Like People, 52 half-hours, color, tape; Mel Tillis Specials, 13 half-hours, color, tape; Lucky Jim Adventure Show, 26/26 half-hours, color, film; Gangway for Comedy, 130/130 half-hour daily strip, color, tape; Stars On Ice, 24/24 half-hours, color, tape; The Seventh Sense, 26/26 half-hours, color, tape.

Ring-Ting-A-Ling Productions

161 West 54th St., Suite 403, New York, NY 10019 (212) 245-8298. Robert Weiner, pres. (Producer of films and TV programs.)

Riviera Productions

6610 Selma Ave., Hollywood, CA 90028 (213) HO 2-8585. F.W. Zens, exec. producer; Leif Rise, assoc. producer, docums., shorts. (Produces filmed series, commercials, features.)

- Punch and Trudy (6); Teletunes (15); Aqua-Lung Adventures (13).

Peter Rodgers Organization, Ltd.

Park Westwood Tower, 969 Hilgard Ave., Los Angeles, CA 90024; (213) 473-4709. Peter S. Rodgers, pres.; Susan M. Warner, secty.-treas.

Roffman, Richard H. Associates

697 West End Avenue, N.Y., NY 10025 (212) RI9-3647. Richard H. Roffman, pres.; Malvina Cohn, John Bowman, v.p.s. (Producers and distributors of shows also do Public Relations and publicity and promotion for shows—also do casting for shows, etc.; also use guests and feature information on numerous cable TV and UHF-TV and AM-radio shows in N.Y.C. area).

Rourke, Jack, Productions

Box 1705, Burbank, CA 91507. Jack Rourke, pres.; William Hagens, v.p.; Jim Rourke, v.p., production; sales; Hank Edwards, v.p., promotion.

Rouzer, Danny, Studio

7022 Melrose Ave., Hollywood, CA 90038 (213) 936-2494. Danny Rouzer, IATSE director of photography. (All types 16mm & 35mm films, Kinescopes, pilots & test commercials, plus video taping and tape to film transfer).

Ruby-Spears Enterprises, Inc.

(A division of The Taft Entertainment Co.) 3459 Cahuenga Blvd., Hollywood, CA 90068 (213) 874-5100. Joseph Ruby, bd. chm.; Kenneth Spears, pres.

- Series: (ABC) Thundarr the Barbarian, The Adventures of Goldie Gold, Heathcliff and Marmaduke.
- Animated Specials: The Puppy Saves the Circus (ABC), Vampire Vabbit (ABC), Miss Switch to the Rescue, Parts 1 & 2 (ABC).

S

SFM Media Service Corp.

1180 Ave. of Americas, New York, NY 10036.

- SFM Holiday Network (package of family feature films), Mobil Showcase Network, Documentary Network, March of Time series, Car Care Central (automotive series).

Sandler Institutional Films, Inc.

7449 Melrose Ave., Hollywood, CA 90046 (213) 876-2021. Allan Sandler, pres. (Production & distribution of educational, institutional films & programming.)

Sextant, Inc.

116 East 68th St., New York, NY 10021 (212) 737-1744. Robert D. Graff, pres.; (TV and motion picture producers.)

- Winston Churchill; The Valiant Years; Franklin D. Roosevelt series; Noah and The Flood; Inside the Movie Kingdom, Young Cassidy.

Show Biz, Inc.

110 21st Ave., So., Nashville, TN 37203 (615) 327-2532. J.R. Dunlap, pres.; Stan Sellers, sr. v.p., mkt.; Dick Montgomery, v.p., sls. mgr.

Showcorporation

P.O. Box 1070, Stamford, CT 06904 (203) 327-9252. C.R. Manby, pres.; John Manby, v.p.) distributor of tape and film series and feature films.)

Silverbach-Lazarus & Co.

9911 West Pico Blvd., Los Angeles, CA (213) 552-2660. Alan Silverbach, chmn; Herb Lazarus, pres. (Producers representatives, consultants, and distributors.)

Slesinger, Stephen, Inc.

1111 N. Westshore Blvd., Tampa, FL 33607 (813) 879-6968 or (813) 837-8773. Shirley A. Lasswell, pres. (Produce films.)

Sonney Amusement Enterprises, Inc.

1654 Cordova St., Los Angeles, CA 90007; Daniel Sonney, pres.; (Producers, distributors of domestic and foreign films.)

Spalla, Rick, Video Productions

Subsidiaries: Hollywood Newsreel Syndicate, Inc., and Rick Spalla Production, Inc.

1622 North Gower St., Hollywood, CA 90028 (213) 469-7307; 301 W. 45th St., New York, NY 10036; 765-4646. Organized 1957. Rick Spalla, pres.; Anthony J. Spalla, v.p.; Jeff Spalla, prod. supvr; (Film and video tape production-producer and packager of TV programs, Commercials and industrial films.)

- Syndicated TV shows: Portrait of a Star; Hollywood Backstage; High Road to Danger. Hollywood Guest Shot; Hobby Nobbing . . . with the Stars; Century of Fashion; The Open Road, Hollywood's Fantastic Artists' and Models' Ball; Hollywood Star Newsreel, and Holiday on Wheels; The Great Getaway; Live two hour Special: California State network hook-up. Miss California International Beauty Pageant (2 hours) (Annual Competition in April).

Spectra Films, Inc.

140 W. 57th St., New York, NY 10019 (212) 757-2509. Fred Pressburger, pres.; Judith Wister, v.p. (Film & tape production)

Aaron Spelling Productions, Inc.

132 So. Rodeo Drive, Beverly Hills, CA 90212 (213) 858-2000. (Production of motion pictures and television) Aaron Spelling, pres.

Sportlite Films

20 N. Wacker Drive, Chicago, Ill. 60606 (302) 236-8488. Alfred D. LeVine, pres.; Jane Gallagher, comptroller; Harry Roehm, unit mgr. (Sales and Rental of Pro Sports action pictures For Libraries, Industry, Govt., Schools, Armed Forces, Library Systems, TV.

- Major League Baseball Incl. World Series Films from '67 thru '80; All Star Games '67 thru '80; Specialty subjects Roberto Clemente, Lou Brock, Johnny Bench + Pitchers Whitey Ford & Bob Gibson ½ Hrs (Color); 16mm Color/Sound or Video. INDY 500 Races from 1958 to 1980 (Each a Separate Film ½ hr.). NGF Instructional (6) Byron Nelson Golf (13 quarters); Women's Golf (4); All American Wrestling Hours B&W/Sd (52); Boxing From Rainbo (26); U.S. Open Golf-1970 to 1981; PGA Tournament 1969–1980; Annual Sports Reviews '69–'74; Horse Racing (13); Play Good Golf (Wright, Revolta, Armour); Center Ice Hockey; Record Breakers NHL. WCT Tennis (1971, 1972, 1973, 1974, 1975, 1976, 1977, 1978, 1979, 1980, 1981). World Cup Soccer (3); Big 10 Football (130); Notre Dame (10); Chicago Bears (13); Marilyn Monroe; Bowling (Ladewig, Nagy Fazio).; Golf Etiquette (1); NFL Video Super Bowls & Bloopers. College Football (8). Texas Rasslin' (13).

Spotlite News, Inc.

15036 Valley Vista Blvd., Sherman Oaks, CA 91403 (213) 986-4622. Algernon G. Walker, pres.; Irene J. Brooks, v.p.; Marcia J. Walker, secty.-treas. (Newsfilm service for commercial clients.) Affiliated with Allend'or Productions Incorporated.

Storyboard, Inc.

(See The Hubley Studios, Inc.)

Survival Anglia Ltd.

420 Lexington Avenue, New York, NY, 10017 (212) 867-6979. John F. Ball, pres.; James T. de Kay, senior v.p.; Victor Simpkins, mgr. business affairs & corporate communications.

- *Properties:* Tales of the Unexpected (66 half-hours, bizarre, mystery-suspense anthology with international stars); Tales of the Unexpected, Series Three (22 half-hours); World of Survival (21 half-hour wildlife episodes hosted by John Forsythe); For a Better World (6 one-hour wildlife/natural history adventures from around the world); Lions of Etosha (one hour special on lions filmed by cinematographer Des Bartlett); Warriors of The Gods (one hour special on Asian elephants by Dieter Plage).

Allen Swift, Inc.

110 West 57th St., New York, NY 10019; Tel: (212) 582-8925. Allen Swift, pres. & creative dir.; Cal A. Stadlen, producer & gen. mgr. (Producers of TV commercials and programs).

Syndicast Services, Inc.

919 Third Ave., New York, NY 10022 (212) 751-3394. Mitchell (Mickey) Johnson, pres.; Leonard V. Koch, exec. v.p.; Sheldon Boden, exec. v.p.; William Koblenzer, v.p. sales; Kathy Antonich, dir. of public relations

- Medix, public affairs health/medical program; The Racers, hosted by Johnny Rutherford; Don Kirshner's Rock Concert, 90-minute rock/variety series; Black Athletes Hall of Fame, annual OTO special hosted by Bill Cosby; Coral Jungle four episodes; Focus On: Hollywood Now, 30-minute magazine format reviewing tv, film, music industries, hosted by CBS personality. David Sheehan; NBC pilot projects; NCAA Football Highlights, three 30-minute specials highlighting 1977 games to air prior to 1978, football season; California Jam II, series of four 60-minute specials on rock concert "happening"; Ebony Affair, weekly 30-minute music/variety series.

T

TVS Television Network

280 Park Ave., New York City 10017 (212) 949-3939. Lee S. Eden, pres.; Bill Madden, v.p. sls.; Richard Hussey, v.p., sports; Steven Hyden, mgr. finance; Ellen Bedell, dir. sta. relations.

D. L. Taffner/Limited

1370 Ave. of Americas, New York City 10019 (212) 245-4680. Donald L. Taffner, pres. (Distributor).

- Benny Hill Show (100 half-hrs. plus feature); Children's Animated Classics (15 1-4 hr. & 90-min.); Hollywood (13 1-hr. documentaries); Kenny Everett Video Show (26 half-hrs.); King of Kensington (65 half-hrs.); Three's Company (130 half-hrs.); Wayne & Shuster (80 half hrs.); World at War (52 half hrs. or 26 hrs., plus ten 1-hr. specials).

Tape-Films, Industries

619 West 54 Street, New York, NY 10019 (212) 977-8600; 941 N. Highland Avenue, Los Angeles, California 90038 (213) 461-3361; 301 E. Erie Street. Chicago, Illinois (312) 951-6700. Charles A. Ahto, v.p./div. mgr. (Duplication and distribution of TV commercials, film and video tape, complete storage facilities).

Telecine Film Studios, Inc.

P.O. Box 613, Park Ridge, IL 60068 (302) 823-1418. Byron L. Friend, pres.; Blanche Sands, office mgr. (Producers of motion pictures, TV series, TV commercials, video-tape.)

- Zoo Parade; Magic Ranch.

Telemated Motion Pictures

77 Bleecker St., N.Y., NY 10012 (212) 682-3434. Saul Taffet, producer-director (Producer of documentaries, commercials, industrial, corporate image, public relations, training, sales promotion, and educational films.

Telemount Pictures, Inc.

P.O. Box 1106, Santa Monica, CA 90406. Henry B. Donovan, pres. and producer. (Producers of Cowboy G-Men film series.)

- 39 Westerns; 13 shorts.

Telepictures Corporation

One Dag Hammarskjold Plaza, New York, NY 10017 (212) 838-1122. Michael J. Solomon, bd. chm. & chief exec. officer; Michael Garin, pres. & chief operating officer; David Lawi, v.p., secty. & treas.; Dick Robertson, v.p.-marketing; Louis Israel, v.p.-gen. sls. mgr.; Jules Stein, pres., TPC Films Intl. (subsidiary); Rocco Viglietta, v.p.-non theatrical and foreign theatrical sales; Ken Levine controller; Jody Shapiro, acct. exec.; Bonnie Adamson, dr. of operations; Patricia Straub, dir., sls. admin. (Distributor)

- *Worldwide Product:* All the Kind Strangers, Angel City, Before and After, Better Late Than Never, Bounty Hunter, Christmas Without Snow, Death of a Princess, Exploring the Unknown, Personal Effects, Hey Abbott, Suicide's Wife, Traffic Jam, Reunion, Last Voyage of The "Valhalla" (2-hour made for TV movies). Clarence Games and the Rock, Golden Moment: An Olympic Love Story, Guyana Tragedy: The Story of Jim Jones, The Sacketts (4-hour made for TV mini-series), Riding Tall, The Savage is Loose (2-hour theatrical motion picture) Forty-Four Days of Christmas, Detective School, One Flight Up, Games People Play, Masters of the Martial Arts, Pepero, Women of Russia (series, numerous episodes, various running times), Falmouth: The Great American Road Race, Hong Kong: Arrival of Boat 845, Lost Treasure of The Concepcion, Third Annual Cheerleading Special (30, 60, 90 minute specials).

Teletest

7243 Santa Monica Blvd., Hollywood, CA 90046 (213) 876-2021. Frank Kratochvil, producer/director. (Producer of test commercials, industrial, corporate communication and public relations films.)

Television Exploitation

1407 Broadway, New York, NY 10018 (212) 563-3105. Robert S. Gettinger, pres. (Distributes feature films.)

- 21 features.

Television Syndication Group, Inc.

366 N. Broadway, Suite 209, Jericho, NY 11753 (516) 935-9024. Jeffrey P. Smith, pres.

Teleworld, Inc.

10 Columbus Circle, New York, NY 10019 (212) 489-9310. Robert Seidelman, pres.; Dalton Danon, vice-pres.; Chad Mason, southern sls. mgr.; Noah Jacobs, eastern sls. mgr.

- Specials: 6 hour long entertainment specials produced and telecast by the BBC, featuring recording artists Helen Reddy, Dionne Warwicke, The Stylistics, Trini Lopez, Freda Payne, Teresa Brewer, Abbe Lane, and comedians George Burns, Frank Gorshin, Marty Allen and Pat Henry. Feature Films: Tele 20 Volume II Package, 20 features including "Hells Angels On Wheels," "The Losers," "And Soon The Darkness" and "Run Angel Run"; Chillers, 29 Horror-sci-fi features including "Corridors of Blood," "Haunted Strangler," "Horror Castle," "Creation of the Humanoids," "Blood of The Vampire," and "Devil Doll"; Action Adventure Classics, 51 features including "Buffalo Bill," "Temple of 1,000 Lights," "Blood For A Silver Dollar," "Assassination in Rome," "Dynamite Joe," and "Espionage in Lisbon."
- Series: Space Maidens and Castaway, 2 new "mini-series" available as 13 half-hour programs. Also may be shown as six one hour or four 90 minute programs; Hopalong Cassidy, 20 first run feature films, 54 one hour programs, 52 half-hour programs.

Telic

317 West 89th St., New York, NY 10024 (212) 595-7078. Woody Siegel, exec. director. (Directors, co-producers, consultants)

Terr, Michael R., Productions

2473 Crestview Dr., Hollywood, CA 90046 (1-213) 654-7506. John Reinhardt, director; Michael Terr, producer. (Film programs and commercials.)

- Wildcat Rider (26); Dancing Age (13); Native Son; Gods of Bali; Black Man in the White World; Lolita Go Go; Here Comes the Sexy; Aliki; Two Kids That Stole A Moon; Songs of Siberia (Musical); The enemies; Jazz at Its Level Best (13 one-half hour color musical); Jazz on Stage (13) one-half hour color-musical; A Tribute to Louis Armstrong and Mahalia Jackson (2½ hrs. Show); Manchurian Massacre (1976); Opium War Atom Boy (90 min.); (1976); Animation; The Open Heart; Mauricette; Cry in The Wind; Our Soviet Sister. New Films: Aba—The Black Superman, Walk the Walk, Deadwood 76, Eegah, Spies-a-Go-Go, The Face of Terror, Fallguy, The Choppers, A Fourth for Marriage.

Thomas, Bob Productions

55 West 42nd St., New York, NY 10036 (212) 221-3602. Robert G. Thomas, pres. (Motion picture and TV producer; motion picture holography; 3-D films)

Time-Life Television

(A Division of Time-Life Films, Inc.) Time & Life Building, New York, NY 10020 (212) 841-2275 Production Office: 12229 Ventura Blvd., Studio City, CA 91604. (213) 506-7700. Austin O. Furst, jr., pres. & chief exec. officer; David Susskind, senior exec. producer; William E. Miller, senior v.p.; Robert J. Peyton, v.p., gen. mgr. for syndication; Robert L. Greenstein, domestic sls mgr.; Chips Barrabee, eastern regional sls. mgr.; Thomas Todd, southern regional sls. mgr.; Jack Garrison, midwestern regional sls. mgr.; Jack Donahue, western regional sls. mgr., Frank Miller, director, international sales; Maria Daniolos, sls. exec.; Wynn Nathan, v.p. television operations; Stephen Scheffer, v.p., programming; Harvey Chertok, v.p. advertising, promotion and publicity. Stephen Elsky, mgr., TV operations.

- Series: The Real McCoys (224 half-hours for comedy strip programming, starring Walter Brennan, Richard Crenna and Kathleen Nolan); Wild, Wild World of Animals (129 half-hours narrated by William Conrad); A Horseman Riding by (13 one-hours, saga of English country life in the Twentieth Century); All Creatures Great and Small (13 one-hours based on the book about an English country-side veterinarian); The Good Life (100 half-hour celebrity interviews with a health and fitness angle-featuring tennis star John Newcomb and Pennie DuPont as hosts; Joyce Davidson Show (Half-hour strip talk/interview program); Food, Wine and Friends (24 half-hours with Robert Carrier in his English mansion where his celebrity-guests enjoy food and wines); World War II: G.I. Diary (25 half-hours narrated by Lloyd Bridges); Doctor Who (98 half-hours of science fiction): The Harold Lloyd World of Comedy (26 first-run half-hours); Monty Python's Flying Circus (39 half-hours); Murder Most English (5 Lord Peter Wimsey detective mysteries; 20 one-hour programs): Vision on (32 half-hours for children): America (26 half-hours or 13 one-hours starring Alistair Cooke); The Goodies (26 comedy half-hours in the Monty Python tradition); The Onedin Line (42 one-hour adventure dramas); Anna Karenina (10 one-hours).
- Features: Danny Wilson's One Hour Family Movies (14); Time-Life Volume 1 (25 movies); Time-Life Volume 2 (23 new titles); Harold Lloyd Film Classics (11 feature films); The Big Nine (9 movies).
- Specials and Mini-Series: An Englishman's Castle (3 one-hour programs of political drama); Moll Flanders (2 100-minute plays based on Daniel Defoe's story starring Julie Foster); The Africans (Profile of today's problem continent in six one hours or 3 two hours or 2 three hours. Hosted by Lou Gossett, Jr.); Americans (13 one-hour profiles including Rosalynn Carter, Woody Hayes, Jodie Foster and others); Ten Who Dared (10 one-hours of adventures with ten of the world's most well-known explorers); Europe, the Mighty Continent (13 one-hour documentaries hosted by Peter Ustinov); Fall of Eagles (13 one-hour dramas of Europe's royal families, narrated by Patricia Neal; Glittering Prizes (Six 75 minutes of critically-acclaimed dramas about life at Cambridge University); BBC Outlook (3 one-hour current documentary specials); The Shirley Bassey Show (6 one-hour specials with well-known guests); The Fight Against Slavery (Docudrama on British slave trade in 6 one-hour programs); The Search for the Nile (6 one-hours narrated by James Mason); The Commanders (7 one-hour profiles of the top military leaders of World War II); Family Classics (10 half-hours of "Ivanhoe"; 9 half-hours of "Little Women"; 6 half-hours of "The Black Tulip"; 13 half-hours of "Last of the Mohicans" and 8 half-hours of "Tom Brown's School Days."); Ripping Yarns (6 half-hour comedy programs in the Monty Python tradition); Fawlty Towers (6 comedy half-hours starring John (Monty Python) Cleese); When the Boat Comes in (26 hour dramas of a miner's family); Window on the World (11 one-hour specials narrated by Burgess Meredith on various subjects); Civilisation (14 one-hours with Dr. Jacob Bronowski as host); Wodehouse Playhouse (13 half-hour comedies); War and Peace (19 one-hours, with Sebastian Cabot as host); Billy Smart's Circus (4 one-hour specials, also available as 4 half-hours); The Six Wives of Henry VIII (6—90-minute dramas); The Fall and Rise of Reginald Perrin (14 half-hours); The Leo Sayer Show (A superstar half-hour TV special); Premiere (6 half-hour dramas); The Dick Van Dyke—Harold Lloyd (90-minute special).

Titus Productions, Inc.

211 E. 51 St., New York, NY 10019 (212) PL 2-6460. Herbert Brodkin, pres.; Robert Buzz Berger, v.p. (Producers of motion pictures, television series and specials.)

Tobias, Herb, and Associates, Inc.

1901 Avenue of the Stars, Suite 840, Century City, Los Angeles, CA 90067 277-6211. (Talent Agency.)

Tower Productions

11541 Landale St., No. Hollywood, CA 91602. H.G. and Barbara J. Rhinelander, owners. (Live programs.)

- Space Patrol (live, plus 300 half hour and 200 15 minute kinescopes).

Transvue TV International Co.

A Division of Transvue Pictures Corp. 14724 Ventura Blvd., Sherman Oaks, CA 91403 (213) 990-5600. Herbert B. Schlosberg, pres.; K. Galloway, secty.

Trident America Programs, Inc.

540 Madison Ave., New York, NY 10022 (212) 355-7070. William F. Storke, pres., Christina Thomas, director of sales. North American distribution of Yorkshire and Tyne Tees

Television (U.K.) product . . . single dramas, documentaries, light entertainment & children's programs.)

Trident Television Associates, Inc.

540 Madison Ave., New York NY 10022 (212) 355-7070. G.E. Ward Thomas, bd. chm.; Arthur Zeiger, pres. & chief exec. officer; Elliott Abrams, vice chm. & chief operating officer; Murray Oken, v.p., domestic sales. (Worldwide television distribution of motion pictures and television programming).

Turn of the Century Fights, Inc.

9 East 40th St., New York, 10016 (212) 532-1711. William D. Cayton, pres.; Jim Jacobs, dir. (Produce and distribute fight films.)

- Knock-Out (now total of 400 subjects, from 1897 thru 1980).
- The Legendary Champions (90-min. feature)

Twentieth Century-Fox Television, A Division of Twentieth Century-Fox Film Corp.

P.O. Box 900, Beverly Hills, CA, 90213 (213) 277-2211. For officers and personnel, see listing under Companies.

- New York Office: 40 West 57th St., New York, NY 10019. Tel.: (212) 977-5500. Joseph Greene, v.p., & gen. sls. mgr.
- Chicago Office: 35 East Wocker Driver, Chicago, IL 60601. Tel.: (312) 372-1589. Stuart Stringfellow, central div. sls. mgr.
- Dallas Office: 11551 Forest Central Drive, Dallas, TX 75243. Tel: (214) 343-9252. Tom Maples, southern division sales manager.
- Series Available:
- Hours: Adventures in Paradise (91); Bracken's World (41); Bus Stop (25); Cade's County (24); Century Theatre—The Fortunes of Nigel (5), A Little Princess (3), The Pathfinder (5), Daniel Boone (165); Five Fingers (16); Follow the Sun (30); Jackie Gleason (100); Hogen (9); Ben Hall (13); Hong Kong (26); How Green Was My Valley (6); Husbands, Wives and Lovers (11); Jack the Ripper (6); Judd for the Defense (50); Lancer (51); Land of the Giants (51); Legend of Custer (17); The Long Hot Summer (26) Lost in Space (83); The Monroes (26); Monte Carlo Show (24); Nashville 99 (4); The New Perry Mason (15); Paper Chase (22); Quiller (13); The Starlost (16); The Time Tunnel (30); 12 O'Clock High (78); Voyage to the Bottom of the Sea (110); Web (6); Young Daniel Boone (8).
- Half-Hours: Adventures in Rainbow Country (26); Anna & the King (13); Arnie (48); Audubon Wildlife Theatre (78); Batman (120); Billy Liar (7); Broken Arrow (72); Circus! (52); Dobie Gillis (147); Doctor Doolittle (17); Double Deckers (17); Fantastic Voyage (17); The Felony Squad (73); The Ghost & Mrs. Muir (50); The Green Hornet (26); The Hardy Boys (17); Journey to the Center of the Earth (17); Julia (86); The Legend of Jesse James (34); The Loner (26); M*A*S*H* (198); That's Hollywood (50); The Incredible World of Adventure (31); Masquerade Party; My Friend Flicka (39); Nanny and the Professor (54); Orson Welles Great Mysteries (26); Peyton Place (514); Return to the Planet of the Apes (animated) (13); Roll Out! (13); Room 222 (113); Run, Joe, Run!; Salty (26); Valentine's Day (34).
- Specials: An Unauthorized Biography: In Search of Howard Hughes; Fox Fanfare; Edith Piaf; Prime of Miss Jean Brodie (3); Tourist (5); TV and the Fabulous 50's.
- Features: Century 5 (35); Century 6 (32); Century 7 (25); Century 8 (25); Century 9 (30); Century 10 (30); Fox 1 (50); Fox II (50); Fox III (50); Golden Century (50); Laurel and Hardy (6); Mark One (11); Planet of the Apes (5); Premiere One (20); Shirley Temple (20).

Twyman Films, Inc.

Box 605, 329 Salem Avenue, Dayton OH 45401 (513) 222-4014. Alan P. Twyman, pres. (Distributor of feature films and shorts from an International Collection.)

- A Paris; Josef von Sternberg's Anatahan; Louis Malle's Calcutta; Jean Luc Godard's Le Gai Savoir; The Fables of Jean de la Fontaine (33) (in French or English); The Food Film; Irregular Stories; King, Murray; The Old Man and the Mouse; Stuck On the Screen; Frederic Rossif's A Wall In Jerusalem; What If . . . ; Lightning.

U

UPA Pictures, Inc.

4440 Lakeside Drive, Burbank, CA, 91505 (213) THornwall 2-7171: Henry G. Saperstein, pres. & exec. producer. (Producers of animated commercials, theatrical films, industrial and TV programs.)

UPA Productions of America

4440 Lakeside Drive, Burbank, CA 91505 (213) 849-6666. Henry G. Saperstein, pres.; (Distributes Mr. Magoo, Dick Tracy, Gerald McBoing-Boing Color Cartoons, family specials, and features, Godzilla, Rodan, sci-fi features etc.

UPITN Corp. (television news agency of United Press International and Independent Television News Ltd.)

77 West 66th Street, New York, NY 10023. (212) 787-0842. Kenneth Coyte, v.p. & chief executive officer; Roy Berke, U.S. manager. (Produces and distributes newsfilm, Roving Report documentary films, industrial and education documentaries, and industrial film clips.

FILM LIBRARY: 321 W. 44th Street, New York, NY 10036 (212) 265-4520. Vincent O'Reilley, library head.

United Artists Television, Inc.

729 Seventh Avenue, New York, NY 10019 (212) 575-3000. (Distributor of feature films; producer and distributor of filmed series and special programs.) For officers and sales directors and regional offices, see listing under Companies.

- Feature Films and cartoons: UA Showcase 11 (30); UA Showcase 10 (30); UA Showcase 9 (29); UA Showcase 8 (30); UA Showcase 7 (30); UA Showcase 6 (30); UA Showcase 5 (31); UA Showcase 3 (27); UA Showcase 2 (21); UA Showcase 1 (19); UA Award (58); UA Box Office (53); UA Premier (44); UA Festival Showcase (44); Charlie Chan/Mr. Wong (12); Joe E. Brown Comedy Time (19) Prime Time Showcase II (24); Prime Time Showcase I (23); Warner Superstars (200); RKO Super (200); AAP Features (26); Warner Bros. Cartoons (327); Popeye Cartoons (234); MGM-11 (30); MGM 10 (30); MGM/8 (25); MGM pre 48: 200 & 1 (200); The Select Group (150); Red Skelton Comedy Time (8); Buster Keaton Slapstick (6); Andy Hardy Theatre (15); Library (310).
- Series: Pink Panther Show (24); Gilligan's Island (98); Rat Patrol (58); Mothers-in-Law (56); Patty Duke Show (104); Outer Limits (49); Popeye (234); Bugs Bunny & His Superstars (327); Yesterday's Newsreel (139); Sports Album (105); East Side/West Side (26); Adventures at Scott Island; Bat Masterson; Bold Venture, Boston Blackie; Case of Dangerous Robin; Cesar's World; Dial 999; Dr. Christian; Everglades; Favorite Story; Harbor Command, Hey Landlord; Highway Patrol; Hollywood & the Stars; I Led 3 Lives; King of Diamonds; Lee Marvin Presents Lawbreaker; Lock-Up; MacKenzies Raiders; Man and the Challenge; Man Called X; Men Into Space; Men of Annapolis; Miami Undercover; Mr. District Attorney; My Mother the Car; New Phil Silvers Show; Ripcord; Rough Riders; Science Fiction Theatre; Sea Hunt, Story of; Tales of the Vikings; Target; This Man Dawson; Tombstone Territory; Troubleshooters; Ultra Man; Unexpected; West Point; Aquanauts; Human Jungle; Stoney Burke; Wolper Specials I and II; Circus Parade

United Artists Television of Canada (A Division of United Artists Television of the Americas, Inc.)

2 Carlton Street, Toronto 2, Ontario, Canada; 363-4033. William Wineberg, gen mgr.

United Artists Television International

729 Seventh Ave., New York, NY 10019 (212) 575-4784. (For officers and branches see listing under Companies.)

United Press International (see also UPITN)

220 E. 42nd St., New York, NY 10017 (212) 850-8600. R.W. Beaton, pres.; H.L. Stevenson, v.p. & ed.; (Company distributes news, audio and film services).

United States Productions, Inc.

47 East 88th St., New York, NY 10028 (212) 876-3774. Francis C. Thayer, pres.; Stowe C. Phelps, exec. prod.; Bert Shapiro, prod.-dir.; David Sawyer, prod.-dir.; Thomas Phipps, script editor. (Producers of theatrical, educational films, documentaries, industrials, commercials.)

Universal Television

A Division of Universal City Studios, Inc. Universal City, CA 91608 (213) 985-4321. (For officers see listing under Companies.)

V

Vavin Inc.

235 East 46th St., New York, NY 10017 (212) 753-3923. Richard de Rochemont, chairman of board and president; NZ moreno, vice-president.

- Produce television films, closed circuit telecasts.

Van Pragg Productions, Inc.

135 E. 55 St., New York, NY 10022 (212) 838-2111. William Van Pragg, exec. dir., Eugene Van Praag, Jeff Glasserow, staff directors.

Veritas Productions, Inc.

1 Laurie Drive, Englewood Cliffs, NJ 07632 (201) 568-4214 or New York City (212) 765-4240. Richard S. Dubelman, pres. (TV and motion picture production)

Versatile Television Productions, Inc.

324 Broadway, Cape Girardeau, MO Edgewater 5-8816. TWX Cape Girardeau 968; Robert O. Hirsch, pres.; Jerry Hollis, mgr. (Producers of animated, live action film and video tape commercials and industrials.

VIACOM Enterprises

1211 Avenue of the Americas, New York, NY 10036 (212) 575-5175. Willard Block, pres. Enterprises; Joe DiSalvo, asst. to pres.; Michael Lambert, v.p. domestic Synd.; Joe Zaleski, v.p. and gen. sls. mgr., domestic synd.; Don Toye, v.p. sls. development; Jack Kelly, v.p. off-network sls.; Peter Yaman, v.p. first run sales; Sal Campo, mgr. pay TV sls.; John Cosgrove, mgr. northeast div.; Bill Stynes, mgr. Mid-Atlantic division; Andrea Cetera, dir. mkt. strategy; Raul Lefcovich, v.p. intl.; James Marrinan, v.p. intl. mktg; Robert Goldfarb, v.p. program planning & development; Barry Weiner, v.p. program packaging; Charlie Tolep, v.p. & gen. mgr., merchandising & licensing; Howard Berk, mktg. mgr; Jerry Kaufer, v.p. creative services; Kate Kelleher, assoc. dir. creative service; Eric Veale, v.p. operations; Bill Theis, dir. operations; Mike Gerber, v.p. business affairs, East Coast.

CBS Studio Center, 4024 North Radford Ave., North Hollywood, CA 91604. (213) 760-5000. Fred Schneier, sr. v.p., program acquisition & motion pictures; Tim Steele, v.p. program development, West Coast; Ron Lightstone, v.p. business affairs.

9720 Wilshire Blvd., Beverly Hills, CA 90212 (213) 278-5644. George Faber, gen., mgr., communications, Don Joannes, mgr., Western division sls.

Domestic Offices: Southeast: 6720 Powers Ferry Rd., N.W., Atlanta, GA 30399 (404) 955-1231, George Milner, mgr.; Southwest: 2 Turtle Creek Village, Dallas, TX 85219 (214) 521-0961, James McCormick, v.p., Southwest Division; Ed Wilson, sls. rep. in Midwest: 625 North Michigan Ave., Chicago, IL 60611 (312) 943-6500, J. Warren Tomassene, mgr., Midwest division, Jeffrey Kinney, sls. rep.; Western: P.O. Box 476, Fairfax, CA 94930; Edward Hewitt, mgr.

- Properties: Air Power (26); All in the Family (207); Alvin Show (26); Andry Griffith Show (249); Assignment: Foreign Legion (26); Beverly Hillbillies (216); Bob Newhart Show (142); Brothers Branigan (39); Call Me Mr. D (51); Colonel Flack (39); Cosmic Christmas (1) *Barter; Circus (26); Devil and Daniel Mouse (1) *Barter; Dick Van Dyke Show (158); December Bride (154); Family Affair (138); Family Feud (half-hour strip); Gomer Pyle (150); Grizzly Adams (38); Grey Ghost (39); Gunsmoke (226); Harlem Globetrotters (22); Have Gun Will Travel (156); Hawaii Five-O; Hogan's Heroes (168); Honeymooners (39); I Love Lucy (179); Intergalactic Thanksgiving (1) *Barter; Lucy-Desi Comedy Hour (13); Marshall Dillon (156); Mary Tyler Moore Show (168); Millionaire (188); Most Important Person (66); My Three Sons (160); Navy Log (102); Our Miss Brooks (127); Ozzie's Girls (24); Perry Mason (245); Petticoat Junction (148); Phil Silvers Show (138); Price is Right (35 new episodes); Rawhide (144); Romie-O and Julie-8 (1) *Barter; Rookies (90); San Francisco Beat (183); Spy! (6); Terrytoons (1200); Trackdown (71); True (33); $25,000 Pyramid (30); Twentieth Century (52); Twilight Zone (151); U.S. Border Patrol (39); Viacom Feature Films: Vols. I, II, III, IV, V, VI & Movie Greats (216 titles); Whirlybirds (111); Whistler (39); Wild Wild West (104).

Victory Television

The Biltmore, Madison Ave. & 43rd St., New York, NY 10017 (212) 687-1516. (Distributors of series and features) James T. Victory, pres.; Ben Okulski, western sales; John Rohrs, northwestern sales.

- Streets of San Francisco (119 one-hour action adventure starring Karl Malden and Michael Douglas); Match Game (Half-hour, first run); Rhoda (110 half-hour sitcom); World of People (200 half-hour episodes).

Video Films Inc.

679 E. Mandoline, Madison Heights, MI 48071 (313) 588-7171. Clifford Hanna, pres., Hrayr Toukhanian, v.p. (Producers of industrial films, tapes, TV commercials.)

- Commercials, Spots, Animation.

Video Programs International

450 Main St., New Rochelle, NY (914) 576-3330. Leonard Kirtman, pres.; Wendy S. Lidell and Robert Zaslow, assoc. producers. (Production and distribution of films and videotapes for all TV media: cable, pay, broadcasting.

Vidistrib, Inc.

4209 Troost Avenue, Studio City, CA, 91604. Tel.: (213) 760-2849. John P. Ballinger, pres.; Rita Cross, v.p.-treas. (TV show sales & distribution syndication, barter, consultation).

Viking Films, Inc.

1600 Broadway, New York, NY 10036 (212) CI 5-8858. Israel M. Berman, pres., exec. producer. (Produce educational, children's subjects, sciences, musicals, cartoons & puppets.)

W

Wade, Roger, Productions Inc.

15 W. 44th St., New York, NY 10036 (212) 575-9111. Roger W. Wade, pres. (Producers of motion pictures for industry, sound slide films, animation commercials, slides.)

Warner Bros. Television

4000 Warner Blvd., Burbank, CA 91522. (213) 843-6000. For officers and personnel, see listing under companies.

- Domestic TV Distribution: 57 Spring Court, Syosset, NY 11791, Peter Affe, v.p. eastern sls. div. (516) 364-9198; Box 999, Covington, LA 70433, Bill Seiler, southeastern sls., dir. (504) 897-2703; 3058 Brockwood Circle, Fort Mitchell, KY. 41017, John Louis, sls. dir. (606) 341-2585; 360 Broadway Ave., Sun Prairie, Wisc., 53590 Robert S. Mitchell, sls. dir. (608) 249-9100; 5161 West 58th Place, Los Angeles, CA 90056, Ed Donaldson, western sls. dir. (213) 299-8005.
- Foreign Distribution: U.K., Eire, Europe, Middle East & Africa: Jack Cook, v.p., 135 Wardour St., London W1V 4AP, England, Australia & New Zealand: Matt Brown, sls. dir., 46 Market St., Box 2661, G.P.L., Sydney, N.S.W. 2001 Australia; Japan, Korea, Taiwan & Guam: Louis K. Mizuno, dir. sls., P.O. Box 762, Tokyo Central Japan; Latin America: Jorge Sanchez, v.p., Latin American sls.; Acapulco 37, Mexico 7 D.F.; Brazil: Lourenberg do Nascimento, Brazil TV mgr., Warner Bros. (South), Inc., Rua Senador Dantas, 19-10 Andar, Rio de Janeiro, RJ, Brazil; France & Colonies, Luxembourg, French Language Station, Switzerland, Monaco, Belgium, French Equatorial Africa & Senegal: Michael Lecourt, TV sls., 20 Rue Troyon, 75 Paris 8, France; Italy, Italian Language Station Switzerland: Christian Livorness, TV sls., 16 Lungotevere Flaminio, Rome 00196, Italy; Spain & Portugal: Xavier Azpiroz, TV sls., Paseo de Gracia, 77, Barcelona 8, Spain.
- Canadian Distribution: Warner Bros., Distributing (Canada) Ltd., 70 Carlton Street, Toronto 2, Ontario, Canada. Max Sheine, mgr. booking dept.; Joan McGregor, admin. asst.
- Domestic Syndication: Roots: The Next Generations (14 hours, color); The Phenomenon of Roots (1 hour special, color); Roots (12 hours, color); Kung Fu (62 hours, color); Wonder Woman (61 hours, color); Pearl (6-hour, multi-part production); The David L. Wolper Specials of the Seventies (30 hours, color); The Waltons (1 hour, color) Welcome Back, Kotter (95 half-hours, color); Chico And The Man (88 half-hours, color); Volume 22 (38 features, all in color); Volume 21 (26 features, all in color); The FBI Story (4 features, all in color); Volume 20 (30 features, all in color); Volume 19 (30 features, 29 in color); Volume 18 (28 features, 25 in color); Volume 17 (27 features, 22 in color); Volume 16 (27 features, 20 in color); Volume 15 (25 features, 18 in color); Volume 14 (27 features, 22 in color); Volume 13 (30 features, 22 in color); Volume 12 (15 features, 13 in color); Volume 11 (15 features, 12 in color); Volume 2-A (22 features, 13 in color) Volume 1-A (23 features, 16 in color); 13 Classic Thrillers (13 features, all in color); Tarzan Features (32 features, 9 in color); The Bowery Boys (48 features); Charlie Chan Film Festival (21 features); Starlite 6 (29 features, 15 in color); Starlite 5 (29 features, 18 in color); Starlite 4 (30 features, 16 in color); Starlite 3 (30 features, 19 in color); Starlite 2 (46 features, 34 in color); Starlite 1 (29 features, 16 in color); The FBI (234 hours, all in color); Maverick (124 hours, western spoof); Tarzan (57 hours, jungle adventures, all in color); F Troop (65 half-hours, 31 in color); Superman (104 live-action half-hours, 52 in color);

Batman/Superman/Aquaman (69 animated half-hours, all in color); Porky Pig & Friends (156 cartoons in color); Bugs Bunny & Friends (100 cartoons in color).

Weiss Global Enterprises

2055 South Saviers Road, Suite 12, Oxnard, CA 93030 (805) 486-4495 Cable: WEISSPICT. Adrian Weiss, pres.; Steven A. Weiss, secty./treas.; Ethel L. Weiss, v.p.; Laurie Weiss, v.p.; Joleen Meibos, oper. mgr.

- Features: Galaxy "14": 14 features including Ginger in the Morning, Molly and Lawless John, Only the Cool; Golden Showmanship "9": 9 features including Cattle Queen of Montana, Slightly Scarlett; Parade "4": 4 features of the 60's, featuring Marshall Thompson and Hedy Lamarr; Impact "120": 120 features produced by Robert L. Lippert, including Baron of Arizona, I Shot Jesse James, King Dinosaur, Rocketship X-M, Sins of Jezebel; Westerns: 60 action features starring Johnny Mack Brown, Harry Carey, Fred Kohler, Jr., Rex Lease, Buddy Roosevelt and Bob Steele; Vintage Flicks: 24 features from the '30's and '40's; One-Hour Documentaries: The Brave Rifles (51 mins.), Our Time in Hell (51 mins.); Those Crazy Americans (54 mins.).
- Serials: Custer's Last Stand, (15 episodes); The Black Coin, (15 episodes); The Clutching Hand, (15 episodes).
- Series: Make Room For Daddy (161 half-hour programs); My Little Margie (126 half-hour shows); Journey! 26 half-hours—all in color); Waterfront (78 half-hour adventures); The Adventure of Jim Bowie (76 half-hours); The Traveler/Northwest Traveler (184 half-hour, true-life adventures); Ski West, (39 half-hour sports programs); Craig Kennedy, Criminologist, (26 half-hour mysteries); Thrill of Your Life, (13 half-hours); Canine Comments (13 quarter-hour short subjects).
- Comedy Shorts: The Chuckle Heads (150 five-minute slapstick comedy shorts).
- Cartoons: Alice by Walt Disney (10 ten-minute cartoons); Krazy Kid Kartunes (4 six-minute cartoons); Nursery Rhymes (6 one and one-half min. cartoons).

William Morris Agency, Inc.

1350 Avenue of the Americas, New York, NY 10019; (212) 586-5100. Beverly Hills Office: 151 El Camino (213) 274-7451. Nashville Office: 2325 Crestmoor Rd., 385-0310. London, Rome and Munich. (Representatives for artists and all other creative talent in all fields of the entertainment world.) Abe Lastfogel, Beverly Hills, and Nat Lefkowitz, New York, chmen. of bd., emeriti; Morris Stoller, Beverly Hills, chm. of bd.; Sam Weisbord, Beverly Hills, pres.; exec. v.p.s: Beverly Hills; Norman Brokaw, Roger Davis, Tony Fantozzi, Stan Kamen, Walter Zifkin; New York: Lee Stevens, Lou Weiss, Larry Lewis, New York, v.p.; treas.

Television Department:

Beverly Hills: Larry Auerbach, Arthur Axelman, Jeralyn Badgley, Alan Berger, Norman Brokaw, Bruce Brown, Lee Cohen, Robert Crestani, Ruth Englehardt, Tony Fantozzi, Bob Goodman, Sam Haskell, Martin Hurwitz, Alan Iezman, Jerry Katzman, Steve Konow, Sol Leon, Debbie Miller, Marc Pariser, Pam Prince, Joe Rivkin, John Rosenmayer, Hal Ross, David Schiff, Michael Shaw, Steve Weiss, Fred Westheimer, Jeff Witjas, Paul Yamamoto.

New York: Lucy Aceto, Leo Bookman, Arthur Fuhrer, Jim Griffin, Leon Memoli, Jonathan Russo, James Sarnoff, Lee Stevens, Lou Weiss, Ron Yatter.

Worldvision Enterprises, Inc.

660 Madison Avenue, New York, NY 10021. (212) 832-3838; Kevin O'Sullivan, pres. & chief executive officer; Neil Delman, exec. v.p.; Hal Golden, senior v.p., domestic sales, Howard Lloyd, senior v.p., western division; Jerry Rettig, senior v.p., creative services; Bert Cohen, senior v.p., intl sls; Al Hartigan, senior v.p.—special projects; Larry Gottlieb, senior v.p. finance; Hal Graham, v.p., special projects, west coast; John Ryan, v.p. eastern division manager; Jim Thomson, v.p. southern division manager; Randy Hanson, sales manager, western division; Ben DeAugusta, v.p., operations; Eugene Moss, v.p.-advertising, promo & public relations; Lucille Shevett, dir. adv & promo; Claudia Young, dir. office mgmt. & personnel; John Barrett, v.p., central division manager; Paul Danylik, account executive, central division; Bill Baffi, account executive, eastern division; Reggie Jester, account executive, southern division; Ed O'Brien, account executive, western division; Martin Weinstein, director, pay-cable programming, Burton Rosenburgh, general sales manager, Evergreen Programs Inc.; Stephen Appel, assistant to the president. Domestic offices: New York, NY 10021, 660 Madison Ave. (212) 832-3838. Chicago, IL 60611, 625 N. Michigan Blvd. (312) 642-2650. Beverly Hills, CA 90212, 9465 Wilshire Blvd. (213) 273-7667. Atlanta, GA 30346, 14 Perimeter Center East (404) 394-7444.

- Little House On The Prairie (166 hours); The Love Boat (150 hours); Barnaby Jones (177 hours); The Man From Atlantis (20½ hours); Holocaust (10 hours); Against the Wind (13 hours); Prime VII (25 features); Prime VI (19 features); Prime V (26 features); Prime IV (26 features); Prime III (16 features); Prime II (16 features); Prime I (10 features); The Don Lane Show (one-hour strip); The Doris Day Show (128 half-hours); Let's Make a Deal (half-hours); Newlywed Game (half-hours); Casper Cartoons (244); Dark Shadows (half-hour strip series); Wonders of the Wild (26 half-hours); The Mod Squad (124 hours); Ben Casey (153 hours); The Next Step Beyond (24 half-hours); Combat (152 hours); The Invaders (43 hours); The Fugitive (120 hours).

Wrather Corporation

270 No. Canon Drive, Beverly Hills, CA 90210 (213) 278-8521. (Producers & distributor of film series.) Subsidiaries include Lassie Television Inc. and Lone Ranger Television, Inc.

- Lassie; Lone Ranger; Lone Ranger Cartoon; Sgt. Preston of the Yukon.

Wright, Carter, Enterprises

6533 Hollywood Blvd., Hollywood CA 90028. (213) HO 9-0944. Carter Wright, pres.; Stephanie Morriott, secty.; June Martin, treas. (Live and film shows, commercials).

Zoetrope Studios

529 Pacific St., San Francisco, CA 94133. (415) 788-7500. Robert R. Spiotta, pres.; Bernard Gersten, v.p. (Motion picture production and rental of production facilities.)

Television

Stations

Channel Allocations

TELEVISION STATIONS

[A listing of television stations in the United States. All stations are authorized by and operate under the approval of the Federal Communications Commission, Washington, DC]

ALABAMA

Anniston

WHMA-TV (Channel 40) CBS

P.O. Box 40, Anniston, AL 36201 (205) 237-8651.

Harry E. Mabry, gen. mgr.; David Atchison, prog. dir.; Bill Hagler, gen. sales mgr.; James A. Hudson, chief eng.

Operation: 1969.

Birmingham

WAPI (Channel 13) NBC

Newhouse Broadcasting Corp., P.O. Box 10502, Birmingham, AL 35202 (205) 933-2720.

Donald D. Wear, vice-pres. and genl. mgr.; Robert L. Williams, sales mgr.; Paul D. Gross, chief eng.; Everett Holle, prog. dir.; James B. Chenoweth, promo. dir.; nat'l. reps., Major Market Television.

Operation: 1949; TV sets in area: 593,000.

WBMG (Channel 42) CBS

Birmingham Television Company, Box 6146, Birmingham, AL 35209 (205) 252-9821.

Roy H. Park, pres.; Hugh M. Smith, exec. v.p. & gen. mgr.; Charles Roundtree, operations dir.; Fred Vinson, chief eng.; Ray White, news dir.; Linda McCarty, traffic dir.

Operation: 1965. TV homes in area: 541,000.

WBRC-TV (Channel 6) ABC

Taft Broadcasting Company, P.O. Box 6, Birmingham, AL 35201 (205) 322-4701.

Richard S. Stephen, gen. mgr.; Ann Bryant, prgm. dir.; Bob Fanning, gen. sales mgr.; Harry L. Lawson, chief eng.; Telerys Agency, nat'l rep.

Operation: 1949; TV sets in area: 695,500.

Dothan

WDHN (Channel 18) ABC

Southeast Alabama Broadcasting Co., P.O. Box 6237, Dothan, AL 36302 (205) 794-2651.

David Antoniak, pres.; Emery McCullough, v.p. & gen. mgr.; Ruth McCullough, asst. mgr.; Elaine Bollinger, prgm. dir.; David Kirkland, prodn. mgr.; Richard Blanton, dir. of eng.

WTVY (Channel 4) CBS, ABC

WTVY, Inc., P.O. Box 1089, Dothan, AL 36301 (205) 792-3195.

Charles Woods, owner; Eldon Klapal, mgr.; Reginald Mitchell, prgm. dir.; Jerry Vann, news dir.; Wayne Register, prodn. mgr.; John Gause, sales mgr.; Glynn Rogers, Sr., dir. of eng.; regional rep., natl. rep. Avery Knodel.

Operation: 1955; TV sets in area: 331,800.

Florence

WOWL-TV (Channel 15)

Television Muscle Shoals, Inc., 840 Cypress Mill Road, P.O. Box 2220, Florence, AL 35630 (205) 764-7711.

Dick Biddle, chm. of bd. & CEO; Jack Worley, pres.; Lincoln Williams, operations mgr.; Anthony Cox, chief eng.; Sara Biddle, community relations dir.; Terry E. Witt, pro. dir.; Rick Adams, news desk mgr.

Operation: 1957; TV homes in area: approx. 234,000.

Huntsville

WAAY-TV (Channel 31)

Rocket City Television, Inc., 1000 Monte Sano Blvd., S.E.; Huntsville, AL 35801 (205) 539-1783.

M.D. Smith IV, pres.; Dan Whitsett, v.p. & sls. mgr.; Robert A. Gay, chief eng.

Operation: 1959.

WHNT-TV (Channel 19) CBS

The Times Alabama Broadcasting, Inc. 960 Monte Sano Blvd., P.O. Box 19, Huntsville, AL 35804 (205) 539-5743.

Thomas M. Percer, exec. v.p. & gen. mgr.; Grady Smith, dir. of sales; Dick Wright, operations mgr.; Jerome Hall, chief eng.; Dan Cooper, news dir.; nat'l. rep: Katz Television.

Operation: 1963.

WAFF-TV (Channel 48) NBC

American Valley Corp., Huntsville, AL 35801 (205) 533-4848.

Ken Gerdes, v.p. & gen. mgr.; Lee Brantley, station operations mgr.; Harvey Libow sls. mgr.; Bishop Ellison, chief eng.

Operation: 1954.

Mobile

WALA-TV (Channel 10) NBC

Universal Communications Corp., 210 Government St., Mobile, AL 36601, P.O. Box 1548 (205) 433-3754. TWX 810 741-2642. FAX-FFB.

Robert F. Finke, v.p. & gen. mgr.; Andy Lee, gen. sls. mgr.; Larry Pate prog. dir., John Long, chief eng.; Jim Briggs, prod. mgr.; Wayne Barnett, news dir.; Ron Gollnick, sports dir.; Stuart Kellogg, pro. mgr.; Karen Smith, traffic mgr.; nat'l rep.: MMT Sales.

Operation: 1953; TV sets in area: 425,900.

WKRG-TV (Channel 5) CBS

WKRG-TV, Inc., 162 St. Louis St., Mobile, AL, 36601 (205) 432-5505.

Zelma K. Giddens, pres. and treas.; Toulmin Greer, secty.; C.P. Persons, Jr., exec. v.p. and gen. mgr.; Thomas W. Diamond, asst. secty.; Curtis Willard, chief eng.; nat'l rep.: Katz Agency.

Operation: 1955; TV sets in area: 422,400.

Montgomery

WCOV-TV (Channel 20) CBS

WCOV, Inc., P.O. Box 2505, Montgomery, AL 36105 (205) 288-7020.

H. Barfield, pres.; Karl Richards, v.p. & gen. mgr.; Phil Witt, chief eng.; nat'l rep.: Meeker Inc.

Operation: 1953; TV sets in area, 122,200.

WKAB-TV (Channel 32) ABC

Montgomery Independent Telecasters, Inc., P.O. Box 3236, Montgomery, AL 36109 (205) 272-5331.

Cy N. Bahakel, pres.; John Rogers, gen. mgr.; Tommy Spain, gen. sls. mgr.; Judy Motowicki, Operations mgr.; Ken Myers, chief eng. nat'l rep., H.R. Television.

Operation: March 24, 1962; TV households: 156,000.

WSFA-TV (Channel 12) NBC

Cosmos Broadcasting Corp., 10 E. Delano, P.O. Box 2566, Montgomery, AL 36105 (205) 281-2900.

Charles Batson, chm of bd. * pres.; Charles Whitehunt, v.p. & gen. mgr.; Robert Ward, gen. sls. mgr., Richard Payne, chief eng.; natl. rep., Harrington, Righter & Parsons

Operation: 1954; TV sets in area: 367,300.

Selma

WSLA (Channel 8) CBS

Central Alabama Broadcasters, Inc., Box 1888, Selma, AL 36701 (205) 875-2240.

Tommy G. Bond, gen. & sales mgr.; George Singleton, operations mgr.

Operation: 1960.

Tuscaloosa

WCFT-TV (Channel 33) NBC, CBS

Service Broadcasters, Inc.

Stan Siegal, gen. mgr.; Jim Caruthers, sls. mgr.; Tommy Ray, operations mgr.; John Fletcher, chief eng., Eva Rozier, promo. dir.; nat'l rep.: Adam Young Inc.

Operation: Oct. 29, 1965; TV households: 79,900.

ALASKA

Anchorage

KENI-TV (Channel 2) NBC, ABC

Midnight Sun Broadcasters, Inc., 4th Ave. Theatre Bldg, 99510; P.O. Box 1160; (907) 272-7461.

Alvin O. Bramstedt, station mgr.; Sandra Moore, prog. dir.; Charles Gray, tech. dir.; nat'l rep.; Simpson/Reilly & Assoc. and Avery Knodel, Inc.

Operation: 1953; TV sets in area: 51,800.

KIMO (Channel 13)

ABC Alaska Broadcasting, Inc. 3910 Seward Highway, Anchorage, AK 99503

Duane L. Triplett, pres. & gen. mgr.; Richard M. Zook, sr. v.p. & station mgr.; Robert G. Underwood, gen. sls. mgr., David Geesin, prgm. dir.; Nat'l rep, HR Television.

KTVA (Channel 11) CBS

Northern Television, Inc., Box 2200, Anchorage, AK; 272-3456.

A.G. Hiebert, chm.; Ron Moore, pres., Anchorage Div.; Bruce Sloan, operations mgr.; nat'l rep., Art Moore, Inc., Adam Young, Inc.

Operation: 1953; TV sets in area: 70,000.

Fairbanks

KFAR-TV (Channel 2) NBC, ABC

Midnight Sun Broadcasters Inc., P.O. 910, Fairbanks, AK 99701.

A.O. Bramstedt, gen. mgr.; Donald Andon, station mgr.; Charles Gray, chief eng.; nat'l rep. S., J.C. Fletcher, N.Y.; William Simpson & Assoc.

Operation: 1955; TV sets in area: 15,500.

KTVF (Channel 11) CBS

Northern Television, Inc., Box 950, Fairbanks, AK 99707 (907) 452-5121.

A.G. Hiebert, chm.; Henry Hove, pres.; nat'l rep., Art Moore, Inc.

Operation: 1955; TV sets in area: 27,000.

Juneau

KINY-TV (Channel 8) NBC

Midnight Sun Broadcasters, Inc., 231 South Franklin St., Juneau, AK 99801 (907) 586-1800.

Kenneth L. Wiley, gen. mgr.; David Moore, chief eng.; nat'l reps., James C. Fletcher, Jr., NY; Simpson and Associates, Pacific Coast.

Operation: 1956.

Sitka

KIFW-TV (Channel 13) CBS

Sitka Broadcasting Co., Box 299 Sitka, AK 99835. (907) 747-3422.

Roy Paschal, pres. & gen. mgr.; Steven L. Rhyner, sta. mgr. & prog. dir.

Operation: 1966.

ARIZONA

Flagstaff

KNAZ-TV (Channel 2) NBC

Grand Canyon TV Co., Box 1843, Flagstaff, AZ 86002 (602) 774-1818.

Dan Modisett, v.p. & gen. mgr.; Tom Monroe, chief engr.; Arvis Duke, sales mgr.

Operation: 1970.

Nogales-Tucson

KZAZ (Channel 11)

Roadrunner Television, A Limited Partnership, 2445 N. Tucson Blvd., Tucson, AZ 85716. 134 Grand Ave., Nogales, AZ 85621 (602) 795-0311.

Gene Adelstein, gen. mgr.; Janice Clark, business mgr.; Harry West, prodn. mgr.; Rich Heatley, chief eng.; Ed Kassel, film dir.; nat'l rep., Adam Young, Inc.

Operation: February, 1967; TV households: 182,000.

Phoenix

KOOL-TV (Channel 10) CBS

Kool Radio-Television Inc., 511 W. Adams St., Phoenix, AZ (602) 257-1234.

Tom Chauncey, chm. & pres.; Homer Lane, exec. vice-pres. & gen. mgr.; Bob Martin, v.p., programming;

Harry Ebbesen, gen. sls. mgr.; Albin Hillstrom, v.p., chief eng.; nat'l rep., Sebtel.
Operation: 1953; TV sets in area: 539,200.

KPAZ (Channel 21)

Trinity Broadcasting of Arizona, Inc., 3351 E. McDowell Rd., Phoenix, AZ 85008 (602) 273-1477.
Paul F. Crouch, pres.
Operation: December, 1969; TV households: 310,000.

KPHO-TV (Channel 5)

Broadcasting Division, Meredith Corporation, 4016 N. Black Canyon, Phoenix, AZ 85017 (602) 248-7474.
William C. McReynolds, v.p. & gen. mgr.; Norris Reichel, gen. sls. mgr.; Don Pauly, sls. mgr.; Ernie Hill, natl. sls. mgr.; Ed Aiken, prgm. mgr.; Linda Stone, promo. mgr.
Operation: 1949; Total TV Homes: in survey area, 569,700; metro area, 311,400.

KTVK (Channel 3) ABC

Arizona Television Company, 3435 North 16th St., P.O. Box 5068, Phoenix, AZ.
Ernest W. McFarland, pres.; Burton LaDow, gen. mgr.; Earl Mehaffey, chief eng.; nat'l rep., Edward Petry Co., Inc.
Operation: 1955; TV homes in area: 480,000.

Phoenix-Mesa

KPNX-TV (Channel 12) NBC

KPNX Broadcasting Co., 1101 N. Central Ave., P.O. Box 711, Phoenix, AZ (602) 258-7333.
C.E. Cooney, pres., gen. mgr.; Bob Allingham, v.p., prog. dir.; Cliff Stevens, v.p., eng.; nat'l rep., Blair Television,
Operation: 1954; TV sets in area: 470,200.

Tucson

KGUN, TV (Channel 9)

May Broadcasting Co., P.O. Box 5707, Tucson, AZ 85703 (602) 792-9933.
Scott Vaughan, v.p. & gen. mgr.; Jack Parris, v.p. & station mgr.; Larry Schnebly, regional sls. mgr.; Mike Sommerfeld, gen. sls. mgr.; John Sawyer, chief acct.; Glenn Hills, chief eng.; nat'l rep. Petry.
Operation: 1956; TV sets in area: 238,970.

KOLD-TV (Channel 13) CBS

Universal Communications Corp., 115 W. Drachman St., Tucson, AZ (602) 624-2511.
Jay Watson, v.p. & gen. mgr.; Paul Cassidy, gen. sls. mgr.; Tom Foos, mgr.-programming/pro.; Phil Dunton, chief eng.; Ken Strobeck, news dir.; natl rep., MMT.
Operation: 1953; TV sets in area, 190,000.

KVOA-TV (Channel 4) NBC

Channel 4-TV, Inc., P.O. Box 5188, 209 W. Elm, Tucson, AZ 85703 (602) 792-2270.
Jon F. Ruby, v.p. & gen. mgr.; Jim Joslyn, station mgr.; Dave Kerrigan, prod. mgr.; Fred Allison, news dir.; Ralph Turk, chief eng.; Renee Bear, controller. nat'l. rep., Blair.
Operation: 1953; TV sets in area: 238,970.

Yuma

KYEL-TV (Channel 13) NBC

Chapman TV of Tuscaloosa, Inc., Box 592, Yuma, AZ 85364 (602) 782-3881.
Cliff Brown, pres. & gen. mgr.; Tom Hagner, sls. mgr.; K.E. Jones, chief eng.; nat'l rep., Avery-Knodel Inc.
Operation: December 1963; TV households, 755,900.

ARKANSAS

Fayetteville-Springdale

KTVP (Channel 29) ABC, CBS

Noark Investment Co., P.O. Box 4150, Fort Smith, AR 72901 (501) 783-4105.
George T. Hernreich, pres.; Robert Hernreich, gen. mgr.; Pat Parker, gen. sls. mgr.; Charles Davis, chief eng.; Karen Littlejohn, sls. secty.
Operation: February, 1969; TV households: 41,300.

Ft. Smith

KFPW-TV (Channel 40) CBS, ABC

KFPW Broadcasting Co., 2415 N. Albert Pike, Ft. Smith, AR 72901 (501) 783-4105.
George Hernreich, owner; Robert Hernreich, gen. mgr.
Operation: 1971.

KFSM-TV (Channel 5) NBC, ABC

Buford TV Inc. of Ft. Smith, 318 N. 13th St., Ft. Smith, AR 72901 (501) 783-3131
Robert Browning, exec. v.p. & gen. mgr.; Gene Graham, gen. sls. mgr.; James Anhalt, chief eng.
Operation: 1956.

KLMN (Channel 24)

MCM Broadcasting Co., c/ Melvin B. Caldwell, Box 1867, Ft. Smith, AR 72902.
David C. McClinton, pres.; Melvin B. Caldwell, exec. v.p. & gen. mgr.; Bill Conrad, secty.-treas.
Operation: 1978.

Jonesboro

KAIT-TV (Channel 8) ABC

George T. Hernreich, P.O. Box 790, Jonesboro, AR 72401.
Robert Hernreich gen. mgr.; David Cunningham, station mgr.; Al Banks, sls. mgr.; nat'l rep., Avery-Knodel.
Operation: July, 1963; TV households: 157,600.

Little Rock

KARK-TV (Channel 4) NBC

Mullins Broadcasting Co., 201 West Third, Little Rock, AR 72203 (501) 376-2481; TWX 910-722-7417.
A subsidiary of Combined Communications Corp., Phoenix, AZ.
David J. Jones, pres. & gen. mgr.; Ray V. Brown, gen. sls. mgr.; Jim Tuley, pgm. dir. David A. Course, promo. dir.; Dean Howard, v.p. & chief eng.; nat'l rep., Blair Television.
Operation: 1954; TV sets in area: 546,310.

KTHV (Channel 11) CBS

Arkansas Television Co., 8th & Izard Sts., P.O. Box 269, Little Rock, AR 72203 (501) 376-1111.

Robert L. Brown, pres. & gen. mgr., KTHV, Channel 11; C.S. Berry, chm. of bd., ARK Television Co.; Marcus George, pres. & treas., ARK Television Co.; Lonnie Gibbons, sales mgr.; Bob Hicks, prog. mgr.; nat'l rep: Katz.
Operation: 1955; TV sets in area: 647,500.

Little Rock-Pine Bluff

KATV (Channel 7) ABC
Leake Industries, Inc. 401 Main Street, Little Rock, AR 72203 (501) 372-7777. Also: Jefferson Square Shopping Center, Pine Bluff, AR 71603 (501) 535-7630; Suite 104, Meyer Bldg., Hot Springs, AR 71901 (501) 624-5777; McIlroy Bank, Fayetteville, AR 72701 (501) 442-0324.
James C. Leake, chmn.; John G. Leake, pres.; Dale Nicholson, v.p. & gen. ad mgr.; Rod Holliday, secty.-treas.; Cal Dring, natl sls. mgr.; Richard Smith, local sls. mgr.; A.R. Garrett, v.p. eng'.; Eric Nelson, prgm. dir.; nat'l rep. Edward Petry & Co., Inc.
Operation: 1953; TV sets in area; 647,500.

CALIFORNIA

Bakersfield

KBAK-TV (Channel 29) ABC
Harriscope Broadcasting Corp., Box 2929, Bakersfield, CA 93303 (805) 327-7955.
Burt I. Harris, pres.; Wayne Killmer, v.p. & gen. mgr.; Bob Banks, chief eng'.; nat'l rep. Katz.
Operation: 1953; TV sets in area: 560,100.

KERO-TV (Channel 23) NBC
McGraw-Hill Bdctg. Co., Inc., 321 21 Street, Bakersfield, CA 93303 (805) 327-1441. P.O. Box 2367.
Norman E. Walt, pres.; Fred Carlton, v.p. & gen. mgr.; Bill Kline, prog. dir.; Norman Hall, dir. of eng'.; Dave Halyaman, news dir.
Operation: 1953; TV sets in area: 280,000.

KPWR-TV (Channel 17) CBS
Gillett Broadcasting Co., P.O. Box 1700, Bakersfield, CA 93302 (805) 327-7511. Dave Moore, operations mgr.; Mike Foster, prod. mgr.
Operation: 1959; TV sets in area: 105,000.

Chico

KHSL-TV (Channel 12) CBS
Golden Empire Broadcasting Co., Box 489, Chico, CA 95927 (916) 342-0141.
C.H. Kinsley Jr., exec. v.p.; Douglas C. Sundby, gen. mgr. & sls. mgr.; Richard Newman, natl. sls. mgr.; Bert Westhoff, operations mgr.; Russell B. Pope, dir. of eng'.; Frank Krug, chief eng. nat'l rep., Avery Knodel, Inc.
Operation: 1953; TV sets in area: 101,460.

El Centro

KECC-TV (Channel 9) CBS
Telebroadcasters of Calif. Inc., Box 29, El Centro, CA 92244 (714) 352-9670.
Richard Nix, gen. mgr.; Greg Weir, prog. dir.; Richard Nix, chief eng.; nat'l rep., UBC sales.
Operation: Dec. 1968; TV Households: 436,300.

Eureka

KIEM (Channel 3) CBS, NBC
California Oregon Radio Co. Box 3E, Eureka, CA (707) 443-3123.
Harvey Ingham, pres. & gen. mgr.; Donald King, operations mgr.; Elmer Haskin, v.p. & gen. sls. mgr.; nat'l rep., Avery-Knodel.
Operation: 1953.

KVIQ (Channel 6)
California Northwest Broadcasting Co., P.O. Box 1019, Eureka, CA 95501 (707) 443-3061.
Carl R. McConnel, pres.; Lynn F. Olsen, gen. mgr.; Sallie Gentry, prog. dir.; Mike Fiest, chief eng.
Operation: 1958.

Fontana (Los Angeles)

KTBN-TV (Channel 40)
International Panorama TV, Inc., P.O. Box A, Santa Ana, CA 92711 (714) 832-2950.
Paul F. Crouch, pres.; Terrence Hickey, prog. dir.; Ben Miller, dir. of eng.; Tim Flynn, dir. of programming; Sam Starr, dir. of sls.
Operation: 1967.

Fresno

KAIL
Trans-America Broadcasting Corp., 6803 West Blvd., Inglewood, CA 90302 (209) 299-9753.
John Lockhart mgr.
Operation: 1961.

KJEO (Channel 47) ABC
Retlaw Broadcasting Co., Box 5455, Fresno, CA (209) 222-2411.
Joseph C. Drilling, pres.; William A. Sawyers, v.p. & gen. mgr.; Walter Davis, local sls. mgr.; Jay Hill, chief eng.; Gary Vautin, prog. & promo. dir.; Andrew Mastoras, v.p.-bus. mgr.; Janice Crosby, v.p.-brdct. stnd.; Frances Hernandez, traffic mgr.; nat'l rep.: Peters, Griffin, Woodward.
Operation: 1953; Sets in area: 577,300.

KMJ-TV (Channel 25) NBC
McClatchy Broadcasting Co., 1544 Van Ness Ave., Fresno, CA 93786 (209) 268-6666.
Eleanor McClatchy, pres.; Charles K. McClatchy, secty.-treas.; L.R. Matushak, Walter Jones, John Hamyln, Verne Bonette, dirs.; Jeff Nagle, mgr.; Edgar Deatherage, sls. mgr.; Ken Waligarski, prog. & prod. mgr.; Gordon Lee, chief engineer.
Operation: 1953; TV sets in area: 261,740.

Hanford-Fresno

KFTV (Channel 21)
Spanish International Communications Corp., 8515 E. Lacey Blvd., Hanford, CA 93230 (209) 584-3362.
Daniel D. Villaneuva, v.p.; Ernesto Balleste, gen. mgr.; Lawrence Beckman, chief eng.
Operation: 1962.

Los Angeles

KABC -TV (Channel 7) ABC
American Broadcasting Companies, Inc., ABC Television Center, Hollywood CA 90027 (213) 557-7777.
Thomas K. Van Amburg, v.p. & gen. mgr.; Norman H. Sloane, dir. of admin.; Randall W. Hooper, dir., network prog. operations; Stephen L. Schaefer, sls. service mgr.; William F. Dryer, dir. of eng.; Martin Kirschen,

dir. of finance & business affairs; Dennis Swanson, news dir.; natl rep., ABC Television Spot Sales, Inc.
Operation: 1949.

KCOP (Channel 13)
KCOP Television Inc., a Chris Craft Industries, station, 915 N. La Brea Ave., Hollywood, CA 90038 (213) 851-1000.
Evan Thompson, pres. & gen. mgr.; Bill Frank, v.p. & station mgr.; Dave Woodcock, gen. sls. mgr.; Steve Friedheim, natl. sls. mgr.; Sally Crawford, local sls. mgr.; Win Korabell, dir. of eng.; Jane Clark, bus. mgr.; Peter Schlesinger, prgm. dir.; Gary Davis, dir. of adv. & promo.
Operation: 1948; TV sets in area: 4,132,100.

KBSC-TV (Channel 52)
Oak Broadcasting Systems, Inc., 1901 Ave. of Stars, Los Angeles, CA 90067 (213) 240-6820.
Andrew Jerrold Perenchio, pres. & gen. mgr.; John Mohr, station mgr.; Bill Morrison, sls. mgr.; Joe Hannigan, chief eng.; Alan Maretsky, traffic mgr.; Joni Penncok, pub. affairs dir.
Operation: 1966.

KHJ-TV (Channel 9)
KHJ-TV (Division of RKO General, Inc.), 5515 Melrose Avenue, Hollywood, CA 90038 (213) 467-5459.
Chuck Velona, gen. mgr., KHJ-TV; Mike Sobel, gen. sls. mgr.; Walt Baker, program dir.; Buck Evans, chief eng.; Alan Timpson, controller, KHJ Division; nat'l rep., Blair.
Operation: 1948; TV sets in area: 3,511,000.

KHOF-TV (Channel 30)
Faith Center, Inc., 1615 S. Glendale Ave., Glendale, CA 91205 (213) 245-7575.
Dr. W. Eugene Scott, pres.; Joe Shackelford, chief eng.; Bruce Henderson, gen. mgr.
Operation: 1969.

KMEX-TV (Channel 34)
Spanish International Bcstg. Co., 5420 Melrose Ave., Los Angeles, CA 90038 (213) 466-8131.
Daniel Villanueva, gen. mgr.; Robert Porter, v.p., eng'.; Louis Sweeney, v.p., nat'l. sls.; Charles Barry, sls. mgr.; rep., Spanish International Network Sales Inc.
Operation: September, 1962; TV households: 1,500,000.

KNBC (Channel 4) NBC
National Broadcasting Co., Inc., 3000 W. Alameda Ave., Burbank, CA 91523 (213) 840-4444.
Thomas M. Straszewski, NBC v.p. & KNBC gen. mgr.; Bud Carey, sta. mgr.; David Ingraham, dir., business affairs; Robert Kennedy, mgr. adv. & promo.; Richard Askin, dir. sales; Petar Sardelich, mgr. production & operations; Warren Baker, dir. programs; Howard D. Sturm, mgr. public affairs; Saundra Willia, mgr. community relations/press & publicity; Jim Brooks, chief engineer; Rep: NBC Spot Sales.
Operation: Jan. 1949.

KNXT (Channel 2) CBS
CBS, Inc. 6121 Sunset Blvd., Los Angeles, CA 90028
James S. Bennett, v.p. & gen. mgr.; Ron Tindiglia, v.p. & station mgr.; Jonathan Rodgers, news dir.; Larry Forsdick, dir. of sls.; Norman L. Bacon, dir. of planning & admin.
Operation: 1950; TV sets in area: 5,028,100.

KSCI (Channel 18)
Global Television, Inc., 1950 Cotner Ave., Los Angeles, CA 90025 (213) 479-8081.
Walter L. Koch, pres. & gen. mgr.; Robert Doane, v.p.; Jeffrey Glasser, prog. dir.; Douglas Lung, chief eng.
Operation: 1977.

KTLA (Channel 5)
Golden West Broadcasters, 5800 Sunset Blvd. Hollywood, CA 90028.
Gene Autry, chm. of bd.; John T. Reynolds, pres. & coo; Anthony B. Cossara, pres., TV div.; Steven Bell, v.p. & gen. mgr., KTLA.
Operation: 1947; TV sets in area: 4,138,000.

KTTV (Channel 11)
5746 Sunset Blvd. Hollywood, CA 90028 (213) 462-7111.
Charles D. Young v.p. & gen. mgr.; Martin Bergman, v.p. & gen. sls. mgr., Ron Tillman, v.p. and programming dir.; Bob Qudeen, nat'l sls. mgr.; Ted Bending, commercial operations dir.; Lew Schatzer, dir. of promo. and pub.; Ray Green, exec. prod.; Lee Pardee, v.p. & chief eng.; Walter R. Kalb, dir. of bus. affairs; Larry Attebery, news dir.; Owned and operated by Metromedia Television, Inc. Rep., Metro TV Sales.
Operation: Jan. 1949.

KWHY-TV (Channel 22)
Coast TV Broadcasting Corp., 5545 Sunset Blvd., Los Angeles, CA 90028 (213) 466-5441.
Wiley D. Bunn, pres.; Robert W. Bunn, gen. mgr.; Mike Wagner, acct. exec.; John H. Nelson, acct. exec.; Eugene A. Harris, David J. Zulli, chief engs.; Rod Buchser, operations mgr.
Operation: 1963.

Modesto

KLOC-TV (Channel 19) Spanish language
KLOC Broadcasting Co., Box 3689, Modesto, CA 95352 (209) 529-2024.
Chester Smith, pres.; Luis Mercado, natl. sls.; David Hornbaker, operations mgr.

Monterey

KMST-TV
Retlaw Broadcasting Co., P.O. Box 1938, Monterey, CA 93940 (408) 649-0460.
Joseph Drilling, pres.; Ben Tucker, gen. mgr.; John Vera, sls. mgr.; Kathryn Pratt, public affairs dir.; James Erickson, prod. mgr.; Ken Warren, chief eng.; Parker Jackson, pro. dir.; Melinda Borum, traffic mgr.; Dan Cullen, news dir.
Operations: February, 1969. TV households: 174,700

Palm Springs

KMIR-TV (Channel 36) NBC
Desert Empire TV Corp., P.O. Box 1506, Palm Springs, CA 92263 (714) 568-3636.
John Conte, pres. & gen. mgr.; William Secrist, prog./prod. mgr.; John Weston, news director. Rep.: Katz Television.
Operation: Oct. 1968; TV households: 54,900.

KESQ-TV (Channel 42, Cable 3) ABC

Drawer 420, Palm Springs, CA. Owned by Esquire, Inc., 488 Madison Ave., New York, NY 10022 (714) 328-8881.

Keith Swinehart, pres. & gen. mgr.; Jack Brembeck, gen. sls. mgr.; Bob Martins, v.p., operations & engineering.

Operation: October 1968; TV households: 63,700.

Redding

KRCR TV (Channel 7) ABC-NBC

Sacramento Valley Television, Inc., 2770 Pioneer Dr., Redding (Shasta Co.) CA 96001 (916) 243-7777.

Richard W. Green, pres. & gen. mgr.; William B. Smullin, v.p., dir. of operations; Doreeta Domke, operations mgr.; nat'l rep., PGW

Sacramento

KCRA-TV (Channel 3) NBC

Kelly Broadcasting Co., 310 10th St., Sacramento, CA 95814 (916) 444-7300.

Robert E. Kelly, partner; Jon S. Kelly, partner; Nina N. Kelly, partner; Don Saraceno, gen. mgr.; nat'l rep.: Blair Television, Inc.

Operation: 1955; TV sets in area: 558,900.

KMUV-TV (Channel 31)

Grayson TV Co., Inc., Box B, 500 Media Pl., Sacramento CA 95813 (916) 929-0300.

Andrew C. Bartalini, pres.; William Crabtree, v.p.; Dr. Samuel Klor, v.p.; Della G. Grayson, secty.; George Artz, secty.; Mike Pappas, gen. mgr.; Ed Verbeski, gen. sls. mgr.; Alvin R. Wilder, chief eng.

Operation: 1974.

KTXL (Channel 40)

Camellia City Telecasters, Inc., P.O. Box 40, Sacramento, CA 95801 (916) 452-8221.

Jack F. Matranga, pres., gen. mgr. & film buyer; Robert Parente, natl. sales mgr.; Robert Nelson, chief eng.; Cal Bollwinkel, operations & prog. mgr.

Operation: 1968.

KXTV (Channel 10) CBS

Great Western Broadcasting Corp., 400 Broadway, Sacramento, CA 95818 (916) 441-4041.

Philip Lombardo, pres.; Joseph E. Lake, v.p. & gen. mgr.; William Bradley, gen. sls. mgr.; Roger Flessing, pgm. mgr.; Judy Law, promo. mgr.; Fred Lindsay, chief eng.; nat'l rep., Peters, Griffin, Woodward, Inc.

Operation: 1955; TV sets in area: 648,500.

Salinas

KSBW-TV (Channel 8) NBC

Blair Broadcasting of California, Inc., P.O. Box 1651, 238 John St. Salinas, CA 93901 (408) 422-6422. (Affiliated with KSBY-TV, San Luis Obispo, CA)

Keith H. Moon, v.p. gen. mgr.; James Magnuson, gen. sls. mgr.; William Fanta, dir. of programming; nat'l reps., Blair.

Operation: 1953; TV homes in area: 166,400.

San Diego

KCST-TV (Channel 39) NBC

KCST-TV, Inc., 8330 Engineer Rd., San Diego, CA 92111 (714) 279-3939.

Bill L. Fox, gen. mgr.; Bruce Stein, nat'l. sls. mgr.; Ric Schwartz, prod. mgr.; Tom Wimberly, chief eng.; Bill Fouch, operations mgr.; Tom Mitchell, news dir.

Operation: 1965.

KFMB-TV (Channel 8) CBS

Midwest Television, Inc., 7677 Engineer Road, San Diego, CA 92101 (714) 292-5362.

August C. Meyer, Jr., pres.; Robert L. Myers, vice-pres. & gen. mgr.; Bill Moylan, v.p., operations; John Weigand, chief eng.; nat'l rep. Peters Griffin, Woodward, Inc.

Operation: 1949, TV sets in area: 1,015,800.

KGTV (Channel 10) ABC

McGraw-Hill Broadcasting Co., P.O. Box 81047, San Diego, CA 92138 (714) 237-1010.

Norm Walt, pres.; Clayton H. Brace, v.p. & gen. mgr.; Leroy Bellwood, dir. of eng.; Howard Oleff, natl. sls. mgr.; Joan Carson, gen. sls. mgr.; Helen Hutchings, dir., business affairs; A. Philip Corvo, prog. dir.; William Strubbe, dir. adv., prom.

Operation: 1953.

XETV (Channel 6) Independent

Bay City Television, Inc., 8253 Ronson Road, Television Heights, San Diego, CA 92111 (714) 279-6666.

Julian M. Kaufman, v.p. & gen. mgr.; Louis J. Meiers, treas./bus. mgr.; Martin Colby, sta. mgr. and gen. sls. mgr.; Ms. Carol Healey, prog. & promo. dir.; Ms. Joan E. O'Laughlin, operation mgr.; Rudolph Regalado, asst. local sls. mgr.; Felipe Fernandez, chief eng.; nat'l rep., Balir Television Assoc.

Operation: 1953; TV sets in area: 582,450.

XEWT-TV (Channel 12) Spanish Language

Televisora de Calimex, S.A., P.O. Box 12, Tijuana, Baja California, Mexico (903) 385-9201.

Jose Marquez, gen. mgr.; Carlos H. Luna, prog. prom. mgr.; Roberto Espinoza, business mgr.

San Francisco

KBHK-TV (Channel 44)

Field Communications Corp., 420 Taylor St., San Francisco, CA 94102 (415) 885-3750.

Bill White, v.p. & gen. mgr.; Walter Adler, gen. sls. mgr.; Kerry Andrews, sls. mgr.; Tom Spitz, prog. mgr.; Mary Lou Manelli, news & pub. affairs mgr.; John Nielsen, eng. mgr.; Jo Russell, prod. mgr.

Operation: January, 1968. TV households: 2,352,700.

KDTV (Channel 14) Spanish language

Bahia de San Francisco TV Co., 2200 Palou Ave., San Francisco 94124, an affiliate of SIN—The National Spanish Television Network.

Reynold V. Anselmo, pres.; Emilio Nicolas, Jr., v.p. & gen. mgr.; Jorge Belon, prod. & programming dir.; Luz Diez, business mgr.; George Ledoux, chief eng.

Operation: 1975.

KEMO-TV (Channel 20) Spanish Language

Leon A. Crosby, 36692 Riviera Dr., Fremont, CA 94124 (415) 285-6420.

Leon Crosby, gen. mgr.; Rene de la Rosa, station mgr.; Walt Conway, sls. mgr.; David Graham, chief eng.

Operation: 1968; TV households, 770,300.

KGO-TV (Channel 7) ABC
American Broadcasting Companies, Inc., 277 Golden Gate Ave., San Francisco, CA 94102 (415) 863-0077.
Russ Coughlan, v.p. & gen. mgr.; Mark Mandala, gen. sls. mgr.; Don Azars, prog. dir.; Dennis Fitch, dir. of adv., prom & publicity; Pete Jacobus, news dir.; Jim Von Striver, chief eng.; Beverly Hayon, Comm. affairs dir.; nat'l rep., ABC Spot Sales.
Operation: 1949; TV homes in area: 1,575,000.

KPIX (Channel 5) CBS
Group W (Westinghouse Broadcasting Co.), 2655 Van Ness Ave., San Francisco, CA 94109 (415) 776-5100.
Pat Polillo, v.p. & gen. mgr.; John Davison, asst. gen. mgr.; Peter E. Schruth, vice chm.; Herbert Farber, gen. sls. mgr.; Bob George, pgm. mgr.; Dick Newton, creative svcs. dir.; Leona Wong, info. svcs. mgr.; Gary Elion, news dir.; Walt Nichol, enge. mgr.; Vic Lai, bus. mgr.; Sandy Johnson, human resources mgr.
Operation: 1948; TV households in area: 1,894,000.

KRON-TV (Channel 4) NBC
Chronicle Broadcasting Co., subsidiary of The Chronicle Publishing Company, 1001 Van Ness Avenue, San Francisco, CA 441-4444.
Paul A. Wischmeyer, pres. and gen. mgr.; F.A. Martin III, v.p. and station mgr.; Stephen J. Levin, v.p./operations; George Davidson, gen. sls. mgr.; Lawrence Pozzi, chief eng.; Chuck Snyder, prgm. mgr.; Jean Harper, news dir.; Dennis Fitch, dir., creative socs; Bill Row, adv. mgr.; nat'l rep., Peters, Griffin & Woodward, Inc.
Operation: 1949; TV households in area: 2,827,100.

KTSF-TV (Channel 26)
Lincoln Television, Inc., 185 Berry St., San Francisco, CA 94107 (415) 495-4995.
Lilliam L. Howell, owner; Richard O'Hearn, gen. mgr.; Drew Pfieffer, prog. mgr.

KTVU (Channel 2)
San Francisco-Oakland Television, No. 2 Jack London Square, Oakland, CA 94607 415-834-1212.
Alan J. Bell, v.p. & gen. mgr.; Richard Goldman, gen. sls. mgr.; Michael Liff, natl. sales mgr.; Jay Sondheim, local sls. mgr.; Thomas A. Breen, prog. dir.; Marlene Holderbaum, contr.; Barry Smith, promo. mgr.; Ian Zellick, dir. of pub. affairs; Jay Finkelman, station mgr.; Ray Jacobs, dir. of operations; Fred Zehnder, news dir.; Ray Swenson, chief eng.
Operation: 1958.

San Jose

KGSC-TV (Channel 36)
Continental Urban TV Corp., c/o Vincent B. Welch, 300 Farragut Bldg., 900 17th St., N.W. Washington, DC 20006; (408) 298-6676. Studio: 1536 Kerley Dr., San Jose, CA 95112. (408) 298-6676.
Sid Connolly, pres. & gen. mgr.; Tom Maguire, gen. sls. mgr.; Sharon Dean, natl. sls. mgr.; LaVoda Brown Heath, reg. sls. mgr.; Frank Foge, chief eng.; Nancy Seiden, bus. mgr.; John Stevens, operations mgr.
Operation: 1967.

KNTV (Channel 11)
Gill Industries, 645 Park Avenue, San Jose, CA 95110; (408) 286-1111.
Robert M. Hosfeldt, v.p. & gen. mgr.; Stewart B. Park, prog. director & film buyer; John Yearwood, gen. sls. mgr.; Lee Schell, news dir.; Lou Bell, chief engr.; Barbara Smith, contr.; nat'l rep., Katz Television.
Operation: 1955.

San Luis Obispo

KSBY-TV (Channel 6)
Blair Broadcasting of California, Inc., Box 1368, San Luis Obispo, CA 93402 (805) 543-0920.
David Cox., v.p./gen. mgr.; Larry Scott, operations mgr.; Ken Reid, chief eng.
Operation: 1953.

Santa Barbara

KEYT (Channel 3) ABC
Key Television Inc., P.O. drawer X, Santa Barbara, CA 93102.
William F. Luton, pres. & treas.; Hal O'Donnell, gen. mgr.; William Luton, Jr., promo. mgr.; William Luton, Jr., gen. sls. mgr.; Renee Foley, program supvr.; William Pierce, news dir.; John Wilson chief engr.; nat'l rep., PGW.
Operation: 1953; TV sets in area: 200,000.

Santa Maria

KCOY-TV (Channel 12)
Central Coast, Broadcasters, Inc., Box 1217, Santa Maria, CA 93454 (805) 922-0505
Robert Burris, v.p. & gen. mgr.; Sue English, oper. mgr.; Dennis Bornhoft, chief eng.; Alan Ross, news dir. Rep.: H.R. Television Co.
Operation: 1964; TV households: 193,000.

Stockton/Sacramento

KOVR (Channel 13)
McClatchy Broadcasting Company, 225 E. Miner Ave., Stockton, CA 95202 (209) 466-6981. Sacramento Studio and Main Business Office: 1216 Arden Way, Sacramento, CA 95815 (916) 927-1313.
Chas. F. Kennedy, pres. & gen. mgr.; Doug Grant, sls. mgr.; Phil Linn, chief eng.; Henry Vrick, prog. mgr.; nat'l rep., Katz Agency.
Operation: 1954; TV sets in area (Sacramento/Stockton Mkt.) 769,700.

Tulare

KMPH (Channel 26)
Pappas Television, Inc., 2600 S. Mooney Blvd., Visalia, CA 93277 (209) 733-2600.
Harry J. Pappas, pres.; Sam H. Horel, v.p. & gen. mgr.; Joseph Shaffer, v.p. & prog. dir. & operations mgr.; Dick Parr, news dir.

COLORADO

Colorado Springs—Pueblo

KKTV (Channel 11) CBS
Capitol of Colorado Corp., P.O. Box 2110, 80901; 3100 N. Nevada Ave., Colorado Springs, CO (303) 634-2844.
T.M. Hederman, pres.; George W. Jeffrey, v.p. & gen. mgr.; nat'l rep. Katz, West.
Operation: 1952; TV households in area: 341,300.

KOAA-TV (Channel 5) NBC
Sangre de Cristo Communications, Inc., P.O. Box 876, 2200 7th Avenue, Pueblo, CO 81002 (303) 544-5782.
John O. Gilbert, pres. & gen. mgr.; Kenneth Renfrow, chief eng.; Jack Sinclair, asst. gen. mgr., production & oper.; Larry Keenan, gen. sls. mgr.; Jim Schaeffer, local sls. mgr.; Helen Solorzano, nat'l sls. mgr.; Charles Claverie, news dir.
Operation: 1953; TV sets in area: 296,700.

KRDO-TV (Channel 13) ABC
Pikes Peak Broadcasting Co., P.O. Box 1457, Colorado Springs, CO 80901 (303) 632-1515.
Harry W. Hoth, Jr., pres.; Neil O. Klocksiem, gen. mgr.; Charles H. Upton, chief eng.
Operation: 1953; TV sets in area: 196,600.

Denver

KBTV (Channel 9) ABC
Combined Communcations Corp., 1089 Bannock St., Denver, CO 80217 (303) 825-5288.
Alvin G. Flanagan, pres., CCC broadcast div.; Charles T. Leasure, pres. & gen. mgr.; Cecil Walker, v.p. & controller; Herb Schubarth, v.p. & chief eng.; Joe Franzgrote, v.p. & gen. sls. mgr.; Carl Akers, v.p., news; Lawrence Deutsch, local sls. mgr.; Don Franzen, prod. mgr.; Marjorie Schmidt, promo. mgr.; Mardee McKinlay, pub. affairs dir.; Ron Scott, news dir.; nat'l rep.: Peters, Griffin, Woodward, Inc.
Operation: 1952; TV sets in area: 713,300.

KMGH-TV (Channel 7) CBS
McGraw-Hill Bdg. Co., Inc., 123 Speer Blvd., Denver, CO 80717 (303) 292-3456.
Norman E. Walt, pres.; Robert A. Hart, Jr., v.p. & gen. mgr.; James Keeley, gen. sls. mgr.; Bob Jones, prog. dir.; Jeff Bell, dir. of adv. & promo.; Elva Gilmer, traffic mgr.; Paul Thompson, news dir.; Eugene Jenkins, chief eng.
Operation: 1953.

KWGN (Channel 2)
WGN of Colorado, Inc., 6160 S. Wabash Way, Englewood, CO (303) 740-2222.
Robert A. Innes, v.p. & gen. mgr.; Mel Smith, prog. mgr.; Tanette Maffei, promo. mgr.; Lorrie Shilling, on-air promo. mgr.; Tarey Thornburg, film dir.; Ron Peters, chief eng.; Jack Brennan, v.p. & dir. of mktd.
Operation: 1952; TV sets in area: 568,400.

KOA-TV (Channel 4) NBC
General Electric Broadcasting Company of Colorado, Inc., 1044 Lincoln Street, P.O. Box 5012 T.A., Denver, CO, 80217 (303) 861-8111.
Brian Cobb, v.p./gen. mgr.; James H. MacDermott, mgr.; Robert C. Biauvelt, mgr. operations; David La France, gen. sls. mgr.; Don Perry, local sls. mgr.; Randy Oswald, natl. sls. mgr.; Kent Lillie, gen. sls. mgr.; Steve Harris, sls. mgr.; Dan Fiorillio, operations mgr.; Dan Akens, news dir.

Durango

KREZ-TV (Channel 6) ABC, NBC, CBS
(Satellite of KREX-TV, Grand Junction. CO)
XYZ TV Inc., Box 789, Grand Junction, CO 81502 (303) 247-0820.
W.G. Walter, resident mgr.
Operation: Nov., 1963.

Grand Junction

KJCT-TV (Channel 8) ABC, P.O. Box 3788, Grand Junction, CO 81502
Harry W. Hoth, pres.; Wesley Ferns, gen. mgr.; Roger Hightower, chief eng.
Operation: 1979; TV households: 49,400

KREX-TV (Channel 5) CBS, NBC
XYZ Television, Inc., Hillcrest Manor, Grand Junction, CO (303) 242-5000.
Carl Q. Anderson, pres. & gen. mgr. nat'l rep.: Seltel, Bill Sullivan, sta. mgr.
Operation: 1954; TV sets in area: 61,000 (ARB).

Montrose

KREY-TV (Channel 10) NBC, CBS, ABC
(Satellite of KREX-TV, Grand Junction, CO)
XYZ Television Inc., Box 49, Montrose, CO 81501 (303) 249-9601.
Carl Q. Anderson, pres. & gen. mgr.; Joe Reichman, sta. mgr.; nat'l rep., Seltel.
Operation: 1956.

Sterling

KTVS (Channel 3) CBS, ABC
(Satellite of KYCU-TV, Cheyenne, WY)
Wyneco Communications, Inc., 2923 E. Lincolnway, Cheyenne, WY 82001.
Carl J. Occhipinti, gen. mgr.; Gene Huston, res. mgr.; Jerry Homer, chief eng.; nat'l. rep.: Katz.
Operation: Dec. 1963.

CONNECTICUT

Hartford

WFSB-TV (Channel 3)
Post-Newsweek Stations, Connecticut, Inc. 3 Constitution Plaza, Hartford, CT 06115 (203) 525-0801.
G. William Ryan, vice pres. & gen. mgr.; Cathy Nierle, bus. mgr.; Dick Ahlea, news dir.; Mike Dorsey, gen. sls. mgr.
Operation: 1957; TV sets in area: 1,151,447.

WHCT (Channel 18)
A Division of Faith Center, Inc., 555 Asylum Street, Hartford, CT 06105 (203) 525-2611.

New Britain

WVIT (Channel 30) NBC
Viacom Broadcasting, Inc., 1211 Ave. of Americas, New York, NY 10036; Hartford: (203) 521-3030, New Britain: (203) 223-3603.
Paul M. Hughes, pres. & gen. mgr.; Lewis Freifeld, dir. sls.
Operation: 1953; TV sets in area, 313,000.

New Haven-Hartford

WTNH-TV (Channel 8) ABC
Capital Cities Communications, Inc., 135 College St., New Haven. CT (203) 777-3611.
Peter K. Orne, gen. mgr.; Marc Edwards, gen. sls. mgr.; Will Mebrane, prog. mgr.; O. Lyle Koch, promo. mgr.; Tom Kirby, news dir.; Bruce Lawrence, sls. mgr.;

Steve Perler, bus. mgr.; Bill Ellison, dir. pub. aff.; Larry Curtis, chief eng.; nat'l rep.: Blair TV, Inc.
Operation: 1948; TV sets in area: 1,949,200.

Waterbury

WATR-TV (Channel 20) NBC

WATR, Inc., 79 Baldwin Avenue, Waterbury, CT (203) 755-1121.
B. Preston Gilmore, pres ; Bob Holczer, mgr.; Jan Alencynowicz, prog. dir.; nat'l rep.: Spot Time, Ltd.
Operation: 1953: TV sets in area: 676,100.

DISTRICT OF COLUMBIA

Washington

WDCA-TV (Channel 20)

Channel 20 Inc., 5202 River Rd., Washington, DC 20016.
Milton Grant. v.p. & gen. mgr.; Joseph Girard, gen. sls. mgr.; Jim Reid, prog. mgr.; Don A. Doughty, dir. of engineering; Dan O'Brian, operations mgr.; Raymond E. Simko, bus. mgr.; rep.: PGW.
Operation: April, 1966; TV households: 1,724,500.

WJLA TV (Channel 7) ABC

Allbritton Communications Co. 4461 Connecticut Avenue, N.W. Washington, DC 20008
Thomas B. Cookerly, pres. & Gen. mgr.; Stan Pederson, dir. of brdcst'.; Len Deibert, mgr. pub. Affairs; Al Norcott, dir. of Engineering; Jim Boaz, dir. of mrktg.; nat'l Rep: Edward Petry & Co.

WRC-TV (Channel 4) NBC

National Broadcasting Co., 4001 Nebraska Ave., N.W., Washington, DC 20016 (202) 686-4000.
John Rohrbeck, v.p. & gen. mgr.; Ann Berk, station mgr.; Ned Paddock, dir. sls.; Allan Horlick, mgr. sls.; Dave Uhrich, dir. prog.; David Nuell, news dir.; Crawford McGill, dir. oper. & eng.; Pete Porrello, dir. emp.; Bill Scholz, dir. plant facilities & purchasing.
Operation: June, 1947; TV households: 4,637,400.

WDVM-TV (Channel 9) CBS

A division of the Evening News Association, Inc. 4001 Brandywine St., NW, Washington, DC 20016
Edwin W. Pfeiffer, v.p./gen. mgr.; Ronald Townsend, station mgr.; Edward P. Jones, prgm. mgr.; Henry K. Yaggi III, dir. of sls.; Gail Neylan, dir. of mrktg.; Lawrence P. Herbster, dir. of bus. affairs; Henry Owen, dir. of eng.; Rep.: MMT Sales.
Operation: 1949; TV households in ADI area: 1,390,300.

WTTG (Channel 5)

Metromedia Television, a division of Metromedia, Inc., 5151 Wisconsin Ave., N.W. Washington DC 20016 (202) 244-5151.
William Carpenter, v.p. & gen. mgr.; Jerry Marcus, v.p. & gen. sls. mgr.; Albert Harmon, v.p. in chg. of eng'.; Stanley Rudick, v.p. and prog. dir.; Kenneth Becker, v.p. &l bus. mgr.; Jill Krasner, dir. of adv. & promo.; Pat Myers, commercial op. mgr.; nat'l rep.; Metro TV Sales.
Operation: 1947; TV sets in area, 2,647,100.

FLORIDA

Daytona Beach-Orlando

WESH-TV (Channel 2)

Cowles Broadcasting Inc., P.O. Box 1551, Daytona Beach, FL 32015 (904)252-2222.
John M. Haberlan, pres. & gen. mgr.; Walter M. Strouse, exec. v.p. and stn. mgr.; Jack Hitchcock, gen. sls. mgr. and asst. station mgr.; Kenneth Smith, oprns. mgr.; nat'l rep.: The Katz Agency.
Operation: 1956; TV households: 1,120,800.

Fort Myers

WBBH-TV (Channel 20) NBC

Broadcasting-Telecasting Services Inc., 3719 Central Ave., Fort Myers, FL 33901 (813) 936-0195.
William N. Dakos, chmn.; Howard L. Hoffman, pres. & gen. mgr.; David McKelvey, 2nd v.p. & chief eng.; Ruth Plumb, compt.; Bill Roof, local sls. mgr.; Steve Grissom, news admins.
Operation: December, 1968; TV households: 213,900.

WEVU (Channel 26) ABC

Caloosa TV Corp., 123 How Lane, New Brunswick, NJ 08903 (813) 597-6038.
Joseph Buerry, gen mgr.; E.F. Shadburne, sls. mgr.; Dona Fussell, prog. dir.
Operation: 1974.

WINK-TV (Channel 11) CBS

Fort Myers Broadcasting Co., 2824 Palm Beach Blvd., Fort Myers, FL 33902, (813) 334-1331.
Edward J. McBride, pres.; Robert F. Doty, v.p. & gen. mgr.; Jr.; Don Locke, gen. sls. mgr.; Clarence Mosley, dir. eng.; Harry Horn, news dir.; Fred Greene, public affairs dir.; nat'l. rep.: Blair Television.
Operation: 1954.

Fort Pierce

WTVX (Channel 34) CBS

Indian River Television Inc., Tr. 1, Box 313, Ft. Pierce, FL 33450.
Ruhl W. Koblegard, pres.; Alto Adams, Jr., v.p.; Ben R. Bryan, secty.; Walter J. Warren, treas.; Jim Holmes, gen. mgr.; Jack Tallcott, sls. mgr.
Operation: April 5, 1966; TV households: 69,400.

Gainesville

WCJB (Channel 20) ABC

Gainesville Television, Inc. P.O. Box 13414, Gainesville, FL 32604 (904) 372-3543.
Alex Skiffington, v.p. & gen. mgr.; William Bryan, chief. eng.; Gregory Johnson, asst. gen. mgr.
Operation: 1971.

Jacksonville

WJKS-TV (Channel 17) NBC

9117 Hogan Rd., Box 17000, Jacksonville, FL 32216 (904) 641-1700.
Gary Adler, v.p. & gen. mgr.; Robert Craig, gen. sls. mgr.; Russ Meyerson, prgm. mgr.; Harold Johnson, chief engr.
Operation: Feb. 1966.

WJXT (Channel 4) CBS

WJXT, operated by the Post-Newsweek Stations, Florida Inc., P.O. Box 5270; 1851 Southampton Road, Jacksonville, FL 32207 (904) 399-4000.

Harry Kalkines, v.p. & gen. mgr.; Barry Barth, dir., bdcst. operations; Leonard L. Mosby, area v.p.; Lynn Fairbanks, gen. sls. mgr.; Christy Birong, op. mgr.; Steve Flanagan, eng. mgr.; natl. rep., Top Market Television, Inc.

Operation: 1949; TV sets in area: 546,000.

WTLV (Channel 12) NBC

Television 12 of Jacksonville, Inc., Television P.O. Box TV-12, Jacksonville, FL 32231 (904) 354-1212.

Jim Kontoloen, gen. mgr.; Harry McClintock, prgm. dir.; Robert Epstein, local sls. mgr.; Karen Heniger, natl. sls. mgr.; Harry Alexander, prod. dir.; Tom Sanders, news dir.; Rosemary Shouldice, promo. dir.; Clyde Smith, chief eng.

Operation: 1957.

Miami

WCIX-TV (Channel 6)

Coral Television Corp., 1111 Brickell Ave., Miami, FL 33131 (305) 377-0811.

Alexander M. Tamler, pres.; Edward Q. Adams, sr. v.p. & gen. mgr.; Harvey Cohen, v.p. & gen. sls. mgr.; Linda Tucker, promo. mgr.; Jack Cowart, dir. of eng.; Jack Willard, bus. mgr.; nat'l. rep: Petry TV.

Operation: Sept. 1967; TV households: 7,423,000.

WCKT (Channel 7) NBC

Sunbeam Television Corp., P.O. Box 1118, Miami, FL 33138 (305) 751-6692.

Edmund N. Ansin, pres.; Robert W. Leider, gen. mgr.; John Fenwick, sls. mgr.; Neil Murphy, contr.; Robert Paasch, Chief eng.; Al Sternberg, op. mgr.; Frank Struzzieri, film mgr.; Charmaine Meyer, creative services mgr.; Wilson Griffith, pub. Affairs dir.; natl rep: Harrington, Righter & Parsons, Inc. Southern: Clem & Lowrance.

Operation: 1956; TV households: 1,065,400.

WHFT-TV (Channel 45)

Lester Sumrall Evangelistic Assn. Inc., Box 12, South Bend, IN 46624 (305) 624-1700.

Lester Sumrall, pres., Robert R. Brown, v.p. & gen. mgr.

Operation: 1975.

WLTV (Channel 23)

Spanish International Communications Corp. 695 N.W. 199th St., Miami FL 33169. (305) 856-2323 (sales) (305) 652-4000 (operations).

Joaquin Blaya, v.p. & gen. mgr.; Harold M. Parks, natl. sls. mgr.; Vivian Quevedo, sta. mgr.; Tony Segredo, sls. mgr.

Operation: 1954.

WPLG (Channel 10) ABC

(Formerly WLBW-TV) Post-Newsweek Stations of FA Inc. 3900 Biscayne Blvd., Miami FL 33137.

Alan Perris, v.p. & gen. mgr.; Tom Weems, chief eng.; Dick Wexo, gen. sls. mgr.; Paul Nugent, natl. sls. mgr.; Rufus Devane, local sls. mgr.; Norman A. Davis, area v.p.

Operation: November 1961; TV households: 1,925,000.

WTVJ (Channel 4) CBS

Wometco Enterprises, Inc., 316 N. Miami Ave., Miami, FL 33128 (305) 377-8241.

Mitchell Wolfson, pres.; Dick Hollands, v.p.—broadcast staff; William R. Brazzil, v.p. & gen. mgr.; Doug Barker, station mgr.; Ralph Renick, v.p. in chg. of news; Bob Bower, chief eng.; Joe Kline, dir. tech. oper.; Tom Fraioli, gen. sls. mgr.; Barry Allentuck, nat. sls. mgr.; Bob Blacher, local sls. mgr.; Dillie Dekle, dir. of prgm. op.; nat'l rep.: MMT Sales, Inc.

Operation: 1949; TV homes in area: 1,390,500.

Orlando

WDBO-TV (Channel 6) CBS

The Outlet Co., P.O. Box 1833, Orlando, FL 32802 (305) 843-0006.

David E. Henderson, pres. bdcst. div.; Arnold F. Schoen, Jr., v.p. & gen. mgr.; Chalmers S. Stromberg, chief eng.; Everett H. Hughes, prog. mgr. & film buyer; Joseph S. Haller, gen. sls. mgr.; nat'l rep.; Petry Television, Inc.

Operation: 1954; TV homes in area: 1,087,500.

WFTV (Channel 9) ABC

Channel Nine of Orlando, 639 W. Central, Orlando, FL 32802 (305) 841-9000.

Walter M. Windsor, gen. mgr.; Clifton L. Conley, gen. sales mgr.; Bruce Baker, local sales dir.; Clifford S. Pine, opns. mgr.; S. Paul Barron, chief eng.; Ramcen Panomore, film dir.; Richard Flynn, bus. mgr.

Palm Beach

WPTV (Channel 5) NBC

Scripps-Howard Bcastg. Co., P.O. Box 510, Palm Beach, FL 33480. Studio: 622 N. Flagler Drive, West Palm Beach, FL (305) 655-5455.

Robert R. Regalbuto, v.p. & gen. mgr.; William J. Brooks, asst. gen. mgr.; Arvo O. Katajisto, op. dir.; Edie Brackeen, gen. sales mgr.; Ed Roos, chief eng.; natl. rep.: Blair Television Associates, Inc.

Operation: Aug. 1954. TV households: 1,152,800.

Panama City

WJHG-TV (Channel 7) ABC

WJHG-TV, P.O. Box 2349, Panama City, FL 32401 (904) 234-2125.

James H. Gray, pres.; Ray H. Holloway, v.p. and gen. mgr.; Roger Jones, sales mgr.; Ron Jones, prod. mgr.; David William, chief eng.; nat'l rep.: Katz.

Operation: 1953; TV homes in area: 293,700.

WMBB (Channel 13) NBC

Octagon Broadcasting Co., Box 1340, Panama City, FL 32401 (904) 769-2312.

Anthony C. Kupris, pres.; John A. Kupris, v.p.; Mary R. Kupris, secty.-treas.; Douglas M. Grimm, gen. mgr.; Jan Moses, prog. dir.; Steve Cook, prod. mgr.; Larry Young, chief eng.

Operation: October, 1973; TV households: 255,200.

Pensacola

WEAR-TV (Channel 3) ABC

Rollins Telecasting, Inc., Box 431, Pensacola, FL 32591 (904) 455-7311.

James C. Roddey, group v.p., media; William P. Eaton, Jr., dir. TV; Milt de Reyna, gen. mgr. & film buyer; James Grimes, gen. sls. mgr.; William Brennan, op. mgr.; Harry Babb, chief eng.

Operation: 1954; TV households: 443,000.

Sarasota

WXLT-TV (Channel 40) ABC

Sarasota-Bradenton Florida TV Co., Inc., P.O. Box 940, Sarasota, FL 33578 (813) 922-0777.

Robert R. Nelson, pres. & gen. mgr.; Lin Rickard, prod. mgr.; Mike Burham, chief eng.; Tom Johnson, sls. mgr.

Operation: 1971.

Tallahassee

WECA-TV (Channel 27) ABC

Allen Communications, Inc., P.O. Box 13327, Tallahassee, FL 32308 (904) 893-3127.

E.C. Allen, pres.; James Matthews, v.p. & gen. mgr.; Kathy Smith, business mgr.; Bill Black, regional sls. mgr.; bus. mgr.; Lynne Stephenson, prog. dir.; Ray Quinn, gen. sls. mgr.

Operation: September, 1976.

Tampa

WFLA-TV (Channel 8) NBC

905 Jackson St., P.O. Box 1410, Tampa, FL 33601 (305) 229-7781.

William B. Faber, pres.; Wesley Quinn, station mgr.; Doug Duperrault, prog. dir.; Jack Jarvis, op. dir.; Larry Kellogg, promo. mgr.; Harry Smith, gen. sls. mgr.; Pat McLaughlin, chief eng.; nat'l rep.; John Blair & Co.

Operation: 1955; TV sets in area: 807,200.

WTVT (Channel 13) CBS

Gaylord Broadcasting Co., 3213 West Kennedy Blvd., Box 22013, Tampa, FL 33622; Transmitter, Route 1, Box 767, Riverview, FL 33569 (305) 876-1313.

Harry M. Apel, v.p., Gaylord Broadcasting Co. and gen. mgr., WTVT; Bill Diaz, gen. sls. mgr.; Dan Boger, prod. mgr.; Walter Rhoads, promo. mgr.; Hugh Smith, news dir.; Lowell Otto, chief eng.; nat'l rep.: Tela Rep.

Operation: 1955; TV sets in area: 2,429,800.

St. Petersburg

WTOG (Channel 44)

Hubbard Broadcasting Inc., 3415 University Ave., St. Paul, MN 55402 (813) 576-4444.

James C. Dowdle, gen. mgr.; James L. Tomlin, gen. sls. mgr.; John Suder, natl. sls. mgr.; Gary L. Simcox, op. mgr.; George Orgera, Jr., chief eng.

Operation: 1968.

WTSP-TV (Channel 10) ABC

WTSP-TV, St. Petersburg/Tampa, FL P.O. Box 10,000, St. Petersburg, FL 33733.

Lawrence M. Clamage, vice pres. & gen. mgr.; John Purcell, v.p., sls.; Jayne Boyd, prgm. mgr.; George Faulder, news dir.; Peter Ford, chief eng. Michael Stopnick, promo. mgr.

W. Palm Beach

WPEC-TV (Channel 12) ABC

Fairfield Drive, West Palm Beach, FL 33407 (305) 848-7211.

Alex W. Dreyfoos, Jr., pres.; Thom Moody, prgm. dir.; Robert C. Weigand, gen. mgr.

Operation: 1955.

GEORGIA

Albany

WALB-TV (Channel 10) NBC

Gray Communications Systems, Inc., P.O. Box 3130, Albany, GA 31708 (912) 883-0154.

James H. Gray, pres.; Dorothy A. Zimmerman, secty; Raymond E. Carow, v.p. and gen. mgr.; Barbara Jones, treas.; Perley E. Eppley, dir. of eng.; nat'l rep.: Katz.

Operation: 1954.

Atlanta

WAGA-TV (Channel 5) CBS

Storer Broadcasting Co., 1551 Briarcliff Rd. N.E., P.O. Box 4207, Atlanta, GA 30302 (404) 875-5551.

Paul Raymon, v.p. gen. mgr.; Jim Ferguson, prog. dir.; Dale Clark, pub. affairs dir.; Gene Adams, chief eng.; Lana Russell, prod. manager; Dick Goss, promo. mgr.; Spencer Koch; nat'l. sls. mgr.; Jack O'Hern, gen. sls. mgr.; Dianne Harnell, local sls. mgr.; Chuck Lipton, mktg. dir.; nat'l rep.: Storer TV sales.

Operation: 1949.

WATL-TV (Channel 36)

Briarcliff Communications Group, Inc., Suite 814, 1800 Peachtree Rd., N.E., Atlanta, GA 30309 (404) 355-0036.

D.R. Jones, chmn.; Edward Herlihy, v.p. & gen. mgr.; Don Hess, prog. dir. Winston Blumberg, sls. mgr.

Operation: 1954.

WANX-TV (Channel 46)

CBN Continental Broadcasting Network, Inc. 1810 Briarcliff Rd., P.O. Box 98097, Atlanta, GA 30359 (404) 325-3103.

Herman Ramsey, v.p. & gen. mgr.; Bruce Fleming, gen. sls. mgr.

Operation: 1971.

WSB-TV (Channel 2) NBC

Cox Broadcasting Corporation, 1601 W. Peachtree Street, N.E., Atlanta, GA 30309 (404) 897-7000.

Clifford M. Kirtland, Jr., pres., Cox; Stanley G. Mouse, exec. v.p., Cox; William A. Schwartz, pres. bdgst. div., Cox; Don Elliot Heald, v.p. & gen. mgr., WSB-TV; Dale Smith, station mgr., WSB-TV; Ed Godfrey, news dir., WSB-TV; A.R. Van Cantfort, prgm. mgr., WSB-TV; John Carpenter, gen. sls. mgr., WSB-TV; William Powers, dir. of engineering, WSB-TV.

Operation: 1948.

WTCG (Channel 17)

Turner Communications Corp., 1018 W. Peachtree St., N.W., Atlanta, GA 30309 (404) 875-7317.

R.E. Turner, pres.; Sidney Pike, v.p., operations; Don Lachowski, gen. sls mgr.; Tom Ashley, natl. sls. mgr.; Gerry Hogan, local sls. mgr.; Gene Wright, chief eng.; Ned Jay, promo. mgr.

Operation: 1967.

WXIA-TV (Channel 11) ABC

Pacific & Southern Broadcasting Co., Inc., 1611 W. Peachtree St., N.E. Atlanta, GA 30309.

Jeff Davidson, pres. & gen. mgr.; Neil Kuvin, prgm. dir.; Howard Kaufman, gen. sls. mgr.; Elliot Bass, natl sls. mgr.; Russ Artz, chief eng.; Dick Williams, new dir.; Ben Perry, community affairs dir.; Jon Beacher, promo. mgr.; nat'l rep.: PGW-TV.

Operation: Sept., 1951, TV households: 1,568,400.

Augusta

WATU-TV (Channel 26) NBC

P.O. Box 1526, Augusta, GA 30903 (803) 827-0026.

Nick Evans, Jr., gen. mgr.; Lee Sheridan, sta. mgr.; Al Van Dinteren, chief eng.

Operation: Dec. 24, 1968; TV households: 183,700.

WJBF (Channel 6) ABC, NBC

Fuqua Television, Inc., Box 1404, Augusta, GA 30903 (803) 722-6664.

J.B. Fuqua, chm. of bd.; John Radeck, pres. & gen. mgr.; John Wright, treas.; Walter Campbell, Jr., v.p. sales; Terry Sams, v.p. prog.; John Jopling, v.p. eng.; James E. Davis, v.p. news; nat'l rep.: H/R Television.

Operation: 1953; TV sets in area: 556,000.

WRDW-TV (Channel 12)

Ziff Davis Broadcasting Company, Drawer 1212, Augusta, GA 30903.

Ralph E. Becker, pres.; Jack West, v.p. & gen. mgr.; William J. Kemple, gen. sls. mgr.; C. Don Smith, chief engineer; Hal Edwards, program mgr.

Operation: Feb. 1954; TV households; 447,500.

Columbus

WRBL-TV (Channel 3) CBS

Columbus Broadcasting Co., Inc. 1350 13th Ave., Columbus, GA 31902 (404) 322-0601.

J. Barnett Woodruff, pres.; George Gingell, v.p. & gen. mgr.; Paul T. Berry, v.p. prgm. & op.; Robert Walton, v.p.-sales; Lacy Worrell, chief eng.

Operation: 1953; TV sets in area: 600,000.

WTVM (Channel 9) ABC

Martin Theatres of Georgia, Inc., 1909 Wynnton Rd., Columbus, GA (404) 324-6471.

Joe Windsor, pres.; Lynn Avery, v.p./gen. mgr., Terry Johnson, news dir.; Bob King, chief eng.; Carroll Ward, prog. dir.; Eddie Bland, prod. mgr.; nat'l rep.: Blair Television.

Operation: 1953: TV sets in area: 719,900.

WYEA-TV (Channel 38) NBC

Eagle Bcstg. Co., P.O. Box 6389, Columbus, GA 31907 (404) 561-3838.

Charles Grisham, pres.; David Rothschild II, v.p. & treas.; Lee R. Redmond, Jr., secty.; Walter McCroba, gen. mgr.; Ken Gerdes, gen. sls. mgr.; Arnett Jones, op. mgr.; Frank Hardman, chief eng.; Al Fleming, news dir.

Operation: 1970.

Macon

WCWB-TV (Channel 41) NBC

Bibb Television, Inc., P.O. Box 4328, Macon, GA 31208 Charles Morris, (912) 746-1455.

Charles Morris, chmn.; Lanier Finch, pres.; Loran Nesmith, sta. mgr.; Joe Spears, chief eng.; Chuca Teel, prod. mgr.

Operation: Dec. 1968; TV households: 126,200.

WMAZ-TV (Channel 13) CBS basic, ABC

Multimedia Broadcasting Company, Box 5008, Macon, GA 31208 (912) 746-7311.

James T. Lynagh, pres.; Albert Sanders, v.p. & gen. mgr.; Elmo Simmons, Jr., gen. sls. mgr.; Wayne Sawyer, admin. coord.; Jim Elrod, dir. opns.; Jim Lawhon, chief engineer, nat'l rep.: Katz Agency, Inc.

Savannah

WJCL (Channel 22) ABC

Lewis Bcstg. Corp., 10001 Abercorn St., Extension, Savannah, GA 31406 (912) 925-0022.

J. Fred Pierce, exec. v.p. & gen. mgr.; Christopher A. Barbieri, v.p., sls. & prgm.; Larry Walker, opns. dir.

Operation: 1970.

WSAV (Channel 3) NBC

WSAV-TV P.O. Box 2429, 1430 East Victory Dr., Savannah, GA 31402 (912) 236-0303.

David R. Bradley, pres.-treas.; Victor A. Modeer, exec. v.p.; Henry H. Bradley, v.p.; David R. Bradley, Jr., v.p.; Tom Matthews, v.p. & gen. mgr.; Randy Peltier, commercial mgr.; Jack Moore, news dir.; Curt Avery, prod. mgr.; Byron Strong, chief eng.; nat'l rep.: AK Television.

Operation: 1956; TV sets in area: 206,500.

WTOC-TV (Channel 11) CBS

Savannah Broadcasting Co., 516 Abercorn St., Savannah, GA 31402 (912) 232-0127.

W.T. Knight, chmn. of bd.; F. Schley Knight, pres.; Esther S. Pruett, asst. secty.-treas.; S.G. Knight, v.p. & treas.; Dwight J. Bruce, v.p. & prog. dir.; Walt Kessel, promo. mgr.; Kyle E. Goodman, chief eng.; nat'l rep.: Katz Agency Inc.

Operations: 1954; TV sets in area: 150,500.

Thomasville-Tallahassee (Fla.)

WCTV (Channel 6) CBS

John H. Phipps Broadcasting Stations, Inc., P.O. Box 3048, Tallahassee FL 32303 (904) 893-6666.

John H. Phipps, chm.; Colin S. Phipps, pres.; Joe Hosford, v.p. & gen. mgr.; Jerry L. Williams, pgm. & op. mgr.; nat'l rep.: Blair TV.

Operation: 1955; TV sets in area: 355,000.

HAWAIIAN ISLANDS

Hilo, Hawaii

KHAW-TV (Channel 11) NBC

(Satellite of KHON-TV, Honolulu)

KHON-TV, Inc., 1170 Auahi St., Honolulu, HI 96814.

Personnel same as KHON.

Operation: Nov. 1961.

KHVO (Channel 13) ABC

(Satellite of KHVH-TV, Honolulu)

Western Telestations Inc., Naniloa Hotel, Hilo, HI 96814 935-8289, Hilo; rep: The Katz Agency; 935-8289.

Operation: May, 1960; TV sets in area.

KPUA-TV (Channel 9) CBS
(Satellite of KGMB-TV, Honolulu)
Heftel Television-Hilo, Inc., P.O. Box 581, 1534 Kapiolani Blvd., Honolulu, HI 96814.
Operation: 1955.

Honolulu, Oahu

KGMB TV (Channel 9) CBS
Lee Enterprises, Incorporated 1534 Kapiolani Blvd., Honolulu, HI 96814 P.O. Box 581, Honolulu, HI 96809 (808) 944-5200.
Richard A. Weiner, gen. mgr.; Robert B. Fox, prod. mgr.; Phil Arnone, prgm. mgr.; Robert W. Sevey, news dir.; James Doney, dir. of community svcs.; Barry Albright, chief eng.; John Funk, sls. mgr.; Ed Wong, contr.; nat'l Rep.: Katz Agency.
Operation: 1952. TV households (State): 261,170.

KHON-TV (Channel 2) Honolulu
KAII-TV WAILUKU SATELLITE
KHAW-TV HILO SATELLITE
KHON-TV, Inc. 615 Piikoi St., Honolulu HI 96814.
William S. Snyder, pres. & gen. mgr.; nat'l rep.: Peters Griffin Woodward.
Operation: 1952; TV sets in area: 293,500.

KIKU-TV (Channel 13)
Mid-Pacific Television Associates, 150-B Puuhale Road, Honolulu, HI 96819
Richard F. Schaller, gen. mgr.; Richard Eaton, pres.; Joanne Ninomiya, gen. mgr.
Operation: July, 1962; TV homes: 148,700.

KITV (Channel 4) ABC
Western Telestations, Inc., 1290 Ala Moana Blvd., Honolulu, HI 96814; 537-3991.
Dick Grimm, v.p. & gen. mgr.; Chuck Hartman, chief eng.; Muneo Hamada, bus. mgr.; Don Rockwell, news dir.; Larry Cazavan, prog. dir.; Charlie Riley, prod. mgr.
Operation: 1954.

Wailuku, Maui

KAII-TV (Channel 7) NBC
(Satellite of KHON-TV, Honolulu) KHON-TV, Inc., 1170 Auahi St., Honolulu, HI 96814.
Personnel: same as KHON-TV, Honolulu.
Operation: Nov. 1958.

KGMV-TV (Channel 3) CBS
(Satellite of KGMB-TV Honolulu, Heftel Broadcasting-Maui, Inc., P.O. Box 1574, Kahului, HI 96732.
Cecil Heftel, pres., treas. & gen. mgr.; Earl McDaniel, gen. mgr.; Julius M. Vetter, chief eng. & sta. mgr.
Operation: 1955.

KMVI-TV (Channel 12) ABC
(Semi-satellite of separately-owned, KITV, Honolulu)
Maui Publishing Co., Ltd., Box 550, Wailuku, Hawaii; 96814.
Operation: Dec., 1955.

IDAHO

Boise

KBCI-TV (Channel 2) CBS
Eugene Television, Inc., 1007 Jefferson St., Boise, ID 83707 (208) 336-5222.
Robert Booth, pres., Glenn Nickell, v.p.; James A. Johntz, Jr., mgr.; Jack Sansman, sls. mgr.; Paul Greene, prog. dir.; nat'l rep.: Katz Television, Art Moore & Associates, Inc. (Pacific Northwest).
Operation: 1953; TV sets in area: 160,000.

KTVB-TV (Channel 7) NBC, ABC
KTVB, Inc. 5407 Fairview Ave., P.O. Box 7 Boise, ID 83701 (208) 375-7277.
Georgia M. Davidson, bd. chm. & dir.; Robert E. Krueger, pres. & gen. mgr.; Sal Celeski, news dir.; nat'l rep.: Avery-Knodel, Inc.

Idaho Falls

KID-TV (Channel 3) CBS
KID Broadcasting Corp., Box 2008, Idaho Falls, ID 83401 (208) 522-5100.
J. Allen Jensen, pres. & gen. mgr.; Dewain Silvester, sta. mgr.; Kim Southwick, op. dir.; Carroll Secrist, chief eng.; nat'l rep.: Peters, Griffin, & Woodward, Inc.—Art Moore & Associates, Inc. (Pacific Northwest).
Operation: 1953; TV sets in area: 131,060.

KIFI (Channel 8)
The Post Company, P.O. Box 2148, Idaho Falls, ID 83401 (208) 523-1171.
Jim Brady, pres. & gen. mgr.; Herman Haefele, op. dir. prgm. dir.; Cecil Cops, chief eng.; nat'l rep.: Avery-Knodel, Inc.; Seattle rep.: Avery-Knodel, Seattle and Portland; Denver rep.: Robert Hix Co., Inc.
Operation: 1961.

Lewiston

KIVI (Channel 6) ABC
Futura Communications, 1866 East Chisholm Dr., Nampa, ID 83651 (208) 336-0500.
Robert V. Hansberger, pres. Michael W. Gee, gen. mgr.; James R. Barto, ops. dir.; Ken Ritchie, local & regional sls. mgr.; Rhea Surmelis, natl. sls.; Jim Barton, bus. mgr.; Brink Chipman, news dir.; Andy Suk, chief eng.; nat'l. rep.: H-R TV, Inc.; regional rep.: Tacher.
Operation: 1974.

KLEW-TV (Channel 3) CBS
NWG Broadcasting Co., 1115 Idaho St., Lewiston, ID 83501.
Gerry Cornwell, mgr.; Marlin Jackson, chief eng.; Vanessa Gerwels, news dir.; nat'l. rep.: H.R.
Operation: 1955.

Pocatello-Blackfoot

KPVI (Channel 6) ABC
Futura Titanium Corp., 1866 E. Chisholm Dr., Nampa, ID 83702 (208) 233-6667.
Robert V. Hansberger, pres.; Michael W. Gee, gen. mgr.; James Barto, ops. mgr.; Brian Hogan, sta. mgr.; Joe Emig, gen. sls. mgr.; Jean Morand, chief eng.; Gregg Licht, news dir.; natl. rep.: Seltel.
Operation: 1974; TV households: 89,900.

Twin Falls

KMVT (Channel 11) NBC, CBS, ABC
The KLIX Corporation, 2217 Elizabeth Blvd. & Eastland Dr., Twin Falls, ID 83301 (208) 733-1280.
Dale G. Moore, pres.; Earl E. Morgenroth, v.p.; Doug

G. Moore, gen. mgr.; George Brown, ops. mgr.; Len Gute, chief eng.; nat'l rep.: Avery-Knodel, Inc., Simpson/Reilly. (Northwest).
Operation: 1955; TV sets in area: 43,860.

ILLINOIS

Champaign

WCIA (Channel 3) CBS

Midwest Television, Inc., 509 S. Neil St., Champaign, IL 61820 (217) 356-8333.

A.C. Meyer, pres.; Guy Main, exec. v.p.; Jack B. Everette, asst. to the pres. and dir. of eng.; Gerald P. Johnson, sls. mgr.; Leonard Davis, natl. sales mgr.; Darrell Blue, prod. mgr.; William J. Helms, prog. mgr.; John Ketterer, film buyer; nat'l rep.: Peters, Griffin, Woodward.
Operation: 1953; TV sets in area: 329,000.

WICD-TV (Channel 15) NBC

Plains Television Corp., 250 Country Fair Dr., Champaign, IL 61820 (217) 352-7673.

Elmer Balaban, pres.; Milton D. Friedland, v.p. & gen. mgr.; Robert B. Farrow, gen. sales mgr.; Joseph Norris, station mgr.; Ed Mason, sls. mgr.; David Boyer, chief eng.; Ken Lotka, op. mgr.; nat'l rep., Katz Television Midwest.
Operation: 1959.

Chicago

WBBM-TV (Channel 2) CBS

CBS, Inc., 630 N. McClurg Court, Chicago, IL 60611 (312) 944-6000.

David L. Nelson, v.p. & gen. mgr.; James Bennett, dir. of bdcst'.; Lilly Eide, dir. of communications; Warner Saunders, dir. of community affairs; Eric Ober, dir. of news; Edward A. Spray, prgm. mgr.; Rhoda Horwitz, sls. traffic mgr.; Dianne Jennings, dir. of research.
Operation: 1946; TV sets in area: 2,679,260.

WCIU-TV (Channel 26) Spanish Intl. Network

Weigel Bdg. Co., Board of Trade, Chicago, IL 60604.

Howard Shapiro, pres.; Bob Ward, sls. mgr.; Peter Zomaya, asst. station mgr.; Rosa Negrete, film dir.; Richard Ashenfelter, chief eng.; Ben Larson, Shirley Harris, Henry Lopez, news directors.
Operation: 1964.

WFLD (Channel 32)

Field Communications Corp., 300 North State St., Chicago, IL 60610.

Derk Zimmerman, v.p. & gen. mgr.; Doug Knight, prog. mgr.; Vincent Barresi, gen. sls. mgr.; William P. Kusack, eng. mgr.; nat'l. rep.: Field Spot Sales.
Operation: Jan., 1966; TV households: 3,138,000.

WGN-TV (Channel 9)

WGN, Continental Broadcasting Company, 2501 W. Bradley Pl., Chicago, IL 60618 (312) 528-2311.

Daniel T. Pecaro, pres. & chief exec. of.; Donald H. Knautz, v.p. assist. to the pres. & dir. of bdcst'. op.; Sheldon Cooper, pres., WGN-TV,; Daniel D. Calibraro, v.p., corporate affairs; Joseph G. Antelo, v.p., dir. of sls.; Jack P. Jacobson, v.p., & ops. mgr.; Paul Davis, news. dir.; John B. Brickhouse, v.p., mgr. of sports; Charlotte O'Brien, mgr. of community affairs; William H. Wills, mgr. of pub. rel.; Harry Q. Trigg, mgr. of film.

WLS-TV (Channel 7) ABC

American Broadcasting Companies Inc., 190 N. State St., Chicago, IL 60601 (312) 263-0880.

Peter B. Desnoes, v.p. & gen. mgr.; Richard Taylor, gen. sls. mgr.; Michael Velazquez, natl. sls. mgr.; Robert Hoffman, local sls. mgr.; Clar W. Heider, admin. asst.; Jeffrey M. McGrath, dir. of programming; Ernst Ledebuhr, contr.; Thomas Kuelbs, news dir.; Sharon McGill, research dir.; Joseph Kresnicka, dir. of eng.; Timothy J. Bennett, adv. promo. dir.; natl. rep.: ABC TV Spot Sales.
Operation: Oct., 1943; TV homes: 2,995,000.

WMAQ-TV (Channel 5) NBC

National Broadcastg. Co., Merchandise Mart, Chicago, IL 60654 (312) 861-5555.

Monte Newman, v.p. & gen. mgr.; Dick Daggett, dir. sls.; Fred DeMarco, sta. mgr.; Rillon Smith, prgm. dir.; Gene Cartwright, dir. op.; Bill Marshall, dir., eng'.; Paul Beavers, News dir.; Carole Cartwright, mgr., Community Programs; Gail Sikevitz, exec. prod., prog. dept.; natl. rep.: NBC Spot Sales.
Operation: 1949; TV sets in area: 3,530,230

WSNS-TV (Channel 44)

Video 44, a Joint Venture, 430 W. Grant Pl., Chicago, IL 60614 (312) 929-1200.

Burt I. Harris, pres., Harriscope (L.A.); Alan Silverman, pres., Essaness Theatre Corp.; Edward L. Morris, v.p. & gen. mgr.; Kevin Murphy, chief eng.; Mark Konrus, sls. mgr.
Operation: April, 1970.

Decatur

WAND (Channel 17) ABC

(Serving Decatur, Springfield & Champaign)

WAND-Television, Inc. (subsidiary of Lin Broadcasting Corp., Inc.), 904 Southside Drive, Decatur, IL 62525 (217) 424-2500.

B. Geoghegan, pres. & gen. mgr.; T.J. Vaughan, v.p. Programming/Op.; Larry Katt, v.p. sls.; Nat'l Rep.: Blair.
Operation: 1953; TV sets in area: 569,506.

Harrisburg

WSIL-TV (Channel 3) ABC Primary

WSIL-TV Holding Co., 21 W. Poplar St., Harrisburg. IL; 62946.

Mackie Nicholes, pres.
Operation: 1953.

Moline

WQAD-TV (Channel 8)

Moline Television Corp., 3003 Park 16th St., Moline, IL 61625 (309) 764-9694.

Arthur M. Swift, exec. v.p. & gen. mgr.; William Snyder, gen. sls. mgr.; Richard Jolliffe, natl. sls. mgr.; Lloyd Bolkcom, pub. svc. mgr. (pub. affairs); Ed Claseman, bus. mgr. & asst. treas.; Ray Krueger, chief eng.; Phil Riske, dir. of news; Ron Klayman, prog. & promo. mgr.
Operation: 1963.

Peoria

WEEK-TV (Channel 25) NBC

Mid-America Television Company, 2907 Springfield Rd., E. Peoria, IL 61611 (309) 699-3961.

Philip Mergener, gen. mgr.; Roger Ottenbach, bdcst. mgr.; Ken Tofanelli, chief. eng.; Vic Burnett, pub. & promo. mgr.; nat'l rep.: Katz Agency.

Operation: 1953; TV sets in area: 325,000.

WMBD (Channel 31) CBS

Midwest Television, Inc., 3131 N. University, Peoria, IL 61611 (309) 688-3131.

August C. Meyer, Jr., pres.; Guy F. Main, exec. v.p.; Gene Robinson, v.p. and gen. mgr.; Gene C. Robinson, stat. mgr.; Llyod Peterson, op. mgr.; Walter Konetsco, chief eng.; nat'l rep.; Petry

Operation: 1958; TV sets in area: 338,300.

WRAU-TV (Channel 19) ABC

Forward of Illinois, Inc. 500 N. Stewart St., Creve Coeur, IL 61611 (309) 698-1919.

Robert E. Rice, pres. & gen. mgr.; William Thorson, sr. v.p. & gen. sls. mgr.; nat'l. rep.: Seltel.

Operation: Oct. 1953; TV households; 333,900.

Quincy

WGEM-TV (Channel 10) NBC

Quincy Broadcasting Co., 513 Hampshire, Quincy, IL 62306 (217) 222-6840.

T.A. Oakley, pres.; Dave Oakley, v.p./gen. mgr.; Fred Colgan, op. mgr.; Ben Stewart, v.p.-sls.; Jim Martens, chief eng.; nat'l rep.: Avery-Knodel, Inc.

Operation: 1953; TV sets in area: 251,300.

Rock Island

WHBF-TV (Channel 4) CBS

Rock Island Broadcasting Co., Telco Bldg., 231 18 St., Rock Island, IL 61201 (309) 786-5441.

Miss Marguerite F. Potter, pres.; Edward E. DeLong III, exec. v.p.; Robert J. Sinnett, v.p. & gen. mgr.; Dewey L. Nelsen, nat. sls. mgr.; Wesley C. Hoots, chief eng.; nat'l rep.: Petry Television, Inc.

Operation: 1950; TV sets in area: 486,000.

Rockford

WIFR-TV (Channel 23) CBS

WIFR Television, Inc., 2523 N. Meridian Rd., Rockford, IL 61105.

Dennis Blake West, pres.; Michael R. Lennon, gen. sls. mgr.; Jack Keefe, news dir.; Aaron Silberman, prog. dir.; Ben Pflederes, chief eng.; Roy Bilsbarrow, cont.; Barbara Schobinger, prog. supvr.

Operation: 1965.

WREX-TV (Channel 13) ABC

WREX-TV, West Auburn at Winnebago Rds., Rockford, IL 61105 (910) 642-0770.

Member of The Gilmore Broadcasting Corp.

James S. Gilmore, owner; John McWeeny, gen. mgr.; Jim Smith, dir. of natl. sls.; Warren A. Anderson, local sls. mgr.; Gene Hines, film mgr.; natl. rep.: Seltel.

Operation: September, 1953; TV households: 401,600.

WTVO (Channel 17) NBC

Winnebago Television Corp., N. Meridian Rd. & State, Rockford, IL 61105 (910) 963-5413.

Harry Balaban, pres.; Hal Froelich, mgr.; Richard McDaniel, prgm. dir.; Tom Anderson, sales mgr.; Al Petzke, chief eng.; Frank Sanders, film dir.; nat'l rep.: Adam Young, Inc.

Operation: 1953; TV sets in area: 580,300

Springfield

WICS (Channel 20) NBC

WICS-TV, Inc., 2680 East Cook St., Springfield, IL 62703 (217) 753-5620.

Larry H. Israel, pres.; Steven Kumble, secty.-treas.; Harry Balaban, secty.; Milton D. Friedland, v.p. & gen. mgr.; Robert R. Swan, gen. sls. mgr.; Jerrold Merritt, chief eng.; Gary Spears, prog. mgr.; Robert Meyer, commercial prodn. mgr.; H.K. Springer, contr.; nat'l rep.: Katz Agency.

Operation: 1953: TV sets in area: 421,000.

INDIANA

Bloomington-Indianapolis

WTTV (Channel 4) Independent

Sarkes Tarzian, Inc., 3490 Bluff Road, Indianapolis, IN 46217 (317) 787-2211.

Sarkes Tarzian, pres.; Elmer C. Snow, gen. mgr., bdcst. div.; Don Tillman, prgm. mgr.; Clyde Dutton, gen. sls. mgr.; Bernie Souers, asst. gen. sls. mgr.; Don Morgan, chief engineer.

Operation: 1949; TV sets in area: 650,000.

Elkhart-South Bend

WSJV (Channel 28) ABC

WSJV, Inc. (owner)—Subsidiary of Quincy Newspapers (Illinois) Inc., Box 1646, Elkhart, IN 46515 (812) 293-8616; South Bend (219) 233-9353.

Thomas A. Oakley, pres.; Don E. Fuller, v.p. & gen. mgr.; Andy Rinaldi, gen. sls. mgr.; nat'l rep., Blair.

Operation: 1954; TV sets in ADI (as of 5/79): 268,000.

Evansville

WEHT-TV (Channel 25) CBS

Gilmore Broadcasting Corp., P.O. Box 395, Evansville, IN 47704 (812) 424-9215.

James S. Gilmore, Jr., pres.; Ernest D. Madden, gen. mgr.; Jerry Kunkel, gen. sls. mgr.; Doug Padgett, contr.; Elmer Chancellor, chief eng.; nat'l rep.: Blair Television.

Operation: 1953; TV sets in area: 290,600.

WFIE (Channel 14) NBC

WFIE, Inc., 1115 Mt. Auburn Rd., Evansville, IN 47712 (812) 426-1414.

Conrad L. Cagle, pres. & gen. mgr.; Ralph S. Jackson, exec. v.p.; Bob West, dir. ops; Bill Hollander, news dir.

Operation: 1953; TV sets in area: 214,000.

WTVW (Channel 7)

Evansville Television, Inc., 405 Carpenter St., Evansville, (812) 422-1121.

E. Berry Smith, gen. mgr.; nat'l rep.: Blair Television.

Operation: 1956; TV sets in area: 247,480.

Fort Wayne

WANE-TV (Channel 15) CBS

Indiana Broadcasting Corp., 2915 W. State Blvd., Fort Wayne, IN 46808 (219) 424-1515.

Philip J. Lombardo, pres.; John Irvin, v.p., gen. mgr.; Don Palmer, fin. mgr.; Ron Mack, prog. dir.; Jim Giffen, chief eng.; nat'l rep.: H-R-P.

Operation: 1954; TV sets in area: 253,000.

WFFT-TV (Channel 55)

Ontario Corp., c/o Robert W. Faull, Box 2255, Fort Wayne, IN 46801 (219) 424-5555.

Van P. Smith, pres.; Robert W. Faull, gen. mgr.; Richard J. Stawicki, prog. dir.; James E. Jump, sls. mgr.; James E. Miller, chief eng.

Operation: December, 1977.

WKJG-TV (Channel 33) NBC

Thirty-Three, Inc., 2633 W. State Blvd., Ft. Wayne, IN 46808 (219) 422-7474.

Anton Hulman, Jr., pres.; Joseph R. Cloutier, exec. v.p. & treas.; Hilliard Gates, gen. mgr.; Calo Mahlock, asst. gen. mgr.; William Kline, sls. mgr.; Eugene A. Chase, chief eng.; nat'l rep., Blair Television.

Operation: 1953; TV sets in area: 280,000.

WPTA (Channel 21) ABC

WPTA-TV, Inc., 3401 Butler Road, Ft. Wayne, IN 46808 (219) 483-0584.

Edwin C. Metcalfe, pres./gen. mgr.; Marvin Gottlieb, v.p. sta. & gen. sls mgr.; Barbara Wigham, v.p. dir. of prgm. & promo.; Bill MacDonald, v.p., sls mgr.; Roger Ball, news dir.; Ray Krueger, chief eng.; nat'l rep.: Blair Television.

Operation: 1957.

Indianapolis

WHMB-TV (Channel 40)

LeSea Bldg., Co., Box 12, South Bend, IN (317) 773-5050.

Lester Sumrall, pres.; Larry W. Vehorn, chief eng.; Wanda Linville, traffic mgr.

Operation: 1971.

WISH-TV (Channel 8) CBS

Corinthion Broadcasting Corp., 1950 North Meridian St., Indianapolis, IN 46624 (317) 924-4381.

Philip Lombardo, pres.; Harold Lewis, fin. v.p.; Bill Stough, v.p. & gen. mgr.; Joe Girard, gen. sls. mgr.; John Demshock, chief eng.; Don Dillion, prog., mgr.; Phil Michael, promo. mgr.; R. Lee Giles, news dir.; Karen Miller, prod. mgr.; Hallie Crombaugh, community affairs dir.; Bruce Ayers, fin. mgr.; nat'l. rep.: PGW.

Operation: 1954; TV sets in area: 893,500.

WRTV (Channel 6) NBC

McGraw-Hill Broadcasting Company, Inc., 1330 North Meridian St., Indianapolis, IN 46202 (317) 635-9788.

Norman Walt, pres.; Jerry Chapman, v.p. & gen. mgr.; Jim Mathis, gen. sls. mgr.; Don Rinker, program mgr.; Robert Flanders, v.p. & dir. of eng.; Tim Hnedak, promo. mgr.; Beverly May, bus. mgr.; Bob Gamble, news dir.; nat'l. rep.; Harrington, Righter and Parsons, Inc.

Operation: 1949.

WTHR (Channel 13) ABC

Video Indiana, Inc., 1401 N. Meridian St., Indianapolis, IN 46202 (317) 639-2311.

Christopher Duffy, v.p. & gen. mgr.; Barbara Gurwitz, pub. affairs dir.; Mel Grossman, gen. sls. mgr.; Tom Rose, prog. dir.; Ron Arendall, mgr. of eng'.; Bob Schaefer, news dir.; John Newcomb, business mgr.; nat'l rep.: John Blair & Co.

Operation: 1957; Ownership change: 1975. TV sets in area: 1,284,200.

Lafayette

WLFI-TV (Channel 18)

RJN Broadcasting, Inc., 2605 Yeager Rd., West Lafayette, IN 47906; (Studio); P.O. Box 18, Lafayette, IN (317) 463-3516.

Richard F. Shively, pres.; Harold Shively, v.p. & gen. mgr.; Dick Wade, sls. mgr.; Ken Gardner, chief eng.; Nina Hart, prog. mgr.

Operation: May, 1953; TV households: 319,900.

South Bend

WHME-TV (Channel 46)

Lester Sumrall Evangelistic Assn. Inc., Box 12, South Bend, IN 46624 (219) 291-8200.

Lester Sumrall, pres.; Steve Sumrall, group v.p.; Larry D. Scott, v.p. & gen. mgr.

Operation: July, 1974.

WNDU-TV (Channel 16) NBC

Michiana Telecasting Corp., Juniper Road, Box 1616, South Bend, IN 46634.

Bazil O'Hagan, pres. & gen. mgr.; Paul Walton, Jr., dir. prog. operations; nat'l rep.; Adam Young, Inc.

Operation: 1955; TV homes in area: 270,000.

WSBT-TV (Channel 22) CBS

WSBT, Inc., 300 W. Jefferson Blvd., South Bend, IN 46601 (219) 233-3141.

Jack E. Douglas, pres. & gen. mgr.; Jerry Bleck, nat. sales mgr.; N.J. Gassensmith, program dir. & film dir.; Robert Bell, chief eng.; nat'l rep.; Katz Agency.

Operation: 1952; Sets in area: 327,200.

Terre Haute

WBAK-TV (Channel 38) ABC

WBAK Television Co., P.O. Box 719, Terre Haute, IN 47808 (812) 238-1515.

Cy N. Bahakel, pres.; Phil Brassie, gen. mgr.; David Pierce, chief eng.; Larry Blackerby, prod. mgr.

Operation: 1973.

WTHI-TV (Channel 10) CBS

Wabash Valley Broadcasting Corp., 918 Ohio St., Terre Haute, IN 47808 (812) 232-9481.

J.R. Cloutier, exec. v.p. & tres.; Russ Arnold, v.p. & gen. mgr.; Dave Bailey, gen. sls. mgr.; Ruth Baker, prgrm. dir.; nat'l rep. Katz Agency.

Operation: 1954; TV sets in area: 449,300.

WTWO (Channel 2) NBC

Indiana Telecasting Corp., P.O. Box 299, Terre Haute, IN 47808.

John T. Gelder, Jr., exec. v.p. & gen. mgr.; Robert B. Beall, asst. gen. mgr. & sls. mgr.; James Underwood, oper. mgr.; nat'l. rep.: Peters, Griffin, Woodward, Inc.

Operation: Sept., 1965.

IOWA

Ames

WOI-TV (Channel 5) ABC

Iowa State Univ. of Science & technology, Ames, IA 50010 (515) 294-5555.

W. Robert Parks, pres.; Robert C. Mulhall, gen. mgr.; Ted Tostlebe, comm. mgr.; Ralph Willey, promo. mgr.; Keith Kethcam, chief eng.; Don Minnaugh, film dir.; nat'l rep.: H-R Television, Inc.

Operation: 1950; TV sets in area: 400,000.

Cedar Rapids

KCRG-TV (Channel 9) ABC

Cedar Rapids Television Co., 2nd Ave. at 5th St. S.E., Cedar Rapids, IA 52401 (319) 398-8422.

Joseph F. Hladky, Jr., pres.; Ed Lasko, v.p. & gen. mgr.; Ron Stack, gen. sls. mgr.; Edna A. Herbst, v.p. & promotion dir.; Walt Alliss, chief eng. & ops. dir.; Paul Dicker, prog. dir.; James Laymon, bus. mgr.; nat'l rep.: H-R.

Operation: 1953; TV homes in area: 493,300.

WMT (Channel 2) CBS

WMT-TV, Broadcast Park, Box 2147, Cedar Rapids, IA (319) 395-6000. WMT, Inc.

Kelly Atherton, pres. & sta. mgr.; Mike Bock, gen. sls mgr.; Dennis Czechamski, dir. TV operation; Bob Burns, chief eng.; nat'l rep.: The Katz Agency, Inc.

Operation: 1953; TV sets in area: 373,800.

Davenport

WOC-TV (Channel 6) NBC

WOC Broadcasting Co., 805 Brady St., Davenport, IA 52808 (319) 324-1661.

D.D. Palmer, pres. & treas.; Lee Marts, gen. mgr.; Harold Heath, sls. mgr.; Coy M. Bullard, tech. dir.; nat'l rep.: Peters, Griffin, Woodward, Inc.

Operation: 1949; TV sets in area: 472,780.

Des Moines

KCCI-TV (Channel 8) CBS

Cowles Communications, Inc., Broadcasting Division, 9th & Pleasant St., Des Moines, IA 50308 (515) 243-4141.

Bill Hippee, v.p. & gen. mgr.; Joe Hudgens, prog. dir.; Nolan Quam, gen. sls. mgr.; Dick Percival, bus. mgr.; Jake Sneller, chief eng.; Guy Koeningsberger, adv. & promo. mgr.; nat'l rep.: The Katz Agency, Inc.

Operation: 1955; TV sets in area: 295,200.

WHO-TV (Channel 13) NBC

WHO Broadcasting Co., 1100 Walnut St., Des Moines, IA 50308 (515) 288-6511.

George C. Carpenter, III, vice pres. & gen. mgr.; William H. Jackson, prgm. dir. & film buyer; Talmage Thompson, gen. sls. mgr.; Phil Tomas, news dir.; Verna M. Smith, promo. mgr.; Charles Myers, Chief Eng.; nat'l rep.: Blair Television.

Operation: 1954; households: 364,840.

Dubuque

KDUB-TV (Channel 40) ABC

Lloyd Hearing Aid Corp., 128 Kishwaukee St., Rockford, IL 61110 (319) 556-4040.

Marvin E. Palmquist, pres. & treas.; Chuck Cyberski, station mgr.; John Sloan, sls. mgr.; Neil Black, prodn. mgr.; John Barton, news dir.; Lloyd Berg, chief eng.

Operation: June, 1970.

Fort Dodge

KVFD-TV (Channel 50) NBC

Northwest Television Co., Television Square, Fort Dodge, IA (319) 576-7606.

Edward Breen, pres. & gen. mgr.; Fred Breen, vice-pres.; Dave Sinclair, secty.; Allan R. Lllmis II, treas.; Richard Johnson, prog. dir.; Gordon De Koster, sls. mgr.; Ray Gales, chief eng.; nat'l rep.: Bernard Howard, Inc.

Operation: 1953; TV sets in area: 143,601.

Kirksville, Mo.-Ottumwa, IA

KTVO (Channel 3) ABC

Broadcast Center, Ottumwa, IA 52501 (515) 682-4535.

Alvina M. Britz, gen. mgr.; Joe Galvan, sls. mgr.; Gerald Dreger, dir. eng.; Dean Weitenhogen, dir. pub. info.; nat'l rep.: PGW.

Operation: 1955; TV sets in area: 146,400.

Mason City

KIMT-TV (Channel 3) CBS

Iowa Television, Inc., Second & Pennsylvania, Mason City, IA 50401; 423-2540.

John C. Shott, v.p. & gen. mgr.; Earl Schoborg, sls. mgr.; John Hegeman, chief eng.; nat'l rep.: Katz Agency.

Operation: 1954; TV sets in area: 449,000.

Sioux City

KCAU-TV (Channel 9) ABC-TV

Forward of Iowa, Inc., 7th & Douglas Sts., Sioux City, IA 51101 (712) 277-2345.

William F. Turner, pres. & gen. mgr.; Jack Gilbert, v.p.-sta. mgr.; Ray Cole, v.p., fin.; Paul J. O'Bryan, asst. v.p. sales mgr.; Bob Dovichak v.p., mktg.; Bob Lawrence, v.p. news; David L. Slater, op. mgr.; William Harling, v.p., prod. & tech. ops.; Jim Henry, promo. dir.; Pat Francis, traffic dir.; Gene Sherman, sports dir.; George Lindblade, creative services dir.; Reps.: Seltel, Ramsland. Bob Hix, Denver.

Operation: 1953; TV homes in area: 248,320.

KMEG-TV (Channel 14) CBS

Medallion Broadcasters Inc., Box 657, 7th & Floyd Blvd., Sioux City, IA 51102 (712) 277-3554.

A. James Ebel, pres.; Robert B. Donovan, v.p. & gen. mgr.; Bruce Lewis, sta. mgr.; Don Welch, gen. sls. mgr.; Lyle Johnson, chief eng.; Greg Funk, bus. mgr.

Operation: Sept., 1967; TV homes: 152,900.

KTIV (Channel 14) NBC

KTIV Television Co., Signol Hill Box 87, Sioux City, IA (712) 258-0545.

Harry Slife, pres. Lyn P. Stoyer, exec. v.p. & gen. mgr.; Chuck Lutz, v.p. prog.; Dave Madsen, prod. dir.; Dave Jennings, chief eng.; nat'l rep.: John Blair Co.

Operation: 1954; TV sets in area: 336,000.

Waterloo

KWWL-TV (Channel 7) NBC

Black Hawk Broadcasting Co., E. 4th & Franklin, Waterloo, IA 50703 (319) 291-1200.

Robert Buckmaster, chmn.; Harry G. Slife, pres.; William L. Bolster, v.p., gen. mgr.; Thomas E. Rodden, secty.-treas.; Thomas L. Trunnell, sect.; Edward M. Tink, v.p. & dir. eng.; James Waterbury, v.p., ops.; nat'l rep., Blair.
Operation: 1953; TV sets in area: 412,700.

KANSAS

Ensign

KTVC (Channel 6) CBS

Southwest Kansas Television Co., Box 157, Dodge City, KS 67801.
Leigh Warner, pres.; W.H. Bailey, gen. mgr.;
Operation: 1957.

Garden City

KGLD-TV (Channel 11) NBC

(Satellite of KARD-TV, Wichita, KS
Part of Kansas State Network)
Kansas State Network, Inc., P.O. Box 333, 833 North Main St., Wichita, KS 67201 (316) 265-5631. Rep.: Telerep.
Operation: Nov. 1958.

KUPK-TV (Channel 13) ABC

Satellite of KAKE-TV, Wichita, KA)
KAKE-TV & Radio Inc., Box 10, Wichita, KA 67201.
Martin Umansky, pres. & gen. mgr.; Bob Surber, v.p. & sta. mgr.; George Sides, news dir.; Paul Hinderliter, chief eng.; John Strauss, prog. dir.
Operation: Oct., 1964.

Goodland

KLOE-TV (Channel 10) CBS

Kays, Inc., Broadcast Plaza Box 569, Goodland, KA 67735 (913) 899-2321.
Kay Melia, gen. mgr.; natl. rep.: Blair.
Operation: 1958.

Great Bend

KCKT (Channel 2) NBC

(Satellite of KARD-TV, Wichita, KA)
Kansas State Network, Inc., P.O. Box 333, Wichita, KA 67201 (316) 265-5631.
Rep.: Tele-Rep.
Operation: Nov., 1954.

Hays

KAYS-TV (Channel 7) CBS, ABC

(Satellite KLOE-TV Goodland, KA)
KAYS Inc., Box 817, Hays, KA, 67601.
Robert E. Schmidt, gen. mgr.; Bernard Brown, sta. mgr.; Dannia Massier, chief engineer; Ken Fitch, prod. mgr.
Operation: Sept., 1958; TV households: 97,000.

Pittsburg

KOAM-TV (Channel 7) NBC

Mid-Continent Telecasting, Inc., P.O. Box 659, Pittsburg, KA 66762 (316) 231-0400.

Lester L. Cox, chm.; Bill Bengston, v.p. & gen. mgr.; Mildred Cox & H. Lang Rogers, v.p.; Fred Hughes, secty.; Louis R. Martin, V.P., prgm'. & op.; Jack Lawrence, chief eng.; nat'l rep., The Katz Agency, Inc.
Operation: 1953.

Topeka

KTSB (Channel 27) NBC

Studio Broadcasting System Div., Highwood Service Inc., P.O. Box 2700, Topeka, KS 66601 (913) 582-4000.
Wm. R. Hirshey, v.p.-bdcst'.; Harry L. Strader, station mgr.; Jim Evers, dir. of prodn. and op.; Scott McIver, dir. of acct'.; Herbert L. Brown, gen. sls. mgr.; William S. Beeman, dir. of eng'.; Bob Totten, dir. of news.
Operation: 1967; TV households: 737,390.

WIBW-TV (Channel 13) CBS

Stauffer Publications, Inc., Box 119, Topeka, KS 66601 (913) 272-3456.
Oscar Stauffer, chmn.; Stanley H. Stauffer, pres.; Jerry Holley, v.p., brdgt.; George Logan, gen. mgr.; Ray C. Senate, natl. sls. mgr.; Paul Winders, op. mgr.; Dick Siley, prog. dir.; natl. rep.: Katz TV.
Operation: November, 1953; TV households: 514,800.

Wichita

KAKE-TV (Channel 10) ABC

KAKE-TV and Radio, Inc., 1500 North West St., Box 10, Wichita, KS (316) 943-4221.
Mark H. Adams, chmn. bd.; Charles E. Jones, treas.; Martin Umansky, pres. & gen. mgr.; Harold Newby, v.p. & chief eng.; John Froome, v.p. and asst. gen. mgr.; Bryce Benedict, v.p. and gen. sales mgr.; Paul Threlfall, v.p. & news mgr.; nat'l rep.: Katz Agency, Inc.
Operation: 1954.

KARD-TV (Channel 3) NBC

(Satellites—KCKT-TV, Great Bend, KS, Ch. 2; KGLD-TV, Garden City, KS, Ch. 11; KOMC-TV, McCook, NB, Ch. 8)
Kansas State Network, Inc., P.O. Box 333; 833 North Main St., Wichita, KS 67201 (316) 265-5631.
Robert C. Waddill, exec. v.p.; Larry Proffitt, gen. sls. mgr.; Jerryanne Hadley, creative services dir.; Don Hain, chief eng.; natl. rep.: Katz
Operation: 1955; TV sets in area: 417,000.

Wichita-Hutchinson

KTVH (Channel 12) CBS

Wichita-Hutchinson Co., Inc., 2815 E. 37th St., North, P.O. Box 12, Wichita, KS 67201 (316) 838-1411; 1800 N. Plum St., Hutchinson, KS (311) 665-5503.
Stuart I. Mackie, pres. & gen. mgr.; Joseph Mahan, gen. sls. mgr.; John Mileham, v.p./admin.; Larry Enter, chief eng.; Ron Scott, v.p./news and op.; Robert Wine, contr.
Operation: 1953; TV sets in area: 375,000.

KENTUCKY

Bowling Green

WBKO (Channel 13) ABC

Bluegrass Media, Inc., 537 E. 10th St., Box 1198, Bowling Green, KY 42101.

Clyde Payne, pres. & gen. mgr.; Randy Odil, sls./sta. mgr.; Gene Prather, prog. dir.; Dave Chumley, chief eng.; Roy Brassfield, news dir.; nat'l. rep.: Avery Knodel Television.

Operation: June, 1962; TV households: 265,600.

Hazard

WKYH-TV (Channel 57)

Hazard Television Co., Inc., P.O. Box 929, Hazard, KY 41701 (606) 436-4444, 436-2522.

William D. Gorman, pres.; Susan Gorman, gen. mgr.; Sharon Shepherd, v.p. & prog. dir.; Rich McKinney, chief eng.; John Edwards, farm editor; Darwin Singleton, news div. & sls.; rep.: Max Goldfarb.

Operations: Oct., 1969.

Lexington

WKYT-TV (Channel 27) CBS

Kentucky Central Television Inc., Box 5037, Lexington, KY 40505; Offices: Rt. 4, Winchester Rd.; (606) 299-0411.

Ralph Gabbard, v.p. & gen. mgr.; Al Taylor, v.p., op.; Jere Pigue, v.p. sls.; Hall Schumacher, dir. of eng'.

Operation: June 1967; TV households: 409,800.

WLEX (Channel 18) NBC

WLEX-TV, Inc., P.O. Box 1457, Russell Cave Road, Lexington, KY 40591 (606) 255-4404.

Harry C. Barfield, pres. & gen. mgr.; Wm. B. Gess, v.p.; John G. Atchinson, secty.; Joe Oliver, v.p. & reg. sls. dir.; Bob Davidson, v.p., local sales; L. Neuzel, chief prog.; Al Scheer, chief eng.; nat'l. rep.: Blair.

Operation: 1955; TV households in area: 411,200.

WTVQ-TV (Channel 62) ABC

STARR-WTVQ-TV Inc., P.O. Box 5590, Lexington, KY 40555 (606) 299-6262.

William L. Service, gen. mgr.; Dave Winters, news dir.; Jerry Fox, prog. mgr.; Jack Lackey, chief eng.; Rep.: Katz TV.

Operation: 1968; TV households: 212,100.

Louisville

WAVE-TV (Channel 3) NBC basic

Orion Broadcasting Inc., P.O. Box 32970, Louisville, KY 40232; 725 S. Floyd Street, Louisville, KY 40203 (502) 585-2201.

Jim Keelor, pres. & gen. mgr.; S. Wheeler Rudd, gen. sls. mgr.; John R. Karem, nat'l. sls. mgr.; Anne Coleman, prgm. dir.; J. Harold Culver, dir. of eng'. & op.; Ed Godfrey, news dir.; David Taylor, prom. supvr.; nat'l. rep.: Katz Agency.

Operation: 1948; TV sets in area: 871,300.

WDRB-TV (Channel 41)

Consolidated Bdg., Co., 1051 E. Main St., Louisville, KY 40206 (502) 584-9701.

Elmer F. Jaspan, pres. & gen. mgr.; Jim Riggs, gen. sls. mgr.; David Tung, ops. mgr.; Bob Cheesman, chief eng.

Operation: 1971.

WHAS-TV (Channel 11) CBS

WHAS, Inc., P.O. Box 1084, Louisville, KY 40201; 582-7840.

Barry Bingham, Sr., bd. chmn.; Barry Bingham, Jr., pres.; Bob Morse, exec. v.p. & gen. mgr.; Donna Zapata, sta. mgr.; Richard Sweeney, prog. dir.; Steve Steinberg, dir. of brdg.; Bill Bratton, dir. eng.; Mike Lerner, promo. dir.; Don Hoover, film dir.; nat'l. rep., Harrington, Righter and Parsons, Inc.

Operation: 1950; TV sets in area: (according to ADI survey area household sets in 1977 Broadcasting yearbook) 782,900.

WLKY-TV (Channel 32) ABC

Combined Communications of Kentucky, Inc., 1918 Mellwood Ave., P.O. Box 6205, Louisville, KY 40206 (502) 893-3671.

Paul Blue, pres. & gen. mgr.; William W. Campbell, v.p. & gen. sls. mgr.; Dick Polgreen, sls. mgr.; Bob Jones, prog. dir.; Tom Becherer, news dir.; Paul Kelley, v.p. & chief eng.; Kathy Baetz, promo. dir.; Lora Bradshaw, community affairs dir.

Operation: September, 1961; households: 610,900.

Paducah

WPSD-TV (Channel 6) NBC

Paducah Newspapers, Inc., P.O. Box 1197, 100 Television Lane, Paducah, KY 42001 (502) 442-8214.

Fred Paxton, pres. & mgring dir.; John Williams, gen. mgr.; nat'l. rep., Blair.

Operation: 1957; TV sets in area: 325,000.

LOUISIANA

Alexandria

KALB-TV (Channel 5) NBC

Lanford Telecasting Co., Inc., 605-11 Washington St., Alexandria, LA 71301 (318) 445-2456.

T.B. Lanford, pres.; Robert E. Miller, gen. mgr.; Berton Chandior, natl. sls. mgr.; Dowell Bushnell, sls. mgr.; Dick Grace, sls. serv. mgr.; Tom Webb, news dir.; Ron Saunders, prgm. dir.; nat'l. rep. The Katz Agency.

Operation: 1954; TV sets in area; 378,500.

Baton Rouge

WAFB-TV (Channel 9) CBS

Guaranty Broadcasting Corp., 844 Government St., Baton Rouge, LA 70821.

George A. Foster, Jr., bd. chmn.; Tom E. Gibbens, pres. & gen. mgr.; Harry P. Gamble, Jr., v.p., and gen. counsel; James D. Russell, secty. & treas.; Ed Lamy, v.p. & prog. dir.; Mervyn Rhys, v.p. & comm. mgr.; Donald K. Allan, v.p. & chief eng., oper. mgr.; Grace McElveen, promo.-mdsing. dir., nat'l. rep.: Katz Television.

Operation: 1953; TV sets in area: 460,000.

WBRZ (Channel 2) ABC

Louisiana Television Broadcasting Corp., P.O. Box 2906, Baton Rouge, La. 708921; 387-2222.

Douglas L. Manship, pres.; Charles Manship, treas.; Richard Manship, secty. & sta. mgr.; Jules L. Mayeux, gen. mgr.; Barbara Bree, prog. mgr.; William R. Yordy, chief eng.; Robert J. Headlee, gen. sls. mgr.; Jim Primeaux, promo. mgr.; John Spain, news dir.; Jamie Politz, pub. affairs & personnel dir.; Pat Cheramie, local sls. mgr.; nat'l. rep.: Blair Television.

Operation: 1955; TV sets TSA; 556,700; TV sets ADI 200,600.

WRBT (Channel 33) NBC

WRBT Television, Inc., P.O. Box 14685, Baton Rouge, LA 70808 (504) 766-3233.

Cyril E. Vetter, pres. & sta. mgr.; Bill Hathorn, gen. sls. mgr.
Operations: 1971.

Lafayette

KATC (Channel 3) ABC
Acadian Television Corp., 1103 Eraste Landry Rd., Lafayette, LA 70501 (318) 232-6111.
Mrs. Frances T. Kurzeg, pres.; William A. Patton, v.p. & gen. mgr.; Roy K. Rhodes, sls. mgr.; reps: The Meeker Co.; James S. Ayers Co., Inc. (Southeast); Busby, Finch & Woods (Southwest).
Operation: 1962; TV households: 357,600.

KLFY-TV (Channel 10) CBS
Texoma Broadcasters, Inc., 2102 Jefferson St., Lafayette, LA 70501 (318) 233-2152.
M.N. Bostick, pres.; Thomas G. Pears, gen. mgr.; Terry Dover, op. mgr.; Keith Townsdin, chief eng.; Dan Coleman, prod. spvsr.; nat'l. rep.: H-R Television, Inc.
Operation: 1955; TV sets in area: 321,000.

Lake Charles

KPLC-TV (Channel 7) NBC
Calcasieu TV & Radio, Inc., 320 Division St., P.O. Box 1488, Lake Charles, LA 70601 (318) 439-9071.
G.R. Chambers, owner & pres.; Rita Guillory, gen. mgr.; Jerry Smithwick, v.p., sls.; Al Smith, v.p., eng.; nat'l. rep.: Katz.
Operation: 1954; TV sets in area: 247,800.

Monroe

KLAA (Channel 14) NBC
Monroe TV Inc., 701 Parkwood Dr., West Monroe, LA 71291 (318) 388-0114.
Kenneth E. Meyer, pres.; Jonathan Wood, mgr., Rodney Evans, chief eng.
Operation: August, 1967; TV households: 92,200.

KNOE-TV (Channel 8) CBS
Noe Enterprises Inc., Knoe Road, P.O. Box 4067, Monroe, LA 71201 (318) 322-8155.
James A. Noe, Jr., pres. & gen. mgr.; Allen Jones, v.p. & natl. sls. mgr.; Jack McCall, v.p., & prog. dir.; Raymond Boyd, v.p. & tech. dir.; Lloyd Voorhees, v.p. & regional/local sls. mgr.; Ansel Smith, v.p., ops.; rep.: Katz.
Operation: 1953; TV sets in area: 400,000.

KTVE (Channel 10) NBC-ABC
(A Gray Communications System Station)
KTVE, Inc., 400 West Main, El Dorado, AR 71730 (318) 323-7716; 206 Catalpa, Monroe, LA.
James S. Tighe, v.p. & gen. mgr.; John Long, chief eng.; Charles Price, reg. sls. mgr.; Helen Johnson, nat'l. traffic dir.; Lou Slaughter, prog. dir.; reps.: The Meeker Company; South: James S. Ayers Co.; Clyde Melville, Texas.
Operation: 1955; TV sets in area: 292,600.

New Orleans

WDSU-TV (Channel 6) NBC
Cosmos Broadcasting Corp., 520 Royal St., New Orleans, LA 70130 (504) 588-9378.
Dixon C. Lourvorn, sr., v.p. & gen. mgr.; Jeff Williams, ast. gen. mgr.; Mimi Mathis, promo. dir.; Terry Schroeder, bus. mgr.; Fred Wuenschel, chief eng.; Arthur Alpert, news dir.; Richard Levy, gen. sls. mgr.; David R. Chase, prog. dir.; nat'l. rep.: Blair Television.

WGNO-TV (Channel 26) ABC, CBS, NBC
Communications Corp. of the South, Inc., 2912 International Trade Mart, Canal St., LA 70112 (504) 522-6211.
Seymour Smith, pres.; Xavier DeSoto, dir. of prgm'.; William Taylor, chief eng.; June Allen, news dir.; rep.; Adam Young Inc.
Operation: 1967; TV households: 425,800.

WVUE-TV (Channel 8) ABC
Gaylord Broadcasting Co., 1025 S. Jefferson Davis Pkwy., New Orleans, LA 70125 (504) 486-6161; TWX 810-951-6090.
Robert Olson, v.p. & gen. mgr.; Jack Long, asst. gen. mgr. & gen. sls. mgr.; Donald B. Wilburn, prgm.; Rich Brase, prom. mgr.; Michael Caruso, chief eng.; nat'l. rep.: TeleRep.
Operation: 1959; TV sets in area: 2,000,000.

WWL-TV (Channel 4) CBS
Loyola University, 6363 St. Charles Ave., New Orleans, LA 70118.
J. Michael Early, v.p., bdcst'. & gen. mgr.; Maurice Guillerman, asst. gen. mgr. & gen. sls. mgr.; John Pela, dir. of creative svcs. & film buyer; Jerrold Whaley, nat'l. sls. mgr.; James Daboval, local sls. mgr.; Phil Johnson, news dir.; Hugh Burney, chief eng.; nat'l. rep.: Katz TV.
Operation: September, 1957; TV households: 684,000.

Shreveport

KSLA-TV (Channel 12) CBS basic
KSLA-TV Inc., P.O. Box 4812, Shreveport, LA 71104 (318) 222-1212.
Winston B. Linam, gen. mgr.; Ben McLaughlin, sls. mgr.; John Renshaw, prog. dir.; Robert Weimar, promo. dir.; John Hitt, chief eng.; nat'l. rep.: Harrington, Righter & Parsons.
Operation: 1954; TV sets in area: 657,800.

KTAL-TV (Channel 6) NBC
KTAL-TV, Inc., 3150 No. Market St., Shreveport, LA 71104 (318) 425-2422. Texas: 3227 Summerhill Rd., Texarkana; (214) 793-1133.
W.E. Hussman, pres.; H. Lee Bryant, v.p. for bdcst'.; Doug Yoder, nat'l. sls. mgr.; Kenny Gardner, sta. mgr.; Jean Byrd, prog. dir.; George Tracy, dir. of eng.
Operation: 1953; TV sets in area: 490,900.

KTBS-TV (Channel 3) ABC
KTBS, Inc., 312 E. Kings Highway, Shreveport, LA 868-3644.
E. Newton Wray, pres. & gen. mgr.; George D. Wray, Jr., vice-pres.; Charles W. Wray, secy-treas.; Edwin Wray, sta. mgr. & promo. mgr.; Marvin L. Perry, Jr., film dir.; Dave Hendricks, chief eng.; nat'l rep.: Katz Agency.
Operation: 1955; TV sets in area: 657,800.

MAINE

Bangor

WABI-TV (Channel 5) CBS
Community Broadcasting Service, 35 Hildreth St., Bangor, ME 04401 (207) 947-8321.

Walter Dickson, pres.; George J. Gonyar, v.p. & gen. mgr.; Dale Carter, chief eng.; nat'l rep.: Seltel; Kettell-Carter, (New England); Andy McDermott, (Canada).
Operation: 1953; TV sets in area: 140,000.

WLBZ-TV (Channel 2) NBC
WLBZ Television, Inc., Box 934, Mt. Hope Ave., Bangor, ME 04401 (207) 942-4822.
Mary R. Thompason, pres.; Don Powers, exec. v.p.; Margo Cobb, station mgr.; Bruce Scott, chief eng.; nat'l. rep.: Katz Agency.
Operation: 1954; TV sets in area: 125,000.

WVII-TV (Channel 7) ABC
Eastern Maine broadcasting System, Inc., 41 Farm Rd. Bangor, ME 04401 (207) 945-6457.
John Pineau, pres. & gen. mgr.; Barbara J. Cyr.; natl./local/regional sls. mgr.; Gary Kasparek, prod. mgr., Jim Rohn, news dir. Rep.: Avery-Knodel, Inc.
Operation: 1965, TV households: 220,000.

Poland Spring

WMTW-TV (Channel 8)
Mt. Washington TV, a divisison of Harron Communications Corp., Inc., Poland Spring, ME 04274; 998-4311. Sales dept., Portland; (207) 733-5664.
Paul Harron, pres.; Al Ritter, v.p. & gen. mgr.; Robert M. Joyce, asst. gen. mgr.; Claus Wiese, prod. mgr.; Parker H. Vincent, chief eng.; nat'l rep.; PGW Reps.
Operation: 1954.

Portland

WCSH-TV (Channel 6) NBC
Maine Radio and TV Co., 579 Congress St., Portland, ME 04101.
Mary R. Thompson, pres.; Jack S. Atwood, gen. mgr.; Don Powers, sta. mgr.; Arthur Leavitt, chief eng.; Bruce McGorrill, sls. mgr., Lew Colby, promo. dir.; George Bowler, film dir.; nat'l rep.: Katz Agency.
Operation: 1953; TV sets in area: 210,000.

WGAN-TV (Channel 13) CBS
Guy Gannett Broadcasting Services, Northport Plaza, Portland, ME 04104 (207) 797-9330.
Jean Gannett Hawley, pres.; Robert L. Gilbertson, exec. v.p.; David King, v.p. & gen. mgr.; Fred Desjardins, chief eng.; Alvin F. Wright, treas. nat'l rep., Blair TV assoc.
Operation: 1954; TV sets in area: 288,000.

Presque Isle

WAGM-TV (Channel 8)
Aroostook Broadcasting Corp., P.O. Box 1149, Presque Isle, ME 04769 (207) 764-4461.
W.L. Dickson, pres.; Norm Johnson, v.p. & gen. mgr.
Operation: 1956.

MARYLAND

Baltimore

WBAL-TV (Channel 11) NBC
The Hearst Corp., Maryland's Broadcast Center, Baltimore, MD 21211 (301) 467-3000.
Franklin C. Snyder, v.p. & gen. mgr.; Hearst Broadcasting Stations; Lawrence Carino, v.p. & gen. mgr.; John Frankenfield, asst. station mgr. & prog. dir.; William J. Katsafanas, gen. sls. mgr.; Jack Gilmore, natl. sls. mgr.; Robert Morgan, reg. sls. mgr ; Eugene E. Keenan, Jr., local sls. mgr.; nat'l. rep.: Blair.
Operation: March, 1948; TV households: 2,613,200.

WBFF (Channel 45)
Chesapeake TV Inc., 3500 Parkdale Ave., Baltimore, MD 31211 (301) 462-4500.
Julian S. Smith, gen. mgr.; Frederick M. Himes, Jr., op. mgr.; Robert Simmons, gen. sls. mgr.; Bruce K. Lumpkin, natl. sls. mgr ; William LeFevre, prog. mgr.; Dwight Weems, prod. mgr.; Augustine Levero, mkt. mgr.; Sharon Christopher, public affairs mgr.; Dennis Winters, chief eng.
Operation: 1971.

WJZ-TV (Channel 13) ABC
Westinghouse Broadcasting Co., Inc., 3725 Malden Ave., Baltimore, MD 21211 (301) 466-0013.
Daniel L. Ritchie, pres. & chief exec. off., Westinghouse Broadcasting Co.; Paul Yates, v.p. & gen. mgr.; H. Brian O'Neill, creative svcs. mgr.; Stan Rudick, prgm. mgr.; Steven Engles, gen. sls. mgr.; Glenn Lahman, chief eng.; Richard Reingold, news dir.; Harold Murphy, bus. mgr.
Operation: 1948; TV sets in area: 2,653,500.

WMAR-TV (Channel 2) CBS
WMAR, Inc., 6400 York Road Baltimore, MD 21212 (301) 377-2222.
Dale Wright, pres. & gen. mgr.; Charles Horich, v.p. & dir. of bdcst'.; Robert O. Donahue, v.p. dir. & of eng.; Peter M. Ballard, corp. secty. & bus. mgr.; Russell A. Jones, treas. & office mgr.; Edwin B. Mich, prod. mgr.; Janet B. Covington, dir. of pub. affairs.
Operation: 1947; TV sets in area: 2,390,600.

Hagerstown

WHAG-TV
Henson Aviation, Inc., P.O. Box 481, Hagerstown, MD 21740 (301) 797-4400.
Arch H. McDonald, v.p. & gen. mgr.; Harold Heagy, oper. & prog. mgr.; Ken Johnson, chief eng.; Greg Suchanek, prod. mgr.; Rep.: Adam Young.
Operation: 1970.

Salisbury

WBOC-TV (Channel 16) ABC, CBS, NBC
Mid Atlantic Communications, Inc. TV Park, Salisbury, MD 21801; (301) 749-1111.
Thomas Draper, pres. & gen mgr.; Jack Ward, operations mgr.; Larry Young, program dir.; William Kenton, sls. mgr.; Vincent Donovan, chief eng.; nat'l rep.; Katz Agency.
Operation: 1954; TV sets in area: 81,970.

MASSACHUSETTS

Adams-Pittsfield

WCDC (Channel 19)
(Satellite of WTEN, Albany, NY)

Knight-Ridder Broadcasting, Inc. 341 Northern Blvd., Albany, NY 12204 (518) 436-4822.

Boston

WBZ TV (Channel 4) NBC

Westinghouse Broadcasting Co., Inc., 1170 Soldiers Field Road, Boston, MA 02134 (617) 787-7000.

Sy Yanoff, v.p. & gen. mgr.; John Spinola, sta. mgr.; Dan Lawlor, gen. sls. mgr.; Jeff Rosser, news dir.; George St. Andre, dir. of eng.; Dick Kurlander, prog. mgr.; Jolen Schmauss, creative services dir.; Terri Park, information services mgr.; Bob Houghton, cont.; Karen Huggins, human resources mgr.; Bob Hayes, traffic mgr.; Bill Ferrick, ops. mgr.

Operation: 1948; TV sets in area: 2,700,000.

WCVB-TV (Channel 5) ABC

Boston Broadcasters, Inc., 5 TV Place, Needham, MA 02192.

Dr. Leo L. Beranek, chrm.; Robert M. Bennett, pres.; Gerald McGavich, v.p. & dir. of sls.; Paul La Camera, v.p., dir. of community svcs.; Clifford Curley, prog. mgr.; Philip Balboni, dir. of pub. affairs; Kate Steed, dir. of adv. prom.

Operation: 1972.

WNAC-TV (Channel 7) ABC

RKO General Broadcasting, Government Center, Boston, MA 02114.

Stephen P. Mathis, gen. mgr.; Peter La Bruzzo, dir. of sls.; Earl C. Davis, dir. of adv. & grom.; Ladislav Hlavaty, dir. of eng.; nat'l rep.: RKO Television Representatives, Inc.

Operation: 1948; TV homes in area: 2,034,200.

WSBK-TV (Channel 38) NBC, ABC, CBS

New Boston TV, Inc., 83 Leo Birmingham Pkwy, Brighton, MA 02135 (617) 783-3838.

Joseph C. Dimino, gen. mgr.; Daniel J. Berkery, sta. mgr.; B.J. Anderson, promo. mgr.; William B. Knight, compt.; John Anderson, dir. of sls. develop.; Stuart Tauber, natl. sls. mgr.; Ron Pulera, local sls. mgr.; Dick Beach, op. mgr. promo. mgr.; Doug Henderson, chief eng.

Operation: 1964.

WXNE-TV (Channel 25)

Christian Broadcasting Network, Inc., 1318 Spratley St., Portsmouth, VA 23705 (617) 449-4200.

M.G. Robertson, pres.; Robert E. Johnson, v.p.; William B. Knight, gen. mgr.; William T. Gregory, chf. eng.

Operation: 1977.

Boston-Cambridge

WLVI-TV (Channel 56)

Field Communications, Box 56, Boston, MA 02101 (617) 288-3200.

Gerald R. Walsh, gen. mgr.; Howard J. Stevens, prog. mgr.; Kevin J. Dunn, gen. sls. mgr.; Robert S. Kraus, sls. mgr.; Andrew M. Seyffert, bus. mgr.; Gail Satz, prod. mgr.; Elizabeth Wrublin, promo. mgr.; Natalie McIver, news & pub. affairs mgr.

Operation: 1953.

Greenfield

WRLP-TV (Channel 32)

Box 2210, Springfield, MA 01101 (413) 785-1201.

William M. Pepin, station mgr.; Paul T. Mello, sales mgr.; Ralph Jay, chief eng.

Operation: 1957.

Springfield-Holyoke

WHYN-TV (Channel 40) ABC

The WHYN Stations Corp., 1300 Liberty Street, Springfield, MA 01101 (413) 785-1911.

Clifford Lefkovich, gen. mgr.; Dick Wylie, natl sls. mgr.; Rep.: Katz Agency.

Operation: 1953; TV sets in area: 1,213,500.

WWLP (Channel 22) NBC

Springfield Television Broadcasting Corp., Box 2210, Springfield, MA 01101 (413) 786-2200.

William L. Putnam, chm.; Kathryn F. Broman, pres.; Roland L. Filiault, treas. & compt.; James H. Ferguson, Jr., exec. v.p.; John Gergie, v.p., eng.; E. Holland Low, natl. sls. mgr.; Daniel T. Sullivan, Jr., local sls. mgr.; W. Buzz Sawyer, v.p. & station mgr.; nat'l rep.: H-R Television—Kettell-Carter (New England).

Operation: 1953; TV sets in area: 864,000.

Worcester

WSMW-TV (Channel 27)

SIBOS Inc., 127 Beverly Rd., Worcester, MA 01605 (617) 852-0027.

Brian A. Higgins, v.p. & gen. mgr.; Robert L. Fouracre, gen. sls. mgr.; Donald J. Reilly, asst. gen. mgr.

Operation: 1970; TV households: 2,416,500.

MICHIGAN

Alpena-Oscoda

WBKB-TV (Channel 11) CBS

Thunder Bay Broadcasting Co., 1390 Bagley St., Alpena, MI 49707 (517) 356-3434.

Thomas Disinger, v.p. & gen. mgr.; William Jackson, dir. of news & pub. affairs; John Rushlow, prod. mgr.; Jeri Foor, traffic mgr.; James Marr, chief eng.

Operation: 1975.

Battle Creek-Kalamazoo

WUHQ-TV (Channel 41) ABC

Channel 41, Inc., 5200 West Dickman Road, Battle Creek, MI 49016 (616) 968-9341.

John W. Lawrence, pres.; William J. Lawrence, Jr., bd. chmn.; Jerry Colvin, exec. v.p. & gen. mgr.; Linda Rosh, prog. & prom. mgr.; Michael J. Laemers, chief eng. & ops. mgr.; Phillip R. Ranns, controller; John Hampel, film dir.

Operation: 1971; Receivers in area: 938,000.

Bay City-Saginaw

WNEM-TV (Channel 5) NBC

WNEM-Television; Broadcasting div./Meredith Corp.; Offices and studios, 107 N. Franklin, Saginaw, MI (517) 755-8191. Sales office & studio: G-3426 Miller Road, Flint, MI (313) 732-2050.

Todd Holmes, v.p. & gen. mgr.; Terry Dolan, prog. dir.; Paul Virciglio, gen. sls. mgr.; Ken Preston, chief eng.; Bill Avery, news dir.; nat'l. rep.: MMT Sales, Inc.

Operation: 1954; TV households in area: 424,550.

Cadillac

WWTV (Channel 9) CBS, ABC

Wilson Communications, Inc. P.O. Box 627; Cadillac, MI 49601 (616) 775-3478.

Ralph C. Wilson, Jr., chmn.; William R. Hirshey, pres.; Fred M. Hutching, v.p. & gen. mgr.; nat'l rep.: Seltel.

Operation: Jan. 1, 1954; TV sets in area: 789,200.

Cheboygan

WTOM (Channel 4) NBC

Midwestern Television Company, Traverse City, MI 49684 (616) 947-7675.

Les Biederman, pres.; Robert L. Dreher, gen. mgr.; William H. Kiker, v.p., chief eng.; Drew McClay, sec.-treas.; R.H. Loomis, bus. mgr ; nat'l rep.: Adam Young, Inc., Andy McDermott (Canada).

Operation: 1959.

Detroit

WGPR-TV (Channel 62)

WGPR, Inc., 3140-6 E. Jefferson Ave., Detroit, MI 48207 (313) 259-8862.

Dr. William V. Banks, pres. & gen. mgr.; James W. Panagos, v.p. & gen. sls. mgr.; Tenicia Gregory, v.p./cont. & sta. mgr.; Joe Spencer, acting prog. dir.; Ulysses W. Boykin, v.p. public affairs.

Operation: 1975.

WJBK-TV (Channel 2) CBS

Storer Broadcasting Company, Number Two Storer Place, Southfield, MI 48075 (313) 557-9000.

Bill Michaels, chm.; Peter Storer, pres.; Terry Lee, ex. v.p.; Jay Watson, v.p., gen. mgr.; Robert J. McBride, sta. mgr.; Mike Von Ende, news dir.; Tom Fox, community affairs dir.; James F. Major, prgm. dir.; Timothy J. Rocha, promo. mgr.

Operation: 1948; TV sets in area: 2,034,000.

WKBD-TV (Channel 50)

Kaiser Bdg. Co., 420 Taylor St., San Francisco, CA 94102 (313) 444-8500.

George H. Williams, v.p. & gen. mgr.; George J. Kapel, gen. sls. mgr.; Les Fine, sls. mgr.; George A. Sperry, prog. mgr.; Curtis Bertsch, business mgr.; Al Martin, dir. of eng.; Toby Cunningham, prod. mgr.; Wayne Studer, prom. mgr.

Operation: 1965.

WDIV-TV (Channel 4) NBC

Post-Newsweek Station, Michigan, Inc. 622 Lafayette Blvd., Detroit, MI 48231 (313) 222-0400.

Amy S. McCombs, v.p. & gen. mgr.; James Snyder, v.p. & news dir.

Operation: 1947; TV sets in area; U.S. total survey area: 3,182,000, total coverage area (inc. Canada): 3,448,000.

WXYZ-TV (Channel 7) ABC

WXYZ, Inc., Broadcast House, 20777 W. Ten Mile Rd., Detroit. MI 48231 (313) 444-1111.

James R. Osborn, v.p. & gen. mr.; Charles Kocher, chief eng.; Francis L. Gannon, gen. sls. mgr.; Jeanne Findlater, prgm. dir.; Jan Coil, natl. sls. service mgr.; Milda Skorupskas, traffic mgr.; Jon Gluck, research dir.; Dave Kenworthy, promotion & pub. mgr.; Kenneth Sims, film dir.; Philip Sweenie, natl. sls. mgr.; John Reynolds, compt.; Barbara Collins, assistant bus. mgr.; nat'l rep.: ABC Television Spot Sales, Inc.

Operation: 1948; TV sets in area: 1,823,423.

Detroit-Windsor (Canada)

CBET (Channel 9) CBS

Canadian Bdg. Corp., 1500 Bronson Ave., Ottawa, Ont. N9A 5K9 (519) 254-2831.

J.L. Hackney, gen. mgr.; D.E. West, prog. dir.; R.J. Kryger, dir. of eng.; F.J. Quinn, dir. of prod.; N.T. Addison, mgr., public relations.

Operation: 1954.

Escanaba

WJMN-TV (Channel 3) NBC

(Satellite of WFRV-TV, Green Bay, WI)

WFRV Inc., Mason at Roosevelt, Green Bay, WI 54305.

Rep.: The Katz Agency.

Operation: 1969.

Flint

WJRT-TV (Channel 12) ABC

Knight-Ridder Broadcasting, Inc., 2302 Lapeer Rd., Flint MI 48503 (313) 233-313.

Albert J. Gillen, pres.; Thomas T. Hartman, v.p.-gen. mgr.; Robert H. Battersby, treasurer-vice president; nat'l rep.; Blair.

Operation: 1958; TV households in area: 1,156,700.

Grand Rapids

WOTV-TV (Channel 8) NBC basic

Time-Life Broadcast, Inc., 120 College Ave., S.E., P.O. Box B, Grand Rapids, MI 49501 (616) 459-4125.

Marvin R. Chauvin, v.p. & gen. mgr.; Russell Vossen, sta. mgr.; Richard A. Groat, gen. sls. mgr.; Rich Bevak, prom. mgr.; Donald P. Gallagher, chief eng.; nat'l rep.: The Katz Agency, Inc.

Operation: 1951; TV sets in area: 532,500.

WZZM (Channel 13)

Wometco West Michigan Television Inc., P.O. Box Z, 645 Three Mile Rd., N W., Grand Rapids, MI 49501.

George U. Lyons, pres. & gen. mgr.; Michael Seagly, ops. mgr.; Willard O. Smith, gen. sls. mgr.

Kalamazoo

WKZO-TV (Channel 3) CBS

Fetzer Television Corporation, 590 W. Maple St., Kalamazoo, MI 49001 (616) 345-2101.

John E. Fetzer, chmn.; Carl E. Lee, pres. & gen. mgr.; Jerry Raffel, natl. sales mgr.; Gordon Anderson, adm. v.p. & mgr. James Garrett, chief eng.; nat'l rep.: Avery-Knodel, Inc.

Operation: 1950; TV sets in area: 785,764.

Lansing

WJIM-TV (Channel 6) CBS

Gross Telecasting, Inc., Country House, E. Saginaw at Howard St., Lansing, MI 48904 (517) 372-8282.

Harold F. Gross, chrm. of bd.; James H. Gross, pres. & gen. mgr.; Carl Onken, tech. oper. dir.; nat'l rep.: Harrington, Righter & Parsons.

Operation: 1950; TV homes in area: 1,063,100.

Lansing-Onondaga

WILX-TV (Channel 10) NBC

Television Corporation of Michigan, Inc., 100 N. Pennsylvania Ave., Lansing, MI 48909 (517) 482-0721.

Edward E. Wilson, pres.; William Hart, gen. mgr. & film buyer; John F. Cundiff, gen. sls. mgr.; Stuart S. Hazard, prod. supvr.; Richard A. Swank, chief eng.; John Tallerico, news dir.; Linda Murphy, office mgr.

Marquette

WLUC-TV (Channel 6)

WLUC, Inc., P.O. Box 460, Marquette, MI 49855.

Thomas L. Gagnum, gen. mgr.; Dan Olsen, reg. & local sls. mgr.; Jack La Forest, prog. dir.; John Truitt, chief eng.; Bruce Saari, indust. film dir.; Don Ranta, film dir.; nat'l rep., Peters, Griffin & Woodward.

Operation: 1956.

Saginaw-Bay City

WEYI-TV (Channel 25) CBS

Ziff-Davis Broadcasting Co., P.O. Box 3265, Saginaw, MI 48605.

Robert B. Farrow, v.p. & gen. mgr.; George Leh, prog. mgr.; John E. Grandon, bus. mgr.; Ernest Hill, v.p. & gen. sls. mgr.; Robert Volz, chief eng.

Operation: 1953.

Traverse City

WGTU (Channel 29) WGTQ (Channel 8) ABC

Michigan Television Network, 201 E. Front St., Traverse City, MI 49684; (616) 946-2900.

Harry Calcutt, pres.; Gene Ellerman, exec. v.p. & gen. mgr.; Pat Mullen, sls. mgr.; Paul Baumgartner, chief eng.; Brad Graham, prog. & news dir.

Operation: 1971.

WPBN-TV (Channel 7) NBC

WPBN-TV, Inc., Midwestern Television Co., Paul Bunyan Bldg., Traverse City, MI 49684 (616) 947-7675.

Les Biederman, pres.; Harry R. Lipson, gen. mgr. & film buyer; Robert L. Dreher, sls. mgr.; William H. Kiker, v.p. & chf. eng.

Operation: 1954.

MINNESOTA

Alexandria

KCMT (Channel 7)

Central Minnesota Television Co., Box 612, Alexandria, MN 56308 (612) 763-5166.

Tom Barnes, Jr., pres.; Glenn Flint, v.p. & gen. mgr.; Kenneth Bechtel, sta. mgr.; John Gunther, sls. mgr.; Carl Iverson, chief eng.; rep. Avery-Knodel, Inc., Harry Hyett Co. (Minn.).

Operation: 1958.

Austin

KAAL-TV (Channel 6) ABC

Minnesota-Iowa Television Co., P.O. Box 577, Austin, MN 55912 (507) 433-8836.

John A. MacGreger, v.p. & gen. mgr.; Terry W. Dolan, v.p./programming; Jerald Jones, v.p./engineering; Jon Janes, news dir.; Harry Sliffe, pres.

Duluth-Superior (Wis.)

KBJR-TV (Channel 6) NBC

RJR Communications, Inc., KBJR Building, Duluth MN 55802 (218) 727-8484.

Robert J. Rich, pres. & gen. mgr.; R.W. Pearson, v.p.; Jack McKenna, prog. dir.; Fred Mognussen, chief eng.; Edwin M. Conrad, pub. aff. dir.; nat'l rep.: Katz Agency.

Operation: 1954; TV sets in area: 336,500.

KDLH-TV (Channel 3) CBS

KDAL, Inc., 425 West Superior St., Duluth, MN 55802 (218) 727-8911.

John LaForge, gen. mgr.; Rudy Luukinen, chief eng.; Joe Golden, sls. mgr.; Earl Henton, prog. dir.; nat'l rep.: Blair (Minneapolis).

Operation: 1954; TV sets in area: 326,090.

WDIO-TV (Channel 10) ABC

Channel 10, Inc., 10 Observation Road, Duluth, MN 55811 (218) 727-6864.

Frank P. Befera, pres. & gen. mgr.; Mike Daley, gen. sls. mgr.; Dave Poirer, prog. mgr.; George W. Woody, chief eng.; Dennis Anderson, news dir.; Walt Strasser, film dir.; Paul Hayden, on-sta. prom.

Operation: 1966.

Mankato

KEYC-TV (Channel 12) CBS

Mankato Broadcasting Corp., P.O. Box 128, Mankato, MN 56001 (507) 625-7905.

Howard J. Brown, pres.; Hap Halligan, v.p. & gen. mgr.; Eugene W. Schulte, v.p.-secty.-treas.; Lloyd Miller, sls. mgr.; Charles Pasek, prog. dir.; David Hooge, chief eng.

Operation: October, 1960; TV households: 202,200.

Minneapolis-St. Paul

KMSP-TV (Channel 9)

United Television, Inc. 6975 York Ave. So., Minneapolis, MN 55435 (612) 925-3300.

Donald Swartz, pres. & gen. mgr.; Roland King, sta. mgr.; Stuart Swartz, gen. sls. mgr.; Joe Carney, op. mgr.; Tony Burden, news dir.; Glenn Smith, dir. of eng.; nat'l rep.: Blair Television.

Operation: 1955; TV sets in area: 962,800.

KSTP-TV (Channel 5) ABC

Hubbard Broadcasting, Inc., 3415 University Ave., Minneapolis, MN 55414 (612) 646-5555.

Stanley S. Hubbard, pres. & gen. mgr.; Gerald D. Deeney, v.p. & treas.; James E. Blake, sls. mgr.; Jeanne Kosek, promo. mgr.; John Degan, prod. supv.; George Merrill, chief eng.; Ralph J. Dolan, oper. mgr.; nat'l rep. Petry Television, Inc.

Operation: 1948.

WCCO-TV (Channel 4) CBS

Midwest Radio-Television, Inc. 50 South 9th St., Minneapolis, MN 55402 (612) 330-2400.

W.T. Doar, Jr., pres.; James M. Rupp. exec. v.p. Loren Lucke, treas.; Paul Arnzen, office mgr.; Robert McGann, natl sls. mgr.; Martha Stanville, adv. & promo. Tom Cousins, community affairs/pub. relations dir.; Ronald Handberg, gen. mgr.; Everett Klosterman, contr. nat'l rep.; TeleRep, Inc.

Operation: 1949; TV sets in area 1,683,400.

WTCN-TV (Channel 11)

WTCN-TV, Inc., 441 Boone Ave. No., Minneapolis, MN 55427 (612) 546-1111.

Raymond J. Schonbak, v.p. & gen. mgr.; Al DeVaney, gen. sls. mgr.; Karl Gensheimer, local sls, mgr.; Art Ludwig, prog. dir.; Cecelia Drazek, prom. dir.; Chuck Biechlin, news dir.; Robert Muchow, chief eng.; natl. rep.: Metromedia TV Sales.

Operation: 1965; TV homes in area: 1,364,000.

Rochester

KTTC-TV (Channel 10) NBC

KTTC Television, Inc., 601 First Ave., S.W., Rochester, MN 55901.

Ed Hutchings, gen. mgr.; Joe St. George, gen. sls. mgr.; Ronald E. Gruber, op. mgr.; C.H. Sanders, chief eng.; natl. rep.: Avery-Knodel.

Operation: July, 1953; TV households: 363,300.

Walker

KNMT (Channel 12) NBC, ABC

Central Minnesota TV Co., Box 612, Alexandria, MN 56308.

Tom Barnes, Jr., pres.; Glenn W. Flint, v.p. & gen. mgr.; Kenneth Bechtel, sta. mgr.; Carl Iverson, chief eng.; rep.: Avery-Knodel Inc.

Operation: 1964.

MISSISSIPPI

Biloxi-Gulfport-Pascagoula

WLOX-TV (Channel 13) ABC

WLOX Broadcasting Co., DedBuys Rd., Biloxi, MS 39531 (601) 896-4778.

T.B. Majure, ex. v.p. & gen. mgr.; Earl Blessey, gen. sls. mgr.; Jas. O. Ray, chief eng.; Ray Williams, pgm. dir.; rep.: Avery-Knodel.

Operation: 1962; TV households: 369,100.

Columbus

WCBI-TV (Channel 4) CBS

Columbus Television, Inc., owner, P.O. Box 271, Columbus, MS 39701 (601) 327-4444.

Tommy Bond, gen. mgr.; Jim White, prod. mgr.; Algie Surratt, chief eng.; J.D. Reed, prog. dir.; nat'l rep.: PGW.

Operation: 1956; TV sets in area: 237,200.

Greenwood

WABG-TV (Channel 6) ABC

Mississippi Telecasting Co., Inc., P.O. Drawer 720, Greenwood, MS 38931 (601) 453-4001.

Cy N. Bahakel, pres.; Gene B. Creasy, gen. mgr.; Ada Mitchell, office mgr.; Robert Taylor, chief eng.

Operation: 1959; TV households: 336,700.

Jackson

WAPT (Channel 16) ABC

Clay Broadcasting of Mississippi, Inc., Box 10297, Jackson, MS 39209 (601) 922-1607.

Ben Strickland, v.p. & gen. mgr.; Joe Root, station mgr.; Danny Bardin, dir. of eng.

Operation: 1970; TV households: 538,600.

WJTV (Channel 12) CBS

Capitol Broadcasting Co., P.O. Box 8887, Jackson, MS 39204 (601) 372-6311.

Owens F. Alexander, exec. v.p. & gen. mgr.; Alan Axtell, gen. sls. mgr.; C.A. Perkins, chief eng.; Gen Ward, prog. dir.; Joe Lesem, news dir.; Phyllis Brooks, dir. of pub. affairs; Glenn Dear, prod. supvr.; nat'l rep.: The Katz Agency, Inc.

Operation: 1954; TV sets in area 434,600.

WLBT (Channel 3) NBC

Communications Improvement, Inc., 715 South Jefferson, Jackson, MS 39205 (601) 948-3333.

Dr. Aaron Shirley, pres.; Blanche Young, sect. & treas.; William H. Dilday, Jr., g. mgr.; Earl K. Moore, gen. coun.; Floyd Kinard, chief eng.; Larry Keeler, prod. mgr.; Bruce Nicholson, film dir.; nat'l rep.; Blair Television.

Operation: 1953; TV sets in area: 342,000.

Laurel-Hattiesburg

WDAM-TV (Channel 7)

Service Broadcasters, Inc., P.O. Box 1978, Hattiesburg, MS 39401.

S.A. Rosenbaum, pres.; Marvin Reuben, exec. v.p. & gen. mgr.; Frank Pilgrim, sls. mgr.; Bobby Smith, chief eng. and oper. mgr.; Ann Daley, prog. mgr.; nat'l rep.: Adam Young VTM.

Operation: 1956; TV set in area 204,500.

WHTV (Channel 24) ABC, NBC

Central TV, Inc., Box 5185, Meridian, MS 39301 (609) 693-2933.

Central TV, Inc., Box 5185, Meridian, MS 39301 (609) 693-2933.

Ronald E. Hale, pres. & gen. mgr.; Bob Horton, sta. mgr.

Operation: June, 1963; TV households: 60,900.

WTOK-TV (Channel 11) CBS basic, NBC, ABC

Southern Television Corp., Southern Bldg., Box 2988, Meridian, MS 39401 (601) 693-1441.

Robert F. Wright, pres. & gen. mgr.; W.B. Crooks, Jr., v.p. & sales mgr. Cecil Germany, nat'l. sls. mgr.; George Shannon, prod dir. & film dir.; Joe Saxon, chief eng.; nat'l. rep.: Katz Agency.

Operation: 1953; TV sets in area: 151,900.

Tupelo

WTVA (Channel 9) NBC

WTWV, Inc., Beech Springs Rd., Box 350, Tupelo, MS 38801 (601) 842-7620.

Frank K. Spain, pres.; Mark Ledbetter, exec. v.p. & gen. mgr.; Charlie H. Hicks, sta. & sales mgr.; Charles Ed Bishop, prog. dir.; Jimmy A. Stembridge, dir. of eng.

Operation: 1957; TV sets in area: 433,900.

MISSOURI

Cape Girardeau

KFVS-TV (Channel 12) CBS

American Hirsch Broadcasting Co., Box 100,310, 324 Broadway, Cape Girardeau, MO 63701 (314) 335-1212.

Walter H. McCrobe, pres.; Larry Harris, v.p. & gen. mgr.; George W. Jeter, secty-treas.; nat'l rep.: Katz Agency.
Operation: 1954.

Columbia

KCBJ-TV (Channel 17) ABC

Channel Seventeen, Inc., 501 Business Loop 70 East, Columbia, MO 65201 (314) 449-0917.
Richard Koenig, pres. & gen. mgr.; Tom Koneg, sls. mgr.; Judith Pulliam, prog. dir.; Howard McCartney, chief eng.
Operation: 1971.

KOMU-TV (Channel 8) NBC

The Curators of the University of Missouri, Highway 63, South, Columbia, MO 65201 (314) 442-1122.
Thomas R. Gray, gen. mgr.; Don Ruggles, sales mgr.; Monte Gurwit, dir. of prog.; William J. Reagan, chief eng.; Dick Nelson, dir. of news; nat'l rep., Avery-Knodel, Inc.
Operation: 1953; TV sets in area: 505,200.

Hannibal, Mo.-Quincy, Ill.

KHQA-TV (Channel 7) CBS, ABC

Lee Enterprises, Inc., 510 Main St., Quincy, IL 62301.
Merritt Milligan, v.p.; John Phillips, gen. mgr.; Robert Fruehe, natl. sales mgr.; Jack Rose, local sls. mgr.; Lee Ellis, tech. dir.; Tom Robey, prgm. dir.; Gene Terry, news dir.; nat'l rep.: H-R TV Inc.
Operation: 1953; TV sets in area: 212,000.

Jefferson City

KRCG-TV (Channel 13) CBS

Box 659, Jefferson City, MO 65102 (314) 896-5144.
Owned by Mid-America Television Co., 114 W. 11th St., Kansas City, MO 64105; (816) 842-3358.
W.A. Bates, pres.; E.J. Schuelein, gen. mgr.-natl. sales; O.L. Jarrett, local-reg. sales mgr.; Warren Polmanteer, chief eng.; Lee Gordon, prgm. dir.; Don Shinkle, news dir.; nat'l rep.: Katz Agency.

Joplin

KODE-TV (Channel 12) ABC

Gilmore Broadcasting Corp. (licensee), 1928 W. 13th St., Joplin, MO (417) 623-7260.
John S. Markward, gen. mgr.; Clyde Helton, gen. sales mgr.; Jim Lobbey, dir. com. affairs; Gary Cooper, promo. mgr.; nat'l rep.: Seltel.
Operation: 1954; TV set sets in area: 403,300 TV households per May '81 ARB.

KTVJ (Channel 16) CBS

Mid-American Bcstg. Inc., P.O. Box 1393, Joplin, MO 64801 (417) 781-2345.
Mike Mitchell, gen. mgr., Ron Bormaster, sls. mgr.; Mel Brooks, chief eng.; Dave Chubb, prod. mgr.; Ron Gross, news dir.
Operation: September, 1967; TV households: 290,600.

Kansas City

KBMA-TV (Channel 41)

Westport TV, Inc., BMA Tower, 1 Penn Valley Park, Kansas City, MO 64108 (816) 756-1441.
Bob Wormington, gen. mgr.; Stuart Powell, gen. sls. mgr.; Dave Klein, sta. mgr.; Harold DeGood, chief eng.
Operation: September, 1970; TV households: 1,225,400.

KCMO-TV (Channel 5) CBS

Meredith Corp., 125 E. 31 St., Kansas City, MO 64108 (913) 677-5555.
Charles M. McAbee, v.p. & gen. mgr.; Gerald Noonan, gen. sls. mgr.; Steve Smith, dir. of eng.; Erwin Parthe, prog. dir.; Michael Casserly, news dir.
Operation: 1953.

KMBC-TV (Channel 9) ABC

Metromedia Inc., 1049 Central St., Kansas City, MO 64105 (816) 421-2650.
R. Kent Replogle, v.p. & gen. mgr.; Robert Harris, v.p. & gen. sls. mgr.; Ray Janz, v.p., local sls. mgr.; Gene Cless, prog. dir.; Gerald Golden, v.p. & bus. mgr.; Ridge Shannon, news dir.; Richard Goggin, prom. mgr.
Operation: August, 1953; TV households; 990,300.

WDAF-TV (Channel 4) NBC

Taft Broadcasting Company, Signal Hill, Kansas City, MO 64108 (816) 753-4567.
Dudley Taft, pres.; Earl Beall, v.p. & gen. mgr.; Randy Smith, gen. sales mgr.; Buddy Turner, operations mgr.; James Schmidt, chief eng.; nat'l rep.: Telerep.
Operation: 1949; TV sets in area: 1,510,400.

Poplar Bluff

KPOB-TV (Channel 15) ABC

(Satellite of WSIL-TV, Harrisburg, IL)
Turner-Farrar Association, 21½ W. Poplar St., Harrisburg, IL Rep.: Adam Young.
Operation: 1961.

St. Louis

KDNL-TV (Channel 30

Evans Broadcasting Corp., 300 Park Ave., New York, NY 10036 (314) 436-3030.
Jack Petrik, pres. & gen. mgr.; Len Ziegel, v.p., sls.; Michael Gilham, gen. sls. mgr.; W.E. Anderson, chief eng.
Operation: 1969.

KMOX-TV (Channel 4) CBS

CBS, Inc. One Memorial Drive, St. Louis, MO 63102 (314) 621-2345.
Allan Cohen, v.p., gen. mgr.; Mike Estivo, dir. planning & admin.; Don Merl, prog. dir.; Joe Orso, dir. sls.; Bob Maier, retail sls. mgr.; Alan Batten, dir. creative svcs.; Wally Wesley, dir. tech. oper.; Fred Burrows, news dir.; nat'l rep.: CBS-TV STNS.
Operation: 1958; est. TV sets in area: 1,037,390.

KPLR (Channel 11)

220 Television, Inc., 4935 Lindell Blvd., St. Louis, MO. 63108 (314) 367-7211.
Harold Koplar, bd. chmn.; Edward Koplar, pres.; Harold Protter, exec. v.p. & gen. mgr.; William Crowell, v.p./sls.; Larry Marcus, v.p. admin. & planning; James Gonsey, dir. of eng.; Gail Brekke, natl. sls. mgr.; B. Thomas Koon, lcl. sls. mgr.; Gentry Trotter, dir. of bdcst. devel.; nat'l. rep.: TeleRep.
Operation: 1956; TV sets in area: 2,500,000.

KSD-TV (Channel 5) NBC

KSD/KSD-TV, Inc., 1111 Olive St., St. Louis, MO 63101 (314) 421-5055.

Michael E. Pulitzer, pres., Ray Karpowicz, v.p.; Tom Bellow, gen. sls. mgr.; Ken Elkins, gen. mgr.; Rebecca Goodrum, prog. mgr.; Harry Honig, dir. adv. & promo.; Fred Steurer, chief eng.; Greg Strauser, natl. sls. mgr.; Don Burnes, dir. info. svcs.; Jay Randolph, sports dir.; George Noory, news dir.; nat'l rep.: Blair TV.

Operation: 1947; TV households in area: 999,100.

KTVI (Channel 2) ABC

Times Mirror Broadcasting, 5915 Berthold Ave., St. Louis, MO 63110 (314) 647-2222.

Paul Wischmeyer, gen. mgr.; John McElfresh, dir. of sales; Roy Wittneben, dir. of eng.; Vic Skaggs, dir. of prog.; nat'l rep.: MMT Sales, Inc.

Operation: 1953; TV sets in area: 1,498,000.

Springfield

KMTC (Channel 27) ABC

Midland TV corp., 2322 S. Cedarbrook, Springfield, MO 65804 (417) 862-7091.

Kenneth E. Meyer, pres. & gen. mgr.; B.D. Blankenship, chief eng.; Ben Kuehnle, op. mgr.; B.H. Alley, prog. dir.

Operation: 1968.

KOLR-TV (Channel 10) CBS

Independent Broadcasting Co., P.O. Box 1716 S.S.S., Springfield, MO 65806 (417) 862-7474.

J.H.G. Cooper, pres.; Kevin McAndrews, v.p. & gen. mgr.; Irving Schwab, secty.; John O. Cooper, treas.; Kevin McAndrews, prog. dir.; Walt Stansbury, chief eng.; Jack Lloyd, gen. sls. mgr.; nat'l rep., Katz.

Operation: 1953; TV sets in area: 405,400.

KYTV (Channel 3) NBC

KY-3, Inc., 999 W. Sunshine, Springfield, MO (417) 866-2766.

Donald S. Moeller, pres. & gen. mgr.; Houston Harte, v.p.; Robert G. Harbut, v.p.; Larry D. Franklin, v.p. & asst. secty.; John P. Zanotti, secty.; A.L. Hall, treas.-asst. secty.

Operation: 1953; TV sets in area: 459,600.

MONTANA

Billings

KTVQ (Channel 2) CBS Primary, NBC

Garryowen Corp., 3203 3rd Ave. N., P.O. Box 2557, Billings, MT (406) 252-5611.

Vic Miller, gen. mgr.; Lee Lareva, opers. dir.; Sarah Etchart, sales mgr.; Don Bradley, film buyer; J.R. Middleton, chief eng.; nat'l rep.: Avery-Knodel, Art Moore.

Operation: 1953; TV sets in area: 94,240.

KULR-TV (Channel 8) ABC

Harriscope Broadcasting, Inc., P.O. Box 2512, Billings, MT 59103.

Burt Harris, pres.; E. "Bob" Merrill, v.p. & gen. mgr.

Operation: 1958; TV households: 86,700.

Butte

KTVM (Channel 6) NBC, CBS

(Satellite of KECI-TV, Missoula, MT)

Eagle Communications Inc., 340 W. Main St., Missoula MT 59801.

Bob Precht, chm.; Michael Stocklin, gen. mgr.; Doug Draeder, chief eng.

KXLF-TV (Channel 4) ABC, CBS

Garryowen Butte TV Inc., 1003 S. Montana St., Butte, MT 59701 (406) 792-9111.

Joseph S. Sample, pres.; Paul Simitzes, gen. mgr.; Diana Grants, natl. sls. mgr.; Ron Tass, local sls. mgr.; Vernon Hogge, chief eng.

Operation: 1953; TV households: 120,050.

Glendive

KXGN-TV (Channel 5) CBS, NBC

Glendive Broadcasting Corp., Broadcast Bldg., 210 South Douglas, Glendive, MT 59330 (406) 365-3377.

Lewis W. Moore, pres.; Dan Frenzel, mgr.; Donald J. Powers, chief eng.; Winnifred Norton, office mgr.

Operation: 1957.

Great Falls

KFBB-TV (Channel 5) ABC, NBC, CBS

Advance Corporation, P.O. Box 1139, Great Falls, MT (406) 453-4377.

Jack Olson, v.p. & gen. mgr.; Stan Whitman, sta. mgr./natl. sls. mgr./prog. dir.; Jack Fisher, local sls. mgr.; Ted Schroeder, ops./prod. mgr.; Bonnie Merrick, business. mgr.; Ron Schlosser, chief eng.; Mack May, news dir.; natl. rep.: Seltel, Inc.

Operation: 1954; TV homes in area: 100,300.

KRTV (Channel 3) NBC, CBS

Garryowen Corp., P.O. Box 1331, Great Falls, MT 59403.

Cliff Ewing, gen. mgr.

Operation: 1958.

Helena

KTVG-TV (Channel 12) NBC

Helena TV, Inc., Box 6125, Helena, MT 59604 (406) 443-5050.

Lynn Koch, owner, gen. mgr.; Frank Coile, chief eng.; rep.: Spot Time.

Operation: Jan., 1958; TV households: 91,200.

Kalispell

KCFW-TV (Channel 9) CBS, NBC

(Satellite of KECI-TV; Missoula, MT)

401 First Ave. E., P.O. Box 857, Kalispell, MT 59901.

Bob Precht, chr.; Mike Stocklin, v.p. & gen. mgr.; Mark Holston, news dir.; Jim Ludwig, sls. mgr.; C.J. Cannaliato, chief eng.

Operation: June, 1968.

Miles City

KYUS-TV (Channel 3)

Custer Bcstg. Corp., Box 760, Miles City, MT 59301 (406) 232-3540.

David G. Rivenes, gen. mgr; Ella M. Rivenes, oper. mgr.; Charles Vick, chief eng. Ella M. Rivenes, continuity.; Ragna Ann Jensen, eng. remotes.

Operation: Sept., 1969.

Missoula

KECI-TV (Channel 13) NBC, Bx CBS

Eagle Communications, 340 West Main, Missoula, MT (406) 721-2063.

Bob Precht, chmn.; Mike Stocklin, gen. mgr.; Jack Schommer, prog. dir.; Joy Hoffman, traffic dir.

TV households in area: 124,000 (ARB).

KPAX-TV (Channel 8) ABC, CBS

Garryowen Corp., Box 3500, Butte, MT 59701.

Pete Friden, gen. mgr.; Sarah Etchart, natl. sls. mgr.; Bill Moller, news dir.; Don Bradley, film buyer; Louis Tischer, chief eng.

Operation: May, 1970.

NEBRASKA

Albion

KCNA-TV (Channel 8) ABC

(Semi-satellite of KHGI-TV, Kearney, NE)

Same personnel as KHGI and Wendy Determan, sta. mgr.; natl. rep.: Katz.

Operation: Dec., 1964.

Grand Island

KGIN-TV (Channel 11) CBS

(Satellite of KOLN-TV, Lincoln, NE)

Cornhusker Television Corp., KOLN-TV, Lincoln, NE Studio: 231 S. Locust St., Grand Island, NE 68801.

Operation: Oct., 1961.

Hastings

KHAS-TV (Channel 5) NBC

Nebraska Television Corp., Highway 281, Box 578, Hastings, NE 68901 (402) 463-1321.

John T. Benson, gen. mgr.; Donald R. Seaton, v.p.; James D. Conway, secty.-treas.; Duane B. Allison, asst. secty.-treas. & chief eng.; Randy Nicholson, film dir.; nat'l rep.; Adam Young, Inc.

Operation: 1956; TV sets in area: 116,000.

Hay Springs

KDUH (Channel 4)

(satellite of KOTA-TV).

Duhamel Entrprises, Inc., P.O. Box 1760, Rapid City, SD 57709.

Mrs. Helen S. Duhamel, chmn., William F. Duhamel, pres. & gen. mgr.; Ralph Davison, bdcst. mgr.; Jerry Dishong, sta. mgr.; Jim Hildahl, chief eng.; nat'l. rep.: Katz Agency.

Operation: 1958.

Hayes Center

KWNB-TV (Channel 6) ABC

(Satellite of KHGI-TV, Kearney, NE)

Same personnel as KHGI-TV; nat'l. rep.: Katz.

Operation: 1956; TV sets in area: 52,000.

Kearney

KHGI-TV (Channel 13) ABC

NTV Network, Kearney, NE (308) 743-2494. (Also Operates KWNB-TV, Hayes Center, NE, KCNA-TV, Albion, NE & KSNB-TV, Superior, NE.)

Frank E. Brosseau, pres.; Wayne W. Lansche, v.p. & gen. sls. mgr.; Jerry Fuehrer, chief eng.; natl. rep.: Katz.

Operation: 1953; TV households in area 615,800.

Lincoln

KOLN-TV/KGIN-TV (Channels 10 & 11) CBS

Cornhusker Television Corp., Box 30350, Lincoln, NE 68503 (402) 434-8251.

John E. Fetzer, pres.; A. James Ebel, v.p. & gen. mgr.; Duane Holman, sls. mgr.; Paul Jensen, asst. mgr. & prog. dir.; Robert Furman, prod. dir.; Ray Taylor, chief eng.; nat'l rep.: Avery-Knodel, Inc.

Operation: 1953; TV sets in area: 294,000.

McCook-Oberlin

KOMC (Channel 8) NBC

(Satellite of KARD-TV, Wichita, KS)

Kansas State Network Inc., P.O. Box 333, Wichita, KS 67201. Rep.: Tele-Rep.

Operation: Oct., 1959.

North Platte

KNOP-TV (Channel 2) NBC

North Platte Television, Inc., Box 749, North Platte, NE 69101. (308) 532-2222.

Ulysses A. Carlini, gen. mgr.; Jim Harmon, news dir.; Mark Himmler, chf. eng.

Operation: 1958.

Omaha

KETV (Channel 7) ABC

Channel 7, 27th & Douglas Sts., Omaha, NE 68131 (402) 345-7777.

Ken Elkins, gen. mgr.; Danny Thompson, gen. sls. mgr.; Gary R. Neilsen, prgm., pub. affairs mgr.; Robert Harrison, bus. mgr.; nat'l. rep.: Peters, Griffin, Woodward, Inc.

Operation: 1957; TV sets in area: 518,600.

KMTV (Channel 3) NBC

May Broadcasting Co., 10714 Mockingbird Dr., Omaha, NE 68131 (402) 592-3333.

Edward W. May, pres.; Owen L. Saddler, bd. chmn.; Jim Shaum, secty.; Norman W. Williams, exec. v.p. & gen. mgr.; Roger C. Ottenbach, v.p. & gen. sls mgr., KMTV; Don Browers, prod. mgr.; Maureen Waldron, promo. mgr.; Larry Steele, chief eng.; Art Jacobson, op. mgr.; Frances M. Rankin, treas.; Richard Roncka, gen. sls. mgr.; nat'l rep.: Petry Television, Inc.

Operation: 1949; TV sets in area: 415,000.

WOWT (Channel 6) CBS

3501 Farnam St., Omaha, NE 68131 (402) 346-6666.

Ed Zachary, pres. & gen. mgr.; Robert Dalton, sales mgr.; John Dixon, prog. dir.; Jim McGaffin, dir. public affairs; Steve Murphy, news dir.; Judy Horan, dir. promo. & merch.; David Kuehm, lcl. sls.; nat'l rep.: PGW Sales, Inc.

Operation: 1949; TV sets in area 573,900.

Scottsbluff

KSTF (Channel 10) CBS, ABC

(Satellite of KYCU-TV, Cheyenne, WY.)

Wyneco Comm., Inc., 2923 E. Lincolnway, Cheyenne, WY (307) 634-7755.
Carl J. Occhipinti, gen. mgr.; Bill Brennan, res. mgr.; Jerry Homer, chief eng.; nat'l rep.: Katz.
Operation: 1955.

Superior

KSNB-TV (Channel 4) ABC
(Satellite of KHGI-TV, Kearney, NE.)
Same personnel as KHGI-TV.; nat'l rep.: Katz.
Operation: October, 1965.

NEVADA

Henderson-Las Vegas

KVVU-TV (Channel 5)
Carson Broadcasting Corp., 1555 E. Flamingo Road, Las Vegas, NV 89109 (702) 565-9755.
Herbert Kaufman, pres.; Forrest Chu, v.p.; Rusty Durante, v.p. & gen. mgr.; Bill Utton, gen. sls. mgr.; Jackie Smith, dir. of eng.
Operation: Sept., 1967; TV households: 155,000.

Las Vegas

KLAS-TV (Channel 8) CBS
3228 Channel 8 Dr., P.O. Box 15047, Las Vegas, NV 89114 (702) 733-8850.
Mark Smith, v.p./Gen. Mgr.; Ed Tabor, gen. sls. mgr.; Tony Malone, loc. sls. mgr.; Tom Franklin, dir. of eng'.; Lisa Nolan, promo. mgr; John Nelson, chief eng.; nat'l rep.: Katz Television.
Operation: 1953; TV sets in area: 213,000.

KORK-TV (Channel 3) NBC
Wester Communications, Inc., P.O. Box 550, Las Vegas, NV 89114 (702) 382-1760.
Donald W. Reynolds, pres.; Gene Spry, v.p. & gen. mgr.; Walter I. Hopper, prog. mgr.; Jim Garner, chief eng.; nat'l rep.: Peters, Griffin & Woodward.
Operation: 1955.

KTNV-TV (Channel 13) ABC
KTNV-TV, 3355 S. Valley View, Las Vegas, NV 89102 (702) 876-1313.
Ed Quinn, v.p. & gen. mgr.; Nancy Hunt, nat'l sls. mgr.; Steve Mauldin, local sls mgr.; Joann Paivanas, prog. coordinator; George Maupin, sports dir.; Michael Williams, production mgr.; Matt James, news dir.; David Beard, chief eng.; William A. Kellogg, controller; Sherri Tolar, promotions mgr.
Operation: 1956; TV households: TSA 213,000, ADI 177,800.

Reno

KCRL-TV (Channel 4) NBC
Circle L. Inc., 1790 Vassar St., Reno, NV 89510 (702) 322-9145.
Charles E. Cord, pres. & gen. mgr.; Jim Elliott, sta. mgr.; Norvel Seyler, chief eng.; Dick DeWitt, news dir.; rep.: Avery-Knodel, Inc.
Operation: 1962; TV households: 111,100.

KOLO-TV (Channel 8) ABC
Nevada Radio-Television, Inc., Box 10,000, Reno, NV 89510 (702) 786-8880.
Donald W. Reynolds, pres.; James C. Herzig. v.p. & gen. mgr.; Jack Hartman, gen. sales mgr.; W.I. Hopper, prog./operations mgr.; Earl Ling, film dir.; Robert Northam, chief eng.; nat'l reps.: Peters, Griffin, Woodward.
Operation: 1953; TV sets in area: 110,000.

KTVN (Channel 2) CBS
Washoe Empire, 300 S. Virginia St., Reno, NV 89505 (702) 786-2212.
Lee D. Hirschland, pres. & gen. mgr.; Will Chism, oper. mgr.; Al Richards, chief eng.; Jackson Fleming, sales mgr.; Ed Pearce, news editor; rep.: H-R Television.
Operation: 1967; TV households: 71,000.

NEW HAMPSHIRE

Hanover

WNNE-TV (Channel 31) NBC
Taft Broadcasting Corp., 4808 San Felipe Rd., Houston, TX 77027 (802) 295-7564.
William R. Loftus, gen. mgr.
Operation: 1978.

Manchester

WMUR-TV (Channel 9) ABC
United Television Co. of New Hampshire, Inc., 1819 Elm St., Manchester, NH 03104 (603) 623-8061.
Richard Eaton, prs.; Sam Phillips, gen. mgr.; Ray Harris, program dir.; nat'l rep.; UBC Sales.
Operation: 1954; TV sets in area: 965,576.

NEW JERSEY

Wildwood

WCMC-TV (Channel 40) NBC
South Jersey Broadcasting Corp., Northfield, NJ (609) 641-1440.
William C. Wotring, Jr., pres. & gen. mgr.; Frank Strohm, sales mgr.; Fred Sharky, prog. dir.; Carlton S. Owen, oper. mgr.; rep.: Dome-Messervey Co. Inc. (Phila.).
Operation: 1966; TV households: 100,200.

NEW MEXICO

Albuquerque

KGGM-TV (Channel 13) CBS
New Mexico Broadcasting Co., Inc., 1414 Coal Ave., S.W., Albuquerque, NM 87104 (505) 243-2285.
Bruce A. Hebenstreit, pres.; Paul Thorne, v.p. & sls. mgr.; Jack Mahoney, gen. mgr.; Brenda Yager, promo. dir.; Dick Doyle, chief eng.; Bud Holgerson, operations mgr.

KMXN-TV (Channel 23)
Spanish Int'l. Communications Corp., 505 Marquette N.W., Suite 1810, Albuquerque, NM 87102 (505) 843-7023.
Eddie L. Pena, chm. of bd.; Manuel A. Cantu, secty.-treas. & gen. mgr.; Manuela M. Bruno, prog. dir.
Operation: 1975.

KOAT-TV (Channel 7) ABC
Telestation, 1377 University Avenue, N.W., Albuquerque, NM (505) 247-1010.

Max A. Sklower, gen. mgr.; Elvin Smith, op. mgr.; James Sharman, chief eng.; Roy Leone, film dir.; nat'l rep.: Blair.

Operation: 1953, TV homes in area: 276,000.

KOB-TV (Channel 4) NBC
Hubbard Broadcasting, Inc., 4 Broadcast Plaza, S.W., Albuquerque, NM (505) 243-4411.

Stanley S. Hubbard, pres.; Jerry Danziger, gen. mgr.; Dave Herman, sls. mgr.; Fern Bibeau, chief eng.; nat'l rep.: Edward Petry Co.

Operation: 1948; TV sets in area: 442,700.

Carlsbad

KAVE-TV (Channel 6) ABC
Box 12277, El Paso, TX 79912.

John Walton, pres.; Jerry Wright, gen. mgr.; Jack Harrel, chief eng.

Operation: 1956; TV sets in area: 15,600.

Clovis

KFDW-TV (Channel 12) CBS
Mel Wheeler, Inc., Box 125 Denton, TX 76201 (505) 762-0221.

Mike Lee, gen. mgr.; Frank Caldwell, chief eng.

Operation: 1956.

Farmington

KMCC-TV (Channel 12) ABC
McAlister Television, Inc., P.O. Box 2805, Lubbock, TX 79408 (806) 745-2828

Jerry Rogers, gen. mgr.; Fred Johnson, chief eng.

Operation: 1956.

Roswell

KBIM-TV (Channel 10) CBS
Holsum, Inc., Box 910, Roswell, NM 88201 (505) 622-2120.

W.C. Taylor, exec. v.p.; Joe Carriere, v.p. & sales mgr.; J.H. King, v.p. oper. & prgm'. mgr.; Frank Tracy, mgr., news dept.; E.C. Rader, chief eng.; rep.: Avery-Knodel, Inc.

Operation: 1966; TV homes: 113,300.

KSWS-TV (Channel 8) NBC
State Telecasting Co., Columbia, SC (Operating satellite to KCBD-TV, Lubbock, TX)

Operation: 1953; TV sets in area: 94,100.

NEW YORK

Albany

WAST (Channel 13) ABC
Viacom Broadcasting Corp., P.O. Box 4035, Albany, NY 12204 (518) 436-4791.

Michael J. Corken, v.p. & gen. mgr.; Tom Burke, sta. mgr.; Frederick E. Lass, Chief eng.; Laurence J. Reilly, dir. of prod. sls.; Cynthia Williams, promo. dir.; Joseph Krone, promo. mgr.; Larry Price, news dir.; Richard Herd, gen. sls. mgr.; Fred Malbone, nat'l. sls. mgr.; Linda Cummings, traffic mgr.

Operation: 1954.

WRGB (Channel 6) CBS
General Electric Bdg. Co., Inc., 1400 Balltown Rd., Schenectady, NY 12309 (518) 385-1385.

James J. Delmonico, v.p. & gen. mgr.; David Lynch, sta. mgr.; Arthur R. Garland, mgr., prog., pub. affairs; Charles B. King, mgr., op.; Donald J. Decker, mgr., news; Terry Walden, mgr., prom.

Operation: 1947.

WTEN (Channel 10)
Knight-Ridder Broadcasting Company, Incorporated, 341 Northern Blvd., Albany, NY 12204 (518) 436-4822.

Binghamton

WMGC-TV (Channel 34) ABC
Pinnacle Communications, Inc., Box 813, Binghamton, NY 13902.

Philip Marcella, pres. & gen. mgr.; Roy Sova, gen. sls. mgr.; rep., Avery-Knodel.

Operation: 1962; TV households: 224,300.

WBNG-TV (Channel 12) CBS
Gateway Communications, Inc., 50 Front Street, Binghamton, NY 13902 (607) 723-7311.

Donald F. Snyder, gen. mgr.; John Mucha, pgm. dir.; Monty Pinker, gen. sales mgr.; Raymond J. McCormack, local sales mgr.; Michael LaMonica, eng.

WICZ-TV (Channel 40)
Stainless, Inc., 70 Henry St., Binghamton, NY 13902 (607) 723-8251.

Jesse Pevear, gen. mgr.; G. Ricciardelli, tech. dir.; nat'l rep.: H-R Television.

Buffalo

WIVB-TV (Channel 4) CBS
Buffalo Broadcasting Co., Inc., 2077 Elmwood Ave., Buffalo, NY 14207; (716) 874-4410.

Leslie G. Arries, Jr., pres.; Quintin E. Renner, prog. dir.; Patricia Sroka, TV sales mgr.; Greg Caroll, creative svs. dir.; Ralph Thompsom, dir. of eng. nat'l rep.: Harrington, Righter & Parsons, Inc.

Operation: 1948; TV households in area: 879,300; U.S. & Canada: 2,228,300.

WGR-TV (Channel 2) NBC
Taft Broadcasting Company, 259 Delaware Ave., Buffalo, NY 14202.

Paul "Dino" Dinovitz, gen. mgr.; Farrell Meisel, prog. mgr.; Jim Conschafter, gen. sls. mgr.; Sue Pearce, bus. mgr.; Ann Tallman, promo. mgr.; Dick Westlund, chief eng.-mgr. of eng.

Operation: 1954; TV sets in area: 1,120,000.

WKBW-TV (Channel 7)
WKBW-TV, Capital Cities Communications, Inc., 7 Broadcast Plaza, Buffalo, NY 14202.

Philip R. Beuth, v.p. & gen. mgr.; Richard Shepard, gen. sales mgr.; Rom Martzolf, program mgr.; Kathleen McGarry, promo. mgr.; Don Holland, chief engineer; rep.: Blair TV.

Operation: 1958.

WUTV (Channel 29)
Whitehaven Entertainment Corp., 951 Whitehaven Road, Buffalo, NY 14072 (716) 773-7531

Herman Pease, pres. & gen. mgr.; Albert M. Benz, dir. of adm.; Ray Felckowski, dir. of eng'.; Robert Jay, gen. sls. mgr.; Donna Rodriquez, prgm. dir.

Operation: December, 1970; TV households: 1,213,100.

Carthage-Watertown

WWNY-TV (Channel 7) CBS, ABC, NBC

The Brockway Co., Box 211, Watertown, NY 13601 (315) 782-2600.

John B. Johnson, pres. & treas.; Catherine C. Johnson, v.p. & secy.; Daniel C. Lesmeiser, mgr. of bdcst.; Tom Bruce, sls. mgr.; David M. Graves, prog. mgr.; Glenn H. Hall, dir. oper.; David P. Cummings, film dir.; Michael R. Yonkovig, dir. of eng.; nat'l rep.: Avery-Knodel, Inc.

Operation: 1954; TV sets in area: 156,920.

Elmira

WENY-TV (Channel 36) ABC

WENY, Inc., P.O. Box 208, Elmira, NY 14902 (607) 739-3636.

Howard L. Green, exec. v.p.; Michael C. Steele, gen. mgr.; C. Robert Edwards, sta. mgr.; John Herrick, prog. mgr.; Carl C. Egolf, dir. of eng.

Operation: Nov., 1969.

WSYE-TV (Channel 18) NBC

(Satellite of WSYR-TV, Syracuse, NY)

Newhouse Broadcasting Corp., 1030 James St., Syracuse, NY 13203.

E.R. Vadebonocoeur, pres. & gen. mgr.; W.V. Rothrum, v.p. & prog. mgr.; W.R Alford, dir. of sales; Donald H. Buies, gen. mgr.; T.A. Green, eng.; nat'l rep.: Harrington, Righter & Parsons, Inc.

Operation: 1956; TV sets in area: 79,000.

New York City

WABC-TV (Channel 7) ABC

American Broadcasting Company, 7 Lincoln Square, New York, NY 10019 (212) 887-7777.

Phil Boyer, v.p. & gen. mgr.; John Curry, Jr., v.p., adv.; Chuck Gingold, pgm. dir.; Cliff Love, editorial dir.; Howard Rothstein, bus. mgr.; Jon Olken, dir. of creative services; Andrew Feinstein, local sls. mgr.

Operation: 1948; TV sets in area: 6,000,000.

WCBS-TV (Channel 2) CBS

CBS Television Stations, Division of Columbia Broadcasting System, Inc., 51 West 52 Street, New York, NY 10019 (212) 765-4321.

Thomas F. Leahy, Sr. broadcast v.p., CBS/Broadcast Group; Neil E. Derrough, pres., CBS Television Stations Division; Edward M. Joyce, v.p. & gen. mgr. WCBS-TV; Peter Temple, sta. mgr., WCBS-TV; John J. Rafferty, dir. of sls.; Laurie Weng, gen. sls. mgr.; Joan Zenker, loc. sls. mgr.; Eglon Simons, nat'l. sls. mgr.; Michael Dainard, retail/developmental sls. mgr.; Renee Isely, dir. of planning & admin.; Steve Cohen, news dir.; Kirk Varner, mgr. adv. & promo.; Ann Levine May, dir. of broadcasting; Toni Johnson, oper. mgr.; William C. Lacey, director of broadcast admin.; Joseph Passarella, dir. of communications; Phil Shimkin, dir. of programming; Gary Cummings, dir. of editorials & station services.

Operation: 1941; TV sets in area: 6,000,000.

WNBC-TV (Channel 4) NBC

WNBC-TV, 30 Rockefeller Plaza, New York, NY 10020 (212) 247-8300.

Albert T. Jerome, v.p. & gen. mgr.; Richard L. Lobo, sta. mgr.; Hilary Hendler, dir. sls.; Karen L. Copeland, dir. programs; Ron Kershaw, dir. news; Peter Sang, mgr., adv. & promo.; Joseph Michaels, dir., editorials; Charles Licaneli, mgr., broadcast standards; Spencer Wiles, mgr., community relations.

Operation: 1941; TV sets in area: 6,000,000.

WNEW-TV (Channel 5)

Metromedia Television, a division of Metromedia, Inc. 205 East 67th Street, New York, NY 10021 (212) 535-1000.

Robert E. O'Connor, v.p. & gen. mgr.; John von Soosten, v.p. & prgm. mgr.; William D. Kelly, v.p. & chief eng.

Operation: 1944; TV households in area: 5,887,330.

WOR-TV (Channel 9)

RKO General Inc., 1440 Broadway, New York, NY 10018 (212) 564-8000.

Robert J. Williamson, v.p. & gen. mgr.; Robert Fennimore, v.p. & gen. sls. mgr.; Peter Mitchell, natl. sls. mgr.; Andrew Duca, dir. of creative svcs.; Michael Nigro, contr.; Richard O'Toole, mrch'. dir.; George Snowden, v.p., prog.

WPIX (Channel 11)

WPIX Inc., 11 WPIX Plaza, New York, NY 10017 (212) 949-1100.

Leavitt J. Pope, pres.; R.N. Hughes, sen. v.p., prog. & community affairs; Gerard Mulderrig, sen. v.p. sls.; Julie Alex, v.p. oper.; Rosemary Amata, asst. cont.; Donald M. Carney, v.p. sports; William L. Cooper, v.p. film; John Corporon, v.p. news; Pat Austin, v.p.-treas.; Laurence Lineham, v.p. sls.; Otis S Freeman, v.p. engr.; John P. Healy, v.p., treas. & cont.; John McGowan, v.p., prog. sls.; Jane Perlman, v.p., research; David H. Polinger, v.p. & asst. to pres.; Gerard Puccio, v.p., sales; Yvonne R. Stewart, v.p., pers. & Off. mgmt.

Operation: 1948; TV sets in area: 8,702,000 (ARB).

New York-Newark (N.J.)

WNJU-TV (Channel 47)

1020 Broad St., Newark, NJ 07102 (201) 643-9100.

Carlos Barba, v.p. & gen. mgr.; Thomas Johansen, gen. sls. mgr.; Felix Suria, dir. of prgm.; Gladys Carames, dir. of pub. affairs; George Kraus, chief eng.; Sylvia Pascual, traffic mgr.; Dominick Cascio, business mgr.

Operation: May, 1965; TV households: 2,300,000.

WWHT-TV (Channels 68/69)

WWHT, Inc., 390 W. Market St., Newark, NJ 07107.

Herb Lefkowitz, gen. mgr.; Joe Giardina, dir. of technical oper.; Lia Afriat, mgr. of programming.

Operation: July, 1977.

New York-Paterson (N.J.)

WXTV (Channel 41)

Spanish International Communications Corp., 250 Park Ave., New York City, 10017.

Edward J. DeGray, pres. & treas.; Leandro Blanco, prog. dir.; rep.: Spanish International Network Inc.

Operation: 1968.

Plattsburg-Burlington (Vt.)

WPTZ (Channel 5) NBC

Rollins Telecasting Inc., P.O. Box 647, Atlanta, GA 30301.

W.P. Eaton, Jr., gen. mgr.; Lincoln Dixon, chief eng.; nat'l rep.: Blair Television.

Operation: 1954; TV homes in area: 252,600 (U.S.); 954,880 (Canada).

Rochester

WHEC-TV (Channel 10) CBS

WHEC, Inc., 191 East Avenue, Rochester, NY 14604 (716) 546-5670.

John P. Decker, pres. & gen. mgr.; Alfred Kronquest, v.p. mktg.; Howard Huff, tech. admin.; nat'l rep.; Katz, Co.

Operation: 1953; TV sets in area, 453,500.

WOKR (Channel 13) ABC

Flower City Television Corp., 4225 W. Henrietta Rd., Rochester, NY 14623 (716) 334-8700.

Lawrence Busse, gen. mgr.; Terry Fauth, prog. dir.; David Pender, cont.; Gordon Sclar, promo. mgr.; Clyde Parker, eng. mgr.; rep.: PGW.

Operation: 1962; TV households: 1,191,800.

WROC-TV (Channel 8) NBC

Ziff-Davis Broadcasting Company, One Park Ave., NYC 10016, Broadcast Center, 201 Humboldt St., Rochester, NY (716) 288-8400.

Francis A. DiProsa, v.p. & gen. mgr.; Lemuel B. Schofield, prog. dir.; Mike Morgan, news dir.; nat'l rep.: The Petry Co.

Operation: 1949; TV sets in area: 479,000.

Syracuse

WIXT-TV (Channel 9) ABC

WIXT-TV, Inc., P.O. Box 9, Syracuse, NY 13214 (315) 446-4780.

Carl Jaquint, v.p. & gen. mgr.; Robert Fishman, v.p., sls.; Ken Rothrum, loc. sls. mgr.; Charles F. Mulvey, chief eng.

Operation: 1962.

WSYR-TV (Channel 3) NBC

Newhouse Broadcasting Corp., 1030 James St., Syracuse, NY 13203 (315) 474-3911.

E.R. Vadeboncoeuer, pres.; Mitzi E. Newhouse, v.p. & secy.; David J. Shurtleff, v.p. & gen. mgr.; E. James McDonald, prog. dir.; A.J. Eicholzer, chief eng.; Louis B. Switzer, promo. mgr.; Norman Cissna, dir. of sls.; Fran Procopio, film ed.; nat'l rep.: Harrington, Righter and Parsons, Inc.

Operation: 1950; TV sets in area: 525,000.

WTVH-TV (Channel 5) CBS basic

Meredith Corp. 980 James St., Syracuse, NY 13203 (315) 474-8511.

Larry Rhodes, v.p. & gen. mgr.; Albert Chismark, dir. of eng.; nat'l rep.: MMT Sales Inc.

Operation: 1948 as WHEN-TV; TV sets in area: 480,000.

Utica

WKTV (Channel 2) NBC

Harron Communications Corp., P.O. Box 2; Utica, NY 13504 (315) 733-0404.

Shell Storrier, secty. treas., & gen. mgr.; Paul Harron, Jr., pres.; Daniel German, program. dir. & promo. mgr.; Frank Guzzetti, film dir.; Marie Zumpano, traffic mgr.; Shaun Fitz-Gerald, merch. mgr.; Merv Ainsworth, chief eng.; nat'l rep.: P.G.W.

Operation: 1949; TV homes in area: 340,100.

WUTR (Channel 20) ABC

Roy H. Park Bdcst'. of Utica-Rome Inc., P.O. Box 20, Utica, NY 13503 (315) 797-5220.

Edward Bizzell, exec. v.p. & gen. mgr.; John S. Mucha, oper. mgr.; Wayne Perry, chief eng.; rep.: Blair Television.

Operation: Feb., 1970.

NORTH CAROLINA

Asheville

WANC-TV (Channel 21)

Thoms Broadcasting, 75 Elk Mt. Scenic Hgwy., Asheville, NC 28804 (704) 254-4663.

Meredith S. Thomas, pres.; Richard Thomas, exec. v.p.; John Randolph, chief eng.

Operation: 1953; TV sets in area: 40,000.

WLOS-TV (Channel 13) ABC

Wometco Skyway Broadcasting Co., Box 2150, Asheville, NC 28802 (704) 255-0013. 7 Laurens Court, P.O. Box 2363, Greenville, SC 29602 (803) 242-6091.

Morton S. Cohn, pres. & gen. mgr.; Dave DeArmond, v.p. & sta. mgr., Bill Pfeiffer, v.p., prog.; Ken Glover, sales mgr.; Kay Jenkins, dir. of sales devel.; Ernie Sigmon, promotion mgr.; C. Lee Davis, chief engineer; Dan Kulinski, business manager; rep.: MMT Sales Inc.

Operation: 1954; TV households in area: 1,254,300 TSA.

Charlotte

WBTV (Channel 3) CBS

Jefferson Pilot Broadcasting Co., One Julian Price Place, Charlotte, NC 28208 (704) 374-3500.

Joseph M. Bryan, chm. bd.; Wallace J. Jorgenson, pres.; James G. Babb, Jr., exec. v.p.; John A. Edgerton, v.p. & man. dir.; Stephen H. Greenwald, news and info. mgr.; John H. Hutchinson, Jr., broadcast oprs. mgr.; Mark D. deCastrique, program oprs. mgr.; Donald L. Baxter, gen. sls. mgr.; George K. Reynolds, tech. op. mgr.; Jim Thacker, sports dir.; Mrs. Tommi L. Jones, v.p./personnel; Joseph B. Young, v.p./research & planning; Gerald W. King, v.p./treas.; George B. Grills, Jr., dir. of eng.; Larry M. Harding, v.p./public aff.

Operation: 1949; TV sets in area: 1,150,540.

WCCB-TV (Channel 18) Independent

Mecklenburg Television Broadcasters, Inc., 1 Television Place, Charlotte, NC 28205 (704) 372-1800.

Cy N. Bahakel, pres.; Don E. Kassner, v.p. engineering; Robert K. Larr, gen. mgr.; Guy Titman, chief engineer; Beverly Poston, pgm. dir.; Tony Alexi, news director; Carla Alden, traffic mgr.; rep.: Katz Television.

Operation: Dec., 1953.

WPCQ (Channel 36)

Westinghouse Broadcasting Co., P.O. Box 18665, Charlotte, NC 28218 (704) 536-3636.

Robert Schuessler, v.p. & gen. mgr.; John Cottingham, sls. mgr.; Nat Tucker, dir. bdcst. op.; Jack Barr, chief eng.

Operation: 1967; TV households: 2,185,700.

WSOC-TV (Channel 9) NBC

Carolina Broadcasting Company, P.O. Box 34665, Charlotte, NC 28234 (704) 372-0930.

Freeman R. Jones, v.p. & gen. mgr.; C.M. Murphy, treas.; Marvin Pridgen, national sales mgr.; Fred Collins, regional sales mgr.; John E. Callaghan, station mgr.; Robert McCourt, pgm. dir.; Lewis Brooks, news dir.; Edie Reno, prom. & merch. mgr.; Charles R. Whitley, chief eng.; nat'l rep.: TeleRep Inc.

Operation: 1957; TV sets in area. 1,130,700.

Durham

WTVD (Channel 11) CBS

Capital Cities Communications Inc., Box 2009, 411 Liberty St., Durham, N.C. 27702 (919) 683-1111.

Richard F. Appleton, v.p. & gen. mgr.; Richard Spinner, gen sls. mgr.; Jim Duncan, pgm. mgr.; Ned Warwick, news dir.; William Higgs, bus. mgr.; Edward Wall, chief eng.; Walter Norflett, dir. of pub. affairs; Bob Johnson, prod. mgr.; Tom Domer, promo. mgr.

Operation: 1954; TV homes in area: 521,400 (ADI).

Durham-Raleigh

WPTF-TV (Channel 28) NBC

Durham Life Broadcasting Service, Inc., Box 3540, Durham, NC 27702 (919) 832-9783.

T.A. Upchurch, chm.; Carl V. Venters, Jr., pres.; C.D. Little, Jr., treas.-cont.; R. Douglas McLarty, v.p. & gen. mgr.; Peter W. Devereaux, gen. sls. mgr.; Wilbur Brann, eng. dir.

Operation: November, 1968; TV households, 517,400.

Greensboro

WFMY-TV (Channel 2) CBS

WFMY Television Corp., Phillips Ave., Greensboro, N.C. 27420 (919) 379-9639.

Mark Conrad, v.p. & gen. mgr.; Dave Alexander, prog. dir.; nat'l. rep.: Blair

Operation: 1949; TV homes in area: 797,800.

Greenville

WNCT-TV (Channel 9) CBS

Roy A. Park Broadcasting, Inc., P.O. Box 898, Greenville, NC 27834 (919) 756-3180.

Roy H. Park, pres. treas.; Mrs. Dorothy D. Park, secretary; T.B. Maxfied, asst. secretary; K.B. Skinner, asst. treas., asst. secretary; Blake C. Lewis, Jr., gen. mgr.; Hank Tribley, gen. mgr.; Dr. John D. Caldwell and William S. Wellons, dir.; John G. Clark, Jr., gen. sls. mgr.; Malcolm Nicholson, prod. mgr.; Ed Fields, prog. mgr.; Heber Adams, chief eng.; nat'l rep.: Blair TV.

Operation: 1953; TV sets in area: 476,300.

Hickory

WHKY-TV (Channel 14)

Catawba Valley Bcstg. Co., Box 1059, Hickory, NC 28601 (704) 322-5726.

Thomas E. Long, sta. mgr., sls. mgr. & chief eng.; Jeffrey B. Long, prog. dir.

Operation: 1968; TV households: 125,700.

High Point

WGHP-TV, Inc.

P.O. Box 2688, High Point, NC 27261 (919) 883-7131.

Eugene H. Bohi, pres. & gen. mgr.; Frank Terry, contr.; Dave Whitaker, v.p. & sta. mgr.; Jim Bennett, exec. dir. news.

Operation: October, 1963; TV households: 1,306,500.

New Bern

WCTI-TV (Channel 12) ABC

Malrite of North Carolina, P.O. Box 2325, New Bern, NC 28560.

B.T. Jenkins, v.p. & gen. mgr.

Raleigh

WRAL-TV (Channel 5) ABC

Capitol Broadcasting Co., P.O. Box 12,000, Raleigh, NC 27605 (919) 828-2511.

James F. Goodman, pres. & chief exec. off.; Louise S. Stephenson, secy.; David H. Witherspoon, v.p. Fred Barber, v.p.; Ray Wilkinson, v.p., John Greene, opr. mgr.; Bud Coggins, gen. sls. mgr.; Walter Voigt, v.p. A.H. Moore, jr., v.p. Lee Poole, dir. of eng.; nat'l rep.: Katz Television American.

Operation: 1956; TV sets in area: 684,700.

Washington

WITN-TV (Channel 7) NBC

WITN-TV Inc., U.S. 17 South, Box Office 468, Washington, NC 27889 (919) 946-3131.

W.R. Roberson Jr., chmn. & chief exec. ofcr.; Dick Paul, pres. & chief op. ofr.; W.E. Bass, v.p.-loc. sls.; Paul Oughton, v.p.-op.; Roy Hardee, v.p.-news; Al Manning, v.p.-eng.; Robert R. Culler, v.p. & gen. sls. mgr.; Frank Brady, local sls. mgr.; John Ratcliffe, prod. mgr.; Fran Williams, asst. secty; Kay Currie, woman's dir., nat'l rep.: Katz Television.

Operation: Sept. 1955; TVHH: 672,000.

Wilmington

WECT-TV (Channel 6) NBC, CBS

Atlantic Telecasting Corp., 322 Shipyard Blvd., Wilmington, NC 28401 (919) 791-8070.

Dan D. Cameron, pres.; Wayne Jackson, sta. mgr.; Bill Elks, oper. dir.; C.D. Martin, Jr., sls. mgr.; Paul A. Brissette, Jr., exec. v.p. & gen. mgr.; Mike Loizides, chief eng.; Ernie Whitmeyer, prod. dir.; nat'l rep.: Adam Young.

Operation: 1954; TV sets in area: 325,000.

WWAY (Channel 3) ABC

Clay Broadcasting Corp., Box 2068, 615 N. Front St. Wilmington, NC (919) 762-8581.

George Diab, pres. Henry E. Sullivan, sls. mgr.; Mitchell Saieed, v.p./gen. mgr.; George Allen, sta. mgr.; Art Madeley, chief engineer; rep., Seltel.

Operation: 1964; TV households: 570,450.

Winston-Salem

WXII-TV (Channel 12) NBC

700 Coliseum Drive, Winston-Salem, NC 27106 (919) 721-9944.

Walter E. Bartlett, pres., Multimedia Broadcasting Co.; Reynard A. Corley, v.p. & gen. mgr.; John Comas, mgr., prgm. & film buyer; Zack Manning, gen. sls. mgr.; Gene Hebert, news dir.; Douglas Bales, promo. mgr.; Henry Hunt, dir. of eng.; nat'l rep.: Katz Television.

Operation: 1953; households: 1,306,500 (TSA-May '79 ARB).

NORTH DAKOTA

Bismarck

KFYR-TV (Channel 5) NBC

Meyer Broadcasting Co., Broadway at Fourth, Bismarck, ND 58501 (701) 223-0900.

William Ekberg, pres. & gen. mgr.; Jerry Hegel, gen. sls. mgr.; Merton Johnsrud, dir. of prog.; Carl Zeller, dir. of eng.; nat'l rep: Blair Television Associates.

Operation: 1953; TV sets in area: 123,900.

KXMB-TV (Channel 12) CBS-ABC

Bismarck Television, Inc., 1811 N. 15th, Bismarck, ND 58501 (701) 223-9197.

Steve Reiten, sta. mgr. & v.p.; John von Rheden, sls. mgr.; Dick Gottschald, news dir.; Keuin Brown, prod. mgr.; John Kranle, chief eng.; natl. rep.: Katz.

Operation: 1955; TV sets in area: 42,205.

Devil's Lake-Grand Forks

WDAZ-TV (Channel 8) NBC

(Satellite of WDAY-TV, Fargo, ND)

WDAY Inc., 207 Fifth St. N., Fargo, ND 58102 (701) 237-6500.

Rep.: Peters, Girffin, Woodward, Inc.; Andy McDermott Ltd. (Canada).

Operation: 1967.

Dickinson

KDIX-TV (Channel 2) CBS, ABC

Dickinson Broadcasting Corp., 119 Second Ave. W., Dickinson, ND 58601 (701) 225-5133.

S.T. Deck, pres.; Bob Stockert, chief eng.; nat'l rep.: Young TV.

Operation: 1956; TV sets in area: 52,000.

KTHI-TV, (Channel 11) ABC

Spokane Television, Inc., Box 1878, Fargo, ND 58102 (701) 237-5211; Box 127, Grand Forks, ND 58201 (701) 772-3481.

John Hrubesky, gen. mgr.; Dave Lunde, sta. & natl. sales mgr.; Tom McGeehan, loc. sls. mgr.; Roger Johnson, chief eng.; David Knudtson, op. mgr.; Beverly Paulson, traffic mgr.; nat'l rep.: HRP.

Fargo

WDAY-TV (Channel 6) NBC

WDAY, Inc., 207 No. 5 St., Fargo, ND 58102 (701) 237-6500.

William Marcil, pres.; Bill Burn, exec. vice-pres.; Scott Park, gen. mgr. TV; Sumner Rassmussen, tech. dir.; Roy Pedersen, promo. mgr.; Norm Schrader, news dir.; nat'l rep.: Peters, Griffin, Woodward, Inc.

Operation: 1953; TV sets in area: 174,700.

Minot

KMOT (Channel 10) NBC, ABC

Meyer Broadcasting Co., Box 1120, Minot, ND 58701 (701) 852-4101.

W.A. Ekberg, pres.; Stan Wilson, sta. mgr.

KXMC-TV (Channel 13) CBS, ABC

KXMC TV, Inc., KXMC Bldg., Minot, ND 58501 (701) 838-2104.

Chester Reiten, pres. & gen. mgr.; Wm. Hurley, v.p.; Henry Buechler, mgr.; Rod Romine, prog. dir.; Duane Aase, chief eng.; nat'l rep.: Katz.

Operation: 1953; TV sets in area: 129,000.

Valley City-Fargo

KXJB-TV (Channel 4) CBS

North America Communications Corp., 1805 E. Main St. West Fargo, ND 58078 (701) 282-0444.

Tom Barnes, Jr., pres. & gen. mgr.; Paul Wickre, gen. sls. mgr. & prog. dir.; Carl Iverson, chief eng.; nat'l rep., H&R Television.

Operation: 1954; TV sets in area: 186,000.

Williston

KUMV-TV (Channel 8)

Box 1287, Williston, ND 58801; Meyer Broadcasting Co., Fourth at Broadway, Bismarck, ND 58501

William A. Ekberg, pres.; Roger Rien, sta. mgr.

Operation: 1957.

OHIO

Akron

WAKR-TV (Channel 23) ABC

Summit Radio Corp., P.O. Box 1590, 853 Copley Rd., Akron, OH 44309 (216) 535-7831.

Roger G. Berk, pres.; Bob Bostiam, v.p. & sta. mgr.; James L. Berk, v.p.; E.F. Devine, v.p. & treas.: "Chip" Fox, sales mgr.; Wm. F. O'Neil, Jr., prod. dir.; Leo Zody, film ed.; Irwin L. Knopp, chief eng.; nat'l rep.: Meeker Television.

Operation: 1953; TV sets in area: 340,000.

Canton

WJAN (Channel 17)

PTL of Heritage Village Church & Missionary Fellowship, Inc., 6600 Atlantic Blvd., NE, Louisville, OH 44641.

James O Bakker, pres.; Arthur T. Lawling, secty. & treas.; Ron Beardmore, gen. mgr.

Operation: January, 1967; TV households: 519,400.

Cincinnati

WCPO-TV (Channel 9) CBS

Scripps-Howard Broadcasting Co., 500 Central Ave., Cincinnati, OH 45202 (513) 721-9900.

Robert D. Gordon, gen. mgr., Rick Reeves, dir. of ops.; Al Schottelkotte, v.p., news dir.; Clint Pace, gen. sls. mgr.; J.B. Chase, nat'l. sls. mgr.; Don Howe, local sls. mgr.; Dwayne Stevens, prod. mgr.; Lee Davis, dir. of community affairs; Bob Doerr, dir. of inform. serv.; Larry Murtaugh, cont.; Burnie Wayner, chief eng.; nat'l rep.: Blair-TV, Inc.

Operation: 1949; TV sets in area: 1,000,000.

WKRC-TV (Channel 12) ABC

Taft Broadcasting Company, 1906 Highland Avenue, Cincinnati, OH 45219 (513) 651-1200.

Dudley S. Taft, pres.; Don Chapin, v.p., sls.; Nick Bolton, v.p. & gen. mgr.; Jim O'Shields, gen. sls. mgr.; Carl Flickinger, natl. sls. mgr.; Ed Herber, v.p., ops.; Ray Owen, chief eng.; Don North, news dir.; John Hill, film dir.

Operation: 1949; TV sets in area: 1,004,100.

WLWT (Channel 5) NBC

James T. Lynagh, gen. mgr.; Karl M. Wiener, v.p./treas.; Joe Lewin, sta. mgr.; Elizabeth W. Mills, communications dir.; Al Snyder, news dir.; William Spiegel, prog. dir.; Edwin Dooley, chief eng.; Mickey Fisher, traf. mgr.; Thomas Storey, film dir.; nat'l rep.: Katz Television.

Operation: 1948; households: 1,700,900 (TSA—May '77 ARB).

WXIX-TV (Channel 19)

Metromedia Television, 485 Lexington Avenue, New York, NY 10017. Studio: 10490 Taconic Terrace, Cincinnati, OH 45215 (513) 772-1919.

Allan Ginsberg, v.p. & gen. mgr.; Joseph Mazza, v.p. & gen. sls. mgr.; Mike Polatschek, lcl. sls. mgr.; Sandra Pastoor, v.p. & prgm. dir.; Patricia Stiphout, promo. mgr.; James D. Parker, v.p. & chief eng.; rep., Metromedia Television Sales.

Operation: 1968; TV households: 546,000.

Cleveland

WEWS (Channel 5) ABC

Scripps-Howard Broadcasting Co., 3001 Euclid Ave., Cleveland, OH 44115 (216) 431-5555.

Jack R. Howard, bd. chm.; Donald L. Perris, pres.; Edward D. Cervenak, v.p. & gen. mgr.; Peter Hlinka, Ohio adv. mgr. Gary Stark, pgm. dir.; James H. Knight, sls. mgr.; Albert A. Varga, chief eng.; rep. Blair-TV, Inc.

Operation: 1947; Total TV households: 1,824,400.

WJKW-TV (Channel 8) CBS

Storer Broadcasting Co., 5800 S. Marglinal Rd., Cleveland OH 44103 (216) 431-8888.

William J. Flynn, v.p. & gen. mgr.; James Clayton, natl. sls. mgr.; Louis Gatfozzi, prog. dir.; Dave Bell, chf. eng.

Operation: 1949.

WKYC-TV (Channel 3) NBC

National Broadcasting Co., 1403 East Sixth Street, Cleveland, Ohio 44114 (216) 344-3333.

Neal Van Ells, v.p. & gen. mgr.; Don Camera, sta. mgr., Marysue A. Gaudio, adv. & promo. mgr.; John Llewellyn, dir. of sls.; Raymond Smith, dir., tech. ops.; Alan Thiel, bus. mgr.; Gregory R. Stehlin, prog. dir.; Lee Bailey, news, exec. prod.; nat'l rep.: NBC Spot Sales.

Operation: 1948; TV sets in area: 1,564,600.

Columbus

WBNS-TV (Channel 10) CBS

Dispatch Printing Co. 770 Twin Rivers Drive, Columbus, OH 43216 (216) 460-3700.

Gene D'Angelo, pres & gen. mgr.; Arnold Routson, v.p. sls.; William Orr, v.p., eng'.; John A. Haldi, vice pres., prgm.; Dale Laackman, prod. dir.; Gerald Cary, treas.; Larry Maisel, News dir.; Jay Scafone, dir. adv. & promo.; nat'l rep.:. Blair TV Inc

Operation: 1949; TV sets in area: 557,400.

WCMH-TV (Channel 4) NBC

Outlet Broadcasting, 3165 Olentangy River Rd., Columbus, OH 43202 (614) 263-5441.

David E. Henderson, pres.; Gary Robinson, v.p. & gen. mgr.; Richard Lowe, gen. sales mgr.; Gary Lowell, nat'l sales mgr.; Dave Miller, prog. dir.; Darryl Moeller, community serv. dir. Dave Smith, prom. mgr.; Richard Townley, news dir.; Bob Dye, chief eng.; Dave Goff, prod. mgr.; Juanita Mitchell, film dir., nat'l rep.: Petry Television Inc.

Operation: 1949; TV homes in area: 1,289,900.

WTVN (Channel 6) ABC

Taft Broadcasting Co., 1271 Dublin Road, Columbus, 43215 (614) 481-6666

Charles Mechen, bd. chm.; Dudley S. Taft, pres.; Ro Grignon, exec. v.p.; Fred von Stade, v.p. & gen. mgr.; George Coles, gen. sls. mgr.; Mrs. Pat Cramer, prog. mgr.; Dave Parr, promo. dir. Sue Swartz, pub. affairs dir.; John R. Owen, chief eng.; nat'l rep.: Harrington, Righter & Parsons, Inc.

Operation: 1949; TV sets in area: 727,000.

Dayton

WDTN (Channel 2) NBC

Grinnell Communications Corporation, 4595 South Dixie Ave., P.O. Box 741, Dayton, OH 45401.

Ray W. Colie, exec. v.p. & gen. mgr.; Don E. Meineke, v.p., sls.; John Lyons, promo. & pub. mgr.; Don Prijatel, sta. mgr.; Roy Poland, chief eng.; nat'l rep.: Petry.

Operation: 1949; TV homes in area: 1,029,500.

WHIO-TV (Channel 7) CBS

Miami Valley Broadcasting Corp., 1414 Wilmington Avenue, Dayton, OH 45401 (513) 254-5311.

Clifford M. Kirtland, bd. chm.; William A. Schwartz, pres.; Stanley G. Mouse, sr., v.p.; Neil Pugh, v.p. & sta. mgr.; John Hanley, cont.; Robert Wells, dir. of info. svcs.; John Clark, prog. dir. Ray Benedict, chief eng.; nat'l rep.: Tele-rep

Operation: 1949; TV sets in area: 437,160.

WKEF (Channel 22) ABC

Springfield Television; Studio: 1731 Soldiers Home Road, Dayton, OH 45418 (513) 263-2662.

George J. Mitchell, pres., Ohio op.; Lon Mirolli, gen. sls. mgr.; Tim Coppock, prod. mgr.; Darrell Hunter, chief eng.; Berl Golub, prog. mgr. & film buyer; Sandy Patton, promo. dir.; Johnny Walker, dir. community affairs/special events; Kevin Kelly, news dir. rep.: H-R Television.

Operation: 1964. TV households: 1,399,000.

Lima

WLIO-TV (Channel 35) NBC, ABC

Lima Communications Corporation, 1424 Rice Ave., Lima, Ohio 45805 (419) 228-8835.

Wayne Current, pres.; Lewis Dickey, v.p., oper. officer; James C. Dages, v.p. & gen. mgr.; Douglas A. Ferguson, sta. mgr. & prog. dir.; Maurice Lamb, chief eng.; Bruce A. Opperman, sls. mgr.; nat'l rep.: The Katz Co.

Operation: 1953; TV sets in area: 293,500.

Lorain-Cleveland

WUAB (Channel 43)

Gaylord Broadcasting Co. of Ohio, 8443 Day Dr., Parma, OH 44129 (216) 845-6043.

Jack J. Moffitt, gen. mgr.; Ron St. Charles, prog. dir.; Nelson Flanagan, dir. of retail svcs.; Edward Fleming, contr.; Robert Hartman, gen. sales mgr.; Harry Sykora, mgr. of engineering; Frank Fletcher, prom. mgr.

Steubenville

WTOV-TV

Ziff-Davis Broadcasting Co. Box 9999, Steubenville, OH 43952.

I. Martin Pompadur, pres.; Ziff Corp.; Ralph E. Becker, pres., Ziff-Davis Broadcasting Co.; Ray J. Chumley, vice pres./gen. mgr.; James McCreary, v.p./contr.; Tom Hansen, prgm. mgr.

Toledo

WDHO-TV (Channel 24) ABC

D.H. Overmyer Telecasting Co., Inc., 300 S. Byrne Rd., Toledo, OH 43615 (419) 535-0024.

William J. Shock, v.p./gen. mgr.; C. Robert Thompson, gen. sls. mgr.; James Bowe, v.p. eng./ops.; rep.: Katz Television.

Operation: 1966; TV households: 1,775,100.

WSPD-TV (Channel 13) NBC

The Storer Broadcasting Co., 136 Huron St., Toledo OH 43604 (419) 255-1313.

Keith McKennedy, v.p. & gen. mgr.; H.W. Ray, asst. gen. mgr.; George Carlino. natl. sls. mgr.; Mark Whelan, loc. sls. mgr.; Tom Foos, promo. mgr.; nat'l rep.: Storer Television Sales, Inc.

Operation: 1948; TV sets in area: 751,100.

WTOL (Channel 11) CBS

Cosmos Broadcasting Corp., P.O. Box 715, Toledo, OH 43601 (419) 244-7411.

Charles A. Batson, pres.; Law Epps, gen. mgr.; Jack Spring, gen. sls. mgr.; Jim Foley, natl/regional sls. mgr.; Jack Sander, lcl. sls. mgr.; Wyatt McDaniels, chief eng.; Lamont McLoughlin, news & pub. aff. dir.; Donald M. Hess, prog. promo. dir.; David MacAllister, prod. mgr.; Shirley Clark, public affairs coordinator; Jim Skinner, film dir.; Carol Webb, traffic mgr.; Joe Gillis, special projects dir.; nat'l rep.: Peters, Griffin, Woodward.

Operation: 1958; TV sets in area: 1,735,300.

Youngstown

WFMJ-TV (Channel 21) NBC

The Vindicator Printing Co., 101 W. Boardman St., Youngstown, OH 44503 (216) 744-8611.

William J. Brown, pres. & gen. mgr.; Mitchell F. Stanley, mgr.; Gene Donahay, TV prog. mgr.; Ralph Barone, film dir.; Seymour Raymond, chief eng.; nat'l rep., Blair TV.

Operation: 1953; TV households in area: 853,200.

WKBN-TV (Channel 27) CBS

WKBN Broadcasting Corp., 3930 Sunset Blvd., Youngstown, OH 44501 (216) 782-1144.

W.P. Williamson, Jr., bd. chm.; Ruth L. Cruikshank, secty.; W.P. Williamson, III, pres. & gen. mgr.; David V. Stewart, treas., sales dir.; Norm Berger, prog. dir.; nat'l rep.: Katz Agency.

Operation: 1953; TV sets in area: 514,000.

WYTV (Channel 33)

WYTV, Inc., 3800 Shady Run Rd., Youngstown, OH 44502.

Geoffrey Pearce, gen. mgr.

Zanesville

WHIZ-TV (Channel 18) NBC, ABC

Southeastern Ohio Television System, Downard Rd., Zanesville, OH 43701 (614) 453-0361.

Allan Land, group mgr.; Robert Hodous, gen. mgr.; Reed Mellon, prog. dir.; Elmer Hartmeyer, chief eng.; Franklin Young, film oper. mgr.; nat'l rep.: Spot Time Ltd.

Operation: 1953; TV sets in area: 80,000.

OKLAHOMA

Ada

KTEN (Channel 10) ABC, CBS, NBC

Eastern Oklahoma Television Co., Inc., P.O. Box 10, Adal OK 74820 (405) 332-3311.

Bill Hoover, pres. & gen. mgr.; James Lallar, chief eng.; Dan Hoover, v.p. & prog. dir.; nat'l rep.: Savalli/Gates.

Operation: 1954; TV sets in area: 178,310.

Ardmore

KXII-TV (Channel 12)

Texoma Broadcasters, Inc., Box 1175, Sherman, TX.

M.N. Bostick, pres.; Richard R. Adams, gen. mgr.; Dennis Lite, chief eng.; Mike Leucke, prod. mgr.; nat'l rep.: Seltel.

Operation: 1956.

Lawton

KSWO-TV (Channel 7) ABC

KSWO Television Co., Inc., P.O. Box 708, Lawton, OK 73502 (405) 355-7000.

R.H. Drewry, pres., gen. mgr.; Larry Patton, sta. & gen. sls. mgr.; Jim Lockerd, chief eng.

Operation: 1953; TV sets in area: 225,000.

Oklahoma City

KOCO-TV (Channel 5) ABC

Gannett Broadcast Group, 1300 E. Britton, Oklahoma City, OK 73113.

Al Parsons, pres. & gen. mgr.; David Loveland, v.p. & gen. sls. mgr.; Ted Newcomb, v.p. & chief eng.; Don Troutt, compt.; Mike Palmer, v.p., bdcst. op.

Operation: 1954; TV households: 654,600.

KTVY (Channel 4) NBC

KTVY, Inc., 500 East Britton Road, Oklahoma City, OK 73114 (405) 478-1212.

Lee Allan Smith, v.p. & gen. mgr.; Dan Bates, asst. gen. mgr.; Bill Thrash, prog. mgr.; Nick Panos, nat'l. sls. mgr.; Jim Whittaker, promo, mgr.; John Bushnell, chief eng.

Operation: 1949; TV sets in area: 537,380.

KWTV (Channel 9) CBS

Griffin Television, Inc. P.O. Box 14159, Oklahoma City, OK 73113 (405) 843-6641.

John Griffin, owner; Jacques DeLier, pres. & gen.

mgr.; Duane Harm, sta. mgr.; Jack Sallaska, prgm. dir.; Troy Brown, chief eng.; Charles Andrews, prom. & pub. dir.; nat'l rep.: HRP.

Operation: 1953; TV sets in area: 739,000.

Sayre

KVIJ-TV

(Satellite of KVII-TV, Amarillo, TX)

Marsh Media, Ltd., Box 13000, Amarillo, TX 79101 (405) 928-3233.

Glen Neal, res. eng. supvr.; rep.: Katz Television.

Operation: 1961.

Tulsa

KOTV (Channel 6) CBS

KOTV, Inc., Third & Frankfort Sts., Tulsa, OK 74120 (918) 582-9233.

Philip J. Lombardo, pres.; Allan Howard, vice pres. & gen. mgr.; Joe Matthews, gen. sales mgr.; Michael Grant, promo. mgr.; Bob Levy, prog. mgr.; nat'l rep.: PGW, Inc.

Operation: 1949; TV sets in area: 742,300.

KJHR-TV (Channel 2) NBC

Scripps-Howard Broadcasting Company, P.O. Box 2, Tulsa, OK 74101 (918) 743-2222.

Ben Hevel, gen. mgr.; Wally Dunham, asst. gen. mgr.; Michael Casserly, news dir.; Charles Pfaff, natl. sls. mgr.; Art Bates, chief eng.; nat'l rep.; Blair Television.

Operation: 1954; TV sets in area: 460,300.

KTUL-TV (Channel 8) ABC

KTUL-TV, Inc., Box 8, Tulsa, OK 74101 (918) 446-3351.

James C. Leake, chmn.; John G. Leake, pres.; Thomas Goodgame, v.p. gen. mgr.; Robert Gregory, v.p., news.; Gary Ricketts, oper. mgr.; Edward Hunt, gen. sls. mgr.; Bill Shock, dir. eng.; Carl Bartholomew, promo. mgr.; nat'l rep.; Edward Petry & Co., Inc.

Operation: 1954; TV homes in area: 481,400.

OREGON

Coos Bay

KCBY-TV (Channel 11) NBC

(Satellite of KVAL-TV, Eugene, OR)

Eugene Television Inc., P.O. Box 1156, Coos Bay, OR 97420 (503) 269-1111.

S.W. McCready, gen. mgr.; Gordon Shaw, oper. mgr.; Irv Brown, chief eng.

Eugene

KEZI-TV (Channel 9) ABC

Liberty Communications, Inc., P.O. Box 7009, Eugene, OR 97401 (503) 343-3301.

Don E. Tykeson, pres.; Jim Putney, gen. mgr.; Dave Larson, opns. mgr.; Dick Walker, lcl. sls. mgr.; Kris Bushnell, natl./reg. sls.; Joe Smelser, prod. mgr.; Dennis Hunt, chief eng.; nat'l rep.: PGW.

Operation: 1960; TV households: 226,200.

KVAL-TV (Channel 13) Primary NBC

Eugene Television, Inc., Blanton Heights, P.O. Box 1313, Eugene, OR 97401 (503) 342-4961.

Glenn E. Nickell, v.p., gen. mgr.; John Doyle, news dir.; Bonnie Cole, natl sales; Gordon Bussey, lcl. sls. mgr.; Gregg McKay, film dir.; Jim Bowen, chief eng.; nat'l rep.: Katz Television, Inc.

Operation: 1954; TV sets in area: 263,800.

Klamath Falls

KOTI-TV (Channel 2)

Oregon Broadcasting Co., P.O. Box 2K, Klamath Falls, OR 97601 (503) 884-8131.

Donald E. Smullin, pres.; Dan Markham, station mgr.; nat'l rep.: PGW; reg. rep.: Art Moore & Associates.

Operation: 1956.

Medford

KTVL-TV (Channel 10) NBC, ABC

Sierra/Cascade Communications, Inc., Box 10, Medford, OR 97501 (503) 773-7373.

J.L. DeArmond, pres.; Ray Johnson, exec. v.p. & gen. mgr.; John Larkin, mgr.; Art Chipman, natl. sales mgr.; Ellis Feinstein, dir. of eng'.; rep.: H-R Television, Simpson Reilly & Associates (Portland & Seattle).

Operation: 1961; TV households: 127,800.

KOBI-TV (Channel 5) ABC, CBS

Oregon Broadcasting Co., P.O. Box 5M, Medford, OR 97501 (503) 779-5555

William B. Smullin, pres.; Patricia D. Smullin, v.p.; Ed Zander, mgr.; Bill Kirk, chief eng.; nat'l rep.: Peters, Griffin, Woodward.

Operation: 1953.

Portland

KATU (Channel 2) ABC

Fisher Broadcasting, Inc., 2153 N.E. Sandy Blvd., Portland, OR 97232 (503) 231-4222.

W.W. Warren, pres.; Thomas R. Dargan, exec. v.p. & gen. mgr.; S.V. Hinman, asst. gen. mgr.; Robert Kalstad, dir. of prog.; Wm. Vandermay, mgr. of eng.; D.B. Ludwig, gen. sls. mgr.; Jim Rogers, natl. sls. mgr.; Joella Werlin, pub. affairs mgr.; rep : Katz Television.

Operation: 1962; TV households: 768,700.

KGW-TV (Channel 8)

King Broadcasting Company, 1501 S.W. Jefferson St., Portland, OR 97201 (503) 226-5000.

Dean H. Woodring, gen. man.

Operation: 1956.

KOIN-TV (Channel 6) CBS

KOIN-TV, Inc., 140 S.W. Columbia St., Portland, OR 97201 (503) 243-6666.

R.M. Schafbuch, gen. mgr.; Howard Kennedy, sta. mgr.; Lee Carlson, gen. sls. mgr.; Jane Hoyt, promo. mgr.; C. Stephen Currie, bdcst. opr. mgr.; Herb Davidson, asst. chief eng.; nat'l rep.: Harrington, Righter & Parsons.

Operation: 1953; TV sets in area: 925,000.

KPTV (Channel 12)

Oregon Television, Inc., P.O. Box 3401, Portland, OR 97208; 735 S.W. 20th Place, Portland, OR (503) 222-9921.

John S. Hansen, pres. & gen. mgr.; Marty Brantley, station mgr.; Gordon White, prog. mgr. & film buyer; Craig A. Miller, promo. mgr.; Gene Phelps, chief eng.; rep.: Tele-Rep, Inc., Tacher & Co. (Seattle).
Operation: 1955; TV households: 789,000.

Roseburg

KPIC (Channel 4) NBC Primary
(Affiliate of KVAL-TV, Eugene, OR)
Southwest Oregon Television Broadcasting Corp., studio & office: 655 West Umpqua, Roseburg, OR 97470.
C.H. Fisher, pres.; Wm. B. Smullin, v.p. & treas.; Richard A. Booth, op. mgr.; S.W. McCready, gen. mgr., sta. mgr.; nat'l rep.: H/R Television, Inc., Art Moore & Associates, Inc. (Pacific Northwest).
Operation: 1956.

PENNSYLVANIA

Allentown

WFMZ-TV (Channel 69)
Maranatha Broadcasting Co., Inc., E. Rock Rd., Allentown, PA 18103 (215) 797-4530.
Richard C. Dean, pres. & gen. mgr.; David G. Hinson, sls. & prog. dir.; David W. Noel, news dir.; Kathy Yelsits, office mgr.; Barry Fisher, chief eng.; sales rep.: Spot Time.
Operation: November, 1976.

Altoona

WOPC (Channel 38) ABC
John R. Powley, Box 191, R.D., Dysart, PA 16636 (814) 943-2607.
John R. Powley, gen. mgr.
Operation: 1974.

WTAJ-TV (Channel 10) CBS
A Gateway Station; 5000 Sixth Avenue, Altoona, PA 16603 (814) 944-2031.
J. Thomas Conners, v.p. & gen. mgr.; George L. Walsh, gen. sls. mgr.; Douglas Parker, prog. mgr.; Jim Turcovsky, prod. mgr.; Richard Kline, chief eng.; Thomas F. Murphy, bus. mgr.; Gary L. Delph, promo. mgr.; George Tyll, news dir.
Operation: 1953.

Erie

WICU (Channel 12) NBC
Great Lakes Communications, Inc., 3514 State St. Erie, PA 16508 (814) 454-5201.
Arthur M. Ingram, pres.; Frederick Kalmbach, gen. mgr.; Clarence Paolella, gen. sls. mgr.; Mike Csop, chief eng.; Eric Johnson, prog. dir.; nat'l rep.: Katz TV Continental; Canadian rep.: Radio/Television Reps., Ltd. (Toronto).
Operation: 1949; TV sets in area: 274,000.

WJET-TV (Channel 24) ABC
Jet Broadcasting Co., Inc., 1635 Ash St., Erie, PA 16503 (814) 864-4902.
Myron Jones, pres.; Pete Coticchia, gen. mgr.; Ken Fanazini, prog. dir.; John Kanzus; rep.; Adam Young Inc.
Operation: 1966; TV households: 530,400.

WSEE (Channel 35) CBS
Gillett Broadcasting of P.A., Inc., 1220 Peach St., Erie, PA 16501 (814) 455-7575.
Jerry Montgomery, pres. & gen. mgr.; Jack Rea, v.p./opns. mgr.; James Penna, chief eng.; nat'l. rep.: Peters, Griffin & Woodward.
Operation: 1954; U.H.F. sets in area: 245,400.

Harrisburg

WHP-TV (Channel 21) CBS-TV
WHP, Inc., P.O. Box 1507, Harrisburg, PA 17105 (717) 238-2100.
Donald D. Wear, pres. & gen. mgr.; Donald C. Yeiser, v.p. sls.; Maurice Gannaway, dir. of eng.; nat'l rep.: Katz.
Operation: 1953; TV households in ADI: 460,400.

WTPA (Channel 27) ABC
Div. of Newhouse Broadcasting Corp.; Syracuse, NY; Studio: 3235 Hoffman St., Harrisburg, PA (814) 236-2727.
E.R. Vadeboncoeur, pres.; Paul S. Abbott, gen. mgr.; Jack Thornsley, sls. mgr.; Paul Baker, prog. dir.; Charles W. Baker, chief eng.; nat'l rep.: Harrington, Righter & Parsons, Inc.
Operation: 1953; TV sets in area: 340,000.

Johnstown

WJAC-TV (Channel 6) NBC
WJAC, Inc., Hickory Lane, Johnstown, PA 15907 (814) 255-5831.
James M. Edwards, Sr., v.p. & gen. mgr.; Richard D. Schrott, gen. sls. mgr.; Christopher J. Taylor, prog. dir.; Arthur Vrooman, chief eng.; nat'l rep.: Harrington, Righter & Parsons.
Operation: 1949; TV sets in area, 1,222,912.

WJNL-TV (Channel 19) CBS
Cover Broadcasting Inc., Cover Hill, Johnstown, PA 15902 (814) 535-8554.
John E. Gelormino, pres. & gen. mgr.; John Kershes-key, chf. eng.
Operation: 1953.

Lancaster

WGAL-TV (Channel 8) NBC
WGAL Television, Inc., 8 West King St., Lancaster, PA 17603 (814) 393-5851.
Harry J. Shaub, Jr., pres. & gen. mgr.; Harold E. Miller, sta. mgr.; Nelson Sears, prog. mgr.; Leroy K. Strine, natl. sls. mgr.; Harold E. Miller, Jr., lcl. sls. mgr.; Robert H. Sima, sls. svc. mgr.; nat'l rep.: The Meeker Co., Inc.
Operation: March, 1949; TV households: 1,267,900.

Lancaster/Lebanon

WLYH-TV (Channel 15)
Television Hill, Lebanon, PA 17042 (814) 273-4551.
Licensee: Gateway Communications, Inc., Suite 612, Executive Building, Cherry Hill, NJ 08034 (609) 667-8100.
Robert W. Eolin, v.p. & gen. mgr.
Operation: May 2, 1957.

Philadelphia

KYW-TV (Channel 3) NBC

Group W (Westinghouse Broadcasting Co.), Independence Mall East, Philadelphia, PA 19106 (215) 238-4700.

Pat Pollio, vice pres. & gen. mgr.; Bill Aber, sta. mgr.; Dave Nelson, prgm. mgr.; Bernie Weber, creative svcs. dir.; Michael Lape, gen. sls. mgr.; Catherine Shaffer, sales mgr.; Robert Fields, chief eng.; nat'l rep.: TVAR.

Operation: 1941; TV sets in area; 2,275,000.

WCAU-TV (Channel 10) CBS

WCAU, Inc., City Line Ave. & Monument Rd., Philadelphia PA 19131 (215) 839-7000.

J.R. Feldman, vice-pres. & gen. mgr.; Richard C. Goldstein, sta. mgr.; Philip Press, sls. dir.; Melvin D. Levine, dir. of labor relations & admin.; Bill Lawlor, news dir.; Thomas Pearson, dir. of bdcst.; John Byrne, dir. of eng.; Gordon W. West, mgr. of film & videotape services; John F. Horstmann, administrator; nat'l rep.: CBS Television National Sales.

Operation: 1948; TV sets in area: 2,373,100.

WKBS-TV (Channel 48)

Field Communications Co., 3201 S. 26th St., Philadelphia, PA 19145 (215) 336-6400.

Vincent F. Barresi, gen. mgr.; John M. Fignar, sls. mgr.; Craig Bachran, sls. mgr.; Joseph Weber, prog. mgr.; Martin S. Jacobs, news & pub. affairs dir.; James Lawless, bus. mgr.; rep.: ITVS.

Operation: 1965; TV households: 2,986,201.

WPHL-TV (Channel 17)

WPHL-TV Inc., 5001 Wynnefield Ave., Philadelphia, PA 19131 (215) 878-1700.

Eugene McCurdy, pres. & gen. mgr.; Zvi Shoubin, prog. dir.; Gordon French, sls. dir.; John Ruzzi, gen. sls. mgr.; Joel Levitt, prod. mgr.; Walt Bundy, chief eng.; Bob DiMattina, dir. of finance.

Operation: July, 1960; TV households: 3,407,700.

WPVI (Channel 6) ABC

Capital Cities Communication, Inc., 4100 City Line Ave., Philadelphia, PA 19131 (215) 878-9700.

Lawrence J. Pollock, v.p. & gen. mgr.; Arnie Kleiner, dir. sls.; Thomas K. Byrnes, gen. sls. mgr.; Art Moore, prom. mgr.; Alan Nesbitt, news dir.; Irwin L. Ross, chief eng.

Operation: 1947; TV sets in area: 4,556,200.

WTAF-TV (Channel 29)

Taft Broadcasting Co., 330 Market St., Philadelphia, PA 19106 (215) 925-2929

Randall E. Smith, gen. mgr.; Elliott Troshinsky, gen. sls. mgr.; Chuck Wing, lcl. sls. mgr.; Gregg Miller, prog. mgr.; Karen Corbin, prom. mgr.; Richard Owens, chief eng.; rep.: Telerep.

Operation: 1965; TV households: 3,037,000.

Pittsburgh

KDKA-TV (Channel 2) CBS

Westinghouse Broadcasting Co., Inc., 1 Gateway Center, Pittsburgh, PA 15222 (412) 392-2200.

Thomas L. Goodgame, v.p. & gen. mgr.; Carolyn Wean, sta. mgr.; Deb Zeyen, prog. mgr.; Joe Berwanger, gen. sls. mgr.; Jack Cvetic, chief eng.; Brad Crum, creative svcs. dir.; Carolyn McClair, info. svcs. dir.

WPXI (Channel 11)

WPXI Corp., 11 Television Hill, Pittsburgh, PA 15230 (412) 237-1100.

Clifford M. Kirtland, Jr., bd. chmn.; William Schwartz, pres. & chief operating officer; Jack P. McCarthy, v.p. & gen. mgr.; Patrick Gmiter, v.p. & sta. mgr.; Donald Cunningham, prgm. dir.; Theodore Lon, contr.; James O. Ross, gen. sls. mgr.; Richard Skrinjar, dir. of adv. & promo.; Robert Wehrman, v.p. & chief eng.; Kenneth Tiven, news dir.

Operation, 1957.

WPGH-TV (Channel 53)

Broadcasting Group/Meredith Corporation, 750 Ivory Avenue, Pittsburgh, PA 15214 (412) 931-8600.

Norris Reichel, v.p./gen. mgr.; Bill Saltzgiver, prgm. mgr.; Mark Albright, dir. eng.; Ed Trimble, gen. sls. mgr.; John Tamerlano, lcl. sls mgr.; Kathy Saunders, natl. sls. mgr.; Harry Tate, promo. mgr.; David A. Case, prod. mgr.; rep.: MMT Sales, Inc.

Operation under Meredith: Oct. 1978; Households (TV): 1,678,900.

WTAE-TV (Channel 4) ABC

The Hearst Corp., 4400 Ardmore Blvd., Pittsburgh, PA 15230 (412) 242-4300.

Franklin C. Snyder, gen. mgr. of bdcst' for Hearst; John G. Conomikes, gen. mgr. of TV for Hearst; Fred Young, v.p. & gen. mgr.; for WTAE-TV; Jack Shenkan, v.p. sales & sta. mgr.; James E. Hurley, dir. of eng.; David Crantz, prom. mgr.; Gary Dreispul, dir. of prog.; Kerry Richards, dir. of creative svcs.; Edward Young, operations mgr.; Ray Coleman, community & station svcs. dir.; Joe Rovitto, news dir.; nat'l rep.: Katz Agency Inc.

Scranton/Wilkes-Barre

WDAU-TV (Channel 22) CBS

Scranton Broadcasters, Inc., 1000 Wyoming Ave., Scranton, PA 18509 (717) 961-2222.

Mrs. M.E. Megargee, pres.; Mrs. M.A. Holcomb, gen. mgr.; Mort Rosenthal, prog. dir.; Doug Holcomb, prom. dir.; nat'l rep : Blair.

Operation: 1953; TV sets in area: 751,710.

WNEP-TV (Channel 16) ABC

N.E.P. Communications, Inc., Wilkes-Barre/Scranton Airport, Avoca, PA 18641 (717) 826-1616.

Thomas P. Shelburne, bd. chmn.; Thomas Shelburne, Jr., pres.; Chester Saxwicki, v.p./eng.; Dean Hinson, v.p. mktg.; Peter P Kozloski, treas.; Jay Newman, news dir.; nat'l rep.: Katz Agency, Inc.

Operation: 1954; TV sets in area: 352,375.

Wilkes-Barre

WBRE-TV (Channel 28) NBC

WBRE-TV, Inc. 62 S. Franklin St., Wilkes-Barre, PA 18703; (717) 823-3101.

Louis G. Baltimore, pres. David M. Baltimore, v.p. & gen. mgr.; F.D. Coslett, gen. sls. mgr.; Charles Sakoski, Sr., chief eng.; nat'l rep.: Edward Petry & Co.

Operation: 1953; TV sets in area: 939,300.

York

WSBA-TV (Channel 43) CBS

Susquehanna Broadcasting Co., 140 E. Market St., York, PA 17401 (717) 843-0043; Station: 2005 S. Queen St., York, PA 17403.

Louis J. Appell, Jr., pres.; Helen P. Appell, v.p.; George N. Appell, treas.; Peter P. Brubaker, v.p./finance; James A. DeBold, sta. mgr.; H. David Berlin, sls. mgr.; Sonia L. Stroliman, prog. dir.; nat'l rep.: Avery-Knodel.

Operation: 1952; TV sets in area 700,000.

RHODE ISLAND

Providence

WJAR-TV (Channel 10) NBC

The Outlet Co., 111 Dorrance St., Providence, RI 02903 (401) 751-5700.

J.S. Sinclair, bd. chmn.; Bruce G. Sundlun, pres. and CEO; David E. Henderson, pres./Outlet Broadcasting; John Sawhill, v.p. & gen. mgr.; Wyatt Mcdaniel, chief eng.; Jim Macfarlan, gen. sls. mgr.; Larry Sylvia, nat'l. sls. mgr.; nat'l. rep.: Petry & Co.

Operation: 1949; TV sets in area: 1,891,000.

WPRI-TV (Channel 12) ABC

Knight-Ridder Broadcasting, Inc., 25 Catamore Boulevard, East Providence, RI 02914 (401) 438-7200.

William E. Wuerch, pres. & gen. mgr.; Gary Masters, gen. mgr.; Anthony Colosimo, nat'l. sls. mgr.; Robert Mc Caughey, reg. sls. mgr.; Edward Passarelli, bus. mgr.; T. Arthur Bone, v.p. & chief eng.; Robert Kinkead, prog. mgr.; Leslie Smith, promo. mgr.; Edwin Hart, news dir.; Joseph Fogarty, admin. asst. to gen. mgr.; nat'l rep.: MMT Sales, Inc.

Operation: 1955; TV Households: 2,737,600 (ARB 5-81).

WTEV (Channel 6) ABC

Telestation WTEV, Television Center, 8th & Spring Sts., New Bedford, MA 02741 (617) 993-2651.

Harry J. Shaub, exec. v.p. & gen. mgr.; Vance L. Eckersley, sta. mgr.; John H. Candioto, sls. mgr.; Mann Reed, reg. sls. mgr.; John Robinson, prog. mgr.; Lee Tanner, oper. mgr.; Peter Mandell, prom. mgr.; Truman Taylor, news dir.; Maurice Wynn, chief eng.; rep.: Meeker Co., Inc.

Operation: 1963; TV households: 533,700.

SOUTH CAROLINA

Anderson

WAIM-TV (Channel 50) ABC, CBS

Wilton E. Hall, 321 Kingsley Rd., Anderson, SC 29621 (803) 226-1511.

Wilton E. Hall, owner; Betty Block, mgr.; Robert Baldwin, dir. of eng.; S.A. Hall, Jr., commercial mgr.; Marshall Gilliard, prod. mgr.; nat'l rep.: Devney.

Operation: 1953; TV sets in area: 172,000.

Augusta

WRDW-TV (Channel 12) CBS

Rust Craft Broadcasting Corp., 1301 George Ave., North Augusta, SC 29621 (803) 824-5432.

Marshall Berkman, pres.; Fred Weber, v.p. James N. Armistead, gen. mgr.; Larry Fischer, sls. mgr.; Lee Sheridan, op. mgr.; Edgar Goforth, chief eng.; nat'l rep.: The Katz Agency.

Charleston

WCBD-TV (Channel 2) ABC

State Telecasting Co. Inc., P.O. Box 879, Charleston, SC 29402.

W. Frank Harden, pres. & gen. mgr.; Virgil V. Evans, v.p.; Grange Cuthbert, asst. gen. mgr.; J.W. (Bill) Evans, III, dir. sls. & mktg.; Truman Kemp, lcl. sls. mgr.; James Abbott, dir. bdcst. ops.; Tessa Hooker, promo. mgr.; Don Feldman, news dir.; Walter Nelson, chief eng.

Operation: 1954.

WCIV (Channel 4) NBC

First Charleston Corp., Highway 703, Mt. Pleasant, SC 29464 (803) 884-8513.

Wm. E. Lucas, pres. & gen. mgr.; Frank Hunt, news mgr.; J.L. Edwards, sales mgr.; Carl Shaw, chief eng.

WCSC-TV (Channel 5) CBS

WCSC, Inc. 485 East Bay St., P.O. Box 186, Charleston, SC 29402 (803) 723-8371.

John M. Rivers, Sr., chmn. & treas. of WCSC, Inc.; John M. Rivers, Jr., pres.; Martha Rivers, 1st v.p. & secy.; Gus Bailey, Jr., v.p., gen. mgr.; WCSC-TV; O. Kendall Hege, v.p., gen. sls. mgr.; Doug Atherton, regl. ls. mgr.; Jack R. Becknell, Jr., v.p., dir. of eng'.; Bob Pettit, film dir.; Eve O. Newitt, promo. mgr.; Pat Sellars, contr., nat'l rep.: Peters, Griffin, Woodward, Inc.

Operation: 1953; TV Households: 139,500 ADI (ARB, May 1977).

Columbia

WIS-TV (Channel 10) NBC

Cosmos Broadcasting Corp., 1111 Bull St., P.O. Box 367, Columbia, SC 29202; (803) 799-1010.

J. Law Epps, pres.; Francis M. Hipp, chm., exec. comm.; Dixon C. Lovvorn, sr. v.p.; Richard M. Coulter, v.p., oper.; Betty M. Robertson, v.p., admin.; James R. Sefert, v.p. & gen. mgr.; WIS-TV; John C. Thorne, v.p., plan. & dev.; Charles A. Whitehurst, v.p.; John L. Sander, v.p.; R. Glenn Hilliard, secty.; Joseph P. Gill, dir. engineering; William D. Ridings, controller; rep.: Harrington, Righter & Parsons, Inc.

Operation: 1953; TV sets in area: 853,400.

WLTX-TV (Channel 19) CBS

Capital Communications, Inc., P.O. Drawer M, Columbia, SC 29250 (803) 776-3600.

Richard T. Laughridge v.p. & gen. mgr.; Donald M. Bryant. prog. dir.; Bill Aull, chief eng.; Richard T. Laughridge, natl. sls. mgr.; nat'l rep.: Katz Television.

Operation: 1953; TV sets in area: 289,800.

WOLO-TV (Channel 25) ABC

Columbia TV Broadcasters Inc., Box 4217, Columbia, SCS 29240 (803) 754-7525.

Donald H. Krauss, gen. mgr.; Tom Meacham, sls. mgr.; Chuck Britt, chief eng.; rep.: Seltel.

Operation: 1961; TV households in area: 215,100.

Florence

WBTW (Channel 13) CBS

Daily Telegraph Printing Co., 3430 N. TV Road, Florence, SC 29501 (803) 662-1565.

Joseph B. Foster, asst. v.p. & sta. mgr.; George L. Sansbury, sls. mgr.; Jim Griffin, news dir.; Russell Smith, prod. mgr.; Bernard E. Moore, techl. mgr.; nat'l rep.: Avery Knodel.

Operation: 1954; TV households: 353,400.

Greenville

WFBC-TV (Channel 4) NBC

Multimedia Broadcasting Co., 505 Rutherford St., Greenville, SC 29602.

Douglas A. Smith, sr. vice pres. & gen. mgr.; Bill Wheless, prgm. dir.; Dave McAtee, sls. dir.; Jim Lynn, chief eng.; Andy Scott, promo. mgr.; David Foster, op. mgr.; nat'l rep.: Katz Agency.

Operation: 1953; TV sets in area: 752,200.

Greenville-Spartanburg-Asheville (NC)

WLOS-TV (Channel 13) ABC

Wometco Skyway Bcstg. Co., Box 2150, Asheville, NC 28802 (704) 254-4321. 333 Wade Hampton Blvd., P.O. Box 2363, Greenville, SC (803) 242-6091.

Mitchell Wolfson, pres.; Louis Wolfson, v.p.; Bill Brazzil, v.p. sales, Wometco bdcst. div.; Morton S. Chon, pres. & gen. mgr.; Dave DeArmond, v.p. sales; Bill Pfeiffer, v.p., prog. oper. mgr.; Loyd B. Leonard, rtl. sls. mgr.; Maurice C. Barkley, reg. sls. mgr.; Kay Jenkins, dir. of sales dev.; Tony Smith, promo. dir.; George Sprague, chief eng.; Larry Stem, bus. & credit mgr.; rep.: Peters, Griffin, Woodward, Inc.

Operation: 1954; TV households in area: 1,144,100.

Spartanburg

WSPA-TV (Channel 7) CBS

Spartan Radiocasting Co., Box 1717, Spartanburg, SC 29304 (803) 576-7777.

Walter J. Brown, pres.; K. James Yager, exec. v.p.; Joe Varholy, gen. mgr.; Walter Vieth, secty.-treas.; David Handy, prgm./stn. mgr.; Gil Anderson, dir. of eng'.; Richard Compton, chief eng.; Tom Ahner, trans. engineer; Jan Haswell, promo. mgr.; Ray Johnson, gen. sales mgr.; Bruce Miller, local/reg. sales mgr.; Al Kamhi, news dir.; nat'l rep.: Seltel.

Operation: 1956; TV sets in area: 2,300,000.

SOUTH DAKOTA

Aberdeen

KABY-TV (Channel 9) NBC

Forum Communications Co., North Highway 281, Box 1520, Aberdeen, SD 57401 (605) 225-9200.

Scott Park, v.p.; Robert S. Kerr, gen. mgr.; Bill Belcher, news dir.; Jim Dunham, natl. sls. mgr.; Earl Miller, chief eng.

Operation: 1958.

Florence

KDLO-TV (Channel 3) CBS

(Satellite of KELO-TV, Sioux Falls, SD)

Midcontinent Broadcasting Co., c/ KELO-TV, Phillips Ave. & 13th St., Sioux Falls, SD 57102 (605) 336-1100.

Joe Floyd, pres.; N.L. Benston, v.p.; E.R. Ruben, secty.-treas.; Evans Nord, v.p. & gen. mgr.; Joe H. Floyd. v.p.; nat'l rep.: H-R Television, Inc., Wayne Evans & Assoc. (Minneapolis).

Operation: 1955; TV sets in area: 375,500.

Lead

KHSD-TV (Channel 11) NBC

Duhamel Broadcasting Enterprises, Box 1760, Rapid City, S.D. 57709.

William F. Duhamel, pres. & gen. mgr.; Ralph Davison, bdcst. mgr.: Don Larsen, dir. of eng.; rep.; Katz.

Operation: November, 1966.

KIVV-TV (Channel 5) ABC

Dakota Broadcasting Co., Inc., Box 677, Rapid City, SD 57709 (605) 394-7777.

Ray Moser, gen. mgr.; Peter Sieler, sta. mgr.; Harvey Sachau, mgr. of eng.; Gerry Fenske, local sls. mgr.; Les Hofland, prod. mgr.; Robert Slocum, contr.; Bill Avery, news dir.; rep.: PGW & Bob Hix, Denver.

Operation: July, 1976.

Rapid City

KEVN-TV (Channel 7) ABC

Dakota Broadcasting Co., Inc., Box 677, Rapid City, SD 57709 (605) 394-7777.

Ray Moser, gen. mgr.; Peter Sieler, sta. mgr.; Harvey Sachau, mgr. of eng.; Gerry Fenske, lcl. sls. mgr.; Les Hofland, prod. mgr.; Robert Slocum, contr.; Bill Avery, news dir.; rep.: PGW & Bob Hix, Denver.

Operation: July, 1976.

KOTA-TV (Channel 3) NBC

KDUH Channel 4, KHSD Channel 11, KSGW Channel 12 (satellites).

Duhamel Broadcasting Enterprises, Duhamel Bldg., 6th & St. Joe Streets, P.O. Box 1760, Rapid City, SD 57709 (605) 342-2000.

Mrs. Helen S. Duhamel, chm.; William F. Duhamel, pres. & gen. mgr.; Ralph Davison, bdcst. mgr.; Dan Larsen, dir. of engineering; Verne Sheppard, farm dir.; nat'l rep: Katz Agency; Omaha rep.. The Soderlund Company.

Operation: 1955; TV sets in area: 80,200.

Reliance

KPLO-TV (Channel 6) CBS

(Satellite of KELO-TV, Sioux Falls, SD)

Midcontinent Broadcasting Co., c/o KELO-TV., Phillips Ave. & 13th St., Sioux Falls, SD 57102 (605) 336-1100.

Joe Floyd, pres.; N.L. Bentson, v.p.; E.R. Ruben, secty.-treas.; Evans A. Nord v.p., gen. mgr.; Joe H. Floyd, v.p.; nat'l rep.: H-R Television, Wayne Evans & Assoc. (Minneapolis).

Operation: 1957; TV sets in area, 375,500.

Sioux Falls

KELO-TV (Channel 11) CBS

Midcontinent Broadcasting Co., Phillips at 13th, Sioux Falls, SD 57102 (605) 336-1100.

Joe Floyd, pres.; N.L. Bentson, v.p.; E.R. Ruben, secty.-treas.; Evans A. Nord vice-pres., gen. mgr.; Joe H. Floyd, v.p.; Tom Sheeley, news, prgm. v.p.; Jack Townsend, sls. devel. dir.; Scott Burri, prod. mgr.; Vince Davi, sls. mrchd'. mgr.; nat'l rep.: H-R Television, Inc., Wayne Evans & Assoc. (Minneapolis).

Operation: 1953; TV sets in area: 375,500.

KSFY-TV (Channel 13)

South Dakota Broadcasting Co., 6th & Dakota, Sioux Falls, SD 57102.

Jim Dunham, v.p. & gen. mgr.; Warren Murray, dir. of eng.; Gar Kellogg, prod. dir.; John Dunnicliff, prom. mgr.; Marge Hokenstad, prog. dir.; Jim Paulson, news dir.; sales rep.: Katz.

Operation: July, 1960; TV households: 309,300.

KXON-TV (Channel 5) ABC

Gillete Broadcasting of South Dakota, Inc., Box 1200 Mitchell, SD 57301; (605) 996-7501.

Darald McElroy, opns. mgr.; Joseph R. Shields, gen. mgr., mkt.; Arnold Monday, chief eng.

Operation: 1960; TV households: 158,800.

TENNESSEE

Chattanooga

WDEF-TV (Channel 12) CBS

Roy H. Park, Broadcasting of Tennessee, Inc., Broadcast Center, 3300 Broad St., Chattanooga, TN 37402 (615) 267-3392.

Roy H. Park, pres.; Louis J. Rocke, gen. mgr.; Joe Killebrew, sls. mgr.; nat'l rep.: Blair Television.

Operation: 1954; TV sets in area: 300,000.

WRCB-TV (Channel 3) NBC

WRCB-TV, Ziff-Davis Broadcasting of Tenn., Inc. 900 Whitehall Road, Chattanooga, TN 37402 (615) 267-5412.

William G. Egan, v.p. & gen. mgr.; Gary Rockey, prog. dir.; Dennis Siewert, sls. mgr.; David Muscari, promo, mgr.

Operation: 1956; TV sets in area: 571,800.

WRIP-TV (Channel 61)

WRIP, Inc., 270 N. Crest Rd., Chattanooga, TN 37404 (404) 866-3855.

Col. Jay Sadow, pres. & gen. mgr.; Jim Hester, optns. mgr.; Joe Lowe, dir. of eng.; Jim Hester, prog. dir.

Operation: 1972.

WTVC (Channel 9) ABC

WTVC, Inc., Box 1150, Chattanooga, TN 37401 (615) 266-4151.

Jane Dowden, pres. & gen. mgr.; Lee Johnson, prog. mgr.; Roger Hale, chief eng.; nat'l rep.: HR TV.

Operation: 1958; TV homes in area: 466,100.

Crossville

WCPT-TV (Channel 55)

WCPT-TV, Inc., Box 608, Crossville, TN 38555; (615) 484-5169.

Carl Smith, pres.; John A. Cunningham, v.p./gen. mgr. James P. Young, sls. mgr.; Rita L. Young, prod. mgr.; Helen L. Cunningham, office mgr.; Wilda Golob, news & pub. affairs dir.; Robin Cunningham, film dir.; John M. Cunningham, chief eng.

Operation: October, 1976.

Jackson

WBBJ-TV (Channel 7) ABC

Jackson Telecasters Inc.; Box 2387, Muse Street, Jackson, TN 38301 (901) 424-4515.

Gerry Quick, gen. mgr. & natl./reg. sls. mgr.; Edward Ward, chief eng.; Barbara Hinson, traffic mgr.; Yvonne Smith, lcl. sls. mgr.; nat'l rep.: H/R.

Operation: 1965; TV homes in area: 348,000.

Johnson City

WJHL-TV (Channel 11) CBS-ABC

Roy H. Park Broadcasting of the Tri-Cities, Inc., 338 East Main St., Johnson City, TN 37602 (615) 926-2151.

Roy H. Park, pres.; W.H. Lancaster, Jr., v.p. & gen. mgr.; John B. Babcock, v.p. op.; K.B. Skinner, treas.; Mrs. Dorothy Park, secy.; Carl E. Dickenson, chief eng.; nat'l rep.: Blair-TV.

Operation: 1953; TV sets in area: 556,100.

Kingsport-Johnson City-Bristol

WKPT-TV (Channel 19) ABC

Holston Valley Broadcasting Corp., 222 Commerce, Kingsport, TN 37660.

George DeVault, Jr., exec. v.p. & gen. mgr.; Ray Walker, v.p. & gen. sls. mgr.; Harold T. Dougherty, v.p. & chief eng.; rep.: Avery-Knodel.

Operation: 1969; TV homes in ADI: 233,000.

Knoxville

WATE-TV (Channel 6) NBC

Nationwide Communications Inc., 1306 Broadway, N.E. Knoxville, TN 37917 (615) 637-9666.

C.W. Fullerton, pres.; Clark Pollock, v.p. & gen. mgr.; Jack W. Lee, gen. mgr.; John Reese, prog. dir.; Bob King, prod. dir.; Don E. Watkins, chief eng.; Jim Mikels, gen. sls. mgr.; nat'l rep.: Blair.

Operation: 1953; TV sets in area: 454,300.

WBIR-TV (Channel 10) CBS

Multimedia Broadcasting Co., 1513 Hutchinson, Knoxville, TN 37901 (615) 637-1010.

Wilson C. Wearn, pres.; Neal Branch, v.p. & gen. mgr.; Peter Fennelly, prog. dir. & film buyer; Robert E. Horton, chief eng.; Lynn Leopper, sls. mgr.; nat'l rep.: The Katz Agency.

Operation: 1956; TV sets in area: 384,800.

WTVK (Channel 26) NBC

South Central Broadcasting Corp., Box 1388, Knoxville, TN 37901; Memorial Park Road on Sharps Ridge, Knoxville, TN 37901 (615) 687-2312.

John D. Engelbrecht, pres.; George F. Stoltz, v.p.; Leighman D. Groves, secty.-treas.; Duane V. Eastvold, gen. mgr.; Jerry Whitener, sls. mgr.; Carl Lawson, chief eng.; nat'l rep., H.R.

Operations: 1953; TV sets in area: 302,700.

Memphis

WHBQ-TV (Channel 13) ABC

RKO General Inc. (NY), 485 So. Highland Ave., P.O. Box 11407, Memphis, TN 38111 (901) 323-7661.

Alex Bonner, gen. mgr.; Jack Bolton, sls. mgr.; Jack Bolton, lcl. sls. mgr.; Robert Lewis, prog. dir.; Larry Doughlan, chief eng.; nat'l rep.: RKO Television Representatives, Inc.

Operation: 1953; TV households in area: 803,200.

WMC-TV (Channel 5) NBC

Scripps-Howard Broadcasting Co., 1960 Union Ave., Memphis, TN 38104 (901) 274-8515.

M.E. Greiner, Jr., v.p. & gen. mgr.; Edw. F. Greaney, Jr., asst gen. mgr. news & public affairs; J.P. Tucker, asst. gen. mgr. sls.; Ron Klayman, dir. bdcst. svcs.; James Zarchin, news dir.; Robert A. Frase, dir. eng.; Jay Perkins, prod. mgr.; Olivia Holmes, promo. mgr.; nat'l rep.: Blair-TV, Inc.

Operation: 1948; TV sets in area: 649,280.

WREG-TV (Channel 3) CBS

New York Times Broadcasting Service, Inc., 803 Channel 3 Drive, Memphis, TN 38103 (901) 525-1313.

Charles B. Brakefield, pres. & gen. mgr.; Jack Michael, v.p. & prog. dir.; Frank Roberts, natl. sales mgr.; Olin Morris, dir. pub. affairs; nat'l rep.: The Katz Agency, Inc.

Operation: 1956; TV sets in area: 515,000.

Nashville

WTVF-TV (Channel 5) CBS

Channel Five Television Co., 474 James Robertson Pkwy., Nashville, TN 37219 (615) 244-5000.

Jock Harris, pres.; Harold Crump, v.p. & gen. mgr.; Thomas Erwin, v.p. & sta. mgr.; Chris Clark, v.p. & news div.; Bill Bumpers, v.p., opers. nat'l rep.: The Katz Agency, Inc.

Operation: 1954; TV sets in area: 813,000.

WNGE (Channel 2) ABC

General Electric Broadcasting Company, Inc., 441 Murfreesboro Road, Nashville, TN 37210 (615) 259-2200.

Edward F. Shadburne, v.p. & gen. mgr.; James C. Anthony, sta. mgr.; Paul Siracuse, gen. sls. mgr.; Buddy Ragan, prog. & promo. mgr.; Robert Hafner, oper. mgr.; Gene Parker, mgr. eng.; nat'l rep.: Blair Television.

Operation: 1953; TV sets in area: 634,500.

WSM-TV (Channel 4) NBC

WSM, Inc., P.O. Box 100, Nashville, TN 37202 (615) 749-2244.

Irving Waugh, pres.; Tom Griscom, v.p. & gen. mgr.; Tommy King, comm. mgr.; Elmer Alley, prog. dir. and film buyer; Lee Whitehurst, chief eng.; nat'l rep.: Edward Petry & Co., Inc.

Operation: 1950; TV homes in area: 546,800.

WZTV (Channel 17)

Reel Broadcasting Co., Inc., Multimedia, Inc., P.O. Box 90764, Nashville, TN 37209 (615) 329-1717.

Robert Zelle, gen. mgr.; Dan F. Maslau, sta. mgr.; Thom Neeson, gen. sls. mgr.; Judy Jenkins, dir. prog. ops.; Ray McInturff, chief eng.

Operation: August, 1968.

TEXAS

Abilene

KRBC-TV (Channel 9) NBC

Abilene Radio & Television Co., 4510 S. 14th St., Abilene, TX (915) 692-4242; TWX: 910-897-5414.

Dale Ackers, pres. & gen. mgr.; Wm. Terry, sta. mgr.; Bryan Mundy, prog. mgr.; Gene Smith, chief eng.; nat'l rep.: Blair Television.

Operation: 1953; TV sets in area: 138,000.

Abilene-Sweetwater

KTXS-TV (Channel 12) ABC, CBS

Texas Key Broadcasters, Inc., U.S. Hwy. 83 Bypass No., Abilene, TX 79604 (915) 677-2281.

Robert P. Jackson, gen. mgr., sales mgr.; J.T. Allen, chief engineer; nat'l rep.: Meeker.

Amarillo

KAMR-TV (Channel 4) NBC

Cannan Communications, Inc., P.O. Box 1224, Wichita Falls, TX 76307 (806) 376-5804.

Darrold A. Cannan, Jr., pres.; Ray Poindexter, v.p. & gen. mgr.; Irl Brown, sls. mgr.; Calvin Handy, bus. mgr.; Ken High, chief eng.; Bruce Mitchell, op. mgr.; Steve Snyder, news dir.; Ruth Goodwin, film dir.; Carolyn York, prom. dir.

Operation: 1953

KFDA-TV (Channel 10) CBS

Bass Broadcasting Company, P.O. Box 1400, Amarillo, TX 79105; (806) 383-2226.

Melvin L. Wheeler, pres.; Earle H. Rast, Jr., v.p. & gen. mgr.; Jack Byrd, oper. mgr.; Ron Slover, news director; Roy McCoy, weather director; Dave Coons, ch. eng.; Larry S. Faler, sales mgr.; Bedford Forrest, farm dir.

Operation: 1953; TV sets in area 222,200.

KVII (Channel 7) ABC

Marsh Media, Ltd., Vaughn Bldg., Box 13000, Amarillo, TX 79101 (806) 373-1787.

James R. McCormick, gen. mgr.; John Patrick, gen. sales mgr.

Operation: 1957.

Austin

KTBC-TV (Channel 7) CBS

The Texas Broadcasting Corp. Box 2223, Austin, TX 78767 (512) 472-2424.

R.K. Evans, pres. & gen. mgr. Gus Stewart, sls. mgr.; Al Mustin, natl/reg. sls. mgr.; Wally Pryor, mgr.; Arthur Riley, chg. eng.

KTVV (Channel 36) NBC

Kingstip Communications, Inc., P.O. Box 490, Austin TX 78767 (512) 476-3636.

Al Tanksley, pres. & gen. mgr.; Larry Cole, gen. sls. mgr.; Bill Hamrick, chief eng.

Operation: 1965.

KVUE (Channel 24) ABC

KVUE-TV, Inc. Box 9927, Austin TX 78766 (512) 459-6521.

Joe Jerkins, v.p. & gen. mgr.; Jim Norton, gen. sls. mgr.; Al Helton, prog. mgr.; Michael J. Wenglar, chief eng.

Operation: 1971.

Beaumont

KBMT-TV (Channel 12) ABC

Texas Telecasting, Inc., Box 1550, Beaumont, TX 77706 (713) 883-7512.

Michael D. McKinnon, pres. & gen. mgr.; Wm. E. Moore, v.p., sta. mgr.; George Gould, v.p., gen. sales mgr., Earl Conner, chief eng.; nat'l rep.: Blair Television.

Operation: 1961.

KFDM-TV (Channel 6) CBS

Belo Broadcasting Corp., P.O. Box 7128, 2955 Interstate 10 East, Beaumont, TX 77706 (713) 892-6622.

Mike Shapiro, pres.; Terry Ford, v.p. & sta. mgr.; Aubrey Jenkins, treas.; Harold Bartlett, chief eng.; nat'l rep., Petry & Co.

Operation: 1955; TV sets in area: 241,000.

Big Spring

KWAB-TV (Channel 4) CBS, ABC

Grayson Entrprises, Inc., 2500 Kentucky Way, Big Spring, TX 79720 (915) 267-6325.

Pat Hales, gen. mgr.; Joe Hull, natl. sls. rep.: Tony Radican, prod. mgr.

Operation: 1956. TV sets in area: 82,200.

Bryan

KBTX-TV (Channel 3) CBS, ABC

Brazos Broadcasting Co., E. 29th St., P.O. Drawer 3730, Bryan, TX 77801 (713) 846-7777.

M.N. Bostick, pres.; Harry L. Gillam, v.p. & sta. mgr.; John M. Lawrence, III, secty.-treas.; nat'l rep.: H-R television.

Operation: May, 1957; TV sets in area: 116,900.

Corpus Christi

KIII (Channel 3) ABC

South Texas Telecasting Co., Inc., 4750 S. Padre Island Dr., Corpus Christi, TX 78404.

Michael D. McKinnon, pres. & gen. mgr.; W.G. Heneger, auditor; Bob White, station & sls. mgr.; Billy Brotherton, sls. mgr.; rep.: Blair Television.

Operation: 1964; TV households: 318,900.

KORO (Channel 28)

Telecorpus, Inc., 102 N. Mesquite, Corpus Christi, TX 78401 (512) 883-2823.

Servando Caballero, gen. mgr.; Felipe Franco, chf. eng.

Operation: 1977.

KRIS-TV (Channel 6) NBC

Gulf Coast Broadcasting Co., Box 840, Corpus Christi, TX 78404 (512) 883-6511.

T. Frank Smith, pres. & gen. mgr.; Van Johnson, 1st v.p. & natl sales.; Jack Farmer, sls. mgr.; Marvin Born, dir. of eng.; nat'l rep.: Peters, Griffin, Woodward, Inc.

Operation: 1956; TV sets in area: 116,780.

KZTV (Channel 10) CBS

K-Six Television, Inc., Wilson Bldg., Corpus Christi, TX; 78404.

Vann M. Kennedy, pres. & gen. mgr.; Eugene Looper, prog. dir.

Operation: 1956.

Dallas-Fort Worth

KDFW-TV (Channel 4) CBS

400 North Griffin St., Dallas, TX 75202 (214) 742-5711.

John J. McCrory, pres.; I.W. Baker, v.p. & dir. of sls.; Gene Cuny, prog. dir.; William B. Honeycutt, dir. of eng'.; national rep.: Harrington, Righter & Parsons.

Operation: 1949; TV sets in area: 1,351,600.

KXAS-TV (Channel 5) NBC

North Texas Broadcasting Corp., 3900 Barnett St., Ft. Worth, TX 76101 (817) 429-1550.

Blake Byrne, pres. & gen. mgr.; Russ Thornton, dir., admin. & devel.; Manny Munoz, dir. of bus. affairs; Patty Parker, gen. sls. mgr.; Sue McDonnell, sta. sls. mgr.; Tom Bedford, eng. dir.

Operation: 1948.

KXTX-TV (Channel 39)

Christian Broadcasting Network, Inc., Box 19307, Dallas TX 75219 (214) 521-3900.

M.G. Robertson, pres.; Bob Johnson, v.p. & gen. mgr.; Roger Baerwolf, sta. mgr.; Harold Wheat, op. mgr.; Dick Bove, gen. sls. mgr.; Art Runyon, chief eng.; Bill Dunn, prod. mgr.; sales rep.: Adam Young, Inc.

Operation: February, 1968; TV households: 2,006,100.

WFAA-TV (Channel 8) ABC

A.H. Belo Corp. (parent organization for The Dallas Morning News, WFAA-KZEW Radio and WFAA-TV), Communications Center, Dallas, TX 75202 (214) 748-9631.

Ward Huey, pres. & CEO, David T. Lane, v.p. & gen. mgr.; Phillip Keller, sta. mgr.; Chris Westerkamp, gen. sls. mgr.; Bob Turner, prod. & oper. mgr.; John Miller, news dir.; Charles Martin, chief eng.; Lee Minard, dir. creative svcs.; Aubrey Jenkins, controller; Petry Television, Inc., nat'l rep.

Operation: 1949; TV sets in area: 1,345,300.

El Paso

KDBC-TV (Channel 4) CBS

Portal Communications, Inc., 2201 Wyoming Ave., El Paso, Texas 79903 (915) 532-6551.

David Scribner, pres.; Ed Sleighel, v.p. & gen. mgr.; William McCormack, bus. mgr.; Charles Maly, prog. dir.; Frank Jordan, chief eng.; Sam Kobren, gen. sls. mgr.; nat'l rep.: H-R.

Operation: 1953; TV sets in area: 180,800.

KTSM-TV (Channel 9) NBC

Tri-State Broadcasting Co., Inc., 801 N. Oregon St., El Paso, TX 79903 (915) 532-5432.

Karl O. Wyler, pres. & gen. mgr.; Jack Rye, exec. v.p. & progs.; John Phelan, v.p., sls.; W.T. McGill, v.p., eng'.; nat'l rep.: Avery-Knodel.

Operation: 1953; TV households in area: 172,500.

KVIA-TV (Channel 13)

KVIA-TV Marsh Media—El Paso, Box 12277, El Paso, TX 79912.

Wayne M. Roy, gen. mgr.; Don Guill, ops-mgr/prog. dir.; Fred Witt, gen. sls. mgr.; George Taylor, chief eng.; natl rep.: Katz Television.

Operation: 1956; TV sets in area: 168,600.

XEJ-TV (Channel 5) (Mexican Border—Juarez)

Television de la Frontera, S.A., Box 9555, El Paso, TX; 79903.

Pedro Meneses Hoyos, pres.; Maria Dolores Heredia, gen. mgr.; Guadalupe Ramirez, sls. mgr.; Ramon Hurtado, prog. mgr.; Hector Noriega, prod. mgr.; Esteban Guerrero, chief eng.; nat'l rep.: Spanish International Network Sales.

Operation: 1954.

XEPM-TV (Channel 2) (Juarez)

XEPM-TV Television Inc., Apdo. No. 271, Juarez, Mexico.

Pedro Meneses, Jr., owner; M.D. Heredia, gen. mgr.; Sergio D. Molinar, Jr., sales mgr.

Operation: 1961.

Fort Worth

KTVT (Channel 11)

Gaylord Broadcasting Co., 4801 W. Freeway, Fort Worth, TX 76107 (817) 738-1951; Branch office: 10111 N. Central Expressway, Dallas, TX 75231.

Jack Berning, v.p. & gen. mgr.; Ocella Allison, asst. gen. mgr., natl. rep.: Telerep.

Harlingen

KGBT-TV (Channel 4) CBS, NBC

Harbenito Broadcasting Company, P.O. Box 711, Harlingen, TX 78551 (512) 423-3910; The Hilton, McAllen, TX (512) 686-8731.

McHenry Tichenor, bd. chmn.; McHenry T. Tichenor, pres.; Mal Kasanoff, exec. v.p.; David Lykes, v.p./station mgr.; Dale Boren, chief studio eng.; nat'l. rep.: Katz.

Operation: 1953. TV sets in area: 124,310.

Houston

KRIV-TV (Channel 26)

Metromedia Television, P.O. Box 22810, 3935 Westheimer, Houston, TX 77027 (713) 626-2610.

M. Jerry Marcus, v.p. & gen. mgr.; Mike Polatschek, gen. sls. mgr.

Operation: August, 1971; TV households: 1,115,900.

KHOU-TV (Channel 11) CBS

Gulf Television Corp., P.O. Box 11, Houston, TX 77001 (713) 526-1111.

Thomas F. Kenney, v.p. & gen. mgr.; Frank Gregg, gen. sls. mgr.; Richard Ramirez, natl. sls. mgr.; Ted Kohn, prog. dir.; Charles Boyer, prod. mgr.; Tom Spaulding, adv. & promo. mgr.; Ian Pearson, news dir.; natl. rep.: Peters, Griffin and Woodward.

Operation: March, 1953; TV households: 1,151,600.

KHTV (Channel 39)

KHTV, 7700 Westpark Drive, P.O. Box 1439, Houston, TX 77001 (713) 781-3930.

WKY Television System, Inc., E.L. Gaylord, pres.; Gene R. Jacobsen, gen. mgr.

Began operating 1/6/67.

KPRC-TV (Channel 2) NBC

Channel Two & VTV Co., 8181 Southwest Freeway, Houston, TX 77027 (713) 771-4631.

Mrs. Oveta Culp Hobby, bd. chmn.; Jack Harris, pres. & gen. mgr.; Jack McGrew, exec. v.p.; Paul Hundorff, oper. mgr.; Tom Reiff, prgm. dir.; Larry Weidman, news dir.; Jeffrey H. Lee, gen. sls. mgr.; nat'l rep., Petry Television, Inc.

Operation: 1949; TV sets in area: 1,035,800.

KTRK-TV (Channel 13) ABC

Capital Cities Communications, Inc.; mailing address: P.O. Box 12, Houston, TX 77001; offices and studios: 3310 Bissonnet, Houston, TX (713) 666-0713; TWX: 910-881-1511; Transmitter: Alameda, TX.

Thomas S. Murphy, pres. & chmn.; Kenneth Johnson, v.p. & gen. mgr.; Jim Mascusi, oper. dir.; Larry Autry, prod. dir.; Nancy K. Duke, bus. mgr.; Jeanne White, pub. svc. dir.; Charles Wolf, prom. dir.; Walt Hawver, news oper. mgr.; Jess Mitchell, chief studio eng.; Betty Milligan, pub. svc. dir.; Paul Bures, sales mgr.; Robert Sutton, chief transmitter eng.; spot sales rep., Blair TV.

Operation: 1954; TV homes in area: 944,000.

Laredo

KGNS-TV (Channel 8) NBC ABC

Gulf Coast Broadcasting Co., Box 2829, Laredo, TX 78041 (512) 727-8888.

Bill McDonald, v.p. & gen. mgr.; Velia Herrera, asst. to gen. mgr.

Operation: 1956; TV sets in area: 23,000

KVTV (Channel 13) CBS

K-SIX TV Inc., Wilson Bldg., 601 Leopard St., Corpus Christi, TX 78401 (512) 723-2923.

Vann M. Kennedy, pres. & gen. mgr.; Richard M. King, sta. mgr.; Nicholas Sanchez, news dir.; Matt Sanderford, chief eng.; Frank Montgomery, resident eng.; sls. rep.: H-R TV.

Operation: December, 1973.

Lubbock

KCBD-TV (Channel 11) NBC

State Telecasting Co., Inc., Box 2190, Lubbock, TX 79408 (806) 744-1414.

W. Robert McKinsey, gen. mgr.; Hank Taylor, prgm. dir.; Frank Lee, dir. of eng'.; Carl Minor, sales mgr.; nat'l rep.: Blair Television.

Operation: 1953: TV sets in area: 224,000.

KLBK-TV (Channel 13) CBS, ABC

Grayson Enterprises, Inc., 7400 University Ave., Lubbock, TX 79408 (806) 744-2345.

W.F. de Tournillon, v.p. & gen. mgr.; Ford Robertson, lcl. sls. mgr.; B.J. Latta, chief eng.; Pat Halos, prog. dir.; nat'l rep.: Meeker Co.

Operation: 1952; TV sets in area: 188,000.

KMCC (Channel 28) ABC

McAlister TV Enterprises Inc., Box 2805, Lubbock, TX 79408 (806) 745-2828.

Bill McAlister, gen. mgr.; Don Turner, sls. mgr.; Bill Baker, chief eng.; A.C. Wimberly, prog. dir.

Operation: 1968; TV homes in area: 122,600.

Lufkin

KTRE-TV (Channel 9) NBC, ABC, CBS

Buford Television Inc. of Lufkin, Box 729, Lufkin, TX 75901 (713) 634-7771.

Lanny Pogue, gen. mgr./mrktg.) Bert Walling, gen. mgr./resources; nat'l rep.: Katz.

Operation: 1955.

Midland

KMID-TV (Channel 2) NBC

Midessa Television Co., Inc. KMID-TV, Midland, TX 79701 (915) 563-2222.

J. Conrad Dunagan, pres.; W.D. Noel, v.p.; T.E. Rodman, secty.-treas.; Ray Herndon, gen. mgr.; John Foster, sta./sls. mgr.; nat'l rep., Avery-Knodel.

Operation: 1953; TV sets in area: 166,700.

Monahans

KMOM-TV (Channel 9) ABC

Grayson Enterprises, Inc., Drawer N. Monahans, TX 79756.

Pat Hales, gen. mgr.; Joe Hull, rep. & natl sls. mgr.; Joe Connally, chief eng.; Tony Radican, prod. mgr.

Operation: 1958; TV homes in area: 143,100.

Odessa

KOSA-TV (Channel 7) CBS

Forward Communications of Texas, Inc., 1211 N. Whitaker, Odessa, TX 79760 (915) 337-8301.

Doyce Elliott, pres. & gen. mgr.; T. Hughes, v.p. & sta. mgr.; Vic Luddington, natl. sls. mgr.; Ed Costello, local sls. mgr.; Bob Stephens, chief eng.; George Duesler, film dir.; nat'l rep.: Seltel.

Operation: 1956; TV homes in area: 119,500.

Port Arthur

KJAC-TV (Channel 4) NBC

Southeast Texas Television Corp., P.O. Box 3257, 1748 Woodworth Blvd., Port Arthur, TX 77640 (915) 985-5557.

Geo Diab, pres.; Roy D. Shotts, v.p.-gen. mgr.; Jack Dahmer, comm. mgr.; Arthur Skipper, chief eng.; nat'l rep.: Katz.

Operation: 1957; TV sets in area: 167,000.

San Angelo

KACB-TV (Channel 3) NBC

(Satellite of KRBC-TV, Abilene)

Abilene Radio & TV Co., 4510 14th St., P.O. Box 178, Abilene, TX 79604 (915) 692-4242.

Dale Ackers, pres.; William Terry, sta. & sls. mgr.; Bryan Mundy, prog. dir.; Gene Smith, chief eng.; rep.: Blair TV.

Operation: 1962.

KCTV (Channel 8) CBS

Jewell Television, P.O. Box 1941, 1011 E., 28th St., San Angelo, TX (915) 655-7383.

Gil Schneider, gen. mgr.; John McGee, gen. sls. mgr.; Gordon Hay, opns. mgr.; Donna Russell, natl. sls.; Eddy Smith, chief eng.

Operation: 1953; TV sets in area: 69,900.

San Antonio

KENS-TV (Channel 5) CBS

Harte-Hanks Television, Inc., Ave. E. & 4th St., P.O. Box TV 5, 78299 (512) 225-5211.

Wayne Kearl, pres.; William Moll, chmn. & gen. mgr.; Elliott Kleeman, v.p. & gen. sls. mgr.; Linda Riss, prog. dir.; Ray McMillan, chief eng.; nat'l Rep.; Katz Agency.

Operation: 1950; TV sets in area: 702,403.

KMOL-TV (Channel 4) NBC

United TV, Inc., P.O. Box 2641, San Antonio, TX 78205 (512) 226-4251.

Edward V. Cheviot, v.p. & gen. mgr.; Morton A. Barrett, gen. sls. mgr.; Kevin Donohue, lcl. sls. mgr.; Jack Carroll, prog. dir.; Don McMullin, news dir.; David Karbach, dir. of eng.; Bob Garcia, film dir.; rep.: Blair Television.

Operation: 1949; TV homes: 442,300.

KSAT-TV (Channel 12)

The Outlet Co., Box 2478, San Antonio, TX 78298 (512) 226-7611.

Joseph S. Sinclair, bd. chmn.; Bruce G. Sundlun, pres.; David E. Henderson, pres. bdcst. div.; James Schiavone, v.p. & gen. mgr.; Rick Andrycha, prog. mgr.

Operation: 1957; TV homes in area: 815,380.

KWEX-TV (Channel 41)

The Spanish International Communications Corp., 411 E. Durango Blvd., San Antonio, TX 78204; (512) 227-4141.

Emilio Nicolas, v.p. & gen. mgr.; Julius Germano, Jr., v.p. & gen. sls. mgr.; Armando Solis, v.p. & bus. mgr.; James Meek, v.p. & chief eng.; Amparo Ortiz, oper. mgr.; Rick Sowa, network oper.

Temple—Waco

KCEN-TV (Channel 6) NBC

Channel 6, Inc., P.O. Box 188, Temple, TX 76501 (817) 773-6868; Waco office, 815 Professional Bldg., Waco, TX 76710 (817) 756-6605; TWX 910-890-4870.

Frank W. Mayborn, pres.; Dan Lesmeister, v.p. & gen. mgr.; Earl Hill, dir. of eng.; nat'l rep.: Blair TV.

Operation: 1953; TV sets in area: 210,900.

Tyler-Longview

KLTV (Channel 7) NBC, CBS, ABC

Channel 7, Inc., Box 957, Tyler, TX 75710 (214) 592-3871.

Robert P. Buford, Gerld B. & Geoffrey R. Buford, owners; Robert P. Buford, pres.; Frank E. Melton, exec. v.p. & gen. mgr.; Bob Way, chief eng.; Gayle Helms, prog. dir.; nat'l rep.: Katz Agency.

Operation: 1954; TV sets in area: 178,500.

Victoria

KXIX (Channel 19) ABC

Victoria Communications Corporation, P.O. Box 1879, Victoria, TX 77901.

Operation: 1969.

Waco

KWTX-TV (Channel 10) CBS, ABC

KWTX Broadcasting Company, 4520 Bosque Blvd., P.O. Box 7528, Waco, TX 76710 (817) 776-1330.

M.N. Bostick, pres. & gen. mgr.; Thomas Stribbling, v.p.; Robert Levy, secty.-treas.; John M. Brinegar, sls. mgr.; Barbara DeVeny, natl-regl. sls. contact; Jesse Dove, chief eng.; Ralph Webb, prog. dir.; nat'l rep.: H-R Television, Inc.

Operation: 1955; TV sets in area: 204,300.

Weslaco

KRGV-TV (Channel 5) NBC, ABC

Mobile Video Tapes, Inc., 900 E. Expressway Avenue, Weslaco, TX 78596 (512) 968-5555.

Douglas L. Manship, pres.; Ray Alexander, gen. mgr.; Jim Cook, local sales mgr.; Sara Skaggs, gen. sls. mgr.; Dina Manship, prgm. dir.; Rick Diaz, news dir.; nat'l rep.: Blair Television.

Operation: 1954; TV sets in area: 178,100.

Wichita Falls

KAUZ-TV (Channel 6) CBS

Wichita Falls Telecasters, P.O. Box 2130, Wichita Falls, TX 76307.

Ray Clymer & W. Erle Whiter, owners; Bill Noble, bus. mgr.; Max Andrews, prog. dir.; Joe George, mktg. dir.; Leon S. Hoeffner, chief eng.; nat'l. rep.: Blair Television.

KFDX-TV (Channel 3) NBC
Clay Broadcasting Corp. of Texas; Box 4000, Wichita Falls, TX 76308 (817) 692-4530.
Lyell Clay, chmn.; George Diab, pres.; John McGee, secty-treas.; Warren Silver, gen. mgr.; Tom Witherspoon, oper. mgr.; Chuck Howard, gen. sales mgr.; William Enloe, dir. of eng.; Peggy Edgar, film ed.; nat'l rep.: Katz.
Operation: 1953; TV sets in area: 250,000.

UTAH

Salt Lake City

KSL-TV (Channel 5) CBS
KSL, Inc., 145 Social Hall Ave., Salt Lake City, UT 84111; 237-2500.
Jay W. Lloyd, pres. & gen. mgr.; William R. Murdock, v.p., sls. & mktg.; Scott R. Clawson, v.p., programming; L.K. Abbott, v.p., ops.; Howard G. Smith, chief eng.; nat'l rep.: Petry
Operation: 1949; TV homes in area: 771,200.

KTVX (Channel 4) ABC
United TV, Inc., 6975 York Ave. S., Minneapolis, MN 55435.
Donald Swartz, pres.; Harold (Hack) Woolley, v.p., gen'l mgr.; Thomas Muir, bus. mgr.; L. Boyd Mullins, sls. mgr.; Dan Rainger, prog. dir.; Les Wells, promo. mgr.; Ron Jacobson, tech. dir.; nat'l rep.: Blair Television.
Operation: 1948; TV sets in area: 427,800.

KUTV (Channel 2) NBC
KUTV Inc., 2185 S. 3600 West, Salt Lake City, UT 84119 (801) 973-3333.
George C. Hatch, pres.; Robert H. Temple, v.p. & gen. mgr.; LaMar Smith, prog. dir.; Dick Thiriot, film buyer, nat'l rep.: Katz.
Operation: 1954; TV set in area: 771,200.

VERMONT

Burlington

WCAX-TV (Channel 3) CBS
Mt. Mansfield Television Inc., 100 Market Sq., Burlington, VT 05451 (802) 862-5761.
S.T. Martin, pres., gen. mgr. & prog. dir.; Phil Waterman, v.p. & gen. sls. mgr.; Stuart Hall, v.p. & prod. mgr.; Peter R. Martin, v.p., news-pub. affairs; Charles Liese, chief eng.; A. Sherman Rutter, film mgr.; nat'l rep., Avery-Knodel, Inc.
Operation: 1954; TV sets in area: U.S. 150,000, Canada 270,000.

WEZF-TV (Channel 22) ABC
International TV Corp., Box 22, Burlington, VT 05401 (802) 658-8022.
Donald G. Martin, pres.; Tom Pierce, gen. mgr.; George Rousseau, station mgr.; Charles Kail, sls. mgr.; Sargent Cathrall, chief eng
Operation: 1968.

VIRGINIA

Bristol

WCYB-TV (Channel 5) NBC, ABC
Appalachian Broadcasting Corporation, Cumberland St., Bristol, VA 24201 (703) 669-4161.
Robert H. Smith, Sr., pres. & gen. mgr.; J. Fey Rogers, v.p. & sls. mgr.; Harry M. Danie., treas.; Charles M. Gore, secty.; R.V Robinson, chief eng.; Charles Lineberger, prgm. op. mgr.; Jim Evans, prod. mgr.; Walter Crockett, news dir.; nat'l rep.: The Meeker Co. & James S. Ayers Co.
Operation: 1956; TV sets in area: 368,870.

Charlottesville

WVIR-TV (Channel 29) NBC
Virginia Broadcasting Corp., P O. Box 751, Charlottesville, VA 22902 (804) 977-7082.
Lloyd T. Smith, Jr., pres.; Harold B. Wright, Jr., v.p. & gen. mgr.; Jim Fernald, natl. sls. mgr.
Operation: March, 1973; TV households: 82,000.

Hampton-Norfolk

WVEC-TV (Channel 13) ABC
Peninsula Broadcasting Corp. P.O. Box 400, Hampton, VA 23369 (804) 722-6331; 110 Third St., Norfolk, VA 23510 (804) 625-1313.
Thomas P. Chisman, pres. & gen. mgr.; J. William Beindorf, gen. sls. mgr.; Matthew L. Kreiner, natl. sls. mgr.; Harrol A. Beauer III, local sls. mgr.; L.W. Kliewer, v.p. oper.; George Hillier, chief eng.; James O. Gray, promotion mgr.; David Whitener, prog. mgr. & film buyer; nat'l rep.: PCW
Operation: 1953; TV sets in area: 406,200.

Harrisonburg

WHSV-TV (Channel 3) ABC, NBC
Shenandoah Valley Television Systems, Inc. P.O. Box TV 3, Rawley Pike, Harrisonburg, VA 22801 (703) 433-9191.
Arthur Hamilton, pres. & gen. mgr.; Sammy Bland, sales mgr.; Zane Bowman, production dir.; Ken McNulty, mgr. news; Ira T. Lowe, dir. of engineering.
Operation: 1953; TV households: 293,500.

Lynchburg

WSET-TV (Channel 13) ABC
WSET, Inc., 2320 Langhorne Rd., P.O. Box 11588, Lynchburg, VA 24506 (804) 528-1313.
Rodger E. Divens pres. & gen. mgr.; Howard Satterwhite, business mgr., Jack Tacketl, v.p. & sls. mgr.; Willis Little, chief eng.; nat'l rep.: Katz Agency.
Operation: 1953; TV households in area: 319,300.

Norfolk

WTKR-TV (Channel 3) CBS
WTKR-TV, Inc., 720 Boush St., Norfolk, VA 23510 (804) 446-100.
William A. Gietz, pres.; John Baggio, sls. mgr.; Joe Perkins, dir. of ops.; Dave Tressel, promo. mgr.; Tom Roland, dir. of community affairs; Bob McCall, dir. of news; Doyle Thompson, dir. of eng.
Operation: 1950; TV sets in area: 467,300.

Norfolk-Portsmouth-Newport News

WAVY-TV (Channel 10) NBC
WAVY Television, Inc. (Division of LIN Broadcasting Corporation), 801 W. Portsmouth, VA 23704 (804) 397-3441.

Ed Frech, pres. & gen. mgr.; Harold Soldinger, v.p. & ass. gen. mgr.; Edward J. Hennessy, gen. sls. mgr.; Waverly Foster, chief eng.; nat'l rep.: Blair Television.
Operation: 1957.

Portsmouth

WYAH-TV (Channel 27)
Christian Broadcasting Network, Inc., Box 111, Portsmouth, VA 23704 (804) 393-2505.
M.G. Robertson, pres.; Robert Johnson, v.p.; Thomas Smith, gen. mgr.; Ken Reichenbach, chief eng.; Charles Tipton, operations mgr.; Bill Malendoski, prod. mgr.; James Vaughan, promo. mgr.
Operation: 1961; TV homes in area: 488,200.

Richmond

WTVR (Channel 6) CBS
Roy H. Park Broadcasting of Virginia, Inc., 3301 W. Broad St., P.O. Box 11064, Richmond, VA 23201 (804) 355-8611.
John R. Mahoney, exec. v.p. & gen. mgr.; Bruce Cynar, natl. sls. mgr.; David B. Totty, lcl./rgl. sales mgr.; John V. Shand, prog. mgr. & dir. of community affairs; Vern Stant, chief eng.; Don Dale, news dir.; Blair-TV, Inc.
Operation: 1948; TV sets in area: 665,900.

WWBT (Channel 12) NBC
Jefferson Pilot Broadcasting Company of Virginia, P.O. Box 12, 5710 Midlothian Pike, Richmond, VA 23201 (804) 233-5461.
W.J. Jorgenson, pres.; Robert L. McRaney, Jr., v.p. & mng. dir.; Grady Pridgen, gen. sales mgr.; Dave Frasier, tech. op. mgr.; John Shreves, prgrm./oper. mgr.; nat'l rep.: Petry.
Operation: 1956; TV sets in area: 498,700.

Richmond-Petersburg

WXEX-TV (Channel 8) ABC
Nationwide Communications, Inc. 230 S. Crater Rd., Petersburg, VA 23803 (804) 733-7876. Box 888, Richmond 23201 643-0166.
Ellis Shook, gen. mgr.; Joe Killebrew, gen. sls. mgr.; Eric Zitron, lcl. sls. mgr.; Gene Walsh, prgm. dir.; Guy Freeman, chief eng.; nat'l rep.: Harrington, Righter & Parsons.
Operation: 1955.

Roanoke

WDBJ-TV (Channel 7) CBS
WDBJ Television, Inc. Mailing Address: P.O. Call Service 7, Roanoke, VA 24022; Street Address: 2001 Colonial Ave., S.W., Roanoke, VA 24015 (703) 344-7000.
C.H. McKeever, pres. & gen. mgr.; Kay Hall, prog. mgr.; Tom Blair, gen. sls. mgr.; George Bassett, dir./On-Air ops.; Lorraine Snebold, viewer promo. mgr.; L.A. Dyson, chief eng.; nat'l rep.; Harrington, Righter & Parsons, Inc.
Operation: 1955; TV homes in area: 945,800.

WSLS-TV (Channel 10) NBC
Roy H. Park Broadcasting of Roanoke, Inc., Church Ave. & Third Street, Roanoke, VA 24011 (703) 344-9226.
Robert H. Teter, exec. v.p. & gen. mgr.; Donald L. Jones, mgr. eng.; Richard K. Burton, gen. sls. mgr.; nat'l rep.: Blair TV.
Operation: 1952; TV sets in area: 680,900.

WASHINGTON

Bellingham

KVOS-TV (Channel 12) CBS
KVOS Television Corp., 1151 Ellis St., Bellingham, WA 98225 (206) 734-4104.
Mitchell Wolfson, pres.; Frank Jamk, v.p. & gen. mgr.; John Price, chief eng.; Doug Davis, sls. mgr.; nat'l rep.: Adam Young, (U.S.); Metrospot (Canada).
Operation: 1953; TV sets in area: 1,000,000.

Pasco

KEPR-TV (Channel 19) CBS
Satellite of KIMA-TV, Yakima, WA)
NWG Broadcasting Co., P.O. Box 702, Yakima, WA 98907 (509) 452-9111.
John J. Noel, Jr., pres.; Dale Hazen, gen. mgr.; Dale Hall, sta. mgr.; Mike Rastovich, dir. of eng.; nat'l rep.: HR Television, Inc.; northwest: Art Moore & Associates, Inc.
Operation: 1954; TV sets in area: 334,400.

Richland-Pasco-Kennewick

KNDU (Channel 25) NBC
(Associated with KNDO, Yakima, WA)
Columbia Empire Broadcasting Corp., 1608 S. 24th Ave., Yakima, WA 98901 P.O. Box 2523, Tri-Cities, WA 99302 (509) 783-6151.
Rep.: Katz.
Operation: 1961.

Seattle

KING-TV (Channel 5) NBC
King Broadcasting Co., 320 Aurora Avenue North, Seattle, WA 98109 (206) 223-5000.
Ancil H. Payne, pres.; Payson Hall, v.p., finance & planning; Jim Kime, v.p., op. & dev.; George V. Willoughby, Jr., v.p. & cor. cnl.; John Orose, treas.; Eric S. Bremner, v.p. & gen. mgr.; Robert Guy, prog. dir.) Don LaCombe, prod. dir.; Linda Clark, oper. dir.; Norm Heffron, news dir.; Emory Bundy, pub. affairs dir.; Nancy Gullick, promo. dir.; Sturges Dorrance, sls. mgr.; Jack Shawcroft, chief eng.; nat'l rep.: Blair TV, Inc.
Operation: 1948; TV sets in area: 941,600.

KIRO-TV (Channel 7) CBS
KIRO, Inc., Broadcast House, Third & Broad, Seattle, WA. 98121 (206) 624-7077.
Kenneth L. Hatch, pres.; Al F. Mladenich, sen. v.p.; Glenn C. Wright, v.p. & sta. mgr.; Michael Poth, local sls. mgr.; Charles Morris, eng. dir.; Judy Law, prog. dir.; natl. rep.: Petry Television, Inc.
Operation: February, 1958; TV households: 976,800.

KOMO-TV (Channel 4) ABC
Fisher Broadcasting Inc., 100 Fourth Ave., N., Seattle WA 98109.
W.W. Warren, chmn. & CEO; John Behnke, pres. & COO; Jay W. Giesa, v.p. & gen. mgr.; Ed Lackner, admin. mgr.; Robert H. Plummer, dir. of eng'.; Pop Hagerty, mchd' mgr.; nat'l rep.: The Katz Agency.
Operation: 1953; TV sets in area: 1,038,500.

Spokane

KHQ-TV (Channel 6) NBC
KHQ, Inc., 4202 S. Regal, Spokane, WA 99203 (509) 448-4666.

J. Birney Blair, pres. & gen. mgr.; Robt. L. Busse, contr.; Richard J. Alegria, lcl. sls. mgr.; John V. Reber, nat'l/rgl. sls. mgr.; Larry W. Gants, promo. mgr.; Homer Mason, prog. dir.; Dee Waymire, eng. dir.; nat'l rep.: Katz Agency.

Operation: 1952; TV sets in area: 528,800 ARB.

KREM-TV (Channel 2) CBS

King Broadcasting Co., S-4103 Regal St., Spokane, WA 99203 (509) 448-2000.

Irwin Starr, v.p. & gen. mgr.; Tony Twibell, sls. mgr.; Bill Hall, op. mgr.; Warren P. Pritchard, chief eng.; nat'l rep.: Blair.

Operation: 1954; TV sets in area: 393,200.

KXLY-TV (Channel 4) CBS

KXLY-TV (Spokane Television, Inc.) W. 500 Boone Ave., Spokane, WA 99201 (509) 328-9084.

Mrs. Morgan Murphy, pres.; Wayne F. McNulty, v.p. & gen. mgr.; Ted Bravakis, sls. mgr.; Ron Simmons, oper. mgr.; T.O. Jorgenson, chief eng.; nat'l rep.: P.G.W.

Operation: 1953; TV sets in area: 358,600.

Tacoma

KSTW (Channel 11)

Gaylord Bdg. Co., Box 11411, Tacoma, WA 98411 (206) 572-5789.

Charles Edwards, v.p. & gen. mgr.; Bob Hamacher, gen. sls. mgr.; Paul Crittenden, chief eng.

Operation: 1953; TV households: 1,038,500.

Yakima

KAPP (Channel 35)

Apple Valley Bcstg. Co., 1610 South 24th Ave., Yakima, WA 98902 (509) 453-0351.

Wayne McNulty, v.p.; R. Scott Marriner, sta. mgr.; Paul Bateman, chief eng.; T.O. Jorgenson, eng. supvr.; Bob Ivers, news dir.; rep.: Peters, Griffin, Woodward, Inc.

Operation: September, 1970; TV households: 148,700.

KIMA-TV (Channel 29) CBS

NWG Broadcasting Co., P.O. Box 702, Yakima, WA 98907 (509) 452-9111. Affiliated stations: KEPR-TV, Pasco, WA; KLEW-TV, Lewiston, ID.

John J. Noel, Jr., pres. Dale Hazen, gen. mgr.; Jim Nolan, prog. dir.; Mike Rastovich, dir. of eng.; nat'l rep.: HR Television, Inc.; Pacific Northwest, Art Moore & Associates, Inc.

Operation: 1953; TV sets in area 768,900.

KNDO (Channel 23) NBC

(Operates associated station KNDU, Richland-Pasco-Kenewick, WA)

Columbia Empire Broadcasting Corp., 1608 S. 24th Ave., Box 10028, Yakima, WA 98909 (509) 248-2300.

Hugh E. Davis, pres.; Scott Hayner, gen. mgr.; Ben Shronshire, natl. sls. mgr.; Ronald Krons, eng'. dir.; Opal Laurent, compt., office mgr.; Larry Baker, news dir.

Operation: 1959.

WEST VIRGINIA

Bluefield

WHIS-TV (Channel 6) NBC

Daily Telegraph Printing Co., Broadcast Center, Bluefield, WV 24701 (304) 327-7114.

Ned Shott, pres.; Al Marra, gen. mgr.; Erlene Branch, prog. dir.; Kenneth Dick, chief eng.; nat'l rep.: Avery-Knodel.

Operation: 1955; TV sets in area: 635,700.

Charleston

WCHS-TV (Channel 8) CBS

WCHS AM-TV Corp., 1111 Virginia St. E., Charleston, WV (304) 346-5358.

O. Wayne Rollins, bd. chm.; R. Randall Rollins, pres.; Robert O. Paxson, gen. mgr.; Richard E. Wagner, sta. mgr.; Dennis Adkins, gen. sls. mgr.; Robert Roush, chief eng.; Kent S. Collins, news dir. nat'l rep.: H.R.

Operation: 1954.

Clarksburg

WBOY-TV (Channel 12)

912 West Pike Street, Clarksburg, WV 26301 (304) 623-3311.

B. Imes, pres.; Skip Simms v.p. & gen. mgr.; Bill Whitfield, secty & group mgr. Joe Wright, chief eng.; nat'l rep.: PGW.

Operation: 1957.

Huntington-Charleston

WOWK-TV (Channel 13) ABC

Gateway Communications, Inc., 625 Fourth Ave., Huntington, WV 25701 (304) 525-7661.

Leo M. MacCourtney, v.p. & gen. mgr.; Jack Dempsey, gen. sls. mgr.; James Maher, lcl. sls. mgr.; Paul N. Dicker, prog. dir.; Jacqui Mullen, news dir.; Robert Willis, promo. dir.; Paul A. Bock, chief eng.; sls. rep.: Blair TV.

Operation: October, 1955; TV households: 831,200.

WSAZ-TV (Channel 3) NBC

Lee Enterprises, Inc., 645 Fifth Avenue, Huntington, WV 25701 (304) 697-4780.

Lloyd G. Schermer, pres. & treas.; George R. Andrick, gen. mgr.; Harold B. Hall, sls. mgr.; William Watts, natl. sls. mgr.; Robert Harris, Charleston operations mgr.; Steve Grzyb, prom. mgr.; George M. Curry, prog. dir.; Bob Brunner, news dir.; John P. Clay, chief eng.

Operation: 1949; TV sets in area: 684,700.

Oak Hill

WOAY-TV (Channel 4) ABC, Primary

Thomas Broadcasting Co., Box 251, Oak Hill, WV 25901 (304) 469-3361.

Robert R. Thomas III, pres.; Curtis W. Butler, gen. mgr.; John Thomas, prod. mgr.; Bob Cannon, oper. mgr.; nat'l. rep.: Spot Time.

Operation: 1954; TV sets in area: 377,820.

Parkersburg

WTAP (Channel 15) NBC

Benedek Broadcasting Inc., 121 W. 7th St., Parkersburg, WV 26102 (304) 485-4588.

Sherman A. Grimm, gen. mgr.; Glenn Wilson, prog. dir.; Charles Helmick, chief engineer; nat'l rep.: Seltel.
Operation: 1953; TV households in area: 70,620.

Weston-Clarksburg

WDTV (Channel 5) CBS

Withers Broadcasting Co. of WV P.O. Box 480, Bridgeport, WV 26330 (304) 842-3558.

W. Russell Withers, Jr., pres.; Dusty Wayman, sta. mgr.; Nick Pelligrin, prog. dir. & traffic dir.; rep.: Avery-Knodel.

Operation: 1960; TV households, 798,200.

Wheeling

WTRF-TV (Channel 7) NBC, ABC

96 16th St., Wheeling, WV 26003 (304) 232-7777.

Charles E. Sherman, pres./gen. mgr.; Robert McFarland, chief eng.; Gary Vorhees, oper. mgr.; Frank Curtis, v.p., asst. gen. mgr./sls. mgr.; C. Kirk Jackson, reg. sls. mgr., mdse. dir.; William Ney, film dir.; Pat Gaughan, pub. affairs dir.; Don Murdock, continuity dir.; Ged Young, promo. dir.; George Carroll, prog. dir.; Mark Davis, news dir.

Operation: 1953; TV sets in area: 523,000.

WISCONSIN

Eau Claire

WEAU-TV (Channel 13) NBC

WEAU Inc., 1907 South Hastings Way, Eau Claire, WI 54701.

V.I. Minahan, pres.; D.L. Nelson, v.p.; Clark Wideman, gen. mgr.; Robert Fallis, chief eng.; nat'l rep.: PGW; Hurley Co. (Minneapolis).

Operation: 1953; TV households: 394,080.

Green Bay

WBAY-TV (Channel 2) CBS

A division of Nationwide Communications Inc., 115 So. Jefferson St., Green Bay, WI 54301 (414) 432-3331.

William M. Dunaway, gen. mgr.; Richard Burnton, gen. sls. mgr.; Richard Millhiser, prog. dir.; nat'l rep.: Blair, Inc.

Operation: 1953; TV sets in area: 651,300.

WFRV-TV (Channel 5) NBC

WFRV Inc., Mason & Roosevelt St., Green Bay, WI 54303 (414) 437-5411.

Robert O. Southard, pres. & gen. mgr.; Ralph S. Jackson, v.p.; Al Lucero, gen. sls. mgr.; Doug Clemensen, dir. opns. & prog.; Harry Hill, chief eng.; nat'l rep.: The Katz Agency.

Operation: 1955; TV households in area: 485,040.

WLUK-TV (Channel 11) ABC

WLUK, Inc., P.O. Box 7711, Green Bay, WI 54303 (414) 494-8711; TWX 910 263-1267.

Thomas A. Hutchison, gen. mgr.; Frank Jonas, gen. sls. mgr.; Monty G. Stock, op.-prgm. mgr.; Jim Steffey, chief eng.; nat'l rep.: Peters, Griffin, Woodward.

Operation: 1954.

La Crosse

WKBT (Channel 8) CBS

Gross Telecasting of Wisconsin, Inc., 141 So. 6th St., La Crosse WI 54601 (608) 782-4678.

Peter S. Good, vice president & gen. mgr.; Edith S. Harris, natl. sls. mgr.; Al Leeman, chief eng.; nat'l rep.: Avery-Knodel, Inc.; Harry S. Hyett Co.—Rameland, Inc. (Minneapolis).

Operation: 1954; TV homes: 241,100 (NSI).

WXOW-TV (Channel 19) ABC

(Satellite of WKOW-TV, Madison, WI)

Liberty Television Inc., P.O. Box 100, Madison, WI 53701.

Terry Shockley, pres.; David Peschau, v.p. & gen. mgr.; Bruce Liljegren, v.p. op.; Jack Brennan, v.p., sls.; Arthur Osburn, chief eng.; rep.: H-R Television; Katz.

Operation: March, 1970; TV households: 151,400.

Madison

WISC-TV (Channel 3) CBS

Television Wisconsin, Inc., 4801 West Beltline Hwy., Madison, WI 53711 (608) 271-4321.

Mrs. Elizabeth Murphy, pres.; George A. Nelson, v.p. & secty.; Stephen Herling, gen. mgr.; H. Kenneth Harley, treas.; Richard S. Nickeson, natl. sls. mgr.; John Hrubesky, prog. dir.; Walter Hariu, chief eng.; nat'l rep.: Peters, Griffin, Woodward, Inc.

Operation: 1956; TV sets in area: 635,000.

WKOW-TV (Channel 27) ABC

Liberty Television, Inc., 5727 Tokay Blvd., Madison WI 53701 (608) 274-1234.

Terry Shockley, pres.; nat'l rep.: Katz Agency.

Operation: 1953; TV households in area: 194,800.

WMTV (Channel 15) NBC

Forward Television Inc., 615 Forward Dr., Madison, WI 53711 (608) 274-1515.

Thomas E. Bolger, pres.; Laurie Leonard, v.p. & sta. mgr.; Fred Sole, opns. mgr.

Operation: 1953; TV households in area: 140,000.

Milwaukee

WISI-TV (Channel 12) ABC

WISN Division—The Hearst Corp., 759 No. 19 St., Milwaukee, WI 53233 (414) 342-8812.

Mickey L. Hooten, v.p. & gen. mgr.; Howard Ritchie, v.p. of sls.; Rick Henry, natl. sls. mgr.; Bill Lind, lcl. sls. mgr.; Eric Stearman, prod. mgr.; Stan Marinoff, dir. of bdcst. ops.; Tom Hauff, dir. of news & pub. affairs; Jeff Nettesheim, dir. of creative; Noel Enders, resident cont.

Operation: 1954; TV sets in area: 675,600.

WITI-TV (Channel 6) CBS

Storer Broadcasting Co., 9001 North Green Bay Road, Milwaukee, WI 53209 (414) 355-6666.

Andrew P. Potos, vice-pres. & gen. mgr.; Raymond Engelhart, bus. mgr.; Allen Haushalter, prog. dir.; Robert Truscott, dir. of eng.; Donald Roering, chief eng.; Ted Sidelko, promo. mgr.

Operation: 1956; TV sets in area: 1,380,400.

WTMJ-TV (Channel 4) NBC

WTMJ Inc., 720 East Capitol Drive, Milwaukee, WI 53201 (414) 332-9611.

Michael McCormick, pres.; Richard Herbst, v.p. & gen. mgr.; Jerry Heilman, gen. sls. mgr.; James Wulliman, mgr. of eng.; Gerald McGrath, prgm. mgr.; Ed Hinshaw, ed. dir. & mgr. of pub. affairs; Wayne Godsey,

mgr. news op.; John Denney, exec. news ed.; James Fitzgerald, film supv.; Rod Synnes, comm. rel. dir.
Operation: 1947; TV households in area: 968,200.

WVTV (Channel 18)

Gaylord Broadcasting Co., P.O. Box 1818, Milwaukee, WI 53201 (414) 442-7050.
Joseph Loughlin gen. mgr.; Mike Schuch, asst. gen. mgr.; Al Leitl, gen. sls. mgr.; Jack Stratmeyer, prod. mgr. Lori Wucherer, promo. mgr.; William Witt, dir. of eng.; Bill Franks, prog. dir.; rep.: Telerep.
Operation: 1959; TV homes: 665,200.

Rhinelander (Wausau)

WAEO-TV (Channel 12) NBC

Northland TV Inc., Box 858, Rhinelander, WI 54501 (715) 369-4700.
Dr. Jasper F. Williams, M.D., bd. chm. John W. Swain, pres.; Carl R. Hilstrom, asst. v.p.; & gen. mgr.; Ned Westman, chief eng.; rep.: Spot Time Ltd.
Operation: 1966; TV homes in area: 230,000.

Wausau

WAOW-TV (Channel 9) ABC

(Outlet of The Wisc. TV Network)
Horizons Communication Corp. of Wis., 1908 Grand Ave., Wausau WI 54401 (715) 842-2251.
Terry Shockley, exec. v.p.; Bruce Liljegren, gen. mgr.; Dennis Lerch, op. mgr.; Don Stenz, chief eng.; rep.: Katz Television.
Operation: 1965; TV households: 170,310.

WSAU-TV (Channel 7) CBS

Forward Communications Corp., 1114 Grand Ave., Wausau, WI 54401 (715) 845-4211.
Wm. F. McCormick, chm.; Richard D. Dudley, pres. & gen mgr. of Forward; Gene R. Anderson, v.p. finance; James W. Frey, v.p./sls.; Bart Kellenhauser, pres. & gen. mgr. WSAU-TV; Jack Crowley, promo. & prog.. mgr.; Don Cochart, chief eng.; nat'l rep.: Meeker Co.
Operation: 1954.

WYOMING

Casper

KTWO-TV (Channel 2) NBC, ABC, CBS

Harriscope Broadcasting Corp., 4200 E. Second Street, Casper, WY 82601 (307) 237-3711.
Burt I. Harris, pres.; Jack Rosenthal, exec. v.p.; Bob Price, v.p. and gen. mgr.; Steve Broomell, chief eng.; Byron Dunbar, film dir.; Dave Borino, prog. dir.; Debbie Hales, traffic dir.; nat'l rep.: Katz; regional rep.: Jonn McGuire, Denver.
Operation: 1957; TV homes in area: 109,200.

Cheyenne

KYCU-TV (Channel 5) CBS, ABC, NBC

Wyneco Communications, Inc., 2923 East Lincolnway, Cheyenne, WY; 82601 (307) 634-7755.
Carl J. Occhipinti, gen. mgr.; Margie O'Brien, natl. sls.; Stuart Seibel, prog. dir.; nat'l rep.: Katz.
Operation: 1954; TV sets in area 82,776.

Riverton-Worland-Thermopolis-Lander

KTNW-TV (Channel 10) NBC

Hi-Ho Broadcasting of Wyoming, 242 Railway Plaza, Rivertow, WY 82501 (307) 856-0875.
Preston Oliver, mgr.; Marilyn Stehle, prog. dir.
Operation: 1980; Households: 17,800.

Public Broadcasting Service Station Affiliates

Alabama
Birmingham .... WBIQ
Cheaha State Park .... WCIQ
Demopolis .... WIIQ
Dozier .... WDIQ
Florence .... WFIQ
Huntsville .... WHIQ
Louisville .... WGIQ
Mobile .... WEIQ
Montgomery .... WAIQ

Alaska
Anchorage .... KAKM
Bethel .... KYUK-TV
Fairbanks .... KUAC-TV

American Samoa
Pago Pago .... KVZK

Arizona
Phoenix .... KAET
Tucson .... KUAT

Arkansas
Arkadelphia .... KETG
Fayetteville .... KAFT
Jonesboro .... KTEJ
Little Rock .... KETS

California
Eureka .... KEET
Fresno .... KMTF
Huntington Beach .... KOCE-TV
Los Angeles .... KCET
Los Angeles .... KLCS
Redding .... KIXE
Sacramento .... KVIE
San Bernardino .... KVCR-TV
San Diego .... KPBS-TV
San Francisco .... KQED
San Francisco .... KQEC
San Jose .... KTEH
San Mateo .... KCSM-TV

Colorado
Denver .... KRMA-TV
Pueblo .... KTSC

Connecticut
Bridgeport .... WEDW
New Haven .... WEDY
Norwich .... WEDN

Delaware
Wilmington .... WHYY-TV

District of Columbia
Washington .... WETA-TV

Florida
Gainesville .... WUFT
Jacksonville .... WJCT
Miami .... WPBT
Miami .... WLRN-TV
Miami .... WTHS
Orlando .... WMFE-TV
Pensacola .... WSRE
Tallahassee .... WFSU-TV
Tampa .... WUSF-TV
Tampa/St. Petersburg .... WEDU

Georgia
Athens .... WGTV
Atlanta .... WETV
Augusta .... WCES-TV
Chatsworth .... WCLP-TV
Cochran .... WDCO-TV
Columbus .... WJSP-TV
Dawson .... WACS-TV
Pelham .... WABW-TV
Savannah .... WVAN-TV
Waycross .... WXGA-TV

Guam
Agana .... KGTF

Hawaii
Honolulu .... KHET
Wailuku .... KMEB-TV

Idaho
Boise .... KAID
Moscow .... KUID-TV
Pocatello .... KBGL

Illinois
Carbondale .... WSIU-TV
Chicago .... WTTW
Chicago .... WXXW
Olney .... WUSI-TV
Peoria .... WTVP
Urbana .... WILL-TV

Indiana
Bloomington .... WTIU
Evansville .... WNIN
Indianapolis .... WFYI
Muncie .... WIPB
St. John .... WCAE
South Bend .... WNIT
Vincennes .... WVUT

Iowa
Council Bluffs .... KBIN
Des Moines .... KDIN-TV
Fort Dodge .... KTIN
Iowa City .... KIIN-TV
Mason City .... KYIN
Red Oak .... KHIN
Sioux City .... KSIN
Waterloo .... KRIN

Kansas
Topeka .... KTWU
Wichita .... KPTS

Kentucky
Ashland .... WKAS
Bowling Green .... WKGR
Covington .... WCVN
Elizabethtown .... WKZT
Hazard .... WKHA
Lexington .... WKLE
Louisville .... WKPC-TV
Louisville .... WKMJ
Madisonville .... WKMA
Morehead .... WKMR
Murray .... WKMU
Owenton .... WKON
Pikeville .... WKPI
Somerset .... WKSO

Louisiana
Baton Rouge .... WLPB-TV
Monroe .... KLTM
New Orleans .... WYES-TV

Maine
Augusta .... WCBB
Biddeford .... WMEG
Calais .... WMED-TV
Orono .... WMEB-TV
Presque Isle .... WMEM-TV

Maryland
Annapolis .... WAPB

Baltimore . . . . WMPB
Hagerstown . . . . WWPB-TV
Salisbury . . . . WCPB

Massachusetts
Boston . . . . WGBH-TV
Boston . . . . WGBX-TV
Springfield . . . . WGBY-TV

Michigan
Alpena . . . . WCML-TV
Detroit . . . . WTVS
East Lansing . . . . WKAR-TV
Grand Rapids . . . . WGVC
Marquette . . . . WNMU-TV
Mt. Pleasant . . . . WCMU-TV
University Center . . . . WUCM-TV

Minnesota
Appleton . . . . KWCM-TV
Austin . . . . KAVT-TV
Duluth-Superior, Wis. . . . . WDSE-TV
Minneapolis-St. Paul . . . . KTCA-TV
St. Paul . . . . KTCI-TV

Mississippi
Biloxi . . . . WMAH
Booneville . . . . WMAE
Bude . . . . WMAU
Greenwood . . . . WMAO
Jackson . . . . WMAA
Meridian . . . . WMAW
Oxford . . . . WMAV
State College . . . . WMAB

Missouri
Kansas City . . . . KCPT
St. Louis . . . . KETC
Springfield . . . . KOZK

Nebraska
Alliance . . . . KTNE-TV
Bassett . . . . KMNE-TV
Hastings . . . . KHNE-TV
Lexington . . . . KLNE-TV
Lincoln . . . . KUON-TV
Merriman . . . . KRNE-TV
Norfolk . . . . KXNE-TV
North Platte . . . . KPNE-TV
Omaha . . . . KYNE-TV

Nevada
Las Vegas . . . . KLVX

New Hampshire
Berlin . . . . WEDB-TV
Durham . . . . WENH
Hanover . . . . WHED-TV
Keene . . . . WEKW-TV
Littleton . . . . WLED-TV

New Jersey
Camden . . . . WNJS
Montclair . . . . WNJM
New Brunswick . . . . WNJB
Trenton . . . . WNJT

New Mexico
Albuquerque . . . . KNME-TV
Las Cruces . . . . KRWG
Portales . . . . KENW

New York
Binghamton . . . . WSKG
Buffao . . . . WNED-TV
Garden City . . . . WLIW
New York . . . . WNYC-TV
New York . . . . WNYE-TV
New York . . . . WNET
Norwood . . . . WNPI-TV
Plattsburgh . . . . WCFE-TV
Rochester . . . . WXXI
Schenectady . . . . WMHT
Syracuse . . . . WCNY-TV
Watertown . . . . WNPE

North Carolina
Asheville . . . . WUNF-TV
Chapel Hill . . . . WUNC-TV
Charlotte . . . . WTVI
Columbia . . . . WUND-TV
Concord . . . . WUNG-TV
Greenville . . . . WUNK-TV
Linville . . . . WUNE-TV
Wilmington . . . . WUNJ-TV
Winston-Salem . . . . WUNL-TV

North Dakota
Fargo-Moorhead, Minn. . . . . KFME
Grand Forks . . . . KGFE

Ohio
Akron . . . . WEAO
Aliance . . . . WNEO
Athens . . . . WOUB-TV
Bowling Green . . . . WBGU-TV
Cambridge . . . . WOUC-TV
Cincinnati . . . . WCET
Cleveland . . . . WVIZ-TV
Columbus . . . . WOSU-TV
Dayton . . . . WPTD
Oxford . . . . WPTO
Portsmouth . . . . WPBO-TV
Toledo . . . . WGTE-TV

Oklahoma
Oklahoma City . . . . KETA
Tulsa . . . . KOED-TV

Oregon
Corvallis . . . . KOAC-TV
La Grande . . . . KTVR
Medford . . . . KSYS
Portland . . . . KOAP-TV
Salem . . . . KVDO-TV

Pennsylvania
Bethlehem-Allentown . . . . WLVT-TV
Erie . . . . WQLN
Hershey . . . . WITF-TV
Pittsburgh . . . . WQED
Pittsburgh . . . . WQEX
Scranton-Wilkes-Barre . . . . WVIA-TV
University Park . . . . WPSX-TV

Puerto Rico
Mayaguez . . . . WIPM-TV
San Juan . . . . WIPR-TV

Rhode Island
Providence . . . . WSBE-TV

South Carolina
Allendale . . . . WEBA-TV
Beuafort . . . . WJWJ-TV
Charleston . . . . WITV
Columbia . . . . WRLK-TV
Florence . . . . WJPM-TV
Greenville . . . . WNTV
Sumter . . . . WRJA-TV

South Dakota
Aberdeen . . . . KDSD-TV
Brookings . . . . KESD-TV
Eagle Butte . . . . KPSD-TV
Lowry . . . . KQSD-TV
Pierre . . . . KTSD-TV
Rapid City . . . . KBHE-TV
Vermillion . . . . KUSD-TV

Tennessee
Chattanooga . . . . WTCI
Knoxville . . . . WSJK-TV
Lexington . . . . WLJT-TV
Memphis . . . . WKNO-TV
Nashville . . . . WDCN-TV

Texas
Austin-San Antonio . . . . KLRN-TV
College Station . . . . KAMU-TV
Corpus Christi . . . . KEDT
Dallas . . . . KERA-TV
Houston . . . . KUHT
Killeen . . . . KNCT

Lubbock .... KTXT-TV
Wichita Falls .... KIDZ-TV

Utah
Provo .... KBYU-TV
Salt Lake City .... KUED

Vermont
Burlington .... WETK
Rutland .... WVER
St. Johnsbury .... WVTB
Windsor .... WVTA

Virginia
Harrisonburg .... WVPT
Norfolk .... WHRO-TV
Norton .... WSVN-TV
Richmond .... WCVE-TV
Richmond .... WCVW
Roanoke .... WBRA-TV

Virgin Islands
Charlotte Amalie .... WTJZ-TV

Washington
Pullman .... KWSU-TV
Seattle .... KCTS-TV
Spokane .... KSPS-TV
Tacoma .... KTPS
Tacoma .... KCPQ-TV
Yakima .... KYVE-TV

West Virginia
Beckley .... WSWP-TV
Huntington .... WMUL-TV
Morgantown .... WWVU

Wisconsin
Green Bay .... WPNE
La Crosse .... WHLA
Madison .... WHA-TV
Menomoine-Colfax .... WHWC
Milwaukee .... WMVS
Milwaukee .... WMVT
Wausau .... WHRM-TV

Advertising Agencies

* **STATION REPRESENTATIVES**

ADVERTISING AGENCIES

A.D. ADAMS ADVERTISING, 19 East 38th St., New York NY 10016 (212) 685-9060.
A.D. ADAMS, pres.; MARK S. OSMANSKI TV director.

WILLIAM HART ADLER, INC., One Concourse Plaza, 4711 Golf Road, Skokie, IL 60076.
WILLIAM HART ADLER, pres. and treas.

THE AITKIN-KYNETT CO., INC., 4 Penn Center, Philadelphia, PA 19103 (215) LOcust 8-6700.
FRANK GALLAGHER; CHARLES C. PENNPACKER; JOHN HALDEMAN, time-buyers; MICHAEL A. WALSH, sr. v.p. & media director; ED BATES, TV dir.

ALLIED ADVERTISING AGENCY, INC., 830 Statler Office Building, Boston, MA 02116 (617) 482-4100.
HOWARD SHAMBAN, TV dir.; MARC F. SHAMBAN, TV time buyer.

ANDERSON DAVID & SEEDS, INC., Colonial Building, Clinton, NY 13323; (212) 853-6141.
JAY C. ANDERSON, pres.; MARK ANDERSON, acct. exec.

ARNOLD & COMPANY, INC., Park Square Building, Boston, MA 02116.
GERARD GOLDEN, pres.; ARNOLD Z. ROSOFF, chairman & treasurer; LEN KARSAVOL, creative dir.; DENNIS LEE, copy chief; MARIE KACHINSKI, media dir.; STAVROS COSMOPULOS, exec. v.p., creative services; FRED SIEGEL, media; PHYLLIS MAYNARD, media; ANNE GOTZ, JOAN BENNETT, GINNY MEANEY, media.

N.W. AYER INCORPORATED, 1345 Ave. of the Americas, New York, NY 10105. (212) 974-7400.
MARCELLA ROSEN, sr. v.p. & media dir.; DAVID LEHMKUHL, group media dir., planning; JANE TWYON, group media dir., planning; RICHARD WELSH, group media dir., network & negotiation;, DIANE ZGONC, group media dir., planning; SUSAN BELL, mgr. spot buying.

BBDO CHICAGO, INC., 410 N. Michigan Ave., Chicago, IL 60611 (312) DElaware 7-7860.
ARTHUR E. MEYERHOFF, chmn. of the bd.; ROGER J. MOHR, pres.; MICHAEL PARADISE, radio/TV dir.

BARBER AND DRULLARD, INC., 633 Delaware Ave., Buffalo, NY 14202 (716) 886-3585.
HENRY L. BARBER, chm; E.T. DRULLARD, pres.

TED BATES ADVERTISING/NEW YORK, 1515 Broadway, New York, NY 10036 (212) 869-3131.
ROBERT E. JACOBY, Chm.; ROBERT A. BRUNS, Pres.; WALTER E. REICHEL, Sr. v.p.-Exec. Dir. Media & Programs Dept.; JOEL M. SEGAL, Sr. V.P.-Dir. TV/Radio Div. Division of TED BATES WORLDWIDE, INC. (same address) with affiliates in Fort Lauderdale, Chicago, Minneapolis, San Francisco, Los Angeles, Canada, Mexico, Argentina, Brazil, Chile, Colombia, Ecuador, Peru, Venezuela, United Kingdom, Norway, Sweden, Denmark, West Germany, Austria, Netherlands, Belgium, France, Spain, Italy, Greece, South Africa, Malaysia, Thailand, Singapore, Hong Kong, Japan, Philippines, Australia, New Zealand.

BAUERLEIN, INC., Hibernia Bank Building, New Orleans, LA 70112 (504) 522-5461.

BEAN ADV. AGENCY, 4003 N. New Jersey St., Indianapolis, IN 46205.
HAROLD BEAN, owner.

BEHRENDS & COMPANY, INC., P.O. Box 2556, 101 W. Green Drive High Point, NC 27261.
RICHARD D. BEHRENDS, pres.; OLIVER S. TRONE, III, exec. v.p.; JOHN H. LOCKHART, JR., senior v.p. & marketing dir.; RICHARD FURMANSKI, v.p. & senior art dir.; ROBERT A. BODLE, JR., v.p. & copy/broadcast dir.; ROBERT P. DONOHOE, v.p., sls. promo.; SYLVIA Y. PASSAVANT, secty-treas.; LEE REITZEL, media dir.

HERBERT S. BENJAMIN ASSOCIATES, INC., 2736 Florida St., Baton Rouge, LA 70802 (504) 387-0611.
LEE HERZBERG, JR., chmn. of the bd/pres.; ALLAN R. BRENT, exec. vice-pres./creative dir.; RICHARD K. GOLDBERGER, Senior v.p.; Margaret S. Webb, v.p./secty.-treas.; ARTHUR E. FULTON, v.p.; J. AUBREY SHAMBURGER, v.p./exec. art dir.; MRS. LENORE USHER, v.p./acct. exec.; GUS WALES, v.p., assoc. creative dir.; MRS. ELAINE LEBOUEF, asst. v.p./media dir.; MRS. CORA WILKINSON, asst. treas.; RICHARD J. BARRA, v.p./manager Lafayette, LA branch office.

WALTER F. BENNETT CO., 20 N. Wacker Drive, Chicago, IL 60606 (312) 372-1131.
WALTER F. BENNETT, principal; FRED DIENERT, mgr., partner; JANE GUNTHER, radio and TV time buyer.
Branch Offices: 1634 Spruce St., Philadelphia, PA 19103; (215) 545-2003; 1717 N. Highland Ave., Hollywood, CA 90028; (213) 464-8291.

BENTON & BOWLES, 909 Third Avenue, New York, NY 10022 (212) 758-6200.
JOHN S. BOWEN, pres. & chief exec. officer; VICTOR G. BLOEDE, chm. of bd. & chm., Benton & Bowles International; GEORGE SIMKO, sr. v.p. & group exec.

TED BERNSTEIN ASSOCIATES INC., 1414 Avenue of the Americas, New York, NY 10019. (212) 421-4660.
HOWARD GOLDSTEIN, time buyer, acct. exec.

RALPH BING ADVERTISING CO., 16109 Selva Dr., San Diego, CA 92128. (714) 487-7444.
RALPH BING, TV dir.

FRANK BLOCK ASSOC., The Chase, 10th Floor, St. Louis, MO 63108. (314) 367-9600.
JAMES A. RINDLER, dir. of media.

BOTSFORD, KETCHUM, INC., 55 Union St., San Francisco, CA 94111.
DENIS HITCHMOUGH, senior v.p., media; MORTY BARAN, dir. broadcast prod.; PATTY MORAN, v.p. mgr. broadcast business affairs.

LEO P. BOTT, JR., ADVERTISING, 111 E. Chestnut St., Chicago, IL 60611 (312) 787-7890.
LEO P. BOTT, JR., TV dir.; ED MORGAN, Timebuyer.

BRAVERMAN-MIRISCH INC., 9255 Sunset Blvd., Los Angeles, CA 90069 (213) 272-8608; 274-5204.
MILLICENT BRAVERMAN, pres.

R.C. BRETH, INC., P.O. Box 929, 310 Pine St., Green Bay, WI 54305; (414) 432-4650.
R.C. BRETH, pres.; ROBERT PATRICKIS, treas.

BURLINGAME/GROSSMAN, INC., 35 E. Wacker Dr., Chicago, IL 60601 (312) 332-1515.
S.J. BRICHTA, TV dir.

LEO BURNETT COMPANY, INC., Prudential Plaza Chicago, IL 60601 (312) 565-5959
C.R. (JACK) KOPP, chm. of the bd. & Chief Exec. officer; MARTIN H. SNITZER, chm of the Exec. committee; JOHN J. KINSELLA, pres.; WAYNE A. (AL) WIGGINS, pres., Leo Burnett Int'l.; A. EUGENE MCKEOUGH, exec. v.p.; NORMAN L. MUSE, exec. v.p.; JOHN C. KRAFT, Senior v.p. & corp. controller; MICHAEL J. RICH, senior v.p.; MICHAEL BRESLIN, senior v.p., gen. counsel.

BYRDE, RICHARD & POUND, INC., 200 Madison Ave., New York, NY 10016 (212) 684-2400.

HAROLD CABOT & CO., INC., 10 High Street, Boston, MA 02110; (617) HAncock 6-7600.
JAMES I. SUMMERS, pres.; JAMES ENGLISH, media director, LOUIS STAMOULIS, dir. Radio-TV.

CALDWELL-VAN RIPER, INC., IBM Bldg., 2827 Rupp Dr., Fort Wayne, IN 46815 (219) 484-4127.
SUSAN M. DILLON, media director.

CAMPBELL-EWALD CO., 30400 Van Dyke, Warren, MI (313) 574-3400.
THOMAS B. ADAMS, chm. of bd.; RICHARD D. O'CONNOR, vice chm.; PAUL L. JOHN, pres.; THOMAS J. GLYNN, exec. v.p. media dir.; THOMAS A. TUCKER, sr. v.p. & gen. mgr.
Branch Offices: Atlanta, Chicago, Los Angeles, New York, San Francisco, Dallas, Kansas City, Washington, Knoxville.

CAMPBELL-MITHUN, INC., Northstar Center, Minneapolis, MN 55402 (612) 339-7383.
STANHOPE E. BLUNT, chm.-exec. comm. & CEO (Mpls.); GEORGE H. GRUENWALD, chm. of bd. & COO (Mpls.); DAVID R. SEIBEL, pres. (Chgo.); RAYMOND L. SACHS, exec. v.p. & genl. admin. mgr. (Mpls.); WILLIAM E. STEIN, exec. v.p. & gen. mgr. (Chgo.); EARL A. HERZOG, v.p., media dir. (Mpls.); ART EDELSTEIN, v.p. media dir. (Chgo.); BOB BOURQUARD, v.p. & assoc. media dir. (Mpls.); DICK MORIARITY, assoc. media dir. (Mpls.); ALAN GERSTEN, assoc. media dir. (Mpls.); ALICE KELLEY, assoc. media dir. (Chgo.); JANET BAK, assoc. media dir. (Chgo.).
Branch Offices: 111 East Wacker Dr., Chicago, IL 60601 (312) 565-3800.

COLE & WEBER, INC., 3100 South 176 St., Seattle, WA 98188 (206) 433-6200.

COMPTON ADVERTISING, INC., 625 Madison Ave., New York, NY 10022.
O. MILTON GOSSETT, pres. & chief exec. officer; C. CAPUANO, v.p. art & TV commercial production; BRUCE COX, v.p. mgr. radio/TV programming. Branch office: Los Angeles, CA; Southfield, MI.

THE CRAMER-KRASSELT CO., 733 N. Van Buren St., Milwaukee, WI 53202 (414) 276-3500.
ROBERT L. CHRISTIANSEN, chmn. of bd. & treas.; TED WING, pres; C. ROSS LITTIG, vice chm. of bd.; DONALD E. POM, media dir.-TV/radio; CAROL WALCZAK, TV/radio time buyer; KATHY CASEY, broadcast services mgr.

CRESWELL, MUNSELL, FULTZ & ZIRBEL, INC., 4211 Signal Ridge Road N.E. Cedar Rapids, IA 52406 (319) 393-0200.
E.W. MUNSELL, pres.; R.P. ZIRBEL, R.D. MEYOCKS, and M.L. WHITE, sr. v.ps.—radio/TV producers, T.F. TAYLOR, v.p. Exec. Prod.; MIKE MURRAY, RICHARD STERN, and BRUCE SKINNER, prods.

CRESWELL, MUNSELL, FULTZ & ZIRBEL INC., 727 E. 2nd St., Des Moines, IA 50306 (515) 247-8550. W. J. FULTZ, exec. v.p., mng. dir..;

CSAKY & WARLICK ADVERTISING, INC., 728 East Main St., Richmond, VA 23212.
TIBOR CSAKY, pres., JUNIUS B. WARLICK, vice pres. & creative director.

CUNNINGHAM & WALSH, CHICAGO INC., 875 Michigan Ave., Chicago, IL 60611 (312) 943-9400.
DONALD P. CARTER, chm. of bd.; JOHN DOHERTY, pres.; DONALD GWALTNEY, sr. v.p., dir.; creative services; CHARLES C. CATCHER, sr. v.p., mgr. creative services & dir.; broadcast prod.; FRED WILLSON, sr. v.p. gen. mgr.

D'ARCY-MACMANUS & MASIUS, INC., 360 Madison Ave., New York, NY 10017. (212) 850-7300.
New York: EUGENE A. PETRILLO, exec. v.p., corporate broadcast dir.; DENNIS ALTMAN, SR. v.p. creative dir.; GEORGE KARALEKAS, SR. v.p. media dir.; Atlanta: REGIS NAPOLITANO, media dir.; WARD WIXON, v.p. creative dir.; Bloomfield Hills: ROBERT C. SPAULDING, v.p. media dir.; RONALD W. MONCHAK, exec. v.p., creative dir.; JOHN PIKE, v.p. broadcast; Chicago: JOSEPH P. ZELLER, SR. v.p. media dir.; LAWRENCE D. SENTEN, exec. v.p. dir. creative services; Los Angeles: FAYE LIZOTTE, regional broadcast buyer; Minneapolis-St. Paul: RONALD D. GESKEY, media dir.; GENE CARR, v.p. creative dir.; St. Louis: FRANK KOPEC, Sr. v.p., media dir.; LARRY STILLMAN, SR. V.P., exec. creative dir.; DON SAGER, v.p. mgr. broadcast; San Francisco: JOE GREENBERG, v.p. media dir.; LEN ALARIA, v.p. creative dir.; Toronto: J. CHARLES AMIS, v.p. dir. media services; PAUL GOTTLIEB, v.p. dir. creative services; Mexico City: JORGE ZAMUDIO, media dir.; PEDRO CARDOSO, media dir.; LUIS OJEDA, v.p. creative services dir.

DAVIS, JOHNSON, MOGUL & COLOMBATTO, INC., 3435 Wilshire Blvd., Los Angeles, CA 90010 (213) 388-3332.
ROBERT DAVIS, pres., odd days; ROBERT J. COLOMBATTO, pres., even days; BRAD BALL, creative dir.; MICHAEL MALONE, media mgr.

DIENER/HAUSER/BATES CO., INC., 116 N. Robertson Blvd., Suite 210, Los Angeles, CA 90048 (213) 855-1900.
EUGENE B. COFSKY, pres. West Coast div., BERNICE LIVINGSTON, v.p.

DIENER/HAUSER/BATES CO., INC., 1400 E. Oakland Park Blvd., Ft. Lauderdale, FL 33334 (305) 564-1234.
BOB KATZ, v.p.

W.B. DONER & CO., 26711 Northwestern, Southfield, MI 48034; (313) 354-9700; also 2305 N. Charles Street, Baltimore, MD 21218 (301) 338-1600.
W.B. DONER, chm. of exec. comm.; HERBERT D. FRIED, chm. of bd.; STANLEY T. BURKOFF, pres.; SEYMOUR (SKIP) ROBERTS, exec. v.p.; M. KEVIN WIXTED, treas.; JAMES DALE, SR., v.p./exec. crea. dir.

DOREMUS & COMPANY (Subsidiary of BBDO Intl., Inc.), 120 Broadway, New York, NY 10005 (212) 964-0700.
FRANKLIN E. SCHAFFER, chairman; G. BARRY MCMENNAMIN, president; WILLIAM B. PALMER, creative dir.; LOUIS CROSSIN, media dir.
Branch offices: Boston, Chicago, San Francisco, Washington, Los Angeles, Miami, Minneapolis, Rockford, London.
Corresp. offices: Akron, Denver, Baltimore, Detroit, Cleveland, Pittsburgh, Dallas, Hartford, Houston, Atlanta, Kansas City, St. Louis.

DOYLE DANE BERNBACH INC., 437 Madison Avenue, New York, NY 10022 (212) 826-2000.
WILLIAM BERNBACH, chm. of exec. committee; JOSEPH R. DALY, chm. of the bd & CEO; ROBERT LEVENSON, vice chm., creative; NEIL R. AUSTRIAN, pres. & COO.

EDELSTEIN-OKERWALL ADVERTISING AGENCY, 2711 Crawford Ave., Evanston, IL 60201 (312) 869-4411.
CONRAD M. OKERWALL, radio & TV dir.

EISNER & ASSOCIATES, INC., 12 W. Madison St., Baltimore, MD 21201; (301) 685-3390.
H W. EISNER, pres.; HAL BRAUNER, media dir.; STEVE EISNER, TV dir.

MIKE FADELL ADVERTISING AGENCY, 6101 York, S. Minneapolis, MN 55410 (612) 922-5606.
EDNA FADELL, TV dir. & timebuyer.

FAHLGREN & FERRISS, INC., 120 E. 4th St., Cincinnati, OH 45202; (513) 241-9200.
ROGER J. CROTTY, v.p. dir. of media; WILLIAM FARSON, secty.

FAIRFAX INC., ADVERTISING, 635 Madison Ave., New York, NY 10022 (212) 350-1800.
STANLEY I. FISHEL, dir. of television; RICHARD MIRENDA, art director; BENNETT KESSLER, copy chief; JONATHAN HORN, media dir.

FOOTE, CONE & BELDING, 200 Park Avenue, New York, NY 10166 (21) 880-9000.
JOHN O'TOOLE, chm.
Offices: Chicago, Los Angeles, New York, San Francisco, Frankfurt-Am-Main, London, Toronto, Mexico City, Sydney, London, Paris, Milan, Copenhagen, Stockholm, Amsterdam, Brussels, Melbourne, Madrid, Puerto Rico, Rome.

FOSTER & DAVIES, INC., Keith Building, Cleveland, OH 44115 (216) 241-0711.

ALBERT FRANK-GUENTHER LAW, INC., 61 Broadway, New York, NY 10006 (212) 248-5200.
JOHN V. MCADAMS, chmn. of bd.; JOHN W. ANGUS, pres.; LAWRENCE BUTNER, sr. v.p. & TV/radio dir.
Branch offices: Chicago, Philadelphia, Los Angeles.

GERBER ADVERTISING AGENCY, 1305 S.W. 12th Ave., Portland, OR 97201 (503) 221-0100.
BETTY CHIMENTI, media dir.; CATHY MACDONALD, assoc. med. dir.

GILLHAM ADVERTISING INC., Fifth Floor, Deseret Plaza Bldg., 15 East First South, Salt Lake City, UT 84111 (801) 328-0281.
LON RICHARDSON, JR., pres.; ARTHUR W. ATACK, exec. v.p.; R.H. OGAARD, sec.; PERRY REEL, media dir.; DOUGLAS NEWMAN, prod. mgr.; JEAN PEARSON, traffic coordinator.

GREY ADVERTISING, INC., 777 Third Ave., New York, NY 10017 (212) 546-2000
EDWARD H. MEYER, pres. & chief exec. officer; ALEC GERSTER, v.p., dir. of media & programming.
Branch offices: Amsterdam-Baarn, Auckland, Brussels, Buenos Aires, Caracas, Chicago, Copenhagen, Detroit, Dusseldorf, Glasgow, Gothenburg, Hong Kong, Johannesburg, London, Los Angeles, Madrid, Malmoe, Milan, Minneapolis/St. Paul, Montevideo, Montreal, New York, Oslo, Paris, Rio De Janeiro, San Francisco, Santiago, Singapore, Stockholm, Sydney, Tokyo, Toronto, Vienna

THE GRISWOLD-ESHLEMAN CO., 12th Floor—55 Public Square, Cleveland, OH 44113; (216) 696-3400.
ROBERT L. BAUMGARDNER, chmn. emeritus; RICHARD E. GERBRACHT, chief exec. officer and chmn.; PATRICK J. MORIN, pres., THOMAS A. LOUNDS, JR., sr. v.p. & mktg. serv. dir.; EDMUND G. NORWICK, JR., dir. bus. media; CYNTHIA BEERBOHM, broadcast buyer; WILLIAM BROWN, broadcast buyer; CATHERINE CRACO, broadcast producer; MEL HORACE, film producer.

GRUBB, GRAHAM & WILDER, INC., Century 21 Plaza, Champaign, IL 61820 (217) 344-4334.
ROBERT F. GRUBB, chmn.; JOHN A. GRUBB, pres.; PATRICK CHELF, media dir.

ART GRUBER ASSOCIATES, ADVERTISING, 4407 East Lake Street, Minneapolis, MN 55406 (612) 721-1661.
ARTHUR R. GRUBER, pres. & treas.; E.M. GRUBER, sec.

HARRIS & LOVE, INC./ADVERTISING, 19 E. 2nd South, Salt Lake City, UT 84111 (801) 532-7333.
DONALD G. WARE, exec. v.p.

HART-CONWAY CO., INC., 300 Triangle Bldg., Rochester, NY 14604 (716) 232-2930.
PETER B. WHITCOMBE, pres.; WILLIAM WESTERFELD, exec. v.p.; PETER TONERY, prod. v.p.

HENDERSON ADVERTISING, INC., 60 Pelham Pointe, Greenville, SC 29615; P. O. Box 2247; Greenville, SC 29602 (Mailing Address) (803) 271-6000.
JAMES M. HENDERSON, chm. of bd. and pres.; FRED C. WALKER, vice chm.; RALPH CALLAHAN, exec. v.p.; LARRY RICHARDSON, exec. v.p./exec. creative dir.; BOB REUSCHLE, sr. v.p./media dir.

HICKS AND GREIST, INC. 522 Fifth Ave., New York, NY 10036 (212) 764-3020.
CHARLES V. SKOOG, JR., chmn.; D. EDWARD RICCHIUTO, pres.; S. PARKER, media dir.; DEMETRI KONTUPOULOS, TV prod. dir.

WM. F. HOLLAND CO., 252 Ludlow Ave., Cincinnati, OH 45220 (513) 221-1252
EARL HOLLAND, pres.

HOUCK ADVERTISING, 3615 Brambleton Ave., S.W., (Mailing Address: P.O. Box 12487, Roanoke, VA 24026 (703) 989-6655.
WILLIAM B. HOUCK, pres.; WILLIAM B. HOUCK, JR., acct. exec.; PRISCILLA D. EVANS, time buyer.

HUTCHINS/YOUNG & RUBICAM INC., 400 Midtown Tower; Rochester, NY 14604. (716) 546-6480.
FRANK M. HUTCHINS, chm.; JAMES N. MOREY, pres. & chief exec. officer; ROBERT H. MAHARRY, v.p., creative dir.; CHARLES E. RELLER, v.p. finance & adm., treasurer; OWEN J. CRUMB, v.p. & gen. mgr., pub. rel. div.; HARVEY J. BOZZI, JR., v.p./director of special projects.

HENRY J. KAUFMAN & ASSOCIATES, INC., 2233 Wisconsin Ave., N.W., Washington, DC 20007 (202) 333-0700.
HENRY J. KAUFMAN, founder chm.; ALFRED H. EDELSON, chmn.; STUART E. KARU, pres. & CEO; Michael G. Carberry, exec. v.p.; J. ROGER VILSACK, Sr. v.p., creative services; IRVING SLOAN, sr. v.p., acct. services; PHILIP N. SABATINO, v.p./exec. AD & assoc. creative dir.; EDWARD F. O'KANE, v.p.; PHILIP I. RABIN, v.p.; BETH A. WRIGHT, v.p., res. dir.; JOAN YONKLER, v.p., media dir.; WILLIAM H. ARENS, treas.; ESTHER C. TRUCKSESS, corporate secty.

KELLY, ZAHRNDT & KELLY INC., 10805 Sunset Office Dr., St. Louis, MO 63127.
ROBERT W. KELLY, dir.; WILLIAM A. WILSON, MARY HOPPER, CHARLES CROWE, DAVE ERICH, acct. execs.

KERR, WEST & GISH, INC., 460 Tenth Circle, North Nashville, TN 37203 (615) 255-1142.
AL KERR, pres.

KM&G INTERNATIONAL INC., Four Gateway Center, Pittsburgh, PA 15222
Pittsburgh: WILLIAM H. GENGE, Chmn. of the Bd.; EDWARD T. PARRACK, v. chmn.; ROBERT H. SAVAGE, pres.; WILLIAM R. PARKER, exec. v.p.; HERBERT D. GORDON, senior v.p. & Dir. media & operations; RAYMOND C. WERNER, senior v.p. and creative dir.; RICHARD SLEEMAN, broadcast services mgr.; JOHN H. MATHER, senior v.p. & dir. of marketing research dept.; EDWARD L. GRAF, exec. v.p., finance & law.
New York office: Ketchum, New York. HENRY J. WASIAK, pres.; WALTER KAPRIELIAN, exec. v.p.; WILLIAM H. LYNN, senior v.p., dir. media/TV programming (N.Y.); PHILIP J. SHIFFMAN, v.p. & dir. media network res. (N.Y.); Ketchum International Inc., JOSEPH M. COOGLE, pres.

KEYES, MARTIN, 841 Mountain Ave., Springfield, NJ (201) 376-7300.
Specialist; ROBYN GREEN-TAYLOR, media director.

KRUPNICK & ASSOCIATES, INC., 135 N. Meramec, St. Louis, MO 63105 (314) 862-9393.
WILLIAM W. QUADE, TV production.

KUTTNER & KUTTNER, INC., 646 N. Michigan Ave., Chicago, IL 60611 (312) 787-7940.

LEWIS & GILMAN, INC., 1700 Market Street, Philadelphia, P 19103 (215) LO8-3775.
R.G. WILDER, pres.; BERNARD OSTROF, sr. v.p. creativ director; WESLEY G. HOTCHKISS, v.p.; C.D. HARRINC TON, V.P., MEDIA DIRECTOR: GLORIA PEDALINC media supervisor; RICHARD BONDS, media supervisor.

THE W.E. LONG ADVERTISING AGENCY, 309 W. Washingto St., Chicago, IL 60606 (312) RA6-4606.
C.J. HEMRICH, gen. mgr.; JOHN PESICK, dir., creativ services; JOE GURA, art dir.; ANITA RAUH, media superv sor.

LOWE & HALL ADVERTISING, INC., 501 East North Street, P.C Box 1824, Greenville, SC (803) 242-5350.
EDWARD E. BROWER, JR., pres.; CAROL C. YOCKEY exec. v.p. and treas.; EDWARD B. BROWER, JR., v.p.; W.T HALL, v.p.; FRANK B. VAN AUKEN, v.p.; DAWN L NEWTON, secty.

MacFARLAND, AVEYARD & CO., 333 N. Michigan Ave., Chi cago, IL 60601; (312) RAndolph 6-9360.
FAYE MacFARLAND, pres.

MALCOLM-HOWARD, 203 N. Wabash Ave., Chicago, IL (312 ANdover 3-0022.
M.G. HOLLAND, pres.; LILLIAN HOLLAND, acct. exec.

McCANN-ERICKSON, INC., 485 Lexington Avenue, New York NY 10017. 697-6000.
STEPHEN LEFF, exec. v.p. dir., media services; WM. F CROASDALE, senior v.p., assoc. dir. R/TV; MARTIN COLE v.p. TV pgm. exec.; MICHAEL YUDIN, v.p., TV pgm. exec. SUE SWENSON, TV pgrm. exec.; CINDY GURMANN, TV pgm. exec.; ROBERT SIMON, v.p., dir. bus. affairs; LEC EGAN, v.p. mgr. bus. affairs; C.J. KETTLER, mgr. networ research; JOHN RHONE, TV pgm. exec., Los Angeles.

EDWARD J. McELROY ADV., INC., 3600 Wilshire Blvd., Lo Angeles, CA 90005 (213) 381-3045.
E.D. BROTSOS, pres. & acct. sup.; PAMELA K. AR SENAULT, media dir.

MCKENZIE, KING & GORDON, 1800 N. Argyle, Hollywood, C. 90028 (213) 466-3421.
MARIE MCKENZIE, pres.; DON MCKENZIE, creative dir. MARIE SELTZER, media dir.

MELDRUM & FEWSMITH, INC., 1220 Huron Rd., Cleveland 15 OH 241-2141.
CHRIS PERRY, v.p. creative dir., BRUCE S. CHILDERS mkt. services.

MELTZER, ARON & LEMEN, INC., Steuart Tower, One Marke Plaza, San Francisco, CA 94105 (415) 543-0850.
STAN COLBERSON, media dir.
Branches: Detroit, 1950 First National Bldg.; New York, 122 Avenue of the Americas; Chicago, 2 No. Riverside Plaza.

MOSS ADVERTISING INC., 633 Third Avenue, New York, NY 10017 (212) MU 7-7377.
JOSEPH H. MOSS, chm.; MURRAY BRAUMAN, pres.

NATIONAL EXPORT ADVERTISING SERVICE, INC., 27 Madison Ave., New York, NY 10016 (212) 686-7060.
Branch: Hughes & Associates, Inc., Banco Popular Center, Hat Rey, Puerto Rico

NEEDHAM & GROHMANN, INC., 30 Rockefeller Plaza, New York NY 10020; 245-6200.
H.A. HEINSIUS, president; WILLIAM R. HOGAN, art dir. JOHN L. GILLESPIE, N. ROBERT LORMAN, ANTHONY E.F. CORNWELL, v.p.'s.

NEEDHAM, HARPER & STEERS ADVERTISING, INC., 909 Third Avenue, New York, NY 10022 (212) 758-7600.
PAUL C. HARPER, JR., chmn. of the bd. & chief exec. offi cer; JOEL P. BAUMWOLL, pres. NH&S New York; BAR RON Z. BIEDERMAN, pres. NH&S/Issues & Images; LOIS KOREY, exec. v.p. & creative dir.; 303 E. Wacker Dr., Chicago, IL 60601 (312) 861-0200. B. BLAIR VEDDER, JR.; chie op. officer; KEITH L. REINHARD, pres. NH&S Chicago JOEL M. HOCHBERG, sr. v.p. & creative dir.; 10889 Wilshire Blvd., Los Angeles, CA 90024 (213) 208-5000. BRADLEY H. ROBERTS, pres. NH&S West; RICHARD KELLEY, sr. v.p. & creative dir.

NOWAK BARLOW JOHNSON, 117 Highbridge St., Fayetteville, NY 13066. (305) 637-9895.
DONALD G. NOWAK, pres. & CEO; Ed Gabriel, audio-visual dir.

PARKSON ADVERTISING AGENCY, INC., 767 Fifth Ave., N.Y., NY 10153 (212) 752-3300.
EDWARD KLETTER, chm. of bd.; PHILIP AGISIM, pres.

HERBERT GRUBER, snr. v.p., broadcast media dir.; RICHARD BARLOW, dir. radio/TV production

PASCOE, STARLING & POLLOCK, INC., Seattle Tower, Seattle, WA 98101 (206) MUtual 2-0353.
WILFRED A. PASCOE, pres.; JACK W. STARLING, v.p.

THE PITLUK GROUP, 100 Richmond by the River, San Antonio, TX 78205; (512) CA3-6131.
LOUIS H. PITLUK, pres.; JACK N. PITLUK, JR., v.p.; PAZ HAUFLER, med. dir.; LINDA HUGHES, med. dir.; ROGER CHRISTIAN, creative dir.

ROCKLIN, IRVING & ASSOCIATES 32 W. Randolph St., Chicago, IL (312) RA 6-2324.
IRVING ROCKLIN, chm. of bd.; LEAH ROCKLIN, pres.; JERROLD ROCKLIN, v.p.; ROBERT ROCKLIN, v.p.; ROBERT SPENCER, TV dir.; PATTY JO WONG, Radio dir.; JOEL MINK, art dir.; CYNTHIA J. BOIVIN, auditor; GEORGE BREYER, comptroller; CAROLYN FISCHER, traffic mgr.

RUDER & FINN INCORPORATED, 110 East 59th St. New York, NY 10022 (212) 593-6400.
DAVID FINN, chmn. of bd.; WILLIAM RUDER, pres.; SUSAN BLANE, television-film dept.

RUMRILL-HOYT, INC., 1895 Mt. Hope Avenue, Rochester, NY 14620 (716) 271-2150.
EUGENE F. NOVAK, pres. & CEO; JOSEPH M. VALERIO, exec. v.p. & gen. mgr.; DONNA BIEDENBACH, R-TV supervisor; WILLIAM J. MACKEY, v.p. and media dir.

LOWE RUNKLE CO., 6801 N. Broadway, Oklahoma City, OK 73114 (405) 848-6800.
HOWARD NEUMANN, pres.; ALAN ATKINS, MICHELE WILLINGHAM, TV prods.

SCOPE COMMUNICATIONS, INC., 56 Post Road West, Westport, CT 06880; (203) 227-7254.
ROBERT ZUCKERMAN, pres.; MARK ZUCKERMAN, v.p.

SHOTWELL AND PARTNERS, INCORPORATED, 1400 Harding Place, Charlotte, NC 28204 (704) 372-8880
GARY A. LUDWICK, creative dir./producer

SILTON/TURNER, INC., 320 Statler Office Building, Boston, MA 02116 (617) 542-9460.
RAMON H. SILTON, pres.; JASON N. SILTON, senior v.p./creative; SALLY L. STITT, senior v.p./media; WILLIAM E. PARANDES, v.p. operations/broadcast production; SCOTT W. BARAN, v.p./communications.

SIMONS MICHELSON ZIEVE INC., 3250 West Big Beaver Road, Troy, MI 48084 (313) 649-5600.
MORTON ZIEVE, JAMES A MICHELSON, officers; JON KATZ, TV dept. dir.

SPIRO & ASSOCIATES, INC., 100 S. Broad St., Phila., PA (215) WAlnut 3-5400.
Herb Smith, v.p., media; LINDA ROSANIO, asst. media dir.

SPITZ ADVERTISING INC., 530 Oak St., Syracuse, NY (315) 479-5581.
WILLIAM SPITZ, president; CHRIS SLATER radio-TV dir.; NICK BIBKO, acct. exec.; LIZ MOORE, time buyer.

ROBERT L. STEVENSON, INC., 16th Floor, 301 Fifth Ave. Bldg., Pittsburgh PA 15222. (412) 281-4901.
RONALD KOEPER pres.

SYNDICAST SERVICES, INC., 919 Third Ave. New York, NY 10022 (212) 751-3394.
MITCHELL (MICKEY) JOHNSON, pres.; LEONARD V. KOCH, exec. v.p.; SHELDON BODEN, exec. v.p.; WILLIAM KOBLENZER, v.p. sales.; KATHY ZANGSAM, dir./prom./adv.

TATHAM-LAIRD & KUDNER, 625 N. Michigan Avenue, Chicago, IL 60611; Phone: 337-4400.
CHICAGO: JEROME F. BIRN, mng. partner & chrmn. mgmt. comm.; CHARLOTTE BEERS, mng. partner & chief operating officer; WILLIAM J. BIENEMANN, mng. partner, mktg. services & business development; DAVID A. FAIRMAN, mng. partner, exec. creative dir.; JOHN J. VALVERDE, partner, chief financial officer; LEONARD F. MATERNA, media dir.; WERNER WOLFF, mgr. broadcast business affairs
TLK Direct Marketing: 605 Third Avenue, New York, NY 10016 (212) 972-9000. DAVID A. HENEBERRY, partner.

TAYLOR-JESSOP ADVERTISING CO., INC., 11 South Portage Path, Akron, OH 44308 (216) 376-2169.
H. TAYLOR PROTHEROE, pres.; ROBERT G. HAWKINS, v.p.; ANNA CONTE. v.p.

J. WALTER THOMPSON COMPANY, 420 Lexington Ave., New York, NY 10017 (212) 867-1000
DON JOHNSON, chmn. of bd. and chief exec. officer; JOHN SHARMAN, chmn. of the exec. committee & chief operating officer; BURT MANNING, vice-chmn; GLEN FORTINBERRY, vice-chmn.; ROBERT E. BUCHANAN, senior v.p. in charge of media; MARIE LUISI, senior v.p. in charge of spot buying syndication & cable; FARLAN MYERS, senior v.p. dir. programming, western division (Los Angeles); JOHN H. P. DAVIS, v.p., dir. of programming, eastern division.
Other offices: Atlanta, Chicago, Detroit, Los Angeles, Miami, New York, San Francisco, Washington.
Service offices: Boston, Charlotte, Cleveland, Dallas, Denver, Houston, Indianapolis, Kansas City, Louisville, Memphis, New Orleans, Oklahoma City, Phoenix, Richmond, St. Louis.
International offices: Toronto, Montreal, Vancouver, Mexico City, Buenos Aires, Sao Paulo, Rio de Janeiro, Santiago, Lima, Caracas, Vienna, Brussels, Copenhagen, Paris, Frankfurt, Milan, Amsterdam, Madrid, Stockholm, Zurich, London, Sydney, Melbourne, Tokyo, Auckland, Manila.

TRACY-LOCKE ADVERTISING INC. Plaza of the Americas, South Tower, P.O. Box 50129, Dallas, TX 75250 (214) 742-3131.
MORRIS HITE, bd. chm.; NORM CAMPBELL, pres.; BOB BELTON, v.p. bdcst prod.; LARRY SPIEGEL, sr. v.p., dir. media & research; WALT ATKINSON, v.p., media serv. dir.; CLAUDE CAYLOR, v.p., media dir.; JUDY MILLER, v.p., media buying director; GINA BURKS, dir. of network; CAROL CASTELLANO, dir. of spot.

VAN BRUNT & CO., ADVERTISING-MARKETING INC., 300 East 42nd Street, New York, NY 10017 (212) 949-1300.
EDWIN VAN BRUNT, chm. bd.; ALBERT VAN BRUNT, pres.; HOWARD KARP, creative dir.; WILLIAM BUCKLEY, dir. of broadcast.

VICTOR VAN DER LINDE CO., INC., 381 Broadway, Westwood, NJ 07675 (201) 664-6830.
A.K. KINGSLEY, v.p.

WARWICK, WELSH & MILLER, INC., 375 Park Avenue, New York, NY 10152, (212) 751-4700.
JOHN P. WARWICK, chm. & pres.; JOHN V. CHERVOKAS, exec. v.p.-creative director; LOU KOHUTH, broadcast business mgr.; FLORENCE HAYLE, casting dir.; JOHN HOWE, senior prod.; LINDA KRAMER, prod.; SUSAN MONAHAN, prod. co-ordinator; PETER ELLERS, assoc. prod.

MORTON DENNIS WAX AND ASSOCIATES, 200 West 51st St., New York, NY 10019 (212) 247-2159
MORTON DENNIS WAX, pres.; SANDRA WAX, v.p./gen. mgr.; DAVID LUDWIG; asst. to pres.; MADELYN FUDEMAN, MARK FELDMAN, IRENE KRAUSS, MICKEY WALLACH, JAMES SLIMAN, DAVID SAMUELS, JIM WEINREBE, acct. execs.

WEBSTER & HARRIS ADVERTISING AGENCY, 1313 Broadway, Lubbock, TX 79401 (806) 747-2588.
REX WEBSTER, LEON HARRIS & DOUG HODEL, acct. execs. partners.

WEISS & GELLER, INC., 880 Third Ave., New York, NY 10022.
MAX A. GELLER chmn. bd. MAX TENDRICH, pres.; ARTHUR ROSS, exec. v.p. creative dir.

WILLIS/CASE/HARWOOD, INC., 3411 Office Park Drive, Dayton, OH; 45439 (513) 299-7394.
WILLIAM GUISEWITE, pres.; SANDRA MAGEE, dir. of media.

YOUNG & RUBICAM, INC., 285 Madison Ave., New York, NY 10017 (212) 953-2000.
EDWARD N. NEY, chm., pres. & chief exec. officer; ALEXANDER KROLL, exec. v.p.; pres. Y&R USA div.; JOSEPH W. OSTROW, exec. v.p., dir. communications svcs.; RICHARD H. LOW, sr. v.p., dir. bdcst, programming & purchasing.
Y&R USA OFFICES: Cedar Rapids, IA.; Cincinnati, OH; Denver, CO, Detroit, MI; Houston, TX; Kansas City, MO; New York, NY; Los Angeles, Oakland, San Diego & San Francisco, CA; Phoenix, AZ; Rochester, NY; St. Petersburg, FL.
Y&R International Offices: Adelaide, Melbourne & Sydney, Australia; Madrid & Barcelona, Spain; Beirut, Lebanon; Berne, Switzerland; Brussels, Belgium; Buenos Aires, Argentina; Copenhagen, Denmark; Frankfurt & Munich, Germany; Stockholm & Gothenburg, Sweden; Hong Kong; London, England; Mexico City, Mexico; Milan, Italy; Montreal & Toronto, Canada; Oslo, Norway; Paris, France; San Juan, Puerto Rico; Santo Domingo, Dominican Republic; Sao Paulo, Brazil; Singapore; Tokyo, Japan; Vienna, Austria.

STATION REPRESENTATIVES

AVERY-KNODEL TELEVISION, 437 Madison Avenue, New York, NY 10022 (212) 421-5600.
J.W. KNODEL, chmn.; ROBERT J. KIZER, pres.; F. ROBERT KALTHOFF, exec. v.p.
Branch Offices: Chicago 60601, 2600 Prudential Plaza, 467-6111, F. ROBERT KALTHOFF; San Francisco 94104, 155 Montgomery Street, 986-8082, JOSEPH FRIEDMAN; Los Angeles 90005, 633 South Shatto Place, 385-6394; Atlanta 30326, 3340 Peachtree Road, N.E., 233-4474, JAMES GOSSETT; Dallas 75219, 3626 North Hall St., Suite 608, 528-1741 WALLIS S. IVY; Detroit, 48084, DAVID BRANGAN; St. Louis 63102, 410 Mansion House Center, 241-5051, FRED W. JOHNSON; Minneapolis, St. Paul: toll free exchange, 929-1300, Edna, MN 55435, 6700 France Ave., Suite 155, MAUREEN KVAM; Denver 80206, 280 Columbine Ave., 321-1086, BOB HIX; Portland 97204, 620 SW 5th, Suite 907, 223-5980, RON HILLBURY; Seattle 98109, P.O. Box 19417, 624-9200, JACK DAVIES; Haddonfield 08033, 537 Mt. Vernon Avenue, TOM WOOD.

BLAIR TELEVISION 717 Fifth Ave., New York, NY 10022 (212) 752-0400.
HARRY B. SMART, chairman/C.E.O., WALTER A. SCHWARTZ, pres.; PATRICK J. DEVLIN, senior v.p./sls. operations; JAMES R. KELLY, Sr. v.p./nat. sls.; ARTHUR C. STRINGER, senior v.p./sls. administration; DAN HARRISON, v.p./sls. mgr., ABC/Blue Team; JAMES R. BLOOM, v.p./sls. mgr, ABC/Green Team; DONALD J. GORMAN, v.p./sls. mgr, ABC/Red Team; WILLIAM E. BREDA, Jr., v.p./Sls. mgr., CBS/Red Team; DAVE HERMAN, v.p./sls. mgr., CBS/Green Team; BRIGGS PALMER, v.p./NY sls. mgr./NBC; STEVEN T. LOSEE, v.p./sls. mgr, NBC/Blue Team; TIMOTHY MCAULIFF, sls. mgr., NBC/Green Team; WILLIAM MURPHY, v.p./sts. mgr., NBC/Red Team.
Branch offices: Chicago, Atlanta, Boston, Cleveland, Dallas, Detroit, Jacksonville, Los Angeles, Minneapolis, Philadelphia, St. Louis, San Francisco.

CBS TELEVISION STATIONS NATIONAL SALES, 51 West 52nd St., New York, NY 10019 (212) 975-4321.
RHODERIC H. PERTH, v.p. & gen. mgr.; GARY MONTANUS, dir. eastern sales; Branch offices: Chicago, 630 North McClurg Court, (312) 944-6000, INGER JENSON, dir., midwest sales; Detroit, 26877 Northwest Hwy., (313) 352-2800, CHET COLLINS, mgr.; San Francisco: 1 Embarcadero Center, (415) 433-0500, FRANKLIN WHEELER, mgr.; Los Angeles: Sunset Blvd., (213) 460-3000, ROBERT BERGLUND; Atlanta: 6 Piedmont Center, (404) 261-2227, SAM STOLWORTH, mgr.

DONALD COOKE, INC., 521 Fifth Ave., New York, NY (212) 682-5844.
DONALD COOKE, pres. & sales-TV mgr.
Branch offices: Chicago, 60645; 3322 West Peterson Ave., 478-5544. HOOPER JONES; Los Angeles, 90211: 111 N. La Cienaga Blvd. 652-1441; CHARLES DWYER; San Francisco: 199 Crestmonte Dr.; 731-36201 DAVID SANDBERG; Kansas City; 1012 Baltimore Bldg.; 471-5502, EUGENE GRAY; St. Louis: 468 Equitable Bldg. 10 S. Broadway, 231-9151; EUGENE GRAY.

CREED ASSOCIATES, INC., Park Square Bldg., Boston, MA 02116 (617) 482-4845.
R.A. KEATING, R.P. LUETTERS, IRENE KELLEHER.

DEVNEY ORGANIZATION, INC. THE, 141 E. 44th St., New York, NY 10017 (212) 661-2460.
E.J. DEVNEY, president.

DORA-CLAYTON AGENCY, INC., P.O. Drawer 7308, Atlanta, GA 30309 (404) 881-0220.
DANIEL A. HAIGHT, ARLETTE G. HAIGHT

GILL-PERNA, INC., 745 Fifth Avenue, New York 10022 (212) 838-4740.
HELEN GILL, pres.; JOHN J. PERNA, JR., exec. v.p.
Branch offices: Chicago: 75 East Wacker Drive, (312) 346-9393. THOMAS J. GRIFFIN, midwest mgr.; Hollywood: 6922 Hollywood Blvd., Suite 508; (213) 462-2371, HUGH WALLACE; San Francisco: 700 Montgomery St. (415) 397-0535, "SANDY" SANDEBERG & WARD GLENN; Detroit; 2915 Book Bldg., (313) 962-7022, MAX GOLDFARB.

INTER-AMERICAN PUBLICATIONS, INC., 41 E. 42nd St., New York, NY 10017 (212) 682-0884.
OCTAVIO GONZALEZ, gen. mgr. & vice pres.

THE KATZ AGENCY, INC., One Dag Hammarskjold Plaza, New York, N.Y 10022 (212) 972-2100.
JAMES L. GREENWALD, pres.; OLIVER T. BLACKWELL, exec. v.p.; SAMUEL T. JONES, v.p., finance & treas.; EDWARD SCHNUR, controller; DICK MENDELSON, v.p., corp. dvpmt.; KENNETH A. MILLS, secy.; MARGARET ALCOTT, asst. secy., BARRY LEWIS, special projects dir., GORDON HASTINGS, v.p., dir. new business; GEORGE FELDMAN, v.p., research planning.
Katz Television: DAVID S. ALLEN, pres.: Katz American: FRANK J. MCCANN, pres.; Katz Continental; RICHARD GOLDSTEIN, v.p. American regnl. mgr.; EDWARD PAPAZIAN, v.p., Continental regnl. mgr.; MICHAEL RAOUNAS, v.p., sls. mgr. American Red; JAMES BELOYIANIS, sls. mgr. American White; LEN GRAZIANO, v.p., sls. mgr. American Blue; ROBERT LEVENSTEIN, v.p. sls. mgr. Continental Silver; JACK MULDERRIG, sls. mgr. Continental Gold.
Branch Offices: Chicago 60611, Equitable Bldg., Suite 3125, 401 N. Michigan Ave.: (312) 670-3200. VICTOR FERRANTE, v.p., regnl. mgr.; Atlanta 30361, 100 Colony Square: (404) 892-3811. LARRY SHRUM, v.p., mgr. American; THOMAS BARRETT, mgr. Continental. Boston 02116, 20 Providence St.: (617) 542-5458, EDWARD O'ROURKE, mgr. Charlotte 28209, Park Seneca Bldg., Suite 209: (704) 523-5505. PAUL QUINN, v.p. mgr. Dallas 75219, 3626 N. Hall St.: (214) 526-7941. J. DAVID RUTLEDGE, v.p., mgr. Detroit (Troy), 1650 W. Big Beaver Road, Troy, 48084, (313) 643-4722. PATRICK MCNEW, v.p., mgr. American; JOHN HOFFMAN, mgr. Continental. Jacksonville 32202, One Independent Square, Suite 2201: (904) 358-2914. L. HERBERT KING, mgr. Los Angeles 90036, 5900 Wilshire Blvd.: (213) 937-5250. GERALD J. JONES, v.p. mgr. Memphis 38112, 2714 Union Ave. Extended, Suite 110; (901) 454-0168. SWAN BURRUS, mgr. Minneapolis 55420, 2850 Metro Drive, Suite 311: (612) 854-0020. TOM HETHERINGTON, sls. mgr. American; JOHN C. ARONSON, mgr. Continental. Philadelphia 19102. 3 Parkway, Suite 622: (215) 567-7950. DAVID CHODIKOFF, mgr. American; JAMES CARMINE, mgr. Continental. St. Louis 63102, 10 Broadway. RICHARD W. EPP, v.p., mgr., American; MICHAEL GRATZ, mgr. Continental. Seattle 98119, North Tower, Suite 305, 100 W. Harrison Plaza; San Francisco 94111, One California Bldg., Suite 2325. DAVID S. ABBEY, v.p. regnl. mgr. (415) 362-7628.

NONA KIRBY CO., INC., 1028 Statler Office Bldg., Boston, MA 02111. (617) 482-6117

ART MOORE INC., 220 W. Mercer, Seattle, WAS 98119 (206) 282-3400.
GREG SMITH, EARLE BAKER, SANDY RUNNION, GREG SWEENEY, RICK STAEB, Branch: Portland, OR 97204, 520 S.W. Yamhill, (503) 228-7284, DARRELL RUTTER, JACK SALING, TEDDI JONES.

NBC SPOT SALES, NATIONAL BROADCASTING COMPANY, INC., 30 Rockefeller Plaza, New York, NY 10020 (212) 247-8300.
RICHARD H. CLOSE, v.p.; BYRON GOODELL, dir. of TV spot sales; RICHARD BERGH, eastern TV mgr. BILL FROMM, dir. of new business & prom.

NATIONAL TIME SALES, 275 Madison Ave., New York, NY 10016 (212) 889-1200.
H. FRANKLIN, sales mgr.
Branch offices: Chicago, 60645: 3322 W. Peterson Ave.; (312) 478-5544; HOOPER JONES; Atlanta; 30309: 1819 Peachtree Rd. N.E.; (404) 355-7961; BERNARD OCHS; Dallas; 75202: 1407 Main St.; (214) 748-2172; JACK RILEY; San Francisco: 94133: 559 Pacific Ave.; SAM POSNER; 111 N. La Cienega Blvd., Beverly Hills,: 90028; CHUCK DWYER; San Jose,: 95129: 4320 Stevens Creek Blvd. (408) 249-5287; GENE BERNALD.

THE BERNARD I. OCHS CO., 1819 Peachtree Rd., N.E., Atlanta, GA 355-6432.
BARNEY OCHS; PAT CASTLEBERRY, sec.

PAN AMERICAN BROADCASTING CO., 9 E. 41st St., New York, NY 10022 (212) 661-6550.
GENE BERNALD, pres.; HENRY FRANKLIN, R.J. DUALSKY.

PETERS, GRIFFIN, WOODWARD, INC., Olympic Tower, 645 Fifth Ave., New York, NY 10022 (212) 826-6000.
JAMES M. KING, pres.; DENNIS K. GILLESPIE, exec. v.p.;

ARTHUR C. ELLIOT, Sr. v.p plnng/development; Colleen Roach, v.p./finance/admin.

EDWARD PETRY & CO., INC., 3 East 54th St., New York, NY 10022 (212) 688-0200.
MARTIN F. CONNELLY, chm. of bd.; BOB MUTH, v. chm.; ART SCOTT, pres.; DAVE ALLEN, exec. v.p.; ED KARLIK, Senior v.p. & dir. of sls.; GEORGE BLINN, v.p. & dir. of sls.; AL ROTHSTEIN, v.p., group sls. mgr.; BROWNING HOLCOMBE, JR., v.p. dir. of manpower development; C.W. SCHELLENGER, senior v.p. of finance & administration; BRUCE FAUSER, group sls. mgr.; ARNOLD ANNEX, group sls. mgr.; TED VAN ERK, dir. of administration; MIKE LAMBERT, v.p. of Programming.
Branches: Chicago, 60611: 410 N. Michigan Ave., Suite 1280, (312) 644-9660. JOHN HEISE, general manager; Atlanta, 30305: 550 Pharr Rd., Suite 560, (404) 266-9650. JOHN DOBSON, sls. mgr.; Dallas, 75240: 1047 One Lincoln Centre, 5400 LBJ Freeway (214) 239-5310. DON TRAVIS, sls. mgr.; Detroit-Troy, 48084: 2855 Coolidge Road, Suite 205, (313) 649-0100. JIM DONNELLY, sales manager; Los Angeles, 90036: 5670 Wilshire Blvd., Suite 2180 (213) 937-4330. BOB MIGGINS, v.p., dir. of west coast operations; San Francisco, 94109: 781 Beach St., (415) 441-2227. ERNIE RIZZUTI, sls. mgr.; Philadelphia, 19103: 2 Penn Center Plaza, Suite 820, (215) 561-6230. JOHN MORROW, sls. mgr.; St. Louis, 63102: 100 North Broadway, (314) 241-2445. L.D. VAN VALKENBURGH, v.p. & sls. mgr.; Boston, 02116: 1016 Statler Bldg., (617) 482-6440. DIANE GOSDANIAN, v.p. & sls. mgr.; Minneapolis, 55402: 706 Second Ave. S., Suite 1138, (612) 339-6606. STU LUTZ, sales manager.

RADIO-TELEVISION REPRESENTATIVES LIMITED, 9 Price Street, Toronto, Ontario. M4W 1Z1 9623221.
ROBERT H. QUINN, pres. & gen. mgr.; WILL HAWKING, v.p.-television & asst. gen. mgr.
Branch offices: Montreal Que.: Montreal, Room 201, 1118 St. Catherine St. W. Phone: 866-3466, MARK BURKO-mgr.; Vancouver, B.C.; 1131 Richards Street; 685-0288; FRANK JOBES; Winnipeg, Manitoba; 802-228 Notre Dame St., 943-0574; A.J. MESSNER.

SELTEL, INC., 750 Third Ave., New York, NY 10017; (212) 697-0944. PETER N. SISAM, chm.; HARRY H. WISE, JR., vice chm.; ARTHUR C. STRINGER, pres./ CEO; EDGAR E. WHITE, exec. v.p./COO; MARY JO ROZELL, v.p.; STEVE BOUTIS, v.p./ fin.; PHILIP L. CORPER, v.p./gen. sls. mgr.
Branch offices: Atlanta: 1819 Peachtree Road, N.E. Atlanta, GA 30309, CAL CASS, mgr.; Boston: 1400 Statler Office Bldg., Boston, MA 02116, SANTO CRUPI, mgr.; Charlotte: 5821 Park Road, Charlotte, NC 28209, CLAUDIA BROOKS, mgr.; Chicago: 401 No. Michigan Ave., Chicago, IL 60611, WILLIAM D. LESLIE, v.p., mgr.; Dallas: Lee Park Bldg., 3511 No. Hall St., Dallas, TX 75219, FRANCES PAINTER, mgr.; Des Moines: 345 Insurance Exchange Bldg., Des Moines, IA, GENE SHAW, mgr.; Detroit: 26111 Evergreen Road, Southfield, MI 48076, CAROL J. CHARRON, mgr.; Los Angeles: 5900 Wilshire Blvd., Los Angeles, CA 90036, ROY S. EDWARDS, JR., sn. v.p.; Minneapolis: 905 Foshay Tower, Minneapolis, MN 55402, BOB BOWDEN, mgr.; Philadelphia: Western Savings Bank Bldg., 1346 Chestnut St., Philadelphia, PA 19107, CATHERINE S. BRADY, mgr.; Portland: 1512 S.W. 18th Ave., Portland OR 97201, RICHARD GOHLMAN, mgr.; St. Louis: 712 No. Second St., St. Louis, MO 63102, RICHARD J. QUIGLEY III, mgr.; San Francisco: 1700 Montgomery St., San Francisco, CA 94111, GEORGE HEMMERLE, v.p./mgr.; Seattle: 1411 Fourth Ave. Bldg., Seattle, WA 98101, MICK TACHER, mgr.

STAUFFER COMMUNICATIONS, Inc., 100% owner of WIBW-TV, Box 119, Topeka, KA 66601 (913) 272-3456.
JERRY HOLLEY, v.p.; GEORGE LOGAN, TV station mgr.; R.C. SENATE, nat'l. sls. mgr. Represented nationally by The Katz Agency, Inc., New York, Chicago, Dallas, Detroit, Atlanta, San Francisco, Los Angeles, Seattle, Charlotte, Cleveland, Memphis, Portland, Boston, Jacksonville, Minneapolis, St. Louis, Philadelphia.

STORER BROADCASTING COMPANY, 1177 Kane Concourse, Miami Beach, FL 33154 (305) 866-0211.
BILL MICHAELS, chm. of bd.; PETER STORER, pres.; TERRY LEE, exec. v.p.; JOHN McCOY, v.p. Law; ARNO MUELLER, v.p. CATV Div.; BOB EDELL, v.p. adv. & p.r.; LEONARD SPRAGG, v.p. engineering; WARREN ZWICKY, v.p. WA counsel, ABIAH H. CHURCH, v.p. asst. secretary; KEN BAGWELL, v.p., television division; NEIL ROCKOFF, v.p. radio division; KEN MOSHER, v.p. treasurer.

STORER TELEVISION SALES, INC., 800 Third Ave., New York, NY 10017 (212) 935-2600.
FRANCIS BARRON, pres.; ANDREW P. POTOS, v.p.-sales; JOSEPH DOWLING, v.p.-research.
Branch Offices: Chicago, 333 N. Michigan Ave., 236-9550, CHARLES LIEBER, midwestern sls. mgr., Detroit: Number Two Storer Place, 557-0300; JAMES E. GOLDSMITH, gen. sls. mgr.; Atlanta, 1375 Peachtree St., N.E. 892-7202, J. WILLIAM MASON, gen. sls. mgr.; San Francisco, 50 California St., Tel.: 981-8860, SHAWN McGREEVY, S.F. mgr.; Los Angeles, 3660 Wilshire Blvd., 380-7320, JOSEPH E. MERTENS, west coast sales mgr.

TELEVISION ADVERTISING REPRESENTATIVES, INC., 90 Park Ave., New York, NY 10016 (212) 867-0100.
KENNETH T. MACDONALD, pres.
Branch offices: Chicago, 60611: 625 N. Michigan Ave., 527-4567, WILLIAM B. CONDON, v.p., midwest mgr.; Detroit 48226: 617 City National Bank Building, 965-6454. ROBERT NEWSHAM, mgr.; San Francisco 94104: 44 Montgomery St., 397-5088, E. JAMES HODGETT, mgr.; Hollywood 90028: 6290 Sunset Blvd., 466-1144, DOUG CLARK, mgr.; Atlanta 30309: 1800 Peachtree Rd., N.W., 351-5300, Robert Schuessler, mgr.

ADAM YOUNG, INC., 3 E. 54th St., New York, NY 10022 (212) 826-6900.
ADAM YOUNG, chm. of bd.; FRANK MARTIN, pres.; MICHAEL T. MEMBRADO, exec. v.p.
Branch Offices: Chicago: 360 N. Michigan Ave., Suite 2001, (312) 346-3017, JERRY BRITT; Atlanta: 1819 Peachtree Road, N.E., (404) 351-9288, NICK IMBORNONE: Dallas; 3626 N. Hall St. (214) 521-1240, JACK RILEY; Detroit: 2100 W. Big Beaver Rd., Troy (313) 362-0030 DONALD R. COTE; Los Angeles: 4929 Wilshire Blvd., (213) 938-2081, DAN MCCARTHY; 7503, GEORGE B. FAULDER; St. Louis: 112 N. 4th St., (314) 231-5143, BRUCE SCHNEIDER; San Francisco: 500 Sansome Street, (415) 986-5366, JOHN CASTLETON.

Programs

* NETWORK
* Syndicated

1981–82 Prime Time Shows

Series titles are listed alphabetically by network, along with time slots, suppliers, production staff heads (executive producer: EP; supervising producer: SP; senior producer: Sr. P; producer: P; co-producer: CP; director: D), cast regulars and semi-regulars.

ABC-TV

Series Title	Day	Hr.	Mins.	Supplier	Production Principals	Cast Regulars & Semi-Regulars
ABC Sunday Night Movie	Sun	9:00	120	Various		
ABC's NFL Monday Night Football	Mon	9:00	120+	ABC Sports	EP: Roone Arledge P: Bob Goodrich D Chet Forte	Frank Gifford, Howard Cosell, Don Meredith, Fran Tarkenton
Barney Miller (Starts in Nov.)	Thu	9:00	30	Four D Prods.	EP: Danny Arnold, Roland Kibbee P: Frank Dungan, Jeff Stein CP: Gary Shaw	Hal Linden, Max Gail, Ron Glass, Steve Landesberg, Ron Carey, James Gregory
Benson	Fri	8:00	30	Witt-Thomas-Harris Prods.	EP: John Rich, Paul Junger Witt, Tony Thomas, Norman Barasch P: Bob Colleary D: John Rich	Robert Guillaume, James Noble, Inga Swenson, Missy Gold, Ethan Phillips, Rene Auberjonois, Caroline McWilliams.
Best Of The West	Thu	8:30	30	Weinberger-Daniels Prods.-Paramount TV	EP: Earl Pomerantz	Joel Higgins, Carlene Watkins, Meeno Peluce, Leonard Frey, Tom Ewell, Valri Bromfield, Tracey Walter
Bosom Buddies	Thu	9:00	30	Miller-Milkis-Boyett Prods.-Paramount TV	EP: Thomas L. Miller, Edward K. Milkis, Robert L. Boyett, Chris Thompson P: Don Van Atta CP: Leonard Ripps, Stu Silver D: Joel Zwick	Tom Hanks, Peter Scolari, Donna Dixon, Telma Hopkins, Holland Taylor, Wendie Jo Sperber
Code Red	Sun	7:00	60	Irwin Allen Prods.-Columbia Pictures TV	P: Irwin Allen	Lorne Greene, Andrew Stevens, Martina Deignan, Julie Adams, Sam J. Jones, Adam Rich
Dynasty	Wed	10:00	60	Aaron Spelling Prods.-Fox-Cat Prods.	EP: Aaron Spelling, Douglas S. Cramer SP: E. Duke Vincent P: Elaine Rich, Ed Ledding	John Forsythe, Linda Evans, James Farentino, Joan Collins, Pamela Sue Martin, Al Corley, John James, Pamela Bellwood, Lee Bergere
Fall Guy, The	Wed	9:00	60	Glen A. Larson Prods.-20th-Fox TV	EP: Glen A. Larson SP: Harker Wade P: Larry Brody CP: Lee Majors	Lee Majors, Doug Barr, Jo Ann Pflug, Heather Thomas
Fantasy Island	Sat	10:00	60	Spelling-Goldberg Prods.-Columbia Pictures TV	EP: Aaron Spelling, Leonard Goldberg SP: Arthur Rowe P: Skip Webster, Don Ingalls CP: Carl Pingitore	Ricardo Montalban, Herve Villechaize, Wendy Schaal
Greatest American Hero, The	Wed	8:00	60	Stephen J. Cannell Prods	EP: Stephen J. Cannell, Juanita Bartlett SP: Jo Swerling Jr. P: Frank Lupo D: Rod Holcomb	William Katt, Robert Culp, Connie Sellecca, Michael Pare, Faye Grant, Brandon Williams
Happy Days	Tue	8:00	30	Miller-Milkis Prods.-Henderson Prod. Co.-Paramount TV	EP: Thomas L. Miller, Edward K. Milkis, Garry K. Marshall SP: Lowell Ganz P: Jerry Paris, Ronny Hallin D: Jerry Paris	Henry Winkler, Tom Bosley, Marion Ross, Anson Williams, Erin Moran, Al Molinaro, Scott Baio, Lynda Goodfriend, Cathy Silvers, Ted McGinley
Hart To Hart	Tue	10:00	60	Spelling-Goldberg Prods.-Rona II	EP: Aaron Spelling, Leonard Goldberg P: Mart Crowley	Robert Wagner, Stefanie Powers, Lionel Stander
King's Crossing	Sat	8:00	60	Lorimar Prods.-MF Prods	EP: Lee Rich, Michael Filerman P: Sally Robinson	Bradford Dillman, Mary Frann, Marilyn Jones, Doran Clark, Daniel Zippi, Linda Hamilton, Beatrice Straight, Dorothy Meyer, Michael Zaslow
Laverne & Shirley	Tue	8:30	30	Miller-Milkis Prods.-Henderson Prod. Co.- Paramount TV	EP: Garry K. Marshall, Thomas L. Miller, Edward K. Milkis SP: Arthur Silver P: Gary Menteer	Penny Marshall, Cindy Williams, Betty Garrett, David L. Lander, Michael McKean, Phil Foster, Eddie Mekka, Leslie Easterbrook

Series Title	Day	Hr.	Mins.	Supplier	Production Principals	Cast Regulars & Semi-Regulars
Love Boat, The	Sat	9:00	60	Aaron Spelling Prods.	EP: Aaron Spelling, Douglas S. Cramer P: Art Baer & Ben Joelson, Harvey Bullock & Ray Allen, Henry Colman.	Gavin MacLeod, Bernie Kopell, Fred Grandy, Ted Lange, Jill Whelan, Lauren Tewes
Maggie	Fri	9:00	30	Ermar Prods.	EP: Erma Bombeck P: Tom Whedon D: John Tracy	Miriam Flynn, James Hampton, Robert Kiger, Garin Bougie, Doris Roberts
Making A Living	Fri	9:30	30	Witt-Thomas Prods.	EP: Paul Junger Witt, Tony Thomas P: Gloria Banta, Greg Antonacci CP: Deborah Leschin D: Jay Sandrich, John Bowab	Louise Lasser, Marian Mercer, Gail Edwards, Ann Jillian, Barrie Youngfellow, Paul Kreppel, Earl Boen
Mork & Mindy	Thu	8:00	30	Henderson Prod. Co.-Miller-Milkis Prods.-Paramount TV	EP: Garry K. Marshall, Tony Marshall SP: Bruce Johnson, Dale McRaven P: Brian Levant	Robin Williams, Pam Dawber, Conrad Janis, Elizabeth Kerr, Robert Donner, Jonathan Winters, Ralph James
Open All Night	Fri	8:30	30	Freeway Prods.	EP: Bernie Brillstein P: Tom Patchett & Jay Tarses D: Will MacKenzie, Tom Patchett	George Dzundza, Susan Tyrrell, Sam Whipple, Bubba Smith, Jay Tarses, Bever-Leigh Banifield
Strike Force	Fri	10:00	60	Aaron Spelling Prods.	EP: Aaron Spelling, Douglas S. Cramer SP: E. Duke Vincent P: Joseph T. Naar, Michael Fisher	Robert Stack, Michael Goodwin, Dorian Harewood, Richard Romanus, Trisha Noble, Herb Edelman
Taxi!	Thu	9:30	30	John Charles Walters Prods.-Paramount TV	EP: James L. Brooks, Ed. Weinberger, Stan Daniels P: Howard Getwitz, Ian Praiser, Ken Estin	Judd Hirsch, Danny DeVito, Tony Danza, Marilu Henner, Christopher Lloyd, Andy Kaufman
That's Incredible	Mon	8:00	60	Alan Landsburg Prods.	EP: Alan Lansburg, Merrill Grant SP: Woody Fraser P: David Yarnell D: Arthur Forrest	John Davidson, Cathy Lee Crosby, Fran Tarkenton
Three's Company	Tue	9:00	30	NRW Co.-TTC Prods.	EP: Michael Ross, Bernie West, George Burditt P: Martin Rips, Joseph Staretski, George Sunga D: Dave Powers	John Ritter, Joyce DeWitt, Priscilla Barnes, Don Knotts, Richard Kline, Jenilee Harison
Today's FBI	Sun	8:00	60	David Gerber Co.	EP: David Gerber P: Sy Salkowitz CP: Christopher Seiter	Mike Connors, Joseph Cali, Carol Potter, Richard Hill, Harold Sylvester
Too Close For Comfort	Tue	9:30	30	Don-El Prods	EP: Arne Sultan, Earl Barret SP: Norman Hopps P: Austin & Irma Kalish D: Russ Petranto, Lee Lockhead	Ted Knight, Nancy Dussault, Deborah Van Valkenburgh, Lydia Cornell, JM J. Bullock, Deena Freeman
20/20	Thu	10:00	60	ABC News	EP: Av Westin	Hugh Downs, Geraldo Rivera, Sylvia Chase, Thomas Hoving

CBS-TV

Series Title	Day	Hr.	Mins.	Supplier	Production Principals	Cast Regulars & Semi-Regulars
Alice	Sun	9:00	30	Warner Bros. TV	EP: Bob Carroll Jr., Madelyn Davis P: Vic Rauseo, Linda Morris CP: Jerry Madden	Linda Lavin, Vic Tayback, Beth Howland, Philip McLeon, Celia Weston, Marvin Kaplan, Dave Madden
Archie Bunker's Place . .2. .	Sun	8:00	30	Tandem Prods.-Ugo Prods.	P: Joseph Gannon	Carroll O'Connor, Danielle Brisebois, Allan Melvin, Denise Miller, Steven Hendrickson, Jason Wingreen, Barbara Meek, Bill Quinn, Anne Meara, Abraham Alvarez
CBS Saturday Night Movies	Sat.	9:00	120	Various		
CBS Tuesday Night Movies	Tue	9:00	120	Various		
Dallas	Fri	10:00	60	Lorimar Prods.	EP: Philip Capice SP: Arthur Bernard Lewis P: Leonard Katzman	Barbara BelGeddes, Patrick Duffy, Linda Gray, Larry Hagman, Susan Howard, Steve Kanaly, Ken Kercheval, Victoria Principal, Charlene Tilton, Leigh McCloskey, Howard Keel, Priscilla Pointer, Jared Martin, Audrey Landers
Dukes Of Hazzard, The	Fri	9:00	60	Lou-Step Prods.-Warner Bros. TV	EP: Paul R. Picard SP: Rod Amateau P: Myles Wilder, Ralph Riskin	Tom Wopat, John Schneider, Catherine Bach, Denver Pyle, Rick Hurst, Ben Jones, James Best, Sorrell Brooke, Waylon Jennings.

Series Title	Day	Hr.	Mins.	Supplier	Production Principals	Cast Regulars & Semi-Regulars
Falcon Crest (Starts Dec. 4)	Fri	10:00	60	Lorimar Prods.-Amanda Prods.-MF Prods.	EP: Michael Filerman, Earl Hamner P: Malcolm Harding	Jane Wyman, Susan Sullivan, Robert Foxworth, Lorenzo Lamas, Abby Dalton, Billy Moses, Jamie Rose, Michael Swan, Nick Ramus, Silvano Gallardo
House Calls	Mon	9:30	30	Alex Winitsky-Arlene Sellers Prods.-Universal TV	EP: Jerome Davis, Arthur U. Gregory SP: Bill & Kathy Greer	Wayne Rogers, Lynn Redgrave, Sharon Gless, Raymond Buktenica, David Wayne, Mark L. Taylor, Deedy Peters, Aneta Corsaut
Incredible Hulk, The	Fri	8:00	60	Universal TV	EP: Kenneth Johnson SP: Nicholas Corea P: Robert Bennett Steinhauer, Karen Harris, Jill Sherman	Bill Bixby, Jack Colvin, Lou Ferrigno
Jeffersons, The	Sun	9:30	30	TAT Communications	EP: David W. Duclon, Ron Leavitt P: Michael G. Moye D: Bob Lally	Isabel Sanford, Sherman Hemsley, Roxie Roker, Franklin Cover, Marla Gibbs, Berlinda Tolbert, Mike Evans
Jessica Novak	Thu	10:00	60	20th-Fox TV	EP: Marc Merson, Jerry Ludwig P: Paul Waigner	Helen Shaver, David Spielberg, Andrew Rubin, Eric Kilpatrick, Nina Wilcox
Knots Landing	Thu	9:00	60	Lorimar Prods.-Roundelay/MF Prods.	EP: David Jacobs, Michael Filerman P: Joseph B. Wallenstein	James Houghton, Kim Lankford, Michele Lee, Constance McCashin, Donna Mills, Don Murray, John Pleshette, Ted Shackelford, Joan Van Ark, Julie Harris, Claudia Lonow, Steve Shaw, Pat Petersen, Tonya Crowe, Bobby Jacoby
Lou Grant	Mon	10:00	60	MTM Enterprises	EP: Gene Reynolds P: Seth Freeman	Edward Asner, Robert Walden, Linda Kelsey, Mason Adams, Jack Bannon, Daryl Anderson, Nancy Marchand
Magnum, P.I.	Thu	8:00	60	Belasarius Prods.-Glen A. Larson Prods.-Universal TV	EP: Don Bellisario P: Rick Weaver, Andrew Schneider, Doug Green	Tom Selleck, John Hillerman, Rogert E. Mosley, Larry Manetti
Mash	Mon	9:00	30	20th-Fox TV	EP: Burt Metcalfe SP: John Rappaport P: Thad Mumford, Dan Wilcox, Dennis Koenig	Alan Alda, Mike Farrell, Harry Morgan, Loretta Swit, David Ogden Stiers, Jamie Farr, William Christopher
Mr. Merlin	Wed	8:00	30	Larry Larry Prods.-Columbia Pictures TV	EP: Larry Rosen, Larry Tucker P: Joel Rogosin	Barnard Hughes, Clark Brandon, Jonathan Prince, Elaine Joyce
Nurse	Wed	9:00	60	Robert Halmi Inc.-Viacom	EP: Robert Halmi SP: Jack Guss P: Camille Marchetta	Michael Learned, Robert Reed, Dennis Boutsikaris, Hattie Winston, Hortensia Colorado, Bonnie Hellman
One Day At A Time	Sun	9:30	30	TAT Communications-Allwhit Inc.	EP: Alan Rafkin, Dick Bensfield, Perry Grant P: Bud Wiser D: Alan Rafkin	Bonnie Franklin, Pat Harrington, Glenn Scarpelli, Valerie Bertinelli, Nanette Fabray, Shelley Fabares
Private Benjamin	Mon	8:00	30	Warner Bros. TV	EP: Bob Carroll Jr., Madelyn Davis P: Leonard B. Kaufman, James Stein, Robert Illes	Lorna Patterson, Eileen Brennan, Hal Williams, Ann Ryerson, Lisa Raggio, Francesca Roberts, Lucy Webb
Shannon	Wed	10:00	60	Universal TV-The Aubrey Co.	EP: James T. Aubrey, James McAdams SP: Alvin Sapinsley P: David J. O'Connell	Kevin Dobson, Charlie Fields, Al Ruscio, Karen Kondazian, Michael Durrell
Simon & Simon	Tue	8:00	60	Universal TV	EP: Philip DeGuere SP: John G. Stephens P: Richard Chapman, Karen Harris	Jameson Parker, Gerald McRaney, Jeannie Wilson, Mary Carver, Eddie Barth
60 Minutes	Sun	7:00	60	CBS News	EP: Don Hewitt Sr. P: Philip Scheffler D: Arthur Bloom	Mike Wallace, Morley Safer, Harry Reasoner, Ed Bradley, Andy Rooney
Trapper John, M.D.	Sun	10:00	60	Frank Glicksman Prods.-Don Brinkley Inc.-20th-Fox TV	EP: Frank Glicksman P: Don Brinkley	Pernell Roberts, Gregory Harrison, Charles Siebert, Christopher Norris, Madge Sinclair, Brian Mitchell, Simon Scott
Two Of Us, The	Mon	8:30	30	Marble Arch Prods.	EP: Martin Starger P: Charlie Hauck D: Peter Bonerz	Peter Cook, Mimi Kennedy, Oliver Clark, Dana Hill, Tim Thomerson
WKRP In Cincinnati	Wed	8:30	30	MTM Enterprises	EP: Hugh Wilson P: Blake Hunter, Peter Torokvei	Gary Sandy, Gordon Jump, Loni Anderson, Richard Sanders, Tim Reid, Frank Bonner, Jan Smithers, Howard Hesseman
Walt Disney	Sat	8:00	60	Walt Disney Prods.	EP: William R. Yates	

NBC-TV

Series Title	Day	Hr.	Mins.	Supplier	Production Principals	Cast Regulars & Semi-Regulars
Angie Dickinson Show, The (Starts Jan. 9)	Sat	10:00	60	Carson Prods.	EP: Carol Evan McKeand, Nigel McKeand	Angie Dickinson

Series Title	Day	Hr.	Mins.	Supplier	Production Principals	Cast Regulars & Semi-Regulars
Barbara Mandrell & The Mandrell Sisters	Sat	8:00	60	Krofft Entertainment	EP: Sid & Marty Krogfft P: Frank Peppiatt D: Jack Regas	Barbara Mandrell, Louise & Irlene Mandrell, Krofft Puppets
Bret Maverick	Tue	9:00	60	Warner Bros. TV-Comanche Prods.	EP: Meta Rosenberg SP: Gordon Dawson P: Charles Floyd Johnson, Geoffrey Fischer	James Garner, Marilyn Hassett, Stuart Margolin, Ed Bruce, John Shearin, Richard Hamilton
Chips	Sun	8:00	60	Rosener TV-MGM-TV	EP: Cy Chermak P: William D. Gordon, James Doherty	Larry Wilcox, Erik Estrada, Bruce Jenner, Robert Pine, Randi Oakes, Brodie Greer, Paul Linke, Lou Wagner
Devlin Connection, The	Fri	10:00	60	Nightwatch Prods. Viacom	EP: John Wilder P: Peter Lefcourt	Rock Hudson, Jack Scalia
Diff'rent Strokes	Thu	9:00	30	Tandem Prods.	EP: Howard Leeds P: Ben Starr, Martin Cohan D: Gerren Keith	Conrad Bain, Gary Coleman, Todd Bridges, Dana Plato
Facts Of Life, The	Wed	9:00	30	TAT Communications	EP: Jack Elinson P: Jerry Mayer	Charlotte Rae, Lisa Whelchel, Kim Fields, Mindy Cohn, Nancy McKeon, Geri Jewell
Father Murphy	Tue	8:00	60	NBC Prods.	EP: Michael Landon P: Kent McCray D: William F. Claxton, Leo Penn	Merlin Olsen, Moses Gunn, Katherine Cannon, Timothy Gibbs
Fitz & Bones (Starts Oct. 24)	Sat	10:00	60	Glen A. Larson Prods.-Universal TV	EP: Glen A. Larson, Lou Shaw, James Parriott P: Ben Kadish	Tom & Dick Smothers, Mike Kellin, Diana Muldaur, Lynnette Mettey, Roger C. Carmel
Flamingo Road	Tue	10:00	60	Lorimar Prods.-MF Prods.	EP: Michael Filerman, Lee Rich SP: Kathy Shelley P: Gareth Davies	John Beck, Woody Brown, Howard Duff, Morgan Fairchild, Mark Harmon, Kevin McCarthy, Cristina Raines, Barbara-Rush, Stella Stevens, Joel Bailey
Flintstones, The	Sun	7:00	30	Hanna-Barbera Prods.	EP: William Hanna, Joseph Barbera	Voices of Henry Corden, Mel Blanc, Jean Vander Pyl, Gay Autterson, Don Messick, John Stephenson
Gimme A Break	Thu	9:30	30	Alan Landsburg Prods.	EP: Mort Lachman P: Geoff Neigher, Chick Mitchell	Nell Carter, Dolph Sweet
Harper Valley	Thu	8:00	30	Universal TV	EP: Bill Dial, Jerry Davis SP: Rod Daniel P: Robert F. O'Neill, Les Sheldon	Barbara Eden, Fannie Flagg, Anne Francine, Mills Watson, George Gobel, Jenn Thompson, Bridget Hanely, Rod McCary
Here's Boomer!	Sun	7:30	30	A.C. Lyles Prods.-Daniel Wilson Prods.-Paramount TV	EP: A.C. Lyles, Daniel Wilson SP: Daniel Wilson P: Fran Sears	Johnny (dog)
Hill Street Blues	Thu	10:00	60	MTM Enterprises	EP: Steven Bochco SP: Gregory Hoblit P: David Anspaugh, Anthony Yerkovich	Daniel J. Travanti, Michael Conrad, Michael Warren, Bruce Weitz, James B. Sikking, Joe Spano, Rene Enriquez, Kiel Martin, Taurean Blacque, Barbara Bosson, Betty Thomas, Ed Marinaro, Charles Haid, Veronica Hamel
Lewis & Clark	Thu	8:30	30	Carson Prods.	EP: Howard West, George Shapiro P: Ed Scharlach, Tom Tenowich	Gabe Kaplan, Guich Koock, Ilene Graff, Amy Linker, Mike McManus, Wendy Holcombe, Clifton James
Little House On The Prairie, The	Mon	8:00	60	NBC Prods.-Ed Friendly	EP: Michael Landon P: Kent McCray D: Michael Landon, Maury Dexter	Michael Landon, Karen Grassle, Melissa Gilbert, Melissa Sue Anderson, Matthew Laborteaux, Jason Bateman, Missy Francis, Lindsay & Sidney Greenbush, Dean Butler, Linwood Boomer, Katherine MacGregor, Richard Bull, Jonathan Gilbert
Love, Sidney	Wed	9:30	30	Warner Bros. TV-R.G. Prods.	EP: George Eckstein SP: Ernest Chambers P: April Kelly	Tony Randall, Swoosie Kurtz, Kaleena Kiff
McClain's Law	Fri	9:00	60	Eric Bercovici Prods.-MGM-TV	EP: Eric Bercovici SP: Richard St. Johns P: Mark Rodgers, Robert H. Justman	James Arness, Marshall Colt, George DiCenzo, Conchata Ferrell, Carl Franklin
NBC Magazine	Fri	8:00	60	NBC News	EP: Tom Wolzien Sr. P: Wallace Westfeldt, Dennis Sullivan, David Schmerler D: George Paul	Betsy Aaron, Douglas Kiker, Jack Perkins, Garrick Utley
NBC Monday Night At Movies	Mon	9:00	120	Various		
NBC Sunday Night At Movies	Sun	9:00	120	Various		
Nashville Palace	Sat	9:00	60	Nashville Palace Inc.	EP: John Aylesworth P: Sam Lovullo D: Bill Davis	Slim Pickens, Lynn Hancock, Dixie Dozen Dancers
Powers Of Matthew Star, The (Starts Dec. 6)	Sun	7:00	60	Harve Bennett Prods.-Paramount TV	EP: Harve Bennett	Peter Barton, Louis Gossett Jr.

Series Title	Day	Hr.	Mins.	Supplier	Production Principals	Cast Regulars & Semi-Regulars
Quincy, M.E.	Wed	10:00	60	Universal TV-Glen A. Larson Prods.	EP: David Moessinger SP: William O. Cairncross, Lester Wm. Berke P: Sam Egan	Jack Klugman, Garry Walberg, John S. Ragin, Robert Ito, Val Bisoglio, Joseph Roman
Real People	Wed	8:00	60	George Schlatter Prods.	EP: George Schlatter P: John Barbour, Bob Wynn D: Dave Caldwell	Sarah Purcell, John Barbour, Skip Stephenson, Bill Rafferty, Fred Willard, Mark Russell, Byron Allen, Peter Billingsley
Television Inside & Out (Starts Dec. 5)	Sat	10:00	60	Miss Ronna Enterprises	P-D: Larry Einhorn	Rona Barrett

Organizations

- NATIONAL GROUPS
- REGIONAL UNITS
- GUILDS & UNIONS

ORGANIZATIONS

Academy of Television Arts and Sciences

4605 Lankershim Blvd., Suite 800, No. Hollywood CA 91602 (213) 506-7880 (Formed in 1948 to advance the arts and sciences of television.) Awards Emmys for nighttime programming.
PRESIDENT
John H. Mitchell
FIRST VICE PRESIDENT
Diana Muldaur
SECOND VICE PRESIDENT
Doug Duitsman
SECRETARY
Art Schneider
TREASURER
Michael Baugh
VICE PRESIDENT—LOS ANGELES
Warren Baker
EXECUTIVE DIRECTOR
Tricia McLeod Robin
AWARDS ADMINISTRATOR
John Leverence
BOARD OF GOVERNORS
Walt Baker, Janice Brandon, Bob Bring, Terry Carter, Don Balluck, Jim Blakeley, Bruce Broughton, Patrick Collins, Dean K. Craig, Don De Mesquita, Friz Freleng, Mike Gavin, Richard Glouner, B. Donald Grant, Bill Hayes, Seymour Klate, Herbert Klynn, Robert F. Lewine, Margaret Loesch, Jerry London, Stan Margulies, Jess Marlow, Jay Michaels, Don Ohlmeyer, Carroll Pratt, Judy Price, Jan Scott, Alan Shayne, Jack Shea, Howard Smith, James A. Smith, Jean Stapleton, Mort Stevens, Howard D. Sturm, Worley Thorne, Ted Voigtlander, Jerry Weiss, Murray Weissman, Dick Wilson, John Rundall, Al Rush, David L. Wolper.

Actor's Equity Association (AAAA-AFL-CIO-CLC)

(Organized May 16, 1913; Membership, 26,000) 165 West 46th St., New York, NY 10036 (212) 869-8530; Jay Moran, 465 California St., San Francisco 91404, Suite 210; Edward Weston, 6430 Sunset Blvd., Hollywood 90028; John Van Eyck, 360 N. Michigan Ave., Chicago 60601.
PRESIDENT
Theodore Bikel
FIRST VICE-PRESIDENT
Barbara Colton
SECOND VICE-PRESIDENT
Nancy Lynch
THIRD VICE-PRESIDENT
Jeanna Belkin
FOURTH VICE-PRESIDENT
Vincent Beck
TREASURER
Robert Fitzsimmons
RECORDING SECRETARY
Carl Harms
EXECUTIVE SECRETARY
Alan Eisenberg
ASSOCIATE EXECUTIVE SECRETARY
Willard Swire
COUNSEL
Cohn, Glickstein, Lurie, Ostrin, Lubel
MIDWEST REGIONAL V.P.
Michael Lloyd
WESTERN REGIONAL V.P.
Joseph Ruskin

Advertising Council, The

825 Third Avenue, New York, NY 10022 (212) 758-0400.
CHAIRWOMAN
Patricia Carbine
PRESIDENT
Robert P. Keim
DIRECTOR OF PUBLIC AFFAIRS
Benjamin S. Greenberg
BRANCH OFFICES
The Advertising Council, 1717 North Highland Ave., Los Angeles CA; The Advertising Council, 1930 Rhode Island Ave., N.W., Washington DC 20036.

Advertising Research Foundation

3E. 54th St., New York, NY 10022 (212) Plaza 751-5656.
CHAIRMAN
William S. Rubens
VICE CHAIRMAN
Michael J. Naples
PRESIDENT
Edgar A. Roll
SECRETARY
Theodore F. Dunn
GENERAL COUNSEL
Gilbert H. Weil, Robert L. Sherman

American Advertising Federation

1225 Connecticut Avenue, N.W., Washington, DC 20036 (202) 659-1800.
CHAIRMAN OF THE BOARD
Chester R. Green
PRESIDENT
Howard H. Bell
EXEC. VICE PRESIDENT
Jonah Gitlitz

American Association of Advertising Agencies

666 Third Avenue, New York, NY 10017 (212) 682-2500.
CHAIRMAN OF THE BOARD
Don Johnston, J. Walter Thompson Co.
VICE-CHAIRMAN
John S. Bowen, Benton & Bowles
SECRETARY-TREASURER
Harold A. Shoup
PRESIDENT
Leonard S. Matthews
WASHINGTON REPRESENTATIVE
Charles F. Adams, Executive V.P.
SENIOR VICE PRESIDENT
Harry Paster
VICE-PRESIDENTS
Donald Ambuhl, Marilyn Bockman, Tony Gruber, Walter Haase, Robert Purcell
EXECUTIVE TREASURER
James D. McMahon

American Cinema Editors

(Organized November 28, 1950; Membership: 250) 4416½ Finley Avenue, Los Angeles, CA 90027 (213) 660-4425.
PRESIDENT
John A. Martinelli
VICE-PRESIDENT
Jerrold L. Ludwig
SECRETARY
Byron "Buzz" Brandt
TREASURER
Ernest V. Milano
BOARD OF DIRECTORS
Bernard Balmuth, David G. Blangsted, Robert N. (Toby) Brown, John F. Burnett, Sheldon Kahn, Frederic L. Knudtson, Rita Roland.

American Federation of Musicians (AFL-CIO)

(Organized October, 1896; Membership: 280,000) 1500 Broadway, New York, NY 10036 (212) 869-1330.
PRESIDENT
Victor W. Fuentealba, 1500 Broadway, New York, NY (212) 869-1330.

VICE-PRESIDENT
David Winstein, 2401 Esplanade Avenue, New Orleans, LA 70119 (504) 947-1700.
CANADA VICE-PRESIDENT
J. Alan Wood, 86 Overlea Blvd., Toronto, Ont., Canada M4H 1C6.
SECRETARY-TREASURER
J. Martin Emerson, 1500 Broadway, New York, NY (212) 869-1330.
EXECUTIVE BOARD
Above Officers and Max L. Arons, 261 West 52nd Street, New York, NY 10019; Mark Tully Massagli, 5020 Stacey Avenue, Las Vegas, NV 89108; Eugene V. Frey, 19 West Court St., Cincinnati, OH 45202; Max Herman, 817 North Vine Street, Hollywood, CA 90038; Harold Dessent, 175 West Washington St., Chicago, IL 60602.

American Federation of Television and Radio Artists (AAAA-AFL-CIO)

(Organized August, 1937; Membership 45,000) 1350 Avenue of the Americas, New York, NY 10019 (212) 265-7700.
PRESIDENT
Bill Hillman
FIRST VICE PRESIDENT
Jackson Beck
SECOND VICE PRESIDENT
Reuben Weiss
VICE-PRESIDENTS
Tom Pettit, Brad Phillips, Bill Cardille, Stan Farber, John Fitzgerald, Ginny Tyler, Bill Burkett
TREASURER
Elizabeth Morgan
RECORDING SECRETARY
Peter Cleaveland
NATIONAL EXECUTIVE DIRECTOR
Sanford I. Wolff
ASSISTANT NATIONAL EXECUTIVE SECRETARY
John C. Hall, Jr.
DIRECTOR, CENTRAL REGION
Herbert Neuer
DIRECTOR, WESTERN REGION
Allan Davis
DIRECTOR, EASTERN REGION
Kenneth Groot
CONTROLLER
Dan Friedenreich
NATIONAL REPRESENTATIVES
Harold Kocin, Walter Grinspan, James Doolittle, Louis Santillana, Jr.
NATIONAL MEMBERSHIP
Agnes Conklin

American Guild of Authors & Composers

40 W. 57th St., New York, NY 10019; PLaza 7-8833. California Office: 6430 Sunset Blvd., Hollywood, CA 90028. (213) HO 2-1108.
PRESIDENT
Ervin Drake

American Guild of Musical Artists, Inc. (AAAA-AFL)

(Organized 1936; Membership 4,000) 1841 Broadway, New York, NY 10023; COlumbus 5-3687-8-9.
PRESIDENT
Gene Boucher
FIRST VICE-PRESIDENT
Richard Gill
SECOND VICE-PRESIDENT
Henry Butler
THIRD VICE-PRESIDENT
Eugenia Hoeflin
FOURTH VICE-PRESIDENT
Muriel Costa-Greenspon
FIFTH VICE-PRESIDENT
Betty Baisch
TREASURER
Lawrence Davidson
RECORDING SECRETARY
Elinor Harper
NATIONAL EXECUTIVE SECRETARY
DeLloyd Tibbs
COUNSEL
Becker & London

ASSISTANT TO THE NAT'L EXEC. SECRETARY
Joan Greenspan, Alan Olsen
FINANCIAL SECRETARY & MEMBERSHIP DEPARTMENT
Catherine Thomas
ADMINISTRATIVE ASSISTANT
Annelise Kamada
DIR. OF PUBLIC RELATIONS
Mildred Grant

CANADA: Graham Spicer, 64 Shuter Street, Toronto, Ontario M5B 1B1, 416-869-1334; CHICAGO: Herbert Neuer, 307 N. Michigan Ave., Chicago, 60601, (312) 372-8081; LOS ANGELES: Dennis Moss, 1308 West Eighth St., Los Angeles 90017 (213) 385-3071; NEW ENGLAND: Robert M. Segal, 11 Beacon Street, Boston, 02108 (617) 742-0208; NEW ORLEANS: Kay Long, 34 San Jose Ave., New Orleans 70121, (504) 835-4180 NORTHWEST: Carolyn Carpp, 2253 Gilman Dr. W., 202, Seattle, WA 98119 (206) 282-0804; PHILADELPHIA: Mark P. Muller, Lafayette Bldg. 8th Floor, 5th and Chestnut Streets, Philadelphia 19106, 8th Floor, SAN FRANCISCO: Harry Polland, Donald Tayer, 100 Bush St., Suite 1500, San Francisco, 94104 (415) 986-4060; TEXAS: Benny Hopper, 4745 Shands Drive, Mesquite, 75149 (214) 279-4720; WASHINGTON, DC: Evelyn Freyman, Suite 210, Chevy Chase Center Bldg., Washington, DC 20015 (202) 657-2560.

American Guild of Variety Artists (AAAA AFL-CIO)

(Organized July 14, 1939; Registered Membership: 35,000; Active Membership: 5,000) 1540 Broadway, New York, NY 10036 (212) 765-0800.
HONORARY PRESIDENT
George Burns
HONORARY FIRST VICE PRESIDENT
Rod McKuen
HONORARY THIRD VICE PRESIDENT
Ann Corio
EXECUTIVE PRESIDENT
Alan Jan Nelson
EXECUTIVE VICE PRESIDENT
Will Aherne
SECRETARY TREASURER
Jack London
REGIONAL VICE PRESIDENTS
Dick Martin, Margie Coate, Ronald Puckett, Stanley Burns, Bob Freed, Tommy Spencer, Harry Lewis, Thomas Merriweather, Tina Marie, Steve Stephens, Ted Monson, Bobby Faye, Don George, Etta Rollins, Ralph Shaw.

American Humane Association

(Organized 1877) Hollywood office: 8480 Beverly Blvd., Hollywood, CA 90048 (213) 653-3394; Denver: 5351 S. Roslyn Street, Englewood, CO 80110. (Liaison with the television and motion picture industry as supervisors of animal action in television and motion picture production.)
PRESIDENT
Joseph Q. Betzendorfer, Jr.
VICE-PRESIDENT
Hans W. Hoel
TREASURER
Charles Ennis
ASSISTANT SECRETARY
Charles W. Friedrichs

American Radio Relay League, Inc.

225 Main Street, Newington, CT 06111 (203) 666-1541.
PRESIDENT
Harry J. Dannals
SECRETARY AND GENERAL MANAGER
Richard L. Baldwin
TREASURER
John Huntoon
COMMUNICATIONS MANAGER
John Lindholm

American Society of Cinematographers, Inc.

(Organized 1919; Membership: 286) 1782 North Orange Drive, Hollywood, CA 90028 (213) 876-5080.
PRESIDENT
Harry Wolf

FIRST VICE-PRESIDENT
Stanley Cortez
SECOND VICE-PRESIDENT
Leonard South
THIRD VICE-PRESIDENT
Milton Krasner
SECRETARY
George Folsey
TREASURER
Linwood G. Duhn

American Society of Composers, Authors and Publishers (ASCAP)

(Organized February 13, 1914; Membership: 17,000 Music Writers, 6,500 Publishers) One Lincoln Plaza, New York, NY 10023 (212) 595-3050; 6430 Sunset Blvd., Hollywood, CA 90028.
PRESIDENT
Hal David
VICE-PRESIDENTS
Irwin Robinson, Arthur Hamilton
SECRETARY
Morton Gould
TREASURER
Leon J. Brettler
ASSISTANT SECRETARY
George Duning
ASSISTANT TREASURER
Ed Murphy
COUNSEL
Bernard Korman
MANAGING DIRECTOR
Paul Marks
WESTERN REGIONAL DIRECTOR
Todd Brabec, Michael Gorfaine, ASCAP, 6430 Sunset Boulevard, Hollywood, CA 90028
SOUTHERN REGIONAL EXECUTIVE DIRECTOR
Connie Bradley, ASCAP, Two Music Square West, Nashville, TN 37203

Artists' Representatives Association, Inc.

1350 Avenue of Americas, New York, NY 10019 (212) 586-5487.
PRESIDENT EMERITUS
Eddie Elkort
HONORARY TREASURER
Joseph Singer
PRESIDENT
David C. Baumgarten
FIRST VICE PRESIDENT
Howard Hausman
VICE PRESIDENTS
Jack Russell, Irwin H. Moss, Lee Salomon
TREASURER
Larry Lewis
SECRETARY
Ben Kuchuk
COUNSEL EMERITUS
Abraham Males
BOARD OF GOVERNORS
Williard Alexander, David C. Baumgarten, Oscar Cohen, Elliott Gunty, Jack Green, Howard Hausman, Ben Kuchuk, Irwin H. Moss, Robert Rexer, Dolores Rosaler, Max Roth, Jack Russell, Charles Ryan, Lee Salomon, Joe Sully.

Associated Actors and Artistes of America (AAAA)-AFL-CIO

(Organized July 18, 1919; Membership: 85,000) 165 West 46th St., New York, NY 10036 (212) 869-0358.
PRESIDENT
Frederick O'Neal
VICE-PRESIDENTS
Chester Migden, DeLloyd Tibbs, Willard Swire, H. O'Neil Shanks, Alan Jan Nelson
TREASURER
Harold M. Hoffman
EXECUTIVE SECRETARY
Sanford I. Wolff
AFFILIATES
Actors' Equity Association, American Federation of Television and Radio Artists, American Guild of Musical Artists, American Guild of Variety Artists, Asociacion Puertorriqueña de Artistas y Tecnicos del Espectaculo, Hebrew Actors Union, Italian Actors Union, Screen Actors Guild, Screen Extras Guild.

Association of Independent Video and Filmmakers, Inc.

625 Broadway, New York, NY 10012 (212) 473-3400. (Membership: over 1,000 independent professionals, including producers, directors, writers and technicians in film and video.)
PRESIDENT
Jane Morrison

Association of National Advertisers, Inc.

155 East 44th St., New York, NY 10017 (212) 697-5950.
PRESIDENT
Peter W. Allport
VICE-PRESIDENTS
Herbert A. Ahigren, William D. Kistler, Harry L. Darling, Samuel Thurm
SECRETARY
James R. Cosgrove

Association of Talent Agents

(Organized July, 1937; Membership: 150) (Official Organization of Talent Agents in Hollywood.) 9255 Sunset Boulevard, Suite 930, Los Angeles, CA 90069 (213) 274-0628.
PRESIDENT
Marvin Faris
FIRST VICE-PRESIDENT
Roger Davis
VICE-PRESIDENT
Sandy Bresler
VICE-PRESIDENT
J. Carter Gibson
SECRETARY-TREASURER
Sandra L. Joseph

Authors' Guild, Inc.

(Membership: 5,800) 234 West 44th St., New York, NY 10036 (212) 398-0838.
PRESIDENT
Anne Edwards
VICE-PRESIDENT
Madeline L'Engle
SECRETARY
Gerold Frank
TREASURER
Barbara W. Tuchman
COUNCIL
Roger Angell, Anne Beattie, Bruce Bliven, Jr., John Brooks, Susan Brownmiller, Robert A. Caro, E. L. Doctorow, Gerold Frank, Mary Gordon, Philip Hamburger, John Hersey, Daniel Hoffman, Elizabeth Janeway, Bel Kaufman, Joseph P. Lash, Madeleine L'Engle, Milton Meltzer, Nancy Milford, Herbert Mitgang, Toni Morrison, Albert Murray, Victor S. Navasky, Sidney Offit, Frederik Pohl, Peter S. Prescott, Nora Sayre, Isaac Bashevis Singer, Louis L. Snyder, Barbara W. Tuchman, Sarah E. Wright.
EXECUTIVE SECRETARY
Peter Heggie

The Authors League of America, Inc.

(Membership: 10,000) Authors League, 234 West 44th St., New York, NY 10036 (212) 391-9198.
PRESIDENT
Harrison E. Salisbury
VICE PRESIDENT
Robert Anderson
SECRETARY
Toni Morrison
TREASURER
Garson Kanin
COUNCIL
Robert Anderson, Roger Angell, John Brooks, Anne Edwards, Jules Feiffer, Frank D. Gilroy, Gerold Frank, Ruth Goodman Goetz, James Goldman, John Hersey, John K. Hutchens, Garson Kanin, Elizabeth Janeway, James Kirkwood, Jerome Lawrence, Madeleine L'Engle, Walter Lord, Terrence McNally, Mary Rodgers, Peter Stone, Barbara W. Tuchman, Herbert Mitgang, Jerome Weidman, E. L. Doctorow
COUNSEL
Irwin Karp

ADMINISTRATIVE ASSISTANT
Shirley S. Beck

Brand Names Foundation, Inc.

477 Madison Avenue, New York, NY 10022 (212) 753-4131; Cable address: Brand Names, New York.

CHAIRMAN OF EXECUTIVE COMMITTEE
William B. Stevens, Jr.
VICE-CHAIRMAN & ASSISTANT TREASURER
Herbert Abramson
PRESIDENT & TREASURER
H. Ford Perine
VICE PRESIDENT
Harold B. Kerr
SECRETARY & ASSISTANT TREASURER
June H. Rene

Broadcast Music, Inc.

320 West 57th Street, New York, NY 10019 (212) 586-2000. 6255 Sunset Boulevard, Hollywood, CA 90028 (213) 465-2111; 10 Music Square, East, Nashville, TN (615) 259-3625; 1325 Remington Road, Schaumberg, IL 60195 (312) 843-7771; 1320 S. Dixie Highway, Coral Gables, FL 33146 (305) 666-6122; 1650 Borel Place, San Mateo, CA 94402 (213) 441-7255; 8401 Westheimer St., Houston, TX 77063 (713) 626-8570; Pan Am Bldg., Hato Rey, Puerto Rico 00917 (809) 754-6490.

BOARD CHAIRMAN
Robert Wells
PRESIDENT
Edward M. Cramer
SENIOR VICE PRESIDENT, PERFORMING RIGHTS ADMINISTRATION
Theodora Zavin
VICE PRESIDENT, FINANCE, AND TREASURER
Edward J. Molinelli
VICE PRESIDENT, NASHVILLE
Frances W. Preston
VICE PRESIDENT, CALIFORNIA
Ronald M. Anton
VICE PRESIDENT, LICENSING
Alan H. Smith
COUNSEL AND SECRETARY
Edward W. Chapin

Broadcast Pioneers

320 West 57th Street, New York, NY 10019 (212) 586-2000.

PRESIDENT
Reid G. Chapman
FIRST VICE-PRESIDENT
Dr. Hal Niven
TREASURER
Sol Paul
EXECUTIVE DIRECTOR
Edward J. DeGray

Broadcasting Foundation of America

52 Vanderbilt Ave., New York, NY 10017; (212) 986-6448. (Organized 1955).

CHAIRMAN OF THE BOARD
George E. Probst
EXECUTIVE DIRECTOR
Calvin W. Stillman
SECRETARY
Calvin Stillman
TREASURER
John Merriam

Catholic Actors Guild of America

(Organized April, 1914; Membership: 700)

1501 Broadway, New York, NY 10036 (212) 398-1868.

PRESIDENT
Jack O'Brian
VICE-PRESIDENTS
Frederick O'Neal, John J. Martin, Lisa Di Julio
CHAIRMAN OF EXECUTIVE BOARD
William J. O'Malley

EXECUTIVE SECRETARY
Michael J. Tucci

Catholic Broadcasting Office

(See Office for Film & Broadcasting)

Catholic Communications Foundation

Ste. 198, Box 9000, Carlsbad, CA 92008 (714) 942-1120. Established by the Catholic Fraternal Benefit Societies to assist the national, regional and local radio and television activities of the Catholic Church in the United States. The Foundation awards communications training scholarships to the Institute for Religious Communications. The CCF's board of directors is composed of Bishops of the Catholic Church, fraternal benefit society executives and representatives from the communications industry.

CHAIRMAN
Most Reverend Anthony G. Bosco, Auxiliary Bishop of Pittsburgh
PRESIDENT
Most Reverend Kenneth Povish, Bishop of Lansing, MI.
EXECUTIVE DIRECTOR
James R. Bastis

Catholic Writers Guild of America, Inc.

(Organized December 12, 1919); 65 E. 89th St., New York, NY 10028.

HONORARY PRESIDENT
Terence Cardinal Cooke, Archbishop of New York
VICE-PRESIDENT
Bernard McMahon
SPIRITUAL DIRECTOR
Most Rev. Philip J. Furlong, D.D.

Clear Channel Broadcasting Service

1776 K St., N.W., Washington, DC 20006. (202) 296-4855.

PRESIDENT
E. W. Wendell, WSM, Nashville, TN
TREASURER
A. M. Herman, WBAP, Fort Worth
SECRETARY
Daniel C. Calibraro, WGN, Chicago
LEGAL COUNSEL
R. Russell Eagan of Kirkland & Ellis, Washington, DC.

Communication Commission of National Council of Churches of Christ in the USA

475 Riverside Dr., New York, NY 10027 (212) 870-2567.

CHAIRPERSON
Sonia Francis
ASSISTANT GENERAL SECRETARY FOR COMMUNICATION, NCC
William F. Fore
EXECUTIVE DIRECTOR FOR BROADCASTING
D. Williams McClurken
DIRECTOR FOR MEDIA RESOURCES
David W. Pomeroy

Corporation for Public Broadcasting

1111 16th St., N.W. Washington, DC 20036 (202) 293-6160.

PRESIDENT
Robben W. Fleming
SPECIAL ASSISTANT TO THE PRESIDENT FOR TELECOMMUNICATIONS
Ronald C. Bornstein
VICE PRESIDENT AND TREASURER
Fred P. Wacker
VICE PRESIDENT AND CONTROLLER
Harvey G. Dickerson

GENERAL COUNSEL AND SECRETARY
Paul A. Mutino

Directors Guild of America, Inc. (DGA)

National office: 7950 Sunset Blvd., Hollywood, CA 90046 (213) 656-1220. 110 West 57th St., New York, NY 10019 (212) 581-0370; Chicago: 40 East Oak St., Chicago, IL 60611 (312) 787-5050.

PRESIDENT
George Schaefer
VICE PRESIDENTS
Tom Donovan, Gilbert Cates, John Avildsen, Jud Taylor, Karl Genus
SECRETARIES
William Beaudine, Jr., Enid Roth
TREASURERS
Sheldon Leonard, Marilyn Jacobs
NATIONAL EXECUTIVE SECRETARY
Michael H. Franklin
BOARD MEMBERS
Alan Alda, John Avildsen, William Beaudine, Jr., Jerome J. Blumenthal, Paul Bogart, Mel Brooks, Gilbert Cates, Hal Cooper, Andrew M. Costikyan, Tom Donovan, Milton Felsen, Karl Genus, Arthur Hiller, Marilyn Jacobs, Norman Jewison, Elia Kazan, Sheldon Leonard, Ira Marvin, Franklin Melton, Chris Montross, Edmund Nadell, Ted Nathanson, John Rich, Enid Roth, Jonathan Sanger, George Schaefer, Franklin Schaffner, Jane Schimel, Jack Shea, Elliott Silverstein, Abby Singer, Jud Taylor

The Dramatists Guild, Inc.

(Membership: 708 Active; 2746 Associate; 284 Subscribing) 234 West 44th St., New York, NY 10036 (212) 398-9366.

PRESIDENT
Stephen Sondheim
VICE PRESIDENT
Richard Lewine
SECRETARY
Sheldon Harnick
EXECUTIVE DIRECTOR
David E. LeVine
COUNSEL
Irwin Karp

Electronic Industries Association

2001 Eye Street, N.W., Washington, DC 20006 (202) 659-2200.

CHAIRMAN OF THE BOARD
Thomas A. Campobasso, Vice President, International and Government Relations, Electronics Operations, Rockwell International Corp., Dallas, TX
PRESIDENT
Peter F. McCloskey
TREASURER
William F. Tait, RCA Corporation, Washington, DC
GENERAL COUNSEL & SECRETARY
Raymond E. Johnson, EIA Headquarters, Washington, DC
VICE PRESIDENTS
Travis Marshall, Motorola, Inc., Washington, DC (Communications Division); William E. Boss, RCA Corporation, Indianapolis (Consumer Electronics Group); Arthur F. Kelly, Jr., Amperex Electronic Corporation, Hicksville, NY (Distributor Products Division); D. Clem Arnold, McDonnell Douglas Electronics Company, St. Charles, MO (Government Division); J. Earl Templeton, Illinois Tool Works, Inc., Chicago (Parts Division); Roger Damm, Motorola, Inc., Phoenix (Solid State Products Division); Roger W. Slinkman, GTE Corporation, Seneca Falls, NY (Tube Division)
CHAIRMAN, ENGINEERING DEPARTMENT
Glenn W. Carter, Dale Electronics, Inc., Columbus, NB
CHAIRMAN, MARKETING SERVICE DEPARTMENT
Paul J. Terhaar, GTE Corporation, Batavia, NY
CHAIRMAN, GOVERNMENT PROCUREMENT RELATIONS COUNCIL
Lawrence Caruso, General Electric Company, Wash., DC
CHAIRMAN, INDUSTRIAL RELATIONS COUNCIL
David Barclay, Hughes Aircraft Company, CA
CHAIRMAN, INTERNATIONAL BUSINESS COUNCIL
John B. Copeland, Motorola, Inc., Washington, DC
CHAIRMAN, LEGISLATIVE AFFAIRS COUNCIL
Susan Clark, Rockwell International, Washington, DC

Episcopal Actors Guild of America, Inc.

(Organized 1926; 750 members) 1 East 29th St., New York, NY 10016 (212) 685-2927.

HONORARY PRESIDENTS PRESIDING BISHOP
The Most Reverend John Maury Allin, D.D.
BISHOP OF NEW YORK
The Right Reverend Paul Moore, Jr., S.T.D.
RETIRED BISHOP OF NEW YORK
The Right Reverend Horace W. B. Donegan, D.D.
PRESIDENT
Ray Heatherton
VICE PRESIDENTS
Rev. Norman J. Catir, Jr., Warden of the Guild Walter Abel, Joan Fontaine, Rex Harrison
RECORDING SECRETARY
Judy Frank
TREASURER
Anthony Mercede
EXECUTIVE SECRETARY
Lon C. Clark
COUNCIL
Myrtle Ash, Dorothy Blackburn, Ralph Bell, Col. C. L. Campbell, Kathleen Claypool, John Connell, Elizabeth Council, Tom Dillon, Alfred Drake, Jose Ferrer, Larry Gates, Eleanor Cody Gould, Bernard Grossman, Nelson Gurnee, Peter Harris, Bernard Hughes, Josephine Hutchinson, Frederic Langford, Robert Lissauer, Mildred Murray, Mildred Natwick, Louis Racnow, Anne Seymour, Jean Stapleton, Donald Sutphin, Lila Tyng, L. Henderson Werth.

Federal Communications Commission

1919 M Street, N.W., Washington, DC 20054.

CHAIRMAN
Charles D. Ferris
COMMISSIONERS
Robert E. Lee, James H. Quello, Abbott M. Washburn, Joseph B. Fogarty, Margita E. White, Tyrone Brown
CHIEF, OFFICE OF PLANS AND POLICY
Carlos V. Roberts
CHIEF, OFFICE OF OPINIONS AND REVIEW
David W. Warren, Jr.
CHAIRMAN, REVIEW BOARD
Leonidas P. B. Emerson
CHIEF JUDGE, OFFICE OF ADMINISTRATIVE LAW JUDGES
Chester F. Naumowicz, Jr.
GENERAL COUNSEL
Robert R. Bruce
CHIEF ENGINEER
William A. Luther
EXECUTIVE DIRECTOR
Richard D. Lichtwardt
CHIEF, BROADCAST BUREAU
Wallace E. Johnson
CHIEF, CABLE TELEVISION BUREAU
James R. Hobson
CHIEF, COMMON CARRIER BUREAU
Walter R. Hinchman
CHIEF, SAFETY AND SPECIAL RADIO SERVICES BUREAU
Charles A. Higginbotham
CHIEF, FIELD OPERATIONS BUREAU
C. Phyll Horne
SECRETARY
William J. Tricarico
CHIEF, CONSUMER ASSISTANCE OFFICE
Belle O'Brien
CHIEF, PUBLIC INFORMATION OFFICE
Samuel M. Sharkey

The Harry Fox Agency

110 East 59th St., New York, NY 10022 (212) 751-1930.

(Music Licensing Bureau)

PRESIDENT
Albert Berman

Girls Friday of Show Business

(See Women in Show Business)

Institute of Electrical and Electronics Engineers, Inc.

345 East 47th Street, New York, NY 10017 (212) 644-7900.

PRESIDENT
Dr. J.J. Suran
EXECUTIVE VICE PRESIDENT
Dr. Leo Young
VICE PRESIDENT—EDUCATIONAL ACTIVITIES
Dr. B.J. Leon
VICE PRESIDENT—PROFESSIONAL ACTIVITIES
Dr. B.O. Weinschel
VICE PRESIDENT—PUBLICATION ACTIVITIES
Dr. R.W. Lucky
VICE PRESIDENT—REGIONAL ACTIVITIES
P.F. Carroll
VICE PRESIDENT—TECHNICAL ACTIVITIES
Dr. C.L. Hogan
SECRETARY/TREASURER
D.S. Brereton
GENERAL MANAGER AND EXECUTIVE DIRECTOR
Eric Herz

International Alliance of Theatrical Stage Employes & Moving Picture Machine Operators of the U.S. and Canada (AFL-CIO)

(Organized nationally, July 17, 1893; internationally, October 1, 1902.) 1515 Broadway, New York, NY 10036 (212) 730-1770.

PRESIDENT
Walter F. Diehl
GENERAL SECRETARY-TREASURER
Harold F. Chadwick
FIRST VICE-PRESIDENT
George J. Flaherty, Hollywood, CA
SECOND VICE-PRESIDENT
A.L. "Pat" Travers, Toronto, Ont.
THIRD VICE-PRESIDENT
Paul G. Robertson, Miami, FL
FOURTH VICE-PRESIDENT
Edward C. Powell, San Rafael, CA
FIFTH VICE-PRESIDENT
Adrian Short, Jr., Bay Village, OH
SIXTH VICE-PRESIDENT
John J. Nolan, Paramus, NJ
SEVENTH VICE-PRESIDENT
J.E. Johnson, Ft. Worth, TX
EIGHTH VICE-PRESIDENT
John J. Ryan, Burtonsville, MD
NINTH VICE-PRESIDENT
Al Parizer, Woodbridge, CT
INTERNATIONAL TRUSTEES
James J. Riley, Cleveland, OH; William L. Daniel, Chamblee, GA; Allan H. Wasson, Grand Junction, CO

The Alliance is comprised of approximately 900 local unions covering the United States, Canada and Hawaii. Following is a list of the New York, Chicago and Hollywood locals:

PRODUCTION

AFFILIATED PROPERTY CRAFTSMEN LOCAL 44 (IATSE—AFL—CIO), HOLLYWOOD

(Organized May 15, 1939) 7429 Sunset Blvd., Hollywood, CA 90046 (213) 876-2320 and 838-7788.

ART DIRECTORS, LOCAL 876 (IATSE) HOLLYWOOD

(Chartered January 7, 1960) 7715 Sunset Blvd., Hollywood, CA 90046 (213) 876-4330.

BROADCASTING STUDIO EMPLOYEES, LOCAL 782 (IATSE) NEW YORK

(Chartered Jan. 5, 1945) 41 Nottingham Rd., Glen Rock, NJ 07452 (201) 445-7372.

FIRST AID EMPLOYEES, LOCAL 767 (IATSE), LOS ANGELES

(Chartered Oct. 30, 1942) 8736 Swinton Ave., Sepulveda, CA 91343 (213) 894-3781.

INTERNATIONAL PHOTOGRAPHERS OF THE MOTION PICTURE INDUSTRIES (Cameramen) IPMPI LOCAL 666, CHICAGO

(Chartered Jan. 1, 1929) Suite 1122, 327 S. La Salle St., Chicago, IL (312) 341-0966.

IPMPI, LOCAL 659 (IATSE), LOS ANGELES

(Organized 1928) 7715 Sunset Blvd., Hollywood, CA 90046 (213) 876-0160.

IPMPI, LOCAL 644 (IATSE), NEW YORK

(Organized Nov. 15, 1926) 250 W. 57th St., New York, NY 10019 (212) 247-3860.

INTERNATIONAL SOUND TECHNICIANS OF THE MOTION PICTURE BROADCAST AND AMUSEMENT INDUSTRIES, LOCAL 695 (IATSE-AFL), LOS ANGELES

(Organized Sept. 15, 1930) 15840 Ventura Blvd., Encino, CA 91436 (213) 872-0452 and 981-0452.

LABORATORY TECHNICIANS, LOCAL 780 (IATSE), CHICAGO

(Chartered Nov. 10, 1944) 327 S. La Salle St., Room 1720, Chicago, IL (312) 922-7105.

LABORATORY TECHNICIANS, LOCAL 683 (IATSE-AFL-CIO), LOS ANGELES

(Organized Sept. 29, 1919) 6721 Melrose Ave., Hollywood, CA 90038 (212) 935-1123.

LABORATORY TECHNICIANS, LOCAL 702 (IATSE-AFL), NEW YORK

(Organized September, 1937) Room 1405, 165 West 46th Street, New York, NY 10036 (212) PLaza 7-5540.

MAKE-UP ARTISTS & HAIR STYLISTS, LOCAL 706 (IATSE), HOLLYWOOD

11519 Chandler Blvd., No. Hollywood, CA 91601.

MAKE-UP ARTISTS AND HAIR STYLISTS, LOCAL 798 (IATSE), NEW YORK

(Chartered Feb. 18, 1949) 1790 Broadway, New York, NY 10019 (212) 354-6016.

MOTION PICTURE COSTUMERS, LOCAL 705 (IATSE-AFL), HOLLYWOOD

(Chartered Nov. 1, 1937) 1427 N. La Brea Ave., Hollywood, CA 90028 (213) 851-0220.

MOTION PICTURE EDITORS GUILD, LOCAL 776 (IATSE), LOS ANGELES

(Chartered Aug. 2, 1944) Secretary's Address: 7715 Sunset Blvd., Hollywood, CA 90046 (213) 876-4770.

MOTION PICTURE FILM EDITORS, LOCAL 771 (IATSE), NEW YORK

(Chartered Aug. 18, 1943) 630 Ninth Ave., New York, NY 10036 (212) 582-3728.

MOTION PICTURE SCRIPT SUPERVISORS AND PRODUCTION OFFICE COORDINATORS LOCAL 161

250 W. 57 St., Room 1202, New York, NY 10019 (212) 245-4562.

MOTION PICTURE SET PAINTERS, LOCAL 729 (IATSE), HOLLYWOOD

(Chartered Aug. 1, 1953) 12754 Ventura Blvd., No. Hollywood, CA 91604 (213) 984-3000.

MOTION PICTURE STUDIO CINETECHNICIANS, LOCAL 789 (IATSE), HOLLYWOOD

(Chartered April 16, 1945) 1635 Vista Del Mar. Hollywood, CA 90028. (213) 462-7288.

MOTION PICTURE SCREEN CARTOONISTS, LOCAL 839 (IATSE), HOLLYWOOD

(Chartered Jan. 18, 1952) 12441 Ventura Blvd., Studio City, CA 91604.

MOTION PICTURE SCREEN CARTOONISTS, LOCAL 841 (IATSE), NEW YORK

(Chartered April 16, 1952) 25 W. 43rd St., New York, NY 10036.

MOTION PICTURE STUDIO ELECTRICAL TECHNICIANS, LOCAL 728 (IATSE), AND M.P.M.O. of U.S. AND CANADA-A.F.L-C.I.O.

(Chartered May 15, 1939) 3400 Barham Blvd., Hollywood, CA 90068.

MOTION PICTURE CRAFTS SERVICE, LOCAL 727 (IATSE), HOLLYWOOD

(Organized May 15, 1939) 12754 Ventura Blvd., Suite 205, North Hollywood, CA 91604.

MOTION PICTURE STUDIO ART CRAFTSMEN. (Illustrators and Matte Artists) LOCAL 790 (IATSE), HOLLYWOOD

(Chartered April 17, 1945) Suite 210, 7715 Sunset Blvd., Los Angeles, CA 90046 (213) 876-2010.

MOTION PICTURE STUDIO GRIPS, LOCAL 80 (IATSE), HOLLYWOOD

(Organized May 15, 1939) 6926 Melrose Ave., Los Angeles, CA 90038 (213) 931-1419.

MOTION PICTURE STUDIO TEACHERS AND WELFARE WORKERS, LOCAL 884 (IATSE) HOLLYWOOD

(Chartered September 1, 1960) 9617 Haas Avenue, Los Angeles, CA 90047 (213) 754-8419.

PUBLICISTS, LOCAL 818 (IATSE), HOLLYWOOD

(Chartered July 11, 1955) 1427 N. La Brea Ave., Hollywood CA 90028

RADIO AND TELEVISION SOUND EFFECTS, LOCAL 844 (IATSE), NEW YORK

(Chartered July 17, 1952), Box 637, Ansonia Station, New York, NY 10023.

SCENIC & TITLE ARTISTS, LOCAL 816 (IATSE) LOS ANGELES

(Chartered March 31, 1949) 7429 Sunset Blvd., Hollywood, CA 90046 (213) 876-1440.

SCRIPT SUPERVISORS, LOCAL 871 (IATSE), HOLLYWOOD

(Chartered January 1, 1958) 7715 Sunset Blvd., Hollywood, CA 90046 (213) 876-4433.

SET DESIGNERS AND MODEL MAKERS, LOCAL 847 (IATSE), HOLLYWOOD

(Chartered Nov. 14, 1952) Suite 210, 7715 Sunset Blvd., Los Angeles 46 CA (213) 876-2010.

STORY ANALYSTS, LOCAL 854 (IATSE), HOLLYWOOD

(Chartered Oct. 18, 1954) 7715 Sunset Blvd., Los Angeles, CA 90046.

MOTION PICTURE STUDIO MECHANICS, LOCAL 476 (IATSE), CHICAGO

(Chartered Feb. 2, 1931) Room 1743; 327 S. LaSalle St., Chicago, IL 60604. (312) 922-5215.

STUDIO MECHANICS, LOCAL 52 (IATSE-AFL), NEW YORK

(Organized 1924) 221 W. 57th St., New York, NY 10019 (212) 765-0741.

STUDIO PROJECTIONISTS, LOCAL 165 (IATSE), HOLLYWOOD

(Chartered May 15, 1939) 6640 Sunset Blvd., Suite 110, Hollywood, CA 90028 (213) 461-2985.

TELEVISION BROADCASTING STUDIO EMPLOYEES, LOCAL 794 (IATSE), NEW YORK

(Chartered June 7, 1945) 144 Aster Ave., Hawthorn, New York, NY 10532.

DISTRIBUTION

FILM EXCHANGE EMPLOYEES, BACK ROOM, LOCAL B-45 (IATSE), CHICAGO

(Chartered May 1, 1937)

FILM EXCHANGE EMPLOYEES, BACK ROOM, LOCAL B-61 (IATSE), LOS ANGELES

(Chartered May 1, 1937) 7715 Sunset Blvd., Room 226, Hollywood, CA 90046.

FILM EXCHANGE EMPLOYEES, FRONT OFFICE, LOCAL F-45 (IATSE), CHICAGO

(Chartered Sept. 4, 1942) c/o G.R. Kuehnl, Apt. 1611, 5455 Sheridan Rd., Chicago, IL 60640.

In addition to the above, there are 34 locals of Back Room Employees and 29 locals of Front Office Employees in the other exchange cities.

MOTION PICTURE HOME OFFICE AND FILM EXCHANGE EMPLOYEES, LOCAL H-63 (IATSE), NEW YORK

(Chartered Mar. 19, 1945) 1650 Broadway, New York, NY 10019 (212) 247-7630.

EXHIBITION

AMUSEMENT AREA EMPLOYEES, LOCAL B-192 (IATSE) LOS ANGELES

(Chartered Oct. 1, 1965) 7715 Sunset Blvd., Room 226, Los Angeles, CA 90046.

EXHIBITION EMPLOYEES, LOCAL 829 (IATSE), NEW YORK

(Chartered December 11, 1950) 150 E. 58 St., New York, NY 10022 (212) 752-4427.

PROJECTIONISTS LOCAL 110 (IATSE), CHICAGO

(Chartered Feb. 4, 1915) 875 N. Michigan Ave. Suite 4160, Chicago, IL 60611 (312) 787-0220.

PROJECTIONISTS LOCAL 150 (IATSE), LOS ANGELES

(Chartered July 16, 1908) 1800 S. Vermont Ave., Los Angeles, CA 90006 (213) 733-1191.

PROJECTIONISTS LOCAL 306 (IATSE), NEW YORK

(Organized July, 1913) 745 7th Ave., New York, NY 10019 (212) 586-5157, and 55 Flatbush Ave., Brooklyn, NY (212) 852-2700.

STAGE EMPLOYEES, LOCAL 4 (IATSE), BROOKLYN

(Chartered April 8, 1888) 2917 Glenwood Rd., Brooklyn, NY 11210 (212) 252-8777.

STAGE EMPLOYEES, LOCAL 2 (IATSE), CHICAGO

(Chartered July 17, 1893) 222 W. Adams St., Room 1345, Chicago, IL 60606.

STAGE EMPLOYEES, LOCAL 33 (IATSE), LOS ANGELES

(Chartered Mar. 1, 1896) 4605 Lankershin Blvd. Suite 833, N. Hollywood, CA 91602 (213) 985-0633-4.

STAGE EMPLOYEES, LOCAL 1 (IATSE), NEW YORK

(Chartered July 17, 1893) 1775 Broadway, New York, NY 10019 (212) 489-7710.

THEATRE EMPLOYEES, LOCAL B-46 (IATSE), CHICAGO

(Chartered May 1, 1937) 875 N. Michigan Ave. Suite 4160, Chicago, IL 60611 (312) 787-0220.

THEATRE EMPLOYEES, LOCAL B-183 (IATSE), NEW YORK

(Chartered May 6, 1942) 235 W. 46th St., Rm. 320, New York, NY 10036 (212) 245-2331, 2332.

THEATRICAL WARDROBE ATTENDANTS, LOCAL 769 (IATSE), CHICAGO

3314 Lake Shore Drive, Apt. 805, Chinago, IL 60657.

THEATRICAL WARDROBE ATTENDANTS, LOCAL 768 (IATSE), LOS ANGELES

(Chartered Dec. 3, 1942) 5909 Melrose Ave. Suite 17, Los Angeles, CA 90038.

THEATRICAL WARDROBE ATTENDANTS, LOCAL 764 (IATSE), NEW YORK

(Chartered Sept. 4, 1942) 1501 Broadway, Room 1604, New York, NY (212) 582-4910-1.

TREASURERS AND TICKET SELLERS & CASHIERS, LOCAL 750 (IATSE), CHICAGO

(Chartered Aug. 1, 1941) 188 W. Randolph St., Room 1920, Chicago, IL 60601.

TREASURERS AND TICKET SELLERS, LOCAL 857 (IATSE), LOS ANGELES

(Chartered June 1, 1955) 6513 Hollywood Blvd., Room 204, Hollywood, CA 90028.

TREASURERS AND TICKET SELLERS, LOCAL 751 (IATSE), NEW YORK

(Chartered Aug. 1, 1941) Hotel Piccadilly, 227 West 45th St., New York, NY 10036 (212) Circle 5-7186.

Outside the three cities covered by the above listings, there are over 900 IATSE locals in the exhibition field.

International Brotherhood of Electrical Workers (AFL)

(Organized November 28, 1891; Membership over 1,000,000) 1125 15th Street, N.W., Washington, DC 20005 (202) 833-7000.

INTERNATIONAL PRESIDENT
Charles H. Pillard, 1125 15th St., N.W., Washington, DC 20005
INTERNATIONAL SECRETARY
Ralph A. Leigon, 1125 15th St., N.W., Washington, DC 20005
INTERNATIONAL TREASURER
Thomas Van Arsdale, Jr., 158-11 Jewel Ave., Flushing, NY 11365
DISTRICT OFFICES
Willowdale, Ontario: 45 Sheppard Ave. East, Suite 401; K. G. Rose
Braintree, MA: 161 Forbes Road, Fourth Floor, John E. Flynn
White Plains, NY: 222 Mamaroneck Ave., J. J. Barry
Cincinnati, OH: 7710 Beading Rd., Suite 9, B. G. Williamson
Birmingham, AL: No. 2 Metroplex Dr., Suite 113, D. H. Waters
Carol Stream, IL: 373 Schmale Road, Suite 201, James P. Conway
Arlington, TX: 2701 Ave. E. East, Suite 412, R. G. Duke
Denver, CO: 2460 West 26th Ave., Suite 264-C, L. C. Farnan
Walnut Creek, CA: 150 N. Wiget Lane, Suite 100, S. R. McCann
Rosemont, IL: 10400 W. Higgins Rd., Suite 400, A. M. Ripp
Springfield, MO: 300 So. Jefferson, Suite 300, J. F. Moore
Chattanooga, TN: Franklin Bldg., Suite 515, M. A. Williams

IBEW LOCAL 349 (FILM)

(Organized April 24, 1904) 1657 N.W. 17th Avenue, Miami, FL 33135 (305) 325-1330.

BUSINESS MANAGER
Art Fernandez

IBEW LOCAL 40 (FILM)

(Organized March 5, 1923) 5643 Vineland Ave., No. Hollywood, CA 91601 (213) 877-1171.

BUSINESS MANAGER
Russell J. Bartley

International Radio and Television Society Inc.

420 Lexington Avenue, New York, NY 10170 (212) 532-4546.

PRESIDENT
Auram Butensky
FIRST VICE PRESIDENT
James H. Rosenfield
VICE PRESIDENTS
Charles A. DeBare, Walter A. Schwartz, Sherril W. Taylor, Arthur A. Watson
SECRETARY
Jerome S. Boros
TREASURER
Alan R. Griffith
EXECUTIVE DIRECTOR
Stephen B. Labunski

Musicians Union Local 47 American Federation of Musicians (AFL-CIO)

(Organized October 30, 1894; Membership: 15,000) 817 Vine Street, Hollywood 38, CA Hollywood 2-2161.

PRESIDENT
Max Herman
VICE PRESIDENT
Vince DiBari
SECRETARY
Marl Young
TREASURER
Bob Manners
TRUSTEES
Johnny Rotella, Abe Most, Lyle (Spud) Murphy
DIRECTORS
Chase Craig, Nellie Lutcher, Frank (Chico) Guerrero, Ray Siegel, Serena Williams, Tibor Zelig
TRIAL BOARD
Clint Neagley, chairman; Beverly Carmen, Bob Karp, Thomas Cortez, Joe R. Holguin, Peggy Gilbert, Tony Horowitz.

National Academy of Television Arts and Sciences, The

(National Office: 110 West 57th St., New York, NY 10019 (212) 586-8424. (Formed to advance the arts and sciences of television.) (Organized 1957.) Awards Emmys for daytime, children's, religious, sports and news documentary programming.

CHAIRMAN OF THE BOARD
Joel Chaseman
PRESIDENT
John Cannon
VICE CHAIRMAN
Lee Polk
VICE PRESIDENT
Jack Moffit
SECRETARY
Richard Rector
TREASURER
Alfred L. Plant
BOARD OF TRUSTEES
Edward Adams, Joel Albert, Rod Burton, Harvey Chertok, David Davidson, George Dessart, Charles Dutcher III, Martha Greenhouse, Keith Gunther, Linda Hobkirk, Ralph Hodges, Dr. Jack Hunter, Arthur Kent, Ann Loring, Clarence McIntosh, Art Pattison, Richard Rector, Danny Royal, James Ruddle, Dick Schneider, Robert Simon, Christine Spencer, Frank Strnad, Bill Stull, Jo Subler and Jo Schmidt.
CHAPTERS
Arizona: David Debarger, Pres., KAET, Arizona State Univ., Tempe, AZ 85281.
Atlanta: Bill Sanders, Pres., Georgia Assoc. of Broadcaster, Suite 604, 6065 Roswell Rd., NE, Atlanta, GA 30328
Boston New England: John Atkinson, Pres., RKO General Building, Government Center, Boston, MA 02114
Chicago: Robert Smith, Pres., WBBM, 639 North McClurg Ct., Chicago, IL 60611
Cleveland: Jack Moffitt Pres., WUAB, 8443 Day Drive, Parma, OH 44129
Columbus/Dayton/Cincinnati: Dick Thrall, Multimedia Broadcasting, 140 West Ninth Street, Cincinnati, Ohio 45202
Detroit: John Schimpf, Pres., Michigan Bell, 1365 Cass Ave., Suite 1916, Detroit, MI 48226
Miami: Frank Gatto, Pres., 326 North Miami Ave., Miami, FL 33128
New York: Dick Schneider, Pres., Cazadero Productions, 139 East 35th St., New York, NY 10016
San Diego: Bill Gruber, Pres., KCOX P.O. Box 19700, San Diego, CA 92119
San Francisco: Dave McLean, Pres., KRON-TV, Chronicle Broadcasting Co., P.O. Box 3412, San Francisco, CA 94119
St. Louis: Irvin Davis, Pres., Clayton Davis and Assoc., 8229 Maryland, St. Louis, MO 63105
Seattle: Don Lacombe, Pres., KING-TV, 320 Aurora Ave North, P.O. Box 24525, Seattle, Washington, 98109
Washington, D.C.: Rich Adams WDVM, 49th Street and Brandwynne NW, Washington, D.C 20016.

National Association of Broadcast Employees and Technicians

(Organized nationally in 1933 as the Association of Technical Employees, name changed in 1940 to NABET internationally in 1952) AFL-CIO, CLC (U.S.A. and Canada) Office of International—80 E. Jackson Blvd., Chicago, IL 60604 (312) 922-2462

International Officers:
PRESIDENT
Edward M. Lynch
VICE-PRESIDENT
James P. Nolan
PRESIDENT, CANADA
Armand Bergeron
SECRETARY-TREASURER, U.S.
Duane R. Corder

National Association of Broadcasters

1771 N St., N.W., Washington, DC 20036 (202) 293-3500. (Organized 1922 to foster and promote the development of the arts of aural and visual broadcasting in all its forms; to protect its members in every lawful and proper manner from injustices and unjust exactions; to do all things necessary and proper to encourage and promote customs and practices which will strengthen and maintain the broadcasting industry to the end that it may best serve the public).

CHAIRMAN
Thomas E. Bolger
PRESIDENT
Vincent T. Wasilewski
RADIO BOARD OF DIRECTORS
Harry Barker, Martin Beck, Fred Conger, Chuck Cooper, Edward Fritts, Cliff Gill, J. William Grimes, Len Hensel, Robert Hilker, Ben Hoberman, Eugene Jackson, Bruce Johnson, Michael Lareau, Jerry Lee, John Lemme, Arnold Lerner, Roy Maple, Stanley McKenzie, Thomas McKinney, Dick Painter, Robert Pricer, Walter Rubens, Martin Rubenstein, Ted Smith, Thom Smith, William Stakelin, Marion Stephenson, Cullie Tarleton, Bob Thomas, J. T. Whitlock, Charles Wright
TELEVISION BOARD OF DIRECTORS
Leslie Arries, Eugene Bohi, William Brazzil, Kathryn Broman, Reid Chapman, Eugene Cowen, Don Curran, William Dilday, Jr., W. Frank Harden, Peter Kenney, Robert King, William Lilley, III, Gert Schmidt, Mark Smith, Walter Windsor

National Association of Television Program Executives, Inc.

310 Madison Avenue, Suite 1207, New York, NY 10017 (212) 661-0270.

PRESIDENT
Steve Currie, KOIN-TV, Portland
FIRST VICE PRESIDENT
Chuck Larsen, WABC-TV, New York
SECOND VICE PRESIDENT
Phil Corvo, KGTV, San Diego
IMMEDIATE PAST-PRESIDENT
Lucille Salhany, Taft Broadcasting, Philadelphia
EXECUTIVE DIRECTOR
George Back, New York
BOARD OF DIRECTORS
Linda Rios-Brook, Stan Marinoff, Gene Swerdloff, Ron Gold, Sandra Pastoor, Michael Garin (Class of 1981). Linda DesMarais, Don Tillman, Dave Sifford (Class of 1982). Gary Dreispul, Brooke Spectorsky, Ron Klayman, John von Soosten (Class of 1983).
DIRECTOR OF PUBLIC RELATIONS
Robert A. Bernstein, March Five, Inc., New York
COUNSEL
Richard Barovick
CONSULTANTS
Dick Block, Neil Harris

National Music Publishers' Association, Inc.

110 East 59th Street, New York, NY 10022 (212) 751-1930.

PRESIDENT
Leonard Feist
VICE-PRESIDENTS
Wesley H. Rose, Ralph Peer II, Michael Stewart
SECRETARY
Leon J. Brettler
TREASURER
Sidney B. Herman

Office for Film and Broadcasting (OFB)

Suite 1300, 1011 First Avenue, New York, NY 10022. (212) 644-1880.

DIRECTOR
Rev. Patrick J. Sullivan S.J.

Office for Film and Broadcasting (OFB): A unit of the Department of Communication of the National Conference of Catholic Bishops and the United States Catholic Conference, was formed in January, 1972, through a reorganization of the former National Catholic Office for Motion Pictures and the National Catholic Office for Radio and Television.

OFB provides a national service of information, for diocesan communications offices throughout the country.

For the motion picture medium, OFB publishes the twice-monthly *Film & Broadcasting Review* which reviews all current nationally released 35mm films and provides information about resources for film utilization and education (16mm films, books, magazines, festivals). The critical reviews are addressed to the moral as well as artistic dimensions of motion pictures and are the result of a consensus based on the reactions of OFB's professional staff and board of consultors. All films reviewed are also classified according to the OFB rating system. Film classifications appear every month as part of the *Review*. In addition, the publication carries information and evaluative studies on trends and issues pertinent to television, with an emphasis on educational material.

The division maintains a 16mm film library of Catholic program material and a consultation service for educational and religious film program directors.

For the broadcast media, OFB is responsible for cooperating with the three major networks (ABC, CBS and NBC) in the production of all regularly scheduled network radio and television programs involving Catholic participation: (See Radio and Television).

For the Catholic press, OFB publishes a weekly film/broadcast service consisting of reviews, information, articles and photos. Special projects are undertaken with individual publications, religious and general.

OFB also plays a liaison role for the NCCB/USCC with the film and broadcasting industries, national media, and religious agencies and organizations. It is a member of OCIC and UNDA, the international Catholic organizations for film and broadcasting, respectively. Consultations and information services are also provided for the Pontifical Commission for Social Communications and the communications offices of national episcopal conferences throughout the world.

Permanent Charities Committee of the Entertainment Industries

Samuel Goldwyn-Permanent Charities Building, 463 N. La Cienega Blvd., Los Angeles, CA 90048 (213) 652-4680. (A donor federation within the entertainment industries supporting community wide charities.)

PRESIDENT
Albert A. Dorskind
FIRST VICE PRESIDENT
Josef Bernay
SECOND VICE PRESIDENT
Paul Masterson
SECRETARY
Sidney P. Solow
TREASURER
George Staff
EXECUTIVE VICE PRESIDENT
William E. Arnold
DIRECTOR OF DEVELOPMENT
Irwin J. Kaplan
CAMPAIGN CHAIRMAN
David Dortort

Producer-Writers Guild of America Pension Plan

310 North San Vicente Boulevard, Suite 305, Los Angeles, CA 90048 (213) 659-6430.

CHAIRMAN
Robert Cortes
VICE CHAIRMAN
Harvey Lehman
SECRETARY
David Harmon
VICE SECRETARY
Irma Kalish

Radio Advertising Bureau, Inc.

485 Lexington Avenue, New York, NY 10017 (212) 599-6666.

CHAIRMAN OF BOARD
George Duncan
PRESIDENT
Miles David
SENIOR VICE PRESIDENT, MARKETING
Richard J. Montesano
VICE PRESIDENT, NATIONAL SALES
William Cummings
SENIOR VICE PRESIDENT, LOCAL SALES
Joseph Vincent
VICE PRESIDENT, ASS'T SECRETARY-TREASURER
William Shriftman

SECRETARY
Cary Simpson
TREASURER
Stephen P. Bellinger

SESAC, INC.

(One of the world's foremost music licensing organizations. A special projects department handles scoring for motion pictures, slide films, syndicated TV series and agency produced commercials; programming for background music and premium albums; and the leasing of masters.)

The Coliseum Tower, 10 Columbus Circle, New York, NY 10019 (212) 586-3450.

CHAIRMAN
A.H. Prager
TREASURER
R.C. Heinecke

Screen Actors Guild (AAAA-APL)

(Organized July, 1933; Membership: 46,000) 7750 Sunset Blvd., Hollywood, CA 90046 876-3030. Branches—New York, 1700 Broadway, New York, NY 10019, John T. McGuire; Chicago, 307 N. Michigan Ave., Herbert H. Neuer; Boston, 11 Beacon Street, Robert M. Segal; Lathrup Village, MI 48076, 28690 Southfield Rd., Jane Ford, Dallas, TX 75204, 3220 Lemmon Ave., Clinta Dayton; San Francisco, CA 94104, 100 Bush St., Don Tayer; FL, 3266 Ponce de Leon Blvd., Coral Gables, FL, Melvin B. Karl.

PRESIDENT
William Schallert
FIRST VICE-PRESIDENT
Kent McCord
SECOND VICE-PRESIDENT
Larry Keith
THIRD VICE-PRESIDENT
Joseph Ruskin
FOURTH VICE-PRESIDENT
Lee Zimmer
FIFTH VICE-PRESIDENT
Regina Waldon
SIXTH VICE-PRESIDENT
Hugh Lampman
VICE PRESIDENTS
Ralph Bell, Kathleen Freeman, Bob Kaliban, Constance B. Minnett, Jessica Walter
RECORDING SECRETARY
Philip Sterling
TREASURER
Gilbert Perkins

NATIONAL EXECUTIVE STAFF

NATIONAL EXECUTIVE SECRETARY
Chester L. Migden
ASSISTANT NATIONAL EXECUTIVE SECRETARIES
Ken Orsatti, James L. Nissen, John T. McGuire, Melvin B. Karl, Michel DeMers, Paulyne Golden
INFORMATION DIRECTOR
Kim Fellner
COUNSEL
Berger, Kahn, Shafton & Moss
Shea, Gould, Climenko & Casey
CONTROLLER
Paulyne Golden

Society of Motion Picture and Television Engineers

(Organized 1916; Membership: 9,000) 862 Scarsdale Ave., Scarsdale, NY 10583 (914) 472-6606)

PRESIDENT
Robert M. Smith
EXECUTIVE VICE PRESIDENT
Charles E. Anderson
PAST PRESIDENT
William D. Hedden
ENGINEERING VICE-PRESIDENT
Roland J. Zavada
EDITORIAL VICE-PRESIDENT
K. Blair Benson
FINANCIAL VICE-PRESIDENT
Joseph A. Flaherty
CONFERENCE VICE-PRESIDENT
Harry Teitelbaum
SECTIONS VICE-PRESIDENT
Leonard F. Coleman
VICE PRESIDENT PHOTO-SCIENCE AFFAIRS
Daan Zwick
VICE PRESIDENT FOR EDUCATIONAL AFFAIRS
Raymond Fielding
VICE PRESIDENT FOR TELEVISION AFFAIRS
Frederick M. Remley
VICE PRESIDENT FOR PHOTO INSTRUMENTATION AFFAIRS
Lincoln L. Endelman
VICE PRESIDENT FOR MOTION-PICTURE AFFAIRS
Fred J. Scobey
SECRETARY
Harold J. Eady
TREASURER
Charles A. Ahto
EXECUTIVE DIRECTOR
Donald F. Breidt
EDITOR
David A. Howell

Station Representatives Association, Inc.

230 Park Avenue, New York, NY 10017 (212) 687-2484.

PRESIDENT
Al Masini
VICE PRESIDENT/RADIO
Bill Fortenbaugh
VICE PRESIDENT/TV
Lloyd Werner
SECRETARY
Robert V. Coppinger
TREASURER
Dave Allen
MANAGING DIRECTOR
M. S. Kellner
DIRECTORS
(Radio board) Charlie Colombo, Bob Coppinger, Lou Faust, Bill Fortenbaugh, Ray Lindau, Peter Moore, George Wolfson. (TV board) Lloyd Werner, Bill Bee, Bob Kizer, Al Masini, Don Robinson, Wally Schwartz, Jay Waters

Television Bureau of Advertising

485 Lexington Avenue, New York, NY 10017 (212) 661-8440. (Organized 1954.); 3340 Peachtree Rd., N.E. Suite 1415, Tower Place, Atlanta GA 30326 (404) 262-7627; 3141 Hood Street, Suite 220, Dallas, TX 75219 (214) 522-4050; 400 N. Michigan Avenue, Chicago, IL 60611 (312) 644-1215; 3155 West Big Beaver Road, Suite 217, Troy, MI 48084 (313) 649-6688; 6380 Wilshire Blvd., Los Angeles, CA 90048 (213) 653-8890.

PRESIDENT
Roger D. Rice
CHAIRMAN OF THE BOARD
Thomas Cookerly
SECRETARY
David E. Henderson
TREASURER
David Johnson
CHAIRMAN EX-OFFICIO
Marvin L. Shapiro
EXECUTIVE VICE PRESIDENT OPERATIONS
George G. Huntington
EXECUTIVE VICE PRESIDENT SALES
Robert A. Lefko
SENIOR VICE PRESIDENT RESEARCH
Harvey Spiegel
SENIOR VICE PRESIDENT NATIONAL SALES
Richard D. Severance
VICE PRESIDENT CENTRAL DIVISION
Dave Michels
VICE PRESIDENT, DETROIT
Richard O'Donnell
VICE PRESIDENT, CREATIVE SERVICES
Beverly Keene
VICE PRESIDENT, WESTERN DIVISION
Robert Fairbanks
VICE PRESIDENT, COMMUNICATIONS
Robert M. Grebe
VICE PRESIDENT, TECHNICAL SERVICES
Walter Vetter

VICE PRESIDENT, LOCAL SALES PROJECTS
Robert Baker
DIRECTOR OF RETAIL
Walter Bills
VICE PRESIDENT, PACIFIC
Arthur Trudeau
VICE PRESIDENT, SOUTHWESTERN SALES
Chuck Hanson
VICE PRESIDENT, SOUTHEASTERN SALES
Perry Bascom

Television Information Office

745 Fifth Avenue, New York, NY 10022 (212) 759-6800.

DIRECTOR
Roy Danish
MANAGER OF CREATIVE SERVICES
Bert Briller
MANAGERS, STATION SERVICES
Louis B. Ames, James Folsom, Henry W. Levinson
MANAGER, RESEARCH SERVICES
James Poteat

Theatre Authority, Inc.

(Organized May 21, 1934) 485 Fifth Avenue, New York, NY 10017 (212) MUrray Hill 2-4215-6.

PRESIDENT
Harold M. Hoffman
FIRST VICE PRESIDENT
Elizabeth Morgan
SECOND VICE PRESIDENT
Willard Swire
THIRD VICE PRESIDENT
John T. McGuire
FOURTH VICE PRESIDENT
Alan Jan Nelson
RECORDING SECRETARY
Sandford I. Wolff
TREASURER
Delloyd Tibbs
COUNSEL
Jacob I. Goodstein
ADVISORY COMMITTEE
Julie Andrews, Fred Astaire, Pearl Bailey, Lucille Ball, Harry Belafonte, Joey Bishop, Sammy Davis, Jr., Maurice Evans, Henry Fonda, Jackie Gleason, Helen Hayes, Charlton Heston, Jerome Hines, Bob Hope, George Jessel, Alan King, Jerry Lewis, George London, Dean Martin, Mary Martin, Gregory Peck, Debbie Reynolds, Dinah Shore, Frank Sinatra, Cornelia Otis Skinner, Kate Smith, Eleanor Steber, Barbra Streisand, Danny Thomas.

United Nations—Radio and Visual Services Division, Department of Public Information

United Nations, New York, NY 10017 (212) 754-6945.

DIRECTOR, RADIO AND VISUAL SERVICES DIVISION
Marcel Martin
CHIEF OF VISUAL SERVICE
Gilbert Lauzun.

Women in Show Business

(A philanthropic organization composed of women in the entertainment industry dedicated to providing funds for reconstructive surgery for children) P.O. Box 2535, North Hollywood, CA 91602 (213) 855-1010.

PRESIDENT
Bea Colgan
EXECUTIVE VICE PRESIDENT/WAYS & MEANS
Rori Lindo
FIRST VICE PRESIDENT/PROGRAM
Marilynn Mitchell
SECOND VICE PRESIDENT/MEMBERSHIP
Joy Pederson
CORRESPONDING/RECORDING SECRETARY
Susan Bierfischer
TREASURER
Judy Collinge
PHILANTHROPY CHAIRMAN
Christie Palmer
PUBLICITY CHAIRMAN
Scherr Lillico
BULLETIN CHAIRMAN
Judie Gold
SOCIAL CHAIRMAN
Kathy Smith
ADVISOR
La Donna Webb

Writers Guild of America

NATIONAL CHAIRMAN
Daniel Taradash

WRITERS GUILD OF AMERICA, EAST, INC.

555 West 57th Street, New York, NY 10019 (212) 245-6180.

PRESIDENT
Allegra Branson
VICE-PRESIDENT
Edward Adler
SECRETARY-TREASURER
Ken Gaughran
EXECUTIVE DIRECTOR
Leonard Wasser

WRITERS GUILD OF AMERICA, WEST, INC.

8955 Beverly Boulevard, Los Angeles, CA 90048 (213) 550-1000.

PRESIDENT
Melville Shavelson
VICE-PRESIDENT
Frank Pierson
SECRETARY-TREASURER
William Ludwig
EXECUTIVE DIRECTOR
Leonard Chassman

State Associations of Broadcasters

ALABAMA BROADCASTERS ASSOCIATION
President: Donald Cooke, WAMI, Opp.
Exec. Dir.: Richard W. Deason, Box 6246, University (35486).

ALASKA BROADCASTERS ASSOCIATION
President: Roy Robinson, KFQD, Anchorage.

ARIZONA BROADCASTERS ASSOCIATION
President: William McReynolds, KPHO-TV, Phoenix.
Exec. Dir.: Ken Heady, Box 654, Scottsdale (85252).

ARKANSAS BROADCASTERS ASSOCIATION
President: Darrel Cunningham, KAIT-TV, Jonesboro.
Exec. Director: Patricia C. Willcox, Cantrell Place, Suite 100, 2311 Biscayne Dr., Little Rock, AR 72207.

CALIFORNIA BROADCASTERS ASSOCIATION
Chairman: Frank W. McLaurin, KSRO, Santa Rosa, CA
Exec. Dir.: Howard J. Smiley, 1107 9th St., Sacramento (95814).

COLORADO BROADCASTERS ASSOCIATION
President: Steve Heater, KSPN, Aspen.
Exec. Dir.: John E. Morrill, 7138 W. Frost Dr., Littleton (80123).

CONNECTICUT BROADCASTERS ASSOCIATION
President: Enzo DeDominicis, WRCQ & WRCH-FM, New Britain.
Exec. Dir.: Bob Meinsen, 15 Highland Park Rd., North Haven, 06473.

FLORIDA ASSOCIATION OF BROADCASTERS
President: Herbert Levin, WQBA, Miami.
Exec. Dir.: Kenneth F. Small, 3131 NW 13th St., Gainesville (32601).

GEORGIA ASSOCIATION OF BROADCASTERS
President: Tom Lloyd, WLAQ Radio, Rome, GA
Exec. Dir.: William Saunders, 6065 Roswell Rd. NE, Atlanta (30328).

HAWAIIAN ASSOCIATION OF BROADCASTERS
President: Richard A. Weiner, KGMB-TV, Honolulu.

IDAHO STATE BROADCASTERS ASSOCIATION
President: David Capps, KEEP, Twin Falls
Exec. Director: John Tischendorf, Box 884, Boise, 83701.

ILLINOIS BROADCASTERS ASSOCIATION
President: Phil Nolan, WIND, Chicago.
Exec. Dir.: James G. Turpin, 726 S. College, Springfield (62704).

INDIANA BROADCASTERS ASSOCIATION INC.
President: G. Earl Metzger, WITZ AM-FM, Jasper.
Exec. V.P.: 300 Board of Trade Bldg. 143 N. Meridian St., Indianapolis (46204).

IOWA BROADCASTERS ASSOCIATION
President: Philip Kelly, WDBQ, Dubuque.
Exec. Secy.: Jack Shelley, 1230 Marston Ave., Ames (50010).

KANSAS ASSOCIATION OF BROADCASTERS
President: Bob Templeton, KAYS/AM, Hays.
Executive Director: Jesse Prisock, 1052 N. Waco, Wichita, KS 67203.

KENTUCKY BROADCASTERS ASSOCIATION
President: Fred Paxton, WPSD-TV, Paducah, KY
Exec. Secy.: J.T. Whitlock, Box 680, Lebanon (40033).

LOUISIANA BROADCASTERS ASSOCIATION
President: Jules Mayeux, WBRZ, Baton Rouge.
Exec. Secy.: John H. Pennybacker, Box 16078, L.S.U., Baton Rouge (70893).

MAINE ASSOCIATION OF BROADCASTERS
President: John Martin, WMTW-TV, Poland Spring.
Exec. Secy.: Norman G Gallant, 160 Bangor St., Augusta (04330).

MARYLAND-D.C.-DELAWARE BROADCASTERS ASSOCIATION
President: Thomas J. Dougherty, Metromedia, Washington.
Exec. Dir.: Robert B. Cochrane, Route 1, Box 125, St. Michaels, MD (21663).

MASSACHUSETTS BROADCASTERS ASSOCIATION
President: Joseph Rayball, WARA, Attleboro.
Exec. Dir.: 42 Curtis Ave., Marlboro (01752). Phone: 617-481-3063.

MICHIGAN ASSOCIATION OF BROADCASTERS
President: Thomas M Girocco, WOTV, Grand Rapids.
Exec. Secy.: Thomas J. Cleary P.O. Box 16015, Lansing (48901).

MINNESOTA BROADCASTERS ASSOCIATION
President: Dick Painter, KYSM, Mankato, MN.
Exec. Secy.: Ray Foslid, 421 E. 1st Ave., Shakopee, MN 55379.

MISSISSIPPI BROADCASTERS ASSOCIATION
President: Chuck Cooper, WKOR, Starkville.
Exec. Secy.: Ray Butterfield, 14104 Big Ridge Rd., Biloxi (39532).

MISSOURI BROADCASTERS ASSOCIATION
President: Charles McAbee Jr., KCMO-TV, Kansas City.
Exec. V.P.: Ted Griffin, 1800 Southwest Blvd., Jefferson City (65101).

MONTANA BROADCASTERS ASSOCIATION
President: Lynn Koch, KGVO-TV, Missoula.
Exec. Dir.: Robert W. Hoene, 213 5th Ave., Helena (59601).

NEBRASKA BROADCASTERS ASSOCIATION
President: John Howard, KGFW, Kearney.
Exec. V.P.: Frank Fogarty, Box 31802, Omaha (68131).

NEVADA BROADCASTERS ASSOCIATION
President: James C. Herzig, KCLO-TV, Reno.
Exec. Secy.: Earl Ling, KOH, Reno (89505).

NEW HAMPSHIRE ASSOCIATION OF BROADCASTERS
President: Bob Connelly, WWNH, Rochester.
Secy.-Treas.: Robert Connelly, WWNH, Rochester Hill Rd., Rochester (03867).

NEW JERSEY BROADCASTERS ASSOCIATION
President: Cathy Clark, WFPG Atlantic City.
Exec. Secy.: Arnold Zucker, Johnston Hall, Rutgers University, New Brunswick (08903).

NEW MEXICO BROADCASTERS ASSOCIATION
President: Jim Gober, KWYK, Farmington.
Exec. Dir.: Dee Schelling, 790-9D Tramway Lane NE Albuquerque, (87122).

NEW YORK STATE BROADCASTERS ASSOCIATION INC.
President: Robert M. Peebles, WROW AM-FM, Albany.
Assn. Secy.: Ellen S. Cody, 9 Herbert Dr., Latham (12110).

NORTH CAROLINA ASSOCIATION OF BROADCASTERS
President: Jerry Oakley, WFBS Radio, Spring Lake.
Exec. Dir.: Wade H. Hargrove, Box 627, Raleigh (27602).

NORTH DAKOTA BROADCASTERS ASSOCIATION
President: Norm Engstrom, WDAY, Fargo.
Exec. Secy.: C. H. Logan, N.D. State U., Fargo (58102).

OHIO ASSOCIATION OF BROADCASTERS
President: Neal Van Ells, WKYC-TV, Cleveland.
Exec. V.P.: Thomas C. Sawyer, 41 S. High St., Columbus (43215).

OKLAHOMA BROADCASTERS ASSOCIATION
President: Allan Muchmore, WBBZ, Ponca City.
Exec. Secy.: Leon Rogers, Box 457, Shawnee (74801).

OREGON ASSOCIATION OF BROADCASTERS
President: Henry Florip, KYJC, Medford.
Gen. Mgr.: Vern Mueller, Box 3236, Eugene (97403).

PENNSYLVANIA ASSOCIATION OF BROADCASTERS
President: Victor C. Diehm, Jr., WAZL, Hazleton.
Executive Director and Counsel: Robert H. Maurer. 407 N. Front St., Harrisburg (17101).

RHODE ISLAND BROADCASTERS ASSOCIATION
President: Richard Rakovan, WPRQ, Providence.
Secy.: Leonard Davey, WJAR-TV, Providence.

ROCKY MOUNTAIN BROADCASTERS ASSOCIATION
President: Joseph J. Tennessen, KFKA, Greeley.
Exec. Dir.: Henry H. Fletcher, Box 220, Pocatello, ID (83201).

SOUTH CAROLINA BROADCASTERS ASSOCIATION
President: Joseph Foster, WBTW-TV, Florence, S.C.
Exec. Mgr.: Dr. Richard M Uray, USC College of Journalism, Columbia, SC (29208).

SOUTH DAKOTA BROADCASTERS ASSOCIATION
President: Mike Edwards, KORN, Mitchell.
Exec. Dir.: Verl Thomson, 4609 W. 12th St., Sioux Falls (57106). (605) 336-7110.

TENNESSEE ASSOCIATION OF BROADCASTERS
President: Dean Osmundson.
Exec. Dir.: Paul Ruhle, Box 2688, Nashville (37219).

TEXAS ASSOCIATION OF BROADCASTERS
President: Mel Z. Gilbert, KSNY, Snyder.
Exec. V.P.: Bonner McLane, P.O. Box 14787, Suite 201, 314 Highland Mall Blvd. Austin, TX 78761.

UTAH BROADCASTERS ASSOCIATION
President: John Miner, KDXU, St. George.
Exec. Secy.: Earl J. Glade Jr., Box 401, Provo (84601).

VERMONT ASSOCIATION OF BROADCASTERS
President: Peter Martin, WCAX-TV, Burlington.
Secretary: Brian Harwood, WRFN, Stowe.

VIRGINIA ASSOCIATION OF BROADCASTERS
President: J. William Poole, WFLS, Fredericksburg.
Exec. Secy.: Peter Easter, Easter Associates, Charlottesville (22902).

WASHINGTON STATE ASSOCIATION OF BROADCASTERS
President: James A. Murphy, 1411 4th Ave. Bldg., Seattle (98101). (206) 622-2991.

WEST VIRGINIA BROADCASTERS ASSOCIATION
President: George R. Andrick, WSAZ-TV, Huntington.
Exec. Dir.: Jack Williams, 1910 Parkview St., Hungtington (25701).

WISCONSIN BROADCASTERS ASSOCIATION
President: Richard Dudley, Forward Communications, Wausau.
Exec. Dir.: Joe wkilleen, Box 11725, Shorewood (53211).

WYOMING ASSOCIATION OF BROADCASTERS
President: Fred Hildebrand, KVOC, Casper.
Exec. Secy.: Ray Lansing, Box 1873, Cheyenne (82001).

The Press

- TRADE PUBLICATIONS
- NEWSPAPER LISTING

TRADE PUBLICATIONS

Quigley Publishing Company

Publishers of Motion Picture Product Digest (25 times a year), Motion Picture Almanac (Annual), Television Almanac (Annual), – Fame. 159 West 53rd St., New York, NY 10019; (212) 247-3100; Cable: Quigpubco, New York.

PRESIDENT AND PUBLISHER
Martin Quigley, Jr.
EDITOR-IN-CHIEF
Richard Gertner
SECRETARY
Michael Mayer
GENERAL MANAGER
Robert McDonald

LONDON BUREAU

William Pay, Manager and London Editor; 15 Samuel Road, Langdon Hills, Basildon, Essex, England. Tel.: 0268-42824.

FOREIGN BUREAUS

ARGENTINA: Natalio Bruski, Calle Paso 240 (8°-C), Buenos Aires.
BELGIUM: Marc Turfkruyer, Van Maerlantstraat, 71 Antwerp.
EGYPT: Ahmed Sami, 4 El Ommara St., Apt. 13, Abbassia, Cairo.
FINLAND: Mrs. Hellevi Rinkihainen, Finnish Film Foundation, Kaisaniemenkatu 3 B 25 SF-00100 Helsinki 10.
GREECE: Rena Velissariou, 8432 Kolokotroni Str., Agia Paraskevi, Attikis, Athens (804).
HOLLAND: Philip de Schaap, 9 Henriette Bosmansstrant, Amsterdam (Zuid).
JAPAN: A.C. Pinder, Whaley-Eaton Service, Central P.O. Box 190, Tokyo, Japan.
NEW ZEALAND: Mark R. Galloway, N.Z. Film Services Ltd., P.O. Box 9340, Courteneay Pl., Wellington, C.3.
PAKISTAN: A.R. Slote, P.O. Box 7426, Karachi. 1-3.
SPAIN: Mariano del Pozo, Conde de Penalver 36, Madrid 6.
SRI LANKA: Chandra Perea, 437 Pethiygoda, Kelanuja, Sri Lanka, Ceylon.
SWITZERLAND: Gerda Goepfert-Klimisch, Film Beld Agentur 6911, Barbengo, Ticino.

QUIGLEY PUBLICATIONS

Motion Picture Product Digest

(Published 25 times a year) 159 West 53rd St., New York, NY 10019; 247-3100; Cable: Quigpubco, New York.

PRESIDENT AND PUBLISHERS
Martin Quigley, Jr.
EDITOR-IN-CHIEF
Richard Gertner
OPERATIONS MANAGER
Robert McDonald

Motion Picture Almanac

(Annually) 159 West 53rd St., New York, NY 10019; 247-3100; Cable: Quigpubco, New York.

EDITOR
Richard Gertner
ASSOCIATE EDITOR
William Pay
ADVERTISING AND DISTRIBUTION
Robert McDonald

Television Almanac

(Annually) 159 West 53rd St., New York, NY 10019; 247-3100; Cable: Quigpubco, New York.

EDITOR
Richard Gertner
ASSOCIATE EDITOR
William Pay
ADVERTISING AND DISTRIBUTION
Robert McDonald

Fame

159 West 53rd St., New York, NY 10019; 247-3100; Cable: Quigpubco, New York.

EDITOR
Richard Gertner
ASSOCIATE EDITOR
William Pay

Advertising Age

(Weekly) 740 Rush St., Chicago, Il. 60611; 649-5200; 220 East 42nd St., New York, NY 10017; (212) 210-0100.

CHAIRMAN
Mrs. G.D. Crain
CHAIRMAN, EXECUTIVE COMMITTEE
S.R. Bernstein
PUBLISHER
Louis F. Demarco
PRESIDENT & EDITOR-IN-CHIEF
Rance Crain

The American Cinematographer

(Monthly on the 1st—Semi-technical) Published by American Society of Cinematographers, Inc., 1782 North Orange Drive, Hollywood, CA 90028; 876-5080.

EDITOR
Herb A. Lightman
ASSISTANT EDITOR
Lyle Dusing
ACCOUNTING
Barbara Prevedel
RESEARCH
Patty Armacost
EDITORIAL ADVISORY BOARD
Milton Krasner, Chairman
Charles G. Clarke
Stanley Cortez
Linwood Dunn
Ernest Laszlo
Lester Shorr

Back Stage

(Weekly dealing with theatrical and nontheatrical films, television shows and commercials, cassettes, radio, and theater, published Friday), 165 W. 46th St., New York, NY 10036, (212) LT 1-1080; 5670 Wilshire Blvd., L.A. CA 90036, (213) 936-5201; 841 North Addison Ave., Elmhurst, IL 60126 (312) 834-7533.

MANAGING EDITOR & PUBLISHER
Allen Zwerdling
ADVERTISING DIRECTOR & PUBLISHER
Ira Eaker
EDITOR
Tom Tolnay

Back Stage Television, Film, Tape & Syndication Director

(Annually, January), 165 W. 46 St., New York, NY 10036, (212) LT 1-1080.

EDITOR
Allen Zwerdling

Billboard

(Weekly dealing with records, music, entertainment artists, video hardware and software, etc., published Saturdays), 9000 Sunset Blvd., Los Angeles, Calif. 90069; 273-7040; 1 Astor Plaza, New York, NY 10036, 764-7300; 150 N. Wacker Drive, Chicago, IL 60606; Central 6-9818; 2160 Patterson Street, Cincinnati, O. 45214; 733 15 St., N.W., Washington 5, D.C.; 393-2580.

PUBLISHER
Lee Zhito
EXECUTIVE VICE PRESIDENT
Jules Perel
MANAGING EDITOR
Adam White
ASSOCIATE PUBLISHERS
Tom Noonan, Bill Wardlow
EDITOR-IN-CHIEF
Gerry Wood

Broadcasting—The News Magazine of the Fifth Estate

(Weekly on Monday; also publishes Broadcasting Yearbook every March.) Washington Headquarters: 1735 DeSales Street, N.W. 20036. (202) 638-1022; New York: 630 Third Ave., 10017 (212) 599-2830; Hollywood: 1680 No. Vine St. 90028. (213) 463-3148.

CHAIRMAN AND EDITOR
Sol Taishoff
PRESIDENT AND PUBLISHER
Lawrence B. Taishoff
DIRECTOR OF SALES & MARKETING
David N. Whitcombe
GENERAL SALES MANAGER (NEW YORK)
Winfield R. Levi
VICE-PRESIDENT-EXECUTIVE EDITOR
Edwin H. James
MANAGING EDITOR
Donald V. West
CHIEF CORRESPONDENT (NEW YORK)
Rufus Crater

Business Screen

(Six times per year, supplement to Back Stage. Magazine established 1938, covers industrial, documentary, in-pant films & tapes primarily, plus industry news.)

PUBLISHERS
Allen Zwerdling
Ira Eaker

Celebrity Bulletin

(A chronicle of the day to day activities of celebrities in New York, Hollywood, London, Paris, and Rome. Published daily.) 171 West 57th Street, New York, NY 10019; Plaza 7-7979

PUBLISHER
Celebrity Service, Inc.

Celebrity Register

(A biographical index, published every five years) 171 West 57th Street, New York, NY 10019; PLaza 7-7979.

PUBLISHER
Celebrity Service, Inc
EDITOR
Earl Blackwell

Contact Book

(A trade directory, published bi-annually) 171 West 57th Street, New York, NY 10019; Plaza 7-7979.

PUBLISHER
Celebrity Service, Inc.
EDITOR
Barbara A. McGurn

Daily Variety

(Motion picture, television, radio, theatre, night clubs daily) 1400 N. Cahuenga Blvd., Hollywood, CA 90028; (213) 469-1141.

EDITOR
Thomas M. Pryor
ADVERTISING MANAGER
Ned Diamant
MARKETING DIRECTOR
Michael Malak
GENERAL MANAGER
F. Phil Turner
PRODUCTION MANAGER
Bob Butler

Electronics

(Design, production and application bi-weekly for engineers and management) McGraw-Hill Publications Co., 1221 Avenue of the Americas, New York, NY 10020. Tel.: (212) 971-1221.

PUBLISHER
Paul W. Reiss
EDITOR-IN-CHIEF
Samuel Weber
MANAGING EDITOR
Arthur Erikson, International
Ray Papece, Technical

Facts, Figures & Film

(TV programmers magazine). 100 Lafayette Drive, Syosset, NY 1-791 (516) 496-3355.

PUBLISHER
Harvey Seslowsky
VICE PRESIDENT-EDITOR-IN-CHIEF
Miss Avra Fliegelman

Film News

Pub. 4 times in the year by Film News Company. (Spring, Summer, Fall, Winter) The international review of AV materials and equipment. Editorial and business office: Box 619, LaSalle, IL 61301

EDITOR
J. Paul Carvico

Film Quarterly

(Quarterly; published Fall, Winter, Spring and Summer. A critical journal of motion pictures and their related arts; successor to *The Quarterly of Film, Radio and Television and the Hollywood Quarterly.*) Editorial, Sales, and Advertising office: University of California Press, Berkeley, CA 94720; (415) 642-6333.

EDITOR
Ernest Callenbach
Published by University of California Press

Hollywood Report

(Quarterly) 917 So. Tremaine Avenue, Hollywood, 90019, CA; WE 9-2345.

EDITOR
Syd Cassyd

Hollywood Reporter, The

(Film, TV, Entertainment, Daily) 6715 Sunset Boulevard, Hollywood, CA 90028; (213) 464-7411; and 1501 Broadway, N.Y., N.Y., 10036; (212) 354-1858.

PUBLISHER & EDITOR-IN-CHIEF
Tichi Wilkerson
BUREAUS
NEW YORK: Martin Gould, 1501 Broadway, N.Y.C. 10036.
PARIS: Jerome Brierre, 7, rue Claude Monet, Boulogne.
ROME: Giovanna Grassi, via Monserrato 149.
LONDON: Kenelm Jenour, 57 Duke St., WI.
TORONTO: Gloria Kissin, 17 Sunfield Rd., Downsview.
HONG KONG: Fred Marshall, P.O. Box 30510, Causeway Bay.

I.A.T.S.E. Official Bulletin

(Quarterly) 1515 Broadway, New York, NY 10036. Tel.: (212) 730-1770.

EDITOR
James J. Riley

International Photographer

(Monthly, on the 5th) 7715 Sunset Boulevard, Hollywood, CA 90046; 876-0160.

PUBLISHER
International Photographer
PUBLISHER'S AGENT
Gerald K. Smith
BUSINESS MANAGER
Mrs. Terry M. Burley
MANAGING EDITOR
George Toscas

International Motion Picture Almanac

(See Quigley Publications)

Journal of the Society of Motion Picture and Television Engineers

(Technical monthly) Editorial office: SMPTE Headquarters, 862 Scarsdale Ave., Scarsdale, NY 10583; (914) 472-6606.

EDITOR
David A. Howell
ADVERTISING MANAGER
Jeffrey B. Friedman

Popular Electronics

(Technical monthly) 1 Park Ave., New York, NY 10016; 725-3568.

PUBLISHING DIRECTOR
Edgar W. Hopper
PUBLISHER
Joseph E. Mesics
PRESIDENT
Phillip B. Korsant
EDITORIAL DIRECTOR
Arthur P. Salsberg
BEVERLY HILLS OFFICE
9025 Wilshire Blvd. 9021

Radio-Electronics

(Monthly dealing with technical aspects for TV, radio and electronic service technicians, high-fidelity enthusiasts and experimenters) 200 Park Ave. S., New York, NY 10003; (212) 777-6400.

PUBLISHER
Larry Steckler
EDITOR-IN-CHIEF
M. Harvey Gernsback
MANAGING EDITOR
Art Kleinman

Ross Reports Television

150 Fifth Ave.; 924-0320.

EDITOR & PUBLISHER
Jerry Leichter

Teletales, The News Letter of the National Academy of Television Receiver Owners of America

(Monthly) 917 So. Tremaine, Hollywood 90019, CA; WE 9-2345.

EDITOR
Syd Cassyd

Television Digest

(Published weekly. Television Digest with Consumer Electronics, TV & CATV Addenda, AM/FM Addenda; Annual Television Factbook, Annual Cable & Station Coverage Atlas, Weekly Satellite Week, Weekly Video Week, Biweekly Public Broadcasting Report, Monthly Early Warnings Report, Daily Communications) 1836

Jefferson Place, N.W., Washington, DC 20036; (202) 872-9200.
EDITOR & PUBLISHER
Albert Warren
EDITORIAL DIRECTOR
David Lachenbruch
EXECUTIVE EDITOR
Dawson B. Nail
SENIOR EDITOR
Paul L. Warren

Television Factbook

(Published annually by Television Digest, Inc.) 1836 Jefferson Place, N.W., Washington, DC 20036.
EDITOR & PUBLISHER
Albert Warren
EDITORIAL DIRECTOR
Mary Appel
MANAGING EDITOR
Marilou B. Legge

Television Index

150 Fifth Avenue, New York, NY 10011; (212) 924-0320.
EDITOR & PUBLISHER
Jerry Leichter

Television/Radio Age

(Bi-weekly trade magazine) 1270 Sixth Ave., New York, NY 10020; PL 7-8400.
PUBLISHER
Sol Paul
EDITOR
Alfred J. Jaffe
EDITORIAL DIRECTOR
Sanford Josephson
MANAGING EDITOR
Ed Rosenthal
ASSISTANT EDITORS
Bob Sobel, George Swisshelm
PRODUCTION DIRECTOR
Marvin Rabach
BRANCH OFFICE
6255 Sunset Blvd., Hollywood, CA 90028; (213) 464-3552

Variety

(General amusements and professional weekly, published Wednesdays) 154 West 46th Street, New York, NY 10036; (212) 582-2700; 1400 N. Cahuenga Blvd., Hollywood, CA 90028; (213) 469-1141; 1050 Potomac St., N.W., Washington, DC 20005; (202) 965-4301; 400 North Michigan Ave., Chicago, IL 60611; (312) 337-4984; PARIS: France, 33 Champs Elysees, Paris 75008, France; Phone 225-08-07; ROME: VARIETY, Via Bissolati, 20, 00187, Rome, Italy; 463-290. LONDON: VARIETY, 49 St. James's St., London SW1A 1JX, England; 493-4561; MADRID, Spain, Calle Lagasca, 104; Phone 276-4262; SYDNEY, N.S.W. 1-7 Albion Place, Phone 61-3124.
PUBLISHER AND EXECUTIVE EDITOR
Syd Silverman
ADVERTISING MANAGER
Mort Bryer

NEWSPAPERS IN PRINCIPAL MARKETS OF THE U.S.

Arranged alphabetically by states, this list gives the names and addresses of newspapers in cities and market areas of over 100,000 population which maintain regular motion picture and entertainment departments.

Alabama

Birmingham, The News, 2200 North 4th Ave., 35202.
Birmingham, Post-Herald, 2200 North 4th Ave., 35202.
Huntsville, Times, 2317 Memorial Pkway., 35807.
Mobile, Mobile Press Register, 304 Government St., 36602.
Montgomery, Advertiser, 107 South Lawrence, 36104.

Arizona

Phoenix, Republic, 120 East Van Buren St., 85004.
Tucson, Arizona Star, 208 North Stone, 87501.

Arkansas

Little Rock, Democrat, Capitol Ave. & Scott, 72203.
Little Rock, Arkansas Gazette, 112 West Third St., 72203.

California

Anaheim, Bulletin, 232 South Lemon St., 92805.
Berkeley, Berkeley Gazette, 2048 Center St., 94704.
Burbank, Daily Review, 220 E. Orange Grove Ave., 91502.
Culver City, Evening Star News, 4043 Irving Place, 90230.
Fresno, Bee, 1559 Van Ness Ave., 93721.
Glendale, News Press, 111 North Isabel, 91209.
Long Beach, Independent Press-Telegram, 604 Pine Ave., 90801.
Huntington Park, Daily Signal, 6414 Rugby Ave., 90255.
Los Angeles, Herald-Examiner, 1111 S. Broadway, 90054.
Los Angeles, Times Mirror Co., Times Mirror Sq., 90053.
Oakland, The Tribune Publishing Co., 401-13th St., 94612.
Pasadena, Star News, 525 East Colorado St., 91109.
Riverside, Press-Enterprise Co., 3512-14th St., 92502.
Sacramento, Bee, 21st & Q Streets, 95813.
Sacramento, Sacramento Union Inc., 301 Capitol Mall, 95812.
San Bernardino, Sun, 399 "D" St., 92401.
San Diego, DBA Union-Tribune, 940 Third Ave., 92112.
San Francisco, Chronicle, 901 Mission St., 94119.
San Francisco, The Examiner, 110 5th Ave., 94119.
San Gabriel Valley, The Tribune, 2037 W. San Bernardino Rd., 91790.
San Jose, Mercury-News, 750 Ridder Park Dr., 95131.
San Pedro, News Pilot, 356 W. 7th Street, 90733.
Santa Ana, Register, 625 North Grand Ave., 92711.
Santa Monica, Evening Outlook, 1540 Third, 90406.
Stockton, Stockton Daily Record, 530 E. Market St., 95202.
Torrance, South Bay Breeze, 5215 Torrance Blvd., 90503.
Van Nuys, Van Nuys News & Valley News & Green Sheet, 14539 Sylvan St., 91408.

Colorado

Denver, The Post, 650 15th St., 80202.
Denver, Rocky Mt. News, 400 West Colfax Ave., 80204.

Connecticut

Bridgeport, Post, 410 State St., 06602.
Hartford, Courant, 285 Broad St., 06101.
Hartford, The Times, 10 Prospect St., 06101.
New Haven, The Journal-Courier, 367 Orange St., 06503.
New Haven, Register, 367 Orange St., 06503.
Waterbury, American-Republican, Inc., 389 Meadow St., 06702.

Delaware

Wilmington, News-Journal Co., 831 Orange St., 19899.

District of Columbia

Washington, Post, 1515 "L" St., N.W., 20005.

Florida

Fort Lauderdale, News, 101 North New River Drive East, 33302.
Jacksonville, Florida Times-Union, One Riverside Ave., 32201.
Miami, Herald, 1 Herald Plaza, 33101.
Miami, The News, 1 Herald Plaza, 33101.
St. Petersburg, Times, P.O. Box 1121, 33731.
Tampa, Tribune, 505 East J.F. Kennedy Blvd., 33602.

Georgia

Atlanta, Journal-Constitution, 10 Forsyth St. N.W., 30302.
Columbus, Ledger-Enquirer Newspapers, 17 West 12th St., 31902.
Macon, Telegraph, 120 Broadway, 31201.
Savannah, Savannah News-Press, 105-111 West Bay St., 31402.

Hawaii

Honolulu, Advertiser, P.O. Box 3110, 96802.
Honolulu, Honolulu Star-Bulletin, Inc. 605 Kapiolani Blvd., 96813.
Honolulu, Hawaii Hochi, 917 Kokea St., 96817.
Honolulu, Hawaii Times, 916 Nuuana Ave., 96807.

Illinois

Chicago, Chicago Today, 445 N. Michigan Ave., 60611.
Chicago, Daily Calumet, 9120 Baltimore, 60617.
Chicago, The News, 401 North Wabash, 60611.
Chicago, The Sun-Times, Field Enterprises, Inc., Newspaper Div., 401 North Wabash, 60611.
Chicago, Tribune, 435 North Michigan Ave., 60611.
Joliet, Herald News, 78 No. Scott St., 60431.
Moline, Daily Dispatch, 1720-5th Ave., 61265.
Peoria, Peoria Journal Star Inc., 1 News Plaza, 61601.
Rock Island, Argus, 1724-4th Ave., 61202.
Rockford, Register-Star, 99 East State St., 61105.
Waukegan, News-Sun, 100 W. Madison St., 60085.

Indiana

Evansville, Courier & Press, 201 N.W. 2nd St., 47701.
Fort Wayne, Journal-Gazette, 701 S. Clinton St., 46802.

Fort Wayne, News Sentinel, 600 West Main St., 46802.
Gary, Post-Tribune, 1065 Broadway, 46402.
Hammond, Times, 417 Fayette St., 46320.
Indianapolis, Star & News, 307 North Pennsylvania St., 46206.
South Bend, Tribune, 225 West Colfax, 46626.

Iowa

Cedar Rapids, Gazette, 500-3rd Ave. S.E., 52401.
Davenport, Times Democrat, 124 E. 2nd St., 52808.
Des Moines, Register & Tribune, 715 Locust St., 50304.
Sioux City, Journal, 5th & Douglas St., 51102.
Waterloo, Courier, 501 Commercial St., 50704.

Kansas

Kansas City, Kansan, 901 North 8th St., 66101.
Topeka, Capital-Journal, 6th & Jefferson, 66607.
Wichita, The Eagle and Beacon, 825 E. Douglas St., 67201.

Kentucky

Louisville, The Courier-Journal & Times, 525 West Broadway, 40202.

Louisiana

Baton Rouge, Advocate, 525 Lafayette St., 70821.
New Orleans, The Times-Picayune, 3800 Howard Ave., 70140.
Shreveport, Journal-Times, 222 Lake St., 71102.

Maryland

Baltimore, News American, Lombard & South Sts., 21230.
Baltimore, Sun, Calvert & Center Sts., 21230.

Massachusetts

Boston Christian Science Monitor, One Norway St., 02115.
Boston, Globe, 135 Morrissey Blvd., 02107.
Boston, Boston-Herald Traveler, 300 Harrison Ave., 02106.
Boston, Record American, 5 Winthrop Square, 02106.
New Bedford, Standard-Times, 555 Pleasant St., 02742.
Springfield, Republican, 1860 Main St., 01101.
Worcester, Worcester Telegram & Gazette, 20 Franklin St., 01601.

Michigan

Ann Arbor, News, 340 E. Huron St., 48106.
Detroit, Free Press, 321 West Lafayette Blvd., 48231.
Detroit, Evening News, 615 Lafayette Blvd., 48231.
Flint, The Journal, 200 East 1st St., 48502.
Grand Rapids, The Press, Press Plaza, Vandenberg Center, 49502.
Kalamazoo, Kalamazoo Gazette, 401 S. Burdick St., 49003.
Lansing, State Journal, 120 East Lenawee St., 48919.
Muskegon, Chronicle, 981 Third St., 49443.
Pontiac, Press, 48 W. Huron St., 48056.
Royal Oak, Tribune, 210 E. Third St., 48068.
Saginaw, News, 203 South Washington Ave., 48605.

Minnesota

Duluth, Herald, 424 West 1st, 55801.
Minneapolis, Star & Tribune, Fifth and Portland Ave., 55415.
St. Paul, Dispatch, 55 East 4th St., 55101.

Mississippi

Biloxi-Gulfport, Herald, De Buys Rd., 39501.
Jackson, Clarion Ledger-News, Box 40, 39205.

Missouri

Kansas City, Star, 1729 Grand Ave., 64108.
St. Louis, Globe-Democrat, 12th Blvd. at Delmar, 63101.
St. Louis, The Post-Dispatch, 1133 Franklin Ave., 63101.
Springfield, Springfield News & Leader, 651 Boonville, 65801.

Nebraska

Lincoln, Journal-Star, 926 "P" Street, 68501.
Omaha, World-Herald, 14th & Dodge Sts., 68102.

Nevada

Las Vegas, Review-Journal, 737 North Main St., 89101.
Las Vegas, Sun, 121 S. Highland Ave., 89104.

New Hampshire

Manchester, Union Leader, 35 Amherst St., 03105.

New Jersey

Camden, Courier-Post, 08101.
Elizabeth, Journal, 295-299 North Broad St., 07207.
Jersey City, Journal, 30 Journal Square, 07306.
Newark, Evening News, 215-221 Market St., 07101.
Newark, Newark Morning Ledger, _ Star Ledger Plaza, 07101.
Passaic-Clifton, Herald-News, 988 Main Ave., 07055.
Paterson, News, News Plaza & Straight St., 07509.
Trenton, Times, 500 Perry St., 08605.

New Mexico

Albuquerque, Journal, 7th & Silver S.W., 87101.

New York

Albany, Times-Union, 645 Albany-Shaker Road, 12201.
Buffalo, The Evening News, 218 Main St., 14240.
Buffalo, Courier-Express, 787 Main St., 14240.
Long Island, Newsday, 500 Stewart Ave., Garden City, 11530.
New York City:
Brooklyn Daily, Inc., 2427 Surf Ave., Brooklyn, 11224.
The News, 220 East 42nd St., 10017.
The Post, 210 South St., 10002.
The Times, 229 West 43rd St., 10036.
Wall St. Journal, 30 Broad St., 10004.
Niagara Falls, Gazette, 310 Niagara St., 14302.
Rochester, Times-Union, 55 Exchange St., 14614.
Syracuse, The Post-Standard, Clinton Square, 13202.
Utica, Observer-Dispatch, 221 Oriskany Plaza, 13503.
Yonkers, The Herald Statesman, Larkin Plaza, 10702.

North Carolina

Charlotte, Observer & News, 600 South Tryon St., 28201.
Durham, Herald, 115-19 Market St., 27702.
Greensboro, The News, 200-04 North Davie St., 27402.
Raleigh, News & Observer, 215 S McDowell St., 27601.
Winston-Salem, Journal-Sentinel, 416-20 North Marshall, 27102.

North Dakota

Fargo, Forum, 101-5th St., 58102.

Ohio

Akron, The Beacon Journal, 44 East Exchange St., 44309.
Canton, The Repository, 500 Market Ave., South, 44702.
Cincinnati, The Enquirer, 617 Vine St., 45201.
Cincinnati, Post & Times-Star, 800 Broadway St., 45202.
Cleveland, The Plain Dealer, 1801 Superior Ave., 44114.
Cleveland, The Press, 901 Lakeside Ave., 44114.
Columbus, The Citizen-Journal, 34 South Third St., 43216.
Columbus, The Dispatch, 34 South Third St., 43216.
Dayton, Journal-Herald, 4th & Ludlow Sts., 45401.
Toledo, The Toledo Blade, 541 Superior St., 43604.
Youngstown, The Vindicator, Vindicator Square, 44501.

Oklahoma

Oklahoma City, Oklahoma Journal, 7430 S.E. 15 St., 73110.
Oklahoma City, Oklahoman, 500 North Broadway, 73125.
Tulsa, The World & Tribune, 315 South Boulder Ave., 74102.

Oregon

Portland, The Oregon Journal, 1320 Southwest Broadway, 97201.
Portland, The Oregonian, 1320 Southwest Broadway, 97201.

Pennsylvania

Allentown, The Call, 6th & Linden Sts., 18105.
Erie, The Times, 20 East 12th St., 16501.
Philadelphia, The Bulletin, 30th & Market Sts., 19101.
Philadelphia, The Inquirer, 400 North Broad St., 19101.
Philadelphia, The News, 400 N. Broad St., 19101.
Pittsburgh, Post Gazette, 50 Blvd. of Allies, 15222.
Pittsburgh, Pittsburgh Press Co., 34 Blvd. of Allies., 15222.
Reading, The Eagle, 345 Penn St., 19601.
Scranton, Tribune, 338 North Washington Ave., 18501.
Scranton, The Times, Penn & Spruce, 18501.

Rhode Island

Providence, The Journal, 75 Fountain St., 02902.

South Carolina

Columbia, The State-Record, P.O. Box 1333, 29202.

Tennessee

Chattanooga, The News-Free Press, 400 E. 11th St., 37401.
Chattanooga, Times, 117 E. 10th St., 37401.
Knoxville, The Journal, 210 W. Church Ave., 37901.
Knoxville, The News-Sentinel, 204 West Church Ave., 37901.
Memphis Press-Scimitar, 495 Union Ave., 38101.
Nashville, The Banner, 1100 Broadway, 37203.
Nashville, The Tennessean, 1100 Broadway, 37203.

Texas

Abilene, Reporter News, North 2nd & Cypress Sts., 79604.
Amarillo, The Globe-Times, 900 Harrison St., 79105.
Austin, American-Statesman, 308 Guadalupe, 78767.
Beaumont, The Enterprise, 380 Walnut St., 77704.
Corpus Christi, The Caller-Times, 820 Lower Broadway, 78401.
Dallas, The News, Communications Center, 75222.
Dallas, The Times Herald, 1101 Pacific, 75202.
El Paso, Times, 401 Mills St., 79999.
Fort Worth, The Press, 507 Jones, 76102.
Fort Worth, The Star-Telegram, 400 West 7th St., 76101.
Houston, The Chronicle, 512-20 Travis St., 77002.
Houston, The Post, 4747 Southwest Freeway, 77001.
Lubbock, The Avalanche-Journal, 8th St. & Avenue "J", 79408.
San Antonio, The Express, Avenue "E" & 3rd St., 78205.
San Antonio, The Light, McCullough & Broadway, 78206.
Waco, Tribune-Herald, 900 Franklin, 76703.
Wichita Falls, The Times, 1301 Lamar St., 76307.

Utah

Salt Lake City, Desert News, 34 East 1st St. South, 84110.
Salt Lake City, The Tribune, 143 South Main St., 84101.

Virginia

Alexandria, Gazette, 717 No. St. Asaph St., 22313.
Arlington, Northern Va. Sun, 3409 Wilson Blvd., 22210.
Newport News-Hampton, The Daily Press, 7505 Warwick Blvd. 23607.
Norfolk, Virginian-Pilot, 150 West Brambleton Ave., 23501.
Richmond, Times-Dispatch, 333 East Grace St., 23213.
Roanoke, Times-World, 201-09 W. Campbell Ave., 24010.

Washington

Seattle, The Post-Intelligencer, 6th & Wall, 98121.
Seattle, The Times, Fairview Ave. N & John, 98111.
Seattle, Daily Journal of Commerce, 83 Columbia St., 98104.
Spokane, Chronicle, W. 927 Riverside, 99210.
Tacoma, The Tribune, 711 Helens Ave., 98401.

Wisconsin

Madison, Capital-Times, 115 South Carroll St., 53701.
Milwaukee, The Journal, 333 West State, 53201.
Milwaukee, Sentinel, 918 N. 4th St., 53201.

The Industry in Great Britain and Ireland

- INDUSTRY DEVELOPMENTS
- PROGRAM CONTRACTORS
- SERVICES
- AD AGENCIES
- COMPANIES AND PERSONNEL
- ORGANIZATIONS
- STUDIOS AND LABORATORIES

BRITISH YEAR IN REVIEW

THE UK television scene during 1981 was dominated by the IBA's (Independent Television Authority's) decisions on the future pattern of commercial television. The Authority, for instance, announced the re-allocation of franchises to run for a period of 8 years from June 1982 together with a new nationwide breakfast-time service. In addition the Broadcasting Act 1980 came into effect giving the Authority responsibility for a new nationwide Fourth Channel television service scheduled to begin in November 1982.

Effectively, the whole population is covered by the IBA's transmissions and 98 per cent of all homes, 20 million in total, had at least one television set. 22 per cent have more than one television set and 62 per cent can receive programmes in colour. Virtually all homes with television can receive all three channels, IBA, BBC1 and BBC2.

The average commercial evening audience during the peak viewing hours of 7.30–10.30 for the year as a whole is 10 million people in 5.8 million homes. During the year 49 per cent of total time spent watching television was spent watching commercial television, 39 per cent watching BBC1 and 12 per cent watching BBC2. Each person, on average, spent a total of 3.4 hours each day watching television.

Such statistics are an indication that even more prosperous times are ahead for commercial television in the U.K. while the British Broadcasting Corporation (BBC) continues to face financial difficulties. The BBC does not transmit advertising and is financed by government grants related to the broadcast receiving licence fees paid by members of the public.

Nevertheless, a healthy competitive situation exists between the BBC and IBA networks. For instance, the BBC claimed at one time of the ceremony, an audience of 25 million for the Royal Wedding on July 29, 1981.

Also, BBC Enterprises—the Corporation's overseas selling organisation—has found new markets and increased trading enabling a record £13 million gross sales in its first trading year as a limited company, an increase of 18 per cent over the previous year's figures. Notable trading increases were also achieved by Film and Video Library Sales and BBC Records and Tapes who had another record year with gross sales well over the £1 million mark for the second year running.

Sales of American programmes to British television are governed, to some extent, by the limitation, by government decree, of foreign material. There is also the British cinema exhibition industry's embargo restricting the televising of any theatrical films to three-year-old productions. Nevertheless, American films and series are among the biggest draws with British viewers.

Nevertheless, a growing number of theatrical films are transmitted with pricews being paid by both the commercial contractors and the BBC rising rapidly. For instance, "Gone With The Wind" was included with a package of films in a $9 million deal with the BBC and a similar arrangement with the commercial networks made "Jaws" available.

However, unless the "free" television networks want to go the way of cinema exhibition, they will need to heed the new ways of distributing video programmes to the home.

Direct-broadcast satellites, "distribution" satellites used in conjunction with cable networks, "subscription" cable services, video-cassette and video disc machines, together with more "off-air" programmes, video games and home computers, all seem likely to lead to more competition for the time and screens of viewers.

Overseas transactions in respect of film and television material — £ million

	1968	1969	1970	1971	1972	1973	1974	1975	1976	1977	1978	1979
Receipts[1]												
Film companies[2]	44.5	40.7	27.6	36.2	31.5	27.4	29.5	45.9	51.1	60.5	85.2	86.1
Television companies[3]	5.1	6.9	5.7	10.6	9.1	10.1	10.4	12.6	18.4	35.7	37.0	41.9
Total	49.6	47.6	33.3	46.8	40.6	37.5	39.9	58.5	69.5	96.2	122.2	128.0
Expenditure[4]												
Film companies[2]	21.4	26.8	15.4	19.6	20.3	21.5	23.0	29.5	31.6	30.1	42.7	51.5
Television companies[3]	4.1	5.2	4.8	4.9	5.0	7.7	9.1	10.8	17.6	16.9	21.7	28.0
Total	25.5	32.0	20.2	24.5	25.3	29.2	32.1	40.3	49.2	47.0	64.4	79.5
Receipts less expenditure												
Film companies[2]	23.1	13.9	12.2	16.6	11.2	5.9	6.5	16.4	19.5	30.4	42.5	34.6
Television companies[3]	2.0	1.7	0.9	5.7	4.1	2.4	1.3	1.8	0.8	18.8	15.3	13.9
Total	24.1	15.6	13.1	22.3	15.3	8.3	7.8	18.2	20.3	49.2	57.8	48.5

[1]Sums receivable from overseas residents. [2]Includes transactions by film companies in respect of rights restricted to television. [3]Includes transactions in respect of BBC sound broadcasting and independent Television News. [4]Sums payable to overseas residents.

Overseas transactions of the BBC and IBA programme contractors[3]: analysis by area £ million

1978

	Total	EC	Other western Europe	North America	Other developed countries[2]	Rest of the world[3]
Receipts[1]	37.0	6.2	2.5	16.8	6.5	5.0
Expenditure[4]	21.7	3.9	1.9	11.8	1.0	3.1

1979

	Total	EC	Other western Europe	North America	Other developed countries[2]	Rest of the world[2]
Receipts[1]	41.9	9.7	3.6	18.3	7.0	3.3
Expenditure[4]	28.0	6.4	1.9	14.8	1.4	3.5

Source: *British Business.*

The Balance of ITV Transmissions (Weekly average, year ended 29th March 1981)

	DURATION		PERCENTAGE	
	hrs	mins	1979–80	1980–81
News and news magazines	10	56	10½	10½
Current affairs and general factual	12	35	12	11¾
Arts	1	12	1¼	1
Religion	2	27	2½	2½
Adult education	2	54	2¾	2¾
School programmes	6	11	6	6¼
Pre-school education	1	29	1½	1¼
Children's informative programmes	2	22	2¼	2¼
'INFORMATIVE'	40	06	38¾	38
Plays, drama, TV movies	24	45	23¾	24¾
Feature films	7	23	7	8¼
'NARRATIVE'	32	08	30¾	33
Children's non-factual	7	45	7¼	7¼
Entertainment and music	13	43	13¼	13¼
Sport	10	07	9¾	8¼
TOTAL ALL PROGRAMMES	104	00	100	100

Miscellaneous transmissions of 11 minutes' duration are not specified above but are included in the total.
Source: *IBA.*

GOVERNMENT UNITS AND PROGRAM CONTRACTORS

HOME OFFICE

Queen Anne's Gate, London SW1H 9AT. Tel.: 01-213-3000.

The Secretary of State for the Home Department is empowered to licence persons or corporate bodies to operate broadcasting stations as laid down in the Wireless Telegraphy Act, 1949. The BBC is incorporated by Royal Charter and operates under this and a Licence and Agreement granted on the 7th July, 1969. The provisions of the Charter and the Licence and Agreement have now been extended to 31st July 1981. The Independent Broadcasting Authority operates under the Independent Broadcasting Acts of 1973 and 1974 which have been extended by the Independent Authority Act 1978 until 31st December 1981 and under a licence which runs until 31st December 1981.

BRITISH BROADCASTING CORPORATION TELEVISION SERVICE

Television Centre, Wood Lane, London, W.12. Telephone: 01-743 8000.

CHAIRMAN OF GOVERNORS: George Howard
ADMINISTRATION
Director General: Ian Trethowan
Managing Director, Television: Alasdair Milne
Deputy Managing Director: Bill Cotton
Controller BBC-1: Alan Hart
Controller BBC-2: Brian Wenham
Chief Accountant: Dick Bates
Controller, Planning and Resource Management: Michael Checkland
Controller Personnel: Roger Chase
Assistant Controller Television Administration: John Keeble
PROGRAMME DEPARTMENTS
Presentation
Head of Presentation Department: Malcolm Walker
Community Programmes
Editor: Michael Fentiman
Drama Group
Head of Group: Shaun Sutton
Head of Plays: Keith Williams
Head of Series and Serials: Graeme McDonald
Light Entertainment Group
Head of Group: James Gilbert
Head of Comedy: John Howard Davies
Head of Variety: Robin Nash
Current Affairs Group
Head of Current Affairs Programmes: John Gau
Outside Broadcasts Group
Head of Group: Cliff Morgan
Head of Sport: Jonathan Martin
Head of Entertainment & Events Programmes: Phil Lewis
Science and Features
Head of Science and Features: Peter Goodchild
Music and Arts
Head of Music and Arts: Humphrey Burton
Documentary Programmes
Head of Documentary Features: Will Wyatt
Children's Programmes
Head of Children's Programmes: Edward Barnes
School Broadcasting
Head of School Broadcasting: J.D.P. Thomas
Continuing Education
Head of Continuing Education: Sheila Innes
Open University
Head of Open University Production Centre: Bob Rowland
Religious Broadcasting
Head of Religious Programmes: Dr. Colin Morris
NEWS AND CURRENT AFFAIRS
Director, News and Current Affairs: Richard Francis
Assistant Director, News and Current Affairs: Alan Protheroe
Editor Television News: Peter Wood
PROGRAMME SERVICES
Studio Management
Head of Studio Management: Harry Cottrell
Scenic Services Group
General Manager, Design and Scenic Services: D. R. Thomas
Scenic Operations
Head of Scenic Construction: Chris Cave
Scenic Construction
Head of Scenic Construction: D.A. Sargeant
Joint Design-Scenic Services Contracts
Design Group
General Manager Television Design: Cliff Hatts
Television Design
Head of Television Design Department: Peter Seddon
Visual Effects
Head of Visual Effects Department: M.J.F. Harris
Graphic Design
Head of Graphic Design: Brian Tregidden
Costume
Head of Costume Department: Miss M.E. McPherson
Make-Up
Head of Make-Up Department: Jack Elliott
Film Operations and Services
General Manager, Film Operations and Services: Ray Mobsby
Head of Film Operations: Peter Jarvis
Script Unit
Head of Television Script Unit: John Scotney
Artists' Bookings
Head of Artists' Contracts: Glyn Martin
Co-Productions
Head of Co-Productions: Tom English
Programme Acquisition
General Manager, Programme Acquisition: Gunnar Rugheimer
ENTERPRISES LTD.
Managing Director, Enterprises: Bryon Parkin
INFORMATION SERVICES
Head of Information Services: Michael Bunce
Chief Press & Publicity Officer: David Stevens
LIAISON UNIT
Head of Television Liaison: Brian Denney
ENGINEERING
Chief Engineer: Bob Longman
Assistant Chief Engineer, Television Operations: Harold Anstey
Assistant Chief Engineer, Television Developments: Ken Kinally
Head of Engineering, Television Projects: Tony Stanley
Head of Engineering, Television Services: Harry Goodings
Head of Engineering, Television Recordings: Douglas Parsons
Head of Engineering, Television Studios: Charles Paton
Head of Engineering, Television Outside Broadcasts: Derek Thorogood
Head of Engineering, Television Network: Norman Taylor
Head of Engineering, Television News: Henry Tarner
INTERNATION UNIT
Head of International Unit: Hugh Sheppard

INDEPENDENT BROADCASTING AUTHORITY

70 Brompton Road, London, SW3 1EY. Tel.: 01-584 7011.

The Authority was created by Parliament in 1955 as the Independent Television Authority to provide a comprehensive television service additional to that of the BBC. Under the terms of the Sound Broadcasting Act 1972, the Authority was renamed the Independent Broadcasting Authority and its functions were extended to cover the provision of Independent Local Radio. In addition, the Broadcasting Act 1980 provided for the setting up of the Channel Four Television Company as a subsidiary of the IBA; it is due to start broadcasting in England, Scotland and Northern Ireland in November, 1982. A separate Welsh Fourth Channel Authority has also been established by this Act. The Authority selects, and appoints the programme companies; supervises the programmes provided by the contractors and their scheduling; controls the advertising; and builds, owns and operates the transmitting stations. Fifteen television programme companies provide programmes in the fourteen independent television regions of the United Kingdom (two companies share the contract for London). As part of its review of the ITV system the Authority, in December 1980, offered a contract to broadcast a new national breakfast-time television service, to start in 1983, to TV-AM. In addition, changes were made to some ITV companies. By May, 1981, 26 Independent Local Radio stations were providing a programme service in 25 areas of the United Kingdom (two companies operate in London). A further 8 stations were due to start broadcasting in the following 12 months. Another 25 stations have been proposed, and these are being considered by the Home Secretary. Both services are financed by the sale of advertising time. The members of the Authority are appointed by the Home Secretary and consist of a chairman and deputy chairman and at present, ten other members, of whom three make the interests of Scotland, Wales and Northern Ireland, respectively, their special care.
Chairman: Lord Thomson of Monifieth
Director General: Sir Brian Young
Deputy Director General: A.W. Pragnell, O.B.E., D.F.C.
Director of Television: C. Shaw
Director of Engineering: T.S. Robson
Director of Radio: John Thompson

Director of Finance: R.D. Downhan
Head of Information: Miss B. Hosking
Head of Advertising Control: H. Theobalds
Dep. Director of Television: D. Glencross
Head of Research: Dr. Ian Haldane
Religious Broadcasting Officer: Christopher Martin
Head of Educational Programme Services: L.C. Taylor
Secretary: Bryan Rook

INDEPENDENT BROADCASTING AUTHORITY GENERAL ADVISORY COUNCIL

Drawn from various walks of life, the Council is appointed to give the IBA impartial advice on the general pattern and content of ITV programmes.
Chairman: Sir Ian Maclennan, KCMG

INDEPENDENT BROADCASTING AUTHORITY ADVERTISING ADVISORY COMMITTEE

The Committee, under the terms of the Independent Broadcasting Authority Act 1973, must be representative of organisations, authorities and persons concerned with standards of conduct of advertising including, in particular, the advertising of medical or surgical goods or services, and the public as consumers.
Chairman: Prof. A.L. Diamond

INDEPENDENT BROADCASTING AUTHORITY SCOTTISH COMMITTEE

Chairman: Rev. W.J. Morris, JP.

INDEPENDENT BROADCASTING AUTHORITY NORTHERN IRELAND COMMITTEE

Chairman: Mrs. J. McIvor

INDEPENDENT BROADCASTING AUTHORITY WELSH COMMITTEE

Chairman: Prof. H. Morris-Jones

INDEPENDENT TELEVISION NEWS LTD.

ITN House, 48 Wells Street, London, W1P 4DE. Telephone: 01-637 2424.

Directors: The Lord Buxton (Chairman), Norman Collins, Frank Copplestone, Bryan Cowgill, Paul Fox, William Hodgson, Alex Mair, David McCall, Daniel Moloney, David Nicholas (Editor), David Plowright, Brian Tesler.

PROGRAM CONTRACTORS

(Appointed by the Independent Television Authority).

ANGLIA TELEVISION LTD.

Head office: Anglia House, Norwich, NR1 3JG. Telephone: Norwich 615151, Telex Telanglia 97424, London Office: Brook House, Park Lane, London, Telephone: 01-408 2288. Telex 25353.

EAST OF ENGLAND TRANSMITTERS:
Chairman: The Marquess Townshend of Raynham.
Directors: The Lord Buxton M.C.DL., Sir John Woolf, Professor Glyn Daniel, R.G. Joice, Desmond E. Longe, MC.DL, D.S. McCall, J.P. Margetson, P. Garner.
Programme Controller: P. Garner.
Sales Director: J.P. Margetson.
Company Secretary: G.M. Rae

ATV NETWORK LTD.

ATV Centre, Birmingham B1 2JP. Tel:021-643-9898; ATV Studio Centre, Eldon Avenue, Borehamwood, Herts. WD6 IJF Tel.: 01-953-6100; ATV House, 17 Great Cumberland Place, London W1A 1AG. Tel.: 01-262-8040.

MIDLANDS FRANCHISE HOLDER
PRESIDENT: Lord Grade
Directors: Jack Gill CBE (Chairman), Lord Windlesham (Managing Director), Leonard Mathews OBE, Francis Essex, Charles Denton, Dennis Basinger, Cecil Clarke, Cliff Baty, Peter Mears, John Madocks, CBE, Ann Spokes
Officers: Leonard Mathews OBE (Senior Resident Director), Francis Essex, Charles Denton (Director of Programmes), Dennis Basinger (Studio Controller—Elstree), Cecil Clarke (Head of Special Drama), Cliff Baty (Financial Director), Richard Creasey (Head of Documentaries), Alan Deeley (Chief Press Officer), Peter Gibson (Head of Staff Relations), Bob Gillman (Head of Regional Development), Philip Grosset (Head of Educational and Religious Broadcasting), Terry Johnston (Head of Regional Programmes), Gerry Kaye (Chief Engineer), Brian Lewis (Film Production Executive), Anthony Lucas (Legal Adviser and Company Secretary), Jean Morton (Head of Audience Relations), Margaret Matheson (Head of Drama), John Scoffield (Head of Light Entertainment), John Terry (Programme Planning and Promotion Controller), Frank Usher (Head of Technical Services), Dorothy Viljoen (Head of Scripts), Billy Wright CBE (Head of Sport and Outside Broadcasting), Peter Mears (Director of Sales), Stanley Smith (General Sales Manager), Barry Spencer (Sales Controller)

BORDER TELEVISION LTD.

The Television Centre, Carlisle, CA1 3NT, Cumberland. Telephone: 0228 25101. London Office: 33 Margaret Street, London, W1N 7LA. Telephone: 01-323 4711.

NORTHWEST ENGLAND—SOUTH SCOTLAND, NORTH NORTHUMBERLAND, AND ISLE OF MAN TRANSMITTERS
Directors: Sir John L. Burgess, O.B.E., T.D., D.L., J.P. (Chairman), James L. Bredin (Managing), B. Blyth, M. Inst. M.S.M. (Sales Director), H.J. Brewis, The Earl of Lonsdale, J.I.M. Smail, O.B.E., M.C., T.D., D.L., Mrs. Moira Shearer Kennedy, D.W. Trimble, MA., R.H. Watts, F.C.I.S. F.C.C.A F.C.M.A. (Dep. Managing Director and Company Secretary), E. Wright, M.A. (Dep. Chairman), J.C. Clucas, F.C.A., Dr. June Paterson-Brown, M.B. Ch.B.
Chief Engineer: H.J. Gower, C. Eng. FIEE.
Assistant Controller of Programmes (Planning): E. Hadwin
Assistant Controller of Programmes (Production): D. Batey
Editor: K. DeVonald
Public Relations Manager: F.J. Bennett

CHANNEL TELEVISION

The Television Centre: Rouge Bouillon St. Helier, Jersey. Telephone: Jersey 73999.

Chairman: E.D. Collas, C.B.E.
Directors: Harold Fielding, J.R. Riley, J. Kay Mouat, G. Le G. Peek, M. Letto, A.E. O'D. Troy.
Managing Director: Kenneth Killip, O.B.E.
Head of Sales: M.J. LeCocq
Operations Manager: F.B. Turner

GRAMPIAN TELEVISION LTD.

Queen's Cross, Aberdeen. AB9 2XJ: Telephone: 0224 53553; 103-105 Marketgait, Dundee. DD1 1QT. Telephone: 0382 21777.

NORTH-EAST SCOTLAND TRANSMITTER
Directors: Iain M. Tennant, J.P. (Chairman); G. Wallace Adam B. Sc.; Robert L. Christie; Fiona J. Lyall, M.B., C.H.B., D.P.H., Alex Mair, M.B.E., A.C.M.A.; The Lord Forbes, K.B.E., D.L., J P.; Neil Paterson; Sir George Sharp, O.B.E., J.P.; Calum A. Macleod, M.A , LLB. (Deputy Chairman), Donald H. Waters, C.A., Neil R. Welling
EXECUTIVES
Chief Executive: Alex Mair, M.B.E.; Operations Manager: Robert Christie; Programme Planning Controller: Alastair Beaton; Publicity and Promotions Executive: Michael J. McLintock, M.A
Head of News and Current Affairs: Edward Brocklebank
Head of Engineering: Alex Ramsay
Company Secretary: Donald H. Waters C.A.
Facilities Executive: Michael Stubbings
Production Executive: Andrew McNeil
Head of Sales and Marketing: Neil R. Welling

GRANADA TELEVISION LTD.

36 Golden Square, London, W.1. Telephone: 01-734 8080. Granada TV Centre, Manchester 3. Telephone: 061-832 7211.

LANCASHIRE TRANSMITTER
Directors: Sir Denis Forman (Chairman & Joint Managing Director), A. Berstein (Deputy Chairman), Sir Paul Bryan, W.R. Carr, L.A W. Diamond, W. Dickson, FCA (Secretary & Financial Director), D. H. Harker, B.L. Heads, D.E. Plowright (Joint Managing Director), A. Quinn, P.R Rennie, M. Scott, Mrs. J. Wooller, L.C. Young.
Granada International: Barrie Heads (Managing Director)

HTV LTD.

London Sales Office: 99 Baker Street, London, W1-M 2AJ. Telephone: 01-486 4311. The Television Centre, Cardiff CF I 9XL. Telephone: Cardiff 21021. Television Centre, Bristol BS4 3HG. Telephone: Bristol 778366 Regent House, Regent Street, Wrexham. Telephone: Wrexham 51777. Scottish Life House, Bridge St., Manchester, Tele.: 061-834 3765.

WALES AND WEST OF ENGLAND TRANSMITTERS
Chairman: Lord Harlech, P.C., K.C.M.G.
Managing Director: R.W. Wordley
Sales Director: C.D. Romaine
Research Executive: I. Platt
Secretary: G.S. Tovey, FCIS, ACA.
Programme Planning: C. Grace

LONDON WEEKEND TELEVISION LTD.

South Bank Television Centre, Kent House, Upper Ground, London SE1 9LT. Telephone: 01-261 3434.

LONDON TRANSMITTER—Friday 7p.m. to Sunday close-down
Chairman: John Freeman
Managing Director: Brian Tesler
General Manager: Vic Gardiner
Director of Programmes: Michael Grade

Finance Director: Peter McNally
Sales Director: Ron Miller
Secretary: Cyril Orr
Controller Staff Relations: Roy van Gelder
Production Controller and Deputy General Manager: Peter Cazaly
Chief Engineer: Roger Appleton
Director of Corporate Affairs: R.J. Potter

RADIO TELEFIS EIREANN
Donnybrook Dublin 4, Ireland: Telephone: (01) 693111. London Office: Ireland House, 150 New Bond Street, London, W1Y OHD Tel.: 01-493 8921 NATIONAL NETWORK OF THE REPUBLIC OF IRELAND.

Chairman: Mr. P. Moriarty
Director-General: G.T. Waters
Deputy Director-General: T.V. Finn
Asst. Director-General: R.K. Gahan
Sales Controller: Donal Stanley
Head of Information & Publications: Louis McRedmond
Director of Television Programmes: Dick Hill
Director of Radio Programmes: Michael Carroll
Controller of Programmes, RTE 1: John Kelleher
Controller of Programmes, RTE 2: Ted Dolan

SCOTTISH TELEVISION LTD.
Cowcaddens, Glasgow, Scotland. Telephone: Glasgow 332 9999.

CENTRAL SCOTLAND TRANSMITTER
Directors: Sir Campbell Fraser (*Chairman*) LL.D. D. Univ., William Brown C.B.E. (*Deputy Chairman and Managing Director*), Gavin Boyd C.B.E., Bill Bryden, Sir Samuel Curran D.L. F.R.S., Mrs. Dorothy Dunnett, Charles A. Fraser M.V.O., Hugh W. Henry (*Sales Director*), Lewis J.M. Hynd O.B.E. (*Company Secretary*), David K. Johnstone (*Director of Programmes*), Mrs. Barbara Leburn M.B.E.J.P., Sir Iain M. Stewart LL.D., Lord Taylor of Gryfe D.L. LL.D., The Earl of Wemyss and March K.T. ILL.D.
Public Relations Manager: John Loch

TELEVISION SOUTH LTD.
Spenser House, 60-61 Buckingham Gate, London, SW1. Tel.: 01-8289898.

SOUTH OF ENGLAND
Managing Director: James Gatward

TELEVISION SOUTHWEST LTD.
Derry's Cross, Plymouth, Devon. Tel.: 0752 663322.

SOUTH WEST ENGLAND TRANSMITTER
Jnt. Man. Directors: Peter Battle, Kevin Goldstein-Jackson

THAMES TELEVISION LTD.
Thames Television House, 306-318 Euston Road, London NW1 3BB. Tel.: 01-287 9494. Teddington Lock, Teddington, Middlesex TW11 9NT. Tel.: 01-977 3252. Norfolk House, Smallbrook Queensway, Birmingham B5 4LF. Tel.: 021-643 9151.

LONDON TRANSMITTER—Monday to Friday
Directors: H.S.L. Dundas (Chairman); B.G. Cowgill (Managing Director); Mrs. M. Baker; H.S.L. Dundas, DSO. DFC.; Sir John Read; J.F. Shaw (Director of Sales): C.S. Wills; Lord Brabourne, Sir Richard Cave, J.T. Davey, Ian Scott, R.M. Denny.
Executives: B.E. Marr (Director of Administration and Company Secretary); R.J. Godfrey (Engineering & Technical Director); Nigel Ryan (Dir. of Programmes), Muir Sutherland (Managing Director Thames Television International); F. Atkinson (Technical Controller); D. Cullimore (Public Relations Director); D. Stevenson (Sales Controller); M. Lawson (Assistant Director of Finance); P. Jones (Controller of Light Entertainment); J. O'Keefe (Industrial Relations); M.J. Frankam (Controller of Drama); D. Hunt (Assistant Director of Finance); G. Lugg (Controller of Programme Planning & Liaison); I. Martin (Controller of Features, Education & Religion); A.C. —Parkinson (Controller, Administration); E.E. Parry (Controller, Programme Services); D. Thornes (Research & Marketing Services Manager); R. Alison (Controller of Sports and Outside Broadcasts); J. Hambley (Controller of Childrens' Programmes).

THAMES INTERNATIONAL
(Wholly-owned subsidiary for programme sales)
Managing Director: Muir Sutherland.

TRIDENT TELEVISION LTD.
Trident House, 8 Grafton Street, London, W1Y 3LA. Telephone: 01-493 1237.

(Parent company controlling Tyne Tees Television Ltd., and Yorkshire Television Ltd.)
Chairman and Managing Director: G.E. Ward Thomas, CBE, DFC
Assistant Managing Directors: Paul Fox; Peter S. Paine, CBE, DFC
Commercial Director:
Directors: Sir Ralph Carr-Ellison, TD; R.H. Dickinson, MA; Sir Richard B. Graham, BT, OBE, DL; Sir James E. Hanson; J.G.S. Linacre, CBE, AFC, DFM; John Wilkinson; P. Wrightson, OBE, Clive Leach

TV-AM
3 Deanery Street, London, W1Y 5LH. Tel.: 01 493 6550.
Chairman: Peter Jay

TYNE TEES TELEVISION LTD.
Head Office: The Television Centre, City Road, Newcastle upon Tyne, NE1 2AL. Telephone: 0632 61-0181. Middlesbrough Office: Corporation House, Corporation Road, Middlesbrough. Tel.: 0642 219181

NORTH EAST TRANSMITTER
Directors: Sir Ralph Carr-Ellison (Chairman) TD, JP; Peter S. Paine DFC, (Managing Director); R.H. Dickinson M.A.; J.P. Graham FCIS, (Company Secretary); G. Oliver Worsley; Viscount Ridley TD, DL; Peter Wrightson, OBE; Sir Maurice Sutherland, Lord Pearl, Professor L.W. Martin, Andy Allan, BA, John Tonge, MBE.
General Manager: John Tonge, M.B.E.

ULSTER TELEVISION LTD.
Havelock House, Ormeau Road, Belfast, BT7 1EB Northern Ireland. Telephone: Belfast 28122. London Office: 19 Marylebone Rd., London, N.W.1. 5JJ. Telephone: 01-486 5211.

NORTHERN IRELAND TRANSMITTER
Directors: J.L. MacQuitty, Q.C., M.A., LL.B. (Chairman), R.B. Henderson, C.B.E., M.A., F.R.T.S. (Deputy Chairman & Managing Director); J.B. McGuckian, B.Sc. (Econ.) (Deputy Chairman); M.R. Hutcheson (Sales Director), J.A. Creagh (Assistant Managing Director); J.B. Waddell (Controller, Local Programmes); Angela, Countess of Antrim, LL.D.; R.E. Benner; O.B.C.; Miss Betty E. Box, O.B.E.; H.R.C. Catherwood, C.S.G. Falloon; Captain O.W.J. Henderson, D.L., F.B.I.M.; G.C. Hutchinson; Major G.B. MacKean, D.L. J.P.; Mrs. Betty MacQuitty, B.Sc. (Econ.); E.M.R. O'Driscoll, LL.D. (Alternate E.J. O'Driscoll)

YORKSHIRE TELEVISION LTD.
The Television Centre, Leeds LS3 1JS. Tel.: 0532 38283; Telex: 557232. Trident House, 8 Grafton Street, London W1X 3LA. 29-30, Old Burlington Street, London, W1X 1LB. Tel.: 01-493-1237; Telex: 25202. Charter Square, Sheffield S1 4HS. Tel.: 0742 23262. 185 Ferensway, Hull HU1 3PH. Tel.: 0482 24488. 2 Saltergate, Lincoln LN2 1DH. Tel.: 0522 30738. Brazennose House, Brazennose Street, Manchester M2 5BP Tel.: 061-834-4228. 8 Bull Ring Lane, Grimsby DN31 1DY. Tel.: 0472 57026.

YORKSHIRE TRANSMITTER
Executives: Kenneth Bellini (Head of Programme Planning); Mrs. Liz Evett (Head of Casting); Frank Smith, (Head of Current Affairs & Documentaries); Brian Harris (Head of Programme Services); Lawrie Higgins (Head of Outside Broadcasts & Sport); Clive Leach (Sales Director); Philip Parker, MIEE (Director of Engineering); David Thorne (Regional Sales Manager); Leslie Thornby, FCIS (Company Secretary); Keith Smith, (Public Relations Manager); Joy Whitby (Head of Children's Programmes); David Cunliffe, Head of Drama); Duncan Wood (Head of Light Entertainment).

TV Program Material—Producers and Distributors

ABACUS PRODUCTIONS LTD.
33 Shelton Street, London WC2; Tel.: 01-240 1277; Cables: ABAFILMS LONDON.
Directors: John Mackey (Chairman), Bob Cuff, Ron Trainer (Jt. MD's).
Entertainment and advertising material provided. Type of film service: Live action (narrative), Live Action (dialogue); miniature and matte process; stop motion; model animation; studios: 2 stages (both with cove; 2 cutting rooms; optical printer; rostrum camera.); 2 Preview Theatres.

ADVISION LTD.
23 Gosfield Street, London, W1P 7HB. Tel.: 01-580 5707. Telex 28668.
Directors: K.S. Hibberd, A. Whetstone, R. Cameron, B. Hibberd, G. Middleton.
(Sound recording facilities: music and film dubbing studios.)

AEGIS PRODUCTIONS LTD.
Thames House, II7 Albert Bridge Road, London, S.W.11. Tel.: 01-228 7766

AEROFILMS
Gate Studios, Station Road, Boreham Wood, Herts: WD6 1EJ; Tel.: 01-207 0666.

AIR-TIME PRODUCTIONS LTD.
50-Frith Street, London, W.1. Tel.: 01-734 9304.
Director: Johnny Fielder.

JOHN ANDERSON HIRE
Court Lane, Iver, Bucks. Tel.: Iver 653737 (production facilities).

GERRY ANDERSON PRODUCTIONS LTD.
Pinewood Studios, Iver Heath, Bucks. Tel.: 0753 652344.
Directors: Gerry Anderson, Keith Shackleton.

EAMONN ANDREWS STUDIOS LTD.
Television Club, 46/48 Harcourt Street, Dublin, Ireland. Tel.: Dublin, 758891.
Directors: Eamonn Andrews, Dermod Cafferky, Lorlan Bourke.
(Tape, disc-recording, filming, Radio & TV programme contractors.)

ANIMALS UNLIMITED
Careys Wood, Smallfield, Surrey: (Animal agency). Tel.: 034 284.
(Animal agency.)

ANIMATED MOTION PICTURES LTD.
72 Wardour Street, London W.1. Tel.: 01-437 8282.
Directors: Nat Miller (Managing), D. Siebenmann.

ANVIL FILM & RECORDING GROUP LTD.
Denham Studios, North Orbital Rd., Denham, Nr. Uxbridge, Middx. Tel.: Denham 833522.
Directors: Ken Cameron, O.B.E., B.Sc. (Chairman), R.I.C.H. Warren (Managing), R.W. Keen, K. Somerville, C. Eng. MIERE.

JAMES ARCHIBALD, 35 Morpeth Mansions, Morpeth Terrace, London, SW1P 1EU. Tel.: 01-828 9691.

ARMADA PRODUCTIONS
86/88 Wardour Street, London WIV 3AT. Tel.: 01-734 6312.
Managing Director: J.D. Dooley.

ATHOS TV AND FILM FACILITIES LTD.
Athos House, Cambridge Rd., Hanwell, London, W.7. Tel.: 01-579 4301-6. Telex: 28905 Ref. 325.
Directors: P.G.A. Bucknall, A.G. Eggleton, The Viscount Portman, D.B. Bucknall, P.S. Hodges.
(Production, 35/16mm cutting rooms, dubbing, recording studios.)

AUSTRALIAN BROADCASTING COMMISSION
54 Portland Place, London, WIN 4DY. Tel.: 01-580 9601. Cables: Austcast, London. Telex: 263897.
Manager for Europe: S. Revill.

AUTOCUE LTD.
Autocue House, 265 Merton Road, London, S.W.18 5JS. Cables: Autocue, London, SW18. Telex: Autocue Telexir 896691G.
Directors: J. Garrett, J.F. Lucas, C.C. Lambert
(Script prompting facilities for film TV and public speaking.)

BALFOUR FILMS (Samaritan Films Ltd.)
76 Brewer Street, London, W1R 4HU. Tel.: 01-437 3877.
Directors: Anne Balfour Fraser, Eric Falk, Timothy Burrill, David Thompson, Brian Little, Frances Cockburn.
(Documentary, advertising and ent. material.)

BASIC FILMS LTD.
68 Lebanom Park, Twickonham TW1 3OQ, Tel: 01-891 0338.
Directors: J.B. Napier-Bell, H.J. Palk; Secretary: M. Napier-Bell.
(Producers of educ. industrial and advt. films.)

BBC TELEVISION ENTERPRISES
Villiers House, The Broadway, London, W.5 2PA. Tel.: 01-743 5588. Cables: Telecentre, London. Telex: 934678 or 265781; Overseas Sales Offices: Westfield Towers, 100 William Street, Sydney, N.S.W. 2011, Australia; Suite 510, Manulife, 55 Bloor Street West, Toronto, Ontario, Canada; Distributors for USA and Latin America: Time-Life Films, Time-Life Building, 1271 Avenue of the Americas, New York, N.Y. 10020, U.S.A.
General Manager: Bryon Parkin.

BIOGRAPHIC FILMS LTD.
90 Dean Street, London, W.1. Tel.: 01-437 9682.
Directors: Keith Learner, Vera E. Linnecar, Nancy Hanna.
(Cartoon, live action, puppets.)

BOULTON-HAWKER FILMS LTD.
Hadleigh, Ipswich, Suffolk 1P7 5BG. Tel.: 0473 82 2235.
Directors: K.P. Boulton, M.J. Boulton, S.J. Boulton, N.H.L. Rea, D.A. Boulton

BRITISH FILMS LTD.
Carlyle House, 235 Vauxhall Bridge Road, London, S.W.1. Tel.: 01-828 7965 Telex: 947165.
(Prestige, advertising and instructional films.)

BRITISH TRANSPORT FILMS
Melbury House, Melbury Terrace, London, NWI 6LP. Tel.: 01-262 3232.

BROCKET PRODUCTIONS LTD.
27-29 Berwick Street, London, W1V 3RF. Tel.: 01-437 0136.
Directors: Matt McCarthy, John Black.

B.D. PRODUCTIONS
Suite 43/44, Mansfield Chambers, 17 St. Ann's Square, Manchester M27PD. Tel.: 061-834 9396.
Derek Burrell-Davis, Mary David.

CAMERA LOCATION SERVICES
16 St. James Close, St. John's Woking, Surrey. Tel.: 04862 4645.
(Researchers, suppliers film and photographic locations.)

CAVALIER FILMS LTD.
34 Deacon's Hill Road, Elstree Herts. Tel.: 01-953 1403.
Chairman & Managing Director: J.E. Rogers, F.R.P.S., M.B.K.S.
(Specialised films, documentary, commercials.)

CBS NEWS
100 Brompton Rd., London, S.W.3. Tel.: 01-584 3366. Telex: 916 319. Cables: Colnews London.
(News, radio and TV coverage.)

CENTRAL OFFICE OF INFORMATION
Hercules Road, London, SE1 7DU. Tel.: 01-928 2345.
Films and Television Division: Director: A.C. White.
Radio Services: Head: K.G. Hicks.

ROGER CHERRILL LTD.
65-66 Dean Street, London W1V 5HD. Tel.: 01-437 7972.
(Post sync Dialogue and Effects.)

CHESS VALLEY FILMS LTD.
Film House, Little Chalfont, Bucks: HP7 9PY. Tel: 02-404 2222.
Directors: Ronald E. Haddock, V.G. Haddock, Heather M. Davies, MBKS.
(Production 35/16mm film and videotape and allied services.)

CHRYSALIS VISUAL PROGRAMMING
12-13 Stratford Place, London, W1N 9AF. Tel.: 01-408 2355. Telex: 21753.

CINE-LINGUAL LTD.
27-29 Berwick Street, London, W1V 3RF. Tel.: 01-437 0136.
Directors: Matt McCarthy, John Black.
(Dubbing, post production & cutting services.)

CINESOUND EFFECTS LIBRARY LTD.
Imperial Studios, Maxwell Road, Elstree Way, Boreham Wood, Herts WD6 IWE. Tel.: Elstree 5837 and 5545, 1587, 4904.
(Sound effects.)

CINEVISTA LTD.
Suite 4, 3 Bateman Street, London, W.1. Tel.: 01-437 7187.
Directors: John Simmons, Cecilia Simmons.
(Entertainment and sponsored material.)

COLOUR FILM SERVICES LTD.
22-5 Portman Close, Baker Street, London, W1H OEP. Tel.: 01-486 2281. 881.
Chairman: Roland Chase.
Managing Director: H. Manley.
(Advertising material.)

COLUMBIA PICTURES TELEVISION
A Division of Columbia Pictures Corporation Ltd. St. Margaret's House, 19/23 Wells Street, London, W1P 3FP. Tel.: 01-637 8444. Telex: 23121. Cables: Columfilm, London.
Vice President, European Zone: Tim Vignoles.
(Production and distribution TV film series.)

STAN COX ASSOCIATES LTD.
84 Wardour street, London, W.1. Tel.: 01-437 9934.

CRANEZOOM LTD.
14a Albert Court, Kensington Gore, London, SW7 2BG. Tel.: 01-589-0204.

CRESTVIEW PRODUCTIONS LTD.
159-165 Grt. Porland Street, London, W.1. Tel.: 01-637 1164. Cables: Helpful, London. W1.
Directors: Ian R. Warren, T.L. Donald.
(TV series.)

CROWN INTERNATIONAL PRODUCTIONS LTD.
145 Wardour Street, London, W1V 3TB. Tel.: 01-437 2945/6.

CTVC
Hillside, Merry Hill Road, Bushey, Watford WD2 1DR. Tel.: 01-950 4426.
(Training centre; production and hire facilities.)

CYGNET GUILD COMMUNICATIONS LTD.
(Film Producers Guild Ltd., Verity Films Ltd., Technical & Scientific Films Ltd., Unit 7 Productions Ltd., Interfilm (London) Ltd., Talkiestrips Ltd., Guild Animation Ltd., Larkin Studios Ltd., Cygnet Films Ltd., Stewart Films Ltd., Stanley Schofield Productions Ltd.) Bushey Studios, Melbourne Road, Bushey, Herts. Tel.: 01-950 5791. Telex: 922 996.
Managing Director: R. Evans.

DATA FILM DISTRIBUTORS LTD.
20 Great Chapel Street, London, W1V 3AQ. Tel.: 01-734 9987.
Directors: W. Goodman, B.V. Goodman, N. Aphale (Secretary).

DE LANE LEA LTD.
75 Dean Street, London, W.1. Tel.: 01-439 1721.
(Entertainment & Advertising films; Sound Recording, Foreign Version Dubbing, editing facilities.)

DE LANE LEA MUSIC LTD.
The Music Centre, Engineers Way, Wembley, Middx: Tel.: 01-903 4611. Telex: 923400.
Contact: Peter Harris.

DE WOLFE LTD.
80-88 Wardour Street, London, W1V 3LF Tel.: 01-437 4933-4.
Directors: James Sylvester de Wolfe, Gordon Chambers.
(Library record music, original scores.)

DIFFUSION D'INFORMATIONS PAR LE FILM S.A.
(Associate of Film Producers Guild Ltd.) 22 Rue de la Corraterie, Geneva, Switzerland.
Directors: N.C.B. Abbott, G. Busby, A. Marks.
(Documentary & industrial films.)

WALT DISNEY PRODUCTIONS LTD.
68 Pall Mall, London, S.W.1 Tel.: 01-839 8010. Telex: 21532.

DORLAND ADVERTISING LTD.
121-141, Westbourne Terrace, London, W.2. Tel: 01-262 5077.

DRAKE AUDIO VISUAL
212 Whitchurch Road, Cardiff, Wales. Tel.: Cardiff 24502/3.
Managing Director: R.G. Drake.
Marketing Director: H. Drake.
Sales Manager: R. Jenkins.

DRUMMER FILMS LTD.
14 Haywood Close, Pinner HA5 3LQ.
Man. Dir.: Martin M. Harris (Producer).

DUCK LANE FILM PRODUCTIONS LTD.
8 Duck Lane, London, W1V1FL Tel.: 01-439 3912. Telex 25554.
Directors: D.R. Andrews, A. O'agviar.

EDUCATIONAL & TELEVISION FILMS LTD.
247a Upper Street, London, N1-1RU. Tel.: 01-226 2298/9.
General Manager: Stanley Forman.

EMI AUDIO VISUAL SERVICES LTD.
5 Dean Street, London, W1V 6QD. Tel.: 01-437 8651. Telex: 8952737.
Managing Director: Donald MacLean.

EMI FILMS LTD.
142 Wardour STreet, London, W1A 3BY. Tel.: 01-437 0444. Telex: 22760.

EPA INTERNATIONAL PROGRAMMERS LTD.
70 Wardour Street, London, W.1. Tel.: 01-439 4351.

EUSTON FILMS LTD.
365 Euston Road, London, N.W.1.

EWART & COMPANY (STUDIO) LTD.
$c3 Wandworth Plain, London, S.W.18. Tel.: 01-874 0131. Cables: Keithewart, London. Telex: 929509.
Managing Director: Keith Ewart.
(Cinema, and television commercials, full production facilities and 2" Broadcast VTR Landline to Post Office Tower.)

EYELINE FILMS LTD.
77 Dean Street, London, W.1. Tel.: 01-734 3391.
Directors: Harold Orton, Jacki Roblin, George H. Brown, Alfred Shaughnessy, Gordon Grimward.
(Ent. advertising material.)

FILM AND TELEVISION MARKETING SERVICES LTD.
4 Tilney Street, London W.1. Tel.: 01-499 2446.
Directors: Richard Blayney, Sheila Brewster, Barbara Norris.
(Dist. of TV programmes and feature films.)

FIVE CITIES FILMS LTD.
Elite Studios. Toller Lane, Bradford 8. Tel.: Bradford 46711/2/3.
Directors: D. Roebuck, S.J. Josephs.
(Commercials advertising and ind. films.)

FLAMINGO FILM PRODUCTIONS
P.O. Box 324, London, SW1X 8LG. Tel.: 01-235 8243.
Directors: R. Kruger, J.S. Kruger.
(Producers of feature films and advertising material.)

FOWLER-CHAPMAN COMPANY LTD.
14a Albert Court, Kensington Gore, London, SW7 2BG. Tel.: 01-589 0204.

FREMANTLE INTERNATIONAL, INC.
30 Dean Street, London W.1. Tel.: 01-437 2673.
European Director: Anthony Gruner.
(Distributors TV programme material.)

JAMES GARRETT & PARTNERS LTD.
5 Queen Street, London, W.1. Tel.: 499 6452. Telex: 261163.
Directors: J.L.M.P. Garrett (Chmn. & Man. Dir.); D.T. Peers (Dep'y. Chmn.); M. Garrett; G.A. Forster.
(Advertising film producers.)

GATEWAY PRODUCTIONS LTD.
Head Office: Waverley Road, Yate, Bristol BS17 5RB. Tel.: 0454 316774. London Office: 470/472 Green Lanes, Palmers Green, London, N. 13. 5XF. Tel.: 882-0177.
Directors: G.R. Heywood, I.F. Jenkins, D.F. Littlechild, G.L. Smart, N. Whalley.
(Producers and Distributors of 16mm. educational, industrial and training films, 8mm. loops, 35mm. filmstrips and Learning Packages.)

GATEWAY FILM PRODUCTIONS
470-472 Green Lanes, Palmers Green, London, N. 13 5XF. Tel.: 01-882 0177.
Chairman: G.R. Heywood.
Directors: G.L. Smart, A.R.P.S., M.B.K.S., Managing N. Whalley, D.F. Littlechild, I.F. Jenkins.
(Industrial, advertising, educational and training films.)

GATEWAY EDUCATIONAL MEDIA
Head Office: Waverley Road, Yate, Bristol BS17 5RB. Tel.: 0454 316774. London Office: 470/472 Green Lanes, Palmers Green, London, N. 13 5XF. Tel.: 01-882 0177.
Directors: G.R. Heywood, I.F. Jenkins, D.F. Littlechild, G.L. Smart, N. Whalley.
(Prod. dist. 16mm. educ., films 8 mm. films lays, 35mm. filmstrips and Learning Packages.)

GENERAL SCREEN ENTERPRISES LTD.
Film House, 97 Oxford Road, Uxbridge, Middx. Tel.: Uxbridge 31931. Telex: 934883.
Managing Director: Fred Chandler.
Scriptwriters: Tony Hill, Tony Church.
(Main titles, trailers, commercials, opticals, special effects.)

G.H.W. PRODUCTIONS LTD.
52 Queen Anne Street, London, W1M 9LA. Tel.: 01-935 1186.
Directors: Peter Rogers, Betty E. Box, O.B.E.; D.G. Truscott.

GLOBAL TELEVISION SERVICES LTD.
159-165, Great Portland Street, London, W1N 6NR. Tel.: 01-637 1164.
Directors: Ian R. Warren, T.L. Donald, I.L. Donald, J. Thompson.
European Purchasing Agent for programmes for Australian Broadcasting Commission and Broadcasting Corporation of New Zealand. World Distributors for HTV, Scottish Television. European Distributors for Canadian Broadcasting Corporation, Societe Radio Canada, Australian Broadcasting Commission and TV I of New Zealand.

BOB GODFREY'S MOVIE EMPORIUM
55 Neal Street, London, W.C.2 Tel: 01-240 1793.
(Specialists in animation and live action for entertainment and advertising.)

G.P.A. (FILMS) LTD.
22 Romily Street, London, W.1. Telephone: 01-734 6994.
Director: Gerry Poulson.

GRANADA TELEVISION INTERNATIONAL LTD.
36 Golden Square, London, W1R 4AH Tel.: 01-734 8080. Telex: 27937.
Managing Director: Barrie Heads.

ELAINE GREENE LTD.
31, Newington Green, London, N16 9PU. Tel.: 01-249 2971.
Directors: Elaine Greene, Ilsa Yardley, Kenneth Thomson.
(Literary agency.)

GUILD SOUND & VISION LTD.
Woodston House, Oundle Road Peterborough PE2 9PZ. Tel.: 0733 63122.
(Distribution and marketing of film and television programmes.)

HALAS AND BATCHELOR ANIMATION LTD.
3/7 Kean St. London, WC2, Tel.: 01-240 3888.
Directors: John Halas, Joy Batchelor Jack King, Harold Whitaker, Rolf Deyhle.

HALLIFORD STUDIOS
Manygate Lane, Shepperton, Middx.: Tel.: Walton-on-Thames 26341. Cables: Hallifilms.
Directors: D.R. Andrews, E.S. Andrews.
Studio Manager: Eugene Andrews.

HAMMER FILM PRODUCTIONS LTD.
EMI Elstree Studios, Boreham, Herts. Tel.: 01-953 1600.
Directors: Brian Lawrence, Roy Skeggs.

HAMPDEN GURNEY STUDIOS, LTD.
Hampden Gurney Street, London, W1H 5AL. Tel.: 01-402 6255. Cables: Jaksonfilm London W1. Telex: 896559 GECOMS-G.

STEWART HARDY FILMS LTD.
Manor Way, Boreham Wood. Herts. Tel.: 01-953 3730/4743.
Directors: R.K. Hardy, J.R.F. Stewart, J.R. Williams, J.D. Messenger.
(Animation, diagram, cartoon films.)

HAWKE STUDIOS LTD.
22 Chenies Street, London, WC1E 7EX. Tel.: 01-580 7452.
(Advertising designers.)

HELIOS TELEFILM PRODUCTIONS LTD.
17 Pembridge Place, London, W.2. Tel.: Bayswater 0497.
Directors: Otto Kreisler, C.H. Bonner, T.A. Moodie.
(Feature and Television films.)

HEMDALE FILM GROUP LTD.
3 Audley Square, South Audley Street, London, W1Y 5DR. Tel.: 01-491 7491.
Directors: J. Daly, D.J. Dawson, J.B. Stutter, A. Goodin.

HEMISPHERE PRODUCTIONS LTD.
105 Mount Street, London, W.1. Tel.: 01-493 5041.

GERARD HOLDSWORTH PRODUCTIONS LTD.
31 Palace Street, London, SW1E 5HW Tel.: 01-828 1671.
Directors: Gerard A. Holdsworth, Mary V. Holdsworth, P.H. Filmer-Sankey, Don Kelly.
(Documentaries, TV program material.)

I.F.A. (SCOTLAND) LTD.
19 North Claremont Street, Glasgow, C3. Scotland. Tel.: 041-332 3620.

IMAGE INTERNATIONAL VIDEO AND FILM PRODUCTIONS LTD.
5 Kendrick Mews, London, S.W.7. Tel.: 01-584 6048.

INCA FILMS LTD.
76 Brewer Street, London, W1R 4HU. Tel.: 01-437 3877.
Directors: Anne Balfour Fraser, Timothy Burrill, Brian Little, Eric Falk, Frances Cockburn.
(Documentary, Advertising and Entertainment material.)

INDEPENDENT TELEVISION NEWS LTD.
ITN House, 48 Wells Street, London, W1P 4DE Tel.: 01-637 2424.
(Production facilities.)

INTERVISION
102 Holland Park Avenue, London, W11 4VA. Tel.: 01-727 1453. Telex: 557363.
Contacts: Mike Tenner, Richard Cooper.

ITC ENTERTAINMENT LTD.
17 Great Cumberland Place, London, W1A 1AG. Tel.: 01-262 8040.
Directors: Lord Grade, Jack Gill, E.J. Kingham, E.S. Birk, Ian R. Jessel, David C. Withers, A.D. Brook.
(Production and world-wide distribution of TV programmes and features.)

ITN LTD.
ITN House, 48 Wells Street, London, W1P 4DE. Tel.: 01-637 2424.
(Colour television film & sound facilities in central London.)

BRIAN JACKSON FILMS LTD.
Hampden Gurney Studios. Hampden Gurney Street, London, W1H 5AL. Tel.: 01-402 6255. Cables: JAKSONFILM London, W1. Telex: 896559 GECOMS-G.

JARAS ENTERTAINMENTS LTD.
21 Cavendish Place, London, W1M 9DL. Tel.: 01-580 5927.

CYRIL JENKINS PRODUCTIONS LTD.
7 Coniston Court, Carlton Drive, London SW15 2BZ. Tel.: 01-788 2733.
Directors: Cyril G. Jenkins, J.M. Jenkins.
(Entertainment, documentary and advertising material.)

KADEK VISION LTD.
Shepperton Studios Centre, Squires Bridge Road, Shepperton, Middx. TW17 0QD.

KRUGER ORGANISATION (FILM & TV DISTRIBUTION) LTD.
P.O. Box 324, London, SW1X 8LG. Tel.: 01-235 8243.
Directors: J.S. Kruger, R. Kruger, H.R. Kruger.

LEE ELECTRIC (LIGHTING) LTD.
128 Wembley Park Drive, Wembley, Middx. Tel.: 01-903 9655 and 01-903 9811.

LEE FILTERS LTD.
Central Way, Walworth Industrial Estate, Andover, Hants. SP10 5AN. Tel.: 0264-66245. Telex: 477259.

LEE SCAFFOLDING LTD.
47 Brunel Road, East Acton, London, W.3. Tel.: 01-749 7305.

LEGION FILM PRODUCTIONS
16 Bloomfield Terrace, London, SW1W 8PG. Tel.: 01-730 9581.

LE VIEN FILMS LTD.
15 Chesterfield Hill, London, W.1. Tel.: 01-629 4545. Cables: Vienfilms London, W.1.
Executive Producer: Jack Le Vien.
(TV films and features.)

LIBRA PRODUCTIONS LTD.
Marl Cottage, Mark Cross, Crowborough, Sussex. Tel.: Rotherfield 2172/2746.
Directors: Kevin Scott (USA), Thomas Kyffin.
(Producers of Television Programmes).

LION TELEVISION SERVICES LTD.
Shepperton Studios, Squires Bridge Road, Shepperton, Middx. Tel.: 09328 60241.
Gen. Manager: Alan J. Mashford

LITTLE KING PRODUCTIONS LTD.
245 Old Marylebone Road, London, NW1 5QT. Tel.: 01-402 8385.
Directors: T.G. Little, D.S.C., Robert King, M.B.E., E.K. King, Dr. Simon C. Nicholas, M.B., B. Chir.
(Production facilities, entertainment and advertising films.)

LORIMAR TELEVISION
16 Berkeley Street, London, W.1. Tel.: 01-493 1566.
Managing Director: Ray Lewis.

MACQUEEN FILM ORGANIZATION LTD.
15 Seal Road, Sevenoaks, Kent. Tel.: 0732 53774.
Managing Director: Nona Lee.
(Advertising, documentary, medical, scientific films & commercials.)

MMA PRESENTATIONS LTD.
16 Neal's Yard, Monmouth Street, London, WC2H 9DP. Tel.: 01-379 3595.
(Music and jingles, radio and TV producers, scripts, conference organisers.)

MARK I PRODUCTIONS LTD.
Lee International Studios, 128 Wembley Park Drive, Wembley, Middx, HA9 8JE. Tel.: 01-902 1262.

MLG FILM AGENCY LTD.
15-17 Old Compton Street, London, W1V 6JR. Tel.: 01-437 9541. Cables: Eaglepix, London, W.1.
Directors: Barry Jacobs, A. Jacobs.

ANTHONY MORRIS LONDON LTD.
6 Goodwins Court, St. Martin's Lane, London, WC2N 4LL. Tel.: 01-836 0576. Cables: AMATEL. Telex: 267664.

MOTION PICTURE MARKETING LTD.
26 York Street, London, W1H 1FR. Tel: 01-935 3565. Telex: 266173.
Directors: Derek Blades (Man.)

THE MOVING PICTURE COMPANY
25 Noel Street, London, W.1. Tel: 01-734 9151.
(Production Facilities.)

CECIL MUSK PRODUCTIONS.
19 Mole House, Kingfisher Close, Walton-on-Thames, KT124LF. Tel.: Walton-on-Thames 20199.
Directors: Edward Cecil Musk, Gretchen Roma Musk.
(Producers programme material.)

NBC NEWS WORLDWIDE INC.
25 St. James' Street, S.W.1. Telephone: 01-839 4571. Telex: 27765.
London Correspondent: Garrick Utley.
General Manager Europe: Irwin Margolis.
London Bureau Chief: Robert McFarland.

NATIONAL SCREEN SERVICE LTD.
Studios: 15 Wadsworth Road, Greenford, Middx. Telex: 934522. Tel.: 01-998 2851. London Office: Nascreno House, Soho Square, London, W1V 6BH. Tel.: 01-437 4851.
Chairman and Chief Executive: Russell Cradick.
Production Controller: Bill Seymour.
(Producers and distributors of publicity accessories; trailers, main titles, opticals, special effects and TV commercials.)

NATIONAL TELEFILM ASSOCIATES INC. (UK) LTD.
28 Berkeley Square, London, W1X 5HA. Tel: 01-499 5945 Telex: 881 2226 NTA LDN.
Foreign Manager: Art Jacobs.

RICHARD NEED PRODUCTIONS LTD.
93-95 Wardour Street, London W.1. Tel: 01-437 1782.
Managing Directors: N.C.B. Abbott, G. Busby.
(Commercials.)

NEW CENTRAL FILM SERVICES (SOHO) LTD.
97-99 Dean Street, London, W.1. Tel.: 01-734 5827-8.
(Editing equipment hire, cutting rooms, production offices cutting room supplies.)

NICHOLAS CARTOON FILMS LTD.
Maidensgrove, Nr. Henley-on-Thames, Oxon. Tel.: Turville Heath 286.
Directors: D.N. Spargo, P.M. Spargo.
(Cartoon.)

NUMO PRODUCTIONS LTD.
3 Richmond Mews, London, W.1. Tel.: 01-439 4017.

ORB PRODUCTIONS LTD.
72 Wardour Street, London, W.1. Tel.: 437 8282. Cables ORBILM, London.
Directors: Nat Miller, Louise Miller.
(TV shorts, feature distribution.)

ORIGINAL ELECTRIC PICTURE COMPANY
15 Pembroke Gardens, Kensington, London, W.8. Tel. 01-603 8192.
Production: Stanley Dubens.
Account Executive: Roger Moss. (Commercials, Documentaries, Video, AV Presentations.)

OVERSEAS FILM AND TELEVISION CENTRE LTD.
Overseas House, 19-23 Ironmonger Row, London, EC1 Tel.: 01-251 2793.
Directors: D.S. Bowden, P. Whale, A. Muscat, K. Minchell, C.G.E. Gover.
(Ent. and adv. producers.)

PACESETTER PRODUCTIONS LTD.
82 Wardour Street, London W1V 3LF. Tel: 01-437 3907.
Managing Director: Ronald C. Spencer.
(Commercials, shorts, doc., ind. and children's films.)

PADDINGTON AND COMPANY LTD.
22 Crawford Place, London, W.1.
Directors: Michael Bond, Ivor Wood, Nicholas Durbridge.

PARAMOUNT TELEVISION
23 Berkeley House, Hay Hill, London, W.1. Tel.: 01-499 5837.
Managing Director: Peter Cary.
(TV film distributors.)

PAXTON RECORDS
Paxton Music Ltd., Borough Green, Sevenoaks, Kent. Tel.: Borough Green 3261.
(Music library on disc and tapes for films, TV and radio.)

PEARL & DEAN GROUP LTD.
Broadwick House, Broadwick Street, London, W1V 1FP. Tel.: 01-434 2200.
Managing Director: J. Simmonds.

THE PICTURE HOUSE
24a, Melville Street, Edinburgh EH3 7NS. Scotland. Tel.: 031-225 7267.
Partners: Vance Chapman, David McCowan Hill.

PHOENIX FILMS LTD.
Manor Road, Goring-On-Thames, RG8 9DP. Tel.: 04-914 2365.
Directors: B.G. Hanson, M.E. Hanson.
(Live-action, cartoon, stop motion.)

PLATO FILMS LTD.
247A Upper Street, London, N1 IRU. Tel.: 01-226 2298/9.
Managing Director: Stanley Forman.
(Entertainment and documentary material. Film Library.)

POLYTEL FILMS LTD.
1 Rockley Road, London, W14 ODL. Tel.: 01-743 3474. Telex: 298816.
(Producers, distributors, film and TV material.)

S. PRESBURY & CO. LTD.
Gloucester House, 19 Charing Cross Road, London, W.C.2. Tel.: 01-930 3601.
Directors: P.S. Presbury (Governing), J. Guy Presbury, R. Salisbury.
(Cinema, TV commercials, Doc. Tech. films.)

PRIMROSE FILM PRODUCTIONS LTD.
61 Regents Park Road, London, N.W.1. Tel.: 01-722 7475 and 01-340 9627.
Directors: Louis Hagen, Anne Hagen.
(Animation, live action, children's programmes.)

PROSPECT STUDIOS LTD.
High Street, Barnes, London, SW13 9LE. Tel.: 01-876 6284-5.

RADIO LUXEMBOURG (LONDON) LTD.
38 Hertford Street, London, W1V 8BA. Tel.: 01-493 5961.
Managing Director: Patrick Cox.
Sales Director: Tim Bradshaw.
(Representatives of European television production group—Societe Internationale de Programmes pour Telediffusion. British representatives of Radio Tele-Luxembourg.)

RANDOM FILM PRODUCTIONS LTD.
25, The Burroughs, Hendon, London, NW4 4AR. Tel.: 01-202 4177.
Directors: F.G. Woosnam-Mills, M.R. Woosnam-Mills, S.A. Tennant.
(Documentaries and Educational.)

RANK ADVERTISING FILMS LTD.
127 Wardour Street, London, W.1. Tel.: 01-439 9531.
Chairman: J. Daly.
Managing Director: D.R. Thomas.
(Production and distribution of cinema and TV commercials, audio-visual programmes.)

RANK PHICOM VIDEO GROUP
Film House, 142 Wardour Street, London, W1V 4BU. Tel.: 01-734 2511. Telex 261237.
Managing Director: G. Sadler.
Managing Director (TV1): I. Reed.
Managing Director (A&V): R. Osborne

RAYANT TELEVISION LTD.
6 Broadwick Street, London, W.1. Tel: 01-437 7422.
Directors: A.G. Lowes, P. Woodford.

REDIFFUSION LTD.
Carlton House, Lower Regent Street, London, SW1Y 4LS Tel.: 01-930 0221.
Managing Director: R.M. Denny

THE ROBERT STIGWOOD GROUP LTD.
67 Brook Street, London W1Y 1YE. Tel: 01-629 9121. Cables: Stigwood, London, W1. Telex: 264267.

REX ROBERTS STUDIOS LTD.
22, Glasthule Road, Dun Laoghaire, Co. Dublin, Ireland. Tel.: Dublin 808305.
(Adv. material.)

SAGITTA PRODUCTIONS LTD.
c/o Pamela Gillis Management, 46 Sheldon Avenue, London, N6 4JR. Tel.: 01-340 7868.
Directors: John Hawkesworth, John Whitney.
(TV and film series.)

SAMUELSON FILM SERVICE LTD.
303-315 Cricklewood Broadway, Edgware Road, London, NW2 6PQ. Tel.: 01-452 8090. Cables: Samcine London. Telex: 21430.
Directors: S. Samuelson C.B.E. (Chairman and Chief Executive), M. Samuelson, (Man. Dir.), D. Samuelson, N.Samuelson.
(Equipment supply.)

SCOTT KYFFIN PRODUCTIONS LTD.
Broomfield Cottage, Garfield Road, Camberley, Surrey.
Chairman: Sir Charles Evans, K.C.B., C.B.E., D.S.O., D.S.C.

SEVEN SEAS FILMS LTD.
83 Duke Street, Grosvenor Square, London, W.1. Tel.: 01-629 0788.
Directors: J. Rayfield, Brian Everett.

SIGNAL FILMS LTD.
13 Hobury Street, London, SW10. Tel.: 01-352 0218. Telex: 21239.
Directors: D.M. Rankin (Managing), D.S.K. Rankin.

SILVERPINE STUDIOS LTD.
Bray, Co. Wicklow, Ireland. Tel.: Dublin 867691/2 and 867073.
(Advt. and Ent. material.)

THE SINCLAIR FILM COMPANY LTD.
Charton Manor, Farningham, Kent. Tel.: 0322 862188
Managing Director: Ian Sinclair.
(Entertainment, documentary and TV films.)

SOUTHERN TELEVISION INTERNATIONAL
159-165 Great Portland Street, London, W1N 6NR. Tel.: 01-637 1164. Telex: 263953.
Contact: Ian Warren.

SOVEREIGN FILMS LTD.
Charton Manor, Farningham, Kent. Tel.: 0322 862188
Directors: Ian Sinclair (Managing), Viscount Throwley.
(Entertainment, documentary, & TV films)

STEWART FILMS LTD.
(A Division of Cygnet Guild Communications, LTD.) 2 Orchard Road, Malvern, Worcs. Tel.: Malvern 4975.
Directors: John R.F. Stewart, R. Evans.
(Shorts, instructional films and audio visual programmes.)

SUPERGROUND LTD.
Suite 4, 60-62 Old Compton Street, London, W.1. Tel.: 01-734 6197.

SURVIVAL ANGLIA LTD.
Brook House, Park Lane, London, W.1. Tel.: 01-408 2288.
Executive Director: M.A. Hay

SWIFT FILM PRODUCTIONS
1 Wool Road, Wimbledon, London, S.W.20. Tel.: 01-946 2040.
Director: T. Peter Hadingham, M.B.K.S.
(Industrial, advertising and documentary & TV 16mm films, filmstrips, AV programmes.)

TALBOT TELEVISION LTD.
30 Dean Street, London, W1V 5AN. Tel.: 01-437 2673.
Director: Paul Talbot.
Chief Enxecutive: Anthony Gruner.
(TV film distributors.)

TARA LTD.
35 Endell Street, London, W.C.2. Tel.: 01-836 0213 and 01-836 9407.
Directors: Jack Saward, Charmian Saward.

TELE-PRODUCTIONS LTD.
Suite 12-16, 175 Piccadilly London, W1V 9DB. Tel.: 01-491 2821
Managing Director: Marcia Glassman.
(Supplies of artistes, voice-overs, commentators, announcers, personalities.)

TELEVISION INTERNATIONAL OPERATIONS LIMITED 9-11 Windmill Street, London W1P 1HF. Tel.: 01-637 2477. Telex: 268208 TVIUK G. Cables: Incolour London.
Managing Director: Ian Reed.
Sales Director: Joe Vick.
Studios: outside broadcasting, mobile VTR. All line and colour standards; full technical crews and prod. VTR recording and editing. Network links. Digital Standards Conversion, cassette duplication (all formats).LTD.

TELEVISION REPORTERS INTERNATIONAL LTD.
EMI Elstree Studios, Borehan Wood, Herts. Tel.: 01-953 1600.
Directors: Robert S. Baker, Roger G. Moore, John Goodman.

THAMES TELEVISION INTERNATIONAL
149 Tottenham Court Road, London, W1P 9LL. Tel.: 01-387 9494.

THORN EMI VIDEO PROGRAMMES LTD.
Thorn House, Upper St. Martin's Lane, London, WC2H 9ED. Tel.: 01-836 2444. Telex 263201 TE1 ENFQ.
Managing Director: Nicholas Bingham.

TOONDER FILM (IRELAND) LTD.
18 Gardiner Place, Dublin, Ireland. Tel.: Dublin 874101.
(Ent. & Advt. material.)

TOWN AND COUNTRY PRODUCTIONS LTD
21 Cheyne Row, London, S.W.3. Tel.: 01-352 7950.
Directors: Bill Latto, Stella Smallwood.

TRANSVIDEO LTD.
St. John's Wood Studios, St. John's Wood Terrace, London, N.W.8. Tel.: 01-722 9255.
Managing Director: M. Silverman.

TRIANGLE FILM PRODUCTIONS LTD.
15 Oslo Court, Prince Albert Road, London, N.W.8 Tel.: 01-722 5656. Cables: Triphilms London N.W.8.

TRIBUNE PRODUCTIONS
(Television Reporters International Ltd.)
EMI Elstree Studios, Boreham Wood, Herts. Tel: 01-953 1600.
Directors: Robert S. Baker, Roger G. Moore, John Goodman.

TRICKFILM STUDIOS LTD.
13 Charlotte Mews, London W.1. Tel.: 01-580 0183.

TRIDENT TELEVISION
Trident House, 8 Grafton Street, London, W1 2PN. Tel.: 01-493 1237.

TRILION VIDEO LTD.
36-44 Brewer Street, London, W.1. Tel.: 01-439 4177.
Managing Director: Bill Hope.

TRIUMPH FILMS INC.
10 Blomfield Villas, London, W.2. Tel.: 01-286 1711.
Director: Peter Hanmer.
(Producers of advertising film and shorts.)

TWICKENHAM FILM STUDIOS LTD.
Twickenham Studios, The Barons, St. Margarets, Twickenham, Middx. Tel.: 01-892 4477.
Directors: A. Baring, G. Coen, G. Humphreys, John T. Davey, FCA, T.L. Kirby, S.J. Mullens.
Secretary: A. Miller.

TYBURN PRODUCTION LTD.
Pinewood Studios, Iver Heath, Bucks. Tel.: IVER 651700. Telex: 847505.
Directors: Kevin Francis, F.H. Toby, J.W. Malins.
Executive Producer: Kevin Francis
Story and Research Editor: Gillian Garrow
Music Supervisor: Philip Markell
(Producers of Theatrical Motion Pictures and TV programming).

UNICORN PICTURES LTD.
Suite 4, 3 Bateman Street, London, W.1. Tel.: 01 437 7187.
Director: John Simmons.

UNITED MOTION PICTURES (LONDON) LTD.
Boston House, 36-38 Fitzroy Square, London, W.1. Tel.: 01-580 1171.
Chief Executive: Trevor Barnard

UPITN INTERNATIONAL NEWS FILM SYNDICATION
ITN House, Wells Street, London, W.1. Tel.: 01-637 2424.
President: Kenneth A. Coyte.
(TV news syndication, current affair programmes and commercial filming worldwide.)

VISCOM PRODUCTION LTD.
25-27 Farringdon Road, London, EUM 3HA. Tel: 01-404 5041. Telegraphic Address: Agavision London, W1. Telex 24474 VISCOM G.
Directors: D. Kingsley (chmn.), Patrick F. Friesner, Geoffrey J. Reeve, I.P.M. Phillips.
(Int. visual communications.)

VIKING FILMS LTD.
6A Rodney Road, Roehampton, London, SW15 5DS. Tel.: 01-789 1532.
Directors: E.A. Humphriss, Monica Ross.
(Studio facilities.)

VISNEWS LTD.
Cumberland Avenue, London, N.W.10 7EH. Tel.: 01-965 7733.
Managing Director: Brian Quinn.
General Manager: Peter Marshall.
Company Secretary/Financial Controller: N.A.J. Bull.
NEWS DIVISION:
Editor in Chief, Robert Kearsley.
Dep. Editor-in-Chief: Nichol Hutton.
BROADCAST FACILITIES:
Chief Engineer: Roy Vitty.
PRODUCTION SERVICES DIVISION:
General Manager: Julian Dinsell.

VISUAL PROGRAMME SYSTEMS LTD.
16 Broadwick Street, London, W1V 1FH. Tel.: 01-734 1787. Telex: 21879.

VOICE OF LONDON LTD.
245 Old Marylebone Road, London, NW1 5RT. Tel.: 01-402 8385.
Studio Manager: Miss Pat Smith.
(Studio facilities.)

WALLACE PRODUCTIONS LTD.
8 Berwick Street, London, W.1. Tel.: 01-734 1888.
Directors: The Hon. M.J. Samuel (Chairman), D.S.K. Rankin (Managing), W.G. Bayley.
(Entertainment & advertising material.)

WARNER BROS. TELEVISION LTD.
Warner House, Pinewood Studios, Iver Heath, Bucks. Tel.: Iver 654 545.

JOSEF WEINBERGER LTD.
10-16 Rathbone Street, London, W1P 2BJ. Tel.: 01-580 2827.
(Representing Chappell Recorded Music Library, London and Telemusic Library, France.)

WESTCHESTER PRODUCTIONS LTD.
Unit N.22, Cricklewood Trading Estate, Claremont Road, London, NW2 1TU. Tel.: 01-450 2584 (Night: 01-452 6289) Cables: Westquip, London, NW2.
(Agents, dist., film and TV equipment.)

DERICK WILLIAMS PRODUCTION LTD.
Charter House, 52 Gloucester Place, London, W.1. Tel.: Welbeck 3521-2 & 0431-2.
Marylebone Studios, 245 Marylebone Road, London, N.W.1. Tel: Ambassador 1881/2.
Directors: Derick Williams (Managing), Jack Harrison, G. Brian Herbert, Douglas A. Duncan.
(Commercials and programmes.)

RICHARD WILLIAMS ANIMATION LTD.
London Studio: 13 Soho Square, London, W1V 5FB. Tel.: 01-437 4455. Cables: Animfilm, London W1V 5FB. Telex: 299556 Anfilm G. Hollywood Studio: 3193 Cahuenga Boulevard, Hollywood, CA 90068, USA. Tel.: (213) 851-8060.
Directors: Richard Williams, Carl Gover.

WILLOUGHBY FILM PRODUCTIONS LTD.
26 Bryanston Square, London, W.1. Tel.: 01-262 0883.
Directors: George W. Willoughby, E.G.G.S. Willoughby.

MAURICE WINNICK (ASSOCIATES) LTD.
52 Gordon Mansions, Torrington Place, London, WC1E 7HG. Tel.: 01-636 4122. Cables: Showbiz London.
Directors: Philip Hindin, Dorothy Hindin, J. Winnick, J. Simmons.

(Promoters of panel and audience participation games. Goodson-Todman rep. in Europe; Theatrical and TV producers.)

WORLDMARK PRODUCTIONS LTD.
Samcine House, 303-315 Cricklewood Broadway, London, NW2 6PQ. Tel.: 01-452 8090. Telex: 21430.
Directors: Tony Maylam, Michael Samuelson.

WORLD MIRROR PRODUCTIONS LTD.
Denham Studios, North Orbital Rd., Denham, Uxbridge, Middx. Tel.: Denham 833522.
Managing Directors: Ken Cameron, R.I.C.H. Warren. (Documentaries, travelogues.)

WORLD WIDE ANIMATION LTD.
21-25, St. Anne's Court, London, W1V 3AW. Tel.: 01-434 1121.
Directors: C. Parish, P. Gilpin.

WORLD WIDE TELEVISION FILM SERVICES LTD.
21-25, St. Anne's Court, London, W1V 3AW. Tel.: 01-434 1121.
Directors: C. Parris, P. Gilpin

WYATT CATTANEO PRODUCTIONS LTD.
22 Charing Cross Rd., London, W.C.2. Tel.: 01-379 6444.

SONNY ZAHL ASSOCIATES LTD.
56-60 Wigmore Street, London, W1H 9DG. Tel.: 01-486-7547.
(incorporating Kavanagh Entertainments for writers.)

ZOOM TELEVISION LTD.
Video Production Centre: 15-19 New Fetter Lane, London, EC4P 4EU. Tel.: 01-353 3641. Telex: 21833. Pinewood Studios, Iver Heath, Bucks. Tel.: 0753 654044. Telex: 848534. Northern Office: Caledonian House, Tatton Street, Knutsford, Cheshire. Tel.: 0565 54533. Systems Engineering Division: Cowley Mill Trading Estate, Longbridge Way, Uxbridge, Middx. UB8 2YG. Tel.: 0895 57981.
Managing Director: P.H. Thompson.

Services for TV Producers

CAMERAS—ACCESSORIES FILM STOCK—TELE-CINE AND EDITING EQUIPMENT

ACMADE INTERNATIONAL
Oakside, (Division of Oakside International Holdings Ltd.), Oxford Road, Uxbridge, Middx. Tel.: Uxbridge 36313. Telex: 8954606.

AGFA-GEVAERT LTD.
Great West Road, Brentford, Middx. Tel.: 01-560 2131.

AMPEX GREAT BRITAIN LTD.
Acre Road, Reading, Berks. Tel.: 0734 85200.

AUTOCUE LTD.
265 Merton Road, London. S.W.18 5JS. Tel: 01-870 0104/7, Cables Autocue London SW18 5JS. TELEX Autocue Telexir 896691 G.

F.W.O. BAUCH LTD.
49 Theobald Street, Boreham Wood, Herts. Tel.: 01-953 0091. Telex: 27502. Cables: Bauch Borehamwood.

R.R. BEARD LTD.
10 Trafalgar Avenue, London, SE15 6NR. Tel.: 01-703 3136 and 9638.

BELL & HOWELL LTD.
Alperton House, Bridgewater Road, Wembley, Middx. HAO IEG. Tel.: 01-902 8812.

BERKEY COLORTRAN (UK)
U.K. Branch of Berkey Photo Inc., P.O. Box 5, Burrell Way, Thetford, Norfolk, IP24 3RB. Tel.: Thetford 2484. Telex: 81294. Cables: Berkmole.

CINESOUND INTERNATIONAL LTD.
Imperial Studios, Maxwell Road, Boreham Wood, Herts. LWD6 IWE. Tel: 01-953 5387.

CINETECHNIC LTD.
169 Oldfield Lane, Greenford, Middlesex. Tel. 01-578 1011 and 35 Briardale Gardens, London, NW3 7PN. Tel.: 01-435 2289.

CINTRON
Grove House, 551 London Road, Isleworth, Middx. TW7 4DS. Tel.: 01-568 0131. Telex: 935054.

EDRIC AUDIO VISUAL LTD.
34-36 Oak End Way, Gerrard's Cross, Bucks. Tel.: (02813) 84646/7/8.

ENGLISH ELECTRIC VALVE COMPANY LTD.
Waterhouse Lane, Chelmsford, Essex. Tel.: 9245-61777. Telegrams: Enelectico, Chelmsford. Telex: 99103.

EVERSHED POWER-OPTICS LTD.
Bridge Wharf, Bridge Road, Chertsey, Surrey KT16 8LJ. (STD 09328) 61181, Telex: 929945 Epochy G. Cable: Lenservo Chertsey.

FRANSCOPE
Grove House, 551 London Road, Isleworth, Middx. TW7 4DS. Tel.: 01-568 0131. Telex: 935054.

FUJI PHOTO FILM (UK) LTD.
99 Baker Street, London, WIM IFB. Tel.: 01 487 5711. Telex: 8812995.

HAYDEN LABORATORIES LTD.
Hayden House, Chiltern Hill, Chalfront St. Peter, Gerrards Cross, Bucks. SL9 9EW. Tel.: 02 813 88447/89221.

HEYDEN & SON LTD.
(Prevost Editing Tables and Film Projectors). 24 Vivian Avenue, London, NW4. Tel: 01-202 0244.

HUMPHRIES FILM LABORATORIES LTD.
71/81 Whitfield Street, London, W1A 2HL. Tel.: 01-636 3636.

INTERNATIONAL VIDEO CORPORATION
10 Portman Road, Reading, Berkshire. RG3 1JR. Tel.: 0734 585421. Telex: 847579.

ITN LTD.
ITN House, 48 Wells Street, London, W1P 4DE. Tel.: 01-637 2424.

HEYDEN/KEM
24 Vivian Avenue, Hendon Central, London, NW4 3XP. Tel.: 01-202 0244. Telex: 28303.

H.A. KINGSTON
34 Charlotte Street, London, W.1. Tel.: Museum 5308.

K.G.M. VIDIAIDS LTD.
Clock Tower Road, Isleworth, Midd. Tel.: 01-568 0151.

KODAK LTD. Motion Picture Sales
P.O. Box 66, Kodak House, Station Road, Hemel Hempstead, Herts. UP1 1JU. Tel: Hemel Hempstead 62331.

LEEVERS RICH EQUIPMENT LTD.
319 Trinity Road, London, SW18 3SL. Tel: 01-874 9054. Telex: 923455. Cables: Leemag. London.

LEE ELECTRIC (LIGHTING) LTD.
128 Wembley Park Drive, Wembley, Middx. Tel.: 01-903 9655 and 01-903 9811.

LIPSNER-SMITH CO. LTD.
Unit 7, Longbridge Way, Cowley Mill Trading Estate, Uxbridge, UB8 2YG. Tel.: Uxbridge 52191.

MARCONI COMMUNICATION SYSTEMS LTD.
Marconi House, Chelmsford, Essex, CMI IPL. Tel.: (0245) 353221.

ERNEST F. MOY LTD.
134 Bayham Street, Camden Town, London, N.W.1. OBB. Tel.: 01-485 5451. Telegraphic Address: Movedor, London, N.W.1.

NEILSON-HORDELL LTD.
Unit 11, Central Trading Estate, Staines, Middx. TW18 4UU. Tel.: 0784 56456 and 54105.

ON-THE-SPOT EQUIPMENT LTD.
208 Kensal Road, Westbourne Park, London, W.10. Tel.: 01-969 6496.

RANK CINTEL
Watton Road, Ware, Herts. Tel.: Ware 3939. Telex: 81415 Cables: Rank Ware.

FREDERICK G. PALMER LTD.
40 Frith Street, London, W.1. Tel.: Gerrard 4796.

PHOTOGRAPHIC ELECTRICAL CO. LTD.
71, Dean Street, London, W1V 6 DE. Tel.: 01-437 4633/4. Cables: Photoelect London. Telex: 25554.

PYE TELECOMMUNICATIONS LTD.
9 Priestley Way, Eldonwall Estate, Edgware Road, London, NW27AF. Tel.: 01-452 6411. Head Office: St. Andrews Road, Cambridge. Tel.: 0223 61222.

RANK FILM EQUIPMENT
P.O. Box 52, Great West Road, Brentford, Middx. Tel.: 01-568 9222.

RANK PRECISION INDUSTRIES LTD.
Broadcast Equipment Group, Rank Cintel, Watton Road, Ware, Herts. Tel.: Ware 3939. Cables: Rank Ware.

RANK TAYLOR HOBSON
P.O. Box 36, Gothlaxton Street, Leicester LE2 OSP. Tel.: 0533 23801. Telex: 34411.

RANK STRAND STUDIO DIVISION
P.O. Box 54, Great West Road, Brentford, Middlesex, TW8 9HR. Tel.: 01-568 9222.

RCA LTD.
Lincoln Way, Windmill Road, Sunbury-on-Thames, Middlesex. Tel.: Sunbury 85511.

ROBERT RIGBY LTD.
Premier Works, 4/6 Northngton Street, London WC1N 2JH. Tel.: 01-405 2944.

RONFORD LTD.
Shepperton Studio Centre, Shepperton, Middx. TW17 0QD. Tel.: Chertsey (09328) 64111/61423.

SAMUELSON FILM SERVICE LTD.
303-315 Cricklewood Broadway, London, NW2 6PQ. Tel.: 01-452 8090. Cables: Samcine London. Telex: 21430.

SHURE ELECTRONICS LTD.
Eccleston Road, Maidstone, Kent, ME15 6AU. Tel.: 0622 59881.

THORN EMI LTD.
Thorn House, Upper St. Martins Lane, London, WC2H 9ED. Tel.: 01-836 2444.

W. VINTEN LTD.
Western Way, Bury St. Edmunds, Suffolk. IP33 3TB. Tel.: Bury St. Edmunds 2121.

WESTREX COMPANY LTD.
152 Coles Green Road, London, N.W.2 7HE. Tel.: 01-452 5401.

FILM & MUSIC LIBRARIES

ABBEY ROAD STUDIOS
3 Abbey Road, London NW8. Tel.: 01-286 1161.

ACME FILM PRODUCTIONS LTD.
174 Wardour Street, London, W1V 3AB. Tel.: 01-437 1572/1889.

ARDMORE & BEECHWOOD LTD.
21 Denmark Street, London, WC2H 8NF. Tel.: 01-836 6699.

BOOSEY & HAWKES MUSIC PUBLISHERS LIMITED
295 Regent Street, London, W1R 8JH. Tel.: 01-580 2060. Cables: Sonorous; London, W.1.R. 8JH.

BOURNE MUSIC LTD.
34/36 Maddox Street, London, W1R 9PD. Tel.: 01-493 6412 and 6583.

BRENT LABORATORIES LTD.
North Circular Road, Cricklewood, London, N.W.2. Tel.: 01-452 4271.

CHARLES BRULL LTD.
18 Tiverton Road, London, NW10 3HL. Tel.: 01-969 3776.

C.T.S. STUDIOS LTD. and DE LANE LEA MUSIC LTD.
THE MUSIC CENTRE
Engineers Way, Wembley, Middlesex. Tel.: 01-903 4611. Telex: 923400.

DE WOLFE LTD.
80-88 Wardour Street, London, W1V 3LF. Tel.: 437 4933-4.

FILM BOOKING OFFICES, LTD.
(Film Library Ltd.)
174 Wardour Street, London, W1V 3AB. Tel.: 01-437 1572.

HARMONIC MUSIC LIBRARY LTD.
26 West Street, London, W.C.2. Tel.: 01-836 1724-5.

ITN LTD.
ITN House, 48 Wells Street, London, W1P 4DE. Tel.: 01-637 2424.

MOZART EDITION (GREAT BRITAIN) LTD.
199 Wardour St., London, W1V 3FA. Tel.: 01-734 3711.

THE MUSIC CENTRE
Engineers Way, Wembley, Middx: Tel: 01-903 4611.

W. PAXTON & CO. LTD.
Paxton Music Ltd., Borough Green, Sevenoaks, Kent. Tel.: Borough Green 3261.

PEER INTERNATIONAL LIBRARY LTD.
8 Denmark Street, London, WC2H 8LT. Tel.: 01-836 4524.

PORTLAND RECORDING STUDIOS LTD.
35 Portland Place, London, W1N 3AG. Tel.: 01-637 2111/4.

SOUTHERN LIBRARY OF RECORDED MUSIC LTD.
8 Denmark Street, London, WC2H 8LT. Tel.: 01-836 4524.

STUDIO FILM AND VIDEO LABORATORIES LTD.
71 Dean Street, London, W.1. Tel.: 437-0831.

JOSEF WEINBERGER LTD.
10-16 Rathbone Street, London, W1P 2BJ. Tel.: 01-580 2827.
(Representing Chappell Recorded Music Library and Telemusic Library, France).

LABORATORIES

BRENT LABORATORIES LTD.
North Circular Road, Cricklewood, London, N.W.2. 7AT. Tel.: 01-452 4271.

COLOUR FILM SERVICES LTD.
22-25 Portman Close, Baker Street, London, W1A 4BE. Tel.: 01-486 2881. Cables: Colorphilm, London.

FILMATIC LABORATORIES LTD.
16 Colville Road, London, W11 2BS. Tel.: 01-229 9347.

HUMPHRIES FILM LABORATORIES LTD.
71/81. Whitfield Street, London, W1A 2HL. Tel.: 01-636-3636.

Managing Director: J.C. Jeffrey.

ITN LTD.
ITN House, 48 Wells Street, London, W1P 4DE. Tel.: 01-637 2424.

HALLIFORD STUDIOS
Manygate Lane, Shepperton, Middx. Tel.: Walton-on-Thames 26341. Cables: Hallifilms.

KAY LABORATORIES LTD.
22 Soho Square, London, W.1. Tel.: 01-437-7811. Telex: 28463.

KAY LABORATORIES VIDEO DIVISION.
22 Soho Square, London, W.1. Tel.: 01-437 7811.
Manager: R.J. Venis.

KENSINGTON FILM CO. LTD.
1 St. Mary Abbots Place, London, W.8. Tel.: 01-603 6693.

KINGLY STUDIOS LTD.
17 Kingly Court, Kingley Street, London, W.1. Tel.: Gerrard 6372.

KODAK LTD.
Motion Picture Sales, Kodak House, Station Road, Hemel Hempstead, Herts. Tel.: 0442 62331.

LEE INTERNATIONAL FILM STUDIOS LTD.
128 Wembley Park Drive, Wembley, Middx. Tel.: 01-902 1262.

LEEVERS-RICH & CO. LTD.
80/82 Wardour Street, London, W1V 3LF. Tel.: 01-437 4161.

MORENO SOUND STUDIOS
35 Broadhurst Gardens, Hampstead, London, N.W.6. Tel.: Maida Vale 1841.

PINEWOOD STUDIOS
Iver Heath, Bucks. Tel.: Iver 651700. Telex: 847505 PINE W G. Cables: PINEWOOD Iver Heath.

PORTLAND RECORDING STUDIOS LTD.
35 Portland Place, London, W1N 3AG. Tel.: 01-637 2111/4.

RANK FILM LABORATORIES LTD.
Denham, Uxbridge, Middx. UB9 5HQ. Tel.: Denham (0895) 832323. Telex: 934704.
Managing Director: J. W. Downer.
Director of Operations: R.W. Login.
Sales Controller: M.A. Levy.

RCA LTD.
Lincoln Way, Windmill Road, Sunbury-on-Thames, Middx. Tel.: Sunbury 85511.

REEDS COLOUR FILM LABORATORIES LTD.
89/91 Wardour Street, London, W1V 4EP. Tel.: 01-437 5548.

SAMUELSON FILM SERVICE LTD.
303-315 Cricklewood Broadway, London, N.W.2. 6PQ. Tel.: 01-452 8090.

STUDIO FILM LABORATORIES LTD.
8-14 Meard Street, London, W1V 3HR. Tel.: 01-437 0831.

TECHNICOLOR LTD.
P.O. Box 7, Bath Road, West Drayton, Middx. UB7 0DB. Tel.: 01-759 5432.

TWICKENHAM FILM STUDIOS, LTD.
St. Margarets, Twickenham, Middlesex. TW1 2AW. Tel.: 01-892 4477. Telex: 8814497.

UNITED MOTION PICTURES (LONDON) LTD.
Boston House, 36/38 Fitzroy Square, London, W.1. Tel.: 01-580 1171.

WORLD WIDE RECORDING STUDIOS
21-25, St. Anne's Court, London, W1V 3AW. Tel.: 01-434 1121.

MARKET RESEARCH

BRITISH MARKET RESEARCH BUREAU LTD.
Saunders House, 53 The Mall, London, W.5. Tel.: 01-567 3060.

EXCHANGE TELEGRAPH COMPANY LTD. (THE)
Extel House, East Harding Street, London, E.C.4. Tel.: 01-353 1080.

MASS-OBSERVATION LTD.
229-243 Shepherds Bush Road, London, W.6. Tel.: 01-748 9172.

MOTION PICTURE RESEARCH
Barnaby House, 19 Rathbone Street, London, W1P 1AF. Tel.: 01-636 6128.

RESEARCH SERVICES LTD.
St. Martin's House, 1‑0 St. Martin's Lane, London, WC2N ½BH. Tel.: 01-240-2450. Telex. 21914.

TELLEX MONITORS LTD
47 Gray's Inn Road, London, WUX 8PR. Tel.: 01-405 7151.

RECORDING STUDIOS

ADVISION SOUND STUDIOS
23 Gosfield Street, London, W1P 7HB. Tel.: 01-580 5707. Telex: 28668.

AIR STUDIOS
214 Oxford Street, London, W.‑. Tel.: 01-637 2758.

EAMONN ANDREWS STUDIOS LTD.
The Television Club, 46/48, Harcourt Street, Dublin, Ireland. Tel: Dublin 758891.

ANVIL FILM & RECORDING GROUP
Denham Studios, North Orbital Road, Denham, Bucks. Tel.: Denham 833522.

AUDIO INTERNATIONAL RECORDING STUDIOS LTD.
18 Rodmarton Street, London, W1H 3FW. Tel.: 01-486 6466.

BRENT LABORATORIES LTD.
North Circular Road, Cricklewood, London, N.W.2. 7AT. Tel.: 01-452 4271.

CBS RECORDING STUDIOS
CBS Records. 31-37 Whitfield Street, London, W1P 5RE. Tel.: 01-636 3434.

CINESOUND EFFECTS LIBRARY LTD.
Imperial Studios, Maxwell Road, Elstree Way, Boreham Wood, Herts, WD6 1WE. Tel.: Elstree 5837 and 5545, 4904, 1587.

CLARENCE STUDIOS LTD.
7 College Approach, Greenwich, London, S.E.10. Tel.: Greenwich 3798 and 4910.

COLOUR FILM SERVICES LTD.
22-25 Portman Close, Baker Street, London, W1A 4GE. Tel.: 01-486 2881.

DE LANE LEA MUSIC LTD.
The Music Centre Engineer's Way, Wembley, Middx: Tel.: 01-903 4611. Telex: 923400.

DELTA SOUND SERVICES LTD.
Shepperton Studio Centre, Squires Bridge Road, Shepperton, Middx. TW17 0QD. Tel.: Chertsey 62045/46.

EMI ELSTREE STUDIOS LTD.
Boreham Wood, Herts. Tel.: 01-953 1600. Telegrams: Emifilms, Boreham Wood. Telex: 922436 EFILMS G.

KAY LABORATORIES SOUND STUDIOS
22 Soho Square, London, W.1. Tel.: 01-437 7811. Telex: 28463.

MERCURY SOUND STUDIO
86 Wardour Street, London, W.1. Tel.: 01-734 0263/4.

MUSIC CENTRE SOUND STUDIOS
(C.T.S. STUDIOS LTD. and DE LANE LEA MUSIC LTD.) Engineers Way, Wembley, Middx. Tel.: 01-903 4611. Telex: 923400.

NATIONAL SCREEN SERVICE LTD.
15 Wadsworth Road, Greenford, Middx. Tel.: 01-998-2851. Cables: Nascreno Greenford, and Nascreno House, 27 Soho Square, London, W.1. Tel.: 01-437 4851-6. Cables: Nascreno London, W.1.

NEW CENTRAL FILM SERVICES (SOHO) LTD.
97-99 Dean Street, London W.1. Tel.: 01-734 5827-8.

PORTLAND RECORDING STUDIOS LTD.
35 Portland Place, London, W1N 3AG Tel.: 01-637 2111.

RANK FILM LABORATORIES LTD.
Denham, Uxbridge, Middlesex. UB9 5HQ Tel.: Denham (0895) 832323.
Managing Director: J.W. Downer.
Director of Operations: R.W. Login.
Sales Controller: M.A. Levy.

STUDIO FILM AND VIDEO LABORATORIES LTD.
71 Dean Street, London, W.1. Tel.: 01-437 0831.

UNITED MOTION PICTURES (LONDON) LTD.
Boston House, 36/38 Fitzroy Square, London, W.1. Tel.: 01-580 1171.

W.F.S. (FILM FACILITIES) LTD.
153 Wardour Street, London, W1V 3TB. Tel.: 01-437 5532.

TITLING—SPECIAL EFFECTS

BIOGRAPHIC FILMS LTD.
90 Dean Street, London, W.1. Tel.: 01-437 9682 and 4879.

BRENT LABORATORIES LTD.
North Circular Road, Cricklewood, London, N.W.2 7AT. Tel.: 01-452 4271.

CINEMAC
174 Wardour Street, London, W.1. Tel.: Gerrard 1091.

CINE-SOUND EFFECTS LTD.
(Soundefex Ltd., Lansdowne Vaults Ltd.) Imperial Studios, Maxwell Road Boreham Wood, Herts. Tel.: 01-953 5837 and 1587, 4904, 5545.

ELECTROCRAFT CONSULTANTS LTD. (Electrocraft Instruments Ltd.)
Head Office: Coombeswell, Farnham Lane, Haslemere, Surrey. Hindhead 5190. Sales Office Works: Liss Mill, Mill Road, Liss, Hants. Tel.: Liss 3444.

FILMTEXT LTD.
37 Kew Road, Richmond Surrez, TW9 2NQ. Tel.: 01-940 5034.

GENERAL SCREEN ENTERPRISES LTD.
97 Oxford Road, Uxbridge, Middx. Tel.: Uxbridge 31931.

HUMPHRIES FILM LABORATORIES LTD.
71/81 Whitfield Street, London, W1A 2HL. Tel.: 01-636 3636.

KAY LABORATORIES LTD.
(Film & Video Labs.)
91/95, Gillespie Road, Highbury, London, N.5. Tel: 01-226 4422. Telex: 28463.
22, Soho Square, London, W.1. Tel: 01-437 7811.
Chairman: David Martineau.
Vice-Chairman: F. Alan Martineau.
Directors: M.D. Kochman, G.R. Hawkes, N.J. Simpson, H.O. Golightly.

KAY LABORATORIES VIDEO DIVISION
22 Soho Square, London, W.1. Tel.: 01-437 7811.
Manager: R.J. Venis

NATIONAL SCREEN SERVICE LTD.
15 Wadsworth Road, Greenford, Middx. Tel: 01-998 2851 and Nascreno House, 27 Soho Square, London, W.1. Tel.: 01-437 4851-6. Telex: 934522.

RANK FILM LABORATORIES LTD.
Denham Uxbridge, Middlesex. UB9 5HQ. Tel.: Denham (0895) 832323.
Managing Director: J.W. Downer.
Director of Operations: R.W. Login.
Sales Controller: M.A. Levy.

RANK PHICOM VIDEO GROUP
Film House, 142 Wardour Street, London, W1V 4BV. Telephone: 01-734 2235. Telex: 261237 Cables: Vidipost London W1.

REEDS COLOUR FILM LABORATORIES LTD.
89/91 Wardour Street, London, W1V 4EP. Tel.: 01-437 5548.

STUDIO FILM AND VIDEO LABORATORIES LTD.
71 Dean Street, London, W.1. Tel.: 01-437 0831.

TECHNICOLOR LTD.
(Subsidiary of Technicolor Inc.) P.O. Box No. 7 Bath Road, West Drayton, Middlesex. UB7 0DB. Tel.: 01-759 5432. Telegraphic and Cable Address: Technicolor, West Drayton. TELEX: 22344.

UNIVERSAL FILM LABORATORY LTD.
Braintree Road, Ruislip, Middx. Tel.: 01-841 5101.

STUDIOS

ATV ELSTREE STUDIOS
Boreham Wood, Elstree, Herts. Tel.: 01-953 6100.
(Owned by ATV Network Ltd.) Four Stages totaling 58,416 sq. ft.)

BRAY STUDIOS
Windsor Road, Windsor, Berks. Tel.: Maidenhead 22111 Telex: 849 462.
Studio Manager: David Goodenough. Contact: Fiona Lalto. 3 Stages Total 13,300 sq. ft.

BUSHEY
(Cygnet Guild Communications Ltd.)
Melbourne Road, Bushey, Herts. Tel.: 01-950 1621.
Studio Manager: Wilf Partington
2 Stages (70′ × 40′, 30′ × 20′)

CINEVIDEO LTD.
245 Old Marylebone Road, London, NW1 5QT. Tel.: 01-724 1363.

CREATIVE FILM MAKERS LTD.
Pottery Lane House, Pottery Lane, Holland Park, London W11 4LZ Tel.: 01-229 5131.

EMI ELSTREE STUDIOS LTD.
Boreham Wood, Herts. Tel.: 01-953 1600. Telegrams: Emifilms, Boreham Wood. Telex: 922436 Eflims G.
Nine sound stages. One 30,000 square feet, four 15,000 square feet, two 8,000 square feet, one 15,750 square feet, one 5,200 square feet.)
Theatres: Three viewing, one dubbing, one looping. RCA sound. Complete Dolby installation in transfers, looping, dubbing and viewing for producing Dolby encoded optical sound tracks.
Managing Director: Andrew Mitchell.

GATEWAY FILM PRODUCTIONS
470-472 Green Lanes, London, NI3 5XF. Tel.: 01-882 0177.

HALLIFORD FILM STUDIOS
Manygate Lane, Shepperton, Middlesex. Tel.: Walton-on-Thames 26341 (5 lines). Cables: Hallifilms, Shepperton.
Two states (60 × 60, 60 × 40) totalling 6,000 sq. ft.
Studio Manager: Eugene Andrews.

ITN LTD.
ITN House, 48 Wells Street, London, WIP 4DE. Telephone: 01-637-2424.
2 Television Studios, 7 EMI colour cameras, VTR and telerecording, telecine, etc.

KAY LABORATORIES SOUND STUDIOS
22, Soho Square, London, W.1. Tel.: 01 437-7811. Telex: 28463.
Managing Director: M.D. Kochman.
Studio Manger: G. Latter.

LEE INTERNATIONAL FILM STUDIOS
128 Wembley Park Drive, Wembley, Middx. Tel: 01-902 1262.

MORENO FILM STUDIOS
35 Broadhurst Gardens, London, N.W.6. Tel.: Maida Vale 0645.
Directors: George Moreno (Managing), Fred Thompson.

NATIONAL FILM STUDIOS OF IRELAND
Ardmore House, Bray, Co. Wicklow, Ireland. Tel.: Dublin 862971. Telex: 30418.

PINEWOOD STUDIOS
Iver Heath, Bucks. Tel.: Iver 651700. Telex: 847505 Pine w G.
Chairman: T.E. Chilton.
Managing Director: C.R. Howard.

A. 165 ft. × 110 ft. × 35 ft. (with tank 40 ft. × 30 ft. × 8 ft.)
B. 110 ft. × 80 ft. × 34 ft.
C. 110 ft. × 80 ft. × 34 ft.
D. 165 ft. × 110 ft. × 35 ft. (with tank 40 ft. × 30 ft. × 8 ft.)
E. 165 ft. × 110 ft. × 35 ft. (with tank 40 ft. × 30 ft. × 8 ft.)
F. 100 ft. × 75 ft. × 35 ft. (with tank 20 ft. × 20 ft. × 8 ft.)
G. 55 ft. × 49 ft. × 23 ft.
H. 90 ft. × 37 ft. × 28 ft.
LARGE PROCESS STAGE: 175 ft. × 28 ft., including West Wing: 68 ft. × 33 ft.
SMALL PROCESS STAGE: 75 ft. × 32 ft. × 17 ft.
SPECIAL EFFECTS STAGE (silent)—89 ft. × 80 ft. × 20 ft. and Glass Matte Dept.
J. 110 ft. × 80 ft. × 30 ft. (dual-purpose, film and TV).
K. 110 ft. × 80 ft. × 30 ft. (dual-purpose, film and TV).
L. 105 ft. × 90 ft. × 31 ft. (dual-purpose, film and TV).
M. 105 ft. × 90 ft. × 31 ft. (dual-purpose, film and TV).
Stage (silent) 336 ft. × 139 ft × 40 ft.

SOUND & PROJECTION DEPARTMENT: Nagras, Sennheisers, Fisher booms. Playback, P.A. Fully staffed transfer section (tape/35mm/cassette/disc.)
Seven theatres, with arc illumination:
Theatre 1: Viewing—35mm Fedi reversible projectors. COMOPT, SEPOPT, SEPMAG. All formats including Scope.
Max. screen width 30ft. 50 seats.
Dubbing—Mono or Dolby stereo. Neve console. 14 dubbers.
Theatre 2: Viewing—35mm Westar reversible projectors. COMOPT, SEPOPT, COMMAG, SEPMAG. All formats including Scope.
Max. screen width 34ft; 50 seats.
Dubbing—Comprehensive stereo facility, utilising 3 6-track recorders for separated dialogue, effects and music. All Dolby formats, including divided surrounds. 60-input console with computer control of faders and panners. 16 dubbers.
Theatres 3 & 4: 15-seater viewing theatres with all formats, COMOPT, SEPOPT, SEPMAG.
Theatre 5: Post-sync and effects theatre, 47′ × 30′ × 18′. All surfaces, splash tank, instant props. 3-track reversible.
Theatre 6: Viewing—35mm Westar, reversible projectors. COMOPT, SEPOPT, SEPMAG. All formats including Scope.
Max. screen width 11ft; 15 seats.
Dubbing, post-sync and effects. 18-input console, 3-track reversible system. Dolby noise reduction.
Theatre 7: Luxury viewing room. 70/35/16mm projection. COMOPT, SEPOPT (including Dolby stereo), COMMAG, SEPMAG. Every screen and soundtrack format, including divided surrounds.
Max. screen width 31ft. 115 super seats. PA, capability for audio-visual. Reception lounge available.
Theatres 1, 2 and 6 are equipped with Westrex variable area photo recorders.
CAMERA DEPARTMENT: Range of 35 mm cameras available, including studio reflexes, Mitchell standard and high speed, Arriflex. Lenses include Varotal and Angenieux zooms, Kowa primes. Wide choice of accessories: crystal motors, geared and fluid heads. Oxberry and Moviola dollies; Academy and Boby cranes.
EDITING: 44 35mm cutting rooms available.
SPECIAL EFFECTS: Fully equipped process projection dept., including front projection and 70m rear and foreground projection. Models of all kinds. Equipped for all effects. Beam splitting cameras for monochrome, color and scope traveling matte.
STILLS: Stills studio. Still cameras to cover all requirements. Dressing rooms. Production and Unit Offices: Many available.
EXTERIOR LOT: 72 acres with formal gardens, lake, woods and concrete service roads. Paddock tank 230 ft. wide narrowing to 110 ft. Backing 240 ft. × 60 ft. Another tank 75 ft. × 225 ft. Both tanks contain approximately 800,000 gallons. Props available and world's largest scenedock. Comprehensive catering dept.
Chairman: T.E. Chilton.

PRODUCTION VILLAGE
(Samuelson Film Service Ltd)
100 Cricklewood Lane, London, NW2 2DS. Tel.: 01-4508969.

PROSPECT STUDIOS LTD.
High Street, Barnes, London, S.W.13. Tel.: 876 6284/5.

RANK PHICOM VIDEO GROUP
142 Wardour Street, London, W1V 4BU. Tel.: 01-734 2235. Telex: 261 237. Cables: Vidipost, London WI.
Managing Director: G. Sadler.
Managing Director (TVI): I. Reed.
Managing Director (ATV): R. Osborne.

SEEZERS TELEVISION STUDIOS
30, Upper Abbey Street, Dublin 1. Tel.: Dublin 48909 and 47079.
Stage sizes: 50 × 70; 20 × 12.

ST. JOHN'S WOOD STUDIOS
St. John's Wood Terrace, London, N.W.8. Tel.: 01-722 9255.
Managing Director: Michael Silverman
Two sound stages: (63 × 32 × 25), (45 × 25× 23).
Facilities for features and commercials, film or videotape.

SHEPPERTON STUDIO CENTRE
Studios Road, Shepperton, Middlesex. TW17 0QD. Tel.: 09328 62611. Telex: 929416 Movies G.
Seven sound and two silent stages.

TELEVISION INTERNATIONAL OPERATIONS LIMITED
9-11 Windmill Street, London W1P 1HF. Tel.: 01-637 2477. Telex: 268208 TVIUK G. Cables: Incolour London.
Managing Director: Ian Reed
Sales Director: Joe Vick.
Studios: outside broadcasting, mobile VTR. All line and colour standards; full technical crews and prod. VTR recording and editing. Network links. Digital Standards Conversion, cassette duplication (all formats).

W.F.S. (FILM FACILITIES) LTD.
153 Wardour Street, London, W1V 3TB Tel.: 01-437 5532.

TWICKENHAM
St. Margaret's, Twickenham. Tel.: 01-892 4477.
Owned by Twickenham Film Studios, Ltd. One stage 7,192 square feet. One stage 5700 square feet. One stage 2000 square feet. RCA Sound. Re-recording and dubbing theatres with 36 input 6 track stereo Rock & Roll.
Directors: G. Coen, G Humphreys, John T. Davey, F.C.A., T.L. Kirkby, M. Landsberger S.J. Mullens.

WORLD WIDE RECORDING STUDIOS
21-25 St. Anne's Court, London, W1V 3AW. Tel.: 01-434 1121.
Director of Recording: P. Gilpin.

ZOOM TELEVISION LTD.
Video Production Centre, 15-19 New Fetter Lane, London, EC4P 4EU. Tel.: 01-353 3641. Telex: 21833.
Director: John Hocking.

NEWS FILM SERVICE

CBS NEWS
100 Brompton Road, London, S.W.3. Tel.: 01-584 3366.

NBC NEWS WORLDWIDE INC.
25 St. James Street, London, SW1. Tel.: 01-839 4571. Telex: 27765.
London Correspondent: Garrick Utley.
London Bureau Chief R. McFarland.

CHARLES H. FOX LTD.
25 Shelton Street, London, WC2H 9HX. Tel.: 01-240 3111.

ROBERT WHITE AND SONS
25 Shelton Street, London, WC2H. 9HX Tel: 01-836 8237.

UPITIN INC.
(International Film Syndication)
ITN House, Welles St., London W1. Tel.: 637 2424.
Pres.: K.A. Coyte.

VISNEWS LTD.
Cumberland Avenue, London N.W.10. Tel.: 965 7733.
Managing Director: Brian Quinn
General Manager: Peter Marshal
NEWS DIVISION:
Editor-in-Chief: Robert Kearsley
Deputy-E-I-C: Nichol Hutton
BROADCAST FACILITIES:
Chief Engineer: Roy Vitty.
PRODUCTION SERVICES DIVISION:
General Manager: Julian Dinsell

COSTUME SUPPLIERS

MORRIS ANGEL & SON LTD.
119-123 Shaftesbury Avenue, London, W.C.2. Tel.: 01-836 5678. (P.B.X.)

BERMAN AND NATHANS LTD.
Head Office: 18 Irving Street, Leicester Square, London, WC2H 7AX. Tel.: 01-839 1651.
Main Store: 40 Camden Street, London, NW1 OEN. Tel: 01-387 0999.

BRITISH ADVERTISING AGENCIES

THE ADVERTISING BUREAU
Gratton House, 9-13, Newton Road, Leeds, L57 4HU. Tel.: Leeds 620351.
Premier House, Southampton Row, London, WC1B 5AL. Tel.: 01-278 1274.
Kent House, 21, Spring Gardens, Manchester, M2. Tel.: 061 832 5459.
T.V. Executive: A.C. Murray.

ALLARDYCE HAMPSHIRE LTD.
10-12 Carlisle Street, London, W1V 2RF. Tel.: 01-437 8644.

APEX ADVERTISING LTD.
47 Cross Road, Coventry. Tel.: Coventry 85151.
T.V. Executive: D.R. Harris.

A.P. SERVICES LTD.
186-8 Shaftesbury Avenue, London, W.C.2. Tel.: Covent Garden 3121.
TV Executive: A.I. Smith.

ARKS LTD.
17 Harcourt Street, Dublin 2. Tel.: Dublin 01-757981.
TV Executive: A.J. O'Leary.

A.T.A. TELEVISION SERVICES (A.T.A. Advertising Ltd.) 33 Queen Anne Street, London, W1M 0AL. Tel.: 01-493 4030.
T.V. Executive: Norman Ward.

BBDO BATTEN, BARTON, DURSTINE & OSBORN LTD.
25/27 St. George Street, London, W1R ONT. Tel.: 01-493 8455.
Head of Television: John C. Henderson.

B.B.P. ADVERTISING (BOSWELL, BARRATT & PHILLIPS LTD.)
B.B.P. House, 117 John Bright Street, Birmingham B1 IBQ.
T.V. Executive: H. Harms.

ALFRED BATES & SON LTD.
130 Fleet Street, London, E.C.4. Telephone: 01-353 8831.

BENTON & BOWLES LTD.
197 Knightsbridge, London, S.W.7. Tel.: Kensington 1444.
Head of Television: W. Adams.

ROBERT BRANDON & PARTNERS LTD.
10 Clifford Street, London, W.1. Tel.: 01-734 0511.

BRIDGE ADVERTISING LTD.
178-202 Great Portland Street, London, W1N 6AQ. Tel.: 01-637-7481/5.
T.V. Executive: W.H. Pearson.

BROCKIE HASLAM AND ALLON LTD.
72-74 Brewer St., London, W.1. Tel.: 01-734 0491.

D.H. BROCKLESBY LTD.
Nicholsons House, Maidenhead, Berkshire. Tel.: Maidenhead 29322. Telex: 84407.
T.V. Executive: Colin Lovejoy.

BROOKS PUBLICITY SERVICES
4, Seaford Court. 220/222, Great Portland Street, London, W1N 5HH. Tel.: 01-388 2117. Telex: 21237G.

L. GRAHAM BROWNE (LONDON) LIMITED.
56/58, Gamage Building, Holborn Circus, London, E.C.1. Tel.: 01-405 6207.

BROWNE'S ADVERTISING LTD.
117 Piccadilly, London, W1V OBL. Tel.: 01-499 5351.
Creative Director: David Knight.

BRUNNING ADVERTISING & MARKETING LTD.
Brunning House, 100 Whitechapel Road, London, E.1. Tel.: 01-247 6525.
T.V. Executive: D. Walker.

BRUNNING ADVERTISING & MARKETING (LEEDS) LTD.
Apson House, 53 Margate, Leeds LS9 7DR. Tel.: Leeds 30054-5.

BRUNNING ADVERTISING & MARKETING (MANCHESTER) LTD.
57 Hilton Street, Manchester, M1 2EJ. Tel.: 061-236 5636.

BUTLER DENNIS & GARLAND LTD.
153-155 Regent Street, London, W1R 7FD. Tel.: 01-437 4377.
Creative Controller: A.D. Dell.

CAVENDISH PUBLICITY SERVICE LTD.
34, Oxford Street (4th floor), London, W.1. Tel.: 01-580 3377.
T.V. Executive: A.G. Howard.

CLARENDON ADVERTISING LTD.
57 Clarendon Road, Leeds 2. Tel.: Leeds 35121.
T.V. Executive: John Somers.

COLLETT, DICKENSON PEARCE & PARTNERS LTD.
Howland House, Howland Street, London, W1P6AJ. Tel.: 01-636 3355.
Media: M. Yershon.
T.V.: P. Levelle.

JOHN COLLINGS & PARTNERS LTD.
Nascreno House, Soho Square, London, W.1. Tel.: Regent 2141.
Media: A.J. Blair.
T.V.: D. Scott-Job.

COMMERCIAL ADVERTISING SERVICE LTD.
Silverwood House, 129 Barlow Moor Road, West Didsbury, Manchester, M.20 8PP. Tel.: 061-434 1211.
T.V. Executives: K. Greenhalgh, J.L. Richman, J.A. Gatcliffe, P. Newman, T. Lee.

CONNELL, MAY AND STEAVENSON LTD.
William Blake House, Marshall St., London, W.1. Tel.: 01-734 9041.

E. ALLAN COOPER & CO. LTD.
Waldorf House, 18 Exeter Street, London, W.C.2. Tel.: 01-836 3104.

S.V. COURT ADVERTISING LTD.
Chronicle House, 72-78 Fleet Street, London, E.C.4 1HY. Tel.: 01-353 2771.
T.V.: Executive: Mrs. Diana Gibson.

W.S. CRAWFORD LTD.
141, Westbourne Terrace, London, W2 6JR. Tel.: 01-723 3227.

DAVIDSON PEARCE, BERRY & SPOTTISWOODE LTD.
67, Brompton Road, London, S.W.3. Tel.: 01-589-4595.
Chairman & Managing Director: C.J. Hawes.

DOMAS LTD. ADVERTISING AND MARKETING
77 Ranelagh Road, Dublin 6, Ireland. Tel.: Dublin 971075-7.

DORLAND ADVERTISING LTD.
121-141 Westbourne Terrace, London, W.2. Tel.: 01-262 5077.

DOYLE DANE BERNBACH LTD.
62-64 Baker Street, London, W.1. Tel.: 01-486 7122.
Media Director: T. Wheeler.

DRYDEN-ALLEN LTD.
4 Hills Road, Cambridge, CB2 1JP. Tel.: Cambridge 65215/6.

DUNCAN ONGLEY ASSOCIATES LTD.
43-44 Windmill Street, London, W.1. Tel.: 01-437 8761.

EVERETTS LTD.
10 Grecoat Place, London, SW1P 1SE. Tel.: 01-799 7822.
T.V. Director: Bernie Kay.

FARMER ADVERTISING LTD.
Stentor House, 16 Moncrieffe Road, Sheffield PS7 1HS.
Chairman: J.C.H. McMillan; Joint.
Managing Directors: J.C. Wells, J. Wardle.
Directors: C.W.R. Guthrie, C.D. Farmer, V. Smith, J.L. Bradshaw, R.C. Milner.

FARMER ADVERTISING (HULL) LTD.
80 Beverley Road, Hull. Tel.: Hull 29001.
T.V. Executive: D.N. Dean.

FOGARTY ADVERTISING LTD.
72 Haddington Road, Dublin 4, Ireland. Tel.: Dublin 681455.

FOOTE, CONE & BELDING LTD.
82 Baker Street, London, W.1. Tel.: 01-935 4426.
T.V. Head: Ian Hoath.

FORBES KEIR LTD.
DuBarry Bldg., 58/64 Lime Street, Liverpool. Tel.: 051-709 6162.
T.V. Directors: J.G.S. Furness, M.I.P.A., L. Hamblet, M.I.P.A., A. McDermott.
T.V. Executive: J.G.S. Furness.

FORDHAM SADLER ADVERTISING LTD.
West End House, Hills Place, London, W.1. Tel.: 01-437 8146.
T.V. Executive: Peter Rose.

GARRATT BAULCOMBE PARTNERS LTD.
Bolton House, 194 Old Brompton Road, London, SW5 OAS. Tel.: 01-370 6781.

FRANK GAYTON ADVERTISING LTD.
156 Upper New Walk, Leicester. Tel.: Leicester 26001.
T.V. Directors: Alan W. Gayton, Nigel Parry Williams.

GEE ADVERTISING LTD.
London, Leicester, Leeds, Manchester. T.V. Department: 19 Grafton Street, London, W.1. Tel.: Hyde Park 6827.

THE GENERAL ADVERTISING CO. OF LONDON LTD.
The Centre, 68 High Street, Weybridge, Surrey, KT13 8JF. Tel.: Weybridge 40251.

G.S. GERRARD LTD.
10 Argyll Street, London, W.1. Tel.: 01-437 6816.
T.V. Executive: W. Purvis.

GODBOLDS LTD.
99 Aldwych, London, W.C.2. Tel: 01-242 9671.
T.V. Director: Geoffrey Godbold.

GOLLEY, SLATER & PARTNERS LTD.
42 Drury Lane, London, W.C.2. Tel.: 01-240 5131.
Media: S. Cullen.

GRAHAM & GILLIES & WARWICK LTD.
Ingestre Court, Ingestre Place, London, W1R 4AL. Tel.: 01-437 9055.

GRANT ADVERTISING LTD.
2 Basil Street, London SW3 IAA. Tel.: 01-584 7050.
T.V. Executive: C. Coast.

GREAT GUNS! LTD.
79 New Cavendish Street, London W1M 8AJ. Tel.: 01-636 4020.
Executive Producer: J.G. Warner.

HERBERT GREAVES LTD.
26 Brown Street, Manchester, W2 1DL. Tel.: 061-832 3841.
T.V. Executives: Derrick H. Greaves, Miss M. Greaves.

JAMES M. GREENWOOD ADVERTISING LTD.
100 Whitechapel Road, London, E.1. Tel.: 01-247 6525.

GREY ADVERTISING LTD.
50-51 Cenduit Street, London, W1R OAR. Tel.: 01-493 6700.

CREIGHTON GRIFFITHS ROYDS LTD.
30, Cathedral Road, Cardiff, Wales. Tel.: Cardiff 397711.
Account Executives: R. Petersen, D.M. Creighton Griffiths, G. Huws.
T.V. Buyer: G.J. Hughes.

JOHN HALIFAX LTD.
84 Northampton Street, Leicester. Tel.: Leicester 50841/2.
T.V. Executives: Gordon A. Penman H.C. Clements.

HARDING, QUANTRILL, O'BRIEN LTD.
Byrom House, 21 Quay St., Manchester, M3 3HT. Tel.: 061-832 2523.

HARWOOD PRESS (ADVERTISING) LTD.
21-51, West Street, Harrow, Middlesex. Tel.: 01-422 5656.
T.V. Executive: R.B.F. Harwood.

HEWLAND, RUDER & FINN INTERNATIONAL LTD.
15 Albemarle Street, London, W.1. Tel.: 01-492 1143.
Film & T.V. Executive: V.A. Stump.

CHARLES F. HIGHAM LTD.
Hesketh House, Portman Square, London, W1H OJH. Tel.: 0-486 5041.
T.V. Executive: Peter King.
Creative Director: Brian Baldwin.

HOBSON BATES & PARTNERS LTD.
Hobson House, 155 Gower Street, London, W.C.1. Tel.: 01-387 4313.
T.V. Production Controller: R. Evans.

CHARLES HOBSON & GREY LTD.
12/13 Conduit Street, London, W.1. Tel.: Hyde Park 6700.
T.V. Executive: M.W. Reynolds.

HECTOR HUGHES & PARTNERS LTD.
Colquhoun House, Broadwick Street, London, W.1. Tel.: 01-437 0724.
T.V. Executives: R.C. Harrold, R. Collins.

HUNTER ADVERTISING LTD.
Royal Exchange Bldgs., Glasgow, C.1. Scotland. Tel.: 041-221 1162.
T.V. Executive: A. Ferguson Hunter.

HUTCHINSON PUBLICITY LTD.
3, Fitzroy Square, London, W.1. Tel.: 01-388 7601.
T.V. Executive: Jeremy Cox.

GEORGE HYNES & PARTNERS LTD.
New Mercury House, 82 Farringdon Street, London, E-C.4. Tel.: 01-353 4030.

INDUSTRIAL PUBLICITY SERVICE LTD.
I.P.S. House, 38 Monmouth Street, London, W.C.2. Tel.: Covent Garden 1902.
T.V. Executive: P.M. Graham.

INTERLINK ADVERTISING LTD.
63/66 St. Martin's Lane, London, WC2N 4JK. Tel.: 01-836 4466.
Managing Director: F.J. Josling, FIPA.

IRISH INTERNATIONAL ADVERTISING & MARKETING LTD.
9 Fitzwilliam Square, Dublin. Tel.: 760231.
Managing Director: Finbar P. Costello.

JAFFE & YOUNG ADVERTISING LTD.
43 South Street, London. W1Y 5PD. Tel.: 01-409 0932.
Managing Director: R.C. Young.

WALTER JUDD LTD.
1A, Bow Lane, London, E.C.4. Tel.: 01-236 4541.

KINGSCOTT LTD.
18/20 St. Andrew Street, London, E.C.4. Tel.: 01-353 6316/9.
T.V. Executive: L.D. Conquest, M.I.P.A.

LINTAS LTD.
Lintas House, New Fetter Lane, London, E.C.4. Tel.: 01-483 8030.
Creative Director: I.M. Drake.

LONGLEYS & HOFFMANN LTD.
60-61 Lionel Street, Birmingham, 83 1JF. Tel.: 01-236 7532.
T.V. Executive: P.J. Niblett.
Time Buyer: R.D. Rawlinson.

LOVELL AND RUPERT CURTIS LTD.
30 Bouverie Street, London, E.C.4. Tel.: 01-353 0871.
T.V. Executive: A.C. Hind, M.I.P.A.

L.P.A. LTD.
2 Basil Street, London, SW3 1AD. Tel.: 01-584 7050.

LUCIEN TELEVISION & SCREEN ADVERTISING LTD.
Hesketh House, Portmen Square, London, W.1. Tel.: 01-486 3161.
T.V. Executive: V.H. Rawlinson.

MACKAY & PARTNERS LTD.
36 Upper Brook Street, London, W.1. Tel.: 01-493 1343.
T.V. Executive: Walter Speight.

BARBOUR MacLAREN ADVERTISING LTD.
101, West Nile Street, Glasgow, C.1. Tel.: 041 332 9211.
T.V. Executives: D. Sanderson, T. Willies.

MacLAREN, DUNKLEY, FRIEDLANDER LTD.
197 Oxford Street, London, W.1. Tel.: 01-437 9000.

McCALLUM ADVERTISING LTD.
7 Queen Street, Edinburgh, EH2 1JP, Scotland. Tel.: 031-266 6441.
T.V. Executive: M. Gilderdale.

McCANN-ERICKSON ADVERTISING LTD.
McCann-Erickson House, 36 Howland Street, London, W.1. Tel.: 01-580 6690.
Executive Creative Director: Barry Day.
Creative Directors: John Booth, Peter Hodgson, Andy Rork.

McCONNELL & CO. (ADVERTISING) LTD.
25 Princes Drive, Colwyn Bay, North Wales. Tel.: Colwyn Bay 2889.
T.V. Executive: E.M. Roberts.

MASIUS, WYNNE-WILLIAMS & D'ARCY-MACMANUS LTD.
2 St. James's Square, London, S.W.1. Tel.: 01-839 3422.
Creative Directors: R. Greene, C. Sharpe.

OGILVY, BENSON & MATHER LTD.
Brettenham House, Lancaster Place, London, W.C.2. E 7EZ. Tel.: 01-836 2466.
T.V. Manager: R. Barker.

MAYFAIR ADVERTISING LTD.
Cliffords Inn, London, E.C.4. Tel.: 01-242 8135.

MEARNS AND GILL ADVERTISING LTD.
7 Carden Place, Aberdeen, Scotland. Tel.: Aberdeen 26311.
T.V. Executives: J.A. Graeme Mearns, J.D. Gill, J. Cardno, G.J. Scott.

PETER A. MENZIES ADVERTISING
2 Newton Place, Glasgow, C.3 7PT. Scotland.

MITCHELLS ADVERTISING SERVICE LTD.
420 Sauchiehall Street, Glasgow, C.2. Scotland. Tel.: Douglas 2241/2.

JOHN NEWTON (Publicity), LTD.
198 Corn Exchange, Manchester 4. Tel.: 061-834 1828.
T.V. Executives: T.D. Farmer, D.F. George.

NATIONAL ADVERTISING CORPORATION LTD.
150 Regent Street, London, W.1. Tel.: 01-437 7866 (20 lines).
T.V. Executive: Mike Storm.

NORMAN, CRAIG AND KUMMEL LTD.
Greater London House, Hampstead Road, London, N.W.1. Tel.: 01-387 9311.

O'KENNEDY-BRINDLEY LTD.
64-6 Lower Gardiner Street, Dublin, Ireland.
T.V. Executive: Niel O'Kennedy.

OMNIFIC ADVERTISING, LTD.
Astral House, 17-19 Maddox Street, London, W1R 0EY. Tel.: 01-499 0477.

OSBORNE LTD.
Hesketh House, Portman Square, London, W1H 9FG. Tel.: 01-486 5877.
Creative Director: Geoff Drew, AMIPA.
T.V. Director: P. King.

J.H. PARKER & CO. LTD.
Northern Bank Chambers, 114 Grafton Street, Dublin 2. Tel.: 771 371.
T.V. Executive: P.J. Dempsey.

PEERS ADVERTISING LTD.
Walton House, High Street, Bromley, Kent: Tel.: 01-464 3661.
Media: A.M. Albinson.

ALFRED PEMBERTON LTD.
Pemberton House, 29 Marylebone Road, London NW1 5JU. Tel.: 01-486 1222.

PERARD FOX AND PARTNERS LTD.
9 Hillgate Street, London, W.8. Tel.: 01-727 3141.
T.V. Executive: Ruth Rose.

PICKERSGILL PUBLICITY SERVICE LTD.
Lavender Walk, Leeds 9. Tel.: Leeds 29137.
T.V. Executives: L. Pickersgill, D. Walls.

REDHEADS ADVERTISING LTD.
11-17 Archbold Terrace, Newcastle-on-Tyne NE2 IDW. Tel.: 0632 815424, and 20 Bride Lane, Fleet Street, London, EC4Y 8DX. Tel.: 01-353 4691.
Directors: C.H.D. Redhead, R.D. Redhead, F.W. Watson.

REID WALKER LTD.
Hutton House, Hutton Street, London, EC4Y 8HP. Tel.: 01-583 9371. Telex: 885857.
T.V. Executive: N.R. Field, F.I.P.A.

RICHMOND TOWERS LTD.
26 Fitzroy Square, London, W1P 6BT. Tel.: 01-388 7421.
T.V. Executive: L. Carpenter.

ROBERTSON & SCOTT (ADVERTISING) LTD.
42 Charlotte Square, Edinburgh EH2 4HJ, Scotland. Tel.: 031-255 6602. Glasgow Office: 124 St. Vincent St., Glasgow. Tel.: 041-221 9021.
Managing Director: D.J. Mackenzie, M.I.P.A.

ROE DOWNTON LTD.
103-109 Wardour Street, London, WIV 3TD. Tel.: 01-439 7111.

F. JOHN ROE ASSOCIATES LTD.
20 St. Ann's Square, Manchester. Tel.: 061 832 5111. & F. John Roe Advertising Ltd. 73 Grosvenor Street, London, W.1. Tel.: 01-493 8501.
Directors: Derek J. Roe, Graham Roe.

ROLES & PARKER LTD.
Eagle Street, London, W.C.1. Tel.: 01-242 7744.
Media: J.V. Standish.

ROYDS LONDON LTD.
Royds House, Mandeville Place, London, W1M 6AE. Tel: 01-935 7733.
T.V., Film & Radio Dept. Director: Eve Samuel.

SAATCHI & SAATCHI COMPTON LTD.
80-84 Charlotte Street, London, W1A 1AO. Tel.: 01-636 5060.

SAWARD, BAKER & CO. LTD.
79 New Cavendish Street, London, W1M 8AJ. Tel.: 01-636 4020.
Media Director: D.J. Clark, AMIPA.
TV, Radio & Cinema Manager: J.G. Warner.

SCIENTIFIC ADVERTISING & MARKETING LTD.
Clifford's Inn, Fleet Street, London, E.C.4. Tel.: 01-242 8844.
T.V. Executives: R.C. Hilling, M.I.P.A., T.G. Hartshorne, M.A.A., M.I.P.A.

SEVERN-SMITH ADVERTISING LTD.
Mason Road, Kidderminster. Tel.: Kidderminster 62876. Bradford Office: Rooley Lane, Bradford 1. Tel.: Bradford 682344. Exeter Office: Southernhay East, Exeter. Tel.: Exeter 70747.
T.V. Executive: N. Tanner.

SHAW & UNDERWOOD (ADVERTISING SERVICES) LTD.
164 High Street, Bordesley, Birmingham B12 OLF. Tel.: 021-772 5001/2.
T.V. Executive: C.I. Randle.

SMEE'S ADVERTISING LTD.
3-5 Duke Street, Manchester Square, London, W.1. Tel.: 01-486 6644.
Media Controller: B.A. Jupp.

A.N. SPICER (ADVERTISING) LTD.
3 The Parsonage, Manchester 3.

THE SPIERS ADVERTISING LTD.
36 Southbourne Crescent, London, NW4 2JY. Tel.: 01-346 4002.
T.V. Executive: B.X. Spiers.

STUART ADVERTISING AGENCY LTD.
22 Red Lion Street, London, WC1R, 4PX. Tel.: 01-831 7444.

TALBOT ADVERTISING
44-5 Fleet Street, London, E.C.4. Tel.: 01-353 5438.
T.V. Executive: P.J. Riley.

J. WALTER THOMPSON CO. LTD.
40 Berkeley Square, London, WIX 6AD. Tel.: 01-629 9496.

OON & HEATH LTD.
1266 Warwick Road, Knowle, Solihull, West Midlands. Tel.: Knowle 4164/7.
Marketing and Television Manager: D.H. Langstone.

UNITED KINGDOM ADVERTISING CO. LTD.
Aspen House, 25 Dover Street, London, W.1. Tel.: 01-491 7621.
T.V. Executive: Mike Storm.

C. VERNON & SONS LTD.
178-202 Great Portland St., London, W.1. Tel.: 01-637 1488.
Head of T.V.: Roy A. Fowler.

C.P. WAKEFIELD LTD.
Wakefield House, 153 Fleet Street, London, E.C.4. Tel.: Fleet Street 4588 (3 lines) & 3521 (4 lines).
T.V. Executive: R.J. Wakefield, M.I.P.A.

WARWICK PARSONS JACKSON LTD.
14-16, Great Pulteney Street, London, W1R 4AR. Tel.: 01-734 8951.
T.V. Executive; Production: G.N. Jackson.
Buying: L. Cowdrey.

WASEY CAMPBELL/EWAND LTD.
Williams House, Eastbourne Terrace, London, W.2 6LD. Tel.: 01-262-3424.

JACK WHITE & ASSOCIATES
Stratford House, Moseley Road, Birmingham, B12 OHU. Tel.: 021-772 4147-8-9.
T.V. Executives: H.J.V. White, M.I.P.A., M.A.A., H.G. Griffiths, A.M.I.P.A.

WHITES OF BRISTOL
Priory House, Marlborough Street, Bristol, BS1 3NY. Tel.: Bristol 23053.

SIDNEY F. WICKS LTD.
21 Newton Street, Manchester M1 1HJ. Tel.: 061-236 0947.
T.V. Executive: David W. Wicks.

WILCOX ADVERTISING LTD.
10A, Chandos Street, London, W1M 9DE. Tel.: Regent 1471-5.
T.V. Executives: W.E. Chester, A. Nicholas.

DAVID WILLIAMS AND KETCHUM LTD.
52, Bedford Row, London, WC1R 4LX. Tel.: 01-242 8123.
T.V. Executive: Frank Gonella.

WTV STUDIOS LTD.
112/114 Thorpe Road, Norwich, NOR 15T. Tel.: Norwich 26721.

CONROY WYKES ADVERTISING LTD.
18-20 St. Andrew Street, London, E.C.4. Tel.: 01-353 0951.
Media: Miss F. Thomas
T.V.: Miss F. Thomas.

YOUNG & RUBICAM LTD.
Greater London House, Hampstead Road, London, N.W.1. Tel.: 01-387 9366.
Creative Director: Dennis Auton.
Manager T.V. Dept.: Gilbert King.

PROFESSIONAL & TRADE ASSOCIATIONS

THE ADVERTISING ASSOCIATION
Abford House, 15 Wilton Road, London, SW1. Tel.: 01-828-2771.
(The Advertising Association is a federation of all trade associations with advertising interests: advertisers, media, agencies and services. It aims to ensure that everyone with a view or a voice in advertising—participants, supporters and critics—should understand the role of advertising in society and the economy, and the part it plays in creating prosperity and full employment.)
Director General: Roger Underhill.

ASSOCIATION OF BROADCASTING AND ALLIED STAFFS
Thorndike House, 70-76 Bell Street, London, NWI 6SP. Tel.: 01-262 0162.
General Secretary: D.A. Hearn.

ASSOCIATION OF CINEMATOGRAPH, TELEVISION AND ALLIED TECHNICIANS
(Affiliated to Trades Union Congress and Labour Party). 2 Soho Square, London, W1V 6DD. Tel.: 01-437 8506.
President: Ron Bowey.
General Secretary: Alan Sapper.
(The Association is the recognized Trade Union for film, television and radio technicians and negotiates salaries and working conditions. It also controls the A.C.T.T. Employment Bureau as an agency exclusively for film and TV technicians.)

ASSOCIATION OF PROFESSIONAL RECORDING STUDIOS LTD.
23 Chestnut Avenue, Chorleywood, Herts, WD3 4HA. Tel.: Rickmansworth 72907.
Secretary: E.L. Masek.

BRITISH ACADEMY OF FILM AND TELEVISION ARTS
195 Piccadilly, London, W1V 9LG. Tel.: 01-734 0022.
President: H.R.H. The Princess Anne.
Vice-President: Sir Richard Attenborough, C.B.E.
Chairman: Timothy Burrill.
Vice-Chairman: Film Otto Plaschkas.
Television: James Callan Jones.
Hon. Treasurer: Martin Schute
Director: Reginald Collin.
The British Academy of Film and Television Arts exists in order to promote, improve and advance original and creative work amongst people engaged in film and television production.

BRITISH ACTORS' EQUITY ASSOCIATION
(Incorporating the Variety Artistes Fed.) 8 Harley Street, London, W1N 2AB. Tel.: 01 637 9311. Cables: Britequity, London.
President: John Barron.
Honorary Treasurer: Milton Johns.
General Secretary: Peter Plouviez.

BRITISH FILM INSTITUTE
127 Charing Cross Road, London WC2H OEA., Tel. 01-437 4355.
Founded 1933. Principal object to encourage the development of the art of the film, to promote its use as a record of contemporary life and manners and to foster public appreciation and study of it from these points of view. The Institute has a similar role in relation to television. The Board of Governors is appointed by the Secretary of State for Education and Science. The National Film Archive and the National Film Theatre are departments of the Institute which publishes Sight and Sound, a quarterly film magazine and the Monthly Film Bulletin, and provides services for education, film availability regions and production.
Chairman: Sir Basil Engholm KCB.
Director: Anthony Smith.
Deputy Director & Secretary. Gerry Rawlinson.

BRITISH FILM PRODUCERS ASSOCIATION LTD.
162 Wardour Street, London, WI. Tel.: 01-734 2142.
President: Kenneth Maidment.
Assistant Secretary: Andrew Patrick.

BRITISH INDUSTRIAL AND SCIENTIFIC FILM ASSOCIATION
26 D'Arblay Street, London, W1V 3FH. Tel.: 01-439 8441.
Director: Colonel K.A.M. Bennett, M.C.

BRITISH KINEMATOGRAPH SOUND AND TELEVISION SOCIETY
110/112 Victoria House, Vernon Place, London, WC1B 4AD. Tel.: 01-242 8400.

President: J. Aldred.
Secretary: William Pay
Founded in 1931, the Society was incorporated in 1946 to service the industries of its title encouraging technical and scientific progress. To further these aims, the Society disseminates to its Members information on technical developments within these industries, arranges technical lectures and demonstrations, and encourages the exchange of ideas. The broad nature of its purpose is made possible by the subscriptions of its Members and by its freedom from political or commercial bias. The *BKSTS Journal* is published monthly and is sent free to all members.

BRITISH MUSIC INFORMATION CENTRE
(Reference library of works by 20th Century British composers.)
10 Stratford Place, London, W1N 9AE. Tel.: 01-499 8567.

Librarian: Roger W. Wright.

THE BRITISH RADIO AND ELECTRONIC EQUIPMENT MANUFACTURERS' ASSOCIATION
Twentieth Century House, 31 Soho Square, London W1V 5DG. Telephone 01-734 7471. Telex 27869.

BRITISH SOCIETY OF CINEMATOGRAPHERS LTD.
(To promote and encourage the pursuit of the highest standards in the craft of motion picture photography)

c/o Technicolor Ltd., Bath Road, Hammondsworth, West Drayton, Middx. Te.: 01-759 5432.
Secretary and Treasurer; Frances Coffey.

CABLE TELEVISION ASSOCIATION OF GREAT BRITAIN
295 Regent Street, London, W1R 7YA. Tel: 01-637 4591.

Secretary: A.W. Brittain, A.C.I.S.

CENTRAL CASTING LTD.
(Licensed annually by the Dept. of Employment.) 2 Lexington Street, London, W1R 3HS. Tel.: 01-437 1881.

Directors: O. Beuselinck, B.J. Kingham, K.L. Maidment.
General Manager: F.T. Kennedy.

CINEMA AND TELEVISION BENEVOLENT FUND
Royalty House, 72 Dean Street, London, W1V 6LT Tel.: 01-437 6567.

Executive Director: Maj. Gen. C.M. Grigg, CBE, MC.
Secretary: H.V. Hughes, F.C.A.
(The Fund gives relief by financial grants and allowances to needy members or ex-members of the film industry or independent TV, and their widows; maintenance and education of orphans and generally to assist those in distress due to sickness, injury, unemployment or old age. Convalescence is available to assist in recovery after illness or operations at "Glebelands," Wokingham, Berks. Admission to Convalescent Home free upon application to the Secretary.)

COMPOSERS GUILD OF GREAT BRITAIN
10 Stratford Place, London, W1N 9AE. Tel.: 01-499 8567.

President: Sir Lennox Berkeley, CBE.
General Secretary: Miss Elizabeth Yeoman.
(Objects: To further artistic and professional interests of its members.)

CONFEDERATION OF ENTERTAINMENT UNIONS
60-62 Clapham Road, London, SW9 0JJ. (Confederation of trades unions engaged in the entertainment industry.)

President: Alan Sapper.
Vice-President: P. Plouviez.
Treasurer: S. Ogden.
Secretary: J. Morton.

ELECTRICAL ELECTRONIC TELECOMMUNICATION AND PLUMBING UNION
Hayes Court, West Common Road, Bromley, BR2 7AU. Tel.: 01-462-7755.
London Area Office: 15 Abbeville Road, London, SW4 9LA. Tel.: 01-675-2211.

(Representing electrical operatives engaged in studio production.)
General Secretary: F.J. Chapple.

THE ENTERTAINMENT AGENTS' ASSOCIATION LTD.
18 Charing Cross Road, London, WC2H OHR. Tel.: 01-240 1724.

President: Bunny Lewis.
Hon. Secretary: Gordon Blackie.
(Founded in 1927 as the Agents' Assoc. and name changed to present one in February, 1966. The Association comprises the principal agencies in all branches of the entertainment industry.)

FILM ARTISTES' ASSOCIATION
F.A.A. House, 61 Marloes Road, London, W.8. Tel.: 01-937 4567-8.

General Secretary: Sean Branningan.

FILMS OF SCOTLAND
32a Rutland Crescent, Edinburgh EH1 2BW Scotland. Tel.: 031-229 3456.

Hon. President: George Singleton, C.B.E., J.P.
Chairman: Andrew Stewart, C.B.E., LL.D
Hon. Treasurer: William Wallace.
Director: James Wilson.
(Objects: Films of Scotland was set up in 1955 to encourage the production of Scottish films in the national interest. Productions are undertaken for public and private sponsors—for cinemas, for television, for video outlets and for non-theatrical purposes. There is a 16 mm lending library with more than a hundred titles—Scottish life as seen by the camera, literally from A to Z.)

GENERAL & MUNICIPAL WORKERS' UNION
Thorne House, Ruxley Ridge, Claygate, Esher, Surrey. Tel.: Esher 62081.

General Secretary: David Basnett.

GUILD OF BRITISH CAMERA TECHNICIANS
303-315 Cricklewood Broadway, London, NW2 6PQ. Tel.: 01-405 3821.

Secretary: Brian M. Rose, MBKS.

GUILD OF FILM ART DIRECTORS
12 Quick Road, Chiswick, London, W.4. Tel.: 01-994 2731.

Secretary: Martin Atkinson.

GUILD OF FILM PRODUCTION ACCOUNTANTS AND FINANCIAL ADMINISTRATORS (GFPA)
Pinewood Studios, Iver, Bucks. SL0 ONH. Tel.: Iver 651700.

Chairman: Cyril Howard
Hon. Secretary: Robin Busby

GUILD OF FILM PRODUCTION EXECUTIVES
Pinewood Studios, Iver Heath, Bucks. Iver Tel.: 651700.

President: Johnny Goodman.
The Guild is an association of many of the more experienced production executives in the UK film industry.

GUILD OF TELEVISION CAMERAMEN
25 Carrholm Road, Leeds, LS7 2NQ.

THE INCORPORATED SOCIETY OF BRITISH ADVERTISERS LTD.
2 Basil Street, London, SW3 1AG. Tel.: 01-584 5221.

(*Objects:* The protection and advancement of the advertising interests of all member companies.)
Director: K.N. Miles.

INDEPENDENT TELEVISION COMPANIES ASSOCIATION, LTD.
Knighton House, 52-66 Mortimer Street, London, W1N 8AN. Tel.: 01-636 6866.

The Association exists to represent and promote the interests of the programme companies appointed by the Independent Television Authority. The members are Anglia Television, ATV Network, Border Television, Channel Television, Grampian Television, Granada Television, HTV, London Weekend Television, Scottish Television, Southern Television, Thames Television, Tyne Tees Television, Ulster Television, Westward Television, Yorkshire Television.

INSTITUTE OF PRACTITIONERS IN ADVERTISING
44 Belgrave Square, London, SW1X 8QS. Tel.: 01-235 7020.

INSTITUTION OF ELECTRONIC AND RADIO ENGINEERS
99 Gower Street, London, WC1E 6AZ. Tel.: 01-388 3071.

MECHANICAL COPYRIGHT PROTECTION SOCIETY LTD.
Elgar House, 41 Streatham High Road, London, SW16 1ER. Tel.: 01-769 4400. Cables and Telegrams: Mecolico, London, S.W.16.

OBJECTS
The Society has the authority of some ten thousand music copyright owners in the U.K. and Commonwealth, the U.S.A., Germany, Italy and other countries to licence the use of their works for all forms of mechanical reproduc-

tion, including sound synchronisation in cinematograph, television and advertising films.
Managing Director: R.W. Montgomery
Company Secretary and Financial Controller: K.R.D. Lowde.
Commercial Operations Controller: G.J. Churchill.

MOTION PICTURE EXPORT ASSOCIATION OF AMERICA INC.
162-170 Wardour Street, London, W.1. Tel.: 01-437 7700.

MUSICIANS' UNION
60-62, Clapham Road, London, SW9 OJJ Cables: Amuse., London, S.W.9. Tel.: 01-582 5566.
General Secretary: John Morton.

NATIONAL ASSOCIATION OF THEATRICAL, TELEVISION AND KINE EMPLOYEES
155, Kennington Park Rd., London, SE11 4JU. Tel.: 735 9068.
General Secretary: J.L. Wilson.

NATIONAL TELEVISION RENTAL ASSOCIATION LTD.
295, Regent Street, London W1R 7YA. Tel.: 01-637 4591.
Secretary: A.W. Brittain, A.C.I.S.

THE PERFORMING RIGHT SOCIETY LTD.
29-33 Berners Street, London, W1P 4AA. Tel.: 01-580 5544.
An Association of Composers, Authors and Publishers of Music.

THE PERSONAL MANAGERS ASSOCIATION LTD.
c/o Jill Foster Ltd., 35 Brompton Road, London, SW3 1DE.
(An association of the Principal Personal managers who represent stars, feature players, authors, writers, producers, directors and technicians.)
Chairman: Peter Dunlop.

PHONOGRAPHIC PERFORMANCE LTD.
Ganton House, 14/22 Ganton Street, London, W1V 1LB. Tel.: 01-437 0311.
(A Company founded by the British Recording Industry to own and exercise the public performance and broadcasting rights in sound recordings in the U.K. and to issue licences for public performances and broadcasting.)
General Manager: H.G.S. Gilbert.
Company Secretary: J.B. Love.

THE RADIO, ELECTRICAL AND TELEVISION RETAILERS' ASSOCIATION (RETRA) LTD.
RETRA House, 57-61 Newington Causeway, London, SE1 6BE. Tel: 01-403 1463.
Secretary: R.T. Edom

THE RADIO INDUSTRY COUNCIL
31 Soho Square, London, W1V 5DG. Tel.: 01-734 7471.
Director: O.P. Sutton, CBE.
Secretary: R.B.S. Purdy, OBE.

RADIO SOCIETY OF GREAT BRITAIN
35 Doughty Street, London, W.C.1N. 2AE. Tel.: 01-837 8688.
General Manager: David A. Evans.

ROYAL TELEVISION SOCIETY
Tavistock House East, Tavistock Square, London, WC1H 9HR, Tel.: 01-387 1970

SCREEN ADVERTISING WORLD ASSOCIATION LTD.
205 Wardour Street, London, W1V 3FA. Tel.: 01-734 7621.

SOCIETY OF AUTHORS
84 Drayton Gardens, London, S.W.10 9SD. Tel.: 01-373 6642.
(An independent trade union representing authors.)

SOCIETY OF AUTHORS BROADCASTING GROUP
(Representing Film, TV and Radio Writers)
(A subsidiary group within the Society of Authors) 84 Drayton Gardens, London, SW10 9SD. Tel.: 01-363 6642.

VARIETY CLUB OF GT. BRITAIN (TENT NO. 36)
Avon House, 3rd Floor, 360 Oxford Street, London, W.1. Tel.: 01-491 4521.
Press Guy: David Jones.

THE WRITERS GUILD OF GREAT BRITAIN
430 Edgward Road, London, W2 1EH Tel.: 01-723 8074-6.
(In 1959 the British Screen and Television Writers Association and the Radio and Television Writers Association were amalgamated to form this Guild. Its object to further and protect the interests of its members and to obtain for screen theatre and book writers terms comparable to those long accepted in other fields.)
President: Eric Paice.
Hon. Treasurer: Alice Robinson.
Gen. Sec.: Ian Rowland Hill, M.A.

EQUIPMENT & APPARATUS MANUFACTURERS

ARMSTRONG AUDIO LTD.
Warlters Road, London, N7 ORZ. Tel.: 01-607 3213.

BELLING AND LEE LTD.
(Electronic Components, RFI Shielding and Suppression, Great Cambridge Road, Enfield, Middlesex. Tel.: 01-363 5393. Telex: 263265.

DECCA RADIO AND TELEVISION LTD.
Neachells Lane, Willenhall, Wolverhampton, West Midlands WV13 3RW.

EMI LTD.
EMI House, 20 Manchester Square, London, W.1. Tel.: 01-486 4488.

ENGLISH ELECTRIC VALVE COMPANY LTD.
Waterhouse Lane, Chelmsford, Essex. Tel: 0245 61777. Telegrams: Enelectico, Chelmsford. Telex: 99103.

GENERAL ELECTRIC CO. LTD. (THE)
1 Stanhope Gate, London, W1A 1EH. Tel.: 01-493 8484.

ITT CONSUMER PRODUCTS (UK) LTD.
Maidstone Road, Sidcup, Kent. DA14 5HT. Tel.: 01-300 7733.

MARCONI COMMUNICATION SYSTEMS LTD.
Marconi House, New Street, Chelmsford, Essex, CMI IPL. Tel.: 0245 353221.

PYE TVT LTD.
(The Broadcasting Company of Philips) P.O. Box 41, Coldhams Lane, Cambridge. CB1 3JU. Tel.: 0223 45115.

RANK AUDIO VISUAL
P.O. Box 70, Great West Road, Brentford, Middx. TW8 9HR. Tel.: 01-568 9222.

RANK CINTEL
Walton Road, Ware, Herts. Tel.: Ware 3939. Telex: 81415.

RANK STRAND STUDIO DIVISION
P.O. Box 54, Great West Road, Brentford, Middlesex. TW8 9HR, Tel: 01-568 9222.

RCA LTD. (Subsidiary of RCA Corporation)
Lincoln Way Windmill Road, Sunbury-on-Thames, Middlesex. Tel.: Sunbury 85511.

ROBERT RIGBY LTD.
Premier Works, 4/6 Northington Street, London, WC1N 2JH. Tel.: 01-405 2944.

THORN EMI LTD.
Thorn House, Upper St. Martins Lane, London, WC2H 9ED. Tel.: 01-836 2444.

The World Market

- MARKET ANALYSES
- STATIONS
- OUTLOOK
- PROGRAM SOURCES

THE WORLD MARKET

ALBANIA

Radiodiffusion TV Albanaise, Tirana, government-owned experimental station, broadcasts sixteen hours weekly to some 55,000 sets, many of them located in public places so that segments of the populace not owning sets can view the programs.

ARGENTINA

Since May 9, 1981, Channel 11 of Buenos Aires, which was the only one of the four existing channels in the city which still had a black and white system, converted to color in the German adopted Pal-N System. Therefore, all the channels actually in Buenos Aires are in color.

In Argentina, although the channels are theoretically private enterprises, they are Government-administrated, and allocated under an auction method such as lately has been Channel 6 of the city of Trenque Lauquen and Channel 13, a new channel in the city of San Juan.

Exception to this method will be Channel ATC (7) of Buenos Aires, which will continue to maintain its actual official regime and will form part of the Official Service of Radiodifussion, which is constituted by one TV channel and one radio station in the capital of the Republic, as well as in the capital of each Province.

This long-awaited law that covers TV and radio, was sanctioned on September 18, 1980, and it has determined, among other dispositions, the private ownership of the TV and radio stations, and issues authorizations for the new ones.

The license permit is granted for a period of 15 years, starting when the channel commences to operate and has an option to extend the period for ten new years, for only one time, which can be extended to 20 in frontier channels.

These permits are granted only to Argentine citizens by birth or naturalized for over 10 years, who have moral qualities, culture and are able to demonstrate the origin of their funds. No permits will be granted to those who are partners or owners of radio stations and may also have connections with Argentine or foreign newspapers. The law also establishes that the maximum publicity time space allowed is of 12 minutes per hour.

The law has no provisions for dubbing, but it establishes a promotional regime, such as tariff exceptions for those films, serials and programs which are dubbed into Spanish within the country by Argentine professionals. The booking must have more than 50 per cent Argentine advertisements daily, and minors are protected between 8 and 22 o'clock, not permitting during this pictures or programs "that might affect the mentality of minors under 18."

Since the beginning of 1981, the devaluation of the Argentine peso with regards to the dollar has been at this writing of 230 per cent. Actually the cost of a 20-inch TV set is around U.S. $800.—and a 14-inch set U.S. $550.

—NAT BRUSKI

BAHRAIN

Bahrain TV, P.O. Box 1075, Manama, Bahrain, operates 1 channel and 1 relay for some 90,000 color sets.

BANGLADESH

Pilot station established in 1964 by government. Now includes originating station at Dacca with relay stations at Chittagong, Natore, Khulna, Sylhet, and Rangpur plus two under construction. Address Bangladesh TV, Television Bhaban, Rampura P.O. Box 456, Dacca.

BRAZIL

Brazil now has some 15,000,000 sets using PAL. The audience is estimated at about 75,000,000.

Brazilian law requires that all owners of radio or TV stations be of Brazilian nationality. A further law stipulates that all foreign films shown on TV be dubbed, and that no station may transmit more than 22 halfhours of foreign films for each 72 hours of weekly transmission time. Stations which transmit less than 12 hours per day have a corresponding lower quota. Seven sound studios, four in Rio de Janeiro and three in Sao Paulo take care of the dubbing of all American and some Italian, Argentine and Mexican films that come into the country.

Professional Association of Radio and TV Station Owners: Sindicato dos Proprietarios de Impresas de Radio Diffusao a Televisao, Avenida Presidente 'Churchill 97 Sala 606, Rio de Janeiro.

BRUNEI

Established 1975 by governmental Radio TV Brunei, Bander Seri Begawan, Brunei. Two channels.

CAMBODIA

Radio Cambodia Television, 28 Ave Preah Mohaksatryany Kossonak, Phnom Penh City, broadcasts daily on Channel 8 with relay to Bokor. Sets are estimated at 35,000.

CHINA (Communist)

At the present time the Chinese People's Republic has three television stations operating in the capital city of Peking and some 37 other stations in the nation servicing close to 1 billion people. TV ownership has shot up from 1 set to each 16,400 people 10 years ago to 1 set for each 280 persons or some 4,000,000 sets.

In American money, black and white sets cost $370 and $1,335 for color; but there is usually a waiting list. China now makes over 1 million sets a year and imports over 1 million annually from Japan. Some broadcasts are in color and local production of color sets began in 1981.

At present most TV studios broadcast hours nightly and at least 25% of their programs is instructional—including lessons in English, French and Japanese.

In addition to some domestic motion pictures shown occasionally on TV, mainland China is open to Western films as well as to those from some of the Communist bloc. As for TV features and documentaries, Peking has already signed exchange contracts with various Western nations.

The Central Broadcasting Administration, Hsin Men, Peking, is now offering to accept commercials on the national TV system charging $625 for a 15-second spot or a full minute for $1,680. Peking Municipal Television station began broadcasting May 16, 1979, as the capital's third station. By 1985 China says it will produce 2.3 million TV sets annually.

CHINA (Taiwan)

Television began in 1962 by the Educational Television Station which later became the private China Television Service in 1971. There are now three networks: Taiwan Television Enterprise (1962), China Television Company (1969) and China Television Service. All three act as private corporations deriving revenues from commercials. These three stations have 24 transmitters with total broadcasts of about 265 hours weekly for some 4.7 million set owners who pay about $1.50 annually for a license. About 80% of the programs are domestic, with most of the imports from the U.S.A. Over one-third of the sets are color. Taiwan has two communication satellite stations. A new rule states that commercials must not exceed 15% total broadcast time. Movie theatre owners complain that the three imported feature films shown weekly on television hurt their box office.

New rules call for no commercials during programs of less than 30 minutes and all commercials to secure government approval. In the future, 80% of all programs must originate in Taiwan with entertainment programs taking up no more than 55% of broadcast time and 45% or more devoted to educational, cultural or news types.

With relay stations it is estimated that 100% of the island's population of 16 million receives TV reception. About 20% of programs are imports, mostly from America, and these are broadcast in the original English with Chinese captions, which has served to develop knowledge of the English language among the populace.

TV showings of full-length Chinese-language, feature films imported from Hong Kong were stopped following retaliatory action by local cinema interests but the government may step in later and permit such showings, except on Sundays and holidays when the theatres reap their largest harvest. Inquiries should be addressed: Ministry of Communications, Taipei, Taiwan, Republic of China, or to: China Radio/TV Assoc. 6-144 Hsin Shen (South) Rd., Section 1, Taipei, Taiwan, Republic of China.

EGYPT

The television of Egypt is attached to the Ministry of Information, which is headed by Dr. Mansour Hassan. The Chairman of the Board of the TV is Mrs. Hemat Moustafa.

The Central station in Cairo transmits about 150 hours a week through two channels covering most of the Egypt areas including Asswan and Luxor.

Egyptian TV follows the European technical transmission with an image of 825 scansions.

The TV receivers in use number around 7 million. The Arabic Department shoots a movie news review weekly and it is sent to 35 stations all over the world with which Arab TV exchange cultural and documentary materials.

FINLAND

Finland has a two-channel television system that differs considerably from the other countries. There is at present in Finland one radio and television operating licence which has been granted to the Finnish Broadcasting Company, Oy Yleisradio Ab. The existence of MTV, Oy Mainos-TV-Reklam Ab, which is a private commercial TV company, is acknowledged in the same licence.

The basis for official relations between the companies were laid down in more detail in the basic agreement concluded in 1976. Television broadcasting in Finland is under parliamentary control. The Administrative Council of the Finnish Broadcasting Company appointed by Parliament is the highest organ exercising programme policy. It has two functions. The Administrative Council exercises programme policy control over both companies. In addition, as its name implies, it is the highest administrative organ in the Finnish Broadcasting Company.

MTV has its own corresponding administrative council.

As a bilingual country Finland also has programmes in Swedish on both tv-channels.

Some 19 hours per week—out of the total of 75—are sold by Oy Yleisradio Ab to MTV. At the end of 1980 a one-year price agreement was signed, according to which MTV's levy to the Finnish Broadcasting Company is at least 154,5 million marks in 1981. MTV produces nearly 40% of its programmes.

Number of color sets is today about 0.8 million (50% of all sets).

The Finnish Broadcasting Company: Head Office, Kesäkatu 2, 00260 Helsinki 26 (Telephone 441141) Broadcasting Center, Yleisradiokeskus, Pasila 00240 Helsinki 24 (Telephone 418811) TV 2 Television Center, Tohloppi, 33270 Tampere 27 (Telephone 445455), MTV, Oy Mainos-TV-Reklam AB, Pasilankatu 44, 00240 Helsinki 24 (Telephone 413300).

GREECE

TV is nationwide now and the most popular and cheap entertainment of the Greek public, harming all other kinds of recreation, movies mostly.

There are two TV stations in Athens, the first operated by the National Radio and Television organization (E.R.T.) and the other by the Information Service of the Armed Forces (U.E.N.E.D.)

TV like sound broadcasting is controlled by the state and responsibility by both services is invested in both the above state units.

The E.R.T. Station operates seven channels with 25 relay stations and some more towers and translators. The U.E.N.E.D. Station operates six channels and has an equal number of relay stations and towers as the E.R.T.

Each station puts on the air an eight-hours program on weekdays and ten to eleven hours on Sundays and holidays, which are mostly videotaped. Some 65–70% of this program is covered by news, educational subjects, local serials, discussion panels, etc. and 30–35% is covered by foreign feature films and feature serials.

The E.R.T. Station is a full member of the E.B.U. and transmits all important events of international interest. It is estimated that the number of the TV sets sold all over the country reached 2,000,000.

TV in Greece is being transformed gradually into color, as of 1979.

TV Commercials: A great number of spots are sold daily by each of the above stations to several advertising agents. The price for one minute varies according to the hour from $500 to $3,000. However, a small percentage of the commercial program offered by the local agencies is covered by foreign films.

Greek Radio Television (ERT): Director General: Constantine Hondros; Assistant Director General: George Stephanakis; Director of Television: Spilios Haramis; Director of TV Programme: George Anastopoulos; Director of TV Production: George Dambassis; Director of News Dept.: E. Bisticas; Director of Public Relations: George Tsoutsias.

Ypiressia Enoplon Dynameon (YENED): Director General Sotirios Vavaroutsos; Assistant Director General: Colonel Souranis; Director of Television: Lieutenant Colonel Photiou; Director of Radio Broadcasting: Vice-Captain Milis; Director of TV Programme: Spyros Skondras; Director of News Dept.: Costas Sismanis.

Agencies: There are several agencies acting on behalf of various foreign TV companies: Hellas Television International, 96 Academias Str. Athens, representing G.M., Screen Gems, Warner Bros, Rai-Sacis, Paramount TV, EMI, Rank, Titanus, Egyptian TV, Spanish TV, World Vision, Romulus, Thames TV, World Fair Communication, Film and TV Marketing-Servicing, etc. Apollo TV International, 46 King Constantine Avenue, Athens, representing United Artists and Betta Film. Yannis Triantafyllis, 8 Klissovis Str. Athens agent of the I.T.C. Dennis Pteropoulos, agent of the N.T.A. Triboukis agent of the German TV, etc.

Local TV Production: Local TV production is meant for local consumption only and this is due to their low budgets on one hand and to the difficulty of finding a market abroad. There are some independent producers now, however, who are aiming at the foreign market, having started negotiations with foreign filmmakers for producing and/or co-producing TV films good enough for foreign distribution. TV Producers Television Enterprise S.A. (TVE) head office 28, Kapodistriou Str. Athens, operating its own studio and offering studio facilities for all types of B/W and color TV programs local or foreign. Chairman: G. Constantinidis. TVEM, 52 Skoufa Str. AthensManagee Eleni Mavili-Photopoulou. Basile Maros, 26 Democritou Str. Athens. Dennis Petropoulos, II Lycavitou Str. Athens. Studio A.T.A., 258 Lenorman Str. Athens, Aris Tsopanellis, 5-7 Kythadineon Str. Athens, DYAS TV Production, Telecolor, Loudogon St.

—RENA VELISSARIOU

GUAM (U.S. Possession)

Guam Television operates one station with two channels for 17 hours daily, servicing an estimated 147,000 sets in the hands of Americans and natives. Owner is Pacific Broadcasting Corp., Agana. Some color is broadcast. The Department of Education, Agana, operates Channel 12, some 14 hours daily.

HOLLAND

The number of TV sets was 4,181,409 on Jan. 1, 1981 (on Jan. 1, 1980 4,110,581). The total Dutch population is about 14,000,000. Practically every Dutch family has a TV set now; two sets per family is no more an exception.

Competition of TV with motion picture theatres is still strong, but the situation of the movie theatres is rather healthy, though less than last year. TV competition is stronger in winter than in summer (weak TV programs). German and Belgian TV can be received well by the Dutch border provinces in particular, but also where a central antennae system exists. The market for black and white sets seems saturated. A color TV (biggest screen of 66 cm) costs about hfl. 2500 (one dollar is hfl. 2.70 now). Various shops sell at lower prices than the "official" ones).

There are two channels, with commercials under the name of Ster-Televisie. Commercials are shown every night on both channels, except on Sunday. A program is never interrupted by a commercial.

The Nederlandse Omroep Stichting (NOS) is the central governmental radio and TV organization, but there are eight main subdivision stations: a Roman-Catholic one (KRO), two political ones (left): VARA and VPRO, two Protestant sections, NCRV and EO, and three neutral sections: AVRO, TROS and VOO (with a big number of popular youth programs, pop and country music in particular). AVRO has the biggest number of members, closely followed by TROS. There are also some minor TV organizations.

Although the NOS has an important part of the playing time, the eight stations mentioned are completely independent, as far as their activities are concerned. Of the educational and religious stations Teleac (series of lessons in foreign language, popular subjects, mathematic courses, etc.) and IKON (religious) are the most important ones.

The approximate cost of a black and white TV set, antennae included, is unchanged, from 250 (a small one) to 1,000 guilders. A yearly contribution of hfl. 135,—is due by the owner of a set, to be paid to the government. This is the fee for a radio set included.

Feature films are programmed about two nights per week, and, in summer, now and then, also on afternoons (all ages films). On Wednesday nights only if there is no football game (Football is very popular here). The film industry organization (Nederlandse Bioscoopbond) has an agreement with the TV authorities that pictures can only be shown on TV 40 months or more after the first release date in motion picture theatres; exceptions are possible. Although showings of motion pictures on TV on weekends are not included in that agreement, the attitude of the Bioscoopbond has become more flexible. The countries of origin of pictures on TV differ, but American, English and German films are shown mainly, also TV series from these countries. Quality of programs is good; also many comments on the political news.

Due to the high standard of living, many people can afford to buy a TV set cash. Buying by installment has never been popular here, though that system makes more clients, gradually, mainly due to the decreasing economical situation. The Dutch stations are in Hilversum, as broadcasting is. Philips, Aristona and Erres are the most important manufacturers of TV sets. Big import from Germany and Japan. Nederlandse Omroep Stichting (NOS): 53, Emmastreaat, P.O. Box 10, 1200 JB Hilversum.

The Nederlandse Bioscoopbond, some time ago, started a law suit against a local cable TV company in Amstelveen trying to obtain a sentence from the judge (in court) about the author rights of a producer of a motion picture shown on cable TV; the legal uncertainty is criticized by the Federation. The judge rejected the point of view of the Federation, but the law suit continues. End of this year a decision of the Highest Court is expected. The main American companies started a similar process against the TV cable company, with the same result.

—PHILIP DE SCHAAP

HONG KONG

Established 1957 by Rediffusion (HK) as closed circuit, converted to conventional TV in 1974. Rediffusion Television Ltd. Kowloon Central P.O. Box 1414, Kowloon, Hong Kong. Additional outlets: Television Broadcasts (HK-TVB), 77 Broadcast Drive, P.O. Box 100, Kowloon, Hong Kong; and, Radio TV Hong Kong, Kowloon Central P.O. Box 200, Hong Kong (governmental unit acting as producer for above).

INDIA

Governmental TV centers operated in New Delhi, Madras, Bombay, Calcutta, Lucknow, Srinagar, Jaipur, Raipur, Amritsar, etc.—20 in all. Sets are estimated at over 2 million. Address: Television India, Mandi House, New Delhi 110001, India.

INDONESIA

The government-owned RRI-TV has been on the air since 1962 and now operates 6 stations in Jakarta, Jogakarta, Bandung, Semarang and Tjiribon, with 95 relays which will be expanded to 125. Broadcasting hours for each average some 40 hours weekly). Programs are 50% entertainment, 25% educational and 25% public service (60% live and 40% canned). Some broadcasts in color. Commercials have been banned since 1981 by government order. Sets are estimated at 898,000. Inquiries to TV Republik Indonesia (TVRI), Yayasan Televisi, Senayan, Jakarta, Indonesia.

IRELAND

Telefis Eireann is the official name of the television service in Ireland.

Television, like sound broadcasting, is run as an independent State Statuary body and responsibility for both services is vested in Radio Eireann, the Broadcasting Authority. This Authority, appointed by the Government, consists of nine members serving on a part-time basis under a chairman selected by the Government. The present chairman is T. Andrews.

The Minister for Posts and Telegraphs, who is responsible for the autonomously operated Authority, reserves to himself the duties of fixing and collecting license fees from domestic viewers, licenses transmitting stations, and has a veto on the appointment and conditions of the Authority's top executive. The Director-General, Kevin McCourt, is employed full-time.

The Television service began in 1962. It has five main high power transmitters and thirteen satellite "transposers," serving areas outside the range of the main transmitters. Its national coverage is estimated at 98 per cent of the population. The national technical standard is a 625 line one, but programs are also broadcast on a 405 line in the Eastern and Northwestern parts of the country.

The Authority, which has the responsibility of providing programs of suitable cultural and national content, transmits 44 hours per week, with occasional additional programs. In addition, educational programs for schools, two and a half hours per week, are broadcast for 33 weeks of the year on behalf of the Department of Education.

Of the 44 hours of weekly programs, 22 are of home origin. The balance are imported mainly from Great Britain and America.

There are about 350,000 receiving sets in operation, of which about 200,000 are able to receive programs from Great Britain's BBC-TV ane Independent TV stations as well as Radio Telefis Eireann.

The national annual license fee is five pounds, of which three pounds fifteen shillings is allocated to the television service, and the balance to the sound radio service. Radio Telefis Eireann also earns considerable revenue from the sale of spot advertising to commercial interests. Revenue from all sources in the past year was 2,868,000 pounds. Expenditure was 2,375,000 pounds.

ISRAEL

Israel television broadcasts daily in Arabic and Hebrew. Arabic programs start at 18:00 and continue till 20:00. Programs include documentaries, round-ups, news, films, variety shows, discussions, games, etc. These programs include, as well, weekly interviews with various individuals of the Arab Israeli population who can convey regards to their relatives in neighboring countries, such as Egypt, Syria, Lebanon, Jordan, through this media.

At 20:00 the Hebrew programs begin and include same variety of broadcast as the Arabic language. Hebrew broadcasts continue up to 23:30 or midnight. The majority of Hebrew programs are subtitled in Arabic; and foreign language broadcasts, including films, are subtitled in both Hebrew and Arabic. Similarly, most of the Arabic programs have Hebrew subtitles.

Besides the above mentioned programs, television broadcasts every morning (except Saturday) educational programs beginning at 8:00 and continuing up till 16:00. This is followed by children's programs which continue till 18:30 and include films, puzzles, games, cartoons, etc.

JAPAN

Non-commercial TV broadcasting began in Tokyo Feb. 1, 1953, by the quasi-governmental Nippon Hoso Kyokai (NHK) followed seven months later by the first commercial TV broadcast of Nippon Television Network (NTV). Broadcast time for NHK, which began with only 866 set-owners paying a fee of Yen 200 monthly, was 4 hours a day compared to six hours by NTV, which was affiliated with the Yomiuri newspapers. By 1960 NHK possessed 60 stations and there were 47 commercial stations broadcasting, while the number of sets had increased to about 7 million, topping the 10 million mark two years later.

Although commercial networks as they are established in the U.S. are legally forbidden, there came about a nationwide grouping of commercial stations with the Tokyo stations acting as pivotal points and providing tieups and programs. Thus in 1959, 16 commercial stations formed Japan News Network which soon spread to other programming areas, with JOKR in Tokyo, followed by three other such groupings centering on NTV, Nippon Educational Television (renamed Asahi National Broadcasting in 1977), and the Fuji TV key stations in Tokyo. It was then apparent that several newspapers with national sales systems, through capital tieups, were virtually in control of the national commercial networks.

Educational broadcasts were begun by NHK in 1959, followed three weeks later by Nippon Educational Television, which later dropped the "educational" pretense and became a regular broadcaster, just as TV Tokyo (JOTX) did later, when it was found impossible to operate a private educational TV station in Japan for financial reasons.

Taking the Tokyo Metropolitan area, which comprises almost 30% of Japan's population of 116 million, we find it served by Channel 1 NHK general network station, Channel 3 NHK educational network station, Channels 4, 6, 8 and 10 operated by NTV, TBS, Fuji, TV Asahi commercial network stations respectively, plus semi-independent TV Tokyo (JOTX) and TVK (UHF) 42.

NHK's general programming lists 8 regional stations throughout the nation, 45 local stations and numerous relay stations in the VHF range to bring general programming to 3,206 outlets (2,699 UHF outlets). NHK educational facilities comprise 8 regional stations, 37 local stations, 2,207 relay stations for a total of 3,145 outlets of which 2,639 are UHF, including use of mini-satellite relays. It must be emphasized that in most cases (but not all) the general and educational services make use of the same facilities.

There are 36 companies operating both radio and TV stations and 58 operating only TV stations. In the VHF sector some 48 companies operate 489 stations, in UHF 46 companies operate 4,115 stations, plus 5 SHF stations, a total of 4,609.

These four "networks" operate in the seven key areas of Japan—Tokyo, Osaka, Nagoya, Fukuoka, Sapporo, Hiroshima and Sendai—which are 4 commercial-station areas (plus NHK). Other stations located in less populous areas use a cross network arrangement by which programs from plural key stations are received. Each network virtually blankets Japan with the exception of ANB which covers about 80% of the country. In addition, the non-commercial NHK network covers the country.

Community TV in Japan is still in its infancy with some 20 stations listed while cable television is confined to a few hotels or apartment houses in the largest cities.

While NHK collects monthly fees from over 29,987,716 registered sets, the Home Ministry estimates the total number is 62,215,000 (including households with more than one) of which 9,423,000 are black & white and 52,792,000 color sets. Some 98% of registered households possess one or more color sets for which they are assessed a flat $3.90 a month versus $2.30 for black-and-white sets. It is estimated that over 92% of all households are equipped with TV. Thus, it is no surprise that NHK is by far the wealthiest broadcasting corporation in the world with 16,560 employees and a 23-story broadcasting center in Tokyo. Revenues run about $1 billion annually and latest figures show a deficit for the year.

Surveys show that the average Japanese spends 3 hours 29 minutes each weekday watching TV, 4 hours 5 minutes on Sundays—a decrease over previous surveys.

In the Tokyo area, which comprises about 30% of the nation's TV audience, a survey says average housewives spend an average of 5 hours 18 minutes daily, an increase from the average of 4.5 hours spent by wives in their 20s as those polled got older until an average of 7 hours daily was reached for those in their 70s. Their favorites were not so much the so-called soap operas as they were the afternoon emceed shows meant for women audiences. Sports features, especially baseball and sumo (traditional Japanese wrestling), were the male choices with golf tournaments and American football taking the place of the once-popular bowling.

Japan imports a number of American TV shows but none has made the Japanese Top Ten listing. Included are "This Is Your World," "Quincy," "Vegas," "Little House on the Prairie," "Hart to Hart," "Muppet Show," "Sesame Street," "Kingston Confidential," and others.

The NHK general network broadcasting hours average 18 hours daily with 32.7% of programs news or commentaries, 16.4% educational, 25.6% cultural and 25.3% entertainment. NHK educational network averages 18 hours daily with 15 news, 30% educational and 19% cultural subjects.

Average daily broadcast time for commercial stations averaged 19 hr. 21 min. since late-night broadcasting is not encouraged to conserve energy. Of this broadcasting average over 17 hours consisted of sponsored programs and 2 hr. of sustaining programs. Although broadcasters claimed that advertising commercials (including a daily average of 241 spot announcements) took only 10% of total broadcasting time of each station on the average, anyone with a stop watch knows that this is not true and over 60% of Japanese polled indicated an opinion that there were far too many commercials, while 50% objected to the ads shown in conjunction with kiddies' programs and thought such programs should carry no ads.

Advertisers on TV are headed by food and beverage makers (24.9%), followed by cosmetics and detergents (14.3%), pharmaceuticals (9.2%), household articles and equipment (8.1%), and electrical appliances (6.5%). In the past year some $3.3 billion was thus spent, an increase of some $432.4 million or almost 15%. Presently there is little room left for more commercials since air time has decreased to save energy so rates are going up. Thus, in the past year, program commercials increased by 2.1% while spot commercials decreased 1.1%.

Commercial stations claimed that 46% of daily broadcasting time was devoted to entertainment, 24% to cultural subjects, 12.9% to news and 3.1% to sports. Sponsors of a 30-minute prime time show pay about $6,500 on a Tokyo station with other spots going for $2,700 per sponsor. Commercial stations took 35.5% of all ad expenditures compared to 31.9% for newspapers.

Japanese movie producers, while curtailing sales of feature films to TV because of possible effect on their own box office admissions, are themselves active in producing TV features for domestic use. Not only are they able to produce them more cheaply (principally because of different union wage scales) but they can do better jobs and make better use of their studio facilities.

Production cost for a one-hour program is only $60,000, about one-fifth of the cost in the U.S. Payments for authoring a 30-minute script is $200 at NHK and as low as $100 at commercial stations. However, writers with a national reputation get much more.

A study of 107 commercial TV stations reveals that although the UHF outlets located in the periphery of the large cities are doing only moderately well as regards operating profits, the older-established VHF stations, particularly those affiliated with networks and in Tokyo or Osaka, enjoyed a prosperous year.

Japan exports very few TV programs of its own but it is the world's largest exporter of TV sets and in addition sells TV equipment, complete broadcasting studios and nationwide setups abroad.

Satellite broadcasts from abroad are used in Japan for momentous new happenings, and normally for sports events such as the Olympic Games or championship boxing matches in which one participant is Japanese. However, overseas heavyweight championship bouts are aired without charge on regular broadcasts.

While a critic would rate Japanese TV in general as somewhat below the not very high American standard, and take umbrage at the great percentage of time taken by commercials on all stations except those of NHK, the average Japanese is satisfied with it. Lately, the government has taken steps to prevent further intensification by stations of audio volume for commercials and limited them to 18% of total program time. In the past year 497 Japanese feature films and 822 imported features have been televised.

The Japanese import a large number of foreign full-length feature films in addition to American TV programs, since Japanese feature films have only limited TV use due to the film industry's desire to protect the theatres. However, the foreign imports are not as popular with TV audiences as their Japanese counterparts (though at theatres last year foreign imports outdrew Japanese films). Thus, except for rare events, full length feature films usually have the highest average viewing rate (14.2%), followed by a growing number of quiz games (13.6%), sports programs 14.5%, samurai dramas 13%, specials 13%, action dramas 12.8%, children's shows 12.8%, cartoons 12.1%, variety shows 11.9%, musical shows 11%, news 10.6%, imported TV shows 8.9%, educational shows 8.8%, in the Tokyo area. Also in the Tokyo area of over 20 million people surveys indicate that 40% of all sets are in use between 6 am and 9 am, 25.9% from 9-12, 47.4% from noon to 1 pm, 36.9% from 1-6 pm, 63.2% from 6 to midnight, 9.6% from midnight to 1 am. The average for the broadcasting day is 44.9%, with prime times of 7-10 pm 77.8% and 7-11 pm at 73.1%.

In 1979 both commercial and non-commercial television stations in the main cities of Japan began audio multiplex broadcasting of a limited number of daily programs. Thus with appliances costing from $70 upward for a set built for multiple reception a listener may enjoy stereophonic sound and/or original dialogue of some imported movies or TV features (such imports are almost invariably dubbed into Japanese language), thus bringing in an era of bilingual broadcasting in a limited way. This is done by fuller utilization of a station's assigned wave length, for example NTV channel 4 in Tokyo is in the frequency band ranging from 170–176 MHz, using 171.25 MHz for video and 175.75 MHz for audio, with multipex audio making use of the gap between.

Addresses of the commercial networks can be found in the World TV Broadcasters list in this volume and additional information can be obtained from National Association of Commercial Broadcasters in Japan, Bungei Shunjusha Building, 3 Kioi-cho, Chiyoda-ku, Tokyo, or from Advertising Association of Broadcasters in Japan, 8-10 Nishi Ginza, Tokyo. For non-commercial TV information one should address Nippon Hoso Kyokai (NHK), 2-2-1 Jinnan, Shibuya-ku, Tokyo 150.

—A.C. PINDER

MADAGASCAR

Television Madagascar, P.O. Box 442, Tanarive, Madagascar, operates a single station and three relays.

MALAYSIA—SINGAPORE

TV Malaysia was established in 1963 and now serves an estimated 554,000 sets on an average of 6 hours daily, half of broadcasts are in color. Originating station is at Kuala Lumpur, serving two networks on the main peninsula. One network extends to Sabah, where there is also a secondary station at Kota Kinabalu, which operates a small network of its own, extending to Sarawak. Inquiries to TV Malaysia (RTM), Broadcasting Dept., Kuala Lumpur 22-10. Sets: 775,000.

TV Singapore went on the air in 1963 and now serves an estimated 561,000 sets on two channels in four different languages. One channel broadcasts 66 hours weekly the other 40 hours; of this total about 82 hours weekly is in color. Inquiries to Radio-TV Singapore, P.O. Box 1902, Singapore 11.

MEXICO

Commercial television was inaugurated in Mexico in 1950 with the creation of Channel 4. Not long after this, Channels 2 and 5 were launched. For five years there was intense competition for viewers and advertising revenue. Finally, in 1955, recognizing the limiting parameters of the market, the three rival channels united to form *TELESISTEMA MEXICANO, S.A.* (TSM).

In 1959, Channel 11, with its cultural mission clearly delineated in its charter, initiated a regular broadcasting schedule under State management. However, financial resources supplied by the Ministry of Public Education severely curb the station's effectiveness.

Involvement in the XIX Olympic Games in 1968 had far-reaching effects on Mexico's television industry. A high capacity microwave network was constructed linking Mexico City and the northern border; a satellite ground station was built. These preparations made possible near blanket coverage of the Republic and the successful on-going transmission of color.

The emergence of channels 13 and 8 toward the end of the decade plunged TSM back into a competitive situation.

In 1972 when the State assumed control of Channel 13, a new concept of "commercial" television had to be adopted.

Striving to attract advertisers by producing the most convincing "numbers," TSM and TELEVISION INDEPENDIENTE DE MEXICO (Channel 8) left themselves open to severe criticism for the "debasement" of public tastes in entertainment. TSM and TIM realized that they could not survive as antagonists. Early in 1973 they joined forces to become a "joint venture" called *TELEVISA, S.A.* The new organization pooled the facilities and managerial talents of Channels 2, 4, 5 and 8 but, within strictly defined limits, maintained autonomy of operation.

In 1976, TELEVISA's annual programming for the four channels totalled 21,423 hours produced for a potential public of more than 25,000,000 viewers distributed among 5,000,000 TV homes.

To effectively transmit Televisa's signal to an estimated 70% of the Republic's territory, 79 repeater stations operate throughout the country. Channel 2 is a national network, Channel 4 reaches two states while Channel 5 covers 18 of the 32 states in the country and Channel 8 serves the large urban metropolitan zone of Valley of Mexico.

Televisa recently opened a market in the United States for the Spanish-speaking population. It currently broadcasts 19 hours per week transmitted directly from Mexico City, and distributed to 11 U.S. cities by Spanish International network.

Beginning in 1972, as a quasi-state enterprise, Channel 13 received a government subsidy. It is now, however, making attempts to free itself from dependence on state financing. Functioning as a national network composed of 26 repeater stations, it produced 4,152 hours during 1976. These programs reach 3,500,000 TV homes.

TELEVISION RURAL DE MEXICO is a sponsored project which extends a broad spectrum of television programming in order to reach isolated rural populations. TRM claims 13,000,000 viewers.

Considerable investment in time and resources is being made by the Federal Government in collaboration with commercial television to conduct a massive educational programme coordinating audio-visual techniques with in class instruction.

With the creation of the microwave network, in 1968, CABLEVISION (a CATV System) was established to provide North American programming, from NBC, CBS and ABC, on Channels 7 and 10 for its Mexico City subscribers.

A recent innovation was *TELSTAR*, a consumer or viewer service channel supplying continuous information on time, weather conditions, community services and short updated news summaries. Operated by *CABLEVISION, TELSTAR* broadcasts on Channel 12.

Any description of television programming must take into consideration the Presidential Decree of 1969 which stipulated that all broadcasting channels must cede 12.5 per cent of their daily transmission time to the government in view of a 25 per cent tax on their gross advertising revenues. In order to make full use of this time, the government in 1977, created the *DIRECCION DE RADIO, TELEVISION Y CINEMATOGRAFIA* in charge of producing programmes for these broadcasts

The chief statutory instrument regulating Mexican television is the 1960 *LEY FEDERAL DE RADIO Y TELEVISION*. This law defines the government's relation to programme content, franchises and permits and all technical matters pertaining to broadcast operation.

MICRONESIA

This Trust Territory administered by the U.S. has MICRONESIA Broadcasting Corp., Taga Hotel, Saipan, Mariana Islands, comprising 2 stations broadcasting 16 hours daily, on a commercial basis.

MONGOLIA

This small member of the Communist bloc established Radio-TV Mongolia in 1970 consisting of a single station with satellite connection with Soviet Russia. Address: Mongolian Radio-TV center, Ulan Bator. Sets number 5,250.

NEW CALEDONIA

Radiodiffusion Television Francais, Noumea, broadcasts about 30 hours weekly on three channels to some 38,000 set owners in this French possession in the Pacific.

NEW ZEALAND

The NZ Broadcasting Service established in 1962 was in 1975 replaced by 2 independently-operated networks which in 1977 were brought under the Broadcasting Corporation of NZ. South Pacific TV, Anzus House, Box 3819, Auckland; and, Avalon TV Center, P.O. Box 30945, Lower Hutt, N.Z. Sets are estimated as over 1 million.

NORTH KOREA

Experimental telecasting, using Russian equipment since 1960s, followed by establishment of a national network of 20 stations with regional stations and relays. Korean Central Broadcasting Committee. Television Section, Pyongyang, Democratic Peoples Republic of (North) Korea.

OMAN

Oman Color TV, Ministry of Information & Culture, P.O. Box 600, Muscar, Oman. Three stations and 1 relay for 95,500 sets.

PAKISTAN

Television came to Pakistan in November, 1964, when the first station at Lahore went on the air. There were hardly 500 sets then. In May, 1967, Pakistan Television Corporation Ltd., a public limited company was formed. It is now running all five TV stations in the country. These stations are at Karachi, Lahore, Rawalpindi-Islamabad, Peshawar and Quetta and transmit four to six hours programmes, including canned foreign films. All these TV stations and eight boosters (three each at Sind and Punjab and one each at NWFP and one in Baluchistan) are equipped with modern technical facilities, including video tape recording, background projection and arrangement for showing both 16 and 35mm films at the stations.

A major breakthrough in the field of TV communication was provided with the opportunity to link Pakistan with the outside world through satellite in the mid-1970s. On July 24, 1981, PTV, for the first time, introduced a new technique of showing an interview of the Pakistani ambassador to Britain (in London) which was taken from Karachi via satellite. Now all the Pakistani viewers can see live programmes from abroad, as all the TV stations have been linked with the National Microwave Network. Some of the Middle East countries' television programs, including Oman, Dubai, Saudi Arabia and Abu Dhabi, are clearly visible in Karachi and surrounding areas with the help of boosters or high antenna. Indian (Amritsar) TV is clearly seen in Lahore and surrounding areas.

Meanwhile about 100 cities and towns of the country are covered by TV telecast. With the installation of boosters about 85% of the population was expected to be able to see television in Pakistan by the end of 1981. Surveys have shown that colour TV has reached most of the well-to-do houses and black and white in most of the average houses throughtout Pakistan. PTC is being run as a commercial service. Advertisements can be telecast in breaks or between the programmes. An average 12 to 15 minutes in an hour is consumed by advertisements on Karachi TV. The pressure on other stations is much lower. The 35% increase in the rate of commercials in 1980 and a further recent increase of 15% during 1981, along with censorship of TV programmes under Islamic laws have affected the business. The latest announcement of not showing women in TV commercials other than domestic ones is also expected to affect business of PTV in the year under review.

The approximate number of receiving sets in the whole of Pakistan is 1,200,000 to 1,400,000, more than 66% of which are licensed sets. The license fee which was Rs. 150 since 1977, has been increased to Rs. 200 since July, 1980, and remains the same in 1981–82.

For the first time colour programmes were introduced on PTV on December 20, 1976. Today there are three full fledged colour studios in operation at Karachi and Islamabad and Lahore; and two colour stations will be completed in the country during the next five years at Quetta and Peshawar. However, Quetta station is likely to start colour transmission by the end of 1982 or in the beginning of 1983. Besides, one re-broadcast centre will be set up at Rojhan and three translator centres at Narowal and Sibi.

The Quetta-Chaman-Khojak TV link will also be completed during the Fifth Plan Period. These projects will not only take the TV signals to the less-developed regions of the country but will also meet the needs of the remote area of the NWFP, Baluchistan, Azad Kashmir and Northern Areas. Most of the canned and domestic programmes are telecast in colour daily from Karachi, Lahore and Islamabad stations. The number of colour sets is over 300,000 and would have increased to more than double but the present ML Government has called it a luxury and imposed 75% duty on import.

Most of the equipment installed in the present TV centre at Quetta has been recovered from the pilot TV centre and has been reconditioned by Pakistani engineers.

Currently about 1,500 creative technicians, including producers, cameramen, designers, editors, engineers and other technical staff are engaged with the PTC. Artistes and performers are not on the permanent staff of the PTV; they are paid daily wages according to their work in the programmes. The highest remuneration aid to a performer is Rs. 19/-per minute and the lowest (minor role) gets Rs. 3/-per minute. About 2,500 outside talents, including composers, singers, actors, announcers, instrumentalists, writers, etc. participate in the PTV programmes. A training institute for the artistes and technicians is likely to be opened in the near future at Karachi, Islamabad and Quetta.

—A.R. SLOTE

PHILIPPINES

Broadcasts began in 1953 by DZAQ-TV then owned by Alto Broadcasting. At present 23 stations are on the air for commercial broadcasting and one government station; most are located in Manila. Number of sets is estimated at over 3,600,000. ABS-CBN was replaced by KBS. Others are Associated Broadcasting. Inter-Island Broadcasting, Metropolitan Broadcasting, Philippines Broadcasting. Radio Mindanao Network, Republic Broadcasting and Southern Broadcasting. More provincial outlets are planned.

PUERTO RICO

Television started in 1954, and there are about 660,000 sets in Puerto Rico. Most of the public places such as restaurants, clubs, have television sets for their patrons and the medium is available in local hotels upon request.

Even though most Puerto Ricans know English, and the number of persons who will acquire a working knowledge of English will increase considerably through intensification of English instruction in public schools and the migration of Puerto Rican families to the United States, and workers who come back from the States with more knowledge of English, television programming is in English and Spanish. WKAQ-TV, owned and operated by the newspaper El Mundo, on channel 4, has 30 per cent programs in English of which 10 per cent have Spanish sub-titles. There are 10 private stations in all.

Kinescope versions of top North American television shows are represented on both stations.

QATAR

Qatar TV Service (governmental), Ministry of Information, P.O. Box 1836 Doha, Qatar. Color since 1973. One station and relay.

SAMOA (AMERICAN)

One TV station serves some 36,000 sets on the American part of this Pacific island as well as a number of sets on the part administered by New Zealand. The station is also used daily for educational instructions to local schools. Address: Television of Samoa, Pago Pago. There are six channels.

SAUDI ARABIA

Established 1957, now 7 government stations and 1 private station. General Directorate of TV, Ministry of Information, P.O. Box 570, Riyadh, Saudi Arabia; ARAMCO-TV, Box 96, Dhahran, Saudi Arabia.

SOUTH KOREA

Commercial TV broadcasting began in 1956, but the station burned three years later and it was not until 1961 when the government established the Korea Broadcasting Service (KBS-TV) network. The commercial network of Tongyang Broadcasting Co. (TBC) began in 1964 and was followed by the commercial Munwha Broadcasting Co. (MBC) in 1969. The American Forces Korea Network (AFKN-TV) was also established to serve American troops, and now comprises 6 stations.

Korean broadcasts are limited to 5.5 hours of evening time on weekdays, 11 hours on Saturday and 17 on Sunday. Monthly fees of about $1.00 are collected and advertisements accepted. KBS had 2 key stations and 51 relays, MBC 8 stations and TBC 2 stations. TV sets were estimated at some 6.3 million.

In 1981 the Republic of Korea "consolidated" all news media and took over the private stations, so daily listings read: KBS-1 TV (7 cities), KBS-2 TV (2), KBS-3 TV-UHF (6), MBC-TV (6). Korean Broadcasting Advertisement Corporation handles all TV and radio advertising.

—A.C. PINDER

SPAIN

The first time Television Espanola—(now Radio-Televisión Espanola RTVE)—transmitted on a regular basis was October 29, 1956, on a scan system of 625 lines at 25 frames per second. That program lasted three hours with a range of 70 kilometres, broadcasting for about 400 sets. The First Channel (VHF) was followed by a Second Channel (UHF) in 1967, complementary at first, but then sort of competitive and with a clearer trend towards "culture." Channel 1 covers now all of Spain, including the Balearic and Canary Islands. Channel 2 is reaching 80 per cent of national territory. According to the Director of Television, Channel 2 has a ceiling of 85 per cent of territory, since the investment costs double when the 70 per cent level of covering is reached.

RTVE is now an autonomous agency under the Ministry of Culture (formerly of Information), Paseo de la Castellana 109, Madrid 16. All programs are controlled by a Censorship Board. Drama, films and episodes of series are classified according to ratings: "general" (no sign on the screen), "older than 14" (a small rhombus), "older than 18" (two small rhombuses).

The total number of sets was 9,573,621 at the beginning of 1981. The production of programs is entirely financed by advertising and tie-in merchandising. Both channels carry ad spots. Due to nocturnal habits of Spaniards, the costliest time comes around 10 P.M. RTVE advertising rules out tobacco—except Cuban cigars—, and welcomes wines and liquors at any time. Very often the screen shows smiling, drinking señoritas.

For some time Spain saw a fight between color systems: German PAL and French SECAM. Several years ago the Government decided for PAL, but later the Ministry of Information declared the French system was also tested before reaching a decision. Fact is that RTVE was already broadcasting in PAL color about 90 per cent of its programs. The decision was final in September 1978, and PAL was the winner, naturally. There are 1,451,188 color sets in the country. Dealers expect sales to soar till the beginning of the Soccer World Championship on June 13, 1982.

Staff and Studios: Top RTVE executives: Fernando Castedo, Director general of the "Autonomous Agency"; José Maria Otero, Secretary General; Miguel Angel Toledano, Director of Spanish TVE; Francisco Guerrero, Director of International Relations.

The brand-new Statute of Television began to apply in 1981. Public opinion voiced soon protests against low ethical levels of programming, mainly in TVE dramas and talk shows. Two serious clangers by Ignacio Gabilondo, Director of Newscasts (about Spain and NATO, and unemployment in Andalusia), produced his immediate dismissal in June.

RTVE Headquarters are at Prado del Rey, Madrid 24, with general offices and nine studios with total area of 3,000 square metres. Additional studios are Chamartin, Madrid, with 3,500 square metres; Miramar, Barcelona, with 1,800 square metres; and Hospitalet, Barcelona, with 1,500 square meters. For out-of-town broadcasts, RTVE has a fleet of mobile units.

Fixed, salaried personnel is about 2,300, with another estimated 5,000 contributing freelancers. An important group of steady contributors is claiming now permanent status before the Labour Courts.

Programming Schedule and Contents: A new planning began on June 20, 1981. Channel 1 broadcasts 69 h 30′ weekly, with accent on week-ends. Same happens with Channel 2, which broadcasts 27 h 45′ weekly. Total of weekly broadcasting time: 97 h 15′. Special programs such as prize boxing bouts, sports championships, bull-fights, royal and political messages, etc., move or remove normal programming.

About 30 per cent of programming consists in filmed foreign production. Along with some choice British series ("Upstairs, Downstairs"), French dramas and Japanese cartoons for kids, many popular Anglo-American productions are eventually shown by TVE, from documentaries to cartoons and series as "The New Avengers", "Eight is Enough", "The Love Boat", "The Benny Hill Show", "Little House on the Prairie", "Tales of the Unknown", "Las Vegas", "Lou Grant", "The Hulk". After the impact of "Roots" and "Holocaust" in 1980, the audience welcomed in 1981 big American series as "Centennial", "Dallas", "The Dream Machine" and similar fare. Price range for half hour of screening time is $1,000–$1,350. One-hour series generally brings twice the half-hour price. Price range for feature films is $4,000–$10,000, dubbed into Spanish at Madrid Studios. Only "Film Society" screenings (Channel 2) accept foreign prints with Spanish subtitles.

Foreign Markets: Through its Foreign Promotion Department, RTVE sells many productions to the Spanish-speaking world. New markets were found from Northern Europe to Greece, and in the Far and Middle East. RTVE has been awarded prizes at the Festivals of Monte Carlo, Montreux, Prague, Berlin and Milan (MIFED). Quality and budgets are growing in Spanish TV. Examples: "Don Quijote de la Mancha" (cartoon series), "Cervantes" (drama series).

The national market for videocassettes is having in 1981 an unexpected bonanza, thanks in part to the "free expression" laws which apply a blind eye to porno tapes. Cable-TV is not yet heard of, but no less than four companies have applied for licenses to set up private TV networks in the near future. It seems that the monopoly of TVE is beginning to end.

—MARIANO DEL POZO

SRI LANKA (Ceylon)

The Government of Sri Lanka intends inaugrating nationwide television in February, 1982. It is intended that a new company, the Sri Lanka Rupavahini Corporation, will run the island republic's TV services. The Corporation's Act will be based on the Sri Lanka Broadcasting Corporations Act with minor amendments.

Rupavahini is the Singhalese and Tamil word for Television. It is intended that the Rupavahini Corporation will provide a daily school's telecast between 10 and 11:30 a.m. on schooldays. In the first year of operation schools' TV telecast will concentrate on General Science, Mathematics, Geography and English language programmes for General Certificate of Education "Ordinary" Level students in the Sinhala and Tamil languages (native languages).

The main Rupavahini Corporation transmission will be from 6 to 10:20 p.m. There will be 20 minute news bulletins in Sinhala, Tamil and English at 7:20, 8:20 and 10:20 p.m., respectively. Before the Sinhalese news bulletin there will be children's programmes and public information/education programmes. There will be entertainment programmes after the Sinhala news. Rupavahini will be a commercial station and commercials will be shown in clusters after the news.

The Rupavahini Corporation equipment will be PAL system "B" in colour but using the high band format. ITN, the private TV station which was taken over by the Government at the request of the majority of its owners due to teething difficulties, will continue as a second Channel, catering largely to the English-speaking minority. It will depend largely on imported products. ITN uses low-band casettes. It is intended that ITN (Independent Television Network) will also function under the Rupavahini Corporation Board but under a separate executive head. Both the Rupavahini Corporation and the ITN will fall under the purview of the Ministry of State which handles Television, Radio, Information Tourism, Wild life and State Printing.

The Independent Television Network hopes to have children's TV programmes in Sinhala and Tamil from January 1, 1982. The Secretary of Ministry of State, Sarath Amunugama, informed that most of the shows and programmes are already being prepared. These programmes will also be used for teaching various subjects to school children of all ages. The usual informative and entertainment programmes telecast every evening now will also continue.

By January, 1982 people all over Sri Lanka will be able to view TV—some of them with powerful booster antennae. Many schools will also be requested to train their children for special TV programmes.

Since the Independent Television Network commenced its transmissions over two years ago they catered mostly to the English-speaking community. There were, however, a weekly Sinhala or Tamil movie and newscasts in Sinhala and Tamil.

Thirty trainees who completed the Canadian International Development Agency (CIDA sponsored course of training in television programming will be the first in TV training to be sponsored by CIDA anywhere in the world. The course of training was of three month's duration, which included script writing, interview techniques, research and production in Television journalism.

The Japanese Government will provide trained and competent Japanese TV personnel to work along with Sri Lankan staff for the first two years of TV operations in Sri Lanka.

Programmes are obtained from various foreign agencies on a telecast and return basis. It operates on a cycle and at times they get the programmes direct from the agency and at times from some other country that has been using them. When ITN completes the telecasting, they return either to the agency or to another country where the agency wants them sent. The fee is between £50 and 200 sterling pounds per hour, depending on the quality of the programmes. These rates are, however, concessionary rates allowed to them in the initial stages as they are in their infancy. These programmes are sent in video casette (U-matic) form and 16 mm film. Locally produced films are obtained through negotiations with State Film Corporation. A royalty of Rs. 5,000 (US $264) for productions over 5 years, and Rs. 700 for ($37) for Tamil films are paid.

They have to get the sponsorship because they cannot telecast them on their own, due to lack of funds. If the sponsorship increases, they state that they can then get better quality films. Films depicting various aspects of a country are given to them free by the missions here in Sri Lanka. They make use of such films when the ITN is short of programmes.

There is a TV Advisory Board which chooses programmes for the ITN. They get the foreign news from VISNEWS or from EUROVISION news. The later is obtained by tapping the Earth Satellite station at Padukka in Sri Lanka.

Japanese Government has donated 69 TV programmes free to the ITN for telecasting. In 1982 February when National TV starts operating, ITN will continue to operate as an independent TV station on channel 12, while the Sri Lanka Rupavahini Corporation (National TV) will operate on Channel 5.

There are more than 88 000 Television sets in use at present two years after TV was introduced.

The license fee was reduced from Rs.250/-(US $14) to Rs.150/-($8) on Black and White sets and Rs.250/-($14/-) on Colour sets for a year.

Minister of State: Hon. Ananda Tissa de Alwis. M.P.: Secretary to the Ministry: Sarath Amunugama; TV Consultant to Ministry: Naren Chitty. Address of Ministry: Ministry of State, 2nd Floor, G C E C Building, Sir Baron Jayatileke Mawatha, Colombo 1. Sri Lanka

Competent Authority on Independent Television Network: M.D.T.L. Guruge. Address of I.T.N. C/O. Sri Lanka Broadcasting Corporation, Torringtonf Square, Colombo 7, Sri Lanka.

Sri Lanka Broadcasting Corporation: Chairman, Eamon Kariyakarawana; Director General of Broadcasting: D.T.L. Guruge; Director of News: Kumar Abeyasinghe. Address: Sri Lanka Broadcasting Corporation, Torringtonf Square, Colombo 7, Sri Lanka.

Liasoning & Promotion of Television Films, Programme Production and Sales: Chandra S. Perera, TV Film Producer-Cinematographer. Representing: N.B.C. News New York and U.P.I.T.N. (London and New York). Address: 437 Pethiyagoda, KELANIYA, SRI LANKA. Cables: "TELECINEX"—Colombo. Phone: 075-7285.

—CHANDRA S. PERERA

SWITZERLAND

The Swiss Broadcasting Corporation (SBC) holds a State concession for sound and television broadcasting. The SBC handles all programme matters, while the PTT enterprises are responsible for the transmitters and relay facilities.

As of April 30, 1981, there were 992,395 TV sets in operation in Switzerland and 1124 transmitters and boosters on the air. The French and German language-stations schedule about 65 hours weekly, each. The Italian language transmitters have about 60 hours of programming the week (without Eurovision—European TV Programme exchanges.)

February 1, 1965 commercial TV was introduced on a limited basis. From January 1, 1974, 20 minutes on weekdays at the utmost, 1t 17,400 francs per minute from 1/1/80. Annual licenses cost 120 francs (138 francs from 1/1/79) for private sets.

About 68.5 per cent of all households have colour TV sets.

—GERDA GOEPPERT-KLIMISCH

TAHITI

Office of Radiodiff-Television Francaise, Papeete, operates four channels 34 hours weekly to serve the estimated 17,350 sets of this

French possession. It is owned by the government and buys its canned programs in Paris.

THAILAND

Thai Television Company (government-owned) opened its first station in 1953 and now operates four stations (two using color) and 8 relays on a commercial basis. Stations are also operated under names of Bangkok Broadcasting & TV, Bangkok Entertainment Co. and Thai Army, all located in Bangkok. Sets are estimated 2.3 million, some color.

In 1980 a ban on broadcasts from 6:30 to 8:00 PM, as a power-saving measure, allegedly improved "family relationships" so at present broadcasts are limited to 8 hours on weekdays, and 16 hours on weekends and holidays.

UNITED ARAB EMIRATES

Established 1969 by Thompson TV International and Abu Dhabi TV Service. In 1971 United Arab Emirates TV Service, P.O. Box 637 Abu Dhabi established separate facilities which now consist of three stations and three relays. Dubai Radio and Color TV, P.O. Box 1695, Dubai, now broadcasts on two channels from 1 station and 1 relay in Arabic, French and English.

VIETNAM

Since 1968 the Government Information Department of the Democratic Republic of Vietnam has operated one non-commercial station in Hanoi. Number of sets is about 510,000.

A second station was added in Saigon (now known as Ho Chi-min City) when South Vietnam was "liberated" and the divided country became one.

WEST GERMANY

Latest statistics on television in Germany:

TV Households: Out of approximately 24,000,000 German households, 23,000,000 have at least one television set. Of these, about 16,000,000 have at least one color set. Some 23,000,000 households have at least one radio. Total TV set sales per year: about $2.2-billion.

Transmission: Only over-the-air TV transmission operates in Germany. Cable and satellite transmission is being planned, but neither form is likely to appear in significant commercial terms until after 1985. After five years of discussion, pilot cable projects are underway in Ludwigshafen, Dusseldorf and Berlin. First German TV satellite won't be launched until 1984, with serious technical consideration of the form getting underway in earnest by about 1986.

TV Networks: Germany has no networks similar to webs in the U.S. In fact, there are no privately-owned commercial TV stations; all telecommunication is state-owned. Basically, the two national TV outlets are: ARD, the national hookup of nine regional stations; and ZDF or second channel, a single nationally-beamed station based in Mainz near Frankfurt. ARD's nine stations each contribute to one joint national broadcasting operation, five regional channels and several radio channels. ARD has a budget of about $1,640-billion; ZDF, $556,000,000. Commercials are aired on both stations in sharply limited timeslots. Stations are supported largely by $6.50 monthly license fees paid by TV set owners.

WORLD TV BROADCASTERS

Europe

ALBANIA
RADIODIFFUSION ALBANAISE
Rue Conference de Peza, 3, Tirana.

AUSTRIA
OSTERREICHISCHER RUNDFUNK GmbH ORF
Argentinierstrasse 30a, 1041 Vienna Postfach 700.

BELGIUM
RADIODIFFUSION-TELEVISION BELGE (R.T.B.) (FRENCH)
Boulevard A. Reyers, 1040, Bruxelles.
BELGISCHE RADIO EN TELEVISIE (B.R.T.) FLEMISH)
Omreopcentrum A. Reyerslaan, 1040, Brussel.

BULGARIA
BULGARIAN TELEVISION
Todor Strashimirov 2, Sofia 4.

CYPRUS
CYPRUS BROADCASTING CORPORATION
P.O. Box 4824, Nicosia.

CZECHOSLOVAKIA
CESKOSLOVENSKA TELEVIZIE
M. Gorkeho 29, Prague 1.

DENMARK
RADIO DENMARK
Television Service, Radio Denmark, Radiohuset, Rosenornsalle, Alle 22, Copenhagen.

FINLAND
FINNISH BROADCASTING CO.
Kesakatu 2, Helsinki 26.

FRANCE
TELEDIFFUSION DE FRANCE
116, avenue du President Kennedy. F-75790 Paris Cedex 16.

GERMANY

GERMAN FEDERAL REPUBLIC
DEUTSCHE RUNDFUNK-UND FERNSEHANSTALTEN
BAYERISCHER RUNDFUNK
Rundfunkplatz 1, 8 Munich 2.
HESSISCHER RUNDFUNK
Frankfurt (Main), Bertramstr. 8.
NORDEUTSCHER RUNDFUNK
Hamburg, 13, Rothenbaumchaussee 132/134.
RADIO BREMEN
Heinrich-Hertz-Str. 13, Bremen, 33.
SAARLANDISCHER RUNDFUNK
66 Saarbrucken, Funkhaus Halberg, Postfach 1050.
SENDER FREIES BERLIN
Berlin 19, Masurenallee 8-14.
SUEDDEUTSCHER RUNDFUNK
7000 Stuttgart 1, Neckarstr. 145.
SUEDWESTFUNK
757 Baden-Baden, Hans-Bredow-Strass.
WESTDEUTSCHER RUNDFUNK
Appellhofplatz 1, 5 Cologne 1. Postfach 101950.
ZWEITES DEUTSCHES FERNSEHEN
Essenheimer Landstrasse, 65 Mainz-Lerchenberg, Postfach 4040.

GIBRALTAR
GIBRALTAR BROADCASTING, LTD.
Television House, Wellington Front.
GIBRALTAR TELEVISION
28 John Mackintosh Square, Gibraltar.

HOLLAND
NEDERLANDSE OMROEP STICHTING (N.O.S.)
P.O. Box 10, Hilversum.

HUNGARY
MAGYAR RADIO ES TELEVIZIO
Television Direction, V. Szabadsagter, 17, Budapest.

IRELAND
RADIO EIREANN TELEFIS
Donnybrook, Dublin 4.

ITALY
RADIO-TELEVISIONE ITALIANA
Viale Mazzini 14, 00195, Rome.

LUXEMBOURG
TELE LUXEMBOURG
Villa Louvigny.

MALTA
THE MALTA TELEVISION SERVICE LTD.
Television House, P O. Box 384 Gwardamanga.

MONACO
TELE MONTE CARLO
16, Blvd. Princesse Charlotte, Monte Carlo.

POLAND
POLSKIE RADIO I TELEWIZJA
Woronicza St., Warsaw.

PORTUGAL
RTP-RADIOTELEVISAO PORTUGUESA S.A.R.L.
Rua de S. Domingos (a Lapa) 26, Lisboa.

ROMANIA
RADIODIFUZIUNEA TELEVIZUNEA ROMANA
Calea Dorobantilor, 191, Bucarest O. Romania.

SPAIN
TELEVISION ESPANOLA
de Prado del Rey, Madrid.

SWEDEN
SVERIGES RADIO
AB S-105 10 Stockholm.

SWITZERLAND
SWISS BROADCASTING CORPORATION
Giacommettistrasse 3, CH-3000 Berne 15.

USSR
SOVJET TELEVISION
Akademika Koroljova 12, Moscow.

YUGOSLAVIA
TELEVISION YOUGOSLAVE
Boriso Kidrica 70/1 Belgrade.

Africa

ALGERIA
RADIODIFFUSION-TELEVISION ALGERIENNE
21 Blvd. des Martyrs, Algiers.

EGYPT
U.A.R. BROADCASTING CORPORATION OF CAIRO
Television Bldg., Cornish Road, P.O. Box 1186, Cairo.

GHANA
GHANA BROADCASTING CORPORATION
P.O. Box 1633, Accra.

KENYA
KENYA BROADCASTING SERVICE
P.O. Box 30456, Nairobi.

MOROCCO
RADIODIFFUSION TELEVISION MAROCAINE
1, Zankat al Brihi, Rabat.

NIGERIA
BROADCASTING COMPANY OF NORTHERN NIGERIA, LTD.
Television House, P.O. Box 250, Kaduna.
NIGERIAN BROADCASTING CORP.
P.M.B. 12005, Victoria Island, Lagos, Nigeria.
WESTERN NIGERIA RADIOVISION SERVICE
Television House, P.O. Box 1460, Ibadan.

SIERRA LEONE
SIERRA LEONE BROADCASTING SERVICE
Private Mail Bag, Freetown.

Asia

NATIONALIST CHINA
TAIWAN TELEVISION ENTERPRISES LTD.
No. 10, Pa Te Road, Sec 3, Taipei.
CHINESE TELEVISION SERVICE
N. 100, Kuang Fu South Road, Taipei.

GUAM
KUAM TELEVISION
Pacific Broadcasting, Box 368, Agana, Guam.

HONG KONG
TELEVISION BROADCASTS, LTD.
77 Broadcast Drive, Hong Kong.
REDIFFUSION (Hong Kong Ltd.)
81 Broadcast Drive, Kowloon, Hong Kong.

INDIA
ALL INDIA TELEVISION
Television Centre, All India Radio, Akashwani Bhawan, 5th Floor, Parliament Street, New Delhi, 110001.

INDONESIA
JAJASAN TELEVISI RI (RRI-TV)
Senajan, Jakarta.

IRAN
TELEVISION OF IRAN
P.O. Box 33-200, Av. Pahlavi, Tehran.

IRAQ
TV-STATION OF THE REP. OF IRAQ
Ministry of Guidance, Broadcasting House, Salhiya, Baghdad.

JAPAN
NIPPON HOSO KYOKAI (NHK)
2-2-1 Jinnan, Shibuya-ku, Tokyo 150.
THE YOMIURI TELEVISION BROADCASTING CORP. (YTV) JCIX-TV
Osaka.
ASAHI BROADCASTING CO. LTD. (ABC) JOBX-TV
Osaka.
KANSAI TELEVISION CORP. (KTV) JONR-TV
Osaka.
MAINICHI BROADCASTING SYSTEM (MBS) JOQR-TV
Osaka.
FUJI TELECASTING CO.
7 Kawada-cho, Ichigaya, Shinjuku, Tokyo.
ASAHI NATIONAL BROADCASTING CO. (TV ASAHI or ANB) formerly Nippon Educational Television (NET)
6-4-10 Roppongi, Tokyo.
NIPPON TELEVISION NETWORK (NTV)
14 Niban-cho, Chiyoda-ku, Tokyo.
TOKYO BROADCASTING SYSTEM (TBS)
5-3-6, Akasaka, Minato-ku, Tokyo.
NATIONAL ASSOCIATION OF COMMERCIAL BROADCASTERS
Bungei Shunju Bldg., 3 Kioi-cho, Chiyoda-Ku, Tokyo.
NIPPON HERALD
1 Shimbashi Ekimae Bldg., 2-20-75, Shimbashi, Tokyo.
INTERNATIONAL MOTION PICTURE
Hattori Bldg, Kyobashi 2-11-4, Chuoku, Tokyo-104.
K.R.C. (TV Section)
26-6 Akasaka, 8-chome, Tokyo.
INTERLINGUAL-INTERNATIONAL INC.
No. 7 Mori Bldg., 2 Nishikubo Tomoe-cho, Shiba, Tokyo 105.
INTERVISION, INC.
205, 15-6, 7-Chome Akasaka, Tokyo 105.
JAPAN INTERNATIONAL ENTERPRISES
17-20 Shimbashi, 6-chome, Tokyo.
MCA JAPAN LTD.
Maison Hirakawa, 2-5-2 Hirakawa-cho, Chiyoda-ku, Tokyo.
NEW JAPAN FILMS INC.
12 Iagura Katamachi, Azabu, Tokyo.
PACIFIC TV CORP.
3, 5-chome, Ginza, Tokyo.
WARNER BROS.
3-2-4 Kyobashi, Chuo-ku, Tokyo 104
JAPAN (Audience Research)
A.C. NIELSEN CO.
1-7-1 Naka-Keguro, Meguro-ku, Tokyo 153.
AUDIENCE STUDIOS (Japan)
Yoneda Bldg., 17-20 Shimbashi, Tokyo.
VIDEO RESEARCH CORP.
2-16-7 Ginza, Chuo-ku, Tokyo 104.
TV TOKYO (Channel 12 JOTX)
4-4 Shiba Koen, Tokyo.
CABLE TELEVISION JCTV
c/o ANB, 6-4/10 Roppongi, Minato-ku, Tokyo.
TVK (UHF) Channel 42
Yokohama.

SOUTH KOREA
KOREA BROADCASTING SYSTEM
33-3 ga, Namsan-Rong, Chung-Ku, Seoul.

KUWAIT
KUWAIT-TV
P.O. Box 621, Kuwait.

LEBANON
COMPAGNIE LIBANAISE DE TELEVISION
P.O. Box 4848, Beirut.

MALAYSIA
TELEVISION MALAYSIA
Dept. of Broadcasting, P.O. Box 1047, Kuala Lumpur.

PHILIPPINES
INTERCONTINENTAL BROADCASTING CORPORATION
P. Guevarra Ave., San Juan Rizal.
KANLAON BROADCASTING SYSTEM, INC.
Broadcast Plaza, Bohol Ave., Quezon City.
REPUBLIC BROADCASTING SYSTEM
E de los Santos Avenue, Diliman, Quezon City.

SAUDI ARABIA
SAUDI ARABIAN TELEVISION
P.O. Box 570, Riyadh.

SINGAPORE
TELEVISION SINGAPORE
P.O. Box 1902, Singapore.

SYRIA
SYRIAN TELEVISION
Omayad Square, Damascus.

THAILAND
THAI TELEVISION CO. LTD.
Bangkhunprom, Bangkok.
THAI ARMY TELEVISION STATION
Pahol Youthin Road, Bangkok.
BANGKOK ENTERTAINMENT
2357 New Petchburi, Bangkok.

TURKEY
TURKISH RADIO-TELEVISION CORP.
Nevzat Tandogan Cad. 2 Kavaklidere, Ankara.

Pacific

AMERICAN SAMOA
TELEVISION OF SAMOA
Department of Education, Pago Pago, 96799.

AUSTRALIA
AUSTRALIAN BROADCASTING COMMISSION
Government owned.
145 Elizabeth St., P.O. Box 487 G.P.O., Sydney, N.S.W. 2001.
AMALGAMATED TELEVISION SERVICES
Pty. Ltd. Television Station A.T.N., Channel 7
Television Centre, Epping, 2121, Sydney, N.S.W.
TELEVISION CORPORATION LIMITED
Television Station TCN, Channel 9
Artarmon Road, Willoughby, 2068, Sydney, N.S.W.
UNITED TELECASTERS LIMITED (CHANNEL A-10)
Epping & Pittwater Rds., North Ryde 2113, N.S.W.
HERALD-SUN T.V.
Pty. Ltd. Television Station HSV, Channel 7
Cnr. Dorcas and Wells St., South Melbourne, Victoria.
GENERAL TELEVISION CORP.
Pty. Ltd. Television Station GTV, Channel 9
22-46 Bendigo Street, Richmond, Melbourne, Victoria.
AUSTRAMA TELEVISION (CHANNEL 0)
ATV C/Ansett Transport Industries Ltd.
489 Swanston Street, Melbourne Victoria.
CANBERRA TELEVISION Ltd.
Television Station CTC, Channel 7
Canberra Television Centre, Black Mountain, Canberra ACT, P.O. Box 633, Canberra.
BRISBANE TV LIMITED
Sir Samuel Griffiths Drive, Mt. Coot-tha, Brisbane, Queensland.
QUEENSLAND TELEVISION LIMITED
Sir Samuel Griffiths Drive, Mt. Coot-tha, Brisbane, Queensland.
TELEVISION BROADCASTER LTD.
Television Station ADS, Channel 7
125 Strangeways Terrace, North Adelaide, South Australia.
SOUTHERN TELEVISION CORP. LTD.
Television Station NWS, Channel 9
202-208 Tynte Street, North Adelaide, South Australia.
TVW LIMITED
Television Station TVW, Channel 7
Osborne Park Road, Tuart Hill, Perth, W.A.
TASMANIAN TELEVISION LIMITED
TVT, Channel A-6
G.P.O. Box 1209M, Hobart (Tas.)

MARIANAS ISLANDS
KUAM TELEVISION
P.O. Box 368, Agana.

NEW CALEDONIA
RADIODIFFUSION TV, NOUMEA
Box 327, Noumea.

NEW ZEALAND
NEW ZEALAND BROADCASTING COUNCIL
P.O. Box 30, 355, Lower Hutt, New Zealand.

TAHITI
TELEVISION FRANCE L
Box 125, Papeete.

Central America

COSTA RICA
TELEVISORA DE COSTA RICA LTDA.
Apt. 3376, San Jose.

CUBA
CMQ-TV
Radiocentre, Habana.
CMBF-TV
Mazon y San Miguel, Habana.

DOMINICAN REPUBLIC
RAHINTEL TELEVISION
P.O. Box 1220, zone 7.

GUATEMALA
RADIO-TELEVISION GUATEMALA, S.A.
Apt. 1367, Guatemala.

HAITI
TELE-HAITI
P.O. Box 1126, Port-au-Prince.

HONDURAS (Rep.)
RADIO Y TELEVISION HONDURENA, S.A.
Tegucigalpa.

MEXICO
TELEVISION DE LA FRONTERA, S.A.
Vincente Guerrero num. 704 Ciudad Juarez, Chih.

RADIO PANAMERICANA, S.A.
Av. Chapultepec num. 18 Mexico, D.F.
TELEVISION DEL NORTE, S.A.
Aparatado Postal num. 1833 Monterrey, N.L.
TELEVIMEX, S.A.
Av. Chapultepec num. 18 Mexico D.F.
RADIO TELEVISION, S.A.
Calle Quinta num. 210 Desps. 14-16 Edificio Barbachano Tiajuana, B.C.
CADENA TELEVISION DEL NORTE, S.A.
Monterrey, N.L. (Dom. Av. Chapultepec num. 18 Mexico, D.F.)
RAFAEL TIJERINA CARRANZA
Aldama num. 2611 Nuevo Laredo, Tams.
TELEVISION TAPATIA, S.A. de C.V.
Circunvalacion num. 265 Guadalajara, Jal.
TELEVISORA DE LA LAGUNA, S.A.
Torreon, Coah. (Dom. Av. Chapultepec num. 18, Mexico, D.F.)
TELEVISION PASO DE CORTES, S.A.
Av. Chapultepec num. 18 Mexico, D.F.
SERGIO R. MOLINAR FERNANDEZ
Apartado Postal num. 271 Ciudad Juarez, Chih.
TELEVISORA DE CHIHUAHUA, S.A.
Chihuahua, Chih. (Dom. Av. Chapultepec num. 18 Mexico, D.F.)
TELEVISORA DE HERMOSILLO, S.A.
Hermosillo, Son. (Dom. Av. Chapultepec num. 18 Mexico, D.F.)
TELEVISORA DE GUADALAJARA, S.A.
Guadalajara, Jal. (Dom. Av. Chapultepec num. 18 Mexico, D.F.)
TELEVISORA DE CALIMEX, S.A.
Tiajuana, B.C. (Dom. Av. Chapultepec num. 18 Mexico. D.F.)
TELEVISORA DE OCCIDENTE, S.A.
Guadalajara, Jal. (Dom. Av. Chapultepec num. 18, Mexico, D.F.)
TELEVISORA DEL NORTE, S.A.
Monterrey, N.L. (Dom. Av. Chapultepec num. 18 Mexico, D.F.)
TELEVISORA DE YUCATAN, S.A.
Merida, Yuc. (Dom. Av. Chapultepec num. 18 Mexico, D.F.)
TELEVISION DE GUANAJUATO, S.A.
Cerro del Zamorano, Qro. (Dom. Av. Chapultepec num. 18 Mexico, D.F.)
TV DE CULIACAN, S.A.
Culiacan, Sin. (Dom. Av. Chapultepec num. 18 Mexico, D.F.)
TELEVISORA DE MEXICALI, S.A.
Apartado Postal num. 1356 Mexicali, B.C. (Dom. Av. Chapultepec num. 18 Mexico, D.F.
TELEVISION DE COLIMA, S.A. de C.V.
Anillo Circunvalacion num. 1080 Nte. Colima, Col.
FELIPE ARRIOLA GANDARA
Av. Alvaro Obregon num. 138 Nogales, Son.
CARLOS FERRAEZ MATOS
Calle Bolivar entre Diaz Miron y Boulevard Miguel Aleman Veracruz, Ver.
TELEVISION GONZALES CAMERENA, S.A.
Av. Chapultepec num. 18 Mexico, D.F.
TELEVISORA DEL GOLFO, S.A.
Ciudad Madero, Tams. (Dom. Av. Chapultepec num. 18 Mexico, D.F.)
TELEVISION DE MEXICO, S.A.
Av. Chapultepec num. 18 Mexico, D.F.
ANTONIO SUAREZ REYES
Av. Punta Banda y Eugenia, Frallion-Amiento Chapultepec Ensenada, B.C.

NETHERLANDS ANTILLES
TELECURACAO
P.O. Box 415, Curacao.

NICARAGUA
TELEVISION DE NICARAGUA, S.A.
Apdo 1505, Managua.

PANAMA
CIRCUITO RPC-TELEVISION
Direccion de Correos y Telecommunicaciones, Ministerio de Gobierno y Justicia, Panama.

TELEVISORA NACIONAL S.A. PANAMA
Ric. Chiari, Panam.

PUERTO RICO
WKAQ-TV
Telemundo Inc., Box W.
WIPM-TV
Dept. of Education
Hato Rey, San Juan.
WAPA-TV
WAPA-TV Broadcasting Corp.
P.O. Box 2050, San Juan.
WORA-TV
Quality Telecasting Corp.
Box 43, Mayaguez.
WIPR-TV
Dept. of Education
Hato Rey, San Juan.
WRIK-TV
Ponce TV Corp.
Darlington Bldg.
WSUR-TV
American Colonial Broadcasting Corp.
Box 4189, San Juan.
WKBM-TV
American Colonial Broadcasting Corp.
Box 4189, San Juan.
WOLE-TV
Western Broadcasting Corp.
Box 415, Aguadilla.

TRINIDAD
TRINIDAD AND TOBAGO TELEVISION CO. LTD.
Television House
12a Maraval Road, Port-of-Spain.

VIRGIN ISLANDS
ISLAND TELERADIOSERVICE, INC.
P.O. Box 1947, St. Thomas.

South America

ARGENTINA
PRIMERA TELEVISION ARGENTINA
Av. Alem 735, Buenos Aires.
COMPANIA ARGENTINA DE TELEVISION, S.A.
Castex 3345, Buenos Aires.
PRODUCCIONES ARGENTINA DE TV
Lima 1261, Buenos Aires.
TELEONCE
Calle Pavon 2444, Buenos Aires.
RIO DE LA PLATA
San Juan, 1160, Buenos Aires.
DIFUSORA MARPLATENSE
Av. Luro 2907, Mar del Plata.
SOCIEDAD DIFUSORA MENDOZA
Garibaldi 7, Mendoza.
TELECOLOR, S.A.
Av. Gral Paz 266, Cordoba.

BRAZIL
TELEVISAO-TUPI
Av. Joao Luiz Alves, 13-Urca, Rio de Janeiro.
TELEVISAO-RIO
Av. Atlantica 4264, Rio de Janeiro.
EMISSORA DE TELEVISAO CONTINENTAL
Rue das Laranjeiras 291, Rio de Janeiro.
TELEVISAO MAYRINK VEIGA
Rio de Janeiro.
RADIO GLOBO DO RIO DE JANEIRO
Rua Irineau Marinho 35, Rio de Janeiro.

CHILE

TV CHILE

Bellavista 0990, Casilla Postal 16.104, Correo 9, Santiago.

COLOMBIA

R. TELEVISORA NACIONAL

Direccion General de la Radiotelevisora National de Colombia
Apartado Nacional 1824, Bogota

ECUADOR

LA VOZ DE LOS ANDES—TELEVISION

Cas: 691, Quito.

PRIMERA TELEVISION ECUATORIANA S.A.

Illingworth 114 y Pichincha, Guayaquil.

PERU

MINISTRY OF EDUCATION

Ministerio de Educacion Publica
Direccion de Cultera, Lima.

TELEVISORA AMERICA

Montero Rosas, Lima.

TELEVISORA PANAMERICANA

Ave. Arequipa 1110, Lima.

TELEVISORA VICTORIA

Av. Tacna 543, Lima.

TELEVISORA EXCELSIOR

Ave. Manco Capac, Lima.

TELEVISORA EL SOL

Jose Galvez 1040, Lima.

TELEVISORA SUR PERUANA

P.O. Box 514, Arequipa.

URUGUAY

TV MONTE CARLO

18 de Julio 1855, Montevideo.

SAETA-TV

Lorenzo Carnelli 1234, Montevideo.

VENEZUELA

TELEVISORA NACIONAL

Apt. 3979, Caracas.

CORPORACION VENEZOLANA DE TELEVISION

Apartado 6674, Caracas.

R. CARACAS TV

Apartado 2057, Caracas.

The Television Code

The Television Code

A NEW CODE of standards and practices for the television industry went into effect March 1, 1952 following a series of developments almost a year in the making.

The code is an aftermath of a meeting held in Washington in June, 1951 at which NBC made public a code of its own. Set up as a standard of operations in television, this code bore an essential resemblance to the instrument earlier adopted by the same network for radio broadcasting. Both NBC codes, a Code of Standards adopted for radio by the National Association of Broadcasters—predecessor organization of the National Association of Radio and Television Broadcasters—in Atlantic City in 1947 and now the industry-wide TV code, drew upon the Motion Picture Production Code for their pattern.

Four subcommittees were active in preparation of the television code throughout the summer of 1951. Their tentative draft was approved by the NARTB board in September and a final polishing job undertaken in Washington shortly before the text was ratified in Chicago, October 19, 1951.

The Television Code provides for a Television Code Review Board which now consists of nine members. All members must be subscribers to the Television Code. The Code Review Board oversees policy direction of the Television Code and recommends to the NAB Television Board of Directors any changes in Code provisions for the latter group's ratification. The interpretation, implementation and enforcement of the Television Code's programming and advertising standards is the basic responsibility of the Code Authority Director.

In those phases that deal with programming and its approaches, the code bears striking similarity to the Motion Picture Production Code, conceived and formulated in 1929 by Martin Quigley, and ratified by the Motion Picture Producers and Distributors Association of America, Inc. late in 1945 when the late Eric Johnston was named president.

Meeting in Bandera TX in February, 1952, TV board members of NAB approved a review board of five and adopted a plan for financing administration of the code as well as an operating budget of $40,000 for the first year of operations.

There have been various revisions of the code over the years. This most recent edition, the 21st, is dated January, 1980.

Under the financing plan, each subscribing TV station originally paid an annual fee based on the highest hourly rate quoted on its rate card, up to a maximum of $1,400. If an NAB member, $250 of this was credited against NAB dues, leaving the balance for defraying code.

In 1974 the TV board adopted a new basis for determining Television Code subscriber fees. Primary basis for the new fee formula is the size of the market being served. Other factors affecting the actual amount of the fee are the independent or affiliate nature of each station, the VHF and UHF status of each station and the intermixed or de-intermixed nature of the market.

For example, the highest fee is charged to a network affiliated VHF station or a network affiliated UHF station in a de-intermixed market when that station serves one of the top three markets. The lowest fee is charged to an independent UHF station serving markets 176-210. Other fees vary in between.

The Code is copyrighted by the National Association of Broadcasters. The Code Authority maintains offices in Washington, New York and Hollywood. Stockton Helffrich is director with offices in New York.

Text of the revised code follows:

PREAMBLE

Television is seen and heard in nearly every American home. These homes include children and adults of all ages, embrace all races and all varieties of philosophic or religious conviction and reach those of every educational background. Television broadcasters must take this pluralistic audience into account in programming their stations. They are obligated to bring their positive responsibility for professionalism and reasoned judgment to bear upon all those involved in the development, production and selection of programs.

The free, competitive American system of broadcasting which offers programs of entertainment, news, general information, education and culture is supported and made possible by revenues from advertising. While television broadcasters are responsible for the programming and advertising on their stations, the advertisers who use television to convey their commercial messages also have a responsibility to the viewing audience. Their advertising messages should be presented in an honest, responsible and tasteful manner. Advertisers should also support the endeavors of broadcasters to offer a diversity of programs that meet the needs and expectations of the total viewing audience.

The viewer also has a responsibility to help broadcasters serve the public. All viewers should make their criticisms and positive suggestions about programming and advertising known to the broadcast licensee. Parents particularly should oversee the viewing habits of their children, encouraging them to watch programs that will enrich their experience and broaden their intellectual horizons.

PROGRAM STANDARDS

I. Principles Governing Program Content

It is in the interest of television as a vital medium to encourage programs that are innovative, reflect a high degree of creative skill, deal with significant moral and social issues and present challenging concepts and other subject matter that relate to the world in which the viewer lives.

Television programs should not only reflect the influence of the established institutions that shape our values and culture, but also expose the dynamics of social change which bear upon our lives.

To achieve these goals, television broadcasters should be conversant with the general and specific needs, interests and aspirations of all the segments of the communities they serve. They should affirmatively seek out responsible representatives of all parts of their communities so that they may structure a broad range of programs that will inform, enlighten, and entertain the total audience.

Broadcasters should also develop programs directed toward advancing the cultural and educational aspects of their communities.

To assure that broadcasters have the freedom to program fully and responsibly, none of the provisions of this Code should be construed as preventing or impeding broadcast of the broad range of material necessary to help broadcasters fulfill their obligations to operate in the public interest.

The challenge to the broadcaster is to determine how suitably to present the complexities of human behavior. For television, this requires exceptional awareness of considerations peculiar to the medium.

Accordingly, in selecting program subjects and themes, great care must be exercised to be sure that treatment and presentation are made in good faith and not for the purpose of sensationalism or to shock or exploit the audience or appeal to prurient interests or morbid curiosity.

Additionally, entertainment programming inappropriate for viewing by a general family audience should not be broadcast during the first hour of network entertainment programming in prime time and in the immediately preceding hour. In the occasional case when an entertainment program in this time period is deemed to be inappropriate for such an audience, advisories should be used to alert viewers. Advisories should also be used when programs in later prime time periods contain material that might be disturbing to significant segments of the audience.

These advisories should be presented in audio and video form at the beginning of the program and when deemed appropriate at a later point in the program. Advisories should also be used responsibly in promotional material in advance of the program. When using an advisory, the broadcaster should attempt to notify publishers of television program listings.

Special care should be taken with respect to the content and treatment of audience advisories so that they do not disserve their intended purpose by containing material that is promotional, sensational or exploitative. Promotional announcements for programs that include advisories should be scheduled on a basis consistent with the purpose of the advisory. (*Latter was effective in September of 1975.*)

II. Responsibility Toward Children

Broadcasters have a special responsibility to children. Programs designed primarily for children should take into account the range of interests and needs of children from instructional and cultural material to a wide variety of entertainment material. In their totality, programs should contribute to the sound, balanced development of children to help them achieve a sense of the world at large and informed adjustments to their society.

In the course of a child's development, numerous social factors and forces, including television, affect the ability of the child to make the transition to adult society.

The child's training and experience during the formative years should include positive sets of values which will allow the child to become a responsible adult, capable of coping with the challenges of maturity.

Children should also be exposed, at the appropriate times, to a reasonable range of the realities which exist in the world sufficient to help them make the transition to adulthood.

Because children are allowed to watch programs designed primarily for adults, broadcasters should take this practice into account in the presentation of material in such programs when children may constitute a substantial segment of the audience.

All the standards set forth in this section apply to both program and commercial material designed and intended for viewing by children.

III. Community Responsibility

1. Television broadcasters and their staffs occupy positions of unique responsibility in their communities and should conscientiously endeavor to be acquainted fully with the community's needs and characteristics in order better to serve the welfare of its citizens.

2. Requests for time for the placement of public service announcements or programs should be carefully reviewed with respect to the character and reputation of the group, campaign or organization involved, the public interest content of the message, and the manner of its presentation.

IV. Special Program Standards

1. Violence, physical or psychological, may only be projected in responsibly handled contexts, not used exploitatively. Programs involving violence should present the consequences of it to its victims and perpetrators.

Presentation of the details of violence should avoid the excessive, the gratuitous and the instructional.

The use of violence for its own sake and the detailed dwelling upon brutality or physical agony, by sight or by sound, are not permissible.

The depiction of conflict, when presented in programs designed primarily for children, should be handled with sensitivity.

2. The treatment of criminal activities should always convey their social and human effects.

The presentation of techniques of crime in such detail as to be instructional or invite imitation shall be avoided.

3. Narcotic addiction shall not be presented except as a destructive habit. The use of illegal drugs or the abuse of legal drugs shall not be encouraged or shown as socially acceptable.

4. The use of gambling devices or scenes necessary to the development of plot or as appropriate background is acceptable only when presented with discretion and in moderation, and in a manner which would not excite interest in, or foster, betting nor be instructional in nature.

5. Telecasts of actual sports programs at which on-the-scene betting is permitted by law shall be presented in a manner in keeping with federal, state and local laws, and should concentrate on the subject as a public sporting event.

6. Special precautions must be taken to avoid demeaning or ridiculing members of the audience who suffer from physical or mental afflictions or deformities.

7. Special sensitivity is necessary in the use of material relating to sex, race, color, age, creed, religious functionaries or rites, or national or ethnic derivation.

8. Subscribers shall not broadcast any material which they determine to be obscene, profane, or indecent.

Above and beyond the requirements of law, broadcasters must consider the family atmosphere in which many of their programs are viewed. There shall be no graphic portrayal of sexual acts by sight or sound. The portrayal of implied sexual acts must be essential to the plot and presented in a responsible and tasteful manner.

Subscribers are obligated to bring positive responsibility and reasoned judgment to bear upon all those involved in the development, production and selection of programs.

9. The presentation of marriage, the family and similarly important human relationships, and material with sexual connotations, shall not be treated exploitatively or irresponsibly, but with sensitivity. Costuming and movements of all performers shall be handled in a similar fashion.

10. The use of liquor and the depiction of smoking in program content shall be de-emphasized. When shown, they should be consistent with plot and character development.

11. The creation of a state of hypnosis by act or detailed demonstration on camera is prohibited, and hypnosis as a form of "parlor game" antics to create humorous situations within a comedy setting is forbidden.

12. Program material pertaining to fortune-telling, occultism, astrology, phrenology, palm-reading, numerology, mind-reading, character-reading, and the like is unacceptable if it encourages people to regard such fields as providing commonly accepted apprasials of life.

13. Professional advice, diagnosis and treatment will be presented in comformity with law and recognized professional standards.

14. Any technique whereby an attempt is made to convey information to the viewer by transmitting messages below the threshold of normal awareness is not permitted.

15. The use of animals, consistent with plot and character delineation, shall be in conformity with accepted standards of humane treatment.

16. Quiz and similar programs that are presented as contests of knowledge, information, skill or luck must, in fact, be genuine contests; and the results must not be controlled by collusion with or between contestants, or by any other action which will favor one contestant against any other.

17. The broadcaster shall be constantly alert to prevent inclusion of elements within a program dictated by factors other than the requirements of the program itself. The acceptance of cash payments or other considerations in return for including scenic properties, the choice and identification of prizes, the selection of music and other creative program elements and inclusion of any identification of commercial products or services, their trade names or advertising slogan within the program are prohibited except in accordance with Sections 317 and 508 of the Communications Act.

18. Contests may not constitute a lottery.

19. No program shall be presented in a manner which through artifice or simulation would mislead the audience as to any material fact. Each broadcaster must exercise reasonable judgment to determine whether a particular

method of presentation would constitute a material deception, or would be accepted by the audience as normal theatrical illusion.

20. A television broadcaster should not present fictional events or other non-news material as authentic news telecasts or announcements, nor should he permit dramatizations in any program which would give the false impression that the dramatized material constitutes news.

21. The standards of this Code covering program content are also understood to include, wherever applicable, the standards contained in the advertising section of the Code.

V. Treatment of News and Public Events

General

Television Code standards relating to the treatment of news and public events are, because of constitutional considerations, intended to be exhortatory. The standards set forth hereunder encourage high standards of professionalism in broadcast journalism. They are not to be interpreted as turning over to others the broadcaster's responsibility as to judgments necessary in news and public events programming.

News

1. A television station's news schedule should be adequate and well-balanced.
2. News reporting should be factual, fair and without bias.
3. A television broadcaster should exercise particular discrimination in the acceptance, placement and presentation of advertising in news programs so that such advertising should be clearly distinguishable from the news content.
4. At all times, pictorial and verbal material for both news and comment should comform to other sections of these standards, wherever such sections are reasonably applicable.
5. Good taste should prevail in the selection and handling of news:

Morbid, sensational or alarming details not essential to the factual report, especially in connection with stories of crime or sex, should be avoided. News should be telecast in such a manner as to avoid panic and unnecessary alarm.

6. Commentary and analysis should be clearly identified as such.
7. Pictorial material should be chosen with care and not presented in a misleading manner.
8. All news interview programs should be governed by accepted standards of ethical journalism, under which the interviewer selects the questions to be asked. Where there is advance agreement materially restricting an important or newsworthy area of questioning, the interviewer will state on the program that such limitation has been agreed upon. Such disclosure should be made if the person being interviewed requires that questions be submitted in advance or if he participates in editing a recording of the interview prior to its use on the air.
9. A television broadcaster should exercise due care in his supervision of content, format, and presentation of newscasts originated by his station, and in his selection of newscasters, commentators, and analysts.

Public Events

1. A television broadcaster has an affirmative responsibility at all times to be informed of public events, and to provide coverage consonant with the ends of an informed and enlightened citizenry.
2. The treatment of such events by a television broadcaster should provide adequate and informed coverage.

VI. Controversial Public Issues

1. Television provides a valuable forum for the expression of responsible views on public issues of a controversial nature. The television broadcaster should seek out and develop with accountable individuals, groups and organizations, programs relating to controversial public issues of import to his fellow citizens; and to give fair representation to opposing sides of issues which materially affect the life or welfare of a substantial segment of the public.
2. Requests by individuals, groups or organizations for time to discuss their views on controversial public issues should be considered on the basis of their individual merits, and in the light of the contribution which the use requested would make to the public interest, and to a well-balanced program structure.
3. Programs devoted to the discussion of controversial public issues should be identified as such. They should not be presented in a manner which would mislead listeners or viewers to believe that the program is purely of an entertainment, news, or other character.
4. Broadcasts in which stations express their own opinions about issues of general public interest should be clearly identified as editorials. They should be unmistakably identified as statements of station opinion and should be appropriately distinguished from news and other program material.

VII. Political Telecasts

1. Political telecasts should be clearly identified as such. They should not be presented by a television broadcaster in a manner which would mislead listeners or viewers to believe that the program is of any other character.

(Ref.: Communications Act of 1934, as amended, Secs. 315 and 317, and FCC Rules and Regulations, Secs. 3.654, 3.657, 3.663, as discussed in NAB's "Political Broadcast Catechism & The Fairness Doctrine.")

VIII. Religious Programs

1. It is the responsibility of a television broadcaster to make available to the community appropriate opportunity for religious presentations.
2. Programs reach audiences of all creeds simultaneously. Therefore, both the advocates of broad or ecumenical religious precepts, and the exponents of specific doctrines, are urged to present their positions in a manner conducive to viewer enlightenment on the role of religion in society.
3. In the allocation of time for telecasts of religious programs the television station should use its best efforts to apportion such time fairly among responsible individuals, groups and organizations.

ADVERTISING STANDARDS

IX. General Advertising Standards

1. This Code establishes basic standards for all television broadcasting. The principles of acceptability and good taste within the Program Standards section govern the presentation of advertising where applicable. In addition, the Code establishes in this section special standards which apply to television advertising.
2. A commercial television broadcaster makes his facilities available for the advertising of products and services and accepts commercial presentations for such advertising. However, a television broadcaster should, in recognition of his responsibility to the public, refuse the facilities of his station to an advertiser where he has good reason to doubt the integrity of the advertiser, the truth of the advertising representations, or the compliance of the advertiser with the spirit and purpose of all applicable legal requirements.
3. Identification of sponsorship must be made in all sponsored programs in accordance with the requirements of the Communications Act of 1934, as amended, and the Rules and Regulations of the Federal Communications Commission.
4. Representations which disregard normal safety precautions shall be avoided.

Children should not be represented, except under proper adult supervision, as being in contact with or demonstrating a product recognized as potentially dangerous to them.

5. In consideration of the customs and attitudes of the communities served, each television broadcaster should refuse his facilities to the advertisement of products and services, or the use of advertising scripts, which the station has good reason to believe would be objectionable to a substantial and responsible segment of the community. These standards should be applied with judgment and

flexibility, taking into consideration the characteristics of the medium, its home and family audience, and the form and content of the particular presentation.

6. The advertising of hard liquor (distilled spirits) is not acceptable.

7. The advertising of beer and wines is acceptable only when presented in the best of good taste and discretion, and is acceptable only subject to Federal and local laws (*See Television Code Interpretation No. 4*).

8. Advertising by institutions or enterprises which in their offers of instruction imply promises of employment or make exaggerated claims for the opportunities awaiting those who enroll for courses is generally unacceptable.

9. The advertising of firearms/ammunition is acceptable provided it promotes the product only as sporting equipment and conforms to recognized standards of safety as well as all applicable laws and regulations. Advertisements of firearms/ammunition by mail order are unacceptable. The advertising of fireworks is acceptable subject to all applicable laws.

10. The advertising of fortune-telling, occultism, astrology, phrenology, palm-reading, numerology, mind-reading, character-reading or subjects of a like nature is not permitted.

11. Because all products of a personal nature create special problems, acceptability of such products should be determined with especial emphasis on ethics and the canons of good taste. Such advertising of personal products as is accepted must be presented in a restrained and obviously inoffensive manner.

12. The advertising of tip sheets and other publications seeking to advertise for the purpose of giving odds or promoting betting is unacceptable.

The lawful advertising of government organizations which conduct legalized lotteries is acceptable provided such advertising does not unduly exhort the public to bet.

The advertising of private or governmental organizations which conduct legalized betting on sporting contests is acceptable provided such advertising is limited to institutional type announcements which do not exhort the public to bet.

13. An advertiser who markets more than one product should not be permitted to use advertising copy devoted to an acceptable product for purposes of publicizing the brand name or other identification of a product which is not acceptable.

14. "Bait-switch" advertising, whereby goods or services which the advertiser has no intention of selling are offered merely to lure the customer into purchasing higher-priced substitutes, is not acceptable.

15. Personal endorsements (testimonials) shall be genuine and reflect personal experience. They shall contain no statement that cannot be supported if presented in the advertiser's own words.

X. Presentation of Advertising

1. Advertising messages should be presented with courtesy and good taste; disturbing or annoying material should be avoided; every effort should be made to keep the advertising message in harmony with the content and general tone of the program in which it appears.

2. The role and capability of television to market sponsors' products are well recognized. In turn, this fact dictates that great care be exercised by the broadcaster to prevent the presentation of false, misleading or deceptive advertising. While it is entirely appropriate to present a product in a favorable light and atmosphere, the presentation must not, by copy or demonstration, involve a material deception as to the characteristics, performance or appearance of the product.

Broadcast advertisers are responsible for making available, at the request of the Code Authority, documentation adequate to support the validity and truthfulness of claims, demonstrations and testimonials contained in their commercial messages.

3. The broadcaster and the advertiser should exercise special caution with the content and presentation of television commercials placed in or near programs designed for children. Exploitation of children should be avoided. Commercials directed to children should in no way mislead as to the product's performance and usefulness.

Commercials, whether live, film or tape, within programs initially designed primarily for children under 12 years of age shall be clearly separated from program material by an appropriate device.

Trade name identification or other merchandising practices involving the gratuitous naming of products is discouraged in programs designed primarily for children.

Appeals involving matters of health which should be determined by physicians should not be directed primarily to children.

4. No children's program personality or cartoon character shall be utilized to deliver commercial messages within or adjacent to the programs in which such a personality or cartoon character regularly appears. This provision shall also apply to lead-ins to commercials when such lead-ins contain sell copy or imply endorsement of the product by program personalities or cartoon characters. (Effective September 1975.)

5. Appeals to help fictitious characters in television programs by purchasing the advertiser's product or service or sending for a premium should not be permitted, and such fictitious characters should not be introduced into the advertising message for such purposes.

6. Commercials for services or over-the-counter products involving health considerations are of intimate and far-reaching importance to the consumer. The following principles should apply to such advertising.

a. Physicians, dentists or nurses or actors representing physicians, dentists or nurses, shall not be employed directly or by implication. These restrictions also apply to persons professionally engaged in medical services (e.g., physical therapists, pharmacists, dental assistants, nurses' aides).

b. Visual representations of laboratory settings may be employed, provided they bear a direct relationship to bona fide research which has been conducted for the product or service. (*See Television Code, X, 11*) In such cases, laboratory technicians shall be identified as such and shall not be employed as spokesmen or in any other way speak on behalf of the product.

c. Institutional announcements not intended to sell a specific product or service to the consumer and public service announcements by non-profit organizations may be presented by accredited physicians, dentists or nurses, subject to approval by the broadcaster. An accredited professional is one who has met required qualifications and has been licensed in his resident state.

7. Advertising should offer a product or service on its positive merits and refrain from discrediting, disparaging or unfairly attacking competitors, competing products, other industries, professions or institutions.

8. A sponsor's advertising messages should be confined within the framework of the sponsor's program structure. A television broadcaster should avoid the use of commercial announcements which are divorced from the program either by preceding the introduction of the program (as in the case of so-called "cow-catcher" announcements) or by following the apparent sign-off of the program (as in the case of so-called trailer or "hitchhike" announcements). To this end, the program itself should be announced and clearly identified, both audio and video, before the sponsor's advertising material is first used, and should be signed off, both audio and video, after the sponsor's advertising material is last used.

9. Since advertising by television is a dynamic technique, a television broadcaster should keep under surveillance new advertising devices so that the spirit and purpose of these standards are fulfilled.

10. A charge for television time to churches and religious bodies is not recommended.

11. Reference to the results of bona fide research, surveys or tests relating to the product to be advertised shall not be presented in a manner so as to create an impression of fact beyond that established by the work that has been conducted.

XI. Advertising of Medical Products

1. The advertising of medical products presents considerations of intimate and far-reaching importance to the consumer because of the direct bearing on his health.

2. Because of the personal nature of the advertising of medical products, claims that a product will effect a cure and the indiscriminate use of such words as "safe," "with-

out risk," "harmless," or terms of similar meaning should not be accepted in the advertising of medical products on television stations.

3. A television broadcaster should not accept advertising material which in his opinion offensively describes or dramatizes distress or morbid situations involving ailments, by spoken word, sound or visual effects.

XII. Contests

1. Contests shall be conducted with fairness to all entrants, and shall comply with all pertinent laws and regulations. Care should be taken to avoid the concurrent use of the three elements which together constitute a lottery—prize, chance and consideration.

2. All contest details, including rules, eligibility requirements, opening and termination dates should be clearly and completely announced and/or shown, or easily accessible to the viewing public, and the winners' names should be released and prizes awarded as soon as possible after the close of the contest.

3. When advertising is accepted which requests contestants to submit items of product identification or other evidence of purchase of products, reasonable facsimiles thereof should be made acceptable unless the award is based upon skill and not upon chance.

4. All copy pertaining to any contest (except that which is required by law) associated with the exploitation or sale of the sponsor's product or service, and all references to prizes or gifts offered in such connection should be considered a part of and included in the total time allowances as herein provided. (*See Television Code, XIV*)

XIII. Premiums and Offers

1. Full details of proposed offers should be required by the television broadcaster for investigation and approved before the first announcement of the offer is made to the public.

2. A final date for the termination of an offer should be announced as far in advance as possible.

3. Before accepting for telecast offers involving a monetary consideration, a television broadcaster should satisfy himself as to the integrity of the advertiser and the advertiser's willingness to honor complaints indicating dissatisfaction with the premium by returning the monetary consideration.

4. There should be no misleading descriptions or visual representations of any premiums or gifts which would distort or enlarge their value in the minds of the viewers.

5. Assurances should be obtained from the advertiser that premiums offered are not harmful to person or property.

6. Premiums should not be approved which appeal to superstition on the basis of "luck-bearing" powers or otherwise.

XIV. Time Standards for Non-Program Material

The following standards for quantity and placement of non-program material are designed to serve the interests of the viewers and reflect as well consideration of the appropriate interests of broadcasters and advertisers within that public interest context.

1. Non-Program Material Definition:

Non-program material, in both prime and all other time, includes billboards, commercials and promotional announcements.

Non-program material also includes:

a. In programs of 90 minutes in length or less, credits in excess of 30 seconds per program except in feature films. In no event should credits exceed 40 seconds in such programs.

The 40-second limitation on credits shall not apply, however, in any section governed by a contract entered into before Oct. 1, 1971.

b. In programs longer than 90 minutes, credits in excess of 50 seconds per program, except in feature films. In no event should credits exceed 60 seconds in such programs.

The only exclusions from the foregoing definition of non-program material are: (1) public service announcements; (2) voiceover credits program information announcements not to exceed 30 seconds; (3) scheduling information regarding unusual special news programs; (4) scheduling information presented in special programs of indeterminate length (e.g., special news, sports or other special events programs) regarding an immediately upcoming program(s) whose regular broadcast timex has been affected by the indeterminate length program, and (5) promotional/informational material used to fulfill time requirements of certain formats two hours in length or longer such as theatrical length motion pictures, mini-series, made-for-television movies, occasional special long-length versions of series programs. In such situations, the broadcaster may, no more than 12 times per year, and at his/her discretion, utilize promotional/informational material to fill the time requirement provided such material does not exceed an average of two minutes per hour. The limit on the amount of such material per program shall be four minutes and, when used, the material shall be generic in nature.

2. Allowable Time for Non-Program Material:

a. Prime Time

1. Prime time is a continuous period of not less than three consecutive hours per broadcast day as designated by the station between the hours of 6:00 P.M. and midnight.

2. In prime time on network-affiliated stations, the amount of non-program material shall not exceed nine minutes 30 seconds in any 60-minute period. When deemed necessary by the broadcaster, an additional 30 seconds per hour may be used for promotional announcements.

b. In all other time, non-program material shall not exceed 16 minutes in any 60-minute period.

c. Children's Programming Time—Defined as those hours other than prime time in which programs initially designed primarily for children under 12 years of age are scheduled.

Within this time period on Saturday and Sunday, non-program material shall not exceed 9 minutes, 30 seconds in any 60-minute period.

Within this time period on Monday through Friday, non-program material shall not exceed 12 minutes in any 60-minute period.

3. Averaging Concept:

In prime time and all other time, programs of 90 minutes in length or longer, the reasonable averaging of the amount of allowable time for non-program material and/or of the number of allowable interruptions is permitted for the purpose of preserving program continuity in the interests of the viewer. In such situations, one or more 60-minute period(s) may contain more than the allowable amount of non-program material and/or more than the allowable number of interruptions providing the remaining 60-minute period(s) contains appropriately less non-program material and/or fewer interruptions so that, on average, each hour of the program is compliant with applicable Television Code standards. Additionally, when the telecast of live programs of indeterminate length, such as sports programs and on-the-spot news coverage, carries over from one time period to another (e.g., prime time to all other time), allowance will be made for the reasonable overage of non-program material in any given hour of the program providing the total amount of non-program material does not exceed the total allowed for the period of time which the program covers.

4. Program Interruptions:

a. Definition: A program interruption is any occurrence of non-program material within the main body of the program.

b. In prime time, the number of program interruptions shall not exceed two within any 30-minute program, or four within any 60-minute program.

Programs longer than 60 minutes shall be pro-rated at two interruptions per half-hour.

The number of interruptions in 60-minute variety shows shall not exceed five.

c. In all other time, the number of interruptions shall not exceed four within any 30-minute program period.

d. In children's weekend programming time, as above defined in 2c, the number of program interruptions shall not exceed two within any 30-minute program or four within any 60-minute program.

e. In both prime time and all other time, the following interruption standard shall apply within programs of 15 minutes or less in length:

5-minute program—1 interruption;
10-minute program—2 interruptions;
15-minute program—2 interruptions.

f. News, weather, sports and special events programs are exempt from the interruption standard because of the nature of such programs.

5. Consecutive Announcements:
In both prime time and all other time. No more than five non-program material announcements shall be scheduled consecutively within programs, and no more than four non-program material announcements shall be scheduled consecutively during station breaks. The consecutive non-program material limitation shall not apply to a single sponsor who wishes to further reduce the number of interruptions in the program.

6. A multiple product announcement is one in which two or more products or services are presented within the framework of a single announcement. A multiple product announcement shall not be scheduled in a unit of time less than 60 seconds, except where integrated so as to appear to the viewer as a single message. A multiple product announcement shall be considered integrated and counted as a single announcement if:

a. the products or services are related and interwoven within the framework of the announcement (related products or services shall be defined as those having a common character, purpose and use); and

b. the voice(s), setting, background and continuity are used consistently throughout so as to appear to the viewer as a single message.

Multiple product announcements of 60 seconds in length or longer not meeting this definition of integration shall be counted as two or more announcements under this section of the Code. This provision shall not apply to retail or service establishments.

7. Reasonable and limited identification of prizes and donors' names where the presentation of contest awards or prizes is a necessary part of program content shall not be included as non-program material as defined above.

8. Programs presenting women's service features, shopping guides, fashion shows, demonstrations and similar material provide a special service to the public in which certain material normally classified as non-program is an informative and necessary part of the program content. Because of this, the time standards may be waived by the Code Authority to a reasonable extent on a case-by-case basis.

9. Gratuitous references in a program to a nonsponsor's product or service should be avoided except for normal guest identification.

10. Stationary backdrops or properties in television presentations showing the sponsor's name or product, the name of his product, his trade-mark or slogan should be used only incidentally and should not obtrude on program interest or entertainment.

Note: From time to time the Code Authority issues advertising guidelines and clarifications expanding on provisions of the Code. Among areas covered are acne, alcoholic beverages, arthritis and rheumatism remedies, bronchitis, comparative advertising, children's premiums and offers, children's TV advertising, disparagement, hallucinogens, hypnosis, lotteries, men-in-white, nonprescription medications, personal products, testimonials, time standards, toys, vegetable oils and margarines, and weight reducing products/services. Copies may be obtained from any NAB Code Authority office.

Time Standards for Independent Stations

1. Non-program elements shall be considered as all-inclusive, with the exception of required credits, legally required station identifications, and "bumpers." Promotion spots and public service announcements, as well as commercials, are to be considered non-program elements.

2. The allowed time for non-program elements, as defined above, shall not exceed seven minutes in a 30-minute period or multiples thereof in prime (prime time is defined as any three contiguous hours between 6:00 P.M. and midnight, local time), or eight minutes in a 30-minute period or multiples thereof during all other times.

3. Where a station does not carry a commercial in a station break between programs, the number of program interruptions shall not exceed four within any 30-minute program, or seven within any 60-minute program, or 10 within any 90-minute program, or 13 in any 120-minute program. Stations which do carry commercials in station breaks between programs shall limit the number of program interruptions to three within any 30-minute program, or six within any 60-minute program, or nine within any 90-minute program, or 12 in any 120-minute program. News, weather, sports, and special events are exempted because of format.

4. Not more than four non-program material announcements as defined above shall be scheduled consecutively. An exception may be made only in the case of a program 60 minutes or more in length, when no more than seven non-program elements may be scheduled consecutively by stations who wish to reduce the number of program interruptions.

5. The conditions of paragraphs three and four shall not apply to live sports programs where the program format dictates and limits the number of program interruptions.

INTERPRETATIONS

Interpretation No. 1
June 7, 1956, Revised June 9, 1958
"Pitch" Programs

The "pitchman" technique of advertising on television is inconsistent with good broadcast practice and generally damages the reputation of the industry and the advertising profession.

Sponsored program-length segments consisting substantially of continuous demonstrations or sales presentation, violate not only the time standards established in the Code but the broad philosophy of improvement implicit in the voluntary Code operation and are not acceptable.

Interpretation No. 2
June 7, 1956
Hollywood Film Promotion

The presentation of commentary or film excerpts from current theatrical releases in some instances may constitute commercial material under the Time Standards for Non-Program Material. Specifically, for example, when such presentation, directly or by inference, urges viewers to attend, it shall be counted against the commercial allowance for the program of which it is a part.

Interpretation No. 3
January 23, 1959
Prize Identification

Aural and/or visual prize identification of up to 10 seconds duration may be deemed "reasonable and limited" under the language of Paragraph 7 of the Time Standards for Non-Program Material. Where such identification is longer than 10 seconds, the entire announcement or visual presentation will be charged against the total commercial time for the program period.

Interpretation No. 4
March 4, 1965
Drinking on Camera

Paragraph 7, Section IX, General Advertising Standards, states that the "advertising of beer and wine is acceptable only when presented in the best of good taste and discretion." This requires that commercials involving beer and wine avoid any representation of on-camera drinking.

Interpretation No. 5
April 8, 1975

The scheduling provisions of Section I (Principles Governing Program Content) shall not apply to programs under contract to a station as of April 8, 1975, all episodes of which were then in existence, if such station is unable, despite reasonable goof faith efforts, to edit such programs to make them appropriate for family viewing or to reschedule them so as not to occupy family viewing periods. This exception shall in no event apply after September 1, 1977. Any such programs excepted from scheduling provisions shall, of course, bear the required advisory notices. (Effective September 1975)

REGULATIONS AND PROCEDURES

The following Regulations and Procedures shall obtain as an integral part of the Television Code of the National Association of Broadcasters:

I. Name

The name of this Code shall be *The Television Code of the National Association of Broadcasters.**

II. Purpose of the Code

The purpose of this Code is cooperatively to mtaintain a level of television programming which gives full conclusion to the educational, informational, cultural, economic, moral and entertainment needs of the American public to the end that more and more people will be better served.

III. Subscribers

Section 1. Eligibility

Any individual, firm or corporation which is engaged in the operation of a television broadcast station or network, or which holds a construction permit for a television broadcast station within the United States or its dependencies, shall, subject to the approval of the Television Board of Directors as hereinafter provided, be eligible to subscribe to the Television Code of the NAB to the extent of one subscription for each such station and/or network which it operates or for which it holds a construction permit; provided, that a non-television member of NAB shall not become eligible via Code subscription to receive any of the member services or to exercise any of the voting privileges of a member.

Section 2. Certification of Subscription

Upon subscribing to the Code, subject to the approval of the Television Board of Directors, there shall be granted forthwith to each such subscribing station authority to use the NAB Television Seal of Good Practice," a copyrighted and registered seal to be provided in the form of a certificate, a slide and/or a film, signifying that the recipient thereof is a subscriber in good standing to the Television Code of the NAB. The seal and its significance shall be appropriately publicized by the NAB.

Section 3. Duration of Subscription

Subscription shall continue in full force and effect until thirty days after the first of the month following receipt of notice of written resignation. Subscription to the Code shall be effective from the date of application subject to the approval of the Television Board of Directors; provided, that the subscription of a television station going on the air for the first time shall, for the first six months of such subscription, be probationary, during which time its subscription can be summarily revoked by an affirmative two-thirds vote of the Television Board of Directors without the usual processes specified below.

Section 4. Suspension of Subscription

Any subscription, and/or the authority to utilize and show the above-noted seal, may be voided, revoked or temporarily suspended for television programming, including commercial copy, which, by theme, treatment by incident, in the judgment of the Television Board constitutes a continuing, willful or gross violation of any of the provisions of the Television Code, by an affirmative two-thirds vote of the Television Board of Directors at a regular or special meeting; provided, however, that the following conditions and procedures shall apply:

A. *Preferring of Charges—Conditions Precedent:*

Prior to the preferring of charges to the Television Board of Directors concerning violation of the Code by a subscriber, the Television Code Review Board (hereinafter provided for) (1) Shall have appropriately, and in good time, informed and advised such subscriber of any and all complaints and information coming to the attention of the Television Code Review Board and relating to the programming of said subscriber, (2) Shall have reported to, and advised, said subscriber by analysis, interpretation, recommendation or otherwise, of the possibility of a violation or breach of the Television Code by the subscriber, and (3) Shall have served upon the subscriber, by Registered Mail a Notice of Intent to prefer charges, at least 20 days prior to the filing of any such charges with the Television Board of Directors. During this period the Television Code Review Board may, within its sole discretion, reconsider its proposed action based upon such written reply as the subscriber may care to make, or upon such action as the subscriber may care to take program-wise, in conformance with the analysis, interpretation, or recommendation of the Television Code Review Board.

(i) Notice of Intent

The Notice of Intent shall include a statement of the grounds and reasons for the proposed charges, including appropriate references to the Television Code.

(ii) Time

In the event that the nature of the program in question is such that time is of the essence, the Television Code Review Board may prefer charges within less than the 20 days above specified, provided that a time certain in which reply may be made is included in its Notice of Intent, and provided that its reasons therefor must be specified in its statement of charges preferred.

B. *The Charges:*

The subscriber shall be advised in writing by Registered Mail of the charges preferred. The charges preferred by the Television Code Review Board to the Television Board of Directors shall include the grounds and reasons therefor, together with specific references to the Television Code. The charges shall contain a statement that the condition precedent, herein before described, have been met.

C. *Hearing:*

The subscriber shall have the right to a hearing and may exercise same by filing an answer within 10 days of the date of such notification.

D. *Waiver:*

Failure to request shall be deemed a waiver of the subscriber's right thereto.

E. *Designation.*

If a hearing is requested by the subscriber, it shall be designated as promptly as possible and at such time and place as the Television Board may specify.

F. *Confidential Status:*

Hearings shall be closed; and all correspondence between a subscriber and the Television Code Review Board and/or the Television Board of Directors concerning specific programming shall be confidential; provided, however, that the confidential status of these procedures may be waived by a subscriber.

G. *Presentation; Representation:*

A subscriber against whom charges have been preferred, and who has exercised his right to a hearing, shall be entitled to effect presentation of his case personally, by agent, by attorney, or by deposition and interrogatory.

H. *Intervention:*

Upon request by the subscriber-respondent or the Television Code Review Board, the Television Board of Directors, in its discretion, may permit the intervention of one or more other subscribers as parties-in-interest.

I. *Transcript:*

A stenographic transcript record shall be taken and shall be certified by the Chairman of the Television Board of Directors to the office of the Secretary of the National Association of Broadcasters, where it shall be maintained. The transcript shall not be open to inspection unless otherwise provided by the party respondent in the proceeding.

J. *Television Code Review Board; Counsel:*

The Television Code Review Board may, at its discretion, utilize the services of an attorney from the staff of the NAB for the purpose of effecting its presentation in a hearing matter.

K. *Order of Procedure:*
At hearings the Television Code Review Board shall open and close.

L. *Cross-Examination:*
The right of cross-examination shall specifically obtain. Where procedure has been by deposition or interrogatory, the use of cross-interrogatories shall satisfy this right.

M. *Presentation:*
Oral and written evidence may be introduced by the subscriber and by the Television Code Review Board. Oral argument may be had at the hearing and written memoranda or briefs may be submitted by the subscriber and by the Television Code Review Board. The Television Board of Directors may admit such evidence as it deems relevant, material and competent, and may determine the nature and length of the oral argument and the written argument or briefs.

N. *Authority of Presiding Officer; of Television Board of Directors:*
The Presiding Officer shall rule upon all interlocutory matters, such as, but not limited to, the admissibility of evidence, the qualifications of witnesses, etc. On all other matters, authority to act shall be vested in a majority of the Television Board unless otherwise provided.

O. *Films, Transcriptions, etc.:*
Films, kinescopes, records, transcriptions, or other mechanical reproductions of television programs, properly identified, shall be accepted into evidence when relevant.

P. *Continuances and Extensions:*
Continuance and extension of any proceeding or for the time of filing or performing any act required or allowed to be done within a specific time may be granted upon request, for a good cause shown. The Board or the Presiding Officer may recess or adjourn a hearing for such time as may be deemed necessary, and may change the place thereof.

Q. *Findings and Conclusions:*
The Television Board of Directors shall decide the case as expeditiously as possible and shall notify the subscriber and the Television Code Review Board, in writing, of the decision. The decision of the Television Board of Directors shall contain findings of fact with conclusions, as well as the reasons or bases therefor. Findings of fact shall set out in detail and with particularity all basic evidentiary facts developed on the record (with appropriate citations to the transcript of record or exhibit relied on for each evidentiary fact) supporting the conclusion reached.

R. *Reconsideration or Reharing:*
A request for reconsideration or rehearing may be filed by parties to the hearing. Requests for reconsideration or rehearing shall state with particularity in what respect the decision or any matter determined therein is claimed to be unjust, unwarranted, or erroneous, and with respect to any finding of fact shall specify the pages of record relied on. If the existence of any newly discovered evidence is claimed, the request shall be accompanied by a verified statement of the facts together with the facts relied on to show that the party, with due diligence, could not have known or discovered such facts at the time of the hearing. The request for a rehearing may seek:

a. Reconsideration
b. Additional oral argument
c. Reopening of the proceedings
d. Amendment of any findings, or
e. Other relief.

S. *Time for Filing:*
Requests for reconsideration or rehearing shall be filed within 10 days after receipt by the respondent of the decision. Opposition thereto may be filed within five days after the filing of the request.

T. *Penalty, Suspension of:*
At the discretion of the Television Board, application of any penalty provided for in the decision may be suspended until the Board makes final disposition of the request for reconsideration or rehearing.

U. *Disqualification:*
Any member of the Television Board may disqualify himself, or upon good cause shown by any interested party, may be disqualified by a majority vote of the Television Board.

Section 5. Additional Procedures
When necessary to the proper administration of the Code, additional rules of procedure will be established from time to time as authorized by the By-Laws of the NAB; in keeping therewith, special consideration shall be given to the procedures for receipt and processing of complaints and to necessary rules to be adopted from time to time, taking into account the source and nature of such complaints; such rules to include precautionary measures such as the posting of bonds to cover costs and expenses of processing same; and further provided that special consideration will be given to procedures insuring the confidential status of proceedings related to Code observance.

Section 6. Amendment and Review
Because of the new and dynamic aspects inherent in television broadcast, the Television Code, as a living, flexible and continuing document, may be amended from time to time by the Television Board of Directors; provided that said Board is specifically charged with review and reconsideration of the entire Code, its appendices and procedures, at least once each year.

Section 7. Termination of Contracts
All subscribers on the air at the time of subscription to the Code shall be permitted that period prior to and including the earliest legal cancellation date to terminate any contracts, then outstanding, calling for program presentations which would not be in conformity with the Television Code, provided, however, that in no event shall such period be longer than 52 weeks.

IV. Affiliate Subscribers

Section 1. Eligibility
Any individual, firm or corporation, which is engaged in the production or distribution, lease, or sale of recorded programs for television presentation, subject to the approval of the Television Code Review Board as hereinafter provided, shall be eligible to become an affiliate subscriber to the Television Code of the NAB.

Section 2. Certification of Subscription
Upon becoming an affiliate subscriber to the Code, subject to the approval of the Television Code Review Board, there shall be granted forthwith to each such affiliate subscriber authority to use a copyrighted and registered seal and declaration, in a manner approved by the Television Code Review Board, identifying the individual firm or corporation as an affiliate subscriber to the Television Code of the NAB. Such authority shall not constitute formal clearance or approval by the Television Code Review Board of specific film programs or other recorded material.

Section 3. Duration of Affiliate Subscription
The affiliate subscription shall continue in full force and effect until 30 days after the first of the month following receipt of a written notice of resignation. The affiliate subscription of the Code shall be effective from the date of application subject to the approval of the Television Code Review Board.

Section 4. Suspension of Affiliate Subscription
Any affiliate subscription and the authority to utilize and show the above-noted seal may be voided, revoked, or temporarily suspended for the sale or distribution for television presentation of any film or other recorded material which by theme, treatment, or incident, in the judgment of the Television Code Review Board, constitutes a continuing, willful or gross violation of any of the provisions of the Television Code, by a majority vote of the Television Code Review Board at any regular or special meeting. The conditions and procedures applicable to subscribers shall not apply to affiliate subscribers.

Section 5. Representation of Affiliate Subscribers
Any affiliate subscriber or group of affiliate subscribers may authorize an individual or association to act for them in connection with their relations with the Television Code Review Board by filing a written notice of such representation with the Board. Such representation, how-

ever, in no way will limit the right of the Television Code Review Board to suspend individual affiliate subscribers in accordance with the provisions of Section 4.

V. Rates

Each subscriber and affiliate subscriber shall pay "administrative" rates in accordance with such schedule, at such time, and under such conditions as may be determined from time to time by the Television Board (see Article VI, section 8, C. Television Board (3) and (4), By-Laws of the NAB); provided, that appropriate credit shall be afforded to a television member of the NAB against the regular dues which he or it pays to NAB.

VI. The Television Code Review Board

Section 1. Composition

There shall be a continuing committee entitled the Television Code Review Board to be composed of not more than nine members, all of whom shall be from subscribers to the Television Code. They shall be appointed by the President of NAB, subject to confirmation by the Television Board, and may include one member from each of the subscribing nationwide television networks. Members of the Television Board shall not be eligible to serve on the Review Board. Due consideration shall be given, in making the appointments, to factors of diversification of geographical location, market size, company representation and network affiliation.

No person shall continue as a member of the Television Code Review Board if the station or entity he represents ceases to subscribe to the Television Code. In such a case a vacancy occurs in the office immediately, and a successor may be appointed to serve out the unexpired term.

All terms shall be for two years, commencing at the close of the annual meeting of the membership following appointment.

A. *Limitation of Service:*

No person shall serve for more than two terms of two years each, consecutively, as a member of the Television Code Review Board; provided, however, this limitation shall not apply to network representatives.

Serving out the unexpired term of a former member shall not constitute a term within the meaning of this section.

B. *Meetings:*

The Television Code Review Board shall meet at least twice in each calendar year on a date to be determined by the Chairman. The chairman, or the Code Authority Director, may, at any time, on at least five days written notice, call a special meeting of the Board.

C. *Quorum:*

For all purposes, a majority of the members of the Television Code Review Board shall constitute a quorum.

Section 2. Authority and Responsibilities

The Television Code Review Board is authorized and directed:

(1) To recommend to the Television Board of Directors amendments to the Television Code; (2) to consider, in its discretion, any appeal from any decision made by the Code Authority Director with respect to any matter which has arisen under the Code, and to suspend, reverse, or modify any such decision; (3) to prefer formal charges, looking toward the suspension or revocation of the authority to show the Code seal, to the Television Board of Directors concerning violations and breaches of the Television Code by a subscriber; (4) to be available to the Code Authority Director for consultation on any and all matters affecting the Television Code.

VII. Code Authority Director

Section 1. Director

There shall be a position designated as the Code Authority Director. This position shall be filled by appointment of the President of NAB, subject to the approval of the Board of Directors.

Section 2. Authority and Responsibilities

The Code Authority Director is authorized and directed: (1) To maintain a continuing review of all programming and advertising material presented over television, especially that of subscribers to the Television Code of NAB; (2) to receive, screen and clear complaints concerning television programming; (3) to define and interpret words and phrases in the Television Code; (4) to develop and maintain appropriate liaison with governmental agencies and with responsible and accountable organizations and institutions; (5) to inform, expeditiously and properly, a subscriber to the Television Code of complaints or commendations, as well as to advise all subscribers concerning the attitudes and desires program-wise of accountable organizations and institutions, and of the American public in general; (6) to review and monitor, if necessary, any certain series of programs, daily programming, or any other program presentations of a subscriber, as well as to request recorded material, or script and copy, with regard to any certain program presented by a subscriber; (7) to reach conclusions and make recommendations or prefer charges to the Television Code Review Board concerning violations and breaches of the Television Code by a subscriber; (8) to recommend to the Code Review Board amendments to the Television Code.

A. *Delegation of Powers and Responsibilities:*

The Code Authority Director shall appoint such executive staff as is needed, consistent with resources, to carry out the above described functions, and may delegate to this staff such responsibilities as he may deem necessary.

*By-Laws of the National Association of Broadcasters, Article VI, section 8, C: "Television Board. The Television board is hereby authorized:—(4) to enact, amend and promulgate standards of practice or codes for its Television members, and to establish such methods to secure observance thereof as it may deem advisable:—."

MOTION PICTURE Product Digest
July 26, 1978
Vol. 6, No. 4
August 9, 1978
Vol. 6, No. 5
reviews: By RICHARD GERTNER
Burt Reynolds Is at His Macho Best
in "Hooper": Tribute to Stuntmen
Hooper
Warner Bros.
Pictures Reviewed in This Issue